AF886638

ATF Federal Explosives Law and Regulations

Plus

ATF State Laws and Published Ordinances Firearms

Table of Contents

Organized Crime Control Act of 1970, Title XI

Regulation of Explosives

Public Law 91-452, Approved October 15, 1970 (as Amended)

[**NOTE:** Any reference to the Internal Revenue Code of 1954 refers to the Internal Revenue Code of 1986 (§2, Public Law 99-514, 100 Stat. 2085, October 22, 1986.)]

Purpose

SEC. 1101. The Congress hereby declares that the purpose of this title is to protect interstate and foreign commerce against interference and interruption by reducing the hazard to persons and property arising from misuse and unsafe or insecure storage of explosive materials. It is not the purpose of this title to place any undue or unnecessary Federal restrictions or burdens on law-abiding citizens with respect to the acquisition, possession, storage, or use of explosive materials for industrial, mining, agricultural, or other lawful purposes, or to provide for the imposition by Federal regulations of any procedures or requirements other than those reasonably necessary to implement and effectuate the provisions of this title.

Sec. 1102. Title 18, United States Code, is amended by adding after chapter 39 the following chapter:

Chapter 40. Importation, Manufacture, Distribution and Storage of Explosive Materials

Editor's Note: The sections of law set out herein were added by Public Law 91-452, Title XI, §1102(a), Oct. 15, 1970, 84 Stat. 952-959, and remain un-changed unless otherwise footnoted.

Sec.

§ 841. Definitions

As used in this chapter

(a) "Person" means any individual, corporation, company, association, firm, partnership, society, or joint stock company.

(b) "Interstate or foreign commerce" means commerce between any place in a State and any place outside of that State, or within any possession of the United States (not including the Canal Zone) or the District of Columbia, and commerce between places within the same State but through any place outside of that State. "State" includes the District of Columbia, the Commonwealth of Puerto Rico, and the possessions of the United States (not including the Canal Zone).

(c) "Explosive materials" means explosives, blasting agents, and detonators.

(d) Except for the purposes of subsections (d), (e), (f), (g), (h), (i), and (j) of section 844 of this title, **"explosives"** means any chemical compound mixture, or device, the primary or common purpose of which is to function by explosion; the term includes, but is not limited to, dynamite and other high explosives, black powder, pellet powder, initiating explosives, detonators, safety fuses, squibs, detonating cord, igniter cord, and igniters. The Attorney General shall publish and revise at least annually in the Federal Register a list of these and any additional explosives which he determines to be within the coverage of this chapter. For the purposes of subsections (d), (e), (f), (g), (h), and (i) of section 844 of this title, the term "explosive" is defined in subsection (j) of such section 844.

(e) "Blasting agent" means any material or mixture, consisting of fuel and oxidizer, intended for blasting, not otherwise defined as an explosive: Provided, That the finished product, as mixed for use or shipment, cannot be detonated by means of a numbered 8 test blasting cap when unconfined.

(f) "Detonator" means any device containing a detonating charge that is used for initiating detonation in an explosive; the term includes, but is not limited to, electric blasting caps of instantaneous and delay types, blasting caps for use with safety fuses and detonating-cord delay connectors.

(g) "Importer" means any person engaged in the business of importing or bringing explosive materials into the United States for purposes of sale or distribution.

(h) "Manufacturer" means any person engaged in the business of manufacturing explosive materials for purposes of sale or distribution or for his own use.

(i) "Dealer" means any person engaged in the business of distributing explosive materials at wholesale or retail.

(j) "Permittee" means any user of explosives for a lawful purpose, who has obtained either a user permit or a limited permit under the provisions of this chapter.

(k) "Attorney General" means the Attorney General of the United States.

(l) "Crime punishable by imprisonment for a term exceeding one year" shall not mean (1) any Federal or State offenses pertaining to antitrust violations, unfair trade practices, restraints of trade, or other similar offenses relating to the regulation of business practices as the Attorney General may be regulation designate, or (2) any State offense (other than one involving a firearm or explosive) classified by the laws of the State as a misdemeanor and punishable by a term of imprisonment of two years or less.

(m) "Licensee" means any importer, manufacturer, or dealer licensed under the provisions of this chapter.

(n) "Distribute" means sell, issue, give, transfer, or otherwise dispose of.

(o) "Convention on the Marking of Plastic Explosives" means the Convention on the Marking of Plastic Explosives for the Purpose of Detection, Done at Montreal on 1 March 1991.

(p) "Detection agent" means any one of the substances specified in this subsection when introduced into a plastic explosive or formulated in such explosive as a part of the manufacturing process in such a manner as to achieve homogeneous distribution in the finished explosive, including—

(1) Ethylene glycol dinitrate (EGDN), C[2]H[4](NO[3])[2], molecular weight 152, when the minimum concentration in the finished explosive is 0.2 percent by mass;

(2) 2,3-Dimethyl-2,3-dinitrobutane (DMNB), C[6]H[12](NO[2])[2], molecular weight 176, when the minimum concentration in the finished explosive is 0.1 percent by mass;

(3) Para-Mononitrotoluene (p-MNT), C[7]H[7]NO[2], molecular weight 137, when the minimum concentration in the finished explosive is 0.5 percent by mass;

(4) Ortho-Mononitrotoluene (o-MNT), C[7]H[7]NO[2], molecular weight 137, when the minimum concentration in the finished explosive is 0.5 percent by mass; and

(5) any other substance in the concentration specified by the Attorney General, after consultation with the Secretary of State and the Secretary of Defense, that has been added to the table in part 2 of the Technical Annex to the Convention on the Marking of Plastic Explosives.

(q) "Plastic explosive" means an explosive material in flexible or elastic sheet form formulated with one or more high explosives which in their pure form has a vapor pressure less than 10-<4> Pa at a temperature of 25 degrees C., is formulated with a binder material, and is as a mixture malleable or flexible at normal room temperature.

(r) "Alien" means any person who is not a citizen or national of the United States.

(s) "Responsible person" means an individual who has the power to direct the management and policies of the applicant pertaining to explosive materials.

(Added Pub. L. 91-452, title XI, Sec. 1102(a), Oct. 15, 1970, 84 Stat. 952; amended Pub. L. 104-132, title VI, Sec. 602, Apr. 24, 1996, 110 Stat. 1288; Pub. L. 107-296, title XI, Sec. 1112(e)(1), (3), 1122(a), Nov. 25, 2002, 116 Stat. 2276, 2280.)

§ 842. Unlawful acts

(a) It shall be unlawful for any person—

(1) to engage in the business of importing, manufacturing, or dealing in explosive materials without a license issued under this chapter;

(2) knowingly to withhold information or to make any false or fictitious oral or written statement or to furnish or exhibit any false, fictitious, or misrepresented identification, intended or likely to deceive for the purpose of obtaining explosive materials, or a license, permit, exemption, or relief from disability under the provisions of this chapter;

(3) other than a licensee or permittee knowingly–

(A) to transport, ship, cause to be transported, or receive any explosive materials; or

(B) to distribute explosive materials to any person other than a licensee or permittee; or

(4) who is a holder of a limited permit—

(A) to transport, ship, cause to be transported, or receive in interstate or foreign commerce any explosive materials; or

(B) to receive explosive materials from a licensee or permittee, whose premises are located outside the State of residence of the limited permit holder, or on more than 6 separate occasions, during the period of the permit, to receive explosive materials from 1 or more licensees or permittees whose premises are located within the State of residence of the limited permit holder.

(b) It shall be unlawful for any licensee or permittee to knowingly distribute any explosive materials to any person other than—

(1) a licensee;

(2) a holder of a user permit; or

(3) a holder of a limited permit who is a resident of the State where distribution is made and in which the premises of the transferor are located.

(c) It shall be unlawful for any licensee to distribute explosive materials to any person who the licensee has reason to believe intends to transport such explosive materials into a State where the purchase, possession, or use of explosive materials is prohibited or which does not permit its residents to transport or ship explosive materials into it or to receive explosive materials in it.

(d) It shall be unlawful for any person knowingly to distribute explosive materials to any individual who:

(1) is under twenty-one years of age;

(2) has been convicted in any court of a crime punishable by imprisonment for a term exceeding one year;

(3) is under indictment for a crime punishable by imprisonment for a term exceeding one year;

(4) is a fugitive from justice;

(5) is an unlawful user of or addicted to any controlled substance (as defined in section 102 of the Controlled Substances Act);

(6) has been adjudicated a mental defective or who has been committed to a mental institution;

(7) is an alien, other than an alien who—

(A) is lawfully admitted for permanent residence (as defined in section 101 (a)(20) of the Immigration and Nationality Act);

(B) is in lawful nonimmigrant status, is a refugee admitted under section 207 of the Immigration and Nationality Act or is in asylum status under section 208 of the Immigration and Nationality Act and—

(i) is a foreign law enforcement officer of a friendly foreign government, as determined by the Secretary in consultation with the Secretary of State, entering the United States on official law enforcement business, and the shipping, transporting, possession, or receipt of explosive materials is in furtherance of this official law enforcement business; or

(ii) is a person having the power to direct or cause the direction of the management and policies of a corporation, partnership, or association licensed pursuant to section 843(a), and the shipping, transporting, possession, or receipt of explosive materials is in furtherance of such power;

(C) is a member of a North Atlantic Treaty Organization (NATO) or other friendly foreign military force, as determined by the Attorney General in consultation with the Secretary of Defense, who is present in the United States under military orders for training or other military purpose authorized by the United States and the shipping, transporting, possession, or receipt of explosive materials is in furtherance of the authorized military purpose; or

(D) is lawfully present in the United States in cooperation with the Director of Central Intelligence, and the shipment transportation, receipt, or possession of the explosive materials is in furtherance of such cooperation;

(8) has been discharged from the armed forces under dishonorable conditions;

(9) having been a citizen of the United States, has renounced the citizenship of that person.

(e) It shall be unlawful for any licensee knowingly to distribute any explosive materials to any person in any State where the purchase, possession, or use by such person of such explosive materials would be in violation of any State law or any published ordinance applicable at the place of distribution.

(f) It shall be unlawful for any licensee or permittee willfully to manufacture, import, purchase, distribute, or receive explosive materials without making such records as the Attorney General may by regulation require, including, but not limited to, a statement of intended use, the name, date, place of birth, social security number or taxpayer identification number, and place of residence of any natural person to whom explosive materials are distributed. If explosive materials are distributed to a corporation or other business entity, such records shall include the identity and principal and local places of business and the name, date, place of birth, and place of residence of the natural person acting as agent of the corporation or other business entity in arranging the distribution.

(g) It shall be unlawful for any licensee or permittee knowingly to make any false entry in any record which he is required to keep pursuant to this section or regulations promulgated under section 847 of this title.

(h) It shall be unlawful for any person to receive, possess, transport, ship, conceal, store, barter, sell, dispose of, or pledge or accept as security for a loan, any stolen explosive materials which are moving as, which are part of, which constitute, or which have been shipped or transported in, interstate or foreign commerce, either before or after such materials were stolen, knowing or having reasonable cause to believe that the explosive materials were stolen.

(i) It shall be unlawful for any person—

(1) who is under indictment for, or who has been convicted in any court of, a crime punishable by imprisonment for a term exceeding one year;

(2) who is a fugitive from justice;

(3) who is an unlawful user of or addicted to any controlled substance (as defined in section 102 of the Controlled Substances Act);

(4) who has been adjudicated as a mental defective or who has been committed to a mental institution;

(5) who is an alien, other than an alien who—

(A) is lawfully admitted for permanent residence (as that term is defined in section 101(a)(20) of the Immigration and Nationality Act);

(B) is in lawful nonimmigrant status, is a refugee admitted under section 207 of the Immigration and Nationality Act or is in asylum status under section 208 of the Immigration and Nationality Act and--

(i) is a foreign law enforcement officer of a friendly foreign government, as determined by the Secretary in consultation with the Secretary of State, entering the United States on official law enforcement business, and the shipping, transporting, possession, or receipt of explosive materials is in furtherance of this official law enforcement business; or

(ii) is a person having the power to direct or cause the direction of the management and policies of a corporation, partnership, or association licensed pursuant to section 843(a), and the shipping, transporting, possession, or receipt of explosive materials is in furtherance of such power;

(C) is a member of a North Atlantic Treaty Organization (NATO) or other friendly foreign military force, as determined by the Attorney General in consultation with the Secretary of Defense, who is present in the United States under military orders for training or other military purpose authorized by the United States and the shipping, transporting, possession, or receipt of explosive materials is in furtherance of the authorized military purpose; or

(D) is lawfully present in the United States in cooperation with the Director of Central Intelligence, and the shipment, transportation, receipt, or possession of the explosive materials is in furtherance of such cooperation;

(6) who has been discharged from the armed forces under dishonorable conditions;

(7) who, having been a citizen of the United States, has renounced the citizenship of that person; to ship or transport any explosive in or affecting interstate or foreign commerce or to receive or possess any explosive which has been shipped or transported in or affecting interstate or foreign commerce.

(j) It shall be unlawful for any person to store any explosive material in a manner not in conformity with regulations promulgated by the Attorney General. In promulgating such regulations, the Attorney General shall take into consideration the class, type, and quantity of explosive materials to be stored, as well as the standards of safety and security recognized in the explosives industry.

(k) It shall be unlawful for any person who has knowledge of the theft or loss of any explosive materials from his stock, to fail to report such theft or loss within twenty-four hours of discovery thereof, to the Attorney General and to appropriate local authorities.

(l) It shall be unlawful for any person to manufacture any plastic explosive that does not contain a detection agent.

(m) **(1)** It shall be unlawful for any person to import or bring into the United States, or export from the United States, any plastic explosive that does not contain a detection agent.

(2) This subsection does not apply to the importation or bringing into the United States, or the exportation from the United States, of any plastic explosive that was imported or brought into, or manufactured in the United States prior to the date of enactment of this subsection [enacted April 24, 1996] by or on behalf of any agency of the United States performing military or police functions (including any military reserve component) or by or on behalf of the National Guard of any State, not later than 15 years after the date of entry into force of the Convention on the Marking of Plastic Explosives, with respect to the United States.

(n) **(1)** It shall be unlawful for any person to ship, transport, transfer, receive, or possess any plastic explosive that does not contain a detection agent.

(2) This subsection does not apply to—

(A) the shipment, transportation, transfer, receipt, or possession of any plastic explosive that was imported or brought into, or manufactured in the United States prior to the date of enactment of this subsection [enacted April 24, 1996] by any person during the period beginning on that date and ending 3 years after that date of enactment; or

(B) the shipment, transportation, transfer, receipt, or possession of any plastic explosive that was imported or brought into, or manufactured in the United States prior to the date of enactment of this subsection [enacted April 24, 1996] by or on behalf of any agency of the United States performing a military or police function (including any military reserve component) or by or on behalf of the National Guard of any State, not later than 15 years after the date of entry into force of the Convention on the Marking of Plastic Explosives, with respect to the United States.

(o) It shall be unlawful for any person, other than an agency of the United States (including any military reserve component) or the National Guard of any State, possessing any plastic explosive on the date of enactment of this subsection [enacted April 24, 1996], to fail to report to the Attorney General within 120 days after such date of enactment the quantity of such explosives possessed, the manufacturer or importer, any marks of identification on such explosives, and such other information as the Attorney General may prescribe by regulation.

(p) Distribution of information relating to explosives, destructive devices, and weapons of mass destruction.

(1) Definitions. In this subsection—

(A) the term "destructive device" has the same meaning as in section 921(a)(4);

(B) the term "explosive" has the same meaning as in section 844(j); and

(C) the term "weapon of mass destruction" has the same meaning as in section 2332a(c)(2).

(2) Prohibition. It shall be unlawful for any person—

(A) to teach or demonstrate the making or use of an explosive, a destructive device, or a weapon of mass destruction, or to distribute by any means information pertaining to, in whole or in part, the manufacture or use of an explosive, destructive device, or weapon of mass destruction, with the intent that the teaching, demonstration, or information be used for, or in furtherance of, an activity that constitutes a Federal crime of violence; or

(B) to teach or demonstrate to any person the making or use of an explosive, a destructive device, or a weapon of mass destruction, or to distribute to any person, by any means, information pertaining to, in whole or in part, the manufacture or use of an explosive, destructive device, or weapon of mass destruction, knowing that such person intends to use the teaching, demonstration, or information for, or in furtherance of, an activity that constitutes a Federal crime of violence.

(Added Pub. L. 91-452, title XI, Sec. 1102(a), Oct. 15, 1970, 84 Stat. 953; amended Pub. L. 100-690, title VI, Sec. 6474(c), (d), Nov. 18, 1988, 102 Stat. 4380; Pub. L. 101-647, title XXXV, Sec. 3521, Nov. 29, 1990, 104 Stat. 4923; Pub. L. 103-322, title XI, Sec. 110508, 10516, Sept. 13, 1994, 108 Stat. 2018, 2020; Pub. L. 104-132, title VI, Sec. 603, title VII, Sec. 707, Apr. 24, 1996, 110 Stat. 1289, 1296; Pub. L. 106-54, Sec. 2(a), Aug. 17, 1999, 113 Stat. 398; Pub. L. 107-296, title XI, Sec. 1112(e)(3), 1122(b), 1123, Nov. 25, 2002, 116 Stat. 2276, 2280, 2283.)

§ 843. Licenses and user permits

(a) An application for a user permit or limited permit or a license to import, manufacture, or deal in explosive materials shall be in such form and contain such information as the Attorney General shall by regulation prescribe, including the names of and appropriate identifying information regarding all employees who will be authorized by the applicant to possess explosive materials, as well as fingerprints and a photograph of each responsible person. Each applicant for a license or permit shall pay a fee to be charged as set by the Attorney General, said fee not to exceed $ 50 for a limited permit and $ 200 for any other license or permit. Each license or user permit shall be valid for not longer than 3 years from the date of issuance and each limited permit shall be valid for not longer than 1 year from the date of issuance. Each license or permit shall be renewable upon the same conditions and subject to the same restrictions as the original license or permit, and upon payment of a renewal fee not to exceed one-half of the original fee.

(b) Upon the filing of a proper application and payment of the prescribed fee, and subject to the provisions of this chapter and other applicable laws, the Attorney General shall issue to such applicant the appropriate license or permit if--

(1) the applicant (or, if the applicant is a corporation, partnership, or association, each responsible person with respect to the applicant) is not a person described in section 842(i);

(2) the applicant has not willfully violated any of the provisions of this chapter or regulations issued hereunder;

(3) the applicant has in a State premises from which he conducts or intends to conduct business;

(4) **(A)** the Secretary verifies by inspection or, if the application is for an original limited permit or the first or second renewal of such a permit, by such other means as the Secretary determines appropriate, that the applicant has a place of storage for explosive materials which meets such standards of public safety and security against theft as the Attorney General by regulations shall prescribe; and

(B) subparagraph (A) shall not apply to an applicant for the renewal of a limited permit if the Secretary has verified, by inspection within the preceding 3 years, the matters described in subparagraph (A) with respect to the applicant; and

(5) the applicant has demonstrated and certified in writing that he is familiar with all published State laws and local ordinances relating to explosive materials for the location in which he intends to do business;

(6) none of the employees of the applicant who will be authorized by the applicant to possess explosive materials is any person described in section 842(i); and

(7) in the case of a limited permit, the applicant has certified in writing that the applicant will not receive explosive materials on more than 6 separate occasions during the 12-month period for which the limited permit is valid.

(c) The Attorney General shall approve or deny an application within a period of 90 days for licenses and permits beginning on the date such application is received by the Attorney General.

(d) The Attorney General may revoke any license or permit issued under this section if in the opinion of the Attorney General the holder thereof has violated any provision of this chapter or any rule or regulation prescribed by the Attorney General under this chapter, or has become ineligible to acquire explosive materials under section 842(d). The Attorney General's action under this subsection may be reviewed only as provided in subsection (e)(2) of this section.

(e) **(1)** Any person whose application is denied or whose license or permit is revoked shall receive a written notice from the Attorney General stating the specific grounds upon which such denial or revocation is based. Any notice of a revocation of a license or permit shall be given to the holder of such license or permit prior to or concurrently with the effective date of the revocation.

(2) If the Attorney General denies an application for, or revokes a license, or permit, he shall, upon request by the aggrieved party, promptly hold a hearing to review his denial or revocation. In the case of a revocation, the Attorney General may upon a request of the holder stay the effective date of the revocation. A hearing under this section shall be at a location convenient to the aggrieved party. The Attorney General shall give written notice of his decision to the aggrieved party within a reasonable time after the hearing. The aggrieved party may, within sixty days after receipt of the Secretary's written decision, file a petition with the United States court of appeals for the district in which he resides or has his principal place of business for a judicial review of such denial or revocation, pursuant to sections 701-706 of title 5, United States Code.

(f) Licensees and holders of user permits shall make available for inspection at all reasonable times their records kept pursuant to this chapter or the regulations issued hereunder, and licensees and permittees shall submit to the Attorney General such reports and information with respect to such records and the contents thereof as he shall by regulations prescribe. The Attorney General may enter during business hours the premises (including places of storage) of any licensee or holder of a user permit, for the purpose of inspecting or examining (1) any records or documents required to be kept by such licensee or permittee, under the provisions of this chapter or regulations issued hereunder, and (2) any explosive materials kept or stored by such licensee or permittee at such premises. Upon the request of any State or any political subdivision thereof, the Attorney General may make available to such State or any political subdivision thereof, any information which he may obtain by reason of the provisions of this chapter with respect to the identification of persons within such State or political subdivision thereof, who have purchased or received explosive materials, together with a description of such explosive materials. The Secretary may inspect the places of storage for explosive materials of an applicant for a limited permit or, at the time of renewal of such permit, a holder of a limited permit, only as provided in subsection (b)(4).

(g) Licenses and user permits issued under the provisions of subsection (b) of this section shall be kept posted and kept available for inspection on the premises covered by the license and permit.

(h) **(1)** If the Secretary receives, from an employer, the name and other identifying information of a responsible person or an employee who will be authorized by the employer to possess explosive materials in the course of employment with the employer, the Secretary shall determine whether the responsible person or employee is one of the persons described in any paragraph of section 842(i). In making the determination, the Secretary may take into account a letter or document issued under paragraph (2).

(2) **(A)** If the Secretary determines that the responsible person or the employee is not one of the persons described in any paragraph of section 842(i), the Secretary shall notify the employer in writing or electronically of the determination and issue, to the responsible person or employee, a letter of clearance, which confirms the determination.

(B) If the Secretary determines that the responsible person or employee is one of the persons described in any paragraph of section 842(i), the Secretary shall notify the employer in writing or electronically of the determination and issue to the responsible person or the employee, as the case may be, a document that—

(i) confirms the determination;

(ii) explains the grounds for the determination;

(iii) provides information on how the disability may be relieved; and

(iv) explains how the determination may be appealed.

(i) Furnishing of samples.

(1) In general. Licensed manufacturers and licensed importers and persons who manufacture or import explosive materials or ammonium nitrate shall, when required by letter issued by the Secretary, furnish—

(A) samples of such explosive materials or ammonium nitrate;

(B) information on chemical composition of those products; and

(C) any other information that the Secretary determines is relevant to the identification of the explosive materials or to identification of the ammonium nitrate.

(2) Reimbursement. The Secretary shall, by regulation, authorize reimbursement of the fair market value of samples furnished pursuant to this subsection, as well as the reasonable costs of shipment.

(Added Pub. L. 91-452, title XI, Sec. 1102(a), Oct. 15, 1970, 84 Stat. 955; amended Pub. L. 107-296, title XI, Sec. 1112(e)(3), 1122(c)-(h), 1124, Nov. 25, 2002, 116 Stat. 2276, 2281, 2282, 2285.)

§ 844. Penalties

(a) Any person who—

(1) violates any of subsections (a) through (i) or (l) through (o) of section 842 shall be fined under this title, imprisoned for not more than 10 years, or both; and

(2) violates subsection (p)(2) of section 842, shall be fined under this title, imprisoned not more than 20 years, or both.

(b) Any person who violates any other provision of section 842 of this chapter shall be fined under this title or imprisoned not more than one year, or both.

(c) **(1)** Any explosive materials involved or used or intended to be used in any violation of the provisions of this chapter or any other rule or regulation promulgated thereunder or any violation of any criminal law of the United States shall be subject to seizure and forfeiture, and all provisions of the Internal Revenue Code of 1986 relating to the seizure, forfeiture, and disposition of firearms, as defined in section 5845(a) of that Code, shall, so far as applicable, extend to seizures and forfeitures under the provisions of this chapter.

(2) Notwithstanding paragraph (1), in the case of the seizure of any explosive materials for any offense for which the materials would be subject to forfeiture in which it would be impracticable or unsafe to remove the materials to a place of storage or would be unsafe to store them, the seizing officer may destroy the explosive materials forthwith. Any destruction under this paragraph shall be in the presence of at least 1 credible witness. The seizing officer shall make a report of the seizure and take samples as the Attorney General may by regulation prescribe.

(3) Within 60 days after any destruction made pursuant to paragraph (2), the owner of (including any person having an interest in) the property so destroyed may make application to the Attorney General for reimbursement of the value of the property. If the claimant establishes to the satisfaction of the Attorney General that—

(A) the property has not been used or involved in a violation of law; or

(B) any unlawful involvement or use of the property was without the claimant's knowledge, consent, or willful blindness, the Attorney General shall make an allowance to the claimant not exceeding the value of the property destroyed.

(d) Whoever transports or receives, or attempts to transport or receive, in interstate or foreign commerce any explosive with the knowledge or intent that it will be used to kill, injure, or intimidate any individual or unlawfully to damage or destroy any building, vehicle, or other real or personal property, shall be imprisoned for not more than ten years, or fined under this title, or both; and if personal injury results to any person, including any public safety officer performing duties as a direct or proximate result of conduct prohibited by this subsection, shall be imprisoned for not more than twenty years or fined under this title, or both; and if death results to any person, including any public safety officer performing duties as a direct or proximate result of conduct prohibited by this subsection, shall be subject to imprisonment for any term of years, or to the death penalty or to life imprisonment.

(e) Whoever, through the use of the mail, telephone, telegraph, or other instrument of interstate or foreign commerce, or in or affecting interstate or foreign commerce, willfully makes any threat, or maliciously conveys false information knowing the same to be false, concerning an attempt or alleged attempt being made, or to be made, to kill, injure, or intimidate any individual or unlawfully to damage or destroy any building, vehicle, or other real or personal property by means of fire or an explosive shall be imprisoned for not more than 10 years or fined under this title, or both.

(f) **(1)** Whoever maliciously damages or destroys, or attempts to damage or destroy, by means of fire or an explosive, any building, vehicle, or other personal or real property in whole or in part owned or possessed by, or leased to, the United States, or any department or agency thereof, or any institution or organization receiving Federal financial assistance, shall be imprisoned for not less than 5 years and not more than 20 years, fined under this title, or both.

(2) Whoever engages in conduct prohibited by this subsection, and as a result of such conduct, directly or proximately causes personal injury or creates a substantial risk of injury to any person, including any public safety officer performing duties, shall be imprisoned for not less than 7 years and not more than 40 years, fined under this title, or both.

(3) Whoever engages in conduct prohibited by this subsection, and as a result of such conduct directly or proximately causes the death of any person, including any public safety officer performing duties, shall be subject to the death penalty, or imprisoned for not less than 20 years or for life, fined under this title, or both.

(g) **(1)** Except as provided in paragraph (2), whoever possesses an explosive in an airport that is subject to the regulatory authority of the Federal Aviation Administration, or in any building in whole or in part owned, possessed, or used by, or leased to, the United States or any department or agency thereof, except with the written consent of the agency, department, or other person responsible for the management of such building or airport, shall be imprisoned for not more than five years, or fined under this title, or both.

(2) The provisions of this subsection shall not be applicable to—

(A) the possession of ammunition (as that term is defined in regulations issued pursuant to this chapter) in an airport that is subject to the regulatory authority of the Federal Aviation Administration if such ammunition is either in checked baggage or in a closed container; or

(B) the possession of an explosive in an airport if the packaging and transportation of such explosive is exempt from, or subject to and in accordance with, regulations of the Pipeline and Hazardous Materials Safety Administration for the handling of hazardous materials pursuant to chapter 51 of title 49.

(h) Whoever—

(1) uses fire or an explosive to commit any felony which may be prosecuted in a court of the United States, or

(2) carries an explosive during the commission of any felony which may be prosecuted in a court of the United States, including a felony which provides for an enhanced punishment if committed by the use of a deadly or dangerous weapon or device shall, in addition to the punishment provided for such felony, be sentenced to imprisonment for 10 years. In the case of a second or subsequent conviction under this subsection, such person shall be sentenced to imprisonment for 20 years. Notwithstanding any other provision of law, the court shall not place on probation or suspend the sentence of any person convicted of a violation of this subsection, nor shall the term of imprisonment imposed under this subsection run concurrently with any other term of imprisonment including that imposed for the felony in which the explosive was used or carried.

(i) Whoever maliciously damages or destroys, or attempts to damage or destroy, by means of fire or an explosive, any building, vehicle, or other real or personal property used in interstate or foreign commerce or in any activity affecting interstate or foreign commerce shall be imprisoned for not less than 5 years and not more than 20 years, fined under this title, or both; and if personal injury results to any person, including any public safety officer performing duties as a direct or proximate result of conduct prohibited by this subsection, shall be imprisoned for not less than 7 years and not more than 40 years, fined under this title, or both; and if death results to any person, including any public safety officer performing duties as a direct or proximate result of conduct prohibited by this subsection, shall also be subject to imprisonment for any term of years, or to the death penalty or to life imprisonment.

(j) For the purposes of subsections (d), (e), (f), (g), (h), and (i) of this section and section 842(p), the term "explosive" means gunpowders, powders used for blasting, all forms of high explosives, blasting materials, fuses (other than electric circuit breakers), detonators, and other detonating agents, smokeless powders, other explosive or incendiary devices within the meaning of paragraph (5) of section 232 of this title, and any chemical compounds, mechanical mixture, or device that contains any oxidizing and combustible units, or other ingredients, in such proportions, quantities, or packing that ignition by fire, by friction, by concussion, by percussion, or by detonation of the compound, mixture, or device or any part thereof may cause an explosion.

(k) A person who steals any explosives materials which are moving as, or are a part of, or which have moved in, interstate or foreign commerce shall be imprisoned for not more than 10 years, fined under this title, or both.

(l) A person who steals any explosive material from a licensed importer, licensed manufacturer, or licensed dealer, or from any permittee shall be fined under this title, imprisoned not more than 10 years, or both.

(m) A person who conspires to commit an offense under subsection (h) shall be imprisoned for any term of years not exceeding 20, fined under this title, or both.

(n) Except as otherwise provided in this section, a person who conspires to commit any offense defined in this chapter shall be subject to the same penalties (other than the penalty of death) as the penalties prescribed for the offense the commission of which was the object of the conspiracy.

(o) Whoever knowingly transfers any explosive materials, knowing or having reasonable cause to believe that such explosive materials will be used to commit a crime of violence (as defined in section 924(c)(3)) or drug trafficking crime (as defined in section 924(c)(2)) shall be subject to the same penalties as may be imposed under subsection (h) for a first conviction for the use or carrying of an explosive material.

(p) Theft reporting requirement.

(1) In general. A holder of a license or permit who knows that explosive materials have been stolen from that licensee or permittee, shall report the theft to the Attorney General not later than 24 hours after the discovery of the theft.

(2) Penalty. A holder of a license or permit who does not report a theft in accordance with paragraph (1), shall be fined not more than $ 10,000, imprisoned not more than 5 years, or both.

(Added Pub. L. 91-452, title XI, Sec. 1102(a), Oct. 15, 1970, 84 Stat. 956; amended Pub. L. 97-298, Sec. 2, Oct. 12, 1982, 96 Stat. 1319; Pub. L. 98-473, title II, Sec. 1014, Oct. 12, 1984, 98 Stat. 2142; Pub. L. 99-514, Sec. 2, Oct. 22, 1986, 100 Stat. 2095; Pub. L. 100-690, title VI, Sec. 6474(a), (b), Nov. 18, 1988, 102 Stat. 4379; Pub. L. 101-647, title XXXV, Sec. 3522, Nov. 29, 1990, 104 Stat. 4924; Pub. L. 103-272, Sec. 5(e)(7), July 5, 1994, 108 Stat. 1374; Pub. L. 103-322, title VI, Sec. 60003(a)(3), title XI, Sec. 110504(b), 110509, 110515(b), 110518(b), title XXXII, Sec. 320106, 320917(a), title XXXIII, Sec. 330016(1)(H), (K), (L), (N), Sept. 13, 1994, 108 Stat. 1969, 2016, 2018, 2020, 2111, 2129, 2147, 2148; Pub. L. 104-132, title VI, Sec. 604, title VII, Sec. 701, 706, 708(a), (c)(3), 724, Apr. 24, 1996, 110 Stat. 1289, 1291, 1295-1297, 1300; Pub. L. 104-294, title VI, Sec. 603(a), Oct. 11, 1996, 110 Stat. 3503; Pub. L. 106-54, Sec. 2(b), Aug. 17, 1999, 113 Stat. 399; Pub. L. 107-296, title XI, Sec. 1112(e)(3), 1125, 1127, Nov. 25, 2002, 116 Stat. 2276, 2285.)

§ 845. Exceptions; relief from disabilities

(a) Except in the case of subsections (l), (m), (n), or (o) of section 842 and subsections (d), (e), (f), (g), (h), and (i) of section 844 of this title, this chapter shall not apply to:

(1) aspects of the transportation of explosive materials via railroad, water, highway, or air that pertain to safety, including security, and are regulated by the United States Department of Transportation or the Department of Homeland Security;

(2) the use of explosive materials in medicines and medicinal agents in the forms prescribed by the official United States Pharmacopeia, or the National Formulary;

(3) the transportation, shipment, receipt, or importation of explosive materials for delivery to any agency of the United States or to any State or political subdivision thereof;

(4) small arms ammunition and components thereof;

(5) commercially manufactured black powder in quantities not to exceed fifty pounds, percussion caps, safety and pyrotechnic fuses, quills, quick and slow matches, and friction primers, intended to be used solely for sporting, recreational, or cultural purposes in antique firearms as defined in section 921(a)(16) of title 18 of the United States Code, or in antique devices as exempted from the term "destructive device" in section 921(a)(4) of title 18 of the United States Code; and

(6) the manufacture under the regulation of the military department of the United States of explosive materials for, or their distribution to or storage or possession by the military or naval services or other agencies of the United States; or to arsenals, navy yards, depots, or other establishments owned by, or operated by or on behalf of, the United States.

(b) **(1)** A person who is prohibited from shipping, transporting, receiving, or possessing any explosive under section 842(i) may apply to the Secretary for relief from such prohibition.

(2) The Secretary may grant the relief requested under paragraph (1) if the Secretary determines that the circumstances regarding the applicability of section 842(i), and the applicant's record and reputation, are such that the applicant will not be likely to act in a manner dangerous to public safety and that the granting of such relief is not contrary to the public interest.

(3) A licensee or permittee who applies for relief, under this subsection, from the disabilities incurred under this chapter as a result of an indictment for or conviction of a crime punishable by imprisonment for a term exceeding 1 year shall not be barred by such disability from further operations under the license or permit pending final action on an application for relief filed pursuant to this section.

(c) It is an affirmative defense against any proceeding involving subsections (l) through (o) of section 842 if the proponent proves by a preponderance of the evidence that the plastic explosive—

(1) consisted of a small amount of plastic explosive intended for and utilized solely in lawful—

(A) research, development, or testing of new or modified explosive materials;

(B) training in explosives detection or development or testing of explosives detection equipment; or

(C) forensic science purposes; or

(2) was plastic explosive that, within 3 years after the date of enactment of the Antiterrorism and Effective Death Penalty Act of 1996 [enacted April 24, 1996], will be or is incorporated in a military device within the territory of the United States and remains an integral part of such military device, or is intended to be, or is incorporated in, and remains an integral part of a military device that is intended to become, or has become, the property of any agency of the United States performing military or police functions (including any military reserve component) or the National Guard of any State, wherever such device is located.

(3) For purposes of this subsection, the term "military device" includes, but is not restricted to, shells, bombs, projectiles, mines, missiles, rockets, shaped charges, grenades, perforators, and similar devices lawfully manufactured exclusively for military or police purposes.

(Added Pub. L. 91-452, title XI, Sec. 1102(a), Oct. 15. 1970, 84 Stat. 958; amended Pub. L. 93-639, Sec. 101, Jan. 4, 1975, 88 Stat. 2217; Pub. L. 104-132, title VI, Sec. 605, Apr. 24, 1996, 110 Stat. 1289; Pub. L. 107-296, title XI, Sec. 1126, Nov. 25, 2002, 116 Stat. 2285; Pub. L. 109-59, title VII, Sec. 7127, Aug. 10, 2005, 119 Stat. 1909.)

§ 846. Additional powers of the Attorney General

(a) The Attorney General is authorized to inspect the site of any accident, or fire, in which there is reason to believe that explosive materials were involved, in order that if any such incident has been brought about by accidental means, precautions may be taken to prevent similar accidents from occurring. In order to carry out the purpose of this subsection, the Attorney General is authorized to enter into or upon any property where explosive materials have been used, are suspected of having been used, or have been found in an otherwise unauthorized location. Nothing in this chapter shall be construed as modifying or otherwise affecting in any way the investigative authority of any other Federal agency. In addition to any other investigatory authority they have with respect to violations of provisions of this, the Federal Bureau of Investigation, together with the Bureau of Alcohol, Tobacco, Firearms, and Explosives, shall have authority to conduct investigations with respect to violations of subsection (d), (e), (f), (g), (h), or (i) of section 844 of this title.

(b) The Attorney General is authorized to establish a national repository of information on incidents involving arson and the suspected criminal misuse of explosives. All Federal agencies having information concerning such incidents shall report the information to the Attorney General pursuant to such regulations as deemed necessary to carry out the provisions of this subsection. The repository shall also contain information on incidents voluntarily reported to the Attorney General by State and local authorities.

(Added Pub. L. 91-452, title XI, Sec. 1102(a), Oct. 15, 1970, 84 Stat. 959; amended Pub. L. 104-208, div. A, title I, Sec. 101(f) (title VI, Sec. 654(a)), Sept. 30, 1996, 110 Stat. 3009-314, 3009-369; Pub. L. 107-296, title XI, Sec. 1112(e)(2), (3), Nov. 25, 2002, 116 Stat. 2276.)

§ 847. Rules and regulations

The administration of this chapter shall be vested in the Attorney General. The Attorney General may prescribe such rules and regulations as he deems reasonably necessary to carry out the provisions of this chapter. The Attorney General shall give reasonable public notice, and afford to interested parties opportunity for hearing, prior to prescribing such rules and regulations.

(Added Pub. L. 91-452, title XI, Sec. 1102(a), Oct. 15, 1970, 84 Stat. 959; amended Pub. L. 107-296, title XI, Sec. 1112(e)(3), Nov. 25, 2002, 116 Stat. 2276.)

§ 848. Effect on State Law

No provision of this chapter shall be construed as indicating an intent on the part of the Congress to occupy the field in which such provision operates to the exclusion of the law of any State on the same subject matter, unless there is a direct and positive conflict between such provision and the law of the State so that the two cannot be reconciled or consistently stand together.

(Added Pub. L. 91-452, title XI, Sec. 1102(a), Oct. 15, 1970, 84 Stat. 959.)

Title 27, Code of Federal Regulations
Part 555—Commerce in Explosives

Section Contents

Subpart H— Exemptions

Subpart I—Unlawful Acts, Penalties, Seizures and Forfeitures

Subpart J—Marking of Plastic Explosives

Subpart K—Storage

Authority: 18 U.S.C. 847.

Source: T.D. ATF–87, 46 FR 40384, Aug. 7, 1981, unless otherwise noted.

Editor's Note: There are several name changes for titles referenced in the regulations, 27 CFR Part 555. The title "Chief, Firearms and Explosives Licensing Center" has been changed to "Chief, National Explosives Licensing Center". Also, the title "Regional director (compliance)" has been changed to "Director, Industry Operations" within the respective ATF Field Divisions. All regulations and rulings contained within this publication were current as of 1/26/07. For more recent information, please refer to the ATF website at www.atf.gov.

Subpart A—Introduction

§ 555.1 Scope of regulations.

(a) In general. The regulations contained in this part relate to commerce in explosives and implement Title XI, Regulation of Explosives (18 U.S.C. Chapter 40; 84 Stat. 952), of the Organized Crime Control Act of 1970 (84 Stat. 922), Pub. L. 103–322 (108 Stat. 1796), Pub. L. 104–132 (110 Stat. 1214), and Pub. L. 107–296 (116 Stat. 2135).

(b) Procedural and substantive requirements. This part contains the procedural and substantive requirements relative to:

(1) The interstate or foreign commerce in explosive materials;

(2) The licensing of manufacturers and importers of, and dealers in, explosive materials;

(3) The issuance of permits;

(4) The conduct of business by licensees and operations by permittees;

(5) The storage of explosive materials;

(6) The records and reports required of licensees and permittees;

(7) Relief from disabilities under this part;

(8) Exemptions, unlawful acts, penalties, seizures, and forfeitures; and

(9) The marking of plastic explosives.

[T.D. ATF–87, 46 FR 40384, Aug. 7, 1981, as amended by T.D. ATF–363, 60 FR 17449, Apr. 6, 1995; T.D. ATF–387, 62 FR 8376, Feb. 25, 1997; ATF No. 1, 68 FR 13780, Mar. 20, 2003]

§ 555.2 Relation to other provisions of law.

The provisions in this part are in addition to, and are not in lieu of, any other provision of law, or regulations, respecting commerce in explosive materials. For regulations applicable to commerce in firearms and ammunition, see Part 478 of this chapter. For regulations applicable to traffic in machine guns, destructive devices, and certain other firearms, see Part 479 of this chapter. For statutes applicable to the registration and licensing of persons engaged in the business of manufacturing, importing or exporting arms, ammunition, or implements of war, see section 38 of the Arms Export Control Act (22 U.S.C. 2778), and regulations of Part 447 of this chapter and in Parts 121 through 128 of Title 22, Code of Federal Regulations. For statutes applicable to nonmailable materials, see 18 U.S.C. 1716 and implementing regulations. For statutes applicable to water quality standards, see 33 U.S.C. 1341.

Subpart B—Definitions

§ 555.11 Meaning of terms.

When used in this part, terms are defined as follows in this section. Words in the plural form include the singular, and vice versa, and words indicating the masculine gender include the feminine. The terms "includes" and "including" do not exclude other things not named which are in the same general class or are otherwise within the scope of the term defined.

Act. 18 U.S.C. Chapter 40.

Adjudicated as a mental defective. (a) A determination by a court, board, commission, or other lawful authority that a person, as a result of marked subnormal intelligence, or mental illness, incompetency, condition, or disease:

(1) Is a danger to himself or to others; or

(2) Lacks the mental capacity to contract or manage his own affairs.

(b) The term will include—

(1) A finding of insanity by a court in a criminal case; and

(2) Those persons found incompetent to stand trial or found not guilty by reason of lack of mental responsibility by any court or pursuant to articles 50a and 76b of the Uniform Code of Military Justice, 10 U.S.C. 850a, 876b.

Alien. Any person who is not a citizen or national of the United States.

Ammunition. Small arms ammunition or cartridge cases, primers, bullets, or smokeless propellants designed for use in small arms, including percussion caps, and 3/32 inch and other external burning pyrotechnic hobby fuses. The term does not include black powder.

Appropriate identifying information. The term means, in relation to an individual:

(a) The full name, date of birth, place of birth, sex, race, street address, State of residence, telephone numbers (home and work), country or countries of citizenship, and position at the employer's business or operations of responsible persons and employees authorized to possess explosive materials;

(b) The business name, address, and license or permit number with which the responsible person or employee is affiliated;

(c) If an alien, INS-issued alien number or admission number; and

(d) Social security number, as optional information (this information is not required but is helpful in avoiding misidentification when a background check is conducted).

Approved storage facility. A place where explosive materials are stored, consisting of one or more approved magazines, conforming to the requirements of this part and covered by a license or permit issued under this part.

Articles pyrotechnic. Pyrotechnic devices for professional use similar to consumer fireworks in chemical composition and construction but not intended for consumer use. Such articles meeting the weight limits for consumer fireworks but not labeled as such and classified by U.S. Department of Transportation regulations in 49 CFR 172.101 as UN0431 or UN0432.

Artificial barricade. An artificial mound or revetted wall of earth of a minimum thickness of three feet, or any other approved barricade that offers equivalent protection.

ATF. (a) Prior to January 24, 2003. The Bureau of Alcohol, Tobacco and Firearms, Department of the Treasury, Washington, DC.

(b) On and after January 24, 2003. The Bureau of Alcohol, Tobacco, Firearms and Explosives, Department of Justice, Washington, DC.

ATF officer. (a) Prior to January 24, 2003. An officer or employee of the Bureau of Alcohol, Tobacco and Firearms (ATF) authorized to perform any function relating to the administration or enforcement of this part.

(b) On and after January 24, 2003. An officer or employee of the Bureau of Alcohol, Tobacco, Firearms and Explosives (ATF) authorized to perform any function relating to the administration or enforcement of this part.

Authority having jurisdiction for fire safety. The fire department having jurisdiction over sites where explosives are manufactured or stored.

Barricaded. The effective screening of a magazine containing explosive materials from another magazine, a building, a railway, or a highway, either by a natural barricade or by an artificial barricade. To be properly barricaded, a straight line from the top of any sidewall of the magazine containing explosive materials to the eave line of any other magazine or building, or to a point 12 feet above the center of a railway or highway, will pass through the natural or artificial barricade.

Blasting agent. Any material or mixture, consisting of fuel and oxidizer, that is intended for blasting and not otherwise defined as an explosive; if the finished product, as mixed for use or shipment, cannot be detonated by means of a number 8 test blasting cap when unconfined. A number 8 test blasting cap is one containing 2 grams of a mixture of 80 percent mercury fulminate and 20 percent potassium chlorate, or a blasting cap of equivalent strength. An equivalent strength cap comprises 0.40–0.45 grams of PETN base charge pressed in an aluminum shell with bottom thickness not to exceed to 0.03 of an inch, to a specific gravity of not less than 1.4 g/cc., and primed with standard weights of primer depending on the manufacturer.

Bulk salutes. Salute components prior to final assembly into aerial shells, and finished salute shells held separately prior to being packed with other types of display fireworks.

Bullet-sensitive explosive materials. Explosive materials that can be exploded by 150-grain M2 ball ammunition having a nominal muzzle velocity of 2700 fps (824 mps) when fired from a .30 caliber rifle at a distance of 100 ft (30.5 m), measured perpendicular. The test material is at a temperature of 70 to 75 degrees F (21 to 24 degrees C) and is placed against a ½ inch (12.4 mm) steel backing plate.

Bureau. (a) Prior to January 24, 2003. The Bureau of Alcohol, Tobacco and Firearms, Department of the Treasury, Washington, DC.

(b) On and after January 24, 2003. The Bureau of Alcohol, Tobacco, Firearms and Explosives, Department of Justice, Washington, DC.

Business premises. When used with respect to a manufacturer, importer, or dealer, the property on which explosive materials are manufactured, imported, stored or distributed. The premises include the property where the records of a manufacturer, importer, or dealer are kept if different than the premises where explosive materials are manufactured, imported, stored or distributed. When used with respect to a user of explosive materials, the property on which the explosive materials are received or stored. The premises includes the property where the records of the users are kept if different than the premises where explosive materials are received or stored.

Chief, Firearms and Explosives Licensing Center. The ATF official responsible for the issuance and renewal of licenses and permits under this part.

Committed to a mental institution. A formal commitment of a person to a mental institution by a court, board, commission, or other lawful authority. The term includes a commitment to a mental institution involuntarily. The term includes commitment for mental defectiveness or mental illness. It also includes commitments for other reasons, such as for drug use. The term does not include a person in a mental institution for observation or a voluntary admission to a mental institution.

Common or contract carrier. Any individual or organization engaged in the business of transporting passengers or goods.

Consumer fireworks. Any small firework device designed to produce visible effects by combustion and which must comply with the construction, chemical composition, and labeling regulations of the U.S. Consumer Product Safety Commission, as set forth in title 16, Code of Federal Regulations, parts 1500 and 1507. Some small devices designed to produce audible effects are included, such as whistling devices, ground devices containing 50 mg or less of explosive materials, and aerial devices containing 130 mg or less of explosive materials. Consumer fireworks are classified as fireworks UN0336, and UN0337 by the U.S. Department of Transportation at 49 CFR 172.101. This term does not include fused setpieces containing components which together exceed 50 mg of salute powder.

Controlled substance. A drug or other substance, or immediate precursor, as defined in section 102 of the Controlled Substances Act, 21 U.S.C. 802. The term includes, but is not limited to, marijuana, depressants, stimulants, and narcotic drugs. The term does not include distilled spirits, wine, malt beverages, or tobacco, as those terms are defined or used in Subtitle E of the Internal Revenue Code of 1986, as amended.

Crime punishable by imprisonment for a term exceeding one year. Any offense for which the maximum penalty, whether or not imposed, is capital punishment or imprisonment in excess of one year. The term does not include (a) any Federal or State offenses pertaining to antitrust violations, unfair trade practices, restraints of trade, or (b) any State offense (other than one involving a firearm or explosive) classified by the laws of the State as a misdemeanor and punishable by a term of imprisonment of two years or less.

Customs officer. Any officer of the Customs Service or any commissioned, warrant, or petty officer of the Coast Guard, or any agent or other person authorized to perform the duties of an officer of the Customs Service.

Dealer. Any person engaged in the business of distributing explosive materials at wholesale or retail.

Detonator. Any device containing a detonating charge that is used for initiating detonation in an explosive. The term includes, but is not limited to, electric blasting caps of instantaneous and delay types, blasting caps for use with safety fuses, detonating-cord delay connectors, and nonelectric instantaneous and delay blasting caps.

Director. (a) Prior to January 24, 2003. The Director, Bureau of Alcohol, Tobacco and Firearms, Department of the Treasury, Washington, DC. (b) On and after January 24, 2003. The Director, Bureau of Alcohol, Tobacco, Firearms and Explosives, Department of Justice, Washington, DC.

Discharged under dishonorable conditions. Separation from the U.S. Armed Forces resulting from a dishonorable discharge or dismissal adjudged by general court-martial. The term does not include any separation from the Armed Forces resulting from any other discharge, e.g., a bad conduct discharge.

Display fireworks. Large fireworks designed primarily to produce visible or audible effects by combustion, deflagration, or detonation. This term includes, but is not limited to, salutes containing more than 2 grains (130 mg) of explosive materials, aerial shells containing more than 40 grams of pyrotechnic compositions, and other display pieces which exceed the limits of explosive materials for classification as "consumer fireworks." Display fireworks are classified as fireworks UN0333, UN0334 or UN0335 by the U.S. Department of Transportation at 49 CFR 172.101. This term also includes fused setpieces containing components which together exceed 50 mg of salute powder.

Distribute. To sell, issue, give, transfer, or otherwise dispose of. The term does not include a mere change of possession from a person to his agent or employee in connection with the agency or employment.

Executed under penalties of perjury. Signed with the required declaration under the penalties of perjury as provided on or with respect to the return, form, or other document or, where no form of declaration is required, with the declaration:

"I declare under the penalties of perjury that this —(insert type of document, such as, statement, application, request, certificate), including the documents submitted in support thereof, has been examined by me and, to the best of my knowledge and belief, is true, correct, and complete".

Explosive actuated device. Any tool or special mechanized device which is actuated by explosives, but not a propellent actuated device.

Explosive materials. Explosives, blasting agents, water gels and detonators. Explosive materials include, but are not limited to, all items in the "List of Explosive Materials" provided for in §555.23.

Explosives. Any chemical compound, mixture, or device, the primary or common purpose of which is to function by explosion. The term includes, but is not limited to, dynamite and other high explosives, black powder, pellet powder, initiating explosives, detonators, safety fuses, squibs, detonating cord, igniter cord, and igniters.

Fireworks. Any composition or device designed to produce a visible or an audible effect by combustion, deflagration, or detonation, and which meets the definition of "consumer fireworks" or "display fireworks" as defined by this section.

Fireworks mixing building. Any building or area used for mixing and blending pyrotechnic compositions except wet sparkler mix.

Fireworks nonprocess building. Any office building or other building or area in a fireworks plant where no fireworks, pyrotechnic compositions or explosive materials are processed or stored.

Fireworks plant. All land and buildings thereon used for or in connection with the assembly or processing of fireworks, including warehouses used with or in connection with fireworks plant operations.

Fireworks plant warehouse. Any building or structure used exclusively for the storage of materials which are neither explosive materials nor pyrotechnic compositions used to manufacture or assemble fireworks.

Fireworks process building. Any mixing building; any building in which pyrotechnic compositions or explosive materials is pressed or otherwise prepared for finished and assembly; or any finishing or assembly building.

Fireworks shipping building. A building used for the packing of assorted display fireworks into shipping cartons for individual public displays and for the loading of packaged displays for shipment to purchasers.

Flash powder. An explosive material intended to produce an audible report and a flash of light when ignited which includes but is not limited to oxidizers such as potassium chlorate or potassium perchlorate, and fuels such as sulfur or aluminum powder.

Fugitive from justice. Any person who has fled from the jurisdiction of any court of record to avoid prosecution for any crime or to avoid giving testimony in any criminal proceeding. The term also includes any person who has been convicted of any crime and has fled to avoid imprisonment.

Hardwood. Oak, maple, ash, hickory, or other hard wood, free from loose knots, spaces, or similar defects.

Highway. Any public street, public alley, or public road, including a privately financed, constructed, or maintained road that is regularly and openly traveled by the general public.

Identification document. A document containing the name, residence address, date of birth, and photograph of the holder and which was made or issued by or under the authority of the United States Government, a State, political subdivision of a State, a foreign government, a political subdivision of a foreign government, an international governmental or an international quasi-governmental organization which, when completed with information concerning a particular individual, is of a type intended or commonly accepted for the purpose of identification of individuals.

Importer. Any person engaged in the business of importing or bringing explosive materials into the United States for purposes of sale or distribution.

Indictment. Includes an indictment or information in any court under which a crime punishable by imprisonment for a term exceeding one year may be prosecuted.

Inhabited building. Any building regularly occupied in whole or in part as a habitation for human beings, or any church, schoolhouse, railroad station, store, or other structure where people are accustomed to assemble, except any building occupied in connection with the manufacture, transportation, storage, or use of explosive materials.

Interstate or foreign commerce. Commerce between any place in a State and any place outside of that State, or within any possession of the United States or the District of Columbia, and commerce between places within the same State but through any place outside of that State.

Licensed dealer. A dealer licensed under this part.

Licensed importer. An importer licensed under this part.

Licensed manufacturer. A manufacturer licensed under this part to engage in the business of manufacturing explosive materials for purposes of sale or distribution or for his own use.

Licensee. Any importer, manufacturer, or dealer licensed under this part.

Limited permit. A permit issued to a person authorizing him to receive for his use explosive materials from a licensee or permittee in his state of residence on no more than 6 occasions during the 12-month period in which the permit is valid. A limited permit does not authorize the receipt or transportation of explosive materials in interstate or foreign commerce.

Magazine. Any building or structure, other than an explosives manufacturing building, used for storage of explosive materials.

Manufacturer. Any person engaged in the business of manufacturing explosive materials for purposes of sale or distribution or for his own use.

Mass detonation (mass explosion). Explosive materials mass detonate (mass explode) when a unit or any part of a larger quantity of explosive material explodes and causes all or a substantial part of the remaining material to detonate or explode.

Mental institution. Includes mental health facilities, mental hospitals, sanitariums, psychiatric facilities, and other facilities that provide diagnoses by licensed professionals of mental retardation or mental illness, including a psychiatric ward in a general hospital.

Natural barricade. Natural features of the ground, such as hills, or timber of sufficient density that the surrounding exposures which require protection cannot be seen from the magazine when the trees are bare of leaves.

Number 8 test blasting cap. (See definition of "blasting agent.")

Permittee. Any user of explosives for a lawful purpose who has obtained either a user permit or a limited permit under this part.

Person. Any individual, corporation, company, association, firm, partnership, society, or joint stock company.

Plywood. Exterior, construction grade (laminated wood) plywood.

Propellant actuated device. Any tool or special mechanized device or gas generator system which is actuated by a propellant or which releases and directs work through a propellant charge.

Pyrotechnic compositions. A chemical mixture which, upon burning and without explosion, produces visible, brilliant displays, bright lights, or sounds.

Railway. Any steam, electric, or other railroad or railway which carries passengers for hire.

Region. A geographical region of the Bureau of Alcohol, Tobacco and Firearms.

Regional director (compliance). The principal regional official responsible for administering regulations in this part.

Renounced U.S. citizenship. (a) A person has renounced his U.S. citizenship if the person, having been a citizen of the United States, has renounced citizenship either—

(1) Before a diplomatic or consular officer of the United States in a foreign state pursuant to 8 U.S C. 1481(a)(5); or

(2) Before an officer designated by the Attorney General when the United States is in a state of war pursuant to 8 U.S.C. 1481(a)(6).

(b) The term will not include any renunciation of citizenship that has been reversed as a result of administrative or judicial appeal.

Responsible person. An individual who has the power to direct the management and policies of the applicant pertaining to explosive materials. Generally, the term includes partners, sole proprietors, site managers, corporate officers and directors, and majority shareholders.

Salute. An aerial shell, classified as a display firework, that contains a charge of flash powder and is designed to produce a flash of light and a loud report as the pyrotechnic effect.

Screen barricade. Any barrier that will contain the embers and debris from a fire or deflagration in a process building, thus preventing propagation of fire to other buildings or areas. Such barriers shall be constructed of metal roofing, ¼ to ½ inch (6 to 13 mm) mesh screen, or equivalent material. The barrier extends from floor level to a height such that a straight line from the top of any side wall of the donor building to the eave line of any exposed building intercepts the screen at a point not less than 5 feet (1.5 m) from the top of the screen. The top 5 feet (1.5 m) of the screen is inclined towards the donor building at an angle of 30 to 45 degrees.

Softwood. Fir, pine, or other soft wood, free from loose knots, spaces, or similar defects.

State. A State of the United States. The term includes the District of Columbia, the Commonwealth of Puerto Rico, and the possessions of the United States.

State of residence. The State in which an individual regularly resides or maintains his home. Temporary stay in a State does not make the State of temporary stay the State of residence.

Theatrical flash powder. Flash powder commercially manufactured in premeasured kits not exceeding 1 ounce and mixed immediately prior to use and intended for use in theatrical shows, stage plays, band concerts, magic acts, thrill shows, and clown acts in circuses.

Unlawful user of or addicted to any controlled substance. A person who uses a controlled substance and has lost the power of self-control with reference to the use of a controlled substance; and any person who is a current user of a controlled substance in a manner other than as prescribed by a licensed physician. Such use is not limited to the use of drugs on a particular day, or within a matter of days or weeks before possession of the explosive materials, but rather that the unlawful use has occurred recently enough to indicate that the individual is actively engaged in such conduct. A person may be an unlawful current user of a controlled substance even though the substance is not being used at the precise time the person seeks to acquire explosive materials or receives or possesses explosive materials. An inference of current use may be drawn from evidence of a recent use or possession of a controlled substance or a pattern of use or possession that reasonably covers the present time, e.g., a conviction for use or possession of a controlled substance within the past year; multiple arrests for such offenses within the past 5 years if the most recent arrest occurred within the past year; or persons found through a drug test to use a controlled substance unlawfully, provided that the test was administered within the past year. For a current or former member of the Armed Forces, an inference of current use may be drawn from recent disciplinary or other administrative action based on confirmed drug use, e.g., court-martial conviction, nonjudicial punishment, or an administrative discharge based on drug use or drug rehabilitation failure.

U.S.C. The United States Code.

User-limited permit. A user permit valid only for a single purchase transaction, a new permit being required for a subsequent purchase transaction.

User permit. A permit issued to a person authorizing him (a) to acquire for his own use explosive materials from a licensee in a State other than the State in which he resides or from a foreign country, and (b) to transport explosive materials in interstate or foreign commerce.

Water gels. Explosives or blasting agents that contain a substantial proportion of water.

(18 U.S.C. 847 (84 Stat. 959); 18 U.S.C. 926 (82 Stat. 1226)

[T.D. ATF–87, 46 FR 40384, Aug. 7, 1981, as amended by T.D. ATF–92, 46 FR 46916, Sept. 23, 1981; T.D. ATF–200, 50 FR 10497, Mar. 15, 1985; T.D. ATF–290, 54 FR 53053, Dec. 27, 1989; T.D. ATF–293, 55 FR 3720, Feb. 5, 1990; T.D. ATF–314, 56 FR 49140, Sept. 27, 1991; T.D. ATF–382, 61 FR 38084, July 23, 1996; T.D. ATF–400, 63 FR 45001, Aug. 24, 1998; T.D. ATF–446, 66 FR 16602, Mar. 27, 2001; ATF No. 1, 68 FR 13780, Mar. 20, 2003]

Subpart C—Administrative and Miscellaneous Provisions

§ 555.21 Forms prescribed.

(a) The Director is authorized to prescribe all forms required by this part. All of the information called for in each form shall be furnished as indicated by the headings on the form and the instructions on or pertaining to the form. In addition, information called for in each form shall be furnished as required by this part.

(b) Requests for forms should be mailed to the ATF Distribution Center, 7943 Angus Court, Springfield, Virginia 22153.

[T.D. ATF–92, 46 FR 46916, Sept. 23, 1981, as amended by T.D. ATF–249, 52 FR 5961, Feb. 27, 1987; T.D. 372, 61 FR 20724, May 8, 1996]

§ 555.22 Alternate methods or procedures; emergency variations from requirements.

(a) Alternate methods or procedures. The permittee or licensee, on specific approval by the Director as provided by this paragraph, may use an alternate method or procedure in lieu of a method or procedure specifically prescribed in this part. The Director may approve an alternate method or procedure, subject to stated conditions, when he finds that:

(1) Good cause is shown for the use of the alternate method or procedure;

(2) The alternate method or procedure is within the purpose of, and consistent with the effect intended by, the specifically prescribed method or procedure and that the alternate method or procedure is substantially equivalent to that specifically prescribed method or procedure; and

(3) The alternate method or procedure will not be contrary to any provision of law and will not result in an increase in cost to the Government or hinder the effective administration of this part. Where the permittee or licensee desires to employ an alternate method or procedure, he shall submit a written application to the regional director (compliance), for transmittal to the Director. The application shall specifically describe the proposed alternate method or procedure and shall set forth the reasons for it. Alternate methods or procedures may not be employed until the application is approved by the Director. The permittee or licensee shall, during the period of authorization of an alternate method or procedure, comply with the terms of the approved application. Authorization of any alternate method or procedure may be withdrawn whenever, in the judgment of the Director, the effective administration of this part is hindered by the continuation of the authorization. As used in this paragraph, alternate methods or procedures include alternate construction or equipment.

(b) Emergency variations from requirements. The Director may approve construction, equipment, and methods of operation other than as specified in this part, where he finds that an emergency exists and the proposed variations from the specified requirements are necessary and the proposed variations:

(1) Will afford security and protection that are substantially equivalent to those prescribed in this part;

(2) Will not hinder the effective administration of this part; and

(3) Will not be contrary to any provisions of law.

Variations from requirements granted under this paragraph are conditioned on compliance with the procedures, conditions, and limitations set forth in the approval of the application. Failure to comply in good faith with the procedures, conditions, and limitations shall automatically terminate the authority for the variations and the licensee or permittee shall fully comply with the prescribed requirements of regulations from which the variations were authorized. Authority for any variation may be withdrawn whenever, in the judgment of the Director, the effective administration of this part is hindered by the continuation of the variation. Where the licensee or permittee desires to employ an emergency variation, he shall submit a written application to the regional director (compliance) for transmittal to the Director. The application shall describe the proposed variation and set forth the reasons for it. Variations may not be employed until the application is approved, except when the emergency requires immediate action to correct a situation that is threatening to life or property. Corrective action may then be taken concurrent with the filing of the application and notification of the Director via telephone.

(c) Retention of approved variations. The licensee or permittee shall retain, as part of his records available for examination by ATF officers, any application approved by the Director under this section.

§ 555.23 List of explosive materials.

The Director shall compile a list of explosive materials, which shall be published and revised at least annually in the Federal Register. The "List of Explosive Materials" (ATF Publication 5400.8) is available at no cost upon request from the ATF Distribution Center, 7943 Angus Court, Springfield, Virginia 22153.

[T.D. ATF–290, 54 FR 53054, Dec. 27, 1989, as amended by T.D. ATF–446, 66 FR 16602, Mar. 27, 2001]

§ 555.24 Right of entry and examination.

(a) Any ATF officer may enter during business hours the premises, including places of storage, of any licensee or holder of a user permit for the purpose of inspecting or examining any records or documents required to be kept under this part, and any facilities in which explosive materials are kept or stored.

(b) Any ATF officer may inspect the places of storage for explosive materials of an applicant for a limited permit or, in the case of a holder of a limited permit, at the time of renewal of such permit.

(c) The provisions of paragraph (b) of this section do not apply to an applicant for the renewal of a limited permit if an ATF officer has, within the preceding 3 years, verified by inspection that the applicant's place of storage for explosive materials meets the requirements of subpart K of this part.
[ATF No. 1, 68 FR 13781, Mar. 20, 2003]

§ 555.25 Disclosure of information.

Upon receipt of written request from any State or any political subdivision of a State, the regional director (compliance) may make available to the State or political subdivision any information which the regional director (compliance) may obtain under the Act with respect to the identification of persons within the State or political subdivision, who have purchased or received explosive materials, together with a description of the explosive materials.

§ 555.26 Prohibited shipment, transportation, receipt, possession, or distribution of explosive materials.

(a) General. No person, other than a licensee or permittee knowingly may transport, ship, cause to be transported, or receive any explosive materials: Provided, That the provisions of this paragraph (a) do not apply to the lawful purchase by a nonlicensee or nonpermittee of commercially manufactured black powder in quantities not to exceed 50 pounds, if the black powder is intended to be used solely for sporting, recreational, or cultural purposes in antique firearms as defined in 18 U.S.C. 921(a)(16), or in antique devices as exempted from the term "destructive device" in 18 U.S.C. 921(a)(4).

(b) Holders of a limited permit. No person who is a holder of a limited permit may—

(1) Transport, ship, cause to be transported, or receive in interstate or foreign commerce any explosive materials;

(2) Receive explosive materials from a licensee or permittee, whose premises are located outside the State of residence of the limited permit holder; or

(3) Receive explosive materials on more than 6 separate occasions, during the period of the permit, from one or more licensees or permittees whose premises are located within the State of residence of the limited permit holder. (See §555.105(b) for the definition of "6 separate occasions.")

(c) Possession by prohibited persons. No person may ship or transport any explosive material in or affecting interstate or foreign commerce or receive or possess any explosive materials which have been shipped or transported in or affecting interstate or foreign commerce who:

(1) Is under indictment or information for, or who has been convicted in any court of, a crime punishable by imprisonment for a term exceeding one year;

(2) Is a fugitive from justice;

(3) Is an unlawful user of or addicted to any controlled substance (as defined in section 102 of the Controlled Substances Act (21 U.S.C. 802) and §555.11);

(4) Has been adjudicated as a mental defective or has been committed to a mental institution;

(5) Is an alien, other than an alien who—

(i) Is lawfully admitted for permanent residence (as that term is defined in section 101(a)(20) of the Immigration and Nationality Act (8 U.S.C. 1101)); or

(ii) Is in lawful nonimmigrant status, is a refugee admitted under section 207 of the Immigration and Nationality Act (8 U.S.C. 1157), or is in asylum status under section 208 of the Immigration and Nationality Act (8 U.S.C. 1158), and—

(A) Is a foreign law enforcement officer of a friendly foreign government, as determined by the Attorney General in consultation with the Secretary of State, entering the United States on official law enforcement business, and the shipping, transporting, possession, or receipt of explosive materials is in furtherance of this official law enforcement business;

(B) Is a person having the power to direct or cause the direction of the management and policies of a corporation, partnership, or association licensed pursuant to section 843(a) of the Act, and the shipping, transporting, possession, or receipt of explosive materials is in furtherance of such power;

(C) Is a member of a North Atlantic Treaty Organization (NATO) or other friendly foreign military force, as determined by the Attorney General in consultation with the Secretary of Defense, (whether or not admitted in a nonimmigrant status) who is present in the United States under military orders for training or other military purpose authorized by the United States, and the shipping, transporting, possession, or receipt of explosive materials is in furtherance of the military purpose; or

(D) Is lawfully present in the United States in cooperation with the Director of Central Intelligence, and the shipment, transportation, receipt, or possession of the explosive materials is in furtherance of such cooperation;

(6) Has been discharged from the armed forces under dishonorable conditions; or

(7) Having been a citizen of the United States, has renounced citizenship.

(d) Distribution to prohibited persons. No person may knowingly distribute explosive materials to any individual who:

(1) Is under twenty-one years of age;

(2) Is under indictment or information for, or who has been convicted in any court of, a crime punishable by imprisonment for a term exceeding one year;

(3) Is a fugitive from justice;

(4) Is an unlawful user of or addicted to any controlled substance (as defined in section 102 of the Controlled Substances Act (21 U.S.C. 802) and §555.11);

(5) Has been adjudicated as a mental defective or has been committed to a mental institution;

(6) Is an alien, other than an alien who—

(i) Is lawfully admitted for permanent residence (as that term is defined in section 101(a)(20) of the Immigration and Nationality Act (8 U.S.C. 1101)); or

(ii) Is in lawful nonimmigrant status, is a refugee admitted under section 207 of the Immigration and Nationality Act (8 U.S.C. 1157), or is in asylum status under section 208 of the Immigration and Nationality Act (8 U.S.C. 1158), and—

(A) Is a foreign law enforcement officer of a friendly foreign government, as determined by the Attorney General in consultation with the Secretary of State, entering the United States on official law enforcement business, and the shipping, transporting, possession, or receipt of explosive materials is in furtherance of this official law enforcement business;

(B) Is a person having the power to direct or cause the direction of the management and policies of a corporation, partnership, or association licensed pursuant to section 843(a) of the Act, and the shipping, transporting, possession, or receipt of explosive materials is in furtherance of such power;

(C) Is a member of a North Atlantic Treaty Organization (NATO) or other friendly foreign military force, as determined by the Attorney General in consultation with the Secretary of Defense, (whether or not admitted in a nonimmigrant status) who is present in the United States under military orders for training or other military purpose authorized by the United States, and the shipping, transporting, possession, or receipt of explosive materials is in furtherance of the military purpose; or

(D) Is lawfully present in the United States in cooperation with the Director of Central Intelligence, and the shipment, transportation, receipt, or possession of the explosive materials is in furtherance of such cooperation;

(7) Has been discharged from the armed forces under dishonorable conditions; or

(8) Having been a citizen of the United States, has renounced citizenship.

(e) See §555.180 for regulations concerning the prohibited manufacture, importation, exportation, shipment, transportation, receipt, transfer, or possession of plastic explosives that do not contain a detection agent.

[ATF No. 1, 68 FR 13781, Mar. 20, 2003]

§ 555.27 Out-of-State disposition of explosive materials.

(a) No nonlicensee or nonpermittee may distribute any explosive materials to any other nonlicensee or nonpermittee who the distributor knows or who has reasonable cause to believe does not reside in the State in which the distributor resides.

(b) The provisions of this section do not apply on and after May 24, 2003.

[ATF No. 1, 68 FR 13782, Mar. 20, 2003]

§ 555.28 Stolen explosive materials.

No person shall receive, conceal, transport, ship, store, barter, sell, or dispose of any stolen explosive materials knowing or having reasonable cause to believe that the explosive materials were stolen.

§ 555.29 Unlawful storage.

No person shall store any explosive materials in a manner not in conformity with this part.

§ 555.30 Reporting theft or loss of explosive materials.

(a) Any licensee or permittee who has knowledge of the theft or loss of any explosive materials from his stock shall, within 24 hours of discovery, report the theft or loss by telephoning 1–800–800–3855 (nationwide toll free number) and on ATF F 5400.5 (formerly Form 4712) in accordance with the instructions on the form. Theft or loss of any explosive materials shall also be reported to appropriate local authorities.

(b) Any other person, except a carrier of explosive materials, who has knowledge of the theft or loss of any explosive materials from his stock shall, within 24 hours of discovery, report the theft or loss by telephoning 1–800–800–3855 (nationwide toll free number) and in writing to the nearest ATF office. Theft or loss shall be reported to appropriate local authorities.

(c) Reports of theft or loss of explosive materials under paragraphs (a) and (b) of this section must include the following information, if known:

(1) The manufacturer or brand name.

(2) The manufacturer's marks of identification (date and shift code).

(3) Quantity (applicable quantity units, such as pounds of explosives, number of detonators, etc.).

(4) Description (dynamite, blasting agents, detonators, etc.) and United Nations (UN) identification number, hazard division number, and classification letter, e.g., 1.1D, as classified by the U.S. Department of Transportation at 49 CFR 172.101 and 173.52.

(5) Size (length and diameter).

(d) A carrier of explosive materials who has knowledge of the theft or loss of any explosive materials shall, within 24 hours of discovery, report the theft or loss by telephoning 1–800–800–3855 (nationwide toll free number). Theft or loss shall also be reported to appropriate local authorities. Reports of theft or loss of explosive materials by carriers shall include the following information, if known:

(1) The manufacturer or brand name.

(2) Quantity (applicable quantity units, such as pounds of explosives, number of detonators, etc.).

(3) Description (United Nations (UN) identification number, hazard division number, and classification letter, e.g., 1.1D) as classified by the U.S. Department of Transportation at 49 CFR 172.101 and 173.52.

[T.D. ATF–87, 46 FR 40384, Aug. 7, 1981, as amended by T.D. ATF–400, 63 FR 45002, Aug. 24, 1998]

§ 555.31 Inspection of site accidents or fires; right of entry.

Any ATF officer may inspect the site of any accident or fire in which there is reason to believe that explosive materials were involved. Any ATF officer may enter into or upon any property where explosive materials have been used, are suspected of having been used, or have been found in an otherwise unauthorized location.

§ 555.32 Special explosive devices.

The Director may exempt certain explosive actuated devices, explosive actuated tools, or similar devices from the requirements of this part. A person who desires to obtain an exemption under this section for any special explosive device, which as designed does not constitute a public safety or security hazard, shall submit a written request to the Director. Each request shall be executed under the penalties of perjury and contain a complete and accurate description of the device, the name and address of the manufacturer or importer, the purpose of and use for which it is intended, and any photographs, diagrams, or drawings as may be necessary to enable the Director to make a determination. The Director may require that a sample of the device be submitted for examination and evaluation. If it is not possible to submit the device, the person requesting the exemption shall advise the Director and designate the place where the device will be available for examination and evaluation.

§ 555.33 Background checks and clearances (effective May 24, 2003).

(a) Background checks.

(1) If the Director receives from a licensee or permittee the names and appropriate identifying information of responsible persons and employees who will be authorized by the employer to possess explosive materials in the course of employment with the employer, the Director will conduct a background check in accordance with this section.

(2) The Director will determine whether the responsible person or employee is one of the persons described in any paragraph of section 842(i) of the Act (see §555.26). In making such determination, the Director may take into account a letter or document issued under paragraph (a)(3) of this section.

(3)(i) If the Director determines that the responsible person or the employee is not one of the persons described in any paragraph of section 842(i) of the Act (see §555.26), the Director will notify the employer in writing or electronically of the determination and issue, to the responsible person or employee, as the case may be, a letter of clearance which confirms the determination.

(ii) If the Director determines that the responsible person or employee is one of the persons described in any paragraph of section 842(i) of the Act (see §555.26), ATF will notify the employer in writing or electronically of the determination and issue to the responsible person or the employee, as the case may be, a document that confirms the determination; explains the grounds for the determination; provides information on how the disability may be relieved; and explains how the determination may be appealed. The employer will retain the notification as part of his permanent records in accordance with §555.121. The employer will take immediate steps to remove the responsible person from his position directing the management or policies of the business or operations as they relate to explosive materials or, as the case may be, to remove the employee from a position requiring the possession of explosive materials. Also, if the employer has listed the employee as a person authorized to accept delivery of explosive materials, as specified in §555.103 or §555.105, the employer must remove the employee from such list and immediately, and in no event later than the second business day after such change, notify distributors of such change.

(b) Appeals and correction of erroneous system information—

(1) In general. A responsible person or employee may challenge the adverse determination set out in the letter of denial, in writing and within 45 days of issuance of the determination, by directing his or her challenge to the basis for the adverse determination, or to the accuracy of the record upon which the adverse determination is based, to the Director. The appeal request must include appropriate documentation or record(s) establishing the legal and/or factual basis for the challenge. Any record or document of a court or other government entity or official furnished in support of an appeal must be certified by the court or other government entity or official as a true copy. In the case of an employee, or responsible person who did not submit fingerprints, such appeal must be accompanied by two properly completed FBI Forms FD–258 (fingerprint card). The Director will advise the individual in writing of his decision and the reasons for the decision.

(2) Employees. The letter of denial, among other things, will advise an employee who elects to challenge an adverse determination to submit the fingerprint cards as described above. The employee also will be advised of the agency name and address that originated the record containing the information causing the adverse determination ("originating agency"). At that time, and where appropriate, an employee is encouraged to apply to the originating agency to challenge the accuracy of the record(s) upon which the denial is based. The originating agency may respond to the individual's application by addressing the individual's specific reasons for the challenge, and by indicating whether additional information or documents are required. If the record is corrected as a result of the application to the originating agency, the individual may so notify ATF which will, in turn, verify the record correction with the originating agency and take all necessary steps to contact the agency responsible for the record system and correct the record. The employee may provide to ATF additional and appropriate documentation or record(s) establishing the legal and/or factual basis for the challenge to ATF's decision to uphold the initial denial. If ATF does not receive such additional documentation or record(s) within 45 days of the date of the decision upholding the initial denial, ATF will close the appeal.

(3) Responsible persons. The letter of denial, among other things, will advise a responsible person of the agency name and address which originated the record containing the information causing the adverse determination ("originating agency"). A responsible person who elects to challenge the adverse determination, where appropriate, is encouraged to apply to the originating agency to challenge the accuracy of the record(s) upon which the denial is based. The originating agency may respond to the individual's application by addressing the individual's specific reasons for the challenge, and by indicating whether additional information or documents are required. If the record is corrected as a result of the application to the originating agency, the individual may so notify ATF which will, in turn, verify the record correction with the originating agency and take all necessary steps to contact the agency responsible for the record system and correct the record. A responsible person may provide additional documentation or records as specified for employees in paragraph (b)(2) of this section.

(Approved by the Office of Management and Budget under control number 1140–0081)

[ATF No. 1, 68 FR 13783, Mar. 20, 2003]

§ 555.34 Replacement of stolen or lost ATF Form 5400.30 (Intrastate Purchase of Explosives Coupon (IPEC)).

When any Form 5400.30 is stolen, lost, or destroyed, the person losing possession will, upon discovery of the theft, loss, or destruction, immediately, but in all cases before 24 hours have elapsed since discovery, report the matter to the Director by telephoning 1–888–ATF-BOMB (nationwide toll free number). The report will explain in detail the circumstances of the theft, loss, or destruction and will include all known facts that may serve to identify the document. Upon receipt of the report, the Director will make such investigation as appears appropriate and may issue a duplicate document upon such conditions as the circumstances warrant.

(Approved by the Office of Management and Budget under control number 1140–0077)

[ATF No. 1, 68 FR 13783, Mar. 20, 2003]

Subpart D—Licenses and Permits

§ 555.41 General.

(a) Licenses and permits issued prior to May 24, 2003.

(1) Each person intending to engage in business as an importer or manufacturer of, or a dealer in, explosive materials, including black powder, must, before commencing business, obtain the license required by this subpart for the business to be operated. Each person who intends to acquire for use explosive materials from a licensee in a State other than the State in which he resides, or from a foreign country, or who intends to transport explosive materials in interstate or foreign commerce, must obtain a permit under this subpart; except that it is not necessary to obtain a permit if the user intends to lawfully purchase:

(i) Explosive materials from a licensee in a State contiguous to the user's State of residence and the user's State of residence has enacted legislation, currently in force, specifically authorizing a resident of that State to purchase explosive materials in a contiguous State; or

(ii) Commercially manufactured black powder in quantities not to exceed 50 pounds, intended to be used solely for sporting, recreational, or cultural purposes in antique firearms or in antique devices.

(2) Each person intending to engage in business as an explosive materials importer, manufacturer, or dealer must file an application, with the required fee (see §555.42), with ATF in accordance with the instructions on the form (see §555.45). A license will, subject to law, entitle the licensee to transport, ship, and receive explosive materials in interstate or foreign commerce, and to engage in the business specified by the license, at the location described on the license. A separate license must be obtained for each business premises at which the applicant is to manufacture, import, or distribute explosive materials except under the following circumstances:

(i) A separate license will not be required for storage facilities operated by the licensee as an integral part of one business premises or to cover a location used by the licensee solely for maintaining the records required by this part.

(ii) A separate license will not be required of a licensed manufacturer with respect to his on-site manufacturing.

(iii) It will not be necessary for a licensed importer or a licensed manufacturer (for purposes of sale or distribution) to also obtain a dealer's license in order to engage in business on his licensed premises as a dealer in explosive materials.

(iv) A separate license will not be required of licensed manufacturers with respect to their on-site manufacture of theatrical flash powder.

(3) Except as provided in paragraph (a)(1) of this section, each person intending to acquire explosive materials from a licensee in a State other than a State in which he resides, or from a foreign country, or who intends to transport explosive materials in interstate or foreign commerce, must file an application, with the required fee (see §555.43), with ATF in accordance with the instructions on the form (see §555.45). A permit will, subject to law, entitle the permittee to acquire, transport, ship, and receive in interstate or foreign commerce explosive materials. Only one permit is required under this part.

(b) Licenses and permits issued on and after May 24, 2003—

(1) In general.

(i) Each person intending to engage in business as an importer or manufacturer of, or a dealer in, explosive materials, including black powder, must, before commencing business, obtain the license required by this subpart for the business to be operated.

(ii) Each person who intends to acquire for use explosive materials within the State in which he resides on no more than 6 separate occasions during the 12-month period in which the permit is valid must obtain a limited permit under this subpart. (See §555.105(b) for definition of "6 separate occasions.")

(iii) Each person who intends to acquire for use explosive materials from a licensee or permittee in a State other than the State in which he resides, or from a foreign country, or who intends to transport explosive materials in interstate or foreign commerce, or who intends to acquire for use explosive materials within the State in which he resides on more than 6 separate occasions during a 12-month period, must obtain a user permit under this subpart.

(iv) It is not necessary to obtain a permit if the user intends only to lawfully purchase commercially manufactured black powder in quantities not to exceed 50 pounds, intended to be used solely for sporting, recreational, or cultural purposes in antique firearms or in antique devices.

(2) Importers, manufacturers, and dealers. Each person intending to engage in business as an explosive materials importer, manufacturer, or dealer must file an application, with the required fee (see §555.42), with ATF in accordance with the instructions on the form (see §555.45). A license will, subject to law, entitle the licensee to transport, ship, and receive explosive materials in interstate or foreign commerce, and to engage in the business specified by the license, at the location described on the license. A separate license must be obtained for each business premises at which the applicant is to manufacture, import, or distribute explosive materials except under the following circumstances:

(i) A separate license will not be required for storage facilities operated by the licensee as an integral part of one business premises or to cover a location used by the licensee solely for maintaining the records required by this part.

(ii) A separate license will not be required of a licensed manufacturer with respect to his on-site manufacturing.

(iii) It will not be necessary for a licensed importer or a licensed manufacturer (for purposes of sale or distribution) to also obtain a dealer's license in order to engage in business on his licensed premises as a dealer in explosive materials. No licensee will be required to obtain a user permit to lawfully transport, ship, or receive explosive materials in interstate or foreign commerce.

(iv) A separate license will not be required of licensed manufacturers with respect to their on-site manufacture of theatrical flash powder.

(3) Users of explosive materials.

(i) A limited permit will, subject to law, entitle the holder of such permit to receive for his use explosive materials from a licensee or permittee in his state of residence on no more than 6 separate occasions during the 12-month period in which the permit is valid. A limited permit does not authorize the receipt or transportation of explosive materials in interstate or foreign commerce. Holders of limited permits who need to receive explosive materials on more than 6 separate occasions during a 12-month period must obtain a user permit in accordance with this subpart.

(ii) Each person intending to acquire explosive materials from a licensee in a State other than a State in which he resides, or from a foreign country, or who intends to transport explosive materials in interstate or foreign commerce, must file an application for a user permit, with the required fee (see §555.43), with ATF in accordance with the instructions on the form (see §555.45). A user permit will, subject to law, entitle the permittee to transport, ship, and receive in interstate or foreign commerce explosive materials. Only one user permit per person is required under this part, irrespective of the number of locations relating to explosive materials operated by the holder of the user permit.

(Approved by the Office of Management and Budget under control number 1140–0083)

[ATF No. 1, 68 FR 13783, Mar. 20, 2003, as amended by ATF 5F, 70 FR 30633, May 27, 2005]

§ 555.42 License fees.

(a) Each applicant shall pay a fee for obtaining a three year license, a separate fee being required for each business premises, as follows:

(1) Manufacturer—$200.

(2) Importer—$200.

(3) Dealer—$200.

(b) Each applicant for a renewal of a license shall pay a fee for a three year license as follows:

(1) Manufacturer—$100.

(2) Importer—$100.

(3) Dealer—$100.

[T.D. ATF–400, 63 FR 45002, Aug. 24, 1998]

§ 555.43 Permit fees.

(a) Each applicant must pay a fee for obtaining a permit as follows:

(1) User—$100 for a three-year period.

(2) User-limited (nonrenewable)—$75.

(3) Limited—$25 for a one-year period.

(b) **(1)** Each applicant for renewal of a user permit must pay a fee of $50 for a three-year period.

(2) Each applicant for renewal of a limited permit must pay a fee of $12 for a one-year period.

[ATF No. 1, 68 FR 13785, Mar. 20, 2003]

§ 555.44 License or permit fee not refundable.

No refund of any part of the amount paid as a license or permit fee will be made where the operations of the licensee or permittee are, for any reason, discontinued during the period of an issued license or permit. However, the license or permit fee submitted with an application for a license or permit will be refunded if that application is denied, withdrawn, or abandoned, or if a license is cancelled subsequent to having been issued through administrative error.

§ 555.45 Original license or permit.

(a) Licenses issued prior to May 24, 2003. Any person who intends to engage in business as an explosive materials importer, manufacturer, or dealer, or who has not timely submitted application for renewal of a previous license issued under this part, shall file with ATF an application for License, Explosives, ATF F 5400.13 with ATF in accordance with the instructions on the form. The application must be executed under the penalties of perjury and the penalties imposed by 18 U.S.C. 844(a). The application is to be accompanied by the appropriate fee in the form of a money order or check made payable to the Bureau of Alcohol, Tobacco and Firearms. ATF F 5400.13 may be obtained from any ATF office. The Chief, Firearms and Explosives Licensing Center, will not approve an application postmarked on or after March 20, 2003, unless it is submitted with a Responsible Person Questionnaire, ATF Form 5400.28. Form 5400.28 must be completed in accordance with the instructions on the form.

(b) Permits issued prior to May 24, 2003. Any person, except as provided in §555.41(a), who intends to acquire explosive materials from a licensee in a state other than the State in which that person resides, or from a foreign country, or who intends to transport explosive materials in interstate or foreign commerce, or who has not timely submitted application for renewal of a previous permit issued under this part, shall file an application for Permit, Explosives, ATF F 5400.16 or Permit, User Limited Special Fireworks, ATF F 5400.21 with ATF in accordance with the instructions on the form. The application must be executed under the penalties of perjury and the penalties imposed by 18 U.S.C. 844(a). The application is to be accompanied by the appropriate fee in the form of a money order or check made payable to the Bureau of Alcohol, Tobacco and Firearms. ATF F 5400.16 and ATF F 5400.21 may be obtained from any ATF office. The Chief, Firearms and Explosives Licensing Center, will not approve an application postmarked on or after March 20, 2003, unless it is submitted with a Responsible Person Questionnaire, ATF Form 5400.28. Form 5400.28 must be completed in accordance with the instructions on the form.

(c) Licenses and permits issued on and after May 24, 2003—

(1) License. Any person who intends to engage in the business as an importer of, manufacturer of, or dealer in explosive materials, or who has not timely submitted an application for renewal of a previous license issued under this part, must file an application for License, Explosives, ATF F 5400.13, with ATF in accordance with the instructions on the form. ATF Form 5400.13 may be obtained by contacting any ATF office. The application must:

(i) Be executed under the penalties of perjury and the penalties imposed by 18 U.S.C. 844(a);

(ii) Include appropriate identifying information concerning each responsible person;

(iii) Include a photograph and fingerprints for each responsible person;

(iv) Include the names of and appropriate identifying information regarding all employees who will be authorized by the applicant to possess explosive materials by submitting ATF F 5400.28 for each employee; and

(v) Include the appropriate fee in the form of money order or check made payable to the Bureau of Alcohol, Tobacco, Firearms and Explosives.

(2) User permit and limited permit. Except as provided in 555.41(b)(1)(iv), any person who intends to acquire explosive materials in the State in which that person resides or acquire explosive materials from a licensee or holder of a user permit in a State other than the State in which that person resides, or from a foreign country, or who intends to transport explosive materials in interstate or foreign commerce, or who has not timely submitted an application for renewal of a previous permit issued under this part, must file an application for Permit, Explosives, ATF F 5400.16 or Permit, User Limited Display Fireworks, ATF F 5400.21 with ATF in accordance with the instructions on the form. ATF Form 5400.16 and ATF Form 5400.21 may be obtained by contacting any ATF office. The application must:

(i) Be executed under the penalties of perjury and the penalties imposed by 18 U.S.C. 844(a);

(ii) Include a photograph, fingerprints, and appropriate identifying information for each responsible person;

(iii) Include the names of and appropriate identifying information regarding all employees who will be authorized by the applicant to possess explosive materials by submitting ATF F 5400.28 for each employee; and

(iv) Include the appropriate fee in the form of money order or check made payable to the Bureau of Alcohol, Tobacco, Firearms and Explosives.

(3) The Chief, Firearms and Explosives Licensing Center, will conduct background checks on responsible persons and employees authorized by the applicant to possess explosive materials in accordance with §555.33. If it is determined that any responsible person or employee is described in any paragraph of section 842(i) of the Act, the applicant must submit an amended application indicating removal or reassignment of that person before the license or permit will be issued.

(Approved by the Office of Management and Budget under control number 1140–0083)

(18 U.S.C. 847 (84 Stat. 959); 18 U.S.C. 926 (82 Stat. 1226))
[T.D. ATF–200, 50 FR 10497, Mar. 15, 1985, as amended by T.D. ATF–400, 63 FR 45002, Aug. 24, 1998; ATF No. 1, 68 FR 13785, Mar. 20, 2003]

§ 555.46 Renewal of license or permit.

(a) If a licensee or permittee intends to continue the business or operation described on a license or permit issued under this part after the expiration date of the license or permit, he shall, unless otherwise notified in writing by the Chief, Firearms and Explosives Licensing Center, execute and file prior to the expiration of his license or permit an application for license renewal, ATF F 5400.14 (Part III), or an application for permit renewal, ATF F 5400.15 (Part III), accompanied by the required fee, with ATF in accordance with the instructions on the form. In the event the licensee or permittee does not timely file a renewal application, he shall file an original application as required by §555.45, and obtain the required license or permit in order to continue business or operations.

(b) A user-limited permit is not renewable and is valid for a single purchase transaction. Applications for all user-limited permits must be filed on ATF F 5400.16 or ATF F 5400.21, as required by §555.45.

(18 U.S.C. 847 (84 Stat. 959); 18 U.S.C. 926 (82 Stat. 1226))
[T.D. ATF–87, 46 FR 40384, Aug. 7, 1981, as amended by T.D. ATF–200, 50 FR 10497, Mar. 15, 1985; T.D. ATF–290, 54 FR 53054, Dec. 27, 1989; T.D. ATF–400, 63 FR 45002, Aug. 24, 1998]

§ 555.47 Insufficient fee.

If an application is filed with an insufficient fee, the application and fee submitted will be returned to the applicant.

(18 U.S.C. 847 (84 Stat. 959); 18 U.S.C. 926 (82 Stat. 1226))
[T.D. ATF–200, 50 FR 10498, Mar. 15, 1985]

§ 555.48 Abandoned application.

Upon receipt of an incomplete or improperly executed application, the applicant will be notified of the deficiency in the application. If the application is not corrected and returned within 30 days following the date of notification, the application will be considered as having been abandoned and the license or permit fee returned.

§ 555.49 Issuance of license or permit.

(a) Issuance of license or permit prior to May 24, 2003. (1) The Chief, Firearms and Explosives Licensing Center, will issue a license or permit if—

(i) A properly executed application for the license or permit is received; and

(ii) Through further inquiry or investigation, or otherwise, it is found that the applicant is entitled to the license or permit.

(2) The Chief, Firearms and Explosives Licensing Center, will approve a properly executed application for a license or permit, if:

(i) The applicant is 21 years of age or over;

(ii) The applicant (including, in the case of a corporation, partnership, or association, any individual possessing, directly or indirectly, the power to direct or cause the direction of the management and policies of the corporation, partnership, or association) is not a person to whom distribution of explosive materials is prohibited under the Act;

(iii) The applicant has not willfully violated any provisions of the Act or this part;

(iv) The applicant has not knowingly withheld information or has not made any false or fictitious statement intended or likely to deceive, in connection with his application;

(v) The applicant has in a State, premises from which he conducts business or operations subject to license or permit under the Act or from which he intends to conduct business or operations;

(vi) The applicant has storage for the class (as described in §555.202) of explosive materials described on the application, unless he establishes to the satisfaction of the Chief, Firearms and Explosives Licensing Center, that the business or operations to be conducted will not require the storage of explosive materials;

(vii) The applicant has certified in writing that he is familiar with and understands all published State laws and local ordinances relating to explosive materials for the location in which he intends to do business; and

(viii) The applicant for a license has submitted the certificate required by section 21 of the Federal Water Pollution Control Act, as amended (33 U.S.C. 1341).

(3) The Chief, Firearms and Explosives Licensing Center, will approve or the regional director (compliance) will deny any application for a license or permit within the 45-day period beginning on the date a properly executed application was received. However, when an applicant for license or permit renewal is a person who is, under the provisions of §555.83 or §555.142, conducting business or operations under a previously issued license or permit, action regarding the application will be held in abeyance pending the completion of the proceedings against the applicant's existing license or permit, or renewal application, or final action by the Director on an application for relief submitted under §555.142, as the case may be.

(4) The license or permit and one copy will be forwarded to the applicant, except that in the case of a user-limited permit, the original only will be issued.

(5) Each license or permit will bear a serial number and this number may be assigned to the licensee or permittee to whom issued for as long as he maintains continuity of renewal in the same region.

(b) Issuance of license or permit on and after May 24, 2003.

(1) The Chief, Firearms and Explosives Licensing Center, will issue a license or permit if:

(i) A properly executed application for the license or permit is received; and

(ii) Through further inquiry or investigation, or otherwise, it is found that the applicant is entitled to the license or permit.

(2) The Chief, Firearms and Explosives Licensing Center, will approve a properly executed application for a license or permit, if:

(i) The applicant (or, if the applicant is a corporation, partnership, or association, each responsible person with respect to the applicant) is not a person described in any paragraph of section 842(i) of the Act;

(ii) The applicant has not willfully violated any provisions of the Act or this part;

(iii) The applicant has not knowingly withheld information or has not made any false or fictitious statement intended or likely to deceive, in connection with his application;

(iv) The applicant has in a State, premises from which he conducts business or operations subject to license or permit under the Act or from which he intends to conduct business or operations;

(v) The applicant has storage for the class (as described in §555.202) of explosive materials described on the application;

(vi) The applicant has certified in writing that he is familiar with and understands all published State laws and local ordinances relating to explosive materials for the location in which he intends to do business;

(vii) The applicant for a license has submitted the certificate required by section 21 of the Federal Water Pollution Control Act, as amended (33 U.S.C. 1341);

(viii) None of the employees of the applicant who will be authorized by the applicant to possess explosive materials is a person described in any paragraph of section 842(i) of the Act; and

(ix) In the case of an applicant for a limited permit, the applicant has certified in writing that the applicant will not receive explosive materials on more than 6 separate occasions during the 12-month period for which the limited permit is valid.

(3) The Chief, Firearms and Explosives Licensing Center, will approve or the regional director (compliance) will deny any application for a license or permit within the 90-day period beginning on the date a properly executed application was received. However, when an applicant for license or permit renewal is a person who is, under the provisions of §555.83 or §555.142, conducting business or operations under a previously issued license or permit, action regarding the application will be held in abeyance pending the completion of the proceedings against the applicant's existing license or permit, or renewal application, or final action by the Director on an application for relief submitted under §555.142, as the case may be.

(4) The license or permit and one copy will be forwarded to the applicant, except that in the case of a user-limited permit, the original only will be issued.

(5) Each license or permit will bear a serial number and this number may be assigned to the licensee or permittee to whom issued for as long as he maintains continuity of renewal in the same region.

(Approved by the Office of Management and Budget under control number 1140–0082)

[ATF No. 1, 68 FR 13785, Mar. 20, 2003]

§ 555.50 Correction of error on license or permit.

(a) Upon receipt of a license or permit issued under this part, each licensee or permittee shall examine the license or permit to insure that the information on it is accurate. If the license or permit is incorrect, the licensee or permittee shall return the license or permit to the Chief, Firearms and Explosives Licensing Center, with a statement showing the nature of the error. The Chief, Firearms and Explosives Licensing Center, shall correct the error, if the error was made in his office, and return the license or permit. However, if the error resulted from information contained in the licensee's or permittee's application for the license or permit, the Chief, Firearms and Explosives Licensing Center, shall require the licensee or permittee to file an amended application setting forth the correct information and a statement explaining the error contained in the application. Upon receipt of the amended application and a satisfactory explanation of the error, the Chief, Firearms and Explosives Licensing Center, shall make the correction on the license or permit and return it to the licensee or permittee.

(b) When the Chief, Firearms and Explosives Licensing Center, finds through any means other than notice from the licensee or permittee that an incorrect license or permit has been issued, (1) the Chief, Firearms and Explosives Licensing Center, may require the holder of the incorrect license or permit to return the license or permit for correction, and (2) if the error resulted from information contained in the licensee's or permittee's application for the license or permit, the Chief, Firearms and Explosives Licensing Center, shall require the licensee or permittee to file an amended application setting forth the correct information, and a

statement satisfactorily explaining the error contained in the application. The Chief, Firearms and Explosives Licensing Center, then shall make the correction on the license or permit and return it to the licensee or permittee.
[T.D. ATF–87, 46 FR 40384, Aug. 7, 1981, as amended by T.D. ATF–290, 54 FR 53054, Dec. 27, 1989]

§ 555.51 Duration of license or permit.

(a) Prior to May 24, 2003. An original license or permit is issued for a period of three years. A renewal license or permit is issued for a period of three years. However, a user-limited permit is valid only for a single purchase transaction.

(b) On and after May 24, 2003.

(1) An original license or user permit is issued for a period of three years. A renewal license or user permit is also issued for a period of three years. However, a user-limited permit is valid only for a single purchase transaction.

(2) A limited permit is issued for a period of one year. A renewal limited permit is also issued for a period of one year.
[ATF No. 1, 68 FR 13786, Mar. 20, 2003]

§ 555.52 Limitations on license or permit.

(a) The license covers the business of explosive materials specified in the license at the licensee's business premises (see §555.41(b)).

(b) The permit is valid with respect to the type of operations of explosive materials specified in the permit.
[T.D. ATF–87, 46 FR 40384, Aug. 7, 1981, as amended by T.D. ATF–387, 62 FR 8376, Feb. 25, 1997; ATF 5F, 70 FR 30633, May 27, 2005]

§ 555.53 License and permit not transferable.

Licenses and permits issued under this part are not transferable to another person. In the event of the lease, sale, or other transfer of the business or operations covered by the license or permit, the successor must obtain the license or permit required by this part before commencing business or operations. However, for rules on right of succession, see §555.59.

§ 555.54 Change of address.

(a) During the term of a license or permit, a licensee or permittee may move his business or operations to a new address at which he intends to regularly carry on his business or operations, without procuring a new license or permit. However, in every case, the licensee or permittee shall—

(1) Give notification of the new location of the business or operations to the Chief, Firearms and Explosives Licensing Center at least 10 days before the move; and

(2) Submit the license or permit to the Chief, Firearms and Explosives Licensing Center. The Chief, Firearms and Explosives Licensing Center will issue an amended license or permit, which will contain the new address (and new license or permit number, if any).

(b) Licensees and permittees whose mailing address will change must notify the Chief, Firearms and Explosives Licensing Center, at least 10 days before the change.

(Paragraph (b) approved by the Office of Management and Budget under control number 1140–0080)
[T.D. ATF–87, 46 FR 40384, Aug. 7, 1981, as amended by T.D. ATF–290, 54 FR 53054, Dec. 27, 1989; ATF No. 1, 68 FR 13786, Mar. 20, 2003]

§ 555.56 Change in trade name.

A licensee or permittee continuing to conduct business or operations at the location shown on his license or permit is not required to obtain a new license or permit by reason of a mere change in trade name under which he conducts his business or operations. However, the licensee or permittee shall furnish his license or permit and any copies furnished with the license or permit for endorsement of the change to the Chief, Firearms and Explosives Licensing Center, within 30 days from the date the licensee or permittee begins his business or operations under the new trade name.
[T.D. ATF–87, 46 FR 40384, Aug. 7, 1981, as amended by T.D. ATF–290, 54 FR 53054, Dec. 27, 1989]

§ 555.57 Change of control, change in responsible persons, and change of employees.

(a) In the case of a corporation or association holding a license or permit under this part, if actual or legal control of the corporation or association changes, directly or indirectly, whether by reason of change in stock ownership or control (in the corporation holding a license or permit or in any other corporation), by operation of law, or in any other manner, the licensee or permittee shall, within 30 days of the change, give written notification executed under the penalties of perjury, to the Chief, Firearms and Explosives Licensing Center. Upon expiration of the license or permit, the corporation or association shall file an ATF F 5400.13 or an ATF F 5400.16 as required by §555.45, and pay the fee prescribed in §555.42(b) or §555.43(b).

(b) For all licenses or permits issued on and after May 24, 2003, each person holding the license or permit must report to the Chief, Firearms and Explosives Licensing Center, any change in responsible persons or employees authorized to possess explosive materials. Such report must be submitted within 30 days of the change and must include appropriate identifying information for each responsible person. Reports relating to newly hired employees authorized to possess explosive materials must be submitted on ATF F 5400.28 for each employee.

(c) Upon receipt of a report, the Chief, Firearms and Explosives Licensing Center, will conduct a background check, if appropriate, in accordance with §555.33.

(d) The reports required by paragraph (b) of this section must be retained as part of a licensee's or permittee's permanent records for the period specified in §555.121.

(Approved by the Office of Management and Budget under control number 1140–0074)
[T.D. ATF–87, 46 FR 40384, Aug. 7, 1981, as amended by T.D. ATF–290, 54 FR 53054, Dec. 27, 1989; ATF No. 1, 68 FR 13786, Mar. 20, 2003]

§ 555.58 Continuing partnerships.

Where, under the laws of the particular State, the partnership is not terminated on death or insolvency of a partner, but continues until the winding up of the partnership affairs is completed, and the surviving partner has the exclusive right to the control and possession of the partnership assets for the purpose of liquidation and settlement, the surviving partner may continue to conduct the business or operations under the license or permit of the partnership. If the surviving partner acquires the business or operations on completion of settlement of the partnership, he shall obtain a license or permit in his own name from the date of acquisition, as provided in §555.45. The rule set forth in this section will also apply where there is more than one surviving partner.

§ 555.59 Right of succession by certain persons.

(a) Certain persons other than the licensee or permittee may secure the right to carry on the same explosive materials business or operations at the same business premises for the remainder of the term of license or permit. These persons are:

(1) The surviving spouse or child, or executor, administrator, or other legal representative of a deceased licensee or permittee; and

(2) A receiver or trustee in bankruptcy, or an assignee for benefit of creditors.

(b) In order to secure the right of succession, the person or persons continuing the business or operations shall submit the license or permit and all copies furnished with the license or permit for endorsement of the succession to the Chief, Firearms and Explosives Licensing Center, within 30 days from the date on which the successor begins to carry on the business or operations.
[T.D. ATF–87, 46 FR 40384, Aug. 7, 1981, as amended by T.D. ATF–290, 54 FR 53054, Dec. 27, 1989]

§ 555.60 Certain continuances of business or operations.

A licensee or permittee who furnishes his license or permit to the Chief, Firearms and Explosives Licensing Center, for correction, amendment, or endorsement, as provided in this subpart, may continue his business or operations while awaiting its return.
[T.D. ATF–87, 46 FR 40384, Aug. 7, 1981, as amended by T.D. ATF–290, 54 FR 53054, Dec. 27, 1989]

§ 555.61 Discontinuance of business or operations.

Where an explosive materials business or operations is either discontinued or succeeded by a new owner, the owner of the business or operations discontinued or succeeded shall, within 30 days, furnish notification of the discontinuance or succession and submit his license or permit and any copies furnished with the license or permit to the Chief, Firearms and Explosives Licensing Center. (See also §555.128.)
[T.D. ATF–87, 46 FR 40384, Aug. 7, 1981, as amended by T.D. ATF–290, 54 FR 53054, Dec. 27, 1989]

§ 555.62 State or other law.

A license or permit issued under this part confers no right or privilege to conduct business or operations, including storage, contrary to State or other law. The holder of a license or permit issued under this part is not, by reason of the rights and privileges granted by that license or permit, immune from punishment for conducting an explosive materials business or operations in violation of the provisions of any State or other law. Similarly, compliance with the provisions of any State or other law affords no immunity under Federal law or regulations.

§ 555.63 Explosives magazine changes.

(a) General.

(1) The requirements of this section are applicable to magazines used for other than temporary (under 24 hours) storage of explosives.

(2) A magazine is considered suitable for the storage of explosives if the construction requirements of this part are met during the time explosives are stored in the magazine.

(3) A magazine is considered suitable for the storage of explosives if positioned in accordance with the applicable table of distances as specified in this part during the time explosives are stored in the magazine.

(4) For the purposes of this section, notification of the regional director (compliance) may be by telephone or in writing. However, if notification of the regional director (compliance) is in writing it must be at least three business days in advance of making changes in construction to an existing magazine or constructing a new magazine, and at least five business days in advance of using any reconstructed magazine or added magazine for the storage of explosives.

(b) Exception. Mobile or portable type 5 magazines are exempt from the requirements of paragraphs (c) and (d) of this section, but must otherwise be in compliance with paragraphs (a) (2) and (3) of this section during the time explosives are stored in such magazines.

(c) Changes in magazine construction. A licensee or permittee who intends to make changes in construction of an existing magazine shall notify the regional director (compliance) describing the proposed changes prior to making any changes. Unless otherwise advised by the regional director (compliance), changes in construction may commence after explosives are removed from the magazine. Explosives may not be stored in a reconstructed magazine before the regional director (compliance) has been notified in accordance with paragraph (a)(4) of this section that the changes have been completed.

(d) Magazines acquired or constructed after permit or license is issued. A licensee or permittee who intends to construct or acquire additional magazines shall notify the regional director (compliance) in accordance with paragraph (a)(4) of this section describing the additional magazines and the class and quantity of explosives to be stored in the magazine. Unless otherwise advised by the regional director (compliance), additional magazines may be constructed, or acquired magazines may be used for the storage of explosives. Explosives must not be stored in a magazine under construction. The regional director (compliance) must be notified that construction has been completed.

[T.D. ATF–87, 46 FR 40384, Aug. 7, 1981, as amended by T.D. ATF–400, 63 FR 45002, Aug. 24, 1998]

Subpart E—License and Permit Proceedings

§ 555.71 Opportunity for compliance.

Except in cases of willfulness or those in which the public interest requires otherwise, and the regional director (compliance) so alleges in the notice of denial of an application or revocation of a license or permit, no license or permit will be revoked or renewal application denied without first calling to the attention of the licensee or permittee the reasons for the contemplated action and affording him an opportunity to demonstrate or achieve compliance with all lawful requirements and to submit facts, arguments, or proposals of adjustment. The notice of contemplated action, ATF F 5400.12, will afford the licensee or permittee 15 days from the date of receipt of the notice to respond. If no response is received within the 15 days, or if after consideration of relevant matters presented by the licensee or permittee, the regional director (compliance) finds that the licensee or permittee is not likely to abide by the law and regulations, he will proceed as provided in §555.74.

[T.D. ATF–87, 46 FR 40384, Aug. 7 1981, as amended by T.D. ATF–446, 66 FR 16602, Mar. 27, 2001]

§ 555.72 Denial of initial application.

Whenever the regional director (compliance) has reason to believe that an applicant for an original license or permit is not eligible to receive a license or permit under the provisions of §555.49, he shall issue a notice of denial on ATF F 5400.11. The notice will set forth the matters of fact and law relied upon in determining that the application should be denied, and will afford the applicant 15 days from the date of receipt of the notice in which to request a hearing to review the denial. If no request for a hearing is filed within that time, a copy of the application, marked *"Disapproved"*, will be returned to the applicant.

§ 555.73 Hearing after initial application is denied.

If the applicant for an original license or permit desires a hearing, he shall file a request with the regional director (compliance) within 15 days after receipt of the notice of denial. The request should include a statement of the reasons for a hearing. On receipt of the request, the regional director (compliance) shall refer the matter to an administrative law judge who shall set a time and place (see §555.77) for a hearing and shall serve notice of the hearing upon the applicant and the regional director (compliance) at least 10 days in advance of the hearing date. The hearing will be conducted in accordance with the hearing procedures prescribed in part 71 of this chapter (see §555.82). Within a reasonable time after the conclusion of the hearing, and as expeditiously as possible, the administrative law judge shall render his recommended decision. He shall certify to the complete record of the proceedings before him and shall immediately forward the complete certified record, together with four copies of his recommended decision, to the regional director (compliance) for decision.

§ 555.74 Denial of renewal application or revocation of license or permit.

If following the opportunity for compliance under §555.71, or without opportunity for compliance under §555.71, as circumstances warrant, the regional director (compliance) finds that the licensee or permittee is not likely to comply with the law or regulations or is otherwise not eligible to continue operations authorized under his license or permit, the regional director (compliance) shall issue a notice of denial of the renewal application or revocation of the license or permit, ATF F 5400.11 or ATF F 5400.10, as appropriate. In either case, the notice will set forth the matters of fact constituting the violations specified, dates, places, and the sections of law and regulations violated. The notice will, in the case of revocation of a license or permit, specify the date on which the action is effective, which date will be on or after the date the notice is served on the licensee or permittee. The notice will also advise the licensee or permittee that he may, within 15 days after receipt of the notice, request a hearing and, if applicable, a stay of the effective date of the revocation of his license or permit.

§ 555.75 Hearing after denial of renewal application or revocation of license or permit.

If a licensee or permittee whose renewal application has been denied or whose license or permit has been revoked desires a hearing, he shall file a request for a hearing with the regional director (compliance). In the case of the revocation of a license or permit, he may include a request for a stay of the effective date of the revocation. On receipt of the request the regional director (compliance) shall advise the licensee or permittee whether the stay of the effective date of the revocation is granted. If the stay of the effective date of the revocation is granted, the regional director (compliance) shall refer the matter to an administrative law judge who shall set a time and place (see §555.77) for a hearing and shall serve notice of the hearing upon the licensee or permittee and the regional director (compliance) at least 10 days in advance of the hearing date. If the stay of the effective date of the revocation is denied, the licensee or permittee may request an immediate hearing. In this event, the regional director (compliance) shall immediately refer the matter to an administrative law judge who shall set a date and place for a hearing, which date shall be no later than 10 days from the date the licensee or permittee requested an immediate hearing. The hearing will be held in accordance with the applicable provisions of part 71 of this chapter. Within a reasonable time after the conclusion of the hearing, and as expeditiously as possible, the administrative law judge shall render his decision. He shall certify to the complete record of the proceeding before him and shall immediately forward the complete certified record, together with two copies of his decision, to the regional director (compliance), serve one copy of his decision on the licensee or permittee or his counsel, and transmit a copy to the attorney for the Government.

§ 555.76 Action by regional director (compliance).

(a) Initial application proceedings. If, upon receipt of the record and the recommended decision of the administrative law judge, the regional director (compliance) decides that the license or permit should be issued, the regional director (compliance) shall cause the application to be approved, briefly stating, for the record, his reasons. If he contemplates that the denial should stand, he shall serve a copy of the administrative law judge's recommended decision on the applicant, informing the applicant of his contemplated action and affording the applicant not more than 10 days in which to submit proposed findings and conclusions or exceptions to the recommended decision with supporting reasons. If the regional director (compliance), after consideration of the record of the hearing and of any proposed findings, conclusions, or exceptions filed with him by the applicant, approves the findings, conclusions and recommended decision of the administrative law judge, the regional director (compliance) shall cause the license or permit to be issued or disapproved the application accordingly. If he disapproves the findings, conclusions, and recommendation of the administrative law judge, in whole or in part, he shall by order make such findings and conclusions as in his opinion are warranted by the law and the facts in the record. Any decision of the regional director (compliance) ordering the disapproval of an initial application for a license or permit shall state the findings and conclusions upon which it is based, including his ruling upon each proposed finding, conclusion, and exception to the administrative law judge's recommended decision, together with a statement of his findings and conclusions, and reasons or basis for his findings and conclusions, upon all material issues of fact, law or discretion presented on the record. A signed duplicate original of the decision will be served upon the applicant and the original copy containing certificate of service will be placed in the official record of the proceedings. If the decision of the regional director (compliance) is in favor of the applicant, he shall issue the license or permit, to be effective on issuance.

(b) Renewal application and revocation proceedings. Upon receipt of the complete certified records of the hearing, the regional director (compliance) shall enter an order confirming the revocation of the license or permit, or disapproving the application, in accordance with the administrative law judge's findings and decision, unless he disagrees with the findings and decision. A signed duplicate original of the order, ATF F 5400.9, will be served upon the licensee or permittee and the original copy containing certificate of service will be placed in the official record of the proceedings. If the regional director (compliance) disagrees with the findings and decision of the administrative law judge, he shall file a petition with the Director for review of the findings and decision, as provided in §555.79. In either case, if the renewal application denial is sustained, a copy of the application marked "Disapproved" will be returned to the applicant. If the renewal application denial is reversed, a license or permit will be issued to become effective on expiration of the license or permit being renewed, or on the date of issuance, whichever is later. If the proceedings involve the revocation of a license or permit which expired before a decision is in favor of the licensee or permittee, the regional director (compliance) shall:

(1) If renewal application was timely filed and a stay of the effective date of the revocation was granted, cause to be issued a license or permit effective on the date of issuance;

(2) If renewal application was not timely filed but a stay of the effective date of the revocation had been granted, request that a renewal application be filed and, following that, cause to be issued a license or permit to be effective on issuance; or

(3) If a stay of the effective date of the revocation had not been granted, request that an application be filed as provided in §555.45, and process it in the same manner as for an application for an original license or permit.

(c) Curtailment of stay of revocation effective date. If, after approval of a request for a stay of the effective date of an order revoking a license or permit but before actions are completed under this subpart, the regional director (compliance) finds that it is contrary to the public interest for the licensee or permittee to continue the operations or activities covered by his license or permit, the regional director (compliance) may issue a notice of withdrawal of the approval, effective on the date of issuance. Notice of withdrawal will be served upon the licensee or permittee in the manner provided in §555.81.

[T.D. ATF–87, 46 FR 40384, Aug. 7, 1981, as amended by T.D. ATF–290, 54 FR 53054, Dec. 27, 1989]

§ 555.77 Designated place of hearing.

The designated place of hearing set as provided in §555.73 or §555.75, will be at the location convenient to the aggrieved party.

§ 555.78 Representation at a hearing.

An applicant, licensee, or permittee may be represented by an attorney, certified public accountant, or other person recognized to practice before the Bureau of Alcohol, Tobacco and Firearms as provided in 31 CFR Part 8, if he has otherwise complied with the applicable requirements of 26 CFR 601.521 through 601.527. The regional director (compliance) shall be represented in proceedings under §§555.73 and 555.75 by an attorney in the office of the chief counsel or regional counsel who is authorized to execute and file motions, briefs, and other papers in the proceedings, on behalf of the regional director (compliance), in his own name as *"Attorney for the Government"*.

[T.D. ATF–87, 46 FR 40384, Aug. 7, 1981, as amended by T.D. ATF–92, 46 FR 46916, Sept. 23, 1981]

§ 555.79 Appeal on petition to the Director.

An appeal to the Director is not required prior to filing an appeal with the U.S. Court of Appeals for judicial review. An appeal may be taken by the applicant, licensee, or permittee to the Director from a decision resulting from a hearing under §555.73 or §555.75. An appeal may also be taken by a regional director (compliance) from a decision resulting from a hearing under §555.75 as provided in §555.76(b). The appeal shall be taken by filing a petition for review on appeal with the Director within 15 days of the service of an administrative law judge's decision or an order. The petition will set forth facts tending to show (a) action of an arbitrary nature, (b) action without reasonable warrant in fact, or (c) action contrary to law and regulations. A copy of the petition will be filed with the regional director (compliance) or served on the applicant, licensee, or permittee, as the case may be. In the event of appeal, the regional director (compliance) shall immediately forward the complete original record, by certified mail, to the Director for his consideration, review, and disposition as provided in subpart I of part 71 of this chapter. When, on appeal, the Director affirms the initial decision of the regional director (compliance) or the administrative law judge, as the case may be, the initial decision will be final.

§ 555.80 Court review.

An applicant, licensee, or permittee may, within 60 days after receipt of the decision of the administrative law judge or the final order of the regional director (compliance) or the Director, file a petition for a judicial review of the decision, with the U.S. Court of Appeals for the district in which he resides or has his principal place of business. The Director, upon notification that a petition has been filed, shall have prepared a complete transcript of the record of the proceedings. The regional director (compliance) or the Director, as the case may be, shall certify to the correctness of the transcript of the record, forward one copy to the attorney for the Government in the review of the case, and file the original record of the proceedings with the original certificate in the U.S. Court of Appeals.

§ 555.81 Service on applicant, licensee, or permittee.

All notices and other formal documents required to be served on an applicant, licensee, or permittee under this subpart will be served by certified mail or by personal delivery. Where service is by personal delivery, the signed duplicate original copy of the formal document will be delivered to the applicant, licensee, or permittee, or, in the case of a corporation, partnership, or association, by delivering it to an officer, manager, or general agent, or to its attorney of record.

§ 555.82 Provisions of part 200 made applicable.

The provisions of subpart G of part 200 of this chapter, as well as those provisions of part 71 relative to failure to appear, withdrawal of an application or surrender of a permit, the conduct of hearings before an administrative law judge, and record of testimony, are hereby made applicable to application, license, and permit proceedings under this subpart to the extent that they are not contrary to or incompatible with this subpart.

§ 555.83 Operations by licensees or permittees after notice of denial or revocation.

In any case where a notice of revocation has been issued and a request for a stay of the effective date of the revocation has not been granted, the licensee or permittee shall not engage in the activities covered by the license or permit pending the outcome of proceedings under this subpart. In any case where notice of revocation has been issued but a stay of the effective date of the revocation has been granted, the licensee or permittee may continue to engage in the activities covered by his license or permit unless, or until, formally notified to the contrary: Provided, That in the event the license or permit would have expired before proceedings under this subpart are completed, timely renewal application must have been filed to continue the license or permit beyond its expiration date. In any case where a notice of denial of a renewal application has been issued, the licensee or permittee may continue to engage in the activities covered by the existing license or permit after the date of expiration of the license or permit until proceedings under this subpart are completed.

Subpart F—Conduct of Business or Operations

§ 555.101 Posting of license or user permit.

A license or user permit issued under this part, or a copy of a license or user permit, will be posted and available for inspection on the business premises at each place where explosive materials are manufactured, imported, or distributed.

[T.D. ATF–87, 46 FR 40384, Aug. 7, 1981. Redesignated by T.D. ATF–487, 68 FR 3748, Jan. 24, 2003. ATF No. 1, 68 FR 13786, Mar. 20, 2003]

§ 555.102 Authorized operations by permittees.

(a) In general. A permit issued under this part does not authorize the permittee to engage in the business of manufacturing, importing, or dealing in explosive materials. Accordingly, if a permittee's operations bring him within the definition of manufacturer, importer, or dealer under this part, he shall qualify for the appropriate license.

(b) Distributions of surplus stocks—

(1) Distributions of surplus stocks prior to May 24, 2003. Permittees are not authorized to engage in the business of sale or distribution of explosive materials. However, permittees may dispose of surplus stocks of explosive materials to other licensees or permittees in accordance with §555.103, and to nonlicensees or to nonpermittees in accordance with §555.105(a)(4).

(2) Distributions of surplus stocks on and after May 24, 2003. Permittees are not authorized to engage in the business of sale or distribution of explosive materials. However, permittees may dispose of surplus stocks of explosive materials to other licensees or permittees in accordance with §555.103 and §555.105.

[T.D. ATF–400, 63 FR 45002, Aug. 24, 1998, as amended by ATF No. 1, 68 FR 13787, Mar. 20, 2003]

§ 555.103 Transactions among licensees/permittees and transactions among licensees and holders of user permits.

(a) Transactions among licensees/permittees prior to May 24, 2003—

(1) General.

(i) A licensed importer, licensed manufacturer or licensed dealer selling or otherwise distributing explosive materials (or a permittee disposing of surplus stock to a licensee or another permittee) who has the certified information required by this section may sell or distribute explosive materials to a licensee or permittee for not more than 45 days following the expiration date of the distributee's license or permit, unless the distributor knows or has reason to believe that the distributee's authority to continue business or operations under this part has been terminated.

(ii) A licensed importer, licensed manufacturer or licensed dealer selling or otherwise distributing explosive materials (or a permittee disposing of surplus stock to another licensee or permittee) must verify the license or permit status of the distributee prior to the release of explosive materials ordered, as required by this section.

(iii) Licensees or permittees desiring to return explosive materials to a licensed manufacturer may do so without obtaining a certified copy of the manufacturer's license.

(iv) Where possession of explosive materials is transferred at the distributor's premises, the distributor must in all instances verify the identity of the person accepting possession on behalf of the distributee before relinquishing possession. Before the delivery at the distributor's premises of explosive materials to an employee of a licensee or permittee, or to an employee of a common or contract carrier transporting explosive materials to a licensee or permittee, the distributor delivering explosive materials must obtain an executed ATF F 5400.8, Explosives Delivery Record, from the employee before releasing the explosive materials. The ATF F 5400.8 must contain all of the information required on the form and required by this part.

Example 1. An ATF F 5400.8 is required when:

a. An employee of the purchaser takes possession at the distributor's premises.

b. An employee of a common or contract carrier hired by the purchaser takes possession at the distributor's premises.

Example 2. An ATF F 5400.8 is not required when:

a. An employee of the distributor takes possession of the explosives for the purpose of transport to the purchaser.

b. An employee of a common or contract carrier hired by the distributor takes possession of the explosives for the purpose of transport to the purchaser.

(2) License/permit verification of individuals.

(i) The distributee must furnish a certified copy (or, in the case of a user-limited, the original) of the license or permit. The certified copy need be furnished only once during the current term of the license or permit. Also, a licensee need not furnish certified copies of licenses to other licensed locations operated by such licensee.

(ii) The distributor may obtain any additional verification as the distributor deems necessary.

(3) License/permit verification of business organizations.

(i) A business organization may (in lieu of furnishing a certified copy of a license) furnish the distributor a certified list which contains the name, address, license number and date of license expiration of each licensed location. The certified list need be furnished only once during the current term of the license or permit. Also, a business organization need not furnish a certified list to other licensed locations operated by such business organization.

(ii) A business organization must, prior to ordering explosive materials, furnish the licensee or permittee a current certified list of the representatives or agents authorized to order explosive materials on behalf of the business organization showing the name, address, and date and place of birth of each representative or agent. A licensee or permittee may not distribute explosive materials to a business organization on the order of a person who does not appear on the certified list of representatives or agents and, if the person does appear on the certified list, the licensee or permittee must verify the identity of such person.

(4) Licensee/permittee certified statement.

(i) A licensee or permittee ordering explosive materials from another licensee or permittee must furnish a current, certified statement of the intended use of the explosive materials, e.g., resale, mining, quarrying, agriculture, construction, sport rocketry, road building, oil well drilling, seismographic research, to the distributor.

(ii) For individuals, the certified statement of intended use must specify the name, address, date and place of birth, and social security number of the distributee.

(iii) For business organizations, the certified statement of intended use must specify the taxpayer identification number, the identity and the principal and local places of **business.**

(iv) The licensee or permittee purchasing explosive materials must revise the furnished copy of the certified statement only when the information is no longer current.

(5) User-limited permit transactions. A user-limited permit issued under the provisions of this part is valid for only a single purchase transaction and is not renewable (see §555.51). Accordingly, at the time a user-limited permittee orders explosive materials, the licensed distributor must write on the front of the user-limited permit the transaction date, his signature, and the distributor's license number prior to returning the permit to the user-limited permittee.

(b) Transactions among licensees/permittees on and after May 24, 2003—

(1) General.

(i) A licensed importer, licensed manufacturer or licensed dealer selling or otherwise distributing explosive materials (or a holder of a user permit disposing of surplus stock to a licensee; a holder of a user permit; or a holder of a limited permit who is within the same State as the distributor) who has the certified information required by this section may sell or distribute explosive materials to a licensee or permittee for not more than 45 days following the expiration date of the distributee's license or permit, unless the distributor knows or has reason to believe that the distributee's authority to continue business or operations under this part has been terminated.

(ii) A licensed importer, licensed manufacturer or licensed dealer selling or otherwise distributing explosive materials (or a holder of a user permit disposing of surplus stock to another licensee or permittee) must verify the license or permit status of the distributee prior to the release of explosive materials ordered, as required by this section.

(iii) Licensees or permittees desiring to return explosive materials to a licensed manufacturer may do so without obtaining a certified copy of the manufacturer's license.

(2) Verification of license/user permit.

(i) Prior to or with the first order of explosive materials, the distributee must provide the distributor certified copy (or, in the case of a user-limited, the original) of the distributee's license or user permit. However, licensees or holders of user permits that are business organizations may (in lieu of a certified copy of a license or user permit) provide the distributor with a certified list that contains the name, address, license or user permit number, and date of the license or user permit expiration of each location.

(ii) The distributee must also provide the distributor with a current list of the names of persons authorized to accept delivery of explosive materials on behalf of the distributee. The distributee ordering explosive materials must keep the list current and provide updated lists to licensees and holders of user permits on a timely basis. A distributor may not transfer possession of explosive materials to any person whose name does not appear on the current list of names of persons authorized to accept delivery of explosive materials on behalf of the distributee. Except as provided in paragraph (b)(3) of this section, in all instances the distributor must verify the identity of the distributee, or the employee of the distributee accepting possession of explosive materials on behalf of the distributee, by examining an identification document (as defined in §555.11) before relinquishing possession.

(iii) A licensee or holder of a user permit ordering explosive materials from another licensee or permittee must provide to the distributor a current, certified statement of the intended use of the explosive materials, e.g., resale, mining, quarrying, agriculture, construction, sport rocketry, road building, oil well drilling, seismographic research, etc.

(A) For individuals, the certified statement of intended use must specify the name, address, date and place of birth, and social security number of the distributee.

(B) For business organizations, the certified statement of intended use must specify the taxpayer identification number, the identity and the principal and local places of business.

(C) The licensee or holder of a user permit purchasing explosive materials must revise the furnished copy of the certified statement only when the information is no longer current.

(3) Delivery of explosive materials by common or contract carrier. When a common or contract carrier will transport explosive materials from a distributor to a distributee who is a licensee or holder of a user permit, the distributor must take the following actions before relinquishing possession of the explosive materials:

(i) Verify the identity of the person accepting possession for the common or contract carrier by examining such person's valid, unexpired driver's license issued by any State, Canada, or Mexico; and

(ii) Record the name of the common or contract carrier (i.e., the name of the driver's employer) and the full name of the driver. This information must be maintained in the distributor's permanent records in accordance with §555.121.

(4) User-limited permit transactions. A user-limited permit issued under the provisions of this part is valid for only a single purchase transaction and is not renewable (see §555.51). Accordingly, at the time a user-limited permittee orders explosive materials, the licensed distributor must write on the front of the user-limited permit the transaction date, his signature, and the distributor's license number prior to returning the permit to the user-limited permittee.

(Approved by the Office of Management and Budget under control number 1140–0079)

[ATF No. 1, 68 FR 13787, Mar. 20, 2003, as amended by ATF No. 2, 68 FR 53512, Sept. 11, 2003]

§ 555.104 Certified copy of license or permit.

Except as provided in §555.49(a), each person issued a license or permit under this part shall be furnished together with his license or permit a copy for his certification. If a person desires an additional copy of his license or permit for certification and for use under §555.103, he shall:

(a) Make a reproduction of the copy of his license or permit and execute the certification on it;

(b) Make a reproduction of his license or permit, enter on the reproduction the statement: "I certify that this is a true copy of a (insert the word license or permit) issued to me to engage in the specified business or operations", and sign his name next to the statement; or

(c) Submit a request, in writing, for certified copies of his license or permit to the Chief, Firearms and Explosives Licensing Center. The request will show the name, trade name (if any), and address of the licensee or permittee and the number of copies of the license or permit desired. There is a fee of $1 for each copy of a license or permit issued by the Chief, Firearms and Explosives Licensing Center under this paragraph. Fee payment must accompany each request for additional copies of a license or permit. The fee must be paid by (1) cash, or (2) money order or check made payable to the Bureau of Alcohol, Tobacco and Firearms.

[T.D. ATF–87, 46 FR 40384, Aug. 7, 1981, as amended by T.D. ATF–290, 54 FR 53054, Dec. 27, 1989]

§ 555.105 Distributions to nonlicensees, nonpermittees, and limited permittees.

(a) Distributions to nonlicensees and nonpermittees prior to May 24, 2003.

(1) This section will apply in any case where distribution of explosive materials to the distributee is not otherwise prohibited by the Act or this part.

(2) Except as provided in paragraph (a)(3) of this section, a licensed importer, licensed manufacturer, or licensed dealer may distribute explosive materials to a nonlicensee or nonpermittee if the nonlicensee or nonpermittee is a resident of the same State in which the licensee's business premises are located, and the nonlicensee or nonpermittee furnishes to the licensee the explosives transaction record, ATF F 5400.4, required by §555.126. Disposition of ATF F 5400.4 will be made in accordance with §555.126.

(3) A licensed importer, licensed manufacturer, or licensed dealer may sell or distribute explosive materials to a resident of a State contiguous to the State in which the licensee's place of business is located if the purchaser's State of residence has enacted legislation, currently in force, specifically authorizing a resident of that State to purchase explosive materials in a contiguous State and the purchaser and the licensee have, prior to the distribution of the explosive materials, complied with all the requirements of paragraphs (a)(2), (a)(5), and (a)(6) of this section applicable to intrastate transactions occurring on the licensee's business premises.

(4) A permittee may dispose of surplus stocks of explosive materials to a nonlicensee or nonpermittee if the nonlicensee or nonpermittee is a resident of the same State in which the permittee's business premises or operations are located, or is a resident of a State contiguous to the State in which the permittee's place of business or operations are located, and if the requirements of paragraphs (a)(2), (a)(3), (a)(5), and (a)(6) of this section are fully met.

(5) A licensed importer, licensed manufacturer, or licensed dealer selling or otherwise distributing explosive materials to a business entity must verify the identity of the representative or agent of the business entity who is authorized to order explosive materials on behalf of the business entity. Each business entity ordering explosive materials must furnish the distributing licensee prior to or with the first order of explosive materials a current certified list of the names of representatives or agents authorized to order explosive materials on behalf of the business entity. The business entity ordering explosive materials is responsible for keeping the certified list current. A licensee may not distribute explosive materials to a business entity on the order of a person whose name does not appear on the certified list.

(6) Where the possession of explosive materials is transferred at the distributor's premises, the distributor must in all instances verify the identity of the person accepting possession on behalf of the distributee before relinquishing possession. Before the delivery at the distributor's premises of explosive materials to an employee of a nonlicensee or nonpermittee, or to an employee of a common or contract carrier transporting explosive materials to a nonlicensee or nonpermittee, the distributor delivering explosive materials must obtain an executed ATF F 5400.8 from the employee before releasing the explosive materials. The ATF F 5400.8 must contain all of the information required on the form and by this part. (See examples in **§555.103(a)).**

(7) A licensee or permittee disposing of surplus stock may sell or distribute commercially manufactured black powder in quantities of 50 pounds or less to a nonlicensee or nonpermittee if the black powder is intended to be used solely for sporting, recreational, or cultural purposes in antique firearms as defined in 18 U.S.C. 921(a)(16), or in antique devices as exempted from the term "destructive device" in 18 U.S.C. 921(a)(4).

(b) Distributions to holders of limited permits on and after May 24, 2003.

(1) This section will apply in any case where distribution of explosive materials to the distributee is not otherwise prohibited by the Act or this part.

(2) A licensed importer, licensed manufacturer or a licensed dealer may distribute explosive materials to a holder of a limited permit if such permittee is a resident of the same State in which the licensee's business premises are located, the holder of the limited permit presents in person or by mail ATF Form 5400.4, Limited Permittee Transaction Report (LPTR), and the licensee completes Form 5400.4 in accordance with §555.126(b). In no event will a licensee distribute explosive materials to a holder of a limited permit unless the holder presents a Form 5400.4 with an original unaltered and unexpired Intrastate Purchase of Explosives Coupon (IPEC), ATF Form 5400.30, affixed. The coupon must bear the name, address, permit number, and the coupon number of the limited permittee seeking distribution of the explosives.

(3) A holder of a limited permit is authorized to receive explosive materials from a licensee or permittee whose premises are located in the same State of residence in which the premises of the holder of the limited permit are located on no more than 6 separate occasions during the one-year period of the permit. For purposes of this section, the term "6 separate occasions" means six deliveries of explosive materials. Each delivery must—

(i) Relate to a single purchase transaction made on one ATF F 5400.4;

(ii) Be referenced on one commercial invoice or purchase order; and

(iii) Be delivered to the holder of the limited permit in one shipment delivered at the same time.

(4) A holder of a user permit may dispose of surplus stocks of explosive materials to a licensee or holder of a user permit, or a holder of a limited permit who is a resident of the same State in which the premises of the holder of the user permit are located. A holder of a limited permit may dispose of surplus stocks of explosive materials to another holder of a limited permit who is a resident of the same State in which the premises of the distributor are located, if the transaction complies with the requirements of paragraph (b)(2) of this section and §555.126(b). A holder of a limited permit may also dispose of surplus stocks of explosive materials to a licensee or holder of a user permit if the disposition occurs in the State of residence of the holder of the limited permit. (See §555.103.)

(5) Each holder of a limited permit ordering explosive materials must furnish the distributing licensee prior to or with the first order of the explosive materials a current list of the names of employees authorized to accept delivery of explosive materials on behalf of the limited permittee. The distributee ordering explosive materials must keep the list current and provide updated lists to licensees and holders of user permits on a timely basis. A licensed importer, licensed manufacturer, licensed dealer, or permittee, selling or otherwise distributing explosive materials to a holder of a limited permit must, prior to delivering the explosive materials, obtain from the limited permittee a current list of persons who are authorized to accept deliveries of explosive materials on behalf of the limited permittee. A licensee or permittee may not deliver explosive materials to a person whose name does not appear on the list.

(6)(i) Delivery at the distributor's premises. Where possession of explosive materials is transferred directly to the distributee at the distributor's premises, the distributor must obtain an executed Form 5400.4 in accordance with §555.126(b) and must in all instances verify the identity of the person accepting possession on behalf of the distributee by examining an identification document (as defined in §555.11) before relinquishing possession.

(ii) Delivery by distributor. Where possession of explosive materials is transferred by the distributor to the distributee away from the distributor's premises, the distributor must obtain an executed Form 5400.4 in accordance with §555.126(b) and must in all instances verify the identity of the person accepting possession on behalf of the distributee by examining an identification document (as defined in §555.11) before relinquishing possession.

(iii) Delivery by common or contract carrier hired by the distributor. Where a common or contract carrier hired by the distributor will transport explosive materials from the distributor to a holder of a limited permit:

(A) The limited permittee must, prior to delivery of the explosive materials, complete the appropriate section on Form 5400.4, affix to the Form 5400.4 one of the six IPECs he has been issued, and provide the form to the distributor in person or by mail.

(B) The distributor must, before relinquishing possession of the explosive materials to the common or contract carrier:

(1) Verify the identity of the person accepting possession for the common or contract carrier by examining such person's valid, unexpired driver's license issued by any State, Canada, or Mexico; and

(2) Record the name of the common or contract carrier (i.e., the name of the driver's employer) and the full name of the driver. This information must be maintained in the distributor's permanent records in accordance with §555.121.

(C) At the time of delivery of the explosive materials, the common or contract carrier, as agent for the distributor, must verify the identity of the person accepting delivery on behalf of the distributee, note the type and number of the identification document (as defined in §555.11) and provide this information to the distributor. The distributor must enter this information in the appropriate section on Form 5400.4.

(iv) Delivery by common or contract carrier hired by the distributee. Where a common or contract carrier hired by the distributee will transport explosive materials from the distributor to a holder of a limited permit:

(A) The limited permittee must, prior to delivery of the explosive materials, complete the appropriate section on Form 5400.4, affix to the Form 5400.4 one of the six IPECs he has been issued, and provide the form to the distributor in person or by mail.

(B) Before the delivery at the distributor's premises to the common or contract carrier who will transport explosive materials to the holder of a limited permit, the distributor must:

(1) Verify the identity of the person accepting possession for the common or contract carrier by examining such person's valid, unexpired driver's license issued by any State, Canada, or Mexico; and

(2) Record the name of the common or contract carrier (i.e., the name of the driver's employer) and the full name of the driver. This information must be maintained in the distributor's permanent records in accordance with §555.121.

(7) A licensee or permittee disposing of surplus stock may sell or distribute commercially manufactured black powder in quantities of 50 pounds or less to a holder of a limited permit, nonlicensee, or nonpermittee if the black powder is intended to be used solely for sporting, recreational, or cultural purposes in antique firearms as defined in 18 U.S.C. 921(a)(16), or in antique devices as exempted from the term "destructive device" in 18 U.S.C. 921(a)(4).

(Approved by the Office of Management and Budget under control number 1140–0075)

[ATF No. 1, 68 FR 13788, Mar. 20, 2003, as amended by ATF No. 2, 68 FR 53513, Sept. 11, 2003]

§ 555.106 Certain prohibited distributions.

(a) A licensee or permittee may not distribute explosive materials to any person except—

(1) A licensee;

(2) A holder of a user permit; or

(3) A holder of a limited permit who is a resident of the State where distribution is made and in which the premises of the transferor are located.

(b) A licensee shall not distribute any explosive materials to any person:

(1) Who the licensee knows is less than 21 years of age;

(2) In any State where the purchase, possession, or use by a person of explosive materials would be in violation of any State law or any published ordinance applicable at the place of distribution;

(3) Who the licensee has reason to believe intends to transport the explosive materials into a State where the purchase, possession, or use of explosive materials is prohibited or which does not permit its residents to transport or ship explosive materials into the State or to receive explosive materials in the State; or

(4) Who the licensee has reasonable cause to believe intends to use the explosive materials for other than a lawful purpose.

(c) A licensee shall not distribute any explosive materials to any person knowing or having reason to believe that the person:

(1) Is, except as provided under §555.142 (d) and (e), under indictment or information for, or was convicted in any court of, a crime punishable by imprisonment for a term exceeding 1 year;

(2) Is a fugitive from justice;

(3) Is an unlawful user of marijuana, or any depressant or stimulant drug, or narcotic drug (as these terms are defined in the Controlled Substances Act, 21 U.S.C. 802);

(4) Was adjudicated as a mental defective or was committed to a mental institution;

(5) Is an alien, other than an alien who—

(i) Is lawfully admitted for permanent residence (as that term is defined in section 101(a)(20) of the Immigration and Nationality Act (8 U.S.C. 1101));

(ii) Is in lawful nonimmigrant status, is a refugee admitted under section 207 of the Immigration and Nationality Act (8 U.S.C. 1157), or is in asylum status under section 208 of the Immigration and Nationality Act (8 U.S.C. 1158), and—

(A) Is a foreign law enforcement officer of a friendly foreign government, as determined by the Attorney General in consultation with the Secretary of State, entering the United States on official law enforcement business, and the shipping, transporting, possession, or receipt of explosive materials is in furtherance of this official law enforcement business;

(B) Is a person having the power to direct or cause the direction of the management and policies of a corporation, partnership, or association licensed pursuant to section 843(a), and the shipping, transporting, possession, or receipt of explosive materials is in furtherance of such power;

(C) Is a member of a North Atlantic Treaty Organization (NATO) or other friendly foreign military force, as determined by the Attorney General in consultation with the Secretary of Defense, (whether or not admitted in a nonimmigrant status) who is present in the United States under military orders for training or other military purpose authorized by the United States, and the shipping, transporting, possession, or receipt of explosive materials is in furtherance of the military purpose; or

(D) Is lawfully present in the United States in cooperation with the Director of Central Intelligence, and the shipment, transportation, receipt, or possession of the explosive materials is in furtherance of such cooperation;

(6) Has been discharged from the armed forces under dishonorable conditions; or

(7) Having been a citizen of the United States, has renounced citizenship.

(d) The provisions of this section do not apply to the purchase of commercially manufactured black powder in quantities not to exceed 50 pounds, intended to be used solely for sporting, recreational, or cultural purposes in antique firearms or in antique devices, if the requirements of §555.105(a)(7) or (b)(7) are fully met.

[T.D. ATF-87, 46 FR 40384, Aug. 7, 1981. Redesignated by T.D. ATF–487, 68 FR 3748, Jan. 24, 2003. ATF No. 1, 68 FR 13790, Mar. 20, 2003]

§ 555.107 Record of transactions.

Each licensee and permittee shall keep records of explosive materials as required by subpart G of this part.

§ 555.108 Importation.

(a) Explosive materials imported or brought into the United States by a licensed importer or holder of a user permit may be released from customs custody to the licensed importer or holder of a user permit upon proof of his status as a licensed importer or holder of a user permit. Proof of status must be made by the licensed importer or holder of a user permit furnishing to the customs officer a certified copy of his license or permit (see §555.103).

(b) A nonlicensee or nonpermittee may import or bring into the United States commercially manufactured black powder in quantities not to exceed 50 pounds. Upon submitting to the customs officer completed ATF F 5400.3, certifying that the black powder is intended to be used solely for sporting, recreational, or cultural purposes in antique firearms or in antique devices, black powder may be released from customs custody. The disposition of the executed ATF F 5400.3 will be in accordance with the instructions on the form.

(c) The provisions of this section are in addition to, and are not in lieu of, any applicable requirement under 27 CFR Part 447.

(d) For additional requirements relating to the importation of plastic explosives into the United States on or after April 24, 1997, see §555.183.

(e) For requirements relating to the marking of imported explosive materials, see §555.109.

[T.D. ATF–87, 46 FR 40384, Aug. 7, 1981, as amended by T.D. ATF–387, 62 FR 8376, Feb. 25, 1997; ATF No. 1, 68 FR 13790, Mar. 20, 2003; ATF 5F, 70 FR 30633, May 27, 2005]

§ 555.109 Identification of explosive materials.

(a) General. Explosive materials, whether manufactured in the United States or imported, must contain certain marks of identification.

(b) Required marks

(1) Licensed manufacturers. Licensed manufacturers who manufacture explosive materials for sale or distribution must place the following marks of identification on explosive materials at the time of manufacture:

(i) The name of the manufacturer; and

(ii) The location, date, and shift of manufacture. Where a manufacturer operates his plant for only one shift during the day, he does not need to show the shift of manufacture.

(2) Licensed importers.

(i) Licensed importers who import explosive materials for sale or distribution must place the following marks of identification on the explosive materials they import:

(A) The name and address (city and state) of the importer; and

(B) The location (city and country) where the explosive materials were manufactured, date, and shift of manufacture. Where the foreign manufacturer operates his plant for only one shift during the day, he does not need to show the shift of manufacture.

(ii) Licensed importers must place the required marks on all explosive materials imported prior to distribution or shipment for use, and in no event later than 15 days after the date of release from Customs custody.

(c) General requirements.

(1) The required marks prescribed in this section must be permanent and legible.

(2) The required marks prescribed in this section must be in the English language, using Roman letters and Arabic numerals.

(3) Licensed manufacturers and licensed importers must place the required marks on each cartridge, bag, or other immediate container of explosive materials that they manufacture or import, as well as on any outside container used for the packaging of such explosive materials.

(4) Licensed manufacturers and licensed importers may use any method, or combination of methods, to affix the required marks to the immediate container of explosive materials, or outside containers used for the packaging thereof, provided the identifying marks are legible, permanent, show all the required information, and are not rendered unreadable by extended periods of storage.

(5) If licensed manufacturers or licensed importers desire to use a coding system and omit printed markings on the container that show all the required information specified in paragraphs (b)(1) and (2) of this section, they must file with ATF a letterhead application displaying the coding that they plan to use and explaining the manner of its application. The Director must approve the application before the proposed coding can be used.

(d) Exceptions —

(1) Blasting caps. Licensed manufacturers or licensed importers are only required to place the identification marks prescribed in this section on the containers used for the packaging of blasting caps.

(2) Alternate means of identification. The Director may authorize other means of identifying explosive materials, including fireworks, upon receipt of a letter application from the licensed manufacturer or licensed importer showing that such other identification is reasonable and will not hinder the effective administration of this part.

(Paragraph (b)(2) approved by the Office of Management and Budget under control number 1140–0055)

[ATF 5F, 70 FR 30633, May 27, 2005]

§ 555.110 Furnishing of samples (Effective on and after January 24, 2003).

(a) In general. Licensed manufacturers and licensed importers and persons who manufacture or import explosive materials or ammonium nitrate must, when required by letter issued by the Director, furnish—

(1) Samples of such explosive materials or ammonium nitrate;

(2) Information on chemical composition of those products; and

(3) Any other information that the Director determines is relevant to the identification of the explosive materials or to identification of the ammonium nitrate.

(b) Reimbursement. The Director will reimburse the fair market value of samples furnished pursuant to paragraph (a) of this section, as well as reasonable costs of shipment.

(Approved by the Office of Management and Budget under control number 1140–0073)

[ATF No. 1, 68 FR 13790, Mar. 20, 2003]

Subpart G—Records and Reports

§ 555.121 General.

(a) **(1)** Licensees and permittees shall keep records pertaining to explosive materials in permanent form (i.e., commercial invoices, record books) and in the manner required in this subpart.

(2) Licensees and permittees shall keep records required by this part on the business premises for five years from the date a transaction occurs or until discontinuance of business or operations by the licensee or permittee. (See also §555.128 for discontinuance of business or operations.)

(b) ATF officers may enter the premises of any licensee or holder of a user permit for the purpose of examining or inspecting any record or document required by or obtained under this part (see §555.24). Section 843(f) of the Act requires licensees and holders of user permits to make all required records available for examination or inspection at all reasonable times. Section 843(f) of the Act also requires licensees and permittees (including holders of limited permits) to submit all reports and information relating to all required records and their contents, as the regulations in this part prescribe.

(c) Each licensee and permittee shall maintain all records of importation, production, shipment, receipt, sale, or other disposition, whether temporary or permanent, of explosive materials as the regulations in this part prescribe. Sections 842(f) and 842(g) of the Act make it unlawful for any licensee or permittee knowingly to make any false entry in, or fail to make entry in, any record required to be kept under the Act and the regulations in this part.

(Approved by the Office of Management and Budget under control number 1512–0373)

[T.D. ATF–87, 46 FR 40384, Aug. 7, 1981, as amended by T.D. ATF–172, 49 FR 14941, Apr. 16, 1984; ATF No. 1, 68 FR 13790, Mar. 20, 2003]

§ 555.122 Records maintained by licensed importers.

(a) Each licensed importer shall take true and accurate physical inventories which will include all explosive materials on hand required to be accounted for in the records kept under this part. The licensed importer shall take a special inventory

(1) At the time of commencing business, which is the effective date of the license issued upon original qualification under this part;

(2) At the time of changing the location of his business to another region;

(3) At the time of discontinuing business; and

(4) At any time the regional director (compliance) may in writing require. Each special inventory is to be prepared in duplicate, the original of which is submitted to the regional director (compliance), and the duplicate retained by the licensed importer. If a special inventory specified by paragraphs (a) (1) through (4) of this section has not been taken during the calendar year, at least one physical inventory will be taken. However, the record of the yearly inventory, other than a special inventory required by paragraphs (a) (1) through (4) of this section, will remain on file for inspection instead of being sent to the regional director (compliance). (See also §555.127.)

(b) Each licensed importer shall, not later than the close of the next business day following the date of importation or other acquisition of explosive materials, enter the following information in a separate record:

(1) Date of importation or other acquisition.

(2) Name or brand name of manufacturer and country of manufacture.

(3) Manufacturer's marks of identification.

(4) Quantity (applicable quantity units, such as pounds of explosives, number of detonators, number of display fireworks, etc.).

(5) Description (dynamite (dyn), blasting agents (ba), detonators (det), display fireworks (df), etc.) and size (length and diameter or diameter only of display fireworks).

(c) Each licensed importer shall, not later than the close of the next business day following the date of distribution of any explosive materials to another licensee or a permittee, enter in a separate record the following information:

(1) Date of disposition.

(2) Name or brand name of manufacturer and country of manufacture.

(3) Manufacturer's marks of identification.

(4) Quantity (applicable quantity units, such as pounds of explosives, number of detonators, number of display fireworks, etc.).

(5) Description (dynamite (dyn), blasting agents (ba), detonators (det), display fireworks (df), etc.) and size (length and diameter or diameter only of display fireworks).

(6) License or permit number of licensee or permittee to whom the explosive materials are distributed.

(d) The regional director (compliance) may authorize alternate records to be maintained by a licensed importer to record his distribution of explosive materials when it is shown by the licensed importer that alternate records will accurately and readily disclose the information required by paragraph (c) of this section. A licensed importer who proposes to use alternate records shall submit a letter application to the regional director (compliance) and shall describe the proposed alternate records and the need for them. Alternate records are not to be employed by the licensed importer until approval is received from the regional director (compliance).

(e) Each licensed importer shall maintain separate records of the sales or other distribution made of explosive materials to nonlicensees or nonpermittees. These records are maintained as prescribed by §555.126.

(Approved by the Office of Management and Budget under control number 1512–0373)

[T.D. ATF–87, 46 FR 40384, Aug. 7, 1981, as amended by T.D. ATF–172, 49 FR 14941, Apr. 16, 1984; T.D. ATF–293, 55 FR 3721, Feb. 5, 1990; T.D. ATF–400, 63 FR 45003, Aug. 24, 1998]

§ 555.123 Records maintained by licensed manufacturers.

(a) Each licensed manufacturer shall take true and accurate physical inventories which will include all explosive materials on hand required to be accounted for in the records kept under this part. The licensed manufacturer shall take a special inventory

(1) At the time of commencing business, which is the effective date of the license issued upon original qualification under this part;

(2) At the time of changing the location of his premises to another region;

(3) At the time of discontinuing business; and

(4) At any other time the regional director (compliance) may in writing require. Each special inventory is to be prepared in duplicate, the original of which is submitted to the regional director (compliance), and the duplicate retained by the licensed manufacturer. If a special inventory required by paragraphs (a) (1) through (4) of this section has not been taken during the calendar year, at least one physical inventory will be taken. However, the record of the yearly inventory, other than a special inventory required by paragraphs (a) (1) through (4) of this section, will remain on file for inspection instead of being sent to the regional director (compliance). (See also §555.127.)

(b) Each licensed manufacturer shall not later than the close of the next business day following the date of manufacture or other acquisition of explosive materials, enter the following information in a separate record:

(1) Date of manufacture or other acquisition.

(2) Manufacturer's marks of identification.

(3) Quantity (applicable quantity units, such as pounds of explosives, number of detonators, number of display fireworks, etc.).

(4) Name, brand name or description (dynamite (dyn), blasting agents (ba), detonators (det), display fireworks (df), etc.) and size (length and diameter or diameter only of display fireworks).

(c) Each licensed manufacturer shall, not later than the close of the next business day following the date of distribution of any explosive materials to another licensee or a permittee, enter in a separate record the following information:

(1) Date of disposition.

(2) Name or brand name of manufacturer or name of importer, as applicable, if acquired other than by his own manufacture.

(3) Manufacturer's marks of identification.

(4) Quantity (applicable quantity units, such as pounds of explosives, number of detonators, number of display fireworks, etc.).

(5) Description (dynamite (dyn), blasting agents (ba), detonators (det), display fireworks (df), etc.) and size (length and diameter or diameter only of display fireworks).

(6) License or permit number of licensee or permittee to whom the explosive materials are distributed.

(d) Each licensed manufacturer who manufactures explosive materials for his own use shall, not later than the close of the next business day following the date of use, enter in a separate record the following information:

(1) Date of use.

(2) Quantity (applicable quantity units, such as pounds of explosives, number of detonators, number of special fireworks, etc.).

(3) Description (dynamite (dyn), blasting agents (ba), detonators (det), display fireworks (df), etc.) and size (length and diameter or diameter only of display fireworks).

Exception: A licensed manufacturer is exempt from the recordkeeping requirements of this subsection if the explosive materials are manufactured for his own use and used within a 24 hour period at the same site.

(e) The regional director (compliance) may authorize alternate records to be maintained by a licensed manufacturer to record his distribution or use of explosive materials when it is shown by the licensed manufacturer that alternate records will accurately and readily disclose the information required by paragraph (c) of this section. A licensed manufacturer who proposes to use alternate records shall submit a letter application to the regional director (compliance) and shall describe the proposed alternate records and the need for them. Alternate records are not to be employed by the licensed manufacturer until approval is received from the regional director (compliance).

(f) Each licensed manufacturer shall maintain separate records of the sales or other distribution made of explosive materials to nonlicensees or nonpermittees. These records are maintained as prescribed by §555.126.

(Approved by the Office of Management and Budget under control number 1512–0373)

[T.D. ATF–87, 46 FR 40384, Aug. 7, 1981, as amended by T.D. ATF–172, 49 FR 14941, Apr. 16, 1984; T.D. ATF–293, 55 FR 3721, Feb. 5, 1990; T.D. ATF–400, 63 FR 45003, Aug. 24, 1998]

§ 555.124 Records maintained by licensed dealers.

(a) Each licensed dealer shall take true and accurate physical inventories which will include all explosive materials on hand required to be accounted for in the records kept under this part. The licensed dealer shall take a special inventory

(1) At the time of commencing business, which is the effective date of the license issued upon original qualification under this part;

(2) At the time of changing the location of his premises to another region;

(3) At the time of discontinuing business; and

(4) At any other time the regional director (compliance) may in writing require. Each special inventory is to be prepared in duplicate, the original of which is submitted to the regional director (compliance), and the duplicate retained by the licensed dealer. If a special inventory required by paragraphs (a)(1) through (4) of this section has not been taken during the calendar year, at least one physical inventory will be taken. However, the record of the yearly inventory, other than a special inventory required by paragraphs (a)(1) through (4) of this section, will remain on file for inspection instead of being sent to the regional director (compliance). (See also §555.127.)

(b) Each licensed dealer shall, not later than the close of the next business day following the date of purchase or other acquisition of explosive materials (except as provided in paragraph (d) of this section), enter the following information in a separate record:

(1) Date of acquisition.

(2) Name or brand name of manufacturer and name of importer (if any).

(3) Manufacturer's marks of identification.

(4) Quantity (applicable quantity units, such as pounds of explosives, number of detonators, number of display fireworks, etc.).

(5) Description (dynamite (dyn), blasting agents (ba), detonators (det), display fireworks (df), etc.) and size (length and diameter or diameter only of display fireworks).

(6) Name, address, and license or permit number of the person from whom the explosive materials are received.

(c) Each licensed dealer shall, not later than the close of the next business day following the date of use (if the explosives are used by the dealer) or the date of distribution of any explosive materials to another licensee or a permittee (except as provided in paragraph (d) of this section), enter in a separate record the following information:

(1) Date of disposition.

(2) Name or brand name of manufacturer and name of importer (if any).

(3) Manufacturer's marks of identification.

(4) Quantity (applicable quantity units, such as pounds of explosives, number of detonators, number of display fireworks, etc.).

(5) Description (dynamite (dyn), blasting agents (ba), detonators (det), display fireworks (df), etc.) and size (length and diameter or diameter only of display fireworks).

(6) License or permit number of licensee or permittee to whom the explosive materials are distributed.

(d) When a commercial record is kept by a licensed dealer showing the purchase or other acquisition information required for the permanent record prescribed by paragraph (b) of this section, or showing the distribution information required for the permanent record prescribed by paragraph (c) of this section, the licensed dealer acquiring or distributing the explosive materials may, for a period not exceeding seven days following the date of acquisition of distribution of the explosive materials, delay making the required entry into the permanent record of acquisition or distribution. However, until the required entry of acquisition or disposition is made in the permanent record, the commercial record must be (1) kept by the licensed dealer separate from other commercial documents kept by the licensee, and (2) readily available for inspection on the licensed premises.

(e) The regional director (compliance) may authorize alternate records to be maintained by a licensed dealer to record his acquisition or disposition of explosive materials, when it is shown by the licensed dealer that alternate records will accurately and readily disclose the required information. A licensed dealer who proposes to use alternate records shall submit a letter application to the regional director (compliance) and shall describe the proposed alternate records and the need for them. Alternate records are not to be employed by the licensed dealer until approval is received from the regional director (compliance).

(f) Each licensed dealer shall maintain separate records of the sales or other distribution made of explosive materials to nonlicensees or nonpermittees. These records are maintained as prescribed by §555.126.

(Approved by the Office of Management and Budget under control number 1512–0373)

[T.D. ATF–87, 46 FR 40384, Aug. 7, 1981, as amended by T.D. ATF–172, 49 FR 14941, Apr. 16, 1984; T.D. ATF–293, 55 FR 3721, Feb. 5, 1990; T.D. ATF–400, 63 FR 45003, Aug. 24, 1998]

§ 555.125 Records maintained by permittees.

(a) Records maintained by permittees prior to May 24, 2003.

(1) Each permittee must take true and accurate physical inventories that will include all explosive materials on hand required to be accounted for in the records kept under this part. The permittee must take a special inventory—

(i) At the time of commencing business, which is the effective date of the permit issued upon original qualification under this part;

(ii) At the time of changing the location of his premises to another region;

(iii) At the time of discontinuing business; and

(iv) At any other time the regional director (compliance) may in writing require. Each special inventory is to be prepared in duplicate, the original of which is submitted to the regional director (compliance) and the duplicate retained by the permittee. If a special inventory required by paragraphs (a)(1)(i) through (iv) of this section has not been taken during the calendar year, a permittee is required to take at least one physical inventory. However, the record of the yearly inventory, other than a special inventory required by paragraphs (a)(1)(i) through (iv) of this section, will remain on file for inspection instead of being sent to the regional director (compliance). (See also §555.127.)

(2) Each permittee must, not later than the close of the next business day following the date of acquisition of explosive materials, enter the following information in a separate record:

(i) Date of acquisition;

(ii) Name or brand name of manufacturer;

(iii) Manufacturer's marks of identification;

(iv) Quantity (applicable quantity units, such as pounds of explosives, number of detonators, number of display fireworks, etc.);

(v) Description (dynamite (dyn), blasting agents (ba), detonators (det), display fireworks (df), etc., and size (length and diameter or diameter only of display fireworks)); and

(vi) Name, address, and license number of the persons from whom the explosive materials are received.

(3) Each permittee must, not later than the close of the next business day following the date of disposition of surplus explosive materials to another permittee or a licensee, enter in a separate record the information prescribed in §555.124(c).

(4) Each permittee must maintain separate records of disposition of surplus stocks of explosive materials to nonlicensees or nonpermittees as prescribed in §555.126.

(5) The regional director (compliance) may authorize alternate records to be maintained by a permittee to record his acquisition of explosive materials, when it is shown by the permittee that alternate records will accurately and readily disclose the required information. A permittee who proposes to use alternate records must submit a letter application to the regional director (compliance) and must describe the proposed alternate records and the need for them. Alternate records are not to be employed by the permittee until approval is received from the regional director (compliance).

(b) Records maintained by permittees on and after May 24, 2003.

(1) Each holder of a user permit must take true and accurate physical inventories that will include all explosive materials on hand required to be accounted for in the records kept under this part. The permittee must take a special inventory—

(i) At the time of commencing business, which is the effective date of the permit issued upon original qualification under this part;

(ii) At the time of changing the location of his premises;

(iii) At the time of discontinuing business; and

(iv) At any other time the regional director (compliance) may in writing require. Each special inventory is to be prepared in duplicate, the original of which is submitted to the regional director (compliance) and the duplicate retained by the permittee. If a special inventory required by paragraphs (b)(1)(i) through (iv) of this section has not been taken during the calendar year, a permittee is required to take at least one physical inventory.

The record of the yearly inventory, other than a special inventory required by paragraphs (b)(1)(i) through (iv) of this section, will remain on file for inspection instead of being sent to the regional director (compliance). (See also §555.127.)

(2) Each holder of a limited permit must take true and accurate physical inventories, at least annually, that will include all explosive materials on hand required to be accounted for in the records kept under this part.

(3) Each holder of a user permit or a limited permit must, not later than the close of the next business day following the date of acquisition of explosive materials, enter the following information in a separate record:

(i) Date of acquisition;

(ii) Name or brand name of manufacturer;

(iii) Manufacturer's marks of identification;

(iv) Quantity (applicable quantity units, such as pounds of explosives, number of detonators, number of display fireworks, etc.);

(v) Description (dynamite (dyn), blasting agents (ba), detonators (det), display fireworks (df), etc., and size (length and diameter or diameter only of display fireworks)); and

(vi) Name, address, and license number of the persons from whom the explosive materials are received.

(4) Each holder of a user permit or a limited permit must, not later than the close of the next business day following the date of disposition of surplus explosive materials to another permittee or a licensee, enter in a separate record the information prescribed in §555.124(c).

(5) When a record book is used as a permittee's permanent record the permittee may delay entry of the required information for a period not to exceed seven days if the commercial record contains all of the required information prescribed by paragraphs (b)(3) and (b)(4) of this section. However, the commercial record may be used instead of a record book as a permanent record provided that the record contains all of the required information prescribed by paragraphs (b)(3) and (b)(4) of this section.

(6) Each holder of a user permit or a limited permit must maintain separate records of disposition of surplus stocks of explosive materials to holders of a limited permit as prescribed in §555.126.

(7) The regional director (compliance) may authorize alternate records to be maintained by a holder of a user permit or a limited permit to record his acquisition of explosive materials, when it is shown by the permittee that alternate records will accurately and readily disclose the required information. A permittee who proposes to use alternate records must submit a letter application to the regional director (compliance) and must describe the proposed alternate records and the need for them. Alternate records are not to be employed by the permittee until approval is received from the regional director (compliance).

(Approved by the Office of Management and Budget under control number 1140–0030)

[ATF No. 1, 68 FR 13790, Mar. 20, 2003]

§ 555.126 Explosives transaction record for distribution of explosive materials prior to May 24, 2003 and Limited Permittee Transaction Report for distribution of explosive materials on and after May 24, 2003.

(a) Explosives transaction record for distribution of explosive materials prior to May 24, 2003.

(1) A licensee or permittee shall not temporarily or permanently distribute explosive materials to any person, other than another licensee or permittee, unless he records the transaction on an explosives transaction record, ATF F 5400.4.

(2) Before the distribution of explosive materials to a nonlicensee or nonpermittee who is a resident of the State in which the licensee or permittee maintains his business premises, or to a nonlicensee or nonpermittee who is not a resident of the State in which the licensee or permittee maintains his business premises and is acquiring explosive materials under §555.105(a)(3), the licensee or permittee distributing the explosive materials shall obtain an executed ATF F 5400.4 from the distributee which contains all of the information required on the form and by the regulations in this part.

(3) Completed ATF F 5400.4 is to be retained by the licensee or permittee as part of his permanent records in accordance with paragraph (a)(4) of this section.

(4) Each ATF F 5400.4 is retained in numerical (by transaction serial number) order commencing with "1" and continuing in regular sequence. When the numbering of any series reaches "1,000,000," the licensee or permittee may recommence the series. The recommenced series is to be given an alphabetical prefix or suffix. Where there is a change in proprietorship, or in the individual, firm, corporate name or trade name, the series in use at the time of the change may be continued.

(5) The requirements of this section are in addition to any other recordkeeping requirement contained in this part.

(6) A licensee or permittee may obtain, upon request, a supply of ATF F 5400.4 from the Director.

(b) Limited Permittee Transaction Report for distribution of explosive materials on and after May 24, 2003.

(1) A licensee or permittee may not distribute explosive materials to any person who is not a licensee or permittee. A licensee or permittee may not distribute explosive materials to a limited permittee unless the distributor records the transaction on ATF Form 5400.4, Limited Permittee Transaction Report.

(2) Before distributing explosive materials to a limited permittee, the licensee or permittee must obtain an executed Form 5400.4 from the limited permittee with an original unaltered and unexpired Intrastate Purchase of Explosives Coupon (IPEC) affixed. Except when delivery of explosive materials is made by a common or contract carrier who is an agent of the limited permittee, the licensee, permittee, or an agent of the licensee or permittee, must verify the identity of the of the holder of the limited permit by examining an identification document (as defined in §555.11) and noting on the Form 5400.4 the type of document presented. The licensee or permittee must complete the appropriate section on Form 5400.4 to indicate the type and quantity of explosive materials distributed, the license or permit number of the seller, and the date of the transaction. The licensee or permittee must sign and date the form and include any other information required by the instructions on the form and the regulations in this part.

(3) One copy of Form 5400.4 must be retained by the distributor as part of his permanent records in accordance with paragraph (b)(4) of this section and for the period specified in §555.121. The distributor must mail the other copy of Form 5400.4 to the Bureau of Alcohol, Tobacco, Firearms and Explosives in accordance with the instructions on the form.

(4) Each Form 5400.4 must be retained in chronological order by date of disposition, or in alphabetical order by name of limited permittee. A licensee may not, however, use both methods in a single recordkeeping system. Where there is a change in proprietorship by a limited permittee, the forms may continue to be filed together after such change.

(5) The requirements of this section are in addition to any other recordkeeping requirement contained in this part.

(Approved by the Office of Management and Budget under control number 1140–0078)

[T.D. ATF–87, 46 FR 40384, Aug. 7, 1981, as amended by T.D. ATF–93, 46 FR 50787, Oct. 15, 1981; T.D. ATF–172, 49 FR 14941, Apr. 16, 1984; T.D. ATF–446, 66 FR 16602, Mar. 27, 2001; ATF No. 1, 68 FR 13791, Mar. 20, 2003]

§ 555.127 Daily summary of magazine transactions.

In taking the inventory required by §§555.122, 555.123, 555.124, and 555.125, a licensee or permittee shall enter the inventory in a record of daily summary transactions to be kept at each magazine of an approved storage facility; however, these records may be kept at one central location on the business premises if separate records of daily transactions are kept for each magazine. Not later than the close of the next business day, each licensee and permittee shall record by manufacturer's name or brand name, the total quantity received in and removed from each magazine during the day, and the total remaining on hand at the end of the day. Quantity entries for display fireworks may be expressed as the number and size of individual display fireworks in a finished state or as the number of packaged display segments or packaged displays. Information as to the number and size of display fireworks contained in any one packaged display segment or packaged display shall be provided to any ATF officer on request. Any discrepancy which might indicate a theft or loss of explosive materials is to be reported in accordance with §555.30.

[T.D. ATF–293, 55 FR 3722, Feb. 5, 1990, as amended by T.D. ATF–400, 63 FR 45003, Aug. 24, 1998]

§ 555.128 Discontinuance of business.

Where an explosive materials business or operations is discontinued and succeeded by a new licensee or new permittee, the records prescribed by this subpart shall appropriately reflect such facts and shall be delivered to the successor. Where discontinuance of the business or operations is absolute, the records required by this subpart must be delivered within 30 days following the business or operations discontinuance to any ATF office located in the region in which the business was located, or to the ATF Out-of-Business Records Center, Spring Mills Office Park, 882 T.J. Jackson Drive, Falling Waters, West Virginia 25419. Where State law or local ordinance requires the delivery of records to other responsible authority, the Chief, Firearms and Explosives Licensing Center may arrange for the delivery of the records required by this subpart to such authority. (See also, §555.61.)

[T.D. ATF–290, 54 FR 53054, Dec. 27, 1989, as amended by T.D. ATF–446a, 66 FR 19089, Apr. 13, 2001; ATF No. 1, 68 FR 13792, Mar. 20, 2003]

§ 555.129 Exportation.

Exportation of explosive materials is to be in accordance with the applicable provisions of section 38 of the Arms Export Control Act (22 U.S.C. 2778) and implementing regulations. However, a licensed importer, licensed manufacturer, or licensed dealer exporting explosive materials shall maintain records showing the manufacture or acquisition of explosive materials as required by this part and records showing the quantity, the manufacturer's name or brand name of explosive materials, the name and address of the foreign consignee of the explosive materials, and the date the explosive materials were exported. See §555.180 for regulations concerning the exportation of plastic explosives.

[T.D. ATF–87, 46 FR 40384, Aug. 7, 1981, as amended by T.D. ATF–387, 62 FR 8377, Feb. 25, 1997]

§ 555.130 [Reserved]

Subpart H—Exemptions

§ 555.141 Exemptions.

(a) General. Except for the provisions of §§555.180 and 555.181, this part does not apply to:

(1) Any aspect of the transportation of explosive materials via railroad, water, highway, or air which is regulated by the U.S. Department of Transportation and its agencies, and which pertains to safety. For example, regulations issued by the Department of Transportation addressing the security risk of aliens transporting explosives by commercial motor or railroad carrier from Canada preclude the enforcement of 18 U.S.C. 842(i)(5) against persons shipping, transporting, receiving, or possessing explosives incident to and in connection with the commercial transportation of explosives by truck or rail from Canada into the United States. Questions concerning this exception should be directed to ATF's Public Safety Branch in Washington, DC.

(2) The use of explosive materials in medicines and medicinal agents in the forms prescribed by the official United States Pharmacopeia or the National Formulary. "The United States Pharmacopeia and The National Formulary," USP and NF Compendia, are available from the United States Pharmacopeial Convention, Inc., 12601 Twinbrook Parkway, Rockville, Maryland 20852.

(3) The transportation, shipment, receipt, or importation of explosive materials for delivery to any agency of the United States or to any State or its political subdivision.

(4) Small arms ammunition and components of small arms ammunition.

(5) The manufacture under the regulation of the military department of the United States of explosive materials for, or their distribution to or storage or possession by, the military or naval services or other agencies of the United States.

(6) Arsenals, navy yards, depots, or other establishments owned by, or operated by or on behalf of, the United States.

(7) The importation, distribution, and storage of fireworks classified as UN0336, UN0337, UN0431, or UN0432 explosives by the U.S. Department of Transportation at 49 CFR 172.101 and generally known as "consumer fireworks" or "articles pyrotechnic."

(8) Gasoline, fertilizers, propellant actuated devices, or propellant actuated industrial tools manufactured, imported, or distributed for their intended purposes.

(9) Industrial and laboratory chemicals which are intended for use as reagents and which are packaged and shipped pursuant to U.S. Department of Transportation regulations, 49 CFR Parts 100 to 177, which do not require explosives hazard warning labels.

(10) Model rocket motors that meet all of the following criteria—

(i) Consist of ammonium perchlorate composite propellant, black powder, or other similar low explosives;

(ii) Contain no more than 62.5 grams of total propellant weight; and

(iii) Are designed as single-use motors or as reload kits capable of reloading no more than 62.5 grams of propellant into a reusable motor casing.

(b) Black powder. Except for the provisions applicable to persons required to be licensed under subpart D, this part does not apply with respect to commercially manufactured black powder in quantities not to exceed 50 pounds, percussion caps, safety and pyrotechnic fuses, quills, quick and slow matches, and friction primers, if the black powder is intended to be used solely for sporting, recreational, or cultural purposes in antique firearms, as defined in 18 U.S.C. 921(a)(16) or antique devices, as exempted from the term "destructive devices" in 18 U.S.C. 921(a)(4).

[T.D. ATF–87, 46 FR 40384, Aug. 7, 1981 as amended by T.D. ATF–87, 46 FR 46916, Sept. 23, 1981; T.D. ATF–293, 55 FR 3722, Feb. 5, 1990; T.D. ATF–387, 62 FR 8377, Feb. 25, 1997; T.D. ATF–400, 63 FR 45003, Aug. 24, 1998; ATF No. 1, 68 FR 13792, Mar. 20, 2003; ATF 6F, 71 FR 46101, Aug. 11, 2006]

§ 555.142 Relief from disabilities (effective January 24, 2003).

(a) Any person prohibited from shipping or transporting any explosive in or affecting interstate or foreign commerce or from receiving or possessing any explosive which has been shipped or transported in or affecting interstate or foreign commerce may make application for relief from disabilities under section 845(b) of the Act.

(b) An application for relief from disabilities must be filed with the Director by submitting ATF Form 5400.29, Application for Restoration of Explosives Privileges, in accordance with the instructions on the form. The application must be supported by appropriate data, including the information specified in paragraph (f) of this section. Upon receipt of an incomplete or improperly executed application for relief, the applicant will be notified of the deficiency in the application. If the application is not corrected and returned within 30 days following the date of notification, the application will be considered abandoned.

(c) (1) The Director may grant relief to an applicant if it is established to the satisfaction of the Director that the circumstances regarding the disability and the applicant's record and reputation are such that the applicant will not be likely to act in a manner dangerous to public safety and that the granting of such relief is not contrary to the public interest.

(2) Except as provided in paragraph (c)(3) of this section, the Director will not grant relief if the applicant—

(i) Has not been discharged from parole or probation for a period of at least 2 years;

(ii) Is a fugitive from justice;

(iii) Is a prohibited alien;

(iv) Is an unlawful user of or addicted to any controlled substance;

(v) Has been adjudicated a mental defective or committed to a mental institution, unless the applicant was subsequently determined by a court, board, commission, or other lawful authority to have been restored to mental competency, to be no longer suffering from a mental disorder, and to have had all rights restored; or

(vi) Is prohibited by the law of the State where the applicant resides from receiving or possessing explosive materials.

(3)(i) The Director may grant relief to aliens who have been lawfully admitted to the United States or to persons who have not been discharged from parole or probation for a period of at least 2 years if he determines that the applicant has a compelling need to possess explosives, such as for purposes of employment.

(ii) The Director may grant relief to the persons identified in paragraph (c)(2) of this section in extraordinary circumstances where the granting of such relief is consistent with the public interest.

(d) A person who has been granted relief under this section is relieved of all disabilities imposed by the Act for the disabilities disclosed in the application. The granting of relief will not affect any disabilities incurred subsequent to the date the application was filed. Relief from disabilities granted to aliens will be effective only so long as the alien retains his or her lawful immigration status.

(e)(1) A licensee or permittee who is under indictment or information for, or convicted of, a crime punishable by imprisonment for a term exceeding one year during the term of a current license or permit, or while he has pending a license or permit renewal application, shall not be barred from licensed or permit operations for 30 days after the date of indictment or information or 30 days after the date upon which his conviction becomes final. Also, if he files his application for relief under this section within such 30 day period, he may further continue licensed or permit operations while his application is pending. A licensee or permittee who does not file an application within 30 days from the date of his indictment or information, or within 30 days from the date his conviction becomes final, shall not continue licensed or permit operations beyond 30 days from the date of his indictment or information or beyond 30 days from the date his conviction becomes final.

(2) In the event the term of a license or permit of a person expires during the 30 day period following the date of indictment of information of during the 30 day period after the date upon which his conviction becomes final or while his application for relief is pending, he shall file a timely application for renewal of his license or permit in order to continue licensed or permit operations. The license or permit application is to show that the applicant has been indicted or under information for, or convicted of, a crime punishable by imprisonment for a term exceeding one year.

(3) A licensee or permittee shall not continue licensed or permit operations beyond 30 days following the date the Director issues notification that the licensee's or permittee's application for removal of the disabilities resulting from an indictment, information or conviction has been denied.

(4) When a licensee or permittee may no longer continue licensed or permit operations under this section, any application for renewal of license of permit filed by the licensee or permittee while his application for removal of disabilities resulting from an indictment, information or conviction is pending, will be denied by the regional director (compliance).

(f)(1) Applications for relief from disabilities must include the following information:

(i) In the case of a corporation, or of any person having the power to direct or control the management of the corporation, information as to the absence of culpability in the offense for which the corporation, or any such person, was indicted, formally accused or convicted;

(ii) In the case of an applicant who is an individual, two properly completed FBI Forms FD–258 (fingerprint card), and a written statement from each of three references who are not related to the applicant by blood or marriage and have known the applicant for at least 3 years, recommending the granting of relief;

(iii) Written consent to examine and obtain copies of records and to receive statements and information regarding the applicant's background, including records, statements and other information concerning employment, medical history, military service, immigration status, and criminal record;

(iv) In the case of an applicant having been convicted of a crime punishable by imprisonment for a term exceeding one year, a copy of the indictment or information on which the applicant was convicted, the judgment of conviction or record of any plea of nolo contendere or plea of guilty or finding of guilt by the court;

(v) In the case of an applicant under indictment, a copy of the indictment or information;

(vi) In the case of an applicant who has been adjudicated a mental defective or committed to a mental institution, a copy of the order of a court, board, commission, or other lawful authority that made the adjudication or ordered the commitment, any petition that sought to have the applicant so adjudicated or committed, any medical records reflecting the reasons for commitment and diagnoses of the applicant, and any court order or finding of a court, board, commission, or other lawful authority showing the applicant's discharge from commitment, restoration of mental competency and the restoration of rights;

(vii) In the case of an applicant who has been discharged from the Armed Forces under dishonorable conditions, a copy of the applicant's Certificate of Release or Discharge from Active Duty (Department of Defense Form 214), Charge Sheet (Department of Defense Form 458), and final court martial order;

(viii) In the case of an applicant who, having been a citizen of the United States, has renounced his or her citizenship, a copy of the formal renunciation of nationality before a diplomatic or consular officer of the United States in a foreign state or before an officer designated by the Attorney General when the United States was in a state of war (see 8 U.S.C. 1481(a)(5) and (6)); and

(ix) In the case of an applicant who is an alien, documentation that the applicant is an alien who has been lawfully admitted to the United States; certification from the applicant including the applicant's INS-issued alien number or admission number, country/countries of citizenship, and immigration status, and certifying that the applicant is legally authorized to work in the United States, or other purposes for which possession of explosives is required; certification from an appropriate law enforcement agency of the applicant's country of citizenship stating that the applicant does not have a criminal record; and, if applicable, certification from a Federal explosives licensee or permittee or other employer stating that the applicant is employed by the employer and must possess explosive materials for purposes of employment. These certifications must be submitted in English.

(2) Any record or document of a court or other government entity or official required by paragraph (f)(1) of this section must be certified by the court or other government entity or official as a true copy.

(Approved by the Office of Management and Budget under control number 1140–0076)

[T.D. ATF–87, 46 FR 40384, Aug. 7, 1981. Redesignated by T.D. ATF–487, 68 FR 3748, Jan. 24, 2003. ATF No. 1, 68 FR 13792, Mar. 20, 2003]

Subpart I—Unlawful Acts, Penalties, Seizures and Forfeitures

§ 555.161 Engaging in business without a license.

Any person engaging in the business of importing, manufacturing, or dealing in explosive materials without a license issued under the Act, shall be fined not more than $10,000 or imprisoned not more than 10 years, or both.

§ 555.162 False statement or representation.

Any person who knowingly withholds information or makes any false or fictitious oral or written statement or furnishes or exhibits any false, fictitious, or misrepresented identification, intended or likely to deceive for the purpose of obtaining explosive materials, or a license, permit, exemption, or relief from disability under the Act, shall be fined not more than $10,000 or imprisoned not more than 10 years, or both.

§ 555.163 False entry in record.

Any licensed importer, licensed manufacturer, licensed dealer, or permittee who knowingly makes any false entry in any record required to be kept under subpart G of this part, shall be fined not more than $10,000 or imprisoned not more than 10 years, or both.

[T.D. ATF–87, 46 FR 40384, Aug. 7, 1981, as amended by T.D. ATF–400, 63 FR 45003, Aug. 24, 1998]

§ 555.164 Unlawful storage.

Any person who stores any explosive material in a manner not in conformity with this part, shall be fined not more than $1,000 or imprisoned not more than one year, or both.

§ 555.165 Failure to report theft or loss.

(a) Any person who has knowledge of the theft or loss of any explosive materials from his stock and fails to report the theft or loss within 24 hours of discovery in accordance with §555.30, shall be fined not more than $1,000 or imprisoned not more than one year, or both.

(b) On and after January 24, 2003, any licensee or permittee who fails to report a theft of explosive materials in accordance with §555.30 will be fined under title 18 U.S.C., imprisoned not more than 5 years, or both.

[T.D. ATF–87, 46 FR 40384, Aug. 7, 1981, as amended by ATF No. 1, 68 FR 13793, Mar. 20, 2003]

§ 555.166 Seizure or forfeiture.

Any explosive materials involved or used or intended to be used in any violation of the Act or of this part or in any violation of any criminal law of the United States are subject to seizure and forfeiture, and all provisions of title 26, U.S.C. relating to the seizure, forfeiture, and disposition of firearms, as defined in 26 U.S.C. 5845(a), will, so far as applicable, extend to seizures and forfeitures under the Act. (See §72.27 of this title for regulations on summary destruction of explosive materials which are impracticable or unsafe to remove to a place of storage.)

[T.D. ATF–87, 46 FR 40384, Aug. 7, 1981, as amended by T.D. ATF–363, 60 FR 17449, Apr. 6, 1995]

Subpart J—Marking of Plastic Explosives

§ 555.180 Prohibitions relating to unmarked plastic explosives.

(a) No person shall manufacture any plastic explosive that does not contain a detection agent.

(b) No person shall import or bring into the United States, or export from the United States, any plastic explosive that does not contain a detection agent. This paragraph does not apply to the importation or bringing into the United States, or the exportation from the United States, of any plastic explosive that was imported or brought into, or manufactured in the United States prior to April 24, 1996, by or on behalf of any agency of the United States performing military or police functions (including any military reserve component) or by or on behalf of the National Guard of any State, not later than 15 years after the date of entry into force of the Convention on the Marking of Plastic Explosives with respect to the United States, i.e., not later than June 21, 2013.

(c) No person shall ship, transport, transfer, receive, or possess any plastic explosive that does not contain a detection agent. This paragraph does not apply to:

(1) The shipment, transportation, transfer, receipt, or possession of any plastic explosive that was imported or brought into, or manufactured in the United States prior to April 24, 1996, by any person during the period beginning on that date and ending on April 24, 1999; or

(2) The shipment, transportation, transfer, receipt, or possession of any plastic explosive that was imported or brought into, or manufactured in the United States prior to April 24, 1996, by or on behalf of any agency of the United States performing a military or police function (including any military reserve component) or by or on behalf of the National Guard of any State, not later than 15 years after the date of entry into force of the Convention on the Marking of Plastic Explosives with respect to the United States, i.e., not later than June 21, 2013.

(d) When used in this subpart, terms are defined as follows:

(1) Convention on the Marking of Plastic Explosives means the Convention on the Marking of Plastic Explosives for the Purposes of Detection, Done at Montreal on 1 March 1991.

(2) "Date of entry into force" of the Convention on the Marking of Plastic Explosives means that date on which the Convention enters into force with respect to the U.S. in accordance with the provisions of Article XIII of the Convention on the Marking of Plastic Explosives. The Convention entered into force on June 21, 1998.

(3) Detection agent means any one of the substances specified in this paragraph when introduced into a plastic explosive or formulated in such explosive as a part of the manufacturing process in such a manner as to achieve homogeneous distribution in the finished explosive, including—

(i) Ethylene glycol dinitrate (EGDN), C[2]H[4](NO[3])[2], molecular weight 152, when the minimum concentration in the finished explosive is 0.2 percent by mass;

(ii) 2,3-Dimethyl-2,3-dinitrobutane (DMNB), C[6]H[12](NO[2])[2], molecular weight 176, when the minimum concentration in the finished explosive is 0.1 percent by mass;

(iii) Para-Mononitrotoluene (p-MNT), C[7]H[7]NO[2], molecular weight 137, when the minimum concentration in the finished explosive is 0.5 percent by mass;

(iv) Ortho-Mononitrotoluene (o-MNT), C[7]H[7]NO[2], molecular weight 137, when the minimum concentration in the finished explosive is 0.5 percent by mass; and

(v) Any other substance in the concentration specified by the Director, after consultation with the Secretary of State and Secretary of Defense, that has been added to the table in Part 2 of the Technical Annex to the Convention on the Marking of Plastic Explosives.

(4) Plastic explosive means an explosive material in flexible or elastic sheet form formulated with one or more high explosives which in their pure form has a vapor pressure less than 10<4>Pa at a temperature of 25 degrees C, is formulated with a binder material, and is as a mixture malleable or flexible at normal room temperature. High explosives, as defined in §555.202(a), are explosive materials which can be caused to detonate by means of a blasting cap when unconfined.

[T.D. ATF–387, 62 FR 8376, Feb. 25, 1997, as amended by T.D. ATF–419, 64 FR 55628, Oct. 14, 1999]

§ 555.181 Reporting of plastic explosives.

All persons, other than an agency of the United States (including any military reserve component) or the National Guard of any State, possessing any plastic explosive on April 24, 1996, shall submit a report to the Director no later than August 22, 1996. The report shall be in writing and mailed by certified mail (return receipt requested) to the Director at P.O. Box 50204, Washington, DC 20091–0204. The report shall include the quantity of plastic explosives possessed on April 24, 1996; any marks of identification on such explosives; the name and address of the manufacturer or importer; the storage location of such explosives, including the city and State; and the name and address of the person possessing the plastic explosives.

(Approved by the Office of Management and Budget under control number 1512–0535)

[T.D. ATF–382, 61 FR 38085, July 23, 1996, as amended by T.D. ATF–387, 62 FR 8377, Feb. 25, 1997]

§ 555.182 Exceptions.

It is an affirmative defense against any proceeding involving §§555.180 and 555.181 if the proponent proves by a preponderance of the evidence that the plastic explosive—

(a) Consisted of a small amount of plastic explosive intended for and utilized solely in lawful—

(1) Research, development, or testing of new or modified explosive materials;

(2) Training in explosives detection or development or testing of explosives detection equipment; or

(3) Forensic science purposes; or

(b) Was plastic explosive that, by April 24, 1999, will be or is incorporated in a military device within the territory of the United States and remains an integral part of such military device, or is intended to be, or is incorporated in, and remains an integral part of a military device that is intended to become, or has become, the property of any agency of the United States performing military or police functions (including any military reserve component) or the National Guard of any State, wherever such device is located. For purposes of this paragraph, the term "military device" includes, but is not restricted to, shells, bombs, projectiles, mines, missiles, rockets, shaped charges, grenades, perforators, and similar devices lawfully manufactured exclusively for military or police purposes.
[T.D. ATF–387, 62 FR 8377, Feb. 25, 1997]

§ 555.183 Importation of plastic explosives on or after April 24, 1997.

Persons filing Form 6 applications for the importation of plastic explosives on or after April 24, 1997, shall attach to the application the following written statement, prepared in triplicate, executed under the penalties of perjury:

(a) "I declare under the penalties of perjury that the plastic explosive to be imported contains a detection agent as required by 27 CFR 555.180(b)"; or

(b) "I declare under the penalties of perjury that the plastic explosive to be imported is a "small amount" to be used for research, training, or testing purposes and is exempt from the detection agent requirement pursuant to 27 CFR 555.182."
[T.D. ATF–387, 62 FR 8377, Feb. 25, 1997]

§ 555.184 Statements of process and samples.

(a) A complete and accurate statement of process with regard to any plastic explosive or to any detection agent that is to be introduced into a plastic explosive or formulated in such plastic explosive shall be submitted by a licensed manufacturer or licensed importer, upon request, to the Director.

(b) Samples of any plastic explosive or detection agent shall be submitted by a licensed manufacturer or licensed importer, upon request, to the Director.

(Paragraph (a) approved by the Office of Management and Budget under control number 1512–0539)
[T.D. ATF–387, 62 FR 8378, Feb. 25, 1997]

§ 555.185 Criminal sanctions.

Any person who violates the provisions of 18 U.S.C. 842(l)–(o) shall be fined under title 18, U.S.C., imprisoned for not more than 10 years, or both.
[T.D. ATF–387, 62 FR 8378, Feb. 25, 1997]

§ 555.186 Seizure or forfeiture.

Any plastic explosive that does not contain a detection agent in violation of 18 U.S.C. 842(l)–(n) is subject to seizure and forfeiture, and all provisions of 19 U.S.C. 1595a, relating to seizure, forfeiture, and disposition of merchandise introduced or attempted to be introduced into the U.S. contrary to law, shall extend to seizures and forfeitures under this subpart. See §72.27 of this chapter for regulations on summary destruction of plastic explosives that do not contain a detection agent.
[T.D. ATF–387, 62 FR 8378, Feb. 25, 1997]

Subpart K—Storage

§ 555.201 General.

(a) Section 842(j) of the Act and §555.29 of this part require that the storage of explosive materials by any person must be in accordance with the regulations in this part. Further, section 846 of this Act authorizes regulations to prevent the recurrence of accidental explosions in which explosive materials were involved. The storage standards prescribed by this subpart confer no right or privileges to store explosive materials in a manner contrary to State or local law.

(b) The Director may authorize alternate construction for explosives storage magazines when it is shown that the alternate magazine construction is substantially equivalent to the standards of safety and security contained in this subpart. Any alternate explosive magazine construction approved by the Director prior to August 9, 1982, will continue as approved unless notified in writing by the Director. Any person intending to use alternate magazine construction shall submit a letter application to the regional director (compliance) for transmittal to the Director, specifically describing the proposed magazine. Explosive materials may not be stored in alternate magazines before the applicant has been notified that the application has been approved.

(c) A licensee or permittee who intends to make changes in his magazines, or who intends to construct or acquire additional magazines, shall comply with §555.63.

(d) The regulations set forth in §§555.221 through 555.224 pertain to the storage of display fireworks, pyrotechnic compositions, and explosive materials used in assembling fireworks and articles pyrotechnic.

(e) The provisions of §555.202(a) classifying flash powder and bulk salutes as high explosives are mandatory after March 7, 1990: Provided, that those persons who hold licenses or permits under this part on that date shall, with respect to the premises covered by such licenses or permits, comply with the high explosives storage requirements for flash powder and bulk salutes by March 7, 1991.

(f) Any person who stores explosive materials shall notify the authority having jurisdiction for fire safety in the locality in which the explosive materials are being stored of the type, magazine capacity, and location of each site where such explosive materials are stored. Such notification shall be made orally before the end of the day on which storage of the explosive materials commenced and in writing within 48 hours from the time such storage commenced.

(Paragraph (f) approved by the Office of Management and Budget under control number 1512–0536)

[T.D. ATF–87, 46 FR 40384, Aug. 7, 1981, as amended by T.D. ATF–293, 55 FR 3722, Feb. 5, 1990; T.D. ATF–400, 63 FR 45003, Aug. 24, 1998]

§ 555.202 Classes of explosive materials.

For purposes of this part, there are three classes of explosive materials. These classes, together with the description of explosive materials comprising each class, are as follows:

(a) High explosives. Explosive materials which can be caused to detonate by means of a blasting cap when unconfined, (for example, dynamite, flash powders, and bulk salutes). See also §555.201(e).

(b) Low explosives. Explosive materials which can be caused to deflagrate when confined (for example, black powder, safety fuses, igniters, igniter cords, fuse lighters, and "display fireworks" classified as UN0333, UN0334, or UN0335 by the U.S. Department of Transportation regulations at 49 CFR 172.101, except for bulk salutes).

(c) Blasting agents. (For example, ammonium nitrate-fuel oil and certain water-gels (see also §555.11).

[T.D. ATF–87, 46 FR 40384, Aug. 7, 1981, as amended by T.D. ATF–293, 55 FR 3722, Feb. 5, 1990; T.D. ATF–400, 63 FR 45003, Aug. 24, 1998]

§ 555.203 Types of magazines.

For purposes of this part, there are five types of magazines. These types, together with the classes of explosive materials, as defined in §555.202, which will be stored in them, are as follows:

(a) Type 1 magazines. Permanent magazines for the storage of high explosives, subject to the limitations prescribed by §§555.206 and 555.213. Other classes of explosive materials may also be stored in type 1 magazines.

(b) Type 2 magazines. Mobile and portable indoor and outdoor magazines for the storage of high explosives, subject to the limitations prescribed by §§555.206, 555.208(b), and 555.213. Other classes of explosive materials may also be stored in type 2 magazines.

(c) Type 3 magazines. Portable outdoor magazines for the temporary storage of high explosives while attended (for example, a "day-box"), subject to the limitations prescribed by §§555.206 and 555.213. Other classes of explosives materials may also be stored in type 3 magazines.

(d) Type 4 magazines. Magazines for the storage of low explosives, subject to the limitations prescribed by §§555.206(b), 555.210(b), and 555.213. Blasting agents may be stored in type 4 magazines, subject to the limitations prescribed by §§555.206(c), 555.211(b), and 555.213. Detonators that will not mass detonate may also be stored in type 4 magazines, subject to the limitations prescribed by §§555.206(a), 555.210(b), and 555.213.

(e) Type 5 magazines. Magazines for the storage of blasting agents, subject to the limitations prescribed by §§555.206(c), 555.211(b), and 555.213.

§ 555.204 Inspection of magazines.

Any person storing explosive materials shall inspect his magazines at least every seven days. This inspection need not be an inventory, but must be sufficient to determine whether there has been unauthorized entry or attempted entry into the magazines, or unauthorized removal of the contents of the magazines.

§ 555.205 Movement of explosive materials.

All explosive materials must be kept in locked magazines meeting the standards in this subpart unless they are:

(a) In the process of manufacture;

(b) Being physically handled in the operating process of a licensee or user;

(c) Being used; or

(d) Being transported to a place of storage or use by a licensee or permittee or by a person who has lawfully acquired explosive materials under §555.106.

§ 555.206 Location of magazines.

(a) Outdoor magazines in which high explosives are stored must be located no closer to inhabited buildings, passenger railways, public highways, or other magazines in which high explosives are stored, than the minimum distances specified in the table of distances for storage of explosive materials in §555.218.

(b) Outdoor magazines in which low explosives are stored must be located no closer to inhibited buildings, passenger railways, public highways, or other magazines in which explosive materials are stored, than the minimum distances specified in the table of distances for storage of low explosives in §555.219, except that the table of distances in §555.224 shall apply to the storage of display fireworks. The distances shown in §555.219 may not be reduced by the presence of barricades.

(c)(1) Outdoor magazines in which blasting agents in quantities of more than 50 pounds are stored must be located no closer to inhabited buildings, passenger railways, or public highways than the minimum distances specified in the table of distances for storage of explosive materials in §555.218.

(2) Ammonium nitrate and magazines in which blasting agents are stored must be located no closer to magazines in which high explosives or other blasting agents are stored than the minimum distances specified in the table of distances for the separation of ammonium nitrate and blasting agents in §555.220. However, the minimum distances for magazines in which explosives and blasting agents are stored from inhabited buildings, etc., may not be less than the distances specified in the table of distances for storage of explosives materials in §555.218.

[T.D. ATF–87, 46 FR 40384, Aug. 7, 1981, as amended by T.D. ATF–293, 55 FR 3722, Feb. 5, 1990; T.D. ATF–400, 63 FR 45003, Aug. 24, 1998]

§ 555.207 Construction of type 1 magazines.

A type 1 magazine is a permanent structure: a building, an igloo or "Army-type structure", a tunnel, or a dugout. It is to be bullet-resistant, fire-resistant, weather-resistant, theft-resistant, and ventilated.

(a) Buildings. All building type magazines are to be constructed of masonry, wood, metal, or a combination of these materials, and have no openings except for entrances and ventilation. The ground around building magazines must slope away for drainage or other adequate drainage provided.

(1) Masonry wall construction. Masonry wall construction is to consist of brick, concrete, tile, cement block, or cinder block and be not less than 6 inches in thickness. Hollow masonry units used in construction must have all hollow spaces filled with well-tamped, coarse, dry sand or weak concrete (at least a mixture of one part cement and eight parts of sand with enough water to dampen the mixture while tamping in place). Interior walls are to be constructed of, or covered with, a nonsparking material.

(2) Fabricated metal wall construction. Metal wall construction is to consist of sectional sheets of steel or aluminum not less than number 14-gauge, securely fastened to a metal framework. Metal wall construction is either lined inside with brick, solid cement blocks, hardwood not less than four inches thick, or will have at least a six inch sand fill between interior and exterior walls. Interior walls are to be constructed of, or covered with, a nonsparking material.

(3) Wood frame wall construction. The exterior of outer wood walls is to be covered with iron or aluminum not less than number 26-gauge. An inner wall of, or covered with nonsparking material will be constructed so as to provide a space of not less than six inches between the outer and inner walls. The space is to be filled with coarse, dry sand or weak concrete.

(4) Floors. Floors are to be constructed of, or covered with, a nonsparking material and shall be strong enough to bear the weight of the maximum quantity to be stored. Use of pallets covered with a nonsparking material is considered equivalent to a floor constructed of or covered with a nonsparking material.

(5) Foundations. Foundations are to be constructed of brick, concrete, cement block, stone, or wood posts. If piers or posts are used, in lieu of a continuous foundation, the space under the buildings is to be enclosed with metal.

(6) Roof. Except for buildings with fabricated metal roofs, the outer roof is to be covered with no less than number 26-guage iron or aluminum, fastened to at least ⅞ inch sheathing.

(7) Bullet-resistant ceilings or roofs. Where it is possible for a bullet to be fired directly through the roof and into the magazine at such an angle that the bullet would strike the explosives within, the magazine is to be protected by one of the following methods:

(i) A sand tray lined with a layer of building paper, plastic, or other nonporous material, and filled with not less than four inches of coarse, dry sand, and located at the tops of inner walls covering the entire ceiling area, except that portion necessary for ventilation.

(ii) A fabricated metal roof constructed of 3⁄16-inch plate steel lined with four inches of hardwood. (For each additional ⅙ inch of plate steel, the hardwood lining may be decreased one inch.)

(8) Doors. All doors are to be constructed of not less than ¼ inch plate steel and lined with at least two inches of hardwood. Hinges and hasps are to be attached to the doors by welding, riveting or bolting (nuts on inside of door). They are to be installed in such a manner that the hinges and hasps cannot be removed when the doors are closed and locked.

(9) Locks. Each door is to be equipped with (i) two mortise locks; (ii) two padlock fastened in separate hasps and staples; (iii) a combination of a mortise lock and a padlock; (iv) a mortise lock that requires two keys to open; or (v) a three-point lock Padlocks must have at least five tumblers and a casehardened shackle of at least ⅜ inch diameter. Padlocks must be protected with not less than ¼ inch steel hoods constructed so as to prevent sawing or lever action on the locks, hasps, and staples. These requirements do not apply to magazine doors that are adequately secured on the inside by means of a bolt, lock, or bar that cannot be actuated from the outside.

(10) Ventilation. Ventilation is to be provided to prevent dampness and heating of stored explosive materials. Ventilation openings must be screened to prevent the entrance of sparks Ventilation openings in side walls and foundations must be offset or shielded for bullet-resistant purposes. Magazines having foundation and roof ventilators with the air circulating between the side walls and the floors and between the side walls and the ceiling must have a wooden lattice lining or equivalent to prevent the packages of explosive materials from being stacked against the side walls and blocking the air circulation.

(11) Exposed metal. No sparking material is to be exposed to contact with the stored explosive materials. All ferrous metal nails in the floor and side walls, which might be exposed to contact with explosive materials, must be blind nailed, countersunk, or covered with a nonsparking lattice work or other nonsparking material.

(b) Igloos, "Army-type structures", tunnels, and dugouts. Igloo, "Army-type structure", tunnel, and dugout magazines are to be constructed of reinforced concrete, masonry, metal, or a combination of these materials. They must have an earthmound covering of not less than 24 inches on the top, sides and rear unless the magazine meets the requirements of paragraph (a)(7) of this section. Interior walls and floors must be constructed of, or covered with, a nonsparking material. Magazines of this type are also to be constructed in conformity with the requirements of paragraph (a)(4) and paragraphs (a)(8) through (11) of this section.

§ 555.208 Construction of type 2 magazines.

A type 2 magazine is a box, trailer, semitrailer, or other mobile facility.

(a) Outdoor magazines—

(1) General. Outdoor magazines are to be bullet-resistant, fire-resistant, weather-resistant, theft-resistant, and ventilated. They are to be supported to prevent direct contact with the ground and, if less than one cubic yard in size, must be securely fastened to a fixed object. The ground around outdoor magazines must slope away for drainage or other adequate drainage provided. When unattended, vehicular magazines must have wheels removed or otherwise effectively immobilized by kingpin locking devices or other methods approved by the Director.

(2) Exterior construction. The exterior and doors are to be constructed of not less than ¼-inch steel and lined with at least two inches of hardwood. Magazines with top openings will have lids with water-resistant seals or which overlap the sides by at least one inch when in a closed position.

(3) Hinges and hasps. Hinges and hasps are to be attached to doors by welding, riveting, or bolting (nuts on inside of door). Hinges and hasps must be installed so that they cannot be removed when the doors are closed and locked.

(4) Locks. Each door is to be equipped with (i) two mortise locks; (ii) two padlocks fastened in separate hasps and staples; (iii) a combination of a mortise lock and a padlock; (iv) a mortise lock that requires two keys to open; or (v) a three-point lock. Padlocks must have at least five tumblers and a case-hardened shackle of at least ⅜-inch diameter. Padlocks must be protected with not less than ¼-inch steel hoods constructed so as to prevent sawing or lever action on the locks, hasps, and staples. These requirements do not apply to magazine doors that are adequately secured on the inside by means of a bolt, lock, or bar that cannot be actuated from the outside.

(b) Indoor magazines—

(1) General. Indoor magazines are to be fire-resistant and theft-resistant. They need not be bullet-resistant and weather-resistant if the buildings in which they are stored provide protection from the weather and from bullet penetration. No indoor magazine is to be located in a residence or dwelling. The indoor storage of high explosives must not exceed a quantity of 50 pounds. More than one indoor magazine may be located in the same building if the total quantity of explosive materials stored does not exceed 50 pounds. Detonators must be stored in a separate magazine (except as provided in §555.213) and the total quantity of detonators must not exceed 5,000.

(2) Exterior construction. Indoor magazines are to be constructed of wood or metal according to one of the following specifications:

(i) Wood indoor magazines are to have sides, bottoms and doors constructed of at least two inches of hardwood and are to be well braced at the corners. They are to be covered with sheet metal of not less than number 26-gauge (.0179 inches). Nails exposed to the interior of magazines must be countersunk.

(ii) Metal indoor magazines are to have sides, bottoms and doors constructed of not less than number 12-gauge (.1046 inches) metal and be lined inside with a nonsparking material. Edges of metal covers must overlap sides at least one inch.

(3) Hinges and hasps. Hinges and hasps are to be attached to doors by welding, riveting, or bolting (nuts on inside of door). Hinges and hasps must be installed so that they cannot be removed when the doors are closed and locked.

(4) Locks. Each door is to be equipped with (i) two mortise locks; (ii) two padlocks fastened in separate hasps and staples; (iii) a combination of a mortise lock and a padlock; (iv) a mortise lock that requires two keys to open; or (v) a three-point lock. Padlocks must have at least five tumblers and a case-hardened shackle of at least ⅜-inch diameter. Padlocks must be protected with not less than ¼-inch steel hoods constructed so as to prevent sawing or lever action on the locks, hasps, and staples. Indoor magazines located in secure rooms that are locked as provided in this subparagraph may have each door locked with one steel padlock (which need not be protected by a steel hood) having at least five tumblers and a case-hardened shackle of at least ⅜-inch diameter, if the door hinges and lock hasp are securely fastened to the magazine.

These requirements do not apply to magazine doors that are adequately secured on the inside by means of a bolt, lock, or bar that cannot be actuated from the outside.

(c) Detonator boxes. Magazines for detonators in quantities of 100 or less are to have sides, bottoms and doors constructed of not less than number 12-gauge (.1046 inches) metal and lined with a nonsparking material. Hinges and hasps must be attached so they cannot be removed from the outside. One steel padlock (which need not be protected by a steel hood) having at least five tumblers and a case-hardened shackle of at least ⅜-inch diameter is sufficient for locking purposes.

§ 555.209 Construction of type 3 magazines.

A type 3 magazine is a "day-box" or other portable magazine. It must be fire-resistant, weather-resistant, and theft-resistant. A type 3 magazine is to be constructed of not less than number 12-gauge (.1046 inches) steel, lined with at least either ½-inch plywood or ½-inch Masonite-type hardboard.

Doors must overlap sides by at least one inch. Hinges and hasps are to be attached by welding, riveting or bolting (nuts on inside).

One steel padlock (which need not be protected by a steel hood) having at least five tumblers and a case-hardened shackle of at least ⅜-inch diameter is sufficient for locking purposes. Explosive materials are not to be left unattended in type 3 magazines and must be removed to type 1 or 2 magazines for unattended storage.

§ 555.210 Construction of type 4 magazines.

A type 4 magazine is a building, igloo or "Army-type structure", tunnel, dugout, box, trailer, or a semitrailer or other mobile magazine.

(a) Outdoor magazines—

(1) General. Outdoor magazines are to be fire-resistant, weather-resistant, and theft-resistant. The ground around outdoor magazines must slope away for drainage or other adequate drainage be provided. When unattended, vehicular magazines must have wheels removed or otherwise be effectively immobilized by kingpin locking devices or other methods approved by the Director.

(2) Construction. Outdoor magazines are to be constructed of masonry, metal-covered wood, fabricated metal, or a combination of these materials. Foundations are to be constructed of brick, concrete, cement block, stone, or metal or wood posts. If piers or posts are used, in lieu of a continuous foundation, the space under the building is to be enclosed with fire-resistant material. The walls and floors are to be constructed of, or covered with, a nonsparking material or lattice work. The doors must be metal or solid wood covered with metal.

(3) Hinges and hasps. Hinges and hasps are to be attached to doors by welding, riveting, or bolting (nuts on inside of door). Hinges and hasps must be installed so that they cannot be removed when the doors are closed and locked.

(4) Locks. Each door is to be equipped with (i) two mortise locks; (ii) two padlocks fastened in separate hasps and staples; (iii) a combination of a mortise lock and a padlock; (iv) a mortise lock that requires two keys to open; or (v) a three-point lock. Padlocks must have at least five tumblers and case-hardened shackle of at least ⅜ inch diameter. Padlocks must be protected with not less than ¼ inch steel hoods constructed so as to prevent sawing or lever action on the locks, hasps, and staples.

These requirements do not apply to magazine doors that are adequately secured on the inside by means of a bolt, lock, or bar that cannot be actuated from the outside.

(b) Indoor magazine—

(1) General. Indoor magazines are to be fire-resistant and theft-resistant. They need not be weather-resistant if the buildings in which they are stored provide protection from the weather. No indoor magazine is to be located in a residence or dwelling. The indoor storage of low explosives must not exceed a quantity of 50 pounds. More than one indoor magazine may be located in the same building if the total quantity of explosive materials stored does not exceed 50 pounds. Detonators that will not mass detonate must be stored in a separate magazine and the total number of electric detonators must not exceed 5,000.

(2) Construction. Indoor magazines are to be constructed of masonry, metal-covered wood, fabricated metal, or a combination of these materials. The walls and floors are to be constructed of, or covered with, a nonsparking material. The doors must be metal or solid wood covered with metal.

(3) Hinges and hasps. Hinges and hasps are to be attached to doors by welding, riveting, or bolting (nuts on inside of door). Hinges and hasps must be installed so that they cannot be removed when the doors are closed and locked.

(4) Locks. Each door is to be equipped with (i) two mortise locks; (ii) two padlocks fastened in separate hasps and staples; (iii) a combination of a mortise lock and padlock; (iv) a mortise lock that requires two keys to open; or (v) a three-point lock.

Padlocks must have at least five tumblers and a case-hardened shackle of at least ⅜ inch diameter. Padlocks must be protected with not less than ¼ inch steel hoods constructed so as to prevent sawing or lever action on the locks, hasps, and staples. Indoor magazines located in secure rooms that are locked as provided in this subparagraph may have each door locked with one steel padlock (which need not be protected by a steel hood) having at least five tumblers and a case-hardened shackle of at least ⅜ inch diameter, if the door hinges and lock hasp are securely fastened to the magazine.

These requirements do not apply to magazine doors that are adequately secured on the inside by means of a bolt, lock, or bar that cannot be actuated from the outside.

§ 555.211 Construction of type 5 magazines.

A type 5 magazine is a building, igloo or "Army-type structure", tunnel, dugout, bin, box, trailer, or a semitrailer or other mobile facility.

(a) Outdoor magazines—

(1) General. Outdoor magazines are to be weather-resistant and theft-resistant. The ground around magazines must slope away for drainage or other adequate drainage be provided. When unattended, vehicular magazines must have wheels removed or otherwise be effectively immobilized by kingpin locking devices or other methods approved by the Director.

(2) Construction. The doors are to be constructed of solid wood or metal.

(3) Hinges and hasps. Hinges and hasps are to be attached to doors by welding, riveting, or bolting (nuts on inside of door). Hinges and hasps must be installed so that they cannot be removed when the doors are closed and locked.

(4) Locks. Each door is to be equipped with (i) two mortise locks; (ii) two padlocks fastened in separate hasps and staples; (iii) a combination of a mortise lock and a padlock; (iv) a mortise lock that requires two keys to open; or (v) a three-point lock. Padlocks must have at least five tumblers and a case-hardened shackle of at least ⅜ inch diameter. Padlocks must be protected with not less than ¼ inch steel hoods constructed so as to prevent sawing or lever action on the locks, hasps, and staples. Trailers, semitrailers, and similar vehicular magazines may, for each door, be locked with one steel padlock (which need not be protected by a steel hood) having at least five tumblers and a case-hardened shackle of at least ⅜ inch diameter, if the door hinges and lock hasp are securely fastened to the magazine and to the door frame.

These requirements do not apply to magazine doors that are adequately secured on the inside by means of a bolt, lock, or bar that cannot be actuated from the outside.

(5) Placards. The placards required by Department of Transportation regulations at 49 CFR part 172, subpart F, for the transportation of blasting agents shall be displayed on all magazines.

(b) Indoor magazines—

(1) General. Indoor magazines are to be theft-resistant. They need not be weather-resistant if the buildings in which they are stored provide protection from the weather. No indoor magazine is to be located in a residence or dwelling. Indoor magazines containing quantities of blasting agents in excess of 50 pounds are subject to the requirements of §555.206 of this subpart.

(2) Construction. The doors are to be constructed of wood or metal.

(3) Hinges and hasps. Hinges and hasps are to be attached to doors by welding, riveting, or bolting (nuts on inside). Hinges and hasps must be installed so that they cannot be removed when the doors are closed and locked.

(4) Locks. Each door is to be equipped with (i) two mortise locks; (ii) two padlocks fastened in separate hasps and staples; (iii) a combination of a mortise lock and a padlock; (iv) a mortise lock that requires two keys to open; or (v) a three-point lock.

Padlocks must have at least five tumblers and a case-hardened shackle of at least ⅜ inch diameter. Padlocks must be protected with not less than ¼ inch steel hoods constructed so as to prevent sawing or lever action on the locks, hasps, and staples. Indoor magazines located in secure rooms that are locked as provided in this subparagraph may have each door locked with one steel padlock (which need not be protected by a steel hood) having at least five tumblers and a case-hardened shackle of at least ⅜ inch diameter, if the door hinges and lock hasps are securely fastened to the magazine and to the door frame.

These requirements do not apply to magazine doors that are adequately secured on the inside by means of a bolt, lock, or bar that cannot be actuated from the outside.

[T.D. ATF–87, 46 FR 40384, Aug. 7, 1981, as amended by T.D. ATF–298, 55 FR 21863, May 30, 1990]

§ 555.212 Smoking and open flames.

Smoking, matches, open flames, and spark producing devices are not permitted:

(a) In any magazine;

(b) Within 50 feet of any outdoor magazine; or

(c) Within any room containing an indoor magazine.

§ 555.213 Quantity and storage restrictions.

(a) Explosive materials in excess of 300,000 pounds or detonators in excess of 20 million are not to be stored in one magazine unless approved by the Director.

(b) Detonators are not to be stored in the same magazine with other explosive materials, except under the following circumstances:

(1) In a type 4 magazine, detonators that will not mass detonate may be stored with electric squibs, safety fuse, igniters, and igniter cord.

(2) In a type 1 or type 2 magazine, detonators may be stored with delay devices and any of the items listed in paragraph (b)(1) of this section.

§ 555.214 Storage within types 1, 2, 3, and 4 magazines.

(a) Explosive materials within a magazine are not to be placed directly against interior walls and must be stored so as not to interfere with ventilation. To prevent contact of stored explosive materials with walls, a nonsparking lattice work or other nonsparking material may be used.

(b) Containers of explosive materials are to be stored so that marks are visible. Stocks of explosive materials are to be stored so they can be easily counted and checked upon inspection.

(c) Except with respect to fiberboard or other nonmetal containers, containers of explosive materials are not to be unpacked or repacked inside a magazine or within 50 feet of a magazine, and must not be unpacked or repacked close to other explosive materials. Containers of explosive materials must be closed while being stored.

(d) Tools used for opening or closing containers of explosive materials are to be of nonsparking materials, except that metal slitters may be used for opening fiberboard containers. A wood wedge and a fiber, rubber, or wooden mallet are to be used for opening or closing wood containers of explosive materials. Metal tools other than nonsparking transfer conveyors are not to be stored in any magazine containing high explosives.

§ 555.215 Housekeeping.

Magazines are to be kept clean, dry, and free of grit, paper, empty packages and containers, and rubbish. Floors are to be regularly swept. Brooms and other utensils used in the cleaning and maintenance of magazines must have no spark-producing metal parts, and may be kept in magazines. Floors stained by leakage from explosive materials are to be cleaned according to instructions of the explosives manufacturer. When any explosive material has deteriorated it is to be destroyed in accordance with the advice or instructions of the manufacturer. The area surrounding magazines is to be kept clear of rubbish, brush, dry grass, or trees (except live trees more than 10 feet tall), for not less than 25 feet in all directions. Volatile materials are to be kept a distance of not less than 50 feet from outdoor magazines. Living foliage which is used to stabilize the earthen covering of a magazine need not be removed.

§ 555.216 Repair of magazines.

Before repairing the interior of magazines, all explosive materials are to be removed and the interior cleaned. Before repairing the exterior of magazines, all explosive materials must be removed if there exists any possibility that repairs may produce sparks or flame. Explosive materials removed from magazines under repair must be

(a) placed in other magazines appropriate for the storage of those explosive materials under this subpart, or

(b) placed a safe distance from the magazines under repair where they are to be properly guarded and protected until the repairs have been completed.

§ 555.217 Lighting.

(a) Battery-activated safety lights or battery-activated safety lanterns may be used in explosives storage magazines.

(b) Electric lighting used in any explosives storage magazine must meet the standards prescribed by the "National Electrical Code," (National Fire Protection Association, NFPA 70–81), for the conditions present in the magazine at any time. All electrical switches are to be located outside of the magazine and also meet the standards prescribed by the National Electrical Code.

(c) Copies of invoices, work orders or similar documents which indicate the lighting complies with the National Electrical Code must be available for inspection by ATF officers.

§ 555.218 Table of distances for storage of explosive materials.

Quantity of Explosives		Distances in feet							
Pounds over	Pounds not over	Inhabited buildings		Public highways with traffic volume 3000 or fewer vehicles/day		Passenger railways-public highways with traffic volume more than 3,000 vehicles/ day		Separation of magazines	
		Barricaded	Unbarricaded	Barricaded	Unbarricaded	Barricaded	Unbarricaded	Barricaded	Unbarricaded
0	5	70	140	30	60	51	102	6	12
5	10	90	180	35	70	64	128	8	16
10	20	110	220	45	90	81	162	10	20
20	30	125	250	50	100	93	186	11	22
30	40	140	280	55	110	103	206	12	24
40	50	150	300	60	120	110	220	14	28
50	75	170	340	70	140	127	254	15	30
75	100	190	380	75	150	139	278	16	32
100	125	200	400	80	160	150	300	18	36
125	150	215	430	85	170	159	318	19	38
150	200	235	470	95	190	175	350	21	42
200	250	255	510	105	210	189	378	23	46
250	300	270	540	110	220	201	402	24	48
300	400	295	590	120	240	221	442	27	54
400	500	320	640	130	260	238	476	29	58
500	600	340	680	135	270	253	506	31	62
600	700	355	710	145	290	266	532	32	64
700	800	375	750	150	300	278	556	33	66
800	900	390	780	155	310	289	578	35	70
900	1,000	400	800	160	320	300	600	36	72
1,000	1,200	425	850	165	330	318	636	39	78
1,200	1,400	450	900	170	340	336	672	41	82
1,400	1,600	470	940	175	350	351	702	43	86
1,600	1,800	490	980	180	360	366	732	44	88
1,800	2,000	505	1,010	185	370	378	756	45	90
2,000	2,500	545	1,090	190	380	408	816	49	98
2,500	3,000	580	1,160	195	390	432	864	52	104
3,000	4,000	635	1,270	210	420	474	948	58	116
4,000	5,000	685	1,370	225	450	513	1,026	61	122
5,000	6,000	730	1,460	235	470	546	1,092	65	130
6,000	7,000	770	1,540	245	490	573	1,146	68	136
7,000	8,000	800	1,600	250	500	600	1,200	72	144
8,000	9,000	835	1,670	255	510	624	1,248	75	150
9,000	10,000	865	1,730	260	520	645	1,290	78	156
10,000	12,000	875	1,750	270	540	687	1,374	82	164
12,000	14,000	885	1,770	275	550	723	1,446	87	174
14,000	16,000	900	1,800	280	560	756	1,512	90	180
16,000	18,000	940	1,880	285	570	786	1,572	94	188
18,00	20,000	975	1,950	290	580	813	1,626	98	196
20,000	25,000	1,055	2,000	315	630	876	1,752	105	210
25,000	30,000	1,130	2,000	340	680	933	1,866	112	224
30,000	35,000	1,205	2,000	360	720	981	1,962	119	238
35,000	40,000	1,275	2,000	380	760	1,026	2,000	124	248
40,000	45,000	1,340	2,000	400	800	1,068	2,000	129	258
45,000	50,000	1,400	2,000	420	840	1,104	2,000	135	270
50,000	55,000	1,460	2,000	440	880	1,140	2,000	140	280
55,000	60,000	1,515	2,000	455	910	1,173	2,000	145	290
60,000	65,000	1,565	2,000	470	940	1,206	2,000	150	300
65,000	70,000	1,610	2,000	485	970	1,236	2,000	155	310
70,000	75,000	1,655	2,000	500	1,000	1,263	2,000	160	320
75,000	80,000	1,695	2,000	510	1,020	1,293	2,000	165	330
80,000	85,000	1,730	2,000	520	1,040	1,317	2,000	170	340
85,000	90,000	1,760	2,000	530	1,060	1,344	2,000	175	350
90,000	95,000	1,790	2,000	540	1,080	1,368	2,000	180	360
95,000	100,000	1,815	2,000	545	1,090	1,392	2,000	185	370
100,000	110,000	1,835	2,000	550	1,100	1,437	2,000	195	390
110,000	120,000	1,855	2,000	555	1,110	1,479	2,000	205	410
120,000	130,000	1,875	2,000	560	1,120	1,521	2,000	215	430
130,000	140,000	1,890	2,000	565	1,130	1,557	2,000	225	450
140,000	150,000	1,900	2,000	570	1,140	1,593	2,000	235	470
150,000	160,000	1,935	2,000	580	1,160	1,629	2,000	245	490
160,000	170,000	1,965	2,000	590	1,180	1,662	2,000	255	510
170,000	180,000	1,990	2,000	600	1,200	1,695	2,000	265	530
180,000	190,000	2,010	2,010	605	1,210	1,725	2,000	275	550
190,000	200,000	2,030	2,030	610	1,220	1,755	2,000	285	570
200,000	210,000	2,055	2,055	620	1,240	1,782	2,000	295	590
210,000	230,000	2,100	2,100	635	1,270	1,836	2,000	315	630
230,000	250,000	2,155	2,155	650	1,300	1,890	2,000	335	670
250,000	275,000	2,215	2,215	670	1,340	1,950	2,000	360	720
275,000	300,000	2,275	2,275	690	1,380	2,000	2,000	385	770

Table: American Table of Distances for Storage of Explosives (December 1910), as Revised and Approved by the Institute of Makers of Explosives—July, 1991.

Notes to the Table of Distances for Storage of Explosives

(1) Terms found in the table of distances for storage of explosive materials are defined in §555.11.

(2) When two or more storage magazines are located on the same property, each magazine must comply with the minimum distances specified from inhabited buildings, railways, and highways, and, in addition, they should be separated from each other by not less than the distances shown for "Separation of Magazines," except that the quantity of explosives contained in cap magazines shall govern in regard to the spacing of said cap magazines from magazines containing other explosives. If any two or more magazines are separated from each other by less than the specified "Separation of Magazines" distances, then such two or more magazines, as a group, must be considered as one magazine, and the total quantity of explosives stored in such group must be treated as if stored in a single magazine located on the site of any magazine of the group, and must comply with the minimum of distances specified from other magazines, inhabited buildings, railways, and highways.

(3) All types of blasting caps in strengths through No. 8 cap should be rated at 1½ lbs. (1.5 lbs.) of explosives per 1,000 caps. For strengths higher than No. 8 cap, consult the manufacturer.

(4) For quantity and distance purposes, detonating cord of 50 or 60 grains per foot should be calculated as equivalent to 9 lbs. of high explosives per 1,000 feet. Heavier or lighter core loads should be rated proportionately.

[T.D. ATF–87, 46 FR 40384, Aug. 7, 1981, as amended by T.D. ATF–400, 63 FR 45003, Aug. 24, 1998; T.D. ATF–446, 66 FR 16602, Mar. 27, 2001; T.D. ATF–446a, 66 FR 19089, Apr. 13, 2001]

§ 555.219 Table of distances for storage of low explosives.

Pounds		From Inhabited building distance (feet)	From public railroad and highway distance (feet)	From above ground magazine (feet)
Over	Not over			
0	1,000	75	75	50
1,000	5,000	115	115	75
5,000	10,000	150	150	100
10,000	20,000	190	190	125
20,000	30,000	215	215	145
30,000	40,000	235	235	155
40,000	50,000	250	250	165
50,000	60,000	260	260	175
60,000	70,000	270	270	185
70,000	80,000	280	280	190
80,000	90,000	295	295	195
90,000	100,000	300	300	200
100,000	200,000	375	375	250
200,000	300,000	450	450	300

Table: Department of Defense Ammunition and Explosives Standards, Table 5–4.1 Extract; 4145.27 M, March 1969

§ 555.220 Table of separation distances of ammonium nitrate and blasting agents from explosives or blasting agents.

Donor weight (pounds)		Minimum separation distance of acceptor from donor when barricaded (feet)		Minimum thickness of artificial barricades (inches)
Over	Not over	Ammonium nitrate	Blasting agent	
0	100	3	11	12
100	300	4	14	12
300	600	5	18	12
600	1,000	6	22	12
1,000	1,600	7	25	12
1,600	2,000	8	29	12
2,000	3,000	9	32	15
3,000	4,000	10	36	15
4,000	6,000	11	40	15
6,000	8,000	12	43	20
8,000	10,000	13	47	20
10,000	12,000	14	50	20
12,000	16,000	15	54	25
16,000	20,000	16	58	25
20,000	25,000	18	65	25
25,000	30,000	19	68	30
30,000	35,000	20	72	30
35,000	40,000	21	76	30
40,000	45,000	22	79	35
45,000	50,000	23	83	35
50,000	55,000	24	86	35
55,000	60,000	25	90	35
60,000	70,000	26	94	40
70,000	80,000	28	101	40
80,000	90,000	30	108	40
90,000	100,000	32	115	40
100,000	120,000	34	122	50
120,000	140,000	37	133	50
140,000	160,000	40	144	50
160,000	180,000	44	158	50
180,000	200,000	48	173	50
200,000	220,000	52	187	60
220,000	250,000	56	202	60
250,000	275,000	60	216	60
275,000	300,000	64	230	60

Table: National Fire Protection Association (NFPA) Official Standard No. 492, 1968

Notes of Table of Separation Distances of Ammonium Nitrate and Blasting Agents from Explosives or Blasting Agents

(1) This table specifies separation distances to prevent explosion of ammonium nitrate and ammonium nitrate-based blasting agents by propagation from nearby stores of high explosives or blasting agents referred to in the table as the "donor." Ammonium nitrate, by itself, is not considered to be a donor when applying this table. Ammonium nitrate, ammonium nitrate-fuel oil or combinations thereof are acceptors. If stores of ammonium nitrate are located within the sympathetic detonation distance of explosives or blasting agents, one-half the mass of the ammonium nitrate is to be included in the mass of the donor.

(2) When the ammonium nitrate and/or blasting agent is not barricaded, the distances shown in the table must be multiplied by six. These distances allow for the possibility of high velocity metal fragments from mixers, hoppers, truck bodies, sheet metal structures, metal containers, and the like which may enclose the "donor." Where explosives storage is in bullet-resistant magazines or where the storage is protected by a bullet-resistant wall, distances and barricade thicknesses in excess of those prescribed in the table in §555.218 are not required.

(3) These distances apply to ammonium nitrate that passes the insensitivity test prescribed in the definition of ammonium nitrate fertilizer issued by the Fertilizer Institute. 1 Ammonium nitrate failing to pass the test must be stored at separation distances in accordance with the table in §555.218.

[1] Definition and Test Procedures for Ammonium Nitrate Fertilizer, Fertilizer Institute 1015–18th St. N.W. Washington, DC 20036.

(4) These distances apply to blasting agents which pass the insensitivity test prescribed in regulations of the U.S. Department of Transportation (49 CFR part 173).

(5) Earth or sand dikes, or enclosures filled with the prescribed minimum thickness of earth or sand are acceptable artificial barricades. Natural barricades, such as hills or timber of sufficient density that the surrounding exposures which require protection cannot be seen from the "donor" when the trees are bare of leaves, are also acceptable.

(6) For determining the distances to be maintained from inhabited buildings, passenger railways, and public highways, use the table in §555.218.

§ 555.221 Requirements for display fireworks, pyrotechnic compositions, and explosive materials used in assembling fireworks or articles pyrotechnic.

(a) Display fireworks, pyrotechnic compositions, and explosive materials used to assemble fireworks and articles pyrotechnic shall be stored at all times as required by this Subpart unless they are in the process of manufacture, assembly, packaging, or are being transported.

(b) No more than 500 pounds (227 kg) of pyrotechnic compositions or explosive materials are permitted at one time in any fireworks mixing building, any building or area in which the pyrotechnic compositions or explosive materials are pressed or otherwise prepared for finishing or assembly, or any finishing or assembly building. All pyrotechnic compositions or explosive materials not in immediate use will be stored in covered, non-ferrous containers.

(c) The maximum quantity of flash powder permitted in any fireworks process building is 10 pounds (4.5 kg).

(d) All dry explosive powders and mixtures, partially assembled display fireworks, and finished display fireworks shall be removed from fireworks process buildings at the conclusion of a day's operations and placed in approved magazines.

[T.D. ATF–293, 55 FR 3722, Feb. 5, 1990, as amended by T.D. ATF–400, 63 FR 45004, Aug. 24, 1998]

§ 555.222 Table of distances between fireworks process buildings and between fireworks process and fireworks nonprocess buildings.

Net weight of fireworks1 (pounds)	Display fireworks2 (feet)	Consumer fireworks3 (feet)
0-100	57	37
101-200	69	37
201-300	77	37
301-400	85	37
401-500	91	37
Above 500	Not permitted4 5	Not permitted4 5

[1] Net weight is the weight of all pyrotechnic compositions, and explosive materials and fuse only.

[2] The distances in this column apply only with natural or artificial barricades. If such barricades are not used, the distances must be doubled.

[3] While consumer fireworks or articles pyrotechnic in a finished state are not subject to regulation, explosive materials used to manufacture or assemble such fireworks or articles are subject to regulation. Thus, fireworks process buildings where consumer fireworks or articles pyrotechnic are being processed shall meet these requirements.

[4] A maximum of 500 pounds of in-process pyrotechnic compositions, either loose or in partially-assembled fireworks, is permitted in any fireworks process building. Finished display fireworks may not be stored in a fireworks process building.

[5] A maximum of 10 pounds of flash powder, either in loose form or in assembled units, is permitted in any fireworks process building. Quantities in excess of 10 pounds must be kept in an approved magazine.

[T.D. ATF–293, 55 FR 3723, Feb. 5, 1990, as amended by T.D. ATF–400, 63 FR 45004, Aug. 24, 1998]

§ 555.223 Table of distances between fireworks process buildings and other specified areas.

Net weight of fireworks1 (pounds)	Display fireworks1 (feet)	Consumer fireworks2 (feet)
0-100	200	25
101-200	200	50
201-300	200	50
301-400	200	50
401-500	200	50
Above 500	Not permitted	Not permitted

Distance from Passenger Railways, Public Highways, Fireworks Plant Buildings used to Store Consumer Fireworks and Articles Pyrotechnic, Magazines and Fireworks Shipping Buildings, and Inhabited Buildings. [3 4 5]

[1] Net weight is the weight of all pyrotechnic compositions, and explosive materials and fuse only.

[2] While consumer fireworks or articles pyrotechnic in a finished state are not subject to regulation, explosive materials used to manufacture or assemble such fireworks or articles are subject to regulation. Thus, fireworks process buildings where consumer fireworks or articles pyrotechnic are being processed shall meet these requirements.

[3] This table does not apply to the separation distances between fireworks process buildings (see § 555.222) and between magazines (see §§ 555.218 and 555.224).

[4] The distances in this table apply with or without artificial or natural barricades or screen barricades. However, the use of barricades is highly recommended.

[5] No work of any kind, except to place or move items other than explosive materials from storage, shall be conducted in any building designated as a warehouse. A fireworks plant warehouse is not subject to § 555.222 or this section, tables of distances.

[T.D. ATF–293, 55 FR 3723, Feb. 5, 1990, as amended by T.D. ATF–400, 63 FR 45004, Aug. 24, 1998]

§ 555.224 Table of distances for the storage of display fireworks (except bulk salutes).

Net weight of fireworks1 (pounds)	Distance between magazine and inhabited building, passenger railway, or public highway3 4 (feet)	Distance between magazines2 3 (feet)
0-1000	150	100
1,001-5,000	230	150
5,001-10,000	300	200
Above 10,000	Use Table § 55.218	

[1] Net weight is the weight of all pyrotechnic compositions, and explosive materials and fuse only.

[2] For the purposes of applying this table, the term "magazine" also includes fireworks shipping buildings for display fireworks.

[3] For fireworks storage magazines in use prior to (30 days from the date of publication of the final rule in the Federal Register), the distances in this table may be halved if properly barricaded between the magazine and potential receptor sites.

[4] This table does not apply to the storage of bulk salutes. Use table at § 555.218.

[T.D. ATF–293, 55 FR 3723, Feb. 5, 1990, as amended by T.D. ATF–400, 63 FR 45004, Aug. 24, 1998]

Questions and Answers (Revised 1/07)
18 U.S.C. Chapter 40 and 27 CFR Part 555

Introduction

The following list of Questions and Answers are intended to aid you in gaining a better understanding of:

18 U.S.C. Chapter 40 – Importation, Manufacture, Distribution and Storage of Explosive Materials and the implementing regulations issued within:

27 CFR Part 555 – Commerce in Explosives

This listing is not all-inclusive. However it contains some of the most frequently asked questions that ATF receives. These questions and answers are intended only as a general overview. To determine how the law and regulations apply to your specific circumstances, you must refer directly to the applicable law and regulation or contact your local ATF Office of Industry Operations. Also, please be aware that both the law and regulations are subject to change. Please contact your local ATF office for the most up-to-date information. You can also find the latest Federal explosives regulations on the ATF website: www.atf.gov.

Unless otherwise stated, these Questions and Answers apply only to Federal law and regulations. States and local jurisdictions have, in many cases, enacted their own requirements relating to explosives. Check with appropriate State or local authorities for information on those requirements. Compliance with Federal law and regulations does not exempt any person from compliance with any State or local requirements.

A Table of Contents and a Subject Index has been included for your convenience. The Index is located at the end of the Questions and Answers section.

Table of Contents

General Questions

1. Who is affected by the Federal explosives law?

The law affects all persons who import, manufacture, deal in, purchase, use, store, or possess explosive materials. It also affects those who ship, transport or cause to be transported, or receive explosive materials. Also, see 18 U.S.C. 845 and 27 CFR 555.141 for exemptions.

2. What changes were brought about by the Safe Explosives Act?

Among other things, the Act mandated that all persons who wish to receive or transport explosive materials must first obtain a Federal explosives license or permit. In addition, the act imposed new restrictions on who may lawfully receive and possess explosive materials (See question 4). All Federal explosive licensees and permittees and their responsible persons and employees authorized to possess explosives are affected by the new requirements and background checks mandated by the Act.

3. Can I obtain general information from ATF on the Internet?

Yes. ATF maintains a website on the Internet at *www.atf.gov*.

4. Does the law make some classes of persons ineligible to receive a Federal license to import, manufacture, or deal in explosive materials or to receive a Federal explosives permit?

Yes. A license or permit will not be issued to any person who:

(a) Is under indictment for, or who has been convicted in any court, of a crime punishable by imprisonment for a term exceeding one year;

(b) Is a fugitive from justice;

(c) Is an unlawful user of or addicted to any controlled substance (as defined in section 102 of the Controlled Substances Act (21 U.S.C. 802));

(d) Has been adjudicated a mental defective or who has been committed to a mental institution;

(e) Is an alien (with certain exceptions);

(f) Has been discharged from the armed forces under dishonorable conditions; or,

(g) Having been a citizen of the United States, has renounced his citizenship. [18 U.S.C. 842(i), 843(b)(1); 27 CFR 555.49(b)(2)(i)]

5. Are there classes of persons to whom the distribution of explosive materials by licensees is prohibited?

Yes. No person shall knowingly distribute explosive materials to any individual listed in Question 4, or to an individual who is under 21 years of age. [18 U.S.C. 842(d); 27 CFR 555.26(d)]

6. What other distributions of explosive materials by licensees and permittees are prohibited?

A licensee or permittee shall not knowingly distribute any explosive materials to any person who:

(a) Is not a licensee [18 U.S.C. 842(b), 27 CFR 555.105, 555.106];

(b) Is not a holder of a user permit [18 U.S.C. 842(b); 27 CFR 555.105, 555.106];

(c) Is not a holder of a limited permit who resides in the same State where distribution is made and in which premises of the transferor are located. [18 U.S.C. 842(b); 27 CFR 555.105, 555.106];

(d) The licensee has reason to believe intends to transport such explosive materials into a State where the purchase, possession, or use of explosive materials is prohibited or which does not permit its residents to transport or ship explosive materials into the State or to receive explosive materials in the State. [18 U.S.C. 842(c); 27 CFR 555.106]

(e) Is in any State where the purchase, possession, or use by such person of such explosive materials would be in violation of any State law or any published ordinance applicable at the place of distribution. [18 U.S.C. 842(e); 27 CFR 555.106(b)(2)]

7. Does Federal law prohibit certain persons from receiving or possessing explosive materials?

Yes. The law prohibits the receipt or possession of explosive materials by any person listed in question 4. [18 U.S.C. 842(i); 27 CFR 555.26, 555.49(b)]

8. May a licensed manufacturer, importer, or dealer distribute explosive materials to nonlicensees and nonpermittees?

No. Every person who receives explosive materials must first obtain a Federal explosives license or permit. Distribution of explosive materials to persons who do not hold a license or permit is unlawful. Also, see 18 U.S.C. 845 and 27 CFR 555.141 for exemptions. [18 U.S.C. 842(a), 842(b); 27 CFR 555.26(a), 555.106]

9. Does Federal law provide penalties for purchasers who give false information at the time of purchasing explosive materials?

Yes. The penalty for knowingly providing false information or misrepresented identification is a maximum 10 years' imprisonment and/or a fine not exceeding $250,000. [18 U.S.C. 842(a)(2), 844(a)]

10. Is the theft of explosive materials, as well as the possession of stolen explosive materials, a Federal crime?

Yes. It is a Federal crime for any person to steal any explosive materials. It is also a Federal crime for any person to receive, possess, transport, ship, conceal, store, barter, sell, dispose of, or pledge or accept as security for a loan any stolen explosive materials. [18 U.S.C. 842(h), 844(k), (l)]

11. Are thefts and losses of explosive materials required to be reported to ATF?

Yes. Any licensee or permittee who has knowledge of the theft or loss of any explosive materials from his or her stock shall, within 24 hours of discovery, report the theft or loss by telephoning 800-461-8841 (Monday-Friday 8:00a.m.– 5:00 p.m. Eastern Time) or 888-283-2662 (after hours and weekends) (nationwide toll free number) and on ATF Form 5400.5, "Report of Theft or Loss – Explosive Materials", in accordance with the instructions on the form. The theft or loss shall also be reported to appropriate local authorities. The same requirements are imposed upon persons other than licensees and permittees, except that nonlicensees and nonpermittees, other than carriers, need not report a theft or loss on Form 5400.5, but must report the theft or loss by telephone, using the same numbers: 800-461-8841 (Monday-Friday 8:00a.m.–5:00 p.m. Eastern Time) or 888-283-2662 (after hours and weekends) and in writing to the nearest ATF office. The theft or loss shall also be reported to appropriate local authorities. Carriers of explosive materials must report a theft or loss by telephone but need not make the report on the ATF form or in writing. See 27 CFR 555.30 for the specific information required to be reported in connection with a theft or loss. [18 U.S.C. 842(k), 18 U.S.C. 844(p); 27 CFR 555.30]

12. My company holds a Federal explosives license and after conducting an inventory of our explosives on hand, we noticed one case of dynamite missing. After double-checking all Daily Summaries of Magazine Transactions, invoices, and delivery sheets, we still cannot reconcile the discrepancy. What are we required to do?

This should be considered a theft or loss of explosive materials. As stated in the answer to Question 11, you must report the theft or loss of explosive materials, within 24 hours of discovery, to ATF by telephone (toll free: 800-461-8841 (Monday-Friday 8:00 a.m.– 5:00 p.m. Eastern Time) or 888-283-2662 (after hours and weekends). ATF Form 5400.5, "Report of Theft or Loss – Explosive Materials", must then be completed and forwarded in accordance with the instructions on the form. [18 U.S.C. 842(k); 27 CFR 555.30]

13. May ATF conduct warrantless inspections of licensees' and permittees' records of explosives materials, stocks of such materials, and magazines?

Except for limited permit holders, any ATF officer may, without a warrant, enter during business hours the premises, including places of storage, of any licensee or permittee for the purpose of inspecting or examining any records or documents required to be kept by the law and regulations and any explosive materials kept or stored at the premises. For inspection purposes, "business hours" includes hours during which business is actually conducted, not just those hours stated on license applications. Any licensee or permittee who refuses to permit the inspection or examination is subject to having his or her license or permit revoked, as well as to denial of an application to renew the license or permit. For limited permit holders, an ATF officer may inspect the places of storage for explosive materials of either an applicant for a limited permit or at the time of renewal of such permit, but in no event shall such inspection occur more than once every three years. [18 U.S.C. 843(b)(4), 18 U.S.C. 843(f)]

14. Will ATF investigate accidents involving explosive materials?

ATF is authorized to inspect the site of any accident or fire where there is reason to believe that explosive materials were involved. Other Federal agencies, or State or local agencies, may also investigate such incidents, depending on the circumstances. [18 U.S.C. 846(a); 27 CFR 555.31]

15. Is black powder subject to regulation under Federal explosives laws?

Black powder is an explosive material for purposes of Federal explosives laws and regulations. However, the law exempts from regulation commercially manufactured black powder in quantities not exceeding 50 pounds (as well as percussion caps, safety and pyrotechnic fuses, quills, quick and slow matches, and friction primers) intended to be used solely for sporting, recreational, or cultural purposes in antique firearms as defined in 18 U.S.C 921(a)(16) or in antique devices exempted from the term "destructive device" in 18 U.S.C. 921(a)(4). However, persons engaged in the business of importing, manufacturing, or dealing in black powder in any quantity must have a Federal explosives license. [18 U.S.C. 841(c), 841(d), 845(a)(5); 27 CFR 555.11: definitions of "explosives" and "explosive materials", 555.141(b)]

16. Is small arms ammunition subject to regulation under Federal explosives laws?

No. The law specifically exempts small arms ammunition and components thereof. (See also Question 81.) [18 U.S.C. 845(a)(4)]

17. Are binary explosives subject to regulation under Federal explosives laws?

Until the compounds are mixed, they are not classified as explosives and, therefore, are not subject to control. However, once mixed, binary explosives are "explosive materials" and are subject to all applicable Federal requirements. A person who mixes or combines compounds of binary explosives for the purpose of sale or distribution or for the person's own business use is a "manufacturer" of explosive materials and must be licensed as a manufacturer under the law. [18 U.S.C. 841(h); 27 CFR 555.11: definition of "manufacturer"]

18. Does ATF have any regulations governing the actual transportation of explosive materials?

Federal explosives laws and regulations generally prohibit any person from transporting explosive materials interstate or intrastate unless the person has a Federal explosives license or permit. Also, the transportation of stolen explosives materials is a Federal crime (see also Question 10). However, the law exempts from regulation under 18 U.S.C. Chapter 40 and 27 CFR Part 555 aspects of the transportation of explosive materials via railroad, water, highway, or air which are regulated by the United States Department of Transportation, and the Department of Homeland Security, and agencies thereof and which pertain to safety and security. [18 U.S.C. 842(a)(3), 842(h), 845(a)(1); 27 CFR 555.26, 555.28, 555.141(a)(1)]

19. Are common or contract carriers required to obtain a Federal explosives license or permit to transport explosive materials?

No. The actual transportation of explosive materials by carriers is subject to Department of Transportation or Department of Homeland Security regulations. [18 U.S.C. 845(a)(1); 27 CFR 555.141(a)(1)]

20. What is the "Explosives List"?

The Explosives List is a comprehensive (but not all-inclusive) listing of explosive materials which have been determined to be within the coverage of Chapter 40. The list is published annually by ATF (the most recent list can be found under the Explosives Laws and Regulations link on the ATF website). [18 U.S.C. 841(d); 27 CFR 555.23]

21. May a person under the age of 21 be lawfully employed by an explosives business and lawfully receive, possess, and use explosive materials on behalf of the business?

Yes. Federal explosives law prohibits any person from distributing explosive materials to persons under 21 years of age. However, it does not prohibit the delivery to or possession of explosive materials by persons under the age of 21 who are receiving or using the materials on behalf of their employers to whom the materials were lawfully sold. [18 U.S.C. 842(d), (i); 27 CFR 555.11 (definition of "distribute"), 555.26, 555.106(b)(1)]

22. ATF regulations require explosive materials to be stored at certain minimum distances from a "public highway". What is a "public highway" for purposes of the regulations?

The term "highway" is defined in 27 CFR 555.11 as "any public street, public alley, or public road, including a privately financed, constructed, or maintained road that is regularly and openly traveled by the general public." Privately financed, constructed, or maintained roads that are marked and barricaded in a manner that prevents access by the general public do not fall within the meaning of the term and would, therefore, be exempt from table of distance requirements. [27 CFR 555.11: definition of "highway"](See also ATF Ruling 2005-2)

23. Is an airport runway or taxiway considered a public highway for purposes of the Table of Distances for storage of explosive materials?

No. However, airport terminals are considered inhabited buildings for Table of Distance requirements.

24. How is shock tube regulated by ATF?

Shock tube contains highly explosive material. However, it may be stored as a low explosive when not attached to a detonator. [27 CFR 555.202(b), 555.213]

25. What is an EX number?

An EX number is a number, preceded by the prefix "EX-", which is issued and used by the Department of Transportation (DOT) to identify an explosive which has been tested and classified by DOT. See U.S. Department of Transportation regulations at 49 CFR 171.8 and 49 CFR 173.56.

26. What is a UN number?

A UN (United Nations) number is used by DOT as a method of identification and classification of products for shipping purposes. UN numbers are different from the hazard class or division designations used by DOT (for example, 1.1, 1.2, 1.3, 1.4, and 1.5). ATF regulations in 27 CFR Part 555 also use UN numbers to help identify certain explosives. [27 CFR 555.11]

27. Can Federal explosives disabilities resulting from a conviction of a crime punishable by imprisonment for a term exceeding one year be removed if the conviction is expunged or set aside or the convicted person has received a pardon for the offense or has had his or her civil rights restored?

A person convicted of, or under indictment for, a "crime punishable by imprisonment for a term exceeding one year" may not lawfully receive or possess explosive materials or be issued a Federal explosives license or permit. The term "crime punishable by imprisonment for a term exceeding one year" does not include offenses pertaining to antitrust violations, unfair trade practices, restraints of trade, or any State offense (other than one involving a firearm or explosive) classified as a misdemeanor and punishable by imprisonment for 2 years or less. There are only 3 means by which Federal explosives disabilities resulting from a conviction of, or indictment for, a "crime punishable by imprisonment for a term exceeding one year" can be removed:

(a) A decision of a court invalidating a conviction on the basis that the conviction was unconstitutional;

(b) In the case of a Federal conviction, a presidential pardon; and

(c) The granting of relief from Federal explosives disabilities by ATF pursuant to the filing of a relief application with the Director. Information on how to apply for relief and contact information for the Relief of Disabilities Section is available on the ATF website. [18 U.S.C. 841(l), 842(d),(i), 845(b); 27 CFR 555.11: definition of "crime punishable by imprisonment for a term exceeding one year", 555.26(c), 555.142]

Licenses and Permits

28. Who needs a Federal explosives license or permit?

All persons who wish to transport, ship, cause to be transported, or receive explosive materials must first obtain a Federal explosives license or permit. Certain exemptions apply. [18 U.S.C. 842(b); 18 U.S.C. 845; 27 CFR 555.26(a), 27 CFR 555.141]

29. Who is eligible for a Federal explosives license or permit?

The Chief, Federal Explosives Licensing Center, will approve a properly completed application for a license or permit on ATF Form 5400.13/5400.16 if the applicant:

(a) Is not a person prohibited from possessing or receiving explosive materials under 18 U.S.C. 842(i) and none of the applicant's "responsible persons" are prohibited under section 842(1); (see also Question 4 in General Q&A);

(b) Has not willfully violated any provision of Chapter 40 or the regulations in 27 CFR Part 555;

(c) Has not knowingly withheld information or has not made any false or fictitious statement intended or likely to deceive, in connection with the application,

(d) Has premises in a State from which he intends to conduct business or operations;

(e) Has storage for the class (as described in 27 CFR 555.202) of explosive materials described on the application;

(f) Has certified in writing that he is familiar with and understands all published State laws and local ordinances relating to explosive materials for the location in which he intends to do business;

(g) Has submitted the certificate required by section 21 of the Federal Water Pollution Control Act, as amended (33 U.S.C. 1341) [18 U.S.C. 843(b); 27 CFR 555.49(b)];

(h) None of the applicant's employees authorized to possess explosives are prohibited persons under 18 U.S.C. 842(i); and

(i) In the case of an applicant for a limited permit, the applicant has certified that the applicant will not receive explosive materials on more than 6 occasions during the 12-month period for which the limited permit is valid.

30. What activities are covered by licenses and permits?

Licenses allow persons to engage in the business of importing, manufacturing, or dealing in explosive materials. Any individual or business entity intending to engage in any of these activities must first obtain a license. A user permit allows the receipt and transportation of explosive materials. A limited permit allows the receipt of explosive materials from a licensee or permittee within the permittee's state of residence only, and on no more than six occasions in the 12-month period during which the limited permit is valid. A limited permit does not authorize the receipt or transportation of explosive materials in interstate or foreign commerce. [27 CFR 555.11: definitions of "importer", "manufacturer", "dealer", "limited permit", and "user permit", 555.41]

A separate license is needed for each business premises where an explosives business or activity is conducted. Only a single user permit is needed by a permittee who uses explosives in more than one location. [27 CFR 555.41]

31. What is a Limited Permit?

A limited permit is for persons who wish to transport, ship, cause to be transported, or receive explosive materials in intrastate commerce only. This permit is designed for the infrequent receipt of explosive materials by intrastate users. The limited permit will allow a purchaser to receive explosive materials on no more than six separate occasions from in-state licensees or permittees during the 12-month period of the permit. The limited permit does not allow the holder to transport, ship, cause to be transported, or receive explosive materials in interstate commerce.

32. What is the duration of a license or permit?

(a) A user license or permit is valid for a period of 3 years.

(b) The user-limited permit is valid only for a single purchase transaction.

(c) Limited permits are valid for no more than six separate receipts of explosive materials during a 12-month period. [27 CFR 555.51]

33. What are the fees for licenses and permits?

Each license applicant must pay a fee of $200 for obtaining a 3-year license, a separate license and fee being required for each business premises. The fee for renewal of a license is $100 for a 3-year license. [27 CFR 555.42]

Each applicant for a user permit must pay a fee of $100 for a 3-year permit, and each applicant for a user-limited permit (nonrenewable) must pay a fee of $75. The fee for renewal of a user permit is $50 for a 3-year permit. [27 CFR 555.43]

Each applicant for a limited permit must pay a fee of $25 for a 1-year limited permit. The fee for renewal of a limited permit is $12 for a 1-year limited permit. [27 CFR 555.43]

34. Will the Government investigate an application for a license or permit?

ATF will investigate any applicant before issuing a license or permit. Additionally, ATF must inspect places of storage and conduct background checks on responsible persons and employee possessors authorized to possess explosives [18 U.S.C. 843(b); 18 U.S.C. 843(h); 27 CFR 555.33, 27 CFR 555.49(b)]

35. What may a licensed explosives dealer do?

A licensed dealer may engage in the business of distributing explosive materials at wholesale or retail [27 CFR 555.11: definition of "dealer"]

36. What may a licensed explosives importer do?

A licensed importer may engage in the business of importing or bringing explosive materials into the United States for purposes of sale or distribution. It is not necessary for a licensed importer to also obtain a dealer's license to engage in business on his or her licensed premises as a dealer in explosive materials (see also Question 52 and 53) [27 CFR 555.11: definition of "importer", 27 CFR 555.41(b)(2)]

37. When is a manufacturer's license required?

A manufacturer's license is required by persons engaged in the business of manufacturing explosive materials for sale, distribution, or for their own business use. For example, persons engaged in the business of providing a blasting service using explosives of their own manufacture would be required to have a manufacturer's license. Persons who manufacture explosives for their personal, non-business use are not required to have a manufacturer's license. However, no person may ship, transport, cause to be transported, or receive explosive materials unless such person holds a license or permit. [27 CFR 555.11: definition of "manufacturer", 555.41(b)] A separate manufacturer's license is not required by a licensed manufacturer for the purpose of on-site manufacture, for example, mixing binary explosives or making blasting agents at a quarry or other job site. It is not necessary for a licensed manufacturer to also obtain a dealer's license to engage in business on his or her licensed premises as a dealer in explosive materials (see also Question 52 and 53) [27 CFR 555.11: definition of "manufacturer", 555.41(b)(2)]

38. How do I apply for a Federal explosives license or permit?

You can request an application for a Federal explosives license or permit from the Federal Explosives Licensing Center at 877-283-3352 or from the ATF Distribution Center at 703-455-7801. As part of the application process, you must complete and submit an ATF Form 5400.13/5400.16, Application for Explosives License or Permit. You must also submit the names, identifying information, fingerprints, and photographs of all responsible persons. In addition, you must submit the names and identifying information of all employees who are authorized to possess explosive materials in the course of their employment on ATF Form 5400.28, Employee Possessor Questionnaire. [27 CFR 555.45(c)]

39. Who is a "responsible person"?

Federal explosives laws define a "responsible person" as an individual who has the power to direct the management and policies of the applicant pertaining to explosive materials. Responsible persons generally include sole proprietors and explosives facility site managers. In the case of a corporation, association, or similar organization, responsible persons generally include only those corporate directors/officers, and stockholders, who have the power to direct management and policies as they pertain to explosive materials.

For example, a corporate vice president whose duties include acquiring and approving contracts with explosives distributors would be considered a responsible person. Other corporate officials whose duties do not include the power to direct the management and policies of the applicant pertaining to explosive materials, for example, a vice president responsible solely for human resources, would not typically be considered a responsible person. Each applicant for a license or permit must assess the corporate and other management responsibilities for all key personnel and determine whether or not these duties place the individual in the position of being a responsible person. [18 U.S.C. 841(s), 27 CFR 555.11: definition of "responsible person"]

40. Who is a "possessor of explosives"?

A possessor of explosives is any employee of a license or permit holder or any employee of an applicant for a license or permit who has or will have actual physical possession of explosive materials or who has or will have constructive possession of explosive materials. For example, persons who physically handle explosive materials would be considered to be actual possessors of explosive materials. This would include employees who directly handle explosive materials as part of the production process; employees who handle explosive materials in order to ship, transport, or sell them; and employees, such as blasters and their helpers who actually use explosive materials. A constructive possessor is any person who has access to explosive materials, without physically handling them. For example, a supervisor at a construction site who keeps keys for storage magazines in which explosives are stored or who directs the use of explosive materials by other employees has constructive possession of explosives.

41. Why is it necessary to provide new and additional information on responsible persons and employee possessors of explosives?

The law requires this information for ATF to conduct background checks on all responsible persons and employee possessors to restrict the availability of explosives to authorized persons only and to reduce the risk of prohibited persons acquiring explosive materials. [18 U.S.C. 843(h); 27 CFR 555.33, 555.45(c)]

42. When will I need to submit the identifying information for my responsible persons and employee possessors of explosives?

(a) All license and permit applicants and any renewal applicants must submit identifying information for responsible persons and employee possessors (and fingerprints and photographs for responsible persons) upon submission of an original or renewal application.

(b) Any new responsible person added after a license or permit has been issued by ATF must be reported to ATF within 30 days. However, the submission of fingerprints and photographs by the new responsible person is required only at the time of any subsequent renewal.

(c) For all licenses and permits (new and renewal), any new employee possessors must be reported to ATF within 30 days of hire on the Employee Possessor Questionnaire form (ATF F 5400.28). [27 CFR 555.45(c), 27 CFR 555.57(b)]

43. How do I get my fingerprints taken?

Fingerprints must be submitted on Fingerprint Identification Cards, FD-258 that have been issued by ATF. The fingerprint cards must contain the following ORI information: WVATF0900; ATF-NATL EXPL LIC, MARTINSBURG WV. These fingerprint cards may be obtained by contacting the Federal Explosives Licensing Center at 877-283-3352 or the ATF Distribution Center at 703-455-7801. The fingerprint cards must be completed by your local law enforcement authority.

44. Will ATF notify me whether or not my responsible persons and employee possessors have passed their background checks?

Yes. A "Notification of Clearance" will be issued directly to all license or permit holders advising whether their responsible persons and employee possessors have been cleared to possess explosive materials, or are or may be prohibited from possessing explosives. These notices must be retained as part of the license or permit holders permanent records. In addition, letters of clearance or denial will be issued directly to responsible persons and employee possessors. [27 CFR 555.33]

45. What notification will I receive if one of my responsible persons or employee possessors does not pass their ATF background check?

If an individual does not pass the background check, a letter will be sent to the licensee or permittee who submitted the individual's name indicating that the individual was denied. A letter will also be sent to that individual explaining the prohibition and outlining appeal and relief procedures, as may be applicable. Unless and until an appeal overturns the denial or relief from disabilities is granted, that individual may not lawfully possess explosives. [27 CFR 555.33]

46. Who will conduct the background checks on applicants, responsible persons, and possessors?

ATF will perform the background checks. If employers wish to require their own background checks as a condition of employment, they may do so. However, such a background check will not be accepted in place of the ATF background check. [27 CFR 555.33]

47. May I sell black powder without a license?

No. Anyone who engages in the business of selling black powder, regardless of quantity, must be licensed as an explosives dealer. [27 CFR 555.41(b)]

48. Is a manufacturer's license required to acquire and mix binary explosives?

If the individual purchasing the binary explosives is engaged in the business of manufacturing explosives, i.e., mixes and uses them in the operation of a commercial business (for example, operating a quarry, or providing the service of removing stumps or boulders from a farm field), then a manufacturer's license is required.

An individual farmer who merely wishes to mix the binary explosives to remove obstacles from his field and provides no other outside service would not need a manufacturer's license.

Please note, however: A Federal explosives license or permit would be required to obtain any explosive device, such as detonators, used to initiate the mixed binary explosives. In addition, transportation of any explosive material, including mixed binary explosives, without a Federal license or permit is prohibited. [27 CFR 555.11: Definition of "manufacturer"; 27 CFR 555.26, 555.41(b)]

49. What is theatrical flash powder and is there a license for its manufacture?

Theatrical flash powder is flash powder commercially manufactured in premeasured kits not exceeding 1 ounce in weight, and mixed immediately prior to use and intended for use in events such as theatrical shows, stage plays, band concerts, magic acts, thrill shows, and clown acts in circuses. A manufacturer's license allows on-site manufacturers to operate nationally on one license issued to their principal place of business. [27 CFR 555.11: definitions of "flash powder" and "theatrical flash powder", 555.41(b)]

50. Is a separate license required for each location where business is conducted?

Yes. A separate license is required for each location where business is conducted. However, a separate license is not required for:

(a) Facilities used only for the storage of explosive materials;

(b) Locations used solely for the storage of records relating to the business; and

(c) Licensed manufacturers' on-site manufacturing. [27 CFR 555.41(b)]

51. Must a person who engages in the business of both manufacturing and importing at the same location have both licenses?

Yes. The licenses for manufacturing and importing allow a person to engage in separate and distinct activities and a separate license is required for each activity. However, a manufacturer or an importer does not need a separate dealer's license to also distribute explosive materials from the licensed premises. [27 CFR 555.41(b)]

52. Does a licensed manufacturer, importer, or dealer need a permit to use explosive materials?

No. No licensee will be required to obtain a user permit to lawfully transport, ship, or receive explosive materials in interstate or foreign commerce. [27 CFR 555.41(b)(2)]

53. Does a Federal license or permit exempt the holder from State or local requirements?

No. A license or permit confers no right or privilege to conduct business or operations, including storage, contrary to State or other law. All legal requirements must be followed, whether Federal, State, or local. [18 U.S.C. 848; 27 CFR 555.62]

54. Who is authorized to import explosive materials?

Any licensed importer is authorized to engage in the business of importing explosive materials for sale, distribution, or their own use. Any licensed manufacturer, dealer, or holder of a user permit may import explosive materials for their own use only. Licensees and user permittees importing explosive materials must provide to the U.S. Customs and Border Protection (CBP) a copy of the license or permit. Note, however, that in the case of certain military explosives or propellant powder or other components of small arms ammunition, Federal firearms regulations require the importer to provide an approved ATF Form 6 to the CBP. [27 CFR 555.41(b)(2), 555.41(b)(3), 447.21, 555.108(a), 555.183, 478.113]

55. How may an employee of an explosives licensee or permittee qualify to accept delivery of explosive materials for the employer?

The employee must be on the current list of representatives or agents authorized to accept delivery of explosive materials on behalf of the employer and be an authorized employee possessor of explosives. [27 CFR 555.103(b), 555.105(b)]

56. When an explosives licensee or permittee sends one of their truck drivers to the distributor's premises to pick up explosive materials that have been purchased by the licensee or permittee, will the driver be required to sign any forms?

No, however the driver is required to furnish the seller with an identification document as defined in 27 CFR Part 555.11. [27 CFR 555.103(b), 555.105(b)]

57. Will a licensee or permittee be notified in advance when the license or user permit needs to be renewed?

Generally, prior to expiration of the license or permit, a licensee or permittee will be notified. The application form must be completed and filed with ATF before expiration of the current license or permit for the renewal to be considered timely. However, if a licensee or permittee does not receive a renewal notification, it is still that licensee's or permittee's responsibility to ensure that an application is filed prior to expiration of the current license or permit. [27 CFR 555.46]

58. I have timely filed my application for renewal of my license (or user's permit) but I have not received my new license (or permit). May I continue in business even though the expiration date shown on my license or permit has passed? If so, how long?

Yes. You may continue to operate the business pursuant to your current license or permit until the application for renewal is acted upon. [5 U.S.C. 558]

59. Can a license or permit be revoked?

Yes. The Director, Industry Operations for the ATF Field Division in which a licensee or permittee is located may revoke a license or permit if the holder has violated any provision of 18 U.S.C. Chapter 40 or its implementing regulations or has become ineligible to receive explosive materials under 18 U.S.C. 842(i). [18 U.S.C. 843(d); 27 CFR 555.71, 555.74]

60. If a Federal explosives licensee or permittee is indicted for or convicted of a "crime punishable by imprisonment for a term exceeding one year", may he or she continue operations under the license or permit?

As stated in the answer to Question 4 in General Q&A, a person under indictment for, or convicted of, a crime punishable by imprisonment for a term exceeding one year is not eligible to be issued a license or permit. However, a licensee or permittee who is indicted for, or convicted of, such a crime during the term of his or her existing license or permit is not barred from licensed or permit operations for 30 days after the date of the indictment or the date the conviction becomes final. If the licensee or permittee files an application for relief from disabilities within such 30-day period, he or she may continue licensed or permit operations while the application is pending. If a relief application is not filed during that period, the licensee or permittee may not continue operations beyond such 30-day period. The right of a licensee to continue licensed or permitted operations beyond such 30-day period is also conditioned on the licensee or permittee timely filing a license or permit renewal application disclosing that the applicant has been indicted for, or convicted of, the crime. A licensee or permittee may not continue operations beyond 30 days following the date the Director issues notification that the relief application has been denied. [18 U.S.C. 845(b); 27 CFR 555.142]

61. May a licensed dealer make a sale to a holder of a limited permit in an adjoining State?

No. Sales may not be made to limited permittees who are out-of-State residents. [18 U.S.C. 842(a); 27 CFR 555.11: definition of "limited permit", 555.41(b)(3)]

Recordkeeping

62. Does a licensee or permittee have to keep records of the acquisition, distribution, and storage of explosive materials?

Yes. Licensees and permittees must keep records of acquisitions, dispositions, and storage of explosive materials. [18 U.S.C. 842(f), 847; 27 CFR 555.107, 555.122-.125, and 555.127, Subpart G]

63. How do licensees and permittees account for explosive quantities in their records?

If acquisitions are recorded by weight, then distribution must also be recorded by weight. If acquisitions are recorded by physical count (e.g., by units), then distribution must also be recorded by physical count. [27 CFR 555.122-.125]

64. Must a licensee or permittee maintain a daily summary of magazine transactions?

Yes. After the initial inventory required by regulations has been taken, the inventory shall be entered in a record of daily transactions. Not later than the close of the next business day, each licensee and permittee shall record by manufacturer's name or brand name the total quantity received in and removed from each magazine during the day and the total remaining on hand at the end of the day. [27 CFR 555.127]

65. Where must a licensee or permittee keep the daily summary of magazine transactions?

The records must either be kept at each magazine or at one central location on the business premises, provided a separate record of daily transactions is maintained for each magazine [27 CFR 555.127]

66. How can I obtain additional copies of ATF Forms?

Forms are available on-line at www.atf.gov/forms/5000.htm#explosives. Requests for forms should be mailed to the ATF Distribution Center, 7943 Angus Court, Springfield, Virginia 22153. You may also have forms mailed to you by submitting an on-line request at www.atf.gov/dcof/index.htm, or by telephoning your request to 703-455-7801. [27 CFR 555.21(b)]

67. Does a purchaser of black powder have to sign any forms at the time of purchase?

If 50 pounds or less of commercially manufactured black powder is being purchased, and the powder is intended to be used solely for sporting, recreational, or cultural purposes in antique firearms as defined in 18 U.S.C. 921(a)(16) or in antique devices exempt from the term "destructive device" in 18 U.S.C. 921(a)(4), no form is required. However, if the black powder is being purchased for any other purpose (regardless of quantity), the purchaser or other transferee must possess a Federal explosives license or permit. [18 U.S.C. 845(a)(5); 18 U.S.C. 926(c); 27 CFR 555.141(b), 555.26(a)]

68. Is there a requirement for licensees and permittees to make an annual inventory of explosive materials on hand?

Yes. An inventory is required to be taken at least once a year. [27 CFR 555.122-.125]

69. When must ATF Form 5400.4, "Limited Permittee Transaction Report (LPTR)", be executed?

Before distribution of explosive materials to a limited permittee, the licensee or permittee must obtain an executed ATF F 5400.4 from the limited permittee with an original unaltered and unexpired Intrastate Purchase of Explosives Coupon (IPEC) attached. Except when delivery of explosive materials is made by a common or contract carrier who is an agent of the limited permittee, the licensee or permittee must verify the identity of the holder of the limited permit by examining an identification document (as defined in 555.11) and noting on the ATF F 5400.4 the type of document presented. The licensee or permittee must complete the appropriate section on ATF F 5400.4 to indicate the type and quantity of explosive materials distributed, the license or permit number of the seller, and the date of the transaction. The licensee or permittee must sign and date the form. [27 CFR 555.126(b)]

70. Do the ATF Forms 5400.4 have to be maintained by the licensee or permittee making the sale?

Yes. One copy of ATF F 5400.4 must be retained by the seller as part of his permanent records in chronological order by date of disposition, or in alphabetical order by name of limited permittee. They must be maintained for a period of five years. [27 CFR 555.126]

71. May I keep computerized records?

Yes. See ATF Ruling 2007-1.

Storage

72. Who must comply with the storage requirements?

Except for those items and activities given exempt status under 18 U.S.C. 845 (also see 27 CFR 555.141), or exempted under 27 CFR 555.32, Special Explosives Devices, all persons who store explosive materials must store them in conformity with the provisions of Subpart K of the regulations, unless the person or the materials are exempt from regulation. [18 U.S.C. 842(j); 27 CFR 555.29, 555.141, 555.201(a)]

73. What are the classes of explosive materials for storage purposes?

There are 3 classes of explosive materials:

(a) High explosives (for example, dynamite, flash powders, and bulk salutes);

(b) Low explosives (for example, black powder, safety fuses, igniters, igniter cords, fuse lighters, and "display fireworks", except for bulk salutes); and

(c) Blasting agents (for example, ammonium nitrate-fuel oil and certain water gels). [27 CFR 555.202]

74. May a person store explosive materials in a residence or dwelling?

No. Storage of explosive materials in a residence or dwelling is prohibited. [27 CFR 555.208(b), 555.210(b), 555.211(b)]

75. What is the "Table of Distances"?

This table lists the minimum acceptable distances separating explosives magazines from inhabited buildings, passenger railroads, public highways, and other explosives magazines. The table is contained in 27 CFR 555.218.

76. When low and high explosives are stored together, how is the distance determined to meet the table of distance requirements?

The table of distances at 27 CFR 555.218 would be applied using the total weight of explosive materials in the magazine. [27 CFR 555.218]

77. Is it necessary to inspect my explosives magazines on a regular basis?

Yes. Any person storing explosives must inspect the magazines at least once every 7 days to determine whether there has been unauthorized entry or attempted entry into the magazines or unauthorized removal of the contents of the magazines. [27 CFR 555.204]

78. What are the requirements for making changes or additions to an approved storage facility?

Making changes in construction to an approved explosives magazine or adding a magazine requires that ATF be notified. However, mobile or portable type 5 magazines and magazines used for the temporary (under 24 hours) storage of explosive materials are exempt from this requirement. See 27 CFR 555.63 for details.

79. Is any type of black powder fuse exempt from storage requirements?

Yes, 3/32-inch and other external burning pyrotechnic hobby fuses are exempt from the requirements of Federal explosives laws and regulations. [18 U.S.C. 845(a)(4-5); 27 CFR 555.11: definition of "ammunition", 555.141(a)(4), 555.141(b)]

80. With the exception of 3/32-inch pyrotechnic safety fuse for use in small arms, must black powder fuses generally be stored in approved explosives magazines?

Yes. Generally igniter fuses, time fuses, blasting fuses, safety fuses, or other black powder fuses by whatever name known, must be stored in approved magazines.

81. Is smokeless powder designed for use in small arms ammunition subject to the explosives storage requirements?

Smokeless propellants designed for use in small arms ammunition are exempt from regulation under 18 U.S.C. Chapter 40 and the regulations in 27 CFR Part 555. However, it should be noted that persons engaged in the business of importing or manufacturing smokeless propellants must have a Federal explosives license. Additionally, smokeless propellant designed for use other than small arms ammunition is not exempt. Therefore, explosives products such as squibs, fireworks, theatrical special effects, or other articles that may be utilizing smokeless propellants are regulated and must be stored accordingly.

82. My office building, in which several company employees work during the day in connection with my explosives business, is located in the general area of my explosives magazine. Do the regulations and the Table of Distances apply to this building as an "inhabited building"?

No. A building such as an office building or repair shop which is part of the premises of an explosives business and is used by the business in connection with the manufacture, transportation, storage, or use of explosive materials is not considered to be an "inhabited building". [27 CFR 555.11: definition of "inhabited building", 555.218]

83. Am I required to notify my State or local authorities about my explosives storage magazines?

Yes. All persons who store explosive materials must notify the fire department having jurisdiction over the site where explosive materials are manufactured or stored. Notification must be made orally by the end of the day on which storage begins and in writing within 48 hours from the time storage began. The notification must include the type of explosive materials, magazine capacity, and the location of each storage site. [27 CFR 555.11: Definition of "authority having jurisdiction for fire safety", 27 CFR 555.201(f)]

84. What is the definition of a "case hardened shackle?"

Case hardening involves putting carbon (or a combination of carbon and nitrogen) into the surface of the steel to make it a high-carbon steel, which can be hardened by heat treatment. Only the outer skin gets hard in this manner. The center is still tough and malleable. This makes for a strong lock with a tough surface.

85. Can detonators be stored with detonating cord?

No. However, products which are manufactured with a detonator attached to the detonating cord as an integral part need not be disassembled and stored separately. [27 CFR 555.213]

86. Are there storage requirements for oxidizers, such as ammonium nitrate?

In general, no. However, when a magazine or bin containing ammonium nitrate is located within the sympathetic detonation distance of other explosives or blasting agents, it must be stored in accordance with the table of distances in 27 CFR 555.220.

87. Are State and local government agencies required to store their explosive materials in conformity with Federal storage regulations?

Yes. There is no exemption in the law or regulations for the storage of explosive materials by any State or political subdivision thereof. [18 U.S.C. 842(j), 845(a)(6); 27 CFR 555.141(a)(3), (a)(5)]

Fireworks

Fireworks are defined in the Federal explosives regulations as any composition or device designed to produce a visible or an audible effect by combustion, deflagration, or detonation. Fireworks are further divided into two broad classifications, consumer fireworks or display fireworks as defined at 27 CFR Part 555.11.

88. Are "consumer fireworks" subject to regulation under the Federal explosives laws?

No. The importation, distribution, and storage of fireworks defined as consumer fireworks are exempted from the provisions of the Federal explosives laws. However, because they contain pyrotechnic compositions classed by ATF as explosive materials, the manufacture of consumer fireworks requires a manufacturer's license. In addition, pyrotechnic compositions used in the manufacture of consumer fireworks must be stored in accordance with regulations in 27 CFR Subpart K. Consumer fireworks are defined as "any small firework device designed to produce visible effects by combustion and which must comply with the construction, chemical composition, and labeling regulations of the U.S. Consumer Product Safety Commission, as set forth in title 16, Code of Federal Regulations, parts 1500 and 1507. Some small devices designed to produce audible effects are included, such as whistling devices, ground devices containing 50 mg or less of explosive materials, and aerial devices containing 130 mg or less of explosive materials. Consumer fireworks are classified as fireworks UN0336 and UN0337 by the U.S. Department of Transportation at 49 CFR 172.101. This term does not include fused set pieces containing components which together exceed 50 mg of salute powder." [27 CFR 555.11: definition of "consumer fireworks"; definition of "licensed manufacturer", 555.141(a)(7)]

89. Are "display fireworks" considered to be explosive materials subject to regulation under Federal explosives laws and regulations?

Yes. Display fireworks include, but are not limited to, salutes containing more than 2 grains (130 mg) of explosive materials, aerial shells containing more than 40 grams of pyrotechnic compositions, and other display pieces which exceed the limits of explosive materials for classification as "consumer fireworks". These fireworks are classified as fireworks UN0333, UN0334, or UN0335 by regulations of the U.S. Department of Transportation at 49 CFR 172.101. Display fireworks also include fused set pieces containing components which together exceed 50 mg of salute powder. [27 CFR 555.11: definition of "display fireworks"]

90. How must display fireworks be stored?

Display fireworks, with the exception of bulk salutes, are considered low explosives and, at a minimum, must be stored in type 4 storage magazines. They may also be stored in type 1 or type 2 magazines. Bulk salutes, which are defined as either salute components prior to final assembly into aerial shells, (or) finished salute shells held separately prior to being packed with other types of display fireworks, are classified as high explosives. As such, bulk salutes may only be stored in type 1 or type 2 magazines specifically constructed for the storage of high explosives. [27 CFR 555.11, 555.202(b), 555.203(d), 555.207, 555.208, 555.210]

91. Are "Articles Pyrotechnic" subject to the requirements of the Federal explosives regulations?

The importation, distribution, and storage of fireworks defined as "Articles Pyrotechnic", are exempt from the Federal explosives laws and regulations. However, because they contain pyrotechnic compositions classed by ATF as explosive materials, the manufacture of items defined as "articles pyrotechnic" requires an ATF manufacturer's license. In addition, pyrotechnic compositions used in the manufacture of articles pyrotechnic must be stored in accordance with regulations in 27 CFR Subpart K. [27 CFR 555.11: definitions of "articles pyrotechnic" and "consumer fireworks", 555.141(a)(7)]

92. Must partially assembled display fireworks be removed from a drying building for overnight storage?

Yes. At the end of a day's manufacturing operations, all dry explosive powders and mixtures and partially assembled and finished display fireworks must be removed from fireworks process buildings and stored in a magazine meeting the storage requirements in 27 CFR Part 555, Subpart K. [27 CFR 555.205, 555.221]

93. What areas of a fireworks manufacturing plant are considered to be "fireworks process buildings?"

Fireworks process buildings include any buildings in which pyrotechnic compositions or explosives materials are mixed, pressed, finished, or assembled. Fireworks process buildings do not include plant warehouses, office buildings, or other buildings and areas in which no fireworks, pyrotechnic compositions, or explosive materials are processed or stored. [27 CFR 555.11: definition of "fireworks process building"]

94. Under what conditions may I temporarily store display fireworks (including low explosives for choreographed shows) on trucks?

See ATF Ruling 2007-2.

95. What types of fireworks require an ATF license or permit in order to be lawfully transported or received?

Any fireworks defined as "display fireworks" in 27 CFR 555.11 may be lawfully received or transported only by persons who hold a valid license or permit. No ATF license or permit is required to receive or transport "consumer fireworks" or "articles pyrotechnic". [18 U.S.C. 842(a)(3); 27 CFR 555.26, 555.141(a)(7)]

Plastic Explosives

96. What is a plastic explosive?

A plastic explosive is defined as "an explosive material in flexible or elastic sheet form formulated with one or more high explosives which in their pure form has a vapor pressure less than 10^{-4} Pa at a temperature of 250 C., is formulated with a binder material, and is as a mixture malleable or flexible at normal room temperature." [18 U.S.C. 841(q); 27 CFR 555.180(d)(4)]

97. What plastic explosives are required to contain detection agents?

All plastic explosives manufactured or imported on or after April 24, 1996, must contain a detection agent. Federal law enforcement agencies and the military may possess unmarked plastic explosives if they meet the requirements of the use-up period described in Question 103. [18 U.S.C. 841(q), 842(n); 27 CFR 555.180]

98. What are the permissible detection agents for marking plastic explosives?

These agents are listed in the law and regulations at 18 U.S.C. 841(p) and 27 CFR 555.180(d)(3).

99. Is it lawful to manufacture plastic explosives that do not contain a detection agent?

No. [18 U.S.C. 842(l); 27 CFR 555.180(a)]

100. Is it lawful to import into the United States plastic explosives that do not contain a detection agent?

No. The importation of plastic explosives into the United States requires that the importer file ATF Form 6 certifying that the imported plastic explosives contain the required detection agent, or is exempted from the marking requirements as provided in the regulations. [18 U.S.C. 842(m); 27 CFR 555.180(b), 555.182, 555.183]

101. Is it lawful to ship, transport, transfer, receive, or possess any plastic explosive that does not contain a detection agent?

No. However, a 15-year use-up period is provided for Federal law enforcement agencies and the military for unmarked plastic explosives imported into or manufactured in the U.S. prior to April 24, 1996. [18 U.S.C. 842(n); 27 CFR 555.180(c)]

102. If a person acquired plastic explosives not containing a detection agent before April 24, 1996, may he or she continue to lawfully possess the explosives?

No. With the exception of the use-up period provided by law for Federal law enforcement agencies or the military, the time period for lawful possession of unmarked plastic explosives terminated on April 24, 1999. [18 U.S.C. 842(n); 27 CFR 555.180(c)]

103. Are police departments exempt from the prohibition against possessing unmarked plastic explosives after April 24, 1999?

No. Police departments and other State or local law enforcement agencies could lawfully possess unmarked plastic explosives acquired on or before April 24, 1996, until April 24, 1999. Such agencies still possessing unmarked plastic explosives should destroy them or abandon them to ATF. Contact the nearest ATF field office for information. [18 U.S.C. 842(n); 27 CFR 555.180(c)(1)]

U.S. Military Explosives

104. Would an ATF license or permit be needed to demilitarize (demil) U.S. military explosives?

As long as the demil operator has a valid Department of Defense contract to perform such operations, the operations would be exempt from 27 CFR Part 555 and no license or permit would be required. However, if title to the explosive materials has passed from the military to the demil operator and the operator intends to resell the explosives on the commercial market, then such operations may be regulated by ATF (e.g., storage, sales, manufacturing) and an ATF license or permit may be needed. Contact the nearest ATF field office for further information. [18 U.S.C. 845(a)(3), (a)(6) and 27 CFR 555.141(a)(3), (a)(5)]

105. Would a civilian contractor who is manufacturing explosive materials pursuant to a government contract for or on behalf of the United States military be entitled to the exemptions from the explosives laws and regulations?

Yes, provided that all the explosive materials in question are manufactured under a government contract. Any explosive materials manufactured in anticipation of receiving a government contract would not qualify for this exemption.

If the contractor manufactures any explosive materials not pursuant to a U.S. military contract, the manufacture and the explosive materials are subject to all requirements of the law and regulations. [18 U.S.C. 845(a)(3), (a)(6); 27 CFR 555.26, 555.41, 555.141(b)]

106. Is an ATF licensee or permittee, whose licensed premises are located on a U.S. military installation, subject to the regulations in 27 CFR Part 555?

All activities conducted outside the scope of a U.S. Government contract are subject to the requirements of Part 555, even if the activities are conducted on property owned by the military. [18 U.S.C. 845(a)(3), (a)(6); 27 CFR 555.26, 555.29, 555.41, 555.141(a)(3), (a)(5)]

Index to Questions and Answers

ATF Explosives Rulings (Revised 1/07)

Table of Contents

1. 27 CFR 55.11: Meaning of Terms (Also § 55.206)

An office or repair shop used in connection with the manufacture, etc. of explosive materials is not an "inhabited building."

ATF Ruling 75-20

ATF has held that a building, such as an office or repair shop, which is a part of the premises of an explosives manufacturer and is used in connection with the manufacture, transportation, storage, or use of explosive materials, is not an "inhabited building." Section 55.11 of 27 CFR defines inhabited building as "any building regularly occupied in whole or in part as a habitation for human beings, or any church, schoolhouse, railroad station, store, or other structure where people are accustomed to assemble, except any building occupied in connection with the manufacture, transportation, storage, or use of explosive materials."

Regulations in 27 CFR §§ 55.206 and 55.218 set forth provisions concerning the location of storage facilities and the minimum distances such storage facilities may be located from, among other things, "inhabited buildings." These provisions are intended to provide protection to persons who inhabit buildings located near premises where explosives are manufactured, stored, etc. However, it is the intent of § 55.11 to exempt buildings used by the explosives industry in connection with the manufacture transportation, storage, or use of explosive materials from the table of distance requirements on "inhabited buildings."
[75 CB 64]

2. 27 CFR 55.207: Construction of Type 1 Magazines (Also § 55.210)

Certain explosives storage facilities meeting standards of construction prescribed by the Department of Defense Explosives Safety Board for such storage are approved by the Bureau.

ATF Ruling 75-21

ATF has held that explosives storage facilities with smooth-finished concrete floors that were constructed under contract for the use of the Department of Defense (DOD) and that are presently being leased to licensees and permittees for the storage of commercial explosives are considered to be in compliance with the requirements for nonsparking floors, as set forth in 27 CFR §§ 55.207(a)(4), 55.207(b), and 55.210, for the storage of all types of fully packaged explosives, pyrotechnics and propellants, with the exception of black powder. Any other such magazines which have smooth finished concrete floors and which meet or exceed DOD construction specifications will also be considered to be in compliance with the requirements of Part 555 with respect to nonsparking floors. It is the responsibility of the licensee or permittee to provide verification that such facilities were manufactured under DOD specifications or that the facilities meet or exceed such specification standards. If the Division Director determines that the concrete floors of type 1 or type 4 magazines do not meet the preceding requirements, he will require such floors to be covered with a nonsparking material, such as epoxy paint or mastic. [75 CB 67]

3. 27 CFR 55.41: Licenses and Permits-General

Certain companies that manufacture explosive materials for use in their own operations are required to obtain licenses as manufacturers of explosive materials.

ATF Ruling 75-31

ATF has held that companies, such as public utility companies engaged in line and facility construction, which manufacture explosives on a regular or continual basis are considered to be engaged in the business of manufacturing explosive materials and must be appropriately licensed as required by 18 U.S.C. 842. The term "manufacturer" is defined in 18 U.S.C. 841(h) as "any person engaged in the business of manufacturing explosive materials for purposes of sale or distribution or for his own use." Although the term "engaged in the business" is not susceptible to a rigid definition within 18 U.S.C. §§ 841-848, it is interpreted to imply an element of continuity or habitual practice; an element clearly present in the operations of companies described herein. Therefore, these companies are considered to be "engaged in the business" and must be licensed as explosives manufacturers. [75 CB 65]

4. 27 CFR 555.109: Identification of Explosive Materials

Methods of marking containers of explosive materials are prescribed.

ATF Ruling 75-35

Editors note: ATF Ruling 75-35 was rendered obsolete pursuant to ATF 5F, 70 Federal Register 30626 (May 27, 2005), and effective July 26, 2005.

5. 27 CFR 55.11: Meaning of Terms- State of Residence

"State of residence" of business entities who use explosive materials; distribution of explosive materials by licensees to out-of-State business entities other than licensees and permittees; and distribution to nonresident employees of such entities are discussed.

Editor's Note: Provisions of ATF Ruling 76-4 were modified, in part, by the Safe Explosives Act.

Effective May 24, 2003, it is unlawful for any person to receive explosive materials unless such person holds an ATF license of permit. It is also unlawful for any licensee or permittee to knowingly distribute explosive materials to any person who does not hold a license or permit. The only relevance remaining in the term "State of residence" is for distribution of explosive materials to, and receipt by, limited permit holders. Pursuant to 18 U.S.C. 842(a)(3) and (a)(4), limited permit holders may, on not more than 3 separate occasions, lawfully receive explosive materials from a licensee or permittee whose premises are located within the state of residence of the limited permit holder. ATF Rule 76-4 continues to apply in determining whether a limited permit holder has acquired a "State of residence" for purposes of receipt of explosives under 18 U.S.C. 842(a)(4)(B).

6. 27 CFR 55.126: Explosives Transaction Record

Under certain conditions, a single Form 5400.4 may be used to cover a series of deliveries.

Editor's Note: The provisions of ATF Rule 76-10 were rendered obsolete by ATF No. 1, 68 FR 13791, Mar. 20, 2003.

7. 27 CFR 55.207: Construction of Type 1 Storage Facilities (Also § 55.208)

Alternate construction standards for storage facilities for explosive materials are prescribed.

ATF Ruling 76-18

Section 842(j) of 18 U.S.C. states: "It shall be unlawful for any person to store any explosive material in a manner not in conformity with regulations promulgated by the Secretary. In promulgating such regulations, the Secretary shall take into consideration the class, type, and quantity of explosive materials to be stored, as well as the standards of safety and security recognized in the explosives industry." The regulations in 27 CFR §§ 55.207 and 55.208 prescribe types of storage facilities for explosive materials and provide (among other things) that such storage facilities shall be bullet resistant.

Section 55.201(b) provides that alternate storage facilities may be authorized for the storage of explosive materials when it is shown that such alternate facilities are or will be constructed in a manner substantially equivalent to the standards of construction contained in the applicable regulations.

The term "bullet-resistant" means resistant to penetration of a bullet of 150 grain M2 ball ammunition having a nominal muzzle velocity of 2700 feet per second fired from a .30 caliber rifle from a distance of 100 feet perpendicular to the wall or door. It has been determined that a wide range of construction criteria meet the bullet-resistant requirements of regulations for construction of storage facilities for explosive materials. In order to promote standards of safety and security in the storage of explosive materials while allowing the industry a wide latitude in the selection of construction materials, it is held that storage facilities (magazines) that are constructed according to the following minimum specifications are bullet-resistant and meet the requirements of the regulations as set forth in 27 CFR Part 55 (All steel and wood dimensions are actual thicknesses. To meet the concrete block and brick dimensions indicated, the manufacturers' represented thicknesses may be used).

(a) Exterior of ⅝ inch steel, lined with an interior of any type of nonsparking material.

(b) Exterior of ½ inch steel, lined with an interior of not less than ⅜ inch plywood.

(c) Exterior of ⅜ inch steel, lined with an interior of two inches of hardwood.

(d) Exterior of ⅜ inch steel, lined with an interior of three inches of softwood or 2¼ inches of plywood.

(e) Exterior of ¼ inch steel, lined with an interior of three inches of hardwood.

(f) Exterior of ¼ inch steel, lined with an interior of five inches of softwood or 5 ¼ inches of plywood.

(g) Exterior of ¼ inch steel, lined with an intermediate layer of two inches of hardwood and an interior lining of 1½ inches of plywood.

(h) Exterior of 3/16 inch steel, lined with an interior of four inches of hardwood.

(i) Exterior of 3/16 inch steel, lined with an interior of seven inches of softwood or 6¾ inches of plywood.

(j) Exterior of 3/16 inch steel, lined with an intermediate layer of three inches of hardwood and an interior lining of ¾ inch of plywood.

(k) Exterior of ⅛ inch steel, lined with an interior of five inches of hardwood.

(l) Exterior of ⅛ inch steel, lined with an interior of nine inches of softwood.

(m) Exterior of ⅛ inch steel, lined with an intermediate layer of four inches of hardwood and an interior lining of ¾ inch plywood.

(n) Exterior of any type of fire-resistant material which is structurally sound, lined with an intermediate layer of four inches of solid concrete block, OR four inches of solid brick OR four inches of solid concrete; AND, an interior lining of ½ inch plywood placed securely against the masonry lining.

(o) Standard eight inch concrete block with voids filled with well-tamped sand/cement mixture.

(p) Standard eight inch solid brick.

(q) Exterior of any type of fire-resistant material which is structurally sound, lined with an intermediate six inch space filled with well-tamped dry sand or well-tamped sand/cement mixture.

(r) Exterior of 1/8 inch steel, lined with a first intermediate layer of ¾ inch plywood, a second intermediate layer of 3⅝ inches of well-tamped dry sand or sand/cement mixture and an interior lining of ¾ inch plywood.

(s) Exterior of any type of fire-resistant material, lined with a first intermediate layer of ¾ inch plywood, a second intermediate layer of 3⅝ inches well-tamped dry sand or sand/cement mixture, a third intermediate layer of ¾ inch plywood, and a fourth intermediate layer of two inches of hardwood OR 14 gauge steel AND an interior lining of ¾ inch plywood.

(t) Eight inch thick solid concrete. [76 CB 106]

8. 27 CFR 55.213: Quantity and Storage Restrictions(Also § 55.208)

Alternate magazine construction standards for storage of electric blasting caps with other explosive materials are prescribed.

ATF Ruling 77-24

Section 842(j) of 18 U.S.C. states: "It shall be unlawful for any person to store any explosive material in a manner not in conformity with regulations promulgated by the Secretary. In promulgating such regulations, the Secretary shall take into consideration the class, type, and quantity of explosive materials to be stored, as well as the standards of safety and security recognized in the explosives industry." The regulations in 27 CFR § 55.213 restrict the storage of blasting caps with other explosive materials. Section 55.201(b) provides that alternate storage magazines may be authorized for the storage of explosive materials when it is shown that such alternate magazines are or will be constructed in a manner substantially equivalent to the standards of construction contained in the applicable regulations.

ATF recognizes that the transportation and storage of explosive materials in the same

vehicle along with electric blasting caps is often desired. The Institute of Makers of Explosives established a recommended standard for such transport in their Safety Library Publication No. 22, dated November 5, 1971 [revised January 1985]. This standard prescribes the minimum construction criteria for:

(a) A container securely attached—

(1) Above the cab of the vehicle (see Figure 1, Appendix A), and

(2) To the vehicle frame under the cargo space (see Figure 2, Appendix A), or

(b) A built-in compartment in the cargo space of the vehicle (see Appendix B).

In addition to motorized vehicles, consideration was also given for the use of similar criteria on portable wheeled trailers being used as magazines under § 55.208(a) of the regulations (see

Appendix E). In order to insure standards of safety and security in the storage of explosive materials while allowing the industry a proper latitude in the construction of magazines, it is held that vehicles used for transporting and for storing explosive materials that are constructed in conformity with the standards listed below, and in compliance with all other safety and security provisions contained in Part 55 (e.g., effectively immobilized when unattended) will meet the requirements of ATF regulations. Even though constructed on the same vehicle, each compartment will be considered as a separate magazine. The two magazines on the vehicle will, however, be considered as one magazine when applying the American Table of Distances [see Table at § 55.218].

Construction Standards For Storage of Electric Blasting Caps (Non Mass-Detonating)

a. The container or compartment must provide for total enclosure of the electric blasting caps.

b. The partition between the explosives storage compartment and the electric blasting cap compartment must be of laminate construction consisting of A/C grade or better exterior plywood, gypsum board [sheetrock] and low carbon steel plates. In order of arrangement, the laminate must conform to the following, with minimum thickness of each lamination as indicated:
½ inch plywood
½ inch gypsum board [sheetrock],
⅛ inch low carbon steel, and,
¼ inch plywood, with the ¼ inch plywood facing the explosives storage compartment. See Appendix C for details of laminate construction. The door to the electric blasting cap compartment must be of metal construction or solid wood covered with metal; the outside walls and top must be of the same construction as the rest of the vehicle or trailer. If high explosives or bullet sensitive explosive materials are stored in the vehicle, then the storage compartment of the vehicle must be constructed so as to be bullet-resistant.

c. As an alternative to the construction requirements shown in paragraph b, a container for use only as illustrated in Appendix A may be used when constructed as follows:

1. The top, lid or door, and the sides and bottom of each container must be of laminate construction consisting of A/C grade or better exterior plywood, solid hardwood, gypsum board [sheetrock], and sheet metal. In order of arrangement, the laminate must conform to the following, with minimum thickness of each lamination as indicated:
¼ inch plywood,
1 inch solid hardwood,
½ inch plywood,
½ inch gypsum board [sheetrock] (OR ¼ inch particle board), and 22 gauge sheet metal, constructed inside to outside in that order.
See Appendix D for details of laminate construction.

2. The hardwood must be fastened together with wood screws, the ½ inch plywood must be fastened to the hardwood with wood screws, the inner ¼ inch plywood must be fastened to the hardwood with adhesive, and the 22 gauge sheet metal must be attached to the exterior of the container with screws.

d. The laminate composite material must be securely bound together by waterproof adhesive or other equally effective means.

e. The steel plates at the joints of laminations must be secured by continuous fillet welds.

f. All interior surfaces of the container or compartment must be constructed so as to prevent contact of contents with any sparking metal.

g. There must be direct access to the container or into a compartment from outside the vehicle.

h. Each container or compartment must have a snug fitting continuous piano-type hinged lid or door equipped with a locking device (or devices).

i. Without permitting direct access to contents under normal conditions, the locking or hinging mechanisms must permit at least one edge of the lid or door to rise or move outward at least ½ inch when subjected to internal pressure.

j. The exterior of the container or compartment must be weather-resistant. [77 CB 191]

9. 18 U.S.C. 842(j): Storage of Explosives

27 CFR 55.208(b)(1), 55.210(b)(1), and 55.211(b)(1): Indoor Storage of Explosives in a Residence or Dwelling

ATF will approve variances to store explosives in a residence or dwelling only upon certain conditions including, but not limited to, receipt of a certification of compliance with State and local law, and documentation that local fire safety officials have received a copy of the certification.

ATF Ruling 2002-3

Section 842(j) of 18 U.S.C. states: "It shall be unlawful for any person to store any explosive material in a manner not in conformity with regulations promulgated by the Secretary."

The regulations in 27 CFR 55.208(b)(1), 55.210(b)(1), and 55.211(b)(1) specify that no indoor magazine is to be located in a residence or dwelling. Section 55.22 specifies that the Director may allow alternate methods or procedures in lieu of a method or procedure specifically prescribed in the regulations. Specifically, section 55.22(a)(3) provides that such "variances" are permissible only in certain circumstances, including where "[t]he alternate method or procedure will not be contrary to any provision of law and will not . . . hinder the effective administration of this part."

ATF has been advised that certain variances previously approved for storage of explosives in residences or dwellings are in violation of State or local zoning law. ATF believes it is important to ensure that approval of variances is in compliance with all State and local provisions.

To obtain a variance for indoor storage of explosives in a residence or dwelling, ATF has determined that a person must submit to ATF a certification signed under penalty of perjury along with the request for the variance. The certification must:

1. State that the proposed alternative storage method will comply with all applicable State and local law;

2. Provide the name, title, address, and phone number of the authority having jurisdiction for fire safety of the locality in which the explosive materials are being stored; and,

3. Demonstrate that the person has mailed or delivered the certification to the authority identified in (2).

When required by the Director, such persons must furnish other documentation as may be necessary to determine whether a variance should be approved.

Held, ATF will approve variances to store explosives in a residence or dwelling only upon certain conditions including, but not limited to, receipt of a certification of compliance with State and local law, and documentation that local fire safety officials have received a copy of the certification.

Date signed: August 23, 2002

10. 18 U.S.C. 842(j): Storage of Explosives 27 CFR 55.208(b)(1), 55.210(b)(1), and 55.211(b)(1): Indoor Storage of Explosives in Business Premises Directly Adjacent to a Residence or Dwelling

ATF Ruling 2002-4

ATF requires approval of variances for indoor storage of explosives in business premises directly adjacent to a residence or dwelling.

Section 842(j) of 18 U.S.C. states: "It shall be unlawful for any person to store any explosive material in a manner not in conformity with regulations promulgated by the Secretary."

The regulations in 27 CFR 55.208(b)(1), 55.210(b)(1), and 55.211(b)(1) specify that no indoor magazine is to be located in a residence or dwelling. Section 55.22 specifies that the Director may allow alternate methods or procedures in lieu of a method or procedure specifically prescribed in the regulations. Specifically, section 55.22(a)(3) provides that such "variances" are permissible only in certain circumstances, including where "[t]he alternate method or procedure will not be contrary to any provision of law and will not ... hinder the effective administration of this part."

ATF has been asked whether businesses that are directly adjacent to living quarters may lawfully store explosive materials in the business premises. The issue presented is whether the premises amount to a "residence or dwelling" within the meaning of the regulations cited above.

Even where the business premises are segregable from the living quarters by the existence of a door or a common wall, the business premises retain their character as a residence or dwelling. Accordingly, indoor storage of explosives in such premises is generally prohibited and can be allowed only pursuant to an approved variance.

Held, ATF requires approval of variances for indoor storage of explosives in business premises directly adjacent to a residence or dwelling. ATF may approve such variances upon receipt of all appropriate certification and other documentation as may be requested.

Date signed: August 23, 2002

11. 18 U.S.C. 842(f): Unlawful Acts, 27 CFR 555.105(b)(6)(iii): Distribution of Explosives to Limited Permittees

Distributors distributing explosive materials to holders of limited permits via common or contract carrier may verify receipt of the explosive materials by telephone, facsimile, e-mail, or other means within three business days of shipment in lieu of requiring the common or contract carrier to verify the identity of the person accepting delivery of the explosives. The distributor shall make a notation on ATF Form 5400.4 indicating whether the shipment was received and the date and time of the contact with the distributee.

ATF Ruling 2003-5

The Bureau of Alcohol, Tobacco, Firearms and Explosives (ATF) has received questions from the explosives industry regarding the requirement under 27 CFR 555.105(b)(6)(iii) that, effective May 24, 2003, a common or contract carrier hired by a Federal explosives licensee or permittee verify the identity of the distributee, note the type and number of the identification document, and provide this information to the distributor. The distributor is required to record this information on ATF Form 5400.4, Limited Permittee Transaction Report (LPTR). Industry members have informed ATF that this requirement places an undue burden on common and contract carriers. Drivers are concerned that verifying the identity of persons accepting delivery of explosive materials by examining an identification document and providing the identification information to the distributor will be overly time consuming. Drivers are also concerned that they could be held personally liable for delivering explosives to persons not authorized to receive them.

ATF imposed the verification requirements of section 555.105(b)(6)(iii) to ensure that when explosive materials are sold by a distributor to a holder of a limited permit and transported by a common or contract carrier hired by the distributor, the explosive materials are delivered only to a person authorized to receive them. ATF continues to believe it is important that sellers of explosive materials verify that such materials are delivered to persons authorized to receive them.

However, it was not ATF's intention to impose an undue burden on common or contract carriers.

Section 555.22, Title 27, CFR, provides that the Director may approve an alternate method or procedure in lieu of a method or procedure specifically prescribed in the regulations when he finds that:

(1) Good cause is shown for the use of the alternate method or procedure;

(2) The alternate method or procedure is within the purpose of, and consistent with the effect intended by, the specifically prescribed method or procedure and that the alternate method or procedure is substantially equivalent to that specifically prescribed method or procedure; and

(3) The alternate method or procedure will not be contrary to any provision of law and will not result in an increase in cost to the Government or hinder the effective administration of Part 555.

ATF finds that there is good cause to authorize a variance to the provisions of section 555.105(b)(6)(iii) due to the undue burden placed on common or contract carriers by the verification requirement. Accordingly, ATF authorizes the following alternate method or procedure to the identification verification requirements of section 555.105(b)(6)(iii):

The distributor shall, no later than three business days after shipment of the explosive materials, contact the distributee by telephone, facsimile, e-mail or any other means to ensure that the shipment has been received. The distributor shall make a notation on ATF Form 5400.4 indicating whether the shipment was received and the date and time of the contact with the distributee.

ATF finds that the above alternate method is consistent with the verification provisions of section 555.105(b)(6)(iii), because it will ensure that delivery has taken place and document the information in the distributor's records.

The alternate method is not contrary to any provision of law, will not increase the costs to ATF, and will not hinder the effective administration of the regulations in 27 CFR Part 555.

Held, pursuant to 27 CFR 555.22, ATF authorizes a variance from the requirements of 27 CFR 555.105(b)(6)(iii) for Federal explosives licensees and permittees making distributions of explosive materials to holders of limited permits via common or contract carrier. As an alternate method or procedure, distributors distributing explosive materials to holders of limited permits via common or contract carrier may verify receipt of the explosive materials by telephone, facsimile, e-mail or other means within three business days of shipment in lieu of requiring the common or contract carrier to verify the identity of the person accepting delivery of the explosives. The distributor shall make a notation on ATF Form 5400.4 indicating whether the shipment was received and the date and time of the contact with the distributee.

Date signed: May 23, 2003.

12. 27 CFR 555.11: Meaning of Terms

ATF provides guidance on three different private roads and whether they are "highways" as defined in 27 CFR 555.11.

ATF Ruling 2005-2

The Federal explosives laws, 18 U.S.C. Chapter 40, require all persons to store explosive materials in a manner in conformity with regulations issued by the Attorney General. 18 U.S.C. 842(j). The Attorney General has delegated his authority to administer and enforce the Federal explosives laws to the Director, ATF. 28 CFR 0.130. Regulations in 27 CFR Part 555 implement the provisions of the Federal explosives laws. Storage regulations in 27 CFR Part 555, Subpart K, provide that outdoor magazines in which high explosives are stored must be located no closer to inhabited buildings, passenger railways, public highways, or other magazines in which high explosives are stored than the minimum distances specified in the table of distances for storage of explosive materials in section 555.218 of the regulations. 27 CFR 555.206.

Section 555.11 of the regulations defines the term "highway" as "any public street, public alley, or public road, including a privately financed, constructed, or maintained road that is regularly and openly traveled by the general public."

In Scenario A, a private road owned by a corporation is used by the general public as an access road to a parking lot owned by the corporation. The road is near an explosives magazine. The road does not have a gate, sign, or other means of restricting access to the road. The road is also used by the general public on a daily basis to gain access to other public streets.

In Scenario B, a company that manufactures display fireworks, a logging company, and an individual who owns buildings utilized to store his collection of automobiles all occupy property to which the only access is a privately owned road. A separate party that leases to these three entities owns the property. The road is located on private property, and a locked gate at the entrance to the road prevents access by the general public. The display fireworks company, the logging company, and the individual storing automobiles all have keys to unlock the gate and travel on the road when needed. The gate is locked at all times, and there is no evidence that the road is open to anyone other than the two businesses and one individual who require access to their property.

In Scenario C, an explosives company maintains explosives magazines in a quarry area that has a roadway traversing through the quarry. The quarry owns the property, and the road is maintained by the quarry. The road has a gate and there are signs advising no trespassing. However, when ATF officials visited the location on several occasions, the gate was left open and members of the public regularly utilized the roadway as a shortcut between two major highways. There were no indications the owner of the property took any steps to prevent members of the public from utilizing the roadway.

Applying the regulatory definition of "highway" to the three scenarios, the road in Scenario A is clearly a highway that is subject to the tables of distance in Part 555. Although it is privately owned, it is regularly and openly traveled by members of the general public without restriction.

The roadway in Scenario B is not a "highway" as defined. Access is restricted at all times and there is no evidence the general public regularly travels on the roadway. Access to the road is provided to only a limited number of persons who have a legal right to travel the road. Accordingly, this road is not regularly and openly traveled by members of the general public.

ATF concludes that the roadway described in Scenario C is a "highway" as defined in 27 CFR 555.11. Although access to the roadway is restricted by a gate and "No trespassing" signs are posted, the gate is not closed at all times. Furthermore, ATF observation indicates that the roadway is regularly and openly traveled by members of the general public. Based on these facts, the roadway is a highway which is subject to the tables of distance in 27 CFR Part 555.

Held, a private road with no gate, signs, or other means of restricting access that is used by the general public as an access road to a parking lot and as access to other public streets is a "highway" as defined in 27 CFR 555.11.

Held further, a private road with a locked gate at the entrance that is locked at all times and used by a limited number of persons leasing or owning property accessed by the road is not a "highway" as defined in 27 CFR 555.11.

Held further, a private roadway traversing a quarry with a gate restricting access and a "no trespassing" sign is a "highway," as defined in 27 CFR 555.11, because the gate is not locked at all times and the general public regularly utilizes the roadway as a shortcut between two public highways.

Date approved: September 8, 2005

13. 27 CFR 555.11: Meaning of Terms

ATF provides guidance on two situations involving structures and whether they are "inhabited buildings" as defined in 27 CFR Part 555.

ATF Ruling 2005-3

The Federal explosives laws, 18 U.S.C. Chapter 40, require all persons to store explosive materials in a manner in conformity with regulations issued by the Attorney General. 18 U.S.C. 842(j). The Attorney General has delegated his authority to administer and enforce the Federal explosives laws to the Director, ATF. 28 CFR 0.130. Regulations in 27 CFR Part 555 implement the provisions of the Federal explosives laws. Storage regulations in 27 CFR Part 555, Subpart K, provide that outdoor magazines in which high explosives are stored must be located no closer to inhabited buildings, passenger railways, public highways, or other magazines in which high explosives are stored than the minimum distances specified in the table of distances for storage of explosive materials in section 555.218 of the regulations. 27 CFR 555.206.

The regulation at 27 CFR 555.11 defines the term "inhabited building" as "any building regularly occupied in whole or in part as a habitation for human beings, or any church, schoolhouse, railroad station, store, or other structure where people are accustomed to assemble, except any building occupied in connection with the manufacture, transportation, storage, or use of explosive materials."

In Scenario A, an explosives licensee leases explosives magazines to an individual who uses the magazines for storage of goods other than explosive materials. The magazines are located adjacent to magazines used by the explosives licensee for the storage of explosive materials. The magazines used by the lessee are not separated by the minimum distances required for the separation of magazines from "inhabited buildings" as required by the regulations in 27 CFR Part 555.

The magazines are visited regularly by the individual who stores property in the magazines, but no additional persons accompany the individual when he is present at the magazine. However, the individual hires contractors to repair equipment stored in the facilities from time to time, and 1-3 employees of the contractor may occasionally be present for short periods of time at the storage site. However, such visits occur no more than 3-5 times per year. In addition, an employee from the water company visits the storage site once a month to read the water meter, an employee from the power company reads the power meter once a month, and other vendors may be present at the site for short periods of time for other legitimate purposes.

In Scenario B, a licensed manufacturer of explosives X leases a unit in an industrial park that shares a common wall with a unit leased by licensed manufacturer Y. Both licensees store explosives in magazines located inside and outside the units. The magazines of Manufacturer X and the building used by Manufacturer Y are not separated by the minimum distance prescribed in 27 CFR Part 555 for the separation of magazines and inhabited buildings. Likewise, the magazines of Manufacturer Y and the building used by Manufacturer X are not separated by the minimum distance prescribed in 27 CFR Part 555 for the separation of magazines and inhabited buildings. In both structures, employees and contractors are regularly present during work hours for purposes of carrying on the manufacturing and distributing businesses of the two licensees. This includes personnel who work in the manufacturing plant, those who work on the loading dock to load and ship explosives products to customers, and those who work in the office taking orders, sending out invoices, and handling other clerical work for the businesses.

Applying the law and regulations to the facts of Scenario A, ATF concludes that the leased structures used by the individual to store items other than explosives are not being used as a habitation for human beings and are not buildings occupied in connection with the manufacture, transportation, storage, or use of explosive materials. Accordingly, the sole issue remaining is whether the structure is one where people are accustomed to assemble.

Noteworthy, the regulation uses the term "people," which is the plural version of "person." Thus, ATF believes that more than one person must "assemble" at the structure for it to be an "inhabited building." In addition, the word "assemble" is defined, in part, as "To bring or gather together in a group or whole." The word "assembly" is defined, in part, as "A group of persons gathered for a common purpose." The American Heritage Dictionary, Second College Edition, Houghton Mifflin Co., 1982. It is clear that the presence of one person at a structure or location cannot be an assembly of any sort. Accordingly, in situations where one person is present at a particular structure, whether on a regular or infrequent basis, such a structure is not an "inhabited building" as defined in 27 CFR 555.11.

Likewise, occasional visits to the storage facility by mail delivery persons or employees of public utility companies for brief periods of time would not be an "assembly" that would make the facility an inhabited building. However, if 2 or more repair persons are present at the facility to make repairs to equipment stored there, such persons would be there for a common purpose, and would have "assembled" at the structure. However, the structure would be an "inhabited building" only if it is a structure where people are accustomed to assemble. The word "accustom" is defined as "To familiarize, as by constant practice, use, or habit: accustomed himself to working long hours." The word "accustomed" is defined as "Usual, characteristic, or normal: worked with her accustomed thoroughness." The American Heritage Dictionary, Second College Edition, Houghton Mifflin Co., 1982. These definitions indicate that a structure will be one where people are accustomed to assemble only if there is some degree of continuity, regularity, or frequency to such assembly.

Infrequent, occasional visits to the storage site by 2 or more repair persons would not make the storage facility an "inhabited building," because such intermittent visits would not be "customary." Only where 2 or more persons are present at the site for a common purpose and on a regular basis would the building fit within the definition of "inhabited building."

To address Scenario B, it is apparent that the building leased by Manufacturer X is exempt from the definition of "inhabited building" as to the magazines of Manufacturer X, and the building leased by Manufacturer Y is likewise exempt as to the magazines of Manufacturer Y. This is because both buildings are occupied in connection with the manufacture, transportation, and storage of explosive materials. Clearly, the employees of both licensees are aware that explosive materials are present on the premises and they assume the risk of any such operation. A more difficult question is presented by the buildings of Manufacturer X and the magazines of Manufacturer Y and vice versa. ATF cannot assume that all employees are cognizant of the activities of their neighbors in the industrial park. Thus, it cannot be assumed that the employees are knowingly assuming the risk of explosive materials stored in magazines owned by the other licensee.

Given the plain language of the regulation, however, ATF does not believe it is appropriate to deny the coverage of the regulatory exemption to adjoining licensees on the basis of an assumption of the risk analysis. The current regulatory definition excludes from the definition of "inhabited building" any building occupied in connection with the manufacture, transportation, storage, or use of explosive materials, regardless of the knowledge of the building's occupants. Accordingly, ATF concludes that the industrial units occupied by Manufacturer X and Manufacturer Y are both exempted from the definition of "inhabited building" as to the magazines of each other as well as to their own magazines.

Held, a structure used to store items other than explosive materials that is visited on a regular basis by one individual is not an "inhabited building" as defined in 27 CFR 555.11, because it is not a structure where people are accustomed to assemble. Where 2 or more persons are present at the structure to repair equipment stored therein and such visits to the site are occasional and infrequent, the structure is not an "inhabited building" because the visits are not "customary." However, where 2 or more persons make regular visits to the structure for a common purpose, the structure is an "inhabited building," and explosives magazines may not be stored closer to the structure than the minimum distances specified in the regulations in 27 CFR Part 555.

Held further, buildings occupied by licensed explosives manufacturers in connection with the manufacture, transportation, storage, or use of explosive materials are not included within the definition of "inhabited building" as to magazines located on their own premises. In addition, buildings occupied by licensed explosives manufacturers in connection with the manufacture, transportation, storage, or use of explosives are not included within the definition of "inhabited building" as to magazines located on property owned by another licensee.

Date approved: November 25, 2005

14. 18 U.S.C. 842(f): Records Required for Explosives Licensees and Permittees

27 CFR 555, Subpart G: Records and Reports

27 CFR 555.22: Alternate Methods or Procedures; Emergency Variations from Requirements

Under specified conditions, approval is granted to utilize computerized records as required records under 27 CFR 555, Subpart G.

ATF Ruling 2007-1

Section 842(f), Title 18, United States Code, makes it unlawful for any licensee or permittee to willfully manufacture, import, purchase, distribute, or receive explosive materials without making such records as the Attorney General may by regulation require, including, but not limited to, a statement of intended use, the name, date, place of birth, social security number or taxpayer identification number, and place of residence of any natural person to whom explosive materials are distributed.

Regulations implementing section 842(f) are in 27 CFR Part 555, Subpart G. The regulations in this subpart specify the records required to be created and maintained by licensed importers (section 555.122), licensed manufacturers (section 555.123), licensed dealers (section 555.124), and permittees (section 555.125). The regulation in section 555.121 provides that licensees and permittees must keep records pertaining to explosive materials in permanent form (i.e., commercial invoices, record books) and in the manner required in Subpart G. In addition, sections 555.122-555.125 specifically allow licensees and permittees to use an alternate record to record the distribution of explosive materials when it is shown that the alternate records would accurately and readily disclose the information required by the regulations. These regulations require licensees and permittees who propose to use alternate records to submit a letter application to ATF describing the proposed alternate records and the need for them. Alternate records are not to be employed until approval from ATF is received.

Regulations at 27 CFR 555.22 allow for the approval and use of an alternate method or procedure in lieu of a method or procedure specifically prescribed in Part 555. ATF may approve an alternate method or procedure when it is found that—

(1) Good cause is shown for the use of the alternate method or procedure;

(2) The alternate method or procedure is within the purpose of, and consistent with the effect intended by, the specifically prescribed method or procedure and that the alternate method or procedure is substantially equivalent to that specifically prescribed method or procedure; and

(3) The alternate method or procedure will not be contrary to any provision of law and will not result in an increase in cost to the Government or hinder the effective administration of 27 CFR Part 555.

With advances in technology and the dramatic decrease in the cost of computers, many businesses rely upon computers to maintain records of their inventory, sales, customer lists, and other business information. Even the smallest home-based business utilizes computers to record and maintain business information. Creating and maintaining records in a computer database, rather than paper form, makes it easier to ensure accuracy of records and makes it less likely that records will be lost or misplaced. In addition, maintaining records via computer generally saves time and money in bookkeeping and auditing expenses. This utilization of computers has allowed companies to automate inventories, utilizing technology such as bar codes or RFID (radio frequency identification) chips. This facilitates better accountability of product overall, reducing the potential of everyday accounting errors. Over the years ATF has seen a significant increase in the number of requests from explosives licensees and permittees for authorization to utilize computerized records rather than paper records of acquisition and distribution and other required records, such as magazine transaction records. ATF routinely approves requests to utilize computerized records, with certain conditions, finding that the use of such records is substantially equivalent to methods of record keeping set forth in the regulations in 27 CFR Part 555, Subpart G.

Several explosives industry members have asked whether computerized records may be maintained without obtaining written approval from ATF if they contain all the required information specified in the regulations and are maintained in a permanent form. Additionally, industry members have questioned whether computer records in combination with paper records may be maintained if they are permanent and contain all the information required by the regulations.

ATF has determined that records of acquisition and disposition, magazine summary records, and the other records required by 27 CFR Part 555, Subpart G, satisfy the standard of permanency and are substantially equivalent to paper records if they meet the following criteria:

1. All data entered into the computer system must be recorded into the database and cannot be capable of being edited or modified at a later date. The software system must retain any correction of errors as an entirely new entry, without deleting or modifying the original entry. The system may allow for entries in a notes column to explain any correction.

2. The system must have a reliable daily memory backup capability to protect the data from accidental deletion or other system failure.

It is also acceptable for licensees/permittees to maintain required records using a combination of a computer program, commercial invoices, and other documents, provided that all of the required information is maintained in the records in permanent form. Any use of a computer for any portion of the required records must comply with the standards outlined above. However, each particular transaction must be self-contained with all the required information in the same recordkeeping medium. As one example, dispositions of explosives by a dealer cannot be separated by keeping the dates of disposition and the manufacturer's name or brand name in the computer, and all the other required information for that disposition on separate written documents.

ATF finds that good cause exists for authorizing the use of a computer to create and maintain the records required by 27 CFR Part 555, Subpart G, as the use of computers is accepted throughout the business community as a reliable, cost-efficient means of maintaining business records. ATF also finds that the use of a computer to maintain required records, contingent upon the requirements outlined above, is consistent with the effect intended by the requirements of Subpart G, as it will result in a permanent, reliable record that will accurately indicate acquisitions and dispositions of explosive materials.

Finally, ATF finds that the use of computer records properly containing all the required information should not hinder the effective administration of the Federal explosives laws or regulations – use of such records generally makes it easier for ATF to conduct inventories of product on hand and to audit required records. Accordingly, ATF concludes that the requirements for approval of an alternate method or procedure in accordance with 27 CFR Part 555, sections 555.22 and 555.122-555.125, are met.

Held, persons holding licenses and permits issued under 18 U.S.C., Chapter 40, may use computers to create and maintain all or any portion of the records required by 18 U.S.C. 842(f) and 27 CFR Part 555, Subpart G, if the following conditions are satisfied:

1. All data entered into the computer system must be recorded into the database and cannot be capable of being edited or modified at a later date. The software system must retain any correction of errors as an entirely new entry, without deleting or modifying the original entry. The system may allow for entries in a notes column to explain any correction.

2. The system must have a reliable daily memory backup capability to protect the data from accidental deletion or other system failure.

Held further, licensees and permittees who wish to use computers to create and maintain all or a portion of their required records in accordance with the requirements set forth in this ruling are not required to obtain advance approval in accordance with 27 CFR 555.22 or 555.121-125.

Held further, licensees and permittees utilizing a combination of a computer program, commercial invoices, and other paper documents as required records must ensure that the required information for a particular transaction is fully contained in the same recordkeeping medium.

Held further, this ATF approved alternate method or procedure for computerized records shall not be withdrawn unless the holder of said variance is so advised by ATF in writing or no longer holds a Federal explosives license or permit.

Date approved: January 18, 2007

15. 18 U.S.C. 842(j): Storage of Explosives

27 CFR 555.210: Construction of Type 4 Magazines

27 CFR 555.215: Housekeeping

27 CFR 555.201: Notification of Local Fire Officials

27 CFR 555.63: Explosives Magazine Changes

27 CFR 555.22: Alternative Methods or Procedures; Emergency Variations from Requirements

Under specified conditions, display fireworks may be temporarily stored in locked and attended motor vehicles at the explosives magazine site(s) and at fireworks display site(s) without meeting the locking requirements of 27 CFR 555.210 provided certain additional security measures are in place. Additionally, allowance per 27 CFR 555.215 is made for the fuel tanks containing volatile materials that may be on the temporary storage vehicles. Finally, slight variation is provided for notification requirements to ATF and local fire officials.

ATF Rule. 2007-2

The Federal explosives laws, 18 U.S.C. Chapter 40, require all persons to store explosive materials in a manner in conformity with regulations issued by the Attorney General. 18 U.S.C. 842(j). The Attorney General has delegated the authority to administer and enforce the Federal explosives laws to the Director, ATF. 28 CFR 0.130. Regulations in 27 CFR Part 555, implement the provisions of the Federal explosives laws.

The regulation at 27 CFR 555.210(a) states, in part, "[o]utdoor magazines are to be fire-resistant, weather-resistant, and theft-resistant." Partly to satisfy the theft-resistant requirement, this section requires that each door be equipped with two mortise locks; two padlocks fastened in separate hasps and staples; a combination of a mortise lock and a padlock; a mortise lock that requires two keys to open; or a three-point lock. Padlocks must have at least five tumblers and a case-hardened shackle of at least

⅜-inch diameter. Padlocks must be protected with not less than ¼-inch steel hoods constructed so as to prevent sawing or lever action on the locks, hasps, and staples.

The regulation at 27 CFR 555.215 states, in part, "[v]olatile materials are to be kept a distance of not less than 50 feet from outdoor magazines."

The regulation at 27 CFR 555.201 requires, in part, that any person storing explosive materials notify local fire authorities orally before the end of the day on which storage of the explosive materials began and in writing within 48 hours from the time such storage began.

The regulation at 27 CFR 555.63 requires that any licensee or permittee who acquires (adds) a storage magazine must notify ATF at least five business days in advance of using any added explosives storage magazine.

Regulations at 27 CFR 555.22 allow for the approval and use of an alternate method or procedure in lieu of a method or procedure specifically prescribed in Part 555. ATF may approve an alternate method or procedure when it is found that—

(1) Good cause is shown for the use of the alternate method or procedure;

(2) The alternate method or procedure is within the purpose of, and consistent with the effect intended by, the specifically prescribed method or procedure and that the alternate method or procedure is substantially equivalent to that specifically prescribed method or procedure; and

(3) The alternate method or procedure will not be contrary to any provision of law and will not result in an increase in cost to the Government or hinder the effective administration of 27 CFR Part 555.

ATF has approved a significant number of variances for temporary storage for a specified amount of time before a display fireworks event, as well as during and after the event until the remaining explosive materials can be placed back into the appropriate storage magazine.

Preparation of display fireworks shows and the transportation of explosive materials to numerous show sites often take place over a period of several days. Preparing and temporarily storing the fireworks for these shows ordinarily take place on delivery trucks and trailers in one storage location where the proprietor already maintains storage of explosives materials with a high degree of security and safety by complying with the provisions of 27 CFR Part 555.

Many display fireworks shows also take several days to prepare at the show site. During preparation and after the show is completed, explosive materials frequently must be temporarily stored. This is often either extra product that was brought to the show or misfires that have been maintained and must be returned to permanent storage.

Allowing flexibility through alternate methods or procedures for specific regulations increases both safety and security at these show sites. These procedures are needed to increase public safety, as well as facilitate smooth operations for the display fireworks industry. The highest risk of incidents involving the accidental ignition of display fireworks is during handling, with the next highest risk being transportation. Providing no flexibility to allow storage in the delivery vehicles would require the industry to dangerously load and unload from storage magazines to vehicles and back into a storage magazine. Additionally, ATF believes that providing this guidance allows for preplanning by the proprietor and consistency of regulatory application nationwide.

One of the major dangers around explosives is fire. Therefore, the regulations require that volatile materials be maintained a distance of not less than 50 feet from outdoor explosives storage magazines. ATF believes that requiring attended storage for display fireworks temporarily stored in vehicles will ensure public safety, in lieu of the 50 foot separation requirement. The attendee should be able to alert the proper authorities if needed to ensure that a fire does not compromise this storage, or may relocate these temporary storage magazines to a safe location away from an identified fire.

Held, ATF will approve alternate methods or procedures for the temporary storage of display fireworks in locked and attended vehicles at explosives magazine site(s), as well as at the fireworks display site(s), under the following conditions:

1. The doors to each storage compartment containing explosive materials must be locked with at least one steel padlock having at least five tumblers and a casehardened shackle of at least ⅜-inch diameter. The padlock does not need to be protected by a steel hood. However, each temporary storage magazine must be attended at all times for security purposes. The vehicle is considered "attended" when an authorized individual is within 100 feet of all temporary storage and has an unobstructed view of the vehicle(s) containing the explosive materials. The individual must remain awake and observant of activities around the vehicle(s).

2. The person who temporarily stores the explosive materials must notify in writing the authority having jurisdiction for fire safety in the locality in which the explosive materials are stored no less than 3 Federal office business days prior to utilizing the additional temporary storage magazine(s).

3. The person who temporarily stores the explosive materials must notify ATF in writing of the location of this storage no less than 3 Federal office business days prior to utilizing the additional temporary storage magazine(s).

All other provisions of 27 CFR Part 555 must be complied with as prescribed.

Held further, this ATF-approved alternate method or procedure for the temporary storage of display fireworks in locked and attended vehicles shall not expire unless the holder of said variance is so advised by ATF or no longer holds a Federal explosives license or permit.

Date approved: January 18, 2007

16. 18 U.S.C. 842(j): Storage of Explosives

27 CFR 555.211(a): Immobilization of Outdoor Type 5 Mobile Storage Magazines

27 CFR 555.215: Housekeeping

27 CFR 555.22: Alternative Methods or Procedures; Emergency Variations from Requirements

Under specified conditions, blasting agents may be stored in mobile type 5 magazines (bulk delivery trucks) without meeting the prescribed immobilization requirements of 27 CFR 555.211.

ATF Rule 2007-3

The Federal explosives laws, 18 U.S.C. Chapter 40, require all persons to store explosive materials in a manner in conformity with regulations issued by the Attorney General. 18 U.S.C. 842(j). The Attorney General has delegated his authority to administer and enforce the Federal explosives laws to the Director, ATF. 28 CFR 0.130. Regulations in 27 CFR

Part 555 implement the provisions of the Federal explosives laws.

The regulation at 27 CFR 555.211(a)(1) states, in part, "[o]utdoor magazines are to be weather-resistant and theft-resistant." This section further states, "[w]hen unattended, vehicular magazines must have wheels removed or otherwise be effectively immobilized by kingpin locking devices or other methods approved by the Director."

The regulation at 27 CFR 555.215 states, in part, "[v]olatile materials are to be kept a distance of not less than 50 feet from outdoor magazines."

The regulation at 27 CFR 555.22 allows for the approval and use of an alternate method or procedure in lieu of a method or procedure specifically prescribed in Part 555. ATF may approve an alternate method or procedure when it is found that—

(1) Good cause is shown for the use of the alternate method or procedure;

(2) The alternate method or procedure is within the purpose of, and consistent with the effect intended by, the specifically prescribed method or procedure and that the alternate method or procedure is substantially equivalent to that specifically prescribed method or procedure; and

(3) The alternate method or procedure will not be contrary to any provision of law and will not result in an increase in cost to the Government or hinder the effective administration of 27 CFR Part 555.

ATF has approved a significant number of variances for an alternate means of immobilizing preloaded bulk delivery vehicles when some additional security measures were put in place.

Bulk delivery vehicles are routinely utilized for on-site delivery of blasting services. Often these trucks contain a blasting agent as defined under 27 CFR 555.11. Utilization of these bulk products delivered on-site and used immediately has increased safety and security by reducing the number of remotely located storage trailers containing packaged blasting agent products. These delivery vehicles generally leave the explosives storage locations at unusual times of day or night, and for safety reasons they are loaded the day before for next day's delivery. These trucks are incapable of being disabled by a kingpin locking device, and the requirement to remove the wheels for immobilization is obviously not feasible. Additionally, there are times when explosive material remains on the vehicle when it returns from use. Most of these products degrade during handling, therefore, removing them from the bulk vehicle would not be an option.

One of the major dangers surrounding explosives is fire. Therefore, the regulations at 27

CFR 555.215 require that volatile materials be maintained a distance of not less than 50 feet from outdoor explosive storage magazines. Because these preloaded storage vehicles contain a fuel tank filled with volatile materials, they must remain in an area protected from fire such as gravel, paved, or closely mowed designated parking area.

ATF believes that the following alternate method of operation is substantially equivalent to the prescribed methods. Increased safety for employees and the public provides good cause for this alternate method. It is not contrary to law and will not result in any increased cost to the Government. Overall, ATF believes allowing for this flexibility assists with the effective administration of 27 CFR Part 555.

Held, ATF will approve alternate methods or procedures for the preloading and temporary storage of bulk blasting agents in delivery vehicles at explosive magazine site(s), when the security and immobilization meets the following criteria:

1. All doors on the vehicle are locked, the ignition key is removed, and the key is secured away from the truck.

2. When the site is not in operation, outer perimeter security is established. This may be by a variety of means such as a locked gate, security guards, fence, natural features, or a combination of these.

3. Each potential access point to explosive materials on a storage vehicle will be secured with a minimum of one padlock that has at least five tumblers and casehardened shackle of at least ⅜-inch diameter.

4. Each vehicle shall be immobilized through the use of a steering wheel locking device, lockable battery disconnect switch, or both.

5. All vehicles preloaded with blasting agents shall be parked in a company designated area not susceptible to fire propagation such as bare dirt, gravel, rock, paving, or closely mowed parking lot.

All other provisions of 27 CFR Part 555 must be complied with as prescribed.

Held further, licensees and permittees who wish to use the alternate method or procedure set forth in this ruling are not required to obtain advance approval in accordance with 27 CFR 555.22.

Held further, this ATF-approved alternate method or procedure for the temporary storage of bulk blasting agent products in locked and properly secured vehicles shall not be withdrawn unless the holder of said variance is so advised by ATF in writing or no longer holds a Federal explosives license or permit.

Date approved: January 26, 2007

APPENDIX A

PERMANENTLY MOUNTED CONTAINERS

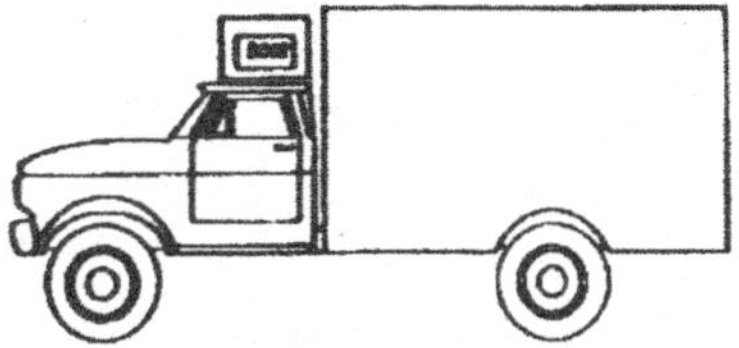

Figure 1

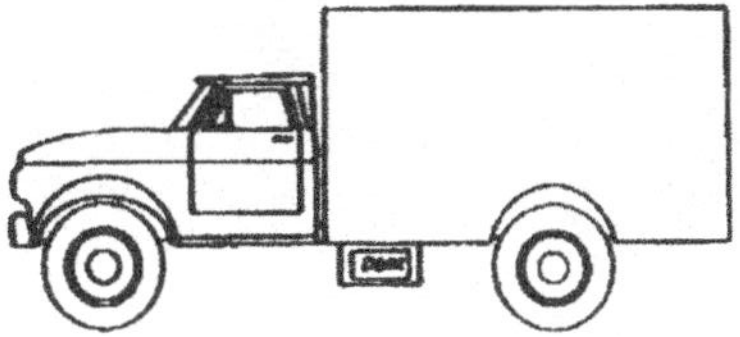

Figure 2

NOTE: The configurations shown in Figures 1 and 2 are equally applicable to multi-axle and "cab-over" vehicles.

[Diagrams: Courtesy of IME]

APPENDIX B

COMPARTMENTS

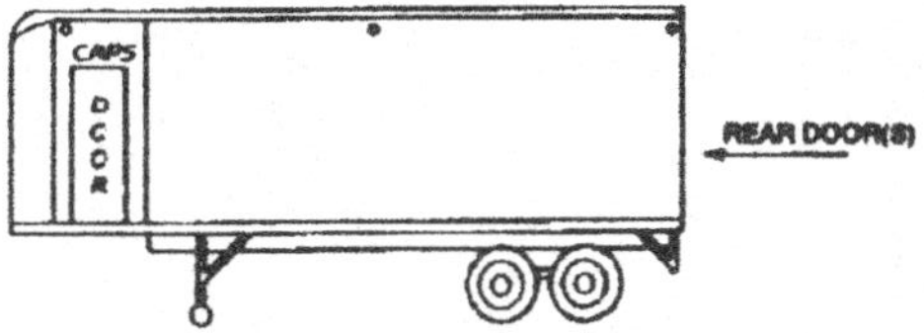

Figure 1

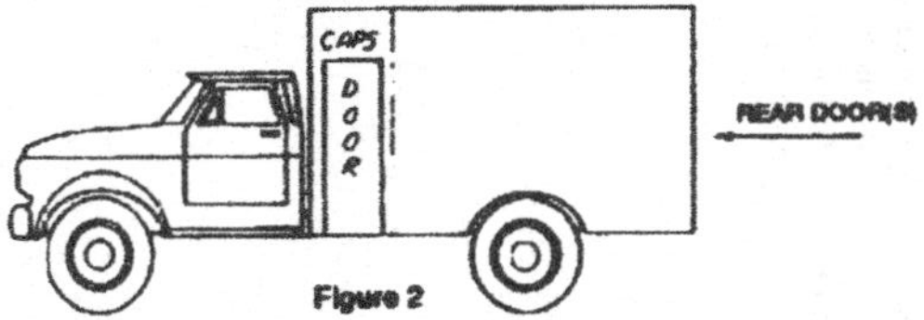

Figure 2

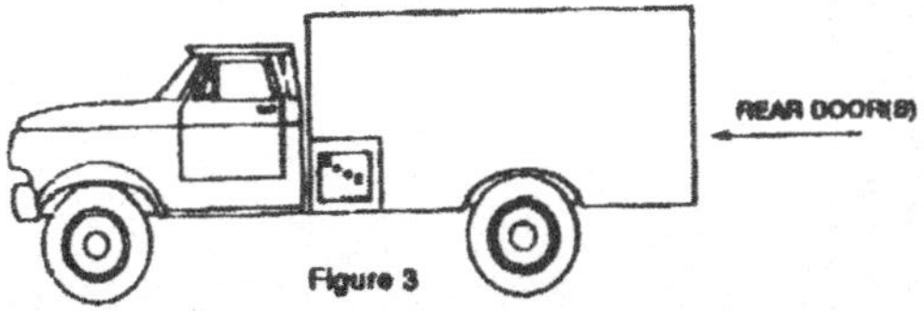

Figure 3

NOTE: The configurations shown in Figures 1 and 2 are equally applicable to multi-axle and "cab-over" vehicles.

[Diagrams: Courtesy of IME]

ATF Rul. 77–24, Appendix A and B

APPENDIX C

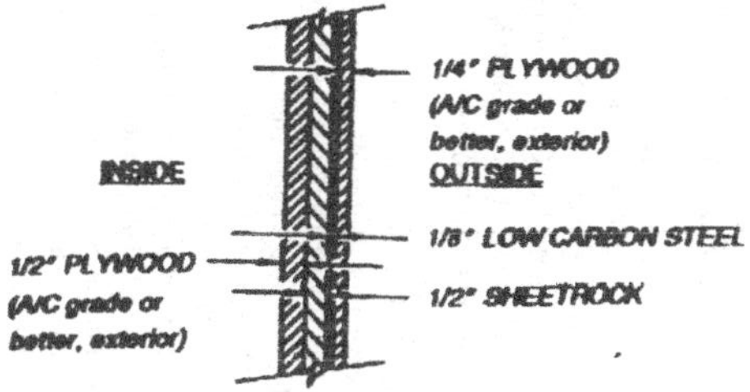

Sketch of laminate construction for container or compartment for electric blasting caps use, as illustrated in Appendix A, B, and E.

APPENDIX D

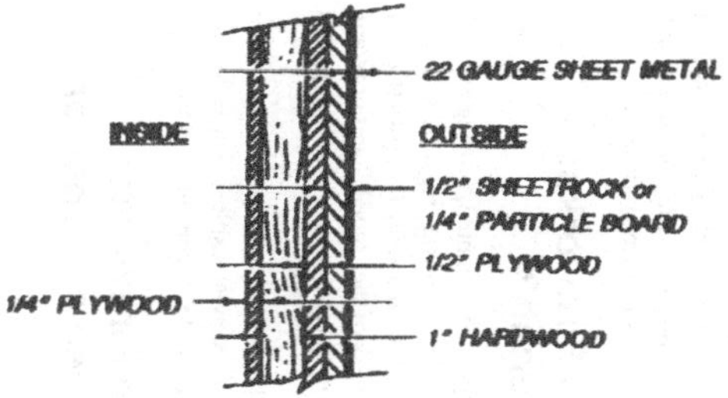

Sketch of laminate construction for container or compartment for electric blasting caps; restricted to use as illustrated in Appendix A.

[Diagrams: Courtesy of IME]

APPENDIX E

PORTABLE WHEELED TRAILERS

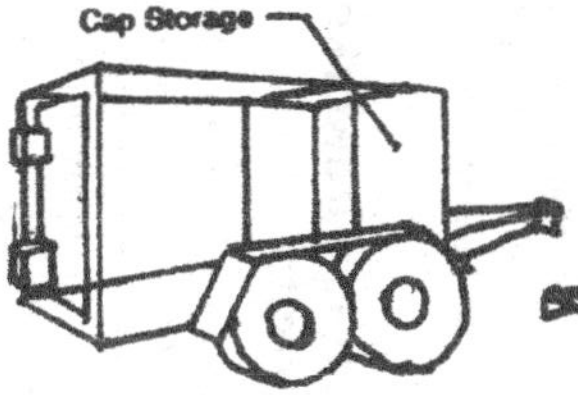

Figure 1

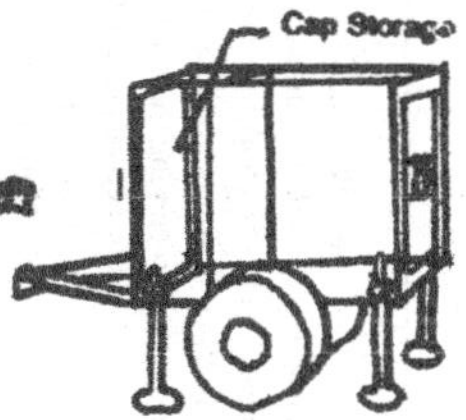

Figure 3

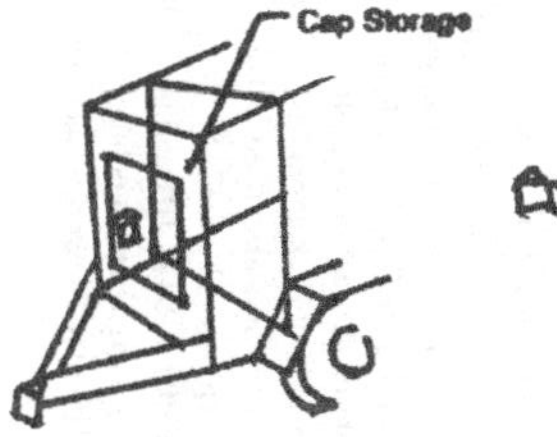

Figure 2

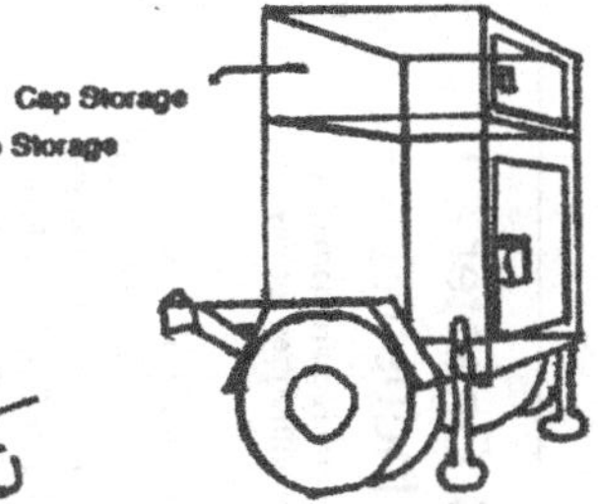

Figure 4

ATF Rul. 77-24, Appendix C – E

General Information (Revised 8/06)

Effect of 18 U.S.C. Chapter 40 On the Fireworks Industry

[Caution! This item discusses Federal requirements only. Please contact your State or local authorities for any additional requirements.]

Title XI of the Organized Crime Control Act of 1970 (18 U.S.C. Chapter 40) establishes controls over explosive materials, including black powder and other pyrotechnic compositions commonly used in fireworks. Part 555 of Title 27, Code of Federal Regulations (CFR), contains the regulations which implement Title XI. Section 555.141(a)(7) exempts "the importation, distribution, and storage of fireworks classified as UN0336, UN0337, UN0431, or UN0432 explosives by the U.S. Department of Transportation at 49 CFR 172.101 and generally known as 'consumer fireworks' or 'articles pyrotechnic.'" Section 555.141(a)(7) does not exempt "display fireworks," as defined in 555.11.

With Respect to Fireworks: Who needs a license?

1. Manufacturers of black powder;
2. Manufacturers of any other explosive material used in manufacturing consumer fireworks or display fireworks; and
3. Importers of, or dealers in, display fireworks.

With Respect to Fireworks: Who needs a permit?

1. All persons transporting, shipping, causing to be transported, or receiving display fireworks, regardless of whether for their own use or for commercial display purposes (Certain exemptions apply, e.g. agencies of the United States or of any State or political subdivisions thereof are exempt from permit requirements); and
2. A person, other than a licensee, transporting, shipping, causing to be transported, or receiving explosive materials for use in manufacturing display fireworks or consumer fireworks.

With Respect to Fireworks: Who may not need a license or permit?

Frequently, persons contracting for display fireworks (e.g., for Fourth of July observances) from a Federal explosives licensee or permittee receive a total service, including the services of a pyrotechnician who transports display fireworks in interstate or intrastate commerce to the site of the display and conducts and supervises the display. In these instances, the customers purchase and receive the contractor's services and not the explosive materials themselves (i.e. the cost of the services includes the contractor's expense in providing the fireworks utilized), and the cost of the services includes the dealer's expense in providing the fireworks utilized. When business is transacted in this manner, the customers purchasing and receiving the services need not obtain Federal explosives licenses or permits under Part 555 as long as they are not transporting, shipping, causing to be transported, or receiving explosive materials. Note: the transportation of explosive materials to the display sites would be authorized by the Federal explosives license or permit of the licensee or permittee providing the services.

With Respect to Fireworks: Types of permits

1. User permit: Allows the permit holder to transport, ship, cause to be transported, and receive display fireworks in interstate or foreign commerce for his or her own use and not for resale. This permit is issued at a cost of $100 for a 3-year period and is renewable at a cost of $50 for a 3-year period.
2. User-limited permit: Identical to the user permit but issued for a single purchase transaction, only. The fee is $75; the permit is nonrenewable.

With Respect to Fireworks: Storage

The law prohibits any person from storing any explosive materials in a manner not in conformity with the regulations promulgated by the Attorney General (18 U.S.C. 842(j)). Pursuant to this section, the Attorney General has prescribed storage regulations in 27 CFR Part 555, Subpart K. Display fireworks must be stored in conformity with the regulations. Display fireworks generally contain perchlorate mixture explosives, potassium chlorate base explosive mixtures, and black powder, which are entered on the List of Explosive Materials with numerous others. (The List, which is not all-inclusive, is annually compiled and readily available without charge from the address set out in 27 CFR 555.23 or online at www.atf.gov.) Display fireworks must be stored as low explosives in magazines meeting, at a minimum, the requirements for type 4 storage magazines prescribed by 27 CFR 555.210 unless they contain other classes of explosives. Bulk salutes must be stored as high explosives in type 1 or type 2 magazines. The net weight of the explosive materials contained in the display fireworks may be used in determining compliance with table of distance requirements. To determine the actual weight of the materials, it may be necessary to contact their manufacturers. The manufacturer of exempt or nonexempt fireworks having stocks of explosive materials on hand to be used in the manufacture of fireworks must store the stocks in conformity with applicable storage requirements.

Explosives Dealer's and User's Guide to Federal Explosives Regulation

Explosives May Not be Distributed by Licensees (Or by Any Person) to Any Person Who:

1. Is under indictment for, or who has been convicted of a crime punishable by imprisonment for a term exceeding one year.

2. Is an unlawful user of, or addicted to, marijuana or any depressant or stimulant drug or narcotic drug (as these terms are defined in section 102 of the Controlled Substances Act).

3. Has been adjudicated as a mental defective or has been committed to a mental institution.

4. Is a fugitive from justice.

5. Is an alien (with certain exceptions)

6. Has been discharged from the armed forces under dishonorable conditions; or

7. Having been a citizen of the United States, has renounced citizenship.

8. Is less than 21 years of age.

Dealers in Explosives Must:

Have a current and valid Federal explosives license.

Have proper storage facilities.

Keep accurate and complete records.

Verify that each buyer has a Federal explosives license or permit.

Verify buyers' identities.

Users of Explosives

Federal permits are required of those who transport, ship, cause to be transported, or receive explosive materials. Among other things, the permittee must keep complete and accurate records of the acquisitions and dispositions of explosives materials. Unless otherwise exempted by law, no person may receive or transport any explosive materials without a permit.

No person shall store any explosive material in a matter not in conformity with applicable regulations.

All persons must report to ATF and local authorities any loss or theft of their explosive materials within 24 hours of discovery.

A Federal license or permit does not confer any right or privilege to violate any state law or local ordinance.

The above summary is general and does not purport to fully convey the Federal explosives law and regulations pertaining to dealers and users.

Black Powder Transactions

Public Law 93-639 (1975) allows nonlicensees/nonpermittees to purchase commercially manufactured black powder, in quantities of 50 pounds or less, solely for sporting, recreational or cultural purposes for use in antique firearms or antique devices. A nonlicensee or nonpermittee purchasing black powder under the exemption need not be a resident of the State in which the dealer is located. Also, the categories of persons to whom the distribution of explosive materials is prohibited do not apply to black powder transactions made under the exemption. Acquisitions of black powder not qualifying under this exemption are subject to the same regulatory requirements that govern any other low explosive.

All persons who distribute black powder, regardless of quantity, must be licensed as explosives dealers and, among other things, must provide adequate storage.

Explosives Security

Through prompt reporting of losses and thefts of explosives and increased emphasis on physical security, explosives licensees and permittees can contribute greatly to efforts by Federal, State and local authorities to reduce the incidence of bombings and other criminal misuse of explosives in the United States. The following actions are of prime importance and in some instances required:

Report . . . any thefts or losses of explosives within 24 hours of discovery, by telephone, to ATF (toll free 1-800-800-3855) and to appropriate local authorities. Because the States and many municipalities have designated specific agencies to investigate the theft or loss of explosives, licensees and permittees are urged to be familiar with State and local reporting procedures and appropriate contact points.

Follow . . . telephone notification with a written report on ATF Form 5400.5, "Report of Theft or Loss—Explosive Materials," to the nearest ATF Division Office, and in accordance with the instructions on the form.

Observe . . . activity around magazines, within business premises, and on job sites, particularly if strangers appear to be loitering in the area in which explosives are being kept. On-site users should take special care to assure that explosives removed from storage for use on the job are either detonated or accounted for and unused items returned to storage.

Review . . . recordkeeping practices to assure that no discrepancies exist and that no figures in reported inventories have been manipulated, and correct any clerical errors promptly. Should any questions arise concerning explosives security procedures or any aspect of explosives regulation coming under the jurisdiction of ATF, do not hesitate to contact ATF.

Note: For Q&A's on regulatory requirements governing recordkeeping and storage, see "Questions and Answers" numbers 62-87.

Additional Information

The flow of useful information is an essential ingredient in the effective administration of regulatory programs. The Bureau of Alcohol, Tobacco, Firearms and Explosives is the Federal agency charged with the responsibility of administering laws impacting the explosives industry. We call your attention to the following publication distributed by ATF that merits your attention:

The Explosives Newsletter

During 1989 ATF developed the Explosives Newsletter, an information service for Federal explosives licensees and permittees which is intended to help explosives industry members better understand the Federal laws under which they must operate. It also includes other items of particular interest to the explosives industry. There is no charge for the Explosives Newsletter; licensees and permittees automatically receive copies when new editions are published.

Explosives industry members having questions on the Federal explosives laws and regulations may address their inquiries to:

Bureau of Alcohol, Tobacco, Firearms and Explosives
Explosives Industry Programs Branch
99 New York Avenue, N.E.
Mailstop 6E405
Washington, DC 20226

Direct e-mail inquiries on general questions or variance requests may be sent to the branch at EIPB@atf.gov.

List of Explosive Materials

Pursuant to the provisions of section 841(d) of title 18, U.S.C., and 27 CFR 555.23, the Director, Bureau of Alcohol, Tobacco, Firearms and Explosives, must revise and publish in the Federal Register at least annually a list of explosives determined to be within the coverage of 18 U.S.C. Chapter 40, Importation, Manufacture, Distribution and Storage of Explosive Materials. This chapter covers not only explosives, but also blasting agents and detonators, all of which are defined as explosive materials in section 841(c) of title 18, U.S.C. Accordingly, the following is the current List of Explosive Materials subject to regulation under 18 U.S.C. Chapter 40. Materials constituting blasting agents are marked by an asterisk. While the list is comprehensive, it is not all-inclusive. The fact that an explosive material may not be on the list does not mean that it is not within the coverage definitions in section 841 of title 18, U.S.C. Explosive materials are listed alphabetically by their common names, followed by chemical names and synonyms in brackets. This revised list is effective as of September 18, 2006.

List of Explosive Materials

A

Acetylides of heavy metals.
Aluminum containing polymeric propellant.
Aluminum ophorite explosive.
Amatex.
Amatol.
Ammonal.
Ammonium nitrate explosive mixtures (cap sensitive).
*Ammonium nitrate explosive mixtures (non-cap sensitive).
Ammonium perchlorate having particle size less than 15 microns.
Ammonium perchlorate composite propellant.
Ammonium perchlorate explosive mixtures.
Ammonium picrate [picrate of ammonia, Explosive D].
Ammonium salt lattice with isomorphously substituted inorganic salts.
*ANFO [ammonium nitrate-fuel oil].
Aromatic nitro-compound explosive mixtures.
Azide explosives.

B

Baranol.
Baratol.
BEAF [1,2-bis (2, 2-difluoro-2-nitroacetoxyethane)].
Black powder.
Black powder based explosive mixtures.
*Blasting agents, nitro-carbo-nitrates, including non-cap sensitive slurry and water gel explosives.
Blasting caps.
Blasting gelatin.
Blasting powder.
BTNEC [bis (trinitroethyl) carbonate].
BTNEN [bis (trinitroethyl) nitramine].
BTTN [1,2,4 butanetriol trinitrate].
Bulk salutes.
Butyl tetryl.

C

Calcium nitrate explosive mixture.
Cellulose hexanitrate explosive mixture.
Chlorate explosive mixtures.
Composition A and variations.
Composition B and variations.
Composition C and variations.
Copper acetylide.
Cyanuric triazide.
Cyclonite [RDX].
Cyclotetramethylenetetranitramine [HMX].
Cyclotol.
Cyclotrimethylenetrinitramine [RDX].

D

DATB [diaminotrinitrobenzene].
DDNP [diazodinitrophenol].
DEGDN [diethyleneglycol dinitrate].
Detonating cord.
Detonators.
Dimethylol dimethyl methane dinitrate composition.
Dinitroethyleneurea.
Dinitroglycerine [glycerol dinitrate].
Dinitrophenol.
Dinitrophenolates.
Dinitrophenyl hydrazine.
Dinitroresorcinol.
Dinitrotoluene-sodium nitrate explosive mixtures.
DIPAM [dipicramide; diaminohexanitrobiphenyl].
Dipicryl sulfone.
Dipicrylamine.
Display fireworks.
DNPA [2,2-dinitropropyl acrylate].
DNPD [dinitropentano nitrile].
Dynamite.

E

EDDN [ethylene diamine dinitrate].
EDNA [ethylenedinitramine].
Ednatol.
EDNP [ethyl 4,4-dinitropentanoate].
EGDN [ethylene glycol dinitrate].
Erythritol tetranitrate explosives.
Esters of nitro-substituted alcohols.
Ethyl-tetryl.
Explosive conitrates.
Explosive gelatins.
Explosive liquids.
Explosive mixtures containing oxygen-releasing inorganic salts and hydrocarbons.
Explosive mixtures containing oxygen-releasing inorganic salts and nitro bodies.
Explosive mixtures containing oxygen-releasing inorganic salts and water insoluble fuels.
Explosive mixtures containing oxygen-releasing inorganic salts and water soluble fuels.
Explosive mixtures containing sensitized nitromethane.

Explosive mixtures containing tetranitromethane (nitroform).
Explosive nitro compounds of aromatic hydrocarbons.
Explosive organic nitrate mixtures.
Explosive powders.

F

Flash powder.
Fulminate of mercury.
Fulminate of silver.
Fulminating gold.
Fulminating mercury.
Fulminating platinum.
Fulminating silver.

G

Gelatinized nitrocellulose.
Gem-dinitro aliphatic explosive mixtures.
Guanyl nitrosamino guanyl tetrazene.
Guanyl nitrosamino guanylidene hydrazine.
Guncotton.

H

Heavy metal azides.
Hexanite.
Hexanitrodiphenylamine.
Hexanitrostilbene.
Hexogen [RDX].
Hexogene or octogene and a nitrated Nmethylaniline.
Hexolites.
HMTD [hexamethylenetriperoxidediamine].
HMX [cyclo-1,3,5,7-tetramethylene 2,4,6,8-tetranitramine; Octogen].
Hydrazinium nitrate/hydrazine/aluminum explosive system.
Hydrazoic acid.

I

Igniter cord.
Igniters.
Initiating tube systems.

K

KDNBF [potassium dinitrobenzofuroxane].

L

Lead azide.
Lead mannite.
Lead mononitroresorcinate.
Lead picrate.
Lead salts, explosive.
Lead styphnate [styphnate of lead, lead trinitroresorcinate].
Liquid nitrated polyol and trimethylolethane.
Liquid oxygen explosives.

M

Magnesium ophorite explosives.
Mannitol hexanitrate.
MDNP [methyl 4,4-dinitropentanoate].
MEAN [monoethanolamine nitrate].
Mercuric fulminate.
Mercury oxalate.
Mercury tartrate.
Metriol trinitrate.
Minol-2 [40% TNT, 40% ammonium nitrate, 20% aluminum].
MMAN [monomethylamine nitrate]; methylamine nitrate.
Mononitrotoluene-nitroglycerin mixture.
Monopropellants.

N

NIBTN [nitroisobutametriol trinitrate].
Nitrate explosive mixtures.
Nitrate sensitized with gelled nitroparaffin.
Nitrated carbohydrate explosive.
Nitrated glucoside explosive.
Nitrated polyhydric alcohol explosives.
Nitric acid and a nitro aromatic compound explosive.
Nitric acid and carboxylic fuel explosive.
Nitric acid explosive mixtures.
Nitro aromatic explosive mixtures.
Nitro compounds of furane explosive mixtures.
Nitrocellulose explosive.
Nitroderivative of urea explosive mixture.
Nitrogelatin explosive.
Nitrogen trichloride.
Nitrogen tri-iodide.
Nitroglycerine [NG, RNG, nitro, glyceryl trinitrate, trinitroglycerine].
Nitroglycide.
Nitroglycol [ethylene glycol dinitrate, EGDN].
Nitroguanidine explosives.
Nitronium perchlorate propellant mixtures.
Nitroparaffins Explosive Grade and ammonium nitrate mixtures.
Nitrostarch.
Nitro-substituted carboxylic acids.
Nitrourea.

O

Octogen [HMX].
Octol [75 percent HMX, 25 percent TNT].
Organic amine nitrates.
Organic nitramines.

P

PBX [plastic bonded explosives].
Pellet powder.
Penthrinite composition.
Pentolite.
Perchlorate explosive mixtures.
Peroxide based explosive mixtures.
PETN [nitropentaerythrite, pentaerythrite tetranitrate, pentaerythritol tetranitrate].
Picramic acid and its salts.
Picramide.
Picrate explosives.
Picrate of potassium explosive mixtures.
Picratol.
Picric acid (manufactured as an explosive).
Picryl chloride.

Picryl fluoride.
PLX [95% nitromethane, 5% ethylenediamine].
Polynitro aliphatic compounds.
Polyolpolynitrate-nitrocellulose explosive gels.
Potassium chlorate and lead sulfocyanate explosive.
Potassium nitrate explosive mixtures.
Potassium nitroaminotetrazole.
Pyrotechnic compositions.
PYX [2,6-bis(picrylamino)] 3,5-dinitropyridine.

R

RDX [cyclonite, hexogen, T4, cyclo-1,3,5,-trimethylene-2,4,6,-trinitramine; hexahydro-1,3,5-trinitro-S-triazine].

S

Safety fuse.
Salts of organic amino sulfonic acid explosive mixture.
Salutes (bulk).
Silver acetylide.
Silver azide.
Silver fulminate.
Silver oxalate explosive mixtures.
Silver styphnate.
Silver tartrate explosive mixtures.
Silver tetrazene.
Slurried explosive mixtures of water, inorganic oxidizing salt, gelling agent, fuel, and sensitizer (cap sensitive).
Smokeless powder.
Sodatol.
Sodium amatol.
Sodium azide explosive mixture.
Sodium dinitro-ortho-cresolate.
Sodium nitrate explosive mixtures.
Sodium nitrate-potassium nitrate explosive mixture.
Sodium picramate.
Special fireworks.
Squibs.
Styphnic acid explosives.

T

Tacot [tetranitro-2,3,5,6-dibenzo-1,3a,4,6a tetrazapentalene].
TATB [triaminotrinitrobenzene].
TATP [triacetonetriperoxide].
TEGDN [triethylene glycol dinitrate].
Tetranitrocarbazole.
Tetrazene [tetracene, tetrazine, 1(5-tetrazolyl)-4-guanyl tetrazene hydrate].
Tetrazole explosives.
Tetryl [2,4,6 tetranitro-N-methylaniline].
Tetrytol.
Thickened inorganic oxidizer salt slurried explosive mixture.
TMETN [trimethylolethane trinitrate].
TNEF [trinitroethyl formal].
TNEOC [trinitroethylorthocarbonate].
TNEOF [trinitroethylorthoformate].
TNT [trinitrotoluene, trotyl, trilite, triton].
Torpex.
Tridite.
Trimethylol ethyl methane trinitrate composition.
Trimethylolthane trinitratenitrocellulose.
Trimonite.
Trinitroanisole.
Trinitrobenzene.
Trinitrobenzoic acid.
Trinitrocresol.
Trinitro-meta-cresol.
Trinitronaphthalene.
Trinitrophenetol.
Trinitrophloroglucinol.
Trinitroresorcinol.
Tritonal.

U

Urea nitrate.

W

Water-bearing explosives having salts of oxidizing acids and nitrogen bases, sulfates, or sulfamates (cap sensitive).
Water-in-oil emulsion explosive compositions.

X

Xanthamonas hydrophilic colloid explosive mixture.

Approved: September 18, 2006
Michael J. Sullivan, Acting Director.

[FR Doc. E6–15850 Filed 9–26–06; 8:45 am]

SPECIAL MESSAGE from ATF

Bureau of Alcohol, Tobacco, Firearms and Explosives

WASHINGTON, DC 20226

I am pleased to provide you with the 29th Edition of State Laws and Published Ordinances – Firearms (ATF P 5300.5). This publication is designed to help you comply with Federal and State firearms laws. Specifically, it will assist you in complying with the Gun Control Act of 1968 (GCA). The GCA prohibits licensees from selling or delivering a firearm to a nonlicensee whose receipt or possession of the firearm would violate State or local laws applicable at the place of sale or delivery. In addition, this book will assist you in making lawful over-the-counter sales of rifles and shotguns to out-of-State residents—transactions that must meet the legal requirements of both your State of residence and the purchaser's State of residence.

State Laws and Published Ordinances – Firearms also includes a "Ready Reference Table" to provide you with additional assistance for your business operations. When deciding whether to make a sale, you should first consult the full text of the State requirements, and if appropriate, contact State and local authorities before making a decision.

As a Federal Firearms Licensee, you play a critical role in ensuring that only eligible persons are able to acquire firearms from you. Your acquisition and disposition records, Firearms Transaction Record ATF Forms 4473, and Reports of Multiple Sales or Other Dispositions of Pistols and Revolvers ATF Form 3310.4, have proven invaluable to law enforcement agencies at all levels of government in tracing firearms involved in crime.

If you have any questions concerning major issues affecting the daily conduct of your business, you may call these numbers with regard to the following:

NICS checks: Contact the FBI NICS Operations Center at 1-877-444-6427.

Your Federal firearms license or renewal application: Contact our Federal Firearms Licensing Center at 1-866-662-2750.

Federal firearms laws or regulations: Contact the ATF Industry Operations field office nearest you. A complete alphabetical listing is located on pages vi–viii of this publication.

Stolen or lost firearms, including National Firearms Act weapons: Contact 1-888-930-9275 (Monday – Friday 8:00 am – 4:30 pm) or 1-800-800-3855 (Evenings, Weekends and Holidays). If you have knowledge of any unlawful activities concerning firearms, contact 1-800-ATF-GUNS.

Interpretation of a State or local law or ordinance: Contact your State police, local law enforcement authority, or your State Attorney General's office (see pages ix–x).

A Special Message from ATF

Reports of Multiple Sales or Other Dispositions of Pistols and Revolvers (ATF Form 3310.4) should be mailed to the address indicated on the form or faxed (1-877-283-0288) to the ATF National Tracing Center and to the designated Chief Law Enforcement Officer where your business is located. (You must retain the third copy as part of your permanent records.)

If you are going out of business with no successor, you are required to ship your firearms records to the ATF Out-of-Business Records Center, 244 Needy Road, Martinsburg, WV 25405.

Materials and ideas for future editions of this publication are welcome and may be sent to the ATF Firearms Programs Division, Attention: Chief, Firearms Programs Division, Room 6N/672, 99 New York Avenue, N.E., Washington, DC 20226.

Kenneth E. Melson
Deputy Director

TABLE OF CONTENTS

Introduction & General Information

State and Local Laws and Ordinances*

* The editors of this publication endeavored, where local firearm regulation has not been preempted by State law, to include ordinances for those major local jurisdictions that make their laws readily accessible to the public. States and many local jurisdictions now make their laws available on the Web; See, e.g., http://www.prairienet.org/~scruffy/f.htm (State statutes) and http://www.spl.org/default.asp?pageID=collection_municodes (municipal codes). However, space and time constraints precluded including every local jurisdiction in this publication. Therefore, users of this publication are encouraged to consult with State and local authorities to ascertain the most current firearms laws and regulations applicable to their particular situation.

Jurisdiction	The jurisdiction's laws in this publication are current through:
Alabama	**2008 First Special Session, including No. 08-277**
Alaska	**2007 Legislative Sessions**
American Samoa	**March 2007**
Arizona	**First Regular Session of the 48th Legislature 2007**
Arkansas	**2008 First Extraordinary Session, June 18, 2008**
California	**2007-2008 Third Extra. Session ch. 7 and ch. 267 of the 2008 Regular Session approved 8/26/08, and Proposition 99 approved by voters 6/3/08**
Colorado	**2008 Supplement**
Connecticut	**2008 Supplement**
Delaware	**76 Del. Laws c. 184 (Nov. 9, 2007)**
District of Columbia	**D.C. Law 17-207, Effective July 18, 2008**
Florida	**Chapters 2008-2 through 2008-63, 2008-71, 2008-72, 2008-92, 2008-93, 2008-100, 2008-105, 2008-116, 2008-123, 2008-128, 2008-131, 2008-145, 2008-150, 2008-247**
Georgia	**2007 Regular Session**
Guam	**P.L. 29-081, May 9, 2008**
Hawaii	**2008 Legislation of the 2008 Regular Session and First Special Session**
Idaho	**2008 Regular Session**
Illinois	**P.A. 95-711, Dec. 15, 2007**
Indiana	**2008 Second Regular Session**
Iowa	**2007 Supplement**
Kansas	**2007 Legislative Session**
Kentucky	**2008 First Extraordinary Session**
Louisiana	**2008 First and Second Extraordinary Sessions**
Maine	**Emergency legislation through Chapter 700 of the Second Regular Session of the 123rd Legislature; current with Ch. 470-560 of the Second Regular Session of the 123rd Legislature**
Maryland	**2008 Regular Session**
Massachusetts	**Act 217 of the 2007 Legislative Session**
Michigan	**Mich. Pub. Acts 300, Oct. 8, 2008**
Minnesota	**Oct. 1, 2007**
Mississippi	**2008 Regular and First Extraordinary Session**
Missouri	**2007 First Extra Session of the Regular Session of the 94th General Assembly**
Montana	**2007 Regular and Special Sessions**
Nebraska	**2008 Second Session and the May 2008 Primary Election**
Nevada	**74th (2007) Session and 24th Special (2008) Session**
New Hampshire	**Chapter 1 of the 2008 Special Session**
New Jersey	**213th Legislature, First Annual Session (P.L. 2008 ch. 97 & J.R. 13)**
New Mexico	**2nd Special Session of the 48th Legislature (2008)**
New York	**Ch. 564, Sep. 4, 2008**
North Carolina	**2008 Regular Session**
North Dakota	**2007 Legislative Session**
No. Mariana Is.	**Sessions to August 2007**
Ohio	**Legislation passed by the 127th Ohio General Assembly and filed with the Secretary of State through November 3, 2008**
Oklahoma	**Emergency Legislation through Chapter 2 of the First Regular Session of the 52nd Legislature**
Oregon	**2007 Regular Session of the 74th Legislative Assembly**
Pennsylvania	**2008 Regular Session and the 2008 Special Sessions**
Puerto Rico	**2008 Sessions**
Rhode Island	**January 2008 Session**
South Carolina	**2007 Regular Session**
South Dakota	**2008 Legislation passed at the 83rd Regular Sessions including Supreme Court Rule 08-09 and the November 2008 General Election**
Tennessee	**2008 Regular Session**
Texas	**2007 Regular Session**
Utah	**2008 Second Special Session**
Vermont	**2007 Adjourned Session (2008)**
Virginia	**2008 Regular Session, Acts 2008, cc. 1 to 884, 2008 Special Session I, cc. 1 and 2, 2008 Special Session II, cc. 1 to 11**
Virgin Islands	**Act 7055 of the 2008 Regular Session**
Washington	**Legislation through 2008**
West Virginia	**2008 Regular and 1st and 2nd Extraordinary Sessions**
Wisconsin	**2007 Wis. Act 41; Dec. 18, 2007**
Wyoming	**2008 Legislation**

FOR QUESTIONS CONCERNING UNLAWFUL ACTIVITIES CONTACT THE ATF CRIMINAL ENFORCEMENT FIELD DIVISIONS BELOW

(Alphabetical by City):

Atlanta Field Division
2600 Century Parkway N.E.
Suite 300
Atlanta, GA 30345-3104
(404) 417-2600

Baltimore Field Division
G.H. Fallon Building
31 Hopkins Plaza, 5th Floor
Baltimore, MD 21201
(443) 965-2000

Boston Field Division
10 Causeway Street
Room 253
Boston, MA 02222-1047
(617) 557-1200

Charlotte Field Division
6701 Carmel Road
Suite 200
Charlotte, NC 28226
(704) 716-1800

Chicago Field Division
525 West Van Buren Street
Suite 600
Chicago, IL 60607
(312) 846-7200

Columbus Field Division
37 West Broad Street
Suite 200
Columbus, OH 43215
(614) 827-8400

Dallas Field Division
1114 Commerce Street
Room 303
Dallas, TX 75242
(469) 227-4300

Denver Field Division
1961 Stout Street, Room 674
Denver, CO 80294
(303) 844-7450

Detroit Field Division
1155 Brewery Park Blvd.
Suite 300
Detroit, MI 48207-2602
(313) 202-3400

Houston Field Division
15355 Vantage Pkwy West
Suite 200
Houston, TX 77032
(281) 372-2900

Kansas City Field Division
2600 Grand Avenue
Suite 200
Kansas City, MO 64108
(816) 559-0700

Los Angeles Field Division
550 N. Brand Blvd.
Suite 800
Glendale, CA 91203
(818) 265-2500

Louisville Field Division
600 Dr. Martin Luther King Jr. Place
Suite 322
Louisville, KY 40202
(502) 753-3400

Miami Field Division
11410 NW 20th Street
Suite 200
Miami, FL 33172
(305) 597-4800

Nashville Field Division
5300 Maryland Way
Suite 200
Brentwood, TN 37027
(615) 565-1400

Newark Field Division
1 Garret Mountain Plaza
Suite 500
Woodland Park, NJ 07424
(973) 413-1179

New Orleans Field Division
One Galleria Blvd., Suite 1700
Metairie, LA 70001
(504) 841-7000

New York Field Division
241 37th Street
3rd Floor
Brooklyn, NY 11232
(718) 650-4000

Philadelphia Field Division
The Curtis Center
601 Walnut Street, Suite 1000 E
Philadelphia, PA 19106
(215) 446-7800

Phoenix Field Division
201 East Washington Street
Suite 940
Phoenix, AZ 85004
(602) 776-5400

San Francisco Field Division
5601 Arnold Road
Suite 400
Dublin, CA 94568-7724
(925) 479-7500

Seattle Field Division
Jackson Federal Building
915 2nd Avenue, Room 790
Seattle, WA 98174
(206) 389-5800

St. Paul Field Division
1870 MINN. World Trade Center
30 East Seventh Street
Suite 1900
St. Paul, MN 55101
(651) 726-0200

Tampa Field Division
400 North Tampa Street
Suite 2100
Tampa, FL 33602
(813) 202-7300

Washington Field Division
1401 H Street, NW
Suite 900
Washington, DC 20226
(202) 648-8010

TO RECEIVE ATF FORMS AND PUBLICATIONS

ATF Distribution Center
PO Box 5950
Springfield, VA 22150-5950
(703) 455-7801

TO OBTAIN ASSISTANCE REGARDING YOUR LICENSE

ATF Federal Firearms Licensing Center (FFLC)
244 Needy Road
Martinsburg, WV 25405
(866) 662-2750

TO SHIP OUT-OF-BUSINESS RECORDS

ATF National Tracing Center
244 Needy Road
Martinsburg, WV 25405
(800) 788-7133

FOR QUESTIONS CONCERNING FEDERAL FIREARMS LAWS, REGULATIONS, PROCEDURES OR POLICIES CONTACT AN ATF INDUSTRY OPERATIONS FIELD OFFICE BELOW

(Alphabetical by City):

ALBUQUERQUE II FO (IO)
201 THIRD STREET SUITE 1550
ALBUQURQUE, NM 87102
(505) 346-6914

ATLANTA V FIELD OFFICE (IO)
2600 CENTURY PARKWAY,
ROOM 300
ATLANTA, GA 30345
(404) 417-2670

BALTIMORE V FIELD OFFICE (IO)
FALLON FEDERAL BUILDING
31 HOPKINS PLAZA, 5TH FLOOR
BALTIMORE, MD 21201
(443) 965-2120

BIRMINGHAM II FIELD OFFICE (IO)
920 18TH STREET NORTH
ROOM 237
BIRMINGHAM, AL 35203
(205) 583-5950

BOSTON V FIELD OFFICE (IO)
10 CAUSEWAY STREET, ROOM 701
BOSTON, MA 02222
(617) 557-1250

BUFFALO II FIELD OFFICE (IO)
598 MAIN STREET, SUITE 201
BUFFALO, NY 14202
(716) 853-5160

CHARLOTTE III FIELD OFFICE (IO)
6701 CARMEL ROAD, SUITE 200
CHARLOTTE, NC 28226
(704) 716-1830

CINCINNATI II FIELD OFFICE (IO)
550 MAIN STREET
ROOM 1516
CINCINNATI, OH 45202
(513) 684-3351

CLEVELAND III FIELD OFFICE (IO)
6745 ENGLE ROAD
SUITE 210
MIDDLEBURG HEIGHTS, OH 44130
(216) 522-3374

COLUMBIA FIELD OFFICE (IO)
1835 ASSEMBLY STREET
COLUMBIA, SC 29201
(803) 765-5722

DALLAS V FIELD OFFICE (IO)
114 COMMERCE STREET
ROOM 303
DALLAS, TX 75242
(469) 227-4415

DENVER III FIELD OFFICE (IO)
1961 STOUT STREET, SUITE 630C
DENVER, CO 80294
(303) 844-7545

DETROIT V FIELD OFFICE (IO)
1155 BREWERY PARK BLVD
SUITE 300
DETROIT, MI 48207
(313) 202-3550

DOWNERS GROVE III FIELD OFFICE (IO)
3250 LACEY ROAD
SUITE 400
DOWNERS GROVE, IL 60515
(630) 725-5290

DUBLIN III FIELD OFFICE (IO)
5601 ARNOLD ROAD, SUITE 400
DUBLIN, CA 94568
(925) 479-7530

FAIRVIEW HEIGHTS FIELD OFFICE (IO)
333 SALEM PLACE
SUITE 225
FAIRVIEW HEIGHTS, IL 62208
(618) 632-0704

FALLS CHURCH III FIELD OFFICE (IO)
7799 LEESBURG PIKE
SUITE 1050, N. TOWER
FALLS CHURCH, VA 22043-2413
(703) 287-1130

FORT PIERCE II FIELD OFFICE (IO)
1660 SW. LUCIE WEST BLVD.
SUITE 400
PORT ST. LUCIE, FL 34986
(772) 924-2809

FORT WORTH II FIELD OFFICE (IO)
6000 WESTERN PLACE, SUITE 400
FT WORTH, TX 76107
(817) 862-2850

GLENDALE III FIELD OFFICE (IO)
550 N. BRAND BLVD., SUITE 700
GLENDALE, CA 91203
(818) 265-2540

GRAND RAPIDS II FIELD OFFICE (IO)
38 WEST FULTON, SUITE 200
GRAND RAPIDS, MI 49503
(616) 301-6100

GREENSBORO II FIELD OFFICE (IO)
1801 STANLEY ROAD
ROOM 300
GREENSBORO, NC 27407-2643
(336) 235-4950

HARRISBURG II FIELD OFFICE (IO)
MARKET SQUARE PLAZA
17 NORTH 2ND STREET, SUITE 1400
HARRISBURG, PA 17101
(717) 231-3400

HARTFORD FIELD OFFICE (IO)
450 MAIN STREET, ROOM 610
HARTFORD, CT 06103
(860) 240-3400

HELENA II FIELD OFFICE (IO)
10 WEST 15TH STREET
SUITE 2400
HELENA, MT 59626
(406) 441-1100

HOUSTON FIELD OFFICE VII (IO)
15355 VANTAGE PARKWAY WEST
SUITE 200
HOUSTON, TX 77032-1965
(281) 372-2950

INDIANAPOLIS II FIELD OFFICE (IO)
151 N. DELAWARE STREET, SUITE 1000
INDIANAPOLIS, IN 46204
(317) 287-3500

JACKSON II FIELD OFFICE (IO)
100 WEST CAPITOL STREET, SUITE 335
JACKSON, MS 39269
(601) 292-4025

JACKSONVILLE III FIELD OFFICE (IO)
5210 BELFORT ROAD
SUITE 350
JACKSONVILLE, FL 32256
(904) 380-5500

KANSAS CITY III & VI FIELD OFFICE (IO)
2600 GRAND AVENUE, SUITE 200
KANSAS CITY, MO 64108
(816) 559-0730

LANSDALE FIELD OFFICE (IO)
100 WEST MAIN STREET
SUITE 300B
LANSDALE, PA 19446
(215) 362-1840

LAS VEGAS III FIELD OFFICE (IO)
600 S. LAS VEGAS BLVD.
SUITE 540
LAS VEGAS, NV 89101
(702) 387-4650

LEXINGTON II FIELD OFFICE (IO)
1040 MONARCH STEET
SUITE 250
LEXINGTON, KY 40513
(859) 219-4508

LITTLE ROCK II FIELD OFFICE (IO)
425 WEST CAPITOL, SUITE 775A
LITTLE ROCK, AR 72201
(501) 324-6457

LOUISVILLE II FIELD OFFICE (IO)
600 DR. MARTIN L KING JR PL.,
SUITE 354
LOUISVILLE, KY 40202
(502) 753-3500

LUBBOCK II FIELD OFFICE (IO)
SENTRY PLAZA III
5214 68TH STREET, SUITE 300
LUBBOCK, TX 79424
(806) 798-1030

MACON II FIELD OFFICE (IO)
1645 FOREST HILL ROAD
SUITE 200
MACON, GA 31210
(478) 405-2520

MC ALLEN II FIELD OFFICE (IO)
1100 EAST LAUREL AVENUE
SUITE 301
MCALLEN, TX 78504
(956) 661-7950

MERRILLVILLE II FIELD OFFICE (IO)
8420 INDIANA STREET 2ND FLOOR
MERRILLVILLE, IN 46410
(219) 753-6310

MIAMI VI FIELD OFFICE (IO)
5225 NW 87TH AVENUE, SUITE 400
MIAMI, FL 33178
(305) 597-4920

MILWAUKEE II FIELD OFFICE
1000 NORTH WATER STREET
SUITE 1400
MILWAUKEE, WI 53202
(414) 727-6200

MOBILE II FIELD OFFICE (IO)
110 BEAUREGARD STREET
SUITE 300
MOBILE, AL 36602
(251) 405-5000

NASHVILLE II FIELD OFFICE (IO)
5300 MARYLAND WAY, SUITE 200
BRENTWOOD, TN 37027
(615) 565-1420

NEW ORLEANS III FIELD OFFICE (IO)
ONE GALLERIA BLVD., SUITE 1700
METAIRIE, LA 70001
(504) 841-7120

NEW YORK VI FIELD OFFICE (IO)
241 37TH STREET
3RD FLOOR
BROOKLYN, NY 11232
(718) 650-4060

NEW JERSEY III FIELD OFFICE (IO)
3 GARRET MOUNTAIN PLAZA, SUITE 202
WEST PATTERSON, NJ 07424
(973) 247-3030

OKLAHOMA CITY I FIELD OFFICE (IO)
55 NORTH ROBINSON, ROOM 229
OKLAHOMA CITY, OK 73102
(405) 297-5073

OMAHA FIELD II OFFICE (IO)
2707 NORTH 108TH STREET, SUITE 202
OMAHA, NE 68164
(402) 952-2635

ORLANDO II FIELD OFFICE (IO)
3452 LAKE LYNDA DRIVE, SUITE 325
ORLANDO, FL 32817
(407) 384-2420

PHOENIX III FIELD OFFICE (IO)
201 EAST WASHINGTON STREET
SUITE 940
PHOENIX, AZ 85004
(602) 776-5400

PITTSBURGH III FIELD OFFICE (IO)
1000 LIBERTY AVENUE
SUITE 1414
PITTSBURGH, PA 15222
(412) 395-0600

PORTLAND III FIELD OFFICE (IO)
1201 N.E. LLOYD BLVD.
ROOM 720
PORTLAND, OR 97232
(503) 331-7830

RICHMOND II FIELD OFFICE (IO)
1011 BOULDER SPRINGS DRIVE
SUITE 300
RICHMOND, VA 23225
(804) 200-4141

ROANOKE II FIELD OFFICE (IO)
310 FIRST STREET, SW
SUITE 500
ROANAKE, VA 24011
(540) 983-6920

SACRAMENTO II FIELD OFFICE (IO)
1325 J STREET, SUITE 1530
SACRAMENTO, CA 95814
(916) 498-5095

SALT LAKE CITY II FIELD OFFICE (IO)
257 EAST 200 SOUTH, SUITE 475
SALT LAKE CITY, UT 84111
(801) 524-7000

SAN ANTONIO II FIELD OFFICE (IO)
8610 BROADWAY, SUITE 400
SAN ANTONIO, TX 78217
(210) 805-2777

SAN DIEGO III FIELD OFFICE (IO)
9449 BALBOA AVENUE, SUITE 200
SAN DIEGO, CA 92123
(858) 966-1030

SANTA ANA II FIELD OFFICE (IO)
34 CIVIC CENTER PLAZA
SUITE 6121
SANTA ANA, CA 92701
(714) 347-9151

SEATTLE II FIELD OFFICE (IO)
915 SECOND AVENUE
7TH FLOOR
SEATTLE, WA 98174
(206) 389-6800

SPOKANE II FIELD OFFICE (IO)
1313 NORTH ATLANTIC STREET
SUITE 4100
SPOKANE, WA 99201
(509) 324-7881

SPRINGFIELD II FIELD OFFICE (IO)
3161 W. WHITE OAKS DRIVE
SUITE 200
SPRINGFIELD, IL 62704
(217) 547-3675

ST. LOUIS III FIELD OFFICE (IO)
ROBERT A. YOUNG FEDERAL BLDG.
1222 SPRUCE STREET, ROOM 6.205
ST. LOUIS, MO 63103
(314) 269-2250

ST. PAUL II FIELD OFFICE (IO)
30 EAST 7TH STREET, SUITE 1700
ST. PAUL, MN 55101
(651) 726-0220

SYRACUSE II FIELD OFFICE (IO)
100 SOUTH CLINTON STREET, ROOM 517
SYRACUSE, NY 13260
(315) 448-0898

TAMPA II FIELD OFFICE (IO)
925 U.S. HIGHWAY 301 SOUTH
TAMPA, FL 33619
(813) 612-2470

TUCSON III FIELD OFFICE (IO)
201 E. WASHINGTON STREET
SUITE 940
PHOENIX, AZ 85004
(602) 776-5400

WILKES BARRE FIELD OFFICE (IO)
7 NORTH WILKESBARRE BLVD.
SUITE 271-M
WILKESBARRE, PA 18702
(570) 820-2210

ATTORNEYS GENERAL

Alabama
Office of the Attorney General
State House
11 South Union Street
Montgomery, AL 36130
(334) 242-7300

Alaska
Office of the Attorney General
P.O. Box 110300
Diamond Courthouse, 4th Floor
Juneau, AK 99811-0300
(907) 465-3600

American Samoa
Office of the Attorney General
American Samoa
Government Executive Office Bldg.
Pago Pago, AS 96799
011 (684) 633-4163

Arizona
Office of the Attorney General
1275 West Washington Street
Phoenix, AZ 85007
(602) 542-4266

Arkansas
Office of the Attorney General
200 Tower Building
323 Center Street
Little Rock, AR 72201-2610
(501) 682-2007

California
Office of the Attorney General
1300 I Street, Suite 1740
Sacramento, CA 95814
(916) 445-9555

Colorado
Office of the Attorney General
Department of Law
1525 Sherman Street, 5th Floor
Denver, CO 80203
(303) 866-4500

Connecticut
Office of the Attorney General
55 Elm Street
Hartford, CT 06141-0120
(860) 808-5318

Delaware
Office of the Attorney General
Carvel State Office Building
820 North French Street
Wilmington, DE 19801
(302) 577-8500

District of Columbia
Office of the Corporation Counsel
1350 Pennsylvania Ave., NW
Suite 409
Washington, DC 20004
(202) 727-3400

Florida
Office of the Attorney General
The Capitol
PL 01
Tallahassee, FL 32399-1050
(850) 414-3990

Georgia
Office of the Attorney General
40 Capitol Square
Atlanta, GA 30334-1300
(404) 656-3300

Guam
Office of the Attorney General
Judicial Center Building
Suite 2-200E
120 West O'Brien Drive
Hagatna, Guam 96910
(671) 475-3409

Hawaii
Office of the Attorney General
425 Queen Street
Honolulu, HI 96813
(808) 586-1282

Idaho
Office of the Attorney General
P.O. Box 83720
Boise, ID 83720-1000
(208) 334-2400

Illinois
Office of the Attorney General
James R. Thompson Center South
100 West Randolph Street
12th Floor
Chicago, IL 60601
(312) 814-3000

Indiana
Office of the Attorney General
Indiana Government Center
402 West Washington Street
5th Floor
Indianapolis, IN 46204
(317) 232-6201

Iowa
Office of the Attorney General
Hoover State Office Building
1305 East Walnut
Des Moines, IA 50319
(515) 281-5164

Kansas
Office of the Attorney General
120 S.W. 10th Avenue, 2nd Floor
Topeka, KS 66612-1597
(785) 296-2215

Kentucky
Office of the Attorney General
State Capitol, Room 118
700 Capitol Avenue
Frankfort, KY 40601
(502) 696-5300

Louisiana
Office of the Attorney General
Department of Justice
P.O. Box 94005
Baton Rouge, LA 70804
(225) 326-6000

Maine
Office of the Attorney General
Six State House Station
Augusta, ME 04333-0006
(207) 626-8800

Maryland
Office of the Attorney General
200 St. Paul Place
Baltimore, MD 21202-2202
(410) 576-6300

Massachusetts
Office of the Attorney General
One Ashburton Place
Boston, MA 02108-1698
(617) 727-2200

Michigan
Office of the Attorney General
P.O. Box 30212
525 West Ottawa Street
Lansing, MI 48909-0212
(517) 373-1110

Minnesota
Office of the Attorney General
State Capitol
Suite 102
St. Paul, MN 55155
(651) 296-3353

Mississippi
Office of the Attorney General
Department of Justice
Post Office Box 220
Jackson, MS 39205-0220
(601) 359-3680

Missouri
Office of the Attorney General
Supreme Court Building
207 West High Street
Jefferson City, MO 65101
(573) 751-3321

Montana
Office of the Attorney General
Justice Building
215 North Sanders
Helena, MT 59620-1401
(406) 444-2026

Nebraska
Office of the Attorney General
State Capitol
Post Office Box 98920
Lincoln, NE 68509-8920
(402) 471-2682

Nevada
Office of the Attorney General
Old Supreme Court Building
100 North Carson Street
Carson City, NV 89701
(775) 684-1100

New Hampshire
Office of the Attorney General
State House Annex
33 Capitol Street
Concord, NH 03301-6397
(603) 271-3658

New Jersey
Office of the Attorney General
Richard J. Hughes Justice Complex
25 Market St., CN 080
Trenton, NJ 08625
(609) 292-8740

New Mexico
Office of the Attorney General
Post Office Drawer 1508
Santa Fe, NM 87504-1508
(505) 827-6000

New York
Office of the Attorney General
Department of Law -The Capitol
2nd Floor
Albany, NY 12224
(518) 474-7330

North Carolina
Office of the Attorney General
Department of Justice
P.O. Box 629
Raleigh, NC 27602-0629
(919) 716-6400

North Dakota
Office of the Attorney General
State Capitol
600 East Boulevard Avenue
Bismarck, ND 58505-0040
(701) 328-2210

No. Mariana Islands
Office of the Attorney General
Administration Building
P.O. Box 10007
Saipan, MP 96950-8907
(670) 664-2341

Ohio
Office of the Attorney General
State Office Tower
30 East Broad Street, 17th Floor
Columbus, OH 43266-0410
(614) 466-4320

Oklahoma
Office of the Attorney General
State Capitol, Room 112
2300 North Lincoln Boulevard
Oklahoma City, OK 73105
(405) 521-3921

Oregon
Office of the Attorney General
Justice Building
1162 Court Street NE
Salem, OR 97301-4096
(503) 378-4732

Pennsylvania
Office of the Attorney General
1600 Strawberry Square
Harrisburg, PA 17120
(717) 787-3391

Puerto Rico
Office of the Attorney General
P.O. Box 9020192
San Juan, PR 00902-0192
(787) 721-2900

Rhode Island
Office of the Attorney General
150 South Main Street
Providence, RI 02903
(401) 274-4400

South Carolina
Office of the Attorney General
Rembert C. Dennis Office Bldg.
P.O. Box 11549
Columbia, SC 29211-1549
(803) 734-3970

South Dakota
Office of the Attorney General
1302 E. Highway 14, Suite 1
Pierre, SD 57501-8501
(605) 773-3215

Tennessee
Office of the Attorney General
500 Charlotte Avenue
Nashville, TN 37243
(615) 741-5860

Texas
Office of the Attorney General
Capitol Station
P.O. Box 12548
Austin, TX 78711-2548
(512) 463-2100

Utah
Office of the Attorney General
State Capitol, Rm 236
Salt Lake City, UT 84114-0810
(801) 538-9600

Vermont
Office of the Attorney General
109 State Street
Montpelier, VT 05609-1001
(802) 828-3173

Virgin Islands
Office of the Attorney General
Department of Justice
G.E.R.S. Complex 488 50C Kronprinsdens
Gade
St. Thomas, VI 00802
(340) 774-5666

Virginia
Office of the Attorney General
900 East Main Street
Richmond, VA 23219
(804) 786-2071

Washington
Office of the Attorney General
P.O. Box 40100
1125 Washington Street, SE
P.O. Box 40100
Olympia, WA 98504-0100
(360) 753-6200

West Virginia
Office of the Attorney General
State Capitol, Room E26
1900 Kanawha Boulevard East
Charleston, WV 25305
(304) 558-2021

Wisconsin
Office of the Attorney General
114 East State Capitol
P.O. Box 7857
Madison, WI 53707-7857
(608) 266-1221

Wyoming
Office of the Attorney General
123 State Capitol Building
Cheyenne, WY 82002
(307) 777-7841

"READY REFERENCE" TABLE

(Providing citations to specific portions of each jurisdiction's laws; please see the endnotes and fully review the sections cited.)

JURISDICTION NAME	PURCHASER WAITING PERIOD	LICENSE: DEALER, MANUFACTURER, ETC.	LOCAL GOVERNMENT LIMITS (PREEMPTION)	LIMITS TO INTERSTATE PURCHASE AND SALE
ALABAMA		13A-11-78 [pistols]; 40-12-143 [handguns] & 40-12-158 [long guns]	11-45-1.1; 11-80-11	
ALASKA			18.65.778 [concealed handguns]; 29.35.145	
AMERICAN SAMOA		46.4222 [import]; 46.4223 [sale]		46.4222 [importation]
ARIZONA			13-3108, but see 15-341	13-3106 & 13-3112(U)(V)(W)
ARKANSAS			14-16-504; 14-54-1411	5-73-125
CALIFORNIA	Penal Code 12072(c)	Penal Code 12070 et seq.; 12086 [firearm manufacture]; 12095 [short-barreled shotguns]; 12250 [machine guns]; 12287 [assault weapons & .50 BMG rifles]; 12305 [destructive devices]	Gov't Code 53071 & 53071.5	Penal Code 12071(b)(8)(C); 12071(b)(3)(A); 12076
COLORADO			18-12-105.6 [firearms in vehicles]; 29-11.7-101 et seq.	12-27-101 to 104
CONNECTICUT	29-37a [2 weeks: long guns]	29-28 [handguns]		
DELAWARE		Title 24,§§ 901 to 905	Title 9, §330(c); Title 22, §111	
DISTRICT OF COLUMBIA	22-4508 [48 hrs: pistols]	7-2504.01 et seq.; 22-4509 & 4510		7-2505.02(b)(1)
FLORIDA	790.0655 [3 days: handguns]		790.33; 790.335 [registration]	790.28
GEORGIA		43-16-1 et seq. [handguns & arms <15"]	16-11-173	10-1-100 & 101
GUAM		60104 & 60115 [register]		
HAWAII	134-2 [14 to 20 days to obtain a license to purchase any handgun]	134-31 et seq.		
IDAHO		Section Repealed		18-3314 & 3315
ILLINOIS	Ch. 720, 5/24-3(A)(g) [72 hrs: concealable; 24 hrs: long guns, stunguns and tasers]		Ch. 430, 65/13.1 [not preempted] Ch. 720, 5/24-10 [affirmative defense]	Ch. 430, 65/3a
INDIANA		35-47-2-14 to 16 & 21	35-47-11	35-47-5-6
IOWA			724.28	
KANSAS				48-1901 to 1904
KENTUCKY			65.870; 237.110(19)	237.020
LOUISIANA		40:1787 [register]	40:1796	40:1801 to 1804
MAINE			Tit. 25, §2011	
MARYLAND	Pub. Safety Art. 5-123 & 124 [7 days: regulated firearms]	Pub. Safety Art. 5-106 et seq. [regulated firearms] & 11-105 [explosives]	Crim. Law Art. 4-209; Pub. Safety Art. 5-104, 133(a) & 134(a) [regulated firearms]	Pub. Safety Art. 5-204 [long guns]
MASSACHUSETTS		Ch. 140, §122 et seq.		
MICHIGAN			123.1101 to 123.1104	3.111 & 3.112
MINNESOTA	624.7132, subd. 4 [5 business days: pistols, assault weapons]		471.633 & 634; 609.67 subd. 6; 624.7131 subd. 12; 624.7132 subd. 16; 624.717; 624.74 subd. 4	624.71
MISSISSIPPI			45-9-51 & 53	
MISSOURI	Section Repealed		21.750	407.500 & 407.505
MONTANA			45-8-351	Section Repealed
NEBRASKA	69-2405 [3 days; handgun]		69-2425 [no preemption]	28-1211
NEVADA		202.440 [local license]		
NEW HAMPSHIRE		159.8 & 159.10 [handguns]	159:26	159:8-a

"READY REFERENCE" TABLE

JURISDICTION NAME	PURCHASER WAITING PERIOD	LICENSE: DEALER, MANUFACTURER, ETC.	LOCAL GOVERNMENT LIMITS (PREEMPTION)	LIMITS TO INTERSTATE PURCHASE AND SALE
NEW JERSEY	2C:58-3 [up to 30 days for permit]	2C:58-1 & 58-2	2C:1-5.d.	
NEW MEXICO				30-7-9
NEW YORK	400.00 (4.a) [up to 6 months for permit]	Penal Law 400.00	Penal Law 400.00.6 [licenses]	Penal Law 265.40
NORTH CAROLINA	14-404 [up to 30 days for handgun permit]		14-409.40; 14-415.23 [concealed handguns]	14-409.10
NORTH DAKOTA			62.1-01-03	
NORTHERN MARIANA IS.		2209; 2210; 2217	2227 [no preemption]	
OHIO			Note after 2923.12: 2004 Ohio Laws File 53 (HB 12) §9 [concealed handguns]	2923.22
OKLAHOMA			Title 21, §1289.24	Title 21, §1288
OREGON			166.170 to 176	166.490
PENNSYLVANIA	6111(a) [48 hrs]	6112 & 6113	6120	6141.1
PUERTO RICO		Tit. 25, §§ 456, 456g, 458		
RHODE ISLAND	11-47-35 & 35.2 [7 days]	11-47-19 [machine gun manufacturers]; 11-47-38 & 39 [retail dealers]	11-47-58	11-47-36 [concealable firearm]
SOUTH CAROLINA		23-31-130 & 150 [pistols]; 23-31-370 [machine guns]	23-31-510 & 520	23-31-10 & 20
SOUTH DAKOTA	23-7-9 [48 hrs: pistols (concealed permit holders exempt)]		7-18A-36; 8-5-13; 9-19-20	23-7-40
TENNESSEE			39-17-1314	
TEXAS			Local Gov't Code 229.001 & 235.021 to 024	Penal Code 46.07
UTAH			76-10-500	76-10-524
VERMONT			Title 24, §2295	Title 13, §§ 4014 & 4015
VIRGINIA			15.2-915 to 915.4; 15.2-1206 to 1209.1	
VIRGIN ISLANDS	466 [48 hrs]	461-462 & 467-468		470 [importation]
WASHINGTON	9.41.090(1) [5 days: pistols]	9.41.100 & 9.41.110	9.41.290	9.41.122 & 9.41.124
WEST VIRGINIA			8-12-5a	
WISCONSIN	175.35 [48 hrs: handguns]		66.0409	175.30
WYOMING			6-8-401	

ENDNOTES

(N.B.: The text of the above-cited provisions should be thoroughly examined in context to ascertain their TRUE effect.)

1. Blank spaces indicate no relevant statutes were located.
2. Jurisdictions include the 50 States, the District of Columbia, American Samoa, Guam, Northern Mariana Islands, Puerto Rico, and Virgin Islands.
3. "PURCHASER WAITING PERIOD" – generally refers to the period between purchaser application for firearms and allowable receipt or delivery. Exceptions exist among the jurisdictions.
4. "LICENSE: DEALER, MANUFACTURER, ETC." – generally means the person must have BOTH a Federal and State license.
5. "LOCAL GOVERNMENT LIMITS (PREEMPTION)" – means that the jurisdiction overrides its subordinate jurisdictions in whole or in part.
6. "LIMITS TO INTERSTATE PURCHASE AND SALE" (also known as "Contiguous State Provisions") – those legislative limits to interstate purchase and sale enacted by jurisdictions based on the GCA are cited, if available.

NOTICE: For an official interpretation of a jurisdiction's law, consult the appropriate government officials of that jurisdiction.

BRADY LAW

STATES IN WHICH THE FBI CONDUCTS NICS CHECKS FOR ALL FIREARMS TRANSACTIONS

Alabama	Kentucky	Ohio
Alaska	Louisiana	Oklahoma
American Samoa	Maine	Puerto Rico
Arizona	Massachusetts	Rhode Island
Arkansas	Minnesota	South Carolina
Delaware	Mississippi	South Dakota
District of Columbia	Missouri	Texas
Georgia	Montana	Vermont
Guam	New Mexico	Virgin Islands
Idaho	New York	West Virginia
Indiana	North Dakota	Wyoming
Kansas	Northern Mariana Island	

STATES THAT ACT AS THE POINT OF CONTACT (POC) FOR ALL FIREARMS TRANSACTIONS

California	Illinois	Tennessee
Colorado	Nevada	Utah
Connecticut	New Jersey	Virginia
Florida	Oregon	
Hawaii	Pennsylvania	

STATES THAT ACT AS A PARTIAL POINT OF CONTACT (POC) FOR NICS CHECKS

Iowa	Permit POC for handguns. FBI for long guns.
Maryland	Maryland State Police for handguns and assault weapons. FBI for long guns and pawn redemptions.
Michigan	Permit POC for handguns. FBI for long guns.
Nebraska	Permit POC for handguns. FBI for long guns.
New Hampshire	New Hampshire Department of Safety for handguns. FBI for long guns.
North Carolina	Permit POC for handguns. FBI for long guns.
Washington	Chief law enforcement officer for handguns. FBI for long guns and pawn redemptions.
Wisconsin	Wisconsin Department of Justice for handguns. FBI for long guns.

The most up-to-date information regarding Point of Contact States can be found on the ATF website at: http://www.atf.gov/firearms/bradylaw/poc_chart.htm

PERMANENT BRADY PERMIT CHART

Note: Notwithstanding the dates set forth below, permits qualify as alternatives to the background check requirements of the Brady law for no more than 5 years from the date of issuance. The permit must be valid under State law in order to qualify as a Brady alternative.

JURISDICTION	QUALIFYING PERMITS
Alabama	None
Alaska	Concealed weapons permits marked NICS-Exempt qualify.
American Samoa	None
Arizona	Concealed weapons permits qualify.
Arkansas	Concealed handgun licenses issued on or after April 1, 1999 qualify.*
California	Entertainment firearms permits only.
Colorado	None
Connecticut	None
Delaware	None*
District of Columbia	None*
Florida	None*
Georgia	Georgia firearms licenses qualify.
Guam	None*
Hawaii	Permits to acquire and licenses to carry qualify.
Idaho	Concealed weapons permits qualify.
Illinois	None
Indiana	None
Iowa	Permits to acquire a handgun and concealed weapons permits qualify.
Kansas	None
Kentucky	Concealed weapons permits issued on or after July 12, 2006 qualify.
Louisiana	None*
Maine	None*
Maryland	None*
Massachusetts	None*
Michigan	Licenses to Purchase a Pistol qualify. Concealed Pistol Licenses (CPLs) issued on or after November 22, 2005, qualify as an alternative to a National Instant Criminal Background Check System (NICS) check. CPLs issued prior to November 22, 2005 and Temporary Concealed Pistol Licenses do not qualify as NICS alternative.
Minnesota	None*
Mississippi	License to carry concealed pistol or revolver issued to individuals under Miss. Stat. Ann. § 45-9-101 qualify. (NOTE: security guard permits issued under Miss. Stat. Ann. § 97-37-7 do not qualify.)
Missouri	None*
Montana	Concealed weapons permits qualify.

Nebraska	Handgun purchase certificates qualify.
New Hampshire	None
New Jersey	None
New Mexico	None
New York	None
North Carolina	Permits to purchase a handgun and concealed handgun permits qualify.
North Dakota	Concealed weapons permits issued on or after December 1, 1999 qualify.*
No. Mariana Islands	None
Ohio	None
Oklahoma	None*
Oregon	None*
Pennsylvania	None
Puerto Rico	None
Rhode Island	None
South Carolina	Concealed weapons permits qualify.
South Dakota	None*
Tennessee	None
Texas	Concealed handgun licenses qualify.
Utah	Concealed weapons permits qualify.
Vermont	None
Virginia	None
Virgin Islands	None
Washington	None*
West Virginia	None
Wisconsin	None
Wyoming	Concealed weapons permits qualify.

*While certain permits issued in these states prior to November 30, 1998 were "grandfathered" as Brady alternatives, none of these grandfathered permits would still be valid under State law as of November 30, 2003.

The most up-to-date Permanent Brady Permit Chart can be found on the ATF website at:
http://www.atf.gov/firearms/bradylaw/permit_chart.htm

RELEVANCY CRITERIA

The following criteria were used in deciding which published State firearms laws and local ordinances are included in this publication.

RELEVANT

Firearms laws and ordinances published by State and local jurisdictions include those which:

1. Prohibit the sale, possession, or receipt of any type of firearm, or allow any exception to the prohibitions;
2. Prohibit the sale, possession, or receipt of any type of firearm to or by any of a particular class of persons (e.g., convicted felons, fugitives from justice, illegal aliens, mental incompetents, unlawful drug users and addicts, persons convicted of misdemeanor crimes of domestic violence);
3. Restrict the sale, possession, or receipt of certain firearms, such as destructive devices (e.g., grenades and other explosive or incendiary devices), fully automatic weapons, short-barreled rifles, short-barreled shotguns, and/or silencers;
4. Require waiting periods or notifications to law enforcement officials before weapons may be delivered to purchasers;
5. Require a permit, license, or a State/local ID card before a weapon may be sold or delivered to a purchaser or recipient;
6. Restrict the age at which it is lawful for a person to purchase or receive a firearm;
7. Restrict the sale or transfer of firearms (e.g., requirement for a State or local license to sell firearms or recordkeeping or notice requirements imposed as a condition of lawful sale of firearms).
8. Require proper storage or safety equipment for firearms; and
9. Regulate gun shows.

NOT RELEVANT

Firearms laws and ordinances published by State and local jurisdictions NOT included are those which:

1. Prohibit the carrying of weapons, either concealed or not, such as legal handguns, long-barreled rifles or shotguns, slingshots, dirks, daggers, etc. on persons or in vehicles, or at other specific locations (e.g., where alcohol is sold, in public areas or buildings), except schools;
2. Prohibit the discharge of weapons within city limits, parks, on private property, or other specific locations;
3. Permit the carrying of concealed weapons, except those concealed weapons provisions that act as alternatives to the Brady background check;
4. Require notification to local police departments or other designated officials AFTER the sale and delivery have been made;
5. Concern the operation of shooting ranges; and
6. Concern the use or possession of weapons in conjunction with non-firearm crimes and the seizure and forfeiture of weapons as a consequence thereof.

UNLAWFUL ACTS

"It shall be unlawful for any licensed importer, licensed manufacturer, licensed dealer, or licensed collector to sell or deliver any firearm to any person in any State where the purchase or possession by such person of such firearm would be in violation of any State law or any published ordinance applicable at the place of sale, delivery or other disposition, unless the licensee knows or has reasonable cause to believe that the purchase or possession would not be in violation of such State law or such published ordinance." [18 U.S.C. § 922(b)(2)] See:, 26 U.S. v. Decker, 335 F.Supp. 1168,1171 (1970), affirmed, 446 F.2d 164 (8th Cir. 1971), and Service Arms Co., Inc. v. U.S., 463 F.Supp. 21 (W.D. Okla. 1978).

FEDERAL AGE RESTRICTIONS

The following is provided due to lower age provisions in some State and local ordinances:

Federal law prohibits Federal firearms licensees from selling or delivering "any firearm or ammunition to any individual who the licensee knows or has reasonable cause to believe is less than eighteen years of age, and, if the firearm is other than a shotgun or rifle, or ammunition for a shotgun or rifle, to any individual who the licensee knows or has reasonable cause to believe is less than twenty-one years of age." [18 U.S.C. § 922(b)(1)]

Where State or local law is MORE restrictive than the Federal law, the State/local law shall apply.

PUBLISHER'S NOTE

Ammunition interchangeable between rifles and handguns (such as .22 caliber rimfire) may be sold to an individual 18 years of age or older, but less than 21, if the licensee is satisfied that the ammunition is being acquired for use in a rifle.

Title 11. Counties and Municipal Corporations

11-45-1.1. Handguns. No incorporated municipality shall have the power to enact any ordinance, rule, or regulation which shall tax, restrict, prevent, or in any way affect the possession or ownership of handguns by the citizens of this state. The entire subject matter of handguns is reserved to the State Legislature. This section shall not be construed to limit or restrict the power of a municipality to adopt ordinances which make the violation of a state handgun law a violation of a municipal ordinance to the same extent as other state law violations, or to limit or restrict the power of a municipal court to exercise concurrent jurisdiction with the district court over violations of state handgun laws which may be prosecuted as breaches of a municipal ordinance.

11-47-12. Gunpowder and explosives storage.
It is the duty of the corporate authorities of every city or town to provide a suitable fireproof building without the limits of the town or city for the storage of gunpowder or other explosive material on such terms as the corporate authorities may prescribe.

11-80-11. Firearms.

(a) No county or municipal corporation, instrumentality, or political subdivision thereof, by ordinance, resolution, or other enactment, shall regulate in any manner gun shows, the possession, ownership, transport, carrying, transfer, sale, purchase, licensing, registration or use of firearms, ammunition, components of firearms, firearms dealers, or dealers in firearm components.

(b)(1) Subsection (a) does not affect the authority a municipality has under law to regulate the discharge of firearms within the limits of the municipality or the authority a county has under law enacted prior August 1, 2000, to regulate the discharge of firearms within the jurisdiction of the county.

(2) Subsection (a) does not affect the authority of the state, a county, or a municipality to assess, enforce, and collect sales taxes, use taxes, and gross receipts taxes in the nature of sales taxes as defined by Section 40-2A-3(8), Code of Alabama 1975, on the retail sale of firearms and ammunition or to assess, enforce, and collect business licenses from firearms or ammunition manufacturers, trade associations, distributors, or dealers for the privilege of engaging in business.

Further, nothing herein shall exempt any business which uses firearms or ammunition in the conduct of its business or any business which leases or sells firearms or ammunition from the provisions of county and municipal planning and zoning laws, as long as the code, ordinance, or regulations are not used to circumvent the intent of subsection (a).

This section shall not be construed to limit or restrict the power of a municipality to adopt or enforce ordinances which make the violation of a state firearm law a violation of a municipal ordinance to the same extent as other state law violations...

Title 13A. Criminal Code

Chapter 7. Offenses Involving Property Damage and Trespass

Article 3. Arson and Explosives

13A-7-44. Criminal possession of explosives.

(a) A person commits the crime of criminal possession of explosives if he possesses, manufactures, buys, sells or transports any explosive, and intends that the explosive be used in the commission of a crime involving violence to another person or destruction of another's property.

(b) Criminal possession of explosives is a Class C felony.

Chapter 11. Offenses Against Public Order and Safety

Article 3. Offenses Relating to Firearms and Weapons

Division 1. General Provisions.

13A-11-57. Selling, giving or lending pistol or knife to minor. Any person who sells, gives or lends to any minor any pistol or bowie knife, or other knife of like kind or description, shall, on conviction, be fined not less than $50.00 nor more than $500.00.

13A-11-58. Sale or purchase of rifles, shotguns and ammunition in adjoining state.

(a) Any resident of Alabama authorized to sell and deliver rifles, shotguns, and ammunition may sell and deliver them to a resident of any state where the sale of the firearms and ammunition is legal. Any purchaser of the firearm or ammunition may take or send it out of the state or have it delivered to his or her place of residence.

(b) Any resident of Alabama who legally purchases rifles, shotguns, and ammunition in any state where the purchase is legal may take delivery of the weapons either in the state where they were purchased or in Alabama.

13A-11-60. Possession or sale of brass or steel teflon-coated handgun ammunition.

(a) Except as provided in subsection (b) of this section, the possession or sale of brass or steel teflon-coated handgun ammunition is illegal anywhere within the State of Alabama. The possession or sale of said ammunition or any ammunition of like kind designed to penetrate bullet-proof vests, shall be unlawful and punishable as provided in subsection (c) of this section.

(b) The provisions of this section shall not apply to state or local law enforcement officers; nor shall it apply to the possession or sale of teflon-coated lead or brass ammunition designed to expand upon contact.

(c) Any person who while armed with a firearm in the commission or attempted commission of any felony, has in his or her immediate possession, teflon-coated ammunition for such firearm, upon conviction of such felony or attempted felony, in addition and consecutive to the punishment prescribed for said felony or attempted felony, shall be punished by the imposition of an additional term of three years in the penitentiary.

(d) Any person violating the provisions of this section shall be guilty of a Class C felony as defined by Section 13A-5-3.

13A-11-61.1 Discharging a firearm into an unoccupied school bus or school building.

(a) No person shall shoot or discharge a firearm into an occupied or unoccupied school bus or school building.

(b) A person who shoots or discharges a firearm into an occupied school bus or school building shall be guilty of a Class B felony.

(c) A person who shoots or discharges a firearm into an unoccupied school bus or school building shall be guilty of a Class C felony.

(d) This section shall not be construed to repeal other criminal laws.

Whenever conduct prescribed by any provision of this section is also prescribed by any other provision of law, the provision which carries the more serious penalty shall be applied.

Division 1A. Rifles and Shotguns.

13A-11-62. Definitions. For purposes of this division, the following terms shall have the following meanings, unless the context clearly indicates otherwise:

(1) Firearm. Definition is same as provided in Section 13A-8-1(4) ["A weapon from which a shot is discharged by gunpowder."].

(2) Rifle. Any weapon designed or redesigned, made or remade, and intended to be fired from the shoulder and designed or redesigned and made or remade to use the energy of the explosive in a fixed metallic cartridge to fire only a single projectile through a rifled bore for each pull of the trigger.

(3) Shotgun. A weapon designed or redesigned, made or remade, and intended to be fired from the shoulder and designed or redesigned and made or remade to use the energy of the explosive in a fixed shotgun shell to fire through a smooth bore either a number of ball shot or a single projectile for each single pull of the trigger.

(4) Short-barreled rifle. A rifle having one or more barrels less than 16 inches in length and any weapon made from a rifle (whether by alteration, modification, or otherwise) if such weapon, as modified, has an overall length of less than 26 inches.

(5) Short-barreled shotgun. A shotgun having one or more barrels less than 18 inches in length and any weapon made from a shotgun (whether by alteration, modification, or otherwise) if such weapon as modified has an overall length of less than 26 inches.

13A-11-63. Short-barreled rifle or shotgun; possession, sale, etc.

(a) A person who possesses, obtains, receives, sells, or uses a short-barreled rifle or a short-barreled shotgun is guilty of a Class C felony.

(b) This section does not apply to a peace officer who possesses, obtains, receives, sells, or uses a short-barreled rifle or a short-barreled shotgun in the course of or in connection with his official duties.

13A-11-64. Identification number, mark or name; altering; possession after alteration. A person who either:

(1) Changes, alters, removes, or obliterates the name of the maker, model, manufacturer's number or other mark or identification of any firearm, or

(2) Possesses, obtains, receives, sells, or uses a firearm after the maker, model, manufacturer's number or other mark or identification has been changed, altered, removed, or obliterated, is guilty of a Class C felony.

13A-11-65. Violation in connection with commission of other felony. Violation of Section 13A-11-63(a) or Section 13A-11-64 in the course of, or in connection with the commission of any other felony shall be a Class B felony, and the punishment imposed therefor shall be in addition to the punishment imposed for the other felony.

13A-11-66.Supplement to other laws--Penalties provided by other laws. This division is supplemental to any other law and the penalties provided herein are in addition to any other penalties provided by law. This division shall not be construed to limit or in any way reduce the minimum and maximum penalties provided in any other law.

Division 2. Pistols.

13A-11-70. Definitions. For the purposes of this division, the following terms shall have the respective meanings ascribed by this section:

(1) Pistol. Any firearm with a barrel less than 12 inches in length.

(2) Crime of violence. Any of the following crimes or an attempt to commit any of them, namely, murder, manslaughter, (except manslaughter arising out of the operation of a vehicle), rape, mayhem, assault with intent to rob, assault with intent to ravish, assault with intent to murder, robbery, burglary, kidnapping and larceny.

(3) Person. Such term includes any firm, partnership, association or corporation.

13A-11-72. Violent felons, drug addicts or drunkards; possession and ownership restrictions--School grounds; possession or carrying on prohibited.

(a) No person who has been convicted in this state or elsewhere of committing or attempting to commit a crime of violence shall own a pistol or have one in his or her possession or under his or her control.

(b) No person who is a drug addict or an habitual drunkard shall own a pistol or have one in his or her possession or under his or her control.

(c) Subject to the exceptions provided by Section 13A-11-74, no person shall knowingly with intent to do bodily harm carry or possess a deadly weapon on the premises of a public school.

(d) Possession of a deadly weapon with the intent to do bodily harm on the premises of a public school in violation of subsection (c) of this section is a Class C felony.

(e) Law enforcement officers are exempt from this section, and persons with pistol permits issued pursuant to Section 13A-11-75, are exempt from the provisions of subsection (c) of this section.

(f) The term "public school" as used in this section applies only to a school composed of grades K-12 and shall include a school bus used for grades K-12.

(g) The term "deadly weapon" as used in this section means a firearm or anything manifestly designed, made, or adapted for the purposes of inflicting death or serious physical injury, and such term includes, but is not limited to, a bazooka, hand grenade, missile, or explosive or incendiary device; a pistol, rifle, or shot-gun; or a switch-blade knife, gravity knife, stiletto, sword, or dagger; or any club, baton, billy, blackjack, bludgeon, or metal knuckles.

13A-11-73. License to carry pistol; generally. No person shall carry a pistol in any vehicle or concealed on or about his person, except on his land, in his own abode or fixed place of business, without a license therefor as hereinafter provided.

13A-11-74. License to carry pistol; exceptions. The provisions of Section 13A-11-73 shall not apply to marshals, sheriffs, prison and jail wardens and their regularly employed deputies, policemen and other law enforcement officers of any state or political subdivision thereof, or to the members of the army, navy or marine corps of the United States or of the national guard, or to the members of the national guard organized reserves or state guard organizations when on duty or going to or from duty, or to the regularly enrolled members of any organization duly authorized to purchase or receive such weapons from the United States or from this state; provided, that such members are at or are going to or from their places of assembly or target practices, or to officers or employees of the United States duly authorized to carry a pistol, or to any person engaged in manufacturing, repairing or dealing in pistols, or the agent or representative of such person possessing, using, or carrying a pistol in the usual or ordinary course of such business, or to any common carrier, except taxi-cabs, licensed as a common carrier, or to any person permitted by law to possess a pistol while carrying it unloaded in a secure wrapper, from the place of purchase to his home or place of business, or to or from a place of repair or in moving from one place of abode or business to another.

13A-11-75. License to carry pistol; issuance; form and content; copies; fee; revocation. The sheriff of a county, upon the application of any person residing in that county, may issue a qualified or unlimited license to such person to carry a pistol in a vehicle or concealed on or about his person within this state for not more than one year from date of issue, if it appears that the applicant has good reason to fear injury to his person or property or has any other proper reason for carrying a pistol, and that he is a suitable person to be so licensed. The license shall be in triplicate, in form to be prescribed by the Secretary of State, and shall bear the name, address, description and signature of the licensee and the reason given for desiring a license. The original thereof shall be delivered to the licensee, the duplicate shall, within seven days, be sent by registered or certified mail to the Director of Public Safety and the triplicate shall be preserved for six years by the authority issuing the same. The fee for issuing such license shall be $1.00 which shall be paid into the county treasury unless otherwise provided by local law. Prior to issuance of a license, the sheriff shall contact available local, state, and federal criminal history data banks to determine whether possession of a firearm by an applicant would be a violation of state or federal law. The sheriff may revoke a license upon proof that the licensee is not a proper person to be licensed.

13A-11-76. Delivery to minors, violent felons, etc. No person shall deliver a pistol to any person under the age of 18 or to one who he has reasonable cause to believe has been convicted of a crime of violence or is a drug addict, an habitual drunkard or of unsound mind.

13A-11-78. License to sell pistol; generally. No retail dealer shall sell or otherwise transfer, or expose for sale or transfer, or have in his possession with intent to sell, or otherwise transfer, any pistol without being licensed as hereinafter provided.

13A-11-79. License to sell pistol; granting; term; conditions; fee. The duly constituted licensing authorities of any city, town or political subdivision of this state may grant licenses in forms prescribed by the secretary of state, effective for not more than one year from date of issue, permitting the licensee to sell pistols at retail within this state subject to the following conditions, in addition to those specified in Section 13A-11-77, for breach of any of which the license shall be forfeited and the licensee subject to punishment as provided in this division The business shall be carried on only in the building designated in the license. The license or a copy thereof, certified by the issuing authority, shall be displayed on the premises where it can easily be read. No pistol shall be sold in violation of any provisions of this division, nor shall a pistol be sold under any circumstances unless the purchaser is personally known to the seller or shall present clear evidence of his identity. A true record in triplicate shall be made of every pistol sold, in a book kept for the purpose, the form of which may be prescribed by the Secretary of State and shall be personally signed by the purchaser and by the person effecting the sale, the caliber, make, model and manufacturer's number of the weapon, the name, address, occupation, color and place of birth of purchaser and a statement signed by the purchaser that he has never been convicted in this state or elsewhere of a crime of violence. One copy shall be sent within six hours by registered or certified mail to the chief of police of the municipality or the sheriff of the county of which the dealer is a resident; the dealer shall within seven days send the duplicate to the Secretary of State; and the dealer shall retain the triplicate for six years. No pistol or imitation thereof or placard advertising the sale thereof shall be displayed in any part of any premises where it can readily be seen from the outside. The fee for issuing said license shall be $.50, which fee shall be paid into the State Treasury.

13A-11-80. Loaning pistols. No person shall make any loan secured by a mortgage, deposit or pledge of a pistol contrary to this division, nor shall any person lend or give a pistol to another or otherwise deliver a pistol contrary to the provisions of this division.

13A-11-81. False information or identification; purchase of pistol; application for license. No person shall, in purchasing or otherwise securing delivery of a pistol or in applying for a license to carry the same give false information or offer false evidence of his identity.

13A-11-83. Applicability; pistols as curiosities or ornaments. This division shall not apply to the purchase, possession or sale of pistols as curiosities or ornaments or to the transportation of such pistols unloaded and in a bag, box or securely wrapped package, but not concealed on the person.

13A-11-84. Penalty---Seizure of pistol--Destruction or use by law enforcement officer.

(a) Every violation of subsection (a) of Section 13A-11-72 or of Sections 13A-11-81 or 13A-11-82 shall be punishable by imprisonment for not more than five years. Every violation of subsection (b) of Section 13A-11-72 or of Sections 13A-11-73, 13A-11-74 and 13A-11-77 through 13A-11-80 shall be punishable by imprisonment for any term less than one year or by a fine of not more than $500.00, or both. The punishment for violating Section 13A-11-78 or 13A-11-79 may include revocation of license.

(b) It shall be the duty of any sheriff, policeman or other peace officer of the State of Alabama, arresting any person charged with violating Sections 13A-11-71 through 13A-11-73, or any one or more of said sections, to seize the pistol or pistols in the possession or under the control of the person or persons charged with violating said section or sections, and to deliver said pistol or pistols to one of the following named persons: if a municipal officer makes the arrest, to the city clerk or custodian of stolen property of the municipality employing the arresting officer; if a county, state or other peace officer makes the arrest, to the sheriff of the county in which the arrest is made. The person receiving the pistol or pistols from the arresting officer shall keep it in a safe place in as good condition as received until disposed of as hereinafter provided. Within five days after the final conviction of any person arrested for violating any of the above-numbered sections, the person receiving possession of the pistol or pistols, seized as aforesaid, shall report the seizure and detention of said pistol or pistols to the district attorney within the county where the pistol or pistols are seized, giving a full description thereof, the number, make and model thereof, the name of the person in whose possession it was found when seized, the person making claim to same or any interest therein, if the name can be ascertained or is known, and the date of the seizure. Upon receipt of the report from the per son receiving possession of the pistol or pistols as aforesaid, it shall be the duty of the district attorney within the county wherein the pistol or pistols were seized to forthwith file a complaint in the circuit court of the proper county, praying that such seized pistol or pistols be declared contraband, be forfeited to the state and be destroyed. Any person, firm or corporation or association of persons in whose possession said pistol or pistols may be seized or who claim to own the same or any interest therein shall be made a party defendant to said complaint, and thereupon such matter shall proceed and be determined in the circuit court of the proper county in the same form and manner, as near as may be, as in the forfeiture and destruction of gaming devices, except as herein otherwise provided. When any judgment of condemnation and forfeiture is made in any case filed under the provisions of this section, the judge making such judgment shall direct therein the destruction of the pistol or pistols by the person receiving possession of said pistol or pistols from the arresting officer in the presence of the clerk or register of the court, unless the judge is of the opinion that the nondestruction thereof is necessary or proper in the ends of justice, in which event and upon recommendation of the district attorney, the judge shall award the pistol or pistols to the sheriff of the county or to the chief of police of the municipality to be used exclusively by the sheriff or the chief of police in the enforcement of law, and the sheriff of the county and the chiefs of police of the municipalities shall keep a permanent record of all pistols awarded to them as provided for herein, to be accounted for as other public property, and said order, in the event that no appeal is taken within 15 days from the rendition thereof, shall be carried out and executed before the expiration of 20 days from the date of the judgment. The court, at its discretion, shall direct in said judgment that the costs of the proceedings be paid by the person in whose possession said pistol or pistols were found when seized, or by any party or parties who claim to own said pistol or pistols, or any interest therein, and who contested the condemnation and forfeiture thereof.

13A-11-85. Out-of-state license; reciprocity.

(a) A person licensed to carry a handgun in any state whose laws recognize and give effect in that state to a license issued under the laws of the State of Alabama shall be authorized to carry a handgun in this state. This section shall apply to a licenseholder from another state only while the licenseholder is not a resident of this state. A licenseholder from another state shall carry the handgun in compliance with the laws of this state.

(b) The Attorney General shall periodically publish a list of states which meet the requirements of subsection (a).

13A-11-224. Keeping gunpowder or explosives in city or town. Any person who keeps on hand, at any one time, within the limits of any incorporated city or town, for sale or for use, more than 50 pounds of gunpowder or other explosives shall, on conviction, be fined not less than $100.00.

Title 40. Revenue and Taxation

Chapter 12. Licenses

40-12-143. Weapons dealers. Persons dealing in pistols, revolvers, maxim silencers, bowie knives, dirk knives, brass knucks or knucks of like kind, whether principal stock in trade or not shall pay the following license tax: in cities and towns of 35,000 inhabitants and over, $150; and in all other places, $100. All persons dealing in pistols, revolvers and maxim silencers shall be required to keep a permanent record of the sale of every pistol, revolver or maxim silencer, showing the date of sale, serial number or other identification marks, manufacturer's name, caliber and type, and also the name and address of the purchaser, which record shall always be open for inspection by any peace officer of the State of Alabama or any municipality thereof. The failure to keep such record shall subject such person to having his license revoked by the probate judge of the county where such license was issued on motion of any district attorney of the State of Alabama.

40-12-158. Shotgun sales.

(a) Each person dealing in shotguns, rifles of .22 caliber or over, metallic ammunition or shotgun shells shall pay a license tax of $25 in cities of 100,000 inhabitants or over; $10 in cities or towns of 7,000 and less than 100,000 inhabitants; and $3 in all other places, whether incorporated or not.

(b) Regularly licensed rolling stores selling any or all of the articles enumerated in this section shall, in addition to the license provided in Section 40-12-174, pay a license tax of $5 to the state and $5 to the county in each county in which they sell or offer such articles for sale.

Title 41 State Government

Chapter 9 Boards and Commissions

41-9-649. Firearm purchase background check -- Mims-Russell Law Enforcement Protection Act.

All transfers or purchases of firearms conducted by a licensed importer, licensed manufacturer, or licensed dealer shall be subject to the national instant criminal background check system (NICS) created by the federal "Brady Handgun Violence Prevention Act" (P.L. No. 103-159), the relevant portion of which is codified at 18 U.S.C.S. § 922(t). To the extent possible, all information from any state or local government agency that is necessary to complete a NICS check shall be provided to the Criminal Justice Information Center. The Criminal Justice Information Center Commission shall promulgate rules and regulations necessary to implement a complete NICS Check. The commission shall also ensure that all information received shall be used solely for the purposes of compliance with NICS and every effort is made to protect the privacy of this information. All proposed rules shall go through the privacy and security committee of the commission which shall seek consultation from the President of the Probate Judges' Association and the Commissioner of the Department of Mental Health and Mental Retardation and consumer advocates as recommended by the commissioner.

[Current through the 2008 First Special Session, including No. 08-277]

ALASKA

ALASKA STAT.

Title 8. Business and Professions

Chapter 76. Pawnbrokers and Secondhand Dealers

8.76.010. Transactions to be entered in book kept at place of business

(a) A person engaged in the business of buying and selling secondhand articles, or lending money on second-hand articles, except a bank, shall maintain a book, in permanent form, in which the person shall enter in legible English at the time of each loan, purchase, or sale

(1) the date of the transaction;

(2) the name of the person conducting the transaction;

(3) the name, age and address of the customer;

(4) a description of the property bought or received in pledge, which includes for any firearm, watch, camera, or optical equipment bought or received in pledge, the name of the maker, the serial, model, or other number, and all letters and marks inscribed;

(5) the price paid or amount loaned;

(6) the signature of the customer.

(b) A person who lends money on secondhand articles under (a) of this section and is located in a municipality that has a population of over 5,000 shall also maintain an electronic record that provides the information required by (a)(1) and (4) of this section for the secondhand articles on which the person lends money. The person shall submit the electronic record as required by the municipal law enforcement agency.

8.76.020. Manner of recording entry. The entries in the book and the electronic record required by as 08.76.010 shall appear in chronological order and, when made in a book, in ink or indelible pencil. Blank lines may not be left between entries. Obliterations, alterations, or erasures may not be made. Corrections shall be made by drawing a line through the entry without destroying its legibility, and, when made in a book, the line shall be drawn in ink. The book shall be open to the inspection of a peace officer at reasonable times.

8.76.030. Criminal liability. A person who violates AS 08.76.010 or 08.76.020 is guilty of a misdemeanor and upon conviction is punishable by a fine of not more than $500, or by imprisonment for not more than six months, or by both.

Title 11. Criminal Law

Chapter 46. Offenses Against Property

11.46.260. Removal of identification marks.

(a) A person commits the crime of removal of identification marks if, with intent to cause interruption to the ownership of another, the person defaces, erases, or otherwise alters or attempts to deface, erase, or otherwise alter any serial number or identification mark placed or inscribed on a propelled vehicle, bicycle, firearm, movable or immovable construction tool or equipment, appliance, merchandise, or other article or its component parts.

(b) Removal of identification marks is

(1) a class C felony if the value of the property on which the serial number or identification mark appeared is $500 or more;

(2) a class A misdemeanor if the value of the property on which the serial number or identification mark appeared is $50 or more but less than $500;

(3) a class B misdemeanor if the value of the property on which the serial number or identification mark appeared is less than $50.

11.46.270. Unlawful possession.

(a) A person commits the crime of unlawful possession if the person possesses a propelled vehicle, bicycle, firearm, movable or immovable construction tool or equipment, appliance, merchandise or other article or its component parts knowing that the serial number or identification mark placed on it by the manufacturer or owner for the purpose of identification has been defaced, erased, or otherwise altered with the intent of causing interruption to the ownership of another.

(b) Unlawful possession is

(1) a class C felony if the value of the property on which the serial number or identification mark appeared is $500 or more;

(2) a class A misdemeanor if the value of the property on which the serial number or identification mark appeared is $50 or more but less than $500;

(3) a class B misdemeanor if the value of the property on which the serial number or identification mark appeared is less than $50.

Chapter 61. Offenses Against Public Order

11.61.190. Misconduct involving weapons in the first degree.

(a) A person commits the crime of misconduct involving weapons in the first degree if the person

(1) uses or attempts to use a firearm during the commission of an offense under AS 11.71.010 - 11.71.040; or

(2) discharges a firearm from a propelled vehicle while the vehicle is being operated and under circumstances manifesting substantial and unjustifiable risk of physical injury to a person or damage to property.

(b) Misconduct involving weapons in the first degree is a class A felony.

11.61.195. Misconduct involving weapons in the second degree.

(a) A person commits the crime of misconduct involving weapons in the second degree if the person knowingly

(1) possesses a firearm during the commission of an offense under AS 11.71.010 - 11.71.040;

(2) violates AS 11.61.200 (a)(1) and is within the grounds of or on a parking lot immediately adjacent to

(A) a public or private preschool, elementary, junior high, or secondary school without the permission of the chief administrative officer of the school or district or the designee of the chief administrative officer; or

(B) an entity, other than a private residence, licensed as a child care facility under AS 47.32, or recognized by the federal government for the care of children; or

(3) discharges a firearm at or in the direction of

(A) a building with reckless disregard for a risk of physical injury to a person; or

(B) a dwelling.

(b) Misconduct involving weapons in the second degree is a class B felony.

11.61.200. Misconduct involving weapons in the third degree.

(a) A person commits the crime of misconduct involving weapons in the third degree if the person

(1) knowingly possesses a firearm capable of being concealed on one's person after having been convicted of a felony or adjudicated a delinquent minor for conduct that would constitute a felony if committed by an adult by a court of this state, a court of the United States, or a court of another state or territory;

(2) knowingly sells or transfers a firearm capable of being concealed on one's person to a person who has been convicted of a felony by a court of this state, a court of the United States, or a court of another state or territory;

(3) manufactures, possesses, transports, sells, or transfers a prohibited weapon;

(4) knowingly sells or transfers a firearm to another whose physical or mental condition is substantially impaired as a result of the introduction of an intoxicating liquor or controlled substance into that other person's body;

(5) removes, covers, alters, or destroys the manufacturer's serial number on a firearm with intent to render the firearm untraceable;

(6) possesses a firearm on which the manufacturer's serial number has been removed, covered, altered, or destroyed, knowing that the serial number has been removed, covered, altered, or destroyed with the intent of rendering the firearm untraceable;

(7) violates AS 11.46.320 and, during the violation, possesses on the person a firearm when the person's physical or mental condition is impaired as a result of the introduction of an intoxicating liquor or controlled substance into the person's body;

(8) violates AS 11.46.320 or 11.46.330 by entering or remaining unlawfully on premises or in a propelled vehicle in violation of a provision of an order issued or filed under AS 18.66.100 - 18.66.180 or issued under former AS 25.35.010(b) or 25.35.020 and, during the violation, possesses on the person a defensive weapon or a deadly weapon, other than an ordinary pocketknife;

(9) communicates in person with another in violation of AS 11.56.740 and, during the communication, possesses on the person a defensive weapon or a deadly weapon, other than an ordinary pocketknife;

(10) resides in a dwelling knowing that there is a firearm capable of being concealed on one's person or a prohibited weapon in the dwelling if the person has been convicted of a felony by a court of this state, a court of the United States, or a court of another state or territory, unless the person has written authorization to live in a dwelling in which there is a concealable weapon described in this paragraph from a court of competent jurisdiction or from the head of the law enforcement agency of the community in which the dwelling is located;

(11) discharges a firearm from a propelled vehicle while the vehicle is being operated in circumstances other than described in AS 11.61.190(a)(2); or

(12) knowingly possesses a firearm that is concealed on the person after having been convicted of a felony or adjudicated a delinquent minor for conduct that would constitute a felony if committed by an adult by a court of this state, a court of the United States, or a court of another state or territory.

(b) It is an affirmative defense to a prosecution

(1) under (a)(1) of this section that

(A) the person convicted of the prior offense on which the action is based received a pardon for that conviction;

(B) the underlying conviction upon which the action is based has been set aside under AS 12.55.085 or as a result of post-conviction proceedings; or

(C) a period of 10 years or more has elapsed between the date of the person's unconditional discharge on the prior offense or adjudication of juvenile delinquency and the date of the violation of (a)(1) of this section, and the prior conviction or adjudication of juvenile delinquency did not result from a violation of AS 11.41 or of a similar law of the United States or of another state or territory;

(2) under (a)(2) or (10) of this section that

(A) the person convicted of the prior offense on which the action is based received a pardon for that conviction;

(B) the underlying conviction upon which the action is based has been set aside under AS 12.55.085 or as a result of post-conviction proceedings; or

(C) a period of 10 years or more has elapsed between the date of the person's unconditional discharge on the prior offense and the date of the violation of (a)(2) or (10) of this section, and the prior conviction did not result from a violation of AS 11.41 or of a similar law of the United States or of another state or territory.

(c) It is an affirmative defense to a prosecution under (a)(3) of this section that the manufacture, possession, transportation, sale, or transfer of the prohibited weapon was in accordance with registration under 26 U.S.C. 5801-5872 (National Firearms Act).

(d) It is an affirmative defense to a prosecution under (a)(11) of this section that the person was using a firearm while hunting, trapping, or fishing in a manner not prohibited by statute or regulation.

(e) The provisions of (a)(3) and (11) of this section do not apply to a peace officer acting within the scope and authority of the officer's employment.

(f) For purposes of (a)(12) of this section, a firearm on a person is concealed if it is covered or enclosed in any manner so that an observer cannot determine that it is a firearm without removing it from that which covers or encloses it or without opening, lifting, or removing that which covers or encloses it. A firearm on a person is not concealed if it is unloaded and is encased in a closed container designed for transporting firearms.

(g) It is an affirmative defense to a prosecution under (a)(12) of this section that

(1) either

(A) the defendant convicted of the prior offense on which the action is based received a pardon for that conviction;

(B) the underlying conviction upon which the action is based has been set aside under AS 12.55.085 or as a result of post-conviction proceedings; or

(C) a period of 10 years or more has elapsed between the date of the defendant's unconditional discharge on the prior offense or adjudication of juvenile delinquency and the date of the violation of (a)(12) of this section, and the prior conviction or adjudication of juvenile delinquency did not result from a violation of AS 11.41 or of a similar law of the United States or of another state or territory; and

(2) at the time of possession, the defendant was

(A) in the defendant's dwelling or on land owned or leased by the defendant appurtenant to the dwelling; or

(B) actually engaged in lawful hunting, fishing, trapping, or other lawful outdoor activity that necessarily involves the carrying of a weapon for personal protection.

(h) As used in this section,

(1) "prohibited weapon" means any

(A) explosive, incendiary, or noxious gas

(i) mine or device that is designed, made, or adapted for the purpose of inflicting serious physical injury or death;

(ii) rocket, other than an emergency flare, having a propellant charge of more than four ounces;

(iii) bomb; or

(iv) grenade;

(B) device designed, made, or adapted to muffle the report of a firearm;

(C) firearm that is capable of shooting more than one shot automatically, without manual reloading, by a single function of the trigger; or

(D) rifle with a barrel length of less than 16 inches, shotgun with a barrel length of less than 18 inches, or firearm made from a rifle or shotgun which, as modified, has an overall length of less than 26 inches;

(2) "unconditional discharge" has the meaning ascribed to it in AS 12.55.185.

(i) Misconduct involving weapons in the third degree is a class C felony.

11.61.210. Misconduct involving weapons in the fourth degree.

(a) A person commits the crime of misconduct involving weapons in the fourth degree if the person

(1) possesses on the person, or in the interior of a vehicle in which the person is present, a firearm when the person's physical or mental condition is impaired as a result of the introduction of an intoxicating liquor or a controlled substance into the person's body in circumstances other than described in AS 11.61.200(a)(7);

(2) discharges a firearm from, on, or across a highway;

(3) discharges a firearm with reckless disregard for a risk of damage to property or a risk of physical injury to a person under circumstances other than those described in AS 11.61.195 (a)(3)(A);

(4) manufactures, possesses, transports, sells, or transfers metal knuckles;

(5) manufactures, sells, or transfers a switchblade or a gravity knife;

(6) knowingly sells a firearm or a defensive weapon to a person under 18 years of age;

(7) other than a preschool, elementary, junior high, or secondary school student, knowingly possesses a deadly weapon or a defensive weapon, without the permission of the chief administrative officer of the school or district or the designee of the chief administrative officer, within the buildings of, on the grounds of, or on the school parking lot of a public or private preschool, elementary, junior high, or secondary school, on a school bus while being transported to or from school or a school-sponsored event, or while participating in a school-sponsored event, except that a person 21 years of age or older may possess

(A) a deadly weapon, other than a loaded firearm, in the trunk of a motor vehicle or encased in a closed container in a motor vehicle;

(B) a defensive weapon;

(C) an unloaded firearm if the person is traversing school premises in a rural area for the purpose of entering public or private land that is open to hunting and the school board with jurisdiction over the school premises has elected to have this exemption apply to the school premises; in this subparagraph, "rural" means a community with a population of 5,500 or less that is not connected by road or rail to Anchorage or Fairbanks or with a population of 1,500 or less that is connected by road or rail to Anchorage or Fairbanks; or

(8) being a preschool, elementary, junior high, or secondary school student, knowingly possesses a deadly weapon or a defensive weapon, within the buildings of, on the grounds of, or on the school parking lot of a public or private preschool, elementary, junior high, or secondary school, on a school bus while being transported to or from school or a school-sponsored event, or while participating in a school-sponsored event, except that a student may possess a deadly weapon, other than a firearm as defined under 18 U.S.C. 921, or a defensive weapon if the student has obtained the prior permission of the chief administrative officer of the school or district or the designee of the chief administrative officer for the possession.

(b) [Repealed]

(c) The provisions of (a)(7) of this section do not apply to a peace officer acting within the scope and authority of the officer's employment.

(d) Misconduct involving weapons in the fourth degree is a class A misdemeanor.

11.61.220. Misconduct involving weapons in the fifth degree.

(a) A person commits the crime of misconduct involving weapons in the fifth degree if the person

(1) is 21 years of age or older and knowingly possesses a deadly weapon, other than an ordinary pocket knife or a defensive weapon,

(A) that is concealed on the person, and, when contacted by a peace officer, the person fails to

(i) immediately inform the peace officer of that possession; or

(ii) allow the peace officer to secure the deadly weapon, or fails to secure the weapon at the direction of the peace officer, during the duration of the contact;

(B) that is concealed on the person within the residence of another person unless the person has first obtained the express permission of an adult residing there to bring a concealed deadly weapon within the residence;

(2) knowingly possesses a loaded firearm on the person in any place where intoxicating liquor is sold for consumption on the premises;

(3) being an unemancipated minor under 16 years of age, possesses a firearm without the consent of a parent or guardian of the minor;

(4) knowingly possesses a firearm

(A) within the grounds of or on a parking lot immediately adjacent an entity, other than a private residence, licensed as a child care facility under AS 47.32 or recognized by the federal government for the care of children, except that a person 21 years of age or older may possess an unloaded firearm in the trunk of a motor vehicle or encased in a closed container of a motor vehicle;

(B) within a

(i) courtroom or office of the Alaska Court System; or

(ii) courthouse that is occupied only by the Alaska Court System and other justice-related agencies;

(C) within a domestic violence or sexual assault shelter that receives funding from the state; or

(5) possesses or transports a switchblade or a gravity knife; or

(6) is less than 21 years of age and knowingly possesses a deadly weapon, other than an ordinary pocket knife or a defensive weapon, that is concealed on the person.

(b) In a prosecution under (a)(6) of this section, it is an affirmative defense that the defendant, at the time of possession, was

(1) in the defendant's dwelling or on land owned or leased by the defendant appurtenant to the dwelling; or

(2) actually engaged in lawful hunting, fishing, trapping, or other lawful outdoor activity that necessarily involves the carrying of a weapon for personal protection.

(c) The provisions of (a)(2) and (4) of this section do not apply to a peace officer acting within the scope and authority of the officer's employment.

(d) In a prosecution under (a)(2) of this section, it is

(1) an affirmative defense that

(A) [Repealed]

(B) the loaded firearm was a concealed handgun as defined in AS 18.65.790; and

(C) the possession occurred at a place designated as a restaurant for the purposes of AS 04.16.049 and the defendant did not consume intoxicating liquor at the place;

(2) a defense that the defendant, at the time of possession, was on business premises

(A) owned by or leased by the defendant; or

(B) in the course of the defendant's employment for the owner or lessee of those premises.

(e) For purposes of this section, a deadly weapon on a person is concealed if it is covered or enclosed in any manner so that an observer cannot determine that it is a weapon without removing it from that which covers or encloses it or without opening, lifting, or removing that which covers or encloses it; a deadly weapon on a person is not concealed if it is an unloaded firearm encased in a closed container designed for transporting firearms.

(f) For purposes of (a)(2) and (e) of this section, a firearm is loaded if the

(1) firing chamber, magazine, clip, or cylinder of the firearm contains a cartridge; and

(2) chamber, magazine, clip, or cylinder is installed in or on the firearm.

(g) Misconduct involving weapons in the fifth degree is a class B misdemeanor.

(h) The provisions of (a)(1) and (6) of this section do not apply to a

(1) peace officer of this state or a municipality of this state acting within the scope and authority of the officer's employment;

(2) peace officer employed by another state or a political subdivision of another state who, at the time of the possession, is

(A) certified as a peace officer by the other state; and

(B) acting within the scope and authority of the officer's employment; or

(3) police officer of this state or a police officer or chief administrative officer of a municipality of this state; in this paragraph, "police officer" and "chief administrative officer" have the meanings given in AS 18.65.290.

(i) In a prosecution

(1) under (a)(4)(B) of this section, it is a defense that the defendant, at the time of possession, was authorized to possess the firearm under a rule of court;

(2) under (a)(4)(C) of this section, it is a defense that the defendant, at the time of possession, was authorized in writing by the administrator of the shelter to possess the firearm.

(j) In (a)(1) of this section, "contacted by a peace officer" means stopped, detained, questioned, or addressed in person by the peace officer for an official purpose.

11.61.250. Unlawful furnishing of explosives.

(a) A person commits the crime of unlawful furnishing of explosives if the person furnishes an explosive substance or device to another knowing that the other intends to use the substance or device to commit a crime.

(b) Unlawful furnishing of explosives is a class C felony.

Chapter 81. General Provisions

11.81.900. Definitions.

(a) For purposes of this title, unless the context requires otherwise, ...

(17) "deadly weapon" means any firearm, or anything designed for and capable of causing death or serious physical injury; including a knife, an axe, a club, metal knuckles, or an explosive; ...

(20) "defensive weapon" means an electric stun gun, or a device to dispense mace or a similar chemical agent, that is not designed to cause death or serious physical injury; ...

(26) "firearm" means a weapon, including a pistol, revolver, rifle, or shotgun, whether loaded or unloaded, operable or inoperable, designed for discharging a shot capable of causing death or serious physical injury; ...

(48) "possess" means having physical possession or the exercise of dominion or control over property; ...

Title 12. Code of Criminal Procedure

Chapter 62. Criminal Justice Information and Records Checks

Article 2. National Crime History Record Check

12.62.400 National criminal history record checks for employment, licensing, and other noncriminal justice purposes. To obtain a national criminal history record check for determining a person's qualifications for a license, permit, registration, employment, or position, a person shall submit the person's fingerprints to the department with the fee established by AS 12.62.160. The department may submit the fingerprints to the Federal Bureau of Investigation to obtain a national criminal history record check of the person for the purpose of evaluating a person's qualifications for ...

(9) a concealed handgun permit under AS 18.65.700 - 18.65.790; ...

Title 29. Municipal Government

Chapter 35. Municipal Powers and Duties

29.35.145. Regulation of Firearms.

(a) The authority to regulate firearms is reserved to the state, and, except as specifically provided by statute, a municipality may not enact or enforce an ordinance regulating the possession, ownership, sale, transfer, use, carrying, transportation, licensing, taxation, or registration of firearms.

(b) Municipalities may enact and enforce ordinances

(1) that are identical to state law and that have the same penalty as provided for by state law;

(2) restricting the discharge of firearms in any portion of their respective jurisdictions where there is a reasonable likelihood that people, domestic animals, or property will be jeopardized; ordinances enacted or enforced under this paragraph may not abridge the right of the individual guaranteed by art. I, sec. 19, Constitution of the State of Alaska, to bear arms in defense of self or others;

(3) restricting the areas in their respective jurisdictions in which firearms may be sold; a business selling firearms may not be treated more restrictively than other businesses located within the same zone; and

(4) prohibiting the possession of firearms in the restricted access area of municipal government buildings; the municipal assembly shall post notice of the prohibition against possession of firearms at each entrance to the restricted access area.

(c) The prohibition on taxation in (a) of this section does not include imposition of a sales tax that is levied on all products sold within a municipality.

(d) This section applies to home rule and general law municipalities.

(e) In this section,

(1) "firearms" includes firearms, or any other element relating to firearms or parts thereof including ammunition and reloading components;

(2) "restricted access area" means the area beyond a secure point where visitors are screened and does not include common areas of ingress and egress open to the general public.

[Current through all 2007 Legislative Sessions]

Anchorage Municipal Code

Title 8. Penal Code

Chapter 8.25. Weapon Crimes

8.25.060. Firearms on school grounds.

A. It is unlawful for any person to knowingly have in his possession or control within the grounds of or on a parking lot immediately adjacent to a public or private preschool, elementary, junior high or high school:

1. A revolver, pistol or other firearm, except a person who is over 21 years of age, and who is not a preschool, elementary school, junior high, or secondary school student, may possess an unloaded firearm in the trunk of a motor vehicle or encased in a closed container in a motor vehicle; ...

B. Subsection A. of this section shall not apply to peace officers or persons who have express authorization of the school district superintendent or his designee or, in the case of a private or religious school, express authorization of the chief administrative officer of that school.

C. In the case of a conviction under subsection A. of this section the weapon shall be disposed of as provided in section 8.05.020.F.

D. Weapons possessed by persons under 18 years of age (minors) in violation of subsection A. of this section shall be seized and may be forfeited to the municipality in accordance with this section. ...

E. Violation of this section shall, upon conviction, be punished by a fine of not more than $10,000.00 or imprisonment for not more than one year, or both such fine and imprisonment.

8.25.090. Sale or furnishing of firearms to minors.

A. A person may not knowingly sell a firearm or defensive weapon to a minor under 18 years of age.

B. Violation of this section shall, upon conviction, be punished by a fine of not more than $10,000.00 or imprisonment for not more than one year, or both such fine and imprisonment.

[Anchorage Municipal Code current through March 31, 2008]

Code of Ordinances, City of Fairbanks

Chapter 46. Offenses

Article VI. Weapons Offenses

46-291. Definitions. The following words, terms and phrases, when used in this article, shall have the meanings ascribed to them in this section, except where the context clearly indicates a different meaning:

Dangerous weapon means a firearm, an airgun or BB gun, a knife other than an ordinary pocketknife having a blade not more than 3½ inches in length, a dagger, slingshot, crossbow, metal knuckles, blackjack, billy or any other instrument by the use of which physical injury may readily be inflicted upon any person, but does not include any instrument or tool which is not commonly used as or considered to be a weapon and which in the circumstances under which it is possessed or used is clearly not intended to be used, or to be available for use, as a weapon.

Firearm means a weapon, including a pistol, revolver, rifle, shotgun or airgun, whether loaded or unloaded, operable or inoperable, designed for discharging a shot or projectile capable of causing death or physical injury. ...

46-292. Failing to maintain a firearm transaction record.

(a) Except as provided in subsection (b) of this section, a pawnbroker, secondhand dealer or other person who, in the course of his business, buys, sells, rents, exchanges or otherwise receives or transfers firearms commits the offense of failing to maintain a firearm transaction record if he fails to record and make available to the city police, in a book permanently kept for the purpose, the following information concerning every receipt or transfer of a firearm in the course of his business:

(1) The name, address, date and place of birth, height, weight and race of the person to whom the firearm was transferred or from whom it was received.

(2) The number and state or other issuing agency of the driver's license or other identification card issued by the federal or any state government, or by any subdivision or agency of either government, bearing a picture of the issuee, which is used to confirm the identity of the person to whom the firearm was transferred or from whom it was received.

(3) The manufacturer, type, model, caliber or gauge, and serial number of the firearm received or transferred.

(4) The name of the employee or other person in the business of receiving or transferring firearms who transferred or received the firearm.

(b) A person does not commit the offense defined by subsection (a) of this section if:

(1) The firearm was received for the sole purpose of repair or customizing and such firearm is returned to the person from whom it was received;

(2) The person to whom he transferred or from whom he received the firearm is an importer, manufacturer, dealer or collector licensed under the provisions of 18 USC 921 - 928; or

(3) He maintains and makes available to the city police with respect to a particular transaction a permanent set of copies of the records of receipt, sale or other disposition of firearms as required by 18 USC 921 - 928.

46-294. Possession of a dangerous weapon on school grounds.

(a) A person other than a peace officer commits the offense of possession of a dangerous weapon on school grounds if he knowingly possesses a dangerous weapon inside a school or on school grounds.

(b) A dangerous weapon is possessed within the meaning of this section if a person has physical possession of the weapon or the exercise of dominion or control over the weapon.

(c) This section does not prohibit a person from transporting and possessing an concealed dangerous weapon upon school grounds with the consent of the school principal or the principal's designee.

(d) "School" as used in this section means either a public school operated by the Fairbanks North Star Borough School District or a private school.

[Fairbanks General Code current through Ord. 5961, effective March 5, 2007]

City and Borough of Juneau Code of Ordinances

Title 42. Penal Code

Chapter 42.20. Offenses Against Public Order

42.20.080. Destructive devices.

(a) Definitions. The following words and phrases when used in this chapter shall have the meanings ascribed in this subsection:

(1) Destructive device includes any of the following weapons except such devices as are designed primarily for emergency or distress signaling purposes:

(A) Any projectile containing any explosive or incendiary material or any other chemical substance, including, but not limited to, that which is commonly known as tracer or incendiary ammunition, except tracer ammunition manufactured for use in shotguns;

(B) Any bomb, grenade, explosive missile, or similar device or any launching device therefor, whether commercially manufactured or not;

(C) Any weapon of a caliber greater than .60 caliber which fires fixed ammunition, or any ammunition therefor other than a shotgun or shotgun ammunition;

(D) Any rocket, rocket-propelled projectile, or similar device of a diameter greater than sixty hundredths of an inch, or any launching device therefor, and any rocket, rocket-propelled projectile or similar device containing any explosive or incendiary material or any other chemical substance, other than the propellant for such device;

(E) Any breakable container which contains a flammable liquid with a flashpoint of one hundred fifty degrees Fahrenheit or less and has a wick or similar device capable of being ignited, other than a device which is commercially manufactured primarily for the purpose of illumination.

(2) Explosive means any material defined as an explosive in the Fire Prevention Code adopted under title 30 of this Code.

(b) Sale, purchase or possession of destructive devices exceptions. Nothing in this chapter shall prohibit the sale to, purchase by, possession of, or use of destructive devices by:

(1) Any peace officer of any recognized governmental agency, while on duty and acting within the scope and course of his employment;

(2) Any member of the Army, Navy, Air Force, or Marine Corps of the United States, or the National Guard, while on duty and acting within the scope and course of his service, nor shall anything in this chapter prohibit the sale to, purchase by, possession by, or use by any person who is a regularly employed and paid officer, employee or member of a fire department or fire protection or firefighting agency of the federal government, the state of Alaska, or a municipal corporation of this state, while on duty and acting within the scope and course of his employment, of any equipment used by such department or agency in the course of fire suppression.

(c) Carrying or placing destructive device. It is unlawful for any person knowingly to:

(1) Carry any explosive or destructive device on any vessel, aircraft, car, or other vehicle that transports passengers for hire;

(2) Place or carry any explosive or destructive device, while on board any such vessel, aircraft, car or other vehicle, in any hand baggage, roll, or other container;

(3) Place any explosive or destructive device in any baggage which is later checked with any common carrier.

(d) Possession of destructive device.

(1) It is unlawful for any person, firm, or corporation purposely, knowingly, recklessly or negligently to possess within the city and borough any destructive device except as provided by this chapter.

(2) Negligent possession shall be possession where the actor is unaware of such possession, but where the ordinary man in the actor's position would have known of such possession or would have been on inquiry to discover such possession. ...

(g) Exceptions. Nothing in subsections (a) - (e) of this section shall be construed to prohibit the legitimate possession or use of explosives by a person licensed by the state of Alaska to handle explosives where such possession and use is in compliance with title 30 of this Code and the proper permits have been obtained, or where explosives are being lawfully transported in inter- or intrastate commerce; nor shall a construction be given which would prohibit the possession of hand loading materials of the type and quantity excepted by title 30 of this Code where such materials are lawfully possessed and used for hand loading purposes.

(h) Penalties. Violation of subsection (e) of this section is a Class A misdemeanor. Violation of any other subsection of this section is a Class B misdemeanor.

42.20.085. Weapons on school grounds.

(a) It is unlawful for any person except an authorized person to:

(1) Possess a deadly weapon, or a defensive weapon while in or upon school grounds, or

(2) Place a deadly weapon or a defensive weapon in any locker, desk, or other place in or upon school grounds.

(b) As used in this section,

(1) Authorized person means:

(A) A peace officer;

(B) A police reserve officer;

(C) A member of the National Guard or the armed forces of the United States in the line of duty;

(D) A person proceeding to, engaging in, or returning from target practice or other authorized activity at a place and time sanctioned by the Superintendent of Schools.

(2) Deadly weapon means any firearm, or anything designed for and capable of causing death or serious physical injury, including metal knuckles, sharp-edged or pointed throwing devices, or any knife equipped with a blade over three inches in length.

(3) Defensive weapon means an electric stun gun, or a device to dispense mace or a similar chemical agent, that is not designed to cause death or serious physical injury. ...

(5) School grounds means any city and borough school building, school administration building or associated playground, parking lot, or any athletic field, while such athletic field is being used for an activity sponsored and supervised by the city and borough school district.

(c) Violation of this section is a Class B misdemeanor.

[City and Borough of Juneau Code of Ordinances current through Ord. No. 2008-21(b)am, adopted July 14, 2008, effective August 14, 2008]

AMERICAN SAMOA

AM. SAMOA CODE ANN.

Title 46. Criminal Justice

Chapter 42. Weapons

46.4201. Definitions. ...

(b) "Deface" means to alter or destroy the manufacturer's or importer's serial number or any distinguishing number or identification mark.

(c) "Explosive weapon" means any explosive, incendiary, or poison gas bomb or similar device designed or adapted for the purpose of inflicting death, serious physical injury, or substantial property damage; or any device designed or adapted for delivering or shooting such a weapon.

(d) "Firearm" means any weapon that is designed or adapted to expel a projectile by the action of an explosive.

(e) "Firearm silencer" means any instrument, attachment, or appliance that is designed or adapted to muffle the noise made by the firing of any firearm.

(f) "Gas gun" means any gas ejective device, weapon, cartridge, container or contrivance other than a gas bomb, that is designed or adapted for the purpose of ejecting any poison gas that will cause death or serious physical injury, but not any device that ejects mace or other repellant or temporary incapacitating substance.

(g) "Intoxicated" means substantially impaired mental or physical capacity resulting from introduction of any substance into the body. ...

(j) "Machine gun" means any firearm that is capable of firing more than 2 shots automatically, without manual reloading, by a single function of the trigger. . .

(l) "Rifle" means any firearm designed or adapted to be fired from the shoulder and to use the energy of the explosive in a fixed metallic cartridge to fire a projectile through a rifled bore by a single function of the trigger.

(m) "Short barrel" means any barrel length of less than 16" for a rifle and 18" for a shotgun, or an overall rifle and shotgun length of less than 26".

(n) "Shotgun" means any firearm designed or adapted to be fired from the shoulder and to use the energy of the explosive in a fixed shot-gun shell to fire a number of shots or single projectile through a smoothbore barrel by a single function of the trigger.

46.4202. Prohibited weapons.

(a) A person commits a crime if he knowingly possesses, manufactures, transports, repairs, or sells:

(1) an explosive weapon;

(2) a machine gun;

(3) a gas gun;

(4) a short-barreled rifle or shotgun;

(5) a firearm silencer;

(8) any other arms, as defined in section 46.4220, for which a valid license from the Commissioner of Public Safety has not been obtained.

(b) A person does not commit a crime under this section if his conduct:

(1) was incident to the performance of official duty by the armed forces, a governmental law enforcement agency, or a penal institution;

(2) was incident to engaging in a lawful commercial or business transaction with an organization listed in paragraph (b)(1); or

(3) was incident to using an explosive weapon reasonably related to a lawful industrial or commercial enterprise;

(4) was incident to displaying the weapon in a public museum or exhibition; or

(5) was incident to dealing with the weapon solely as a curio, ornament, or keepsake, or to using it in a manner reasonably related to a lawful dramatic performance; but if the weapon is a type described in paragraph (a)(1), (3), (4), or (5), it must be in a nonfunctioning condition that it cannot readily be made operable. No machine gun may be possessed, manufactured, transported, repaired, or sold as a curio, ornament, or keepsake even if it is inoperable and cannot be made readily operable.

(c) The defendant has the burden of injecting the issue of an exemption under subsection (b).

(d) A crime under paragraph (a)(1), (2), (3), (4), or (5) is a class C felony; a crime under paragraph (a)(6), (7), or (8) is a class A misdemeanor.

46.4203. Unlawful use of weapons.

(a) A person commits the crime of unlawful use of weapons if he knowingly:

(1) carries concealed on or about his person a knife, a firearm, a blackjack or any other weapon readily capable of lethal use; ...

(5) possesses or discharges a firearm or projectile weapon while intoxicated;

(b) Exemptions.

(1) Paragraphs (a)(1), (3), (4), (6), (7) and (8) do not apply to or affect any of the following:

(A) peace officers, or any person summoned by these officers to assist in making arrests or preserving the peace while actually engaged in assisting the officer;

(B) wardens, superintendents and keepers of prisons, jails and other institutions for the detention of persons accused or convicted of crime;

(C) members of the armed forces while performing their official duty.

(c) The defendant has the burden of injecting the issue of an exemption under subsection (b).

(d) Unlawful use of weapons is a class D felony unless committed under paragraph (a)(5), (6), (7), or (8), then it is a class B misdemeanor.

46.4204. Defacing firearm.

(a) A person commits the crime of defacing a firearm if he knowingly defaces any firearm.

(b) Defacing firearm is a class A misdemeanor.

46.4205. Possession of a defaced firearm.

(a) A person commits the crime of possession of a defaced firearm if he knowingly possesses a firearm which does not have the manufacturer's or importer's serial number engraved or cast on the receiver or frame of the firearm.

(b) Possession of a defaced firearm is a class B misdemeanor.

46.4206. Unlawful transfer of weapons.

(a) A person commits the crime of unlawful transfer of weapons if he:

(1) knowingly sells, leases, loans, gives away, or delivers a firearm or ammunition for a firearm to any person, who, under the provisions of 46.4207, is not lawfully entitled to possess it;

(2) Knowingly sells, leases, loans, gives away, or delivers a knife, rifle, shotgun or blackjack to a person less than 18 years old without the consent of the child's custodial parent or guardian, or recklessly sells, leases, loans, gives away, or delivers any other firearm to a person less than 18 years old; provided, that this does not prohibit delivery of those weapons to any peace officer or member of the armed forces while performing his official duty; or

(3) recklessly sells, leases, loans, gives away, or delivers a firearm or ammunition for a firearm to a person who is intoxicated.

(b) Unlawful transfer of weapons under paragraph (a)(1) is a class D felony; unlawful transfer of weapons under paragraphs (a)(2) and (3) is a class A misdemeanor.

46.4207. Unlawful possession of firearms and firearm ammunition.

(a) A person commits the crime of unlawful possession of a firearm or firearm ammunition if he has any firearm or firearm ammunition in his possession, and

(1) he has been convicted of a dangerous felony or confined therefor in this territory or elsewhere during the 5-year period immediately preceding the date of that possession; or

(2) he is a fugitive from justice, an habitual drunkard, a drug addict, or is currently adjudged mentally incompetent.

(b) Unlawful possession of a firearm or firearm ammunition is a class C felony.

46.4220. Definition of "arms". As used in 46.4220 through 46.4234, "arms" includes guns, rifles, pistols, air rifles, air pistols, gas rifles, gas pistols, ammunition, shells cartridges, gunpowder, dynamite, nitroglycerine, blasting powder, fireworks, and all other firearms and explosives and materials for the manufacture of the same.

46.4221. License - Required when.

(a) It is unlawful of any person, whether permanently or temporarily resident within American Samoa or whether on shore or on board any vessel, anchored, moored, or docked in any harbor in American Samoa, to have in his possession any arms without first having obtained a license from the Commissioner of Public Safety.

(b) A license to possess arms shall not be issued by the Commissioner of Public Safety unless the application therefor has been approved by the attorney general, and that such approval shall be given only after a background investigation has been conducted on the applicant; and that the:

(1) applicant is not a convicted felon; and

(2) applicant does not have any mental disorder or any disease which may endanger the public if a license to possess arm(s) is issued to him; and

(3) applicant is not a member of any organization that advocates the overthrowing of the government of American Samoa or that of the United States.

(c) A license shall be issued only for this ownership and possession of 12, 16, 20 and 410 gauge shotguns and shotgun shells and 22 caliber rifles and their ammunitions.

(d) Licenses issued prior to the enactment of subsection (c) remain valid. No additional licenses shall be issued for renewals of existing licenses provided in 46.4227 and transfers of arms validly licensed provided in 46.4229(b).

46.4222. License - Required for import.

(a) It is unlawful of any person to import arms into American Samoa without having obtained a license therefor from the Commissioner of Public Safety.

(b) A license to import arms shall not be issued by the Commissioner of Public Safety unless the application has been approved by the Attorney General.

(c) Unless otherwise authorized, only those shotguns and rifles referred to in section 46. 221(c) ASCA may be imported with licenses.

(d) The customs officers may confiscate any guns that are being imported into the Territory in violation of law. Confiscated guns must be surrendered to the custody of the Commissioner of Public Safety within 5 days of confiscation.

46.4223. License - Required for the sale of arms.

(a) It is unlawful for any person to sell or in any other way transfer the right of possession of any arms without having obtained from the Commissioner of Public Safety a license to sell arms. The application for such license shall contain such information as may be required by the Commissioner of Public Safety.

(b) A license to sell arms shall not be issued by the Treasurer unless the application for the license has been approved by the Governor or his designated representative. No license shall be issued for the sale of arms other than shotguns and .22 caliber rifles as set out in 46.221(c) and ammunition therefor.

46.4224. License - Information required.

(a) Every person who obtains a license to possess, import, or sell arms shall, upon the written request of the Governor or his designated representative, furnish such information concerning such arms as may be reasonably required.

(b) Each license issued shall specify the number, quantity, and description of the arms which may be possessed, imported, or sold, or otherwise transferred under it.

46.4225. License - Possession required when carrying arms. Every person to whom a license to possess arms is issued, shall when carrying such arms or any part thereof, have with him the license to possess such arms, and shall produce the same for inspection upon demand of any officer or official of the government.

46.4226. License - Revocation. Any license issued under authority of this title may be altered or revoked by the Governor or his designated representative at any time for good cause.

46.4227. License - Renewal.

(a) Licenses to possess arms shall expire on 10 January of the year following their issue. Each holder of a license to possess arms shall, between the 1st and the 10th of January of each year, submit his license to possess arms for the previous year, together with the annual license fee, to the Commissioner of Public Safety.

(b) The Commissioner of Public Safety may renew the license with or without examining the arms for which the license is to be issued; but the holder of the license shall, upon the demand of the Commissioner of Public Safety, submit the arms to him for examination.

46.4228. Marking arms for identification. Each person to whom a license to possess arms is issued shall, upon receipt of such arms, produce at the office of the Commissioner of Public Safety his license to possess arms, together with the arms specified in said license. Such arms shall be examined and compared with the license and, if found to correspond therewith, shall be marked with such letter as may be designated by the Commissioner of Public Safety and also marked with a number indicating the order of the license, and registration as specified in the license, unless the arm has plainly visible and distinctive serial number stamped on it. Such arms when duly marked shall be redelivered to the licensee, together with license. If the provisions of this section are not complied with, the license shall be revoked, and the arms may be confiscated as though no license had been issued.

46.4229. Sales to persons without licenses - Grandfather clause.

(a) No person shall sell or otherwise transfer any arms to any person who does not hold a valid and existing license to possess the particular firearms to be sold.

(b) Arms no longer permitted to be licensed but for which current, valid licenses were issued prior to the effective date of section 46.4221(c) may, in the discretion of the Commissioner of Public Safety and in the manner provided in this chapter, be transferred to persons obtaining licenses therefor.

46.4233. Authorized possession and use of arms without license.

(a) This chapter does not prohibit the possession and use of arms and other police weapons by any member of the police force, armed forces of the United States or employees if the government of the United States and law enforcement officers of other states or territories if these arms are properly issued by the issuing authorities and are brought into the Territory in the course of performance of official duties.

(b) The Governor or his designated representative may authorize the pulenu'u or police of any village to possess and use arms in connection with his official duties without first obtaining a license therefor.

(c) The Governor may enter into reciprocal agreements with states whose law enforcement officers may be assigned on official duty in the Territory to permit these law enforcement officers to carry firearms without registration.

46.4234. Violation - Penalty.

(a) Any person who violates any of the provisions of this chapter or who refuses to obey any lawful order issued under the authority of this chapter is guilty of a class A misdemeanor and shall, upon conviction, be sentenced accordingly, and any arms involved may be confiscated by the government. ...

[Current through March 2007]

ARIZONA
ARIZ. REV. STAT.

Title 13. Criminal Code

Chapter 9. Probation and Restoration of Civil Rights

13-904. Suspension of civil rights and occupational disabilities.

A. A conviction for a felony suspends the following civil rights of the person sentenced:

1. The right to vote.

2. The right to hold public office of trust or profit.

3. The right to serve as a juror.

4. During any period of imprisonment any other civil rights the suspension of which is reasonably necessary for the security of the institution in which the person sentenced is confined or for the reasonable protection of the public.

5. The right to possess a gun or firearm.

B. Persons sentenced to imprisonment shall not thereby be rendered incompetent as witnesses upon the trial of a criminal action or proceeding, or incapable of making and acknowledging a sale or conveyance of property.

C. A person sentenced to imprisonment is under the protection of the law, and any injury to his person, not authorized by law, is punishable in the same manner as if such person was not convicted and sentenced.

D. The conviction of a person for any offense shall not work forfeiture of any property, except if a forfeiture is expressly imposed by law. All forfeitures to the state, unless expressly imposed by law, are abolished.

E. A person shall not be disqualified from employment by this state or any of its agencies or political subdivisions, nor shall a person whose civil rights have been restored be disqualified to engage in any occupation for which a license, permit or certificate is required to be issued by this state solely because of a prior conviction for a felony or misdemeanor within or without this state. A person may be denied employment by this state or any of its agencies or political subdivisions or a person who has had his civil rights restored may be denied a license, permit or certificate to engage in an occupation by reason of the prior conviction of a felony or misdemeanor if the offense has a reasonable relationship to the functions of the employment or occupation for which the license, permit or certificate is sought.

F. Subsection E of this section is not applicable to any law enforcement agency.

G. Any complaints concerning a violation of subsection E of this section shall be adjudicated in accordance with the procedures set forth in title 41, chapter 6 and title 12, chapter 7, article 6.

H. A person who is adjudicated delinquent under section 8-341 for a felony does not have the right to carry or possess a gun or firearm.

13-905. Restoration of civil rights; persons completing probation.

A. A person who has been convicted of two or more felonies and whose period of probation has been completed may have any civil rights which were lost or suspended by his felony conviction restored by the judge who discharges him at the end of the term of probation. ...

C. If the person was convicted of a dangerous offense under section 13-604, the person may not file for the restoration of his right to possess or carry a gun or firearm. If the person was convicted of a serious offense as defined in section 13-604 the person may not file for the restoration of his right to possess or carry a gun or firearm for ten years from the date of his discharge from probation. If the person was convicted of any other felony offense, the person may not file for the restoration of his right to possess or carry a gun or firearm for two years from the date of his discharge from probation.

13-906. Applications by persons discharged from prison.

A. Upon proper application, a person who has been convicted of two or more felonies and who has received an absolute discharge from imprisonment may have any civil rights which were lost or suspended by his conviction restored by the superior court judge by whom the person was sentenced or his successors in office from the county in which he was originally sentenced. ...

C. If the person was convicted of a dangerous offense under section 13-604, the person may not file for the restoration of his right to possess or carry a gun or firearm. If the person was convicted of a serious offense as defined in section 13-604 the person may not file for the restoration of his right to possess or carry a gun or firearm for ten years from the date of his absolute discharge from imprisonment. If the person was convicted of any other felony offense, the person may not file for the restoration of his right to possess or carry a gun or firearm for two years from the date of his absolute discharge from imprisonment.

13-907. Setting aside judgment of convicted person on discharge; application; release from disabilities; exceptions.

A. Except as otherwise provided in this section, every person convicted of a criminal offense, on fulfillment of the conditions of probation or sentence and discharge by the court, may apply to the judge, justice of the peace or magistrate who pronounced sentence or imposed probation or such judge, justice of the peace or magistrate's successor in office to have the judgment of guilt set aside. The convicted person shall be informed of this right at the time of discharge.

B. The application to set aside the judgment may be made by the convicted person or by the convicted person's attorney or probation officer authorized in writing.

C. If the judge, justice of the peace or magistrate grants the application, the judge, justice of the peace or magistrate shall set aside the judgment of guilt, dismiss the accusations or information and order that the person be released from all penalties and disabilities resulting from the conviction other than those imposed by:

1. The department of transportation pursuant to section 28-3304, 28-3306, 28-3307, 28-3308 or 28-3319, except that the conviction may be used as a conviction if such conviction would be admissible had it not been set aside and may be pleaded and proved in any subsequent prosecution of such person by the state or any of its subdivisions for any offense or used by the department of transportation in enforcing section 28-3304, 28-3306, 28-3307, 28-3308, 28-3319 as if the judgment of guilt had not been set aside. ...

13-909. Restoration of civil rights; persons completing probation for federal offenses.

A. A person who has been convicted of two or more felonies and whose period of probation has been completed may have any civil rights which were lost or suspended by his felony conviction in a United States district court restored by the presiding judge of the superior court in the county in which he now resides, upon filing of an affidavit of discharge from the judge who discharged him at the end of the term of probation.

B. Upon proper application, a person who has been discharged from probation either prior to or after adoption of this chapter may have any civil rights which were lost or suspended by his felony conviction restored by an application filed with the clerk of the superior court in the county in which he now resides. The clerk of the superior court shall process the application upon request of the person involved or his attorney.

C. If the person was convicted of an offense which would be a dangerous offense under section 13-604, the person may not file for the restoration of his right to possess or carry a gun or firearm. If the person was convicted of an offense which would be a serious offense as defined in section 13-604 the person may not file for the restoration of his right to possess or carry a gun or firearm for ten years from the date of his discharge from probation. If the person was convicted of any other felony offense, the person may not file for the restoration of his right to possess or carry a gun or firearm for two years from the date of his discharge from probation.

13-910. Applications by persons discharged from federal prison.

A. Upon proper application, a person who has been convicted of two or more felonies and who has received an absolute discharge from imprisonment in a federal prison may have any civil rights which were lost or suspended by his conviction restored by the presiding judge of the superior court in the county in which he now resides.

B. A person who is subject to the provisions of subsection A of this section may file, no sooner than two years from the date of his absolute discharge, an application for restoration of civil rights that shall be accompanied by a certificate of absolute discharge from the director of the federal bureau of prisons, unless it is shown to be impossible to obtain such certificate. Such application shall be filed with the clerk of the superior court in the county in which the person now resides, and such clerk shall be responsible for processing applications for restoration of civil rights upon request of the person involved or his attorney.

C. If the person was convicted of an offense which would be a dangerous offense under section 13-604, the person may not file for the restoration of his right to possess or carry a gun or firearm. If the person was convicted of an offense which would be a serious offense as defined in section 13-604 the person may not file for the restoration of his right to possess or carry a gun or firearm for ten years from the date of his absolute discharge from imprisonment. If the person was convicted of any other felony offense, the person may not file for the restoration of his right to possess or carry a gun or firearm for two years from the date of his absolute discharge from imprisonment.

13-912. Restoration of civil rights for first offenders; exception.

A. Any person who has not previously been convicted of any other felony shall automatically be restored any civil rights that were lost or suspended by the conviction if the person both:

1. Completes a term of probation or receives an absolute discharge from imprisonment.

2. Pays any fine or restitution imposed.

B. This section does not apply to a person's right to possess weapons as defined in section 13-3101 unless the person applies to a court pursuant to section 13-905 or 13-906.

13-912.01. Restoration of civil rights; persons adjudicated delinquent.

A. A person who was adjudicated delinquent and whose period of probation has been completed may have his right to possess or carry a gun or firearm restored by the judge who discharges the person at the end of his term of probation.

B. A person who was adjudicated delinquent and who has been discharged from probation, on proper application, may have his right to carry or possess a gun or firearm restored by the judge of the juvenile court in the county where the person was adjudicated delinquent or his successors. The clerk of the superior court shall process the application on the request of the person involved or the person's attorney. The applicant shall serve a copy of the application on the county attorney.

C. If the person's adjudication was for a dangerous offense under section 13-604, a serious offense as defined in section 13-604, burglary in the first degree, burglary in the second degree or arson, the person may not file for the restoration of his right to possess or carry a gun or firearm until the person attains thirty years of age. If the person's adjudication was for any other felony offense, the person may not file for the restoration of his right to possess or carry a gun or firearm for two years from the date of his discharge.

Chapter 31. Weapons and Explosives

13-3101. Definitions.

A. In this chapter, unless the context otherwise requires:

1. "Deadly weapon" means anything that is designed for lethal use. The term includes a firearm.

2. "Deface" means to remove, alter or destroy the manufacturer's serial number.

3. "Explosive" means any dynamite, nitroglycerine, black powder or other similar explosive material, including plastic explosives. Explosive does not include ammunition or ammunition components such as primers, percussion caps, smokeless powder, black powder and black powder substitutes used for hand loading purposes.

4. "Firearm" means any loaded or unloaded handgun, pistol, revolver, rifle, shotgun or other weapon that will expel, is designed to expel or may readily be converted to expel a projectile by the action of an explosive. Firearm does not include a firearm in permanently inoperable condition.

5. Improvised explosive device" means a device that incorporates explosives or destructive, lethal, noxious, pyrotechnic or incendiary chemicals and that is designed to destroy, disfigure, terrify or harass.

6. "Occupied structure" means any building, object, vehicle, watercraft, aircraft or place with sides and a floor that is separately securable from any other structure attached to it, that is used for lodging, business, transportation, recreation or storage and in which one or more human beings either are or are likely to be present or so near as to be in equivalent danger at the time the discharge of a firearm occurs. Occupied structure includes any dwelling house, whether occupied, unoccupied or vacant.

7. "Prohibited possessor" means any person:

(a) Who has been found to constitute a danger to himself or to others pursuant to court order under section 36-540, and whose court ordered treatment has not been terminated by court order.

(b) Who has been convicted within or without this state of a felony or who has been adjudicated delinquent for a felony and whose civil right to possess or carry a gun or firearm has not been restored.

(c) Who is at the time of possession serving a term of imprisonment in any correctional or detention facility.

(d) Who is at the time of possession serving a term of probation pursuant to a conviction for a domestic violence offense as defined in section 13-3601 or a felony offense, parole, community supervision, work furlough, home arrest or release on any other basis or who is serving a term of probation or parole pursuant to the interstate compact under title 31, chapter 3, article 4.

(e) Who is an undocumented alien or a nonimmigrant alien traveling with or without documentation in this state for business or pleasure or who is studying in this state and who maintains a foreign residence abroad. This subsection does not apply to:

(i) nonimmigrant aliens who possess a valid hunting license or permit that is lawfully issued by a state in the United States.

(ii) nonimmigrant aliens who enter the United States to participate in a competitive target shooting event or to display firearms at a sports or hunting trade show that is sponsored by a national, state or local firearms trade organization, devoted to the competitive use or other sporting use of firearms.

(iii) certain diplomats.

(iv) officials of foreign governments or distinguished foreign visitors who are designated by the United States Department of State.

(v) persons who have received a waiver from the United States Attorney General.

8. "Prohibited weapon":

(A) Includes the following:

(I) An item that is a bomb, grenade, rocket having a propellant charge of more than four ounces or mine and that is explosive, incendiary or poison gas.

(II) A device that is designed made or adapted to muffle the report of a firearm.

(III) A firearm that is capable of shooting more than one shot automatically, without manual reloading, by a single function of the trigger.

(IV) A rifle with a barrel length of less than sixteen inches, or shotgun with a barrel length of less than eighteen inches, or any firearm that is made from a rifle or shotgun and that, as modified, has an overall length of less than twenty-six inches.

(V) An instrument, including a nunchaku, that consists of two or more sticks, clubs, bars or rods to be used as handles, connected by a rope, cord, wire or chain, in the design of a weapon used in connection with the practice of a system of self-defense.

(VI) A breakable container that contains a flammable liquid with a flash point of one hundred fifty degrees Fahrenheit or less and that has a wick or similar device capable of being ignited.

(VII) A chemical or combination of chemicals, compounds or materials, including dry ice, that is possessed or manufactured for the purpose of generating a gas to cause a mechanical failure, rupture or bursting or an explosion or detonation of the chemical or combination of chemicals, compounds or materials.

(VIII) An improvised explosive device.

(IX) Any combination of parts or materials that is designed and intended for use in making or converting a device into an item set forth in. item (I), (VI) or (VIII) of this subdivision.

(B) Does not include:

(I) Any fireworks that are imported, distributed or used in compliance with state laws or local ordinances.

(II) Any propellant, propellant actuated devices or propellant actuated industrial tools that are manufactured, imported or distributed for their intended purposes.

(III) A device that is commercially manufactured primarily for the purpose of illumination.

B. The items set forth in subsection A, paragraph 8, subdivision (a), items (I), (II), (III) and (IV) of this section do not include any firearms or devices that are registered in the national firearms registry and transfer records of the United States treasury department or any firearm that has been classified as a curio or relic by the United States treasury department.

13-3102. Misconduct involving weapons; defenses; classification; definitions.

A. A person commits misconduct involving weapons by knowingly:

1. Carrying a deadly weapon without a permit pursuant to section 13-3112 except a pocket knife concealed on his person; or

2. Carrying a deadly weapon without a permit pursuant to section 13-3112 concealed within immediate control of any person in or on a means of transportation; or

3. Manufacturing, possessing, transporting, selling or transferring a prohibited weapon , except that if the violation involves dry ice, a person commits misconduct involving weapons by knowingly possessing the dry ice with the intent to cause injury to or death of another person or to cause damage to the property of another person; or

4. Possessing a deadly weapon or prohibited weapon if such person is a prohibited possessor; or

5. Selling or transferring a deadly weapon to a prohibited possessor; or

6. Defacing a deadly weapon; or

7. Possessing a defaced deadly weapon knowing the deadly weapon was defaced; or

8. Using or possessing a deadly weapon during the commission of any felony offense included in chapter 34 of this title; or

9. Discharging a firearm at an occupied structure in order to assist, promote or further the interests of a criminal street gang, a criminal syndicate or a racketeering enterprise; or

10. Unless specifically authorized by law, entering any public establishment or attending any public event and carrying a deadly weapon on his person after a reasonable request by the operator of the establishment or the sponsor of the event or the sponsor's agent to remove his weapon and place it in the custody of the operator of the establishment or the sponsor of the event for temporary and secure storage of the weapon pursuant to section 13-3102.01; or

11. Unless specifically authorized by law, entering an election polling place on the day of any election carrying a deadly weapon; or

12. Possessing a deadly weapon on school grounds; or

13. Unless specifically authorized by law, entering a nuclear or hydroelectric generating station carrying a deadly weapon on his person or within the immediate control of any person; or

14. Supplying, selling or giving possession or control of a firearm to another person if the person knows or has reason to know that the other person would use the firearm in the commission of any felony; or

15. Using, possessing or exercising control over a deadly weapon in furtherance of any act of terrorism as defined in section 13-2301 or possessing or exercising control over a deadly weapon knowing or having reason to know that it will be used to facilitate any act of terrorism as defined in section 13-2301.

B. Subsection A, paragraph 1 of this section shall not apply to a person in his dwelling, on his business premises or on real property owned or leased by that person.

C. Subsection A, paragraphs 1, 2, 3, 7, 10, 11, 12 and 13 of this section shall not apply to:

1. A peace officer or any person summoned by any peace officer to assist and while actually assisting in the performance of official duties; or

2. A member of the military forces of the United States or of any state of the United States in the performance of official duties; or

3. A warden, deputy warden or correctional officer of the state department of corrections; or

4. A person specifically licensed, authorized or permitted pursuant to a statute of this state or of the United States.

D. Subsection A, paragraphs 3 and 7 of this section shall not apply to:

1. The possessing, transporting, selling or transferring of weapons by a museum as a part of its collection or an educational institution for educational purposes or by an authorized employee of such museum or institution, if:

(a) Such museum or institution is operated by the United States or this state or a political subdivision of this state, or by an organization described in 26 United States Code section 170(c) as a recipient of a charitable contribution; and

(b) Reasonable precautions are taken with respect to theft or misuse of such material.

2. The regular and lawful transporting as merchandise; or

3. Acquisition by a person by operation of law such as by gift, devise or descent or in a fiduciary capacity as a recipient of the property or former property of an insolvent, incapacitated or deceased person.

E. Subsection A, paragraph 3 of this section shall not apply to the merchandise of an authorized manufacturer of or dealer in prohibited weapons, when such material is intended to be manufactured, possessed, transported, sold or transferred solely for or to a dealer, a regularly constituted or appointed state, county or municipal police department or police officer, a detention facility, the military service of this or another state or the United States, a museum or educational institution or a person specifically licensed or permitted pursuant to federal or state law.

F. Subsection A, paragraph 1 of this section shall not apply to a weapon or weapons carried in a belt holster that is wholly or partially visible, or carried in a scabbard or case designed for carrying weapons that is wholly or partially visible or carried in luggage. Subsection A, paragraph 2 of this section shall not apply to a weapon or weapons carried in a case, holster, scabbard, pack or luggage that is carried within a means of transportation or within a storage compartment, map pocket, trunk or glove compartment of a means of transportation.

G. Subsection A, paragraph 10 of this section shall not apply to shooting ranges or shooting events, hunting areas or similar locations or activities.

H. Subsection A, paragraph 3 of this section shall not apply to a weapon described in section 13-3101, subsection A, paragraph 8, subdivision(A), item (V) , if such weapon is possessed for the purposes of preparing for, conducting or participating in lawful exhibitions, demonstrations, con-tests or athletic events involving the use of such weapon. Subsection A, paragraph 12 of this section shall not apply to a weapon if such weapon is possessed for the purposes of preparing for, conducting or participating in hunter or firearm safety courses.

I. Subsection A, paragraph 12 of this section shall not apply to the possession of a:

1. Firearm that is not loaded and that is carried within a means of transportation under the control of an adult provided that if the adult leaves the means of transportation the firearm shall not be visible from the outside of the means of transportation and the means of transportation shall be locked.

2. Firearm for use on the school grounds in a program approved by a school.

J. The operator of the establishment or the sponsor of the event or the employee of the operator or sponsor or the agent of the sponsor, including a public entity or public employee, is not liable for acts or omissions pursuant to subsection A, paragraph 10 of this section unless the operator, sponsor, employee or agent intended to cause injury or was grossly negligent.

K. Misconduct involving weapons under subsection A, paragraph 9, 14 or 15 of this section is a class 3 felony. Misconduct involving weapons under subsection A, paragraph 3, 4, 8 or 13 of this section is a class 4 felony. Misconduct involving weapons under subsection A, paragraph 12 of this section is a class 1 misdemeanor unless the violation occurs in connection with conduct that violates section 13-2308, subsection A, paragraph 5, section 13-2312, subsection C, section 13-3409 or section 13-3411, in which case the offense is a class 6 felony. Misconduct involving weapons under subsection A, paragraph 5, 6 or 7 of this section is a class 6 felony. Misconduct involving weapons under subsection A, paragraph 1, 2, 10 or 11 of this section is a class 1 misdemeanor.

L. For the purposes of this section:

1. "Public establishment" means a structure, vehicle or craft that is owned, leased or operated by this state or political subdivision of this state.

2. "Public event" means a specifically named or sponsored event of limited duration that is either conducted by a public entity or conducted by a private entity with a permit or license granted by a public entity. Public event does not include an unsponsored gathering of people in a public place.

3. "School" means a public or nonpublic kindergarten program, common school or high school.

4. "School grounds" means in, or on the grounds of, a school.

13-3102.01. Storage of deadly weapons; definitions.

A. If an operator of a public establishment or a sponsor of a public event requests that a person carrying a deadly weapon remove the weapon, the operator or sponsor shall provide temporary and secure storage. The storage shall be readily accessible on entry into the establishment or event and allow for the immediate retrieval of the weapon on exit from the establishment or event.

B. This section does not apply to the licensed premises of any public establishment or public event with a license issued pursuant to title 4.

C. The operator of the establishment or the sponsor of the event or the employee of the operator or sponsor or the agent of the sponsor, including a public entity or public employee, is not liable for acts or omissions pursuant to this section unless the operator, sponsor, employee or agent intended to cause injury or was grossly negligent.

D. For the purposes of this section, "public establishment" and "public event" have the same meanings prescribed in section 13-3102.

13-3103. Misconduct involving explosives; classification

A. A person commits misconduct involving explosives by knowingly:

1. Keeping or storing a greater quantity than fifty pounds of explosives in or upon any building or premises within a distance of one-half mile of the exterior limits of a city or town, except in vessels, railroad cars or vehicles receiving and keeping them in the course of and for the purpose of transportation; or

2. Keeping or storing percussion caps or any blasting powder within two hundred feet of a building or premises where explosives are kept or stored; or

3. Selling, transporting or possessing explosives without having plainly marked, in a conspicuous place on the box or package containing the explosive, its name, explosive character and date of manufacture.

4. This section shall not apply to any person who legally keeps, stores or transports explosives, percussion caps or blasting powder as a part of their business.

B. Misconduct involving explosives is a class 1 misdemeanor.

13-3105. Forfeiture of weapons and explosives.

A. Upon the conviction of any person for the violation of any felony in this state in which a deadly weapon, dangerous instrument or explosive was used, displayed or unlawfully possessed by such person, the court shall order the article forfeited and sold, destroyed or otherwise properly disposed.

B. Upon the conviction of any person for the violation of section 13-2904, subsection A, paragraph 6 or section 13-3102, subsection A, paragraph 1, 2, 8 or 10, the court may order the forfeiture of the deadly weapon or dangerous instrument involved in the offense.

C. If at any time the court finds pursuant to rule 11 of the Arizona rules of criminal procedure that a person who is charged with a violation of this title is incompetent, the court shall order that any deadly weapon, dangerous instrument or explosive used, displayed or unlawfully possessed by the person during the commission of the alleged offense be forfeited and sold, destroyed or otherwise properly disposed.

13-3106. Firearm purchase in other states.

A person residing in this state, or a corporation or other business entity maintaining a place of business in this state, may purchase or otherwise obtain firearms anywhere in the United States if such purchase or acquisition fully complies with the laws of this state and the state in which the purchase or acquisition is made and the purchaser and seller, prior to the sale or delivery for sale, have complied with all the requirements of the federal gun control act of 1968, Public Law 90-618, section 922, subsection (c) and the code of federal regulations, volume 26, section 178.96, subsection (c).

13-3108. Firearms regulated by state; state preemption; violation; classification.

A. Except as provided in subsection C of this section, a political subdivision of this state shall not enact any ordinance, rule or tax relating to the transportation, possession, carrying, sale, transfer or use of firearms or ammunition or any firearm or ammunition components in this state.

B. A political subdivision of this state shall not require the licensing or registration of firearms or ammunition or any firearm or ammunition components or prohibit the ownership, purchase, sale or transfer of firearms or ammunition or any firearm or ammunition components.

C. This section does not prohibit a political subdivision of this state from enacting and enforcing any ordinance or rule pursuant to state law, to implement or enforce state law or relating to any of the following:

1. Imposing any privilege or use tax on the retail sale, lease or rental of, or the gross proceeds or gross income from the sale, lease or rental of, firearms or ammunition or any firearm or ammunition components at a rate that applies generally to other items of tangible personal property.

2. Prohibiting a minor who is unaccompanied by a parent, grandparent or guardian or a certified hunter safety instructor or certified firearms safety instructor acting with the consent of the minor's parent, grandparent or guardian from knowingly possessing or carrying on the minor's person, within the minor's immediate control or in or on a means of transportation a firearm in any place that is open to the public or on any street or highway or on any private property except private property that is owned or leased by the minor or the minor's parent, grandparent or guardian. Any ordinance or rule that is adopted pursuant to this paragraph shall not apply to a minor who is fourteen, fifteen, sixteen or seventeen years of age and who is engaged in any of the following:

(a) Lawful hunting or shooting events or marksmanship practice at established ranges or other areas where the discharge of a firearm is not prohibited.

(b) Lawful transportation of an unloaded firearm for the purpose of lawful hunting.

(c) Lawful transportation of an unloaded firearm between the hours of 5:00 a.m. and 10:00 p.m. for the purpose of shooting events or marksmanship practice at established ranges or other areas where the discharge of a firearm is not prohibited.

(d) Any activity that is related to the production of crops, livestock, poultry, livestock products, poultry products or ratites or storage of agricultural commodities.

3. The use of land and structures, including a business relating to firearms or ammunition or their components or a shooting range in the same manner as other commercial businesses. Notwithstanding any other law, this paragraph does not authorize a political subdivision to regulate the sale or transfer of firearms on property it owns, leases, operates or controls in a manner that is different than or inconsistent with state law. For the purposes of this paragraph, a use permit or other contract that provides for the use of property owned, leased, operated or controlled by a political subdivision shall not be considered a sale, conveyance or disposition of property.

4. Regulating employees or independent contractors of the political subdivision who are acting within the course and scope of their employment or contract.

5. Limiting firearms possession in parks or preserves of one square mile or less in area to persons who possess a concealed weapons permit issued pursuant to section 13-3112. The political subdivision shall post reasonable notice at each park or preserve. The notice shall state the following: "Carrying a firearm in this park is limited to persons who possess a permit issued pursuant to section 13-3112." In parks or preserves that are more than one square mile in area, a political subdivision may designate developed or improved areas in which the political subdivision may limit firearms possession to persons who possess a concealed weapons permit issued pursuant to section 13-3112. The political subdivision shall post reasonable notice at each designated developed or improved area. The notice shall state the following: "Carrying a firearm in this developed or improved area is limited to persons with a permit issued pursuant to section 13-3112." For the purposes of this paragraph, "developed or improved area" means an area of property developed for public recreation or family activity, including picnic areas, concessions, playgrounds, amphitheaters, racquet courts, swimming areas, golf courses, zoos, horseback riding facilities and boat landing and docking facilities. Developed or improved area does not include campgrounds, trails, paths or roadways except trails, paths and roadways directly associated with and adjacent to designated developed or improved areas. Any notice that is required by this paragraph shall be conspicuously posted at all public entrances and at intervals of one-fourth mile or less where the park, preserve or developed or improved area has an open perimeter. Any limitation imposed by a political subdivision pursuant to this paragraph shall not apply to a person:

(a) Engaged in a permitted firearms or hunter safety course conducted in a park by a certified hunter safety instructor or certified firearms safety instructor.

(b) At a properly supervised range, as defined in section 13-3107, at a permitted shooting event, at a permitted firearms show or in a permitted hunting area.

(c) Legally transporting, carrying, storing or possessing a firearm in a vehicle.

(d) Going directly to or from an area where the person is lawfully engaged in hunting, marksmanship practice or recreational shooting.

(e) Traversing a trailhead area in order to gain access to areas where the possession of firearms is not limited.

(f) Using trails, paths or roadways to go directly to or from an area where the possession of firearms is not limited and where no reasonable alternate means of access is available.

6. Limiting or prohibiting the discharge of firearms in parks and preserves except:

(a) As allowed pursuant to chapter 4 of this title.

(b) On a properly supervised range as defined in section 13-3107.

(c) In an area recommended as a hunting area by the Arizona game and fish department and approved and posted as required by the political subdivision's chief law enforcement officer. Any such area may be closed when deemed unsafe by the political subdivision's chief law enforcement officer or the director of the Arizona game and fish department.

(d) To control nuisance wildlife by permit from the Arizona game and fish department or the United States fish and wildlife service.

(e) By special permit of the chief law enforcement officer of the political subdivision.

(f) As required by an animal control officer in performing duties specified in section 9-499.04 and title 11, chapter 7, article 6.

(g) In self defense or defense of another person against an animal attack if a reasonable person would believe that deadly physical force against the animal is immediately necessary and reasonable under the circumstances to protect oneself or the other person.

D. A violation of any ordinance established pursuant to subsection c, paragraph 6 of this section is a class 2 misdemeanor unless the political subdivision designates a lesser classification by ordinance.

13-3109. Sale or gift of firearm to minor; classification.

A. Except as provided in subsection C of this section, a person who sells or gives to a minor, without written consent of the minor's parent or legal guardian, a firearm, ammunition or a toy pistol by which dangerous and explosive substances may be discharged is guilty of a class 6 felony.

B. Nothing in this section shall be construed to require reporting sales of firearms, nor shall registration of firearms or firearms sales be required.

C. The temporary transfer of firearms and ammunition by firearms safety instructors, hunter safety instructors, competition coaches or their assistants shall be allowed if the minor's parent or guardian has given consent for the minor to participate in activities such as firearms or hunting safety courses, firearms competition or training. With the consent of the minor's parent or guardian, the temporary transfer of firearms and ammunition by an adult accompanying minors engaged in hunting or formal or informal target shooting activities shall be allowed for those purposes.

13-3111. Minors prohibited from carrying or possessing firearms; exceptions; seizure and forfeiture; penalties; classification.

A. Except as provided in subsection B, an unemancipated person who is under eighteen years of age and who is unaccompanied by a parent, grandparent or guardian, or a certified hunter safety instructor or certified firearms safety instructor acting with the consent of the unemancipated person's parent or guardian, shall not knowingly carry or possess on his person, within his immediate control, or in or on a means of transportation a firearm in any place that is open to the public or on any street or highway or on any private property except private property owned or leased by the minor or the minor's parent, grandparent or guardian.

B. This section does not apply to a person who is fourteen, fifteen, sixteen or seventeen years of age and who is any of the following:

1. Engaged in lawful hunting or shooting events or marksmanship practice at established ranges or other areas where the discharge of a firearm is not prohibited.

2. Engaged in lawful transportation of an unloaded firearm for the purpose of lawful hunting.

3. Engaged in lawful transportation of an unloaded firearm between the hours of 5:00 a.m. and 10:00 p.m. for the purpose of shooting events or marksmanship practice at established ranges or other areas where the discharge of a firearm is not prohibited.

4. Engaged in activities requiring the use of a firearm that are related to the production of crops, livestock, poultry, livestock products, poultry products, or ratites or in the production or storage of agricultural commodities.

C. If the minor is not exempt under subsection B and is in possession of a firearm, a peace officer shall seize the firearm at the time the violation occurs.

D. In addition to any other penalty provided by law, a person who violates subsection A shall be subject to the following penalties:

1. If adjudicated a delinquent juvenile for an offense involving an unloaded firearm, a fine of not more than two hundred fifty dollars, and the court may order the suspension or revocation of the person's driver license until the person reaches eighteen years of age. If the person does not have a driver license at the time of the adjudication, the court may direct that the department of transportation not issue a driver license to the person until the person reaches eighteen years of age.

2. If adjudicated a delinquent juvenile for an offense involving a loaded firearm, a fine of not more than five hundred dollars, and the court may order the suspension or revocation of the person's driver license until the person reaches eighteen years of age. If the person does not have a driver license at the time of the adjudication, the court may direct that the department of transportation not issue a driver license to the person until the person reaches eighteen years of age.

3. If adjudicated a delinquent juvenile for an offense involving a loaded or unloaded firearm, if the person possessed the firearm while the person was the driver or an occupant of a motor vehicle, a fine of not more than five hundred dollars and the court shall order the suspension or revocation of the person's driver license until the person reaches eighteen years of age. If the person does not have a driver license at the time of adjudication, the court shall direct that the department of transportation not issue a driver license to the person until the person reaches eighteen years of age. If the court finds that no other means of transportation is available, the driving privileges of the child may be restricted to travel between the child's home, school and place of employment during specified periods of time according to the child's school and employment schedule.

E. Firearms seized pursuant to subsection C shall be held by the law enforcement agency responsible for the seizure until the charges have been adjudicated or disposed of otherwise or the person is convicted. Upon adjudication or conviction of a person for a violation of this section, the court shall order the firearm forfeited. However, the law enforcement agency shall return the firearm to the lawful owner if the identity of that person is known.

F. If the court finds that the parent or guardian of a minor found responsible for violating this section knew or reasonably should have known of the minor's unlawful conduct and made no effort to prohibit it, the parent or guardian is jointly and severally responsible for any fine imposed pursuant to this section or for any civil actual damages resulting from the unlawful use of the firearm by the minor.

G. This section is supplemental to any other law imposing a criminal penalty for the use or exhibition of a deadly weapon. A minor who violates this section may be prosecuted and adjudicated delinquent for any other criminal conduct involving the use or exhibition of the deadly weapon.

H. A person who violates subsection A is guilty of a class 6 felony.

13-3113. Adjudicated delinquents; firearm possession; violation; classification. A person who was previously adjudicated delinquent for an offense that would be a felony if committed by an adult and who possesses, uses or carries a firearm within ten years from the date of his adjudication or his release or escape from custody is guilty of a class 5 felony for a first offense and a class 4 felony for a second or subsequent offense if the person was previously adjudicated for an offense that if committed as an adult would constitute:

1. Burglary in the first degree.

2. Burglary in the second degree.

3. Arson.

4. Any felony offense involving the use or threatening exhibition of a deadly weapon or dangerous instrument.

5. A serious offense as defined in section 13-604.

13-3115. Forensics firearms identification system

The department of public safety is authorized to establish and maintain a forensics firearms identification system designed to provide investigative information on criminal street gangs and the unlawful use of firearms.

13-3117. Remote stun guns; sales records; use; classification; definitions.

A. It is unlawful for a person or entity to do any of the following:

1. Sell an authorized remote stun gun without keeping an accurate sales record as to the identity of the purchaser with the manufacturer of the authorized remote stun gun. The identification that is required by this paragraph shall be verified with a government issued identification. This requirement does not apply to secondary sales.

2. Knowingly use or threaten to use a remote stun gun or an authorized remote stun gun against a law enforcement officer who is engaged in the performance of the officer's official duties.

B. This section does not:

1. Preclude the prosecution of any person for the use of a remote stun gun or an authorized remote stun gun during the commission of any criminal offense.

2. Preclude any justification defense under chapter 4 of this title.

C. The regulation of remote stun guns and authorized remote stun guns is a matter of statewide concern.

D. A violation of:

1. Subsection A, paragraph 1 is a petty offense.

2. Subsection A, paragraph 2 is a class 4 felony.

E. For the purposes of this section:

1. "Authorized remote stun gun" means a remote stun gun that has all of the following:

(a) An electrical discharge that is less than one hundred thousand volts and less than nine joules of energy per pulse.

(b) A serial or identification number on all projectiles that are discharged from the remote stun gun.

(c) An identification and tracking system that, on deployment of remote electrodes, disperses coded material that is traceable to the purchaser through records that are kept by the manufacturer on all remote stun guns and all individual cartridges sold.

(d) A training program that is offered by the manufacturer.

2. "Remote stun gun" means an electronic device that emits an electrical charge and that is designed and primarily employed to incapacitate a person or animal either through contact with electrodes on the device itself or remotely through wired probes that are attached to the device or through a spark, plasma, ionization or other conductive means emitting from the device.

13-3118. Possession or storage of firearms; restrictions prohibited; exceptions.

(a) Except for the legislature, this state and any agency or political subdivision of this state shall not enact or implement any law, rule or ordinance relating to the possession or storage of firearms other than as provided in statute.

(b) This section does not prohibit:

(1) a state, county or municipal judicial department, law enforcement agency or prosecutorial agency from prohibiting a deadly weapon pursuant to section 13-3102(a)(10).

(2) a political subdivision of this state from enacting any rule or ordinance requiring a business that obtains a secondhand firearm by purchase, trade or consignment to retain the firearm for a period of not more than ten days at its place of business or another storage location that is approved by the applicable law enforcement agency.

Title 15. Education

Chapter 3. Local Governance of Schools

Article 3. Powers and Duties of School District Governing Boards

15-341. General powers and duties; immunity; delegation.

A. The governing board shall: ...

25. Notwithstanding section 13-3108, prescribe and enforce policies and procedures that prohibit a person from carrying or possessing a weapon on school grounds unless the person is a peace officer or has obtained specific authorization from the school administrator. ...

Title 44. Trade and Commerce

Chapter 11. Regulations Concerning Particular Businesses

Article 3. Pawnbrokers

44-1627. Licensing; requirements.

A. A person shall not act as a pawnbroker until licensed by the sheriff of the county in which the person regularly conducts business.

B. A pawnbroker shall obtain a separate license for each pawnshop owned by that pawnbroker.

C. A pawnbroker license may not be sold or transferred without the approval of the sheriff or the sheriff's designee.

D. A pawnbroker shall not conduct business at a location other than a licensed location except for firearms transactions that are permitted by a federally licensed firearms dealer at an organized gun show. ...

[Current through the First Regular Session of the 48th Legislature 2007]

ARKANSAS
ARK. CODE

Title 5. Criminal Offenses

Chapter 73. Weapons

Subchapter 1. Possession and Use Generally

5-73-101. Definitions. As used in this chapter:

(1) "Blasting agent" means any material or mixture consisting of fuel and oxidizer intended for blasting if finished product as mixed for use or shipment cannot be detonated by means of a No. 8 test blasting cap when unconfined;

(2) "Contraband" means any explosive material that was used with the knowledge and consent of the owner to facilitate a violation of this subchapter, as well as any explosive material possessed under circumstances prohibited by law;

(3) "Destruction Device" means:

(A) Any of the following:

(i) Any explosive, incendiary or poison gas;

(ii) Bomb;

(iii) Grenade;

(iv) Rocket having a propellant charge of more than four ounces (4 ozs);

(v) Missile having an explosive or incendiary charge of more than one-quarter ounce;

(vi) Mine; or

(vii) Similar device; and

(B) Any combination of parts either designed or intended for use in converting any device into a destructive device as defined in subdivision (3)(A) and from which a destructive device may be readily assembled for use as a weapon;

(4)(A) "Detonator" means any device containing any initiating or primary explosive that is used for initiating detonation.

(B) A detonator may not contain more than ten (10) grams of total explosives by weight, excluding ignition or delay charges, and may include, without limitation, electric blasting caps of instantaneous and delay types, blasting caps for use with safety fuses, detonating cord delay connectors, and non-instantaneous and delay blasting caps that use detonating cord, shock tube, or any other replacement for electric leg wires;

(5) "Distribute" means to sell, issue, give, transfer, or otherwise dispose of explosive material;

(6) "Explosive Material" means an explosive, blasting agent, or detonator;

(7)(A) "Explosive" means any chemical compound mixture or device, the primary or common purpose of which is to function by explosion.

(B) "Explosive" includes, without limitation:

(i) Dynamite and other high explosive;

(ii) Black powder;

(iii) Pellet powder;

(iv) An initiating explosive;

(v) A detonator;

(vi) A safety fuse;

(viii) A squib;

(viii) A detonating cord;

(ix) An igniter cord;

(x) An igniter;

(xi) Any material determined to be within the scope of Title 18 U. S. C. § 841; and

(xii) Any material classified as an explosive other than consumer fireworks, 1.4 (Class C, Common), by the United States Department of Transportation.

(8) "Instrument of crime" means anything manifestly designed, made, adapted, or commonly used for a criminal purpose.

(9) "Minor" means any person under eighteen (18) years of age; and

(10) "Violent felony conviction" means a conviction for any felony offense against the person which is codified in § 5-101 et seq., § 5-12-101 et seq., § 5-13-201 et seq., § 5-13-301 et seq., and § 5-14-201 et seq., or any other offense containing as an element of the offense one (1) of the following;

(A) The use of physical force;

(B) The use or threatened use of serious physical force;

(C) The infliction of physical harm; or

(D) The creation of a substantial risk of serious physical harm.

5-73-102. Possessing instrument of crime.

(a) A person commits the offense of possessing an instrument of crime if he possesses any instrument of crime with a purpose to employ it criminally.

(b) Possessing an instrument of crime is a Class A misdemeanor.

5-73-103. Possession of firearms by certain persons.

(a) Except as provided in subsection (d) of this section or unless authorized by and subject to such conditions as prescribed by the Governor, or his or her designee, or the Bureau of Alcohol, Tobacco, Firearms and Explosives of the United States Department of Justice, or other bureau or office designated by the United States Department of Justice, no person shall possess or own any firearm who has been:

(1) Convicted of a felony;

(2) Adjudicated mentally ill; or

(3) Committed involuntarily to any mental institution.

(b)(1) A determination by a jury or a court that a person committed a felony constitutes a conviction for purposes of subsection (a) of this section even though the court suspended imposition of sentence or placed the defendant on probation.

(2) However, the determination by a jury or court that the person committed a felony does not constitute a conviction for purposes of subsection (a) of this section if the per-son is subsequently granted a pardon explicitly restoring the ability to possess a firearm.

(c)(1) A person who violates this section commits a Class B felony if:

(A) The person has a prior violent felony conviction; or

(B) The person's current possession of a firearm involves the commission of another crime; or

(C) The person has been previously convicted under this section or a similar provision from another jurisdiction.

(2) A person who violates this section commits a Class D felony if he or she has been previously convicted of a felony and his or her present conduct or the prior felony conviction does not fall within subdivision (c)(1) of this section.

(3) Otherwise, the person commits a Class A misdemeanor.

(d) The Governor may restore without granting a pardon, the right of a convicted felon or an adjudicated delinquent to own and possess a firearm upon the recommendation of the chief law enforcement officer in the jurisdiction in which the person resides, so long as the underlying felony or delinquency adjudication:

(1) Did not involve the use of a weapon; and

(2) Occurred more than eight (8) years ago.

5-73-104. Criminal use of prohibited weapons.

(a) A person commits the offense of criminal use of prohibited weapons if, except as authorized by law, he or she uses, possesses, makes, repairs, sells, or otherwise deals in any:

(1) Bomb;

(2) Machine gun;

(3) Sawed-off shotgun or rifle;

(4) Firearm specially made or specially adapted for silent discharge;

(5) Metal knuckles; or

(6) Other implement for the infliction of serious physical injury or death.

(b) It is a defense to prosecution under this section that:

(1) The defendant was a law enforcement officer, prison guard, or member of the armed forces acting in the course and scope of his or her duty at the time he or she used or possessed the prohibited weapon; or

(2) The defendant used, possessed, made, repaired, sold, or otherwise dealt in any article enumerated in subsection (a) of this chapter under circumstances negating any likelihood that the weapon could be used as a weapon.

(c) (1) Criminal use of prohibited weapons is a Class B felony if the weapon is a bomb, machine gun, or firearm specially made or specially adapted for silent discharge. Otherwise, it is a Class D felony.

(2) Otherwise, criminal use of prohibited weapons is a Class D Felony.

5-73-105. Legitimate manufacture, repair, and transportation of prohibited weapons.

Section 5-73-104 shall not be construed to prohibit the manufacture, repair, transportation, or sale of the weapons enumerated in § 5-73-104 to or for an authorized representative of:

(1) The armed forces; or

(2) Any law enforcement agency.

5-73-106. Defacing a firearm.

(a) A person commits the offense of defacing a firearm when he or she knowingly removes, defaces, mars, covers, alters, or destroys the manufacturer's serial number or identification mark of a firearm.

(b) Defacing a firearm is a Class D felony.

5-73-107. Possession of a defaced firearm.

(a) A person commits the offense of possession of a defaced firearm if he or she knowingly possesses a firearm with a manufacturer's serial number or other identification mark required by law that has been removed, defaced, marred, altered, or destroyed.

(b) It is a defense to a prosecution under this section that the person reported the possession to the police or other governmental agency prior to arrest or the issuance of an arrest warrant or summons.

(c)(1) Possession of a defaced firearm is a Class D felony.

(2) However, possession of a defaced firearm is a Class A misdemeanor if the manufacturer's serial number or other identification mark required by law is merely covered or obstructed, but still retrievable.

5-73-108. Criminal acts involving explosives.

(a)(1) A person commits the offense of criminal possession of explosive material or a destructive device if the person:

(A) Sells, possesses, manufactures, transfers, or transports explosive material or a destructive device; and

(B) Either:

(i) Has the purpose of using that explosive material or destructive device to commit an offense; or

(ii) Knows or should know that another person intends to use that explosive material or destructive device to commit an offense.

(2) Criminal possession of explosive material or a destructive device is a Class B felony.

(b)(1) A person commits the offense of criminal distribution of explosive material if he or she knowingly distributes explosive material to any individual who:

(A) Has pleaded guilty or nolo contendere to or been found guilty of a crime in state or federal court punishable by imprisonment for a term exceeding one (1) year;

(B) Is under indictment or has been formally charged for a crime punishable by imprisonment for a term exceeding one (1) year;

(C) Is a fugitive from justice;

(D) Is an unlawful user of or addicted to any controlled substance; or

(E) Has been adjudicated mentally incompetent.

(2) Criminal distribution of explosive material is a Class C felony.

(c)(1) A person commits the offense of possession of stolen explosive material if he or she:

(A) Receives, possesses, transports, ships, conceals, stores, barters, sells, disposes of, or pledges or accepts as security for a loan any stolen explosive materials; and

(B) Knows or has reasonable cause to believe that the explosive material was stolen.

(2) Possession of stolen explosive material is a Class C felony.

(d) It is a Class A misdemeanor for any person to store any explosive material in a manner not in conformity with the Arkansas Fire Prevention Code.

(e) A person who commits theft of any explosive material with the intent to cause harm to a person or property is guilty of a Class B felony.

(f) Any explosive material determined to be contraband is subject to seizure by a law enforcement officer and to being destroyed in conformity with the Arkansas Fire Prevention Code.

5-73-109. Furnishing a deadly weapon to a minor.

(a) A person commits the offense of furnishing a deadly weapon to a minor if he or she sells, barters, leases, gives, rents, or otherwise furnishes a firearm or other deadly weapon to a minor without the consent of a parent, guardian, or other person responsible for general supervision of the minor's welfare.

(b)(1) Furnishing a deadly weapon to a minor is a Class A misdemeanor.

(2) However, furnishing a deadly weapon to a minor is a Class B felony if the deadly weapon is:

(A) A handgun;

(B) A sawed-off or short-barreled shotgun, as defined in § 5-1-102;

(C) A sawed-off or short-barreled rifle, as defined in § 5-1-102;

(D) A firearm that has been specially made or specially adapted for silent discharge;

(E) A machine gun;

(F) An explosive or incendiary device, as defined in § 5-71-301;

(G) Metal knuckles;

(H) A defaced firearm, as defined in § 5-73-107; or

(I) Another implement for the infliction of serious physical injury or death that serves no common lawful purpose.

5-73-110. Disarming minors and mentally defective or irresponsible persons -- Disposition of property seized.

(a) Subject to constitutional limitation, nothing in this section and §§ 5-73-101 -- 5-73-109 shall be construed to prohibit a law enforcement officer from disarming, without arresting, a minor or person who reasonably appears to be mentally defective or otherwise mentally irresponsible, when that person is in possession of a deadly weapon.

(b) Property seized pursuant to subsection (a) of this section may be:

(1) Returned to the parent, guardian, or other person entrusted with care and supervision of the person so disarmed; or

(2) Delivered to the custody of a court having jurisdiction to try criminal offenses, in which case the court shall:

(A) Treat the property as contraband under §§ 5-5-101 and 5-5-102; or

(B) Issue an order requiring that at a certain time the parent, guardian, or person entrusted with the care and supervision of the person disarmed show cause why the seized property should not be so treated.

(c) Notice of the show cause proceedings may be given in the manner provided for service of criminal summons under Rule 6.3 of Arkansas Rules of Criminal Procedure.

5-73-119. Handguns -- Possession by minor or possession on school property.

(a)(1) No person in this state under the age of eighteen (18) years of age shall possess a handgun.

(2)(A) A violation of subdivision (a)(1) of this section is Class A misdemeanor.

(B) A violation of subdivision (a)(1) of this section is a Class D felony if the person has previously:

(i) Been adjudicated delinquent for a violation of subdivision (a)(1) of this section;

(ii) Been adjudicated delinquent for any offense that would be a felony if committed by an adult; or

(iii) Pleaded guilty or nolo contendere to, or been found guilty of, a felony in circuit court while under the age of eighteen (18) years of age.

(b)(i) No person in this state shall possess a firearm:

(A) Upon the developed property of the public or private schools, K-12; or

(B) In or upon any school bus; or

(C) At a designated bus stop as identified on the route lists published by a school district each year.

(2)(A) A violation of subdivision (b)(1) of this section shall be a Class D felony.

(B) No sentence imposed for a violation of subdivision (b)(1) of this sections shall be suspended or probated or treated as a first offense under § 16-93-301 et seq.

(c)(1) No person in this state shall possess a handgun upon the property of any private institution of higher education or a publicly supported institutions of higher education in this state on or about his or her person, in a vehicle occupied by him or her, or otherwise readily available for use with a purpose to employ the handgun as a weapon against a person.

(2) A violation of subdivision (c)(1) of this section shall be a Class D felony.

(d) A "handgun" means a firearm, capable of firing rimfire ammunition or centerfire ammunition and designed or constructed to be fired with one (1) hand.

(e) It is a defense to prosecution under this section that at the time of the act of possessing a handgun or firearm:

(1) The person is in his own dwelling or place of business or on property in which he or she has a possessory or proprietary interest, except upon the property of a public or private institution of higher learning;

(2) The person is a law enforcement officer, correctional officer, or member of the armed forces acting in the course and scope of his or her official duties;

(3) The person is assisting a law enforcement officer, correctional officer, or member of the armed forces acting in the course and scope of his or her official duties pursuant to the direction or request of the law enforcement officer, correctional officer, or member of the armed forces;

(4) The person is a licensed security guard acting in the course and scope of his or her duties;

(5) The person is hunting game with a handgun or firearm that may be hunted with a handgun or firearm under the rules and regulations of the Arkansas State Game and Fish Commission or is en route to or from a hunting area for the purpose of hunting game with a handgun or firearm;

(6) The person is a certified law enforcement officer;

(7) The person is on a journey, unless the person is eighteen (18) years of age or less; or

(8) The person is participating in a certified hunting safety course sponsored by the commission or a firearm safety course recognized and approved by the commission or by a state or national non-profit organization qualified and experienced in firearm safety;

(9) The person is participating in a school-approved educational course or sporting activity involving the use of firearms; or

(10) The person is a minor engaged in lawful marksmanship competition or practice or other lawful recreational shooting under the supervision of his or her parent, legal guardian, or other person twenty-one (21) years of age or older standing in loco parentis, or is traveling to or from this activity with an unloaded handgun or firearm accompanied by his or her parent, legal guardian, or other person twenty-one (21) years of age or older standing in loco parentis.

5-73-125. Interstate sale and purchase of shotguns, rifles, and ammunition.

(a) The sale of shotguns and rifles and ammunition in this state to residents of adjacent states is authorized pursuant to regulations issued by the Secretary of the Treasury under the Federal Gun Control Act of 1968, 18 U.S.C. §921 et seq., as the act is in effect on March 4, 1969.

(b) A resident of this state is permitted to purchase a rifle, shotgun, or ammunition in an adjacent state as expressly authorized pursuant to the regulations issued under the Federal Gun Control Act of 1968, 18 U.S.C. §921 et seq., as the act is in effect on March 4, 1969.

5-73-129. Furnishing a handgun or a prohibited weapon to a felon.

(a) A person commits the offense of furnishing a handgun to a felon if he or she sells, barters, leases, gives, rents, or otherwise furnishes a handgun to a person who he or she knows has been found guilty of or pleaded guilty or nolo contendere to a felony.

(b) A person commits the offense of furnishing a prohibited weapon to a felon if he or she sells, barters, leases, gives, rents, or otherwise furnishes:

(1) A sawed-off shotgun or rifle;

(2) A firearm that has been specially made or specially adapted for silent discharge;

(3) A machine gun;

(4) A bomb;

(5) Metal knuckles;

(6) A defaced firearm, as defined in § 5-73-107; or

(7) Other implement for the infliction of serious physical injury or death that serves no common lawful purpose, to a person who has been found guilty of or who has pleaded guilty or nolo contendere to a felony.

(c) Furnishing a handgun or a prohibited weapon to a felon is a Class B felony.

5-73-131. Possession or use of weapons by incarcerated persons.

(a) A person commits the offense of possession or use of weapons by incarcerated persons if, without approval of custodial authority he or she uses, possesses, makes, repairs, sells, or otherwise deals in any weapon, including, but not limited to, any bomb, firearm, knife, or other implement for the infliction of serious physical injury or death and that serves no common lawful purpose, while incarcerated in the Department of Correction, the Department of Community Correction, or a county or municipal jail or detention facility.

(b) Possession or use of weapons by incarcerated persons is a Class D felony.

(c) This section is not applicable to possession of a weapon by an incarcerated person before he or she completes the standard booking and search procedures in a jail facility after arrest.

5-73-132. Sale, rental, or transfer of firearm to person prohibited from possessing firearms.

(a) A person shall not sell, rent, or transfer a firearm to any person who he or she knows is prohibited by state or federal law from possessing the firearm.

(b)(1) Violation of this section is a Class A misdemeanor, unless the firearm is:

(A) A handgun;

(B) A sawed-off or short-barreled shotgun, as defined in § 5-1-102;

(C) A sawed-off or short-barreled rifle, as defined in § 5-1-102;

(D) A firearm that has been specially made or specially adapted for silent discharge;

(E) A machine gun;

(F) An explosive or incendiary device, as defined in § 5-71-301;

(G) A defaced firearm, as defined in § 5-73-107; or

(H) Other implement for the infliction of serious physical injury or death that serves no common lawful purpose.

(2) If the firearm is listed in subdivision (b)(1) of this section, a violation of this section is a Class B felony.

5-73-133. Possession of a taser stun gun.

(a) As used in this section, "taser stun gun" means any device that:

(1) Is powered by an electrical charging unit such as a battery; and

(2) Either:

(A) Emits an electrical charge in excess of twenty thousand (20,000) volts; or

(B) Is otherwise capable of incapacitating a person by an electrical charge.

(b)(1) No person who is eighteen (18) years of age or under may purchase or possess a taser stun gun.

(2) No person shall sell, barter, lease, give, rent, or otherwise furnish a taser stun gun to a person who is eighteen (18) years of age or under.

(c) Any law enforcement officer using a taser stun gun shall be properly trained in the use of the taser stun gun and informed of any danger or risk of serious harm and injury that may be caused by the use of the taser stun gun on a person.

(d)(1) A person who violates subdivision (b)(1) of this section is deemed guilty of an unclassified misdemeanor punishable by a fine of not less than five hundred dollars ($500) nor more than one thousand dollars ($1,000).

(2) A person who violates subdivision (b)(2) of this section is deemed guilty of a Class B felony.

Subchapter 2. Uniform Machine Gun Act

5-73-201. Title.

This subchapter may be cited as the "Uniform Machine Gun Act.".

5-73-202 Definitions.

As used in this sub-chapter:

(1) "Crime of violence" means any of the following crimes or an attempt to commit any of them:

(A) Murder;

(B) Manslaughter;

(C) Kidnapping;

(D) Rape;

(E) Mayhem;

(F) Assault to do great bodily harm;

(G) Robbery;

(H) Burglary;

(I) Housebreaking;

(J) Breaking and entering; and

(K) Larceny.

(2) "Machine gun" means a weapon of any description by whatever name known, loaded or unloaded, from which more than five (5) shots or bullets may be rapidly, or automatically, or semi-automatically, discharged from a magazine, by a single function of the firing device; and

(3) "Person" includes a firm, partnership, association, or corporation.

5-73-203. Uniformity of interpretation.

This subchapter shall be so interpreted and construed as to effectuate its general purpose to make uniform the law of those states which enact it.

5-73-204. Possession or use for offensive or aggressive purposes unlawful.

Possession or use of a machine gun for offensive or aggressive purpose is declared to be a crime punishable by imprisonment in the state penitentiary for a term of not less than ten (10) years.

5-73-205. Presumption of offensive or aggressive purpose.

(a) Possession or use of a machine gun is presumed to be for an offensive or aggressive purpose:

(1) When the machine gun is on premises not owned or rented for bona fide permanent residence or business occupancy by the person in whose possession the machine gun may be found;

(2) When in the possession of or used by an unnaturalized foreign-born person or a person who has been convicted of a crime of violence in any court of record, state or federal, of the United States of America, its territories or insular possessions;

(4) When empty or loaded pistol shells of 30 (.30 in. or 7.63 mm.) or larger caliber which have been or are susceptible of use in the machine gun are found in the immediate vicinity of the machine gun.

(b) A machine gun is exempt from the presumption of offensive or aggressive purpose if:

(1) The machine gun has been registered to a corporation in the business of manufacturing ammunition or a representative of the corporation under the National Firearms Act, 26 U.S.C. § 5801 et seq., or the Gun Control , 18 U.S.C. §921 et seq.;

(2) The machine gun is being used primarily to test ammunition in a non-offensive and non-aggressive manner by the corporation or the corporation's representative that the machine gun is registered to; and

(3) The corporation or the corporation's representative is not prohibited from the possession of a firearm by any state or federal law.

5-73-206. Evidence of possession or use.

The presence of a machine gun in any room, boat, or vehicle is evidence of the possession or use of the machine gun by each person occupying the room, boat, or vehicle where the weapon is found.

5-73-207. Manufacture for military, non-aggressive, or nonoffensive use. Nothing contained in this subchapter prohibits or interferes with:

(1) The manufacture for and sale of machine guns to the military forces or the peace officers of the United States or of any political subdivision of the United States, or the transportation required for that purpose;

(2) The possession of a machine gun for scientific purpose, or the possession of a machine gun not usable as a weapon and possessed as a curiosity, ornament, or keepsake; or

(3) The possession of a machine gun other than one adapted to use pistol cartridges of 30 (.30 in. or 7.63 mm.) or larger caliber, for a purpose manifestly not aggressive or offensive.

5-73-208. Registration by manufacturers.

(a) Every manufacturer shall keep a register of all machine guns manufactured or handled by the manufacturer.

(b) This register shall show:

(1) The model and serial number, date of manufacture, sale, loan, gift, delivery, or receipt, of every machine gun, the name, address, and occupation of the person to whom the machine gun was sold, loaned, given, or delivered, or from whom it was received; and

(2) The purpose for which it was acquired by the person to whom the machine gun was sold, loaned, given, or delivered, or from whom received.

(c) Upon demand every manufacturer shall permit any marshal, sheriff, or police officer to inspect the manufacturer's entire stock of machine guns, parts, and supplies therefor, and shall produce the register, required by this section, for inspection.

(d) A violation of any provision of this section is punishable by a fine of not less than .... hundred dollars.

Subchapter 3. Concealed Handguns

5-73-305. Criminal penalty.

Any person who knowingly submits a false answer to any question on an application for a license issued pursuant to this subchapter, or who knowingly submits a false document when applying for a license issued pursuant to this subchapter upon conviction is guilty of a Class B misdemeanor.

5-73-306. Prohibited places.

No license to carry a concealed handgun issued pursuant to this subchapter authorizes any person to carry a concealed handgun into:

...

(14) Any school, college, community college, or university campus building or event, unless for the purpose of participating in an authorized firearms-related activity;...

5-73-307. List of license holders.

(a) The Department of Arkansas State Police shall maintain an automated listing of license holders and this information shall be available on-line, upon request, at any time, to any law enforcement agency through the Arkansas Crime Information Center.

(b) Nothing in this subchapter shall be construed to require or allow the registration, documentation, or providing of a serial number with regard to any firearm.

5-73-308. License -- Issuance or denial.

(a)(1)(A) The Director of the Department of Arkansas State Police may deny a license if within the preceding five (5) years the applicant has been found guilty of one (1) or more crimes of violence constituting a misdemeanor or for the offense of carrying a weapon.

(B) The director may revoke a license if the licensee has been found guilty of one (1) or more crimes of violence within the preceding three (3) years.

(2) Subdivision (a)(1) of this section does not apply to a misdemeanor that has been expunged or for which the imposition of sentence was suspended.

(3) Upon notification by any law enforcement agency or a court and subsequent written verification, the director shall suspend a license or the processing of an application for a license if the licensee or applicant is arrested or formally charged with a crime that would disqualify the licensee or applicant from having a license under this subchapter until final disposition of the case.

(b)(1) The director may deny a license if the sheriff or chief of police, if applicable, of the applicant's place of residence submits an affidavit that the applicant has been or is reasonably likely to be a danger to himself or herself or others or to the community at large as the result of the applicant's mental or psychological state, as demonstrated by past patterns of behavior or participation in an incident involving unlawful violence or threats of unlawful violence, or if the applicant is under a criminal investigation at the time of applying for a license.

(2) Within one hundred twenty (120) days after the date of receipt of the items listed in § 5-73-311(a), the director shall:

(A) Issue the license; or

(B) Deny the application based solely on the ground that the applicant fails to qualify under the criteria listed in this subchapter.

(3)(A) If the director denies the application, the director shall notify the applicant in writing, stating the grounds for denial.

(B) The decision of the director is final.

5-73-309. License -- Requirements.

The Director of the Department of Arkansas State Police shall issue a license to carry a concealed handgun if the applicant:

(1) Is a citizen of the United States;

(2)(A) Is a resident of the state and has been a resident continuously for ninety (90) days or longer immediately preceding the filing of the application.

(B) However, subdivision (2)(A) of this section does not apply to any:

(i) Retired city, county, state, or federal law enforcement officer; or

(ii) Active duty military personnel who submit documentation of their active duty status;

(3) Is twenty-one (21) years of age or older;

(4) Does not suffer from a mental or physical infirmity that prevents the safe handling of a handgun and has not threatened or attempted suicide;

(5)(A) Has not been convicted of a felony in a court of this state, of any other state, or of the United States without having been pardoned for conviction and had firearms possession rights restored.

(B) A record of a conviction that has been sealed or expunged under Arkansas law does not render an applicant ineligible to receive a concealed handgun license if:

(i) The applicant was sentenced prior to March 13, 1995; or

(ii) The order sealing or expunging the applicant's record of conviction complies with § 16-90-605;

(6) Is not subject to any federal, state, or local law that makes it unlawful to receive, possess, or transport any firearm, and has had his or her background check successfully completed through the Department of Arkansas State Police and the Federal Bureau of Investigation's National Instant Check System;

(7)(A) Does not chronically or habitually abuse a controlled substance to the extent that his or her normal faculties are impaired.

(B) It is presumed that an applicant chronically and habitually uses a controlled substance to the extent that his or her faculties are impaired if the applicant has been voluntarily or involuntarily committed to a treatment facility for the abuse of a controlled substance or has been found guilty of a crime under the provisions of the Uniform Controlled Substances Act, § 5-64-101 et seq., or a similar law of any other state or the United States relating to a controlled substance within the three-year period immediately preceding the date on which the application is submitted;

(8)(A) Does not chronically or habitually use an alcoholic beverage to the extent that his or her normal faculties are impaired.

(B) It is presumed that an applicant chronically and habitually uses an alcoholic beverage to the extent that his or her normal faculties are impaired if the applicant has been voluntarily or involuntarily committed as an alcoholic to a treatment facility or has been convicted of two (2) or more offenses related to the use of alcohol under a law of this state or similar law of any other state or the United States within the three-year period immediately preceding the date on which the application is submitted;

(9) Desires a legal means to carry a concealed handgun to defend himself or herself;

(10) Has not been adjudicated mentally incompetent;

(11) Has not been voluntarily or involuntarily committed to a mental institution or mental health treatment facility;

(12) Is not a fugitive from justice or does not have an active warrant for his or her arrest;

(13) Has satisfactorily completed a training course as prescribed and approved by the director; and

(14) Signs a statement of allegiance to the United States Constitution and the Arkansas Constitution.

5-73-314. Lost or destroyed license - Change of address.

(a) Within thirty (30) days after the changing of a permanent address, or within thirty (30) days after having a license or handgun lost or disposed of, the licensee shall notify the Director of the Department of Arkansas State Police in writing of the change or loss or disposition..

(b) If a concealed handgun license is lost or destroyed, the person to whom the license was issued shall comply with the provisions of subsection (a) of this section and may obtain a duplicate license or substitute license upon:

(1) Payment to the Department of Arkansas State police a fee established by the director under the Arkansas Administrative procedure Act § 25-15-201 et seq.; and

(2) Furnishing a notarized statement to the department that the handgun or license has been lost or disposed of

5-73-315. Possession of license - Identification of licensee.

(a) Any licensee possessing a valid license issued pursuant to this subchapter may carry a concealed handgun.

(b) The licensee shall

(1) Carry the license, together with valid identification, at any time when the licensee is carrying a concealed handgun; and

(2) Display both the license and proper identification upon demand by a law enforcement officer.

5-73-317. Rules and regulations.

The director is authorized to promulgate rules and regulations to permit the efficient administration of this subchapter.

Title 14. Local Government

Chapter 16. Powers of Counties Generally

14-16-504. Regulation by local unit of government.

(a) As used in this section, "local unit of government" means a city, town, or county.

(b)(1)(A) A local unit of government shall not enact any ordinance or regulation pertaining to, or regulate in any other manner, the ownership, transfer, transportation, carrying, or possession of firearms, ammunition for firearms, or components of firearms, except as otherwise provided in state or federal law.

(B) This shall not prevent the enactment of an ordinance regulating or forbidding the unsafe discharge of a firearm.

(c)(1) Notwithstanding subsection (b) of this section, the governing body of a local unit of government, following the proclamation by the Governor of a state of emergency, may enact an emergency ordinance regulating the transfer, transportation, or carrying of firearms or components of firearms.

(2) Such emergency ordinance shall not be effective for a period of more than twenty (20) days and shall be enacted by a two-thirds (2/3) majority of the governing body.

Chapter 54. Powers of Municipalities Generally

14-54-1411. Firearms and ammunition. [Same as §14-16-504, above.]

[Current through the 2008 First Extraordinary Session , June 18, 2008]

CALIFORNIA
CAL. CODE

California Penal Code

Part I. Of Crimes and Punishment

Title 15. Miscellaneous Crimes

Chapter 1. Schools

626.9. Possession of firearm in school zone or on grounds of public or private university or college; Exceptions

(a) This section shall be known, and may be cited, as the Gun-Free School Zone Act of 1995.

(b) Any person who possesses a firearm in a place that the person knows, or reasonably should know, is a school zone, as defined in paragraph (1) of subdivision (e), unless it is with the written permission of the school district superintendent, his or her designee, or equivalent school authority, shall be punished as specified in subdivision (f).

(c) Subdivision (b) does not apply to the possession of a firearm under any of the following circumstances:

(1) Within a place of residence or place of business or on private property, if the place of residence, place of business, or private property is not part of the school grounds and the possession of the firearm is otherwise lawful.

(2) When the firearm is an unloaded pistol, revolver, or other firearm capable of being concealed on the person and is in a locked container or within the locked trunk of a motor vehicle.

This section does not prohibit or limit the otherwise lawful transportation of any other firearm, other than a pistol, revolver, or other firearm capable of being concealed on the person, in accordance with state law.

(3) When the person possessing the firearm reasonably believes that he or she is in grave danger because of circumstances forming the basis of a current restraining order issued by a court against another person or persons who has or have been found to pose a threat to his or her life or safety. This subdivision may not apply when the circumstances involve a mutual restraining order issued pursuant to Division 10 (commencing with Section 6200) of the Family Code absent a factual finding of a specific threat to the person's life or safety. Upon a trial for violating subdivision (b), the trier of a fact shall determine whether the defendant was acting out of a reasonable belief that he or she was in grave danger.

(4) When the person is exempt from the prohibition against carrying a concealed firearm pursuant to subdivision (b), (d), (e), or (h) of Section 12027.

(d) Except as provided in subdivision (b), it shall be unlawful for any person, with reckless disregard for the safety of another, to discharge, or attempt to discharge, a firearm in a school zone, as defined in paragraph (1) of subdivision (e).

The prohibition contained in this subdivision does not apply to the discharge of a firearm to the extent that the conditions of paragraph (1) of subdivision (c) are satisfied.

(e) As used in this section, the following definitions shall apply:

(1) "School zone" means an area in, or on the grounds of, a public or private school providing instruction in kindergarten or grades 1 to 12, inclusive, or within a distance of 1,000 feet from the grounds of the public or private school.

(2) "Firearm" has the same meaning as that term is given in Section 12001.

(3) "Locked container" has the same meaning as that term is given in subdivision (c) of Section 12026.1.

(4) "Concealed firearm" has the same meaning as that term is given in Sections 12025 and 12026.1.

(f)(1) Any person who violates subdivision (b) by possessing a firearm in, or on the grounds of, a public or private school providing instruction in kindergarten or grades 1 to 12, inclusive, shall be punished by imprisonment in the state prison for two, three, or five years.

(2) Any person who violates subdivision (b) by possessing a firearm within a distance of 1,000 feet from the grounds of a public or private school providing instruction in kindergarten or grades 1 to 12, inclusive, shall be punished as follows:

(A) By imprisonment in the state prison for two, three, or five years, if any of the following circumstances apply:

(i) If the person previously has been convicted of any felony, or of any crime made punishable by Chapter 1 (commencing with Section 12000) of Title 2 of Part 4.

(ii) If the person is within a class of persons prohibited from possessing or acquiring a firearm pursuant to Section 12021 or 12021.1 of this code or Section 8100 or 8103 of the Welfare and Institutions Code.

(iii) If the firearm is any pistol, revolver, or other firearm capable of being concealed upon the person and the offense is punished as a felony pursuant to Section 12025.

(B) By imprisonment in a county jail for not more than one year or by imprisonment in the state prison for two, three, or five years, in all cases other than those specified in subparagraph (A).

(3) Any person who violates subdivision (d) shall be punished by imprisonment in the state prison for three, five, or seven years.

(g)(1) Every person convicted under this section for a misdemeanor violation of subdivision (b) who has been convicted previously of a misdemeanor offense enumerated in Section 12001.6 shall be punished by imprisonment in a county jail for not less than three months, or if probation is granted or if the execution or imposition of sentence is suspended, it shall be a condition thereof that he or she be imprisoned in a county jail for not less than three months.

(2) Every person convicted under this section of a felony violation of subdivision (b) or (d) who has been convicted previously of a misdemeanor offense enumerated in Section 12001.6, if probation is granted or if the execution of sentence is suspended, it shall be a condition thereof that he or she be imprisoned in a county jail for not less than three months.

(3) Every person convicted under this section for a felony violation of subdivision (b) or (d) who has been convicted previously of any felony, or of any crime made punishable by Chapter 1 (commencing with Section 12000) of Title 2 of Part 4, if probation is granted or if the execution or imposition of sentence is suspended, it shall be a condition thereof that he or she be imprisoned in a county jail for not less than three months.

(4) The court shall apply the three-month minimum sentence specified in this subdivision, except in unusual cases where the interests of justice would best be served by granting probation or suspending the execution or imposition of sentence without the minimum imprisonment required in this subdivision or by granting probation or suspending the execution or imposition of sentence with conditions other than those set forth in this subdivision, in which case the court shall specify on the record and shall enter on the minutes the circumstances indicating that the interests of justice would best be served by this disposition.

(h) Notwithstanding Section 12026, any person who brings or possesses a loaded firearm upon the grounds of a campus of, or buildings owned or operated for student housing, teaching, research, or administration by, a public or private university or college, that are contiguous or are clearly marked university property, unless it is with the written permission of the university or college president, his or her designee, or equivalent university or college authority, shall be punished by imprisonment in the state prison for two, three, or four years. Notwithstanding subdivision (k), a university or college shall post a prominent notice at primary entrances on noncontiguous property stating that firearms are prohibited on that property pursuant to this subdivision.

(i) Notwithstanding Section 12026, any person who brings or possesses a firearm upon the grounds of a campus of, or buildings owned or operated for student housing, teaching, research, or administration by, a public or private university or college, that are contiguous or are clearly marked university property, unless it is with the written permission of the university or college president, his or her designee, or equivalent university or college authority, shall be punished by imprisonment in the state prison for one, two, or three years. Notwithstanding subdivision (k), a university or college shall post a prominent notice at primary entrances on noncontiguous property stating that firearms are prohibited on that property pursuant to this subdivision.

(j) For purposes of this section, a firearm shall be deemed to be loaded when there is an unexpended cartridge or shell, consisting of a case that holds a charge of powder and a bullet or shot, in, or attached in any manner to, the firearm, including, but not limited to, in the firing chamber, magazine, or clip thereof attached to the firearm. A muzzle-loader firearm shall be deemed to be loaded when it is capped or primed and has a powder charge and ball or shot in the barrel or cylinder.

(k) This section does not require that notice be posted regarding the proscribed conduct.

(l) This section does not apply to a duly appointed peace officer as defined in Chapter 4.5 (commencing with Section 830) of Title 3 of Part 2, a full-time paid peace officer of another state or the federal government who is carrying out official duties while in California, any person summoned by any of these officers to assist in making arrests or preserving the peace while he or she is actually engaged in assisting the officer, a member of the military forces of this state or of the United States who is engaged in the performance of his or her duties, a person holding a valid license to carry the firearm pursuant to Article 3 (commencing with Section 12050) of Chapter 1 of Title 2 of Part 4, or an armored vehicle guard, engaged in the performance of his or her duties, as defined in subdivision (e) of Section 7521 of the Business and Professions Code.

(m) This section does not apply to a security guard authorized to carry a loaded firearm pursuant to Section 12031.

(n) This section does not apply to an existing shooting range at a public or private school or university or college campus.

(o) This section does not apply to an honorably retired peace officer authorized to carry a concealed or loaded firearm pursuant to subdivision (a) or (i) of Section 12027 or paragraph (1) or (8) of subdivision (b) of Section 12031.

626.95. Penalty for specified weapons offenses on playgrounds or youth centers

(a) Any person who is in violation of paragraph (2) of subdivision (a), or subdivision (b), of Section 417, or Section 12025 or 12031, upon the grounds of or within a playground, or a public or private youth center during hours in which the facility is open for business, classes, or school-related programs, or at any time when minors are using the facility, knowing that he or she is on or within those grounds, shall be punished by imprisonment in the state prison for one, two, or three years, or in a county jail not exceeding one year.

(b) State and local authorities are encouraged to cause signs to be posted around playgrounds and youth centers giving warning of prohibition of the possession of firearms upon the grounds of or within playgrounds or youth centers.

(c) For purposes of this section, the following definitions shall apply:

(1) "Playground" means any park or recreational area specifically designed to be used by children that has play equipment installed, including public grounds designed for athletic activities such as baseball, football, soccer, or basketball, or any similar facility located on public or private school grounds, or on city or county parks.

(2) "Youth center" means any public or private facility that is used to host recreational or social activities for minors while minors are present.

(d) It is the Legislature's intent that only an actual conviction of a felony of one of the offenses specified in this section would subject the person to firearms disabilities under the federal Gun Control Act of 1968 (P.L. 90-618; 18 U.S.C. Sec. 921).

Part 4. Prevention of Crimes and Apprehension of Criminals

Title 2. Control of Deadly Weapons

Chapter 1. Firearms

Article 1. General Provisions

12000. Citation of chapter This chapter shall be known and may be cited as "The Dangerous Weapons Control Law."

12001. Definitions

(a)(1) As used in this title, the terms "pistol," "revolver," and "firearm capable of being concealed upon the person" shall apply to and include any device designed to be used as a weapon, from which is expelled a projectile by the force of any explosion, or other form of combustion, and that has a barrel less than 16 inches in length. These terms also include any device that has a barrel 16 inches or more in length which is designed to be interchanged with a barrel less than 16 inches in length.

(2) As used in this title, the term "handgun" means any "pistol," "revolver," or "firearm capable of being concealed upon the person."

(b) As used in this title, "firearm" means any device, designed to be used as a weapon, from which is expelled through a barrel a projectile by the force of any explosion or other form of combustion.

(c) As used in Sections 12021, 12021.1, 12070, 12071, 12072, 12073, 12078, 12101, and 12801 of this code, and Sections 8100, 8101, and 8103 of the Welfare and Institutions Code, the term "firearm" includes the frame or receiver of the weapon.

(d) For the purposes of Sections 12025 and 12031, the term "firearm" also shall include any rocket, rocket propelled projectile launcher, or similar device containing any explosive or incendiary material whether or not the device is designed for emergency or distress signaling purposes.

(e) For purposes of Sections 12070, 12071, and paragraph (8) of subdivision (a), and subdivisions (b), (c), (d), and (f) of Section 12072, the term "firearm" does not include an unloaded firearm that is defined as an "antique firearm" in Section 921(a)(16) of Title 18 of the United States Code.

(f) Nothing shall prevent a device defined as a "handgun," "pistol," "revolver," or "firearm capable of being concealed upon the person" from also being found to be a short-barreled shotgun or a short-barreled rifle, as defined in Section 12020.

(g) For purposes of Sections 12551 and 12552, the term "BB device" means any instrument that expels a projectile, such as a BB or a pellet, not exceeding 6mm caliber, through the force of air pressure, gas pressure, or spring action, or any spot marker gun.

(h) As used in this title, "wholesaler" means any person who is licensed as a dealer pursuant to Chapter 44 (commencing with Section 921) of Title 18 of the United States Code and the regulations issued pursuant thereto who sells, transfers, or assigns firearms, or parts of firearms, to persons who are licensed as manufacturers, importers, or gunsmiths pursuant to Chapter 44 (commencing with Section 921) of Title 18 of the United States Code, or persons licensed pursuant to Section 12071, and includes persons who receive finished parts of firearms and assemble them into completed or partially completed firearms in furtherance of that purpose.

"Wholesaler" shall not include a manufacturer, importer, or gunsmith who is licensed to engage in those activities pursuant to Chapter 44 (commencing with Section 921) of Title 18 of the United States Code or a person licensed pursuant to Section 12071 and the regulations issued pursuant thereto. A wholesaler also does not include those persons dealing exclusively in grips, stocks, and other parts of firearms that are not frames or receivers thereof.

(i) As used in Section 12071 or 12072, "application to purchase" means any of the following:

(1) The initial completion of the register by the purchaser, transferee, or person being loaned the firearm as required by subdivision (b) of Section 12076.

(2) The initial completion and transmission to the department of the record of electronic or telephonic transfer by the dealer on the purchaser, transferee, or person being loaned the firearm as required by subdivision (c) of Section 12076.

(j) For purposes of Section 12023, a firearm shall be deemed to be "loaded" whenever both the firearm and the unexpended ammunition capable of being discharged from the firearm are in the immediate possession of the same person.

(k) For purposes of Sections 12021, 12021.1, 12025, 12070, 12072, 12073, 12078, 12101, and 12801 of this code, and Sections 8100, 8101, and 8103 of the Welfare and Institutions Code, notwithstanding the fact that the term "any firearm" may be used in those sections, each firearm or the frame or receiver of the same shall constitute a distinct and separate offense under those sections.

(l) For purposes of Section 12020, a violation of that section as to each firearm, weapon, or device enumerated therein shall constitute a distinct and separate offense.

(m) Each application that requires any firearms eligibility determination involving the issuance of any license, permit, or certificate pursuant to this title shall include two copies of the applicant's fingerprints on forms prescribed by the Department of Justice. One copy of the fingerprints may be submitted to the United States Federal Bureau of Investigation.

(n) As used in this chapter, a "personal handgun importer" means an individual who meets all of the following criteria:

(1) He or she is not a person licensed pursuant to Section 12071.

(2) He or she is not a licensed manufacturer of firearms pursuant to Chapter 44 (commencing with Section 921) of Title 18 of the United States Code.

(3) He or she is not a licensed importer of firearms pursuant to Chapter 44 (commencing with Section 921) of Title 18 of the United States Code and the regulations issued pursuant thereto.

(4) He or she is the owner of a handgun.

(5) He or she acquired that handgun outside of California.

(6) He or she moves into this state on or after January 1, 1998, as a resident of this state.

(7) He or she intends to possess that handgun within this state on or after January 1, 1998.

(8) The handgun was not delivered to him or her by a person licensed pursuant to Section 12071 who delivered that firearm following the procedures set forth in Section 12071 and subdivision (c) of Section 12072.

(9) He or she, while a resident of this state, had not previously reported his or her ownership of that handgun to the Department of Justice in a manner prescribed by the department that included information concerning him or her and a description of the firearm.

(10) The handgun is not a firearm that is prohibited by subdivision (a) of Section 12020.

(11) The handgun is not an assault weapon, as defined in Section 12276 or 12276.1.

(12) The handgun is not a machinegun, as defined in Section 12200.

(13) The person is 18 years of age or older.

(o) For purposes of paragraph (6) of subdivision (n):

(1) Except as provided in paragraph (2), residency shall be determined in the same manner as is the case for establishing residency pursuant to Section 12505 of the Vehicle Code.

(2) In the case of members of the Armed Forces of the United States, residency shall be deemed to be established when he or she was discharged from active service in this state.

(p) As used in this code, "basic firearms safety certificate" means a certificate issued by the Department of Justice pursuant to Article 8 (commencing with Section 12800) of Chapter 6 of Title 2 of Part 4, prior to January 1, 2003.

(q) As used in this code, "handgun safety certificate" means a certificate issued by the Department of Justice pursuant to Article 8 (commencing with Section 12800) of Chapter 6 of Title 2 of Part 4, as that article is operative on or after January 1, 2003.

(r) As used in this title, "gunsmith" means any person who is licensed as a dealer pursuant to Chapter 44 (commencing with Section 921) of Title 18 of the United States Code and the regulations issued pursuant thereto, who is engaged primarily in the business of repairing firearms, or making or fitting special barrels, stocks, or trigger mechanisms to firearms, or the agent or employee of that person.

(s) As used in this title, "consultant-evaluator" means a consultant or evaluator who, in the course of his or her profession is loaned firearms from a person licensed pursuant to Chapter 44 (commencing with Section 921) of Title 18 of the United States Code and the regulations issued pursuant thereto, for his or her research or evaluation, and has a current certificate of eligibility issued to him or her pursuant to Section 12071.

12001.5. Short-barreled shotguns or rifles Except as expressly provided in Section 12020, and solely in accordance with Section 12020, no person may manufacture, import into this state, keep for sale, offer for sale, give, lend, or possess any short-barreled shotgun or short-barreled rifle, as defined in Section 12020, and nothing else in this chapter shall be construed as authorizing the manufacture, importation into the state, keeping for sale, offering for sale, or giving, lending, or possession of any short-barreled shotgun or short-barreled rifle, as defined in Section 12020.

12001.6. Offenses involving violent use of firearm As used in this chapter, an offense which involves the violent use of a firearm includes any of the following:

(a) A violation of paragraph (2) or (3) of subdivision (a) of Section 245 or a violation of subdivision (d) of Section 245.

(b) A violation of Section 246.

(c) A violation of paragraph (2) of subdivision (a) of Section 417.

(d) A violation of subdivision (c) of Section 417.

Article 2. Unlawful Carrying and Possession of Weapons

12020. Manufacture, importation, sale, possession; Exceptions

(a) Any person in this state who does any of the following is punishable by imprisonment in a county jail not exceeding one year or in the state prison:

(1) Manufactures or causes to be manufactured, imports into the state, keeps for sale, or offers or exposes for sale, or who gives, lends, or possesses any cane gun or wallet gun, any undetectable firearm, any firearm which is not immediately recognizable as a firearm, any camouflaging firearm container, any ammunition which contains or consists of any flechette dart, any bullet containing or carrying an explosive agent, any ballistic knife, any multiburst trigger activator, any nunchaku, any short-barreled shotgun, any short-barreled rifle, any metal knuckles, any belt buckle knife, any leaded cane, any zip gun, any shuriken, any unconventional pistol, any lipstick case knife, any cane sword, any shobi-zue, any air gauge knife, any writing pen knife, any metal military practice hand grenade or metal replica hand grenade, or any instrument or weapon of the kind commonly known as a blackjack, slungshot, billy, sandclub, sap, or sandbag.

(2) Commencing January 1, 2000, manufactures or causes to be manufactured, imports into the state, keeps for sale, or offers or exposes for sale, or who gives, or lends, any large-capacity magazine.

(3) Carries concealed upon his or her person any explosive substance, other than fixed ammunition.

(4) Carries concealed upon his or her person any dirk or dagger.

However, a first offense involving any metal military practice hand grenade or metal replica hand grenade shall be punishable only as an infraction unless the offender is an active participant in a criminal street gang as defined in the Street Terrorism and Enforcement and Prevention Act (Chapter 11 (commencing with Section 186.20) of Title 7 of Part 1). A bullet containing or carrying an explosive agent is not a destructive device as that term is used in Section 12301.

(b) Subdivision (a) does not apply to any of the following:

(1) The sale to, purchase by, or possession of short-barreled shotguns or short-barreled rifles by police departments, sheriffs' offices, marshals' offices, the California Highway Patrol, the Department of Justice, or the military or naval forces of this state or of the United States for use in the discharge of their official duties or the possession of short-barreled shotguns and short-barreled rifles by peace officer members of a police department, sheriff's office, marshal's office, the California Highway Patrol, or the Department of Justice when on duty and the use is authorized by the agency and is within the course and scope of their duties and the peace officer has completed a training course in the use of these weapons certified by the Commission on Peace Officer Standards and Training.

(2) The manufacture, possession, transportation or sale of short-barreled shotguns or short-barreled rifles when authorized by the Department of Justice pursuant to Article 6 (commencing with Section 12095) of this chapter and not in violation of federal law.

(3) The possession of a nunchaku on the premises of a school which holds a regulatory or business license and teaches the arts of self-defense.

(4) The manufacture of a nunchaku for sale to, or the sale of a nunchaku to, a school which holds a regulatory or business license and teaches the arts of self-defense.

(5) Any antique firearm. For purposes of this section, "antique firearm" means any firearm not designed or redesigned for using rimfire or conventional center fire ignition with fixed ammunition and manufactured in or before 1898 (including any matchlock, flintlock, percussion cap, or similar type of ignition system or replica thereof, whether actually manufactured before or after the year 1898) and also any firearm using fixed ammunition manufactured in or before 1898, for which ammunition is no longer manufactured in the United States and is not readily available in the ordinary channels of commercial trade.

(6) Tracer ammunition manufactured for use in shotguns.

(7) Any firearm or ammunition that is a curio or relic as defined in Section 478.11 of Title 27 of the Code of Federal Regulations and which is in the possession of a person permitted to possess the items pursuant to Chapter 44 (commencing with Section 921) of Title 18 of the United States Code and the regulations issued pursuant thereto. Any person prohibited by Section 12021, 12021.1, or 12101 of this code or Section 8100 or 8103 of the Welfare and Institutions Code from possessing firearms or ammunition who obtains title to these items by bequest or intestate succession may retain title for not more than one year, but actual possession of these items at any time is punishable pursuant to Section 12021, 12021.1, or 12101 of this code or Section 8100 or 8103 of the Welfare and Institutions Code. Within the year, the person shall transfer title to the firearms or ammunition by sale, gift, or other disposition. Any person who violates this paragraph is in violation of subdivision (a).

(8) Any other weapon as defined in subsection (e) of Section 5845 of Title 26 of the United States Code and which is in the possession of a person permitted to possess the weapons pursuant to the federal Gun Control Act of 1968 (Public Law 90-618), as amended, and the regulations issued pursuant thereto. Any person prohibited by Section 12021, 12021.1, or 12101 of this code or Section 8100 or 8103 of the Welfare and Institutions Code from possessing these weapons who obtains title to these weapons by bequest or intestate succession may retain title for not more than one year, but actual possession of these weapons at any time is punishable pursuant to Section 12021, 12021.1, or 12101 of this code or Section 8100 or 8103 of the Welfare and Institutions Code. Within the year, the person shall transfer title to the weapons by sale, gift, or other disposition. Any person who violates this paragraph is in violation of subdivision (a). The exemption provided in this subdivision does not apply to pen guns.

(9) Instruments or devices that are possessed by federal, state, and local historical societies, museums, and institutional collections which are open to the public, provided that these instruments or devices are properly housed, secured from unauthorized handling, and, if the instrument or device is a firearm, unloaded.

(10) Instruments or devices, other than short-barreled shotguns or short-barreled rifles, that are possessed or utilized during the course of a motion picture, television, or video production or entertainment event by an authorized participant therein in the course of making that production or event or by an authorized employee or agent of the entity producing that production or event.

(11) Instruments or devices, other than short-barreled shotguns or short-barreled rifles, that are sold by, manufactured by, exposed or kept for sale by, possessed by, imported by, or lent by persons who are in the business of selling instruments or devices listed in subdivision (a) solely to the entities referred to in paragraphs (9) and (10) when engaging in transactions with those entities.

(12) The sale to, possession of, or purchase of any weapon, device, or ammunition, other than a short-barreled rifle or short-barreled shotgun, by any federal, state, county, city and county, or city agency that is charged with the enforcement of any law for use in the discharge of their official duties, or the possession of any weapon, device, or ammunition, other than a short-barreled rifle or short-barreled shotgun, by peace officers thereof when on duty and the use is authorized by the agency and is within the course and scope of their duties.

(13) Weapons, devices, and ammunition, other than a short-barreled rifle or short-barreled shotgun, that are sold by, manufactured by, exposed or kept for sale by, possessed by, imported by, or lent by, persons who are in the business of selling weapons, devices, and ammunition listed in subdivision (a) solely to the entities referred to in paragraph (12) when engaging in transactions with those entities.

(14) The manufacture for, sale to, exposing or keeping for sale to, importation of, or lending of wooden clubs or batons to special police officers or uniformed security guards authorized to carry any wooden club or baton pursuant to Section 12002 by entities that are in the business of selling wooden batons or clubs to special police officers and uniformed security guards when engaging in transactions with those persons.

(15) Any plastic toy hand grenade, or any met-al military practice hand grenade or metal replica hand grenade that is a relic, curio, memorabilia, or display item, that is filled with a permanent in-ert substance or that is otherwise permanently altered in a manner that prevents ready modifi-cation for use as a grenade.

(16) Any instrument, ammunition, weapon, or device listed in subdivision (a) that is not a firearm that is found and possessed by a person who meets all of the following:

(A) The person is not prohibited from possessing firearms or ammunition pursuant to Section 12021 or 12021.1 or paragraph (1) of subdivision (b) of Section 12316 of this code or Section 8100 or 8103 of the Welfare and Institutions Code.

(B) The person possessed the instrument, ammunition, weapon, or device no longer than was necessary to deliver or transport the same to a law enforcement agency for that agency's disposition according to law.

(C) If the person is transporting the listed item, he or she is transporting the listed item to a law enforcement agency for disposition according to law.

(17) Any firearm, other than a short-barreled rifle or short-barreled shotgun, that is found and possessed by a person who meets all of the following:

(A) The person is not prohibited from possessing firearms or ammunition pursuant to Section 12021 or 12021.1 or paragraph (1) of subdivision (b) of Section 12316 of this code or Section 8100 or 8103 of the Welfare and Institutions Code.

(B) The person possessed the firearm no longer than was necessary to deliver or transport the same to a law enforcement agency for that agency's disposition according to law.

(C) If the person is transporting the firearm, he or she is transporting the firearm to a law enforcement agency for disposition according to law.

(D) Prior to transporting the firearm to a law enforcement agency, he or she has given prior notice to that law enforcement agency that he or she is transporting the firearm to that law enforcement agency for disposition according to law.

(E) The firearm is transported in a locked container as defined in subdivision (d) of Section 12026.2.

(18) The possession of any weapon, device, or ammunition, by a forensic laboratory or any authorized agent or employee thereof in the course and scope of his or her authorized activities.

(19) The sale of, giving of, lending of, importation into this state of, or purchase of, any large-capacity magazine to or by any federal, state, county, city and county, or city agency that is charged with the enforcement of any law, for use by agency employees in the discharge of their official duties whether on or off duty, and where the use is authorized by the agency and is within the course and scope of their duties.

(20) The sale to, lending to, transfer to, purchase by, receipt of, or importation into this state of, a large-capacity magazine by a sworn peace officer as defined in Chapter 4.5 (commencing with Section 830) of Title 3 of Part 2 who is authorized to carry a firearm in the course and scope of his or her duties.

(21) The sale or purchase of any large-capacity magazine to or by a person licensed pursuant to Section 12071.

(22) The loan of a lawfully possessed large-capacity magazine between two individuals if all of the following conditions are met:

(A) The person being loaned the large-capacity magazine is not prohibited by Section 12021, 12021.1, or 12101 of this code or Section 8100 or 8103 of the Welfare and Institutions Code from possessing firearms or ammunition.

(B) The loan of the large-capacity magazine occurs at a place or location where the possession of the large-capacity magazine is not otherwise prohibited and the person who lends the large-capacity magazine remains in the accessible vicinity of the person to whom the large-capacity magazine is loaned.

(23) The importation of a large-capacity magazine by a person who lawfully possessed the large-capacity magazine in the state prior to January 1, 2000, lawfully took it out of the state, and is returning to the state with the large-capacity magazine previously lawfully possessed in the state.

(24) The lending or giving of any large-capacity magazine to a person licensed pursuant to Section 12071, or to a gunsmith, for the purposes of maintenance, repair, or modification of that large-capacity magazine.

(25) The return to its owner of any large-capacity magazine by a person specified in paragraph (24).

(26) The importation into this state of, or sale of, any large-capacity magazine by a person who has been issued a permit to engage in those activities pursuant to Section 12079, when those activities are in accordance with the terms and conditions of that permit.

(27) The sale of, giving of, lending of, importation into this state of, or purchase of, any large-capacity magazine, to or by entities that operate armored vehicle businesses pursuant to the laws of this state.

(28) The lending of large-capacity magazines by the entities specified in paragraph (27) to their authorized employees, while in the course and scope of their employment for purposes that pertain to the entity's armored vehicle business.

(29) The return of those large-capacity magazines to those entities specified in paragraph (27) by those employees specified in paragraph (28).

(30)(A) The manufacture of a large-capacity magazine for any federal, state, county, city and county, or city agency that is charged with the enforcement of any law, for use by agency employees in the discharge of their official duties whether on or off duty, and where the use is authorized by the agency and is within the course and scope of their duties.

(B) The manufacture of a large-capacity magazine for use by a sworn peace officer as defined in Chapter 4.5 (commencing with Section 830) of Title 3 of Part 2 who is authorized to carry a firearm in the course and scope of his or her duties.

(C) The manufacture of a large-capacity magazine for export or for sale to government agencies or the military pursuant to applicable federal regulations.

(31) The loan of a large-capacity magazine for use solely as a prop for a motion picture, television, or video production.

(32) The purchase of a large-capacity magazine by the holder of a special weapons permit issued pursuant to Section 12095, 12230, 12250, 12286, or 12305, for any of the following purposes:

(A) For use solely as a prop for a motion picture, television, or video production.

(B) For export pursuant to federal regulations.

(C) For resale to law enforcement agencies, government agencies, or the military, pursuant to applicable federal regulations.

(c)(1) As used in this section, a "short-barreled shotgun" means any of the following:

(A) A firearm which is designed or redesigned to fire a fixed shotgun shell and having a barrel or barrels of less than 18 inches in length.

(B) A firearm which has an overall length of less than 26 inches and which is designed or redesigned to fire a fixed shotgun shell.

(C) Any weapon made from a shotgun (whether by alteration, modification, or otherwise) if that weapon, as modified, has an overall length of less than 26 inches or a barrel or barrels of less than 18 inches in length.

(D) Any device which may be readily restored to fire a fixed shotgun shell which, when so restored, is a device defined in subparagraphs (A) to (C), inclusive.

(E) Any part, or combination of parts, designed and intended to convert a device into a device defined in subparagraphs (A) to (C), inclusive, or any combination of parts from which a device defined in subparagraphs (A) to (C), inclusive, can be readily assembled if those parts are in the possession or under the control of the same person.

(2) As used in this section, a "short-barreled rifle" means any of the following:

(A) A rifle having a barrel or barrels of less than 16 inches in length.

(B) A rifle with an overall length of less than 26 inches.

(C) Any weapon made from a rifle (whether by alteration, modification, or otherwise) if that weapon, as modified, has an overall length of less than 26 inches or a barrel or barrels of less than 16 inches in length.

(D) Any device which may be readily restored to fire a fixed cartridge which, when so restored, is a device defined in subparagraphs (A) to (C), inclusive.

(E) Any part, or combination of parts, designed and intended to convert a device into a device defined in subparagraphs (A) to (C), inclusive, or any combination of parts from which a device defined in subparagraphs (A) to (C), inclusive, may be readily assembled if those parts are in the possession or under the control of the same person. ...

(4) As used in this section, a "wallet gun" means any firearm mounted or enclosed in a case, resembling a wallet, designed to be or capable of being carried in a pocket or purse, if the firearm may be fired while mounted or enclosed in the case.

(5) As used in this section, a "cane gun" means any firearm mounted or enclosed in a stick, staff, rod, crutch, or similar device, designed to be, or capable of being used as, an aid in walking, if the firearm may be fired while mounted or enclosed therein.

(6) As used in this section, a "flechette dart" means a dart, capable of being fired from a firearm, that measures approximately one inch in length, with tail fins that take up approximately five-sixteenths of an inch of the body. ...

(9) As used in this section, a "camouflaging firearm container" means a container which meets all of the following criteria:

(A) It is designed and intended to enclose a firearm.

(B) It is designed and intended to allow the firing of the enclosed firearm by external controls while the firearm is in the container.

(C) It is not readily recognizable as containing a firearm.

"Camouflaging firearm container" does not include any camouflaging covering used while engaged in lawful hunting or while going to or returning from a lawful hunting expedition.

(10) As used in this section, a "zip gun" means any weapon or device which meets all of the following criteria:

(A) It was not imported as a firearm by an importer licensed pursuant to Chapter 44 (commencing with Section 921) of Title 18 of the United States Code and the regulations issued pursuant thereto.

(B) It was not originally designed to be a firearm by a manufacturer licensed pursuant to

Chapter 44 (commencing with Section 921) of Title 18 of the United States Code and the regulations issued pursuant thereto.

(C) No tax was paid on the weapon or device nor was an exemption from paying tax on that weapon or device granted under Section 4181 and Subchapters F (commencing with Section 4216) and G (commencing with Section 4221) of Chapter 32 of Title 26 of the United States Code, as amended, and the regulations issued pursuant thereto.

(D) It is made or altered to expel a projectile by the force of an explosion or other form of combustion. ...

(12) As used in this section, an "unconventional pistol" means a firearm that does not have a rifled bore and has a barrel or barrels of less than 18 inches in length or has an overall length of less than 26 inches. ...

(20) As used in this section, a "rifle" means a weapon designed or redesigned, made or remade, and intended to be fired from the shoulder and designed or redesigned and made or remade to use the energy of the explosive in a fixed cartridge to fire only a single projectile through a rifled bore for each single pull of the trigger.

(21) As used in this section, a "shotgun" means a weapon designed or redesigned, made or remade, and intended to be fired from the shoulder and designed or redesigned and made or remade to use the energy of the explosive in a fixed shotgun shell to fire through a smooth bore either a number of projectiles (ball shot) or a single projectile for each pull of the trigger.

(22) As used in this section, an "undetectable firearm" means any weapon which meets one of the following requirements:

(A) When, after removal of grips, stocks, and magazines, it is not as detectable as the Security Exemplar, by walk-through metal detectors calibrated and operated to detect the Security Exemplar.

(B) When any major component of which, when subjected to inspection by the types of X-ray machines commonly used at airports, does not generate an image that accurately depicts the shape of the component. Barium sulfate or other compounds may be used in the fabrication of the component.

(C) For purposes of this paragraph, the terms "firearm," "major component," and "Security Exemplar" have the same meanings as those terms are defined in Section 922 of Title 18 of the United States Code.

All firearm detection equipment newly installed in nonfederal public buildings in this state shall be of a type identified by either the United States Attorney General, the Secretary of Transportation, or the Secretary of the Treasury, as appropriate, as available state-of-the-art equipment capable of detecting an undetectable firearm, as defined, while distinguishing innocuous metal objects likely to be carried on one's person sufficient for reasonable passage of the public.

(23) As used in this section, a "multiburst trigger activator" means one of the following devices:

(A) A device designed or redesigned to be attached to a semiautomatic firearm which allows the firearm to discharge two or more shots in a burst by activating the device.

(B) A manual or power-driven trigger activating device constructed and designed so that when attached to a semiautomatic firearm it increases the rate of fire of that firearm. ...

(25) As used in this section, "large-capacity magazine" means any ammunition feeding device with the capacity to accept more than 10 rounds, but shall not be construed to include any of the following:

(A) A feeding device that has been permanently altered so that it cannot accommodate more than 10 rounds.

(B) A .22 caliber tube ammunition feeding device.

(C) A tubular magazine that is contained in a lever-action firearm.

(d) Knives carried in sheaths which are worn openly suspended from the waist of the wearer are not concealed within the meaning of this section.

12020.3. Fines based on color of firearm

Any person who, for commercial purposes, purchases, sells, manufacturers, ships, transports, distributes, or receives a firearm, where the coloration of the entire exterior surface of the firearm is bright orange or bright green, either singly, in combi-nation, or as the predominant color in combina-tion with other colors in any pattern, is liable for a civil fine in an action brought by the city attor-ney of the city or the district attorney for the county of not more than ten thousand dollars ($10,000).

12020.5. Advertising unlawful weapons

It shall be unlawful for any person, as defined in Section 12277, to advertise the sale of any weapon or device whose possession is prohibited by Section 12020, 12220, 12280, 12303, 12320, 12321, 12355, or 12520 in any newspaper, magazine, circular, form letter, or open publication that is published, distributed, or circulated in this state, or on any billboard, card, label, or other advertising medium, or by means of any other advertising device.

12021. Possession of firearm by convicted felon or other specified persons; Exemptions

(a)(1) Any person who has been convicted of a felony under the laws of the United States, of the State of California, or any other state, government, or country, or of an offense enumerated in subdivision (a), (b), or (d) of Section 12001.6, or who is addicted to the use of any narcotic drug, and who owns, purchases, receives, or has in his or her possession or under his or her custody or control any firearm is guilty of a felony.

(2) Any person who has two or more convictions for violating paragraph (2) of subdivision (a) of Section 417 and who owns, purchases, receives, or has in his or her possession or under his or her custody or control any firearm is guilty of a felony.

(b) Notwithstanding subdivision (a), any person who has been convicted of a felony or of an offense enumerated in Section 12001.6, when that conviction results from certification by the juvenile court for prosecution as an adult in an adult court under Section 707 of the Welfare and Institutions Code, and who owns or has in his or her possession or under his or her custody or control any firearm is guilty of a felony.

(c)(1) Except as provided in subdivision (a) or paragraph (2) of this subdivision, any person who has been convicted of a misdemeanor violation of Section 71, 76, 136.1, 136.5, or 140, subdivision (d) of Section 148, Section 171b, 171c, 171d, 186.28, 240, 241, 242, 243, 244.5, 245, 245.5, 246.3, 247, 273.5, 273.6, 417, 417.6, 422, 626.9, 646.9, 12023, or 12024, subdivision (b) or (d) of Section 12034, Section 12040, subdivision (b) of Section 12072, subdivision (a) of former Section 12100, Section 12220, 12320, or 12590, or Section 8100, 8101, or 8103 of the Welfare and Institutions Code, any firearm-related offense pursuant to Sections 871.5 and 1001.5 of the Welfare and Institutions Code, or of the conduct punished in paragraph (3) of subdivision (g) of Section 12072, and who, within 10 years of the conviction, owns, purchases, receives, or has in his or her possession or under his or her custody or control, any firearm is guilty of a public offense, which shall be punishable by imprisonment in a county jail not exceeding one year or in the state prison, by a fine not exceeding one thousand dollars ($1,000), or by both that imprisonment and fine. The court, on forms prescribed by the Department of Justice, shall notify the department of persons subject to this subdivision. However, the prohibition in this paragraph may be reduced, eliminated, or conditioned as provided in paragraph (2) or (3).

(2) Any person employed as a peace officer described in Section 830.1, 830.2, 830.31, 830.32, 830.33, or 830.5 whose employment or livelihood is dependent on the ability to legally possess a firearm, who is subject to the prohibition imposed by this subdivision because of a conviction under Section 273.5, 273.6, or 646.9, may petition the court only once for relief from this prohibition. The petition shall be filed with the court in which the petitioner was sentenced. If possible, the matter shall be heard before the same judge who sentenced the petitioner. Upon filing the petition, the clerk of the court shall set the hearing date and shall notify the petitioner and the prosecuting attorney of the date of the hearing. Upon making each of the following findings, the court may reduce or eliminate the prohibition, impose conditions on reduction or elimination of the prohibition, or otherwise grant relief from the prohibition as the court deems appropriate:

(A) Finds by a preponderance of the evidence that the petitioner is likely to use a firearm in a safe and lawful manner.

(B) Finds that the petitioner is not within a prohibited class as specified in subdivision (a), (b), (d), (e), or (g) or Section 12021.1, and the court is not presented with any credible evidence that the petitioner is a person described in Section 8100 or 8103 of the Welfare and Institutions Code.

(C) Finds that the petitioner does not have a previous conviction under this subdivision no matter when the prior conviction occurred.

In making its decision, the court shall consider the petitioner's continued employment, the interest of justice, any relevant evidence, and the totality of the circumstances. The court shall require, as a condition of granting relief from the prohibition under this section, that the petitioner agree to participate in counseling as deemed appropriate by the court. Relief from the prohibition shall not relieve any other person or entity from any liability that might otherwise be imposed. It is the intent of the Legislature that courts exercise broad discretion in fashioning appropriate relief under this paragraph in cases in which relief is warranted. However, nothing in this paragraph shall be construed to require courts to grant relief to any particular petitioner. It is the intent of the Legislature to permit persons who were convicted of an offense specified in Section 273.5, 273.6, or 646.9 to seek relief from the prohibition imposed by this subdivision.

[Publisher's Note: On June 4, 2004, the Bureau of Alcohol, Tobacco, Firearms, and Explosives (ATF) advised that individuals who may have had their convictions of a misdemeanor crime of domestic violence set aside or dismisssed pursuant to California Penal Code §1203.4 still remain prohibited from acquiring or possessing firearms under federal law, and firearms prohibition relief afforded to peace officers pursuant to §12021 (c)(2) or §1203.4 is ineffective (http://caag.state.ca.us/firearms/infobuls/).]

(3) Any person who is subject to the prohibition imposed by this subdivision because of a conviction of an offense prior to that offense being added to paragraph (1) may petition the court only once for relief from this prohibition. The petition shall be filed with the court in which the petitioner was sentenced. If possible, the matter shall be heard before the same judge that sentenced the petitioner. Upon filing the petition, the clerk of the court shall set the hearing date and notify the petitioner and the prosecuting attorney of the date of the hearing. Upon making each of the following findings, the court may reduce or eliminate the prohibition, impose conditions on reduction or elimination of the prohibition, or otherwise grant relief from the prohibition as the court deems appropriate:

(A) Finds by a preponderance of the evidence that the petitioner is likely to use a firearm in a safe and lawful manner.

(B) Finds that the petitioner is not within a prohibited class as specified in subdivision (a), (b), (d), (e), or (g) or Section 12021.1, and the court is not presented with any credible evidence that the petitioner is a person described in Section 8100 or 8103 of the Welfare and Institutions Code.

(C) Finds that the petitioner does not have a previous conviction under this subdivision, no matter when the prior conviction occurred.

In making its decision, the court may consider the interest of justice, any relevant evidence, and the totality of the circumstances. It is the intent of the Legislature that courts exercise broad discretion in fashioning appropriate relief under this paragraph in cases in which relief is warranted. However, nothing in this paragraph shall be construed to require courts to grant relief to any particular petitioner.

(4) Law enforcement officials who enforce the prohibition specified in this subdivision against a person who has been granted relief pursuant to paragraph (2) or (3) shall be immune from any liability for false arrest arising from the enforcement of this subdivision unless the person has in his or her possession a certified copy of the court order that granted the person relief from the prohibition. This immunity from liability shall not relieve any person or entity from any other liability that might otherwise be imposed.

(d)(1) Any person who, as an express condition of probation, is prohibited or restricted from owning, possessing, controlling, receiving, or purchasing a firearm and who owns, purchases, receives, or has in his or her possession or under his or her custody or control, any firearm but who is not subject to subdivision (a) or (c), is guilty of a public offense, which shall be punishable by imprisonment in a county jail not exceeding one year or in the state prison, by a fine not exceeding one thousand dollars ($1,000), or by both that imprisonment and fine. The court, on forms provided by the Department of Justice, shall notify the department of persons subject to this subdivision. The notice shall include a copy of the order of probation and a copy of any minute order or abstract reflecting the order and conditions of probation.

(2) For any person who is subject to subdivision (a), (b), or (c), the court shall, at the time judgment is imposed, provide on a form supplied by the Department of Justice, a notice to the defendant prohibited by this section from owning, purchasing, receiving, possessing, or having under his or her custody or control, any firearm. The notice shall inform the defendant of the prohibition regarding firearms and include a form to facilitate the transfer of firearms. Failure to provide the notice shall not be a defense to a violation of this section.

(e) Any person who (1) is alleged to have committed an offense listed in subdivision (b) of Section 707 of the Welfare and Institutions Code, an offense described in subdivision (b) of Section 1203.073, or any offense enumerated in paragraph (1) of subdivision (c), or any offense described in subdivision (a) of Section 12025, subdivision (a) of Section 12031, or subdivision (a) of Section 12034, and (2) is subsequently adjudged a ward of the juvenile court within the meaning of Section 602 of the Welfare and Institutions Code because the person committed an offense listed in subdivision (b) of Section 707 of the Welfare and Institutions Code, an offense described in subdivision (b) of Section 1203.073, any offense enumerated in paragraph (1) of subdivision (c), or any offense described in subdivision (a) of Section 12025, subdivision (a) of Section 12031, or subdivision (a) of Section 12034, shall not own, or have in his or her possession or under his or her custody or control, any firearm until the age of 30 years. A violation of this subdivision shall be punishable by imprisonment in a county jail not exceeding one year or in the state prison, by a fine not exceeding one thousand dollars ($1,000), or by both that imprisonment and fine. The juvenile court, on forms prescribed by the Department of Justice, shall notify the department of persons subject to this subdivision. Notwithstanding any other law, the forms required to be submitted to the department pursuant to this subdivision may be used to determine eligibility to acquire a firearm.

(f) Subdivision (a) shall not apply to a person who has been convicted of a felony under the laws of the United States unless either of the following criteria is satisfied:

(1) Conviction of a like offense under California law can only result in imposition of felony punishment.

(2) The defendant was sentenced to a federal correctional facility for more than 30 days, or received a fine of more than one thousand dollars ($1,000), or received both punishments.

(g)(1) Every person who purchases or receives, or attempts to purchase or receive, a firearm knowing that he or she is prohibited from doing so by a temporary restraining order or injunction issued pursuant to Section 527.6 or 527.8 of the Code of Civil Procedure, a protective order as defined in Section 6218 of the Family Code, a protective order issued pursuant to Section 136.2 or 646.91 of this code, or by a protective order issued pursuant to Section 15657.03 of the Welfare and Institutions Code, is guilty of a public offense, which shall be punishable by imprisonment in a county jail not exceeding one year or in the state prison, by a fine not exceeding one thousand dollars ($1,000), or by both that imprisonment and fine.

(2) Every person who owns or possesses a firearm knowing that he or she is prohibited from doing so by a temporary restraining order or injunction issued pursuant to Section 527.6 or 527.8 of the Code of Civil Procedure, a protective order as defined in Section 6218 of the Family Code, a protective order issued pursuant to Section 136.2 or 646.91 of this code, or by a protective order issued pursuant to Section 15657.03 of the Welfare and Institutions Code, is guilty of a public offense, which shall be punishable by imprisonment in a county jail not exceeding one year, by a fine not exceeding one thousand dollars ($1,000), or by both that imprisonment and fine.

(3) The Judicial Council shall provide notice on all protective orders that the respondent is prohibited from owning, possessing, purchasing, receiving, or attempting to purchase or receive a firearm while the protective order is in effect. The order shall also state that the firearm shall be relinquished to the local law enforcement agency for that jurisdiction or sold to a licensed gun dealer, and that proof of surrender or sale shall be filed within a specified time of receipt of the order. The order shall state the penalties for a violation of the prohibition. The order shall also state on its face the expiration date for relinquishment.

(4) If probation is granted upon conviction of a violation of this subdivision, the court shall impose probation consistent with the provisions of Section 1203.097.

(h)(1) A violation of subdivision (a), (b), (c), (d), or (e) is justifiable where all of the following conditions are met:

(A) The person found the firearm or took the firearm from a person who was committing a crime against him or her.

(B) The person possessed the firearm no longer than was necessary to deliver or transport the firearm to a law enforcement agency for that agency's disposition according to law.

(C) If the firearm was transported to a law enforcement agency, it was transported in accordance with paragraph (18) of subdivision (a) of Section 12026.2.

(D) If the firearm is being transported to a law enforcement agency, the person transporting the firearm has given prior notice to the law enforcement agency that he or she is transporting the firearm to the law enforcement agency for disposition according to law.

(2) Upon the trial for violating subdivision (a), (b), (c), (d), or (e), the trier of fact shall determine whether the defendant was acting within the provisions of the exemption created by this subdivision.

(3) The defendant has the burden of proving by a preponderance of the evidence that he or she comes within the provisions of the exemption created by this subdivision.

(i) Subject to available funding, the Attorney General, working with the Judicial Council, the California Alliance Against Domestic Violence, prosecutors, and law enforcement, probation, and parole officers, shall develop a protocol for the implementation of the provisions of this section. The protocol shall be designed to facilitate the enforcement of restrictions on firearm ownership, including provisions for giving notice to defendants who are restricted, provisions for informing those defendants of the procedures by which defendants shall dispose of firearms when required to do so, provisions explaining how defendants shall provide proof of the lawful disposition of firearms, and provisions explaining how defendants may obtain possession of seized firearms when legally permitted to do so pursuant to this section or any other provision of law. The protocol shall be completed on or before January 1, 2005.

12021.1. Possession of firearms by persons previously convicted of specified violent crimes

(a) Notwithstanding subdivision (a) of Section 12021, any person who has been previously convicted of any of the offenses listed in subdivision (b) and who owns or has in his or her possession or under his or her custody or control any firearm is guilty of a felony. A dismissal of an accusatory pleading pursuant to Section 1203.4a involving an offense set forth in subdivision (b) does not affect the finding of a previous conviction. If probation is granted, or if the imposition or execution of sentence is suspended, it shall be a condition of the probation or suspension that the defendant serve at least six months in a county jail.

(b) As used in this section, a violent offense includes any of the following:

(1) Murder or voluntary manslaughter.

(2) Mayhem.

(3) Rape.

(4) Sodomy by force, violence, duress, menace, or threat of great bodily harm.

(5) Oral copulation by force, violence, duress, menace, or threat of great bodily harm.

(6) Lewd acts on a child under the age of 14 years.

(7) Any felony punishable by death or imprisonment in the state prison for life.

(8) Any other felony in which the defendant inflicts great bodily injury on any person, other than an accomplice, that has been charged and proven, or any felony in which the defendant uses a firearm which use has been charged and proven.

(9) Attempted murder.

(10) Assault with intent to commit rape or robbery.

(11) Assault with a deadly weapon or instrument on a peace officer.

(12) Assault by a life prisoner on a noninmate.

(13) Assault with a deadly weapon by an inmate.

(14) Arson.

(15) Exploding a destructive device or any explosive with intent to injure.

(16) Exploding a destructive device or any explosive causing great bodily injury.

(17) Exploding a destructive device or any explosive with intent to murder.

(18) Robbery.

(19) Kidnapping.

(20) Taking of a hostage by an inmate of a state prison.

(21) Attempt to commit a felony punishable by death or imprisonment in the state prison for life.

(22) Any felony in which the defendant personally used a dangerous or deadly weapon.

(23) Escape from a state prison by use of force or violence.

(24) Assault with a deadly weapon or force likely to produce great bodily injury.

(25) Any felony violation of Section 186.22.

(26) Any attempt to commit a crime listed in this subdivision other than an assault.

(27) Any offense enumerated in subdivision (a), (b), or (d) of Section 12001.6.

(28) Carjacking.

(29) Any offense enumerated in subdivision (c) of Section 12001.6 if the person has two or more convictions for violating paragraph (2) of subdivision (a) of Section 417.

(c) Any person previously convicted of any of the offenses listed in subdivision (b) which conviction results from certification by the juvenile court for prosecution as an adult in adult court under the provisions of Section 707 of the Welfare and Institutions Code, who owns or has in his or her possession or under his or her custody or control any firearm is guilty of a felony. If probation is granted, or if the imposition or execution of sentence is suspended, it shall be a condition of the probation or suspension that the defendant serve at least six months in a county jail.

(d) The court shall apply the minimum sentence as specified in subdivisions (a) and (c) except in unusual cases where the interests of justice would best be served by granting probation or suspending the imposition or execution of sentence without the imprisonment required by subdivisions (a) and (c), or by granting probation or suspending the imposition or execution of sentence with conditions other than those set forth in subdivisions (a) and (c), in which case the court shall specify on the record and shall enter on the minutes the circumstances indicating that the interests of justice would best be served by the disposition.

12029. Weapons constituting nuisances; Confiscation and destruction Except as provided in Section 12020, blackjacks, slungshots, billies, nunchakus, sandclubs, sandbags, shurikens, metal knuckles, short-barreled shotguns or short-barreled rifles as defined in Section 12020, and any other item which is listed in subdivision (a) of Section 12020 and is not listed in subdivision (a) of Section 12028 are nuisances, and the Attorney General, district attorney, or city attorney may bring an action to enjoin the manufacture of, importation of, keeping for sale of, offering or exposing for sale, giving, lending, or possession of, any of the foregoing items. These weapons shall be subject to confiscation and summary destruction whenever found within the state. These weapons shall be destroyed in the same manner as other weapons described in Section 12028, except that upon the certification of a judge or of the district attorney that the ends of justice will be subserved thereby, the weapon shall be preserved until the necessity for its use ceases.

12035. Criminal storage of a firearm

(a) As used in this section, the following definitions apply:

(1) "Locking device" means a device that is designed to prevent the firearm from functioning and when applied to the firearm, renders the firearm inoperable.

(2) "Loaded firearm" has the same meaning as set forth in subdivision (g) of Section 12031.

(3) "Child" means a person under 18 years of age.

(4) "Great bodily injury" has the same meaning as set forth in Section 12022.7.

(5) "Locked container" has the same meaning as set forth in subdivision (d) of Section 12026.2.

(b)(1) Except as provided in subdivision (c), a person commits the crime of "criminal storage of a firearm of the first degree" if he or she keeps any loaded firearm within any premises that are under his or her custody or control and he or she knows or reasonably should know that a child is likely to gain access to the firearm without the permission of the child's parent or legal guardian and the child obtains access to the firearm and thereby causes death or great bodily injury to himself, herself, or any other person.

(2) Except as provided in subdivision (c), a person commits the crime of "criminal storage of a firearm of the second degree" if he or she keeps any loaded firearm within any premises that are under his or her custody or control and he or she knows or reasonably should know that a child is likely to gain access to the firearm without the permission of the child's parent or legal guardian and the child obtains access to the firearm and thereby causes injury, other than great bodily injury, to himself, herself, or any other person, or carries the firearm either to a public place or in violation of Section 417.

(c) Subdivision (b) shall not apply whenever any of the following occurs:

(1) The child obtains the firearm as a result of an illegal entry to any premises by any person.

(2) The firearm is kept in a locked container or in a location that a reasonable person would believe to be secure.

(3) The firearm is carried on the person or within such a close proximity thereto that the individual can readily retrieve and use the firearm as if carried on the person.

(4) The firearm is locked with a locking device that has rendered the firearm inoperable.

(5) The person is a peace officer or a member of the armed forces or National Guard and the child obtains the firearm during, or incidental to, the performance of the person's duties.

(6) The child obtains, or obtains and discharges, the firearm in a lawful act of self-defense or defense of another person, or persons.

(7) The person who keeps a loaded firearm on any premise that is under his or her custody or control has no reasonable expectation, based on objective facts and circumstances, that a child is likely to be present on the premises.

(d) Criminal storage of a firearm is punishable as follows:

(1) Criminal storage of a firearm in the first degree, by imprisonment in the state prison for 16 months, or two or three years, by a fine not exceeding ten thousand dollars ($10,000), or by both that imprisonment and fine; or by imprisonment in a county jail not exceeding one year, by a fine not exceeding one thousand dollars ($1,000), or by both that fine and imprisonment.

(2) Criminal storage of a firearm in the second degree, by imprisonment in a county jail not exceeding one year, by a fine not exceeding one thousand dollars ($1,000), or by both that imprisonment and fine.

(e) If the person who allegedly violated this section is the parent or guardian of a child who is injured or who dies as the result of an accidental shooting, the district attorney shall consider, among other factors, the impact of the injury or death on the person alleged to have violated this section when deciding whether to prosecute an alleged violation. It is the Legislature's intent that a parent or guardian of a child who is injured or who dies as the result of an accidental shooting shall be prosecuted only in those instances in which the parent or guardian behaved in a grossly negligent manner or where similarly egregious circumstances exist. This subdivision shall not otherwise restrict, in any manner, the factors that a district attorney may consider when deciding whether to prosecute alleged violations of this section.

(f) If the person who allegedly violated this section is the parent or guardian of a child who is injured or who dies as the result of an accidental shooting, no arrest of the person for the alleged violation of this section shall occur until at least seven days after the date upon which the accidental shooting occurred.

In addition to the limitation contained in this subdivision, a law enforcement officer shall consider the health status of a child who suffers great bodily injury as the result of an accidental shooting prior to arresting a person for a violation of this section, if the person to be arrested is the parent or guardian of the injured child.

The intent of this subdivision is to encourage law enforcement officials to delay the arrest of a parent or guardian of a seriously injured child while the child remains on life-support equipment or is in a similarly critical medical condition.

(g)(1) The fact that the person who allegedly violated this section attended a firearm safety training course prior to the purchase of the firearm that is obtained by a child in violation of this section shall be considered a mitigating factor by a district attorney when he or she is deciding whether to prosecute the alleged violation.

(2) In any action or trial commenced under this section, the fact that the person who allegedly violated this section attended a firearm safety training course prior to the purchase of the firearm that is obtained by a child in violation of this section, shall be admissible.

(h) Every person licensed under Section 12071 shall post within the licensed premises the notice required by paragraph (7) of subdivision (b) of that section, disclosing the duty imposed by this section upon any person who keeps a loaded firearm.

12036. Child access to firearms

(a) As used in this section, the following definitions shall apply:

(1) "Locking device" means a device that is designed to prevent the firearm from functioning and when applied to the firearm, renders the firearm inoperable.

(2) "Child" means a person under the age of 18 years.

(3) "Off-premises" means premises other than the premises where the firearm was stored.

(4) "Locked container" has the same meaning as set forth in subdivision (d) of Section 12026.2.

(b) A person who keeps a pistol, revolver, or other firearm capable of being concealed upon the person, loaded or unloaded, within any premise that is under his or her custody or control and he or she knows or reasonably should know that a child is likely to gain access to that firearm without the permission of the child's parent or legal guardian and the child obtains access to that firearm and thereafter carries that firearm off-premises, shall be punished by imprisonment in a county jail not exceeding one year, by a fine not exceeding one thousand dollars ($1,000), or by both that imprisonment and fine.

(c) A person who keeps any firearm within any premises that is under his or her custody or control and he or she knows or reasonably should know that a child is likely to gain access to the firearm without the permission of the child's parent or legal guardian and the child obtains access to the firearm and thereafter carries that firearm off-premises to any public or private preschool, elementary school, middle school, high school, or to any school-sponsored event, activity, or performance whether occurring on school grounds or elsewhere, shall be punished by imprisonment in a county jail not exceeding one year, by a fine not exceeding five thousand dollars ($5,000), or by both that imprisonment and fine.

(d) A pistol, revolver, or other firearm capable of being concealed upon the person that a child gains access to and carries off-premises in violation of this section shall be deemed "used in the commission of any misdemeanor as provided in this code or any felony" for the purpose of subdivision (b) of Section 12028 regarding the authority to confiscate firearms and other deadly weapons as a nuisance.

(e) This section shall not apply if any one of the following circumstances exists:

(1) The child obtains the firearm as a result of an illegal entry into any premises by any person.

(2) The firearm is kept in a locked container or in a location that a reasonable person would believe to be secure.

(3) The firearm is locked with a locking device that has rendered the firearm inoperable.

(4) The firearm is carried on the person within such a close range that the individual can readily retrieve and use the firearm as if carried on the person.

(5) The person is a peace officer or a member of the Armed Forces or National Guard and the child obtains the firearm during, or incidental to, the performance of the person's duties.

(6) The child obtains, or obtains and discharges, the firearm in a lawful act of self-defense or defense of another person or persons.

(7) The person who keeps a firearm has no reasonable expectation, based on objective facts and circumstances, that a child is likely to be present on the premises.

(f) If the person who allegedly violated this section is the parent or guardian of a child who is injured or who dies as the result of an accidental shooting, the district attorney shall consider, among other factors, the impact of the injury or death on the person alleged to have violated this section when deciding whether to prosecute the alleged violation. It is the Legislature's intent that a parent or guardian of a child who is injured or who dies as the result of an accidental shooting shall be prosecuted only in those instances in which the parent or guardian behaved in a grossly negligent manner or where similarly egregious circumstances exist. This subdivision shall not otherwise restrict, in any manner, the factors that a district attorney may consider when deciding whether to prosecute alleged violations of this section.

(g) If the person who allegedly violated this section is the parent or guardian of a child who is injured or who dies as the result of an accidental shooting, no arrest of the person for the alleged violation of this section shall occur until at least seven days after the date upon which the accidental shooting occurred.

In addition to the limitation contained in this subdivision, a law enforcement officer shall consider the health status of a child who suffers great bodily injury as the result of an accidental shooting prior to arresting a person for a violation of this section, if the person to be arrested is the parent or guardian of the injured child.

The intent of this subdivision is to encourage law enforcement officials to delay the arrest of a parent or guardian of a seriously injured child while the child remains on life-support equipment or is in a similarly critical medical condition.

(h)(1) The fact that the person who allegedly violated this section attended a firearm safety training course prior to the purchase of the firearm that is obtained by a child in violation of this section shall be considered a mitigating factor by a district attorney when he or she is deciding whether to prosecute the alleged violation.

(2) In any action or trial commenced under this section, the fact that the person who allegedly violated this section attended a firearm safety training course prior to the purchase of the firearm that is obtained by a child in violation of this section, shall be admissible.

(i) Every person licensed under Section 12071 shall post within the licensed premises the notice required by paragraph (7) of subdivision (b) of that section, disclosing the duty imposed by this section upon any person who keeps any firearm.

Article 3 Licenses to Carry Pistols and Revolvers

12050. Issuance; Revocation; Amendment

(a)(1)(A) The sheriff of a county, upon proof that the person applying is of good moral character, that good cause exists for the issuance, and that the person applying satisfies any one of the conditions specified in subparagraph (D) and has completed a course of training as described in subparagraph (E), may issue to that person a license to carry a pistol, revolver, or other firearm capable of being concealed upon the person in either one of the following formats:

(i) A license to carry concealed a pistol, revolver, or other firearm capable of being concealed upon the person.

(ii) Where the population of the county is less than 200,000 persons according to the most recent federal decennial census, a license to carry loaded and exposed in that county a pistol, revolver, or other firearm capable of being concealed upon the person.

(B) The chief or other head of a municipal police department of any city or city and county, upon proof that the person applying is of good moral character, that good cause exists for the issuance, and that the person applying is a resident of that city and has completed a course of training as described in subparagraph (E) may issue to that person a license to carry a pistol, revolver, or other firearm capable of being concealed upon the person in either one of the following formats:

(i) A license to carry concealed a pistol, revolver, or other firearm capable of being concealed upon the person.

(ii) Where the population of the county in which the city is located is less than 200,000 persons according to the most recent federal decennial census, a license to carry loaded and exposed in that county a pistol, revolver, or other firearm capable of being concealed upon the person.

(C) The sheriff of a county or the chief or other head of a municipal police department of any city or city and county, upon proof that the person applying is of good moral character, that good cause exists for the issuance, and that the person applying is a person who has been deputized or appointed as a peace officer pursuant to subdivision (a) or (b) of Section 830.6 by that sheriff or that chief of police or other head of a municipal police department, may issue to that person a license to carry concealed a pistol, revolver, or other firearm capable of being concealed upon the person. Direct or indirect fees for the issuance of a license pursuant to this subparagraph may be waived. The fact that an applicant for a license to carry a pistol, revolver, or other firearm capable of being concealed upon the person has been deputized or appointed as a peace officer pursuant to subdivision (a) or (b) of Section 830.6 shall be considered only for the purpose of issuing a license pursuant to this subparagraph, and shall not be considered for the purpose of issuing a license pursuant to subparagraph (A) or (B). ...

(b) A license may include any reasonable restrictions or conditions which the issuing authority deems warranted, including restrictions as to the time, place, manner, and circumstances under which the person may carry a pistol, revolver, or other firearm capable of being concealed upon the person.

(c) Any restrictions imposed pursuant to subdivision (b) shall be indicated on any license issued.

(d) A license shall not be issued if the Department of Justice determines that the person is within a prohibited class described in Section 12021 or 12021.1 of this code or Section 8100 or 8103 of the Welfare and Institutions Code.

(e)(1) The license shall be revoked by the local licensing authority if at any time either the local licensing authority is notified by the Department of Justice that a licensee is within a prohibited class described in Section 12021 or 12021.1 of this code or Section 8100 or 8103 of the Welfare and Institutions Code, or the local licensing authority determines that the person is within a prohibited class described in Section 12021 or 12021.1 of this code or Section 8100 or 8103 of the Welfare and Institutions Code.

(2) If at any time the Department of Justice determines that a licensee is within a prohibited class described in Section 12021 or 12021.1 of this code or Section 8100 or 8103 of the Welfare and Institutions Code, the department shall immediately notify the local licensing authority of the determination.

(3) If the local licensing authority revokes the license, the Department of Justice shall be notified of the revocation pursuant to Section 12053. The licensee shall also be immediately notified of the revocation in writing.

(f)(1) A person issued a license pursuant to this section may apply to the licensing authority for an amendment to the license to do one or more of the following:

(A) Add or delete authority to carry a particular pistol, revolver, or other firearm capable of being concealed upon the person.

(B) Authorize the licensee to carry concealed a pistol, revolver, or other firearm capable of being concealed upon the person.

(C) If the population of the county is less than 200,000 persons according to the most recent federal decennial census, authorize the licensee to carry loaded and exposed in that county a pistol, revolver, or other firearm capable of being concealed upon the person.

(D) Change any restrictions or conditions on the license, including restrictions as to the time, place, manner, and circumstances under which the person may carry a pistol, revolver, or other firearm capable of being concealed upon the person.

(2) When the licensee changes his or her address, the license shall be amended to reflect the new address and a new license shall be issued pursuant to paragraph (3).

(3) If the licensing authority amends the license, a new license shall be issued to the licensee reflecting the amendments.

(4)(A) The licensee shall notify the licensing authority in writing within 10 days of any change in the licensee's place of residence.

(B) If the license is one to carry concealed a pistol, revolver, or other firearm capable of being concealed upon the person, then it may not be revoked solely because the licensee changes his or her place of residence to another county if the licensee has not breached any conditions or restrictions set forth in the license or has not fallen into a prohibited class described in Section 12021 or 12021.1 of this code or Section 8100 or 8103 of the Welfare and Institutions Code. However, any license issued pursuant to subparagraph (A) or (B) of paragraph (1) of subdivision (a) shall expire 90 days after the licensee moves from the county of issuance if the licensee's place of residence was the basis for issuance of the license.

(C) If the license is one to carry loaded and exposed a pistol, revolver, or other firearm capable of being concealed upon the person, the license shall be revoked immediately if the licensee changes his or her place of residence to another county.

(5) An amendment to the license does not extend the original expiration date of the license and the license shall be subject to renewal at the same time as if the license had not been amended.

(6) An application to amend a license does not constitute an application for renewal of the license.

(g) Nothing in this article shall preclude the chief or other head of a municipal police department of any city from entering an agreement with the sheriff of the county in which the city is located for the sheriff to process all applications for licenses, renewals of licenses, and amendments to licenses, pursuant to this article.

12051. Applications; Form and contents; Punishment for false statements

(a)(1) The standard application form for licenses described in paragraph (3) shall require information from the applicant including, but not limited to, the name, occupation, residence and business address of the applicant, his or her age, height, weight, color of eyes and hair, and reason for desiring a license to carry the weapon. Applications for licenses shall be filed in writing, and signed by the applicant. Any license issued upon the application shall set forth the licensee's name, occupation, residence and business address, his or her age, height, weight, color of eyes and hair, the reason for desiring a license to carry the weapon, and shall, in addition, contain a description of the weapon or weapons authorized to be carried, giving the name of the manufacturer, the serial number, and the caliber. The license issued to the licensee may be laminated.

(2) Applications for amendments to licenses shall be filed in writing and signed by the applicant, and shall state what type of amendment is sought pursuant to subdivision (f) of Section 12050 and the reason for desiring the amendment.

(3)(A) Applications for amendments to licenses, applications for licenses, amendments to licenses, and licenses shall be uniform throughout the state, upon forms to be prescribed by the Attorney General. The Attorney General shall convene a committee composed of one representative of the California State Sheriffs' Association, one representative of the California Police Chiefs' Association, and one representative of the Department of Justice to review, and as deemed appropriate, revise the standard application form for licenses. The committee shall meet for this purpose if two of the committee's members deem that necessary. The application shall include a section summarizing the statutory provisions of state law that result in the automatic denial of a license.

(B) The forms shall contain a provision whereby the applicant attests to the truth of statements contained in the application.

(C) An applicant shall not be required to complete any additional application or form for a license, or to provide any information other than that necessary to complete the standard application form described in subparagraph (A), except to clarify or interpret information provided by the applicant on the standard application form.

(D) The standard application form described in subparagraph (A) is deemed to be a local form expressly exempt from the requirements of the Administrative Procedures Act, Chapter 3.5 (commencing with Section 11340) of Part 1 of Division 3 of Title 2 of the Government Code.

(b) Any person who files an application required by subdivision (a) knowing that statements contained therein are false is guilty of a misdemeanor.

(c) Any person who knowingly makes a false statement on the application regarding any of the following shall be guilty of a felony:

(1) The denial or revocation of a license, or the denial of an amendment to a license, issued pursuant to Section 12050.

(2) A criminal conviction.

(3) A finding of not guilty by reason of insanity.

(4) The use of a controlled substance.

(5) A dishonorable discharge from military service.

(6) A commitment to a mental institution.

(7) A renunciation of United States citizenship.

Article 4. Licenses to Sell Firearms

12070. Penalty for unlicensed sale, lease or transfer; Exceptions

(a) No person shall sell, lease, or transfer firearms unless he or she has been issued a license pursuant to Section 12071. Any person violating this section is guilty of a misdemeanor.

(b) Subdivision (a) does not include any of the following:

(1) The sale, lease, or transfer of any firearm by a person acting pursuant to operation of law, a court order, or pursuant to the Enforcement of Judgments Law (Title 9 (commencing with Section 680.010) of Part 2 of the Code of Civil Procedure), or by a person who liquidates a personal firearm collection to satisfy a court judgment.

(2) A person acting pursuant to subdivision (e) of Section 186.22a or subdivision (c) of Section 12028.

(3) The sale, lease, or transfer of a firearm by a person who obtains title to the firearm by intestate succession or by bequest or as a surviving spouse pursuant to Chapter 1 (commencing with Section 13500) of Part 2 of Division 8 of the Probate Code, provided the person disposes of the firearm within 60 days of receipt of the firearm.

(4) The infrequent sale, lease, or transfer of firearms.

(5) The sale, lease, or transfer of used firearms other than pistols, revolvers, or other firearms capable of being concealed upon the person, at gun shows or events, as specified in Section 12071, by a person other than a licensee or dealer, provided the person has a valid federal firearms license and a current certificate of eligibility issued by the Department of Justice, as specified in Section 12071, and provided all the sales, leases, or transfers fully comply with subdivision (d) of Section 12072. However, the person shall not engage in the sale, lease, or transfer of used firearms other than pistols, revolvers, or other firearms capable of being concealed upon the person at more than 12 gun shows or events in any calendar year and shall not sell, lease, or transfer more than 15 used firearms other than pistols, revolvers, or other firearms capable of being concealed upon the person at any single gun show or event. In no event shall the person sell more than 75 used firearms other than pistols, revolvers, or other firearms capable of being concealed upon the person in any calendar year.

A person described in this paragraph shall be known as a "Gun Show Trader."

The Department of Justice shall adopt regulations to administer this program and shall recover the full costs of administration from fees assessed applicants.

As used in this paragraph, the term "used firearm" means a firearm that has been sold previously at retail and is more than three years old.

(6) Deliveries, sales, or transfers of firearms between or to importers and manufacturers of firearms licensed to engage in business pursuant to Chapter 44 (commencing with Section 921) of Title 18 of the United States Code and the regulations issued pursuant thereto.

(7) The sale, delivery, or transfer of firearms by manufacturers or importers licensed pursuant to Chapter 44 (commencing with Section 921) of Title 18 of the United States Code and the regulations issued pursuant thereto to dealers or wholesalers.

(8) Deliveries and transfers of firearms made pursuant to Section 12028, 12028.5, or 12030.

(9) The loan of a firearm for the purposes of shooting at targets, if the loan occurs on the premises of a target facility which holds a business or regulatory license or on the premises of any club or organization organized for the purposes of practicing shooting at targets upon established ranges, whether public or private, if the firearm is at all times kept within the premises of the target range or on the premises of the club or organization.

(10) Sales, deliveries, or transfers of firearms by manufacturers, importers, or wholesalers licensed pursuant to Chapter 44 (commencing with Section 921) of Title 18 of the United States Code and the regulations issued pursuant thereto to persons who reside outside this state who are licensed pursuant to Chapter 44 (commencing with Section 921) of Title 18 of the United States Code and the regulations issued pursuant thereto, if the sale, delivery, or transfer is in accordance with Chapter 44 (commencing with Section 921) of Title 18 of the United States Code and the regulations issued pursuant thereto.

(11) Sales, deliveries, or transfers of firearms by persons who reside outside this state and are licensed outside this state pursuant to Chapter 44 (commencing with Section 921) of Title 18 of the United States Code and the regulations issued pursuant thereto to wholesalers, manufacturers, or importers, if the sale, delivery, or transfer is in accordance with Chapter 44 (commencing with Section 921) of Title 18 of the United States Code and the regulations issued pursuant thereto.

(12) Sales, deliveries, or transfers of firearms by wholesalers to dealers.

(13) Sales, deliveries, or transfers of firearms by persons who reside outside this state to persons licensed pursuant to Section 12071, if the sale, delivery, or transfer is in accordance with Chapter 44 (commencing with Section 921) of Title 18 of the United States Code, and the regulations issued pursuant thereto.

(14) Sales, deliveries, or transfers of firearms by persons who reside outside this state and are licensed pursuant to Chapter 44 (commencing with Section 921) of Title 18 of the United States Code and the regulations issued pursuant thereto to dealers, if the sale, delivery, or transfer is in accordance with Chapter 44 (commencing with Section 921) of Title 18 of the United States Code and the regulations issued pursuant thereto.

(15) The delivery, sale, or transfer of an unloaded firearm by one wholesaler to another wholesaler if that firearm is intended as merchandise in the receiving wholesaler's business.

(16) The loan of an unloaded firearm or the loan of a firearm loaded with blank cartridges for use solely as a prop for a motion picture, television, or video production or entertainment or theatrical event.

(17) The delivery of an unloaded firearm that is a curio or relic, as defined in Section 478.11 of Title 27 of the Code of Federal Regulations, by a person licensed as a collector pursuant to Chapter 44 (commencing with Section 921) of Title 18 of the United States Code and the regulations issued pursuant thereto with a current certificate of eligibility issued pursuant to Section 12071 to a dealer.

(c) (1) As used in this section, "infrequent" means:

(A) For pistols, revolvers, and other firearms capable of being concealed upon the person, less than six transactions per calendar year. For this purpose, "transaction" means a single sale, lease, or transfer of any number of pistols, revolvers, or other firearms capable of being concealed upon the person.

(B) For firearms other than pistols, revolvers, or other firearms capable of being concealed upon the person, occasional and without regularity.

(2) As used in this section, "operation of law" includes, but is not limited to, any of the following:

(A) The executor or administrator of an estate, if the estate includes firearms.

(B) A secured creditor or an agent or employee thereof when the firearms are possessed as collateral for, or as a result of, a default under a security agreement under the Commercial Code.

(C) A levying officer, as defined in Section 481.140, 511.060, or 680.260 of the Code of Civil Procedure.

(D) A receiver performing his or her functions as a receiver, if the receivership estate includes firearms.

(E) A trustee in bankruptcy performing his or her duties, if the bankruptcy estate includes firearms.

(F) An assignee for the benefit of creditors performing his or her functions as an assignee, if the assignment includes firearms.

(G) A transmutation of property between spouses pursuant to Section 850 of the Family Code.

(H) Firearms received by the family of a police officer or deputy sheriff from a local agency pursuant to Section 50081 of the Government Code.

(I) The transfer of a firearm by a law enforcement agency to the person who found the firearm where the delivery is to the person as the finder of the firearm pursuant to Article 1 (commencing with Section 2080) of Chapter 4 of Division 3 of the Civil Code.

12071.Application for local license to sell firearms at retail; Certificate of eligibility; Requirements and exemptions; Forfeiture of license; Centralized list of licensees; Inspection of dealers; Fee; Request for information

(a)(1) As used in this chapter, the term "licensee," "person licensed pursuant to Section 12071," or "dealer" means a person who has all of the following:

(A) A valid federal firearms license.

(B) Any regulatory or business license, or licenses, required by local government.

(C) A valid seller's permit issued by the State Board of Equalization.

(D) A certificate of eligibility issued by the Department of Justice pursuant to paragraph (4).

(E) A license issued in the format prescribed by paragraph (6).

(F) Is among those recorded in the centralized list specified in subdivision (e).

(2) The duly constituted licensing authority of a city, county, or a city and county shall accept applications for, and may grant licenses permitting, licensees to sell firearms at retail within the city, county, or city and county. The duly constituted licensing authority shall inform applicants who are denied licenses of the reasons for the denial in writing.

(3) No license shall be granted to any applicant who fails to provide a copy of his or her valid federal firearms license, valid seller's permit issued by the State Board of Equalization, and the certificate of eligibility described in paragraph (4).

(4) A person may request a certificate of eligibility from the Department of Justice and the Department of Justice shall issue a certificate to an applicant if the department's records indicate that the applicant is not a person who is prohibited from possessing firearms.

(5) The department shall adopt regulations to administer the certificate of eligibility program and shall recover the full costs of administering the program by imposing fees assessed to applicants who apply for those certificates.

(6) A license granted by the duly constituted licensing authority of any city, county, or city and county, shall be valid for not more than one year from the date of issuance and shall be in one of the following forms:

(A) In the form prescribed by the Attorney General.

(B) A regulatory or business license that states on its face "Valid for Retail Sales of Firearms" and is endorsed by the signature of the issuing authority.

(C) A letter from the duly constituted licensing authority having primary jurisdiction for the applicant's intended business location stating that the jurisdiction does not require any form of regulatory or business license or does not otherwise restrict or regulate the sale of firearms.

(7) Local licensing authorities may assess fees to recover their full costs of processing applications for licenses.

(b) A license is subject to forfeiture for a breach of any of the following prohibitions and requirements:

(1)(A) Except as provided in subparagraphs (B) and (C), the business shall be conducted only in the buildings designated in the license.

(B) A person licensed pursuant to subdivision (a) may take possession of firearms and commence preparation of registers for the sale, delivery, or transfer of firearms at gun shows or events, as defined in Section 478.100 of Title 27 of the Code of Federal Regulations, or its successsor, if the gun show or event is not conducted from any motorized or towed vehicle. A person conducting business pursuant to this subparagraph shall be entitled to conduct business as authorized herein at any gun show or event in the state without regard to the jurisdiction within this state that issued the license pursuant to subdivision (a), provided the person complies with (i) all applicable laws, including, but not limited to, the waiting period specified in subparagraph (A) of paragraph (3), and (ii) all applicable local laws, regulations, and fees, if any.

A person conducting business pursuant to this subparagraph shall publicly display his or her license issued pursuant to subdivision (a), or a facsimile thereof, at any gun show or event, as specified in this subparagraph.

(C) A person licensed pursuant to subdivision (a) may engage in the sale and transfer of firearms other than pistols, revolvers, or other firearms capable of being concealed upon the person, at events specified in subdivision (g) of Section 12078, subject to the prohibitions and restrictions contained in that subdivision.

A person licensed pursuant to subdivision (a) also may accept delivery of firearms other than pistols, revolvers, or other firearms capable of being concealed upon the person, outside the building designated in the license, provided the firearm is being donated for the purpose of sale or transfer at an auction or similar event specified in subdivision (g) of Section 12078.

(D) The firearm may be delivered to the purchaser, transferee, or person being loaned the firearm at one of the following places:

(i) The building designated in the license.

(ii) The places specified in subparagraph (B) or (C).

(iii) The place of residence of, the fixed place of business of, or on private property owned or lawfully possessed by, the purchaser, transferee, or person being loaned the firearm.

(2) The license or a copy thereof, certified by the issuing authority, shall be displayed on the premises where it can easily be seen.

(3) No firearm shall be delivered:

(A) Within 10 days of the application to purchase, or, after notice by the department pursuant to subdivision (d) of Section 12076, within 10 days of the submission to the department of any correction to the application, or within 10 days of the submission to the department of any fee required pursuant to subdivision (e) of Section 12076, whichever is later.

(B) Unless unloaded and securely wrapped or unloaded and in a locked container.

(C) Unless the purchaser, transferee, or person being loaned the firearm presents clear evidence of his or her identity and age to the dealer.

(D) Whenever the dealer is notified by the Department of Justice that the person is in a prohibited class described in Section 12021 or 12021.1 of this code or Section 8100 or 8103 of the Welfare and Institutions Code. The dealer shall make available to the person in the prohibited class a prohibited notice and transfer form, provided by the department, stating that the person is prohibited from owning or possessing a firearm, and that the person may obtain from the department the reason for the prohibition.

(4) No pistol, revolver, or other firearm or imitation thereof capable of being concealed upon the person, or placard advertising the sale or other transfer thereof, shall be displayed in any part of the premises where it can readily be seen from the outside.

(5) The licensee shall agree to and shall act properly and promptly in processing firearms transactions pursuant to Section 12082.

(6) The licensee shall comply with Sections 12073, 12076, and 12077, subdivisions (a) and (b) and paragraph (1) of subdivision (f) of Section 12072, and subdivision (a) of Section 12316.

(7) The licensee shall post conspicuously within the licensed premises the following warnings in block letters not less than one inch in height:

(A) "IF YOU KEEP A LOADED FIREARM WITHIN ANY PREMISES UNDER YOUR CUSTODY OR CONTROL, AND A PERSON UNDER 18 YEARS OF AGE OBTAINS IT AND USES IT, RESULTING IN INJURY OR DEATH, OR CARRIES IT TO A PUBLIC PLACE, YOU MAY BE GUILTY OF A MISDEMEANOR OR A FELONY UNLESS YOU STORED THE FIREARM IN A LOCKED CONTAINER OR LOCKED THE FIREARM WITH A LOCKING DEVICE, TO KEEP IT FROM TEMPORARILY FUNCTIONING."

(B) "IF YOU KEEP A PISTOL, REVOLVER, OR OTHER FIREARM CAPABLE OF BEING CONCEALED UPON THE PERSON, WITHIN ANY PREMISES UNDER YOUR CUSTODY OR CONTROL, AND A PERSON UNDER 18 YEARS OF AGE GAINS ACCESS TO THE FIREARM, AND CARRIES IT OFF-PREMISES, YOU MAY BE GUILTY OF A MISDEMEANOR, UNLESS YOU STORED THE FIREARM IN A LOCKED CONTAINER, OR LOCKED THE FIREARM WITH A LOCKING DEVICE, TO KEEP IT FROM TEMPORARILY FUNCTIONING."

(C) "IF YOU KEEP ANY FIREARM WITHIN ANY PREMISES UNDER YOUR CUSTODY OR CONTROL, AND A PERSON UNDER 18 YEARS OF AGE GAINS ACCESS TO THE FIREARM, AND CARRIES IT OFF-PREMISES TO A SCHOOL OR SCHOOL-SPONSORED EVENT, YOU MAY BE GUILTY OF A MISDEMEANOR, INCLUDING A FINE OF UP TO FIVE THOUSAND DOLLARS ($5,000), UNLESS YOU STORED THE FIREARM IN A LOCKED CONTAINER, OR LOCKED THE FIREARM WITH A LOCKING DEVICE."

(D) "DISCHARGING FIREARMS IN POORLY VENTILATED AREAS, CLEANING FIREARMS, OR HANDLING AMMUNITION MAY RESULT IN EXPOSURE TO LEAD, A SUBSTANCE KNOWN TO CAUSE BIRTH DEFECTS, REPRODUCTIVE HARM, AND OTHER SERIOUS PHYSICAL INJURY. HAVE ADEQUATE VENTILATION AT ALL TIMES. WASH HANDS THOROUGHLY AFTER EXPOSURE."

(E) "FEDERAL REGULATIONS PROVIDE THAT IF YOU DO NOT TAKE PHYSICAL POSSESSION OF THE FIREARM THAT YOU ARE ACQUIRING OWNERSHIP OF WITHIN 30 DAYS AFTER YOU COMPLETE THE INITIAL BACKGROUND CHECK PAPERWORK, THEN YOU HAVE TO GO THROUGH THE BACKGROUND CHECK PROCESS A SECOND TIME IN ORDER TO TAKE PHYSICAL POSSESSION OF THAT FIREARM."

(F) "NO PERSON SHALL MAKE AN APPLICATION TO PURCHASE MORE THAN ONE PISTOL, REVOLVER, OR OTHER FIREARM CAPABLE OF BEING CONCEALED UPON THE PERSON WITHIN ANY 30-DAY PERIOD AND NO DELIVERY SHALL BE MADE TO ANY PERSON WHO HAS MADE AN APPLICATION TO PURCHASE MORE THAN ONE PISTOL, REVOLVER, OR OTHER FIREARM CAPABLE OF BEING CONCEALED UPON THE PERSON WITHIN ANY 30-DAY PERIOD."

(8)(A) Commencing April 1, 1994, and until January 1, 2003, no pistol, revolver, or other firearm capable of being concealed upon the person shall be delivered unless the purchaser, transferee, or person being loaned the firearm presents to the dealer a basic firearms safety certificate.

(B) Commencing January 1, 2003, no dealer may deliver a handgun unless the person receiving the handgun presents to the dealer a valid handgun safety certificate. The firearms dealer shall retain a photocopy of the handgun safety certificate as proof of compliance with this requirement.

(C) Commencing January 1, 2003, no handgun may be delivered unless the purchaser, transferee, or person being loaned the firearm presents documentation indicating that he or she is a California resident. Satisfactory documentation shall include a utility bill from within the last three months, a residential lease, a property deed, or military permanent duty station orders indicating assignment within this state, or other evidence of residency as permitted by the Department of Justice. The firearms dealer shall retain a photocopy of the documentation as proof of compliance with this requirement.

(D) Commencing January 1, 2003, except as authorized by the department, no firearms dealer may deliver a handgun unless the recipient performs a safe handling demonstration with that handgun. The demonstration shall commence with the handgun unloaded and locked with the firearm safety device with which it is required to be delivered, if applicable. While maintaining muzzle awareness, that is, the firearm is pointed in a safe direction, preferably down at the ground, and trigger discipline, that is, the trigger finger is outside of the trigger guard and along side of the handgun frame, at all times, the handgun recipient shall correctly and safely perform the following:

(i) If the handgun is a semiautomatic pistol:

(I) Remove the magazine.

(II) Lock the slide back. If the model of firearm does not allow the slide to be locked back, pull the slide back, visually and physically check the chamber to ensure that it is clear.

(III) Visually and physically inspect the chamber, to ensure that the handgun is unloaded.

(IV) Remove the firearm safety device, if applicable. If the firearm safety device prevents any of the previous steps, remove the firearm safety device during the appropriate step.

(V) Load one bright orange, red, or other readily identifiable dummy round into the magazine. If no readily identifiable dummy round is available, an empty cartridge casing with an empty primer pocket may be used.

(VI) Insert the magazine into the magazine well of the firearm.

(VII) Manipulate the slide release or pull back and release the slide.

(VIII) Remove the magazine.

(IX) Visually inspect the chamber to reveal that a round can be chambered with the magazine removed.

(X) Lock the slide back to eject the bright orange, red, or other readily identifiable dummy round. If the handgun is of a model that does not allow the slide to be locked back, pull the slide back and physically check the chamber to ensure that the chamber is clear. If no readily identifiable dummy round is available, an empty cartridge casing with an empty primer pocket may be used.

(XI) Apply the safety, if applicable.

(XII) Apply the firearm safety device, if applicable. This requirement shall not apply to an Olympic competition pistol if no firearms safety device, other than a cable lock that the department has determined would damage the barrel of the pistol, has been approved for the pistol, and the pistol is either listed in paragraph (2) of subdivision (h) of Section 12132 or is subject to paragraph (3) of subdivision (h) of Section 12132.

(ii) If the handgun is a double-action revolver:

(I) Open the cylinder.

(II) Visually and physically inspect each chamber, to ensure that the revolver is unloaded.

(III) Remove the firearm safety device. If the firearm safety device prevents any of the previous steps, remove the firearm safety device during the appropriate step.

(IV) While maintaining muzzle awareness and trigger discipline, load one bright orange, red, or other readily identifiable dummy round into a chamber of the cylinder and rotate the cylinder so that the round is in the next-to-fire position. If no readily identifiable dummy round is available, an empty cartridge casing with an empty primer pocket may be used.

(V) Close the cylinder.

(VI) Open the cylinder and eject the round.

(VII) Visually and physically inspect each chamber to ensure that the revolver is unloaded.

(VIII) Apply the firearm safety device, if applicable. This requirement shall not apply to an Olympic competition pistol if no firearms safety device, other than a cable lock that the department has determined would damage the barrel of the pistol, has been approved for the pistol, and the pistol is either listed in paragraph (2) of subdivision (h) of Section 12132 or is subject to paragraph (3) of subdivision (h) of Section 12132.

(iii) If the handgun is a single-action revolver:

(I) Open the loading gate.

(II) Visually and physically inspect each chamber, to ensure that the revolver is unloaded.

(III) Remove the firearm safety device required to be sold with the handgun. If the firearm safety device prevents any of the previous steps, remove the firearm safety device during the appropriate step.

(IV) Load one bright orange, red, or other readily identifiable dummy round into a chamber of the cylinder, close the loading gate and rotate the cylinder so that the round is in the next-to-fire position. If no readily identifiable dummy round is available, an empty cartridge casing with an empty primer pocket may be used.

(V) Open the loading gate and unload the revolver.

(VI) Visually and physically inspect each chamber to ensure that the revolver is unloaded.

(VII) Apply the firearm safety device, if applicable. This requirement shall not apply to an Olympic competition pistol if no firearms safety device, other than a cable lock that the department has determined would damage the barrel of the pistol, has been approved for the pistol, and the pistol is either listed in paragraph (2) of subdivision (h) of Section 12132 or is subject to paragraph (3) of subdivision (h) of Section 12132.

(E) The recipient shall receive instruction regarding how to render that handgun safe in the event of a jam.

(F) The firearms dealer shall sign and date an affidavit stating that the requirements of subparagraph (D) have been met. The firearms dealer shall additionally obtain the signature of the handgun purchaser on the same affidavit. The firearms dealer shall retain the original affidavit as proof of compliance with this requirement.

(G) The recipient shall perform the safe handling demonstration for a department certified instructor.

(H) No demonstration shall be required if the dealer is returning the handgun to the owner of the handgun.

(I) Department certified instructors who may administer the safe handling demonstration shall meet the requirements set forth in subdivision (j) of Section 12804.

(J) The persons who are exempt from the requirements of subdivision (b) of Section 12801, pursuant to Section 12807, are also exempt from performing the safe handling demonstration.

(9) Commencing July 1, 1992, the licensee shall offer to provide the purchaser or transferee of a firearm, or person being loaned a firearm, with a copy of the pamphlet described in Section 12080 and may add the cost of the pamphlet, if any, to the sales price of the firearm.

(10) The licensee shall not commit an act of collusion as defined in Section 12072.

(11) The licensee shall post conspicuously within the licensed premises a detailed list of each of the following:

(A) All charges required by governmental agencies for processing firearm transfers required by Sections 12076, 12082, and 12806.

(B) All fees that the licensee charges pursuant to Sections 12082 and 12806.

(12) The licensee shall not misstate the amount of fees charged by a governmental agency pursuant to Sections 12076, 12082, and 12806.

(13) Except as provided in subparagraphs (B) and (C) of paragraph (1) of subdivision (b), all firearms that are in the inventory of the licensee shall be kept within the licensed location. The licensee shall report the loss or theft of any firearm that is merchandise of the licensee, any firearm that the licensee takes possession of pursuant to Section 12082, or any firearm kept at the licensee's place of business within 48 hours of discovery to the appropriate law enforcement agency in the city, county, or city and county where the licensee's business premises are located.

(14) Except as provided in subparagraphs (B) and (C) of paragraph (1) of subdivision (b), any time when the licensee is not open for business, all inventory firearms shall be stored in the licensed location. All firearms shall be secured using one of the following methods as to each particular firearm:

(A) Store the firearm in a secure facility that is a part of, or that constitutes, the licensee's business premises.

(B) Secure the firearm with a hardened steel rod or cable of at least one-eighth inch in diameter through the trigger guard of the firearm. The steel rod or cable shall be secured with a hardened steel lock that has a shackle. The lock and shackle shall be protected or shielded from the use of a boltcutter and the rod or cable shall be anchored in a manner that prevents the removal of the firearm from the premises.

(C) Store the firearm in a locked fireproof safe or vault in the licensee's business premises.

(15) The licensing authority in an unincorporated area of a county or within a city may impose security requirements that are more strict or are at a higher standard than those specified in paragraph (14).

(16) Commencing January 1, 1994, the licensee shall, upon the issuance or renewal of a license, submit a copy of the same to the Department of Justice.

(17) The licensee shall maintain and make available for inspection during business hours to any peace officer, authorized local law enforcement employee, or Department of Justice employee designated by the Attorney General, upon the presentation of proper identification, a firearms transaction record.

(18)(A) On the date of receipt, the licensee shall report to the Department of Justice in a format prescribed by the department the acquisition by the licensee of the ownership of a pistol, revolver, or other firearm capable of being concealed upon the person.

(B) The provisions of this paragraph shall not apply to any of the following transactions:

(i) A transaction subject to the provisions of subdivision (n) of Section 12078.

(ii) The dealer acquired the firearm from a wholesaler.

(iii) The dealer is also licensed as a secondhand dealer pursuant to Article 4 (commencing with Section 21625) of Chapter 9 of Division 8 of the Business and Professions Code.

(iv) The dealer acquired the firearm from a person who is licensed as a manufacturer or importer to engage in those activities pursuant to Chapter 44 (commencing with Section 921) of Title 18 of the United States Code and any regulations issued pursuant thereto.

(v) The dealer acquired the firearm from a person who resides outside this state who is licensed pursuant to Chapter 44 (commencing with Section 921) of Title 18 of the United States Code and any regulations issued pursuant thereto.

(19) The licensee shall forward in a format prescribed by the Department of Justice, information as required by the department on any firearm that is not delivered within the time period set forth in Section 478.102(c) of Title 27 of the Code of Federal Regulations.

(20)(A) Firearms dealers may require any agent who handles, sells, or delivers firearms to obtain and provide to the dealer a certificate of eligibility from the department pursuant to paragraph (4) of subdivision (a). The agent or employee shall provide on the application, the name and California firearms dealer number of the firearms dealer with whom he or she is employed.

(B) The department shall notify the firearms dealer in the event that the agent or employee who has a certificate of eligibility is or becomes prohibited from possessing firearms.

(C) If the local jurisdiction requires a background check of the agents or employees of the firearms dealer the agent or employee shall obtain a certificate of eligibility pursuant to subparagraph (A).

(D) Nothing in this paragraph shall be construed to preclude a local jurisdiction from conducting an additional background check pursuant to Section 11105 or prohibiting employment based on criminal history that does not appear as part of obtaining a certificate of eligibility, provided however, that the local jurisdiction may not charge a fee for the additional criminal history check.

(E) The licensee shall prohibit any agent who the licensee knows or reasonably should know is within a class of persons prohibited from possessing firearms pursuant to Section 12021 or 12021.1 of this code, or Section 8100 or 8103 of the Welfare and Institutions Code, from coming into contact with any firearm that is not secured and from accessing any key, combination, code, or other means to open any of the locking devices described in clause (ii) of subparagraph (G) of this paragraph.

(F) Nothing in this paragraph shall be construed as preventing a local government from enacting an ordinance imposing additional conditions on licensees with regard to agents.

(G) For purposes of this section, the following definitions shall apply:

(i) An "agent" is an employee of the licensee.

(ii) "Secured" means a firearm that is made inoperable in one or more of the following ways:

(I) The firearm is inoperable because it is secured by a firearms safety device listed on the department's roster of approved firearms safety devices pursuant to subdivision (d) of Section 12088 of this chapter.

(II) The firearm is stored in a locked gun safe or long-gun safe which meets the standards for department-approved gun safes set forth in Section 12088.2.

(III) The firearm is stored in a distinct locked room or area in the building that is used to store firearms that can only be unlocked by a key, a combination, or similar means.

(IV) The firearm is secured with a hardened steel rod or cable that is at least one-eighth of an inch in diameter through the trigger guard of the firearm. The steel rod or cable shall be secured with a hardened steel lock that has a shackle. The lock and shackle shall be protected or shielded from the use of a boltcutter and the rod or cable shall be anchored in a manner that prevents the removal of the firearm from the premises.

(c)(1) As used in this article, "clear evidence of his or her identity and age" means either of the following:

(A) A valid California driver's license.

(B) A valid California identification card issued by the Department of Motor Vehicles.

(2) As used in this section, a "secure facility" means a building that meets all of the following specifications:

(A) All perimeter doorways shall meet one of the following:

(i) A windowless steel security door equipped with both a dead bolt and a doorknob lock.

(ii) A windowed metal door that is equipped with both a dead bolt and a doorknob lock. If the window has an opening of five inches or more measured in any direction, the window shall be covered with steel bars of at least 1/2-inch diameter or metal grating of at least 9 gauge affixed to the exterior or interior of the door.

(iii) A metal grate that is padlocked and affixed to the licensee's premises independent of the door and doorframe.

(B) All windows are covered with steel bars.

(C) Heating, ventilating, air-conditioning, and service openings are secured with steel bars, metal grating, or an alarm system.

(D) Any metal grates have spaces no larger than six inches wide measured in any direction.

(E) Any metal screens have spaces no larger than three inches wide measured in any direction.

(F) All steel bars shall be no further than six inches apart.

(3) As used in this section, "licensed premises," "licensed place of business," "licensee's place of business," or "licensee's business premises" means the building designated in the license.

(4) For purposes of paragraph (17) of subdivision (b):

(A) A "firearms transaction record" is a record containing the same information referred to in subdivision (a) of Section 478.124, Section 478.124a, and subdivision (e) of Section 478.125 of Title 27 of the Code of Federal Regulations.

(B) A licensee shall be in compliance with the provisions of paragraph (17) of subdivision (b) if he or she maintains and makes available for inspection during business hours to any peace officer, authorized local law enforcement employee, or Department of Justice employee designated by the Attorney General, upon the presentation of proper identification, the bound book containing the same information referred to in Section 478.124a and subdivision (e) of Section 478.125 of Title 27 of the Code of Federal Regulations and the records referred to in subdivision (a) of Section 478.124 of Title 27 of the Code of Federal Regulations.

(d) Upon written request from a licensee, the licensing authority may grant an exemption from compliance with the requirements of paragraph (14) of subdivision (b) if the licensee is unable to comply with those requirements because of local ordinances, covenants, lease conditions, or similar circumstances not under the control of the licensee.

(e)(1) Except as otherwise provided in this paragraph, the Department of Justice shall keep a centralized list of all persons licensed pursuant to subparagraphs (A) to (E), inclusive, of paragraph (1) of subdivision (a),. The department may remove from this list any person who knowingly or with gross negligence violates this article. Up-on removal of a dealer from this list, notification shall be provided to local law enforcement and licensing authorities in the jurisdiction where the dealer's business is located.

(2) The department shall remove from the centralized list any person whose federal firearms license has expired or has been revoked.

(3) Information compiled from the list shall be made available, upon request, for the following purposes only:

(A) For law enforcement purposes.

(B) When the information is requested by a person licensed pursuant to Chapter 44 (commencing with Section 921) of Title 18 of the United States Code for determining the validity of the license for firearm shipments.

(C) When information is requested by a person promoting, sponsoring, operating, or otherwise organizing a show or event as defined in Section 478.100 of Title 27 of the Code of Federal Regulations, or its successor, who possesses a valid certificate of eligibility issued pursuant to Section 12071.1, if that information is requested by the person to determine the eligibility of a prospective participant in a gun show or event to conduct transactions as a firearms dealer pursuant to subparagraph (B) of paragraph (1) of subdivision (b).

(4) Information provided pursuant to paragraph (3) shall be limited to information necessary to corroborate an individual's current license status as being one of the following:

(A) A person licensed pursuant to subparagraphs (A) to (E), inclusive, of paragraph (1) of subdivision (a).

(B) A person licensed pursuant to Chapter 44 (commencing with Section 921) of Title 18 of the United States Code and who is not subject to the requirement that he or she be licensed pursuant to subparagraphs (A) to (E), inclusive, of paragraph (1) of subdivision (a).

(f) The Department of Justice may inspect dealers to ensure compliance with this article. The department may assess an annual fee, not to exceed one hundred fifteen dollars ($115), to cover the reasonable cost of maintaining the list described in subdivision (e), including the cost of inspections. Dealers whose place of business is in a jurisdiction that has adopted an inspection program to ensure compliance with firearms law shall be exempt from that portion of the department's fee that relates to the cost of inspections. The applicant is responsible for providing evidence to the department that the jurisdiction in which the business is located has the inspection program.

(g) The Department of Justice shall maintain and make available upon request information concerning the number of inspections conducted and the amount of fees collected pursuant to subdivision (f), a listing of exempted jurisdictions, as defined in subdivision (f), the number of dealers removed from the centralized list defined in subdivision (e), and the number of dealers found to have violated this article with knowledge or gross negligence.

(h) Paragraph (14) or (15) of subdivision (b) shall not apply to a licensee organized as a nonprofit public benefit or mutual benefit corporation organized pursuant to Part 2 (commencing with Section 5110) or Part 3 (commencing with Section 7110) of Division 2 of the Corporations Code, if both of the following conditions are satisfied:

(1) The nonprofit public benefit or mutual benefit corporation obtained the dealer's license solely and exclusively to assist that corporation or local chapters of that corporation in conducting auctions or similar events at which firearms are auctioned off to fund the activities of that corporation or the local chapters of the corporation.

(2) The firearms are not pistols, revolvers, or other firearms capable of being concealed upon the person.

12071.1 Requirement of certificate of eligibility to organize gun show

(a) No person shall produce, promote, sponsor, operate, or otherwise organize a gun show or event, as specified in subparagraph (B) of paragraph (1) of subdivision (b) of Section 12071, unless that person possesses a valid certificate of eligibility from the Department of Justice. Unless the department's records indicate that the applicant is a person prohibited from possessing firearms, a certificate of eligibility shall be issued by the Department of Justice to an applicant provided the applicant does all of the following:

(1) Certifies that he or she is familiar with the provisions of this section and Section 12071.4.

(2) Ensures that liability insurance is in effect for the duration of an event or show in an amount of not less than one million dollars ($1,000,000).

(3) Provides an annual list of the gun shows or events that the applicant plans to promote, produce, sponsor, operate, or otherwise organize during the year for which the certificate of eligibility is issued, including the date, time, and location of the gun shows or events.

(b) If during that year the information required by paragraph (3) of subdivision (a) changes, or additional gun shows or events will be promoted, produced, sponsored, operated, or otherwise organized by the applicant, the producer shall notify the Department of Justice no later than 30 days prior to the gun show or event.

(c) As used in this section, a "licensed gun show producer" means a person who has been issued a certificate of eligibility by the Department of Justice pursuant to subdivision (a). No regulations shall be required to implement this subdivision.

(d) The Department of Justice shall adopt regulations to administer the certificate of eligibility program under this section and shall recover the full costs of administering the program by fees assessed applicants who apply for certificates. A licensed gun show producer shall be assessed an annual fee of eighty-five dollars ($85) by the department.

(e)(1) A willful failure by a gun show producer to comply with any of the requirements of this section, except for the posting of required signs, shall be a misdemeanor punishable by a fine not to exceed two thousand dollars ($2,000), and shall render the producer ineligible for a gun show producer license for one year from the date of the conviction.

(2) The willful failure of a gun show producer to post signs as required by this section shall be a misdemeanor punishable by a fine not to exceed one thousand dollars ($1,000) for the first offense and not to exceed two thousand dollars ($2,000) for the second or subsequent offense, and with respect to the second or subsequent offense, shall render the producer ineligible for a gun show producer license for one year from the date of the conviction.

(3) Multiple violations charged pursuant to paragraph (1) arising from more than one gun show or event shall be grounds for suspension of a producer's certificate of eligibility pending adjudication of the violations.

(f) Prior to the commencement of a gun show or event, the producer thereof shall, upon written request, within 48 hours, or a later time specified by the requesting law enforcement agency, make available to the requesting law enforcement agency with jurisdiction over the facility, a complete and accurate list of all persons, entities, and organizations that have leased or rented, or are known to the producer to intend to lease or rent, any table, display space, or area at the gun show or event for the purpose of selling, leasing, or transferring firearms.

The producer shall thereafter, upon written request, for every day the gun show or event operates, within 24 hours, or a later time specified by the requesting law enforcement agency, make available to the requesting law enforcement agency with jurisdiction over the facility, an accurate, complete, and current list of the persons, entities, and organizations that have leased or rented, or are known to the producer to intend to lease or rent, any table, display space, or area at the gun show or event for the purpose of selling, leasing, or transferring firearms.

This subdivision applies to persons, entities, and organizations whether or not they participate in the entire gun show or event, or only a portion thereof.

(g) The information that may be requested by the law enforcement agency with jurisdiction over the facility, and that shall be provided by the producer upon request, may include, but is not limited to, the following information relative to a vendor who offers for sale firearms manufactured after December 31, 1898: his or her complete name, and a driver's license or identification card number.

(h) The producer and facility manager shall prepare an annual event and security plan and schedule that shall include, at a minimum, the following:

(1) The type of shows or events including, but not limited to, antique or general firearms.

(2) The estimated number of vendors offering firearms for sale or display.

(3) The estimated number of attendees.

(4) The number of entrances and exits at the gun show or event site.

(5) The location, dates, and times of the shows or events.

(6) The contact person and telephone number for both the producer and the facility.

(7) The number of sworn peace officers employed by the producer or the facilities manager who will be present at the show or event.

(8) The number of nonsworn security personnel employed by the producer or the facility's manager who will be present at the show or event.

(i) The annual event and security plan shall be submitted by either the producer or the facility's manager to the Department of Justice and the law enforcement agency with jurisdiction over the facility. Not later than 15 days prior to the commencement of the gun show or event, the producer shall submit to the department, the law enforcement agency with jurisdiction over the facility site, and the facility's manager a revised event and security plan if significant changes have been made since the annual plan was submitted, including a revised list of vendors that the producer knows, or reasonably should know, will be renting tables, space, or otherwise participating in the gun show or event. The event and security plan shall be approved by the facility's manager prior to the event or show after consultation with the law enforcement agency with jurisdiction over the facility. No gun show or event shall commence unless the requirements of this subdivision are met.

(j) The producer shall be responsible for informing prospective gun show vendors of the requirements of this section and of Section 12071.4 that apply to vendors.

(k) The producer shall, within seven calendar days of the commencement of the show or event, but not later than noon on Friday for a show or event held on a weekend, submit a list of all prospective vendors and designated firearms transfer agents who are licensed firearms dealers to the Department of Justice for the purpose of determining whether these prospective vendors and designated firearms transfer agents possess valid licenses and are thus eligible to participate as licensed dealers at the show or event. The department shall examine its records and if it determines that a dealer's license is not valid, it shall notify the show or event producer of that fact prior to the commencement of the show or event.

(l) If a licensed firearms dealer fails to cooperate with a producer or fails to comply with the applicable requirements of this section or Section 12071.4, that person shall not be allowed to participate in that show or event.

(m) If a producer fails to comply with subdivision (j) or (k), the gun show or event shall not commence until those requirements are met.

(n) All producers shall have written contracts with all gun show vendors selling firearms at the show or event.

(o) The producer shall require that signs be posted in a readily visible location at each public entrance to the show containing, but not limited to, the following notices:

(1) This gun show follows all federal, state, and local firearms and weapons laws without exception.

(2) All firearms carried onto the premises by members of the public will be checked, cleared of any ammunition, secured in a manner that prevents them from being operated, and an identification tag or sticker will be attached to the firearm prior to the person being allowed admittance to the show.

(3) No member of the public under the age of 18 years shall be admitted to the show unless accompanied by a parent, grandparent, or legal guardian.

(4) All firearms transfers between private parties at the show shall be conducted through a licensed dealer in accordance with applicable state and federal laws.

(5) Persons possessing firearms on this facility must have in their immediate possession government-issued photo identification, and display it upon request to any security officer or any peace officer, as defined in Section 830.

(p) The show producer shall post, in a readily visible location at each entrance to the parking lot at the show, signage that states: "The transfer of firearms on the parking lot of this facility is a crime."

(q) It is the intent of the Legislature that the certificate of eligibility program established pursuant to this section be incorporated into the certificate of eligibility program established pursuant to Section 12071 to the maximum extent practicable.

12071.4. Gun Show Enforcement and Security Act

(a) This section shall be known, and may be cited as, the Gun Show Enforcement and Security Act of 2000.

(b) All gun show or event vendors shall certify in writing to the producer that they:

(1) Will not display, possess, or offer for sale any firearms, knives, or weapons for which possession or sale is prohibited.

(2) Acknowledge that they are responsible for knowing and complying with all applicable federal, state, and local laws dealing with the possession and transfer of firearms.

(3) Will not engage in activities that incite or encourage hate crimes.

(4) Will process all transfers of firearms through licensed firearms dealers as required by state law.

(5) Will verify that all firearms in their possession at the show or event will be unloaded, and that the firearms will be secured in a manner that prevents them from being operated except for brief periods when the mechanical condition of a firearm is being demonstrated to a prospective buyer.

(6) Have complied with the requirements of subdivision (e).

(7) Will not display or possess black powder, or offer it for sale.

(c) All firearms transfers at the gun show or event shall be in accordance with applicable state and federal laws.

(d) Except for purposes of showing ammunition to a prospective buyer, ammunition at a gun show or event may be displayed only in closed original factory boxes or other closed containers.

(e) Prior to the commencement of a gun show or event, each vendor shall provide to the producer all of the following information relative to the vendor, the vendor's employees, and other persons, compensated or not, who will be working or otherwise providing services to the public at the vendor's display space if firearms manufactured after December 31, 1898, will be offered for sale:

(1) His or her complete name.

(2) His or her driver's license or state-issued identification card number.

(3) His or her date of birth.

The producer shall keep the information at the show's or event's onsite headquarters for the duration of the show or event, and at the producer's regular place of business for two weeks after the conclusion of the show or event, and shall make the information available upon request to any sworn peace officer for purposes of the officer's official law enforcement duties.

(f) Vendors and employees of vendors shall wear name tags indicating first and last name.

(g) No person at a gun show or event, other than security personnel or sworn peace officers, shall possess at the same time both a firearm and ammunition that is designed to be fired in the firearm. Vendors having those items at the show for sale or exhibition are exempt from this prohibition.

(h) No member of the public who is under the age of 18 years shall be admitted to, or be permitted to remain at, a gun show or event unless accompanied by a parent or legal guardian. Any member of the public who is under the age of 18 shall be accompanied by his or her parent, grandparent, or legal guardian while at the show or event.

(i) Persons other than show or event security personnel, sworn peace officers, or vendors, who bring firearms onto the gun show or event premises shall sign in ink the tag or sticker that is attached to the firearm prior to being allowed admittance to the show or event, as provided for in subdivision (j).

(j) All firearms carried onto the premises of a gun show or event by members of the public shall be checked, cleared of any ammunition, secured in a manner that prevents them from being operated, and an identification tag or sticker shall be attached to the firearm, prior to the person being allowed admittance to the show. The identification tag or sticker shall state that all firearms transfers between private parties at the show or event shall be conducted through a licensed dealer in accordance with applicable state and federal laws. The person possessing the firearm shall complete the following information on the tag before it is attached to the firearm:

(1) The gun owner's signature.

(2) The gun owner's printed name.

(3) The identification number from the gun owner's government-issued photo identification.

(k) All persons possessing firearms at the gun show or event shall have in his or her immediate possession, government-issued photo identification, and display it upon request, to any security officer, or any peace officer.

(l) Unless otherwise specified, a first violation of this section is an infraction. Any second or subsequent violation is a misdemeanor. Any person who commits an act which he or she knows to be a violation of this section is guilty of a misdemeanor for a first offense.

12072. Restrictions on transfer of firearms; Act of collusion; Punishment for violations

(a)(1) No person, corporation, or firm shall knowingly supply, deliver, sell, or give possession or control of a firearm to any person within any of the classes prohibited by Section 12021 or 12021.1.

(2) No person, corporation, or dealer shall sell, supply, deliver, or give possession or control of a firearm to any person whom he or she has cause to believe to be within any of the classes prohibited by Section 12021 or 12021.1 of this code or Section 8100 or 8103 of the Welfare and Institutions Code.

(3)(A) No person, corporation, or firm shall sell, loan, or transfer a firearm to a minor, nor sell a handgun to an individual under 21 years of age.

(B) Subparagraph (A) shall not apply to or affect those circumstances set forth in subdivision (p) of Section 12078.

(4) No person, corporation, or dealer shall sell, loan, or transfer a firearm to any person whom he or she knows or has cause to believe is not the actual purchaser or transferee of the firearm, or to any person who is not the person actually being loaned the firearm, if the person, corporation, or dealer has either of the following:

(A) Knowledge that the firearm is to be subsequently loaned, sold, or transferred to avoid the provisions of subdivision (c) or (d).

(B) Knowledge that the firearm is to be subsequently loaned, sold, or transferred to avoid the requirements of any exemption to the provisions of subdivision (c) or (d).

(5) No person, corporation, or dealer shall acquire a firearm for the purpose of selling, transferring, or loaning the firearm, if the person, corporation, or dealer has either of the following:

(A) In the case of a dealer, intent to violate subdivision (b) or (c).

(B) In any other case, intent to avoid either of the following:

(i) The provisions of subdivision (d).

(ii) The requirements of any exemption to the provisions of subdivision (d).

(6) The dealer shall comply with the provisions of paragraph (18) of subdivision (b) of Section 12071.

(7) The dealer shall comply with the provisions of paragraph (19) of subdivision (b) of Section 12071.

(8) No person shall sell or otherwise transfer his or her ownership in a pistol, revolver, or other firearm capable of being concealed upon the person unless the firearm bears either:

(A) The name of the manufacturer, the manufacturer's make or model, and a manufacturer's serial number assigned to that firearm.

(B) The identification number or mark assigned to the firearm by the Department of Justice pursuant to Section 12092.

(9)(A) No person shall make an application to purchase more than one pistol, revolver, or other firearm capable of being concealed upon the person within any 30-day period.

(B) Subparagraph (A) shall not apply to any of the following:

(i) Any law enforcement agency.

(ii) Any agency duly authorized to perform law enforcement duties.

(iii) Any state or local correctional facility.

(iv) Any private security company licensed to do business in California.

(v) Any person who is properly identified as a full-time paid peace officer, as defined in Chapter 4.5 (commencing with Section 830) of Title 3 of Part 2, and who is authorized to, and does carry a firearm during the course and scope of his or her employment as a peace officer.

(vi) Any motion picture, television, or video production company or entertainment or theatrical company whose production by its nature involves the use of a firearm.

(vii) Any person who may, pursuant to Section 12078, claim an exemption from the waiting period set forth in subdivision (c) of this section.

(viii) Any transaction conducted through a licensed firearms dealer pursuant to Section 12082.

(ix) Any person who is licensed as a collector pursuant to Chapter 44 (commencing with Section 921) of Title 18 of the United States Code and the regulations issued pursuant thereto and who has a current certificate of eligibility issued to him or her by the Department of Justice pursuant to Section 12071.

(x) The exchange of a pistol, revolver, or other firearm capable of being concealed upon the person where the dealer purchased that firearm from the person seeking the exchange within the 30-day period immediately preceding the date of exchange or replacement.

(xi) The replacement of a pistol, revolver, or other firearm capable of being concealed upon the person when the person's pistol, revolver, or other firearm capable of being concealed upon the person was lost or stolen, and the person reported that firearm lost or stolen prior to the completion of the application to purchase to any local law enforcement agency of the city, county, or city and county in which he or she resides.

(xii) The return of any pistol, revolver, or other firearm capable of being concealed upon the person to its owner.

(b) No person licensed under Section 12071 shall supply, sell, deliver, or give possession or control of a pistol, revolver, or firearm capable of being concealed upon the person to any person under the age of 21 years or any other firearm to a person under the age of 18 years.

(c) No dealer, whether or not acting pursuant to Section 12082, shall deliver a firearm to a person, as follows:

(1) Within 10 days of the application to purchase, or, after notice by the department pursuant to subdivision (d) of Section 12076, within 10 days of the submission to the department of any correction to the application, or within 10 days of the submission to the department of any fee required pursuant to subdivision (e) of Section 12076, whichever is later.

(2) Unless unloaded and securely wrapped or unloaded and in a locked container.

(3) Unless the purchaser, transferee, or person being loaned the firearm presents clear evidence of his or her identity and age, as defined in Section 12071, to the dealer.

(4) Whenever the dealer is notified by the Department of Justice that the person is in a prohibited class described in Section 12021 or 12021.1 of this code or Section 8100 or 8103 of the Welfare and Institutions Code.

(5)(A) Commencing April 1, 1994, and until January 1, 2003, no pistol, revolver, or other firearm capable of being concealed upon the person shall be delivered unless the purchaser, transferee, or person being loaned the firearm presents to the dealer a basic firearms safety certificate.

(B) Commencing January 1, 2003, no handgun shall be delivered unless the purchaser, transferee, or person being loaned the handgun presents a handgun safety certificate to the dealer.

(6) No pistol, revolver, or other firearm capable of being concealed upon the person shall be delivered whenever the dealer is notified by the Department of Justice that within the preceding 30-day period the purchaser has made another application to purchase a pistol, revolver, or other firearm capable of being concealed upon the person and that the previous application to purchase involved none of the entities specified in subparagraph (B) of paragraph (9) of subdivision (a).

(d) Where neither party to the transaction holds a dealer's license issued pursuant to Section 12071, the parties to the transaction shall complete the sale, loan, or transfer of that firearm through a licensed firearms dealer pursuant to Section 12082.

(e) No person may commit an act of collusion relating to Article 8 (commencing with Section 12800) of Chapter 6. For purposes of this section and Section 12071, collusion may be proven by any one of the following factors:

(1) Answering a test applicant's questions during an objective test relating to firearms safety.

(2) Knowingly grading the examination falsely.

(3) Providing an advance copy of the test to an applicant.

(4) Taking or allowing another person to take the basic firearms safety course for one who is the applicant for a basic firearms safety certificate or a handgun safety certificate.

(5) Allowing another to take the objective test for the applicant, purchaser, or transferee.

(6) Using or allowing another to use one's identification, proof of residency, or thumbprint.

(7) Allowing others to give unauthorized assistance during the examination.

(8) Reference to unauthorized materials during the examination and cheating by the applicant.

(9) Providing originals or photocopies of the objective test, or any version thereof, to any person other than as authorized by the department.

(f)(1) No person who is licensed pursuant to Chapter 44 (commencing with Section 921) of Title 18 of the United States Code shall deliver, sell, or transfer a firearm to a person who is licensed pursuant to Chapter 44 (commencing with Section 921) of Title 18 of the United States Code and whose licensed premises are located in this state unless:

(A) Commencing July 1,2008, a person who is licensed pursuant to Chapter 44 (commencing with Section 921) of Title 18 of the United States Code may not deliver, sell, or transfer a firearm to a person in California who is licensed pursuant to Chapter 44 (commencing with Section 921) of Title 18 of the United States Code unless, prior to delivery, the person intending to deliver, sell, or transfer the firearm obtains a verification number via the internet for the intended delivery, sale, or transfer, from the department. If Internet service is unavailable to either the department or the licensee due to a technical or other malfunction, or a federal firearms licensee who is located outside of California does not possess a computer or have Internet access, alternate means of communication, including facsimile or telephone, shall be made available for a licensee to obtain a verification number in order to comply with this section. This Subdivision shall not apply to the delivery, sale, or transfer of a short-barreled rifle, or short-barreled shotgun, as defined in Section 12020, or to a machinegun as defined in Section 12200, or to an assault weapon as defined in Sections 12276, 12276.1, and 12276.5.

(B) For every identification number request received pursuant to this Section, the Department shall determine whether the intended recipient is on the centralized list of firearms dealers pursuant to this Section, or the centralized list of exempted federal firearms licensees pursuant to Subdivision (a) of Section 12083, or the centralized list of firearms manufacturers pursuant to Subdivision (f) of Section 12086.

(C) If the Department finds that the intended recipient is on one of these lists, the Department shall issue to the inquiring party, a unique identification number for the intended delivery, sale, or transfer. in addition to the unique verification number, the Department may provide to the inquiring party information necessary for determining the eligibility of the intended recipient to receive the firearm. The person intending to deliver, sell, or transfer the firearm shall provide the unique verification number to the recipient along with the firearm upon delivery, in a manner to be determined by the Department.

(D) If the Department finds that the intended recipient is not on one of these lists, the Department shall notify the inquiring party that the intended recipient is ineligible to receive the firearm.

(E) The Department shall prescribe the manner in which the verification numbers may be requested via the Internet, or by alternate means of communication, such as by facsimile or telephone, including all required enrollment information and procedures.

(2)(A) On or after January 1, 1998, within 60 days of bringing a pistol, revolver, or other firearm capable of being concealed upon the person into this state, a personal handgun importer shall do one of the following:

(i) Forward by prepaid mail or deliver in person to the Department of Justice, a report prescribed by the department including information concerning that individual and a description of the firearm in question.

(ii) Sell or transfer the firearm in accordance with the provisions of subdivision (d) or in accordance with the provisions of an exemption from subdivision (d).

(iii) Sell or transfer the firearm to a dealer licensed pursuant to Section 12071.

(iv) Sell or transfer the firearm to a sheriff or police department.

(B) If the personal handgun importer sells or transfers the pistol, revolver, or other firearm capable of being concealed upon the person pursuant to subdivision (d) of Section 12072 and the sale or transfer cannot be completed by the dealer to the purchaser or transferee, and the firearm can be returned to the personal handgun importer, the personal handgun importer shall have complied with the provisions of this paragraph.

(C) The provisions of this paragraph are cumulative and shall not be construed as restricting the application of any other law. However, an act or omission punishable in different ways by this section and different provisions of the Penal Code shall not be punished under more than one provision.

(D)(i) On and after January 1, 1998, the department shall conduct a public education and notification program regarding this paragraph to ensure a high degree of publicity of the provisions of this paragraph.

(ii) As part of the public education and notification program described in this subparagraph, the department shall do all of the following:

(I) Work in conjunction with the Department of Motor Vehicles to ensure that any person who is subject to this paragraph is advised of the provisions of this paragraph, and provided with blank copies of the report described in clause (i) of subparagraph (A) at the time that person applies for a California driver's license or registers his or her motor vehicle in accordance with the Vehicle Code.

(II) Make the reports referred to in clause (i) of subparagraph (A) available to dealers licensed pursuant to Section 12071.

(III) Make the reports referred to in clause (i) of subparagraph (A) available to law enforcement agencies.

(IV) Make persons subject to the provisions of this paragraph aware of the fact that reports referred to in clause (i) of subparagraph (A) may be completed at either the licensed premises of dealers licensed pursuant to Section 12071 or at law enforcement agencies, that it is advisable to do so for the sake of accuracy and completeness of the reports, that prior to transporting a pistol, revolver, or other firearm capable of being concealed upon the person to a law enforcement agency in order to comply with subparagraph (A), the person should give prior notice to the law enforcement agency that he or she is doing so, and that in any event, the pistol, revolver, or other firearm capable of being concealed upon the person should be transported unloaded and in a locked container.

(iii) Any costs incurred by the department to implement this paragraph shall be absorbed by the department within its existing budget and the fees in the Dealers' Record of Sale Special Account allocated for implementation of this subparagraph pursuant to Section 12076.

(3) Where a person who is licensed as a collector pursuant to Chapter 44 (commencing with Section 921) of Title 18 of the United States Code and the regulations issued pursuant thereto, whose licensed premises are within this state, acquires a pistol, revolver, or other firearm capable of being concealed upon the person that is a curic or relic, as defined in Section 478.11 of Title 27 of the Code of Federal Regulations, outside of this state, takes actual possession of that firearm outside of this state pursuant to the provisions of subsection (j) of Section 923 of Title 18 of the United States Code, as amended by Public Law 104-208, and transports that firearm into this state, within five days of that licensed collector transporting that firearm into this state, he or she shall report to the department in a format prescribed by the department his or her acquisition of that firearm.

(4)(A) It is the intent of the Legislature that a violation of paragraph (2) or (3) shall not constitute a "continuing offense" and the statute of limitations for commencing a prosecution for a violation of paragraph (2) or (3) commences on the date that the applicable grace period specified in paragraph (2) or (3) expires.

(B) Paragraphs (2) and (3) shall not apply to a person who reports his or her ownership of a pistol, revolver or other firearm capable of being concealed upon the person after the applicable grace period specified in paragraph (2) or (3) expires if evidence of that violation arises only as the result of the person submitting the report described in paragraph (2) or (3).

(g)(1) Except as provided in paragraph (2), (3), or (5), a violation of this section is a misdemeanor.

(2) If any of the following circumstances apply, a violation of this section is punishable by imprisonment in the state prison for two, three, or four years.

(A) If the violation is of paragraph (1) of subdivision (a).

(B) If the defendant has a prior conviction of violating the provisions, other than paragraph (9) of subdivision (a), of this section or former Section 12100 of this code or Section 8101 of the Welfare and Institutions Code.

(C) If the defendant has a prior conviction of violating any offense specified in subdivision (b) of Section 12021.1 or of a violation of Section 12020, 12220, or 12520, or of former Section 12560.

(D) If the defendant is in a prohibited class described in Section 12021 or 12021.1 of this code or Section 8100 or 8103 of the Welfare and Institutions Code.

(E) A violation of this section by a person who actively participates in a "criminal street gang" as defined in Section 186.22.

(F) A violation of subdivision (b) involving the delivery of any firearm to a person who the dealer knows, or should know, is a minor.

(3) If any of the following circumstances apply, a violation of this section shall be punished by imprisonment in a county jail not exceeding one year or in the state prison, or by a fine not to exceed one thousand dollars ($1,000), or by both that fine and imprisonment.

(A) A violation of paragraph (2), (4), or (5) of subdivision (a).

(B) A violation of paragraph (3) of subdivision (a) involving the sale, loan, or transfer of a pistol, revolver, or other firearm capable of being concealed upon the person to a minor.

(C) A violation of subdivision (b) involving the delivery of a pistol, revolver, or other firearm capable of being concealed upon the person.

(D) A violation of paragraph (1), (3), (4), (5), or (6) of subdivision (c) involving a pistol, revolver, or other firearm capable of being concealed upon the person.

(E) A violation of subdivision (d) involving a pistol, revolver, or other firearm capable of being concealed upon the person.

(F) A violation of subdivision (e).

(4) If both of the following circumstances apply, an additional term of imprisonment in the state prison for one, two, or three years shall be imposed in addition and consecutive to the sentence prescribed.

(A) A violation of paragraph (2) of subdivision (a) or subdivision (b).

(B) The firearm transferred in violation of paragraph (2) of subdivision (a) or subdivision (b) is used in the subsequent commission of a felony for which a conviction is obtained and the prescribed sentence is imposed.

(5)(A) A first violation of paragraph (9) of subdivision (a) is an infraction punishable by a fine of fifty dollars ($50).

(B) A second violation of paragraph (9) of subdivision (a) is an infraction punishable by a fine of one hundred dollars ($100).

(C) A third or subsequent violation of paragraph (9) of subdivision (a) is a misdemeanor.

(D) For purposes of this paragraph each application to purchase a pistol, revolver, or other firearm capable of being concealed upon the person in violation of paragraph (9) of subdivision (a) shall be deemed a separate offense.

12073. Dealers required to keep register or record of electronic or telephonic transfer; Violation a misdemeanor

(a) As required by the Department of Justice, every dealer shall keep a register or record of electronic or telephonic transfer in which shall be entered the information prescribed in Section 12077.

(b) This section shall not apply to any of the following transactions:

(1) The delivery, sale, or transfer of an unloaded firearm that is not a handgun by a dealer to another dealer upon proof of compliance with the requirements of paragraph (1) of subdivision (f) of Section 12072.

(2) The delivery, sale, or transfer of an unloaded firearm by a dealer to another dealer if that firearm is intended as merchandise in the receiving dealer's business upon proof of compliance with the requirements of paragraph (1) of subdivision (f) of Section 12072.

(3) The delivery, sale, or transfer of an unloaded firearm by a dealer to a person licensed as an importer or manufacturer pursuant to Chapter 44 (commencing with Section 921) of Title 18 of the United States Code and any regulations issued pursuant thereto.

(4) The delivery, sale, or transfer of an unloaded firearm by a dealer who sells, transfers, or delivers the firearm to a person who resides outside this state who is licensed pursuant to Chapter 44 (commencing with Section 921) of Title 18 of the United States Code and any regulations issued pursuant thereto.

(5) The delivery, sale, or transfer of an unloaded firearm by a dealer to a wholesaler if that firearm is being returned to the wholesaler and is intended as merchandise in the wholesaler's business.

(6) The delivery, sale, or transfer of an unloaded firearm that is not a handgun by a dealer to himself or herself.

(7) The loan of an unloaded firearm by a dealer who also operates a target facility which holds a business or regulatory license on the premises of the building designated in the license or whose building designated in the license is on the premises of any club or organization organized for the purpose of practicing shooting at targets upon established ranges, whether public or private, to a person at that target facility or club or organization, if the firearm is kept at all times within the premises of the target range or on the premises of the club or organization.

(8) The delivery of an unloaded firearm by a dealer to a gunsmith for service or repair.

(9) The return of an unloaded firearm to the owner of that firearm by a dealer, if the owner initially delivered the firearm to the dealer for service or repair.

(10) The loan of an unloaded firearm by a dealer to a person who possesses a valid entertainment firearms permit issued pursuant to Section 12081, for use solely as a prop in a motion picture, television, video, theatrical, or other entertainment production or event.

(11) The loan of an unloaded firearm by a dealer to a consultant-evaluator, if the loan does not exceed 45 days from the date of delivery of the firearm by the dealer to the consultant-evaluator.

(c) A violation of this section is a misdemeanor.

12074. Preparation and furnishing of register; Standards for electronic transfer of applicant information

(a) The register shall be prepared by and obtained from the State Printer and shall be furnished by the State Printer only to dealers on application at a cost to be determined by the Department of General Services for each 100 leaves in quadruplicate, one original and three duplicates for the making of carbon copies. The original and duplicate copies shall differ in color, and shall be in the form provided by this article.

(b) Where the electronic transfer of applicant information is used, the Department of Justice shall develop the standards for all appropriate electronic and telephonic equipment and telephone numbers to effect the transfer of information to the department.

12075. Information furnished to Department of Justice; Nontransferability of register The State Printer upon issuing a register shall forward to the Department of Justice the name and business address of the dealer together with the series and sheet numbers of the register. The register shall not be transferable. If the dealer moves his business to a different location he shall notify the department of such fact in writing within 48 hours.

12076. Format for submission of information; Contents and retention of records; Examination of records relating to purchaser; Fees; Discretionary acts by department

(a)(1) Before January 1, 1998, the Department of Justice shall determine the method by which a dealer shall submit firearm purchaser information to the department and the information shall be in one of the following formats:

(A) Submission of the register described in Section 12077.

(B) Electronic or telephonic transfer of the information contained in the register described in Section 12077.

(2) On or after January 1, 1998, electronic or telephonic transfer, including voice or facsimile transmission, shall be the exclusive means by which purchaser information is transmitted to the department.

(3) On or after January 1, 2003, except as permitted by the department, electronic transfer shall be the exclusive means by which information is transmitted to the department. Telephonic transfer shall not be permitted for information regarding sales of any firearms.

(b)(1) Where the register is used, the purchaser of any firearm shall be required to present clear evidence of his or her identity and age, as defined in Section 12071, to the dealer, and the dealer shall require him or her to sign his or her current legal name and affix his or her residence address and date of birth to the register in quadruplicate. The salesperson shall affix his or her signature to the register in quadruplicate as a witness to the signature and identification of the purchaser. Any person furnishing a fictitious name or address or knowingly furnishing any incorrect information or knowingly omitting any information required to be provided for the register and any person violating any provision of this section is guilty of a misdemeanor, provided however, that any person who is prohibited from obtaining a firearm pursuant to Section 12021 or 12021.1 of this code, or Section 8100 or 8103 of the Welfare and Institutions Code who knowingly furnishes a fictitious name or address or knowingly furnishes any incorrect information or knowingly omits any information required to be provided for the register shall be punished by imprisonment in a county jail not exceeding one year or imprisonment in the state prison for a term of 8, 12, or 18 months.

(2) The original of the register shall be retained by the dealer in consecutive order. Each book of 50 originals shall become the permanent register of transactions that shall be retained for not less than three years from the date of the last transaction and shall be available for the inspection of any peace officer, Department of Justice employee designated by the Attorney General, or agent of the federal Bureau of Alcohol, Tobacco, Firearms, and Explosives upon the presentation of proper identification, but no information shall be compiled therefrom regarding the purchasers or other transferees of firearms that are not pistols, revolvers, or other firearms capable of being concealed upon the person.

(3) Two copies of the original sheet of the register, on the date of the application to purchase, shall be placed in the mail, postage prepaid, and properly addressed to the Department of Justice in Sacramento.

(4) If requested, a photocopy of the original shall be provided to the purchaser by the dealer.

(5) If the transaction is a private party transfer conducted pursu-ant to Section 12082, a photocopy of the original shall be provided to the seller by the dealer, upon request. The dealer shall redact all of the purchaser's personal information, as required pursuant to paragraph (1) of subdivision (b) and paragraph (1) of subdivision (c) of Section 12077, from the seller's copy, and the seller's personal information from the purchaser's copy.

(c)(1) Where the electronic or telephonic transfer of applicant information is used, the purchaser shall be required to present clear evidence of his or her identity and age, as defined in Section 12071, to the dealer, and the dealer shall require him or her to sign his or her current legal name to the record of electronic or telephonic transfer. The salesperson shall affix his or her signature to the record of electronic or telephonic transfer as a witness to the signature and identification of the purchaser. Any person furnishing a fictitious name or address or knowingly furnishing any incorrect information or knowingly omitting any information required to be provided for the electronic or telephonic transfer and any person violating any provision of this section is guilty of a misdemeanor, provided however, that any person who is prohibited from obtaining a firearm pursuant to Section 12021 or 12021.1 of this code, or Section 8100 or 8103 of the Welfare and Institutions Code who knowingly furnishes a fictitious name or address or knowingly furnishes any incorrect information or knowingly omits any information required to be provided for the register shall be punished by imprisonment in a county jail not exceeding one year or imprisonment in the state prison for a term of 8, 12, or 18 months.

(2) The record of applicant information shall be transmitted to the Department of Justice in Sacramento by electronic or telephonic transfer on the date of the application to purchase.

(3) The original of each record of electronic or telephonic transfer shall be retained by the dealer in consecutive order. Each original shall become the permanent record of the transaction that shall be retained for not less than three years from the date of the last transaction and shall be provided for the inspection of any peace officer, Department of Justice employee designated by the Attorney General, or agent of the federal Bureau of Alcohol, Tobacco, Firearms, and Explosives upon the presentation of proper identification, but no information shall be compiled therefrom regarding the purchasers or other transferees of firearms that are not pistols, revolvers, or other firearms capable of being concealed upon the person.

(4) If requested, a copy of the record of electronic or telephonic transfer shall be provided to the purchaser by the dealer.

(5) If the transaction is a private party transfer conducted pursu-ant to Section 12082, a copy shall be provided to the seller by the dealer, upon request. The dealer shall redact all of the purchaser's personal information, as required pursuant to paragraph (1) of subdivision (b) and paragraph (1) of subdivision (c) of Section 12077, from the seller's copy, and the seller's personal information from the purchaser's copy.

(d)(1) The department shall examine its records, as well as those records that it is authorized to request from the State Department of Mental Health pursuant to Section 8104 of the Welfare and Institutions Code, in order to determine if the purchaser is a person described in Section 12021, 12021.1, or subparagraph (A) of paragraph (9) of subdivision (a) of Section 12072 of this code or Section 8100 or 8103 of the Welfare and Institutions Code.

(2) To the extent that funding is available, the Department of Justice may participate in the National Instant Criminal Background Check System (NICS), as described in subsection (t) of Section 922 of Title 18 of the United States Code, and, if that participation is implemented, shall notify the dealer and the chief of the police department of the city or city and county in which the sale was made, or if the sale was made in a district in which there is no municipal police department, the sheriff of the county in which the sale was made, that the purchaser is a person prohibited from acquiring a firearm under federal law.

(3) If the department determines that the purchaser is a person described in Section 12021, 12021.1, or subparagraph (A) of paragraph (9) of subdivision (a) of Section 12072 of this code or Section 8100 or 8103 of the Welfare and Institutions Code, it shall immediately notify the dealer and the chief of the police department of the city or city and county in which the sale was made, or if the sale was made in a district in which there is no municipal police department, the sheriff of the county in which the sale was made, of that fact.

(4) If the department determines that the copies of the register submitted to it pursuant to paragraph (3) of subdivision (b) contain any blank spaces or inaccurate, illegible, or incomplete information, preventing identification of the purchaser or the pistol, revolver, or other firearm to be purchased, or if any fee required pursuant to subdivision (e) is not submitted by the dealer in conjunction with submission of copies of the register, the department may notify the dealer of that fact. Upon notification by the department, the dealer shall submit corrected copies of the register to the department, or shall submit any fee required pursuant to subdivision (e), or both, as appropriate and, if notification by the department is received by the dealer at any time prior to delivery of the firearm to be purchased, the dealer shall withhold delivery until the conclusion of the waiting period described in Sections 12071 and 12072.

(5) If the department determines that the information transmitted to it pursuant to subdivision (c) contains inaccurate or incomplete information preventing identification of the purchaser or the pistol, revolver, or other firearm capable of being concealed upon the person to be purchased, or if the fee required pursuant to subdivision (e) is not transmitted by the dealer in conjunction with transmission of the electronic or telephonic record, the department may notify the dealer of that fact. Upon notification by the department, the dealer shall transmit corrections to the record of electronic or telephonic transfer to the department, or shall transmit any fee required pursuant to subdivision (e), or both, as appropriate, and if notification by the department is received by the dealer at any time prior to delivery of the firearm to be purchased, the dealer shall withhold delivery until the conclusion of the waiting period described in Sections 12071 and 12072.

(e) The Department of Justice may require the dealer to charge each firearm purchaser a fee not to exceed fourteen dollars ($14), except that the fee may be increased at a rate not to exceed any increase in the California Consumer Price Index as compiled and reported by the California Department of Industrial Relations. The fee shall be no more than is necessary to fund the following:

(1)(A) The department for the cost of furnishing this information.

(B) The department for the cost of meeting its obligations under paragraph (2) of subdivision (b) of Section 8100 of the Welfare and Institutions Code.

(2) Local mental health facilities for state-mandated local costs resulting from the reporting requirements imposed by Section 8103 of the Welfare and Institutions Code.

(3) The State Department of Mental Health for the costs resulting from the requirements imposed by Section 8104 of the Welfare and Institutions Code.

(4) Local mental hospitals, sanitariums, and institutions for state-mandated local costs resulting from the reporting requirements imposed by Section 8105 of the Welfare and Institutions Code.

(5) Local law enforcement agencies for state-mandated local costs resulting from the notification requirements set forth in subdivision (a) of Section 6385 of the Family Code.

(6) Local law enforcement agencies for state-mandated local costs resulting from the notification requirements set forth in subdivision (c) of Section 8105 of the Welfare and Institutions Code.

(7) For the actual costs associated with the electronic or telephonic transfer of information pursuant to subdivision (c).

(8) The Department of Food and Agriculture for the costs resulting from the notification provisions set forth in Section 5343.5 of the Food and Agricultural Code.

(9) The department for the costs associated with subparagraph (D) of paragraph (2) of subdivision (f) of Section 12072.

(10) The department for the costs associated with funding Department of Justice firearms-related regulatory and enforcement activities related to the sale, purchase, loan, or transfer of firearms pursuant to this chapter.

The fee established pursuant to this subdivision shall not exceed the sum of the actual processing costs of the department, the estimated reasonable costs of the local mental health facilities for complying with the reporting requirements imposed by paragraph (2) of this subdivision, the costs of the State Department of Mental Health for complying with the requirements imposed by paragraph (3) of this subdivision, the estimated reasonable costs of local mental hospitals, sanitariums, and institutions for complying with the reporting requirements imposed by paragraph (4) of this subdivision, the estimated reasonable costs of local law enforcement agencies for complying with the notification requirements set forth in subdivision (a) of Section 6385 of the Family Code, the estimated reasonable costs of local law enforcement agencies for complying with the notification requirements set forth in subdivision (c) of Section 8105 of the Welfare and Institutions Code imposed by paragraph (6) of this subdivision, the estimated reasonable costs of the Department of Food and Agriculture for the costs resulting from the notification provisions set forth in Section 5343.5 of the Food and Agricultural Code, the estimated reasonable costs of the department for the costs associated with subparagraph (D) of paragraph (2) of subdivision (f) of Section 12072, and the estimated reasonable costs of department firearms-related regulatory and enforcement activities related to the sale, purchase, loan, or transfer of firearms pursuant to this chapter.

(f)(1) The Department of Justice may charge a fee sufficient to reimburse it for each of the following but not to exceed fourteen dollars ($14), except that the fee may be increased at a rate not to exceed any increase in the California Consumer Price Index as compiled and reported by the California Department of Industrial Relations:

(A) For the actual costs associated with the preparation, sale, processing, and filing of forms or reports required or utilized pursuant to Section 12078.

(B) For the actual processing costs associated with the submission of a Dealers' Record of Sale to the department.

(C) For the actual costs associated with the preparation, sale, processing, and filing of reports utilized pursuant to subdivision (l) of Section 12078 or paragraph (18) of subdivision (b) of Section 12071, or clause (i) of subparagraph (A) of paragraph (2) of subdivision (f) of Section 12072, or paragraph (3) of subdivision (f) of Section 12072.

(D) For the actual costs associated with the electronic or telephonic transfer of information pursuant to subdivision (c).

(2) If the department charges a fee pursuant to subparagraph (B) of paragraph (1) of this subdivision, it shall be charged in the same amount to all categories of transaction that are within that subparagraph.

(3) Any costs incurred by the Department of Justice to implement this subdivision shall be reimbursed from fees collected and charged pursuant to this subdivision. No fees shall be charged to the dealer pursuant to subdivision (e) for implementing this subdivision.

(g) All money received by the department pursuant to this section shall be deposited in the Dealers' Record of Sale Special Account of the General Fund, which is hereby created, to be available, upon appropriation by the Legislature, for expenditure by the department to offset the costs incurred pursuant to this section, paragraph (1) and subparagraph (D) of paragraph (2) of subdivision (f) of Section 12072, Sections 12083 and 12099, subdivision (c) of Section 12131, Sections 12234, 12289, and 12289.5, and subdivisions (f) and (g) of Section 12305.

(h) Where the electronic or telephonic transfer of applicant information is used, the department shall establish a system to be used for the submission of the fees described in subdivision (e) to the department.

(i)(1) Only one fee shall be charged pursuant to this section for a single transaction on the same date for the sale of any number of firearms that are not pistols, revolvers, or other firearms capable of being concealed upon the person or for the taking of possession of those firearms.

(2) In a single transaction on the same date for the delivery of any number of firearms that are pistols, revolvers, or other firearms capable of being concealed upon the person, the department shall charge a reduced fee pursuant to this section for the second and subsequent firearms that are part of that transaction.

(j) Only one fee shall be charged pursuant to this section for a single transaction on the same date for taking title or possession of any number of firearms pursuant to paragraph (18) of subdivision (b) of Section 12071 or subdivision (c) or (i) of Section 12078.

(k) Whenever the Department of Justice acts pursuant to this section as it pertains to firearms other than pistols, revolvers, or other firearms capable of being concealed upon the person, the department's acts or omissions shall be deemed to be discretionary within the meaning of the California Tort Claims Act pursuant to Division 3.6 (commencing with Section 810) of Title 1 of the Government Code.

(l) As used in this section, the following definitions apply:

(1) "Purchaser" means the purchaser or transferee of a firearm or a person being loaned a firearm.

(2) "Purchase" means the purchase, loan, or transfer of a firearm.

(3) "Sale" means the sale, loan, or transfer of a firearm.

(4) "Seller" means, if the transaction is being conducted pursuant to Section 12082, the person selling, loaning, or transferring the firearm.

12076.5. Firearms Safety and Enforcement Special Fund established

(a) The Firearms Safety and Enforcement Special Fund is hereby established in the State Treasury and shall be administered by the Department of Justice. Notwithstanding Section 13340 of the Government Code, all moneys in the fund are continuously appropriated to the Department of Justice without regard to fiscal years for the purpose of implementing and enforcing the provisions of Article 8 (commencing with Section 12800), as added by the Statutes of 2001, enforcing the provisions of this title, and for the establishment, maintenance and upgrading of equipment and services necessary for firearms dealers to comply with Section 12077.

(b) The Department of Justice may require firearms dealers to charge each person who obtains a firearm a fee not to exceed five dollars ($5) for each transaction. Revenues from this fee shall be deposited in the Firearms Safety and Enforcement Special Fund.

12077. Form and contents of register and record of electronic transfer

(a) The Department of Justice shall prescribe the form of the register and the record of electronic transfer pursuant to Section 12074.

(b)(1) For handguns, information contained in the register or record of electronic transfer shall be the date and time of sale, make of firearm, peace officer exemption status pursuant to subdivision (a) of Section 12078 and the agency name, dealer waiting period exemption pursuant to subdivision (n) of Section 12078, dangerous weapons permit holder waiting period exemption pursuant to subdivision (r) of Section 12078, curio and relic waiting period exemption pursuant to subdivision (t) of Section 12078, California Firearms Dealer number issued pursuant to Section 12071, for transactions occurring prior to January 1, 2003, the purchaser's basic firearms safety certificate number issued pursuant to Sections 12805 and 12809, for transactions occurring on or after January 1, 2003, the purchaser's handgun safety certificate number issued pursuant to Article 8 (commencing with Section 12800), manufacturer's name if stamped on the firearm, model name or number, if stamped on the firearm, if applicable, serial number, other number (if more than one serial number is stamped on the firearm), any identification number or mark assigned to the firearm pursuant to Section 12092, caliber, type of firearm, if the firearm is new or used, barrel length, color of the firearm, full name of purchaser, purchaser's complete date of birth, purchaser's local address, if current address is temporary, complete permanent address of purchaser, identification of purchaser, purchaser's place of birth (state or country), purchaser's complete telephone number, purchaser's occupation, purchaser's sex, purchaser's physical description, all legal names and aliases ever used by the purchaser, yes or no answer to questions that prohibit purchase including, but not limited to, conviction of a felony as described in Section 12021 or an offense described in Section 12021.1, the purchaser's status as a person described in Section 8100 of the Welfare and Institutions Code, whether the purchaser is a person who has been adjudicated by a court to be a danger to others or found not guilty by reason of insanity, whether the purchaser is a person who has been found incompetent to stand trial or placed under conservatorship by a court pursuant to Section 8103 of the Welfare and Institutions Code, signature of purchaser, signature of salesperson (as a witness to the purchaser's signature), salesperson's certificate of eligibility if he or she has obtained a certificate of eligibility, name and complete address of the dealer or firm selling the firearm as shown on the dealer's license, the establishment number, if assigned, the dealer's complete business telephone number, any information required by Section 12082, any information required to determine whether or not paragraph (6) of subdivision (c) of Section 12072 applies, and a statement of the penalties for any person signing a fictitious name or address or for knowingly furnishing any incorrect information or for knowingly omitting any information required to be provided for the register.

(2) Effective January 1, 2003, the purchaser shall provide his or her right thumbprint on the register in a manner prescribed by the department. No exception to this requirement shall be permitted except by regulations adopted by the department.

(3) The firearms dealer shall record on the register or record of electronic transfer the date that the handgun is delivered.

(c)(1) For firearms other than handguns, information contained in the register or record of electronic transfer shall be the date and time of sale, peace officer exemption status pursuant to subdivision (a) of Section 12078 and the agency name, auction or event waiting period exemption pursuant to subdivision (g) of Section 12078, California Firearms Dealer number issued pursuant to Section 12071, dangerous weapons permit holder waiting period exemption pursuant to subdivision (r) of Section 12078, curio and relic waiting period exemption pursuant to paragraph (1) of subdivision (t) of Section 12078, full name of purchaser, purchaser's complete date of birth, purchaser's local address, if current address is temporary, complete permanent address of purchaser, identification of purchaser, purchaser's place of birth (state or country), purchaser's complete telephone number, purchaser's occupation, purchaser's sex, purchaser's physical description, all legal names and aliases ever used by the purchaser, yes or no answer to questions that prohibit purchase, including, but not limited to, conviction of a felony as described in Section 12021 or an offense described in Section 12021.1, the purchaser's status as a person described in Section 8100 of the Welfare and Institutions Code, whether the purchaser is a person who has been adjudicated by a court to be a danger to others or found not guilty by reason of insanity, whether the purchaser is a person who has been found incompetent to stand trial or placed under conservatorship by a court pursuant to Section 8103 of the Welfare and Institutions Code, signature of purchaser, signature of salesperson (as a witness to the purchaser's signature), salesperson's certificate of eligibility if he or she has obtained a certificate of eligibility, name and complete address of the dealer or firm selling the firearm as shown on the dealer's license, the establishment number, if assigned, the dealer's complete business telephone number, any information required by Section 12082, and a statement of the penalties for any person signing a fictitious name or address or for knowingly furnishing any incorrect information or for knowingly omitting any information required to be provided for the register.

(2) Effective January 1, 2003, the purchaser shall provide his or her right thumbprint on the register in a manner prescribed by the department. No exception to this requirement shall be permitted except by regulations adopted by the department.

(3) The firearms dealer shall record on the register or record of electronic transfer the date that the handgun is delivered.

(d) Where the register is used, the following shall apply:

(1) Dealers shall use ink to complete each document.

(2) The dealer or salesperson making a sale shall ensure that all information is provided legibly. The dealer and salespersons shall be informed that incomplete or illegible information will delay sales.

(3) Each dealer shall be provided instructions regarding the procedure for completion of the form and routing of the form. Dealers shall comply with these instructions which shall include the information set forth in this subdivision.

(4) One firearm transaction shall be reported on each record of sale document. For purposes of this subdivision, a "transaction" means a single sale, loan, or transfer of any number of firearms that are not handguns.

(e) The dealer or salesperson making a sale shall ensure that all required information has been obtained from the purchaser. The dealer and all salespersons shall be informed that incomplete information will delay sales.

(f) Effective January 1, 2003, the purchaser's name, date of birth, and driver's license or identification number shall be obtained electronically from the magnetic strip on the purchaser's driver's license or identification and shall not be supplied by any other means except as authorized by the department. This requirement shall not apply in either of the following cases:

(1) The purchaser's identification consists of a military identification card.

(2) Due to technical limitations, the magnetic stripe reader is unable to obtain the required information from the purchaser's identification. In those circumstances, the firearms dealer shall obtain a photocopy of the identification as proof of compliance.

(3) In the event that the dealer has reported to the department that the dealer's equipment has failed, information pursuant to this subdivision shall be obtained by an alternative method to be determined by the department.

(g) As used in this section, the following definitions shall control:

(1) "Purchaser" means the purchaser or transferee of a firearm or the person being loaned a firearm.

(2) "Purchase" means the purchase, loan, or transfer of a firearm.

(3) "Sale" means the sale, loan, or transfer of a firearm.

12077.5. Firearms eligibility check; Fees; Application; Immunity of department; Requiring another person to obtain eligibility check or notification prohibited; Statements on application

(a) An individual may request that the Department of Justice perform a firearms eligibility check for that individual. The applicant requesting the eligibility check shall provide the information required by subdivision (c) of Section 12077 to the department, in an application specified by the department.

(b) The department shall charge a fee of twenty dollars ($20) for performing the eligibility check authorized by this section, but not to exceed the actual processing costs of the department. After the department establishes fees sufficient to reimburse the department for processing costs, fees charged may increase at a rate not to exceed the legislatively approved cost-of-living adjustment for the department's budget or as otherwise increased through the Budget Act.

(c) An applicant for the eligibility check pursuant to subdivision (a) shall complete the application, have it notarized by any licensed California Notary Public, and submit it by mail to the department. Upon receipt of a notarized application and fee, the department shall do all of the following:

(1) Examine its records, and the records it is authorized to request from the State Department of Mental Health pursuant to Section 8104 of the Welfare and Institutions Code, to determine if the purchaser is a person described in Section 12021 or 12021.1 of this code, or Section 8100 or 8103 of the Welfare and Institutions Code.

(2) Notify the applicant by mail of its determination of whether the applicant is a person described in Section 12021 or 12021.1 of this code, or Section 8100 or 8103 of the Welfare and Institutions Code. The department's notification shall state either "eligible to possess firearms as of the date the check was completed" or "ineligible to possess firearms as of the date the check was completed."

(d) If the department determines that the information submitted to it in the application contains any blank spaces, or inaccurate, illegible, or incomplete information, preventing identification of the applicant, or if the required fee is not submitted, the department shall not be required to perform the firearms eligibility check.

(e) The department shall make applications to conduct a firearms eligibility check as described in this section available to licensed firearms dealers and on the department's Web site.

(f) The department shall be immune from any liability arising out of the performance of the firearms eligibility check, or any reliance upon the firearms eligibility check.

(g) No person or agency may require or request another person to obtain a firearms eligibility check or notification of a firearms eligibility check pursuant to this section. A violation of this subdivision is a misdemeanor.

(h) The department shall include on the application specified in subdivision (a) and the notification of eligibility specified in subdivision (c) the following statements:

"No person or agency may require or request another person to obtain a firearms eligibility check or notification of firearms eligibility check pursuant to Section 12077.5 of the Penal Code. A violation of these provisions is a misdemeanor."

"If the applicant for a firearms eligibility check purchases, transfers, or receives a firearm through a licensed dealer as required by law, a waiting period and background check are both required."

12078. Transactions and persons exempt from operation of specified provisions

(a)(1) The waiting periods described in Sections 12071 and 12072 shall not apply to deliveries, transfers, or sales of firearms made to persons properly identified as full-time paid peace officers as defined in Chapter 4.5 (commencing with Section 830) of Title 3 of Part 2, provided that the peace officers are authorized by their employer to carry firearms while in the performance of their duties. Proper identification is defined as verifiable written certification from the head of the agency by which the purchaser or transferee is employed, identifying the purchaser or transferee as a peace officer who is authorized to carry firearms while in the performance of his or her duties, and authorizing the purchase or transfer. The certification shall be delivered to the dealer at the time of purchase or transfer and the purchaser or transferee shall identify himself or herself as the person authorized in the certification. The dealer shall keep the certification with the record of sale. On the date that the delivery, sale, or transfer is made, the dealer delivering the firearm shall transmit to the Department of Justice an electronic or telephonic report of the transaction as is indicated in subdivision (b) or (c) of Section 12077.

(2) Subdivision (b) of Section 12801 and the preceding provisions of this article do not apply to deliveries, transfers, or sales of firearms made to authorized law enforcement representatives of cities, counties, cities and counties, or state or federal governments for exclusive use by those governmental agencies if, prior to the delivery, transfer, or sale of these firearms, written authorization from the head of the agency authorizing the transaction is presented to the person from whom the purchase, delivery, or transfer is being made. Proper written authorization is defined as verifiable written certification from the head of the agency by which the purchaser or transferee is employed, identifying the employee as an individual authorized to conduct the transaction, and authorizing the transaction for the exclusive use of the agency by which he or she is employed. Within 10 days of the date a handgun is acquired by the agency, a record of the same shall be entered as an institutional weapon into the Automated Firearms System (AFS) via the California Law Enforcement Telecommunications System (CLETS) by the law enforcement or state agency. Those agencies without access to AFS shall arrange with the sheriff of the county in which the agency is located to input this information via this system.

(3) Subdivision (b) of Section 12801 and the preceding provisions of this article do not apply to the loan of a firearm made by an authorized law enforcement representative of a city, county, or city and county, or the state or federal government to a peace officer employed by that agency and authorized to carry a firearm for the carrying and use of that firearm by that peace officer in the course and scope of his or her duties.

(4) Subdivision (b) of Section 12801 and the preceding provisions of this article do not apply to the delivery, sale, or transfer of a firearm by a law enforcement agency to a peace officer pursuant to Section 10334 of the Public Contract Code. Within 10 days of the date that a handgun is sold, delivered, or transferred pursuant to Section 10334 of the Public Contract Code to that peace officer, the name of the officer and the make, model, serial number, and other identifying characteristics of the firearm being sold, transferred, or delivered shall be entered into the Automated Firearms System (AFS) via the California Law Enforcement Telecommunications System (CLETS) by the law enforcement or state agency that sold, transferred, or delivered the firearm. Those agencies without access to AFS shall arrange with the sheriff of the county in which the agency is located to input this information via this system.

(5) Subdivision (b) of Section 12801 and the preceding provisions of this article do not apply to the delivery, sale, or transfer of a firearm by a law enforcement agency to a retiring peace officer who is authorized to carry a firearm pursuant to Section 12027.1. Within 10 days of the date that a handgun is sold, delivered, or transferred to that retiring peace officer, the name of the officer and the make, model, serial number, and other identifying characteristics of the firearm being sold, transferred, or delivered shall be entered into the Automated Firearms System (AFS) via the California Law Enforcement Telecommunications System (CLETS) by the law enforcement or state agency that sold, transferred, or delivered the firearm. Those agencies without access to AFS shall arrange with the sheriff of the county in which the agency is located to input this information via this system.

(6) Subdivision (d) of Section 12072 and subdivision (b) of Section 12801 do not apply to sales, deliveries, or transfers of firearms to authorized representatives of cities, cities and counties, counties, or state or federal governments for those governmental agencies where the entity is acquiring the weapon as part of an authorized, voluntary program where the entity is buying or receiving weapons from private individuals. Any weapons acquired pursuant to this paragraph shall be disposed of pursuant to the applicable provisions of Section 12028 or 12032.

(7) Subdivision (d) of Section 12072 and subdivision (b) of Section 12801 shall not apply to the sale, loan, delivery, or transfer of a firearm made by an authorized law enforcement representative of a city, county, city and county, state, or the federal government to any public or private nonprofit historical society, museum, or institutional collection or the purchase or receipt of that firearm by that public or private nonprofit historical society, museum, or institutional collection if all of the following conditions are met:

(A) The entity receiving the firearm is open to the public.

(B) The firearm prior to delivery is deactivated or rendered inoperable.

(C) The firearm is not subject to Section 12028, 12028.5, 12030, or 12032.

(D) The firearm is not prohibited by other provisions of law from being sold, delivered, or transferred to the public at large.

(E) Prior to delivery, the entity receiving the firearm submits a written statement to the law enforcement representative stating that the firearm will not be restored to operating condition, and will either remain with that entity, or if subsequently disposed of, will be transferred in accordance with the applicable provisions of this article and, if applicable, Section 12801.

(F) Within 10 days of the date that the firearm is sold, loaned, delivered, or transferred to that entity, the name of the government entity delivering the firearm, and the make, model, serial number, and other identifying characteristics of the firearm and the name of the person authorized by the entity to take possession of the firearm shall be reported to the department in a manner prescribed by the department.

(G) In the event of a change in the status of the designated representative, the entity shall notify the department of a new representative within 30 days.

(8) Subdivision (d) of Section 12072 and subdivision (b) of Section 12801 shall not apply to the sale, loan, delivery, or transfer of a firearm made by any person other than a representative of an authorized law enforcement agency to any public or private nonprofit historical society, museum, or institutional collection if all of the following conditions are met:

(A) The entity receiving the firearm is open to the public.

(B) The firearm is deactivated or rendered inoperable prior to delivery.

(C) The firearm is not of a type prohibited from being sold, delivered, or transferred to the public.

(D) Prior to delivery, the entity receiving the firearm submits a written statement to the person selling, loaning, or transferring the firearm stating that the firearm will not be restored to operating condition, and will either remain with that entity, or if subsequently disposed of, will be transferred in accordance with the applicable provisions of this article and, if applicable, Section 12801.

(E) If title to a handgun is being transferred to the public or private nonprofit historical society, museum, or institutional collection, then the designated representative of that public or private historical society, museum or institutional collection within 30 days of taking possession of that handgun, shall forward by prepaid mail or deliver in person to the Department of Justice, a single report signed by both parties to the transaction, that includes information identifying the person representing that public or private historical society, museum, or institutional collection, how title was obtained and from whom, and a description of the firearm in question, along with a copy of the written statement referred to in subparagraph (D). The report forms that are to be completed pursuant to this paragraph shall be provided by the Department of Justice.

(F) In the event of a change in the status of the designated representative, the entity shall notify the department of a new representative within 30 days.

(b)(1) Section 12071, subdivisions (c) and (d) of Section 12072, and subdivision (b) of Section 12801 shall not apply to deliveries, sales, or transfers of firearms between or to importers and manufacturers of firearms licensed to engage in that business pursuant to Chapter 44 (commencing with Section 921) of Title 18 of the United States Code and the regulations issued pursuant thereto.

(2) Subdivision (b) of Section 12801 shall not apply to the delivery, sale, or transfer of a handgun to a person licensed pursuant to Section 12071, where the licensee is receiving the handgun in the course and scope of his or her activities as a person licensed pursuant to Section 12071.

(c)(1) Subdivision (d) of Section 12072 shall not apply to the infrequent transfer of a firearm that is not a handgun by gift, bequest, intestate succession, or other means by one individual to another if both individuals are members of the same immediate family.

(2) Subdivision (d) of Section 12072 shall not apply to the infrequent transfer of a handgun by gift, bequest, intestate succession, or other means by one individual to another if both individuals are members of the same immediate family and all of the following conditions are met:

(A) The person to whom the firearm is transferred shall, within 30 days of taking possession of the firearm, forward by prepaid mail or deliver in person to the Department of Justice, a report that includes information concerning the individual taking possession of the firearm, how title was obtained and from whom, and a description of the firearm in question. The report forms that individuals complete pursuant to this paragraph shall be provided to them by the Department of Justice.

(B) The person taking title to the firearm shall first obtain a handgun safety certificate.

(C) The person receiving the firearm is 18 years of age or older.

(3) As used in this subdivision, "immediate family member" means any one of the following relationships:

(A) Parent and child.

(B) Grandparent and grandchild.

(d)(1) Subdivision (d) of Section 12072 shall not apply to the infrequent loan of firearms between persons who are personally known to each other for any lawful purpose, if the loan does not exceed 30 days in duration and, when the firearm is a handgun, commencing January 1, 2003, the individual being loaned the handgun has a valid handgun safety certificate.

(2) Subdivision (d) of Section 12072, and subdivision (b) of Section 12801 shall not apply to the loan of a firearm where all of the following conditions exist:

(A) The person loaning the firearm is at all times within the presence of the person being loaned the firearm.

(B) The loan is for a lawful purpose.

(C) The loan does not exceed three days in duration.

(D) The individual receiving the firearm is not prohibited from owning or possessing a firearm pursuant to Section 12021 or 12021.1 of this code, or by Section 8100 or 8103 of the Welfare and Institutions Code.

(E) The person loaning the firearm is 18 years of age or older.

(F) The person being loaned the firearm is 18 years of age or older.

(e) Section 12071, subdivisions (c) and (d) of Section 12072, and subdivision (b) of Section 12801 shall not apply to the delivery of a firearm to a gunsmith for service or repair, or to the return of the firearm to its owner by the gunsmith.

(f) Subdivision (d) of Section 12072 and subdivision (b) of Section 12801 shall not apply to the sale, delivery, or transfer of firearms by persons who reside in this state to persons who reside outside this state who are licensed pursuant to Chapter 44 (commencing with Section 921) of Title 18 of the United States Code and the regulations issued pursuant thereto, if the sale, delivery, or transfer is in accordance with Chapter 44 (commencing with Section 921) of Title 18 of the United States Code and the regulations issued pursuant thereto.

(g)(1) Subdivision (d) of Section 12072 shall not apply to the infrequent sale or transfer of a firearm, other than a handgun, at auctions or similar events conducted by nonprofit mutual or public benefit corporations organized pursuant to the Corporations Code.

As used in this paragraph, the term "infrequent" shall not be construed to prohibit different local chapters of the same nonprofit corporation from conducting auctions or similar events, provided the individual local chapter conducts the auctions or similar events infrequently. It is the intent of the Legislature that different local chapters, representing different localities, be entitled to invoke the exemption created by this paragraph, notwithstanding the frequency with which other chapters of the same nonprofit corporation may conduct auctions or similar events.

(2) Subdivision (d) of Section 12072 shall not apply to the transfer of a firearm other than a handgun, if the firearm is donated for an auction or similar event described in paragraph (1) and the firearm is delivered to the nonprofit corporation immediately preceding, or contemporaneous with, the auction or similar event.

(3) The waiting period described in Sections 12071 and 12072 shall not apply to a dealer who delivers a firearm handgun at an auction or similar event described in paragraph (1), as authorized by subparagraph (C) of paragraph (1) of subdivision (b) of Section 12071. Within two business days of completion of the application to purchase, the dealer shall forward by prepaid mail to the Department of Justice a report of the same as is indicated in subdivision (c) of Section 12077. If the electronic or telephonic transfer of applicant information is used, within two business days of completion of the application to purchase, the dealer delivering the firearm shall transmit to the Department of Justice an electronic or telephonic report of the same as is indicated in subdivision (c) of Section 12077.

(h) Subdivision (d) of Section 12072 and subdivision (b) of Section 12801 shall not apply to the loan of a firearm to a person 18 years of age or older for the purposes of shooting at targets if the loan occurs on the premises of a target facility that holds a business or regulatory license or on the premises of any club or organization organized for the purposes of practicing shooting at targets upon established ranges, whether public or private, if the firearm is at all times kept within the premises of the target range or on the premises of the club or organization.

(i)(1) Subdivision (d) of Section 12072 shall not apply to a person who takes title or possession of a firearm that is not a handgun by operation of law if the person is not prohibited by Section 12021 or 12021.1 of this code or Section 8100 or 8103 of the Welfare and Institutions Code from possessing firearms.

(2) Subdivision (d) of Section 12072 shall not apply to a person who takes title or possession of a handgun by operation of law if the person is not prohibited by Section 12021 or 12021.1 of this code or Section 8100 or 8103 of the Welfare and Institutions Code from possessing firearms and all of the following conditions are met:

(A) If the person taking title or possession is neither a levying officer as defined in Section 481.140, 511.060, or 680.210 of the Code of Civil Procedure, nor a person who is receiving that firearm pursuant to subparagraph (G), (I), or (J) of paragraph (2) of subdivision (u), the person shall, within 30 days of taking possession, forward by prepaid mail or deliver in person to the Department of Justice, a report of information concerning the individual taking possession of the firearm, how title or possession was obtained and from whom, and a description of the firearm in question. The reports that individuals complete pursuant to this paragraph shall be provided to them by the department.

(B) If the person taking title or possession is receiving the firearm pursuant to subparagraph (G) of paragraph (2) of subdivision (u), the person shall do both of the following:

(i) Within 30 days of taking possession, forward by prepaid mail or deliver in person to the department, a report of information concerning the individual taking possession of the firearm, how title or possession was obtained and from whom, and a description of the firearm in question. The reports that individuals complete pursuant to this paragraph shall be provided to them by the department.

(ii) Prior to taking title or possession of the firearm, the person shall obtain a handgun safety certificate.

(C) Where the person receiving title or possession of the handgun is a person described in subparagraph (I) of paragraph (2) of subdivision (u), on the date that the person is delivered the firearm, the name and other information concerning the person taking possession of the firearm, how title or possession of the firearm was obtained and from whom, and a description of the firearm by make, model, serial number, and other identifying characteristics, shall be entered into the Automated Firearms System (AFS) via the California Law Enforcement Telecommunications System (CLETS) by the law enforcement or state agency that transferred or delivered the firearm. Those agencies without access to AFS shall arrange with the sheriff of the county in which the agency is located to input this information via this system.

(D) Where the person receiving title or possession of the handgun is a person described in subparagraph (J) of paragraph (2) of subdivision (u), on the date that the person is delivered the firearm, the name and other information concerning the person taking possession of the firearm, how title or possession of the firearm was obtained and from whom, and a description of the firearm by make, model, serial number, and other identifying characteristics, shall be entered into the AFS via the CLETS by the law enforcement or state agency that transferred or delivered the firearm. Those agencies without access to AFS shall arrange with the sheriff of the county in which the agency is located to input this information via this system. In addition, that law enforcement agency shall not deliver that handgun to the person referred to in this subparagraph unless, prior to the delivery of the same, the person presents proof to the agency that he or she is the holder of a handgun safety certificate.

(3) Subdivision (d) of Section 12072 shall not apply to a person who takes possession of a firearm by operation of law in a representative capacity who subsequently transfers ownership of the firearm to himself or herself in his or her individual capacity. In the case of a handgun, the individual shall obtain a handgun safety certificate prior to transferring ownership to himself or herself, or taking possession of a handgun in an individual capacity.

(j) Subdivision (d) of Section 12072 and subdivision (b) of Section 12801 shall not apply to deliveries, transfers, or returns of firearms made pursuant to Section 12021.3, 12028, 12028.5, or 12030.

(k) Section 12071, subdivision (c) of Section 12072, and subdivision (b) of Section 12801 shall not apply to any of the following:

(1) The delivery, sale, or transfer of unloaded firearms that are not handguns by a dealer to another dealer upon proof of compliance with the requirements of paragraph (1) of subdivision (f) of Section 12072.

(2) The delivery, sale, or transfer of unloaded firearms by dealers to persons who reside outside this state who are licensed pursuant to Chapter 44 (commencing with Section 921) of Title 18 of the United States Code and the regulations issued pursuant thereto.

(3) The delivery, sale, or transfer of unloaded firearms to a wholesaler if the firearms are being returned to the wholesaler and are intended as merchandise in the wholesaler's business.

(4) The delivery, sale, or transfer of unloaded firearms by one dealer to another dealer if the firearms are intended as merchandise in the receiving dealer's business upon proof of compliance with the requirements of paragraph (1) of subdivision (f) of Section 12072.

(5) The delivery, sale, or transfer of an unloaded firearm that is not a handgun by a dealer to himself or herself.

(6) The loan of an unloaded firearm by a dealer who also operates a target facility that holds a business or regulatory license on the premises of the building designated in the license or whose building designated in the license is on the premises of any club or organization organized for the purposes of practicing shooting at targets upon established ranges, whether public or private, to a person at that target facility or that club or organization, if the firearm is at all times kept within the premises of the target range or on the premises of the club or organization.

(l) A person who is exempt from subdivision (d) of Section 12072 or is otherwise not required by law to report his or her acquisition, ownership, or disposal of a handgun or who moves out of this state with his or her handgun may submit a report of the same to the Department of Justice in a format prescribed by the department.

(m) Subdivision (d) of Section 12072 and subdivision (b) of Section 12801 shall not apply to the delivery, sale, or transfer of unloaded firearms to a wholesaler as merchandise in the wholesaler's business by manufacturers or importers licensed to engage in that business pursuant to Chapter 44 (commencing with Section 921) of Title 18 of the United States Code and the regulations issued pursuant thereto, or by another wholesaler, if the delivery, sale, or transfer is made in accordance with Chapter 44 (commencing with Section 921) of Title 18 of the United States Code.

(n)(1) The waiting period described in Section 12071 or 12072 shall not apply to the delivery, sale, or transfer of a handgun by a dealer in either of the following situations:

(A) The dealer is delivering the firearm to another dealer and it is not intended as merchandise in the receiving dealer's business.

(B) The dealer is delivering the firearm to himself or herself and it is not intended as merchandise in his or her business.

(2) In order for this subdivision to apply, both of the following shall occur:

(A) If the dealer is receiving the firearm from another dealer, the dealer receiving the firearm shall present proof to the dealer delivering the firearm that he or she is licensed pursuant to Section 12071 by complying with paragraph (1) of subdivision (f) of Section 12072.

(B) Whether the dealer is delivering, selling, or transferring the firearm to himself or herself or to another dealer, on the date that the application to purchase is completed, the dealer delivering the firearm shall forward by prepaid mail to the Department of Justice a report of the same and the type of information concerning the purchaser or transferee as is indicated in subdivision (b) of Section 12077. Where the electronic or telephonic transfer of applicant information is used, on the date that the application to purchase is completed, the dealer delivering the firearm shall transmit an electronic or telephonic report of the same and the type of information concerning the purchaser or transferee as is indicated in subdivision (b) of Section 12077.

(o) Section 12071 and subdivisions (c) and (d) of Section 12072 shall not apply to the delivery, sale, or transfer of firearms regulated pursuant to Section 12020, Chapter 2 (commencing with Section 12200), or Chapter 2.3 (commencing with Section 12275), if the delivery, sale, or transfer is conducted in accordance with the applicable provisions of Section 12020, Chapter 2 (commencing with Section 12200), or Chapter 2.3 (commencing with Section 12275).

(p)(1) Paragraph (3) of subdivision (a) and subdivision (d) of Section 12072 shall not apply to the loan of a firearm that is not a handgun to a minor, with the express permission of the parent or legal guardian of the minor, if the loan does not exceed 30 days in duration and is for a lawful purpose.

(2) Paragraph (3) of subdivision (a) of Section 12072, subdivision (d) of Section 12072, and subdivision (b) of Section 12801 shall not apply to the loan of a handgun to a minor by a person who is not the parent or legal guardian of the minor if all of the following circumstances exist:

(A) The minor has the written consent of his or her parent or legal guardian that is presented at the time of, or prior to the time of, the loan, or is accompanied by his or her parent or legal guardian at the time the loan is made.

(B) The minor is being loaned the firearm for the purpose of engaging in a lawful, recreational sport, including, but not limited to, competitive shooting, or agricultural, ranching, or hunting activity, or a motion picture, television, or video production, or entertainment or theatrical event, the nature of which involves the use of a firearm.

(C) The duration of the loan does not exceed the amount of time that is reasonably necessary to engage in the lawful, recreational sport, including, but not limited to, competitive shooting, or agricultural, ranching, or hunting activity, or a motion picture, television, or video production, or entertainment or theatrical event, the nature of which involves the use of a firearm.

(D) The duration of the loan does not, in any event, exceed 10 days.

(3) Paragraph (3) of subdivision (a), and subdivision (d), of Section 12072, and subdivision (b) of Section 12801 shall not apply to the loan of a handgun to a minor by his or her parent or legal guardian if both of the following circumstances exist:

(A) The minor is being loaned the firearm for the purposes of engaging in a lawful, recreational sport, including, but not limited to, competitive shooting, or agricultural, ranching, or hunting activity, or a motion picture, television, or video production, or entertainment or theatrical event, the nature of which involves the use of a firearm.

(B) The duration of the loan does not exceed the amount of time that is reasonably necessary to engage in the lawful, recreational sport, including, but not limited to, competitive shooting, or agricultural, ranching, or hunting activity, or a motion picture, television, or video production, or entertainment or theatrical event, the nature of which involves the use of a firearm.

(4) Paragraph (3) of subdivision (a), and subdivision (d), of Section 12072 shall not apply to the transfer or loan of a firearm that is not a handgun to a minor by his or her parent or legal guardian.

(5) Paragraph (3) of subdivision (a), and subdivision (d), of Section 12072 shall not apply to the transfer or loan of a firearm that is not a handgun to a minor by his or her grandparent who is not the legal guardian of the minor if the transfer is done with the express permission of the parent or legal guardian of the minor.

(6) Subparagraph (A) of paragraph (3) of subdivision (a) of Section 12072 shall not apply to the sale of a handgun if both of the following requirements are satisfied:

(A) The sale is to a person who is at least 18 years of age.

(B) The firearm is an antique firearm as defined in paragraph (16) of subsection (a) of Section 921 of Title 18 of the United States Code.

(q) Subdivision (d) of Section 12072 shall not apply to the loan of a firearm that is not a handgun to a licensed hunter for use by that licensed hunter for a period of time not to exceed the duration of the hunting season for which that firearm is to be used.

(r) The waiting period described in Section 12071 or 12072 shall not apply to the delivery, sale, or transfer of a firearm to the holder of a special weapons permit issued by the Department of Justice issued pursuant to Section 12095, 12230, 12250, or 12305. On the date that the application to purchase is completed, the dealer delivering the firearm shall transmit to the Department of Justice an electronic or telephonic report of the same as is indicated in subdivision (b) or (c) of Section 12077.

(s)(1) Subdivision (d) of Section 12072 and subdivision (b) of Section 12801 shall not apply to the infrequent loan of an unloaded firearm by a person who is neither a dealer as defined in Section 12071 nor a federal firearms licensee pursuant to Chapter 44 of Title 18 of the United States Code, to a person 18 years of age or older for use solely as a prop in a motion picture, television, video, theatrical, or other entertainment production or event.

(2) Subdivision (d), and paragraph (1) of subdivision (f), of Section 12072, and subdivision (b) of Section 12801 shall not apply to the loan of an unloaded firearm by a person who is not a dealer as defined in Section 12071 but who is a federal firearms licensee pursuant to Chapter 44 of Title 18 of the United States Code, to a person who possesses a valid entertainment firearms permit issued pursuant to Section 12081, for use solely as a prop in a motion picture, television, video, theatrical, or other entertainment production or event. The person loaning the firearm pursuant to this paragraph shall retain a photocopy of the entertainment firearms permit as proof of compliance with this requirement.

(3) Subdivision (b) of Section 12071, subdivision (c) of, and paragraph (1) of subdivision (f) of, Section 12072, and subdivision (b) of Section 12801 shall not apply to the loan of an unloaded firearm by a dealer as defined in Section 12071, to a person who possesses a valid entertainment firearms permit issued pursuant to Section 12081, for use solely as a prop in a motion picture, television, video, theatrical, or other entertainment production or event. The dealer shall retain a photocopy of the entertainment firearms permit as proof of compliance with this requirement.

(4) Subdivision (b) of Section 12071, Subdivision (c) and Paragraph (1) of Subdivision (f) of Section 12072, and Subdivision (b) of Section 12801 shall not apply to the loan of an unloaded firearm to a consultant-evaluator by a person licensed pursuant to Section 12071 if the loan does not exceed 45 days from the date of delivery. At the time of the loan, the consultant-evaluator shall provide the following information, which the dealer shall retain for two years:

(a) A photocopy of a valid, current, government-issued identification to determine the consultant-evaluator's identity, including, but not limited to, a California driver's license, identification card, or passport.

(b) A photocopy of the consultant-evaluator's valid, current certificate of eligibility.

(c) A letter from the person licensed as an importer, manufacturer, or dealer pursuant to Chapter 44 (commencing with Section 921) of Title 18 of the United States Code, with whom the consultant-evaluator has a bona fide business relationship. The letter shall detail the bona fide business purposes for which the firearm is being loaned and confirm that the consultant-evaluator is being loaned the firearm as part of a bona fide business relationship.

(d) The signature of the consultant-evaluator on a form indicating the date the firearm is loaned and the last day the firearm may be returned.

(t)(1) The waiting period described in Section 12071 or 12072 shall not apply to the sale, delivery, loan, or transfer of a firearm that is a curio or relic, as defined in Section 478.11 of Title 27 of the Code of Federal Regulations, or its successor, by a dealer to a person who is licensed as a collector pursuant to Chapter 44 (commencing with Section 921) of Title 18 of the United States Code and the regulations issued pursuant thereto who has a current certificate of eligibility issued to him or her by the Department of Justice pursuant to Section 12071. On the date that the delivery, sale, or transfer is made, the dealer delivering the firearm shall transmit to the Department of Justice an electronic or telephonic report of the transaction as is indicated in subdivision (b) or (c) of Section 12077.

(2) Subdivision (d) of Section 12072 shall not apply to the infrequent sale, loan, or transfer of a firearm that is not a handgun, which is a curio or relic manufactured at least 50 years prior to the current date, but not including replicas thereof, as defined in Section 478.11 of Title 27 of the Code of Federal Regulations, or its successor.

(u) As used in this section:

(1) "Infrequent" has the same meaning as in paragraph (1) of subdivision (c) of Section 12070.

(2) "A person taking title or possession of firearms by operation of law" includes, but is not limited to, any of the following instances wherein an individual receives title to, or possession of, firearms:

(A) The executor or administrator of an estate if the estate includes firearms.

(B) A secured creditor or an agent or employee thereof when the firearms are possessed as collateral for, or as a result of, a default under a security agreement under the Commercial Code.

(C) A levying officer, as defined in Section 481.140, 511.060, or 680.260 of the Code of Civil Procedure.

(D) A receiver performing his or her functions as a receiver if the receivership estate includes firearms.

(E) A trustee in bankruptcy performing his or her duties if the bankruptcy estate includes firearms.

(F) An assignee for the benefit of creditors performing his or her functions as an assignee, if the assignment includes firearms.

(G) A transmutation of property consisting of firearms pursuant to Section 850 of the Family Code.

(H) Firearms passing to a surviving spouse pursuant to Chapter 1 (commencing with Section 13500) of Part 2 of Division 8 of the Probate Code.

(I) Firearms received by the family of a police officer or deputy sheriff from a local agency pursuant to Section 50081 of the Government Code.

(J) The transfer of a firearm by a law enforcement agency to the person who found the firearm where the delivery is to the person as the finder of the firearm pursuant to Article 1 (commencing with Section 2080) of Chapter 4 of Division 3 of the Civil Code.

12079. Permit for possession, transportation or sale of large capacity magazines

(a) Upon a showing that good cause exists, the Department of Justice may issue permits for the possession, transportation, or sale between a person licensed pursuant to Section 12071 and an out-of-state client, of large capacity magazines.

(b) For purposes of this section, "large capacity magazine" shall have the same meaning as that set forth in paragraph (25) of subdivision (c) of Section 12020.

12080. Distribution and sale of pamphlet summarizing state firearms laws

(a) The Department of Justice shall prepare a pamphlet which summarizes California firearms laws as they pertain to persons other than law enforcement officers or members of the armed services.

(b) The pamphlet shall include the following matters:

(1) Lawful possession.

(2) Licensing procedures.

(3) Transportation and use of firearms.

(4) Acquisition of hunting licenses.

(5) The safe handling and use of firearms.

(6) Various methods of safe storage and child proofing of firearms.

(7) The availability of firearms safety programs and devices.

(8) The responsibilities of firearms ownership.

(9) The operation of various types of firearms.

(10) The lawful use of deadly force.

(c) The department shall offer copies of the pamphlet at actual cost to firearms dealers licensed pursuant to Section 12071 who shall have copies of the most current version available for sale to retail purchasers or transferees of firearms. The cost of the pamphlet, if any, may be added to the sale price of the firearm. Other interested parties may purchase copies directly from the Department of General Services. The pamphlet shall declare that it is merely intended to provide a general summary of laws applicable to firearms and is not designed to provide individual guidance for specific areas. Individuals having specific questions shall be directed to contact their local law enforcement agency or private counsel.

(d) The Department of Justice or any other public entity shall be immune from any liability arising from the drafting, publication, or dissemination of the pamphlet or any reliance upon it. All receipts from the sale of these pamphlets shall be deposited as reimbursements to the support appropriation for the Department of Justice.

[Publisher's Note: The pamphlet is available online at http://caag.state.ca.us/firearms/index.html.**]**

12081. Loans of unloaded firearms for use as props exempt from specified transfer requirements

(a) Any person who is at least 21 years of age may apply for an entertainment firearms permit from the Department of Justice that authorizes the permitholder to possess firearms loaned to him or her for use solely as a prop in a motion picture, television, video, theatrical, or other entertainment production or event. Upon receipt of an initial or renewal application submitted as specified in subdivision (b), the department shall examine its records, records the department is authorized to request from the State Department of Mental Health pursuant to Section 8104 of the Welfare and Institutions Code, and records of the National Instant Criminal Background Check System as described in subsection (t) of Section 922 of Title 18 of the United States Code, in order to determine if the applicant is prohibited from possessing or receiving firearms. The department shall issue an entertainment firearms permit only if the records indicate that the applicant is not prohibited from possessing or receiving firearms pursuant to any federal, state, or local law.

(b)(1) Requests for entertainment firearms permits shall be made on application forms prescribed by the Department of Justice that require applicant information, including, but not limited to, the following:

(A) Complete name.

(B) Residential and mailing address.

(C) Telephone number.

(D) Date of birth.

(E) Place of birth.

(F) Country of citizenship and, if other than United States, alien number or admission number.

(G) Valid driver's license number or valid identification card number issued by the California Department of Motor Vehicles.

(H) Social security number.

(I) Signature.

(2) All applications must be submitted with the appropriate fee as specified in subdivision (c).

(3) An initial application for an entertainment firearms permit shall require the submission of fingerprint images and related information in a manner prescribed by the department, for the purpose of obtaining information as to the existence and nature of a record of state or federal level convictions and state or federal level arrests for which the department establishes that the individual was released on bail or on his or her own recognizance pending trial as needed to determine whether the applicant may be issued the permit. Requests for federal level criminal offender record information received by the Department of Justice pursuant to this section shall be forwarded by the department to the Federal Bureau of Investigation.

(4) The Department of Justice shall review the criminal offender record information specified in subdivision (l) of Section 11105 for entertainment firearms permit applicants.

(5) The Department of Justice shall review subsequent arrests, pursuant to Section 11105.2, to determine the continuing validity of the permit as specified in subdivision (d) for all entertainment firearms permitholders.

(6) Any person who furnishes a fictitious name or address or knowingly furnishes any incorrect information or knowingly omits any information required to be provided on this application is guilty of a misdemeanor.

(c) The Department of Justice shall recover the full costs of administering the program by assessing the following application fees:

(1) For the initial application: one hundred four dollars ($104). Of this sum, fifty-six dollars ($56) shall be deposited into the Fingerprint Fee Account, and forty-eight dollars ($48) shall be deposited into the Dealer Record of Sale Account.

(2) For each annual renewal application: twenty-nine dollars ($29), which shall be deposited into the Dealer Record of Sale Account.

(d) The implementation of subdivisions (a), (b), and (c) by the department is exempt from the Administrative Procedure Act (Chapter 3.5 (commencing with Section 11340) of Part 1 of Division 3 of Title 2 of the Government Code).

(e) The department shall annually review and shall adjust the fees specified in subdivision (c), if necessary, to fully fund, but not to exceed the actual costs of, the permit program provided for by this section, including enforcement of the program.

(f) An entertainment firearms permit issued by the Department of Justice shall be valid for one year from the date of issuance. If at any time during that year the permitholder becomes prohibited from possessing or receiving firearms pursuant to any federal, state, or local law, his or her entertainment firearms permit shall be no longer valid.

12082. Procedure for completing sale, loan, or transfer through dealer; Adoption of regulations; Violations by dealer

(a) A person shall complete any sale, loan, or transfer of a firearm through a person licensed pursuant to Section 12071 in accordance with this section in order to comply with subdivision (d) of Section 12072. The seller or transferor or the person loaning the firearm shall deliver the firearm to the dealer who shall retain possession of that firearm. The dealer shall then deliver the firearm to the purchaser or transferee or the person being loaned the firearm, if it is not prohibited, in accordance with subdivision (c) of Section 12072. If the dealer cannot legally deliver the firearm to the purchaser or transferee or the person being loaned the firearm, the dealer shall forthwith, without waiting for the conclusion of the waiting period described in Sections 12071 and 12072, return the firearm to the transferor or seller or the person loaning the firearm. The dealer shall not return the firearm to the seller or transferor or the person loaning the firearm when to do so would constitute a violation of subdivision (a) of Section 12072. If the dealer cannot legally return the firearm to the transferor or seller or the person loaning the firearm, then the dealer shall forthwith deliver the firearm to the sheriff of the county or the chief of police or other head of a municipal police department of any city or city and county who shall then dispose of the firearm in the manner provided by Sections 12028 and 12032. The purchaser or transferee or person being loaned the firearm may be required by the dealer to pay a fee not to exceed ten dollars ($10) per firearm, and no other fee may be charged by the dealer for a sale, loan, or transfer of a firearm conducted pursuant to this section, except for the applicable fee that the Department of Justice may charge pursuant to Section 12076. Nothing in these provisions shall prevent a dealer from charging a smaller fee. The fee that the department may charge is the fee that would be applicable pursuant to Section 12076, if the dealer was selling, transferring, or delivering a firearm to a purchaser or transferee or a person being loaned a firearm, with any other parties being involved in the transaction."

(b) The Attorney General shall adopt regulations under this section to do all of the following:

(1) Allow the seller or transferor or the person loaning the firearm, and the purchaser or transferee or the person being loaned the firearm, to complete a sale, loan, or transfer through a dealer, and to allow those persons and the dealer to comply with the requirements of this section and

Sections 12071, 12072, 12076, and 12077 and to preserve the confidentiality of those records.

(2) Where a personal handgun importer is selling or transferring a pistol, revolver, or other firearm capable of being concealed upon the person to comply with clause (ii) of subparagraph (A) of paragraph (2) of subdivision (f) of Section 12072, to allow a personal handgun importer's ownership of the pistol, revolver, or other firearm capable of being concealed upon the person being sold or transferred to be recorded in a manner that if the firearm is returned to that personal handgun importer because the sale or transfer cannot be completed, the Department of Justice will have sufficient information about that personal handgun importer so that a record of his or her ownership can be maintained in the registry provided by subdivision (c) of Section 11106.

(3) Ensure that the register or record of electronic transfer shall state the name and address of the seller or transferor of the firearm or the person loaning the firearm and whether or not the person is a personal handgun importer in addition to any other information required by Section 12077.

(c) Notwithstanding any other provision of law, a dealer who does not sell, transfer, or keep an inventory of handguns is not required to process private party transfers of handguns.

(d) A violation of this section by a dealer is a misdemeanor.

12083. Centralized list of exempted federal firearms licensees; Annual fee; Records on file; On-site inspection; Adoption of regulations

(a) Commencing January 1, 2008, the Department of Justice shall keep a centralized list of persons who identify themselves as being licensed pursuant to Chapter 44 (commencing with Section 921) of Title 18 of the United States Code as a dealer, pawnbroker, importer or manufacturer of firearms whose licensed premises are within this state and who declare to the department an exemption from the firearms dealer licensing requirements of Section 12070. The list shall be known as the centralized list of exempted federal firearms licensees. To qualify for placement on the centralized list, an applicant shall do all of the following:

(1) Possess a valid federal firearms license pursuant to Chapter 44 (commencing with Section 921) of Title 18 of the United States Code as a dealer, pawnbroker, importer, or manufacturer of firearms.

(2) Maintain eligibility under California law to possess firearms by possessing a current, valid certificate of eligibility pursuant to Section 12071.

(3) Maintain with the department a signed declaration enumerating the applicant's statutory exemptions from licensing requirements of Section 12070. Any person furnishing a fictitious name, knowingly furnishing any incorrect information, or knowingly omitting any information for the declaration shall be guilty of a misdemeanor.

(b) Commencing January 1, 2008, the department shall assess an annual fee of one hundred fifteen dollars ($115) to cover its costs of maintaining the centralized list of exempted federal firearms licensees prescribed by subdivision (a), conducting inspections in accordance with this section, and for the cost of maintaining the firearm shipment verification number system described in subdivision (f) of Section 12072. The department may increase the fee at a rate not to exceed the increase in the California Consumer Price Index as compiled and reported by the Department of Industrial Relations. The fees collected shall be deposited in the Dealers' Record of Sale Special Account.

(c)(1) Any person licensed pursuant to Chapter 44 (commencing with Section 921) of Title 18 of the United States Code as a dealer, pawnbroker, importer, or manufacturer of firearms whose licensed premises are within this state shall not import or receive firearms from any source unless listed on the centralized list of firearms dealers pursuant to Section 12071, or the centralized list of exempted federal firearms licensees pursuant to subdivision (a), or the centralized list of firearms manufacturers pursuant to subdivision (f) of Section 12086.

(2) A violation of this subdivision is a misdemeanor.

(d)(1) All persons on the centralized list of exempted federal firearms licensees prescribed by subdivision (a) shall record and keep on file for three years, the verification number that shall accompany firearms received from other federal firearms licensees pursuant to subdivision (f) of Section 12072.

(2) A violation of this subdivision is cause for immediate removal from the centralized list.

(e) Information compiled from the list described in subdivision (a) shall be made available for the following purposes:

(1) Requests from local, state, and federal law enforcing agencies and the duly constituted city, county, and city and county licensing authorities.

(2) When the information is requested by a person licensed pursuant to Chapter 44 (commencing with Section 921) of Title 18 of the United States Code for determining the validity of the license for firearm shipments.

(f) The department may conduct onsite inspections at the business premises of a person on the centralized list described in subdivision (a) to determine compliance with firearms laws pursuant to Article 4 (commencing with Section 12070) of Chapter 1 of Title 2 of Part 4 of the Penal Code. The department shall work in consultation with the Bureau of Alcohol, Tobacco, Firearms, and Explosives to ensure that licensees are not subject to duplicative inspections. During the inspection the following firearm records shall be made available for review:

(1) Federal records referred to in subdivision (a) of Section 478.125 of Title 27 of the Code of Federal Regulations and the bound book containing the same information referred to in Section 478.124a and subdivision (e) of Section 478.125 of Title 27 of the Code of Federal Regulations.

(2) Verification numbers issued pursuant to subdivision (f) of Section 12072.

(3) Any other records requested by the department to determine compliance with this article.

(g) The department may remove from the centralized list described in subdivision (a), any person who violates this article.

(h) The department may adopt regulations as necessary to carry out the provisions of this section, subdivision (f) of Section 12072, and Section 12071. The department shall work in consultation with the Bureau of Alcohol, Tobacco, Firearms, and Explosives to ensure that state regulations are not duplicative of federal regulations.

12085 Prohibition of unlicensed manufacture of firearms in state; Exception; Disposition of records on cessation of business; Violation

(a) Commencing July 1, 1999, no person, firm, or corporation licensed to manufacture firearms pursuant to Chapter 44 (commencing with Section 921) of Title 18 of the United States Code may manufacture firearms within this state unless licensed pursuant to Section 12086.

(b) Subdivision (a) does not apply to a person licensed to manufacture firearms pursuant to Chapter 44 (commencing with Section 921) of Title 18 of the United States Code who manufactures fewer than 100 firearms in a calendar year within this state.

(c) If a person, firm, or corporation required to be licensed pursuant to Section 12086 ceases operations, then the records required pursuant to paragraphs (6) and (10) of subdivision (c) of Section 12086 shall be forwarded to the federal Bureau of Alcohol, Tobacco, and Firearms within three days of the closure of business.

(d) A violation of this section is a misdemeanor.

(e)(1) As used in this section and Section 12086, the term "firearm" includes the frame or receiver of the weapon.

(2) As used in this section and Section 12086, the term "firearm" includes the unfinished frame or receiver of a weapon that can be readily converted to the functional condition of a finished frame or receiver.

(3) For purposes of this section and Section 12086, the term "firearm" does not include an unloaded firearm that is defined as an "antique firearm" in paragraph (16) of subsection (a) of Section 921 of Title 18 of the United States Code.

12086. Granting of licenses to manufacture firearms in state; Requirements for licensees; "Secure facility"; Revocation of license

(a)(1) As used in this section, "licensee" means a person, firm, or corporation that satisfies both of the following:

(A) Has a license issued pursuant to paragraph (2) of subdivision (b).

(B) Is among those recorded in the centralized list specified in subdivision (f).

(2) As used in this section, "department" means the Department of Justice.

(b)(1) The Department of Justice shall accept applications for, and shall grant licenses permitting, the manufacture of firearms within this state. The department shall inform applicants who are denied licenses of the reasons for the denial in writing.

(2) No license shall be granted by the department unless and until the applicant presents proof that he or she has all of the following:

(A) A valid license to manufacture firearms issued pursuant to Chapter 44 (commencing with Section 921) of Title 18 of the United States Code.

(B) Any regulatory or business license, or licenses, required by local government.

(C) A valid seller's permit or resale certificate issued by the State Board of Equalization, if applicable.

(D) A certificate of eligibility issued by the Department of Justice pursuant to paragraph (4) of subdivision (a) of Section 12071.

(3) The department shall adopt regulations to administer this section and Section 12085 and shall recover the full costs of administering the program by collecting fees from license applicants. Recoverable costs shall include, but not be limited to, the costs of inspections and maintaining a centralized list of licensed firearm manufacturers. The fee for licensed manufacturers who produce fewer than 500 firearms in a calendar year within this state shall not exceed two hundred fifty dollars ($250) per year or the actual costs of inspections and maintaining a centralized list of firearm manufacturers and any other duties of the department required pursuant to this section and Section 12085, whichever is less.

(4) A license granted by the department shall be valid for no more than one year from the date of issuance and shall be in the form prescribed by the Attorney General.

(c) A licensee shall comply with the following prohibitions and requirements:

(1) The business shall be conducted only in the buildings designated in the license.

(2) The license or a copy thereof, certified by the department, shall be displayed on the premises where it can easily be seen.

(3) Whenever a licensee discovers that a firearm has been stolen or is missing from the licensee's premises, the licensee shall report the loss or theft within 48 hours of the discovery to all of the following:

(A) The Department of Justice, in a manner prescribed by the department.

(B) The federal Bureau of Alcohol, Tobacco, and Firearms.

(C) The police department in the city or city and county where the building designated in the license is located.

(D) If there is no police department in the city or city and county where the building designated in the license is located, the sheriff of the county where the building designated in the license is located.

(4)(A) The licensee shall require that each employee obtain a certificate of eligibility pursuant to paragraph (4) of subdivision (a) of Section 12071, which shall be renewed annually, prior to being allowed to come into contact with any firearm.

(B) The licensee shall prohibit any employee who the licensee knows or reasonably should know is within a class of persons prohibited from possessing firearms pursuant to Section 12021 or 12021.1 of this code, or Section 8100 or 8103 of the Welfare and Institutions Code, from coming into contact with any firearm.

(5)(A) Each firearm the licensee manufactures in this state shall be identified with a unique serial number stamped onto the firearm utilizing the method of compression stamping.

(B) Licensed manufacturers who produce fewer than 500 firearms in a calendar year within this state may serialize long guns only by utilizing a method of compression stamping or by engraving the serial number onto the firearm.

(C) The licensee shall stamp the serial number onto the firearm within one business day of the time the receiver or frame is manufactured.

(D) The licensee shall not use the same serial number for more than one firearm.

(6)(A) The licensee shall record the type, model, caliber, or gauge, and serial number of each firearm manufactured or acquired, and the date of the manufacture or acquisition, within one business day of the manufacture or acquisition.

(B) The licensee shall maintain permanently within the building designated in the license the records required pursuant to subparagraph (A).

(C) Backup copies of the records described in subparagraph (A), whether electronic or hard copy, shall be made at least once a month. These backup records shall be maintained in a facility separate from the one in which the primary records are stored.

(7)(A) The licensee shall allow the department to inspect the building designated in the license to ensure compliance with the requirements of this section.

(B) The licensee shall allow any peace officer, authorized law enforcement employee, or Department of Justice employee designated by the Attorney General, upon the presentation of proper identification, to inspect facilities and records during business hours to ensure compliance with the requirements of this section.

(8) The licensee shall store in a secure facility all firearms manufactured and all barrels for firearms manufactured.

(9)(A) The licensee shall notify the chief of police or other head of the municipal police department in the city or city and county where the building designated in the license is located that the licensee is manufacturing firearms within that city or city and county and the location of the licensed premises.

(B) If there is no police department in the city or city and county where the building designated in the license is located, the licensee shall notify the sheriff of the county where the building designated in the license is located that the licensee is manufacturing firearms within that county and the location of the licensed premises.

(10) For at least 10 years, the licensee shall maintain records of all firearms that are lost or stolen, as prescribed by the department.

(d) Except as otherwise provided in subdivision (e), as used in this section, a "secure facility" means that the facility satisfies all of the following: ...

(e) For purposes of this section, any licensed manufacturer who produces fewer than 500 firearms in a calendar year within this state may maintain a "secure facility" by complying with all of the requirements described in subdivision (d), or may design a security plan that is approved by the Department of Justice or the federal Bureau of Alcohol, Tobacco, and Firearms.

(1) If a security plan is approved by the federal Bureau of Alcohol, Tobacco, and Firearms, the approved plan, along with proof of approval, shall be filed with the Department of Justice and the local police department. If there is no police department, the filing shall be with the county sheriff's office.

(2) If a security plan is approved by the Department of Justice, the approved plan, along with proof of approval, shall be filed with the local police department. If there is no police department, the filing shall be with the county sheriff's office.

(f)(1) Except as otherwise provided in this subdivision, the Department of Justice shall maintain a centralized list of all persons licensed pursuant to paragraph (2) of subdivision (b). The centralized list shall be provided annually to each police department and county sheriff within the state.

(2) Except as provided in paragraph (3), the license of any licensee who violates this section may be revoked.

(3) The license of any licensee who knowingly or with gross negligence violates this section or violates this section three times shall be revoked, and that person, firm, or corporation shall become permanently ineligible to obtain a license pursuant to this section.

(g)(1) Upon the revocation of the license, notification shall be provided to local law enforcement authorities in the jurisdiction where the licensee's business is located and to the federal Bureau of Alcohol, Tobacco, and Firearms.

(2) The department shall make information concerning the location and name of a licensee available, upon request, for the following purposes only:

(A) Law enforcement.

(B) When the information is requested by a person licensed pursuant to Chapter 44 (commencing with Section 921) of Title 18 of the United States Code for determining the validity of the license for firearm shipments.

(3) Notwithstanding paragraph (2), the department shall make the name and business address of a licensee available to any person upon written request.

(h) The Department of Justice shall maintain and make available upon request information concerning the number of inspections conducted and the amount of fees collected pursuant to paragraph (3) of subdivision (b), the number of licensees removed from the centralized list described in subdivision (f), and the number of licensees found to have violated this section.

Article 4.5. Firearms Safety Devices

12087. Citation of article This article shall be known and may be cited as the "Aroner-Scott-Hayden Firearms Safety Act of 1999."

12087.6. Definitions As used in this article:

(a) "Firearms safety device" means a device other than a gun safe that locks and is designed to prevent children and unauthorized users from firing a firearm. The device may be installed on a firearm, be incorporated into the design of the firearm, or prevent access to the firearm.

(b) "Gun safe" means a locking container that fully contains and secures one or more firearms, and that meets the standards for gun safes adopted pursuant to Section 12088.2.

(c) "Long-gun safe" means a locking container designed to fully contain and secure a rifle as defined in paragraph (20) of subdivision (c) of Section 12020 or a shotgun as defined in paragraph (21) of subdivision (c) of Section 12020, that has a locking system consisting of either a mechanical combination lock or an electronic combination lock that has at least 1,000 possible unique combinations consisting of a minimum of three numbers, letters, or symbols per combination, and that is not listed on the roster maintained pursuant to subdivision (d) of Section 12088.

12088.1. Requirement of safety device upon sale or transfer; Exemptions; Requirement of warning language or labels

(a) All firearms sold or transferred in this state by a licensed firearms dealer, including private transfers through a dealer, and all firearms manufactured in this state, shall include or be accompanied by a firearms safety device that is listed on the Department of Justice's roster of approved firearms safety devices and that is identified as appropriate for that firearm by reference to either the manufacturer and model of the firearm, or to the physical characteristics of the firearm that match those listed on the roster for use with the device.

(b) All firearms sold or transferred in this state by a licensed firearms dealer, including private transfers through a dealer, and all firearms manufactured in this state shall be accompanied with warning language or labels as described in Section 12088.3.

(c)(1) All long-gun safes commercially sold or transferred in this state, or manufactured in this state for sale in this state, that do not meet the standards for gun safes adopted pursuant to Section 12088.2 shall be accompanied by the following warning:

"WARNING: This gun safe does not meet the safety standards for gun safes specified in California Penal Code Section 12088.2. It does not satisfy the requirements of Penal Code Section 12088.1, which mandates that all firearms sold in California be accompanied by a firearms safety device or proof of ownership, as required by law, of a gun safe that meets the Section 12088.2 minimum safety standards developed by the California Attorney General."

(2) This warning shall be conspicuously displayed in its entirety on the principal display panel of the gun safe's package, on any descriptive materials that accompany the gun safe, and on a label affixed to the front of the gun safe.

(3) This warning shall be displayed in both English and Spanish in conspicuous and legible type in contrast by typography, layout, or color with other printed matter on the package or descriptive materials in a manner consistent with Part 1500.121 of Title 16 of the Code of Federal Regulations, or successor regulations thereto.

(d) The sale or transfer of a firearm shall be exempt from subdivision (a) if both of the following apply:

(1) The purchaser or transferee owns a gun safe that meets the standards set forth in Section 12088.2. Gun safes shall not be required to be tested, and therefore may meet the standards without appearing on the Department of Justice roster.

(2) The purchaser or transferee presents an original receipt for purchase of the gun safe, or other proof of purchase or ownership of the gun safe as authorized by the Attorney General, to the firearms dealer. The dealer shall maintain a copy of this receipt or proof of purchase with the dealers' record of sales of firearms.

(e) The sale or transfer of a firearm shall be exempt from subdivision (a) if all of the following apply:

(1) The purchaser or transferee purchases an approved safety device no more than 30 days prior to the day the purchaser or transferee takes possession of the firearm.

(2) The purchaser or transferee presents the approved safety device to the firearms dealer when picking up the firearm.

(3) The purchaser or transferee presents an original receipt to the firearms dealer which shows the date of purchase, the name, and the model number of the safety device.

(4) The firearms dealer verifies that the requirements in (1) to (3), inclusive, have been satisfied.

(5) The firearms dealer maintains a copy of the receipt along with the dealers' record of sales of firearms.

12088.15. Prohibited activities; Penalties for violation; Civil action **(a)** No person shall keep for commercial sale, offer, or expose for commercial sale, or commer-cially sell any firearms safety device that is not listed on the roster maintained pursuant to sub-division (d) of Section 12088, or that does not comply with the standards for firearms safety de-vices adopted pursuant to Section 12088.2.

(b) No person may distribute as part of an organized firearm safety program, with or without consideration, any firearm safety device that is not listed on the roster maintained pursuant to subdivision (d) of Section 12088 or does not comply with the standards for firearms safety devices adopted pursuant to Section 12088.2.

(c) No long-gun safe may be manufactured in this state for sale in this state that does not comply with the standards for gun safes adopted pursuant to Section 12088.2, unless the long-gun safe is labeled by the manufacturer consistent with the requirements of Section 12088.1.

(d)(1) Any person who keeps for commercial sale, offers, or exposes for commercial sale, or who commercially sells a long-gun safe that does not comply with the standards for gun safes adopted pursuant to Section 12088.2, and who knows or has reason to know, that the long-gun safe in question does not meet the standards for gun safes adopted pursuant to Section 12088.2, is in violation of this section, and is punishable as provided in subdivision (e), unless the long-gun safe is labeled pursuant to Section 12088.1.

(2) Any person who keeps for commercial sale, offers, or exposes for commercial sale, or who commercially sells a long-gun safe that does not comply with the standards for gun safes adopted pursuant to Section 12088.2, and who removes or causes to be removed from the long-gun safe, the label required pursuant to Section 12088.1, is in violation of this section, and is punishable as provided in subdivision (e).

(e) A violation of this section is punishable by a civil fine of up to five hundred dollars ($500). A second violation of this section that occurs within five years of the date of a previous offense is punishable by a civil fine of up to one thousand dollars ($1,000) and, if the violation is committed by a licensed firearms dealer, the dealer shall be ineligible to sell firearms in this state for 30 days. A third or subsequent violation that occurs within five years of two or more previous offenses is punishable by a civil fine of up to five thousand dollars ($5,000) and, if the violation is committed by a licensed firearms dealer, the firearms dealer shall be permanently ineligible to sell firearms in this state.

(f) The Attorney General, a district attorney, or a city attorney may bring a civil action for a violation of the provisions of this section.

12088.3. Warning labels

(a) The packaging of any firearm and any descriptive materials that accompany any firearm sold or transferred in this state, or delivered for sale in this state, by any licensed manufacturer, or licensed dealer, shall bear a label containing the following warning statement:

WARNING

Children are attracted to and can operate firearms that can cause severe injuries or death. Prevent child access by always keeping guns locked away and unloaded when not in use. If you keep a loaded firearm where a child obtains and improperly uses it, you may be fined or sent to prison.

A yellow triangle containing an exclamation mark shall appear immediately before the word "Warning" on the label.

(b) If the firearm is sold or transferred without accompanying packaging, the warning label or notice shall be affixed to the firearm itself by a method to be prescribed by regulation of the Attorney General.

(c) The warning statement required under subdivisions (a) and (b) shall be:

(1) Displayed in its entirety on the principal display panel of the firearm's package, and on any descriptive materials that accompany the firearm.

(2) Displayed in both English and Spanish in conspicuous and legible type in contrast by typography, layout, or color with other printed matter on that package or descriptive materials in a manner consistent with Part 1500.121 of Title 16, of the Code of Federal Regulations, or successor regulations thereto.

12088.4 Nonconformity of gunsafe or safety device with required standards

If at any time the Attorney General determines that a gun safe or firearms safety device subject to the provisions of this art-icle and sold after January 1, 2002, does not conform with the standards required by subdivi-sion (a) of Section 12088.1 or Section 12088.2, the Attorney General may order the recall and replacement of the gun safe or firearms safety device, or order that the gun safe or firearm safety device be brought into conformity with those requirements. If the firearms safety device cannot be separated from the firearm without damaging the firearm, the Attorney General may order the recall and replacement of the firearm. If the firearms safety device can be separated and reattached to the firearm without damaging the firearm, the licensed manufacturer or li-censed firearms dealer shall immediately pro-vide a conforming replacement as instructed by the Attorney General.

12088.6. Fines and other sanctions for violations

Any violation of Section 12088.1 or Section 12088.3 is punishable by a fine of one thousand dollars ($1,000). On the second violation of any of those sections, the licensed firearm manufacturer shall be ineligible to manufacture, or the licensed fire-arm dealer shall be ineligible to sell, firearms in this state for 30 days, and shall be punished by a fine of one thousand dollars ($1,000). On the third violation of any of those sections, a firearm manufacturer shall be permanently ineligible to manufacture firearms in this state. On the third violation of any of those sections, a licensed fire-arm dealer shall be permanently ineligible to sell firearms in this state.

12088.8. Exemptions

(a) This article does not apply to the commerce of any firearm defined as an "antique firearm" in paragraph (16) of subsection (a) of Section 921 of Title 18 of the United States Code.

(b) This article shall not apply to the commerce of any firearm intended to be used by a salaried, full-time peace officer as defined in Chapter 4.5 (commencing with Section 830) of Title 3 of Part 2 for purposes of law enforcement. Nothing in this article shall preclude local governments, local agencies, or state law enforcement agencies from requiring their peace officers to store their firearms in gun safes or attach firearms safety devices to those firearms.

12088.9. Transaction fee to support costs of program; creation of Firearm Safety Account

(a) The Department of Justice may require each dealer to charge each firearm purchaser or transferee a fee not to exceed one dollar ($1) for each firearm transaction. The fee shall be for the purpose of supporting department program costs related to this act, including the establishment, maintenance, and upgrading of related data base systems and public rosters.

Article 5. Obliteration of Identification Marks

12090. Tampering with marks on firearms; Permission

Any person who changes, alters, removes or obliterates the name of the maker, model, manufacturer's number, or other mark of identification, including any distinguishing number or mark assigned by the Department of Justice on any pistol, revolver, or any other firearm, without first having secured written permission from the department to make such change, alteration or removal shall be punished by imprisonment in the state prison.

12091. Changing identification marks

Possession of any pistol or revolver upon which the name of the maker, model, manufacturer's number or other mark of identification has been changed, altered, removed, or obliterated, shall be presumptive evidence that the possessor has changed, altered, removed, or obliterated the same. **[Publisher's Note:** This section has been held unconstitutional in part in the case of *In re Christopher K.*, 110 Cal.Rptr.2d 914, 91 Cal.App.4th 853 (App. 4 Dist. 2001).**]**

12092. Assigning number or mark; Restoration of number or mark

The Department of Justice upon request may assign a distinguishing number or mark of identification to any firearm whenever it is without a manufacturer's number, or other mark of identification or whenever the manufacturer's number or other mark of identification or the distinguishing number or mark assigned by the department has been destroyed or obliterated.

12093. Placing number or mark on pistol or revolver

Any person may place or stamp on any pistol, revolver, or other firearm any number or identifying indicium, provided the number or identifying indicium does not change, alter, remove, or obliterate the manufacturer's name, number, model, or other mark of identification. This section does not prohibit restoration by the owner of the name of the maker, model, or of the original manufacturer's number or other mark of identification when such restoration is authorized by the department, nor prevent any manufacturer from placing in the ordinary course of business the name of the maker, model, manufacturer's number, or other mark of identification upon a new firearm.

12094. Possession or disposition of pistol, revolver, or other firearm without correct identification mark; Exceptions

(a) Any person with knowledge of any change, alteration, removal, or obliteration described herein, who buys, receives, disposes of, sells, offers for sale, or has in his or her possession any pistol, revolver, or other firearm which has had the name of the maker, model, or the manufacturer's number or other mark of identification including any distinguishing number or mark assigned by the Department of Justice changed, altered, removed, or obliterated is guilty of a misdemeanor.

(b) Subdivision (a) does not apply to any of the following:

(1) The acquisition or possession of a firearm described in subdivision (a) by any member of the military forces of this state or of the United States, while on duty and acting within the scope and course of his or her employment.

(2) The acquisition or possession of a firearm described in subdivision (a) by any peace officer described in Chapter 4.5 (commencing with Section 830) of Title 3 of Part 2, while on duty and acting within the scope and course of his or her employment.

(3) The acquisition or possession of a firearm described in subdivision (a) by any employee of a forensic laboratory, while on duty and acting within the scope and course of his or her employment.

(4) The possession and disposition of a firearm described in subdivision (a) by a person who meets all of the following:

(A) He or she is not prohibited from possessing firearms or ammunition pursuant to Section 12021 or 12021.1 or paragraph (1) of subdivision (b) of Section 12316 of this code, or Section 8100 or 8103 of the Welfare and Institutions Code.

(B) The person possessed the firearm no longer than was necessary to deliver the same to a law enforcement agency for that agency's disposition according to law.

(C) If the person is transporting the firearm, he or she is transporting the firearm to a law enforcement agency in order to deliver the firearm to the law enforcement agency for the agency's disposition according to law.

(D) If the person is transporting the firearm to a law enforcement agency, he or she has given prior notice to the law enforcement agency that he or she is transporting the firearm to that law enforcement agency for that agency's disposition according to law.

(E) The firearm is transported in a locked container as defined in subdivision (d) of Section 12026.2.

Article 6. Permits

12095. Permits for short-barreled shotguns or rifles

(a) If it finds that it does not endanger the public safety, the Department of Justice may issue permits initially valid for a period of one year, and renewable annually thereafter, for the manufacture, possession, transportation, or sale of short-barreled shotguns or short-barreled rifles upon a showing that good cause exists for the issuance thereof to the applicant for the permit. No permit shall be issued to a person who is under 18 years of age.

(b) Good cause, for the purposes of this section, shall be limited to only the following:

(1) The permit is sought for the manufacture, possession, or use with blank cartridges, of a short-barreled rifle or short-barreled shotgun, solely as props for a motion picture, television, or video production or entertainment event.

(2) The permit is sought for the manufacture of, exposing for sale, keeping for sale, sale of, importation or lending of short-barreled rifles or short-barreled shotguns to the entities listed in paragraph (1) of subdivision (b) of Section 12020 by persons who are licensed as dealers or manufacturers under the provisions of Chapter 53 (commencing with Section 5801) of Title 26 of the United States Code, as amended, and the regulations issued pursuant thereto.

12096. Permit applications, forms, and fees

Applications for permits shall be filed in writing, signed by the applicant if an individual, or by a member or officer qualified to sign if the applicant is a firm or corporation, and shall state the name, business in which engaged, business address, and a full description of the use to which the short-barreled shotguns or short-barreled rifles are to be put.

Applications and permits shall be uniform throughout the state on forms prescribed by the Department of Justice.

Each applicant for a permit shall pay at the time of filing his or her application a fee determined by the Department of Justice not to exceed the application processing costs of the Department of Justice. A permit granted pursuant to this article may be renewed one year from the date of issuance, and annually thereafter, upon the filing of a renewal application and the payment of a permit renewal fee not to exceed the application processing costs of the Department of Justice. After the department establishes fees sufficient to reimburse the department for processing costs, fees charged shall increase at a rate not to exceed the legislatively approved annual cost-of-living adjustments for the department's budget.

12097. Permit placement; Identifying numbers

(a) Every person, firm, or corporation to whom a permit is issued shall keep it on his or her person or at the place where the short-barreled shotguns or short-barreled rifles are kept. The permit shall be open to inspection by any peace officer or any other person designated by the authority issuing the permit.

(b) Every short-barreled shotgun or short-barreled rifle possessed pursuant to the provisions of this article shall bear a unique identifying number. If a weapon does not bear a unique identifying number, the Department of Justice shall assign a number which shall be placed or stamped on that weapon.

12098. Revocation of permit

Permits issued in accordance with this article may be revoked by the issuing authority at any time when it appears that the need for the short-barreled shotguns or short-barreled rifles has ceased or that the holder of the permit has used the short-barreled shotguns or short-barreled rifles for purposes other than those allowed by the permit or that the holder of the permit has not exercised great care in retaining custody of any weapons possessed under the permit.

12099. Annual inspection for security and safe storage by Department of Justice

(a) Except as provided in subdivision (b), the Department of Justice shall, for every person, firm, or corporation to whom a permit is issued pursuant to this article, annually conduct an inspection for security and safe storage purposes, and to reconcile the inventory of short-barreled shotguns and short-barreled rifles.

(b) A person, firm, or corporation with an inventory of fewer than five devices that require any Department of Justice permit shall be subject to an inspection for security and safe storage purposes, and to reconcile inventory, once every five years, or more frequently if determined by the department.

Article 7. Juveniles

12101. Possession of concealable firearm or live ammunition by minor

(a) (1) A minor shall not possess a pistol, revolver, or other firearm capable of being concealed upon the person.

(2) Paragraph (1) shall not apply if one of the following circumstances exists:

(A) The minor is accompanied by his or her parent or legal guardian, and the minor is actively engaged in, or is in direct transit to or from, a lawful, recreational sport, including, but not limited to, competitive shooting, or agricultural, ranching, or hunting activity, or a motion picture, television, or video production, or entertainment or theatrical event, the nature of which involves this use of a firearm.

(B) The minor is accompanied by a responsible adult, the minor has the prior written consent of his or her parent or legal guardian, and the minor is actively engaged in, or is in direct transit to or from, a lawful, recreational sport, including, but not limited to, competitive shooting, or agricultural, ranching, or hunting activity, or a motion picture, television, or video production, or entertainment or theatrical event, the nature of which involves the use of a firearm.

(C) The minor is at least 16 years of age, the minor has the prior written consent of his or her parent or legal guardian and the minor is actively engaged in, or is in direct transit to or from, a lawful recreational sport, including, but not limited to, competitive shooting, or agricultural, ranching, or hunting activity, or a motion picture, television, or video production, or entertainment or theatrical event, the nature of which involves the use of a firearm.

(D) The minor has the prior written consent of his or her parent or legal guardian, the minor is on lands owned or lawfully possessed by his or her parent or legal guardian, and the minor is actively engaged in, or is in direct transit to or from, a lawful, recreational sport, including, but not limited to, competitive shooting, or agricultural, ranching, or hunting activity, or a motion picture, television, or video production, or entertainment or theatrical event, the nature of which involves the use of a firearm.

(b) **(1)** A minor shall not possess live ammunition.

(2) Paragraph (1) shall not apply if one of the following circumstances exists:

(A) The minor has the written consent of his or her parent or legal guardian to possess live ammunition.

(B) The minor is accompanied by his or her parent or legal guardian.

(C) The minor is actively engaged in, or is going to or from, a lawful, recreational sport, including, but not limited to, competitive shooting, or agricultural, ranching, or hunting activity the nature of which involves the use of a firearm.

(c) Every minor who violates this section shall be punished as follows:

(1) By imprisonment in the state prison or in a county jail if one of the following applies:

(A) The minor has been found guilty previously of violating this section.

(B) The minor has been found guilty previously of an offense specified in subdivision (b) of Section 12021.1 or in Section 12020, 12220, 12520, or 12560.

(C) The minor has been found guilty of a violation of paragraph (1) of subdivision (a).

(2) Violations of this section other than those violations specified in paragraph (1) shall be punishable as a misdemeanor.

(d) In a proceeding to enforce this section brought pursuant to Article 14 (commencing with Section 601) of Chapter 2 of Part 1 of the Welfare and Institutions Code, the court may require the custodial parent or legal guardian of a minor who violates this section to participate in classes on parenting education that meet the requirements established in Section 16507.7 of the Welfare and Institutions Code.

(e) As used in this section, "responsible adult" means a person at least 21 years of age who is not within a class of persons prohibited from owning or possessing firearms by virtue of Section 12021 or 12021.1 of this code, or Section 8100 or 8103 of the Welfare and Institutions Code.

(f) It is not the intent of the Legislature in enacting the amendments to this section or to Section 12078 to expand or narrow the application of current statutory or judicial authority as to the rights of minors to be loaned or to possess live ammunition or a firearm for the purpose of self-defense or the defense of others.

Chapter 1.3. Unsafe Handguns

12125. Possession of concealable firearm or live ammunition by minor

(a) Commencing January 1, 2001, any person in this state who manufactures or causes to be manufactured, imports into the state for sale, keeps for sale, offers or exposes for sale, gives, or lends any unsafe handgun shall be punished by imprisonment in a county jail not exceeding one year.

(b) This section shall not apply to any of the following:

(1) The manufacture in this state, or importation into this state, of any prototype pistol, revolver, or other firearm capable of being concealed upon the person when the manufacture or importation is for the sole purpose of allowing an independent laboratory certified by the Department of Justice pursuant to Section 12130 to conduct an independent test to determine whether that pistol, revolver, or other firearm capable of being concealed upon the person is prohibited by this chapter, and, if not, allowing the department to add the firearm to the roster of pistols, revolvers, and other firearms capable of being concealed upon the person that may be sold in this state pursuant to Section 12131.

(2) The importation or lending of a pistol, revolver, or other firearm capable of being concealed upon the person by employees or authorized agents of entities determining whether the weapon is prohibited by this section.

(3) Firearms listed as curios or relics, as defined in Section 478.11 of Title 27 of the Code of Federal Regulations.

(4) The sale or purchase of any pistol, revolver or other firearm capable of being concealed upon the person, if the pistol, revolver, or other firearm is sold to, or purchased by, the Department of Justice, any police department, any sheriff's official, any marshal's office, the Youth and Adult Correctional Agency, the California Highway Patrol, any district attorney's office, or the military or naval forces of this state or of the United States for use in the discharge of their official duties. Nor shall anything in this section prohibit the sale to, or purchase by, sworn members of these agencies of any pistol, revolver, or other firearm capable of being concealed upon the person.

(c) Violations of subdivision (a) are cumulative with respect to each handgun and shall not be construed as restricting the application of any other law. However, an act or omission punishable in different ways by this section and other provisions of law shall not be punished under more than one provision, but the penalty to be imposed shall be determined as set forth in Section 654.

12126. "Unsafe handgun"; "Chamber load indicator"; "Magazine disconnect mechanism"; "Semiautomatic pistol"

As used in this chapter, "unsafe handgun" means any pistol, revolver, or other firearm capable of being concealed upon the person, as defined in subdivision (a) of Section 12001, for which any of the following is true:

(a) For a revolver:

(1) It does not have a safety device that, either automatically in the case of a double-action firing mechanism, or by manual operation in the case of a single-action firing mechanism, causes the hammer to retract to a point where the firing pin does not rest upon the primer of the cartridge.

(2) It does not meet the firing requirement for handguns pursuant to Section 12127.

(3) It does not meet the drop safety requirement for handguns pursuant to Section 12128.

(b) For a pistol:

(1) It does not have a positive manually operated safety device, as determined by standards relating to imported guns promulgated by the federal Bureau of Alcohol, Tobacco, and Firearms.

(2) It does not meet the firing requirement for handguns pursuant to Section 12127.

(3) It does not meet the drop safety requirement for handguns pursuant to Section 12128.

(4) Commencing January 1, 2006, for a center-fire semiautomatic pistol that is not already listed on the roster pursuant to Section 12131, it does not have either a chamber load indicator, or a magazine disconnect mechanism.

(5) Commencing January 1, 2007, for all center fire semiautomatic pistols that are not already listed on the roster pursuant to Section 12131, it does not have both a chamber load indicator and if it has a detachable magazine, a magazine disconnect mechanism.

(6) Commencing January 1, 2006, for all rimfire semiautomatic pistols that are not already listed on the roster pursuant to Section 12131, it does not have a magazine disconnect mechanism, if it has a detachable magazine.

(7) Commencing January 1, 2010, for all semiautomatic pistols that are not already listed on the roster pursuant to section 12131, it is not designed and equipped with a microscopic array of characters that identify the make, model, and serial number of the pistol, etched or otherwise imprinted in two or more places on the interior surface or internal working parts of the pistol, and that are transferred by imprinting on each cartridge case when the firearm is fired, provided that the Department of Justice certifies that the technology used to create the imprint is available to more than one manufacturer unencumbered by any patent restrictions. The attorney general may also approve a method of equal or greater reliability and effectiveness in identifying the specific serial number of a firearm from spent cartridge casings discharged by that firearm than that which is set forth in this paragraph, to be thereafter required as otherwise set forth by this paragraph where the Attorney General certifies that this new method is also unencumbered by any patent restrictions. Approval by the attorney general shall include notice of that fact via regulations adopted by the Attorney General for purposes of implementing that method for purposes of this paragraph. The microscopic array of characters required by this section shall not be considered the name of the maker, model, manufacturer's number, or other mark of identification, including any distinguishing number or mark assigned by the Department of Justice, within the meaning of Sections 12090 and 12094.

(c) As used in this section, a "chamber load indicator" means a device that plainly indicates that a cartridge is in the firing chamber. A device satisfies this definition if it is readily visible, has incorporated or adjacent explanatory text or graphics, or both, and is designed and intended to indicate to a reasonably foreseeable adult user of the pistol, without requiring the user to refer to a user's manual or any other resource other than the pistol itself, whether a cartridge is in the firing chamber.

(d) As used in this section, a "magazine disconnect mechanism" means a mechanism that prevents a semiautomatic pistol that has a detachable magazine from operating to strike the primer of ammunition in the firing chamber when a detachable magazine is not inserted in the semiautomatic pistol.

(e) As used in this section, a "semiautomatic pistol" means a pistol, as defined in subdivision (a) of Section 12001, the operating mode of which uses the energy of the explosive in a fixed cartridge to extract a fired cartridge and chamber a fresh cartridge with each single pull of the trigger.

12127. "Firing requirement for handguns"

(a) As used in this chapter, the "firing requirement for handguns" means a test in which the manufacturer provides three handguns of the make and model for which certification is sought to an independent testing laboratory certified by the Attorney General pursuant to Section 12130. These handguns may not be refined or modified in any way from those that would be made available for retail sale if certification is granted. The magazines of a tested pistol shall be identical to those that would be provided with the pistol to a retail customer. The laboratory shall fire 600 rounds from each gun, stopping after each series of 50 rounds has been fired for 5 to 10 minutes to allow the weapon to cool, stopping after each series of 100 rounds has been fired to tighten any loose screws and clean the gun in accordance with the manufacturer's instructions, and stopping as needed to refill the empty magazine or cylinder to capacity before continuing. The ammunition used shall be of the type recommended by the handgun manufacturer in the user manual, or if none is recommended, any standard ammunition of the correct caliber in new condition that is commercially available. A handgun shall pass this test if each of the three test guns meets both of the following:

(1) Fires the first 20 rounds without a malfunction that is not due to ammunition that fails to detonate.

(2) Fires the full 600 rounds with no more than six malfunctions that are not due to ammunition that fails to detonate and without any crack or breakage of an operating part of the handgun that in creases the risk of injury to the user.

(b) If a pistol or revolver fails the requirements of either paragraph (1) or (2) of subdivision (a) due to ammunition that fails to detonate, the pistol or revolver shall be retested from the beginning of the "firing requirement for handguns" test. A new model of the pistol or revolver that failed due to ammunition that fails to detonate may be submitted for the test to replace the pistol or revolver that failed.

(c) As used in this section, "malfunction" means a failure to properly feed, fire, or eject a round, or failure of a pistol to accept or eject the magazine, or failure of a pistol's slide to remain open after the magazine has been expended.

12128. "Drop safety requirement for handguns"

As used in this chapter, the "drop safety requirement for handguns" means that at the conclusion of the firing requirements for handguns described in Section 12127, the same certified independent testing laboratory shall subject the same three handguns of the make and model for which certification is sought, to the following test:

A primed case (no powder or projectile) shall be inserted into the chamber. For pistols, the slide shall be released, allowing it to move forward under the impetus of the recoil spring, and an empty magazine shall be inserted. For both pistols and revolvers, the weapon shall be placed in a drop fixture capable of dropping the pistol from a drop height of 1m + 1cm (39.4 + 0.4 in.) onto the largest side of a slab of solid concrete having minimum dimensions of 7.5 x 15 x 15 cm (3 x 6 x 6 in.). The drop distance shall be measured from the lowermost portion of the weapon to the top surface of the slab. The weapon shall be dropped from a fixture and not from the hand. The weapon shall be dropped in the condition that it would be in if it were dropped from a hand (cocked with no manual safety applied). If the design of a pistol is such that upon leaving the hand a "safety" is automatically applied by the pistol, this feature shall not be defeated. An approved drop fixture is a short piece of string with the weapon attached at one end and the other end held in an air vise until the drop is initiated.

The following six drops shall be performed:

(a) Normal firing position with barrel horizontal.

(b) Upside down with barrel horizontal.

(c) On grip with barrel vertical.

(d) On muzzle with barrel vertical.

(e) On either side with barrel horizontal.

(f) If there is an exposed hammer or striker, on the rearmost point of that device, otherwise on the rearmost point of the weapon.

The primer shall be examined for indentations after each drop. If indentations are present, a fresh primed case shall be used for the next drop.

The handgun shall pass this test if each of the three test guns does not fire the primer.

12129. Certification by manufacturers and sellers that firearms are not unsafe

Every person who is licensed as a manufacturer of firearms pursuant to Chapter 44 (commencing with Section 921) of Title 18 of the United States Code who manufactures firearms in this state, and every person who imports into the state for sale, keeps for sale, or offers or exposes for sale any firearm, shall certify under penalty of perjury and any other remedy provided by law that every model, kind, class, style, or type of pistol, revolver, or other firearm capable of being concealed upon the person that he or she manufactures or im-ports, keeps, or exposes for sale is not an un-safe handgun as prohibited by this chapter.

12130. Testing requirement; Certification of laboratories; Test report; Requirements for center-fire semiautomatic pistols and rimfire semiautomatic pistols

(a) Any pistol, revolver, or other firearm capable of being concealed upon the person manufactured in this state, imported into the state for sale, kept for sale, or offered or exposed for sale, shall be tested within a reasonable period of time by an independent laboratory certified pursuant to subdivision (b) to determine whether that pistol, revolver, or other firearm capable of being concealed upon the person meets or exceeds the standards defined in Section 12126.

(b) On or before October 1, 2000, the Department of Justice shall certify laboratories to verify compliance with the standards defined in Section 12126. The department may charge any laboratory that is seeking certification to test any pistol, revolver, or other firearm capable of being concealed upon the person pursuant to this chapter a fee not exceeding the costs of certification.

(c) The certified testing laboratory shall, at the manufacturer's or importer's expense, test the firearm and submit a copy of the final test report directly to the Department of Justice along with a prototype of the weapon to be retained by the department. The department shall notify the manufacturer or importer of its receipt of the final test report and the department's determination as to whether the firearm tested may be sold in this state.

(d)(1) Commencing January 1, 2006, no center-fire semiautomatic pistol may be submitted for testing pursuant to this chapter if it does not have either a chamber load indicator as defined in subdivision (c) of Section 12126, or a magazine disconnect mechanism as defined in subdivision (d) of Section 12126 if it has a detachable magazine.

(2) Commencing January 1, 2007, no center-fire semiautomatic pistol may be submitted for testing pursuant to this chapter if it does not have both a chamber load indicator as defined in subdivision (c) of Section 12126 and a magazine disconnect mechanism as defined in subdivision (d) of Section 12126.

(3) Commencing January 1, 2006, no rimfire semiautomatic pistol may be submitted for testing pursuant to this chapter if it has a detachable magazine, and does not have a magazine disconnect mechanism as defined in subdivision (d) of Section 12126.

12131. (First of two; Operative term contingent) Roster of firearms determined not to be unsafe; Fee for maintenance of roster

(a) On and after January 1, 2001, the Department of Justice shall compile, publish, and thereafter maintain a roster listing all of the pistols, revolvers, and other firearms capable of being concealed upon the person that have been tested by a certified testing laboratory, have been determined not to be unsafe handguns, and may be sold in this state pursuant to this title. The roster shall list, for each firearm, the manufacturer, model number, and model name.

(b)(1) The department may charge every person in this state who is licensed as a manufacturer of firearms pursuant to Chapter 44 (commencing with Section 921) of Title 18 of the United States Code, and any person in this state who manufactures or causes to be manufactured, imports into the state for sale, keeps for sale, or offers or exposes for sale any pistol, revolver, or other firearm capable of being concealed upon the person in this state, an annual fee not exceeding the costs of preparing, publishing, and maintaining the roster pursuant to subdivision (a) and the costs of research and development, report analysis, firearms storage, and other program infrastructure costs necessary to implement this chapter.

(2) Any pistol, revolver, or other firearm capable of being concealed upon the person that is manufactured by a manufacturer who manufactures or causes to be manufactured, imports into the state for sale, keeps for sale, or offers or exposes for sale any pistol, revolver, or other firearm capable of being concealed upon the person in this state, and who fails to pay any fee required pursuant to paragraph (1), may be excluded from the roster.

(3) If a purchaser has initiated a transfer of a handgun that is listed on the roster as not unsafe, and prior to the completion of the transfer, the handgun is removed from the roster of not unsafe handguns because of failure to pay the fee required to keep that handgun listed on the roster, the handgun shall be deliverable to the purchaser if the purchaser is not otherwise prohibited from purchasing or possessing the handgun. However, if a purchaser has initiated a transfer of a handgun that is listed on the roster as not unsafe, and prior to the completion of the transfer, the handgun is removed from the roster pursuant to subdivision (f), the handgun shall not be deliverable to the purchaser.

(c) The Attorney General may annually retest up to 5 percent of the handgun models that are listed on the roster described in subdivision (a).

(d) The retesting of a handgun model pursuant to subdivision (c) shall conform to the following:

(1) The Attorney General shall obtain from retail or wholesale sources, or both, three samples of the handgun model to be retested.

(2) The Attorney General shall select the certified laboratory to be used for the retesting.

(3) The ammunition used for the retesting shall be of a type recommended by the manufacturer in the user manual for the handgun. If the user manual for the handgun model makes no ammunition recommendation, the Attorney General shall select the ammunition to be used for the retesting. The ammunition shall be of the proper caliber for the handgun, commercially available, and in new condition.

(e) The retest shall be conducted in the same manner as the testing prescribed in Sections 12127 and 12128.

(f) If the handgun model fails retesting, the Attorney General shall remove the handgun model from the roster maintained pursuant to subdivision (a).

(g) A handgun model removed from the roster pursuant to subdivision (f) may be reinstated on the roster if all of the following are met:

(1) The manufacturer petitions the Attorney General for reinstatement of the handgun model.

(2) The manufacturer pays the Department of Justice for all of the costs related to the reinstatement testing of the handgun model, including the purchase price of the handguns, prior to reinstatement testing.

(3) The reinstatement testing of the handguns shall be in accordance with subdivisions (d) and (e).

(4) The three handgun samples shall be tested only once for reinstatement. If the sample fails it may not be retested.

(5) If the handgun model successfully passes testing for reinstatement, and if the manufacturer of the handgun is otherwise in compliance with this chapter, the Attorney General shall reinstate the handgun model on the roster maintained pursuant to subdivision (a).

(6) The manufacturer shall provide the Attorney General with the complete testing history for the handgun model.

(7) Notwithstanding subdivision (c), the Attorney General may, at any time, further retest any handgun model that has been reinstated to the roster.

12131. (Second of two; Operative date contingent) Roster of firearms determined not to be unsafe; Fee for maintenance of roster; Annual retest

(a) On and after January 1, 2001, the Department of Justice shall compile, publish, and thereafter maintain a roster listing all of the pistols, revolvers, and other firearms capable of being concealed upon the person that have been tested by a certified testing laboratory, have been determined not to be unsafe handguns, and may be sold in this state pursuant to this title. The roster shall list, for each firearm, the manufacturer, model number, and model name.

(b) (1) The department may charge every person in this state who is licensed as a manufacturer of firearms pursuant to Chapter 44 (commencing with Section 921) of Title 18 of the United States Code, and any person in this state who manufactures or causes to be manufactured, imports into the state for sale, keeps for sale, or offers or exposes for sale any pistol, revolver, or other firearm capable of being concealed upon the person in this state, an annual fee not exceeding the costs of preparing, publishing, and maintaining the roster pursuant to subdivision (a) and the costs of research and development, report analysis, firearms storage, and other program infrastructure costs necessary to implement this chapter.

(2) Any pistol, revolver, or other firearm capable of being concealed upon the person that is manufactured by a manufacturer who manufactures or causes to be manufactured, imports into the state for sale, keeps for sale, or offers or exposes for sale any pistol, revolver, or other firearm capable of being concealed upon the person in this state, and who fails to pay any fee required pursuant to paragraph (1), may be excluded from the roster.

(3) If a purchaser has initiated a transfer of a handgun that is listed on the roster as not unsafe, and prior to the completion of the transfer, the handgun is removed from the roster of not unsafe handguns because of failure to pay the fee required to keep that handgun listed on the roster, the handgun shall be deliverable to the purchaser if the purchaser is not otherwise prohibited from purchasing or possessing the handgun. However, if a purchaser has initiated a transfer of a handgun that is listed on the roster as not unsafe, and prior to the completion of the transfer, the handgun is removed from the roster pursuant to subdivision (f), the handgun shall not be deliverable to the purchaser.

(c) The Attorney General may annually retest up to 5 percent of the handgun models that are listed on the roster described in subdivision (a).

(d) The retesting of a handgun model pursuant to subdivision (c) shall conform to the following:

(1) The Attorney General shall obtain from retail or wholesale sources, or both, three samples of the handgun model to be retested.

(2) The Attorney General shall select the certified laboratory to be used for the retesting.

(3) The ammunition used for the retesting shall be of a type recommended by the manufacturer in the user manual for the handgun. If the user manual for the handgun model makes no ammunition recommendation, the Attorney General shall select the ammunition to be used for the retesting. The ammunition shall be of the proper caliber for the handgun, commercially available, and in new condition.

(e) The retest shall be conducted in the same manner as the testing prescribed in Sections 12127 and 12128.

(f) If the handgun model fails retesting, the Attorney General shall remove the handgun model from the roster maintained pursuant to subdivision (a).

(g) A handgun model removed from the roster pursuant to subdivision (f) may be reinstated on the roster if all of the following are met:

(1) The manufacturer petitions the Attorney General for reinstatement of the handgun model.

(2) The manufacturer pays the Department of Justice for all of the costs related to the reinstatement testing of the handgun model, including the purchase price of the handguns, prior to reinstatement testing.

(3) The reinstatement testing of the handguns shall be in accordance with subdivisions (d) and (e).

(4) The three handgun samples shall be tested only once for reinstatement. If the sample fails it may not be retested.

(5) If the handgun model successfully passes testing for reinstatement, and if the manufacturer of the handgun is otherwise in compliance with this chapter, the Attorney General shall reinstate the handgun model on the roster maintained pursuant to subdivision (a).

(6) The manufacturer shall provide the Attorney General with the complete testing history for the handgun model.

(7) Notwithstanding subdivision (c), the Attorney General may, at any time, further retest any handgun model that has been reinstated to the roster.

12131.5. Listing of firearms differing only cosmetically from firearms on roster

(a) A firearm shall be deemed to satisfy the requirements of subdivision (a) of Section 12131 if another firearm made by the same manufacturer is already listed and the unlisted firearm differs from the listed firearm only in one or more of the following features:

(1) Finish, including, but not limited to, bluing, chrome-plating, oiling, or engraving.

(2) The material from which the grips are made.

(3) The shape or texture of the grips, so long as the difference in grip shape or texture does not in any way alter the dimensions, material, linkage, or functioning of the magazine well, the barrel, the chamber, or any of the components of the firing mechanism of the firearm.

(4) Any other purely cosmetic feature that does not in any way alter the dimensions, material, linkage, or functioning of the magazine well, the barrel, the chamber, or any of the components of the firing mechanism of the firearm.

(b) Any manufacturer seeking to have a firearm listed under this section shall provide to the Department of Justice all of the following:

(1) The model designation of the listed firearm.

(2) The model designation of each firearm that the manufacturer seeks to have listed under this section.

(3) A statement, under oath, that each unlisted firearm for which listing is sought differs from the listed firearm only in one or more of the ways identified in subdivision (a) and is in all other respects identical to the listed firearm.

(c) The department may, in its discretion and at any time, require a manufacturer to provide to the department any model for which listing is sought under this section, to determine whether the model complies with the requirements of this section.

12132. Chapter inapplicable to certain sales, loans and transfers; Pistols designed for Olympic target shooting

This chapter shall not apply to any of the following:

(a) The sale, loan, or transfer of any firearm pursuant to Section 12082 in order to comply with subdivision (d) of Section 12072.

(b) The sale, loan, or transfer of any firearm that is exempt from the provisions of subdivision (d) of Section 12072 pursuant to any applicable exemption contained in Section 12078, if the sale, loan, or transfer complies with the requirements of that applicable exemption to subdivision (d) of Section 12072.

(c) The sale, loan, or transfer of any firearm as described in paragraph (3) of subdivision (b) of Section 12125.

(d) The delivery of a pistol, revolver, or other firearm capable of being concealed upon the person to a person licensed pursuant to Section 12071 for the purposes of the service or repair of that firearm.

(e) The return of a pistol, revolver, or other firearm capable of being concealed upon the person by a person licensed pursuant to Section 12071 to its owner where that firearm was initially delivered in the circumstance set forth in subdivisions (a), (d), (f) or (j).

(f) The delivery of a pistol, revolver, or other firearm capable of being concealed upon the person to a person licensed pursuant to Section 12071 for the purpose of a consignment sale or as collateral for a pawnbroker loan.

(g) The sale, loan, or transfer of any pistol, revolver, or other firearm capable of being concealed upon the person listed as a curio or relic, as defined in Section 178.11 of the Code of Federal Regulations.

(h)(1) The Legislature finds a significant public purpose in exempting pistols that are designed expressly for use in Olympic target shooting events. Therefore, those pistols that are sanctioned by the International Olympic Committee and by USA Shooting, the national governing body for international shooting competition in the United States, and that are used for Olympic target shooting purposes at the time that the act adding this subdivision is enacted, and that fall within the definition of "unsafe handgun" pursuant to paragraph (3) of subdivision (b) of Section 12126 shall be exempt, as provided in paragraphs (2) and (3).

(2) This chapter shall not apply to any of the following pistols, because they are consistent with the significant public purpose expressed in paragraph (1):

MANUFACTURER	MODEL	CALIBER
ANSCHUTZ	FP	.22LR
BENELLI	MP90	.22LR
BENELLI	MP90	.32 S&W LONG
BENELLI	MP95	.22LR
BENELLI	MP95	.32 S&W LONG
DRULOV	FP	.22LR
GREEN	ELECTROARM	.22LR
HAMMERLI	100	.22LR
HAMMERLI	101	.22LR
HAMMERLI	102	.22LR
HAMMERLI	162	.22LR
HAMMERLI	280	.22LR
HAMMERLI	280	.32 S&W LONG
HAMMERLI	FP10	.22LR
HAMMERLI	MP33	.22LR
HAMMERLI	SP20	.22LR
HAMMERLI	SP20	.32 S&W LONG
MORINI	CM102E	.22LR
MORINI	22M	.22LR
MORINI	32M	.32 S&W LONG
MORINI	CM80	.22LR
PARDINI	GP	.22 SHORT
PARDINI	GPO	.22 SHORT
PARDINI	GP-SCHUMANN	.22 SHORT
PARDINI	HP	.32 S&W LONG
PARDINI	K22	.22LR
PARDINI	MP	.32 S&W LONG
PARDINI	PGP75	.22LR
PARDINI	SP	.22LR
PARDINI	SPE	.22LR
SAKO	FINMASTER	.22LR
STEYR	FP	.22LR
VOSTOK	IZH NO. 1	.22LR
VOSTOK	MU55	.22LR
VOSTOK	TOZ35	.22LR
WALTHER	FP	.22LR
WALTHER	GSP	.22LR
WALTHER	GSP	.32 S&W LONG
WALTHER	OSP	.22 SHORT
WALTHER	OSP-2000	.22 SHORT

(3) The department shall create a program that is consistent with the purpose stated in paragraph (1) to exempt new models of competitive firearms from this chapter. The exempt competitive firearms may be based on recommendations by USA Shooting consistent with the regulations contained in the USA Shooting Official Rules or may be based on the recommendation or rules of any other organization that the department deems relevant.

(i) The sale, loan, or transfer of any semiautomatic pistol that is to be used solely as a prop during the course of a motion picture, television, or video production by an authorized participant therein in the course of making that production or event or by an authorized employee or agent of the entity producing that production or event.

(j) The delivery of a pistol, revolver, or other firearm capable of being concealed upon the person to a person licensed pursuant to Section 12071 where the firearm is being loaned by the licensee to a consultant-evaluator.

(k) The delivery of a pistol, revolver, or other firearm capable of being concealed upon the person by a person licensed pursuant to Section 12071 where the firearm is being loaned by the licensee to a consultant-evaluator.

(l) The return of a pistol, revolver, or other firearm capable of being concealed upon the person to a person licensed pursuant to Section 12071 where it was initially delivered pursuant to subdivision (k).

12133. Exemption from chapter of certain single-action revolvers

(a) The provisions of this chapter shall not apply to a single-action revolver that has at least a 5-cartridge capacity with a barrel length of not less than three inches, and meets any of the following specifications:

(1) Was originally manufactured prior to 1900 and is a curio or relic, as defined in Section 478.11 of Title 27 of the Code of Federal Regulations.

(2) Has an overall length measured parallel to the barrel of at least 7 1/2 inches when the handle, frame or receiver, and barrel are assembled.

(3) Has an overall length measured parallel to the barrel of at least 7 1/2 inches when the handle, frame or receiver, and barrel are assembled and that is currently approved for importation into the United States pursuant to the provisions of paragraph (3) of subsection (d) of Section 925 of Title 18 of the United States Code.

(b) The provisions of this chapter shall not apply to a single-shot pistol with a barrel length of not less than six inches and that has an overall length of at least 101/2 inches when the handle, frame or receiver, and barrel are assembled.

Chapter 2. Machineguns

Article 1. General Provisions

12200. Definition The term "machinegun" as used in this chapter means any weapon which shoots, is designed to shoot, or can readily be restored to shoot, automatically, more than one shot, without manual reloading, by a single function of the trigger. The term shall also include the frame or receiver of any such weapon, any part designed and intended solely and exclusively, or combination of parts designed and intended, for use in converting a weapon into a machinegun, and any combination of parts from which a machinegun can be assembled if such parts are in the possession or under the control of a person. The term also includes any weapon deemed by the federal Bureau of Alcohol, Tobacco, and Firearms as readily convertible to a machinegun under Chapter 53 (commencing with Section 5801) of Title 26 of the United States Code.

12201. Exceptions to application of chapter

Nothing in this chapter shall affect or apply to any of the following:

(a) The sale to, purchase by, or possession of machineguns by police departments, sheriffs' offices, marshals' offices, district attorneys' offices, the California Highway Patrol, the Department of Justice, the Department of Corrections for use by the department's Special Emergency Response Teams and Law Enforcement Liaison/Investigations Unit, or the military or naval forces of this state or of the United States for use in the discharge of their official duties, provided, however, that any sale to these entities be transacted by a person who is permitted pursuant to Section 12230 and licensed pursuant to Section 12250.

(b) The possession of machineguns by regular, salaried, full-time peace officer members of a police department, sheriff's office, marshal's office, district attorney's office, the California Highway Patrol, the Department of Justice, or the Department of Corrections for use by the department's Special Emergency Response Teams and Law Enforcement Liaison/Investigations Unit when on duty and if the use is within the scope of their duties.

Article 2. Unlawful Possession of Machine Guns

12220.Possession, sale, and manufacture

(a) Any person, firm, or corporation, who within this state possesses or knowingly transports a machinegun except as authorized by this chapter, is guilty of a public offense and upon conviction thereof shall be punished by imprisonment in the state prison, or by a fine not to exceed ten thousand dollars ($10,000), or by both such fine and imprisonment.

(b) Any person, firm, or corporation who within this state intentionally converts a firearm into a machinegun, or who sells, or offers for sale, or knowingly manufactures a machinegun, except as authorized by this chapter, is punishable by imprisonment in the state prison for four, six, or eight years.

Article 3. Permits

12230. Permits

The Department of Justice may issue permits for the possession, manufacture, and transportation or possession, manufacture, or transportation of machineguns, upon a satisfactory showing that good cause exists for the issuance thereof to the applicant for the permit, but no permit shall be issued to a person who is under 18 years of age.

12231. Permit applications; Filing

Applications for permits shall be filed in writing, signed by the applicant if an individual, or by a member or officer qualified to sign if the applicant is a firm or corporation, and shall state the name, business in which engaged, business address and a full description of the use to which the firearms are to be put.

Applications and permits shall be uniform throughout the state on forms prescribed by the Department of Justice.

Each applicant for a permit shall pay at the time of filing his or her application a fee determined by the Department of Justice not to exceed the application processing costs of the Department of Justice. A permit granted pursuant to this article may be renewed one year from the date of issuance, and annually thereafter, upon the filing of a renewal application and the payment of a permit renewal fee not to exceed the application processing costs of the Department of Justice. After the department establishes fees sufficient to reimburse the department for processing costs, fees charged shall increase at a rate not to exceed the legislatively approved annual cost-of-living adjustments for the department's budget.

12232. Keeping; Inspection

Every person, firm or corporation to whom a permit is issued shall keep it on his person or at the place where the firearms are kept. The permit shall be open to inspection by any peace officer or any other person designated by the authority issuing the permit.

12233.. Revocation of permits

Permits issued in accordance with this chapter may be revoked by the issuing authority at any time when it appears that the need for the firearms has ceased or that the holder of the permit has used the firearms for purposes other than those allowed by the permit or that the holder of the permit has not exercised great care in retaining custody of any weapons possessed under the permit.

12234. Annual inspection for security and safe storage by Department of Justice

(a) Except as provided in subdivision (b), the Department of Justice shall, for every person, firm, or corporation to whom a permit is issued pursuant to this article, annually conduct an inspection for security and safe storage purposes, and to reconcile the inventory of machine guns.

(b) A person, firm, or corporation with an inventory of fewer than five devices that require any Department of Justice permit shall be subject to an inspection for security and safe storage purposes, and to reconcile inventory, once every five years, or more frequently if determined by the department.

Article 4. Licenses to Sell Machineguns

12250. Grant of license; Conditions of retention; Application; Fees

(a) The Department of Justice may grant licenses in a form to be prescribed by it effective for not more than one year from the date of issuance, to permit the sale at the place specified in the license of machineguns subject to all of the following conditions, upon breach of any of which the license shall be revoked:

(1) The business shall be carried on only in the place designated in the license.

(2) The license or a certified copy thereof must be displayed on the premises in a place where it may easily be read.

(3) No machinegun shall be delivered to any person not authorized to receive the same under the provisions of this chapter.

(4) A complete record must be kept of sales made under the authority of the license, showing the name and address of the purchaser, the descriptions and serial numbers of the weapons purchased, the number and date of issue of the purchaser's permit, if any, and the signature of the purchaser or purchasing agent. This record shall be open to the inspection of any peace officer or other person designated by the Attorney General.

(b) Applications for licenses shall be filed in writing, signed by the applicant if an individual or by a member or officer qualified to sign if the applicant is a firm or corporation, and shall state the name, business in which engaged, business address and a full description of the use to which the firearms are to be put.

Applications and licenses shall be uniform throughout the state on forms prescribed by the Department of Justice.

Each applicant for a license shall pay at the time of filing his or her application a fee determined by the Department of Justice not to exceed the application processing costs of the Department of Justice. A license granted pursuant to this article may be renewed one year from the date of issuance, and annually thereafter, upon the filing of a renewal application and the payment of a license renewal fee not to exceed the application processing costs of the Department of Justice. After the department establishes fees sufficient to reimburse the department for processing costs, fees charged shall increase at a rate not to exceed the legislatively approved annual cost-of-living adjustments for the department's budget.

12251. Possession of machinegun as nuisance; Surrender and destruction

It shall be a public nuisance to possess any machinegun in violation of this chapter, and the Attorney General, any district attorney or any city attorney may bring an action before the superior court to enjoin the possession of any such machinegun.

Any such machinegun found to be in violation of this chapter shall be surrendered to the Department of Justice, and the department shall destroy such machinegun so as to render it unusable and unrepairable as a machinegun, except upon the filing of a certificate with the department by a judge or district attorney stating that the preservation of such machinegun is necessary to serve the ends of justice.

Chapter 2.3. Roberti-Roos Assault Weapons Control Act of 1989

Article 1. General Provisions

12275. Citation of chapter This chapter shall be known as the Roberti-Roos Assault Weapons Control Act of 1989 and the .50 Caliber BMG Regulation Act of 2004.

12275.5. Legislative findings and declarations

(a) The Legislature hereby finds and declares that the proliferation and use of assault weapons poses a threat to the health, safety, and security of all citizens of this state. The Legislature has restricted the assault weapons specified in Section 12276 based upon finding that each firearm has such a high rate of fire and capacity for firepower that its function as a legitimate sports or recreational firearm is substantially outweighed by the danger that it can be used to kill and injure human beings. It is the intent of the Legislature in enacting this chapter to place restrictions on the use of assault weapons and to establish a registration and permit procedure for their lawful sale and possession. It is not, however, the intent of the Legislature by this chapter to place restrictions on the use of those weapons which are primarily designed and intended for hunting, target practice, or other legitimate sports or recreational activities.

(b) The Legislature hereby finds and declares that the proliferation and use of .50 BMG rifles, as defined in Section 12278, poses a clear and present terrorist threat to the health, safety, and security of all residents of, and visitors to, this state, based upon findings that those firearms have such a high capacity for long distance and highly destructive firepower that they pose an unacceptable risk to the death and serious injury of human beings, destruction or serious damage of vital public and private buildings, civilian, police and military vehicles, power generation and transmission facilities, petrochemical production and storage facilities, and transportation infrastructure. It is the intent of the Legislature in enacting this chapter to place restrictions on the use of these rifles and to establish a registration and permit procedure for their lawful sale and possession.

12276. "Assault Weapon" As used in this chapter, "assault weapon" shall mean the following designated semiautomatic firearms:

(a) All of the following specified rifles:

(1) All AK series including, but not limited to, the models identified as follows:

(A) Made in China AK, AKM, AKS, AK47, AK47S, 56, 56S, 84S, and 86S.

(B) Norinco 56, 56S, 84S, and 86S.

(C) Poly Technologies AKS and AK47.

(D) MAADI AK47 and ARM.

(2) UZI and Galil.

(3) Beretta AR-70.

(4) CETME Sporter.

(5) Colt AR-15 series.

(6) Daewoo K-1, K-2, Max 1, Max 2, AR 100, and AR 110C.

(7) Fabrique Nationale FAL, LAR, FNC, 308 Match, and Sporter.

(8) MAS 223.

(9) HK-91, HK-93, HK-94, and HK-PSG-1.

(10) The following MAC types:

(A) RPB Industries Inc. sM10 and sM11.

(B) SWD Incorporated M11.

(11) SKS with detachable magazine.

(12) SIG AMT, PE-57, SG 550, and SG 551.

(13) Springfield Armory BM59 and SAR-48.

(14) Sterling MK-6.

(15) Steyer AUG.

(16) Valmet M62S, M71S, and M78S.

(17) Armalite AR-180.

(18) Bushmaster Assault Rifle.

(19) Calico M-900.

(20) J&R ENG M-68.

(21) Weaver Arms Nighthawk.

(b) All of the following specified pistols:

(1) UZI.

(2) Encom MP-9 and MP-45.

(3) The following MAC types:

(A) RPB Industries Inc. sM10 and sM11.

(B) SWD Incorporated M-11.

(C) Advance Armament Inc. M-11.

(D) Military Armament Corp. Ingram M-11.

(4) Intratec TEC-9.

(5) Sites Spectre.

(6) Sterling MK-7.

(7) Calico M-950.

(8) Bushmaster Pistol.

(c) All of the following specified shotguns:

(1) Franchi SPAS 12 and LAW 12.

(2) Striker 12.

(3) The Streetsweeper type S/S Inc. SS/12.

(d) Any firearm declared by the court pursuant to Section 12276.5 to be an assault weapon that is specified as an assault weapon in a list promulgated pursuant to Section 12276.5.

(e) The term "series" includes all other models that are only variations, with minor differences, of those models listed in subdivision (a), regardless of the manufacturer.

(f) This section is declaratory of existing law, as amended, and a clarification of the law and the Legislature's intent which bans the weapons enumerated in this section, the weapons included in the list promulgated by the Attorney General pursuant to Section 12276.5, and any other models which are only variations of those weapons with minor differences, regardless of the manufacturer. The Legislature has defined assault weapons as the types, series, and models listed in this section because it was the most effective way to identify and restrict a specific class of semiautomatic weapons.

12276.1. Further definition of "assault weapon"

(a) Notwithstanding Section 12276, "assault weapon" shall also mean any of the following:

(1) A semiautomatic, centerfire rifle that has the capacity to accept a detachable magazine and any one of the following:

(A) A pistol grip that protrudes conspicuously beneath the action of the weapon.

(B) A thumbhole stock.

(C) A folding or telescoping stock.

(D) A grenade launcher or flare launcher.

(E) A flash suppressor.

(F) A forward pistol grip.

(2) A semiautomatic, centerfire rifle that has a fixed magazine with the capacity to accept more than 10 rounds.

(3) A semiautomatic, centerfire rifle that has an overall length of less than 30 inches.

(4) A semiautomatic pistol that has the capacity to accept a detachable magazine and any one of the following:

(A) A threaded barrel, capable of accepting a flash suppressor, forward handgrip, or silencer.

(B) A second handgrip.

(C) A shroud that is attached to, or partially or completely encircles, the barrel that allows the bearer to fire the weapon without burning his or her hand, except a slide that encloses the barrel.

(D) The capacity to accept a detachable magazine at some location outside of the pistol grip.

(5) A semiautomatic pistol with a fixed magazine that has the capacity to accept more than 10 rounds.

(6) A semiautomatic shotgun that has both of the following:

(A) A folding or telescoping stock.

(B) A pistol grip that protrudes conspicuously beneath the action of the weapon, thumbhole stock, or vertical handgrip.

(7) A semiautomatic shotgun that has the ability to accept a detachable magazine.

(8) Any shotgun with a revolving cylinder.

(b) The Legislature finds a significant public purpose in exempting pistols that are designed expressly for use in Olympic target shooting events. Therefore, those pistols that are sanctioned by the International Olympic Committee and by USA Shooting, the national governing body for international shooting competition in the United States, and that are used for Olympic target shooting purposes at the time the act adding this subdivision is enacted, and that would otherwise fall within the definition of "assault weapon" pursuant to this section are exempt, as provided in subdivision (c).

(c) "Assault weapon" does not include either of the following:

(1) Any antique firearm.

(2) Any of the following pistols, because they are consistent with the significant public purpose expressed in subdivision (b):

MANUFACTURER	MODEL	CALIBER
BENELLI	MP90	.22LR
BENELLI	MP90	.32 S&W LONG
BENELLI	MP95	.22LR
BENELLI	MP95	.32 S&W LONG
HAMMERLI	280	.22LR
HAMMERLI	280	.32 S&W LONG
HAMMERLI	SP20	.22LR
HAMMERLI	SP20	.32 S&W LONG
PARDINI	GPO	.22 SHORT
PARDINI	GP-SCHUMANN	.22 SHORT
PARDINI	HP	.32 S&W LONG
PARDINI	MP	.32 S&W LONG
PARDINI	SP	.22LR
PARDINI	SPE	.22LR
WALTHER	GSP	.22LR
WALTHER	GSP	.32 S&W LONG
WALTHER	OSP	.22 SHORT
WALTHER	OSP-2000	.22 SHORT

(3) The Department of Justice shall create a program that is consistent with the purposes stated in subdivision (b) to exempt new models of competitive pistols that would otherwise fall within the definition of "assault weapon" pursuant to this section from being classified as an assault weapon. The exempt competitive pistols may be based on recommendations by USA Shooting consistent with the regulations contained in the USA Shooting Official Rules or may be based on the recommendation or rules of any other organization that the department deems relevant.

(d) The following definitions shall apply under this section:

(1) "Magazine" shall mean any ammunition feeding device.

(2) "Capacity to accept more than 10 rounds" shall mean capable of accommodating more than 10 rounds, but shall not be construed to include a feeding device that has been permanently altered so that it cannot accommodate more than 10 rounds.

(3) "Antique firearm" means any firearm manufactured prior to January 1, 1899.

(e) This section shall become operative January 1, 2000.

12276.5. Description of assault weapons for identification purposes; Promulgation of list

(a) The Attorney General shall prepare a description for identification purposes, including a picture or diagram, of each assault weapon listed in Section 12276, and any firearm declared to be an assault weapon pursuant to this section, and shall distribute the description to all law enforcement agencies responsible for enforcement of this chapter. Those law enforcement agencies shall make the description available to all agency personnel.

(b)(1) Until January 1, 2007, the Attorney General shall promulgate a list that specifies all firearms designated as assault weapons in Section 12276 or declared to be assault weapons pursuant to this section. The Attorney General shall file that list with the Secretary of State for publication in the California Code of Regulations. Any declaration that a specified firearm is an assault weapon shall be implemented by the Attorney General who, within 90 days, shall promulgate an amended list which shall include the specified firearm declared to be an assault weapon. The Attorney General shall file the amended list with the Secretary of State for publication in the California Code of Regulations. Any firearm declared to be an assault weapon prior to January 1, 2007, shall remain on the list filed with the Secretary of State.

(2) Chapter 3.5 (commencing with Section 11340) of Division 3 of Title 2 of the Government Code, pertaining to the adoption of rules and regulations, shall not apply to any list of assault weapons promulgated pursuant to this section.

(c) The Attorney General shall adopt those rules and regulations that may be necessary or proper to carry out the purposes and intent of this chapter.

12277. "Person" As used in this chapter, "person" means an individual, partnership, corporation, limited liability company, association, or any other group or entity, regardless of how it was created.

12278. ".50 BMG rifle"; ".50 BMG cartridge"; "Antique firearm"

(a) As used in this chapter, a ".50 BMG rifle" means a center fire rifle that can fire a .50 BMG cartridge and is not already an assault weapon pursuant to Section 12276, 12276.1, or 12276.5, or a machinegun, as defined in Section 12200.

(b) As used in this chapter, a ".50 BMG cartridge" means a cartridge that is designed and intended to be fired from a center fire rifle and that meets all of the following criteria:

(1) It has an overall length of 5.54 inches from the base to the tip of the bullet.

(2) The bullet diameter for the cartridge is from .510 to, and including, .511 inch.

(3) The case base diameter for the cartridge is from .800 inch to, and including, .804 inch.

(4) The cartridge case length is 3.91 inches.

(c) A ".50 BMG rifle" does not include any "antique firearm," nor any curio or relic as defined in Section 178.11 of Title 27 of the Code of Federal Regulations.

Article 2. Unlawful Activities

12280. Manufacture, distribution, sale, or possession of weapon; Distinct and separate offense for each weapon where multiple weapons are involved in specified prohibited activity; Enhancement of punishment for another crime; Exceptions

(a)(1) Any person who, within this state, manufactures or causes to be manufactured, distributes, transports, or imports into the state, keeps for sale, or offers or exposes for sale, or who gives or lends any assault weapon or any .50 BMG rifle, except as provided by this chapter, is guilty of a felony, and upon conviction shall be punished by imprisonment in the state prison for four, six, or eight years.

(2) In addition and consecutive to the punishment imposed under paragraph (1), any person who transfers, lends, sells, or gives any assault weapon or any .50 BMG rifle to a minor in violation of paragraph (1) shall receive an enhancement of one year.

(3) Except in the case of a first violation involving not more than two firearms as provided in subdivisions (b) and (c), for purposes of this section, if more than one assault weapon or .50 BMG rifle is involved in any violation of this section, there shall be a distinct and separate offense for each.

(b) Any person who, within this state, possesses any assault weapon, except as provided in this chapter, shall be punished by imprisonment in a county jail for a period not exceeding one year, or by imprisonment in the state prison. However, a first violation of these provisions is punishable by a fine not exceeding five hundred dollars ($500) if the person was found in possession of no more than two firearms in compliance with subdivision (c) of Section 12285 and the person meets all of the following conditions:

(1) The person proves that he or she lawfully possessed the assault weapon prior to the date it was defined as an assault weapon pursuant to Section 12276, 12276.1 or 12276.5.

(2) The person has not previously been convicted of a violation of this section.

(3) The person was found to be in possession of the assault weapon within one year following the end of the one-year registration period established pursuant to subdivision (a) of Section 12285.

(4) The person relinquished the firearm pursuant to Section 12288, in which case the assault weapon shall be destroyed pursuant to Section 12028.

(c) Any person who, within this state, possesses any .50 BMG rifle, except as provided in this chapter, shall be punished by a fine of one thousand dollars ($1,000), imprisonment in a county jail for a period not to exceed one year, or by both that fine and imprisonment. However, a first violation of these provisions is punishable by a fine not exceeding five hundred dollars ($500) if the person was found in possession of no more than two firearms in compliance with subdivision (a) of Section 12285 and the person meets the conditions set forth in paragraphs (1), (2), and (3):

(1) The person proves that he or she lawfully possessed the .50 BMG rifle prior to January 1, 2005.

(2) The person has not previously been convicted of a violation of this section.

(3) The person was found to be in possession of the .50 BMG rifle within one year following the end of the .50 BMG rifle registration period established pursuant to subdivision (a) of Section 12285.

(4) Firearms seized pursuant to this subdivision from persons who meet all of the conditions set forth in paragraphs (1), (2), and (3) shall be returned unless the court finds in the interest of public safety, after notice and hearing, that the .50 BMG rifle should be destroyed pursuant to Section 12028. Firearms seized from persons who do not meet the conditions set forth in paragraphs (1), (2), and (3) shall be destroyed pursuant to Section 12028.

(d) Notwithstanding Section 654 or any other provision of law, any person who commits another crime while violating this section may receive an additional, consecutive punishment of one year for violating this section in addition and consecutive to the punishment, including enhancements, which is prescribed for the other crime.

(e) Subdivisions (a), (b), and (c) shall not apply to the sale to, purchase by, importation of, or possession of assault weapons or a .50 BMG rifle by the Department of Justice, police departments, sheriffs' offices, marshals' offices, the Department of Corrections and Rehabilitation, the Department of the California Highway Patrol, district attorneys' offices, Department of Fish and Game, Department of Parks and Recreation, or the military or naval forces of this state or of the United States, or any federal law enforcement agency for use in the discharge of their official duties.

(f)(1) Subdivisions (b) and (c) shall not prohibit the possession or use of assault weapons or a .50 BMG rifle by sworn peace officer members of those agencies specified in subdivision (e) for law enforcement purposes, whether on or off duty.

(2) Subdivisions (a), (b), and (c) shall not prohibit the delivery, transfer, or sale of an assault weapon or a .50 BMG rifle to, or the possession of an assault weapon or a .50 BMG rifle by, a sworn peace officer member of an agency specified in subdivision (e), if the peace officer is authorized by his or her employer to possess or receive the assault weapon or the .50 BMG rifle. Required authorization is defined as verifiable written certification from the head of the agency, identifying the recipient or possessor of the assault weapon as a peace officer and authorizing him or her to receive or possess the specific assault weapon. For this exemption to apply, in the case of a peace officer who possesses or receives the assault weapon prior to January 1, 2002, the officer shall register the assault weapon pursuant to Section 12285 on or before April 1, 2002; in the case of a peace officer who possesses or receives the assault weapon on or after January 1, 2002, the officer shall register the assault weapon pursuant to Section 12285 not later than 90 days after possession or receipt. In the case of a peace officer who possesses or receives a .50 BMG rifle on or before January 1, 2005, the officer shall register the .50 BMG rifle on or before April 30, 2006. In the case of a peace officer who possesses or receives a .50 BMG rifle after January 1, 2005, the officer shall register the .50 BMG rifle not later than one year after possession or receipt. The peace officer must include with the registration, a copy of the authorization required pursuant to this paragraph.

(3) Nothing in this section shall be construed to limit or prohibit the delivery, transfer, or sale of an assault weapon or a .50 BMG rifle to, or the possession of an assault weapon or a .50 BMG rifle by, a member of a federal law enforcement agency provided that person is authorized by the employing agency to possess the assault weapon or .50 BMG rifle.

(g) Subdivision (b) shall not apply to the possession of an assault weapon during the 90-day period immediately after the date it was specified as an assault weapon pursuant to Section 12276.5, or during the one-year period after the date it was defined as an assault weapon pursuant to Section 12276.1, if all of the following are applicable:

(1) The person is eligible under this chapter to register the particular assault weapon.

(2) The person lawfully possessed the particular assault weapon prior to the date it was specified as an assault weapon pursuant to Section 12276.5, or prior to the date it was defined as an assault weapon pursuant to Section 12276.1.

(3) The person is otherwise in compliance with this chapter.

(h) Subdivisions (a), (b), and (c) shall not apply to the manufacture by persons who are issued permits pursuant to Section 12287 of assault weapons or .50 BMG rifles for sale to the following:

(1) Exempt entities listed in subdivision (e).

(2) Entities and persons who have been issued permits pursuant to Section 12286 or 12287.

(3) Entities outside the state who have, in effect, a federal firearms dealer's license solely for the purpose of distribution to an entity listed in paragraphs (4) to (6), inclusive.

(4) Federal military and law enforcement agencies.

(5) Law enforcement and military agencies of other states.

(6) Foreign governments and agencies approved by the United States State Department.

(i) Subdivision (a) shall not apply to a person who is the executor or administrator of an estate that includes an assault weapon or a .50 BMG rifle registered under Section 12285 or that was possessed pursuant to paragraph (1) of subdivision (f) that is disposed of as authorized by the probate court, if the disposition is otherwise permitted by this chapter.

(j) Subdivisions (b) and (c) shall not apply to a person who is the executor or administrator of an estate that includes an assault weapon or a .50 BMG rifle registered under Section 12285 or that was possessed pursuant to paragraph (1) of subdivision (f) if the assault weapon or .50 BMG rifle is possessed at a place set forth in paragraph (1) of subdivision (c) of Section 12285 or as authorized by the probate court.

(k) Subdivision (a) shall not apply to either of the following:

(1) A person who lawfully possesses and has registered an assault weapon or .50 BMG rifle pursuant to this chapter who lends that assault weapon or .50 BMG rifle to another if all the following apply:

(A) The person to whom the assault weapon or .50 BMG rifle is lent is 18 years of age or over and is not in a class of persons prohibited from possessing firearms by virtue of Section 12021 or 12021.1 of this code or Section 8100 or 8103 of the Welfare and Institutions Code.

(B) The person to whom the assault weapon or .50 BMG rifle is lent remains in the presence of the registered possessor of the assault weapon or .50 BMG rifle.

(C) The assault weapon or .50 BMG rifle is possessed at any of the following locations:

(i) While on a target range that holds a regulatory or business license for the purpose of practicing shooting at that target range.

(ii) While on the premises of a target range of a public or private club or organization organized for the purpose of practicing shooting at targets.

(iii) While attending any exhibition, display, or educational project that is about firearms and that is sponsored by, conducted under the auspices of, or approved by a law enforcement agency or a nationally or state recognized entity that fosters proficiency in, or promotes education about, firearms.

(2) The return of an assault weapon or .50 BMG rifle to the registered possessor, or the lawful possessor, which is lent by the same pursuant to paragraph (1).

(l) Subdivisions (b) and (c) shall not apply to the possession of an assault weapon or .50 BMG rifle by a person to whom an assault weapon or .50 BMG rifle is lent pursuant to subdivision (k).

(m) Subdivisions (a), (b), and (c) shall not apply to the possession and importation of an assault weapon or a .50 BMG rifle into this state by a nonresident if all of the following conditions are met:

(1) The person is attending or going directly to or coming directly from an organized competitive match or league competition that involves the use of an assault weapon or a .50 BMG rifle.

(2) The competition or match is conducted on the premises of one of the following:

(A) A target range that holds a regulatory or business license for the purpose of practicing shooting at that target range.

(B) A target range of a public or private club or organization that is organized for the purpose of practicing shooting at targets.

(3) The match or competition is sponsored by, conducted under the auspices of, or approved by, a law enforcement agency or a nationally or state recognized entity that fosters proficiency in, or promotes education about, firearms.

(4) The assault weapon or .50 BMG rifle is transported in accordance with Section 12026.1 or 12026.2.

(5) The person is 18 years of age or over and is not in a class of persons prohibited from possessing firearms by virtue of Section 12021 or 12021.1 of this code or Section 8100 or 8103 of the Welfare and Institutions Code.

(n) Subdivisions (b) and (c) shall not apply to any of the following persons:

(1) A person acting in accordance with Section 12286 or 12287.

(2) A person who has a permit to possess an assault weapon or a .50 BMG rifle issued pursuant to Section 12286 or 12287 when he or she is acting in accordance with Section 12285, 12286, or 12287.

(o) Subdivisions (a), (b), and (c) shall not apply to any of the following persons:

(1) A person acting in accordance with Section 12285.

(2) A person acting in accordance with Section 12286, 12287, or 12290.

(p) Subdivisions (b) and (c) shall not apply to the registered owner of an assault weapon or a .50 BMG rifle possessing that firearm in accordance with subdivision (c) of Section 12285.

(q) Subdivision (a) shall not apply to the importation into this state of an assault weapon or a .50 BMG rifle by the registered owner of that assault weapon or a .50 BMG rifle if it is in accordance with the provisions of subdivision (c) of Section 12285.

(r) Subdivision (a) shall not apply during the first 180 days of the 2005 calendar year to the importation into this state of a .50 BMG rifle by a person who lawfully possessed that .50 BMG rifle in this state prior to January 1, 2005.

(s) Subdivision (c) shall not apply to the possession of a .50 BMG rifle that is not defined or specified as an assault weapon pursuant to this chapter, by any person prior to May 1, 2006, if all of the following are applicable:

(1) The person is eligible under this chapter to register that .50 BMG rifle.

(2) The person lawfully possessed the .50 BMG rifle prior to January 1, 2005.

(3) The person is otherwise in compliance with this chapter.

(t) Subdivisions (a), (b), and (c) shall not apply to the sale of assault weapons or .50 BMG rifles by persons who are issued permits pursuant to Section 12287 to any of the following:

(1) Exempt entities listed in subdivision (e).

(2) Entities and persons who have been issued permits pursuant to Section 12286 or 12287.

(3) Federal military and law enforcement agencies.

(4) Law enforcement and military agencies of other states.

(5) Foreign governments and agencies approved by the United States State Department.

(6) Officers described in subdivision (f) who are authorized to possess assault weapons or .50 BMG rifles pursuant to subdivision (f).

(u) As used in this chapter, the date a firearm is an assault weapon is the earliest of the following:

(1) The effective date of an amendment to Section 12276 that adds the designation of the specified firearm.

(2) The effective date of the list promulgated pursuant to Section 12276.5 that adds or changes the designation of the specified firearm.

(3) The operative date of Section 12276.1, as specified in subdivision (d) of that section.

[Publisher's Note: 2003 Cal. Stat. 499 (SB 238), §6.5 provides: "It is the intent of the Legislature in amending Section 12280 of the Penal Code to delete the exemption allowing retired peace officers to obtain an assault weapon from their employing agency upon retirement. These amendments are intended to make Section 12280 of the Penal Code consistent with the holding in Silveira v. Lockyer (2003) 312 F.3d 1052 by the Ninth Circuit Court of Appeals, which held that exemption to be unconstitutional. The amendments deleting the exemption are therefore declaratory of existing law."**]**

12281.SKS rifles

(a) Any person who, or firm, company, or corporation that, operated a retail or other commercial firm, company, or corporation, and manufactured, distributed, transported, imported, possessed, possessed for sale, offered for sale, or transferred, for commercial purpose, an SKS rifle in California between January 1, 1992, and December 19, 1997, shall be immune from criminal prosecution under Section 12280. The immunity provided in this subdivision shall apply retroactively to any person who, or firm, company, or corporation that, is or was charged by complaint or indictment with a violation of Section 12280 for conduct related to an SKS rifle, whether or not the case of that person, firm, company, or corporation is final.

(b) Any person who possessed, gave, loaned, or transferred an SKS rifle in California between January 1, 1992, and December 19, 1997, shall be immune from criminal prosecution under Section 12280. The immunity provided in this subdivision shall apply retroactively to any person who was charged by complaint or indictment with a violation of Section 12280 for conduct related to an SKS rifle, whether or not the case of that person is final.

(c) Any SKS rifle in the possession of any person who, or firm, company, or corporation that, is described in subdivision (a) or (b), shall not be subject to seizure by law enforcement for violation of Section 12280 prior to January 1, 2000.

(d) Any person, firm, company, or corporation, convicted under Section 12280 for conduct relating to an SKS rifle, shall be permitted to withdraw his or her plea of guilty or nolo contendere, or to reopen his or her case and assert the immunities provided in this section, if the court determines that the allowance of the immunity is in the interests of justice. The court shall interpret this section liberally to the benefit of the defendant.

(e) The Department of Justice shall notify all district attorneys on or before January 31, 1999, of the provisions of this section. The department shall identify all criminal prosecutions in the state for conduct related to SKS rifles within 90 days of the effective date of this section. In all cases so identified by the Attorney General, the district attorneys shall inform defense counsel, or the defendant if the defendant is in propria persona, in writing, of the provisions of this section within 120 days of the effective date of this section.

(f)(1) Any person, firm, company, or corporation that is in possession of an SKS rifle shall do one of the following on or before January 1, 2000:

(A) Relinquish the SKS rifle to the Department of Justice pursuant to subdivision (h).

(B) Relinquish the SKS rifle to a law enforcement agency pursuant to Section 12288.

(C) Dispose of the SKS rifle as permitted by Section 12285.

(2) Any person who has obtained title to an SKS rifle by bequest or intestate succession shall be required to comply with subparagraph (A) or (B) of paragraph (1) of this subdivision unless he or she otherwise complies with paragraph (1) of subdivision (b) of Section 12285.

(3) Any SKS rifle relinquished to the department pursuant to this subdivision shall be in a manner prescribed by the department.

(4) The department shall conduct a public education and notification program as described in Section 12289, commencing no later than January 1, 1999.

(g) Any person who complies with subdivision (f) shall be exempt from the prohibitions set forth in subdivision (a) or (b) of Section 12280 for those acts by that person associated with complying with the requirements of subdivision (f).

(h)(1) The department shall purchase any SKS rifle relinquished pursuant to subdivision (f) from funds appropriated for this purpose by the act amending this section in the 1997-98 Regular Session of the Legislature or by subsequent budget acts or other legislation. The department shall adopt regulations for this purchase program that include, but are not limited to, the manner of delivery, the reimbursement to be paid, and the manner in which persons shall be informed of the state purchase program.

(2) Any person who relinquished possession of an SKS rifle to a law enforcement agency pursuant to Section 12288 prior to the effective date of the purchase program set forth in paragraph (1) shall be eligible to be reimbursed from the purchase program. The procedures for reimbursement pursuant to this paragraph shall be part of the regulations adopted by the department pursuant to paragraph (1).

(i) Notwithstanding paragraph (11) of subdivision (a) of Section 12276, an "SKS rifle" under this section means all SKS rifles commonly referred to as "SKS Sporter" versions, manufactured to accept a detachable AK-47 magazine and imported into this state and sold by a licensed gun dealer, or otherwise lawfully possessed in this state by a resident of this state who is not a licensed gun dealer, between January 1, 1992, and December 19, 1997.

(j) Failure to comply with subdivision (f) is a public offense punishable by imprisonment in the state prison, or in a county jail, not exceeding one year.

(k) In addition to the regulations required pursuant to subdivision (h), emergency regulations for the purchase program described in subdivision (h) shall be adopted pursuant to Chapter 3.5 (commencing with Section 11340) of Part 1 of Division 3 of Title 2 of the Government Code.

12282. Possession of assault weapon as public nuisance; Civil action; Destruction of weapon

(a) Except as provided in Section 12280, possession of any assault weapon, as defined in Section 12276, 12276.1, or 12276.5, or of any .50 BMG rifle, as defined in Section 12278, in violation of this chapter is a public nuisance, solely for purposes of this section and subdivision (d) of Section 12028. The Attorney General, any district attorney, or any city attorney, may, in lieu of criminal prosecution, bring a civil action or reach a civil compromise in any superior court to enjoin the possession of the assault weapon or .50 BMG rifle that is a public nuisance.

(b) Upon motion of the Attorney General, district attorney, or city attorney, a superior court may impose a civil fine not to exceed three hundred dollars ($300) for the first assault weapon or .50 BMG rifle deemed a public nuisance pursuant to subdivision (a) and up to one hundred dollars ($100) for each additional assault weapon or .50 BMG rifle deemed a public nuisance pursuant to subdivision (a).

(c) Any assault weapon or .50 BMG rifle deemed a public nuisance under subdivision (a) shall be destroyed in a manner so that it may no longer be used, except upon a finding by a court, or a declaration from the Department of Justice, district attorney, or city attorney stating that the preservation of the assault weapon or .50 BMG rifle is in the interest of justice.

(d) Upon conviction of any misdemeanor or felony involving the illegal possession or use of an assault weapon, the assault weapon shall be deemed a public nuisance and disposed of pursuant to subdivision (d) of Section 12028.

Article 3. Registration and Permits

12285. Registration procedure; Fee; Sale or transfer to licensed gun dealer; Conditions for possession; Persons excluded from registration or possession; Forgiveness period

(a)(1) Any person who lawfully possesses an assault weapon, as defined in Section 12276, prior to June 1, 1989, shall register the firearm by January 1, 1991, and any person who lawfully possessed an assault weapon prior to the date it was specified as an assault weapon pursuant to Section 12276.5 shall register the firearm within 90 days with the Department of Justice pursuant to those procedures that the department may establish. Except as provided in subdivision (a) of Section 12280, any person who lawfully possessed an assault weapon prior to the date it was defined as an assault weapon pursuant to Section 12276.1, and which was not specified as an assault weapon under Section 12276 or 12276.5, shall register the firearm within one year of the effective date of Section 12276.1, with the department pursuant to those procedures that the department may establish. The registration shall contain a description of the firearm that identifies it uniquely, including all identification marks, the full name, address, date of birth, and thumbprint of the owner, and any other information that the department may deem appropriate. The department may charge a fee for registration of up to twenty dollars ($20) per person but not to exceed the actual processing costs of the department. After the department establishes fees sufficient to reimburse the department for processing costs, fees charged shall increase at a rate not to exceed the legislatively approved annual cost-of-living adjustment for the department's budget or as otherwise increased through the Budget Act. The fees shall be deposited into the Dealers' Record of Sale Special Account.

(2) Except as provided in subdivision (a) of Section 12280, any person who lawfully possesses any .50 BMG rifle prior to January 1, 2005, that is not specified as an assault weapon under Section 12276 or 12276.5 or defined as an assault weapon pursuant to Section 12276.1, shall register the .50 BMG rifle with the department no later than April 30, 2006, pursuant to those procedures that the department may establish. The registration shall contain a description of the firearm that identifies it uniquely, including all identification marks, the full name, address, date of birth, and thumbprint of the owner, and any other information that the department may deem appropriate. The department may charge a fee for registration of twenty-five dollars ($25) per person to cover the actual processing and public education campaign costs of the department. The fees shall be deposited into the Dealers' Record of Sale Special Account. Data-processing costs associated with modifying the department's data system to accommodate .50 caliber BMG rifles shall not be paid from the Dealers Record of Sale Special Account.

(b)(1) Except as provided in paragraph (2), no assault weapon possessed pursuant to this section may be sold or transferred on or after January 1, 1990, to anyone within this state other than to a licensed gun dealer, as defined in subdivision (c) of Section 12290, or as provided in Section 12288. Any person who (A) obtains title to an assault weapon registered under this section or that was possessed pursuant to paragraph (1) of subdivision (f) of Section 12280 by bequest or intestate succession, or (B) lawfully possessed a firearm subsequently declared to be an assault weapon pursuant to Section 12276.5, or subsequently defined as an assault weapon pursuant to Section 12276.1, shall, within 90 days, render the weapon permanently inoperable, sell the weapon to a licensed gun dealer, obtain a permit from the Department of Justice in the same manner as specified in Article 3 (commencing with Section 12230) of Chapter 2, or remove the weapon from this state. A person who lawfully possessed a firearm that was subsequently declared to be an assault weapon pursuant to Section 12276.5 may alternatively register the firearm within 90 days of the declaration issued pursuant to subdivision (f) of Section 12276.5.

(2) A person moving into this state, otherwise in lawful possession of an assault weapon, shall do one of the following:

(A) Prior to bringing the assault weapon into this state, that person shall first obtain a permit from the Department of Justice in the same manner as specified in Article 3 (commencing with Section 12230) of Chapter 2.

(B) The person shall cause the assault weapon to be delivered to a licensed gun dealer, as defined in subdivision (c) of Section 12290, in this state in accordance with Chapter 44 (commencing with Section 921) of Title 18 of the United States Code and the regulations issued pursuant thereto. If the person obtains a permit from the Department of Justice in the same manner as specified in Article 3 (commencing with Section 12230) of Chapter 2, the dealer shall redeliver that assault weapon to the person. If the licensed gun dealer, as defined in subdivision (c) of Section 12290, is prohibited from delivering the assault weapon to a person pursuant to this paragraph, the dealer shall possess or dispose of the assault weapon as allowed by this chapter.

(3) Except as provided in paragraph (4), no .50 BMG rifle possessed pursuant to this section may be sold or transferred on or after January 1, 2005, to anyone within this state other than to a licensed gun dealer, as defined in subdivision (c) of Section 12290, or as provided in Section 12288. Any person who obtains title to a .50 BMG rifle registered under this section or that was possessed pursuant to paragraph (1) of subdivision (f) of Section 12280 by bequest or intestate succession shall, within 180 days of receipt, render the weapon permanently inoperable, sell the weapon to a licensed gun dealer, obtain a permit from the Department of Justice in the same manner as specified in Article 3 (commencing with Section 12230) of Chapter 2, or remove the weapon from this state.

(4) A person moving into this state, otherwise in lawful possession of a .50 BMG rifle, shall do one of the following:

(A) Prior to bringing the .50 BMG rifle into this state, that person shall first obtain a permit from the Department of Justice in the same manner as specified in Article 3 (commencing with Section 12230) of Chapter 2.

(B) The person shall cause the .50 BMG rifle to be delivered to a licensed gun dealer, as defined in subdivision (c) of Section 12290 in this state in accordance with Chapter 44 (commencing with Section 921) of Title 18 of the United States Code and the regulations issued pursuant thereto. If the person obtains a permit from the Department of Justice in the same manner as specified in Article 3 (commencing with Section 12230) of Chapter 2, the dealer shall redeliver that .50 BMG rifle to the person. If the licensed gun dealer, as defined in subdivision (c) of Section 12290 is prohibited from delivering the . 50 caliber BMG rifle to a person pursuant to this paragraph, the dealer shall dispose of the .50 BMG rifle as allowed by this chapter.

(c) A person who has registered an assault weapon or registered a .50 BMG rifle under this section may possess it only under any of the following conditions unless a permit allowing additional uses is first obtained under Section 12286:

(1) At that person's residence, place of business, or other property owned by that person, or on property owned by another with the owner's express permission.

(2) While on the premises of a target range of a public or private club or organization organized for the purpose of practicing shooting at targets.

(3) While on a target range that holds a regulatory or business license for the purpose of practicing shooting at that target range.

(4) While on the premises of a shooting club which is licensed pursuant to the Fish and Game Code.

(5) While attending any exhibition, display, or educational project which is about firearms and which is sponsored by, conducted under the auspices of, or approved by a law enforcement agency or a nationally or state recognized entity that fosters proficiency in, or promotes education about, firearms.

(6) While on publicly owned land if the possession and use of a firearm described in Section 12276, 12276.1, 12276.5, or 12278, is specifically permitted by the managing agency of the land.

(7) While transporting the assault weapon or .50 BMG rifle between any of the places mentioned in this subdivision, or to any licensed gun dealer, as defined in subdivision (c) of Section 12290, for servicing or repair pursuant to subdivision (b) of Section 12290, if the assault weapon is transported as required by Section 12026.1.

(d) No person who is under the age of 18 years, no person who is prohibited from possessing a firearm by Section 12021 or 12021.1, and no person described in Section 8100 or 8103 of the Welfare and Institutions Code may register or possess an assault weapon or .50 BMG rifle.

(e) The department's registration procedures shall provide the option of joint registration for assault weapons or .50 BMG rifle owned by family members residing in the same household.

(f) For 90 days following January 1, 1992, a forgiveness period shall exist to allow persons specified in subdivision (b) of Section 12280 to register with the Department of Justice assault weapons that they lawfully possessed prior to June 1, 1989.

(g)(1) Any person who registered a firearm as an assault weapon pursuant to the provisions of law in effect prior to January 1, 2000, where the assault weapon is thereafter defined as an assault weapon pursuant to Section 12276.1, shall be deemed to have registered the weapon for purposes of this chapter and shall not be required to reregister the weapon pursuant to this section.

(2) Any person who legally registered a firearm as an assault weapon pursuant to the provisions of law in effect prior to January 1, 2005, where the assault weapon is thereafter defined as a .50 caliber BMG rifle pursuant to Section 12278, shall be deemed to have registered the weapon for purposes of this chapter and shall not be required to reregister the weapon pursuant to this section.

(h) Any person who registers his or her assault weapon during the 90-day forgiveness period described in subdivision (f), and any person whose registration form was received by the Department of Justice after January 1, 1991, and who was issued a temporary registration prior to the end of the forgiveness period, shall not be charged with a violation of subdivision (b) of Section 12280, if law enforcement becomes aware of that violation only as a result of the registration of the assault weapon. This subdivision shall have no effect upon persons charged with a violation of subdivision (b) of Section 12280 of the Penal Code prior to January 1, 1992, provided that law enforcement was aware of the violation before the weapon was registered.

12286. Permit requirement Any person who lawfully acquired an assault weapon before June 1, 1989, or a .50 BMG rifle before January 1, 2005, and wishes to use it in a manner different than specified in subdivision (c) of Section 12285, who lawfully acquired an assault weapon between June 1, 1989, and January 1, 1990, and wishes to keep it after January 1, 1990, or who wishes to acquire an assault weapon after January 1, 1990, or a .50 BMG rifle after January 1, 2005, shall first obtain a permit from the Department of Justice in the same manner as specified in Article 3 (commencing with Section 12230) of Chapter 2.

12287. Issuance of permits for manufacture or sale of assault weapons or .50 BMG rifles for possession by specified persons and entities

(a) The Department of Justice may, upon a finding of good cause, issue permits for the manufacture of assault weapons or .50 BMG rifles for the sale to, purchase by, or possession of assault weapons or .50 BMG rifles by, any of the following:

(1) The agencies listed in subdivision (e), and the officers described in subdivision (f) of Section 12280.

(2) Entities and persons who have been issued permits pursuant to this section or Section 12286.

(3) Entities outside the state who have, in effect, a federal firearms dealer's license solely for the purpose of distribution to an entity listed in paragraphs (4) to (6), inclusive.

(4) Federal law enforcement and military agencies.

(5) Law enforcement and military agencies of other states.

(6) Foreign governments and agencies approved by the United States State Department.

(b) Application for the permits, the keeping and inspection thereof, and the revocation of permits shall be undertaken in the same manner as specified in Article 3 (commencing with Section 12230) of Chapter 2.

12288. Relinquishment of weapon to peace officers

Any individual may arrange in advance to relinquish an assault weapon or a .50 BMG rifle to a police or sheriff's department. The assault weapon or .50 BMG rifle shall be transported in accordance with Section 12026.1.

12289.5.Annual inspection for security and safe storage by Department of Justice

(a) Except as provided in subdivision (b), the Department of Justice shall, for every person, firm, or corporation to whom a permit is issued pursuant to this article, annually conduct an inspection for security and safe storage purposes, and to reconcile the inventory of assault weapons.

(b) A person, firm, or corporation with an inventory of fewer than five devices that require any Department of Justice permit shall be subject to an inspection for security and safe storage purposes, and to reconcile inventory, once every five years, or more frequently if determined by the department.

Article 4. Licensed Gun Dealers

12290. Transportation, display or sale of weapons; Service or repair; "Licensed gun dealer"

(a) Any licensed gun dealer, as defined in subdivision (c), who lawfully possesses an assault weapon or .50 BMG rifle pursuant to Section 12285, in addition to the uses allowed in Section 12285, may transport the firearm between dealers or out of the state if that person is permitted pursuant to the National Firearms Act, display it at any gun show licensed by a state or local governmental entity, sell it to a resident outside the state, or sell it to a person who has been issued a permit pursuant to Section 12286. Any transporting allowed by this section must be done as required by Section 12026.1.

(b)(1) Any licensed gun dealer, as defined in subdivision (c), may take possession of any assault weapon or .50 BMG rifle for the purposes of servicing or repair from any person to whom it is legally registered or who has been issued a permit to possess it pursuant to this chapter.

(2) Any licensed gun dealer, as defined in subdivision (c), may transfer possession of any assault weapon or .50 BMG rifle received pursuant to paragraph (1), to a gunsmith for purposes of accomplishing service or repair of the same. Transfers are permissible only to the following persons:

(A) A gunsmith who is in the dealer's employ.

(B) A gunsmith with whom the dealer has contracted for gunsmithing services. In order for this subparagraph to apply, the gunsmith receiving the assault weapon or .50 BMG rifle shall hold all of the following:

(i) A dealer's license issued pursuant to Chapter 44 (commencing with Section 921) of Title 18 of the United States Code and the regulations issued pursuant thereto.

(ii) Any business license required by a state or local governmental entity.

(c) The term "licensed gun dealer," as used in this article, means a person who is licensed pursuant to Section 12071 and who has a permit to sell assault weapons or .50 BMG rifles pursuant to Section 12287.

Chapter 2.5. Destructive Devices

12301. "Destructive device"; "Explosive"

(a) The term "destructive device," as used in this chapter, shall include any of the following weapons:

(1) Any projectile containing any explosive or incendiary material or any other chemical substance, including, but not limited to, that which is commonly known as tracer or incendiary ammunition, except tracer ammunition manufactured for use in shotguns.

(2) Any bomb, grenade, explosive missile, or similar device or any launching device therefor.

(3) Any weapon of a caliber greater than 0.60 caliber which fires fixed ammunition, or any ammunition therefor, other than a shotgun (smooth or rifled bore) conforming to the definition of a "destructive device" found in subsection (b) of Section 479.11 of Title 27 of the Code of Federal Regulations, shotgun ammunition (single projectile or shot), antique rifle, or an antique cannon. For purposes of this section, the term "antique cannon" means any cannon manufactured before January 1, 1899, which has been rendered incapable of firing or for which ammunition is no longer manufactured in the United States and is not readily available in the ordinary channels of commercial trade. The term "antique rifle" means a firearm conforming to the definition of an "antique firearm" in Section 479.11 of Title 27 of the Code of Federal Regulations.

(4) Any rocket, rocket-propelled projectile, or similar device of a diameter greater than 0.60 inch, or any launching device therefor, and any rocket, rocket-propelled projectile, or similar device containing any explosive or incendiary material or any other chemical substance, other than the propellant for that device, except those devices as are designed primarily for emergency or distress signaling purposes.

(5) Any breakable container which contains a flammable liquid with a flashpoint of 150 degrees Fahrenheit or less and has a wick or similar device capable of being ignited, other than a device which is commercially manufactured primarily for the purpose of illumination.

(6) Any sealed device containing dry ice (CO_2) or other chemically reactive substances assembled for the purpose of causing an explosion by a chemical reaction.

(b) The term "explosive," as used in this chapter, shall mean any explosive defined in Section 12000 of the Health and Safety Code.

12302. Sale to, or purchase, possession, transportation, storage, or use of, by law enforcement officers, military personnel, or firefighters

Nothing in this chapter shall prohibit the sale to, purchase by, or possession, transportation, storage, or use of, destructive devices or explosives by:

(a) Any peace officer listed in Section 830.1 or 830.2, or any peace officer in the Department of Justice authorized by the Attorney General, while on duty and acting within the scope and course of his or her employment.

(b) Any member of the Army, Navy, Air Force, or Marine Corps of the United States, or the National Guard, while on duty and acting within the scope and course of his or her employment.

Nothing in this chapter prohibits the sale to, or the purchase, possession, transportation, storage, or use by any person who is a regularly employed and paid officer, employee, or member of a fire department or fire protection or firefighting agency of the federal government, the State of California, a city, county, city and county, district, or other public or municipal corporation or political subdivision of this state, while on duty and acting within the scope and course of his or her employment, of any equipment used by that department or agency in the course of fire suppression.

12303. Possession of destructive device; Punishment

Any person, firm, or corporation who, within this state, possesses any destructive device, other than fixed ammunition of a caliber greater than .60 caliber, except as provided by this chapter, is guilty of a public offense and upon conviction thereof shall be punished by imprisonment in the county jail for a term not to exceed one year, or in state prison, or by a fine not to exceed ten thousand dollars ($10,000) or by both such fine and imprisonment.

12303.6. Sale or transportation of destructive device; Punishment

Any person, firm, or corporation who, within this state, sells, offers for sale, or knowingly transports any destructive device, other than fixed ammunition of a caliber greater than .60 caliber, except as provided by this chapter, is guilty of a felony and is punishable by imprisonment in the state prison for two, three or four years.

12304. Sale, possession, or transportation of fixed ammunition; Punishment; Subsequent conviction

Any person, firm or corporation who, within this state, sells, offers for sale, possesses or knowingly transports any fixed ammunition of a caliber greater than .60 caliber, except as provided in this chapter, is guilty of a public offense and upon conviction thereof shall be punished by imprisonment in the county jail for a term not to exceed six months or by a fine not to exceed one thousand dollars ($1,000), or by both such fine and imprisonment.

A second or subsequent conviction shall be punished by imprisonment in the county jail for a term not to exceed one year, or by imprisonment in the state prison, or by a fine not to exceed three thousand dollars ($3,000), or by both such fine and imprisonment.

12305. Permit for manufacture, sale or use of destructive device; Inspection for security and safe storage

(a) Every dealer, manufacturer, importer, and exporter of any destructive device, or any motion picture or television studio using destructive devices in the conduct of its business, shall obtain a permit for the conduct of that business from the Department of Justice.

(b) Any person, firm, or corporation not mentioned in subdivision (a) shall obtain a permit from the Department of Justice in order to possess or transport any destructive device. No permit shall be issued to any person who meets any of the following criteria:

(1) Has been convicted of any felony.

(2) Is addicted to the use of any narcotic drug.

(3) Is a person in a class prohibited by Section 8100 or 8103 of the Welfare and Institutions Code or Section 12021 or 12021.1 of this code.

(c) Applications for permits shall be filed in writing, signed by the applicant if an individual, or by a member or officer qualified to sign if the applicant is a firm or corporation, and shall state the name, business in which engaged, business address and a full description of the use to which the destructive devices are to be put.

(d) Applications and permits shall be uniform throughout the state on forms prescribed by the Department of Justice.

(e) Each applicant for a permit shall pay at the time of filing his or her application a fee not to exceed the application processing costs of the Department of Justice. A permit granted pursuant to this article may be renewed one year from the date of issuance, and annually thereafter, upon the filing of a renewal application and the payment of a permit renewal fee not to exceed the application processing costs of the Department of Justice. After the department establishes fees sufficient in amount to cover processing costs, the amount of the fees shall only increase at a rate not to exceed the legislatively approved cost-of-living adjustment for the department.

(f) Except as provided in subdivision (g), the Department of Justice shall, for every person, firm, or corporation to whom a permit is issued pursuant to this article, annually conduct an inspection for security and safe storage purposes, and to reconcile the inventory of destructive devices.

(g) A person, firm, or corporation with an inventory of fewer than five devices that require any Department of Justice permit shall be subject to an inspection for security and safe storage purposes, and to reconcile inventory, once every five years, or more frequently if determined by the department.

12307. Nuisances; Injunction; Destruction of device

The possession of any destructive device in violation of this chapter shall be deemed to be a public nuisance and the Attorney General or district attorney of any city, county, or city and county may bring an action before the superior court to enjoin the possession of any destructive device.

Any destructive device found to be in violation of this chapter shall be surrendered to the Department of Justice, or to the sheriff or chief of police, if the sheriff or chief of police has elected to perform the services required by this section. The department, sheriff, or chief of police shall destroy the destructive device so as to render it unusable and unrepairable as a destructive device, except upon the filing of a certificate with the department by a judge or district attorney stating that the preservation of the destructive device is necessary to serve the ends of justice.

12312. Possession of materials with intent to make destructive device or explosive; Punishment

Every person who possesses any substance, material, or any combination of substances or materials, with the intent to make any destructive device or any explosive without first obtaining a valid permit to make such destructive device or explosive, is guilty of a felony, and is punishable by imprisonment in the state prison for two, three, or four years.

Chapter 2.6. Ammunition

12316.Sale of ammunition to minor or person under specified age; Ownership or possession of ammunition by person prohibited from having firearm; Carrying of ammunition onto school grounds

(a)(1) Any person, corporation, or dealer who does either of the following shall be punished by imprisonment in a county jail for a term not to exceed six months, or by a fine not to exceed one thousand dollars ($1,000), or by both the imprisonment and fine:

(A) Sells any ammunition or reloaded ammunition to a person under 18 years of age.

(B) Sells any ammunition or reloaded ammunition designed and intended for use in a handgun to a person under 21 years of age. As used in this subparagraph, "ammunition" means handgun ammunition as defined in subdivision (a) of Section 12323. Where ammunition or reloaded ammunition may be used in both a rifle and a handgun, it may be sold to a person who is at least 18 years of age, but less than 21 years of age, if the vendor reasonably believes that the ammunition is being acquired for use in a rifle and not a handgun.

(2) Proof that a person, corporation, or dealer, or his or her agent or employee, demanded, was shown, and acted in reasonable reliance upon, bona fide evidence of majority and identity shall be a defense to any criminal prosecution under this subdivision. As used in this subdivision, "bona fide evidence of majority and identity" means a document issued by a federal, state, county, or municipal government, or subdivision or agency thereof, including, but not limited to, a motor vehicle operator's license, California state identification card, identification card issued to a member of the armed forces, or other form of identification that bears the name, date of birth, description, and picture of the person.

(b)(1) No person prohibited from owning or possessing a firearm under Section 12021 or 12021.1 of this code or Section 8100 or 8103 of the Welfare and Institutions Code shall own, possess, or have under his or her custody or control, any ammunition or reloaded ammunition.

(2) For purposes of this subdivision, "ammunition" shall include, but not be limited to, any bullet, cartridge, magazine, clip, speed loader, autoloader, or projectile capable of being fired from a firearm with a deadly consequence.

(3) A violation of this subdivision is punishable by imprisonment in a county jail not to exceed one year or in the state prison, by a fine not to exceed one thousand dollars ($1,000), or by both the fine and imprisonment.

(c) Unless it is with the written permission of the school district superintendent, his or her designee, or equivalent school authority, no person shall carry ammunition or reloaded ammunition onto school grounds, except sworn law enforcement officers acting within the scope of their duties or persons exempted under sub-paragraph (A) of paragraph (1) of subdivision (a) of Section 12027. This subdivision shall not apply to a duly appointed peace officer as defined in Chapter 4.5 (commencing with Section 830) of Title 3 of Part 2, a full-time paid peace officer of another state or the federal government who is carrying out official duties while in California, any person summoned by any of these officers to assist in making an arrest or preserving the peace while he or she is actually engaged in assisting the officer, a member of the military forces of this state or of the United States who is engaged in the performance of his or her duties, a person holding a valid license to carry the firearm pursuant to Article 3 (commencing with Section 12050) of Chapter 1 of Title 2 of Part 4, or an armored vehicle guard, who is engaged in the performance of his or her duties, as defined in subdivision (e) of Section 7521 of the Business and Professions Code.

A violation of this subdivision is punishable by imprisonment in a county jail for a term not to exceed six months, a fine not to exceed one thousand dollars ($1,000), or both the imprisonment and fine.

(d)(1) A violation of paragraph (1) of subdivision (b) is justifiable where all of the following conditions are met:

(A) The person found the ammunition or reloaded ammunition or took the ammunition or reloaded ammunition from a person who was committing a crime against him or her.

(B) The person possessed the ammunition or reloaded ammunition no longer than was necessary to deliver or transport the ammunition or reloaded ammunition to a law enforcement agency for that agency's disposition according to law.

(C) The person is prohibited from possessing any ammunition or reloaded ammunition solely because that person is prohibited from owning or possessing a firearm only by virtue of Section 12021.

(2) Upon the trial for violating paragraph (1) of subdivision (b), the trier of fact shall determine whether the defendant is subject to the exemption created by this subdivision.

(3) The defendant has the burden of proving by a preponderance of the evidence that he or she is subject to the exemption provided by this subdivision.

12320. Possession; Punishment

Any person, firm, or corporation who, within this state knowingly possesses any handgun ammunition designed primarily to penetrate metal or armor is guilty of a public offense and upon conviction thereof shall be punished by imprisonment in the state prison, or in the county jail for a term not to exceed one year, or by a fine not to exceed five thousand dollars ($5,000), or by both such fine and imprisonment.

12321. Manufacture, transport or sale; Punishment

Any person, firm, or corporation who, within this state, manufactures, imports, sells, offers to sell, or knowingly transports any handgun ammunition designed primarily to penetrate metal or armor is guilty of a felony and upon conviction thereof shall be punished by imprisonment in state prison, or by a fine not to exceed five thousand dollars ($5,000), or by both such fine and imprisonment.

12322. Exemptions

Nothing in this chapter shall apply to or affect either of the following:

(a) The sale to, purchase by, possession of, or use of any ammunition by any member of the Army, Navy, Air Force, or Marine Corps of the United States, or the National Guard, while on duty and acting within the scope and course of his or her employment, or any police agency or forensic laboratory or any person who is the holder of a valid permit issued pursuant to Section 12305.

(b) The possession of handgun ammunition designed primarily to penetrate metal or armor by a person who found the ammunition, if he or she is not prohibited from possessing firearms or ammunition pursuant to Section 12021, 12021.1, or paragraph (1) of subdivision (b) of Section 12316 of this code or Section 8100 or 8103 of the Welfare and Institutions Code and is transporting the ammunition to a law enforcement agency for disposition according to law.

12323. Definitions

As used in this chapter, the following definitions shall apply:

(a) "Handgun ammunition" means ammunition principally for use in pistols, revolvers, and other firearms capable of being concealed upon the person, as defined in subdivision (a) of Section 12001, notwithstanding that the ammunition may also be used in some rifles.

(b) "Handgun ammunition designed primarily to penetrate metal or armor" means any ammunition, except a shotgun shell or ammunition primarily designed for use in rifles, that is designed primarily to penetrate a body vest or body shield, and has either of the following characteristics:

(1) Has projectile or projectile core constructed entirely, excluding the presence of traces of other substances, from one or a combination of tungsten alloys, steel, iron, brass, beryllium copper, or depleted uranium, or any equivalent material of similar density or hardness.

(2) Is primarily manufactured or designed, by virtue of its shape, cross-sectional density, or any coating applied thereto, including, but not limited to, ammunition commonly known as "KTW ammunition," to breach or penetrate a body vest or body shield when fired from a pistol, revolver, or other firearm capable of being concealed upon the person.

(c) "Body vest or shield" means any bullet-resistant material intended to provide ballistic and trauma protection for the wearer or holder.

(d) "Rifle" shall have the same meaning as defined in paragraph (20) of subdivision (c) of Section 12020.

12324. Deactivated ammunition

Nothing in this chapter shall prohibit the possession, importation, sale, attempted sale, or transport of ammunition from which the propellant has been removed and the primer has been permanently deactivated.

12325. Manufacture of ammunition pursuant to government contract

Nothing in this chapter shall prohibit the manufacture of ammunition under contracts approved by agencies of the state or federal government.

Chapter 5. Firearm Devices

Article 1. General Provisions

12500. "Silencer" The term "silencer" as used in this chapter means any device or attachment of any kind designed, used, or intended for use in silencing, diminishing, or muffling the report of a firearm. The term "silencer" also includes any combination of parts, designed or redesigned, and intended for use in assembling a silencer or fabricating a silencer and any part intended only for use in such assembly or fabrication.

12501. Exceptions from proscription on possession of silencer

Section 12520 shall not apply to, or affect, any of the following:

(a) The sale to, purchase by, or possession of silencers by agencies listed in Section 830.1, or the military or naval forces of this state or of the United States for use in the discharge of their official duties.

(b) The possession of silencers by regular, salaried, full-time peace officers who are employed by an agency listed in Section 830.1, or by the military or naval forces of this state or of the United States when on duty and when the use of silencers is authorized by the agency and is within the course and scope of their duties.

(c) The manufacture, possession, transportation, or sale or other transfer of silencers to an entity described in subdivision (a) by dealers or manufacturers registered under Chapter 53 (commencing with Section 5801) of Title 26 of the United States Code, and the regulations issued pursuant thereto.

Article 2. Unlawful Possession of Firearm Silencers

12520. Punishment for possession

Any person, firm, or corporation who within this state possesses a silencer is guilty of a felony and upon conviction thereof shall be punished by imprisonment in the state prison or by a fine not to exceed ten thousand dollars ($10,000) or by both.

Chapter 6. Miscellaneous

Article 1. BB Devices and Imitation Firearms

12550. Definitions As used in this article, the following definitions apply:

(a) "BB device" is defined in subdivision (g) of Section 12001.

(b) "Firearm" is defined in subdivision (b) of Section 12001.

(c) "Imitation firearm" means any BB device, toy gun, replica of a firearm, or other device that is so substantially similar in coloration and overall appearance to an existing firearm as to lead a reasonable person to perceive that the device is a firearm.

12551. Sale of BB device to minor

Every person who sells to a minor any BB device is guilty of a misdemeanor.

12552. Furnishing weapon to minor

(a) Every person who furnishes any BB device to any minor, without the express or implied permission of the parent or legal guardian of the minor, is guilty of a misdemeanor.

(b) As used in this section, "furnishes" means any of the following:

(1) A loan.

(2) A transfer that does not involve a sale.

12553. Alteration of imitation firearm or device as misdemeanor

(a)(1) Any person who changes, alters, removes, or obliterates any coloration or markings that are required by any applicable state or federal law or regulation, for any imitation firearm, or device described in subdivision (c) of Section 12555, in any way that makes the imitation firearm or device look more like a firearm is guilty of a misdemeanor.

(2) This subdivision shall not apply to manufacturer, importer, or distributor of imitation firearms or to the lawful use in theatrical productions, including motion pictures, television, and stage productions.

(b) Any manufacturer, importer, or distributor of imitation firearms that fails to comply with any applicable federal law or regulation governing the marking of a toy, look-alike or imitation firearm as defined by federal law or regulation is guilty of a misdemeanor.

12554. Conspicuous advisory required; Penalty for failure to comply

(a) Any imitation firearm manufactured after July 1, 2005, shall, at the time of offer for sale in this state, be accompanied by a conspicuous advisory in writing as part of the packaging, but not necessarily affixed to the imitation firearm to the effect that the product may be mistaken for a firearm by law enforcement officers or others, that altering the coloration or markings required by state or federal law or regulations so as to make the product look more like a firearm is dangerous, and may be a crime, and that brandishing or displaying the product in public may cause confusion and may be a crime.

(b) Any manufacturer, importer, or distributor that fails to comply with this advisory for any imitation firearm manufactured after July 1, 2005, shall be liable for a civil fine for each action brought by a city attorney or district attorney of not more than one thousand dollars ($1,000) for the first action, five thousand dollars ($5,000) for the second action, and ten thousand dollars ($10,000) for the third action and each subsequent action.

12555. Authorized purposes; Liability for civil fine

(a) Any person who, for commercial purposes, purchases, sells, manufactures, ships, transports, distributes, or receives, by mail order or in any other manner, an imitation firearm except as authorized by this section shall be liable for a civil fine in an action brought by the city attorney or the district attorney of not more than ten thousand dollars ($10,000) for each violation.

(b) The manufacture, purchase, sale, shipping, transport, distribution, or receipt, by mail or in any other manner, of imitation firearms is authorized if the device is manufactured, purchased, sold, shipped, transported, distributed, or received for any of the following purposes:

(1) Solely for export in interstate or foreign commerce.

(2) Solely for lawful use in theatrical productions, including motion picture, television, and stage productions.

(3) For use in a certified or regulated sporting event or competition.

(4) For use in military or civil defense activities, or ceremonial activities.

(5) For public displays authorized by public or private schools.

(c) As used in this section, "imitation firearm" does not include any of the following:

(1) A nonfiring collector's replica that is historically significant, and is offered for sale in conjunction with a wall plaque or presentation case.

(2) A BB device, as defined in subdivision (g) of Section 12001.

(3) A device where the entire exterior surface of the device is white, bright red, bright orange, bright yellow, bright green, bright blue, bright pink, or bright purple, either singly or as the predominant color in combination with other colors in any pattern, as provided by federal regulations governing imitation firearms, or where the entire device is constructed of transparent or translucent materials which permits unmistakable observation of the device's complete contents, as provided by federal regulations governing imitation firearms.

12556. Open display or exposure in public place prohibited; Punishment for violation; Exceptions

(a) No person may openly display or expose any imitation firearm, as defined in Section 12550, in a public place.

(b) Violation of this section, except as provided in subdivision (c), is an infraction punishable by a fine of one hundred dollars ($100) for the first offense, and three hundred dollars ($300) for a second offense.

(c) A third or subsequent violation of this section is punishable as a misdemeanor.

(d) Subdivision (a) shall not apply to the following, when the imitation firearm is:

(1) Packaged or concealed so that it is not subject to public viewing.

(2) Displayed or exposed in the course of commerce, including commercial film or video productions, or for service, repair, or restoration of the imitation firearm.

(3) Used in a theatrical production, a motion picture, video, television, or stage production.

(4) Used in conjunction with a certified or regulated sporting event or competition.

(5) Used in conjunction with lawful hunting, or lawful pest control activities.

(6) Used or possessed at certified or regulated public or private shooting ranges.

(7) Used at fairs, exhibitions, expositions, or other similar activities for which a permit has been obtained from a local or state government.

(8) Used in military, civil defense, or civic activities, including flag ceremonies, color guards, parades, award presentations, historical reenactments, and memorials.

(9) Used for public displays authorized by public or private schools or displays that are part of a museum collection.

(10) Used in parades, ceremonies, or other similar activities for which a permit has been obtained from a local or state government.

(11) Displayed on a wall plaque or in a presentation case.

(12) Used in areas where the discharge of a firearm is lawful.

(13) A device where the entire exterior surface of the device is white, bright red, bright orange, bright yellow, bright green, bright blue, bright pink, or bright purple, either singly or as the predominant color in combination with other colors in any pattern, or where the entire device is constructed of transparent or translucent materials which permits unmistakable observation of the device's complete contents. Merely having an orange tip as provided in federal law and regulations does not satisfy this requirement. The entire surface must be colored or transparent or translucent.

(e) For purposes of this section, the term "public place" means an area open to the public and includes streets, sidewalks, bridges, alleys, plazas, parks, driveways, front yards, parking lots, automobiles, whether moving or not, and buildings open to the general public, including those that serve food or drink, or provide entertainment, and the doorways and entrances to buildings or dwellings.

(f) Nothing in this section shall be construed to preclude prosecution for a violation of Section 171b, 171.5, or 626.10.

Article 8. Handgun Safety Certificate

12800. Legislative intent It is the intent of the Legislature in enacting this article to require that persons who obtain handguns have a basic familiarity with those firearms, including, but not limited to, the safe handling and storage of those firearms. It is not the intent of the Legislature to require a handgun safety certificate for the mere possession of a firearm.

12801. Definitions

(a) As used in this article, the following definitions shall apply:

(1) "Department" means the Department of Justice.

(2) "DOJ Certified Instructor" or "certified instructor" means a person designated as a handgun safety instructor by the Department of Justice pursuant to subdivision (d) of Section 12804.

(b) No person shall do either of the following:

(1) Purchase or receive any handgun, except an antique firearm, as defined in paragraph (16) of subsection (a) of Section 921 of Title 18 of the United States Code, without a valid handgun safety certificate.

(2) Sell, deliver, loan, or transfer any handgun, except an antique firearm, as defined in paragraph (16) of subsection (a) of Section 921 of Title 18 of the United States Code, to any person who does not have a valid handgun safety certificate.

(c) Any person who violates subdivision (b) is guilty of a misdemeanor.

(d) The provisions of this section are cumulative, and shall not be construed as restricting the application of any other law. However, an act or omission punishable in different ways by different provisions of this code shall not be punished under more than one provision.

12802. Prohibited acts

(a) No person may commit an act of collusion as specified in Section 12072.

(b) Any person who alters, counterfeits, or falsifies a handgun safety certificate, or who uses or attempts to use any altered, counterfeited, or falsified handgun safety certificate to purchase a handgun is guilty of a misdemeanor.

(c) The provisions of this section are cumulative and shall not be construed as restricting the application of any other law. However, an act or omission punishable in different ways by this section and different provisions of this code shall not be punished under more than one provision.

12804. Development of instruction manual; Audiovisual materials; written objective test; Certified instructors

(a) The department shall develop an instruction manual in English and in Spanish by October 1, 2002. The department shall make the instructional manual available to firearms dealers licensed pursuant to Section 12071, who shall make it available to the general public. Essential portions of the manual may be included in the pamphlet described in Section 12080.

(b) The department shall develop audiovisual materials in English and in Spanish by March 1, 2003, to be issued to instructors certified by the department.

(c)(1) The department shall develop a written objective test, in English and in Spanish, and prescribe its content, form, and manner, to be administered by an instructor certified by the department. If the person taking the test is unable to read, the examination shall be administered orally. The test shall cover, but not be limited to, all of the following: ...

(e) If a dealer licensed pursuant to Section 12071 or his or her employee, or where the managing officer or partner is certified as an instructor pursuant to this article, he or she shall also designate a separate room or partitioned area for a person to take the objective test, and maintain adequate supervision to assure that no acts of collusion occur while the objective test is being administered.

(f) The department shall solicit input from any reputable association or organization, including any law enforcement association that has as one of its objectives the promotion of firearms safety, in the development of the handgun safety certificate instructional materials.

(g) The department shall develop handgun safety certificates to be issued by instructors certified by the department, to those persons who have complied with this article.

(h) The department shall be immune from any liability arising from implementing this section.

(i) The department shall update test materials related to this article every five years.

(j) Department Certified Instructor applicants shall have a certification to provide training from one of the following organizations as specified, or any entity found by the department to give comparable instruction in firearms safety, or the applicant shall have similar or equivalent training to that provided by the following, as determined by the department: ...

12805. Passing the objective test; Fees; Enforcement activities

(a) An applicant for a handgun safety certificate shall successfully pass the objective test referred to in paragraph (1) of subdivision (c) of Section 12804, with a passing grade of at least 75 percent. Any person receiving a passing grade on the objective test shall immediately be issued a handgun safety certificate by the instructor. ...

12806. Handgun safety certificate

(a) A handgun safety certificate shall include, but not be limited to, the following information:

(1) A unique handgun safety certificate identification number.

(2) The holder's full name.

(3) The holder's date of birth.

(4) The holder's driver's license or identification number.

(5) The holder's signature.

(6) The signature of the issuing instructor.

(7) The date of issuance.

(b) The handgun safety certificate shall expire five years after the date that it was issued by the certified instructor.

12807. Persons exempted from handgun safety certificate requirement

(a) The following persons, properly identified, are exempted from the handgun safety certificate requirement in subdivision (b) of Section 12801:

(1) Any active or honorably retired peace officer, as defined in Chapter 4.5 (commencing with Section 830) of Title 3 of Part 2.

(2) Any active or honorably retired federal officer or law enforcement agent.

(3) Any reserve peace officer, as defined in Section 832.6.

(4) Any person who has successfully completed the course of training specified in Section 832.

(5) A firearms dealer licensed pursuant to Section 12071, who is acting in the course and scope of his or her activities as a person licensed pursuant to Section 12071.

(6) A federally licensed collector who is acquiring or being loaned a handgun that is a curio or relic, as defined in Section 478.11 of Title 27 of the Code of Federal Regulations, who has a current certificate of eligibility issued to him or her by the department pursuant to Section 12071.

(7) A person to whom a handgun is being returned, where the person receiving the firearm is the owner of the firearm.

(8) A family member of a peace officer or deputy sheriff from a local agency who receives a firearm pursuant to Section 50081 of the Government Code.

(9) Any individual who has a valid concealed weapons permit issued pursuant to Section 12050.

(10) An active, or honorably retired member of the United States Armed Forces, the National Guard, the Air National Guard, the active reserve components of the United States, where individuals in those organizations are properly identified. For purposes of this section, proper identification includes the Armed Forces Identification Card, or other written documentation certifying that the individual is an active or honorably retired member.

(11) Any person who is authorized to carry loaded firearms pursuant to subdivision (c) or (d) of Section 12031.

(12) Persons who are the holders of a special weapons permit issued by the department pursuant to Section 12095, 12230, 12250,or 12305.

(b) The following persons who take title or possession of a handgun by operation of law in a representative capacity, until or unless they transfer title ownership of the handgun to themselves in a personal capacity, are exempted from the handgun safety certificate requirement in subdivision (b) of Section 12801:

(1) The executor or administrator of an estate.

(2) A secured creditor or an agent or employee thereof when the firearms are possessed as collateral for, or as a result of, or an agent or employee thereof when the firearms are possessed as collateral for, or as a result of, a default under a security agreement under the Commercial Code.

(3) A levying officer, as defined in Section 481.140, 511.060, or 680.260 of the Code of Civil Procedure.

(4) A receiver performing his or her functions as a receiver.

(5) A trustee in bankruptcy performing his or her duties.

(6) An assignee for the benefit of creditors performing his or her functions as an assignee.

12809. Operative date of article Except for the provisions of Section 12804, this article shall become operative on January 1, 2003.

California Family Code

Division 10. Prevention of Domestic Violence

Part 1. Short Title and Definitions

Part 4. Protective Orders and Other Domestic Violence Prevention Orders

Chapter 3. Registration and Enforcement of Orders

6389. Person subject to protective order prohibited from owning or possessing firearm; Relinquishment of firearm; Employment exemption

(a) A person subject to a protective order, as defined in Section 6218, shall not own, possess, purchase, or receive a firearm while that protective order is in effect. Every person who owns, possesses, purchases or receives, or attempts to purchase or receive a firearm while the protective order is in effect is punishable pursuant to subdivision (g) of Section 12021 of the Penal Code.

(b) On all forms providing notice that a protective order has been requested or granted, the Judicial Council shall include a notice that, upon service of the order, the respondent shall be ordered to relinquish possession or control of any firearms and not to purchase or receive or attempt to purchase or receive any firearms for a period not to exceed the duration of the restraining order.

(c) (1) Upon issuance of a protective order, as defined in Section 6218, the court shall order the respondent to relinquish any firearm in the respondent's immediate possession or control, or subject to the respondent's immediate possession or control.

(2) The relinquishment ordered pursuant to Paragraph (1) shall occur by immediately surrendering the firearm in a safe manner, upon request of any law enforcement officer, to the control of the officer, after being served with the protective order. Alternatively, if no request is made by a law enforcement officer, the relinquishment shall occur within 24 hours of being served with the order, by either surrendering the firearm in a safe manner to the control of local law enforcement officials, or by selling the firearm to a licensed gun dealer, as specified in Section 12071 of the Penal Code. The law enforcement officer or licensed gun dealer taking possession of the firearm pursuant to this Subdivision shall issue a receipt to the person relinquishing the firearm at the time of relinquishment. A person ordered to relinquish any firearm pursuant to this Subdivision shall file with the court that issued the protective order, within 48 hours after being served with the order, the receipt showing the firearm was surrendered to a local law enforcement agency or sold to a licensed gun dealer. Failure to timely file a receipt shall constitute a violation of the protective order.

(3) The application forms for protective orders adopted by the Judicial Council and approved by the Department of Justice shall be amended to require the petitioner to describe the number, types, and locations of any firearms presently known by the petitioner to be possessed or controlled by the respondent.

(4) It is recommended that every law enforcement agency in the state develop, adopt, and implement written policies and standards for law enforcement officers who request immediate relinquishment of firearms.

(d) If the respondent declines to relinquish possession of any firearm based on the assertion of the right against self-incrimination, as provided by the Fifth Amendment to the United States Constitution and Section 15 of Article I of the California Constitution, the court may grant use immunity for the act of relinquishing the firearm required under this section.

(e) A local law enforcement agency may charge the respondent a fee for the storage of any firearm pursuant to this section. This fee shall not exceed the actual cost incurred by the local law enforcement agency for the storage of the firearm. For purposes of this subdivision, "actual cost" means expenses directly related to taking possession of a firearm, storing the firearm, and surrendering possession of the firearm to a licensed dealer as defined in Section 12071 of the Penal Code or to the respondent.

(f) The restraining order requiring a person to relinquish a firearm pursuant to subdivision (c) shall state on its face that the respondent is prohibited from owning, possessing, purchasing, or receiving a firearm while the protective order is in effect and that the firearm shall be relinquished to the local law enforcement agency for that jurisdiction or sold to a licensed gun dealer, and that proof of surrender or sale shall be filed with the court within a specified period of receipt of the order. The order shall also state on its face the expiration date for relinquishment. Nothing in this section shall limit a respondent's right under existing law to petition the court at a later date for modification of the order.

(g) The restraining order requiring a person to relinquish a firearm pursuant to subdivision (c) shall prohibit the person from possessing or controlling any firearm for the duration of the order. At the expiration of the order, the local law enforcement agency shall return possession of any surrendered firearm to the respondent, within five days after the expiration of the relinquishment order, unless the local law enforcement agency determines that (1) the firearm has been stolen, (2) the respondent is prohibited from possessing a firearm because the respondent is in any prohibited class for the possession of firearms, as defined in Sections 12021 and 12021.1 of the Penal Code and Sections 8100 and 8103 of the Welfare and Institutions Code, or (3) another successive restraining order is used against the respondent under this section. If the local law enforcement agency determines that the respondent is the legal owner of any firearm deposited with the local law enforcement agency and is prohibited from possessing any firearm, the respondent shall be entitled to sell or transfer the firearm to a licensed dealer as defined in Section 12071 of the Penal Code. If the firearm has been stolen, the firearm shall be restored to the lawful owner upon his or her identification of the firearm and proof of ownership.

(h) The court may, as part of the relinquishment order, grant an exemption from the relinquishment requirements of this section for a particular firearm if the respondent can show that a particular firearm is necessary as a condition of continued employment and that the current employer is unable to reassign the respondent to another position where a firearm is unnecessary. If an exemption is granted pursuant to this subdivision, the order shall provide that the firearm shall be in the physical possession of the respondent only during scheduled work hours and during travel to and from his or her place of employment. In any case involving a peace officer who as a condition of employment and whose personal safety depends on the ability to carry a firearm, a court may allow the peace officer to continue to carry a firearm, either on duty or off duty, if the court finds by a preponderance of the evidence that the officer does not pose a threat of harm. Prior to making this finding, the court shall require a mandatory psychological evaluation of the peace officer and may require the peace officer to enter into counseling or other remedial treatment program to deal with any propensity for domestic violence.

(i) During the period of the relinquishment order, a respondent is entitled to make one sale of all firearms that are in the possession of a local law enforcement agency pursuant to this section. A licensed gun dealer, who presents a local law enforcement agency with a bill of sale indicating that all firearms owned by the respondent that are in the possession of the local law enforcement agency have been sold by the respondent to the licensed gun dealer, shall be given possession of those firearms, at the location where a respondent's firearms are stored, within five days of presenting the local law enforcement agency with a bill of sale.

(j) The disposition of any unclaimed property under this section shall be made pursuant to Section 1413 of the Penal Code.

(k) The return of a firearm to any person pursuant to subdivision (g) shall not be subject to the requirements of subdivision (d) of Section 12072 of the Penal Code.

(l) If the respondent notifies the court that he or she owns a firearm that is not in his or her immediate possession, the court may limit the order to exclude that firearm if the judge is satisfied the respondent is unable to gain access to that firearm while the protective order is in effect.

(m) Any respondent to a protective order who violates any order issued pursuant to this section shall be punished under the provisions of subdivision (g) of Section 12021 of the Penal Code.

California Welfare and Institutions Code

Division 8. Miscellaneous

Chapter 3. Firearms

8100. Possession of firearm by patient with mental disorder; Waiting period for patients threatening physical violence

(a) A person shall not have in his or her possession or under his or her custody or control, or purchase or receive, or attempt to purchase or receive, any firearms whatsoever or any other deadly weapon, if on or after January 1, 1992, he or she has been admitted to a facility and is receiving inpatient treatment and, in the opinion of the attending health professional who is primarily responsible for the patient's treatment of a mental disorder, is a danger to self or others, as specified by Section 5150, 5250, or 5300, even though the patient has consented to that treatment. A person is not subject to this subdivision once he or she is discharged from the facility.

(b)(1) A person shall not have in his or her possession or under his or her custody or control, or purchase or receive, or attempt to purchase or receive, any firearms whatsoever or any other deadly weapon for a period of six months whenever, on or after January 1, 1992, he or she communicates to a licensed psychotherapist, as defined in subdivisions (a) to (e), inclusive, of Section 1010 of the Evidence Code, a serious threat of physical violence against a reasonably identifiable victim or victims. The six-month period shall commence from the date that the licensed psychotherapist reports to the local law enforcement agency the identity of the person making the communication. The prohibition provided for in this subdivision shall not apply unless the licensed psychotherapist notifies a local law enforcement agency of the threat by that person. The person, however, may own, possess, have custody or control over, or receive or purchase any firearm if a superior court, pursuant to paragraph (3) and upon petition of the person, has found, by a preponderance of the evidence, that the person is likely to use firearms or other deadly weapons in a safe and lawful manner.

(2) Upon receipt of the report from the local law enforcement agency pursuant to subdivision (c) of Section 8105, the Department of Justice shall notify by certified mail, return receipt requested, a person subject to this subdivision of the following:

(A) That he or she is prohibited from possessing, having custody or control over, receiving, or purchasing any firearm or other deadly weapon for a period of six months commencing from the date that the licensed psychotherapist reports to the local law enforcement agency the identity of the person making the communication. The notice shall state the date when the prohibition commences and ends.

(B) That he or she may petition a court, as provided in this subdivision, for an order permitting the person to own, possess, control, receive, or purchase a firearm.

(3) Any person who is subject to paragraph (1) may petition the superior court of his or her county of residence for an order that he or she may own, possess, have custody or control over, receive, or purchase firearms. At the time the petition is filed, the clerk of the court shall set a hearing date and notify the person, the Department of Justice, and the district attorney. The people of the State of California shall be the respondent in the proceeding and shall be represented by the district attorney. Upon motion of the district attorney, or upon its own motion, the superior court may transfer the petition to the county in which the person resided at the time of the statements, or the county in which the person made the statements. Within seven days after receiving notice of the petition, the Department of Justice shall file copies of the reports described in Section 8105 with the superior court. The reports shall be disclosed upon request to the person and to the district attorney. The district attorney shall be entitled to a continuance of the hearing to a date of not less than 14 days after the district attorney is notified of the hearing date by the clerk of the court. The court, upon motion of the petitioner establishing that confidential information is likely to be discussed during the hearing that would cause harm to the person, shall conduct the hearing in camera with only the relevant parties present, unless the court finds that the public interest would be better served by conducting the hearing in public. Notwithstanding any other provision of law, declarations, police reports, including criminal history information, and any other material and relevant evidence that is not excluded under Section 352 of the Evidence Code, shall be admissible at the hearing under this paragraph. If the court finds by a preponderance of the evidence that the person would be likely to use firearms in a safe and lawful manner, the court shall order that the person may have custody or control over, receive, possess, or purchase firearms. A copy of the order shall be submitted to the Department of Justice. Upon receipt of the order, the department shall delete any reference to the prohibition against firearms from the person's state summary criminal history information.

(c) "Discharge," for the purposes of this section, does not include a leave of absence from a facility.

(d) "Attending health care professional," as used in this section, means the licensed health care professional primarily responsible for the person's treatment who is qualified to make the decision that the person has a mental disorder and has probable cause to believe that the person is a danger to self or others.

(e) "Deadly weapon," as used in this section and in Sections 8101, 8102, and 8103, means any weapon, the possession or concealed carrying of which is prohibited by Section 12020 of the Penal Code.

(f) "Danger to self," as used in subdivision (a), means a voluntary person who has made a serious threat of, or attempted, suicide with the use of a firearm or other deadly weapon.

(g) A violation of subdivision (a) of, or paragraph (1) of subdivision (b) of, this section shall be a public offense, punishable by imprisonment in the state prison, or in a county jail for not more than one year, by a fine not exceeding one thousand dollars ($1,000), or by both that imprisonment and fine.

(h) The prohibitions set forth in this section shall be in addition to those set forth in Section 8103.

(i) Any person admitted and receiving treatment prior to January 1, 1992, shall be governed by this section, as amended by Chapter 1090 of the Statutes of 1990, until discharged from the facility.

8101. Giving deadly weapon to mental patient; Punishment

(a) Any person who shall knowingly supply, sell, give, or allow possession or control of a deadly weapon to any person described in Section 8100 or 8103 shall be punishable by imprisonment in the state prison, or in a county jail for a period of not exceeding one year, by a fine of not exceeding one thousand dollars ($1,000), or by both the fine and imprisonment.

(b) Any person who shall knowingly supply, sell, give, or allow possession or control of a firearm to any person described in Section 8100 or

8103 shall be punished by imprisonment in the state prison for two, three, or four years.

(c) "Deadly weapon," as used in this section has the meaning prescribed by Section 8100.

8103. Certificate for possession required for specified classes of people; Prohibition against possession for specified persons; Violation as felony

(a)(1) No person who after October 1, 1955, has been adjudicated by a court of any state to be a danger to others as a result of a mental disorder or mental illness, or who has been adjudicated to be a mentally disordered sex offender, shall purchase or receive, or attempt to purchase or receive, or have in his or her possession, custody, or control any firearm or any other deadly weapon unless there has been issued to the person a certificate by the court of adjudication upon release from treatment or at a later date stating that the person may possess a firearm or any other deadly weapon without endangering others, and the person has not, subsequent to the issuance of the certificate, again been adjudicated by a court to be a danger to others as a result of a mental disorder or mental illness.

(2) The court shall immediately notify the Department of Justice of the court order finding the individual to be a person described in paragraph (1). The court shall also notify the Department of Justice of any certificate issued as described in paragraph (1).

(b)(1) No person who has been found, pursuant to Section 1026 of the Penal Code or the law of any other state or the United States, not guilty by reason of insanity of murder, mayhem, a violation of Section 207, 209, or 209.5 of the Penal Code in which the victim suffers intentionally inflicted great bodily injury, carjacking or robbery in which the victim suffers great bodily injury, a violation of Section 451 or 452 of the Penal Code involving a trailer coach, as defined in Section 635 of the Vehicle Code, or any dwelling house, a violation of paragraph (1) or (2) of subdivision (a) of Section 262 or paragraph (2) or (3) of subdivision (a) of Section 261 of the Penal Code, a violation of Section 459 of the Penal Code in the first degree, assault with intent to commit murder, a violation of Section 220 of the Penal Code in which the victim suffers great bodily injury, a violation of Section 12303.1, 12303.2, 12303.3, 12308, 12309, or 12310 of the Penal Code, or of a felony involving death, great bodily injury, or an act which poses a serious threat of bodily harm to another person, or a violation of the law of any other state or the United States that includes all the elements of any of the above felonies as defined under California law, shall purchase or receive, or attempt to purchase or receive, or have in his or her possession or under his or her custody or control any firearm or any other deadly weapon.

(2) The court shall immediately notify the Department of Justice of the court order finding the person to be a person described in paragraph (1).

(c)(1) No person who has been found, pursuant to Section 1026 of the Penal Code or the law of any other state or the United States, not guilty by reason of insanity of any crime other than those described in subdivision (b) shall purchase or receive, or attempt to purchase or receive, or shall have in his or her possession, custody, or control any firearm or any other deadly weapon unless the court of commitment has found the person to have recovered sanity, pursuant to Section 1026.2 of the Penal Code or the law of any other state or the United States.

(2) The court shall immediately notify the Department of Justice of the court order finding the person to be a person described in paragraph (1). The court shall also notify the Department of Justice when it finds that the person has recovered his or her sanity.

(d)(1) No person found by a court to be mentally incompetent to stand trial, pursuant to Section 1370 or 1370.1 of the Penal Code or the law of any other state or the United States, shall purchase or receive, or attempt to purchase or receive, or shall have in his or her possession, custody, or control any firearm or any other deadly weapon, unless there has been a finding with respect to the person of restoration to competence to stand trial by the committing court, pursuant to Section 1372 of the Penal Code or the law of any other state or the United States.

(2) The court shall immediately notify the Department of Justice of the court order finding the person to be mentally incompetent as described in paragraph (1). The court shall also notify the Department of Justice when it finds that the person has recovered his or her competence.

(e)(1) No person who has been placed under conservatorship by a court, pursuant to Section 5350 or the law of any other state or the United States, because the person is gravely disabled as a result of a mental disorder or impairment by chronic alcoholism shall purchase or receive, or attempt to purchase or receive, or shall have in his or her possession, custody, or control any firearm or any other deadly weapon while under the conservatorship if, at the time the conservatorship was ordered or thereafter, the court which imposed the conservatorship found that possession of a firearm or any other deadly weapon by the person would present a danger to the safety of the person or to others. Upon placing any person under conservatorship, and prohibiting firearm or any other deadly weapon possession by the person, the court shall notify the person of this prohibition.

(2) The court shall immediately notify the Department of Justice of the court order placing the person under conservatorship and prohibiting firearm or any other deadly weapon possession by the person as described in paragraph (1). The notice shall include the date the conservatorship was imposed and the date the conservatorship is to be terminated. If the conservatorship is subsequently terminated before the date listed in the notice to the Department of Justice or the court subsequently finds that possession of a firearm or any other deadly weapon by the person would no longer present a danger to the safety of the person or others, the court shall immediately notify the Department of Justice.

(3) All information provided to the Department of Justice pursuant to paragraph (2) shall be kept confidential, separate, and apart from all other records maintained by the Department of Justice, and shall be used only to determine eligibility to purchase or possess firearms or other deadly weapons. Any person who knowingly furnishes that information for any other purpose is guilty of a misdemeanor. All the information concerning any person shall be destroyed upon receipt by the Department of Justice of notice of the termination of conservatorship as to that person pursuant to paragraph (2).

(f)(1) No person who has been (A) taken into custody as provided in Section 5150 because that person is a danger to himself, herself, or to others, (B) assessed within the meaning of Section 5151, and (C) admitted to a designated facility within the meaning of Sections 5151 and 5152 because that person is a danger to himself, herself, or others, shall own, possess, control, receive or purchase, or attempt to own, possess, control, receive, or purchase any firearm for a period of five years after the person is released from the facility. A person described in the preceding sentence, however, may own, possess, control, receive, or purchase, or attempt to own, possess, control, receive, or purchase any firearm if the superior court has pursuant to paragraph (5), found that the People of the State of California have not met their burden pursuant to paragraph (6).

(2) For each person subject to this subdivision, the facility shall immediately, on the date of admission, submit a report to the Department of Justice, on a form prescribed by the Department of Justice, containing information that includes, but is not limited to, the identity of the person and the legal grounds upon which the person was admitted to the facility.

Any report prescribed by this subdivision shall be confidential, except for purposes of the court proceedings described in this subdivision and for determining the eligibility of the person to own, possess, control, receive, or purchase a firearm.

(3) Prior to, or concurrent with, the discharge, the facility shall inform a person subject to this subdivision that he or she is prohibited from owning, possessing, controlling, receiving, or purchasing any firearm for a period of five years. Simultaneously, the facility shall inform the person that he or she may request a hearing from a court, as provided in this subdivision, for an order permitting the person to own, possess, control, receive, or purchase a firearm. The facility shall provide the person with a form for a request for a hearing. The Department of Justice shall prescribe the form. Where the person requests a hearing at the time of discharge, the facility shall forward the form to the superior court unless the person states that he or she wil submit the form to the superior court.

(4) The Department of Justice shall provide the form upon request to any person described in paragraph (1). The Department of Justice shall also provide the form to the superior court in each county. A person described in paragraph (1) may make a single request for a hearing at any time during the five-year period. The request for hearing shall be made on the form prescribed by the department or in a document that includes equivalent language.

(5) Any person who is subject to paragraph (1) who has requested a hearing from the superior court of his or her county of residence for an order that he or she may own, possess, control, receive, or purchase firearms shall be given a hearing. The clerk of the court shall set a hearing date and notify the person, the Department of Justice, and the district attorney. The People of the State of California shall be the plaintiff in the proceeding and shall be represented by the district attorney. Upon motion of the district attorney, or on its own motion, the superior court may transfer the hearing to the county in which the person resided at the time of his or her detention, the county in which the person was detained, or the county in which the person was evaluated or treated. Within seven days after the request for a hearing, the Department of Justice shall file copies of the reports described in this section with the superior court. The reports shall be disclosed upon request to the person and to the district attorney. The court shall set the hearing within 30 days of receipt of the request for a hearing. Upon showing good cause, the district attorney shall be entitled to a continuance not to exceed 14 days after the district attorney was notified of the hearing date by the clerk of the court. If additional continuances are granted, the total length of time for continuances shall not exceed 60 days. The district attorney may notify the county mental health director of the hearing who shall provide information about the detention of the person that may be relevant to the court and shall file that information with the superior court. That information shall be disclosed to the person and to the district attorney. The court, upon motion of the person subject to paragraph (1) establishing that confidential information is likely to be discussed during the hearing that would cause harm to the person, shall conduct the hearing in camera with only the relevant parties present, unless the court finds that the public interest would be better served by conducting the hearing in public. Notwithstanding any other law, declarations, police reports, including criminal history information, and any other material and relevant evidence that is not excluded under Section 352 of the Evidence Code, shall be admissible at the hearing under this section.

(6) The people shall bear the burden of showing by a preponderance of the evidence that the person would not be likely to use firearms in a safe and lawful manner.

(7) If the court finds at the hearing set forth in paragraph (5) that the people have not met their burden as set forth in paragraph (6), the court shall order that the person shall not be subject to the five-year prohibition in this section on the ownership, control, receipt, possession or purchase of firearms. A copy of the order shall be submitted to the Department of Justice. Upon receipt of the order, the Department of Justice shall delete any reference to the prohibition against firearms from the person's state mental health firearms prohibition system information.

(8) Where the district attorney declines or fails to go forward in the hearing, the court shall order that the person shall not be subject to the five-year prohibition required by this subdivision on the ownership, control, receipt, possession, or purchase of firearms. A copy of the order shall be submitted to the Department of Justice. Upon receipt of the order, the Department of Justice shall, within 15 days, delete any reference to the prohibition against firearms from the person's state mental health firearms prohibition system information.

(9) Nothing in this subdivision shall prohibit the use of reports filed pursuant to this section to determine the eligibility of persons to own, possess, control, receive, or purchase a firearm if the person is the subject of a criminal investigation, a part of which involves the ownership, possession, control, receipt, or purchase of a firearm.

(g)(1) No person who has been certified for intensive treatment under Section 5250, 5260, or 5270.15 shall own, possess, control, receive, or purchase, or attempt to own, possess, control, receive, or purchase any firearm for a period of five years.

Any person who meets the criteria contained in subdivision (e) or (f) who is released from intensive treatment shall nevertheless, if applicable, remain subject to the prohibition contained in subdivision (e) or (f).

(2) For each person certified for intensive treatment under paragraph (1), the facility shall immediately submit a report to the Department of Justice, on a form prescribed by the department, containing information regarding the person, including, but not limited to, the legal identity of the person and the legal grounds upon which the person was certified. Any report submitted pursuant to this paragraph shall only be used for the purposes specified in paragraph (2) of subdivision (f).

(3) Prior to, or concurrent with, the discharge of each person certified for intensive treatment under paragraph (1), the facility shall inform the person of that information specified in paragraph (3) of subdivision (f).

(4) Any person who is subject to paragraph (1) may petition the superior court of his or her county of residence for an order that he or she may own, possess, control, receive, or purchase firearms. At the time the petition is filed, the clerk of the court shall set a hearing date and notify the person, the Department of Justice, and the district attorney. The People of the State of California shall be the respondent in the proceeding and shall be represented by the district attorney. Upon motion of the district attorney, or on its own motion, the superior court may transfer the petition to the county in which the person resided at the time of his or her detention, the county in which the person was detained, or the county in which the person was evaluated or treated. Within seven days after receiving notice of the petition, the Department of Justice shall file copies of the reports described in this section with the superior court. The reports shall be disclosed upon request to the person and to the district attorney. The district attorney shall be entitled to a continuance of the hearing to a date of not less than 14 days after the district attorney was notified of the hearing date by the clerk of the court. The district attorney may notify the county mental health director of the petition, and the county mental health director shall provide information about the detention of the person that may be relevant to the court and shall file that information with the superior court. That information shall be disclosed to the person and to the district attorney. The court, upon motion of the person subject to paragraph (1) establishing that confidential information is likely to be discussed during the hearing that would cause harm to the person, shall conduct the hearing in camera with only the relevant parties present, unless the court finds that the public interest would be better served by conducting the hearing in public. Notwithstanding any other provision of law, any declaration, police reports, including criminal history information, and any other material and relevant evidence that is not excluded under Section 352 of the Evidence Code, shall be admissible at the hearing under this section. If the court finds by a preponderance of the evidence that the person would be likely to use firearms in a safe and lawful manner, the court may order that the person may own, control, receive, possess, or purchase firearms. A copy of the order shall be submitted to the Department of Justice. Upon receipt of the order, the Department of Justice shall delete any reference to the prohibition against firearms from the person's state mental health firearms prohibition system information.

(h) For all persons identified in subdivisions (f) and (g), facilities shall report to the Department of Justice as specified in those subdivisions, except facilities shall not report persons under subdivision (g) if the same persons previously have been reported under subdivision (f).

Additionally, all facilities shall report to the Department of Justice upon the discharge of persons from whom reports have been submitted pursuant to subdivision (f) or (g). However, a report shall not be filed for persons who are discharged within 31 days after the date of admission.

(i) Every person who owns or possesses or has under his or her custody or control, or purchases or receives, or attempts to purchase or receive, any firearm or any other deadly weapon in violation of this section shall be punished by imprisonment in the state prison or in a county jail for not more than one year.

(j) "Deadly weapon," as used in this section, has the meaning prescribed by Section 8100.

California Government Code

Title 5. Local Agencies

Division 2. Cities, Counties, and Other Agencies

Part 1. Powers and Duties Common to Cities, Counties, and Other Agencies

Chapter 1. General

Article 4. Miscellaneous

53071. Firearms

It is the intention of the Legislature to occupy the whole field of regulation of the registration or licensing of commercially manufactured firearms as encompassed by the provisions of the Penal Code, and such provisions shall be exclusive of all local regulations, relating to registration or licensing of commercially manufactured firearms, by any political subdivision as defined in Section 1721 of the Labor Code.

53071.5. Preemption of legislative regulation of imitation firearms

By the enforcement of this section, the Legislature occupies the whole field of regulation of the manufacture, sale, or possession of imitation firearms, as defined in Section 12550 of the Penal Code, and that section shall preempt and be exclusive of all regulations relating to the manufacture, sale, or possession of imitation firearms, including regulations governing the manufacture, sale, or possession of BB devices and air rifles described in subdivision (g) of Section 12001 of the Penal Code.

Publisher's Notes

Attorney General Opinion 94-212 (online at http://caag.state.ca.us/opinions/index.htm) provides:

- With regard to firearms registration and licensing, the Legislature has explicitly preempted the adoption of local ordinances, citing Government Code section 53071.
- As to firearms possession at one's residence, business, or other property, state law has preempted the field.
- The state has so thoroughly occupied the field that the Attorney General has no doubt that regulating firearms sales is beyond the reach of local governments; cities and counties have been charged with the execution of the state's program for the licensing of firearms dealers, but their role is ministerial in nature.
- The Attorney General concluded that the language of Penal Code sections 12026 and 12304, construed together, precludes a local entity from prohibiting the sale of handgun ammunition.
- A city ordinance which requires ammunition vendors within the city to record and maintain identification information with respect to each purchaser is not preempted.
- As to firearms use, the state has not preempted the field.

The Department of Justice, Firearms Division prepares a pamphlet that summarizes California firearms laws, and it is available online at http://caag.state.ca.us/firearms/index.html.
[Current through 2007-2008 Third Extra. Session ch. 7 and ch. 267 of the 2008 Regular Session approved 8/26/08, and Proposition 99 approved by voters 6/3/08]

Publisher's Note

California Municipal and County Codes and Ordinances may be accessed online at http://www.igs.berkeley.edu/library/calcodes and http://california.lp.findlaw.com/ca01_codes/municode.html.

Anaheim Municipal Code

Title 4. Business Regulation

Chapter 4.99 Retail Sale of Concealable Firearms

4.99.010. Duly Constituted Licensing Authority. The Finance Director of the City of Anaheim, or his or her designated representative, is designated the City's licensing authority, and is authorized to issue licenses for the retail sale of concealable firearms pursuant to Article 4 of the California Penal Code, commencing with Section 12070. The term "Finance Director" as used herein shall include any representative designated by the Finance Director pursuant to this section.

4.99.030. License Required. No person shall conduct, engage in or carry on the business of selling, leasing, transferring, advertising, offering or exposing for sale, lease or transfer concealable firearms without having first obtained a license from the Finance Director. The license shall not be transferable. The initial license or any renewal thereof shall be valid from the date of issuance through December 31st of the same calendar year. Notwithstanding any other provision of this section, no license shall be required by the City pursuant to this chapter for any activity which is exempt from such licensing by any provision of Section 12070 of the Penal Code or any other provision of state law.

4.99.050. Issuance, Denial and Revocation of License.

.010. The Finance Director shall issue a license to an applicant who satisfies all the requirements contained in Penal Code Sections 12070, et seq., and shall deny a license to an applicant who fails to satisfy any one or more of those requirements.

.020. Any license issued pursuant to this chapter may be revoked by the Finance Director for breach of any of the conditions set forth in Penal Code Section 12071(a).

.030. If an application for a license is denied, the Finance Director shall notify the applicant by certified mail of such denial within thirty days of the date of receipt of the application.

.040. If a license issued pursuant to this chapter is revoked, the Finance Director shall notify the licensee by certified mail of such revocation within ten days of the date of revocation.

.050. Any applicant who is denied a license or any licensee whose license is revoked may appeal such action to the City Manager or his designee within fifteen days after receipt of notice of denial or revocation by filing a written notice of appeal and paying any appeal fee as prescribed by resolution of the City Council within such fifteen day period. The City Manager or his designee shall review the circumstances of such denial or revocation and shall render a written decision. The action of the City Manager or his designee shall be final and conclusive.

Title 6. Public Health and Safety

Chapter 6.32. Dangerous Weapons

6.32.010. "Dangerous or Deadly Weapons" Defined. The term DANGEROUS OR DEADLY WEAPONS as used in Sections 6.32.010 through 6.32.050 of this chapter includes, but is not limited to, any dangerous or deadly weapon within the meaning of any law of this State restricting the use thereof; **...** ; and any firearm other than:

.010. One carried pursuant to valid permit issued by duly authorized governmental authority; or

.020. Any ordinary rifle or shotgun lawfully carried for purposes of hunting or other lawful purpose.

6.32.020. Loitering or Hiding With Concealed Dangerous or Deadly Weapons - Prohibited. It shall be unlawful for any person, while carrying concealed upon his person any dangerous or deadly weapon, to loaf or loiter upon any public street, sidewalk or alley or to wander about from place to place with no lawful business thereby to perform, or to hide, lurk or loiter upon or about the premises of another. . .

6.32.050. Dangerous or Deadly Weapons in Automobiles For Improper Purposes - Prohibited. It shall be unlawful for any person to have in his possession, in any automobile, any dangerous or deadly weapon, but this restriction shall not be deemed to prohibit the carrying of ordinary tools or equipment carried in good faith for uses of honest work, trade or business, or for the purpose of legitimate sport or recreation.

6.32.110. Exception for Bona Fide Educational and Recreational Programs. Nothing contained in Chapter 6.32 shall prohibit the use of the weapons specified therein in connection with a bona fide educational or recreational program under the supervision of a qualified instructor; provided that such program shall have been approved either by the Board of Trustees of any School District or the City Council of the City of Anaheim.

[Current through Anaheim Ord. 6130 passed April 22, 2008]

Bakersfield Municipal Code

Title 9 Public Peace, Morals and Welfare

VIII. Weapons Offenses

Chapter 9.52. Weapons in Public Buildings

Weapons in owner's place of business or residence allowed.

[Current through Ordinance 4518 and the August 2008 code supplement]

Fremont Municipal Code

Title III. Public Safety, Welfare and Morals

Chapter 3. Firearms

3-3100. Sale of weapons to minors. Except as otherwise provided in section 3-3103 of this Code, no persons shall sell, give, loan, or in any way furnish to any person under the age of sixteen years, any gun, revolver, pistol, firearm, spring gun, air gun, BB gun, sling, slingshot, bow and arrow, or other device designed or intended to discharge, or capable of discharging, any dangerous missile.

3-3101. Sale of ammunition to minors. Except as otherwise provided in section 3-3103 of this Code, no person shall sell, give, loan, or in any way furnish, to any person under the age of sixteen years, any cartridge, shell, ammunition, or device containing any explosive substance, designed or intended to be used in, or fired from any gun, revolver, pistol, or other firearm.

3-3102. Possession and use of weapons by minors. Except as otherwise provided in section 3-3103 of this Code, no person under the age of sixteen years shall fire, discharge, shoot, or operate, or participate in the firing, discharging, shooting, or operating, or have in his possession, care, custody, or control, any gun, revolver, pistol, firearm, spring gun, air gun, BB gun, sling, slingshot, bow and arrow, or device designed or intended to discharge, or capable of discharging any dangerous missile, or any cartridge, shell, ammunition, or device containing any explosive substance, designed or intended to be used in or fired from any gun, revolver, pistol, or other firearm.

3-3103. Prohibition does not apply to licensed hunters. Nothing in section 3-3100 3-3101 or 3-3102 of this Code shall be deemed or construed to prohibit the selling, giving, loaning, or furnishing of any article mentioned in sections 3-3100 and 3-3102 of this Code, to any person under the age of sixteen years who is the holder of a valid hunting license issued to such person in accordance with the provisions of the Fish and Game Code of the state; nor prohibit any such licensee under the age of sixteen years from using or possessing any article mentioned in section 3-3102 of this Code. Nothing in this chapter shall be deemed or construed to prohibit any person under the age of sixteen years from firing, discharging, shooting, or operating any article mentioned in section 3-3102 of this Code, when such person is the holder of a valid state hunting license and is engaged in hunting any wild game or predatory bird or animal which may be lawfully hunted and killed, or is lawfully engaged in shooting at any inanimate target or trapshooting device while accompanied by, and under the direct care and control of some responsible adult person, at an established shooting range with a safe background.

Title V. Businesses, Professions and Trades
Chapter 8. Permit Requirements for Firearms Dealers

5-8100. Citation of chapter. This chapter may be referred to as the "Firearms Dealer Permit Ordinance."

5-8101. Purpose and findings. The declared purpose of this chapter is to provide uniform regulations applicable to all firearm dealers in the city. ...

5-8102. Definitions. The following words and phrases, whenever used in this chapter, are defined as follows:

(a) Firearms shall mean any device, designed to be used as a weapon or modified to be used as a weapon, from which a projectile is expelled through a barrel by the force of an explosion or other form of combustion.

(b) Firearms dealer shall mean a person engaged in the business of selling, transferring, leasing, or advertising for sale, transfer, or lease any firearm with the exception of a Bureau of Alcohol, Tobacco and Firearms Federal Firearms License of Type 03 (collectors of curios and relics) and Type 01 (limited to those who engage only in gunsmith activities and does not conduct any sales transactions). Included in this definition are persons involved solely in the warehousing and storage of firearms.

(c) Gunsmith shall mean a person whose occupation is to design, modify, make or repair firearms. For the sales of firearms, refer to "firearms dealer" under section 5-8102(b) of this chapter.

(d) Engaged in business shall mean the conduct of a business by the selling, transferring, or leasing of any firearm; or the preparation for such conduct of business as evidenced by the securing of applicable federal and state licenses; or the holding of one's self out as engaged in the business of selling, transferring, or leasing of firearms in quantity, in series, or in individual transactions, or in any other manner indicative of trade.

(e) Person shall mean any natural person, firm, corporation, organization, company association, business trust, joint-stock company, partnership, joint venture, club, or the agent, servant, manager, officer, employee, or lessee of any of them, and any municipal, political, or governmental corporation, district, body, or agency, other than the city.

(f) Chief of police shall mean the chief of police of the city or his/her designee charged with the administration of this chapter, subject to the administrative direction of the city manager.

(g) Collector of curios and relics shall mean a person who is a collector of any antique firearm or any firearm which is a curio or relic as defined in Section 178.11 et seq. of Title 27 of the Code of Federal Regulations and who only possesses a valid Bureau of Alcohol, Tobacco and Firearms Federal Firearms License of Type 03 (collectors of curios and relics).

(h) Applicant shall mean the firearms dealer as defined in this chapter who is making application to the city for a firearms dealers permit or gunsmith permit as required by section 12071 (a)(1) of the California Penal Code.

5-8103. Permit required.

(a) Firearms dealers: No person, partnership, cooperative, corporation, firm, or association will engage in the business of operating or managing any business in the city, which sells, transfers, leases, or offers or advertises for sale, transfer, or lease, any firearm without first obtaining a firearms dealers permit from the chief of police. Application for any such permit shall be made in writing to the chief of police in such form as shall be prescribed by him/her. The applicant shall provide all information requested, including proof of compliance with all applicable federal, state, and local laws, or the application will not be deemed complete. Persons who are collectors of antique firearms as described in section 5-8102(g) of this chapter shall not be required to obtain and maintain a firearms dealers permit.

(b) Gunsmiths: No person, partnership, cooperative, corporation, firm, or association will engage in the business of operating or managing a business which designs, modifies, makes or repairs firearms in the city without first obtaining a gunsmith permit from the chief of police. Application for any such permit shall be made in writing to the chief of police in such form as shall be prescribed by him/her. The applicant shall provide all information requested, including proof of compliance with all applicable federal, state, and local laws, or the application will not be deemed complete.

5-8104. Application fee.

The application shall be accompanied by a nonrefundable fee as set forth in the City of Fremont Master Fee Resolution. The fee shall include an amount necessary to cover the full cost of processing, background, issuance, inspection, enforcement, and hearing procedures.

5-8105. Administration of chapter; application and investigation for permits.

(a) The chief of police shall administer this chapter.

(b) The chief of police shall conduct an appropriate investigation to determine the eligibility of each owner, partner, employee, manager, officer, agent, lessee, or other responsible person engaged in the business for which the applicant is seeking a firearms dealer permit or gunsmith permit. This investigation shall be conducted for the protection of the public safety and to determine if a permit may be issued. The applicant, owner, partner, employee, manager, officer, agent, lessee, or other responsible person engaged in the business shall fully cooperate by providing necessary authorization for access to private records and information.

5-8106. Permit approval, renewal, suspension, conditions and revocation criteria.

The chief of police may issue or renew a permit only if the following criteria are met. The chief of police is hereby authorized to adopt all necessary administrative regulations to carry out the intent of this section including, but not limited to, the imposition of additional criteria. The chief of police may deny a new permit or suspend, revoke or impose reasonable conditions for issuance or renewal of an existing permit for failure to meet the following criteria:

(a) Applicants must be at least twenty-one years old.

(b) The applicant must hold all required federal, state and local licenses and authorizations.

(c) The applicant has no prior revocation, or denial of similar permit, for good cause within the last two years.

(d) The applicant has provided accurate statements of fact on his/her application.

(e) The applicant must have paid all applicable fees as provided in the City of Fremont Master Fee Resolution.

(f) The applicant is free of a conviction for any of the following offenses:

(1) Any offense that would disqualify the applicant from owning or possessing a firearm under federal, state and local laws.

(2) Any offense related to the manufacturing, sale, possession, use, or registration of any firearm or dangerous or deadly weapon.

(3) Any offense involving the use of force or violence against the person of another.

(4) Any offense involving the manufacturing, sale, possession, or use of any controlled substance as defined in the California State Health and Safety Code.

(g) The applicant is not an unlawful user of any controlled substance as defined by the California State Health and Safety Code, or an excessive user of alcohol to the extent that it would impair his/her ability to be a dealer in firearms.

(h) The applicant is not a mental patient in a hospital or institution, or on leave of absence from any hospital or institution, or adjudicated by a court to be a danger to others as a result of a mental disorder or illness, or a mentally disordered sex offender, or suffers from any psychological disturbance which would impair his/her ability to be a firearms dealer.

(i) The applicant proposes to operate the business in a manner which complies with all federal, state, and local laws.

(j) The applicant operates his or her business:

(1) Within a zoning district where the proposed use is permitted by zoning regulations; or

(2) Within a zoning district where applicant has obtained required permits or approvals as prescribed by this Code and general law.

(k) The applicant can provide evidence of possessory interest in the property at which he/she proposes to conduct business.

5-8107. Security requirements; chief of police authority to adopt administrative regulations. Firearms security regulations shall be adopted as provided in this section.

(a) A primary objective of this chapter is to discourage theft from the premises of firearm dealers, and to provide a safe and secure facility for storage and sale of firearms. To this end the permittee shall comply with all firearms dealer regulations prescribed by the chief of police. The chief of police is hereby authorized to adopt all necessary administrative regulations to carry out the intent of this chapter.

(b) The chief of police shall adopt firearms security regulations requiring, but not limited to, the following:

(1) The provision of secure locks, windows and doors, adequate lighting and an alarm security system in accordance with permitted city security regulations and/or specified by the chief of police.

(2) The storage of all firearms on the premises out of reach of customers in secure, locked locations, so that the access to the firearms is controlled by the dealer and/or owner, partner, employee, manager, officer, agent, lessee, or other responsible person engaged in the business and to the exclusion of others. Storage of ammunition, gunpowder or other flammable or explosive material shall be in accordance with the Uniform Fire Code and the city's hazardous material storage [management] ordinance (Title III, Chapter 12).

Sec. 5-8108. Permit forms.

All permits issued pursuant to this chapter shall be in a form prescribed by the attorney general of the state..

5-8109. Permit duration.

(a) All permits issued pursuant to this chapter shall expire one year after the date of issuance, unless suspended or revoked prior to the expiration date.

(b) All permits issued pursuant to this chapter shall be subject to renewal after one year. The permittee shall file a renewal application and pay all applicable fees as specified in the City of Fremont Master Fee Resolution. Such application will then be reviewed by the chief of police for compliance with the provisions of this chapter. A permit will be issued by the chief of police for a renewal application which meets the provisions specified in this chapter. The permit will be valid for one year.

(c) Applications for renewal of an existing valid permit must be completed and received by the chief of police, in completed form, no later than thirty days prior to the expiration of the current permit.

5-8111. Permit conditions.

(a) All permits issued pursuant to this chapter shall be subject to the following conditions. The violation of any of the listed conditions will be grounds for suspension or revocation of the permit by the chief of police.

(1) No firearms dealer permit will be issued to a person engaged in the business of selling, transferring, leasing, or advertising for sale, transfer, or lease any firearm from a residence.

(2) The business shall be conducted only in the building located at the street address indicated on the permit.

(3) The sales of firearms shall be conducted only by the person(s) listed on the firearms dealer permit issued by the city, known as the "permittee." If the permittee is to be assisted by another person such as an owner, partner, employee, manager, officer, agent, lessee, or other responsible person engaged in the business of the sale of firearms, these individuals must be reviewed under the requirements of subsections 5-8106(a), (e), (f), (g), and (h) of this chapter.

(4) The permittee shall comply with sections 12073, 12074, 12076, 12077, and 12082 and subdivision (b) of section 12072 of the California State Penal Code, to the extent that the provisions remain in effect.

(5) The permittee shall comply with all sections of this chapter.

(b) Any permit pursuant to this chapter shall be subject to additional conditions as the chief of police deems reasonably related to the purpose of this chapter.

5-8112. Permit denial, suspension, conditioning, revocation, denial of renewal.

(a) The chief of police may deny a new permit application or suspend, modify, revoke or deny renewal of an existing permit for failure of the applicant or permittee to satisfy any of the criteria provided in section 5-8106 (permit approval, renewal, suspension, conditions and revocation criteria), section 5-8107 (security requirements; chief of police authority to adopt administrative regulations), section 5-8110 (permit transition period), section 5-8111 (permit conditions), section 5-8115 (permit liability insurance), and section 5-8116 (inspection authority) of this chapter.

(b) Any decision to deny a new permit application, suspend, revoke, modify or deny renewal of an existing permit shall be given to the applicant or permittee in writing, addressed to the permittee or applicant at the address on the application, and shall set forth the factual and legal grounds for the decision.

5-8115. Permit liability insurance.

(a) No firearms dealer permit shall be issued or renewed pursuant to this chapter unless there is in full force and effect prior to issuance of a permit or renewal a policy of insurance in such form that the city attorney and city risk manager deems proper, executed by an insurance company approved by the city attorney whereby the applicant or permittee is insured against liability for damage to property and for injury to or death of any person as a result of the sale, transfer or lease, or advertising for sale, transfer or lease, or advertising for lease, any firearm. The minimum liability limits shall not be less than one million dollars for damages to or destruction of property in any one incident, and one million dollars for the death or injury of any one person provided, however, that additional amounts may be required by the city attorney if deemed necessary. Person(s) engaged in the business of operating or managing a gunsmith business which designs, modifies, makes or repairs firearms will not be required to obtain and maintain liability insurance as discussed in this section.

(b) The policy of insurance shall name the city, its officers, agents and employees as additional insureds. Applicants and permittees shall indemnify, defend and hold harmless the city, its officers, agents, and employees, from claims arising from the negligence of the applicant or permittee. Such policy of insurance shall contain an endorsement providing that the policy will not be canceled until notice in writing has been given to the chief of police sixty days prior to cancellation.

(c) A federally licensed firearms dealer possessing a home occupation permit for the sale of firearms from his/her residence shall be exempted from this requirement.

5-8116. Inspection authority.

The chief of police or his/her designee shall have the right to enter any firearms dealer establishment from time to time during regular business hours to make reasonable inspections to ascertain whether the provisions of this chapter are being complied with. A warrant shall be obtained whenever required by law.

5-8117. Compliance.

Any person engaging in the business of selling, transferring, or leasing, or advertising for sale, transfer, or lease, or offering for sale, transfer, or lease, any firearm on the effective date of this chapter shall have sixty days from the effective date to comply with the provisions of this chapter.

5-8118. Temporary suspension.

(a) If a firearm dealer violates any federal, state or local county or city law, the chief of police may immediately suspend the right of the dealer to sell firearms.

(b) The temporary suspension will not exceed three days. If the violation results in a criminal charge filed by a federal, state, or county district attorney such permit to sell firearms may be suspended until the case is adjudicated in a court of law.

(c) Notice of suspension and rights to appeal and costs shall be the same as provided in section 5-8112 (permit denial, suspension, conditioning, revocation, denial of renewal), section 5-8113 (appeal of adverse action) and section 5-8114 (costs of appeal and attorney fees) of this chapter.

[Codified through Ordinance No. 10-2008, May 27, 2008]

Municipal Code and Charter of Fresno

Article 26 Crimes Against Public Health and Safety

9-2601. Use and Possession of Weapons.

(a) No person shall discharge a firearm into or within the city.

(b) No person shall propel a missile by, or possess a sling shot, slung shot, bow or other instrument commonly used or which is designed for the throwing of any shot, bullet, rock, stone, arrow, or other missile.

(c) No person shall hunt wildlife with any weapon described in this section.

(d) No person under the age of eighteen years shall have in his or her possession any firearm within the city limits unless he or she, at all times during the possession, (1) has on his or her person the written consent of his or her parent or legal guardian, or (2) is accompanied by his or her parent or legal guardian, or (3) is participating in and going to and from an organized, lawful recreational or competitive shooting activity or lawful hunting activity. As used in this subdivision (d), "firearm" means any firearm except (1) any pistol, revolver or firearm capable of being concealed upon the person as described in California Penal Code Section 12001, or (2) a B-B gun, pellet gun or spot-marking gun as described in California Penal Code Section 12001.1, or (3) any other firearm the possession of which by minors is regulated by State law.

(e) This section shall not be deemed to make punishable the act of carrying or discharging a weapon in the lawful discharge of his or her duties by a public officer, or private person duly licensed therefor.

(f) This section shall not apply to the keeping of weapons at a place of business or residence by a person eighteen years of age or older lawfully in possession of the property.

(g) This section shall not apply to the keeping or use of weapons on the premises of any shooting gallery, practice range, skeet field, archery range, or similar place conducted at a fixed location and with regard to which adequate safeguards have been provided to protect persons and property from injury.

(h) This section shall not apply to an organized hunt for the purpose of ridding premises larger than one acre of vermin, when the setting of traps or poison is impractical and the Chief of Police has approved the time, place, and method of the hunt.

(i) No parent, guardian, or any adult person having the control, custody, or charge of any person under the age of eighteen years, shall knowingly permit, allow, or let said person to use or possess a weapon in violation of this section.

(j) The Chief of Police is authorized to seize and impound any weapon which is kept, possessed or used in violation of this section, or the missile thereof, and to hold the same for a period of thirty days and until thereafter claimed by its owner. Such articles may not be returned to any person under the age of eighteen years but may be returned to his or her guardian or parent.

(k) This section shall not be deemed to make punishable an act or acts which are allowed or prohibited by any law of the State.

[Fresno Municipal Code Codified through Ordinance No. 2007-105, enacted Jan. 28, 2008]

Fresno County Ordinance Code

Title 10 Public Peace, Morals and Welfare

Chapter 10.44 Possession and Discharge

10.44.010 By minors.

It is unlawful for any person under the age of fourteen years, unless accompanied by an adult person, to carry on his person or have in his possession any pistol, rifle, gun or other firearm, unless such minor is regularly enlisted in some military organization of this state or the United States or is taking a course in military instruction. (Ord. 371-N § 1 (part), 1974.)

[Fresno County Ordinance Code current through Ordinance 08-019, April 8, 2008]

Glendale Municipal Code

Title 9. Public Peace and Welfare

Chapter 9.24. Weapons

9.24.055. Firearms and weapons generally - Persons under age eighteen not to possess firearms, ammunition, etc. No person under the age of eighteen years shall have in his or her possession, care, custody or control in the city, any gun, revolver, pistol, spring or air gun, or firearm of any description, or any cartridge, shell or other device containing any explosive, or any cartridge, shell or other device designed and intended for use in connection with any gun, revolver, pistol or firearm of any description, or any ammunition of any description containing any explosive.

[Glendale Municipal Code current through 2008]

Huntington Beach Municipal Code

Title 9. Public Peace, Morals and Welfare

Chapter 9.80. Concealed Weapons

9.80.010. Prohibited. 9.80.040. Nuisance. [Huntington Beach Municipal Code current through 2008]

Long Beach Municipal Code

Title 5. Regulation of Businesses, Trades and Professions

Chapter 5.36. Firearms Dealer

5.36.010. Permit-Required. No person shall engage in, carry on or conduct the business of a firearms dealer without first having obtained a permit in accordance with this Chapter and complying with any and all requirements and regulations set forth in this Chapter.

5.36.020. Permit application. Each person, before obtaining a permit to carry on the business of selling firearms, shall make a written application on forms furnished by the Chief of Police, which shall be signed by the applicant. Every applicant must provide all documents required pursuant to Section 12071 of the California Penal Code.

Title 9. Public Peace, Morals and Welfare

Chapter 9.52. Firearms

9.52.010. Sale of ammunition prohibited. No person as principal, agent or otherwise, shall sell, exchange, give or loan any BB shot, pellet, or other missile designed or intended for use in any spring gun, air gun, air pistol or other pistol or gun, or any cartridge, shell or other device containing any explosive and designed or intended for use in any gun, revolver, pistol, or firearm of any description to any person under the age of eighteen years.

9.52.020. Possession prohibited. No person, as principal, agent or otherwise, under the age of eighteen years, shall have in his or her possession, care, custody or control, any gun, revolver, pistol, spring gun, air gun or firearm of any description, or any cartridge, shell or other device containing any explosive, and designed or intended for use in any gun, revolver, pistol or firearm of any description, or any explosive ammunition of any description whatsoever.

[Current through Long Beach Ordinance ORD-08-0019, passed July 15, 2008]

Los Angeles County Code

Title 7. Business Licenses

Division 2. Specific Businesses

Chapter 7.46. Gun Dealers

7.46.010. Gun dealer defined. "Gun dealer" means any person, firm or corporation who sells or otherwise transfers to the public any pistol, revolver, rifle, shotgun, or other firearm, including those persons required by Section 12070(a) of the California Penal Code to obtain a license under California Penal Code Section 12071. "Gun dealer" also includes any person, firm or corporation who purchases, takes in trade or accepts on consignment from the public any firearm of any type, including, but not limited to pistols, revolvers, rifles and shotguns.

7.46.020. License - Required. Every gun dealer shall first procure a license and pay an annual license fee in the amount set forth in Section 7.14.010 of this title, under the appropriate heading.

7.46.030. License - Granted subject to specific conditions. If a license is granted under this chapter, it shall be subject to the conditions set forth in Penal Code Section 12071, for breach of any of which the license shall be subject to forfeiture.

7.46.040. Records and reporting requirement. In addition to properly maintaining all records and documents required by state and federal laws, a gun dealer, upon purchasing, taking in trade or accepting on consignment from the public any firearm of any type, shall comply with buy-form requirements as enumerated in Part 7 of Chapter 7.76 of this title, beginning with Section 7.76.310. A gun dealer shall also be required to report every sale, lease, or other transfer of a firearm to the sheriff on a form prescribed by the sheriff.

7.46.050. Inspection. Upon the request of any designated sheriff representative or any peace officer, a gun dealer shall (1) furnish all records pertaining to the gun dealer's transactions, including, but not limited to, all records required to be maintained by law, and (2) shall permit an inspection of those portions of the licensed premises where the firearms and firearm accessories are located.

7.46.060. Conditions for granting license. No license or renewal license shall be issued under this chapter unless, in addition to the requirements set forth in Section 7.46.030 and Section 7.46.040 above, the gun dealer satisfies each of the following conditions:

A. The gun dealer, and all officers, employees, and agents of said gun dealer, are at least 21 years of age;

B. Neither the gun dealer, nor any officer, employee, or agent of said gun dealer, has had a similar type license previously revoked or denied within the immediately preceding two years;

C. Neither the gun dealer, nor any officer, employee, or agent of said gun dealer, has been convicted of:

1. Any offense disqualifying said individual from owning or possessing a firearm under applicable federal, state, or local laws,

2. Any offense relating to the manufacture, sale, possession, use, or registration of any firearm or dangerous or deadly weapon,

3. Any offense involving the use of force or violence upon the person of another,

4. Any offense involving theft, fraud, dishonesty, or deceit,

5. Any offense involving the manufacture, sale, possession, or use of any controlled substance as defined by the California Health and Safety Code, as said definition now reads or may hereafter be amended to read;

D. The gun dealer has a fixed place of business where all licensed activities will be conducted as required by Section 7.04.060 of this code. The storing of all firearms and munitions shall occur at said fixed place of business. Under no circumstance may the address of the fixed place of business be either a United States Post Office box or a private commercial mailbox. The gun dealer shall provide evidence as owner, lessee or other legal occupant of said fixed place of business. The license shall specify the Post Office address of said fixed place of business;

E. The gun dealer's fixed place of business shall not be located in any area or district that is zoned for residential use;

F. The gun dealer has agreed to indemnify, defend and hold harmless the county of Los Angeles, its officers, agents and employees, from claims arising from the negligent or intentional acts of said gun dealer;

G. In connection with every firearm sold, leased, or otherwise transferred by a gun dealer, said gun dealer must also sell or otherwise provide a trigger lock or similar device reviewed and approved by the sheriff that is designed to prevent the unintentional discharge of the firearm;

H. The gun dealer has obtained a policy of insurance as provided in Section 7.46.070 below; and

I. The gun dealer has complied with the security requirements as provided in Section 7.46.080 below.

7.46.070. Liability insurance.

A. No license or renewal license shall be issued under this chapter unless the gun dealer carries and maintains in full force and effect a policy of insurance, as described in this subsection, in a form approved by the county of Los Angeles and executed by an insurance company admitted to do business in the state of California. This policy of insurance shall insure the gun dealer against liability for damage to property and for injury to or death of any person as a result of the sale, lease, or transfer, or the offering for sale, lease, or transfer, of a firearm. The minimum liability limits shall not be less than $1,000,000.00 for each incident of damage to property or incident of injury or death to a person. The policy shall name the county of Los Angeles as an additional insured.

B. The policy of insurance shall contain an endorsement providing that said policy shall not be canceled until notice in writing has been given to the office of the treasurer and tax collector at least 30 days prior to the time the cancellation becomes effective.

C. If at any time the gun dealer's policy of insurance expires, said gun dealer's license under this chapter will automatically be suspended pursuant to Section 7.08.240 and Section 7.08.250 of this code.

7.46.080. Security requirements.

A. No license or renewal license shall be issued under this chapter unless the gun dealer adheres to security measures as required by the sheriff. These security measures shall include, but not be limited to, the following:

1. The provision of adequate lighting, secure locks, windows, and doors, and fire and theft alarms, as each such item is specified and approved by both the sheriff and the fire department; and

2. The storing of all firearms and munitions on the premises out of reach of customers in secure, locked facilities, so that access to firearms and munitions shall be controlled by the gun dealer or employees of the gun dealer to the exclusion of all others.

B. Upon written request by the gun dealer, the sheriff may approve alternative security measures which he/she determines will provide equivalent or superior security to the premises as the measures required under subsection A above.

7.46.095. Officers, employees and agents of gun dealers defined. Any reference in this chapter to an officer, employee or agent of a gun dealer shall apply only to those persons who directly participate in firearm sale transactions.

7.46.100. Penalty. Any gun dealer violating the provisions of this chapter is guilty of a misdemeanor, punishable by a fine not to exceed $1,000.00, or imprisonment for a term not to exceed six months, or both. This penalty is in addition to all other penalties provided by law, and to the immediate revocation of the gun dealer's license granted under this chapter.

Title 13. Public Peace, Morals and Welfare

Division 8. Weapons

Chapter 13.67. Prohibition on the Sale of Firearms and Ammunition on County Property

13.67.030. Prohibition. The sale of firearms and/or ammunition on county property is prohibited.

13.67.040. Definitions. As used in this chapter, the words hereinafter defined are used as so defined unless it is apparent from the context that a different meaning is intended.

A. County Property. "County property" includes real property owned, leased, subleased, or otherwise assigned by the county, or real property subject to the use and control of the county. It includes real property of the county in the possession of a public or private entity under contract with the county. By way of example, it includes all public buildings and the surrounding grounds owned or leased by the county in the unincorporated and incorporated portions of the county, such as the Los Angeles County Fairgrounds in the city of Pomona.

B. Firearm. "Firearm" includes all devices described in California Penal Code Section 12001, including, but not limited to, any gun, pistol, revolver, rifle or any device, designed or modified to be used as a weapon, from which is expelled through a barrel a projectile by the force of an explosion or other form of combustion. "Firearm" does not include any instrument as referred to in Government Code Section 53071.5.

C. Ammunition. "Ammunition" is any ammunition as defined in California Penal Code Section 12316(b)(2).

D. Peace Officer. "Peace officer" is any person who is a peace officer as defined in Title 3, Part 2, Chapter 4.5 of the California Penal Code (Sections 830, et seq.).

E. Sale. A "sale" is any transaction, with or without the exchange of consideration, which transfers ownership, title, possession, or control of any firearm, or gives, loans, leases, or delivers a firearm. A "sale" includes the act of placing an order for any of the aforementioned transfers. The act of displaying a firearm shall not constitute a sale for purposes of this chapter.

13.67.050. Exceptions. Section 13.67.030 shall not apply to the sale of any firearm by a peace officer when on duty and the sale of such firearm is within the scope of his/her duties. In addition, said section shall not apply to: (A) either the sale of ammunition, or the rental of any firearm, for on-site recreational use, at the Whittier Narrows Trap and Skeet Range within the Whittier Narrows Regional Park; or (B) the sale of shotguns, as defined in California Penal Code section 12020(c)(21), but no other firearms, at the aforementioned Whittier Narrows Trap and Skeet Range during the remaining term of that certain Lease Agreement No. 59348, originally entered into between the Country of Los Angeles and Pachmayr, Limited; or (C) to the public administrator in the distribution of a private estate or to the sale of firearms by its auctioneer to fulfill its obligations under state law.

13.67.060. Penalty. Every violation of this chapter shall constitute a misdemeanor.

[Los Angeles County Code current through Ordinance 2008-0046Z]

Los Angeles Municipal Code

Chapter V. Public Safety and Protection

Article 5. Firearms - Dangerous Weapons

55.09. Ammunition Sales.

(a) The term "firearm ammunition", as used in this section, shall include any ammunition for use in pistols, revolvers, rifles, shot guns, or any other device designed to be used as a weapon from which is expelled a projectile by the force of explosion or other form of combustion.

(b) Except as specified in Subsection (c) and (d), no person, including retail gun dealers, shall sell, give, lend or transfer ownership of any firearm ammunition during the period of seven days prior to the first day of January of each year or on the first day of January of each year, or during the seven days prior to the Fourth of July of each year or on the Fourth of July of each year.

(c) Nothing in this section shall prohibit the sale or transfer of ownership of firearm ammunition by a wholesale gun dealer to another wholesale gun dealer or by a wholesale gun dealer to a retail gun dealer, or of ammunition requested by Section 12324 of the Penal Code of the State of California.

(d) Nothing in this section shall prohibit sale to or acquisition by any person described in Section 12302 or 12322 of the Penal Code of the State of California.

(e) The provisions of this section shall not apply to the sale of any firearm ammunition when such sale is prohibited by Sections 12303, 12303.6, 12304 or 12321 of the Penal Code of the State of California, or is otherwise prohibited by Chapter 2.5, Title 2, Part 4, of the Penal Code of the State of California.

55.11. Records of Ammunition Sales.

(a) The term "firearm ammunition" shall mean any ammunition for use in a pistol, revolver or other firearm, as that term is defined in Section 103.314 of this Code.

(b) The term "vendor", as used in this section, shall mean any person who is engaged in the retail sale of firearm ammunition and retail firearms dealers.

(c) No vendor shall sell or otherwise transfer ownership of any firearm ammunition without at the time of purchase recording the following information on a form to be prescribed by the Board of Police Commissioners:

1. the date of the transaction,

2. the name, address and date of birth of the transferee,

3. the transferee's driver's license or other identification number and the state in which it was issued,

4. the brand, type and amount of ammunition transferred,

5. the transferee's signature, and

6. the name of the sales person who processed the transaction.

The vendor shall also at the time of purchase or transfer obtain the right thumbprint of the purchaser or transferee on the above form.

(d) The records required by this section shall be maintained on the premises of the vendor for a period of not less than two (2) years from the date of the recorded transfer. Said records shall be subject to inspection at any time during normal business hours in accordance with the provisions of Section 103.14 of this Code.

(e) No person shall knowingly make a false entry in, or fail to make a required entry in, or fail to obtain the required thumbprint, or fail to maintain in the required manner records prepared in accordance herewith. No person shall refuse to permit a police department employee to examine any record prepared in accordance with this section during any inspection conducted pursuant to this section, or refuse to permit the use of any record or information therefrom by the police department.

(f) The requirements of this section shall not apply to:

1. any person described in Section 12302 or 12322 of the Penal Code of the State of California.

2. any off-duty peace officer who displays proper agency identification which identifies him or her as an active peace officer.

3. any person who has been issued a permit to carry a concealed weapon under the authority of Section 12050 of the Penal Code of the State of California.

4. any security guard licensed under the authority of Section 12033 of the Penal Code of the State of California.

5. any firearms dealer who has been issued a Federal Firearms License, a Certificate of Eligibility by the State of California and a Seller of Firearms Police Permit by the City of Los Angeles.

(g) Violation of this section shall constitute a misdemeanor.

55.12.1. Disposal of Firearms and Ammunition.

(a) Definitions.

1. "Firearm," as used in this section, means any pistol, rifle, shotgun or other device designed to be used as a weapon, from which a projectile is expelled through a barrel by the force of an explosion or any other form of combustion, or any device that is capable of being altered so as to expel a projectile in that manner. The term firearm includes the frame of such weapon.

2. "Ammunition," as used in this section, means any projectile designed to be expelled through the barrel of a firearm by the force of any explosion or other form of combustion, including, but not limited to, cartridge cases, primers, bullets, or propellant powder.

(b) Disposal of a firearm and/or ammunition in the City of Los Angeles shall only be accomplished in either of the following two ways:

1. By selling, giving away or otherwise transferring the firearm and/or ammunition in accordance with all applicable provisions of the California Penal Code, Welfare and Institutions Code, and this Code; or

2. By surrendering the firearm and/or ammunition to a California law enforcement agency for the purpose of destruction.

(c) Any person in the City of Los Angeles who disposes of a firearm and/or ammunition in any manner other than as required by this section, including, but not limited to, by depositing those item(s) in a public trash receptacle or leaving in any public place, is guilty of a misdemeanor.

55.13. Sale of Ammunition Clips and Similar Devices.

(a) No person shall sell or otherwise transfer to any other person a clip, magazine, belt, drum, feed strip or similar device which has a capacity of, or which can be readily restored or converted to accept, more than ten rounds of ammunition.

(b) The provisions of this section shall not apply to:

1. the sale or transfer of such devices to any person described in Section 12302 or 12322 of the Penal Code of the State of California, or

2. the sale or transfer of ownership of such a device by a wholesale gun dealer to another wholesale gun dealer or by a wholesale gun dealer to a retail gun dealer, or

3. any entity or establishment engaged in the business of motion picture, television, video, theatrical, or re-enactment production, provided the clip, magazine, belt, drum, feed strip or similar device is used solely as a prop in the course of motion picture, television, video, theater, or re-enactment production by an authorized participant therein or by an authorized agent or employee of the entity or establishment, and only if the entity or establishment properly secures such clip, magazine, belt, drum, feed strip or similar device from unauthorized use, or

4. any federal, state or local historical society, museum or institutional collection which is open to the public provided any such clip, magazine, belt, drum, feed strip or similar device is properly housed, secured from unauthorized handling and is unloaded.

55.14. Sale or Purchase of More than One Handgun within A Thirty Day Period Prohibited.

(a) The term "dealer" shall mean a retail firearms dealer licensed by the City of Los Angeles.

(b) No person shall make application to a dealer, as provided in Section 12072 of the California Penal Code, to purchase a pistol, revolver or other firearm capable of being concealed upon the person, herein referred to as "handgun", within thirty (30) days of making a prior application for the purchase of a handgun within the State of California. In computing the thirty (30) day period, the date the application is made shall be counted as the first day.

(c) No dealer shall transfer the title of any handgun to any person whom the dealer knows has made application to purchase more than one handgun within the State of California within a thirty (30) day period prior thereto, nor shall any dealer process a handgun transaction between unlicensed parties pursuant to California Penal Code Section 12072 if the dealer knows that the one receiving the handgun has made application to purchase more than one handgun within the State of California within a thirty (30) day period prior thereto.

(d) No dealer shall transfer the title of any handgun to any person or process a handgun transaction between unlicensed parties pursuant to California Penal Code Section 12072 if the dealer prior to transfer has received notice from the California Department of Justice, herein referred to as the "DOJ," that DOJ has received an application for that person to purchase a handgun within the State of California within a thirty (30) day period prior thereto. For the purposes of this subsection, the date of application of purchase shall be the date the DOJ receives the Dealer Record of Sale or receives the transmission of required information by any medium of communication authorized under state law.

(e) For the purposes of this section, the redemption of a handgun pledged to a pawnbroker shall not be deemed the sale, purchase or transfer of title of that handgun. However, a dealer may not rely on this exemption unless the purchaser provides proof that the transaction was the redemption of a handgun pawned to a pawnbroker by the purchaser.

(f) Each dealer shall maintain records of each sale of a handgun to include the name of the purchaser, type of handgun, date of sale, and any other information required by the Board of Police Commissioners. Such records shall be maintained on the premises of the dealer for a period of not less than two (2) years and shall be subject to inspection at any time during normal business hours.

(g) Each dealer shall post a sign in a conspicuous place with letters at least one inch high stating the obligations and restrictions of purchasers and dealers under this ordinance, pursuant to direction by the Board of Police Commissioners.

(h) The dealer shall require the purchaser, at the time of application for purchase, to indicate, on a form prescribed by the Board of Police Commissioners, whether he or she has or has not made an application for the purchase of a handgun within the last thirty (30) days, and which if any of the exemptions set forth in Subsection (i) is applicable to the purchaser. If any of the exemptions is checked, the restrictions of this section shall not apply, provided all necessary information is supplied to the dealer. The form shall be signed by the purchaser under penalty of perjury. Such form shall be maintained on the premises of the dealer for a period of not less than two (2) years from the date of signing and shall be subject to inspection at any time during normal business hours.

(i) The provisions of this section shall not apply to the following:

1. Any person who is licensed to sell firearms pursuant to Article 4, commencing with Section 12070, of the Penal Code of the State of California.

2. Any law enforcement agency.

3. Any agency duly authorized to perform law enforcement duties.

4. Any state or local correctional facility.

5. Any private security company licensed to do business in the State of California.

6. Any person who is properly identified as a full-time paid peace officer, as defined in Section 830.1, 830.2, 830.4, or 830.5 of the Penal Code of the State of California, and who is authorized to, and does, carry a firearm during the course and scope of his or her employment as a peace officer.

7. Any antique firearm, as defined in Paragraph (16) of Subsection (a) of Section 921 of Title 18 of the United States Code.

8. Any person or entity who purchases firearms in a collector's series or a bulk purchase from an estate sale.

9. Any motion picture, television, or video production company, or entertainment or theatrical company whose production involves the use of a handgun, and which secures such handgun from unauthorized use.

10. Any person who is exempt from the waiting period in Subdivision (c) of Section 12072 of the Penal Code of the State of California.

11. Any person who is exempt from the provisions of Subdivision (d) of Section 12072 of the Penal Code of the State of California.

12. Any person or entity conducting a transaction through a law enforcement agency pursuant to Section 12084 of the Penal Code of the State of California.

13. Any person or entity conducting a transaction described in Subdivision (k) of Section 12078 of the Penal Code of the State of California.

14. Any person who is licensed as a collector pursuant to Chapter 44, (commencing with Section 921) of Title 18 of the United States Code and the regulations issued pursuant thereto, and who has a current certificate of eligibility issued to him or her by the Department of Justice pursuant to Section 12071 of the Penal Code of the State of California.

15. Any person or entity acquiring a handgun by bequest or intestate succession.

16. Any person or entity replacing a single handgun where the handgun was lost or stolen, and the person reported that firearm lost or stolen to the Los Angeles Police Department or appropriate law enforcement agency and presents the dealer with evidence of such reporting.

17. Any seller's exchange or replacement of a handgun for a handgun purchased from that seller within the thirty (30) day period immediately preceding the date of exchange or replacement.

(j) Violation of Subsection (b) shall be punishable as an infraction subject to a fine of $50.00 for the first violation of that subsection, as an infraction subject to a fine of $100.00 for the second violation of the subsection and as a misdemeanor for the third or subsequent violations of that subsection. Each handgun applied for shall be deemed a separate violation.

(k) If any provision of this section is for any reason held to be invalid or unconstitutional by the decision of any court of competent jurisdiction within the State of California, such decision shall not affect the validity of the remaining provisions and the City declares that it would have adopted the remaining provisions irrespective of the fact that a provision or provisions are declared invalid or unconstitutional.

55.15. Fingerprinting of Firearms Purchasers.

(a) As used in this section, the term "firearm" shall mean any pistol, revolver, shotgun, rifle, or other firearm, as that term is defined in Section 103.314 of this Code.

(b) As used in this section, the term "dealer" shall mean any person who is engaged in the retail sale of firearms and any retail firearms dealer, as that term is defined in Section 103.314 of this Code.

(c) No dealer shall sell or otherwise transfer ownership of any firearm without obtaining the right thumbprint of the purchaser or transferee on the California Department of Justice Dealer Record of Sale of Firearm worksheet signed by the purchaser or transferee. The thumbprint shall be recorded at the lower right hand corner on the back side of the worksheet at the time that the worksheet is signed by the purchaser or transferee.

(d) The thumbprint record required by this section shall be maintained on the premises of the dealer throughout the time period during which the Dealer Record of Sale is required to be maintained by the dealer. Such record shall be subject to inspection at any time during normal business hours in accordance with the provisions of Section 103.14 of this Code.

(e) No person shall knowingly fail to obtain a required thumbprint, or knowingly fail to maintain the record of a thumbprint required by this section. No person shall refuse to permit a law enforcement employee to examine any record prepared in accordance with this section during any inspection conducted pursuant to this section, or refuse to permit the use of any such record or information in such record by a law enforcement agency.

(f) The requirements of this section shall not apply when the purchaser or transferee is any of the following:

(1) Any person described in Section 12302 or 12322 of the California Penal Code;

(2) Any off-duty peace officer who displays proper agency identification which identifies him or her as an active peace officer;

(3) Any person who has been issued a permit to carry a concealed weapon under the authority of Section 12050 of the California Penal Code;

(4) Any security guard licensed under the authority of Section 12033 of the California Penal Code;

(5) Any firearms dealer who has been issued a Federal Firearms License, a Certificate of Eligibility by the State of California, and a Seller of Firearms Permit by the City of Los Angeles.

(g) Any violation of this section shall constitute a misdemeanor.

55.16. Sale of Ultracompact Firearms Prohibited.

(a) As used in this section, the term "ultracompact firearm" shall mean any pistol, revolver, or other firearm, which has an overall length of six and three quarter inches (6.75") or less or an overall height of four and one half inches (4.5") or less, as measured with the magazine detached.

(b) As used in this section, the term "dealer" shall mean any person who is engaged in the retail sale of firearms and any retail firearms dealer, as that term is defined in Section 103.314 of this code.

(c) No dealer or other person shall sell, lend, give, transfer ownership of, or otherwise transfer to any other person any ultracompact firearm.

(d) No dealer shall process, pursuant to Penal Code Section 12072(d), any ultracompact firearm transaction between persons not licensed as firearms dealers or not possessing Federal Firearms Licenses.

(e) No dealer or other person shall sell, lend, give, transfer ownership of, or otherwise transfer to any other person any firearm accessory designed to enable or facilitate the carrying of any ultracompact firearm, including but not limited to shoulder holsters, ankle holsters, belt holsters, and other devices designed for the purpose of aiding the concealment of an ultracompact firearm, unless such accessory is designed so that it can be used with a firearm larger than an ultracompact firearm.

(f) The provisions of Subsections (c), (d), and (e) of this section shall not apply when the purchaser or transferee is any of the following:

(1) Any law enforcement agency;

(2) Any agency duly authorized to perform law enforcement duties;

(3) Any state or local correctional facility;

(4) Any private security company licensed to do business in the State of California or employee thereof authorized to carry a firearm within the course and scope of his or her employment with such company;

(5) Any person properly identified as a full-time paid peace officer, as defined in California Penal Code Sections 830.1, 830.2, 830.4, or 830.5, and who is authorized to carry and does carry a firearm within the course and scope of his or her employment as a peace officer;

(6) Any person properly identified as a retired peace officer;

(7) Any person described in Section 12302 or 12322 of the California Penal Code;

(8) Any person with a valid license to carry a concealed weapon issued pursuant to California Penal Code section 12050;

(9) The purchaser of an antique firearm, as defined in Paragraph (16) of Subsection (a) of Section 921 of the United States Code;

(10) The purchaser of a firearm in a collector's series, provided that such firearm falls within the definition of a curio or relic contained in Section 178.11 of Title 27 of the Code of Federal Regulations;

(11) A motion picture, television, or video production company, or an entertainment or theatrical company whose production involves the use of such handgun, provided that the firearm is borrowed or rented for such purpose, that the firearm is secured from unauthorized use during its use for such purpose, and that the firearm is returned to its owner within a reasonable period of time after it is no longer needed for such use;

(12) Any person who is exempt from the provisions of Subsection (d) of Section 12072 of the California Penal Code;

(13) Any person or entity conducting a transaction described in Subdivision (k) of Section 12078 of the California Penal Code;

(14) Any person licensed as a collector pursuant to Chapter 44 (commencing with Section 921) of Title 18 of the United States Code and regulations issued pursuant thereto, and who has a current certificate of eligibility issued to him or her by the state Department of Justice pursuant to Section 12071 of the California Penal Code;

(15) Any person or entity acquiring a firearm by bequest or by intestate succession;

(16) A person redeeming a firearm pledged to a pawnbroker, provided that such person provides proof that he or she is the person who pledged the firearm.

(g) Every dealer shall maintain separate records as to any sales exempted pursuant to Subsection (f) of this section, showing each sale of an ultracompact firearm, including as to each sale the name of the purchaser, type of handgun, date of sale, and documentation of any exemption pursuant to Penal Code Section 12050 or any other exemption set forth in Subsection (f). Such records shall be maintained on the premises of the dealer for a period of not less than two years and shall be subject to inspection at any time during normal business hours in accordance with the provisions of Section 103.14 of this code.

(h) Any violation of this section shall constitute a misdemeanor.

55.18. Sale of Large Caliber Firearms Prohibited.

(a) As used in this section, the term "large caliber firearm" shall mean any firearm, as defined in Section 103.314 of this Code, capable of firing a centerfire cartridge of .50 caliber or larger either by designation or by actual measurement. The term "large caliber firearm" shall include any rifle or handgun.

(b) As used in this section, the term "rifle" shall mean any firearm that is designed or redesigned, made or remade, and intended to be fired from the shoulder and is designed to fire only a single projectile through a rifled bore for each single pull of the trigger. The term "rifle" shall not include any shotgun.

(c) As used in this section, the term "handgun" shall mean any firearm with a barrel less than 16 inches in length. The term "handgun" shall include any pistol, revolver, or concealable firearm as such terms are defined in the California Penal Code.

(d) No person shall sell, give, transfer ownership of, transfer, offer for sale, or display for sale any large caliber firearm.

(e) The provisions of this section shall not apply to any sale or transfer of a firearm which is prohibited under state law, of any destructive device as defined in Section 12301 of the California Penal Code, of any assault weapon as defined in the California Penal Code, or of any .50 BMG rifle as defined in the California Penal Code. The provisions of this section shall not apply to any offer for sale or display in any periodical, solicitation by mail, or use of the internet by a seller or offeror located outside the City of Los Angeles.

(f) The provisions of Subsection (d) of this section shall not apply where the purchaser or transferee is any of the following:

(1) A law enforcement agency;

(2) An agency duly authorized to perform law enforcement duties;

(3) A state or local correctional facility;

(4) A person described in Section 12302 or 12322 of the California Penal Code, when acting within the course and scope of his or her law enforcement duties;

(5) A federal law enforcement officer, when acting in the course and scope of his or her law enforcement duties;

(6) A person who is properly identified as a full-time paid peace officer, as defined in Section 830.1, 830.2, 830.4, or 830.5 of the California Penal Code, and who is authorized to, and does, carry a firearm during the course of his or her employment as a peace officer, and is acting within the course and scope of such employment;

(7) A firearms dealer who has been issued a Federal Firearms License, a Certificate of Eligibility by the State of California, and a permit by the City of Los Angeles to engage in the retail sale of firearms;

(8) A purchaser of a curio or collector firearm. A firearm shall be deemed curio or collector only if it falls within one of these categories:

(A) It was manufactured prior to 1899;

(B) It is classified as a curio or relic pursuant to 27 Code of Federal Regulations section 178.11, and the purchaser maintains a current federal firearms collector license; or

(C) It is a muzzle-loading firearm;

(9) A federal, state, or local historical society, museum, or institutional collection that is open to the public, provided that the large caliber firearm is used for display purposes, is secured from unauthorized use, and is unloaded;

(10) A person acting pursuant to a valid entertainment firearms permit issued by the State of California; or

(11) A person who obtains title to a large caliber firearm by bequest or intestate succession.

(g) Any violation of this section shall constitute a misdemeanor.

[Los Angeles Municipal Code current through Sept. 8, 2008]

Oakland Municipal and Planning Codes

Title 5. Business Taxes and Regulations

Chapter 5.26. Firearms Dealers

5.26.010. Title. This chapter shall be known as the firearms dealer permit ordinance.

5.26.030. Definitions. The following words and phrases, whenever used in this chapter, shall be construed as defined in this section:

"Engaged in the business" means the conduct of a business by the selling, transferring, or leasing of any firearm; or the preparation for such conduct of business as evidenced by the securing of applicable federal or state licenses; or the holding of one's self out as engaged in the business of selling, transferring, or leasing of any firearm, or the selling, transferring, or leasing of firearms in quantity, in series, or in individual transactions, or in any other manner indicative of trade.

"Firearm" means any device, designed to be used as a weapon or modified to be used as a weapon, from which is expelled through a barrel a projectile by the force of an explosion or other form of combustion.

"Firearms dealer" means a person engaged in the business of selling, transferring, or leasing, or advertising for sale, transfer, or lease, or offering or exposing for sale, transfer, or lease, any firearm.

"Person" means natural person, association, partnership, firm, or corporation.

5.26.040. Permit required. It is unlawful for any person, partnership, cooperative, corporation, firm, or association to engage in the business of operating or managing any business which sells, transfers, leases, or offers or advertises for sale, transfer, or lease, any firearm without first obtaining a firearms dealer permit from the Chief of Police of the Oakland Police Department.

5.26.050. Application - Forms, fees. An applicant for a permit under this chapter shall file with the Chief of Police a sworn application in writing, on a form to be furnished by the city. The applicant shall provide all information requested, including proof of compliance with all applicable federal, state, and local laws when required by the Chief of Police, or the application will not be deemed complete. The application shall be accompanied by a nonrefundable fee as set forth in the city municipal license/permit fee schedule. To the extent practicable, the fee amount shall reflect the cost of enforcing the requirements of this chapter.

5.26.080. Security. In order to discourage the theft of firearms stored on the premises of a firearms dealer, any business licensed under this chapter must adhere to security measures as required by the Chief of Police. Security measures shall include but not be limited to:

A. The provision of secure locks, windows and doors, adequate lighting, and alarms as specified by the Chief of Police;

B. Storing of all firearms on the premises out of the reach of customers in secure, locked facilities, so that access to firearms shall be controlled by the dealer or employees of the dealer, to the exclusion of all others.

5.26.110. Permit - Assignment. The assignment or attempt to assign any permit issued pursuant to this chapter is unlawful and any such assignment or attempt to assign a permit shall render the permit null and void.

5.26.120. Permit - Conditions. Any permit issued pursuant to this chapter shall be subject to all of the following conditions, the breach of any of which shall be sufficient cause for revocation of the permit by the Chief of Police:

A. The business shall be carried on only in the building located at the street address shown on the license.

B. The permittee shall comply with Sections 12073, 12074, 12076, 12077 and 12082 and subdivision (b) of Section 12072 of the California Penal Code, to the extent that the provisions remain in effect. Any permit issued pursuant to this chapter shall be subject to such additional conditions as the Chief of Police finds are reasonably related to the purpose of this chapter.

C. The permittee shall not sell, lease or otherwise transfer a firearm without also selling or otherwise providing with each firearm a trigger lock or similar device approved by the Chief of Police that is designed to prevent the unintentional discharge of firearm. The trigger lock or similar device required to be sold pursuant to this subsection must be attached to each firearm prior to completion of the sales transaction.

D. The permittee shall post conspicuously within the licensed premises the following warning in block letters not less than one inch in height as per Penal Code Section 12071(b)(11):
IF YOU LEAVE A LOADED FIREARM WHERE A CHILD OBTAINS AND IMPROPERLY USES IT, YOU MAY BE FINED OR SENT TO PRISON

E. Persons Under the Age of Eighteen Excluded from Establishments Displaying Firearms. No person who maintains or operates any place of business in which firearms are kept, displayed or offered in any manner, sold, furnished or transferred shall permit the entry into the premises of persons under the age of 18 years, unless all firearms and firearms accessories are completely and wholly kept, displayed or offered within a separate room or enclosure to which persons under the age of eighteen years are excluded. Each entrance to such a room or enclosure shall be signposted in block letters not less than one inch in height to the effect that firearms are kept, displayed or offered in such room or enclosure and that persons under the age of eighteen are excluded.

A person who maintains or operates any place of business that openly displays firearms without providing for separate enclosures therefor, must prevent entry into the premises of persons prohibited by age under state law from purchasing firearms. The entrance to the establishment shall be signposted in block letters not less than one inch in height to the effect that firearms are kept, displayed or offered on the premises and that persons under the age of eighteen are excluded.

The person who maintains or operates any place of business in which firearms are kept, displayed or offered in any manner, sold, furnished or transferred, and his employees, agents, and persons acting under his authority, are each and together separately and jointly responsible for requiring bona fide evidence of identity of persons to prevent the entry of persons not permitted to purchase a firearm under state law by reason of age. Bona fide evidence of identity of the person is a document issued by a federal, state, county, or municipal government, or subdivision or agency therefor, including, but not limited to, a motor vehicle operator's license or an identification card issued to a member of the Armed Forces, which contains the name, date of birth, description, and picture of the person.

F. Persons Under the age of Twenty-one Excluded from Establishments Displaying Concealable Firearms. No person who maintains or operates any place of business in which concealable firearms are kept, displayed or offered in any manner, sold, furnished or transferred shall permit the entry into the premises of persons under the age of twenty-one years, unless all concealable firearms and concealable firearms accessories are completely and wholly kept, displayed or offered within a separate room or enclosure to which persons under the age of twenty-one years are excluded. Each entrance to such a room or enclosure shall be signposted in block letters not less than one inch in height to the effect that firearms are kept, displayed or offered in such room or enclosure and that persons under the age of twenty-one (21) are excluded.

A person who maintains or operates any place of business that openly displays concealable firearms without providing for separate enclosures therefor, must prevent entry into the premises of persons prohibited by age under state law from purchasing concealable firearms. The entrance to the establishment shall be signposted in block letters not less than one inch in height to the effect that firearms are kept, displayed or offered on the premises and that persons under the age of twenty-one (21) are excluded.

The person who maintains or operates any place of business in which concealable firearms are kept, displayed or offered in any manner, sold, furnished or transferred, and his employees, agents, and persons acting under his authority, are each and together separately and jointly responsible for requiring bona fide evidence of identity of persons to prevent the entry of persons not permitted to purchase concealable firearms under state law by reason of age. Bona fide evidence of identity of the person is a document issued by a federal, state, county, or municipal government, or subdivision or agency therefor, including, but not limited to, a motor vehicle operator's license or an identification card issued to a member of the Armed Forces, which contains the name, date of birth, description, and picture of the person.

G. Persons Prohibited from Possessing Firearms Excluded from Establishments Displaying Firearms. Persons prohibited from possessing or purchasing firearms pursuant to California Penal Code Section 12021 shall not enter into or loiter about any firearms dealership, defined as a place of business in which firearms are openly kept, displayed or offered in any manner, sold, furnished or transferred pursuant to Penal Code Section 12070. Each entrance to such an establishment shall be signposted in block letters not less than one inch in height to the effect that persons prohibited from possessing firearms pursuant to Penal Code Section 12021 are excluded from the premises.

When a firearms dealer displays or offers for sale firearms within a separate room or enclosure that segregates the firearms and firearms related accessories (including but not limited to, ammunition clips, bullets and holsters) from other general merchandise, each entrance to such a separate room or enclosure shall be signposted in block letters not less than one inch in height to the effect that persons prohibited from possessing firearms pursuant to Penal Code Section 12021 are excluded from entering the separate rooms or enclosures.

Any dealer engaging in the business of selling, transferring, or leasing, or advertising for sale, transfer, or lease, or offering or exposing for sale, transfer, or lease, any firearm within the city who knowingly violates the provisions of Section 5.26.120G, or fails to adhere to the notice provisions of Section 5.26.120G, shall be subject to the penalty provisions of the Oakland Municipal Code, Chapter 5.26, including but not limited to suspension and/or revocation of his or her permit to sell firearms.

Any person prohibited from possessing firearms pursuant to Penal Code Section 12021 who enters into or loiters about firearms dealership in Oakland that has been properly posted pursuant to this section is guilty of a misdemeanor.

5.26.130. Permit - Grounds for revocation. In addition to any provisions constituting grounds for denial shall also constitute grounds for revocation.

5.26.150. Permit - Liability insurance. No permit shall be issued or continued pursuant to this chapter unless there is in full force and effect a policy of insurance in such form as the City Attorney deems proper, executed by an insurance company approved by the City Attorney whereby the applicant or permittee is insured against liability for damage to property and for injury to or death of any person as a result of the sale, transfer or lease, or advertising for sale, transfer, or lease, or offering or exposing for sale, transfer, or lease, any firearm. The minimum liability limits shall not be less than one million dollars ($1,000,000.00) for damage to or destruction of property in any one incident, and one million dollars ($1,000,000.00) for the death or injury to any one person; provided, however, that additional amounts may be required by the City Attorney if deemed necessary.

Such policy of insurance shall contain an endorsement providing that the policy will not be canceled until notice in writing has been given to the city, addressed in care of the Chief of Police, 455 - 7th Street, Oakland, California, 94607, at least thirty (30) days immediately prior to the time such cancellation becomes effective. Further, such policy of insurance shall name the city, its officers, agents, and employees as additional insureds. Additionally, applicants and permittees shall indemnify, defend, and hold harmless the city, its officers, agents, and employees, from claims arising from the negligence of the applicant or permittee.

5.26.160. Permit - Authority to inspect. Any and all investigating officials of the city shall have the right to enter the building designated in the permit from time to time during regular business hours to make reasonable inspections to observe and enforce compliance with building, mechanical, fire, electrical, plumbing, or health regulations, and provisions of this chapter. A police investigator may conduct compliance inspections to insure conformance to all federal, state, and local law, and all provisions of this chapter.

5.26.170 Compliance. Any person engaging in the business of selling, transferring, or leasing, or advertising for sale, transfer, or lease, or offering or exposing for sale, transfer, or lease, any firearm on the effective date of this chapter shall have a period of sixty (60) days after such effective date to comply with the provisions of this chapter.

5.26.180. Temporary suspension of permit to sell firearms.

A. If the dealer violates any federal, state and local county or city law, the Chief of Police may immediately suspend the right of the dealer to sell firearms. This temporary suspension will not exceed three days, if the violation results in a criminal charge filed in court by a federal, state, or county District Attorney such permit to sell firearm may be suspended until the case is adjudicated in a court of law.

B. Notice of suspension shall be mailed to the person(s) who made application for the permit and shall be delivered to the address listed on the permit.

Title 9. Public Peace, Morals and Welfare

Chapter 9.20. Ammunition Sales Registration

9.20.010. Title. This chapter shall be known as the ammunition sales registration ordinance.

9.20.020. Findings and purpose. The City Council finds that the state of California has not preempted the enactment of a municipal ordinance requiring the registration of ammunition sales in Oakland, and further finds that the availability of such ammunition sales records can aid the police department in investigations of major crimes involving the use of firearms. The intent of the City Council in enacting the ordinance codified in this chapter is to implement a requirement that ammunition vendors within the city record and maintain records with respect to each individual purchase of ammunition, so that the purchaser can be traced in the event the ammunition were used in the commission of criminal activity.

9.20.030. Definitions. As used in this chapter:

"Ammunition" means a projectile designated to be expelled by force of an explosion, and intended for use in any firearm, including pistols, revolvers, rifles, and shotguns.

"Vendor" means any individual, person, gun dealer, store, firm or corporation selling ammunition within the city of Oakland.

9.20.040. Record of ammunition sales. Every vendor who sells ammunition in the city shall maintain a record of ammunition sales as prescribed by this chapter. The record shall be maintained on the vendor's premises, on forms supplied by, or approved by, the Oakland Police Department (OPD). An ammunition purchaser must provide to the vendor and the vendor shall record the following information:

A. The date of the transaction;

B. The purchaser's name, address and date of birth;

C. The purchaser's valid driver's license number or other identification number from a valid photographic I.D. such as a passport;

D. The brand of ammunition purchased;

E. The type and amount of ammunition purchased; and

F. The purchaser's signature and vendor's initials.

The information required to be recorded shall be maintained in chronological order by date of sale of the ammunition and shall be retained on the business premises of the vendor for a period not less than two years following the date of the recorded sale of the ammunition.

Oakland police officers may enter a vendor's premises during regular business hours for the purpose of examining, inspecting or copying records required by this chapter.

9.20.050. Violation - Penalty. It is unlawful for any vendor engaged in the retail sale of ammunition in Oakland to knowingly make a false entry in, or fail to make appropriate entry, or fail to properly maintain any such record, or refuse to immediately provide the ammunition sales log for inspection to a police officer upon request.

Violation of any provision of this chapter shall be a misdemeanor, subject to fines and penalties as provided by law. Additionally, failure to abide by the requirements of this chapter shall be grounds for the revocation of a vendor's firearms dealer license, pursuant to Chapter 5.26 of this code.

Chapter 9.36. Weapons

Article II. Firearms and Weapons Violence Prevention

9.36.050. Title. This article shall be known as the firearms and weapons violence prevention ordinance.

9.36.070. Definitions. The following words and phrases, wherever used in this article, shall be construed as defined in this section:

A. "Firearms" means any device, designed to be used as a weapon or modified to be used as a weapon, from which is expelled through a barrel a projectile by the force of an explosion or other form of combustion.

B. "Projectile weapon" means any device or instrument used as a weapon which launches or propels a projectile by means other than the force of an explosion or other form of combustion with sufficient force to cause injury to persons or property. A projectile weapon shall include, but not be limited to, air gun, air pistol, air rifle, gas-operated gun, BB gun, pellet gun, flare gun, dart gun, bow, cross-bow, slingshot, wrist rocket, blow gun, paint gun, or other similar device or instrument.

9.36.090. Enforcement. Violations of this article shall result in arrest as a misdemeanor. The District Attorney shall review the circumstances surrounding the violation and shall charge the violation either as an infraction or as a misdemeanor, except that:

A. Violation of this article for a second or subsequent offense shall be chargeable as a misdemeanor only, and the penalty for conviction of the same shall be punishable by a fine of not more than one thousand dollars ($1,000.00) or by imprisonment in the county jail for a period of not more than one year, or by both.

B. Violation of this article occurring within one thousand five hundred (1,500) feet of a day care center, school or school yard, whether public or private, shall be a misdemeanor, and the penalty for conviction of the same shall be punishable by a fine of not more than one thousand dollars ($1,000.00) or by imprisonment in the county jail for a period of not more than one year, or by both.

C. A person shall be guilty of a separate offense for each and every firing of a projectile weapon or discharge of a firearm, and shall be punished accordingly.

D. Juveniles arrested pursuant to this section shall be subject to Section 602 of the Welfare and Institutions Code.

9.36.100. Parental responsibility for minors. Any parent or legal guardian, of a person over the age of eighteen (18), is also guilty of an offense punishable in accordance with Section 9.36.090 if he or she knows or reasonably should know that a minor is likely to gain access to a firearm or a projectile weapon kept within any premises or vehicle which is under his or her custody or control, and a minor obtains and fires or discharges the firearm or projectile weapon within the city, in violation of Section 9.36.080.

9.36.130. Projectile weapons - Possession of by minors.

A. It is unlawful for any person under the age of eighteen (18) to have in his or her possession within the city limits of Oakland any projectile weapon, as defined in Section 9.36.070. Violation of this provision shall be punishable in the manner provided in Section 9.36.090D.

B. It is unlawful for any parent or legal guardian, or any person over the age of eighteen (18) years, to sell, give or loan to any minor in the city under the age of eighteen (18) years, or to allow such minor to possess, any device or instrument capable of launching a projectile, and/or the projectiles specifically intended to be launched by said device or instrument, as defined hereinabove. Violation of this provision shall be punishable in the manner provided in Section 9.36.090.

C. Any device or instrument capable of launching a projectile, and/or the projectiles specifically intended to be launched by said device or instrument, which is in possession of a minor in violation of this article, is declared to be a nuisance, and shall be surrendered to the Police Department of the city. The Police Department, except upon the certificate of a judge of a court of record or of the District Attorney of Alameda county that the preservation thereof is necessary or proper to the ends of justice, shall destroy any such device or instrument.

9.36.131 Theft or loss of firearms - Reporting of stolen and/or lost firearms required.

A. Any person owning a firearm or in possession of a firearm is required to report the theft or loss of such firearm to the Oakland Police Department when:

1. owner resides in Oakland, and/or

2. the theft or loss of the firearm occurs in Oakland.

B. A person subject to the reporting requirements in subsection A is required to report the theft or loss of a firearm within forty eight (48) hours of knowing that the firearm has been stolen or lost.

C. A person who has experienced the theft or loss of a firearm between August 1, 1992 and July 30, 2002 and who otherwise meets the reporting requirements in subsection A is required to report the loss or theft of such firearm to the Oakland Police Department within sixty (60) days of the effective date of the Amendment under which this new section was adopted.

9.36.141 Penalty - Misdemeanor.

A. Failure to report within forty-eight (48) hours the theft or loss of a firearm pursuant to this subsection B of Section 9.36.131, when the owner or person in possession knew or should have known of the theft or loss, shall be a misdemeanor subjecting the owner to prosecution.

B. Failure to report firearms theft or loss within the timeframe set forth within subsection C of section 9.36.131 shall be a misdemeanor subjecting the owner to prosecution.

Article III. Prohibition on the Sale of Saturday Night Specials, Also Known as Junk Guns

9.36.150. Title. This article shall be known as the city Saturday night special/junk gun sales prohibition and may be so cited.

9.36.160. Purpose and intent. The purpose of this article is to ensure the health, safety, and general welfare of city citizens by eliminating the sale of cheaply made, inadequately designed and poorly manufactured handguns in the city.

9.36.170. Saturday night special defined.

A. Except as provided in subsection B of this section, term "Saturday night special" as used in this article means any of the following:

1. A pistol, revolver, or firearm capable of being concealed upon the person, as those terms are defined in the California Penal Code Section 12001(a), which contains a frame, barrel, breechblock, cylinder or slide that is not completely fabricated of heat treated carbon steel, forged alloy or other material of equal or higher tensile strength;

2. A semiautomatic pistol which:

a. It is not originally equipped by the manufacturer with a locked-breech action, and

b. Is chambered for cartridges developing maximum permissible breech pressures above twenty-four thousand one hundred (24,100) Copper Units of Pressure as standardized by the Sporting Arms and Ammunition Manufacturers Institute,

c. For purpose of subsection (A)(2)(b) of this section, "semiautomatic pistol" means a firearm, as defined in California Penal Code Section 12001(b), which is designed to be held and fired with one hand, and which does the following upon discharge: (i) fires the cartridge in the chamber, (ii) ejects the fired cartridge case, and (iii) loads a cartridge from the magazine into the chamber. "Semiautomatic pistol" shall not include any assault weapons designated in California Penal Code Section 12276;

3. A pistol, revolver, or firearm capable of being concealed upon the person, as those terms are defined in California Penal Code Section 12001(a), which:

a. Uses an action mechanism which is substantially identical in design to any action mechanism manufactured in or before 1898 that was originally chambered for rimfire ammunition developing maximum safe permissible breech pressures below nineteen thousand (19,000) Copper Units of Pressure as standardized by the Sporting Arms and Ammunition Manufacturers Institute, and

b. Is chambered to fire either centerfire ammunition or rimfire ammunition developing maximum permissible breech pressures above nineteen thousand (19,000) Copper Units of Pressure as standardized by the Sporting Arms and Ammunition Manufacturers Institute, and

c. Is not originally equipped by the manufacturer with a nondetachable trigger guard, or

d. If rimfire, is equipped with a barrel of less than twenty (20) bore diameters in overall length protruding from the frame,

e. For purposes of this subsection (A)(3), "action mechanism" means the mechanism of a firearm by which it is loaded, locked, fired and unloaded commonly known as the cycle of operation.

B. Exclusions. The term "Saturday night special" does not include any of the following:

1. Any pistol which is an antique or relic firearm or other weapon falling within the specifications of paragraphs (5), (7) and (8) of subsection (b) of California Penal Code Section 12020; or

2. Any pistol for which the propelling force is classified as pneumatic, that is, of, or related to, compressed air or any other gases not directly produced by combustion; or

3. Children's pop guns or toys; or

4. An "unconventional pistol" as defined in California Penal Code Section 12020(c)(12); or

5. Any pistol which has been modified to either: render it permanently inoperable, or permanently to make it a device no longer classified as a Saturday night special.

9.36.180. Roster of Saturday night specials. On or before January 1, 1997, the Chief of Police or his or her designee shall compile, publish, and thereafter maintain a roster of Saturday night specials. The roster shall list those firearms, by manufacturer and model number, which the Chief of Police or his or her designee determines fit the definition of Saturday night special set forth in Section 9.36.170A.

9.36.190. Notification.

A. Upon completion of a list of firearms to be placed on the roster for the first time, the Police Chief or his or her designee shall endeavor to send written notification to:

1. The manufacturer of every firearm on said list; and

2. Every dealer within the city who is licensed to sell and transfer firearms pursuant to Section 12071 of the Penal Code of the state of California and Chapter 5.26 of this code.

B. Such notification shall do the following:

1. Identify the model number of the firearm which has been classified as a Saturday night special within the meaning of Section 9.36.170A; and

2. Advise the recipient that the recipient may apply for reconsideration of the classification of the firearm as a Saturday night special; and

3. Advise the recipient that the burden of proving a firearm does not constitute a Saturday night special within the meaning of Section 9.36.170A shall be on the recipient.

9.36.220. Publication of the roster. The Chief of Police or his or her designee shall place on the roster those firearms which have been determined to constitute a Saturday night special within the meaning of Section 9.36.170A. The Chief of Police or his or her designee shall cause the roster to be published in the following manner:

A. Notification of the roster's completion shall be published at least once in the official newspaper as designated by the city and circulated in the city within fifteen (15) days after its completion; and

B. A copy of the roster, certified as a true and correct copy thereof, shall be filed in the office of the City Clerk of the city; and

C. A copy of the roster, certified as a true and correct copy thereof, shall be distributed to every dealer within the city who is licensed to sell and transfer firearms pursuant to Section 12071 of the Penal Code of the state of California and Chapter 5.26 of the municipal code.

9.36.250. Sale prohibited. After January 1, 1997, no wholesale or retail firearms dealer as licensed by the city in Chapter 5.26 of the municipal code shall sell, offer or display for sale, give, lend or transfer ownership of, any firearm listed on the roster of Saturday night specials. This section shall not preclude a wholesale or retail gun dealer from processing firearm transactions between unlicensed parties pursuant to Section 12072(d) of the Penal Code of the state of California.

9.36.260. Exemptions. Nothing in this article relative to the sale of Saturday night specials shall prohibit the disposition of any firearm by sheriffs, constables, marshals, police officers, or other duly appointed peace officers in the performance of their official duties, nor to persons who are authorized by the United States Federal Government for use in the performance of their official duties; nor shall anything in this article prohibit the use of any firearm by the above-mentioned persons in the performance of their official duties.

9.36.270. Penalty. Any person violating any of the provisions of this article shall be guilty of a misdemeanor. Any person convicted of a misdemeanor under the provisions of this article shall be punishable by a fine of not more than one thousand dollars ($1,000.00) or by imprisonment for a period not exceeding six months, or by both such fine and imprisonment. Each such person shall be guilty of a separate offense for each and every day during any portion of which any violation of any provision of this article is committed, continued or permitted by such person and shall be punishable accordingly. In addition, any person found to be in violation of this Article shall be considered in noncompliance with the requirements of Chapter 5.26 of this code, and subject to the suspension and or revocation of a firearms dealer permit.

Article V. Prohibition on the Sale of Compact Handguns

9.36.400. Title. This article shall be known as the City of Oakland Compact Handgun Sales Limitation Act.

9.36.410. Purpose and intent. The purpose and intent of this article is to provide for the health, safety, and general welfare of the citizens of Oakland by prohibiting the sale of compact firearms in Oakland.

9.36.420. Definitions.

A. "Dealer" means a retail firearms dealer licensed by the city.

B. "Compact handgun" means a pistol, revolver, and any handgun designed to be concealed upon the person that has a length of six and three quarter inches (6.75") or less or a height of four and one-half inches (4.5") or less, measured with the magazine detached.

9.36.430. Prohibition on the sale of compact handguns. As of the effective date of this article:

A. No firearms dealer in Oakland shall transfer the title of any compact handgun as defined herein to any person. This section shall not preclude a wholesale or retail gun dealer from processing firearms transactions between unlicensed parties pursuant to Section 12072(d) of the Penal Code of the state of California.

B. For the purposes of this section, the redemption of a compact handgun covered by this article pledged to a pawnbroker prior to the effective date of this ordinance shall not be deemed the sale or transfer of title of that handgun. However, a dealer may not rely on this exemption unless the transaction involved the redemption of a handgun pawned to a pawnbroker by the purchaser.

C. Each dealer shall post a sign in a conspicuous place with letters at least one inch high stating the obligations and restrictions of dealers under this ordinance, pursuant to direction by the Oakland Police Department.

D. The provisions of this section shall not apply to the following:

1. Any law enforcement agency;

2. Any agency duly authorized to perform law enforcement duties;

3. Any state or local correctional facility;

4. Any private security company licensed to do business in the state of California;

5. Any person who is properly identified as a full-time paid peace officer, as defined in Section 830.1, 830.2, 830.4, or 830.5 of the Penal Code of the state of California, and who is authorized to carry a firearm during the course and scope of his or her employment as a peace officer;

6. Any antique firearm, as defined in paragraph (16) of subsection (a) of Section 921 of Title 18 of the United States Code;

7. Any motion picture, television, or video production company, or entertainment or theatrical company whose production involves the use of an ultracompact firearm, and which secures such from unauthorized use;

8. Any person who is exempt from the provisions of subdivision (d) of Section 12072 of the Penal Code of the state of California;

9. Any person or entity conducting a transaction described in subdivision (k) of Section 12078 of the Penal Code of the state of California;

10. Any person who is licensed as a collector pursuant to Chapter 44, (commencing with Section 921) of Title 18 of the United States Code and the regulations issued pursuant thereto, and who has a current certificate of eligibility issued to him or her by the Department of Justice pursuant to Section 12071 of the Penal Code of the state of California;

11. Any person or entity acquiring a compact handgun by bequest or intestate succession.

9.36.440. Penalties. Violation of this article by a firearms dealer shall subject the dealer to civil penalties as provided for in Chapter 1.8 and 1.12 of the OMC, and in addition shall subject the dealer to suspension or revocation of the dealer's firearms dealer permit. Each instance wherein a compact handgun is sold, conveyed, transferred or distributed in violation of this article shall be deemed a distinct and separate offense.

[Oakland Municipal and Planning Codes current through Ordinance 12871, passed May 20, 2008]

Riverside County Code

Title 9. Public Peace, Morals and Welfare

Chapter 9.20. Firearms

Article 2. Use and Discharge by Minors

9.20.100. Firearm defined. The word "firearm," as used in this article, includes the following: cannon, gun, pistol, revolver, automatic pistol, rifle, shotgun, air gun, pellet gun and any other weapon designed to discharge one or more projectiles propelled by the expansion of gas.

9.20.110 Discharge unlawful - Exceptions. It is unlawful for any person under the age of eighteen (18) years to shoot or discharge a firearm in the unincorporated territory of the county unless such person: (a) holds and has in his or her possession a permit issued by the county as hereinafter provided; or (b) is in the company of a parent, guardian or other adult person having responsibility for his or her conduct; or (c) holds and has in his or her possession a current valid hunting license issued pursuant to law.

[Riverside County Code current through Ordinances 348.4585, 348.4586, 348.4587, 348.4588 and 880, passed May 13, 2008]

Sacramento City Code

Title 5. Business Licenses and Regulations

Chapter 5.64. Firearm and Ammunition Sales

5.64.010. Definitions.

"Applicant," when the applicant is other than a natural person, includes any officer, director, employee or agent of the applicant who may be engaged in the business of selling, leasing or otherwise transferring any firearm or firearm ammunition.

"Engaging in the business" means the conduct of a business by the selling, leasing or transferring of any firearm or firearm ammunition; or the preparation for such conduct of business as evidenced by the securing of applicable federal or state licenses; or the holding of one's self out as engaged in the business of selling, leasing or otherwise transferring of any firearm or firearm ammunition, or the selling, leasing or transferring of firearms in quantity, in series or in individual transactions, or in any other manner indicative of trade.

"Firearm" means any device, designed to be used as a weapon or modified to be used as a weapon, that expels a projectile through a barrel by the force of an explosion or other form of combustion.

"Firearm ammunition" means any cartridge or encasement containing a bullet or projectile, propellant or explosive charge, and a primer which is used in the operation of a firearm.

5.64.020. License requirement and penalty for violation.

A. It is unlawful for any person, firm, corporation or dealer engaging in the business of selling, leasing or otherwise transferring any firearm or firearm ammunition, to sell, lease or otherwise transfer any firearm or firearm ammunition, without first having obtained a license from the chief of police. Gunsmiths, crafts makers and firearms collectors shall not be required to obtain such a license unless they sell, lease or transfer firearms. The chief of police shall make available application forms requiring applicants to provide the information set forth in Section 5.64.030 of this chapter, and shall collect a nonrefundable application fee from each applicant. Upon receiving the application fee, the chief of police shall issue a receipt to the applicant showing that such application fee has been paid. The issuance of such a receipt shall not authorize an applicant to engage in the sale, lease, or other transfer of firearms or firearm ammunition until the necessary license has been lawfully issued.

B. The application fee shall be established by resolution of the city council and shall be the exclusive source of city funds from which recovery of all costs associated with this chapter may be obtained, including but not limited to, processing applications, monitoring licensees, and enforcing the provisions of this chapter.

C. Any person, firm, corporation or dealer engaging in the business of selling, leasing or otherwise transferring any firearm or firearm ammunition to sell, lease, or otherwise transfer any firearm or firearm ammunition without first having obtained a license from the Sacramento police department or, once having obtained a license, violates any of the conditions set forth in Section 5.64.110 of this chapter or continues to engage in the business after such license is revoked or suspended, shall be in violation of this chapter and guilty of a misdemeanor, punishable by imprisonment in the county jail not exceeding six months, or by fine not exceeding one thousand dollars ($1,000.00), or by both. The chief of police shall prescribe a form on which the chief of police shall inform the California Department of Justice that such violators do not meet the state requirements for gun licensees or dealers pursuant to Penal Code Section 12071(a)(1).

5.64.080. Nonassignability of license. The assignment of any license issued pursuant to this chapter is unlawful and any such assignment shall render the license null and void. In addition, the attempt to transfer or assign any license issued pursuant to this chapter shall be further grounds for revocation.

5.64.090. Security. Each business licensed under this chapter shall meet the requirements of a "secure facility" pursuant to Section 12071(c)(3) of the California Penal Code.

5.64.100. Application for employee work permit.

A. Each employee who engages in the sale, lease or other transfer of firearms or firearm ammunition must obtain either of the following: (1) a valid employee work permit from the chief of police; or (2) a valid certificate of eligibility issued by the California Department of Justice. The work permit requirement does not apply to employees at gun shows that are properly licensed under state and federal law.

B. The application for an employee work permit for an employee engaged in the sale, lease or other transfer of firearms or firearm ammunition shall be submitted to the chief of police on a form provided by the chief of police. Prior to submitting such application, a nonrefundable fee as established by resolution of the city council shall be paid to the chief of police to defray, in part, the cost of investigation and report required by this section. The chief of police shall issue a receipt to the employee showing that such permit application fee has been paid. The issuance of such a receipt shall not authorize an employee to engage in the sale, lease or other transfer or firearms or firearm ammunition until the necessary work permit has been lawfully granted.

C. The application to the chief of police shall be signed under penalty of perjury and set forth the following information:

1. The name, age and address of the employee;

2. The business name and address of the employer;

3. All convictions of the applicant for any misdemeanor or felony;

4. All information relating to licenses or permits relating to firearms or other weapons sought by the applicant from other jurisdictions, including, but not limited to the date(s) of application(s) and whether each such application resulted in issuance of a license or permit;

5. All information relating to suspension(s) or revocation(s) of licenses or permits relating to firearms, including but not limited to the date and circumstances of the suspension(s) or revocation(s);

6. Other reasonable conditions deemed appropriate by the chief of police.

D. The chief of police shall take the fingerprints of the employee and shall confirm the information submitted.

E. The chief of police shall act upon and either approve or deny the application for an employee work permit at the earlier of the following: (1) within one hundred eighty (180) days of receipt of the application; or (2) within fifteen (15) days of receipt of fingerprint verification by the State Department of Justice.

F. The chief of police shall deny the issuance of an employee work permit when any of the following conditions specified in Sections 5.64.060(D) through (G) of this chapter exist.

G. Duration and renewal of work permit. Unless revoked on an earlier date, all employee work permits issued pursuant to this chapter shall expire one year after the date of issuance; provided, however, that such work permits may be renewed by the chief of police for additional one-year periods upon the approval of an application for renewal by the chief of police and payment of the renewal fee established by resolution of the city council. Such renewal application shall set forth the information listed herein, and must be received by the chief of police, in completed form, no later than forty-five (45) days prior to the expiration of the current employee work permit.

H. Nonassignability. The assignment of any employee work permit issued pursuant to this chapter is unlawful and any such assignment shall render the license null and void. In addition, the attempt to transfer or assign any employee work permit issued pursuant to this chapter shall be further grounds for revocation.

I. Grounds for Revocation of Work Permit. In addition to any other provisions of this sec-tion, any circumstances constituting grounds for denial of an employee work permit shall also constitute grounds for revocation of an employee work permit.

5.64.110. License - Conditions. In addition to all other requirements and conditions stated in this chapter, each licensee shall be subject to all of the following conditions, the breach of any of which shall be sufficient cause of revocation of the license by the chief of police:

A. The business shall be carried on only in the building located at the street address shown on the license, except for gun shows as permitted under state and federal law.

B. The licensee shall comply with Sections 12073, 12074, 12076, 12077 and 12082 of the California Penal Code, as those sections now read or may hereafter be amended to read.

C. The licensee shall verify that each employee engaged in the sale, lease or other transfer of any firearm or firearm ammunition has obtained either of the following: (1) a valid employee work permit from the chief of police; or (2) a valid certificate of eligibility issued by the California Department of Justice. An employee shall not engage in the foregoing activities if a work permit is denied or revoked by the chief of police. A licensee who permits an employee to work without a valid work permit or valid certificate of eligibility shall be in violation of this section.

D. The licensee shall not sell, lease or otherwise transfer any firearm, except for leases or transfers for theatrical purposes, without also selling or otherwise providing with each firearm a nonreturnable trigger lock. If the use of a trigger lock on a particular type of firearm is infeasible, then the licensee shall do the following: (1) demonstrate such infeasibility to the satisfaction of the chief of police; and (2) make available for sale a similar device, which is approved by the chief of police, that is designed to prevent the unintentional discharge of the firearm.

E. The licensee shall not sell, offer to expose for sale, give, lease or otherwise transfer to any person, any ammunition feeding device with the capacity to accept more than ten (10) rounds of ammunition.

F. The licensee shall not sell, lease or otherwise transfer any firearm to any person whom the licensee reasonably believes is within any of the classes prohibited by California Penal Code Sections 12021 or 12021.1 or California Welfare and Institutions Code Sections 8100 to 8103, inclusive, as those sections now read or may hereafter be amended to read.

G. The licensee shall comply with all federal, state, and local laws, and all provisions of this chapter.

Any license issued pursuant to this title shall be subject to such additional conditions as the chief of police finds are reasonably related to the purpose of this chapter.

5.64.120. Grounds for revocation. In addition to any other provisions of this chapter, any circumstances constituting grounds for denial of a license shall also constitute grounds for revocation of a license.

5.64.150. Authority to inspect. Any and all investigating officials of the city shall have the right to enter the building and entire premises designated in the license from time to time during regular business hours to make reasonable inspections and to observe and enforce compliance with building, mechanical, fire, electrical, or plumbing regulations to the extent provided by law. Police department employees, as designated by the chief of police, may conduct compliance inspections to insure conformance with all federal, state, and local laws, and all provisions of this chapter.

5.64.160 Compliance by existing dealers. Any person lawfully engaging in the business of selling, leasing, or otherwise transferring any firearm or firearm ammunition on the effective date of the ordinance codified in this chapter shall have a period of ninety (90) days after such effective date to comply with the provisions of this chapter.

5.64.170. Delivery to revenue manager. Upon granting said license, the police department shall forward said license to the revenue manager, who shall issue said license to the applicant upon payment of the license fee, payable in advance.

5.64.180. Rules and regulations. The chief of police is authorized to promulgate such rules and regulations after noticed hearing as will enable the police department to effectuate the purposes of this chapter. Failure to comply with such rules and regulations, or with any other requirements imposed by this chapter, shall constitute grounds for revocation of licenses issued hereunder.

Title 9. Public Peace, Morals and Welfare

Chapter 9.32. Weapons and Explosives

Article I. In General

9.32.010. "Dangerous or deadly weapon" defined. The term "dangerous or deadly weapon" includes, but is not limited to ... any firearm other than one carried pursuant to a valid permit, issued by a duly authorized governmental authority, or any ordinary rifle or shotgun lawfully carried for purposes of hunting or other lawful sport.

9.32.090. Prohibiting sale of air guns to minors.

A. It is unlawful for any person to sell, give or loan to any person under eighteen (18) years of age, any BB gun, air gun, pellet gun, or spring gun, within the city.

B. No person, under the age of eighteen (18) years, shall have in his or her possession or control any BB gun, air gun, pellet gun, or spring gun, within this city, except while in immediate custody of his or her parents as hereinafter provided.

C. The chief of police is authorized to establish training programs for minors, and also to license the establishment of shooting ranges and galleries in proper locations.

D. Any person under the age of eighteen (18) years may, while in custody of and under immediate control of his or her parent or guardian, use a BB gun, air gun, pellet gun, or spring gun, to shoot at an inanimate target on range or in a shooting gallery which has been authorized by the chief of police.

E. Any person desiring to operate and maintain a shooting range for BB guns, air guns, pellet guns, or spring guns, within the city, shall first apply to the chief of police for a permit, on an application form to be furnished by the chief of police. If, upon investigation, it appears the range can be maintained at the location specified in the application, without danger to the public, then and in that event the chief of police may issue a permit.

9.32.110. Explosives generally - Storage for sale in the city. There shall not be kept within the city more than fifty (50) pounds of powder, or other explosive substance, by any one person, and every person keeping such articles shall keep all they have in one metal case, with metal cover and handles, plainly marked "POWDER," which shall be kept near the front door, and shall display conspicuously upon the front of the building a sign plainly marked "GUNPOWDER KEPT HERE."

9.32.120. Explosives generally - Applicability of two preceding sections. Nothing in the two preceding sections shall apply to metallic cartridges or percussion caps.

9.32.150. Public grounds - Discharge of firearms, sales, shows, etc. No person shall, in or upon any public grounds, discharge any cannon or firearm, expose for sale any goods, wares or merchandise or erect or maintain any booth, stand or show; except, in accordance with a permit from the city manager.

9.32.180 Reporting of stolen and/or lost firearms required.

It is unlawful for any person who owns or possesses a firearm to knowingly or negligently fail to report the theft or loss of such firearm to the Sacramento police department within forty-eight (48) hours of the time he or she knew or should have known the firearm has been stolen or lost, when either the owner or possessor resides in the city, or the theft or loss of the firearm occurs in the city. As used in this section, "firearm" means any device, designed to be used as a weapon or modified to be used as a weapon, which expels a projectile through a barrel by the force of an explosion or other form of combustion. Any person violating this section is guilty of a misdemeanor.

[Sacramento City Code current through Ordinance 2008-036, Aug., 2008]

San Diego County Code

Title 2. Licenses, Business Regulations and Business Taxes

Division 1. Business Regulations

Chapter 12. Firearms Dealers

21.1201. Purpose and Intent. California Penal Code sections 12070-12083 regulate the sale of firearms and provide that the duly constituted licensing authority of a county shall accept applications for and may grant licenses permitting licensees to sell firearms at retail within its jurisdiction. This chapter appoints the Sheriff as the licensing authority for retail firearms sales in the unincorporated area of the County and implements these sections of the Penal Code applicable to the County.

21.1202. Definitions., For the purposes of this chapter the following definitions shall apply:

(a) "Firearm" means any device defined as a firearm in Penal Code section 12001.

(b) "Firearms dealer" means a person who meets the requirements of Penal Code section 12071(a)(1) and to whom a license has been issued pursuant to this chapter.

(c) "Firearms dealer employee" means a person who works for a firearms dealer as an agent, employee or representative.

21.1203. License Required. The licensing authority for firearm dealers in the unincorporated area of the County is the Sheriff. It shall be unlawful for any person to sell, lease or transfer a firearm in the unincorporated area of the County without a license issued by the Sheriff pursuant to Penal Code section 12071. This section shall not apply to a firearms transaction excluded under Penal Code section 12070. A person requesting a license under this chapter shall submit an application on a form provided by the Sheriff. A firearms dealer's license is governed by sections 21.101-21.117 and any additional conditions in this chapter.

21.1204. Minimum Age for Firearms Dealer. The minimum age to be a firearms dealer is 21.

21.1205. Additional Grounds for Denial of License. In addition to the grounds for denial of a license under section 21.108 the Sheriff may deny a firearms dealer's license if the dealer fails to meet any requirement under Penal Code section 12071 or if the Sheriff determines the applicant:

(a) Is disqualified from owning or possessing a firearm under federal, State or County laws or regulations.

(b) Has within five years preceding the date of the application been convicted of any offense relating to the manufacture, sale, possession, use or registration of any firearm or any "dangerous weapon" or "deadly weapon," as those terms are defined under federal law, or the law of any state.

21.1206. Clearance for Firearms Dealer Employee.

(a) No person shall be employed as an employee of a firearms dealer without a written clearance from the Sheriff. An employee of a firearms dealer is required to renew the written clearance from the Sheriff annually.

(b) A person seeking to be employed by a firearms dealer shall submit fingerprints and complete a form provided by the Sheriff. A person requesting clearance under this section shall be subject to investigation under section 21.107. The Sheriff may deny the clearance based upon any of the grounds in sections 21.108 and 21.1205. The Sheriff shall issue or deny the clearance in writing within 30 days after a completed request for clearance has been submitted. The Sheriff may suspend or revoke a person's clearance under this section on any of the grounds for which a license or permit may be suspended or revoked under section 21.112.

(c) It shall be unlawful for a firearms dealer to employ any person who does not have a current clearance from the Sheriff to be a firearms dealer employee.

21.1207. No Transactions at Unauthorized Locations. Except as provided in Penal Code section 12071(b)(1)(B) and (C), a firearms dealer shall only conduct firearms transactions at the location specified in the license.

21.1208. Transaction Records and Inspection. A firearms dealer shall maintain all firearms transaction records required by federal and State law and have them available for inspection. During business hours, a firearms dealer shall allow any Sheriff's Department employee and any peace officer to inspect firearms transaction records, firearms, firearm accessories and all places where transaction records, firearms and firearm accessories are stored.

21.1209. – Compliance with Requirement for Secondhand Goods Dealers. A firearms dealer licensed under this chapter who acquires a secondhand or used firearm for resale shall comply with sections 21.701-21.706 and State law requirements for acquisition and sales of secondhand goods.

[San Diego County Code current through September 5, 2008]

San Diego Municipal Code

Chapter 3. Business Regulations, Business Taxes, Permits and Licenses

Article 3. Police Regulated Business Regulations

Division 42. Regulation of Firearm Dealers

33.4201. Permit Required.

(a) It is unlawful for any person to engage in business as a Firearm Dealer without a Firearm Dealer permit issued by the Chief of Police.

(b) It is unlawful for any person to sell, deliver, or otherwise transfer any firearm in violation of state or federal law or without the permit required by Section 33.4201.

33.4202. Definitions.

For the purpose of this Division, the following words mean:

"Chief of Police" means the Chief of Police or the Chief's designated representative.

"Firearm" means any device designed to be used as a weapon, from which a projectile is expelled through a barrel by the force of any explosion or other form of combustion. A "Firearm" includes any device defined as a firearm in California Penal Code Section 12002.

"Firearm Dealer" means any person who obtains a Federal Firearms Dealers License for the business of:

(a) Selling, transferring, or leasing any new or used Firearms at wholesale or retail; or

(b) advertising for sale, transfer, or lease any new or used Firearms at wholesale or retail; or

(c) offering or exposing for sale, transfer, or lease, any new or used Firearms at wholesale or retail.

"Person" has the same meaning as that in San Diego Municipal Code section 11.0210.

33.4203. Application for Permit.

(a) The Chief of Police shall prescribe the application form for a Firearm Dealer permit.

(b) Any person proposing to engage in business as a Firearm Dealer within the City of San Diego shall secure the applicable federal license and shall have thirty (30) calendar days after receipt of such federal license to apply for a City permit in compliance with the provisions of this Division.

33.4204. Issuance or Denial of Permit. Except as otherwise provided in this Code, upon completion of the background investigation of the applicant, the Chief of Police or designated representative shall issue the permit unless:

(a) The applicant has knowingly made false or misleading statements of a material fact or omitted a material fact in the application; or

(b) The operation of the Firearm Dealer business as proposed in the application for the permit will violate any applicable building, fire, health or zoning requirement set forth in this Code; or

(c) The applicant is under twenty one (21) years of age; or

(d) The applicant had a similar permit or license previously revoked or denied for good cause within one year immediately preceding the date of the filing of the application, or

(e) The applicant has not been licensed as a Firearms Dealer as required federal law and the State of California; or

(f) The applicant, within five (5) years immediate preceding the date of the filing of the application has been convicted in a court of competent jurisdiction of any of the following offenses:

(1) Any offense involving the use of force or violence upon the person of another; or

(2) Any offense of theft, embezzlement, or receiving of stolen property; or

(3) Any felony offense involving the sale, manufacture, possession or use of any controlled substance as defined by the California Health and Safety Code; or

(4) Any offense committed in another state which, if committed in this state, would have been punishable as one of the offenses specified above and immediately preceding; or

(g) The applicant is under indictment for, or has been convicted of, any violation of any federal, state, or local law relating to manufacture, sale, transfer, lease, registration, use or possession of any firearm or ammunition; or

(h) The applicant fails to remove the authority of any officer, agent or employee to act on behalf of the applicant in the Firearms Dealer business within five (5) working days after the applicant receives written notification by certified mail or personal delivery from the Chief of Police, that:

(1) any officer, employee, or agent of the applicant is under indictment for, or has been convicted of, any violation of federal, state or local law relating to the manufacture, sale, transfer, lease, registration, use or possession of any firearm or ammunition; or

(2) any officer, employee or agent of the applicant is a person in a prohibited class described in section 12021 or 12021.1 of the California Penal Code or section 8100 or 8103 of the Welfare and Institutions Code; or

(i) The applicant is a person in a prohibited class described in section 12021 or 12021.1 of the California Penal Code or section 8100 or 8103 of the Welfare and Institutions Code; or

(j) The applicant has failed to provide evidence of a possessory interest, such as the interest of an owner, tenant, lessee or sublessee, in the property where the proposed business will be conducted; or

(k) The applicant has failed to obtain a zoning use certificate required by this Code.

33.4206. Right of Inspection. Firearm Dealers shall have their places of business open to Police Department inspection during all hours of operations. Firearm Dealers shall maintain all Firearms records and documents, Firearms and storage places in a manner and place accessible for inspection by Police Department personnel.

33.4207. Permit Not Transferable. A Firearm Dealer permit may be issued only to a specific person to conduct business as a dealer at a specific location and at gun shows in accordance with California Penal Code section 12071. It is unlawful for any person to transfer a Firearm Dealer permit to another person or from one location to another without prior written approval of the Chief of Police. Any attempted transfer shall be ineffective.

33.4208. Permit Restricted to One Location. The Firearm Dealer business may be carried on only in the location designated in the permit. Only one Firearm Dealer permit may be issued per location.

33.4209. Display of Permit Required. The Firearm Dealer permit or a copy thereof, certified by the issuing authority, shall be displayed on the premises where it can be easily read.

33.4210. Business and Security Regulations.

(a) All Firearm Dealers and officers, employees or agents of the Firearm Dealers, shall comply with all provisions of California Penal Code section 12071, which include business regulations and the building specifications for Firearm security.

(b) All sellers of Firearms shall protect such Firearms from theft during business hours in the following manner:

(1) All Firearms shall be in locked cabinets, a secure rack, or a storage area so that access to firearms is controlled by the dealer or an employee, to the exclusion of all others.

(2) The Firearm Dealer, agent, or employee shall be present when a prospective buyer or seller is handling any Firearm.

33.4212. Provision of Trigger Locks.

(a) It is unlawful for any firearm dealer to sell, give, lend or transfer ownership of any firearm without also providing with the firearm a trigger locking or similar device approved by the Chief of Police. The trigger locking or similar device shall be designed to prevent the unintentional discharge of the firearm. The trigger locking or similar device shall be attached to the firearm prior to the completion of the sale, gift, lending or transfer transaction, and the firearm must leave the premises of the firearms dealer with the trigger locking or similar device attached. If a trigger locking or similar device cannot be attached because the firearm lacks a trigger guard, a lockable bag or box shall be used in lieu of a trigger locking or similar device. Nothing in this section shall prevent the firearms dealer from recovering the cost of the trigger locking or similar device by charging a price for the lock, either separately or as part of the price of the firearm.

(b) It is unlawful for any firearm dealer sell, give, lend or transfer ownership of any firearm, without providing with the firearm printed material, approved by the Chief of Police, that advises the user of safe firearm storage practices.

(c) In connection with any sale, gift, loan or transfer of a firearm, the firearm dealer shall obtain from the person receiving the firearm a signed acknowledgment that a trigger locking or similar device required by section 33.4212(a) of this Municipal Code, and the printed material required by section 33.4212(b), has been provided with the firearm. The acknowledgment shall be in a form approved by the Chief of Police. The firearm dealer shall retain signed acknowledgments of receipt in the same manner and to the same extent as required by State law for other firearm transaction records.

(d) A firearm dealer who has fully complied with the provisions of this section relating to trigger locking or similar devices shall not be presumed to have made any representation to the transferee regarding the safety or appropriateness of the use of the trigger locking or similar device, nor shall the firearm dealer be liable in any civil action brought against the firearm dealer, to the extent such liability would be based solely upon the act of furnishing the trigger locking or similar device to a person in compliance with this section.

(e) The requirements of section 33.4212 shall be in force and effect, and shall apply to all firearms dealers now or hereafter licensed, ninety days after the date on which the Chief of Police sends written notice to currently-licensed firearms dealers of the approved trigger locking or similar devices, printed safety material and form of acknowledgment.

(f) The provisions of this section shall not apply to temporary lending transaction in which the firearm does not leave the premises of the firearms dealer.

(g) The requirements in sections 33.4212 (a), (b), and (c) shall not apply to firearms that are curios or relics, as defined by federal law, Title 27 Code of Federal Regulations section 178.

(h) The requirements in sections 33.4212(a), (b) and (c) shall not apply to firearms dealers when they are involved in private party transfers conducted pursuant to California Penal Code section 12082.

Chapter 5. Public Safety, Morals and Welfare

Article 3. Firearms - Dangerous Weapons - Explosives - Hazardous Trades

53.15.2. Possession of Air Guns, Pointed Missiles, Etc. by Minors.

(a) Except as provided in this section, it is unlawful for any person under the age of eighteen to have in his or her possession, upon any public street or sidewalk or in any public gathering place within the corporate limits of the City any air gun, sharp pointed missile, dart or arrow, described in Municipal Code section 53.15.

(b) This section shall not apply to any minor engaged in supervised or otherwise lawful activity involving such weapons, or who is going to or returning from a place where the minor was engaged in such supervised or otherwise lawful activity.

53.16. Penalties for Firearms and Other Weapons Offenses. ...

(c) Any parent or legal guardian, or person over the age of eighteen, is also guilty of a misdemeanor, if:

(1) he or she possesses on any premises or within a vehicle under his or her custody or control, any of the firearms identified in Municipal Code sections 53.10, and

(2) he or she knows or reasonably should know that a minor is likely to gain access to such firearm, and

(3) a minor obtains and fires or discharges such firearm in violation of sections 53.10.

(d) A violation of Municipal Code sections 53.10, 53.15 or 53.15.1 occurring within 1,500 feet of a private or public day care center, school or school grounds, is a misdemeanor, and the penalty for conviction of the same is punishable by a fine of not more than $1,000, or by imprisonment in the county jail for not more than one year, or both. A violation of sections 53.10, 53.15 or 53.15.1 may also be prosecuted in accordance with Chapter 1 of the Municipal Code.

53.31. Assault Weapons, Nuisance, Temporary Custody.

(a) The term "assault weapon," as used in this Section, shall include:

(1) Any semiautomatic action, centerfire rifle or carbine which accepts a detachable magazine with a capacity of twenty (20) rounds or more, with a barrel of more than sixteen (16) inches, including but not limited to the following firearms or their copies: AR 15 semiautomatic assault rifles, Uzi semiautomatic assault rifles or carbines, Ingram Mac-10 semiautomatic assault carbines, Ingram Mac-11 semiautomatic assault carbines, Heckler and Koch 93 semiautomatic assault rifles, Heckler and Koch 91 semiautomatic assault rifles, AK-47 semiautomatic assault rifles, AKM-47 semiautomatic assault rifles, all Avtomat Kalashnikov weapons, M1-A semiautomatic assault rifles, M-14 semiautomatic assault rifles, Thompson semiautomatic carbines and any other semiautomatic carbines manufactured by Auto Ordnance;

(2) Any shotgun with a barrel of more than eighteen (18) inches and a folding stock or magazine capacity of more than six (6) rounds;

(3) Any weapon which may be readily restored to an operable assault weapon, as defined in paragraphs 1 and 2 above; and

(4) Any part, or combination of parts, designed or intended to convert a weapon into an assault weapon, as defined above in paragraphs 1 and 2, or any combination of parts from which an assault weapon may be readily assembled, if those parts are in the possession or under the control of the same person.

(b) As used in this Section, the term "semiautomatic" means a weapon which fires a single projectile for each single pull of the trigger and which employs a magazine.

(c) The term "assault weapon" does not include any of the following:

(1) Any of the above generally and specifically described weapons which is a "machine gun" as that term is defined by Section 12200 of the Penal Code of the State of California; any pistol, revolver or other firearm which is capable of being concealed upon one's person, as defined and regulated by the provisions of Sections 12021 and 12025 of the Penal Code of the State of California;

(2) Any of the following: weapons which do not use fixed ammunition, weapons which were manufactured prior to 1898, manually operated bolt action weapons, lever action weapons, slide action weapons, single-shot weapons. Multiple-barrel weapons, revolving cylinder weapons, semiautomatic weapons which use exclusively Mannlicher-style clips, semiautomatic weapons manufactured prior to 1954, rimfire weapons that employ a tubular magazine;

(3) Any assault weapon which is an antique or relic firearm or other weapon falling within the specifications of paragraphs (5) (7) and (8) of Subsection (b) of Section 12020 of the Penal Code of the State of California;

(4) Any short-barreled rifle or shotgun as defined in Subsection (c) of Penal Code Section 12020; or

(5) Any assault weapon which has been modified either to render it permanently inoperable or to permanently make it a device no longer defined as an assault weapon.

(d) A sheriff, undersheriff, deputy sheriff, marshal, deputy marshal, or police officer, as defined in subdivision (a) of Section 830.1, a member of the University of California Police Department, as defined in subdivision (d) of Section 830.2, and a member of a California State University Police Department, as defined in subdivision (e) of Section 830.2, during the performance of his or her duties within the City of San Diego may take temporary custody of any assault weapon in plain sight or discovered pursuant to a consensual search as necessary for the protection of the peace officer or other persons of the City of San Diego.

(e) The provisions of subsection (d) shall not apply to assault weapons in the possession of any person specifically authorized to possess concealed or loaded firearms within the City limits of San Diego pursuant to the Dangerous Weapons' Control Law, California Penal Code Section 1200 et seq.

(f) Upon taking custody of an assault weapon, the officer shall give the owner or person who possessed the assault weapon a receipt. The receipt shall describe the assault weapon and list any identification or serial number on the assault weapon. The receipt shall indicate where the firearm can be recovered and the date after which the owner or possessor can recover the assault weapon. No assault weapon shall be held less than forty-eight (48) hours. If an assault weapon is not retained for use as evidence related to criminal charges or is not retained because it was illegally possessed, the assault weapon shall be made available to the owner or person who was in lawful possession forty-eight (48) hours after the seizure or as soon thereafter as possible, but no later than seventy-two (72) hours after the seizure.

(g) Any assault weapon which has been taken into custody which has been stolen shall be restored to the lawful owner, as soon as its use for evidence has been served, upon his or her identification of the firearm and proof of ownership.

(h) Any assault weapon taken into custody and held by a police, university police, or sheriff's department or by a marshal's office, for longer than twelve (12) months and not recovered by the owner or person who has lawful possession at the time it was taken into custody, shall be considered a nuisance and sold or destroyed as provided in subdivision (c) of Section 12028.

[San Diego Municipal Code current as of May 2008]

San Francisco Police Code

Article 9. Miscellaneous Conduct Regulations

613. Regulating Sale of Firearms. It shall be unlawful for any person, firm, corporation or dealer engaging in the business of selling, leasing or otherwise transferring any firearm, firearms ammunition, or firearms ammunition component to sell, lease or otherwise transfer any firearm, firearms ammunition or firearms ammunition component without first having obtained a license from the San Francisco Police Department. The Department shall make available application forms requiring applicants to provide the information set forth in Section 613.2, and shall collect a nonrefundable application fee from each applicant.

The Chief of Police shall recommend to the Board of Supervisors, on or before April 1, 1994, a fee which shall be sufficient to recover all costs associated with regulating the sale of firearms under this Article, including but not limited to, processing applications, monitoring licensees, and enforcing the provisions of this Article. The fee shall be set by the Board of Supervisors.

613.1. Definitions.

(a) "Firearm" shall mean any device, designed to be used as a weapon or modified to be used as a weapon, that expels a projectile by the force of an explosion or other form of combustion.

(b) "Firearm ammunition" shall mean any cartridge or encasement containing a bullet or projectile, propellant or explosive charge, and a primer which is used in the operation of a firearm.

(c) "Firearm ammunition component" shall mean any cartridge or encasement, bullet or projectile, primer or propellant or explosive material used in the manufacture of ammunition.

(d) "Firearm capable of being concealed upon the person" shall mean any such firearm as defined in California Penal Code Section 12001(a).

(e) "Transfer" shall include, but shall not be limited to, the redemption of a pawned or pledged firearm by any individual including the individual who pawned or pledged the firearm.

(f) "Ultracompact firearm" shall mean any pistol, revolver, handgun or other firearm that is 6.75 inches or less in length or 4.5 inches or less in height, measured with the magazine detached, but shall not include any such firearm that is an unsafe handgun defined by Section 12126 of the California Penal Code.

613.2. Application Form and Background Check Requirements.

(a) The application for a license to engage in the business of selling, leasing or otherwise transferring any firearm, firearms ammunition, or firearms ammunition component shall be signed under penalty of perjury and shall set forth:

(1) The name, age and address of the applicant, as well as the name, age and address of all persons who will have access to or control of workplace firearms, firearm ammunition, or firearm ammunition components, including but not limited to, the applicant's employees, agents and or supervisors, if any;

(2) The address of the location for which the license is required, together with the business name of such location, if any;

(3) All convictions of the applicant for any of the offenses listed in Section 613.3(e);

(4) All information relating to licenses or permits relating to firearms or other weapons sought by the applicant from other jurisdictions, including, but not limited to date of application and whether each such application resulted in issuance of a license;

(5) All information relating to revocations of licenses or permits relating to firearms, including but not limited to date and circumstances of revocation;

(6) Applicant's agreement to indemnify, defend, and hold harmless the City, its officers, agents, and employees, from and against any and all claims, losses, costs, damages and liabilities of any kind, arising in any manner out of the applicant's negligence or intentional or willful misconduct;

(7) Applicant's understanding that the City shall have the right to enter the building designated in the license from time to time during regular business hours to make reasonable inspections and to investigate and enforce compliance with building, mechanical, fire, electrical, plumbing, or health regulations, provisions of this Article, and all other applicable federal, state, and local laws.

(b) Each application must be accompanied by evidence that the applicant has satisfied the insurance requirements stated in Section 613.13 of this Article.

(c) All persons listed on the application form as having access to, or control of, workplace firearms, firearm ammunition, or firearm ammunition components shall obtain a certificate of eligibility under Penal Code Section 12071 from the state Department of Justice. A copy of the certificate of eligibility for each such person shall be submitted with the application.

(d) The Chief of Police shall conduct a criminal history background check on the applicant and on all other persons listed on the application form as having access to, or control of, workplace firearms, firearm ammunition, or firearm ammunition components, including but not limited to, the applicant's employees, agents and/or supervisors, if any, and shall determine whether such persons have been convicted of any offenses described in subsection (e) of Section 613.3, or are among the persons described in subsections (f) or (g) of Section 613.3. Where the Chief of Police determines that one or more of the applicants, employees, agents or supervisors have been convicted of an offense described in subsection (e) of Section 613.3, or are among the persons described in subsections (f) or (g) of Section 613.3, the applicant shall have 21 days from the mailing of written notification from the Chief of Police to provide evidence in a form acceptable to the Chief of Police that such persons have been removed or reassigned so that they no longer have access to or control of workplace firearms, firearm ammunition, or firearm ammunition components. In the event that an applicant fails to comply with this subsection, the Chief of Police shall deny the license.

(e) As used in this Section, the term "applicant" when the applicant is other than a natural person shall include any officer, director, employee or agent of the applicant who has access to, or control of, workplace firearms, firearm ammunition, or firearm ammunition components.

613.3. Denial of License. The Chief of Police shall deny the issuance of a license when any of the following conditions exist:

(a) The applicant is under the age of 21 years.

(b) The applicant is not licensed as required by all applicable federal, State and local laws.

(c) The applicant has had a firearms license previously revoked or denied for good cause within the immediately preceding two years.

(d) The applicant has made a false or misleading statement of a material fact, or omission of a material fact, in the application for a license. If a license is denied on this ground, the applicant shall be prohibited from reapplying for a license for a period of two years.

(e) The applicant has been convicted of:

(1) Any offense so as to disqualify the applicant from owning or possessing a firearm under applicable federal, State, and local laws, including but not limited to the offenses listed in California Penal Code Section 12021;

(2) Any offense relating to the manufacture, sale, possession, use, or registration of any firearm or dangerous or deadly weapon;

(3) Any offense involving the use of force or violence upon the person of another;

(4) Any offense involving theft, fraud, dishonesty, or deceit, including but not limited to any of the offenses listed in Title 7 (Crimes Against Public Justice) and title 13, Chapters 4 (Forgery and Counterfeiting), 5 (Larceny), 6 (Embezzlement), 7 (Extortion), 8 (False Personation), 13 and 14 (Fraud) of the California Penal Code;

(5) Any offense involving the manufacture, sale, possession or use of any controlled substance as defined by the California Health and Safety Code as said definition now reads or may hereafter be amended to read.

(f) The applicant is within the classes of persons defined in California Welfare and Institutions Code Sections 8100 or 8103.

(g) The applicant is (1) currently, or has been within the past two years, an unlawful user of any controlled substance as defined by the California Health and Safety Code as said definition now reads or may hereafter be amended to read; or (2) an excessive user of alcohol, to the extent that such use would impair his or her fitness to be a dealer in firearms.

(h) The operation of the business as proposed would not comply with all applicable federal, State, and local laws.

(i) The applicant, or an officer, employee, or agent thereof, proposes to operate in the following locations:

(1) Within any RH, RM, RC, NC or RED zoning district, or within 1,000 feet of the exterior limits of any such district;

(2) Within 1,000 feet of a public or private day care center or day care home, or within 1,000 feet of any elementary, junior high or high school whether public or private;

(3) On or within 1,500 feet of the exterior limits of any other premises used as a place of business by a dealer in firearms;

(4) Within 1,000 feet of a community center, church, neighborhood center, recreational center, whether public or private, where regularly scheduled activities are conducted for people under 18 years of age.

(j) The applicant, or an officer, employee or agent thereof does not have, and/or cannot provide evidence of a possessory interest in the property at which the proposed business will be conducted.

(k) Any other ground for denial exists under any applicable provision of federal, State or local law.

(l) The applicant fails to comply with the requirements of subsections (c) or (d) of Section 613.2.

As used in this Section, the term "applicant" when the applicant is other than a natural person shall include any officer, director, employee or agent of the applicant who has access to, or control of, workplace firearms, firearm ammunition or firearm ammunition components.

613.8. Nonassignability of License. The assignment or attempt to assign any license issued pursuant to this Article is unlawful and any such assignment or attempt to assign a license shall render the license null and void.

613.9. Security. In order to discourage the theft of firearms stored on the premises of a fire-arms dealer, each business licensed under this Article must adhere to security measures as re-quired by the Chief of Police. Security measures shall include but not be limited to:

(a) Provision of secure locks, windows and doors, adequate lighting, and alarms as specified by the Chief of Police;

(b) Storage of all firearms on the premises out of the reach of customers in secure, locked facilities, so that access to firearms shall be controlled by the dealer or employees of the dealer, to the exclusion of all others.

613.10. License--Conditions. In addition to all other requirements and conditions stated in this Article, each license shall be subject to all of the following conditions, the breach of any of which shall be sufficient cause for revocation of the license by the Chief of Police:

(a) The business shall be carried on only in the building located at the street address shown on the license, except as otherwise authorized under Section 12071(b)(1) of the California Penal Code.

(b) The licensee shall comply with Sections 12073, 12074, 12076, 12077 and 12082 of the California Penal Code, to the extent that the provisions remain in effect.

(c) The licensee shall not deliver any pistol or revolver to a purchaser earlier than 10 days after the application for the purchase, lease or transfer, unless otherwise provided by State or federal law.

(d) The licensee shall not deliver any firearm to a purchaser, lessee or other transferee unless the firearm is unloaded and securely wrapped or unloaded in a locked container.

(e) The licensee shall not deliver any firearm, firearm ammunition, or firearm ammunition component to a purchaser, lessee or other transferee unless the purchaser, lessee or other transferee presents clear evidence of his or her identity and age to the seller. As used in this Section, "clear evidence of his or her identity and age" includes, but is not limited to, a motor vehicle operator's license, a State identification card, an armed forces identification card, an employment identification card which contains the bearer's signature and photograph, or any similar documentation which provides the seller reasonable assurance of the identity and age of the purchaser.

(f) The licensee shall not display in any part of the premises where it can be readily seen from outside the premises, any firearm, firearm ammunition or imitation thereof, or placard advertising the sale or other transfer thereof, other than a sign identifying the name of the business.

(g) The licensee shall not sell, lease or otherwise transfer to any person any ammunition that:

(1) Serves no sporting purpose;

(2) Is designed to expand upon impact and utilize the jacket, shot or materials embedded within the jacket or shot to project or disperse barbs or other objects that are intended to increase the damage to a human body or other target (including, but not limited to, Winchester Black Talon, Speer Gold Dot, Federal Hydra-

Shok, Hornady XTP, Eldorado Starfire, Hollow Point Ammunition and Remington Golden Sabre ammunition; or

(3) Is designed to fragment upon impact (including, but not limited to, Black Rhino bullets and Glaser Safety Slugs).

This subsection does not apply to conventional hollow-point ammunition with a solid lead core when the purchase is made for official law enforcement purposes and the purchaser is authorized to make such a purchase by the director of a public law enforcement agency such as the Chief of the San Francisco Police Department or the Sheriff of the City and County of San Francisco.

(h) The licensee shall post within the licensee's premises a notice stating the following:

"THE CALIFORNIA PENAL CODE PROHIBITS THE SALE OF FIREARMS OR FIREARMS AMMUNITION TO PERSONS UNDER THE AGE OF 18, AND FURTHER GENERALLY PROHIBITS THE SALE OF A PISTOL, REVOLVER, OR FIREARM CAPABLE OF BEING CONCEALED UPON THE PERSON TO ANY PERSON UNDER THE AGE OF 21."

The posted notice shall be in a conspicuous location, shall be in 36 point type block letters in black ink on a white background, and shall be located so that the notice can easily and clearly be seen by all prospective purchasers of firearms and firearm ammunition.

(i) The licensee shall not sell, lease or otherwise transfer any ultracompact firearm except as authorized by Section 613.10-2 or any 50 caliber firearm or 50 caliber cartridge except as authorized by Section 613.10-1.

(j) Any license issued pursuant to this Article shall be subject to such additional conditions as the Chief of Police finds are reasonably related to the purpose of this Article.

(k) The licensee shall comply with the requirements of Section 613.10-3 and shall, in addition, post the appropriate notice or notices, as specified below, in a conspicuous location at the entrance of the licensee's premises (or at the entrance to the separate room or, enclosure pursuant to Section 613.10-3(c)). Such notice shall be in 36 point type block letters in black ink on a white background.

(1) Licensees that sell, lease or otherwise transfer firearms, other than firearms capable of being concealed on the person, shall post a notice at the entrance to the premises (or at the entrance to the separate room or enclosure pursuant to Section 613.10-3(c)) stating the following:

"THE SAN FRANCISCO POLICE CODE REQUIRES THAT FIREARMS DEALERS PROHIBIT ENTRY BY PERSONS UNDER AGE 18, AND FURTHER PROHIBITS ENTRY BY (1) PERSONS CONVICTED OF A VIOLENT OFFENSE WHO ARE PROHIBITED FROM POSSESSING FIREARMS PURSUANT TO CALIFORNIA PENAL CODE SECTIONS 12021 OR 12021.1; AND (2) PERSONS WHO ARE CURRENTLY PROHIBITED FROM POSSESSING FIREARMS BECAUSE THEY HAVE BEEN ADJUDICATED AS MENTALLY DISORDERED, NOT GUILTY BY REASON OF INSANITY OR INCOMPETENT TO STAND TRIAL."

(2) Licensees that sell, lease or otherwise transfer firearms capable of being concealed on the person shall post a notice at the entrance to the premises (or at the entrance to the separate room or enclosure containing such firearms pursuant to Section 613.10-3(c)) stating the following:

"THE SAN FRANCISCO POLICE CODE REQUIRES THAT FIREARMS DEALERS PROHIBIT ENTRY BY PERSONS UNDER AGE 21, AND FURTHER PROHIBITS ENTRY BY (1) PERSONS CONVICTED OF A VIOLENT OFFENSE WHO ARE PROHIBITED FROM POSSESSING FIREARMS PURSUANT TO CALIFORNIA PENAL CODE SECTIONS 12021 OR 12021.1; AND (2) PERSONS WHO ARE CURRENTLY PROHIBITED FROM POSSESSING FIREARMS BECAUSE THEY HAVE BEEN ADJUDICATED AS MENTALLY DISORDERED, NOT GUILTY BY REASON OF INSANITY OR INCOMPETENT TO STAND TRIAL."

(3) Licensees that sell, lease or otherwise transfer firearms capable of being concealed on the person, but who keep such firearms in a separate room or enclosure in accordance with Section 613.10-3(c) shall post the notice required by paragraph (1) at the entrance to the premises or separate room or enclosure containing firearms that are not capable of being concealed on the person, and shall post the notice required by paragraph (2) at the entrance to the separate room or enclosure containing firearms capable of being concealed on the person.

(l) The licensee shall notify the Chief of Police of the name, age and address of, and submit a certificate of eligibility under Penal Code Section 12071 from the State Department of Justice for, any person not listed on the licensee's application under Section 613.2(a)(1) who will be given access to, or control of, workplace firearms, firearm ammunition, or firearm ammunition components. The licensee shall submit the required information and certificate within 10 days of such person being employed or otherwise being given access to, or control over workplace firearms, firearm ammunition, or firearm ammunition components.

(m) Within the first five business days of April and October of each year, licensees shall cause a physical inventory to be taken that includes a listing of each firearm held by the licensee by make, model, and serial number, together with a listing of each firearm the licensee has sold since the last inventory period. In addition, the inventory shall include a listing of each firearm lost or stolen that is required to be reported pursuant to Penal Code Section 12071(b)(13). Licensees shall maintain a copy of the inventory on the premises for which the license was issued. Immediately upon completion of the inventory, licensees shall forward a copy of the inventory to the address specified by the Chief of Police, by such means as specified by the Chief of Police. With each copy of the inventory, licensees shall include an affidavit signed by the licensee (or, if the licensee is not a natural person, by an officer, general manager, or other principal of the licensee) stating under penalty of perjury that within the first five business days of that April or October, as the case may be, the signer personally confirmed the presence of the firearms reported on the inventory.

613.10-1. Sale or Transfer of 50 Caliber Firearms and Cartridges Restricted.

(a) Findings...

(b) Purpose and Intent. The purpose and intent of this Section is to protect the health, safety, security and general welfare of the citizens of the City and County of San Francisco and the City's law enforcement personnel by reducing the risk of personal injury, death or property damage caused by 50 caliber firearms. Specifically with respect to regulating 50 caliber handguns, it is the purpose and intent of this Section to protect of the citizens of the City and County of San Francisco and the City's law enforcement personnel by reducing the risk of personal injury, death or property damage caused by persons using 50 caliber handguns. It is not the intent of this Section to address the problem of handgun safety, as addressed, for example in Sections 12125 through 12133 of the California Penal Code, or to otherwise regulate 50 caliber handguns based on consumer product safety considerations for the person using the handgun.

(c) Definitions.

(1) As used in this section, the term "50 caliber firearm" shall mean any firearm, as defined in Section 613.1 of this Article, capable of firing a centerfire 50 caliber cartridge.

(2) As used in this section, the term "50 caliber cartridge" shall mean a firearm ammunition cartridge in 50 caliber, either by designation or actual measurement, or any metric equivalent, including but not limited to a .50 BMG cartridge, that is capable of being fired from a centerfire rifle or a handgun. The term "50 caliber cartridge" does not include any memorabilia or display item that is filled with a permanent inert substance or that is otherwise permanently altered in a manner that prevents ready modification for use as live ammunition.

(3) As used in this section, the term "rifle" shall mean any firearm that is designed or redesigned, made or remade, and intended to be fired from the shoulder, and which is designed or redesigned, made or remade to use the energy in a fixed cartridge to fire only a single projectile through a rifled bore for each single pull of the trigger. The term "rifle" shall not include any shotgun.

(4) As used in this section, the term "handgun" shall mean any firearm with a barrel less than 16 inches in length. The term "handgun" shall include any pistol, revolver, or concealable firearm as such terms are defined in the California Penal Code.

(d) Sale or Transfer of 50 Caliber Firearms and Cartridges Restricted. No person shall sell, give, transfer, offer for sale or display for sale any 50 caliber firearm or 50 caliber cartridge except as authorized by paragraph (e) of this Section.

(e) Exceptions.

(1) The provisions of Subsection (d) of this Section shall not apply to:

(A) the sale or other transfer of a 50 caliber firearm or 50 caliber cartridge which is prohibited under state law, including, but not limited to, the sale or transfer of any assault weapon as defined in the California Penal Code;

(B) the sale or other transfer of a 50 caliber firearm which is an antique firearm, as defined in paragraph (16) of subsection (a) of Section 921 of Title 18 of the United States Code;

(C) the delivery of a 50 caliber firearm to a licensed gunsmith, as defined in California Penal Code Section 12001(r), or to a person licensed pursuant to California Penal Code Section 12071, for purposes of service or repair, or to the return of the firearm to its owner by the gunsmith or the licensee following the completion of service or repairs;

(D) the return of a 50 caliber firearm to its owner by a person licensed pursuant to this Article where the firearm was initially delivered to the licensee for the purpose of a consignment sale or as collateral for a pawnbroker loan; or

(E) the offer for sale or display in any periodical, solicitation by mail or use of the internet of a 50 caliber firearm or 50 caliber cartridge by any person or entity where the seller or offeror is located outside of the City and County of San Francisco.

(F) the loan of an unloaded 50 caliber firearm to a person for use solely as a prop in a motion picture, television, video, theatrical or other entertainment production or event, provided that such person is in compliance with Section 12081 of the California Penal Code.

(2) The provisions of Subsection (d) of this Section shall not apply to the sale or other transfer of a 50 caliber firearm or 50 caliber cartridge where the purchaser or transferee is any of the following:

(A) A law enforcement agency;

(B) An agency duly authorized to perform law enforcement duties;

(C) A state or local correctional facility;

(D) A federal law enforcement officer, provided such law enforcement officer is authorized to acquire a 50 caliber firearm and/or 50 caliber cartridges and does so while acting within the course and scope of his or her employment:

(E) A person described in Sections 12302 or 12322(a) of the California Penal Code; provided such person is authorized to acquire a 50 caliber firearm and/or 50 caliber cartridges and does so while acting within the course and scope of his or her employment;

(F) A person who is properly identified as a full-time paid peace officer as defined in Sections 830.1, 830.2, 830.4, or 830.5 of the California Penal Code, provided such peace officer is authorized to acquire a 50 caliber firearm and/or 50 caliber cartridge and does so while acting within the course and scope of his or her employment;

(G) A firearms dealer who has been issued a Federal Firearms License, and who is in compliance with the requirements of Section 12071 of the California Penal Code,

(H) A person or entity conducting a transaction described in subdivision (k) of Section 12078 of the California Penal Code:

(I) A person acquiring a 50 caliber firearm or 50 caliber cartridge by bequest or intestate succession, or otherwise by operation of law, provided that such firearms or cartridges are transferred to a law enforcement agency or to a person licensed pursuant to California Penal Code Section 12071 within 12 months from the date that the person obtains title;

(J) A non-profit entity that is authorized to destroy firearms, and which has agreed to destroy the firearm being transferred;

(K) A federal, state, or local historical society, museum, or institutional collection that is open to the public; provided that the 50 caliber firearm is used for display purposes, is unloaded, and is secured from unauthorized uses.

(f) Penalties.

(1) Violation of this Section shall be punishable as a misdemeanor. In addition, in the case of a violation of this section by a firearms dealer licensed under this Article, each violation of this Section shall constitute grounds for suspension or revocation of the licensee's firearms dealer license.

(2) Each transaction in violation of this Section shall be deemed a distinct and separate violation.

(g) Sunset Provision. If the State of California enacts legislation prohibiting or restricting the sale or transfer of 50 caliber rifles, 50 caliber handguns, or 50 caliber cartridges covered by this section, the requirements of this Section shall be suspended as of the date such legislation goes into effect with respect to any requirements of this Section that apply to the firearms or cartridges prohibited or restricted by state law. If the State of California subsequently repeals any such legislation, the suspension shall expire by operation of law and the previously suspended provisions shall become operative.

613.10-2. Sale of Ultracompact Firearms Restricted.

(a) Findings. ...

(b) Purpose and Intent. The purpose and intent of this Section is to protect the health, safety, and general welfare of the citizens of the City and County of San Francisco by reducing the potential for death or injury to citizens and law enforcement personnel attributable to ultracompact firearms. It is not the intent of this Section to address the problem of handgun safety, as addressed, for example, in Sections 12125 through 12133 of the California Penal Code, or to otherwise regulate ultracompact firearms based on consumer product safety considerations for the person using the firearm.

(c) Sale of Ultracompact Firearms Restricted. No person licensed pursuant to this Article shall sell, lease or otherwise transfer any ultracompact firearm except as authorized by paragraph (d) of this Section. Nothing in this section shall preclude any person licensed pursuant to this Article from processing firearms transactions between unlicensed parties pursuant to subdivision (d) of Section 12072 of the Penal Code of the State of California.

(d) Exceptions. The requirements of this Section shall not apply to the sale, lease or other transfer of an ultracompact firearm in the following circumstances:

(1) To any law enforcement agency;

(2) To any agency duly authorized to perform law enforcement duties;

(3) To any state or local correctional facility;

(4) To a federal law enforcement officer, provided such law enforcement officer is authorized to acquire an ultracompact firearm and does so while acting within the course and scope of his or her employment;

(5) To a private security company licensed to do business in the State of California;

(6) To a person described in Sections 12302 or 12322(a) of the California Penal Code, provided such person is authorized to acquire an ultracompact firearm and does so while acting within the course and scope of his or her employment;

(7) To any person who is properly identified as a full-time paid peace officer, as defined in Section 830.1, 830.2, 830.4, or 830.5 of the California Penal Code, provided such peace officer is authorized to acquire an ultracomplact firearm and does so while acting within the course and scope of his or her employment;

(8) To the sale, lease or other transfer of any antique firearm, as defined in paragraph (16) of subsection (a) of Section 921 of Title 18 of the United States Code.

(9) To the loan of an ultracompact firearm to a person for use solely as a prop in a motion picture, television, or video, theatrical or other entertainment production or event, provided that such person is in compliance with Section 12081 of the California Penal Code;

(10) To any person or entity conducting a transaction that is exempt from the provisions of subdivision (d) of Section 12072 of the California Penal Code;

(11) To any person or entity conducting a transaction described in subdivision (k) of Section 12078 of the Penal Code of the State of California.

(12) To a firearms dealer who has been issued a Federal Firearms License, and who is in compliance with the requirements of Section 12071 of the California Penal Code;

(13) To any person or entity acquiring an ultracompact firearm by bequest, intestate succession or otherwise by operation of law;

(14) To a non-profit entity that is authorized to destroy firearms, and which has agreed to destroy the firearm being transferred.

(15) To a federal, state, or local historical society, museum, or institutional collection that is open to the public, provided that the ultracompact firearm is used for display purposes, is unloaded, and is secured from unauthorized use;

(16) To the delivery of an ultracompact firearm to a licensed gunsmith, as defined in California Penal Code Section 12001(r), or to a person licensed pursuant to California Penal Code Section 12071, for purposes of service or repair, or to the return of the firearm to its owner by the gunsmith or by licensee following the completion of service or repairs;

(17) To the return of an ultracompact firearm to its owner by a person licensed pursuant to this Article where the firearm was initially delivered to the licensee for the purpose of a consignment sale or as collateral for a pawnbroker loan.

(e) Penalties.

(1) Violation of this section shall be punishable as a misdemeanor. In addition, each violation of this Section shall constitute grounds for suspension or revocation of the licensee's firearms dealer license under this Article.

(2) Each transaction in violation of this Section shall be deemed a distinct and separate violation.

613.10-3. Persons Prohibited From Purchasing Firearms Not Permitted On Licensees' Premises.

(a) No person who is prohibited from possessing or purchasing firearms pursuant to California Penal Code Sections 12021 or 12021.1 or Subsections (a), (b), (c) or (d) of California Welfare and Institutions Code Sections 8103 shall enter into any place of business that is licensed pursuant to this Article to engage in the business of selling, leasing or otherwise transferring firearms and which displays the notices required by Section 613.10(r), except in accordance with paragraph (c) of this Section.

(b)(1) No person licensed pursuant to this Article to sell, lease or otherwise transfer firearms shall permit the entry onto the premises that are the subject of the license by any person who the licensee knows or has reason to know is prohibited from possessing or purchasing firearms pursuant to California Penal Code Sections 12021 or 12021.1, or Subsections (a), (b), (c) or (d), of California Welfare and Institutions Code Section 8103, except in accordance with paragraph (c) of this Section.

(2) No person licensed pursuant to this Article to sell, lease or otherwise transfer firearms shall permit the entry onto the premises that are the subject of the license by any person under 18 years of age except in accordance with paragraph (c) of this section. No person licensed pursuant to this Article to sell, lease or otherwise transfer firearms and who keeps or displays for sale, lease or other transfer firearms capable of being concealed on the person shall permit the entry onto the premises that are the subject of the license by any person under 21 years of age, except in accordance with paragraph (c) of this section.

(3) The licensee and any of his or her agents, employees or other persons acting under the licensee's authority shall be responsible for requiring clear evidence of age and identity of persons to prevent the entry of persons not permitted to purchase a firearm under state law by reason of age. Clear evidence of age and identity includes, but is not limited to, a motor vehicle operator's license, a State identification card, an armed forces identification card, or an employment identification card which contains the bearer's signature, photograph, and age, or any similar documentation which provides reasonable assurance of the identity and age of the individual.

(c) Exceptions.

(1) It shall not be a violation of this section for any person who is otherwise prohibited pursuant to subsection (a) from entering or being present on the premises to enter or be present on the premises if the firearms and related accessories (including, but not limited to, ammunition, ammunition clips and holsters) are kept or displayed within a separate room or enclosure that separates such firearms and related accessories from other merchandise, and such persons are excluded from the separate room or enclosure.

(2) It shall not be a violation of this section for any person who the licensee is otherwise required to keep from entering or being present on the premises pursuant to subsection (b) to enter or be present on the premises if the firearms and related accessories (including, but not limited to ammunition, ammunition clips and holsters) are kept or displayed within a separate room or enclosure that separates such firearms and related accessories from other merchandise, and such persons are excluded from the separate room or enclosure. Where a licensee keeps or displays weapons capable of being concealed on the person in a separate roam or enclosure that separates such firearms and related accessories, including, but not limited to, ammunition, ammunition clips and holsters) from other merchandise (including other firearms), it shall not be a violation of this section for persons at least 18 years old but less than 21years old to enter or be present on the premises if such persons are excluded from the separate room or enclosure containing firearms capable of being concealed on the person.

(d) Penalty for violation.

(1) Any person who is prohibited from possessing or purchasing firearms pursuant to California Penal Code Sections 12021 or 12021.1 or Subsections (a), (b), (c) or (d) of California Welfare and Institutions Code Section 8103 and who knowingly enters into any place of business that is licensed pursuant to this Article to engage in the business of selling, leasing or otherwise transferring firearms in violation of subsection (a) shall be guilty of a misdemeanor.

(2) Any licensee or other person acting under the authority of a licensee, including, but not limited to, employees or agents of a licensee, who knowingly allows a person to enter the licensee's premises in violation of this section shall be guilty of a misdemeanor.

613.11. License - Grounds For Revocation. In addition to any other provisions of this Article, any circumstances constituting grounds for denial of a license shall also constitute grounds for revocation of a license.

613.13. License - Liability Insurance. No license shall be issued or reissued pursuant to this Article unless there is in full force and effect a policy of insurance in such form as the City Attorney and the City's Risk Manager deem proper, executed by an insurance company approved by the City's Risk Manager, whereby the applicant or licensee is insured against liability for damage to property and for injury to or death of any person as a result of the sale, transfer or lease, or advertising for sale, transfer or lease, or offering or exposing for sale, transfer or lease, of any firearm, firearm ammunition or firearm ammunition component. The minimum liability limits shall not be less than $1,000,000 for damage to or destruction of property in any one incident, and $1,000,000 for the death or injury to any one person; provided, however, that additional amounts may be required by the City Attorney or City's Risk Manager if deemed necessary to protect the interests of the City and its residents.

Such policy of insurance shall contain an endorsement providing that the policy will not be canceled until notice in writing has been given to the City, address in care of the Chief of Police, 850 Bryant Street, San Francisco, California, 94103, at 30 days immediately prior to the time such cancellation becomes effective. Further, such policy of insurance shall name the City, its officers, agents, and employees as additional insureds.

In addition, applicants and licensees shall agree in writing that they shall indemnity, defend, and hold harmless the City, its officers, agents, and employees, from claims arising out of the negligence or the intentional or willful misconduct of the applicant or licensee.

613.14. License - Authority to Inspect. Any and all investigating officials of the City shall have the right to enter the building designated in the license from time to time during regular business hours to make reasonable inspections and to observe and enforce compliance with building, mechanical, fire, electrical, plumbing, or health regulations, and provisions of this Article. A police investigator may conduct compliance inspections to insure conformance with all federal, state, and local laws, and all provisions of this Article.

613.16. Temporary Suspension of License.

(a) If the Chief of Police has reason to believe that a licensee has (1) violated any federal, state or local law relating to the sale, lease, transfer, use or possession of firearms, firearm ammunition or firearms ammunition components, or (2) has committed any of the offenses set forth in Section 613.3(e), the Chief of Police may immediately suspend for a period not to exceed three days the right of the licensee to sell, lease or otherwise transfer firearms, firearm ammunition or firearm ammunition components.

However, if the licensee is charged by a federal, state, or local prosecuting attorney with a violation of any such law the Chief of Police may suspend such license until the charges are dismissed or the licensee is found not guilty in a court of law. If the Chief of Police suspends a license pursuant to this provision, the licensee shall be provided an opportunity to present evidence to the Chief or his or her designee that the pending charges are without legal merit. The Chief's decision regarding whether to suspend a license pending the outcome of such charges shall be appealable to the Board of Permit Appeals, at which proceeding the Chief shall have the burden of proof to justify his decision.

(b) Notice of suspension shall be mailed to the person(s) who applied for the license and shall be delivered to the address listed on the license.

613.17. Delivery to Tax Collector. Upon granting said license, the Police Department shall forward said license to the Tax Collector who shall issue said license to the applicant upon payment of the license fee, payable in advance.

613.18. Rules and Regulations. The Chief of Police is hereby authorized to promulgate such rules and regulations after noticed hearing as will enable the Police Department to effectuate the purposes of this ordinance. Failure to comply with such rules and regulations, or with any other requirements imposed by this ordi-nance, shall constitute grounds for revocation of licenses issued hereunder.

613.19. Penalties. Any person violating any provision of this Article shall be guilty of a misdemeanor or an infraction. The complaint charging such violation shall specify whether the violation is a misdemeanor or infraction, which decision shall be that of the District Attorney. If charged as an infraction, upon conviction, the violator shall be punished by a fine of not more than $100 for each provision violated. If charged as a misdemeanor, upon conviction, the violator shall be punished by a fine of not less than $500 or more than $1,000 for each provision violated or by imprisonment in the County Jail for a period of not more than six months, or by both such fine and imprisonment. In any accusatory pleading charging a violation of this Section, if the defendant has been previously convicted of a violation of this Section, each such previous violation and conviction shall be charged in the accusatory pleading. Any person violating any provision of this Section a second or subsequent time within a 10-year period shall be guilty of a misdemeanor and shall be punished by a fine of not less than $500 and not more than $1,000 for each provision violated, or by imprisonment in the County Jail for a period of not more than six months, or by both such fine and imprisonment.

614. Exceptions. Sections 613 to 613.20 inclusive, of this Article, shall not apply to wholesale dealers in their business intercourse with retail dealers, nor to wholesale or retail dealers in the regular or ordinary transport of unloaded firearms as merchandise to other wholesale or retail dealers by mail, express or other mode of shipment to points outside the City and County of San Francisco.

615. Records of Ammunition Sales.

(a) Definitions.

(1) "Firearm ammunition," as used in this Section, shall include any ammunition for use in any pistol or revolver, or semiautomatic rifle or assault weapon, but shall not include ammunition for shotguns that contains shot that is No. 4 or smaller.

(2) "Semiautomatic rifle," as used in this Section, shall mean any repeating rifle which utilizes a portion of the energy of a firing cartridge to extract the fired cartridge case and chamber the next round, and which requires a separate pull of the trigger to fire each cartridge.

(3) "Assault weapon," as used in this Section, shall mean any of the weapons designated in California Penal Code Section 12276 or 12276.1,

(4) "Vendor" as used in this Section, shall mean any person who is engaged in the sale of firearm ammunition, including any retail firearms dealer.

(b) No vendor shall sell or otherwise transfer ownership of any firearm ammunition without at the time of purchase recording the following information on a form to be prescribed by the Chief of Police: (1) the name of the vendor (including the name of the specific individual) transferring ownership to the transferee; (2) the place where the transfer occurred; (3) the date and time of the transfer; (4) the name, address and date of birth of the transferee; (5) the transferee's driver's license number, or other identification number, and the state in which it was issued; (6) the brand, type and amount of ammunition transferred; and (7) the transferee's signature.

(c) The records required by this Section shall be maintained on the premises of the vendor for a period of not less than two years from the date of the recorded transfer. Said records shall be subject to inspection at any time during normal business hours.

(d) No person shall knowingly make a false entry in, or fail to make a required entry in, or fail to maintain in the required manner records prepared in accordance herewith. No person shall refuse to permit a police department employee to examine any record prepared in accordance with this Section during any inspection conducted pursuant to this Section.

(e) Penalties.

(1) First Conviction. Any person violating any provision of this Section shall be guilty of an infraction. Upon conviction of the infraction, the violator shall be punished by a fine of not less than $50 nor more than $100.

(2) Subsequent Convictions. In any accusatory pleading charging a violation of this Section, if the defendant has been previously convicted of a violation of this Section, each such previous violation and conviction shall be charged in the accusatory pleading. Any person violating any provision of this Section a second time within a 90-day period shall be guilty of a misdemeanor and shall be punished by a fine of not less than $300 and not more than $400 for each provision violated, or by imprisonment in the County Jail for a period of not more than six months, or by both such fine and imprisonment. Any person violating any provision of this Section, a third time, and each subsequent time, within a 30-day period shall be guilty of a misdemeanor and shall be punished by a fine of not less than $400 and not more than $500 for each provision violated, or by imprisonment in the County Jail for a period of not more than six months, or by both such fine and imprisonment. ...

616. Reporting the Loss or Theft of Firearms.

(a) Any person that owns or is otherwise in possession of a firearm shall report the theft or loss of such firearm to the San Francisco Police Department within 48 hours of becoming aware of the theft or loss whenever (1) the owner resides in San Francisco, or (2) the theft or loss of the firearm occurs in San Francisco.

(b) The failure of an owner or person in possession of a firearm to report the theft or loss of the firearms within 48 hours of when the owner or person in possession becomes aware or should have become aware of the theft or loss shall be punishable in accordance with Section 613.19.

617. Prohibition Against Possession or Sale of Firearms or Ammunition on Property Controlled by the City and County of San Francisco.

A. Legislative Findings.

(1) The national and statewide statistical information available from numerous sources overwhelmingly demonstrates that the incidence of gunshot fatalities and injuries has reached alarming and thus, unacceptable proportions; and

(2) Government at all levels has a substantial interest in protecting the people from those who acquire guns illegally and use them to commit crimes resulting in injury or death of their victims; and

(3) Recent events throughout the City and County have generated additional fears of random usage of guns to commit violence on unsuspecting residents, children and adults alike; and

(4) The Board of Supervisors finds that crimes and injuries committed with the use of a firearm are prevalent in San Francisco, with local statistics showing an increase in homicides by use of a gun increasing from 39 gun homicides out of 63 total homicides in 2001, which is a 63 percent increase, to 68 gun homicides out of 85 total homicides in 2006, which is an 80 percent increase. Local statistics also show that San Francisco Police Department has seized 1,158 guns in 2005, and 1,104 guns in 2006. In 2007, there have already been 25 gun homicides.

(5) In the City and County of San Francisco, the number of nonfatal injuries from guns has steadily increased. While, in 2001, 81 patients were admitted to SF General Trauma Center for serious injuries resulting from gun shots, the number of patients admitted for serious injuries rose to 228 by 2006. Similarly, the total number of shootings that resulted in nonfatal injuries documented by SFPD was 269 in 2005, 303 in 2006, and in 2007 this number has already reached 105 by May 10; and

(6) Gun crimes in and around schools and on buses carrying students to and from school have become increasingly common; and

(7) In 2003 and 2004, 52 percent of the City's gun violence victims were under the age of 25.

(8) Homicides committed with handguns are the leading cause of firearms related injuries and death in California; and

(9) The widespread availability of illegally obtained firearms has resulted in a significant rise in the number of shooting incidents across the County; and

(10) The Board of Supervisors has authority over the management and control of City and County property, and it may regulate, by ordinance, the manner in which the property of the City and County is accessed and used by members of the public; and

(11) Prohibiting the possession or sale of firearms and/or ammunition on City and County property will promote the public health and safety by contributing to the reduction in the presence of firearms and the potential for gunshot fatalities and injuries in the county. It will increase the confidence of members of the public that they are not at risk of injury from firearms when they seek to use the property and facilities of the City and County. In particular, this Board of Supervisors finds that an enormous number of the general public utilizes the parks, playgrounds and squares of San Francisco. This Board finds that prohibiting the possession or sale of firearms and ammunition on City and County property will help to ensure the safety of the general public and specifically children who are among the most vulnerable in our society; and

(12) The California Supreme Court has ruled that State Law does not preempt local laws banning the possession and sale of firearms and ammunition on their property. In Nordyke v. King (2002) 27 Cal.4th 875, the Supreme Court upheld an Alameda County ordinance banning the possession of firearms and ammunition on county owned property and in Great Western v. County of Los Angeles (2002) 27 Cal.4th 853, the Supreme Court upheld a Los Angeles County Ordinance prohibiting all sales of firearms and ammunition on county property. These rulings uphold the legal ability of the Board of Supervisors to ban the possession and sale of firearms and ammunition on City and County property.

B. Legislative Intent. With passage of this ordinance, the City and County seeks to ensure that its property and facilities are used in a manner consistent with promoting the health, safety and welfare of all of its residents.

C. Definitions.

(1) City and County Property.

(a) As used in this section, the term "City property" means real property, including any buildings thereon, owned or leased by the City and County of San Francisco (hereinafter "City"), and in the City's possession or in the possession of a public or private entity under contract with the City to perform a public purpose including but not limited to the following property: recreational and park property including but not limited to Golden Gate Park, the San Francisco Zoo, Hilltop Park and San Francisco's parks and playgrounds, plazas including but not limited to United Nations Plaza and Hallidie Plaza, community centers such as Ella Hill Hutch Community Center, and property of the Department of Recreations and Parks, the Port, and the Public Utilities Commission.

The term "City property" does not include any "local public building" as defined in Penal Code Section 171b(c), where the state regulates possession of firearms pursuant to Penal Code Section 171b.

The term "City property" also does not include the public right-of-way owned by the City and County of San Francisco including any area across, along, on, over, upon, and within the dedicated public alleys, boulevards, courts, lanes, roads, sidewalks, streets, and ways within the City or any property owned by the City that is outside the territorial limits of the City and County of San Francisco.

(2) Firearms. As used in this section the term "firearm" is any gun, pistol, revolver, rifle or any device, designed or modified to be used as a weapon, from which is expelled through a barrel a projectile by the force of an explosion or other form of combustion. "Firearm" does not include imitation firearms or BB guns and air rifles as defined in Government Code Section 53071.5.

(3) Ammunition. "Ammunition" is any ammunition as defined in California Penal Code Section 12316(b)(2).

C. Possession or Sale of Firearms or Ammunition on County Property Prohibited. No person shall:

(1) Bring onto or possess on county property a firearm, loaded or unloaded, or ammunition for a firearm.

(2) Sell on county property a firearm, loaded or unloaded, or ammunition for a firearm.

D. Exceptions, Ban on Possession. Section C.(1) above shall not apply to the following:

(1) A peace officer, retired peace officer or person assisting a peace officer when authorized to carry a concealed weapon under

Penal Code Section 12027(a) or a loaded firearm under Penal Code Section 12031(b)(1) and under 18 U.S.C. 926B or 926C.

(2) Members of the armed forces when on duty or other organizations when authorized to carry a concealed weapon under Penal Code Section 12027(c) or a loaded firearm under Penal Code Section 12031(b)(4).

(3) Military or civil organizations carrying unloaded weapons while parading or when going to and from their organizational meetings when authorized to carry a concealed weapon under Penal Code Section 12027(d).

(4) Guards or messengers of common carriers, banks and other financial institutions when authorized to carry a concealed weapon under Penal Code Section 12027(e) and armored vehicle guards when authorized to carry a loaded weapon under Penal Code Section 12031(b)(7).

(5) Persons who are at a target range.

(6) Honorably retired Federal officers or agents of Federal law enforcement agencies when authorized to carry a concealed weapon under Penal Code Section 12027(i) or a loaded weapon under Penal Code Section 12031(b)(8).

(7) The public administrator in the distribution of a private estate or to the sale of firearms by its auctioneer to fulfill its obligation under State Law.

(8) Patrol special police officers, animal control officers or zookeepers, and harbor police officers, when authorized to carry a loaded firearm under Penal Code Section 12031(c).

(9) A guard or messenger of a common carrier, bank or other financial institution; a guard of a contract carrier operating an armored vehicle; a licensed private investigator, patrol operator or alarm company operator; a uniformed security guard or night watch person employed by a public agency; a uniformed security guard or uniformed alarm agent; a uniformed employee of private patrol operator or private investigator when any of the above are authorized to carry a loaded firearm under Penal Code Section 12031(d).

(10) Any authorized participant in a motion picture, television or video production or entertainment event when the participant lawfully uses a firearm as part of that production or event.

E. Exception, Ban on Sale. Section C.(2) above shall not apply to the following:

(1) Purchase or sale of a firearm or ammunition for a firearm by a federal, state or local law enforcement agency or by any other Federal, State or local governmental entity.

(2) The public administrator in the distribution of a private estate or to the sale of firearms by its auctioneer to fulfill its obligation under state law.

(3) Sale of ammunition at a target range for use at the target range.

F. Penalty for Violation. Any person who violates any of the provisions of this Section 617(c) shall be deemed guilty of a misdemeanor and upon conviction shall be punished by a fine not to exceed $1,000.00 or by imprisonment in the county jail not to exceed six months, or by both.

G. Severability. If any provision, clause or word of this Section or the application thereof to any person or circumstance is held invalid, such invalidity shall not affect any other provision, clause, word or application of this Section which can be given effect without the invalid provision, clause or word, and to this end the provisions of this chapter are declared to be severable.

Article 13. Miscellaneous Regulations for Professions and Trades

840. Carrying Firearms, Etc. - Regulations Governing. Any person who in the course of employment or his livelihood carries a firearm or any other deadly or dangerous weapons as defined by Section 1291-A, Article 17, Chapter VIII, Part II of the San Francisco Municipal Code, concealed or unconcealed, shall register with the Chief of Police in writing by signed statement, stating the name, occupation, residence and business address of the registrant, his age, height, weight, color of eyes and hair, and reason desired to carry such weapon, and shall submit therewith two sets of fingerprints and one photograph.

841. Carrying Firearms, etc.--Requirements. Any person carrying a firearm or any other deadly or dangerous weapon as defined by said Section 1291-A, Article 17, Chapter VIII, of the San Francisco Municipal Code, in the City and County of San Francisco, must:

(1) Be at least 21 years of age;

(2) Be a citizen of the United States;

(3) Be of good moral character;

(4) Be able to manifest a knowledge of Sections 197, 834, 835, 837 and 847 of the Penal Code of the State of California.

842. Carrying Firearms, Etc. - Penalty for Violation. Any person who shall violate any of the provisions of Section 840 to 842 of this Article shall be deemed guilty of a misdemeanor and shall be punished by a fine of not exceeding $500, or by imprisonment in the County Jail for not exceeding six months, or by both such fine and imprisonment.

Article 35. Firearm Strict Liability Act

3501. Definitions.

(a) "Firearm" shall have the same meaning as in San Francisco Police Code Section 613.1(a).

(b) "Dealer" means any person engaged in the business of selling firearms at wholesale or retail and specifically includes pawnbrokers who take or receive firearms as security for the payment or repayment of money.

(c) "Importer" means any person engaged in the business of importing or bringing firearms into the United States for sale or distribution.

(d) "Manufacturer" means any person in business to manufacture or assemble a firearm or ammunition for sale or distribution.

(e) "Law enforcement agency" means a federal, state or local law enforcement agency, state militia or an agency of the United States government.

(f) "Law enforcement official" means any officer or agent of an agency defined in paragraph (e) of this section who is authorized to use a firearm in the course of his or her work.

(g) "Internal personalized safety feature" means any internal locking device or other mechanical or electrical device integral to the frame of the firearm that prevents any unauthorized use of the firearm. Such mechanical or electrical devices can include but are not limited to devices that use computer microchips, radio signals or user fingerprints as a means to "recognize" an authorized user. A trigger lock or other external device shall not be considered an internal personalized safety feature.

3502. Imposition of Strict Liability.

(a) Each manufacturer, importer and/or dealer of a firearm shall be held strictly liable in tort, without regard to fault or proof of defect, for all direct and consequential damages arising from bodily injury or death where the bodily injury or death results from the discharge within the jurisdiction of the City and County of San Francisco of any firearm manufactured, imported, distributed, sold, leased or otherwise transferred by the manufacturer, importer and/or dealer, except that no liability shall be imposed pursuant to this subsection for a discharge that occurs prior to the effective date of this section.

(b) Exemptions and Limitations.

(1) No action may be commenced pursuant to this section by any person who is injured or killed by the discharge of a firearm while such person is committing or attempting to commit a crime (whether or not such crime is actually charged), or while such person is attempting to evade arrest by a law enforcement official. This exemption shall be in the nature of an affirmative defense, and shall be proven by a preponderance of the evidence.

(2) No action may be commenced pursuant to this section by any person injured or killed by the discharge of a firearm by a law enforcement official.

(3) This section shall not limit in scope any cause of action, other than that provided by this section, available to a person injured by or killed by a firearm.

(4) Nothing in this section shall prevent a manufacturer, importer or dealer from seeking whole or partial indemnity or contribution for any liability incurred under this section from any third party wholly or partially responsible for the injury or death.

(5) No action may be commenced pursuant to this section by any person for a self-inflicted injury.

(6) No action may be commenced pursuant to this section where the firearm was equipped with an internal personalized safety feature at that time of its first retail sale.

(7) If any manufacturer, importer or dealer has purchased and has in effect at the time of the injury an insurance policy that covers any and all damages, including but not limited to bodily injury or death resulting from the discharge of the specific firearm involved in the incident, the liability imposed under this section as to that manufacturer, mporter or dealer shall not exceed the total amount of coverage available under said policy provided that the total coverage available under the policy shall not be less than $100,000 per incident.

(8) No action may be commenced pursuant to this section where the firearm involved is either (a) a shotgun without a magazine or having a fixed magazine of four or less rounds or (b) a rifle without a magazine or having a fixed magazine of four or less rounds.

Article 45. Firearms and Weapons Violence Prevention Ordinance

4500. Title. This Article shall be known as the Firearms and Weapons Violence Prevention Ordinance.

4501. Definitions. The following words and phrases, whenever used in this Article, shall be construed as defined in this Section:

(a) "Firearm" means any device, designed to be used as a weapon or modified to be used as a weapon, that expels a projectile by the force of an explosion or other form of combustion.

(b) "Projectile weapon" means any device or instrument used as a weapon which launches or propels a projectile by means other than the force of an explosion or other form of combustion with sufficient force to cause injury to persons or property. A projectile weapon shall include, but not be limited to, air gun, air pistol, air rifle, gas operated gun, BB gun, pellet gun, flare gun, dart gun, bow, cross-bow, slingshot, wrist rocket, blow gun, paint gun, or other similar device or instrument.

4503. Enforcement.

(a) Except as otherwise provided in this Section, any person violating any provision of this Article shall be guilty of a misdemeanor or an infraction. The complaint charging such violation shall specify whether the violation is a misdemeanor or infraction, which decision shall be that of the District Attorney. If charged as an infraction, upon conviction, the violator shall be punished by a fine of not more than $100 for each provision violated. If charged as a misdemeanor, upon conviction, the violator shall be punished by a fine of not more than $500 for each provision violated or by imprisonment in the County Jail for a period of not more than six months, or by both such fine and imprisonment.

(b) Any person violating any provision of this Article a second or subsequent time within a 10-year period shall be guilty of a misdemeanor and shall be punished by a fine of not more than $1,000 for each violation, or by imprisonment in the County Jail for a period of not more than six months, or by both such fine and imprisonment.

(c) Any person violating any provision of this Article within 1,500 feet of a day care center, school or school yard, whether public or private, shall be guilty of a misdemeanor and shall be punished by a fine of not more than $1,000 for each such violation, or by imprisonment in the County Jail for a period of not more than six months, or by both such fine and imprisonment.

(d) A person shall be guilty of a separate offense for each and every discharge of a firearm or firing of a projectile weapon, and shall be punished accordingly.

(e) Juveniles arrested pursuant to this Section shall be subject to Section 602 of the Welfare and Institutions Code.

4504. Parental Responsibility for Minors. Any parent or legal guardian, or a person over the age of 18, is also guilty of an offense punishable in accordance with Section 4503 if he or she knows or reasonably should know that a minor is likely to gain access to a firearm or a projectile weapon kept within any premises or vehicle which is under his or her custody or control, and a minor obtains and fires or discharges the firearm or projectile weapon within the City and County of San Francisco, in violation of Section 4502.

4505. Firearms and Projectile Weapons; Confiscation and Disposal of. Any firearm or projectile weapon discharged within the City and County of San Francisco in violation of the provisions of Section 4502 is hereby declared to be a nuisance, and shall be surrendered to the Police Department of the City and County of San Francisco. The Chief of Police, except upon the certificate of a judge of a court of record, or of the District Attorney that the preservation thereof is necessary or proper to the ends of justice, shall destroy or cause to be destroyed such firearms and projectile weapons, provided, however, that in the event any such firearm or projectile weapon is determined to have been stolen, the same shall not be destroyed but shall be returned to the lawful owner as soon as its use as evidence has been served, upon identification of the firearm or projectile weapon and proof of ownership thereof.

4506. Firearms And Projectile Weapons; Exceptions. ...

(b) Use of firearms and projectile weapons may be permissible when integral to the pursuit of specific competitive and sporting events, including but not limited to events such as target and skeet shooting, upon issuance of a permit from the Chief of Police to persons conducting the event or engaged in the business of providing the location at which the event is to take place. The Chief of Police shall formulate criteria for the application, issuance, and renewal of such permits, and may require as a condition of approval the posting of any bond, or proof of adequate liability insurance.

4507. Firearms and Projectile Weapons; Possession of by Minors.

(a) It shall be unlawful for any person under the age of 18 to have in his or her possession within the City and County of San Francisco any firearm or projectile weapon, as defined in Section 4501. Violation of this provision shall be punishable in the manner provided in Section 4503.

(b) It shall be unlawful for any parent or legal guardian, or any person over the age of 18 years, to sell, give or otherwise transfer to any minor in the City and County of San Francisco under the age of 18 years, or to allow such minor to possess, any firearm or projectile weapon, as defined in Section 4501. Violation of this provision shall be punishable in the manner provided in Section 4503.

(c) Any firearm or projectile weapon, which is in possession of a minor in violation of this Article, is hereby declared to be a nuisance, and shall be surrendered to the Police Department of the City and County of San Francisco and disposed of in accordance with the provisions of Section 4505 above.

4512. Handguns Located in a Residence To Be Kept in a Locked Container or Disabled with a Trigger Lock.

(a) Prohibition. No person shall keep a handgun within a residence owned or controlled by that person unless the handgun is stored in a locked container or disabled with a trigger lock that has been approved by the California Department of Justice.

(b) Definitions.

(1) "Residence." As used in this Section, "residence" is any structure intended or used for human habitation including but not limited to houses, condominiums, rooms, in law units, motels, hotels, SRO's, time-shares, recreational and other vehicles where human habitation occurs.

(2) "Locked container." As used in this Section, "locked container" means a secure container which is fully enclosed and locked by a padlock, key lock, combination lock or similar locking device.

(3) "Handgun." As used in this Section, "handgun" means any pistol, revolver, or other firearm that is capable of being concealed upon the person, designed to be used as a weapon, capable of expelling a projectile by the force of any explosion or other form of combustion, and has a barrel less than 16 inches in length.

(4) "Trigger lock." As used in this Section, a "trigger lock" means a trigger lock that is listed in the California Department of Justice's list of approved firearms safety devices and that is identified as appropriate for that handgun by reference to either the manufacturer and model of the handgun or to the physical characteristics of the handgun that match those listed on the roster for use with the device under Penal Code Section 12088(d).

(c) Exceptions. This Section shall not apply in the following circumstances:

(1) The handgun is carried on the person of an individual over the age of 18.

(2) The handgun is under the control of a person who is a peace officer under Penal Code Section 830.

(d) Lost or Stolen Handguns. In order to encourage reports to law enforcement agencies of lost or stolen handguns pursuant to San Francisco Police Code Section 616, a person who files a report with a law enforcement agency notifying the agency that a handgun has been lost or stolen shall not be subject to prosecution for violation of Section 4512(a) above.

(e) Penalty. Every violation of this Section shall constitute a misdemeanor and upon conviction shall be punished by a fine not to exceed $1,000.00 or by imprisonment in the county jail not to exceed six months, or by both.

(f) Severability. If any provision, clause or word of this chapter or the application thereof to any person or circumstance is held invalid, such invalidity shall not affect any other provision, clause, word or application of this Section which can be given effect without the invalid provision, clause or word, and to this end the provisions of this Section are declared to be severable.

[San Francisco Police Code current through Ordinance No. 113-08, June 30, 2008]

San José Municipal Code

Title 10. Public Peace, Morals and Welfare

Chapter 10.32. Weapons

Part 2. Weapons and Minors

10.32.020. Restrictions on weapons for persons under the age of eighteen years. No person shall sell, dispose of or give to any minor under the age of eighteen years, any pistol, air-gun or other weapon capable of receiving and discharging any charge, cartridge or explosive, without the written request of his parent or guardian; and no minor under the age of eighteen years shall have in his possession any such pistol or weapon unless by the written consent of his parent or guardian.

Part 3. Concealable Firearms Dealers

10.32.030. License required. No person shall engage in the business of selling or transferring, or shall advertise for sale, or offer or expose for sale, any pistol, revolver or other firearm capable of being concealed upon a person without first obtaining and keeping in current effect the requisite license as provided under this part. A license shall be required for each location where selling or transferring of such firearms takes place.

10.32.040. Application for license. Each person applying for a license under this part shall submit an application and a questionnaire response in support of such application to the chief of police. Such application shall be accompanied by a fee (in the form of a check or cash) as set forth in the schedule of fees established by resolution of council, said fee to defray the cost of processing such application. This fee shall be in addition to fees for processing noncriminal fingerprint cards. Each application shall specify only one location where the sale or transfer of concealable firearms should take place. If a licensee changes his or her place or business, an application for the new location shall be submitted, accompanied by a fee (in the form of a check or cash) as set forth in the schedule of fees established by resolution of council, said fee to defray the cost of processing such application. Such application shall be considered an initial application and not an application for renewal.

10.32.050. Approval by chief of police. The chief of police shall have the authority to approve or not approve applications under this part. For the purposes of considering applications for licenses under this part, the chief of police shall apply the minimum written standards hereinbelow set forth in Section 10.32.060. If an applicant meets all such standards, the chief of police shall issue a license.

10.32.060. Procedural requirements. Applicant must comply with the following requirements:

A. Possess valid and current federal firearms permit;

B. Pay permit fee to police department;

C. Pay fingerprinting fee to police department;

D. Complete personal history questionnaire on form provided by the San Jose police department;

E. Be fingerprinted at police department;

F. Be photographed at police department;

G. Be interviewed at police department by officer assigned to concealable firearms unit;

H. Sign authorization for release of pertinent records.

10.32.065. Business site standards.

A. Business location must be in compliance with city of San José zoning codes and regulations.

B. Business site must be a permanent building having four solid walls and a roof; constructed of wood, metal, cement or like materials and resting on a foundation of cement, stone, brick or metal or similar materials commonly used in the construction of foundations for permanent buildings.

C. Building doors and windows must be alarmed with a system of any manufacture which will cause an audible alarm to sound on the exterior of the building or a silent alarm to a centralized monitored facility signaling unauthorized entry during nonbusiness hours.

D. Business inventory of an explosive or flammable nature must be stored in compliance with federal, state of California and city of San José fire codes.

E. Inventory of firearms must conform to the type of federal firearms permit issued to the permittee.

F. Business must conform to Penal Code Section 12071, state of California, i.e.:

1. The business shall be carried on only in the building designated in the license.

2. The license or a copy thereof (when issued), certified by the issuing authority, shall be displayed on the premises where it can easily be seen.

3. No pistol or revolver shall be delivered:

a. Within fifteen days of the application for the purchase, and when delivered shall be unloaded and securely wrapped; nor

b. Unless the purchaser either is personally known to the seller or shall present clear evidence of his identity.

4. No pistol or revolver, or imitation thereof, or placard advertising the sale or other transfer thereof, shall he displayed in any part of the premises where it can readily be seen from the outside.

10.32.066 Approval of application - Fitness standards. Any application submitted under Section 10.32.060 shall be approved if the applicant:

A. Fulfills all requirements of Section 10.32.060 of this part; and

B. Is not under indictment for, nor has ever been convicted of a crime involving fraud, dishonesty, deceit or violence; and

C. Has not done any unlawful act involving firearms with the intent to injure seriously another person; and

D. Has not done any act involving dishonesty, fraud or deceit with the intent substantially to benefit him or herself or another, or substantially to injure another; and

E. Is not a fugitive from justice; and

F. Is not an unlawful user of or addicted to marijuana or any depressant or stimulant drug, narcotic drug, or excessive use of alcohol, to the extent that such use would impair his or her fitness to deal in concealable firearms; and

G. Has not been adjudicated as a mental defective, has not been committed to a mental institution, or does not suffer from any severe psychological disturbance which would seriously impair his or her fitness to deal in concealable firearms; and

H. Has not willfully violated any provisions of this chapter; and

I. Has not willfully made false statements as to any material fact in application for this license; and

J. Has not done any act which if done by a licensee would be grounds for suspension or revocation of license; and

K. Meets all business site standards under Section 10.32.065.

10.32.067. Maintenance and submission of records. Each licensed dealer shall maintain records of importation, shipment, receipt, sale or other disposition of firearms and ammunition, and shall make such records available at all reasonable times, and shall submit to the licensing authority such reports and information upon reasonable request. The licensing authority may enter during business hours the premises (includeing places of storage) of any firearms or ammunitions dealer for the purpose of inspecting or examining (1) any records or documents required to be kept; (2) any firearms or ammunition kept or stored at such premises.

10.32.100. Licenses nonassignable. Except as otherwise hereinafter provided, no license issued under this part may be sold, transferred or assigned by the licensee or by operation of law, to any other person or persons; and any such sale, transfer or assignment, or attempted sale, transfer or assignment, be deemed to constitute a voluntary surrender of such license and such license shall thereafter be deemed terminated and void.

10.32.110. Display of license. Every license issued under this part shall be displayed on the premises where the concealable firearms are sold in such a fashion that it can be easily read by patrons.

10.32.150 Weapons - Possession, sale and offering for sale prohibited. No person shall possess, sell, or offer for sale any device which is designed:

A. To propel any flechette or dart by means of compressed gas; and

B. To include in its operation the following:

1. Mechanical energy from an external source is supplied to the device;

2. The device compresses gas; and

3. The compressed gas propels the flechette or dart.

Title 20. Zoning

Chapter 20.80. Specific Use Regulations

20.80.720. Prohibited uses. The following are not permitted as home occupations: ...

C. Firearm sales and service. ...

[San José Municipal Code current through Ord. 28338, June 24, 2008]

Santa Ana Municipal Code

Chapter 10. Crimes and Miscellaneous Law Enforcement Provisions

Article II. Minors

10-176. Possession of firearms or ammunition by a minor.

(a) No minor shall possess any firearm or ammunition in a public place.

(b) No person shall sell, exchange, give or loan, or cause or permit to be sold, exchanged, given or loaned, any firearm or ammunition to any person under the age of eighteen (18) years.

(c) For the purpose of this section, the term "firearm" means any weapon which will or is designed to or may readily be converted to expel a projectile by the action of an explosive.

(d) For the purpose of this section, the term ammunition means any unexpended cartridge or shell, consisting of a case which holds a charge of powder and a bullet or shot, or any device containing any explosive designed and intended for use in any firearm, or any device containing any explosive.

(e) Nothing in this section shall prohibit the possession of any firearm by a minor at a bona fide firing range or in an automobile vehicle while enroute to or from hunting any game animal that may be legally hunted under the provisions of the Fish and Game Code, provided said minor has the written permission of his parent or guardian to have such firearm or is accompanied by his parent or guardian while he has such firearm in his possession.

Article IX. Licensing of Firearms Sales

10-500. Purpose and scope.

(a) The purposes of this article are as follows:

(1) To implement Article 4 (commencing with section 12070) of Title 2 of Part 4 of the Penal Code of the State of California.

(2) To provide procedures for the grant or denial of licenses for the sale, lease, or transfer of firearms and the suspension or revocation of the same.

(3) To regulate the sale, lease or transfer of firearms as necessary and appropriate for the protection of the public health, safety and welfare.

(b) This article applies to all sales, leases and transfers of firearms for which a license is required by Section 12070 of the Penal Code of the State of California.

10-501. Definitions. As used in this article the following terms shall have the following meanings:

Chief of police means the chief of police of the City of Santa Ana and any employee within the Santa Ana Police Department who is designated by the chief of police to act as his representative in the implementation of this article.

Penal Code means the Penal Code of the State of California, as it may from time to time be amended.

License and **licensee** mean the license required by Section 12070 of the Penal Code of the State of California and the person holding such a license, unless the context shows that a different type of license is meant.

Other terms used in this article shall have the same meanings as set forth in the Penal Code.

10-502. Chief of police as licensing authority. The chief of police is the duly authorized licensing authority for licenses granted pursuant to Sections 12070 and 12071 of the Penal Code.

10-503. Fixed place of business required. Each licensee must have a fixed place of business and possess all permits, licenses, certificates, variances or other instruments of approval or evidences that any conditions exist for the sale of firearms as required by any other section of this Code or by any statute or code provisions of the state. Except as may otherwise be provided by the Penal Code, all sales of firearms may only be made at said fixed location.

10-504. Branch establishments; multiple places of business; separate licenses required. A separate firearm seller's license fee must be paid and a separate firearm seller's license obtained for each branch establishment of the business transacted and carried on within the city.

10-505. Overlapping business. If in addition to the business of selling firearms any person shall engage in, conduct, manage, or carry on at the same time and on the same premises any other business requiring any permit, license, certificate, variance or other instrument of approval or evidence that any conditions exist as required by any other section of this Code or by statute or code provisions of the state, such person shall comply with all of the provisions affecting each business.

10-506. Licensee responsible for the conduct of the business. It shall be the duty of the licensee to see that no firearms are sold or offered for sale in violation of this article or the Penal Code.

10-507. Retention and maintenance of sales information and sales records; inspection.

(a) All firearm and ammunition sales information and sales records shall be retained for a period of three (3) years and shall be maintained and stored within the licensed location during normal business hours. All such records shall be maintained in an orderly manner, consistent with federal, state and local laws and shall be located so that all records are immediately and readily accessible for inspection in a specific space allocated for record retention.

(b) All sales information and records relating to sales of firearms and/or ammunition shall be made available to any police officer or other duly authorized representative of the chief of police upon request. Stock in trade shall likewise be made available for inspection as appropriate to allow for comparison of such stock with business records. No licensee shall attempt to dissuade or impair said officers' or duly authorized representative's inspection.

10-508. Evidence of doing business as a seller of firearms. When any person shall by the use of signs, circulars, cards, telephone books, newspapers, or trade publications, advertise, holdout, or represent that he or she is a seller of firearms, or when any person holds an active license, certificate, or permit issued by a governmental agency indicating that he or she is a seller of firearms within the city and such person fails to deny in a sworn statement given to the chief of police that he or she is not a seller of firearms within the city, after being requested to do so by the chief of police, then these facts shall be considered prima facie evidence that such person is a seller of firearms within the city.

10-509. Rules and regulations. The chief of police shall adopt and enforce, by rules and regulations, security requirements for the protection from theft of firearms and ammunition sold and maintained by the licensee. Such rules may require that licensee provide burglar alarm systems, separate storage areas for ammunition and other measures designed to prevent the theft of such merchandise from the premises of licensee. The chief of police shall also have the power to adopt other rules and regulations not inconsistent with the provisions of this article as may be necessary or desirable to aid in the enforcement of the provisions of this article.

10-511. Application for license.

(a) An application for a license shall be filed with the chief of police on forms furnished by the chief of police. Such application shall be accompanied by a fee in an amount established by resolution of the city council, and shall contain the following information: ...

(b) The application submitted shall be accompanied by copies of the following licenses, permits, and certificates:

(1) A valid federal firearms license issued in the name of applicant.

(2) A valid seller's permit issued by the State Board of Equalization in the name of applicant issued for the city.

(3) A valid certificate of eligibility issued by the California Department of Justice.

(4) A valid city business license issued for the retail sale of firearms.

(5) A statement of zoning code compliance in accordance with subsection (c) of this section

10-519. Use of license by unlicensed person. No person holding a license required by this article shall permit any other person to use the license, the licensed premises or the licensee's name for the purpose of evading any provision of this article.

10-520. Duplicates. A duplicate license may be issued to replace any license previously issued hereunder which has been lost or destroyed upon the licensee filing a statement of such fact, and, at the time of filing such statement, paying a duplicate license fee equal to ten (10) percent of the original fee for the firearm seller's license.

10-521. Posting. Every licensee shall keep the license prominently posted in a conspicuous and public place upon the premises specified in the license where such business is carried on. All other federal, state, and city licenses, certificates, and permits relative to conducting the business of firearm sales shall likewise be posted together with the firearm seller's license.

10-522. Prohibition of display of suspended or revoked license. No person shall reproduce, duplicate, copy or alter an original firearm seller's license for the purpose of circumventing the requirement of this section by giving a false or misleading impression that any duplication of an original license is valid and that the appropriate fee therefor has been paid; neither shall any person exhibit a suspended or revoked license.

10-523. Transferability.

(a) Licenses shall not be transferable except as provided in this section.

(b) When a business for which a license has been issued is sold or transferred to one (1) of the transferees listed in this section, the chief of police may endorse a change of ownership on such license upon written application by the transferee. Only the following transferees or successors shall be entitled to such transfer of permit.

(1) Duly qualified representatives of licensees regularly appointed by courts of competent jurisdiction, assignees for the benefit of creditors, and spouses or children of deceased licensees;

(2) The surviving partner or partners of a dissolved partnership;

(3) A new partnership consisting of the members of a predecessor partnership, plus their spouse or spouses of any such members;

(4) A new corporation organized and controlled by an individual or unincorporated licensee for the purpose of acquiring, and which has acquired the assets of the business previously owned and operated by such licensee;

(5) Upon dissolution of a closely held corporation, the stockholders to whom the assets are distributed.

(c) A transferee or successor entitled to a transfer of a license must file an application for such transfer with the chief of police within thirty (30) days after he assumes [ownership or control] of the business. The chief of police may require such evidence of the transfer of ownership or control as he deems necessary. A fee equal to twenty (20) percent of the original fee for the license shall accompany the application.

(d) The chief of police shall transfer the license to an applicant entitled to such transfer unless he finds that the transfer will adversely affect the peace, health, safety, and general welfare of the public or that the transferee does not possess the qualifications required of original applicants.

10-524. Change of location. Where no conflict exists with the city's zoning regulations, as evidenced by a new statement of zoning code compliance pursuant to subsection 10-511(c), a change of location may be endorsed on a firearm seller's license by the chief of police upon written application by a licensee, accompanied by a change of location fee equal to twenty (20) percent of the original fee for the firearm seller's license shall accompany the application.

10-525. No refund upon termination of business. The license fee paid under the terms of this article is not refundable upon termination of a business or for any unused portion or term of a license period.

10-526. Right of entry. The chief of police or any police officer shall have the power and authority to enter the premises of any business engaging in firearms sales, leases or transfers for which a license is required during business hours, free of any charge, and at such other times as may be reasonable following due process of law, and request a display of the required license together with any other federal, state, and city licenses, certificates, and permits relative to conducting the business of firearm sales. No person having custody or control of such place of business shall fail to comply with any such request.

10-527. Penalty for violation. Every violation of the provisions of this article shall be deemed to be a misdemeanor, and upon conviction thereof shall be punishable as provided for in section 1-8 of this Code. Each day any violation of any said provision of this article shall constitute a separate offense.

[Santa Ana Municipal Code Codified through Ord. No. 2739, enacted April 2, 2007]

Stockton Municipal Code

Chapter 4. Public Safety Code

Part II. Firearms and Fireworks

Division 3. Sale of Firearms and Ammunition to Minors

4-044. Sale of Firearms and/or Ammunition to Minors. It shall be unlawful, in the City of Stockton, to sell firearms and/or ammunition of **any kind to minors under the age** of fourteen (14) years, unless accompanied by parent or guardian.

Division 4. License to Sell Firearms

4-045. Sale of Concealed Firearms. It shall be unlawful to sell pistols, revolvers, or other firearms capable of being concealed upon the person, at retail, within the City of Stockton, without a license issued in the manner **herein provided for.**

4-046. License to Sell. The City Manager is hereby authorized to issue licenses to sell, at retail, within the City of Stockton, pistols, revolvers, and other firearms capable of being concealed upon the person. Such licenses shall be issued in the form prescribed by the Attorney General and shall be effective for not more than one (1) year from date of issue.

4-046.1. License - Chief of Police to Approve. The said licenses shall be issued only after a report in writing concerning said prospective licensee has been made by the Chief of Police to the City Manager. Said licenses shall not be issued to any person who

(a) has been convicted of a felony and has served a term of imprisonment therefor in a State or Federal prison.

(b) has been convicted of any violation of the Dangerous Weapons Control Law of the State of California, or

(c) is not of good moral character.

4-046.2. License - Revocation. Any of the reasons which would prevent the original issuance of said license shall also make the revocation of said license mandatory by the City Manager.

[Current through Stockton Ord. 018-07 effective Sept. 13, 2007]

COLORADO
Colo. Rev. Stat.

Title 12. Professions and Occupations General

Article 26. Firearms - Dealers

12-26-101. Definitions. As used in this article, unless the context otherwise requires:

(1)(a) "Firearms" means a pistol, revolver, or other weapon of any description, loaded or unloaded, from which any shot, bullet, or other missile can be discharged, the length of the barrel of which, not including any revolving, detachable, or magazine breech, does not exceed twelve inches.

(b) "Firearms" does not include firearms, as defined in paragraph (a) of this subsection (1), for which ammunition is not sold or which there is reasonable ground for believing are not capable of being effectually used.

12-26-102. Retail dealers - record - inspection. Every individual, firm, or corporation engaged, within this state, in the retail sale, rental, or exchange of firearms, pistols, or revolvers shall keep a record of each pistol or revolver sold, rented, or exchanged at retail. The record shall be made at the time of the transaction in a book kept for that purpose and shall include the name of the person to whom the pistol or revolver is sold or rented or with whom exchanged; his age, occupation, residence, and, if residing in a city, the street and number therein where he resides; the make, caliber, and finish of said pistol or revolver, together with its number and serial letter, if any; the date of the sale, rental, or exchange of said pistol or revolver; and the name of the employee or other person making such sale, rental, or exchange. The record book shall be open at all times to the inspection of any duly authorized police officer.

12-26-103. Record - failure to make - penalty. Every individual, firm, or corporation who fails to keep the record provided for in section 12-26-102 or who refuses to exhibit such record when requested by a police officer and any purchaser, lessee, or exchanger of a pistol or revolver who, in connection with the making of such record, gives false information is guilty of a misdemeanor and, upon conviction thereof, shall be punished by a fine of not less than twenty-five dollars nor more than one hundred dollars, or by imprisonment in the county jail for not more than one year, or by both such fine and imprisonment.

Article 26.1. Background checks at gun shows - penalty

12-26.1-101. Background checks at gun shows - penalty.

(1) Before a gun show vendor transfers or attempts to transfer a firearm at a gun show, he or she shall:

(a) require that a background check, in accordance with section 24-33.5-424, C.R.S., be conducted of the prospective transferee; and

(b) obtain approval of a transfer from the Colorado Bureau of Investigation after a background check has been requested by a licensed gun dealer, in accordance with section 24-33.5-424, C.R.S.

(2) A gun show promoter shall arrange for the services of one or more licensed gun dealers on the premises of the gun show to obtain the background checks required by this article.

(3) If any part of a firearm transaction takes place at a gun show, no firearm shall be transferred unless a background check has been obtained by a licensed gun dealer.

(4) Any person violating the provisions of this section commits a class 1 misdemeanor and shall be punished as provided in section 18-1.3-501, C.R.S.

12-26.1-102. Records - penalty.

(1) A licensed gun dealer who obtains a background check on a prospective transferee shall record the transfer, as provided in section 12-26-102, C.R.S., and retain the records, as provided in section 12-26-103, C.R.S., in the same manner as when conducting a sale, rental, or exchange at retail.

(2) Any individual who gives false information in connection with the making of such records commits a class 1 misdemeanor and shall be punished as provided in section 18-1.3-501, C.R.S.

12-26.1-103. Fees imposed by licensed gun dealers. For each background check conducted at a gun show, a licensed gun dealer may charge a fee not to exceed ten dollars.

12-26.1-104. Posted notice - penalty.

(1) A gun show promoter shall post prominently a notice, in a form to be prescribed by the executive director of the department of public safety or his or her designee, setting forth the requirement for a background check as provided in this article.

(2) Any person violating the provisions of this section commits a class 1 misdemeanor and shall be punished as provided in section 18-1.3-501, C.R.S.

12-26.1-105. Exemption. The provisions of this article shall not apply to the transfer of an antique firearm, as defined in 18 U.S.C. sec. 921(a)(16), as amended, or a curio or relic, as defined in 27 C.F.R. sec. 178.11, as amended.

12-26.1-106. Definitions. As used in this article, unless the context otherwise requires:

(1) "Collection" means a trade, barter, or in-kind exchange for one or more firearms.

(2) "Firearm" means any handgun, automatic, revolver, pistol, rifle, shotgun, or other instrument or device capable or intended to be capable of discharging bullets, cartridges, or other explosive charges.

(3) "Gun show" means the entire premises provided for an event or function, including but not limited to parking areas for the event or function, that is sponsored to facilitate, in whole or in part, the purchase, sale, offer for sale, or collection of firearms at which:

(a) twenty-five or more firearms are offered or exhibited for sale, transfer, or exchange; or

(b) not less than three gun show vendors exhibit, sell, offer for sale, transfer, or exchange firearms.

(4) "Gun show promoter" means a person who organizes or operates a gun show.

(5) "Gun show vendor" means any person who exhibits, sells, offers for sale, transfers or exchanges, any firearm at a gun show, regardless of whether the person arranges with a gun show promoter for a fixed location from which to exhibit, sell, offer for sale, transfer, or exchange any firearm.

(6) "Licensed gun dealer" means any person who is a licensed importer, licensed manufacturer, or dealer licensed pursuant to 18 U.S.C. sec. 923, as amended, as a federally licensed firearms dealer.

Article 27. Firearms - Purchase in Contiguous State

12-27-101. Legislative declaration - nonresident.

(1) It is declared by the general assembly that it is lawful for a licensed importer, licensed manufacturer, licensed dealer, or a licensed collector (licensed under the federal "Gun Control Act of 1968") whose place of business is in this state to sell or deliver a rifle or shotgun to a resident of a state contiguous to this state, subject to the following restrictions and requirements:

(a) The purchaser's state of residence must permit such sale or delivery by law.

(b) The sale must fully comply with the legal conditions of sale in both such contiguous states.

(c) The purchaser and the licensee must have complied, prior to the sale or delivery for sale of the rifle or shotgun, with all of the requirements of section 922 (c) of the federal "Gun Control Act of 1968" applicable to interstate transactions other than those at the licensee's business premises.

12-27-102. Legislative declaration - residents.

(1) It is declared by the general assembly that it is lawful for a resident of this state, otherwise qualified, to purchase or receive delivery of a rifle or shotgun in a state contiguous to this state, subject to the following restrictions and requirements:

(a) The sale must fully comply with the legal conditions of sale in both such contiguous states;

(b) The purchaser and the licensee must have complied, prior to the sale or delivery for sale of the rifle or shotgun, with all of the requirements of section 922 (c) of the federal "Gun Control Act of 1968", applicable to interstate transactions other than at the licensee's business premises.

12-27-103. Definitions. As used in this article, unless the context otherwise requires:

(1) "A state contiguous to this state" means any state having a common border with this state.

(2) All other terms shall be construed as such terms are defined in the federal "Gun Control Act of 1968".

12-27-104. Article does not apply - when.

(1) The provisions of this article do not apply to:

(a) Transactions between licensed importers, licensed manufacturers, licensed dealers, and licensed collectors;

(b) The loan or rental of a firearm to any person for temporary use for lawful sporting purposes;

(c) A person who is participating in any organized rifle or shotgun match or contest, or is engaged in hunting, in a state other than his state of residence and whose rifle or shotgun has been lost or stolen or has become inoperative in such other state, and who purchases a rifle or shotgun in such other state from a licensed dealer if such person presents to such dealer a sworn statement:

(I) That his rifle or shotgun was lost or stolen or became inoperative while participating in such a match or contest, or while engaged in hunting, in such other state; and

(II) Identifying the chief law enforcement officer of the locality in which such person resides, to whom such licensed dealer shall forward such statement by registered mail.

Title 18. Criminal Code

Article 12. Offenses Relating to Firearms and Weapons

Part 1. Firearms and Weapons – General

18-12-101. Definitions – peace officer affirmative defense.

(1) As used in this article, unless the context otherwise requires:

(a) "Adult" means any person eighteen years of age or older. ...

(b) "Bomb" means any explosive or incendiary device or Molotov cocktail as defined in section 9-7-103, C.R.S., or any chemical device which causes or can cause an explosion, which is not specifically designed for lawful and legitimate use in the hands of its possessor.

(c) "Firearm silencer" means any instrument, attachment, weapon, or appliance for causing the firing of any gun, revolver, pistol, or other firearm to be silent or intended to lessen or muffle the noise of the firing of any such weapon.

(d) "Gas gun" means a device designed for projecting gas-filled projectiles which release their contents after having been projected from the device and includes projectiles designed for use in such a device. ...

(e.5) "Handgun" means a pistol, revolver, or other firearm of any description, loaded or unloaded, from which any shot, bullet, or other missile can be discharged, the length of the barrel of which, not including any revolving, detachable, or magazine breech, does not exceed twelve inches.

(e.7) "Juvenile" means any person under the age of eighteen years. ...

(g) "Machine gun" means any firearm, whatever its size and usual designation, that shoots automatically more than one shot, without manual reloading, by a single function of the trigger.

(h) "Short rifle" means a rifle having a barrel less than sixteen inches long or an overall length of less than twenty-six inches.

(i) "Short shotgun" means a shotgun having a barrel or barrels less than eighteen inches long or an overall length of less than twenty-six inches.

(i.5) "Stun gun" means a device capable of temporarily immobilizing a person by the infliction of an electrical charge. ...

(2) It shall be an affirmative defense to any provision of this article that the act was committed by a peace officer in the lawful discharge of his duties.

18-12-102. Possessing a dangerous or illegal weapon - affirmative defense.

(1) As used in this section, the term "dangerous weapon" means a firearm silencer, machine gun, short shotgun, short rifle, or ballistic knife.

(2) As used in this section, the term "illegal weapon" means a blackjack, gas gun, metallic knuckles, gravity knife, or switchblade knife.

(3) A person who knowingly possesses a dangerous weapon commits a class 5 felony. Each subsequent violation of this subsection (3) by the same person shall be a class 4 felony.

(4) A person who knowingly possesses an illegal weapon commits a class 1 misdemeanor.

(5) It shall be an affirmative defense to the charge of possessing a dangerous weapon, or to the charge of possessing an illegal weapon, that the person so accused was a peace officer or member of the armed forces of the United States or Colorado National Guard acting in the lawful discharge of his duties, or that said person has a valid permit and license for possession of such weapon.

18-12-103. Possession of a defaced firearm. A person commits a class 1 misdemeanor if he knowingly and unlawfully possesses a firearm, the manufacturer's serial number of which, or other distinguishing number or identification mark, has been removed, defaced, altered, or destroyed, except by normal wear and tear.

18-12-103.5. Defaced firearms - contraband - destruction.

(1) After a judgment of conviction under section 18-12-103 or 18-12-104 has become final, any defaced firearm upon which the judgment was based shall be deemed to be contraband, the possession of which is contrary to the public peace, health, and safety.

(2) Defaced firearms which are deemed to be contraband shall be placed in the possession of the Colorado bureau of investigation or of a local law enforcement agency designated by the Colorado bureau of investigation and shall be destroyed or rendered permanently inoperable.

18-12-104. Defacing a firearm. A person commits a class 1 misdemeanor if such person knowingly removes, defaces, covers, alters, or destroys the manufacturer's serial number or any other distinguishing number or identification mark of a firearm.

18-12-105.5. Unlawfully carrying a weapon - unlawful possession of weapons - school, college, or university grounds.

(1) A person commits a class 6 felony if such person knowingly and unlawfully and without legal authority carries, brings, or has in such person's possession a deadly weapon as defined in section 18-1-901 (3) (e) in or on the real estate and all improvements erected thereon of any public or private elementary, middle, junior high, high, or vocational school or any public or private college, university, or seminary, except for the purpose of presenting an authorized public demonstration or exhibition pursuant to instruction in conjunction with an organized school or class, for the purpose of carrying out the necessary duties and functions of an employee of an educational institution that require the use of a deadly weapon, or for the purpose of participation in an authorized extracurricular activity or on an athletic team.

(2) [Deleted]

(3) It shall not be an offense under this section if:

(a) The weapon is unloaded and remains inside a motor vehicle while upon the real estate of any public or private college, university, or seminary; or

(b) The person is in that person's own dwelling or place of business or on property owned or under that person's control at the time of the act of carrying; or

(c) The person is in a private automobile or other private means of conveyance and is carrying a weapon for lawful protection of that person's or another's person or property while traveling; or

(d) The person, at the time of carrying a concealed weapon, held a valid written permit to carry a concealed weapon issued pursuant to section 18-12-105.1, as said section existed prior to its repeal; except that it shall be an offense under this section if the person was carrying a concealed handgun in violation of the provisions of section 18-12-214 (3); or

(d.5) The weapon involved was a handgun and the person held a valid permit to carry a concealed handgun or a temporary emergency permit issued pursuant to part 2 of this article; except that it shall be an offense under this section if the person was carrying a concealed handgun in violation of the provisions of section 18-12-214 (3); or

(e) The person is a peace officer, as described in section 16-2.5-101, C.R.S., when carrying a weapon in conformance with the policy of the employing agency as provided in section 16-2.5-101 (2), C.R.S.; or

(f) [Deleted]

(g) [Deleted]

(h) The person has possession of the weapon for use in an educational program approved by a school which program includes, but shall not be limited to, any course designed for the repair or maintenance of weapons.

18-12-106. Prohibited use of weapons.

(1) A person commits a class 2 misdemeanor if:

(a) He knowingly and unlawfully aims a firearm at another person; or

(b) Recklessly or with criminal negligence he discharges a firearm or shoots a bow and arrow; or

(c) He knowingly sets a loaded gun, trap, or device designed to cause an explosion upon being tripped or approached, and leaves it unattended by a competent person immediately present; or

(d) The person has in his or her possession a firearm while the person is under the influence of intoxicating liquor or of a controlled substance, as defined in section 12-22-303 (7), C.R.S. Possession of a permit issued under section 18-12-105.1, as it existed prior to its repeal, or possession of a permit or a temporary emergency permit issued pursuant to part 2 of this article is no defense to a violation of this subsection (1). ...

18-12-108. Possession of weapons by previous offenders.

(1) A person commits the crime of possession of a weapon by a previous offender if the person knowingly possesses, uses, or carries upon his or her person a firearm as described in section 18-1-901 (3) (h) or any other weapon that is subject to the provisions of this article subsequent to the person's conviction for a felony, or subsequent to the person's conviction for attempt or conspiracy to commit a felony, under Colorado or any other state's law or under federal law.

(2)(a) Except as otherwise provided by paragraphs (b) and (c) of this subsection (2), a person commits a class 6 felony if the person violates subsection (1) of this section.

(b) A person commits a class 5 felony, as provided by section 18-12-102, if the person violates subsection (1) of this section and the weapon is a dangerous weapon, as defined in section 18-12-102 (1).

(c) A person commits a class 5 felony if the person violates subsection (1) of this section and the person's previous conviction was for burglary, arson, or any felony involving the use of force or the use of a deadly weapon and the violation of subsection (1) of this section occurs as follows:

(I) From the date of conviction to ten years after the date of conviction, if the person was not incarcerated; or

(II) From the date of conviction to ten years after the date of release from confinement, if such person was incarcerated or, if subject to supervision imposed as a result of conviction, ten years after the date of release from supervision.

(d) Any sentence imposed pursuant to this subsection (2) shall run consecutively with any prior sentences being served by the offender.

(3) A person commits the crime of possession of a weapon by a previous offender if the person knowingly possesses, uses, or carries upon his or her person a firearm as described in section 18-1-901 (3) (h) or any other weapon that is subject to the provisions of this article subsequent to the person's adjudication for an act which, if committed by an adult, would constitute a felony, or subsequent to the person's adjudication for attempt or conspiracy to commit a felony, under Colorado or any other state's law or under federal law.

(4)(a) Except as otherwise provided by paragraphs (b) and (c) of this subsection (4), a person commits a class 6 felony if the person violates subsection (3) of this section.

(b) A person commits a class 5 felony, as provided by section 18-12-102, if the person violates subsection (3) of this section and the weapon is a dangerous weapon, as defined in section 18-12-102 (1).

(c) A person commits a class 5 felony if the person commits the conduct described in subsection (3) of this section and the person's previous adjudication was based on an act that, if committed by an adult, would constitute burglary, arson, or any felony involving the use of force or the use of a deadly weapon and the violation of subsection (3) of this section occurs as follows:

(I) From the date of adjudication to ten years after the date of adjudication, if the person was not committed to the department of institutions, or on or after July 1, 1994, to the department of human services; or

(II) From the date of adjudication to ten years after the date of release from commitment, if such person was committed to the department of institutions, or on or after July 1, 1994, to the department of human services or, if subject to supervision imposed as a result of an adjudication, ten years after the date of release from supervision.

(d) Any sentence imposed pursuant to this subsection (4) shall run consecutively with any prior sentences being served by the offender.

(5) A second or subsequent offense under paragraphs (b) and (c) of subsection (2) and paragraphs (b) and (c) of subsection (4) of this section is a class 4 felony.

(6)(a) Upon the discharge of any inmate from the custody of the department of corrections, the department shall provide a written advisement to such inmate of the prohibited acts and penalties specified in this section. The written advisement, at a minimum, shall include the written statement specified in paragraph (c) of this subsection (6).

(b) Any written stipulation for deferred judgment and sentence entered into by a defendant pursuant to section 18-1.3-102, shall contain a written advisement of the prohibited acts and penalties specified in this section. The written advisement, at a minimum, shall include the written statement specified in paragraph (c) of this subsection (6).

(c) The written statement shall provide that:

(I)(A) A person commits the crime of possession of a weapon by a previous offender in violation of this section if the person knowingly possesses, uses, or carries upon his or her person a firearm as described in section 18-1-901 (3) (h), or any other weapon that is subject to the provisions of this title subsequent to the person's conviction for a felony, or subsequent to the person's conviction for attempt or conspiracy to commit a felony, or subsequent to the person's conviction for a misdemeanor crime of domestic violence as defined in 18 U.S.C. sec. 921 (a) (33) (A), or subsequent to the person's conviction for attempt or conspiracy to commit such misdemeanor crime of domestic violence; and

(B) For the purposes of this paragraph (c), "felony" means any felony under Colorado law, federal law, or the laws of any other state; and

(II) A violation of this section may result in a sentence of imprisonment or fine, or both.

(d) The act of providing the written advisement described in this subsection (6) or the failure to provide such advisement may not be used as a defense to any crime charged and may not provide any basis for collateral attack on, or for appellate relief concerning, any conviction.

18-12-108.5. Possession of handguns by juveniles - prohibited - exceptions - penalty.

(1)(a) Except as provided in this section, it is unlawful for any person who has not attained the age of eighteen years knowingly to have any handgun in such person's possession.

(b) Any person possessing any handgun in violation of paragraph (a) of this subsection (1) commits the offense of illegal possession of a handgun by a juvenile.

(c)(I) Illegal possession of a handgun by a juvenile is a class 2 misdemeanor.

(II) For any second or subsequent offense, illegal possession of a handgun by a juvenile is a class 5 felony.

(d) Any person under the age of eighteen years who is taken into custody by a law enforcement officer for an offense pursuant to this section shall be taken into temporary custody in the manner described in section 19-2-508, C.R.S.

(2) This section shall not apply to:

(a) Any person under the age of eighteen years who is:

(I) In attendance at a hunter's safety course or a firearms safety course; or

(II) Engaging in practice in the use of a firearm or target shooting at an established range authorized by the governing body of the jurisdiction in which such range is located or any other area where the discharge of a firearm is not prohibited; or

(III) Engaging in an organized competition involving the use of a firearm or participating in or practicing for a performance by an organized group under 501 (c) (3) as determined by the federal internal revenue service which uses firearms as a part of such performance; or

(IV) Hunting or trapping pursuant to a valid license issued to such person pursuant to article 4 of title 33, C.R.S.; or

(V) Traveling with any handgun in such person's possession being unloaded to or from any activity described in subparagraph (I), (II), (III), or (IV) of this paragraph (a);

(b) Any person under the age of eighteen years who is on real property under the control of such person's parent, legal guardian, or grandparent and who has the permission of such person's parent or legal guardian to possess a handgun;

(c) Any person under the age of eighteen years who is at such person's residence and who, with the permission of such person's parent or legal guardian, possesses a handgun for the purpose of exercising the rights contained in section 18-1-704 or section 18-1-704.5.

(3) For the purposes of subsection (2) of this section, a handgun is "loaded" if:

(a) There is a cartridge in the chamber of the handgun; or

(b) There is a cartridge in the cylinder of the handgun, if the handgun is a revolver; or

(c) The handgun, and the ammunition for such handgun, is carried on the person of a person under the age of eighteen years or is in such close proximity to such person that such person could readily gain access to the handgun and the ammunition and load the handgun.

(4) Repealed.

18-12-108.7. Unlawfully providing or permitting a juvenile to possess a handgun – penalty – unlawfully providing a firearm other than a handgun to a juvenile – penalty.

(1)(a) Any person who intentionally, knowingly, or recklessly provides a handgun with or without remuneration to any person under the age of eighteen years in violation of section 18-12-108.5 or any person who knows of such juvenile's conduct which violates section 18-12-108.5 and fails to make reasonable efforts to prevent such violation commits the crime of unlawfully providing a handgun to a juvenile or permitting a juvenile to possess a handgun.

(b) Unlawfully providing a handgun to a juvenile or permitting a juvenile to possess a handgun in violation of this subsection (1) is a class 4 felony.

(2)(a) Any person who intentionally, knowingly, or recklessly provides a handgun to a juvenile or permits a juvenile to possess a handgun, even though such person is aware of a substantial risk that such juvenile will use a handgun to commit a felony offense, or who, being aware of such substantial risk, fails to make reasonable efforts to prevent the commission of the offense, commits the crime of unlawfully providing or permitting a juvenile to possess a handgun. A person shall be deemed to have violated this paragraph (a) if such person provides a handgun to or permits the possession of a handgun by any juvenile who has been convicted of a crime of violence, as defined in section 18-1.3-406 or any juvenile who has been adjudicated a juvenile delinquent for an offense which would constitute a crime of violence, as defined in section 18-1.3-406, if such juvenile were an adult.

(b) Unlawfully providing a handgun to a juvenile or permitting a juvenile to possess a handgun in violation of this subsection (2) is a class 4 felony.

(3) With regard to firearms other than handguns, no person shall sell, rent, or transfer ownership or allow unsupervised possession of a firearm with or without remuneration to any juvenile without the consent of the juvenile's parent or legal guardian. Unlawfully providing a firearm other than a handgun to a juvenile in violation of this subsection (3) is a class 1 misdemeanor.

(4) It shall not be an offense under this section if a person believes that a juvenile will physically harm the person if the person attempts to disarm the juvenile or prevent the juvenile from committing a violation of section 18-12-108.5.

18-12-109. Possession, use, or removal of explosives or incendiary devices - possession of components thereof - chemical, biological, and nuclear weapons - persons exempt - hoaxes.

(1) As used in this section:

(a)(I) "Explosive or incendiary device" means:

(A) Dynamite and all other forms of high explosives, including, but not limited to, water gel, slurry, military C-4 (plastic explosives), blasting agents to include nitro-carbon-nitrate, and ammonium nitrate and fuel oil mixtures, cast primers and boosters, R.D.X., P.E.T.N., electric and nonelectric blasting caps, exploding cords commonly called detonating cord or detcord or primacord, picric acid explosives, T.N.T. and T.N.T. mixtures, and nitroglycerin and nitroglycerin mixtures;

(B) Any explosive bomb, grenade, missile, or similar device; and

(C) Any incendiary bomb or grenade, fire bomb, or similar device, including any device, except kerosene lamps, which consists of or includes a breakable container including a flammable liquid or compound and a wick composed of any material which, when ignited, is capable of igniting such flammable liquid or compound and can be carried or thrown by one individual acting alone.

(II) "Explosive or incendiary device" shall not include rifle, pistol, or shotgun ammunition, or the components for handloading rifle, pistol, or shotgun ammunition.

(b)(I) "Explosive or incendiary parts" means any substances or materials or combinations thereof which have been prepared or altered for use in the creation of an explosive or incendiary device. Such substances or materials may include, but shall not be limited to, any:

(A) Timing device, clock, or watch which has been altered in such a manner as to be used as the arming device in an explosive;

(B) Pipe, end caps, or metal tubing which has been prepared for a pipe bomb;

(C) Mechanical timers, mechanical triggers, chemical time delays, electronic time delays, or commercially made or improvised items which, when used singly or in combination, may be used in the construction of a timing delay mechanism, booby trap, or activating mechanism for any explosive or incendiary device.

(II) "Explosive or incendiary parts" shall not include rifle, pistol, or shotgun ammunition, or the components for handloading rifle, pistol, or shotgun ammunition, or any signaling device customarily used in operation of railroad equipment.

(2) Any person who knowingly possesses, controls, manufactures, gives, mails, sends, or causes to be sent an explosive or incendiary device commits a class 4 felony.

(2.5) Any person who knowingly possesses, controls, manufacturers, gives, mails, sends, or causes to be sent a chemical, biological, or radiological weapon commits a class 3 felony.

(3) Subsection (2) of this section shall not apply to the following persons:

(a) A peace officer while acting in his official capacity transporting or otherwise handling explosives or incendiary devices;

(b) A member of the armed forces of the United States or Colorado National Guard while acting in his official capacity;

(c) An authorized employee of the office of active and inactive mines in the division of reclamation, mining, and safety while acting within the scope of his or her employment;

(d) A person possessing a valid permit issued under the provisions of article 7 of title 9, C.R.S., or an employee of such permittee acting within the scope of his employment;

(e) A person who is exempt from the necessity of possessing a permit under the provisions of section 9-7-106 (5), C.R.S., or an employee of such exempt person acting within the scope of his employment.

(f) A person or entity authorized to use chemical, biological, or radiological materials in their lawful business operations while using the chemical, biological, or radiological materials in the course of legitimate business activities. Authorized users shall include clinical, environmental, veterinary, agricultural, public health, or radiological laboratories and entities otherwise licensed to possess radiological materials.

(4) Any person who knowingly uses or causes to be used or gives, mails, sends, or causes to be sent an explosive or incendiary device or a chemical, biological, or radiological weapon or materials in the commission of or in an attempt to commit a felony commits a class 2 felony.

(5) Any person who removes or causes to be removed or carries away any explosive or incendiary device from the premises where said explosive or incendiary device is kept by the lawful user, vendor, transporter, or manufacturer thereof, without the consent or direction of the lawful possessor, commits a class 4 felony. A person convicted of this offense shall be subjected to a mandatory minimum sentence of two years in the department of corrections.

(5.5) Any person who removes or causes to be removed or carries away any chemical, biological, or radiological weapon from the premises where said chemical, biological, or radiological weapon is kept by the lawful user, vendor, transporter, or manufacturer thereof, without the consent or direction of the lawful possessor, commits a class 3 felony. A person convicted of this offense shall be subject to a mandatory minimum sentence of four years in the department of corrections.

(6) Any person who possesses any explosive or incendiary parts commits a class 4 felony.

(6.5) Any person who possesses any chemical weapon, biological weapon, or radiological weapon parts commits a class 3 felony.

(7) Any person who manufactures or possesses or who gives, mails, sends, or causes to be sent any false, facsimile, or hoax explosive or incendiary device or chemical, biological, or radiological weapon to another person or places any such purported explosive or incendiary device or chemical, biological, or radiological weapon in or upon any real or personal property commits a class 5 felony.

(8) Any person possessing a valid permit issued under the provisions of article 7 of title 9, C.R.S., or an employee of such permittee acting within the scope of his employment, who knowingly dispenses, distributes, or sells explosive or incendiary devices to a person who is not authorized to possess or control such explosive or incendiary device commits a class 4 felony.

18-12-111. Unlawful purchase of firearms.

(1) Any person who knowingly purchases or otherwise obtains a firearm on behalf of or for transfer to a person whom the transferor knows or reasonably should know is ineligible to possess a firearm pursuant to federal or state law commits a class 4 felony.

(2)(a) Any person who is a licensed dealer, as defined in 18 U.S.C. sec. 921 (a) (11), shall post a sign displaying the provisions of subsection (1) of this section in a manner that is easily readable. The person shall post such sign in an area that is visible to the public at each location from which the person sells firearms to the general public.

(b) Any person who violates any provision of this subsection (2) commits a class 2 petty offense and, upon conviction thereof, shall be punished by a fine of two hundred fifty dollars.

Part 2. Permits to Carry Concealed Handguns

18-12-201. Legislative declaration.

(1) The general assembly finds that:

(a) There exists a widespread inconsistency among jurisdictions within the state with regard to the issuance of permits to carry concealed handguns and identification of areas of the state where it is lawful to carry concealed handguns;

(b) This inconsistency among jurisdictions creates public uncertainty regarding the areas of the state in which it is lawful to carry concealed handguns;

(c) Inconsistency results in the arbitrary and capricious denial of permits to carry concealed handguns based on the jurisdiction of residence rather than the qualifications for obtaining a permit;

(d) The criteria and procedures for the lawful carrying of concealed handguns historically has been regulated by state statute and should be consistent throughout the state to ensure the consistent implementation of state law; and

(e) It is necessary that the state occupy the field of regulation of the bearing of concealed handguns since the issuance of a concealed handgun permit is based on a person's constitutional right of self-protection and there is a prevailing state interest in ensuring that no citizen is arbitrarily denied a concealed handgun permit and in ensuring that the laws controlling the use of the permit are consistent throughout the state.

(2) Based on the findings specified in subsection (1) of this section, the general assembly hereby concludes that:

(a) The permitting and carrying of concealed handguns is a matter of statewide concern; and

(b) It is necessary to provide statewide uniform standards for issuing permits to carry concealed handguns for self-defense.

(3) In accordance with the findings and conclusions specified in subsections (1) and (2) of this section, the general assembly hereby instructs each sheriff to implement and administer the provisions of this part 2. The general assembly does not delegate to the sheriffs the authority to regulate or restrict the issuance of permits provided for in this part 2 beyond the provisions of this part 2. An action or rule that encumbers the permit process by placing burdens on the applicant beyond those sworn statements and specified documents detailed in this part 2 or that creates restrictions beyond those specified in this part 2 is in conflict with the intent of this part 2 and is prohibited.

18-12-202. Definitions. As used in this part 2, unless the context otherwise requires:

(1) "Bureau" means the Colorado bureau of investigation within the department of public safety. ...

(4) "Handgun" means a handgun as defined in section 18-12-101 (1) (e.5); except that the term does not include a machine gun as defined in section 18-12-101 (1) (g). ...

(6) "Permit" means a permit to carry a concealed handgun issued pursuant to the provisions of this part 2; except that "permit" does not include a temporary emergency permit issued pursuant to section 18-12-209. ...

18-12-203. Criteria for obtaining a permit.

(1) Beginning May 17, 2003, except as otherwise provided in this section, a sheriff shall issue a permit to carry a concealed handgun to an applicant who:

(a) Is a legal resident of the state of Colorado. For purposes of this part 2, a person who is a member of the armed forces and is stationed pursuant to permanent duty station orders at a military installation in this state, and a member of the person's immediate family living in Colorado, shall be deemed to be a legal resident of the state of Colorado.

(b) Is twenty-one years of age or older;

(c) Is not ineligible to possess a firearm pursuant to section 18-12-108 or federal law;

(d) Has not been convicted of perjury under section 18-8-503, in relation to information provided or deliberately omitted on a permit application submitted pursuant to this part 2;

(e)(I) Does not chronically and habitually use alcoholic beverages to the extent that the applicant's normal faculties are impaired.

(II) The prohibition specified in this paragraph (e) shall not apply to an applicant who provides an affidavit, signed by a professional counselor who is licensed pursuant to article 43 of title 12, C.R.S., and specializes in alcohol addiction, stating that the applicant has been evaluated by the counselor and has been determined to be a recovering alcoholic who has refrained from using alcohol for at least three years.

(f) Is not an unlawful user of or addicted to a controlled substance as defined in section 18-18-102 (5).Whether an applicant is an unlawful user of or addicted to a controlled substance shall be determined as provided in federal law and regulations.

(g) Is not subject to:

(I) A protection order issued pursuant to section 18-1-1001 or section 19-2-707, C.R.S., that is in effect at the time the application is submitted; or

(II) A permanent protection order issued pursuant to article 14 of title 13, C.R.S.; or

(III) A temporary protection order issued pursuant to article 14 of title 13, C.R.S., that is in effect at the time the application is submitted;

(h) Demonstrates competence with a handgun by submitting:

(I) Evidence of experience with a firearm through participation in organized shooting competitions or current military service;

(II) Evidence that, at the time the application is submitted, the applicant is a certified instructor;

(III) Proof of honorable discharge from a branch of the United States armed forces within the three years preceding submittal of the application;

(IV) Proof of honorable discharge from a branch of the United States armed forces that reflects pistol qualifications obtained within the ten years preceding submittal of the application;

(V) A certificate showing retirement from a Colorado law enforcement agency that reflects pistol qualifications obtained within the ten years preceding submittal of the application; or

(VI) A training certificate from a handgun training class obtained within the ten years preceding submittal of the application. The applicant shall submit the original training certificate or a photocopy thereof that includes the original signature of the class instructor. In obtaining a training certificate from a handgun training class, the applicant shall have discretion in selecting which handgun training class to complete.

(2) Regardless of whether an applicant meets the criteria specified in subsection (1) of this section, if the sheriff has a reasonable belief that documented previous behavior by the applicant makes it likely the applicant will present a danger to self or others if the applicant receives a permit to carry a concealed handgun, the sheriff may deny the permit.

(3)(a) The sheriff shall deny, revoke, or refuse to renew a permit if an applicant or a permittee fails to meet one of the criteria listed in subsection (1) of this section and may deny, revoke, or refuse to renew a permit on the grounds specified in subsection (2) of this section.

(b) Following issuance of a permit, if the issuing sheriff has a reasonable belief that a permittee no longer meets the criteria specified in subsection (1) of this section or that the permittee presents a danger as described in subsection (2) of this section, the sheriff shall suspend the permit until such time as the matter is resolved and the issuing sheriff determines that the permittee is eligible to possess a permit as provided in this section.

(c) If the sheriff suspends or revokes a permit, the sheriff shall notify the permittee in writing, stating the grounds for suspension or revocation and informing the permittee of the right to seek a second review by the sheriff, to submit additional information for the record, and to seek judicial review pursuant to section 18-12-207.

18-12-204. Permit contents - validity - carrying requirements.

(1)(a) Each permit shall bear a color photograph of the permittee and shall display the signature of the sheriff who issues the permit. In addition, the sheriffs of this state shall ensure that all permits issued pursuant to this part 2 contain the same items of information and are the same size and the same color.

(b) A permit is valid for a period of five years after the date of issuance and may be renewed as provided in section 18-12-211. A permit issued pursuant to this part 2, including a temporary emergency permit issued pursuant to section 18-12-209, is effective in all areas of the state, except as otherwise provided in section 18-12-214.

(2)(a) A permittee, in compliance with the terms of a permit, may carry a concealed handgun as allowed by state law. The permittee shall carry the permit, together with valid photo identification, at all times during which the permittee is in actual possession of a concealed handgun and shall produce both documents upon demand by a law enforcement officer. Failure to produce a permit upon demand by a law enforcement officer raises a rebuttable presumption that the person does not have a permit. Failure to carry and produce a permit and valid photo identification upon demand as required in this subsection (2) is a class 1 petty offense. A charge of failure to carry and produce a permit and valid photo identification upon demand pursuant to this subsection (2) shall be dismissed by the court if, at or before the permittee's scheduled court appearance, the permittee exhibits to the court a valid permit and valid photo identification, both of which were issued to the permittee prior to the date on which the permittee was charged with failure to carry and produce a permit and valid photo identification upon demand.

(b) The provisions of paragraph (a) of this subsection (2) apply to temporary emergency permits issued pursuant to section 18-12-209.

(3)(a) A person who may lawfully possess a handgun may carry a handgun under the following circumstances without obtaining a permit and the handgun shall not be considered concealed:

(I) The handgun is in the possession of a person who is in a private automobile or in some other private means of conveyance and who carries the handgun for a legal use, including self-defense; or

(II) The handgun is in the possession of a person who is legally engaged in hunting activities within the state.

(b) The provisions of this subsection (3) shall not be construed to authorize the carrying of a handgun in violation of the provisions of section 18-12-105 or 18-12-105.5.

18-12-205. Sheriff - application - procedure - background check

(1)(a) To obtain a permit, a person shall submit a permit application on a statewide standardized form developed by the sheriffs and available from each sheriff. The permit application form shall solicit only the following information from the applicant:

(I) The applicant's full name, date of birth, and address;

(II) The applicant's birth name, if different from the name provided pursuant to subparagraph (I) of this paragraph (a), and any other names the applicant may have used or by which the applicant may have been known;

(III) The applicant's home address or addresses for the ten-year period immediately preceding submittal of the application;

(IV) Whether the applicant is a resident of this state as of the date of application and whether the applicant has a valid driver's license or other state-issued photo identification or military order proving residence; and

(V) Whether the applicant meets the criteria for obtaining a permit specified in section 18-12-203(1).

(b) The permit application form shall not require the applicant to waive or release a right or privilege, including but not limited to waiver or release of privileged or confidential information contained in medical records.

(2)(a) An applicant shall complete the permit application form and return it, in person, to the sheriff of the county or city and county in which the applicant resides, to the sheriff of the county or city and county in which the applicant maintains a secondary residence or owns or leases real property used by the applicant in a business, or to the sheriff that previously issued a permit to the applicant. The applicant shall sign the completed permit application form in person before the sheriff. The applicant shall provide his or her signature voluntarily upon a sworn oath that the applicant knows the contents of the permit application and that the information contained in the permit application is true and correct. An applicant who knowingly and intentionally makes a false or misleading statement on a permit application or deliberately omits any material information requested on the application commits perjury as described in section 18-8-503. Upon conviction, the applicant shall be punished as provided in section 18-1.3-501.In addition, the applicant shall be denied the right to obtain or possess a permit, and the sheriff shall revoke the applicant's permit if issued prior to conviction.

(b) An applicant shall also submit to the sheriff a permit fee not to exceed one hundred dollars for processing the permit application. The sheriff shall set the amount of the permit fee as provided in subsection (5) of this section. In addition, the applicant shall submit an amount specified by the director of the bureau, pursuant to section 24-72-306, C.R.S., for processing the applicant's fingerprints through the bureau and through the federal bureau of investigation. Neither the permit fee nor the fingerprint processing fee shall be refundable in the event the sheriff denies the applicant's permit application or suspends or revokes the permit subsequent to issuance.

(3) In addition to the items specified in subsection (2) of this section, an applicant, when submitting the completed permit application, shall submit the following items to the sheriff:

(a) Documentary evidence demonstrating competence with a handgun as specified in section 18-12-203 (1) (h); and

(b) A full frontal view color photograph of the applicant's head taken within the thirty days immediately preceding submittal of the permit application; except that the applicant need not submit a photograph if the sheriff photographs the applicant for purposes of issuing a permit. Any photograph submitted shall show the applicant's full head, including hair and facial features, and the depiction of the applicant's head shall measure one and one-eighth inches wide and one and one-fourth inches high.

(4)(a) The sheriff shall witness an applicant's signature on the permit application as provided in subsection (2) of this section and verify that the person making application for a permit is the same person who appears in any photograph submitted and the same person who signed the permit application form. To verify the applicant's identity, the applicant shall present to the sheriff the applicant's valid Colorado driver's license or valid Colorado or military photo identification.

(b) After verifying the applicant's identity, the sheriff shall take two complete sets of the applicant's fingerprints. The sheriff shall submit both sets of fingerprints to the bureau, and the sheriff shall not retain a set of the applicant's fingerprints.

(c) After receipt of a permit application and the items specified in this section, the sheriff shall verify that the applicant meets the criteria specified in section 18-12-203 (1) and is not a danger as described in section 18-12-203 (2).The verification at a minimum shall include requesting the bureau to conduct a search of the national instant criminal background check system and a search of the state integrated criminal justice information system to determine whether the applicant meets the criteria specified in section 18-12-203 (1).In addition, if the applicant resides in a municipality or town, the sheriff shall consult with the police department of the municipality or town in which the applicant resides, and the sheriff may consult with other local law enforcement agencies.

(5) The sheriff in each county or city and county in the state shall establish the amount of the new and renewal permit fees within his or her jurisdiction. The amount of the new and renewal permit fees shall comply with the limits specified in paragraph (b) of subsection (2) of this section and section 18-12-211 (1), respectively. The fee amounts shall reflect the actual direct and indirect costs to the sheriff of processing permit applications and renewal applications pursuant to this part 2.

18-12-206. Sheriff - issuance or denial of permits - report

(1) Within ninety days after the date of receipt of the items specified in section 18-12-205, a sheriff shall:

(a) Approve the permit application and issue the permit; or

(b) Deny the permit application based solely on the ground that the applicant fails to qualify under the criteria listed in section 18-12-203 (1) or that the applicant would be a danger as described in section 18-12-203 (2). If the sheriff denies the permit application, he or she shall notify the applicant in writing, stating the grounds for denial and informing the applicant of the right to seek a second review of the application by the sheriff, to submit additional information for the record, and to seek judicial review pursuant to section 18-12-207.

(2) If the sheriff does not receive the results of the fingerprint checks conducted by the bureau and by the federal bureau of investigation within ninety days after receiving a permit application, the sheriff shall determine whether to grant or deny the permit application without considering the fingerprint check information. If, upon receipt of the information, the sheriff finds that the permit was issued or denied erroneously, based on the criteria specified in section 18-12-203 (1) and (2), the sheriff shall either revoke or issue the permit, whichever is appropriate.

(3)(a) Each sheriff shall maintain a list of the persons to whom he or she issues permits pursuant to this part 2. Upon request by another criminal justice agency for law enforcement purposes, the sheriff may, at his or her discretion, share information from the list of permittees with a law enforcement agency for the purpose of determining the validity of a permit. A database maintained pursuant to this subsection (3) and any database operated by a state agency that includes permittees shall be searchable only by name.

(b)(I) Notwithstanding the provisions of paragraph (a) of this subsection (3), on and after July 1, 2011, a sheriff shall not share information from the list of permittees with a law enforcement agency for the purpose of creating a statewide database of permittees, and any law enforcement agency that receives information concerning permittees from a sheriff shall not use the information to create or maintain a statewide database of permittees. Any information concerning a permittee that is included in a statewide database pursuant to paragraph (a) of this subsection (3) shall be removed from the database no later than July 1, 2011.

(II) Prior to the repeal in subparagraph (I) of this paragraph (b), the state auditor's office shall conduct a performance audit of the statewide database of permittees as provided in section 2-3-118, C.R.S.

(c) Except for suspected violations of sections 18-12-105 and 18-12-105.5, a peace officer may not use or search a database of permittees maintained by a law enforcement agency to establish reasonable suspicion for a traffic stop, or when contacting an individual, to justify probable cause for a search or seizure of a person or a person's vehicle or property.

(4) Each sheriff shall annually prepare a report specifying, at a minimum, the number of permit applications received during the year for which the report was prepared, the number of permits issued during the year, the number of permits denied during the year, the reasons for denial, the number of revocations during the year, and the reasons for the revocations. The report shall not include the name of a person who applies for a permit, regardless of whether the person receives or is denied a permit. Each sheriff shall submit the report on or before March 1, 2004, and on or before March 1 each year thereafter, to the members of the general assembly. In addition, each sheriff shall provide a copy of the annual report prepared pursuant to this subsection (4) to a member of the public upon request.

18-12-208. Colorado bureau of investigation - duties

(1) Upon receipt of a permit applicant's fingerprints from a sheriff pursuant to section 18-12-205 (4) or upon a sheriff's request pursuant to section 18-12-211 (1), the bureau shall process the full set of fingerprints to obtain any available state criminal justice information or federal information pursuant to section 16-21-103 (5), C.R.S., and shall report any information received to the sheriff. In addition, within ten days after receiving the fingerprints, the bureau shall forward one set of the fingerprints to the federal bureau of investigation for processing to obtain any available state criminal justice information or federal information.

(2) The bureau shall use the fingerprints received pursuant to this part 2 solely for the purposes of:

(a) Obtaining information for the issuance or renewal of permits; and

(b) Notifying an issuing sheriff that a permittee has been arrested for or charged with an offense that would require revocation or suspension of the permit or that a permittee has been convicted of such an offense.

(3) On or before January 15, 2004, and on or before January 15 each year thereafter until January 15, 2007, the bureau shall provide to the general assembly a list of the jurisdictions in which the sheriff provides to the bureau the names of persons to whom the sheriff issues permits.

18-12-210. Maintenance of permit - address change - invalidity of permit.

(1) Within thirty days after a permittee changes the address specified on his or her permit or within three business days after his or her permit is lost, stolen, or destroyed, the permittee shall notify the issuing sheriff of the change of address or permit loss, theft, or destruction. Failure to notify the sheriff pursuant to this subsection (1) is a class 1 petty offense.

(2) If a permit is lost, stolen, or destroyed, the permit is automatically invalid. The person to whom the permit was issued may obtain a duplicate or substitute therefor upon payment of fifteen dollars to the issuing sheriff and upon submission of a notarized statement to the issuing sheriff that the permit has been lost, stolen, or destroyed.

(3) The provisions of this section apply to temporary emergency permits issued pursuant to section 18-12-209.

18-12-213. Reciprocity.

(1) A permit to carry a concealed handgun or a concealed weapon that is issued by a state that recognizes the validity of permits issued pursuant to this part 2 shall be valid in this state in all respects as a permit issued pursuant to this part 2 if the permit is issued to a person who is:

(a) Twenty-one years of age or older; and

(b)(I) A resident of the state that issued the permit, as demonstrated by the address stated on a valid picture identification that is issued by the state that issued the permit and is carried by the permit holder; or

(II) A resident of Colorado for no more than ninety days, as determined by the date of issuance on a valid picture identification issued by Colorado and carried by the permit holder.

(2) For purposes of this section, a "valid picture identification" means a driver's license or a state identification issued in lieu of a driver's license.

18-12-211. Renewal of permits

(1) Within one hundred twenty days prior to expiration of a permit, the permittee may obtain a renewal form from the issuing sheriff and renew the permit by submitting to the issuing sheriff a completed renewal form, a notarized affidavit stating that the permittee remains qualified pursuant to the criteria specified in section 18-12-203 (1) (a) to (1) (g), and the required renewal fee not to exceed fifty dollars, as set by the sheriff pursuant to section 18-12-205 (5).The renewal form shall meet the requirements specified in section 18-12-205 (1) for an application. The sheriff shall verify pursuant to section 18-12-205 (4) that the permittee meets the criteria specified in section 18-12-203 (1) (a) to (1) (g) and is not a danger as described in section 18-12-203 (2) and shall either renew or deny the renewal of the permit in accordance with the provisions of section 18-12-206 (1).If the sheriff denies renewal of a permit, the permittee may seek a second review of the renewal application by the sheriff and may submit additional information for the record. The permittee may also seek judicial review as provided in section 18-12-207.

(2) A permittee who fails to file a renewal form on or before the permit expiration date may renew the permit by paying a late fee of fifteen dollars in addition to the renewal fee established pursuant to subsection (1) of this section. No permit shall be renewed six months or more after its expiration date, and the permit shall be deemed to have permanently expired. A person whose permit has permanently expired may reapply for a permit, but the person shall submit an application for a permit and the fee required pursuant to section 18-12-205. A person who knowingly and intentionally files false or misleading information or deliberately omits material information required under this section is subject to criminal prosecution for perjury under section 18-8-503.

18-12-214. Authority granted by permit-carrying restrictions.

(1)(a) A permit to carry a concealed handgun authorizes the permittee to carry a concealed handgun in all areas of the state, except as specifically limited in this section. ...

(2) A permit issued pursuant to this Part 2 does no authorize a person to carry a concealed handgun into a place where the carrying of firearms is prohibited by federal law.

(3) A permit issued pursuant to this Part 2 does not authorize a person to carry a concealed handgun onto the real property, or into any improvements erected thereon, of a public elementary, middle, junior high, or high school; except that:

(a) A permittee may have a handgun on the real property of the public school so long as the handgun remains in his or her vehicle and, if the permittee is not in the vehicle, the handgun is in a compartment within the vehicle and the vehicle is locked.

(b) A permittee who is employed or retained by contract by a school district as a school security officer may carry a concealed handgun onto the real property, or into any improvement erected thereon, of a public elementary, middle, junior high, or high school while the permittee is on duty.

(c) A permittee may carry a concealed handgun on undeveloped real property owned by a school district that is used for hunting or other shooting sports. ...

18-12-216. Permits issued prior to May 17, 2003.

(1) A permit issued pursuant to section 18-12-105.1, as it existed prior to its repeal, shall permanently expire on June 30, 2007, or on the expiration date specified on the permit, whichever occurs first. ...

Title 24. Government - State

Article 33.5. Public Safety

Part 4. Colorado Bureau of Investigation

24-33.5-424. National instant criminal background check system - state point of contact - grounds for denial of firearm transfer - appeal - rule-making - unlawful acts - repeal.

(1) For purposes of this section:

(a) "18 U.S.C. sec. 922 (t)" means 18 U.S.C. sec. 922 (t) as it exists as of March 7, 2000, or as it may be amended.

(b) "Firearm" has the same meaning as set forth in 18 U.S.C. sec. 921 (a) (3), as amended.

(c) "NICS system" means the national instant criminal background check system created by Public Law 103-159, known as the federal "Brady Handgun Violence Prevention Act", the relevant portion of which is codified at 18 U.S.C. sec. 922 (t).

(d) "Transfer" means the sale or delivery of any firearm in this state by a transferor to a transferee. "Transfer" shall include redemption of a pawned firearm by any person who is not licensed as a federal firearms licensee by the federal bureau of alcohol, tobacco, and firearms or any of its successor agencies. "Transfer" shall not include the return or replacement of a firearm that had been delivered to a federal firearms licensee for the sole purpose of repair or customizing.

(e) "Transferee" means any person who is not licensed as a federal firearms licensee by the federal bureau of alcohol, tobacco, and firearms or any of its successor agencies, in accordance with the federal "Gun Control Act of 1968", chapter 44 of title 18 U.S.C., as amended, and to whom a transferor wishes to sell or deliver a firearm.

(f) "Transferor" means any licensed importer, licensed manufacturer, or licensed dealer as defined in 18 U.S.C. sec. 921 (a) (9), (a) (10), and (a) (11), as amended, respectively.

(2) The bureau is hereby authorized to serve as a state point of contact for implementation of 18 U.S.C. sec. 922 (t), all federal regulations and applicable guidelines adopted pursuant thereto, and the NICS system.

(3)(a) The bureau, acting as the state point of contact for implementation of 18 U.S.C. sec. 922 (t), shall transmit a request for a background check in connection with the prospective transfer of a firearm to the NICS system and may also search other databases. The bureau shall deny a transfer of a firearm to a prospective transferee if the transfer would violate 18 U.S.C. sec. 922 (g) or (n) or result in the violation of any provision of state law, including but not limited to section 18-12-108 (4) (c), C.R.S., involving acts which, if committed by an adult, would constitute a burglary, arson, or any felony involving the use of force or the use of a deadly weapon.

(b)(I) In addition to the grounds for denial specified in paragraph (a) of this subsection (3), the bureau shall deny a transfer of a firearm if at any time the bureau transmits the request or searches other databases, information indicates that the prospective transferee:

(A) Has been arrested for or charged with a crime for which the prospective transferee if convicted, would be prohibited under state or federal law from purchasing, receiving, or possessing a firearm and either there has been no final disposition of the case or the final disposition is not noted in the other databases; or

(B) Is the subject of an indictment, an information, or a felony complaint alleging that the prospective transferee has committed a crime punishable by imprisonment for a term exceeding one year as defined in 18 U.S.C. sec. 921 (a) (20), as amended, and either there has been no final disposition of the case or the final disposition is not noted in the other databases.

(II) This paragraph (b) shall be repealed, effective July 1, 2010.

(c) The bureau is authorized to cooperate with federal, state, and local law enforcement agencies to perform or assist any other law enforcement agency in performing any firearm retrievals, and to assist in the prosecution of any rescinded transfers.

(4) Pursuant to section 16-21-103 (4) (c), C.R.S., and section 19-1-304 (1) (b.8), C.R.S., the bureau shall receive and process information concerning final case disposition data of any cases prosecuted in a court in this state within seventy-two hours after the final disposition of the case for purposes of carrying out its duties under this section.

(5)(a) Upon denial of a firearm transfer, the bureau shall notify the transferor and send notice of the denial to the NICS system, pursuant to 18 U.S.C. sec. 922 (t). In addition, the bureau shall immediately send notification of such denial and the basis for the denial to the federal, state, and local law enforcement agencies having jurisdiction over the area in which the transferee resides and in which the transferor conducts any business.

(b) Upon denial of a firearm transfer, the transferor shall provide the transferee with written information prepared by the bureau concerning the procedure by which the transferee may request a review of the denial and of the instant criminal background check records that prompted the denial. The bureau shall render a final administrative decision regarding denial within thirty days after receiving information from the transferee that demonstrates the transfer was improperly denied.

(c) In the case of any transfer denied pursuant to paragraph (b) of subsection (3) of this section, the inability of the transferee to obtain the final disposition of a case that is no longer pending shall not constitute the basis for the continued denial of the transfer so long as the transferee provides a letter signed and verified by the clerk of the court or his or her designee that indicates that no final case disposition information is available. Upon presentation of such letter to the bureau, the bureau shall reverse the denial.

(d) If the bureau reverses a denial, the bureau shall immediately request that the agency that provided the records prompting the denial make a permanent change to such records if necessary to reflect accurate information. In addition, the bureau shall provide immediate notification of such reversal to all agencies and entities that had been previously notified of a denial pursuant to paragraph (a) of this subsection (5).

(6) If in the course of conducting any background check pursuant to this section, whether the firearms transaction is approved or denied, the bureau obtains information that indicates the prospective transferee is the subject of an outstanding warrant, the bureau shall immediately provide notification of such warrant to the federal, state, and local law enforcement agencies having jurisdiction over the area in which the transferee resides and in which the transferor conducts any business.

(7)(a) The executive director or his or her designee shall adopt such rules as are necessary to:

(I) Carry out the duties of the bureau as the state point of contact, as those duties are set forth in federal law, and assist in implementing 18 U.S.C. sec. 922 (t), all federal regulations and applicable guidelines adopted pursuant thereto, and the NICS system; and

(II) Ensure the proper maintenance, confidentiality, and security of all records and data provided pursuant to this section.

(b) The rules adopted pursuant to paragraph (a) of this subsection (7) shall include, but need not be limited to:

(I) Procedures whereby a prospective transferee whose transfer is denied may request a review of the denial and of the instant criminal background check records that prompted the denial;

(II) Procedures regarding retention of records obtained or created for purposes of this section or for implementation of 18 U.S.C. sec. 922 (t); except that the bureau shall not retain a record for more than forty-eight hours after the day on which the bureau approves the transfer;

(III) Procedures and forms adopted by the bureau that request information from and establish proper identification of a prospective transferee and that may correspond with any firearms transaction record required by 18 U.S.C. sec. 922(t). Such procedures and forms shall not preclude any person from making a lawful firearm transfer under this section.

(IV) Procedures for carrying out the duties under this section, including at a minimum:

(A) That the bureau shall be open for business at least twelve hours per day every calendar day, except Christmas day and Thanksgiving day, in order to transmit the requests for a background check to the NICS system and search other databases;

(B) That the bureau shall provide a toll-free telephone number for any person calling from within the state that is operational every day that the office is open for business for the purpose of responding to requests from transferors in accordance with this section; and

(C) That the bureau shall employ and train personnel at levels that ensure prompt processing of the reasonably anticipated volume of inquiries received under this section.

(8) Nothing in this section shall be construed to create any civil cause of action for damages in addition to that which is available under the "Colorado Governmental Immunity Act", article 10 of this title.

(9) No act performed by the bureau or its agents in carrying out their lawful duties under this section shall be construed to be a violation of any provision of title 18, C.R.S.

(10)(a) It is unlawful for:

(I) Any person, in connection with the acquisition or attempted acquisition of a firearm from any transferor, to willfully make any false or fictitious oral or written statement or to furnish or exhibit any false, fictitious, or misrepresented identifcation that is intended or likely to deceive such transferor with respect to any fact material to the lawfulness of the sale or other disposition of such firearm under federal or state law;

(II) Any transferor knowingly to request criminal history record information or a background check under false pretenses or knowingly to disseminate criminal history record information to any person other than the subject of such information;

(III) Any agent or employee or former agent or employee of the bureau knowingly to violate the provisions of this section.

(b) Any person who violates the provisions of paragraph (a) of this subsection (10) commits a class 1 misdemeanor and shall be punished as provided in section 18-1.3-501, C.R.S.

(11) Any transferor who complies with the provisions of this section shall not be subject to any civil or criminal liability or regulatory sanction that may arise from the lawful transfer or lawful denial of the transfer of a firearm.

Title 29. Government - Local Miscellaneous

Article 11.7. Regulation of Firearms

29-11.7-101. Legislative declaration.

(1) The general assembly hereby finds that:

(a) Section 3 of article II of the state constitution, the article referred to as the state bill of rights, declares that all persons have certain inalienable rights, which include the right to defend their lives and liberties;

(b) Section 13 of article II of the state constitution protects the fundamental right of a person to keep and bear arms and implements section 3 of article II of the state constitution;

(c) The general assembly recognizes a duty to protect and defend the fundamental civil rights set forth in paragraphs (a) and (b) of this subsection (1);

(d) There exists a widespread inconsistency among jurisdictions within the state with regard to firearms regulations;

(e) This inconsistency among local government laws regulating lawful firearm possession and ownership has extraterritorial impact on state citizens and the general public by subjecting them to criminal and civil penalties in some jurisdictions for conduct wholly lawful in other jurisdictions;

(f) Inconsistency among local governments of laws regulating the possession and ownership of firearms results in persons being treated differently under the law solely on the basis of where they reside, and a person's residence in a particular county or city or city and county is not a rational classification when it is the basis for denial of equal treatment under the law;

(g) This inconsistency places citizens in the position of not knowing when they may be violating the local laws and therefore being unable to avoid violating the law and becoming subject to criminal and other penalties.

(2) Based on the findings specified in subsection (1) of this section, the general assembly concludes that:

(a) The regulation of firearms is a matter of statewide concern;

(b) It is necessary to provide statewide laws concerning the possession and ownership of a firearm to ensure that law-abiding persons are not unfairly placed in the position of unknowingly committing crimes involving firearms.

29-11.7-102. Firearms database-prohibited.

(1) A local government, including a law enforcement agency, shall not maintain a list or other form of record or database of:

(a) Persons who purchase or exchange firearms or who leave firearms for repair or sale on consignment;

(b) Persons who transfer firearms, unless the persons are federally licensed firearms dealers;

(c) The descriptions, including serial numbers, of firearms purchased, transferred, exchanged, or left for repair or sale on consignment.

29-11.7-103. Regulation - type of firearm – prohibited. A local government may not enact an ordinance, regulation, or other law that prohibits the sale, purchase, or possession of a firearm that a person may lawfully sell, purchase, or possess under state or federal law. Any such ordinance, regulation, or other law enacted by a local government prior to March 18, 2003, is void and unenforceable.

29-11.7-104. Regulation - carrying - posting. A local government may enact an ordinance, regulation, or other law that prohibits the open carrying of a firearm in a building or specific area within the local government's jurisdiction. If a local government enacts an ordinance, regulation, or other law that prohibits the open carrying of a firearm in a building or specific area, the local government shall post signs at the public entrances to the building or specific area informing persons that the open carrying of firearms is prohibited in the building or specific area.

[Current through the 2008 Supplement]

CONNECTICUT

CONN. GEN. STAT.

Title 29. Public Safety and State Police

Chapter 529. Division of State Police

29-27. "Pistol" and "revolver" defined. The term "pistol" and the term "revolver", as used in sections 29-28 to 29-38, inclusive, mean any firearm having a barrel less than twelve inches in length.

29-28. Permit for sale at retail of pistol or revolver. Permit to carry pistol or revolver. Confidentiality of name and address of permit holder. Permits for out-of-state residents.

(a) No person who sells ten or more pistols or revolvers in a calendar year or is a federally-licensed firearm dealer shall advertise, sell, deliver, or offer or expose for sale or delivery, or have in such person's possession with intent to sell or deliver, any pistol or revolver at retail without having a permit therefor issued as provided in this subsection. The chief of police or, where there is no chief of police, the warden of the borough or the first selectman of the town, as the case may be, may, upon the application of any person, issue a permit in such form as may be prescribed by the Commissioner of Public Safety for the sale at retail of pistols and revolvers within the jurisdiction of the authority issuing such permit. No permit for the sale at retail of any pistol or revolver shall be issued unless the applicant holds a valid eligibility certificate for a pistol or revolver issued pursuant to section 29-36f or a valid state permit to carry a pistol or revolver issued pursuant to subsection (b) of this section and the applicant submits documentation sufficient to establish that local zoning requirements have been met for the location where the sale is to take place except that any person selling or exchanging a pistol or revolver for the enhancement of a personal collection or for a hobby or who sells all or part of such person's personal collection of pistols or revolvers shall not be required to submit such documentation for the location where the sale or exchange is to take place.

(b) Upon the application of any person having a bona fide residence or place of business within the jurisdiction of any such authority, such chief of police, warden or selectman may issue a temporary state permit to such person to carry a pistol or revolver within the state, provided such authority shall find that such applicant intends to make no use of any pistol or revolver which such applicant may be permitted to carry under such permit other than a lawful use and that such person is a suitable person to receive such permit. No state or temporary state permit to carry a pistol or revolver shall be issued under this subsection if the applicant (1) has failed to successfully complete a course approved by the Commissioner of Public Safety in the safety and use of pistols and revolvers including, but not limited to, a safety or training course in the use of pistols and revolvers available to the public offered by a law enforcement agency, a private or public educational institution or a firearms training school, utilizing instructors certified by the National Rifle Association or the Department of Environmental Protection and a safety or training course in the use of pistols or revolvers conducted by an instructor certified by the state or the National Rifle Association, (2) has been convicted of a felony or of a violation of subsection (c) of section 21a-279, section 53a-58, 53a-61, 53a-61a, 53a-62, 53a-63, 53a-96, 53a-175, 53a-176, 53a-178 or 53a-181d, (3) has been convicted as delinquent for the commission of a serious juvenile offense, as defined in section 46b-120, (4) has been discharged from custody within the preceding twenty years after having been found not guilty of a crime by reason of mental disease or defect pursuant to section 53a-13, (5) has been confined in a hospital for persons with psychiatric disabilities, as defined in section 17a-495, within the preceding twelve months by order of a probate court, (6) is subject to a restraining or protective order issued by a court in a case involving the use, attempted use or threatened use of physical force against another person, (7) is subject to a firearms seizure order issued pursuant to subsection (d) of section 29-38c after notice and hearing, (8) is prohibited from shipping, transporting, possessing or receiving a firearm pursuant to 18 USC 922(g)(4), (9) is an alien illegally or unlawfully in the United States, or (10) is less than twenty-one years of age. Nothing in this section shall require any person who holds a valid permit to carry a pistol or revolver on October 1, 1994, to participate in any additional training in the safety and use of pistols and revolvers. Upon issuance of a temporary state permit to the applicant, the local authority shall forward the original application to the commissioner. Not later than sixty days after receiving a temporary state permit, an applicant shall appear at a location designated by the commissioner to receive the state permit. Said commissioner may then issue, to any holder of any temporary state permit, a state permit to carry a pistol or revolver within the state. Upon issuance of the state permit, the commissioner shall make available to the permit holder a copy of the law regarding the permit holder's responsibility to report the loss or theft of a firearm and the penalties associated with the failure to comply with such law. Upon issuance of the state permit, the commissioner shall forward a record of such permit to the local authority issuing the temporary state permit. The commissioner shall retain records of all applications, whether approved or denied. The copy of the state permit delivered to the permittee shall be laminated and shall contain a full-face photograph of such permittee. A person holding a state permit issued pursuant to this subsection shall notify the issuing authority within two business days of any change of such person's address. The notification shall include the old address and the new address of such person.

(c) No issuing authority may require any sworn member of the Department of Public Safety or an organized local police department to furnish such sworn member's residence address in a permit application. The issuing authority shall allow each such sworn member who has a permit to carry a pistol or revolver issued by such authority, to revise such member's application to include a business or post office address in lieu of the residence address. The issuing authority shall notify each such member of the right to revise such application.

(d) Notwithstanding the provisions of sections 1-210 and 1-211, the name and address of a person issued a permit to sell at retail pistols and revolvers pursuant to subsection (a) of this section or a state or a temporary state permit to carry a pistol or revolver pursuant to subsection (b) of this section, or a local permit to carry pistols and revolvers issued by local authorities prior to October 1, 2001, shall be confidential and shall not be disclosed, except (1) such information may be disclosed to law enforcement officials acting in the performance of their duties, (2) the issuing authority may disclose such information to the extent necessary to comply with a request made pursuant to section 29-33 for verification that such state or temporary state permit is still valid and has not been suspended or revoked, and the local authority may disclose such information to the extent necessary to comply with a request made pursuant to section 29-33 for verification that a local permit is still valid and has not been suspended or revoked, and (3) such information may be disclosed to the Commissioner of Mental Health and Addiction Services to carry out the provisions of subsection (c) of section 17a-500.

(e) The issuance of any permit to carry a pistol or revolver does not thereby authorize the possession or carrying of a pistol or revolver in any premises where the possession or carrying of a pistol or revolver is otherwise prohibited by law or is prohibited by the person who owns or exercises control over such premises.

(f) Any bona fide resident of the United States having no bona fide residence or place of business within the jurisdiction of any local authority in the state, but who has a permit or license to carry a pistol or revolver issued by the authority of another state or subdivision of the United States, may apply directly to the Commissioner of Public Safety for a permit to carry a pistol or revolver in this state. All provisions of subsections (b), (c), (d) and (e) of this section shall apply to applications for a permit received by the commissioner under this subsection.

Sec. 3(a) A person is guilty of firearms trafficking if such person, knowingly and intentionally, directly or indirectly, causes one or more firearms that such person owns, is in possession of or is in control of to come into the possession of or control of another person whom such person knows or has reason to believe is prohibited from owning or possessing any firearm under state or federal law.

(b) Any person who violates any provision of this section shall be guilty of a class C felony if such person, on or after the effective date of this section, sells, delivers or otherwise transfers five or fewer firearms, and a class B felony if such person, on or after the effective date of this section, sells, delivers or otherwise transfers more than five firearms.

(c) For the purposes of this section, "firearm" means "firearm" as defined in section 53a-3 of the general statutes, but does not include a rifle or shotgun or an antique firearm as defined in subsection (b) of section 29-37a of the general statutes.

29-28a. Application for permit. Notice of decision to applicant.

(a) Requests for temporary state permits under section 29-28 shall be submitted to the chief of police, or, where there is no chief of police, to the warden of the borough or the first selectman of the town, as the case may be, on application forms prescribed by the Commissioner of Public Safety. Upon written request by any person for a temporary state permit not on a prescribed application form, or upon request by any person for such application form, the local authority shall supply such forms. When any such request is made in person at the office of the local authority, the local authority shall supply such application form immediately. When any such request is made in any other manner, the local authority shall supply such application form not later than one week after receiving such request. If such application form is not supplied within the time limits required by this section, the request therefor shall constitute a sufficient application. If any local authority fails to supply an application form upon the request of any person, such person may request an application form from the Commissioner of Public Safety or any barracks of the division of state police, and the time limits and procedures set forth in this section for handling requests for such forms shall be applicable.

(b) The local authority shall, not later than eight weeks after a sufficient application for a temporary state permit has been made, inform the applicant that such applicant's request for a temporary state permit has been approved or denied. The local authority shall forward a copy of the application indicating approval or denial of the temporary state permit to the Commissioner of Public Safety. If the local authority has denied the application for a temporary state permit, no state permit may be issued. The commissioner shall, not later than eight weeks after receiving an application indicating approval from the local authority, inform the applicant in writing that the applicant's application for a state permit has been approved or denied, or that the results of the national criminal history records check have not been received. If grounds for denial become known after a temporary state permit has been obtained, the temporary state permit shall be immediately revoked pursuant to section 29-32.

29-30. Fees for pistol and revolver permits. Expiration and renewal of permits.

(a) The fee for each permit originally issued under the provisions of subsection (a) of section 29-28 for the sale at retail of pistols and revolvers shall be one hundred dollars and for each renewal thereof one hundred dollars. The fee for each state permit originally issued under the provisions of subsection (b) of section 29-28 for the carrying of pistols and revolvers shall be seventy dollars plus sufficient funds as required to be transmitted to the Federal Bureau of Investigation to cover the cost of a national criminal history records check. The local authority shall forward sufficient funds for the national criminal history records check to the commissioner no later than five business days after receipt by the local authority of the application for the temporary state permit. Thirty-five dollars shall be retained by the local authority. Upon approval by the local authority of the application for a temporary state permit, thirty-five dollars shall be sent to the commissioner. The fee to renew each state permit originally issued under the provisions of subsection (b) of section 29-28, as amended by this act, shall be thirty-five dollars. Upon deposit of such fees in the General Fund, ten dollars of each fee shall be credited within thirty days to the appropriation for the Department of Public Safety to a separate non-lapsing account for the purposes of the issuance of permits under subsections (a) and (b) of section 29-28.

(b) A local permit originally issued before October 1, 2001, whether for the sale at retail of pistols and revolvers or for the carrying of pistols and revolvers, shall expire five years after the date it becomes effective and each renewal thereof shall expire five years after the expiration date of the permit being renewed. On and after October 1, 2001, no local permit for the carrying of pistols and revolvers shall be renewed.

(c) A state permit originally issued under the provisions of section 29-28 for the carrying of pistols and revolvers shall expire five years after the date such permit becomes effective and each renewal thereof shall expire five years after the expiration date of the state permit being renewed and such renewal shall not be contingent on the renewal or issuance of a local permit. A temporary state permit issued for the carrying of pistols and revolvers shall expire sixty days after the date it becomes effective, and may not be renewed.

(d) The renewal fee required pursuant to subsection (a) of this section shall apply for each renewal which is requested not earlier than thirty-one days before, and not later than thirty-one days after, the expiration date of the state permit being renewed.

(e) No fee or portion thereof paid under the provisions of this section for issuance or renewal of a state permit shall be refundable except if such permit for which the fee or portion thereof was paid was not issued or renewed. The portion of the fee expended on the national criminal history records check for any such permit that was not issued or renewed shall not be refunded.

(f) The issuing authority shall send a notice of the expiration of a state permit to carry a pistol or revolver, issued pursuant to section 29-28, to the holder of such permit, by first class mail, not less than ninety days before such expiration, and shall enclose with such notice a form for the renewal of said state permit. A state permit to carry a pistol or revolver, issued pursuant to section 29-28, shall be valid for a period of ninety days after the expiration date, except this provision shall not apply to any permit to carry a pistol or revolver which has been revoked or for which revocation is pending, pursuant to section 29-32.

29-31. Display of permit to sell. Record of sales. No sale of any pistol or revolver shall be made except in the room, store or place described in the permit for the sale of pistols and revolvers, and such permit or a copy thereof certified by the authority issuing the same shall be exposed to view within the room, store or place where pistols or revolvers are sold or offered or exposed for sale, and no sale or delivery of any pistol or revolver shall be made unless the purchaser or person to whom the same is to be delivered is personally known to the vendor of such pistol or revolver or the person making delivery thereof or unless the person making such purchase or to whom delivery thereof is to be made provides evidence of his identity. The vendor of any pistol or revolver shall keep a record of each pistol or revolver sold in a book kept for that purpose, which record shall be in such form as is prescribed by the Commissioner of Public Safety and shall include the date of the sale, the caliber, make, model and manufacturer's number of such pistol or revolver and the name, address and occupation of the purchaser thereof, and shall be signed by the purchaser and by the person making the sale, each in the presence of the other, and shall be preserved by the vendor of such pistol or revolver for at least six years.

29-32. Revocation of permit. Notification. Confiscation. Penalty for failure to surrender permit.

(a) For the purposes of this section, "conviction" means the entry of a judgment of conviction by any court of competent jurisdiction.

(b) Any state permit or temporary state permit for the carrying of any pistol or revolver may be revoked by the Commissioner of Public Safety for cause and shall be revoked by said commissioner upon conviction of the holder of such permit of a felony or of any misdemeanor specified in subsection (b) of section 29-28 or upon the occurrence of any event which would have disqualified the holder from being issued the state permit or temporary state permit pursuant to subsection (b) of section 29-28. Upon the revocation of any state permit or temporary state permit, the person whose state permit or temporary state permit is revoked shall be notified in writing and such state permit or temporary state permit shall be forthwith delivered to the commissioner. Any law enforcement authority shall confiscate and immediately forward to the commissioner any state permit or temporary state permit that is illegally possessed by any person. The commissioner may revoke the state permit or temporary state permit based upon the commissioner's own investigation or upon the request of any law enforcement agency. Any person who fails to surrender any permit within five days of notification in writing of revocation thereof shall be guilty of a class C misdemeanor.

(c) Any local permit for the carrying of a pistol or revolver issued prior to October 1, 2001, may be revoked by the authority issuing the same for cause, and shall be revoked by the authority issuing the same upon conviction of the holder of such permit of a felony or of any misdemeanor specified in subsection (b) of section 29-28 or upon the occurrence of any event which would have disqualified the holder from being issued such local permit. Upon the revocation of any local permit, the person whose local permit is revoked shall be notified in writing and such permit shall be forthwith delivered to the authority issuing the same. Upon the revocation of any local permit, the authority issuing the same shall forthwith notify the commissioner. Upon the revocation of any permit issued by the commissioner, the commissioner shall forthwith notify any local authority which the records of the commissioner show as having issued a currently valid local permit to the holder of the permit revoked by the commissioner. Any person who fails to surrender such permit within five days of notification in writing or revocation thereof shall be guilty of a class C misdemeanor.

29-32b. Board of Firearms Permit Examiners. Appeals to board. Hearings.

(a) There shall be established a Board of Firearms Permit Examiners, within the Department of Public Safety for administrative purposes only, hereinafter referred to as the board, to be comprised of seven members appointed by the Governor to serve during his term and until their successors are appointed and qualify. With the exception of public members, the members shall be appointed from nominees of the Commissioner of Public Safety, the Connecticut State Association of Chiefs of Police, the Commissioner of Environmental Protection, The Connecticut State Rifle and Revolver Association, Inc., and Ye Connecticut Gun Guild, Inc., and each of said organizations shall be entitled to representation on the board. At least one member of the board shall be a lawyer licensed to practice in this state, who shall act as chairman of the board during the hearing of appeals brought under this section.

(b) Any person aggrieved by any refusal to issue or renew a permit or certificate under the provisions of section 29-28 or 29-36f, or by any limitation or revocation of a permit or certificate issued under any of said sections, or by a refusal or failure of any issuing authority to furnish an application as provided in section 29-28a, may, within ninety days after receipt of notice of such refusal, limitation or revocation, or refusal or failure to supply an application as provided in section 29-28a, and without prejudice to any other course of action open to such person in law or in equity, appeal to the board. On such appeal the board shall inquire into and determine the facts, de novo, and unless it finds that such a refusal, limitation or revocation, or such refusal or failure to supply an application, as the case may be, would be for just and proper cause, it shall order such permit or certificate to be issued, renewed or restored, or the limitation removed or modified, as the case may be. If the refusal was for failure to document compliance with local zoning requirements, under subsection (a) of section 29-28, the board shall not issue a permit.

(c) Any person aggrieved by the action of an issuing authority may file with the board a clear and concise statement of the facts on which he relies for relief, and shall state the relief sought by the appellant. The receipt by the board of the appellant's statement shall initiate the appeals process, and no appeal may be rejected for mere lack of formality. The board shall, within ten days next following receipt of the appeal, set a time and place at which the appeal shall be heard. The board, while such appeal is pending, may request such additional information from the appellant and from the issuing authority as it deems reasonably necessary to conduct a fair and impartial hearing, and shall require of the issuing authority from whose decision or action the appeal is being sought a statement in writing setting forth the reasons for such failure, refusal, revocation or limitation. Failure or refusal of the issuing authority to furnish such written statement, or to supply the appellant with an application, at least ten days prior to the hearing shall be cause for the board to grant the relief sought, forthwith and without further hearing.

(d) The board shall hold hearings at such times and places as it in its discretion reasonably determines to be required, but not less than once every ninety days, and shall give reasonable notice of the time and place of the hearing to the appellant and to the issuing authority. The board shall have the power to compel attendance at its sessions.

(e) All appeals hearings shall be conducted in an informal manner, but otherwise according to the rules of evidence, and all witnesses shall be sworn by the chairman. The board shall cause a verbatim transcript of the hearing to be kept in such manner as it may determine, and shall furnish such transcript to any party appealing its decision as hereinafter set forth. The statements of witnesses made under oath shall be privileged. Decisions of the board shall be by majority vote and shall be communicated in writing to the appellant and to the issuing authority within twenty days after the rendering of the decision. If any issuing authority neglects or refuses to comply with a decision of the board within ten days after notice of the board's decision has been given to such issuing authority, the board shall apply to the Superior Court for a writ of mandamus to enforce the board's decision.

(f) Any person aggrieved by the decision of the board may appeal therefrom in accordance with the provisions of section 4-183.

(g) The board shall serve without compensation, but its members shall be entitled to reasonable subsistence and travel allowances in the performance of their duties.

29-33. Sale, delivery or transfer of pistols and revolvers. Procedure. Penalty.

(a) No person, firm or corporation shall sell, deliver or otherwise transfer any pistol or revolver to any person who is prohibited from possessing a pistol or revolver as provided in section 53a-217c.

(b) On and after October 1, 1995, no person may purchase or receive any pistol or revolver unless such person holds a valid permit to carry a pistol or revolver issued pursuant to subsection (b) of section 29-28, a valid permit to sell at retail a pistol or revolver issued pursuant to subsection (a) of section 29-28 or a valid eligibility certificate for a pistol or revolver issued pursuant to section 29-36f or is a federal marshal, parole officer or peace officer.

(c) No person, firm or corporation shall sell, deliver or otherwise transfer any pistol or revolver except upon written application on a form prescribed and furnished by the Commissioner of Public Safety. Such person, firm or corporation shall insure that all questions on the application are answered properly prior to releasing the pistol or revolver and shall retain the application, which shall be attached to the federal sale or transfer document, for at least twenty years or until such vendor goes out of business. Such application shall be available for inspection during normal business hours by law enforcement officials. No sale, delivery or other transfer of any pistol or revolver shall be made unless the person making the purchase or to whom the same is delivered or transferred is personally known to the person selling such pistol or revolver or making delivery or transfer thereof or provides evidence of his identity in the form of a motor vehicle operator's license, identity card issued pursuant to section 1-1h or valid passport. No sale, delivery or other transfer of any pistol or revolver shall be made until the person, firm or corporation making such transfer obtains an authorization number from the Commissioner of Public Safety. Said commissioner shall perform the national instant criminal background check and make a reasonable effort to determine whether there is any reason that would prohibit such applicant from possessing a pistol or revolver as provided in section 53a-217c. If the commissioner determines the existence of such a reason, the commissioner shall deny the sale and no pistol or revolver shall be sold, delivered or otherwise transferred by such person, firm or corporation to such applicant.

(d) No person, firm or corporation shall sell, deliver or otherwise transfer any pistol or revolver, other than at wholesale, unless such pistol or revolver is equipped with a reusable trigger lock, gun lock or gun locking device appropriate for such pistol or revolver, which lock or device shall be constructed of material sufficiently strong to prevent it from being easily disabled and have a locking mechanism accessible by key or by electronic or other mechanical accessory specific to such lock or device to prevent unauthorized removal. No pistol or revolver shall be loaded or contain therein any gunpowder or other explosive or any bullet, ball or shell when such pistol or revolver is sold, delivered or otherwise transferred.

(e) Upon the sale, delivery or other transfer of any pistol or revolver, the person making the purchase or to whom the same is delivered or transferred shall sign a receipt for such pistol or revolver which shall contain the name and address of such person, the date of sale, the caliber, make, model and manufacturer's number and a general description of such pistol or revolver, the identification number of such person's permit to carry pistols or revolvers, issued pursuant to subsection (b) of section 29-28, permit to sell at retail pistols or revolvers, issued pursuant to subsection (a) of said section, or eligibility certificate for a pistol or revolver, issued pursuant to section 29-36f, if any, and the authorization number designated for the transfer by the Department of Public Safety. The person, firm or corporation selling such pistol or revolver or making delivery or transfer thereof shall give one copy of the receipt to the person making the purchase of such pistol or revolver or to whom the same is delivered or transferred, shall retain one copy of the receipt for at least five years, and shall send, by first class mail, or electronically transmit, within forty-eight hours of such sale, delivery or other transfer, one copy of the receipt to the Commissioner of Public Safety and one copy of the receipt to the chief of police or, where there is no chief of police, the warden of the borough or the first selectman of the town, as the case may be, of the town in which the transferee resides.

(f) The provisions of this section shall not apply to antique pistols or revolvers. An antique pistol or revolver, for the purposes of this section, means any pistol or revolver which was manufactured in or before 1898 and any replica of such pistol or revolver provided such replica is not designed or redesigned for using rimfire or conventional centerfire fixed ammunition except rimfire or conventional centerfire fixed ammunition which is no longer manufactured in the United States and not readily available in the ordinary channel of commercial trade.

(g) The provisions of this section shall not apply to the sale, delivery or transfer of pistols or revolvers between (1) a federally-licensed firearm manufacturer and a federally-licensed firearm dealer, (2) a federally-licensed firearm importer and a federally-licensed firearm dealer, or (3) federally-licensed firearm dealers.

(h) If the court finds that a violation of this section is not of a serious nature and that the person charged with such violation (1) will probably not offend in the future, (2) has not previously been convicted of a violation of this section, and (3) has not previously had a prosecution under this section suspended pursuant to this subsection, it may order suspension of prosecution. The court shall not order suspension of prosecution unless the accused person has acknowledged that he understands the consequences of the suspension of prosecution. Any person for whom prosecution is suspended shall agree to the tolling of any statute of limitations with respect to such violation and to a waiver of his right to a speedy trial. Such person shall appear in court and shall be released to the custody of the Court Support Services Division for such period, not exceeding two years, and under such conditions as the court shall order. If the person refuses to accept, or, having accepted, violates such conditions, the court shall terminate the suspension of prosecution and the case shall be brought to trial. If such person satisfactorily completes his period of probation, he may apply for dismissal of the charges against him and the court, on finding such satisfactory completion, shall dismiss such charges. If the person does not apply for dismissal of the charges against him after satisfactorily completing his period of probation, the court, upon receipt of a report submitted by the Court Support Services Division that the person satisfactorily completed his period of probation, may on its own motion make a finding of such satisfactory completion and dismiss such charges. Upon dismissal, all records of such charges shall be erased pursuant to section 54-142a. An order of the court denying a motion to dismiss the charges against a person who has completed his period of probation or terminating the participation of a defendant in such program shall be a final judgment for purposes of appeal.

(i) Any person who violates any provision of this section shall be guilty of a class D felony, except that any person who sells, delivers or otherwise transfers a pistol or revolver in violation of the provisions of this section, knowing that such pistol or revolver is stolen or that the manufacturer's number or other mark of identification on such pistol or revolver has been altered, removed or obliterated, shall be guilty of a class B felony, and any pistol or revolver found in the possession of any person in violation of any provision of this section shall be forfeited.

29-34. False statement or information in connection with sale or transfer of pistol or revolver prohibited. Sale or transfer to person under twenty-one years of age prohibited. Temporary transfers. Penalties.

(a) No person shall make any false statement or give any false information connected with any purchase, sale, delivery or other transfer of any pistol or revolver. Any person violating any provision of this subsection shall be guilty of a class D felony.

(b) No person shall sell, barter, hire, lend, give, deliver or otherwise transfer to any person under the age of twenty-one years any pistol or revolver, except that a pistol or revolver may be temporarily transferred to any person only for the use by such person in target shooting or on a firing or shooting range, provided such use is otherwise permitted by law and is under the immediate supervision of a person eligible to possess a pistol or revolver. Any person violating any provision of this subsection shall be guilty of a class D felony for which one year of the sentence imposed may not be suspended or reduced by the court.

(c) Any pistol or revolver found in the possession of any person in violation of any provision of this section shall be forfeited.

29-35. Carrying of pistol or revolver without permit prohibited. Exceptions.

(a) No person shall carry any pistol or revolver upon his or her person, except when such person is within the dwelling house or place of business of such person, without a permit to carry the same issued as provided in section 29-28. The provisions of this subsection shall not apply to the carrying of any pistol or revolver by any parole officer or peace officer of this state, or parole officer or peace officer of any other state while engaged in the pursuit of official duties, or federal marshal or federal law enforcement agent, or to any member of the armed forces of the United States, as defined in section 27-103, or of this state, as defined in section 27-2, when on duty or going to or from duty, or to any member of any military organization when on parade or when going to or from any place of assembly, or to the transportation of pistols or revolvers as merchandise, or to any person transporting any pistol or revolver while contained in the package in which it was originally wrapped at the time of sale and while transporting the same from the place of sale to the purchaser's residence or place of business, or to any person removing such person's household goods or effects from one place to another, or to any person while transporting any such pistol or revolver from such person's place of residence or business to a place or individual where or by whom such pistol or revolver is to be repaired or while returning to such person's place of residence or business after the same has been repaired, or to any person transporting a pistol or revolver in or through the state for the purpose of taking part in competitions, taking part in formal pistol or revolver training, repairing such pistol or revolver or attending any meeting or exhibition of an organized collectors' group if such person is a bona fide resident of the United States and is permitted to possess and carry a pistol or revolver in the state or subdivision of the United States in which such person resides, or to any person transporting a pistol or revolver to and from a testing range at the request of the issuing authority, or to any person transporting an antique pistol or revolver, as defined in section 29-33. For the purposes of this subsection, "formal pistol or revolver training" means pistol or revolver training at a locally approved or permitted firing range or training facility, and "transporting a pistol or revolver" means transporting a pistol or revolver that is unloaded and, if such pistol or revolver is being transported in a motor vehicle, is not readily accessible or directly accessible from the passenger compartment of the vehicle or, if such pistol or revolver is being transported in a motor vehicle that does not have a compartment separate from the passenger compartment, such pistol or revolver shall be contained in a locked container other than the glove compartment or console. Nothing in this section shall be construed to prohibit the carrying of a pistol or revolver during formal pistol or revolver training or repair.

(b) The holder of a permit issued pursuant to section 29-28 shall carry such permit upon one's person while carrying such pistol or revolver.

29-36. Alteration of firearm identification mark, number or name.

(a) No person shall remove, deface, alter or obliterate the name of any maker or model or any maker's number or other mark of identification on any firearm as defined in section 53a-3. The possession of any firearm upon which any identifying mark, number or name has been removed, defaced, altered or obliterated shall be prima facie evidence that the person owning or in possession of such firearm has removed, defaced, altered or obliterated the same.

(b) Any person who violates any provision of this section shall be fined not more than one thousand dollars or imprisoned not more than five years or both and any firearm found in the possession of any person in violation of said provision shall be forfeited.

29-36f. Eligibility certificate for pistol or revolver.

(a) Any person who is twenty-one years of age or older may apply to the Commissioner of Public Safety for an eligibility certificate for a pistol or revolver.

(b) The Commissioner of Public Safety shall issue an eligibility certificate unless said commissioner finds that the applicant:

(1) has failed to successfully complete a course approved by the Commissioner of Public Safety in the safety and use of pistols and revolvers including, but not limited to, a safety or training course in the use of pistols and revolvers available to the public offered by a law enforcement agency, a private or public educational institution or a firearms training school, utilizing instructors certified by the National Rifle Association or the Department of Environmental Protection and a safety or training course in the use of pistols or revolvers conducted by an instructor certified by the state or the National Rifle Association;

(2) has been convicted of a felony or of a violation of subsection (c) of section 21a-279, section 53a-58, 53a-61, 53a-61a, 53a-62, 53a-63, 53a-96, 53a-175, 53a-176, 53a-178 or 53a-181d;

(3) has been convicted as delinquent for the commission of a serious juvenile offense, as defined in section 46b-120;

(4) has been discharged from custody within the preceding twenty years after having been found not guilty of a crime by reason of mental disease or defect pursuant to section 53a-13;

(5) has been confined in a hospital for persons with psychiatric disabilities, as defined in section 17a-495, within the preceding twelve months by order of a probate court;

(6) is subject to a restraining or protective order issued by a court in a case involving the use, attempted use or threatened use of physical force against another person;

(7) is subject to a firearms seizure order issued pursuant to subsection (d) of section 29-38c after notice and hearing;

(8) is prohibited from shipping, transporting, possessing or receiving a firearm pursuant to 18 USC 922(g)(4); or (9) is an alien illegally or unlawfully in the United States.

29-36g. Application for eligibility certificate. Criminal history records check. Deadline for approval or denial of application. Form of certificate. Change of address. Confidentiality of name and address of certificate holder. Scope of certificate.

(a) Requests for eligibility certificates under section 29-36f shall be submitted to the Commissioner of Public Safety on application forms prescribed by the commissioner. No eligibility certificate for a pistol or revolver shall be issued under the provisions of said section unless the applicant for such certificate gives to the Commissioner of Public Safety, upon the commissioner's request, full information concerning the applicant's criminal record and relevant information concerning the applicant's mental health history. The commissioner shall require each applicant to submit to state and national criminal history records checks. The commissioner shall take a full description of such applicant. The commissioner shall take the fingerprints of such applicant or conduct any other method of positive identification required by the State Police Bureau of Identification or the Federal Bureau of Investigation. The commissioner shall record the date the fingerprints were taken in the applicant's file and shall conduct criminal history records checks in accordance with section 29-17a. The commissioner shall, within sixty days of receipt of the national criminal history records check from the Federal Bureau of Investigation, either approve the application and issue the eligibility certificate or deny the application and notify the applicant of the reason for such denial in writing.

(b)(1) With respect to any application for an eligibility certificate filed with the Commissioner of Public Safety on or before July 1, 1995, the commissioner shall, not later than October 1, 1995,

(A) approve the application and issue the eligibility certificate,

(B) issue a temporary eligibility certificate, or

(C) deny the application and notify the applicant of the reason for such denial in writing.

(2) With respect to any application for an eligibility certificate filed with the Commissioner of Public Safety after July 1, 1995, the commissioner shall, within ninety days,

(A) approve the application and issue the eligibility certificate,

(B) issue a temporary eligibility certificate, or

(C) denies the application and notify the applicant of the reason for such denial in writing.

(3) A temporary certificate issued under this subsection shall be valid until such time as the commissioner either approves or denies the application.

(c) An eligibility certificate for a pistol or revolver shall be of such form and content as the commissioner may prescribe, shall be signed by the certificate holder and shall contain an identification number, the name, address, place and date of birth, height, weight and eye color of the certificate holder and a full-face photograph of the certificate holder.

(d) A person holding an eligibility certificate issued by the commissioner shall notify the commissioner within two business days of any change of his address. The notification shall include his old address and his new address.

(e) Notwithstanding the provisions of sections 1-210 and 1-211, the name and address of a person issued an eligibility certificate for a pistol or revolver under the provisions of section 29-36f shall be confidential and shall not be disclosed, except (1) such information may be disclosed to law enforcement officials acting in the performance of their duties, (2) the Commissioner of Public Safety may disclose such information to the extent necessary to comply with a request made pursuant to section 29-33 for verification that such certificate is still valid and has not been suspended or revoked, and (3) such information may be disclosed to the Commissioner of Mental Health and Addiction Services to carry out the provisions of subsection (c) of section 17a-500.

(f) An eligibility certificate for a pistol or revolver shall not authorize the holder thereof to carry a pistol or revolver upon his person in circumstances for which a permit to carry a pistol or revolver issued pursuant to subsection (b) of section 29-28 is required under section 29-35.

29-36i. Revocation of eligibility certificate.

(a) Any eligibility certificate for a pistol or revolver shall be revoked by the Commissioner of Public Safety upon the occurrence of any event which would have disqualified the holder from being issued the certificate pursuant to section 29-36f.

(b) Upon the revocation of any eligibility certificate, the person whose eligibility certificate is revoked shall be notified in writing and such certificate shall be forthwith delivered to the Commissioner of Public Safety. Any person who fails to surrender such certificate within five days of notification in writing of revocation thereof shall be guilty of a class C misdemeanor.

29-36k. Transfer or surrender of firearms by persons ineligible to possess same. Penalty.

(a) Not later than two business days after the occurrence of any event that makes a person ineligible to possess a pistol or revolver or other firearm, such person shall (1) transfer in accordance with section 29-33 all pistols and revolvers which such person then possesses to any person eligible to possess a pistol or revolver and transfer in accordance with any applicable state and federal laws all other firearms to any person eligible to possess such other firearms by obtaining an authorization number for the sale or transfer of the firearm from the Commissioner of Public Safety, and submit a sale or transfer of firearms form to said commissioner within two business days, or (2) deliver or surrender such pistols and revolvers and other firearms to the Commissioner of Public Safety. The commissioner shall exercise due care in the receipt and holding of such pistols and revolvers and other firearms.

(b) Such person, or such person's legal representative, may, at any time up to one year after such delivery or surrender, transfer such pistols and revolvers in accordance with the provisions of section 29-33 to any person eligible to possess a pistol or revolver and transfer such other firearms in accordance with any applicable state and federal laws to any person eligible to possess such other firearms. Upon notification in writing by the transferee and such person, the Commissioner of Public Safety shall within ten days deliver such pistols and revolvers or other firearms to the transferee. If, at the end of such year, such pistols and revolvers or other firearms have not been so transferred, the commissioner shall cause them to be destroyed.

(c) Any person who fails to transfer or surrender any such pistols and revolvers and other firearms as provided in this section shall be subject to the penalty provided for in section 53a-217 or 53a-217c.

29-36l. Verification of eligibility of persons to receive or possess firearms. State database. Instant criminal background check. Immunity of seller or transferor. Authorization number required.

(a) The Commissioner of Public Safety shall establish a state database within one year of October 1, 1994, that any person, firm or corporation who sells or otherwise transfers pistols or revolvers may access, by telephone or other electronic means in addition to the telephone, for information to be supplied immediately, on whether a permit to carry a pistol or revolver, issued pursuant to subsection (b) of section 29-28, a permit to sell at retail a pistol or revolver, issued pursuant to subsection (a) of section 29-28, or an eligibility certificate for a pistol or revolver, issued pursuant to section 29-36f, is valid and has not been revoked or suspended.

(b) Upon establishment of the database, the commissioner shall notify each person, firm or corporation holding a permit to sell at retail pistols or revolvers issued pursuant to subsection (a) of section 29-28 of the existence and purpose of the system and the means to be used to access the database.

(c) The Department of Public Safety shall establish days and hours during which the telephone number or other electronic means shall be operational for purposes of responding to inquiries, taking into consideration the normal business hours of retail firearm businesses.

(d)(1) The Department of Public Safety shall be the point of contact for initiating a background check through the National Instant Criminal Background Check System (NICS), established under section 103 of the Brady Handgun Violence Prevention Act, on individuals purchasing firearms.

(2) The Department of Public Safety, Department of Mental Health and Addiction Services and Judicial Department shall, in accordance with state and federal law regarding confidentiality, enter into a memorandum of understanding with the Federal Bureau of Investigation for the purpose of implementing the National Instant Criminal Background Check System in the state. The Department of Public Safety shall report the name, date of birth and physical description of any person prohibited from possessing a firearm pursuant to 18 USC 922(g) or (n) to the National Instant Criminal Background Check System Index, Denied Persons Files.

(e) Any person, firm or corporation that contacts the Department of Public Safety to access the database established under this section and determine if a person is eligible to receive or possess a firearm shall not be held civilly liable for the sale or transfer of a firearm to a person whose receipt or possession of such firearm is unlawful or for refusing to sell or transfer a firearm to a person who may lawfully receive or possess such firearm if such person, firm or corporation relied in good faith, on the information provided to such person, firm or corporation by said department, unless the conduct of such person, firm or corporation was unreasonable or reckless.

(f) Any person, firm or corporation that sells, delivers or otherwise transfers any firearm pursuant to section 29-33 or section 29-37a shall contact the Department of Public Safety to access the database established under this section and receive an authorization number for such sale, delivery or transfer. The provisions of this subsection shall not apply to: (1) Any sale, delivery or transfer of an antique firearm manufactured in or before 1898, including any firearm with a matchlock, flintlock, percussion cap or similar type of ignition system manufactured in or before 1898; (2) any sale, delivery or transfer of any replica of any firearm described in subdivision (1) of this subsection if such replica uses rimfire or conventional centerfire fixed ammunition which is no longer manufactured in the United States and which is not readily available in the ordinary channels of commercial trade; (3) transactions between persons who are licensed as firearms importers or collectors, manufacturers or dealers pursuant to 18 USC 921 et seq.; (4) the transfer of firearms to and from gunsmiths for purposes of repair only; and (5) any sale, delivery or transfer of any firearm to any agency of the United States, the state of Connecticut or any local government.

29-36m. Regulations. The Commissioner of Public Safety shall adopt regulations in accordance with the provisions of chapter 54 to carry out the provisions of sections 18-81i, 29-27, 29-28, subsection (a) of section 29-30, section 29-32, subsection (b) of section 29-32b, sections 29-33, 29-34, 29-36f to 29-36l, inclusive, subsection (a) of section 29-37, subsections (a) and (b) of section 53-202d and section 53a-217c.

29-36n. Protocol concerning transfer or surrender of pistols and revolvers.

(a) The Commissioner of Public Safety, in conjunction with the Chief State's Attorney and the Connecticut Police Chiefs Association, shall develop a protocol to ensure that persons who become ineligible to possess a pistol or revolver have, in accordance with section 29-36k, transferred such pistol or revolver to a person eligible to possess such pistol or revolver or have delivered or surrendered such pistol or revolver to said commissioner.

(b) The Commissioner of Public Safety, in conjunction with the Chief State's Attorney and the Connecticut Police Chiefs Association, shall update the protocol developed pursuant to subsection (a) of this section to reflect the provisions of sections 29-7h, 29-28, 29-28a, 29-29, 29-30, 29-32 and 29-35, subsections (b) and (e) of section 46b-15, subsections (c) and (d) of section 46b-38c and sections 53-202a, 53-202l, 53-202m and 53a-217 and shall include in such protocol specific instructions for the transfer of pistols and revolvers when the assistance of more than one law enforcement agency is necessary to effect the requirements of section 29-36k.

29-37. Penalties.

(a) Any person violating any provision of section 29-28 or 29-31 shall be fined not more than five hundred dollars or imprisoned not more than three years or both, and any pistol or revolver found in the possession of any person in violation of any of said provisions shall be forfeited.

(b) Any person violating any provision of subsection (a) of section 29-35 may be fined not more than one thousand dollars and shall be imprisoned not less than one year or more than five years, and, in the absence of any mitigating circumstances as determined by the court, one year of the sentence imposed may not be suspended or reduced by the court. The court shall specifically state the mitigating circumstances, or the absence thereof, in writing for the record. Any pistol or revolver found in the possession of any person in violation of any provision of subsection (a) of section 29-35 shall be forfeited.

(c) Any person violating any provision of subsection (b) of section 29-35 shall have committed an infraction and shall be fined thirty-five dollars.

29-37a. Sale or delivery at retail of firearm other than pistol or revolver. Procedure.

(a) No person, firm or corporation may deliver, at retail, any firearm, as defined in section 53a-3, other than a pistol or revolver, to any person unless such person makes application on a form prescribed and furnished by the Commissioner of Public Safety, which shall be attached by the vendor to the federal sale or transfer document and filed and retained by the vendor for at least twenty years or until such vendor goes out of business. Such application shall be available for inspection during normal business hours by law enforcement officials. No sale or delivery of any firearm shall be made until the expiration of two weeks from the date of the application, and until the person, firm or corporation making such sale, delivery or transfer has insured that such application has been completed properly and has obtained an authorization number from the Commissioner of Public Safety for such sale, delivery or transfer. The Department of Public Safety shall make every effort, including performing the national instant criminal background check, to determine if the applicant is eligible to receive such firearm. If it is determined that the applicant is ineligible to receive such firearm, the Commissioner of Public Safety shall immediately notify the person, firm or corporation to whom such application was made and no such firearm shall be sold or delivered to such applicant by such person, firm or corporation. When any firearm is delivered in connection with the sale or purchase, such firearm shall be enclosed in a package, the paper or wrapping of which shall be securely fastened, and no such firearm when delivered on any sale or purchase shall be loaded or contain any gunpowder or other explosive or any bullet, ball or shell.

(b) Upon the delivery of the firearm, the purchaser shall sign in triplicate a receipt for such firearm which shall contain the name and address of such purchaser, the date of sale, caliber, make, model and manufacturer's number and a general description thereof. Not later than twenty-four hours after such delivery, the vendor shall send by first class mail or electronically transfer one receipt to the Commissioner of Public Safety and one receipt to the chief of police or, where there is no chief of police, the warden of the borough or the first selectman, of the town in which the purchaser resides, and shall retain one receipt, together with the original application, for at least five years. The waiting period specified in subsection (a) of this section during which delivery may not be made and the provisions of this subsection shall not apply to any federal marshal, parole officer or peace officer, or to the delivery at retail of (1) any firearm to a holder of a valid state permit to carry a pistol or revolver issued under the provisions of section 29-28 or a valid eligibility certificate issued under the provisions of section 29-36f, (2) any firearm to an active member of the armed forces of the United States or of any reserve component thereof, (3) any firearm to a holder of a valid hunting license issued pursuant to chapter 490, or (4) antique firearms. For the purposes of this section, "antique firearm" means any firearm which was manufactured in or before 1898 and any replica of such firearm provided such replica is not designed or redesigned for using rimfire or conventional centerfire fixed ammunition except rimfire or conventional centerfire fixed ammunition which is no longer manufactured in the United States and not readily available in the ordinary channel of commercial trade.

29-37b. Retail dealer to equip pistols and revolvers with gun locking device and provide written warning at time of sale. Penalty.

(a) Each person, firm or corporation which engages in the retail sale of any pistol or revolver, at the time of sale of any such pistol or revolver, shall (1) equip such pistol or revolver with a reusable trigger lock, gun lock or gun locking device appropriate for such firearm, which lock or device shall be constructed of material sufficiently strong to prevent it from being easily disabled and have a locking mechanism accessible by key or by electronic or other mechanical accessory specific to such lock or device to prevent unauthorized removal, and (2) provide to the purchaser thereof a written warning which shall state in block letters not less than one inch in height: "UNLAWFUL STORAGE OF A LOADED FIREARM MAY RESULT IN IMPRISONMENT OR FINE."

(b) Each such person, firm or corporation shall conspicuously post and at all times display the warning specified in subsection (a) of this section in block letters not less than three inches in height.

(c) Any person, firm or corporation which violates any provision of this section shall be fined not less than five hundred dollars for each violation.

29-37d. Firearms dealer to install burglar alarm system on premises of its establishment. Exceptions. On and after July 1, 1993, each business organization which engages in the retail sale of firearms, as defined in section 53a-3, as a regular course of trade or business, shall have a burglar alarm system installed on the premises of its establishment in which ten or more firearms are stored and kept for sale. Such alarm system shall be directly connected to the local police department or monitored by a central station and shall activate upon unauthorized entry or interruption to such system. For the purposes of this section, "business organization" means a sole proprietorship, partnership, firm, corporation or other form of business or legal entity. The provisions of this section shall not apply to any person who (1) sells or exchanges a firearm for the enhancement of a personal collection or as a hobby, (2) sells all or part of a personal collection of firearms, or (3) sells firearms from his own residence and keeps for sale not more than ten firearms.

29-37e. False statement or information in connection with sale or transfer of firearm other than pistol or revolver prohibited.

(a) No person shall make any false statement or give any false information connected with any purchase, sale, delivery or other transfer of any firearm other than a pistol or revolver. Any person violating any provision of this subsection shall be guilty of a class D felony.

(b) Any firearm found in the possession of any person in violation of this section shall be forfeited.

29-37f. Qualifications of retail store employees who sell firearms. No person, firm or corporation that engages in the retail sale of goods, where the principal part of such trade or business is the retail sale of goods other than firearms, shall employ a person to sell firearms in a retail store unless such person (1) is at least eighteen years of age, (2) has submitted to state and national criminal history records checks and such checks indicate that such person has not been convicted of a felony or a violation specified in subdivision (2) of subsection (b) of section 29-36f, and (3) has successfully completed a course or testing approved by the Commissioner of Public Safety in firearms safety and statutory procedures relating to the sale of firearms. The sale of firearms by such person, firm or corporation shall be accomplished only by an employee qualified pursuant to this section. Any employer who employs a person to sell firearms in violation of the provisions of this section shall be liable for a civil penalty of not more than ten thousand dollars per day for each violation. The Attorney General shall institute a civil action to recover such penalty.

29-37g. Gun show requirements.

(a) For the purposes of this section,

(1) "gun show" means any event (A) at which fifty or more firearms are offered or exhibited for sale, transfer or exchange to the public and (B) at which two or more persons are exhibiting one or more firearms for sale, transfer or exchange to the public; and

(2) "gun show promoter" means any person who organizes, plans, promotes or operates a gun show.

(b) Not later than thirty days before commencement of a gun show, the gun show promoter shall notify the chief of police or, where there is no chief of police, the warden of the borough or the first selectman of the town in which the gun show is to take place of the date, time, duration and location of the gun show.

(c) No person, firm or corporation shall sell, deliver or otherwise transfer a firearm at a gun show until such person, firm or corporation has complied with the provisions of section 29-36l.

29-37i. Responsibilities re storage of loaded firearms with respect to minors.

No person shall store or keep any loaded firearm on any premises under his control if he knows or reasonably should know that a minor is likely to gain access to the firearm without the permission of the parent or guardian of the minor unless such person (1) keeps the firearm in a securely locked box or other container or in a location which a reasonable person would believe to be secure or (2) carries the firearm on his person or within such close proximity thereto that he can readily retrieve and use it as if he carried it on his person. For the purposes of this section, "minor" means any person under the age of sixteen years.

29-37j. Purchase of firearm with intent to transfer it to person prohibited from purchasing or possessing.

(a) Any person who purchases a firearm, as defined in section 53a-3, pursuant to section 29-33 or 29-37a with the intent to transfer such firearm to any other person who the transferor knows or has reason to believe is prohibited from purchasing or otherwise receiving such a firearm pursuant to section 29-33 or 29-37a shall be fined not more than one thousand dollars or imprisoned not more than five years or both.

(b) Any person prohibited from purchasing or otherwise receiving or possessing a firearm and who solicits, employs or assists any person in violating the provisions of subsection (a) of this section shall be guilty of a class B misdemeanor. If the violation of subsection (a) of this section involves a transfer of more than one firearm, such person shall be guilty of a class A misdemeanor. Each transfer shall constitute a separate offense.

(c) Any person convicted of violating the provisions of subsection (a) or (b) of this section and who was convicted of a felony within the prior five-year period shall be guilty of a class D felony.

29-38. Weapons in vehicles.

(a) Any person who knowingly has, in any vehicle owned, operated or occupied by such person, any weapon, any pistol or revolver for which a proper permit has not been issued as provided in section 29-28 or any machine gun which has not been registered as required by section 53-202, shall be fined not more than one thousand dollars or imprisoned not more than five years or both, and the presence of any such weapon, pistol or revolver, or machine gun in any vehicle shall be prima facie evidence of a violation of this section by the owner, operator and each occupant thereof. The word "weapon", as used in this section, means any BB gun, any blackjack, any metal or brass knuckles, any police baton or nightstick, any dirk knife or switch knife, any knife having an automatic spring release device by which a blade is released from the handle, having a blade of over one and one-half inches in length, any stiletto, any knife the edged portion of the blade of which is four inches or over in length, any martial arts weapon or electronic defense weapon, as defined in section 53a-3, or any other dangerous or deadly weapon or instrument.

(b) The provisions of this section shall not apply to:

(1) Any officer charged with the preservation of the public peace while engaged in the pursuit of such officer's official duties;

(2) any security guard having a baton or nightstick in a vehicle while engaged in the pursuit of such guard's official duties;

(3) any person enrolled in and currently attending a martial arts school, with official verification of such enrollment and attendance, or any certified martial arts instructor, having any such martial arts weapon in a vehicle while traveling to or from such school or to or from an authorized event or competition;

(4) any person having a BB gun in a vehicle provided such weapon is unloaded and stored in the trunk of such vehicle or in a locked container other than the glove compartment or console; and

(5) any person having a knife, the edged portion of the blade of which is four inches or over in length, in a vehicle if such person is

(A) any member of the armed forces of the United States, as defined in section 27-103, or any reserve component thereof, or of the armed forces of this state, as defined in section 27-2, when on duty or going to or from duty,

(B) any member of any military organization when on parade or when going to or from any place of assembly,

(C) any person while transporting such knife as merchandise or for display at an authorized gun or knife show,

(D) any person while lawfully removing such person's household goods or effects from one place to another, or from one residence to another,

(E) any person while actually and peaceably engaged in carrying any such knife from such person's place of abode or business to a place or person where or by whom such knife is to be repaired, or while actually and peaceably returning to such person's place of abode or business with such knife after the same has been repaired,

(F) any person holding a valid hunting, fishing or trapping license issued pursuant to chapter 490 or any salt water fisherman while having such knife in a vehicle for lawful hunting, fishing or trapping activities, or

(G) any person participating in an authorized historic reenactment.

29-38b. Determination of commitment status of person who applies for or seeks renewal of firearm permit or certificate. Report on status of application.

(a) The Commissioner of Public Safety, in fulfilling his obligations under sections 29-28 to 29-38a, inclusive, and section 53-202d, shall verify that any person who, on or after October 1, 1998, applies for or seeks renewal of a permit to sell at retail a pistol or revolver, a permit to carry a pistol or revolver, an eligibility certificate for a pistol or revolver or a certificate of possession for an assault weapon has not been confined in a hospital for persons with psychiatric disabilities, as defined in section 17a-495, within the preceding twelve months by order of a probate court, by making an inquiry to the Department of Mental Health and Addiction Services in such a manner so as to only receive a report on the commitment status of the person with respect to whom the inquiry is made including identifying information in accordance with the provisions of subsection (b) of section 17a-500.

(b) If the Commissioner of Public Safety determines pursuant to subsection (a) of this section that a person has been confined in a hospital for persons with psychiatric disabilities, as defined in section 17a-495, within the preceding twelve months by order of a probate court, said commissioner shall report the status of such person's application for or renewal of a permit to sell at retail a pistol or revolver, a permit to carry a pistol or revolver, an eligibility certificate for a pistol or revolver or a certificate of possession for an assault weapon to the Commissioner of Mental Health and Addiction Services for the purpose of fulfilling his responsibilities under subsection (c) of section 17a-500.

29-38c. Seizure of firearms of person posing risk of imminent personal injury to self or others.

(a) Upon complaint on oath by any state's attorney or assistant state's attorney or by any two police officers, to any judge of the Superior Court, that such state's attorney or police officers have probable cause to believe that (1) a person poses a risk of imminent personal injury to himself or herself or to other individuals, (2) such person possesses one or more firearms, and (3) such firearm or firearms are within or upon any place, thing or person, such judge may issue a warrant commanding a proper officer to enter into or upon such place or thing, search the same or the person and take into such officer's custody any and all firearms. Such state's attorney or police officers shall not make such complaint unless such state's attorney or police officers have conducted an independent investigation and have determined that such probable cause exists and that there is no reasonable alternative available to prevent such person from causing imminent personal injury to himself or herself or to others with such firearm.

(b) A warrant may issue only on affidavit sworn to by the complainant or complainants before the judge and establishing the grounds for issuing the warrant, which affidavit shall be part of the seizure file. In determining whether grounds for the application exist or whether there is probable cause to believe they exist, the judge shall consider: (1) Recent threats or acts of violence by such person directed toward other persons; (2) recent threats or acts of violence by such person directed toward himself or herself; and (3) recent acts of cruelty to animals as provided in subsection (b) of section 53-247 by such person. In evaluating whether such recent threats or acts of violence constitute probable cause to believe that such person poses a risk of imminent personal injury to himself or herself or to others, the judge may consider other factors including, but not limited to (A) the reckless use, display or brandishing of a firearm by such person, (B) a history of the use, attempted use or threatened use of physical force by such person against other persons, (C) prior involuntary confinement of such person in a hospital for persons with psychiatric disabilities, and (D) the illegal use of controlled substances or abuse of alcohol by such person. If the judge is satisfied that the grounds for the application exist or that there is probable cause to believe that they exist, such judge shall issue a warrant naming or describing the person, place or thing to be searched. The warrant shall be directed to any police officer of a regularly organized police department or any state police officer. It shall state the grounds or probable cause for its issuance and it shall command the officer to search within a reasonable time the person, place or thing named for any and all firearms. A copy of the warrant shall be given to the person named therein together with a notice informing the person that such person has the right to a hearing under this section and the right to be represented by counsel at such hearing.

(c) The applicant for the warrant shall file a copy of the application for the warrant and all affidavits upon which the warrant is based with the clerk of the court for the geographical area within which the search will be conducted no later than the next business day following the execution of the warrant. Prior to the execution and return of the warrant, the clerk of the court shall not disclose any information pertaining to the application for the warrant or any affidavits upon which the warrant is based. The warrant shall be executed and returned with reasonable promptness consistent with due process of law and shall be accompanied by a written inventory of all firearms seized.

(d) Not later than fourteen days after the execution of a warrant under this section, the court for the geographical area where the person named in the warrant resides shall hold a hearing to determine whether the seized firearms should be returned to the person named in the warrant or should continue to be held by the state. At such hearing the state shall have the burden of proving all material facts by clear and convincing evidence. If, after such hearing, the court finds by clear and convincing evidence that the person poses a risk of imminent personal injury to himself or herself or to other individuals, it may order that the firearm or firearms seized pursuant to the warrant issued under subsection (a) of this section continue to be held by the state for a period not to exceed one year, otherwise the court shall order the seized firearm or firearms to be returned to the person named in the warrant. If the court finds that the person poses a risk of imminent personal injury to himself or herself or to other individuals, it shall give notice to the Department of Mental Health and Addiction Services which may take such action pursuant to chapter 319i as it deems appropriate.

(e) Any person whose firearm or firearms have been ordered seized pursuant to subsection (d) of this section, or such person's legal representative, may transfer such firearm or firearms in accordance with the provisions of section 29-33 or other applicable state or federal law, to any person eligible to possess such firearm or firearms. Upon notification in writing by such person, or such person's legal representative, and the transferee, the head of the state agency holding such seized firearm or firearms shall within ten days deliver such firearm or firearms to the transferee.

29-38d. Interstate transportation of firearms through state.

(a) The provisions of sections 29-35 and 29-38 shall not apply to the interstate transportation of firearms through this state in accordance with 18 USC 926A and 927, as amended from time to time, by any person who is not otherwise prohibited from shipping, transporting, receiving or possessing a firearm. Such person may transport a firearm for any lawful purpose from any place where such person may lawfully possess and carry such firearm through this state to any other place where such person may lawfully possess and carry such firearm provided such transportation is in accordance with subsection (b) of this section.

(b) During the transportation of a firearm through this state as authorized in subsection (a) of this section, such firearm shall be unloaded and neither such firearm nor any ammunition being transported shall be readily accessible or directly accessible from the passenger compartment of the vehicle. If the vehicle does not have a compartment separate from the passenger compartment, such firearm shall be unloaded and such firearm and any ammunition being transported shall be contained in a locked container other than the glove compartment or console.

(c) No person who is transporting a firearm through this state in accordance with this section may use or carry such firearm or sell, deliver or otherwise transfer such firearm while in this state.

29-38e. Statewide firearms trafficking task force. Composition. Duties.

(a) There shall be within the Division of State Police, within the Department of Public Safety, a state-wide firearms trafficking task force for the effective cooperative enforcement of the laws of this state concerning the distribution and possession of firearms.

(b) The task force shall be comprised of municipal and state law enforcement officers and may include federal law enforcement officers. Such task force shall be authorized to conduct any investigation authorized by this section at any place within the state as may be deemed necessary.

(c) The task force may request and may receive from any federal, state or local agency, cooperation and assistance in the performance of its duties, including the temporary assignment of personnel which may be necessary to carry out the performance of its functions.

(d) The task force may enter into mutual assistance and cooperation agreements with other states pertaining to firearms law enforcement matters extending across state boundaries, and may consult and exchange information and personnel with agencies of other states with reference to firearms law enforcement problems of mutual concern.

(e) The Commissioner of Public Safety may appoint a commanding officer and such other personnel as the commissioner deems necessary for the duties of the task force, within available appropriations.

(f) The task force shall: (1) Review the problem of illegal trafficking in firearms and its effects, including its effects on the public, and implement solutions to address the problem; (2) identify persons illegally trafficking in firearms and focus resources to prosecute such persons; (3) track firearms which were sold or distributed illegally and implement solutions to remove such firearms from persons illegally in possession of them; and (4) coordinate its activities with other law enforcement agencies within and without the state.

29-38f. Statewide Firearms Trafficking Task Force Policy Board. Composition. Duties.

(a) There shall be a State-Wide Firearms Trafficking Task Force Policy Board within the Division of State Police, within the Department of Public Safety, for administrative purposes only, consisting of the Commissioner of Public Safety, the Chief State's Attorney, the agent in Connecticut in charge of the federal Bureau of Alcohol, Tobacco and Firearms, the president of the Connecticut Police Chiefs Association and five chiefs of police designated by said association, each to serve for a term of one year, provided one such chief of police shall be from a municipality with a population of one hundred thousand or more.

(b) The policy board shall direct the formulation of policies and operating procedures of the task force.

(c) The policy board may apply for and administer any federal, state, local or private appropriations or grant funds made available for the operation of the task force.

(d) The receipts from the sale of seized firearms pursuant to section 54-36e shall be deposited in the General Fund and credited to a separate, nonlapsing forfeit firearms account which shall be established by the Comptroller. All moneys in the account are deemed to be appropriated and shall be expended for the purposes established in section 29-38e.

Title 53. Crimes

Chapter 943. Offenses Against Public Peace and Safety

53-202. Machine guns.

(a) As used in this section:

(1) "Machine gun" shall apply to and include a weapon of any description, loaded or unloaded, which shoots, is designed to shoot or can be readily restored to shoot automatically more than one projectile, without manual reloading, by a single function of the trigger, and shall also include any part or combination of parts designed for use in converting a weapon into a machine gun and any combination of parts from which a machine gun can be assembled if such parts are in the possession of or under the control of a person.

(2) "Crime of violence" shall apply to and include any of the following named crimes or an attempt to commit any of the same: Murder, manslaughter, kidnapping, sexual assault and sexual assault with a firearm, assault in the first or second degree, robbery, burglary, larceny and riot in the first degree.

(3) "Projectile" means any size bullet that when affixed to any cartridge case may be propelled through the bore of a machine gun.

(b) Any person who possesses or uses a machine gun in the perpetration or attempted perpetration of a crime of violence shall be imprisoned not less than ten years nor more than twenty years.

(c) Any person who possesses or uses a machine gun for an offensive or aggressive purpose shall be fined not more than one thousand dollars or imprisoned not less than five years nor more than ten years or be both fined and imprisoned.

(d) The possession or use of a machine gun shall be presumed to be for an offensive or aggressive purpose:

(1) When the machine gun is on premises not owned or rented, for bona fide permanent residence or business occupancy, by the person in whose possession the machine gun was found; or

(2) when in the possession of, or used by, an unnaturalized foreign-born person, or a person who has been convicted of a crime of violence in any state or federal court of record of the United States of America, its territories or insular possessions; or

(3) when the machine gun is of the kind described in subsection (g) hereof and has not been registered as therein required; or

(4) when empty or loaded projectiles of any caliber which have been or are susceptible of use in the machine gun are found in the immediate vicinity thereof.

(e) The presence of a machine gun in any room, boat or vehicle shall be presumptive evidence of the possession or use of the machine gun by each person occupying such room, boat or vehicle.

(f) Each manufacturer shall keep a register of all machine guns manufactured or handled by the manufacturer. Such register shall show the model and serial number, and the date of manufacture, sale, loan, gift, delivery or receipt, of each machine gun, the name, address and occupation of the person to whom the machine gun was sold, loaned, given or delivered, or from whom it was received and the purpose for which it was acquired by the person to whom the machine gun was sold, loaned, given or delivered. Upon demand, any manufacturer shall permit any marshal or police officer to inspect such manufacturer's entire stock of machine guns, and parts and supplies therefor, and shall produce the register, herein required, for inspection. Any person who violates any provision of this subsection shall be fined not more than two thousand dollars.

(g) Each machine gun in this state adapted to use projectiles of any caliber shall be registered in the office of the Commissioner of Public Safety within twenty-four hours after its acquisition and, thereafter, annually, on July first. Blanks for registration shall be prepared by said commissioner and furnished upon application. To comply with this subsection, the application as filed shall show the model and serial number of the gun, the name, address and occupation of the person in possession, and from whom and the purpose for which the gun was acquired. The registration data shall not be subject to inspection by the public. Any person who fails to register any gun as required hereby shall be presumed to possess the same for an offensive or aggressive purpose. The provisions of this subsection shall not apply to any machine gun which has been registered under the provisions of subsection (f) and which is still in the actual possession of the manufacturer.

(h) No provision of this section shall apply to:

(1) The manufacture of machine guns for sale or transfer to the United States government, to any state, territory or possession of the United States or to any political subdivision thereof or to the District of Columbia;

(2) the possession of a machine gun rendered inoperable by welding of all critical functioning parts and possessed as a curiosity, ornament or keepsake; or

(3) a machine gun acquired, transferred or possessed in accordance with the National Firearms Act, as amended, provided such machine gun shall be subject to the provisions of subsection (g) of this section.

53-202a. Assault weapons: Definition.

(a) As used in this section and sections 53-202b to 53-202k, inclusive, "assault weapon" means:

(1) Any selective-fire firearm capable of fully automatic, semiautomatic or burst fire at the option of the user or any of the following specified semiautomatic firearms: Algimec Agmi; Armalite AR-180; Australian Automatic Arms SAP Pistol; Auto-Ordnance Thompson type; Avtomat Kalashnikov AK-47 type; Barrett Light-Fifty model 82A1; Beretta AR-70; Bushmaster Auto Rifle and Auto Pistol; Calico models M-900, M-950 and 100-P; Chartered Industries of Singapore SR-88; Colt AR-15 and Sporter; Daewoo K-1, K-2, Max-1 and Max-2; Encom MK-IV, MP-9 and MP-45; Fabrique Nationale FN/FAL, FN/LAR, or FN/FNC; FAMAS MAS 223; Feather AT-9 and Mini-AT; Federal XC-900 and XC-450; Franchi SPAS-12 and LAW-12; Galil AR and ARM; Goncz High-Tech Carbine and High-Tech Long Pistol; Heckler & Koch HK-91, HK-93, HK-94 and SP-89; Holmes MP-83; MAC-10, MAC-11 and MAC-11 Carbine type; Intratec TEC-9 and Scorpion; Iver Johnson Enforcer model 3000; Ruger Mini-14/5F folding stock model only; Scarab Skorpion; SIG 57 AMT and 500 series; Spectre Auto Carbine and Auto Pistol; Springfield Armory BM59, SAR-48 and G-3; Sterling MK-6 and MK-7; Steyr AUG; Street Sweeper and Striker 12 revolving cylinder shotguns; USAS-12; UZI Carbine, Mini-Carbine and Pistol; Weaver Arms Nighthawk; Wilkinson "Linda" Pistol;

(2) A part or combination of parts designed or intended to convert a firearm into an assault weapon, as defined in subdivision (1) of this subsection, or any combination of parts from which an assault weapon, as defined in subdivision (1) of this subsection, may be rapidly assembled if those parts are in the possession or under the control of the same person;

(3) Any semiautomatic firearm not listed in subdivision (1) of this subsection that meets the following criteria:

(A) A semiautomatic rifle that has an ability to accept a detachable magazine and has at least two of the following:

(i) A folding or telescoping stock;

(ii) A pistol grip that protrudes conspicuously beneath the action of the weapon;

(iii) A bayonet mount;

(iv) A flash suppressor or threaded barrel designed to accommodate a flash suppressor; and

(v) A grenade launcher; or

(B) A semiautomatic pistol that has an ability to accept a detachable magazine and has at least two of the following:

(i) An ammunition magazine that attaches to the pistol outside of the pistol grip;

(ii) A threaded barrel capable of accepting a barrel extender, flash suppressor, forward handgrip or silencer;

(iii) A shroud that is attached to, or partially or completely encircles, the barrel and that permits the shooter to hold the firearm with the nontrigger hand without being burned;

(iv) A manufactured weight of fifty ounces or more when the pistol is unloaded; and

(v) A semiautomatic version of an automatic firearm; or

(C) A semiautomatic shotgun that has at least two of the following:

(i) A folding or telescoping stock;

(ii) A pistol grip that protrudes conspicuously beneath the action of the weapon;

(iii) A fixed magazine capacity in excess of five rounds; and

(iv) An ability to accept a detachable magazine; or

(4) A part or combination of parts designed or intended to convert a firearm into an assault weapon, as defined in subdivision (3) of this subsection, or any combination of parts from which an assault weapon, as defined in subdivision (3) of this subsection, may be rapidly assembled if those parts are in the possession or under the control of the same person.

(b) As used in this section and sections 53-202b to 53-202k, inclusive, the term "assault weapon" does not include any firearm modified to render it permanently inoperable.

53-202b. Sale or transfer of assault weapon prohibited. Class C felony.

(a)(1) Any person who, within this state, distributes, transports or imports into the state, keeps for sale, or offers or exposes for sale, or who gives any assault weapon, except as provided by sections 29-37j and 53-202a to 53-202k, inclusive, and subsection (h) of section 53a-46a, shall be guilty of a class C felony and shall be sentenced to a term of imprisonment of which two years may not be suspended or reduced.

(2) Any person who transfers, sells or gives any assault weapon to a person under eighteen years of age in violation of subdivision (1) of this subsection shall be sentenced to a term of imprisonment of six years, which shall not be suspended or reduced and shall be in addition and consecutive to the term of imprisonment imposed under subdivision (1) of this subsection.

(b) The provisions of subsection (a) of this section shall not apply to:

(1) The sale of assault weapons to the Department of Public Safety, police departments, the department of correction or the military or naval forces of this state or of the United States for use in the discharge of their official duties;

(2) A person who is the executor or administrator of an estate that includes an assault weapon for which a certificate of possession has been issued under section 53-202d which is disposed of as authorized by the probate court, if the disposition is otherwise permitted by sections 29-37j and 53-202a to 53-202k, inclusive, and subsection (h) of section 53a-46a;

(3) The transfer by bequest or intestate succession of an assault weapon for which a certificate of possession has been issued under section 53-202d.

53-202c. Possession of assault weapon prohibited. Class D felony.

(a) Except as provided in section 53-202e, any person who, within this state, possesses any assault weapon, except as provided in sections 29-37j, 53-202a to 53-202k, inclusive, and 53-202o and subsection (h) of section 53a-46a, shall be guilty of a class D felony and shall be sentenced to a term of imprisonment of which one year may not be suspended or reduced; except that a first-time violation of this subsection shall be a class A misdemeanor if (1) the person presents proof that he lawfully possessed the assault weapon prior to October 1, 1993, and (2) the person has otherwise possessed the firearm in compliance with subsection (d) of section 53-202d.

(b) The provisions of subsection (a) of this section shall not apply to the possession of assault weapons by members or employees of the Department of Public Safety, police departments, the department of correction or the military or naval forces of this state or of the United States for use in the discharge of their official duties; nor shall anything in sections 29-37j and 53-202a to 53-202k, inclusive, and subsection

(h) of section 53a-46a prohibit the possession or use of assault weapons by sworn members of these agencies when on duty and the use is within the scope of their duties.

(c) The provisions of subsection (a) of this section shall not apply to the possession of an assault weapon by any person prior to July 1, 1994, if all of the following are applicable:

(1) The person is eligible under sections 29-37j and 53-202a to 53-202k, inclusive, and subsection (h) of section 53a-46a to apply for a certificate of possession for the assault weapon by July 1, 1994;

(2) The person lawfully possessed the assault weapon prior to October 1, 1993; and

(3) The person is otherwise in compliance with sections 29-37j and 53-202a to 53-202k, inclusive, and subsection (h) of section 53a-46a.

(d) The provisions of subsection (a) of this section shall not apply to a person who is the executor or administrator of an estate that includes an assault weapon for which a certificate of possession has been issued under section 53-202d, if the assault weapon is possessed at a place set forth in subdivision (1) of subsection (d) of section 53-202d or as authorized by the Probate Court.

53-202d. Certificate of possession of assault weapon. Certificate of transfer of assault weapon to gun dealer. Circumstances where possession of assault weapon authorized.

(a) Any person who lawfully possesses an assault weapon, as defined in section 53-202a, prior to October 1, 1993, shall apply by October 1, 1994, or, if such person is a member of the military or naval forces of this state or of the United States and is unable to apply by October 1, 1994, because he or she is or was on official duty outside of this state, shall apply within ninety days of returning to the state to the Department of Public Safety, for a certificate of possession with respect to such assault weapon. The certificate shall contain a description of the firearm that identifies it uniquely, including all identification marks, the full name, address, date of birth and thumbprint of the owner, and any other information as the department may deem appropriate. The department shall adopt regulations in accordance with the provisions of chapter 54 not later than January 1, 1994, to establish procedures with respect to the application for and issuance of certificates of possession pursuant to this section. Notwithstanding the provisions of sections 1-210 and 1-211, the name and address of a person issued a certificate of possession shall be confidential and shall not be disclosed, except such records may be disclosed to (1) law enforcement agencies, and (2) the Commissioner of Mental Health and Addiction Services to carry out the provisions of subsection (c) of section 17a-500.

(b) No assault weapon possessed pursuant to this section may be sold or transferred on or after January 1, 1994, to any person within this state other than to a licensed gun dealer, as defined in subsection (d) of section 53-202f, or as provided in section 53-202e, or by bequest or intestate succession. Any person who obtains title to an assault weapon for which a certificate of possession has been issued under this section by bequest or intestate succession shall, within ninety days of obtaining title, apply to the Department of Public Safety for a certificate of possession as provided in subsection (a) of this section, render the weapon permanently inoperable, sell the weapon to a licensed gun dealer or remove the weapon from the state. Any person who moves into the state in lawful possession of an assault weapon, shall, within ninety days, either render the weapon permanently inoperable, sell the weapon to a licensed gun dealer or remove the weapon from this state, except any person who is a member of the military or naval forces of this state or of the United States, is in lawful possession of an assault weapon and has been transferred into the state after October 1, 1994, may, within ninety days of arriving in the state, apply to the Department of Public Safety for a certificate of possession with respect to such assault weapon.

(c) If an owner of an assault weapon sells or transfers the weapon to a licensed gun dealer, he shall, at the time of delivery of the weapon, execute a certificate of transfer and cause the certificate to be mailed or delivered to the Commissioner of Public Safety. The certificate shall contain:

(1) The date of sale or transfer;

(2) the name and address of the seller or transferor and the licensed gun dealer, their social security numbers or motor vehicle operator license numbers, if applicable;

(3) the licensed gun dealer's federal firearms license number and seller's permit number;

(4) a description of the weapon, including the caliber of the weapon and its make, model and serial number; and

(5) any other information the commissioner prescribes.

The licensed gun dealer shall present his motor vehicle operator's license or social security card, federal firearms license and seller's permit to the seller or transferor for inspection at the time of purchase or transfer. The Commissioner of Public Safety shall maintain a file of all certificates of transfer at his central office.

(d) A person who has been issued a certificate of possession of an assault weapon under this section may possess it only under the following conditions:

(1) At that person's residence, place of business or other property owned by that person, or on property owned by another with the owner's express permission;

(2) While on the premises of a target range of a public or private club or organization organized for the purpose of practicing shooting at targets;

(3) While on a target range which holds a regulatory or business license for the purpose of practicing shooting at that target range;

(4) While on the premises of a licensed shooting club;

(5) While attending any exhibition, display or educational project which is about firearms and which is sponsored by, conducted under the auspices of, or approved by a law enforcement agency or a nationally or state recognized entity that fosters proficiency in, or promotes education about, firearms; or

(6) While transporting the assault weapon between any of the places mentioned in this subsection, or to any licensed gun dealer, as defined in subsection (d) of section 53-202f, for servicing or repair pursuant to subsection (c) of section 53-202f, provided the assault weapon is transported as required by section 53-202f.

53-202e. Relinquishment of assault weapon to law enforcement agency.

Any individual may arrange in advance to relinquish an assault weapon to a police department or the Department of Public Safety. The assault weapon shall be transported in accordance with the provisions of section 53-202f.

53-202f. Transportation of assault weapon. Authorized actions of gun dealer.

(a) While transporting an assault weapon between any of the places mentioned in subdivisions (1) to (6), inclusive, of subsection (d) of section 53-202d, no person shall carry a loaded assault weapon concealed from public view or knowingly have, in any motor vehicle owned, operated or occupied by him (1) a loaded assault weapon, or (2) an unloaded assault weapon unless such weapon is kept in the trunk of such vehicle or in a case or other container which is inaccessible to the operator of or any passenger in such vehicle. Any person who violates the provisions of this subsection shall be fined not more than five hundred dollars or imprisoned not more than three years or both.

(b) Any licensed gun dealer, as defined in subsection (d) of this section, who lawfully possesses an assault weapon pursuant to section 53-202d, in addition to the uses allowed in section 53-202d, may transport the assault weapon between dealers or out of the state, display it at any gun show licensed by a state or local governmental entity or sell it to a resident outside the state. Any transporting of the assault weapon allowed by this subsection must be done as required by subsection (a) of this section.

(c)(1) Any licensed gun dealer, as defined in subsection (d) of this section, may take possession of any assault weapon for the purposes of servicing or repair from any person to whom has been issued a certificate of possession for such weapon pursuant to sections 29-37j and 53-202a to 53-202k, inclusive, and subsection (h) of section 53a-46a.

(2) Any licensed gun dealer may transfer possession of any assault weapon received pursuant to subdivision (1) of this subsection, to a gunsmith for purposes of accomplishing service or repair of the same. Transfers are permissible only to the following persons:

(A) A gunsmith who is in the dealer's employ;

(B) A gunsmith with whom the dealer has contracted for gunsmithing services, provided the gunsmith receiving the assault weapon holds a dealer's license issued pursuant to Chapter 44, commencing with Section 921, of Title 18 of the United States Code and the regulations issued pursuant thereto.

(d) The term "licensed gun dealer", as used in sections 29-37j and 53-202a to 53-202k, inclusive, and subsection (h) of section 53a-46a means a person who has a federal firearms license and a permit to sell firearms pursuant to section 29-28.

53-202g. Report of theft of assault weapon.

(A) Any person who lawfully possesses an assault weapon under Sections 29-37j and 53-202a to 53-202k, inclusive, and subsection (h) of Section 53a-46a or a firearm, as defined in Section 53a-3, that is lost or stolen from such person shall report the loss or theft to the organized local police department for the town in which the loss or theft occurred or, if such town does not have an organized local police department, to the state police troop having jurisdiction for such town within seventy-two hours of when such person discovered or should have discovered the loss or theft. Such department or troop shall forthwith forward a copy of such report to the commissioner of public safety. the provisions of this subsection shall not apply to the loss or theft of an antique firearm as defined in subsection (b) of section 29-37a.

(b) Any person who fails to make a report required by subsection (a) of this section within the prescribed time period shall commit an infraction and be fined not more than ninety dollars for a first offense and be guilty of a class d felony for any subsequent offense, except that, if such person intentionally fails to make such report within the prescribed time period, such person shall be guilty of a class c felony. any person who violates subsection (a) of this section for the first offense shall not lose such person's right to hold or obtain any firearm permit under the general statutes.

53-202h. Temporary transfer or possession of assault weapon for transport to out-of-state event. The provisions of subsection (a) of section 53-202b and subsection (a) of section 53-202c shall not apply to the temporary transfer or possession of an assault weapon, for which a certificate of possession has been issued pursuant to section 53-202d, for purposes of transporting such weapon to and from any shooting competition or exhibition, display or educational project which is about firearms and which is sponsored by, conducted under the auspices of, or approved by a law enforcement agency or a nationally or state recognized entity that fosters proficiency in, or promotes education about, firearms, which competition, exhibition, display or educational project is held outside this state.

53-202i. Circumstances in which manufacture or transportation of assault weapons not prohibited. Nothing in sections 29-37j and 53-202a to 53-202k, inclusive, and subsection (h) of section 53a-46a shall be construed to prohibit any person, firm or corporation engaged in the business of manufacturing assault weapons in this state from manufacturing or transporting assault weapons in this state for sale within this state in accordance with subdivision (1) of subsection (b) of section 53-202b or for sale outside this state.

53-202j. Commission of a class A, B or C felony with an assault weapon: Eight-year nonsuspendable sentence. Any person who commits any class A, B or C felony and in the commission of such felony uses, or is armed with and threatens the use of, or displays, or represents by his words or conduct that he possesses an assault weapon, as defined in section 53-202a, shall be imprisoned for a term of eight years, which shall not be suspended or reduced and shall be in addition and consecutive to any term of imprisonment imposed for conviction of such felony.

53-202k. Commission of a class A, B or C felony with a firearm: Five-year nonsuspendable sentence. Any person who commits any class A, B or C felony and in the commission of such felony uses, or is armed with and threatens the use of, or displays, or represents by his words or conduct that he possesses any firearm, as defined in section 53a-3, except an assault weapon, as defined in section 53-202a, shall be imprisoned for a term of five years, which shall not be suspended or reduced and shall be in addition and consecutive to any term of imprisonment imposed for conviction of such felony.

53-202l. Armor piercing and incendiary .50 caliber ammunition: Definition. Sale or transfer prohibited. Class D felony.

(a) For the purposes of this section:

(1) "Armor piercing .50 caliber bullet" means any .50 caliber bullet that is (A) designed for the purpose of, (B) held out by the manufacturer or distributor as, or (C) generally recognized as having a specialized capability to penetrate armor or bulletproof glass, including, but not limited to, such bullets commonly designated as "M2 Armor-Piercing" or "AP", "M8 Armor-Piercing Incendiary" or "API", "M20 Armor-Piercing Incendiary Tracer" or "APIT", "M903 Caliber .50 Saboted Light Armor Penetrator" or "SLAP", or "M962 Saboted Light Armor Penetrator Tracer" or "SLAPT".

(2) "Incendiary .50 caliber bullet" means any .50 caliber bullet that is (A) designed for the purpose of, (B) held out by the manufacturer or distributor as, or (C) generally recognized as having a specialized capability to ignite upon impact, including, but not limited to, such bullets commonly designated as "M1 Incendiary", "M23 Incendiary", "M8 Armor-Piercing Incendiary" or "API", or "M20 Armor-Piercing Incendiary Tracer" or "APIT".

(b) Any person who knowingly distributes, transports or imports into the state, keeps for sale or offers or exposes for sale or gives to any person any ammunition that is an armor piercing .50 caliber bullet or an incendiary .50 caliber bullet shall be guilty of a class D felony, except that a first-time violation of this subsection shall be a class A misdemeanor.

(c) The provisions of subsection (b) of this section shall not apply to the following:

(1) The sale of such ammunition to the Department of Public Safety, police departments, the Department of Correction or the military or naval forces of this state or of the United States for use in the discharge of their official duties;

(2) A person who is the executor or administrator of an estate that includes such ammunition that is disposed of as authorized by the Probate Court; or

(3) The transfer by bequest or intestate succession of such ammunition.

(d) If the court finds that a violation of this section is not of a serious nature and that the person charged with such violation (1) will probably not offend in the future, (2) has not previously been convicted of a violation of this section, and (3) has not previously had a prosecution under this section suspended pursuant to this subsection, it may order suspension of prosecution in accordance with the provisions of subsection (h) of section 29-33 of the general statutes.

53-202m. Circumstances when assault weapons exempt from limitations on transfers and registration requirements. Notwithstanding any provision of the general statutes, sections 53-202a to 53-202k, inclusive, of the general statutes and section 2 of this act, shall not be construed to limit the transfer or require the registration of an assault weapon as defined in subdivision (3) or (4) of subsection (a) of section 53-202a of the general statutes provided such firearm was legally manufactured prior to September 13, 1994.

53-202n. Possession of specified assault weapon permitted under certain circumstances. Notice requirement.

(a) For the purposes of subsection (a) of section 53-202c, this section and section 53-202o, "specified assault weapon" means any of the following firearms: Auto-Ordnance Thompson type, Avto-mat Kalashnikov AK-47 type, or MAC-10, MAC-11 and MAC-11 Carbine type.

(b) The provisions of subsection (a) of section 53-202c of the general statutes shall not apply to any person who

(1) in good faith purchased or otherwise obtained title to a specified assault weapon on or after October 1, 1993, and prior to May 8, 2002, in compliance with any state and federal laws concerning the purchase or transfer of firearms,

(2) is not otherwise disqualified or prohibited from possessing such specified assault weapon, and

(3) has notified the Department of Public Safety in accordance with subsection (c) of this section prior to October 1, 2003, that he or she possesses such specified assault weapon.

(c) A person complies with the notice requirement of subdivision (3) of subsection (b) of this section if such person provides the Department of Public Safety with: (1) A copy of the proof of purchase for such specified assault weapon, and (2) one of the following: (A) A copy of state form DPS-3 with respect to such specified assault weapon, (B) a copy of federal ATF Form 4473 with respect to such specified assault weapon, or (C) a sworn affidavit from such person that such specified assault weapon was purchased in compliance with any state and federal laws concerning the purchase or transfer of firearms; except that, if such person does not have a copy of the proof of purchase for such specified assault weapon, such person may satisfy the requirement of subdivision (1) of this subsection by, not later than January 1, 2003, providing such information as the department may require on a form prescribed by the department together with a sworn affidavit from such person that such specified assault weapon was purchased in compliance with any state and federal laws concerning the purchase or transfer of firearms.

(d) Any person who is a member of the military or naval forces of this state or of the United States and is unable to meet the notice requirements of subdivision (3) of subsection (b) and of subsection (c) of this section by October 1, 2003, because such person is or was on official duty outside this state, may file such notice within ninety days of returning to the state.

(e) As proof that a person has complied with the notice requirement of this section and that such notice has been received by the Department of Public Safety, the department shall issue a certificate of possession for such specified assault weapon. Such certificate shall contain a description of the firearm that identifies it uniquely, including all identification marks, and the full name, address and date of birth of the owner.

53-202o. Affirmative defense in prosecution for possession of specified assault weapon.

(a) In any prosecution for a violation of section 53-202c of the general statutes based on the possession by the defendant of a specified assault weapon, it shall be an affirmative defense that the defendant

(1) in good faith purchased or otherwise obtained title to such specified assault weapon on or after October 1, 1993, and prior to May 8, 2002, in compliance with any state and federal laws concerning the purchase or transfer of firearms,

(2) is not otherwise disqualified or prohibited from possessing such specified assault weapon, and

(3) has possessed such specified assault weapon in compliance with subsection (d) of section 53-202d of the general statutes.

(b) In any such prosecution, if such defendant proves such affirmative defense by a preponderance of the evidence, the specified assault weapon shall be returned to such defendant upon such defendant notifying the Department of Public Safety in accordance with subdivision (3) of subsection (b) and of subsection (c) of section 3 of this act and obtaining a certificate of possession, provided such notification is made not later than October 1, 2003.

53-202aa. Firearms trafficking: Class C or Class B felony.

(a) A person is guilty of firearms trafficking if such person, knowingly and intentionally, directly or indirectly, causes one or more firearms that such person owns, is in possession of or is in control of to come into the possession of or control of another person whom such person knows or has reason to believe is prohibited from owning or possessing any firearm under state or federal law.

(b) Any person who violates any provision of this section shall be guilty of a class C felony if such person, on or after October 1, 2007, sells, delivers or otherwise transfers five or fewer firearms, and a class B felony if such person, on or after October 1, 2007, sells, delivers or otherwise transfers more than five firearms.

(c) For the purposes of this section, "firearm" means "firearm" as defined in section 53a-3, but does not include a rifle or shotgun or an antique firearm as defined in subsection (b) of section 29-37a.

53-205. Shotguns, rifles and muzzleloaders in vehicles and snowmobiles. No person shall carry or possess in any vehicle or snowmobile any shotgun or rifle or muzzleloader of any gauge or caliber while such shotgun or rifle or muzzleloader contains in the barrel, chamber or magazine any loaded shell or cartridge capable of being discharged or when such muzzleloader has a percussion cap in place or when the powder pan of a flint lock contains powder. Muzzleloader as used in this section means a rifle or shotgun, incapable of firing a self-contained cartridge and which must be loaded at the muzzle end. The enforcement officers of the Department of Environmental Protection are empowered to enforce this section. The provisions of this section shall not apply to members of the military departments of the government or state while on duty or while traveling to or from assignments, or to enforcement officers, security guards or other persons employed to protect public or private property while in the performance of such duties. Any person who violates any provision of this section shall be fined not less than ten nor more than one hundred dollars or be imprisoned not more than thirty days or be both fined and imprisoned.

53-206b. Unlawful training in use of firearms, explosive or incendiary devices or techniques capable of causing injury. Class C felony.

(a) As used in this section:

(1) "Civil disorder" means a public disturbance involving acts of violence by a group of three or more persons which causes an immediate danger of or results in damage to the property of or injury to any other person.

(2) "Explosive or incendiary device" means

(A) dynamite and all other forms of high explosives,

(B) any explosive bomb, grenade, missile or similar device, and

(C) any incendiary bomb or grenade, fire bomb or similar device, including any device which

(i) consists of or includes a breakable container which contains a flammable liquid or compound and a wick composed of any material which, when ignited, is capable of igniting such flammable liquid or compound, and

(ii) can be carried or thrown by an individual.

(3) "Firearm" means a firearm as defined in section 53a-3.

(b) No person shall

(1) teach or demonstrate to any person the use, application or making of any firearm, explosive or incendiary device, or technique capable of causing injury or death to a person, knowing or intending that such firearm, explosive, incendiary device or technique will be unlawfully employed for use in, or in furtherance of, a civil disorder; or

(2) assemble with one or more persons for the purpose of training with, practicing with or being instructed in the use of any firearm, explosive or incendiary device, or technique capable of causing injury or death to a person, intending to employ unlawfully such firearm, explosive, incendiary device or technique for use in, or in furtherance of, a civil disorder.

(c) Any person who violates any provision of this section shall be guilty of a class C felony.

(d) Nothing in this section shall make unlawful any act of any peace officer, as defined in section 53a-3, performed in the lawful discharge of his official duties.

53-206c. Sale, carrying and brandishing of facsimile firearms prohibited. Class B misdemeanor.

(a) For the purposes of this section:

(1) "Facsimile of a firearm" means (A) any nonfunctional imitation of an original firearm which was manufactured, designed and produced since 1898, or (B) any nonfunctional representation of a firearm other than an imitation of an original firearm, provided such representation could reasonably be perceived to be a real firearm. Such term does not include any look-a-like, nonfiring, collector replica of an antique firearm developed prior to 1898, or traditional BB or pellet-firing air gun that expels a metallic or paint-contained projectile through the force of air pressure.

(2) "Firearm" means firearm as defined in section 53a-3.

(b) No person shall give, offer for sale or sell any facsimile of a firearm. The provisions of this subsection shall not apply to any facsimile of a firearm, which, because of its distinct color, exaggerated size or other design feature, cannot reasonably be perceived to be a real firearm.

(c) Except in self defense, no person shall carry, draw, exhibit or brandish a facsimile of a firearm or simulate a firearm in a threatening manner, with intent to frighten, vex or harass another person.

(d) No person shall draw, exhibit or brandish a facsimile of a firearm or simulate a firearm in the presence of a peace officer, firefighter, emergency medical technician or paramedic engaged in the performance of his duties knowing or having reason to know that such peace officer, firefighter, emergency medical technician or paramedic is engaged in the performance of his duties, with intent to impede such person in the performance of such duties.

(e) Any person who violates any provision of this section shall be guilty of a class B misdemeanor.

53-206d. Carrying a loaded firearm while under influence prohibited.

(a)(1) No person shall carry a pistol, revolver, machine gun, shotgun, rifle or other firearm, which is loaded and from which a shot may be discharged, upon his person (A) while under the influence of intoxicating liquor or any drug, or both, or (B) while the ratio of alcohol in the blood of such person is ten-hundredths of one per cent or more of alcohol, by weight.

(2) Any person who violates any provision of this subsection shall be guilty of a class B misdemeanor. ...

Title 53a. Penal Code

Chapter 950. General Provisions

53a-3. Definitions. Except where different meanings are expressly specified, the following terms have the following meanings when used in this title:

(1) "Person" means a human being, and, where appropriate, a public or private corporation, a limited liability company, an unincorporated association, a partnership, a government or a governmental instrumentality;

(2) "Possess" means to have physical possession or otherwise to exercise dominion or control over tangible property; ...

(6) "Deadly weapon" means any weapon, whether loaded or unloaded, from which a shot may be discharged, or a switchblade knife, gravity knife, billy, blackjack, bludgeon, or metal knuckles. The definition of "deadly weapon" in this subdivision shall be deemed not to apply to section 29-38 or 53-206;

(7) "Dangerous instrument" means any instrument, article or substance which, under the circumstances in which it is used or attempted or threatened to be used, is capable of causing death or serious physical injury, and includes a "vehicle" as that term is defined in this section and includes a dog that has been commanded to attack, except a dog owned by a law enforcement agency of the state or any political subdivision thereof or of the federal government when such dog is in the performance of its duties under the direct supervision, care and control of an assigned law enforcement officer; ...

(15) "Machine gun" means a weapon of any description, irrespective of size, by whatever name known, loaded or unloaded, from which a number of shots or bullets may be rapidly or automatically discharged from a magazine with one continuous pull of the trigger and includes a submachine gun;

(16) "Rifle" means a weapon designed or redesigned, made or remade, and intended to be fired from the shoulder and designed or redesigned and made or remade to use the energy of the explosive in a fixed metallic cartridge to fire only a single projectile through a rifled bore for each single pull of the trigger;

(17) "Shotgun" means a weapon designed or redesigned, made or remade, and intended to be fired from the shoulder and designed or redesigned and made or remade to use the energy of the explosive in a fixed shotgun shell to fire through a smooth bore either a number of ball shot or a single projectile for each single pull of the trigger;

(18) "Pistol" or "revolver" means any firearm having a barrel less than twelve inches;

(19) "Firearm" means any sawed-off shotgun, machine gun, rifle, shotgun, pistol, revolver or other weapon, whether loaded or unloaded from which a shot may be discharged.

(20) "Electronic defense weapon" means a weapon which by electronic impulse or current is capable of immobilizing a person temporarily, but is not capable of inflicting death or serious physical injury, including a stun gun or other conductive energy device.

Chapter 952. Offenses

Part XXI. Miscellaneous Offenses

53a-211. Possession of a sawed-off shotgun or silencer: Class D felony.

(a) A person is guilty of possession of a sawed-off shotgun or a silencer when he owns, controls or possesses any sawed-off shotgun that has a barrel of less than eighteen inches or an overall length of less than twenty-six inches or when he owns, controls or possesses any silencer designed to muffle the noise of a firearm during discharge.

(b) The provisions of this section shall not apply to persons, firms, corporations or museums licensed or otherwise permitted by federal or state law to possess, control or own sawed-off shotguns or silencers.

(c) Possession of a sawed-off shotgun or a silencer is a class D felony.

53a-212. Stealing a firearm: Class D felony.

(a) A person is guilty of stealing a firearm when, with intent to deprive another of his firearm or to appropriate the same to himself or a third party, he wrongfully takes, obtains or withholds a firearm, as defined in subdivision (l9) of section 53a-3.

(b) Stealing a firearm is a class D felony.

53a-216. Criminal use of firearm or electronic defense weapon: Class D felony.

(a) A person is guilty of criminal use of a firearm or electronic defense weapon when he commits any class A, B or C or unclassified felony as defined in section 53a-25 and in the commission of such felony he uses or threatens the use of a pistol, revolver, machine gun, shotgun, rifle or other firearm or electronic defense weapon. No person shall be convicted of criminal use of a firearm or electronic defense weapon and the underlying felony upon the same transaction but such person may be charged and prosecuted for both such offenses upon the same information.

(b) Criminal use of a firearm or electronic defense weapon is a class D felony for which five years of the sentence imposed may not be suspended or reduced by the court.

53a-217. Criminal possession of a firearm or electronic defense weapon: Class D felony.

(a) A person is guilty of criminal possession of a firearm or electronic defense weapon when such person possesses a firearm or electronic defense weapon and

(1) has been convicted of a felony,

(2) has been convicted as delinquent for the commission of a serious juvenile offense, as defined in section 46b-120,

(3) knows that such person is subject to

(A) a restraining or protective order of a court of this state that has been issued against such person, after notice and an opportunity to be heard has been provided to such person, in a case involving the use, attempted use or threatened use of physical force against another person, or

(B) a foreign order of protection, as defined in section 46b-15a, as amended by this act, that has been issued against such person in a case involving the use, attempted use or threatened use of physical force against another person,

(4) knows that such person is subject to a firearms seizure order issued pursuant to subsection (d) of section 29-38c after notice and an opportunity to be heard has been provided to such person, or

(5) is prohibited from shipping, transporting, possessing or receiving a firearm pursuant to 18 USC 922(g)(4). For the purposes of this section, "convicted" means having a judgment of conviction entered by a court of competent jurisdiction.

(b) Criminal possession of a firearm or electronic defense weapon is a class D felony, for which two years of the sentence imposed may not be suspended or reduced by the court.

53a-217a. Criminally negligent storage of a firearm: Class D felony.

(a) A person is guilty of criminally negligent storage of a firearm when he violates the provisions of section 29-37i and a minor obtains the firearm and causes the injury or death of himself or any other person. For the purposes of this section, "minor" means any person under the age of sixteen years.

(b) The provisions of this section shall not apply if the minor obtains the firearm as a result of an unlawful entry to any premises by any person.

(c) Criminally negligent storage of a firearm is a class D felony.

53a-217b. Possession of a weapon on school grounds: Class D felony.

(a) A person is guilty of possession of a weapon on school grounds when, knowing that such person is not licensed or privileged to do so, such person possesses a firearm or deadly weapon, as defined in section 53a-3, (1) in or on the real property comprising a public or private elementary or secondary school or (2) at a school-sponsored activity as defined in subsection (h) of section 10-233a.

(b) The provisions of subsection (a) of this section shall not apply to the otherwise lawful possession of a firearm

(1) by a person for use in a program approved by school officials in or on such school property or at such school-sponsored activity,

(2) by a person in accordance with an agreement entered into between school officials and such person or such person's employer,

(3) by a peace officer, as defined in subdivision (9) of section 53a-3, while engaged in the performance of such peace officer's official duties, or

(4) by a person while traversing such school property for the purpose of gaining access to public or private lands open to hunting or for other lawful purposes, provided such firearm is not loaded and the entry on such school property is permitted by the local or regional board of education.

(c) Possession of a weapon on school grounds is a class D felony.

53a-217c. Criminal possession of a pistol or revolver: Class D felony.

(a) A person is guilty of criminal possession of a pistol or revolver when such person possesses a pistol or revolver, as defined in section 29-27, and

(1) has been convicted of a felony or of a violation of subsection (c) of section 21a-279 or section 53a-58, 53a-61, 53a-61a, 53a-62, 53a-63, 53a-96, 53a-175, 53a-176, 53a-178 or 53a-181d,

(2) has been convicted as delinquent for the commission of a serious juvenile offense, as defined in section 46b-120,

(3) has been discharged from custody within the preceding twenty years after having been found not guilty of a crime by reason of mental disease or defect pursuant to section 53a-13,

(4) has been confined in a hospital for persons with psychiatric disabilities, as defined in section 17a-495, within the preceding twelve months by order of a probate court,

(5) knows that such person is subject to

(A) a restraining or protective order of a court of this state that has been issued against such person, after notice and an opportunity to be heard has been provided to such person, in a case involving the use, attempted use or threatened use of physical force against another person, or

(B) a foreign order of protection, as defined in section 46b-15a, as amended by this act, that has been issued against such person in a case involving the use, attempted use or threatened use of physical force against another person,

(6) knows that such person is subject to a firearms seizure order issued pursuant to subsection (d) of section 29-38c after notice and an opportunity to be heard has been provided to such person,

(7) is prohibited from shipping, transporting, possessing or receiving a firearm pursuant to 18 USC 922(g)(4), or (8) is an alien illegally or unlawfully in the United States. For the purposes of this section, "convicted" means having a judgment of conviction entered by a court of competent jurisdiction.

(b) Criminal possession of a pistol or revolver is a class D felony.

[Current through the 2008 Supplement]

Bridgeport Municipal Code

Title 9 Public Peace and Welfare

Chapter 9.16. Weapons

9.16.010. Permit to carry certain weapons – Issuance authorized. Pursuant to Section 53-206 of the General Statutes, the mayor or chief of police may, for good and sufficient reason, issue a permit to any proper person to carry any of the weapons or instruments described in such section.

9.16.020 Permit to carry certain weapons - Fee. A fee of five dollars ($5.00) shall be charged for each permit issued pursuant to Section 9.16.010.

9.16.030 Permit to carry certain weapons-Records-Disposition of fees.
The board of police commissioners shall provide suitable permit books in which permits issued pursuant to Section 9.16.010 shall be numbered consecutively and in duplicate. The mayor shall turn over to the chief of police any permit fees received by him. The chief of police shall, on the first and sixteenth day of each month, turn over to the city treasurer the amount collected for received by him for such permits and deliver to the comptroller the duplicate copies of the permits so issued.

9.16.040 Permit to carry certain weapons - Term - Revocation. The permit required by Section 9.16.010 shall be for the period of one year from the date of its issuance and may be revoked by the chief of police during such year at his discretion.

9.16.050 Air rifles prohibited.

A. It is unlawful and is prohibited for any person to store, sell, offer or expose for sale at retail or have in possession with intent to sell at retail or with intent to use within the limits of the city any air gun, rifle or pistol, spring gun or pistol or any implement not a firearm which impels with force a pellet of any kind.

B. The term "at retail" as used in this section shall mean any sale of less than one dozen such articles, or any sale to anyone not a bona fide dealer therein.

C. The chief of police is authorized, empowered and directed to seize, remove and destroy any air gun or any implement not a firearm which impels with force a pellet of any kind which shall be used or discharged within the city or which shall be held, stored or possessed in violation of this section.

9.16.060 Resale of city-owned firearms-Prohibited.

A. All firearms which become the property of the city, either through purchase, seizure, donation or any other means will not be re-sold.

B. All firearms will be destroyed, at such time as they are no longer necessary as evidence in a criminal proceeding or for such other use the chief of police determines reasonable and appropriate, so that they cannot be re-sold or re-used by a secondary user or purchaser. (

[Current through September 17, 2007]

The Municipal Code of Hartford

Chapter 21. Licenses and Permits Generally

Article II. Pistol Permits

Division 1. Generally

21-31. Definitions. The following words, terms and phrases, when used in this article, shall have the meanings ascribed to them in this section, except where the context clearly indicates a different meaning:

Ammunition means any projectile or other device which will or is designed to or may readily be converted to be expelled from any pistol or revolver.

Antique pistol or revolver means any pistol or revolver which was manufactured in or before 1898 and any replica of such pistol or revolver provided such replica is not designed or redesigned for using rimfire or conventional centerfire fixed ammunition except rimfire or conventional centerfire fixed ammunition which is no longer manufactured in the United States and not readily available in the ordinary channel of commercial trade.

Chief of police means the chief of police of the city police department.

Permit includes the word "license."

Person means any individual, corporation, company, association, firm, partnership, society, joint stock company, or organization of any kind.

Pistol and revolver mean any firearm having a barrel less than twelve (12) inches in length.

Transfer includes to sell, assign, lease, loan, give away, or otherwise cause the lawful title or rightful possession of a firearm to vest in another.

Division 2. Sale or Transfer of Pistols and Revolvers

21-51. Permit - Required. It shall be unlawful for any person to advertise, sell, deliver, or offer or expose for sale or delivery, or have in his possession with intent to sell or deliver, or loan, exchange, give away or otherwise transfer the lawful ownership or possession of any pistol or revolver at retail without having a permit therefor issued as provided in this division.

21-52. Same - Application. An application for a permit to sell or transfer pistols and revolvers shall be made in writing to the chief of police on such suitable forms provided or approved by him, setting forth the name and social security number of the applicant, the applicant's residence or registered address, and the address where business is to be conducted. The applicant must be a bona fide resident of, or have a place of business in the city in order to be eligible for a permit. The application for a permit shall also state that the applicant (including in the case of a corporation, partnership, or association, any individual possessing, directly or indirectly, the power to direct or cause the direction of the management and policies of the corporation, partnership or association) is eighteen (18) years of age or older, that he is not under indictment for nor has been convicted in any court of a felony, is not a fugitive from justice, is not addicted to the use of narcotics, has not been a patient in a mental institution within the past five (5) years, and is not mentally retarded.

21-53. Same - Issuance.

(a) The chief of police shall issue a permit to sell or transfer pistols and revolvers to applicants who have satisfactorily complied with the requirements of this division. It shall be the duty of the chief of police to refuse the permit to any applicant who fails, refuses or is unable to comply with all of the requirements specified in this division.

(b) The chief of police shall notify the applicant not later than eight (8) weeks after a sufficient application has been submitted that the request for a permit has been approved or denied.

21-54. Same - Expiration. A permit for the sale or transfer of pistols and revolvers shall expire five (5) years after the date it becomes effective; and renewal thereof shall expire five (5) years after the expiration date of the permit being renewed.

21-55. Same - Fee. The fee for each permit issued under the provisions of this division shall be fifteen dollars ($15.00).

21-56. Same - Revocation. The chief of police shall have the authority to revoke a permit issued under this division upon determining that the permit holder has violated any provision of this division or is no longer able to fully comply with all of the requirements specified in this division, and the money paid for such permit shall be forfeited to the city.

21-57. Same - Display at place of sale. No sale of any pistol or revolver shall be made except in the room, store or place described in the permit for the sale of pistols and revolvers, and such permit or a copy thereof certified by the authority issuing the same shall be exposed to view within the room, store or place where pistols or revolvers are sold or offered or exposed for sale.

21-58. Required records.

(a) Content of records. Any seller of pistols and revolvers other than a manufacturer selling to a bona fide wholesaler or a retailer or a wholesaler selling to a bona fide retailer shall keep a record of all such pistols and revolvers sold, leased, loaned, given away or otherwise transferred. Such record shall contain the following information:

(1) The name, social security number, age, address, and permit number of the transferor;

(2) The name, social security number, age, address, and permit number of the transferee;

(3) The date of the sale;

(4) The name of the manufacturer, the caliber, make, model and serial number of the pistol or revolver.

(b) Inspection. Records of transfers shall be open for inspection by any duly authorized law enforcement official or by the chief of police or city clerk of the city at all reasonable times.

(c) Evidence of identification of transferee. The person to whom the pistol or revolver is to be delivered must provide evidence of his identification. Such evidence must include a picture identification, as well as information as to the bearer's age and address. The record of transfer shall be signed by the transferor and the transferee, each in the presence of the other.

21-59. Application to purchase.

(a) Required; delivery; waiting period. No person shall transfer a pistol or revolver at retail or otherwise transfer except upon written application on a form prescribed and furnished by the commissioner of public safety in triplicate. A copy of the application is to be mailed by first class mail on the day of receipt to the chief of police and one (1) to the commissioner of public safety. No sale or delivery of any pistol or revolver shall be made until the expiration of two (2) weeks from the date of the mailing of such copies.

(b) Exceptions. The waiting period specified in subsection (a) of this section during which delivery may be made shall not apply to the holder of a valid state permit to carry pistols and revolvers, nor to any federal marshal, sheriff, parole officer or peace officer. The provisions of this section shall not apply to antique pistols or revolvers.

(c) Prohibited acts. No person shall sell at retail, deliver or otherwise transfer any pistol or revolver to any alien. No person shall make any false statement or give any false information connected with any purchase or other transfer of any pistol or revolver. No person shall sell or otherwise transfer any pistol or revolver to any other person under the age of eighteen (18) years of age.

21-60. Report of sale or transfer. Any transferor of pistols and revolvers shall, upon selling or otherwise transferring a pistol or revolver, make a report of the sale or gift, which report shall contain the date of sale or transfer, name, age, address, occupation, physical description of purchaser or donee, the purpose for which purchased, the kind, description, including serial number of the pistol or revolver, and the consideration paid therefor, the city and state permit number, if any, and/or the driver's license number, if any. Such report of sale shall be open for inspection by any duly authorized law enforcement official or by the chief of police or the city clerk at all reasonable times. In addition, the information contained in this section shall be provided by the seller to the chief of police no later than the last business day of each calendar month representing all sales, lease, loans, gifts, or other transfers of any pistol or revolver by the seller.

21-61. Statement of intent. The city's experience is that many firearms used in violent crimes are weapons which have been stolen. Existing state law provides for a statewide firearms tracking task force within the division of state police. Requiring owners of firearms to promptly report to the police the loss or theft of firearms will enable the Hartford Police Department to timely investigate and hopefully solve the incidents of initial theft and may well result in taking firearms off the street before they can be used to perpetrate a violent crime. In addition, this legislation provides the city with a vehicle to recover costs expended in connection with police services from owners who fail to report stolen firearms. The section is within the city's police powers authority and its power to protect the public interest.

21-62.Owner to report loss or theft of firearm. In the city, any person who is the lawful owner of a firearm and any permitted firearm's owner shall report the loss or theft thereof from premises in the city or from their person to the Hartford Police Department within seventy-two (72) hours of becoming aware of said theft or loss.

Sec. 21-63. Failure to report loss or theft of firearm. In the event that a lost or stolen firearm is determined to have been used to aid or abet the commission of a felony in the city and it shall become known to the Hartford Police that the owner of said firearm failed to report its loss or theft within seventy-two (72) hours of becoming aware of such loss or theft to the organized local police department in the municipality, or the Connecticut State Police, then the Corporation Counsel for the City of Hartford, on a showing of good cause, and consistent with state law, may sue the said owner to recover the costs to the city of police services and other expenses associated with the investigation and prosecution of said felony.

Division 3. Permit to Carry Pistol or Revolver

21-71. Required. It shall be unlawful for any person to have in his possession in the city any pistol or revolver, except when such person is within his dwelling house or place of business, unless a permit to carry a pistol or revolver has been issued in accordance with the provisions of this division.

21-72. Application.

(a) Documents required generally. Every person applying for a permit to carry a pistol or revolver in the city shall provide the following documents to the Hartford police department, records division, between 8:00 a.m. and 6:00 p.m., Monday through Friday, or 8:00 a.m. to 12:00 noon on Saturday: ...

21-73. Issuance, fee. Upon satisfactorily completing all aspects of the application requirements for a permit to carry a pistol or revolver, the applicant will be scheduled to meet with the commander of the records unit of the police department. At that time, a fee of fifteen dollars ($15.00) in cash, check or money order made payable to the city shall be paid and a permit issued. If, at any point in the process an application is denied, the payment to the Federal Bureau of Investigation shall be nonrefundable, however, no other charges for processing the application shall be made.

21-74. Exceptions to division. The provisions of this division shall not apply to the carrying of any pistol or revolver by any sheriff, parole officer or peace officer of any other state while engaged in the pursuit of his official duties, or federal marshal or federal law enforcement agent, or to any member of the armed forces of the United States, or of this state, when on duty or going to or from duty, or to any member of any military organization when on parade or when going to or from any place of assembly, or to the transportation of pistols or revolvers as merchandise, or to any person carrying any pistol or revolver while contained in the package in which it was originally wrapped at the time of sale and while carrying the same from the place of sale to the purchaser's residence or place of business, or to any person removing his household goods or effects from one (1) place to another, or to any person while carrying any such pistol or revolver from his place of residence or business to a place or person where or by whom such pistol or revolver is to be repaired or while returning to his place or residence or business after the same has been repaired, or to any person carrying a pistol or revolver in or through the state for the purpose of taking part in competitions or attending any meeting or exhibition of an organized collectors' group if such person is a bona fide resident of the United States having a permit or license to carry a firearm issued by the authority of any other state or subdivision of the United States, or to any person carrying a pistol or revolver to and from a testing range at the request of the issuing authority, or to any person carrying an antique pistol or revolver.

Chapter 25. Offenses - Miscellaneous

25-18. Facsimiles of firearms.

(a) For the purposes of this section:

(1) "Facsimile of a firearm" means

(A) any nonfunctional imitation of an original firearm which was manufactured, designed and produced since 1898, or

(B) any nonfunctional representation of a firearm other than an imitation of an original firearm, provided such representation could reasonably be perceived to be a real firearm.

Such term does not include any look-a-like, nonfiring, collector replica of an antique firearm developed prior to 1898, or traditional BB. Or pellet-firing air gun that expels a metallic or paint-contained projectile through the force of air pressure.

(2) "Firearm" means firearm as defined in Connecticut General Statutes section 53a-3.

(b) No person shall give, offer for sale or sell any facsimile of a firearm. The provisions of this subsection shall not apply to any facsimile of a firearm, which, because of its distinct color, exaggerated size or other design feature, cannot reasonable be perceived to be a real firearm.

(c) Except in self defense, no person shall carry, draw, exhibit or brandish a facsimile of a firearm or simulate a firearm in a threatening manner, with intent to frighten, vex or harass another person.

(d) No person shall draw, exhibit or brandish a facsimile of a firearm or simulate a firearm in the presence of a peace officer, firefighter, emergency medical technician or paramedic engaged in the performance of his duties knowing or having reason to know that such peace officer, firefighter, emergency medical technician or paramedic is engaged in the performance of such duties.

[Hartford City Code codified through Ordinance No. 09-08, enacted April 28, 2008]

Code of Ordinances of the City of New Haven

Chapter 18. Offenses and Miscellaneous Provisions

18-12. Firearms and ammunition - Sales to children. No person shall sell to any child under the age of sixteen (16) years, without the written consent of the parent or guardian of such child, any cartridge or fixed ammunition of which any fulminate is a component part, or any gun, pistol, or other mechanical contrivance arranged for the explosion of such cartridge, or of any fulminate.

18-12.1. Same - Sale of pistols and revolvers.

(a) No person shall advertise, sell, offer or expose for sale, or have in his possession with intent to sell, any pistol or revolver at retail unless such person shall have obtained:

(1) A federal license as a dealer in firearms or ammunition from the bureau of alcohol, tobacco and firearms;

(2) A state permit for the sale at retail of pistols and revolvers within the city; and

(3) A state permit to engage in or conduct business as a seller within the State of Connecticut for the place of business in which such a sale of any pistol or revolver at retail shall occur from the state tax commissioner.

(b) No sale of any pistol or revolver at retail by any person qualified to conduct such a sale under subsection (a) shall be conducted in a private dwelling, no part of which is open to the general public.

(c) All sales of any pistol or revolver at retail by any person qualified to conduct such a sale under subsection (a) shall be conducted in premises located on property zoned as a business district or in premises for which a variance has been granted for the sale of pistols or revolvers at retail.

(d) For the purposes of this section:

(1) The term "sale of any pistol or revolver at retail" means any transfer of title, exchange or barter, in any manner or by any means whatsoever, of any pistol or revolver for a consideration for any purpose other than resale in the regular course of business;

(2) The term "pistol or revolver" means any firearm having a barrel less than twelve (12) inches in length.

[City of New Haven Code of Ordinances codified through Ord. No. 1554, enacted June 4, 2007]

DELAWARE
DEL. CODE

Title 9. Counties

Part I. Provisions Affecting All Counties

Chapter 3. County Governments Generally

330. General powers and duties.

(c) The county governments shall enact no law or regulation prohibiting, restricting or licensing the ownership, transfer, possession or transportation of firearms or components of firearms or ammunition except that the discharge of a firearm may be regulated; provided any law, ordinance or regulation incorporates the justification defenses as found in Title 11 of the Delaware Code.

Title 11. Crimes and Criminal Procedure

Part I. Delaware Criminal Code

Chapter 2. General Provisions Concerning Offenses

222. General definitions. When used in this Criminal Code:

(5) "Deadly weapon" includes a firearm, as defined in paragraph (11) of this section, a bomb, a knife of any sort (other than an ordinary pocketknife carried in a closed position), switchblade knife, billy, blackjack, bludgeon, metal knuckles, slingshot, razor, bicycle chain or ice pick or any dangerous instrument, as defined in paragraph (4) of this section, which is used, or attempted to be used, to cause death or serious physical injury. For the purpose of this definition, an ordinary pocketknife shall be a folding knife having a blade not more than 3 inches in length.

(12) "Firearm" includes any weapon from which a shot, projectile or other object may be discharged by force of combustion, explosive, gas and/or mechanical means, whether operable or inoperable, loaded or unloaded. It does not include a BB gun.

Chapter 5. Specific Offenses

Subchapter VII. Offenses Against Public Health, Order and Decency

Subpart E. Offenses Involving Deadly Weapons and Dangerous Instruments

1444. Possessing a destructive weapon; class E felony.

(a) A person is guilty of possessing a destructive weapon when the person sells, transfers, buys, receives or has possession of a bomb, bombshell, firearm silencer, sawed-off shotgun, machine gun or any other firearm or weapon which is adaptable for use as a machine gun.

(b) Possessing a destructive weapon is a class E felony. This section does not apply to members of the military forces or to members of a police force in this State duly authorized to carry a weapon of the type described; nor shall the provisions contained herein apply to authorized and certified (by an accredited state enforcement agency) state and federal wildlife biologists possessing firearm silencers for the purposes of wildlife disease or wildlife population control, or persons possessing machine guns for scientific or experimental research and development purposes, which machine guns have been duly registered under the National Firearms Act of 1968 (26 U.S.C. §5801 et seq.).

(c) The term "shotgun" as used in this section means a weapon designed or redesigned, made or remade, and intended to be fired from the shoulder and designed or redesigned and made or remade to use the energy of the explosive in a fixed shotgun shell to fire through a smooth bore either a number of projectiles (ball shot) or a single projectile for each pull of the trigger. The term "sawed-off shotgun" as used in this section means a shotgun having 1 or more barrels less than 18 inches in length or any weapon made from a shotgun (whether by alteration, modification or otherwise) if such weapon as modified has an overall length of less than 26 inches.

1445. Unlawfully dealing with a dangerous weapon; unclassified misdemeanor. A person is guilty of unlawfully dealing with a dangerous weapon when:

(1) The person possesses, sells or in any manner has control of:

a. A weapon which by compressed air or by spring discharges or projects a pellet, slug or bullet, except a BB or air gun which does not discharge or project a pellet or slug larger than a BB shot; or

b. A pellet, slug or bullet, intending that it be used in any weapon prohibited by paragraph (1)a. of this section; or

(2) The person sells, gives or otherwise transfers to a child under 16 years of age a BB or air gun or spear gun or BB shot, unless the person is that child's parent or guardian, or unless the person first receives the permission of said parent or guardian; or

(3) Being a parent, the person permits the person's child under 16 years of age to have possession of a firearm or a BB or air gun or spear gun unless under the direct supervision of an adult; or

(4) The person sells, gives or otherwise transfers to a child under 18 years of age a firearm or ammunition for a firearm, unless the person is that child's parent or guardian, or unless the person first receives the permission of said parent or guardian; or

(5) The person sells, gives or otherwise transfers a firearm to any person knowing that said person intends to commit any felony, class A misdemeanor or drug related criminal offense while in possession of said firearm.

Unlawfully dealing with a firearm or dangerous weapon is an unclassified misdemeanor, unless the person is convicted under paragraph (4) of this section, in which case it is a class G felony, or unless the person is convicted under paragraph (5) of this section, in which case it is a class E felony.

1448. Possession and purchase of deadly weapons by persons prohibited; penalties.

(a) Except as otherwise provided herein, the following persons are prohibited from purchasing, owning, possessing or controlling a deadly weapon or ammunition for a firearm within the State:

(1) Any person having been convicted in this State or elsewhere of a felony or a crime of violence involving physical injury to another, whether or not armed with or having in possession any weapon during the commission of such felony or crime of violence;

(2) Any person who has ever been committed for a mental disorder to any hospital, mental institution or sanitarium, unless the person possesses a certificate of a medical doctor or psychiatrist licensed in this State that the person is no longer suffering from a mental disorder which interferes or handicaps the person from handling deadly weapons;

(3) Any person who has been convicted for the unlawful use, possession or sale of a narcotic, dangerous drug or central nervous system depressant or stimulant as those terms were defined prior to the effective date of the Uniform Controlled Substances Act in June 1973 or of a narcotic drug or controlled substance as defined in Chapter 47 of Title 16;

(4) Any person who, as a juvenile, has been adjudicated as delinquent for conduct which, if committed by an adult, would constitute a felony, unless and until that person has reached their 25th birthday;

(5) Any juvenile, if said deadly weapon is a handgun, unless said juvenile possesses said handgun for the purpose of engaging in lawful hunting, instruction, sporting or recreational activity while under the direct or indirect supervision of an adult. For the purpose of this subsection, a handgun shall be defined as any pistol, revolver or other firearm designed to be readily capable of being fired when held in 1 hand;

(6) Any person who is subject to a Family Court protection from abuse order (other than an ex parte order), but only for so long as that order remains in effect or is not vacated or otherwise terminated, except that this paragraph shall not apply to a contested order issued solely upon § 1041(1)d., e., or h. of Title 10, or any combination thereof;

(7) Any person who has been convicted in any court of any misdemeanor crime of domestic violence. For purposes of this paragraph, the term "misdemeanor crime of domestic violence" means any misdemeanor offense that:

a. Was committed by a member of the victim's family, as "family" is defined in § 901(12) of Title 10 (regardless, however, of the state of residence of the parties); by a former spouse of the victim; by a person who co-habitated with the victim at the time of the offense; or by a person with a child in common with the victim; and

b. Is an offense as defined under §601, §602, §603, §611, §614, §621, §625, §628, §763, §765, §766, §767, §781, §785 or §791 of this title, or any similar offense when committed or prosecuted in another jurisdiction; or

(8) Any person who, knowing that he or she is the defendant or co-defendant in any criminal case in which that person is alleged to have committed any felony under the laws of this State, the United States or any other state or territory of the United States, becomes a fugitive from justice by failing to appear for any scheduled court proceeding pertaining to such felony for which proper notice was provided or attempted. It is no defense to a prosecution under this paragraph that the person did not receive notice of the scheduled court proceeding.

(9) Except for "antique firearms", and validly seized deadly weapons or ammunition from a person prohibited as a result of a felony conviction under paragraph (a)(1) or (a)(3) of this section, or validly seized under paragraph (a)(8) of this section may be disposed of by the law enforcement agency holding the weapon or ammunition, after the exhaustion of any right of direct appeal, pursuant to § 2311 of this title.

a. "Antique firearm" means any firearm not designed or redesigned for using rimfire or conventional centerfire ignition with fixed ammunition and manufactured in or before 1898 and also any firearm using fixed ammunition manufactured in or before 1898, for which ammunition is no longer manufactured in the United States and is not readily restored to a firing condition.

b. A person prohibited under this section has the burden of proving that the subject firearm is an antique firearm as defined in paragraph (a)(9)a. of this section subject to an exemption under this section and § 2311 of this title.

(b) Any prohibited person as set forth in subsection (a) of this section who knowingly possesses, purchases, owns or controls a deadly weapon or ammunition for a firearm while so prohibited shall be guilty of possession of a deadly weapon or ammunition for a firearm by a person prohibited.

(c) Possession of a deadly weapon by a person prohibited is a class F felony, unless said deadly weapon is a firearm or ammunition for a firearm, in which case it is a class D felony. As used herein, the word "ammunition" shall mean 1 or more rounds of fixed ammunition designed for use in and capable of being fired from a pistol, revolver, shotgun or rifle but shall not mean inert rounds or expended shells, hulls or casings.

(d) Any person who is a prohibited person solely as the result of a conviction for an offense which is not a felony shall not be prohibited from purchasing, owning, possessing or controlling a deadly weapon or ammunition for a firearm if 5 years have elapsed from the date of conviction.

(e)(1) Notwithstanding any provision of this Section or Code to the contrary, any person who is a prohibited person as described in this Section and who knowingly possesses, purchases, owns, or controls a firearm or destructive weapon while so prohibited shall receive a minimum sentence of:

a. one (1) year at Level V, if the person has previously been convicted of a violent felony;

b. three (3) years at Level V, if the person does so within ten (10) years of the date of conviction for any violent felony or the date of termination of all periods of incarceration or confinement imposed pursuant to said conviction, whichever is the later date; or

c. five (5) years at Level V, if the person has been convicted on two or more separate occasions of any violent felony.

(2) Any person who is a prohibited person as described in this section because of a conviction for a violent felony and who, while in possession or control of a firearm in violation of this section, negligently causes serious physical injury to or the death of another person through the use of such firearm, shall be guilty of a class B felony and shall receive a minimum sentence of:

a. Four years at Level V; or

b. Six years at Level V, if the person causes such injury or death within 10 years of the date of conviction for any violent felony or the date of termination of all periods of incarceration or confinement imposed pursuant to said conviction, whichever is the later date; or

c. Ten years at Level V, if the person has been convicted on 2 or more separate occasions of any violent felony.

d. Nothing in this paragraph shall be deemed to be a related or included offense of any other provision of this Code. Nothing in this paragraph shall be deemed to preclude prosecution or sentencing under any other provision of this Code nor shall this paragraph be deemed to repeal any other provision of this Code.

(3) Any sentence imposed pursuant to this subsection shall not be subject to the provisions of § 4215 of this title. For the purposes of this subsection, "violent felony" means any felony so designated by § 4201(c) of this title, or any offense set forth under the laws of the United States, any other state or any territory of the United States which is the same as or equivalent to any of the offenses designated as a violent felony by § 4201(c) of this title.

(4) Any sentence imposed for a violation of this subsection shall not be subject to suspension and no person convicted for a violation of this subsection shall be eligible for good time, parole or probation during the period of the sentence imposed.

(f)(1) Upon conviction, any person who is a prohibited person as described in paragraph (a)(5) of this section and who is 14 years of age or older shall, for a first offense, receive a minimum sentence of 6 months of Level V incarceration, and shall receive a minimum sentence of 1 year of Level V incarceration for a second and subsequent offense, which shall not be subject to suspension. Any sentence imposed pursuant to this subsection shall not be subject to §§ 4205(b) and 4215 of this title.

(2) The penalties prescribed by this subsection and subsection (g) of this section shall be imposed regardless of whether or not the juvenile is determined to be amenable to the rehabilitative process of the Family Court pursuant to § 1010(c) of Title 10 or any successor statute.

(g) In addition to the penalties set forth in subsection (f) of this section herein, a person who is a prohibited person as described in paragraph (a)(5) of this section and who is 14 years of age or older shall, upon conviction of a first offense, be required to view a film and/or slide presentation depicting the damage and destruction inflicted upon the human body by a projectile fired from a gun, and shall be required to meet with, separately or as part of a group, a victim of a violent crime, or with the family of a deceased victim of a violent crime. The Division of Youth Rehabilitative Service, with the cooperation of the Office of the Chief Medical Examiner and the Violent Crimes Compensation Board, shall be responsible for the implementation of this subsection.

1448A. Criminal history record checks for sales of firearms.

a. No licensed importer, licensed manufacturer or licensed dealer shall sell or deliver from inventory any firearm, as defined in § 222(11) of this title, to another person, other than a licensed importer, licensed manufacturer, licensed dealer or licensed collector, until the licensee has:

(1) Obtained from the potential buyer or transferee a nonexpired identification issued by a governmental agency indicating the address, sex and date of birth of the buyer or transferee and bearing a photograph of the transferee obtained from the potential buyer or transferee, a secondary form of identification which indicates the same name and address of the buyer or transferee as noted on the government issued identification, and has inspected said identifications and has determined that they accurately reflect the identify of the buyer or transferee. For purposes of this section, an original gas, electric, telephone or other utility bill may qualify as a secondary form of identification, if it contains the requisite information; and

(2) Obtained a completed consent form from the potential buyer or transferee, which form shall have been promulgated by the State Bureau of Identification (SBI) and provided by the licensed importer, licensed manufacturer or licensed dealer, which shall include the name, address, birth date, gender, race, and Social Security number, driver's license number or other identification number of such potential buyer or transferee; and

(3) Requested, by means of a toll-free telephone call pursuant to subsection (e) of this section, the SBI to conduct a criminal history and involuntary commitment of an adult record check; and

(4) Received a unique approval number for that inquiry from the SBI, and has recorded the date and approval number on the consent form; and

(5) In the event that the potential buyer or transferee is prohibited from purchasing a firearm, within 7 days of the notification of said prohibition, the licensee shall deliver to the State Bureau of Identification a copy of all consent forms executed by the potential buyer, along with the following information:

a. The name, address and telephone number of the licensee and of any agent or employee of the licensee who witnessed the prohibited attempted sale or transfer; and

b. The number, type, descriptions and serial numbers of any firearms which the buyer or the transferee attempted to purchase or otherwise acquire.

Any information received by the State Bureau of Identification pursuant to this para-graph may be provided to any law enforcement agency for the purpose of investigating and prosecuting any criminal conduct.

(b) Upon receipt of a request for a criminal history and involuntary commitment of an adult record check, the SBI during the licensee's call or by return call, shall:

(1) Review its criminal history and involuntary commitment of an adult records to determine if the potential buyer or transferee is prohibited from receipt or possession of a firearm pursuant to § 1448 of this title or federal law; and

(2) Inform the licensee making the inquiry either:

a. That its records demonstrate that the potential buyer or transferee is so prohibited; or

b. Provide the licensee with a unique approval number.

(c) In the event of electronic failure or similar emergency beyond the control of the SBI, the SBI shall immediately notify the requesting licensee of the reason for, and estimated length of, such delay. After such notification, the SBI shall no later than the end of the 3rd business day following a request for a criminal history and involuntary commitment of an adult record check of the licensee, either:

(1) Inform the licensee that its records demonstrate that the potential buyer or transferee is prohibited from receipt or possession of a firearm pursuant to § 1448 of this title or federal law; or

(2) Provide the licensee with a unique approval number. Unless notified by the end of the 3rd business day following a request for a records check that the potential buyer or transferee is so prohibited, and without regard to whether the licensee has received a unique approval number, the licensee may complete the sale or delivery and shall not be deemed in violation of this section with respect to such sale or delivery.

(d)(1) Any records containing any of the information set forth in paragraph (a)(1) pertaining to a potential buyer or transferee who is not found to be prohibited from receipt or possession of a firearm by reason of § 1448 of this title or federal law shall be confidential and may only be disclosed by any officer or employee of the SBI to other law enforcement agencies. The SBI and any other law enforcement agencies shall destroy any such records after it communicates the corresponding approval number to the licensee and such records shall be destroyed within 60 days after the day of receipt of the licensee's request.

(2) Notwithstanding contrary provisions of this subsection, the SBI shall maintain a log of dates of requests for criminal history record checks and unique approval numbers corresponding to such dates for a period of not longer than 1 year.

(3) Nothing in this section shall be construed to allow the State to maintain records containing the names of licensees who receive unique approval numbers or to maintain records of firearm transactions, including the names or other identification of licensees and potential buyers or transferees, involving persons not prohibited by § 1448 of this title and federal law from the receipt or possession of firearms. However, the SBI may retain whatever information it receives including, but not limited to, the identifying information of potential buyers or transferees, if the SBI has probable cause to believe the potential buyer or transferee is committing a crime.

(e) The SBI shall establish a toll-free telephone number which shall be operational between the hours of 9:00 a.m. and 9:00 p.m., Monday through Saturday, and 9:00 a.m. to 5:00 p.m. Sunday for purposes of responding to inquiries as described in this section from licensed manufacturers, licensed importers and licensed dealers. The foregoing notwithstanding, the telephone number need not be operational on Christmas Day, Thanksgiving Day or on Easter Sunday. The SBI shall employ and maintain such personnel as are necessary to administer the provisions of this section.

(f) Any person who is denied the right to receive or purchase a firearm as a result of the procedures established by this section may request an amendment of any errors in the record pertaining to the person by petitioning the SBI. If the SBI fails to amend the record within 30 days, the person requesting the amendment may petition the Superior Court in the county of residence for a writ of mandamus directing the SBI to amend the record. The Court shall award the petitioner all reasonable attorney fees and other costs, if it determines that SBI willfully refused to amend the record. If the record as corrected demonstrates that such person is not prohibited from receipt or possession of a firearm by § 1448 of this title or federal law, the SBI shall destroy any records it maintains which contain any information derived from the criminal history and involuntary commitment of an adult records check set forth in paragraphs (a)(3) and (a)(4) of this section.

(g) The SBI shall promulgate regulations to ensure the identity, confidentiality and security of all records and data provided pursuant to this section.

(h) A licensed importer, licensed manufacturer or licensed dealer is not required to comply with the provisions of this section in the event of:

(1) Unavailability of telephone service at the licensed premises due to:

a. The failure of the entity which provides telephone service in the state, region, or other geographical area in which the licensee is located; or

b. The interruption of telephone service by reason of hurricane, tornado, flood, natural disaster, or other act of God, or war, invasion, insurrection, riot, or other bona fide emergency, or other reason beyond the control of the licensee; or

(2) Failure of the SBI reasonably to comply with the requirements of subsections (b) and (c) of this section. Within 72 hours of the normalization of telephone service, the licensed importer, licensed manufacturer or licensed dealer shall communicate to SBI the identifying data as set forth in subsection (a)(4) of this section for each sale or delivery of a firearm during the unavailability of telephone service.

(i) Compliance with the provisions of this section shall be a complete defense to any claim or cause of action under the laws of this State for liability for damages arising from the importation or manufacture, or the subsequent sale or transfer to any person who has been convicted in any court of a crime punishable by imprisonment for a term exceeding 1 year, of any firearm which has been shipped or transported in interstate or foreign commerce.

(j) The provisions of this section shall not apply to:

(1) Any firearm (including any firearm with a matchlock, flintlock, percussion cap, or similar type of ignition system) manufactured in or before 1898;

(2) Any replica of any firearm described in paragraph (1) of this subsection if such replica:

a. Is not designed or redesigned to use rimfire or conventional centerfire fixed ammunition; or

b. Uses rimfire or conventional centerfire fixed ammunition which is no longer manufactured in the United States and which is not readily available in the ordinary channels of commercial trade;

(3) Any shotgun, which is defined as a firearm designed or intended to be fired from the shoulder and designed or made to fire through a smooth bore either a number of ball shot or a single projectile for each single pull of the trigger;

(4) The return, by a licensed pawnbroker, of a firearm to the person from whom it was received;

(5) Transactions in which the potential buyer or transferee holds a valid concealed deadly weapons license pursuant to § 1441 of this title; and

(6) Transactions involving a law enforcement officer as defined by § 222(14) of this title.

(k) Any licensed dealer, licensed manufacturer, licensed importer or employee thereof who willfully and intentionally requests a criminal history record check from the SBI for any purpose other than compliance with subsection (a) of this section, or willfully and intentionally disseminates any criminal history record information to any person other than the subject of such information or discloses to any person the unique identification number shall be guilty of a class A misdemeanor.

(l) Any person who, in connection with the purchase, transfer, or attempted purchase or transfer of a firearm pursuant to subsection (a) of this section, willfully and intentionally makes any materially false oral or written statement or willfully and intentionally furnishes or exhibits any false identification intended or likely to deceive the licensee shall be guilty of a class G felony.

(m) Any licensed dealer, licensed manufacturer, licensed importer or employee thereof who willfully and intentionally sells or delivers a firearm in violation of this section shall be guilty of a class A misdemeanor. Second or subsequent offenses by an individual shall be a class G felony.

(n) The SBI shall provide to the judiciary committees of the Senate and House of Representatives an annual report including the number of inquiries made pursuant to this section for the prior calendar year. Such report shall include, but not be limited to, the number of inquiries received from licensees, the number of inquiries resulting in a determination that the potential buyer or transferee was prohibited from receipt or possession of a firearm pursuant to §1448 of this title or federal law, and the estimated costs of administering this section.

(o) This section shall become effective 6 months from July 20, 1990 or at such time as the SBI has notified all licensed importers, licensed manufacturers and licensed dealers in writing that the procedures and toll-free number described in this section are operational, whichever shall occur first.

(p) Violations of this section shall be in the exclusive jurisdiction of the Superior Court.

(q) Notwithstanding Chapter 89 of this title, Chapter 10 of Title 29, and other Delaware laws, the SBI is authorized and directed to release records and data required by this section. The SBI shall not release or disclose criminal records or data except as specified in subsections (b) and (c) of this section.

1450. Receiving a stolen firearm; class F felony. A person is guilty of receiving a stolen firearm if the person intentionally receives, retains or disposes of a firearm of another person with intent to deprive the owner of it or to appropriate it, knowing that it has been acquired under circumstances amounting to theft, or believing that it has been so acquired. Receiving a stolen firearm is a class F felony. Knowledge that a firearm has been acquired under circumstances amounting to theft may be presumed in the case of a person who acquires it for a consideration which the person knows is substantially below its reasonable value.

1451. Theft of a firearm; class F felony.

(a) A person is guilty of theft of a firearm when the person takes, exercises control over or obtains a firearm of another person intending to deprive the other person of it or appropriate it.

(b) Theft of a firearm is a class F felony.

1454. Giving a firearm to person prohibited; class F felony. A person is guilty of giving a firearm to certain persons prohibited when the person sells, transfers, gives, lends or otherwise furnishes a firearm to a person knowing that said person is a person prohibited as is defined in § 1448 of this title.

Giving a firearm to certain persons prohibited is a class F felony.

1455. Engaging in a firearms transaction on behalf of another; class F felony; class C felony. A person is guilty of engaging in a firearms transaction on behalf of another when the person purchases or obtains a firearm on behalf of a person not qualified to legally purchase, own or possess a firearm in this State or for the purpose of selling, giving or otherwise transferring a firearm to a person not legally qualified to purchase, own or possess a firearm in this State.

Engaging in a firearms transaction on behalf of another is a class F felony for the first offense, and a class C felony for each subsequent like offense.

1456. Unlawfully permitting a minor access to a firearm; class A misdemeanor.

(a) A person is guilty of unlawfully permitting a minor access to a firearm when the person intentionally or recklessly stores or leaves a loaded firearm within the reach or easy access of a minor and where the minor obtains the firearm and uses it to inflict serious physical injury or death upon the minor or any other person.

(b) It shall be an affirmative defense to a prosecution under this section if:

(1) The firearm was stored in a locked box or container or in a location which a reasonable person would have believed to be secure from access to a minor; or

(2) The minor obtains the firearm as the result of an unlawful entry by any person; or

(3) The serious physical injuries or death to the minor or any other person results from a target or sport shooting accident or hunting accident.

(c) Unlawfully permitting a minor access to a firearm is a class A misdemeanor.

1457. Possession of a weapon in a Safe School and Recreation Zone; class D, E, or F: class A or B misdemeanor.

(a) Any person who commits any of the offenses described in subsection (b) of this section, or any juvenile who possesses a firearm or other deadly weapon, and does so while in or on a "Safe School and Recreation Zone" shall be guilty of the crime of possession of a weapon in a Safe School and Recreation Zone.

(b) The underlying offenses in Title 11 shall be:

(1) Section 1442. Carrying a concealed deadly weapon; class G felony; class E felony.

(2) Section 1444. Possessing a destructive weapon; class E felony.

(3) Section 1446. Unlawfully dealing with a switchblade knife; unclassified misdemeanor.

(4) Section 1448. Possession and purchase of deadly weapons by persons prohibited; class F felony.

(5) Section 1452. Unlawfully dealing with knuckles-combination knife; class B misdemeanor.

(6) Section 1453. Unlawfully dealing with martial arts throwing star; class B misdemeanor.

(c) For the purpose of this section, "Safe School and Recreation Zone" shall mean:

(1) Any building, structure, athletic field, sports stadium or real property owned, operated, leased or rented by any public or private school including, but not limited to, any kindergarten, elementary, secondary or vocational-technical school or any college or university, within 1,000 feet thereof; or

(2) Any motor vehicle owned, operated, leased or rented by any public or private school including, but not limited to, any kindergarten, elementary, secondary, or vocational-technical school or any college or university; or

(3) Any building or structure owned, operated, leased or rented by any county or municipality, or by the State, or by any board, agency, commission, department, corporation or other entity thereof, or by any private organization, which is utilized as a recreation center, athletic field or sports stadium.

(d) Nothing in this section shall be construed to preclude or otherwise limit a prosecution of or conviction for a violation of this chapter or any other provision of law. A person may be convicted both of the crime of possession of a weapon in a Safe School and Recreation Zone and of the underlying offense as defined elsewhere by the laws of the State.

(e) It shall not be a defense to a prosecution for a violation of this section that the person was unaware that the prohibited conduct took place on or in a Safe School and Recreation Zone.

(f) It shall be an affirmative defense to a prosecution for a violation of this section that the weapon was possessed pursuant to an authorized course of school instruction, or for the purpose of engaging in any school-authorized sporting or recreational activity. The affirmative defense established in this section shall be proved by a preponderance of the evidence. Nothing herein shall be construed to establish an affirmative defense with respect to a prosecution for any offense defined in any other section of this chapter.

(g) It is an affirmative defense to prosecution for a violation of this section that the prohibited conduct took place entirely within a private residence, and that no person under the age of 18 was present in such private residence at any time during the commission of the offense. The affirmative defense established in this section shall be proved by the defendant by a preponderance of the evidence. Nothing herein shall be construed to establish an affirmative defense with respect to a prosecution for an offense defined in any other section of this chapter.

(h) This section shall not apply to any law enforcement or police officer, or to any security officer as defined in Chapter 13 of Title 24.

(i) For purposes of this section only, "deadly weapon" shall include any object described in § 222(5) or § 222(11) of this title or BB guns.

(j) The penalty for possession of a weapon in a Safe School and Recreation Zone shall be:

(1) If the underlying offense is a class B misdemeanor, the crime shall be a class A misdemeanor;

(2) If the underlying offense is an unclassified misdemeanor, the crime shall be a class B misdemeanor;

(3) If the underlying offense is a class E, F, or G felony, the crime shall be one grade higher than the underlying offense.

(4) In the event that an elementary or secondary school student possesses a firearm or other deadly weapon in a Safe School and Recreation Zone, in addition to any other penalties contained in this section, the student shall be expelled by the local School Board or charter school board of directors for a period of not less than 180 days unless otherwise provided for in federal or state law.

1459. Possession of a weapon with a removed, obliterated or altered serial number.

(a) No person shall knowingly transport, ship, possess or receive any firearm with the knowledge that the importer's or manufacturer's serial number has been removed, obliterated or altered in a manner that has disguised or concealed the identity or origin of the firearm.

(b) This section shall apply to a firearm manufactured prior to 1973.

(c) Possessing, transporting, shipping or receiving a firearm with a removed, obliterated or altered serial number pursuant to this section is a class D felony.

Part II. Criminal Procedure Generally

Chapter 43. Sentencing, Probation, Parole and Pardons

Subchapter VI. Clemency

4364. Effect of Pardon; Restoration of Civil Rights. Except as otherwise provided by the Delaware Constitution, or expressly by any provision of the Delaware Code or any court rule, the granting of an unconditional pardon by the Governor shall have the effect of fully restoring all civil rights to the person pardoned. Such civil rights include, but are not limited to, the right to vote, the right to serve on a jury if selected, the right to purchase or possess deadly weapons and the right to seek and hold public office provided however, that this section shall not limit or affect the Governor's authority to place lawful conditions upon the granting of a pardon. Notwithstanding the granting of a pardon or any provision of this section, no person who shall be convicted of embezzlement of the public money, bribery, perjury or other infamous crime, shall be eligible to a seat in either House of the General Assembly, or capable of holding any office of trust, honor or profit under this State.

Title 22. Municipalities

Chapter 1. General Provisions

111. Limitation on firearm regulations. The municipal governments shall enact no law, ordinance or regulation prohibiting, restricting or licensing the ownership, transfer, possession or transportation of firearms or components of firearms or ammunition except that the discharge of a firearm may be regulated; provided any law, ordinance or regulation incorporates the justification defenses as found in Title 11. Nothing contained herein shall be construed to invalidate municipal ordinances existing before July 4, 1985, and any ordinance enacted after July 4, 1985, is hereby repealed. Notwithstanding the provisions of this section to the contrary, the City of Wilmington may, in addition to the nature and extent of regulation permitted by this section, enact any law or ordinance governing the possession or concealment of a paintball gun within its corporate limits as it deems necessary to protect the public safety.

Title 24. Professions and Occupations

Chapter 9. Deadly Weapons Dealers

901. License requirement. No person shall engage in the business of selling any pistol or revolver, or stiletto, steel or brass knuckles, or other deadly weapon made especially for the defense of one's person without first having obtained a license therefor, which license shall be known as "special license to sell deadly weapons." No person licensed or unlicensed shall possess, sell or offer for sale any switchblade knife.

This section shall not apply to toy pistols, pocket knives or knives used for sporting purposes and in the domestic household, or surgical instruments or tools of any kind.

902. Application and fee for license; duration; renewal. Whoever desires to engage in the business of selling any of the articles referred to in the first paragraph of § 901 of this title shall apply to the Department of State to obtain a license to conduct such business and shall pay an application fee of $50 to the Department. The license shall entitle the holder thereof to conduct such business until June 1 next succeeding its date. An application for renewal of such license shall be accompanied by a payment of $50 to the Department.

903. Sale to persons under 21 or intoxicated persons. No person shall sell to a person under the age of 21 or any intoxicated person any of the articles referred to in the first paragraph of § 901 of this title.

904. Records.

(a) Any person desiring to engage in the business described in this chapter shall keep and maintain in the place of business at all times a record. In such record the businessperson shall enter the date of the sale, the name and address of the person purchasing any deadly weapon, the number and kind of deadly weapon so purchased, the age of the purchaser, and the mode of identification bearing a picture which shall include but it is not limited to a driver's license. The record shall at all times be open for inspection by any judge, justice of the peace, police officer, constable or other peace officer of this State.

(b) Any person engaging in the business described in this chapter shall keep and maintain a list of current employees including their names, former names used, dates of birth, physical descriptions and social security numbers. The required employee list and all attachments thereto shall be considered confidential but shall, nevertheless, be open for inspection by any police officer of this State or of any political subdivision of this State, within their respective jurisdiction, at any time, at the licensee's primary place of business and during the licensee's regular business hours. No person licensed under this chapter shall knowingly allow any employee who is a person prohibited from possessing a deadly weapon pursuant to § 1448 of Title 11 to facilitate a sale of a deadly weapon. All employers licensed to do business pursuant to this chapter shall, prior to employment and at least once during each calendar year thereafter, perform a telephonic criminal history record check of each employee utilizing the procedures set forth in § 1448A of Title 11 and shall make and maintain a record thereof using the State Bureau of Identification Criminal History Record Information and Mental Health Information Consent Form (Form 544). A copy of each such form shall be attached to the above required employee list for inspection upon the valid request of a police officer of this State or of any political subdivision of this State, within their respective jurisdiction.

904A. Criminal history checks for sales between unlicensed persons.

(a) For purposes of this section, "licensed firearm dealer" means any person licensed as a deadly weapons dealer pursuant to Chapter 9 of Title 24 and 18 U.S.C. § 921 et seq.

(b) As a condition of its license, any dealer holding a license pursuant to this chapter shall facilitate the transfer of a firearm, as that term is defined in § 222 of Title 11, from any person who is not so licensed, upon the request of said unlicensed seller pursuant to the following procedure:

(1) The prospective buyer and seller shall jointly appear at the place of business of the dealer, during said dealer's regular hours of business, and shall inform the dealer of their desire to avail themselves of the advantages of the procedure set forth herein.

(2) The dealer shall then subject the prospective buyer to a criminal history record check pursuant to the terms of § 1448A of Title 11.

(3) In the event that said record check reveals that the prospective buyer is prohibited from possessing, purchasing or owning a firearm pursuant to § 1448 of Title 11, the dealer shall so inform both parties of that fact and the transfer shall not take place.

(4) Any dealer who is asked to facilitate the transfer of a firearm pursuant to the terms of this section, may charge a reasonable fee for said service, said fee not to exceed $20 per criminal history check performed pursuant to this procedure.

(5) Failure or refusal on the part of the dealer to facilitate the transfer of a firearm pursuant to the procedures set forth herein shall be adequate cause to suspend the license of said dealer for a period not to exceed 30 days per occurrence.

Title 28 Sports and Amusements

Chapter 8. Purchase of Rifles and Shotguns in States Contiguous to this State

801. Definitions.

(a) As used in this chapter, the term "a state contiguous to this State" shall mean any state having a common border with this State.

(b) As used in this chapter, all other terms shall be given the meaning prescribed in 18 U.S.C. § 921 (the Gun Control Act of 1968), and the regulations duly promulgated thereunder as presently enacted or promulgated and as hereafter modified. (28 Del. C. 1953, § 801; 57 Del. Laws, c. 393.)

802. Lawful acts.

It shall be lawful for a person residing in this State, including a corporation or other business entity maintaining a place of business in this State, to purchase or otherwise obtain a rifle or shotgun in a state contiguous to this State and to receive or transport such rifle or shotgun into this State, subject, however, to such other laws of the State or its political subdivision as may be applicable and subject to § 102 of the Gun Control Act of 1968, 18 U.S.C. § 922. (28 Del. C. 1953, § 802; 57 Del. Laws, c. 393.)

803. Not applicable to federal licensees.

This chapter shall not apply or be construed to affect in any way the purchase, receipt or transportation of rifles and shotguns by federally licensed firearms manufacturers, importers, dealers or collectors. (28 Del. C. 1953, § 803; 57 Del. Laws, c. 393.)

[Current through 76 Del. Laws c. 184 (Nov. 9, 2007)]

Code of Ordinances, City of Dover

Chapter 70. Offenses and Miscellaneous Provisions

Sec. 70-2. Weapons generally.

(a) *Prohibited.* It shall be unlawful for any person to have in his possession, carry or use a revolver or pistol of any description, a shotgun or rifle which may be used for the explosion of bullets and cartridges, or any air gun, BB gun, gas-operated gun or spring-operated gun, or any instrument, toy or weapon commonly known as a "peashooter," "slingshot" or "beany," or any bow made for the purpose of throwing or projecting missiles of any kind by any means whatsoever, or any knife, whether that instrument is called by any name set forth above or by any other name.

(b) *License.* The weapons prohibited in subsection (a) of this section shall be permitted upon the grant of a license therefor, if required, in the following situations:

(1) *On owner's property.* Possession within the possessor's own domicile or business.

(2) *Other license or permit.* Possession, carrying or use in conformity with a license issued by the state for that weapon or a permit issued by the chief of police.

(3) *Shooting ranges.* Possession or use at licensed shooting galleries or ranges when the instrument can be fired, discharged or operated in a manner that shall not endanger persons or property, and also in a manner that shall prevent the projectile from traversing any grounds or space outside the limits of the gallery or range.

(4) *Unloaded and cased.* Carrying of any type of gun whatsoever, when unloaded and properly cased, to or from any licensed gallery or range or to or from an area where hunting is allowed by law. This subsection shall not limit the permitted carrying of guns allowed by subsection (b)(5) of this section.

(5) *Officers.* Carrying, wearing, possessing and using, by United States marshals, sheriffs, constables and their deputies, and any regular, special or ex officio police officer, or any other law enforcement officer while on duty, or as shall be necessary in the proper discharge of their duties.

(6) *State license.* Possession, carrying or use in conformity with a license issued by the state for that purpose in areas approved by the chief of police, and only then when used in such a manner that shall not endanger persons or property.

(7) *Defense.* Persons exercising their legal right of self-defense or defense of property or others.

(c) *Discharge.* It shall be unlawful for any person to discharge any gun, pistol, revolver or other firearm within the city, except lawfully pursuant to subsections (b)(2), (3) and (5)–(7) of this section, and those persons excepted by law.

(d) *Business display.* It shall be unlawful for any pawnbroker, secondhand dealer or other person who engages in business in the city to display, or to place on exhibition in any show window or other window facing upon any street, any pistol, revolver or other firearm with a barrel of less than 12 inches in length, or any switchblade knife, or any brass or metal knuckles, or any club loaded with lead or other weight, or any blackjack or billyclub.

(e) *Furnishing to certain persons.* It shall be unlawful for any person to purchase from, or sell, loan or furnish any weapon to, any person under the influence of any alcoholic beverage or any narcotic drug, stimulant or depressant, to any person in a condition of agitation and excitability or to a minor under the age of 18 years.

(f) *Forfeiture.* Every person convicted of a violation of this section or section 70-3, or any state law relating to weapons shall forfeit to the city the dangerous or deadly weapon concealed or displayed. Upon a finding of guilt, it shall then be the duty of the court, after the expiration of a reasonable time, to declare by written order that the chief of police shall either dispose of the weapon or retain the weapon for official police department use.

(g) *Definition.* As used in subsection (e) and (f) of this section, the term "weapon" shall mean any revolver or pistol, shotgun or rifle which may be used for the explosion of bullets and cartridges, or any air gun, BB gun, or any instrument, toy or weapon commonly known as a "peashooter," "slingshot" or "beany," or any bow made for the purpose of throwing or projecting missiles of any kind, or any knife.

[Dover Code codified through Ordinances of 2008-25 adopted June 9, 2008]

New Castle County Code

Chapter 22. Offenses and Miscellaneous Provisions

Article 3. Weapons

22.03.001. Possession by persons under sixteen (16) years of age. Possession of firearms, air pressure guns, including BB guns, bows and metal tipped arrows, by anyone sixteen (16) years of age or younger is prohibited, unless under direct supervision of a person twenty-one (21) years of age or over.

Sec. 22.03.009. Stun guns and taser guns.

A. It shall be unlawful for any person to manufacture, make, deliver, transport, trade, give, sell, purchase or possess a stun gun or taser gun.

B. The term "stun gun," as used in this Section, means any battery-powered, pulsed electrical device of high voltage and low or no amperage which, when pressed against a person, can disrupt the central nervous system and cause temporary loss of voluntary muscle control of such person.

C. The term "taser gun," as used in this Section, means any device contained in a package which permits it to be hand held, containing an electrical supply unit, and into which an expendable plastic cassette may be inserted and which operates by shooting or projecting a ball, net or dart carrying fine wires from the package to the target and which, upon hitting a person, can send out current capable of disrupting the person's nervous system in such a manner as to render the person incapable of functioning normally.

D. Exempt from this Section shall be any law enforcement officer while performing his or her lawful duties within this County and those selling to any law enforcement officer while performing his or her lawful duties within this County.

E. A conviction of violation of this Section shall be punishable as provided in Section 1.01.009 of this Code.

[New Castle County Code codified through Ordinance No. 08-040, enacted July 8, 2008]

Wilmington City Code

Chapter 5. Businesses

Article II. Business Licenses

Division 2. Specific Businesses and Activities

5-79. Retailers of firearms. No person shall receive a license as a retailer of firearms unless he complies with the provisions of all applicable ordinances and laws.

Chapter 36. Miscellaneous Offenses and Provisions

Article V. Offenses Involving Public Safety

Division 2. Weapons and Related Offenses

36-156. Armor-piercing bullets.

(a) Definitions. The following words, terms and phrases, when used in this section, shall have the meanings ascribed to them in this subsection, except where the context clearly indicates a different meaning:

(1) *Armor-piercing bullet* means any bullet which is coated with a nonstick fluoropolymer finish, such as the registered trademark finishes, Teflon, Halon, Halar, Flvon, Soreflon or Algoflon.

(2) *Bullet* means a round or elongated missile designed to be fired from a firearm.

(3) *Bulletproof vest* means any commercially available, soft, lightweight body armor, usually consisting of several layers of a polyaramid fiber tradenamed "Kevlar."

(b) Offenses. It shall be unlawful for any person to bring into the city or to manufacture, sell, distribute, possess or use armor-piercing bullets or any other bullets similarly coated with a nonstick fluoropolymer finish. It shall further be unlawful for any person which is in the business of manufacturing firearms to possess the component parts of any armor-piercing bullet.

(c) Penalties. Any person who violates the provisions of this section shall upon conviction be subject to a minimum fine of $100.00 and a maximum fine of $500.00, or to imprisonment not exceeding 90 days, or both, for each offense.

36-157. Firearms dealers; storage, display.

(a) The following words, terms and phrases, when used in this section, shall have the meanings ascribed to them in this subsection, except where the context clearly indicates a different meaning:

(1) *Dealer* means any individual, firm, association, partnership or corporation engaged in the business of selling firearms, purchasing firearms for resale or conducting a gunsmith or firearms repair business. Whenever used in any cause prescribing and imposing a penalty, the term "dealer," as applied to any partnership or association, shall mean the partners or members thereof, and as applied to any corporation, shall include the officers thereof.

(2) *Firearms* means any rifle, revolver, pistol or shotgun capable of propelling a projectile by means of an explosive material or charge.

(b) During the hours they are not regularly open for business, dealers shall store all firearms in accordance with the following requirements:

(1) No firearms shall be displayed in windows.

(2) All firearms must be placed in an approved safe, vault or properly secured storeroom. Any dealer may comply with the requirements of this section by providing an approved steel safe wherein any firearms may be stored and locked during nonbusiness hours.

(c) Before promulgating any regulations designed to carry out the intent and purpose of this section, the department of licenses and inspections shall consult with the police department.

(d) No dealer shall receive a permit to store firearms unless he complies with the provisions of this section and regulations issued pursuant thereto.

(e) Any dealer who violates the provisions of this section shall be advised in writing by the department of licenses and inspections of the nature of the violation, and shall be required to comply with the provisions of this section within the period indicated in such notice; provided, that in no case shall the time permitted for such compliance exceed 60 days. Each day that any dealer fails to comply with the requirements of this section or to make the changes indicated in any notice of violation, after the period allowed for such compliance has expired, shall constitute a separate violation of this section.

(f) Any person who violates the provisions of this section shall, upon summary conviction, be subject to a minimum fine of $100.00 and a maximum fine of $500.00 or to imprisonment not exceeding 90 days, or both, for such offense.

36-158. Certain firearms prohibited.

(a) The following words, terms and phrases, when used in this section, shall have the meanings ascribed to them in this subsection, except where the context clearly indicates a different meaning:

(1) *Converted starter pistol* means a starter pistol which has been altered to fire a projectile with sufficient force to cause death or physical injury.

(2) *Short-barreled rifle* means a rifle having one or more barrels less than 16 inches in length and any weapon made from a rifle, whether by alteration, modification or otherwise, if such weapon, as modified, has an overall length of less than 26 inches.

(3) *Smooth-bore shot revolver* means a revolver with a smooth-bore having been reamed out so that it can be used to fire shot-shell.

(4) *Zip gun* means any weapon or instrument not originally designed to be a firearm which has been made or altered to discharge a projectile with sufficient force to cause death or physical injury.

(b) It shall be unlawful for any person to manufacture, make, deliver, transport, trade, give, sell or possess a smooth-bore shot revolver, short-barreled rifle, zip gun, or converted starter pistol.

(c) Any law enforcement officer while performing his lawful duties within the city shall be exempted from the effect of this section. For purposes of this section, the term "law enforcement officer" includes police officers, the attorney general, the attorney general's deputies and investigators, the sheriff, and the sheriff's deputies, prison guards, constables and bailiffs.

(d) A conviction of violation of this section shall be punishable by a fine of not less than $500.00 and not more than $2,500.00 or by both such fine and imprisonment not exceeding six months. The minimum sentence of a $500.00 fine shall not be subject to suspension or reduction for any reason.

36-161. Stun guns, etc.

(a) It shall be unlawful for any person to manufacture, make, deliver, transport, trade, give, sell, or possess a converted tear gas gun, stun gun or taser gun.

(b) The following words, terms and phrases, when used in this section, shall have the meanings ascribed to them in this subsection, except where the context clearly indicates a different meaning:

(1) *Converted tear gas gun* means a tear gas gun which has been altered to fire a projectile with sufficient force to cause death or physical injury.

(2) *Stun gun* means any battery-powered, pulsed electrical device of high voltage and low or no amperage which, when pressed against a person, can disrupt the central nervous system and cause temporary loss of voluntary muscle control of such person.

(3) *Taser gun* means any device contained in a package which permits it to be handheld, containing an electrical supply unit, and into which an expendable plastic cassette may be inserted and which operates by shooting or projecting a ball, net, or dart carrying fine wires from the package to the target and which, upon hitting a person can send out current capable of disrupting the person's nervous system in such a manner as to render the person incapable of functioning normally.

(c) Any law enforcement officer while performing his lawful duties within the city shall be exempted from the effect of this section. For purposes of this section, the term "law enforcement officer" shall include police officers, the attorney general, the attorney general's deputies and investigators, the sheriff, and the sheriff's deputies, prison guards, constables and bailiffs.

(d) A conviction of violation of this section shall be punishable by a fine of not less than $500.00 and not more than $2,500.00 or by both such fine and imprisonment not exceeding six months. The minimum sentence of $500.00 fine shall not be subject to suspension or reduction for any reason.

Sec. 36-165. Paint pellet guns.

(a) It shall be unlawful for any person to use or discharge a paint pellet gun in the city, usually used in connection with the game known as "paint ball" which is similar to the childhood game of "capture the flag," unless said discharge occurs in a licensed, supervised recreational facility during the facility's hours of operation. For purposes of this sub-section, a paint pellet gun or paint ball gun is defined to include any instrument which releases a projectile, including but not limited to paint balls, which consist of a thin plastic shell or capsule filled with a water-soluble paint, propelled by spring mechanism, compressed gas, explosive charge, or any combination thereof, regardless of speed or distance at which the projectile is capable of being released. While they may not be lethal, they can cause serious injury.

(b) It shall be unlawful for any person to conceal a paint pellet gun or paint ball gun in the city, without regard to whether the gun is discharged or contains any projectiles.

(c) It shall be unlawful for any minor to possess a paint pellet gun or paint ball gun in the city, without regard to whether the gun is discharged or contains any projectiles. For purposes of this subsection a minor is defined as any person who has not reached the age of 18 as of the date of the offense.

(d) A conviction of violation of this section shall be punishable by a fine of not less than $500.00.

[Wilmington City Code codified through Ord. No. 08-053, enacted July 10, 2008]

DISTRICT OF COLUMBIA

D.C. CODE

Publisher's Note

The District of Columbia Official Code 2001 Edition (West) changed the numbering from the previous 1981 edition.

Division I. Government of District

Title 1. Government Organization

Chapter 3. Specified Governmental Authority

Subchapter II. Regulatory Authority

Part C. General

1-303.43. Regulations relative to firearms, explosives, and weapons The Council of the District of Columbia is hereby authorized and empowered to make, and the Mayor of the District of Columbia is hereby authorized and empowered to enforce, all such usual and reasonable police regulations, in addition to those already made under §§ 1-303.01 to 1-303.03 as the Council may deem necessary for the regulation of firearms, projectiles, explosives, or weapons of any kind in the District of Columbia.

Title 5. Police, Firefighters, and Chief Medical Examiner

Chapter 1. Metropolitan Police

Subchapter XVII. Miscellaneous

5-133.16. Transfer of ammunition feeding devices prohibited Except as provided in § 7-2507.05, and § 22-4517, the Metropolitan Police Department shall not transfer any ammunition feeding device in its possession to any person or entity other than a law enforcement officer or governmental agency for law enforcement purposes.

Title 7. Human Health Care and Safety

Subtitle J. Public Safety

Chapter 25. Firearms Control
Unit A. Firearms Control Regulations

Subchapter I. General Provisions

7-2501.01. Definitions. As used in this chapter the term:

(1) "Acts of Congress" means:

(A) Chapter 32 of Title 22;

(B) Omnibus Crime Control and Safe Streets Act of 1968, as amended (Title VII, Unlawful Possession or Receipt of Firearms (82 Stat. 1236; 18 U.S.C. Appendix)); and

(C) An Act to Amend Title 18, United States Code, To Provide for Better Control of the Interstate Traffic in Firearms Act of 1968 (82 Stat. 1213; 18 U.S.C. § 921 et seq.).

(2) "Ammunition" means cartridge cases, shells, projectiles (including shot), primers, bullets (including restricted pistol bullets), propellant powder, or other devices or materials designed, redesigned, or intended for use in a firearm or destructive device.

(3) "Antique firearm" means:

(A) Any firearm (including any firearm with a matchlock, flintlock, percussion cap, or similar type of ignition system) manufactured in or before 1898; and

(B) Any replica of any firearm described in subparagraph (A) if such replica:

(i) Is not designed or redesigned for using rimfire or conventional centerfire fixed ammunition; or

(ii) Uses rimfire or conventional ammunition which is no longer manufactured in the United States and which is not readily available in the ordinary channels of commercial trade.

(4) "Chief" means the Chief of Police of the Metropolitan Police Department of the District of Columbia or his designated agent.

(5) "Crime of violence" means a crime of violence as defined in § 22-4501, committed in any jurisdiction, but does not include larceny or attempted larceny.

(6) "Dealer's license" means a license to buy or sell, repair, trade, or otherwise deal in firearms, destructive devices, or ammunition as provided for in subchapter IV of Unit A of this unit.

(7) "Destructive device" means:

(A) An explosive, incendiary, or poison gas bomb, grenade, rocket, missile, mine, or similar device;

(B) Any device by whatever name known which will, or is designed or redesigned, or may be readily converted or restored to expel a projectile by the action of an explosive or other propellant through a smooth bore barrel, except a shotgun;

(C) Any device containing tear gas or a chemically similar lacrimator or sternutator by whatever name known;

(D) Any device designed or redesigned, made or remade, or readily converted or restored, and intended to stun or disable a person by means of electric shock;

(E) Any combination of parts designed or intended for use in converting any device into any destructive device; or from which a destructive device may be readily assembled: Provided, that the term shall not include:

(i) Any pneumatic, spring, or BB gun which expels a single projectile not exceeding .18 inch in diameter;

(ii) Any device which is neither designed nor redesigned for use as a weapon;

(iii) Any device originally a weapon which has been redesigned for use as a signaling, line throwing, or safety device; or

(iv) Any device which the Chief finds is not likely to be used as a weapon.

(8) "District" means District of Columbia.

(9) "Firearm" means any weapon which will, or is designed or redesigned, made or remade, readily converted or restored, and intended to, expel a projectile or projectiles by the action of an explosive; the frame or receiver of any such device; or any firearm muffler or silencer: Provided, that such term shall not include:

(A) Antique firearms; or

(B) Destructive devices;

(C) Any device used exclusively for line throwing, signaling, or safety, and required or recommended by the Coast Guard or Interstate Commerce Commission; or

(D) Any device used exclusively for firing explosive rivets, stud cartridges, or similar industrial ammunition and incapable for use as a weapon.

(10) "Machine gun" means any firearm which shoots, is designed to shoot, or can be readily converted or restored to shoot:

(A) Automatically, more than 1 shot by a single function of the trigger;

(B) Semiautomatically, more than 12 shots without manual reloading.

(11) "Organization" means any partnership, company, corporation, or other business entity, or any group or association of 2 or more persons united for a common purpose.

(12) "Pistol" means any firearm originally designed to be fired by use of a single hand.

(13) "Registration certificate" means a certificate validly issued pursuant to this chapter evincing the registration of a firearm pursuant to this chapter.

(13a) "Restricted pistol bullet" means any bullet designed for use in a pistol which, when fired from a pistol with a barrel of 5 inches or less in length, is capable of penetrating commercially available body armor with a penetration resistance equal to or greater than that of 18 layers of kevlar.

(14) "Rifle" means a grooved bore firearm using a fixed metallic cartridge with a single projectile and designed or redesigned, made or remade, and intended to be fired from the shoulder.

(15) "Sawed-off shotgun" means a shotgun having a barrel of less than 20 inches in length; or a firearm made from a shotgun if such firearm as modified has an overall length of less than 26 inches or any barrel of less than 20 inches in length.

(16) "Shotgun" means a smooth bore firearm using a fixed shotgun shell with either a number of ball shot or a single projectile, and designed or redesigned, made or remade, and intended to be fired from the shoulder.

(17) "Short barreled rifle" means a rifle having any barrel less than 16 inches in length, or a firearm made from a rifle if such firearm as modified has an overall length of less than 26 inches or any barrel of less than 16 inches.

(18) "Weapons offense" means any violation in any jurisdiction of any law which involves the sale, purchase, transfer in any manner, receipt, acquisition, possession, having under control, use, repair, manufacture, carrying, or transportation of any firearm, ammunition, or destructive device.

Subchapter II. Firearms and Destructive Devices

7-2502.01. Registration requirements.

(a) Except as otherwise provided in this unit, no person or organization in the District of Columbia ("District") shall receive, possess, control, transfer, offer for sale, sell, give, or deliver any destructive device, and no person or organization in the District shall possess or control any firearm, unless the person or organization holds a valid registration certificate for the firearm. A registration certificate may be issued:

(1) To an organization if:

(A) The organization employs at least 1 commissioned special police officer or employee licensed to carry a firearm whom the organization arms during the employee's duty hours; and

(B) The registration is issued in the name of the organization and in the name of the president or chief executive officer of the organization;

(2) In the discretion of the Chief of Police, to a police officer who has retired from the Metropolitan Police Department; or

(3) In the discretion of the Chief of Police, to the Fire Marshal and any member of the Fire and Arson Investigation Unit of the Fire Prevention Bureau of the Fire Department of the District of Columbia, who is designated in writing by the Fire Chief, for the purpose of enforcing the arson and fire safety laws of the District of Columbia.

(b) Subsection (a) of this section shall not apply to:

(1) Any law enforcement officer or agent of the District or the United States, or any law enforcement officer or agent of the government of any state or subdivision thereof, or any member of the armed forces of the United States, the National Guard or organized reserves, when such officer, agent, or member is authorized to possess such a firearm or device while on duty in the performance of official authorized functions;

(2) Any person holding a dealer's license: Provided, that the firearm or destructive device is:

(A) Acquired by such person in the normal conduct of business;

(B) Kept at the place described in the dealer's license; and

(C) Not kept for such person's private use or protection, or for the protection of his business;

(3) With respect to firearms, any nonresident of the District participating in any lawful recreational firearm-related activity in the District, or on his way to or from such activity in another jurisdiction: Provided, that such person, whenever in possession of a firearm, shall upon demand of any member of the Metropolitan Police Department, or other bona fide law enforcement officer, exhibit proof that he is on his way to or from such activity, and that his possession or control of such firearm is lawful in the jurisdiction in which he resides: Provided further, that such weapon shall be unloaded, securely wrapped, and carried in open view.

7-2502.02. Registration of certain firearms prohibited.

(a) A registration certificate shall not be issued for a:

(1) Sawed-off shotgun;

(2) Machine gun;

(3) Short-barreled rifle; or

(4) Pistol not validly registered to the current registrant in the District prior to September 24, 1976, except that the provisions of this section shall not apply to any organization that employs at least 1 commissioned special police officer or other employee licensed to carry a firearm and that arms the employee with a firearm during the employee's duty hours or to a police officer who has retired from the Metropolitan Police Department.

(b) Nothing in this section shall prevent a police officer who has retired from the Metropolitan Police Department from registering a pistol.

7-2502.03. Qualifications for registration; information required for registration.

(a) No registration certificate shall be issued to any person (and in the case of a person between the ages of 18 and 21, to the person and his signatory parent or guardian) or organization unless the Chief determines that such person (or the president or chief executive in the case of an organization):

(1) Is 21 years of age or older: Provided, that the Chief may issue to an applicant between the ages of 18 and 21 years old, and who is otherwise qualified, a registration certificate if the application is accompanied by a notarized statement of the applicant's parent or guardian:

(A) That the applicant has the permission of his parent or guardian to own and use the firearm to be registered; and

(B) The parent or guardian assumes civil liability for all damages resulting from the actions of such applicant in the use of the firearm to be registered: Provided further, that such registration certificate shall expire on such person's 21st birthday;

(2) Has not been convicted of a crime of violence, weapons offense, or of a violation of this chapter;

(3) Is not under indictment for a crime of violence or a weapons offense;

(4) Has not been convicted within 5 years prior to the application of any:

(A) Violation in any jurisdiction of any law restricting the use, possession, or sale of any narcotic or dangerous drug; or

(B) A violation of § 22-407, regarding threats to do bodily harm, or § 22-404, regarding assaults and threats, or any similar provision of the law of any other jurisdiction so as to indicate a likelihood to make unlawful use of a firearm;

(5) Within the 5-year period immediately preceding the application, has not been acquitted of any criminal charge by reason of insanity or has not been adjudicated a chronic alcoholic by any court: Provided, that this paragraph shall not apply if such person shall present to the Chief, with the application, a medical certification indicating that the applicant has recovered from such insanity or alcoholic condition and is capable of safe and responsible possession of a firearm;

(6) Within the 5 years immediately preceding the application, has not been voluntarily or involuntarily committed to any mental hospital or institution: Provided, that this paragraph shall not apply, if such person shall present to the Chief, with the application, a medical certification that the applicant has recovered from whatever malady prompted such commitment;

(7) Does not appear to suffer from a physical defect which would tend to indicate that the applicant would not be able to possess and use a firearm safely and responsibly;

(8) Has not been adjudicated negligent in a firearm mishap causing death or serious injury to another human being;

(9) Is not otherwise ineligible to possess a pistol under § 22-4503;

(10) Has not failed to demonstrate satisfactorily a knowledge of the laws of the District of Columbia pertaining to firearms and the safe and responsible use of the same in accordance with tests and standards prescribed by the Chief: Provided, that once this determination is made with respect to a given applicant for a particular type of firearm, it need not be made again for the same applicant with respect to a subsequent application for the same type of firearms: Provided, further, that this paragraph shall not apply with respect to any firearm reregistered pursuant to § 7-2502.06; and

(11) Has vision better than or equal to that required to obtain a valid driver's license under the laws of the District of Columbia: Provided, that current licensure by the District of Columbia of the applicant to drive, shall be prima facie evidence that such applicant's vision is sufficient and: Provided further, that this determination shall not be made with respect to persons applying to reregister any firearm pursuant to § 7-2502.06.

(b) Every person applying for a registration certificate shall provide on a form prescribed by the Chief:

(1) The full name or any other name by which the applicant is known;

(2) The present address and each home address where the applicant has resided during the 5-year period immediately preceding the application;

(3) The present business or occupation and any business or occupation in which the applicant has engaged during the 5-year period immediately preceding the application and the addresses of such businesses or places of employment;

(4) The date and place of birth of the applicant;

(5) The sex of the applicant;

(6) Whether (and if so, the reasons) the District, the United States or the government of any state or subdivision of any state has denied or revoked the applicant's license, registration certificate, or permit pertaining to any firearm;

(7) A description of the applicant's role in any mishap involving a firearm, including the date, place, time, circumstances, and the names of the persons injured or killed;

(8) The intended use of the firearm;

(9) The caliber, make, model, manufacturer's identification number, serial number, and any other identifying marks on the firearm;

(10) The name and address of the person or organization from whom the firearm was obtained, and in the case of a dealer, his dealer's license number;

(11) Where the firearm will generally be kept;

(12) Whether the applicant has applied for other registration certificates issued and outstanding;

(13) Such other information as the Chief determines is necessary to carry out the provisions of this unit.

(c) Every organization applying for a registration certificate shall:

(1) With respect to the president or chief executive of such organization, comply with the requirements of subsection (b) of this section; and

(2) Provide such other information as the Chief determines is necessary to carry out the provisions of this unit.

7-2502.04. Fingerprints and photographs of applicants; application in person required.

(a) The Chief may require any person applying for a registration certificate to be fingerprinted if, in his judgment, this is necessary to conduct an efficient and adequate investigation into the matters described in § 7-2502.03 and to effectuate the purpose of this unit: Provided, that any person who has been fingerprinted by the Chief within 5 years prior to submitting the application need not, in the Chiefs discretion, be fingerprinted again if he offers other satisfactory proof of identity.

(b) Each applicant, other than an organization, shall submit with the application 2 full-face photographs of himself, 13/4 by 17/8 inches in size which shall have been taken within the 30-day period immediately preceding the filing of the application.

(c) Every applicant (or in the case of an organization, the president or chief executive, or a person authorized in writing by him), shall appear in person at a time and place prescribed by the Chief, and may be required to bring with him the firearm for which a registration certificate is sought, which shall be unloaded and securely wrapped, and carried in open view.

7-2502.05. Application signed under oath; fees.

(a) Each applicant (the president or chief executive in the case of an organization) shall sign an oath or affirmation attesting to the truth of all the information required by § 7-2502.03.

(b) Each application required by this subchapter shall be accompanied by a nonrefundable fee to be established by the Mayor: Provided, that such fee shall, in the judgment of the Mayor, reimburse the District for the cost of services provided under this subchapter.

7-2502.06. Time for filing registration applications.

(a) An application for a registration certificate shall be filed (and a registration certificate issued) prior to taking possession of a firearm from a licensed dealer or from any person or organization holding a registration certificate therefor. In all other cases, an application for registration shall be filed immediately after a firearm is brought into the District. It shall be deemed compliance with the preceding sentence if such person personally communicates with the Metropolitan Police Department (as determined by the Chief to be sufficient) and provides such information as may be demanded: Provided, that such person files an application for a registration certificate within 48 hours after such communication.

(b) Any firearm validly registered under prior regulations must be registered pursuant to this unit in accordance with procedures to be promulgated by the Chief. An application to register such firearm shall be filed pursuant to this unit within 60 days of September 24, 1976.

7-2502.07. Issuance of registration certificate; time period; corrections.

(a) Upon receipt of a properly executed application for registration certificate, the Chief, upon determining through inquiry, investigation, or otherwise, that the applicant is entitled and qualified under the provisions of this unit, thereto, shall issue a registration certificate. Each registration certificate shall be in duplicate and bear a unique registration certificate number and such other information as the Chief determines is necessary to identify the applicant and the firearm registered. The duplicate of the registration certificate shall be delivered to the applicant and the Chief shall retain the original.

(b) The Chief shall approve or deny an application for a registration certificate within a 60-day period beginning on the date the Chief receives the application, unless good cause is shown, including nonreceipt of information from sources outside the District government: Provided, that in the case of an application to register a firearm validly registered under prior regulations, the Chief shall have 365 days after the receipt of such application to approve or deny such application. The Chief may hold in abeyance an application where there is a revocation proceeding pending against such person or organization.

(c) Upon receipt of a registration certificate, each applicant shall examine same to ensure that the information thereon is correct. If the registration certificate is incorrect in any respect, the person or organization named thereon shall return it to the Chief with a signed statement showing the nature of the error. The Chief shall correct the error, if it occurred through administrative error. In the event the error resulted from information contained in the application, the applicant shall be required to file an amended application setting forth the correct information, and a statement explaining the error in the original application. Each amended application shall be accompanied by a fee equal to that required for the original application.

(d) In the event the Chief learns of an error in a registration certificate other than as provided in subsection (c) of this section, he may require the holder to return the registration certificate for correction. If the error resulted from information contained in the application, the person or organization named therein shall be required to file an amended application as provided in subsection (c) of this section.

(e) Each registration certificate issued by the Chief shall be accompanied by a statement setting forth the registrant's duties under this unit.

(f) In the discretion of the Chief of Police, a registration certificate may be issued to a retired police officer who is a resident of the District of Columbia for a pistol and ammunition which conforms to the Metropolitan Police Department General Orders and policies.

(g) When the retired police officer ceases to be a resident of the District of Columbia the registration certificate expires.

(h) Nothing in this unit shall create an entitlement to a registration certificate for a retired police officer. If the Chief of Police denies a retired police officer's registration certificate application, the Chief of Police shall state the reasons for the denial in writing.

(i) The District of Columbia shall not incur any liability by reason of the issuance or denial of a certificate, nor for any use made of the registered firearm.

7-2502.08. Duties of registrants. Each person and organization holding a registration certificate, in addition to any other requirements imposed by this unit, or the acts of Congress, shall:

(1) Notify the Chief in writing of:

(A) The loss, theft, or destruction of the registration certificate or of a registered firearm (including the circumstances, if known) immediately upon discovery of such loss, theft, or destruction;

(B) A change in any of the information appearing on the registration certificate or required by § 7-2502.03;

(C) The sale, transfer or other disposition of the firearm not less than 48 hours prior to delivery, pursuant to such sale, transfer or other disposition, including:

(i) Identification of the registrant, the firearm and the serial number of the registration certificate;

(ii) The name, residence, and business address and date of birth of the person to whom the firearm has been sold or transferred; and

(iii) Whether the firearm was sold or how it was otherwise transferred or disposed of.

(2) Return to the Chief, the registration certificate for any firearm which is lost, stolen, destroyed, or otherwise transferred or disposed of, at the time he notifies the Chief of such loss, theft, destruction, sale, transfer, or other disposition.

(3) Have in his possession, whenever in possession of a firearm, the registration certificate for such firearm, and exhibit the same upon the demand of a member of the Metropolitan Police Department, or other law enforcement officer.

7-2502.09. Revocation of registration certificate. A registration certificate shall be revoked if:

(1) Any of the criteria in § 7-2502.03 are not currently met;

(2) The registered firearm has become an unregisterable firearm under the terms of § 7-2502.02, or a destructive device;

(3) The information furnished to the Chief on the application for a registration certificate proves to be intentionally false; or

(4) There is a violation or omission of the duties, obligations or requirements imposed by § 7-2502.08.

7-2502.10. Procedure for denial and revocation of registration certificate.

(a) If it appears to the Chief that an application for a registration certificate should be denied or that a registration certificate should be revoked, the Chief shall notify the applicant or registrant of the proposed denial or revocation, briefly stating the reason or reasons therefor. Service may be made by delivering a copy of the notice to the applicant or registrant personally, or by leaving a copy thereof at the place of residence identified on the application or registration with some person of suitable age and discretion then residing therein, or by mailing a copy of the notice by certified mail to the residence address identified on the application or certificate, in which case service shall be complete as of the date the return receipt was signed. In the case of an organization, service may be made upon the president, chief executive, or other officer, managing agent or person authorized by appointment or law to receive such notice as described in the preceding sentence at the business address of the organization identified in the application or registration certificate. The person serving the notice shall make proof thereof with the Chief in a manner prescribed by him. In the case of service by certified mail, the signed return receipt shall be filed with the Chief together with a signed statement showing the date such notice was mailed; and if the return receipt does not purport to be signed by the person named in the notice, then specific facts from which the Chief can determine that the person who signed the receipt meets the appropriate qualifications for receipt of such notice set out in this subsection. The applicant or registrant shall have 15 days from the date the notice is served in which to submit further evidence in support of the application or qualifications to continue to hold a registration certificate, as the case may be: Provided, that if the applicant does not make such a submission within 15 days from the date of service, the applicant or registrant shall be deemed to have conceded the validity of the reason or reasons stated in the notice, and the denial of revocation shall become final.

(b) Within 10 days of the date upon which the Chief receives such a submission, he shall serve upon the applicant or registrant in the manner specified in subsection (a) of this section notice of his final decision. The Chief's decision shall become effective at the expiration of the time within which to file a notice of appeal pursuant to the District of Columbia Administrative Procedure Act (D.C. Code, § 2-501 et seq.) or, if such a notice of appeal is filed, at the time the final order or judgment of the District of Columbia Court of Appeals becomes effective.

(c) Within 7 days of a decision unfavorable to the applicant or registrant becoming final, the applicant or registrant shall:

(1) Peaceably surrender to the Chief the firearm for which the registration certificate was revoked in the manner provided in § 7-2507.05; or

(2) Lawfully remove such firearm from the District for so long as he has an interest in such firearm; or

(3) Otherwise lawfully dispose of his interest in such firearm.

7-2502.11. Information prohibited from use as evidence in criminal proceedings. No information obtained from a person under this subchapter or retained by a person in order to comply with any section of this subchapter, shall be used as evidence against such person in any criminal proceeding with respect to a violation of this unit, occurring prior to or concurrently with the filing of the information required by this subchapter: Provided, that this section shall not apply to any violation of § 22-2402 or § 7-2507.04.

Subchapter III. Estates Containing Firearms

7-2503.01. Rights and responsibilities of executors and administrators.

(a) The executor or administrator of an estate containing a firearm shall notify the Chief of the death of the decedent within 30 days of his appointment or qualification, whichever is earlier.

(b) Until the lawful distribution of such firearm to an heir or legatee or the lawful sale, transfer, or disposition of the firearm by the estate, the executor or administrator of such estate shall be charged with the duties and obligations which would have been imposed by this unit upon the decedent, if the decedent were still alive: Provided, that such executor or administrator shall not be liable to the criminal penalties of § 7-2507.06.

Subchapter IV. Licensing of Firearms Businesses

7-2504.01. Manufacture of firearms, destructive devices or ammunition prohibited; requirement for dealer's license.

(a) No person or organization shall manufacture any firearm, destructive device or parts thereof, or ammunition, within the District: Provided, that persons holding registration certificates may engage in handloading, reloading, or custom loading ammunition for his registered firearms: Provided further, that such person may not handload, reload, or custom load ammunition for others.

(b) No person or organization shall engage in the business of selling, purchasing, or repairing any firearm, destructive device, parts therefor, or ammunition, without first obtaining a dealer's license, and no licensee shall engage in the business of selling, purchasing, or repairing firearms which are unregisterable under § 7-2502.02, destructive devices, or parts therefor, except pursuant to a valid work or purchase order, for those persons specified in § 7-2502.01(b)(1).

(c) Any license issued pursuant to this section shall be issued by the Metropolitan Police Department as a Public Safety endorsement to a basic business license under the basic business license system as set forth in subchapter I-A of Chapter 28 of Title 47.

7-2504.02. Qualifications for dealer's license; application; fee.

(a) Any person eligible to register a firearm under this chapter and who, if a registrant, has not previously failed to perform any of the duties imposed by this unit; and, any person eligible under the acts of Congress to engage in such business, may obtain a dealer's license, or a renewal thereof, which shall be valid for a period of not more than 1 year from the date of issuance. The license required by this unit, shall be in addition to any other license or licensing procedure required by law.

(b) Each application for a dealer's license and each application for renewal thereof shall be made on a form prescribed by the Chief, shall be sworn to or affirmed by the applicant, and shall contain:

(1) The information required by § 7-2502.03 (a);

(2) The address where the applicant conducts or intends to conduct his business;

(3) Whether the applicant, prior to September 24, 1976, held a license to deal in deadly weapons in the District; and

(4) Such other information as the Chief may require, including fingerprints and photographs of the applicant, to carry out the purposes of this unit.

(c) Each application for a dealer's license, or renewal shall be accompanied by a fee established by the Mayor: Provided, that such fee shall in the judgment of the Mayor, reimburse the District for the cost of services provided under this subchapter.

(d) Any license issued pursuant to this section shall be issued as a Public Safety endorsement to a basic business license under the basic business license system as set forth in subchapter I-A of Chapter 28 of Title 47.

7-2504.03. Issuance of dealer's license; time period; corrections.

(a) Upon receipt of a properly executed application for a dealer's license, or renewal thereof, the Chief, upon determining through further inquiry, investigation, or otherwise, that the applicant is entitled and qualified under the provisions of this unit thereto, shall issue a dealer's license. Each dealer's license shall be in duplicate and bear a unique dealer's license number, and such other information as the Chief determines is necessary to identify the applicant and premises. The duplicate of the dealer's license shall be delivered to the applicant and the Chief shall retain the original.

(b) The Chief shall approve or deny an application for a registration certificate within a 60-day period beginning on the date the Chief receives the application, unless good cause is shown, including nonreceipt of information from sources outside the District government. The Chief may hold in abeyance an application where there is any firearms revocation proceeding pending against such person.

(c) Upon receipt of a dealer's license, each applicant shall examine the same to ensure that the information thereon is correct. If the dealer's license is incorrect in any respect, the person named thereon shall return the same to the Chief with a signed statement showing the nature of the error. The Chief shall correct the error, if it occurred through administrative error. In the event the error resulted from information contained in the application, the applicant shall be required to file an amended application explaining the error in the original application. Each amended application shall be accompanied by a fee equal to that required for the original application.

(d) In the event the Chief learns of an error in a dealer's license, other than as provided in subsection (c) of this section, he may require the holder to return the dealer's license for correction. If the error resulted from information contained in the application, the person named shall be required to file an amended application as provided in subsection (c) of this section.

(e) Each dealer's license issued by the Chief shall be accompanied by a statement setting forth a dealer's duties under this unit.

7-2504.04. Duties of licensed dealers; records required.

(a) Each person holding a dealer's license, in addition to any other requirements imposed by this unit, the acts of Congress, and other law, shall:

(1) Display the dealer's license in a conspicuous place on the premises;

(2) Notify the Chief in writing:

(A) Of the loss, theft, or destruction of the dealer's license (including the circumstances, if known) immediately upon the discovery of such loss, theft, or destruction;

(B) Of a change in any of the information appearing on the dealer's license or required by § 7-2504.02 immediately upon the occurrence of any such change;

(3) Keep at the premises identified in the dealer's license a true and current record in book form of:

(A) The name, address, home phone, and date of birth of each employee handling firearms, ammunition, or destructive devices;

(B) Each firearm or destructive device received into inventory or for repair including the:

(i) Serial number, caliber, make, model, manufacturer's number (if any), dealer's identification number (if any), registration certificate number (if any) of the firearm, and similar descriptive information for destructive devices;

(ii) Name, address, and dealer's license number (if any) of the person or organization from whom the firearm or destructive device was purchased or otherwise received;

(iii) Consideration given for the firearm or destructive device, if any;

(iv) Date and time received by the licensee and in the case of repair, returned to the person holding the registration certificate; and

(v) Nature of the repairs made;

(C) Each firearm or destructive device sold or transferred including the:

(i) Serial number, caliber, make, model, manufacturer's number or dealer's identification number, and registration certificate number (if any) of the firearm or similar information for destructive devices;

(ii) Name, address, registration certificate number or license number (if any) of the person or organization to whom transferred;

(iii) The consideration for transfer; and

(iv) Time and date of delivery of the firearm or destructive device to the transferee;

(D) Ammunition received into inventory including the:

(i) Brand and number of rounds of each caliber or gauge;

(ii) Name, address, and dealer's license or registration number (if any) of the person or organization from whom received;

(iii) Consideration given for the ammunition; and

(iv) Date and time of the receipt of the ammunition;

(E) Ammunition sold or transferred including:

(i) Brand and number of rounds of each caliber or gauge;

(ii) Name, address and dealer's license number (if any) of the person or organization to whom sold or transferred;

(iii) If the purchaser or transferee is not a licensee, the registration certificate number of the firearm for which the ammunition was sold or transferred;

(iv) The consideration for the sale and transfer; and

(v) The date and time of sale or transfer.

(b) The records required by subsection (a) of this section shall upon demand be exhibited during normal business hours to any member of the Metropolitan Police Department.

(c) Each person holding a dealer's license shall, when required by the Chief in writing, submit on a form and for the periods of time specified, any record information required to be maintained by subsection (a) of this section, and any other information reasonably obtainable therefrom.

7-2504.05. Revocation of dealer's license. A dealer's license shall be revoked if:

(1) Any of the criteria in § 7-2504.04 is not currently met; or

(2) The information furnished to the Chief on the application for a dealer's license proves to be intentionally false; or

(3) There is a violation or omission of the duties, obligations, or requirements imposed by § 7-2504.04.

7-2504.06. Procedure for denial and revocation of dealer's license.

(a) If it appears to the Chief that an application for a dealer's license should be denied or that a dealer's license should be revoked, the Chief shall notify the applicant or registrant of the proposed denial or revocation briefly stating the reason or reasons therefor. Service may be made as provided for in § 7-2502.10(a). The applicant or dealer shall have 15 days from the date of service in which to submit further evidence in support of the application or qualifications to continue to hold a dealer's license, as the case may be: Provided, that if the applicant or dealer does not make such a submission within 15 days from the date of service, the applicant or dealer shall be deemed to have conceded the validity of the reason or reasons stated in the notice, and the denial or revocation shall become final.

(b) Within 10 days of the date upon which the Chief receives such a submission, the Chief shall serve upon the applicant or registrant in the manner provided in § 7-2502.10(a) notice of his final decision. The Chief's decision shall become effective at the expiration of the time within which to file a notice of appeal pursuant to the District of Columbia Administrative Procedure Act (D.C. Code, § 2-501 et seq.) or, if such a notice of appeal is filed, at the time the final order or judgment of the District of Columbia Court of Appeals becomes effective.

(c) Within 45 days of a decision becoming effective, which is unfavorable to a licensee or to an applicant for a dealer's license, the licensee or applicant shall:

(1) If he is eligible to register firearms pursuant to this chapter, register such firearms in his inventory as are capable of registration pursuant to this unit;

(2) Peaceably surrender to the Chief any firearms in his inventory which he does not register, and all destructive devices in his inventory in the manner provided for in § 7-2507.05;

(3) Lawfully remove from the District any firearm in his inventory which he does not register and all destructive devices and ammunition in his inventory for so long as he has an interest in them; or

(4) Otherwise lawfully dispose of any firearms in his inventory which he does not register and all destructive devices and ammunition in his inventory.

7-2504.07. Display of firearms or ammunition by dealers; security; employees of dealers.

(a) No licensed dealer shall display any firearm or ammunition in windows visible from a street or sidewalk. All firearms, destructive devices, and ammunition shall be kept at all times in a securely locked place affixed to the premises except when being shown to a customer, being repaired, or otherwise being worked on.

(b) No licensee shall knowingly employ any person in his establishment if such person would not be eligible to register a firearm under this unit.

7-2504.08. Identification number on firearm required before sale. No licensee shall sell or offer for sale any firearm which does not have imbedded into the metal portion of such firearm a unique manufacturer's identification number or serial number, unless the licensee shall have imbedded into the metal portion of such firearm a unique dealer's identification number.

7-2504.09. Certain information obtained from or retained by dealers not to be used as evidence in criminal proceedings. No information obtained from or retained by a licensed dealer to comply with this subchapter shall be used as evidence against such licensed dealer in any criminal proceeding with respect to a violation of this unit occurring prior to or concurrently with the filing of such information: Provided, that this section shall not apply to any violation of § 22-2402, or of § 7-2507.04.

Subchapter V. Sale and Transfer of Firearms, Destructive Devices, and Ammunition

7-2505.01. Sales and transfers prohibited. No person or organization shall sell, transfer or otherwise dispose of any firearm, destructive device or ammunition in the District except as provided in § 7-2502.10(c), § 7-2505.02, or § 7-2507.05.

7-2505.02. Permissible sales and transfers.

(a) Any person or organization eligible to register a firearm may sell or otherwise transfer ammunition or any firearm, except those which are unregisterable under § 7-2502.02, to a licensed dealer.

(b) Any licensed dealer may sell or otherwise transfer:

(1) Ammunition, excluding restricted pistol bullets, and any firearm or destructive device which is lawfully a part of such licensee's inventory, to any nonresident person or business licensed under the acts of Congress and the jurisdiction where such person resides or conducts such business;

(2) Ammunition, including restricted pistol bullets, and any firearm or destructive device which is lawfully a part of such licensee's inventory to:

(A) Any other licensed dealer;

(B) Any law enforcement officer or agent of the District or the United States of America when such officer or agent is on duty, and acting within the scope of his duties when acquiring such firearm, ammunition, or destructive device, if the officer or agent has in his possession a statement from the head of his agency stating that the item is to be used in such officer's or agent's official duties.

(c) Any licensed dealer may sell or otherwise transfer a firearm except those which are unregisterable under § 7-2502.02, to any person or organization possessing a registration certificate for such firearm: Provided, that if the Chief denies a registration certificate, he shall so advise the licensee who shall thereupon: (1) Withhold delivery until such time as a registration certificate is issued, or, at the option of the purchaser; (2) declare the contract null and void, in which case consideration paid to the licensee shall be returned to the purchaser: Provided further, that this subsection shall not apply to per-sons covered by subsection (b) of this section.

(d) Except as provided in subsections (b) and (e) of this section, no licensed dealer shall sell or otherwise transfer ammunition unless:

(1) The sale or transfer is made in person; and

(2) The purchaser exhibits, at the time of sale or other transfer, a valid registration certificate, or in the case of a nonresident, proof that the weapon is lawfully possessed in the jurisdiction where such person resides;

(3) The ammunition to be sold or transferred is of the same caliber or gauge as the firearm described in the registration certificate, or other proof in the case of nonresident; and

(4) The purchaser signs a receipt for the ammunition which (in addition to the other records required under this unit) shall be maintained by the licensed dealer for a period of 1 year from the date of sale.

(e) Any licensed dealer may sell ammunition to any person holding an ammunition collector's certificate on September 24, 1976: Provided, that the collector's certificate shall be exhibited to the licensed dealer whenever the collector purchases ammunition for his collection: Provided further, that the collector shall sign a receipt for the ammunition, which shall be treated in the same manner as that required under paragraph (4) of subsection (d) of this section.

Subchapter VI. Possession of Ammunition

7-2506.01. Persons permitted to possess ammunition. No person shall possess ammunition in the District of Columbia unless:

(1) He is a licensed dealer pursuant to subchapter IV of this unit;

(2) He is an officer, agent, or employee of the District of Columbia or the United States of America, on duty and acting within the scope of his duties when possessing such ammunition;

(3) He is the holder of the valid registration certificate for a firearm of the same gauge or caliber as the ammunition he possesses; except, that no such person shall possess restricted pistol bullets; or

(4) He holds an ammunition collector's certificate on September 24, 1976.

Subchapter VII. Miscellaneous Provisions

7-2507.01. Security mortgages, deposits, or pawns with firearms, destructive devices, or ammunition prohibited; loan or rental of firearms, destructive devices, or ammunition prohibited.

(a) No firearm, destructive device, or ammunition shall be security for, or be taken or received by way of any mortgage, deposit, pledge, or pawn.

(b) No person may loan, borrow, give, or rent to or from another person, any firearm, destructive device, or ammunition.

7-2507.02. Firearms required to be unloaded and disassembled or locked. Except for law enforcement personnel described in § 7-2502.01(b)(1), each registrant shall keep any firearm in his possession unloaded and disassembled or bound by a trigger lock or similar device unless such firearm is kept at his place of business, or while being used for lawful recreational purposes within the District of Columbia.

7-2507.03. Firing ranges.

Any person operating a firing range in the District, shall in addition to any other requirement imposed by law, register with the Chief, on a form prescribed by him, which shall include the business name of the range, the location, the names and home addresses of the owners and principal officers, the types of weapons fired there, the number and types of weapons normally stored there, the days and hours of operation, and such other information as the Chief shall require.

7-2507.04. False information; forgery or alteration.

(a) It shall be unlawful for any person purchasing any firearm or ammunition, or applying for any registration certificate or dealer's license under this unit, or in giving any information pursuant to the requirements of this unit, to knowingly give false information or offer false evidence of identity.

(b) It shall be unlawful for anyone to forge or alter any application, registration certificate, or dealer's license submitted, retained or issued under this unit.

7-2507.05. Voluntary surrender of firearms, destructive devices, or ammunition; immunity from prosecution; determination of evidentiary value of firearm.

(a) If a person or organization within the District voluntarily and peaceably delivers and abandons to the Chief any firearm, destructive device or ammunition at any time, such delivery shall preclude the arrest and prosecution of such person on a charge of violating any provision of this unit with respect to the firearm, destructive device, or ammunition voluntarily delivered. Delivery under this section may be made at any police district, station, or central headquarters, or by summoning a police officer to the person's residence or place of business. Every firearm and destructive device to be delivered and abandoned to the Chief under this section shall be unloaded and securely wrapped in a package, and, in the case of delivery to a police facility, the package shall be carried in open view. No person who delivers and abandons a firearm, destructive device, or ammunition under this section, shall be required to furnish identification, photographs, or fingerprints. No amount of money shall be paid for any firearm, destructive device, or ammunition delivered and abandoned under this section.

(b) Whenever any firearm, destructive device, or any ammunition is surrendered under this section or pursuant to § 7-2502.10(c)(1), the Chief shall inquire of the United States Attorney and the Corporation Counsel for the District whether such firearm is needed as evidence; provided, that if the same is not needed as evidence, it shall be destroyed.

7-2507.06. Penalties. Any person convicted of a violation of any provision of this unit shall be fined not more than $1,000 or imprisoned for not more than 1 year, or both; except that:

(1) A person who knowingly or intentionally sells, transfers, or distributes a firearm, destructive device, or ammunition to a person under 18 years of age shall be fined not more than $10,000 or imprisoned for not more than 10 years, or both.

(2)(A) Except as provided in subparagraph (B) of this paragraph, any person who is convicted a second time for possessing an unregistered firearm shall be fined not more than $5,000 or imprisoned not more than 5 years, or both.

(B) A person who in the person's dwelling place, place of business, or on other land possessed by the person, possesses a pistol, or firearm that could otherwise be registered, shall be fined not more than $1,000 or imprisoned not more than 1 year, or both.

7-2507.06a. Seizure and forfeiture of conveyances.

(a) For the purposes of this section, the term "owner" means a person with an ownership interest in the specific conveyance sought to be forfeited. The term "owner" does not include:

(1) A person with only a general unsecured interest in, or claim against, the conveyance;

(2) A bailee; or

(3) A nominee who exercises no dominion or control over the conveyance.

(b) Any conveyance, including vehicles and vessels in which any person or persons transport, possess, or conceal any firearm, as that term is defined in § 7-2501.01, or in any manner use to facilitate a violation of § 7-2502.02 or § 22-4503 or § 22-4504, shall be seized and forfeited to the District of Columbia, provided that:

(1) No conveyance used by any person as a duly licensed common carrier in the course of transacting business as a licensed common carrier is subject to forfeiture under this section unless it appears that the owner or other person in charge of the conveyance is a consenting party or has knowledge of a violation of this section; and

(2) The forfeiture of a conveyance encumbered by a bona fide security interest is subject to the interest of the secured party if the secured party neither had knowledge of, nor consented to, the illegal act giving rise to forfeiture.

(c) An innocent owner's interest in a conveyance which has been seized shall not be forfeited under this section.

(1) A person is an innocent owner if he or she establishes, by a preponderance of the evidence:

(A) That he or she did not know that a person or persons in the conveyance was transporting, possessing, or concealing any firearm or that the conveyance was involved in or was being used in the commission of any illegal act involving any firearm; or

(B) That, upon receiving knowledge of the presence of any illegal firearm in or on the conveyance or that the conveyance was being used in the commission of an illegal act involving a forfeiture, he or she took action to terminate the presence in or on the conveyance of the person, persons, or firearms.

(2)(A) A claimant who establishes a lack of knowledge under subsection (c)(1)(A) of this section shall be considered an innocent owner unless the government, in rebuttal, establishes the existence of facts and circumstances that should have created a suspicion that the conveyance was being or would be used for an illegal purpose. In that case, the claimant must establish that, in light of such facts and circumstances, he or she did all that reasonably could be expected to prevent the use of the conveyance in the commission of any such illegal act.

(B) A person who willfully blinds himself or herself to a fact shall be considered to have had knowledge of that fact.

(d) Except as otherwise expressly provided by this section, all seizures and forfeitures of conveyances under this section shall follow the procedures set forth in § 48-905.02.

7-2507.08. Construction of unit. Nothing in this unit shall be construed, or applied to necessarily require, or excuse noncompliance with any provision of any federal law. This unit and the penalties prescribed in § 7-2507.06, for violations of this unit, shall not supersede but shall supplement all statutes of the District and the United States in which similar conduct is prohibited or regulated.

Unit B. Strict Liability for Illegal Sale and Distribution of Firearms

7-2531.01. Definitions. For the purposes of this unit, the term:

(1) "Dealer" means:

(A) Any person engaged in the business of selling firearms at wholesale or retail;

(B) Any person engaged in the business of repairing firearms or of making or fitting special barrels, stocks, or trigger mechanisms to firearms; or

(C) Any person who is a pawnbroker who takes or receives by way of pledge or pawn, any firearm as security for the payment or repayment of money.

(2) "Engaged in the business" means:

(A) A person who devotes time, attention, and labor to dealing in firearms as a regular course of trade or business with the principal objective of livelihood and profit through the repetitive purchase and resale of firearms. The term "engaged in business" shall not include a person who makes occasional sales, exchanges, or purchases of firearms for the enhancement of a personal collection or for a hobby, or who sells all or part of this personal collection of firearms; or

(B) A person who devotes time, attention, and labor to importing firearms as a regular course of trade or business with the principal objective of livelihood and profit through the sale or distribution of the firearms imported.

(3) "Firearm" shall have the same meaning as in § 7-2501.01.

(4) "Illegal sale" means:

(A) Failure to establish proof of the purchaser's residence in a jurisdiction where the purchase of the weapon is legal or ignoring proof of the purchaser's residence in the District of Columbia;

(B) Failure to comply with District of Columbia registration and waiting requirements prior to delivery of the firearm to the purchaser when proof of District of Columbia residence is provided;

(C) Failure to maintain full, complete, and accurate records of firearm sales as required by local, state, and federal law; or

(D) Knowingly and willfully maintaining false records with the intent to misrepresent the name and address of persons purchasing firearms, or the type of firearm sold to those persons.

(5) "Importer" means any person engaged in the business of importing or bringing firearms or ammunition into the United States for purposes of sale or distribution.

(6) "Law enforcement agency" means a federal, state, or local law enforcement agency, state militia, or an agency of the United States government.

(7) "Law enforcement officer" means any employee or agent of a law enforcement agency who is authorized to use a firearm in the course of employment.

(8) "Manufacturer" means any person in business to manufacture or assemble a firearm or ammunition for sale or distribution.

(9) "Pawnbroker" means any person whose business or occupation includes the taking or receiving, by way of pledge or pawn, of any firearm as security for the payment or repayment of money.

7-2531.02. Liability.

(a) Any manufacturer, importer, or dealer of a firearm who can be shown by a preponderance of the evidence to have knowingly and willfully engaged in the illegal sale of a firearm shall be held strictly liable in tort, without regard to fault and without regard to either: (1) an intent to interfere with a legally protected interest; or (2) a breach of duty to exercise reasonable care, for all direct and consequential damages that arise from bodily injury or death if the bodily injury or death proximately results from the discharge of the firearm in the District of Columbia, regardless of whether or not the person operating the firearm is the original, illegal purchaser.

(b) Any individual who can be shown by a preponderance of the evidence to have knowingly and willfully engaged in the illegal sale, loan, lease, or rental of a firearm for money or anything of value shall be held strictly liable in tort, without regard to fault and without regard to either: (1) an intent to interfere with a legally protected interest; or (2) a breach of duty to exercise reasonable care, for all direct and consequential damages that arise from bodily injury or death if the bodily injury or death proximately results from the discharge of the firearm in the District of Columbia regardless of whether or not the person operating the firearm is the original, illegal purchaser.

(c) Nothing in this unit shall relieve from liability any person who commits a crime, is negligent, or who might otherwise be liable for acts committed with the firearm.

7-2531.03. Exemptions.

(a) No firearm originally distributed to a law enforcement agency or a law enforcement officer shall provide the basis for liability under this unit.

(b) No action may be brought pursuant to this unit by a person who can be shown by a preponderance of the evidence to have committed a self-inflicted injury or by a person injured by a firearm while committing a crime, attempting to commit a crime, engaged in criminal activity, or engaged in a delinquent act.

(c) No action may be brought pursuant to this unit by a person who can be shown by a preponderance of the evidence to be engaged in the sale or distribution of illegal narcotics.

(d) No action may be brought pursuant to this unit by a person who either: (1) assumed the risk of the injury that occurred; or (2) negligently contributed to the injury that occurred.

Unit C. Assault Weapons Manufacturing Strict Liability

7-2551.01. Definitions. For the purposes of this unit, the term:

(1) "Assault weapon" includes the following weapons:

(A) Norinco, Mitchell, and Poly Technologies Avtomat Kalashnikovs (all models);

(B) Action Arms Israeli Military Industries UZI and Galil;

(C) Beretta AR-70 (SC-70);

(D) Colt AR-15 and CAR-15;

(E) Fabrique Nationale FN/FAL, FN/LAR, and FNC;

(F) MAC 10 and MAC 11;

(G) Steyr AUG;

(H) INTRATEC TEC-9; and

(I) Street Sweeper and Striker 12.

(2) "Handgun" means a firearm with a barrel-less than 12 inches in length at the time of manufacture.

(3) "Dealer" and "importer" shall have the same meaning as in 18 U.S.C. § 921.

(4) "Machine gun" shall have the same meaning as in paragraph (10) of § 7-2501.01.

(5) "Manufacturer" means any person in business to manufacture or assemble a firearm or ammunition for sale or distribution.

(6) "Law enforcement agency" means a federal, state, or local law enforcement agency, state militia, or an agency of the United States government.

(7) "Law enforcement officer" means any officer or agent of an agency defined in paragraph (6) of this section who is authorized to use a handgun or machine gun in the course of his or her work.

7-2551.02. Liability. Any manufacturer, importer, or dealer of an assault weapon or machine gun shall be held strictly liable in tort, without regard to fault or proof of defect, for all direct and consequential damages that arise from bodily injury or death if the bodily injury or death proximately results from the discharge of the assault weapon or machine gun in the District of Columbia.

7-2551.03. Exemptions.

(a) No assault weapon originally distributed to a law enforcement agency or a law enforcement officer shall provide the basis for liability under this unit.

(b) No action may be brought pursuant to this unit by a person injured by an assault weapon while committing a crime.

(c) This section shall not operate to limit in scope any cause of action, other than that provided by this unit, available to a person injured by an assault weapon.

(d) Any defense that is available in a strict liability action shall be available as a defense under this unit.

(e) Recovery shall not be allowed under this unit for a self-inflicted injury that results from a reckless, wanton, or willful discharge of an assault weapon.

Division IV. Criminal Law and Procedure and Prisoners

Title 22. Criminal Offenses and Penalties

Subtitle VI. Regulation and Possession of Weapons

Chapter 45. Weapons and Possession of Weapons

22-4501. Definitions.

(a) "Pistol," as used in this chapter, means any firearm with a barrel less than 12 inches in length.

(b) "Sawed-off shotgun," as used in this chapter, means any shotgun with a barrel less than 20 inches in length.

(c) "Machine gun," as used in this chapter, means any firearm which shoots automatically or semiautomatically more than 12 shots without reloading.

(d) "Person," as used in this chapter, includes individual, firm, association, or corporation.

(e) "Sell" and "purchase" and the various derivatives of such words, as used in this chapter, shall be construed to include letting on hire, giving, lending, borrowing, and otherwise transferring.

(f) "Crime of violence," as used in this chapter, means any of the following crimes, or an attempt to commit any of the same, namely: murder, manslaughter, an act of terrorism, manufacture or possession of a weapon of mass destruction, use, dissemination, or detonation of a weapon of mass destruction, first degree sexual abuse, second degree sexual abuse, or child sexual abuse, mayhem, maliciously disfiguring another, abduction, kidnapping, burglary, robbery, housebreaking, any assault with intent to kill, commit first degree sexual abuse, second degree sexual abuse, or child sexual abuse, or robbery, assault with a dangerous weapon, assault with intent to commit any offense punishable by imprisonment in the penitentiary, arson, or extortion or blackmail accompanied by threats of violence or aggravated assault.

(g) "Dangerous crime," as used in this chapter, means distribution of or possession with intent to distribute a controlled substance, if the offense is punishable by imprisonment for more than 1 year. For the purposes of this definition, the term "controlled substance" means any substance defined as such in the District of Columbia Code or any Act of Congress.

(h) "Playground," as used in this chapter, means any facility intended for recreation, open to the public, and with any portion of the facility that contains 1 or more separate apparatus intended for the recreation of children, including, but not limited to, sliding boards, swingsets, and teeterboards.

(i) "Video arcade," as used in this chapter, means any facility legally accessible to persons under 18 years of age, intended primarily for the use of pinball and video machines for amusement, and which contains a minimum of 10 pinball or video machines.

(j) "Youth center," as used in this chapter, means any recreational facility or gymnasium (including any parking lot appurtenant thereto), intended primarily for use by persons under 18 years of age, which regularly provides athletic, civic, or cultural activities.

22-4502.01. Gun free zones; enhanced penalty.

(a) All areas within 1000 feet of an appropriately identified public or private day care center, elementary school, vocational school, secondary school, college, junior college, or university, or any public swimming pool, playground, video arcade, youth center, or public library, or in and around public housing as defined in 42 U.S.C.S. § 1437a(b), the development or administration of which is assisted by the United States Department of Housing and Urban Development, or in or around housing that is owned, operated, or financially assisted by the District of Columbia Housing Authority, or an event sponsored by any of the above entities shall be declared a gun free zone. For the purposes of this subsection, the term "appropriately identified" means that there is a sign that identifies the building or area as a gun free zone.

(b) Any person illegally carrying a gun within a gun free zone shall be punished by a fine up to twice that otherwise authorized to be imposed, by a term of imprisonment up to twice that otherwise authorized to be imposed, or both.

(c) The provisions of this section shall not apply to a person legally licensed to carry a firearm in the District of Columbia who lives or works within 1000 feet of a gun free zone or to members of the Army, Navy, Air Force, or Marine Corps of the United States; the National Guard or Organized Reserves when on duty; the Post Office Department or its employees when on duty; marshals, sheriffs, prison, or jail wardens, or their deputies; policemen or other duly-appointed law enforcement officers; officers or employees of the United States duly authorized to carry such weapons; banking institutions; public carriers who are engaged in the business of transporting mail, money, securities, or other valuables; and licensed wholesale or retail dealers.

22-4503. Unlawful possession of pistol.

(a) No person shall own or keep a pistol, or have a pistol in his or her possession or under his or her control, within the District of Columbia, if:

(1) Such person is a drug addict;

(2) Such person has been convicted in the District of Columbia or elsewhere of a felony;

(3) Such person has been convicted of violating § 22-2701, § 22-2722, or §§ 22-3502 to 22-3506; or

(4) Such person is not licensed under § 22-4510 to sell weapons, and such person has been convicted of violating this chapter.

(b) No person shall keep a pistol for, or intentionally make a pistol available to, such a person, knowing that such person has been so convicted or that such person is a drug addict. Whoever violates this section shall be punished as provided in § 22-4515, unless the violation occurs after such person has been convicted of a violation of this section, in which case such person shall be imprisoned for not more than 10 years.

22-4507. Certain sales of pistols prohibited. No person shall within the District of Columbia sell any pistol to a person who he or she has reasonable cause to believe is not of sound mind, or is forbidden by § 22-4503 to possess a pistol, or, except when the relation of parent and child or guardian and ward exists, is under the age of 21 years.

22-4508. Transfers of firearms regulated. No seller shall within the District of Columbia deliver a pistol to the purchaser thereof until 48 hours shall have elapsed from the time of the application for the purchase thereof, except in the case of sales to marshals, sheriffs, prison or jail wardens or their deputies, policemen, or other duly appointed law enforcement officers, and, when delivered, said pistol shall be securely wrapped and shall be unloaded. At the time of applying for the purchase of a pistol the purchaser shall sign in duplicate and deliver to the seller a statement containing his or her full name, address, occupation, color, place of birth, the date and hour of application, the caliber, make, model, and manufacturer's number of the pistol to be purchased and a statement that the purchaser is not forbidden by § 22-4503 to possess a pistol. The seller shall, within 6 hours after such application, sign and attach his or her address and deliver 1 copy to such person or persons as the Chief of Police of the District of Columbia may designate, and shall retain the other copy for 6 years. No machine gun, sawed-off shotgun, or blackjack shall be sold to any person other than the persons designated in § 22-4514 as entitled to possess the same, and then only after permission to make such sale has been obtained from the Chief of Police of the District of Columbia. This section shall not apply to sales at wholesale to licensed dealers.

22-4509. Dealers of weapons to be licensed. No retail dealer shall within the District of Columbia sell or expose for sale or have in his or her possession with intent to sell, any pistol, machine gun, sawed-off shotgun, or blackjack without being licensed as provided in § 22-4510. No wholesale dealer shall, within the District of Columbia, sell, or have in his or her possession with intent to sell, to any person other than a licensed dealer, any pistol, machine gun, sawed-off shotgun, or blackjack.

22-4510. Licenses of weapons dealers; records; by whom granted; conditions.

(a) The Mayor of the District of Columbia may, in his or her discretion, grant licenses and may prescribe the form thereof, effective for not more than 1 year from date of issue, permitting the licensee to sell pistols, machine guns, sawed-off shotguns, and blackjacks at retail within the District of Columbia subject to the following conditions in addition to those specified in § 22-4509, for breach of any of which the license shall be subject to forfeiture and the licensee subject to punishment as provided in this chapter:

(1) The business shall be carried on only in the building designated in the license.

(2) The license or a copy thereof, certified by the issuing authority, shall be displayed on the premises where it can be easily read.

(3) No pistol shall be sold: (A) if the seller has reasonable cause to believe that the purchaser is not of sound mind or is forbidden by § 22-4503 to possess a pistol or is under the age of 21 years; and (B) unless the purchaser is personally known to the seller or shall present clear evidence of his or her identity. No machine gun, sawed-off shotgun, or blackjack shall be sold to any person other than the persons designated in § 22-4514 as entitled to possess the same, and then only after permission to make such sale has been obtained from the Chief of Police of the District of Columbia.

(4) A true record shall be made in a book kept for the purpose, the form of which may be prescribed by the Mayor, of all pistols, machine guns, and sawed-off shotguns in the possession of the licensee, which said record shall contain the date of purchase, the caliber, make, model, and manufacturer's number of the weapon, to which shall be added, when sold, the date of sale.

(5) A true record in duplicate shall be made of every pistol, machine gun, sawed-off shotgun, and blackjack sold, said record to be made in a book kept for the purpose, the form of which may be prescribed by the Mayor of the District of

Columbia and shall be personally signed by the purchaser and by the person effecting the sale, each in the presence of the other and shall contain the date of sale, the name, address, occupation, color, and place of birth of the purchaser, and, so far as applicable, the caliber, make, model, and manufacturer's number of the weapon, and a statement by the purchaser that the purchaser is not forbidden by § 22-4503 to possess a pistol. One copy of said record shall, within 7 days, be forwarded by mail to the Chief of Police of the District of Columbia and the other copy retained by the seller for 6 years.

(6) No pistol or imitation thereof or placard advertising the sale thereof shall be displayed in any part of said premises where it can readily be seen from the outside. No license to sell at retail shall be granted to anyone except as provided in this section.

(b) Any license issued pursuant to this section shall be issued by the Metropolitan Police Department as a Public Safety endorsement to a basic business license under the basic business license system as set forth in subchapter I-A of Chapter 28 of Title 47 of the District of Columbia Code.

22-4511. False information in purchase of weapons prohibited. No person shall, in purchasing a pistol or in applying for a license to carry the same, or in purchasing a machine gun, sawed-off shotgun, or blackjack within the District of Columbia, give false information or offer false evidence of his or her identity.

22-4512. Alteration of identifying marks of weapons prohibited. No person shall within the District of Columbia change, alter, remove, or obliterate the name of the maker, model, manufacturer's number, or other mark or identification on any pistol, machine gun, or sawed-off shotgun. Possession of any pistol, machine gun, or sawed-off shotgun upon which any such mark shall have been changed, altered, removed, or obliterated shall be prima facie evidence that the possessor has changed, altered, removed, or obliterated the same within the District of Columbia; provided, however, that nothing contained in this section shall apply to any officer or agent of any of the departments of the United States or the District of Columbia engaged in experimental work.

22-4513. Exceptions. Except as provided in § 22-4502 and § 22-4514(b), this chapter shall not apply to toy or antique pistols unsuitable for use as firearms.

22-4514. Possession of certain dangerous weapons prohibited; exceptions.

(a) No person shall within the District of Columbia possess any machine gun, sawed-off shotgun, or any instrument or weapon of the kind commonly known as a blackjack, slungshot, sand club, sandbag, switchblade knife, or metal knuckles, nor any instrument, attachment, or appliance for causing the firing of any firearm to be silent or intended to lessen or muffle the noise of the firing of any firearms; provided, however, that machine guns, or sawed-off shotguns, and blackjacks may be possessed by the members of the Army, Navy, Air Force, or Marine Corps of the United States, the National Guard, or Organized Reserves when on duty, the Post Office Department or its employees when on duty, marshals, sheriffs, prison or jail wardens, or their deputies, policemen, or other duly-appointed law enforcement officers, including any designated civilian employee of the Metropolitan Police Department, or officers or employees of the United States duly authorized to carry such weapons, banking institutions, public carriers who are engaged in the business of transporting mail, money, securities, or other valuables, wholesale dealers and retail dealers licensed under § 22-4510.

(b) No person shall within the District of Columbia possess, with intent to use unlawfully against another, an imitation pistol, or a dagger, dirk, razor, stiletto, or knife with a blade longer than 3 inches, or other dangerous weapon.

(c) Whoever violates this section shall be punished as provided in § 22-4515 unless the violation occurs after such person has been convicted in the District of Columbia of a violation of this section, or of a felony, either in the District of Columbia or in another jurisdiction, in which case such person shall be imprisoned for not more than 10 years.

22-4515. Penalties. Any violation of any provision of this chapter for which no penalty is specifically provided shall be punished by a fine of not more than $1,000 or imprisonment for not more than 1 year, or both.

22-4515a. Manufacture, transfer, use, possession, or transportation of Molotov cocktails, or other explosives for unlawful purposes, prohibited; definitions; penalties.

(a) No person shall within the District of Columbia manufacture, transfer, use, possess, or transport a Molotov cocktail. As used in this subsection, the term "Molotov cocktail" means: (1) a breakable container containing flammable liquid and having a wick or a similar device capable of being ignited; or (2) any other device designed to explode or produce uncontained combustion upon impact; but such term does not include a device lawfully and commercially manufactured primarily for the purpose of illumination, construction work, or other lawful purpose.

(b) No person shall manufacture, transfer, use, possess, or transport any device, instrument, or object designed to explode or produce uncontained combustion, with the intent that the same may be used unlawfully against any person or property.

(c) No person shall, during a state of emergency in the District of Columbia declared by the Mayor pursuant to law, or during a situation in the District of Columbia concerning which the President has invoked any provision of Chapter 15 of Title 10, United States Code, manufacture, transfer, use, possess, or transport any device, instrument, or object designed to explode or produce uncontained combustion, except at his or her residence or place of business.

(d) Whoever violates this section shall:

(1) for the first offense, be sentenced to a term of imprisonment of not less than 1 and not more than 5 years;

(2) for the second offense, be sentenced to a term of imprisonment of not less than 3 and not more than 15 years; and

(3) for the third or subsequent offense, be sentenced to a term of imprisonment of not less than 5 years and not more than 30 years. In the case of a person convicted of a third or subsequent violation of this section, Chapter 402 of Title 18, United States Code (Federal Youth Corrections Act) shall not apply. For purposes of imprisonment following revocation of release authorized by § 24-403.01(b)(7), the third or subsequent conviction for an offense defined by this section is a Class A felony.

22-4517. Dangerous articles; definition; taking and destruction; procedure.

(a) As used in this section, the term "dangerous article" means:

(1) Any weapon such as a pistol, machine gun, sawed-off shotgun, blackjack, slingshot, sandbag, or metal knuckles; or

(2) Any instrument, attachment, or appliance for causing the firing of any firearms to be silent or intended to lessen or muffle the noise of the firing of any firearms.

(b) A dangerous article unlawfully owned, possessed, or carried is hereby declared to be a nuisance.

(c) When a police officer, in the course of a lawful arrest or lawful search, or when a designated civilian employee of the Metropolitan Police Department in the course of a lawful search, discovers a dangerous article which the officer reasonably believes is a nuisance under subsection (b) of this section the officer shall take it into his or her possession and surrender it to the Property Clerk of the Metropolitan Police Department.

(d)(1) Within 30 days after the date of such surrender, any person may file in the office of the Property Clerk of the Metropolitan Police Department a written claim for possession of such dangerous article. Upon the expiration of such period, the Property Clerk shall notify each such claimant, by registered mail addressed to the address shown on the claim, of the time and place of a hearing to determine which claimant, if any, is entitled to possession of such dangerous article. Such hearing shall be held within 60 days after the date of such surrender.

(2) At the hearing the Property Clerk shall hear and receive evidence with respect to the claims filed under paragraph (1) of this subsection. Thereafter he or she shall determine which claimant, if any, is entitled to possession of such dangerous article and shall reduce his or her decision to writing. The Property Clerk shall send a true copy of such written decision to each claimant by registered mail addressed to the last known address of such claimant.

(3) Any claimant may, within 30 days after the day on which the copy of such decision was mailed to such claimant, file an appeal in the Superior Court of the District of Columbia. If the claimant files an appeal, he or she shall at the same time give written notice thereof to the Property Clerk. If the decision of the Property Clerk is so appealed, the Property Clerk shall not dispose of the dangerous article while such appeal is pending and, if the final judgment is entered by such court, he or she shall dispose of such dangerous article in accordance with the judgment of such court. The Superior Court of the District of Columbia is authorized to determine which claimant, if any, is entitled to possession of the dangerous article and to enter a judgment ordering a disposition of such dangerous article consistent with subsection (f) of this section.

(4) If there is no such appeal, or if such appeal is dismissed or withdrawn, the Property Clerk shall dispose of such dangerous article in accordance with subsection (f) of this section.

(5) The Property Clerk shall make no disposition of a dangerous article under this section, whether in accordance with his or her own decision or in accordance with the judgment of the Superior Court of the District of Columbia, until the United States Attorney for the District of Columbia certifies to the Property Clerk that such dangerous article will not be needed as evidence.

(e) A person claiming a dangerous article shall be entitled to its possession only if: (1) such person shows, on satisfactory evidence, that such person is the owner of the dangerous article or is the accredited representative of the owner, and that the ownership is lawful; (2) such person shows on satisfactory evidence that at the time the dangerous article was taken into possession by a police officer or a designated civilian employee of the Metropolitan Police Department, it was not unlawfully owned and was not unlawfully possessed or carried by the claimant or with his or her knowledge or consent; and (3) the receipt of possession by the claimant does not cause the article to be a nuisance. A representative is accredited if such person has a power of attorney from the owner.

(f) If a person claiming a dangerous article is entitled to its possession as determined under subsections (d) and (e) of this section, possession of such dangerous article shall be given to such person. If no person so claiming is entitled to its possession as determined under subsections (d) and (e) of this section, or if there be no claimant, such dangerous article shall be destroyed. In lieu of such destruction, any such serviceable dangerous article may, upon order of the Mayor of the District of Columbia, be transferred to and used by any federal or District Government law-enforcing agency, and the agency receiving same shall establish property responsibility and records of these dangerous articles.

(g) The Property Clerk shall not be liable in damages for any action performed in good faith under this section.

Title 24. Prisoners and Their Treatment

Chapter 2. Prisons and Prisoners

Subchapter VII. Correctional Treatment Facility

24-261.02a. Registration of firearms for private operator

(a) In order to register firearms, the private operator shall follow the following procedures:

(1) To register for interim approval, the private operator shall provide the Chief of the Metropolitan Police Department ("Chief of Police") with the serial numbers and storage places of firearms in the private operator's possession in the District of Columbia. If the Chief of Police determines that the information provided is satisfactory, he or she shall issue interim approval to the private operator for the weapons identified and held in the private operator's possession. The interim approval shall be valid for 90 days, during which time the private operator shall complete the actions necessary to register for permanent approval.

(2) (A) To register for permanent approval, the private operator shall provide the Chief of Police with the following information:

(i) The names and such other identifying information as the Chief of Police may require, of all private correctional officers who will be authorized by the private operator to carry and use firearms in the course of their assigned duties;

(ii) Records or other evidence acceptable to the Chief of Police to demonstrate that each private correctional officer authorized to carry and use firearms has received instructions about all applicable rules of the Department of Corrections or the Federal Bureau of Prisons regarding the use of force and deadly force in the course of his or her duties;

(iii) Records or other evidence acceptable to the Chief of Police to demonstrate that each private correctional officer authorized to carry and use firearms has successfully completed the training required by § 24-261.02(d); and

(iv) A sworn affidavit signed by each private correctional officer authorized to carry and use firearms attesting that he or she has read and understands all applicable rules of the Department of Corrections or the Federal Bureau of Prisons regarding the use of force and deadly force in the course of his or her duties.

(B) The Chief of Police, upon determining that the information submitted in accordance with this paragraph is satisfactory, shall issue permanent registration approval to the private operator for the firearms in the private operator's possession in the District of Columbia.

(b) A private operator who is issued firearms registration approval pursuant to this section shall be subject to the duties and revocation provisions set forth in §§ 7-2502.08 and 7-2502.09, and other applicable rules and laws of the District of Columbia. A private operator shall notify the Chief of Police whenever any private correctional officer authorized to carry and use firearms leaves the private operator's employment at a facility in the District or otherwise ceases to be authorized to carry and use firearms.

(c) Nothing in § 24-261.02 or this section shall be construed to allow any private correctional officer or any other person to remove any weapon registered to the private operator from the premises and grounds of the private operator's facility except in the performance of assigned duties and in accordance with laws and rules of the District and federal governments.

[Current through D.C. Law 17-207, Effective July 18, 2008]

FLORIDA
FLA. STAT.

Title XLVI. Crimes

Chapter 790. Weapons and Firearms

790.001. Definitions As used in this chapter, except where the context otherwise requires:

(1) "Antique firearm" means any firearm manufactured in or before 1918 (including any matchlock, flintlock, percussion cap, or similar early type of ignition system) or replica thereof, whether actually manufactured before or after the year 1918, and also any firearm using fixed ammunition manufactured in or before 1918, for which ammunition is no longer manufactured in the United States and is not readily available in the ordinary channels of commercial trade.

(2) "Concealed firearm" means any firearm, as defined in subsection (6), which is carried on or about a person in such a manner as to conceal the firearm from the ordinary sight of another person.

(3)(a) "Concealed weapon" means any dirk, metallic knuckles, slungshot, billie, tear gas gun, chemical weapon or device, or other deadly weapon carried on or about a person in such a manner as to conceal the weapon from the ordinary sight of another person.

(b) "Tear gas gun" or "chemical weapon or device" means any weapon of such nature, except a device known as a "self-defense chemical spray." "Self-defense chemical spray" means a device carried solely for purposes of lawful self-defense that is compact in size, designed to be carried on or about the person, and contains not more than two ounces of chemical.

(4) "Destructive device" means any bomb, grenade, mine, rocket, missile, pipebomb, or similar device containing an explosive, incendiary, or poison gas and includes any frangible container filled with an explosive, incendiary, explosive gas, or expanding gas, which is designed or so constructed as to explode by such filler and is capable of causing bodily harm or property damage; any combination of parts either designed or intended for use in converting any device into a destructive device and from which a destructive device may be readily assembled; any device declared a destructive device by the Bureau of Alcohol, Tobacco, and Firearms; any type of weapon which will, is designed to, or may readily be converted to expel a projectile by the action of any explosive and which has a barrel with a bore of one-half inch or more in diameter; and ammunition for such destructive devices, but not including shotgun shells or any other ammunition designed for use in a firearm other than a destructive device. "Destructive device" does not include:

(a) A device which is not designed, redesigned, used, or intended for use as a weapon;

(b) Any device, although originally designed as a weapon, which is redesigned so that it may be used solely as a signaling, line-throwing, safety, or similar device;

(c) Any shotgun other than a short-barreled shotgun; or

(d) Any nonautomatic rifle (other than a short-barreled rifle) generally recognized or particularly suitable for use for the hunting of big game.

(5) "Explosive" means any chemical compound or mixture that has the property of yielding readily to combustion or oxidation upon application of heat, flame, or shock, including but not limited to dynamite, nitroglycerin, trinitrotoluene, or ammonium nitrate when combined with other ingredients to form an explosive mixture, blasting caps, and detonators; but not including:

(a) Shotgun shells, cartridges, or ammunition for firearms;

(b) Fireworks as defined in s. 791.01;

(c) Smokeless propellant powder or small arms ammunition primers, if possessed, purchased, sold, transported, or used in compliance with s. 552.241;

(d) Black powder in quantities not to exceed that authorized by chapter 552, or by any rules adopted thereunder by the Department of Financial Services, when used for, or intended to be used for, the manufacture of target and sporting ammunition or for use in muzzle-loading flint or percussion weapons.

The exclusions contained in paragraphs (a)-(d) do not apply to the term "explosive" as used in the definition of "firearm" in subsection (6).

(6) "Firearm" means any weapon (including a starter gun) which will, is designed to, or may readily be converted to expel a projectile by the action of an explosive; the frame or receiver of any such weapon; any firearm muffler or firearm silencer; any destructive device; or any machine gun. The term "firearm" does not include an antique firearm unless the antique firearm is used in the commission of a crime.

(7) "Indictment" means an indictment or an information in any court under which a crime punishable by imprisonment for a term exceeding 1 year may be prosecuted.

(8) "Law enforcement officer" means:

(a) All officers or employees of the United States or the State of Florida, or any agency, commission, department, board, division, municipality, or subdivision thereof, who have authority to make arrests;

(b) Officers or employees of the United States or the State of Florida, or any agency, commission, department, board, division, municipality, or subdivision thereof, duly authorized to carry a concealed weapon;

(c) Members of the Armed Forces of the United States, the organized reserves, state militia, or Florida National Guard, when on duty, when preparing themselves for, or going to or from, military duty, or under orders;

(d) An employee of the state prisons or correctional systems who has been so designated by the Department of Corrections or by a warden of an institution;

(e) All peace officers;

(f) All state attorneys and United States attorneys and their respective assistants and investigators.

(9) "Machine gun" means any firearm, as defined herein, which shoots, or is designed to shoot, automatically more than one shot, without manually reloading, by a single function of the trigger.

(10) "Short-barreled shotgun" means a shotgun having one or more barrels less than 18 inches in length and any weapon made from a shotgun (whether by alteration, modification, or otherwise) if such weapon as modified has an overall length of less than 26 inches.

(11) "Short-barreled rifle" means a rifle having one or more barrels less than 16 inches in length and any weapon made from a rifle (whether by alteration, modification, or otherwise) if such weapon as modified has an overall length of less than 26 inches.

(12) "Slungshot" means a small mass of metal, stone, sand, or similar material fixed on a flexible handle, strap, or the like, used as a weapon.

(13) "Weapon" means any dirk, knife, metallic knuckles, slungshot, billie, tear gas gun, chemical weapon or device, or other deadly weapon except a firearm or a common pocketknife, plastic knife, or blunt-bladed table knife.

(14) "Electric weapon or device" means any device which, through the application or use of electrical current, is designed, redesigned, used, or intended to be used for offensive or defensive purposes, the destruction of life, or the infliction of injury.

(15) "Dart-firing stun gun" means any device having one or more darts that are capable of delivering an electrical current.

(16) "Readily accessible for immediate use" means that a firearm or other weapon is carried on the person or within such close proximity and in such a manner that it can be retrieved and used as easily and quickly as if carried on the person.

(17) "Securely encased" means in a glove compartment, whether or not locked; snapped in a holster; in a gun case, whether or not locked; in a zippered gun case; or in a closed box or container which requires a lid or cover to be opened for access.

(18) "Sterile area" means the area of an airport to which access is controlled by the inspection of persons and property in accordance with federally approved airport security programs.

(19) "Ammunition" means an object consisting of all of the following:

(a) A fixed metallic or nonmetallic hull or casing containing a primer.

(b) One or more projectiles, one or more bullets, or shot.

(c) Gunpowder.

All of the specified components must be present for an object to be ammunition.

790.065. Sale and delivery of firearms

(1) A licensed importer, licensed manufacturer, or licensed dealer may not sell or deliver from her or his inventory at her or his licensed premises any firearm to another person, other than a licensed importer, licensed manufacturer, licensed dealer, or licensed collector, until she or he has:

(a) Obtained a completed form from the potential buyer or transferee, which form shall have been promulgated by the Department of Law Enforcement and provided by the licensed importer, licensed manufacturer, or licensed dealer, which shall include the name, date of birth, gender, race, and social security number or other identification number of such potential buyer or transferee and has inspected proper identification including an identification containing a photograph of the potential buyer or transferee.

(b) Collected a fee from the potential buyer for processing the criminal history check of the potential buyer. The fee shall be established by the Department of Law Enforcement and may not exceed $8 per transaction. The Department of Law Enforcement may reduce, or suspend collection of, the fee to reflect payment received from the Federal Government applied to the cost of maintaining the criminal history check system established by this section as a means of facilitating or supplementing the National Instant Criminal Background Check System. The Department of Law Enforcement shall, by rule, establish procedures for the fees to be transmitted by the licensee to the Department of Law Enforcement. All such fees shall be deposited into the Department of Law Enforcement Operating Trust Fund, but shall be segregated from all other funds deposited into such trust fund and must be accounted for separately. Such segregated funds must not be used for any purpose other than the operation of the criminal history checks required by this section. The Department of Law Enforcement, each year prior to February 1, shall make a full accounting of all receipts and expenditures of such funds to the President of the Senate, the Speaker of the House of Representatives, the majority and minority leaders of each house of the Legislature, and the chairs of the appropriations committees of each house of the Legislature. In the event that the cumulative amount of funds collected exceeds the cumulative amount of expenditures by more than $2.5 million, excess funds may be used for the purpose of purchasing soft body armor for law enforcement officers.

(c) Requested, by means of a toll-free telephone call, the Department of Law Enforcement to conduct a check of the information as reported and reflected in the Florida Crime Information Center and National Crime Information Center systems as of the date of the request.

(d) Received a unique approval number for that inquiry from the Department of Law Enforcement, and recorded the date and such number on the consent form. However, if the person purchasing, or receiving delivery of, the firearm is a holder of a valid concealed weapons or firearms license pursuant to the provisions of s. 790.06 or holds an active certification from the Criminal Justice Standards and Training Commission as a "law enforcement officer," a "correctional officer," or a "correctional probation officer" as defined in s. 943.10(1), (2), (3), (6), (7), (8), or (9), the provisions of this subsection do not apply.

(2) Upon receipt of a request for a criminal history record check, the Department of Law Enforcement shall, during the licensee's call or by return call, forthwith:

(a) Review any records available to determine if the potential buyer or transferee:

1. Has been convicted of a felony and is prohibited from receipt or possession of a firearm pursuant to s. 790.23;

2. Has been convicted of a misdemeanor crime of domestic violence, and therefore is prohibited from purchasing a firearm;

3. Has had adjudication of guilt withheld or imposition of sentence suspended on any felony or misdemeanor crime of domestic violence unless 3 years have elapsed since probation or any other conditions set by the court have been fulfilled or expunction has occurred; or

4. Has been adjudicated mentally defective or has been committed to a mental institution by a court and as a result is prohibited by federal law from purchasing a firearm.

a. As used in this subparagraph, "adjudicated mentally defective" means a determination by a court that a person, as a result of marked subnormal intelligence, or mental illness, incompetency, condition, or disease, is a danger to himself or herself or to others or lacks the mental capacity to contract or manage his or her own affairs. The phrase includes a judicial finding of incapacity under s. 744.331(6)(a), an acquittal by reason of insanity of a person charged with a criminal offense, and a judicial finding that a criminal defendant is not competent to stand trial.

b. As used in this subparagraph, "committed to a mental institution" means involuntary commitment, commitment for mental defectiveness or mental illness, and commitment for substance abuse. The phrase includes involuntary inpatient placement as defined in s. 394.467, involuntary outpatient placement as defined in s. 394.4655, involuntary assessment and stabilization under s. 397.6818, and involuntary substance abuse treatment under s. 397.6957, but does not include a person in a mental institution for observation or discharged from a mental institution based upon the initial review by the physician or a voluntary admission to a mental institution.

c. In order to check for these conditions, the department shall compile and maintain an automated database of persons who are prohibited from purchasing a firearm based on court records of adjudications of mental defectiveness or commitments to mental institutions. Clerks of court shall submit these records to the department within 1 month after the rendition of the adjudication or commitment. Reports may be submitted in an automated format. The reports must, at a minimum, include the name, along with any known alias or former name, the sex, and the date of birth of the subject. The department shall delete any mental health record from the database upon request of an individual when 5 years have elapsed since the individual's restoration to capacity by court order after being adjudicated an incapacitated person under s. 744.331, or similar laws of any other state; or, in the case of an individual who was previously committed to a mental institution under chapter 394, or similar laws of any other state, when the individual produces a certificate from a licensed psychiatrist that he or she has not suffered from disability for at least 5 years prior to the date of request for removal of the record. When the department has received a subsequent record of an adjudication of mental defectiveness or commitment to a mental institution for such individual, the 5-year timeframe shall be calculated from the most recent adjudication of incapacitation or commitment.

d. The department is authorized to disclose the collected data to agencies of the Federal Government and other states for use exclusively in determining the lawfulness of a firearm sale or transfer. The department is also authorized to disclose any collected data to the Department of Agriculture and Consumer Services for purposes of determining eligibility for issuance of a concealed weapons or concealed firearms license and for determining whether a basis exists for revoking or suspending a previously issued license pursuant to s. 790.06(10). When a potential buyer or transferee appeals a nonapproval based on these records, the clerks of court and mental institutions shall, upon request by the department, provide information to help determine whether the potential buyer or transferee is the same person as the subject of the record. Photographs and any other data that could confirm or negate identity must be made available to the department for such purposes, notwithstanding any other provision of state law to the contrary. Any such information that is made confidential or exempt from disclosure by law shall retain such confidential or exempt status when transferred to the department.

(b) Inform the licensee making the inquiry either that records demonstrate that the buyer or transferee is so prohibited and provide the licensee a nonapproval number, or provide the licensee with a unique approval number.

(c) 1. Review any records available to it to determine whether the potential buyer or transferee has been indicted or has had an information filed against her or him for an offense that is a felony under either state or federal law, or, as mandated by federal law, has had an injunction for protection against domestic violence entered against the potential buyer or transferee under s. 741.30, has had an injunction for protection against repeat violence entered against the potential buyer or transferee under s. 784.046, or has been arrested for a dangerous crime as specified in s. 907.041(4)(a) or for any of the following enumerated offenses:

a. Criminal anarchy under ss. 876.01 and 876.02.

b. Extortion under s. 836.05.

c. Explosives violations under s. 552.22(1) and (2).

d. Controlled substances violations under chapter 893.

e. Resisting an officer with violence under s. 843.01.

f. Weapons and firearms violations under this chapter.

g. Treason under s. 876.32.

h. Assisting self-murder under s. 782.08.

i. Sabotage under s. 876.38.

j. Stalking or aggravated stalking under s. 784.048.

If the review indicates any such indictment, information, or arrest, the department shall provide to the licensee a conditional nonapproval number.

2. Within 24 working hours, the department shall determine the disposition of the indictment, information, or arrest and inform the licensee as to whether the potential buyer is prohibited from receiving or possessing a firearm. For purposes of this paragraph, "working hours" means the hours from 8 a.m. to 5 p.m. Monday through Friday, excluding legal holidays.

3. The office of the clerk of court, at no charge to the department, shall respond to any department request for data on the disposition of the indictment, information, or arrest as soon as possible, but in no event later than 8 working hours.

4. The department shall determine as quickly as possible within the allotted time period whether the potential buyer is prohibited from receiving or possessing a firearm.

5. If the potential buyer is not so prohibited, or if the department cannot determine the disposition information within the allotted time period, the department shall provide the licensee with a conditional approval number.

6. If the buyer is so prohibited, the conditional nonapproval number shall become a nonapproval number.

7. The department shall continue its attempts to obtain the disposition information and may retain a record of all approval numbers granted without sufficient disposition information. If the department later obtains disposition information which indicates:

a. That the potential buyer is not prohibited from owning a firearm, it shall treat the record of the transaction in accordance with this section; or

b. That the potential buyer is prohibited from owning a firearm, it shall immediately revoke the conditional approval number and notify local law enforcement.

8. During the time that disposition of the indictment, information, or arrest is pending and until the department is notified by the potential buyer that there has been a final disposition of the indictment, information, or arrest, the conditional nonapproval number shall remain in effect.

(3) In the event of scheduled computer downtime, electronic failure, or similar emergency beyond the control of the Department of Law Enforcement, the department shall immediately notify the licensee of the reason for, and estimated length of, such delay. After such notification, the department shall forthwith, and in no event later than the end of the next business day of the licensee, either inform the requesting licensee if its records demonstrate that the buyer or transferee is prohibited from receipt or possession of a firearm pursuant to Florida and Federal law or provide the licensee with a unique approval number. Unless notified by the end of said next business day that the buyer or transferee is so prohibited, and without regard to whether she or he has received a unique approval number, the licensee may complete the sale or transfer and shall not be deemed in violation of this section with respect to such sale or transfer.

(4)(a) Any records containing any of the information set forth in subsection (1) pertaining to a buyer or transferee who is not found to be prohibited from receipt or transfer of a firearm by reason of Florida and federal law which records are created by the Department of Law Enforcement to conduct the criminal history record check shall be confidential and exempt from the provisions of s. 119.07(1) and may not be disclosed by the Department of Law Enforcement or any officer or employee thereof to any person or to another agency. The Department of Law Enforcement shall destroy any such records forthwith after it communicates the approval and nonapproval numbers to the licensee and, in any event, such records shall be destroyed within 48 hours after the day of the response to the licensee's request.

(b) Notwithstanding the provisions of this subsection, the Department of Law Enforcement may maintain records of NCIC transactions to the extent required by the Federal Government, and may maintain a log of dates of requests for criminal history records checks, unique approval and nonapproval numbers, license identification numbers, and transaction numbers corresponding to such dates for a period of not longer than 2 years or as otherwise required by law.

(c) Nothing in this chapter shall be construed to allow the State of Florida to maintain records containing the names of purchasers or transferees who receive unique approval numbers or to maintain records of firearm transactions.

(d) Any officer or employee, or former officer or employee of the Department of Law Enforcement or law enforcement agency who intentionally and maliciously violates the provisions of this subsection commits a felony of the third degree punishable as provided in s. 775.082 or s. 775.083.

(5) The Department of Law Enforcement shall establish a toll-free telephone number which shall be operational 7 days a week with the exception of Christmas Day and New Year's Day, for a period of 12 hours a day beginning at 9 a.m. and ending at 9 p.m., for purposes of responding to inquiries as described in this section from licensed manufacturers, licensed importers, and licensed dealers. The Department of Law Enforcement shall employ and train such personnel as are necessary expeditiously to administer the provisions of this section.

(6) Any person who is denied the right to receive or purchase a firearm as a result of the procedures established by this section may request a criminal history records review and correction in accordance with the rules promulgated by the Department of Law Enforcement.

(7) It shall be unlawful for any licensed dealer, licensed manufacturer, or licensed importer willfully and intentionally to request criminal history record information under false pretenses, or willfully and intentionally to disseminate criminal history record information to any person other than the subject of such information. Any person convicted of a violation of this subsection commits a felony of the third degree punishable as provided in s. 775.082 or s. 775.083.

(8) The Department of Law Enforcement shall promulgate regulations to ensure the identity, confidentiality, and security of all records and data provided pursuant to this section.

(9) This section shall become effective at such time as the Department of Law Enforcement has notified all licensed importers, licensed manufacturers, and licensed dealers in writing that the procedures and toll-free number described in this section are operational. This section shall remain in effect only during such times as the procedures described in subsection (2) remain operational.

(10) A licensed importer, licensed manufacturer, or licensed dealer is not required to comply with the requirements of this section in the event of:

(a) Unavailability of telephone service at the licensed premises due to the failure of the entity which provides telephone service in the state, region, or other geographical area in which the licensee is located to provide telephone service to the premises of the licensee due to the location of said premises; or the interruption of telephone service by reason of hurricane, tornado, flood, natural disaster, or other act of God, war, invasion, insurrection, riot, or other bona fide emergency, or other reason beyond the control of the licensee; or

(b) Failure of the Department of Law Enforcement to comply with the requirements of subsections (2) and (3).

(11) Compliance with the provisions of this chapter shall be a complete defense to any claim or cause of action under the laws of any state for liability for damages arising from the importation or manufacture, or the subsequent sale or transfer to any person who has been convicted in any court of a crime punishable by imprisonment for a term exceeding 1 year, of any firearm which has been shipped or transported in interstate or foreign commerce. The Department of Law Enforcement, its agents and employees shall not be liable for any claim or cause of action under the laws of any state for liability for damages arising from its actions in lawful compliance with this section.

(12)(a) Any potential buyer or transferee who willfully and knowingly provides false information or false or fraudulent identification commits a felony of the third degree punishable as provided in s. 775.082 or s. 775.083.

(b) Any licensed importer, licensed manufacturer, or licensed dealer who violates the provisions of subsection (1) commits a felony of the third degree punishable as provided in s. 775.082 or s. 775.083.

(c) Any employee or agency of a licensed importer, licensed manufacturer, or licensed dealer who violates the provisions of subsection (1) commits a felony of the third degree punishable as provided in s. 775.082 or s. 775.083.

(d) Any person who knowingly acquires a firearm through purchase or transfer intended for the use of a person who is prohibited by state or federal law from possessing or receiving a firearm commits a felony of the third degree, punishable as provided in s. 775.082 or s. 775.083.

(13) This section does not apply to employees of sheriff's offices, municipal police departments, correctional facilities or agencies, or other criminal justice or governmental agencies when the purchases or transfers are made on behalf of an employing agency for official law enforcement purposes.

(14) This section is repealed effective October 1, 2009.

790.0655. Purchase and delivery of handguns; mandatory waiting period; exceptions; penalties

(1)(a) There shall be a mandatory 3-day waiting period, which shall be 3 days, excluding weekends and legal holidays, between the purchase and the delivery at retail of any handgun. "Purchase" means the transfer of money or other valuable consideration to the retailer. "Handgun" means a firearm capable of being carried and used by one hand, such as a pistol or revolver. "Retailer" means and includes every person engaged in the business of making sales at retail or for distribution, or use, or consumption, or storage to be used or consumed in this state, as defined in s. 212.02(13).

(b) Records of handgun sales must be available for inspection by any law enforcement agency, as defined in s. 934.02, during normal business hours.

(2) The 3-day waiting period shall not apply in the following circumstances:

(a) When a handgun is being purchased by a holder of a concealed weapons permit as defined in s. 790.06.

(b) To a trade-in of another handgun.

(3) It is a felony of the third degree, punishable as provided in s. 775.082, s. 775.083, or s. 775.084:

(a) For any retailer, or any employee or agent of a retailer, to deliver a handgun before the expiration of the 3-day waiting period, subject to the exceptions provided in subsection (2).

(b) For a purchaser to obtain delivery of a handgun by fraud, false pretense, or false representation.

790.115. Possessing or discharging weapons or firearms at a school-sponsored event or on school property prohibited; penalties; exceptions

(1) A person who exhibits any sword, sword cane, firearm, electric weapon or device, destructive device, or other weapon as defined in s. 790.001(13), including a razor blade, box cutter, or common pocket knife, except as authorized in support of school-sanctioned activities, in the presence of one or more persons in a rude, careless, angry, or threatening manner and not in lawful self-defense, at a school-sponsored event or on the grounds or facilities of any school, school bus, or school bus stop, or within 1,000 feet of the real property that comprises a public or private elementary school, middle school, or secondary school, during school hours or during the time of a sanctioned school activity, commits a felony of the third degree, punishable as provided in s. 775.082, s. 775.083, or s. 775.084. This subsection does not apply to the exhibition of a firearm or weapon on private real property within 1,000 feet of a school by the owner of such property or by a person whose presence on such property has been authorized, licensed, or invited by the owner.

(2)(a) A person shall not possess any firearm, electric weapon or device, destructive device, or other weapon as defined in s. 790.001(13), including a razor blade or box cutter, except as authorized in support of school-sanctioned activities, at a school-sponsored event or on the property of any school, school bus, or school bus stop; however, a person may carry a firearm:

1. In a case to a firearms program, class or function which has been approved in advance by the principal or chief administrative officer of the school as a program or class to which firearms could be carried;

2. In a case to a career center having a firearms training range; or

3. In a vehicle pursuant to s. 790.25(5); except that school districts may adopt written and published policies that waive the exception in this subparagraph for purposes of student and campus parking privileges.

For the purposes of this section, "school" means any preschool, elementary school, middle school, junior high school, secondary school, career center, or postsecondary school, whether public or nonpublic.

(b) A person who willfully and knowingly possesses any electric weapon or device, destructive device, or other weapon, as defined in s. 790.001(13), including a razor blade or box cutter, except as authorized in support of school-sanctioned activities, in violation of this subsection commits a felony of the third degree, punishable as provided in s. 775.082, s. 775.083, or s. 775.084.

(c)(1) A person who willfully and knowingly possesses any firearm in violation of this subsection commits a felony of the third degree, punishable as provided in s. 775.082, s. 775.083, or s. 775.084.

(2) A person who stores or leaves a loaded firearm within the reach or easy access of a minor who obtains the firearm and commits a violation of subparagraph 1. commits a misdemeanor of the second degree, punishable as provided in s. 775.082 or s. 775.083; except that this does not apply if the firearm was stored or left in a securely locked box or container or in a location which a reasonable person would have believed to be secure, or was securely locked with a firearm-mounted push-button combination lock or a trigger lock; if the minor obtains the firearm as a result of an unlawful entry by any person; or to members of the Armed Forces, National Guard, or State Militia, or to police or other law enforcement officers, with respect to firearm possession by a minor which occurs during or incidental to the performance of their official duties.

(d) A person who discharges any weapon or firearm while in violation of paragraph (a), unless discharged for lawful defense of himself or herself or another or for a lawful purpose, commits a felony of the second degree, punishable as provided in s. 775.082, s. 775.083, or s. 775.084.

(e) The penalties of this subsection shall not apply to persons licensed under s. 790.06. Persons licensed under s. 790.06 shall be punished as provided in s. 790.06(12), except that a licenseholder who unlawfully discharges a weapon or firearm on school property as prohibited by this subsection commits a felony of the second degree, punishable as provided in s. 775.082, s. 775.083, or s. 775.084.

(3) This section does not apply to any law enforcement officer as defined in s. 943.10(1), (2), (3), (4), (6), (7), (8), (9), or (14).

(4) Notwithstanding s. 985.24, s. 985.245, or s. 985.25(1), any minor under 18 years of age who is charged under this section with possessing or discharging a firearm on school property shall be detained in secure detention, unless the state attorney authorizes the release of the minor, and shall be given a probable cause hearing within 24 hours after being taken into custody. At the hearing, the court may order that the minor continue to be held in secure detention for a period of 21 days, during which time the minor shall receive medical, psychiatric, psychological, or substance abuse examinations pursuant to s. 985.18, and a written report shall be completed.

790.151. Using firearm while under the influence of alcoholic beverages, chemical substances, or controlled substances; penalties

(1) As used in ss. 790.151-790.157, to "use a firearm" means to discharge a firearm or to have a firearm readily accessible for immediate discharge.

(2) For the purposes of this section, "readily accessible for immediate discharge" means loaded and in a person's hand.

(3) It is unlawful and punishable as provided in subsection (4) for any person who is under the influence of alcoholic beverages, any chemical substance set forth in s. 877.111, or any substance controlled under chapter 893, when affected to the extent that his or her normal faculties are impaired, to use a firearm in this state.

(4) Any person who violates subsection (3) commits a misdemeanor of the second degree, punishable as provided in s. 775.082 or s. 775.083.

(5) This section does not apply to persons exercising lawful self-defense or defense of one's property.

790.161. Making, possessing, throwing, projecting, placing, or discharging any destructive device or attempt so to do, felony; penalties A person who willfully and unlawfully makes, possesses, throws, projects, places, discharges, or attempts to make, possess, throw, project, place, or discharge any destructive device:

(1) Commits a felony of the third degree, punishable as provided in s. 775.082 or s. 775.084.

(2) If the act is perpetrated with the intent to do bodily harm to any person, or with the intent to do property damage, or if the act results in a disruption of governmental operations, commerce, or the private affairs of another person, commits a felony of the second degree, punishable as provided in s. 775.082 or s. 775.084.

(3) If the act results in bodily harm to another person or in property damage, commits a felony of the first degree, punishable as provided in s. 775.082 or s. 775.084.

(4) If the act results in the death of another person, commits a capital felony, punishable as provided in s. 775.082. In the event the death penalty in a capital felony is held to be unconstitutional by the Florida Supreme Court or the United States Supreme Court, the court having jurisdiction over a person previously sentenced to death for a capital felony shall cause such person to be brought before the court, and the court shall sentence such person to life imprisonment if convicted of murder in the first degree or of a capital felony under this subsection, and such person shall be ineligible for parole. No sentence of death shall be reduced as a result of a determination that a method of execution is held to be unconstitutional under the State Constitution or the Constitution of the United States.

790.1612. Authorization for governmental manufacture, possession, and use of destructive devices The governing body of any municipality or county and the Division of State Fire Marshal of the Department of Financial Services have the power to authorize the manufacture, possession, and use of destructive devices as defined in s. 790.001(4).

790.17. Furnishing weapons to minors under 18 years of age or persons of unsound mind and furnishing firearms to minors under 18 years of age prohibited

(1) A person who sells, hires, barters, lends, transfers, or gives any minor under 18 years of age any dirk, electric weapon or device, or other weapon, other than an ordinary pocketknife, without permission of the minor's parent or guardian, or sells, hires, barters, lends, transfers, or gives to any person of unsound mind an electric weapon or device or any dangerous weapon, other than an ordinary pocketknife, commits a misdemeanor of the first degree, punishable as provided in s. 775.082 or s. 775.083.

(2)(a) A person may not knowingly or willfully sell or transfer a firearm to a minor under 18 years of age, except that a person may transfer ownership of a firearm to a minor with permission of the parent or guardian. A person who violates this paragraph commits a felony of the third degree, punishable as provided in s. 775.082, s. 775.083, or s. 775.084.

(b) The parent or guardian must maintain possession of the firearm except pursuant to s. 790.22.

790.174. Safe storage of firearms required

(1) A person who stores or leaves, on a premise under his or her control, a loaded firearm, as defined in s. 790.001, and who knows or reasonably should know that a minor is likely to gain access to the firearm without the lawful permission of the minor's parent or the person having charge of the minor, or without the supervision required by law, shall keep the firearm in a securely locked box or container or in a location which a reasonable person would believe to be secure or shall secure it with a trigger lock, except when the person is carrying the firearm on his or her body or within such close proximity thereto that he or she can retrieve and use it as easily and quickly as if he or she carried it on his or her body.

(2) It is a misdemeanor of the second degree, punishable as provided in s. 775.082 or s. 775.083, if a person violates subsection (1) by failing to store or leave a firearm in the required manner and as a result thereof a minor gains access to the firearm, without the lawful permission of the minor's parent or the person having charge of the minor, and possesses or exhibits it, without the supervision required by law:

(a) In a public place; or

(b) In a rude, careless, angry, or threatening manner in violation of s. 790.10.

This subsection does not apply if the minor obtains the firearm as a result of an unlawful entry by any person.

(3) As used in this act, the term "minor" means any person under the age of 16.

790.175. Transfer or sale of firearms; required warnings; penalties

(1) Upon the retail commercial sale or retail transfer of any firearm, the seller or transferor shall deliver a written warning to the purchaser or transferee, which warning states, in block letters not less than 1/4 inch in height:

"IT IS UNLAWFUL, AND PUNISHABLE BY IMPRISONMENT AND FINE, FOR ANY ADULT TO STORE OR LEAVE A FIREARM IN ANY PLACE WITHIN THE REACH OR EASY ACCESS OF A MINOR UNDER 18 YEARS OF AGE OR TO KNOWINGLY SELL OR OTHERWISE TRANSFER OWNERSHIP OR POSSESSION OF A FIREARM TO A MINOR OR A PERSON OF UNSOUND MIND."

(2) Any retail or wholesale store, shop, or sales outlet which sells firearms must conspicuously post at each purchase counter the following warning in block letters not less than 1 inch in height:

"IT IS UNLAWFUL TO STORE OR LEAVE A FIREARM IN ANY PLACE WITHIN THE REACH OR EASY ACCESS OF A MINOR UNDER 18 YEARS OF AGE OR TO KNOWINGLY SELL OR OTHERWISE TRANSFER OWNERSHIP OR POSSESSION OF A FIREARM TO A MINOR OR A PERSON OF UNSOUND MIND."

(3) Any person or business knowingly violating a requirement to provide warning under this section commits a misdemeanor of the second degree, punishable as provided in s. 775.082 or s. 775.083.

790.18. Sale or transfer of arms to minors by dealers It is unlawful for any dealer in arms to sell or transfer to a minor any firearm, pistol, Springfield rifle or other repeating rifle, bowie knife or dirk knife, brass knuckles, slungshot or electric weapon or device. A person who violates this section commits a felony of the second degree, punishable as provided in s. 775.082, s. 775.083, or s. 775.084.

790.22. Use of BB guns, air or gas-operated guns, or electric weapons or devices by minor under 16; limitation; possession of firearms by minor under 18 prohibited; penalties

(2) Any adult responsible for the welfare of any child under the age of 16 years who knowingly permits such child to use or have in his or her possession any BB gun, air or gas-operated gun, electric weapon or device, or firearm in violation of the provisions of subsection (1) of this section commits a misdemeanor of the second degree, punishable as provided in s. 775.082 or s. 775.083.

(3) A minor under 18 years of age may not possess a firearm, other than an unloaded firearm at his or her home, unless:

(a) The minor is engaged in a lawful hunting activity and is:

1. At least 16 years of age; or

2. Under 16 years of age and supervised by an adult.

(b) The minor is engaged in a lawful marksmanship competition or practice or other lawful recreational shooting activity and is:

(1) At least 16 years of age; or

(2) Under 16 years of age and supervised by an adult who is acting with the consent of the minor's parent or guardian.

(c) The firearm is unloaded and is being transported by the minor directly to or from an event authorized in paragraph (a) or paragraph (b).

(4)(a) Any parent or guardian of a minor, or other adult responsible for the welfare of a minor, who knowingly and willfully permits the minor to possess a firearm in violation of subsection (3) commits a felony of the third degree, punishable as provided in s. 775.082, s. 775.083, or s. 775.084.

(5)(a) A minor who violates subsection (3) commits a misdemeanor of the first degree; for a first offense, may serve a period of detention of up to 3 days in a secure detention facility; and, in addition to any other penalty provided by law, shall be required to perform 100 hours of community service; and:

(6) Any firearm that is possessed or used by a minor in violation of this section shall be promptly seized by a law enforcement officer and disposed of in accordance with s. 790.08(1)-(6).

(7) The provisions of this section are supplemental to all other provisions of law relating to the possession, use, or exhibition of a firearm.

790.221. Possession of short-barreled rifle, short-barreled shotgun, or machine gun; penalty

(1) It is unlawful for any person to own or to have in his or her care, custody, possession, or control any short-barreled rifle, short-barreled shotgun, or machine gun which is, or may readily be made, operable; but this section shall not apply to antique firearms.

(2) A person who violates this section commits a felony of the second degree, punishable as provided in s. 775.082, s. 775.083, or s. 775.084.

(3) Firearms in violation hereof which are lawfully owned and possessed under provisions of federal law are excepted.

790.23. Felons and delinquents; possession of firearms, ammunition, or electric weapons or devices unlawful

(1) It is unlawful for any person to own or to have in his or her care, custody, possession, or control any firearm, ammunition, or electric weapon or device, or to carry a concealed weapon, including a tear gas gun or chemical weapon or device, if that person has been:

(a) Convicted of a felony in the courts of this state;

(b) Found, in the courts of this state, to have committed a delinquent act that would be a felony if committed by an adult and such person is under 24 years of age;

(c) Convicted of or found to have committed a crime against the United States which is designated as a felony;

(d) Found to have committed a delinquent act in another state, territory, or country that would be a felony if committed by an adult and which was punishable by imprisonment for a term exceeding 1 year and such person is under 24 years of age; or

(e) Found guilty of an offense that is a felony in another state, territory, or country and which was punishable by imprisonment for a term exceeding 1 year.

(2) This section shall not apply to a person convicted of a felony whose civil rights and firearm authority have been restored.

(3) Except as otherwise provided in subsection(4), any person who violates this section com-mits a felony of the second degree, punishable as provided in s. 775.082, s. 775.083, or s. 775.084.

(4) Notwithstanding the provisions of S. 874.04, if the offense described in subsection (1) has been committed by a person who has previously qualified or currently qualifies for the penalty enhancements provided for in S. 874.04, the offense is a felony of the first degree, punishable by a term of years not exceeding life or as provided in S. 775.082, S. 775.083, or S. 775.084.

790.233. Possession of firearm or ammunition prohibited when person is subject to an injunction against committing acts of domestic violence; penalties

(1) A person may not have in his or her care, custody, possession, or control any firearm or ammunition if the person has been issued a final injunction that is currently in force and effect, restraining that person from committing acts of domestic violence, and that has been issued under s. 741.30.

(2) A person who violates subsection (1) commits a misdemeanor of the first degree, punishable as provided in s. 775.082 or s. 775.083.

(3) It is the intent of the Legislature that the disabilities regarding possession of firearms and ammunition are consistent with federal law. Accordingly, this section shall not apply to a state or local officer as defined in s. 943.10(14), holding an active certification, who receives or possesses a firearm or ammunition for use in performing official duties on behalf of the officer's employing agency, unless otherwise prohibited by the employing agency.

790.235. Possession of firearm or ammunition by violent career criminal unlawful; penalty

(1) Any person who meets the violent career criminal criteria under s. 775.084(1)(d), regardless of whether such person is or has previously been sentenced as a violent career criminal, who owns or has in his or her care, custody, possession, or control any firearm, ammunition, or electric weapon or device, or carries a concealed weapon, including a tear gas gun or chemical weapon or device, commits a felony of the first degree, punishable as provided in s. 775.082, s. 775.083, or s. 775.084. A person convicted of a violation of this section shall be sentenced to a mandatory minimum of 15 years' imprisonment; however, if the person would be sentenced to a longer term of imprisonment under s. 775.084 (4)(d), the person must be sentenced under that provision. A person convicted of a violation of this section is not eligible for any form of discretionary early release, other than pardon, executive clemency, or conditional medical release under s. 947.149.

(2) For purposes of this section, the previous felony convictions necessary to meet the violent career criminal criteria under s. 775.084(1)(d) may be convictions for felonies committed as an adult or adjudications of delinquency for felonies committed as a juvenile. In order to be counted as a prior felony for purposes of this section, the felony must have resulted in a conviction sentenced separately, or an adjudication of delinquency entered separately, prior to the current offense, and sentenced or adjudicated separately from any other felony that is to be counted as a prior felony.

(3) This section shall not apply to a person whose civil rights and firearm authority have been restored.

790.25. Lawful ownership, possession, and use of firearms and other weapons

(1) Declaration of Policy. The Legislature finds as a matter of public policy and fact that it is necessary to promote firearms safety and to curb and prevent the use of firearms and other weapons in crime and by incompetent persons without prohibiting the lawful use in defense of life, home, and property, and the use by United States or state military organizations, and as otherwise now authorized by law, including the right to use and own firearms for target practice and marksmanship on target practice ranges or other lawful places, and lawful hunting and other lawful purposes.

(2) Uses Not Authorized.

(a) This section does not authorize carrying a concealed weapon without a permit, as prohibited by ss. 790.01 and 790.02.

(b) The protections of this section do not apply to the following:

1. A person who has been adjudged mentally incompetent, who is addicted to the use of narcotics or any similar drug, or who is a habitual or chronic alcoholic, or a person using weapons or firearms in violation of ss. 790.07-790.115, 790.145-790.19, 790.22-790.24;

2. Vagrants and other undesirable persons as defined in s. 856.02 **[Publisher's Note:** s. 856.02 was repealed by s. 3, ch. 72-133**]**;

3. A person in or about a place of nuisance as defined in s. 823.05, unless such person is there for law enforcement or some other lawful purpose.

(3) Lawful Uses. The provisions of ss. 790.053 and 790.06 do not apply in the following instances, and, despite such sections, it is lawful for the following persons to own, possess, and lawfully use firearms and other weapons, ammunition, and supplies for lawful purposes:

(a) Members of the Militia, National Guard, Florida State Defense Force, Army, Navy, Air Force, Marine Corps, Coast Guard, organized reserves, and other armed forces of the state and of the United States, when on duty, when training or preparing themselves for military duty, or while subject to recall or mobilization;

(b) Citizens of this state subject to duty in the Armed Forces under s. 2, Art. X of the State Constitution, under chapters 250 and 251, and under federal laws, when on duty or when training or preparing themselves for military duty;

(c) Persons carrying out or training for emergency management duties under chapter 252;

(d) Sheriffs, marshals, prison or jail wardens, police officers, Florida highway patrol officers, game wardens, revenue officers, forest officials, special officers appointed under the provisions of chapter 354, and other peace and law enforcement officers and their deputies and assistants and full-time paid peace officers of other states and of the Federal Government who are carrying out official duties while in this state;

(e) Officers or employees of the state or United States duly authorized to carry a concealed weapon;

(f) Guards or messengers of common carriers, express companies, armored car carriers, mail carriers, banks, and other financial institutions, while actually employed in and about the shipment, transportation, or delivery of any money, treasure, bullion, bonds, or other thing of value within this state;

(g) Regularly enrolled members of any organization duly authorized to purchase or receive weapons from the United States or from this state, or regularly enrolled members of clubs organized for target, skeet, or trap shooting, while at or going to or from shooting practice; or regularly enrolled members of clubs organized for modern or antique firearms collecting, while such members are at or going to or from their collectors' gun shows, conventions, or exhibits;

(h) A person engaged in fishing, camping, or lawful hunting or going to or returning from a fishing, camping, or lawful hunting expedition;

(i) A person engaged in the business of manufacturing, repairing, or dealing in firearms, or the agent or representative of any such person while engaged in the lawful course of such business;

(j) A person firing weapons for testing or target practice under safe conditions and in a safe place not prohibited by law or going to or from such place;

(k) A person firing weapons in a safe and secure indoor range for testing and target practice;

(l) A person traveling by private conveyance when the weapon is securely encased or in a public conveyance when the weapon is securely encased and not in the person's manual possession;

(m) A person while carrying a pistol unloaded and in a secure wrapper, concealed or otherwise, from the place of purchase to his or her home or place of business or to a place of repair or back to his or her home or place of business;

(n) A person possessing arms at his or her home or place of business;

(o) Investigators employed by the several public defenders of the state, while actually carrying out official duties, provided such investigators:

1. Are employed full time;

2. Meet the official training standards for firearms established by the Criminal Justice Standards and Training Commission as provided in s. 943.12(5) and the requirements of ss. 493.6108(1)(a) and 943.13(1)-(4); and

3. Are individually designated by an affidavit of consent signed by the employing public defender and filed with the clerk of the circuit court in the county in which the employing public defender resides.

(p) Investigators employed by the capital collateral regional counsel, while actually carrying out official duties, provided such investigators:

1. Are employed full time;

2. Meet the official training standards for firearms as established by the Criminal Justice Standards and Training Commission as provided in s. 943.12(1) and the requirements of ss. 493.6108(1)(a) and 943.13(1)-(4); and

3. Are individually designated by an affidavit of consent signed by the capital collateral regional counsel and filed with the clerk of the circuit court in the county in which the investigator is headquartered.

(5) Possession in Private Conveyance Notwithstanding subsection (2), it is lawful and is not a violation of s. 790.01 for a person 18 years of age or older to possess a concealed firearm or other weapon for self-defense or other lawful purpose within the interior of a private conveyance, without a license, if the firearm or other weapon is securely encased or is otherwise not readily accessible for immediate use. Nothing herein contained prohibits the carrying of a legal firearm other than a handgun anywhere in a private conveyance when such firearm is being carried for a lawful use. Nothing herein contained shall be construed to authorize the carrying of a concealed firearm or other weapon on the person. This subsection shall be liberally construed in favor of the lawful use, ownership, and possession of firearms and other weapons, including lawful self-defense as provided in s. 776.012.

790.27. Alteration or removal of firearm serial number or possession, sale, or delivery of firearm with serial number altered or removed prohibited; penalties

(1)(a) It is unlawful for any person to knowingly alter or remove the manufacturer's or importer's serial number from a firearm with intent to disguise the true identity thereof.

(b) Any person violating paragraph (a) is guilty of a felony of the third degree, punishable as provided in s. 775.082, s. 775.083, or s. 775.084.

(2)(a) It is unlawful for any person to knowingly sell, deliver, or possess any firearm on which the manufacturer's or importer's serial number has been unlawfully altered or removed.

(b) Any person violating paragraph (a) is guilty of a misdemeanor of the first degree, punishable as provided in s. 775.082 or s. 775.083.

(3) This section shall not apply to antique firearms.

790.28. Purchase of rifles and shotguns in contiguous states A resident of this state may purchase a rifle or shotgun in any state contiguous to this state if he or she conforms to applicable laws and regulations of the United States, of the state where the purchase is made, and of this state.

790.31. Armor-piercing or exploding ammunition or dragon's breath shotgun shells, bolo shells, or flechette shells prohibited

(1) As used in this section, the term:

(a) "Armor-piercing bullet" means any bullet which has a steel inner core or core of equivalent hardness and a truncated cone and which is designed for use in a handgun as an armor-piercing or metal-piercing bullet.

(b) "Exploding bullet" means any bullet that can be fired from any firearm, if such bullet is designed or altered so as to detonate or forcibly break up through the use of an explosive or deflagrant contained wholly or partially within or attached to such bullet. The term does not include any bullet designed to expand or break up through the mechanical forces of impact alone or any signaling device or pest control device not designed to impact on any target.

(c) "Handgun" means a firearm capable of being carried and used by one hand, such as a pistol or revolver.

(d) "Dragon's breath shotgun shell" means any shotgun shell that contains exothermic pyrophoric misch metal as the projectile and that is designed for the sole purpose of throwing or spewing a flame or fireball to simulate a flamethrower.

(e) "Bolo shell" means any shell that can be fired in a firearm and that expels as projectiles two or more metal balls connected by solid metal wire.

(f) "Flechette shell" means any shell that can be fired in a firearm and that expels two or more pieces of fin-stabilized solid metal wire or two or more solid dart-type projectiles.

(2)(b) Any person who possesses an armor-piercing bullet or exploding bullet with knowledge of its armor-piercing or exploding capabilities loaded in a handgun, or who possesses a dragon's breath shotgun shell, bolo shell, or flechette shell with knowledge of its capabilities loaded in a firearm, is guilty of a felony of the third degree, punishable as provided in s. 775.082, s. 775.083, or s. 775.084.

(3) This section does not apply to:

(a) The possession of any item described in subsection (1) by any law enforcement officer, when possessed in connection with the performance of his or her duty as a law enforcement officer, or law enforcement agency.

(b) The manufacture of items described in subsection (1) exclusively for sale or delivery to law enforcement agencies.

(c) The sale or delivery of items described in subsection (1) to law enforcement agencies.

(2) Limited Exception; County Waiting-Period Ordinances

(a) Any county may have the option to adopt a waiting-period ordinance requiring a waiting period of up to, but not to exceed, 3 working days between the purchase and delivery of a handgun. For purposes of this subsection, "purchase" means payment of deposit, payment in full, or notification of intent to purchase. Adoption of a waiting-period ordinance, by any county, shall require a majority vote of the county commission on votes on waiting-period ordinances. This exception is limited solely to individual counties and is limited to the provisions and restrictions contained in this subsection.

(b) Ordinances authorized by this subsection shall apply to all sales of handguns to individuals by a retail establishment except those sales to individuals exempted in this subsection. For purposes of this subsection, "retail establishment" means a gun shop, sporting goods store, pawn shop, hardware store, department store, discount store, bait or tackle shop, or any other store or shop that offers handguns for walk-in retail sale but does not include gun collectors shows or exhibits, or gun shows.

(c) Ordinances authorized by this subsection shall not require any reporting or notification to any source outside the retail establishment, but records of handgun sales must be available for inspection, during normal business hours, by any law enforcement agency as defined in s. 934.02.

(d) The following shall be exempt from any waiting period:

(1) Individuals who are licensed to carry concealed firearms under the provisions of s. 790.06 or who are licensed to carry concealed firearms under any other provision of state law and who show a valid license;

(2) Individuals who already lawfully own another firearm and who show a sales receipt for another firearm; who are known to own another firearm through a prior purchase from the retail establishment; or who have another firearm for trade-in;

(3) A law enforcement or correctional officer as defined in s. 943.10;

(4) A law enforcement agency as defined in s. 934.02;

(5) Sales or transactions between dealers or between distributors or between dealers and distributors who have current federal firearms licenses; or

(6) Any individual who has been threatened or whose family has been threatened with death or bodily injury, provided the individual may lawfully possess a firearm and provided such threat has been duly reported to local law enforcement.

(3) Policy and Intent.

(a) It is the intent of this section to provide uniform firearms laws in the state; to declare all ordinances and regulations null and void which have been enacted by any jurisdictions other than state and federal, which regulate firearms, ammunition, or components thereof; to prohibit the enactment of any future ordinances or regulations relating to firearms, ammunition, or components thereof unless specifically authorized by this section or general law; and to require local jurisdictions to enforce state firearms laws.

(b) As created by chapter 87-23, Laws of Florida, this section shall be known and may be cited as the "Joe Carlucci Uniform Firearms Act."

790.33. Field of regulation of firearms and ammunition preempted

(1) PREEMPTION. –Except as expressly provided by general law, the Legislature hereby declares that it is occupying the whole field of regulation of firearms and ammunition, including the purchase, sale, transfer, taxation, manufacture, ownership, possession, and transportation thereof, to the exclusion of all existing and future county, city, town, or municipal ordinances or regulations relating thereto. Any such existing ordinances are hereby declared null and void. This subsection shall not affect zoning ordinances which encompass firearms businesses along with other businesses. Zoning ordinances which are designed for the purpose of restricting or prohibiting the sale, purchase, transfer, or manufacture of firearms or ammunition as a method of regulating firearms or ammunition are in conflict with this subsection and are prohibited.

(2) LIMITED EXCEPTION; COUNTY WAITING-PERIOD ORDINANCES.

(a) Any county may have the option to adopt a waiting-period ordinance requiring a waiting period of up to, but not to exceed, 3 working days between the purchase and delivery of a handgun. For purposes of this subsection, "purchase" means payment of deposit, payment in full, or notification of intent to purchase. Adoption of a waiting-period ordinance, by any county, shall require a majority vote of the county commission on votes on waiting-period ordinances. This exception is limited solely to individual counties and is limited to the provisions and restrictions contained in this subsection.

(b) Ordinances authorized by this subsection shall apply to all sales of handguns to individuals by a retail establishment except those sales to individuals exempted in this subsection. For purposes of this subsection, "retail establishment" means a gun shop, sporting goods store, pawn shop, hardware store, department store, discount store, bait or tackle shop, or any other store or shop that offers handguns for walk-in retail sale but does not include gun collectors shows or exhibits, or gun shows.

(c) Ordinances authorized by this subsection shall not require any reporting or notification to any source outside the retail establishment, but records of handgun sales must be available for inspection, during normal business hours, by any law enforcement agency as defined in s. 934.02.

(d) The following shall be exempt from any waiting period:

1. Individuals who are licensed to carry concealed firearms under the provisions of s. 790.06 or who are licensed to carry concealed firearms under any other provision of state law and who show a valid license;

2. Individuals who already lawfully own another firearm and who show a sales receipt for another firearm; who are known to own another firearm through a prior purchase from the retail establishment; or who have another firearm for trade-in;

3. A law enforcement or correctional officer as defined in s. 943.10;

4. A law enforcement agency as defined in s. 934.02;

5. Sales or transactions between dealers or between distributors or between dealers and distributors who have current federal firearms licenses; or

6. Any individual who has been threatened or whose family has been threatened with death or bodily injury, provided the individual may lawfully possess a firearm and provided such threat has been duly reported to local law enforcement.

(3) POLICY AND INTENT.

(a) It is the intent of this section to provide uniform firearms laws in the state; to declare all ordinances and regulations null and void which have been enacted by any jurisdictions other than state and federal, which regulate firearms, ammunition, or components thereof; to prohibit the enactment of any future ordinances or regulations relating to firearms, ammunition, or components thereof unless specifically authorized by this section or general law; and to require local jurisdictions to enforce state firearms laws.

(b) As created by chapter 87-23, Laws of Florida, this section shall be known and may be cited as the "Joe Carlucci Uniform Firearms Act."

790.335. Prohibition of registration of firearms

(1) Legislative Findings and Intent.

(2) Prohibitions. No state governmental agency or local government, special district, or other political subdivision or official, agent, or employee of such state or other governmental entity or any other person, public or private, shall knowingly and willfully keep or cause to be kept any list, record, or registry of privately owned firearms or any list, record, or registry of the owners of those firearms.

(3) Exceptions. The provisions of this sec-tion shall not apply to:

(a) Records of firearms that have been used in committing any crime.

(b) Records relating to any person who has been convicted of a crime.

(c) Records of firearms that have been reported stolen that are retained for a period not in excess of 10 days after such firearms are recovered. Official documentation recording the theft of a recovered weapon may be maintained no longer than the balance of the year entered, plus 2 years.

(d) Firearm records that must be retained by firearm dealers under federal law, including copies of such records transmitted to law enforcement agencies. However, no state governmental agency or local government, special district, or other political subdivision or official, agent, or employee of such state or other governmental entity or any other person, private or public, shall accumulate, compile, computerize, or otherwise collect or convert such written records into any form of list, registry, or database for any purpose.

(e)(1) Records kept pursuant to the recordkeeping provisions of s. 790.065; however, nothing in this section shall be construed to authorize the public release or inspection of records that are made confidential and exempt from the provisions of s. 119.07(1) by s. 790.065(4)(a).

(2) Nothing in this paragraph shall be construed to allow the maintaining of records containing the names of purchasers or transferees who receive unique approval numbers or the maintaining of records of firearm transactions.

(f) Firearm records, including paper pawn transaction forms and contracts on firearm transactions, required by chapters 538 and 539.

(1) Electronic firearm records held pursuant to chapter 538 may only be kept by a secondhand dealer for 30 days after the date of the purchase of the firearm by the secondhand dealer.

(2) Electronic firearm records held pursuant to chapter 539 may only be kept by a pawnbroker for 30 days after the expiration of the loan that is secured by a firearm or 30 days after the date of purchase of a firearm, whichever is applicable.

(3) Except as required by federal law, any firearm records kept pursuant to chapter 538 or chapter 539 shall not, at any time, be electronically transferred to any public or private entity, agency, business, or enterprise, nor shall any such records be copied or transferred for purposes of accumulation of such records into lists, registries, or databases.

(4) Notwithstanding subparagraph 3., secondhand dealers and pawnbrokers may electronically submit firearm transaction records to the appropriate law enforcement agencies as required by chapters 538 and 539; however, the law enforcement agencies may not electronically submit such records to any other person or entity and must destroy such records within 60 days after receipt of such records.

(5) Notwithstanding subparagraph 3., secondhand dealers and pawnbrokers may electronically submit limited firearms records consisting solely of the manufacturer, model, serial number, and caliber of pawned or purchased firearms to a third-party private provider that is exclusively incorporated, exclusively owned, and exclusively operated in the United States and that restricts access to such information to only appropriate law enforcement agencies for legitimate law enforcement purposes. Such records must be destroyed within 30 days by the third-party provider. As a condition of receipt of such records, the third-party provider must agree in writing to comply with the requirements of this section. Any pawnbroker or secondhand dealer who contracts with a third-party provider other than as provided in this act or electronically transmits any records of firearms transactions to any third-party provider other than the records specifically allowed by this paragraph commits a felony of the second degree, punishable as provided in s. 775.082 or s. 775.083.

(g) Records kept by the Department of Law Enforcement of NCIC transactions to the extent required by federal law and a log of dates of requests for criminal history record checks, unique approval and nonapproval numbers, license identification numbers, and transaction numbers corresponding to such dates.

(h) Records of an insurer that, as a condition to providing insurance against theft or loss of a firearm, identify such firearm. Such records may not be sold, commingled with records relating to other firearms, or transferred to any other person or entity. The insurer may not keep a record of such firearm more than 60 days after the policy of insurance expires or after notification by the insured that the insured is no longer the owner of such firearm.

(i) Lists of customers of a firearm dealer retained by such dealer, provided that such lists do not disclose the particular firearms purchased. Such lists, or any parts thereof, may not be sold, commingled with records relating to other firearms, or transferred to any other person or entity.

(j) Sales receipts retained by the seller of firearms or by a person providing credit for such purchase, provided that such receipts shall not serve as or be used for the creation of a database for registration of firearms.

(k) Personal records of firearms maintained by the owner of such firearms.

(l) Records maintained by a business that stores or acts as the selling agent of firearms on behalf of the lawful owner of the firearms.

(m) Membership lists of organizations comprised of firearm owners.

(n) Records maintained by an employer or contracting entity of the firearms owned by its officers, employees, or agents, if such firearms are used in the course of business performed on behalf of the employer.

(o) Records maintained pursuant to s. 790.06 by the Department of Agriculture and Consumer Services of a person who was a licensee within the prior 2 years.

(p) Records of firearms involved in criminal investigations, criminal prosecutions, criminal appeals, and postconviction motions, civil proceedings relating to the surrender or seizure of firearms including protective injunctions, Baker Act commitments, and sheriff's levies pursuant to court judgments, and voluntary surrender by the owner or custodian of the firearm.

(q) Paper documents relating to firearms involved in criminal cases, criminal investigations, and criminal prosecutions, civil proceedings relating to the surrender or seizure of firearms including protective injunctions, Baker Act commitments, and sheriff's levies pursuant to court judgments, and voluntary surrender by the owner or custodian of the firearm.

(r) Noncriminal records relating to the receipt, storage or return of firearms, including, but not limited to, records relating to firearms impounded for storage or safekeeping, receipts proving that a firearm was returned to the rightful owner and supporting records of identification and proof of ownership, or records relating to firearms impounded pursuant to levies or court orders, provided, however, that such records shall not be compiled, sorted, or otherwise arranged into any lists, indexes, or registries of firearms or firearms owners.

(4) Penalties.

(a) Any person who violates a provision of this section commits a felony of the third degree, punishable as provided in s. 775.082 or s. 775.083.

(b) Except as required by the provisions of s. 16, Art. I of the State Constitution or the Sixth Amendment to the United States Constitution, no public funds shall be used to defend the unlawful conduct of any person charged with a violation of this section, unless the charges against such person are dismissed or such person is determined to be not guilty at trial. Notwithstanding this paragraph, public funds may be expended to provide the services of the office of public defender or court-appointed conflict counsel as provided by law.

(c) The governmental entity, or the designee of such governmental entity, in whose service or employ a list, record, or registry was compiled in violation of this section may be assessed a fine of not more than $5 million, if the court determines that the evidence shows that the list, record, or registry was compiled or maintained with the knowledge or complicity of the management of the governmental entity. The Attorney General may bring a civil cause of action to enforce the fines assessed under this paragraph.

(d) The state attorney in the appropriate jurisdiction shall investigate complaints of criminal violations of this section and, where evidence indicates a violation may have occurred, shall prosecute violators.

(5) Construction. This section shall be construed to effectuate its remedial and deterrent purposes. This section may not be construed to grant any substantive, procedural privacy right or civil claim to any criminal defendant, and a violation of this section may not be grounds for the suppression of evidence in any criminal case.

[Current through Chapters 2008-2 through 2008-63, 2008-71, 2008-72 ,2008-92, 2008-93, 2008-100, 2008-105, 2008-116, 2008-123, 2008-128, 2008-131, 2008-145, 2008-150, 2008-247]

Code of Broward County

Chapter 18. Law Enforcement

Article VI. Firearms, Five-Day Waiting Period

18-96. Waiting period; prohibition. There shall be a mandatory five-day waiting period, which shall be five days, excluding weekends and legal holidays, in connection with the sale of firearms occurring within Broward County when the sale is a transfer of money or other valuable consideration, and any part of the sale transacttion is conducted on property to which the public has the right of access. Some examples of properties to which the public has a right of access are: gun shows, firearm exhibits, wholesale and retail stores, and flea markets. No person shall transfer or receive a firearm to or from another person for five days from the hour of such sale, excluding weekends and legal holidays, when the sale is a transfer of money or other valuable consideration, and any part of the sale transacttion is conducted on property to which the public has the right of access. An uninterrupted, continuous, and cumulative aggregate of 120 hours must elapse between such sale and receipt of the firearm, excluding the hours of weekends and legal holidays. A person who violates the prohibition of this Section is guilty of a violation of a county ordinance, punishable as provided in § 125.69, F.S. and the violation shall be prosecuted in the same manner as misdemeanors are prosecuted.

18-97. Criminal history records check; requirement; prohibition.

(a) Legislative intent. It is the intent of the Board of County Commissioners, in the exercise of its constitutionally-granted authority, to ensure that no firearm is sold, offered for sale, or transferred where any part of the transaction is conducted on property to which the public has a right of access unless a national criminal history background check is conducted. As of the dates of the enactment and effectiveness of this Ordinance, Florida statutory law only allows such checks to be performed for and by licensed dealers, licensed importers, and licensed manufacturers. Therefore, a buyer or seller who is not otherwise a licensed dealer, licensed importer, or licensed manufacturer must use a licensed dealer, licensed importer, or licensed manufacturer as an intermediary to ensure such checks are conducted.

(b) Requirement; prohibition. There shall be a mandatory national criminal history records check done in connection with the sale of firearms occurring within Broward County. No person shall transfer or receive a firearm when any part of the sale transaction is conducted on property to which the public has the right of access until all procedures and requirements of § 790.065, F.S. have been complied with by a person statutorily authorized to conduct background information checks, and an approval number has been obtained and documented, as provided by § 790.065, F.S. In case of repeal or amendment of § 790.065, F.S., no person shall transfer or receive a firearm by sale when any part of the sale transaction is conducted on property to which the public has the right of access until all procedures, requirements, and prohibitions set forth in other federal or state laws relating to background checks have been complied with by persons selling or buying firearms. A person who violates the prohibition of this Section is guilty of a violation of a county ordinance, punishable as provided in § 125.69, F.S. and the violation shall be prosecuted in the same manner as misdemeanors are prosecuted.

18-98. Non-applicability to holders of Florida concealed weapon permits. Pursuant to the Florida Constitution, when purchasing a firearm, holders of a Florida concealed weapon permit or license, as prescribed by general law, shall not be subject to the prohibitions and penalties of this article.

18-99. Penalty. Violation of a prohibition of this article shall be punishable by a fine not to exceed $500.00 or by imprisonment in the county jail not to exceed 60 days or by both such fine and imprisonment. Each violation of a prohibition of this article relating to a specific firearm shall constitute a separate and individual violation. Nothing contained herein shall be construed to preempt the imposition of higher penalties imposed by state or federal law.

[Broward County Code current through Ordinance No. 2008-27, enacted June 24, 2008]

Code of Martin County

Chapter 75. Firearms

75.1. Short title. This chapter shall be known and may be cited as the "Martin County Cooling-Off Period Ordinance."

75.2. Definitions. As used in this chapter, unless the context otherwise indicates:

Purchase means the payment of a deposit, payment in full, or notification of intent to purchase.

Retail establishment means gun shop, sporting goods store, pawnshop, hardware store, department store, discount store, bait or tackle shop, or any other store that offers handguns for walk-in retail sale, but does not include gun collector shows or exhibits, or gun shows.

Working day means any day, excluding Saturday, Sunday and legal holidays as defined in F.S. ch. 683.

75.3. Cooling-off period between purchase and delivery of handguns. A waiting period of three working days, measured from the time of purchase of a handgun, must expire before delivery of the handgun is made to an individual by a retail establishment selling said handgun. This section shall apply to all sales of handguns to individuals by a retail establishment except as specifically exempted herein.

75.4. Exemptions. The provisions of this section [chapter] shall not apply to:

A. Individuals who are licensed to carry concealed firearms under the provisions of F.S. § 790.06, who are licensed to carry concealed firearms under any other provision of State law and who show a valid license.

B. Individuals who already lawfully own another firearm and who show a sales receipt for another firearm, who are known to own another firearm through a prior purchase from the retail establishment, or who have another firearm for trade-in.

C. Any law enforcement or correctional officer as defined in F.S. § 943.10.

D. Any law enforcement agency as defined in F.S. § 934.02.

E. Sales or transactions between dealers or between distributors or between dealers and distributors who have current federal firearms licenses.

F. Any individual who has been threatened or whose family has been threatened with death or bodily injury, provided the individual may lawfully possess a firearm and provided such threat has been duly reported to local law enforcement.

75.5. Inspection of records. Records of handgun sales by retail establishments shall be available for inspection, during normal business hours, by any law enforcement agency as defined in F.S. § 934.02.

[Code of Martin County current through Ord. No. 793, enacted March 18, 2008]

Code of Miami- Dade County

Chapter 21. Offenses and Miscellaneous Provisions

Article III. Weapons

Division 1. In General

21-16. Sale, loan, etc., weapons to intoxicated persons, etc. It shall be unlawful for any person to sell, loan or furnish any firearm as defined in Section 21-20.1(a) to any person whom he knows or has reasonable cause to believe is under the influence of alcohol or any narcotic, drug, stimulant, or depressant, or who is of unsound mind, or who is a member of any subversive organization. In addition to all other penalties, such unlawful sale, loan or furnishing shall be grounds for revocation of any license issued by the County to such person.

21-17. Possession of weapons by felons, intoxicated persons, etc. It shall be unlawful for any person who has been convicted of a felony, or who is under the influence of alcohol or a narcotic or drug to wear or have about his person or in any vehicle in which he is an occupant any firearm or other dangerous or deadly weapon.

21-19.1. Sale of Saturday night specials in Miami-Dade County prohibited. It shall be unlawful for any person to sell or otherwise transfer any Saturday night special as defined in Section 21-20.1(b) or offer or expose for sale or transfer any such Saturday night special.

(a) Notwithstanding the foregoing provisions of this section, the sale of Saturday night specials to law enforcement agencies or for authorized military use shall not be subject to the limitations herein set forth.

(b) This section shall not apply to wholesale dealers in their business intercourse with retail dealers nor to retail dealers in their business intercourse with other retail dealers nor to wholesale or retail dealers in the regular or ordinary transportation of unloaded firearms, merchandise by mail, express or other mode of shipment, to points outside the County, nor to sales or transfer of firearms that do not use a self-containing cartridge.

21-20. Registration of sales and transfers required; penalty.

(a) Register. Every person engaged in the business of selling, leasing, or otherwise transferring firearms as defined in Section 21-20.1(a), of the size capable of being concealed on the person, whether such seller, lessor or transferor is a retail dealer, a pawnbroker, or otherwise, shall keep a register in which shall be entered the date and time of sale, lease or transfer; the name of the salesman making the sale, lease or transfer; the place where sold, leased or transferred; the make, model, manufacturer's number, caliber or other marks of identification of such firearm; the purchaser's name, address, color, sex, age, weight, height, complexion, color of hair, birthplace, place of residence, length of residence, occupation, employer, the purpose for which the gun is desired, the date and time of delivery.

(b) Purchaser and seller to sign. The person to whom such firearm is sold, leased or otherwise transferred shall sign and the dealer shall require him to sign his name and affix his address to the register in duplicate and the salesman shall affix his signature in duplicate as a witness to the signature of the person to whom such firearm is sold, leased or transferred.

(c) Waiting period required for handgun. The person to whom such handgun is sold, leased or otherwise transferred shall wait, and the dealer shall require him to wait, a period of seventy-two (72) hours prior to such person acquiring possession of such handgun.

(d) Fictitious name prohibited. Any person signing a fictitious name or address in such register is guilty of a violation of this section.

(e) Exceptions. This section shall not apply to wholesale dealers in their business intercourse with retail dealers nor to retail dealers in their business intercourse with other retail dealers, nor to wholesale or retail dealers in the regular or ordinary transportation of unloaded firearms, merchandise by mail, express or other mode of shipment, to points outside the County, nor to sales or transfer of firearms that do not use a self-containing cartridge.

(f) Penalty. Every person who is convicted for a violation of Section 21-20(c) shall be punished upon a first conviction by imprisonment for not less than six (6) months or by a fine of not less than one thousand dollars ($1,000.00) or by both such fine and imprisonment, in the discretion of the court and on a second and subsequent conviction shall be punished by imprisonment for not less than one (1) year or by a fine of not less than one thousand dollars ($1,000.00) or by both such fine and imprisonment, in the discretion of the court.

Division 2. License to Sell Firearms

21-20.1. Definitions.

(a) The word "firearm" as used in this division shall be construed to mean any firearm, weapon, revolver, pistol, autoloading pistol, modified rifle or shotgun, or any similar mechanism by whatever name known, which is designed to expel a projectile through a gun barrel by the action of any explosive, having the size, length or dimensions which make it capable of being concealed upon the person, and originally designed or altered to be used by one (1) hand and having a barrel length of less than sixteen (16) inches, and an overall length of less than twenty-six and one-half (26 1/2) inches, but the word firearm shall not be construed to mean guns that do not use self-contained cartridges.

(b) The term "Saturday Night Special" shall mean any firearm which is 32 caliber or smaller, whose barrel is less than three (3) inches, except those whose frame is an investment cast or forged steel, or investment cast or forged high tensile alloy.

(c) The word "sale" includes transfer, assignment, pledge, lease, loan, barter, or gift.

21-20.2. License - Required to sell.

(a) It shall be unlawful for any person who, without being licensed as provided in this act [division], to engage in the business of selling or otherwise transferring any handgun or to advertise for sale, or offer or expose for sale or transfer any handgun defined in Section 21-20.1(a) or to engage in the business of repairing handguns. This section applies to persons in the firearms business or in the business of gunsmithing, and does not apply to:

(1) Sales or trades by an unlicensed person to a person licensed hereunder, nor to

(2) Isolated sales, transfers or trades between unlicensed persons who are not engaged in the firearms business or in the business of gunsmithing.

(b) The provisions of this section shall not apply to gun shows, conferences or conventions which are staged under the auspices of a duly recognized nonprofit state or national organization.

21-20.3. Same - Granted by Clerk of Commission. The Clerk of the Board of County Commissioners may grant licenses in the form prescribed by the County Commission permitting the licensee to sell said handguns within the County on the premises named therein.

21-20.4. Application for license; content. Applicants for licenses to sell must file their applications in duplicate copies in the form prescribed by the County Commission, with the Clerk of the County Commission. All applications must be signed and verified by all persons in whose names the license may be issued. The application shall state the full name, age, residence, present and previous occupations of each person so signing the same, and shall also specify the complete name, address and location of the place of business, as well as other facts as may be required to show the good character, competency and integrity of each person so signing.

21-20.4.1. Demonstration of knowledge of County, State and federal laws applicable to sale of firearms required. As part of the application procedure for a license to sell firearms, the Miami-Dade County Commission shall set up two (2) different type licenses. The first license shall be described as the dealer license, which shall be issued to the owner, corporation or partner, or general manager; and the employee license, which shall be issued to any employee who is engaged primarily in retail sale of firearms. The County Commission shall appoint a four-member committee consisting of a member from each of the following: Miami-Dade Police Department, Trail Glade Range or park employee, County Attorney's Office, and a fourth member from the local Miami-Dade County gun dealers. It shall be the duty of this committee to prepare a guide manual from which questions on a short written examination will be taken so that the applicant, whether he be a dealer or employee, shall have to show his knowledge of federal, State and Miami-Dade County laws by a testing procedure which shall be multiple choice or true and false, drawn up by this committee, based on the guide book. If the applicant shall have difficulty with the written or spoken English language, which seriously impairs his ability to demonstrate his knowledge of such laws, special provisions shall be made through an interpreter or otherwise, for the administration of the required written examination to such person. In the event of failure, the applicant may retake the test seven (7) days from date of examination, however, no persons shall be allowed to take this test more than three (3) times in a twelve-month period.

21-20.5. Investigation of application; fingerprinting.

(a) The County Commission may require, upon initial application, for applicants for licenses to be fingerprinted, and may require them to attach to their initial applications their photographs. The Director of the Miami-Dade Police Department shall make a records check of each applicant. The Director of the Miami-Dade Police Department shall make a complete investigation only when so directed by the County Commission.

(b) Such fingerprinting shall be searched for any previous criminal record and shall be placed on file by the Director of the Miami-Dade Police Department of the County. No such fingerprinting may be inspected by any person, other than a peace officer, except upon order of a judge of a court of record on such notice, if any, to the person who has been fingerprinted as the judge may determine.

21-20.6. Issuance of license.

(a) Upon completion of the following:

(1) An investigation showing the statements on the application to be true; and

(2) The taking and passing of the required written examination on gun control laws and applicable [applicability] to the sale of firearms in Miami-Dade County;

The Director of the Miami-Dade Police Department shall inquire into the good character, competency and integrity of the applicant. In making his determination on the character of the applicant, the Director shall be governed by the following provisions:

(i) He shall not consider a past arrest of the applicant which did not result in a conviction; provided, however, allegations which are the basis for any pending criminal charges may be considered if the charges are pending when the application is considered.

(ii) In the case of subsequent applications of the same applicant, the Director shall only consider acts or omissions on the part of the applicant which have occurred subsequent to the date of the issuance of prior permits to the applicant.

If the Director receives testimony regarding the applicant, witnesses shall be sworn and the rules of evidence applicable to quasi-judicial proceedings shall govern. If the Director is satisfied as to the applicant's good character, competency and integrity, a license to sell shall be issued to the applicant within sixty (60) days of application, and the committee should also assign a permanent license number to each applicant. If the Director is not satisfied as to the applicant's good character, competency and integrity, the application shall be denied. Any applicant or other person dissatisfied with the decision of the Director may, within ten (10) days from the date of action by the Director, apply to the County Commission to review the action of the Director. In such event, witnesses shall be sworn and the rules of evidence applicable to quasi-judicial proceedings shall govern. The determination of the Board shall be final unless overturned by a court of competent jurisdiction.

(b) No license may be issued to any person who has been convicted of a felony in this state or elsewhere or who is under eighteen (18) years of age.

(c) No license may be issued solely in the name of a corporation, firm, partnership, company or other fictitious entity, but the corporate name may be listed on the license.

(d) Annual renewal for either dealer or employee shall be a duplicate simple form listing only name, address, company name and address, Miami-Dade County license number, and notarized signature.

21-20.7. File of applications and licenses. After disposition of the application, the original shall be maintained in the files of the County Commission. The duplicate copy shall be filed with the Director of the Miami-Dade Police Department. Attached thereto shall be information as to the final disposition of the application.

21-20.8. Conditions; revocation.

(A) A license issued in accordance with the provisions of this division is subject to the following conditions, for breach of any of which the license is subject to revocation:

(1) The business shall be carried on only in the premises designated in the license.

(2) The license or a copy certified by the issuing authority shall be displayed in a conspicuous place on the premises in which the business is conducted where it can be easily read.

(3) No handgun, or imitation thereof, shall be placed in any display window of the premises.

(4) No handgun shall be delivered:

(a) Until seventy-two (72) hours after the sale has been completed, unless the purchaser is the possessor of a valid license to carry firearms issued within the State, which license to carry shall be displayed at time of sale, except where an operable firearm is traded for another operable firearm as defined in Section 21-20.8 of this chapter.

(b) Unless the purchaser is either personally known to the licensee, or presents conclusive evidence of his identity, proper identification as required by Omnibus Crime Bill of 1968, said evidence to be noted on the sales record.

(c) Unless the delivery is made to the identical individual who has purchased the firearm, by proof of dealer receipt.

(d) The provisions of this section shall not apply to duly employed federal, State, County or municipal law enforcement officers.

(5) A true record of every sale of firearms as, defined in this division shall be kept by the licensee in accordance with the provisions of Section 21-20, Article III of the Miami-Dade County Code, on a form prescribed by the Director of the Miami-Dade Police Department, and a copy of the record of each sale shall be mailed to the Director of the Miami-Dade Police Department within seventy-two (72) hours of the delivery of the firearm.

All firearms with overall length of twenty-six and one-half (26 1/2) inches shall be excluded from this requirement.

(B) The following shall also constitute grounds for revocation of the license:

(1) Conviction of the licensee of any felony.

(2) Any grounds for which the license could have been refused when applied for.

(3) Conviction for violation of Section 20-20.17 of this Code.

21-20.9. Inspection of sales record book. A sales record book, as required by the Federal Firearms Act, shall be open at all reasonable hours for the inspection of any police or peace officer.

21-20.10. Licenses nontransferable; expiration date; renewals.

(a) Licenses are not transferable. It shall be unlawful for any licensee to assign, sell, lease or in any manner transfer or attempt to sell, lease or in any manner to attempt to transfer said license to sell or any rights or duties thereunder, to any person, persons, firm, partnership, corporation, company or other fictitious entity.

(b) If, during the period of the existence of the license, a person so licensed ceases to engage in the said business or ceases to sell firearms, said license may be surrendered to the County Commission who may order the removal of said person's name from the license and thereby relieve said person from future responsibility under the license as of the date of the removal of the same. All copies of said license shall be corrected.

(c) Any licensed employee may work at any other licensed dealer premises in Miami-Dade County without notice of transfer to the County Commission.

(d) All licenses issued under this division shall expire one (1) year after issuance. Every application for a license to sell, including subsequent successive renewals, may be investigated as provided for in Section 21-20.5 of this division. County must, via certified mail, give thirty (30) days written notice of expiration and send simplified renewal forms.

(e) Licenses may be revoked for cause by the County Commission at any time. The licensee or licensees shall be entitled to an immediate hearing upon the merits to obtain reinstatement of said license.

21-20.11. Review of orders denying license, revocation. The Circuit Court shall review orders denying a license to sell and others denying reinstatement of a revoked license.

21-20.12. Fees, charges. The initial fee for a license, whether it be dealer or employee, to sell firearms shall be determined by the Manager and shall be established by administrative order of the Manager and approved by the Board of County Commissioners, which sum must accompany the application; provided, however, that the said sum is returnable if the application is subsequently denied. Renewal application shall have a license fee determined by the Manager, as provided for in this section whether it be dealer or employee. Fees collected for licenses issued as provided in this division may be paid into the County Treasury.

21-20.13. Sale of firearm with serial number defaced prohibited. It shall be unlawful to sell any firearm whose serial number has been altered, changed, disfigured or defaced.

21-20.14. Sale or delivery of firearms to certain classes of persons. It shall be unlawful to sell or deliver any firearm to any person who the seller has reasonable grounds to believe is under the influence of intoxicating liquor, narcotic drugs or barbiturates, or hallucinogens, or is addicted to the use of any narcotic drug or barbiturate, or hallucinogens, or is a habitual alcoholic, or is of unsound mind; or has been convicted of a felony; or is a fugitive from justice; or is a member of a subversive organization.

21-20.14.1. Three working days waiting period.

(a) Definitions.

Purchase means payment of deposit, payment in full, or notification of intent to purchase.

Retail establishment means gun shop, sporting goods store, pawn shop, hardware store, department store, discount store, bait or tackle shop, or any other store or shop that offers handguns for walk-in retail sale but does not include gun collectors shows or exhibits, or gun shows.

(b) Applicability. This section shall apply to all sales of handguns to individuals by retail establishments unless exempted in subsection (d) hereof.

(c) Prohibition. It shall be unlawful and a violation of this section for any retail establishment to deliver a handgun to a purchaser within three (3) working days after the purchase of the handgun. It shall be unlawful and a violation of this section for any individual to take possession of a handgun within three (3) working days after the purchase of the handgun. Records of handgun sales by retail establishments shall be available for inspection during normal business hours by any law enforcement agency as defined in Section 934.02, Florida Statutes.

(d) Exceptions. The following shall be exempted from the provisions of this section.

(1) Individuals who are licensed to carry concealed firearms under the provisions of Section 790.06, Florida Statutes, or who are licensed to carry concealed firearms under any other provisions of State law and who show a valid license;

(2) Individuals who already lawfully own another firearm and who show a sales receipt for another firearm; who are known to own another firearm through a prior purchase from the retail establishment; or who have another firearm for trade-in;

(3) A law enforcement or correctional officer as defined in Section 943.10, Florida Statutes;

(4) A law enforcement agency as defined in Section 943.02, Florida Statutes;

(5) Sales or transactions between dealers or between distributors or between dealers and distributors who have current federal firearms licenses; or

(6) Any individual who has been threatened or whose family has been threatened with death or bodily injury, provided the individual may lawfully possess a firearm and provided such threat has been duly reported to Miami-Dade Police Department or any municipal police department within Miami-Dade County, Florida.

(e) Penalties. Every person who is convicted for a violation of this section shall be punished upon a first conviction by imprisonment for not less than six (6) months or by a fine of not less than one thousand dollars ($1,000.00) or by both such fine and imprisonment, in the discretion of the court and on a second and subsequent conviction shall be punished by imprisonment for not less than one (1) year or by a fine of not less than one thousand dollars ($1,000.00) or by both such fine and imprisonment, in the discretion of the court.

21-20.15. Penalty. Every person who is convicted of a violation of this division shall be punished by a fine not to exceed five hundred dollars ($500.00) or imprisonment in the County Jail for not more than thirty (30) days, or by both such fine and imprisonment; for a second conviction of a violation of this division such person shall be punished by a fine not to exceed one thousand dollars ($1,000.00) or imprisonment in the County Jail not more than twelve (12) months, or by both such fine and imprisonment.

21-20.16. Handgun purchaser's instruction and qualification procedure.

(a) Instruction and qualification procedure required. It shall be unlawful for any person to acquire a handgun in Miami-Dade County from a licensed dealer, unless such person has received safety instruction and otherwise qualified, pursuant to this section, or unless he is specifically exempted from the operation of this section.

(b) In order to qualify under this section the purchaser must demonstrate his knowledge of federal, State and Miami-Dade County laws by passing a short written examination consisting of questions prescribed herein by the County Commission. The purchaser must also demonstrate his ability to safely handle the handgun to be purchased pursuant to guidelines prescribed herein by the County Commission. Persons who have successfully completed the above procedures in the course of purchasing a handgun shall not be required to repeat such procedures in any given year in order to purchase another handgun of the same type and caliber previously purchased. If the purchaser shall have difficulty with the written or spoken English language, which seriously impairs his ability to demonstrate his knowledge of such laws and his ability to safely handle said handgun, special provisions shall be made through an interpreter or otherwise, for the administration of the required written examination and qualification procedure to such person. In the event of failure, the applicant may retake the test seven (7) days from the date of examination, however no person shall be allowed to take this test more than three (3) times in a twelve-month period as a result of such failure.

(c) The transferor shall administer unto the purchaser safety procedures as prescribed by the County Commission prior to the purchaser acquiring ownership or possession of the handgun, and shall execute the prescribed form that said procedure has been imparted to the purchaser.

The above prescribed procedure shall be administered only by duly licensed dealers or salesmen under this chapter.

The signature of the purchaser upon the prescribed form shall serve as a release of the transferor from all civil liability as a result of carrying out this section.

(d) The County Commission shall appoint a four (4) member committee consisting of a member from each of the following: Miami-Dade Police Department, Trail Glade Range or park employee, County Attorney's Office, and a fourth member from the local Miami-Dade County gun dealers.

The committee shall from time to time advise the County Commission of needed changes in the prescribed procedure and form. Such changes shall become effective upon the consideration and approval of the County Commission.

(e) This section shall not apply to:

(1) Law enforcement officers or agents of any state of the United States, or any political subdivision, municipal corporation, department or agency of either, members of the organized militia of any state or the armed forces of the United States, or law enforcement officers of any political subdivision, municipal corporation, department or agency of either, while engaged in the discharge of their official duties.

(2) Wholesale dealers in their business intercourse with retail dealers or retail dealers in their business intercourse with other retail dealers or to wholesale or retail dealers in the regular or ordinary transportation of any unloaded firearms, merchandise by mail, express or other mode of shipment to points outside the country.

(3) Nonresidents of the United States having proper authorization from his or her consulate, acting consulate, commercial attache, or such other authorized representative.

(4) This section applies to persons in the firearms business or in the business of gunsmithing, and does not apply to:

(a) Sales or trades by an unlicensed person or a person licensed hereunder, nor to

(b) Isolated sales, transfers or trades between unlicensed persons who are not engaged in the firearms business or in the business of gunsmithing.

(5) The provisions of this section shall not apply to gun shows, conferences or conventions which are staged under the auspices of a duly recognized nonprofit, State or national organization.

21-20.17. Unlawful to sell handguns to persons who have not qualified.

(a) It shall be unlawful for a licensed dealer to sell, deliver, transfer, or furnish any handgun to any person in Miami-Dade County unless the purchaser has qualified under Section 21-20.16, or

(b) Unless the purchaser exhibits evidence that he is exempt from the requirements of Section 21-20.16(b). Every person who is convicted for violation of this section shall be punished as provided in Section 21-20.15 of this Code.

21-20.18. Five-day waiting period and criminal history records check on firearms sales.

(a) Definitions. For purposes of this section, the following terms shall be defined as follows:

(1) Any part of the transaction means any part of the sales transaction, including but not limited to, the offer of sale, negotiations, the agreement to sell, the transfer of consideration, or the transfer of the firearm.

(2) Antique firearms means any firearm (including any firearm with a matchlock, flintlock, percussion cap, or similar type of ignition system) manufactured in or before 1898; and any replica of any such firearm if such replica (i) is not designed or redesigned for using rimfire or conventional centerfire fixed ammunition, or (ii) uses rimfire or conventional centerfire fixed ammunition which is no longer manufactured in the United States and which is not readily available in the ordinary channels of commercial trade.

(3) Firearm means any weapon which will, is designed to, or may readily be converted to expel a projectile by the action of an explosive; the frame or receiver of any such weapon; and firearm muffler or firearm silencer; any destructive device; or any machine gun. Such term does not include an antique firearm.

(4) Property to which the public has the right of access means any real or personal property to which the public has a right of access, including property owned by either public or private individuals, firms and entities and expressly includes, but is not limited to, flea markets, gun shows and firearms exhibitions.

(5) Sale means the transfer of money or other valuable consideration.

(b) Application and enforcement of section. Law enforcement officers shall have the right to enforce the provisions of this section against any person found violating these provisions within their jurisdiction.

(c) Sale and delivery of firearms; mandatory five-day waiting period. There shall be a mandatory five-day waiting period, which shall be five full days, excluding weekends and legal holidays, between the hour of the sale and the hour of the delivery of any firearm when any part of the transaction is conducted within Miami-Dade County on property to which the public has the right of access.

(d) Sale and delivery of firearms; mandatory criminal records check. No person, whether licensed or unlicensed, shall sell, offer for sale, transfer or deliver any firearm to another person when any part of the transaction is conducted on property to which the public has the right of access within Miami-Dade County until all procedures specified under section 790.065, Florida Statutes, have been complied with by a person authorized by that section to conduct a criminal history check of background information as specified in that section, and the approval number set forth by that section has been obtained and documented. Upon the repeal of section 790.065, Florida Statutes, no person, whether licensed or unlicensed, shall sell, offer for sale, transfer or deliver any firearm to another person when any part of the transaction is conducted on property to which the public has the right of access until all procedures specified under any other state or federal law which requires a national criminal history information or national criminal history check on potential buyer or transferee of firearms have been complied with by any person authorized by law to conduct the required national criminal history or background records check and any required approval under such state or federal law or rule has been obtained. "Person" for purposes of this subsection shall include any person, including, but not limited to a licensed importer, licensed manufacturer or licensed dealer and any unlicensed person.

In the case of a seller who is not a licensed importer, licensed manufacturer or licensed dealer, compliance with section 790.065 or its state or federal successor shall be achieved by the seller requesting that a licensed importer, licensed manufacturer or licensed dealer complete all the requirements of section 790.065 or its state or federal successor. Licensed importers, manufacturers and dealers may charge a reasonable fee of an unlicensed seller to cover costs associated with completing the requirements of section 790.065.

(e) Exemptions. Holders of a concealed weapons permit as prescribed by state law and holders of an active certification from the Criminal Justice Standards and Training Commission as a law enforcement officer, a correctional officer, or a correctional probation officer as set forth in state law shall not be subject to the provisions of this section.

Sales to a licensed importer, licensed manufacturer or licensed dealer shall not be subject to the provisions of this section.

(f) Penalties. Any person violating any provision of this section shall be punished by a fine not to exceed five hundred dollars ($500.00) or by imprisonment not to exceed sixty (60) days in the County Jail, or by both such fine and imprisonment. Nothing contained herein shall be construed to preempt the imposition of any higher penalties imposed by state or federal law.

(g) Reporting of information. To the fullest extent permissible by law, all information acquired in relation to a violation of this section shall be reported to appropriate federal and state officials.

21-20.19. Locking devices required for firearms.

(a) Definitions. For purposes of this section, the following terms shall be defined as follows:

(1) Locking device means a device that when installed on a firearm and secured by means of a key or a mechanically or electronically operated combination lock prevents the firearm from being discharged without first deactivating or removing the device.

(b) If a person stores or leaves a firearm at any location where the person knows or reasonably should know that a minor might gain access to the firearm, the person shall secure the firearm with a locking device or in a securely locked box or container except when it is carried on his or her body or is located within such close proximity that the person can retrieve the firearm and prevent access to it by a minor.

(c) A violation of this section is a breach of a duty of safety owed by the person who owns or possesses the firearm to all minors who might gain access to it and to the general public.

(d) Any person which is convicted of violating this sub-section shall be punished by a fine not to exceed five hundred dollars ($500.00) or by imprisonment in the County jail for not more than sixty (60) days or by both such fine and imprisonment.

(e) Applicability. This section shall apply to the incorporated and unincorporated areas of Miami-Dade County.

[Code of Miami-Dade County current through Ordinance No. 08-86, enacted July 1, 2008]

Palm Beach County Code

Chapter 28. Weapons

Article II. Firearms

28-21. Definitions. As used in this article:

Business day means a 24-hour day (beginning at 12:01 a.m.), excluding weekends and legal holidays.

FDLE means the Florida Department of Law Enforcement.

Firearm means any weapon, including a starter gun or handgun, which will or is designed to or may readily be converted to expel a projectile by the action of an explosive; the frame or receiver of any such weapon; any firearm muffler or firearm silencer; any destructive device; or any machine gun.

Handgun means a firearm capable of being carried and used by one (1) hand, such as a pistol or revolver.

Licensed dealer has the same meaning as in Title 27, Code of Federal Regulations, Part 178.

Licensed importer has the same meaning as in Title 27, Code of Federal Regulations, Part 178.

Licensed manufacturer has the same meaning as in Title 27, Code of Federal Regulations, Part 178.

NICS means the National Instant Criminal Background Check System established pursuant to Title 18, United States Code, Section 922.

Person includes, but is not limited to, any individual, corporation, company, association, firm, partnership, society, or joint stock company.

Sale means the transfer of money or other valuable consideration for any firearm when any part of the transaction is conducted on property to which the public has the right of access, which includes, but is not limited to, flea markets, gun shows and gun exhibitions.

Shotgun means a shoulder-held firearm with a smooth bore designed primarily to fire shells containing multiple projectiles (shotshells).

Shotshell means a metal, plastic or cardboard case containing a primer, propellant and multiple projectiles intended to be fired from a shotgun.

Structure means that which is three (3) feet or more in height which is built or constructed or erected or tied down having a fixed location on the ground such as buildings, homes and mobile homes.

28-23. Mandatory waiting period; Criminal history records check requirement.

(a) No person shall deliver a firearm to another person, other than a licensed dealer, licensed manufacturer or licensed importer, in connection with the sale, as defined in section 28-1, of such firearm for a period of five (5) business days from the date of the sale.

(b) No person shall deliver a firearm to another person, other than a licensed dealer, licensed manufacturer or licensed importer, in connection with the sale, as defined in section 28-1, of such firearm until the seller, either directly or through a licensed dealer, licensed manufacturer or licensed importer, has:

(1) Requested a national criminal history records check of the potential purchaser or transferee of the firearm from FDLE or NICS, and

(2) Received an approval/identification number from FDLE or NICS signifying the potential purchaser or transferee of the firearm would not be prohibited by state or federal law from receiving or possessing a firearm.

(c) The provisions of this section shall not apply to holders of a concealed weapons permit issued pursuant to general law of the State of Florida when purchasing a firearm.

28-24. Enforcement. It shall be the duty and responsibility of law enforcement officers within their respective jurisdictions to enforce this article and all related state laws.

28-25. Penalty. Any person convicted of a violation of any of provision of this article shall be punished in accordance with section 125.69(1), Florida Statutes, as may be amended, by a fine not to exceed five hundred dollars ($500.00) per violation or by imprisonment in the county jail not to exceed sixty (60) days or by both such fine and imprisonment. Each violation of this article shall constitute a separate offense. In addition to the sanctions contained herein, the county shall be authorized to take any other appropriate legal action, including, but not limited to, seeking cease and desist orders, instituting other administrative actions, and requesting temporary and permanent injunctions to enforce the provisions of this article. It is the purpose of this article to provide additional and cumulative remedies.

28-26. Applicability. This article shall be applicable in both the unincorporated and incorporated areas of the county to the extent permitted by law.

[Palm Beach County Code current Ord. No. 08-008, adopted March 24, 2008]

GEORGIA

GA. CODE

Title 10. Commerce and Trade

Chapter 1. Selling and Other Trade Practices

Article 6. Interstate Purchase of Rifles and Shotguns

10-1-100. Out of State Purchase of Rifles and Shotguns by Residents. Residents of the State of Georgia may purchase rifles and shotguns in any state of the United States, provided such residents con-form to applicable provisions of statutes and reg-ulations of the United States, of the State of Georgia, and of the state in which the purchase is made.

10-1-101. Nonresident May Purchase Rifles and Shotguns in Georgia. Residents of any state of the United States may purchase rifles and shotguns in the State of Georgia, provided such residents conform to applicable provisions of statutes and regulations of the United States, of the State of Georgia, and of the state in which such persons reside.

Title 16. Crimes and Offenses

Chapter 11. Offenses Against Public Order and Safety

Article 4. Dangerous Instrumentalities and Practices

Part 1. General Provisions

16-11-101.1. Furnishing pistol or revolver to person under the age of 18 years

(a) For the purposes of this Code section, the term:

(1) "Minor" means any person under the age of 18 years.

(2) "Pistol or revolver" means a pistol or revolver as defined in subsection (a) of Code Section 16-11-132.

(b) It shall be unlawful for a person intentionally, knowingly, or recklessly to sell or furnish a pistol or revolver to a minor, except that it shall be lawful for a parent or legal guardian to permit possession of a pistol or revolver by a minor for the purposes specified in subsection (c) of Code Section 16-11-132 unless otherwise expressly limited by subsection (c) of this Code section.

(c)(1) It shall be unlawful for a parent or legal guardian to permit possession of a pistol or revolver by a minor if the parent or legal guardian knows of a minor's conduct which violates the provisions of Code Section 16-11-132 and fails to make reasonable efforts to prevent any such violation of Code Section 16-11-132.

(2) Notwithstanding any provisions of subsection (c) of Code Section 16-11-132 or any other law to the contrary, it shall be unlawful for any parent or legal guardian intentionally, knowingly, or recklessly to furnish to or permit a minor to possess a pistol or revolver if such parent or legal guardian is aware of a substantial risk that such minor will use a pistol or revolver to commit a felony offense or if such parent or legal guardian who is aware of such substantial risk fails to make reasonable efforts to prevent commission of the offense by the minor.

(3) In addition to any other act which violates this subsection, a parent or legal guardian shall be deemed to have violated this subsection if such parent or legal guardian furnishes to or permits possession of a pistol or revolver by any minor who has been convicted of a forcible felony or forcible misdemeanor, as defined in Code Section 16-1-3, or who has been adjudicated delinquent under the provisions of Article 1 of Chapter 11 of Title 15 for an offense which would constitute a forcible felony or forcible misdemeanor, as defined in Code Section 16-1-3, if such minor were an adult.

(d) Upon conviction of a violation of subsection (b) or (c) of this Code section, a person shall be guilty of a felony and punished by a fine not to exceed $5,000.00 or by imprisonment for not less than three nor more than five years, or both.

16-11-113. Offense of transferring firearm to individual other than actual buyer. Any person who attempts to solicit, persuade, encourage, or entice any dealer to transfer or otherwise convey a firearm other than to the actual buyer, as well as any other person who willfully and intentionally aids or abets such person, shall be guilty of a felony. This Code section shall not apply to a federal law enforcement officer or a peace officer, as defined in Code Section 16-1-3, in the performance of his or her official duties or other person under such officer's direct supervision.

Part 2. Possession of Dangerous Weapons

16-11-120. Short title. This part shall be known and may be cited as the "Georgia Firearms and Weapons Act."

16-11-121. Definitions. As used in this part, the term:

(1) "Dangerous weapon" means any weapon commonly known as a "rocket launcher," "bazooka," or "recoilless rifle" which fires explosive or nonexplosive rockets designed to injure or kill personnel or destroy heavy armor, or similar weapon used for such purpose. The term shall also mean a weapon commonly known as a "mortar" which fires high explosive from a metallic cylinder and which is commonly used by the armed forces as an antipersonnel weapon or similar weapon used for such purpose. The term shall also mean a weapon commonly known as a "hand grenade" or other similar weapon which is designed to explode and injure personnel or similar weapon used for such purpose.

(2) "Machine gun" means any weapon which shoots or is designed to shoot, automatically, more than six shots, without manual reloading, by a single function of the trigger.

(3) "Person" means any individual, partnership, company, association, or corporation.

(4) "Sawed-off rifle" means a weapon designed or redesigned, made or remade, and intended to be fired from the shoulder; and designed or redesigned, made or remade, to use the energy of the explosive in a fixed metallic cartridge to fire only a single projectile through a rifle bore for each single pull of the trigger; and which has a barrel or barrels of less than 16 inches in length or has an overall length of less than 26 inches.

(5) "Sawed-off shotgun" means a shotgun or any weapon made from a shotgun whether by alteration, modification, or otherwise having one or more barrels less than 18 inches in length or if such weapon as modified has an overall length of less than 26 inches.

(6) "Shotgun" means a weapon designed or redesigned, made or remade, and intended to be fired from the shoulder; and designed or redesigned, and made or remade, to use the energy of the explosive in a fixed shotgun shell to fire through a smooth bore either a number of ball shot or a single projectile for each single pull of the trigger.

(7) "Silencer" means any device for silencing or diminishing the report of any portable weapon such as a rifle, carbine, pistol, revolver, machine gun, shotgun, fowling piece, or other device from which a shot, bullet, or projectile may be discharged by an explosive.

16-11-122. Possession of sawed-off shotgun or rifle, machine gun, silencer, or dangerous weapon prohibited. No person shall have in his possession any sawed-off shotgun, sawed-off rifle, machine gun, dangerous weapon, or silencer except as provided in Code Section 16-11-124.

16-11-123. Unlawful possession of firearms or weapons. A person commits the offense of unlawful possession of firearms or weapons when he or she knowingly has in his or her possession any sawed-off shotgun, sawed-off rifle, machine gun, dangerous weapon, or silencer, and, upon conviction thereof, he or she shall be punished by imprisonment for a period of five years.

16-11-124. Exemptions from application of part. This part shall not apply to:

(1) A peace officer of any duly authorized police agency of this state or of any political subdivision thereof, or a law enforcement officer of any department or agency of the United States who is regularly employed and paid by the United States, this state, or any such political subdivision, or an employee of the Department of Corrections of this state who is authorized in writing by the commissioner of corrections to transfer or possess such firearms while in the official performance of his duties;

(2) A member of the National Guard or of the armed forces of the United States to wit: the army, navy, marine corps, air force, or coast guard who while serving therein, possesses such firearm in the line of duty;

(3) Any sawed-off shotgun, sawed-off rifle, machine gun, dangerous weapon, or silencer which has been modified or changed to the extent that it is inoperative. Examples of the requisite modification include weapons with their barrel or barrels filled with lead, hand grenades filled with sand, or other nonexplosive materials;

(4) Possession of a sawed-off shotgun, sawed-off rifle, machine gun, dangerous weapon, or silencer by a person who is authorized to possess the same because he has registered the sawed-off shotgun, sawed-off rifle, machine gun, dangerous weapon, or silencer in accordance with the dictates of the National Firearms Act, 68A Stat. 725 (26 U.S.C. Sections 5841-5862); and

(5) A security officer employed by a federally licensed nuclear power facility or a licensee of such facility, including a contract security officer, who is trained and qualified under a security plan approved by the United States Nuclear Regulatory Commission or other federal agency authorized to regulate nuclear facility security; provided, however that this exemption shall apply only while such security officer is acting in connection with his or her official duties on the premises of such nuclear power facility or on properties outside the facility property pursuant to a written agreement entered into with the local law enforcement agency having jurisdiction over the facility. The exemption under this paragraph does not include the possession of silencers.

16-11-125.Burden of proof as to exemptions. In any complaint, accusation, or indictment and in any action or proceeding brought for the enforcement of this part it shall not be necessary to negative any exception, excuse, proviso, or exemption contained in this part, and the burden of proof of any such exception, excuse, proviso, or exemption shall be upon the defendant.

Part 3. Carrying and Possession of Firearms

16-11-127.1. Carrying weapons within school safety zones, at school functions, or on school property

(a) As used in this Code section, the term:

(1) "School safety zone" means in, on, or within 1,000 feet of any real property owned by or leased to any public or private elementary school, secondary school, or school board and used for elementary or secondary education and in, on, or within 1,000 feet of the campus of any public or private technical school, vocational school, college, university, or institution of post-secondary education.

(2) "Weapon" means and includes any pistol, revolver, or any weapon designed or intended to propel a missile of any kind, or any dirk, bowie knife, switchblade knife, ballistic knife, any other knife having a blade of two or more inches, straight-edge razor, razor blade, spring stick, knuckles whether made from metal, thermoplastic, wood, or other similar material, blackjack, any bat, club, or other bludgeon-type weapon, or any flailing instrument consisting of two or more rigid parts connected in such a manner as to allow them to swing freely, which may be known as a nun chahka, nun chuck, nunchaku, shuriken, or fighting chain, or any disc, of whatever configuration, having at least two points or pointed blades which is designed to be thrown or propelled and which may be known as a throwing star or oriental dart, or any weapon of like kind, and any stun gun or taser as defined in subsection (a) of Code Section 16-11-106. This paragraph excludes any of these instruments used for classroom work authorized by the teacher.

(b) Except as otherwise provided in sub-section (c) of this Code section, it shall be unlawful for any person to carry to or to possess or have under such person's control while within a school safety zone or at a school building, school function, or school property or on a bus or other transportation furnished by the school any weapon or explosive compound, other than fireworks the possession of which is regulated by Chapter 10 of Title 25. Any person who violates this subsection shall be guilty of a felony and, upon conviction thereof, be punished by a fine of not more than $10,000.00, by imprisonment for not less than two nor more than ten years, or both; provided, however, that upon conviction of a violation of this subsection involving a firearm as defined in paragraph (2) of subsection (a) of Code Section 16-11-131, or a dangerous weapon or machine gun as defined in Code Section 16-11-121, such person shall be punished by a fine of not more than $10,000.00 or by imprisonment for a period of not less than five nor more than ten years, or both. A child who violates this subsection shall be subject to the provisions of Code Section 15-11-63.

(c) The provisions of this Code section shall not apply to:

(1) Baseball bats, hockey sticks, or other sports equipment possessed by competitors for legitimate athletic purposes;

(2) Participants in organized sport shooting events or firearm training courses;

(3) Persons participating in military training programs conducted by or on behalf of the armed forces of the United States or the Georgia Department of Defense;

(4) Persons participating in law enforcement training conducted by a police academy certified by the Georgia Peace Officer Standards and Training Council or by a law enforcement agency of the state or the United States or any political subdivision thereof;

(5) The following persons, when acting in the performance of their official duties or when en route to or from their official duties:

(A) A peace officer as defined by Code Section 35-8-2;

(B) A law enforcement officer of the United States government;

(C) A prosecuting attorney of this state or of the United States;

(D) An employee of the Georgia Department of Corrections or a correctional facility operated by a political subdivision of this state or the United States who is authorized by the head of such correctional agency or facility to carry a firearm;

(E) A person employed as a campus police officer or school security officer who is authorized to carry a weapon in accordance with Chapter 8 of Title 20; and

(F) Medical examiners, coroners, and their investigators who are employed by the state or any political subdivision thereof;

(6) A person who has been authorized in writing by a duly authorized official of the school to have in such person's possession or use as part of any activity being conducted at a school building, school property, or school function a weapon which would otherwise be prohibited by this Code section. Such authorization shall specify the weapon or weapons which have been authorized and the time period during which the authorization is valid;

(7) A person who is licensed in accordance with Code Section 16-11-129 or issued a permit pursuant to Code Section 43-38-10, when such person carries or picks up a student at a school building, school function, or school property or on a bus or other transportation furnished by the school or any weapon legally kept within a vehicle in transit through a designated school zone by any person other than a student;

(8) A weapon which is in a locked compartment of a motor vehicle or one which is in a locked container in or a locked firearms rack which is on a motor vehicle which is being used by an adult over 21 years of age to bring to or pick up a student at a school building, school function, or school property or on a bus or other transportation furnished by the school, or when such vehicle is used to transport someone to an activity being conducted on school property which has been authorized by a duly authorized official of the school; provided, however, that this exception shall not apply to a student attending such school;

(9) Persons employed in fulfilling defense contracts with the government of the United States or agencies thereof when possession of the weapon is necessary for manufacture, transport, installation, and testing under the requirements of such contract;

(10) Those employees of the State Board of Pardons and Paroles when specifically designated and authorized in writing by the members of the State Board of Pardons and Paroles to carry a weapon;

(11) The Attorney General and those members of his or her staff whom he or she specifically authorizes in writing to carry a weapon;

(12) Probation supervisors employed by and under the authority of the Department of Corrections pursuant to Article 2 of Chapter 8 of Title 42, known as the "Statewide Probation Act," when specifically designated and authorized in writing by the director of the Division of Probation;

(13) Public safety directors of municipal corporations;

(14) State and federal trial and appellate judges;

(15) United States attorneys and assistant United States attorneys;

(16) Clerks of the superior courts;

(17) Teachers and other school personnel who are otherwise authorized to possess or carry weapons, provided that any such weapon is in a locked compartment of a motor vehicle or one which is in a locked container in or a locked firearms rack which is on a motor vehicle; or

(18) Constables of any county of this state.

(d)(1) This Code section shall not prohibit any person who resides or works in a business or is in the ordinary course transacting lawful business or any person who is a visitor of such resident located within a school safety zone from carrying, possessing, or having under such person's control a weapon within a school safety zone; provided, however, it shall be unlawful for any such person to carry, possess, or have under such person's control while at a school building or school function or on school property, a school bus, or other transportation furnished by the school any weapon or explosive compound, other than fireworks the possession of which is regulated by Chapter 10 of Title 25.

(2) Any person who violates this subsection shall be subject to the penalties specified in subsection (b) of this Code section.

(3) This subsection shall not be construed to waive or alter any legal requirement for possession of weapons or firearms otherwise required by law.

(e) It shall be no defense to a prosecution for a violation of this Code section that:

(1) School was or was not in session at the time of the offense;

(2) The real property was being used for other purposes besides school purposes at the time of the offense; or

(3) The offense took place on a school vehicle.

(f) In a prosecution under this Code section, a map produced or reproduced by any municipal or county agency or department for the purpose of depicting the location and boundaries of the area on or within 1,000 feet of the real property of a school board or a private or public elementary or secondary school that is used for school purposes or within 1,000 feet of any campus of any public or private technical school, vocational school, college, university, or institution of postsecondary education, or a true copy of the map, shall, if certified as a true copy by the custodian of the record, be admissible and shall constitute prima-facie evidence of the location and boundaries of the area, if the governing body of the municipality or county has approved the map as an official record of the location and boundaries of the area. A map approved under this Code section may be revised from time to time by the governing body of the municipality or county. The original of every map approved or revised under this subsection or a true copy of such original map shall be filed with the municipality or county and shall be maintained as an official record of the municipality or county. This subsection shall not preclude the prosecution from introducing or relying upon any other evidence or testimony to establish any element of this offense. This subsection shall not preclude the use or admissibility of a map or diagram other than the one which has been approved by the municipality or county.

(g) A county school board may adopt regulations requiring the posting of signs designating the areas within 1,000 feet of school boards and private or public elementary and secondary schools as "Weapon-free and Violence-free School Safety Zones."

16-11-128. Carrying pistol without license.

(a) A person commits the offense of carrying a pistol without a license when he has or carries on or about his person, outside of his home, motor vehicle, or place of business, any pistol or revolver without having on his person a valid license issued by the judge of the probate court of the county in which he resides, provided that no permit shall be required for persons with a valid hunting or fishing license on their person or for persons not required by law to have hunting licenses who are engaged in legal hunting, fishing, or sport shooting when the persons have the permission of the owner of the land on which the activities are being conducted; provided, further, that the pistol or revolver, whenever loaded, shall be carried only in an open and fully exposed manner.

(b) Upon conviction of the offense of carrying a pistol without a license, a person shall be punished as follows:

(1) For the first offense, he shall be guilty of a misdemeanor; and

(2) For the second offense, and for any subsequent offense, he is guilty of a felony, and, upon conviction thereof, shall be imprisoned for not less than one year nor more than five years.

(c) On and after October 1, 1996, a person licensed to carry a handgun in any state whose laws recognize and give effect within such state to a license issued pursuant to this part shall be authorized to carry a handgun in this state, but only while the licensee is not a resident of this state; provided, however, that such license holder shall carry the handgun in compliance with the laws of this state.

16-11-129. License to carry pistol or revolver; temporary renewal permit; mandamus

(a) Application for license or renewal license; term. The judge of the probate court of each county may, on application under oath and on payment of a fee of $15.00, issue a license or renewal license valid for a period of five years to any person whose domicile is in that county or who is on active duty with the United States armed forces and who is not a domiciliary of this state but who either resides in that county or on a military reservation located in whole or in part in that county at the time of such application, which license or renewal license shall authorize that person to carry any pistol or revolver in any county of this state notwithstanding any change in that person's county of residence or state of domicile. Applicants shall submit the application for a license or renewal license to the judge of the probate court on forms prescribed and furnished free of charge to persons wishing to apply for the license or renewal license. An applicant who is not a United States citizen shall provide sufficient personal identifying data, including without limitation his or her place of birth and United State issued alien or admission number, as the Georgia Bureau of Investigation may prescribe by rule or regulation. An applicant who is nonimmigrant status shall provide proof of his or her qualifications for an exception to the federal firearm prohibition pursuant to 18 U.S.C. Section 922(y). Forms shall be designed to elicit information from the applicant pertinent to his or her eligibility under this Code section, including citizenship, but shall not require data which is nonpertinent or irrelevant such as serial numbers or other identification capable of being used as a de facto registration of firearms owned by the applicant. The Department of Public Safety shall furnish application forms and license forms required by this Code section. The forms shall be furnished to each judge of each probate court within the state at no cost.

(b) Licensing exceptions. No license or renewal license shall be granted to:

(1) Any person who is prohibited from possessing firearms pursuant to 18.U.S.C. Section 922;

(1.1) Any person under 21 years of age;

(2) Any person who is a fugitive from justice or against whom proceedings are pending for any felony, forcible misdemeanor, or violation of Code Section 16-11-126, 16-11-127, or 16-11-128 until such time as the proceedings are adjudicated;

(3) Any person who has been convicted of a felony by a court of this state or any other state; by a court of the United States including its territories, possessions, and dominions; or by a court of any foreign nation and has not been pardoned for such felony by the President of the United States, the State Board of Pardons and Paroles, or the person or agency empowered to grant pardons under the constitution or laws of such state or nation or any person who has been convicted of a forcible misdemeanor and has not been free of all restraint or supervision in connection therewith for at least five years or any person who has been convicted of a violation of Code Section 16-11-126, 16-11-127, or 16-11-128 and has not been free of all restraint or supervision in connection therewith for at least three years, immediately preceding the date of the application;

(4) Any individual who has been hospitalized as an inpatient in any mental hospital or alcohol or drug treatment center within five years of the date of his or her application. The probate judge may require any applicant to sign a waiver authorizing any mental hospital or treatment center to inform the judge whether or not the applicant has been an inpatient in any such facility in the last five years and authorizing the superintendent of such facility to make to the judge a recommendation regarding whether a license to carry a pistol or revolver should be issued. When such a waiver is required by the probate judge, the applicant shall pay to the probate judge a fee of $3.00 for reimbursement of the cost of making such a report by the mental health hospital, alcohol or drug treatment center, or the Department of Human Resources, which the probate judge shall remit to the hospital, center, or department. The judge shall keep any such hospitalization or treatment information confidential. It shall be at the discretion of the probate judge, considering the circumstances surrounding the hospitalization and the recommendation of the superintendent of the hospital or treatment center where the individual was a patient, to issue the license or renewal license; or

(5)(A) Any person, the provisions of paragraph (3) of this subsection notwithstanding, who has been convicted of an offense arising out of the unlawful manufacture, distribution, possession, or use of a controlled substance or other dangerous drug.

(B) As used in this paragraph, the term:

(i) "Controlled substance" means any drug, substance, or immediate precursor included in the definition of controlled substances in paragraph (4) of Code Section 16-13-21.

(ii) "Convicted" means a plea of guilty, a finding of guilt by a court of competent jurisdiction, the acceptance of a plea of nolo contendere, or the affording of first offender treatment by a court of competent jurisdiction irrespective of the pendency or availability of an appeal or an application for collateral relief.

(iii) "Dangerous drug" means any drug defined as such in Code Section 16-13-71; or

(6) Any person not lawfully present in the United States.

(c) Fingerprinting. Following completion of the application for a license or renewal license, the judge of the probate court shall require the applicant to proceed to an appropriate law enforcement agency in the county with the completed application. The appropriate local law enforcement agency in each county shall then capture the fingerprints of the applicant for a license or renewal license to carry a pistol or revolver, place the fingerprint required by subsection (f) of this Code section on a blank license form which has been furnished to the law enforcement agency by the judge of the probate court, and place the name of the applicant on the blank license form. The law enforcement agency shall be entitled to a fee of $5.00 from the applicant for its services in connection with the application.

(d) Investigation of applicant; issuance of license; renewal.

(1) For both license applications and requests for license renewals, the judge of the probate court shall within two business days following the receipt of the application or request direct the law enforcement agency to request a fingerprint based criminal history records check from the Georgia Crime Information Center and Federal Bureau of Investigation for purposes of determining the suitability of the applicant and return an appropriate report to the judge of the probate court. Fingerprints shall be in such form and of such quality as prescribed by the Georgia Crime Information Center and under standards adopted by the Federal Bureau of Investigation. The Georgia Bureau of Investigation may charge such fee as is necessary to cover the cost of the records search.

(2) For both license applications and requests for license renewals, the judge of the probate court shall within two business days following the receipt of the application or request also direct the law enforcement agency to conduct a background check using the Federal Bureau of Investigation's National Instant Criminal Background Check System and return an appropriate report to the probate judge.

(3) When a person who is not a United States citizen applies for a license or renewal of a license under this Code section, the judge of the probate court shall direct the law enforcement agency to conduct a search of the records maintained by the United States Bureau of Immigration and Customs Enforcement. As a condition to the issuance of a license or the renewal of a license, an applicant who is in nonimmigrant status shall provide proof of his or her qualifications for an exception to the federal firearm prohibition pursuant to 18 U.S.C. Section 922(y).

(4) The law enforcement agency shall report to the judge of the probate court within 30 days, by telephone and in writing, of any findings relating to the applicant which may bear on his or her eligibility for a license or renewal license under the terms of this Code section. When no derogatory information is found on the applicant bearing on his or her eligibility to obtain a license or renewal license, a report shall not be required. The law enforcement agency shall return the application and the blank license form with the fingerprint thereon directly to the judge of the probate court within such time period. Not later than ten days after the judge of the probate court receives the report from the law enforcement agency concerning the suitability of the applicant for a firearms license, the judge of the probate court shall issue such applicant a license or renewal license to carry any pistol or revolver unless facts establishing ineligibility have been reported or unless the judge determines such applicant has not met all the qualifications, is not of good moral character, or has failed to comply with any of the requirements contained in this Code section. The judge of the probate court shall date stamp the report from the law enforcement agency to show the date on which the report was received by the judge of the probate court.

(e) Revocation, loss, or damage to license. If, at any time during the period for which the license was issued, the judge of the probate court of the county in which the license was issued shall learn or have brought to his or her attention in any manner any reasonable ground to believe the licensee is not eligible to retain the license, the judge may, after notice and hearing, revoke the license of the person upon adjudication of falsification of application, mental incompetency, chronic alcohol or narcotic usage, conviction of any felony or forcible misdemeanor, or for violation of Code Section 16-11-126, 16-11-127, or 16-11-128. It shall be unlawful for any person to possess a license which has been revoked, and any person found in possession of any such revoked license, except in the performance of his or her official duties, shall be guilty of a misdemeanor. It shall be required that any license holder under this Code section have in his or her possession his or her valid license whenever he or she is carrying a pistol or revolver under the authority granted by this Code section, and his or her failure to do so shall be prima facie evidence of a violation of Code Section 16-11-128. Loss of any license issued in accordance with this Code section or damage to the license in any manner which shall render it illegible shall be reported to the judge of the probate court of the county in which it was issued within 48 hours of the time the loss or damage becomes known to the license holder. The judge of the probate court shall thereupon issue a replacement for and shall take custody of and destroy a damaged license; and in any case in which a license has been lost, he or she shall issue a cancellation order and notify by telephone and in writing each of the law enforcement agencies whose records were checked before issuance of the original license. The judge shall charge the fee specified in subsection (k) of Code Section 15-9-60 for such services.

(f) License specifications. Licenses issued as prescribed in this Code section shall be printed on durable but lightweight card stock, and the completed card shall be laminated in plastic to improve its wearing qualities and to inhibit alterations. Measurements shall be 3 1/4 inches long, and 2 1/4 inches wide. Each shall be serially numbered within the county of issuance and shall bear the full name, residential address, birth date, weight, height, color of eyes, sex, and a clear print of the right index finger of the licensee. If the right index fingerprint cannot be secured for any reason, the print of another finger may be used but such print shall be marked to identify the finger from which the print is taken. The license shall show the date of issuance, the expiration date, and the probate court in which issued and shall be signed by the licensee and bear the signature or facsimile thereof of the judge. The seal of the court shall be placed on the face before the license is laminated. The reverse side of the license shall have imprinted thereon in its entirety Code Section 16-11-127.

(g) Alteration or counterfeiting of license; penalty. A person who deliberately alters or counterfeits such a license card commits a felony and, upon conviction thereof, shall be punished by imprisonment for a period of not less than one nor more than five years.

(h) Licenses for former law enforcement officers. Except as otherwise provided in Code Section 16-11-130, any person who has served as a law enforcement officer for at least ten of the 12 years immediately preceding the retirement of such person as a law enforcement officer shall be entitled to be issued a license as provided for in this Code section without the payment of any of the fees provided for in this Code section. Such person must comply with all the other provisions of this Code section relative to the issuance of such licenses. As used in this subsection, the term "law enforcement officer" means any peace officer who is employed by the United States government or by the State of Georgia or any political subdivision thereof and who is required by the terms of his or her employment, whether by election or appointment, to give his or her full time to the preservation of public order or the protection of life and property or the prevention of crime. Such term shall include conservation rangers.

(i) Temporary renewal licenses.

(1) Any person who holds a license under this Code section to carry a pistol or revolver may, at the time he applies for a renewal of the license, also apply for a temporary renewal license if less than 90 days remain before expiration of the license he then holds or if his previous license has expired within the last 30 days.

(2) Unless the judge of the probate court knows or is made aware of any fact which would make the applicant ineligible for a five-year renewal license, the judge shall at the time of application issue a temporary renewal license to the applicant.

(3) Such a temporary renewal license shall be in the form of a paper receipt indicating the date on which the court received the renewal application and shall show the name, address, sex, age, and race of the applicant and that the temporary renewal license expires 90 days from the date of issue.

(4) During its period of validity the temporary renewal permit, if carried on or about the holder's person together with the holder's previous license, shall be valid in the same manner and for the same purposes as a five-year license.

(5) A $1.00 fee shall be charged by the probate court for issuance of a temporary renewal license.

(6) A temporary renewal license may be revoked in the same manner as a five-year license

(j) When an eligible applicant who is a United States citizen fails to receive a license, temporary permit, or renewal license within the time period required by this Code section and the application or request has been properly filed, the applicant may bring an action in mandamus or other legal proceeding in order to obtain a license, temporary license, or renewal license, and such applicant shall be entitled to recover his or her costs in such action, including reasonable attorney's fees.

16-11-130. Exemptions from Code Sections 16-11-126 through 16-11-128

(a) Code Sections 16-11-126 through 16-11-128 shall not apply to or affect any of the following persons if such persons are employed in the offices listed below or when authorized by federal or state law, regulations, or order:

(1) Peace officers, as such term is defined in paragraph (11) of Code Section 16-1-3, and retired peace officers so long as they remain certified whether employed by the state or a political subdivision of the state or another state or a political subdivision of another state but only if such other state provides a similar privilege for the peace officers of this state;

(2) Wardens, superintendents, and keepers of correctional institutions, jails, or other institutions for the detention of persons accused or convicted of an offense;

(3) Persons in the military service of the state or of the United States;

(4) Persons employed in fulfilling defense contracts with the government of the United States or agencies thereof when possession of the weapon is necessary for manufacture, transport, installation, and testing under the requirements of such contract;

(5) District attorneys, investigators employed by and assigned to a district attorney's office, assistant district attorneys, attorneys or investigators employed by the Prosecuting Attorneys' Council of the State of Georgia, and any retired district attorney, assistant district attorney, district attorneys investigator, or attorney or investigator retired from the Prosecuting Attorneys' Council of the State of Georgia, if such employee is retired in good standing and is receiving benefits under Title 47 or is retired in good standing and receiving benefits from a county or municipal retirement system;

(6) State court solicitors-general; investigators employed by and assigned to a state court solicitor general's office; assistant state court solicitors general; the corresponding personnel of any city court expressly continued in existence as a city court pursuant to Article VI, Section X, Paragraph I, subparagraph (5) of the Constitution; and the corresponding personnel of any civil court expressly continued as a civil court pursuant to said provision of the Constitution;

(7) Those employees of the State Board of Pardons and Paroles when specifically designated and authorized in writing by the members of the State Board of Pardons and Paroles to carry a weapon;

(8) The Attorney General and those members of his or her staff whom he or she specifically authorizes in writing to carry a weapon;

(9) Chief probation officers, probation officers, intensive probation officers, and surveillance officers employed by and under the authority of the Department of Corrections pursuant to Article 2 of Chapter 8 of Title 42, known as the "Statewide Probation Act," when specifically designated and authorized in writing by the director of Division of Probation;

(10) Public safety directors of municipal corporations;

(11) Explosive ordnance disposal technicians, as such term is defined by Code Section 16-7-80, and persons certified as provided in Code Section 35-8-13 to handle animals trained to detect explosives, while in the performance of their duties;

(12) State and federal trial and appellate judges, full-time and permanent part-time judges of municipal and city courts, and former state trial and appellate judges retired from their respective offices under state retirement;

(13) United States Attorneys and Assistant United States Attorneys;

(14) County medical examiners and coroners and their sworn officers employed by county government; and

(15) Clerks of the superior courts.

(b) Code Sections 16-11-126 through 16-11-128 shall not apply to or affect persons who at the time of their retirement from service with the Department of Corrections were chief probation officers, probation officers, intensive probation officers, or surveillance officers, when specifically designated and authorized in writing by the director of Division of Probation.

(c) Code Sections 16-11-126 through 16-11-128 shall not apply to or affect any:

(1) Sheriff, retired sheriff, deputy sheriff, or retired deputy sheriff if such retired deputy sheriff is receiving benefits under the Peace Officers' Annuity and Benefit Fund provided under Chapter 17 of Title 47;

(2) Member of the Georgia State Patrol or agent of the Georgia Bureau of Investigation or retired member of the Georgia State Patrol or agent of the Georgia Bureau of Investigation if such retired member or agent is receiving benefits under the Employees' Retirement System;

(3) Full-time law enforcement chief executive engaging in the management of a county, municipal, state, state authority, or federal law enforcement agency in the State of Georgia, including any college or university law enforcement chief executive that is registered or certified by the Georgia Peace Officer Standards and Training Council; or retired law enforcement chief executive that formerly managed a county, municipal, state, state authority, or federal law enforcement agency in the State of Georgia, including any college or university law enforcement chief executive that was registered or certified at the time of his or her retirement by the Georgia Peace Officer Standards and Training Council, if such retired law enforcement chief executive is receiving benefits under the Peace Officers' Annuity and Benefit Fund provided under Chapter 17 of Title 47 or is retired in good standing and receiving benefits from a county, municipal, State of Georgia, state authority, or federal retirement system; or

(4) Police officer of any county, municipal, state, state authority, or federal law enforcement agency in the State of Georgia, including any college or university police officer that is registered or certified by the Georgia Peace Officer Standards and Training Council, or retired police officer of any county, municipal, state, state authority, or federal law enforcement agency in the State of Georgia, including any college or university police officer that was registered or certified at the time of his or her retirement by the Georgia Peace Officer Standards and Training Council, if such retired employee is receiving benefits under the Peace Officers' Annuity and Benefit Fund provided under Chapter 17 of Title 47 or is retired in good standing and receiving benefits from a county, municipal, State of Georgia, state authority, or federal retirement system.

In addition, any such sheriff, retired sheriff, deputy sheriff, retired deputy sheriff, active or retired law enforcement chief executive, or other law enforcement officer referred to in this subsection shall be authorized to carry a pistol or revolver on or off duty anywhere within the state and the provisions of Code Sections 16-11-126 through 16-11-128 shall not apply to the carrying of such firearms.

(d) A prosecution based upon a violation of Code Section 16-11-126, 16-11-127, or 16-11-128 need not negative any exemptions.

16-11-131. Possession of firearms by convicted felons and first offender probationers.

(a) As used in this Code section, the term:

(1) "Felony" means any offense punishable by imprisonment for a term of one year or more and includes conviction by a court-martial under the Uniform Code of Military Justice for an offense which would constitute a felony under the laws of the United States.

(2) "Firearm" includes any handgun, rifle, shotgun, or other weapon which will or can be converted to expel a projectile by the action of an explosive or electrical charge.

(b) Any person who is on probation as a first offender pursuant to Article 3 of Chapter 8 of Title 42 or who has been convicted of a felony by a court of this state or any other state; by a court of the United States including its territories, possessions, and dominions; or by a court of any foreign nation and who receives, possesses, or transports any firearm commits a felony and, upon conviction thereof, shall be imprisoned for not less than one nor more than five years; provided, however, that if the felony as to which the person is on probation or has been previously convicted is a forcible felony, then upon conviction of receiving, possessing, or transporting a firearm, such person shall be imprisoned for a period of five years.

(b.1) Any person who is prohibited by this Code section from possessing a firearm because of conviction of a forcible felony or because of being on probation as a first offender for a forcible felony pursuant to this Code section and who attempts to purchase or obtain transfer of a firearm shall be guilty of a felony and shall be punished by imprisonment for not less than one nor more than five years.

(c) This Code section shall not apply to any person who has been pardoned for the felony by the President of the United States, the State Board of Pardons and Paroles, or the person or agency empowered to grant pardons under the constitutions or laws of the several states or of a foreign nation and, by the terms of the pardon, has expressly been authorized to receive, possess, or transport a firearm.

(d) A person who has been convicted of a felony, but who has been granted relief from the disabilities imposed by the laws of the United States with respect to the acquisition, receipt, transfer shipment, or possession of firearms by the secretary of the United States Department of the Treasury pursuant to 18 U.S.C. Section 925, shall, upon presenting to the Board of Public Safety proof that the relief has been granted and it being established from proof submitted by the applicant to the satisfaction of the Board of Public Safety that the circumstances regarding the conviction and the applicant's record and reputation are such that the acquisition, receipt, transfer, shipment, or possession of firearms by the person would not present a threat to the safety of the citizens of Georgia and that the granting of the relief sought would not be contrary to the public interest, be granted relief from the disabilities imposed by this Code section. A person who has been convicted under federal or state law of a felony pertaining to antitrust violations, unfair trade practices, or restraint of trade shall, upon presenting to the Board of Public Safety proof, and it being established from said proof, submitted by the applicant to the satisfaction of the Board of Public Safety that the circumstances regarding the conviction and the applicant's record and reputation are such that the acquisition, receipt, transfer, shipment, or possession of firearms by the person would not present a threat to the safety of the citizens of Georgia and that the granting of the relief sought would not be contrary to the public interest, be granted relief from the disabilities imposed by this Code section. A record that the relief has been granted by the board shall be entered upon the criminal history of the person maintained by the Georgia Crime Information Center and the board shall maintain a list of the names of such persons which shall be open for public inspection.

(e) As used in this Code section, the term "forcible felony" means any felony which involves the use or threat of physical force or violence against any person and further includes, without limitation, murder; felony murder; burglary; robbery; armed robbery; kidnapping; hijacking of an aircraft or motor vehicle; aggravated stalking; rape; aggravated child molestation; aggravated sexual battery; arson in the first degree; the manufacturing, transporting, distribution, or possession of explosives with intent to kill, injure, or intimidate individuals or destroy a public building; terroristic threats; or acts of treason or insurrection.

(f) Any person placed on probation as a first offender pursuant to Article 3 of Chapter 8 of Title 42 and subsequently discharged without court adjudication of guilt pursuant to Code Section 42-8-62 shall, upon such discharge, be relieved from the disabilities imposed by this Code section.

16-11-132. Possession of pistol or revolver by person under the age of 18 years

(a)(1) For the purposes of this Code section, the term "pistol" or "revolver" means a firearm of any description, loaded or unloaded, from which any shot, bullet, or other missile can be discharged where the length of the barrel, not including any revolving, detachable, or magazine breech, does not exceed 12 inches; provided, however, that the term pistol or revolver shall not include a gun which discharges shot of .46 centimeters or less in diameter.

(2) For the purposes of this Code section, a pistol or revolver is considered loaded if:

(A) There is a cartridge in the chamber or cylinder of the pistol or revolver;

(B) The person is carrying on his or her body or attached to his or her clothing the pistol or revolver and the ammunition for such pistol or revolver; or

(C) The pistol or revolver and the ammunition for such pistol or revolver are in such close proximity to such person that such person could readily gain access to the pistol or revolver and the ammunition and load the pistol or revolver.

(b) Notwithstanding any other provisions of this part and except as otherwise provided in this Code section, it shall be unlawful for any person under the age of 18 years to possess or have under such person's control a pistol or revolver. A person convicted of a first violation of this subsection shall be guilty of a misdemeanor and shall be punished by a fine not to exceed $1,000.00 or by imprisonment for not more than 12 months, or both. A person convicted of a second or subsequent violation of this sub-section shall be guilty of a felony and shall be punished by a fine of $5,000.00 or by imprisonment for a period of three years, or both.

(c) Except as otherwise provided in subsection (d) of this Code section, the provisions of subsection (b) of this Code section shall not apply to:

(1) Any person under the age of 18 years who is:

(A) Attending a hunter education course or a firearms safety course;

(B) Engaging in practice in the use of a firearm or target shooting at an established range authorized by the governing body of the jurisdiction where such range is located;

(C) Engaging in an organized competition involving the use of a firearm or participating in or practicing for a performance by an organized group under 26 U.S.C. Section 501(c)(3) which uses firearms as a part of such performance;

(D) Hunting or fishing pursuant to a valid license if such person has in his or her possession such a valid hunting or fishing license if required; is engaged in legal hunting or fishing; has permission of the owner of the land on which the activities are being conducted; and the pistol or revolver, whenever loaded, is carried only in an open and fully exposed manner; or

(E) Traveling to or from any activity described in subparagraphs (A) through (D) of this paragraph if the pistol or revolver in such person's possession is not loaded;

(2) Any person under the age of 18 years who is on real property under the control of such person's parent, legal guardian, or grandparent and who has the permission of such person's parent or legal guardian to possess a pistol or revolver; or

(3) Any person under the age of 18 years who is at such person's residence and who, with the permission of such person's parent or legal guardian, possesses a pistol or revolver for the purpose of exercising the rights authorized in Code Section 16-3-21 or 16-3-23.

(d) Subsection (c) of this Code section shall not apply to any person under the age of 18 years who has been convicted of a forcible felony or forcible misdemeanor, as defined in Code Section 16-1-3, or who has been adjudicated delinquent under the provisions of Article 1 of Chapter 11 of Title 15 for an offense which would constitute a forcible felony or forcible misdemeanor, as defined in Code Section 16-1-3, if such person were an adult.

16-11-151. Prohibited training

(a) As used in this Code section, the term "dangerous weapon" has the same meaning as found in paragraph (1) of Code Section 16-11-121.

(b) It shall be unlawful for any person to:

(1) Teach, train, or demonstrate to any other person the use, application, or making of any illegal firearm, dangerous weapon, explosive, or incendiary device capable of causing injury or death to persons either directly or through a writing or over or through a computer or computer network if the person teaching, training, or demonstrating knows, has reason to know, or intends that such teaching, training, or demonstrating will be unlawfully employed for use in or in furtherance of a civil disorder, riot, or insurrection; or

(2) Assemble with one or more persons for the purpose of being taught, trained, or instructed in the use of any illegal firearm, dangerous weapon, explosive, or incendiary device capable of causing injury or death to persons if such person so assembling knows, has reason to know, or intends that such teaching, training, or instruction will be unlawfully employed for use in or in furtherance of a civil disorder, riot, or insurrection.

(c) Any person who violates any provision of subsection (b) of this Code section shall be guilty of a felony and, upon conviction thereof, shall be punished by a fine of not more than $5,000.00 or by imprisonment for not less than one nor more than five years, or both.

16-11-160. Use of machine guns, sawed-off rifles, sawed-off shotguns, or firearms with silencers during commission of certain offenses; enhanced criminal penalties.

(a)(1) It shall be unlawful for any person to possess or to use a machine gun, sawed-off rifle, sawed-off shotgun, or a firearm equipped with a silencer, as those terms are defined in Code Section 16-11-121, during the commission or the attempted commission of any of the following offenses:

(A) Aggravated assault as defined in Code Section 16-5-21;

(B) Aggravated battery as defined in Code Section 16-5-24;

(C) Robbery as defined in Code Section 16-8-40;

(D) Armed robbery as defined in Code Section 16-8-41;

(E) Murder or felony murder as defined in Code Section 16-5-1;

(F) Voluntary manslaughter as defined in Code Section 16-5-2;

(G) Involuntary manslaughter as defined in Code Section 16-5-3;

(H) Sale, possession for sale, transportation, manufacture, offer for sale, or offer to manufacture controlled substances in violation of any provision of Article 2 of Chapter 13 of this title, the 'Georgia Controlled Substances Act';

(I) Terroristic threats or acts as defined in Code Section 16-11-37;

(J) Arson as defined in Code Section 16-7-60, 16-7-61, or 16-7-62 or arson of lands as defined in Code Section 16-7-63;

(K) Influencing witnesses as defined in Code Section 16-10-93; and

(L) Participation in criminal gang activity as defined in Code Section 16-15-4.

(2)(A) As used in this paragraph, the term 'bulletproof vest' means a bullet-resistant soft body armor providing, as a minimum standard, the level of protection known as 'threat level I,' which means at least seven layers of bullet-resistant material providing protection from at least three shots of 158-grain lead ammunition fired from a .38 caliber handgun at a velocity of 850 feet per second.

(B) It shall be unlawful for any person to wear a bulletproof vest during the commission or the attempted commission of any of the following offenses:

(i) Any crime against or involving the person of another in violation of any of the provisions of this title for which a sentence of life imprisonment may be imposed;

(ii) Any felony involving the manufacture, delivery, distribution, administering, or selling of controlled substances or marijuana as provided in Code Section 16-13-30; or

(iii) Trafficking in cocaine, illegal drugs, marijuana, or methamphetamine as provided in Code Section 16-13-31.

(b) Any person who violates paragraph (1) of subsection (a) of this Code section shall be guilty of a felony, and, upon conviction thereof, shall be punished by confinement for a period of ten years, such sentence to run consecutively to any other sentence which the person has received. Any person who violates paragraph (2) of subsection (a) of this Code section shall be guilty of a felony, and, upon conviction thereof, shall be punished by confinement for a period of one to five years, such sentence to run consecutively to any other sentence which the person has received.

(c) Upon the second or subsequent conviction of a person under this Code section, the person shall be punished by life imprisonment. Notwithstanding any other law to the contrary, the sentence of any person which is imposed for violating this Code section a second or subsequent time shall not be suspended by a court or a probationary sentence imposed in lieu thereof.

(d) The punishment prescribed for the violation of subsections (a) and (c) of this Code section shall not be probated or suspended as is provided by Code Section 17-10-7.

(e) Any crime committed in violation of this Code section shall be considered a separate offense.

Part 5. Brady Law Regulations

16-11-171. Definitions. As used in this part, the term:

(1) "Center" means the Georgia Crime Information Center within the Georgia Bureau of Investigation.

(2) "Dealer" means any person licensed as a dealer pursuant to 18 U.S.C. Section 921, et seq., or Chapter 16 of Title 43.

(3) "Firearm means any weapon or the frame or receiver of any such weapons, any firearm muffler or firearm silencer, or any destructive device as defined in 18 U.S.C. Section 912(a)(3).

(4) "Involuntarily hospitalized" means hospitalized as an inpatient in any mental health facility pursuant to Code Section 37-3-81 or hospitalized as an inpatient in any mental health facility as a result of being adjudicated mentally incompetent to stand trial or being adjudicated not guilty by reason of insanity at the time of the crime pursuant to Part 2 of Article 6 of Title 17.

(5) "NICS" means the National Instant Criminal Background Check System created by the federal "Brady Handgun Violence Prevention Act" (P. L. No. 103-159).

16-11-172. Transfers or purchases of firearms subject to the NICS; Information concerning persons who have been involuntarily hospitalized to be forwarded to the FBI; penalties for breach of confidentiality; exceptions.

(a) All transfers or purchases of firearms conducted by a licensed importer, licensed manufacturer, or licensed dealer shall be subject to the NICS. To the extent possible, the center shall provide to the NICS all necessary criminal history information and wanted person records in order to complete an NICS check.

(b) The center shall forward to the Federal Bureau of Investigation information concerning persons who have been involuntarily hospitalized as defined in this part for the purpose of completing an NICS check.

(c) Any government official who willfully or intentionally compromises the identity, confidentiality, and security of any records and data pursuant to this part shall be guilty of a felony and fined no less than $5,000.00 and shall be subject to automatic dismissal from his or her employment.

(d) The provisions of this part shall not apply to:

(1) Any firearm, including any handgun with a matchlock, flintlock, percussion cap, or similar type of ignition system, manufactured in or before 1898;

(2) Any replica of any firearm described in paragraph (1) of this subsection if such replica is not designed or redesigned to use rimfire or conventional centerfire fixed ammunition or uses rimfire or conventional centerfire fixed ammunition which is no longer manufactured in the United States and which is not readily available in the ordinary channels of commercial trade; and

(3) Any firearm which is a curio or relic as defined by 27 C.F.R. 178.11.

16-11-173.Legislative findings; preemption of local regulation and lawsuits; exceptions.

(a)(1) It is declared by the General Assembly that the regulation of firearms is properly an issue of general, statewide concern.

(2) The General Assembly further declares that the lawful design, marketing, manufacture, and sale of firearms and ammunition to the public is not unreasonably dangerous activity and does not constitute a nuisance per se.

(b)(1) No county or municipal corporation, by zoning or by ordinance, resolution, or other enactment, shall regulate in any manner gun shows; the possession, ownership, transport, carrying, transfer, sale, purchase, licensing, or registration of firearms or components of firearms; firearms dealers; or dealers in firearms components.

(2) The authority to bring suit and right to recover against any firearms or ammunition manufacturer, trade association, or dealer by or on behalf of any governmental unit created by or pursuant to an Act of the General Assembly or the Constitution, or any department, agency, or authority thereof, for damages, abatement, or injunctive relief resulting from or relating to the lawful design, manufacture, marketing, or sale of firearms or ammunition to the public shall be reserved exclusively to the state. This paragraph shall not prohibit a political subdivision or local government authority from bringing an action against a firearms or ammunition manufacturer or dealer for breach of contract or express warranty as to firearms or ammunition purchased by the political subdivision or local government authority.

(c) A county or municipal corporation may regulate the transport, carrying, or possession of firearms by employees of the local unit of government in the course of their employment with such local unit of government.

(d) Nothing contained in this Code section shall prohibit municipalities or counties by ordinance, resolution, or other enactment, from requiring the ownership of guns by heads of households within the political subdivision.

(e) Nothing contained in this Code section shall prohibit municipalities or counties, by ordinance, resolution, or other enactment, from reasonably limiting or prohibiting the discharge of firearms within the boundaries of the municipal corporation or county.

Title 43, Professions and Businesses

Chapter 16. Firearms Dealers

43-16-2. License requirement; applicability of chapter to casual sales. Any person, firm, retail dealer, wholesale dealer, pawnbroker, or corporation who shall sell, dispose of, or offer for sale or cause or permit to be sold, disposed of, or offered for sale any pistol, revolver, or short-barreled firearm of less than 15 inches in length, whether the same shall be his own property or whether he shall sell the same as an agent or employee of another, shall obtain from the department a license permitting the sale of such pistols, revolvers, and firearms. Nothing in this chapter shall apply to or prohibit the casual sales of the articles referred to in this Code section between individuals or bona fide gun collectors.

43-16-3.Affidavit of applicant for license. Any person, firm, retail dealer, wholesale dealer, pawnbroker, or corporation who makes application for a license under this chapter must accompany such application with an affidavit of the applicant sworn to before an officer authorized by law to administer oaths, stating that the applicant is a citizen of the United States, has reached the age of 21 years, and has not been convicted of a felony.

43-16-4.Surety bond. There shall be filed with the application for a license under this chapter a bond for the sum of $1,000.00, payable to the State of Georgia and conditioned upon the faithful performance of all provisions of this chapter, signed by the applicant as principal, and with a surety company or two individuals as securities. Such bond must be approved by the department.

43-16-5.License fees. All annual license fees described by this chapter shall be paid to the department on or before July 1 of each year. The department shall issue its receipt for every payment. The annual license payment to acquire such license shall be $ 25.00 for the owner of any establishment which sells any firearms listed in Code Section 43-16-2. The annual employee license fee shall be $ 3.00.

43-16-7.License to be displayed. Every recipient of a license to sell any firearms listed in Code Section 43-16-2 shall keep such license conspicuously displayed on his business premises.

43-16-8. Revocation of license for nonpayment of fee. Should any licensee fail or neglect to pay his annual license fee on or before July 1 of every year, the department shall notify him that his license will be revoked. Unless the fee is paid in full before August 1 of the same year, the department shall revoke the license.

43-16-9. Reinstatement of license. The owner of any establishment or employee thereof whose license for selling such firearms has been revoked for failure to pay the annual license fee may make application to the department for reinstatement. Such application shall be accompanied by a fee of $ 10.00, in addition to the regular license fee required. If the department shall find the applicant guilty only of default in payment of annual license fees, the license may be immediately reinstated.

43-16-10. Authority to revoke license for fraud, unethical practice or crime. The department shall have the power to revoke any license granted by it under this chapter to any person, firm, retail dealer, wholesale dealer, pawnbroker, or corporation, or any agent or employee thereof, found by the Board of Public Safety to be guilty of fraud or willful misrepresentation, or found guilty under the laws of this state of any crime involving moral turpitude, or found guilty of violating Code Section 16-11-101.

43-16-10.1. Record-keeping requirements.

(a) As a condition of any license issued pursuant to this chapter, each licensee shall be required to keep a record of the acquisition and disposition of firearms as provided in this Code section.

(b) The record required by subsection (a) of this Code section shall be identical in form and context to the firearms acquisition and disposition record required by Part 178 of Chapter 1 of Title 27 of the Code of Federal Regulations as it exists on July 1, 1988.

(c) The record required by subsection (a) of this Code section shall be maintained on the licensed premises and shall be open to the inspection of any duly authorized law enforcement officer during the ordinary hours of business or at any reasonable time. The record of each acquisition or disposition of a firearm shall be maintained for a period of not less than five years.

(d) The failure of a licensee to keep and maintain the records required by this Code section shall be grounds for revocation of the license.

43-16-11. Revocation proceedings governed by Administrative Procedure Act. All proceedings for the revocation of licenses issued under this chapter shall be governed by Chapter 13 of Title 50, the "Georgia Administrative Procedure Act."

43-16-12. Violations relating to regulation of dealers in pistols and short barreled firearms. Any person, firm, or corporation who violates this chapter shall be guilty of a misdemeanor.

[Current through 2007 Regular Session]

GUAM

GUAM CODE

Title 9. Crimes and Corrections

Chapter 71. The Guam Gun-Free School Zone Act of 2004

71.10. Title. This Chapter shall be known, and may be cited, as "The Guam Gun-Free School Zone Act of 2004".

71.20. Definitions. As used in this Chapter, the following definitions shall apply:

(a) 'School zone' means an area in, or on the grounds of, a public or private school providing instruction in early childhood, kindergarten or grades 1 to 12, inclusive.

(b) 'Firearm' shall mean as defined in 10 GCA §60100.

(c) 'Concealed firearm' shall mean as defined in 9 GCA §60108(e).

71.30. Person Not Allowed to Possess Firearms. Any person who possesses a firearm in a place that the person knows, or reasonably should know, is a school zone, as defined in paragraph (a) of Subdivision §71.20, shall be punished as specified in Subdivision §71.60.

71.50. Firearms Prohibited on University or College Property.

(a) It shall be unlawful for any person to bring or possess a loaded firearm upon the grounds of a campus of, or buildings owned or operated for student housing, teaching, research, or administration by, a public or private university or college, that are contiguous or are clearly marked university property, unless it is with the written permission of the university or college president, his or her designee, or equivalent university or college authority. Notwithstanding §71.80, a university or college shall post a prominent notice at primary entrances on noncontiguous property stating that firearms are prohibited on that property pursuant to this Section.

(b) It shall be unlawful for any person to bring or possess a firearm upon the grounds of a campus of, or buildings owned or operated for student housing, teaching, research, or administration by, a public or private university or college, that are contiguous or are clearly marked university property, unless it is with the written permission of the university or college president, his or her designee, or equivalent university or college authority. Notwithstanding Section §71.80, a university or college shall post a prominent notice at primary entrances on noncontiguous property stating that firearms are prohibited on that property pursuant to this Section.

71.60. Punishment. Any person who violates §71.30, §71.40, or §71.50 of this Act shall be guilty of a felony of the third degree and any person who is convicted of an offense pursuant to §71.30, §71.40, or §71.50 shall be sentenced as follows:

(a) For a first offense, the Court shall impose a sentence of imprisonment of no more than three (3) years, a fine of not less than One Thousand Dollars ($1,000.00), and mandatory community service of no less than one hundred and fifty (150) hours.

(b) In cases where the person has been convicted of felonies under any provision of this Chapter, the person shall be sentenced to a term of imprisonment which shall not be less than five (5) years and in addition, may be fined not more than Fifteen Thousand Dollars ($15,000.00). The sentence, if for a term of years, shall include a special parole term of not less than one (1) year in addition to such term of imprisonment. Imposition or execution of such sentence shall not be suspended, and probation shall not be granted. Sentence in these cases must also include mandatory community service of no less than one hundred fifty (150) hours unless the term of imprisonment is for life.

(c) The Court shall apply any minimum sentence, fine or community service specified in this Section, except in unusual cases where the interests of justice would best be served by granting probation or suspending the execution or imposition of sentence without the minimum imprisonment, fine or community service required in this Subdivision or by granting probation or suspending the execution or imposition of sentence with conditions other than those set forth in this Section, in which case the Court shall specify on the record and shall enter on the minutes the circumstances indicating that the interests of justice would best be served by this disposition.

71.70. What Constitutes a Loaded Firearm. For purposes of this Chapter, a firearm shall be deemed to be loaded when there is an unexpended cartridge or shell, consisting of a case that holds a charge of powder and a bullet or shot, in, or attached in any manner to, the firearm, including, but not limited to, in the firing chamber, magazine, or clip thereof attached to the firearm. A muzzle-loader firearm shall be deemed to be loaded when it is capped or primed and has a powder charge and ball or shot in the barrel or cylinder.

71.80. Notice.

(a) The Guam Department of Education and other entities covered by this Chapter shall post permanent signs with large visible lettering stating at a minimum, 'Warning this is a Gun-Free Zone' at the main entrances of the covered facilities within their control on or before January 1, 2005. This Section does not require that notice be posted regarding the proscribed conduct for the purposes of prosecution of any violation of this Act.

(b) The Guam Police Department within sixty (60) days of the effective date of this Act shall implement a public relations campaign to inform the general public of its provisions.

71.81. Not Applicable to Peace Officers and Military. This Chapter does not apply to a duly appointed peace officer as defined in §5.55, Article 2, Chapter 5, Title 8, Guam Code Annotated, a full-time paid peace officer of another state or the Federal government who is carrying out official duties while in Guam, any person summoned by any of these officers to assist in making arrests or preserving the peace while he or she is actually engaged in assisting the officer, a member of the military forces of Guam or of the United States who is engaged in the performance of his or her duties, or an armored vehicle guard engaged in the performance of his or her duties.

71.82. Not Applicable to Security Guards. This Chapter does not apply to an on-duty security guard authorized to carry a loaded firearm, provided the security guard is an employee of an entity contracted by the school for security purposes.

71.83. Not Applicable to Existing Shooting Ranges. This Chapter does not apply to an existing shooting range at a public or private school or university or college campus.

Title 10. Health and Safety

Chapter 60. Firearms

[Note: References to "Director of Public Safety" changed to "Chief of Police", and references to "Department of Public Safety" changed to "Guam Police Department" pursuant to P.L. 17-78:1, which repealed §5102 GC providing for the Department of Public Safety and reenacted §5102 establishing the Guam Police Department.**]**

60100. Definitions. As used in this Chapter:

(a) *Firearm* means any weapon, the operating force of which is an explosive. This definition includes pistols, revolvers, rifles, shotguns, machine guns, automatic rifles, noxious gas projectors, mortars, bombs, cannon and submachine guns. The specific mention of certain weapons does not exclude from the definition other weapons operated by explosives.

(b) *Pistol or revolver* means any firearm of any shape whatever and designed to be fired with one hand with a barrel less than twelve inches (12") in length and capable of discharging loaded ammunition or any noxious gas.

(c) *Rifle* means a firearm designed, made, redesigned or remade and intended to be fired from the shoulder and to fire only a single projectile through a rifled bore for each single pull of the trigger, and shall include any such firearm which may be readily restored to fire loaded ammunition, and any antique rifle or long gun, regardless of the type of charge or projectile used, which is capable of being used as a weapon.

(d) *Shotgun* means any firearm designed, made, redesigned or remade and intended to be fired from the shoulder and to fire through a smooth barrel either a number of projectiles (ball shot) or a single projectile, and shall include any such firearm which may be readily restored to fire any of the above, and shall also include any firearm of any age designed and capable of firing the above-mentioned projectiles.

(e) The terms *Chief of Police* and *department* mean the Chief of Police and the Guam Police Department, respectively.

(f) *Body armor* means a commercially available, soft, lightweight material with penetration resistance equal to or greater than that of eighteen (18) layers of KEVLAR.

(g) *Restricted bullet* means a bullet that, as determined by the Chief of Police, when fired from a pistol or revolver with a barrel five inches (5") or less in length, is capable of penetrating body armor.

(h) *Putative Owner* means a person who were it not for the requirements of this Chapter, would be the owner of a firearm, or one who acquires possession of a firearm, or one who acquires possession of a firearm by theft.

(i) *Lawful Owner* means a person whose right to a firearm have been perfected in accordance with §60103 of this title.

60101. Chapter Not Applicable. This Chapter shall not apply to:

(a) Law enforcement officers when using firearms authorized by the Chief of Police in their official duty.

(b) Persons in the Armed Forces of the United States engaged in official duty.

(c) Devices which are not designed or redesigned for use as a weapon. Any device, although originally designed as a firearm, which has been redesigned, or was designed initially for use as a signaling, pyrotechnic, line-throwing, safety, industrial or similar device, surplus ordnance sold, loaned or given by the Secretary of the Army pursuant to the provisions of §§ 4684(2), 4685 or 4686 of Title 10 of the United States Code is not covered by this Chapter.

(d) Those firearms that can not fire fixed ammunition but are loaded through the muzzle or cylinder with a combination of ball and black powder.

60102. Ownership, Etc., of Certain Firearms Prohibited. The manufacture, possession, sale, barter, trade, gift, transfer or acquisition of any machine guns, sub-machine guns, automatic rifles or any other firearm not a rifle having a barrel length of sixteen (16) inches or greater or not a shotgun having a barrel length of eighteen (18) inches or greater or a revolver or pistol having a barrel length of more than twelve (12) inches is prohibited. Mufflers, silencers or devices for deadening the sound of discharged firearms are also prohibited. Any person violating this Section shall be guilty of a felony which shall be punishable for a term of imprisonment of not less than three (3) years and a fine of not less than One Thousand Dollars ($1,000). Imposition of sentence shall not be suspended and the offender shall not be eligible for parole nor work release until the term of imprisonment prescribed herein has been completed nor may probation be imposed in lieu of this portion of the offender's sentence. Provided, however, that in the case of an offender not previously convicted of a felony, the court may sentence the offender to not more than two (2) years imprisonment and the provisions of this Section prohibiting probation, suspension, parole or work release shall not be applicable to such offender.

60103. Ownership, Etc., Permitted. Any person who qualifies under this Chapter may lawfully own, possess, use or carry any rifle, shotgun, pistol or revolver not prohibited by §60102 subject to the conditions and penalties provided in this Chapter.

60104. Dealer: Register: Dealer, etc., must register. Any person who is in the business of selling, buying, renting or trading any firearm shall register with the Department of Revenue and Taxation under the terms and conditions set forth in this Chapter and the rules and regulations adopted hereunder before engaging in any of the above activities. The term in the business of shall mean any person, natural or legal, who engages in the above activities for profit or who so engages on behalf of others, but shall not include private sales, loans, gifts or transfers of lawfully possessed firearms which have been registered and possessed under the provisions of §§60103 and 60110 of this Chapter.

60106. Identification Card: Required. No person shall own, possess, use, carry or acquire any firearm which is lawful under §60103 unless he shall have first obtained from the Department an identification card evidencing his right to such ownership, possession, use or carrying, concealed or otherwise, as stated upon the face of the card; except that any person may use and possess a firearm at any shooting gallery licensed pursuant to §60105, et seq., without first obtaining a permit as otherwise required by this Section. Said card shall be in the possession of the holder and on his person whenever he is using or has in his possession a firearm.

60107. Same: Contents. Identification cards shall have on their face all the following:

(1) The name, address, sex, height and weight of the holder, his birth date, Social Security number, if any, or Guam I.D. number, if any, and the expiration date of the card which shall be three (3) years after the holder's next birthday.

(2) A photograph of the holder taken immediately prior to issuance.

(3) An indication of the type of firearm which may be owned, used, carried, possessed or acquired by the holder, and whether or not the holder may carry the firearm concealed.

(4) A number, unique to the holder, which shall also be assigned to all registration records concerning firearms possessed by the holder.

60108. Same: Restrictions.

(a) No identification card shall issue unless the Department is satisfied that the applicant may lawfully possess, use, carry, concealed or otherwise, own or acquire the type of firearm stated in the application and upon the face of the identification card.

(b) No person shall be issued an identification card:

(1) Who has been convicted by any court of the United States, a state, territory, possession, trust territory or political subdivision thereof of any felony; or

(2) Who is an alien, except temporary permits may be issued to aliens for use only at target ranges operated by persons possessing permits therefor and who are citizens, or only for use at authorized sporting events and except for official representatives of foreign governments in their official capacities; or

(3) Who is presently charged by information or indictment with any crime stated in (b)(1) above; or

(4) Who has been adjudicated incompetent, has been committed to any mental institution; or

(5) Who is under the age of eighteen (18) years; or

(6) Who has been convicted of any violation of the Uniform Controlled Dangerous Substances Act or any misdemeanor where personal injury or use of firearms was an element or factor of the offense unless the Chief of Police has determined that the offense was committed more than ten (10) years previously and that the applicant would not endanger the public safety by receiving an identification card; or

(7) Who, in the determination of the Chief of Police appears to suffer from a physical or mental disease or defect which would adversely affect the safe use of the firearm applied for, unless the person obtains a certificate from a licensed physician stating that the physical or mental disease or defect would not adversely affect his safe operation of the firearm or the public safety.

(c) An applicant for an identification card shall apply therefore on a form supplied by the Department and shall provide such information as may be necessary to afford the Department reasonable opportunity to ascertain the facts required to appear upon the card and facts relevant to the applicant's eligibility for a card, and facts necessary to determine whether the applicant may carry a concealed weapon if such permission is sought. The applicant is required to submit to a national criminal background check conducted by the Federal Bureau of Investigation to process such background checks.

A fingerprint card, or electronic fingerprint data, if so utilized, shall be transmitted to Guam's State Identification Bureau as classified by the Federal Bureau of Investigation for processing of fingerprints.

(d) If the application is not denied, the identification card shall issue within thirty (30) days, except where application has been made to carry a concealed weapon, in which case the card shall issue within sixty (60) days.

(e) For purposes of this Chapter, the term concealed when used in connection with any firearm whatsoever shall mean a firearm which is carried on a person or within a vehicle in such a manner that it cannot be seen by the naked eye, but would be available for use by the person concealing or attempting to conceal the firearm or any other person aware of the firearm's existence. A firearm is not concealed when it is within a locked portion of any vehicle, unloaded, with the ammunition stored outside of the firearm or any clip or ammunition storage chamber attached to the firearm, and outside of the immediate reach of the person so placing the weapon or any other person knowing of its existence. A firearm is concealed if, among other things, it is hidden beneath any article of clothing so that only the shape is visible, but not the firearm itself.

60110. Registration. Any person purchasing, receiving by gift, device or otherwise, acquiring or otherwise coming into permanent possession of a firearm, the possession of which is permitted by this chapter, shall register the same with the Department within three (3) working days after acquiring said firearm on the forms specified by the Department. Such facts and information shall be given so as to enable the Department to record for identification purposes the firearm so registered. It shall be unlawful for any person to own or possess any firearm which has not been registered. No firearm may be registered by the Department unless the person presenting the firearm also displays current identification card evidencing his eligibility to own, possess, use or carry the firearm presented for inspection as to the facts required for registration. Any firearm registration which expires on or after March 1, 1988 or which is thereafter issued under this chapter shall be permanent for as long as the registrant retains the firearm. The Chief of Police shall promulgate rules and regulations establishing a permanent firearms identification card and a reasonable fee to cover the cost incurred.

60111. Transfer of Firearm: Private. Upon the transfer of any firearm from one person not a dealer, manufacturer, wholesaler or repairer to another not a dealer, manufacturer, wholesaler or repairer, other than on loan, a new registration must be obtained by the new owner or one permanently possessing the firearm, indicating the prior owner or possessor, and his identification card number, as well as all information pertinent to the new owner or possessor.

60112. Same: Pawnbroker. Upon transfer to a pawnbroker, and upon retransfer from the pawnbroker to either the owner or to another person, a new registration must be obtained.

60113. Loss, Destruction or Theft of Identification Card or Firearm.

(a) Upon the destruction, mutilation, loss or theft of any identification card, the holder shall report the event to the Department and may obtain a duplicate card upon the payment of the fee required therefor and upon proof of loss, mutilation, destruction or theft. If any portion of the original card remains, the holder must surrender it to the Department before receiving a duplicate.

(b) Upon the loss, theft, damage beyond repair to any firearm registered under this Chapter, the holder of the identification card applicable to the weapon in question shall report the event to the Department, along with the nature of the disposition of the firearm, if lost, destroyed or damaged beyond repair, and the same information shall be recorded upon the registration records kept by the Department applicable to the firearm involved.

(c) The holder of any identification card or any person in whose name a firearm is registered shall report to the Department any change in the facts stated upon the card and any change in the facts which would effect the holder's eligibility to continue to possess a card.

(d) The holder shall report any of the events required to be reported by Subsection (a) and (b) above within three (3) working days of their occurrence and any of the events required to be reported by Subsection (c) within ten (10) days of their occurrence.

60114. Identification Card: Possession.

(a) No person may use or possess a firearm regardless of ownership of the firearm, without possession of an identification card evidencing his eligibility to possess, use or carry such firearms.

(b) No person shall transfer, loan, give, sell either as a part of a business or individually, or permit the use of any firearm unless the person receiving the firearm has an identification card evidencing his eligibility to receive, possess, use and carry the type of firearm which he is to receive, borrow, use, buy or possess or carry.

(c) No person shall use, display, or carry with the intent to use or display, an identification card which has been suspended or revoked, or for which a duplicate has been issued, or which has been defaced or altered. No person shall use any of the above for the purpose of obtaining any firearm.

(d) No person shall willfully alter, remove or obliterate the name of the make, model, manufacturer's number or other mark of identity of any firearm or ammunition. Possession of a firearm or ammunition upon which any mark of identity has been altered, removed or obliterated shall be presumptive evidence that the possessor has altered, removed or obliterated the same.

(e) If any person, in complying with any of the requirements of this Chapter, knowingly gives false information or knowingly offers false evidence of any of the facts required by the Department or by this Chapter, he shall be guilty of unsworn falsification and punished therefore as provided by §52.30 of the Criminal and Correctional Code. Addition-ally, upon conviction, such person shall not be eligible for an identification for a period of ten (10) years, and any existing identification card issued to such person shall be revoked, and any firearms possessed pursuant to such identification card forfeited.

(f) The registered owner shall report to the Guam Police Department the loss, theft, sale or transfer of a firearm within three (3) working days thereof.

60115. Dealer: Registration. Dealers, etc., must register. Any person who is a dealer, manufacturer, wholesaler or retailer of firearms or ammunition within the territory of Guam who must register pursuant to §60104 shall do so by supplying the Department with information required by it relative to the name of the business, name or names of the owners, whether the business is a corporation, sole proprietorship, partnership or any other form, relative to the types of firearms sold, manufactured, bought or repaired by such business, to the security of the storage areas of the business wherein the firearms are stored and any other information deemed necessary by the Department. Upon receipt of the necessary information, the Department shall issue a registration card, which shall be prominently displayed at the place of business of the person or firm so registered and copies to be displayed at each location where firearms are sold.

60116. Conduct of Dealer's Business. Any person or firm licensed pursuant to §60104 shall, in the conduct of his business:

(1) Display no firearms or ammunition in any place where they can be seen from outside the premises.

(2) Keep all firearms and ammunition in a securely locked place at all times except when they are actually being shown to a customer or when actually being repaired or worked on.

(3) Permit only persons who are holders of identification cards making them eligible to purchase, possess and use the firearms sold by the registrant or repaired by the repair facility to have access to the firearms or ammunition, except that nothing in this Section shall be construed to require a customer to show an identification card prior to his engaging in a transaction for which an identification card is required by this Chapter.

(4) Sell only to persons who have a valid identification card making them eligible to purchase, own, possess, use and carry the firearm or ammunition which is the subject of the transaction.

(5) Keep a continuing and up-to-date inventory of all firearms and ammunition in his possession and report the loss, theft, damage or destruction of the same to the Department immediately upon the discovery of the event.

(6) Deliver a written warning to the purchaser or transferee of a firearm a written warning which states in block letters not less than one-fourth inch (1/4") in height: It is unlawful and punishable by imprisonment and a fine for any adult to store or leave a firearm without placing it in a secure locked cabinet or closet, or without a trigger lock in place.

(7) Conspicuously post at each purchase counter the following warning in block letters not less than one (1) inch in height: It is unlawful to store or leave a firearm without placing it in a secure locked cabinet or closet, or without a trigger lock in place.

60117. Repair. No person, other than the owner or possessor of a firearm, may accept any firearm for repair without having first been shown a valid identification card showing the person delivering the firearm to such person accepting it for repair to be eligible to possess the firearm in question. No firearm shall be returned to any person after repair or other work done on it without the repairer having seen the identification card of the one to whom the firearm is to be given.

60118. Private Sale or Transfer. Any person engaging in a private sale, loan or other transfer of a firearm shall be shown by the person receiving the firearm an identification card evidencing the receiver's right to own, possess, use and carry such firearm. It shall be unlawful for a person to transact a private sale, loan, gift or transfer without having seen the valid identification card of the person receiving the firearm.

60120. Rules and Regulations. The Chief of Police may adopt, pursuant to the Administrative Adjudication Act, rules and regulations to implement this Chapter. Nevertheless, the absence of such rules and regulations shall not affect the implementation of this Chapter.

60121. Prohibitions and Penalties.

(a) Any person who knowingly owns or possesses an unregistered firearm shall be punishable by imprisonment of not less than one (1) year and a fine of not less than One Thousand Dollars ($1,000).

(b) For any violation of this Chapter or the rules and regulations adopted hereunder where another penalty is not set forth specifically, any person, corporation, firm or business found guilty shall be guilty of a misdemeanor. In the case of a corporation, business or other firm convicted of a violation of this Chapter, as a legal person, then the individual actually per-forming the acts forbidden or omitting the acts required shall be guilty in addition to the corporation, firm or business as an entity.

(c) Any person who carries a concealed firearm without a valid identification card evidencing permission to carry a concealed weapon shall be guilty of a felony which shall be punishable for a term of imprisonment of not less than three (3) years and a fine of not less than One Thousand Dollars ($1,000).

(d) Any person who knowingly removes, obliterates or alters the identifying marks of a firearm shall be guilty of a felony which shall be punishable for a term of imprisonment of not less than three (3) years and a fine of not less than One Thousand Dollars ($1,000).

(e) Any person purchasing, possessing, using or carrying a firearm without an applicable identification card shall be guilty of a felony which shall be punishable for a term of imprisonment of not less than three (3) years and a fine of not less than One Thousand Dollars ($1,000).

(f) Any person, legal or natural, who transfers, gives, returns, sells or loans any firearm to any person without a valid identification card is guilty of a felony which shall be punishable for a term of imprisonment of not less than three (3) years and a fine of not less than One Thousand Dollars ($1,000). In the case of a violation by a legal person, the natural persons liable for punishment under this Section, in addition to said legal person, shall be the person or persons performing the forbidden acts or omitting the required acts and any responsible officer of the legal person who knew or should have known of the acts or omissions committed.

(g)(1) Any person legal or natural who transfers, gives, returns, sells, or loans any firearm to any minor is guilty of a felony which shall be punishable by a term of imprisonment not less than five (5) years and a fine of not less than Five Thousand Dollars ($5,000). In the case of a violation by a legal person or persons performing the forbidden acts, or omitting the required acts and any responsible officer who knew or should have known of the act as omission committed shall be liable for punishment under this section.

(2) This section does not apply when a minor is being instructed by his parent or guardian in the use of weapons or when that minor, properly licensed, is hunting with his parent or guardian who is also properly licensed and complying with the provisions of Department of Agriculture Hunting Regulations §§15300.00 through 15300. 29.

(h) Any person who negligently entrusts a firearm to a minor is guilty of a felony, which shall be punishable by a term of imprisonment of not less than one (1) year and a fine of not less than One Thousand Dollars ($1,000).

(i) Imposition of sentence pursuant to the provisions of this Section shall not be suspended nor shall probation be granted and the offender shall not be eligible for parole nor work release until the term of imprisonment prescribed herein has been completed.

(j) Notwithstanding provisions of this Section to the contrary, in the case of an offender not previously convicted of a felony, the court may sentence the offender to not more than two (2) years imprisonment for an offense established by the provisions of Subsections (a), (b), (c), (d) or (e) of this Section. The provisions of Subsection (f) of this Section prohibiting probation, suspension, parole or work release shall not be applicable to such an offender.

60121.1. Same: Restricted Bullets. Any person who imports, manufactures or sells a restricted bullet on Guam, except as specifically authorized by the Chief of Police for purposes of public safety shall be guilty of a felony of the third degree and if the person holds a business license or is registered under the provisions of this Chapter, such business license and such registration shall be subject to revocation.

60121.2. Same: Same. Any person who possesses or carries a restricted bullet not in accordance with the regulations promulgated by the Chief of Police shall be guilty of a felony of the third degree.

60122. Fees.

(a) The fee for each identification card, duplicate or renewal thereof shall be Five Dollars ($5.00) and shall be paid to the Treasurer of Guam for the account of the Department of Revenue and Taxation before any identification card may be issued or renewed.

(b) The annual registration fee for every dealer, shooting gallery, manufacturer, wholesaler, retailer and repairer shall be One Hundred Dollars ($100) and shall be payable to the Treasurer of Guam for the account of the Department of Revenue and Taxation before any such registration may be issued, renewed or a duplicate card issued.

(c) The fee for registration of every firearm by an individual shall be Two Dollars ($2.00) payable to the Treasurer of Guam for the account of the Department of Revenue and Taxation before any such registration may be issued or a duplicate issued.

(d) Registrations for dealers, shooting galleries, retailers, wholesalers, manufacturers and repairers shall be renewed annually at the same time as the business license is renewed and shall be prorated on the same scale as in the business license for the establishment involved.

60123. Revalidation of Previous Weapon Registration. All registrations of weapons issued under any prior law are hereby declared revalidated as of the date of enactment of this Act and shall be valid until the next birthday of the holder, at which time an identification card must be obtained and the firearms registered according to the provisions of this Chapter. This Section does not apply to firearms owned, possessed, used or carried in a manner declared illegal by any previous Act.

60124. Registration of Prohibited Firearms. Any firearm which cannot be legally owned, possessed, used or carried pursuant to this Act but which was legally possessed, owned, used or carried prior to the effective date of this Act may not be registered hereafter. However, the owner or possessor may retain said weapons until the time specified in §60123, at which time he must dispose of the firearm. If he cannot do so without loss to himself, the Department is authorized to purchase the firearm from the legal owner based upon the fair market value at the time of last eligibility to possess or own as stated in §60123, and keep or dispose of the firearm as the Chief of Police determines.

60125. Firearms Forfeited: When. Upon the conviction of any violation of this Chapter or of any crime wherein the use of firearms was an element or part or was involved in the commission thereof, the firearms involved in the violation shall be declared forfeit by the court convicting such person and be given to the Department for its use, sale or destruction. In addition, the court shall either revoke the applicable identification card or business registration or suspend the same for a period of not less than six (6) months nor for more than two (2) years. In the case of a conviction of a business, the weapons shall be forfeited only where the registration is revoked. In the case of suspension of a business registration no firearms shall be sold.

60126. Registry. The Department shall maintain a confidential registry, open only to law enforcement officials, of all firearms registered for the life of the weapon on Guam and for two years thereafter. However, once a firearm has been transferred and the registration cancelled in the name of the transferor, then the records of the transferor may be destroyed.

60127. New Residents. Any United States citizen who is in legal possession of a firearm or ammunition thereof pursuant to the laws of any state, territory, possession or Trust Territory of the United States shall have a period of thirty (30) days after arriving on Guam as a new resident in which to apply for an identification card and during that period may continue to use the firearm and, for that purpose, the gun is legally in his possession until an identification card is issued or denied him. Thereafter, all provisions of this Chapter apply in full to such person and he must purchase, possess or use firearms only in compliance therewith. The Section does not apply to any firearm which may not be legally possessed or used, carried or purchased under this Chapter. With respect to such firearms the provisions of §60123 shall apply except that under this Section the new resident has thirty (30) days in which to comply. Further, this Section does not apply to persons transiting Guam. Such transiting persons shall not have in their possession any firearm on Guam without first obtaining permission therefor from the Chief of Police.

60128. Exceptions. The provisions of Civil Code §§1714.2, and 1716.1; and subitems (40) and (41) of §5107, Title 6, Guam Code Annotated, shall not apply:

(a) If the firearm is stored in a secure, locked cabinet or closet in the home of its lawful owner.

(b) If the minor obtains the firearm as a result of the unlawful entry of the home of the owner of the firearm.

(c) To members of the Armed Forces, National Guard, police or other law enforcement officials with respect to firearm possession by a minor which occurs during or incidental to the performance of their official duties.

(d) If the firearm was properly secured by the owner with a trigger lock.

61101. Permit to Receive Explosives.

No person shall receive or possess any explosives without first having secured a permit therefore from the Director of Revenue and Taxation, and it is unlawful for any person to sell, give away or deliver explosives to any person who does not present such a permit.

61106. Dealers in Explosives: License.

No person shall be licensed to engage in the business of manufacturing, transporting, storing, selling, transferring or otherwise disposing of explosives, except upon certification by the Director of Revenue and Taxation that such person, his agents and employees, are qualified, capable and possess facilities to conduct such business in a safe manner.

Title 11. Finance and Taxation

Chapter 72. Commercial Licenses

72156. Explosive Vendors.

An endorsement for explosive vending must be obtained by any person who engages in the business of selling any explosives. Such endorsement is:

(a) Certification by the Guam Fire Chief that such person, his agents and employees, are qualified, capable and possess facilities to conduct such business in a safe manner.

[Current through P.L. 29-081, May 09, 2008]

HAWAII

HAW. REV. STAT.

Title 10. Public Safety and Internal Security

Chapter 134. Firearms, Ammunition and Dangerous Weapons

Part I. General Regulations

134-1. Definitions. As used in this chapter, unless the context indicates otherwise:

"Acquire" means gain ownership of.

"Antique pistol or revolver" means any pistol or revolver manufactured before 1899 and any replica thereof if it either is not designed or redesigned for using rimfire or conventional centerfire fixed ammunition or is designed or redesigned to use rimfire or conventional centerfire fixed ammunition that is no longer manufactured in the United States and is not readily available in the ordinary channels of commercial trade.

"Assault pistol" means a semiautomatic pistol which accepts a detachable magazine and which has two or more of the following characteristics:

(1) An ammunition magazine which attaches to the pistol outside of the pistol grip;

(2) A threaded barrel capable of accepting a barrel extender, flash suppressor, forward hand grip, or silencer;

(3) A shroud which is attached to or partially or completely encircles the barrel and which permits the shooter to hold the firearm with the second hand without being burned;

(4) A manufactured weight of fifty ounces or more when the pistol is unloaded;

(5) A centerfire pistol with an overall length of twelve inches or more; or

(6) It is a semiautomatic version of an automatic firearm;

but does not include a firearm with a barrel sixteen or more inches in length, an antique pistol as defined in section 134-1 or a curio or relic as those terms are used in 18 U.S.C. § 921(16) or 27 C.F.R. 178.11.

"Automatic firearm" means any firearm that shoots, is designed to shoot, or can be readily modified to shoot automatically more than one shot, without a manual reloading, by a single function of the trigger. This term shall also include the frame or receiver of any such firearm, any part designed and intended solely and exclusively, or any combination of parts designed and intended, for use in converting a firearm into an automatic firearm, and any combination of parts from which an automatic firearm can be assembled if the parts are in the possession or under the control of a single person.

"Chief of police" means the chief of police of the counties of Hawaii, Maui, Kauai, or the city and county of Honolulu.

"Crime of violence" means any offense, as defined in Title 37, that involves injury or threat of injury to the person of another.

"Electric gun" means any portable device that is electrically operated to project a missile or electromotive force. It does not include any electric livestock prod used in animal husbandry and any automatic external defibrillator used in emergency medical situations.

"Firearm" means any weapon, for which the operating force is an explosive, including but not limited to pistols, revolvers, rifles, shotguns, automatic firearms, noxious gas projectors, mortars, bombs, and cannon.

"Firearm loaded with ammunition" and **"loaded firearm"** means a firearm with ammunition present within the firing chamber, revolving cylinder, or within a magazine which is inserted in a firearm.

"Fugitive from justice" means any person

(1) who has fled from any state, territory, the District of Columbia, or possession of the United States, to avoid prosecution for a felony or to avoid giving testimony in any criminal proceeding or

(2) who has fled from any country other than the United States and is avoiding lawful extradition back to that country.

"Pistol" or **"revolver"** means any firearm of any shape with a barrel less than sixteen inches in length and capable of discharging loaded ammunition or any noxious gas.

"Public highway" shall have the same meaning as defined in section 264- 1(a).

"Semiautomatic" means the mode of operation by which a firearm uses the energy of the explosive in a fixed cartridge to extract a fired cartridge and chamber a fresh cartridge with each single pull of a trigger.

134-2. Permits to acquire.

(a) No person shall acquire the ownership of a firearm, whether usable or unusable, serviceable or unserviceable, modern or antique, registered under prior law or by a prior owner or unregistered, either by purchase, gift, inheritance, bequest, or in any other manner, whether procured in the State or imported by mail, express, freight, or otherwise, until the person has first procured from the chief of police of the county of the person's place of business or, if there is no place of business, the person's residence or, if there is neither place of business nor residence, the person's place of sojourn, a permit to acquire the ownership of a firearm as prescribed in this section. When title to any firearm is acquired by inheritance or bequest, the foregoing permit shall be obtained before taking possession of a firearm; provided that upon presentation of a copy of the death certificate of the owner making the bequest, any heir or legatee may transfer the inherited or bequested firearm directly to a dealer licensed under section 134-31 or licensed by the United States Department of Justice with-out complying with the requirements of this section.

(b) The permit application form shall include the applicant's name, address, sex, height, weight, date of birth, place of birth, country of citizenship, Social Security number, alien or admission number, and information regarding the applicant's mental health history and shall require the fingerprinting and photographing of the applicant by the police department of the county of registration; provided that where fingerprints and photograph are already on file with the department, these may be waived.

(c) An applicant for a permit shall sign a waiver at the time of application, allowing the chief of police of the county issuing the permit access to any records that have a bearing on the mental health of the applicant. The permit application form and the waiver form shall be prescribed by the attorney general and shall be uniform throughout the State.

(d) The chief of police of the respective counties may issue permits to acquire firearms to citizens of the United States of the age of twenty-one years or more, or duly accredited official representatives of foreign nations, or duly commissioned law enforcement officers of the State who are aliens; provided that any law enforcement officer who is the owner of a firearm and who is an alien shall transfer ownership of the firearm within forty-eight hours after termination of employment from a law enforcement agency. The chief of police of each county may issue permits to aliens of the age of eighteen years or more for use of rifles and shotguns for a period not exceeding sixty days, upon a showing that the alien has first procured a hunting license under chapter 183D, part II. The chief of police of each county may issue permits to aliens of the age of twenty-one years or more for use of firearms for a period not exceeding six months, upon a showing that the alien is in training for a specific organized sport-shooting contest to be held within the permit period. The attorney general shall adopt rules, pursuant to chapter 91, as to what constitutes sufficient evidence that an alien is in training for a sport-shooting contest. Notwithstanding any provision of the law to the contrary and upon joint application, the chief of police may issue permits to acquire firearms jointly to spouses who otherwise qualify to obtain permits under this section.

(e) The permit application form shall be signed by the applicant and by the issuing authority. One copy of the permit shall be retained by the issuing authority as a permanent official record. Except for sales to dealers licensed under section 134-31, or dealers licensed by the United States Department of Justice, or law enforcement officers, or where a license is granted under section 134-9, or where any firearm is registered pursuant to section 134-3(a), no permit shall be issued to an applicant earlier than fourteen calendar days after the date of the application; provided that a permit shall be issued or the application denied before the twentieth day from the date of application. Permits issued to acquire any pistol or revolver shall be void unless used within ten days after the date of issue. Permits to acquire a pistol or revolver require a separate application and permit for each transaction. Permits issued to acquire any rifle or shotgun shall entitle the permittee to make sub-sequent purchases of rifles or shotguns for a period of one year from the date of issue without a separate application and permit for each acquisition, subject to the disqualifications under section 134-7 and subject to revocation under section 134-13; provided that if a permittee is arrested for committing a felony or any crime of violence of for the illegal sale of any drug, the permit shall be impounded and shall be surrendered to the issuing authority. The issuing authority shall perform an inquiry on an applicant who is a citizen of the United States by using the National Instant Criminal Background Check System before any determination to issue a permit or to deny an application is made. If the applicant is not a citizen of the United States and may be eligible to acquire a firearm under this chapter, the issuing authority shall perform an inquiry on the applicant, by using the National Instant Criminal Background Check System, to include a check of the Immigration and Customs Enforcement databases, before any determination to issue a permit or to deny an application is made.

(f) In all cases where a pistol or revolver is acquired from another person within the State, the permit shall be signed in ink by the person to whom title to the pistol or revolver is transferred and shall be delivered to the person who is transferring title to the firearm, who shall verify that the person to whom the firearm is to be transferred is the person named in the permit and enter on the permit in the space provided the following information: name of the person to whom the title to the firearm was transferred; names of the manufacturer and importer; model; type of action; caliber or gauge; and serial number as applicable. The person who is transferring title to the firearm shall sign the permit in ink and cause the permit to be delivered or sent by registered mail to the issuing authority within forty-eight hours after transferring the firearm.

In all cases where receipt of a firearm is had by mail, express, freight, or otherwise from sources without the State, the person to whom the permit has been issued shall make the prescribed entries on the permit, sign the permit in ink, and cause the permit to be delivered or sent by registered mail to the issuing authority within forty-eight hours after taking possession of the firearm.

In all cases where a rifle or shotgun is acquired from another person within the State, the person who is transferring title to the rifle or shotgun shall submit, within forty-eight hours after transferring the firearm, to the authority which issued the permit to acquire, the following information, in writing: name of the person who transferred the firearm, name of the person to whom the title to the firearm was transferred; names of the manufacturer and importer; model; type of action; caliber or gauge; and serial number as applicable.

(g) Effective July 1, 1995, no person shall be issued a permit under this section for the acquisition of a pistol or revolver unless the person, at any time prior to the issuance of the permit, has completed:

(1) An approved hunter education course as authorized under section 183D-28;

(2) A firearms safety or training course or class available to the general public offered by a law enforcement agency of the State or of any county;

(3) A firearms safety or training course offered to law enforcement officers, security guards, investigators, deputy sheriffs, or any division or subdivision of law enforcement or security enforcement by a state or county law enforcement agency; or

(4) A firearms training or safety course or class conducted by a state certified or National Rifle Association certified firearms instructor or a certified military firearms instructor that provides, at a minimum, a total of at least two hours of firing training at a firing range and a total of at least four hours of classroom instruction, which may include a video, that focuses on:

(A) The safe use, handling, and storage of firearms and firearm safety in the home; and

(B) Education on the firearm laws of the State.

An affidavit signed by the certified firearms instructor who conducted or taught the course, providing the name, address, and phone number of the instructor and attesting to the successful completion of the course by the applicant shall constitute evidence of certified successful completion under this paragraph.

(h) No person shall sell, give, lend, or deliver into the possession of another any firearm except in accordance with this chapter.

(i) No fee shall be charged for permits, or applications for permits, under this section, except for a single fee chargeable by and payable to the issuing county, for individuals applying for their first permit, in an amount equal to the fee actually charged by the Federal Bureau of Investigation to the issuing police department for a fingerprint check in connection with that application or permit. In the case of a joint application, the fee provided for in this section may be charged to each person to whom no previous permit has been issued.

134-2.5. Permits for motion picture films or television program production.

(a) Upon a finding that public safety is not endangered, the chief of police of the appropriate county may issue permits, initially valid for a period of one year and renewable annually thereafter, for the possession, transportation, or use, with blank cartridges, of firearms or explosives solely as props for motion picture films or television program production upon a showing that good cause exists for the issuance of a permit to the applicant and upon sufficient proof of a federal firearms license and a state film permit required under section 201-3. No permit shall be issued to a person who is under twenty years of age or who is disqualified under section 134-7.

(b) Applications for permits shall be in writing, signed by the individual applicant or by a member or officer qualified to sign if the applicant is a firm or corporation, and shall state the name, business in which engaged, business address, and a full description of the use to which the firearms or explosives are to be put, including the names of the persons who will actually use the props. The application shall also require the fingerprinting and photographing of the applicant. Applications and permits shall be uniform throughout the State on forms prescribed by the attorney general.

(c) The attorney general shall establish rules pursuant to chapter 91 concerning security requirements for storing and transporting firearms or explosives for which permits are issued. Permits shall be issued only upon a showing of the applicant's ability to meet these security requirements.

(d) A fee of $50 should be charged for each permit issued under this section.

(e) Every applicant to whom a permit is issued shall keep it on the applicant's person or at the place where the firearms or explosives are stored. The permit, firearms and explosives, shall be available for inspection by any law enforcement officer or any other person designated by the respective chief of police.

(f) Every firearm or explosive for which a permit is issued shall bear a unique identifying number. If the firearm or explosive does not bear a unique identifying number, the chief of police of the appropriate county shall assign a number that shall be stamped or placed thereon.

(g) The chief of police of the respective county shall revoke permits issued under this section any time it appears that the holder of the permit has used the firearms or explosives for purposes other than those allowed by the permit or that the holder of the permit has not exercised great care in retaining custody of any firearm or explosive possessed under the permit.

134-3. Registration, mandatory, exceptions.

(a) Every person arriving in the State who brings or by any other manner causes to be brought into the State a firearm of any description, whether usable or unusable, serviceable or unserviceable, modern or antique, shall register the firearm within three days after arrival of the person or of the firearm, whichever arrives later, with the chief of police of the county of the person's place of business or, if there is no place of business, the person's residence or, if there is neither a place of business nor residence, the person's place of sojourn. A nonresident alien may bring firearms not otherwise prohibited by law into the State for a continuous period not to exceed ninety days; provided that the person meets the registration requirement of this section and the person possesses:

(1) A valid Hawaii hunting license procured under chapter 183D, part II, or a commercial or private shooting preserve permit issued pursuant to section 183D-34;

(2) A written document indicating the person has been invited to the State to shoot on private land; or

(3) Written notification from a firing range or target shooting business indicating that the person will actually engage in target shooting.

The nonresident alien shall be limited to a nontransferable registration of not more than ten firearms for the purpose of the above activities.

(b) Every person who acquires a firearm pursuant to section 134-2 shall register the firearm in the manner prescribed by this section within five days of acquisition. The registration shall be on forms prescribed by the attorney general, which shall be uniform throughout the State, and shall include the following information: name of the manufacturer and importer; model; type of action; caliber or gauge; serial number; and source from which receipt was obtained, including the name and address of the prior registrant. If the firearm has no serial number, the permit number shall be entered in the space provided for the serial number, and the permit number shall be engraved upon the receiver portion of the firearm prior to registration. All registration data that would identify the individual registering the firearm by name or address shall be confidential and shall not be disclosed to anyone, except as may be required for processing the registration or as may be required by a law enforcement agency for the lawful performance of its duties or as may be required by order of a court.

(c) Dealers licensed under section 134-31 or dealers licensed by the United States Department of Justice shall register firearms pursuant to this section on registration forms pre-scribed by the attorney general and shall not be required to have the firearms physically inspected by the chief of police at the time of registration.

(d) Registration shall not be required for:

(1) Any device that is designed to fire loose black powder or that is a firearm manufactured before 1899.

(2) Any device not designed to fire or made incapable of being readily restored to a firing condition; or

(3) All unserviceable firearms and destructive devices registered with the Bureau of Alcohol, Tobacco, and Firearms of the United States Department of Justice pursuant to Title 27, Code of Federal Regulations.

(e) No fee shall be charged for the registration.

134-3.5 Disclosure for firearm permit and registration purposes. A health care provider or public health authority shall disclose health information, including protected health care information, relating to an individual's mental health history, to the appropriate county chief of police in response to a request for the information from the chief of police; provided that:

(1) The information shall be used only for the purpose of evaluating the individual's fitness to acquire or own a firearm; and

(2) The individual has signed a waiver permitting release of the health information for that purpose.

134-4. Transfer, possession of firearms.

(a) No transfer of any rifle having a barrel length of sixteen inches or over or any shotgun having a barrel length of eighteen inches or over, whether usable or unusable, serviceable or unserviceable, modern or antique, registered under prior law or by a prior owner, or unregistered shall be made to any person under the age of eighteen years, except as provided by section 134-5.

(b) No person shall possess any firearm that is owned by another, regardless of whether the owner has consented to possession of the firearm, without a permit from the chief of police of the appropriate county, except as provided in subsection (c) and section 134-5.

(c) Any lawfully acquired rifle or shotgun may be lent to an adult for use within the State for a period not to exceed fifteen days without a permit; provided that where the rifle or shotgun is to be used outside of the State, the loan may be for a period not to exceed seventy-five days.

(d) No person shall knowingly lend a firearm to any person who is prohibited from ownership or possession of a firearm under section 134-7.

(e) After July 1, 1992, no person shall bring or cause to be brought into the State an assault pistol. No assault pistol may be sold or transferred on or after July 1, 1992, to anyone within the State other than to a dealer licensed under section 134-32 or the chief of police of any county except that any person who obtains title by bequest or intestate succession to an assault pistol registered within the State shall, within ninety days, render the weapon permanently inoperable, sell or transfer the weapon to a licensed dealer or the chief of police of any county, or remove the weapon from the State.

134-5. Possession by licensed hunters and minors; target shooting; game hunting.

(a) Any person of the age of sixteen years, or over or any person under the age of sixteen years while accompanied by an adult, may carry and use any lawfully acquired rifle or shotgun and suitable ammunition while actually engaged in hunting or target shooting or while going to and from the place of hunting or target shooting; provided that the person has procured a hunting license under chapter 183D, part II. A hunting license shall not be required for persons engaged in target shooting.

(b) A permit shall not be required when any lawfully acquired firearm is lent to a person, including a minor, upon a target range or similar facility for purposes of target shooting; provided that the period of the loan does not exceed the time in which the person actually engages in target shooting upon the premises.

(c) A person may carry unconcealed and use a lawfully acquired pistol or revolver while actually engaged in hunting game mammals, if that pistol or revolver and its suitable ammunition are acceptable for hunting by rules adopted pursuant to section 183D-3 and if that person is licensed pursuant to part II of chapter 183D. The pistol or revolver may be transported in an enclosed container, as defined in section 134-E, in the course of going to and from the place of the hunt, notwithstanding section 134-F.

134-7. Ownership or possession prohibited, when; penalty.

(a) No person who is a fugitive from justice or is a person prohibited from possessing firearms or ammunition under federal law shall own, possess, or control any firearm or ammunition therefor.

(b) No person who is under indictment for, or has waived indictment for, or has been bound over to the circuit court for, or has been convicted in this State or elsewhere of having committed a felony, or any crime of violence, or an illegal sale of any drug shall own, possess, or control any firearm or ammunition therefor.

(c) No person who:

(1) Is or has been under treatment or counseling for addiction to, abuse of, or dependence upon any dangerous, harmful, or detrimental drug, intoxicating compound as defined in section 712-1240, or intoxicating liquor;

(2) Has been acquitted of a crime on the grounds of mental disease, disorder, or defect pursuant to section 704-411; or

(3) Is or has been diagnosed as having a significant behavioral, emotional, or mental disorders [disorder] as defined by the most current diagnostic manual of the American Psychiatric Association or for treatment for organic brain syndromes; shall own, possess, or control any firearm or ammunition therefor, unless the person has been medically documented to be no longer adversely affected by the addiction, abuse, dependence, mental disease, disorder, or defect.

(d) No person who is less than twenty-five years old and has been adjudicated by the family court to have committed a felony, two or more crimes of violence, or an illegal sale of any drug shall own, possess or control any firearm or ammunition therefor.

(e) No minor who:

(1) Is or has been under treatment for addiction to any dangerous, harmful, or detrimental drug, intoxicating compound as defined in section 712-1240, or intoxicating liquor;

(2) Is a fugitive from justice; or

(3) Has been determined not to have been responsible for a criminal act or has been committed to any institution on account of a mental disease, disorder, or defect; shall own, possess, or control any firearm or ammunition therefor, unless the minor has been medically documented to be no longer adversely affected by the addiction, mental disease, disorder, or defect.

For the purposes of enforcing this section, and notwithstanding section 571-84 or any other law to the contrary, any agency within the State shall make its records relating to family court adjudications available to law enforcement officials.

(f) No person who has been restrained pursuant to an order of any court, including an ex parte order as provided in this subsection, from contacting, threatening, or physically abusing any person, shall possess, control, or transfer ownership of any firearm or ammunition therefor, so long as the protective order, restraining order, or any extension is in effect, unless the order, for good cause shown, specifically permits the possession of a firearm and ammunition. The restraining order or order of protection shall specifically include a statement that possession, control, or transfer of ownership of a firearm or ammunition by the person named in the order is prohibited. Such person shall relinquish possession and control of any firearm and ammunition owned by that person to the police department of the appropriate county for safekeeping for the duration of the order or extension thereof. In the case of an ex parte order, the affidavit or statement under oath that forms the basis for the order shall contain a statement of the facts that support a finding that the person to be restrained owns, intends to obtain or to transfer ownership, or possesses a firearm, and that the firearm may be used to threaten, injure, or abuse any person. The ex parte order shall be effective upon service pursuant to section 586-6. At the time of service of a restraining order involving firearms and ammunition issued by any court, the police officer may take custody of any and all firearms and ammunition in plain sight, those discovered pursuant to a consensual search, and those firearms surrendered by the person restrained. If the person restrained is the registered owner of a firearm and knows the location of the firearm but refuses to surrender the firearm or refuses to disclose the location of the firearm, the person restrained shall be guilty of a misdemeanor. In any case, when a police officer is unable to locate the firearms and ammunition either registered under this chapter or known to the person granted protection by the court, the police officer shall apply to the court for a search warrant pursuant to chapter 803 for the limited purpose of seizing the firearm and ammunition.

For the purposes of this subsection, good cause shall not be based solely upon the consideration that the person subject to restraint pursuant to an order of any court, including an ex parte order as provided for in this subsection, is required to possess or carry firearms or ammunition during the course of the person's employment. Good cause consideration may include but not be limited to the protection and safety of the person to whom a restraining order is granted.

(g) Any person disqualified from ownership, possession, control, or the right to transfer ownership of firearms and ammunition under this section shall surrender or dispose of all firearms and ammunition in compliance with section 134-7.3.

(h) Any person violating subsection (a) or (b) shall be guilty of a class C felony; provided that any felon violating subsection (b) shall be guilty of a class B felony. Any person violating subsection (c), (d), (e), (f), or (g) shall be guilty of a misdemeanor.

134-7.3. Seizure of firearms upon disqualification.

(a) If any applicant is denied a permit, the chiefs of police of the respective counties shall send, by certified mail, a notice setting forth the reasons for the denial and may require that the applicant voluntarily surrender all firearms and ammunition to the chief of police where the applicant resides or dispose of all firearms and ammunition. If an applicant fails to voluntarily surrender or dispose of all firearms and ammunition within thirty days from the date notice was mailed, the chief of police may seize all firearms and ammunition.

(b) Any person disqualified from ownership, possession, or control of firearms and ammunition under section 134-7 shall voluntarily surrender all firearms and ammunition to the chief of police where the person resides or dispose of all firearms and ammunition. If any person fails to voluntarily surrender or dispose of all firearms and ammunition within thirty days from the date of disqualification, the chief of police may seize all firearms and ammunition.

(c) For the purposes of this section, "dispose" means selling the firearms to a gun dealer licensed under section 134-31, transferring ownership of the firearms to any person who meets the requirements of section 134-2, or surrendering all firearms to the chief of police where the person resides for storage or disposal; provided, for a person subject to section 134-7(f), "dispose" shall not include transferring ownership of the firearms to any person who meets the requirements of section 134-2.

(d) The chief of police of the respective counties shall adopt procedures to implement and administer the provisions of this section by December 31, 2001.

134-7.5. Seizure of firearms in domestic abuse situations; requirements; return of.

(a) Any police officer who has reasonable grounds to believe that a person has recently assaulted or threatened to assault a family or household member may seize all firearms and ammunition that the police officer has reasonable grounds to believe were used or threatened to be used in the commission of the offense. The police officer may seize any firearms or ammunition that are in plain view of the officer or were discovered pursuant to a consensual search, as necessary for the protection of the officer or any family or household member. Firearms seized under this section shall be taken to the appropriate county police department for safekeeping or as evidence.

(b) Upon taking possession of a firearm or ammunition, the officer shall give the owner or person who was in lawful possession of the firearm or ammunition a receipt identifying the firearm or ammunition and indicating where the firearm or ammunition can be recovered.

(c) The officer taking possession of the firearm or ammunition shall notify the person against whom the alleged assault or threatened assault was inflicted of remedies and services available to victims of domestic violence, including the right to apply for a domestic abuse restraining order.

(d) The firearm or ammunition shall be made available to the owner or person who was in lawful possession of the firearm or ammunition within seven working days after the seizure when:

(1) The firearm or ammunition are not retained for use as evidence;

(2) The firearm or ammunition are not retained because they are possessed illegally;

(3) The owner or person who has lawful possession of the firearm or ammunition is not restrained by an order of any court from possessing a firearm or ammunition; and

(4) No criminal charges are pending against the owner or person who has lawful possession of the firearm or ammunition when a restraining order has already issued.

134-8. Ownership, etc., of automatic firearms, silencers, etc., prohibited; penalties.

(a) The manufacture, possession, sale, barter, trade, gift, transfer, or acquisition of any of the following is prohibited: assault pistols, except as provided by section 134-4(e); automatic firearms; rifles with barrel lengths less than sixteen inches; shotguns with barrel lengths less than eighteen inches; cannons; mufflers, silencers, or devices for deadening or muffling the sound of discharged firearms; hand grenades, dynamite, blasting caps, bombs, or bombshells, or other explosives; or any type of ammunition or any projectile component thereof coated with teflon or any other similar coating designed primarily to enhance its capability to penetrate metal or pierce protective armor; and any type of ammunition or any projectile component thereof designed or intended to explode or segment upon impact with its target.

(b) Any person who installs, removes, or alters a firearm part with the intent to convert the firearm to an automatic firearm shall be deemed to have manufactured an automatic firearm in violation of subsection (a).

(c) The manufacture, possession, sale, barter, trade, gift, transfer, or acquisition of detachable ammunition magazines with a capacity in excess of ten rounds which are designed for or capable of use with a pistol is prohibited. This subsection shall not apply to magazines originally designed to accept more than ten rounds of ammunition which have been modified to accept no more than ten rounds and which are not capable of being readily restored to a capacity of more than ten rounds.

(d) Any person violating subsection (a) or (b) shall be guilty of a class C felony and shall be imprisoned for a term of five years without probation. Any person violating subsection (c) shall be guilty of a misdemeanor except when a detachable magazine prohibited under this section is possessed while inserted into a pistol in which case the person shall be guilty of a class C felony.

134-9. Licenses to carry.

(a) In an exceptional case, when an applicant shows reason to fear injury to the applicant's person or property, the chief of police of the appropriate county may grant a license to an applicant who is a citizen of the United States of the age of twenty-one years or more or to a duly accredited official representative of a foreign nation of the age of twenty-one years or more to carry a pistol or revolver and ammunition therefor concealed on the person within the county where the license is granted. Where the urgency or the need has been sufficiently indicated, the respective chief of police may grant to an applicant of good moral character who is a citizen of the United States of the age of twenty-one years or more, is engaged in the protection of life and property, and is not prohibited under section 134-7 from the ownership or possession of a firearm, a license to carry a pistol or revolver and ammunition therefor unconcealed on the person within the county where the license is granted. The chief of police of the appropriate county, or the chief's designated representative, shall perform an inquiry on an applicant by using the National Instant Criminal Background Check System, to include a check of the Immigration and Customs Enforcement databases where the applicant is not a citizen of the United States, before any determination to grant a license is made. Unless renewed, the license shall expire one year from the date of issue.

(b) The chief of police of each county shall adopt procedures to require that any person granted a license to carry a concealed weapon on the person shall:

(1) Be qualified to use the firearm in a safe manner;

(2) Appear to be a suitable person to be so licensed;

(3) Not to be prohibited under section 134-7 from the ownership or possession of a firearm; and

(4) Not have been adjudged insane or not appear to be mentally deranged.

(c) No person shall carry concealed or unconcealed on the person a pistol or revolver without being licensed to do so under this section or in compliance with sections 134-5(c) or 134-E.

(d) A fee of $10 shall be charged for each license and shall be deposited in the treasury of the county in which the license is granted.

134-10. Alteration of identification marks prohibited. No person shall willfully alter, remove, or obliterate the name of the make, model, manufacturer's number, or other mark of identity of any firearm or ammunition. Possession of a firearm or ammunition upon which any mark of identity has been altered, removed, or obliterated shall be presumptive evidence that the possessor has altered, removed, or obliterated the mark of identity.

134-10.5. Storage of firearm; responsibility with respect to minors. No person shall store or keep any firearm on any premises under the person's control if the person knows or reasonably should know that a minor is likely to gain access to the firearm without the permission of the parent or guardian of the minor, unless the person:

(1) Keeps the firearm in a securely locked box or other container or in a location that a reasonable person would believe to be secure; or

(2) Carries the firearm on the person or within such close proximity thereto that the person readily can retrieve and use it as if it were carried on the person.

For purposes of this section, "minor" means any person under the age of sixteen years.

134-11. Exemptions.

(a) Sections 134-A to 134-G and 134-7 to 134-9, except section 134-7(f), shall not apply:

(1) To state and county law enforcement officers; provided that such persons are not convicted of an offense involving abuse of a family and household member under section 709-906;

(2) To members of the armed forces of the State and of the United States and mail carriers while in the performance of their respective duties if those duties require them to be armed;

(3) To regularly enrolled members of any organization duly authorized to purchase or receive the weapons from the United States or from the State, provided the members are either at, or going to or from, their places of assembly or target practice;

(4) To persons employed by the State, or subdivisions thereof, or the United States while in the performance of their respective duties or while going to and from their respective places of duty if those duties require them to be armed;

(5) To aliens employed by the State, or subdivisions thereof, or the United States while in the performance of their respective duties or while going to and from their respective places of duty if those duties require them to be armed;

(6) To police officers on official assignment in Hawaii from any state which by compact permits police officers from Hawaii while on official assignment in that state to carry firearms without registration. The governor of the State or the governor's duly authorized representative may enter into compacts with other states to carry out this paragraph.

(b) Sections 134-2 and 134-3 shall not apply to such firearms or ammunition that are a part of the official equipment of any federal agency.

(c) Sections134-A to 134-G, 134-8, and 134-9 shall not apply to the possession, transportation, or use, with blank cartridges, of any firearm or explosive solely as props for motion picture film or television program production when authorized by the chief of police of the appropriate county pursuant to section 134-2.5 and not in violation of federal law.

134-12.5. Forfeiture of firearms, ammunition, deadly or dangerous weapons, and switchblade knives; when. All firearms, ammunition, deadly or dangerous weapons, and switchblade knives possessed, used in violation of this chapter or the Hawaii penal code shall be forfeited to the State according to the provisions of chapter 712A and shall be destroyed or, if not destroyed, transferred to the chief of police of the county in which the violation took place for use by and under control of the police department.

134-13. Revocation of permits. All permits and licenses provided for under this part may be revoked, for good cause, by the issuing authority or by the judge of any court.

134-14. Report. Within ten days after the last day of each month, each of the authorities authorized in this chapter to issue or revoke permits and licenses shall make a report to the department of the attorney general of all permits and licenses issued or revoked by the authority as of the last day of the preceding month. The report shall be in the manner and form as the attorney general may prescribe.

134-15. Restriction of materials for manufacture of pistols or revolvers.

(a) It shall be unlawful for any person, includeing a licensed manufacturer, licensed importer, or licensed dealer, to possess, sell, or deliver any pistol or revolver the frame or receiver of which is a die casting of zinc alloy which has a melting temperature of less than 800 degrees Fahrenheit.

(b) This section shall not apply to any pistol or revolver duly registered prior to July 1, 1975 pursuant to section 134-3 or to any antique pistol or revolver.

134-16. Restriction on possession, sale, gift, or delivery of electric guns.

(a) It shall be unlawful for any person, including a licensed manufacturer, licensed importer, or licensed dealer, to possess, offer for sale, hold for sale, sell, give, lend, or deliver any electric gun.

(b) Any electric gun in violation of subsection (a) shall be confiscated and disposed of by the chief of police.

(c) This section shall not apply to law enforcement officers of county police and sheriff departments of this State, or vendors providing electric guns to those entities; provided that electric guns shall at all times remain in the custody and control of the county police or sheriff departments.

(d) The county police and sheriff departments of this State shall maintain records regarding every electric gun in their custody and control. Such records shall report every instance of usage of the electric guns; in particular, records shall be maintained in a similar manner as for those of discharging of firearms. The county police and sheriff departments shall annually report to the legislature regarding these records twenty days before the beginning of each session.

134-17. Penalties.

(a) If any person gives false information or offers false evidence of the person's identity in complying with any of the requirements of this part, that person shall be guilty of a misdemeanor, provided, however that if any person intentionally gives false information or offers false evidence concerning their psychiatric or criminal history in complying with any of the requirements of this part, that person shall be guilty of a class C felony.

(b) Any person who violates section 134-3(a) shall be guilty of a petty misdemeanor.

(c) Any person who violates section 134-2, 134-4, 134-10, 134-15, or 134-16(a) shall be guilty of a misdemeanor. Any person who violates section 134-3(b) shall be guilty of a petty misdemeanor and the firearm shall be confiscated as contraband and disposed of, if the firearm is not registered within five days of the person receiving notice of the violation.

134-18. Qualified immunity for physicians, psychologists or psychiatrists who provide information on permit applicants. There shall be no civil liability for any physician, psychologist or psychiatrist who provides information or renders an opinion in response to an inquiry made for purposes of issuing a firearm permit under section 134-2 or for purposes of investigating the continuing mental health of the holder of a valid firearm permit provided that the physician, psychologist, or psychiatrist acted without malice.

134-23. Place to keep loaded firearms other than pistols and revolvers; penalty.

(a) Except as provided in section 134-5, all firearms shall be confined to the possessor's place of business, residence, or sojourn; provided that it shall be lawful to carry unloaded firearms in an enclosed container from the place of purchase to the purchaser's place of business, residence, or sojourn, or between these places upon change of place of business, residence, or sojourn, or between these places and the following:

(1) A place of repair;

(2) A target range;

(3) A licensed dealer's place of business;

(4) An organized, scheduled firearms show or exhibit;

(5) A place of formal hunter or firearm use training or instruction; or

(6) A police station.

"Enclosed container" means a rigidly constructed receptacle, or a commercially manufactured gun case, or the equivalent thereof that completely encloses the firearm.

(b) Any person violating this section by carrying or possessing a loaded firearm other than a pistol or revolver shall be guilty of a class B felony.

134-24. Place to keep unloaded firearms other than pistols and revolvers; penalty.

(a) Except as provided in section 134-5, all firearms shall be confined to the possessor's place of business, residence, or sojourn; provided that it shall be lawful to carry unloaded firearms in an enclosed container from the place of purchase to the purchaser's place of business, residence, or sojourn, or between these places upon change of place of business, residence, or sojourn, or between these places and the following:

(1) A place of repair;

(2) A target range;

(3) A licensed dealer's place of business;

(4) An organized, scheduled firearms show or exhibit;

(5) A place of formal hunter or firearm use training or instruction; or

(6) A police station.

"Enclosed container" means a rigidly constructed receptacle, or a commercially manufactured gun case, or the equivalent thereof that completely encloses the firearm.

(b) Any person violating this section by carrying or possessing an unloaded firearm other than a pistol or revolver shall be guilty of a class C felony.

134-25. Place to keep pistol or revolver; penalty.

(a) Except as provided in sections 134-5 and 134-9, all firearms shall be confined to the possessor's place of business, residence, or sojourn; provided that it shall be lawful to carry unloaded firearms in an enclosed container from the place of purchase to the purchaser's place of business, residence, or sojourn, or between these places upon change of place of business, residence, or sojourn, or between these places and the following:

(1) A place of repair;

(2) A target range;

(3) A licensed dealer's place of business;

(4) An organized, scheduled firearms show or exhibit;

(5) A place of formal hunter or firearm use training or instruction; or

(6) A police station.

"Enclosed container" means a rigidly constructed receptacle, or a commercially manufactured gun case, or the equivalent thereof that completely encloses the firearm.

(b) Any person violating this section by carrying or possessing a loaded or unloaded pistol or revolver shall be guilty of a class B felony.

134-27. Place to keep ammunition; penalty.

(a) Except as provided in sections 134-5 and 134-9, all ammunition shall be confined to the possessor's place of business, residence, or sojourn; provided that it shall be lawful to carry ammunition in an enclosed container from the place of purchase to the purchaser's place of business, residence, or sojourn, or between these places upon change of place of business, residence, or sojourn, or between these places and the following:

(1) A place of repair;

(2) A target range;

(3) A licensed dealer's place of business;

(4) An organized, scheduled firearms show or exhibit;

(5) A place of formal hunter or firearm use training or instruction; or

(6) A police station.

"Enclosed container" means a rigidly constructed receptacle, or a commercially manufactured gun case, or the equivalent thereof that completely encloses the ammunition.

(b) Any person violating this section shall be guilty of a misdemeanor.

Part II. Firearms, Dealers' Licenses

134-31. License to sell and manufacture firearms; fee. Any person desiring to engage in the business to sell and manufacture firearms for sale in the State either at wholesale or retail, shall annually file an application for a license therefor with the director of finance of each county of the State. The annual fee for the issuance of such license shall be $10 and shall be payable to said director of finance. A license issued hereunder shall expire on June 30 next following the date of issuance of the license unless sooner terminated. Application for renewal of license shall be filed on or before June 30 of each year.

134-32. License to sell and manufacture firearms; conditions. Every license issued pursuant to this part shall be issued and shall be regarded as having been accepted by the licensee subject to the following conditions:

(1) That the licensee at all times shall comply with all provisions of law relative to the sale of firearms.

(2) That the license during any time of national emergency or crisis, as defined in section 134-34, may be canceled or suspended.

(3) That all firearms in the possession and control of any licensee at any time of national emergency or crisis, as defined in section 134-34, may be seized and held in possession or purchased by or on the order of the governor until such time as the national emergency or crisis has passed, or until such time as the licensee and the government of the United States or the government of the State may agree upon some other disposition of the same.

(4) That all firearms in the possession and control of the licensee or registered pursuant to section 134-3(c) by the licensee shall be subject to physical inspection by the chief of police of each county during normal business hours at the licensee's place of business.

(5) That the license may be revoked for a violation of any of the conditions of this section.

134-33. Punishment for violations of section 134-32. Any person who manufactures or sells any firearms within the State without having a valid license so to do, or who being a holder of a license violates any of the terms or conditions of the same, shall be fined not less than $100 nor more than $1,000 or imprisoned not less than three months nor more than one year.

Title 37 Hawaii Penal Code

Chapter 707 Offenses Against the Person

Part III. Criminal Assaults and Related Offenses

707-714.5. Criminally negligent storage of a firearm.

(1) A person commits the offense of criminally negligent storage of a firearm if the person violates section 134-10.5 and a minor obtains the firearm. For purposes of this section, "minor" means any person under the age of sixteen years.

(2) This section shall not apply if the minor obtains the firearm as a result of an unlawful entry to any premises by any person.

(3) Criminally negligent storage of a firearm is a misdemeanor.

[Current through 2008 Legislation of the 2008 Regular Session and First Special Session]

Title 18. Crimes and Punishments

Chapter 33. Firearms, Explosives and Other Deadly Weapons

18-3302. Issuance of licenses to carry concealed weapons

(1) The sheriff of a county, on behalf of the state of Idaho, shall, within ninety (90) days after the filing of an application by any person who is not disqualified from possessing or receiving a firearm under state or federal law, issue a license to the person to carry a weapon concealed on his person within this state. For licenses issued before July 1, 2006, a license shall be valid for four (4) years from the date of issue. For licenses issued on or after July 1, 2006, a license shall be valid for five (5) years from the date of issue. The citizen's constitutional right to bear arms shall not be denied to him, unless he:

(a) Is ineligible to own, possess or receive a firearm under the provisions of state or federal law; or

(b) Is formally charged with a crime punishable by imprisonment for a term exceeding one (1) year; or

(c) Has been adjudicated guilty in any court of a crime punishable by imprisonment for a term exceeding one (1) year; or

(d) Is a fugitive from justice; or

(e) Is an unlawful user of, or addicted to, marijuana or any depressant, stimulant, or narcotic drug, or any other controlled substance as defined in 21 U.S.C. 802; or

(f) Is currently suffering or has been adjudicated as follows, based on substantial evidence:

(i) Lacking mental capacity as defined in section 18-210, Idaho Code; or

(ii) Mentally ill as defined in section 66-317, Idaho Code; or

(iii) Gravely disabled as defined in section 66-317, Idaho Code; or

(iv) An incapacitated person as defined in section 15-5-101(a), Idaho Code; or

(g) Is or has been discharged from the armed forces under dishonorable conditions; or

(h) Is or has been adjudicated guilty of or received a withheld judgment or suspended sentence for one (1) or more crimes of violence constituting a misdemeanor, unless three (3) years has elapsed since disposition or pardon has occurred prior to the date on which the application is submitted; or

(i) Has had entry of a withheld judgment for a criminal offense which would disqualify him from obtaining a concealed weapon license; or

(j) Is an alien illegally in the United States; or

(k) Is a person who having been a citizen of the United States, has renounced his or her citizenship; or

(l) Is under twenty-one (21) years of age; or

(m) Is free on bond or personal recognizance pending trial, appeal or sentencing for a crime which would disqualify him from obtaining a concealed weapon license; or

(n) Is subject to a protection order issued under chapter 63, title 39, Idaho Code, that restrains the person from harassing, stalking or threatening an intimate partner of the person or child of the intimate partner or person, or engaging in other conduct that would place an intimate partner in reasonable fear of bodily injury to the partner or child.

The license application shall be in triplicate, in a form to be prescribed by the director of the Idaho state police, and shall ask the name, address, description and signature of the licensee, date of birth, social security number, military status, and the driver's license number or state identification card number of the licensee if used for identification in applying for the license. The application shall indicate that provision of the social security number is optional. The license application shall contain a warning substantially as follows:

CAUTION: Federal law and state law on the possession of weapons and firearms differ. If you are prohibited by federal law from possessing a weapon or a firearm, you may be prosecuted in federal court. A state permit is not a defense to a federal prosecution.

The sheriff shall require any person who is applying for original issuance of a license to submit his fingerprints in addition to the other information required in this subsection. Within five (5) days after the filing of an application, the sheriff shall forward the application and fingerprints to the Idaho state police for a records check of state and national files. The Idaho state police shall conduct a national fingerprint-based records check and return the results to the sheriff within seventy-five (75) days. The sheriff shall not issue a license before receiving the results of the records check and must deny a license if the applicant is disqualified under any of the criteria listed in paragraphs (a) through (n) of subsection (1) of this section.

The license will be in a form substantially similar to that of the Idaho driver's license. It will bear the signature, name, address, date of birth, picture of the licensee, expiration date and the driver's license number or state identification card number of the licensee if used for identification in applying for the license. Upon issuing a license under the provisions of this section, the sheriff will notify the Idaho state police on a form or in a manner prescribed by the state police. Information relating to an applicant or licensee received or maintained pursuant to this section by the sheriff or Idaho state police is confidential and exempt from disclosure under section 9-338, Idaho Code.

(2) The fee for original issuance of a license shall be twenty dollars ($20.00) paid to the sheriff for the purpose of enforcing the provisions of this chapter. The sheriff may collect any additional fees necessary to cover the cost of processing fingerprints lawfully required by any state or federal agency or department, and the cost of materials for the license lawfully required by any state agency or department, which costs shall be paid to the state.

(3) The fee for renewal of the license shall be fifteen dollars ($ 15.00). The sheriff may collect any additional fees necessary to cover the processing costs lawfully required by any state or federal agency or department, and the cost of materials for the license lawfully required by any state agency or department, which costs shall be paid to the state. If a licensee applying for renewal has not previously been required to submit fingerprints, the sheriff shall require the licensee to do so and may collect any additional fees necessary to cover the cost of processing fingerprints lawfully required by any state or federal agency or department.

(4) Every license that is not, as provided by law, suspended, revoked or disqualified in this state shall be renewable at any time during the ninety (90) day period before its expiration or within ninety (90) days after the expiration date. Renewal notices shall be mailed out ninety (90) days prior to the expiration date of the license. The sheriff shall require the licensee applying for renewal to complete an application. The sheriff shall submit the application to the Idaho state police for a records check of state and national databases. The Idaho state police shall conduct the records check and return the results to the sheriff within thirty (30) days. The sheriff shall not issue a renewal before receiving the results of the records check and must deny a license if the applicant is disqualified under any of the criteria listed in subsection (1), paragraphs (a) through (n) of this section. A renewal license shall be valid for a period of five (5) years. A license so renewed shall take effect on the expiration date of the prior license. A licensee renewing ninety-one (91) days or more after the expiration date of the license shall pay a late renewal penalty of ten dollars ($10.00) in addition to the renewal fee. The fee shall be paid to the sheriff for the purpose of enforcing the provisions of this chapter.

(5) Notwithstanding the requirements of this section, the sheriff of the county of the applicant's residence may issue a temporary emergency license for good cause pending review under subsection (1) of this section.

(6) A city, county or other political subdivision of this state shall not modify the requirements of this section, nor may a political subdivision ask the applicant to voluntarily submit any information not required in this section. A civil action may be brought to enjoin a wrongful refusal to issue a license or a wrongful modification of the requirements of this section. The civil action may be brought in the county in which the application was made or in Ada county at the discretion of the petitioner. Any person who prevails against a public agency in any action in the courts for a violation of subsections (1) through (5) of this section, shall be awarded costs, including reasonable attorney's fees incurred in connection with the legal action.

(7) Except in the person's place of abode or fixed place of business, or on property in which the person has any ownership or leasehold interest, a person shall not carry a concealed weapon without a license to carry a concealed weapon. For the purposes of this section, a concealed weapon means any dirk, dirk knife, bowie knife, dagger, pistol, revolver, or any other deadly or dangerous weapon. The provisions of this section shall not apply to any lawfully possessed shotgun or rifle.

(8) A county sheriff, deputy sheriff, or county employee who issues a license to carry a concealed weapon under this section shall not incur any civil or criminal liability as the result of the performance of his duties under this section.

(9) While in any motor vehicle, inside the limits or confines of any city, a person shall not carry a concealed weapon on or about his person without a license to carry a concealed weapon. This shall not apply to any firearm located in plain view whether it is loaded or unloaded. A firearm may be concealed legally in a motor vehicle so long as the weapon is disassembled or unloaded.

(10) In implementing the provisions of this section on behalf of the state of Idaho, the sheriff shall make applications readily available at the office of the sheriff or at other public offices in his jurisdiction.

(11) The sheriff of a county may issue a license to carry a concealed weapon to those individuals between the ages of eighteen (18) and twenty-one (21) years who in the judgment of the sheriff warrant the issuance of the license to carry a concealed weapon. Such issuance shall be subject to limitations which the issuing authority deems appropriate. Licenses issued to individuals between the ages of eighteen (18) and twenty-one (21) years shall be easily distinguishable from regular licenses.

(12) The requirement to secure a license to carry a concealed weapon under this section shall not apply to the following persons:

(a) Officials of a county, city, state of Idaho, the United States, peace officers, guards of any jail, court appointed attendants or any officer of any express company on duty;

(b) Employees of the adjutant general and military division of the state where military membership is a condition of employment when on duty;

(c) Criminal investigators of the attorney general's office, criminal investigators of a prosecuting attorney's office, prosecutors and their deputies;

(d) Any person outside the limits of or confines of any city while engaged in lawful hunting, fishing, trapping or other lawful outdoor activity;

(e) Any publicly elected Idaho official;

(f) Retired peace officers or detention deputies with at least ten (10) years of service with the state or a political subdivision as a peace officer or detention deputy and who have been certified by the peace officer standards and training council;

(g) Any person who has a valid permit from a state or local law enforcement agency or court authorizing him to carry a concealed weapon. A permit issued in another state will only be considered valid if the permit is in the licensee's physical possession.

(13) When issuing a license pursuant to this section, the sheriff may require the applicant to demonstrate familiarity with a firearm and shall accept any of the following, provided the applicant may select which one:

(a) Completion of any hunter education or hunter safety course approved by the department of fish and game or a similar agency of another state; or

(b) Completion of any national rifle association firearms safety or training course, or any national rifle association hunter education course; or

(c) Completion of any firearms safety or training course or class available to the general public offered by a law enforcement agency, community college, college, university, or private or public institution or organization or firearms training school, utilizing instructors certified by the national rifle association or the Idaho state police; or

(d) Completion of any law enforcement firearms safety or training course or class offered for security guards, investigators, special deputies, or any division or subdivision of a law enforcement agency or security enforcement agency; or

(e) Presents evidence of equivalent experience with a firearm through participation in organized shooting competition or military service; or

(f) Is licensed or has been licensed to carry a firearm in this state or a county or municipality, unless the license has been revoked for cause; or

(g) Completion of any firearms training or training or safety course or class conducted by a state certified or national rifle association certified firearms instructor.

(14) A person carrying a concealed weapon in violation of the provisions of this section shall be guilty of a misdemeanor.

(15) The sheriff of the county where the license was issued or the sheriff of the county where the person resides shall have the power to revoke a license subsequent to a hearing in accordance with the provisions of chapter 52, title 67, Idaho Code, for any of the following reasons:

(a) Fraud or intentional misrepresentation in the obtaining of a license; or

(b) Misuse of a license, including lending or giving a license to another person, or duplicating a license, or using a license with the intent to unlawfully cause harm to a person or property; or

(c) The doing of an act or existence of a condition which would have been grounds for the denial of the license by the sheriff; or

(d) The violation of any of the terms of this section; or

(e) The applicant is adjudicated guilty of or receives a withheld judgment for a crime which would have disqualified him from initially receiving a license.

(16) A person twenty-one (21) years of age or older issued a license to carry a concealed weapon is exempt from any requirement to undergo a records check at the time of purchase or transfer of a firearm from a federally licensed firearms dealer. However, a temporary emergency license issued under subsection (5) of this section shall not exempt the holder of the license from any records check requirement. Temporary emergency licenses shall be easily distinguishable from regular licenses.

(17) The attorney general is authorized to negotiate reciprocal agreements with other states related to the recognition of licenses to carry concealed weapons. The Idaho state police shall keep a copy and maintain a record of all such agreements, which shall be made available to the public.

(18) The provisions of this section are hereby declared to be severable and if any provision of this section or the application of such provision to any person or circumstance is declared invalid for any reason, such declaration shall not affect the validity of remaining portions of this section.

18-3302A. Sale of weapons to minors It shall be unlawful to directly or indirectly sell to any minor under the age of eighteen (18) years any weapon without the written consent of the parent or guardian of the minor. Any person violating the provisions of this section shall be guilty of a misdemeanor and shall be punished by a fine not in excess of one thousand dollars ($1,000), by imprisonment in the county jail for a term not in excess of six (6) months, or by both such fine and imprisonment. As used in this section, "weapon" shall mean any dirk, dirk knife, bowie knife, dagger, pistol, revolver or gun.

18-3302C. Prohibited conduct

Any person obtaining a license under the provisions of section 18-3302, Idaho Code, shall not:

(1) Carry a concealed weapon in a courthouse, juvenile detention facility or jail, public or private school, except as provided in subsection (4)(f) of section 18-3302D, Idaho Code; or

(2) Provide information on the application for a permit to carry a concealed weapon knowing the same to be untrue. Any person violating the provisions of this section shall be guilty of a misdemeanor.

18-3302D. Possessing weapons or firearms on school property

(1)(a) It shall be unlawful and is a misdemeanor for any person to possess a firearm or other deadly or dangerous weapon while on the property of a school or in those portions of any building, stadium or other structure on school grounds which, at the time of the violation, were being used for an activity sponsored by or through a school in this state or while riding school provided transportation.

(b) The provisions of this section regarding the possession of a firearm or other deadly or dangerous weapon on school property shall also apply to students of schools while attending or participating in any school sponsored activity, program or event regardless of location.

(2) Definitions. As used in this section:

(a) "Deadly or dangerous weapon" means any weapon as defined in 18 U.S.C. section 930;

(b) "Firearm" means any firearm as defined in 18 U.S.C. section 921;

(c) "Minor" means a person under the age of eighteen (18) years;

(d) "Possess" means to bring an object, or to cause it to be brought, onto the property of a public or private elementary or secondary school, or onto a vehicle being used for school provided transportation, or to exercise dominion and control over an object located anywhere on such property or vehicle. For purposes of subsection (1)(b) of this section, "possess" shall also mean to bring an object onto the site of a school sponsored activity, program or event, regardless of location, or to exercise dominion and control over an object located anywhere on such a site;

(e) "School" means a private or public elementary or secondary school.

(3) Right to search students or minors. For purposes of enforcing the provisions of this section, employees of a school district shall have the right to search all students or minors, including their belongings and lockers, that are reasonably believed to be in violation of the provisions of this section, or applicable school rule or district policy, regarding the possessing of a firearm or other deadly or dangerous weapon.

(4) The provisions of this section shall not apply to the following persons:

(a) A peace officer;

(b) A person who lawfully possesses a firearm or deadly or dangerous weapon as an appropriate part of a program, an event, activity or other circumstance approved by the board of trustees or governing board;

(c) A person or persons complying with the provisions of section 19-202A, Idaho Code;

(d) Any adult over eighteen (18) years of age and not enrolled in a public or private elementary or secondary school who has lawful possession of a firearm or other deadly or dangerous weapon, secured and locked in his vehicle in an unobtrusive, nonthreatening manner;

(e) A person who lawfully possesses a firearm or other deadly or dangerous weapon in a private vehicle while delivering minor children, students or school employees to and from school or a school activity;

(f) Notwithstanding the provisions of section 18-3302C, Idaho Code, a person or an employee of the school or school district who is authorized to carry a firearm with the permission of the board of trustees of the school district or the governing board.

(5) Penalties. Persons who are found guilty of violating the provisions of this section may be sentenced to a jail term of not more than one (1) year or fined an amount not in excess of one thousand dollars ($1,000) or both. If a violator is a student and under the age of eighteen (18) years, the court may place the violator on probation and suspend the juvenile detention or fine or both as long as the violator is enrolled in a program of study recognized by the court that, upon successful completion, will grant the violator a general equivalency diploma (GED) or a high school diploma or other educational program authorized by the court. Upon successful completion of the terms imposed by the court, the court shall discharge the offender from serving the remainder of the sentence. If the violator does not complete, is suspended from, or otherwise withdraws from the program of study imposed by the court, the court, upon receiving such information, shall order the violator to commence serving the sentence provided for in this section.

18-3302E. Possession of a weapon by a minor

(1) It shall be unlawful for any person under the age of eighteen (18) years to possess or have in possession any weapon, as defined in section 18-3302A, Idaho Code, unless he:

(a) Has the written permission of his parent or guardian to possess the weapon; or

(b) Is accompanied by his parent or guardian while he has the weapon in his possession.

(2) Any minor under the age of twelve (12) years in possession of a weapon shall be accompanied by an adult.

(3) Any person who violates the provisions of this section is guilty of a misdemeanor.

18-3302F. Prohibition of possession of certain weapons by a minor

(1) It shall be unlawful for any person under the age of eighteen (18) years to possess or have in possession any handgun.

(2) Except as provided by federal law, a minor under the age of eighteen (18) years may not possess the following:

(a) A sawed-off rifle or sawed-off shotgun; or

(b) A full automatic weapon.

(3) Any person who violates the provisions of subsection (2) (a) of this section is guilty of a misdemeanor.

(4) Any person who violates the provisions of subsection (2) (b) of this section is guilty of a felony.

(5) For purposes of this section:

(a) "Full automatic weapon" means any firearm which fires, is designed to fire, or can be readily restored to fire, automatically more than one (1) bullet, or other missile without reloading, by a single function of the trigger.

(b) "Handgun" means a pistol, revolver, or other firearm of any description, loaded or unloaded, from which any shot, bullet, or other missile can be discharged, the length of the barrel of which, not including any revolving, detachable, or magazine breech, does not exceed twelve (12) inches. Excluded from this definition are handguns firing a metallic projectile, such as a BB or pellet, through the force of air pressure, CO2 pressure, or spring action or any spot marker gun.

(6) Any person who provides a handgun to a minor when the possession of the handgun by the minor is a violation of the provisions of this section is guilty of a misdemeanor.

18-3302G. Exceptions The provisions of section 18-3302E, Idaho Code, regarding the possession of a weapon by a minor or section 18-3302F, Idaho Code, regarding possession of handguns by minors shall not apply to any of the following:

(1) Patrons firing at lawfully operated target concessions at amusement parks and similar locations provided that the firearms to be used are firmly chained or affixed to the counters;

(2) Any person in attendance at a hunter's safety course or a firearm's safety course;

(3) Any person engaging in practice or any other lawful use of a firearm at an established range or any other area where the discharge of a firearm is not prohibited by state or local law;

(4) Any person engaging in an organized competition involving the use of a firearm, or participating in or practicing for such competetion;

(5) Any minor under eighteen (18) years of age who is on real property with the permission of the owner, licensee, or lessee of the property and who has the permission of a parent or legal guardian or the owner, licensee, or lessee to possess a firearm not otherwise in violation of the law;

(6) Any resident or nonresident hunters with a valid hunting license or other persons who are lawfully engaged in hunting; and

(7) Any person traveling to or from any activity described in subsection (2), (3), (4), (5) or (6) of this section with an unloaded firearm in his possession.

18-3302I. Threatening violence on school grounds

(1)(a) Any person, including a student, who willfully threatens on school grounds by word or act to use a firearm or other deadly or dangerous weapon to do violence to any other person on school grounds is guilty of a misdemeanor.

(b) The threats prohibited by this section encompass only those statements or acts where the speaker or actor intends to communicate a serious expression of an intent to commit an act of unlawful violence to a particular individual or group of individuals. The prosecution is not required to prove that the defendant actually intended to carry out the threat.

(2) Definitions. As used in this section:

(a) "Deadly or dangerous weapon" means a weapon, device, instrument, material or substance that is used for, or is readily capable of, causing death or serious bodily injury;

(b) "Firearm" means any weapon, whether loaded or unloaded, from which a shot, projectile or other object may be discharged by force of combustion, explosive, gas and/or mechanical means, regardless of whether such weapon is operable;

(c) "On school grounds" means in, or on the property of, a public or private elementary or secondary school.

18-3308. Selling explosives, ammunition or firearms to minors No person, firm, association or corporation shall sell or give to any minor under the age of sixteen (16) years any powder, commonly called gunpowder, of any description, or any dynamite or other explosive, or any shells or fixed ammunition of any kind, except shells loaded for use in shotguns and for use in rifles of twenty-two (22) caliber or smaller, or any firearms of any description, without the written consent of the parents or guardian of such minor first had and obtained. Any person, firm, association or corporation violating any of the provisions of this section shall be deemed guilty of a misdemeanor.

18-3310. Shipping loaded firearms Every person who ships, or causes to be shipped, or delivers or causes to be delivered, to any railroad, express or stage company, or to any other common carrier, for shipment as baggage or otherwise, any loaded pistol, revolver, rifle, shotgun or other firearm, is guilty of a misdemeanor.

18-3311. Keeping gunpowder or other explosives in towns Every person who makes or keeps gunpowder, nitroglycerin, or other highly explosive substance, within any city or town, or who carries the same through the streets there-of, in any quantity or manner prohibited by law, or by any ordinance of such city or town, is guilty of a misdemeanor.

18-3312. Injuring another by careless handling and discharge of firearms Any person who handles, uses or operates any firearm in a careless, reckless or negligent manner, or without due caution and circumspection, whereby the same is fired or discharged and maims, wounds or injures any other person or persons, is guilty of a misdemeanor, and upon conviction thereof shall be punished by a fine of not more than one thousand dollars ($ 1,000), or by imprisonment in the county jail for not more than six (6) months, or by both such fine and imprisonment.

18-3314. Resident's purchase of firearm in contiguous state Residents of the state of Idaho may purchase rifles and shotguns in a state contiguous to Idaho, provided that such residents conform to the applicable provisions of the federal Gun Control Act of 1968, and regulations thereunder, as administered by the United States secretary of the treasury, and provided further, that such residents conform to the provisions of law applicable to such a purchase in Idaho and in the contiguous state in which the purchase is made.

18-3315. Resident of contiguous state - purchase of firearm in Idaho Residents of a state contiguous to the state of Idaho may purchase rifles and shotguns in Idaho, provided that such residents conform to the applicable provisions of the federal Gun Control Act of 1968, and regulations thereunder, as administered by the United States secretary of the treasury, and provided further, that such residents conform to the provisions of law applicable to such purchase in Idaho and in the state in which such persons reside.

18-3316. Unlawful possession of a fire-arm

(1) A person who previously has been convicted of a felony who purchases, owns, posses-ses, or has under his custody or control any firearm shall be guilty of a felony and shall be imprisoned in the state prison for a period of time not to exceed five (5) years and by a fine not to exceed five thousand dollars ($5,000).

(2) For the purpose of subsection (1) of this section, "convicted of a felony" shall include a person who has entered a plea of guilty, nolo contendere or has been found guilty of any of the crimes enumerated in section 18-310, Idaho Code, or to a comparable felony crime in another state, territory, commonwealth, or other jurisdiction of the United States.

(3) For the purpose of subsection (1) of this section, "firearm" shall include any weapon from which a shot, projectile or other object may be discharged by force of combustion, explosive, gas and/or mechanical means, whether operable or inoperable.

(4) Subsection (1) of this section shall not apply to a person whose conviction has been nullified by expungement, pardon, setting aside the conviction or other comparable procedure by the jurisdiction where the felony conviction occurred; or whose civil right to bear arms either specifically or in combination with other civil rights has been restored by any other provision of Idaho law.

18-3318. Definitions Definitions as used in sections 18-3319, 18-3320 and 18-3321, Idaho Code:

(1) "Bomb" means any chemical or mixture of chemicals contained in such a manner that it can be made to explode with fire or force, and combined with the method or mechanism intended to cause its explosion. The term includes components of a bomb only when the individual charged has taken steps to place the components in proximity to each other, or has partially assembled components from which a completed bomb can be readily assembled. "Bomb" does not include: rifle, pistol or shotgun ammunition and their components; fireworks; boating, railroad and other safety flares or propellants used in model rockets or similar hobby activities.

(2) "Destructive device" means:

(a) Any explosive, incendiary or poisonous gas:

(i) Bomb;

(ii) Grenade;

(iii) Rocket having a propellant charge of more than four (4) ounces;

(iv) Missile having an explosive or incendiary charge of more than one-fourth (1/4) ounce;

(v) Mine;

(vi) Similar device

(b) Any type of weapon, by whatever name known, which will, or which may be imminently converted to, expel a projectile by the action of an explosive or other propellant, the barrel or barrels of which have a bore of more than .700 inches in diameter, except rifled and unrifled shotguns or shotgun shells.

(c) Components of a destructive device only when the individual charged has taken steps to place the components in proximity to each other, or has partially assembled components from which a completed destructive device can be readily assembled.

(d) The term "destructive device" shall not include:

(i) Any device which is neither designed nor redesigned for use as a weapon;

(ii) Any device which, although originally designed for use as a weapon, has been redesigned for use as a signaling, pyrotechnic, line throwing, safety or similar device;

(iii) Otherwise lawfully owned surplus military ordnance;

(iv) Antiques or reproductions thereof and rifles held for sporting, recreational, investment or display purposes;

(v) Rifle, pistol or shotgun ammunition and their components.

(3) "Shrapnel" means any metal, ceramic, glass, hard plastic or other material of sufficient hardness to puncture human skin when propelled by force of the bomb or destructive device to which it is attached or in which it is contained.

18-3319. Unlawful possession of bombs or destructive devices

(1) Any person who knowingly, intentionally, or recklessly possesses or controls a bomb or destructive device for a purpose unlawful pursuant to title 18, Idaho Code, is guilty of a felony, punishable by up to a five thousand dollar ($5,000) fine and five (5) years in prison.

(2) Any person who knowingly possesses an assembled bomb or assembled destructive device and who:

(a) Has been convicted of a felony; or

(b) Has been found guilty of any crime where such conviction results in the person being prohibited from possessing or owning firearms; or

(c) Is in possession or control of any substance or paraphernalia in violation of section 37-2732B, 37-2734A or 37-2734B, Idaho Code, or the felony provisions of section 37-2732, Idaho Code;

is guilty of a felony, punishable by up to a five thousand dollar ($5,000) fine and five (5) years in prison.

(3) Any person who possesses a bomb or destructive device which by its design will propel shrapnel is guilty of a felony, punishable by up to a five thousand dollar ($5,000) fine and five (5) years in prison.

18-3320. Unlawful use of destructive device or bomb Any person who knowingly, intentionally, or recklessly:

(1) Conspires to use, uses or causes to be used a destructive device or bomb in the commission of or an attempt to commit a felony; or

(2) With the intent to injure the person or property of another, transports a bomb or destructive device; or

(3) Injures another or conspires or attempts to injure another in his person or property through the use of a destructive device or bomb is guilty of a felony, punishable by up to a twenty-five thousand dollar ($25,000) fine and life in prison.

18-3320A. Disposal of destructive devices or bombs Any destructive device or bomb that has been lawfully seized by a law enforcement agency may be destroyed in a reasonable manner. An official record listing the destructive device or bomb destroyed and the location of destruction shall be kept on file at the office of the seizing agency. In the event of such destruction, a photograph, videotape, or similar record of the device or bomb shall be preserved for evidentiary purposes. The destruction of a destructive device or bomb before a preliminary hearing, trial, or both shall not be a bar to prosecution for any violation of law.

18-3321. Persons exempt Unless the intent to injure the person or property of another has been established, the provisions in section 18-3319, Idaho Code, shall not apply to:

(1) Any public safety officer or member of the armed forces of the United States or national guard while acting in his official capacity;

(2) Any person possessing a valid permit issued under the provisions of the international fire code, sections 41-253 and 41-254, Idaho Code, or any employee of such permittee acting within the scope of his employment;

(3) Any person possessing a valid license as an importer, wholesaler, or display operator under the provisions of the Idaho fireworks act, sections 39-2602, 39-2606, 39-2607, 39-2608, 39-2609, 39-2610, 39-2611 and 39-2612, Idaho Code;

(4) A device which falls within the definition of a bomb or destructive device when used on property owned or otherwise in the control of the person using the device;

(5) Those licensed or permitted by the federal government to use or possess a bomb or destructive device.

(6) Those persons who possess a destructive device properly registered and taxed under the provisions of the national firearms act, as amended, as to possession of destructive devices properly registered to such persons.

Chapter 85. Idaho Criminal Gang Enforcement Act

18-8505. Supplying firearms to a criminal gang

(1) A person commits the offense of supplying firearms to a criminal gang if the person knows an individual is a gang member and supplies, sells or gives possession or control of any firearm to that gang member.

(2) Subsection (1) of this section shall not apply to a person who is convicted as a principal to the offense committed by the recipient of the firearm.

(3) A person convicted of a violation of this section shall be imprisoned for a term not to exceed ten (10) years or be fined an amount not to exceed fifty thousand dollars ($ 50,000), or both.

[Current through 2008 Regular Session]

ILLINOIS
ILL. COMP. STAT.

Chapter 430. Public Safety

Act 65. Firearms Owners Identification Card Act

65/0.01. Short title This Act may be cited as the Firearm Owners Identification Card Act.

65/1. Declaration of policy It is hereby declared as a matter of legislative determination that in order to promote and protect the health, safety and welfare of the public, it is necessary and in the public interest to provide a system of identifying persons who are not qualified to acquire or possess firearms, firearm ammunition, stun guns, and tasers within the State of Illinois by the establishment of a system of Firearm Owner's Identification Cards, thereby establishing a practical and workable system by which law enforcement authorities will be afforded an opportunity to identify those persons who are prohibited by Section 24-3.1 of the "Criminal Code of 1961", as amended, from acquiring or possessing firearms and firearm ammunition and who are prohibited by this Act from acquiring stun guns and tasers.

65/1.1. Firearm defined; Firearm ammunition defined For purposes of this Act:

"Has been adjudicated as a mental defective" means the person is the subject of a determination by a court, board, commission or other lawful authority that a person, as a result of marked subnormal intelligence, or mental illness, mental impairment, incompetency, condition, or disease:

(1) is a danger to himself, herself, or to others;

(2) lacks the mental capacity to manage his or her own affairs;

(3) is not guilty in a criminal case by reason of insanity, mental disease or defect;

(4) is incompetent to stand trial in a criminal case;

(5) is not guilty by reason of lack of mental responsibility pursuant to Articles 50a and 72b of the Uniform Code of Military Justice, 10 U.S.C. 850a, 876b

"Counterfeit" means to copy or imitate, without legal authority, with intent to deceive.

"Federally licensed firearm dealer" means a person who is licensed as a federal firearms dealer under Section 923 of the federal Gun Control Act of 1968 (18 U.S.C. 923).

"Firearm" means any device, by whatever name known, which is designed to expel a projectile or projectiles by the action of an explosion, expansion of gas or escape of gas; excluding, however:

(1) any pneumatic gun, spring gun, paint ball gun or BB gun which either expels a single globular projectile not exceeding .18 inch in diameter and which has a maximum muzzle velocity of less than 700 feet per second or breakable paint balls containing washable marking colors;

(2) any device used exclusively for signaling or safety and required or recommended by the United States Coast Guard or the Interstate Commerce Commission;

(3) any device used exclusively for the firing of stud cartridges, explosive rivets or similar industrial ammunition; and

(4) an antique firearm (other than a machine-gun) which, although designed as a weapon, the Department of State Police finds by reason of the date of its manufacture, value, design, and other characteristics is primarily a collector's item and is not likely to be used as a weapon.

"Firearm ammunition" means any self-contained cartridge or shotgun shell, by whatever name known, which is designed to be used or adaptable to use in a firearm; excluding, however:

(1) any ammunition exclusively designed for use with a device used exclusively for signaling or safety and required or recommended by the United States Coast Guard or the Interstate Commerce Commission; and

(2) any ammunition designed exclusively for use with a stud or rivet driver or other similar industrial ammunition.

"Gun show" means an event or function:

(1) at which the sale and transfer of firearms is the regular and normal course of business and where 50 or more firearms are displayed, offered, or exhibited for sale, transfer, or exchange; or

(2) at which not less than 10 gun show vendors display, offer, or exhibit for sale, sell, transfer, or exchange firearms.

"Gun show" includes the entire premises provided for an event or function, including parking areas for the event or function, that is sponsored to facilitate the purchase, sale, transfer, or exchange of firearms as described in this Section.

"Gun show" does not include training or safety classes, competitive shooting events, such as rifle, shotgun, or handgun matches, trap, skeet, or sporting clays shoots, dinners, banquets, raffles, or any other event where the sale or transfer of firearms is not the primary course of business.

"Gun show promoter" means a person who organizes or operates a gun show.

"Gun show vendor" means a person who exhibits, sells, offers for sale, transfers, or exchanges any firearms at a gun show, regardless of whether the person arranges with a gun show promoter for a fixed location from which to exhibit, sell, offer for sale, transfer, or exchange any firearm.

"Sanctioned competitive shooting event" means a shooting contest officially recognized by a national or state shooting sport association, and includes any sight-in or practice conducted in conjunction with the event.

"Stun gun or taser" has the meaning ascribed to it in Section 24-1 of the Criminal Code of 1961.

65/2. Firearm Owner's Identification Card required; exceptions

(a)(1) No person may acquire or possess any firearm, stun gun, or taser within this State without having in his or her possession a Firearm Owner's Identification Card previously issued in his or her name by the Department of State Police under the provisions of this Act.

(2) No person may acquire or possess firearm ammunition within this State without having in his or her possession a Firearm Owner's Identification Card previously issued in his or her name by the Department of State Police under the provisions of this Act.

(b) The provisions of this Section regarding the possession of firearms, firearm ammunition, stun guns, and tasers do not apply to:

(1) United States Marshals, while engaged in the operation of their official duties;

(2) Members of the Armed Forces of the United States or the National Guard, while engaged in the operation of their official duties;

(3) Federal officials required to carry firearms, while engaged in the operation of their official duties;

(4) Members of bona fide veterans organizations which receive firearms directly from the armed forces of the United States, while using the firearms for ceremonial purposes with blank ammunition;

(5) Nonresident hunters during hunting season, with valid nonresident hunting licenses and while in an area where hunting is permitted; however, at all other times and in all other places these persons must have their firearms unloaded and enclosed in a case;

(6) Those hunters exempt from obtaining a hunting license who are required to submit their Firearm Owner's Identification Card when hunting on Department of Natural Resources owned or managed sites;

(7) Nonresidents while on a firing or shooting range recognized by the Department of State Police; however, these persons must at all other times and in all other places have their firearms unloaded and enclosed in a case;

(8) Nonresidents while at a firearm showing or display recognized by the Department of State Police; however, at all other times and in all other places these persons must have their firearms unloaded and enclosed in a case;

(9) Nonresidents whose firearms are unloaded and enclosed in a case;

(10) Nonresidents who are currently licensed or registered to possess a firearm in their resident state;

(11) Unemancipated minors while in the custody and immediate control of their parent or legal guardian or other person in loco parentis to the minor if the parent or legal guardian or other person in loco parentis to the minor has a currently valid Firearm Owner's Identification Card;

(12) Color guards of bona fide veterans organizations or members of bona fide American Legion bands while using firearms for ceremonial purposes with blank ammunition;

(13) Nonresident hunters whose state of residence does not require them to be licensed or registered to possess a firearm and only during hunting season, with valid hunting licenses, while accompanied by, and using a firearm owned by, a person who possesses a valid Firearm Owner's Identification Card and while in an area within a commercial club licensed under the Wildlife Code [520 ILCS 5/1.1 et seq.] where hunting is permitted and controlled, but in no instance upon sites owned or managed by the Department of Natural Resources;

(14) Resident hunters who are properly authorized to hunt and, while accompanied by a person who possesses a valid Firearm Owner's Identification Card, hunt in an area within a commercial club licensed under the Wildlife Code where hunting is permitted and controlled; and

(15) A person who is otherwise eligible to obtain a Firearm Owner's Identification Card under this Act and is under the direct supervision of a holder of a Firearm Owner's Identification Card who is 21 years of age or older while the person is on a firing or shooting range or is a participant in a firearms safety and training course recognized by a law enforcement agency or a national, statewide shooting sports organization.

(c) The provisions of this Section regarding the acquisition and possession of firearms, firearm ammunition, stun guns, and tasers do not apply to law enforcement officials of this or any other jurisdiction, while engaged in the operation of their official duties.

65/3. Transfer of firearms; records; exceptions

(a) Except as provided in Section 3a, no person may knowingly transfer, or cause to be transferred, any firearm, firearm ammunition, stun gun, or taser to any person within this State unless the transferee with whom he deals displays a currently valid Firearm Owner's Identification Card which has previously been issued in his name by the Department of State Police under the provisions of this Act. In addition, all firearm, stun gun, and taser transfers by federally licensed firearm dealers are subject to Section 3.1.

(a-5) Any person who is not a federally licensed firearm dealer and who desires to transfer or sell a firearm while that person is on the grounds of a gun show must, before selling or transferring the firearm, request the Department of State Police to conduct a background check on the prospective recipient of the firearm in accordance with Section 3.1.

(b) Any person within this State who transfers or causes to be transferred any firearm, stun gun, or taser shall keep a record of such transfer for a period of 10 years from the date of transfer. Such record shall contain the date of the transfer; the description, serial number or other information identifying the firearm if no serial number is available; and, if the transfer was completed within this State, the transferee's Firearm Owner's Identification Card number. On or after January 1, 2006, the record shall contain the date of application for transfer of the firearm. On demand of a peace officer such transferor shall produce for inspection such record of transfer. If the transfer or sale took place at a gun show, the record shall include the unique identification number. Failure to record the unique identification number is a petty offense.

(b-5) Any resident may purchase ammunition from a person outside of Illinois. Any resident purchasing ammunition outside the State of Illinois must provide the seller with a copy of his or her valid Firearm Owner's Identification Card and either his or her Illinois driver's license or Illinois State Identification Card prior to the shipment of the ammunition. The ammunition may be shipped only to an address on either of those 2 documents.

(c) The provisions of this Section regarding the transfer of firearm ammunition shall not apply to those persons specified in paragraph (b) of Section 2 of this Act.

65/3a. Reciprocal rights in Iowa, Missouri, Indiana, Wisconsin and Kentucky

(a) Any resident of Illinois who has obtained a firearm owner's identification card pursuant to this Act and who is not otherwise prohibited from obtaining, possessing or using a firearm may purchase or obtain a rifle or shotgun or ammunition for a rifle or shotgun in Iowa, Missouri, Indiana, Wisconsin or Kentucky.

(b) Any resident of Iowa, Missouri, Indiana, Wisconsin or Kentucky or a non-resident with a valid non-resident hunting license, who is 18 years of age or older and who is not prohibited by the laws of Illinois, the state of his domicile, or the United States from obtaining, possessing or using a firearm, may purchase or obtain a rifle, shotgun or ammunition for a rifle or shotgun in Illinois.

(b-5) Any non-resident who is participating in a sanctioned competitive shooting event, who is 18 years of age or older and who is not prohibited by the laws of Illinois, the state of his or her domicile, or the United States from obtaining, possessing, or using a firearm, may purchase or obtain a shotgun or shotgun ammunition in Illinois for the purpose of participating in that event. A person may purchase or obtain a shotgun or shotgun ammunition under this subsection only at the site where the sanctioned competitive shooting event is being held.

(c) Any transaction under this Section is subject to the provisions of the Gun Control Act of 1968 (18 U.S.C. 922 (b)(3)).

65/3.1. Dial up system

(a) The Department of State Police shall provide a dial up telephone system or utilize other existing technology which shall be used by any federally licensed firearm dealer, gun show promoter, or gun show vendor who is to transfer a firearm, stun gun, or taser under the provisions of this Act. The Department of State Police may utilize existing technology which allows the caller to be charged a fee not to exceed $2. Fees collected by the Department of State Police shall be deposited in the State Police Services Fund and used to provide the service.

(b) Upon receiving a request from a federally licensed firearm dealer, gun show promoter, or gun show vendor, the Department of State Police shall immediately approve, or within the time period established by Section 24-3 of the Criminal Code of 1961 regarding the delivery of firearms, stun guns, and tasers notify the inquiring dealer, gun show promoter, or gun show vendor of any objection that would disqualify the transferee from acquiring or possessing a firearm, stun gun, or taser. In conducting the inquiry, the Department of State Police shall initiate and complete an automated search of its criminal history record information files and those of the Federal Bureau of Investigation, including the National Instant Criminal Background Check System, and of the files of the Department of Human Services relating to mental health and developmental disabilities to obtain any felony conviction or patient hospitalization information which would disqualify a person from obtaining or require revocation of a currently valid Firearm Owner's Identification Card.

(c) If receipt of a firearm would not violate Section 24-3 of the Criminal Code of 1961, federal law, or this Act the Department of State Police shall:

(1) assign a unique identification number to the transfer; and

(2) provide the licensee, gun show promoter, or gun show vendor with the number.

(d) Approvals issued by the Department of State Police for the purchase of a firearm are valid for 30 days from the date of issue.

(e)(1) The Department of State Police must act as the Illinois Point of Contact for the National Instant Criminal Background Check System.

(2) The Department of State Police and the Department of Human Services shall, in accordance with State and federal law regarding confidentiality, enter into a memorandum of understanding with the Federal Bureau of Investigation for the purpose of implementing the National Instant Criminal Background Check System in the State. The Department of State Police shall report the name, date of birth, and physical description of any person prohibited from possessing a firearm pursuant to the Firearm Owners Identification Card Act or 18 U.S.C. 922(g) and (n) to the National Instant Criminal Background Check System Index, Denied Persons Files.

(f) The Department of State Police shall promulgate rules not inconsistent with this Section to implement this system.

65/3.2. List of prohibited projectiles; notice to dealers Prior to January 1, 2002, the Department of State Police shall list on the Department's World Wide Web site all firearm projectiles that are prohibited under Sections 24-2.1, 24-2.2, and 24-3.2 of the Criminal Code of 1961, together with a statement setting forth the sentence that may be imposed for violating those Sections. The Department of State Police shall, prior to January 1, 2002, send a list of all firearm projectiles that are prohibited under Sections 24-2.1, 24-2.2, and 24-3.2 of the Criminal Code of 1961 to each federally licensed firearm dealer in Illinois registered with the Department.

65/3.3. Report to the local law enforcement agency

The Department of State Police must report the name and address of a person to the local law enforcement agency where the person resides if the person attempting to purchase a firearm is disqualified from purchasing a firearm because of information obtained under Section 3.1 [430 ILCS 65/3.1] that would disqualify the person from obtaining a Firearm Owner's Identification Card under any of subsections (c) through (n) of Section 8 of this Act [430 ILCS 65/8].

65/4. Application for Firearm Owner's Identification Card

(a) Each applicant for a Firearm Owner's Identification Card must:

(1) Make application on blank forms prepared and furnished at convenient locations throughout the State by the Department of State Police, or by electronic means, if and when made available by the Department of State Police; and

(2) Submit evidence to the Department of State Police that:

(i) He or she is 21 years of age or over, or if he or she is under 21 years of age that he or she has the written consent of his or her parent or legal guardian to possess and acquire firearms and firearm ammunition and that he or she has never been convicted of a misdemeanor other than a traffic offense or adjudged delinquent, provided, however, that such parent or legal guardian is not an individual prohibited from having a Firearm Owner's Identification Card and files an affidavit with the Department as prescribed by the Department stating that he or she is not an individual prohibited from having a Card;

(ii) He or she has not been convicted of a felony under the laws of this or any other jurisdiction;

(iii) He or she is not addicted to narcotics;

(iv) He or she has not been a patient in a mental institution within the past 5 years and he or she has not been adjudicated as a mental defective ;

(v) He or she is not mentally retarded;

(vi) He or she is not an alien who is unlawfully present in the United States under the laws of the United States;

(vii) He or she is not subject to an existing order of protection prohibiting him or her from possessing a firearm;

(viii) He or she has not been convicted within the past 5 years of battery, assault, aggravated assault, violation of an order of protection, or a substantially similar offense in another jurisdiction, in which a firearm was used or possessed;

(ix) He or she has not been convicted of domestic battery or a substantially similar offense in another jurisdiction committed on or after the effective date of this amendatory Act of 1997;

(x) He or she has not been convicted within the past 5 years of domestic battery or a substantially similar offense in another jurisdiction committed before the effective date of this amendatory Act of 1997;

(xi) He or she is not an alien who has been admitted to the United States under a nonimmigrant visa (as that term is defined in Section 101 (a)(26) of the Immigration and Nationality Act (8 U.S.C. 1101(a)(26))), or that he or she is an alien who has been lawfully admitted to the United States under a non-immigrant visa if that alien is:

(1) admitted to the United States for lawful hunting or sporting purposes;

(2) an official representative of a foreign government who is:

(A) accredited to the United States Government or the Government's mission to an international organization having its headquarters in the United States; or

(B) en route to or from another country to which that alien is accredited;

(3) an official of a foreign government or distinguished foreign visitor who has been so designated by the Department of State;

(4) a foreign law enforcement officer of a friendly foreign government entering the United States on official business; or

(5) one who has received a waiver from the Attorney General of the United States pursuant to 18 U.S.C. 922(y)(3);

(xii) He or she is not a minor subject to a peti-tion filed under Section 5-520 of the Juvenile Court Act of 1987 alleging that the minor is a de-linquent minor for the commission of an offense that if committed by an adult would be a felony; and

(xiii) He or she is not an adult who had been adjudicated a delinquent minor under the Juvenile Court Act of 1987 for the commission of an offense that if committed by an adult would be a felony; and

(3) Upon request by the Department of State Police, sign a release on a form prescribed by the Department of State Police waiving any right to confidentiality and requesting the disclosure to the Department of State Police of limited mental health institution admission information from another state, the District of Columbia, any other territory of the United States, or a foreign nation concerning the applicant for the sole purpose of determining whether the applicant is or was a patient in a mental health institution and disqualified because of that status from receiving a Firearm Owner's Identification Card. No mental health care or treatment records may be requested. The information received shall be destroyed within one year of receipt.

(a-5) Each applicant for a Firearm Owner's Identification Card who is over the age of 18 shall furnish to the Department of State Police either his or her driver's license number or Illinois Identification Card number.

(a-10) Each applicant for a Firearm Owner's Identification Card, who is employed as an armed security officer at a nuclear energy, storage, weapons, or development facility regulated by the Nuclear Regulatory Commission and who is not an Illinois resident, shall furnish to the Department of State Police his or her driver's license number or state identification card number from his or her state of residence. The Department of State Police may promulgate rules to enforce the provisions of this subsection (a-10).

(b) Each application form shall include the following statement printed in bold type: "Warning: Entering false information on an application for a Firearm Owner's Identification Card is punishable as a Class 2 felony in accordance with subsection (d-5) of Section 14 of the Firearm Owners Identification Card Act."

(c) Upon such written consent, pursuant to Section 4, paragraph (a)(2)(i), the parent or legal guardian giving the consent shall be liable for any damages resulting from the applicant's use of firearm or firearm ammunition.

65/5. Approval or denial of application; fees The Department of State Police shall either approve or deny all applications within 30 days from the date they are received, and every applicant found qualified pursuant to Section 8 of this Act by the Department shall be entitled to a Firearm Owner's Identification Card upon the payment of a $10 fee. $6 of each fee derived from the issuance of Firearm Owner's Identification Cards, or renewals thereof, shall be deposited in the Wildlife and Fish Fund in the State Treasury; $1 of such fee shall be deposited in the State Police Services Fund and $3 of such fee shall be deposited in the Firearm Owner's Notification Fund. Monies in the Firearm Owner's Notification Fund shall be used exclusively to pay for the cost of sending notices of expiration of Firearm Owner's Identification Cards under Section 13.2 of this Act. Excess monies in the Firearm Owner's Notification Fund shall be used to ensure the prompt and efficient processing of applications received under Section 4 of this Act.

65/6. Contents of Firearm Owner's Identification Card

(a) A Firearm Owner's Identification Card, issued by the Department of State Police at such places as the Director of the Department shall specify, shall contain the applicant's name, residence, date of birth, sex, physical description, recent photograph and signature such other personal identifying information as may be required by the Director. Each Firearm Owner's Identification Card must have the expiration date boldly and conspicuously displayed on the face of the card. Each Firearm Owner's Identification Card must have printed on it the following: "CAUTION - This card does not permit bearer to UNLAWFULLY carry or use firearms." Before December 1, 2002, the Department may use a person's digital photograph and signature from his or her Illinois driver's license or Illinois Identification Card, if available. On and after December 1, 2002, the Department shall use a person's digital photograph and signature from his or her Illinois driver's license or Illinois Identification Card, if available. The Department shall decline to use a person's digital photograph or signature if the digital photograph or signature is the result of or associated with fraudulent or erroneous data, unless otherwise provided by law.

(b) A person applying for a Firearm Owner's Identification Card shall consent to the Department of State Police using the applicant's digital driver's license or Illinois Identification Card photograph, if available, and signature on the applicant's Firearm Owner's Identification Card. The Secretary of State shall allow the Department of State Police access to the photograph and signature for the purpose of identifying the applicant and issuing to the applicant a Firearm Owner's Identification Card.

(c) The Secretary of State shall conduct a study to determine the cost and feasibility of creating a method of adding an identifiable code, background, or other means on the driver's license or Illinois Identification Card to show that an individual is not disqualified from owning or possessing a firearm under State or federal law. The Secretary shall report the findings of this study 12 months after the effective date of this amendatory Act of the 92nd General Assembly.

65/6.1. Altered, forged or counterfeit Firearm Owner's Identification Cards

(a) Any person who forges or materially alters a Firearm Owner's Identification Card or who counterfeits a Firearm Owner's Identification Card commits a Class 2 felony.

(b) Any person who knowingly possesses a forged or materially altered Firearm Owner's Identification Card with the intent to use it commits a Class 2 felony. A person who possesses a Firearm Owner's Identification Card with knowledge that it is counterfeit commits a Class 2 felony.

65/7. Period of validity Except as provided in Section 8 of this Act, a Firearm Owner's Identification Card issued under the provisions of this Act shall be valid for the per-son to whom it is issued for a period of 10 years from the date of issuance.

65/8. Denial or revocation The Department of State Police has authority to deny an application for or to revoke and seize a Firearm Owner's Identification Card previously issued under this Act only if the Department finds that the applicant or the person to whom such card was issued is or was at the time of issuance:

(a) A person under 21 years of age who has been convicted of a misdemeanor other than a traffic offense or adjudged delinquent;

(b) A person under 21 years of age who does not have the written consent of his parent or guardian to acquire and possess firearms and firearm ammunition, or whose parent or guardian has revoked such written consent, or where such parent or guardian does not qualify to have a Firearm Owner's Identification Card;

(c) A person convicted of a felony under the laws of this or any other jurisdiction;

(d) A person addicted to narcotics;

(e) A person who has been a patient of a mental institution within the past 5 years or has been adjudicated as a mental defective ;

(f) A person whose mental condition is of such a nature that it poses a clear and present danger to the applicant, any other person or persons or the community;

For the purposes of this Section, "mental condition" means a state of mind manifested by violent, suicidal, threatening or assaultive behavior.

(g) A person who is mentally retarded;

(h) A person who intentionally makes a false statement in the Firearm Owner's Identification Card application;

(i) An alien who is unlawfully present in the United States under the laws of the United States;

(i-5) An alien who has been admitted to the United States under a non-immigrant visa (as that term is defined in Section 101(a)(26) of the Immigration and Nationality Act (8 U.S.C. 1101(a)(26))), except that this subsection (i-5) does not apply to any alien who has been lawfully admitted to the United States under a non-immigrant visa if that alien is:

(1) admitted to the United States for lawful hunting or sporting purposes

(2) an official representative of a foreign government who is:

(A) accredited to the United States Government or the Government's mission to an international organization having its headquarters in the United States; or

(B) en route to or from another country to which that alien is accredited;

(3) an official of a foreign government or distinguished foreign visitor who has been so designated by the Department of State;

(4) a foreign law enforcement officer of a friendly foreign government entering the United States on official business; or

(5) one who has received a waiver from the Attorney General of the United States pursuant to 18 U.S.C. 922(y)(3);

(j) A person who is subject to an existing order of protection prohibiting him or her from possessing a firearm;

(k) A person who has been convicted within the past 5 years of battery, assault, aggravated assault, violation of an order of protection, or a substantially similar offense in another jurisdiction, in which a firearm was used or possessed;

(l) A person who has been convicted of domestic battery or a substantially similar offense in another jurisdiction committed on or after January 1, 1998;

(m) A person who has been convicted within the past 5 years of domestic battery or a substantially similar offense in another jurisdiction committed before January 1, 1998;

(n) A person who is prohibited from acquiring or possessing firearms or firearm ammunition by any Illinois State statute or by federal law.

(o) A minor subject to a petition filed under Section 5-520 of the Juvenile Court Act of 1987 alleging that the minor is a delinquent minor for the commission of an offense that if committed by an adult would be a felony; or

(p) An adult who had been adjudicated a delinquent minor under the Juvenile Court Act of 1987 for the commission of an offense that if committed by an adult would be a felony.

65/8.1. Circuit Clerk to notify Department of State Police

(a) The Circuit Clerk shall, in the form and manner required by the Supreme Court, notify the Department of State Police of all final dispositions of cases for which the Department has received information reported to it under Section 2.1 of the Criminal Identification Act [20 ILCS 2630/2.1].

(b) Upon adjudication of any individual as a mental defective, as defined in Section 1.1 [430 ILCS 65/1.1], the court shall direct the circuit court clerk to immediately notify the Department of State Police, Firearm Owner's Identification (FOID) department, and shall forward a copy of the court order to the Department.

65/9. Grounds for denial or revocation

Every person whose application for a Firearm Owner's Identification Card is denied, and every holder of such a Card before his Card is revoked or seized, shall receive a written notice from the Department of State Police stating specifically the grounds upon which his application has been denied or upon which his Identification Card has been revoked.

65/10. Appeal of denial or revocation

(a) Whenever an application for a Firearm Owner's Identification Card is denied, whenever the Department fails to act on an application within 30 days of its receipt, or whenever such a Card is revoked or seized as provided for in Section 8 of this Act [430 ILCS 65/8], the aggrieved party may appeal to the Director of the Department of State Police for a hearing upon such denial, revocation or seizure, unless the denial, revocation, or seizure was based upon a forcible felony, stalking, aggravated stalking, domestic battery, any violation of the Illinois Controlled Substances Act [720 ILCS 570/100 et seq.], the Methamphetamine Control and Community Protection Act [720 ILCS 646/1 et seq,], or the Cannabis Control Act [720 ILCS 550/1 et seq.] that is classified as a Class 2 or greater felony, any felony violation of Article 24 of the Criminal Code of 1961 [720 ILCS 5/24-1 et seq.], or any adjudication as a delinquent minor for the commission of an offense that if committed by an adult would be a felony, in which case the aggrieved party may petition the circuit court in writing in the county of his or her residence for a hearing upon such denial, revocation, or seizure.

(b) At least 30 days before any hearing in the circuit court, the petitioner shall serve the relevant State's Attorney with a copy of the petition. The State's Attorney may object to the petition and present evidence. At the hearing the court shall determine whether substantial justice has been done. Should the court determine that substantial justice has not been done, the court shall issue an order directing the Department of State Police to issue a Card.

(c) Any person prohibited from possessing a firearm under Sections 24-1.1 or 24-3.1 of the Criminal Code of 1961 [720 ILCS 5/24-1.1 or 720 ILCS 5/24-3.1] or acquiring a Firearm Owner's Identification Card under Section 8 of this Act [430 ILCS 65/8] may apply to the Director of the Department of State Police or petition the circuit court in the county where the petitioner resides, whichever is applicable in accordance with subsection (a) of this Section, requesting relief from such prohibition and the Director or court may grant such relief if it is established by the applicant to the court's or Director's satisfaction that:

(0.05) when in the circuit court, the State's Attorney has been served with a written copy of the petition at least 30 days before any such hearing in the circuit court and at the hearing the State's Attorney was afforded an opportunity to present evidence and object to the petition;

(1) the applicant has not been convicted of a forcible felony under the laws of this State or any other jurisdiction within 20 years of the applicant's application for a Firearm Owner's Identification Card, or at least 20 years have passed since the end of any period of imprisonment imposed in relation to that conviction,

(2) the circumstances regarding a criminal conviction, where applicable, the applicant's criminal history and his reputation are such that the applicant will not be likely to act in a manner dangerous to public safety; and

(3) granting relief would not be contrary to the public interest.

(d) When a minor is adjudicated delinquent for an offense which if committed by an adult would be a felony, the court shall notify the Department of State Police.

(e) The court shall review the denial of an application or the revocation of a Firearm Owner's Identification Card of a person who has been adjudicated delinquent for an offense that if committed by an adult would be a felony if an application for relief has been filed at least 10 years after the adjudication of delinquency and the court determines that the applicant should be granted relief from disability to obtain a Firearm Owner's Identification Card. If the court grants relief, the court shall notify the Department of State Police that the disability has been removed and that the applicant is eligible to obtain a Firearm Owner's Identification Card.

65/11. Judicial review of administrative decision

All final administrative decisions of the Department under this Act shall be subject to judicial review under the provisions of the Administrative Review Law [735 ILCS 5/3-101 et seq.], and all amendments and modifications thereof, and the rules adopted pursuant thereto. The term "administrative decision" is defined as in Section 3-101 of the Code of Civil Procedure [735 ILCS 5/3-101].

The Director of State Police shall submit a report to the General Assembly on March 1 of each year, beginning March 1, 1991, listing all final decisions by a court of this State upholding, reversing, or reversing in part any administrative decision made by the Department of State Police.

65/12. Transfer of firearms upon death of owner or incident to legal proceedings The provisions of this Act shall not apply to the passing or transfer of any firearm or firearm ammunition upon the death of the owner thereof to his heir or legatee or to the passing or transfer of any firearm or firearm ammunition incident to any legal proceeding or action until 60 days after such passing or transfer.

65/13. Acquisition or possession otherwise prohibited Nothing in this Act shall make lawful the acquisition or possession of firearms or fire-arm ammunition which is otherwise prohibited by law.

65/13.1. Municipal ordinance not affected The provisions of any ordinance enacted by any municipality which requires registration or imposes greater restrictions or limitations on the acquisition, possession and transfer of firearms than are imposed by this Act, are not invalidated or affected by this Act.

65/13.2. Notice of expiration The Department of State Police shall, 60 days prior to the expiration of a Firearm Owner's Identification Card, forward by first class mail to each person whose card is to expire a notification of the expiration of the card and an application which may be used to apply for renewal of the card. It is the obligation of the holder of a Firearm Owner's Identification Card to notify the Department of State Police of any address change since the issuance of the Fire-arm Owner's Identification Card.

65/13.3. Municipal ordinance submission Within 6 months after the effective date of this amendatory Act of the 92nd General Assembly, every municipality must submit to the Department of State Police a copy of every ordinance adopted by the municipality that regulates the acquisition, possession, sale, or transfer of fire-arms within the municipality and must submit, 30 days after adoption, every such ordinance adopted after its initial submission of ordinances under this Section. The Department of State Police shall compile these ordinances and publish them in a form available to the public free of charge and shall periodically update this compilation of ordinances in a manner prescribed by the Director of State Police.

65/14. Sentence

(a) A violation of paragraph (1) of subsection (a) of Section 2, when the person's Firearm Owner's Identification Card is expired but the person is not otherwise disqualified from renewing the card, is a Class A misdemeanor.

(b) Except as provided in subsection (a) with respect to an expired card, a violation of paragraph (1) of subsection (a) of Section 2 is a Class A misdemeanor when the person does not possess a currently valid Firearm Owner's Identification Card, but is otherwise eligible under this Act. A second or subsequent violation is a Class 4 felony.

(c) A violation of paragraph (1) of subsection (a) of Section 2 is a Class 3 felony when:

(1) the person's Firearm Owner's Identification Card is revoked or subject to revocation under Section 8; or

(2) the person's Firearm Owner's Identification Card is expired and not otherwise eligible for renewal under this Act; or

(3) the person does not possess a currently valid Firearm Owner's Identification Card, and the person is not otherwise eligible under this Act.

(d) A violation of subsection (a) of Section 3 is a Class 4 felony. A third or subsequent conviction is a Class 1 felony.

(d-5) Any person who knowingly enters false information on an application for a Firearm Owner's Identification Card, who knowingly gives a false answer to any question on the application, or who knowingly submits false evidence in connection with an application is guilty of a Class 2 felony.

(e) Except as provided by Section 6.1 of this Act, any other violation of this Act is a Class A misdemeanor.

Chapter 720. Criminal Offenses

Act 5. Criminal Code of 1961

Article 16. Theft and Related Offenses

5/16-16. Possession of a stolen firearm

(a) A person commits possession of a stolen firearm when he or she, not being entitled to the possession of a firearm, possesses or delivers the firearm, knowing it to have been stolen or converted. It may be inferred that a person who possesses a firearm with knowledge that its serial number has been removed or altered has knowledge that the firearm is stolen or converted.

(b) Possession of a stolen firearm is a Class 2 felony.

5/16-16.1. Aggravated possession of a stolen firearm

(a) A person commits aggravated possession of a stolen firearm when he or she:

(1) Not being entitled to the possession of not less than 2 and not more than 5 firearms, possesses or delivers those firearms at the same time or within a one year period, knowing the firearms to have been stolen or converted.

(2) Not being entitled to the possession of not less than 6 and not more than 10 firearms, possesses or delivers those firearms at the same time or within a 2 year period, knowing the firearms to have been stolen or converted.

(3) Not being entitled to the possession of not less than 11 and not more than 20 firearms, possesses or delivers those firearms at the same time or within a 3 year period, knowing the firearms to have been stolen or converted.

(4) Not being entitled to the possession of not less than 21 and not more than 30 firearms, possesses or delivers those firearms at the same time or within a 4 year period, knowing the firearms to have been stolen or converted.

(5) Not being entitled to the possession of more than 31 firearms, possesses or delivers those firearms at the same time or within a 5 year period, knowing the firearms to have been stolen or converted.

(b) It may be inferred that a person who possesses a firearm with knowledge that its serial number has been removed or altered has knowledge that the firearm is stolen or converted.

(c) Sentence.

(1) A person who violates paragraph (1) of subsection (a) of this Section commits a Class 1 felony.

(2) A person who violates paragraph (2) of subsection (a) of this Section commits a Class X felony for which he or she shall be sentenced to a term of imprisonment of not less than 6 years and not more than 30 years.

(3) A person who violates paragraph (3) of subsection (a) of this Section commits a Class X felony for which he or she shall be sentenced to a term of imprisonment of not less than 6 years and not more than 40 years.

(4) A person who violates paragraph (4) of subsection (a) of this Section commits a Class X felony for which he or she shall be sentenced to a term of imprisonment of not less than 6 years and not more than 50 years.

(5) A person who violates paragraph (5) of subsection (a) of this Section commits a Class X felony for which he or she shall be sentenced to a term of imprisonment of not less than 6 years and not more than 60 years.

Article 24. Deadly Weapons

5/24-1. Unlawful use of weapons

(a) A person commits the offense of unlawful use of weapons when he knowingly:

(1) (Effective until January 1, 2009) Sells, manufactures, purchases, possesses or carries any bludgeon, black-jack, slung-shot, sand-club, sand-bag, metal knuckles, throwing star, or any knife, commonly referred to as a switchblade knife, which has a blade that opens automatically by hand pressure applied to a button, spring or other device in the handle of the knife, or a ballistic knife, which is a device that propels a knifelike blade as a projectile by means of a coil spring, elastic material or compressed gas; or

(1) (As amended by P.A. 95-809, effective January 1, 2009) Sells, manufactures, purchases, possesses or carries any bludgeon, black-jack, slung-shot, sand-club, sand-bag, metal knuckles or other knuckle weapon regardless of its composition, throwing star, or any knife, commonly referred to as a switchblade knife, which has a blade that opens automatically by hand pressure applied to a button, spring or other device in the handle of the knife, or a ballistic knife, which is a device that propels a knifelike blade as a projectile by means of a coil spring, elastic material or compressed gas; or

(2) Carries or possesses with intent to use the same unlawfully against another, a dagger, dirk, billy, dangerous knife, razor, stiletto, broken bottle or other piece of glass, stun gun or taser or any other dangerous or deadly weapon or instrument of like character; or

(3) Carries on or about his person or in any vehicle, a tear gas gun projector or bomb or any object containing noxious liquid gas or substance, other than an object containing a non-lethal noxious liquid gas or substance designed solely for personal defense carried by a person 18 years of age or older; or

(4) Carries or possesses in any vehicle or concealed on or about his person except when on his land or in his own abode or fixed place of business any pistol, revolver, stun gun or taser or other firearm, except that this subsection (a)(4) does not apply to or affect transportation of weapons that meet one of the following conditions:

(i) are broken down in a non-functioning state; or

(ii) are not immediately accessible; or

(iii) are unloaded and enclosed in a case, firearm carrying box, shipping box, or other container by a person who has been issued a currently valid Firearm Owner's Identification Card; or

(5) Sets a spring gun; or

(6) Possesses any device or attachment of any kind designed, used or intended for use in silencing the report of any firearm; or

(7) Sells, manufactures, purchases, possesses or carries:

(i) a machine gun, which shall be defined for the purposes of this subsection as any weapon, which shoots, is designed to shoot, or can be readily restored to shoot, automatically more than one shot without manually reloading by a single function of the trigger, including the frame or receiver of any such weapon, or sells, manufactures, purchases, possesses, or carries any combination of parts designed or intended for use in converting any weapon into a machine gun, or any combination or parts from which a machine gun can be assembled if such parts are in the possession or under the control of a person;

(ii) any rifle having one or more barrels less than 16 inches in length or a shotgun having one or more barrels less than 18 inches in length or any weapon made from a rifle or shotgun, whether by alteration, modification, or otherwise, if such a weapon as modified has an overall length of less than 26 inches; or

(iii) any bomb, bomb-shell, grenade, bottle or other container containing an explosive substance of over one-quarter ounce for like purposes, such as, but not limited to, black powder bombs and Molotov cocktails or artillery projectiles; or

(8) Carries or possesses any firearm, stun gun or taser or other deadly weapon in any place which is licensed to sell intoxicating beverages, or at any public gathering held pursuant to a license issued by any governmental body or any public gathering at which an admission is charged, excluding a place where a showing, demonstration or lecture involving the exhibition of unloaded firearms is conducted.

This subsection (a)(8) does not apply to any auction or raffle of a firearm held pursuant to a license or permit issued by a governmental body, nor does it apply to persons engaged in firearm safety training courses; or

(9) Carries or possesses in a vehicle or on or about his person any pistol, revolver, stun gun or taser or firearm or ballistic knife, when he is hooded, robed or masked in such manner as to conceal his identity; or

(10) Carries or possesses on or about his person, upon any public street, alley, or other public lands within the corporate limits of a city, village or incorporated town, except when an invitee thereon or therein, for the purpose of the display of such weapon or the lawful commerce in weapons, or except when on his land or in his own abode or fixed place of business, any pistol, revolver, stun gun or taser or other firearm, except that this subsection (a)(10) does not apply to or affect transportation of weapons that meet one of the following conditions:

(i) are broken down in a non-functioning state; or

(ii) are not immediately accessible; or

(iii) are unloaded and enclosed in a case, firearm carrying box, shipping box, or other container by a person who has been issued a currently valid Firearm Owner's Identification Card.

A "stun gun or taser", as used in this paragraph (a) means (i) any device which is powered by electrical charging units, such as, batteries, and which fires one or several barbs attached to a length of wire and which, upon hitting a human, can send out a current capable of disrupting the person's nervous system in such a manner as to render him incapable of normal functioning or (ii) any device which is powered by electrical charging units, such as batteries, and which, upon contact with a human or clothing worn by a human, can send out current capable of disrupting the person's nervous system in such a manner as to render him incapable of normal functioning; or

(11) Sells, manufactures or purchases any explosive bullet. For purposes of this paragraph

(a) "explosive bullet" means the projectile portion of an ammunition cartridge which contains or carries an explosive charge which will explode upon contact with the flesh of a human or an animal. "Cartridge" means a tubular metal case having a projectile affixed at the front thereof and a cap or primer at the rear end thereof, with the propellant contained in such tube between the projectile and the cap; or

(12) (Blank); or

(13) (Added by P.A. 95-885, effective January 1, 2009) Carries or possesses on or about his or her person while in a building occupied by a unit of government, a billy club, other weapon of like character, or other instrument of like character intended for use as a weapon. For the purposes of this Section, "billy club" means a short stick or club commonly carried by police officers which is either telescopic or constructed of a solid piece of wood or other man-made material.

(b) (Effective until January 1, 2009) Sentence. A person convicted of a violation of subsection 24-1(a)(1) through (5), subsection 24-1(a)(10), or subsection 24-1(a)(11) commits a Class A misdemeanor. A person convicted of a violation of subsection 24-1(a)(8) or 24-1(a)(9) commits a Class 4 felony; a person convicted of a violation of subsection 24-1(a)(6) or 24-1(a)(7)(ii) or (iii) commits a Class 3 felony. A person convicted of a violation of subsection 24-1(a)(7)(i) commits a Class 2 felony and shall be sentenced to a term of imprisonment of not less than 3 years and not more than 7 years, unless the weapon is possessed in the passenger compartment of a motor vehicle as defined in Section 1-146 of the Illinois Vehicle Code [625 ILCS 5/1-146], or on the person, while the weapon is loaded, in which case it shall be a Class X felony. A person convicted of a second or subsequent violation of subsection 24-1(a)(4), 24-1(a)(8), 24-1(a)(9), or 24-1(a)(10) commits a Class 3 felony. The possession of each weapon in violation of this Section constitutes a single and separate violation.

(b) (As amended by P.A. 95-885, effective January 1, 2009) Sentence. A person convicted of a violation of subsection 24-1(a)(1) through (5), subsection 24-1(a)(10), subsection 24-1(a)(11), or subsection 24-1(a)(13) commits a Class A misdemeanor. A person convicted of a violation of subsection 24-1(a)(8) or 24-1(a)(9) commits a Class 4 felony; a person convicted of a violation of subsection 24-1(a)(6) or 24-1(a)(7)(ii) or (iii) commits a Class 3 felony. A person convicted of a violation of subsection 24-1(a)(7)(i) commits a Class 2 felony and shall be sentenced to a term of imprisonment of not less than 3 years and not more than 7 years, unless the weapon is possessed in the passenger compartment of a motor vehicle as defined in Section 1-146 of the Illinois Vehicle Code [625 ILCS 5/1-146], or on the person, while the weapon is loaded, in which case it shall be a Class X felony. A person convicted of a second or subsequent violation of subsection 24-1(a)(4), 24-1(a)(8), 24-1(a)(9), or 24-1(a)(10) commits a Class 3 felony. The possession of each weapon in violation of this Section constitutes a single and separate violation.

(c) Violations in specific places.

(1) A person who violates subsection 24-1(a)(6) or 24-1(a)(7) in any school, regardless of the time of day or the time of year, in residential property owned, operated or managed by a public housing agency or leased by a public housing agency as part of a scattered site or mixed-income development, in a public park, in a courthouse, on the real property comprising any school, regardless of the time of day or the time of year, on residential property owned, operated or managed by a public housing agency or leased by a public housing agency as part of a scattered site or mixed-income development, on the real property comprising any public park, on the real property comprising any courthouse, in any conveyance owned, leased or contracted by a school to transport students to or from school or a school related activity, or on any public way within 1,000 feet of the real property comprising any school, public park, courthouse, or residential property owned, operated, or managed by a public housing agency or leased by a public housing agency as part of a scattered site or mixed-income development commits a Class 2 felony and shall be sentenced to a term of imprisonment of not less than 3 years and not more than 7 years.

(1.5) A person who violates subsection 24-1(a)(4), 24-1(a)(9), or 24-1(a)(10) in any school, regardless of the time of day or the time of year, in residential property owned, operated, or managed by a public housing agency or leased by a public housing agency as part of a scattered site or mixed-income development, in a public park, in a courthouse, on the real property comprising any school, regardless of the time of day or the time of year, on residential property owned, operated, or managed by a public housing agency or leased by a public housing agency as part of a scattered site or mixed-income development, on the real property comprising any public park, on the real property comprising any courthouse, in any conveyance owned, leased, or contracted by a school to transport students to or from school or a school related activity, or on any public way within 1,000 feet of the real property comprising any school, public park, courthouse, or residential property owned, operated, or managed by a public housing agency or leased by a public housing agency as part of a scattered site or mixed-income development commits a Class 3 felony.

(2) A person who violates subsection 24-1(a)(1), 24-1(a)(2), or 24-1(a)(3) in any school, regardless of the time of day or the time of year, in residential property owned, operated or managed by a public housing agency or leased by a public housing agency as part of a scattered site or mixed-income development, in a public park, in a courthouse, on the real property comprising any school, regardless of the time of day or the time of year, on residential property owned, operated or managed by a public housing agency or leased by a public housing agency as part of a scattered site or mixed-income development, on the real property comprising any public park, on the real property comprising any courthouse, in any conveyance owned, leased or contracted by a school to transport students to or from school or a school related activity, or on any public way within 1,000 feet of the real property comprising any school, public park, courthouse, or residential property owned, operated, or managed by a public housing agency or leased by a public housing agency as part of a scattered site or mixed-income development commits a Class 4 felony. "Courthouse" means any building that is used by the Circuit, Appellate, or Supreme Court of this State for the conduct of official business.

(3) Paragraphs (1), (1.5), and (2) of this subsection (c) shall not apply to law enforcement officers or security officers of such school, college, or university or to students carrying or possessing firearms for use in training courses, parades, hunting, target shooting on school ranges, or otherwise with the consent of school authorities and which firearms are transported unloaded enclosed in a suitable case, box, or transportation package.

(4) For the purposes of this subsection (c), "school" means any public or private elementary or secondary school, community college, college, or university.

(d) The presence in an automobile other than a public omnibus of any weapon, instrument or substance referred to in subsection (a)(7) is prima facie evidence that it is in the possession of, and is being carried by, all persons occupying such automobile at the time such weapon, instrument or substance is found, except under the following circumstances: (i) if such weapon, instrument or instrumentality is found upon the person of one of the occupants therein; or (ii) if such weapon, instrument or substance is found in an automobile operated for hire by a duly licensed driver in the due, lawful and proper pursuit of his trade, then such presumption shall not apply to the driver.

(e) Exemptions. Crossbows, Common or Compound bows and Underwater Spearguns are exempted from the definition of ballistic knife as defined in paragraph (1) of subsection (a) of this Section.

5/24-1.1. Unlawful Use or Possession of Weapons by Felons or Persons in the Custody of the Department of Corrections Facilities

(a) It is unlawful for a person to knowingly possess on or about his person or on his land or in his own abode or fixed place of business any weapon prohibited under Section 24-1 of this Act or any firearm or any firearm ammunition if the person has been convicted of a felony under the laws of this State or any other jurisdiction. This Section shall not apply if the person has been granted relief by the Director of the Department of State Police under Section 10 of the Firearm Owners Identification Card Act.

(b) It is unlawful for any person confined in a penal institution, which is a facility of the Illinois Department of Corrections, to possess any weapon prohibited under Section 24-1 of this Code or any firearm or firearm ammunition, regardless of the intent with which he possesses it.

(c) It shall be an affirmative defense to a violation of subsection (b), that such possession was specifically authorized by rule, regulation, or directive of the Illinois Department of Corrections or order issued pursuant thereto.

(d) The defense of necessity is not available to a person who is charged with a violation of subsection (b) of this Section.

(e) Sentence. Violation of this Section by a person not confined in a penal institution shall be a Class 3 felony for which the person, if sentenced to a term of imprisonment, shall be sentenced to no less than 2 years and no more than 10 years and any second or subsequent violation shall be a Class 2 felony for which the person shall be sentenced to a term of imprisonment of not less than 3 years and not more than 14 years. Violation of this Section by a person not confined in a penal institution who has been convicted of a forcible felony, a felony violation of Article 24 of this Code or of the Firearm Owners Identification Card Act, stalking or aggravated stalking, or a Class 2 or greater felony under the Illinois Controlled Substances Act, the Cannabis Control Act, or the Methamphetamine Control and Community Protection Act is a Class 2 felony for which the person shall be sentenced to not less than 3 years and not more than 14 years. Violation of this Section by a person who is on parole or mandatory supervised release is a Class 2 felony for which the person, if sentenced to a term of imprisonment, shall be sentenced to not less than 3 years and not more than 14 years. Violation of this Section by a person not confined in a penal institution is a Class X felony when the firearm possessed is a machine gun. Any person who violates this Section while confined in a penal institution, which is a facility of the Illinois Department of Corrections, is guilty of a Class 1 felony, if he possesses any weapon prohibited under Section 24-1 of this Code regardless of the intent with which he possesses it, a Class X felony if he possesses any firearm, firearm ammunition or explosive, and a Class X felony for which the offender shall be sentenced to not less than 12 years and not more than 50 years when the firearm possessed is a machine gun. A violation of this Section while wearing or in possession of body armor as defined in Section 33F-1 is a Class X felony punishable by a term of imprisonment of not less than 10 years and not more than 40 years. The possession of each firearm or firearm ammunition in violation of this Section constitutes a single and separate violation.

5/24-1.6. Aggravated unlawful use of a weapon

(a) A person commits the offense of aggravated unlawful use of a weapon when he or she knowingly:

(1) Carries on or about his or her person or in any vehicle or concealed on or about his or her person except when on his or her land or in his or her abode or fixed place of business any pistol, revolver, stun gun or taser or other firearm; or

(2) Carries or possesses on or about his or her person, upon any public street, alley, or other public lands within the corporate limits of a city, village or incorporated town, except when an invitee thereon or therein, for the purpose of the display of such weapon or the lawful commerce in weapons, or except when on his or her own land or in his or her own abode or fixed place of business, any pistol, revolver, stun gun or taser or other firearm; and

(3) One of the following factors is present:

(A) the firearm possessed was uncased, loaded and immediately accessible at the time of the offense; or

(B) the firearm possessed was uncased, unloaded and the ammunition for the weapon was immediately accessible at the time of the offense; or

(C) the person possessing the firearm has not been issued a currently valid Firearm Owner's Identification Card; or

(D) the person possessing the weapon was previously adjudicated a delinquent minor under the Juvenile Court Act of 1987 for an act that if committed by an adult would be a felony; or

(E) the person possessing the weapon was engaged in a misdemeanor violation of the Cannabis Control Act, in a misdemeanor violation of the Illinois Controlled Substances Act, or in a misdemeanor violation of the Methamphetamine Control and Community Protection Act; or

(F) the person possessing the weapon is a member of a street gang or is engaged in street gang related activity, as defined in Section 10 of the Illinois Streetgang Terrorism Omnibus Prevention Act; or

(G) the person possessing the weapon had a order of protection issued against him or her within the previous 2 years; or

(H) the person possessing the weapon was engaged in the commission or attempted commission of a misdemeanor involving the use or threat of violence against the person or property of another; or

(I) the person possessing the weapon was under 21 years of age and in possession of a handgun as defined in Section 24-3, unless the person under 21 is engaged in lawful activities under the Wildlife Code or described in subsection 24-2(b)(1), (b)(3), or 24-2(f).

(b) "Stun gun or taser" as used in this Section has the same definition given to it in Section 24-1 of this Code.

(c) This Section does not apply to or affect the transportation or possession of weapons that:

(i) are broken down in a non-functioning state; or

(ii) are not immediately accessible; or

(iii) are unloaded and enclosed in a case, firearm carrying box, shipping box, or other container by a person who has been issued a currently valid Firearm Owner's Identification Card.

(d) Sentence. Aggravated unlawful use of a weapon is a Class 4 felony; a second or subsequent offense is a Class 2 felony for which the person shall be sentenced to a term of imprisonment of not less than 3 years and not more than 7 years. Aggravated unlawful use of a weapon by a person who has been previously convicted of a felony in this State or another jurisdiction is a Class 2 felony for which the person shall be sentenced to a term of imprisonment of not less than 3 years and not more than 7 years. Aggravated unlawful use of a weapon while wearing or in possession of body armor as defined in Section 33F-1 by a person who has not been issued a valid Firearms Owner's Identification Card in accordance with Section 5 of the Firearm Owners Identification Card Act is a Class X felony. The possession of each firearm in violation of this Section constitutes a single and separate violation.

5/24-1.7. Armed habitual criminal

(a) A person commits the offense of being an armed habitual criminal if he or she receives, sells, possesses, or transfers any firearm after having been convicted a total of 2 or more times of any combination of the following offenses:

(1) a forcible felony as defined in Section 2-8 of this Code [720 ILCS 5/2-8];

(2) unlawful use of a weapon by a felon; aggravated unlawful use of a weapon; aggravated discharge of a firearm; vehicular hijacking; aggravated vehicular hijacking; aggravated battery of a child; intimidation; aggravated intimidation; gunrunning; home invasion; or aggravated battery with a firearm; or

(3) any violation of the Illinois Controlled Substances Act [720 ILCS 570/100 et seq.] or the Cannabis Control Act [720 ILCS 550/1 et seq.] that is punishable as a Class 3 felony or higher.

(b) Sentence. Being an armed habitual criminal is a Class X felony.

5/24-2. Exemptions

(a) (Effective until January 1, 2009) Subsections 24-1(a)(3), 24-1(a)(4) and 24-1(a)(10) and Section 24-1.6 [720 ILCS 5/24-1 and 720 ILCS 5/24-1.6] do not apply to or affect any of the following:

(a) (As amended by P.A. 95-885, effective January 1, 2009) Subsections 24-1(a)(3), 24-1(a)(4), 24-1(a)(10), and 24-1(a)(13) and Section 24-1.6 do not apply to or affect any of the following:

(1) Peace officers, and any person summoned by a peace officer to assist in making arrests or preserving the peace, while actually engaged in assisting such officer.

(2) Wardens, superintendents and keepers of prisons, penitentiaries, jails and other institutions for the detention of persons accused or convicted of an offense, while in the performance of their official duty, or while commuting between their homes and places of employment.

(3) Members of the Armed Services or Reserve Forces of the United States or the Illinois National Guard or the Reserve Officers Training Corps, while in the performance of their official duty.

(4) Special agents employed by a railroad or a public utility to perform police functions, and guards of armored car companies, while actually engaged in the performance of the duties of their employment or commuting between their homes and places of employment; and watchmen while actually engaged in the performance of the duties of their employment.

(5) Persons licensed as private security contractors, private detectives, or private alarm contractors, or employed by an agency certified by the Department of Professional Regulation, if their duties include the carrying of a weapon under the provisions of the Private Detective, Private Alarm, Private Security, Fingerprint Vendor, and Locksmith Act of 2004 [225 ILCS 446/28], while actually engaged in the performance of the duties of their employment or commuting between their homes and places of employment, provided that such commuting is accomplished within one hour from departure from home or place of employment, as the case may be. Persons exempted under this subdivision (a)(5) shall be required to have completed a course of study in firearms handling and training approved and supervised by the Department of Professional Regulation as prescribed by Section 28 of the Private Detective, Private Alarm, Private Security, Fingerprint Vendor, and Locksmith Act of 2004 [225 ILCS 446/28], prior to becoming eligible for this exemption. The Department of Professional Regulation shall provide suitable documentation demonstrating the successful completion of the prescribed firearms training. Such documentation shall be carried at all times when such persons are in possession of a concealable weapon.

(6) Any person regularly employed in a commercial or industrial operation as a security guard for the protection of persons employed and private property related to such commercial or industrial operation, while actually engaged in the performance of his or her duty or traveling between sites or properties belonging to the employer, and who, as a security guard, is a member of a security force of at least 5 persons registered with the Department of Professional Regulation; provided that such security guard has successfully completed a course of study, approved by and supervised by the Department of Professional Regulation, consisting of not less than 40 hours of training that includes the theory of law enforcement, liability for acts, and the handling of weapons. A person shall be considered eligible for this exemption if he or she has completed the required 20 hours of training for a security officer and 20 hours of required firearm training, and has been issued a firearm control card by the Department of Professional Regulation. Conditions for the renewal of firearm control cards issued under the provisions of this Section shall be the same as for those cards issued under the provisions of the Private Detective, Private Alarm, Private Security, Fingerprint Vendor, and Locksmith Act of 2004 [225 ILCS 446/1 et seq.]. Such firearm control card shall be carried by the security guard at all times when he or she is in possession of a concealable weapon.

(7) Agents and investigators of the Illinois Legislative Investigating Commission authorized by the Commission to carry the weapons specified in subsections 24-1(a)(3) and 24-1(a)(4) [720 ILCS 5/24-1], while on duty in the course of any investigation for the Commission.

(8) Persons employed by a financial institution for the protection of other employees and property related to such financial institution, while actually engaged in the performance of their duties, commuting between their homes and places of employment, or traveling between sites or properties owned or operated by such financial institution, provided that any person so employed has successfully completed a course of study, approved by and supervised by the Department of Professional Regulation, consisting of not less than 40 hours of training which includes theory of law enforcement, liability for acts, and the handling of weapons. A person shall be considered to be eligible for this exemption if he or she has completed the required 20 hours of training for a security officer and 20 hours of required firearm training, and has been issued a firearm control card by the Department of Professional Regulation. Conditions for renewal of firearm control cards issued under the provisions of this Section shall be the same as for those issued under the provisions of the Private Detective, Private Alarm, Private Security, Fingerprint Vendor, and Locksmith Act of 2004 [225 ILCS 446/1 et seq.]. Such firearm control card shall be carried by the person so trained at all times when such person is in possession of a concealable weapon. For purposes of this subsection, "financial institution" means a bank, savings and loan association, credit union or company providing armored car services.

(9) Any person employed by an armored car company to drive an armored car, while actually engaged in the performance of his duties.

(10) Persons who have been classified as peace officers pursuant to the Peace Officer Fire Investigation Act [20 ILCS 2910/0.01 et seq.].

(11) Investigators of the Office of the State's Attorneys Appellate Prosecutor authorized by the board of governors of the Office of the State's Attorneys Appellate Prosecutor to carry weapons pursuant to Section 7.06 of the State's Attorneys Appellate Prosecutor's Act [725 ILCS 210/7.06].

(12) Special investigators appointed by a State's Attorney under Section 3-9005 of the Counties Code [55 ILCS 5/3-9005].

(12.5) Probation officers while in the performance of their duties, or while commuting between their homes, places of employment or specific locations that are part of their assigned duties, with the consent of the chief judge of the circuit for which they are employed.

(13) Court Security Officers while in the performance of their official duties, or while commuting between their homes and places of employment, with the consent of the Sheriff.

(13.5) A person employed as an armed security guard at a nuclear energy, storage, weapons or development site or facility regulated by the Nuclear Regulatory Commission who has completed the background screening and training mandated by the rules and regulations of the Nuclear Regulatory Commission.

(14) Manufacture, transportation, or sale of weapons to persons authorized under subdivisions (1) through (13.5) of this subsection to possess those weapons.

(b) Subsections 24-1(a)(4) and 24-1(a)(10) and Section 24-1.6 [720 ILCS 5/24-1 and 720 ILCS 5/24-1.6] do not apply to or affect any of the following:

(1) Members of any club or organization organized for the purpose of practicing shooting at targets upon established target ranges, whether public or private, and patrons of such ranges, while such members or patrons are using their firearms on those target ranges.

(2) Duly authorized military or civil organizations while parading, with the special permission of the Governor.

(3) Hunters, trappers or fishermen with a license or permit while engaged in hunting, trapping or fishing.

(4) Transportation of weapons that are broken down in a non-functioning state or are not immediately accessible.

(c) Subsection 24-1(a)(7) [720 ILCS 5/24-1] does not apply to or affect any of the following:

(1) Peace officers while in performance of their official duties.

(2) Wardens, superintendents and keepers of prisons, penitentiaries, jails and other institutions for the detention of persons accused or convicted of an offense.

(3) Members of the Armed Services or Reserve Forces of the United States or the Illinois National Guard, while in the performance of their official duty.

(4) Manufacture, transportation, or sale of machine guns to persons authorized under subdivisions (1) through (3) of this subsection to possess machine guns, if the machine guns are broken down in a non-functioning state or are not immediately accessible.

(5) Persons licensed under federal law to manufacture any weapon from which 8 or more shots or bullets can be discharged by a single function of the firing device, or ammunition for such weapons, and actually engaged in the business of manufacturing such weapons or ammunition, but only with respect to activities which are within the lawful scope of such business, such as the manufacture, transportation, or testing of such weapons or ammunition. This exemption does not authorize the general private possession of any weapon from which 8 or more shots or bullets can be discharged by a single function of the firing device, but only such possession and activities as are within the lawful scope of a licensed manufacturing business described in this paragraph.

During transportation, such weapons shall be broken down in a non-functioning state or not immediately accessible.

(6) The manufacture, transport, testing, delivery, transfer or sale, and all lawful commercial or experimental activities necessary thereto, of rifles, shotguns, and weapons made from rifles or shotguns, or ammunition for such rifles, shotguns or weapons, where engaged in by a person operating as a contractor or subcontractor pursuant to a contract or subcontract for the development and supply of such rifles, shotguns, weapons or ammunition to the United States government or any branch of the Armed Forces of the United States, when such activities are necessary and incident to fulfilling the terms of such contract.

The exemption granted under this subdivision (c)(6) shall also apply to any authorized agent of any such contractor or subcontractor who is operating within the scope of his employment, where such activities involving such weapon, weapons or ammunition are necessary and incident to fulfilling the terms of such contract. During transportation, any such weapon shall be broken down in a non-functioning state, or not immediately accessible.

(d) Subsection 24-1(a)(1) [720 ILCS 5/24-1] does not apply to the purchase, possession or carrying of a black-jack or slung-shot by a peace officer.

(e) Subsection 24-1(a)(8) [720 ILCS 5/24-1] does not apply to any owner, manager or authorized employee of any place specified in that subsection nor to any law enforcement officer.

(f) Subsection 24-1(a)(4) and subsection 24-1(a)(10) and Section 24-1.6 [720 ILCS 5/24-1 and 720 ILCS 5/24-1.6] do not apply to members of any club or organization organized for the purpose of practicing shooting at targets upon established target ranges, whether public or private, while using their firearms on those target ranges.

(g) Subsections 24-1(a)(11) and 24-3.1(a)(6) [720 ILCS 5/24-1 and 720 ILCS 5/24-3.1] do not apply to:

(1) Members of the Armed Services or Reserve Forces of the United States or the Illinois National Guard, while in the performance of their official duty.

(2) Bonafide collectors of antique or surplus military ordinance.

(3) Laboratories having a department of forensic ballistics, or specializing in the development of ammunition or explosive ordinance.

(4) Commerce, preparation, assembly or possession of explosive bullets by manufacturers of ammunition licensed by the federal government, in connection with the supply of those organizations and persons exempted by subdivision (g)(1) of this Section, or like organizations and persons outside this State, or the transportation of explosive bullets to any organization or person exempted in this Section by a common carrier or by a vehicle owned or leased by an exempted manufacturer.

(g-5) Subsection 24-1(a)(6) does not apply to or affect persons licensed under federal law to manufacture any device or attachment of any kind designed, used, or intended for use in silencing the report of any firearm, firearms, or ammunition for those firearms equipped with those devices, and actually engaged in the business of manufacturing those devices, firearms, or ammunition, but only with respect to activities that are within the lawful scope of that business, such as the manufacture, transportation, or testing of those devices, firearms, or ammunition. This exemption does not authorize the general private possession of any device or attachment of any kind designed, used, or intended for use in silencing the report of any firearm, but only such possession and activities as are within the lawful scope of a licensed manufacturing business described in this subsection (g-5). During transportation, those devices shall be detached from any weapon or not immediately accessible.

(h) An information or indictment based upon a violation of any subsection of this Article need not negative any exemptions contained in this Article. The defendant shall have the burden of proving such an exemption.

(i) Nothing in this Article shall prohibit, apply to, or affect the transportation, carrying, or possession, of any pistol or revolver, stun gun, taser, or other firearm consigned to a common carrier operating under license of the State of Illinois or the federal government, where such transportation, carrying, or possession is incident to the lawful transportation in which such common carrier is engaged; and nothing in this Article shall prohibit, apply to, or affect the transportation, carrying, or possession of any pistol, revolver, stun gun, taser, or other firearm, not the subject of and regulated by subsection 24-1(a)(7) or subsection 24-2(c) of this Article [720 ILCS 5/24-1 or 720 ILCS 5/24-2], which is unloaded and enclosed in a case, firearm carrying box, shipping box, or other container, by the possessor of a valid Firearm Owners Identification Card.

5/24-2.1. Unlawful use of firearm projectiles

(a) A person commits the offense of unlawful use of firearm projectiles when he or she knowingly manufactures, sells, purchases, possesses, or carries any armor piercing bullet, dragon's breath shotgun shell, bolo shell, or flechette shell.

For the purposes of this Section:

"Armor piercing bullet" means any handgun bullet or handgun ammunition with projectiles or projectile cores constructed entirely (excluding the presence of traces of other substances) from tungsten alloys, steel, iron, brass, bronze, beryllium copper or depleted uranium, or fully jacketed bullets larger than 22 caliber designed and intended for use in a handgun and whose jacket has a weight of more than 25% of the total weight of the projectile, and excluding those handgun projectiles whose cores are composed of soft materials such as lead or lead alloys, zinc or zinc alloys, frangible projectiles designed primarily for sporting purposes, and any other projectiles or projectile cores that the U.S. Secretary of the Treasury finds to be primarily intended to be used for sporting purposes or industrial purposes or that otherwise does not constitute "armor piercing ammunition" as that term is defined by federal law.

The definition contained herein shall not be construed to include shotgun shells.

"Dragon's breath shotgun shell" means any shotgun shell that contains exothermic pyrophoric mesh metal as the projectile and is designed for the purpose of throwing or spewing a flame or fireball to simulate a flame-thrower.

"Bolo shell" means any shell that can be fired in a firearm and expels as projectiles 2 or more metal balls connected by solid metal wire.

"Flechette shell" means any shell that can be fired in a firearm and expels 2 or more pieces of fin-stabilized solid metal wire or 2 or more solid dart-type projectiles.

(b) Exemptions. This Section does not apply to or affect any of the following:

(1) Peace officers.

(2) Wardens, superintendents and keepers of prisons, penitentiaries, jails and other institutions for the detention of persons accused or convicted of an offense.

(3) Members of the Armed Services or Reserve Forces of the United States or the Illinois National Guard while in the performance of their official duties.

(4) Federal officials required to carry firearms, while engaged in the performance of their official duties.

(5) United States Marshals, while engaged in the performance of their official duties.

(6) Persons licensed under federal law to manufacture, import, or sell firearms and firearm ammunition, and actually engaged in any such business, but only with respect to activities which are within the lawful scope of such business, such as the manufacture, transportation, or testing of such bullets or ammunition.

This exemption does not authorize the general private possession of any armor piercing bullet, dragon's breath shotgun shell, bolo shell, or flechette shell, but only such possession and activities which are within the lawful scope of a licensed business described in this paragraph.

(7) Laboratories having a department of forensic ballistics or specializing in the development of ammunition or explosive ordnance.

(8) Manufacture, transportation or sale of armor piercing bullets, dragon's breath shotgun shells, bolo shells, or flechette shells to persons specifically authorized under paragraphs (1) through (7) of this subsection to possess such bullets or shells.

(c) An information or indictment based upon a violation of this Section need not negate any exemption herein contained. The defendant shall have the burden of proving such an exemption.

(d) Sentence. A person convicted of unlawful use of armor piercing bullets shall be guilty of a Class 3 felony.

5/24-2.2. Manufacture, sale or transfer of bullets or shells represented to be armor piercing bullets, dragon's breath shotgun shells, bolo shells, or flechette shells

(a) Except as provided in subsection (b) of this Section, it is unlawful for any person to knowingly manufacture, sell, offer to sell, or transfer any bullet or shell which is represented to be an armor piercing bullet, a dragon's breath shotgun shell, a bolo shell, or a flechette shell as defined in Section 24-2.1 of this Code.

(b) Exemptions. This Section does not apply to or affect any person authorized under Section 24-2.1 to manufacture, sell, purchase, possess, or carry any armor piercing bullet or any dragon's breath shotgun shell, bolo shell, or flechette shell with respect to activities which are within the lawful scope of the exemption therein granted.

(c) An information or indictment based upon a violation of this Section need not negate any exemption herein contained. The defendant shall have the burden of proving such an exemption and that the activities forming the basis of any criminal charge brought pursuant to this Section were within the lawful scope of such exemption.

(d) Sentence. A violation of this Section is a Class 4 felony.

5/24-3. Unlawful Sale of Firearms

(A) A person commits the offense of unlawful sale of firearms when he or she knowingly does any of the following:

(a) Sells or gives any firearm of a size which may be concealed upon the person to any person under 18 years of age.

(b) Sells or gives any firearm to a person under 21 years of age who has been convicted of a misdemeanor other than a traffic offense or adjudged delinquent.

(c) Sells or gives any firearm to any narcotic addict.

(d) Sells or gives any firearm to any person who has been convicted of a felony under the laws of this or any other jurisdiction.

(e) Sells or gives any firearm to any person who has been a patient in a mental hospital within the past 5 years.

(f) Sells or gives any firearms to any person who is mentally retarded.

(g) Delivers any firearm of a size which may be concealed upon the person, incidental to a sale, without withholding delivery of such firearm for at least 72 hours after application for its purchase has been made, or delivers any rifle, shotgun or other long gun, or a stun gun or taser, incidental to a sale, without withholding delivery of such rifle, shotgun or other long gun, or a stun gun or taser for at least 24 hours after application for its purchase has been made. However, this paragraph (g) does not apply to: (1) the sale of a firearm to a law enforcement officer if the seller of the firearm knows that the person to whom he or she is selling the firearm is a law enforcement officer or the sale of a firearm to a person who desires to purchase a firearm for use in promoting the public interest incident to his or her employment as a bank guard, armed truck guard, or other similar employment; (2) a mail order sale of a firearm to a nonresident of Illinois under which the firearm is mailed to a point outside the boundaries of Illinois; (3) the sale of a firearm to a nonresident of Illinois while at a firearm showing or display recognized by the Illinois Department of State Police; or (4) the sale of a firearm to a dealer licensed as a federal firearms dealer under Section 923 of the federal Gun Control Act of 1968 (18 U.S.C. 923). For purposes of this paragraph (g), "application" means when the buyer and seller reach an agreement to purchase a firearm.

(h) While holding any license as a dealer, importer, manufacturer or pawnbroker under the federal Gun Control Act of 1968, manufactures, sells or delivers to any unlicensed person a handgun having a barrel, slide, frame or receiver which is a die casting of zinc alloy or any other nonhomogeneous metal which will melt or deform at a temperature of less than 800 degrees Fahrenheit. For purposes of this paragraph, (1) "firearm" is defined as in the Firearm Owners Identification Card Act; and (2) "handgun" is defined as a firearm designed to be held and fired by the use of a single hand, and includes a combination of parts from which such a firearm can be assembled.

(i) Sells or gives a firearm of any size to any person under 18 years of age who does not possess a valid Firearm Owner's Identification Card.

(j) Sells or gives a firearm while engaged in the business of selling firearms at wholesale or retail without being licensed as a federal firearms dealer under Section 923 of the federal Gun Control Act of 1968 (18 U.S.C. 923). In this paragraph (j):

A person "engaged in the business" means a person who devotes time, attention, and labor to engaging in the activity as a regular course of trade or business with the principal objective of livelihood and profit, but does not include a person who makes occasional repairs of firearms or who occasionally fits special barrels, stocks, or trigger mechanisms to firearms.

"With the principal objective of livelihood and profit" means that the intent underlying the sale or disposition of firearms is predominantly one of obtaining livelihood and pecuniary gain, as opposed to other intents, such as improving or liquidating a personal firearms collection; however, proof of profit shall not be required as to a person who engages in the regular and repetitive purchase and disposition of firearms for criminal purposes or terrorism.

(k) Sells or transfers ownership of a firearm to a person who does not display to the seller or transferor of the firearm a currently valid Firearm Owner's Identification Card that has previously been issued in the transferee's name by the Department of State Police under the provisions of the Firearm Owners Identification Card Act. This paragraph (k) does not apply to the transfer of a firearm to a person who is exempt from the requirement of possessing a Firearm Owner's Identification Card under Section 2 of the Firearm Owners Identification Card Act. For the purposes of this Section, a currently valid Firearm Owner's Identification Card means (i) a Firearm Owner's Identification Card that has not expired or (ii) if the transferor is licensed as a federal firearms dealer under Section 923 of the federal Gun Control Act of 1968 (18 U.S.C. 923), an approval number issued in accordance with Section 3.1 of the Firearm Owners Identification Card Act shall be proof that the Firearm Owner's Identification Card was valid.

(B) Paragraph (h) of subsection (A) does not include firearms sold within 6 months after enactment of Public Act 78-355 (approved August 21, 1973, effective October 1, 1973), nor is any firearm legally owned or possessed by any citizen or purchased by any citizen within 6 months after the enactment of Public Act 78-355 subject to confiscation or seizure under the provisions of that Public Act. Nothing in Public Act 78-355 shall be construed to prohibit the gift or trade of any firearm if that firearm was legally held or acquired within 6 months after the enactment of that Public Act.

(C) Sentence.

(1) Any person convicted of unlawful sale of firearms in violation of any of paragraphs (c) through (h) of subsection (A) commits a Class 4 felony.

(2) Any person convicted of unlawful sale of firearms in violation of paragraph (b) or (i) of subsection (A) commits a Class 3 felony.

(3) Any person convicted of unlawful sale of firearms in violation of paragraph (a) of subsection (A) commits a Class 2 felony.

(4) Any person convicted of unlawful sale of firearms in violation of paragraph (a), (b), or (i) of subsection (A) in any school, on the real property comprising a school, within 1,000 feet of the real property comprising a school, at a school related activity, or on or within 1,000 feet of any conveyance owned, leased, or contracted by a school or school district to transport students to or from school or a school related activity, regardless of the time of day or time of year at which the offense was committed, commits a Class 1 felony. Any person convicted of a second or subsequent violation of unlawful sale of firearms in violation of paragraph (a), (b), or (i) of subsection (A) in any school, on the real property comprising a school, within 1,000 feet of the real property comprising a school, at a school related activity, or on or within 1,000 feet of any conveyance owned, leased, or contracted by a school or school district to transport students to or from school or a school related activity, regardless of the time of day or time of year at which the offense was committed, commits a Class 1 felony for which the sentence shall be a term of imprisonment of no less than 5 years and no more than 15 years.

(5) Any person convicted of unlawful sale of firearms in violation of paragraph (a) or (i) of subsection (A) in residential property owned, operated, or managed by a public housing agency or leased by a public housing agency as part of a scattered site or mixed-income development, in a public park, in a courthouse, on residential property owned, operated, or managed by a public housing agency or leased by a public housing agency as part of a scattered site or mixed-income development, on the real property comprising any public park, on the real property comprising any courthouse, or on any public way within 1,000 feet of the real property comprising any public park, courthouse, or residential property owned, operated, or managed by a public housing agency or leased by a public housing agency as part of a scattered site or mixed-income development commits a Class 2 felony.

(6) Any person convicted of unlawful sale of firearms in violation of paragraph (j) of subsection (A) commits a Class A misdemeanor. A second or subsequent violation is a Class 4 felony.

(7) Any person convicted of unlawful sale of firearms in violation of paragraph (k) of subsection (A) commits a Class 4 felony. A third or subsequent conviction for a violation of paragraph (k) of subsection (A) is a Class 1 felony.

(D) For purposes of this Section:

"School" means a public or private elementary or secondary school, community college, college, or university.

"School related activity" means any sporting, social, academic, or other activity for which students' attendance or participation is sponsored, organized, or funded in whole or in part by a school or school district.

(E) A prosecution for a violation of paragraph (k) of subsection (A) of this Section may be commenced within 6 years after the commission of the offense. A prosecution for a violation of this Section other than paragraph (g) of subsection (A) of this Section may be commenced within 5 years after the commission of the offense defined in the particular paragraph.

5/24-3A. Gunrunning

(a) A person commits gunrunning when he or she transfers 3 or more firearms in violation of any of the paragraphs of Section 24-3 of this Code.

(b) Sentence. A person who commits gunrunning:

(1) is guilty of a Class 1 felony;

(2) is guilty of a Class X felony for which the sentence shall be a term of imprisonment of not less than 8 years and not more than 40 years if the transfer is of not less than 11 firearms and not more than 20 firearms;

(3) is guilty of a Class X felony for which the sentence shall be a term of imprisonment of not less than 10 years and not more than 50 years if the transfer is of more than 20 firearms.

A person who commits gunrunning by transferring firearms to a person who, at the time of the commission of the offense, is under 18 years of age is guilty of a Class X felony.

5/24-3.1. Unlawful possession of firearms and firearm ammunition

(a) A person commits the offense of unlawful possession of firearms or firearm ammunition when:

(1) He is under 18 years of age and has in his possession any firearm of a size which may be concealed upon the person; or

(2) He is under 21 years of age, has been convicted of a misdemeanor other than a traffic offense or adjudged delinquent and has any firearms or firearm ammunition in his possession; or

(3) He is a narcotic addict and has any firearms or firearm ammunition in his possession; or

(4) He has been a patient in a mental hospital within the past 5 years and has any firearms or firearm ammunition in his possession; or

(5) He is mentally retarded and has any firearms or firearm ammunition in his possession; or

(6) He has in his possession any explosive bullet.

For purposes of this paragraph "explosive bullet" means the projectile portion of an ammunition cartridge which contains or carries an explosive charge which will explode upon contact with the flesh of a human or an animal. "Cartridge" means a tubular metal case having a projectile affixed at the front thereof and a cap or primer at the rear end thereof, with the propellant contained in such tube between the projectile and the cap.

(b) Sentence. Unlawful possession of firearms, other than handguns, and firearm ammunition is a Class A misdemeanor. Unlawful possession of handguns is a Class 4 felony. The possession of each firearm or firearm ammunition in violation of this Section constitutes a single and separate violation.

(c) Nothing in paragraph (1) of subsection (a) of this Section prohibits a person under 18 years of age from participating in any lawful recreational activity with a firearm such as, but not limited to, practice shooting at targets upon established public or private target ranges or hunting, trapping, or fishing in accordance with the Wildlife Code or the Fish and Aquatic Life Code.

5/24-3.3. Unlawful Sale or Delivery of Firearms on the Premises of Any School,

Unlawful sale or delivery of firearms on the premises of any school, regardless of the time of day or the time of year, or any conveyance owned, leased or contracted by a school to transport students to or from school or a school related activity, or residential property owned, operated or manag-ed by a public housing agency. Any person 18 years of age or older who sells, gives or delivers any firearm to any person under 18 years of age in any school, regardless of the time of day or the time of year or residential property owned, operated or managed by a public housing agency or leased by a public housing agency as part of a scattered site or mixed-income development, on the real property comprising any school, regardless of the time of day or the time of year or residential property owned, operated and managed by a public housing agency or leased by a public housing agency as part of a scattered site or mixed-income development commits a Class 3 felony. School is defined, for the purposes of this Section, as any public or private elementary or secondary school, community college, college or university. This does not apply to peace officers or to students carrying or possessing firearms for use in school training courses, parades, target shooting on school ranges, or otherwise with the consent of school authorities and which firearms are transported unloaded and enclosed in a suitable case, box or transportation package.

5/24-3.4. Unlawful sale of firearms by liquor licensee

(a) It shall be unlawful for any person who holds a license to sell at retail any alcoholic liquor issued by the Illinois Liquor Control Commission or local liquor control commissioner under the Liquor Control Act of 1934 [235 ILCS 5/1-1 et seq.] or an agent or employee of the licensee to sell or deliver to any other person a firearm in or on the real property of the establishment where the licensee is licensed to sell alcoholic liquors unless the sale or delivery of the firearm is otherwise lawful under this Article and under the Firearm Owners Identification Card Act.

(b) Sentence. A violation of subsection (a) of this Section is a Class 4 felony.

5/24-3.5. Unlawful purchase of a firearm

(a) For purposes of this Section, "firearms transaction record form" means a form:

(1) executed by a transferee of a firearm stating: (i) the transferee's name and address (including county or similar political subdivision); (ii) whether the transferee is a citizen of the United States; (iii) the transferee's State of residence; and (iv) the date and place of birth, height, weight, and race of the transferee; and

(2) on which the transferee certifies that he or she is not prohibited by federal law from transporting or shipping a firearm in interstate or foreign commerce or receiving a firearm that has been shipped or transported in interstate or foreign commerce or possessing a firearm in or affecting commerce.

(b) A person commits the offense of unlawful purchase of a firearm who knowingly purchases or attempts to purchase a firearm with the intent to deliver that firearm to another person who is prohibited by federal or State law from possessing a firearm.

(c) A person commits the offense of unlawful purchase of a firearm when he or she, in purchasing or attempting to purchase a firearm, intentionally provides false or misleading information on a United States Department of the Treasury, Bureau of Alcohol, Tobacco and Firearms firearms transaction record form.

(d) Exemption. It is not a violation of subsection (b) of this Section for a person to make a gift or loan of a firearm to a person who is not prohibited by federal or State law from possessing a firearm if the transfer of the firearm is made in accordance with Section 3 of the Firearm Owners Identification Card Act [430 ILCS 65/3].

(e) Sentence.

(1) A person who commits the offense of unlawful purchase of a firearm:

(A) (Effective until January 1, 2009) is guilty of a Class 4 felony for purchasing or attempting to purchase one firearm;

(A) (As amended by P.A. 95-882, effective January 1, 2009) is guilty of a Class 2 felony for purchasing or attempting to purchase one firearm;

(B) (Effective until January 1, 2009) is guilty of a Class 3 felony for purchasing or attempting to purchase not less than 2 firearms and not more than 5 firearms at the same time or within a one year period;

(B) (As amended by P.A. 95-882, effective January 1, 2009) is guilty of a Class 1 felony for purchasing or attempting to purchase not less than 2 firearms and not more than 5 firearms at the same time or within a one year period;

(C) (Effective until January 1, 2009) is guilty of a Class 2 felony for purchasing or attempting to purchase not less than 6 firearms and not more than 10 firearms at the same time or within a 2 year period;

(C) (As amended by P.A. 95-882, effective January 1, 2009) is guilty of a Class X felony for which the offender shall be sentenced to a term of imprisonment of not less than 9 years and not more than 40 years for purchasing or attempting to purchase not less than 6 firearms at the same time or within a 2 year period.

(D) is guilty of a Class 1 felony for purchasing or attempting to purchase not less than 11 firearms and not more than 20 firearms at the same time or within a 3 year period;

(E) is guilty of a Class X felony for which the person shall be sentenced to a term of imprisonment of not less than 6 years and not more than 30 years for purchasing or attempting to purchase not less than 21 firearms and not more than 30 firearms at the same time or within a 4 year period;

(F) is guilty of a Class X felony for which the person shall be sentenced to a term of imprisonment of not less than 6 years and not more than 40 years for purchasing or attempting to purchase not less than 31 firearms and not more than 40 firearms at the same time or within a 5 year period;

(G) is guilty of a Class X felony for which the person shall be sentenced to a term of imprisonment of not less than 6 years and not more than 50 years for purchasing or attempting to purchase more than 40 firearms at the same time or within a 6 year period.

(2) In addition to any other penalty that may be imposed for a violation of this Section, the court may sentence a person convicted of a violation of subsection (c) of this Section to a fine not to exceed $ 250,000 for each violation.

(f) A prosecution for unlawful purchase of a firearm may be commenced within 6 years after the commission of the offense.

5/24-3.6. Unlawful use of a firearm in the shape of a wireless telephone

(a) For the purposes of this Section, "wireless telephone" means a device that is capable of transmitting or receiving telephonic communications without a wire connecting the device to the telephone network.

(b) A person commits the offense of unlawful use of a firearm in the shape of a wireless telephone when he or she manufactures, sells, transfers, purchases, possesses, or carries a firearm shaped or designed to appear as a wireless telephone.

(c) This Section does not apply to or affect the sale to or possession of a firearm in the shape of a wireless telephone by a peace officer.

(d) Sentence. Unlawful use of a firearm in the shape of a wireless telephone is a Class 4 felony.

5/24-4. Register of sales by dealer

(a) Any seller of firearms of a size which may be concealed upon the person, other than a manufacturer selling to a bona fide wholesaler or retailer or a wholesaler selling to a bona fide retailer, shall keep a register of all firearms sold or given away.

(b) Such register shall contain the date of the sale or gift, the name, address, age and occupation of the person to whom the weapon is sold or given, the price of the weapon, the kind, description and number of the weapon, and the purpose for which it is purchased and obtained.

(c) Such seller on demand of a peace officer shall produce for inspection the register and allow such peace officer to inspect such register and all stock on hand.

(d) Sentence. Violation of this Section is a Class B misdemeanor.

5/24-5. Defacing identification marks of firearms

(a) Any person who shall knowingly or intentionally change, alter, remove or obliterate the name of the importer's or manufacturer's serial number of any firearm commits a Class 2 felony.

(b) A person who possesses any firearm upon which any such importer's or manufacturer's serial number has been changed, altered, removed or obliterated commits a Class 3 felony.

(c) Nothing in this Section shall prevent a person from making repairs, replacement of parts, or other changes to a firearm if those repairs, replacement of parts, or changes cause the removal of the name of the maker, model, or other marks of identification other than the serial number on the firearm's frame or receiver

(d) A prosecution for a violation of this Section may be commenced within 6 years after the commission of the offense.

5/24-7. Weapons offenses; community service In addition to any other sentence that may be imposed, a court shall order any person convicted of a violation of this Article to perform community service for not less than 30 and not more than 120 hours, if community service is available in the jurisdiction and is funded and approved by the county board of the county where the offense was committed. In addition, whenever any person is placed on supervision for an alleged offense under this Article, the supervision shall be conditioned upon the performance of the community service.

This Section does not apply when the court imposes a sentence of incarceration.

5/24-8. Firearm tracing

(a) Upon recovering a firearm from the possession of anyone who is not permitted by federal or State law to possess a firearm, a local law enforcement agency shall use the best available information, including a firearms trace when necessary, to determine how and from whom the person gained possession of the firearm. Upon recovering a firearm that was used in the commission of any offense classified as a felony or upon recovering a firearm that appears to have been lost, mislaid, stolen, or otherwise unclaimed, a local law enforcement agency shall use the best available information, including a firearms trace when necessary, to determine prior ownership of the firearm.

(b) Local law enforcement shall, when appropriate, use the National Tracing Center of the Federal Bureau of Alcohol, Tobacco and Firearms in complying with subsection (a) of this Section.

(c) Local law enforcement agencies shall use the Illinois Department of State Police Law Enforcement Agencies Data System (LEADS) Gun File to enter all stolen, seized, or recovered firearms as prescribed by LEADS regulations and policies.

5/24-9. Firearms; Child Protection

(a) Except as provided in subsection (c), it is unlawful for any person to store or leave, within premises under his or her control, a firearm if the person knows or has reason to believe that a minor under the age of 14 years who does not have a Firearm Owners Identification Card is likely to gain access to the firearm without the lawful permission of the minor's parent, guardian, or person having charge of the minor, and the minor causes death or great bodily harm with the firearm, unless the firearm is:

(1) secured by a device or mechanism, other than the firearm safety, designed to render a firearm temporarily inoperable; or

(2) placed in a securely locked box or container; or

(3) placed in some other location that a reasonable person would believe to be secure from a minor under the age of 14 years.

(b) Sentence. A person who violates this Section is guilty of a Class C misdemeanor and shall be fined not less than $1,000. A second or sub-sequent violation of this Section is a Class A misdemeanor.

(c) Subsection (a) does not apply:

(1) if the minor under 14 years of age gains access to a firearm and uses it in a lawful act of self-defense or defense of another; or

(2) to any firearm obtained by a minor under the age of 14 because of an unlawful entry of the premises by the minor or another person.

(d) For the purposes of this Section, "firearm" as the meaning ascribed to it in Section 1.1 of the Firearm Owners Identification Card Act.

24-9.5. Handgun safety devices

(a) It is unlawful for a person licensed as a federal firearms dealer under Section 923 of the federal Gun Control Act of 1968 (18 U.S.C. 923) to offer for sale, sell, or transfer a handgun to a person not licensed under that Act, unless he or she sells or includes with the handgun a device or mechanism, other than the firearm safety, designed to render the handgun temporarily inoperable or inaccessible. This includes but is not limited to:

(1) An external device that is:

(i) attached to the handgun with a key or combination lock; and

(ii) designed to prevent the handgun from being discharged unless the device has been deactivated.

(2) An integrated mechanical safety, disabling, or locking device that is:

(i) built into the handgun; and

(ii) designed to prevent the handgun from being discharged unless the device has been deactivated.

(b) Sentence. A person who violates this Section is guilty of a Class C misdemeanor and shall be fined not less than $1,000. A second or subsequent violation of this Section is a Class A misdemeanor.

(c) For the purposes of this Section, "handgun" has the meaning ascribed to it in clause (h)(2) of subsection (A) of Section 24-3 of this Code.

(d) This Section does not apply to:

(1) the purchase, sale, or transportation of a handgun to or by a federally licensed firearms dealer or manufacturer that provides or services a handgun for:

(i) personnel of any unit of the federal government;

(ii) members of the armed forces of the United States or the National Guard;

(iii) law enforcement personnel of the State or any local law enforcement agency in the State while acting within the scope of their official duties; and

(iv) an organization that is required by federal law governing its specific business or activity to maintain handguns and applicable ammunition;

(2) a firearm modified to be permanently inoperative;

(3) the sale or transfer of a handgun by a federally licensed firearms dealer or manufacturer described in item (1) of this subsection (d);

(4) the sale or transfer of a handgun by a federally licensed firearms dealer or manufacturer to a lawful customer outside the State; or

(5) an antique firearm.

24-10. Municipal ordinance regulating firearms; affirmative defense to a violation It is an affirmative defense to a violation of a municipal ordinance that prohibits, regulates, or restricts the private ownership of firearms if the individual who is charged with the violation used the firearm in an act of self-defense or defense of another as defined in Sections 7-1 and 7-2 of this Code when on his or her land or in his or her abode or fixed place of business.

[Current through P.A. 95-959, July 18, 2008]

Code of Ordinances, City of Aurora

Chapter 29. Offenses – Miscellaneous

Article III. Offenses Against or Endangering the Person

29-43. Unlawful use of weapons.

(a) A person commits the offense of unlawful use of weapons when he knowingly: ...

(4) Carries or possesses in any vehicle or concealed on or about his person except when on his land or in his own abode or fixed place of business any pistol, revolver, stun gun or taser or other firearm; or ...

(6) Possesses any device or attachment of any kind designed, used or intended for use in silencing the report of any firearm; or

(7) Sells, manufactures, purchases, possesses or carries a machine gun, which shall be defined for the purposes of this subsection as any weapon which shoots, is designed to shoot, or can be readily restored to shoot, automatically more than one (1) shot without manually reloading by a single function of the trigger, including the frame or receiver of any such weapon; or sells, manufactures, purchases, possesses or carries any combination of parts designed or intended for use in converting any weapon into a machine gun, or any combination of parts from which a machine gun can be assembled if such parts are in the possession or under the control of a person, or any rifle having one (1) or more barrels less than sixteen (16) inches in length or a shotgun having one (1) or more barrels less than eighteen (18) inches in length, or a rifle or shotgun which does not have a fixed stock capable of being fired from the shoulder, or any weapon made from a rifle or shotgun, whether by alteration, modification or otherwise, if such a weapon as modified has an overall length of less than twenty-six (26) inches, or any bomb, bomb-shell, grenade, bottle or other container containing an explosive substance of over one-quarter ounce for like purposes, such as, but not limited to, black powder bombs and Molotov cocktails or artillery projectiles; or ...

(10) Carries or possesses on or about his person, upon any public street, alley or other public lands within the corporate limits of the city, except when the invitee thereon or therein, for the purpose of the display of such weapon or the lawful commerce in weapons, or except when on his land or in his own abode or fixed place of business, any pistol, revolver, stun gun or taser or other firearm. A "stun gun or taser," as used in this paragraph (a) means (i) any device which is powered by electrical charging units, such as batteries, and which fires one (1) or several barbs attached to a length of wire and which, upon hitting a human, can send out a current capable of disrupting the person's nervous system in such a manner as to render him incapable of normal functioning; or (ii) any device which is powered by electrical charging units, such as batteries, and which, upon contact with a human or clothing worn by a human, can send out a current capable of disrupting the person's nervous system in such a manner as to render him incapable of normal functioning; or

(11) Sells, manufactures or purchases any explosive bullet. For purposes of this subsection "explosive bullet" means the projectile portion of an ammunition cartridge which contains or carries an explosive charge which will explode upon contact with the flesh of a human or an animal. "Cartridge" means a tubular metal case having a projectile affixed at the front thereof and a cap or primer at the rear end thereof, with the propellant contained in such tube between the projectile and the cap; or

(12) Carries or possesses on or about his person any bludgeon, blackjack, sling-shot, sand-club, sand-bag, metal knuckles, throwing star, butterfly knife, nunchaku, switchblade knife, ballistic knife, tear gas gun projector bomb or any object containing noxious liquid gas, pistol or revolver or other firearm, bomb, grenade, bottle or other container containing an explosive substance of over one-quarter ounce, or cartridge while in the building or on the grounds of any elementary or secondary school, community college, college or university or while on the grounds of any public parks, playground forest preserve or residential property owned, operated or managed by a public housing agency or while on any public way within one thousand (1,000) feet of the real property comprising any school or public park or residential property owned, operated, or managed by a public housing agency. This subsection shall not apply to law enforcement officers or security officers of such school, college, university, park, forest preserve or public housing agency, nor to students carrying or possessing firearms for use in training courses, parades, hunting, target shooting on school ranges, or otherwise with the consent of school authorities and which firearms are transported unloaded and enclosed in a suitable case, box or transportation package.

(b) A person violating subsection (a) of this section shall be guilty of a Misdemeanor III.

(c) The presence in an automobile other than a public omnibus of any weapon, instrument or substance referred to in paragraph (a)(7) is a prima facie evidence that it is in the possession of, and is being carried by, all persons occupying such automobile at the time such weapon, instrument or substance is found, except under the following circumstances:

(1) If such weapon, instrument or substance is found upon the person of one (1) of the occupants therein; or

(2) If such weapon, instrument or substance is found in an automobile operated for hire by a duly licensed driver in the due, lawful and proper pursuit of his trade, then such presumption shall not apply to the driver.

(d) Upon conviction of a violation of this section, any weapon seized shall be confiscated by the trial court. Any weapon so confiscated shall be transferred to the police department for disposition.

29-44. Exemptions.

(a) Section 29-43, paragraphs (a)(3), (a)(4) and (a)(10) do not apply to or affect any of the following:

(1) Peace officers or any person summoned by any peace officers to assist in making arrests or preserving the peace while he is actually engaged in assisting such officer.

(2) Wardens, superintendents and keepers of prisons, penitentiaries, jails and other institutions for the detention of persons accused or convicted of an offense, while in the performance of their official duty, or while commuting between their homes and place of employment.

(3) Members of the Armed Services or Reserve Forces of the United States or the Illinois National Guard or the Reserve Officers Training Corps, while in the performance of their official duty.

(4) Special agents employed by a railroad or a public utility to perform police functions, or guards of armored car companies while actually engaged in the performance of the duties of their employment or commuting between their homes and places of employment; watchmen, while actually engaged in the performance of the duties of their employment.

(5) Persons licensed as private security contractors, private detectives, private alarm contractors or employed by an agency certified by the department of registration and education if their duties include the carrying of a weapon under the provisions of the Private Detective, Private Alarm and Private Security Act of 1983 [225 ILCS 445/1 et seq.], while actually engaged in the performance of the duties of their employment or commuting between their homes and places of employment, provided that such commuting is accomplished within one (1) hour from departure from home or place of employment, as the case may be. Persons exempted under this paragraph shall be required to have completed a course of study in firearms handling and training approved and supervised by the department of registration and education as prescribed by section 28 of the Private Detective, Private Alarm and Private Security Act [225 ILCS 445/28], prior to becoming eligible for this exemption. The department of registration and education shall provide suitable documentation demonstrating the successful completion of the prescribed firearms training. Such documentation shall be carried at all times when such persons are in possession of a concealable weapon.

(6) Any person regularly employed in a commercial or industrial operation for the production of persons employed and private property related to such commercial or industrial operation while actually engaged in the performance of their duty or traveling between sites or properties belonging to the employer of such security guards, and who as such security guards are members of a security force of five (5) persons or more registered with the department of registration and education, provided that such security guard has successfully completed a course of study, approved by and supervised by the department of registration and education, consisting of not less than forty (40) hours of training which shall include theory of law enforcement, liability for acts and the handling of weapons. The department of registration and education shall provide suitable documentation to demonstrate the successful completion of such course. Such documentation shall be carried by the security guard at all times when he is in possession of a concealable weapon.

(7) Agents and investigators of the state legislative investigating commission authorized by the commission to carry the weapons specified in section 29-43, paragraphs (a)(3) and (a)(4), while on duty in the course of any investigation for the commission.

(8) Persons employed by a financial institution for the protection of other employees and property related to such financial institution, while actually engaged in the performance of their duties, commuting between their homes and places of employment, or traveling between sites or properties owned or operated by such financial institution, provided that any person so employed has successfully completed a course of study, approved by and supervised by the department of registration and education, consisting of not less than forty (40) hours of training which includes theory of law enforcement, liability for acts, and the handling of weapons. The department of registration and education shall provide suitable documentation to demonstrate the successful completion of such course, and such documentation shall be carried by the person so trained at all times when such person is in possession of a concealable weapon. For purposes of this paragraph, "financial institution" means a bank, savings and loan association, credit union or company armored car services.

(9) Any person employed by an armored car company to drive an armored car while actually engaged in the performance of his duties.

(10) Persons who have been classified as peace officers pursuant to "an act relating to fire protection, amending certain acts herein named," [20 ILCS 2910/1] enacted by the 82nd General Assembly of the State of Illinois.

(11) Investigators of the office of the state's attorneys appellate prosecutor authorized by the board of governors of the office of the state's attorneys appellate prosecutor to carry weapons pursuant to section 7.06 of the State's Attorneys Appellate Service Commission Act [725 ILCS 210/7.06].

(12) Manufacture, transportation or sale of weapons to persons authorized under paragraphs (1) through (11) of this subsection to possess those weapons.

(b) Section 29-43, paragraphs (a)(4) and (a)(10) do not apply to or affect any of the following:

(1) Members of any club or organization organized for the purpose of practicing shooting at targets upon established target ranges, whether public or private, patrons of such ranges while such members or patrons are using their firearms on those target ranges.

(2) Duly authorized military or civil organizations while parading, with the special permission of the governor.

(3) Licensed hunters, trappers or fishermen while engaged in hunting, trapping or fishing.

(4) Transportation of weapons broken down in a nonfunctioning state or not immediately accessible.

(c) Section 29-43, paragraph (a)(7) does not apply to or affect any of the following:

(1) Peace officers while in the performance of their official duties.

(2) Wardens, superintendents and keepers of prisons, penitentiaries, jails and other institutions for the detention of persons accused or convicted of an offense.

(3) Members of the Armed Services or Reserve Forces of the United States or the Illinois National Guard, while in the performance of their official duty.

(4) Manufacture, transportation or sale of machine guns to persons authorized under paragraphs (1) through (3) of this subsection to possess machine guns, if the machine guns are broken down in a nonfunctioning state or not immediately accessible.

(5) Persons licensed under federal law to manufacture any weapon from which eight (8) or more shots or bullets can be discharged by a single function of the firing device, or ammunition for such weapons, and actually engaged in the business of manufacturing such weapons or ammunition, but only with respect to activities which are within the lawful scope of such business, such as the manufacture, transportation or testing of such weapons or ammunition. This exemption does not authorize the general private possession of any weapon from which eight (8) or more shots or bullets can be discharged by a single function of the firing device but only such possession and activities which are within the lawful scope of a licensed manufacturing business described in this paragraph. During transportation, such weapons shall be broken down in a nonfunctioning state or not immediately accessible.

(6) The manufacture, transport, testing, delivery, transfer or sale, and all lawful commercial or experimental activities necessary thereto, of rifles, shotguns and weapons made from rifles or shotguns, or ammunition for such rifles, shotguns or weapons, where engaged in by a person operating as a contractor or subcontractor pursuant to a contract or subcontract for the development and supply of such rifles, shotguns, weapons or ammunition to the United States government or any branch of the Armed Forces of the United States, when such activities are necessary and incident to fulfilling the terms of such contract.

The exemption granted under this subsection shall also apply to any authorized agent of any such contractor or subcontractor who is operating within the scope of his employment, where such activities involving such weapon, weapons or ammunition are necessary and incident to fulfilling the terms of such contract.

During transportation, any such weapon shall be broken down in a nonfunctioning state, or not immediately accessible.

(d) Section 29-43, paragraph (a)(1) does not apply to the purchase, possession or carrying of a blackjack, sling-shot or nunchaku by a peace officer, nor does it apply to the possession of nunchaku pursuant to instruction in conjunction with an organized school or class, or as part of an authorized public demonstration, exhibition or tournament sponsored by an organized school or class. When transporting nunchaku for a public demonstration, exhibition or tournament, they shall be transported in a closed container which is not immediately accessible.

(e) Section 29-43, paragraph (a)(8) does not apply to any owner, manager or authorized employee of any place specified in that subsection nor to any law enforcement officer.

(f) Section 29-43, paragraphs (a)(4) and (a)(10) do not apply to members of any club or organization organized for the purpose of practicing shooting at targets upon established target ranges, whether public or private, while using their firearms on those target ranges.

(g) Section 29-43, paragraph (a)(11) does not apply to:

(1) Members of the Armed Services or Reserve Forces of the United States or the Illinois National Guard, while in the performance of their official duty.

(2) Bona fide collectors of antique or surplus military ordinance.

(3) Laboratories having a department of forensic ballistics, or specializing in the development of ammunition or explosive ordinance.

(4) Commerce, preparation, assembly or possession of explosive bullets by manufacturers of ammunition licensed by the federal government, in connection with the supply of those organizations and persons exempted by paragraph (g)(1) of this section, or like organizations and persons outside this state, or the transportation of explosive bullets to any organization or person exempted in this section by a common carrier or by a vehicle owned or leased by an exempted manufacturer.

(h) A charge based upon a violation of any subsection need not negative any exemptions contained in this section. The defendant shall have the burden of proving such an exemption.

(i) Nothing in this section shall prohibit, apply to or affect the transportation, carrying or possession of any pistol or revolver, stun gun, taser or other firearm consigned to a common carrier operating under license of the state or the federal government, where such transportation, carrying or possession is incident to the lawful transportation in which such common carrier is engaged; and nothing in this section shall prohibit, apply to or affect the transportation, carrying or possession of any pistol, revolver, stun gun, taser or other firearm, not the subject of and regulated by subsection 29-43, paragraph (a)(7), which is unloaded and enclosed in a case, firearm carrying box, shipping box or other container, by the possessor of a valid firearm owners identification card.

29-46. Possession of ammunition and firearms by underage persons.

(a) No underage person shall acquire, possess or transport any firearm or firearm ammunition within the city, except for the purpose of firearm instruction and any other lawful purpose while in the custody and immediate control of their parent, legal guardian or other person in loco parentis to the underage person so long as said parent, legal guardian or other person in loco parentis to the underage person has a currently valid State of Illinois Firearm Owner's Identification Card.

(b) This section shall not apply to:

(1) Any underage person who was honorably discharged from the United States military and can otherwise own a firearm under state and federal law; or

(2) The acquisition, possession or transportation of a shotgun or rifle by a person eighteen (18) years old or older who has a valid State of Illinois Firearm Owner's Identification Card and has successfully completed a hunter safety education course certified by the State of Illinois.

(c) No person shall sell, loan or furnish to any underage person any firearm or firearm ammunition within the city, except as provided for in paragraph (a) above.

(d) A person violating subsection (a) or (c) of this section shall be guilty of a Misdemeanor III.

(e) Upon conviction of a violation of this section, any weapons seized shall be confiscated by the trial court. Any weapon so confiscated shall be transferred to the police department for disposition.

(f) For the purpose of this section:

(1) Firearm, rifle and shotgun have the meanings ascribed to them in section 29-49 of this Code.

(2) Firearm ammunition means any bullet, self-contained cartridge, shotgun shell, or projectile, by whatever name known, which is designed to be used or adaptable to use in a firearm; excluding, however, any ammunition exclusively designed for use with a device used exclusively for signaling or safety and required or recommended by the United States Coast Guard or the Interstate Commerce Commission, or any ammunition designed exclusively for use with a stud or rivet driver or other similar industrial ammunition.

(3) Underage person means a person under the age of twenty-one (21).

29-47. Air and spring guns.

(a) No person shall possess an air rifle, air gun, air pistol, spring gun, spring pistol, B-B gun, pellet gun or any other implement that is not a firearm and which impels a pellet constructed of hard plastic, steel, lead or other hard material with a force that reasonably may be expected to cause bodily harm, under the conditions specified for firearms in subsections (a)(4), (a)(10) or (a)(12) of section 29-43 of this article. ...

(c) Any person violating this section shall be guilty of a misdemeanor, the fine for which shall be a minimum of two hundred dollars ($200.00) and a maximum of five hundred dollars ($500.00) for each separate violation.

29-49. Transfer or possession of assault ammunition or assault weapons.

(a) No person shall sell, offer or display for sale, give, lend, transfer ownership of, acquire or possess any assault weapon or large capacity ammunition feeding device.

(b) This section shall not apply to:

(1) Any law enforcement officer of this or any other municipality or state of the United States, members of the armed forces of the United States, or the organized militia of this or any other state to the extent that any such person is otherwise authorized to possess an assault weapon or large capacity ammunition feeding device and is acting within the scope of his or her duties;

(2) The transportation of assault weapons or large capacity ammunition feeding devices to persons exempt from this section, as set forth above, by federally licensed manufacturers and dealers, if the weapons or devices are broken down in a nonfunctioning state and not immediately accessible during transportation; or

(3) The government of the United States or the government of any municipality, county, or state.

(4) Any person who, while in their home and in an act of self-defense, uses either an assault weapon owned by a person mentioned in paragraph (1) above or such a weapon for which a certificate of ownership has been issued but for which they are not the certificate holder.

(c) Any assault weapon or large capacity ammunition feeding device possessed, sold or transferred in violation of subsection (a) above is hereby declared to be contraband and shall be seized and appropriately disposed of by the police department.

(d) Any person found in violation of this section shall be guilty of a Misdemeanor IV.

(e) Any person who, prior to the effective date of this section, was legally in possession of an assault weapon or large capacity ammunition feeding device prohibited by this section, shall have ninety (90) days from the effective date of this section to do any of the following without being subject to prosecution hereunder:

(1) Remove the assault weapon or large capacity ammunition feeding device from within the limits of the city;

(2) Modify the assault weapon either to render it permanently inoperable or to permanently make it a device no longer defined as an assault weapon;

(3) Surrender the assault weapon or large capacity ammunition feeding device to the police chief or his designee for appropriate disposal; or

(4) Obtain a certificate of ownership from the Aurora Police Department in accordance with subsection (g) below.

(f) This section shall not apply to:

(1) An antique firearm as defined below;

(2) Any semiautomatic rifle with a fixed stock that cannot accept a detachable magazine or clip that holds more than eight (8) rounds of ammunition;

(3) Any semiautomatic shotgun that cannot hold more than five (5) rounds of ammunition in a fixed or detachable magazine;

(4) Any firearm that can only be operated manually by bolt, pump, lever or slide action.

(5) Any magazine, belt, drum, feed strip or similar device originally manufactured to accept more than fifteen (15) rounds of ammunition but which has been fitted with a permanent block so as to hold no more than fifteen (15) rounds of ammunition.

(g)(1) Persons who, prior to the effective date of this section, lawfully possess assault weapons as defined herein may, within ninety (90) days of the effective date of this section, apply to the Aurora Police Department for a certificate of ownership for said assault weapons.

(a) No certificate of ownership shall, however, be granted for the following weapons:

Avtomat Kalashnikov AK-47 type (including all models of the Norinco, Mitchell and Poly Tech)

Franchi SPAS-12 and LAW-12

MAC-11 carbine type

Street Sweeper and Striker 12 revolving cylinder shotguns

USAS-12

Uzi carbine and mini-carbine

Any assault weapon with a folding or telescoping stock which is not designed to be fired from the Shoulder

Any pistol which meets the definition of an assault weapon except as set forth in paragraph b. below

(b) A certificate of ownership may be granted for any pistol originally manufactured to accept, in the pistol grip, a magazine designed to contain 16-17 rounds of ammunition but which does not extend below the pistol grip and does not otherwise meet the definition of an assault weapon.

(2) The Aurora Police Department shall issue a certificate of ownership for each applicable weapon owned by the applicant if the applicant has a currently valid State of Illinois Firearm Owners Identification card. Said certificate shall not be transferable, except upon the death of the certificate holder and then only to his heir or legatee.

(3) The Aurora Police Department shall not issue a certificate of ownership, and shall revoke any existing certificate of ownership, if the applicant or certificate holder is prohibited, under state or federal law, from possessing the assault weapon for which the certificate of ownership has been applied or issued.

(4) All assault weapons for which a certificate of ownership has been issued shall be kept unloaded and stored in a locked environment when being transported or kept within the City of Aurora.

(5) The City of Aurora shall not enact any ordinance imposing greater regulations upon assault weapons for which a certificate of ownership has been issued.

(h) For the purpose of this section:

(1) Antique firearm means:

(a) Any firearm, including any firearm with a matchlock, flintlock, percussion cap or similar type of ignition system, manufactured in or before 1898; and

(b) Any replica of any firearm described in sub-paragraph (a) if the replica:

(1) Is not designed or redesigned for using rim-fire or conventional centerfire fixed ammunition; or

(2) Uses rimfire or conventional centerfire fixed ammunition which is not readily available in the ordinary channels of commercial trade.

(2) Assault weapon means:

(a) Any of the firearms (or types, replicas, or duplicates in any caliber of the firearms regardless of manufacturer) known as:

A.A. Arms AP-9

Algimec Agmi

Armalite AR-180

Australian Automatic Arms SAP Pistol

Australian Automatic Arms SAR

Auto-Ordnance Thompson type

Avtomat Kalashnikov AK-47 type (including all models of the Norinco, Mitchell, and Poly Tech)

Barrett Light-Fifty model 82A1

Beretta AR-70 (SC-70)

Beretta BM-59

Bushmaster Auto Rifle and Auto Pistol

Calico models M-900, M-950 and 100-P

Chartered Industries of Singapore SR-88

Claridge High Tech

Colt AR-15 and Sporter

Daewoo K-1, K-2, Max-1 and Max-2

Encom MK-IV, MP-9 and MP-45

Fabrique Nationale FN/FAL, FN/LAR or FN/FNC

FAMAS MAS 223

Feather AT-9 and Mini-AT

Federal XC-900 and XC-450

Franchi SPAS-12 and LAW-12

Galil AR and ARM

Heckler & Koch HK-91, HK-93, HK-94, PSG-1, SP-89

Holmes Mp-83

Intratec TEC-9, TEC-DC-9, TEC-22 and Scorpion

Iver Johnson Enforce Model 3000

Iver Johnson PM30 Paratrooper

M14S type

MAC 10, M-10, MAC-11 and MAC-11 Carbine type

Ruger Mini-14/5F (folding stock only) and Mini-14/5RF

Scarab Skorpion

SIG 57 AMT, SIG 550SP, SIG 551SP, SIG PE-57 types and 500 series

Spectre Auto Carbine and Auto Pistol

Springfield Armory BM59, SAR-48 and G-3

Sterling MK-6 and MK-7 and SAR types

Steyr AUG

Street Sweeper and Striker 12 revolving cylinder shotguns

SWD M-10, M-11, M-11/9, M-12

Valmet M63, M71S, M76 or M78

USAS-12

UZI Carbine, Mini-Carbine and Pistol

Weaver Arms Nighthawk

Wilkinson "Linda" Pistol

(b) Any semiautomatic rifle that has a fixed magazine that can hold more than eight (8) rounds or an ability to accept a detachable magazine and has at least two (2) of the following:

(1) A folding or telescoping stock;

(2) A pistol grip that protrudes beneath the action of the weapon and which is separate and apart from the stock;

(3) A bayonet mount;

(4) A flash suppressor;

(5) A barrel with a threaded muzzle;

(6) A grenade launcher.

(c) Any semiautomatic pistol that has an ability to accept a detachable magazine and has at least two (2) of the following:

(1) An ammunition magazine that attaches to the pistol outside of the pistol grip;

(2) A barrel having a threaded muzzle;

(3) A shroud that is attached to, or partially or completely encircles, the barrel and which permits the shooter to hold the firearm with the non-trigger hand without being burned;

(4) A manufactured weight of fifty (50) ounces or more when the pistol is unloaded;

(5) A semiautomatic version of an automatic firearm

(d) Any semiautomatic shotgun that has either a fixed magazine with a capacity in excess of five (5) rounds or an ability to accept a detachable magazine and, in addition, has at least one (1) of the following:

(1) A folding or telescoping stock;

(2) A pistol grip that protrudes beneath the action of the firearm and which is separate and apart from the stock;

(3) Firearm means any device, by whatever name known, which is designed to expel a projectile or projectiles by the action of an explosion, expansion of gas or escape of gas; excluding, however:

(a) Any pneumatic gun, spring gun, paint ball gun or BB gun which either expels a single globular projectile not exceeding .18 inch in diameter and which has a maximum muzzle velocity of less than seven hundred (700) feet per second or breakable paint balls containing washable marking colors;

(b) Any device used exclusively for signaling or safety and required or recommended by the United States Coast Guard or the Interstate Commerce Commission; or

(c) Any device used exclusively for the firing of stud cartridges, explosive rivets or similar industrial ammunition.

(4) Large capacity ammunition feeding device means:

(a) Any magazine, belt, drum, feed strip or similar device that has a capacity of, or can be readily restored or converted to accept, more than fifteen (15) rounds of ammunition; or

(b) Any combination of parts from which a device described in the above clause can be assembled;

The term does not include an attached tubular device designed to accept and capable of operating with only .22 caliber rimfire ammunition.

(5) Pistol means any firearm designed to be fired by the use of a single hand, as well as any combination of parts from which a pistol can be assembled.

(6) Rifle means a firearm designed or redesigned, made or remade, and intended to be fired from the shoulder and designed or redesigned and made or remade to use the energy of the explosive in a fixed metallic cartridge to fire only a single projectile through a rifled bore for each single pull of the trigger.

(7) Semiautomatic means a firearm that utilizes a portion of the energy of a firing cart-ridge to extract the fired cartridge case and chamber the next round, and which requires a separate pull of the trigger to fire each cartridge.

(8) Shotgun means a firearm designed or redesigned, made or remade, and intended to be fired from the shoulder and designed or redesigned and made or remade to use the energy of the explosive in a fixed shotgun shell to fire through a smooth or rifled bore either a number of ball shot or a single projectile for each single pull of the trigger.

29-50. Unlawful storage of ammunition and firearms.

(a) No person shall keep firearm ammunition or a loaded firearm within any premises or vehicle under his or her custody or control which he or she knows, or reasonably should know, a minor is likely to be able to gain access to without permission of the minor's parent or legal guardian, unless the firearm ammunition is in a locked container, the loaded firearm is equipped with a locking device, or said firearm is in a locked container.

(b) For the purposes of this section:

(1) Minor means a person under the age of eighteen (18) years;

(2) Firearm has the meaning ascribed to it in section 29-49 of this Code;

(3) Loaded firearm means a firearm in which there is an unexpended cartridge or shell, consisting of a case that holds a charge of powder or a bullet or shot, in or attached in any manner to the firearm, including, but not limited to, in the firing chamber, magazine or clip thereof attached to the firearm; except that a muzzleloader firearm shall be deemed loaded when it is capped or pinned and has a powder charge and ball or shot in the barrel or cylinder;

(4) Locking device means a device attached to a firearm other than the safety that temporarily prevents the firearm from functioning.

(c) A person violating this section shall be guilty of a Misdemeanor II.

29-51. Replica firearms prohibited.

(a) "Replica firearm" means any toy, lookalike and imitation firearm having the appearance, shape, and/or configuration of any original firearm which was manufactured, designed, and produced since 1898 or any device, object or facsimile made of plastic, wood, metal or any other material that a person could reasonably perceive as an actual firearm. Such term shall not, however, include:

(1) Non-firing collector replica antique firearms, which look authentic and may be a scale model but are not intended as toys modeled on real firearms designed, manufactured, and produced prior to 1898;

(2) Decorative, ornamental, and miniature objects having the appearance, shape and/or configuration of a firearm, including those intended to be displayed on a desk or worn on bracelets, necklaces, key chains, provided that the objects measure no more than thirty-eight (38) millimeters in height by seventy (70) millimeters in length, the length measurement excluding any gun stock length measurement.

(b) It shall be unlawful for any person to sell, manufacture, purchase, possess or carry any replica firearm within the corporate limits of the City of Aurora unless such replica firearm contains, or has affixed to it, one of the markings set forth in subsection (c) of this section or unless this prohibition does not apply pursuant to subsection (d) of this section.

(c) The following markings are approved for replica firearms:

(1) A blaze orange (Federal Standard 595a, February, 1987, color number 12199, issued by the United States General Services Administration) or orange color brighter than that specified by the federal standard color number, solid plug permanently affixed to the muzzle end of the barrel as an integral part of the entire device and recessed no more than six (6) millimeters from the muzzle end of the barrel.

(2) A blaze orange (Federal Standard 595a, February, 1987, color number 12199, issued by the United States General Services Administration) or orange color brighter than that specified by the Federal Standard color number, marking permanently affixed to the exterior surface of the barrel, covering the circumference of the barrel from the muzzle end for a depth of a least six (6) millimeters.

(3) Construction of the replica firearm entirely of transparent or translucent materials which permits unmistakable observation of the replica firearm's complete contents.

(4) Coloration of the entire exterior surface of the replica firearm in white, bright red, bright orange, bright yellow, bright green, bright blue, bright pink, or bright purple, either singly or as the predominant color in combination with other colors in any pattern.

(d) The possession and use of a replica firearm is permitted:

(1) If the device is solely for use and is being used in theatrical productions, including motion picture, television and stage productions.

(2) If the replica firearm is in the possession and control of a carnival and is offered for use to carnival customers on a temporary basis. For the purpose of this section, "carnival" means an enterprise which offers amusement or entertainment to the public by means of one or more amusement attractions or rides.

(e) Penalties. Any person violating this section, upon conviction thereof, shall be fined not less than five hundred dollars ($500.00) nor more than one thousand dollars ($1,000.00) for each offense, or imprisoned for a period not to exceed six (6) months, or both such fine and imprisonment. As an alternative to, or in addition to, any such fine, the violator may be made to perform community service.

[Code of Ordinances, City of Aurora codified through Ordinance No. 008-66, enacted June 24, 2008]

City of Chicago Municipal Code

Title 4. Businesses, Occupations and Consumer Protection

Chapter 4-144. Weapons

Article I. Deadly Weapons

4-144-010. License - Required. It shall be unlawful for any person to engage in the business of selling, or to sell or give away, any pistol, revolver or other firearm, dagger, stiletto, billie, derringer bowie knife, dirk, stun gun or taser, as defined in Section 24-1 of the Illinois Criminal Code, 720 ILCS 5/24-1, or other deadly weapon which can be carried or concealed on the person, without securing a weapons dealer license. The license required by this chapter shall be in addition to any other license required by law.

4-144-020. License - Application. An application for a weapons dealer license shall be made in conformity with the general requirements of this Code relating to applications for licenses. The director of business affairs and licensing shall approve said application before a license shall be issued.

4-144-030. License - Fee. The annual fee for a weapons dealer license shall be as set forth in Section 4-5-010.

4-144-040. Daily report required - Sales or gifts. Every person dealing in the aforementioned deadly weapons shall make out and deliver to the superintendent of police every day before the hour of twelve noon, a legible and correct report of every sale or gift made under authority of his license during the preceding 24 hours, which report shall contain the date of such sale or gift, the name of the purchaser or donee with his or her address and age, the number, kind, description and price of such weapon, the number of the purchaser's permit, and the purpose given by such person for the purchase of such weapon, which report shall be substantially in the following form:

Number of permit
Number of weapon

Name of purchaser
Address of purchaser
Age of purchaser
Kind or description of weapon
For what purpose purchased
Price

4-144-050. Register required. Every person dealing in the aforementioned deadly weapons or ammunition at retail, within the city, shall keep a register of all such weapons and ammunition sold, loaned, rented or given away by him. Such register shall contain the date of the sale, loaning, renting or gift, the number of the permit, the number of the weapon, the name and age of the person to whom the weapon or ammunition is sold, loaned, rented or given, the quantity of ammunition, the price of each item, and the purpose for which it is purchased or obtained. The said register shall be in the following form (see diagram for Section 4-144-050)[omitted]. Such register shall be kept open for the inspection of the police at all reasonable times during business hours.

4-144-060. Restrictions on sales or gifts. It shall be unlawful for any person to sell, barter or give away to any person within the city, any deadly weapon mentioned in Section 4-144-010, except to licensed dealers and to persons who have secured a permit for the purchase of such articles from the superintendent of police as hereinafter required. This section shall not apply to sales made of such articles which are to be delivered or furnished outside the city.

4-144-061. Sale of certain handgun ammunition prohibited. Except as allowed by subsection (e) of Section 8-20-170 of this Code, it shall be unlawful for any person to sell, offer for sale, expose for sale, barter or give away to any person within the city, any ammunition of the following calibers and types:

.45 automatic
.380 automatic
.38 special
.357 magnum
.25 caliber
.22 caliber, including .22 long
9 millimeter

Any other ammunition, regardless of the designation by the manufacturer, distributor or seller, that is capable of being used as a substitute for any of the foregoing.

4-144-062. Sale of handguns without child-proofing or safety devices prohibited. Except as allowed by subsection (e) of Section 8-20-170 of this Code, it shall be unlawful for any person to sell, barter or give away to any person any handgun which does not contain:

(1) A safety mechanism to hinder the use of the handgun by unauthorized users. Such devices shall include, but shall not be limited to, trigger locks, combination handle locks, and solenoid use-limitation devices, and

(2) A load indicator device that provides reasonable warning to potential users such that users even unfamiliar with the weapon would be forewarned and would understand the nature of the warning.

Safety mechanism means a design adaption or nondetachable accessory that lessons the likelihood of unanticipated use of the handgun by other than the owner of the handgun and those specifically authorized by the owner to use the handgun.

A trigger lock means a device that when locked in place by means of a key, prevents a potential user from pulling the trigger of the handgun without first removing the trigger lock by use of the trigger lock's key.

A combination handle lock means a device that is part of the handgun which precludes the use of the handgun unless the combination tumblers are properly aligned.

A solenoid use-limitation device means a device which precludes, by use of a solenoid, the firing of the handgun unless a magnet of the appropriate strength is placed in proximity to the handle of the weapon.

A load indicator means a device which plainly indicates that a bullet is placed in the handgun in a way that pulling the trigger or otherwise handling the handgun may result in detonation.

4-144-070. Permit required when - Issuance conditions. It shall be unlawful for any person to purchase any deadly weapon mentioned in Section 4-144-010 which can be concealed on the person, without first securing from the superintendent of police a permit so to do. Before any such permit is granted, an application in writing shall be made therefor, setting forth in such application the name, address, age, height, weight, complexion, nationality and other elements of identification of the person desiring such permit, and the applicant shall present such evidence of good character as the superintendent of police at his discretion may require.

The superintendent of police shall refuse such permit to any person under 18 years of age, any narcotic addict, any person who has been convicted of a felony under the laws of this state or any other jurisdiction within five years from release from penitentiary or within five years of conviction if penitentiary sentence has not been imposed, and any person who has been released from a mental institution or from the custody of the Illinois Youth Commission within the last five years, or is mentally retarded. Otherwise, in case he shall be satisfied that the applicant is of good moral character, it shall be the duty of the superintendent of police to grant such permit.

4-144-080. Sales display restrictions. It shall be unlawful for any person to exhibit for sale in show cases or show windows, on counters or in any other public manner, any deadly weapon mentioned in Section 4-144-010, or to display any signs, posters, cartoons, or display cards suggesting the sale of any such deadly weapons, or any ammunition whose sale is prohibited pursuant to Section 4-144-061 of this Code.

Article II. Gunsmiths

4-144-100. License - Required. It shall be unlawful for any person to engage in the business of repairing any pistol, revolver, derringer or other firearm which can be concealed on the person without securing a weapons dealer license so to do.

4-144-130. Daily report required - Repairs. Every person licensed under this chapter shall make out and submit to the superintendent of police every day, before twelve noon, a legible and correct report of each firearm received for repair during the preceding 24 hours, which report shall contain the date, name, physical description, age, address and occupation of the owner of such firearm, the type of weapon, its make, and the serial number and bore length of such weapon, which report shall be substantially in the following form:

Date
Name of owner
Physical description of owner
Age of owner
Address of owner
Occupation of owner
Type of weapon
Make of weapon
Serial number
Bore and length of weapon

Article III. Air Rifles and Toy Weapons

4-144-140. License required when. It shall be unlawful for any person to engage in the business of selling or to sell or to give away any air rifle or air gun, or any toy firearms or other toy in the nature of a firearm in which any explosive substance can be used, without securing a weapons dealer license, and no person having secured such license shall sell or give away any such weapon to any person within the city who has not secured a permit from the superintendent of police to purchase such weapon in the manner hereinafter provided.

4-144-145 Replica air guns – Sale or transfer prohibited.

(a) It shall be unlawful for any person to engage in the business of selling or to sell, exhibit for sale, give away or otherwise transfer any replica air gun in the city of Chicago. For purposes of this article, "replica air gun" means and includes any air gun, air pistol, air rifle, spring gun, spring pistol, BB gun, pellet gun or any other implement that a person could reasonably perceive as an actual firearm but that is not a firearm, and that is capable of firing or discharging a projectile constructed of hard plastic, steel, lead or other hard materials with a force that reasonably is expected to cause bodily harm.

4-144-170. Daily report required. Every person licensed under this chapter shall make out and deliver to the superintendent of police every day, before the hour of twelve noon, a legible and correct report of every sale or gift made under authority of said license to sell the kind of weapons or other articles named in Section 4-144-140 during the preceding 24 hours, which report shall contain the date of such sale or gift, the name of the purchaser or donee with his or her address and age, the number, kind, description and price of such weapon or other article, the number of the purchaser's permit, and the purpose for the purchase of such weapon or other article, which report shall be substantially in the following form:

Number of permit
Number of weapon or article
Name of purchaser
Address of purchaser
Age of purchaser
Kind or description of weapon or other article
For what purpose purchased
Price

4-144-180. Permit - Required. It shall be unlawful for any person to purchase any air rifle or air gun, or any toy firearms or other toy in the nature of a firearm in which any explosive substance can be used, without first securing from the superintendent of police a permit so to do. Before any such permit is granted, an application in writing shall be made therefor, setting forth in such application, the name, address, age, height, weight, complexion, nationality and other elements of identification of such person desiring such permit. Such application shall also contain a recommendation from two persons who shall appear to be taxpayer residing within the city that the permit shall issue.

4-144-190. Replica firearms and pellet guns.

(a) It shall be unlawful for any person to purchase, possess, use, sell, give away or otherwise transfer, or to engage in the business of selling or to exhibit for sale, a replica firearm, paint pellet or paint pellet gun in the City of Chicago, except as provided in subsection (c) of this section.

(b) For the purposes of this chapter, the following terms shall have the following meanings:

"Paint pellet" means a pellet or projectile of paint which explodes upon impact.

"Paint pellet gun" means any firearm, toy firearm or toy in the nature of a firearm which is powered by compressed gas and which fires paint pellets.

"Replica firearm" means any device, object or facsimile made of plastic, wood, metal or any other material, that a person could reasonably perceive as an actual firearm but that is incapable of being fired or discharged, except that the term shall not include any replica of an antique firearm, as defined in Section 8-20-030(b) of this Code. Each such replica firearm shall have as an integral part, permanently affixed, a blaze orange plug inserted in the barrel of such replica firearm. Such plug shall be recessed no more than six millimeters from the muzzle end of the barrel of such firearm.

(c) The manufacture, marketing, distribution, sale and possession of replica firearms are permitted if the devices are manufactured, marketed, distributed, sold or held (1) solely for subsequent transportation in intrastate, interstate or foreign commerce, or (2) solely for use in theatrical productions, including motion picture, television and stage productions. Such devices shall not be displayed to the general public or sold for other use in the city. The use or possession of a paint pellet or paint pellet gun is permitted if the use or possession is solely within premises licensed as a public place of amusement; or if the use or possession is solely for the purpose of transporting the paint pellet or paint pellet gun to or from those premises by the licensee or agent or employee of the licensee, or by a common carrier, for purposes of initial delivery, repair or disposal of the paint pellet or paint pellet gun.

(d) Any person who violates the provisions of this section, upon conviction thereof, shall be fined not less than $100.00 nor more than $500.00 for each offense. Any such violation may also be punishable as a misdemeanor by incarceration in a penal institution other than a penitentiary for up to six months under the procedures set forth in Section 1-2-1.1 of the Illinois Municipal Code as amended, and in the Illinois Code of Criminal Procedure, Illinois Revised Statutes, Chapter 38, Sections 100-1 et seq. (1985), as amended, in a separate proceeding. All actions seeking the imposition of fines only shall be filed as quasi-criminal actions subject to the provisions of the Illinois Code of Civil Procedure, Illinois Revised Statutes, Chapter 110, Section 1-101, et seq. (1985), as amended. Each purchase, use, sale, gift or transfer of any such replica firearm, paint pellet or paint pellet gun shall be deemed a separate and distinct offense, and each day a person unlawfully engages in the business of selling or exhibits for sale any such replica firearm, paint pellet or paint pellet gun shall be deemed a separate and distinct offense.

4-144-195. Replica military style weapons. It shall be unlawful for any person to purchase, possess, use, sell, give away or otherwise transfer, or to engage in the business of selling or to exhibit for sale, any replica rocket propelled grenade launcher, bazooka, artillery piece, grenade, mine, bomb, or items similar to weapons designed and manufactured for military purposes or replicas thereof, in the City of Chicago. Any person who violates this section shall be fined not less than $100.00 nor more than $500.00 for each offense.

4-144-200. Granting of permit - Conditions. It shall be the duty of the superintendent of police to refuse such permit to any person having been convicted of any crime, and any minor. Otherwise, if the applicant is of good moral character, the superintendent of police shall grant such permit upon the payment of a fee of $1.00.

4-144-210. Sale or transfer to minors prohibited. It is unlawful for any dealer to sell, lend, rent, give or otherwise transfer an air rifle to any person under the age of 18 years where the dealer knows the person to be under 18 years of age, or where such dealer has failed to make reasonable inquiry relative to the age of such person and such person is under 18 years of age.

It is unlawful for any person to sell, lend or otherwise transfer any air rifle to any person under 18 years of age.

4-144-220. Sales display restrictions. It shall be unlawful for any person to exhibit for sale in show cases, or show windows, on counters, or in any public manner, any air rifle or air gun, or any toy firearm or other toy in the nature of a firearm in which any explosive substance can be used, or to display any signs, posters, cartoons or display cards suggesting the sale of any such weapon or firearm.

4-144-230. Alteration restricted. No person shall alter any air rifle, air gun, toy firearm or toy in the nature of a firearm in such a way that it can fire any type of projectile other than that which it was designed by its manufacturer to fire.

4-144-240. License - Revocation conditions. When the license of any said licensee shall be revoked, no other such license shall be issued to such licensee for a period of three years thereafter.

Article IV. Violation of Chapter Provisions

4-144-250. Violation - Penalties. Any person violating Section 4-144-010 or Section 4-144-060 Section 4-144-061 of this chapter shall be fined not less than $500.00 nor more than $1,000.00 for a first offense and $1,000.00 for each subsequent offense. Any person violating any other provision of this chapter shall be fined not less than $250.00 nor more than $500.00 for a first offense and not less than $500.00 nor more than $1,000.00 for each subsequent offense. Each purchase, sale or gift of any weapon or article mentioned in this chapter shall be deemed a separate offense.

4-144-260. License - Revocation. In case the mayor shall determine that a licensee has violated any provision of this chapter, he shall revoke the weapons dealer license issued to such person, and the money paid for such license shall be forfeited to the city. No other such license shall be issued to such licensee for a period of three years thereafter.

Title 8. Offenses Affecting Public Peace, Morals and Welfare

Chapter 8-16. Offenses By or Against Minors

8-16-090. Firearms for minors. No person shall sell, loan, or furnish to any minor any gun, pistol, or other firearm, or any toy gun, toy pistol, or other toy firearm in which any explosive substance can be used, within the city; except that minors may be permitted, with the consent of their parents or guardians, to use firearms on the premises of a duly licensed shooting gallery, gun club, or rifle club, or to shoot game birds in accordance with the provisions of Section 8-24-050 of this Code.

Chapter 8-20. Weapons

Article I. Possession of Firearms

8-20-010. Unlawful to carry - Exceptions. It shall be unlawful for any person to carry or maintain in any vehicle or about his or her person except when on his or her property or in his or her residence or fixed place of business, any rifle, shotgun, or other firearm; provided, that this section shall not apply to:

(1) Peace officers or any person summoned by any such officers to assist in making arrests or preserving the peace while he is actually engaged in assisting such officer;

(2) Wardens, superintendents and keepers of prisons, penitentiaries, jails, and other institutions for the detention of persons accused or convicted of an offense, while in the performance of their official duty or commuting between their homes and places of employment;

(3) Members of the Armed Services or Reserve Forces of the United States or the Illinois National Guard or the Reserve Officers Training Corps, while in the performance of their official duty;

(4) Special agents employed by a railroad to perform police functions, or employees of a detective agency, watchman-guard or patrolman agency, licensed by the State of Illinois, while actually engaged in the performance of the duties of their employment or commuting between their homes and places of employment;

(5) Agents and investigators of the Illinois Crime Investigating Commission authorized by the Commission to carry weapons, while on duty in the course of any investigation for the Commission;

(6) Manufacture, transportation when the weapons are not immediately accessible to any person, or sale of weapons to persons authorized under law to possess them;

(7) Members of any club or organization organized for the purpose of practicing shooting at targets upon established target ranges, whether public or private, while such members are using their firearms on such target ranges;

(8) Duly authorized military or civil organizations while parading, with the special permission of the Governor;

(9) Licensed hunters or fishermen while engaged in hunting or fishing;

(10) Transportation of weapons broken down in a nonfunctioning state.

8-20-015. Unlawful firearm or laser sight accessory in motor vehicle - Impoundment.

(a) The owner of record of any motor vehicle that contains an unregistered firearm, a firearm that is not broken down in a nonfunctioning state, or a laser sight accessory shall be liable to the city for an administrative penalty of $500.00 plus any towing and storage fees applicable under Section 9-92-080. Any such vehicle shall be subject to seizure and impoundment pursuant to this section. The subsection shall not apply: (1) if the vehicle used in the violation was stolen at the time and the theft was reported to the appropriate police authorities within 24 hours after the theft was discovered or reasonably should have been discovered; (2) if the vehicle is operating as a common carrier and the violation occurs without the knowledge of the person in control of the vehicle; or (3) if the owner proves that the presence of the firearm was permissible pursuant to paragraphs (1) through (10) of Section 8-20-010 or, in the case of an unregistered firearm, the firearm was exempt from registration under subsection (b) of Section 8-20-040; or (4) if the owner proves the presence of a laser sight accessory was permissible pursuant to Section 8-20-165. ...

8-20-020. Violation - Penalty. Any person violating the provisions of Section 8-20-010 shall be fined $500.00.

Article II. Registration of Firearms

8-20-030. Definitions. As used in this Title 8:

(a) "Ammunition" means cartridge cases, shells, projectiles (including shot), primers, bullets, propellant powder, or other devices or materials designed, or intended for use in a firearm or destructive device.

(b) "Antique firearms" means any firearm, including, but not limited to, any firearm with a matchlock, flintlock, percussion cap or similar type of ignition system, which is incapable of being fired or discharged; or any firearm manufactured before 1898 for which cartridge ammunition is not commercially available and is possessed as a curiosity or ornament or for its historical significance or value.

(c) "Corrections officer" means wardens, superintendents and keepers of prisons, penitentiaries, jails or other institutions for the detention of persons accused or convicted of an offense.

(d) "Crime of violence" is defined as any felony committed while armed with a weapon as defined in Chapter 38, Section 33 A-1 of Illinois Revised Statutes, as amended.

(e) "Duty-related firearm" shall mean any weapon which is authorized by any law enforcement agency to be utilized by their personnel in the performance of their official duties.

(f) "Firearm" means any weapon which will, or is designed to or restored to, expel a projectile or projectiles by the action of any explosive; the frame or receiver of any such device; or any firearm muffler or silencer. Provided, that such term shall not include:

(1) antique firearm;

(2) any device used exclusively for line-throwing, signaling, or safety and required or recommended by the United States Coast Guard or Interstate Commerce Commission; or

(3) any device used exclusively for firing explosives, rivets, stud cartridges, or any similar industrial ammunition incapable of use as a weapon.

(g) "Fragmenting bullet" means a bullet that is designed or modified to shatter on impact, or any other bullet that is designed or modified so that more than 50 percent of the mass of the bullet is likely to fragment inside a human or animal target.

(h) "Assault weapon" means any of the following weapons:

(1) Assault Rifles

AK 47 type	86S type
AK 47S type	86S7 type
AK 74 type	87S type
AKS type	Galil type
AKM type	Type 56 type
AKMS type	Type 565 type
84S1 type	Valmet M76 type
Arm type	Valmet M78 type
84S1 type	M76 counter sniper type
84S3 type	FAL type
HK91 type	L1A1A type
HK93 type	SAR 48 type
HK94 type	AUG type
G3SA type	FNC type
K1 type	Uzi carbine
K2 type	Algimec AGMI type
AR100 type	AR180 type
M24S type	MAS 223 type
SIG 550SP type	Beretta BM59 type
SIG 551SP type	Beretta AR70 type

Australian Automatic Arms
CIS SR88 type
SAR type
SKS type with detachable magazine
Colt AR-15
Springfield Armory SAR-48
Springfield Armory BM-59
Bushmaster Auto Rifle
Auto-Ordinance Thompson M1
Ruger Mini 14/5F
Federal XC-900 and XC-450
Feather AT-9 Auto Carbine
Goncz High Tech Carbine
Auto-Ordinance Thompson 1927A1
Iver Johnson PM30 P Paratrooper

(2) Assault Pistols

Uzi type
Heckler & Koch Sp-89 type
Australian Automatic Arms SAP type
Spectre Auto type
Sterling Mark 7 type; and

(3) Any weapon that the superintendent of police defines by regulation as an assault weapon because the design or operation of such weapon is inappropriate for lawful use.

(i) "Assault ammunition" means any ammunition magazine having a capacity of more than twelve (12) rounds of ammunition.

(j) "Disc projectile ammunition" means any ammunition which is composed of multiple disc shaped objects stacked together to form a single round of ammunition, including but not limited to the following types of ammunition: (i) Magdisc type; and (ii) Shatterdisc type.

(k) "Handgun" means a firearm designed to be held and fired by the use of a single hand, and includes a combination of parts from which such firearm can be assembled.

(l) "Machine gun" means any firearm from which eight or more shots or bullets may be discharged by a single function of the firing device.

(m) "Mayor" means the mayor of the City of Chicago or his or her designee.

(n) "Metal piercing bullet" means any bullet that is manufactured with other than a lead or lead alloy core, or ammunition of which the bullet itself is wholly composed of, or machined from, a metal or metal alloy other than lead, or any other bullet that is manufactured to defeat or penetrate bullet resistant properties of soft body armor or any other type of bullet resistant clothing which meets the minimum requirements of the current National Institute for Justice Standards for "Ballistic Resistance of Police Body Armor".

(o) "Organization" means partnership, company, corporation or other business entity, or any group or association of two or more persons united for a common purpose.

(p) "Peace officer" means any person who by virtue of his office or public employment is vested by law with a duty to maintain public order or make arrests for offenses, whether that duty extends to all offenses or is limited to specific offenses.

(q) "Person" shall mean any individual, corporation, company, association, firm, partnership, society, joint stock company or organization of any kind.

(r) "Sawed-off shotgun" means a shotgun having a barrel of less than 18 inches in length or a firearm made from a shotgun if such firearm is modified and has an overall length of less than 26 inches or a barrel of less than 18 inches in length.

(s) "Security personnel" means special agents employed by a railroad or public utility to perform police functions: guards of armored car companies, watchmen, security guards and persons regularly employed in a commercial or industrial operation for the protection of persons employed by or property related to such commercial or industrial operation.

(t) "Short-barreled rifle" means a rifle having any barrel less than 16 inches in length, or a modified firearm if such firearm as modified has an overall length of less than 26 inches or any barrel of less than 16 inches.

(u) "Superintendent" means the superintendent of the Chicago Police Department or his designated representative.

(v) "Safety mechanism" means a design adaption or nondetachable accessory that lessens the likelihood of unanticipated use of the handgun by other than the owner of the handgun and those specifically authorized by the owner to use the handgun.

(w) "Trigger lock" means a device that when locked in place by means of a key, prevents a potential user from pulling the trigger of the handgun without first removing the trigger lock by use of the trigger lock's key.

(x) "Combination handle lock" means a device that is part of the handgun which precludes the use of the handgun unless the combination tumblers are properly aligned.

(y) "Solenoid use-limitation device" means a device which precludes, by use of solenoid, the firing of the handgun unless a magnet of the appropriate strength is placed in proximity to the handle of the weapon.

(z) "Load indicator" means a device which plainly indicates that a bullet is placed in the handgun in a way that pulling the trigger or otherwise handling the handgun may result in detonation.

(aa) "Laser sight accessory" means a laser sighting device which is either integrated into a firearm or capable of being attached to a firearm.

8-20-040. Registration of firearms.

(a) All firearms in the City of Chicago shall be registered in accordance with the provisions of this chapter. It shall be the duty of a person owning or possessing a firearm to cause such firearm to be registered. No person shall within the City of Chicago, possess, harbor, have under his control, transfer, offer for sale, sell, give, deliver, or accept any firearm unless such person is the holder of a valid registration certificate for such firearm. No person shall, within the City of Chicago, possess, harbor, have under his control, transfer, offer for sale, sell, give, deliver, or accept any firearm which is unregisterable under the provisions of this chapter.

(b) This section shall not apply to:

(1) Firearms owned or under the direct control or custody of any federal, state or local governmental authority maintained in the course of its official duties;

(2) Duty-related firearms owned and possessed by peace officers who are not residents of the City of Chicago;

(3) Duty-related firearms owned or possessed by corrections officers; provided, that such corrections officers are not residents of the City of Chicago;

(4) Firearms owned, manufactured or processed by licensed manufacturers of firearms, bulk transporters or licensed sellers of firearms at wholesale or retail, provided that such persons have, in addition to any other license required by law, a valid deadly weapons dealer license issued under Chapter 4-144 of this Code;

(5) Any nonresident of the City of Chicago participating in any lawful recreational firearm-related activity in the city, or on his way to or from such activity in another jurisdiction; provided, that such weapon shall be unloaded and securely wrapped and that his possession or control of such firearm is lawful in the jurisdiction in which he resides;

(6) Peace officers, while in the course of their official duties, who possess and control any firearm or ammunition issued by their department, bureau or agency in the normal course of business;

(7) Private security personnel who possess or control any firearm or ammunition within the City of Chicago; provided, that such firearms shall be owned and maintained by the security firm employing such personnel and shall be registered by the security firm in accordance with this chapter;

(8) Those persons summoned by a peace officer to assist in making an arrest or preserving the peace while actually engaged in assisting the peace officer.

8-20-050. Unregisterable firearms. No registration certificate shall be issued for any of the following types of firearms:

(a) Sawed-off shotgun, machine gun, or short-barreled rifle;

(b) Firearms other than handguns, owned or possessed by any person in the City of Chicago prior to the effective date of this chapter which are not validly registered prior to the effective date of this chapter;

(c) Handguns, except:

(1) Those validly registered to a current owner in the City of Chicago prior to the effective date of this chapter, and which contain each of the following:

(i) A safety mechanism to hinder the use of the handgun by unauthorized users. Such devices shall include, but shall not be limited to, trigger locks, combination handle locks, and solenoid use-limitation devices; and

(ii) A load indicator device that provides reasonable warning to potential users such that even users unfamiliar with the weapon would be forewarned and would understand the nature of the warning;

(2) Those owned by peace officers who are residents of the City of Chicago,

(3) Those owned by security personnel,

(4) Those owned by private detective agencies licensed under Chapter 111.2601, et seq., Illinois Revised Statutes;

(d) Firearm muffler or silencer;

(e) Assault weapons, as defined in Section 8-20-030, unless they are owned by a person who is entitled to own them under Section 8-24-025.

Any person who receives through inheritance any firearm validly registered pursuant to this chapter will be eligible to reregister such firearm within 60 days after obtaining possession or title, provided such person shall be qualified to do so in accordance with this chapter.

8-20-060. Prerequisites to registration - Application for registration.

(a) No registration certificate shall be issued to any person unless such person:

(1) Shall possess a valid Illinois Firearm Owner's Identification Card in accordance with Chapter 38 Section 83-4 of the Illinois Revised Statutes as amended;

(2) Has not been convicted of a crime of violence, as defined herein as weapons offense, or a violation of this chapter; and

(3) Has not been convicted within the five years prior to the application of any:

(i) Violation of any law relating to the use, possession or sale of any narcotic or dangerous drug, or

(ii) Violation of Chapter 38 Section 12-2(a)(1) of the Illinois Revised Statutes, as amended, for aggravated assault or any similar provision of the law of any other jurisdiction; and

(4) Has vision better than or equal to that required to obtain a valid driver's license under the standards established by the Illinois Vehicle Code Chapter 95-1/2, Section 506-4 Illinois Revised Statutes, as amended; and

(5) Is not otherwise ineligible to possess a firearm under any federal, state or local law, statute or ordinance.

(b) All applicants for a registration certificate under this chapter shall file with the superintendent on a form provided, a sworn application in writing.

8-20-070 Fingerprints.

When necessary to establish the identity of any applicant or registrant, such applicant or registrant shall be required to submit to fingerprinting in accordance with procedures and regulations prescribed by the superintendent.

8-20-080 Application fees.

(a) A nonrefundable fee in the amount indicated in subsection (d) of this section shall accompany each initial registration.

(b) A nonrefundable fee in the amount indicated in subsection (d) of this section shall accompany each reregistration application.

(c) The registration fee shall not be applicable to (1) any duty-related handgun of a peace officer domiciled in the City of Chicago, or (2) to any duty- related handgun(s) owned by a resident of the City of Chicago who retired from the Chicago Police Department in good standing and without any disciplinary charges pending, and who is, or is eligible to become, an annuitant of the Policemen's Annuity and Benefit Fund of the City of Chicago, but only if the handgun(s) is registered in that person's name at the time of separation from active duty in the Chicago Police Department.

(d) Registration fees for firearms shall be as follows:

1 firearm	$20.00
2 - 10 firearms	25.00
More than ten firearms	35.00

8-20-090 Filing time.

(a) A registration certificate shall be obtained prior to any person taking possession of a firearm from any source.

(b) Any firearm currently registered must be reregistered pursuant to this chapter and in accordance with rules, regulations and procedures prescribed by the superintendent. An application to reregister such firearm shall be filed within 180 days from the effective date of this chapter; provided, however, that this section shall not apply to law enforcement officers during their tenure of continuous active duty.

8-20-100 Investigations.

Upon receipt of an application for registration of a firearm, the superintendent of police shall investigate the information contained in said application to determine whether the application and firearm meet the requirements for registration under this chapter. Failure by the applicant or registrant to respond to investigation inquiries shall be sufficient grounds for denial or revocation.

8-20-110 Issuance of registration certificate.

(a) Upon receipt of a properly executed application for a registration certificate and the report of the superintendent, the mayor, upon determining that the applicant has complied with the provisions of this chapter, shall authorize the issuance of the registration certificate. Each registration certificate shall be in triplicate and bear a unique registration certificate number and contain such other information as may be necessary to identify the applicant and the firearm registered. The original of the registration certificate shall be retained by the superintendent; the mayor and applicant shall each receive a copy.

(b) The mayor shall approve or deny any application for a registration certificate within a 120-day period beginning on the date the superintendent receives the application unless good cause is shown. In the case of an application to reregister a firearm currently registered, the mayor shall have 365 days after receipt of such application to approve or deny such application unless good cause is shown.

(c) Any application for registration or renewal shall be held in abeyance when there is a criminal proceeding for a crime of violence, or an offense involving a weapon, or a proceeding to revoke firearm registration pending against the applicant until such proceeding has terminated. In the case of a renewal of registration the then-current registration shall be deemed continuing until the termination of such proceedings.

(d) Upon receipt of a registration certificate, each applicant shall examine the same to insure that the information thereon is correct. If the registration certificate is incorrect in any respect, the registrant thereon shall return it to the superintendent with a signed statement showing the nature of the error. The superintendent shall correct the error, if it occurred through administrative error.

In the event that the error resulted from incorrect information contained in the application, the applicant shall be required to file an amended application setting forth the correct information and a statement explaining the error in the original application. Each amended application shall be accompanied by a fee of $2.00.

(e) Each registration certificate authorized to be issued by the mayor shall be accompanied by a statement setting forth the registrant's duties under this chapter.

8-20-120 Revocation – Denial.

A registration certificate shall be revoked or an application for registration or reregistration shall be denied by the mayor when she finds that:

(a) Any of the criteria in Section 8-20-060(a) of this chapter are not currently met; or

(b) The registered firearm is or has become an unregisterable firearm under the terms of Section 8-20-050 of this chapter; or

(c) The information furnished to the superintendent on the application for registration certificate proves to be false; or

(d) The applicant or registrant has violated any of the provisions of this chapter.

8-20-130 Procedures for denial or revocation.

(a) If it is determined that an application for registration or reregistration should be denied or that a registration certificate should be revoked, the mayor shall notify the applicant or registrant in writing of the proposed denial or revocation, briefly stating the reason or reasons therefore.

(b) The applicant or registrant, within ten days after receiving notice of the proposed denial or revocation, may file with the department of administrative hearings a written request for a hearing before the department of administrative hearings.

(c) Within ten days of receipt of a request for hearing, the department of administrative hearings shall give notice of a hearing to be held not less than five days after service of the notice on the person requesting the hearing.

At the hearing, the applicant or registrant may submit further evidence in support of the application for initial registration or to continue to hold a registration certificate as the case may be.

The department of administrative hearings shall issue a written finding stating the reasons for denial or revocation thereof and shall serve a copy of said findings upon the applicant or registrant and all parties appearing or represented at the hearing.

(d) If the applicant or registrant does not request a hearing or submit further evidence within ten days after receiving notification of the proposed denial or revocation, it shall be deemed that the applicant or registrant has conceded the validity of the reason or reasons stated in the notice and the denial or revocation shall become final.

(e) Within three days after notification of a decision unfavorable to the applicant or registrant and all time for appeal in accordance with paragraph 8-20-130(a) through (d) having expired, the applicant or registrant shall:

(1) Peaceably surrender to the Chicago Police Department the firearm for which the applicant was denied or the registration certificate was revoked; or

(2) Remove such firearm from the City of Chicago; or

(3) Otherwise lawfully dispose of his interest in such firearm.

(f) The applicant or registrant shall submit to the superintendent evidence of the disposition of nonregisterable firearms in accordance with paragraph 8-20-130(e)(2) and (3). Such evidence shall be submitted on forms and in the manner prescribed by the superintendent.

8-20-140. Additional duties of registrant. Each person holding a registration certificate shall:

(a) Immediately notify the Chicago Police Department on a form prescribed by the superintendent of:

(1) The loss, theft or destruction of the registration certificate or of a registered firearm immediately upon discovery of such loss, theft, or destruction;

(2) A change in any of the information appearing on the registration certificate;

(3) The sale, transfer or other disposition of the firearm not less than 48 hours prior to delivery.

(b) Immediately return to the superintendent his copy of the registration certificate for any firearm which is lost, stolen, destroyed or otherwise disposed of.

(c) Each registrant shall keep any firearm in his possession unloaded and disassembled or bound by a trigger lock or similar device, unless such firearm is in his possession at his place of residence or business or while being used for lawful recreational purposes within the City of Chicago; provided, this paragraph shall not apply to law enforcement personnel as defined in paragraph 8-20-030(k) and provided further this paragraph shall not apply to security personnel as defined in paragraph 8-20-030(n), while in the course of their employment.

8-20-150. Exhibition of registration. Any person carrying or having in his possession or under his custody or control any firearm, shall have on his person or within his immediate custody a valid registration certificate for such firearm issued hereunder, which shall be exhibited for inspection to any peace officer upon demand. Failure of any such person to so exhibit his registration certificate shall be presumptive evidence that he is not authorized to possess such firearm.

Failure of any person to exhibit a registration certificate for any firearm in his possession, custody or control shall also be cause for the confiscation of such firearms and revocation of any registration certificates issued therefore under this chapter.

8-20-160. Possession of ammunition. No person shall possess ammunition in the City of Chicago unless:

(a) He is a person exempted pursuant to Section 8-20-040 of this chapter; or

(b) He is the holder of a valid registration certificate for a firearm of the same gauge or caliber as the ammunition possessed, and has the registration certificate in his possession while in possession of the ammunition; or

(c) He is a licensed weapons dealer pursuant to Chapter 4-144 or a licensed shooting gallery or gun club pursuant to Chapter 4-149 of this Code.

8-20-165 Possession of laser sight accessories. No person shall sell, offer, or display for sale, give, lend, transfer ownership of, acquire or possess any laser sight accessory in the City of Chicago provided, that this section shall not apply to any members of the armed forces of the United States, or the organized militia of this or any other state, and peace officers as defined in this Code to the extent that any such person is otherwise authorized to acquire or possess a laser sight accessory and is acting within the scope of his or her duties.

8-20-170. Permissible sales and transfers of firearms and ammunition.

(a) No firearm may be sold or otherwise transferred within the City of Chicago except through a licensed weapons dealer as defined in Chapter 4-144 of the Municipal Code of the City of Chicago.

(b) No ammunition may be sold or otherwise transferred within the City of Chicago except through a licensed shooting gallery or gun club as defined in Chapter 4-149, a licensed weapons dealer as defined in Chapter 4-144 or as otherwise allowed by the Municipal Code of the City of Chicago.

(c) No firearm or ammunition shall be security for, or be taken or received by way of any mortgage, deposit, pledge or pawn.

(d) No person may loan, borrow, give or rent to or from another person, any firearm or ammunition except in accordance with this chapter.

(e) A peace officer may additionally sell or transfer any lawfully held firearm or ammunition to another peace officer in accordance with the other provisions of this chapter.

(f) Except as allowed by subsection (e) of this section, no person may sell, offer for sale, barter or transfer within the city any ammunition listed in Section 4-144-061 of this Code.

8-20-175. Firearm Owners Identification Card Act - Compliance required. No person shall sell, give away or otherwise transfer any firearm, as defined in Section 8-20-030, without complying with Sections 3 and 3.1 of the Firearm Owners Identification Card Act, Illinois Revised Statutes, Chapter 38, paragraphs 83-3 and 83-3.1.

8-20-180. False information - Forgery - Alteration.

(a) It shall be unlawful for any person purchasing any firearm or ammunition, or applying for any registration certificate under this chapter, or, in giving any information pursuant to the requirements of this chapter, to knowingly give false information or offer false evidence of identity.

(b) It shall be unlawful for anyone to forge or alter any application or registration certificate submitted, retained or issued under this chapter.

8-20-200 Renewal of registration.

(a) Every registrant must renew his registration certificate annually. Applications for renewal shall be made by such registrants 60 days prior to the expiration of the current registration certificate.

(b) The application for renewal shall include the payment of a renewal fee as follows:

1 firearm	$20.00
2 - 10 firearms	25.00
More than ten firearms	35.00

(c) Failure to comply with the requirement for renewal of registration of a firearm shall cause that firearm to become unregisterable.

(d) All terms, conditions and requirements of this chapter for registration of firearms shall be applicable to renewal or registration of such firearms.

(e) The renewal fee shall not be applicable to duty-related handguns of peace officers domiciled in the City of Chicago.

8-20-230. Authority of the superintendent. The superintendent shall have the authority to promulgate rules and regulations for the implementation of this chapter and to prescribe all forms and the information required thereon.

8-20-240. Acquisition or possession prohibited by law. Nothing in this chapter shall make lawful the acquisition or possession of firearms or firearm ammunition which is otherwise prohibited by law.

8-20-241. Firearm used illegally - Penalty. The owner of an unregistered firearm that is used in any criminal act shall be subject to a fine of $500.00 for each such use, regardless of whether the owner participated in, aided or abetted the criminal act. A fine under this section shall be in addition to any other penalty imposed on the criminal act or use of the firearm.

8-20-250. Violation - Penalty. Any person who violates any provision of this chapter, where no other penalty is specifically provided, shall upon conviction for the first time, be fined not less than $300.00, nor more than $500.00; or be incarcerated for not less than ten days nor more than 90 days or both. Any subsequent conviction for a violation of this chapter shall be punishable by a fine of $500.00 and by incarceration for a term of not less than 90 days, nor more than six months.

Chapter 8-24. Firearms and Other Weapons

8-24-020. Carrying dangerous weapons. ...

No person shall sell, manufacture, purchase, possess or carry any weapon from which eight or more shots or bullets may be discharged by a single function of the firing device. ...

No person shall carry concealed on or about his person a pistol, revolver, derringer or other firearm or dagger, dirk, stiletto, bowie knife, commando knife, any blade of which is released by a spring mechanism, including knives as known "switch-blades" or any other type or kind of knife, any blade of which is more than two and one-half inches in length, ordinary razor or other dangerous weapon except that no person 18 years of age or under shall carry concealed on or about his person, any knife, the blade of which is two inches in length or longer. This provision shall not apply, however, to the following officers while engaged in the discharge of their official duties: sheriffs, coroners, constables, policemen, or other duly constituted police officers and warden, superintendents and keepers of prisons, penitentiaries, jails and other institutions for the detention of persons accused or convicted of crime; nor to the following employees or agents while engaged in the discharge of duties of their employment: conductors, baggage men, messengers, drivers, watchmen, special agents and policemen employed by railroads or express companies; nor to persons lawfully summoned by an officer to assist in making arrests or preserving the peace, while so engaged in assisting such officer.

Any person violating the provisions of this section shall be fined $200.00 for each offense, or shall be punished by imprisonment for a period of six months, or by both such fine and imprisonment.

In addition to all other penalties, weapons used in violation of this section shall be forfeited to, and confiscated by, the city.

8-24-025. Assault weapons or ammunition - Sale Prohibited - Exceptions.

(a) No person shall sell, offer or display for sale, give, lend, transfer ownership of, acquire or possess any assault weapon or assault ammunition, as those terms are defined in Chapter 8-20 of this code. This section shall not apply to any officer, agent, or employee of this or any other municipality or state or of the United States, members of the armed forces of the United States, or the organized militia of this or any other state, and peace officers as defined in this code to the extent that any such person is otherwise authorized to acquire or possess an assault weapon or assault ammunition and is acting within the scope of his or her duties. In addition, this section shall not apply to the acquisition or possession of assault ammunition by persons employed to provide security for armored carriers or mobile check cashing services while in the course of such duties, while commuting directly to or from the person's place of employment, and while at the person's home, if the assault ammunition (1) is acquired or possessed for the use with a weapon that the person has been authorized to carry under Section 28 of the Illinois Private Detective, Private Alarm and Private Security Act of 1983; and (2) consists of an ammunition magazine that has a capacity of 15 or fewer rounds of ammunition.

(b) Any assault weapon or assault ammunition possessed, sold or transferred in violation of subsection (a) is hereby declared to be contraband and shall be seized and disposed of in accordance with the provisions of Section 8-20-220.

(c) Any person found in violation of this section shall be sentenced to not more than six months imprisonment or fined not less than $500.00 and not more than $1,000.00 or both.

(d) Any person who, prior to the effective date of the ordinance codified in this section, was legally in possession of an assault weapon or assault ammunition prohibited by this section shall have 14 days from the effective date of the ordinance codified in this section to do any of the following without being subject to prosecution hereunder:

(1) To remove the assault weapon or ammunition from within the limits of the city of Chicago; or

(2) To modify the assault weapon either to render it permanently inoperable or to permanently make it a device no longer defined as an assault weapon; or

(3) To surrender the assault weapon or ammunition to the superintendent of police or his designee for disposal in accordance with Section 8-20-220.

8-24-026. Fragmenting bullets and metal piercing bullets - Sale prohibited - Exceptions.

(a) No person shall manufacture, sell, offer or display for sale, give, lend, transfer ownership of, acquire or possess any fragmenting bullets, metal piercing bullets, or disc projectile ammunition. This section shall not apply to any officer, agent, or employee of this or any other municipality or state or of the United States, members of the armed forces of the United States, or the organized militia of this or any other state, and peace officers as defined in this code to the extent that any such person is otherwise authorized to acquire or possess fragmenting bullets, metal piercing bullets, or disc projectile ammunition and is acting within the scope of his or her duties.

(b) Any fragmenting bullets, metal piercing bullets, or disc projectile ammunition manufactured, possessed, sold or transferred in violation of subsection (a) are hereby declared to be contraband and shall be seized and disposed of in accordance with the provisions of Section 8-2-220.

(c) Any person found in violation of this section shall be sentenced to not more than six months imprisonment or fined $500.00, or both.

8-24-027 Disguised firearms prohibited.

(a) No person shall purchase, acquire, sell, offer or expose for sale, or possess any firearm that is designed, constructed, modified or disguised to resemble any other object.

(b) Any person who violates subsection (a) of this section shall be guilty of a misdemeanor, and shall be subject to incarceration for not less than 30 days and not more than 180 days for each offense. Each day of a continuing violation, and each purchase, acquisition, sale, offering or exposing for sale, or possession of a different firearm described in subsection (a) shall constitute a separate and distinct offense.

(c) Nothing in this section suspends, repeals or alters any other provision of this Code which limits, restricts or prohibits the purchase, acquisition, sale, offering or exposure for sale, or possession of a firearm.

8-24-060. Violation - Penalty. Any person violating any of the provisions of this chapter, where no other penalty is specifically provided, shall be fined not more than $200.00 for each offense.

[Municipal Code of Chicago codified through Council Journal of March 12, 2008]

Code of the City of Peoria

Chapter 20. Offenses and Miscellaneous Provisions

Article III. Minors

Division 1. Generally

20-51. Selling, etc., weapons to prohibited. No person shall sell, give, loan, hire, barter, furnish or offer to sell, give, loan, hire, barter or furnish, to any minor within the city, any gun, pistol, revolver, fowling piece or toy firearm, in which any explosive substance can be used; or any Bowie knife, dirk, dagger or other deadly weapon of a like character.

Article VI. Weapons

Division 2. Concealable Deadly Weapons

20-181. License to sell, etc.

(a) It shall be unlawful for any person to engage in the business of selling, or to sell or give away, any pistol, revolver, dagger, stiletto, billie, derringer, Bowie knife, dirk or other deadly weapon, which can be concealed on the person, without securing a license so to do.

(b) An application for the license required by subsection (a) of this section shall be made in conformity with the general requirements of article I of chapter 18 of this Code, relating to applications for licenses.

(c) The annual fee for a deadly weapon license shall be $20.00.

(d) In case the city manager shall determine that an applicant for a license under this section has violated any provision of this section, he shall revoke the license of such person for the selling of such weapons, and the money paid for such license shall be forfeited to the city. No other license shall be issued to such licensee for a period of three years thereafter.

20-182. Required certificate and thumbprint.

(a) It shall be unlawful for any person to sell, barter or give away to any person within the city any deadly weapon mentioned in section 20-181, except to licensed dealers, without first obtaining from the person receiving such deadly weapon a signed statement in which he states that he is over 21 years of age; that he is not under indictment for, nor has been convicted of, a crime punishable by imprisonment for a term of one year or more; that he is not a fugitive from justice; nor an unlawful user or addicted to a depressant, stimulant or narcotic drug; nor that he has been adjudicated mentally defective or has been committed to a mental institution; and which statement shall contain a thumbprint from the person receiving such weapon. Such thumbprint shall be the right-hand thumbprint unless circumstances prevent, in which case it shall be the left-hand thumbprint. All such information required by this section shall be entered upon forms provided by the superintendent of police for that purpose.

(b) Any person violating the provisions of this section shall be punished as provided in section 1-5 of this Code.

20-183. Record of weapons sold, loaned, etc., to be kept; form; inspection.

(a) Every person dealing in the weapons mentioned in section 20-181 at retail within the city shall keep a record of all such weapons sold, loaned, rented or given away by him. The record required herein shall be made at the time of the transaction, in a book kept for that purpose, and shall include the name of the person to whom such weapon is sold, loaned, rented or given; his age; date of birth; legal residence; social security number, if any; driver's license number, if any; the kind and description of the weapon; if a firearm, the make, caliber and finish thereof, together with the number or serial letter thereof, if any; his state firearm owner's identification number; the date of the sale, loaning, rental or gift and the name of the employee or other person making such sale, rental, loan or gift.

(b) Such book shall be kept open for the inspection of the police at all reasonable times during business hours.

(c) Any person violating the provisions of this section shall be punished as provided in section 1-5 of this Code.

20-184. Report of sales, etc., to superintendent of police. Every person dealing in the deadly weapons referred to in section 20-181 shall deliver daily reports to the superintendent of police. The reports shall be on forms provided by the superintendent of police and shall be a legible and correct report of every sale or gift made under authority of his license during the preceding 24 hours and shall set forth the information required in section 20-183 and the certificate required by section 20-182.

20-185. Restriction on sale. It shall be unlawful for any person to sell, barter or give away; to any person within the city; any deadly weapon mentioned in section 20-181, to any person known to him to be under 21 years of age or of unsound mind or under indictment or a drug addict or a fugitive from justice or who has been convicted of a crime of violence.

20-186. Exhibiting firearms. It shall be unlawful for any person to exhibit or display any firearm capable of being concealed on the person in any display window on the exterior of a business establishment during hours the establishment is not opened for business.

20-187. Registration.

(a) All permanent residents of the city and those persons who reside in the city continuously for more than ten days who have in their possession any pistol, revolver or gun which may be concealed on the person shall register such gun or pistol with the superintendent of police, setting forth the caliber, make, model and manufacturer's number of the weapon. No fee shall be required for such registration.

(b) For the purpose of this article only, the term "reside" shall mean either the physical presence of a person at a location for the reason of shelter or lodging, or the keeping of a dwelling by a person for the purpose of his shelter.

(c) Any person who does not register a deadly weapon as provided shall be fined not less than $50.00 nor more than as provided in section 1-5 for failure to register such revolver, gun or pistol.

[Peoria City Code codified through Ordinance No. 16295, enacted June 24, 2008]

Code of Ordinances of the City of Rockford

Chapter 5. Business Permits, Taxes and Licenses

Article III. Miscellaneous Business Regulations

Division 9. Weapons Dealers

5-489. License required. It shall be unlawful for any individual, firm, corporation, company or association to engage in the business of repairing or selling, or to repair, sell or give away to any person within the city, a new or used handgun without first having secured a license from the City of Rockford. The term "engaged in the business" means devoting time, attention and labor to engaging in such activity as a regular course of trade or business with the principal objective of profit.

5-490. Record required; inspection. It shall be the duty of every licensee hereunder to keep a permanent register of all weapons sold, repaired or given away by such licensee for a period of ten (10) years from the date of transaction, the register to be in substantially the following form:

(1) Date of transaction;

(2) Serial Number of weapon;

(3) To whom sold or given;

(4) Age and residence of purchaser;

(5) Kind and description of weapon;

(6) The number of purchaser's Firearm Owner's Identification card.

Entries in this register shall be made at the time of the sale, repair, or gift, and the register shall be kept open for inspection by any member of the police department at all reasonable times.

5-491. Report required. Any licensee making any sale, repair or gift of any weapon referred to in section 7-260 shall report such sale or gift by mailing or delivering the report to the chief of police within seven (7) calendar days thereafter, upon a form to be furnished by the city, the report to contain the following information:

(1) The date of transaction;

(2) Serial number of weapon;

(3) To whom sold or given;

(4) For whom repaired;

(5) Age and residence of purchaser;

(6) Residence of person requesting weapon to be repaired;

(7) Kind and description of weapon;

(8) The number of purchaser's firearm owner's identification card.

Chapter 19. Offenses – Miscellaneous

Article II. Weapons

19-52. Furnishing bullets, pellets, arrows, etc., to minors.

It shall be unlawful for any person to sell, give or deliver any ammunition, ball, bullet, pellet, steel-tipped arrow or other missile or projectile to any person under the age of 18 years and it shall be unlawful for any person under the age of 18 years to obtain any ammunition, ball, bullet, pellet, steel-tipped arrow or other missile or projectile by purchase, sale or gift, or in any other manner.

23-26. Sale of weapons.

Whoever being licensed under the provisions of this article shall sell or offer for sale at retail any pistols, revolvers, derringers, bowie knives, dirks or other deadly weapons of similar character shall be required to obtain a license as required in chapter 5, article III, division 12 for dealers in weapons, and shall be governed by all the regulations and provisions of said division.

[Code of Ordinances of the City of Rockford codified through Ordinance No. 2008-104-O, enacted May 27, 2008]

Code of Ordinances of the City of Springfield

Title IX. General Regulations

Chapter 98. Nuisances

98.06. Chronic nuisances.

(a) Any property within the City of Springfield which becomes a chronic nuisance Property is in violation of this chapter and is subject to its remedies. Any person in charge who permits property under his or her ownership or control to be a chronic nuisance property shall be in violation of this chapter and subject to its remedies.

(b) Definitions:

(1) Chronic nuisance property. Chronic nuisance property is property upon which two or more of the following criminal activities have occurred during any 60-day period, as a result of any two separate factual events that have been independently investigated by a law enforcement agency:

a. Unlawful use of weapons, 720 ILCS 5/24-1;

b. Unlawful use or possession of weapons by felons or persons in the custody of the department of corrections facilities, 720 ILCS 5/24-1.1;

c. Aggravated discharge of a firearm, 720 ILCS 5/24-1.2;

d. Reckless discharge of a firearm, 720 ILCS 5/24-1.5;

e. Unlawful sale of firearms, 720 ILCS 5/24- 3;

f. Unlawful possession of firearms and firearm ammunition, 720 ILCS 5/24-3.1;

g. Manufacture or delivery of a controlled substance, 720 ILCS 570/401;

h. Controlled Substance Act, 720 ILCS 570/401; Controlled substance trafficking, 720 ILCS 570/401.1, chemical breakdown of illicit controlled substance, 720 ILCS 570/401.5; Possession unauthorized by this act, 720 ILCS 570/402; look alike substances; manufacture, distribution, advertisement or possession; 720 ILCS 570/404; calculated criminal drug conspiracy, 720 ILCS 570/405; criminal drug conspiracy, 720 ILCS 570/405.1; street gang criminal drug conspiracy, 720 ILCS 570/405.2; criminal synthetic drug manufacturing conspiracy, 720 ILCS 405.3; miscellaneous violations, 720 ILCS 570/406; permitting unlawful use of a building, 720 ILCS 570/406.1; delivery of controlled, counterfeit or look alike substances; 720 ILCS/407;

i. Cannabis Control Act, 720 ILCS 550/1; possession of cannabis, 720 ILCS 550/4; manufacture or delivery of cannabis, 720 ILCS 550/5; cannabis trafficking, 720 ILCS 550/5.1; delivery of cannabis on school grounds, 720 ILCS 550/5.2; casual delivery of cannabis as possession, 720 ILCS 550/6; persons under 18 years of age; delivery; enhancement of penalty, 720 ILCS 550/7; unauthorized production or possession of sativa plant; punishment, 720 ILCS 550/8; calculated criminal cannabis conspiracy; 720 ILCS 550/9;

j. Use of Intoxicating Compounds Act, 720 ILCS 690/.01; use prohibited, 720 ILCS 690/1; sale or delivery of intoxicating compounds, 720 ILCS 690/2;

k. Prostitution, 720 ILCS 5/11-14 or Section 133.04 of the Springfield City Code, 1988, as amended;

l. Solicitation of a sexual act, 720 ILCS 5/11-14.1;

m. Soliciting for a prostitute, 720 ILCS 5/11-15 or Section 133.05 of the Springfield City Code, 1988, as amended;

n. Soliciting for a juvenile prostitute, 720 ILCS 5/11-15.1;

o. Pandering, 720 ILCS 5/11-16;

p. Keeping a place of prostitution, 720 ILCS 5/11-17 or Section 133.06 of the Springfield City Code, 1988, as amended;

q. Keeping a place of juvenile prostitution,720 ILCS 5/11-17.1;

r. Patronizing a prostitute, 720 ILCS 5/11-18 or Section 133.07 of the Springfield City Code, 1988, as amended;

s. Patronizing a juvenile prostitute, 720 ILCS 5/11-18.1;

t. Pimping, 720 ILCS 5/11-19 or Section 133.08 of the Springfield City Code, 1988, as amended;

u. Juvenile pimping, 720 ILCS 5/11-19.1;

v. Disorderly conduct, 720 ILCS 5/26-1;

w. Mob action, 720 ILCS 5/25-1;

x. Public indecency, 720 ILCS 5/11-9;

y. Assault 720 ILCS 5/12-1; aggravated assault, 720 ILCS 5/12-2;

z. Battery, 720 ILCS 5/12-3; battery of an unborn child, 720 ILCS 5/12-3.1; domestic battery, 720 ILCS 5/12-3.2; aggravated battery, 720 ILCS 5/12-4; heinous battery, 720 ILCS 5/12-4.1; aggravated battery with a firearm 720 ILCS 5/12-4.2; aggravated battery of a child, 720 ILCS 5/12-4.3; aggravated battery of an unborn child, 720 ILCS 5/12-4.4; aggravated battery of a senior citizen, 720 ILCS 5/12-4.6;

aa. Unlawful possession or consumption of alcohol by a minor, 235 ILCS 5/6-20;

bb. Garbage, rubbish, brush, weeds, and solid waste, section 170.17.01 of the Springfield City Code, 1988, as amended; receptacles to be provided, section 170.17.02 of the Springfield City Code, 1988, as amended; or solid waste to be placed in receptacles, section 170.17.03 of the Springfield City Code, 1988, as amended.

cc. Accumulation of rubbish or garbage, disposal of rubbish, rubbish storage facilities, disposal of garbage, garbage facilities, containers, all as per Section 307 of the International Property Maintenance Code as adopted by chapter 170.

(2) Control. The ability to regulate, restrain, dominate, counteract or govern conduct that occurs on property.

(3) Owner. Any person, agent, firm or corporation having any legal or equitable interest in the property. Owner includes, but is not limited to a mortgagee in possession in whom is vested all or part of the legal title to the property or all or part of the beneficial ownership and the right to the present use and enjoyment of the premises; or an occupant who can control what occurs on the property.

(4) Permit. To suffer, allow, consent to, acquiesce by failure to prevent, or expressly ascent or agree to the doing of an act.

(5) Person. Any natural person, association, partnership or corporation capable of owning or using property in the City of Springfield.

(6) Person in charge. Any person in actual or constructive possession of a property, including but not limited to an owner, occupant of property under his or her domain, ownership or control.

(7) Property. Any real property, including land which is affixed, incidental or pertinent to the land, including but not limited to any premises, room, house, building, or structure or any separate part or portion thereof.

(c) Commencement of action. When the chief of police receives a report documenting the occurrence of a second nuisance activity at or within a property and determines that the property has become a chronic nuisance property, the chief of police shall:

(1) Notify the person in charge in writing that the property has been determined to be a chronic nuisance property. The notice shall contain the following information: the street address or legal description sufficient for identification of the property; a statement that the chief of police has determined the property to be chronic nuisance property with a concise description of the nuisance activities leading to his/her findings; and a demand that the person in charge respond within ten days to the chief of police and propose a course of action that the chief of police agrees will abate the nuisance activities giving rise to the violation.

(2) Service shall be made either personally or by first class mail, postage pre-paid, return receipt requested, addressed to the person in charge at the address of the property believed to be a chronic nuisance property, or such other place which is likely to give the person in charge notice of the determination by the chief of police.

(3) A copy of the notice shall be served on the owner at such address as shown on the tax rolls of the county in which the property is located, and/or the occupant, at the address of the property, if these persons are different than the person in charge, and shall be made either personally or by first class mail, postage pre-paid.

(4) A copy of the notice shall also be posted at the property after ten days has elapsed from the service or mailing of the notice to the person in charge if the person in charge has not contacted the chief of police.

(5) The failure of any person to receive notice that the property may be a chronic nuisance property shall not invalidate or otherwise affect the proceedings under this chapter.

(6) After the notification, but prior to the commencement of legal proceedings by the city pursuant to this chapter, a person in charge stipulates with the chief of police that the person in charge will pursue a course of action the parties agree will abate the nuisance activities giving rise to the violation, the chief of police may agree to postpone legal proceedings for a period of not less than ten nor more than 30 days. If the agreed course of action does not result in the abatement of the nuisance activity or if no agreement concerning abatement is reached within 30 days, the chief of police shall request authorization for the corporation counsel to commence a legal proceeding to abate the nuisance.

(7) Concurrent with the notification procedures set forth herein, the chief of police shall send copies of the notice, as well as, any other documentation which supports legal proceedings to the corporation counsel.

(8) When a person in charge makes a response to the chief of police as required above, any conduct or statements made in connection with the furnishing of that response shall not constitute an admission that any nuisance activities have or are occurring. This subsection does not require the exclusion of any evidence which is otherwise admissible or offered for any other purpose.

(d) Abatement of nuisance. The corporation counsel of the City of Springfield, Illinois, may commence an action to abate a public nuisance as described above. Upon being satisfied by affidavits or other sworn evidence that an alleged public nuisance exists, the court may without notice or bond enter a temporary restraining order or a preliminary injunction to enjoin any defendant from maintaining such nuisance and may enter an order restraining any defendant from removing or interfering with all property used in connection with the public nuisance.

(e) Burden of proof.

(1) In an action seeking closure of a chronic nuisance property, the city shall have the initial burden of showing by a preponderance of the evidence that the property is a chronic nuisance property.

(2) It is a defense to an action seeking the closure of chronic nuisance property that the owner of the property at the time in question could not in the exercise of reasonable care or diligence, determine that the property had become a public nuisance property, or could not, in spite of the exercise of reasonable care and diligence, control the conduct leading to the findings that the property is a chronic nuisance property.

(3) In establishing the amount of any civil penalty requested, the court may consider any of the following factors and shall state those found applicable:

a. The actions or lack of action taken by the person in charge to mitigate or correct the problem at the property;

b. Whether the problem at the property was repeated or continuous;

c. The magnitude or gravity of the problem;

d. The cooperation of the person in charge with the city; or

e. The cost of the city investigating and correcting or attempting to correct the condition.

(f) Remedies.

(1) In the event a court determines property to be a chronic nuisance property, the court may order that the property be closed and secured against all use and occupancy for a period of not less than 30 days, but not more than 180, or the court may employ any other remedy deemed by it to be appropriate to abate the nuisance.

(2) In addition to the remedy provided in subsection (1) above, the court may impose upon the owner of the property a civil penalty in the amount of up to $100 per day, payable to the City of Springfield, for each day the owner had actual knowledge that the property was a public nuisance property and permitted the property to remain a chronic nuisance property.

(3) In determining what remedy or remedies shall employ, the court may consider evidence of other conduct which has occurred on the property, including, but not limited to:

a. The disturbance of neighbors.

b. The recurrence of loud and obnoxious noises.

(g) Emergency closing procedures.

(1) In the event that it is determined that the property is an immediate threat to the public safety and welfare, the city may apply to the court for such interim relief, as is deemed by the chief of police to be appropriate. In such an event, the notification provision set forth in subsection (c) above need not be complied with, however, the city shall make a diligent effort to notify the person in charge prior to a court hearing.

(2) In the event that the court finds the property constitutes a chronic nuisance property as defined in this section, the court may order the remedy set out above. In addition, in the event that it also finds the person in charge had knowledge of activities or conditions of the property constituting or violating this chapter and permitted the activities to occur, the court may assess a civil fine as provided above.

(3) The court may authorize the City of Springfield to physically secure the property against use or occupancy in the event the owner fails to do so within the time specified by the court. In the event that the city is authorized to secure the property, all reasonable costs incurred by the city to affect a closure shall be made and assessed as a lien against the property. If used herein, "costs" means these costs actually incurred by the city for the physical securing of the property, as well as, tenant relocation costs.

(4) The City of Springfield Office of Public Health affecting the closure shall prepare a statement of cost and the City of Springfield shall thereafter submit said statement to the court for its review. If no objection of the statement is made within the period described by the court, a lien in said amount may be recorded against said property.

(5) Any person who is assessed the cost of closure and/or civil penalty by the court shall be personally liable for the payment thereof by the city.

(6) A tenant is entitled to their reasonable relocation costs, as those are determined by the court if, without actual notice, the tenant moved into the property, after either:

a. The owner or tenant received notice as described herein of the police chief's determination as described above; or

b. Unknown owner or other agent received notice of an action brought pursuant to this section.

[Springfield City Code codified through Ordinance No. 363-06-08, adopted June 3, 2008]

INDIANA

Ind. Code

Title 35. Criminal Law and Procedure

Article 47. Weapons and Instruments of Violence

Chapter 1. Definitions

35-47-1-1. Applicability of definitions in chapter. The definitions in this chapter apply throughout this article.

35-47-1-3. "Dealer". "Dealer" means any person who holds himself out as a buyer and seller of handguns on a regular and continuing basis.

35-47-1-5. "Firearm". "Firearm" means any weapon:

(1) that is:

(A) capable of expelling; or

(B) designed to expel; or

(2) that may readily be converted to expel;

a projectile by means of an explosion.

35-47-1-5.5. "Gun show". "Gun show" has the meaning set forth in 27 C.F.R. 478.100.

35-47-1-6. "Handgun". "Handgun" means any firearm:

(1) designed or adapted so as to be aimed and fired from one (1) hand, regardless of barrel length; or

(2) any firearm with:

(A) a barrel less than sixteen (16) inches in length; or

(B) an overall length of less than twenty-six (26) inches.

35-47-1-7. "Proper person". "Proper person" means a person who:

(1) does not have a conviction for resisting law enforcement under IC 35-44-3-3 within five (5) years before the person applies for a license or permit under this chapter;

(2) does not have a conviction for a crime for which the person could have been sentenced for more than one (1) year;

(3) does not have a conviction for a crime of domestic violence (as defined in IC 35-41-1-6.3), unless a court has restored the person's right to possess a firearm under IC 35-47-4-7;

(4) is not prohibited by a court order from possessing a handgun;

(5) does not have a record of being an alcohol or drug abuser as defined in this chapter;

(6) does not have documented evidence which would give rise to a reasonable belief that he has a propensity for violent or emotionally unstable conduct;

(7) does not make a false statement of material fact on the person's application;

(8) does not have a conviction for any crime involving an inability to safely handle a handgun;

(9) does not have a conviction for violation of the pro-visions of this article within five (5) years of his application; or

(10) does not have an adjudication as a delinquent child for an act that would be a felony if committed by an adult, if the person applying for a license or permit under this chapter is less than twenty-three (23) years of age.

35-47-1-8. "Proper reason". "Proper reason" means for the defense of oneself or the state of Indiana.

35-47-1-9. "Retail". "Retail" means the sale of handguns singly or in small quantities to one who intends to be the ultimate user thereof.

35-47-1-10. "Sawed-off shotgun". "Sawed-off shotgun" means:

(1) a shotgun having one (1) or more barrels less than eighteen (18) inches in length; and

(2) any weapon made from a shotgun (whether by alteration, modification, or otherwise) if the weapon as modified has an overall length of less than twenty-six (26) inches.

35-47-1-11. "Shotgun". "Shotgun" means a weapon designed or redesigned, made or remade, and intended to be fired from the shoulder and designed or redesigned and made or remade to use the energy of the explosive in a fixed shotgun shell to fire through a smooth bore either a number of ball shot or a single projectile for each single pull of the trigger.

35-47-1-12. "Superintendent". "Superintendent" refers to the superintendent of the Indiana state police department.

35-47-1-13. "Wholesale". "Wholesale" means the sale of handguns singly or in bulk lots to one lawfully licensed to deal in handguns, or the sale of a handgun to a governmental law enforcement agency for issue to its employees.

Chapter 2. Regulation of Handguns

35-47-2-1. License required to carry handgun.

(a) Except as provided in subsection (b) and section 2 of this chapter, a person shall not carry a handgun in any vehicle or on or about the person's body, except in the person's dwelling, on the person's property or fixed place of business, without a license issued under this chapter being in the person's possession.

(b) Unless the person's right to possess a firearm has been restored under IC 35-47-4-7, a person who has been convicted of domestic battery under IC 35-42-2-1.3 may not possess or carry a handgun in any vehicle or on or about the person's body in the person's dwelling or on the person's property or fixed place of business.

35-47-2-2. Persons excepted. Section 1 of this chapter does not apply to:

(1) marshals;

(2) sheriffs;

(3) the commissioner of the department of correction or persons authorized by him in writing to carry firearms;

(4) judicial officers;

(5) law enforcement officers;

(6) members of the armed forces of the United States or of the national guard or organized reserves while they are on duty;

(7) regularly enrolled members of any organization duly authorized to purchase or receive such weapons from the United States or from this state who are at or are going to or from their place of assembly or target practice;

(8) employees of the United States duly authorized to carry handguns;

(9) employees of express companies when engaged in company business;

(10) any person engaged in the business of manufacturing, repairing, or dealing in firearms or the agent or representative of any such person having in his possession, using, or carrying a handgun in the usual or ordinary course of that business; or

(11) any person while carrying a handgun unloaded and in a secure wrapper from the place of purchase to his dwelling or fixed place of business, or to a place of repair or back to his dwelling or fixed place of business, or in moving from one dwelling or business to another.

35-47-2-3. Issuance of licenses.

(a) A person desiring a license to carry a handgun shall apply:

(1) to the chief of police or corresponding law enforcement officer of the municipality in which the applicant resides;

(2) if that municipality has no such officer, or if the applicant does not reside in a municipality, to the sheriff of the county in which the applicant resides after the applicant has obtained an application form prescribed by the superintendent; or

(3) if the applicant is a resident of another state and has a regular place of business or employment in Indiana, to the sheriff of the county in which the applicant has a regular place of business or employment.

The superintendent and local law enforcement agencies shall allow an applicant desiring to obtain or renew a license to carry a handgun to submit an application electronically under this chapter if funds are available to establish and maintain an electronic application system.

(b) The law enforcement agency which accepts an application for a handgun license shall collect the following application fees:

(1) From a person applying for a four (4) year handgun license, a ten dollar ($10) application fee, five dollars ($5) of which shall be refunded if the license is not issued.

(2) From a person applying for a lifetime handgun license who does not currently possess a valid Indiana handgun license, a fifty dollar ($50) application fee, thirty dollars ($30) of which shall be refunded if the license is not issued.

(3) From a person applying for a lifetime handgun license who currently possesses a valid Indiana handgun license, a forty dollar ($40) application fee, thirty dollars ($30) of which shall be refunded if the license is not issued.

Except as provided in subsection (h), the fee shall be deposited into the law enforcement agency's firearms training fund or other appropriate training activities fund and used by the agency to train law enforcement officers in the proper use of firearms or in other law enforcement duties, or to purchase firearms or firearm related equipment, or both for the law enforcement officers employed by the law enforcement agency. The state board of accounts shall establish rules for the proper accounting and expenditure of funds collected under this subsection.

(c) The officer to whom the application is made shall ascertain the applicant's name, full address, length of residence in the community, whether the applicant's residence is located within the limits of any city or town, the applicant's occupation, place of business or employment, criminal record, if any, and convictions (minor traffic offenses excepted), age, race, sex, nationality, date of birth, citizenship, height, weight, build, color of hair, color of eyes, scars and marks, whether the applicant has previously held an Indiana license to carry a handgun and, if so, the serial number of the license and year issued, whether the applicant's license has ever been suspended or revoked, and if so, the year and reason for the suspension or revocation, and the applicant's reason for desiring a license. The officer to whom the application is made shall conduct an investigation into the applicant's official records and verify thereby the applicant's character and reputation, and shall in addition verify for accuracy the information contained in the application, and shall forward this information together with the officer's recommendation for approval or disapproval and one (1) set of legible and classifiable fingerprints of the applicant to the superintendent.

(d) The superintendent may make whatever further investigation the superintendent deems necessary. Whenever disapproval is recommended, the officer to whom the application is made shall provide the superintendent and the applicant with the officer's complete and specific reasons, in writing, for the recommendation of disapproval.

(e) If it appears to the superintendent that the applicant:

(1) has a proper reason for carrying a handgun;

(2) is of good character and reputation;

(3) is a proper person to be licensed; and

(4) is:

(A) a citizen of the United States; or

(B) not a citizen of the United States but is allowed to carry a firearm in the United States under federal law;

the superintendent shall issue to the applicant a qualified or an unlimited license to carry any handgun lawfully possessed by the applicant. The original license shall be delivered to the licensee. A copy shall be delivered to the officer to whom the application for license was made. A copy shall be retained by the superintendent for at least four (4) years in the case of a four (4) year license. The superintendent may adopt guidelines to establish a records retention policy for a lifetime license. A four (4) year license shall be valid for a period of four (4) years from the date of issue. A lifetime license is valid for the life of the individual receiving the license. The license of police officers, sheriffs or their deputies, and law enforcement officers of the United States government who have been honorably retired by a lawfully created pension board or its equivalent after twenty (20) or more years of service, shall be valid for the life of these individuals. However, a lifetime license is automatically revoked if the license holder does not remain a proper person.

(f) At the time a license is issued and delivered to a licensee under subsection (e), the superintendent shall include with the license information concerning handgun safety rules that:

(1) neither opposes nor supports an individual's right to bear arms; and

(2) is:

(A) recommended by a nonprofit educational organization that is dedicated to providing education on safe handling and use of firearms;

(B) prepared by the state police department; and

(C) approved by the superintendent.

The superintendent may not deny a license under this section because the information required under this subsection is unavailable at the time the superintendent would otherwise issue a license. The state police department may accept private donations or grants to defray the cost of printing and mailing the information required under this subsection.

(g) A license to carry a handgun shall not be issued to any person who:

(1) has been convicted of a felony;

(2) has had a license to carry a handgun suspended, unless the person's license has been reinstated;

(3) is under eighteen (18) years of age;

(4) is under twenty-three (23) years of age if the person has been adjudicated a delinquent child for an act that would be a felony if committed by an adult; or

(5) has been arrested for a Class A or Class B felony, or any other felony that was committed while armed with a deadly weapon or that involved the use of violence, if a court has found probable cause to believe that the person committed the offense charged.

In the case of an arrest under subdivision (5), a license to carry a handgun may be issued to a person who has been acquitted of the specific offense charged or if the charges for the specific offense are dismissed. The superintendent shall prescribe all forms to be used in connection with the administration of this chapter.

(h) If the law enforcement agency that charges a fee under subsection (b) is a city or town law enforcement agency, the fee shall be deposited in the law enforcement continuing education fund established under IC 5-2-8-2.

(i) If a person who holds a valid license to carry a handgun issued under this chapter:

(1) changes the person's name;

(2) changes the person's address; or

(3) experiences a change, including an arrest or a conviction, that may affect the person's status as a proper person (as defined in IC 35-47-1-7) or otherwise disqualify the person from holding a license;

the person shall, not later than thirty (30) days after the date of a change described under subdivision (3), and not later than sixty (60) days after the date of the change described under subdivision (1) or (2), notify the superintendent, in writing, of the event described under subdivision (3) or, in the case of a change under subdivision (1) or (2), the person's new name or new address.

(j) The state police shall indicate on the form for a license to carry a handgun the notification requirements of subsection (i).

(k) The state police department shall adopt rules under IC 4-22-2 to implement an electronic application system under subsection (a). Rules adopted under this section must require the superintendent to keep on file one (1) set of classifiable and legible fingerprints from every person who has received a license to carry a handgun so that a person who applies to renew a license will not be required to submit an additional set of fingerprints.

First of 2 versions of this section

35-47-2-4. Qualified licenses – Unlimited licenses [as amended by P.L.1-2007].

(a) Licenses to carry handguns shall be either qualified or unlimited, and are valid for:

(1) four (4) years from the date of issue in the case of a four (4) year license; or

(2) the life of the individual receiving the license in the case of a lifetime license.

A qualified license shall be issued for hunting and target practice. The superintendent may adopt rules imposing limitations on the use and carrying of handguns under a license when handguns are carried by a licensee as a condition of employment. Unlimited licenses shall be issued for the purpose of the protection of life and property.

(b) In addition to the application fee, the fee for:

(1) a qualified license shall be:

(A) five dollars ($5) for a four (4) year qualified license;

(B) twenty-five dollars ($25) for a lifetime qualified license from a person who does not currently possess a valid Indiana handgun license; or

(C) twenty dollars ($20) for a lifetime qualified license from a person who currently possesses a valid Indiana handgun license; and

(2) an unlimited license shall be:

(A) thirty dollars ($30) for a four (4) year unlimited license;

(B) seventy-five dollars ($75) for a lifetime unlimited license from a person who does not currently possess a valid Indiana handgun license; or

(C) sixty dollars ($60) for a lifetime unlimited license from a person who currently possesses a valid Indiana handgun license.

The superintendent shall charge a twenty dollar ($20) fee for the issuance of a duplicate license to replace a lost or damaged license. These fees shall be deposited in accordance with subsection (e).

(c) Licensed dealers are exempt from the payment of fees specified in subsection (b) for a qualified license or an unlimited license.

(d) The following officers of this state or the United States who have been honorably retired by a lawfully created pension board or its equivalent after at least twenty (20) years of service or because of a disability are exempt from the payment of fees specified in subsection (b):

(1) Police officers.

(2) Sheriffs or their deputies.

(3) Law enforcement officers.

(4) Correctional officers.

(e) Fees collected under this section shall be deposited in the state general fund.

Second of 2 versions of this section

35-47-2-4. Qualified licenses – Unlimited licenses [as amended by P.L.155-2007].

(a) Licenses to carry handguns shall be either qualified or unlimited, and are valid for:

(1) four (4) years from the date of issue in the case of a four (4) year license; or

(2) the life of the individual receiving the license in the case of a lifetime license.

A qualified license shall be issued for hunting and target practice. The superintendent may adopt rules imposing limitations on the use and carrying of handguns under a license when handguns are carried by a licensee as a condition of employment. Unlimited licenses shall be issued for the purpose of the protection of life and property.

(b) In addition to the application fee, the fee for:

(1) a qualified license shall be:

(A) five dollars ($5) for a four (4) year qualified license;

(B) twenty-five dollars ($25) for a lifetime qualified license from a person who does not currently possess a valid Indiana handgun license; or

(C) twenty dollars ($20) for a lifetime qualified license from a person who currently possesses a valid Indiana handgun license; and

(2) an unlimited license shall be:

(A) thirty dollars ($30) for a four (4) year unlimited license;

(B) seventy-five dollars ($75) for a lifetime unlimited license from a person who does not currently possess a valid Indiana handgun license; or

(C) sixty dollars ($60) for a lifetime unlimited license from a person who currently possesses a valid Indiana handgun license.

The superintendent shall charge a twenty dollar ($20) fee for the issuance of a duplicate license to replace a lost or damaged license. These fees shall be deposited in accordance with subsection (e).

(c) Licensed dealers are exempt from the payment of fees specified in subsection (b) for a qualified license or an unlimited license.

(d) The following officers of this state or the United States who have been honorably retired by a lawfully created pension board or its equivalent after at least twenty (20) years of service or because of a disability are exempt from the payment of fees specified in subsection (b):

(1) Police officers.

(2) Sheriffs or their deputies.

(3) Law enforcement officers.

(4) Correctional officers.

(e) Fees collected under this section shall be deposited in the state general fund.

(f) The superintendent may not issue a lifetime qualified license or a lifetime unlimited license to a person who is a resident of another state. The superintendent may issue a four (4) year qualified license or a four (4) year unlimited license to a person who is a resident of another state and who has a regular place of business or employment in Indiana as described in section 3(a)(3) [IC 35-47-2-3(a)(3)] of this chapter.

35-47-2-5. Suspension or revocation of licenses.

(a) The superintendent may suspend or revoke any license issued under this chapter if he has reasonable grounds to believe that the person's license should be suspended or revoked.

(b) Documented evidence that a person is not a "proper person" to be licensed as defined by IC 35-47-1-7, or is prohibited under section 3(g)(5) [IC 35-47-2-3(g)(5)] of this chapter from being issued a license, shall be grounds for immediate suspension or revocation of a license previously issued under this chapter. However, if a license is suspended or revoked based solely on an arrest under section 3(g)(5) of this chapter, the license shall be reinstated upon the acquittal of the defendant in that case or upon the dismissal of the charges for the specific offense.

(c) A person who fails to promptly return his license after written notice of suspension or revocation commits a Class A misdemeanor. The observation of a handgun license in the possession of a person whose license has been suspended or revoked constitutes a sufficient basis for the arrest of that person for violation of this subsection.

(d) The superintendent shall establish rules under IC 4-22-2 concerning the procedure for suspending or revoking a person's license.

35-47-2-6. Time for review of applications.

(a) Every initial application for any license under this chapter shall be granted or rejected within sixty (60) days after the application is filed.

(b) The period during which an application for the renewal of an existing license may be filed begins one hundred eighty (180) days before the expiration of the existing license. If the application for renewal of an existing license is filed within thirty (30) days of its expiration, the existing license is automatically extended until the application for renewal is passed upon.

35-47-2-7. Prohibited sales or transfers.

(a) Except an individual acting within a parent-minor child or guardian-minor protected person relationship or any other individual who is also acting in compliance with IC 35-47-10, a person may not sell, give, or in any other manner transfer the ownership or possession of a handgun or assault weapon (as defined in IC 35-50-2-11) to any person under eighteen (18) years of age.

(b) It is unlawful for a person to sell, give, or in any manner transfer the ownership or possession of a handgun to another person who the person has reasonable cause to believe:

(1) has been:

(A) convicted of a felony; or

(B) adjudicated a delinquent child for an act that would be a felony if committed by an adult, if the person seeking to obtain ownership or possession of the handgun is less than twenty-three (23) years of age;

(2) is a drug abuser;

(3) is an alcohol abuser; or

(4) is mentally incompetent.

35-47-2-8. Persons to whom sale regulations apply. The regulation of the sale of handguns imposed by this chapter shall apply equally to an occasional sale, trade, or transfer between individual per-sons and to retail transactions between dealers and individual persons.

35-47-2-14. License requirements for retail dealer - Failure to display license. A retail dealer who:

(1) sells;

(2) trades;

(3) transfers;

(4) exposes for sale, trade, or transfer; or

(5) possesses with intent to sell, trade, or transfer; any handgun without being licensed under sections 15 and 16 [IC 35-47-2-15 and IC 35-47-2-16] of this chapter and without displaying his license at all times commits a Class B misdemeanor.

35-47-2-15. Issuance of retail handgun dealer's license - Investigation and fingerprinting.

(a) A person desiring a retail handgun dealer's license shall apply to the sheriff of the county in which he resides, or if he is a resident of another state and has a regular place of business in Indiana, then to the sheriff of the county in which he has a regular place of business.

The applicant shall state his name, full address, occupation, sex, race, age, place of birth, date of birth, nationality, height, weight, build, color of eyes, color of hair, complexion, scars and marks, and any criminal record (minor traffic offenses excepted). The officer to whom the application is made shall verify the application and search his records concerning the applicant's character and reputation.

(b) The officer to whom the application is made shall send to the superintendent:

(1) the verified application;

(2) the results of the officer's investigation; and

(3) the officer's recommendation for approval or disapproval of the application; in as many copies as the superintendent shall designate, and one (1) set of legible and classifiable fingerprints of the applicant. The superintendent may make whatever further investigation he deems necessary. Whenever disapproval is recommended by the officer to whom the application was made, he shall provide the superintendent and the applicant with his complete reasons for the disapproval in writing. If the officer to whom the application is made recommends approval, he shall instruct the applicant in the proper method of taking legible and classifiable fingerprints. If it appears to the superintendent that the applicant is of good character and reputation and a proper person to be licensed, he shall issue to the applicant a retail handgun dealer's license which shall be valid for a period of two (2) years from the date of issue. The fee for the license shall be twenty dollars ($20), which shall be deposited with the officer to whom the application is made, who shall in turn forward it to the superintendent for deposit with the treasurer of state when the application is approved by the superintendent. In the event that the application is disapproved by the superintendent, the fee shall be returned to the applicant along with the complete reasons, in writing, for the disapproval.

(c) No retail dealer's license shall be issued to any person who has been:

(1) convicted of a felony; or

(2) adjudicated a delinquent child for an act that would be a felony if committed by an adult, if the person applying for the retail dealer's license is less than twenty-three (23) years of age; in Indiana or any other state or country.

(d) A retail dealer's license shall permit the licensee to sell handguns at retail within this state subject to the conditions specified in this chapter. The license may be suspended or revoked in accordance with applicable law, and the licensee may be subject to punishment as provided in this chapter.

35-47-2-16. License-designated business site - Display of license - Restrictions on sale of handguns - Display, sale or transfer at gun shows.

(a) A retail dealer's business shall be carried on only in the site designated in the license. A separate license shall be required for each separate retail outlet. Whenever a licensed dealer moves his place of business, he shall promptly notify the superintendent, who shall at once issue an amended license certificate valid for the balance of the license period. This subsection does not apply to sales at wholesale.

(b) The license, certified by the issuing authority, shall be displayed on the business premises in a prominent place where it can be seen easily by prospective customers.

(c) No handgun shall be sold:

(1) in violation of any provision of this chapter; or

(2) under any circumstances unless the purchaser is personally known to the seller or presents clear evidence of his identity.

(d) Notwithstanding subsection (a), a retail dealer may display, sell, or transfer handguns at a gun show in accordance with this chapter and federal law.

35-47-2-17. False information - Confiscation and disposal of handgun. No person, in purchasing or otherwise securing delivery of a handgun or in applying for a license to carry a handgun, shall give false in-formation or offer false evidence of identity. In addition to any penalty provided by this chapter, any handgun obtained through false information shall be subject to confiscation and disposition as provided in this chapter. Upon notice of a violation of this section by the superintendent, it shall be the duty of the sheriff or chief of police or corresponding officer of the jurisdiction in which the purchaser resides to confiscate the firearm and retain it as evidence pending trial for the offense.

35-47-2-18. Alteration, removal or obliteration of identifying marks prohibited - Possession of such handguns prohibited. No person shall:

(1) change, alter, remove, or obliterate the name of the maker, model, manufacturer's serial number, or other mark of identification on any handgun; or

(2) possess any handgun on which the name of the maker, model, manufacturer's serial number, or other mark of identification has been changed, altered, removed, or obliterated; except as provided by applicable United States statute.

35-47-2-19. Firearms exempted. This chapter does not apply to any firearm not designed to use fixed cartridges or fixed ammunition, or any firearm made before January 1, 1899.

35-47-2-20. Effect of full or conditional pardon.

(a) A full pardon from the governor of Indiana for:

(1) a felony other than a felony that is included in IC 35-42; or

(2) a violation of this chapter; removes any disability under this chapter imposed because of that offense, if fifteen (15) years have elapsed between the time of the offense and the application for a license under this chapter.

(b) A conditional pardon described in IC 11-9-2-4 for:

(1) a felony; or

(2) a violation of this chapter; removes a disability under this chapter if the superintendent determines after an investigation that circumstances have changed since the pardoned conviction was entered to such an extent that the pardoned person is likely to handle handguns in compliance with the law.

35-47-2-21. Foreign licenses.

(a) Retail dealers' licenses issued by other states or foreign countries will not be recognized in Indiana except for sales at wholesale.

(b) Licenses to carry handguns, issued by other states or foreign countries, will be recognized according to the terms thereof but only while the holders are not residents of Indiana.

35-47-2-22. Use of false or altered handgun license unlawful. . It is unlawful for any person to use, or to attempt to use, a false, counterfeit, spurious, or altered handgun-carrying license to obtain a handgun contrary to the provisions of this chapter.

35-47-2-23. Violation of chapter.

(a) A person who violates section 3, 4, 5, 14, 15, or 16 [IC 35-47-2-3, IC 35-47-2-4, IC 35-47-2-5, IC 35-47-2-14, IC 35-47-2-15, or IC 35-47-2-16] of this chapter commits a Class B misdemeanor.

(b) A person who violates section 7, 17, or 18 [IC 35-47-2-7, IC 35-47-2-17, or IC 35-47-2-18] of this chapter commits a Class C felony.

(c) A person who violates section 1 [IC 35-47-2-1] of this chapter commits a Class A misdemeanor. However, the offense is a Class C felony:

(1) if the offense is committed:

(A) on or in school property;

(B) within one thousand (1,000) feet of school property; or

(C) on a school bus; or

(2) if the person:

(A) has a prior conviction of any offense under:

(i) this subsection; or

(ii) subsection (d); or

(B) has been convicted of a felony within fifteen (15) years before the date of the offense.

(d) A person who violates section 22 of this chapter commits a Class A misdemeanor. However, the offense is a Class D felony if the person has a prior conviction of any offense under this subsection or subsection (c), or if the person has been convicted of a felony within fifteen (15) years before the date of the offense.

35-47-2-24. Burden of proof – When prosecution dismissed.

(a) In an information or indictment brought for the enforcement of any provision of this chapter, it is not necessary to negate any exemption specified under this chapter, or to allege the absence of a license required under this chapter. The burden of proof is on the defendant to prove that he is exempt under section 2 [IC 35-47-2-2] of this chapter, or that he has a license as required under this chapter.

(b) Whenever a person who has been arrested or charged with a violation of section 1 [IC 35-47-2-1] of this chapter presents a valid license to the prosecuting attorney or establishes that he is exempt under section 2 of this chapter, any prosecution for a violation of section 1 of this chapter shall be dismissed immediately, and all records of an arrest or proceedings following arrest shall be destroyed immediately.

Chapter 2.5. Sale of Handguns

35-47-2.5-1. Applicability.

(a) This chapter does not apply to the following:

(1) Transactions between persons who are licensed as firearms importers or collectors or firearms manufacturers or dealers under 18 U.S.C. 923.

(2) Purchases by or sales to a law enforcement officer or agent of the United States, the state, or a county or local government.

(3) Indiana residents licensed to carry handguns under IC 35-47-2-3.

(b) Notwithstanding any other provision of this chapter, the state shall participate in the NICS if federal funds are available to assist the state in participating in the NICS. If:

(1) the state participates in the NICS; and

(2) there is a conflict between:

(A) a provision of this chapter; and

(B) a procedure required under the NICS;

the procedure required under the NICS prevails over the conflicting provisions of this chapter.

35-47-2.5-2. "Dealer" defined. As used in this chapter, "dealer" includes any person licensed under 18 U.S.C. 923.

35-47-2.5-2.5. "NICS" defined.
As used in this chapter, "NICS" refers to the National Instant Criminal Background Check System maintained by the Federal Bureau of Investigation in accordance with the federal Brady Handgun Violence Prevention Act (18 U.S.C. 9321 et. seq.).

35-47-2.5-3. Completion of Form 4473.

(a) A person purchasing a handgun from a dealer shall complete and sign Bureau of Alcohol, Tobacco, Firearms and Explosives Form 4473.

35-47-2.5-4. Dealer's responsibilities.

(a) A dealer may not sell, rent, trade, or transfer from the dealer's inventory a handgun to a person until the dealer has done all of the following:

(1) Obtained from the prospective purchaser a completed and signed Form 4473, as specified in section 3 of this chapter.

(2) Contacted NICS:

(A) by telephone; or

(B) electronically;

to request a background check on the prospective purchaser.

(3) Received authorization from NICS to transfer the handgun to the prospective purchaser.

(b) The dealer shall record the NICS transaction number on Form 4473 and retain Form 4473 for auditing purposes.

35-47-2.5-5. Photographic identification required - Other documentation of residence.

(a) To establish personal identification and residence in Indiana for purposes of this chapter, a dealer must require a prospective purchaser to present one (1) photographic identification form issued by a governmental agency of the state or by the United States Department of Defense, or other documentation of residence.

(b) Except when photographic identification was issued by the United States Department of Defense, other documentation of residence must show an address identical to that shown on the photographic identification form or as amended by proper notice of change of address filed with the issuing authority. Suitable other documentation of residence includes:

(1) evidence of currently paid personal property tax or real estate tax, a current lease, utility, or telephone bill, a voter registration card, a bank check, a passport, an automobile registration, or a hunting or fishing license;

(2) other current identification allowed as evidence of residency by 27 CFR 178.124 and United States Alcohol, Tobacco, and Firearms Ruling 79-7; or

(3) other documentation of residence, determined to be acceptable by the state police department, that corroborates that the prospective purchaser currently resides in Indiana.

(c) If the photographic identification was issued by the United States Department of Defense, permanent orders may be used as documentation of residence.

35-47-2.5-12. False statement a felony. A person who knowingly or intentionally makes a materially false statement on Form 4473 completed under section 3 [IC 35-47-2.5-3] of this chapter commits a Class D felony.

35-47-2.5-13. Violation by dealer a misdemeanor. Except as otherwise provided in this chapter, a dealer who knowingly or intentionally sells, rents, trades, or transfers a handgun in violation of this chapter commits a Class A misdemeanor.

35-47-2.5-14. Exemptions for children and resale of guns.

(a) This section does not apply to a person who provides a handgun to the following:

(1) A child who is attending a hunters safety course or a firearms safety course or an adult who is supervising the child during the course.

(2) A child engaging in practice in using a firearm for target shooting at an established range or in an area where the discharge of a firearm is not prohibited or is supervised by:

(A) a qualified firearms instructor; or

(B) an adult who is supervising the child while the child is at the range.

(3) A child engaging in an organized competition involving the use of a firearm or participating in or practicing for a performance by an organized group under Section 501(c)(3) of the Internal Revenue Code that uses firearms as a part of a performance or an adult who is involved in the competition or performance.

(4) A child who is hunting or trapping under a valid license issued to the child under IC 14-22.

(5) A child who is traveling with an unloaded firearm to or from an activity described in this section.

(6) A child who:

(A) is on real property that is under the control of the child's parent, an adult family member of the child, or the child's legal guardian; and

(B) has permission from the child's parent or legal guardian to possess a firearm.

(b) A person who purchases a handgun with the intent to:

(1) resell or otherwise provide the handgun to another person who the person knows or has reason to believe is ineligible for any reason to purchase or otherwise receive from a dealer a handgun; or

(2) transport the handgun out of the state to be resold or otherwise provided to another person who the transferor knows is ineligible to purchase or otherwise receive a firearm; commits a Class D felony.

(c) If the violation of this section involves a transfer of more than one (1) handgun, the offense is a Class C felony.

35-47-2.5-15. Violation of exemptions a felony.

(a) A person who is ineligible to purchase or otherwise receive or possess a handgun in Indiana who knowingly or intentionally solicits, employs, or assists any person in violating section 14 of this chapter commits a Class D felony.

(b) If the violation involves a transfer of more than one (1) handgun, the offense is a Class C felony.

Chapter 3. Disposal of Confiscated Weapons

35-47-3-1. Disposal of confiscated weapons. All firearms confiscated pursuant to statute shall, upon conviction of the person for the offense for which the confiscation was made, be disposed of in accordance with this chapter.

35-47-3-2. Firearms not required to be registered - Return to rightful owner - Disposal by law enforcement agency.

(a) This section applies only to firearms which are not required to be registered in the National Firearms Registration and Transfer Record.

(b) Firearms shall be returned to the rightful owner at once following final disposition of the cause if a return has not already occurred under the terms of IC 35-33-5. If the rightful ownership is not known the law enforcement agency holding the firearm shall make a reasonable attempt to ascertain the rightful ownership and cause the return of the firearm. However, nothing in this chapter shall be construed as requiring the return of firearms to rightful owners who have been convicted for the misuse of firearms. In such cases, the court may provide for the return of the firearm in question or order that the firearm be at once delivered:

(1) except as provided in subdivision (2), to the sheriff's department of the county in which the offense occurred; or

(2) to the city or town police force that confiscated the firearm, if:

(A) a member of the city or town police force confiscated the firearm; and

(B) the city or town has a population of more than two thousand five hundred (2,500) and less than two hundred fifty thousand (250,000).

(c) The receiving law enforcement agency shall dispose of firearms under subsection (b), at the discretion of the law enforcement agency, not more than one hundred twenty (120) days following receipt by use of any of the following procedures:

(1) Public sale of the firearms to the general public as follows:

(A) Notice of the sale shall be:

(i) posted for ten (10) days in the county courthouse in a place readily accessible to the general public; and

(ii) advertised in the principal newspaper of the county for two (2) days in an advertisement that appears in the newspaper at least five (5) days prior to the sale.

(B) Disposition of the firearm shall be by public auction in a place convenient to the general public, with disposition going to the highest bidder. However, no firearm shall be transferred to any bidder if that bidder is not lawfully eligible to receive and possess firearms according to the laws of the United States and Indiana.

(C) All handguns transferred under this subdivision shall also be transferred according to the transfer procedures set forth in this article.

(D) Money collected pursuant to the sales shall first be used to defray the necessary costs of administering this subdivision with any surplus to be:

(i) deposited into the receiving law enforcement agency's firearms training fund, if the law enforcement agency is a county law enforcement agency, or into a continuing education fund established under IC 5-2-8-2, if the law enforcement agency is a city or town law enforcement agency; and

(ii) used by the agency exclusively for the purpose of training law enforcement officers in the proper use of firearms or other law enforcement duties, if the law enforcement agency is a county law enforcement agency, or for law enforcement purposes, if the law enforcement agency is a city or town law enforcement agency.

(2) Sale of the firearms to a licensed firearms dealer as follows:

(A) Notice of the sale must be:

(i) posted for ten (10) days in the county courthouse in a place readily accessible to the general public; and

(ii) advertised in the principal newspaper of the county for two (2) days in an advertisement that appears in the newspaper at least five (5) days before the sale.

(B) Disposition of the firearm shall be by auction with disposition going to the highest bidder who is a licensed firearms dealer.

(C) Money collected from the sales shall first be used to defray the necessary costs of administering this subdivision and any surplus shall be:

(i) deposited into the receiving law enforcement agency's firearms training fund or other appropriate training activities fund; and

(ii) used by the agency exclusively for the purpose of training law enforcement officers in the proper use of firearms or other law enforcement duties.

(3) Sale or transfer of the firearms to another law enforcement agency.

(4) Release to the state police department laboratory or other forensic laboratory administered by the state or a political subdivision (as defined in IC 36-1-2-13) for the purposes of research, training, and comparison in conjunction with the forensic examination of firearms evidence.

(5) Destruction of the firearms.

(d) Notwithstanding the requirement of this section mandating disposal of firearms not more than one hundred twenty (120) days following receipt, the receiving law enforcement agency may at its discretion hold firearms it may receive until a sufficient number has accumulated to defray the costs of administering this section if a delay does not exceed one hundred eighty (180) days from the date of receipt of the first firearm in the sale lot. In any event, all confiscated firearms shall be disposed of as promptly as possible.

(e) When a firearm is delivered to the state police department laboratory or other forensic laboratory under subsection (c)(4) and the state police department laboratory or other forensic laboratory determines the laboratory has no further need for the firearm in question, the laboratory shall return the firearm to the law enforcement agency for disposal under subsection (c).

35-47-3-3. Firearms required to be registered - Return to rightful owner - Delivery to law enforcement agency.

(a) This section applies to firearms that are required to be registered in the National Firearms Registration and Transfer Record.

(b) Firearms shall be returned to the rightful owner at once following final disposition of the cause, if such return has not already occurred under the terms of IC 35-33-5, and if such owner remains lawfully entitled to possess such firearms according to applicable United States and Indiana statutes. If rightful ownership is not known, the law enforcement agency holding the firearm shall make a reasonable and diligent effort to ascertain the rightful ownership and cause the return of the firearm being held, providing the owner remains lawfully entitled to possess such firearms.

(c) Firearms that are not returnable under this section shall be at once delivered to:

(1) the sheriff's department of the county in which the offense occurred, unless subdivision (2) applies; or

(2) the city or town police force that confiscated the firearm, if:

(A) a member of the city or town police force confiscated the firearm; and

(B) the city or town has a population of more than two thousand five hundred (2,500) and less than two hundred fifty thousand (250,000); following final disposition of the cause.

(d) When firearms are sent to a law enforcement agency under subsection (c), the law enforcement agency may upon request release the firearms to the state police department laboratory or other forensic laboratory administered by the state or a political subdivision (as defined in IC 36-1-2-13) for the purposes of research, training, and comparison in conjunction with the forensic examination of firearms evidence.

(e) The receiving law enforcement agency or laboratory shall cause the registry of such firearms in the United States National Firearms Registration and Transfer Record within thirty (30) days following receipt from the court.

(f) The court may order such firearms as are not returnable destroyed, specifying the exact manner of destruction and requiring the receiving law enforcement agency or laboratory to make due return to the ordering court the time, date, method of destruction, and disposition of the remains of the destroyed firearm.

(g) No portion of this section shall be construed as requiring the receiving law enforcement agency or laboratory to retain firearms which are inoperable or unserviceable, or which the receiving law enforcement agency or laboratory may choose to transfer as public property in the ordinary course of lawful commerce and exchange.

35-47-3-4. Unlawful disposal of confiscated firearms. A person who knowingly or intentionally:

(1) delivers a confiscated firearm to a person convicted of a felony:

(A) involving use of a firearm; and

(B) which is the basis of the confiscation;

(2) delivers a confiscated firearm to another with knowledge that there is a rightful owner to whom the firearm must be returned; or

(3) fails to deliver a confiscated firearm to the sheriff's department, a city or town police force, the state police department laboratory or a forensic laboratory under this chapter, the state under IC 14-22-39-6, or for disposition after a determination that the rightful owner of the firearm cannot be ascertained or is no longer entitled to possess the confiscated firearm; commits a Class D felony.

Chapter 4. Miscellaneous Provisions

35-47-4-1. Sale or delivery of deadly weapon to intoxicated person unlawful. A person who sells, barters, gives, or delivers any deadly weapon to any per-son at the time in a state of intoxication, knowing him to be in a state of intoxication, or to any per-son who is in the habit of becoming intoxicated, and knowing him to be a person who is in the habit of becoming intoxicated, commits a Class B misdemeanor.

35-47-4-2. Making loan secured by mortgage, deposit or pledge of handgun unlawful. A person who makes a loan secured by a:

(1) mortgage;

(2) deposit; or

(3) pledge; of a handgun commits a Class B misdemeanor.

35-47-4-3. Pointing a firearm.

(a) This section does not apply to a law enforcement officer who is acting within the scope of the law enforcement officer's official duties or to a person who is justified in using reasonable force against another person under:

(1) IC 35-41-3-2; or

(2) IC 35-41-3-3.

(b) A person who knowingly or intentionally points a firearm at another person commits a Class D felony. However, the offense is a Class A misdemeanor if the firearm was not loaded.

35-47-4-5. Possession of firearm by serious violent felon.

(a) As used in this section, "serious violent felon" means a person who has been convicted of:

(1) committing a serious violent felony in:

(A) Indiana; or

(B) any other jurisdiction in which the elements of the crime for which the conviction was entered are substantially similar to the elements of a serious violent felony; or

(2) attempting to commit or conspiring to commit a serious violent felony in:

(A) Indiana as provided under IC 35-41-5-1 or IC 35-41-5-2; or

(B) any other jurisdiction in which the elements of the crime for which the conviction was entered are substantially similar to the elements of attempting to commit or conspiring to commit a serious violent felony.

(b) As used in this section, "serious violent felony" means:

(1) murder (IC 35-42-1-1);

(2) voluntary manslaughter (IC 35-42-1-3);

(3) reckless homicide not committed by means of a vehicle (IC 35-42-1-5);

(4) battery as a:

(A) Class A felony (IC 35-42-2-1(a)(5));

(B) Class B felony (IC 35-42-2-1(a)(4)); or

(C) Class C felony (IC 35-42-2-1(a)(3));

(5) aggravated battery (IC 35-42-2-1.5);

(6) kidnapping (IC 35-42-3-2);

(7) criminal confinement (IC 35-42-3-3);

(8) rape (IC 35-42-4-1);

(9) criminal deviate conduct (IC 35-42-4-2);

(10) child molesting (IC 35-42-4-3);

(11) sexual battery as a Class C felony (IC 35-42-4-8);

(12) robbery (IC 35-42-5-1);

(13) carjacking (IC 35-42-5-2);

(14) arson as a Class A felony or Class B felony (IC 35-43-1-1(a));

(15) burglary as a Class A felony or Class B felony (IC 35-43-2-1);

(16) assisting a criminal as a Class C felony (IC 35-44-3-2);

(17) resisting law enforcement as a Class B felony or Class C felony (IC 35-44-3-3);

(18) escape as a Class B felony or Class C felony (IC 35-44-3-5);

(19) trafficking with an inmate as a Class C felony (IC 35-44-3-9);

(20) criminal gang intimidation (IC 35-45-9-4);

(21) stalking as a Class B felony or Class C felony (IC 35-45-10-5);

(22) incest (IC 35-46-1-3);

(23) dealing in or manufacturing cocaine or a narcotic drug (IC 35-48-4-1);

(24) dealing in methamphetamine (IC 35-48-4-1.1);

(25) dealing in a schedule I, II, or III controlled substance (IC 35-48-4-2);

(26) dealing in a schedule IV controlled substance (IC 35-48-4-3); or

(27) dealing in a schedule V controlled substance (IC 35-48-4-4).

(c) A serious violent felon who knowingly or intentionally possesses a firearm commits unlawful possession of a firearm by a serious violent felon, a Class B felony.

35-47-4-6. Possession of firearm by domestic batterer.

(a) A person who has been convicted of domestic battery under IC 35-42-2-1.3 and who knowingly or intentionally possesses a firearm commits unlawful possession of a firearm by a domestic batterer, a Class A misdemeanor.

(b) It is a defense to a prosecution under this section that the person's right to possess a firearm has been restored under IC 35-47-4-7.

35-47-4-7. Restoration of right to possess firearm by person who has been convicted of domestic violence - Procedure.

(a) Notwithstanding IC 35-47-2, IC 35-47-2.5, the restoration of the right to serve on a jury under IC 33-28-5-18, or the restoration of the right to vote under IC 3-7-13-5, and except as provided in subsections (b), (c), and (f), a person who has been convicted of a crime of domestic violence may not possess a firearm after the person's release from imprisonment or lawful detention.

(b) Not earlier than five (5) years after the date of conviction, a person who has been convicted of a crime of domestic violence may petition the court for restoration of the person's right to possess a firearm. in determining whether to restore the person's right to possess a firearm, the court shall consider the following factors:]

(1) whether the person has been subject to:

(a) a protective order;

(b) a no contact order;

(c) a workplace violence restraining order; or

(d) any other court order that prohibits the person from possessing a firearm.

(2) whether the person has successfully completed a substance abuse program, if applicable.

(3) whether the person has successfully completed a parenting class, if applicable.

(4) whether the person still presents a threat to the victim of the crime.

(5) whether there is any other reason why the person should not possess a firearm, including whether the person failed to satisfy a specified condition under subsection (c) or whether the person has committed a subsequent offense.

(c) the court may condition the restoration of a person's right to possess a firearm upon the person's satisfaction of specified conditions.

(d) if the court denies a petition for restoration of the right to possess a firearm, the person may not file a second or subsequent petition until one (1) year has elapsed after the filing of the most recent petition.]

(e) a person has not been convicted of a crime of domestic violence for purposes of subsection (a) if the conviction has been expunged or if the person has been pardoned.

(f) the right to possess a firearm shall be restored to a person whose conviction is reversed on appeal or on postconviction review at the earlier of the following:

(1) at the time the prosecuting attorney states on the record that the charges that gave rise to the conviction will not be refiled.

(2) ninety (90) days after the final disposition of the appeal or the postconviction proceeding.

Chapter 5. Prohibited Instruments of Violence

35-47-5-4.1. Dealing in sawed-off shotguns prohibited - Exceptions.

(a) A person who:

(1) manufactures;

(2) causes to be manufactured;

(3) imports into Indiana;

(4) keeps for sale;

(5) offers or exposes for sale; or

(6) gives, lends, or possesses; any sawed-off shotgun commits dealing in a sawed-off shotgun, a Class D felony.

(b) The presence of a weapon referred to in subsection (a) in a motor vehicle (as defined under IC 9-13-2-105(a)) except for school buses and a vehicle operated in the transportation of passengers by a common carrier (as defined in IC 8-2.1-17-4) creates an inference that the weapon is in the possession of the persons occupying the motor vehicle. However, the inference does not apply to all the persons occupying the motor vehicle if the weapon is found upon, or under the control of, one (1) of the occupants. In addition, the inference does not apply to a duly licensed driver of a motor vehicle for hire who finds the weapon in the licensed driver's motor vehicle in the proper pursuit of the licensed driver's trade.

(c) This section does not apply to a law enforcement officer who is acting in the course of the officer's official duties or to a person who manufactures or imports for sale or sells a sawed-off shotgun to a law enforcement agency.

35-47-5-5. Firearms exempted. This chapter does not apply to any firearm not designed to use fixed cartridges or fixed ammunition, or any firearm made before January 1, 1899.

35-47-5-6. Interstate firearms sales.

(a) Any resident of Indiana:

(1) who is eighteen (18) years of age or older; and

(2) who is not prohibited by law from obtaining, possessing, or using a firearm; may purchase or obtain a rifle or shotgun in Ohio, Kentucky, Michigan, or Illinois.

(b) Any resident of Ohio, Kentucky, Michigan, or Illinois:

(1) who is eighteen (18) years of age or older; and

(2) who is not prohibited by the laws of Indiana, his domicile, or the United States from obtaining, possessing, or using a firearm; may purchase or obtain a rifle, shotgun, or ammunition for a rifle or a shotgun in Indiana.

(c) Any transaction under this section is subject to the provisions of the Gun Control Act of 1968 (82 Stat. 1213, 18 U.S.C. 0.922(B)(3)).

35-47-5-8. Ownership or possession of machine gun prohibited. A person who owns or possesses a machine gun commits a Class C felony.

35-47-5-9. Operation of loaded machine gun prohibited. A person who operates a loaded machine gun commits a Class B felony.

35-47-5-10. Exemptions to machine guns prohibitions. The provisions of section 8 or 9 [IC 35-47-5-8 or IC 35-47-5-9] of this chapter shall not be construed to apply to any of the following:

(1) Members of the military or naval forces of the United States, National Guard of Indiana, or Indiana State Guard, when on duty or practicing.

(2) Machine guns kept for display as relics and which are rendered harmless and not usable.

(3) Any of the law enforcement officers of this state or the United States while acting in the furtherance of their duties.

(4) Persons lawfully engaged in the display, testing, or use of fireworks.

(5) Agencies of state government.

(6) Persons permitted by law to engage in the business of manufacturing, assembling, conducting research on, or testing machine guns, bombs, airplanes, tanks, armored vehicles, or ordnance equipment or supplies while acting within the scope of such business.

(7) Persons possessing, or having applied to possess, machine guns under applicable United States statutes. Such machine guns must be transferred as provided in this article.

(8) Persons lawfully engaged in the manufacture, transportation, distribution, use, or possession of any material, substance, or device for the sole purpose of industrial, agricultural, mining, construction, educational, or any other lawful use.

35-47-5-11. Use of armor-piercing handgun ammunition prohibited - Exception.

(a) As used in this section, "armor-piercing handgun ammunition" means a cartridge that:

(1) can be fired in a handgun; and

(2) will, upon firing, expel a projectile that has a metal core and an outer coating of plastic.

(b) A person who knowingly or intentionally:

(1) manufactures;

(2) possesses;

(3) transfers possession of; or

(4) offers to transfer possession of; armor-piercing handgun ammunition commits a Class C felony.

(c) This section does not apply to nylon coated ammunition, plastic shot capsules, or ammunition designed to be used in rifles or shotguns.

(d) This section does not apply to a law enforcement officer who is acting in the course of the officer's official duties or to a person who manufactures or imports for sale or sells armor-piercing handgun ammunition to a law enforcement agency.

Chapter 9. Possession of Firearms on School Property and School Buses

35-47-9-1. Security guards – School functions. This chapter does not apply to the following:

(1) A:

(A) federal;

(B) state; or

(C) local;

law enforcement officer.

(2) A person who has been employed or authorized by: (A) a school; or (B) another person who owns or operates property being used by a school for a school function; to act as a security guard, perform or participate in a school function, or participate in any other activity authorized by a school.

(3) A person who: (A) may legally possess a firearm; and (B) possesses the firearm in a motor vehicle that is being operated by the person to transport another person to or from a school or a school function.

35-47-9-2. Possession of firearm on school property or bus. A person who possesses a firearm:

(1) in or on school property;

(2) in or on property that is being used by a school for a school function; or

(3) on a school bus;

commits a Class D felony.

Chapter 10. Children and Handguns

35-47-10-1. Applicability. This chapter does not apply to the following:

(1) A child who is attending a hunters safety course or a firearms safety course or an adult who is supervising the child during the course.

(2) A child engaging in practice in using a firearm for target shooting at an established range or in an area where the discharge of a firearm is not prohibited or supervised by:

(A) a qualified firearms instructor; or

(B) an adult who is supervising the child while the child is at the range.

(3) A child engaging in an organized competition involving the use of a firearm or participating in or practicing for a performance by an organized group under Section 501(c)(3) of the Internal Revenue Code that uses firearms as a part of a performance or an adult who is involved in the competition or performance.

(4) A child who is hunting or trapping under a valid license issued to the child under IC 14-22.

(5) A child who is traveling with an unloaded firearm to or from an activity described in this section.

(6) A child who:

(A) is on real property that is under the control of the child's parent, an adult family member of the child, or the child's legal guardian; and

(B) has permission from the child's parent or legal guardian to possess a firearm.

(7) A child who:

(A) is at the child's residence; and

(B) has the permission of the child's parent, an adult family member of the child, or the child's legal guardian to possess a firearm.

35-47-10-2. "Adult" defined. As used in this chapter, "adult" means a person who is at least eighteen (18) years of age.

35-47-10-3. "Child" defined. As used in this chapter, "child" means a person who is less than eighteen (18) years of age.

35-47-10-4. "Loaded" defined. As used in this chapter, "loaded" means having any of the following:

(1) A cartridge in the chamber or cylinder of a firearm.

(2) Ammunition in close proximity to a firearm so that a person can readily place the ammunition in the firearm.

35-47-10-5. Dangerous possession of firearm. A child who knowingly, intentionally, or recklessly:

(1) possesses a firearm for any purpose other than a purpose described in section 1 of this chapter; or

(2) provides a firearm to another child with or without remuneration for any purpose other than a purpose described in section 1 [IC 35-47-10-1] of this chapter;

commits dangerous possession of a firearm, a Class A misdemeanor. However, the offense is a Class C felony if the child has a prior conviction under this section.

35-47-10-6. Provision of firearm by adult to child. An adult who knowingly, intentionally, or recklessly provides a firearm to a child for any purpose other than those described in section 1 [IC 35-47-10-1] of this chapter, with or without remuneration, commits dangerous control of a firearm, a Class C felony. However, the offense is a Class B felony if the adult has a prior conviction under this section.

Chapter 11: Local Regulations of Firearms

35-47-11-1. Applicability.

(a) Section 2 [IC 35-47-11-2] of this chapter applies to all units (as defined in IC 36-1-2-23). All other sections of this chapter apply to all units other than townships.

(b) This chapter applies only if a statute expressly grants a legislative body the authority to adopt an emergency ordinance under this chapter.

(c) This chapter does not affect the validity of an ordinance adopted before, and in effect on, January 1, 1994.

35-47-11-2. Regulation of ownership, sale, transfer, etc. of firearms or ammunition. Notwithstanding IC 36-1-3, a unit may not regulate in any manner the ownership, possession, sale, transfer, or transportation of firearms (as defined in IC 35-47-1-5) or ammunition except as follows:

(1) This chapter does not apply to land, buildings, or other real property owned or administered by a unit, except highways (as defined in IC 8- 23-1-23) or public highways (as defined in IC 8-2.1-17-14).

(2) Notwithstanding the limitation in this section, a unit may use the unit's planning and zoning powers under IC 36-7-4 to prohibit the sale of firearms within two hundred (200) feet of a school by a person having a business that did not sell firearms within two hundred (200) feet of a school before April 1, 1994.

(3) Notwithstanding the limitation in this section, a legislative body of a unit other than a township may adopt an emergency ordinance or a unit other than a township may take other action allowed under section 6 of this chapter to regulate the sale of firearms anywhere within the unit for a period of not more than seventy-two (72) hours after the regulatory action takes effect.

35-47-11-3. Emergency ordinance - Adoption for disaster. The legislative body of a unit may adopt an emergency ordinance under this chapter if:

(1) a disaster (as defined in IC 10-14-3-1) has occurred or is likely to occur in the unit; and

(2) a local disaster emergency has been declared in the unit under IC 10-14-3-29.

35-47-11-4. Emergency ordinances - Adoption procedure. Notwithstanding any other law, if the conditions described under section 3 [IC 35-47-11-3] of this chapter are present within a unit, the legislative body of the unit may adopt an emergency ordinance under this chapter:

(1) without complying with the public notice and public meeting provisions of:

(A) IC 5-14-1.5; or

(B) any other statute;

(2) on the same day that the ordinance is presented to the legislative body; and

(3) by a majority vote of the members of the legislative body.

35-47-11-5. Time of taking effect - Expiration. An emergency ordinance adopted under section 4 [IC 35-47-11-4] of this chapter:

(1) takes effect on the date and at the time of the adoption of the ordinance; and

(2) expires the earlier of:

(A) seventy-two (72) hours after the time of the adoption of the ordinance; or

(B) a time specified in the emergency ordinance.

35-47-11-6. Restriction on firearms for 72 hours. If:

(1) the conditions described under section 3 [IC 35-47-11-3] of this chapter are present within a unit;

(2) an unsuccessful attempt is made to convene the legislative body for the purpose of adopting an emergency ordinance under this chapter; and

(3) in the case of a municipality, an unsuccessful attempt is made to convene the works board to act under this chapter as if the works board were the legislative body; the executive of a municipality or the presiding officer of a county executive may declare a restriction on the sale of firearms anywhere within the unit for a period of not more than seventy-two (72) hours after the restriction is declared. A declaration under this section has the same effect as an ordinance adopted under section 4 [IC 35-47-11-4] of this chapter and becomes effective and expires as provided in section 5 [IC 35-47-11-5] of this chapter.

Article 47.5. Controlled Explosives

Chapter 2. Definitions

35-47.5-2-1. Applicability. The definitions in this chapter apply throughout this article.

35-47.5-2-4. Destructive device.

(a) "Destructive device" means:

(1) an explosive, incendiary, or overpressure device that is configured as a:

(A) bomb;

(B) grenade;

(C) rocket with a propellant charge of more than four (4) ounces;

(D) missile having an explosive or incendiary charge of more than one-quarter (1/4) ounce;

(E) mine;

(F) Molotov cocktail; or

(G) device that is substantially similar to an item described in clauses (A) through (F);

(2) a type of weapon that may be readily converted to expel a projectile by the action of an explosive or other propellant through a barrel that has a bore diameter of more than one-half (1/2) inch; or

(3) a combination of parts designed or intended for use in the conversion of a device into a destructive device.

(b) The term does not include the following

(1) A pistol, rifle, shotgun, or weapon suitable for sporting or personal safety purposes or ammunition.

(2) A device that is neither designed nor redesigned for use as a weapon.

(3) A device that, although originally designed for use as a weapon, is redesigned for use as a signaling, pyrotechnic, line throwing, safety, or similar device.

(4) A surplus military ordnance sold, loaned, or given by authority of the appropriate official of the United States Department of Defense.

35-47.5-2-5. Detonator. "Detonator" means a device containing a detonating charge that is used to initiate detonation in an explosive, including the following:

(1) Electric blasting caps.

(2) Blasting caps for use with safety fuses.

(3) Detonating cord delay connectors.

(4) Blasting caps for use with a shock tube.

(5) Improvised devices designed to function as a detonator.

35-47.5-2-6. Distribute. "Distribute" means the actual, constructive, or attempted transfer from one (1) person to another.

35-47.5-2-7. Explosives. "Explosives" means a chemical compound or other substance or mechanical system intended to produce an explosion capable of causing injury to persons or damage to property or containing oxidizing and combustible units or other ingredients in such proportions or quantities that ignition, fire, friction, concussion, percussion, or detonation may produce an explosion capable of causing injury to persons or damage to property, including the substances designated in IC 35-47.5-3. The term does not include the following:

(1) A model rocket and model rocket engine designed, sold, and used to propel recoverable aero models.

(2) A paper cap in which the explosive content does not average more than twenty-five hundredths (0.25) grains of explosive mixture per paper cap for toy pistols, toy cannons, toy canes, toy guns, or other devices using paper caps unless the paper cap is used as a component of a destructive device.

35-47.5-2-8. Hoax device. "Hoax device" or "replica" means a device or article that has the appearance of a destructive device or detonator.

35-47.5-2-9. Incendiary. "Incendiary" means a flammable liquid or compound with a flash point not greater than one hundred fifty (150) degrees Fahrenheit, as determined by a Tagliabue or an equivalent closed cup device, including gasoline, kerosene, fuel oil, or a derivative of these substances.

35-47.5-2-12. Property. "Property" means real or personal property of any kind, including money, choses in action, and other similar interests in property.

35-47.5-2-13. Regulated explosive.

(a) "Regulated explosive" includes:

(1) a destructive device; and

(2) an explosive.

(b) The term does not include the following:

(1) An explosive in a manufactured article that is designed and packaged in a manner that is likely to prevent an explosion resulting in property damage or personal injury. A manufactured article to which this subdivision applies includes fixed ammunition for small arms, a firework, and a safety fuse match.

(2) Gasoline, kerosene, naphtha, turpentine, or benzine.

(3) An explosive that is being transported on or in a vessel, railroad car, or highway vehicle in conformity with the regulations adopted by the United States Department of Transportation.

(4) A blasting explosive that is transported or used for agricultural purposes and that is in a quantity that does not exceed two hundred (200) pounds.

(5) Ammonium nitrate or other explosive compounds kept for mining purposes at coal mines regulated under IC 14-34.

Chapter 5. Offenses Relating to Regulated Explosives

35-47.5-5-1. Applicability. Sections 2, 3, 4, 5, and 6 [IC 35-47.5-5-2, IC 35-47.5-5-3, IC 35-47.5-5-4, IC 35-47.5-5-5, and IC 35-47.5-5-6] of this chapter do not apply to the following:

(1) A person authorized to manufacture, possess, transport, distribute, or use a destructive device or detonator under the laws of the United States, as amended, or under Indiana law when the person is acting in accordance with the laws, regulations, and rules issued under federal or Indiana law.

(2) A person who is issued a permit for blasting or surface coal mining by the director of the department of natural resources under IC 14-34 when the person is acting under the laws and rules of Indiana and any ordinances and regulations of the political subdivision or authority of the state where blasting or mining operations are being performed.

(3) Fireworks (as defined in IC 22-11-14-1) and a person authorized by the laws of Indiana and of the United States to manufacture, possess, distribute, transport, store, exhibit, display, or use fireworks.

(4) A law enforcement agency, a fire service agency, the department of homeland security, or an emergency management agency of Indiana, an agency or an authority of a political subdivision of the state or the United States, and an employee or authorized agent of the United States while in performance of official duties.

(5) A law enforcement officer, a fire official, or an emergency management official of the United States or any other state if that person is attending training in Indiana.

(6) The armed forces of the United States or of Indiana.

(7) Research or educational programs conducted by or on behalf of a college, university, or secondary school that are:

(A) authorized by the chief executive officer of the educational institution or the officer's designee; or

(B) conducted under the policy of the educational institution;

and conducted in accordance with the laws of the United States and Indiana.

(8) The use of explosive materials in medicines and medicinal agents in forms prescribed by the most recent published edition of the official United States Pharmacopoeia or the National Formulary.

(9) Small arms ammunition and reloading components of small arms ammunition.

(10) Commercially manufactured black powder in quantities not to exceed fifty (50) pounds, percussion caps, safety and pyrotechnic fuses, quills, quick and slow matches, and friction primers intended to be used solely for sporting, recreational, or cultural purposes in antique firearms or antique devices.

(11) An explosive that is lawfully possessed for use in legitimate agricultural or business activities.

35-47.5-5-2. Unauthorized possession, manufacture, transportation or distribution of destructive devices. A person who knowingly or intentionally: (1) possesses; (2) manufactures; (3) transports; (4) distributes; (5) possesses with the intent to distribute; or (6) offers to distribute;

a destructive device, unless authorized by law, commits a Class C felony.

35-47.5-5-3 Possession, manufacture, transportation or distribution by convicted felon. A person who has been convicted of a felony by an Indiana court or a court of any other state, the United States, or another country and knowingly or intentionally: (1) possesses; (2) manufactures; (3) transports; (4) distributes; (5) possesses with the intent to distribute; or (6) offers to distribute;

a regulated explosive commits a Class C felony. However, the offense is a Class B felony if the person has a prior unrelated conviction for an offense under this section.

35-47.5-5-4. Distribution to convicted felon. A person who knowingly or intentionally distributes a regulated explosive to a person who has been convicted of a felony by an Indiana court or a court of another state, the United States, or an-other country commits a Class C felony.

35-47.5-5-5. Distribution to minor. A person who knowingly or intentionally distributes or offers to distribute: (1) a destructive device; (2) an explosive; or (3) a detonator;

to a person who is less than eighteen (18) years of age commits a Class B felony.

35-47.5-5-6. Possession, manufacture, transportation or distribution of hoax device. A person who: (1) manufactures; (2) possesses; (3) transports; (4) distributes; or (5) uses;

a hoax device or replica with the intent to cause another to believe that the hoax device or replica is a destructive device or detonator commits a Class D felony.

35-47.5-5-8. Possession, transportation, receipt, placing or detonation of destructive device or explosive to kill, injure or intimidate. A person who: (1) possesses; (2) transports; (3) receives; (4) places; or (5) detonates;

a destructive device or explosive with the knowledge or intent that it will be used to kill, injure, or intimidate an individual or to destroy property commits a Class A felony.

35-47.5-5-11. Violation of rules regarding use of regulated explosives. A person who recklessly violates a rule regarding the use of a regulated explosive adopted by the commission under IC 35-47.5-4-4.5 commits a Class A misdemeanor. However, the offense is a Class D felony if the violation of the rule proximately causes bodily injury or death.

[Current through the 2008 Second Regular Session]

Revised Code of the Consolidated City & County of Indianapolis/Marion County

Title II. Public Order and Safety

Chapter 451. Weapons

451-4. Display of dangerous weapons. No pistol, revolver or other dangerous weapon of a similar character, which may be concealed and carried upon the person, shall be displayed for sale where it can be seen in or through any window of any structure fronting on any street or alley in the city.

451-5. Unlawful disposition of dangerous weapons. No person shall sell, give, barter, exchange, lend or otherwise dispose of, or place in the possession of any known or suspected habitual user of narcotics or any known or suspected criminal or a person with criminal purpose, any type of machine gun, sawed-off shotgun, pistol or revolver, or ammunition therefor, or any knucks, billy, sandbag, dagger, dirk, bowie knife or stiletto, or any spring gun, sword cane or any other dangerous weapon of any similar character, which may be carried or concealed on or about the person and which are commonly used and fit to be used unlawfully to inflict harm on or to any person; or any tools, devices or jimmies commonly used for burglary. However, ordinary pocketknives with blades not exceeding five (5) inches in length and so known and sold in legitimate trade shall not be included in the terms of this section, and the provisions of this section shall not apply to any military forces, peace officers or other persons so excepted by law for the possession, use or disposal of any such things.

[Revised Code of the Consolidated City & County of Indianapolis/Marion County codified through Ord. 11, 2008]

Municipal Code of the City of South Bend

Chapter 13. Offenses and Miscellaneous Provisions

Article 10. Selected Weapons and Instruments of Violence Prohibited

Division 1. Loaded Assault Weapons Prohibited.

13-95. Definitions. The following definitions shall apply throughout this Article:

(a) Ammunition shall mean any ammunition cartridge, shell or other device containing explosive or incendiary material and designed and intended for use in any firearm.

(b) Assault Weapons shall mean and include:

(1) Any center fire rifle with automatic action or semi-automatic action or revolving cylinder weapon, or carbine which accepts a detachable magazine with a capacity of over fifteen (15) rounds;

(2) Any semiautomatic shotgun with a magazine capacity of more than nine (9) rounds;

(3) Firearms as defined in Indiana Code § 35-47-1-5, which have threads, lugs, or other characteristics which are designed to facilitate the direct attachment of a silencer, bayonet, grenade launcher, flash suppressor, or folding stock to the firearm; and

(4) Any part or combination of parts designed to facilitate the attachment of a silencer, bayonet, grenade launcher, flash suppressor, or folding stock to a firearm; and

(5) A detachable magazine, drum, belt, feed strip, or similar device which has a capacity of, or can be readily restored or converted to accept more than fifteen (15) rounds of ammunition; and

(6) Any combination of parts which are designed and intended solely and exclusively for assembling a firearm defined to be an assault weapon as defined above, or from which an assault weapon could be assembled into an operable assault weapon if such parts are located in the possession or under the control of one (I) person.

(c) Automatic shall describe the mechanically self-acting of a weapon in which the spent cartridge is ejected, and the weapon reloaded and fired, by the action of the gas generated in firing or by the force of the recoil, resulting in the weapon continuing to fire as long as the trigger is depressed.

(d) Chambered shall mean ready to fire.

(e) Community Correction Officer shall mean any person employed by a governmental agency to oversee individuals who are in custody for alleged violation or convicted of a violation of the law.

(f) Dealer shall have the meaning set forth in Indiana Code § 35-47-1-3.

(g) Firearms shall have the meaning set forth in Indiana Code § 35-47-1-5.

(h) Magazine shall mean a storage place for ammunition which acts as a receptacle for cartridges which are to be fed into the breech or chamber of a rifle, shotgun, or other weapon.

(i) Minor shall mean any person who is under eighteen (18) years of age.

(j) Person shall mean any individual, corporation, company, association, firm, partnership, club or society.

(k) Rifle shall mean a weapon designed or redesigned, made or remade, and usually fired from the shoulder which has spiral grooves cut in the bore to use the energy of ammunition.

(l) Secured shall mean out of reach and in a safe condition.

(m) Security guard services shall mean an entity or individual that engages in the business of providing security guard service to the public for hire.

(n) Semi-automatic shall mean any weapon which fires a single projectile for each single pull of the trigger and which employs a magazine.

(o) Shooting or firing range shall mean a place for target practice with firearms.

(p) Shotgun shall have the meanings set forth in Indiana Code § 35-47-1-11.

13-96. Exceptions.

(a) This division shall not apply to any person listed as an "expected person" in Indiana Code § 35-47-2-2; individuals regulated and approved by the Bureau of Alcohol, Tobacco and Firearms (BATF); community correction officers, and individuals properly licensed or authorized by a BATF stamp to use such weapons in a City licensed shooting or firing range properly equipped and zoned for the use of such weapons. Such individuals shall be required to provide valid proof of their status as an exception to this Division upon reasonable request.

(b) The definition of "assault weapon" as set forth in Section 13-91 [sic] shall not include any of the following:

(1) Weapons that do not use fixed ammunition, weapons that were in production prior to 1898, manually operated bolt-action weapons, lever-action weapons, slide-action weapons, single-shot weapons, semiautomatic weapons which use exclusively manalicher-style clips semiautomatic weapons manufactured prior to 1954, and rimfire weapons that employ a tubular magazine.

(2) Any antique or relic firearm;

(3) Any assault weapon which has been modified either to render it permanently inoperable or to permanently make it a device no longer defined as an assault weapon.

(c) Any assault weapon may be possessed by any federal, state, county or local historical society, museum or educational institution of higher learning which is open to the public, provided such assault weapon is properly housed and secured from unauthorized handling and is always kept unloaded.

13-97. Restrictions on the possession and manufacturing of assault weapons.

(a) No person may manufacture any assault weapon, as defined herein, within the City of South Bend.

(b) No person may possess any loaded assault weapon, as defined herein, within the City of South Bend, except as specifically provided herein.

(c) No person may manufacture or possess any magazine with a capacity of more than fifteen (15) rounds for assault weapons within the City of South Bend, except as specifically provided herein.

13-98. Requirements for transporting and possessing unloaded assault weapons.

(a) Any person who owns any assault weapon as defined herein must:

(1) Keep a flag safety (i.e., plastic tie or chamber plug) properly in place so that the chamber is plugged resulting in the firing pin being blocked at all times;

(2) Keep said assault weapon unloaded at all times; and

(3) Keep said assault weapon in a gun slip, glove, or case so that it is inoperable and secured at all times with a gun tie or padlock; except when using such weapon at a City licensed shooting or firing range as specifically provided herein.

(b) Any person who owns any assault weapon must keep all of its ammunition and magazines separate from the assault weapon, and in a secured container, except as specifically provided herein.

13-99. Penalties; confiscation and destruction of assault weapons.

(a) Any person who violates any of the provisions of this Division shall be fined a minimum of two hundred and fifty dollars ($250.00) and up to a maximum of two thousand five hundred dollars ($2,500.00) for each and every violation.

(b) The owner of an assault weapon found with a seated magazine shall be fined two hundred fifty dollars ($250.00), and if a round is chambered an additional one hundred dollars ($100.00) shall be assessed for each round in the chamber and the magazine.

(c) Any loaded assault weapon or weapons, and/or any magazine with a capacity of more than fifteen (15) rounds found in violation of this division shall be seized and confiscated.

(d) Additionally, any person found in violation of this division shall be required to attend and successfully complete a minimum of two (2) educational seminars on firearms safety and perform a minimum of twenty (20) hours of community service for each violation. Said seminars shall be conducted by the South Bend Police Department and/or the City's duly authorized agent.

(e) Penalties up to a total of two thousand five hundred dollars ($2,500.00) for all citations, seminar fees and fines may be imposed upon an adult when found in violation of this division for each and every violation.

Division 2. Access to Firearms by Minors Prohibited

13-103. Definitions. The definitions set forth in Division I of this Article, where applicable, shall be binding on the interpretation and enforcement of this division.

13-104. Parent and legal guardian required to properly store firearms/ammunition; dealers to conspicuously display signs.

(a) No person, including but not limited to parent(s) or legal guardian(s), shall store, leave, or give a loaded or unloaded firearm(s) and ammunition used by such firearm(s) in any place where the person knows, or reasonably should know based on the totality of the circumstances, that a minor is able to gain access to such firearm(s) and/or ammunition.

(b) All persons purchasing firearm(s) within the city, must provide written proof that he or she has successfully completed a National Rifle Association (or comparable agency) firearms safety instruction program on the proper use and storage of such firearm(s). Written proof of such training must be turned into the front desk of the

South Bend Police Department within sixty (60) days from the date of said purchase.

(c) Any licensed firearm dealer doing business in the city must conspicuously post, at every purchase counter in every store, shop, or sales outlet, signs with the following warning in block letters not less than one (1) inch in height:

"IT IS A VIOLATION OF CITY OF SOUTH BEND CODE § 13-100 TO STORE OR LEAVE A FIREARM AND AMMUNITION WHERE CHILDREN UNDER THE AGE OF EIGHTEEN (18) YEARS OF AGE CAN OBTAIN ACCESS."

"ALL PERSONS PURCHASING FIREARMS ARE REQUIRED TO PROVIDE TO THE SOUTH BEND POLICE DEPARTMENT WRITTEN PROOF OF SUCCESSFULLY COMPLETING AN APPROVED FIREARMS SAFETY INSTRUCTION SESSION ON THE PROPER USE AND STORAGE OF SUCH FIREARMS WITHIN SIXTY (60) DAYS OF DATE OF PURCHASE."

Such warnings shall also be distributed by such licensed firearm dealer to each firearm purchaser at the time of the sale of a firearm.

(d) This section shall not apply when:

(1) A minor's access to a firearm and its ammunition is under the supervision or control of a responsible adult for purposes of lawful hunting or instruction in firearms safety, care, handling, or marksmanship;

(2) A minor has access to a firearm and its ammunition as a result of an unlawful entry into the place in which the firearm was found;

(3) A minor obtains a firearm and its ammunition in a lawful act of self-defense or defense of another person or persons within a domicile; or

(4) A minor, without permission of the lawful possessor of a firearm and its ammunition, obtains the firearm from the possessor's body, when the possessor is unable to prevent the removal of said firearm from his or her person.

13-105. Penalties/ citations/ educational training sessions required for parents, guardians, dealers and minors found in violation.

(a) Any parent or guardian whose firearm gets in the wrongful possession of a minor shall be considered in violation of this division. Said parent or guardian shall be issued an ordinance violation citation with a fine of one hundred dollars ($100.00) for each offense, and said firearm shall be confiscated.

(b) Any person who fails to provide written proof that he or she has successfully completed a National Rifle Association (or comparable agency) approved firearms safety instruction program on the proper and storage of such firearm(s) required by Section 13-100 [sic] shall be considered in violation of this division. An ordinance violation citation shall be issued with a fine of fifty dollars ($50.00), and the firearm shall be confiscated.

... Firearms confiscated under this paragraph shall be returned to the owner only upon said owner showing proof that he or she is the lawful owner and that said person has since successfully completed a National Rifle Association (or comparable agency) on firearms safety instruction program addressing the proper use and storage of firearms.

(c) Any licensed firearm dealer who fails to properly display the signs required in Section 13-100 [sic] shall be considered in violation of that section. Each day and each location required to have such signage which is not in compliance with Section 13-100 [sic], shall be considered a separate offense for which a separate citation may be issued. Said licensed firearm dealer shall be issued an ordinance violation citation with fines of twenty-five dollars ($25.00) for the first violation, and fifty dollar ($50.00) fines for all subsequent violations. ...

(d) In addition to adult violators being issued ordinance violation citation(s), any such person found in violation of this division shall be required to attend and successfully complete a minimum of two (2) educational seminars on firearms safety and perform a minimum of twenty (20) hours of community service for each violation. Said seminars shall be conducted by the South Bend Police Department and/or the City's duly authorized agent. All costs related to said seminars shall be paid by the violator.

(e) Any firearm, found in the possession of a minor, shall be confiscated, and the South Bend Police Department shall be immediately contacted. Said minor shall be required to attend and successfully complete a minimum of two (2) educational seminars specifically focusing on minors and safety awareness. Said seminars shall be conducted by the South Bend Police Department and/or the City's duly authorized agent. All costs related to said safety awareness seminars shall be paid by said minor's parent(s) or guardian(s). Additionally, said minor shall be required to write, in his or her own handwriting, a minimum of a two-page report on safety awareness and perform a minimum of twenty (20) hours of community service for each violation, with credit being given against the community service hours for reports deemed acceptable by the department of the City's agent. Said report must be turned into the principal of the minor's school within seventy-two (72) hours of successfully completing the last seminar. Copies of such reports with the name of the minor being held confidential, shall be sent to the Common Council's Health and Public Safety Committee on the last business day of the month when said seminar was held. Quarterly Committee reports shall be given on the progress of such youth educational safety awareness seminars.

(f) Penalties up to a total of two thousand five hundred dollars ($2,500.00) for all citations, seminar fees and fines may be imposed upon an adult when found in violation of this division for each and every violation.

[Municipal Code of the City of South Bend codified through Ordinance No. 9830-08, enacted Feb. 25, 2008]

IOWA

IOWA CODE

Title XVI. Criminal Law and Procedure

Chapter 724. Weapons

724.1. Offensive weapons. An offensive weapon is any device or instrumentality of the following types:

1. A machine gun. A machine gun is a firearm which shoots or is designed to shoot more than one shot, without manual reloading, by a single function of the trigger.

2. A short-barreled rifle or short-barreled shotgun. A short-barreled rifle or short-barreled shotgun is a rifle with a barrel or barrels less than sixteen inches in length or a shotgun with a barrel or barrels less than eighteen inches in length, as measured from the face of the closed bolt or standing breech to the muzzle, or any rifle or shotgun with an overall length less than twenty-six inches.

3. Any weapon other than a shotgun or muzzle loading rifle, cannon, pistol, revolver or musket, which fires or can be made to fire a projectile by the explosion of a propellant charge, which has a barrel or tube with the bore of more than six-tenths of an inch in diameter, or the ammunition or projectile therefor, but not including antique weapons kept for display or lawful shooting.

4. A bomb, grenade, or mine, whether explosive, incendiary, or poison gas; any rocket having a propellant charge of more than four ounces; any missile having an explosive charge of more than one-quarter ounce; or any device similar to any of these.

5. A ballistic knife. A ballistic knife is a knife with a detachable blade which is propelled by a spring-operated mechanism, elastic material, or compressed gas.

6. Any part or combination of parts either designed or intended to be used to convert any device into an offensive weapon as described in subsections 1 to 5 of this section, or to assemble into such an offensive weapon, except magazines or other parts, ammunition, or ammunition components used in common with lawful sporting firearms or parts including but not limited to barrels suitable for refitting to sporting firearms.

7. Any bullet or projectile containing any explosive mixture or chemical compound capable of exploding or detonating prior to or upon impact, or any shot shell or cartridge containing exothermic pyrophoric misch metal as a projectile which is designed to throw or project a flame or fireball to simulate a flamethrower.

Notwithstanding section 724.2, no person is authorized to possess in this state a shot shell or cartridge intended to project a flame or fireball of the type described in this section.

8. Any mechanical device specifically constructed and designed so that when attached to a firearm silences, muffles or suppresses the sound when fired. However, this subsection does not apply to a mechanical device possessed and used by a person solely for the purpose of shooting a deer pursuant to an approved city special deer population control plan if the person has a valid federal permit to possess and use the mechanical device.

9. An offensive weapon or part or combination of parts therefor shall not include the following:

a. An antique firearm. An antique firearm is any firearm (including any firearm with a matchlock, flintlock, percussion cap, or similar type of ignition system) manufactured in or before 1898 or any firearm which is a replica of such a firearm if such replica is not designed or redesigned for using conventional rimfire or centerfire ammunition or which uses only rimfire or centerfire fixed ammunition which is no longer manufactured in the United States and which is not readily available in the ordinary channels of commercial trade.

b. A collector's item. A collector's item is any firearm other than a machine gun that by reason of its date of manufacture, value, design, and other characteristics is not likely to be used as a weapon. The commissioner of public safety shall designate by rule firearms which the commissioner determines to be collector's items and shall revise or update the list of firearms at least annually.

c. Any device which is not designed or redesigned for use as a weapon; any device which is designed solely for use as a signaling, pyrotechnic, line-throwing, safety, or similar device; or any firearm which is unserviceable by reason of being unable to discharge a shot by means of an explosive and is incapable of being readily restored to a firing condition.

724.2. Authority to possess offensive weapons. Any of the following is authorized to possess an offensive weapon when the person's duties or lawful activities require or permit such possession:

1. Any peace officer.

2. Any member of the armed forces of the United States or of the national guard.

3. Any person in the service of the United States.

4. A correctional officer, serving in an institution under the authority of the Iowa department of corrections.

5. Any person who under the laws of this state and the United States, is lawfully engaged in the business of supplying those authorized to possess such devices.

6. Any person, firm or corporation who under the laws of this state and the United States is lawfully engaged in the improvement, invention or manufacture of firearms.

7. Any museum or similar place which possesses, solely as relics, offensive weapons which are rendered permanently unfit for use.

8. A resident of this state who possesses an offensive weapon which is a curio or relic firearm under the federal Firearms Act, 18 U.S.C. ch. 44,solely for use in the official functions of a historical reenactment organization of which the person is a member, if the offensive weapon has been permanently rendered unfit for the firing of live ammunition. The offensive weapon may, however, be adapted for the firing of blank ammunition.

9. A nonresident who possesses an offensive weapon which is a curio or relic firearm under the federal Firearms Act, 18 U.S.C. ch. 44, solely for use in official functions in this state of a historical reenactment organization of which the person is a member, if the offensive weapon is legally possessed by the person in the person's state of residence and the offensive weapon is at all times while in this state rendered incapable of firing live ammunition. A nonresident who possesses an offensive weapon under this subsection while in this state shall not have in the person's possession live ammunition. The offensive weapon may, however, be adapted for the firing of blank ammunition.

724.3. Unauthorized possession of offensive weapons. Any person, other than a person authorized herein, who knowingly possesses an offensive weapon commits a class "D" felony.

724.4A. Weapons free zones - enhanced penalties.

1. As used in this section, "weapons free zone" means the area in or on, or within one thousand feet of, the real property comprising a public or private elementary or secondary school, or in or on the real property comprising a public park. A weapons free zone shall not include that portion of a public park designated as a hunting area under section 461A.42.

2. Notwithstanding sections 902.9 and 903.1, a person who commits a public offense involving a firearm or offensive weapon, within a weapons free zone, in violation of this or any other chapter shall be subject to a fine of twice the maximum amount which may otherwise be imposed for the public offense.

724.4B. Carrying weapons on school grounds - penalty - exceptions.

1. A person who goes armed with, carries, or transports a firearm of any kind, whether concealed or not, on the grounds of a school commits a class "D" felony. For the purposes of this section, "school" means a public or non-public school as defined in section 280.2.

2. Subsection 1 does not apply to the following:

a. A person listed under section 724.4, subsection 4, paragraphs "b" through "f" or "j".

b. A person who has been specifically authorized by the school to go armed, carry, or transport a firearm on the school grounds, including for purposes of conducting an instructional program regarding firearms.

724.5. Duty to carry permit to carry weapons. A person armed with a revolver, pistol, or pocket billy concealed upon the person shall have in the person's immediate possession the permit provided for in section 724.4, subsection 4, paragraph "i", and shall produce the permit for inspection at the request of a peace officer. Failure to so produce a permit is a simple misdemeanor.

724.6 Professional permit to carry weapons.

1. A person may be issued a permit to carry weapons when the person's employment in a private investigation business or private security business licensed under chapter 80A, or a person's employment as a peace officer, correctional officer, security guard, bank messenger or other person transporting property of a value requiring security, or in police work, reasonably justifies that person going armed. The permit shall be on a form prescribed and published by the commissioner of public safety, shall identify the holder, and shall state the nature of the employment requiring the holder to go armed. A permit so issued, other than to a peace officer, shall authorize the person to whom it is issued to go armed anywhere in the state, only while engaged in the employment, and while going to and from the place of the employment. A permit issued to a certified peace officer shall authorize that peace officer to go armed anywhere in the state at all times. Permits shall expire twelve months after the date when issued except that permits issued to peace officers and correctional officers are valid through the officer's period of employment unless otherwise canceled. When the employment is terminated, the holder of the permit shall surrender it to the issuing officer for cancellation.

2. Notwithstanding subsection 1, fire fighters, as defined in section 411.1, subsection 9, airport fire fighters included under section 97B.49B, emergency rescue technicians, and emergency medical care providers, as defined in section 147A.1, shall not, as a condition of employment, be required to obtain a permit under this section. However, the provisions of this subsection shall not apply to a person designated as an arson investigator by the chief fire officer of a political subdivision.

724.7 Nonprofessional permit to carry weapons. Any person who can reasonably justify going armed may be issued a nonprofessional permit to carry weapons. Such permits shall be on a form prescribed and published by the commissioner of public safety, which shall be readily distinguishable from the professional permit, and shall identify the holder thereof, and state the reason for the issuance of the permit, and the limits of the authority granted by such permit. All permits so issued shall be for a definite period as established by the issuing officer, but in no event shall exceed a period of twelve months.

724.8. Persons eligible for permit to carry weapons. No person shall be issued a professional or nonprofessional permit to carry weapons unless:

1. The person is eighteen years of age or older.

2. The person has never been convicted of a felony.

3. The person is not addicted to the use of alcohol or any controlled substance.

4. The person has no history of repeated acts of violence.

5. The issuing officer reasonably determines that the applicant does not constitute a danger to any person.

6. The person has never been convicted of any crime defined in chapter 708, except "assault" as defined in section 708.1 and "harassment" as defined in section 708.7.

724.9 Firearm training program.

A training program to qualify persons in the safe use of firearms shall be provided by the issuing officer of permits, as provided in section 724.11. The commissioner of public safety shall approve the training program, and the county sheriff or the commissioner of public safety conducting the training program within their respective jurisdictions may contract with a private organization or use the services of other agencies, or may use a combination of the two, to provide such training. Any person eligible to be issued a permit to carry weapons may enroll in such course. A fee sufficient to cover the cost of the program may be charged each person attending. Certificates of completion, on a form prescribed and published by the commissioner of public safety, shall be issued to each person who successfully completes the program. No person shall be issued either a professional or nonprofessional permit unless the person has received a certificate of completion or is a certified peace officer. No peace officer or correctional officer, except a certified peace officer, shall go armed with a pistol or revolver unless the officer has received a certificate of completion, provided that this requirement shall not apply to persons who are employed in this state as peace officers on January 1, 1978 until July 1, 1978, or to peace officers of other jurisdictions exercising their legal duties within this state.

724.10. Application for permit to carry weapons - criminal history check required. A person shall not be issued a permit to carry weapons unless the person has completed and signed an application on a form to be prescribed and published by the commissioner of public safety. The application shall state the full name, driver's license or nonoperator's identification card number, residence, and age of the applicant, and shall state whether the applicant has ever been convicted of a felony, whether the person is addicted to the use of alcohol or any controlled substance, and whether the person has any history of mental illness or repeated acts of violence. The applicant shall also display an identification card that bears a distinguishing number assigned to the card holder, the full name, date of birth, sex, residence address, and a brief description and colored photograph of the card holder. The sheriff shall conduct immediately a criminal history check concerning each applicant by obtaining criminal history data from the department of public safety. A person who knowingly makes a false statement of material fact on the application commits a class "D" felony.

724.15. Annual permit to acquire pistols or revolvers.

1. Any person who acquires ownership of any pistol or revolver shall first obtain an annual permit. An annual permit shall not be issued to any person unless:

a. The person is twenty-one years of age or older.

b. The person has never been convicted of a felony.

c. The person is not addicted to the use of alcohol or a controlled substance.

d. The person has no history of repeated acts of violence.

e. The person has never been convicted of a crime defined in chapter 708, except "assault" as defined in section 708.1 and "harassment" as defined in section 708.7.

f. The person has never been adjudged mentally incompetent.

2. Any person who acquires ownership of a pistol or revolver shall not be required to obtain an annual permit if:

a. The person transferring the pistol or revolver and the person acquiring the pistol or revolver are licensed firearms dealers under federal law;

b. The pistol or revolver acquired is an antique firearm, a collector's item, a device which is not designed or redesigned for use as a weapon, a device which is designed solely for use as a signaling, pyrotechnic, line-throwing, safety, or similar device, or a firearm which is unserviceable by reason of being unable to discharge a shot by means of an explosive and is incapable of being readily restored to a firing condition; or

c. The person acquiring the pistol or revolver is authorized to do so on behalf of a law enforcement agency.

d. The person has obtained a valid permit to carry weapons, as provided in section 724.11.

e. The person transferring the pistol or revolver and the person acquiring the pistol or revolver are related to one another within the second degree of consanguinity or affinity unless the person transferring the pistol or revolver knows that the person acquiring the pistol or revolver would be ineligible to obtain a permit.

3. The annual permit to acquire pistols or revolvers shall authorize the permit holder to acquire one or more pistols or revolvers during the period that the permit remains valid. If the issuing officer determines that the applicant has become disqualified under the provisions of subsection 1, the issuing officer may immediately invalidate the permit.

724.16. Annual permit to acquire required - transfer prohibited.

1. Except as otherwise provided in section 724.15,subsection 2, a person who acquires ownership of a pistol or revolver without a valid annual permit to acquire pistols or revolvers or a person who transfers ownership of a pistol or revolver to a person who does not have in the person's possession a valid annual permit to acquire pistols or revolvers is guilty of an aggravated misdemeanor.

2. A person who transfers ownership of a pistol or revolver to a person that the transferor knows is prohibited by section 724.15 from acquiring ownership of a pistol or revolver commits a class "D" felony.

724.16A. Trafficking in stolen weapons. A person who knowingly transfers or acquires possession, or who facilitates the transfer, of a stolen firearm commits a class "D" felony for a first offense and a class "C" felony for second and subsequent offenses or if the weapon is used in the commission of a public offense. However, this section shall not apply to a person purchasing stolen firearms through a buy-back program sponsored by a law enforcement agency if the firearms are returned to their rightful owners or destroyed.

724.17. Application for annual permit to acquire - criminal history check required. The application for an annual permit to acquire pistols or revolvers may be made to the sheriff of the county of the applicant's residence and shall be on a form prescribed and published by the commissioner of public safety. The application shall state the full name of the applicant, the driver's license or nonoperator's identification card number of the applicant, the residence of the applicant, and the age of the applicant. The applicant shall also display an identification card that bears a distinguishing number assigned to the cardholder, the full name, date of birth, sex, residence address, and brief description and colored photograph of the card-holder, or other identification as specified by rule of the department of public safety. The sheriff shall conduct a criminal history check concerning each applicant by obtaining criminal history data from the department of public safety. A person who knowingly makes a false statement of material fact on the application commits a class "D" felony.

724.18. Procedure for making application for annual permit to acquire. A person may personally request the sheriff to mail an application for an annual permit to acquire pistols or revolvers, and the sheriff shall immediately forward to such person an application for an annual permit to acquire pistols or revolvers. A person shall upon completion of the application personally deliver such application to the sheriff who shall note the period of validity on the application and shall immediately issue the annual permit to acquire pistols or revolvers to the applicant. For the purposes of this section the date of application shall be the date on which the sheriff received the completed application.

724.19. Issuance of annual permit to acquire. The annual permit to acquire pistols or revolvers shall be issued to the applicant immediately upon completion of the application unless the applicant is disqualified under the provisions of section 724.15 and shall be on a form prescribed and published by the commissioner of public safety. The permit shall contain the name of the permittee, the residence of the permittee, and the effective date of the permit.

724.20. Validity of annual permit to acquire pistols or revolvers. The permit shall be valid throughout the state and shall be valid three days after the date of application and shall be invalid one year after the date of application.

724.21. Giving false information when acquiring weapon. A person who gives a false name or presents false identification, or otherwise knowingly gives false material information to one from whom the person seeks to acquire a pistol or revolver, commits a class "D" felony.

724.22. Persons under twenty-one - sale, loan, gift, making available - possession.

1. Except as provided in subsection 3, a person who sells, loans, gives, or makes available a rifle or shotgun or ammunition for a rifle or shotgun to a minor commits a serious misdemeanor for a first offense and a class "D" felony for second and subsequent offenses.

2. Except as provided in subsections 4 and 5, a person who sells, loans, gives, or makes available a pistol or revolver or ammunition for a pistol or revolver to a person below the age of twenty-one commits a serious misdemeanor for a first offense and a class "D" felony for second and subsequent offenses.

3. A parent, guardian, spouse who is eighteen years of age or older, or another with the express consent of the minor's parent or guardian or spouse who is eighteen years of age or older may allow a minor to possess a rifle or shotgun or the ammunition therefor which may be lawfully used.

4. A person eighteen, nineteen, or twenty years of age may possess a firearm and the ammunition therefor while on military duty or while a peace officer, security guard or correctional officer, when such duty requires the possession of such a weapon or while the person receives instruction in the proper use thereof from an instructor who is twenty-one years of age or older.

5. A parent or guardian or spouse who is twenty-one years of age or older, of a person fourteen years of age but less than twenty-one may allow the person to possess a pistol or revolver or the ammunition therefor for any lawful purpose while under the direct supervision of the parent or guardian or spouse who is twenty-one years of age or older, or while the person receives instruction in the proper use thereof from an instructor twenty-one years of age or older, with the consent of such parent, guardian or spouse.

6. For the purposes of this section, caliber .22 rimfire ammunition shall be deemed to be rifle ammunition.

7. Access to loaded firearms by children restricted - penalty. It shall be unlawful for any person to store or leave a loaded firearm which is not secured by a trigger lock mechanism, placed in a securely locked box or container, or placed in some other location which a reasonable person would believe to be secure from a minor under the age of fourteen years, if such person knows or has reason to believe that a minor under the age of fourteen years is likely to gain access to the firearm without the lawful permission of the minor's parent, guardian, or person having charge of the minor, the minor lawfully gains access to the firearm without the consent of the minor's parent, guardian, or person having charge of the minor, and the minor exhibits the firearm in a public place in an unlawful manner, or uses the firearm unlawfully to cause injury or death to a person. This subsection does not apply if the minor obtains the firearm as a result of an unlawful entry by any person. A violation of this subsection is punishable as a serious misdemeanor.

724.25. Felony and antique firearm defined.

1. As used in sections 724.8,subsection 2, and 724.26, the word "felony" means any offense punishable in the jurisdiction where it occurred by imprisonment for a term exceeding one year, but does not include any offense, other than an offense involving a firearm or explosive, classified as a misdemeanor under the laws of the state and punishable by a term of imprisonment of two years or less.

2. As used in this chapter an "antique firearm" means any firearm (including any firearm with a matchlock, flintlock, percussion cap, or similar type of ignition system) manufactured in or before 1898. An antique firearm also means a replica of a firearm so described if the replica is not designed or redesigned for using rimfire or conventional centerfire fixed ammunition or if the replica uses rimfire or conventional centerfire fixed ammunition which is no longer manufactured in the United States and which is not readily available in the ordinary channels of commercial trade.

724.26. Possession, receipt, transportation, or dominion and control of firearms and offensive weapons by felons. A person who is convicted of a felony in a state or federal court, or who is adjudicated delinquent on the basis of conduct that would constitute a felony if committed by an adult, and who knowingly has under the person's dominion and control or possession, receives, or transports or causes to be transported a firearm or offensive weapon is guilty of a class "D" felony.

724.27. Offenders' rights restored. The provisions of sections 724.8, subsection 2, 724.15, subsection 1, paragraphs "b" and "e", and section 724.26 shall not apply to a person who is eligible to have the person's civil rights regarding fire-arms restored under section 914.7 and who is pardoned or has had the person's civil rights restored by the President of the United States or the chief executive of a state and who is expressly authorized by the President of the United States or such chief executive to receive, transport, or possess firearms or destructive devices.

724.28. Prohibition of regulation by political subdivisions. A political subdivision of the state shall not enact an ordinance regulating the ownership, possession, legal transfer, lawful transportation, registration, or licensing of firearms when the ownership, possession, transfer, or transportation is otherwise lawful under the laws of this state. An ordinance regulating firearms in violation of this section existing on or after April 5, 1990, is void.

724.29. Firearm devices. A person who sells or offers for sale a manual or power-driven trigger activating device constructed and designed so that when attached to a firearm increases the rate of fire of the firearm is guilty of an aggravated misdemeanor.

Title XVI. Criminal Law and Procedure

Subtitle 2. Criminal Procedure

Chapter 809. Disposition of Seized Property

809.21 Sale of certain ammunition and firearms. Ammunition and firearms which are not illegal and which are not offensive weapons as defined by section 724.1 may be sold by the department of public safety at public auction. The department of public safety may sell at public auction forfeited legal weapons received from the director of the department of natural resources, except that rifles and shotguns shall be retained by the department of natural resources for disposal according to its rules. The sale of ammunition or firearms pursuant to this section shall be made only to federally licensed firearms dealers or to persons who have a permit to purchase the firearms. Persons who have not obtained a permit may bid on firearms at the public auction. However, persons who bid without a permit must post a fifty percent of purchase price deposit with the commissioner of public safety on any winning bid. No transfer of firearms may be made to a person bidding without a permit until such time as the person has obtained a permit. If the person is unable to produce a permit within two weeks from the date of the auction the person shall forfeit the fifty percent deposit to the department of public safety. All proceeds of a public auction pursuant to this section, less department expenses reasonably incurred, shall be deposited in the general fund of the state. The department of public safety shall be reimbursed from the proceeds for the reasonable expenses incurred in selling the property at the auction.

[Current through 2007 Supplement]

KANSAS
KAN. STAT.

Chapter 21. Crimes and Punishment

Article 42. Crimes Against the Public Safety

21-4202. Aggravated weapons violation.

(a) An aggravated weapons violation is a violation of any of the provisions of K.S.A. 21-4201 and amendments thereto by a person who:

(1) Within five years preceding such violation has been convicted of a nonperson felony under the laws of Kansas or any other jurisdiction or has been released from imprisonment for such nonperson felony; or

(2) has been convicted of a person felony pursuant to the Kansas laws or in any other jurisdiction which is substantially the same as such crime or has been released from imprisonment for such crime, and has not had the conviction of such crime expunged or been pardoned for such crime.

(b)(1) Aggravated weapons violation is a severity level 9, nonperson felony for a violation of subsections (a)(1) through (a)(5) or subsection (a)(9) of K.S.A. 21-4201 and amendments thereto.

(2) Aggravated weapons violation is a severity level 8, nonperson felony for a violation of subsections (a)(6), (a)(7) and (a)(8) of K.S.A. 21-4201 and amendments thereto.

21-4203. Criminal disposal of firearms.

(a) Criminal disposal of firearms is knowingly:

(1) Selling, giving or otherwise transferring any firearm with a barrel less than 12 inches long to any person under 18 years of age;

(2) selling, giving or otherwise transferring any firearms to any person who is both addicted to and an unlawful user of a controlled substance;

(3) selling, giving or otherwise transferring any firearm to any person who, within the preceding five years, has been convicted of a felony, other than those specified in subsection (b), under the laws of this or any other jurisdiction or has been released from imprisonment for a felony and was found not to have been in possession of a firearm at the time of the commission of the offense;

(4) selling, giving or otherwise transferring any firearm to any person who, within the preceding 10 years, has been convicted of a felony to which this subsection applies, but was not found to have been in the possession of a firearm at the time of the commission of the offense, or has been released from imprisonment for such a crime, and has not had the conviction of such crime expunged or been pardoned for such crime;

(5) selling, giving or otherwise transferring any firearm to any person who has been convicted of a felony under the laws of this or any other jurisdiction and was found to have been in possession of a firearm at the time of the commission of the offense; or

(6) selling, giving or otherwise transferring any firearm to any person who is or has been a mentally ill person subject to involuntary commitment for care and treatment, as defined in K.S.A. 59-2946, and amendments thereto, or a person with an alcohol or substance abuse problem subject to involuntary commitment for care and treatment as defined in K.S.A. 59-29b46, and amendments thereto, and such person has not received a certificate of restoration pursuant to K.S.A. 2006 Supp. 75-7c26, and amendments thereto.

(b) Subsection (a)(4) shall apply to a felony under K.S.A. 21-3401, 21-3402, 21-3403, 21-3404, 21-3410 21-3411, 21-3414, 21-3415, 21-3419, 21-3420 21-3421, 21-3427, 21-3502, 21-3506, 21-3518 21-3716, 65-4127a or 65-4127b, or 65-4160 through 65-4164, or K.S.A. 2006 Supp. 21-3442, and amendments thereto, or a crime under a law of another jurisdiction which is substantially the same as such felony.

(c) Criminal disposal of firearms is a class A nonperson misdemeanor.

21-4204. Criminal possession of a firearm.

(a) Criminal possession of a firearm is:

(1) Possession of any firearm by a person who is both addicted to and an unlawful user of a controlled substance;

(2) possession of any firearm by a person who has been convicted of a person felony or a violation of any provision of the uniform controlled substances act under the laws of Kansas or a crime under a law of another jurisdiction which is substantially the same as such felony or violation, or was adjudicated a juvenile offender because of the commission of an act which if done by an adult would constitute the commission of a person felony or a violation of any provision of the uniform controlled substances act, and was found to have been in possession of a firearm at the time of the commission of the offense;

(3) possession of any firearm by a person who, within the preceding five years has been convicted of a felony, other than those specified in subsection (a)(4)(A), under the laws of Kansas or a crime under a law of another jurisdiction which is substantially the same as such felony, has been released from imprisonment for a felony or was adjudicated as a juvenile offender because of the commission of an act which if done by an adult would constitute the commission of a felony, and was found not to have been in possession of a firearm at the time of the commission of the offense;

(4) possession of any firearm by a person who, within the preceding 10 years, has been convicted of:

(A) A felony under K.S.A. 21-3401, 21-3402, 21-3403, 21-3404, 21-3410, 21-3411, 21-3414, 21-3415, 21-3419, 21-3420, 21-3421, 21-3427, 21-3502, 21-3506, 21-3518, 21-3716, 65-4127a or 65-4127b, or 65-4160 through 65-4164, or K.S.A. 2006 Supp.21-3442, and amendments thereto, or a crime under a law of another jurisdiction which is substantially the same as such felony, has been released from imprisonment for such felony, or was adjudicated as a juvenile offender because of the commission of an act which if done by an adult would constitute the commission of such felony, was found not to have been in possession of a firearm at the time of the commission of the offense, and has not had the conviction of such crime expunged or been pardoned for such crime; or

(B) a nonperson felony under the laws of Kansas or a crime under the laws of another jurisdiction which is substantially the same as such nonperson felony, has been released from imprisonment for such nonperson felony or was adjudicated as a juvenile offender because of the commission of an act which if done by an adult would constitute the commission of a nonperson felony, and was found to have been in possession of a firearm at the time of the commission of the offense;

(5) possession of any firearm by any person, other than a law enforcement officer, in or on any school property or grounds upon which is located a building or structure used by a unified school district or an accredited nonpublic school for student instruction or attendance or extracurricular activities of pupils enrolled in kindergarten or any of the grades 1 through 12 or at any regularly scheduled school sponsored activity or event;

(6) refusal to surrender or immediately remove from school property or grounds or at any regularly scheduled school sponsored activity or event any firearm in the possession of any person, other than a law enforcement officer, when so requested or directed by any duly authorized school employee or any law enforcement officer; or

(7) possession of any firearm by a person who is or has been a mentally ill person subject to involuntary commitment for care and treatment, as defined in K.S.A. 59-2946, and amendments thereto, or persons with an alcohol or substance abuse problem subject to involuntary commitment for care and treatment as defined in K.S.A. 59-29b46, and amendments thereto.

(b) Subsection (a)(5) shall not apply to:

(1) Possession of any firearm in connection with a firearms safety course of instruction or firearms education course approved and authorized by the school;

(2) any possession of any firearm specifically authorized in writing by the superintendent of any unified school district or the chief administrator of any accredited nonpublic school;

(3) possession of a firearm secured in a motor vehicle by a parent, guardian, custodian or someone authorized to act in such person's behalf who is delivering or collecting a student; or

(4) possession of a firearm secured in a motor vehicle by a registered voter who is on the school grounds, which contain a polling place for the purpose of voting during polling hours on an election day.

(c) Subsection (a)(7) shall not apply to a person who has received a certificate of restoration pursuant to K.S.A. 2006 Supp. 75-7c26, and amendments thereto.

(d) Violation of subsection (a)(1) or (a)(5) is a class B nonperson select misdemeanor; violation of subsection (a)(2), (a)(3) (a)(4) or (a)(7) is a se-verity level 8, nonperson felony; violation of sub-section (a)(6) is a class A nonperson misdemeanor.

21-4204a. Criminal possession of firearm by a juvenile.

(a) Criminal possession of a firearm by a juvenile is knowingly possessing a firearm with a barrel less than 12 inches long by any person less than 18 years of age.

(b) Criminal possession of a firearm by a juvenile is a class A nonperson misdemeanor. A second or subsequent violation is a severity level 8, nonperson felony.

(c) It shall be a defense to a prosecution of criminal possession of a firearm by a juvenile if such person less than 18 years of age was:

(1) In attendance at a hunter's safety course or a firearms safety course;

(2) engaging in practice in the use of such firearm or target shooting at an established range authorized by the governing body of the jurisdiction in which such range is located;

(3) engaging in an organized competition involving the use of such firearm, or participating in or practicing for a performance by an organization exempt from federal income tax pursuant to section 501(c)(3) of the internal revenue code of 1986 which uses firearms as a part of such performance;

(4) hunting or trapping pursuant to a valid license issued to such person pursuant to article 9 of chapter 32 of the Kansas Statutes Annotated and amendments thereto;

(5) traveling with any such firearm in such person's possession being unloaded to or from any activity described in paragraphs (1) through (4), only if such firearm is secured, unloaded and outside the immediate access of such person;

(6) on real property under the control of such person's parent, legal guardian or grandparent and who has the permission of such parent, legal guardian or grandparent to possess such firearm; or

(7) at such person's residence and who, with the permission of such person's parent or legal guardian, possesses such firearm for the purpose of exercising the rights contained in K.S.A. 21-3211, 21-3212 or 21-3213 and amendments thereto.

(d) This section shall be part of and supplemental to the Kansas criminal code.

21-4206. Confiscation and disposition of weapons; use of proceeds of sale.

(1) Upon conviction of a violation or upon adjudication as a juvenile offender for a violation of K.S.A. 21-4201, 21-4202, 21-4204, 21-4204a or 21-4219, and amendments thereto, any weapon seized in connection therewith shall remain in the custody of the trial court.

(2) Any stolen weapon so seized and detained, when no longer needed for evidentiary purposes, shall be returned to the person entitled to possession, if known. All other confiscated weapons when no longer needed for evidentiary purposes, shall in the discretion of the trial court, be: (a) destroyed; (b) forfeited to the law enforcement agency seizing the weapon for use within such agency, for sale to a properly licensed federal firearms dealer, for trading to a properly licensed federal firearms dealer for other new or used firearms or accessories for use within such agency or for trading to another law enforcement agency for that agency's use; or (c) forfeited to the Kansas bureau of investigation for law enforcement, testing, comparison or destruction by the Kansas bureau of investigation forensic laboratory.

(3) If weapons are sold as authorized by subsection (2), the proceeds of the sale shall be credited to the asset seizure and forfeiture fund of the seizing agency.

Chapter 48. Militia, Defense and Public Safety

Article 19. Sale and Purchase of Certain Firearms

48-1901. Definitions. As used in this act, unless the context otherwise requires, the phrase "a state contiguous to this state" means any state having a common border with Kansas. All other words and phrases used in this act shall have the meanings respectively ascribed to them in the federal gun control act of 1968.

48-1902. Sale or delivery of rifle or shotgun to resident of contiguous state; restrictions. It is hereby declared to be lawful for an importer, manufacturer, dealer or collector licensed under the federal gun control act of 1968, whose place of business is in this state, to sell or deliver a rifle or shotgun to a resident of a state contiguous to this state, subject to the following restrictions and requirements:

(a) The purchaser's state of residence must permit such sale or delivery by law;

(b) The sale must fully comply with the legal conditions of sale in both such states; and

(c) Prior to the sale or delivery for sale of the rifle or shotgun, the purchaser and the licensee must have complied with all of the requirements of section 922 (c) of the federal gun control act of 1968, applicable to interstate transactions other than at the licensee's business premises.

48-1903. Purchase or receipt of rifle or shotgun in contiguous state by resident of Kansas; restrictions. It is hereby declared to be lawful for a resident of this state to purchase or receive delivery of a rifle or shotgun in a state contiguous to this state, subject to the following restrictions and requirements:

(a) The sale must fully comply with the legal conditions of sale in both such states; and

(b) Prior to the sale or delivery for sale of the rifle or shotgun, the purchaser and the licensee must have complied with all of the requirements of section 922 (c) of the federal gun control act of 1968, applicable to interstate transactions other than at the licensee's business premises.

48-1904. Nonapplication of act, when. The provisions of this act do not apply to:

(a) Transactions between importers, manufacturers, dealers and collectors licensed under the federal gun control act of 1968;

(b) the loan or rental of a firearm to any person for temporary use for lawful sporting purposes; and

(c) a person who is engaged in hunting or in participating in any organized rifle or shotgun match or contest in a state other than his or her state of residence, and whose rifle or shotgun has been lost or stolen or has become inoperative in such other state, and who purchases a rifle or shotgun in such other state from a licensed dealer, if such person presents to such dealer a sworn statement:

(1) That his or her rifle or shotgun was lost or stolen or became inoperative while hunting or participating in any such match or contest in such other state; and

(2) Identifying the chief law enforcement officer of the political subdivision in which such person resides, to whom such licensed dealer shall forward such statement by registered mail.

[Current through 2007 Legislative Session Supplement]

Code of the City of Topeka

Chapter 54. Criminal Code

Article IV. Offenses Against Public Safety

Division 2. Weapons

54-105. Furnishing weapons to minors. It shall be unlawful for any person to sell, give, loan or otherwise furnish any pistol or revolver by which a cartridge may be exploded, or any dirk, Bowie knife, knucks, slingshot or other dangerous weapons to any minor.

Article VI. Possession and Use of Certain Types of Ammunition

54-191. Definitions. The term "polymer tipped ammunition" means a projectile which may be used in a handgun or other firearm and which is constructed, in whole or in part of carbon-based plastics (polymers), including but not limited to ammunition which is known as a "Black Rhino" or "Rhino-Ammo" bullets or shells.

54-192. Unlawful. It shall be unlawful for any person to sell, possess or use any polymer tipped ammunition within the city limits.

54-193. Penalty. Any person convicted of violating the provisions of section 54-192 shall be punished by a fine not to exceed $499.00 and/or imprisonment not to exceed 179 days. A person so convicted shall not be eligible for parole until they have paid a fine of $499.00 or served an imprisonment of 179 days.

[Code of the City of Topeka current through Ord. No. 19129, adopted July 15, 2008]

Wichita City Code

Title 5. Public Safety and Morals

Chapter 5.88. Weapons

5.88.010. Unlawful use of weapons.

(1) Unlawful use of a weapon is knowingly:

(a) Selling, manufacturing, purchasing, possessing or carrying any bludgeon, sand club, metal knuckles or throwing star, or any knife, commonly referred to as a switch-blade, which, having the appearance of a pocket knife, also has a blade that opens automatically by hand pressure applied to a button, spring or other device in the handle of the knife, or by other mechanical contrivance, or any knife having a blade that opens or falls or is ejected into position by the force of gravity or by an outward, downward or centrifugal thrust or movement;

(b) Carrying concealed on one's person, or possessing with intent to use the same unlawfully against another, a dagger, dirk, billy, blackjack, slingshot, nightstick, nun-chucks, sap gloves, tomahawk, dangerous knife, straight-edged razor, stiletto or any other dangerous or deadly instrument of like character, except that an ordinary pocket knife with no blade more than four inches in length shall not be construed to be a dangerous knife or a dangerous or deadly weapon or instrument;

(c) Carrying unconcealed on one's person or in any vehicle under one's immediate control, with intent to use the same unlawfully against another, a dagger, dirk, billy, blackjack, slingshot, nightstick, nunchucks, sap gloves, tomahawk, dangerous knife, straight-edged razor, stiletto or any other dangerous or deadly instrument of like character, except that an ordinary pocket knife with no blade more than four inches in length shall not be construed to be a dangerous knife or a dangerous or deadly weapon or instrument;

(d) Carrying any pistol, revolver or other firearm concealed on one's person, while on property open to the public, except when on one's land or in one's abode or fixed place of business;

(e) Carrying on one's person any unconcealed, loaded firearm, while on property open to the public, except when on one's land or in one's abode or fixed place of business;

(f) Carrying in any vehicle under one's immediate control, while on property open to the public, any loaded firearm, except when on one's land or in one's abode or fixed place of business;

(g) Carrying in any air, land or water vehicle an unloaded firearm that is not encased in a container which completely encloses the firearm;

(h) Carrying a loaded or unloaded firearm in a courtroom or within City Hall;

(i) Drawing a pistol, revolver, knife or any other deadly weapon upon any person.

(j) As used in this section, "throwing star" means any instrument, without handles, consisting of a metal plate having three or more radiating points with one or more sharp edges and designed in the shape of a polygon, trefoil, cross, star, diamond or other geometric shape, manufactured for use as a weapon for throwing.

(2) Subsections (1)(a), (b), (c), (d), (e), (f) and (g), (h), and (i) of this subsection shall not apply to or affect any of the following:

(a) Law enforcement officers, or any person summoned by any such officers to assist in making arrests or preserving the peace while actually engaged in assisting such officer;

(b) Wardens, superintendents, directors, security personnel and keepers of prisons, penitentiaries, jails and other institutions for the detention of persons accused or convicted of a crime, while acting within the scope of their authority;

(c) Members of the armed services or reserve forces of the United States or the Kansas National Guard while in the performance of their official duty; or

(d) Manufacture of, transportation to, or sale of weapons to a person authorized under (a) through (c) of this subsection to possess such weapons.

(e) Qualified law enforcement officers or qualified retired law enforcement officers pursuant to the Law Enforcement Officers Safety Act of 2004, 18 U.S.C. 926B and 18 U.S.C. 926C and amendments thereto.

(3) Subsection (1)(d), (e) and (f) of this section shall not apply to or affect the following:

(a) Watchmen, while actually engaged in the performance of the duties of their employment.

(b) Private detectives licensed by the state to carry the firearm involved while actually engaged in the duties of their employment;

(c) Detectives or special agents regularly employed by railroad companies or other corporations to perform full-time security or investigative service, while actually engaged in the duties of their employment; or

(d) The State Fire Marshal, the State Fire Marshal's deputies or any member of a fire department authorized to carry a firearm pursuant to K.S.A. 31-157 and amendments thereto, while engaged in an investigation in which such fire marshal, deputy or member is authorized to carry a firearm pursuant to K.S.A. 31-157 and amendments thereto.

(e) Special deputy sheriffs described in K.S.A. 2001 Supp. 19-827, and amendments thereto, who have satisfactorily completed the basic course of instruction required for permanent appointment as a part-time law enforcement officer under K.S.A. 74-5607a and amendments thereto.

(4) Subsection (1)(i) of this section shall not apply to or affect historical reenactors and actors when engaged in performances and demonstrations. Provided, however, this subsection shall only apply to those performances and demonstrations which have been approved in advance in writing by the city manager or his designee.

(5) Subsection (1) (d), (e), (f), and (g) shall not apply to any person who sells, purchases, possesses or carries a firearm, device or attachment which has been rendered unserviceable by steel weld in the chamber and marriage weld of the barrel to the receiver and which has been registered in the national firearms registration and transfer record in compliance with 26 U.S.C. 5841 et seq. in the name of such person and, if such person transfers such firearm, device or attachment to another person, has been so registered in the transferee's name by the transferor.

(6) Subsections 1(d), 1(e), 1 (f), and 1 (g) shall not apply to any person authorized to carry a concealed firearm pursuant to the Personal and Family Protection Act, K.S.A. 75-7c01 through K.S.A. 75-7c18, and amendments thereto.

(7) Subsection 1 (d) shall not apply to licensed hunters or fishermen, while engaged in hunting or fishing;

(8) It shall be a defense that the defendant is within an exemption.

(9) Any person who violates any of the provisions of this section within the corporate limits of the city shall be guilty of a misdemeanor, and upon conviction thereof shall be punished by a fine not to exceed two thousand five hundred dollars or by imprisonment for not more than one year, or by both such fine and imprisonment.

(10) In addition to the penalty for violation of any of the provisions of this section, it shall be the duty of the municipal court judge:

(a) To order any weapon seized in connection with such violation which is not a firearm to be forfeited to the city and the same shall be destroyed or caused to be destroyed by the chief of police whenever the weapon is no longer needed for evidence;

(b) To order any weapon seized in connection with such violation when no longer needed for evidentiary purposes, shall, in the discretion of the trial court, be:

(i) Destroyed;

(ii) Forfeited to the Wichita Police Department for use within the police department, for sale to a properly licensed federal firearms dealer or for trading to a properly licensed federal firearms dealer by the police department for other new or used firearms or accessories for the Police Department's use; or

(iii) Forfeited to the Kansas Bureau of Investigation for law enforcement, testing, comparison or destruction by the Kansas Bureau of Investigation forensic laboratory.

If weapons are sold as authorized above, the proceeds from any such sale shall be credited to the asset seizure and forfeiture fund of the Wichita Police Department. All transactions involving weapons disposed of under this subsection must have the prior approval of the city manager. All sales of weapons are subject to review by the city council;

(c) Any stolen weapon confiscated in connection with any violation of this section other than subdivision (a) of this subsection shall be returned to the person entitled to possession, if known, when the same is no longer needed for evidence. All other weapons shall be disposed of as provided in subsection (9)(a) and (b) of this section.

Chapter 5.89. Firearms

5.89.010. Definitions. For the purposes of this chapter, the following terms shall have the meaning ascribed to them in this section:

(a) "Minor" means a person who is under the age of eighteen years.

(b) "Firearm" means any loaded or unloaded pistol, revolver, rifle, shotgun or other weapon which will or is designed to or may readily be converted to expel a projectile by the action of an explosion, expanding gases or other combustion. Air rifles, air pistols and BB guns are included in this definition only if capable of expelling projectiles by the sudden release of compressed gas. This term shall not include a firearm which has been rendered unserviceable by steel weld in the chamber and marriage weld of the barrel to the receiver and which has been registered in the national firearms registration and transfer record in compliance with 26 U.S.C. 5841, et seq., and any amendments thereto.

5.89.020. Possession of a firearm by a minor prohibited - Exceptions.

(a) Unless otherwise specifically provided herein, it shall be unlawful for a minor to possess any firearm within the city of Wichita, except when the minor is in the presence of and under the direct supervision of a parent, stepparent, grandparent, stepgrandparent, or legal guardian.

(b) Any minor who is not in the presence of and under the direct supervision of his or her parent, stepparent, grandparent, stepgrandparent, or legal guardian may only possess a firearm in the city of Wichita under the following circumstances:

1. During a hunter education class held pursuant to K.S.A. 32-920 and conducted by a Kansas hunter education instructor who is certified by the Kansas Department of Wildlife and Parks, provided said possession is under the supervision of the instructor;

2. During a firearms instructional or safety training class taught by an instructor certified by the National Rifle Association or other nationally recognized hunting, target or sports shooting organization, provided said possession is under the supervision of the instructor;

3. While transporting an unloaded firearm to and from an excursion for lawful hunting of game birds or animals, provided:

(a) The minor is in possession of a valid hunting license, if said license is required by state or federal law for the purposes of the hunting excursion;

(b) The minor is in possession of a valid hunter education certificate issued to said minor;

(c) The firearm, during transportation, is unloaded and is stored in a case, scabbard, or other container which completely encloses the firearm.

5.89.040. Stolen weapons. Any stolen firearm confiscated in connection with any violation of this chapter shall be returned to the person entitled to possession, if known, when the same is no longer needed for evidence.

5.89.050. Penalty. Any minor violating the provisions of this chapter shall be dealt with in accordance with the Kansas Juvenile Offenders Code, K.S.A. 38-1601, et seq., and any amendments thereto.

[Code of the City of Wichita current through Ord. No. 47-884, adopted May 13, 2008]

KENTUCKY

KY. REV. STAT.

Title XIX. Public Safety and Morals

Chapter 237. Firearms and Destructive Devices

237.020 Right of Kentucky residents to buy firearms in contiguous states - Reciprocity as to residents of contiguous states.

(1) Residents of the Commonwealth of Kentucky who are citizens of the United States shall have the right to purchase or otherwise acquire rifles, shotguns, handguns, and any other firearms which they are permitted to purchase or otherwise acquire under federal law and the Kentucky revised statutes from properly licensed dealers, manufacturers, importers, or collectors, and unlicensed individual persons in Kentucky or in any other state or nation outside of the Commonwealth of Kentucky.

(2) Residents of states other than the Commonwealth of Kentucky who are citizens of the United States shall have the right to purchase or otherwise acquire rifles, shotguns, handguns, and any other firearms which they are permitted to purchase or otherwise acquire under federal law and the Kentucky revised statutes from properly licensed dealers, manufacturers, importers, or collectors and from unlicensed individual persons in the Commonwealth of Kentucky.

(3) Citizens of countries other than the United States shall have the right to purchase or otherwise acquire rifles, shotguns, handguns, and any other firearms which they are permitted to purchase or otherwise acquire under federal law and the Kentucky revised statutes from properly licensed dealers, manufacturers, importers, or collectors and from unlicensed individual persons

(4) All such sales shall conform to the requirements of federal law, the Kentucky Revised Statutes, applicable local ordinances, and the law of the purchaser's state.

237.030 Definitions for KRS 237.040 and 237.050.

(1) "Destructive device" means any explosive, incendiary, or poison gas bomb, grenade, mine, rocket, missile, or similar device and includes the unassembled components from which such a device can be made.

(2) "Booby trap device" includes any device, or substance designed to surreptitiously or covertly take life, endanger life or destroy or damage property and shall not include firearms.

237.040 Criminal possession of destructive device or booby trap device. A person is guilty of criminal possession of a destructive device or a booby trap device when he possesses, manufactures, or transports such substance or device with:

(1) Intent to use that device to commit an offense against the laws of this state, a political subdivision thereof, or of the United States; or

(2) Knowledge that some other person intends to use that device to commit an offense against the laws of this state, a political subdivision thereof, or of the United States.

(3) Mere possession without substantial evidence of the requisite intent is insufficient to bring action under KRS 237.030 to 237.050.

237.050 Exemptions. KRS 237.030 to 237.050 shall not apply to:

(1) Destructive devices or booby trap devices which are possessed by the government of the United States, this state, or a political subdivision thereof;

(2) Any device which is lawfully possessed under the gun control act of 1968, the organized crime control act of 1971, or any other law of the United States or this state, unless a crime is committed therewith;

(3) Nonlethal devices placed on the premises of the owner or the lawful occupant thereof for his own self-protection or the protection of the said property;

(4) The setting of traps suitable and legal for the taking of game by persons licensed or permitted to do so by the game laws of the commonwealth;

(5) Inert devices which cannot readily be restored to operating condition; or

(6) The acquisition, possession, use, or control of firearms.

237.060 Definitions for KRS 237.060 to 237.090 and certain other sections. The following definitions apply in KRS 237.060 to 237.090 and KRS 197.170, 218A.992, 244.125, 244.990, and 514.110, unless the context otherwise requires:

(1) "Handgun" means any pistol or revolver originally designed to be fired by the use of a single hand, or any other firearm originally designed to be fired by the use of a single hand.

(2) "Firearm" means any weapon which will expel a projectile by the action of an explosive.

(3) "Licensed gun dealer" means a person who has a federal firearms license and any business license required by a state or local government entity.

(4) "Loaded" with respect to a firearm means:

(a) There is ammunition in the chamber of the firearm; or

(b) There is ammunition in the cylinder of the firearm; or

(c) There is ammunition in the magazine of a firearm, if the magazine is attached to the firearm.

(5) "Juvenile" means a person who has not attained his eighteenth birthday.

(6) "Ammunition" means loaded ammunition designed for use in any firearm.

(7) "Armor-piercing ammunition" means a projectile or projectile core which may be used in a handgun and which is constructed entirely (excluding the presence of traces of other substances) from one (1) or a combination of tungsten alloys, steel, iron, brass, bronze, beryllium copper, or depleted uranium. "Armor piercing ammunition" does not include shotgun shot required by federal or state environmental or game regulations for hunting purposes, a frangible projectile designed for target shooting, a projectile which the Secretary of the Treasury of the United States finds is primarily intended to be used for sporting purposes, or any other projectile or projectile core which the Secretary of the Treasury of the United States finds is intended to be used for industrial purposes, including a charge used in an oil and gas well perforating device.

(8) "Flanged ammunition" means ammunition with a soft lead core and having sharp flanges which are designed to expand on impact.

237.070 Prohibition against sale or transfer of firearm to convicted felon.

(1) No person shall knowingly sell or transfer a firearm to any person prohibited from possessing it by KRS 527.040.

(2) Any person who violates the provisions of subsection (1) of this section is guilty of a Class A misdemeanor.

(3) Any firearm transferred in violation of this section shall be subject to forfeiture and shall be disposed of pursuant to KRS 237.090.

237.080 Prohibition against manufacture, sale, delivery, transfer, or importation of armor-piercing ammunition - Exceptions.

(1) It shall be unlawful for any person to knowingly manufacture, sell, deliver, transfer, or import armor-piercing ammunition.

(2) Subsection (1) of this section shall not apply to members of the Armed Forces of the United States or law enforcement officers within the scope of their duties, nor shall it prohibit licensed gun dealers from possessing armor-piercing ammunition for the purpose of receiving and transferring it to members of the Armed Forces of the United States, or law enforcement officers for use within the scope of their duties.

(3) A violation of subsection (1) of this section shall be a Class D felony for the first offense and a Class C felony for each subsequent offense.

(4) Any armor-piercing ammunition transferred, sold, or offered for sale, in violation of this section is contraband and shall be seized and summarily forfeited to the state and shall be disposed of pursuant to KRS 237.090.

237.090 Disposition of forfeited firearm or ammunition. Any firearm or ammunition forfeited pursuant to KRS 237.060 to 237.090 shall, upon order of a court of competent jurisdiction, be disposed of or retained as provided in KRS 500.090.

237.095. Persons barred by federal law from purchase of firearms - Duty to notify courts and law enforcement agencies of purchase or attempt to purchase - Protocol for providing notice - Duty to notify petitioner - Immunity from liability.

(1) Upon receiving notice that a person barred from purchasing a firearm under 18 U.S.C. sec. 922(g)(8) has purchased or attempted to purchase a firearm, any agency with the responsibility of entering domestic violence records into the Law Information Network of Kentucky shall notify:

(a) The court in the jurisdiction where the domestic violence order was issued under KRS 403.750; and

(b) The law enforcement agencies, as designated by the Department of Kentucky State Police, that have jurisdiction in the county where the domestic violence order was issued and in the county of the victim's residence if different from the county where the domestic violence order was issued.

(2) The Department of Kentucky State Police shall develop a protocol for providing notice to the required court and law enforcement agencies under subsection (1) of this section. Within the protocol, the Department of Kentucky State Police shall designate which local law enforcement agencies are to receive notice in each county. A minimum of one (1) law enforcement agency shall be designated in each county.

(3) When a designated law enforcement agency for the county where the domestic violence order was issued or where the victim resides receives notice under subsection (1)(b) of this section, that agency shall make reasonable efforts to ensure that the petitioner who obtained the domestic violence order is notified that the respondent has purchased or attempted to purchase a firearm.

(4) Any person carrying out responsibilities under this section shall be immune from civil liability for good faith conduct in carrying out those responsibilities.

(5) This section shall apply only to domestic violence orders issued, or reissued, on or after July 14, 2000, through July 15, 2002.

237.100. Notification of purchase of firearm or attempt to purchase firearm -- Immunity.

(1) Upon receipt of notice that a person barred from purchasing a firearm under 18 U.S.C. sec. 922(g)(8) has purchased or attempted to purchase a firearm, the Justice and Public Safety Cabinet shall make a reasonable effort to provide notice to the petitioner who obtained the domestic violence order issued under KRS 403.750 that the respondent to the order has attempted to purchase a firearm. The Justice and Public Safety Cabinet may contract with a private entity in order to provide notification.

(2) The notification shall be limited to a petitioner who has:

(a) Received a domestic violence protective order issued or reissued under KRS 403.750 on or after July 15, 2002;

(b) Received a domestic violence protective order that involves a respondent who is prohibited by 18 U.S.C. sec. 922(g)(8) from possessing a firearm; and

(c) Provided the Justice and Public Safety Cabinet or the entity with a request for notification.

(3) Any person carrying out responsibilities under this section shall be immune from civil liability for good faith conduct in carrying out those responsibilities. Nothing in this subsection shall limit liability for negligence.

237.110 License to carry concealed deadly weapon - Criteria - Suspension or revocation - Prohibitions - Reciprocity - Reports.

(1) The Department of Kentucky State Police is authorized to issue and renew licenses to carry concealed firearms or other deadly weapons, or a combination thereof, to persons qualified as provided in this section.

(2) An original or renewal license issued pursuant to this section shall:

(a) Be valid throughout the Commonwealth and, except as provided in this section or other specific section of the Kentucky Revised Statutes or federal law, permit the holder of the license to carry firearms, ammunition, or other deadly weapons, or a combination thereof at any location in the Commonwealth.

(b) Unless revoked as provided by law, be valid for a period of five (5) years from the date of issuance;

(c) Authorize the holder of the license to carry a concealed firearm or other deadly weapon, or a combination thereof, on or about his or her person; and

(d) Authorize the holder of the license to carry ammunition for a firearm on or about his or her person.

(3) Prior to the issuance of an original or renewal license to carry a concealed deadly weapon, the Department of Kentucky State Police shall conduct a background check to ascertain whether the applicant is eligible, under 18 U.S.C. sec. 922(g) and (n), any other applicable federal law, and state law, to purchase, receive, or possess a firearm or ammunition, or both. The background check shall include:

(a) A state records check covering the items specified in this subsection, together with any other requirements of this section;

(b) A federal records check, which shall include a National Instant Criminal Background Check System (NICS) check;

(c) A federal immigration alien query if the person is an alien who has been lawfully admitted to the United States by the United States government or an agency thereof; and

(d) In addition to the immigration alien query, if the applicant has not been lawfully admitted to the United States under permanent resident status, the Department of Kentucky State Police shall, if a doubt exists relating to an alien's eligibility to purchase a firearm, consult with the United States Department of Homeland Security, United States Department of Justice, United States Department of State, or other federal agency to confirm whether the alien is eligible to purchase a firearm in the United States, bring a firearm into the United States, or possess a firearm in the United States under federal law.

(4) The Department of Kentucky State Police shall issue an original or renewal license if the applicant:

(a) Is not prohibited from the purchase, receipt, or possession of firearms, ammunition, or both pursuant to 18 U.S.C. 922(g), 18 U.S.C. 922(n), or applicable federal or state law;

(b) 1. Is a citizen of the United States who is a resident of this Commonwealth and has been a resident for six (6) months or longer immediately preceding the filing of the application;

2. Is a citizen of the United States who is a member of the Armed Forces of the United States who is on active duty, who is at the time of application assigned to a military posting in Kentucky, and who has been assigned to a posting in the Commonwealth for six (6) months or longer immediately preceding the filing of the application;

3. Is lawfully admitted to the United States by the United States government or an agency thereof, is permitted by federal law to purchase a firearm, and has been a resident of this Commonwealth for six (6) months or longer immediately preceding the filing of the application; or

4. Is lawfully admitted to the United States by the United States government or an agency thereof, is permitted by federal law to purchase a handgun, is, at the time of the application, assigned to a military posting in Kentucky, and has been assigned to a posting in the Commonwealth for six (6) months or longer immediately preceding the filing of the application;

(c) Is twenty-one (21) years of age or older;

(d) Has not been committed to a state or federal facility for the abuse of a controlled substance or been convicted of a misdemeanor violation of KRS Chapter 218A or similar laws of any other state relating to controlled substances, within a three (3) year period immediately preceding the date on which the application is submitted;

(e) Does not chronically and habitually use alcoholic beverages as evidenced by the applicant having two (2) or more convictions for violating KRS 189A.010 within the three (3) years immediately preceding the date on which the application is submitted, or having been committed as an alcoholic pursuant to KRS Chapter 222 or similar laws of another state within the three (3) year period immediately preceding the date on which the application is submitted;

(f) Does not owe a child support arrearage which equals or exceeds the cumulative amount which would be owed after one (1) year of nonpayment, if the Department of Kentucky State Police has been notified of the arrearage by the Cabinet for Health and Family Services;

(g) Has complied with any subpoena or warrant relating to child support or paternity proceedings. If the Department of Kentucky State Police has not been notified by the Cabinet for Health and Family Services that the applicant has failed to meet this requirement, the Department of Kentucky State Police shall assume that paternity and child support proceedings are not an issue;

(h) Has not been convicted of a violation of KRS 508.030 or 508.080 within the three (3) years immediately preceding the date on which the application is submitted. The commissioner of the Department of Kentucky State Police may waive this requirement upon good cause shown and a determination that the applicant is not a danger and that a waiver would not violate federal law; and

(i) Demonstrates competence with a firearm by successful completion of a firearms safety course offered or approved by the Department of Criminal Justice Training. The firearms safety course shall:

1. Be not more than eight (8) hours in length;

2. Include instruction on handguns, the safe use of handguns, the care and cleaning of handguns, and handgun marksmanship principles;

3. Include actual range firing of a handgun in a safe manner, and the firing of not more than twenty (20) rounds at a full-size silhouette target, during which firing, not less than eleven (11) rounds must hit the silhouette portion of the target; and

4. Include information on and a copy of laws relating to possession and carrying of firearms, as set forth in KRS Chapters 237 and 527, and the laws relating to the use of force, as set forth in KRS Chapter 503.

(5) A legible photocopy of the certificate of completion issued by the Department of Criminal Justice Training shall constitute evidence of qualification under subsection (4)(i) of this section.

(6) (a) Peace officers who are currently certified as peace officers by the Kentucky Law Enforcement Council pursuant to KRS 15.380 to 15.404 and peace officers who are retired and are members of the Kentucky Employees Retirement System, State Police Retirement System, or County Employees Retirement System or other retirement system operated by or for a city, county, or urban-county in Kentucky shall be deemed to have met the training requirement.

(b) Current and retired peace officers of the following federal agencies shall be deemed to have met the training requirement:

1. Any peace officer employed by a federal agency specified in KRS 61.365;

2. Any peace officer employed by a federal civilian law enforcement agency not specified above who has successfully completed the basic law enforcement training course required by that agency;

3. Any military peace officer of the United States Army, Navy, Marine Corps, or Air Force, or a reserve component thereof, or of the Army Reserve or Air Force Reserve who has successfully completed the military law enforcement training course required by that branch of the military; and

4. Any member of the United States Coast Guard serving in a peace officer role who has successfully completed the law enforcement training course specified by the United States Coast Guard.

(b) A license which has expired shall be void and shall not be valid for any purpose other than surrender to the sheriff in exchange for a renewal license.

(15) The licensee shall carry the license at all times the licensee is carrying a concealed firearm or other deadly weapon and shall display the license upon request of a law enforcement officer. Violation of the provisions of this subsection shall constitute a noncriminal violation with a penalty of twenty-five dollars ($25), payable to the clerk of the District Court, but no court costs shall be assessed.

(16) Except as provided in KRS 527.020, no license issued pursuant to this section shall authorize any person to carry a concealed firearm into:

(a) Any police station or sheriff's office;

(b) Any detention facility, prison, or jail;

(c) Any courthouse, solely occupied by the Court of Justice courtroom, or court proceeding;

(d) Any meeting of the governing body of a county, municipality, or special district; or any meeting of the General Assembly or a committee of the General Assembly, except that nothing in this section shall preclude a member of the body, holding a concealed deadly weapon license, from carrying a concealed deadly weapon at a meeting of the body of which he or she is a member;

(e) Any portion of an establishment licensed to dispense beer or alcoholic beverages for consumption on the premises, which portion of the establishment is primarily devoted to that purpose;

(f) Any elementary or secondary school facility without the consent of school authorities as provided in KRS 527.070, any child-caring facility as defined in KRS 199.011, any day-care center as defined in KRS 199.894, or any certified family child-care home as defined in KRS 199.8982, except however, any owner of a certified child-care home may carry a concealed firearm into the owner's residence used as a certified child-care home;

(g) An area of an airport to which access is controlled by the inspection of persons and property; or

(h) Any place where the carrying of firearms is prohibited by federal law.

(17) The owner, business or commercial lessee, or manager of a private business enterprise, day-care center as defined in KRS 199.894 or certified or licensed family child-care home as defined in KRS 199.8982, or a health-care facility licensed under KRS Chapter 216B, except facilities renting or leasing housing, may prohibit persons holding concealed deadly weapon licenses from carrying concealed deadly weapons on the premises and may prohibit employees, not authorized by the employer, holding concealed deadly weapons licenses from carrying concealed deadly weapons on the property of the employer. If the building or the premises are open to the public, the employer or business enterprise shall post signs on or about the premises if carrying concealed weapons is prohibited. Possession of weapons, or ammunition, or both in a vehicle on the premises shall not be a criminal offense so long as the weapons, or ammunition, or both are not removed from the vehicle or brandished while the vehicle is on the premises. A private but not a public employer may prohibit employees or other persons holding a concealed deadly weapons license from carrying concealed deadly weapons, or ammunition, or both in vehicles owned by the employer, but may not prohibit employees or other persons holding a concealed deadly weapons license from carrying concealed deadly weapons, or ammunition, or both in vehicles owned by the employee, except that the Justice and Public Safety Cabinet may prohibit an employee from carrying any weapons, or ammunition, or both other than the weapons, or ammunition, or both issued or authorized to be used by the employee of the cabinet, in a vehicle while transporting persons under the employee's supervision or jurisdiction. Carrying of a concealed weapon, or ammunition, or both in a location specified in this subsection by a license holder shall not be a criminal act but may subject the person to denial from the premises or removal from the premises, and, if an employee of an employer, disciplinary measures by the employer.

(20)(a) A person who has a valid license issued by another state of the United States to carry a concealed deadly weapon in that state may, subject to provisions of Kentucky law, carry a concealed deadly weapon in Kentucky, and his or her license shall be considered as valid in Kentucky.

237.990 Penalties.

(1) Any person who violates any of the provisions of KRS 237.030 to 237.050 shall be guilty of a Class D felony.

(2) Any person who violates any of the provisions of KRS 237.030 to 237.050, and in so doing uses any destructive device or booby trap device to avoid detection by law enforcement or other government personnel or to avoid theft or detection by any other person, of any controlled substance as set forth in KRS Chapter 218A and held in violation of KRS 218A.140, shall be guilty of a Class C felony.

Title L. Kentucky Penal Code

Chapter 527. Offenses Relating to Firearms and Weapons

527.010 Definitions for chapter. The following definitions apply in this chapter unless the context otherwise requires:

(1) "Booby trap device" shall have the same meaning as set forth in KRS 237.030.

(2) "Deface" means to remove, deface, cover, alter, or destroy the manufacturer's serial number or any other distinguishing number or identification mark.

(3) "Destructive device" shall have the same meaning as set forth in KRS 237.030.

(4) "Firearm" means any weapon which will expel a projectile by the action of an explosive.

(5) "Handgun" means any pistol or revolver originally designed to be fired by the use of a single hand, or any other firearm originally designed to be fired by the use of a single hand.

527.030 Defacing a firearm.

(1) A person is guilty of defacing a firearm when he intentionally defaces a firearm.

(2) Defacing a firearm is a Class A misdemeanor.

527.040 Possession of firearm by convicted felon - Exceptions.

(1) A person is guilty of possession of a firearm by a convicted felon when he possesses, manufactures, or transports a firearm when he has been convicted of a felony, as defined by the laws of the jurisdiction in which he was convicted, in any state or federal court and has not:

(a) Been granted a full pardon by the Governor or by the President of the United States;

(b) Been granted relief by the United States Secretary of the Treasury pursuant to the Federal Gun Control Act of 1968, as amended.

(2) Possession of a firearm by a convicted felon is a Class D felony unless the firearm possessed is a handgun in which case it is a Class C felony.

(3) The provisions of this section shall apply to any youthful offender convicted of a felony offense under the laws of this Commonwealth. The exceptions contained in KRS 527.100 prohibiting possession of a handgun by a minor shall not apply to this section.

(4) The provisions of this section with respect to handguns, shall apply only to persons convicted after January 1, 1975, and with respect to other firearms, to persons convicted after July 15, 1994.

527.050 Possession of defaced firearm.

(1) A person is guilty of possession of a defaced firearm when he knowingly possesses a defaced firearm unless he makes a report to the police or other appropriate government agency of such possession prior to arrest or authorization of a warrant by a court.

(2) Possession of a defaced firearm is a Class A misdemeanor.

527.060 Forfeiture. Upon the conviction of any person for the violation of any law of this Commonwealth in which a deadly weapon was used, displayed or unlawfully possessed by such person the court shall order the weapon forfeited to the state and sold, destroyed or otherwise disposed of in accordance with KRS 500.090.

527.070 Unlawful possession of a weapon on school property - Posting of sign - Exemptions.

(1) A person is guilty of unlawful possession of a weapon on school property when he knowingly deposits, possesses, or carries, whether openly or concealed, for purposes other than instructional or school-sanctioned ceremonial purposes, or the purposes permitted in subsection (3) of this section, any firearm or other deadly weapon, destructive device, or booby trap device in any public or private school building or bus, on any public or private school campus, grounds, recreation area, athletic field, or any other property owned, used, or operated by any board of education, school, board of trustees, regents, or directors for the administration of any public or private educational institution. The provisions of this section shall not apply to institutions of postsecondary or higher education.

(2) Each chief administrator of a public or private school shall display about the school in prominent locations, including, but not limited to, sports arenas, gymnasiums, stadiums, and cafeterias, a sign at least six (6) inches high and fourteen (14) inches wide stating:

UNLAWFUL POSSESSION OF A WEAPON ON SCHOOL PROPERTY IN KENTUCKY IS A FELONY PUNISHABLE BY A MAXIMUM OF FIVE (5) YEARS IN PRISON AND A TEN THOUSAND DOLLAR ($10,000) FINE.

Failure to post the sign shall not relieve any person of liability under this section.

(3) The provisions of this section prohibiting the unlawful possession of a weapon on school property shall not apply to:

(a) An adult who possesses a firearm, if the firearm is contained within a vehicle operated by the adult and is not removed from the vehicle, except for a purpose permitted herein, or brandished by the adult, or by any other person acting with expressed or implied consent of the adult, while the vehicle is on school property;

(b) Any pupils who are members of the reserve officers training corps or pupils enrolled in a course of instruction or members of a school club or team, to the extent they are required to carry arms or weapons in the discharge of their official class or team duties;

(c) Any peace officer or police officer authorized to carry a concealed weapon pursuant to KRS 527.020;

(d) Persons employed by the Armed Forces of the United States or members of the National Guard or militia when required in the discharge of their official duties to carry arms or weapons;

(e) Civil officers of the United States in the discharge of their official duties. Nothing in this section shall be construed as to allow any person to carry a concealed weapon into a public or private elementary or secondary school building;

(f) Any other persons, including, but not limited to, exhibitors of historical displays, who have been authorized to carry a firearm by the board of education or board of trustees of the public or private institution;

(g) A person hunting during the lawful hunting season on lands owned by any public or private educational institution and designated as open to hunting by the board of education or board of trustees of the educational institution;

(h) A person possessing unloaded hunting weapons while traversing the grounds of any public or private educational institution for the purpose of gaining access to public or private lands open to hunting with the intent to hunt on the public or private lands, unless the lands of the educational institution are posted prohibiting the entry; or

(i) A person possessing guns or knives when conducting or attending a "gun and knife show" when the program has been approved by the board of education or board of trustees of the educational institution.

(4) Unlawful possession of a weapon on school property is a Class D felony.

527.100 Possession of handgun by minor.

(1) A person is guilty of possession of a handgun by a minor when, being under the age of eighteen (18) years, he possesses, manufactures, or transports a handgun as defined by KRS 527.010, except when the person is:

(a) In attendance at a hunter's safety course or a firearms safety course;

(b) Engaging in practice in the use of a firearm, or target shooting at an established firing range, or any other area where the discharge of a firearm is not prohibited;

(c) Engaging in an organized competition involving the use of a firearm, or participating in or practicing for a performance by a group organized under Section 501(c)(3) of the Internal Revenue Code or any successor thereto which uses firearms as a part of the performance;

(d) Hunting or trapping pursuant to a valid license issued to him pursuant to the statutes or administrative regulations of this Commonwealth;

(e) Traveling to or from any activity described in paragraphs (a) to (d) of this subsection with any unloaded handgun in his possession;

(f) On real property which is under the control of an adult and has the permission of that adult and his parent or legal guardian to possess a handgun; or

(g) At his residence and with the permission of his parent or legal guardian possesses a handgun and is justified under the principles of justification set forth in KRS Chapter 503 in using physical force or deadly physical force.

(2) For the purposes of subsection (1) of this section, a handgun is "loaded" if:

(a) There is a cartridge in the chamber of the handgun; or

(b) There is a cartridge in the cylinder of the handgun, if the handgun is a revolver; or

(c) There is a cartridge in the magazine of a semiautomatic handgun, if the magazine is attached to the handgun; or

(d) The handgun and the ammunition for the handgun, are carried on the person of one under the age of eighteen (18) years or are in such close proximity to him that he could readily gain access to the handgun and the ammunition and load the handgun.

(3) Possession of a handgun by a minor is a Class A misdemeanor for the first offense and a Class D felony for each subsequent offense.

527.110 Unlawfully providing handgun to juvenile or permitting juvenile to possess handgun.

(1) A person is guilty of unlawfully providing a handgun to a juvenile or permitting a juvenile to possesses a handgun when he:

(a) Intentionally, knowingly, or recklessly provides a handgun, with or without remuneration, in violation of KRS 527.040, 527.100, or 600.020 to any person he knows or has reason to believe is under the age of eighteen (18) years; or

(b) Is the parent or legal guardian of a juvenile and intentionally, knowingly, or recklessly provides a handgun to the juvenile or permits the juvenile to possess a handgun knowing that there is a substantial risk that the juvenile will use a handgun to commit a felony offense; or, with knowledge that the juvenile has been convicted of a crime of violence as defined in KRS 439.3401 or has been adjudicated a public offender of an offense which would constitute a crime of violence as defined in KRS 439.3401, provides a handgun to that juvenile.

(2) Unlawfully providing a handgun to a juvenile or permitting a juvenile to possess a handgun is a Class D felony.

[Current through 2008 First Extraordinary Session]

LOUISIANA

LA. REV. STAT.

Title 14. Criminal Law

67.15. Theft of a firearm

A. Theft of a firearm is the misappropriation or taking of a firearm which belongs to another, either without the consent of the other to the misappropriation or taking or by means of fraudulent conduct, practices, or representations. An intent to deprive the other permanently of the firearm is essential.

B. For purposes of this Section, "firearm" means a shotgun or rifle, or a pistol, revolver, or other handgun.

C.(1) For a first offense, the penalty for theft of a firearm shall be imprisonment with or without hard labor for not less than two years nor more than ten years, without the benefit of probation, parole, or suspension of sentence and a fine of one thousand dollars.

(2) For a second offense, the penalty for theft of a firearm shall be imprisonment with or without hard labor for not less than five years nor more than fifteen years, without the benefit of probation, parole, or suspension of sentence and a fine of two thousand dollars.

(3) For a third and subsequent offense, the penalty for theft of a firearm shall be imprisonment at hard labor for not less than fifteen years nor more than thirty years, without the benefit of probation, parole, or suspension of sentence and a fine of five thousand dollars.

69.1. Illegal possession of stolen firearms

A. Illegal possession of stolen firearms is the intentional possessing, procuring, receiving, or concealing of a firearm which has been the subject of any robbery or theft under circumstances which indicate that the offender knew or should have known that the firearm was the subject of a robbery or theft.

B. Whoever commits the crime of illegal possession of firearms shall be punished as follows:

(1) For a first offense, the penalty shall be imprisonment, with or without hard labor, for not less than one year nor more than five years.

(2) For second and subsequent offenses, the penalty shall be imprisonment, with or without hard labor, for not less than two years nor more than ten years.

91. Unlawful sales of weapons to minors

A. Unlawful sales of weapons to minors is the selling or otherwise delivering for value of any firearm or other instrumentality customarily used as a dangerous weapon to any person under the age of eighteen. Lack of knowledge of the minor's age shall not be a defense.

B. Whoever commits the crime of unlawful sales of weapons to minors shall be fined not more than three hundred dollars or imprisoned for not more than six months, or both.

95. Illegal carrying of weapons

A. Illegal carrying of weapons is:

(1) The intentional concealment of any firearm, or other instrumentality customarily used or intended for probable use as a dangerous weapon, on one's person; or

(2) The ownership, possession, custody or use of any firearm, or other instrumentality customarily used as a dangerous weapon, at any time by an enemy alien; or

(3) The ownership, possession, custody or use of any tools, or dynamite, or nitroglycerine, or explosives, or other instrumentality customarily used by thieves or burglars at any time by any person with the intent to commit a crime; or

(4) The manufacture, ownership, possession, custody or use of any switchblade knife, spring knife or other knife or similar instrument having a blade which may be automatically unfolded or extended from a handle by the manipulation of a button, switch, latch or similar contrivance.

(5)(a) The intentional possession or use by any person of a dangerous weapon on a school campus during regular school hours or on a school bus. "School" means any elementary, secondary, high school, or vo-tech school in this state and "campus" means all facilities and property within the boundary of the school property. "School bus" means any motor bus being used to transport children to and from school or in connection with school activities.

(b) The provisions of this Paragraph shall not apply to:

(i) A peace officer as defined by R.S. 14:30(B) in the performance of his official duties.

(ii) A school official or employee acting during the normal course of his employment or a student acting under the direction of such school official or employee.

(iii) Any person having the written permission of the principal or school board and engaged in competition or in marksmanship or safety instruction.

B. (1) Whoever commits the crime of illegal carrying of weapons shall be fined not more than five hundred dollars, or imprisoned for not more than six months, or both.

(2) Whoever commits the crime of illegal carrying of weapons with any firearm used in the commission of a crime of violence as defined in R.S. 14:2(13), shall be fined not more than two thousand dollars, or imprisoned, with or without hard labor, for not less than one year nor more than two years, or both. Any sentence issued pursuant to the provisions of this Paragraph and any sentence issued pursuant to a violation of a crime of violence as defined in R.S. 14:2(13) shall be served consecutively.

C. On a second conviction, the offender shall be imprisoned with or without hard labor for not more than five years.

D. On third and subsequent convictions, the offender shall be imprisoned with or without hard labor for not more than ten years without benefit of parole, probation, or suspension of sentence.

E. If the offender uses, possesses, or has under his immediate control any firearm, or other instrumentality customarily used or intended for probable use as a dangerous weapon, while committing or attempting to commit a crime of violence or while in the possession of or during the sale or distribution of a controlled dangerous substance, the offender shall be fined not more than ten thousand dollars and imprisoned at hard labor for not less than five nor more than ten years without the benefit of probation, parole, or suspension of sentence. Upon a second or subsequent conviction, the offender shall be imprisoned at hard labor for not less than twenty years nor more than thirty years without the benefit of probation, parole, or suspension of sentence.

F. The enhanced penalty upon second, third, and subsequent convictions shall not be applicable in cases where more than five years have elapsed since the expiration of the maximum sentence, or sentences, of the previous conviction or convictions, and the time of the commission of the last offense for which he has been convicted; the sentence to be imposed in such event shall be the same as may be imposed upon a first conviction.

G. (1) The provisions of this Section except Paragraph (4) of Subsection A shall not apply to sheriffs and their deputies, state and city police, constables and town marshals, or persons vested with police power when in the actual discharge of official duties. These provisions shall not apply to sheriffs and their deputies and state and city police who are not actually discharging their official duties, provided that such persons are full time, active, and certified by the Council on Peace Officer Standards and Training and have on their persons valid identification as duly commissioned law enforcement officers.

(2) The provisions of this Section except Paragraph (4) of Subsection A shall not apply to any law enforcement officer who is retired from full-time active law enforcement service with at least twelve years service upon retirement, nor shall it apply to any enforcement officer of the office of state parks, in the Department of Culture, Recreation and Tourism who is retired from active duty as an enforcement officer, provided that such retired officers have on their persons valid identification as retired law enforcement officers, which identification shall be provided by the entity which employed the officer prior to his or her public retirement. The retired law enforcement officer must be qualified annually in the use of firearms by the Council on Peace

Officer Standards and Training and have proof of such qualification. This exception shall not apply to such officers who are medically retired based upon any mental impairment.

(3)(a) The provisions of this Section except Paragraph (4) of Subsection A shall not apply to active or retired reserve or auxiliary law enforcement officers qualified annually by the Council on Peace Officer Standards and Training and who have on their person valid identification as active or retired reserve law or auxiliary municipal police officers. The active or retired reserve or auxiliary municipal police officer shall be qualified annually in the use of firearms by the Council on Peace Officer Standards and Training and have proof of such certification.

(b) For the purposes of this Paragraph, a reserve or auxiliary municipal police officer shall be defined as a volunteer, non-regular, sworn member of a law enforcement agency who serves with or without compensation and has regular police powers while functioning as such agency's representative, and who participates on a regular basis in agency activities including, but not limited to those pertaining to crime prevention or control, and the preservation of the peace and enforcement of the law.

H. The provisions of this Section shall not prohibit active justices or judges of the supreme court, courts of appeal, district courts, parish courts, juvenile courts, family courts, city courts, and traffic courts, constables, coroners, district attorneys and designated assistant district attorneys, United States Attorneys and Assistant United States Attorneys and investigators, and justices of the peace from possessing and concealing a handgun on their person when the justice or judge, constable, coroner, district attorneys and designated assistant district attorneys, United States Attorneys and Assistant United States Attorneys and investigators, or Justices of the peace are certified by the Council on Peace Officer Standards and Training.

I. The provisions of this Section shall not prohibit the carrying of a concealed handgun by a person who is a college or university police officer under the provisions of R.S. 17:1805 and who is carrying a concealed handgun in accordance with the provisions of that statute.

J. The provisions of this Section shall not prohibit the ownership of rescue knives by commissioned full-time law enforcement officers. The provisions of this Section shall not prohibit the carrying of rescue knives by commissioned full-time law enforcement officers who are in the actual discharge of their official duties. The provisions of this Section shall not prohibit the sale of rescue knives to commissioned full-time law enforcement officers. The provisions of this Section shall not prohibit the ownership or possession of rescue knives by merchants who own or possess the knives solely as inventory to be offered for sale to commissioned full-time law enforcement officers. As used in this Subsection, a "rescue knife" is a folding knife, which can be readily and easily opened with one hand and which has at least one blade which is designed to be used to free individuals who are trapped by automobile seat belts, or at least one blade which is designed for a similar purpose. No blade of a rescue knife shall exceed five inches in length.

K. (1) The provisions of this Section shall not prohibit a retired justice, or judge from the supreme court, courts of appeal, district courts, parish courts, juvenile courts, family courts, and city courts from possessing and concealing a handgun on their person provided that such retired justice of judge is certified by the Council of Peace Officers and Training and has on their person a valid identification showing proof of their status as a retired justice or judge.

(2) The retired justice or judge shall be qualified annually in the use of firearms by the Council of Peace Officers and Training and have proof of such certification. However, this Subsection shall not apply to a retired justice or judge who is medically retired based upon any medical impairment.

95.1. Possession of firearm or carrying concealed weapon by a person convicted of certain felonies

A. It is unlawful for any person who has been convicted of a crime of violence as defined in R.S. 14:2(13) which is a felony or simple burglary, burglary of a pharmacy, burglary of an inhabited dwelling, unauthorized entry of an inhabited dwelling, felony illegal use of weapons or dangerous instrumentalities, manufacture or possession of a delayed action incendiary device, manufacture or possession of a bomb, or any violation of the Uniform Controlled Substances Law which is a felony, or any crime which is defined as a sex offense in R.S. 15:541(14.1), or any crime defined as an at-tempt to commit one of the above enumerated offenses under the laws of this state, or who has been convicted under the laws of any other state or of the United States or of any foreign government or country of a crime which, if committed in this state, would be one of the above enumerated crimes, to possess a firearm or carry a concealed weapon.

B. Whoever is found guilty of violating the provisions of this Section shall be imprisoned at hard labor for not less than ten nor more than fifteen years without the benefit of probation, parole, or suspension of sentence and be fined not less than one thousand dollars nor more than five thousand dollars.

C. Except as otherwise specifically provided, this Section shall not apply to the following cases:

(1) The provisions of this Section prohibiting the possession of firearms and carrying concealed weapons by persons who have been convicted of certain felonies shall not apply to any person who has not been convicted of any felony for a period of ten years from the date of completion of sentence, probation, parole, or suspension of sentence.

(2) Upon completion of sentence, probation, parole, or suspension of sentence the convicted felon shall have the right to apply to the sheriff of the parish in which he resides, or in the case of Orleans Parish the superintendent of police, for a permit to possess firearms. The felon shall be entitled to possess the firearm upon the issuing of the permit.

(3) The sheriff or superintendent of police, as the case may be, shall immediately notify the Department of Public Safety, in writing, of the issuance of each permit granted under this Section.

95.1.1. Illegally supplying a felon with a firearm

A. Illegally supplying a felon with a firearm is the intentional giving, selling, donating, providing, lending, delivering, or otherwise transferring a firearm to any person known by the offender to be a person convicted of a felony and prohibited from possessing a firearm as provided for in R.S. 14:95.1.

B. Whoever commits the crime of illegally supplying a felon with a firearm shall be imprisoned for not more than five years and may be fined not less than one thousand dollars nor more than five thousand dollars. At least one year of the sentence imposed shall be served without benefit of parole, probation, or suspension of sentence.

95.2. Carrying a firearm, or dangerous weapon, by a student or nonstudent on school property, at school-sponsored functions or firearm-free zone

A. Carrying a firearm, or dangerous weapon as defined in R.S. 14:2, by a student or nonstudent on school property, at a school sponsored function, or in a firearm-free zone is unlawful and shall be defined as possession of any firearm or dangerous weapon, on one's person, at any time while on a school campus, on school transportation, or at any school sponsored function in a specific designated area including but not limited to athletic competitions, dances, parties, or any extracurricular activities, or within one thousand feet of any school campus.

B. For purposes of this Section, the following words have the following meanings:

(1) "School" means any elementary, secondary, high school, vocational-technical school, college, or university in this state.

(2) "Campus" means all facilities and property within the boundary of the school property.

(3) "School bus" means any motor bus being used to transport children to and from school or in connection with school activities.

(4) "Nonstudent" means any person not registered and enrolled in that school or a suspended student who does not have permission to be on the school campus.

C. The provisions of this Section shall not apply to:

(1) A federal, state, or local law enforcement officer in the performance of his official duties.

(2) A school official or employee acting during the normal course of his employment or a student acting under the direction of such school official or employee.

(3) Any person having the written permission of the principal.

(4) The possession of a firearm occurring within one thousand feet of school property and entirely on private property, or entirely within a private residence, or in accordance with a concealed handgun permit issued pursuant to R.S. 40:1379.1.

(5) Any constitutionally protected activity which cannot be regulated by the state, such as a firearm contained entirely within a motor vehicle.

(6) Any student carrying a firearm to or from a class, in which he is duly enrolled, that requires the use of the firearm in the class.

(7) A student enrolled or participating in an activity requiring the use of a firearm including but not limited to any ROTC function under the authorization of a university.

(8) A student who possesses a firearm in his dormitory room or while going to or from his vehicle or any other person with permission of the administration.

D. (1) Whoever commits the crime of carrying a firearm, or a dangerous weapon as defined in R.S. 14:2, by a student or nonstudent on school property, at a school-sponsored function, or in a firearm-free zone shall be imprisoned at hard labor for not more than five years.

(2) Whoever commits the crime of carrying a firearm, or a dangerous weapon as defined in R.S. 14:2, on school property or in a firearm-free zone with the firearm or dangerous weapon being used in the commission of a crime of violence as defined in R.S. 14:2(13) on school property or in a firearm-free zone, shall be fined not more than two thousand dollars, or imprisoned, with or without hard labor, for not less than one year nor more than five years, or both. Any sentence issued pursuant to the provisions of this Paragraph and any sentence issued pursuant to a violation of a crime of violence as defined in R.S. 14:2(13) shall be served consecutively. Upon commitment to the Department of Public Safety and Corrections after conviction for a crime committed on school property, at a school-sponsored function or in a firearm-free zone, the department shall have the offender evaluated through appropriate examinations or tests conducted under the supervision of the department. Such evaluation shall be made within thirty days of the order of commitment.

E. Lack of knowledge that the prohibited act occurred on or within one thousand feet of school property shall not be a defense.

F. (1) School officials shall notify all students and parents of the impact of this legislation and shall post notices of the impact of this Section at each major point of entry to the school. These notices shall be maintained as permanent notices.

(2)(a) If a student is detained by the principal or other school official for violation of this Section or the school principal or other school official confiscates or seizes a firearm or concealed weapon from a student while upon school property, at a school function, or on a school bus, the principal or other school official in charge at the time of the detention or seizure shall immediately report the detention or seizure to the police department or sheriff's department where the school is located and shall deliver any firearm or weapon seized to that agency.

(b) The confiscated weapon shall be disposed of or destroyed as provided by law.

(3) If a student is detained pursuant to Paragraph (2) of this Subsection for carrying a concealed weapon on campus, the principal shall immediately notify the student's parents.

(4) If a person is arrested for carrying a concealed weapon on campus by a university or college police officer, the weapon shall be given to the sheriff, chief of police, or other officer to whom custody of the arrested person is transferred as provided by R.S. 17:1805(B).

G. Any principal or school official in charge who fails to report the detention of a student or the seizure of a firearm or concealed weapon to a law enforcement agency as required by Paragraph F(2) of this Section within seventy-two hours of notice of the detention or seizure may be issued a misdemeanor summons for a violation hereof and may be fined not more than five hundred dollars or sentenced to not more than forty hours of community service, or both. Upon successful completion of the community service or payment of the fine, or both, the arrest and conviction shall be set aside as provided for in Code of Criminal Procedure Article 894(B).

95.6. Firearm-free zone; notice; signs; crime; penalties

A. A "firearm-free zone" is an area inclusive of any school campus and within one thousand feet of any such school campus, and within a school bus.

B. The provisions of this Section shall not apply to:

(1) A federal, state, or local law enforcement building.

(2) A military base.

(3) A commercial establishment which is permitted by law to have firearms or armed security.

(4) Private premises where a firearm is kept pursuant to law.

(5) Any constitutionally protected activity within the firearm-free zone, such as a firearm contained entirely within a motor vehicle.

C. For purposes of this Section:

(1) "School" means any public or private elementary, secondary, high school, or vocational-technical school, college, or university in this state.

(2) "School campus" means all facilities and property within the boundary of the school property.

(3) "School bus" means any motor bus being used to transport children to and from school or in connection with school activities.

D. The local governing authority which has jurisdiction over zoning matters in which each firearm-free zone is located shall publish a map clearly indicating the boundaries of each firearm-free zone in accordance with the specifications in Subsection A. The firearm-free zone map shall be made an official public document and placed with the clerk of court for the parish or parishes in which the firearm-free zone is located.

E. The state superintendent of education, with the approval of the State Board of Elementary and Secondary Education, and the commissioner of higher education, with the approval of the Board of Regents, shall develop a method by which to mark firearm-free zones, including the use of signs or other markings suitable to the situation. Signs or other markings shall be located in a visible manner on or near each school and on and in each school bus indicating that such area is a firearm-free zone and that such zone extends to one thousand feet from the boundary of school property. The state Department of Education shall assist each approved school with the posting of notice as required in this Subsection.

F. (1) It is unlawful for any person to cover, remove, deface, alter, or destroy any sign or other marking identifying a firearm-free zone as provided in this Section.

(2) Whoever violates the provisions of this Subsection shall be fined not more than one thousand dollars or imprisoned for not more than six months, or both.

95.7. Possession of or dealing in firearms with obliterated number or mark

A. No person shall intentionally receive, possess, carry, conceal, buy, sell, or transport any firearm from which the serial number or mark of identification has been obliterated.

B. This Section shall not apply to any firearm which is an antique or war relic and is inoperable or for which ammunition is no longer manufactured in the United States and is not readily available in the ordinary channels of commercial trade, or which was originally manufactured without such a number.

C. Whoever violates the provisions of this Section shall be fined not more than one thousand dollars and may be imprisoned for not more than six months.

95.8. Illegal possession of a handgun by a juvenile

A. It is unlawful for any person who has not attained the age of seventeen years knowingly to possess any handgun on his person. Any person possessing any handgun in violation of this Section commits the offense of illegal possession of a handgun by a juvenile.

B. (1) On a first conviction, the offender shall be fined not more than one hundred dollars and imprisoned for not less than ninety days and not more than six months.

(2) On a second conviction, the offender shall be fined not more than five hundred dollars and imprisoned with or without hard labor for not more than two years.

(3) On a third or subsequent conviction, the offender shall be fined not more than one thousand dollars and imprisoned at hard labor for not more than five years.

(4) A juvenile adjudicated delinquent under this Section, having been previously found guilty or adjudicated delinquent for any crime of violence as defined by R.S. 14:2(13), or attempt or conspiracy to commit any such offense, shall upon a first or subsequent conviction be fined not less than five hundred dollars and not more than one thousand dollars and shall be imprisoned with or without hard labor for not less than six months and not more than five years. At least ninety days shall be served without benefit of probation, parole, or suspension of sentence.

C. The provisions of this Section shall not apply to any person under the age of seventeen years who is:

(1) Attending a hunter's safety course or a firearms safety course.

(2) Engaging in practice in the use of a firearm or target shooting at an established range.

(3) Hunting or trapping pursuant to a valid license issued to him pursuant to the laws of this state.

(4) Traveling to or from any activity described in Paragraph (1), (2), or (3) of this Subsection while in possession of an unloaded gun.

(5) On real property with the permission of his parent or legal guardian and with the permission of the owner or lessee of the property.

(6) At such person's residence and who, with the permission of such person's parent or legal guardian, possesses a handgun.

(7) Possessing a handgun with the written permission of such person's parent or legal guardian; provided that such person carries on his person a copy of such written permission.

D. For the purposes of this Section "handgun" means a firearm as defined in R.S. 14:37.2, provided however, that the barrel length shall not exceed twelve inches.

319. Sale of toy pistols prohibited; penalty; exceptions No person shall sell or offer to sell any toy pistol constructed so as to accommodate blank powder cartridges, blank cartridges, or shells used in firing or discharging toy pistols.

Whoever violates this Section shall be fined not less than ten dollars nor more than fifty dollars, or imprisoned for not more than ten days, or both.

Paper cap pistols and other toy pistols not constructed so as to accommodate blank powder cartridges or shells are not included within the provisions of this Section.

Title 40. Public Health and Safety

Chapter 6. Department of Public Safety

1379.3. Statewide permits for concealed handguns; application procedures; definitions

A. (1) Notwithstanding any other provision of law to the contrary, the deputy secretary of public safety services of the Department of Public Safety and Corrections shall issue a concealed handgun permit to any citizen who qualifies for a permit under the provisions of this Section and may promulgate rules and adopt regulations regarding concealed handgun permits in accordance with the Administrative Procedure Act. The permit shall contain a permit number, expiration date, photograph, and the name, address, and date of birth of the permittee. ...

B. A concealed handgun permit issued pursuant to this Section shall grant authority to a citizen to carry a concealed handgun on his person.

J. For the purposes of this Section, the following terms shall have the meanings ascribed herein:

(1) "Handgun" means a type of firearm commonly referred to as a pistol or revolver originally designed to be fired by the use of a single hand and which is designed to fire or is capable of firing fixed cartridge ammunition. The term "handgun" shall not include shotguns or rifles that have been altered by having their stocks or barrels cut or shortened. ...

T.(1) Possession of a current and valid concealed handgun permit issued pursuant to this Section shall constitute sufficient evidence of the background check required pursuant to 18 USC 922(t).

Chapter 9. Weapons

Part I. Machine Guns

1751. Definitions For purposes of this Part, "machine gun" includes all firearms of any calibre, commonly known as machine rifles, machine guns, and submachine guns, capable of automatically discharging more than eight cartridges successfully without reloading, in which the ammunition is fed to the gun from or by means of clips, disks, belts, or some other separable mechanical device. "Manufacturer" includes all persons manufacturing machine guns; "Merchant" includes all persons dealing with machine guns as merchandise.

1752. Handling of machine guns unlawful; exceptions No person shall sell, keep or offer for sale, loan or give away, purchase, possess, carry, or transport any machine gun within this state, except that:

(1) All duly appointed peace officers may purchase, possess, carry, and transport machine guns.

(2) This Part does not apply to the Army, Navy, or Marine Corps of the United States, the National Guard, and organizations authorized by law to purchase or receive machine guns from the United States or from this state. The members of such Corps, National Guard, and organizations may possess, carry, and transport machine guns while on duty.

(3) Persons possessing war relics may purchase and possess machine guns which are relics of any war in which the United States was involved, may exhibit and carry the machine guns in the parades of any military organization, and may sell, offer to sell, loan, or give the machine guns to other persons possessing war relics.

(4) Guards or messengers employed by common carriers, banks, and trust companies, and pay-roll guards or messengers may possess and carry machine guns while actually employed in and about the shipment, transportation, or delivery, or in the guarding of any money, treasure, bullion, bonds, or other thing of value. Their employers may purchase or receive machine guns and keep them in their possession when the guns are not being used by their guards or messengers.

(5) Manufacturers and merchants may sell, keep or offer for sale, loan or give away, purchase, possess, and transport machine guns in the same manner as other merchandise except as otherwise provided in this Part. Common carriers may possess and transport unloaded machine guns as other merchandise.

1753. Transfers of possessions permitted in certain cases; method No manufacturer or merchant shall permit any machine gun to pass from his possession to the possession of any person other than:

(1) A manufacturer or a merchant.

(2) A common carrier for shipment to a manufacturer or merchant.

(3) A duly authorized agent of the government of the United States or of this state, acting in his official capacity.

(4) A person authorized to purchase a machine gun under the provisions of paragraphs (1) and (4) of R.S. 40:1752.

Manufacturers or merchants shall not deliver a machine gun to any of the persons authorized to purchase it under the provisions of paragraphs (1) and (4) of R.S. 40:1752 unless the person presents a written permit to purchase and possess a machine gun, signed by the sheriff of the parish in which the manufacturer or merchant has his place of business or delivers the machine gun. The manufacturer or merchant shall retain the written permit and keep it on file in his place of business. Each sheriff shall keep a record of all permits issued by him.

1754. Registers to be kept; inspection thereof Every manufacturer or merchant shall keep a register of all machine guns manufactured or handled by him. This register shall show:

(1) The date of the sale, loan, gift, delivery, or receipt of any machine gun;

(2) The name, address, and occupation of the person to whom the machine gun was sold, loaned, given, or delivered, or from whom it was received; and

(3) The purpose for which the person, to whom the machine gun was sold, loaned, given, or delivered, purchased or obtained it.

Upon demand, every manufacturer or merchant shall permit any sheriff or deputy sheriff or any police officer to inspect his entire stock of machine guns, and parts and supplies therefor, and shall produce the register required in this Section and all written permits to purchase or possess a machine gun, which he has retained and filed in his place of business.

1755. Penalty

A. Any manufacturer who:

(1) Passes possession of or delivers a machine gun to any person in violation of R.S. 40:1753; or

(2) Fails to keep an accurate register, as required in R.S. 40:1754; or

(3) Fails to produce or account for a sheriff's permit for each machine gun sold by him for which a permit is necessary under the provisions of R.S. 40:1753, shall be imprisoned at hard labor for not less than one year nor more than five years.

B. Any person who violates R.S. 40:1752 shall be imprisoned at hard labor for not less than one year nor more than ten years.

C. Whoever, having been convicted of murder, armed or simple robbery, aggravated or simple burglary, or aggravated battery, or an attempt to commit any one of those crimes, thereafter violates any of the provisions of this Part shall be imprisoned at hard labor for not less than three years nor more than ten years.

Part II. Registration

1781. Definitions For the purpose of this Part, the following terms have the meanings ascribed to them in this Section:

(1) "Dealer" means any person not a manufacturer or importer engaged in this state in the business of selling any firearm. The term includes wholesalers, pawnbrokers, and other persons dealing in used firearms.

(2) "Department" means the Department of Public Safety.

(3) "Firearm" means a shotgun having a barrel of less than eighteen inches in length; a rifle having a barrel of less than sixteen inches in length; any weapon made from either a rifle or a shotgun if said weapon has been modified to have an overall length of less than twenty-six inches; any other firearm, pistol, revolver, or shotgun from which the serial number or mark of identification has been obliterated, from which a shot is discharged by an explosive, if that weapon is capable of being concealed on the person; or a machine gun, grenade launcher, flame thrower, bazooka, rocket launcher, excluding black powder weapons, or gas grenade; and includes a muffler or silencer for any firearm, whether or not the firearm is included within this definition. Pistols and revolvers and those rifles and shotguns which have not previously been defined in this Paragraph as firearms from which serial numbers or marks of identification have not been obliterated are specifically exempt from this definition.

(4) "Importer" means any person who imports or brings into the state any firearm.

(5) "Machine gun" means any weapon, including a submachine gun, which shoots or is designed to shoot automatically more than one shot without manual reloading, by a single function of the trigger.

(6) "Manufacturer" means any person who is engaged in this state in the manufacture, assembling, alteration, or repair of any firearm.

(7) "Muffler" or "silencer" includes any device for silencing or diminishing the report of any portable weapon such as a rifle, carbine, pistol, revolver, machine gun, submachine gun, shotgun, fowling piece, or other device from which a shot, bullet, or projectile may be discharged by an explosive and is not limited to mufflers and silencers for firearms as defined in this Section.

(8) "Transfer" includes the sale, assignment, pledge, lease, loan, gift, or other disposition of any firearm.

1782. Exemptions from Part This Part does not apply to the following persons and things:

(1) Sheriffs or equivalent municipal officers in municipalities of over ten thousand, when they are acting in their official capacity.

(2) The arms, accoutrements, and equipment of the military and naval forces of the United States or of other officers of the United States authorized by law to possess weapons of any kind.

(3) The arms, accoutrements, and equipment of the militia

(4) Any firearm which is unserviceable and which is transferred as a curiosity or ornament.

1783. Registration with department of public safety Every person possessing any firearm shall register with the department the number or other mark identifying the firearm, together with his name, address, and place of business or employment, the place where the firearm is usually kept, and, if the person is other than a natural person, the name and home address of the executive officer thereof having control of the firearm and the name and home address of the person having actual possession thereof.

1784. Application to possess or transfer; use of information or evidence No person shall continue to possess or shall transfer any firearm without the prior approval of the department. Interested persons shall file written applications in duplicate on application forms issued in blank for those purposes by the department. In the case of transfers of any firearm, applications shall be filed by both the proposed vendor and the proposed vendee.

The applications shall set forth, in the original and duplicate, the manufacturer's number or other mark identifying the firearm. Both the original and duplicate shall be forwarded to the department. If approved, the original shall be returned to the applicant.

No information or evidence obtained from an application, registration or records required to be submitted or retained by a natural person in order to comply with any provision of this Part or regulations issued thereunder shall, except as provided by the laws on perjury or false swearing, be used, directly or indirectly, as evidence against that person in a criminal proceeding with respect to a violation of law occurring prior to or concurrently with the filing of the application or registration, or the compiling of the records containing the information or evidence.

1785. Possession or dealing in unregistered or illegally transferred weapons No person shall receive, possess, carry, conceal, buy, sell, or transport any firearm which has not been registered or transferred in accordance with this Part.

1786. Seizure and forfeiture of weapons; disposition thereof. Any firearm possessed or transferred in violation of the provisions of this Part may be seized by the department. Upon seizure, it is forfeited to the state and may be disposed of by the department as follows:

No firearm shall be sold at public sale; the department may order the firearm destroyed or may sell it at private sale to any political subdivision of the state or to any officer thereof, or may retain it for its own use, or may transfer it without charge to any executive department or independent establishment of the state for use by it.

1787. Importation, manufacture, or dealing in without registration Upon first engaging in business, every importer, manufacturer, and dealer in firearms shall register with the department his name or style, principal place of business and other places of business in this state. No person required to register under the provisions of this Section shall import, manufacture, or deal in any firearm without having registered as required by this Section.

1788. Identification with number or other mark; obliteration or alteration of number or mark

A. Each manufacturer, importer, and dealer in any firearm shall identify it with a number or other identification mark approved by the department and shall mark or stamp or otherwise place the number or mark thereon in a manner approved by the department.

B. No one shall obliterate, remove, change, or alter this number or mark. Whenever, in a trial for a violation of this Sub-section, the defendant is shown to have or to have had possession of any firearm upon which the number or mark was obliterated, removed, changed, or altered, that possession is sufficient evidence to authorize conviction unless the defendant explains it to the satisfaction of the court.

1789 Records of importers, manufacturers, or dealers Importers, manufacturers, and dealers shall keep such books and records and render such returns in relation to the transactions in firearms specified in this Part as the department requires.

1790. Rules and regulations; importation of firearms The department may prescribe such rules and regulations as are necessary for carrying out the provisions of this Part.

Under regulations prescribed by the department, any firearm may be imported or brought into this state or possessed or transferred when the purpose thereof is shown to be lawful.

1791. Penalty Upon the first violation of any provision of this Part the penalty shall be a fine of not less than five hundred dollars nor more than two thousand dollars and imprisonment with or without hard labor for not less than one nor more than five years. For any subsequent violation of this Part the penalty shall be a fine of not less than two thousand dollars nor more than five thousand dollars and imprisonment at hard labor for not less than five years nor more than ten years.

1792. Possession of unidentifiable firearm; particular penalties; identification of source of firearm

A. No person shall intentionally receive, possess, carry, conceal, buy, sell, transfer, or transport any firearm which has been illegally obtained or from which the serial number or individual identifying mark, as required by R.S. 40:1788, has been intentionally obliterated, altered, removed, or concealed.

B. The provisions of this Section shall not apply to any firearm which is an antique or war relic and is inoperable or for which ammunition is no longer manufactured in the United States and is not readily available in the ordinary channels of commercial trade, or which was originally manufactured without such a number.

C. Except as otherwise provided in this Section, whoever violates the provisions of this Section may be fined not more than one thousand dollars and shall be imprisoned at hard labor for five years without benefit of probation, parole, or suspension of sentence.

D. Prior to sentencing for a conviction under this Section, the defendant shall be given the opportunity to identify the source of the firearm upon which the conviction was based. If the defendant reveals the identity of the source of the weapon and the identity of the source is confirmed by the prosecutor or the court, the defendant shall be fined not more than one thousand dollars or imprisoned for not more than six months, or both.

E. Nothing in this Section shall be construed to prevent the prosecution of an individual who obtained the firearm by theft, robbery, deception, or by other unlawful means from the lawful owner of the firearm.

F. Any illegally obtained firearm or a firearm from which the serial number or identifying mark required by R.S. 40:1788 has been obliterated or altered is hereby declared to be contraband and shall be seized by the law enforcement agency of jurisdiction. If it is determined that a person other than the owner was responsible for removing, altering, or obliterating the serial number or identifying mark, the firearm shall be returned to its lawful owner or may be disposed of according to law but only after a new serial number has been permanently fixed on the firearm. If a new serial number is not so affixed, the firearm shall be destroyed by the law enforcement agency in possession of the firearm.

Part II-A. Miscellaneous Provisions

1796 Preemption of state law

A. No governing authority of a political subdivision shall enact after July 15, 1985, any ordinance or regulation more restrictive than state law concerning in any way the sale, purchase, possession, ownership, transfer, transportation, license, or registration of firearms, ammunition, or components of firearms or ammunition, however this Section shall not apply to the levy and collection of sales and use taxes, license fees and taxes and permit fees, nor shall it affect the authority of political subdivisions to prohibit the possession of a weapon or firearm in certain commercial establishments and public buildings.

B. Nothing in this Section shall prohibit a local governing authority in a high risk area from developing a plan with federally licensed firearms manufacturers, dealers, or importers to secure the inventory of firearms and ammunition of those licensees in order to prevent looting of the licensee's premises in a state of emergency or disaster. Such plan shall be renewed on a periodic basis. The information contained in the plan shall be deemed security procedures as defined in R.S. 44:3.1 and shall be released only to the sheriffs of the parishes or police chiefs of the municipalities in which the declared state of emergency or disaster exists.

C. For purposes of this Section:

(1) "Declared emergency or disaster" means an emergency or disaster declared by the governor or parish president pursuant to the provisions of the Louisiana Homeland Security and Emergency Assistance and Disaster Act.

(2) "High-risk area" means the parishes of Assumption, Calcasieu, Cameron, Iberia, Jefferson, Lafourche, Orleans, Plaquemines, St. Bernard, St. Charles, St. James, St. John, St. Martin, St. Mary, St. Tammany, Tangipahoa, Terrebonne, and Vermilion.

Part III. Purchase of Rifles and Shotguns

1801. Declaration of policy It is declared that it is in the public interest to authorize residents of this state to purchase or otherwise obtain rifles and shotguns or ammunition in states contiguous to this state in compliance with such other laws of this state or its political subdivisions as may be applicable and in compliance with Section 102 of the Gun Control Act of 1968, Public Law 90-618, 18 U.S.C.§ 921 et seq. and it is the declared intention of this state that the sale of shotguns and rifles and the sale of ammunition in this state to residents of adjacent states is hereby authorized pursuant to regulations issued under the Gun Control Act of 1968.

1802. Definitions As used in this Part:

(1) "A state contiguous to this state" shall mean any state having a common border with this state.

(2) All other terms shall be given the meaning prescribed in 18 U.S.C.§ 921 (the Gun Control Act of 1968, Public Law 90-618) and the regulations duly promulgated thereunder as presently enacted or promulgated and as hereafter modified.

1803. Purchase of rifle or shotgun or ammunition It shall be lawful for any person residing in this state, including any corporation or other business entity maintaining a place of business in this state, to purchase or otherwise obtain a rifle or shotgun or ammunition in any state which is contiguous to this state and to receive or transport such rifle or shotgun or ammunition into this state and to permit any person residing in a contiguous state to purchase shotguns, rifles or ammunition in this state and to receive or transport such rifles, shotguns or ammunition in this state.

1804. Application This Part shall not apply or be construed to affect in any way the purchase, receipt or transportation of rifles or shot-guns or ammunition by federally licensed fire-arms manufacturers, importers, dealers or collectors except to permit such purchase, receipt or transportation.

Part IV. Armor-Piercing Bullets

1810. Definitions As used in this Part, "armor-piercing bullet" shall mean any bullet, except a shotgun shell or ammunition primarily designed for use in rifles, that:

(1) Has a steel inner core or core of equivalent density and hardness, truncated cone, and is designed for use in a pistol or revolver as a body armor or metal piercing bullet; or

(2) Has been primarily manufactured or designed, by virtue of its shape, cross-sectional density, or any coating applied thereto, to breach or penetrate body armor when fired from a handgun.

1811. Prohibitions

A. No person shall import, manufacture, sell, purchase, possess, or transfer armor-piercing bullets.

B. Whoever violates the provisions of this Section shall be fined not more than one thousand dollars or imprisoned with or without hard labor for not more than one year, or both.

1812. Exemptions The provisions of this Part shall not apply to:

(1) Law enforcement officers and employees acting in the lawful performance of their duties.

(2) Law enforcement or other authorized agencies conducting a firearms training course, operating a forensic ballistics laboratory, or specializing in the development of ammunition or explosive ordinance.

(3) Department of Corrections officials and employees authorized to carry firearms while engaged in the performance of their official duties.

(4) Members of the armed services or reserve forces of the United States or Louisiana National Guard while engaged in the performance of their official duties.

(5) Federal officials authorized to carry firearms while engaged in the performance of their official duties.

(6) The lawful manufacture, importation, sale, purchase, possession, or transfer of armor-piercing bullets exclusively to or for persons authorized by law to possess such bullets.

(7) A bona fide collector licensed by the Department of Public Safety.

[Current through 2008 Regular Session]

MAINE

ME. REV. STAT.

Title 15. Court Procedure - Criminal

Chapter 15. Possession of Firearms by Felons

393. Possession of firearms prohibited for certain persons.

1. POSSESSION PROHIBITED. A person may not own, possess or have under that person's control a firearm or crossbow, unless that person has obtained a permit under this section, if that person:

A-1. Has been convicted of committing or found not criminally responsible by reason of insanity:

(1) A crime in this State that is punishable by imprisonment for a term of one year or more;

(2) A crime under the laws of the United States that is punishable by imprisonment for a term exceeding one year;

(3) A crime under the laws of any other state that, in accordance with the laws of that jurisdiction, is punishable by a term of imprisonment exceeding one year. This subparagraph does not include a crime under the laws of another state that is classified by the laws of that state as a misdemeanor and is punishable by a term of imprisonment of 2 years or less;

(4) A crime under the laws of any other state that, in accordance with the laws of that jurisdiction, does not come within subparagraph (3) but is elementally substantially similar to a crime in this State that is punishable by a term of imprisonment for one year or more; or

(5) A crime under the laws of the United States, this State or any other state or the Passamaquoddy Tribe or Penobscot Nation in a proceeding in which the prosecuting authority was required to plead and prove that the person committed the crime with the use of:

(a) A firearm or crossbow against a person; or

(b) Any other dangerous weapon;

B. [Repealed 2001, c. 549, §2]

C. Has been adjudicated in this State or under the laws of the United States or any other state to have engaged in conduct as a juvenile that, if committed by an adult, would have been a disqualifying conviction:

(1) Under paragraph A-1, subparagraphs (1) to (4) and bodily injury to another person was threatened or resulted; or

(2) [Deleted]

(3) Under paragraph A-1, subparagraph (5);

D. Is subject to an order of a court of the United States or a state, territory, commonwealth or tribe that restrains that person from harassing, stalking or threatening an intimate partner, as defined in 18 United States Code, Section 921(a), of that person or a child of the intimate partner of that person, or from engaging in other conduct that would place the intimate partner in reasonable fear of bodily injury to the intimate partner or the child, except that this paragraph applies only to a court order that was issued after a hearing for which that person received actual notice and at which that person had the opportunity to participate and that:

(1) Includes a finding that the person represents a credible threat to the physical safety of an intimate partner or a child; or

(2) By its terms, explicitly prohibits the use, attempted use or threatened use of physical force against an intimate partner or a child that would reasonably be expected to cause bodily injury; or

E. Has been:

(1) Committed involuntarily to a hospital pursuant to an order of the district court under Title 34-B, Section 3864 because the person was found to present a likelihood of serious harm, as defined under Title 34-B, Section 3801, Subsection 4, Paragraphs a to c;

(2) Found not criminally responsible by reason of insanity with respect to a criminal charge; or

(3) Found not competent to stand trial with respect to a criminal charge. For the purposes of this subsection, a person is deemed to have been convicted upon the acceptance of a plea of guilty or nolo contendere or a verdict or finding of guilty, or of the equivalent in a juvenile case, by a court of competent jurisdiction.

For the purposes of this subsection, a person is deemed to have been found not criminally responsible by reason of insanity upon the acceptance of a plea of not criminally responsible by reason of insanity or a verdict or finding of not criminally responsible by reason of insanity, or of the equivalent in a juvenile case, by a court of competent jurisdiction.

1-A. LIMITED PROHIBITION FOR NONVIOLENT JUVENILE OFFENSES. A person who has been adjudicated in this State or under the laws of the United States or any other state to have engaged in conduct as a juvenile that, if committed by an adult, would have been a disqualifying conviction under subsection 1, paragraph A-1 but is not an adjudication under subsection 1, paragraph C may not own or have in that person's possession or control a firearm or crossbow for a period of 3 years following completion of any disposition imposed or until that person reaches 18 years of age, whichever is later.

2. APPLICATION AFTER 5 YEARS. A person subject to the provisions of subsection 1, paragraph A-1 or C as a result of a conviction or adjudication may, after the expiration of 5 years from the date that the person is finally discharged from the sentences imposed as a result of the conviction or adjudication, apply to the Commissioner for a permit to carry a firearm or crossbow. That person may not be issued a permit to carry a concealed firearm pursuant to Title 25, chapter 252. A permit issued pursuant to this subsection is valid for 4 years from the date of issue unless sooner revoked for cause by the Commissioner. For purposes of this subsection, "firearm" does not include a firearm defined under 18 United States Code, Section 921(3).

3. CONTENTS. An application under subsection 2 must be on a form prepared by the Commissioner of Public Safety. The application must include the following: the applicant's full name; all aliases; date and place of birth; place of legal residence; occupation; make, model and serial number of the firearm or crossbow sought to be possessed; date, place and nature of conviction; sentence imposed; place of incarceration; name and address of probation or parole officer; date of discharge or release from prison or jail or termination of probation; the reason for the request; and any other information determined by the Commissioner to be of assistance. The application must be accompanied by certified or attested copies of the indictment, information or complaint, judgment and commitment and discharge that are the subject of the conviction.

4. NOTIFICATION, OBJECTION AND HEARING. Upon receipt of an application, the Commissioner shall determine if it is in proper form. If the application is proper, the Commissioner shall within 30 days notify in writing the sentencing or presiding judge, the Attorney General, the district attorney for the county where the applicant resides, the district attorney for the county where the conviction occurred, the law enforcement agency that investigated the crime, the chief of police and sheriff in the municipality and county where the crime occurred and the chief of police and sheriff in the municipality where the applicant resides as of the filing of the application. The commissioner may direct any appropriate investigation to be carried out. If, within 30 days of the sending of notice, any person so notified objects in writing to the issuance of a permit, a permit may not be issued. The Commissioner may deny an application even if no objection is filed.

4-A. APPLICATION FOR RELIEF. EXCEPT AS OTHERWISE PROVIDED, A person subject to the federal prohibition against possession of firearms pursuant to 18 United States Code, Section 922(g)(4) as a result of being adjudicated a mental defective or committed to any psychiatric hospital pursuant to Title 34-b, section 3863 and who has not been committed to a psychiatric hospital pursuant to an order of the district court pursuant to Title 34-b, section 3864 may, after the expiration of 5 years from the date of final discharge from commitment, apply to the Commissioner for relief from the disability.

Relief is not available under this subsection for a person found not criminally responsible by reason of insanity or incompetent to stand trial in a criminal case or a person adjudged by a probate court to lack the capacity to contract or manage the person's own affairs.

a. An application under this subsection must be on a form developed by the Commissioner. The application must include the applicant's full name; all aliases; date and place of birth; place of legal residence; occupation; make and model of the firearm sought to be possessed; reason for the request; date, place and docket number of commitment; name of institution to which applicant was committed; names of providers that provided mental health treatment for the applicant; date of discharge from commitment; release for all mental health records; and any other information determined by the Commissioner to be of assistance. The application must be accompanied by certified or attested copies of the commitment from which the applicant seeks relief and the report of an independent psychologist or psychiatrist licensed to practice in this state specifically addressing the factors set forth in paragraph e. The Commissioner may establish a roster of psychologists and psychiatrists qualified and interested in doing these evaluations. The psychologist or psychiatrist must be available for cross-examination. The psychologist or psychiatrist listed on the roster is an employee for the purposes of the Maine tort claims act for evaluations under this paragraph.

b. The Commissioner has the independent authority to establish the following, to be paid by the applicant:

(1) application fee; and

(2) fees for evaluations required by paragraph a.

c. Upon receipt of a completed application, the Commissioner shall notify persons who received notice of the commitment pursuant to Title 34-b, section 3864, subsection 3, paragraph a, subparagraph (2) and the district attorney, chief of police and sheriff in the municipality and county where the applicant resides of the filing of the application, with a request to provide to the commissioner any information relevant to the factors in paragraph e.

d. Upon receipt of a completed application, the Commissioner shall review the application and determine whether the person has made a prima facie showing of the elements of paragraph e. If the Commissioner determines that the person has made a prima facie showing, the Commissioner shall schedule a hearing.

e. The burden of proof is on the applicant to prove, by clear and convincing evidence, that the circumstances that led to the involuntary commitment to a hospital have changed, that the applicant is not likely to act in a manner dangerous to public safety and that granting the application for relief will not be contrary to the public interest.

f. If the Commissioner finds by clear and convincing evidence that the circumstances that led to the involuntary commitment have changed, that the applicant is not likely to act in a manner dangerous to public safety and that granting the application for relief will not be contrary to the public interest, the Commissioner may grant relief.

g. Notwithstanding any other provision of law, and except as indicated in this paragraph, all applications for relief pursuant to this subsection and documents made a part of the application, refusals and any information of record collected by the Commissioner during the process of determining whether an applicant qualifies for relief are confidential and may not be made available for public inspection or copying unless:

(1) the applicant waives this confidentiality in writing or on the record of any hearing; or

(2) a court of record so orders. Proceedings relating to the grant or denial of relief are not public proceedings under Title 1, chapter 13.

The Commissioner shall make a permanent record, in the form of a summary, of the final decision regarding each application. The summary must include the name of the applicant and indicate whether the application for relief was granted or denied. The information contained in this summary is available for public inspection.

h. An applicant may appeal the denial of an application for relief under this subsection within 30 days of receipt of the written notice of decision by filing a complaint in the district court for de novo review in the district where the department of public safety has its principal office. Hearings are closed unless otherwise agreed to by the applicant. A party aggrieved by a decision of the district court may not appeal as of right. The time for taking the appeal and the manner and any conditions for the taking of the appeal are as the supreme judicial court provides by rule.

5. APPEAL. Any person to whom a permit under subsection 2 has been denied may. file a petition for review pursuant to Title 5, Chapter 375, Subchapter 7.

6. FILING FEE. The commissioner may establish a reasonable filing fee not to exceed $25 to defray costs of processing applications.

7. DEFINITIONS. As used in this section, unless the context otherwise indicates, the following terms have the following meanings.

A. "Firearm" has the same meaning as in Title 17-A, section 2, subsection 12-A.

B. "Not criminally responsible by reason of mental disease or defect" has the same meaning as used in Title 17-A, section 39 and includes the former finding in this State under former provisions of section 103 of "not guilty by reason of mental disease or defect excluding responsibility" as well as any comparable finding under the laws of the United States or any other state.

C. "State" means the State of Maine and "state" means any other state of the United States and includes the District of Columbia, the Commonwealth of Puerto Rico and the possessions of the United States.

D. "Use of a dangerous weapon" has the same meaning as in Title 17-A, section 2, subsection 9, paragraph A.

E. "Commissioner" means the Commissioner of public safety or the Commissioner's designee.

8. PENALTY. A violation of subsection 1, paragraph A-1 or C is a Class C crime. A violation of subsection 1, paragraph D or E is a Class D crime. A violation of subsection 1-A by a person at least 18 years of age is a Class C crime.

For the purposes of this subsection, a person is deemed to have been convicted upon the acceptance of a plea of guilty or nolo contendere or a verdict or finding of guilty, or the equivalent in a juvenile case, by a court of competent jurisdiction.

9. PRIMA FACIE EVIDENCE. Notwithstanding any other law or rule of evidence, a copy of a court abstract provided by a court to the Department of Public Safety, State Bureau of Identification pursuant to Title 34-b, Section 3864, Subsection 12, if certified by the custodian of the records of that bureau, or the custodian's designee, is admissible in a criminal prosecution brought pursuant to this section as prima facie evidence that the person identified in the abstract has been involuntarily committed by the court issuing the abstract and has been provided the notice required in Title 34-B, Section 3864, Subsection 5, Paragraph a-1 and Title 34-b, Section 3864, Subsection 13.

10. SUBPOENA POWER. The Commissioner is authorized to issue a subpoena in the name of the Commissioner in accordance with Title 5, Section 9060, except that this authority applies to any stage of an investigation under this section and is not limited to an adjudicatory hearing. If a witness refuses to obey a subpoena or to give any evidence relevant to proper inquiry by the Commissioner, the attorney general may petition the superior court in the county where the refusal occurred to find the witness in contempt. The attorney general shall cause to be served on that witness an order requiring the witness to appear before the superior court to show cause why the witness should not be adjudged in contempt. The court shall, in a summary manner, hear the evidence and, if it is such as to warrant the court in doing so, punish that witness in the same manner and to the same extent as for contempt committed before the superior court or with reference to the process of the superior court.

11. RULES. The Commissioner may adopt rules to implement the provisions of Subsections 2 to 4-a. Rules adopted pursuant to this subsection are routine technical rules as defined by Title 5, chapter 375, Subchapter 2-a.

Chapter 17. Miscellaneous Provisions

455. Record of sales of firearms.

1. FORMS. A dealer may not:

A. Sell, let or loan a firearm to a person without making a copy of the form a dealer must keep as prescribed by 18 United States Code, Section 923. The copy must be made and marked as 'STATE COPY' before the firearm is delivered; or

B. Refuse to show or refuse to allow inspection of a copy of the form described in paragraph A to a sheriff, deputy sheriff, police officer, constable, game warden or prosecuting attorney.

A person who violates this subsection commits a civil violation for which a fine of $50 may be adjudged.

2. FALSE OR FICTITIOUS NAME. A person may not give a false or fictitious name to a dealer. A person who violates this subsection commits a civil violation for which a fine of $50 may be adjudged.

3. EXCEPTION. This section does not apply to a wholesaler who sells only to other dealers or to a manufacturer who sells only at wholesale.

455-A. Warning requirement upon sales of firearms.

1. POSTING OF CONSPICUOUS WARNING. Except as provided in subsection 1-A, any commercial retail sales outlet that sells firearms shall conspicuously post at each purchase counter where firearms may be purchased the following warning in block letters not less than one inch in height:

"ENDANGERING THE WELFARE OF A CHILD IS A CRIME. IF YOU LEAVE A FIREARM AND AMMUNITION WITHIN EASY ACCESS OF A CHILD, YOU MAY BE SUBJECT TO FINE, IMPRISONMENT OR BOTH.

KEEP FIREARMS AND AMMUNITION SEPARATE.

KEEP FIREARMS AND AMMUNITION LOCKED UP.

USE TRIGGER LOCKS."

1-A. POSTING OF WARNINGS AT GUN SHOWS. The warning sign as described in subsection 1 must be posted at all entrances of an organized gun show.

2. VIOLATION. Any person who fails to post the warning in compliance with subsection 1, commits a civil violation for which a civil forfeiture of not more than $200 may be adjudged.

Title 17-A. Maine Criminal Code

Chapter 1. Preliminary

2. Definitions. As used in this code, unless a different meaning is plainly required, the following words and variants thereof have the following meanings. ...

9. Dangerous weapon.

A. "Use of a dangerous weapon" means the use of a firearm or other weapon, device, instrument, material or substance, whether animate or inanimate, which, in the manner it is used or threatened to be used is capable of producing death or serious bodily injury.

B. "Armed with a dangerous weapon" means in actual possession, regardless of whether the possession is visible or concealed, of:

(1) A firearm;

(2) Any device designed as a weapon and capable of producing death or serious bodily injury; or

(3) Any other device, instrument, material or substance, whether animate or inanimate, which, in the manner it is intended to be used by the actor, is capable of producing or threatening death or serious bodily injury. For purposes of this definition, the intent may be conditional.

C. When used in any other context, "dangerous weapon" means a firearm or any device designed as a weapon and capable of producing death or serious bodily injury.

D. For purposes of this subsection, proof that a thing is presented in a covered or open manner as a dangerous weapon gives rise to a permissible inference under the Maine Rules of Evidence, Rule 303 that it, in fact, is a dangerous weapon. ...

12-A. "Firearm" means any weapon, whether loaded or unloaded, which is designed to expel a projectile by the action of an explosive and includes any such weapon commonly referred to as a pistol, revolver, rifle, gun, machine gun or shotgun. Any weapon which can be made into a firearm by the insertion of a firing pin, or other similar thing, or by repair, is a firearm. ...

Chapter 23. Offenses Against the Family

554. Endangering the welfare of a child

1. A person is guilty of endangering the welfare of a child if that person: ...

B. Knowingly sells, furnishes, gives away or offers to sell, furnish or give away to a child under 16 years of age any intoxicating liquor, cigarettes, tobacco, air rifles, gunpowder, smokeless powder or ammunition for firearms; ...

2. It is an affirmative defense to prosecution under this section that: ...

C. The defendant was the parent, foster parent, guardian or an adult approved by the parent, foster parent or guardian who furnished a child under 16 years of age an air rifle, gunpowder, smokeless powder or ammunition for a firearm for use in a supervised manner.

3. Endangering the welfare of a child is a Class D crime, except that a violation of subsection 1, paragraph B-2 is a Class C crime.

554-A. Unlawful transfer of a firearm to a minor.

1. As used in this section, the following terms have the following meanings.

A. "Transfer" means to sell, furnish, give, lend, deliver or otherwise provide, with or without consideration.

B. "Minor" means a person under 16 years of age.

2. Except as provided in section 554-B, a person is guilty of unlawfully transferring a firearm to a minor if that person, who is not the parent, foster parent or guardian of the minor, knowingly transfers a firearm to a minor.

3. It is an affirmative defense to a prosecution under subsection 2 that:

A. The actor reasonably believed the person receiving the firearm had attained 16 years of age. A reasonable belief can not be based solely upon the physical appearance of the person or upon the oral representation of that person as to that person's age; or

B. The transfer of the firearm to the minor was approved by the parent, foster parent or guardian of the minor.

4. Unlawful transfer of a firearm to a minor is a Class D crime.

554-B. Unlawful transfer of handgun to minor

1. As used in this section, the following terms have the following meanings.

A. "Handgun" means a firearm that has a short stock and is designed to be held and fired by the use of a single hand, or any combination of parts from which a handgun can be assembled.

B. "Minor" means a person under 18 years of age.

C. "Transfer" means to sell, furnish, give, lend, deliver or otherwise provide, with or without consideration.

2. A person is guilty of unlawfully transferring a handgun to a minor if that person knowingly transfers a handgun to a person who the transferor knows or has reasonable cause to believe is a minor.

3. This section does not apply to:

A. A temporary transfer of a handgun to a minor:

(1) With the prior written consent of the minor's parent or guardian and that parent or guardian is not prohibited by federal, state or local law from possessing a firearm; or

(2) In the course of employment, target practice, hunting or instruction in the safe and lawful use of a handgun.

The minor may transport an unloaded handgun in a locked container directly from the place of transfer to a place at which an activity described in this subparagraph is to take place and directly from the place at which such an activity took place to the transferor;

B. A minor who is a member of the United States Armed Forces or the National Guard who possesses or is armed with a handgun in the line of duty;

C. A transfer by inheritance of title to, but not possession of, a handgun to a minor; or

D. The transfer of a handgun to a minor when the minor takes the handgun in self-defense or in defense of another person against an intruder into the residence of the minor or a residence in which the minor is an invited guest.

4. The State may not permanently confiscate a handgun that is transferred to a minor in circumstances in which the transferor is not in violation of this section and if the possession of the handgun by the minor subsequently becomes unlawful because of the conduct of the minor. When that handgun is no longer required by the State for the purposes of investigation or prosecution, the handgun must be returned to the lawful owner.

5. The following penalties apply.

A. A person who violates this section commits a Class D crime, except as provided in paragraph B.

B. A person who violates this section and, at the time of the offense, has one or more prior convictions under this section or for engaging in substantially similar conduct to that contained in this section in another jurisdiction. Section 9-A governs the use of prior convictions when determining a sentence. Violation of this paragraph is a Class C crime.

Chapter 29. Forgery and Related Offenses

705. Criminal simulation

1. A person is guilty of criminal simulation if:

A. With intent to defraud, he makes or alters any property so that it appears to have an age, rarity, quality, composition, source or authorship which it does not in fact possess; or with knowledge of its true character and with intent to defraud, he transfers or possesses property so simulated; or ...

D. With intent to defraud and to prevent identification:

(1) He alters, removes or obscures the manufacturer's serial number or any other distinguishing identification number, mark or symbol upon any automobile, snowmobile, outboard motor, motorboat, aircraft or any other vehicle or upon any machine, firearm or other object; or

(2) He possesses any such object or any such item after that number has been altered, removed or obscured.

2. Criminal simulation is a Class E crime.

Chapter 43. Weapons

1051. Possession of machine gun

1. A person is guilty of possession of a machine gun if, without authority to do so, he knowingly possesses a machine gun.

2. As used in this chapter, "machine gun" means a weapon of any description, by whatever name known, loaded or unloaded, which is capable of discharging a number of projectiles in rapid succession by one manual or mechanical action on the trigger or firing mechanism.

3. Possession of a machine gun is a Class D crime.

1052. Right to possess, carry or transport machine gun Any law enforcement officer of the State of Maine, any law enforcement officer of another state or a territory of the United States, members of the Armed Forces, Maine National Guard and Maine State Guard may possess a machine gun if the possession or carrying of such weapon is in the discharge of his official duties and has been authorized by his appointing authority.

Machine guns manufactured, acquired, transferred or possessed in accordance with the National Firearms Act, as amended, shall be exempt from this chapter.

1053. Confiscation and seizure of machine gun Any machine gun possessed in violation of section 1051 is declared to be contraband and is subject to forfeiture to the State. Any law enforcement officer shall have the power to seize the same with due process.

When a machine gun is seized as provided, the officer seizing the same shall immediately file with the judge before whom such warrant is returnable, a libel against the machine gun, setting forth the seizure and describing the machine gun and the place of seizure in a sufficient manner to reasonably identify it, that it was possessed in violation of law and pray for a decree of forfeiture thereof. Such judge shall fix a time for the hearing of such libel and shall issue his monition and notice of same to all persons interested, citing them to appear at the time and place appointed to show cause why such machine gun should not be declared forfeited, by causing true and attested copies of said libel and monition to be posted in 2 public and conspicuous places in the town and place where such machine gun was seized, 10 days at least before said libel is returnable. In addition, a true and attested copy of the libel and monition shall be served upon the person from whom said machine gun was seized and upon the owner thereof, if their whereabouts can be readily ascertained 10 days at least before said libel is returnable. In lieu of forfeiture proceedings, title to such seized machine gun may be transferred in writing to the State of Maine by the owner thereof. If title to and ownership in the machine gun is transferred to the State, a receipt for the machine gun shall be given to the former owner by the law enforcement officer who seized the machine gun.

1054. Forfeiture of machine gun If no claimant for a machine gun seized under the authority of section 1053 appears, the judge shall, on proof of notice, declare the same to be forfeited to the State. If any person appears and claims such machine gun, as having a right to the possession thereof at the time when the same was seized, he shall file with the judge a claim in writing stating specifically the right so claimed, the foundation thereof, the item so claimed, any exemption claimed, the time and place of the seizure and the name of the law enforcement officer who seized the machine gun, and in it declare that it was not possessed in violation of this chapter, and state his business and place of residence and sign and make oath to the same before said judge. If any person so makes claim, he shall be admitted as a party to the process, and the libel, and may hear any pertinent evidence offered by the libelant or claimant. If the judge is, upon hearing, satisfied that said machine gun was not possessed in violation of this chapter, and that claimant is entitled to the custody thereof, he shall give an order in writing, directed to the law enforcement officer having seized the same, commanding him to deliver to the claimant the machine gun to which he is so found to be entitled, within 48 hours after demand. If the judge finds the claimant not entitled to possess the machine gun, he shall render judgment against him for the libellant for costs, to be taxed as in civil cases before such judge, and issue execution thereon, and shall declare such machine gun forfeited to the State. The claimants may appear and shall recognize with sureties as on appeals in civil actions from a judge. The judge may order that the machine gun remain in the custody of the seizing law enforcement officer, pending the disposition of the appeal. All machine guns declared forfeited to the State, or title to which have been transferred to the State in lieu of forfeiture proceedings shall be turned over to the Chief of the Maine State Police. If said machine gun is found to be of a historic, artistic, scientific or educational value, the State Police may retain the machine gun for an indefinite period of time. Any other machine gun declared forfeited and in possession of the State Police shall be destroyed by a means most convenient to the Chief of the State Police.

1056. Possession of armor-piercing ammunition

1. A person is guilty of possession of armor-piercing ammunition if, without authority to do so, the person knowingly possesses armor-piercing ammunition other than as part of a bona fide collection.

2. As used in this chapter, "armor-piercing ammunition" means a projectile or projectile core that may be used in a handgun and that is constructed entirely, excluding the presence of traces of other substances, from one or a combination of tungsten alloys, steel, iron, brass, beryllium copper or depleted uranium, including but not limited to ammunition commonly known as KTW ammunition. "Armor-piercing ammunition" does not include shotgun shot required by federal or state environmental or game laws, rules or regulations for hunting purposes, a frangible projectile designed for target shooting or any projectile or projectile core found by the United States Secretary of the Treasury or the secretary's delegate, pursuant to 27 Code of Federal Regulations, Section 178.148 or Section 178.149, to be:

A. Primarily intended to be used for sporting purposes; or

B. Used for industrial purposes, including a charge used in an oil and gas well perforating device.

3. Possession of armor-piercing ammunition is a Class C crime.

4. This section does not apply to members of the United States Armed Forces, the United States Reserve Forces or the National Guard, or to law enforcement officers or agencies or forensic laboratories, in the course of duty or employment.

Chapter 47. General Sentencing Provisions

1158-A. Forfeiture of firearms

1. As part of every sentence imposed, except as provided in subsection 2, a court shall order that a firearm must be forfeited to the State if:

A. That firearm constitutes the basis for conviction under:

(1) Title 15, section 393;...

B. The State pleads and proves that the firearm is used by the defendant or an accomplice during the commission of any murder or Class A, Class B or Class C crime or any Class D crime defined in chapter 9, 11 or 13; or

C. The defendant, with the approval of the state, consents to the forfeiture of the firearm.

2. Except as provided in subsection 3, a court may not order the forfeiture of a firearm otherwise qualifying for forfeiture under subsection 1 if another person can satisfy the court by a preponderance of the evidence and prior to the imposition of the defendant's sentence that:

A. Other than in the context of either subsection 1, paragraph A, subparagraph (1) or subsection 1, paragraph B relative to murder or any other unlawful homicide crime in which the firearm used is a handgun, the other person, at the time of the commission of the crime, had a right to possess the firearm to the exclusion of the defendant;

B. In the context of subsection 1, paragraph A, subparagraph (1), the other person, at the time of the commission of the crime, had a right to possess the firearm to the exclusion of the defendant; or

C. In the context of subsection 1, paragraph B relating to murder or any other unlawful homicide crime in which the firearm used is a handgun, the other person, at the time of the commission of the crime, was the rightful owner from whom the handgun had been stolen and the other person was not a principal or accomplice in the commission of the crime.

3. If another person satisfies subsection 2, paragraph B, a court shall nonetheless order the forfeiture of a firearm otherwise qualifying for forfeiture under subsection 1, paragraph A, subparagraph (1) if the State can satisfy the court by a preponderance of the evidence both that the other person knew or should have known that the defendant was a prohibited person under Title 15, section 393 and that the other person intentionally, knowingly or recklessly allowed the defendant to possess or have under the defendant's control the firearm.

4. The Attorney General shall adopt rules governing the disposition to state, county and municipal agencies of firearms forfeited under this section. A handgun not excepted under subsec-tion 2, paragraph C must be destroyed by the State.

5. As used in this section, "handgun" means a firearm, including a pistol or revolver, that has a short stock and is designed to be held and fired by the use of a single hand.

Title 19-A. Domestic Relations

Chapter 101. Protection From Abuse

4006. Hearings

1. FULL HEARING. Within 21 days of the filing of a complaint, a hearing must be held at which the plaintiff must prove the allegation of abuse by a preponderance of the evidence. If a request for temporary, emergency or interim relief is denied, the hearing must be held as soon as practicable within the 21-day period.

2. TEMPORARY ORDERS. The court may enter temporary orders authorized under subsection 5 that it considers necessary to protect the plaintiff or minor child from abuse, on good cause shown in an ex parte proceeding, which the court shall hear and determine as expeditiously as practicable after the filing of a complaint. Immediate and present danger of abuse to the plaintiff or minor child constitutes good cause. An order remains in effect pending a hearing pursuant to subsection 1.

2-A. TEMPORARY ORDERS; POSSESSION OF DANGEROUS WEAPONS. The court may direct the defendant not to possess a firearm or other dangerous weapon for the duration of the temporary order if the complaint demonstrates:

A. Abuse that involves a firearm or other dangerous weapon; or

B. A heightened risk of immediate abuse to the plaintiff or a minor child. In determining whether a heightened risk of immediate abuse is present, the court shall consider, but is not limited to consideration of, whether:

(1) The temporary order of protection is not likely to achieve its purpose in the absence of such a condition;

(2) The defendant has violated orders of protection;

(3) Past or present abuse to a victim resulted in injury;

(4) The abuse occurred in public; and

(5) The abuse includes:

(a) Threats of suicide or homicide;

(b) Killing or threatening to kill pets;

(c) An escalation of violence;

(d) Stalking behavior or extreme obsession;

(e) Sexual violence;

(f) Excessive alcohol or drug use; and

(g) Abuse against a pregnant victim.

If the court prohibits the defendant from possessing a firearm or other dangerous weapon in a temporary order and if the defendant moves for dissolution or modification of an order pursuant to subsection 7, the court must hear and decide the motion as expeditiously as possible and must issue a written decision on the motion within 24 hours after a hearing on that motion.

If the court prohibits the defendant from possessing a dangerous weapon other than a firearm in a temporary order, the court shall specify the type of weapon the defendant is prohibited from possessing.

If the court prohibits the defendant from possessing a firearm or other dangerous weapon in a temporary order, the court shall direct the defendant to relinquish, within 24 hours after service of the order on the defendant or such earlier time as the court specifies in the order, all firearms and specified dangerous weapons in the possession of the defendant to a law enforcement officer or other individual for the duration of the order. If the weapons are relinquished to an individual other than a law enforcement officer, the defendant must file, within 24 hours after such relinquishment, with the court or local law enforcement agency designated in the order a written statement that contains the name and address of the individual holding the weapons and a description of all weapons held by that individual. The court may subsequently issue a search warrant authorizing a law enforcement officer to seize any firearms and other dangerous weapons at any location if there is probable cause to believe such firearms or dangerous weapons have not been relinquished by the defendant. ...

4007. Relief.

1.PROTECTION ORDER; CONSENT AGREEMENT. The court, after a hearing and upon finding that the defendant has committed the alleged abuse, may grant a protective order or, upon making that finding, approve a consent agreement to bring about a cessation of abuse. This subsection does not preclude the parties from voluntarily requesting a consent agreement without a finding of abuse. The court may enter a finding that the defendant represents a credible threat to the physical safety of the plaintiff or a minor child residing in the plaintiff's household. Relief granted under this section may include: ...

A-1. Directing the defendant not to possess a firearm or other dangerous weapon for the duration of the order; ...

1-A. NO POSSESSION OF FIREARM OR DANGEROUS WEAPONS FOR DURATION OF ORDER. If the court prohibits the defendant from possessing a dangerous weapon other than a firearm, the court shall specify the type of weapon the defendant is prohibited from possessing.

If the court prohibits the defendant from possessing a firearm or other dangerous weapon, the court shall direct the defendant to relinquish, within 24 hours after service of the order on the defendant or such earlier time as the court specifies in the order, all firearms and specified dangerous weapons in the possession of the defendant to a law enforcement officer or other individual for the duration of the order. If the weapons are relinquished to an individual other than a law enforcement officer, the defendant must file, within 24 hours after such relinquishment, with the court or local law enforcement agency designated in the order a written statement that contains the name and address of the individual holding the weapons and a description of all weapons held by that individual. The court may subsequently issue a search warrant authorizing a law enforcement officer to seize any firearms and other dangerous weapons at any location if there is probable cause to believe such firearms or dangerous weapons have not been relinquished by the defendant. ...

Chapter 252-A. Firearms Regulations

2011. State preemption.

1. PREEMPTION. The State intends to occupy and preempt the entire field of legislation concerning the regulation of firearms, components, ammunition and supplies. Except as provided in subsection 3, any existing or future order, ordinance, rule or regulation in this field of any political subdivision of the State is void.

2. REGULATION RESTRICTED. Except as provided in subsection 3, no political subdivision of the State, including, but not limited to, municipalities, counties, townships and village corporations, may adopt any order, ordinance, rule or regulation concerning the sale, purchase, purchase delay, transfer, ownership, use, possession, bearing, transportation, licensing, permitting, registration, taxation or any other matter pertaining to firearms, components, ammunition or supplies.

3. EXCEPTION. This section does not prohibit an order, ordinance, rule or regulation of any political subdivision which, with the exception of appropriate civil penalty provisions, conforms exactly with any applicable provision of state law or which regulates the discharge of firearms within a jurisdiction.

4. LAW ENFORCEMENT AGENCY. Nothing in this section limits the power of any law enforcement agency to regulate the type and use of firearms issued or authorized by that agency for use by its employees. For the purposes of this section "law enforcement agency" has the same meaning as set forth in section 3701.

2012. Sale of firearms to include safety brochure

1. DEFINITIONS. As used in this section, unless the context otherwise indicates, the following terms have the following meanings.

A. "Basic firearm safety brochure" means a brochure, produced by a national, nonprofit membership organization that provides a comprehensive voluntary safety program including the training of people in the safe handling and use of firearms or by any other organization that contains the following information relating to firearms:

(1) Rules for safe handling, storage and use of firearms;

(2) Nomenclature and descriptions of various types of firearms; and

(3) Responsibilities of firearm ownership.

B. "Firearm" has the same meaning as in Title 17-A, section 2, subsection 12-A.

C. "Firearm dealer" means a person who is licensed as a dealer under 18 United States Code, Section 923, or who is required to be licensed as a dealer under that section.

2. REQUIREMENT. A firearm dealer must:

A. Include a basic firearm safety brochure with every firearm sold at retail in this State, except that the brochure need not be supplied by the firearm dealer if the firearm manufacturer provides a basic firearm safety brochure with the firearm. The dealer may collect a charge for the brochure, which may not be greater than the dealer's cost to obtain the brochure;

B. Offer to demonstrate to the purchaser the use of a trigger locking device; and

C. Post in a conspicuous place information relating to the availability of known local voluntary firearm safety programs.

3. NO LIABILITY. Organizations that produce basic firearm safety brochures for distribution to firearm dealers for subsequent distribution to purchasers of firearms and firearm dealers are not liable for injuries resulting from the accidental discharge of nondefective firearms purchased from any dealer.

[Current through emergency legislation through Chapter 700 of the Second Regular Session of the 123rd Legislature ; current with Chs. 470-560 of the Second Regular Session of the 123rd Legislature]

Criminal Law Article

Title 1. General Provisions

Subtitle 1. Definitions

1-101. Definitions

(a) In general. In this article the following words have the meanings indicated. ...

(g) Minor. "Minor" means an individual under the age of 18 years. ...

Title 4. Weapon Crimes

Subtitle 2. Handguns

... 4-209. Regulation of weapons and ammunition

(a) State preemption. Except as otherwise provided in this section, the State preempts the right of a county, municipal corporation, or special taxing district to regulate the purchase, sale, taxation, transfer, manufacture, repair, ownership, possession, and transportation of:

(1) A handgun, rifle, or shotgun; and

(2) Ammunition for and components of a handgun, rifle, or shotgun.

(b) Exceptions.

(1) A county, municipal corporation, or special taxing district may regulate the purchase, sale, transfer, ownership, possession, and transportation of the items listed in subsection (a) of this section:

(i) With respect to minors;

(ii) With respect to law enforcement officials of the subdivision; and

(iii) Except as provided in paragraph (2) of this subsection, within 100 yards of or in a park, church, school, public building, and other place of public assembly.

(2) A county, municipal corporation, or special taxing district may not prohibit the teaching of or training in firearms safety, or other educational or sporting use of the items listed in subsection (a) of this section.

(c) Preexisting local laws. To the extent that a local law does not create an inconsistency with this section or expand existing regulatory control, a county, municipal corporation, or special taxing district may exercise its existing authority to amend any local law that existed on or before December 31, 1984. ...

Subtitle 3. Assault Pistols and Detachable Magazines

4-301. "Assault pistol" defined In this subtitle, "assault pistol" means any of the following firearms or a copy regardless of the producer or manufacturer:

(1) AA Arms AP-9 semiautomatic pistol;

(2) Bushmaster semiautomatic pistol;

(3) Claridge HI-TEC semiautomatic pistol;

(4) D Max Industries semiautomatic pistol;

(5) Encom MK-IV, MP-9, OR MP-45 semiautomatic pistol;

(6) Heckler and Koch semiautomatic SP-89 pistol;

(7) Holmes MP-83 semiautomatic pistol;

(8) Ingram MAC 10/11 semiautomatic pistol and variations including the Partisan Avenger and the SWD Cobray;

(9) Intratec TEC-9/DC-9 semiautomatic pistol in any centerfire variation;

(10) P.A.W.S. type semiautomatic pistol;

(11) Skorpion semiautomatic pistol;

(12) Spectre double action semiautomatic pistol (Sile, F.I.E., Mitchell);

(13) UZI semiautomatic pistol;

(14) Weaver Arms semiautomatic Nighthawk pistol; or

(15) Wilkinson semiautomatic "Linda" pistol.

4-302. Scope of subtitle This subtitle does not apply to:

(1) If acting within the scope of official business, personnel of the United States government or a unit of that government, members of the armed forces of the United States or of the national guard, or law enforcement personnel of the State or a local unit in the State;

(2) A firearm modified to render it permanently inoperative;

(3) Purchases, sales, and transport to or by a licensed firearms dealer or manufacturer who is:

(i) Providing or servicing an assault pistol or detachable magazine for a law enforcement unit or for personnel exempted under item (1) of this section; or

(ii) Acting to sell or transfer an assault pistol detachable magazine to a licensed firearm dealer in another state;

(4) Organizations that are required or authorized by federal law governing their specific business or activity to maintain assault pistols and applicable ammunition and detachable magazines;

(5) The receipt of an assault pistol or detachable magazine by inheritance if the decedent lawfully possessed the assault pistol; or

(6) The receipt of an assault pistol or detachable magazine by a personal representative of an estate for purposes of exercising the powers and duties of a personal representative of an estate.

4-303. Assault pistols - prohibited

(a) In general. Except as provided in subsection (b) of this section, a person may not:

(1) Transport an assault pistol into the State; or

(2) Possess, sell, offer to sell, transfer, purchase, or receive an assault pistol.

(b) Exception. A person who lawfully possessed an assault pistol before June 1, 1994 and who registered the assault pistol with the Secretary of the State Police before August 1, 1994 may continue to possess the assault pistol.

4-304. Same - seizure and disposition A law enforcement unit may seize as contraband and dispose of according to regulation an assault pistol transported, sold, transferred, purchased, received, or possessed in violation of this subtitle.

4-305. Detachable magazines - prohibited

(a) Scope of section. This section does not apply to a .22 caliber rifle with a tubular magazine.

(b) Prohibited. A person may not manufacture, sell, offer for sale, purchase, receive, or transfer a detachable magazine that has a capacity of more than 20 rounds of ammunition for a firearm.

4-306. Penalties

(a) In general. A person who violates this subtitle is guilty of a misdemeanor and on conviction is subject to imprisonment not exceeding 3 years or a fine not exceeding $5,000 or both.

(b) Use in a felony or crime of violence. ...

Subtitle 4. Uniform Machine Gun Act

4-401. Definitions

(a) In general. In this subtitle the following words have the meanings indicated. ...

(c) Machine gun. "Machine gun" means a loaded or unloaded weapon that is capable of automatically discharging more than one shot or bullet from a magazine by a single function of the firing device.

4-402. Possession of machine gun

(a) Evidence of possession. The presence of a machine gun in a room, boat, or vehicle is evidence of the possession or use of the machine gun by each person occupying the room, boat, or vehicle.

(b) Exceptions. This subtitle does not prohibit or interfere with:

(1) The manufacture, sale, and transportation of a machine gun for or to a military force or peace officer of the United States, a state, or a political subdivision of a state;

(2) The possession of a machine gun for a scientific purpose;

(3) The possession, as a curiosity, ornament, or keepsake, of a machine gun that cannot be used as a weapon; or

(4) The possession of a machine gun for a purpose that is manifestly not aggressive or offensive.

(c) Seizure and confiscation.

(1) A court may issue a warrant to search for and seize a machine gun possessed in violation of this subtitle under the same procedure as for issuance of a warrant for stolen property.

(2) On application by the State's Attorney, a court may order the confiscation or destruction of a legally seized machine gun or the transfer of the machine gun to a peace officer of the State or a political subdivision of the State.

4-403. Registration of machine gun

(a) Manufacturer registration.

(1) A manufacturer of a machine gun shall keep a register of each machine gun manufactured or handled by the manufacturer.

(2) The register shall contain:

(i) The method of manufacture and serial number of the machine gun;

(ii) The date of manufacture, sale, loan, gift, delivery, and receipt of the machine gun from the manufacturer; and

(iii) The name, address, and occupation of the person to whom the machine gun was sold, loaned, given or delivered, or from whom the machine gun was received, and the purpose for which the machine gun was acquired.

(3) A person who violates this subsection is guilty of a misdemeanor and on conviction is subject to a fine not exceeding $100.

(b) Inspection of manufacturer stock.

(1) On demand, a manufacturer of a machine gun shall allow a marshal, sheriff, or police officer to inspect the manufacturer's entire stock of machine guns, parts, and supplies and the register required under subsection (a) of this section.

(2) A person who violates paragraph (1) of this subsection is guilty of a misdemeanor and on conviction is subject to a fine not exceeding $100.

(c) Registration of possession.

(1) A person who acquires a machine gun shall register the machine gun with the secretary of the state police:

(i) Within 24 hours after acquiring the machine gun; and

(ii) In each succeeding year during the month of May.

(2) The Secretary of the State Police shall prepare and, on request of an applicant, furnish an application form for registration under this subsection.

(3) An application for registration shall contain:

(i) The make, model, serial number, caliber, type, barrel length, finish, and country of origin of the machine gun;

(ii) The name, address, race, gender, date of birth, Maryland driver's license number, and occupation of the person in possession of the machine gun; and

(iii) The name of the person from whom the machine gun was acquired and the purpose for acquiring the machine gun.

(4) Each application for registration filed with the Secretary of the State Police shall be accompanied by a nonrefundable registration fee of $10.

(5) Registration data provided under this section is not open to public inspection. ...

4-405. Use of machine gun for aggressive purpose

(a) Presumption of offensive or aggressive purpose. Possession or use of a machine gun is presumed to be for an offensive or aggressive purpose when:

(1) The machine gun:

(i) Is on premises not owned or rented for bona fide permanent residence or business occupancy by the person in whose possession the machine gun is found;

(ii) Is in the possession of, or used by, an unnaturalized foreign-born person or a person who has been convicted of a crime of violence in any state or federal court of the United States; or

(iii) Is not registered as required under § 4-403 of this subtitle; or

(2) Empty or loaded shells that have been used or are susceptible of being used in the machine gun are found in the immediate vicinity of the machine gun.

(b) Prohibited. A person may not possess or use a machine gun for an offensive or aggressive purpose.

(c) Penalty. A person who violates this section is guilty of a misdemeanor and on conviction is subject to imprisonment not exceeding 10 years.

(d) Statute of limitations and in banc review. A person who violates this section is subject to § 5-106(b) of the Courts Article.

4-406. Uniformity This subtitle shall be interpreted and construed as to effectuate its general purpose to make uniform the law of those states that enact it.

4-407 Short title This subtitle may be cited as the Uniform Machine Gun Act.

Subtitle 5. Destructive Devices

4-501 Definitions

(a) In general. In this subtitle the following words have the meanings indicated.

(b) Destructive device.

(1) "Destructive device" means explosive material, incendiary material, or toxic material that is:

(i) Combined with a delivery or detonating apparatus so as to be capable of inflicting injury to persons or damage to property; or

(ii) Deliberately modified, containerized, or otherwise equipped with a special delivery, activation, or detonation component that gives the material destructive characteristics of a military ordnance.

(2) "Destructive device" includes a bomb, grenade, mine, shell, missile, flamethrower, poison gas, Molotov cocktail, pipe bomb, and petroleum-soaked ammonium nitrate.

(c) Explosive material.

(1) "Explosive material" means material that explodes when detonated and has a destructive capability.

(2) "Explosive material" includes:

(i) Explosives as defined in § 11-101 of the Public Safety Article; and

(ii) Dynamite for construction work, ammonium nitrate, natural gas in pipelines or storage tanks, ether, and cannisterized oxygen for health care facilities.

(3) "Explosive material" does not include items excluded from explosives in § 11-101 of the Public Safety Article when the items are used in their original configuration.

(d) Incendiary material.

(1) "Incendiary material" means a flammable or combustible liquid.

(2) "Incendiary material" includes gasoline, acetone, benzene, butane, jet fuel, fuel oil, kerosene, and diesel fuel.

(e) Toxic material.

(1) "Toxic material" means material that is capable of causing death or serious bodily injury almost immediately on being absorbed through the skin, inhaled, or ingested.

(2) "Toxic material" includes:

(i) Nerve gas, mustard gas, cyanide gas, chlorine gas, sulphuric acid, or their precursors; and

(ii) A biological substance containing a disease organism or microorganism.

4-502. Scope of subtitle This subtitle does not apply to:

(1) A member of the armed forces of the United States or of the national guard or law enforcement personnel of the United States, the State, or a political subdivision of the State while acting within the scope of official duties;

(2) An officer or employee of the United States, the State, or a political subdivision of the State who is authorized to handle a destructive device within the scope of official duties and who is acting within the scope of those duties;

(3) A person authorized by law to possess explosive material, incendiary material, or toxic material who is acting within the scope of authority if the possession of the material is specifically regulated or licensed by law; or

(4) A person who possesses smokeless or black gunpowder under Title 11, Subtitle 1 of the Public Safety Article and uses the gunpowder for loading or reloading small arms ammunition, antique firearms, or replicas of antique firearms.

4-503 Manufacture or possession of destructive device

(a) Prohibited. A person may not knowingly:

(1) Manufacture, transport, possess, control, store, sell, distribute, or use a destructive device; or

(2) Possess explosive material, incendiary material, or toxic material with intent to create a destructive device.

(b) Penalty.

(1) A person who violates this section is guilty of a felony and on conviction is subject to imprisonment not exceeding 25 years or a fine not exceeding $250,000 or both.

(2) A sentence imposed under this subsection may be separate from and consecutive to or concurrent with a sentence for a crime based on the act or acts establishing the violation of this section.

(3) In addition to any other penalty authorized by law, if the person convicted or found to have committed a delinquent act under this section is a minor, the court may order the Motor Vehicle Administration to initiate an action, under the motor vehicle laws, to suspend the driving privilege of the minor for a specified period not to exceed:

(i) For a first violation, 6 months; and

(ii) For each subsequent violation, 1 year or until the person is 21 years old, whichever is longer.

(c) Restitution.

(1) In addition to any penalty provided in subsection (b) of this section, a person convicted or found to have committed a delinquent act under this section may be ordered by the court to pay restitution to:

(i) The State, county, municipal corporation, bicounty agency, county board of education, public authority or special taxing district for actual costs reasonably incurred due to a violation of this section, including the search for, removal of, and damages caused by a destructive device; and

(ii) The owner or tenant of a property for the actual value of any goods, services, or income lost as a result of the evacuation of the property or damage sustained due to a violation of this section.

(2)(i) If a person convicted or found to have committed a delinquent act under this section is a minor, the court may order the minor, the minor's parent, or both to pay the restitution described in paragraph (1) of this subsection.

(ii) Except as otherwise provided in this section, the provisions of title 11, subtitle 6 of the Criminal Procedure Article apply to an order of restitution under this paragraph.

(3) This subsection does not limit the right of a person to restitution under title 11, subtitle 6 of the Criminal Procedure Article.

Title 5. Controlled Dangerous Substances, Prescriptions, and Other Substances

Subtitle 6. Crimes - Controlled Dangerous Substances and Paraphernalia

Part III. Related and Derivative Crimes

5-622. Firearm crimes

(a) "Firearm" defined. In this section, "firearm" includes:

(1) A handgun, antique firearm, rifle, shotgun, short-barreled shotgun, and short-barreled rifle, as those words are defined in § 4-201 of this article;

(2) A machine gun, as defined in § 4-401 of this article; and

(3) A regulated firearm, as defined in § 5-101 of the Public Safety Article.

(b) Prohibited. A person may not possess, own, carry, or transport a firearm if that person has been convicted of:

(1) A felony under this title;

(2) A crime under the laws of another state or of the United States that would be a felony under this title if committed in this State;

(3) Conspiracy to commit a crime referred to in paragraphs (1) and (2) of this subsection; or

(4) An attempt to commit a crime referred to in paragraphs (1) and (2) of this subsection.

(c) Penalty. A person who violates this section is guilty of a felony and on conviction is subject to imprisonment not exceeding 5 years or a fine not exceeding $10,000 or both.

Public Safety Article

Title 5. Firearms

Subtitle 1. Regulated Firearms

5-101. Definitions

(a) In this subtitle the following words have the meanings indicated.

(b) Antique firearm. "Antique firearm" has the meaning stated in § 4-201 of the Criminal Law Article.

(c) Crime of violence. "Crime of violence" means:

(1) abduction;

(2) arson in the first degree;

(3) assault in the first or second degree;

(4) burglary in the first, second, or third degree;

(5) carjacking and armed carjacking;

(6) escape in the first degree;

(7) kidnapping;

(8) voluntary manslaughter;

(9) maiming as previously proscribed under former Article 27, § 386 of the Code;

(10) mayhem as previously proscribed under former Article 27, § 384 of the Code;

(11) murder in the first or second degree;

(12) rape in the first or second degree;

(13) robbery;

(14) robbery with a dangerous weapon;

(15) sexual offense in the first, second, or third degree;

(16) an attempt to commit any of the crimes listed in items (1) through (15) of this subsection; or

(17) assault with intent to commit any of the crimes listed in items (1) through (15) of this subsection or a crime punishable by imprisonment for more than 1 year.

(d) Dealer. "Dealer" means a person who is engaged in the business of:

(1) selling, renting, or transferring firearms at wholesale or retail; or

(2) repairing firearms.

(e) Dealer's license. "Dealer's license" means a State regulated firearms dealer's license.

(f) Designated law enforcement agency. "Designated law enforcement agency" means a law enforcement agency that the Secretary designates to process applications to purchase regulated firearms for secondary sales.

(g) Disqualifying crime. "Disqualifying crime" means:

(1) a crime of violence;

(2) a violation classified as a felony in the State; or

(3) a violation classified as a misdemeanor in the State that carries a statutory penalty of more than 2 years.

(h) Firearm.

(1) "Firearm" means:

(i) a weapon that expels, is designed to expel, or may readily be converted to expel a projectile by the action of an explosive; or

(ii) the frame or receiver of such a weapon.

(2) "Firearm" includes a starter gun.

(i) Firearm applicant. "Firearm applicant" means a person who makes a firearm application.

(j) Firearm application. "Firearm application" means an application to purchase, rent, or transfer a regulated firearm.

(k) Fugitive from justice. "Fugitive from justice" means a person who has fled to avoid prosecution or giving testimony in a criminal proceeding.

(l) Habitual drunkard. "Habitual drunkard" means a person who has been found guilty of any three crimes under § 21-902(a), (b), or (c) of the Transportation Article, one of which occurred in the past year.

(m) Habitual user. "Habitual user" means a person who has been found guilty of two controlled dangerous substance crimes, one of which occurred in the past 5 years.

(n) Handgun.

(1) "Handgun" means a firearm with a barrel less than 16 inches in length.

(2) "Handgun" includes signal, starter, and blank pistols.

(o) Licensee. "Licensee" means a person who holds a dealer's license.

(p) Regulated firearm. "Regulated firearm" means:

(1) a handgun; or

(2) a firearm that is any of the following specific assault weapons or their copies, regardless of which company produced and manufactured that assault weapon:

(i) American Arms Spectre da Semiautomatic carbine;

(ii) AK-47 in all forms;

(iii) Algimec AGM-1 type semi-auto;

(iv) AR 100 type semi-auto;

(v) AR 180 type semi-auto;

(vi) Argentine L.S.R. semi-auto;

(vii) Australian Automatic Arms SAR type semi-auto;

(viii) Auto-Ordnance Thompson M1 and 1927 semi-automatics;

(ix) Barrett light .50 cal. semi-auto;

(x) Beretta AR70 type semi-auto;

(xi) Bushmaster semi-auto rifle;

(xii) Calico models M-100 and M-900;

(xiii) CIS SR 88 type semi-auto;

(xiv) Claridge HI TEC C-9 carbines;

(xv) Colt AR-15, CAR-15, and all imitations except Colt AR-15 Sporter H-BAR rifle;

(xvi) Daewoo MAX 1 and MAX 2, aka AR 100, 110C, K-1, and K-2;

(xvii) Dragunov Chinese made semi-auto;

(xviii) Famas semi-auto (.223 caliber);

(xix) Feather AT-9 semi-auto;

(xx) FN LAR and FN FAL assault rifle;

(xxi) FNC semi-auto type carbine;

(xxii) F.I.E./Franchi LAW 12 and SPAS 12 assault shotgun;

(xxiii) Steyr-AUG-SA semi-auto;

(xxiv) Galil models AR and ARM semi-auto;

(xxv) Heckler and Koch HK-91 A3, HK-93 A2, HK-94 A2 and A3;

(xxvi) Holmes model 88 shotgun;

(xxvii) Avtomat Kalashnikov semiautomatic rifle in any format;

(xxviii) Manchester Arms "Commando" MK-45, MK-9;

(xxix) Mandell TAC-1 semi-auto carbine;

(xxx) Mossberg model 500 Bullpup assault shotgun;

(xxxi) Sterling Mark 6;

(xxxii) P.A.W.S. carbine;

(xxxiii) Ruger mini-14 folding stock model (.223 caliber);

(xxxiv) SIG 550/551 assault rifle (.223 caliber);

(xxxv) SKS with detachable magazine;

(xxxvi) AP-74 Commando type semi-auto;

(xxxvii) Springfield Armory BM-59, SAR-48, G3, SAR-3, M-21 sniper rifle, M1A, excluding the M1 Garand;

(xxxviii) Street sweeper assault type shotgun;

(xxxix) Striker 12 assault shotgun in all formats;

(xl) Unique F11 semi-auto type;

(xli) Daewoo USAS 12 semi-auto shotgun;

(xlii) UZI 9mm carbine or rifle;

(xliii) Valmet M-76 and M-78 semi-auto;

(xliv) Weaver Arms "Nighthawk" semi-auto carbine; or

(xlv) Wilkinson Arms 9mm semi-auto "Terry".

(q) Rent. "Rent" means the temporary transfer for consideration of a regulated firearm that is taken from the property of the owner of the regulated firearm.

(r) Secondary sale. "Secondary sale" means a sale of a regulated firearm in which neither party to the sale:

(1) is a licensee;

(2) is licensed by the federal government as a firearms dealer;

(3) devotes time, attention, and labor to dealing in firearms as a regular course of trade or business with the principal objective of earning a profit through the repeated purchase and resale of firearms; or

(4) repairs firearms as a regular course of trade or business.

(s) Secretary. "Secretary" means the Secretary of State Police or the Secretary's designee.

(t) Straw purchase. "Straw purchase" means a sale of a regulated firearm in which a person uses another, known as the straw purchaser, to:

(1) complete the application to purchase a regulated firearm;

(2) take initial possession of the regulated firearm; and

(3) subsequently transfer the regulated firearm to the person.

5-102. Scope of subtitle This subtitle does not apply to:

(1) the transfer or possession of a regulated firearm or detachable magazine:

(i) for testing or experimentation authorized by the Secretary; and

(ii) by a federally licensed gun manufacturer, dealer, or importer;

(2) the sale, transfer, or possession of an antique firearm;

(3) an unserviceable firearm sold, transferred, or possessed as a curio or museum piece;

(4) law enforcement personnel of any unit of the federal government, members of the armed forces of the United States or the National Guard, or law enforcement personnel of the State or any local agency in the State, while those personnel or members are acting within the scope of their official duties;

(5) a regulated firearm modified to render it permanently inoperative;

(6) purchases, sales, and transportation to or by a federally licensed gun manufacturer, dealer, or importer;

(7) an organization that is required or authorized by federal law governing its specific business or activity to maintain firearms;

(8) the receipt of a regulated firearm by inheritance, if the heir forwards to the Secretary a completed application to purchase or transfer that regulated firearm; or

(9) a signal pistol or other visual distress signal that the United States Coast Guard approves as a marine safety device.

5-103. Effect of subtitle This subtitle does not affect:

(1) a sale or transfer for bona fide resale in the ordinary course of business of a licensee; or

(2) a sale, rental, transfer, or the use of a regulated firearm by a person authorized or required to do so as part of the person's duties as a member of:

(i) an official police force or other law enforcement agency;

(ii) the armed forces of the United States, including all official reserve organizations; or

(iii) the Maryland National Guard.

5-104. Preemption by State This subtitle supersedes any restriction that a local jurisdiction in the State imposes on a sale of a regulated firearm, and the State preempts the right of any local jurisdiction to regulate the sale of a regulated firearm.

5-105. Regulations The Secretary shall adopt regulations to carry out this subtitle.

5-106. Dealer's license

(a) Required. A person must lawfully possess a dealer's license issued by the Secretary before the person engages in the business of selling, renting, or transferring regulated firearms.

(b) One license for each place of business. One dealer's license is required for each place of business where regulated firearms are sold.

5-107. Application for dealer's license required

(a) In general.

(1) An applicant for a dealer's license shall:

(i) submit to the Secretary an application on the form that the Secretary provides; and

(ii) pay to the Secretary an application fee of $50, payable to the Comptroller.

(2) A refund or proration of the application fee is prohibited.

(b) Required information. An application for a dealer's license shall contain:

(1) the applicant's name, address, Social Security number, place and date of birth, height, weight, race, eye and hair color, and signature;

(2) a clear and recognizable photograph of the applicant, unless the photograph has been submitted with a prior year's application;

(3) a set of the applicant's fingerprints, unless the fingerprints have been submitted with a prior year's application; and

(4) a statement by the applicant that the applicant:

(i) is a citizen of the United States;

(ii) is at least 21 years old;

(iii) has never been convicted of a disqualifying crime;

(iv) has never been convicted of a violation classified as a common law crime and received a term of imprisonment of more than 2 years;

(v) is not a fugitive from justice;

(vi) is not a habitual drunkard;

(vii) is not addicted to a controlled dangerous substance or is not a habitual user; and

(viii) has never spent more than 30 consecutive days in a medical institution for treatment of a mental disorder, unless a physician's certificate issued within 30 days before the date of application is attached to the application, certifying that the applicant is capable of possessing a regulated firearm without undue danger to the applicant or to another.

(c) Required warning. Each application for a dealer's license shall contain the following statement: "Any false information supplied or statement made in this application is a crime which may be punished by imprisonment for a period of not more than 3 years, or a fine of not more than $5,000 or both."

(d) Application of corporation. If an applicant is a corporation, a corporate officer who is a resident of the State shall complete and execute the application.

5-108. Criminal history records check

(a) "Central Repository' defined. In this section, "Central Repository" means the Criminal Justice Information System Central Repository of the Department of Public Safety and Correctional Services.

(b) Application required. The Secretary shall apply to the Central Repository for a State and national criminal history records check for each applicant for a dealer's license.

(c) Contents of application. As part of the application for a criminal history records check, the Secretary shall submit to the Central Repository:

(1) two complete sets of the applicant's legible fingerprints taken on forms approved by the Director of the Central Repository and the Director of the Federal Bureau of Investigation;

(2) the fee authorized under § 10-221(b)(7) of the Criminal Procedure Article for access to Maryland criminal history records; and

(3) the mandatory processing fee required by the Federal Bureau of Investigation for a national criminal history records check.

(d) Information forwarded to applicant and State Police. In accordance with §§ 10-201 through 10-234 of the Criminal Procedure Article, the Central Repository shall forward to the applicant and the Secretary a printed statement of the applicant's criminal history record information.

(e) Restrictions on information. Information obtained from the Central Repository under this section:

(1) is confidential and may not be disseminated; and

(2) shall be used only for the licensing purpose authorized by this section.

(f) Subject may contest contents. The subject of a criminal history records check under this section may contest the contents of the printed statement issued by the Central Repository as provided in § 10-223 of the Criminal Procedure Article.

5-109. Investigation of applicant for dealer's license The Secretary shall conduct an investigation to determine the truth or falsity of the information supplied and the statements made in an application for a dealer's license.

5-110. Disapproval of dealer's license application

(a) Grounds. The Secretary shall disapprove an application for a dealer's license if:

(1) the Secretary determines that the applicant supplied false information or made a false statement;

(2) the Secretary determines that the application is not properly completed; or

(3) the Secretary receives a written notification from the applicant's licensed attending physician that the applicant suffers from a mental disorder and is a danger to the applicant or to another.

(b) Notice. If the Secretary disapproves an application for a dealer's license, the Secretary shall notify the applicant in writing of the disapproval.

(c) Effect of disapproval. A person whose application for a dealer's license has been disapproved may not engage in the business of selling, renting, or transferring regulated firearms, unless the disapproval has been subsequently withdrawn by the Secretary or overruled by a court in accordance with subsection (d) of this section.

(d) Appeal.

(1) An applicant who is aggrieved because the Secretary has disapproved the application for a dealer's license may appeal to the circuit court of the county where the applicant's place of business is to be located.

(2) The appeal must be filed not later than 30 days after the Secretary mails notification of disapproval to the applicant.

(3) If the appeal is properly and timely filed, the court shall affirm or reverse the disapproval of the Secretary depending on whether the court finds that:

(i) the applicant supplied false information or made a false statement; or

(ii) the application was not properly completed.

(4) The Secretary or the applicant may appeal the decision of the circuit court to the Court of Special Appeals.

5-111. Term of dealer's license

(a) In general. Unless a dealer's license is renewed for a 1-year term as provided in this section, a dealer's license expires on the first June 30 after its effective date.

(b) Applications for renewal.

(1) Before a dealer's license expires, the licensee periodically may renew it for an additional 1-year term, if the licensee:

(i) is otherwise entitled to be licensed;

(ii) pays to the Secretary a renewal fee of $25, payable to the Comptroller; and

(iii) submits to the Secretary a renewal application on the form that the Secretary provides

(2) A refund or proration of the renewal fee is prohibited.

5-112. Nontransferability of dealer's license; new place of business

(a) Nontransferability of dealer's license. A dealer's license is not transferable.

(b) Notice to Secretary of new place of business. Before moving a place of business, a licensee shall inform the Secretary and surrender the dealer's license.

(c) New dealer's license for new place of business. If a cause to revoke the dealer's license does not exist, the Secretary shall issue a new dealer's license without charge covering the new place of business for the rest of the term of the surrendered dealer's license.

5-113. Display of dealer's license

(a) Required. A licensee shall display conspicuously the dealer's license and any other license required by law at the licensee's place of business.

(b) Identification of licensee and location. The dealer's license shall identify the licensee and the location of the licensee's place of business.

5-114. Suspensions and revocations – Grounds; notice; effect

(a) Suspensions. The Secretary shall suspend a dealer's license if the licensee:

(1) is under indictment for a crime of violence; or

(2) is arrested for a violation of this subtitle that prohibits the purchase or possession of a regulated firearm.

(b) Revocations. The Secretary shall revoke a dealer's license if:

(1) it is discovered that false information has been supplied or false statements have been made in an application required by this subtitle; or

(2) the licensee:

(i) is convicted of a disqualifying crime;

(ii) is convicted of a violation classified as a common law crime and receives a term of imprisonment of more than 2 years;

(iii) is a fugitive from justice;

(iv) is a habitual drunkard;

(v) is addicted to a controlled dangerous substance or is a habitual user;

(vi) has spent more than 30 consecutive days in a medical institution for treatment of a mental disorder, unless the licensee produces a physician's certificate, issued after the last institutionalization and certifying that the licensee is capable of possessing a regulated firearm without undue danger to the licensee or to another;

(vii) has knowingly or willfully manufactured, offered to sell, or sold a handgun not on the handgun roster in violation of § 5-406 of this title; or

(viii) has knowingly or willfully participated in a straw purchase of a regulated firearm.

(c) Notice. If the Secretary suspends or revokes a dealer's license, the Secretary shall notify the licensee in writing of the suspension or revocation.

(d) Effect of suspension or revocation. A person whose dealer's license is suspended or revoked may not engage in the business of selling, renting, or transferring regulated firearms, unless the suspension or revocation has been subsequently withdrawn by the Secretary or overruled by a court in accordance with § 5-116 of this subtitle.

5-115. Suspensions and revocations-Hearings

(a) Right to hearing.

(1) A person whose dealer's license is suspended or revoked and who is aggrieved by the action of the Secretary may request a hearing by writing to the Secretary within 30 days after the Secretary forwards notice to the applicant under § 5-114(c) of this subtitle.

(2) The Secretary shall grant the hearing within 15 days after receiving the request.

(b) Application of contested case provisions. The hearing shall be held in accordance with Title 10, Subtitle 2 of the State Government Article.

5-116. Judicial review

(a) Stay of revocation. A revocation may not take effect while an appeal is pending.

(b) Application of contested case provisions. Any subsequent judicial review shall be held in accordance with Title 10, Subtitle 2 of the State Government Article.

5-117. Application for regulated firearm required A person must submit a firearm application in accordance with this subtitle before the person purchases, rents, or transfers a regulated firearm.

5-118. Firearm application

(a) In general. A firearm applicant shall:

(1) submit to a licensee or designated law enforcement agency a firearm application on the form that the Secretary provides; and

(2) pay to the licensee or designated law enforcement agency an application fee of $10.

(b) Required information. A firearm application shall contain:

(1) the firearm applicant's name, address, Social Security number, place and date of birth, height, weight, race, eye and hair color, signature, driver's or photographic identification soundex number, occupation, and regulated firearm information for each regulated firearm to be purchased, rented, or transferred;

(2) the date and time that the firearm applicant delivered the completed firearm application to the prospective seller or transferor; and

(3) a statement by the firearm applicant under the penalty of perjury that the firearm applicant:

(i) is at least 21 years old;

(ii) has never been convicted of a disqualifying crime;

(iii) has never been convicted of a violation classified as a common law crime and received a term of imprisonment of more than 2 years;

(iv) is not a fugitive from justice;

(v) is not a habitual drunkard;

(vi) is not addicted to a controlled dangerous substance or is not a habitual user;

(vii) has never spent more than 30 consecutive days in a medical institution for treatment of a mental disorder, unless a physician's certificate issued within 30 days before the date of application is attached to the application, certifying that the firearm applicant is capable of possessing a regulated firearm without undue danger to the firearm applicant or to another;

(viii) is not a respondent against whom a current non ex parte civil protective order has been entered under § 4-506 of the Family Law Article;

(ix) if under the age of 30 years at the time of application, has not been adjudicated delinquent by a juvenile court for an act that would be a disqualifying crime if committed by an adult; and

(x) subject to § 5-119 of this subtitle, has completed a certified firearms safety training course that the Police Training Commission conducts without charge or that meets the standards that the Police Training Commission establishes under § 3-207 of this article.

(c) Required warning. Each firearm application shall contain the following statement: "Any false information supplied or statement made in this application is a crime which may be punished by imprisonment for a period of not more than 3 years, or a fine of not more than $5,000, or both."

(d) Firearm application of corporation. If the firearm applicant is a corporation, a corporate officer who is a resident of the State shall complete and execute the firearm application.

5-119. Exemption from certified firearms training course requirement A firearm applicant is not required to complete a certified firearms training course required under §§ 5-118 and 5-134 of this subtitle if the firearm applicant:

(1) has already completed a certified firearms training course required under §§ 5-118 and 5-134 of this subtitle;

(2) is a law enforcement officer of the State or any local law enforcement agency in the State;

(3) is a member, retired member, or honorably discharged member of the armed forces of the United States or the National Guard;

(4) is a member of an organization that is required by federal law governing its specific business or activity to maintain handguns and applicable ammunition; or

(5) holds a permit to carry a handgun under Subtitle 3 of this title.

5-120. Copies of firearm application; fees

(a) Copy to Secretary.

(1) On receipt of a firearm application, a licensee or designated law enforcement agency shall promptly forward one copy of it to the Secretary by certified mail or facsimile machine.

(2) The copy of the firearm application forwarded to the Secretary shall contain the name, address, and signature of the prospective seller, lessor, or transferor.

(b) Other copies.

(1) The prospective seller, lessor, or transferor shall keep one copy of the firearm application for not less than 3 years.

(2) The firearm applicant is entitled to the remaining copy of the firearm application.

(c) Fees.

(1) Except as provided in paragraph (2) of this subsection, the licensee or designated law enforcement agency shall forward the $10 application fee with the firearm application to the Secretary.

(2) A licensee or designated law enforcement agency that uses a facsimile machine to forward the firearm application to the Secretary shall:

(i) be billed $10 for each firearm application forwarded to the Secretary during the month; and

(ii) pay the total application fee by the fifteenth day of the following month.

5-121. Investigation of firearm applicant

(a) Secretary to conduct investigation. On receipt of a firearm application, the Secretary shall conduct an investigation promptly to determine the truth or falsity of the information supplied and statements made in the firearm application.

(b) Request for assistance. In conducting an investigation under this subsection, the Secretary may request the assistance of the Police Commissioner of Baltimore City, the chief of police in any county maintaining a police force, or the sheriff in a county not maintaining a police force.

5-122. Disapproval of firearm application

(a) Grounds. The Secretary shall disapprove a firearm application if:

(1) the Secretary determines that the firearm applicant supplied false information or made a false statement;

(2) the Secretary determines that the firearm application is not properly completed; or

(3) the Secretary receives written notification from the firearm applicant's licensed attending physician that the firearm applicant suffers from a mental disorder and is a danger to the firearm applicant or to another.

(b) Notice.

(1) If the Secretary disapproves a firearm application, the Secretary shall notify the prospective seller, lessor, or transferor in writing of the disapproval within 7 days after the date that the executed firearm application is forwarded to the Secretary by certified mail or facsimile machine.

(2) After notifying the prospective seller, lessor, or transferor under paragraph (1) of this subsection, the Secretary shall notify the prospective purchaser, lessee, or transferee in writing of the disapproval.

(3) The date when the prospective seller, lessor, or transferor forwards the executed firearm application to the Secretary by certified mail or by facsimile machine is the first day of the 7-day period allowed for notice of disapproval to the prospective seller, lessor, or transferor.

5-123. Time for licensee to complete transactions

(a) Seven-day waiting period. A licensee may not sell, rent, or transfer a regulated firearm until after 7 days following the time a firearm application is executed by the firearm applicant, in triplicate, and the original is forwarded by the prospective seller or transferor to the Secretary.

(b) Completion required in 90 days. A licensee shall complete the sale, rental, or transfer of a regulated firearm within 90 days after the firearm application was stamped by the Secretary as not being disapproved.

(c) Incomplete transactions.

(1) If the sale, rental, or transfer of a regulated firearm is not completed within 90 days after the firearm application was stamped by the Secretary as not being disapproved, a licensee shall return the firearm application to the Secretary within 7 days.

(2) The Secretary shall void a firearm application returned under paragraph (1) of this subsection as an incomplete sale, rental, or transfer.

(d) Notification of completed transaction.

(1)(i) A licensee who sells, rents, or transfers a regulated firearm in compliance with this subtitle shall forward a copy of the written notification of the completed transaction to the Secretary within 7 days after delivery of the regulated firearm.

(ii) The notification shall contain an identifying description of the regulated firearm, including its caliber, make, model, any manufacturer's serial number, and any other special or peculiar characteristic or marking by which the regulated firearm may be identified.

(2) The Secretary shall maintain a permanent record of all notifications received of completed sales, rentals, and transfers of regulated firearms in the State.

5-124. Secondary transactions

(a) Seven-day waiting period.

(1) A person who is not a licensee may not sell, rent, transfer, or purchase a regulated firearm until after 7 days following the time a firearm application is executed by the firearm applicant, in triplicate, and the original is forwarded by a licensee to the Secretary.

(2) As an alternative to completing a secondary sale of a regulated firearm through a licensee, a prospective seller, lessor, or transferor and a prospective purchaser, lessee, or transferee may complete the transaction through a designated law enforcement agency.

(b) Processing fee. A firearm applicant for a secondary sale of a regulated firearm through a licensee shall pay to the licensee a processing fee not exceeding $20.

(c) Completion required in 90 days. A person shall complete the sale, rental, or transfer of a regulated firearm within 90 days after the firearm application was stamped by the Secretary as not being disapproved.

(d) Incomplete transactions.

(1) If the sale, rental, or transfer of a regulated firearm is not completed within 90 days after the firearm application was stamped by the Secretary as not being disapproved, a person shall return the firearm application to the Secretary within 7 days.

(2) The Secretary shall void a firearm application returned under paragraph (1) of this subsection as an incomplete sale, rental, or transfer.

(e) Notification of completed transaction.

(1)(i) A person who sells, rents, or transfers a regulated firearm in compliance with this subtitle shall forward a copy of the written notification of the completed transaction to the Secretary within 7 days after delivery of the regulated firearm.

(ii) The notification shall contain an identifying description of the regulated firearm, including its caliber, make, model, any manufacturer's serial number, and any other special or peculiar characteristic or marking by which the regulated firearm may be identified.

(2) The Secretary shall maintain a permanent record of all notifications received of completed sales, rentals, and transfers of regulated firearms in the State.

5-125. Approve, on hold, and disapproved applications

(a) Approved applications. An approved firearm application is valid only for the purchase, rental, or transfer of the regulated firearm listed in the firearm application.

(b) On hold and disapproved applications. A licensee or other person may not sell, rent, or transfer a regulated firearm to a firearm applicant whose firearm application is placed on hold because of an open disposition of criminal proceedings against the firearm applicant or disapproved, unless the hold or disapproval has been subsequently withdrawn by the Secretary or overruled by a court in accordance with § 5-127 of this subtitle.

5-126. Hearings

(a) Right to hearing.

(1) A firearm applicant who is aggrieved by the action of the Secretary may request a hearing by writing to the Secretary within 30 days after the Secretary forwards notice to the firearm applicant under § 5-122 of this subtitle.

(2) The Secretary shall grant the hearing within 15 days after receiving the request.

(b) Application of contested case provisions. The hearing shall be held in accordance with Title 10, Subtitle 2 of the State Government Article.

(c) Venue. The hearing shall be held in the county of the legal residence of the firearm applicant.

5-127. Judicial review Any subsequent judicial review shall be held in accordance with Title 10, Subtitle 2 of the State Government Article.

5-128. Purchases within 30 days - in general.

(a) Scope of subsection. Subsection (b) of this section does not apply to:

(1) a law enforcement agency;

(2) an agency authorized to perform law enforcement duties;

(3) a State or local correctional facility;

(4) a private security company licensed to do business in the State;

(5) the purchase of an antique firearm;

(6) a purchase by a licensee;

(7) the exchange or replacement of a regulated firearm by a seller for a regulated firearm purchased from the seller by the same person seeking the exchange or replacement within 30 days immediately before the exchange or replacement; or

(8) a person whose regulated firearm is stolen or irretrievably lost and who considers it essential that the regulated firearm be replaced immediately, if:

(i) the person provides the licensee with a copy of the official police report or an official summary of the report, a copy of which shall be attached to the firearm application;

(ii) the official police report or official summary of the report contains the name and address of the regulated firearm owner, a description of the regulated firearm, the location of the loss or theft, the date of the loss or theft, and the date when the loss or theft was reported to the law enforcement agency; and

(iii) the loss or theft occurred within 30 days before the person's attempt to replace the regulated firearm, as reflected by the date of loss or theft on the official police report or official summary of the report.

(b) One purchase limit. A person may not purchase more than one regulated firearm in a 30-day period.

(c) On hold and disapproved applications. A licensee or other person may not sell, rent, or transfer a regulated firearm to a firearm applicant whose firearm application is placed on hold because of an open disposition of criminal proceedings against the firearm applicant or disapproved, unless the hold or disapproval has been subsequently withdrawn by the Secretary or overruled by a court in accordance with § 5-127 of this subtitle.

(d) Penalty. A person who violates this section is guilty of a misdemeanor and on conviction is subject to imprisonment not exceeding 3 years or a fine not exceeding $5,000 or both.

5-129. Purchases within 30 days - Multiple purchases allowed.

(a) Requirements. Notwithstanding § 5-128 (b) of this subtitle, a person may purchase more than one regulated firearm in a 30-day period if:

(1) the person applies for and the Secretary approves a multiple purchase; and

(2)(i) the purchase of the regulated firearms is for a private collection or a collector series;

(ii) the purchase of the regulated firearms is a bulk purchase from an estate sale;

(iii)1. the purchase of not more than two regulated firearms is a multiple purchase to take advantage of a licensee's discounted price available only for a multiple purchase; and

2. the purchaser is prohibited from purchasing a regulated firearm during the following 30-day period unless approved under item (i) or (ii) of this item; or

(iv) the purchase is for other purposes similar to items (i) through (iii) of this item.

(b) Application.

(1) The application for a multiple purchase shall:

(i) list the regulated firearms to be purchased;

(ii) state the purpose of the purchase of more than one regulated firearm in a 30-day period;

(iii) be witnessed by a licensee or designated law enforcement agency; and

(iv) be signed under the penalty of perjury by the firearm applicant.

(2) The application for a multiple purchase of regulated firearms shall be attached to a completed firearm application and forwarded to the Secretary by a licensee or designated law enforcement agency.

(c) Background investigation. On receipt of the firearm application and the application for a multiple purchase, the Secretary shall conduct a background investigation as required in § 5-121 of this subtitle.

(d) Penalty. A person who violates this section is guilty of a misdemeanor and on conviction is subject to imprisonment not exceeding 3 years or a fine not exceeding $5,000 or both.

5-130. Gun shows

(a) "Gun show' defined. In this section, "gun show" means any organized gathering open to the public at which any firearm is displayed.

(b) Scope of section. Subsections (c) through (h) of this section do not apply to a licensee.

(c) Temporary transfer permit required. A person must obtain a temporary transfer permit issued by the Secretary before the person displays a regulated firearm for sale or transfer from a table or fixed display at a gun show.

(d) Application for temporary transfer permit.

(1) An applicant for a temporary transfer permit shall:

(i) submit to the Secretary an application on the form that the Secretary provides; and

(ii) pay to the Secretary a fee of $10 for each calendar year.

(2) Each additional temporary transfer permit during the same calendar year shall be issued without charge.

(e) Required information. The application for a temporary transfer permit shall contain any information that is necessary for the Secretary to conduct a computer background investigation.

(f) Required warning. Each application for a temporary transfer permit shall contain the following statement: "Any false information supplied or statement made in this application is a crime which may be punished by imprisonment for a period of not more than 3 years or a fine not more than $5,000 or both.".

(g) Investigation of application; grounds for disapproval.

(1) The Secretary shall conduct an investigation to determine the truth or falsity of the information supplied and the statements made in the application for a temporary transfer permit.

(2) If there is no reason to disapprove the application for a temporary transfer permit, the Secretary shall issue the permit within 7 days after the date of application.

(3) The Secretary shall disapprove an application for a temporary transfer permit if the Secretary determines that:

(i) the applicant supplied false information or made a false statement; or

(ii) the application is not properly completed.

(4) If the Secretary disapproves an application for a temporary transfer permit, the Secretary shall notify the applicant in writing of the disapproval.

(h) Label and display.

(1) A temporary transfer permit shall be clearly labeled "temporary" and shall include the statement: "This is not a license to engage in the business of selling firearms.".

(2) The temporary transfer permit shall be placed in public view as part of any display of a regulated firearm.

(i) Five permit limit.

(1) A person may not receive more than five temporary transfer permits during a single calendar year.

(2) To display a regulated firearm for sale, trade, or transfer at more than five gun shows in a calendar year, a person shall obtain a dealer's license under this subtitle.

(j) Sale or transfer of regulated firearm. A sale or transfer of a regulated firearm from a table or fixed display at a gun show is governed by §§ 5-103, 5-104, 5-117 through 5-129, and 5-136 of this subtitle.

5-131. Handgun identification requirements

(a) Definitions.

(1) In this section the following words have the meanings indicated.

(2) "Manufacturer" means a person who possesses a federal license to engage in the business of manufacturing firearms or ammunition for sale or distribution.

(3) "Projectile" means the part of handgun ammunition that is expelled through the barrel of the handgun by an explosion.

(4) "Shell casing" means the part of handgun ammunition that contains the primer and propellent powder to discharge the projectile.

(b) Manufacturer requirements. A manufacturer that ships or transports a handgun for sale, rental, or transfer in the State shall include in the box with the handgun in a separate, sealed container:

(1) a shell casing of a projectile discharged from the handgun; and

(2) additional information that the Secretary requires to identify the type of handgun and shell casing.

(c) Actions by dealer.

(1) On receipt of a handgun from a manufacturer, the dealer shall confirm to the Department of State Police that the manufacturer has complied with subsection (b) of this section.

(2) On the sale, rental, or transfer of the handgun, the dealer shall forward the sealed container to the Department of State Police Crime Laboratory.

(d) Crime Laboratory database. On receipt of a shell casing and information as required in subsection (b) of this section, the Department of State Police Crime Laboratory shall enter the information in each relevant database.

5-132. Handgun safety devices

(a) Definitions.

(1) In this section the following words have the meanings indicated.

(2) "Authorized user" means the owner of a handgun or a person authorized by the owner to possess and use the handgun.

(3) "External safety lock" means an external device that is:

(i) attached to a handgun with a key or combination lock; and

(ii) designed to prevent a handgun from being discharged unless the device has been deactivated.

(4) "Handgun" does not include a signal, starter, or blank pistol.

(5) "Handgun Roster Board" means the Handgun Roster Board established under § 5-404 of this title.

(6) "Integrated mechanical safety device" means a disabling or locking device that is:

(i) built into a handgun; and

(ii) designed to prevent the handgun from being discharged unless the device has been deactivated.

(7) "Personalized handgun" means a handgun manufactured with incorporated design technology that:

(i) allows the handgun to be fired only by the authorized user; and

(ii) prevents any of the safety characteristics of the handgun from being readily deactivated.

(b) Scope of section. This section does not apply to:

(1) the purchase, sale, or transportation of a handgun to or by a federally licensed gun dealer or manufacturer that provides or services a handgun for:

(i) personnel of any unit of the federal government;

(ii) members of the armed forces of the United States or the National Guard;

(iii) law enforcement personnel of the State or any local law enforcement agency in the State while acting within the scope of their official duties; and

(iv) an organization that is required by federal law governing its specific business or activity to maintain handguns and applicable ammunition;

(2) a firearm modified to be permanently inoperative;

(3) the sale or transfer of a handgun by a federally licensed gun dealer or manufacturer covered under item (1) of this subsection;

(4) the sale or transfer of a handgun by a federally licensed gun dealer or manufacturer to a lawful customer outside the State; or

(5) an antique firearm.

(c) Restriction on sale, rent, or transfer of handguns.

(1) A dealer may not sell, offer for sale, rent, or transfer in the State a handgun manufactured on or before December 31, 2002, unless the handgun is sold, offered for sale, rented, or transferred with an external safety lock.

(2) On or after January 1, 2003, a dealer may not sell, offer for sale, rent, or transfer in the State a handgun manufactured on or after January 1, 2003, unless the handgun has an integrated mechanical safety device.

(d) Report.

(1) The Handgun Roster Board annually shall:

(i) review the status of personalized handgun technology; and

(ii) on or before July 1, report its findings to the Governor and, in accordance with § 2-1246 of the State Government Article, to the General Assembly.

(2) In reviewing the status of personalized handgun technology under paragraph (1) of this subsection, the Handgun Roster Board shall consider:

(i) the number and variety of models and calibers of personalized handguns that are available for sale;

(ii) each study, analysis, or other evaluation of personalized handguns conducted or commissioned by:

1. the National Institute of Justice;

2. a federal, State, or local law enforcement laboratory; or

3. any other entity with an expertise in handgun technology; and

(iii) any other information that the Handgun Roster Board considers relevant.

5-133. Restrictions on possession of regulated firearms

(a) Preemption by State. This section supersedes any restriction that a local jurisdiction in the State imposes on the possession by a private party of a regulated firearm, and the State preempts the right of any local jurisdiction to regulate the possession of a regulated firearm.

(b) Possession of regulated firearm prohibited. A person may not possess a regulated firearm if the person:

(1) has been convicted of a disqualifying crime;

(2) has been convicted of a violation classified as a common law crime and received a term of imprisonment of more than 2 years;

(3) is a fugitive from justice;

(4) is a habitual drunkard;

(5) is addicted to a controlled dangerous substance or is a habitual user;

(6) suffers from a mental disorder as defined in § 10-101(f)(2) of the Health - General Article and has a history of violent behavior against the person or another, unless the person has a physician's certificate that the person is capable of possessing a regulated firearm without undue danger to the person or to another;

(7) has been confined for more than 30 consecutive days to a facility as defined in § 10-101 of the Health - General Article, unless the person has a physician's certificate that the person is capable of possessing a regulated firearm without undue danger to the person or to another;

(8) is a respondent against whom a current non ex parte civil protective order has been entered under § 4-506 of the Family Law Article; or

(9) if under the age of 30 years at the time of possession, has been adjudicated delinquent by a juvenile court for an act that would be a disqualifying crime if committed by an adult.

(c) Penalty for possession by convicted felon.

(1) A person may not possess a regulated firearm if the person was previously convicted of:

(i) a crime of violence; or

(ii) a violation of § 5-602, § 5-603, § 5-604, § 5-605, § 5-606, § 5-607, § 5-608, § 5-609, § 5-612, § 5-613, or § 5-614 of the Criminal Law Article.

(2) A person who violates this subsection is guilty of a felony and on conviction is subject to imprisonment for not less than 5 years, no part of which may be suspended.

(3) A person sentenced under paragraph (1) of this subsection may not be eligible for parole.

(4) Each violation of this subsection is a separate crime.

(d) Possession by person under age 21 years prohibited; exceptions.

(1) Except as provided in paragraph (2) of this subsection, a person who is under the age of 21 years may not possess a regulated firearm or ammunition solely designed for a regulated firearm.

(2) Unless a person is otherwise prohibited from possessing a regulated firearm, this subsection does not apply to:

(i) the temporary transfer or possession of a regulated firearm or ammunition solely designed for a regulated firearm if the person is:

1. under the supervision of another who is at least 21 years old and who is not prohibited by State or federal law from possessing a firearm; and

2. acting with the permission of the parent or legal guardian of the transferee or person in possession;

(ii) the transfer by inheritance of title, and not of possession, of a regulated firearm;

(iii) a member of the armed forces of the United States or the National Guard while performing official duties;

(iv) the temporary transfer or possession of a regulated firearm or ammunition solely designed for a regulated firearm if the person is:

1. participating in marksmanship training of a recognized organization; and

2. under the supervision of a qualified instructor;

(v) a person who is required to possess a regulated firearm for employment and who holds a permit under Subtitle 3 of this title; or

(vi) the possession of a firearm or ammunition for self-defense or the defense of others against a trespasser into the residence of the person in possession or into a residence in which the person in possession is an invited guest.

5-134. Restrictions on sale, rental, or transfer of regulated firearms

(a) Preemption by State. This section supersedes any restriction that a local jurisdiction in the State imposes on the transfer by a private party of a regulated firearm, and the State preempts the right of any local jurisdiction to regulate the transfer of a regulated firearm.

(b) Sale, rental or transfer of regulated firearm prohibited. A dealer or other person may not sell, rent, or transfer a regulated firearm to a purchaser, lessee, or transferee who the dealer or other person knows or has reasonable cause to believe:

(1) is under the age of 21 years;

(2) has been convicted of a disqualifying crime;

(3) has been convicted of a conspiracy to commit a felony;

(4) has been convicted of a violation classified as a common law crime and received a term of imprisonment of more than 2 years;

(5) is a fugitive from justice;

(6) is a habitual drunkard;

(7) is addicted to a controlled dangerous substance or is a habitual user;

(8) suffers from a mental disorder as defined in § 10-101(f)(2) of the Health - General Article, and has a history of violent behavior against the purchaser, lessee, or transferee or another, unless the purchaser, lessee, or transferee possesses a physician's certificate that the recipient is capable of possessing a regulated firearm without undue danger to the purchaser, lessee, or transferee or to another;

(9) has been confined for more than 30 consecutive days to a facility as defined in § 10-101 of the Health - General Article, unless the purchaser, lessee, or transferee possesses a physician's certificate that the recipient is capable of possessing a regulated firearm without undue danger to the purchaser, lessee, or transferee or to another;

(10) is a respondent against whom a current non ex parte civil protective order has been entered under § 4-506 of the Family Law Article;

(11) if under the age of 30 years at the time of the transaction, has been adjudicated delinquent by a juvenile court for an act that would be a disqualifying crime if committed by an adult;

(12) is visibly under the influence of alcohol or drugs;

(13) is a participant in a straw purchase; or

(14) subject to subsection (c) of this section for a transaction under this subsection that is made on or after January 1, 2002, has not completed a certified firearms safety training course conducted free of charge by the Police Training Commission or that meets standards established by the Police Training Commission under § 3-207 of this article.

(c) Exemption from certified firearms training course requirement. A person is not required to complete a certified firearms safety training course under subsection (b)(14) of this section and § 5-118(b)(3)(x) of this subtitle if the person:

(1) has already completed a certified firearms safety training course required under subsection (b)(14) of this section and § 5-118(b)(3)(x) of this subtitle;

(2) is a law enforcement officer of the State or any local law enforcement agency in the State;

(3) is a member, retired member, or honorably discharged member of the armed forces of the United States or the National Guard;

(4) is a member of an organization that is required by federal law governing its specific business or activity to maintain handguns and applicable ammunition; or

(5) has been issued a permit to carry a handgun under Subtitle 3 of this title.

(d) Sale, rental, or transfer of regulated firearm to minor prohibited.

(1) A person may not sell, rent, or transfer:

(i) ammunition solely designed for a regulated firearm to a person who is under the age of 21 years; or

(ii)1. a firearm other than a regulated firearm to a minor;

2. ammunition for a firearm to a minor;

3. pepper mace, which is an aerosol propelled combination of highly disabling irritant based products and is also known as oleo-resin capsicum (O.C.) spray, to a minor; or

4. another deadly weapon to a minor.

(2) A person who violates this subsection is guilty of a misdemeanor and on conviction is subject to imprisonment not exceeding 1 year or a fine not exceeding $1,000 or both.

5-135. Regulated firearms subject to seizure A regulated firearm that is sold, rented, transferred, possessed, received, or purchased in violation of this subtitle may be:

(1) seized by a law enforcement agency as contraband; and

(2) after a finding of guilt, disposed of in accordance with Title 13, Subtitle 2 of the Criminal Procedure Article.

5-136. Straw purchases

(a) Scope of section.

(1) This section does not apply to a person who purchases a regulated firearm as a gift if:

(i) the regulated firearm is a gift to a resident of the State; and

(ii)1. both the purchaser and recipient of the gift comply with the requirements of this subtitle that relate to the possession, sale, rental, receipt, transfer or purchase of a regulated firearm; or

2. if the gift is in the form of a gift certificate, only the recipient of the gift need comply with the requirements of this subtitle that relate to the possession, sale, rental, receipt, transfer, or purchase of a regulated firearm.

(2) If the regulated firearm is a gift to the purchaser's spouse, parent, grandparent, grandchild, sibling, or child, the recipient shall:

(i) complete an application to purchase or transfer a regulated firearm; and

(ii) forward the application to the Secretary within 5 days after receipt of the regulated firearm.

(3) The Secretary shall waive the $10 application fee required under § 5-118(a)(2) of this subtitle for a gift purchased in accordance with this subsection.

(b) Prohibited. A person may not knowingly or willfully participate in a straw purchase of a regulated firearm.

5-137. Out-of-state purchases

(a) Requirements for purchase. A person who seeks to own a regulated firearm and purchases the regulated firearm from an out-of-state federally licensed gun importer, manufacturer, or dealer shall:

(1) have the federally licensed importer, manufacturer, or dealer ship the regulated firearm to a licensee for processing; and

(2) comply with §§ 5-103, 5-104, 5-117 through 5-129, and 5-136 of this subtitle.

(b) Waiver of requirements. If a person purchases a regulated firearm for use within the scope of the person's official duties, the Secretary may waive the 7-day waiting period under § 5-124 of this subtitle for:

(1) law enforcement personnel of any unit of the federal government;

(2) members of the armed forces of the United States or the National Guard; or

(3) law enforcement personnel of the State or any local agency in the State.

5-138. Sale transfer, or disposal of stolen regulated firearm prohibited A person may not possess, sell, transfer, or otherwise dispose of a stolen regulated firearm if the person knows or has reasonable cause to believe that the regulated firearm has been stolen.

5-139. False information or misstatement in application

(a) Prohibited. A person may not knowingly give false information or make a material misstatement in a firearm application or in an application for a dealer's license.

(b) Penalty. A person who violates this section is guilty of a misdemeanor and on conviction is subject to imprisonment not exceeding 3 years or a fine not exceeding $5,000 or both.

5-140. Transporting regulated firearm for unlawful sale or trafficking

(a) Prohibited. A dealer or other person may not transport a regulated firearm into the State for the purpose of unlawfully selling or trafficking of the regulated firearm.

(b) Penalty. A person who violates this section is guilty of a misdemeanor and on conviction is subject to imprisonment not exceeding 10 years or a fine not exceeding $25,000 or both

(c) Separate crime. Each violation of this section is a separate crime.

5-141. Knowing participation in straw purchase

(a) Prohibited. A dealer or other person may not be a knowing participant in a straw purchase of a regulated firearm to a minor or to a person prohibited by law from possessing a regulated firearm.

(b) Penalty. A person who violates this section is guilty of a misdemeanor and on conviction is subject to imprisonment not exceeding 10 years or a fine not exceeding $25,000 or both.

(c) Separate crime. Each violation of this section is a separate crime.

5-142. Removal or alteration of identification mark or number on firearm

(a) Prohibited. A person may not obliterate, remove, change, or alter the manufacturer's identification mark or number on a firearm.

(b) Presumption. If on trial for a violation of this section possession of the firearm by the defendant is established, the defendant is presumed to have obliterated, removed, changed, or altered the manufacturer's identification mark or number on the firearm.

5-143. Knowing participation in violation of subtitle

(a) Prohibited. Except as otherwise provided in this subtitle, a dealer or other person may not knowingly participate in the illegal sale, rental, transfer, purchase, possession, or receipt of a regulated firearm in violation of this subtitle.

(b) Penalty. A person who violates this section is guilty of a misdemeanor and on conviction is subject to imprisonment not exceeding 5 years or a fine not exceeding $10,000 or both.

(c) Separate crime. Each violation of this section is a separate crime.

Subtitle 2. Rifles and Shotguns

5-201. Definitions

(a) In general. In this subtitle the following words have the meanings indicated.

(b) Rifle. "Rifle" has the meaning stated in § 4-201 of the Criminal Law Article.

(c) Short-barreled rifle. "Short-barreled rifle" has the meaning stated in § 4-201 of the Criminal Law Article.

(d) Short-barreled shotgun. "Short-barreled shotgun" has the meaning stated in § 4-201 of the Criminal Law Article.

(e) Shotgun. "Shotgun" has the meaning stated in § 4-201 of the Criminal Law Article.

5-202. Scope of subtitle. This subtitle does not apply to a short-barreled rifle or short-barreled shotgun that is:

(1) an antique firearm as defined in § 4-201 of the Criminal Law Article;

(2) a device designed or redesigned for use other than as a weapon;

(3) a device designed or redesigned for use as a signaling, pyrotechnic, line throwing, safety, or similar device; or

(4) a firearm that cannot:

(i) discharge a projectile by an explosive; and

(ii) be readily restored to a firing condition.

5-203. Possession of short-barreled rifle or short-barreled shotgun

(a) Prohibited. A person may not possess a short-barreled rifle or short-barreled shotgun unless:

(1) the person, while on official business is:

(i) a member of the law enforcement personnel of the federal government, the State, or a political subdivision of the State;

(ii) a member of the armed forces of the United States or the National Guard while on duty or traveling to or from duty;

(iii) a member of the law enforcement personnel of another state or a political subdivision of another state, while temporarily in this State;

(iv) a warden or correctional officer of a correctional facility in the State; or

(v) a sheriff or a temporary or full-time deputy sheriff; or

(2) the short-barreled shotgun or short-barreled rifle has been registered with the federal government in accordance with federal law.

(b) Burden of proof. In a prosecution under this section, the defendant has the burden of proving the lawful registration of the short-barreled shotgun or short-barreled rifle.

(c) Penalty. A person who violates this section is guilty of a misdemeanor and on conviction is subject to imprisonment not exceeding 5 years or a fine not exceeding $5,000 or both.

5-204. Purchasers of rifles and shotguns.

(a) "Adjacent state" defined In this section, "adjacent state" means Delaware, Pennsylvania, Virginia, or West Virginia.

(b) Resident of this State in adjacent state. If a resident of this State is eligible to purchase a rifle or shotgun under the laws of an adjacent state, the resident may purchase a rifle or shotgun from a federally licensed gun dealer in the adjacent state.

(c) Resident of adjacent state in this State. If a resident of an adjacent state is eligible to purchase a rifle or shotgun under the laws of this State, the resident may purchase a rifle or shotgun from a federally licensed gun dealer in this State.

5-205. Possession by person with mental disorder

(a) Prohibited. Unless the person possesses a physician's certificate that the person is capable of possessing a rifle or shotgun without undue danger to the person or to another, a person may not possess a rifle or shotgun if the person:

(1) suffers from a mental disorder as defined in § 10-101(f)(2) of the Health - General Article and has a history of violent behavior against the person or another; or

(2) has been confined for more than 30 consecutive days in a facility as defined in § 10-101 of the Health - General Article.

(b) Penalty. A person who violates this section is guilty of a misdemeanor and on conviction is subject to imprisonment not exceeding 3 years or a fine not exceeding $1,000 or both.

Subtitle 4. Handgun Roster

5-401. Definitions

(a) In general. In this subtitle the following words have the meanings indicated.

(b) Board. "Board" means the Handgun Roster Board.

(c) Handgun. "Handgun" has the meaning stated in § 4-201 of the Criminal Law Article.

(d) Handgun roster. "Handgun roster" means the roster of authorized handguns compiled by the Board under § 5-405 of this subtitle.

(e) Secretary. "Secretary" means the Secretary of State Police or the Secretary's designee.

5-402. Application of subtitle

(a) Manufacture and sale of weapons. This subtitle does not affect a person's right to:

(1) manufacture, sell, or offer to sell a rifle or other weapon that is not defined as a handgun in § 4-201 of the Criminal Law Article;

(2) manufacture a prototype handgun model required for design, development, testing, and approval by the Board; and

(3) manufacture in this State a handgun that is not on the handgun roster by a federally licensed gun manufacturer who is also licensed as a regulated firearms dealer in this State for direct sale to a unit of:

(i) the federal government;

(ii) a state other than this State;

(iii) a local government in a state other than this State; or

(iv) a law enforcement agency in a state other than this State.

(b) Strict liability.

(1) A person is not strictly liable for damages for injuries to another that result from the criminal use of a firearm by a third person.

(2) Paragraph (1) of this subsection does not apply if the person conspired with the third person to commit the criminal act in which the firearm was used or willfully aided, abetted, or caused the commission of the criminal act in which the firearm was used.

(3) This subtitle does not otherwise negate, limit, or modify the doctrine of negligence or strict liability that relates to abnormally dangerous products or activities and defective products.

5-403. Regulations The Secretary shall adopt regulations necessary to carry out this subtitle.

5-404. Handgun Roster Board

(a) Established. There is a Handgun Roster Board in the Department of State Police. ....

5-405. Duties and procedures of Board

(a) Establishment and publication of roster The Board shall:

(1) compile and maintain a handgun roster of authorized handguns that are useful for legitimate sporting, self-protection, or law enforcement purposes;

(2) annually publish the handgun roster in the Maryland Register; and

(3) semiannually send a copy of the handgun roster to all persons who hold a State regulated firearm dealer's license under Subtitle 1 of this title.

(b) Criteria for placement on roster. The Board shall consider carefully each of the following characteristics of a handgun without placing undue weight on any one characteristic in determining whether any handgun should be placed on the handgun roster:

(1) concealability;

(2) ballistic accuracy;

(3) weight;

(4) quality of materials;

(5) quality of manufacture;

(6) reliability as to safety;

(7) caliber;

(8) detectability by the standard security equipment that is commonly used at an airport or courthouse and that is approved by the Federal Aviation Administration for use at airports in the United States; and

(9) utility for legitimate sporting activities, self-protection, or law enforcement.

(c) Placement process.

(1) The Board may place a handgun on the handgun roster on its own initiative.

(2) The Board shall place a handgun on the handgun roster on the successful petition of any person subject to subsections (d) and (e) of this section, unless a court, after all appeals are exhausted, has made a finding that the decision of the Board shall be affirmed.

(3) A petition to place a handgun on the handgun roster shall be submitted to the Board in writing in the form and manner that the Board requires.

(4) A person who petitions for placement of a handgun on the handgun roster has the burden of proving to the Board that the handgun should be placed on the handgun roster.

(d) Action of Board on petition.

(1) Within 45 days after receipt of a petition to place a handgun on the handgun roster, the Board shall:

(i) deny the petition in writing, stating the reasons for denial; or

(ii) approve the petition and publish a description of the handgun in the Maryland Register, including notice that any objection to the handgun's inclusion on the handgun roster shall be filed with the Board within 30 days.

(2) If the Board fails to deny or approve a petition within the time required under paragraph (1) of this subsection, the petition shall be considered denied.

(e) Notice of denial; hearing; appeal.

(1) If the Board denies a petition to place a handgun on the handgun roster, the Board shall notify the petitioner by certified mail, return receipt requested.

(2) The petitioner may request a hearing within 15 days after the date that the Board's denial letter is received.

(3)(i) If the petitioner requests a hearing under paragraph (2) of this subsection, within a reasonable time not to exceed 90 days after receiving the request, the Board shall:

1. hold a hearing on the petition; and

2. issue a written final decision on the petition.

(ii) The Board shall provide notice of the hearing in accordance with Title 10, Subtitle 2 of the State Government Article.

(iii) At a hearing held under this paragraph, the petitioner has the burden of proving to the Board that the handgun should be placed on the handgun roster because the handgun is useful for legitimate sporting activities, self-protection, or law enforcement purposes.

(4) Any party of record who is aggrieved may appeal within 30 days after a final decision of the Board in accordance with Title 10, Subtitle 2 of the State Government Article.

(f) Effect of section. This section does not require the Board to test any handgun or have any handgun tested at the expense of the Board.

5-406. Manufacture or sale of handguns.

(a) Prohibitions

(1) Except as provided in § 5-402 of this subtitle, a person may not manufacture for distribution or sale a handgun that is not included on the handgun roster in the State.

(2) A person may not sell or offer for sale in the State a handgun manufactured after January 1, 1985, that is not included on the handgun roster.

(3) A person may not manufacture, sell, or offer for sale a handgun on which the manufacturer's identification mark or number is obliterated, removed, changed, or otherwise altered.

(b) Injunction authorized. The Secretary may seek an order from a circuit court to permanently or temporarily enjoin the willful and continuous manufacture, sale, or offer for sale, in violation of this section, of a handgun that is not included on the handgun roster.

(c) Penalties.

(1) A person who manufactures a handgun for distribution or sale in violation of this section is guilty of a misdemeanor and on conviction is subject to a fine not exceeding $10,000 for each violation.

(2) A person who sells or offers to sell a handgun in violation of this section is guilty of a misdemeanor and on conviction is subject to a fine not exceeding $2,500 for each violation.

(3) For purposes of this subsection, each handgun manufactured, sold, or offered for sale in violation of this subsection is a separate violation.

Title 11. Explosives

Subtitle 1. Licenses to Engage in Business as Manufacturer or Dealer or to Possess Explosives

11-101. Definitions

(a) In general. In this subtitle the following words have the meanings indicated.

(b) Dealer.

(1) "Dealer" means a person who is engaged in the business of buying or selling explosives.

(2) "Dealer" does not include a manufacturer.

(c) Explosives.

(1) "Explosives" means gunpowder, powders for blasting, high explosives, blasting materials, fuses other than electric circuit breakers, detonators and other detonating agents, smokeless powder, and any chemical compound or mechanical mixture that contains oxidizing and combustible units or other ingredients in such proportions, quantities, or packing that ignition by fire, friction, concussion, percussion, or detonation of any part of the compound or mixture may and is intended to cause an explosion.

(2) "Explosives" includes bombs and destructive devices designed to operate by chemical, mechanical, or explosive action.

(3)"Explosives" does not include fixed ammunition for small arms, small arms ammunition primers, small arms percussion caps, safety and pyrotechnic fuses, quills, quick and slow matches, friction primers, fireworks, or common matches when used in their original configuration.

(d) Explosives for use in firearms. "Explosives for use in firearms" means:

(1) smokeless powder for loading or reloading small arms ammunition; or

(2) black powder for loading or reloading small arms ammunition, antique arms, or replicas of antique arms.

(e) Local licensing authority. "Local licensing authority" means the sheriff or chief of police of the county or community where the applicant for a license resides or has a regular place of business.

(f) Manufacturer. "Manufacturer" means a person who manufactures or otherwise produces explosives.

11-102. Scope of subtitle

(a) Transportation of explosives. This subtitle does not apply to explosives while being transported on vessels, vehicles, or railroad cars, or while being held for delivery, if the transportation or delivery is subject to and conforms with regulations adopted by the United States Department of Transportation or United States Coast Guard.

(b) Safety signals. This subtitle does not apply to the receipt, possession, and use of signals required for the safe operation of vessels, motor vehicles, railroad cars, or aircraft by their operators. ...

11-105. License required; exceptions

(a) In general. Except as otherwise provided in this subtitle, a person shall obtain a license issued under this subtitle before the person engages in business as a manufacturer or dealer, possesses explosives other than explosives for use in firearms, or possesses or stores explosives for use in firearms in the State.

(b) License to engage in business as dealer required.

(1) A person shall obtain a license to engage in business as a dealer under this subtitle before the person engages in the business of loading or reloading small arms ammunition in the State. ...

(c) Exceptions - Armed forces and others handling explosives. This section does not apply to the armed forces of the United States, the National Guard, the State Guard, or officers or employees of the United States, the State, or a local subdivision of the State who are authorized to handle explosives in the performance of their duties.

(d) Exceptions - Possession of explosives for use in firearms.

(1) Subject to paragraph (2) of this subsection, a person need not obtain a license to possess or store up to 5 pounds of smokeless powder for the loading or reloading of small arms ammunition, and up to 5 pounds of black powder for the loading or reloading of small arms ammunition or for use in the loading of antique arms or replicas of antique arms, if the smokeless powder and black powder are stored in their original shipping containers and are possessed only for personal use in firearms.

(2) A person may not possess or store explosives for use in firearms in any quantity in multifamily dwellings, apartments, dormitories, hotels, schools, other public buildings, or buildings or structures open for public use.

(3) Notwithstanding paragraph (2) of this subsection, the State Fire Marshal may issue a permit to allow temporary possession of explosives for use in firearms in a building or structure open for public use. ...

11-112. Records and reports of manufacturers and dealers

(a) Records.

(1) Each manufacturer and each dealer shall keep, for all explosives shipped, purchased, or sold, a record that includes:

(i) the name and address of each consignee, buyer, or seller of the explosives;

(ii) the date of each shipment, purchase, or sale; and

(iii) the amount and description of the explosives.

(2) Each record kept under this subsection shall at all times be open for inspection by agents of the licensing authority and by federal, State, and local law enforcement officers.

(3)(i) Subject to subparagraph (ii) of this paragraph, each manufacturer and each dealer shall provide a copy of each record kept under this subsection to the State Fire Marshal in the form that the State Fire Marshal requires.

(ii) A record kept under this subsection shall be provided on request, but need not be filed more than once in each calendar month.

(b) Reports to licensing authorities.

(1) Subject to paragraph (2) of this subsection, each manufacturer shall file with the licensing authority of each state, other than this State, to which explosives have been shipped by the manufacturer, a report that includes:

(i) the name of each buyer to whom explosives have been shipped in that state; and

(ii) the amount and description of the explosives.

(2) A report required under paragraph (1) of this subsection shall be filed on request, but need not be filed more than once in each calendar month.

(3) In like manner, each manufacturer shall file with the State Fire Marshal a report that includes:

(i) the name of each buyer of explosives in this State; and

(ii) the amount and description of the explosives.

11-113. Reports of theft of explosives Each theft or other unauthorized taking of explosives from a licensee under this subtitle shall be reported by the licensee to the State Fire Marshal:

(1) immediately by telephone; and

(2) by a written report in the form required by the State Fire Marshal.

11-114. Prohibited acts; penalty - In general

(a) Engaging in business as manufacturer or dealer without license prohibited. Except as otherwise provided in this subtitle, a person may not engage in business as a manufacturer or dealer in the State unless the person is licensed under this subtitle.

(b) Possession of explosives other than explosives for use in firearms without license prohibited. Except as otherwise provided in this subtitle, a person may not possess explosives other than explosives for use in firearms in the State unless the person is licensed under this subtitle.

(c) Sale to unlicensed persons prohibited. Except as otherwise provided in this subtitle, a dealer may not sell, barter, give, or dispose of explosives other than explosives for use in firearms to a person unless the person is licensed under this subtitle. ...

(f) Violation of regulations prohibited. A person may not violate a regulation adopted under this subtitle.

(g) Penalty. Except as otherwise provided in § 11-116 of this subtitle, a person who violates this section is guilty of a misdemeanor and on conviction is subject to imprisonment not exceeding 5 years or a fine not exceeding $5,000 or both.

11-115. Prohibited acts - Explosives for use in fire-arms; required reports

(a) Prohibitions on possession or storage of explosives for use in firearms.

(1) A person may not possess at any time or store in any one place more than 5 pounds of smokeless powder or more than 5 pounds of black powder for use in firearms unless the person is licensed under this subtitle.

(2) A person may not engage in the business of loading or reloading small arms ammunition unless the person is licensed to engage in business as a dealer under this subtitle.

(3) Except as otherwise provided in this subtitle, a person may not possess or store explosives for use in firearms in any quantity in multifamily dwellings, apartments, dormitories, hotels, schools or other public buildings, or buildings or structures open for public use.

(b) Prohibition on sale of explosives for use in firearms. A dealer may not sell, barter, give, or dispose of more than 5 pounds of black powder or more than 5 pounds of smokeless power for use in firearms to any one person at any one time unless the person is licensed under this subtitle.

(c) Failure to file reports or records prohibited. A person may not fail to file reports or records required under § 11-112 of this subtitle.

(d) Failure to file report of theft prohibited. A person may not fail to file a report of theft of explosives required under § 11-113 of this subtitle.

(e) Penalty. A person who violates this section is guilty of a misdemeanor and on conviction is subject to imprisonment not exceeding 6 months or a fine not exceeding $500 or both.

11-116. Additional penalties

(b) For violation of § 11-114(c) or conspiracy to violate § 11-114(b)

(1) Except as otherwise provided in paragraph (2) of this subsection, a person who violates § 11-114(c) of this subtitle or who conspires to violate § 11-114(b) of this subtitle is guilty of a felony and on conviction is subject to imprisonment not exceeding 20 years or a fine not exceeding $10,000 or both.

(2) Paragraph (1) of this subsection does not apply to a person who had probable cause to believe that the explosives involved would be used for a purpose other than the violation of:

(i) Title 3, Subtitle 1 or Subtitle 5, Title 5, Subtitle 1, Subtitle 2, Subtitle 3, or Subtitle 4, § 6-602, § 7-402, or § 12-701 of this article;

(ii) Title 1, Subtitle 3, Title 3, Subtitle 7, or § 4-123.1 of the Agriculture Article;

(iii) Title 19, Subtitle 2 or Subtitle 3 of the Business Regulation Article;

(iv) Title 14, Subtitle 29, § 11-810, or § 14-1317 of the Commercial Law Article;

(v) § 3-218, § 3-305(c)(2), § 3-409(a) or (c), § 3-803(b), § 3-807(i), § 3-808(d), § 3-811(c), § 8-801, § 8-802, § 9-602(e), § 11-702(d)(8), § 11-703(d)(5)(iii), § 11-706(b)(8), § 11-708(d)(7)(ii), § 11-711(h)(2), § 11-712(c)(6)(ii), § 11-714(c)(6), § 11-715(g)(2), § 11-716(h)(2), § 11-723(b)(8), or § 11-726 of the Correctional Services Article;

(vi) the Criminal Law Article other than Title 8, Subtitle 2, Part II or § 10-614;

(vii) Title 5, Subtitle 10A of the Environment Article;

(viii) § 5-503 of the Family Law Article;

(ix) Title 20, Subtitle 7 or § 21-259.1 of the Health-General Article;

(x) § 8-713.1, § 8-724.1, § 8-725.5, § 8-725.6, § 8-726.1, § 8-738.1, § 8-740.1, or § 10-411(a) or (d), as it relates to Harford County, of the Natural Resources Article;

(xi) § 14-127 of the Real Property Article;

(xii) Article 2B, Title 22 or § 18-104 of the Code;

(xiii) Article 24, § 11-512, § 11-513, or § 11-514 of the Code;

(xiv) § 109 of the Code of Public Local Laws of Caroline County;

(xv) § 4-103 of the Code of Public Local Laws of Carroll County; or

(xvi) § 8A-1 of the Code of Public Local Laws of Talbot County.

[Current through 2008 Regular Session]

Annapolis Municipal Code

Title 11. Public Peace, Morals And Welfare

Chapter 11.44. Weapons

11.12.120 Public peace and order.
A violation of any provision of this chapter is declared a municipal infraction and for each violation the person is subject to a fine as established by resolution of the City Council.
(Ord. O-5-04 § 1 (part), 2004: Ord. O-1-99 § 1 (part))

11.44.030. Firearms and ammunition - Register of purchasers. Each person engaged in the business of selling or exchanging any firearms, except shotguns and air and cat rifles, or ammunition, shall keep a register of the name and address of the person purchasing any firearms, cartridges or other ammunition, noting the make, caliber and date of purchase. The register shall be open to the inspection of the police at all times.

11.44.040. Firearms and ammunition - Registration of sellers. Each person engaged in the business of selling or exchanging any kind of firearms or ammunition shall register his name and place of business with the City Clerk.

11.44.050. Firearms and ammunition - Minors under eighteen years.

A. A person, whether a licensed dealer or not, may not sell, barter or give away any firearms, other deadly weapons or any ammunition to any minor under the age of eighteen years, except with the express written permission of a parent or guardian of the minor.

B. This section does not apply to a member of any organized militia in the state, when the member is engaged in supervised training, marksmanship activities or any other performance of the member's official duty. The restrictions or limitations contained in this section also do not apply to any adult or qualified supervisor or instructor of a recognized organization engaged in the instruction of marksmanship.

[Annapolis Municipal Code current through Ord. O-17-08 Revised, passed September 8, 2008]

Anne Arundel County Code

Article 9. Crimes and Civil Offenses, and Fines

Title1. Crimes

Subtitle 6. Weapons

9-1-604. Explosives.

(a) Except in a place for blasting or in the course of transportation, a person may not store or possess more than 100 pounds of gunpowder or any quantity of dynamite, nitroglycerine, or other explosive in the County.

(b) A person who violates any provision of this section is guilty of a misdemeanor and on conviction is subject to a fine not exceeding $300. If a corporation violates any provision of this section, the president of the corporation or any officer or member of its board of directors may be prosecuted for the violation.

Article 18. Zoning

Title 10. Requirements for Conditional Uses

18-10-119. Home occupations.
A home occupation shall comply with all of the following requirements.

(1) A home occupation shall be located and conducted entirely in a principal dwelling unit and shall be incidental and secondary to the use of the structure as a dwelling.

(2) A home occupation may not change the character of the dwelling unit and may not exceed 25% of the total floor area.

(3) Home occupations are limited to the following: ...

(x) Repair and maintenance of firearms, including handguns, rifles, shotguns, and antique firearms, as those terms are defined in the Criminal Law Article, § 4-201, of the State Code;...

(4) The operator of a home occupation shall be a resident of the dwelling unit in which the occupation is located.

(5) No more than one nonresident may be employed in the home occupation.

(6) The sale or rental of goods or products other than those produced on the premises by the home occupation is prohibited.

(7) Outside storage is prohibited.

[Anne Arundel County Code current through Sept. 2008]

Baltimore City Code

Article 19. Police Ordinances

Subtitle 59. Weapons

Part 1. Firearms - In General

59-4. Toy cartridge pistols.

(a) Sale, etc., prohibited. It shall not be lawful for any person or persons to sell, give away, or dispose of in any manner, what is known as "the toy cartridge pistol" within the limits of the City of Baltimore under a penalty of $10 for each and every offense, the same to be collected as other fines and penalties are collected.

Part 2. Firearms - Access by Minors

59-11. Definitions.

(a) In general. In this Part, the following terms have the meanings indicated.

(b) Ammunition. "Ammunition" means any cartridge, shell, or other device that contains explosive or incendiary material and is designed or intended for use in any firearm.

(c) Child safety lock. "Child safety lock" means:

(1) a device that, when locked in place, prevents the trigger from being moved and can itself be removed only by using a key or combination; or

(2) any other device that:

(i) when locked in place, otherwise renders the firearm inoperable and can itself be removed only by using a key or combination; and

(ii) has been approved for this purpose by the Police Commissioner.

(d) Firearm. "Firearm" means any pistol, revolver, rifle, shotgun, short-barreled rifle, short-barreled shotgun, or other firearm, except an inoperable antique firearm.

(e) Minor. "Minor" means any person under the age of 18.

59-12. Access by minors prohibited.

(a) Prohibited conduct. Except as provided in subsection (b) of this section, a person may not leave a loaded firearm, or an unloaded firearm that is in close proximity to ammunition, in any location where the person knows or reasonably should know that an unsupervised minor might gain access to the firearm.

(b) Exceptions. Subsection (a) of this section does not apply if:

(1) the minor's access to the firearm is supervised by a person 21 years old or older;

(2) the firearm is in a locked gun cabinet or similar locked location;

(3) the firearm is secured with a child safety lock;

(4) the minor obtained access to the firearm as the result of an unlawful entry to the premises; or

(5) the firearm is in the possession or control of a law enforcement officer while the officer is engaged in official duties.

59-13. Safety locks.

(a) Dealers must provide.

(1) A licensed firearm dealer may not sell, lease, or otherwise transfer a firearm without an accompanying child safety lock suitable for that firearm.

(2) The dealer must provide the child safety lock to the recipient of the firearm when transferring the firearm. The dealer may charge for the child safety lock.

(b) Notices.

(1) A licensed firearm dealer who sells, leases, or otherwise transfers a firearm must post conspicuously in the dealer's place of business:

(i) a notice of the prohibition in § 59-12 of this Part against leaving a firearm where an unsupervised minor can obtain access to it; and

(ii) a notice of the prohibition in subsection (a) of this section against the transfer of a firearm without an accompanying child safety lock.

(2) If the transaction occurs outside the dealer's place of business, or if the dealer does not maintain a place of business in a commercial establishment, the dealer must provide the required notices in writing when transferring the firearm.

59-14. Rules and regulations. The Police Commissioner may adopt rules and regulations to carry out this Part, including but not limited to rules or regulations governing the wording, size, and placement of the notices required by this Part.

59-16. Penalties. Any person who violates any provision of this Part or of a rule or regulation adopted under this Part is guilty of a misdemeanor and, on conviction, is subject to a fine of $1,000 or to imprisonment for 1 year or both.

[Baltimore City Code current through June 30, 2009]

Baltimore County Code, 2003

Article 17. Miscellaneous Provisions and Offenses

Title 2. Firearms and Weapons

17-2-102. Minors.

(a) Prohibited - Purchase of firearm.

(1) A minor may not purchase, trade, acquire in any manner, use, possess, or attempt to use or possess a gun, pistol, rifle, shotgun, or any other type of firearm, unless the minor has filed a statement of possession or use with the Police Department.

(2) The statement of possession or use shall be retained by the Police Department.

(3)(i) The statement of possession or use shall be endorsed by the parents or guardians of the minor, attesting to the knowledge of the parents or guardians of the acquisition, use, possession, or prospective acquisition, use or possession of any firearms.

(ii) The endorsement executed by the parents or guardians of a minor shall provide clearly and without exception or qualification that:

1. Any negligence of the minor in the use or possession of a firearm shall be imputed to the parents or guardians; and

2. The parents or guardians shall be jointly and severally liable with the minor for any civil damages caused by the minor's negligence in the use or possession of a firearm.

(4) The statements required under this section shall be signed and sworn to before a person authorized to administer oaths.

(b) Same - Sale of firearm. A person may not sell, give, or transfer a firearm to a minor unless the statements required under subsection (a) of this section have been filed with the Police Department.

(c) Limitation. This section does not apply to antique or unserviceable firearms sold, transferred, or held as curios or museum pieces.

(d) Serial numbers not required. This section may not be construed to require the registration or listing of firearms by serial number or in any other manner.

17-2-103. Penalty. A person who violates any provision of § 17-2-101 or § 17-2-102 of this title is guilty of a misdemeanor.

[Baltimore County Code current through Bill No.120-08]

Montgomery County Code

Chapter 57. Weapons

57-1. Definitions. In this Chapter, the following words and phrases have the following meanings:

Child safety handgun box: A secure, lockable box designed to hold the handgun being transferred that:

(1) requires a key or combination to remove;

(2) renders the handgun inoperable when locked; and

(3) is approved by Executive regulation under method (2).

Child safety handgun device: A child safety handgun lock or child safety handgun box.

Child safety handgun lock: A device that when locked in place prevents movement of the trigger of the handgun being transferred without first removing the lock by use of a key or combination. "Child safety handgun lock" also includes any other device that can be attached to a handgun and:

(1) requires a key or combination to remove;

(2) renders the handgun inoperable when locked in place; and

(3) is approved by Executive regulation under method (2).

Crime of violence: Murder, voluntary manslaughter, rape, mayhem, kidnapping, robbery, burglary, housebreaking, arson, assault with intent to murder, ravish or rob, assault with deadly weapon or assault with intent to commit any offense punishable by imprisonment for more than one (1) year.

Firearm dealer: A person required by State or federal law to obtain a:

(1) regulated firearms dealer's license; or

(2) temporary transfer permit to display a regulated firearm at a gun show.

Fixed ammunition: Any ammunition composed of a projectile or projectiles, a casing, an explosive charge and a primer, all of which shall be contained as one (1) unit. Cartridges designed, made and intended to be used exclusively (i) in a device for signaling and safety purposes required or recommended by the United States Coast Guard or (ii) for industrial purposes, shall not be considered fixed ammunition. Curios or relics, as defined in regulations promulgated by the United States Secretary of the Treasury pursuant to 18 United States Code, section 921(A)(13), shall not be considered fixed ammunition.

Fugitive from justice: Any person for whom criminal proceedings have been instituted, warrant issued or indictment presented to the grand jury, who has fled from a sheriff or other peace officer within this state, or who has fled from any state, territory, District of Columbia or possession of the United States, to avoid prosecution for crime of violence or to avoid giving testimony in any criminal proceeding involving a felony or treason.

Gun or firearm: Any rifle, shotgun, revolver, pistol, air gun, air rifle or any similar mechanism by whatever name known which is designed to expel a projectile through a gun barrel by the action of any explosive, gas, compressed air, spring or elastic.

(1) The term "antique firearm" means (a) any firearm (including any firearm with a match-lock, flintlock, percussion cap, or similar type of ignition system) manufactured in or before 1898; and (b) any replica of any firearm described in subparagraph (a) if such replica (i) is not designed or redesigned or using rimfire or conventional centerfire fixed ammunition, or (ii) uses rimfire or conventional centerfire fixed ammunition which is no longer manufactured in the United States and which is not readily available in the ordinary channels of commercial trade.

(2) "Handgun" means any pistol, revolver or other firearm capable of being concealed on the person, including a short-barreled shotgun and a short-barreled rifle as these terms are defined below. "Handgun" does not include a shotgun, rifle, or antique firearm.

(3) "Rifle" means a weapon designed or redesigned, made or remade, and intended to be fired from the shoulder and designed or redesigned and made or remade to use the energy of the explosive in a fixed metallic cartridge to fire only a single projectile through a rifled bore for each single pull of the trigger.

(4) The term "short-barreled rifle" means a rifle having one (1) or more barrels less than sixteen (16) inches in length and any weapon made from a rifle (whether by alternation, modification or otherwise) if such weapon, as modified, has an overall length of less than twenty-six (26) inches.

(5) The term "short-barreled shotgun" means a shotgun having one (1) or more barrels less than eighteen (18) inches in length and any weapon made from a shotgun (whether by alteration, modification or otherwise) if such weapon as modified has an overall length of less than twenty-six (26) inches.

(6) "Shotgun" means a weapon designed or redesigned, made or remade, and intended to be fired from the shoulder and designed or redesigned and made or remade to use the energy of the explosive in a fixed shotgun shell to fire through a smooth bore either a number of ball shot or a single projectile for each single pull of the trigger.

Gun shop: An establishment where a handgun, rifle, or shotgun, or ammunition or major component of these guns is sold or transferred. "Gun shop" does not include an area of an establishment that is separated by a secure, physical barrier from all areas where any of these items is located.

Gun show: Any organized gathering where a gun is displayed for sale.

Minor: An individual younger than 18 years old.

Pistol or **revolver:** Any gun with a barrel less than twelve (12) inches in length that uses fixed ammunition.

Place of public assembly: A "place of public assembly" is a government owned park identified by the Maryland-National Capital Park and Planning Commission; place of worship; elementary or secondary school; public library; government owned or operated recreational facility; or multipurpose exhibition facility, such as a fairgrounds or conference center. A place of public assembly includes all property associated with the place, such as a parking lot or grounds of a building.

Sell or purchase: Such terms and the various derivatives of such words shall be construed to include letting on hire, giving, lending, borrowing or otherwise transferring.

Sporting use: "Sporting use" of a firearm and ammunition means hunting or target shooting in compliance with all federal, State, and local laws. Sporting use includes:

(a) participation in a managed hunt sponsored by a government agency; and

(b) the sale or other transfer of ammunition by a sporting club for immediate, on-site use at the club. ...

Vehicle: Any motor vehicle, as defined in the Transportation Article of the Annotated Code of Maryland, trains, aircraft and vessels.

57-7. Access to guns by minors.

(a) A person must not give, sell, rent, lend, or otherwise transfer any rifle or shotgun or any ammunition or major component for these guns in the County to a minor. This subsection does not apply when the transferor is at least 18 years old and is the parent, guardian, or instructor of the minor, or in connection with a regularly conducted or supervised program of marksmanship or marksmanship training.

(b) An owner, employee, or agent of a gun shop must not allow a minor to, and a minor must not, enter the gun shop unless the minor is accompanied by a parent or other legal guardian at all times when the minor is in the gun shop.

(c) This section must be construed as broadly as possible within the limits of State law to protect minors.

57-8. Child safety handgun devices and handguns.

(a) Findings. ...

(b) Child safety handgun device.

(1) A firearm dealer who sells, leases, or otherwise transfers a handgun in the County must provide to the recipient of the handgun a child safety handgun device for the handgun at the time of the transfer. The dealer may charge for the child safety handgun device.

(2) A person who purchases or otherwise receives a handgun from a firearm dealer (or any transferor who would be a firearm dealer if the transfer occurred in the State) after October 8, 1997 must obtain a child safety handgun device for the handgun:

(A) at the time of a transfer in the County; or

(B) before entering the County with the handgun if the transfer occurred outside the County and the transferee resides in the County.

(c) Notices.

(1) A firearm dealer who sells, leases, or otherwise transfers a handgun must post conspicuously in the dealer's place of business a notice of:

(A) the requirement in subsection (b) for a child safety handgun device; and

(B) the prohibition in State law of storing or leaving a loaded firearm in a location where an unsupervised child can gain access to the firearm.

(2) If the firearm dealer transferring a handgun does not maintain a place of business in a commercial establishment, the dealer must provide the notices required by paragraph (1) in writing when transferring the handgun.

(d) Enforcement. The Department of Health and Human Services and any other department designated by the County Executive enforces this section.

(f) Regulations. The Executive may adopt regulations under method (2) to implement this Section.

57-9. Unlawful ownership or possession of firearms. A person must not possess, exercise control over, use, carry, transport, or keep a rifle, shotgun, or pistol, if the person:

(a) is an unlawful user of , addicted to, or is under treatment for an addiction to, marijuana or any depressant or stimulant drug or narcotic drug (as defined in Maryland Criminal Law Code Annotated, sections 1-101, 5-101, 5-401, 5-404, and 5-604); or

(b) has been convicted in any court of a crime of violence, trafficking in narcotics, a criminal violation of any of the provisions of Maryland Public Safety Code Annotated, sections 5-101 to 5-138, 5-142, or any federal firearms control law; or

(c) is a fugitive from justice; or

(d) has been confined to any hospital or institution for treatment of a mental disorder or for mental illness unless a licensed physician has by affidavit stated that the physician is familiar with the person's history of mental illness and that in the physician's opinion the person is not disabled by such illness in a manner which should prevent the person from possessing a rifle or a shotgun; or

(e) has been confined to any hospital or institution for treatment of alcoholism unless a licensed physician has by affidavit stated that the physician is familiar with the person's history of alcoholism and that, in the physician's opinion, the person is no longer suffering from a disability in such a manner which should prevent the person from possessing a rifle or shotgun. (1981 L.M.C., ch. 42, § 1; 2001 L.M.C., ch.11, § 1; 2004 L.M.C., ch. 22, §1.)

57-11. Firearms in or near places of public assembly.

(a) A person must not sell, transfer, possess, or transport a handgun, rifle, or shotgun, or ammunition for these firearms, in or within 100 yards of a place of public assembly.

(b) This section does not:

(1) prohibit the teaching of firearms safety or other educational or sporting use in the areas described in subsection (a);

(2) apply to a law enforcement officer, or a security guard licensed to carry the firearm;

(3) apply to the possession of a firearm or ammunition in the person's own home;

(4) apply to the possession of one firearm, and ammunition for the firearm, at a business by either the owner or one authorized employee of the business;

(5) apply to the possession of a handgun by a person who has received a permit to carry the handgun under State law; or

(6) apply to separate ammunition or an unloaded firearm:

(A) transported in an enclosed case or in a locked firearms rack on a motor vehicle; or

(B) being surrendered in connection with a gun turn-in or similar program approved by a law enforcement agency.

(c) This section does not prohibit a gun show at a multipurpose exhibition facility if:

(1) the facility's intended and actual primary use is firearms sports (hunting or target, trap, or skeet shooting) or education (firearms training); or

(2) no person who owns or operates the facility or promotes or sponsors the gun show received financial or in-kind support from the County (as defined in Section 57-13(a)) during the preceding 5 years, or after December 1, 2001, whichever is shorter; and

(A) no other public activity is allowed at the place of public assembly during the gun show; and

(B) if a minor may attend the gun show:

(i) the promoter or sponsor of the gun show provides to the Chief of Police, at least 30 days before the show:

(a) photographic identification, fingerprints, and any other information the Police Chief requires to conduct a background check of each individual who is or works for any promoter or sponsor of the show and will attend the show; and

(b) evidence that the applicant will provide adequate professional security personnel and any other safety measure required by the Police Chief, and will comply with this Chapter; and

(ii) the Police Chief does not prohibit the gun show before the gun show is scheduled to begin because:

(a) the promoter or sponsor has not met the requirements of clause (i); or

(b) the Police Chief has determined that an individual described in clause (i)(a) is not a responsible individual.

(d) Notwithstanding subsection (a), a gun shop owned and operated by a firearms dealer licensed under Maryland or federal law on January 1, 1997, may conduct regular, continuous operations after that date in the same permanent location under the same ownership if the gun shop:

(1) does not expand its inventory (the number of guns or rounds of ammunition displayed or stored at the gun shop at one time) or square footage by more than 10 percent, or expand the type of guns (handgun, rifle, or shotgun) or ammunition offered for sale since January 1, 1997;

(2) has secure locks on all doors and windows;

(3) physically secures all ammunition and each firearm in the gun shop (such as in a locked box or case, in a locked rack, or with a trigger lock);

(4) has adequate security lighting;

(5) has a functioning alarm system connected to a central station that notifies the police; and

(6) has liability insurance coverage of at least $1,000,000.

57-12. Sale of fixed ammunition.

(a) Legislative intent. The purpose of this section is to provide support to state and local law enforcement officials in their efforts against crime and violence by placing controls on the flow of dangerous ammunition, in addition to those provided by federal law, and to encourage compliance with the state police department's program of voluntary firearm registration. It is not the purpose of this section to place any undue or unnecessary restrictions or burdens on law-abiding citizens with respect to the acquisition, possession, or use of firearms appropriate to the purpose of hunting, trapshooting, target shooting, personal protection, or any other lawful activity, or to discourage or eliminate the private ownership or use of firearms by law-abiding citizens for lawful purposes. It is not the purpose of this section to create, nor does it permit the creation of, any separate system of county registration of firearms or ammunition, or the levying of any county fee in connection with any registration of firearms or ammunition. It is specifically not the intent of this section to serve as a revenue generating measure.

(b) Registration of ammunition dealers. Any ammunition dealer (as defined in 18 United States Code, section 921 et seq.) who conducts business in Montgomery County is required to register with the Montgomery County department of police by maintaining on file with that department, at all times, a valid, current copy of his federal ammunition dealer's license.

(c) Conditions for sale. No ammunition dealer may sell fixed ammunition to any other person, unless:

(1) The sale is made in person;

(2) The purchaser exhibits, at the time of sale, a valid registration certificate or, in the case of a nonresident, proof that the firearm is lawfully possessed in the jurisdiction where the purchaser resides;

(3) The fixed ammunition to be sold is of the same caliber or gauge as the firearm described in the registration certificate, or other proof in the case of a nonresident; and

(4) The purchaser signs a receipt for the ammunition which shall be maintained by the licensed dealer for a period of one (1) year from the date of sale.

(d) Exceptions. The provisions of this section shall not apply to the sale of fixed ammunition:

(1) Which is suitable for use only in rifles or shotguns generally available in commerce, or to the sale of component parts of these types of ammunition;

(2) To any person licensed to possess fixed ammunition under an act of Congress and the law of the jurisdiction where the person resides or conducts business; or

(3) To any law enforcement officer of federal, state, local or any other governmental entity, if the officer has in his possession a statement from the head of his agency stating that the fixed ammunition is to be used in the officer's official duties.

(e) Penalties. Any ammunition dealer who sells fixed ammunition in violation of the provisions of this section shall be guilty of a class C violation, pursuant to section 1-19 of the Montgomery County Code, punishable only by a civil penalty in the amount of fifteen dollars ($15.00).

(f) Exception for incorporated municipalities. This section shall not be effective in any incorporated municipality which by law has authority to enact a law on the same subject. If any such incorporated municipality adopts this section and requests the county to enforce the adopted provisions thereof within its corporate limits, the county may thereafter administer and enforce the same within the incorporated municipality. The county executive is authorized to enter into agreements with incorporated municipalities to enforce and administer the provisions so adopted and to collect the administrative costs of implementation from such municipalities.

[Publisher's note: This section was held to be invalid by the Court of Appeals in *Montgomery County, Maryland, et al. v. Atlantic Guns, Inc., et al.*, 302 Md. 540, 489 A.2d 1114 (1985), and interpreted in *Washington Suburban Sanitary Commission v. Elgin*, 53 Md. App. 452, 454 A.2d 408 (1983).**]**

57-13. Use of public funds.

(a) The County must not give financial or in-kind support to any organization that allows the display and sale of guns at a facility owned or controlled by the organization. Financial or in-kind support means any thing of value that is not generally available to similar organizations in the County, such as a grant, special tax treatment, bond authority, free or discounted services, or a capital improvement constructed by the County.

(b) An organization referred to in subsection (a) that receives direct financial support from the County must repay the support if the organization allows the display and sale of guns at the organization's facility after receiving the County support. The repayment must include the actual, original value of the support, plus reasonable interest calculated by a method specified by the Director of Finance.

[Publisher's note: 2001 L.M.C., ch. 11, §2 provides that §57-13 applies to (1) support that an organization receives from the county after Dec. 1, 2001 and (2) the display of a gun for sale at the facility after Dec. 1, 2001, and that §57-13 expires on Dec. 1, 2011.**]**

57-14. Exemptions from Chapter. Nothing in this Chapter applies to the purchase, ownership or possession of bona fide antique guns which are incapable of use as a gun. Except as provided in Sections 57-7 and 57-11, nothing in this Chapter prohibits the owner or tenant of any land from carrying or discharging a firearm on that land for the purpose of killing predatory animals which prey on, damage or destroy property, livestock, or crops.

57-15. Penalty. Any violation of this Chapter or a condition of an approval certificate issued under this Chapter is a Class A violation to which the maximum penalties for a class A violation apply. Any violation of Section 57-8 is a Class A civil violation.

[Montgomery County Code current as of Aug. 31, 2008]

Code of the Town of Cheverly

Sec. 20-5. Supplying firearm materials or dangerous weapons to underage persons.

It shall be unlawful for any person, licensed dealer or otherwise, to sell, barter or give away shotguns, rifles, or any long guns whatsoever or any ammunition, powder, shot or shells for any dangerous weapon, covered by section 20-4 preceding, to any person under the age of eighteen (18) years. It shall be unlawful for any person, licensed dealer or otherwise, to sell, barter or give away handguns of all types except those classified as antiques to any person under the age of twenty-one (21) years.

(Ord. No. 14-77, 10-13-77)

[Codified throughOrd. No. O-1-09, adopted Jan. 8, 2009.(Supplement No. 33)]

Municipal Code of Town of La Plata

143-5 Weapons.

No pawnbroker, swap shop or secondhand dealer shall receive as a pledge or purchase any revolver, pistol, blackjack or sawed-off shotgun, and no pawnbroker shall display in his window or shop any such weapons for sale.

[Codified through Ordinance No. 08-23, adopted Nov. 25, 2008. (Supplement No. 12)]

Charter and Code Takoma Park

14.16.010 Exemption from County weapons law.

Pursuant to the authority conferred by Article 23A, Section 2B of the Annotated Code of Maryland and by Section 1-203 of the Montgomery County Code, the City exempts itself from the provisions of Chapter 57, Weapons, Section 57-5A, Child Safety Handgun Devices and Handguns, and Section 57-7A, Firearms In or Near Places of Public Assembly, of the Montgomery County Code.

(Ord. 2001-27 § 2(1), 2001)

14.16.020 Definitions.

"Child safety handgun box" means a secure, lockable box designed to hold the handgun being transferred that:

1. Requires a key or combination to remove; and

2. Renders the handgun inaccessible when locked.

"Child safety handgun device" means a child safety handgun lock, child safety handgun box, or integrated mechanical safety device.

"Child safety handgun lock" means a device that when locked in place prevents movement of the trigger of the handgun being transferred without first removing the lock by use of a key or combination. "Child safety handgun lock" also includes any other external device that is:

1. Attached to a handgun with a key or combination lock; and

2. Designed to prevent a handgun from being discharged unless the device has been deactivated or removed.

"Gun" or "firearm" means a handgun, rifle, shotgun, short-barreled rifle, short-barreled shotgun, or any other firearm, whether loaded or unloaded, except it does not include an antique firearm.

1. Antique firearm" means:

a. Any firearm (including any firearm with a matchlock, flintlock, percussion cap, or similar type of ignition system) manufactured in or before 1898; and

b. Any replica of any firearm described in subparagraph (a), if such replica:

i. Is not designed or redesigned for using rimfire or conventional centerfire fixed ammunition, or

ii. Uses rimfire or conventional centerfire fixed ammunition which is no longer manufactured in the United States and which is not readily available in the ordinary channels of commercial trade.

2. "Handgun" means any pistol, revolver, or other firearm capable of being concealed on the person, including a short-barreled shotgun and a short-barreled rifle. "Handgun" does not include a shotgun, rifle, or antique firearm.

3. 'Rifle" means a weapon designed or redesigned, made or remade, and intended to be fired from the shoulder and designed or redesigned and made or remade to use the energy of the explosive in a fixed metallic cartridge to fire only a single projectile through a rifle bore for each single pull of the trigger.

4. "Short-barreled rifle" means a rifle having one or more barrels less than 16" in length and any weapon made from a rifle (whether by alteration, modification or otherwise) if such weapon as modified has an overall length of less than 26".

5. "Short-barreled shotgun" means a shotgun having one or more barrels less than 18" in length and any weapon made from a shotgun (whether by alteration, modification or otherwise) if such weapon as modified has an overall length of less than 26".

6. "Shotgun" means a weapon designed or redesigned, made or remade, and intended to be fired from the shoulder and designed or redesigned and made or remade to use the energy of the explosive in a fixed shotgun shell to fire through a smooth bore either a number of ball shot or a single projectile for each single pull of the trigger.

"Integrated mechanical safety device" means a disabling or locking device that is built into a handgun and is designed to prevent the handgun from being discharged unless the device has been deactivated.

"Law enforcement officer" means:

1. A duly appointed member of a police force or other agency of the United States, of a State or the District of Columbia, or of a County, municipality, or other political subdivision, who is responsible for the prevention and detection of crime and the enforcement of the laws of the United States, a State or the District of Columbia, or a County, municipality, or other political subdivision;

2. Any military or militia personnel directed by the appropriate authority to keep law and order.

"Minor" means an individual younger than 18 years old.

"Place of public assembly" means:

1. A place to which the general public has access and a right to resort for business, entertainment, or other lawful purpose, but is not limited to a place devoted solely to the uses of the public.

2. A place of public assembly includes, but is not limited to:

a. A public park or other public grounds;

b. A place of worship;

c. A school;

d. A public building, including its grounds and curtilage;

e. The front or immediate area or parking lot of any store, shop, restaurant, tavern, shopping center, or other place of business; and

f. A public parking lot.

3. A place of public assembly includes all property associated with the place and located within 100 yards of the place.

(Ord. 2001-27 § 2(2), 2001)

14.16.030 Discharge of firearms prohibited. . . .

14.16.040 Child safety handgun devices.

A. Findings. The unintentional discharge of handguns often causes accidental death or injury to children. Additional safeguards are needed to protect children from injury or death from the unintentional discharge of loaded and unlocked handguns. Requiring all handguns in the City to have and use handgun safety devices can prevent unintentional injuries and fatalities to children.

B. It is unlawful for any person, other than a law enforcement officer, to carry, transport or keep a handgun in the City without a child safety handgun device installed on the handgun.

(Ord. 2001-27 § 2(4), 2001)

14.16.050 Prohibition of firearms in residences that run programs for minors.

A. Except as provided in this section, no person shall use, carry, transport, keep or leave any firearm in any location where the person knows or reasonably should know that a minor could gain access to the firearm in a residence that is used for programs or activities primarily serving minors during the times in which such activities or programs meet. "Programs or activities primarily serving minors" include day care, child care, recreation programs, sports programs, day camps, club meetings, tutoring programs, and other organized and regular educational activities.

B. This section does not:

1. Prohibit the teaching of firearms safety training for educational purposes in the areas described in subsection (A) of this section;

2. Prohibit historic demonstrations using weapons or replicas of weapons for educational purposes in the areas described in subsection (A) of this section;

3. Apply to a law enforcement officer or to a security guard licensed to carry a firearm when the law enforcement officer or security guard is in the course of his or her employment or duty or is traveling to or from the place of employment or duty;

4. Apply to an unloaded firearm in a locked case or in a child safety handgun box; or

5. Apply to activities serving only the person's own minor child or children in the person's own residence, such as babysitting of the child or home schooling.

(Ord. 2001-27 § 2(5), 2001)

14.16.060 Prohibition of firearms in places of public assembly.

A. A person must not purchase, sell, transfer, possess, or transport firearms or ammunition or components for firearms in a place of public assembly.

B. This section does not:

1. Prohibit the teaching of firearms safety training or other educational use in a place of public assembly;

2. Apply to a law enforcement officer;

3. Apply to a security guard licensed to carry a firearm when the security guard is in the course of his or her employment or is traveling to or from the place of employment;

4. Apply to the possession of firearms or ammunition in a person's own home;

5. Apply to the possession of one firearm, and ammunition for the firearm at a business by either the owner or an authorized employee of the business;

6. Apply to the possession of a handgun by a person who has received a permit to carry the handgun under State law; or

7. Apply to separate ammunition or an unloaded firearm:

a. Transported in an enclosed case or in a locked firearms rack on a motor vehicle; or

b. Being surrendered in connection with a gun turn-in or similar program approved by a law enforcement agency.

(Ord. 2001-27 § 2(6), 2001)

14.16.070 Penalty.

Any violation of this chapter is a Class A misdemeanor offense, and on conviction is subject to a fine of $1,000.00 or a term of imprisonment of not more than 180 days, or both.

[Ord. 2001-27 §§ 2(1), (2), (5), (7), 2001]

Chapter 140. Licenses - Sale of Firearms

121. Definitions; Application for License or Identification Card; Exceptions. As used in sections 122 to 131P, inclusive, the following words shall, unless the context clearly requires otherwise, have the following meanings:

"Ammunition", cartridges or cartridge cases, primers (igniter), bullets or propellant powder designed for use in any firearm, rifle or shotgun. The term "ammunition" shall also mean tear gas cartridges, chemical mace or any device or instrument which contains or emits a liquid, gas, powder or any other substance designed to incapacitate.

"Assault weapon", shall have the same meaning as a semiautomatic assault weapon as defined in the federal Public Safety and Recreational Firearms Use Protection Act, 18 U.S.C. section 921(a)(30) as appearing in such section on September 13, 1994, and shall include, but not be limited to, any of the weapons, or copies or duplicates of the weapons, of any caliber, known as: (i) Avtomat Kalashnikov (AK) (all models); (ii) Action Arms Israeli Military Industries UZI and Galil; (iii) Beretta Ar70 (SC-70); (iv) Colt AR-15; (v) Fabrique National FN/FAL, FN/LAR and FNC; (vi) SWD M-10, M-11, M-11/9 and M-12; (vi) Steyr AUG; (vii) INTRATEC TEC-9, TEC-DC9 and TEC-22; and (viii) revolving cylinder shotguns, such as, or similar to, the Street Sweeper and Striker 12; provided, however, that the term assault weapon shall not include: (i) any of the weapons, or replicas or duplicates of such weapons, specified in appendix A to 18 U.S.C. section 922 as appearing in such appendix on September 13, 1994, as such weapons were manufactured on October 1, 1993; (ii) any weapon that is operated by manual bolt, pump, lever or slide action; (iii) any weapon that has been rendered permanently inoperable or otherwise rendered permanently unable to be designated a semiautomatic assault weapon; (iv) any weapon that was manufactured prior to the year 1899; (v) any weapon that is an antique or relic, theatrical prop or other weapon that is not capable of firing a projectile and which is not intended for use as a functional weapon and cannot be readily modified through a combination of available parts into an operable assault weapon; (vi) any semiautomatic rifle that cannot accept a detachable magazine that holds more than five rounds of ammunition; or (vii) any semiautomatic shotgun that cannot hold more than five rounds of ammunition in a fixed or detachable magazine.

"Conviction", a finding or verdict of guilt or a plea of guilty, whether or not final sentence is imposed.

"Firearm", a pistol, revolver or other weapon of any description, loaded or unloaded, from which a shot or bullet can be discharged and of which the length of the barrel or barrels is less than 16 inches or 18 inches in the case of a shotgun as originally manufactured; provided, however, that the term firearm shall not include any weapon that is: (i) constructed in a shape that does not resemble a handgun, short-barreled rifle or short-barreled shotgun including, but not limited to, covert weapons that resemble key-chains, pens, cigarette-lighters or cigarette-packages; or (ii) not detectable as a weapon or potential weapon by x-ray machines commonly used at airports or walk-through metal detectors.

"Gunsmith", any person who engages in the business of repairing, altering, cleaning, polishing, engraving, blueing or performing any mechanical operation on any firearm, rifle, shotgun or machine gun.

"Imitation firearm", any weapon which is designed, manufactured or altered in such a way as to render it incapable of discharging a shot or bullet.

"Large capacity feeding device", (i) a fixed or detachable magazine, box, drum, feed strip or similar device capable of accepting, or that can be readily converted to accept, more than ten rounds of ammunition or more than five shotgun shells; or (ii) a large capacity ammunition feeding device as defined in the federal Public Safety and Recreational Firearms Use Protection Act, 18 U.S.C. section 921(a)(31) as appearing in such appendix on September 13, 1994. The term "large capacity feeding device" shall not include an attached tubular device designed to accept, and capable of operating only with, .22 caliber ammunition.

"Large capacity weapon", any firearm, rifle or shotgun: (i) that is semiautomatic with a fixed large capacity feeding device; (ii) that is semiautomatic and capable of accepting, or readily modifiable to accept, any detachable large capacity feeding device; (iii) that employs a rotating cylinder capable of accepting more than ten rounds of ammunition in a rifle or firearm and more than five shotgun shells in the case of a shotgun or firearm; or (iv) that is an assault weapon. The term "large capacity weapon" shall be a secondary designation and shall apply to a weapon in addition to its primary designation as a firearm, rifle or shotgun and shall not include: (i) any weapon that was manufactured in or prior to the year 1899; (ii) any weapon that operates by manual bolt, pump, lever or slide action; (iii) any weapon that is a single-shot weapon; (iv) any weapon that has been modified so as to render it permanently inoperable or otherwise rendered permanently unable to be designated a large capacity weapon; or (v) any weapon that is an antique or relic, theatrical prop or other weapon that is not capable of firing a projectile and which is not intended for use as a functional weapon and cannot be readily modified through a combination of available parts into an operable large capacity weapon.

"Length of barrel" or "barrel length", that portion of a firearm, rifle, shotgun or machine gun through which a shot or bullet is driven, guided or stabilized and shall include the chamber.

"Licensing authority", the chief of police or the board or officer having control of the police in a city or town, or persons authorized by them.

"Machine gun", a weapon of any description, by whatever name known, loaded or unloaded, from which a number of shots or bullets may be rapidly or automatically discharged by one continuous activation of the trigger, including a submachine gun.

"Purchase" and **"sale"** shall include exchange; the word "purchaser" shall include exchanger; and the verbs "sell" and "purchase", in their different forms and tenses, shall include the verb exchange in its appropriate form and tense.

"Rifle", a weapon having a rifled bore with a barrel length equal to or greater than 16 inches and capable of discharging a shot or bullet for each pull of the trigger.

"Sawed-off shotgun", any weapon made from a shotgun, whether by alteration, modification or otherwise, if such weapon as modified has one or more barrels less than 18 inches in length or as modified has an overall length of less than 26 inches.

"Semiautomatic", capable of utilizing a portion of the energy of a firing cartridge to extract the fired cartridge case and chamber the next round, and requiring a separate pull of the trigger to fire each cartridge.

"Shotgun", a weapon having a smooth bore with a barrel length equal to or greater than 18 inches with an overall length equal to or greater than 26 inches, and capable of discharging a shot or bullet for each pull of the trigger.

"Violent crime", shall mean any crime punishable by imprisonment for a term exceeding one year, or any act of juvenile delinquency involving the use or possession of a deadly weapon that would be punishable by imprisonment for such term if committed by an adult, that: (i) has as an element the use, attempted use or threatened use of physical force or a deadly weapon against the person of another; (ii) is burglary, extortion, arson or kidnapping; (iii) involves the use of explosives; or (iv) otherwise involves conduct that presents a serious risk of physical injury to another.

"Weapon", any rifle, shotgun or firearm.

Where the local licensing authority has the power to issue licenses or cards under this chapter, but no such licensing authority exists, any resident or applicant may apply for such license or firearm identification card directly to the colonel of state police and said colonel shall for this purpose be the licensing authority.

The provisions of sections 122 to 129D, inclusive, and sections 131, 131A, 131B and 131E shall not apply to:

(A) any firearm, rifle or shotgun manufactured in or prior to the year 1899;

(B) any replica of any firearm, rifle or shotgun described in clause (A) if such replica: (i) is not designed or redesigned for using rimfire or conventional centerfire fixed ammunition; or (ii) uses rimfire or conventional centerfire fixed ammunition which is no longer manufactured in the United States and which is not readily available in the ordinary channels of commercial trade; and

(C) manufacturers or wholesalers of firearms, rifles, shotguns or machine guns.

121A. Certificate of Identification of Firearm as Prima Facie Evidence. A certificate by a ballistics expert of the department of the state police or of the city of Boston of the result of an examination made by him of an item furnished him by any police officer, signed and sworn to by such expert, shall be prima facie evidence of his findings as to whether or not the item furnished is a firearm, rifle, shotgun, machine gun, sawed off shotgun or ammunition, as defined by section one hundred and twenty-one, provided that in order to qualify as an expert under this section he shall have previously qualified as an expert in a court proceeding.

122. Licenses; Application to Commissioner after Refusal of License; Fees; Penalty for Improper Issuance. The chief of police or the board or officer having control of the police in a city or town, or persons authorized by them, may, after an investigation into the criminal history of the applicant to determine eligibility for a license under this section, grant a license to any person except an alien, a minor, a person who has been adjudicated a youthful offender, as defined in section fifty-two of chapter one hundred and nineteen, including those who have not received an adult sentence or a person who has been convicted of a felony or of the unlawful use, possession or sale of narcotic or harmful drugs, to sell, rent or lease firearms, rifles, shotguns or machine guns, or to be in business as a gunsmith. Every license shall specify the street and number of the building where the business is to be carried on, and the license shall not protect a licensee who carries on his business in any other place. The licensing authority to whom such application is made shall cause one copy of said applicant's fingerprints to be forwarded to the department of the state police, who shall within a reasonable time thereafter advise such authority in writing of any criminal record of the applicant. The taking of fingerprints shall not be required in issuing a renewal of a license, if the fingerprints of said applicant are on file with the department of the state police. The licensing authority to whom such application is made shall cause one copy of such application to be forwarded to the executive director of the criminal history systems board. Any person refused a license under this section may within ten days thereafter apply to the colonel of state police for such license, who may direct that said licensing authorities grant said license, if, after a hearing, he is satisfied there were no reasonable grounds for the refusal to grant such license and that the applicant was not barred by the provisions of law from holding such a license. The fee for an application for a license issued under this section shall be $100, which shall be payable to the licensing authority and shall not be prorated or refunded in case of revocation or denial. The licensing authority shall retain $25 of the fee; $50 of the fee shall be deposited into the general fund of the commonwealth; and $25 of the fee shall be deposited in the Firearms Fingerprint Identity Verification Trust Fund. A person licensed to sell, rent or lease firearms, rifles, shotguns or machine guns shall not be assessed any additional fee for a gunsmith's license. Whoever knowingly issues a license in violation of this section shall be punished by imprisonment for not less than six months nor more than two years in a jail or house of correction.

122A. Recording of Licenses; Notice to Criminal History Systems Board. The licensing authority, under section one hundred and twenty-two, shall record all is-sued licenses in books, forms or electronic files kept for that purpose, and upon the granting of any such license or renewal thereof or renewal of an expired license shall send notice thereof to the criminal history systems board in a manner prescribed by the executive director of said criminal history systems board; provided, however, that said executive director shall promulgate rules and regulations to ensure the prompt collection, exchange, dissemination, and distribution of such license information. The executive director of the criminal history systems board, upon the application of the licensee, at a price not in excess of the cost thereof, shall furnish said licensee with the necessary sales record books to be kept by him as provided in section one hundred and twenty-three.

122B. License to Sell Ammunition; Penalties. No person shall sell ammunition in the commonwealth unless duly licensed. The chief of police or the board or officer having control of the police in a city or town, or persons authorized by them, may, after an investigation into the criminal history of the applicant to determine eligibility to be licensed under this section, grant a license to any person, except an alien, a minor, a person who has been adjudicated a youthful offender, as defined in section fifty-two of chapter one hundred and nineteen, including those who have not received an adult sentence or a person who has been convicted of a felony in any state or federal jurisdiction, or of the unlawful use, possession or sale of narcotic or harmful drugs, to sell ammunition. Every license shall specify the street and number, if any, of the building where the business is to be carried on. The licensing authority to whom such application is made shall cause one copy of the application to be forwarded to the executive director of the criminal history systems board, who shall within a reasonable time thereafter advise such authority in writing of any criminal record disqualifying the applicant. The fee for an application for a license to sell ammunition shall be $100, which shall be payable to the licensing authority and shall not be prorated or refunded in case of revocation or denial. The licensing authority shall retain $25 of the fee; $50 of the fee shall be deposited into the general fund of the commonwealth; and $25 of the fee shall be deposited in the Firearms Fingerprint Identity Verification Trust Fund. The licensing authority to whom such application is made shall cause one copy of any approved application to be forwarded to the executive director of the criminal history systems board. ...

The secretary of the executive office of public safety may establish such rules and regulations as he may deem necessary to carry out the provisions of this section.

Any person refused a license under this section or once issued a license under this section has had said license suspended or revoked may obtain a judicial review of such refusal, suspension or revocation by filing within thirty days of such refusal, suspension or revocation a petition for review thereof in the district court having jurisdiction in the city or town in which the applicant filed for such license, and a justice of said court, after a hearing, may direct that a license be issued the applicant if satisfied there was no reasonable ground for refusing such license and that the applicant was not prohibited by law from holding the same.

Whoever not being licensed, as hereinbefore provided, sells ammunition within the commonwealth shall be punished by a fine of not less than five hundred nor more than one thousand dollars or by imprisonment for not less than six months nor more than two years.

123. Conditions of Licenses Granted Under § 122. A license granted under section one hundred and twenty-two shall be expressed to be and shall be subject to the following conditions:

First, that the provisions in regard to the nature of the license and the building in which the business may be carried on under it shall be strictly adhered to.

Second, that every licensee shall, before delivery of a firearm, rifle or shotgun, make or cause to be made a true, legible entry in a sales record book to be furnished by the executive director of the criminal history systems board and to be kept for that purpose, specifying the complete description of the firearm, rifle or shotgun, and designation as a large capacity weapon, if applicable, including the make, serial number, if any, type of firearm, rifle or shotgun, and designation as a large capacity weapon, if applicable, whether sold, rented or leased, the date of each sale, rental or lease, the license to carry firearms number or permit to purchase number and the identification card number in the case of a firearm or the identification card number or the license to carry firearms number in the case of a rifle or shotgun, the sex, residence and occupation of the purchaser, renter or lessee, and shall before delivery, as aforesaid, require the purchaser, renter or lessee personally to write in said sales record book his full name. Said book shall be open at all times to the inspection of the police.

Third, that the license or a copy thereof, certified by the official issuing the same, shall be displayed on the premises in a position where it can easily be read.

Fourth, that no firearm, rifle or shotgun, or machine gun shall be displayed in any outer window of said premises or in any other place where it can readily be seen from the outside.

Fifth, that the licensee shall submit a record of all sales, rentals and leases forthwith at the time of such sale, rental or lease via electronic communication link to the executive director of the criminal history systems board.

Sixth, that every firearm, rifle or shotgun shall be unloaded when delivered.

Seventh, that no delivery of a firearm shall be made to any person not having a license to carry firearms issued under the provisions of section one hundred and thirty-one nor shall any delivery of a rifle or shotgun or ammunition be made to any minor nor to any person not having a license to carry firearms issued under the provisions of section one hundred and thirty-one or a firearm identification card issued under the provisions of section one hundred and twenty-nine B nor shall any large capacity firearm or large capacity feeding device therefor be delivered to any person not having a Class A license to carry firearms issued under section 131 nor shall any large capacity rifle or shotgun or large capacity feeding device therefor be delivered to any person not having a Class A or Class B license to carry firearms issued under said section 131; provided, however, that delivery of a firearm by a licensee to a person possessing a valid permit to purchase said firearm issued under the provisions of section one hundred and thirty-one A and a valid firearm identification card issued under section one hundred and twenty-nine B may be made by the licensee to the purchaser's residence or place of business, subject to the restrictions imposed upon such permits as provided under section 131A.

Eighth, that no firearm shall be sold, rented or leased to a minor or a person who has not a permit then in force to purchase, rent or lease the same issued under section one hundred and thirty-one A, and a firearm identification card issued under the provisions of section one hundred and twenty-nine B, or unless such person has a license to carry firearms issued under the provisions of section one hundred and thirty-one; nor shall any rifle or shotgun be sold, rented or leased to a person who has not a valid firearm identification card as provided for in section one hundred and twenty-nine B, or has a license to carry firearms as provided in section one hundred and thirty-one; that no large capacity firearm nor large capacity feeding device therefor shall be sold, rented, leased or transferred to any person not having (i) a Class A license to carry firearms issued under section 131 or (ii) a proper permit issued under section 131A and a firearm identification card issued under section 129B; that no large capacity rifle or shotgun nor large capacity feeding device therefor shall be sold to any person not having a Class A or Class B license to carry firearms issued under said section 131; and that no machine gun shall be sold, rented or leased to any person who has not a license to possess the same issued under section one hundred and thirty-one.

Ninth, that upon the sale, rental or lease of a firearm, subject to a permit to purchase issued under the provisions of section one hundred and thirty-one A, the licensee under section one hundred and twenty-two shall take up such permit to purchase and shall endorse upon it the date and place of said sale, rental or lease, and shall transmit the same to the executive director of the criminal history systems board; and that upon the sale, rental or lease of a machine gun shall endorse upon the license to possess the same the date and place of said sale, rental or lease, and shall within seven days transmit a notice thereof to said executive director. In case of a sale under the provisions of section one hundred and thirty-one E the licensee under section one hundred and twenty-two shall write in the sales record book the number of the license to carry firearms issued the purchaser under the provisions of section one hundred and thirty-one, or the number of the firearm identification card issued the purchaser under the pro-visions of section one hundred and twenty-nine B, whichever is applicable under the provisions of condition Eighth of this section.

Tenth, that this license shall be subject to forfeiture as provided in section one hundred and twenty-five for breach of any of its conditions, and that, if the licensee hereunder is convicted of a violation of any such conditions, this license shall thereupon become void.

Eleventh, that the second, fifth, eighth and ninth conditions shall not apply to a gunsmith with regard to repair or remodeling or servicing of firearms, rifles or shotguns unless said gunsmith has manufactured a firearm, rifle or shotgun for the purchaser, but said gunsmith shall keep records of the work done by him together with the names and addresses of his customers. Such records shall be kept open for inspection by the police at all times.

Twelfth, that any licensee shall keep records of each sale, rental or lease of a rifle or shotgun, specifying the description of said rifle or shotgun, together with the name and address of the purchaser, renter or lessee, and the date of such transaction.

Thirteenth, that the current validity of any firearm identification card, license to carry firearms or permit to purchase, rent or lease firearms presented, and that the person presenting said card, license or permit is the lawful holder thereof, shall be verified by the licensee prior to any sale, rental or lease of a rifle, shotgun, firearm or large capacity feeding device; and, upon being presented with such card or license that is expired, suspended or revoked, the licensee shall notify the licensing authority of the presentment of such expired, suspended or revoked card, license or permit; and further, the licensee may take possession of such card or license provided that, in such case, such licensee shall: (i) issue a receipt, in a form provided by the executive director of the criminal history systems board, to the holder thereof which shall state that the holder's card or license is expired, suspended or revoked, was taken by such licensee and forwarded to the licensing authority by whom it was issued and such receipt shall be valid for the date of issuance for the purpose of providing immunity from prosecution under section 10 of chapter 269 for unlawfully possessing a firearm, rifle or shotgun or large capacity weapon; (ii) notify the cardholder or licensee of his requirement to renew said card or license; and (iii) forward such expired card or license to the licensing authority forthwith; provided, however, that such licensee shall be immune from civil and criminal liability for good faith compliance with the provisions herein.

Fourteenth, that the licensee shall conspicuously post at each purchase counter the following warning in bold type not less than one inch in height: "IT IS UNLAWFUL TO STORE OR KEEP A FIREARM, RIFLE, SHOTGUN OR MACHINE GUN IN ANY PLACE UN-LESS THAT WEAPON IS EQUIPPED WITH A TAMPER-RESISTANT SAFETY DEVICE OR IS STORED OR KEPT IN A SECURELY LOCKED CONTAINER.", and that such licensee shall provide said warning, in writing, to the purchaser or transferee of any firearm, rifle, shotgun or machine gun in bold type not less than one-quarter inch in height.

Fifteenth, that all licensees shall maintain a permanent place of business that is not a residence or dwelling wherein all transactions described in this section shall be conducted and wherein all records required to be kept under this section shall be so kept.

Sixteenth, that no licensee shall sell, lease, rent, transfer or deliver or offer for sale, lease, rent, transfer or delivery to any person any assault weapon or large capacity feeding device that was not otherwise lawfully possessed on September 13, 1994.

Seventeenth, that any licensee from whom a rifle, shotgun, firearm or machine gun is lost or stolen shall report such loss or theft to the licensing authority and the executive director of the criminal history systems board forthwith. Such report shall include a complete description of the weapon, including the make, model, serial number and caliber and whether such weapon is a large capacity weapon.

Eighteenth, that no licensee shall sell, rent, lease, transfer or deliver or offer for sale, lease, transfer or delivery any firearm, to any purchaser in the commonwealth unless such sale is to a business entity that is primarily a firearm wholesaler and the sale, by its terms, prohibits the purchaser from reselling such firearm to a firearm retailer or consumer in the commonwealth if such firearm has a frame, barrel, cylinder, slide or breechblock that is composed of: (i) any metal having a melting point of less than 900 degrees Fahrenheit; (ii) any metal having an ultimate tensile strength of less than 55,000 pounds per square inch; or (iii) any powdered metal having a density of less than 7.5 grams per cubic centimeter. This clause shall not apply to any make and model of firearm for which a sample of three firearms in new condition all pass the following test: Each of the three samples shall fire 600 rounds, stopping every 100 rounds to tighten any loose screws and to clean the gun if required by the cleaning schedule in the user manual, and as needed to refill the empty magazine or cylinder to capacity before continuing. For any firearm that is loaded in a manner other than via a detachable magazine, the tester shal also pause every 50 rounds for ten minutes. The ammunition used shall be the type recommended by the firearm manufacturer in its user manual or, if none is recommended, any standard ammunition of the correct caliber in new condition. A firearm shall pass this test if it fires the first 20 rounds without a malfunction, fires the full 600 rounds with not more than six malfunctions and completes the test without any crack or breakage of an operating part of the firearm. The term "crack" or "breakage" shall not include a crack or breakage that does not increase the danger of injury to the user. For purposes of evaluating the results of this test, malfunction shall mean any failure to feed, chamber, fire, extract or eject a round or any failure to accept or eject a magazine or any other failure which prevents the firearm, without manual intervention beyond that needed for routine firing and periodic reloading, from firing the chambered round or moving a new round into position so that the firearm is capable of firing the new round properly. "Malfunction" shall not include a misfire caused by a faulty cartridge the primer of which fails to detonate when properly struck by the firearm's firing mechanism.

Nineteenth, that no licensee shall sell, rent, lease, transfer or deliver or offer for sale, lease, transfer or delivery any firearm to any purchaser in the commonwealth unless such sale is to a business entity that is primarily a firearms wholesaler, and the sale, by its terms, prohibits such purchaser from reselling such firearm to a firearm retailer or consumer in the common-wealth if such firearm is prone to accidental discharge which, for purposes of this clause, shall mean any make and model of firearm for which a sample of five firearms in new condition all undergo, and none discharge during, the following test: Each of the five sample firearms shall be: (a) test loaded; (b) set so that the firearm is in a condition such that pulling the trigger and taking any action that must simultaneously accompany the pulling of the trigger as part of the firing procedure would fire the handgun; and (c) dropped onto a solid slab of concrete from a height of one meter from each of the following positions: (i) normal firing position; (ii) upside down; (iii) on grip; (iv) on the muzzle; (v) on either side; and (vi) on the exposed hammer or striker or, if there is no exposed hammer or striker, the rear-most part of the firearm. If the firearm is designed so that its hammer or striker may be set in other positions, each sample firearm shall be tested as above with the hammer or striker in each such position but otherwise in such condition that pulling the trigger, and taking any action that must simultaneously accompany the pulling of the trigger as part of the firing procedure, would fire the firearm. Alternatively, the tester may use additional sample firearms of the same make and model, in a similar condition, for the test of each of these hammer striker settings.

Twentieth, that no licensee shall sell, rent, lease, transfer or deliver or offer for sale, lease, transfer or delivery, any firearm to any purchaser in the commonwealth unless such sale is to a business entity that is primarily a firearm wholesaler, and the sale, by its terms, prohibits the purchaser from reselling such firearm to a firearm retailer or consumer in the commonwealth if such firearm is prone to: (i) firing more than once per pull of the trigger; or (ii) explosion during firing.

Twenty-first, that no licensee shall sell, rent, lease, transfer or deliver or offer for sale, lease, transfer or delivery any firearm to any purchaser in the commonwealth unless such sale is to a business entity that is primarily a firearm wholesaler and the sale, by its terms, prohibits the purchaser from reselling such firearm to a firearm retailer or consumer in the commonwealth if such firearm has a barrel less than three inches in length, unless the licensee discloses in writing, prior to the transaction, to the prospective buyer, lessee, deliveree or transferee the limitations of the accuracy of the particular make and model of the subject firearm, by disclosing the make and model's average group diameter test result at seven yards, average group diameter test result at 14 yards and average group diameter test result at 21 yards. For purposes of this clause, "average group diameter test result" shall mean the arithmetic mean of three separate trials, each performed as follows on a different sample firearm in new condition of the make and model at issue. Each firearm shall fire five rounds at a target from a set distance and the largest spread in inches between the centers of any of the holes made in a test target shall be measured and recorded. This procedure shall be repeated two more times on the firearm. The arithmetic mean of each of the three recorded results shall be deemed the result of the trial for that particular sample firearm. The ammunition used shall be the type recommended by the firearm manufacturer in its user manual or, if none is recommended, any standard ammunition of the correct caliber in new condition.

Clauses Eighteenth to Twenty-first, inclusive, of the first paragraph shall not apply to: (a) any firearm lawfully owned or possessed under a license issued under this chapter on or before October 21, 1998; or (b) any firearm designated by the secretary of public safety, with the advice of the gun control advisory board established pursuant to section 1311/2 of chapter 140, as a firearm solely designed and sold for formal target shooting competition. The secretary of public safety shall compile a list, on a bi-annual basis, of firearms designated as formal target shooting firearms in accordance with this paragraph. Such list shall be made available for distribution by the executive office of public safety.

No licensee shall sell any rifle or shotgun, contrary to the provisions of section one hundred and thirty or section 131E.

No person licensed under the provisions of section 122 or section 122B shall sell, rent, lease, transfer or deliver any rifle, shotgun or firearm or ammunition or ammunition feeding device contrary to the provisions of section 130 or section 131E; and no such licensee shall sell, rent, lease, transfer or deliver any rifle, shotgun or firearm or ammunition or ammunition feeding device to any person who does not have in his possession the required firearm identification card or proof of exemption therefrom, license to carry firearms or permit to purchase, rent or lease firearms and who does not present such card, proof, license or permit to the licensee in person at the time of purchase, rental or lease. No person licensed under the provisions of section 122 or section 122B shall fill an order for such weapon, ammunition or ammunition feeding device that was received by mail, facsimile, telephone or other telecommunication unless such transaction or transfer includes the in-person presentation of the required card, proof, license or permit as required herein prior to any sale, delivery or any form of transfer of possession of the subject weapon, ammunition or ammunition feeding device. Transactions between persons licensed under section 122 or between federally licensed dealers shall be exempt from the provisions of this paragraph.

The licensing authority shall enter, one time per calendar year, during regular business hours, the commercial premises owned or leased by any licensee, wherein such records required to be maintained under this section are stored or maintained, and inspect, in a reasonable manner, such records and inventory for the purpose of enforcing the provisions of this section. If such records and inventory contain evidence of violations of this section, the inspecting officer shall produce and take possession of copies of such records and, in the event that the licensee subject to inspection does not possess copying equipment, the inspecting officer shall arrange to have copied, in a reasonable time and manner, such records that contain evidence of such violations and the costs for such copying shall be assessed against the owner of such records. Licensees found to be in violation of this section shall be subject to the suspension or permanent revocation of such license issued under section 122 and to the provisions of section 128. Nothing herein shall prohibit the licensing authority or the department of state police from conducting such inspections pursuant to a valid search warrant issued by a court of competent jurisdiction.

Notwithstanding the provisions of this section, a person licensed under the provisions of section one hundred and twenty-two, or section one hundred and twenty-two B, may sell or transfer firearms, rifles, shotguns, machine guns or ammunition at any regular meeting of an incorporated collectors club or at a gun show open to the general public; provided, however, that all other provisions of this section are complied with and that such sale or transfer is in conformity with federal law or regulations applicable to the transfer or sale of firearms, rifles, shotguns, machine guns or ammunition, including the restrictions imposed upon firearm identification cards issued under section 129B, licenses to carry firearms issued under section 131 and permits to purchase, lease or rent firearms issued under section 131A.

124. Term of Licenses. Licenses issued under sections one hundred and twenty-two and one hundred and twenty-two B shall expire three years from the date of issuance.

125. Suspension or Revocation of Licenses. The officials authorized to issue a license under section one hundred and twenty-two, after due notice to the licensee and reasonable opportunity for him to be heard, may declare his license forfeited, or may suspend his license for such period of time as they may deem proper, upon satisfactory proof that he has violated or permitted a violation of any condition thereof or has violated any provision of this chapter, or has been convicted of a felony. The pendency of proceedings before a court shall not suspend or interfere with the power to declare a forfeiture. If the license is declared forfeited, the licensee shall be disqualified to receive a license for one year after the expiration of the term of the license so forfeited. The executive director of the criminal history systems board shall be notified in writing of any forfeiture under this section.

126. Sign or Advertisement Evidence That Firearms Are Kept for Sale. If there is exposed from, maintained in or permitted to remain on any vehicle or premises any placard, sign or advertisement purporting or designed to announce that firearms, rifles, shotguns or machine guns are kept in or upon such vehicle or premises or that an occupant of any vehicle or premises is a gunsmith, it shall be prima facie evidence that firearms, rifles, shotguns or machine guns are kept in or upon such vehicle or premises for sale or that the occupant is engaged in business as a gunsmith.

127. Transfer of Licenses. The officials authorized to issue a license under section one hundred and twenty-two may transfer licenses from one location to another within the city or town in which the licenses are in force, but such transfer shall be granted only to the original licensee and upon the same terms and conditions upon which the license was originally granted. The executive director of the criminal history systems board shall be notified in writing of any transfers made under this section.

128. Penalty for Violation of License or Selling without License. Any licensee under a license described in section one hundred and twenty-three, and any employee or agent of such a licensee, who violates any provision of said section required to be expressed in the second, fourth, sixth, seventh, eighth, ninth, sixteenth, eighteenth, nineteenth, twentieth or twenty-first condition of said license, and except as provided in section one hundred and twenty-eight A, any person who, without being licensed as herein before provided, sells, rents or leases a firearm, rifle, shotgun or ma-chine gun, or is engaged in business as a gun-smith, shall be punished by a fine of not less than $1,000 nor more than $10,000, or by imprisonment for not less than one year nor more than ten years, or by both such fine and imprisonment.

Evidence that a person sold or attempted to sell a machine gun without being licensed under section one hundred and twenty-three shall, in a prosecution under this section, constitute prima facie evidence that such person is engaged in the business of selling machine guns.

128A. Exceptions to Applicability of § 128. The provisions of section one hundred and twenty-eight shall not apply to any person who, without being licensed as provided in section one hundred and twenty-two, sells or transfers a firearm, rifle or shotgun to a person licensed under said section one hundred and twenty-two, or to a federally licensed firearms dealer or to a federal, state or local historical society, museum or institutional collection open to the public. The provisions of section one hundred and twenty-eight shall not apply to any resident of the commonwealth who, without being licensed as provided in section one hundred and twenty-two, sells or transfers to other than a federally licensed firearms dealer or organization named above not more than four firearms, including rifles and shotguns in any one calendar year; provided, however, that the seller has a firearm identification card or a license to carry firearms, is an exempt person under the conditions of clauses (n), (o), (r) and (s) of the fourth paragraph of section one hundred and twenty-nine C, or is permitted to transfer ownership under the conditions of section one hundred and twenty-nine D and the purchaser has, in the case of sale or transfer of a firearm, a permit to purchase issued under the provisions of section one hundred and thirty-one A and a firearm identification card issued under section one hundred and twenty-nine B, or has such permit to purchase and is an exempt person under the provisions of section one hundred and twenty-nine C, or has been issued a license to carry firearms under the provisions of section one hundred and thirty, or in the case of sale or transfer of a rifle or shotgun, the purchaser has a firearm identification card or a license to carry firearms or is an exempt person as hereinbefore stated; and provided, further, that such resident reports within seven days, in writing to the executive director of the criminal history systems board on forms furnished by said executive director, the names and addressees of the seller and the purchaser of any such large capacity feeding device, firearm, rifle or shotgun, together with a complete description of the firearm, rifle or shotgun, including its designation as a large capacity weapon, if applicable, the calibre, make and serial number and the purchaser's license to carry firearms number, permit to purchase number and identifying number of such documentation as is used to establish exempt person status in the case of a firearm or the purchaser's license to carry number or firearm identification card number or said document identity number, in the case of a rifle or shotgun.

128B. Reports by Residents and Nonresidents Purchasing or Obtaining Firearms from Unlicensed Source. Any resident of the commonwealth who purchases or obtains a firearm, rifle or shotgun or machine gun from any source within or without the commonwealth, other than from a licensee under section one hundred and twenty-two or a person authorized to sell firearms under section one hundred and twenty-eight A, and any non-resident of the commonwealth who purchases or obtains a firearm, rifle, shotgun or machine gun from any source within or without the commonwealth, other than such a licensee or person, and receives such firearm, rifle, shotgun or machine gun, within the commonwealth shall within seven days after receiving such firearm, rifle, shotgun or machine gun, report, in writing, to the executive director of the criminal history systems board the name and address of the seller or donor and the buyer or donee, together with a complete description of the firearm, rifle, shotgun or machine gun, including the caliber, make and serial number. Whoever violates any provision of this section shall for the first offense be punished by a fine of not less than $500 nor more than $1,000 and for any subsequent offense by imprisonment in the state prison for not more than ten years.

129. Penalty for Giving False Name, Information. Whoever in purchasing, renting or hiring a firearm, rifle, shotgun or machine gun, or in making application for any form of license or permit issued in connection therewith, or in requesting that work be done by a gunsmith, gives a false or fictitious name or address or knowingly offers or gives false information concerning the date or place of birth, his citizenship status, occupation, or criminal record, shall for the first offense be punished by a fine of not less than five hundred nor more than one thousand dollars, or by imprisonment for not more than one year, or both; and for a second or subsequent offense, shall be punished by imprisonment for not less than two and one half years nor more than five years in the state prison.

129B. Firearm Identification Card; Application; Suspension or Revocation; Appeal; Investigation of Criminal Record and Mental Health of Applicant. A firearm identification card shall be issued and possessed subject to the following conditions and restrictions:

(1) Any person residing or having a place of business within the jurisdiction of the licensing authority or any person residing in an area of exclusive federal jurisdiction located within a city or town may submit to the licensing authority an application for a firearm identification card, or renewal of the same, which the licensing authority shall issue, unless the applicant:

(i) has ever, in a court of the commonwealth, been convicted or adjudicated a youthful offender or delinquent child, both as defined in section 52 of chapter 119, for the commission of: (a) a felony; (b) a misdemeanor punishable by imprisonment for more than two years; (c) a violent crime as defined in section 121; (d) a violation of any law regulating the use, possession, ownership, transfer, purchase, sale, lease, rental, receipt or transportation of weapons or ammunition for which a term of imprisonment may be imposed; or (e) a violation of any law regulating the use, possession or sale of controlled substances, as defined in section 1 of chapter 94C including, but not limited to, a violation under said chapter 94C; provided, however, that except for the commission of a violent crime or a crime involving the trafficking of controlled substances, if the applicant has been so convicted or adjudicated or released from confinement, probation or parole supervision for such conviction or adjudication, whichever is last occurring, not less than five years immediately preceding such application, such applicant's right or ability to possess a non-large capacity rifle or shotgun shall be deemed restored in the commonwealth with respect to such conviction or adjudication and such conviction or adjudication shall not disqualify such applicant for a firearm identification card;

(ii) has, in any other state or federal jurisdiction, been convicted or adjudicated a youthful offender or delinquent child for the commission of: (a) a felony; (b) a misdemeanor punishable by imprisonment for more than two years; (c) a violent crime as defined in section 121; (d) a violation of any law regulating the use, possession, ownership, transfer, purchase, sale, lease, rental, receipt or transportation of weapons or ammunition for which a term of imprisonment may be imposed; or (e) a violation of any law regulating the use, possession or sale of controlled substances, as defined in section 1 of chapter 94C; provided, however, that, except for the commission of a violent crime or a crime involving the trafficking of weapons or controlled substances, if the applicant has been so convicted or adjudicated or released from confinement, probation or parole supervision for such conviction or adjudication, whichever is last occurring, not less than five years immediately preceding such application, and such applicant's right or ability to possess a rifle or shotgun has been fully restored in the jurisdiction wherein the subject conviction or adjudication was entered, such conviction or adjudication shall not disqualify such applicant for a firearm identification card;

(iii) has been confined to any hospital or institution for mental illness, unless the applicant submits with his application an affidavit of a registered physician attesting that such physician is familiar with the applicant's mental illness and that in such physician's opinion the applicant is not disabled by such an illness in a manner that should prevent the applicant from possessing a firearm, rifle or shotgun;

(iv) is or has been under treatment for or confinement for drug addiction or habitual drunkenness, unless such applicant is deemed to be cured of such condition by a licensed physician, in which case he may make application for such card after the expiration of five years from the date of such confinement or treatment and upon presentation of an affidavit issued by such physician to the effect that such physician knows the applicant's history of treatment and that in such physician's opinion the applicant is deemed cured;

(v) is at the time of the application less than 15 years of age;

(vi) is at the time of the application more than 15 but less than 18 years of age, unless the applicant submits with his application a certificate of his parent or guardian granting the applicant permission to apply for a card;

(vii) is an alien;

(viii) is currently subject to: (a) an order for suspension or surrender issued pursuant to section 3B or 3C of chapter 209A or a similar order issued by another jurisdiction; or (b) a permanent or temporary protection order issued pursuant to chapter 209A or a similar order issued by another jurisdiction; or

(ix) is currently the subject of an outstanding arrest warrant in any state or federal jurisdiction.

(2) Within seven days of the receipt of a completed application for a card, the licensing authority shall forward one copy of the application and one copy of the applicant's fingerprints to the colonel of state police, who shall, within 30 days, advise the licensing authority, in writing, of any disqualifying criminal record of the applicant arising from within or without the commonwealth and whether there is reason to believe that the applicant is disqualified for any of the foregoing reasons from possessing a card; provided, however, that the taking of fingerprints shall not be required in issuing the renewal of a card if the renewal applicant's fingerprints are on file with the department of state police. In searching for any disqualifying history of the applicant, the colonel shall utilize, or cause to be utilized, files maintained by the department of mental health, department of probation and statewide and nationwide criminal justice, warrant and protection order information systems and files including, but not limited to, the National Instant Criminal Background Check System. If the information available to the colonel does not indicate that the possession of a non-large capacity rifle or shotgun by the applicant would be in violation of state or federal law, he shall certify such fact, in writing, to the licensing authority within such 30 day period.

(3) The licensing authority may not prescribe any other condition for the issuance of a firearm identification card and shall, within 40 days from the date of application, either approve the application and issue the license or deny the application and notify the applicant of the reason for such denial in writing; provided, however, that no such card shall be issued unless the colonel has certified, in writing, that the information available to him does not indicate that the possession of a rifle or shotgun by the applicant would be in violation of state or federal law.

(4) A firearm identification card shall be revoked or suspended by the licensing authority or his designee upon the occurrence of any event that would have disqualified the holder from being issued such card or from having such card renewed or for a violation of a restriction provided under this section. Any revocation or suspension of a card shall be in writing and shall state the reasons therefor. Upon revocation or suspension, the licensing authority shall take possession of such card and receipt for fee paid for such card, and the person whose card is so revoked or suspended shall take all action required under the provisions of section 129D. No appeal or post-judgment motion shall operate to stay such revocation or suspension. Notices of revocation and suspension shall be forwarded to the executive director of the criminal history systems board and the commissioner of probation and shall be included in the criminal justice information system. A revoked or suspended card may be reinstated only upon the termination of all disqualifying conditions.

(5) Any applicant or holder aggrieved by a denial, revocation or suspension of a firearm identification card, unless a hearing has previously been held pursuant to chapter 209A, may, within either 90 days after receipt of notice of such denial, revocation or suspension or within 90 days after the expiration of the time limit in which the licensing authority is required to respond to the applicant, file a petition to obtain judicial review in the district court having jurisdiction in the city or town wherein the applicant filed for or was issued such card. A justice of such court, after a hearing, may direct that a card be issued or reinstated to the petitioner if the justice finds that such petitioner is not prohibited by law from possessing such card.

(6) A firearm identification card shall not entitle a holder thereof to possess: (i) a large capacity firearm or large capacity feeding device therefor, except under a Class A license issued to a shooting club as provided under section 131 or under the direct supervision of a holder of a Class A license issued to an individual under section 131 at an incorporated shooting club or licensed shooting range; or (ii) a non-large capacity firearm or large capacity rifle or shotgun or large capacity feeding device therefor, except under a Class A license issued to a shooting club as provided under section 131 or under the direct supervision of a holder of a Class A or Class B license issued to an individual under section 131 at an incorporated shooting club or licensed shooting range. A firearm identification card shall not entitle a holder thereof to possess any rifle or shotgun that is, or in such manner that is, otherwise prohibited by law. A firearm identification card shall be valid for the purpose of purchasing and possessing chemical mace, pepper spray or other similarly propelled liquid, gas or powder designed to temporarily incapacitate. Except as otherwise provided herein, a firearm identification card shall not be valid for the use, possession, ownership, transfer, purchase, sale, lease, rental or transportation of a rifle or shotgun if such rifle or shotgun is a large capacity weapon as defined in section 121.

(7) A firearm identification card shall be in a standard form provided by the executive director of the criminal history systems board in a size and shape equivalent to that of a license to operate motor vehicles issued by the registry of motor vehicles pursuant to section 8 of chapter 90 and shall contain an identification number, name, address, photograph, fingerprint, place and date of birth, height, weight, hair color, eye color and signature of the cardholder and shall be marked "Firearm Identification Card". If a firearm identification card is issued for the sole purpose of purchasing or possessing chemical mace, pepper spray or other similarly propelled liquid, gas or powder designed to temporarily incapacitate, such card shall clearly state that such card is valid for such limited purpose only. The application for such card shall be made in a standard form provided by the executive director of the criminal history systems board which shall require the applicant to affirmatively state, under the pains and penalties of perjury, that he is not disqualified on any of the grounds enumerated in clauses (i) to (ix), inclusive, from being issued such card.

(8) Any person who knowingly files an application containing false information shall be punished by a fine of not less than $500 nor more than $1,000 or by imprisonment for not less than six months nor more than two years in a house of correction, or by both such fine and imprisonment.

(9) A firearm identification card shall be valid, unless revoked or suspended, for a period of not more than 6 years from the date of issue, except that if the cardholder applied for renewal before the card expired, the card shall remain valid for a period of 90 days after the stated expiration date on the card, unless the application for renewal is denied. A card issued on February 29 shall expire on March 1. The executive director of the criminal history systems board shall send by first class mail to the holder of a firearm identification card, a notice of the expiration of the card not less than 90 days before its expiration, and shall enclose with the notice a form for the renewal of the card. The executive director of the criminal history systems board shall include in the notice all pertinent information about the penalties that may be imposed if the firearm identification card is not renewed within the 90 days before expiration.

(9A) Except as provided in clause (9B), the fee for an application for a firearm identification card shall be $100, which shall be payable to the licensing authority and shall not be prorated or refunded in the case of revocation or denial. The licensing authority shall retain $25 of the fee; $50 of the fee shall be deposited in the General Fund; and $25 of the fee shall be deposited in the Firearms Fingerprint Identity Verification Trust Fund. Notwithstanding any general or special law to the contrary, licensing authorities shall deposit quarterly that portion of the firearm identification card application fee which is to be deposited into the General Fund, not later than January 1, April 1, July 1 and October 1 of each year.

(9B) The application fee for a firearm identification card issued for the sole purpose of purchasing or possessing chemical mace, pepper spray or other similarly propelled liquid, gas or powder designed to temporarily incapacitate shall be $25, which shall be payable to the licensing authority and shall not be prorated or refunded in the case of revocation or denial. The licensing authority shall retain 50 per cent of the fee and the remaining portion shall be deposited in the General Fund. Notwithstanding any general or special law to the contrary, licensing authorities shall deposit quarterly that portion of the firearm identification card application fee which is to be deposited into the General Fund, not later than January 1, April 1, July 1 and October 1 of each year. There shall be no application fee for the renewal of a firearm identification card issued under this clause.

A firearm identification card issued under this clause shall display, in clear and conspicuous language, that the card shall be valid only for the purpose of purchasing or possessing chemical mace, pepper spray or other similarly propelled liquid, gas or powder designed to temporarily incapacitate.

(10) Any person over the age of 70 shall be exempt from the requirement of paying a renewal fee for a firearm identification card.

(11) A cardholder shall notify, in writing, the licensing authority that issued such card, the chief of police into whose jurisdiction such cardholder moves and the executive director of the criminal history systems board of any change of address. Such notification shall be made by certified mail within 30 days of its occurrence. Failure to so notify shall be cause for revocation or suspension of such card.

(12) Notwithstanding the provisions of section 10 of chapter 269, any person in possession of a non-large capacity rifle or shotgun whose firearm identification card issued under this section is invalid for the sole reason that it has expired, meaning after 90 days beyond the stated expiration date on the card, but who shall not be disqualified from renewal upon application therefor under this section, shall be subject to a civil fine of not less than $500 nor more than $5,000 and the provisions of said section 10 of said chapter 269 shall not apply; provided, however, that the exemption from the provisions of said section 10 of said chapter 269 provided herein shall not apply if: (i) such firearm identification card has been revoked or suspended, unless such revocation or suspension was caused by failure to give notice of a change of address as required under this section; (ii) revocation or suspension of such firearm identification card is pending, unless such revocation or suspension was caused by failure to give notice of a change of address as required under this section; or (iii) an application for renewal of such firearm identification card has been denied. Any law enforcement officer who discovers a person to be in possession of a rifle or shotgun after such person's firearm identification card has expired or has been revoked or suspended solely for failure to give notice of a change of address shall confiscate any rifle or shotgun and such expired, meaning after 90 days beyond the stated expiration date on the card, or suspended card then in possession, and such officer shall forward such card to the licensing authority by whom it was issued as soon as practicable. Any confiscated weapon shall be returned to the owner upon the renewal or reinstatement of such expired or suspended card within one year of such confiscation or such weapon may be otherwise disposed of in accordance with the provisions of section 129D. Pending the issuance of a renewed firearm identification card, a receipt for the fee paid, after five days following issuance, shall serve as a valid substitute and any rifle or shotgun so confiscated shall be returned, unless the applicant is disqualified. The provisions of this paragraph shall not apply if such person has a valid license to carry firearms issued under section 131 or 131F.

(13) Upon issuance of a firearm identification card under this section, the licensing authority shall forward a copy of such approved application and card to the executive director of the criminal history systems board, who shall inform the licensing authority forthwith of the existence of any disqualifying condition discovered or occurring subsequent to the issuance of a firearm identification card under this section.

(14) Nothing in this section shall authorize the purchase, possession or transfer of any weapon, ammunition or feeding device that is, or in such manner that is, prohibited by state or federal law.

(15) The secretary of the executive office of public safety, or his designee, may promulgate regulations to carry out the purposes of this section.

129C. Firearm Identification Card; Restrictions on Possession, Transfer of Firearms; Reports; Persons and Uses Exempted from Provisions of Section; Purchases by Certain Exempt Persons. No person, other than a licensed dealer or one who has been issued a license to carry a pistol or revolver or an exempt person as hereinafter described, shall own or possess any firearm, rifle, shotgun or ammunition unless he has been issued a firearm identification card by the licensing authority pursuant to the provisions of section one hundred and twenty-nine B.

No person shall sell, give away, loan or otherwise transfer a rifle or shotgun or ammunition other than (a) by operation of law, or (b) to an exempt person as hereinafter described, or (c) to a licensed dealer, or (d) to a person who displays his firearm identification card, or license to carry a pistol or revolver.

A seller shall, within seven days, report all such transfers to the executive director of the criminal history systems board according to the provisions set forth in section one hundred and twenty-eight A, and in the case of loss, theft or recovery of any firearm, rifle, or shotgun or machine gun a similar report shall be made forthwith to both the executive director of the criminal history systems board and the licensing authority in the city or town where the owner resides. Failure to so report shall be cause for suspension or permanent revocation of such person's firearm identification card or license to carry firearms, or both, and shall be punished by a fine of not less than $200 nor more than $1,000 for a first offense and by a fine of not less than $1,000 nor more than $5,000 for a second offense.

The provisions of this section shall not apply to the following exempted persons and uses:

(a) Any device used exclusively for signaling or distress use and required or recommended by the United States Coast Guard or the Interstate Commerce Commission, or for the firing of stud cartridges, explosive rivets or similar industrial ammunition;

(b) Federally licensed firearms manufacturers or wholesale dealers, or persons employed by them or by licensed dealers, or on their behalf, when possession of firearms, rifles or shotguns is necessary for manufacture, display, storage, transport, installation, inspection or testing;

(c) To a person voluntarily surrendering a firearm, rifle or shotgun and ammunition therefor to a licensing authority, the colonel of the state police or his designee if prior written notice has been given by said person to the licensing authority or the colonel of the state police, stating the place and approximate time of said surrender;

(d) The regular and ordinary transport of firearms, rifles or shotguns as merchandise by any common carrier;

(e) Possession by retail customers for the purpose of firing at duly licensed target concessions at amusement parks, piers and similar locations, provided that the firearms, rifles or shotguns to be so used are firmly chained or affixed to the counter and that the proprietor is in possession of a firearm identification card or license to carry firearms;

(f) Possession of rifles and shotguns and ammunition therefor by nonresident hunters with valid nonresident hunting licenses during hunting season;

(g) Possession of rifles and shotguns and ammunition therefor by residents or nonresidents while on a firing or shooting range;

(h) Possession of rifles and shotguns and ammunition therefor by nonresidents traveling in or through the commonwealth, providing that any rifles or shotguns are unloaded and enclosed in a case;

(i) Possession of rifles and shotguns by nonresidents while at a firearm showing or display organized by a regularly existing gun collectors' club or association;

(j) Any new resident moving into the commonwealth, any resident of the commonwealth returning after having been absent from the commonwealth for not less than 180 consecutive days or any resident of the commonwealth upon being released from active service with any of the armed services of the United States with respect to any firearm, rifle or shotgun and any ammunition therefor then in his possession, for 60 days after such release, return or entry into the commonwealth;

(k) Any person under the age of fifteen with respect to the use of a rifle or shotgun by such person in hunting or target shooting, provided that such use is otherwise permitted by law and is under the immediate supervision of a person holding a firearm identification card or a license to carry firearms, or a duly commissioned officer, noncommissioned officer or enlisted member of the United States Army, Navy, Marine Corps, Air Force or Coast Guard, or the National Guard or military service of the commonwealth or reserve components thereof, while in the performance of his duty;

(l) The possession or utilization of any rifle or shotgun during the course of any television, movie, stage or other similar theatrical production, or by a professional photographer or writer for examination purposes in the pursuit of his profession, providing such possession or utilization is under the immediate supervision of a holder of a firearm identification card or a license to carry firearms;

(m) The temporary holding, handling or firing of a firearm for examination, trial or instruction in the presence of a holder of a license to carry firearms, or the temporary holding, handling or firing of a rifle or shotgun for examination, trial or instruction in the presence of a holder of a firearm identification card, or where such holding, handling or firing is for a lawful purpose;

(n) The transfer of a firearm, rifle or shotgun upon the death of any owner to his heir or legatee shall be subject to the provisions of this section, provided that said heir or legatee shall within one hundred and eighty days of such transfer, obtain a firearm identification card or a license to carry firearms if not otherwise an exempt person who is qualified to receive such or apply to the licensing authority for such further limited period as may be necessary for the disposition of such firearm, rifle or shotgun;

(o) Persons in the military or other service of any state or of the United States, and police officers and other peace officers of any jurisdiction, in the performance of their official duty or when duly authorized to possess them;

(p) Carrying or possession by nonresidents of so-called black powder rifles, shotguns, and ammunition therefor as described in such paragraphs (A) and (B) of the third paragraph of section 121, and the carrying or possession of conventional rifles, shotguns, and ammunition therefor by nonresidents who meet the requirements for such carrying or possession in the state in which they reside;

(q) [Deleted]

(r) Possession by a veteran's organization chartered by the Congress of the United States, chartered by the commonwealth or recognized as a nonprofit tax-exempt organization by the Internal Revenue Service and possession by the members of any such organization when on official parade duty or ceremonial occasions.

(s) Possession by federal, state and local historical societies, museums, and institutional collections open to the public, provided such firearms, rifles or shotguns are unloaded, properly housed and secured from unauthorized handling;

(t) the possession of firearms, rifles, shotguns, machine guns and ammunition, by banks or institutional lenders, or their agents, servants or employees, when the same are possessed as collateral for a secured commercial transaction or as a result of a default under a secured commercial transaction.

(u) Any nonresident who is eighteen years of age or older at the time of acquiring a rifle or shotgun from a licensed firearms dealer; provided, however, that such nonresident must hold a valid firearms license from his state of residence; provided, further, that the licensing requirements of such nonresident's state of residence are as stringent as the requirements of the commonwealth for a firearm identification card, as determined by the colonel of the state police who shall, annually, publish a list of those states whose requirements comply with the provisions of this clause.

Any person, exempted by clauses (o), (p) and (q), purchasing a rifle or shotgun or ammunition therefor shall submit to the seller such full and clear proof of identification, including shield number, serial number, military or governmental order or authorization, military or other official identification, other state firearms license, or proof of nonresidence, as may be applicable.

Nothing in this section shall permit the sale of rifles or shotguns or ammunition therefor to a minor under the age of eighteen in violation of section one hundred and thirty nor may any firearm be sold to a person under the age of 21 nor to any person who is not licensed to carry firearms under section one hundred and thirty-one unless he presents a valid firearm identification card and a permit to purchase issued under section one hundred and thirty-one A, or presents such permit to purchase and is a properly documented exempt person as herein before described.

Nothing in this section shall permit the sale or transfer of any large capacity rifle or shotgun or large capacity feeding device therefor to any person not in possession of a Class A or Class B license to carry firearms issued under section 131, or of any large capacity firearm or large capacity feeding device therefor to any person not in possession of a Class A license to carry firearms issued under section 131.

The possession of a firearm identification card issued under section one hundred and twenty-nine B shall not entitle any person to carry a firearm in violation of section ten of chapter two hundred and sixty-nine and, the possession of a firearm identification card issued under section 129B shall not entitle any person to possess any large capacity rifle or shotgun or large capacity feeding device therefor in violation of subsection (m) of said section 10 of said chapter 269.

Any person who, while not being within the limits of his own property or residence, or such person whose property or residence is under lawful search, and who is not exempt under this section, shall on demand of a police officer or other law enforcement officer, exhibit his license to carry firearms, or his firearm identification card or receipt for fee paid for such card, or, after January first, nineteen hundred and seventy, exhibit a valid hunting license issued to him which shall bear the number officially inscribed of such license to carry or card if any. Upon failure to do so such person may be required to surrender to such officer said firearm, rifle or shotgun which shall be taken into custody as under the provisions of section one hundred and twenty-nine D, except that such firearm, rifle or shotgun shall be returned forthwith upon presentation within thirty days of said license to carry firearms, firearm identification card or receipt for fee paid for such card or hunting license as herein before described. Any person subject to the conditions of this paragraph may, even though no firearm, rifle or shotgun was surrendered, be required to produce within thirty days said license to carry firearms, firearm identification card or receipt for fee paid for such card, or said hunting license, failing which the conditions of section one hundred and twenty-nine D will apply. Nothing in this section shall prevent any person from being prosecuted for any violation of this chapter.

129D. Surrender of Firearms upon Revocation, Suspension or Denial of Application for License or Firearm Identification Card; Rights of Owners; Disposition; Regulations. Upon revocation, suspension or denial of an application for a firearm identification card pursuant to the conditions of section one hundred and twenty-nine B, or of any firearms license if said firearms identification card is not then in force or of any machine gun license, the person whose application was so revoked, suspended or denied shall without delay deliver or surrender, to the licensing authority where he resides, all firearms, rifles, shotguns and machine guns and ammunition which he then possesses unless an appeal is pending. Such person, or his legal representative, shall have the right, at any time up to one year after said delivery or surrender, to transfer such firearms, rifles, shotguns and machine guns and ammunition to any licensed dealer or any other person legally permitted to purchase or take possession of such firearms, rifles, shotguns and machine guns and ammunition and upon notification in writing by the purchaser or transferee and the former owner, the licensing authority shall within ten days deliver such firearms, rifles, shotguns and machine guns and ammunition to the transferee or purchaser and due care shall be observed by the licensing authority in the receipt and holding of any such firearm, rifle, shotgun or machine gun and ammunition.

The licensing authority, after taking possession of any firearm, rifle, shotgun, machine gun or ammunition by any means, may transfer possession of such weapon for storage purposes to a federally and state licensed dealer of such weapons and ammunition who operates a bonded warehouse on the licensed premises that is equipped with a safe for the secure storage of firearms and a weapon box or similar container for the secure storage of other weapons and ammunition; provided, however, that the licensing authority shall not transfer to such dealer possession of any weapon that is or may be evidence in any current or pending criminal case concerning a violation of any general or special law, rule or regulation governing the use, possession or ownership of such weapon. Any such dealer that takes possession of a weapon under the provisions of this section shall: (i) inspect such weapon; (ii) issue to the owner a receipt indicating the make, model, caliber, serial number and condition of each weapon so received; and (iii) store and maintain all weapons so received in accordance with such regulations, rules or guidelines as the secretary of the executive office of public safety may establish under this section. The owner shall be liable to such dealer for reasonable storage charges and may dispose of any such weapon as provided under this section by transfer to a person lawfully permitted to purchase or take possession of such weapon.

Firearms, rifles, shotguns or machine guns and ammunition not disposed of after delivery or surrender according to the provisions of this section shall be sold at public auction by the colonel of the state police to the highest bidding person legally permitted to purchase and possess said firearms, rifles, shotguns or machine guns and ammunition and the proceeds shall be remitted to the state treasurer. Any such weapon that is stored and maintained by a licensed dealer as provided under this section may be so auctioned at the direction of: (i) the licensing authority at the expiration of one year following initial surrender or delivery to such licensing authority; or (ii) the dealer then in possession, if the storage charges for such weapon have been in arrears for 90 days; provided, however, that in either case, title shall pass to the licensed dealer for the purpose of transferring ownership to the auctioneer; and provided further, that in either case, after deduction and payment for storage charges and all necessary costs associated with such surrender and transfer, all surplus proceeds, if any, shall be immediately returned to the owner of such weapon.

The secretary of the executive office of public safety may make and promulgate such rules and regulations as are necessary to carry out the provisions of this section.

130. Penalty for Selling Firearms to Minor or Alien. Whoever sells or furnishes a rifle, shotgun or ammunition to any alien eighteen years of age or older who does not hold a permit card issued to him under section one hundred and thirty-one H or, except as provided in this section or section one hundred and thirty-one E, whoever sells or furnishes any alien or any person under eighteen years of age a rifle, shotgun, machine gun or ammunition, or whoever sells or furnishes to any person under 21 years of age a firearm or large capacity rifle or shotgun or ammunition therefor shall have his license to sell firearms, rifles, shotguns, machine guns and or ammunition revoked and shall not be entitled to apply for such license for ten years from the date of such revocation and shall be punished by a fine of not less than $1,000 nor more than $10,000, or by imprisonment in a state prison for not more than ten years or by imprisonment in a house of correction for not more than two and one-half years, or by both such fine and imprisonment. Nothing in this section or section one hundred and thirty-one E shall be construed to prohibit a parent or guardian from allowing his child or ward, who has not attained age fifteen, the supervised use of a rifle or shotgun or ammunition therefor, according to the provisions of section one hundred and twenty-nine C, nor from furnishing his child or ward, who has attained age fifteen, with a rifle or shotgun that is not a large capacity weapon or ammunition; provided, however, that said child or ward, being fifteen years of age or older, has been issued a valid firearm identification card or alien permit to possess a rifle or shotgun which is in his possession. Nothing in this section shall be construed to prohibit an instructor from furnishing rifles or shotguns or ammunition therefor to pupils; provided, however, that said instructor has the consent of a parent or guardian of a pupil under the age of eighteen years.

130B. Firearm Licensing Review Board

(a) There shall be a firearm licensing review board, established within the criminal history systems board, in this section called the board, comprised of 7 members, 1 of whom shall be a member of the criminal history systems board appointed by the executive director and who shall be the chair, 1 of whom shall be the secretary of public safety or his designee, 1 of whom shall be the colonel of state police or his designee, 1 of whom shall be appointed by the Massachusetts Chiefs of Police Association, 1 of whom shall be the attorney general or his designee, 1 whom shall be an attorney with litigation experience in firearm licensing cases and appointed by the governor from a list of qualified persons submitted to the governor by the Massachusetts Bar Association, and 1 of whom shall be a retired member of the judiciary and appointed by the governor.

(b) An applicant for a firearm identification card or license to carry who has been convicted of or adjudicated a delinquent child or youthful offender by reason of an offense or offenses punishable by 2 1/2 years imprisonment or less when committed under the laws of the commonwealth which was not: (a) an assault or battery on a family member or household member, as defined by section 1 of chapter 209A, except that the determination to be made under clause (e) of said section 1 of said chapter 209A shall be made by the review board, may, after the passage of 5 years from conviction, adjudication as a youthful offender or a delinquent child or release from confinement, commitment, probation or parole supervision for such conviction or adjudication, whichever is last occurring, file a petition for review of eligibility with the firearm licensing review board.

(c) The petitioner shall provide to the board a copy of a completed firearm identification card or license to carry application, which application shall have previously been submitted to the licensing authority or be submitted to the licensing authority contemporaneously with the petition filed with the board. The petitioner shall have the burden to prove his suitability to receive a firearm identification card or a license to carry by clear and convincing evidence. The board shall set a reasonable filing fee to file the petition.

(d) If the board determines, by 2/3rds vote, that: (i) the sole disqualifier for the petitioner is any conviction or adjudication as a youthful offender or a delinquent child for an offense or offenses punishable by 2 1/2 years imprisonment or less when committed under the laws of the commonwealth, arising out of a single incident and which does not otherwise disqualify the petitioner under subclauses (a), (d) or (e) of clause (i) or clauses (ii) to (ix), inclusive, of paragraph (1) of section 129B or subclauses (a), (d) or (e) of clause (i) or clauses (ii) to (vii), inclusive, of paragraph (d) of section 131, and which was not an assault or battery on a family member or household members, as defined by section 1 of chapter 209A, except that the determination to be made under clause (e) of said section 1 of said chapter 209A shall be made by the board; (ii) 5 years has passed since such conviction or adjudication or release from confinement, commitment, probation or parole supervision for such conviction or adjudication, whichever is last occurring; and (iii) by clear and convincing evidence, that the petitioner is a suitable person to be a firearm identification card or license to carry holder, the board shall determine that the petitioner's right or ability to possess a firearm is fully restored in the commonwealth with respect to such conviction or adjudication and that such conviction or adjudication shall not prohibit such petitioner from applying to a licensing authority for a firearm identification card or license to carry. The board shall make a determination on a petition within 60 days after receipt of the petition.

(e) The board shall hold hearings at such times and places as in its discretion it reasonably determines to be required, but not less than once every 90 days, and shall give reasonable notice of the time and place of the hearing to the petitioner. The board shall have the power to compel attendance of witnesses at hearings.

(f) All hearings shall be conducted in an informal manner, but otherwise according to the rules of evidence, and all witnesses shall be sworn by the chair. If requested by the petitioner and payment for stenographic services, as determined by the board, accompanies such request, the board shall cause a verbatim transcript of the hearing to be made. The board's decisions and findings of facts therefore shall be communicated in writing to the petitioner and to the licensing authority to whom the petitioner has applied or intends to apply within 20 days of rendering a decision.

(g) Members of the board shall serve without compensation, but shall be entitled to reasonable subsistence and travel allowances in the performance of their duties.

131. License to Carry Firearm; Penalty for Violation. All licenses to carry firearms shall be designated Class A or Class B, and the issuance and possession of any such license shall be subject to the following conditions and restrictions:

(a) A Class A license shall entitle a holder thereof to purchase, rent, lease, borrow, possess and carry: (i) firearms, including large capacity firearms, and feeding devices and ammunition therefor, for all lawful purposes, subject to such restrictions relative to the possession, use or carrying of firearms as the licensing authority deems proper; and (ii) rifles and shotguns, including large capacity weapons, and feeding devices and ammunition therefor, for all lawful purposes; provided, however, that the licensing authority may impose such restrictions relative to the possession, use or carrying of large capacity rifles and shotguns as it deems proper. A violation of a restriction imposed by the licensing authority under the provisions of this paragraph shall be cause for suspension or revocation and shall, unless otherwise provided, be punished by a fine of not less than $1,000 nor more than $10,000; provided, however, that the provisions of section 10 of chapter 269 shall not apply to such violation.

The colonel of state police may, after an investigation, grant a Class A license to a club or facility with an on-site shooting range or gallery, which club is incorporated under the laws of the commonwealth for the possession, storage and use of large capacity weapons, ammunition therefor and large capacity feeding devices for use with such weapons on the premises of such club; provided, however, that not less than one shareholder of such club shall be qualified and suitable to be issued such license; and provided further, that such large capacity weapons and ammunition feeding devices may be used under such Class A club license only by such members that possess a valid firearm identification card issued under section 129B or a valid Class A or Class B license to carry firearms, or by such other persons that the club permits while under the direct supervision of a certified firearms safety instructor or club member who, in the case of a large capacity firearm, possesses a valid Class A license to carry firearms or, in the case of a large capacity rifle or shotgun, possesses a valid Class A or Class B license to carry firearms. Such club shall not permit shooting at targets that depict human figures, human effigies, human silhouettes or any human images thereof, except by public safety personnel performing in line with their official duties.

No large capacity weapon or large capacity feeding device shall be removed from the premises except for the purposes of: (i) transferring such firearm or feeding device to a licensed dealer; (ii) transporting such firearm or feeding device to a licensed gunsmith for repair; (iii) target, trap or skeet shooting on the premises of another club incorporated under the laws of the commonwealth and for transporting thereto; (iv) attending an exhibition or educational project or event that is sponsored by, conducted under the supervision of or approved by a public law enforcement agency or a nationally or state recognized entity that promotes proficiency in or education about semiautomatic weapons and for transporting thereto and therefrom; (v) hunting in accordance with the provisions of chapter 131; or (vi) surrendering such firearm or feeding device under the provisions of section 129D. Any large capacity weapon or large capacity feeding device kept on the premises of a lawfully incorporated shooting club shall, when not in use, be secured in a locked container, and shall be unloaded during any lawful transport. The clerk or other corporate officer of such club shall annually file a report with the colonel of state police and the executive director of the criminal history systems board listing all large capacity weapons and large capacity feeding devices owned or possessed under such license. The colonel of state police or his designee, shall have the right to inspect all firearms owned or possessed by such club upon request during regular business hours and said colonel may revoke or suspend a club license for a violation of any provision of this chapter or chapter 269 relative to the ownership, use or possession of large capacity weapons or large capacity feeding devices.

(b) A Class B license shall entitle a holder thereof to purchase, rent, lease, borrow, possess and carry: (i) non-large capacity firearms and feeding devices and ammunition therefor, for all lawful purposes, subject to such restrictions relative to the possession, use or carrying of such firearm as the licensing authority deems proper; provided, however, that a Class B license shall not entitle the holder thereof to carry or possess a loaded firearm in a concealed manner in any public way or place; and provided further, that a Class B license shall not entitle the holder thereof to possess a large capacity firearm, except under a Class A club license issued under this section or under the direct supervision of a holder of a valid Class A license at an incorporated shooting club or licensed shooting range; and (ii) rifles and shotguns, including large capacity rifles and shotguns, and feeding devices and ammunition therefor, for all lawful purposes; provided, however, that the licensing authority may impose such restrictions relative to the possession, use or carrying of large capacity rifles and shotguns as he deems proper. A violation of a restriction provided under this paragraph, or a restriction imposed by the licensing authority under the provisions of this paragraph, shall be cause for suspension or revocation and shall, unless otherwise provided, be punished by a fine of not less than $1,000 nor more than $10,000; provided, however, that the provisions of section 10 of chapter 269 shall not apply to such violation.

A Class B license shall not be a valid license for the purpose of complying with any provision under this chapter governing the purchase, sale, lease, rental or transfer of any weapon or ammunition feeding device if such weapon is a large capacity firearm or if such ammunition feeding device is a large capacity feeding device for use with a large capacity firearm, both as defined in section 121.

(c) Either a Class A or Class B license shall be valid for the purpose of owning, possessing, purchasing and transferring non-large capacity rifles and shotguns, and for purchasing and possessing chemical mace, pepper spray or other similarly propelled liquid, gas or powder designed to temporarily incapacitate, consistent with the entitlements conferred by a firearm identification card issued under section 129B.

(d) Any person residing or having a place of business within the jurisdiction of the licensing authority or any law enforcement officer employed by the licensing authority or any person residing in an area of exclusive federal jurisdiction located within a city or town may submit to such licensing authority or the colonel of state police, an application for a Class A or Class B license to carry firearms, or renewal of the same, which such licensing authority or said colonel may issue if it appears that the applicant is a suitable person to be issued such license, and that the applicant has good reason to fear injury to his person or property, or for any other reason, including the carrying of firearms for use in sport or target practice only, subject to such restrictions expressed or authorized under this section, unless the applicant:

(i) has, in any state or federal jurisdiction, been convicted or adjudicated a youthful offender or delinquent child for the commission of (a) a felony; (b) a misdemeanor punishable by imprisonment for more than two years; (c) a violent crime as defined in section 121; (d) a violation of any law regulating the use, possession, ownership, transfer, purchase, sale, lease, rental, receipt or transportation of weapons or ammunition for which a term of imprisonment may be imposed; or (e) a violation of any law regulating the use, possession or sale of controlled substances as defined in section 1 of chapter 94C.

(ii) has been confined to any hospital or institution for mental illness, unless the applicant submits with his application an affidavit of a registered physician attesting that such physician is familiar with the applicant's mental illness and that in such physician's opinion the applicant is not disabled by such an illness in a manner that should prevent such applicant from possessing a firearm;

(iii) is or has been under treatment for or confinement for drug addiction or habitual drunkenness, unless such applicant is deemed to be cured of such condition by a licensed physician, and such applicant may make application for such license after the expiration of five years from the date of such confinement or treatment and upon presentment of an affidavit issued by such physician stating that such physician knows the applicant's history of treatment and that in such physician's opinion the applicant is deemed cured;

(iv) is at the time of the application less than 21 years of age;

(v) is an alien;

(vi) is currently subject to: (A) an order for suspension or surrender issued pursuant to section 3B or 3C of chapter 209A or a similar order issued by another jurisdiction; or (B) a permanent or temporary protection order issued pursuant to chapter 209A or a similar order issued by another jurisdiction; or

(vii) is currently the subject of an outstanding arrest warrant in any state or federal jurisdiction.

(e) Within seven days of the receipt of a completed application for a license to carry or possess firearms, or renewal of same, the licensing authority shall forward one copy of the application and one copy of the applicant's fingerprints to the colonel of state police, who shall within 30 days advise the licensing authority, in writing, of any disqualifying criminal record of the applicant arising from within or without the commonwealth and whether there is reason to believe that the applicant is disqualified for any of the foregoing reasons from possessing a license to carry or possess firearms. In searching for any disqualifying history of the applicant, the colonel shall utilize, or cause to be utilized, files maintained by the department of probation and statewide and nationwide criminal justice, warrant and protection order information systems and files including, but not limited to, the National Instant Criminal Background Check System. The colonel shall inquire of the commissioner of the department of mental health relative to whether the applicant is disqualified from being so licensed. If the information available to the colonel does not indicate that the possession of a firearm or large capacity firearm by the applicant would be in violation of state or federal law, he shall certify such fact, in writing, to the licensing authority within said 30 day period.

The licensing authority may also make inquiries concerning the applicant to: (i) the executive director of the criminal history systems board relative to any disqualifying condition and records of purchases, sales, rentals, leases and transfers of weapons or ammunition concerning the applicant; (ii) the commissioner of probation relative to any record contained within the department of probation or the statewide domestic violence record keeping system concerning the applicant; and (iii) the commissioner of the department of mental health relative to whether the applicant is a suitable person to possess firearms or is not a suitable person to possess firearms. The director or commissioner to whom the licensing authority makes such inquiry shall provide prompt and full cooperation for that purpose in any investigation of the applicant.

The licensing authority shall, within 40 days from the date of application, either approve the application and issue the license or deny the application and notify the applicant of the reason for such denial in writing; provided, however, that no such license shall be issued unless the colonel has certified, in writing, that the information available to him does not indicate that the possession of a firearm or large capacity firearm by the applicant would be in violation of state or federal law.

(f) A license issued under this section shall be revoked or suspended by the licensing authority, or his designee, upon the occurrence of any event that would have disqualified the holder from being issued such license or from having such license renewed. A license may be revoked or suspended by the licensing authority if it appears that the holder is no longer a suitable person to possess such license. Any revocation or suspension of a license shall be in writing and shall state the reasons therefor. Upon revocation or suspension, the licensing authority shall take possession of such license and the person whose license is so revoked or suspended shall take all actions required under the provisions of section 129D. No appeal or post-judgment motion shall operate to stay such revocation or suspension. Notices of revocation and suspension shall be forwarded to the executive director of the criminal history systems board and the commissioner of probation and shall be included in the criminal justice information system. A revoked or suspended license may be reinstated only upon the termination of all disqualifying conditions, if any.

Any applicant or holder aggrieved by a denial, revocation or suspension of a license, unless a hearing has previously been held pursuant to chapter 209A, may, within either 90 days after receiving notice of such denial, revocation or suspension or within 90 days after the expiration of the time limit during which the licensing authority is required to respond to the applicant, file a petition to obtain judicial review in the district court having jurisdiction in the city or town wherein the applicant filed for, or was issued, such license. A justice of such court, after a hearing, may direct that a license be issued or reinstated to the petitioner if such justice finds that there was no reasonable ground for denying, suspending or revoking such license and that the petitioner is not prohibited by law from possessing same.

(g) A license shall be in a standard form provided by the executive director of the criminal history systems board in a size and shape equivalent to that of a license to operate motor vehicles issued by the registry of motor vehicles pursuant to section 8 of chapter 90 and shall contain a license number which shall clearly indicate whether such number identifies a Class A or Class B license, the name, address, photograph, fingerprint, place and date of birth, height, weight, hair color, eye color and signature of the licensee. Such license shall be marked "License to Carry Firearms" and shall clearly indicate whether the license is Class A or Class B. The application for such license shall be made in a standard form provided by the executive director of the criminal history systems board, which form shall require the applicant to affirmatively state under the pains and penalties of perjury that such applicant is not disqualified on any of the grounds enumerated above from being issued such license.

(h) Any person who knowingly files an application containing false information shall be punished by a fine of not less than $500 nor more than $1,000 or by imprisonment for not less than six months nor more than two years in a house of correction, or by both such fine and imprisonment.

(i) A license to carry or possess firearms shall be valid, unless revoked or suspended, for a period of not more than 6 years from the date of issue and shall expire on the anniversary of the licensee's date of birth occurring not less than 5 years but not more than 6 years from the date of issue, except that if the licensee applied for renewal before the license expired, the license shall remain valid for a period of 90 days beyond the stated expiration date on the license, unless the application for renewal is denied. Any renewal thereof shall expire on the anniversary of the licensee's date of birth occurring not less than 5 years but not more than 6 years from the effective date of such license. Any license issued to an applicant born on February 29 shall expire on March 1. The fee for the application shall be $100, which shall be payable to the licensing authority and shall not be prorated or refunded in case of revocation or denial. The licensing authority shall retain $25 of the fee; $50 of the fee shall be deposited into the general fund of the commonwealth and not less than $50,000 of the funds deposited into the General Fund shall be allocated to the Firearm Licensing Review Board, established in section 130B, for its operations and that any funds not expended by said board for its operations shall revert back to the General Fund; and $25 of the fee shall be deposited in the Firearms Fingerprint Identity Verification Trust Fund. For law enforcement officials, or local, state, or federal government entities acting on their behalf, the fee for the application shall be set at $25, which shall be payable to the licensing authority and shall not be prorated or refunded in case of revocation or denial. The licensing authority shall retain $12.50 of the fee, and $12.50 of the fee shall be deposited into the general fund of the commonwealth. Notwithstanding any general or special law to the contrary, licensing authorities shall deposit such portion of the license application fee into the Firearms Record Keeping Fund quarterly, not later than January 1, April 1, July 1 and October 1 of each year. Notwithstanding any general or special law to the contrary, licensing authorities shall deposit quarterly such portion of the license application fee as is to be deposited into the General Fund, not later than January 1, April 1, July 1 and October 1 of each year. For the purposes of section 10 of chapter 269, an expired license to carry firearms shall be deemed to be valid for a period not to exceed 90 days beyond the stated date of expiration, unless such license to carry firearms has been revoked.

Any person over the age of 70 and any law enforcement officer applying for a license to carry firearms through his employing agency shall be exempt from the requirement of paying a renewal fee for a Class A or Class B license to carry.

(j)(1) No license shall be required for the carrying or possession of a firearm known as a detonator and commonly used on vehicles as a signaling and marking device, when carried or possessed for such signaling or marking purposes.

(2) No license to carry shall be required for the possession of an unloaded large capacity rifle or shotgun or an unloaded feeding device therefor by a veteran's organization chartered by the Congress of the United States, chartered by the commonwealth or recognized as a nonprofit tax-exempt organization by the Internal Revenue Service, or by the members of any such organization when on official parade duty or during ceremonial occasions. For purposes of this subparagraph, an "unloaded large capacity rifle or shotgun" and an "unloaded feeding device therefor" shall include any large capacity rifle, shotgun or feeding device therefor loaded with a blank cartridge or blank cartridges, so-called, which contain no projectile within such blank or blanks or within the bore or chamber of such large capacity rifle or shotgun.

(k) Whoever knowingly issues a license in violation of this section shall be punished by a fine of not less than $500 nor more than $1,000 or by imprisonment for not less than six months nor more than two years in a jail or house of correction, or by both such fine and imprisonment.

(l) The executive director of the criminal history systems board shall send by first class mail to the holder of each such license to carry firearms, a notice of the expiration of such license not less than 90 days prior to such expiration and shall enclose therein a form for the renewal of such license. The taking of fingerprints shall not be required in issuing the renewal of a license if the renewal applicant's fingerprints are on file with the department of the state police. Any licensee shall notify, in writing, the licensing authority who issued said license, the chief of police into whose jurisdiction the licensee moves and the executive director of the criminal history systems board of any change of address. Such notification shall be made by certified mail within 30 days of its occurrence. Failure to so notify shall be cause for revocation or suspension of said license.

(m) Notwithstanding the provisions of section 10 of chapter 269, any person in possession of a fireatm, rifle or shotgun whose license issued under this section is invalid for the sole reason that it has expired, meaning after 90 days beyond the stated expiration date on the license, but who shall not be disqualified from renewal upon application therefor under this section, shall be subject to a civil fine of not less than $500 nor more than $5,000 and the provisions of section 10 of chapter 269 shall not apply; provided, however, that the exemption from the provisions of said section 10 of said chapter 269 provided herein shall not apply if: (i) such license has been revoked or suspended, unless such revocation or suspension was caused by failure to give notice of a change of address as required under this section; (ii) revocation or suspension of such license is pending, unless such revocation or suspension was caused by failure to give notice of a change of address as required under this section; or (iii) an application for renewal of such license has been denied. Any law enforcement officer who discovers a person to be in possession of a firearm, rifle or shotgun after such person's license has expired, meaning after 90 days beyond the stated expiration date on the license, has been revoked or suspended, solely for failure to give notice of a change of address, shall confiscate such firearm, rifle or shotgun and the expired or suspended license then in possession and such officer, [sic] shall forward such license to the licensing authority by whom it was issued as soon as practicable. The officer shall, at the time of confiscation, provide to the person whose firearm, rifle or shotgun has been confiscated, a written inventory and receipt for all firearms, rifles or shotguns confiscated and the officer and his employer shall exercise due care in the handling, holding and storage of these items. Any confiscated weapon shall be returned to the owner upon the renewal or reinstatement of such expired or suspended license within one year of such confiscation or may be otherwise disposed of in accordance with the provisions of section 129D. The provisions of this paragraph shall not apply if such person has a valid license to carry firearms issued under section 131F.

(n) Upon issuance of a license to carry or possess firearms under this section, the licensing authority shall forward a copy of such approved application and license to the executive director of the criminal history systems board, who shall inform the licensing authority forthwith of the existence of any disqualifying condition discovered or occurring subsequent to the issuance of a license under this section.

(o) No person shall be issued a license to carry or possess a machine gun in the commonwealth, except that a licensing authority or the colonel of state police may issue a machine gun license to:

(i) a firearm instructor certified by the municipal police training committee for the sole purpose of firearm instruction to police personnel;

(ii) a bona fide collector of firearms upon application or upon application for renewal of such license.

(p) The executive director of the criminal history systems board shall promulgate regulations in accordance with chapter 30A to establish criteria for persons who shall be classified as bona fide collectors of firearms.

(q) Nothing in this section shall authorize the purchase, possession or transfer of any weapon, ammunition or feeding device that is, or in such manner that is, prohibited by state or federal law.

(r) The secretary of the executive office of public safety or his designee may promulgate regulations to carry out the purposes of this section.

131¾. Roster of Large Capacity Rifles, Shotguns, Firearms, and Feeding Devices; Publication; Petitions to Amend. The secretary of public safety shall, with the advice of the gun control advisory board established pursuant to the provisions of section 131½, compile and publish a roster of large capacity rifles, shotguns, firearms and feeding devices, all as defined in section 121, and such weapons referred to in clauses Eighteenth to Twenty-first, inclusive, of section 123.

The secretary shall, not less than three times annually, publish the roster in newspapers of general circulation throughout the commonwealth, and shall send a copy thereof to all dealers licensed in the commonwealth under the provisions of said section 122 of said chapter 140; and further, the licensing authority shall furnish said roster to all cardholders and licensees upon initial issuance and upon every renewal of the same.

The secretary may amend the roster upon his own initiative or with the advice of said board. A person may petition the secretary to place a weapon on, or remove a weapon from, the roster, subject to the provisions of this section. A person who so petitions shall give the reasons why the roster should be so amended.

A petition to amend the roster shall be submitted in writing to the secretary and shall be in the form and manner prescribed by the secretary. Upon receipt of the petition to place a weapon on the roster, the secretary shall, within 45 days of receipt of the petition, either notify the petitioner by certified mail that the petition is denied, or it shall modify the roster. An addition to the roster shall be effective on the date it is included in the next publication in newspapers of general circulation as provided under this section.

The secretary may promulgate rules and regulations relative to the appeal of a decision on a petition to modify the roster and any other regulations consistent with the provisions of this section and section 2SS of chapter 29, sections 11 and 14 of chapter 131, sections 121, 122, 122B, 123, 128, 128A, 128B, 129B, 129C, 129D, 130, 131, 131A, 131E, 131F and 131K of chapter 140 to effectuate the purposes of each said section.

131A. Permits to Purchase, Rent or Lease Firearms; Penalty; Fee. A licensing authority under section one hundred and thirty-one, upon the application of a person qualified to be granted a license thereunder by such authority, may grant to such a person, other than a minor, a permit to purchase, rent or lease a firearm if it appears that such purchase, rental or lease is for a proper purpose, and may revoke such permit at will. The colonel of the state police or a person authorized by him, upon the application of a person licensed under section one hundred and thirty-one F, may grant to such licensee, other than a minor, a permit to purchase, rent or lease a firearm, rifle or shotgun, or to purchase ammunition therefor, if it appears that such purchase, rental or lease is for a proper purpose, and may revoke such permit at will. Such permits shall be issued on forms furnished by the executive director of the criminal history systems board, shall be valid for not more than ten days after issue, and a copy of every such permit so issued shall within one week thereafter be sent to the said executive director. The licensing authority may impose such restrictions relative to the caliber and capacity of the firearm to be purchased, rented or leased as he deems proper. Whoever knowingly issues a permit in violation of this section shall be punished by a fine of not less than five hundred nor more than one thousand dollars and by imprisonment for not less than six months nor more than two years in a jail or house of correction.

The fee for the permits shall be $100, which shall be payable to the licensing authority and shall not be prorated or refunded in case of revocation or denial. The licensing authority shall retain $25 of the fee; $50 of the fee shall be deposited into the general fund of the commonwealth; and $25 of the fee shall be deposited in the Firearms Fingerprint Identity Verification Trust Fund.

131B. Penalty for Loan of Money on Firearm, Rifle, Shotgun or Machine Gun; Exceptions. Whoever loans money secured by mortgage, deposit or pledge of a firearm, rifle, shotgun or machine gun shall be punished by a fine of not more than five hundred dollars or by imprisonment for not more than one year, or by both; provided, however that nothing herein shall prohibit a bank or other institutional lender from loaning money secured by a mortgage, deposit, or pledge of a firearm, rifle, shotgun or machine gun to a manufacturer, wholesaler, or dealer of firearms, rifles, or shotguns. The provisions of section one hundred and twenty-three shall not be applicable to any such mortgage, deposit or pledge unless or until the lender takes possession of the collateral upon default or the collateral is removed from the premises of the debtor.

131C. Carrying Firearm in Vehicle; Penalty.

(a) No person carrying a loaded firearm under a Class A license issued under section 131 or 131F shall carry the same in a vehicle unless such firearm while carried therein is under the direct control of such person. Whoever violates the provisions of this subsection shall be punished by a fine of $500.

(b) No person carrying a firearm under a Class B license issued under section 131 or 131F shall possess the same in a vehicle unless such weapon is unloaded and contained within the locked trunk of such vehicle or in a locked case or other secure container. Whoever violates the provisions of this subsection shall be punished by a fine of $500.

(c) No person possessing a large capacity rifle or shotgun under a Class A or Class B license issued under section 131 or 131F shall possess the same in a vehicle unless such weapon is unloaded and contained within the locked trunk of such vehicle or in a locked case or other secure container. Whoever violates the provisions of this subsection shall be punished by a fine of not less than $500 nor more than $5,000.

(d) The provisions of this section shall not apply to (i) any officer, agent or employee of the commonwealth or any state or the United States; (ii) any member of the military or other service of any state or of the United States; (iii) any duly authorized law enforcement officer, agent or employee of any municipality of the commonwealth; provided, however, that any such person described in clauses (i) to (iii), inclusive, is authorized by a competent authority to carry or possess the weapon so carried or possessed and is acting within the scope of his duties.

(e) A conviction of a violation of this section shall be reported forthwith by the court or magistrate to the licensing authority who shall immediately revoke the card or license of the person so convicted. No new such card or license may be issued to any such person until one year after the date of revocation.

131E. Purchase by Licensed Resident; Purchase for Use of Another Prohibited; Penalty. Any resident of the commonwealth may purchase firearms, rifles, shotguns and ammunition feeding devices from any dealer licensed under section 122, or from such person as shall be qualified under section 128A, or ammunition from a licensee under section 122B, subject to the following conditions and restrictions:

(a) rifles, shotguns and feeding devices therefor may be so purchased only upon presentment of: (i) a valid firearm identification card issued under section 129B; or (ii) a valid Class A or Class B license to carry firearms issued under section 131; or (iii) valid proof of exempt status under section 129C; provided, however, that large capacity rifles and shotguns and large capacity feeding devices therefor may be so purchased only upon presentment of a Class A or Class B license to carry firearms issued under said section 131; and provided further, that no rifle or shotgun or ammunition or ammunition feeding device therefor shall be sold to any person less than 18 years of age; and provided further, that no large capacity rifle or shotgun or large capacity feeding device therefor shall be sold to any person less than 21 years of age;

(b) firearms and feeding devices therefor may be so purchased only upon presentment of: (i) a valid Class A or Class B license to carry firearms issued under section 131; or (ii) a valid firearm identification card issued under section 129B together with a valid permit to purchase a firearm issued under section 131A; or (iii) a valid permit to purchase a firearm issued under section 131A together with valid proof of exempt status under section 129C; provided, however, that large capacity firearms and large capacity feeding devices therefor may be so purchased only upon presentment of: (i) a valid Class A license to carry firearms issued under section 131; or (ii) a valid firearm identification card issued under section 129B together with a valid and proper permit to purchase a firearm issued under section 131A; or (iii) a valid and proper permit to purchase a firearm issued under section 131A together with valid proof of exempt status under section 129C; and provided further, that neither a firearm identification card issued under section 129B, nor proof of exempt status under section 129C, shall be valid for the purpose of purchasing any firearm or ammunition feeding device therefor without being presented together with a valid and proper permit to purchase issued under section 131A; and provided further, that an alien permit to possess a rifle or shotgun shall not be valid for the purpose of purchasing firearms or ammunition or ammunition feeding devices therefor; and provided further, that no firearm or ammunition or ammunition feeding device therefor shall be sold to any person less than 21 years of age.

Any person who uses said license to carry firearms or firearm identification card for the purpose of purchasing a firearm, rifle or shotgun for the unlawful use of another, or for resale to or giving to an unlicensed person, shall be punished by a fine of not less than one thousand nor more than fifty thousand dollars, or by imprisonment for not less than two and one-half years nor more than ten years in a state prison, or by both such fine and imprisonment. A conviction of a violation of this section shall be reported forthwith by the court to the licensing authority which issued the license or firearm identification card, which shall immediately revoke the license or firearm identification card of such person. No new license or firearm identification card under section one hundred and twenty-nine B or section one hundred and thirty-one shall be issued to any such person within two years after the date of said revocation.

131F. Temporary License to Carry Firearms Issued to Non-Residents. A Class A or Class B temporary license to carry firearms or feeding devices or ammunition therefor, within the commonwealth, may be issued by the colonel of state police, or persons authorized by him, to a nonresident or any person not falling within the jurisdiction of a local licensing authority or to an alien that resides outside the commonwealth for purposes of firearms competition and subject to such terms and conditions as said colonel may deem proper; provided, however, that no license shall be issued to a person who:

(i) has, in any state or federal jurisdiction, been convicted or adjudicated a youthful offender or delinquent child for the commission of (a) a felony; (b) a misdemeanor punishable by imprisonment for more than two years; (c) a violent crime as defined in section 121; (d) a violation of any law regulating the use, possession, ownership, transfer, purchase, sale, lease, rental, receipt or transportation of weapons or ammunition for which a term of imprisonment may be imposed; or (e) a violation of any law regulating the use, possession or sale of controlled substances, as defined in section 1 of chapter 94C;

(ii) has been confined to any hospital or institution for mental illness, unless the applicant submits with his application an affidavit of a registered physician attesting that such physician is familiar with the applicant's mental illness and that in such physician's opinion the applicant is not disabled by such an illness in a manner that should prevent such applicant from possessing a firearm;

(iii) is or has been under treatment for or confinement for drug addiction or habitual drunkenness, unless such applicant is deemed to be cured of such condition by a licensed physician, and such applicant may make application for said license after the expiration of five years from the date of such confinement or treatment and upon presentment of an affidavit issued by such physician stating that such physician knows the applicant's history of treatment and that in such physician's opinion the applicant is deemed cured;

(iv) is currently subject to: (A) an order for suspension or surrender issued pursuant to section 3B or 3C of chapter 209A or a similar order issued by another jurisdiction; or (B) a permanent or temporary protection order issued pursuant to chapter 209A or a similar order issued by another jurisdiction; or

(v) is currently the subject of an outstanding arrest warrant in any state or federal jurisdiction.

Such license shall be valid for a period of one year but the colonel may renew such license, if in his discretion, such renewal is necessary.

The colonel may also issue such license, subject to such terms and conditions as he deems proper, to any resident of the commonwealth for the purposes of sports competition.

A temporary license issued under this section shall be marked "Temporary License to Carry Firearms", shall clearly indicate whether it is Class A or Class B and shall not be used to purchase firearms in the commonwealth as provided under section 131E. Neither a large capacity firearm nor large capacity feeding device therefor may be carried unless such person has been issued a Class A license; provided, however, that the colonel may permit a Class A or Class B licensee to possess large capacity rifles or shotguns or both, and such entitlement shall be clearly indicated on such license. The fee for an application for the license shall be $100, which shall be payable to the licensing authority and shall not be prorated or refunded in case of revocation or denial. The licensing authority shall retain $25 of the fee; $50 of the fee shall be deposited into the general fund of the commonwealth; and $25 of the fee shall be deposited in the Firearms Fingerprint Identity Verification Trust Fund. A license issued under the provisions of this section to a non-resident who is in the employ of a bank, public utility corporation, or a firm engaged in the business of transferring monies, or business of similar nature, or a firm licensed as a private detective under the provisions of chapter one hundred and forty-seven, and whose application is endorsed by his employer, or who is a member of the armed services and is stationed within the territorial boundaries of the commonwealth and has the written consent of his commanding officer, may be issued for any term not to exceed two years, and said licenses shall expire in accordance with the provisions of section one hundred and thirty-one.

A license, otherwise in accordance with provisions of this section, may be issued to a nonresident employee, whose application is endorsed by his employer, of a federally licensed Massachusetts manufacturer of machine guns to possess within the commonwealth a machine gun for the purpose of transporting or testing relative to the manufacture of machine guns, and the license shall be marked "temporary license to possess a machine gun" and may be issued for any term not to exceed two years and shall expire in accordance with the provisions of section one hundred and thirty-one.

131F½. Temporary License to Carry Firearms During Certain Television, Movie, or Stage Productions. Notwithstanding the provisions of subsection (a) of section ten of chapter two hundred and sixty-nine of the General Laws or any other law to the contrary, the carrying or possession of a firearm and blank ammunition therefor, during the course of any television, movie, stage or other similar theatrical production, by a person within such production, shall be authorized; provided, however, that such carrying or possession of such firearm shall be under the immediate supervision of a person licensed to carry firearms.

131G. Certain Nonresidents Authorized to Carry Firearms In or Through the Commonwealth. Any person who is not a resident of the commonwealth may carry a pistol or revolver in or through the commonwealth for the purpose of taking part in a pistol or revolver competition or attending any meeting or exhibition of any organized group of firearm collectors or for the purpose of hunting; provided, that such person is a resident of the United States and has a permit or license to carry firearms issued under the laws of any state, district or territory thereof which has licensing requirements which prohibit the issuance of permits or licenses to persons who have been convicted of a felony or who have been convicted of the unlawful use, possession or sale of narcotic or harmful drugs; provided, further, that in the case of a person traveling in or through the commonwealth for the purpose of hunting, he has on his person a hunting or sporting license issued by the commonwealth or by the state of his destination. Police officers and other peace officers of any state, territory or jurisdiction within the United States duly authorized to possess firearms by the laws thereof shall, for the purposes of this section, be deemed to have a permit or license to carry firearms as described in this section.

131H. Permit to Own or Possess Firearms by Aliens. No alien shall own or have in his possession or under his control a firearm except as pro-vided in section one hundred and thirty-one F or a rifle or shotgun except as provided in this section or section one hundred and thirty-one F. The colonel of the state police may, after an investigation, issue a permit to an alien to own or have in his possession or under his control a rifle or shotgun; subject to such terms and conditions as said colonel may deem proper. The fee for the permit shall be $100, which shall be payable to the licensing authority and shall not be prorated or refunded in case of revocation or denial. The licensing authority shall retain $25 of the fee; $50 of the fee shall be deposited into the general fund of the commonwealth; and $25 of the fee shall be deposited in the Firearms Fingerprint Identity Verification Trust Fund. Upon issuing such permit said colonel shall so notify, in writing, the chief of police or the board or officer having control of the police in the city or town in which such alien resides. Each such permit card shall expire at twelve midnight on December thirty-first next succeeding the effective date of said permit, and shall be revocable for cause by said colonel. In case of revocation, the fee for such permit shall not be prorated or refunded. Whenever any such permit is revoked, said colonel shall give notification as hereinbefore provided. The permit issued to an alien under this section shall be subject to sections one hundred and twenty-nine B and one hundred and twenty-nine C except as otherwise provided by this section.

Violation of any provision of this section shall be punished by a fine of not less than five hundred nor more than one thousand dollars, and by imprisonment for not more than six months in a jail or house of correction. If, in any prosecution for violation of this section, the defendant alleges that he has been naturalized, or alleges that he is a citizen of the United States, the burden of proving the same shall be upon him. Any firearm, rifle or shotgun owned by an alien or in his possession or under his control in violation of this section shall be forfeited to the commonwealth. Any such firearm, rifle or shotgun may be the subject of a search warrant as provided in chapter two hundred and seventy-six.

The director of law enforcement of the department of fisheries, wildlife and environmental law enforcement, deputy directors of enforcement, chiefs of enforcement, deputy chiefs of enforcement, environmental police officers and deputy environmental police officers, wardens as defined in section one of chapter one hundred and thirty-one, and members of the state police in areas over which they have jurisdiction, and all officers qualified to serve criminal process shall arrest, without a warrant, any person found with a firearm, rifle or shotgun in his possession if they have reason to believe that he is an alien and if he does not have in his possession a valid permit as provided in this section.

131I. Penalty for Making or Possessing Altered, Forged or Counterfeit License to Carry a Firearm or Firearm Identification Card. Whoever falsely makes, alters, forges or counterfeits or procures or assists another to falsely make, alter, forge or counterfeit a license to carry a firearm or a firearm identification card, or whoever forges or without authority uses the signature, facsimile of the signature, or validating signature stamp of the licensing authority or its designee, or whoever possesses, utters, publishes as true or in any way makes use of a falsely made, altered, forged or counterfeited license to carry a firearm or a firearm identification card, shall be punished by imprisonment in a state prison for not more than five years or in a jail or house of correction for not more than two years, or by a fine of not less than five hundred dollars, or both such fine and imprisonment.

131J. Sale or Possession of Electrical Weapons; Penalty. No person shall possess a portable device or weapon from which an electrical current, impulse, wave or beam may be directed, which current, impulse, wave or beam is designed to incapacitate temporarily, injure or kill, except: (1) a federal, state or municipal law enforcement officer, or member of a special reaction team in a state prison or designated special operations or tactical team in a county correctional facility, acting in the discharge of his official duties who has completed a training course approved by the secretary of public safety in the use of such a device or weapon designed to incapacitate temporarily; or (2) a supplier of such devices or weapons designed to incapacitate temporarily, if possession of the device or weapon is necessary to the supply or sale of the device or weapon within the scope of such sale or supply enterprise. No person shall sell or offer for sale such device or weapon, except to federal, state or municipal law enforcement agencies. A device or weapon sold under this section shall include a mechanism for tracking the number of times the device or weapon has been fired. The secretary of public safety shall adopt regulations governing who may sell or offer to sell such devices or weapons in the commonwealth and governing law enforcement training on the appropriate use of portable electrical weapons. Whoever violates this section shall be punished by a fine of not less than $500 nor more than $1,000 or by imprisonment in the house of correction for not less than 6 months nor more than 2-1/2 years, or by both such fine and imprisonment. A law enforcement officer may arrest without a warrant any person whom he has probable cause to believe has violated this section.

131K. Weapons Without Safety Device; Liability of Manufacturers, Importers and Sellers; Exceptions. Any firearm or large capacity weapon, both as defined in section 121, sold within the commonwealth without a safety device designed to prevent the discharge of such weapon by unauthorized users and approved by the colonel of state police including, but not limited to, mechanical locks or devices designed to recognize and authorize, or otherwise allow the firearm to be discharged only by its owner or authorized user, by solenoid use-limitation devices, key activated or combination trigger or handle locks, radio frequency tags, automated fingerprint identification systems or voice recognition, provided, that such device is commercially available, shall be defective and the sale of such a weapon shall constitute a breach of warranty under section 2-314 of chapter 106 and an unfair or deceptive trade act or practice under section 2 of chapter 93A. Any entity responsible for the manufacture, importation or sale as an inventory item or consumer good, both as defined in section 9-102 of chapter 106, of such a weapon that does not include or incorporate such a device shall be individually and jointly liable to any person who sustains personal injury or property damage resulting from the failure to include or incorporate such a device. If death results from such personal injury, such entities shall be liable in an amount including, but not limited to, that provided under chapter 229. Contributory or comparative negligence shall not be valid defenses to an action brought under this section in conjunction with section 2 of chapter 93A or section 2-314 of chapter 106 or both; provided, however, that nothing herein shall prohibit such liable parties from maintaining an action for indemnification or contribution against each other or against the lawful owner or other authorized user of said weapon. Any disclaimer, limit or waiver of the liability provided under this section shall be void.

No entity responsible for the manufacture, importation or sale of such a weapon shall be liable to any person for injuries caused by the discharge of such weapon that does not include or incorporate a safety device as required under this section if such injuries were: (i) self-inflicted, either intentionally or unintentionally, unless such injuries were self-inflicted by a person less than 18 years of age; (ii) inflicted by the lawful owner or other authorized user of said weapon; (iii) inflicted by any person in the lawful exercise of self-defense; or (iv) inflicted upon a co-conspirator in the commission of a crime.

This section shall not apply to any weapon distributed to an officer of any law enforcement agency or any member of the armed forces of the United States or the organized militia of the commonwealth; provided, however, that such person is authorized to acquire, possess or carry such a weapon for the lawful performance of his official duties; and provided further, that any such weapon so distributed is distributed solely for use in connection with such duties. This section shall not apply to any firearm manufactured in or prior to the year 1899, or to any replica of such a firearm if such replica is not designed or redesigned for using rimfire or conventional centerfire fixed ammunition.

131L. Storage of Firearms, Rifles or Shotguns; Requirements; Penalties for Violation.

(a) It shall be unlawful to store or keep any firearm, rifle or shotgun including, but not limited to, large capacity weapons, or machine gun in any place unless such weapon is secured in a locked container or equipped with a tamper-resistant mechanical lock or other safety device, properly engaged so as to render such weapon inoperable by any person other than the owner or other lawfully authorized user. For purposes of this section, such weapon shall not be deemed stored or kept if carried by or under the control of the owner or other lawfully authorized user.

(b) A violation of this section shall be punished, in the case of a firearm, rifle or shotgun that is not a large capacity weapon, by a fine of not less than $500 nor more than $5,000 or by imprisonment for not more than one year, or by both such fine and imprisonment, and in the case of a large capacity weapon or machine gun, by a fine of not less than $1,000 nor more than $10,000 or by imprisonment for not less than one year nor more than ten years, or by both such fine and imprisonment.

(c) A violation of this section shall be punished, in the case of a rifle or shotgun that is not a large capacity weapon and such weapon was stored or kept in a place where a person under the age of 18 who does not possess a valid firearm identification card issued under section 129B may have access without committing an unforeseeable trespass, by a fine of not less than $1,000 nor more than $10,000 or by imprisonment for not less than one year nor more than ten years, or by both such fine and imprisonment.

(d) A violation of this section shall be punished, in the case of a rifle or shotgun that is a large capacity weapon, firearm or machine gun was stored or kept in a place where a person under the age of 18 may have access, without committing an unforeseeable trespass, by a fine of not less than $5,000 nor more than $10,000 or by imprisonment for not less than two and one-half years, nor more than ten years, or by both such fine and imprisonment.

(e) A violation of the provisions of this section shall be evidence of wanton or reckless conduct in any criminal or civil proceeding if a person under the age of 18 who was not a trespasser or was a foreseeable trespasser acquired access to a weapon, unless such person possessed a valid firearm identification card issued under section 129B and was permitted by law to possess such weapon, and such access results in the personal injury to or the death of any person.

(f) This section shall not apply to the storage or keeping of any firearm, rifle or shotgun with matchlock, flintlock, percussion cap or similar type of ignition system manufactured in or prior to the year 1899, or to any replica of any such firearm, rifle or shotgun if such replica is not designed or redesigned for using rimfire or conventional centerfire fixed ammunition.

131M. Sale and Possession of Assault Weapons. No person shall sell, offer for sale, transfer or possess an assault weapon or a large capacity feeding device that was not otherwise lawfully possessed on September 13, 1994. Whoever not being licensed under the provisions of section 122 violates the provisions of this section shall be punished, for a first offense, by a fine of not less than $1,000 nor more than $10,000 or by imprisonment for not less than one year nor more than ten years, or by both such fine and imprisonment, and for a second offense, by a fine of not less than $5,000 nor more than $15,000 or by imprisonment for not less than five years nor more than 15 years, or by both such fine and imprisonment.

The provisions of this section shall not apply to: (i) the possession by a law enforcement officer for purposes of law enforcement; or (ii) the possession by an individual who is retired from service with a law enforcement agency and is not otherwise prohibited from receiving such a weapon or feeding device from such agency upon retirement.

131N. Covert Weapons. No person shall sell, offer for sale, transfer or possess any weapon, capable of discharging a bullet or shot, that is: (i) constructed in a shape that does not resemble a handgun, short-barreled rifle or short-barreled shotgun including, but not limited to, covert weapons that resemble keychains, pens, cigarette lighters or cigarette packages; or (ii) not detectable as a weapon or potential weapon by x-ray machines commonly used at airports or walk-through metal detectors. Whoever violates the provisions of this section shall be punished, for a first offense, by a fine of not less than $1,000 nor more than $10,000 or by imprisonment for not less than one year nor more than ten years, or by both such fine and imprisonment, and for a second offense, by a fine of not less than $5,000 nor more than $15,000 or by imprisonment for not less than five years nor more than 15 years, or by both such fine and imprisonment.

Chapter 148. Fire Prevention

35. Possession of Bombs and Explosives; Penalty. No person shall have in his possession or under his control any bomb or other high explosive, as defined by the rules and regulations made under section nine, contrary to the provisions of this chapter or of any rule or regulation made thereunder. Whoever violates this section shall be punished by a fine of not more than one thousand dollars, or by imprisonment for not more than two and one half years, or both, and any bomb or explosive found in his possession or under his control on such violation shall be forfeited to the commonwealth. Any officer qualified to serve criminal process may arrest without a warrant any person violating this section.

Chapter 209A. Abuse Prevention

3B. Suspension of Firearm License and Surrender of Firearms With Temporary or Emergency Order; Petition for Review. Upon issuance of a temporary or emergency order under section four or five of this chapter, the court shall, if the plaintiff demonstrates a substantial likelihood of immediate danger of abuse, order the immediate suspension and surrender of any license to carry firearms and or firearms identification card which the defendant may hold and order the defendant to surrender all firearms, rifles, shotguns, machine guns and ammunition which he then controls, owns or possesses in accordance with the provisions of this chapter and any license to carry firearms or firearms identification cards which the defendant may hold shall be surrendered to the appropriate law enforcement officials in accordance with the provisions of this chapter and, said law enforcement official may store, transfer or otherwise dispose of any such weapon in accordance with the provisions of section 129D of chapter 140; provided however, that nothing herein shall authorize the transfer of any weapons surrendered by the defendant to anyone other than a licensed dealer. Notice of such suspension and ordered surrender shall be appended to the copy of abuse prevention order served on the defendant pursuant to section seven. Law enforcement officials, upon the service of said orders, shall immediately take possession of all firearms, rifles, shotguns, machine guns, ammunition, any license to carry firearms and any firearms identification cards in the control, ownership, or possession of said defendant. Any violation of such orders shall be punishable by a fine of not more than five thousand dollars, or by imprisonment for not more than two and one-half years in a house of correction, or by both such fine and imprisonment.

Any defendant aggrieved by an order of surrender or suspension as described in the first sentence of this section may petition the court which issued such suspension or surrender order for a review of such action and such petition shall be heard no later than ten court business days after the receipt of the notice of the petition by the court. If said license to carry firearms or firearms identification card has been suspended upon the issuance of an order issued pursuant to section four or five, said petition may be heard contemporaneously with the hearing specified in the second sentence of the second paragraph of section four. Upon the filing of an affidavit by the defendant that a firearm, rifle, shotgun, machine gun or ammunition is required in the performance of the defendant's employment, and upon a request for an expedited hearing, the court shall order said hearing within two business days of receipt of such affidavit and request but only on the issue of surrender and suspension pursuant to this section.

Chapter 266. Crimes Against Property

102A. Possession of Infernal Machine; Notice of Seizure. Whoever, other than a police or other law enforcement officer acting in the discharge of his official duties, has in his possession or under his control an infernal machine or a similar instrument, contrivance or device shall be punished by imprisonment in the state prison for not more than ten years or in jail for not more than two and one half years, or by a fine of not more than one thousand dollars, or by both such fine and imprisonment, and the said machine, instrument, contrivance or device shall be forfeited to the commonwealth. The term "infernal machine", as used in this section, shall include any device for endangering life or doing unusual damage to property, or both, by fire or, explosion, whether or not contrived to ignite or explode automatically and whether or not disguised so as to appear harmless. Notice of the seizure of any such machine, instrument, contrivance or device shall be sent forthwith to the commissioner of public safety and the article seized shall be subject to his order.

102A½. Possession of Hoax Device.

(a) Whoever possesses, transports, uses or places or causes another to knowingly or unknowingly possess, transport, use or place any hoax device or hoax substance with the intent to cause anxiety, unrest, fear or personal discomfort to any person or group of persons shall be punished by imprisonment in a house of correction for not more than two and one-half years or by imprisonment in the state prison for not more than five years or by a fine of not more than $5,000, or by both such fine and imprisonment.

(b) For the purposes of this section, the term "hoax device" shall mean any device that would cause a person reasonably to believe that such device is an infernal machine. For the purposes of this section, the term "infernal machine" shall mean any device for endangering life or doing unusual damage to property, or both, by fire or explosion, whether or not contrived to ignite or explode automatically. For the purposes of this section, the words "hoax substance" shall mean any substance that would cause a person reasonably to believe that such substance is a harmful chemical or biological agent, a poison, a harmful radioactive substance or any other substance for causing serious bodily injury, endangering life or doing unusual damage to property, or both.

(c) This section shall not apply to any law enforcement or public safety officer acting in the lawful discharge of official duties.

(d) The court shall, after a conviction, conduct a hearing to ascertain the extent of costs incurred, damages and financial loss suffered by local, county or state public safety agencies and the amount of property damage caused as a result of the violation of this section. A person found guilty of violating this section shall, in all cases, upon conviction, in addition to any other punishment, be ordered to make restitution to the local, county or state government for any costs incurred, damages and financial loss sustained as a result of the commission of the offense. Restitution shall be imposed in addition to incarceration or fine; however, the court shall consider the defendant's present and future ability to pay in its determinations regarding a fine. In determining the amount, time and method of payment of restitution, the court shall consider the financial resources of the defendant and the burden restitution will impose on the defendant.

102B. Possession of Molotov Cocktail or Similar Device; Exceptions; Penalties. Whoever makes, sells, uses or has in his possession or under his control a bottle or other breakable container containing a flammable liquid into which has been fixed or placed a wick or similar device, and which bottle or container when ignited and thrown will cause a fire or explosion, shall be punished by a fine of not more than one thousand dollars or imprisonment for not more than five years in the state prison or by imprisonment for not more than two and one-half years in a house of correction or both such fine and imprisonment.

The provisions of this section shall not apply to flares, lanterns, fireworks or other such devices used for signal or illumination purposes, or for any other lawful purpose.

Whoever violates any provision of this section may be arrested without a warrant and held in custody in jail or otherwise until a complaint be made against him for such offense unless previously admitted to bail, which complaint shall be made as soon as practicable and in any case within twenty-four hours, Sundays and legal holidays excepted.

Chapter 269. Crimes Against Public Peace

10. Penalty for Unlawfully Carrying Dangerous Weapons, Possessing Machine Gun, etc.

(a) Whoever, except as provided or exempted by statute, knowingly has in his possession; or knowingly has under his control in a vehicle; a firearm, loaded or unloaded, as defined in section one hundred and twenty-one of chapter one hundred and forty without either:

(1) being present in or on his residence or place of business; or

(2) having in effect a license to carry firearms issued under section one hundred and thirty-one of chapter one hundred and forty; or

(3) having in effect a license to carry firearms issued under section one hundred and thirty-one F of chapter one hundred and forty; or

(4) having complied with the provisions of sections one hundred and twenty-nine C and one hundred and thirty-one G of chapter one hundred and forty; or

(5) having complied as to possession of an air rifle or BB gun with the requirements imposed by section twelve B; and whoever knowingly has in his possession; or knowingly has under control in a vehicle; a rifle or shotgun, loaded or unloaded, without either:

(1) being present in or on his residence or place of business; or

(2) having in effect a license to carry firearms issued under section one hundred and thirty-one of chapter one hundred and forty; or

(3) having in effect a license to carry firearms issued under section one hundred and thirty-one F of chapter one hundred and forty; or

(4) having in effect a firearms identification card issued under section one hundred and twenty-nine B of chapter one hundred and forty; or

(5) having complied with the requirements imposed by section one hundred and twenty-nine C of chapter one hundred and forty upon ownership or possession of rifles and shotguns; or

(6) having complied as to possession of an air rifle or BB gun with the requirements imposed by section twelve B; shall be punished by imprisonment in the state prison for not less than two and one-half years nor more than five years, or for not less than 18 months nor more than two and one-half years in a jail or house of correction. The sentence imposed on such person shall not be reduced to less than 18 months, nor suspended, nor shall any person convicted under this subsection be eligible for probation, parole, work release, or furlough or receive any deduction from his sentence for good conduct until he shall have served 18 months of such sentence; provided, however, that the commissioner of correction may on the recommendation of the warden, superintendent, or other person in charge of a correctional institution, grant to an offender committed under this subsection a temporary release in the custody of an officer of such institution for the following purposes only: to attend the funeral of a relative; to visit a critically ill relative; or to obtain emergency medical or psychiatric service unavailable at said institution. Prosecutions commenced under this subsection shall neither be continued without a finding nor placed on file.

No person having in effect a license to carry firearms for any purpose, issued under section one hundred and thirty-one or section one hundred and thirty-one F of chapter one hundred and forty shall be deemed to be in violation of this section.

The provisions of section eighty-seven of chapter two hundred and seventy-six shall not apply to any person seventeen years of age or older, charged with a violation of this subsection, or to any child between ages fourteen and seventeen so charged, if the court is of the opinion that the interests of the public require that he should be tried as an adult for such offense instead of being dealt with as a child.

The provisions of this subsection shall not affect the licensing requirements of section one hundred and twenty-nine C of chapter one hundred and forty which require every person not otherwise duly licensed or exempted to have been issued a firearms identification card in order to possess a firearm, rifle or shotgun in his residence or place of business.

(b) Whoever, except as provided by law, carries on his person, or carries on his person or under his control in a vehicle, any stiletto, dagger, or a device or case which enables a knife with a locking blade to be drawn at a locked position, any ballistic knife, or any knife with a detachable blade capable of being propelled by any mechanism, dirk knife, any knife having a double-edged blade, or a switch knife, or any knife having an automatic spring release device by which the blade is released from the handle, having a blade of over one and one-half inches, or a slung shot, blowgun, blackjack, metallic knuckles or knuckles of any substance which could be put to the same use with the same or similar effect as metallic knuckles, nunchaku, zoobow, also known as klackers or kung fu sticks, or any similar weapon consisting of two sticks of wood, plastic or metal connected at one end by a length of rope, chain, wire or leather, a shuriken or any similar pointed starlike object intended to injure a person when thrown, or any armband, made with leather which has metallic spikes, points or studs or any similar device made from any other substance or a cestus or similar material weighted with metal or other substance and worn on the hand, or a manrikigusari or similar length of chain having weighted ends; or whoever, when arrested upon a warrant for an alleged crime, or when arrested while committing a breach or disturbance of the public peace, is armed with or has on his person, or has on his person or under his control in a vehicle, a billy or other dangerous weapon other than those herein mentioned and those mentioned in paragraph (a), shall be punished by imprisonment for not less than two and one-half years nor more than five years in the state prison, or for not less than six months nor more than two and one-half years in a jail or house of correction, except that, if the court finds that the defendant has not been previously convicted of a felony, he may be punished by a fine of not more than fifty dollars or by imprisonment for not more than two and one-half years in a jail or house of correction.

(c) Whoever, except as provided by law possesses a machine gun, as defined in section one hundred and twenty-one of chapter one hundred and forty, without permission under section one hundred and thirty-one of said chapter one hundred and forty; or whoever owns, possesses or carries on his person, or carries on his person or under his control in a vehicle, a sawed-off shotgun, as defined in said section one hundred and twenty-one of said chapter one hundred and forty, shall be punished by imprisonment in the state prison for life, or for any term of years provided that any sentence imposed under the provisions of this paragraph shall be subject to the minimum requirements of paragraph (a).

(d) Whoever, after having been convicted of any of the offenses set forth in paragraph (a), (b) or (c) commits a like offense or any other of the said offenses, shall be punished by imprisonment in the state prison for not less than five years nor more than seven years; for a third such offense, by imprisonment in the state prison for not less than seven years nor more than ten years; and for a fourth such offense, by imprisonment in the state prison for not less than ten years nor more than fifteen years. The sentence imposed upon a person, who after a conviction of an offense under paragraph (a), (b) or (c) commits the same or a like offense, shall not be suspended, nor shall any person so sentenced be eligible for probation or receive any deduction from his sentence for good conduct.

(e) Upon conviction of a violation of this section, the firearm or other article shall, unless otherwise ordered by the court, be confiscated by the commonwealth. The firearm or article so confiscated shall, by the authority of the written order of the court be forwarded by common carrier to the colonel of the state police, who, upon receipt of the same, shall notify said court or justice thereof. Said colonel may sell or destroy the same, except that any firearm which may not be lawfully sold in the commonwealth shall be destroyed, and in the case of a sale, after paying the cost of forwarding the article, shall pay over the net proceeds to the commonwealth.

(f) The court shall, if the firearm or other article was lost by or stolen from the person lawfully in possession of it, order its return to such person.

(g) Whoever, within this commonwealth, produces for sale, delivers or causes to be delivered, orders for delivery, sells or offers for sale, or fails to keep records regarding, any rifle or shotgun without complying with the requirement of a serial number, as provided in section one hundred and twenty-nine B of chapter one hundred and forty, shall for the first offense be punished by confinement in a jail or house of correction for not more than two and one-half years, or by a fine of not more than five hundred dollars.

(h)(1) Whoever owns, possesses or transfers a firearm, rifle, shotgun or ammunition without complying with the provisions of section 129C of chapter 140 shall be punished by imprisonment in a jail or house of correction for not more than two years or by a fine of not more than $500. Whoever commits a second or subsequent violation of this paragraph shall be punished by imprisonment in a house of correction for not more than two years or by a fine of not more than $1,000 or both. Any officer authorized to make arrests may arrest without a warrant any person whom the officer has probable cause to believe has violated this paragraph.

(2) Any person who leaves a firearm, rifle, shotgun or ammunition unattended with the intent to transfer possession of such firearm, rifle, shotgun or ammunition to any person not licensed under section 129C of chapter 140 or section 131 of chapter 140 for the purpose of committing a crime or concealing a crime shall be punished by imprisonment in a house of correction for not more than 2-1/2 years or in state prison for not more than 5 years.

(i) Whoever knowingly fails to deliver or surrender a revoked or suspended license to carry or possess firearms or machine guns issued under the provisions of section one hundred and thirty-one or one hundred and thirty-one F of chapter one hundred and forty, or firearm identification card, or receipt for the fee for such card, or a firearm, rifle, shotgun or machine gun, as provided in section one hundred and twenty-nine D of chapter one hundred and forty, unless an appeal is pending, shall be punished by imprisonment in a jail or house of correction for not more than two and one-half years or by a fine of not more than one thousand dollars.

(j) Whoever, not being a law enforcement officer, and notwithstanding any license obtained by him under the provisions of chapter one hundred and forty, carries on his person a firearm as hereinafter defined, loaded or unloaded or other dangerous weapon in any building or on the grounds of any elementary or secondary school, college or university without the written authorization of the board or officer in charge of such elementary or secondary school, college or university shall be punished by a fine of not more than one thousand dollars or by imprisonment for not more than one year, or both.

For the purpose of this paragraph, "firearm" shall mean any pistol, revolver, rifle or smoothbore arm from which a shot, bullet or pellet can be discharged by whatever means.

Any officer in charge of an elementary or secondary school, college or university or any faculty member or administrative officer of an elementary or secondary school, college or university failing to report violations of this paragraph shall be guilty of a misdemeanor and punished by a fine of not more than five hundred dollars.

(k) [Deleted]

(l) The provisions of this section shall be fully applicable to any person proceeded against under section seventy-five of chapter one hundred and nineteen and convicted under section eighty-three of chapter one hundred and nineteen, provided, however, that nothing contained in this section shall impair, impede, or affect the power granted any court by chapter one hundred and nineteen to adjudicate a person a delinquent child, including the power so granted under section eighty-three of said chapter one hundred and nineteen.

(m) Notwithstanding the provisions of paragraph (a) or (h), any person not exempted by statute who knowingly has in his possession, or knowingly has under his control in a vehicle, a large capacity weapon or large capacity feeding device therefor who does not possess a valid Class A or Class B license to carry firearms issued under section 131 or 131F of chapter 140, except as permitted or otherwise provided under this section or chapter 140, shall be punished by imprisonment in a state prison for not less than two and one-half years nor more than ten years. The possession of a valid firearm identification card issued under section 129B shall not be a defense for a violation of this subsection; provided, however, that any such person charged with violating this paragraph and holding a valid firearm identification card shall not be subject to any mandatory minimum sentence imposed by this paragraph. The sentence imposed upon such person shall not be reduced to less than one year, nor suspended, nor shall any person convicted under this subsection be eligible for probation, parole, furlough, work release or receive any deduction from his sentence for good conduct until he shall have served such minimum term of such sentence; provided, however, that the commissioner of correction may, on the recommendation of the warden, superintendent or other person in charge of a correctional institution or the administrator of a county correctional institution, grant to such offender a temporary release in the custody of an officer of such institution for the following purposes only: (i) to attend the funeral of a spouse or next of kin; (ii) to visit a critically ill close relative or spouse; or (iii) to obtain emergency medical services unavailable at such institution. Prosecutions commenced under this subsection shall neither be continued without a finding nor placed on file. The provisions of section 87 of chapter 276 relative to the power of the court to place certain offenders on probation shall not apply to any person 17 years of age or over charged with a violation of this section.

This paragraph shall not apply to the possession of a large capacity weapon or large capacity feeding device by (i) any officer, agent or employee of the commonwealth or to obtain emergency medical services unavailable at such institution. Prosecutions commenced under this subsection shall neither be continued without a finding nor placed on file. The provisions of section 87 of chapter 276 relative to the power of the court to place certain offenders on probation shall not apply to any person 17 years of age or over charged with a violation of this section.

The provisions of any other state or the United States, including any federal, state or local law enforcement personnel; (ii) any member of the military or other service of any state or the United States; (iii) any duly authorized law enforcement officer, agent or employee of any municipality of the commonwealth; (iv) any federal, state or local historical society, museum or institutional collection open to the public; provided, however, that any such person described in clauses (i) to (iii), inclusive, is authorized by a competent authority to acquire, possess or carry a large capacity semiautomatic weapon and is acting within the scope of his duties; or (v) any gunsmith duly licensed under the applicable federal law.

(n) Whoever violates paragraph (a) or paragraph (c), by means of a loaded firearm, loaded sawed off shotgun or loaded machine gun shall be further punished by imprisonment in the house of correction for not more than 2-1/2 years, which sentence shall begin from and after the expiration of the sentence for the violation of paragraph (a) or paragraph (c).

(o) For purposes of this section, "loaded" shall mean that ammunition is contained in the weapon or within a feeding device attached thereto.

For purposes of this section, "ammunition" shall mean cartridges or cartridge cases, primers (igniter), bullets or propellant powder designed for use in any firearm, rifle or shotgun.

10A. Sale, etc., of Silencers for Firearms. Any person, other than a federally licensed firearms manufacturer, an authorized agent of the municipal police training committee, or a duly authorized sworn law enforcement officer while acting within the scope of official duties and under the direct authorization of the police chief or his designee, or the colonel of the state police, who sells or keeps for sale, or offers, or gives or disposes of by any means other than submitting to an authorized law enforcement agency, or uses or possesses any instrument, attachment, weapon or appliance for causing the firing of any gun, revolver, pistol or other firearm to be silent or intended to lessen or muffle the noise of the firing of any gun, revolver, pistol or other firearm shall be punished by imprisonment for not more than five years in state prison or for not more than two and one-half years in a jail or house of correction. Nothing contained herein shall be construed to prohibit a federally licensed firearms manufacturer from selling such instrument, attachment, weapon or appliance to authorized law enforcement agencies for law enforcement purposes or to the municipal police training committee for law enforcement training. Upon conviction of a violation of this section, the instrument, attachment or other article shall be confiscated by the commonwealth and forwarded, by the authority of the written order of the court, to the colonel of the state police, who shall destroy said article.

10E. Unlawful Sale or Distribution of Multiple Firearms; Penalties; Mandatory Sentences; Eligibility for Probation or Parole. Whoever, except as provided by law, in a single transaction or occurrence or in a series of transactions within a twelve month period, knowingly or intentionally distributes, sells, or transfers possession of a quantity of firearms, rifles, shotguns, machine guns, or any combination thereof, shall, if the quantity of firearms, rifles, shotguns, machine guns, or any combination thereof is:

(1) Three or more, but less than ten, be punished by a term of imprisonment of not more than ten years in the state prison. No sentence imposed under the provisions of this paragraph shall be for less than a mandatory minimum term of imprisonment of three years and a fine of not more than fifty thousand dollars may be imposed but not in lieu of the mandatory minimum term of imprisonment, as established herein.

(2) Ten or more, but less than twenty, be punished by a term of imprisonment of not more than ten years in the state prison. No sentence imposed under the provisions of this paragraph shall be for less than a mandatory minimum term of imprisonment of five years and a fine of not more than one hundred thousand dollars may be imposed but not in lieu of the mandatory minimum term of imprisonment, as established herein.

(3) Twenty or more, be punished by a term of imprisonment not less than ten years up to life imprisonment in the state prison. No sentence imposed under the provisions of this paragraph shall be for less than a mandatory minimum term of imprisonment of ten years and a fine of not more than one hundred and fifty thousand dollars may be imposed but not in lieu of the mandatory minimum term of imprisonment, as established herein.

A prosecution commenced under this section shall not be placed on file or continued without a finding, and the sentence imposed upon a person convicted of violating any provision of said section shall not be reduced to less than the mandatory minimum term of imprisonment as established in said section, nor shall any sentence of imprisonment imposed upon any person be suspended or reduced until such person shall have served said mandatory minimum term of imprisonment.

A person convicted of violating any provision of this section shall not, until he shall have served the mandatory minimum term of imprisonment established herein, be eligible for probation, parole, furlough, work release, or receive any deduction from his sentence for good conduct under sections one hundred and twenty-nine, one hundred and twenty-nine C and one hundred and twenty-nine D of chapter one hundred and twenty-seven; provided, however, that the commissioner of corrections may, on the recommendation of the warden, superintendent, or other person in charge of the correctional institution, grant to said offender a temporary release in the custody of an officer of such institution for the following purposes: to attend the funeral of a relative, to visit a critically ill relative, or to obtain emergency medical or psychiatric services unavailable at said institution. The provisions of section eighty-seven of chapter two hundred and seventy-six shall not apply to any person, seventeen years of age or over, charged with a violation of said sections, or to any child between the age of fourteen and seventeen, so charged, if the court is of the opinion that the interests of the public require that he shall be tried for such offense instead of being dealt with as a child.

10F. Penalties for Illegal Sale of Large Capacity Weapons or Feeding Devices.

(a) Any person who sells, keeps for sale, or offers or exposes for sale, gives or otherwise transfers any large capacity weapon or large capacity feeding device, both as defined in section 121 of chapter 140, to a person 18 years of age or over, except as permitted under this section or chapter 140, shall be punished by imprisonment in a state prison for not less than two and one-half years nor more than ten years. Any person who commits a second or subsequent such crime shall be punished by imprisonment in a state prison for not less than five years nor more than 15 years. The sentence imposed upon such person shall not be reduced to less than two and one-half years for a first offense, nor less than five years for a second or subsequent such offense, nor suspended, nor shall any person convicted under this subsection be eligible for probation, parole, furlough, work release or receive any deduction from his sentence for good conduct until he shall have served such minimum term of such sentence; provided, however, that the commissioner of correction may, on the recommendation of the warden, superintendent or other person in charge of a correctional institution or the administrator of a county correctional institution, grant to such offender a temporary release in the custody of an officer of such institution for the following purposes only: (i) to attend the funeral of a spouse or next of kin; (ii) to visit a critically ill close relative or spouse; or (iii) to obtain emergency medical services unavailable at such institution. Prosecutions commenced under this subsection shall neither be continued without a finding nor placed on file. The provisions of section 87 of chapter 276 relative to the power of the court to place certain offenders on probation shall not apply to any person 17 years of age or over charged with a violation of this subsection.

(b) Any person who transfers, sells, lends or gives a large capacity weapon or large capacity feeding device to a person under the age of 18, except as permitted under the provisions of chapter 140, shall be punished by imprisonment in a state prison for not less than five nor more than 15 years. The sentence imposed upon such person shall not be reduced to less than five years, nor suspended, nor shall any person convicted under this subsection be eligible for probation, parole, furlough, work release or receive any deduction from his sentence for good conduct until he has served five years of such sentence; provided, however, that the commissioner of corrections may, on the recommendation of the warden, superintendent or other person in charge of a correctional institution or the administrator of a county correctional institution, grant to such offender a temporary release in the custody of an officer of such institution for the following purposes only: (i) to attend the funeral of a spouse or next of kin; (ii) to visit a critically ill close relative or spouse; or (iii) to obtain emergency medical services unavailable at such institution. Prosecutions commenced under this subsection shall neither be continued without a finding nor placed on file. The provisions of section 87 of chapter 276 relative to the power of the court to place certain offenders on probation shall not apply to any person 17 years of age or over charged with a violation of this subsection.

10H. Possession of Firearms While Intoxicated or Under the Influence of Narcotics or Other Drugs; Penalties. Whoever, having in effect a license to carry firearms issued under section 131 or 131F of chapter 140, carries on his person, or has under his control in a vehicle, a loaded firearm, as defined in section 121 of said chapter 140, while under the influence of intoxicating liquor or marijuana, narcotic drugs, depressants or stimulant substances, all as defined in section 1 of chapter 94C, or the vapors of glue shall be punished by a fine of not more than $ 5,000 or by imprisonment in the house of correction for not more than two and one-half years, or by both such fine and imprisonment.

11. Posters Setting Forth Certain Statutes. The state secretary shall, annually, cause to be printed, in English and in such other languages as he may deem necessary, and in large letters so as to be easily read, for use as a poster, section one hundred and thirty-one of chapter one hundred and forty and sections ten, twelve B, and fourteen of this chapter. Sufficient copies of the said posters shall be sent to the clerks and to the superintendents of schools in all cities and towns for their use as herein provided. The city or town clerks shall cause posters received by them to be displayed in such places as they may select, and in such numbers, according to the population of the city or town, as its clerk may deem expedient. The superintendents of schools shall cause the posters received by them to be distributed among the schools within their jurisdiction, and in such numbers as they may deem necessary. The cost of preparing and printing the posters and of distributing them to the various cities and towns shall be paid by the commonwealth, and the cost of placing or affixing them in each city or town shall be paid by that city or town.

11A. Definitions. For the purposes of this section and sections eleven B, eleven C and eleven D, the following words shall have the following meanings:

"Firearm", a firearm as defined in section one hundred and twenty-one of chapter one hundred and forty, or a rifle or shotgun.

"Serial number", the number stamped or placed upon a firearm by the manufacturer in the original process of manufacture.

"Identification number", the number stamped or placed upon a firearm by the colonel of the state police under authority of section eleven D.

11B. Commission of Felony While Possessing Firearm with Removed, Defaced, etc., Serial or Identification Number. Whoever, while in the commission or attempted commission of a felony, has in his possession or under his control a firearm the serial number or identification number of which has been removed, defaced, altered, obliterated or mutilated in any manner shall be punished by imprisonment in the state prison for not less than two and one half nor more than five years, or in a jail or house of correction for not less than six months nor more than two and one half years. Upon a conviction of a violation of this section, said firearm or other article, by the authority of the written order of the court, shall be forwarded to the colonel of the state police, who shall cause said weapon to be destroyed.

11C. Removal or mutilation of serial or identification numbers of firearms; receiving such firearm; possession or control as evidence; destruction. Whoever, by himself or another, removes, defaces, alters, obliterates or mutilates in any manner the serial number or identification number of a firearm, or in any way participates therein, and whoever receives a firearm with knowledge that its serial number or identification number has been removed, defaced, altered obliterated or mutilated in any manner, shall be punished by a fine of not more than two hundred dollars or by imprisonment for not less than one month nor more than two and one half years. Possession or control of a firearm the serial number or identification number of which has been removed, defaced, altered, obliterated or mutilated in any manner shall be prima facie evidence that the person having such possession or control is guilty of a violation of this section; but such prima facie evidence may be rebutted by evidence that such person had no knowledge whatever that such number had been removed, defaced, altered, obliterated or mutilated, or by evidence that he had no guilty knowledge thereof. Upon a conviction of a violation of this section said firearm or other article shall be forwarded, by the authority of the written order of the court, to the colonel of the state police, who shall cause said firearm or other article to be destroyed.

11E. Firearms Sold, etc., by Manufacturers and Dealers to Have Serial Identification Numbers; Penalty. All firearms, rifles and shotguns of new manufacture, manufactured or delivered to any licensed dealer within the commonwealth shall bear serial numbers permanently inscribed on a visible metal area of said firearm, rifle or shotgun, and the manufacturer of said firearm, rifle or shotgun shall keep records of said serial numbers and the dealer, distributor or person to whom the firearm, rifle or shotgun was sold or delivered.

No licensed dealer shall order for delivery, cause to be delivered, offer for sale or sell within the commonwealth any newly manufactured firearm, rifle or shotgun received directly from a manufacturer, wholesaler or distributor not so inscribed with a serial number nor shall any licensed manufacturer or distributor of firearms, rifles or shotguns deliver or cause to be delivered within the commonwealth any firearm, rifle or shotgun not complying with this section.

No licensed manufacturer within the commonwealth shall produce for sale within the United States, its territories or possessions any firearm, rifle or shotgun not complying with paragraph one of this section. Whoever violates this section shall be punished by a fine of five hundred dollars. Each such violation shall constitute a separate offense.

12A. Sale, etc., to Minor Under Eighteen of Air Rifle or BB Gun Penalized; Exceptions. Whoever sells to a minor under the age of eighteen or whoever, not being the parent, guardian or adult teacher or instructor, furnishes to a minor under the age of eighteen an air rifle or so-called BB gun, shall be punished by a fine of not less than fifty nor more than two hundred dollars or by imprisonment for not more than six months.

12B. Possession by Minor Under Eighteen of Air Rifle or BB Gun Regulated; Discharge, etc. of Such Weapon Regulated; Penalty. No minor under the age of eighteen shall have an air rifle or so-called BB gun in his possession while in any place to which the public has a right of access unless he is accompanied by an adult or unless he is the holder of a sporting or hunting license and has on his person a permit from the chief of police of the town in which he resides granting him the right of such possession. No person shall discharge a BB shot, pellet or other object from an air rifle or so-called BB gun into, from or across any street, alley, public way or railroad or railway right of way, and no minor under the age of eighteen shall discharge a BB shot, pellet or other object from an air rifle or BB gun unless he is accompanied by an adult or is the holder of a sporting or hunting license. Whoever violates this section shall be punished by a fine of not more than one hundred dollars, and the air rifle or BB gun or other weapon shall be confiscated. Upon a conviction of a violation of this section the air rifle or BB gun or other weapon shall, by the written authority of the court, be forwarded to the colonel of the state police, who may dispose of said article in the same manner as prescribed in section ten.

12D. Carrying Rifle or Shotgun Having Shells or Cartridges Therein on Public Way Prohibited; Exceptions; Penalty.

(a) Except as exempted or provided by law, no person shall carry on his person on any public way a loaded rifle or shotgun having cartridges or shells in either the magazine or chamber thereof. For purposes of this section, "loaded shotgun or loaded rifle" shall mean any shotgun or rifle having ammunition in either the magazine or chamber thereof, such ammunition including a live cartridge, primer (igniter), bullet or propellant powder designed for use in any firearm, rifle or shotgun and, in the case of a muzzle loading or black powder shotgun or rifle, so-called, a shotgun or rifle containing powder in the flash pan or in the bore or chamber or containing a percussion cap, shot or ball; provided, however, that "loaded shotgun or loaded rifle" shall not include a shotgun or rifle loaded with a blank cartridge, so-called, which contains no projectile within such blank or within the bore or chamber of such shotgun or rifle.

Whoever violates the provisions of this subsection shall be punished by a fine of not less than $ 500 nor more than $ 5,000 or by imprisonment in the house of correction for not more than two years, or by both such fine and imprisonment, and may be arrested without a warrant; provided, however, that if such rifle or shotgun is a large capacity weapon, as defined in section 121 of chapter 140, such person shall be punished by a fine of not less than $ 1,000 nor more than $ 10,000 or by imprisonment for not less than one year nor more than ten years, or by both such fine and imprisonment, and may be arrested without a warrant.

(b) Except as exempted or provided by law, no person shall carry on his person on any public way an unloaded rifle or shotgun, unless such rifle or shotgun is enclosed in a case.

Whoever violates the provisions of this subsection shall be punished by a fine of not less than $100 nor more than $1,000, and may be arrested without a warrant; provided, however, that if such unloaded rifle or shotgun is a large capacity weapon and is carried simultaneously with a fully or partially loaded large capacity feeding device, such person shall be punished by a fine of not less than $1,000 nor more than $10,000 or by imprisonment for not less than one year nor more than ten years, or by both such fine and imprisonment, and may be arrested without a warrant.

This subsection shall not apply to drills, parades, military reenactments or other commemorative ceremonies, color guards or memorial service firing squads, so-called, as permitted by law.

(c) Upon a conviction of a violation of any provision of this section, such rifle or shotgun shall be confiscated by the commonwealth and, upon written order of the court, such weapon shall be forwarded to the colonel of the state police, who may dispose of such weapon in the manner prescribed in section 10.

(d) The provisions of this section shall not apply to the carrying of a loaded or unloaded rifle or shotgun on a public way by (i) any officer, agent or employee of the commonwealth or any other state or the United States, including any federal, state or local law enforcement personnel; (ii) any member of the military or other service of any state or the United States, including members of the national guard, reserves and junior reserve officer training corps; (iii) any duly authorized law enforcement officer, agent or employee of any municipality of the commonwealth; provided, however, that any such person described in clauses (i) to (iii), inclusive, shall be authorized by a competent authority to so carry a loaded or unloaded rifle or shotgun on a public way and such person is acting within the scope of his duties or training; or (iv) a person who is lawfully engaged in hunting and is the holder of a valid hunting or sporting license issued pursuant to chapter 131. This section shall not apply to the operation of a shooting gallery, licensed and defined under the provisions of section 56A of chapter 140, nor to persons using the same.

[Current through Act 377 of the 2008 Legislative Session]

City of Boston Municipal Code

Chapter XVI. Prohibitions, Penalties and Permits

16-5A. Regulating the Sale and Marketing of Replica Firearms in the City of Boston

16-5A.1. Definitions. As used herein, the following words shall have the following meanings:

a. Commissioner shall mean the Commissioner of the Boston Police Department or his designee.

b. Department shall mean the Boston Police Department.

c. Market shall mean to display for sale.

d. Replica firearm shall mean any toy, imitation, or facsimile pistol, revolver, shotgun, rifle, air rifle, B-B gun, pellet gun, machine gun or other similar simulated weapon which because of its color, size, shape or other characteristics, can reasonably be perceived to be a real firearm from which a shot or bullet can be fired and which, pursuant to Subsection 16-5A.4, has been determined to pose a threat to public safety.

e. Sell shall mean to exchange or deliver.

1. For money or its equivalent; or

2. As a promotion or an inducement to buy.

16-5A.2. Regulation. Except as provided in Subsection 16-5A.3 herein, no person or entity shall sell or market any replica firearm in the City of Boston.

16-5A.3. Exceptions. Notwithstanding the foregoing, the sale of replica firearms shall be permitted if the sale of such replica firearms is solely for purposes of or for transportation in intrastate, interstate or foreign commerce.

16-5A.4. Enforcement.

a. The Commissioner shall identify those replica firearms which he deems to pose a threat to public safety and shall make a list thereof. Such list shall include, where available, the following information: the product name, the manufacturer and the model number. The Commissioner shall post the list in the Department and shall make available copies thereof to the general public. The Commissioner shall update the list from time to time as he deems warranted. Any individual may provide information to the Commissioner regarding replica firearms to be added to the list.

b. Any person who observes the sale or marketing of a replica firearm which he believes to be a violation of this section may register a written or oral complaint with the Department. A Boston Police Officer, in response to each complaint, shall investigate the complaint forthwith in person and determine whether there is such a violation.

c. Any Boston Police Officer who observes the sale or marketing of a replica firearm which he believes to be a violation of this section shall investigate forthwith and determine whether there is such a violation.

d. If a Boston Police Officer determines that there is a violation of this section, he shall order the individual or entity to remove immediately from sale or marketing all the replica firearms found to be in violation. In addition, the Department shall impose the following penalties on the individual or entity violating the section:

1. A written citation or warning for the first violation;

2. A one hundred ($100.00) dollar fine for the second violation; and

3. A two hundred ($200.00) dollar fine for the second and any subsequent violations.

For purposes of penalties to be imposed under this section, if a Boston Police Officer determines that a sale or marketing of replica firearms has occurred, that determination shall constitute one violation, regardless of the number of replica firearms involved. A Boston Police Officer shall return, unannounced, to the site of the violation within the thirty (30) day period following the determination of a violation to ensure that the individual or entity has not resumed the sale or marketing of replica firearms.

16-5A.5. Appeal Procedure. Any individual who is aggrieved by an action taken by a Boston Police Officer under this section may, within ten (10) days of such action, file an appeal, in writing, to the Commissioner. After notice to such individual, the Commissioner shall hold a hearing, after which he shall issue a decision in which he affirms, annuls or modifies an action taken by a Boston Police Officer, giving his reasons therefor. The Commissioner shall send the decision to the individual by first class mail within ten (10) days after the hearing. The decision shall be a final administrative decision. The individual shall have thirty (30) days from the date of the written decision to seek judicial review in the Suffolk County Superior Court.

Chapter XVIII. Fees and Charges

18-1.1. "A" Fees and Charges. ...

3. Ammunition, License to Sell. The fee for a license to sell ammunition granted by the Police Commissioner under Section 122B of Chapter 140 of the General Laws shall be two hundred twenty-five ($225.00) dollars.

4. Ammunition, License to Store. The fee for an annual license to store ammunition granted by the Chief of the Fire Department under Section 13 of Chapter 148 of the General Laws shall be sixty ($60.00) dollars. ...

18-1.6. "F" Fees and Charges. ...

15. Firearm Identification Card. The fee for the issuance by the Police Commissioner of a Firearm Identification Card under Section 129B of Chapter 140 of the General Laws shall be fifteen ($15.00) dollars.

16. Firearms, License to Carry or Possess. The fee for a license to carry firearms in the Commonwealth or to possess therein a machine gun granted by the Police Commissioner under Section 131 of Chapter 140 of the General Laws shall be fifty ($50.00) dollars, excepting Police Officers and retired Police Officers of the City of Boston who shall be exempt from the requirements of paying said fee.

16A. Firearms, License to Possess Assault Weapon. The fee for a license to possess an assault weapon or weapons in the City of Boston granted by the Police Commissioner of the City of Boston, or his designee, pursuant to Section 6 of Chapter 596 of the Acts of 1989, shall be twenty-five ($25.00) dollars. Such license shall be for five (5) years. A person having more than one (1) assault weapon shall be required to apply for only one (1) license, which license shall be applicable to all of the assault weapons which that person lists with the Police Commissioner, or his designee, at the time of application.

17. Firearms, License to Sell, Rent, or Lease. The fee for a license to sell, rent or lease firearms, rifles, shotguns, or machine guns granted by the Police Commissioner under Section 122 of Chapter 140 of the General Laws shall be three hundred ($300.00) dollars.

18. Firearms, Permit to Purchase, Rent or Lease. The fee for a permit to purchase, rent, or lease a firearm granted by the Police Commissioner under Section 131A of Chapter 140 of the General Laws shall be fifty ($50.00) dollars. ....

18-1.7. "G" Fees and Charges. ...

13. Gunsmith's License. The fee for a license to be in business as a gunsmith granted by the Police Commissioner under Section 122 of Chapter 140 of the General Laws shall be, in the case of a person licensed to sell, rent, or lease firearms, rifles, shotguns, or machine guns, fifty ($50.00) dollars and in the case of any other persons one hundred ($100.00) dollars.

Chapter XIX. Schools

19-1.1. Possession, Detection, of Weapons in Schoolhouses. No person other than a sworn officer of the Commonwealth, a member of the military service on active duty, or a constable, officer or official of the City or County having a license to carry firearms shall in any schoolhouse have possession of any dangerous weapon, which term shall include any firearm, and any item described in paragraphs (b) or (c) of Section 10 of Chapter 269 of the General Laws, unless authorized by the Superintendent of Schools or the Superintendent's designee. If the Superintendent shall determine public safety requires, a system for detection of weapons may be established in any schoolhouse and no person shall be admitted to, or enter, a schoolhouse without submitting to such detection system. Any person violating the provisions hereof shall be punished by a fine of two hundred ($200.00) dollars.

[City of Boston Municipal Code current through Ord. 2007 C. 12, passed 12/12/07]

The Municipal Code of the City of Cambridge

Title 9. Public Peace, Morals and Welfare

Chapter 9.16. Weapons

9.16.020. Handgun transactions - Council findings. The City Council finds and declares that:

A. Handguns play a major role in the commission of homicide, aggravated assault, armed robbery and accidental injury and death;

B. Handguns should therefore not be readily available within the City;

C. In order to promote and protect the public health, safety and welfare, to preserve peace and good order, and to maintain the internal police of the City, it is necessary to regulate certain handgun transactions.

9.16.030. Handgun transactions - Handgun defined. As used in this section and Sections 9.16.020, 9.16.040 and 9.16.050 of this chapter, "handgun" means any firearm, as defined in Section 121 of Chapter 140 of the General Laws, of which the length of barrel is less than ten inches.

9.16.040. Handgun transactions - Prohibited - Penalty. Any person who sells, rents or leases a handgun shall be punished by a fine of not more than three hundred dollars. Each such sale, rental or lease shall constitute a separate offense.

9.16.050. Handgun transactions - Exemptions from provisions. This section and Sections 9.16.020, 9.16.040 and 9.16.050 of this chapter shall not apply to:

A. Any person who holds a license to sell, rent or lease firearms from the Chief of Police under section 121 of Chapter 140 of the General Laws, if the license was first issued on or before May 15, 1986; or

B. Any person to whom any licensed mentioned in subsection A of this section is transferred; or

C. Any person or circumstances mentioned in Section 128A of Chapter 140 of the General laws.

9.16.090. Replica firearms.

A. Definitions. As used in this section, the following words shall have the following meanings:

1. "Chief" means the Chief of the Cambridge Police Department or his designee.

2. "Department" means the Cambridge Police Department.

3. "Market" means to display for sale.

4. "Replica firearm" means any toy, imitation or facsimile pistol, revolver, shotgun, rifle, air rifle, B-B gun, pellet gun, machine gun or other similar simulated weapon which because of its color, size, shape or other characteristics, can reasonably be perceived to be a real firearm from which a shot or bullet can be fired and which, pursuant to subsection D of this section, has been determined to pose a threat to public safety.

5. means to exchange or deliver:

a. For money or its equivalent, or

b. As a promotion or an inducement to buy.

B. Regulation. Except as provided in subsection C of this section, no person or entity shall sell or market any replica firearm in the City.

C. Exceptions. Notwithstanding the foregoing, the sale of replica firearms shall be permitted if the sale of such replica firearms is solely for purposes of or for transportation in intrastate, interstate or foreign commerce.

D. Enforcement.

1. The Chief shall identify those replica firearms which he deems to pose a threat to public safety and shall make a list thereof. Such list shall include, where available, the following information: the product name, the manufacturer and the model number. The Chief shall post the list in the Department and shall make available copies thereof to the general public. The Chief shall update the list from time to time as he deems warranted. Any individual may provide information to the Chief regarding replica firearms to be added to the list.

2. Any person who observes the sale or marketing of a replica firearm which he believes to be a violation of this section may register a written or oral complaint with the Department. A Cambridge police officer, in response to each complaint, shall investigate the complaint forthwith in person and determine whether there is such a violation.

3. Any Cambridge police officer who observes the sale or marketing of a replica firearm which he believes to be a violation of this section shall investigate forthwith and determine whether there is such a violation.

4. If a Cambridge police officer determines that there is a violation of this section, he shall order the individual or entity to remove immediately from sale or marketing all the replica firearms found to be in violation. In addition, the Department shall impose the following penalties on the individual or entity violating this section:

a. A written citation or warning for the first violation;

b. A one hundred dollar fine for the second violation; and

c. A two hundred dollar fine for the third and any subsequent violations.

5. For purposes of penalties to be imposed under this section, if a Cambridge police officer determines that a sale or marketing of replica firearms has occurred, that determination shall constitute one violation, regardless of the number of replica firearms involved. A Cambridge police officer shall return, unannounced, to the site of the violation within the thirty day period following the determination of a violation to ensure that the individual or entity has not resumed the sale or marketing of replica firearms.

E. Appeal Procedure. Any individual who is aggrieved by an action taken by a Cambridge police officer under this section may, within ten days of such action, file an appeal, in writing, to the Chief. After notice to such individual, the Chief shall hold a hearing, after which he shall issue a decision in which he affirms, annuls or modifies an action taken by a Cambridge police officer, giving his reasons therefor. The Chief shall send the decision to the individual by first class mail within ten days after the hearing. The decision shall be a final administrative decision. The individual shall have thirty days from the date of the written decision to seek judicial review in the Third District Court of Eastern Middlesex County.

[Cambridge City Code current through Ord. 1311 passed Dec. 10, 2007]

Revised Ordinances of the City of Worcester

Part I. Regulatory Ordinances

C - Public Safety

Chapter 11. Licenses
Chief of Police

9. Firearm Sales

(a) No person shall sell any firearm without a license issued by the chief of police pursuant to the provisions of G.L. c. 140, §122. The fee shall be twenty dollars per year. Such licenses shall be issued consistent with the provisions of G.L. c. 140, §§ 121 to 131I.

(b) The penalty for each violation of this section shall be three hundred dollars.

10. Ammunition Sales

(a) No person shall sell any ammunition for any firearm without a license issued by the chief of police pursuant to the provisions of G.L. c. 140, §122B. The fee shall be twenty-five dollars. The fee for a renewal license shall be one dollar per year.

(b) The penalty for each violation of this section shall be three hundred dollars.

[Revised Ordinances of the City of Worcester current as of Sept. 9, 2008]

MICHIGAN
MICH. COMP. LAWS

Chapter 3. Federal and Interstate Relations - Purchase of Rifles and Shotguns

3.111. Rifles and shotguns; purchases by residents. Residents of this state may purchase rifles and shotguns in any state contiguous thereto if they conform to the federal gun control act of 1968 and the regulations issued thereunder, as administered by the secretary of the treasury and the laws of the state where the purchase is made.

3.112. Rifles and shotguns; purchases by nonresidents. Residents of a contiguous state may purchase rifles and shotguns in this state if they conform to the federal gun control act of 1968 and the regulations issued thereunder as administered by the secretary of the treasury and the laws of the state wherein the purchaser resides.

[Publisher's Note: The above statutes were originally enacted as procedural/enabling statutes so as conform state law with the Gun Control Act of 1968 (GCA). In 1986 the GCA was amended to permit long gun transfers from Federal Firearms Licensees to residents of any state, whether contiguous or not, if made at the seller's licensed business premises and only if in full compliance with the legal conditions of sale in both the buyer's and seller's states of residence. The Michigan Attorney General's Office has determined that these provisions do not imply restrictions to sales beyond contiguous states by persons properly licensed by the Federal government to deal in firearms.**]**

Chapter 8. Statutes

8.3t. Firearm; defined; exception. The word "firearm", except as otherwise specifically defined in the statutes, shall be construed to include any weapon from which a dangerous projectile may be propelled by using explosives, gas or air as a means of propulsion, except any smooth bore rifle or handgun designed and manufactured exclusively for propelling BB's not exceeding .177 calibre by means of spring, gas or air.

Chapter 28. Michigan State Police - Firearms.

28.421. Definitions. As used in this act:

(a) "Felony" means that term as defined in section 1 of chapter I of the code of criminal procedure, 1927 PA 175, MCL761.1, or a violation of a law of the United States or another state that is designated as a felony or that is punishable by death or by imprisonment for more than 1 year.

(b) "Firearm" means a weapon from which a dangerous projectile may be propelled by an explosive, or by gas or air. Firearm does not include a smooth bore rifle or handgun designed and manufactured exclusively for propelling by a spring, or by gas or air, BB's not exceeding .177 caliber.

(c) "Misdemeanor" means a violation of a penal law of this state or violation of a local ordinance substantially corresponding to a violation of a penal law of this state that is not a felony or a violation of an order, rule, or regulation of a state agency that is punishable by imprisonment or a fine that is not a civil fine, or both.

(d) "Peace officer" means, except as otherwise provided in this act, an individual who is employed as a law enforcement officer, as that term is defined under section 2 of the commission on law enforcement standards act, 1965 PA 203, MCL 28.602, by this state or another state, a political subdivision of this state or another state, or the United States, and who is required to carry a firearm in the course of his or her duties as a law enforcement officer.

(e) "Pistol" means a loaded or unloaded firearm that is 30 inches or less in length, or a loaded or unloaded firearm that by its construction and appearance conceals it as a firearm.

(f) "Purchaser" means a person who receives a pistol from another person by purchase or gift.

(g) "Reserve peace officer", "auxiliary officer", or "reserve officer" means, except as otherwise provided in this act, an individual authorized on a voluntary or irregular basis by a duly authorized police agency of this state or a political subdivision of this state to act as a law enforcement officer, who is responsible for the preservation of the peace, the prevention and detection of crime, and the enforcement of the general criminal laws of this state, and who is otherwise eligible to possess a firearm under this act.

(h) "Retired police officer" or "retired law enforcement officer" means an individual who was a certified police officer or certified law enforcement officer as those terms are defined under section 2(k) of the commission on the law enforcement standards act, 1965 PA 203, MCL 28.602, and retired in good standing from his or her employment as a police officer or law enforcement officer.

(i) "Seller" means a person who sells or gives a pistol to another person.

28.421a. Concealed pistol licenses; issuance; creation of standardized system. It is the intent of the legislature to create a standardized system for issuing concealed pistol licenses to prevent criminals and other violent individuals from obtaining a license to carry a concealed pistol, to allow law abiding residents to obtain a license to carry a concealed pistol, and to prescribe the rights and responsibilities of individuals who have obtained a license to carry a concealed pistol. It is also the intent of the legislature to grant an applicant the right to know why his or her application for a concealed pistol license is denied and to create a process by which an applicant may appeal that denial.

28.422. License to purchase, carry, or transport pistol; issuance; qualifications; applications; sale of pistol; exemptions; basic pistol safety brochure; forging application; implementation during business hours.

Sec. 2.(1) [Until January 7, 2009]Except as provided in subsection (2), a person shall not purchase, carry, or transport a pistol in this state without first having obtained a license for the pistol as prescribed in this section.

(2) A person who brings a pistol into this state who is on leave from active duty with the armed forces of the United States or who has been discharged from active duty with the armed forces of the United States shall obtain a license for the pistol within 30 days after his or her arrival in this state.

(3) The commissioner or chief of police of a city, township, or village police department that issues licenses to purchase, carry, or transport pistols, or his or her duly authorized deputy, or the sheriff or his or her duly authorized deputy, in the parts of a county not included within a city, township, or village having an organized police department, in discharging the duty to issue licenses shall with due speed and diligence issue licenses to purchase, carry, or transport pistols to qualified applicants residing within the city, village, township, or county, as applicable unless he or she has probable cause to believe that the applicant would be a threat to himself or herself or to other individuals, or would commit an offense with the pistol that would violate a law of this or another state or of the United States. An applicant is qualified if all of the following circumstances exist:

(a) The person is not subject to an order or disposition for which he or she has received notice and an opportunity for a hearing, and which was entered into the law enforcement information network pursuant to any of the following:

(i) Section 464a(1) of the mental health code, 1974 PA 258, MCL 330.1464a.

(ii) Section 5107 of the estates and protected individuals code, 1998 PA 386, MCL 700.5107, or section 444a of former 1978 PA 642.

(iii) Section 2950(9) of the revised judicature act of 1961, 1961 PA 236, MCL 600.2950.

(iv) Section 2950a(7) of 1961 PA 236, MCL 600.2950a.

(v) Section 14 of 1846 RS 84, MCL 552.14.

(vi) Section 6b(5) of chapter V of the code of criminal procedure, 1927 PA 175, MCL 765.6b, if the order has a condition imposed pursuant to section 6b(3) of chapter V of 1927 PA 175, MCL 765.6b.

(vii) Section 16b(1) of chapter IX of 1927 PA 175, MCL 769.16b.

(b) The person is 18 years of age or older or, if the seller is licensed pursuant to section 923 of title 18 of the United States Code, 18 USC 923, is 21 years of age or older.

(c) The person is a citizen of the United States and is a legal resident of this state.

(d) A felony charge against the person is not pending at the time of application.

(e) The person is not prohibited from possessing, using, transporting, selling, purchasing, carrying, shipping, receiving, or distributing a firearm under section 224f of the Michigan penal code, 1931 PA 328, MCL 750.224f.

(f) The person has not been adjudged insane in this state or elsewhere unless he or she has been adjudged restored to sanity by court order.

(g) The person is not under an order of involuntary commitment in an inpatient or outpatient setting due to mental illness.

(h) The person has not been adjudged legally incapacitated in this state or elsewhere. This subdivision does not apply to a person who has had his or her legal capacity restored by order of the court.

(i) The person correctly answers 70% or more of the questions on a basic pistol safety review questionnaire approved by the basic pistol safety review board and provided to the individual free of charge by the licensing authority. If the person fails to correctly answer 70% or more of the questions on the basic pistol safety review questionnaire, the licensing authority shall inform the person of the questions he or she answered incorrectly and allow the person to attempt to complete another basic pistol safety review questionnaire. The person shall not be allowed to attempt to complete more than 2 basic pistol safety review questionnaires on any single day. The licensing authority shall allow the person to attempt to complete the questionnaire during normal business hours on the day the person applies for his or her license.

(4) Applications for licenses under this section shall be signed by the applicant under oath upon forms provided by the director of the department of state police. Licenses to purchase, carry, or transport pistols shall be executed in triplicate upon forms provided by the director of the department of state police and shall be signed by the licensing authority. Three copies of the license shall be delivered to the applicant by the licensing authority.

(5) Upon the sale of the pistol, the seller shall fill out the license forms describing the pistol sold, together with the date of sale, and sign his or her name in ink indicating that the pistol was sold to the licensee. The licensee shall also sign his or her name in ink indicating the purchase of the pistol from the seller. The seller may retain a copy of the license as a record of the sale of the pistol. The licensee shall return 2 copies of the license to the licensing authority within 10 days following the purchase of the pistol.

(6) One copy of the license shall be retained by the licensing authority as an official record for a period of 6 years. The other copy of the license shall be forwarded by the licensing authority within 48 hours to the director of the department of state police. A license is void unless used within 10 days after the date of its issue.

(7) This section does not apply to the purchase of pistols from wholesalers by dealers regularly engaged in the business of selling pistols at retail, or to the sale, barter, or exchange of pistols kept as relics or curios not made for modern ammunition or permanently deactivated. This section does not prevent the transfer of ownership of pistols that are inherited if the license to purchase is approved by the commissioner or chief of police, sheriff, or their authorized deputies, and signed by the personal representative of the estate or by the next of kin having authority to dispose of the pistol.

(8) The licensing authority shall provide a basic pistol safety brochure to each applicant for a license under this section before the applicant answers the basic pistol safety review questionnaire. A basic pistol safety brochure shall contain, but is not limited to providing, information on all of the following subjects:

(a) Rules for safe handling and use of pistols.

(b) Safe storage of pistols.

(c) Nomenclature and description of various types of pistols.

(d) The responsibilities of owning a pistol.

(9) The basic pistol safety brochure shall be supplied in addition to the safety pamphlet required by section 9b.

(10) The basic pistol safety brochure required in subsection (8) shall be produced by a national nonprofit membership organization that provides voluntary pistol safety programs that include training individuals in the safe handling and use of pistols.

(11) A person who forges any matter on an application for a license under this section is guilty of a felony, punishable by imprisonment for not more than 4 years or a fine of not more than $2,000.00, or both.

(12) A licensing authority shall implement this section during all of the licensing authority's normal business hours and shall set hours for implementation that allow an applicant to use the license within the time period set forth in subsection (6).

(1) [Eff January 7, 2009] Except as provided in subsection (2), a person shall not purchase, carry, possess, or transport a pistol in this state without first having obtained a license for the pistol as prescribed in this section.

(2) A person who brings a pistol into this state who is on leave from active duty with the armed forces of the United States or who has been discharged from active duty with the armed forces of the United States shall obtain a license for the pistol within 30 days after his or her arrival in this state.

(3) The commissioner or chief of police of a city, township, or village police department that issues licenses to purchase, carry, possess, or transport pistols, or his or her duly authorized deputy, or the sheriff or his or her duly authorized deputy, in the parts of a county not included within a city, township, or village having an organized police department, in discharging the duty to issue licenses shall with due speed and diligence issue licenses to purchase, carry, possess, or transport pistols to qualified applicants residing within the city, village, township, or county, as applicable unless he or she has probable cause to believe that the applicant would be a threat to himself or herself or to other individuals, or would commit an offense with the pistol that would violate a law of this or another state or of the United States. An applicant is qualified if all of the following circumstances exist:

(a) The person is not subject to an order or disposition for which he or she has received notice and an opportunity for a hearing, and which was entered into the law enforcement information network pursuant to any of the following:

(i) Section 464a(1) of the mental health code, 1974 PA 258, MCL 330.1464a.

(ii) Section 5107 of the estates and protected individuals code, 1998 PA 386, MCL 700.5107, or section 444a of former 1978 PA 642.

(iii) Section 2950(10) of the revised judicature act of 1961, 1961 PA 236, MCL 600.2950.

(iv) Section 2950a(7) of 1961 PA 236 MCL 600.2950a.

(v) Section 14 of 1846 RS 84, MCL 552.14.

(vi) Section 6b(5) of chapter V of the code of criminal procedure, 1927 PA 175, MCL 765.6b, if the order has a condition imposed under section 6b(3) of chapter V of 1927 PA 175, MCL 765.6b.

(vii) Section 16b(1) of chapter IX of 1927 PA 175, MCL 769.16b.

(b) The person is 18 years of age or older or, if the seller is licensed pursuant to section 923 of title 18 of the United States Code, 18 USC 923, is 21 years of age or older.

(c) The person is a citizen of the United States and is a legal resident of this state.

(d) A felony charge against the person is not pending at the time of application.

(e) The person is not prohibited from possessing, using, transporting, selling, purchasing, carrying, shipping, receiving, or distributing a firearm under section 224f of the Michigan penal code, 1931 PA 328, MCL 750.224f.

(f) The person has not been adjudged insane in this state or elsewhere unless he or she has been adjudged restored to sanity by court order.

(g) The person is not under an order of involuntary commitment in an inpatient or outpatient setting due to mental illness.

(h) The person has not been adjudged legally incapacitated in this state or elsewhere. This subdivision does not apply to a person who has had his or her legal capacity restored by order of the court.

(i) The person correctly answers 70% or more of the questions on a basic pistol safety review questionnaire approved by the basic pistol safety review board and provided to the individual free of charge by the licensing authority. If the person fails to correctly answer 70% or more of the questions on the basic pistol safety review questionnaire, the licensing authority shall inform the person of the questions he or she answered incorrectly and allow the person to attempt to complete another basic pistol safety review questionnaire. The person shall not be allowed to attempt to complete more than 2 basic pistol safety review questionnaires on any single day. The licensing authority shall allow the person to attempt to complete the questionnaire during normal business hours on the day the person applies for his or her license.

(4) Applications for licenses under this section shall be signed by the applicant under oath upon forms provided by the director of the department of state police. Licenses to purchase, carry, possess, or transport pistols shall be executed in quadruplicate upon forms provided by the director of the department of state police and shall be signed by the licensing authority. Four copies of the license shall be delivered to the applicant by the licensing authority. A license is void unless used within 10 days after the date it is issued.

(5) If an individual purchases or otherwise acquires a pistol, the seller shall fill out the license forms describing the pistol, together with the date of sale or acquisition, and sign his or her name in ink indicating that the pistol was sold to or otherwise acquired by the purchaser. The purchaser shall also sign his or her name in ink indicating the purchase or other acquisition of the pistol from the seller. The seller may retain a copy of the license as a record of the transaction. The purchaser shall receive 3 copies of the license. The purchaser shall return 2 copies of the license to the licensing authority within 10 days after the date the pistol is purchased or acquired. The return of the copies to the licensing authority may be made in person or may be made by first-class mail or certified mail sent within the 10-day period to the proper address of the licensing authority. A purchaser who fails to comply with the requirements of this subsection is responsible for a state civil infraction and may be fined not more than $250.00. If a purchaser is found responsible for a state civil infraction under this subsection, the court shall notify the department of state police of that determination.

(6) Within 48 hours after receiving the license copies returned under subsection (5), the licensing authority shall forward 1 copy of the license to the department of state police. The licensing authority shall retain the other copy of the license as an official record for not less than 6 years. Within 10 days after receiving the license copies returned under subsection (5), the licensing authority shall electronically enter the information into the pistol entry database as required by the department of state police if it has the ability to electronically enter that information. If the licensing authority does not have that ability, the licensing authority shall provide that information to the department of state police in a manner otherwise required by the department of state police. Any licensing authority that provided pistol descriptions to the department of state police under former section 9 of this act shall continue to provide pistol descriptions to the department of state police under this subsection. The purchaser has the right to obtain a copy of the information placed in the pistol entry database under this subsection to verify the accuracy of that information. The licensing authority may charge a fee not to exceed $1.00 for the cost of providing the copy. The licensee may carry, use, possess, and transport the pistol for 30 days beginning on the date of purchase or acquisition only while he or she is in possession of his or her copy of the license. However, the person is not required to have the license in his or her possession while carrying, using, possessing, or transporting the pistol after this period.

(7) This section does not apply to the purchase of pistols from wholesalers by dealers regularly engaged in the business of selling pistols at retail, or to the sale, barter, or exchange of pistols kept as relics or curios not made for modern ammunition or permanently deactivated. This section does not prevent the transfer of ownership of pistols that are inherited if the license to purchase is approved by the commissioner or chief of police, sheriff, or their authorized deputies, and signed by the personal representative of the estate or by the next of kin having authority to dispose of the pistol.

(8) The licensing authority shall provide a basic pistol safety brochure to each applicant for a license under this section before the applicant answers the basic pistol safety review questionnaire. A basic pistol safety brochure shall contain, but is not limited to providing, information on all of the following subjects:

(a) Rules for safe handling and use of pistols.

(b) Safe storage of pistols.

(c) Nomenclature and description of various types of pistols.

(d) The responsibilities of owning a pistol.

(9) The basic pistol safety brochure shall be supplied in addition to the safety pamphlet required by section 9b.

(10) The basic pistol safety brochure required in subsection (8) shall be produced by a national nonprofit membership organization that provides voluntary pistol safety programs that include training individuals in the safe handling and use of pistols.

(11) A person who forges any matter on an application for a license under this section is guilty of a felony, punishable by imprisonment for not more than 4 years or a fine of not more than $2,000.00, or both.

(12) A licensing authority shall implement this section during all of the licensing authority's normal business hours and shall set hours for implementation that allow an applicant to use the license within the time period set forth in subsection (6).

28.422a. Individual licensed under § 28.425b; sales record; exemption; materially false statement as felony; penalty; rules.

Sec. 2a. (1) [Until January 7, 2009]An individual who is licensed under section 5b to carry a concealed pistol is not required to obtain a license under section 2 to purchase, carry, or transport a pistol.

(2) If an individual licensed under section 5b purchases a pistol, the seller shall complete a sales record in triplicate on a form provided by the department of state police. The record shall include the individual's concealed weapon license number. The individual purchasing the pistol shall sign the record. The seller shall retain 1 copy of the record, provide 1 copy to the individual purchasing the pistol, and forward the original to the department of state police within 10 days following the purchase.

(3) This section does not apply to a person or entity exempt under section 2(7).

(4) An individual who makes a material false statement on a sales record under this section is guilty of a felony punishable by imprisonment for not more than 4 years or a fine of not more than $2,500.00, or both.

(5) The department of state police may promulgate rules to implement this section.

[(1) Eff January 7, 2009, the section heading reads: "Individual licensed under MCL 28.425b; completion of record by seller; duties of purchaser; noncompliance as state civil infraction; penalty; forwarding copy of record to state police; exemption; materially false statement as felony; penalty; rules." and this section provides:]An individual who is licensed under section 5b to carry a concealed pistol is not required to obtain a license under section 2 to purchase, carry, possess, or transport a pistol.

(2) If an individual licensed under section 5b purchases or otherwise acquires a pistol, the seller shall complete a record in quadruplicate on a form provided by the department of state police. The record shall include the purchaser's concealed weapon license number. The purchaser shall sign the record. The seller may retain 1 copy of the record. The purchaser shall receive 3 copies of the record and forward 2 copies to the police department of the city, village, or township in which the purchaser resides, or, if the purchaser does not reside in a city, village, or township having a police department, to the county sheriff, within 10 days following the purchase or acquisition. The return of the copies to the police department or county sheriff may be made in person or may be made by first-class mail or certified mail sent within the 10-day period to the proper address of the police department or county sheriff. A purchaser who fails to comply with the requirements of this subsection is responsible for a state civil infraction and may be fined not more than $250.00. If a purchaser is found responsible for a state civil infraction under this subsection, the court shall notify the department of state police and the concealed weapon licensing board that issued the license of that determination.

(3) Within 48 hours after receiving the record copies returned under subsection (2), the police department or county sheriff shall forward 1 copy of the record to the department of state police. The police department or county sheriff shall retain the other copy of the record as an official record for not less than 6 years. Within 10 days after receiving the record copies returned under subsection (2), the police department or county sheriff shall electronically enter the information into the pistol entry database as required by the department of state police if it has the ability to electronically enter that information. If the police department or county sheriff does not have that ability, the police department or county sheriff shall provide that information to the department of state police in a manner otherwise required by the department of state police. Any police department or county sheriff that provided pistol descriptions to the department of state police under former section 9 of this act shall continue to provide pistol descriptions to the department of state police under this subsection. The purchaser has the right to obtain a copy of the information placed in the pistol entry database under this subsection to verify the accuracy of that information. The police department or county sheriff may charge a fee not to exceed $1.00 for the cost of providing the copy. The purchaser may carry, use, possess, and transport the pistol for 30 days beginning on the date of purchase or acquisition only while he or she is in possession of his or her copy of the record. However, the person is not required to have the record in his or her possession while carrying, using, possessing, or transporting the pistol after this period.

(4) This section does not apply to a person or entity exempt under section 2(7).

(5) An individual who makes a material false statement on a sales record under this section is guilty of a felony punishable by imprisonment for not more than 4 years or a fine of not more than $2,500.00, or both.

(6) The department of state police may promulgate rules to implement this section.]

28.424. Restoration of rights by concealed weapons licensing board; application; fee; determination; circumstances; judicial review.

(1) A person who is prohibited from possessing, using, transporting, selling, purchasing, carrying, shipping, receiving, or distributing a firearm under section 224f(2) of the Michigan penal code, Act No. 328 of the Public Acts of 1931, being section 750.224f of the Michigan Compiled Laws, may apply to the concealed weapons licensing board in the county in which he or she resides for restoration of those rights.

(2) Not more than 1 application may be submitted under subsection (1) in any calendar year. The concealed weapons licensing board may charge a fee of not more than $10.00 for the actual and necessary expenses of each application.

(3) The concealed weapons licensing board shall, by written order of the board, restore the rights of a person to possess, use, transport, sell, purchase, carry, ship, receive, or distribute a firearm if the board determines, by clear and convincing evidence, that all of the following circumstances exist:

(a) The person properly submitted an application for restoration of those rights as provided under this section.

(b) The expiration of 5 years after all of the following circumstances:

(i) The person has paid all fines imposed for the violation resulting in the prohibition.

(ii) The person has served all terms of imprisonment imposed for the violation resulting in the prohibition.

(iii) The person has successfully completed all conditions of probation or parole imposed for the violation resulting in the prohibition.

(c) The person's record and reputation are such that the person is not likely to act in a manner dangerous to the safety of other persons.

(4) If the concealed weapons licensing board pursuant to subsection (3) refuses to restore a right under this section, the person may petition the circuit court for review of that decision.

28.425. Concealed pistol application kits.

(1) County sheriffs, local police agencies, and county clerks shall provide concealed pistol application kits during normal business hours and free of charge to individuals who wish to apply for licenses to carry concealed pistols. ...

28.425a. Concealed weapon licensing board; membership; quorum; voting; clerk; authority and duties; panel; investigation of license applicant; temporary license; compilation of firearms laws by legislative service bureau; distribution; statement.

(1) Each county shall have a concealed weapon licensing board. ...

(9) The legislative service bureau shall compile the firearms laws of this state, including laws that apply to carrying a concealed pistol, and shall provide copies of the compilation to each concealed weapon licensing board in this state for distribution under this subsection. A concealed weapon licensing board shall distribute a copy of the compilation to each individual who applies for a license to carry a concealed pistol at the time the application is submitted. The concealed weapon licensing board shall require the applicant to sign a written statement acknowledging that he or she has received a copy of the compilation. An individual is not eligible to receive a license to carry a concealed pistol until he or she has signed the statement.

28.425b. License application; fee; verification of requirements; determination; circumstances for issuance; fingerprints; issuance or denial; temporary license; definitions.

(1) To obtain a license to carry a concealed pistol, an individual shall apply to the concealed weapon licensing board in the county in which that individual resides. The application shall be filed with the county clerk during the county clerk's normal business hours. The application shall be on a form provided by the director of the department of state police and shall allow the applicant to designate whether the applicant seeks a temporary license. The application shall be signed under oath by the applicant. The oath shall be administered by the county clerk or his or her representative. The application shall contain all of the following information:

(a) The applicant's legal name and date of birth and the address of his or her primary residence. If the applicant resides in a city, village, or township that has a police department, the name of the police department. Information received under this subdivision is confidential, is not subject to disclosure under the freedom of information act, 1976 PA 442, MCL 15.231 to 15.246, and shall not be disclosed to any person except for purposes of this act for law enforcement purposes.

(b) A statement by the applicant that the applicant meets the criteria for a license under this act to carry a concealed pistol.

(c) A statement by the applicant authorizing the concealed weapon licensing board to access any record, including any medical record, pertaining to the applicant's qualifications for a license to carry a concealed pistol under this act. The applicant may request that information received by the concealed weapon licensing board under this subdivision be reviewed in a closed session. If the applicant requests that the session be closed, the concealed weapon licensing board shall close the session only for purposes of this subdivision. The applicant and his or her representative have the right to be present in the closed session. Medical records and personal identifying information received by the concealed weapon licensing board under this subdivision is confidential, is not subject to disclosure under the freedom of information act, 1976 PA 442, MCL 15.231 to 15.246, and shall not be disclosed to any person except for purposes of this act or for law enforcement purposes or if the applicant is convicted of a felony involving a pistol.

(d) A statement by the applicant regarding whether he or she has a history of mental illness that would disqualify him or her under subsection (7)(j) to (l) from receiving a license to carry a concealed pistol, and authorizing the concealed weapon licensing board to access the mental health records of the applicant relating to his or her mental health history. ...

(e) A statement by the applicant regarding whether he or she has ever been convicted in this state or elsewhere for any of the following:

(i) any felony;

(ii) any misdemeanor listed under subsection (7)(h) or (i), if the applicant was convicted of violating that misdemeanor in the 8 years immediately preceding the date of the application.

(f) A statement by the applicant whether he or she has been dishonorably discharged from the United States armed forces.

(g) If the applicant seeks a temporary license, the facts supporting the issuance of that temporary license.

(h) The names, residential addresses, and telephone numbers of 2 individuals who are references for the applicant. Information received under this subdivision is confidential, is not subject to disclosure under the Freedom of Information Act, 1976 PA 442, MCL 15.231 to 15.246, and shall not be disclosed to any person except for purposes of this act and for law enforcement purposes.

(i) A passport-quality photograph of the applicant provided by the applicant at the time of application.

(j) A certificate stating that the applicant has completed the training course prescribed by this act.

(2) The application form shall contain a conspicuous warning that the application is executed under oath and that intentionally making a material false statement on the application is a felony punishable by imprisonment for not more than 4 years or a fine of not more than $2,500.00, or both.

(3) An individual who intentionally makes a material false statement on an application under subsection (1) is guilty of a felony punishable by imprisonment for not more than 4 years or a fine of not more than $2,500.00, or both.

(4) The concealed weapon licensing board shall retain a copy of each application for a license to carry a concealed pistol as an official record. One year after the expiration of a concealed pistol license, the county clerk may destroy the record and maintain only a name index of the record.

(5) Each applicant shall pay a nonrefundable fee of $105.00 by any method of payment accepted by that county for payments of other fees and penalties. Except for a local police agency as provided in subsection (9), a unit of local government, an agency of a unit of local government, or an agency or department of this state shall not charge an additional fee, assessment, or other amount in connection with a license under this section. The fee shall be payable to the county. The county treasurer shall deposit $41.00 of each fee collected under this section in the general fund of the county and credit $26.00 of that deposit to the credit of the county clerk and $15.00 of that deposit to the credit of the county sheriff and forward the balance to the state treasurer. The state treasurer shall deposit the balance of the fee in the general fund to the credit of the department of state police. The department of state police shall use the money received under this act to process the fingerprints and to reimburse the federal bureau of investigation for the costs associated with processing fingerprints submitted under this act. The balance of the money received under this act shall be credited to the department of state police.

(6) The county sheriff on behalf of the concealed weapon licensing board shall verify the requirements of subsection (7)(d), (e), (f), (h), (i), (j), (k), (l), and (m) through the law enforcement information network and report his or her finding to the concealed weapon licensing board. If the applicant resides in a city, village, or township that has a police department, the concealed weapon licensing board shall contact that city, village, or township police department to determine only whether that city, village, or township police department has any information relevant to the investigation of whether the applicant is eligible under this act to receive a license to carry a concealed pistol.

(7) The concealed weapon licensing board shall issue a license to an applicant to carry a concealed pistol within the period required under this act after the applicant properly submits an application under subsection (1) and the concealed weapon licensing board determines that all of the following circumstances exist:

(a) The applicant is 21 years of age or older.

(b) The applicant is a citizen of the United States or is an alien lawfully admitted into the United States, is a resident of this state, and has resided in this state for not less than 6 months.

(c) The applicant has knowledge and has had training in the safe use and handling of a pistol by the successful completion of a pistol safety training course or class that meets the requirements of section 5j, and that is available to the general public and presented by a law enforcement agency, junior or community college, college, or public or private institution or organization or firearms training school.

(d) The applicant is not the subject of an order or disposition under any of the following:

(i) Section 464a of the mental health code, 1974 PA 258, MCL 330.1464a.

(ii) Former section 444a of the revised probate code, 1978 PA 642, MCL 700.444a, or section 5107 of the estates and protected individuals code, 1998 PA 386, MCL 700.5107.

(iii) Sections 2950 and 2950a of the revised judicature act of 1961, 1961 PA 236, MCL 600.2950 and 600.2950a.

(iv) Section 6b of chapter V of the code of criminal procedure, 1927 PA 175, MCL 765.6b, if the order has a condition imposed pursuant to section 6b(3) of chapter V of the code of criminal procedure, 1927 PA 175, MCL 765.6b.

(v) Section 16b of chapter IX of the code of criminal procedure, 1927 PA 175, MCL 769.16b.

(e) The applicant is not prohibited from possessing, using, transporting, selling, purchasing, carrying, shipping, receiving, or distributing a firearm under section 224f of the Michigan penal code, 1931 PA 328, MCL 750.224f.

(f) The applicant has never been convicted of a felony in this state or elsewhere, and a felony charge against the applicant is not pending in this state or elsewhere at the time he or she applies for a license described in this section.

(g) The applicant has not been dishonorably discharged from the United States armed forces.

(h) The applicant has not been convicted of a misdemeanor violation of any of the following in the 8 years immediately preceding the date of application: ...

(i) The applicant has not been convicted of a misdemeanor violation of any of the following in the 3 years immediately preceding the date of application unless the misdemeanor violation is listed under subdivision (h): ...

(j) The applicant has not been found guilty but mentally ill of any crime and has not offered a plea of not guilty of, or been acquitted of, any crime by reason of insanity.

(k) The applicant has never been subject to an order of involuntary commitment in an inpatient or outpatient setting due to mental illness.

(l) The applicant does not have a diagnosed mental illness at the time the application is made regardless of whether he or she is receiving treatment for that illness.

(m) The applicant is not under a court order of legal incapacity in this state or elsewhere.

(n) Issuing a license to the applicant to carry a concealed pistol in this state is not detrimental to the safety of the applicant or to any other individual. A determination under this subdivision shall be based on clear and convincing evidence of repeated violations of this act, crimes, personal protection orders or injunctions, or police reports or other clear and convincing evidence of the actions of, or statements of, the applicant that bear directly on the applicant's ability to carry a concealed pistol.

(8) Upon entry of a court order or conviction of 1 of the enumerated prohibitions for using, transporting, selling, purchasing, carrying, shipping, receiving or distributing a firearm in this section the department of state police shall immediately enter the order or conviction into the law enforcement information network. For purposes of this act, information of the court order or conviction shall not be removed from the law enforcement information network, but may be moved to a separate file intended for the use of the county concealed weapon licensing boards, the courts, and other government entities as necessary and exclusively to determine eligibility to be licensed under this act.

(9) An individual, after submitting an application and paying the fee prescribed under subsection (5), shall request and have classifiable fingerprints taken by the county sheriff or a local police agency if that local police agency maintains fingerprinting capability. If the individual requests that classifiable fingerprints be taken by a local police agency, the individual shall also pay to that local police agency a fee of $15.00 by any method of payment accepted by the unit of local government for payments of other fees and penalties. The county sheriff or local police agency shall take the fingerprints within 5 business days after the request.

(10) The fingerprints shall be taken, under subsection (9), on forms and in a manner prescribed by the department of state police. The fingerprints shall be immediately forwarded to the department of state police for comparison with fingerprints already on file with the department of state police. The department of state police shall forward the fingerprints to the federal bureau of investigation. Within 10 days after receiving a report of the fingerprints from the federal bureau of investigation, the department of state police shall provide a copy to the submitting sheriff's department or local police agency as appropriate and the clerk of the appropriate concealed weapon licensing board. Except as provided in subsection (14), the concealed weapon licensing board shall not issue a concealed pistols license until it receives the fingerprint comparison report prescribed in this subsection. The concealed weapon licensing board may deny a license if an individual's fingerprints are not classifiable by the federal bureau of investigation.

(11) The concealed weapon licensing board shall deny a license to an applicant to carry a concealed pistol if the applicant is not qualified under subsection (7) to receive that license.

(12) A license to carry a concealed pistol that is issued based upon an application that contains a material false statement is void from the date the license is issued.

(13) Subject to subsections (10) and (14), the concealed weapon licensing board shall issue or deny issuance of a license within 45days after the concealed weapon licensing board receives the fingerprint comparison report provided under subsection (10). If the concealed weapon licensing board denies issuance of a license to carry a concealed pistol, the concealed weapon licensing board shall within 5 business days do both of the following:

(a) Inform the applicant in writing of the reasons for the denial. Information under this subdivision shall include all of the following:

(i) A statement of the specific and articulable facts supporting the denial.

(ii) Copies of any writings, photographs, records, or other documentary evidence upon which the denial is based.

(b) Inform the applicant in writing of his or her right to appeal the denial to the circuit court as provided in section5d.

(14) If the fingerprint comparison report is not received by the concealed weapon licensing board within 60 days after the fingerprint report is forwarded to the department of state police by the federal bureau of investigation, the concealed weapon licensing board shall issue a temporary license to carry a concealed pistol to the applicant if the applicant is otherwise qualified for a license. A temporary license issued under this section is valid for 180 days or until the concealed weapon licensing board receives the fingerprint comparison report provided under subsection (10) and issues or denies issuance of a license to carry a concealed pistol as otherwise provided under this act. Upon issuance or the denial of issuance of the license to carry a concealed pistol to an applicant who received a temporary license under this section, the applicant shall immediately surrender the temporary license to the concealed weapon licensing board that issued that temporary license.

(15) If an individual licensed under this act to carry a concealed pistol moves to a different county within this state, his or her license remains valid until it expires or is otherwise suspended or revoked under this act. A license to carry a concealed pistol that is lost, stolen, or defaced may be replaced by the issuing county clerk for a replacement fee of $10.00.

(16) If a concealed weapons licensing board suspends or revokes a license issued under this act, the license is forfeited and shall be returned to the concealed weapon licensing board forthwith.

(17) An applicant or an individual licensed under this act to carry a concealed pistol may be furnished a copy of his or her application under this section upon request and the payment of a reasonable fee.

(18) This section does not prohibit the concealed weapon licensing board from making public and distributing to the public at no cost lists of individuals who are certified as qualified instructors as prescribed under section 5(j).

(19) As used in this section:

(a) "Convicted" means a final conviction, the payment of a fine, a plea of guilty or nolo contendere if accepted by the court, or a finding of guilt for a criminal law violation or a juvenile adjudication or disposition by the juvenile division of probate court or family division of circuit court for a violation that if committed by an adult would be a crime.

(b) "Felony" means that term as defined in section 1 of chapter I of the code of criminal procedure, 1927 PA 175, MCL 761.1, or a violation of a law of the United States or another state that is designated as a felony or that is punishable by death or by imprisonment for more than 1 year.

(c) "Mental illness" means a substantial disorder of thought or mood that significantly impairs judgment, behavior, capacity to recognize reality, or ability to cope with the ordinary demands of life, and includes, but is not limited to, clinical depression.

(d) "Misdemeanor" means a violation of a penal law of this state or violation of a local ordinance substantially corresponding to a violation of a penal law of this state that is not a felony or a violation of an order, rule, or regulation of a state agency that is punishable by imprisonment or a fine that is not a civil fine, or both.

(e) "Treatment" means care or any therapeutic service, including, but not limited to, the administration of a drug, and any other service for the treatment of a mental illness.

28.425g. Pistol subject to seizure and forfeiture. A pistol carried in violation of this act is subject to seizure and forfeiture in the same manner that property is subject to seizure and forfeiture under sections 4701 to 4709 of the revised judicature act of 1961, 1961 PA 236, MCL 600.4701 to 600.4709. This section does not apply if the violation is a state civil infraction under section 5f unless the individual fails to present his or her license within the 45-day period described in that section.

28.425h. Expiration of license issued under former law; renewal license.

(1) An individual who is licensed to carry a concealed pistol on the effective date of the amendatory act that added this section may carry a concealed pistol under that license until the license expires or the individual's authority to carry a concealed pistol under that license is otherwise terminated, whichever occurs first.

(2) An individual who is licensed under this act to carry a concealed pistol on the effective date of the amendatory act that added this section may apply for a renewal license upon the expiration of that license as provided in section 5l.

28.425k. Acceptance of license as implied consent to submit to chemical analysis of breath, blood, or urine.

(1) Acceptance of a license issued under this act to carry a concealed pistol constitutes implied consent to submit to a chemical analysis under this section. This section also applies to individuals listed in section 12a(a) to (f).

(2) An individual shall not carry a concealed pistol while he or she is under the influence of alcoholic liquor or a controlled substance or while having a bodily alcohol content prohibited under this section. A person who violates this section is responsible for a state civil infraction or guilty of a crime as follows: ...

(9) As used in this section:

(a) "Alcoholic liquor" means that term as defined in section 105 of the Michigan liquor control code of 1998, 1998 PA 58, MCL 436.1105.

(b) "Controlled substance" means that term as defined in section 7104 of the public health code, 1978 PA 368, MCL 333.7401.

28.425l. License; validity; duration; renewal; waiver of educational requirements.

(1) A license to carry a concealed pistol issued before July 1, 2003 is valid for 3 years and a license to carry a concealed pistol issued on or after July 1, 2003 but before July 1, 2006 is valid for 5 years, and a license to carry a concealed pistol issued on or after July 1, 2006 is valid until the applicant's date of birth that falls not less than 4 years or more than 5 years after the license is issued. A license that is renewed before the immediately preceding license term expires, but not more than 1 year before the immediately preceding license term expires, shall bear an issue date that is identical to the expiration date of the preceding license. A renewal of a license under section 5b shall, except as provided in this section, be issued in the same manner as an original license issued under section 5b.

(2) The concealed weapon licensing board shall issue or deny issuance of a renewal license within 60 days after the application for renewal is properly submitted. The county clerk shall issue the applicant a receipt for his or her renewal application at the time the application is submitted. The receipt shall contain all of the following:

(a) The name of the applicant.

(b) The date and time the receipt is issued.

(c) The amount paid.

(d) A statement that the receipt is for a license renewal.

(e) A statement of whether the applicant qualifies for an extension under subsection (3).

(f) The name of the county in which the receipt is issued.

(g) An impression of the county seal.

(3) If the concealed weapon licensing board fails to deny or issue a renewal license to the person within 60 days as required under subsection (2), the expiration date of the current license is extended by 180 days or until the renewal license is issued, whichever occurs first. This subsection does not apply unless the person pays the renewal fee at the time the renewal application is submitted and the person has submitted a receipt from a police agency that confirms that a background check has been requested by the applicant.

(4) A person carrying a concealed pistol after the expiration date of his or her license pursuant to an extension under subsection (3) shall keep the receipt issued by the county clerk under subsection (2) and his or her expired license in his or her possession at all times that he or she is carrying the pistol. For the purposes of this act, the receipt is considered to be part of the license to carry a concealed pistol until a renewal license is issued or denied. Failing to have the receipt and expired license in possession while carrying a concealed pistol or failing to display the receipt to a peace officer upon request is a violation of this act.

(5) The educational requirements under section 5b(7)(c) are waived for an applicant who is a retired police officer or retired law enforcement officer.

(6) The educational requirements under section 5b(7)(c) for an applicant who is applying for a renewal of a license under this act are waived except that the applicant shall certify that he or she has completed at least 3 hours' review of the training described under section 5b(7)(c) and has had at least 1 hour of firing range time in the 6 months immediately preceding the subsequent application.

28.425o. Premises on which carrying concealed weapon prohibited; violation.

(1) Subject to subsection (4), an individual licensed under this act to carry a concealed pistol, or who is exempt from licensure under section 12a(f), shall not carry a concealed pistol on the premises of any of the following:

(a) A school or school property except that a parent or legal guardian of a student of the school is not precluded from carrying a concealed pistol while in a vehicle on school property, if he or she is dropping the student off at the school or picking up the child from the school. As used in this section, "school" and "school property" mean those terms as defined in section 237a of the Michigan penal code, 1931 PA 328, MCL 750.237a. ...

(h) A dormitory or classroom of a community college, college, or university. ...

(3) As used in subsection (1), "premises" does not include parking areas of the places identified under subsection (1).

(4) Subsection (1) does not apply to any of the following:

(a) An individual licensed under this act who is a retired police officer or retired law enforcement officer. The concealed weapon licensing board may require a letter from the law enforcement agency stating that the retired police officer or law enforcement officer retired in good standing.

(b) An individual who is licensed under this act and who is employed or contracted by an entity described under subsection (1) to provide security services and is required by his or her employer or the terms of a contract to carry a concealed firearm on the premises of the employing or contracting entity.

(c) An individual who is licensed as a private investigator or private detective under the private detective license act, 1965 PA 285, MCL 338.821 to 338.851.

(d) Any of the following who is licensed under this act while on duty and in the course of his or her employment:

(i) A corrections officer of a county sheriff's department.

(ii) A motor carrier officer or capitol security officer of the department of state police.

(iii) A member of a sheriff's posse.

(iv) An auxiliary officer or reserve officer of a police or sheriff's department.

(v) A parole or probation officer of the department of corrections.

(5) An individual who violates this section is responsible for a state civil infraction or guilty of a crime as follows:

(a) Except as provided in subdivisions (b) and (c), the individual is responsible for a state civil infraction and may be fined not more than $500.00. The court shall order the individual's license to carry a concealed pistol suspended for 6months.

(b) For a second violation, the individual is guilty of a misdemeanor punishable by a fine of not more than $1,000.00. The court shall order the individual's license to carry a concealed pistol revoked.

(c) For a third or subsequent violation, the individual is guilty of a felony punishable by imprisonment for not more than 4 years or a fine of not more than $5,000.00, or both. The court shall order the individual's license to carry a concealed pistol revoked.

28.426. Issuance of license; conditions. A license shall not be issued to an applicant under section 2 or 5b unless both of the following apply:

(a) The issuing agency has determined through the federal national instant criminal background check system (NICS) that the applicant is not prohibited under federal law from possessing or transporting a firearm.

(b) If the applicant is not a United States citizen, the issuing agency has verified through the United States immigration and customs enforcement databases that the applicant is not an illegal alien or a nonimmigrant alien.

28.428. Revocation of licenses; grounds, hearing; suspension; order; notice.

(1) The concealed weapon licensing board that issued a license to an applicant to carry a concealed pistol may revoke that license if the board determines that the individual committed any violation of this act other than a violation of section 5f(4) or if the board determines that the individual is not eligible under this act to receive a license to carry a concealed pistol. If the board determines that the individual has been found responsible for 3 or more state civil infraction violations of this act during the license period, the board shall conduct a hearing and may suspend the individual's license for not more than 1 year. ...

28.430. Theft of firearm; report required; failure to report theft as civil violation; penalty.

(1) A person who owns a firearm shall, within 5 days after he or she knows his or her firearm is stolen, report the theft to a police agency having jurisdiction over that theft.

(2) A person who fails to report the theft of a firearm as required under subsection (1) is responsible for a civil violation and may be fined not more than $500.00.

28.431. Review of criminal histories; report; rules. Before January 1, 1995, the director of the state police shall, if sufficient money is appropriated by the legislature, by rules promulgated by the department of state police, provide a system for the expeditious review of the criminal histories of individuals who purchase firearms and file with the legislature a written report of his or her findings and conclusions.

28.432. Inapplicability of §§ 28.422 and 28.429; citation as "Janet Kukuk act".

Sec. 12. (1) [Until January 7, 2009] Sections 2 and 9 do not apply to any of the following:

(a) A police or correctional agency of the United States or of this state or any subdivision of this state.

(b) The United States army, air force, navy, or marine corps.

(c) An organization authorized by law to purchase or receive weapons from the United States or from this state.

(d) The national guard, armed forces reserves, or other duly authorized military organization.

(e) A member of an entity or organization described in subdivisions (a) to (d) for a pistol while engaged in the course of his or her duties with that entity or while going to or returning from those duties.

(f) A United States citizen holding a license to carry a pistol concealed upon his or her person issued by another state.

(g) The regular and ordinary transportation of a pistol as merchandise by an authorized agent of a person licensed to manufacture firearms or a licensed dealer.

(h) Purchasing, owning, carrying, possessing, using, or transporting an antique firearm. As used in this subdivision, "antique firearm" means that term as defined in section 231a of the Michigan penal code, 1931 PA 328, MCL 750.231a.

(i) An individual carrying, possessing, using, or transporting a pistol belonging to another individual, if the other individual's pistol is properly licensed and inspected under this act and the individual carrying, possessing, using, or transporting the pistol has obtained a license under section 5b to carry a concealed pistol.

(2) The amendatory act that added subdivision (h) shall be known and may be cited as the "Janet Kukuk act".

(1) [Eff January 7, 2009] Section 2 does not apply to any of the following:

(a) A police or correctional agency of the United States or of this state or any subdivision of this state.

(b) The United States army, air force, navy, or marine corps.

(c) An organization authorized by law to purchase or receive weapons from the United States or from this state.

(d) The national guard, armed forces reserves, or other duly authorized military organization.

(e) A member of an entity or organization described in subdivisions (a) through (d) for a pistol while engaged in the course of his or her duties with that entity or while going to or returning from those duties.

(f) A United States citizen holding a license to carry a pistol concealed upon his or her person issued by another state.

(g) The regular and ordinary transportation of a pistol as merchandise by an authorized agent of a person licensed to manufacture firearms or a licensed dealer.

(h) Purchasing, owning, carrying, possessing, using, or transporting an antique firearm. As used in this subdivision, "antique firearm" means that term as defined in section 231a of the Michigan penal code, 1931 PA 328, MCL 750.231a.

(i) An individual carrying, possessing, using, or transporting a pistol belonging to another individual, if the other individual's possession of the pistol is authorized by law and the individual carrying, possessing, using, or transporting the pistol has obtained a license under section 5b to carry a concealed pistol or is exempt from licensure as provided in section 12a .

(2) The amendatory act that added subsection (1)(h) shall be known and may be cited as the "Janet Kukuk act".

28.432a. Persons to whom requirements inapplicable. The requirements of this act for obtaining a license to carry a concealed pistol do not apply to any of the following:

(a) A peace officer of a duly authorized police agency of the United States or of this state or a political subdivision of this state, who is regularly employed and paid by the United States or this state or a subdivision of this state, except a township constable.

(b) A constable who is trained and certified under the commission on law enforcement standards act, 1965 PA 203, MCL 28.601 to 28.616, while engaged in his or her official duties or going to or coming from his or her official duties, and who is regularly employed and paid by a political subdivision of this state.

(c) A person regularly employed by the department of corrections and authorized in writing by the director of the department of corrections to carry a concealed pistol during the performance of his or her duties or while going to or returning from his or her duties.

(d) A member of the United States army, air force, navy, or marine corps while carrying a concealed pistol in the line of duty.

(e) A member of the national guard, armed forces reserves, or other duly authorized military organization while on duty or drill or while going to or returning from his or her place of assembly or practice or while carrying a concealed pistol for purposes of that military organization.

(f) A resident of another state who is licensed by that state to carry a concealed pistol.

(g) The regular and ordinary transportation of a pistol as merchandise by an authorized agent of a person licensed to manufacture firearms.

(h) A person while carrying a pistol unloaded in a wrapper or container in the trunk of his or her vehicle or, if the vehicle does not have a trunk, from transporting that pistol unloaded in a locked compartment or container that is separated from the ammunition for that pistol from the place of purchase to his or her home or place of business or to a place of repair or back to his or her home or place of business, or in moving goods from 1 place of abode or business to another place of abode or business.

(i) A peace officer or law enforcement officer from Canada.

28.434. Unlawful possession; weapon forfeited to state; disposal; immunity.

(1) Subject to section 5g, all pistols, weapons, or devices carried or possessed contrary to this act are declared forfeited to the state, and shall be turned over to the director of the department of state police or his or her designated representative, for disposal under this section.

(2) The director of the department of state police shall dispose of firearms under this section by 1 of the following methods:

(a) By conducting a public auction in which firearms received under this section may be purchased at a sale conducted in compliance with section 4708 of the revised judicature act of 1961, 1961 PA 236, MCL 600.4708, by individuals authorized by law to possess those firearms.

(b) By destroying them.

(c) By any other lawful manner prescribed by the director of the department of state police.

(3) Before disposing of a firearm under this section, the director of the department of state police shall do both of the following:

(a) Determine through the law enforcement information network whether the firearm has been reported lost or stolen. If the firearm has been reported lost or stolen and the name and address of the owner can be determined, the director of the department of state police shall provide 30 days' written notice of his or her intent to dispose of the firearm under this section to the owner, and allow the owner to claim the firearm within that 30-day period if he or she is authorized to possess the firearm.

(b) Provide 30 days' notice to the public on the department of state police website of his or her intent to dispose of the firearm under this section. The notice shall include a description of the firearm and shall state the firearm's serial number, if the serial number can be determined. The department of state police shall allow the owner of the firearm to claim the firearm within that 30-day period if he or she is authorized to possess the firearm. The 30-day period required under this subdivision is in addition to the 30-day period required under subdivision (a).

(4) The department of state police is immune from civil liability for disposing of a firearm in compliance with this section.

28.435. Sale of firearms by federally licensed firearms dealer; sale of trigger lock or secured container; exceptions; brochure or pamphlet; statement of compliance; notice of liability; action by political subdivision against firearm or ammunition producer prohibited; rights of state attorney general; exceptions; effect of subsections (9) through (11); violation; penalties; definitions.

(1) Except as provided in subsection (2), a federally licensed firearms dealer shall not sell a firearm in this state unless the sale includes 1of the following:

(a) A commercially available trigger lock or other device designed to disable the firearm and prevent the discharge of the firearm.

(b) A commercially available gun case or storage container that can be secured to prevent unauthorized access to the firearm.

(2) This section does not apply to any of the following:

(a) The sale of a firearm to a police officer or a police agency.

(b) The sale of a firearm to a person who presents to the federally licensed firearms dealer 1 of the following:

(i) A trigger lock or other device designed to disable the firearm and prevent the discharge of the firearm together with a copy of the purchase receipt for the federally licensed firearms dealer to keep. A separate trigger lock or device and a separate purchase receipt shall be required for each firearm purchased.

(ii) A gun case or storage container that can be secured to prevent unauthorized access to the firearm together with a copy of the purchase receipt for the federally licensed firearms dealer to keep. A separate gun case or storage container and a separate purchase receipt shall be required for each firearm purchased.

(c) The sale of an antique firearm. As used in this subdivision, "antique firearm" means that term as defined in section 231a of the Michigan penal code, 1931 PA 328, MCL 750.231a.

(d) The sale or transfer of a firearm if the seller is not a federally licensed firearms dealer.

(3) A federally licensed firearms dealer shall not sell a firearm in this state unless the firearm is accompanied with, free of charge, a brochure or pamphlet that includes safety information on the use and storage of the firearm in a home environment.

(4) Upon the sale of a firearm, a federally licensed firearms dealer shall sign a statement and require the purchaser to sign a statement stating that the sale is in compliance with subsections (1), (2), and (3).

(5) A federally licensed firearms dealer shall retain a copy of the signed statements prescribed in subsection (4) and, if applicable, a copy of the receipt prescribed in subsection (2)(b), for at least 6 years.

(6) A federally licensed firearms dealer in this state shall post in a conspicuous manner at the entrances, exits, and all points of sale on the premises where firearms are sold a notice that says the following: "You may be criminally and civilly liable for any harm caused by a person less than 18 years of age who lawfully gains unsupervised access to your firearm if unlawfully stored."...

(14) Beginning September 1, 2000, a person who violates this section is guilty of a crime as follows:

(a) Except as provided in subdivision (b) or (c), the person is guilty of a misdemeanor punishable by imprisonment for not more than 93 days or a fine of not more than $500.00, or both.

(b) For a second conviction, the person is guilty of a misdemeanor punishable by imprisonment for not more than 1 year or a fine of not more than $1,000.00, or both.

(c) For a third or subsequent conviction, the person is guilty of a felony punishable by imprisonment for not more than 2 years or a fine of not more than $5,000.00, or both.

(15) As used in this section:

(a) "Federally licensed firearms dealer" means a person licensed under section 923 of title 18 of the United States Code, 18 U.S.C. 923.

(b) "Firearm or ammunition" includes a component of a firearm or ammunition.

(c) "Person" means an individual, partnership, corporation, association, or other legal entity.

(d) "Political subdivision" means a county, city, village, township, charter township, school district, community college, or public university or college.

(e) "Produce" means to manufacture, construct, design, formulate, develop standards for, prepare, process, assemble, inspect, test, list, certify, give a warning or instructions regarding, market, sell, advertise, package, label, distribute, or transfer.

Chapter 123. Local Governmental Affairs - Firearms and Ammunition

123.1101. Definitions. As used in this act:

(a) "Local unit of government" means a city, village, township, or county.

(b) "Pistol" means that term as defined in section 222 of the Michigan penal code, Act No. 328 of the Public Acts of 1931, being section 750. 222 of the Michigan Compiled Laws.

123.1102. Regulation of pistols or other firearms. A local unit of government shall not impose special taxation on, enact or enforce any ordinance or regulation pertaining to, or regulate in any other manner the ownership, registration, purchase, sale, transfer, transportation, or possession of pistols or other firearms, ammunition for pistols or other firearms, or components of pistols or other firearms, except as otherwise provided by federal law or a law of this state.

123.1103. Permissible prohibitions or regulation. This act does not prohibit a local unit of government from doing either of the following:

(a) Prohibiting or regulating conduct with a pistol or other firearm that is a criminal offense under state law.

(b) Prohibiting or regulating the transportation, carrying, or possession of pistols and other firearms by employees of that local unit of government in the course of their employment with that local unit of government.

123.1104. Prohibiting discharge of pistol or other firearm. This act does not prohibit a city or a charter township from prohibiting the discharge of a pistol or other firearm within the jurisdiction of that city or charter township.

Chapter 750. Michigan Penal Code

Chapter XXXVII. Firearms

750.222. Definitions. As used in this chapter:

(a) "Alcoholic liquor" means that term as defined in section 105 of the Michigan liquor control code of 1998, 1998 PA 58, MCL 436.1105.

(b) "Barrel length" means the internal length of a firearm as measured from the face of the closed breech of the firearm when it is unloaded, to the forward face of the end of the barrel.

(c) "Controlled substance" means a controlled substance or controlled substance analogue as those terms are defined in section 7104 of the public health code, 1978 PA 368, MCL 333.7104.

(d) "Firearm" means a weapon from which a dangerous projectile may be propelled by an explosive, or by gas or air. Firearm does not include a smooth bore rifle or handgun designed and manufactured exclusively for propelling by a spring, or by gas or air, BB's not exceeding .177 caliber.

(e) "Pistol" means a loaded or unloaded firearm that is 30 inches or less in length, or a loaded or unloaded firearm that by its construction and appearance conceals itself as a firearm.

(f) "Purchaser" means a person who receives a pistol from another person by purchase, gift, or loan.

(g) "Seller" means a person who sells, furnishes, loans, or gives a pistol to another person.

(h) "Shotgun" means a firearm designed or redesigned, made or remade, and intended to be fired from the shoulder and designed or redesigned and made or remade to use the energy of the explosive in a fixed shotgun shell to fire through a smooth bore either a number of ball shot or a single projectile for each single function of the trigger.

(i) "Short-barreled shotgun" means a shotgun having 1 or more barrels less than 18 inches in length or a weapon made from a shotgun, whether by alteration, modification, or otherwise, if the weapon as modified has an overall length of less than 26 inches.

(j) "Rifle" means a firearm designed or redesigned, made or remade, and intended to be fired from the shoulder and designed or redesigned and made or remade to use the energy of the explosive in a fixed metallic cartridge to fire only a single projectile through a rifled bore for each single pull of the trigger.

(k) "Short-barreled rifle" means a rifle having 1 or more barrels less than 16 inches in length or a weapon made from a rifle, whether by alteration, modification, or otherwise, if the weapon as modified has an overall length of less than 26 inches.

750.223. Selling firearms and ammunition; violations; penalties; "licensed dealer" defined.

(1) A person who knowingly sells a pistol without complying with section 2 of Act No. 372 of the Public Acts of 1927, as amended, being section 28.422 of the Michigan Compiled Laws, is guilty of a misdemeanor, punishable by imprisonment for not more than 90 days, or a fine of not more than $100.00, or both.

(2) A person who knowingly sells a firearm more than 30 inches in length to a person under 18 years of age is guilty of a misdemeanor, punishable by imprisonment for not more than 90 days, or a fine of not more than $500.00, or both. A second or subsequent violation of this subsection is a felony punishable by imprisonment for not more than 4 years, or a fine of not more than $2,000.00, or both. It is an affirmative defense to a prosecution under this subsection that the person who sold the firearm asked to see and was shown a driver's license or identification card issued by a state that identified the purchaser as being 18 years of age or older.

(3) A seller shall not sell a firearm or ammunition to a person if the seller knows that either of the following circumstances exists:

(a) The person is under indictment for a felony. As used in this subdivision, "felony" means a violation of a law of this state, or of another state, or of the United States that is punishable by imprisonment for 4 years or more.

(b) The person is prohibited under section 224f from possessing, using, transporting, selling, purchasing, carrying, shipping, receiving, or distributing a firearm.

(4) A person who violates subsection (3) is guilty of a felony, punishable by imprisonment for not more than 10 years, or by a fine of not more than $5,000.00, or both.

(5) As used in this section, "licensed dealer" means a person licensed under section 923 of chapter 44 of title 18 of the United States Code who regularly buys and sells firearms as a commercial activity with the principal objective of livelihood and profit.

750.224. Weapons; manufacture, sale, or possession as felony; exceptions; "muffler" or "silencer" defined.

(1) A person shall not manufacture, sell, offer for sale, or possess any of the following:

(a) A machine gun or firearm that shoots or is designed to shoot automatically more than 1 shot without manual reloading, by a single function of the trigger.

(b) A muffler or silencer.

(c) A bomb or bombshell.

(d) A blackjack, slungshot, billy, metallic knuckles, sand club, sand bag, or bludgeon.

(e) A device, weapon, cartridge, container, or contrivance designed to render a person temporarily or permanently disabled by the ejection, release, or emission of a gas or other substance.

(2) A person who violates subsection (1) is guilty of a felony, punishable by imprisonment for not more than 5 years, or a fine of not more than $2,500.00, or both.

(3) Subsection (1) does not apply to any of the following:

(a) A self-defense spray or foam device as defined in section 224d.

(b) A person manufacturing firearms, explosives, or munitions of war by virtue of a contract with a department of the government of the United States.

(c) A person licensed by the secretary of the treasury of the United States or the secretary's delegate to manufacture, sell, or possess a machine gun, or a device, weapon, cartridge, container, or contrivance described in subsection (1).

(4) As used in this chapter, "muffler" or "silencer" means 1 or more of the following:

(a) A device for muffling, silencing, or deaden-ing the report of a firearm.

(b) A combination of parts, designed or redesigned, and intended for use in assembling or fabricating a muffler or silencer.

(c) A part, designed or redesigned, and intended only for use in assembling or fabricating a muffler or silencer.

750.224a. Portable device or weapon directing electrical current, impulse, wave, or beam; sale or possession prohibited; exceptions; use of electro-muscular disruption technology; violation; penalty; definitions.

(1) Except as otherwise provided in this section, a person shall not sell, offer for sale, or possess in this state a portable device or weapon from which an electrical current, impulse, wave, or beam may be directed, which current, impulse, wave, or beam is designed to incapacitate temporarily, injure, or kill.

(2) This section does not prohibit any of the following:

(a) The possession and reasonable use of a device that uses electro-muscular disruption technology by any of the following individuals, if the individual has been trained in the use, effects, and risks of the device, and is using the device while performing his or her official duties:

(i) A peace officer.

(ii) An employee of the department of corrections who is authorized in writing by the director of the department of corrections to possess and use the device.

(iii) A local corrections officer authorized in writing by the county sheriff to possess and use the device.

(iv) An individual employed by a local unit of government that utilizes a jail or lockup facility who has custody of persons detained or incarcerated in the jail or lockup facility and who is authorized in writing by the chief of police, director of public safety, or sheriff to possess and use the device.

(v) A probation officer.

(vi) A court officer.

(vii) A bail agent authorized under section 167b.

(viii) A licensed private investigator.

(ix) An aircraft pilot or aircraft crew member.

(x) An individual employed as a private security police officer. As used in this subparagraph, "private security police" means that term as defined in section 2 of the private security business and security alarm act, 1968 PA 330, MCL 338.1052.

(b) Possession solely for the purpose of delivering a device described in subsection (1) to any governmental agency or to a laboratory for testing, with the prior written approval of the governmental agency or law enforcement agency and under conditions determined to be appropriate by that agency.

(3) A manufacturer, authorized importer, or authorized dealer may demonstrate, offer for sale, hold for sale, sell, give, lend, or deliver a device that uses electro-muscular disruption technology to a person authorized to possess a device that uses electro-muscular disruption technology and may possess a device that uses electro-muscular disruption technology for any of those purposes.

(4) A person who violates this section is guilty of a felony punishable by imprisonment for not more than 4 years or a fine of not more than $2,000.00, or both.

(5) As used in this section:

(a) "A device that uses electro-muscular disruption technology" means a device to which all of the following apply:

(i) The device is capable of creating an electro-muscular disruption and is used or intended to be used as a defensive device capable of temporarily incapacitating or immobilizing a person by the direction or emission of conducted energy.

(ii) The device contains an identification and tracking system that, when the device is initially used, dispenses coded material traceable to the purchaser through records kept by the manufacturer.

(iii) The manufacturer of the device has a policy of providing the identification and tracking information described in subparagraph (ii) to a police agency upon written request by that agency.

(b) "Local corrections officer" means that term as defined in section 2 of the local corrections officers training act, 2003 PA 125, MCL 791.532.

(c) "Peace officer" means any of the following:

(i) A police officer or public safety officer of this state or a political subdivision of this state, including motor carrier officers appointed under section 6d of 1935 PA 59, MCL 28.6d, and security personnel employed by the state under section 6c of 1935 PA 59, MCL 28.6c.

(ii) A sheriff or a sheriff's deputy.

(iii) A police officer or public safety officer of a junior college, college, or university who is authorized by the governing board of that junior college, college, or university to enforce state law and the rules and ordinances of that junior college, college, or university.

(iv) A township constable.

(v) A marshal of a city, village, or township.

(vi) A conservation officer of the department of natural resources or the department of environmental quality.

(vii) A law enforcement officer of another state or of a political subdivision of another state or a junior college, college, or university in another state, substantially corresponding to a law enforcement officer described in subparagraphs (i) to (vi).

(viii) A federal law enforcement officer.

750.224b. Short-barreled shotgun or rifle; manufacture, sale, or possession as felony; penalty; exceptions; applicability of § 776.20.

(1) A person shall not manufacture, sell, offer for sale, or possess a short-barreled shotgun or a short-barreled rifle.

(2) A person who violates this section is guilty of a felony punishable by imprisonment for not more than 5 years, or a fine of not more than $2,500.00, or both.

(3) The provisions of this section shall not apply to the sale, offering for sale or possession of a short-barreled rifle or a short-barreled shotgun which the secretary of the treasury of the United States of America, or his delegate, pursuant to U.S.C. title 26, section 5801 through 5872, or U.S.C. title 18, sections 921 through 928, has found to be a curio, relic, antique, museum piece or collector's item not likely to be used as a weapon, but only if the person selling, offering for sale or possessing the firearm has also fully complied with the provisions of sections 2 and 9 of Act No. 372 of the Public Acts of 1927, as amended, being sections 28.422 and 28.429 of the Michigan Compiled Laws.

The provisions of section 20 of chapter 16 of Act No. 175 of the Public Acts of 1927, as added by Act No. 299 of the Public Acts of 1968, being section 776.20 of the Michigan Compiled Laws, are applicable to this subsection.

750.224c. Armor piercing ammunition; manufacture, distribution, sale, or use prohibited; exceptions; violation as felony; pen-

alty; definitions; exemption of projectile or projectile core; rule.

(1) Except as provided in subsection (2), a person shall not manufacture, distribute, sell, or use armor piercing ammunition in this state. A person who willfully violates this section is guilty of a felony, punishable by imprisonment for not more than 4 years, or by a fine of not more than $2,000.00, or both.

(2) This section does not apply to either of the following:

(a) A person who manufactures, distributes, sells, or uses armor piercing ammunition in this state, if that manufacture, distribution, sale, or use is not in violation of chapter 44 of title 18 of the United States Code.

(b) A licensed dealer who sells or distributes armor piercing ammunition in violation of this section if the licensed dealer is subject to license revocation under chapter 44 of title 18 of the United States Code for that sale or distribution.

(3) As used in this section:

(a) "Armor piercing ammunition" means a projectile or projectile core which may be used in a pistol and which is constructed entirely, excluding the presence of traces of other substances, of tungsten alloys, steel, iron, brass, bronze, beryllium copper, or a combination of tungsten alloys, steel, iron, brass, bronze, or beryllium copper. Armor piercing ammunition does not include any of the following:

(i) Shotgun shot that is required by federal law or by a law of this state to be used for hunting purposes.

(ii) A frangible projectile designed for target shooting.

(iii) A projectile that the director of the department of state police finds is primarily intended to be used for sporting purposes.

(iv) A projectile or projectile core that the director of the department of state police finds is intended to be used for industrial purposes.

(b) "Licensed dealer" means a person licensed under chapter 44 of title 18 of the United States Code to deal in firearms or ammunition.

(4) The director of the department of state police shall exempt a projectile or projectile core under subsection (3)(a)(iii) or (iv) if that projectile or projectile core is exempted under chapter 44 of title 18 of the United States Code. The director of state police shall exempt a projectile or projectile core under subsection (3)(a)(iii) or (iv) only by a rule promulgated in compliance with the administrative procedures act of 1969, Act No. 306 of the Public Acts of 1969, being sections 24.201 to 24.328 of the Michigan Compiled Laws.

750.224e. Conversion of semiautomatic firearm to fully automatic firearm; prohibited acts; penalty; applicability; "fully automatic firearm", "licensed collector", and "semiautomatic firearm defined.

(1) A person shall not knowingly do any of the following:

(a) Manufacture, sell, distribute, or possess or attempt to manufacture, sell, distribute, or possess a device that is designed or intended to be used to convert a semiautomatic firearm into a fully automatic firearm.

(b) Demonstrate to another person or attempt to demonstrate to another person how to manufacture or install a device to convert a semiautomatic firearm into a fully automatic firearm.

(2) A person who violates subsection (1) is guilty of a felony punishable by imprisonment for not more than 4 years, or a fine of not more than $2,000.00, or both.

(3) This section does not apply to any of the following:

(a) A police agency of this state, or of a local unit of government of this state, or of the United States.

(b) An employee of an agency described in subdivision (a), if the manufacture, sale, distribution, or possession or attempted manufacture, sale, distribution, or possession or demonstration or attempted demonstration is in the course of his or her official duties as an employee of that agency.

(c) The armed forces.

(d) A member or employee of the armed forces, if the manufacture, sale, distribution, or possession or attempted manufacture, sale, distribution, or possession or demonstration or attempted demonstration is in the course of his or her official duties as a member or employee of the armed forces.

(e) A licensed collector who possesses a device that is designed or intended to be used to convert a semiautomatic firearm into a fully automatic firearm that was lawfully owned by that licensed collector before the effective date of the amendatory act that added this section. This subdivision does not permit a licensed collector who lawfully owned a device that is designed or intended to be used to convert a semiautomatic firearm into a fully automatic firearm before the effective date of the amendatory act that added this section to sell or distribute or attempt to sell or distribute that device to another person after the effective date of the amendatory act that added this section.

(4) As used in this section:

(a) "Fully automatic firearm" means a firearm employing gas pressure or force of recoil to mechanically eject an empty cartridge from the firearm after a shot, and to load the next cartridge from the magazine, without renewed pressure on the trigger for each successive shot.

(b) "Licensed collector" means a person who is licensed under chapter 44 of title 18 of the United States Code to acquire, hold, or dispose of firearms as curios or relics.

(c) "Semiautomatic firearm" means a firearm employing gas pressure or force of recoil to mechanically eject an empty cartridge from the firearm after a shot, and to load the next cartridge from the magazine, but requiring renewed pressure on the trigger for each successive shot.

750.224f. Possession of firearm by person convicted of felony; circumstances; penalty; applicability of section to expunged or set aside conviction; "felony" and "specified felony" defined.

(1) Except as provided in subsection (2), a person convicted of a felony shall not possess, use, transport, sell, purchase, carry, ship, receive, or distribute a firearm in this state until the expiration of 3 years after all of the following circumstances exist:

(a) The person has paid all fines imposed for the violation.

(b) The person has served all terms of imprisonment imposed for the violation.

(c) The person has successfully completed all conditions of probation or parole imposed for the violation.

(2) A person convicted of a specified felony shall not possess, use, transport, sell, purchase, carry, ship, receive, or distribute a firearm in this state until all of the following circumstances exist:

(a) The expiration of 5 years after all of the following circumstances exist:

(i) The person has paid all fines imposed for the violation.

(ii) The person has served all terms of imprisonment imposed for the violation.

(iii) The person has successfully completed all conditions of probation or parole imposed for the violation.

(b) The person's right to possess, use, transport, sell, purchase, carry, ship, receive, or distribute a firearm has been restored pursuant to section 4 of Act No. 372 of the Public Acts of 1927, being section 28.424 of the Michigan Compiled Laws.

(3) A person who possesses, uses, transports, sells, purchases, carries, ships, receives, or distributes a firearm in violation of this section is guilty of a felony, punishable by imprisonment for not more than 5 years, or a fine of not more than $5,000.00, or both.

(4) This section does not apply to a conviction that has been expunged or set aside, or for which the person has been pardoned, unless the expunction, order, or pardon expressly provides that the person shall not possess a firearm.

(5) As used in this section, "felony" means a violation of a law of this state, or of another state, or of the United States that is punishable by imprisonment for 4 years or more, or an attempt to violate such a law.

(6) As used in subsection (2), "specified felony" means a felony in which 1 or more of the following circumstances exist:

(i) An element of that felony is the use, attempted use, or threatened use of physical force against the person or property of another, or that by its nature, involves a substantial risk that physical force against the person or property of another may be used in the course of committing the offense.

(ii) An element of that felony is the unlawful manufacture, possession, importation, exportation, distribution, or dispensing of a controlled substance.

(iii) An element of that felony is the unlawful possession or distribution of a firearm.

(iv) An element of that felony is the unlawful use of an explosive.

(v) The felony is burglary of an occupied dwelling, or breaking and entering an occupied dwelling, or arson.

750.227a. Pistols; unlawful possession by licensee. Any person licensed in accordance with law to carry a pistol because he is engaged in the business of protecting the person or property of another, except peace officers of the United States, the state or any subdivision of the state railroad policemen appointed and commissioned under the provisions of Act No. 114 of the Public Acts of 1941, being sections 470.51 to 470.61 of the Compiled Laws of 1948 or those in the military service of the United States, who shall have a pistol in his possession while not actually engaged in the business of protecting the person or property of another, except in his dwelling house or on other land possessed by him, is guilty of a felony. This section shall not be construed to prohibit such person from carrying an unloaded pistol to or from his place of employment by the most direct route.

750.227c. Transporting or possessing loaded firearm in or upon vehicle; violation as misdemeanor; penalty; applicability to person violating § 312.10(1)(g).

(1) Except as otherwise permitted by law, a person shall not transport or possess in or upon a sailboat or a motor vehicle, aircraft, motorboat, or any other vehicle propelled by mechanical means, a firearm, other than a pistol, which is loaded.

(2) A person who violates this section is guilty of a misdemeanor, punishable by imprisonment for not more than 2 years, or a fine of not more than $2,500.00, or both.

(3) This section does not apply to a person who violates section 10(1)(g) of chapter II of Act No. 286 of the Public Acts of 1929, as amended, being section 312.10 of the Michigan Compiled Laws.

750.227d. Transporting or possessing firearm in or upon motor vehicle or self-propelled vehicle designed for land travel; conditions; violation as misdemeanor; penalty.

(1) Except as otherwise permitted by law, a person shall not transport or possess in or upon a motor vehicle or any self-propelled vehicle designed for land travel a firearm, other than a pistol, unless the firearm is unloaded and is 1 or more of the following:

(a) Taken down.

(b) Enclosed in a case.

(c) Carried in the trunk of the vehicle.

(d) Inaccessible from the interior of the vehicle.

(2) A person who violates this section is guilty of a misdemeanor, punishable by imprisonment for not more than 90 days, or a fine of not more than $100.00, or both.

750.228. [Repealed January 7, 2009] Failure to have pistol inspected; applicability; penalty.

(1) Except as provided in subsection (2), a person who fails to have his or her pistol inspected as required under section 9 of Act No. 372 of the Public Acts of 1927, being section 28.429 of the Michigan Compiled Laws, is guilty of a misdemeanor punishable by imprisonment for not more than 90 days, or a fine of not more than $100.00, or both.

(2) Subsection (1) does not apply to a person who obtained a pistol in violation of section 9 of Act No. 372 of the Public Acts of 1927, before the effective date of the 1990 amendatory act that added this subsection, who has not been convicted of that violation, and who has his or her pistol inspected as required under section 9 of Act No. 372 of the Public Acts of 1927 within 90 days after the effective date of the 1990 amendatory act that added this subsection.

750.229. Pistols accepted in pawn, by second-hand dealer or junk dealer. Any pawnbroker who shall accept a pistol in pawn, or any second-hand or junk dealer, as defined in Act No. 350 of the Public Acts of 1917, who shall accept a pistol and offer or display the same for resale, shall be guilty of a misdemeanor.

750.230. Firearms; altering, removing, or obliterating marks of identity; presumption. A person who shall willfully alter, remove, or obliterate the name of the maker, model, manufacturer's number, or other mark of identity of a pistol or other firearm, shall be guilty of a felony, punishable by imprisonment for not more than 2 years or fine of not more than $1,000.00. Possession of a firearm upon which the number shall have been altered, removed, or obliterated, other than an antique firearm as defined by section 231a (2)(a) or (b), shall be presumptive evidence that the possessor has altered, removed, or obliterated the same.

750.231. Sections 750.224, 750.224a, 750.224b, 750.226a, 750.227, 750.227c, and 750.227d inapplicable to certain persons and organizations.

(1) Except as provided in subsection (2) sections 224, 224a, 224b, 224d, 226a, 227, 227c, and 227d do not apply to any of the following:

(a) A peace officer of an authorized police agency of the United States, of this state, or of a political subdivision of this state, who is regularly employed and paid by the United States, this state, or a political subdivision of this state.

(b) A person who is regularly employed by the state department of corrections and who is authorized in writing by the director of the department of corrections to carry a concealed weapon while in the official performance of his or her duties or while going to or returning from those duties.

(c) A person employed by a private vendor that operates a youth correctional facility authorized under section 20g of PA 232, MCL 791.220g, who meets the same criteria established by the director of the state department of corrections for departmental employees described in subdivision (b) and who is authorized in writing by the director of the department of corrections to carry a concealed weapon while in the official performance of his or her duties or while going to or returning from those duties.

(d) A member of the United States army, air force, navy, or marine corps or the United States coast guard while carrying weapons in the line of or incidental to duty.

(e) An organization authorized by law to purchase or receive weapons from the United States or from this state.

(f) A member of the national guard, armed forces reserve, the United States coast guard reserve, or any other authorized military organization while on duty or drill, or in going to or returning from a place of assembly or practice, while carrying weapons used for a purpose of the national guard, armed forces reserve, the United States coast guard reserve, or other duly authorized military organization.

(g) A security employee employed by the state and granted limited arrest powers under section 6c of 1935 PA 59, MCL 28.6c.

(h) A motor carrier officer appointed under section 6d of 1935 PA 59, MCL 28.6d.

(2) As applied to section 224a(1) only, subsection (1) is not applicable to an individual included under subsection (1)(a), (b), or (c) unless he or she has been trained on the use, effects, and risks of using a portable device or weapon described in section 224a(1).

750.231a. Exceptions to § 750.227(2); definitions.

(1) Subsection (2) of section 227 does not apply to any of the following:

(a) To a person holding a valid license to carry a pistol concealed upon his or her person issued by his or her state of residence except where the pistol is carried in nonconformance with a restriction appearing on the license.

(b) To the regular and ordinary transportation of pistols as merchandise by an authorized agent of a person licensed to manufacture firearms.

(c) To a person carrying an antique firearm as defined in subsection (2), completely unloaded in a closed case or container designed for the storage of firearms in the trunk of a vehicle.

(d) To a person while transporting a pistol for a lawful purpose that is licensed by the owner or occupant of the motor vehicle in compliance with section 2 of 1927 PA 372, MCL 28.422, and the pistol is unloaded in a closed case designed for the storage of firearms in the trunk of the vehicle.

(e) To a person while transporting a pistol for a lawful purpose that is licensed by the owner or occupant of the motor vehicle in compliance with section 2 of 1927 PA 372, MCL 28.422, and the pistol is unloaded in a closed case designed for the storage of firearms in a vehicle that does not have a trunk and is not readily accessible to the occupants of the vehicle.

(2) As used in this section:

(a) "Antique firearm" means either of the following:

(i) A firearm not designed or redesigned for using rimfire or conventional center fire ignition with fixed ammunition and manufactured in or before 1898, including a matchlock, flintlock, percussion cap, or similar type of ignition system or replica of such a firearm, whether actually manufactured before or after 1898.

(ii) A firearm using fixed ammunition manufactured in or before 1898, for which ammunition is no longer manufactured in the United States and is not readily available in the ordinary channels of commercial trade.

(b) "Lawful purpose" includes the following:

(i) While en route to or from a hunting or target shooting area.

(ii) While transporting a pistol en route to or from his or her home or place of business and place of repair.

(iii) While moving goods from 1 place of abode or business to another place of abode or business.

(iv) While transporting a licensed pistol en route to or from a law enforcement agency for the purpose of having a safety inspection performed on the pistol as is required by section 9 of 1927 PA 372, MCL 28.429, or for the purpose of having a law enforcement official take possession of the weapon.

(v) While en route to or from his or her abode or place of business and a gun show or places of purchase or sale.

(vi) While en route to or from his or her abode to a public shooting facility or public land where discharge of firearms is permitted by law, rule, regulation, or local ordinance.

(vii) While en route to or from his or her abode to a private property location where the pistol is to be used as is permitted by law, rule, regulation, or local ordinance.

750.231b. Sale and safety inspection; persons exempt. Sections 223 and 228 do not apply to a duly authorized police or correctional agency of the United States or of the state or any subdivision thereof, nor to the army, air force, navy or marine corps of the United States, nor to organizations authorized by law to purchase or receive weapons from the United States or from this state, nor to the national guard, armed forces reserves or other duly authorized military organizations, nor to a member of such agencies or organizations for weapons used by him for the purposes of such agencies or organizations, nor to a person holding a license to carry a pistol concealed upon his person issued by another state, nor to the regular and ordinary transportation of pistols as merchandise by an authorized agent of a person licensed to manufacture firearms.

750.232. Purchasers of firearms; registration. Any person engaged in any way or to any extent in the business of selling at retail, guns, pistols, other firearms or silencers for fire-arms who shall fail or neglect to keep a register in which shall be entered the name, age, occupation and residence (if residing in the city with the street number of such residence) of each and every purchaser of such guns, pistols, other firearms or silencers for firearms together with the number or other mark of identification, if any, on such gun, pistol, other firearms or silencer for firearms, which said register shall be open to the inspection of all peace officers at all times, shall be guilty of a misdemeanor.

750.232a. Obtaining pistol in violation of § 28.422; intentionally making material false statement on application for license to purchase pistol; using or attempting to use false identification or identification of another person to purchase firearm; penalties.

(1) Except as provided in subsection (2), a person who obtains a pistol in violation of section 2 of Act No. 372 of the Public Acts of 1927, as amended, being section 28.422 of the Michigan Compiled Laws, is guilty of a misdemeanor, punishable by imprisonment for not more than 90 days or a fine of not more than $100.00, or both.

(2) Subsection (1) does not apply to a person who obtained a pistol in violation of section 2 of Act No. 372 of the Public Acts of 1927 before the effective date of the 1990 amendatory act that added this subsection, who has not been convicted of that violation, and who obtains a license as required under section 2 of Act No. 372 of the Public Acts of 1927 within 90 days after the effective date of the 1990 amendatory act that added this subsection.

(3) A person who intentionally makes a material false statement on an application for a license to purchase a pistol under section 2 of Act No. 372 of the Public Acts of 1927, as amended, is guilty of a felony, punishable by imprisonment for not more than 4 years, or a fine of not more than $2,000.00, or both.

(4) A person who uses or attempts to use false identification or the identification of another person to purchase a firearm is guilty of a misdemeanor, punishable by imprisonment for not more than 90 days or a fine of not more than $100.00, or both.

750.233. Firearm; intentionally aiming without malice.

(1) A person who intentionally but without malice points or aims a firearm at or toward another person is guilty of a misdemeanor punishable by imprisonment for not more than 93 days or a fine of not more than $500.00, or both.

(2) This section does not apply to a peace officer of this state or another state, or of a local unit of government of this state or another state, or of the United States, performing his or her duties as a peace officer. As used in this section, "peace officer" means that term as defined in section 215.

750.234f. Possession of firearm by a person less than 18 years of age; exceptions; violation as misdemeanor; penalty.

(1) Except as provided in subsection (2), an individual less than 18 years of age shall not possess a firearm in public except under the direct supervision of an individual 18 years of age or older.

(2) Subsection (1) does not apply to an individual less than 18 years of age who possesses a firearm in accordance with part 401 (wildlife conservation) of the natural resources and environmental protection act, Act No. 451 of the Public Acts of 1994, being sections 324.40101 to 324.40119 of the Michigan Compiled Laws, or part 435 (hunting and fishing licensing) of Act No. 451 of the Public Acts of 1994, being sections 324.43501 to 324.43561 of the Michigan Compiled Laws. However, an individual less than 18 years of age may possess a firearm without a hunting license while at, or going to or from, a recognized target range or trap or skeet shooting ground if, while going to or from the range or ground, the firearm is enclosed and securely fastened in a case or locked in the trunk of a motor vehicle.

(3) An individual who violates this section is guilty of a misdemeanor, punishable by imprisonment for not more than 90 days, or a fine of not more than $100.00, or both.

750.235a. Parent of minor guilty of misdemeanor; conditions; penalty; defense; definitions.

(1) The parent of a minor is guilty of a misdemeanor if all of the following apply:

(a) The parent has custody of the minor.

(b) The minor violates this chapter in a weapon free school zone.

(c) The parent knows that the minor would violate this chapter or the parent acts to further the violation.

(2) An individual convicted under subsection (1) may be punished by 1 or more of the following:

(a) A fine of not more than $2,000.00.

(b) Community service for not more than 100 hours.

(c) Probation.

(3) It is a complete defense to a prosecution under this section if the defendant promptly notifies the local law enforcement agency or the school administration that the minor is violating or will violate this chapter in a weapon free school zone.

(4) As used in this section:

(a) "Minor" means an individual less than 18 years of age.

(b) "School" means a public, private, denominational, or parochial school offering developmental kindergarten, kindergarten, or any grade from 1 through 12.

(c) "School property" means a building, playing field, or property used for school purposes to impart instruction to children or used for functions and events sponsored by a school, except a building used primarily for adult education or college extension courses.

(d) "Weapon free school zone" means school property and a vehicle used by a school to transport students to or from school property.

750.237. Liquor or other drug; possession or use of firearm by person under influence.

(1) An individual shall not carry, have in possession or under control, or use in any manner or discharge a firearm under any of the following circumstances:

(a) The individual is under the influence of alcoholic liquor, a controlled substance, or a combination of alcoholic liquor and a controlled substance.

(b) The individual has an alcohol content of 0.08 or more grams per 100 milliliters of blood, per 210 liters of breath, or per 67 milliliters of urine.

(c) Because of the consumption of alcoholic liquor, a controlled substance, or a combination of alcoholic liquor and a controlled substance, the individual's ability to use a firearm is visibly impaired.

(2) Except as provided in subsections (3) and (4), an individual who violates subsection (1) is guilty of a misdemeanor punishable by imprisonment for not more than 93 days or a fine of not more than $100.00 for carrying or possessing a firearm, or both, and not more than $500.00 for using or discharging a firearm, or both.

(3) An individual who violates subsection (1) and causes a serious impairment of a body function of another individual by the discharge or use in any manner of the firearm is guilty of a felony punishable by imprisonment for not more than 5 years or a fine of not less than $1,000.00 or more than $5,000.00, or both. ...

(4) An individual who violates subsection (1) and causes the death of another individual by the discharge or use in any manner of a firearm is guilty of a felony punishable by imprisonment for not more than 15 years or a fine of not less than $2,500.00 or more than $10,000.00, or both. ...

(9) This section does not prohibit the individual from being charged with, convicted of, or sentenced for any other violation of law arising out of the same transaction as the violation of this section in lieu of being charged with, convicted of, or sentenced for the violation of this section.

750.237a. Individuals engaging in proscribed conduct; violation; penalties; definitions.

(1) An individual who engages in conduct proscribed under section 224, 224a, 224b, 224c, 224e, 226, 227, 227a, 227f, 234a, 234b, or 234c, or who engages in conduct proscribed under section 223(2) for a second or subsequent time, in a weapon free school zone is guilty of a felony punishable by 1 or more of the following:

(a) Imprisonment for not more than the maximum term of imprisonment authorized for the section violated.

(b) Community service for not more than 150 hours.

(c) A fine of not more than 3 times the maximum fine authorized for the section violated.

(2) An individual who engages in conduct proscribed under section 223(1), 224d, 226a, 227c, 227d, 231c, 232a(1), or (4), 233, 234, 234e, 234f, 235, 236, or 237, or who engages in conduct proscribed under section 233(2) for the first time, in a weapon free school zone is guilty of a misdemeanor punishable by 1 or more of the following:

(a) Imprisonment for not more than the maximum term of imprisonment authorized for the section violated or 93 days, whichever is greater.

(b) Community service for not more than 100 hours.

(c) A fine of not more than $2,000.00 or the maximum fine authorized for the section violated, whichever is greater.

(3) Subsections (1) and (2) do not apply to conduct proscribed under a section enumerated in those subsections to the extent that the proscribed conduct is otherwise exempted or authorized under this chapter.

(4) Except as provided in subsection (5), an individual who possesses a weapon in a weapon free school zone is guilty of a misdemeanor punishable by 1 or more of the following:

(a) Imprisonment for not more than 93 days.

(b) Community service for not more than 100 hours.

(c) A fine of not more than $2,000.00.

(5) Subsection (4) does not apply to any of the following:

(a) An individual employed by or contracted by a school if the possession of that weapon is to provide security services for the school.

(b) A peace officer.

(c) An individual licensed by this state or another state to carry a concealed weapon.

(d) An individual who possesses a weapon provided by a school or a school's instructor on school property for purposes of providing or receiving instruction in the use of that weapon.

(e) An individual who possesses a firearm on school property if that possession is with the permission of the school's principal or an agent of the school designated by the school's principal or the school board.

(f) An individual who is 18 years of age or older who is not a student at the school and who possesses a firearm on school property while transporting a student to or from the school if any of the following apply:

(i) The individual is carrying an antique firearm, completely unloaded, in a wrapper or container in the trunk of a vehicle while en route to or from a hunting or target shooting area or function involving the exhibition, demonstration or sale of antique firearms.

(ii) The individual is carrying a firearm unloaded in a wrapper or container in the trunk of the person's vehicle, while in possession of a valid Michigan hunting license or proof of valid membership in an organization having shooting range facilities, and while en route to or from a hunting or target shooting area.

(iii) The person is carrying a firearm in a wrapper or container in the trunk of the person's vehicle from the place of purchase to his or her home or place of business or to a place of repair or back to his or her home or place of business, or in moving goods from one place of abode or business to another place of abode or business.

(iv) The person is carrying an unloaded firearm in the passenger compartment of a vehicle that does not have a trunk, if the person is otherwise complying with the requirements of subparagraph (ii) or (iii) and the wrapper or container is not readily accessible to the occupants of the vehicle.

(6) As used in this section:

(a) "Antique firearm" means either of the following:

(i) A firearm not designed or redesigned for using rimfire or conventional center fire ignition with fixed ammunition and manufactured in or before 1898, including a matchlock, flintlock, percussion cap, or similar type of ignition system or a replica of such a firearm, whether actually manufactured before or after the year 1898.

(ii) A firearm using fixed ammunition manufactured in or before 1898, for which ammunition is no longer manufactured in the United States and is not readily available in the ordinary channels of commercial trade.

(b) "School" means a public, private, denominational, or parochial school offering developmental kindergarten, kindergarten, or any grade from 1 through 12.

(c) "School property" means a building, playing field, or property used for school purposes to impart instruction to children or used for functions and events sponsored by a school, except a building used primarily for adult education or college extension courses.

(d) "Weapon free school zone" means school property and a vehicle used by a school to transport students to or from school property.

750.239. Forfeiture of weapons. All pistols, weapons or devices carried, possessed or used contrary to this chapter are hereby declared forfeited to the state, and shall be turned over to the commissioner of the Michigan state police or his designated representative, for such disposition as the commissioner may prescribe.

Chapter 752. Crimes and Offenses - Spring, Gas, or Air Operated Handguns

752.891. Use or possession of BB handgun by minor. No person under 18 years of age shall use or possess any handgun designed and manufactured exclusively for propelling BB's not exceeding .177 calibre by means of spring, gas or air, outside the curtilage of his domicile unless he is accompanied by a person over 18 years of age.

752.892. Penalty. Any person who violates the provisions of this act is guilty of a misdemeanor.

[Current through Mich. Pub. Acts 348, Dec. 23, 2008]

MINNESOTA

MINN. STAT.

Chapter 97B. Hunting

97B.021. POSSESSION OF FIREARMS BY PER-SONS UNDER AGE 16

Subdivision 1. Restrictions.

(a) Except as provided in this subdivision, a person under the age of 16 may not possess a firearm, unless accompanied by a parent or guardian.

(b) A person under age 16 may possess a firearm without being accompanied by a parent or guardian:

(1) on land owned by, or occupied as the principal residence of, the person or the person's parent or guardian;

(2) while participating in an organized target shooting program with adult supervision;

(3) while the person is participating in a firearms safety program or traveling to and from class; or

(4) if the person is age 14 or 15 and has a firearms safety certificate.

Subd. 2. Seizure of unlawfully possessed firearms. A law enforcement officer shall seize a firearm used in violation of this section. The officer must tag the seized firearm with the name and address of the person from whom it was taken and give the person a receipt. The firearm shall be placed in the custody of the conservation officer in charge of the area where the seizure was made.

Subd. 3. Return or forfeiture of seized firearms. A firearm seized under this section must be returned to the person from whom it was seized when the person presents a firearms safety certificate to the conservation officer. The person must present the certificate within 90 days after the beginning of the first firearms training course in the county after the firearm was seized. If the person does not present a certificate, the firearm is contraband and forfeited to the state, and shall be disposed of as prescribed by the commissioner.

Chapter 242. Corrections; Youth

242.31. RESTORATION OF CIVIL RIGHTS; POSSESSION OF FIREARMS

Subdivision 1. Restoration. Whenever a person who has been committed to the custody of the commissioner of corrections upon conviction of a crime following certification under the provisions of section 260B.125 is finally discharged by order of the commissioner, that discharge shall restore the person to all civil rights. The commissioner shall file a copy of the order with the district court of the county in which the conviction occurred.

Subd. 2. Order of discharge. Whenever a person described in subdivision 1 has been placed on probation by the court pursuant to section 609.135 and, after satisfactory fulfillment of it, is discharged from probation, the court shall issue an order of discharge pursuant to subdivision 2a and section 609.165.

This order restores the defendant to civil rights.

Subd. 2a. Crimes of violence; ineligibility to possess firearms. The order of discharge must provide that a person who has been convicted of a crime of violence, as defined in section 624.712, subdivision 5, is not entitled to ship, transport, possess, or receive a firearm for the remainder of the person's lifetime. Any person who has received such a discharge and who thereafter has received a relief of disability under United States Code, title 18, section 925, or whose ability to possess firearms has been restored under section 609.165, subdivision 1d, shall not be subject to the restrictions of this subdivision.

Chapter 245. Department of Human Services

245.041. PROVISION OF FIREARMS BACKGROUND CHECK INFORMATION Notwithstanding section 253B.23, subdivision 9, the commissioner of human services shall provide commitment information to local law enforcement agencies on an individual request basis by means of electronic data transfer from the department of human services through the Minnesota crime information system for the sole purpose of facilitating a firearms background check under section 624. 7131, 624.7132, or 624.714. The information to be provided is limited to whether the person has been committed under chapter 253B and, if so, the type of commitment.

Chapter 260B. Delinquency

260B.245. EFFECT OF JUVENILE COURT PROCEEDINGS.

Subdivision 1. Effect. ...

(b) A person who was adjudicated delinquent for, or convicted as an extended jurisdiction juvenile of, a crime of violence as defined in section 624.712, subdivision 5, is not entitled to ship, transport, possess, or receive a firearm for the remainder of the person's lifetime. A person who has received a relief of disability under United States Code, title 18, section 925, or whose ability to possess firearms has been restored under section 609.165, subdivision 1d, is not subject to the restrictions of this subdivision.

Chapter 471. Municipal Rights, Powers, Duties

471.633. FIREARMS The legislature preempts all authority of a home rule charter or statutory city including a city of the first class, county, town, municipal corporation, or other govern-mental subdivision, or any of their instrumentalities, to regulate firearms, ammunition, or their respective components to the complete exclusion of any order, ordinance or regulation by them except that:

(a) a governmental subdivision may regulate the discharge of firearms; and

(b) a governmental subdivision may adopt regulations identical to state law.

Local regulation inconsistent with this section is void.

471.634. DEFINITION For purposes of section 471.633, the terms "municipal corporation" and "governmental subdivision," or instrumentality thereof, do not include school districts and other entities composed exclusively of school districts when school boards or school administrators are regulating school grounds, school facilities, school transportation services, school programs, or the conduct of students at any activities conducted under the direct or indirect supervision or control of the school board or administration.

Chapter 609. Criminal Code

609.11. MINIMUM SENTENCES OF IMPRISONMENT ...

Subd. 5. Firearm.

(a) Except as otherwise provided in paragraph (b), any defendant convicted of an offense listed in subdivision 9 in which the defendant or an accomplice, at the time of the offense, had in possession or used, whether by brandishing, displaying, threatening with, or otherwise employing, a firearm, shall be committed to the commissioner of corrections for not less than three years, nor more than the maximum sentence provided by law. Any defendant convicted of a second or subsequent offense in which the defendant or an accomplice, at the time of the offense, had in possession or used a firearm shall be committed to the commissioner of corrections for not less than five years, nor more than the maximum sentence provided by law.

(b) Any defendant convicted of violating section 609.165 or 624.713, subdivision 1, clause (b), shall be committed to the commissioner of corrections for not less than five years, nor more than the maximum sentence provided by law. ...

Subd. 9. Applicable offenses. The crimes for which mandatory minimum sentences shall be served as provided in this section are: ... possession or other unlawful use of a firearm in violation of section 609.165, subdivision 1b, or 624.713, subdivision 1, clause (b), a felony violation of chapter 152; or any attempt to commit any of these offenses. ...

609.165 RESTORATION OF CIVIL RIGHTS; POSSESSION OF FIREARMS

Subdivision 1. Restoration. When a person has been deprived of civil rights by reason of conviction of a crime and is thereafter discharged, such discharge shall restore the person to all civil rights and to full citizenship, with full right to vote and hold office, the same as if such conviction had not taken place, and the order of discharge shall so provide.

Subd. 1a. Certain convicted felons ineligible to possess firearms. The order of discharge must provide that a person who has been convicted of a crime of violence, as defined in section 624.712, subdivision 5, is not entitled to ship, transport, possess, or receive a firearm for the remainder of the person's lifetime. Any person who has received such a discharge and who thereafter has received a relief of disability under United States Code, title 18, section 925, or whose ability to possess firearms has been restored under section 609.165, subdivision 1d, shall not be subject to the restrictions of this subdivision.

Subd. 1b. Violation and penalty.

(a) Any person who has been convicted of a crime of violence, as defined in section 624.712, subdivision 5, and who ships, transports, possesses, or receives a firearm, commits a felony and may be sentenced to imprisonment for not more than 15 years or to payment of a fine of not more than $30,000, or both.

(b) A conviction and sentencing under this section shall be construed to bar a conviction and sentencing for a violation of section 624.713, subdivision 2.

(c) The criminal penalty in paragraph (a) does not apply to any person who has received a relief of disability under United States Code, title 18, section 925, or whose ability to possess firearms has been restored under subdivision 1d.

Subd. 1c. [Repealed]

Subd. 1d. Judicial restoration of ability to possess a firearm by a felon. A person prohibited by state law from shipping, transporting, possessing, or receiving a firearm because of a conviction or a delinquency adjudication for committing a crime of violence may petition a court to restore the person's ability to possess, receive, ship, or transport firearms and otherwise deal with firearms.

The court may grant the relief sought if the person shows good cause to do so and the person has been released from physical confinement.

If a petition is denied, the person may not file another petition until three years have elapsed without the permission of the court.

Subd. 2. Discharge. The discharge may be:

(1) By order of the court following stay of sentence or stay of execution of sentence; or

(2) Upon expiration of sentence.

Subd. 3. Applicability. This section does not apply to a forfeiture of and disqualification for public office as provided in section 609.42, subdivision 2.

609.2242 DOMESTIC ASSAULT ...

Subd. 3. Domestic assaults; firearms.

(a) When a person is convicted of a violation of this subsection or section 609.221, 609.222, 609. 223, or 609.224, or 609.2247, the court shall determine and make written findings on the record as to whether:

(1) the assault was committed against a family or household member, as defined in section 518B.01, subdivision 2;

(2) the defendant owns or possesses a firearm; and

(3) the firearm was used in any way during the commission of the assault.

(b) If the court determines that the assault was of a family or household member, and that the offender owns or possesses a firearm and used it in any way during the commission of the assault, it shall order that the firearm be summarily forfeited under section 609.5316, subdivision 3.

(c) When a person is convicted of assaulting a family or household member and is determined by the court to have used a firearm in any way during commission of the assault, the court may order that the person is prohibited from possessing any type of firearm for any period longer than three years or for the remainder of the person's life. A person who violates this paragraph is guilty of a gross misdemeanor. At the time of the conviction, the court shall inform the defendant whether and for how long the defendant is prohibited from possessing a firearm and that it is a gross misdemeanor to violate this paragraph. The failure of the court to provide this information to a defendant does not affect the applicability of the firearm possession prohibition or the gross misdemeanor penalty to that defendant.

(d) Except as otherwise provided in paragraph (c), when a person is convicted of a violation of this section or section 609.224 and the court determines that the victim was a family or household member, the court shall inform the defendant that the defendant is prohibited from possessing a pistol for three years from the date of conviction and that it is a gross misdemeanor offense to violate this prohibition. The failure of the court to provide this information to a defendant does not affect the applicability of the pistol possession prohibition or the gross misdemeanor penalty to that defendant.

(e) Except as otherwise provided in paragraph (c), a person is not entitled to possess a pistol if the person has been convicted after August 1, 1992, of domestic assault under this section or assault in the fifth degree under section 609.224 and the assault victim was a family or household member as defined in section 518B.01, subdivision 2, unless three years have elapsed from the date of conviction and, during that time, the person has not been convicted of any other violation of this section or section 609.224. Property rights may not be abated but access may be restricted by the courts. A person who possesses a pistol in violation of this paragraph is guilty of a gross misdemeanor. ....

609.66. DANGEROUS WEAPONS

Subdivision 1. Misdemeanor and gross misdemeanor crimes.

(a) Whoever does any of the following is guilty of a crime and may be sentenced as provided in paragraph (b):

(1) recklessly handles or uses a gun or other dangerous weapon or explosive so as to endanger the safety of another; or

(2) intentionally points a gun of any kind, capable of injuring or killing a human being and whether loaded or unloaded, at or toward another; or ...

(5) possesses any other dangerous article or substance for the purpose of being used unlawfully as a weapon against another; or

(6) outside of a municipality and without the parent's or guardian's consent, furnishes a child under 14 years of age, or as a parent or guardian permits the child to handle or use, outside of the parent's or guardian's presence, a firearm or airgun of any kind, or any ammunition or explosive.

Possession of written evidence of prior consent signed by the minor's parent or guardian is a complete defense to a charge under clause (6).

(b) A person convicted under paragraph (a) may be sentenced as follows:

(1) if the act was committed in a public housing zone, as defined in section 152.01, subdivision 19, a school zone, as defined in section 152.01, subdivision 14a, or a park zone, as defined in section 152.01, subdivision 12a, to imprisonment for not more than one year or to payment of a fine of not more than $3,000, or both; or

(2) otherwise, including where the act was committed on residential premises within a zone described in clause (1) if the offender was at the time an owner, tenant, or invitee for a lawful purpose with respect to those residential premises, to imprisonment for not more than 90 days or to payment of a fine of not more than $700, or both.

Subd. 1a. Felony crimes; silencers prohibited; reckless discharge.

(a) Except as otherwise provided in subdivision 1h, whoever does any of the following is guilty of a felony and may be sentenced as provided in paragraph (b):

(1) sells or has in possession any device designed to silence or muffle the discharge of a firearm;

(2) intentionally discharges a firearm under circumstances that endanger the safety of another; or

(3) recklessly discharges a firearm within a municipality.

(b) A person convicted under paragraph (a) may be sentenced as follows:

(1) if the act was a violation of paragraph (a), clause (2), or if the act was a violation of paragraph (a), clause (1) or (3), and was committed in a public housing zone, as defined in section 152.01, subdivision 19, a school zone, as defined in section 152.01, subdivision 14a, or a park zone, as defined in section 152.01, subdivision 12a, to imprisonment for not more than five years or to payment of a fine of not more than $10,000, or both; or

(2) otherwise, to imprisonment for not more than two years or to payment of a fine of not more than $5,000, or both.

Subd. 1b. Felony; furnishing to minors. Whoever, in any municipality of this state, furnishes a minor under 18 years of age with a firearm, airgun, ammunition, or explosive without the prior consent of the minor's parent or guardian or of the police department of the municipality is guilty of a felony and may be sentenced to imprisonment for not more than ten years or to payment of a fine of not more than $20,000, or both. Possession of written evidence of prior consent signed by the minor's parent or guardian is a complete defense to a charge under this subdivision.

Subd. 1c. Felony; furnishing a dangerous weapon. Whoever recklessly furnishes a person with a dangerous weapon in conscious disregard of a known substantial risk that the object will be possessed or used in furtherance of a felony crime of violence is guilty of a felony and may be sentenced to imprisonment for not more than ten years or to payment of a fine of not more than $20,000, or both.

Subd. 1d. Possession on school property; penalty.

(a) Except as provided under paragraphs (c) and (e), whoever possesses, stores, or keeps a dangerous weapon or uses or brandishes a replica firearm or a BB gun while knowingly on school property is guilty of a felony and may be sentenced to imprisonment for not more than two years or to payment of a fine of not more than $5,000, or both.

(b) Whoever possesses, stores, or keeps a replica firearm or a BB gun on school property is guilty of a gross misdemeanor.

(c) Notwithstanding paragraph (a) or (b), it is a misdemeanor for a person authorized to carry a firearm under the provisions of a permit or otherwise to carry a firearm on or about the person's clothes or person in a location the person knows is school property. Notwithstanding section 609.531, a firearm carried in violation of this paragraph is not subject to forfeiture.

(d) As used in this subdivision:

(1) "BB gun" means a device that fires or ejects a shot measuring .18 of an inch or less in diameter;

(2) "dangerous weapon" has the meaning given it in section 609.02, subdivision 6;

(3) "replica firearm" has the meaning given it in section 609.713; and

(4) "school property" means:

(i) a public or private elementary, middle, or secondary school building and its improved grounds, whether leased or owned by the school;

(ii) a child care center licensed under chapter 245A during the period children are present and participating in a child care program;

(iii) the area within a school bus when that bus is being used by a school to transport one or more elementary, middle, or secondary school students to and from school-related activities, including curricular, cocurricular, noncurricular, extracurricular, and supplementary activities; and

(iv) that portion of a building or facility under the temporary, exclusive control of a public or private school, a school district, or an association of such entities where conspicuous signs are prominently posted at each entrance that give actual notice to persons of the school-related use.

(e) This subdivision does not apply to:

(1) Active licensed peace officer;

(2) Military personnel, or students participating in military training, who are on-duty, performing official duties;

(3) Persons authorized to carry a pistol under section 624.714 while I a motor vehicle or outside a motor vehicle to directly place a firearm in, or retrieve it from, the trunk or rear area of the vehicle;

(4) Persons who keep or store in a motor vehicle pistols in accordance with sections 624.714 or 624.715 or other firearms in accordance with section 97B.045;

(5) Firearm safety or marksmanship courses or activities conducted on school property;

(6) Possession of dangerous weapons, BB guns, or replica firearms by a ceremonial color guard;

(7) A gun or knife show held on school property;

(8) Possession of dangerous weapons, BB guns, or replica firearms with written permission of the principal or other person having general control and supervision of the school or the director of a child care center; or

(9) Persons who are on unimproved property owned or leased by a child care center, school, or school district unless the person knows that a student is currently present on the land for a school-related activity.

(f) Notwithstanding section 471.634, a school district or other entity composed exclusively of school districts may not regulate firearms, ammunition, or their respective components, when possessed or carried by nonstudents or nonemployees, in a manner that is inconsistent with this subdivision.

Subd. 1e. Felony; drive-by shooting.

Subd. 1f. Gross misdemeanor; transferring a firearm without background check. A person, other than a federally licensed firearms dealer, who transfers a pistol or semiautomatic military-style assault weapon to another without complying with the transfer requirements of section 624.7132, is guilty of a gross misdemeanor if the transferee possesses or uses the weapon within one year after the transfer in furtherance of a felony crime of violence, and if:

(1) the transferee was prohibited from possessing the weapon under section 624.713 at the time of the transfer; or

(2) it was reasonably foreseeable at the time of the transfer that the transferee was likely to use or possess the weapon in furtherance of a felony crime of violence.

Subd. 1g. Felony; possession in courthouse or certain state buildings. ...

Subd. 2. Exceptions. Nothing in this section prohibits the possession of the articles mentioned by museums or collectors of art or for other lawful purposes of public exhibition.

Subd. 1h. Silencers; authorized for law enforcement purposes. Notwithstanding subdivision 1a, paragraph (a), clause (1), licensed peace officers may use devices designed to silence or muffle the discharge of a firearm for tactical emergency response operations. Tactical emergency response operations include execution of high risk search and arrest warrants, incidents of terrorism, hostage rescue, and any other tactical deployments involving high risk circumstances. The chief law enforcement officer of a law enforcement agency that has the need to use silencing devices must establish and enforce a written policy governing the use of the devices.

609.663. DISPLAY OF HANDGUN AMMUNITION It is a petty misdemeanor to display centerfire metallic case handgun ammunition for sale to the public in a manner that makes the ammunition directly accessible to persons under the age of 18 years, other than employees or agents of the seller, unless the display is under observation of the seller or the seller's employee or agent, or the seller takes reasonable steps to exclude underage persons from the immediate vicinity of the display. Ammunition displayed in an enclosed display case or behind a counter is not directly accessible. This section does not apply to ammunition suitable for big game hunting.

609.666. NEGLIGENT STORAGE OF FIREARMS

Subdivision 1. Definitions. For purposes of this section, the following words have the meanings given.

(a) "Firearm" means a device designed to be used as a weapon, from which is expelled a projectile by the force of any explosion or force of combustion.

(b) "Child" means a person under the age of 18 years.

(c) "Loaded" means the firearm has ammunition in the chamber or magazine, if the magazine is in the firearm, unless the firearm is incapable of being fired by a child who is likely to gain access to the firearm.

Subd. 2. Access to firearms. A person is guilty of a gross misdemeanor who negligently stores or leaves a loaded firearm in a location where the person knows, or reasonably should know, that a child is likely to gain access, unless reasonable action is taken to secure the firearm against access by the child.

Subd. 3. Limitations. Subdivision 2 does not apply to a child's access to firearms that was obtained as a result of an unlawful entry.

609.667. FIREARMS; REMOVAL OR ALTERATION OF SERIAL NUMBER Whoever commits any of the following acts may be sentenced to imprisonment for not more than five years or to payment of a fine of not more than $10,000, or both:

(1) obliterates, removes, changes, or alters the serial number or other identification of a firearm;

(2) receives or possesses a firearm, the serial number or other identification of which has been obliterated, removed, changed, or altered; or

(3) receives or possesses a firearm that is not identified by a serial number.

As used in this section, "serial number or other identification" means the serial number and other information required under United States Code, title 26, section 5842, for the identification of firearms.

609.67. MACHINE GUNS AND SHORT-BARRELED SHOTGUNS

Subdivision 1. Definitions.

(a) "Machine gun" means any firearm designed to discharge, or capable of discharging automatically more than once by a single function of the trigger.

(b) "Shotgun" means a weapon designed, redesigned, made or remade which is intended to be fired from the shoulder and uses the energy of the explosive in a fixed shotgun shell to fire through a smooth bore either a number of ball shot or a single projectile for each single pull of the trigger.

(c) "Short-barreled shotgun" means a shotgun having one or more barrels less than 18 inches in length and any weapon made from a shotgun if such weapon as modified has an overall length less than 26 inches.

(d) "Trigger activator" means a removable manual or power driven trigger activating device constructed and designed so that, when attached to a firearm, the rate at which the trigger may be pulled increases and the rate of fire of the firearm increases to that of a machine gun.

(e) "Machine gun conversion kit" means any part or combination of parts designed and intended for use in converting a weapon into a machine gun, and any combination of parts from which a machine gun can be assembled, but does not include a spare or replacement part for a machine gun that is possessed lawfully under section 609.67, subdivision 3.

Subd. 2. Acts prohibited. Except as otherwise provided herein, whoever owns, possesses, or operates a machine gun, any trigger activator or machine gun conversion kit, or a short-barreled shotgun may be sentenced to imprisonment for not more than five years or to payment of a fine of not more than $10,000, or both.

Subd. 3. Uses permitted. The following persons may own or possess a machine gun or short-barreled shotgun provided the provisions of subdivision 4 are complied with:

(1) law enforcement officers for use in the course of their duties;

(2) chief executive officers of correctional facilities and other personnel thereof authorized by them and persons in charge of other institutions for the retention of persons convicted or accused of crime, for use in the course of their duties;

(3) persons possessing machine guns or short-barreled shotguns which, although designed as weapons, have been determined by the superintendent of the bureau of criminal apprehension or the superintendent's delegate by reason of the date of manufacture, value, design or other characteristics to be primarily collector's items, relics, museum pieces or objects of curiosity, ornaments or keepsakes, and are not likely to be used as weapons;

(4) manufacturers of ammunition who possess and use machine guns for the sole purpose of testing ammunition manufactured for sale to federal and state agencies or political subdivisions;

(5) dealers and manufacturers who are federally licensed to buy and sell, or manufacture machine guns or short-barreled shotguns and who either use the machine guns or short-barreled shotguns in peace officer training under courses approved by the board of peace officer standards and training, or are engaged in the sale of machine guns or short-barreled shotguns to federal and state agencies or political subdivisions; and

(6) persons employed by the Minnesota National Guard as security guards, for use in accordance with applicable federal military regulations.

Subd. 4. Report required.

(a) A person owning or possessing a machine gun or short-barreled shotgun as authorized by subdivision 3, clause (1), (2), (3), or (4) shall, within ten days after acquiring such ownership or possession, file a written report with the bureau of criminal apprehension, showing the person's name and address; the person's official title and position, if any; a description of the machine gun or short-barreled shotgun sufficient to enable identification thereof; the purpose for which it is owned or possessed; and such further information as the bureau may reasonably require.

(b) A dealer or manufacturer owning or having a machine gun or short-barreled shotgun as authorized by subdivision 3, clause (5) shall, by the tenth day of each month, file a written report with the bureau of criminal apprehension showing the name and address of the dealer or manufacturer and the serial number of each machine gun or short-barreled shotgun acquired or manufactured during the previous month.

Subd, 5. Exceptions. This section does not apply to members of the armed services of either the United States or the state of Minnesota for use in the course of their duties or to security guards employed by the Minnesota National Guard for use in accordance with applicable federal military regulations.

Subd. 6. Preemption. Laws 1977, chapter 255, supersedes all local ordinances, rules and regulations.

Chapter 624. Crimes, Other Provisions

624.71. GUN CONTROL, APPLICATION OF FEDERAL LAW

Subdivision 1. Notwithstanding any other law to the contrary, it shall be lawful for any federally licensed importer, manufacturer, dealer, or collector to sell and deliver firearms and ammunition to a resident of a contiguous state in any instance where such sale and delivery is lawful under the federal Gun Control Act of 1968 (Public Law Number 90-618).

Subd. 2. Notwithstanding any other law to the contrary, it shall be lawful for a resident of Minnesota to purchase firearms and ammunition in a contiguous state in any instance where such sale and delivery is lawful under the federal Gun Control Act of 1968 (Public Law Number 90-618).

624.711 DECLARATION OF POLICY It is not the intent of the legislature to regulate shotguns, rifles and other long guns of the type commonly used for hunting and not defined as pistols or semiautomatic military-style assault weapons, or to place costs of administration upon those citizens who wish to possess or carry pistols or semiautomatic military-style assault weapons lawfully, or to confiscate or otherwise restrict the use of pistols or semiautomatic military-style assault weapons by law-abiding citizens.

624.712. DEFINITIONS

Subdivision 1. Scope. As used in sections 624.711 to 624.717, the terms defined in this section shall have the meanings given them.

Subd. 2. Pistol. "Pistol" includes a weapon designed to be fired by the use of a single hand and with an overall length less than 26 inches, or having a barrel or barrels of a length less than 18 inches in the case of a shotgun or having a barrel of a length less than 16 inches in the case of a rifle (a) from which may be fired or ejected one or more solid projectiles by means of a cartridge or shell or by the action of an explosive or the igniting of flammable or explosive substances; or (b) for which the propelling force is a spring, elastic band, carbon dioxide, air or other gas, or vapor.

"Pistol" does not include a device firing or ejecting a shot measuring .18 of an inch, or less, in diameter and commonly known as a "BB gun," a scuba gun, a stud gun or nail gun used in the construction industry or children's pop guns or toys.

Subd. 3. Antique firearm. "Antique firearm" means any firearm, including any pistol, with a matchlock, flintlock, percussion cap, or similar type of ignition system, manufactured before 1899 and any replica of any firearm described herein if such replica is not designed or redesigned, made or remade, or intended to fire conventional rimfire or conventional centerfire ammunition, or uses conventional rimfire or conventional centerfire ammunition which is not readily available in the ordinary channels of commercial trade.

Subd. 4. Saturday night special pistol. "Saturday night special pistol" means a pistol other than an antique firearm or a pistol for which the propelling force is carbon dioxide, air or other vapor, or children's pop guns or toys, having a frame, barrel, cylinder, slide or breech-block:

(a) of any material having a melting point (liquids) of less than 1,000 degrees Fahrenheit, or

(b) of any material having an ultimate tensile strength of less than 55,000 pounds per square inch, or

(c) of any powdered metal having a density of less than 7.5 grams per cubic centimeter.

Subd. 5. Crime of violence. "Crime of violence" means: felony convictions of the following offenses: sections 609.185 (murder in the first degree); 609.19 (murder in the second degree); 609.195 (murder in the third degree); 609.20 (manslaughter in the first degree); 609.205 (manslaughter in the second degree); 609.215 (aiding suicide and aiding attempted suicide); 609.221 (assault in the first degree); 609.222 (assault in the second degree); 609.223 (assault in the third degree); 609.2231 (assault in the fourth degree); 609.229 (crimes committed for the benefit of a gang); 609.235 (use of drugs to injure or facilitate crime); 609.24 (simple robbery); 609.245 (aggravated robbery); 609.25 (kidnapping); 609.255 (false imprisonment); 609.342 (criminal sexual conduct in the first degree); 609.343 (criminal sexual conduct in the second degree); 609.344 (criminal sexual conduct in the third degree); 609.345 (criminal sexual conduct in the fourth degree); 609.377 (malicious punishment of a child); 609.378 (neglect or endangerment of a child); 609.486

(commission of crime while wearing or possessing a bullet-resistant vest); 609.52 (involving theft of a firearm, theft involving the intentional taking or driving of a motor vehicle without the consent of the owner or authorized agent of the owner, theft involving the taking of property from a burning, abandoned, or vacant building, or from an area of destruction caused by civil disaster, riot, bombing, or the proximity of battle, and theft involving the theft of a controlled substance, an explosive, or an incendiary device); 609.561 (arson in the first degree); 609.562 (arson in the second degree); 609.582, subdivision 1, 2, or 3 (burglary in the first through third degrees); 609.66, subdivision 1e (drive-by shooting); 609.67 (unlawfully owning, possessing, operating a machine gun or short-barreled shotgun); 609.71 (riot); 609.713 (terroristic threats); 609.749 (harassment and stalking); 609.855, subdivision 5 (shooting at a public transit vehicle or facility); and chapter 152 (drugs, controlled substances); and an attempt to commit any of these offenses.

Subd. 6. Transfer. "Transfer" means a sale, gift, loan, assignment or other delivery to another, whether or not for consideration, of a pistol or semiautomatic military-style assault weapon or the frame or receiver of a pistol or semiautomatic military-style assault weapon.

Subd. 7. Semiautomatic military-style assault weapon. "Semiautomatic military-style assault weapon" means:

(1) any of the following firearms:

(i) Avtomat Kalashnikov (AK-47) semiautomatic rifle type;

(ii) Beretta AR-70 and BM-59 semiautomatic rifle types;

(iii) Colt AR-15 semiautomatic rifle type;

(iv) Daewoo Max-1 and Max-2 semiautomatic rifle types;

(v) Famas MAS semiautomatic rifle type;

(vi) Fabrique Nationale FN-LAR and FN-FNC semiautomatic rifle types;

(vii) Galil semiautomatic rifle type;

(viii) Heckler & Koch HK-91, HK-93, and HK-94 semiautomatic rifle types;

(ix) Ingram MAC-10 and MAC-11 semiautomatic pistol and carbine types;

(x) Intratec TEC-9 semiautomatic pistol type;

(xi) Sigarms SIG 550SP and SIG 551SP semiautomatic rifle types;

(xii) SKS with detachable magazine semiautomatic rifle type;

(xiii) Steyr AUG semiautomatic rifle type;

(xiv) Street Sweeper and Striker-12 revolving-cylinder shotgun types;

(xv) USAS-12 semiautomatic shotgun type;

(xvi) Uzi semiautomatic pistol and carbine types; or

(xvii) Valmet M76 and M78 semiautomatic rifle types;

(2) any firearm that is another model made by the same manufacturer as one of the firearms listed in clause (1), and has the same action design as one of the listed firearms, and is a redesigned, renamed, or renumbered version of one of the firearms listed in clause (1), or has a slight modification or enhancement, including but not limited to a folding or retractable stock; adjustable sight; case deflector for left-handed shooters; shorter barrel; wooden, plastic, or metal stock; larger clip size; different caliber; or a bayonet mount; and

(3) any firearm that has been manufactured or sold by another company under a licensing agreement with a manufacturer of one of the firearms listed in clause (1) entered into after the effective date of Laws 1993, chapter 326, to manufacture or sell firearms that are identical or nearly identical to those listed in clause (1), or described in clause (2), regardless of the company of production or country of origin.

The weapons listed in clause (1), except those listed in items (iii), (ix), (x), (xiv), and (xv), are the weapons the importation of which was barred by the Bureau of Alcohol, Tobacco, and Firearms of the United States Department of the Treasury in July 1989.

Except as otherwise specifically provided in paragraph (d), a firearm is not a "semiautomatic military-style assault weapon" if it is generally recognized as particularly suitable for or readily adaptable to sporting purposes under United States Code, title 18, section 925, paragraph (d) (3), or any regulations adopted pursuant to that law.

Subd. 8. Included weapons. By August 1, 1993, and annually thereafter, the superintendent of the bureau of criminal apprehension shall publish a current authoritative list of the firearms included within the definition of "semiautomatic military-style assault weapon" under this section. Dealers, purchasers, and other persons may rely on the list in complying with this chapter.

Subd. 9. Business day. "Business day" means a day on which state offices are open for normal business and excludes weekends and legal holidays.

Subd. 10. Crime punishable by imprisonment for a term exceeding one year. "Crime punishable by imprisonment for a term exceeding one year" does not include:

(1) any federal or state offense pertaining to antitrust violations, unfair trade practices, restraints of trade, or other similar offenses relating to the regulation of business practices; or

(2) any state offense classified by the laws of this state or any other state as a misdemeanor and punishable by a term of imprisonment of two years or less.

What constitutes a conviction of a crime shall be determined in accordance with the law of the jurisdiction in which the proceedings were held. Any conviction which has been expunged, or set aside, or for which a person has been pardoned or has had civil rights restored shall not be considered a conviction for purposes of this definition, unless such pardon, expungement, or restoration of civil rights expressly provides that the person may not ship, transport, possess, or receive firearms.

Subd. 11. Commissioner. "Commissioner" means the commissioner of public safety unless otherwise indicated.

624.713. CERTAIN PERSONS NOT TO POSSESS FIREARMS

Subdivision 1. Ineligible persons. The following persons shall not be entitled to possess a pistol or semiautomatic military-style assault weapon or, except for clause (a), any other firearm:

(a) a person under the age of 18 years except that a person under 18 may carry or possess a pistol or semiautomatic military-style assault weapon (i) in the actual presence or under the direct supervision of the person's parent or guardian, (ii) for the purpose of military drill under the auspices of a legally recognized military organization and under competent supervision, (iii) for the purpose of instruction, competition, or target practice on a firing range approved by the chief of police or county sheriff in whose jurisdiction the range is located and under direct supervision; or (iv) if the person has successfully completed a course designed to teach marksmanship and safety with a pistol or semiautomatic military-style assault weapon and approved by the commissioner of natural resources;

(b) except as otherwise provided in clause (i), a person who has been convicted of, or adjudicated delinquent or convicted as an extended jurisdiction juvenile for committing, in this state or elsewhere, a crime of violence. For purposes of this section, crime of violence includes crimes in other states or jurisdictions which would have been crimes of violence as herein defined if they had been committed in this state;

(c) a person who is or has ever been confined in Minnesota or elsewhere as a person who is mentally ill, mentally retarded, or mentally ill and dangerous to the public as defined in section 253B.02, to a treatment facility, or who has ever been found incompetent to stand trial or not guilty by reason of mental illness, unless the person possesses a certificate of a medical doctor or psychiatrist licensed in Minnesota, or other satisfactory proof that the person is no longer suffering from this disability;

(d) a person who has been convicted in Minnesota or elsewhere of a misdemeanor or gross misdemeanor violation of chapter 152, or a person who is or has ever been hospitalized or committed for treatment for the habitual use of a controlled substance or marijuana, as defined in sections 152.01 and 152.02, unless the person possesses a certificate of a medical doctor or psychiatrist licensed in Minnesota, or other satisfactory proof, that the person has not abused a controlled substance or marijuana during the previous two years;

(e) a person who has been confined or committed to a treatment facility in Minnesota or elsewhere as chemically dependent as defined in section 253B.02, unless the person has completed treatment. Property rights may not be abated but access may be restricted by the courts;

(f) a peace officer who is informally admitted to a treatment facility pursuant to section 253B.04 for chemical dependency, unless the officer possesses a certificate from the head of the treatment facility discharging or provisionally discharging the officer from the treatment facility. Property rights may not be abated but access may be restricted by the courts;

(g) a person, including a person under the jurisdiction of the juvenile court, who has been charged with committing a crime of violence and has been placed in a pretrial diversion program by the court before disposition, until the person has completed the diversion program and the charge of committing the crime of violence has been dismissed;

(h) except as otherwise provided in clause (i), a person who has been convicted in another state of committing an offense similar to the offense described in section 609.224, subdivision 3, against a family or household member or section 609.2242, subdivision 3, unless three years have elapsed since the date of conviction and, during that time, the person has not been convicted of any other violation of section 609.224, subdivision 3, or 609.2242, subdivision 3, or a similar law of another state;

(i) a person who has been convicted in this state or elsewhere of assaulting a family or household member and who was found by the court to have used a firearm in any way during commission of the assault is prohibited from possessing any type of firearm for the period determined by the sentencing court; or

(j) a person who:

(1) has been convicted in any court of a crime punishable by imprisonment for a term exceeding one year;

(2) is a fugitive from justice as a result of having fled from any state to avoid prosecution for a crime or to avoid giving testimony in any criminal proceeding;

(3) is an unlawful user of any controlled substance as defined in chapter 152;

(4) has been judicially committed to a treatment facility in Minnesota or elsewhere as a person who is mentally ill, mentally retarded, or mentally ill and dangerous to the public as defined in section 253B.02;

(5) is an alien who is illegally or unlawfully in the United States;

(6) has been discharged from the armed forces of the United States under dishonorable conditions; or

(7) has renounced the person's citizenship having been a citizen of the United States.

(k) a person who has been convicted of the following offenses at the gross misdemeanor level, unless three years have elapsed since the date of conviction and, during that time, the person has not been convicted of any other violation of these sections: section 609.229 (crimes committed for the benefit of a gang); 609.2231, subdivision 4 (assaults motivated by bias); 609.255 (false imprisonment); 609.378 (neglect or endangerment of a child); 609.582, subdivision 4 (burglary in the fourth degree); 609.665 (setting a spring gun); 609.71 (riot); or 609.749 (harassment and stalking). For purposes of this paragraph, the specified gross misdemeanor convictions include crimes committed in other states or jurisdictions which would have been gross misdemeanors if conviction occurred in this state.

A person who issues a certificate pursuant to this subdivision in good faith is not liable for damages resulting or arising from the actions or misconduct with a firearm committed by the individual who is the subject of the certificate.

The prohibition in this subdivision relating to the possession of firearms other than pistols and semiautomatic military-style assault weapons does not apply retroactively to persons who are prohibited from possessing a pistol or semiautomatic military-style assault weapon under this subdivision before August 1, 1994.

The lifetime prohibition on possessing, receiving, shipping, or transporting firearms for persons convicted or adjudicated delinquent of a crime of violence in clause (b), applies only to offenders who are discharged from sentence or court supervision for a crime of violence on or after August 1, 1993.

Subd. 1a. Ineligible to receive, ship, transport. A person presently charged with a crime punishable by imprisonment for a term exceeding one year shall not be entitled to receive, ship, or transport any pistol or semiautomatic military-style assault weapon. A violation of this subdivision is a gross misdemeanor.

Subd. 2. Penalties.

(a) A person named in subdivision 1, clause (a), who possesses a pistol or semiautomatic military-style assault weapon is guilty of a felony and may be sentenced to imprisonment for not more than five years or to payment of a fine of not more than $10,000, or both.

(b) A person named in subdivision 1, clause (b), who possesses any type of firearm is guilty of a felony and may be sentenced to imprisonment for not more than 15 years or to payment of a fine of not more than $30,000, or both. This paragraph does not apply to any person who has received a relief of disability under United States Code, title 18, section 925, or whose ability to possess firearms has been restored under section 609.165, subdivision 1d.

(c) A person named in any other clause of subdivision 1 who possesses any type of firearm is guilty of a gross misdemeanor.

Subd. 3. Notice.

(a) When a person is convicted of, or adjudicated delinquent or convicted as an extended jurisdiction juvenile for committing, a crime of violence as defined in section 624.712, subdivision 5, the court shall inform the defendant that the defendant is prohibited from possessing a pistol or semiautomatic military-style assault weapon for the remainder of the person's lifetime, and that it is a felony offense to violate this prohibition. The failure of the court to provide this information to a defendant does not affect the applicability of the pistol or semiautomatic military-style assault weapon possession prohibition or the felony penalty to that defendant.

(b) When a person, including a person under the jurisdiction of the juvenile court, is charged with committing a crime of violence and is placed in a pretrial diversion program by the court before disposition, the court shall inform the defendant that: (1) the defendant is prohibited from possessing a pistol or semiautomatic military-style assault weapon until the person has completed the diversion program and the charge of committing a crime of violence has been dismissed; (2) it is a gross misdemeanor offense to violate this prohibition; and (3) if the defendant violates this condition of participation in the diversion program, the charge of committing a crime of violence may be prosecuted. The failure of the court to provide this information to a defendant does not affect the applicability of the pistol or semiautomatic military-style assault weapon possession prohibition or the gross misdemeanor penalty to that defendant.

624.7131. TRANSFEREE PERMIT; PENALTY

Subdivision 1. Information. Any person may apply for a transferee permit by providing the following information in writing to the chief of police of an organized full time police department of the municipality in which the person resides or to the county sheriff if there is no such local chief of police:

(a) the name, residence, telephone number and driver's license number or nonqualification certificate number, if any, of the proposed transferee;

(b) the sex, date of birth, height, weight and color of eyes, and distinguishing physical characteristics, if any, of the proposed transferee;

(c) a statement that the proposed transferee authorizes the release to the local police authority of commitment information about the proposed transferee maintained by the commissioner of human services, to the extent that the information relates to the proposed transferee's eligibility to possess a pistol or semiautomatic military-style assault weapon under section 624.713, subdivision 1; and

(d) a statement by the proposed transferee that the proposed transferee is not prohibited by section 624.713 from possessing a pistol or semiautomatic military-style assault weapon.

The statements shall be signed and dated by the person applying for a permit. At the time of application, the local police authority shall provide the applicant with a dated receipt for the application. The statement under clause (c) must comply with any applicable requirements of Code of Federal Regulations, title 42, sections 2.31 to 2.35, with respect to consent to disclosure of alcohol or drug abuse patient records.

Subd. 2. Investigation. The chief of police or sheriff shall check criminal histories, records and warrant information relating to the applicant through the Minnesota crime information system and the national criminal record repository and shall make a reasonable effort to check other available state and local record keeping systems. The chief of police or sheriff shall obtain commitment information from the commissioner of human services as provided in section 245.041.

Subd. 3. Forms. Chiefs of police and sheriffs shall make transferee permit application forms available throughout the community. There shall be no charge for forms, reports, investigations, notifications, waivers or any other act performed or materials provided by a government employee or agency in connection with application for or issuance of a transferee permit.

Subd. 4. Grounds for disqualification. A determination by the chief of police or sheriff that the applicant is prohibited by section 624.713 from possessing a pistol or semiautomatic military-style assault weapon shall be the only basis for refusal to grant a transferee permit.

Subd. 5. Granting of permits. The chief of police or sheriff shall issue a transferee permit or deny the application within seven days of application for the permit. The chief of police or sheriff shall provide an applicant with written notification of a denial and the specific reason for the denial. The permits and their renewal shall be granted free of charge.

Subd. 6. Permits valid statewide. Transferee permits issued pursuant to this section are valid statewide and shall expire after one year. A transferee permit may be renewed in the same manner and subject to the same provisions by which the original permit was obtained, except that all renewed permits must comply with the standards adopted by the commissioner under section 624.7151. Permits issued pursuant to this section are not transferable. A person who transfers a permit in violation of this subdivision is guilty of a misdemeanor.

Subd. 7. Permit voided. The transferee permit shall be void at the time that the holder becomes prohibited from possessing a pistol under section 624.713, in which event the holder shall return the permit within five days to the issuing authority. Failure of the holder to return the permit within the five days is a misdemeanor unless the court finds that the circumstances or the physical or mental condition of the permit holder prevented the holder from complying with the return requirement.

Subd. 8. Hearing upon denial. Any person aggrieved by denial of a transferee permit may appeal the denial to the district court having jurisdiction over the county or municipality in which the denial occurred.

Subd. 9. Permit to carry. A valid permit to carry issued pursuant to section 624.714 constitutes a transferee permit for the purposes of this section and section 624.7132.

Subd. 10. Transfer report not required. A person who transfers a pistol or semiautomatic military-style assault weapon to a person exhibiting a valid transferee permit issued pursuant to this section or a valid permit to carry issued pursuant to section 624.714 is not required to file a transfer report pursuant to section 624.7132, subdivision 1.

Subd. 11. Penalty. A person who makes a false statement in order to obtain a transferee permit knowing or having reason to know the statement is false is guilty of a gross misdemeanor.

Subd. 12. Local regulation. This section shall be construed to supersede municipal or county regulation of the issuance of transferee permits.

624.7132. REPORT OF TRANSFER

Subdivision 1. Required information. Except as provided in this section and section 624.7131, every person who agrees to transfer a pistol or semiautomatic military-style assault weapon shall report the following information in writing to the chief of police of the organized full-time police department of the municipality where the proposed transferee resides or to the appropriate county sheriff if there is no such local chief of police:

(a) the name, residence, telephone number and driver's license number or nonqualification certificate number, if any, of the proposed transferee;

(b) the sex, date of birth, height, weight and color of eyes, and distinguishing physical characteristics, if any, of the proposed transferee;

(c) a statement that the proposed transferee authorizes the release to the local police authority of commitment information about the proposed transferee maintained by the commissioner of human services, to the extent that the information relates to the proposed transferee's eligibility to possess a pistol or semiautomatic military-style assault weapon under section 624.713, subdivision 1;

(d) a statement by the proposed transferee that the transferee is not prohibited by section 624.713 from possessing a pistol or semiautomatic military-style assault weapon; and

(e) the address of the place of business of the transferor.

The report shall be signed and dated by the transferor and the proposed transferee. The report shall be delivered by the transferor to the chief of police or sheriff no later than three days after the date of the agreement to transfer, excluding weekends and legal holidays. The statement under clause (c) must comply with any applicable requirements of Code of Federal Regulations, title 42, sections 2.31 to 2.35, with respect to consent to disclosure of alcohol or drug abuse patient records.

Subd. 2. Investigation. Upon receipt of a transfer report, the chief of police or sheriff shall check criminal histories, records and warrant information relating to the proposed transferee through the Minnesota crime information system and the national criminal record repository and shall make a reasonable effort to check other available state and local record keeping systems. The chief of police or sheriff shall obtain commitment information from the commissioner of human services as provided in section 245.041.

Subd. 3. Notification. The chief of police or sheriff shall notify the transferor and proposed transferee in writing as soon as possible if the chief or sheriff determines that the proposed transferee is prohibited by section 624.713 from possessing a pistol or semiautomatic military-style assault weapon. The notification to the transferee shall specify the grounds for the disqualification of the proposed transferee and shall set forth in detail the transferee's right of appeal under subdivision 13.

Subd. 4. Delivery. Except as otherwise provided in subdivision 7 or 8, no person shall deliver a pistol or semiautomatic military-style assault weapon to a proposed transferee until five business days after the date the agreement to transfer is delivered to a chief of police or sheriff in accordance with subdivision 1 unless the chief of police or sheriff waives all or a portion of the seven day waiting period. The chief of police or sheriff may waive all or a portion of the five business day waiting period in writing if the chief of police or sheriff finds that the transferee requires access to a pistol or semiautomatic military-style assault weapon because of a threat to the life of the transferee or of any member of the household of the transferee.

No person shall deliver a pistol or semiautomatic military-style assault weapon to a proposed transferee after receiving a written notification that the chief of police or sheriff has determined that the proposed transferee is prohibited by section 624.713 from possessing a pistol or semiautomatic military-style assault weapon.

If the transferor makes a report of transfer and receives no written notification of disqualification of the proposed transferee within five business days after delivery of the agreement to transfer, the pistol or semiautomatic military-style assault weapon may be delivered to the transferee.

Subd. 5. Grounds for disqualification. A determination by the chief of police or sheriff that the proposed transferee is prohibited by section 624.713 from possessing a pistol or semiautomatic military-style assault weapon shall be the sole basis for a notification of disqualification under this section.

Subd. 6. Transferee permit. If a chief of police or sheriff determines that a transferee is not a person prohibited by section 624.713 from possessing a pistol or semiautomatic military-style assault weapon, the transferee may, within 30 days after the determination, apply to that chief of police or sheriff for a transferee permit, and the permit shall be issued.

Subd. 7. [Repealed]

Subd. 8. Report not required. If the proposed transferee presents a valid transferee permit issued under section 624.7131 or a valid permit to carry issued under section 624.714, the transferor need not file a transfer report.

Subd. 9. Number of pistols or semiautomatic military-style assault weapons. Any number of pistols or semiautomatic military-style assault weapons may be the subject of a single transfer agreement and report to the chief of police or sheriff. Nothing in this section or section 624.7131 shall be construed to limit or restrict the number of pistols or semiautomatic military-style assault weapons a person may acquire.

Subd. 10. Restriction on records. If, after a determination that the transferee is not a person prohibited by section 624.713 from possessing a pistol or semiautomatic military-style assault weapon, a transferee requests that no record be maintained of the fact of who is the transferee of a pistol or semiautomatic military-style assault weapon, the chief of police or sheriff shall sign the transfer report and return it to the transferee as soon as possible. Thereafter, no government employee or agency shall maintain a record of the transfer that identifies the transferee, and the transferee shall retain the report of transfer.

Subd. 11. Forms; cost. Chiefs of police and sheriffs shall make transfer report forms available throughout the community. There shall be no charge for forms, reports, investigations, notifications, waivers or any other act performed or materials provided by a government employee or agency in connection with a transfer.

Subd. 12. Exclusions. Except as otherwise provided in section 609.66, subdivision 1f, this section shall not apply to transfers of antique firearms as curiosities or for their historical significance or value, transfers to or between federally licensed firearms dealers, transfers by order of court, involuntary transfers, transfers at death or the following transfers:

(a) a transfer by a person other than a federally licensed firearms dealer;

(b) a loan to a prospective transferee if the loan is intended for a period of no more than one day;

(c) the delivery of a pistol or semiautomatic military-style assault weapon to a person for the purpose of repair, reconditioning or remodeling;

(d) a loan by a teacher to a student in a course designed to teach marksmanship or safety with a pistol and approved by the commissioner of natural resources;

(e) a loan between persons at a firearms collectors exhibition;

(f) a loan between persons lawfully engaged in hunting or target shooting if the loan is intended for a period of no more than 12 hours;

(g) a loan between law enforcement officers who have the power to make arrests other than citizen arrests; and

(h) a loan between employees or between the employer and an employee in a business if the employee is required to carry a pistol or semiautomatic military-style assault weapon by reason of employment and is the holder of a valid permit to carry a pistol.

Subd. 13. Appeal. A person aggrieved by the determination of a chief of police or sheriff that the person is prohibited by section 624.713 from possessing a pistol or semiautomatic military-style assault weapon may appeal the determination as provided in this subdivision. The district court shall have jurisdiction of proceedings under this subdivision.

On review pursuant to this subdivision, the court shall be limited to a determination of whether the proposed transferee is a person prohibited from possessing a pistol or semiautomatic military-style assault weapon by section 624.713.

Subd. 14. Transfer to unknown party.

(a) No person shall transfer a pistol or semiautomatic military-style assault weapon to another who is not personally known to the transferor unless the proposed transferee presents evidence of identity to the transferor.

(b) No person who is not personally known to the transferor shall become a transferee of a pistol or semiautomatic military-style assault weapon unless the person presents evidence of identity to the transferor.

(c) The evidence of identity shall contain the name, residence address, date of birth, and photograph of the proposed transferee; must be made or issued by or under the authority of the United States government, a state, a political subdivision of a state, a foreign government, a political subdivision of a foreign government, an international governmental or an international quasi-governmental organization; and must be of a type commonly accepted for the purpose of identification of individuals.

(d) A person who becomes a transferee of a pistol or semiautomatic military-style assault weapon in violation of this subdivision is guilty of a misdemeanor.

Subd. 15. Penalties.

(a) Except as otherwise provided in paragraph (b), a person who does any of the following is guilty of a gross misdemeanor:

(1) transfers a pistol or semiautomatic military-style assault weapon in violation of subdivisions 1 to 13;

(2) transfers a pistol or semiautomatic military-style assault weapon to a person who has made a false statement in order to become a transferee, if the transferor knows or has reason to know the transferee has made the false statement;

(3) knowingly becomes a transferee in violation of subdivisions 1 to 13; or

(4) makes a false statement in order to become a transferee of a pistol or semiautomatic military-style assault weapon knowing or having reason to know the statement is false.

(b) A person who does either of the following is guilty of a felony:

(1) transfers a pistol or semiautomatic military-style assault weapon to a person under the age of 18 in violation of subdivisions 1 to 13; or

(2) transfers a pistol or semiautomatic military-style assault weapon to a person under the age of 18 who has made a false statement in order to become a transferee, if the transferor knows or has reason to know the transferee has made the false statement.

Subd. 16. Local regulation. This section shall be construed to supersede municipal or county regulation of the transfer of pistols.

624.714. CARRYING OF WEAPONS WITHOUT PERMIT; PENALTIES

Subd. 1a. Permit required; penalty. A person, other than a peace officer, as defined in section 626.84, subdivision 1, who carries, holds, or possesses a pistol in a motor vehicle, snowmobile, or boat, or on or about the person's clothes or the person, or otherwise in possession or control in a public place, as defined in section 624.7181, subdivision 1, paragraph (c), without first having obtained a permit to carry the pistol is guilty of a gross misdemeanor. A person who is convicted a second or subsequent time is guilty of a felony.

Subd. 1b. Display of permit; penalty.

(a) The holder of a permit to carry must have the permit card and a driver's license, state identification card, or other government-issued photo identification in immediate possession at all times when carrying a pistol and must display the permit card and identification document upon lawful demand by a peace officer, as defined in section 626.84, subdivision 1. A violation of this paragraph is a petty misdemeanor. The fine for a first offense must not exceed $25. Notwithstanding section 609.531, a firearm carried in violation of this paragraph is not subject to forfeiture.

(b) A citation issued for violating paragraph (a) must be dismissed if the person demonstrates, in court or in the office of the arresting officer, that the person was authorized to carry the pistol at the time of the alleged violation.

(c) Upon the request of a peace officer, a permit holder must write a sample signature in the officer's presence to aid in verifying the person's identity.

(d) Upon the request of a peace officer, a permit holder shall disclose to the officer whether or not the permit holder is currently carrying a firearm.

Subd. 2. Where application made; authority to issue permit; criteria; scope.

(a). Applications by Minnesota residents for permits to carry shall be made to the county sheriff where the applicant resides. Nonresidents, as defined in section 171.01, subdivision 42, may apply to any sheriff. ...

Subd. 7. Permit card contents; expiration; renewal.

(a) Permits to carry must be on an official, standardized permit card adopted by the commissioner of public safety, containing only the name, residence, and driver's license number or state identification card number of the permit holder, if any.

(b) The permit card must also identify the issuing sheriff and state the expiration date of the permit. The permit card must clearly display a notice that a permit, if granted, is void and must be immediately returned to the sheriff if the permit holder becomes prohibited by law from possessing a firearm.

(c) A permit to carry a pistol issued under this section expires five years after the date of issue. It may be renewed in the same manner and under the same criteria which the original permit was obtained, ...

Subd. 7a. Change of address; loss or destruction of permit.

(a) Within 30 days after changing permanent address, or within 30 days of having lost or destroyed the permit card, the permit holder must notify the issuing sheriff of the change, loss, or destruction. Failure to provide notification as required by this subdivision is a petty misdemeanor. The fine for a first offense must not exceed $25. Notwithstanding section 609.531, a firearm carried in violation of this paragraph is not subject to forfeiture.

(b) After notice is given under paragraph (a), a permit holder may obtain a replacement permit card by paying $10 to the sheriff. The request for a replacement permit card must be made on an official, standardized application adopted for this purpose under section 624.7151, and, except in the case of an address change, must include a notarized statement that the permit card has been lost or destroyed.

Subd. 9. Carrying pistols about one's premises or for purposes of repair, target practice. A permit to carry is not required of a person:

(a) To keep or carry about the person's place of business, dwelling house, premises or on land possessed by the person a pistol;

(b) To carry a pistol from a place of purchase to the person's dwelling house or place of business, or from the person's dwelling house or place of business to or from a place where repairing is done, to have the pistol repaired;

(c) To carry a pistol between the person's dwelling house and place of business;

(d) To carry a pistol in the woods or fields or upon the waters of this state for the purpose of hunting or of target shooting in a safe area; or

(e) To transport a pistol in a motor vehicle, snowmobile or boat if the pistol is unloaded, contained in a closed and fastened case, gunbox, or securely tied package.

Subd. 10. False representations. A person who gives or causes to be given any false material information in applying for a permit to carry, knowing or having reason to know the information is false, is guilty of a gross misdemeanor.

Subd. 11. No limit on number of pistols. A person shall not be restricted as to the number of pistols the person may carry. ...

Subd. 13. Exemptions; adult correctional facility officers. A permit to carry a pistol is not required of any officer of a state adult correctional facility when on guard duty or otherwise engaged in an assigned duty. ...

Subd. 18. Employers; public colleges and universities.

(a) An employer, whether public or private, may establish policies that restrict the carry or possession of firearms by its employees while acting in the course and scope of employment. Employment related civil sanctions may be invoked for a violation.

(b) A public postsecondary institution regulated under chapter 136F or 137 may establish policies that restrict the carry or possession of firearms by its students while on the institution's property. Academic sanctions may be invoked for a violation.

(c) Notwithstanding paragraphs (a) and (b), an employer or a postsecondary institution may not prohibit the lawful carry or possession of firearms in a parking facility or parking area. ...

Subd. 22. Short title; construction; severability. This section may be cited as the Minnesota Citizens' Personal Protection Act of 2003. The legislature of the state of Minnesota recognizes and declares that the second amendment of the United States Constitution guarantees the fundamental, individual right to keep and bear arms. The provisions of this section are declared to be necessary to accomplish compelling state interests in regulation of those rights. The terms of this section must be construed according to the compelling state interest test. The invalidation of any provision of this section shall not invalidate any other provision.

Subd. 23. Exclusivity. This section sets forth the complete and exclusive criteria and procedures for the issuance of permits to carry and establishes their nature and scope. No sheriff, police chief, governmental unit, government official, government employee, or other person or body acting under color of law or governmental authority may change, modify, or supplement these criteria or procedures, or limit the exercise of a permit to carry.

624.7141. TRANSFER TO INELIGIBLE PERSON

Subdivision 1. Transfer prohibited. A person is guilty of a gross misdemeanor who intentionally transfers a pistol or semiautomatic military-style assault weapon to another if the person knows that the transferee:

(1) has been denied a permit to carry under section 624.714 because the transferee is not eligible under section 624.713 to possess a pistol or semiautomatic military-style assault weapon;

(2) has been found ineligible to possess a pistol or semiautomatic military-style assault weapon by a chief of police or sheriff as a result of an application for a transferee permit or a transfer report; or

(3) is disqualified under section 624.713 from possessing a pistol or semiautomatic military-style assault weapon.

Subd. 2. Felony. A violation of this section is a felony if the transferee possesses or uses the weapon within one year after the transfer in furtherance of a felony crime of violence.

Subd.3. Subsequent eligibility. This section is not applicable to a transfer to a person who became eligible to possess a pistol or semiautomatic military-style assault weapon under section 624.713 after the transfer occurred but before the transferee used or possessed the weapon in furtherance of any crime.

624.7142. CARRYING WHILE UNDER THE INFLUENCE OF ALCOHOL OR A CONTROLLED SUBSTANCE

Subdivision 1. Acts prohibited. A person may not carry a pistol on or about the person's clothes or person in a public place:

(1) when the person is under the influence of a controlled substance, as defined in section 152.01, subdivision 4;

(2) when the person is under the influence of a combination of any two or more of the elements named in clauses (1) and (4);

(3) when the person is knowingly under the influence of any chemical compound or combination of chemical compounds that is listed as a hazardous substance in rules adopted under section 182.655 and that affects the nervous system, brain, or muscles of the person so as to impair the person's clearness of intellect or physical control;

(4) when the person is under the influence of alcohol;

(5) when the person's alcohol concentration is 0.10 or more; or

(6) when the person's alcohol concentration is less than 0.10, but more than 0.04. ...

Subd. 6. Penalties.

(a) A person who violates a prohibition under subdivision 1, clauses (1) to (5), is guilty of a misdemeanor. A second or subsequent violation is a gross misdemeanor.

(b) A person who violates subdivision 1, clause (6), is guilty of a misdemeanor.

(c) In addition to the penalty imposed under paragraph (a), if a person violates subdivision 1, clauses (1) to (5), the person's authority to carry a pistol in a public place on or about the person's clothes or person under the provisions of a permit or otherwise is revoked and the person may not reapply for a period of one year from the date of conviction.

(d) In addition to the penalty imposed under paragraph (b), if a person violates subdivision 1, clause (6), the person's authority to carry a pistol in a public place on or about the person's clothes or person under the provisions of a permit or otherwise is suspended for 180 days from the date of conviction.

(e) Notwithstanding section 609.531, a firearm carried in violation of subdivision 1, clause (6), is not subject to forfeiture. ....

624.715. EXEMPTIONS; ANTIQUES AND ORNAMENTS Sections 624.713 and 624.714 shall not apply to antique firearms which are carried or possessed as curiosities or for their historical significance or value.

624.7151 STANDARDIZED FORMS

By December 1, 1992, the commissioner shall adopt statewide standards governing the form and contents, as required by sections 624.7131 to 624.714, of every application for a pistol transferee permit, pistol transferee permit, report of transfer of a pistol, application for a permit to carry a pistol, and permit to carry a pistol that is granted or renewed on or after January 1, 1993. Every application for a pistol transferee permit, pistol transferee permit, report of transfer of a pistol, application for a permit to carry a pistol, and permit to carry a pistol that is received, granted, or renewed by a police chief or county sheriff on or after January 1, 1993, must meet the statewide standards adopted by the commissioner. Notwithstanding the previous sentence, neither failure of the Department of Public Safety to adopt standards nor failure of the police chief or county sheriff to meet them shall delay the timely processing of applications nor invalidate permits issued on other forms meeting the requirements of sections 624.7131 to 624.714.

624.716. SATURDAY NIGHT SPECIALS PROHIBITED; PENALTY Any federally licensed firearms dealer who sells a Saturday Night Special Pistol, or any person who manufactures or assembles a Saturday Night Special Pistol in whole or in part, shall be guilty of a gross misdemeanor.

624.7161. FIREARMS DEALERS; CERTAIN SECURITY MEASURES REQUIRED

Subdivision 1. Definitions.

(a) For purposes of this section, the following terms have the meanings given.

(b) "Firearms dealer" means a dealer federally licensed to sell pistols who operates a retail business in which pistols are sold from a permanent business location other than the dealer's home.

(c) "Small firearms dealer" means a firearms dealer who operates a retail business at which no more than 50 pistols are displayed for sale at any time.

(d) "Large firearms dealer" means a firearms dealer who operates a retail business at which more than 50 pistols are displayed for sale at any time.

Subd. 2. Security measures required. After business hours when the dealer's place of business is unattended, a small firearms dealer shall place all pistols that are located in the dealer's place of business in a locked safe or locked steel gun cabinet, or on a locked, hardened steel rod or cable that runs through the pistol's trigger guards. The safe, gun cabinet, rod, or cable must be anchored to prevent its removal from the premises.

Subd. 3. Security standards. The commissioner of public safety shall adopt standards specifying minimum security requirements for small and large firearms dealers. By January 1, 1993, all firearms dealers shall comply with the standards. The standards may provide for:

(1) alarm systems for small and large firearms dealers;

(2) site hardening and other necessary and effective security measures required for large firearms dealers;

(3) a system of inspections, during normal business hours, by local law enforcement officials for compliance with the standards; and

(4) other reasonable requirements necessary and effective to reduce the risk of burglaries at firearms dealers' business establishments.

624.7162. FIREARMS DEALERS; SAFETY REQUIREMENTS

Subdivision 1. Firearms dealers. For purposes of this section, a firearms dealer is any person who is federally licensed to sell firearms from any location.

Subd. 2. Notice required. In each business location where firearms are sold by a firearms dealer, the dealer shall post in a conspicuous location the following warning in block letters not less than one inch in height: "IT IS UNLAWFUL TO STORE OR LEAVE A LOADED FIREARM WHERE A CHILD CAN OBTAIN ACCESS."

Subd. 3. Fine. A person who violates the provisions of this section is guilty of a petty misdemeanor and may be fined not more than $200.

624.717. LOCAL REGULATION Sections 624.711 to 624.716 shall be construed to supersede municipal or county regulation of the carrying or possessing of pistols and the regulation of Saturday Night Special Pistols.

624.719. POSSESSION OF FIREARM BY NONRESIDENT ALIEN A nonresident alien may not possess a firearm except to take game as a nonresident under the game and fish laws. A firearm possessed in violation of this section is contraband and may be confiscated.

624.74. METAL-PENETRATING BULLETS

Subdivision 1. Intent. This section is designed to give law enforcement officers performing their official duties a reasonable degree of protection from penetration of quality body armor. It is not the intent of this section to restrict the availability of ammunition for personal defense, sporting, or hunting purposes.

Subd. 2. Definition. For purposes of this section, "metal-penetrating bullet" means a handgun bullet of 9 mm, .25, .32, .357, .38, .41, .44, or .451 caliber which is comprised of a hardened core equal to the minimum of the maximum attainable hardness by solid red metal alloys which purposely reduces the normal expansion or mushrooming of the bullet's shape upon impact. "Metal-penetrating bullet" excludes any bullet composed of copper or brass jacket with lead or lead alloy cores and any bullet composed of lead or lead alloys. ...

Subd. 4. Local regulation. This section shall be construed to supersede any municipal or county regulation of ammunition, including its component parts.

[Current through 2008 Legislative Session]

Minnesota Rules

Department of Public Safety
Driver and Vehicle Services Division

Chapter 7416. Firearms Permits

7416.0100 APPLICATION FOR A HANDGUN TRANSFEREE PERMIT An application for a handgun transferee permit must be made on a form en-titled "Minnesota Uniform Firearm Application/ Receipt Transferee Permit or Report of Transfer for Firearms." A facsimile of the form is reproduced at part 7416.9911.

7416.0200 PISTOL TRANSFEREE PERMIT A pistol transferee permit must be issued on a form en-titled "Minnesota State Permit to Acquire Hand-guns From Federal Firearms Dealers." A facsimile of the form is reproduced at part 7416.9940.

7416.0300 REPORT OF TRANSFER OF A HANDGUN A report of transfer of a handgun must be made on a form entitled "Minnesota Uniform Firearm Application/Receipt Transferee Permit or Report of Transfer for Firearms." A facsimile of the form is reproduced at part 7416.9911.

7416.0400 APPLICATION FOR A PERMIT TO CARRY A PISTOL An application for a permit to carry a pistol must be made on a form entitled "Minnesota Uniform Firearm Application/Receipt, Carry Permit for Handgun in Public Place." A facsimile of the form is reproduced at part 7416.9931.

7416.0500 PERMIT TO CARRY A PISTOL A permit to carry a pistol must be issued on a form en-titled "Minnesota State Permit to Carry a Hand-gun." The permit, when issued, must be wallet sized and must be covered by plastic or some other material to protect against tampering or alteration of the permit. A facsimile of the form is reproduced at part 7416.9950.

[Current through June 16, 2008]

State Patrol Division

Chapter 7500. Explosives, Blasting Agents, Firearms

7500.0100 DEFINITIONS

Subpart 1. Scope. The terms used in this chapter have the meanings given them in this part. ...

Subp. 6. Explosives. "Explosives" means a chemical compound, mixture, or device, the primary or common purpose of which is to function by explosion. The term includes dynamite and other high explosives, detonators, safety fuses, squibs, detonating cord, igniter cord, and igniters.

For purposes of this subpart, there are three classes of explosive materials. These classes, together with the description of explosive materials comprising each class, are as follows:

A. High explosives are explosive materials that can be caused to detonate by means of a blasting cap when unconfined; for example, dynamite.

B. Low explosives are explosive materials that can be caused to deflagrate when confined, for example: black powder, safety fuses, igniters, igniter cord, fuse lighters, and special fireworks defined as class B explosives by United States Department of Transportation regulations in Code of Federal Regulations, title 49, part 173.

C. Blasting agents include, for example, ammonium nitrate-fuel oil and certain water gels. ...

Subp. 16. Small arms ammunition. "Small arms ammunition" means a shotgun, rifle, pistol, or revolver cartridge; or cartridge for propellant-actuated power devices and industrial guns. This term does not include military-type ammunition containing explosive bursting charges and spotting or pyrotechnic projectiles.

Subp. 17. Small arms ammunition primers. "Small arms ammunition primers" means small percussion-sensitive explosive charges, encased in a cup, used to ignite propellant powder.

Subp. 18. Smokeless propellant. "Smokeless propellant" means a solid propellant, called a smokeless powder in the trade, used in small arms ammunition, cannons, rockets, propellant-actuated power devices, or other devices. ....

7500.3400 SMALL ARMS AMMUNITION

Subpart 1. Restricted scope. Parts 7500.3400 to 7500.3600 do not apply to in-process storage and intraplant transportation during manufacture of small arms ammunition, small arms primers, and smokeless propellants.

This part does not apply to the transportation of small arms ammunition governed by Code of Federal Regulations, title 49, parts 100 to 199 (1983).

Subp. 2. Quantity limitations. No quantity limitations are imposed on the storage of small arms ammunition in warehouses, retail stores, and other general occupancies, except those imposed by limitation of storage facilities and consistency with public safety.

Subp. 3. Separate storage. Small arms ammunition must be separated from flammable liquids, flammable solids as classified by the Department of Transportation in Code of Federal Regulations, title 49, parts 100 to 199 (1983), and oxidizing materials by a fire-resistive wall of one-hour rating or by a distance of 25 feet.

Small arms ammunition must not be stored together with explosives or blasting agents, unless the storage facility is adequate as described in Code of Federal Regulations, title 27, sections 55.201 to 55.220 (1982).

7500.3500 SMOKELESS PROPELLANTS

Subpart 1. Restricted scope. This part does not apply to the transportation of smokeless propellants governed by Code of Federal Regulations, title 49, parts 100 to 199 (1983).

Subp. 1a. Approved container required. Smokeless propellants must be stored in shipping containers approved by the Department of Transportation in Code of Federal Regulations, title 49, parts 100 to 199 (1983).

Subp. 1b. Transportation of certain quantities. Quantities of smokeless propellants of 25 pounds or less in shipping containers approved by the Department of Transportation in Code of Federal Regulations, title 49, parts 100 to 199 (1983), may be transported in a passenger vehicle.

Quantities in excess of 25 pounds but not exceeding 50 pounds that are transported in a passenger vehicle must be in a portable magazine having wooden walls of at least one-inch nominal thickness.

Transportation of quantities in excess of 50 pounds is prohibited in vehicles transporting passengers.

Transportation of quantities in excess of 50 pounds in other than passenger vehicles must comply with Department of Transportation regulations in Code of Federal Regulations, title 49, parts 100 to 199 (1983). In addition, warning placards must be prominently displayed when more than 250 pounds are being transported.

Subp. 2. Storage of propellants intended for personal use. Smokeless propellants intended for personal use in quantities not to exceed 20 pounds may be stored in residences. Quantities in excess of 20 but not exceeding 50 pounds must be stored in a wooden box or cabinet having walls of at least one-inch nominal thickness.

Subp. 3. Commercial displays and stocks. Not more than 20 pounds of smokeless propellants, in containers of one-pound maximum capacity, may be displayed in commercial establishments. Commercial stocks of smokeless propellants greater than 20 pounds but not more than 100 pounds must be stored in approved wooden boxes having walls of at least one-inch nominal thickness. Not more than 50 pounds are permitted in any one box.

Commercial stocks in quantities not to exceed 750 pounds must be stored in storage cabinets having wooden walls of at least one-inch nominal thickness. Not more than 400 pounds are permitted in any one cabinet.

Subp. 4. Quantities over 750 pounds. Quantities in excess of 750 pounds must be stored in magazines constructed and located as specified in part 7500.0800.

7500.3600 SMALL ARMS AMMUNITION PRIMERS

Subpart 1. Restricted scope. This section does not apply to the transportation of small arms ammunition primers governed by Code of Federal Regulations, title 49, parts 100 to 199 (1983).

Subp. 2. Transportation and storage. Small arms ammunition primers must not be transported or stored except in the original shipping container approved by the Department of Transportation in Code of Federal Regulations, title 49, parts 100 to 199 (1983).

Subp. 3. Truck or rail transportation. Truck or rail transportation of small arms ammunition primers must comply with the Department of Transportation regulations in Code of Federal Regulations, title 49, parts 100 to 199 (1983).

Subp. 4. Quantity limitations. Not more than 25,000 small arms ammunition primers may be transported in a passenger vehicle.

Not more than 10,000 small arms ammunition primers may be stored in residences.

Not more than 10,000 small arms ammunition primers may be displayed in commercial establishments.

Subp. 5. Separate storage. Small arms ammunition primers must be separated from flammable liquids, flammable solids as classified by the Department of Transportation in Code of Federal Regulations, title 49, parts 100 to 199 (1983), and oxidizing materials by a fire-resistive wall of one-hour rating or by a distance of 25 feet.

Subp. 6. Storage of large quantities. Quantities of small arms ammunition primers in excess of 1,000 000 must be stored in magazines in accordance with part 7500.0800.

7500.5100 DEFINITIONS

Subpart 1. Scope. For the purpose of parts 7500.5100 to 7500.5600, the following terms have the meanings given them.

Subp. 2. Approved machine gun or short-barrelled shotgun. "Approved machine gun or short-barrelled shotgun" means a machine gun or short-barrelled shotgun that, although designed as a weapon, has been determined by the superintendent as not likely to be used as a weapon and that has been determined by the superintendent to appear on the National Firearms Act Curios and Relics List, as provided by United States Code, title 18, chapter 44, and as issued by the Department of the Treasury, Bureau of Alcohol, Tobacco and Firearms (Washington, D.C.).

Subp. 3. Bureau. "Bureau" means the Minnesota Bureau of Criminal Apprehension.

Subp. 4. Report form. "Report form" means the official form created by the bureau on which the data specified in Minnesota Statutes, section 609.67, subdivision 4 is to be reported to the superintendent.

Subp. 5. Superintendent. "Superintendent" means the superintendent of the bureau or a delegate.

[Current through June 16, 2008]

Chapter 7504. Firearms Dealers; Security Standards

7504.0100 DEFINITIONS

Subpart 1. Scope. For purposes of this chapter, the terms in subparts 2 to 6 have the meanings given them.

Subp. 2. Firearms dealer. "Firearms dealer" means a dealer federally licensed to sell pistols who operates a retail business in which pistols are sold from a permanent business location other than the dealer's home. For the purposes of this chapter, a dealer's home does not include the following:

A. a building located on property that is zoned commercial;

B. a business location where the square footage used for the business of selling firearms exceeds the square footage used for the dealer's residence; or

C. a building located on the same property as the dealer's home that is not attached to the dealer's home.

Subp. 3. Small firearms dealer. "Small firearms dealer" means a firearms dealer who operates a retail business at which no more than 50 pistols are displayed for sale at any time.

Subp. 4. Large firearms dealer. "Large firearms dealer" means a firearms dealer who operates a retail business at which more than 50 pistols are displayed for sale at any time.

Subp. 5. Pistol. "Pistol" includes a weapon designed to be fired by the use of a single hand and with an overall length less than 26 inches, or having a barrel or barrels of a length less than 18 inches in the case of a shotgun or having a barrel of a length less than 16 inches in the case of a rifle (1) from which may be fired or ejected one or more solid projectiles by means of a cartridge or shell or by the action of an explosive or the igniting of flammable or explosive substances; or (2) for which the propelling force is a spring, elastic band, carbon dioxide, air or other gas, or vapor. Pistol does not include a device firing or ejecting a shot measuring .18 of an inch or less in diameter and commonly known as a BB gun, a scuba gun, a stud gun or nail gun used in the construction industry, or children's pop guns or toys.

Subp. 6. Displayed for sale. "Displayed for sale" means a pistol available for sale to customers that is either displayed or stored at the dealer's place of business. This does not include pistols temporarily stored at the place of business for repair or servicing.

7504.0200 Security measures for small firearms dealers. After business hours when the dealer's place of business is unattended, a small firearms dealer shall place all pistols that are located in the dealer's place of business in a locked safe or locked steel gun cabinet, or on a locked, hardened steel rod or cable that runs through the pistols' trigger guards.

A. The door to a safe must be recessed or flush and made of at least seven gauge steel. The body of a safe must be made of hot rolled steel of at least 12 gauge thickness. A safe must have an Underwriters Laboratory (UL) listed Group 2 combination lock.

B. A rod or cable used to secure a pistol must be hardened steel. The rod or cable must be at least one-fourth inch in diameter. The rod or cable must be secured with a hardened steel lock which has a shackle that is protected or shielded from attack by a bolt cutter.

C. No more than five pistols may be affixed to any one rod or cable.

D. The safe, gun cabinet, rod, or cable must be anchored to prevent its removal from the premises.

E. The door to a gun cabinet must be made of at least 14 gauge steel. The door may be recessed, flush, or overlapping. If the door is flush or overlapping, it must be designed to conceal the location of the locking bolts and hinges from the outside of the cabinet. The door must be reinforced and must be attached to the body by one continuous hinge or at least two hinges that are located either inside or outside the body. If the hinges are located outside the body of the safe, the safe must have an interior locking system consisting of permanent or moveable locking pins securing the door from the inside when it is in the closed position. The body of a gun cabinet must be made of hot rolled steel of at least 14 gauge thickness, and must be continuously welded to create a single, solid structure. A gun cabinet must have either a UL-listed group two combination lock or a UL-listed key lock that is encased in a high security, drill-resistant lock body. A key lock must use a restricted key that can only be duplicated by a factory-authorized source.

7504.0300 SECURITY MEASURES FOR LARGE FIREARMS DEALERS A large firearms dealer shall comply with the requirements of items A to H or the requirements of part 7504.0200, items A to D, except that a large firearms dealer shall not use a gun cabinet.

A. The dealer shall install vehicle-resistant barriers to prevent the penetration of the dealer's place of business by a motor vehicle. The barriers must protect any areas that are accessible to vehicles and that have a free run distance of 50 feet or more.

B. The dealer shall secure each perimeter doorway according to sub-item (1), (2), or (3).

(1) The dealer may use a windowless steel security door equipped with both a dead bolt and a doorknob lock.

(2) The dealer may use a windowed metal door that is equipped with both a dead bolt and a doorknob lock. The window must be made of one-half inch polycarbonate or glass reinforced with metal mesh. If the window has an opening of five inches or more measured in any direction, then the window must be covered with steel bars or metal grating affixed to the interior of the door.

(3) The dealer may use a metal grate that is padlocked and affixed to the premises independent of the door and door frame.

C. A dealer shall cover all windows with steel bars that are anchored internally to the wall joists.

D. A dealer shall secure a room where pistols are stored after hours with a locked steel door or metal grating.

E. A dealer shall not display pistols within four feet of a window unless the window is covered by metal screen that is anchored internally to the wall joists.

F. A dealer shall secure heating, ventilating, air conditioning, and service openings with steel bars, metal grating, or an alarm system as described in part 7504.0400.

G. A dealer shall illuminate each perimeter doorway so that the doorway is clearly visible after dark from a distance of 100 feet.

H. Metal grates and grating must have spaces no larger than six inches wide along any diagonal. Metal screen must have spaces no larger than three inches wide along any diagonal. Steel bars must be no more than six inches apart on center.

7504.0400 ELECTRONIC SECURITY FOR ALL FIREARMS DEALERS A firearms dealer shall install and maintain an electronic security system that meets the requirements of items A to I.

A. The system must emit an audible alarm at the dealer's place of business when triggered.

B. The system must transmit a silent alarm when triggered. The silent alarm must be transmitted directly to a public safety answering point where this service is available. If no local public safety answering point provides an alarm-monitoring service, the silent alarm must automatically transmit a violation signal to a UL-approved monitoring station that must notify an appropriate law enforcement agency within two minutes of receiving the violation signal.

C. A large firearms dealer's system must include a backup silent alarm that transmits the alarm as provided in item B if the primary transmission system fails.

D. The system must include a line cut alarm unless the incoming phone lines are hardened by routing them through underground conduit or similar protective barrier. The line cut alarm must be triggered when the phone line is cut, and must emit an audible alarm.

E. The system must monitor all exterior doors, windows, and other entry points, including but not limited to heating, ventilating, air conditioning, and customer and service entry points.

F. The system must use motion and heat sensors to monitor pistol storage areas and alarm control boards.

G. The system must monitor all interior doors that provide access to pistol storage areas.

H. All components of the system must be UL-approved.

I. All components of the system must be independently tested and certified to comply with this part at least once per year. The test must be conducted by an alarm system installation or monitoring firm or a person approved by the alarm system manufacturer or distributor. Written certification must be available for inspection by a local law enforcement authority.

7504.0500 INSPECTION BY LAW ENFORCEMENT A firearms dealer shall make its place of business available to local law enforcement officials during normal business hours for the purpose of verifying compliance with this chapter. A local law enforcement official is not required to give advance notice of an inspection.

7504.0600 EXEMPTIONS Upon written request from a firearms dealer, the commissioner of public safety shall grant an exemption from compliance with a requirement of this chapter if the following conditions are met:

A. the request identifies the requirement from which the dealer wants to be exempt;

B. the request identifies security measures used in lieu of complying with the requirement;

C. the requirement is not specifically set out in statute; and

D. the commissioner determines that the security measures will provide a degree of security similar to the degree of security provided by the requirement or will cause a delay in the unauthorized entry into the dealer's business premises equivalent to the delay provided by complying with the requirement.

[Current through June 16, 2008]

MISSISSIPPI

Miss. Code

Title 45. Public Safety and Good Order

Chapter 9. Weapons

45-9-51. Prohibition against adoption of certain ordinances Subject to the provisions of Section 45-9-53, no county or municipality may adopt any ordinance that restricts or requires the possession, transportation, sale, transfer or ownership of firearms or ammunition or their components.

45-9-53. Exceptions

(1) This Section and Section 45-9-51 do not affect the authority that a county or municipality may have under another law:

(a) To require citizens or public employees to be armed for personal or national defense, law enforcement, or another lawful purpose;

(b) To regulate the discharge of firearms within the limits of the county or municipality. A county or municipality may not apply a regulation relating to the discharge of firearms or other weapons in the extraterritorial jurisdiction of the county or municipality or in an area annexed by the county or municipality after September 1, 1981, if the firearm or other weapon is:

(i) A shotgun, air rifle or air pistol, BB gun or bow and arrow discharged:

1. On a tract of land of ten (10) acres or more and more than one hundred fifty (150) feet from a residence or occupied building located on another property; and

2. In a manner not reasonably expected to cause a projectile to cross the boundary of the tract; or

(ii) A center fire or rim fire rifle or pistol or a muzzle-loading rifle or pistol of any caliber discharged:

1. On a tract of land of fifty (50) acres or more and more than three hundred (300) feet from a residence or occupied building located on another property; and

2. In a manner not reasonably expected to cause a projectile to cross the boundary of tract;

(c) To regulate the use of property or location of businesses for uses therein pursuant to fire code, zoning ordinances, or land-use regulations, so long as such codes, ordinances and regulations are not used to circumvent the intent of Section 45-9-51 or subparagraph (e) of this section;

(d) To regulate the use of firearms in cases of insurrection, riots and natural disasters in which the city finds such regulation necessary to protect the health and safety of the public. However, the provisions of this section shall not apply to the lawful possession of firearms in the home, place of business or in transit to and from the home or place of business;

(e) To regulate the storage or transportation of explosives in order to protect the health and safety of the public, with the exception of black powder which is exempt up to twenty-five (25) pounds per private residence and fifty (50) pounds per retail dealer;

(f) To regulate the carrying of a firearm at: (i) a public park or at a public meeting of a county, municipality or other governmental body; (ii) a political rally, parade or official political meeting; or (iii) a nonfirearm-related school, college or professional athletic event; or

(g) To regulate the receipt of firearms by pawnshops.

(2) The exception provided by subsection (1)(f) of this section does not apply if the firearm was in or carried to and from an area designated for use in a lawful hunting, fishing or other sporting event and the firearm is of the type commonly used in the activity.

45-9-101. License to carry stun gun, concealed pistol or revolver.

(1)(a) The Department of Public Safety is authorized to issue licenses to carry stun guns, concealed pistols or revolvers to persons qualified as provided in this section. Such licenses shall be valid throughout the state for a period of five (5) years from the date of issuance. Any person possessing a valid license issued pursuant to this section may carry a stun gun, concealed pistol or concealed revolver.

(b) The licensee must carry the license, together with valid identification, at all times in which the licensee is carrying a stun gun, concealed pistol or revolver and must display both the license and proper identification upon demand by a law enforcement officer. A violation of the provisions of this paragraph (b) shall constitute a noncriminal violation with a penalty of Twenty-five Dollars ($25.00) and shall be enforceable by summons. ...

(9) Within thirty (30) days after the changing of a permanent address, or within thirty (30) days after having a license lost or destroyed, the licensee shall notify the Department of Public Safety in writing of such change or loss. Failure to notify the Department of Public Safety pursuant to the provisions of this subsection shall constitute a noncriminal violation with a penalty of Twenty-five Dollars ($25.00) and shall be enforceable by a summons.

(10) In the event that a stun gun, concealed pistol or revolver license is lost or destroyed, the person to whom the license was issued shall comply with the provisions of subsection (9) of this section and may obtain a duplicate, or substitute thereof, upon payment of Fifteen Dollars ($15.00) to the Department of Public Safety, and furnishing a notarized statement to the department that such license has been lost or destroyed. ...

(20) The provisions of this section shall be under the supervision of the Commissioner of Public Safety. The commissioner is authorized to promulgate reasonable rules and regulations to carry out the provisions of this section.

Title 75. Regulation of Trade, Commerce and Investments

Chapter 67. Loans

Article 7. Mississippi Pawnshop Act

75-67-305. Information required to be recorded on pawn ticket; detailed recording of transactions required

(1) At the time of making the pawn or purchase transaction, the pawnbroker shall enter upon the pawn ticket a record of the following information which shall be typed or written in ink and in the English language:

(a) A clear and accurate description of the property, including the following:

(i) Brand name;

(ii) Model number;

(iii) Serial number;

(iv) Size;

(v) Color, as apparent to the untrained eye; ...

(viii) In the case of firearms, the type of action, caliber or gauge, number of barrels, barrel length and finish; and

(ix) Any other unique identifying marks, numbers, names or letters;

(b) The name, residence address and date of birth of pledgor or seller;

(c) Date of pawn or purchase transaction;

(d) Driver's license number or social security number or Mississippi identification card number, as defined in Section 45-35-1, Mississippi Code of 1972, of the pledgor or seller or identification information verified by at least two (2) forms of identification, one (1) of which shall be a photographic identification;

(e) Description of the pledgor including approximate height, sex and race;

(f) Amount of cash advanced;

(g) The maturity date of the pawn transaction and the amount due; and

(h) The monthly rate and pawn charge. Such rates and charges shall be disclosed using the requirements prescribed in Regulation Z (Truth in Lending) of the rules and regulations of the Board of Governors of the Federal Reserve.

(2) Each pawn or purchase transaction document shall be consecutively numbered and entered in a corresponding log or record book. Separate logs or record books for pawn and purchase transactions shall be kept.

(3) Records may be in the form of traditional hard copies, computer printouts or magnetic media if readily accessible for viewing on a screen with the capability of being promptly printed upon request.

(4) Every licensee shall maintain a record which indicates the total number of accounts and the total dollar value of all pawn transactions outstanding as of December 31 of each year.

Title 97. Crimes

Chapter 37. Weapons and Explosives

97-37-1. Deadly weapons; carrying while concealed; use or attempt to use; penalties

(1) Except as otherwise provided in Section 45-9-101, any person who carries, concealed in whole or in part, any ... pistol, revolver, or any rifle with a barrel of less than sixteen (16) inches in length, or any shotgun with a barrel of less than eighteen (18) inches in length, machine gun or any fully automatic firearm or deadly weapon, or any muffler or silencer for any firearm, whether or not it is accompanied by a firearm, or uses or attempts to use against another person any imitation firearm, shall upon conviction be punished as follows:

(a) By a fine of not less than One Hundred Dollars ($100.00) nor more than Five Hundred Dollars ($500.00), or by imprisonment in the county jail for not more than six (6) months, or both, in the discretion of the court, for the first conviction under this section.

(b) By a fine of not less than One Hundred Dollars ($100.00) nor more than Five Hundred Dollars ($500 00), and imprisonment in the county jail for not less than thirty (30) days nor more than six (6) months, for the second conviction under this section.

(c) By confinement in the custody of the department of corrections for not less than one (1) year nor more than five (5) years, for the third or subsequent conviction under this section.

(d) By confinement in the custody of the department of corrections for not less than one (1) year nor more than ten (10) years for any person previously convicted of any felony who is convicted under this section.

(2) It shall not be a violation of this section for any person over the age of eighteen (18) years to carry a firearm or deadly weapon concealed in whole or in part within the confines of his own home or his place of business, or any real property associated with his home or business or within any motor vehicle.

(3) It shall not be a violation of this section for any person to carry a firearm or deadly weapon concealed in whole or in part if the possessor of the weapon is then engaged in a legitimate weapon-related sports activity or is going to or returning from such activity. For purposes of this subsection, "legitimate weapon-related sports activity" means hunting, fishing, target shooting or any other legal sports activity which normally involves the use of a firearm or other weapon.

97-37-3. Deadly weapons; forfeiture of weapon; return upon dismissal or acquittal; confiscated firearms may be sold at auction; proceeds of sale used to purchase bullet-proof vests for seizing law enforcement agency

(1) Any weapon used in violation of Section 97-37-1, or used in the commission of any other crime, shall be seized by the arresting officer, may be introduced in evidence, and in the event of a conviction, shall be ordered to be forfeited, and shall be disposed of as ordered by the court having jurisdiction of such offense. In the event of dismissal or acquittal of charges, such weapon shall be returned to the accused from whom it was seized.

(2)(a) If the weapon to be forfeited is merchantable, the court may order the weapon forfeited to the seizing law enforcement agency.

(b) A weapon so forfeited to a law enforcement agency may be sold at auction as provided by Sections 19-3-85 and 21-39-21 to a federally-licensed firearms dealer, with the proceeds from such sale at auction to be used to buy bullet-proof vests for the seizing law enforcement agency.

97-37-5. Unlawful for convicted felon to possess any firearms, or other weapons or devices; penalties; exceptions

(1) It shall be unlawful for any person who has been convicted of a felony under the laws of this state, any other state, or of the United States to possess any firearm or any bowie knife, dirk knife, butcher knife, switchblade knife, metallic knuckles, blackjack, or any muffler or silencer for any firearm unless such person has received a pardon for such felony, has received a relief from disability pursuant to Section 925(c) of Title 18 of the U.S. Code, or has received a certificate of rehabilitation pursuant to subsection (3) of this section.

(2) Any person violating this section shall be guilty of a felony and, upon conviction thereof, shall be fined not more than Five Thousand Dollars ($5,000.00), or committed to the custody of the State Department of Corrections for not less than one(1) year nor more than ten (10) years, or both.

(3) A person who has been convicted of a felony under the laws of this state may apply to the court in which he was convicted for a certificate of rehabilitation. The court may grant such certificate in its discretion upon a showing to the satisfaction of the court that the applicant has been rehabilitated and has led a useful, productive and law-abiding life since the completion of his sentence and upon the finding of the court that he will not be likely to act in a manner dangerous to public safety.

97-37-7. Deadly weapons; persons permitted to carry weapons; bond; permit to carry weapon; grounds for denying application for permit; required weapons training course; reciprocal agreements

(1)(a) It shall not be a violation of Section 97-37-1 or any other statute for pistols, firearms or other suitable and appropriate weapons to be carried by duly constituted bank guards, company guards, watchmen, railroad special agents or duly authorized representatives who are not sworn law enforcement officers, agents or employees of a patrol service, guard service, or a company engaged in the business of transporting money, securities or other valuables, while actually engaged in the performance of their duties as such, provided that such persons have made a written application and paid a nonrefundable permit fee of One Hundred Dollars ($100.00) to the Department of Public Safety.

(b) No permit shall be issued to any person who has ever been convicted of a felony under the laws of this or any other state or of the United States. To determine an applicant's eligibility for a permit, the person shall be fingerprinted. If no disqualifying record is identified at the state level, the fingerprints shall be forwarded by the Department of Public Safety to the Federal Bureau of Investigation for a national criminal history record check. The department shall charge a fee which includes the amounts required by the Federal Bureau of Investigation and the department for the national and state criminal history record checks and any necessary costs incurred by the department for the handling and administration of the criminal history background checks. In the event a legible set of fingerprints, as determined by the Department of Public Safety and the Federal Bureau of Investigation, cannot be obtained after a minimum of three (3) attempts, the Department of Public Safety shall determine eligibility based upon a name check by the Mississippi Highway Safety Patrol and a Federal Bureau of Investigation name check conducted by the Mississippi Safety Patrol at the request of the Department of Public Safety.

(c) A person may obtain a duplicate of a lost or destroyed permit upon payment of a Fifteen Dollar ($15.00) replacement fee to the Department of Public Safety, if he furnishes a notarized statement to the department that the permit has been lost or destroyed.

(d)(i) No less than ninety (90) days prior to the expiration date of a permit, the Department of Public Safety shall mail to the permit holder written notice of expiration together with the renewal form prescribed by the department. The permit holder shall renew the permit on or before the expiration date by filing with the department the renewal form, a notarized affidavit stating that the permit holder remains qualified, and the renewal fee of Fifty Dollars ($50.00); provided, however, that honorably retired law enforcement officers shall be exempt from payment of the renewal fee. A permit holder who fails to file a renewal application on or before its expiration date shall pay a late fee of Fifteen Dollars ($15.00).

(ii) Renewal of the permit shall be required every four (4) years. The permit of a qualified renewal applicant shall be renewed upon receipt of the completed renewal application and appropriate payment of fees.

(iii) A permit cannot be renewed six (6) months or more after its expiration date, and such permit shall be deemed to be permanently expired; the holder may reapply for an original permit as provided in this section.

(2) It shall not be a violation of this or any other statute for pistols, firearms or other suitable and appropriate weapons to be carried by Department of Wildlife, Fisheries and Parks law enforcement officers, railroad special agents who are sworn law enforcement officers, investigators employed by the Attorney General, district attorneys, legal assistants to district attorneys, criminal investigators employed by the district attorneys, investigators or probation officers employed by the Department of Corrections, employees of the State Auditor who are authorized by the State Auditor to perform investigative functions, or any deputy fire marshal or investigator employed by the State Fire Marshal, while engaged in the performance of their duties as such, or by fraud investigators with the Department of Human Services, or by judges of the Mississippi Supreme Court, Court of Appeals, circuit, chancery, county, justice and municipal courts. Before any person shall be authorized under this subsection to carry a weapon, he shall complete a weapons training course approved by the Board of Law Enforcement Officer Standards and Training. Before any criminal investigator employed by a district attorney shall be authorized under this section to carry a pistol, firearm or other weapon, he shall have complied with Section 45-6-11 or any training program required for employment as an agent of the Federal Bureau of Investigation. A law enforcement officer, as defined in Section 45-6-3, shall be authorized to carry weapons in courthouses in performance of his official duties. This section shall in no way interfere with the right of a trial judge to restrict the carrying of firearms in the court-room.

(3) It shall not be a violation of this or any other statute for pistols, firearms or other suitable and appropriate weapons, to be carried by any out-of-state, full-time commissioned law enforcement officer who holds a valid commission card from the appropriate out-of-state law enforcement agency and a photo identification. The provisions of this subsection shall only apply if the state where the out-of-state officer is employed has entered into a reciprocity agreement with the state that allows full-time commissioned law enforcement officers in Mississippi to lawfully carry or possess a weapon in such other states. The Commissioner of Public Safety is authorized to enter into reciprocal agreements with other states to carry out the provisions of this subsection.

97-37-11. Deadly weapons; dealers to keep record of weapons sold Every merchant or dealer or pawnbroker that sells bowie-knifes, dirk-knifes, pistols, brass or metallic knuckles or slungshots, shall keep a record of all sales of such weapons sold, showing the description of the weapons, the name of the purchaser, and the description of weapons date of sale. This record to be opened to public inspection at any time to persons desiring to see it. The dealer who violates this section shall be guilty of a misdemeanor, and upon conviction shall be fined not less than five dollars ($5.00) nor more than twenty-five dollars ($25.00).

97-37-13. Deadly weapons; weapons and cartridges not to be given to minor or intoxicated person It shall not be lawful for any person to sell, give or lend to any minor under eighteen (18) years of age or person intoxicated, knowing him to be a minor under eighteen (18) years of age or in a state of intoxication, any deadly weapon, or other weapon the carrying of which concealed is prohibited, or pistol cartridge; and, on conviction thereof, he shall be punished by a fine not more than One Thousand Dollars ($1,000.00), or imprisoned in the county jail not exceeding one (1) year, or both.

97-37-14. Possession of handgun by minor; act of delinquency; exceptions

(1) Except as otherwise provided in this section, it is an act of delinquency for any person who has not attained the age of eighteen (18) years knowingly to have any handgun in such person's possession.

(2) This section shall not apply to:

(a) Any person who is:

(i) In attendance at a hunter's safety course or a firearms safety course; or

(ii) Engaging in practice in the use of a firearm or target shooting at an established range authorized by the governing body of the jurisdiction in which such range is located or any other area where the discharge of a firearm is not prohibited; or

(iii) Engaging in an organized competition involving the use of a firearm, or participating in or practicing for a performance by an organized group under 501(c)(3) as determined by the federal internal revenue service which uses firearms as a part of such performance; or

(iv) Hunting or trapping pursuant to a valid license issued to such person by the Department of Wildlife, Fisheries and Parks or as otherwise allowed by law; or

(v) Traveling with any handgun in such person's possession being unloaded to or from any activity described in subparagraph (i), (ii), (iii) or (iv) of this paragraph (a) and paragraph (b).

(b) Any person under the age of eighteen (18) years who is on real property under the control of an adult and who has the permission of such adult to possess a handgun.

(3) This section shall not apply to any person who uses a handgun or other firearm to lawfully defend himself from imminent danger at his home or place of domicile and any such person shall not be held criminally liable for such use of a handgun or other firearm.

(4) For the purposes of this section, "handgun" means a pistol, revolver or other firearm of any description, loaded or unloaded, from which any shot, bullet or other missile can be discharged, the length of the barrel of which, not including any revolving, detachable or magazine breech, is less than sixteen (16) inches.

97-37-15. Parent or guardian not to permit minor son to have or carry weapon; penalty Any parent, guardian or custodian who shall knowingly suffer or permit any child under the age of eighteen (18) years to have or to own, or to carry concealed, in whole or in part, any weapon the carrying of which concealed is prohibited, shall be guilty of a misdemeanor, and, on conviction, shall be fined not more than One Thousand Dollars ($1,000.00), and shall be imprisoned not more than six (6) months in the county jail. The provisions of this section shall not apply to a minor who is exempt from the provisions of Section 97-37-14.

97-37-17. Possession of weapons by students; aiding or encouraging

(1) The following definitions apply to this section:

(a) "Educational property" shall mean any public or private school building or bus, public or private school campus, grounds, recreational area, athletic field, or other property owned, used or operated by any local school board, school, college or university board of trustees, or directors for the administration of any public or private educational institution or during a school related activity, and shall include the facilities and property of the Oakley and Columbia juvenile training schools operated by the Department of Human Services; provided, however, that the term "educational property" shall not include any sixteenth section school land or lieu land on which is not located a school building, school campus, recreational area or athletic field.

(b) "Student" shall mean a person enrolled in a public or private school, college or university, or a person who has been suspended or expelled within the last five (5) years from a public or private school, college or university, or a person in the custody of the Oakley or Columbia juvenile training schools operated by the Department of Human Services, whether the person is an adult or a minor.

(c) "Switchblade knife" shall mean a knife containing a blade or blades which open automatically by the release of a spring or a similar contrivance.

(d) "Weapon" shall mean any device enumerated in subsection (2) or (4) of this section.

(2) It shall be a felony for any person to possess or carry, whether openly or concealed, any gun, rifle, pistol or other firearm of any kind, or any dynamite cartridge, bomb, grenade, mine or powerful explosive on educational property. However, this subsection does not apply to a BB gun, air rifle or air pistol. Any person violating this subsection shall be guilty of a felony and, upon conviction thereof, shall be fined not more than Five Thousand Dollars ($5,000.00), or committed to the custody of the State Department of Corrections for not more than three (3) years, or both.

(3) It shall be a felony for any person to cause, encourage or aid a minor who is less than eighteen (18) years old to possess or carry, whether openly or concealed, any gun, rifle, pistol or other firearm of any kind, or any dynamite cartridge, bomb, grenade, mine or powerful explosive on educational property. However, this subsection does not apply to a BB gun, air rifle or air pistol. Any person violating this subsection shall be guilty of a felony and, upon conviction thereof, shall be fined not more than Five Thousand Dollars ($5,000.00), or committed to the custody of the State Department of Corrections for not more than three (3) years, or both.

(4) It shall be a misdemeanor for any person to possess or carry, whether openly or concealed, any BB gun, air rifle, air pistol, bowie knife, dirk, dagger, slingshot, leaded cane, switchblade knife, blackjack, metallic knuckles, razors and razor blades (except solely for personal shaving), and any sharp-pointed or edged instrument except instructional supplies, unaltered nail files and clips and tools used solely for preparation of food, instruction and maintenance on educational property. Any person violating this subsection shall be guilty of a misdemeanor and, upon conviction thereof, shall be fined not more than One Thousand Dollars ($1,000.00), or be imprisoned not exceeding six (6) months, or both.

(5) It shall be a misdemeanor for any person to cause, encourage or aid a minor who is less than eighteen (18) years old to possess or carry, whether openly or concealed, any BB gun, air rifle, air pistol, bowie knife, dirk, dagger, slingshot, leaded cane, switchblade, knife, blackjack, metallic knuckles, razors and razor blades (except solely for personal shaving) and any sharp-pointed or edged instrument except instructional supplies, unaltered nail files and clips and tools used solely for preparation of food, instruction and maintenance on educational property. Any person violating this subsection shall be guilty of a misdemeanor and, upon conviction thereof, shall be fined not more than One Thousand Dollars ($1,000.00), or be imprisoned not exceeding six (6) months, or both.

(6) It shall not be a violation of this section for any person to possess or carry, whether openly or concealed, any gun, rifle, pistol or other firearm of any kind on educational property if:

(a) The person is not a student attending school on any educational property;

(b) The firearm is within a motor vehicle; and

(c) The person does not brandish, exhibit or display the firearm in any careless, angry or threatening manner.

(7) This section shall not apply to:

(a) A weapon used solely for educational or school-sanctioned ceremonial purposes, or used in a school-approved program conducted under the supervision of an adult whose supervision has been approved by the school authority;

(b) Armed forces personnel of the United States, officers and soldiers of the militia and National Guard, law enforcement personnel, any private police employed by an educational institution, State Militia or Emergency Management Corps and any guard or patrolman in a state or municipal institution, and any law enforcement personnel or guard at a state juvenile training school, when acting in the discharge of their official duties;

(c) Home schools as defined in the compulsory school attendance law, Section 37-13-91;

(d) Competitors while participating in organized shooting events;

(e) Any person as authorized in Section 97-37-7 while in the performance of his official duties;

(f) Any mail carrier while in the performance of his official duties; or

(g) Any weapon not prescribed by Section 97-37-1 which is in a motor vehicle under the control of a parent, guardian or custodian, as defined in Section 43-21-105, which is used to bring or pick up a student at a school building, school property or school function.

(8) All schools shall post in public view a copy of the provisions of this section.

97-37-31. Silencers on firearms; armor piercing ammunition; manufacture, sale, possession or use unlawful It shall be unlawful for any person, persons, corporation or manufacturing establishment, not duly authorized under federal law, to make, manufacture, sell or possess any instrument or device which, if used on firearms of any kind, will arrest or muffle the report of said firearm when shot or fired or armor piercing ammunition as defined in federal law. Any person violating this section shall be guilty of a misdemeanor and, upon conviction, shall be fined not more than Five Hundred Dollars ($500.00), or imprisoned in the penitentiary not more than thirty (30) days, or both. All such instruments or devices shall be registered with the Department of Public Safety and any law enforcement agency in possession of such instruments or devices shall submit an annual inventory of such instruments and devices to the Department of Public Safety. The Commissioner of Public Safety shall document the information required by this section.

97-37-33. Toy pistols; sale of pistol or cartridges prohibited; cap pistols excepted If any person shall sell, or offer, or expose for sale any toy pistol, or cartridges, or other contrivance by which such pistols are fired or made to cause an explosion, he shall be guilty of a misdemean-or, and, upon conviction, shall be punished by a fine of not less than five dollars nor more than twenty-five dollars, or by imprisonment in the county jail not less than three days nor more than thirty days, or both.

It is expressly provided, however, that nothing herein shall be construed to prohibit the sale, or offering, or exposure for sale of any toy cap pistols, or other devices, in which paper caps manufactured in accordance with United States Interstate Commerce Commission regulations for packing or shipping of toy paper caps are used or exploded and the sale of such toy cap pistols is hereby declared to be permissible.

97-37-35. Stolen firearms; possession, receipt, acquisition or disposal; offense; punishment

(1) It is unlawful for any person knowingly or intentionally to possess, receive, retain, acquire or obtain possession or dispose of a stolen firearm or attempt to possess, receive, retain, acquire or obtain possession or dispose of a stolen firearm.

(2) It is unlawful for any person knowingly or intentionally to sell, deliver or transfer a stolen firearm or attempt to sell, deliver or transfer a stolen firearm.

(3) Any person convicted of violating this section shall be guilty of a felony and shall be punished as follows:

(a) For the first conviction, punishment by commitment to the Department of Corrections for five (5) years;

(b) For the second and subsequent convictions, the offense shall be considered trafficking in stolen firearms punishable by commitment to the Department of Corrections for not less than fifteen (15) years.

(c) For a conviction where the offender possesses two (2) or more stolen firearms, the offense shall be considered trafficking in stolen firearms punishable by commitment to the Department of Corrections for not less than fifteen (15) years.

(4) Any person who commits or attempts to commit any other crime while in possession of a stolen firearm shall be guilty of a separate felony of possession of a stolen firearm under this section and, upon conviction thereof, shall be punished by commitment to the Department of Corrections for five (5) years, such term to run consecutively and not concurrently with any other sentence of incarceration.

97-37-37. Enhanced Penalty for use of firearm during commission of felony

Section 1. (1) Except to the extent that a greater minimum sentence is otherwise provided by any other provision of law, any person who uses or displays a firearm during the commission of any felony shall, in addition to the punishment provided for such felony, be sentenced to an additional term of imprisonment in the custody of the Department of Corrections of five (5) years, which sentence shall not be reduced or suspended.

(2) Except to the extent that a greater minimum sentence is otherwise provided by any other provision of law, any convicted felon who uses or displays a firearm during the commission of any felony shall, in addition to the punishment provided for such felony, be sentenced to an additional term of imprisonment in the custody of the department of corrections of ten (10) years, to run consecutively, not concurrently, which sentence shall not be reduced or suspended.

[Current through 2008 Regular and 2008 First Extraordinary Session]

MISSOURI

MO. REV. STAT.

Title III. Legislative Branch

Chapter 21. General Assembly

21.750. Firearms legislation preemption by general assembly, exceptions--limitation on civil recovery against firearms or ammunitions manufacturers, when, exception

(1) The general assembly hereby occupies and preempts the entire field of legislation touching in any way firearms, components, ammunition and supplies to the complete exclusion of any order, ordinance or regulation by any political subdivision of this state. Any existing or future orders, ordinances or regulations in this field are hereby and shall be null and void except as provided in subsection 3 of this section.

(2) No county, city, town, village, municipality, or other political subdivision of this state shall adopt any order, ordinance or regulation concerning in any way the sale, purchase, purchase delay, transfer, ownership, use, keeping, possession, bearing, transportation, licensing, permit, registration, taxation other than sales and compensating use taxes or other controls on firearms, components, ammunition, and supplies except as provided in subsection 3 of this section.

(3) Nothing contained in this section shall prohibit any ordinance of any political subdivision which conforms exactly with any of the provisions of sections 571.010 to 571.070, RSMo, with appropriate penalty provisions, or which regulates the open carrying of firearms readily capable of lethal use or the discharge of firearms within a jurisdiction, provided such ordinance complies with the provisions of section 252.243, RSMo.

(4) The lawful design, marketing, manufacture, distribution, or sale of firearms or ammunition to the public is not an abnormally dangerous activity and does not constitute a public or private nuisance.

(5) No county, city, town, village or any other political subdivision nor the state shall bring suit or have any right to recover against any firearms or ammunition manufacturer, trade association or dealer for damages, abatement or injunctive relief resulting from or relating to the lawful design, manufacture, marketing, distribution, or sale of firearms or ammunition to the public. This subsection shall apply to any suit pending as of October 12, 2003, as well as any suit which may be brought in the future. Provided, however, that nothing in this section shall restrict the rights of individual citizens to recover for injury or death caused by the negligent or defective design or manufacture of firearms or ammunition.

(6) Nothing in this section shall prevent the state, a county, city, town, village or any other political subdivision from bringing an action against a firearms or ammunition manufacturer or dealer for breach of contract or warranty as to firearms or ammunition purchased by the state or such political subdivision.

Title XXVI. Trade and Commerce

Chapter 407. Merchandising Practices

407.500. Missouri residents may purchase rifles and shotguns in contiguous states, when Residents of the state of Missouri may purchase rifles and shotguns in a state contiguous to the state of Missouri, provided that such residents conform to the applicable provisions of the Federal Gun Control Act of 1968, and regulations thereunder, as administered by the United States Secretary of the Treasury, and provided further that such residents conform to the provisions of law applicable to such purchase in the state of Missouri and in the contiguous state in which the purchase is made.

407.505. Residents of contiguous states may purchase rifles and shotguns in Missouri, when Residents of a state contiguous to the state of Missouri may purchase rifles and shotguns in the state of Missouri, provided that such residents conform to the applicable provisions of the Federal Gun Control Act of 1968, and regulations thereunder, as administered by the United States Secretary of the Treasury, and provided further that such residents conform to the provisions of law applicable to such purchase in the state of Missouri and in the state in which such persons reside.

Title XXXVIII. Crimes and Punishment; ...

Chapter 571. Weapons Offenses

571.010. Chapter definitions. As used in this chapter, the following terms shall mean:

(1) "Antique, curio or relic firearm," any firearm so defined by the National Gun Control Act, 18 U.S.C. Title 26, Section 5845, and the United States Treasury/Bureau of Alcohol Tobacco and Firearms, 27 CFR Section 178.11:

(a) Antique firearm is any firearm not designed or redesigned for using rim fire or conventional center fire ignition with fixed ammunition and manufactured in or before 1898, said ammunition not being manufactured any longer; this includes any matchlock, wheel-lock, flintlock, percussion cap or similar type ignition system, or replica thereof;

(b) Curio or relic firearm is any firearm deriving value as a collectable weapon due to its unique design, ignition system, operation or at least fifty years old, associated with a historical event, renown personage or major war; ...

(3) "Blasting agent", any material or mixture, consisting of fuel and oxidizer that is intended for blasting, but not otherwise defined as an explosive under this section, provided that the finished product, as mixed for use of shipment, cannot be detonated by means of a numbered 8 test blasting cap when unconfined;

(4) "Concealable firearm", any firearm with a barrel less than sixteen inches in length, measured from the face of the bolt or standing breech;

(5) "Deface," to alter or destroy the manufacturer's or importer's serial number or any other distinguishing number or identification mark;

(6) "Detonator", any device containing a detonating charge that is used for initiating detonation in an explosive, including but not limited to, electric blasting caps of instantaneous and delay types, non-electric blasting caps for use with safety fuse or shock tube and detonating-cord delay connectors;

(7) "Explosive weapon," any explosive, incendiary, or poison gas bomb or similar device designed or adapted for the purpose of inflicting death, serious physical injury, or substantial property damage; or any device designed or adapted for delivering or shooting such a weapon. For the purposes of this subdivision, the term "explosive" shall mean any chemical compound mixture or device, the primary or common purpose of which is to function by explosion, including but not limited to, dynamite and other high explosives, pellet powder, initiating explosives, detonators, safety fuses, squibs, detonating cords, igniter cords, and igniters or blasting agents;

(8) "Firearm," any weapon that is designed or adapted to expel a projectile by the action of an explosive;

(9) "Firearm silencer," any instrument, attachment, or appliance that is designed or adapted to muffle the noise made by the firing of any firearm;

(10) "Gas gun," any gas ejection device, weapon, cartridge, container or contrivance other than a gas bomb, that is designed or adapted for the purpose of ejecting any poison gas that will cause death or serious physical injury, but not any device that ejects a repellant or temporary incapacitating substance;

(11) "Intoxicated," substantially impaired mental or physical capacity resulting from introduction of any substance into the body; ...

(14) "Machine gun," any firearm that is capable of firing more than one shot automatically, without manual reloading, by a single function of the trigger;

(15) "Projectile weapon," any bow, crossbow, pellet gun, slingshot or other weapon that is not a firearm, which is capable of expelling a projectile that could inflict serious physical injury or death by striking or piercing a person;

(16) "Rifle," any firearm designed or adapted to be fired from the shoulder and to use the energy of the explosive in a fixed metallic cartridge to fire a projectile through a rifled bore by a single function of the trigger;

(17) "Short barrel," a barrel length of less than sixteen inches for a rifle and eighteen inches for a shotgun, both measured from the face of the bolt or standing breech, or an overall rifle or shotgun length of less than twenty-six inches;

(18) "Shotgun," any firearm designed or adapted to be fired from the shoulder and to use the energy of the explosive in a fixed shot-gun shell to fire a number of shot or a single projectile through a smooth bore barrel by a single function of the trigger; ...

571.020. Possession - manufacture - transport - repair - sale of certain weapons a crime – exceptions - penalties

1. A person commits a crime if such person knowingly possesses, manufactures, transports, repairs, or sells:

(1) An explosive weapon;

(2) An explosive, incendiary or poison substance or material with the purpose to possess, manufacture or sell an explosive weapon;

(3) A machine gun;

(4) A gas gun;

(5) A short barreled rifle or shotgun;

(6) A firearm silencer;

(7) A switchblade knife;

(8) A bullet or projectile which explodes or detonates upon impact because of an independent explosive charge after having been shot from a firearm; or

(9) Knuckles.

2. A person does not commit a crime pursuant to this section if his conduct:

(1) Was incident to the performance of official duty by the armed forces, national guard, a governmental law enforcement agency, or a penal institution; or

(2) Was incident to engaging in a lawful commercial or business transaction with an organization enumerated in subdivision (1) of this section; or

(3) Was incident to using an explosive weapon in a manner reasonably related to a lawful industrial or commercial enterprise; or

(4) Was incident to displaying the weapon in a public museum or exhibition; or

(5) Was incident to dealing with the weapon solely as a curio, ornament, or keepsake, or to using it in a manner reasonably related to a lawful dramatic performance; but if the weapon is a type described in subdivision or (4) of subsection 1 of this section it must be in such a nonfunctioning condition that it cannot readily be made operable. No short barreled rifle, short barreled shotgun, machine gun, or firearm silencer may be possessed, manufactured, transported, repaired or sold as a curio, ornament, or keepsake, unless such person is an importer, manufacturer, deal-er, or collector licensed by the Secretary of the Treasury pursuant to the Gun Control Act of 1968, U.S.C., Title 18, or unless such firearm is an "antique firearm" as defined in subsection 3 of section 571.080, or unless such firearm has been designated a "collectors item" by the Secretary of the Treasury pursuant to the U.S.C., Title 26, Section 5845(a).

3. A crime pursuant to subdivision (1), (2), (3), (4), (5) or (6) of subsection 1 of this section is a class C felony; a crime pursuant to subdivision (7), (8) or (9) of subsection 1 of this section is a class A misdemeanor.

571.030. Unlawful use of weapons - exceptions - penalties

1. A person commits the crime of unlawful use of weapons if he or she knowingly:

(1) Carries concealed upon or about his or her person a knife, a firearm, a blackjack or any other weapon readily capable of lethal use; or ...

(5) Possesses or discharges a firearm or projectile weapon while intoxicated; or ...

(10) Carries a firearm, whether loaded or unloaded, or any other weapon readily capable of lethal use into any school, onto any school bus, or onto the premises of any function or activity sponsored or sanctioned by school officials or the district school board.

2. Subdivisions (1), (3), (4), (6), (7), (8), (9) and (10) of subsection 1 of this section shall not apply to or affect any of the following:

(1) All state, county and municipal peace officers who have completed the training required by the police officer standards and training commission pursuant to sections 590.030 to 590.050, RSMo, and possessing the duty and power of arrest for violation of the general criminal laws of the state or for violation of ordinances of counties or municipalities of the state, whether such officers are on or off duty, and whether such officers are within or outside of the law enforcement agency's jurisdiction, or all qualified retired peace officers, as defined in subsection 10 of this section, and who carry the identification defined in subsection 11 of this section, or any person summoned by such officers to assist in making arrests or preserving the peace while actually engaged in assisting such officer;

(2) Wardens, superintendents and keepers of prisons, penitentiaries, jails and other institutions for the detention of persons accused or convicted of crime;

(3) Members of the armed forces or national guard while performing their official duty;

(4) Those persons vested by article V, section 1 of the Constitution of Missouri with the judicial power of the state and those persons vested by article III of the Constitution of the United States with the judicial power of the United States, the members of the federal judiciary;

(5) Any person whose bona fide duty is to execute process, civil or criminal;

(6) Any federal probation officer or federal flight deck officer as defined under the federal flight deck officer program, 49 U.S.C. Section 44921;

(7) Any state probation or parole officer, including supervisors and members of the board of probation and parole;

(8) Any corporate security advisor meeting the definition and fulfilling the requirements of the regulations established by the board of police commissioners under section 84.340, RSMo; and

(9) Any coroner, deputy coroner, medical examiner, or assistant medical examiner.

3. Subdivisions (1), (5), (8) and (10) of subsection 1 of this section do not apply when the actor is transporting such weapons in a nonfunctioning state or in an unloaded state when ammunition is not readily accessible or when such weapons are not readily accessible. Subdivision (1) of subsection 1 of this section does not apply to any person twenty-one years of age or older transporting a concealable firearm in the passenger compartment of a motor vehicle, so long as such concealable firearm is otherwise lawfully possessed, nor when the actor is also in possession of an exposed firearm or projectile weapon for the lawful pursuit of game, or is in his or her dwelling unit or upon premises over which the actor has possession, authority or control, or is traveling in a continuous journey peaceably through this state. Subdivision (10) of subsection 1 of this section does not apply if the firearm is otherwise lawfully possessed by a person while traversing school premises for the purposes of transporting a student to or from school, or possessed by an adult for the purposes of facilitation of a school-sanctioned firearm-related event.

4. Subdivisions (1), (8), and (10) of subsection 1 of this section shall not apply to any person who has a valid concealed carry endorsement issued pursuant to sections 571.101 to 571.121 or a valid permit or endorsement to carry concealed firearms issued by another state or political subdivision of another state.

5. Subdivisions (3), (4), (5), (6), (7), (8), (9), and (10) of subsection 1 of this section shall not apply to persons who are engaged in a lawful act of defense pursuant to section 563.031, RSMo.

6. Nothing in this section shall make it unlawful for a student to actually participate in school-sanctioned gun safety courses, student military or ROTC courses, or other school-sponsored firearm-related events, provided the student does not carry a firearm or other weapon readily capable of lethal use into any school, onto any school bus, or onto the premises of any other function or activity sponsored or sanctioned by school officials or the district school board.

7. Unlawful use of weapons is a class D felony unless committed pursuant to subdivision (6), (7), or (8) of subsection 1 of this section, in which cases it is a class B misdemeanor, or subdivision (5) or (10) of subsection 1 of this section, in which case it is a class A misdemeanor if the firearm is unloaded and a class D felony if the firearm is loaded, or subdivision (9) of subsection 1 of this section, in which case it is a class B felony, except that if the violation of subdivision (9) of subsection 1 of this section results in injury or death to another person, it is a class A felony. ...

571.045 Defacing firearm, penalty.

1. A person commits the crime of defacing a firearm if he knowingly defaces any firearm.

2. Defacing a firearm is a class A misdemeanor.

571.050. Possession of defaced firearm, penalty

1. A person commits the crime of possession of a defaced firearm if he knowingly possesses a firearm which is defaced.

2. Possession of a defaced firearm is a class B misdemeanor.

571.060. Unlawful transfer of weapons, penalty

1. A person commits the crime of unlawful transfer of weapons if he:

(1) Knowingly sells, leases, loans, gives away or delivers a firearm or ammunition for a firearm to any person who, under the provisions of section 571.070, is not lawfully entitled to possess such;

(2) Knowingly sells, leases, loans, gives away or delivers a blackjack to a person less than eighteen years old without the consent of the child's custodial parent or guardian, or recklessly, as defined in section 562.016, RSMo, sells, leases, loans, gives away or delivers any firearm to a person less than eighteen years old without the consent of the child's custodial parent or guardian; provided, that this does not prohibit the delivery of such weapons to any peace officer or member of the armed forces or national guard while performing his official duty; or

(3) Recklessly, as defined in section 562.016, RSMo, sells, leases, loans, gives away or delivers a firearm or ammunition for a firearm to a person who is intoxicated.

2. Unlawful transfer of weapons under subdivision (1) of subsection 1 of this section is a class D felony; unlawful transfer of weapons under subdivisions (2) and (3) of subsection 1 of this section is a class A misdemeanor.

571.070. Possession of concealable firearm unlawful for certain persons - penalty

1. A person commits the crime of unlawful possession of a firearm if such person knowingly has any firearm in his or her possession and:

(1) Such person has been convicted of a felony under the laws of this state, or of a crime under the laws of any state or of the United States which, if committed within this state, would be a felony; or

(2) Such person is a fugitive from justice, is habitually in an intoxicated or drugged condition, or is currently adjudged mentally incompetent.

2. Unlawful possession of a firearm is a class C felony.

571.080 Transfer of concealable firearms

A person commits the crime of transfer of a concealable firearm if such person violates 18 U.S.C. Section 922(b) or 18 U.S.C. Section 922(x).

571.095. Confiscation of firearms and ammunition, when -- exceptions

Upon conviction for or attempting to commit a felony in violation of any law perpetrated in whole or in part by the use of a firearm, the court may, in addition to the penalty provided by law for such offense, order the confiscation and disposal or sale or trade to a licensed firearms dealer of firearms and ammunition used in the commission of the crime or found in the possession or under the immediate control of the defendant at the time of his or her arrest. The proceeds of any sale or gains from trade shall be the property of the police department or sheriff's department responsible for the defendant's arrest or the confiscation of the firearms and ammunition. If such firearms or ammunition are not the property of the convicted felon, they shall be returned to their rightful owner if he or she is known and was not a participant in the crime. Any proceeds collected under this section shall be deposited with the municipality or by the county treasurer into the county sheriff's revolving fund established in section 50.535, RSMo.

571.107. Endorsement does not authorize concealed firearms, where - penalty for violation.

1. A concealed carry endorsement issued pursuant to sections 571.101 to 571.121 or a concealed carry endorsement or permit issued by another state or political subdivision of another state shall authorize the person in whose name the permit or endorsement is issued to carry concealed firearms on or about his or her person or vehicle throughout the state. No driver's license or nondriver's license containing a concealed carry endorsement issued pursuant to sections 571.101 to 571.121 or a concealed carry endorsement or permit issued by another state or political subdivision of another state shall authorize any person to carry concealed firearms into: ...

(10) Any higher education institution or elementary or secondary school facility without the consent of the governing body of the higher education institution or a school official or the district school board. Possession of a firearm in a vehicle on the premises of any higher education institution or elementary or secondary school facility shall not be a criminal offense so long as the firearm is not removed from the vehicle or brandished while the vehicle is on the premises; ...

2. Carrying of a concealed firearm in a location specified in subdivisions (1) to (17) of subsection 1 of this section by any individual who holds [a] concealed carry endorsement issued pursuant to sections 571.101 to 571.121 shall not be a criminal act but may subject the person to denial to the premises or removal from the premises. If such person refuses to leave the premises and a peace officer is summoned, such person may be issued a citation for an amount not to exceed one hundred dollars for the first offense. If a second citation for a similar violation occurs within a six-month period, such person shall be fined an amount not to exceed two hundred dollars and his or her endorsement to carry concealed firearms shall be suspended for a period of one year. If a third citation for a similar violation is issued within one year of the first citation, such person shall be fined an amount not to exceed five hundred dollars and shall have his or her concealed carry endorsement revoked and such person shall not be eligible for a concealed carry endorsement for a period of three years. Upon conviction of charges arising from a citation issued pursuant to this subsection, the court shall notify the sheriff of the county which issued the certificate of qualification for a concealed carry endorsement and the department of revenue. The sheriff shall suspend or revoke the certificate of qualification for a concealed carry endorsement and the department of revenue shall issue a notice of such suspension or revocation of the concealed carry endorsement and take action to remove the concealed carry endorsement from the individual's driving record. The director of revenue shall notify the licensee that he or she must apply for a new license pursuant to chapter 302, RSMo, which does not contain such endorsement. A concealed carry endorsement suspension pursuant to sections 571.101 to 571.121 shall be reinstated at the time of the renewal of his or her driver's license. The notice issued by the department of revenue shall be mailed to the last known address shown on the individual's driving record. The notice is deemed received three days after mailing.

[Current through 2007 First Extra Session of the Regular Session of the 94th General Assembly]

MONTANA

MONT. CODE

Title 45. Crimes

Chapter 6. Offenses Against Property

Part 3. Theft and Related Offenses

45-6-326. Obscuring the identity of a machine.

(1) A person commits the offense of obscuring the identity of a machine if he:

(a) removes, defaces, covers, alters, destroys, or otherwise obscures the manufacturer's serial number or any other distinguishing identification number or mark upon any machine, vehicle, electrical device, or firearm with the purpose to conceal, misrepresent, or transfer any such machine, vehicle, electrical device, or firearm; or

(b) possesses with the purpose to conceal, misrepresent, or transfer any machine, vehicle, device, or firearm knowing that the serial number or other identification number or mark has been removed or otherwise obscured.

(2) A person convicted of obscuring the identity of a machine shall be fined not to exceed $500 or be imprisoned in the county jail for a term not to exceed 6 months, or both.

(3) The fact of possession or transfer of any such machine, vehicle, electrical device, or firearm creates a presumption that the person knew the serial number or other identification number or mark had been removed or otherwise obscured.

Chapter 8. Offenses Against Public Order

Part 3. Weapons

45-8-301. Uniformity of interpretation. Sections 45-8-302 through 45-8-305 and 45-8-307 must be interpreted and construed to effectuate their general purpose to make uniform the law of those states that enact them.

45-8-302. Definitions. In 45-8-303 through 45-8-305 and 45-8-307, the following definitions apply:

(1) "Crime of violence" means any of the following crimes or an attempt to commit any of the crimes: any forcible felony, robbery, burglary, and criminal trespass.

(2) "Machine gun" means a firearm designed to discharge more than one shot by a single function of the trigger.

(3) "Person" includes a firm, partnership, association, or corporation.

45-8-304. Possession or use of machine gun for offensive purpose. Possession or use of a machine gun for offensive or aggressive purpose is hereby declared to be a crime punishable by imprisonment in the state penitentiary for a term of not less than 10 years.

45-8-305. Presumption of offensive or aggressive purpose. Possession or use of a machine gun must be presumed to be for an offensive or aggressive purpose when the machine gun is in the possession of or used by a person who has been convicted of a crime of violence in any court of record, state or federal, in the United States of America or its territories or insular possessions.

45-8-307. Exceptions. Sections 45-8-301 through 45-8-305 and this section do not prohibit or interfere with:

(1) the manufacture of machine guns for and sale of machine guns to the military forces or the peace officers of the United States or of any political subdivision of the United States or transportation required for that purpose;

(2) the possession of a machine gun for a scientific purpose or the possession of a machine gun not usable as a weapon and possessed as a curiosity, ornament, or keepsake;

(3) the possession of a machine gun for a purpose manifestly not aggressive or offensive.

45-8-313. Unlawful possession of firearm by convicted person.

(1) A person commits the offense of unlawful possession of a firearm by a convicted person if the person purposely or knowingly purchases or possesses a firearm after the person has been convicted of:

(a) a felony for which the person received an additional sentence under 46-18-221; or

(b) an offense under the law of another state or of the United States that is equivalent to an offense that when committed in Montana is subject to an additional sentence under 46-18-221.

(2) A person convicted of unlawful possession of a firearm by a convicted person shall be imprisoned in a state prison for not less than 2 years or more than 10 years.

(3) A person who has been issued a permit under 45-8-314 may not be convicted of a violation of this section.

45-8-314. Lifetime firearms supervision of certain convicted persons.

(1) For the purposes of rehabilitation and public protection, a person convicted of an offense referred to in 45-8-313 shall, as part of the sentence imposed, be sentenced to life supervision by the state for the purpose of restricting the person's right to purchase and possess firearms. Active supervision by a probation or parole officer is not required but may be imposed by the court. "Supervision" means that the person may not violate 45-8-313 and must comply with other state and federal law restrictions on the purchase and possession of firearms.

(2)(a) A person subject to subsection (1) may apply to the district court for the county in which the person resides for a permit to purchase and possess one or more firearms. The person shall show good cause for the possession of each firearm sought to be purchased and possessed. The grant or denial of the application does not prevent the person from making another application, except that if an application is denied, another application may not be made for the next 12 months.

(b) The application must contain the following information:

(i) the person's full name and any past or present aliases;

(ii) the person's date and place of birth;

(iii) the person's address;

(iv) the person's occupation;

(v) the make and model of each firearm sought to be purchased and possessed;

(vi) the date and place of each conviction of an offense referred to in 45-8-313, the name of the offense, the state and county in which the offense occurred, the sentence imposed, the place or places of incarceration, and the date of discharge from supervision for the last offense;

(vii) the name and business address of the person's last probation or parole officer; and

(viii) any other information considered necessary by the court.

(c) The person shall, at the time of filing the application with the court, mail a copy to the county attorney and county sheriff.

(d) The county attorney or county sheriff may file a written objection with the court. If no objection is filed, the court may grant the permit if it finds that the person has shown good cause to purchase and possess the firearm or firearms listed in the application. If an objection is filed, a hearing must be held within 60 days after the filing of the objection. If the court first finds that the person has shown good cause to purchase and possess the firearm or firearms listed in the application and that, but for the objection, the court would have granted a permit, the court shall decide whether the objection is valid and overrides the good cause showing and requires denial of the permit.

45-8-315. Definition. "Concealed weapon" means any weapon mentioned in 45-8-316 through 45-8-318 and 45-8-321 through 45-8-328 that is wholly or partially covered by the clothing or wearing apparel of the person carrying or bearing the weapon, except that for purposes of 45-8-321 through 45-8-328, concealed weapon means a handgun or a knife with a blade 4 or more inches in length that is wholly or partially covered by the clothing or wearing apparel of the person carrying or bearing the weapon.

45-8-332. Definitions.

(1) "Destructive device", as used in this chapter, includes but is not limited to the following weapons:

(a) a projectile containing an explosive or incendiary material or any other similar chemical substance, including but not limited to that which is commonly known as tracer or incendiary ammunition, except tracer ammunition manufactured for use in shotguns;

(b) a bomb, grenade, explosive missile, or similar device or a launching device therefor;

(c) a weapon of a caliber greater than .60 caliber which fires fixed ammunition or any ammunition therefor, other than a shotgun or shotgun ammunition;

(d) a rocket, rocket-propelled projectile, or similar device of a diameter greater than 0.60 inch or a launching device therefor and a rocket, rocket-propelled projectile, or similar device containing an explosive or incendiary material or any other similar chemical substance other than the propellant for the device, except devices designed primarily for emergency or distress signaling purposes;

(e) a breakable container which contains a flammable liquid with a flashpoint of 150 degrees Fahrenheit or less and which has a wick or similar device capable of being ignited, other than a device which is commercially manufactured primarily for the purpose of illumination.

(2) "Explosive," as used in this chapter, means any explosive defined in rules adopted by the department of justice pursuant to 50-3-102(3).

45-8-333. Reckless or malicious use of explosives. Every person who shall recklessly or maliciously use, handle, or have in his or her possession any blasting powder, giant or Hercules powder, giant caps, or other highly explosive substance whereby any human being is intimidated, terrified or endangered shall be guilty of a misdemeanor.

45-8-334. Possession of a destructive device.

(1) A person who, with the purpose to commit a felony, has in his possession any destructive device on a public street or highway, in or near any theater, hall, school, college, church, hotel, other public building, or private habitation, in, on, or near any aircraft, railway passenger train, car, vessel engaged in carrying passengers for hire, or other public place ordinarily passed by human beings is guilty of the offense of possession of a destructive device.

(2) A person convicted of the offense of possession of a destructive device shall be imprisoned in the state prison for a period of not more than 10 years or be fined an amount of not more than $50,000, or both.

45-8-335. Possession of explosives.

(1) A person commits the offense of possession of explosives if he possesses, manufactures, transports, buys, or sells an explosive compound, flammable material, or timing, detonating, or similar device for use with an explosive compound or incendiary device and:

(a) has the purpose to use such explosive, material, or device to commit an offense; or

(b) knows that another has the purpose to use such explosive material, or device to commit an offense.

(2) A person convicted of the offense of possession of explosives shall be imprisoned in the state prison for any term not to exceed 20 years or be fined an amount not to exceed $ 50,000, or both.

45-8-336. Possession of a silencer.

(1) A person commits the offense of possession of a silencer if he possesses, manufactures, transports, buys, or sells a silencer and has the purpose to use it to commit an offense or knows that another person has such a purpose.

(2) A person convicted of the offense of possession of a silencer is punishable by imprisonment in the state prison for a term of not less than 5 years or more than 30 years or a fine of not less than $1,000 or more than $20,000 or by both such fine and imprisonment.

45-8-337. Possession of unregistered silencer or of bomb or similar device prima facie evidence of unlawful purpose. Possession of a silencer that is not registered under federal law or of a bomb or similar device charged or filled with one or more explosives is prima facie evidence of a purpose to use the same to commit an offense.

45-8-340. Sawed-off firearm - penalty.

(1) A person commits the offense of possession of a sawed-off firearm if he knowingly possesses a rifle or shotgun that when originally manufactured had a barrel length of:

(a) 16 inches or more and an overall length of 26 inches or more in the case of a rifle; or

(b) 18 inches or more and an overall length of 26 inches or more in the case of a shotgun; and

(c) the firearm has been modified in a manner so that the barrel length, overall length, or both, are less than specified in subsection (1)(a) or (1)(b).

(2) The barrel length is the distance from the muzzle to the rear-most point of the chamber.

(3) This section does not apply to firearms possessed:

(a) by a peace officer of this state or one of its political subdivisions;

(b) by an officer of the United States government authorized to carry weapons;

(c) by a person in actual service as a national guardsman;

(d) by a person called to the aid of one of the persons named in subsections (3)(a) through (3)(c);

(e) for educational or scientific purposes in which the firearms are incapable of being fired;

(f) by a person who has a valid federal tax stamp for the firearm, issued by the bureau of alcohol, tobacco, and firearms; or

(g) by a bona fide collector of firearms if the firearm is a muzzleloading, sawed-off firearm manufactured before 1900.

(4) A person convicted of the offense of possession of a sawed-off firearm must be fined not less than $200 or more than $500 or imprisoned in the county jail for not less than 5 days or more than 6 months, or both, upon a first conviction. If a person has one or more prior convictions under this section or one or more prior felony convictions under a law of this state, another state, or the United States, he must be fined an amount not to exceed $1,000 or imprisoned in the state prison for a term not to exceed 5 years, or both.

45-8-344. Use of firearms by children under fourteen prohibited - exceptions. It is unlawful for a parent, guardian, or other person having charge or custody of a minor child under the age of 14 years to permit the minor child to carry or use in public any firearms, except when the child is accompanied by a person having charge or custody of the child or under the supervision of a qualified firearms safety instructor or an adult who has been authorized by the parent or guardian.

45-8-345. Criminal liability of parent or guardian - prosecution.

(1) Any parent, guardian, or other person violating the provisions of 45-8-344 shall be guilty of a misdemeanor.

(2) The county attorney, on complaint of any person, must prosecute violations of 45-8-344.

45-8-351. Restriction on local government regulation of firearms.

(1) Except as provided in subsection (2), no county, city, town, consolidated local government, or other local government unit may prohibit, register, tax, license, or regulate the purchase, sale or other transfer (including delay in purchase, sale, or other transfer), ownership, possession, transportation, use, or unconcealed carrying of any weapon, including a rifle, shotgun, handgun, or concealed handgun.

(2)(a) For public safety purposes, a city or town may regulate the discharge of rifles, shotguns, and handguns. A county, city, town, consolidated local government, or other local government unit has power to prevent and suppress the carrying of concealed or unconcealed weapons to a public assembly, publicly owned building, park under its jurisdiction, or school, and the possession of firearms by convicted felons, adjudicated mental incompetents, illegal aliens, and minors.

(b) Nothing contained herein shall allow any government to prohibit the legitimate display of firearms at shows or other public occasions by collectors and others, nor shall anything contained herein prohibit the legitimate transportation of firearms through any jurisdiction, whether in airports or otherwise.

45-8-360. Establishment of individual licensure. In consideration that the right to keep and bear arms is protected and reserved to the people in Article II, section 12, of the Montana constitution, a person who has not been convicted of a violent, felony crime and who is lawfully able to own or to possess a firearm under the Montana constitution is considered to be individually licensed and verified by the state of Montana within the meaning of the provisions regarding individual licensure and verification in the federal Gun-Free School Zones Act.

45-8-361. Possession or allowing possession of weapon in school building – exceptions - penalties - seizure and forfeiture or return authorized - definitions.

(1) A person commits the offense of possession of a weapon in a school building if the person purposely and knowingly possesses, carries, or stores a weapon in a school building.

(2) A parent or guardian of a minor commits the offense of allowing possession of a weapon in a school building if the parent or guardian purposely and knowingly permits the minor to possess, carry, or store a weapon in a school building.

(3)(a) Subsection (1) does not apply to law enforcement personnel.

(b) The trustees of a district may grant persons and entities advance permission to possess, carry, or store a weapon in a school building.

(4)(a) A person convicted under this section shall be fined an amount not to exceed $500, imprisoned in the county jail for a term not to exceed 6 months, or both. The court shall consider alternatives to incarceration that are avail-able in the community.

(b)(i) A weapon in violation of this section may be seized and, upon conviction of the person possessing or permitting possession of the weapon, may be forfeited to the state or returned to the lawful owner.

(ii) If a weapon seized under the provisions of this section is subsequently determined to have been stolen or otherwise taken from the owner's possession without permission, the weapon must be returned to the lawful owner.

(5) As used in this section:

(a) "school building" means all buildings owned or leased by a local school district that are used for instruction or for student activities. The term does not include a home school provided for in 20-5-109.

(b) "weapon" means any type of firearm, a knife with a blade 4 or more inches in length, a sword, a straight razor, a throwing star, nunchucks, or brass or other metal knuckles. The term also includes any other article or instrument possessed with the purpose to commit a criminal offense.

Part 4. Montana Street Terrorism Enforcement and Prevention Act

45-8-406. Supplying of firearms to criminal street gang.

(1) A person commits the offense of supplying firearms to a criminal street gang if the person purposely or knowingly supplies, sells, or gives possession or control of any firearm to another, and the person has actual knowledge that the other person will use the firearm to commit an offense enumerated in 45-8-405 while actively participating in any criminal street gang whose members engage in a pattern of criminal street gang activity.

(2) Subsection (1) does not apply to a person who is convicted as a principal to the offense committed by the person to whom the firearm was supplied, sold, or given.

(3) A person convicted of the offense of supplying firearms to a criminal street gang shall be imprisoned in the county jail for a term not to exceed 1 year or be fined an amount not to exceed $1,000, or both.

[Current through 2007 Regular and Special Sessions]

NEBRASKA
NEB. REV. STAT.

Chapter 28. Crimes and Punishment

Article 12. Offenses Against Public Health and Safety

28-1201. Terms, defined For purposes of sections 28-1201 to 28-1212, unless the context otherwise requires:

(1) Firearm shall mean any weapon which is designed to or may readily be converted to expel any projectile by the action of an explosive or frame or receiver of any such weapon;

(2) Fugitive from justice shall mean any person who has fled or is fleeing from any peace officer to avoid prosecution or incarceration for a felony;

(3) Juvenile shall mean any person under the age of eighteen years; **...**

(6) Machine gun shall mean any firearm, whatever its size and usual designation, that shoots automatically more than one shot, without manual reloading, by a single function of the trigger;

(7) Short rifle shall mean a rifle having a barrel less than sixteen inches long or an overall length of less than twenty-six inches; and

(8) Short shotgun shall mean a shotgun having a barrel or barrels less than eighteen inches long or an overall length of less than twenty-six inches.

28-1202. Carrying concealed weapon; penalty; affirmative defense

(1)(a) Except as otherwise provided in this section, any person who carries a weapon or weapons concealed on or about his or her person such as a revolver, pistol, bowie knife, dirk or knife with a dirk blade attachment, brass or iron knuckles, or any other deadly weapon commits the offense of carrying a concealed weapon.

(b) It is an affirmative defense that the defendant was engaged in any lawful business, calling, or employment at the time he or she was carrying any weapon or weapons and the circumstances in which such person was placed at the time were such as to justify a prudent person in carrying the weapon or weapons for the defense of his or her person, property, or family.

(2) This section does not apply to a person who is the holder of a valid permit issued under the Concealed Handgun Permit Act if the concealed weapon the defendant is carrying is a handgun as defined in section 69-2429.

(3) Carrying a concealed weapon is a Class I misdemeanor.

(4) In the case of a second or subsequent conviction under this section, carrying a concealed weapon is a Class IV felony.

28-1203. Transportation or possession of machine guns, short rifles, or short shotguns; penalty; exception

(1) Any person or persons who shall transport or possess any machine gun, short rifle, or short shotgun commits a Class IV felony.

(2) The provisions of this section shall not be held to prohibit any act by peace officers, members of the United States armed services, or members of the National Guard of this state, in the lawful discharge of their duties, or persons qualified under the provisions of federal law relating to the short rifle, short shotgun, or machine gun.

28-1204. Unlawful possession of a revolver; exceptions; penalty

(1) Any person under the age of eighteen years who possesses a pistol, revolver, or any other form of short-barreled hand firearm commits the offense of unlawful possession of a revolver.

(2) The provisions of this section shall not apply to the issuance of such firearms to members of the armed forces of the United States, active or reserve, National Guard of this state, or Reserve Officers Training Corps, when on duty or training, or to the temporary loan of pistols, revolvers, or any other form of short-barreled firearms for instruction under the immediate supervision of a parent or guardian or adult instructor.

(3) Unlawful possession of a revolver is a Class III misdemeanor.

28-1204.01. Unlawful transfer of a firearm to a juvenile; exceptions; penalty; county attorney; duty

(1) Any person who knowingly and intentionally does or attempts to sell, provide, loan, deliver, or in any other way transfer the possession of a firearm to a juvenile commits the offense of unlawful transfer of a firearm to a juvenile. The county attorney shall have a copy of the petition served upon the owner of the firearm, if known, in person or by registered or certified mail at his or her last-known address.

(2) This section shall not apply to the transfer of a firearm other than the types specified in section 28-1204 to a juvenile:

(a) From a person related to such juvenile within the second degree of consanguinity or affinity if the transfer of physical possession of such firearm does not occur until such time as express permission has been obtained from the juvenile's parent or guardian;

(b) For a legitimate and lawful sporting purpose; or

(c) Who is under direct adult supervision in an appropriate educational program.

(3) This section shall apply to the transfer of any firearm described in section 28-1204, except as specifically provided in subsection (2) of section 28-1204.

(4) Unlawful transfer of a firearm to a juvenile is a Class IV felony.

28-1204.02. Confiscation of firearm; disposition Any firearm in the possession of a person in violation of section 28-1204 or 28-1204.01 shall be confiscated by a peace officer or other authorized law enforcement officer. Such firearm shall be held by the agency employing such officer until it no longer is required as evidence.

28-1204.04. Unlawful possession of a firearm on school grounds; penalty; exceptions; confiscation of certain firearms; disposition

(1) Any person who possesses a firearm in a school, on school grounds, in a school-owned vehicle, or at a school-sponsored activity or athletic event shall be guilty of the offense of unlawful possession of a firearm on school grounds. Unlawful possession of a firearm on school grounds is a Class II misdemeanor. This subsection shall not apply to (a) the issuance of firearms to or possession by members of the armed forces of the United States, active or reserve, National Guard of this state, or Reserve Officers Training Corps or peace officers or other duly authorized law enforcement officers when on duty or training, (b) firearms which may lawfully be possessed by the person receiving instruction, for instruction under the immediate supervision of an adult instructor, or (c) firearms contained within a private vehicle operated by a nonstudent adult which are not loaded and (i) are encased or (ii) are in a locked firearm rack that is on a motor vehicle. For purposes of this subsection, encased shall mean enclosed in a case that is expressly made for the purpose of containing a firearm and that is completely zipped, snapped, buckled, tied, or otherwise fastened with no part of the firearm exposed.

(2) Any firearm possessed in violation of subsection (1) of this section in a school, on school grounds, in a school-owned vehicle, or at a school-sponsored activity or athletic event shall be confiscated without warrant by a peace officer or may be confiscated without warrant by school administrative or teaching personnel. Any firearm confiscated by school administrative or teaching personnel shall be delivered to a peace officer as soon as practicable.

(3) Any firearm confiscated by or given to a peace officer pursuant to subsection (2) of this section shall be declared a common nuisance and shall be held by the peace officer prior to his or her delivery of the firearm to the property division of the law enforcement agency which employs the peace officer. The property division of such law enforcement agency shall hold such firearm for as long as the firearm is needed as evidence. After the firearm is no longer needed as evidence it shall be destroyed in such manner as the court may direct.

(4) Whenever a firearm is confiscated and held pursuant to this section or section 28-1204.02, the peace officer who received such firearm shall cause to be filed within ten days after the confiscation a petition for destruction of such firearm. The petition shall be filed in the district court of the county in which the confiscation is made. The petition shall describe the firearm held, state the name of the owner, if known, allege the essential elements of the violation which caused the confiscation, and conclude with a prayer for disposition and destruction in such manner as the court may direct. At any time after the confiscation of the firearm and prior to court disposition, the owner of the firearm seized may petition the district court of the county in which the confiscation was made for possession of the firearm. The court shall release the firearm to such owner only if the claim of ownership can reasonably be shown to be true and either (a) the owner of the firearm can show that the firearm was taken from his or her property or place of business unlawfully or without the knowledge and consent of the owner and that such property or place of business is different from that of the person from whom the firearm was confiscated or (b) the owner of the firearm is acquitted of the charge of unlawful possession of a revolver in violation of section 28-1204, unlawful transfer of a firearm to a juvenile, or unlawful possession of a firearm on school grounds. No firearm having significant antique value or historical significance as determined by the Nebraska State Historical Society shall be destroyed. If a firearm has significant antique value or historical significance, it shall be sold at auction and the proceeds deposited in the permanent school fund.

28-1205. Using a deadly weapon to commit a felony; penalty; separate and distinct offense

(1) Any person who uses a firearm, a knife, brass or iron knuckles, or any other deadly weapon to commit any felony which may be prosecuted in a court of this state or who unlawfully possesses a firearm, a knife, brass or iron knuckles, or any other deadly weapon during the commission of any felony which may be prosecuted in a court of this state commits the offense of using a deadly weapon to commit a felony.

(2)(a) Use of a deadly weapon other than a firearm to commit a felony is a Class III felony.

(b) Use of a deadly weapon which is a firearm to commit a felony is a Class II felony.

(3) The crimes defined in this section shall be treated as separate and distinct offenses from the felony being committed, and sentences imposed under this section shall be consecutive to any other sentence imposed.

28-1206. Possession of a deadly weapon by a felon or a fugitive from justice; penalty

(1) Any person who possesses any firearm or brass or iron knuckles and who has previously been convicted of a felony or who is a fugitive from justice commits the offense of possession of a deadly weapon by a felon or a fugitive from justice.

(2) Such felony conviction may have been had in any court in the United States, the several states, territories, or possessions, or the District of Columbia.

(3)(a) Possession of a deadly weapon other than a firearm by a felon or a fugitive from justice is a Class IV felony.

(b) Possession of a deadly weapon which is a firearm by a felon or a fugitive from justice is a Class III felony.

28-1207. Possession of a defaced firearm; penalty

(1) Any person who knowingly possesses, receives, sells, or leases, other than by delivery to law enforcement officials, any firearm from which the manufacturer's identification mark or serial number has been removed, defaced, altered, or destroyed, commits the offense of possession of a defaced firearm.

(2) Possession of a defaced firearm is a Class IV felony.

28-1208. Defacing a firearm; penalty

(1) Any person who intentionally removes, defaces, covers, alters, or destroys the manufacturer's identification mark or serial number or other distinguishing numbers on any firearm commits the offense of defacing a firearm.

(2) Defacing a firearm is a Class IV felony.

28-1209. Failure to register tranquilizer guns; penalty

(1) Any person who fails or neglects to register any gun or other device designed, adapted or used for projecting darts or other missiles containing tranquilizers or other chemicals or compounds which will produce unconsciousness or temporary disability in live animals, with the county sheriff of the county in which the owner of the gun or device resides, commits the offense of failure to register tranquilizer guns.

(2) Failure to register tranquilizer guns is a Class III misdemeanor.

28-1210. Failure to notify the sheriff of the sale of tranquilizer guns; penalty; record

(1) Any person, partnership, limited liability company, or corporation selling any gun or other device as described in section 28-1209 which fails to immediately notify the sheriff of the county of the sale and giving the name and address of the purchaser thereof and the make and number of the gun or device commits the offense of failure to notify the sheriff of the sale of tranquilizer guns.

(2) The sheriff shall keep a record of such sale with the information furnished pursuant to this section.

(3) Failure to notify the sheriff of the sale of tranquilizer guns is a Class III misdemeanor.

28-1211. Firearms; purchase, sell, trade, or convey; conditions The State of Nebraska herewith permits its residents, not otherwise precluded by any applicable laws, to purchase, sell, trade, convey, deliver, or transport rifles, shotguns, ammunition, reloading components or firearm accessories in Nebraska and in states contiguous to Nebraska. This authorization is enacted to implement for this state the permissive firearms sales and delivery provisions in section 922(b), (3)(A) of Public Law 90-618 of the 90th Congress, Second Session. In the event that presently enacted federal restrictions on the purchase of rifles, shotguns, ammunition, reloading components, or firearm accessories are repealed by the United States Congress or set aside by courts of competent jurisdiction, this section shall in no way be interpreted to prohibit or restrict the purchase of shotguns, rifles, ammunition, reloading components, or firearm accessories by residents of Nebraska otherwise competent to purchase same in contiguous or other states.

28-1212.03. Stolen firearm; prohibited acts; violation; penalty Any person who possesses, receives, retains, or disposes of a stolen firearm knowing that it has been or believing that it has been stolen shall be guilty of a Class IV felony unless the firearm is possessed, received, retained, or disposed of with intent to restore it to the owner.

28-1213. Explosives, destructive devices, other terms; defined For purposes of sections 28-1213 to 28-1239, unless the context otherwise requires:

(1) Person means any individual, corporation, company, association, firm, partnership, limited liability company, society, or joint-stock company;

(2) Business enterprise means any corporation, partnership, limited liability company, company, or joint-stock company;

(3) Explosive materials means explosives, blasting agents, and detonators;

(4) Explosives means any chemical compound, mixture, or device, the primary or common purpose of which is to function by explosion, including, but not limited to, dynamite and other high explosives, black powder, pellet powder, initiating explosives, detonators, safety fuses, squibs, detonating cord, ignited cord, igniters, display fireworks as defined in section 28-1241, and firecrackers or devices containing more than one hundred thirty milligrams of explosive composition, but does not include common fireworks as defined in such section, gasoline, kerosene, naphtha, turpentine, benzine, acetone, ethyl ether, benzol, fixed ammunition and primers for small arms, safety fuses, or matches; ...

(7)(a) Destructive devices means:

(i) Any explosive, incendiary, chemical or biological poison, or poison gas (A) bomb, (B) grenade, (C) rocket having a propellant charge of more than four ounces, (D) missile having an explosive or incendiary charge of more than one-quarter ounce, (E) mine, (F) booby trap, (G) Molotov cocktail, (H) bottle bomb, (I) vessel or container intentionally caused to rupture or mechanically explode by expanding pressure from any gas, acid, dry ice, or other chemical mixture, or (J) any similar device, the primary or common purpose of which is to explode and to be used as a weapon against any person or property; or

(ii) Any combination of parts either designed or intended for use in converting any device into a destructive device as defined in subdivision (7)(a)(i) of this section from which a destructive device may be readily assembled.

(b) The term destructive device does not include (i) any device which is neither designed nor redesigned for use as a weapon to be used against person or property, (ii) any device, although originally designed for use as a weapon, which is redesigned for use as a signaling, pyrotechnic, line-throwing, safety, or similar device, (iii) surplus ordnance sold, loaned, or given by the Secretary of the Army pursuant to 10 U.S.C. 4684(2), 4685, or 4686, as such sections existed on March 7, 2006, (iv) any other device which the Nebraska State Patrol finds is not likely to be used as a weapon or is an antique, or (v) any other device possessed under circumstances negating an intent that the device be used as a weapon against any person or property;

(8) Federal permittee means any lawful user of explosive materials who has obtained a federal user permit under 18 U.S.C. chapter 40, as such chapter existed on March 7, 2006;

(9) Federal licensee means any importer, manufacturer, or dealer in explosive materials who has obtained a federal importers', manufacturers', or dealers' license under 18 U.S.C. chapter 40, as such chapter existed on March 7, 2006; and

(10) Smokeless propellants means solid propellants commonly called smokeless powders in the trade and used in small arms ammunition.

28-1214. Explosives control; applicability of act

(1) Sections 28-1213 to 28-1239 shall apply to persons engaged in the manufacture, ownership, possession, storage, use, transportation, purchase, sale, or gift of explosive materials, except as may be otherwise indicated herein.

(2) Sections 28-1213 to 28-1239 shall not apply to explosive materials while being transported in conformity with federal law or regulations, nor, except as may be otherwise provided in such sections, to the ownership, possession, storage, use, transportation, purchase, or sale of explosive materials by the armed forces of the United States, the National Guard, other reserve components of the armed forces of the United States, and the duly constituted police and firefighting forces of the United States and of the state and its political subdivisions in the lawful discharge of their official duties.

28-1215. Unlawful possession of explosive materials, first degree; exception; penalty

(1) Except as provided in subsection (2) of this section, any person who is ineligible to obtain a permit from the Nebraska State Patrol and who possesses or stores explosive materials commits the offense of unlawful possession of explosive materials in the first degree.

(2) Subsection (1) of this section shall not be applicable to any person transporting explosive materials in accordance with section 28-1235 or to any person who has obtained a permit from the Nebraska State Patrol to store or use such explosive materials or, in the case of a business enterprise, a permit to purchase such explosive materials.

(3) Unlawful possession of explosive materials in the first degree is a Class IV felony.

28-1216. Unlawful possession of explosive materials, second degree; penalty

(1) Except as provided in subsection (2) of this section, any person who is eligible to obtain a permit from the Nebraska State Patrol or has valid educational, industrial, commercial, agricultural, or other legitimate need for a permit and who possesses or stores explosive materials without such a permit commits the offense of unlawful possession of explosive materials in the second degree.

(2) The exclusions provided in subsection (2) of section 28-1215 shall be applicable to this section.

(3) Unlawful possession of explosive materials in the second degree is a Class I misdemeanor.

28-1217. Unlawful sale of explosives; penalty

(1) Any person who knowingly and intentionally sells, transfers, issues, or gives any explosive materials to any person who does not display a valid permit issued by the Nebraska

State Patrol authorizing the storage or use of such explosive materials or, in the case of a business enterprise, a permit to purchase such explosive materials commits the offense of unlawful sale of explosives.

(2) Unlawful sale of explosives is a Class IV felony.

28-1220. Possession of a destructive device; penalty; permit or license for explosive materials; no defense

(1) Any person who has in his possession a destructive device, as defined in subdivision (7) of section 28-1213, commits the offense of possession of a destructive device.

(2) A permit or license issued under any state or federal law to possess, own, use, distribute, sell, manufacture, store, or handle in any manner explosive materials shall not be a defense to the crime of possession of a destructive device as defined in this section.

(3) Possession of a destructive device is a Class IV felony.

28-1239. Explosives control; exceptions to sections; enumerated In addition to the exceptions provided in sections 28-1213 to 28-1239, such sections shall not apply to:

(1) The use of explosive materials in medicines and medicinal agents in forms prescribed by the official United States Pharmacopoeia or the National Formulary;

(2) The sale, transportation, shipment, receipt, or importation of explosive materials for delivery to any agency of the United States or to any state or political subdivision thereof;

(3) Small arms ammunition and components thereof;

(4) The storage or possession of or dealing in black powder used for recreation purposes by a sportsperson;

(5) The storage or possession of or dealing in smokeless propellants, percussion caps, primers, and other components used by a sportsperson in the reloading of small arms ammunition;

(6) Bona fide war trophies capable of exploding and innocently found explosive materials possessed under circumstances negating an intent to use the same unlawfully, but the owner thereof shall surrender such items forthwith to any nationally certified hazardous device technician or military explosive ordnance expert upon demand by a law enforcement officer or agency or fire department; and

(7) The storage in minimum amounts necessary for lawful educational purposes of explosive materials to be used in the natural science laboratories of any state-accredited school system.

Chapter 69. Personal Property

Article 24. Handguns

69-2402. Terms, defined For purposes of sections 69-2401 to 69-2425:

(1) Antique handgun or pistol shall mean any handgun or pistol, including those with a matchlock, flintlock, percussion cap, or similar type of ignition system, manufactured in or before 1898 and any replica of such a handgun or pistol if such replica (a) is not designed or redesigned for using rimfire or conventional centerfire fixed ammunition or (b) uses rimfire or conventional centerfire fixed ammunition which is no longer manufactured in the United States and which is not readily available in the ordinary channels of commercial trade;

(2) Criminal history record check shall include a check of the criminal history records of the Nebraska State Patrol and a check of the Federal Bureau of Investigation's National Instant Criminal Background Check System; and

(3) Handgun shall mean any firearm with a barrel less than sixteen inches in length or any firearm designed to be held and fired by the use of a single hand.

69-2403. Sale, lease, rental, and transfer; certificate required; exceptions Except as provided in section 69-2409, a person shall not purchase, lease, rent, or receive transfer of a handgun until he or she has obtained a certificate in accordance with section 69-2404. Except as provided in section 69- 2409, a person shall not sell, lease, rent, or transfer a handgun to a person who has not obtained a certificate. The certificate shall not be required if:

(1) The person acquiring the handgun is a licensed firearms dealer under federal law;

(2) The handgun is an antique handgun;

(3) The person acquiring the handgun is authorized to do so on behalf of a law enforcement agency;

(4) The transfer is a temporary transfer of a handgun and the transferee remains (a) in the line of sight of the transferor or (b) within the premises of an established shooting facility; or

(5) The transfer is between a person and his or her spouse, sibling, parent, child, aunt, uncle, niece, nephew, or grandparent.

69-2404. Certificate; application; fee Any person desiring to purchase, lease, rent, or receive transfer of a handgun shall apply with the chief of police or sheriff of the applicant's place of residence for a certificate. The application may be made in person or by mail. The application form and certificate shall be made on forms approved by the Superintendent of Law Enforcement and Public Safety. The application shall include the applicant's full name, social security number, address, date of birth, and country of citizenship. If the applicant is not a United States citizen, the application shall include the applicant's place of birth and his or her alien admission number. If the application is made in person, the applicant shall also present a current Nebraska motor vehicle operator's license, state identification card, or military identification card, or if the application is made by mail, the application form shall describe the license or card used for identification and be notarized by a notary public who has verified the identification of the applicant through such a license or card. An applicant shall receive a certificate if he or she is twenty-one years of age or older and is not prohibited from purchasing or possessing a handgun by 18 U.S.C. 922. A fee of five dollars shall be charged for each application for a certificate to cover the cost of a criminal history record check.

69-2405. Application; chief of police or sheriff; duties; immunity Upon the receipt of an application for a certificate, the chief of police or sheriff shall issue a certificate or deny a certificate and furnish the applicant the specific reasons for the denial in writing. The chief of police or sheriff shall be permitted up to three days in which to conduct an investigation to determine whether the applicant is prohibited by law from purchasing or possessing a handgun. If the certificate or denial is mailed to the applicant, it shall be mailed to the applicant's address by first-class mail within the three-day period. If it is determined that the purchase or possession of a handgun by the applicant would be in violation of applicable federal, state, or local law, the chief of police or sheriff shall deny the certificate. In computing the three-day period, the day of receipt of the application shall not be included and the last day of the three-day period shall be included. The three-day period shall expire at 11:59 p.m. of the third day unless it is a Saturday, Sunday, or legal holiday in which event the period shall run until 11:59 p.m. of the next day which is not a Saturday, Sunday, or legal holiday. No later than the end of the three-day period the chief of police or sheriff shall issue or deny such certificate and, if the certificate is denied, furnish the applicant the specific reasons for denial in writing. No civil liability shall arise to any law enforcement agency if such law enforcement agency complies with sections 69-2401, 69-2403 to 69-2408, and 69-2409.01.

69-2406. Certificate; denial or revocation; appeal; filing fee Any person who is denied a certificate, whose certificate is revoked, or who has not been issued a certificate upon expiration of the three-day period may appeal within ten days of receipt of the denial or revocation to the county court of the county of the applicant's place of residence. The applicant shall file with the court the specific reasons for the denial or revocation by the chief of police or sheriff and a filing fee of ten dollars in lieu of any other filing fee required by law. The court shall issue its decision within thirty days of the filing of the appeal.

69-2407. Certificate; contents; term; revocation A certificate issued in accordance with section 69-2404 shall contain the holder's name, social security number, address, and date of birth and the effective date of the certificate. A certificate shall authorize the holder to acquire any number of handguns during the period that the certificate is valid. The certificate shall be valid throughout the state and shall become invalid three years after its effective date. If the chief of police or sheriff who issued the certificate determines that the applicant has become disqualified for the certificate under section 69-2404, he or she may immediately revoke the certificate and require the holder to surrender the certificate immediately. Revocation may be appealed pursuant to section 69-2406.

69-2408. False information on application; other violations; penalties; confiscation of handgun Any person who willfully provides false information on an application form for a certificate under section 69-2404 shall, upon conviction, be guilty of a Class IV felony, and any person who intentionally violates any other provision of sections 69-2401, 69-2403 to 69-2407, and 69-2409.01 shall, upon conviction, be guilty of a Class I misdemeanor. As a part of the judgment of conviction, the court may order the confiscation of the handgun.

69-2409. Automated criminal history files; legislative intent; system implementation; Nebraska State Patrol; superintendent; duties; purchase, lease, rental, or transfer; election It is the intent of the Legislature that the Nebraska State Patrol implement an expedited program of upgrading Nebraska's automated criminal history files to be utilized for, among other law enforcement purposes, an instant criminal history record check on handgun purchasers when buying a handgun from a licensed importer, manufacturer, or dealer so that such instant criminal history record check may be implemented as soon as possible on or after January 1, 1995.

The patrol's automated arrest and conviction records shall be reviewed annually by the Superintendent of Law Enforcement and Public Safety who shall report the status of such records within thirty days of such review to the Governor and the Clerk of the Legislature. The instant criminal history record check system shall be implemented by the patrol on or after January 1, 1995, when, as determined by the Superintendent of Law enforcement and Public Safety, eighty-five percent of the Nebraska arrest and conviction records since January 1, 1965, available to the patrol are included in the patrol's automated system. Not less than thirty days prior to implementation and enforcement of the instant check system, the patrol shall send written notice to all licensed importers, manufacturers, and dealers outlining the procedures and toll-free number described in sections 69-2410 to 69-2423.

Upon implementation of the instant criminal history record check system, a person who desires to purchase, lease, rent, or receive transfer of a handgun from a licensed importer, manufacturer, or dealer may elect to obtain such handgun either under sections 69-2401, 69-2403 to 69-2408, and 69-2409.01 or under sections 69-2409.01 and 69-2410 to 69-2423.

69-2410. Importer, manufacturer, or dealer; sale or delivery; duties No importer, manufacturer, or dealer licensed pursuant to 18 U.S.C. 923 shall sell or deliver any handgun to another person other than a licensed importer, manufacturer, dealer, or collector until he or she has:

(1)(a) Inspected a valid certificate issued to such person pursuant to sections 69-2401, 69-2403 to 69-2408, and 69-2409.01; and

(b) Inspected a valid identification containing a photograph of such person which appropriately and completely identifies such person; or

(2)(a) Obtained a completed consent form from the potential buyer or transferee, which form shall be established by the Nebraska State Patrol and provided by the licensed importer, manufacturer, or dealer. The form shall include the name, address, date of birth, gender, race, social security number or other identification number, and country of citizenship of such potential buyer or transferee. If the potential buyer or transferee is not a United States citizen, the completed consent form shall contain the potential buyer's or transferee's place of birth and his or her alien admission number;

(b) Inspected a valid identification containing a photograph of the potential buyer or transferee which appropriately and completely identifies such person;

(c) Requested by toll-free telephone call or other electromagnetic communication that the Nebraska State Patrol conduct a criminal history record check; and

(d) Received a unique approval number for such inquiry from the Nebraska State Patrol indicating the date and number on the consent form.

69-2411. Request for criminal history record check; Nebraska State Patrol; duties; fee

(1) Upon receipt of a request for a criminal history record check, the Nebraska State Patrol shall as soon as possible during the licensee's telephone call or by return telephone call:

(a) Check its criminal history records and check the Federal Bureau of Investigation's National Instant Criminal Background Check System to determine if the potential buyer or transferee is prohibited from receipt or possession of a handgun pursuant to state or federal law; and

(b) Either (i) inform the licensee that its records demonstrate that the potential buyer or transferee is prohibited from receipt or possession of a handgun or (ii) provide the licensee with a unique approval number.

(2) In the event of electronic failure or similar emergency beyond the control of the Nebraska State Patrol, the patrol shall immediately notify a requesting licensee of the reason for and estimated length of such delay. In any event, no later than the end of the next business day the Nebraska State Patrol shall either (a) inform the licensee that its records demonstrate that the potential buyer or transferee is prohibited from receipt or possession of a handgun or (b) provide the licensee with a unique approval number. If the licensee is not informed by the end of the next business day that the potential buyer is prohibited from receipt or possession of a handgun, and regardless of whether the unique approval number has been received, the licensee may complete the sale or delivery and shall not be deemed to be in violation of sections 69-2410 to 69-2423 with respect to such sale or delivery.

(3) A fee of three dollars shall be charged for each request of a criminal history record check required pursuant to section 69-2410, which amount shall be transmitted monthly to the Nebraska State Patrol. Such amount shall be for the purpose of covering the costs of the criminal history record check.

69-2412. Records; confidentiality; destruction

(1) Any records which are created by the Nebraska State Patrol to conduct the criminal history record check containing any of the information set forth in subdivision (2)(a) of section 69-2410 pertaining to a potential buyer or transferee who is not prohibited from receipt or transfer of a handgun by reason of state or federal law shall be confidential and may not be disclosed by the patrol or any officer or employee thereof to any person. The Nebraska State Patrol shall destroy any such records as soon as possible after communicating the unique approval number, and in any event, such records shall be destroyed within forty-eight hours after the date of receipt of the licensee's request.

(2) Notwithstanding the provisions of this section, the Nebraska State Patrol shall only maintain a log of dates of requests for criminal history record checks and unique approval numbers corresponding to such dates for not to exceed one year.

(3) Nothing in this section shall be construed to allow the state to maintain records containing the names of licensees who receive unique approval numbers or to maintain records of handgun transactions, including the names or other identification of licensees and potential buyers or transferees including persons not otherwise prohibited by law from the receipt or possession of handguns.

69-2413. Nebraska State Patrol; toll-free telephone number; personnel The Nebraska State Patrol shall establish a toll-free telephone number which shall be operational seven days a week between 8 a.m. and 10 p.m. for purposes of responding to requests under section 69-2410. The Nebraska State Patrol shall employ and train such personnel as is necessary to expeditiously administer the provisions of sections 69-2410 to 69-2423.

69-2414. Records; amendment; procedure Any person who is denied the right to purchase or receive a handgun as a result of procedures established by sections 69-2410 to 69-2423 may request amendment of the record pertaining to him or her by petitioning the Nebraska State Patrol. If the Nebraska State Patrol fails to amend the record within seven days, the person requesting the amendment may petition the county court of the county in which he or she resides for an order directing the patrol to amend the record. If the person proves by a preponderance of the evidence that the record should be amended, the court shall order the record be amended. If the record demonstrates that such person is not prohibited from receipt or possession of a handgun by state or federal law, the Nebraska State Patrol shall destroy any records it maintains which contain any information derived from the criminal history record check.

69-2416. Licensed importer, manufacturer, or dealer; compliance not required; when A licensed importer, manufacturer, or dealer shall not be required to comply with the provisions of subdivision (2) of section 69-2410 and sections 69-2411 to 69-2423 in the event of:

(1) Unavailability of telephone service at the licensed premises due to (a) the failure of the entity which provides telephone service in the state, region, or other geographical area in which the licensee is located to provide telephone service to the premises due to the location of such premises or (b) the interruption of telephone service by reason of hurricane, flood, natural disaster, other act of God, war, riot, or other bona fide emergency or reason beyond the control of the licensee; or

(2) Failure of the Nebraska State Patrol to comply reasonably with the requirements of sections 69-2410 to 69- 2423.

69-2417. Nebraska State Patrol; licensee; liability defense; when Compliance with sections 69-2410 to 69-2423 shall be a defense by the Nebraska State Patrol and the licensee transferring a handgun in any cause of action under the laws of this state for liability for damages arising from the importation or manufacture, or the subsequent sale or transfer, of any handgun which has been shipped or transported in interstate or foreign commerce to any person who has been convicted in any court of any crime punishable by a term of more than one year.

69-2418. Instant criminal history record check; requirements; exemptions Sections 69-2410 to 69-2423 shall not apply to:

(1) Any antique handgun or pistol;

(2) Any firearm which is a curio or relic as defined in 27 C.F.R. 478.11.

69-2419. Criminal history records; prohibited acts; violation; penalty Any licensed importer, manufacturer, or dealer who knowingly and intentionally requests a criminal history record check from the Nebraska State Patrol for any purpose other than compliance with sections 69-2410 to 69-2423 or knowingly and intentionally disseminates any criminal history record check information to any person other than the subject of such information shall be guilty of a Class I misdemeanor.

69-2420. False statement; false identification; prohibited acts; violation; penalty Any person who, in connection with the pur-chase, transfer, or attempted purchase of a handgun pursuant to sections 69-2410 to 69-2423, knowingly and intentionally makes any materially false oral or written statement or knowingly and intentionally furnishes any false identification intended or likely to deceive the licensee shall be guilty of a Class IV felony.

69-2421. Sale or delivery; violation; penalty Any licensed importer, manufacturer, or dealer who knowingly and intentionally sells or delivers a handgun in violation of sections 69-2401 to 69-2425 shall be guilty of a Class IV felony.

69-2422. Obtaining handgun for prohibited transfer; violation; penalty For purposes of sections 69-2401 to 69-2425, any person who knowingly and intentionally obtains a handgun for the purposes of transferring it to a person who is prohibited from receipt or possession of a handgun by state or federal law shall be guilty of a Class IV felony.

69-2424. Rules and regulations The Nebraska State Patrol shall adopt and promulgate rules and regulations to carry out sections 69-2401 to 69-2425.

69-2425. City or village ordinance; not preempted Any city or village ordinance existing on September 6, 1991, shall not be preempted by sections 69-2401 to 69- 2425.

69-2426. Dealers of firearms; distribution of information; Firearm Information Fund; created

(1) Dealers of firearms shall distribute to all purchasers information developed by the Department of Health and Human Services regarding the dangers of leaving loaded firearms unattended around children.

(2) There is hereby created the Firearm Information Fund. Private contributions shall be credited by the State Treasurer to such fund for the implementation of the provisions of this section.

[Current through the 2008 Special Session and the November 2008 General Election]

Lincoln Municipal Code

Title 9. Public Peace and Welfare

Chapter 9.36. Weapons

9.36.020. Minors Not to be Furnished with Firearms, Ammunition or Weapons.

(a) It shall be unlawful for any person to sell, loan, or furnish to any minor any gun, fowling piece, or other firearm, any ammunition or component thereof, or any pocket knife having a blade more than three and one-half inches in length; provided, it shall be lawful to sell, loan, or furnish shotguns or rifles, of a type commonly used for hunting, and any ammunition or component thereof for the same, to persons eighteen years of age or older.

(b) This section shall not apply to delivery or transfer of rifles or shotguns or ammunition or components thereof to a juvenile:

(1) By the juvenile's parent or legal guardian for a legitimate and lawful sporting purpose; or

(2) Who is under direct adult supervision in an appropriate educational or competitive shooting program.

9.36.025. Firearms Offered for Sale at Retail; Access Restricted. It shall be unlawful for any person to display for sale at retail any firearm or firearms, or any ammunition for firearms or reloading components thereof, without such firearm or firearms, or ammunition for firearms or reloading components thereof, being secured so as to cause them to be inaccessible without the assistance of authorized sales personnel of the retailer.

9.36.030. Report of Sale of Firearms. Any person, firm, association, or corporation dealing in firearms of any type shall, on the same day of the sale of any firearm, except a shotgun or a rifle of a type commonly used for hunting, report the sale to the Police Department on forms as prescribed and furnished by the Police Department. The report shall contain all the information requested thereon.

9.36.090. Transporting Explosives; Port of Entry; Routes; Penalty. It shall be unlawful for any person, firm, or corporation to convey, or transport through any street, avenue, alley, or other public place within the city, any dynamite, nitroglycerine, gunpowder, guncotton, TNT, or any other explosive material, including fireworks of every nature or description, without first having stopped at a port of entry hereinafter designated, and having notified the Police Department of the city of their intention to move said vehicle within or through the city and requesting a police escort. Such vehicle, or vehicles, shall follow such route, or routes, as may be designated to them by such police escort. The City Council shall, by resolution, designate ports of entry at which all such vehicles shall stop.

9.36.100 Unlawful Possession of Firearms.

(a) It shall be unlawful for any person to possess any firearm within the corporate limits or on any property of the City of Lincoln outside the corporate limits when that person has been convicted of any one of the following offenses within the last ten years:

Stalking in violation of Neb. Rev. Stat. § 28-311.03 or any other comparable or similar state statute from another state; Violation of a protection order as set forth in Neb. Rev. Stat. § 42-924 or Violation of a foreign protection order as set forth in Neb. Rev. Stat. § 42-931; False imprisonment in the second degree in violation of Neb. Rev. Stat. § 28-315; Sexual assault in the third degree in violation of Neb. Rev. Stat. § 28-320; Impersonating a peace officer in violation of Neb. Rev. Stat. § 28-610; or, Impersonating police officer in violation of Lincoln Municipal Code § 9.08.060; Debauching a Minor in violation of Neb. Rev. Stat. § 28-805; Obstructing government operations in violation of Neb. Rev. Stat. § 28-901; Resisting arrest in violation of Neb. Rev. Stat. § 28-904; Resisting officer in violation of Lincoln Municipal Code § 9.08.030; Obstructing a peace officer in violation of Neb. Rev. Stat. § 28-906; Interfering with an officer making an arrest in violation of Lincoln Municipal Code § 9.08.020; Carrying concealed weapon in violation of Neb. Rev. Stat. § 28-1202; Criminal child enticement in violation of Neb. Rev. Stat. § 28-311; Implements for escape in violation of Neb. Rev. Stat. § 28-913; Unlawful possession of explosives, second degree in violation of Neb. Rev. Stat. § 28-1216; Use of explosives without a permit in violation of Neb. Rev. Stat. § 28-1218; Concealing the death of another person in violation of Neb. Rev. Stat. § 28-1302; Minors not to be furnished with firearms, ammunition, or weapons in violation of Lincoln Municipal Code § 9.36.020; Discharge of firearms unlawful in violation of Lincoln Municipal Code § 9.36.010; Assault in the third degree in violation of Neb. Rev. Stat. § 28-310; Assault and battery, menacing threats in violation of Lincoln Municipal Code §9.12.010; Unlawful intrusion in violation of Neb. Rev. Stat. § 28-311.08; Violation of custody in violation of Neb. Rev. Stat. § 28-316; Domestic assault in violation of Neb. Rev. Stat. § 28-323; Criminal trespass in the first degree in violation of Neb. Rev. Stat. § 28-520; Contributing to the delinquency of a child in violation of Neb. Rev. Stat. § 28-709; Public indecency in violation of Neb. Rev. Stat. § 28-806; Public indecency or indecent exposure in violation of Lincoln Municipal Code § 9.16.180; Operating a motor vehicle or vessel to avoid arrest in violation of Neb. Rev. Stat. § 28-905; Fleeing in a vehicle to avoid arrest in violation of Lincoln Municipal Code § 10.14.280; Any violation of the Uniform Controlled Substances Act as set forth in Neb. Rev. Stat. §§ 28-401 to 28-456.01; Toxic compounds, unlawful use in violation of Lincoln Municipal Code §9.16.110; Criminal attempt in violation of *Neb. Rev. Stat.* § 28-201 for any of the state crimes set forth in this subsection (a).

(b) It shall be unlawful for any person to possess any firearm within the corporate limits or on any property of the City of Lincoln outside the corporate limits when that person has been convicted of two or more of the following offenses within the last ten years: Driving under the influence of alcoholic liquor or drugs in violation of Neb. Rev. Stat. § 60-6,196; Driving under the influence of alcoholic liquor or drugs in violation of Lincoln Municipal Code § 10.16.030; Implied consent to submit to chemical test, refusal in violation of Neb. Rev. Stat. § 60-6-197; Chemical test refusal in violation of Lincoln Municipal Code §10.16.040; or any conviction under a law of another state or municipality if at the time of the conviction under said law the offence for which the person was convicted would have been a violation of Neb. Rev. Stat. §§ 60-6,196 or 60-6,197.

(c) The provisions of this section shall not apply to (1) the issuance of firearms or the possession by members of the Armed Forces of the United States, active or reserve, the National Guard of this state, or Reserve Officers Training Corps, when on duty or training; or (2) a peace officer as defined by Neb. Rev. Stat. § 28-109(14).

9.36.110 Firearms in Unattended Motor Vehicle; Unlawful. It shall be unlawful for any person to keep a firearm in an unattended motor vehicle for a period in excess of twenty-four hours.

The provisions of this section shall not apply to members of the Armed Forces of the United States, active or reserve, the National Guard of this state, or Reserve Officers Training Corps, when on duty or training, or peace officers or other duly authorized law enforcement officers, nor shall it apply to vehicles containing firearms that are parked in locked enclosures or buildings, such as garages or other storage facilities.

9.36.120 Firearm; Defined. For purposes of Sections 9.36.100 and 9.36.110 of the Lincoln Municipal Code, "firearm" shall mean any weapon which is designed to or may be readily converted to expel any projectile by action of an explosive or frame or receiver of any such weapon including, but not limited to, any pistol, revolver, shotgun, or rifle.

[Lincoln Municipal Code current through Oct. 21, 2008]

Omaha Municipal Code

Chapter 19 Occupation Taxes

Article II. Schedule

19-51. Firearm dealers. All applicants for a permit to engage in the business of buying, selling, renting, pawning, pledging or trading any firearms, as provided in section 19-371 et seq., shall pay a permit fee in the amount of $48 00 per year. Such fee shall be prorated at $4.00 per month or any portion thereof when an application is made other than at the start of a calendar year

Article XI. Firearm Dealers
Division 1. Generally

Sec. 19-361. Definition. For the purposes of this article, the words "firearm dealer" shall mean any person who is engaged in the business of buying, selling, renting, pawning, pledging or trading any firearm, but shall not include any person having an established place of business who is engaged in the business of buying any firearms solely for resale to a person who possesses a federal firearms dealer's license and has an established place of business.

19-362. Record of transactions. All persons who shall engage in the business of a firearm dealer shall keep a book in which shall be legibly written in ink, at the time of any purchase, sale, rental, pawn, pledge or trade of a firearm, an accurate account, description, and identification and serial number, in the English language, of the firearm, the amount of money loaned, paid, or allowed in trade thereon or therefor, when the same was received, and the name, residence and description of the person selling, buying, renting, or trading the same, which book as well as the firearm purchased or taken in trade shall be at all reasonable times open to the inspection of the chief of police, or any member of the police department.

19-363. Daily reports. It shall be the duty of every firearm dealer to make and mail to the chief of police by certified mail every day, for the book required in section 19-362, a list of all firearms or other valuable things received or deposited, purchased, mortgaged, or taken in trade during the previous day, together with the time received, purchased, or taken in trade or in pledge, or from whom the same were purchased or taken in trade; provided the report required by this section shall not apply to Sundays or holidays, but the report of Monday of each week or the first business day thereafter shall cover the business transactions of the Saturday previous, and also the Sunday or holidays previous if any such transactions shall take place on said last-mentioned day; provided that no persons shall be required to furnish such description of any firearm purchased from manufacturers or wholesale dealers having an established place of business or of firearms purchased at open sale from any bankrupt stock, or from any other person doing business and having an established place of business, but such goods must be accompanied by a bill of sale or other evidence of open and legitimate purchase, and must be shown to the chief of police or any member of the police department when demanded.

19-364. Retention of firearms. No firearm received on deposit, purchased or taken in trade by any firearm dealer shall be sold or permitted to be taken from the place of business of such person for the period of ten days after the mailing to the chief of police of the copy and statement as required by section 19-363 requiring a list of all firearms or other valuable things received or deposited, purchased, mortgaged, or taken in trade during the previous day.

Division 2. Permit

19-371. Required. It shall be unlawful for any person to engage in business as a firearm dealer without first obtaining a permit to do so from the permits and inspections division.

19-372. Federal license required. No permit shall be issued under the provisions of this division to an applicant who does not possess a valid, current license issued by the Alcohol, Tobacco, Tax and Firearms Division of the Internal Revenue Service of the Department of the Treasury of the United States.

19-373. Bond. The applicant for a permit required by the provisions of this division shall execute a continuing bond to the city in the sum of $2,000.00 with good and sufficient corporate surety, approved and filed as required by the ordinances of the city and conditioned for the faithful compliance with the provisions of this division.

19-374. Fee. The fee for a permit required by the provisions of this division shall be as provided in section 19-51 of this Code.

19-375. Approval. Before any permit is issued under the provisions of this division, the application therefor shall be approved by the chief of police.

19-376. Business location. No person operating under the provisions of this division shall be allowed to do business in more than one place under one permit, and every permit shall state the place where such business is to be carried on.

19-377. Transfer. Permits issued under the provisions of this division shall not be transferable, nor shall they be assigned.

19-378. Renewal. Any permit issued under the provisions of this division shall be renewed on or before December 31 next succeeding the date of issuance, and annually thereafter.

19-379. Suspension or revocation - Grounds. Any permit issued under the provisions of this division may be suspended or revoked for the violation by the permittee of:

(a) Any provision of this division or other applicable provision of this Code, state law or city ordinance, rule or regulation; or

(b) Any other provision of this Code, state law or city ordinance classified as a misdemeanor or felony.

19-380. Same - Hearing. Before any permit issued under the provisions of this division is suspended or revoked, the permittee shall be given notice of the date and time for a hearing to show cause, if any there be, why such permit should not be revoked. Such hearing shall be within seven days of the date of the notice. At the time and place set for the show-cause hearing, the police chief shall hear all the testimony as to whether or not grounds for the revocation or suspension of the permit exist.

19-381. Same - Action at hearing. If, upon the conclusion of the hearing thereon, it shall appear that grounds for the suspension or revocation of a permit issued under the provisions of this division exist, the police chief shall so find and impose such suspension or revocation as he may deem proper under the circumstances.

19-382. Same - Appeals. A finding, and the action taken thereon, by the police chief at the conclusion of the hearing on the suspension or revocation of a permit issued under the provisions of this division may be appealed by the permittee to the city council upon written application to the council therefor. Upon hearing such appeal, the council may reverse, change or modify the finding or action of the police chief. A permit shall remain in effect pending the outcome of such appeal.

Division 3. Firearms Exhibitions

19-383. Permit required. It shall be unlawful for any person to promote or sponsor a firearms exhibition without first obtaining a permit to do so from the permits and inspections division.

19-384. Federal license required. No permit shall be issued under the provisions of this division to an applicant who does not possess a valid, current license issued by the Alcohol, Tobacco, Tax and Firearms Division of the Internal Revenue Service of the Department of the Treasury of the United States.

19-385. Date of application for permit. No permit for a firearm exhibition shall be issued unless applied for more than five days before the date of such exhibition.

19-386. Permit fee. The fee for a firearms exhibition sponsor's permit shall be $10.00, and this fee shall be waived for retail and/or wholesale firearms dealers currently licensed to do business in the city.

19-387. Duration of exhibition. No firearms exhibition shall continue for more than three consecutive days.

19-388. Approval of permit. Before any permit is issued under the provisions of this division, the application therefor shall be approved by the police chief or someone under his direction.

19-389. Location. No person operating under the provisions of this division shall be allowed to do business in more than one place under one permit.

19-390. Transfer of permit. Permits issued under the provisions of this division shall not be transferable, nor shall they be assigned.

19-391. Duties of promoter. It shall be the duty of the person promoting or sponsoring a firearms exhibition to provide:

(a) Security personnel at each entrance to the exhibition, who shall check every firearm brought into the exhibition by the general public for the purpose of determining that the firearms are not loaded.

(b) Twenty-four-hour-per-day security at the site of the exhibition.

(c) A written notice to each exhibitor stating that all applicable federal, state, and local laws and ordinances must be observed.

(d) To the chief of police or his agent(s), a list of the names and addresses of all exhibitors.

19-392. Duties of exhibitors. Any person who is designated as an exhibitor under section 19-391(d), shall keep an accurate record of all firearms sold, purchased, bartered or traded for, regardless of their antiquity or value or whether said firearms are functioning or nonfunctioning. The record shall be kept in the English language and shall list the serial number, if any, the calibre or gauge, the manufacturer's name, and the name and address of the buyer or seller. Said record shall be made available to the police department, at the request of the police department, during the time of the firearms exhibition and for a period of two years thereafter.

19-392.1. Exhibition or sale of ammunition. It shall be unlawful to exhibit or sell any ammunition in an assembled state at a firearms exhibition.

19-392.2. Possession of concealable weapons. Notwithstanding any other provision of this Code, an exhibitor who is a nonresident of this city may possess concealable firearms without registering them with the chief of police. Such possession shall be legal only at the site of the exhibition, while in direct route to or from the exhibition and during the time of exhibition.

Chapter 20 Offenses

Article VII. Weapons

Division 1. Generally

20-191. Definitions. For the purposes of this article, the following words and phrases shall have the meanings respectively ascribed to them:

Concealable firearm: A firearm having a barrel less than 18 inches in length.

Firearm: Any instrument which releases a projectile by means of an explosive charge.

Machine gun: Any firearm which shoots, is designed to shoot, or can be readily restored to shoot automatically more than one shot without manual reloading by a single function of the trigger. The term shall also include the frame or receiver of any such weapon, any combination of parts designed and intended for use in converting a weapon into a machine gun, and any combination of parts from which a machine gun can be assembled if such parts are in the possession or under the control of a person.

Paint ball gun: Any gun which releases a paint ball propelled by spring mechanism or compressed gas.

Stun gun: Any hand-held electronic device that is powered by an internal power source, such as batteries, and that is capable of introducing an electrical current into the body of a person which when introduced shall be capable of disrupting a person's central nervous system and rendering him/her temporarily incapable of normal functioning. The electrical current may be introduced into the human body by means of direct pressure to the body from fixed electrodes on the electronic device and/or one or more electrodes attached to a length of wire and which, upon being fired from a firearm or mechanical device, strikes the human body.

Weapon: A firearm, stun gun or any other instrument the use of which is intended or likely to cause death or bodily injury.

20-193. Confiscation. As a part of the judgment of conviction of any person under this article, confiscation and destruction of the weapon shall be ordered by the court provided, in cases where the gun in question was reported stolen to the appropriate law enforcement agency, it shall be ordered returned to the lawful innocent owner if claimed by such person prior to the conviction.

Claimed, as used herein means reporting a missing weapon by serial number or other identifiable marking to the appropriate law enforcement agency and then possessing a legal registration for the same weapon.

20-195. Possession or transportation of firearms.

(a) It shall be unlawful for any person to knowingly or purposely transport in any conveyance or in any other manner, or to possess off his own premises, any rifle, shotgun, air gun, air rifle, paint ball gun or machine gun unless the same is unloaded and contained in any enclosed gun case, or unloaded and broken down. The removal of the bolt from any such firearm or carrying the same in a holster type gun case without further breaking down such firearm shall not be deemed to be in compliance with the requirements of this section.

(b) The provisions of this section shall not be applicable to:

(1) Authorized law enforcement officers;

(2) The armed forces of the United States, including ROTC units and the National Guard in the performance of their duties;

(3) The carrying of unloaded and uncased rifles in parades or using rifles in ceremonials;

(4) The possession of rifles, machine guns, or shotguns at shows or exhibits; or

(5) Any other lawful use, purpose or activity, including but not limited to skeet and trap shooting, target shooting at rifle ranges, hunter safety instruction conducted by qualified instructors, when such do not endanger public safety or are detrimental to public welfare; provided, however:

a. The prior approval of the police chief shall have been obtained for such use, purpose or activity; and

b. With respect to skeet and trap shooting ranges located or to be located in public parks, the final determination of the safety of the same shall be made by the city council.

20-198. Sale of firearm to minor; penalty.

(a) It shall be unlawful for any person to sell or otherwise furnish or deliver any firearm, component parts or ammunition to any individual who is known, or there is reason to believe that the individual is less than 18 years of age, and if the firearm, component parts or ammunition is a concealable firearm, to any person who has not reached the age of 21.

(b) Any person convicted under the provisions of this section shall be punished by a mandatory fine of $500.00 and by a mandatory imprisonment of six months.

20-199. Display of firearms or ammunition. It shall be unlawful for any person purposely or knowingly to exhibit for sale or distribution in a display window or any other place which can be seen from a public thoroughfare any firearm or ammunition therefor.

20-200. Permit required for purchasing or renting firearm. It shall be unlawful for any person to sell or rent a concealable firearm to any person who has not obtained a written permit from the chief of police as provided for in this article.

20-201. Pawning firearm. It shall be lawful for any person to pawn, pledge or store a concealable firearm or to accept a firearm in pawn or as a pledge of or for storage from a person who has not registered it as provided for in this article. No such firearm shall be returned to any person until such person has registered it as provided for in this article.

20-202. Report of sales and rentals of firearms. Any person engaged in the sale, pawning, exchange, loan, rental or delivery of firearms must preserve and furnish daily to the chief of police a record of such transaction indicating the date of the transaction, a description of the firearm, including number, color, make, caliber, and all other identifying marks, and the name, address, age, height, complexion, color of hair, color of eyes, weight, and apparent deformities or peculiarities of the person with whom such transaction was had.

20-203. Possession of machine guns.

(a) It shall be unlawful for any person to possess any machine gun.

(b) It shall be an affirmative defense under this section that the accused is a member of one or more of the following classes:

(1) Authorized law enforcement officers in the performance of their official duties;

(2) The armed forces of the United States, including ROTC units and the National Guard in the performance of their duties;

(3) Any person or persons or corporation licensed by the federal government as a dealer in machine guns under the provisions of the National Firearms Act; or

(4) Any person lawfully purchasing and possessing a machine gun under applicable federal law.

20-204. Unlawful possession of concealable firearm; exceptions.

(a) Any person who has not reached the age of 21 who possesses a concealable firearm as defined in this article commits the offense of unlawful possession of a concealable firearm.

(b) The provisions of this section shall not apply to the issuance of such firearms to members of the armed forces of the United States, active or reserve, national guard of the state, or reserve officer training corps, when on duty or training, or to the temporary loan of pistols, revolvers, or any other form of concealable firearms for instruction under the immediate supervision of a parent or guardian or adult instructor.

20-205. Penalties. Any person who has not reached the age of 21 at the time of the offense convicted under the provisions of sections 20-203 and 20-204 shall be punished as follows:

(a) For a first conviction, any person so offending shall be punished by a fine of not exceeding $500.00, and by mandatory imprisonment of not less than 30 days and not more than six months.

(b) For any second subsequent conviction, any persons so offending shall be punished by a fine not exceeding $500.00 and by mandatory imprisonment of not less than six months.

Division 2. Firearm Registration

20-251. Required.

(a) It shall be unlawful for any person to own, have possession of, or maintain control over any concealable firearm which has not been registered to said person with the chief of police in accordance with this division, except when such possession or control is with the knowledge and express consent of the person in whose name such concealable firearm is registered.

(b) A corporation, including a body corporate created by Nebraska statute, may register a concealable firearm in its corporate name. However, the corporation may consent to a person possessing or controlling the corporation's registered concealable firearm only if that person:

(1) Is a part-time or full-time employee of the corporation;

(2) Is acting within the scope of his or her employment with the corporation; and,

(3) Possesses a current identification card issued pursuant to section 20-208 upon satisfaction of the requirements of section 20-207.

20-252. Confiscation of firearm upon conviction of violation. As a part of the judgment of conviction of any person under this division, confiscation of a concealable firearm may be ordered by the court.

20-253. Application.

(a) Any person desiring to register a concealable firearm shall make an application to the chief of police stating therein that he or she holds the qualifications to register a concealable firearm in accordance with this section.

(b) The chief of police will conduct an investigation to determine if the applicant is qualified to register the firearm. A concealable firearm may not be registered to any person who:

(1) Is currently the subject of an active protection order.

(2) Has provided false information on the registration request.

(3) Has a conviction of an offense listed below which has not been pardoned or set aside under state or federal law:

a. Any felony;

b. Carrying a concealed weapon or being a minor in possession of a concealable firearm;

c. Within the previous five years, assault, child abuse, or violation of any provision of

Chapter 20, article VII of the Omaha Municipal Code.

(4) Has been convicted of any charge of domestic violence, including stalking or harassment.

(5) Has a record of mental disorder which would show the applicant to be a danger to self or to others.

(6) Is a fugitive from justice in this state or any other jurisdiction.

(7) Has been dishonorably discharged from the United States Armed Forces.

(8) Is a user of, or addicted to, unlawful controlled substances.

(9) Is not a citizen of the United States.

(10) Is under the age of 21 years.

(c) A registration fee of $10.00 shall accompany each application made to the chief of police for each concealable firearm. The fee will be applied to the administrative costs of processing the application.

20-254. Issuance or denial of permit. Upon the filing of an application for registration of a concealable firearm, the chief of police shall issue a permit or deny a permit and furnish the applicant the specific reasons for the denial in writing. A failure to furnish the applicant a written denial shall constitute issuance of a permit.

20-255. Consideration of application by chief of police; custody of firearm pending decision on permit. The chief of police shall be permitted seven days in which to consider an application for registration of a concealable firearm. The chief of police may take custody of the concealable firearm in the applicant's possession while the application is under consideration whenever he determines that it is in the best interests of public safety to do so.

20-256. Disposition of firearm upon denial of permit. Any person whose application for registration of a concealable firearm is denied shall have ten days in which to provide for proper registration or other lawful disposition of the concealable firearm. During this time the chief of police shall maintain custody of the concealable firearm. If the applicant fails to provide for the proper registration or other lawful disposition of the concealable firearm within this time, it shall be presumed that the concealable firearm is an unregistered concealable firearm and the chief of police may apply to the municipal court for an order of confiscation.

20-257. Revocation.

A firearm registration may be revoked when:

(a) The applicant or registrant fails to hold all of the qualifications set forth in section 20-253 of the Omaha Municipal Code; or

(b) The information furnished on the application for a firearm registration was false or misleading or no longer continues to be true.

In the event revocation occurs under the provision of the Omaha Municipal Code, the chief of police shall furnish the registrant with the specific reasons for the revocation in writing

.20-258. Appeals.

Any person aggrieved by the denial or revocation of any registration issued under the provisions of this division may appeal therefrom to the administrative appeals board by complying with all of the provision of section 2-171 et seq. of this Code.

[Omaha Municipal Code codified through Ord. No. 38203, enacted July 29, 2008]

NEVADA

Nev. Rev. Stat.

Title 15. Crimes and Punishment

Chapter 202. Crimes Against Public Heath and Safety

Weapons

Dangerous Weapons and Firearms

202.253. Definitions. As used in NRS 202. 253 to 202.369, inclusive, and section 1 of this act:

1. "Explosive or incendiary device" means any explosive or incendiary material or substance that has been constructed, altered, packaged or arranged in such a manner that its ordinary use would cause destruction or injury to life or property.

2. "Firearm" means any device designed to be used as a weapon from which a projectile may be expelled through the barrel by the force of any explosion or other form of combustion.

3. "Firearm capable of being concealed upon the person" applies to and includes all firearms having a barrel less than 12 inches in length.

4. "Motor vehicle" means every vehicle that is self-propelled.

202.254. Private person authorized to obtain background check on person who wishes to obtain firearm from him; fee.

1. A private person who wishes to transfer a firearm to another person may, before he transfers the firearm, request that the central repository for Nevada records of criminal history perform a background check on the person who wishes to acquire the firearm.

2. The person who requests the information pursuant to subsection 1 shall provide the central repository with identifying information about the person who wishes to acquire the firearm.

3. Upon receiving a request from a private person pursuant to subsection 1 and the identifying information required pursuant to subsection 2, the central repository shall within 5 business days after receiving the request:

(a) Perform a background check on the person who wishes to acquire the firearm; and

(b) Notify the person who requests the information whether the information available to the central repository indicates that the receipt of a firearm by the person who wishes to acquire the firearm would violate a state or federal law.

4. If the person who requests the information does not receive notification from the central repository regarding his request within 5 business days after making the request, he may presume that the receipt of a firearm by the person who wishes to acquire the firearm would not violate a state or federal law.

5. The central repository may charge a reasonable fee for performing a background check and notifying a person of the results of the background check pursuant to this section.

6. The failure of a person to request the central repository to perform a background check pursuant to this section before transferring a firearm to another person does not give rise to any civil cause of action.

202.257. Possession of firearm when under influence of alcohol, controlled substance or other intoxicating substance; administration of evidentiary test; penalty; forfeiture of firearm.

1. It is unlawful for a person who:

(a) Has a concentration of alcohol of 0.10 or more in his blood or breath; or

(b) Is under the influence of any controlled substance, or is under the combined influence of intoxicating liquor and a controlled substance, or any person who inhales, ingests, applies or otherwise uses any chemical, poison or organic solvent, or any compound or combination of any of these, to a degree which renders him incapable of safely exercising actual physical control of a firearm, to have in his actual physical possession any firearm. This prohibition does not apply to the actual physical possession of a firearm by a person who was within his personal residence and had the firearm in his possession solely for self-defense.

2. Any evidentiary test to determine whether a person has violated the provisions of subsection 1 must be administered in the same manner as an evidentiary test that is administered pursuant to NRS 484.383 to 484.3947, inclusive, except that submission to the evidentiary test is required of any person who is directed by a police officer to submit to the test. If a person to be tested fails to submit to a required test as directed by a police officer, the officer may direct that reasonable force be used to the extent necessary to obtain the samples of blood from the person to be tested, if the officer has reasonable cause to believe that the person to be tested was in violation of this section.

3. Any person who violates the provisions of subsection 1 is guilty of a misdemeanor.

4. A firearm is subject to forfeiture pursuant to NRS 179.1156 to 179.119, inclusive, only if, during the violation of subsection 1, the firearm is brandished, aimed or otherwise handled by the person in a manner which endangered others.

5. As used in this section, the phrase "concentration of alcohol of 0.10 or more in his blood or breath" means 0.10 gram or more of alcohol per 100 milliliters of the blood of a person or per 210 liters of his breath.

202.260. Unlawful possession, manufacture or disposition of explosive or incendiary device: Penalty; exceptions.

1. A person who unlawfully possesses, manufactures or disposes of any explosive or incendiary device with the intent to destroy life or property is guilty of a category B felony and shall be punished by imprisonment in the state prison for a minimum term of not less than 1 year and a maximum term of not more than 6 years, and may be further punished by a fine of not more than $5,000.

2. This section does not prohibit a person from possessing, manufacturing or using any material, component, substance or device as required for the performance of his duties related to mining, agriculture, construction or any other valid occupational purpose or if the person is authorized by a governmental entity which has lawful control over such matters to use those items in the performance of his duties.

3. For the purposes of this section "dispose of" means give, give away, loan, offer, offer for sale, sell or transfer.

202.263. Unlawful manufacture, possession, sale, advertisement or transportation of hoax bomb: Penalty; exceptions.

1. A person shall not manufacture, purchase, possess, sell, advertise for sale or transport a hoax bomb if the person knows or should know that the hoax bomb is to be used to make a reasonable person believe that the hoax bomb is an explosive or incendiary device.

2. A person who violates subsection 1 is guilty of a gross misdemeanor.

3. This section does not prohibit:

(a) The purchase, possession, sale, advertising for sale, transportation or use of a military artifact, if the military artifact is harmless or inert, unless the military artifact is used to make a reasonable person believe that the military artifact is an explosive or incendiary device.

(b) The authorized manufacture, purchase, possession, sale, transportation or use of any material, substance or device by a member of the Armed Forces of the United States, a fire department or a law enforcement agency if the person is acting lawfully while in the line of duty.

(c) The manufacture, purchase, possession, sale, transportation or use of any material, substance or device that is permitted by a specific statute.

4. As used in this section "hoax bomb" means:

(a) An inoperative facsimile or imitation of an explosive or incendiary device; or

(b) A device or object that appears to be or to contain an explosive or incendiary device.

202.265. Possession of dangerous weapon on property or in vehicle of school; penalty; exceptions.

1. Except as otherwise provided in this section, a person shall not carry or possess, while on the property of the of Nevada System of Higher Education or a private or public school or while in a vehicle of a private or public school:

(a) An explosive or incendiary device;

(b) A dirk, dagger or switchblade knife;

(c) A nunchaku or trefoil;

(d) A blackjack or billy club or metal knuckles; or

(e) A pistol, revolver or other firearm.

2. Any person who violates subsection 1 is guilty of a gross misdemeanor.

3. This section does not prohibit the possession of a weapon listed in subsection 1 on the property of a private or public school by a:

(a) Peace officer;

(b) School security guard; or

(c) Person having written permission from the president of a branch or facility of the Nevada System of Higher Education or the principal of the school to carry or possess the weapon.

4. For the purposes of this section:

(a) "Firearm" includes:

(1) Any device used to mark the clothing of a person with paint or any other substance; and

(2) Any device from which a metallic projectile, including any ball bearing or pellet, may be expelled by means of spring, gas, air or other force.

(b) "Nunchaku" has the meaning ascribed to it in NRS 202.350.

(c) "Switchblade knife" has the meaning ascribed to it in NRS 202.350.

(d) "Trefoil" has the meaning ascribed to it in NRS 202.350.

(e) "Vehicle" has the meaning ascribed to "school bus" in NRS 484.148.

202.273. Unlawful manufacture or sale of certain metal-penetrating bullets: Exceptions; penalty.

1. Except as provided in subsection 2, it is unlawful to manufacture or sell any metal-penetrating bullet capable of being fired from a handgun.

2. A person may manufacture and sell metal-penetrating bullets pursuant to an agreement with a law enforcement agency for the sale of such bullets to that agency.

3. A person who violates the provisions of this section is guilty of a gross misdemeanor.

4. As used in this section, "metal-penetrating bullet" means a bullet whose core:

(a) Reduces the normal expansion of the bullet upon impact; and

(b) Is at least as hard as the maximum hardness attainable using solid red metal alloys, and which can be used in a handgun. The term does not include any bullet with a copper or brass jacket and a core of lead or a lead alloy, or a bullet made of lead or lead alloys.

202.275. Possession, manufacture or disposition of short-barreled rifle or short-barreled shotgun: Penalty; exceptions.

1. Except as otherwise provided in subsection 3, a person who knowingly or willfully possesses, manufactures or disposes of any short-barreled rifle or short-barreled shotgun is guilty of a category D felony and shall be punished as provided in NRS 193.130.

2. For purposes of this section:

(a) "Short-barreled rifle" means:

(1) A rifle having one or more barrels less than 16 inches in length; or

(2) Any weapon made from a rifle, whether by alteration, modification or other means, with an overall length of less than 26 inches.

(b) "Short-barreled shotgun" means:

(1) A shotgun having one or more barrels less than 18 inches in length; or

(2) Any weapon made from a shotgun, whether by alteration, modification or other means, with an overall length of less than 26 inches.

3. This section does not prohibit:

(a) The possession or use of any short-barreled rifle or short-barreled shotgun by any peace officer when authorized to do so in the performance of official duties; or

(b) The possession of any short-barreled rifle or short- barreled shotgun by a person who is licensed as a firearms importer, manufacturer, collector or dealer by the United States Department of the Treasury, or by a person to whom such a rifle or shotgun is registered with the United States Department of the Treasury;

(c) The possession of any short-barreled rifle or short-barreled shotgun that has been determined to be a collector's item pursuant to 26 U.S.C. chapter 53 or a curio or relic pursuant to 18 U.S.C. chapter 44.

202.277. Changing, altering, removing or obliterating serial number of firearm prohibited; possession of firearm with serial number changed, altered, removed or obliterated prohibited; penalties.

1. A person shall not intentionally change, alter, remove or obliterate the serial number upon any firearm. Any person who violates the provisions of this subsection is guilty of a category C felony and shall be punished as provided in NRS 193.130.

2. A person shall not knowingly possess a firearm on which the serial number has been intentionally changed, altered, removed or obliterated. Any person who violates the provisions of this subsection is guilty of a category D felony and shall be punished as provided in NRS 193.130.

202.300. Use or possession of firearm by child under age of 18 years; unlawful to aid or permit child to commit violation; penalties; child 14 years of age or older authorized to possess firearm under certain circumstances.

1. Except as otherwise provided in this section, a child under the age of 18 years shall not handle or have in his possession or under his control, except while accompanied by or under the immediate charge of his parent or guardian or an adult person authorized by his parent or guardian to have control or custody of the child, any firearm of any kind for hunting or target practice or for other purposes. A child who violates this subsection commits a delinquent act and the court may order the detention of the child in the same manner as if the child had committed an act that would have been a felony if committed by an adult.

2. A person who aids or knowingly permits a child to violate subsection 1:

(a) Except as otherwise provided in paragraph (b), for the first offense, is guilty of a misdemeanor.

(b) For a first offense, if the person knows or has reason to know that there is a substantial risk that the child will use the firearm to commit a violent act, is guilty of a category C felony and shall be punished as provided in NRS 193.130.

(c) For a second or any subsequent offense, is guilty of a category B felony and shall be punished by imprisonment in the state prison for a minimum term of not less than 1 year and a maximum term of not more than 6 years, and may be further punished by a fine of not more than $5,000.

3. A person does not aid or knowingly permit a child to violate subsection 1 if:

(a) The firearm was stored in a securely locked container or at a location which a reasonable person would have believed to be secure;

(b) The child obtained the firearm as a result of an unlawful entry by any person in or upon the premises where the firearm was stored;

(c) The injury or death resulted from an accident which was incident to target shooting, sport shooting or hunting; or

(d) The child gained possession of the firearm from a member of the military or a law enforcement officer, while the member or officer was performing his official duties.

4. The provisions of subsection 1 do not apply to a child who is a member of the Armed Forces of the United States.

5. Except as otherwise provided in subsection 8, a child who is 14 years of age or older, who has in his possession a valid license to hunt, may handle or have in his possession or under his control, without being accompanied by his parent or guardian or an adult person authorized by his parent or guardian to have control or custody of him:

(a) A rifle or shotgun that is not a fully automatic firearm, if the child is not otherwise prohibited by law from possessing the rifle or shotgun and the child has the permission of his parent or guardian to handle or have in his possession or under his control the rifle or shotgun; or

(b) A firearm capable of being concealed upon the person, if the child has the written permission of his parent or guardian to handle or have in his possession or under his control such a firearm and the child is not otherwise prohibited by law from possessing such a firearm, and the child is traveling to the area in which he will be hunting or returning from that area and the firearm is not loaded, or the child is hunting pursuant to that license.

6. Except as otherwise provided in subsection 8, a child who is 14 years of age or older may handle or have in his possession or under his control a rifle or shotgun that is not a fully automatic firearm if the child is not otherwise prohibited by law from possessing the rifle or shotgun, without being accompanied by his parent or guardian or an adult person authorized by his parent or guardian to have control or custody of him, if the child has the permission of his parent or guardian to handle or have in his possession or under his control the rifle or shotgun and the child is:

(a) Attending a course of instruction in the responsibilities of hunters or a course of instruction in the safe use of firearms;

(b) Practicing the use of a firearm at an established firing range or at any other area where the discharge of a firearm is permitted;

(c) Participating in a lawfully organized competition or performance involving the use of a firearm;

(d) Within an area in which the discharge of firearms has not been prohibited by local ordinance or regulation and he is engaging in a lawful hunting activity in accordance with chapter 502 of NRS for which a license is not required;

(e) Traveling to or from any activity described in paragraph (a), (b), (c) or (d), and the firearm is not loaded;

(f) On real property that is under the control of an adult, and the child has the permission of that adult to possess the firearm on the real property; or

(g) At his residence.

7. Except as otherwise provided in subsection 8, a child who is 14 years of age or older may handle or have in his possession or under his control, for the purpose of engaging in any of the activities listed in paragraphs (a) to (g), inclusive, of subsection 6, a firearm capable of being concealed upon the person, without being accompanied by his parent or guardian or an adult person authorized by his parent or guardian to have control or custody of him, if the child:

(a) Has the written permission of his parent or guardian to handle or have in his possession or under his control such a firearm for the purpose of engaging in such an activity; and

(b) Is not otherwise prohibited by law from possessing such a firearm.

8. A child shall not handle or have in his possession or under his control a loaded firearm if he is:

(a) An occupant of a motor vehicle;

(b) Within any residence, including his residence, or any building other than a facility licensed for target practice, unless possession of the firearm is necessary for the immediate defense of the child or another person; or

(c) Within an area designated by a county or municipal ordinance as a populated area for the purpose of prohibiting the discharge of weapons, unless he is within a facility licensed for target practice.

9. For the purposes of this section, a firearm is loaded if:

(a) There is a cartridge in the chamber of the firearm;

(b) There is a cartridge in the cylinder of the firearm, if the firearm is a revolver; or

(c) There is a cartridge in the magazine and the magazine is in the firearm or there is a cartridge in the chamber, if the firearm is a semiautomatic firearm.

202.310. Sale of firearms to minors; penalty. Any person in this state who sells or barters to a child who is under the age of 18 years, with reckless disregard of whether the child is under the age of 18 years, or with knowledge or reason to know that the child is under the age of 18 years, a pistol, revolver or a firearm capable of being concealed upon the person is guilty of a category B felony and shall be punished by imprisonment in the state prison for a minimum term of not less than 1 year and a maximum term of not more than 6 years, and may be further punished by a fine of not more than $5,000.

202.350. Manufacture, importation, possession or use of dangerous weapon or silencer; carrying concealed weapon without permit; penalties; issuance of permit to carry concealed weapon; exceptions.

1. Except as otherwise provided in this section and NRS 202.3653 to 202.369, inclusive, and section 1 of this act, a person within this state shall not: ...

(d) Carry concealed upon his person any:

(1) Explosive substance, other than ammunition or any components thereof;

(2) Dirk, dagger or machete;

(3) Pistol, revolver or other firearm, or other dangerous or deadly weapon; or

(4) Knife which is made an integral part of a belt buckle.

2. Except as otherwise provided in NRS 202.275 and 212.185, a person who violates any of the provisions of:

(a) Paragraph (a) or (c) or subparagraph (2) or (4) of paragraph (d) of subsection 1 is guilty:

(1) For the first offense, of a gross misdemeanor.

(2) For any subsequent offense, of a category D felony and shall be punished as provided in NRS 193.130.

(b) Paragraph (b) or subparagraph (1) or (3) of paragraph (d) of subsection 1 is guilty of a category C felony and shall be punished as provided in NRS 193.130.

3. Except as otherwise provided in this subsection, the sheriff of any county may, upon written application by a resident of that county showing the reason or the purpose for which a concealed weapon is to be carried, issue a permit authorizing the applicant to carry in this state the concealed weapon described in the permit. The sheriff shall not issue a permit to a person to carry a switchblade knife. This subsection does not authorize the sheriff to issue a permit to a person to carry a pistol, revolver or other firearm.

4. Except as otherwise provided in subsection 5, this section does not apply to:

(a) Sheriffs, constables, marshals, peace officers, correctional officers employed by the department of corrections, special police officers, police officers of this state, whether active or honorably retired, or other appointed officers.

(b) Any person summoned by any peace officer to assist in making arrests or preserving the peace while the person so summoned is actually engaged in assisting such an officer.

(c) Any full-time paid peace officer of an agency of the United States or another state or political subdivision thereof when carrying out official duties in the State of Nevada.

(d) Members of the Armed Forces of the United States when on duty.

5. The exemptions provided in subsection 4 do not include a former peace officer who is retired for disability unless his former employer has approved his fitness to carry a concealed weapon.

6. The provisions of paragraph (b) of subsection 1 do not apply to any person who is licensed, authorized or permitted to possess or use a machine gun or silencer pursuant to federal law. The burden of establishing federal licensure, authorization or permission is upon the person possessing the license, authorization or permission.

7. This section shall not be construed to prohibit a qualified law enforcement officer or a qualified retired law enforcement officer from carrying a concealed weapon in this State if he is authorized to do so pursuant to 18 U.S.C. § 926B or 196C.

8. As used in this section:

(a) "Concealed weapon" means a weapon described in this section that is carried upon a person in such a manner as not to be discernible by ordinary observation.

(b) "Honorably retired" means retired in Nevada after completion of 10 years of creditable service as a member of the public employees' retirement system. A former peace officer is not "honorably retired" if he was discharged for cause or resigned before the final disposition of allegations of serious misconduct.

(c) "Machine gun" means any weapon which shoots, is designed to shoot or can be readily restored to shoot more than one shot, without manual reloading, by a single function of the trigger. ...

(e) "Qualified law enforcement officer" has the meaning ascribed to it in 18 U.S.C. § 926B(c).

(f) "Qualified retired law enforcement officer" has the meaning ascribed to it in 18 U.S.C. § 926C(c).

(g) "Silencer" means any device for silencing, muffling or diminishing the report of a firearm, including any combination of parts, designed or redesigned, and intended for use in assembling or fabricating a silencer or muffler, and any part intended only for use in such assembly or fabrication. ...

202.357. Electronic stun device: Use prohibited except for self-defense; possession by certain persons prohibited; sale, gift or other provision to certain persons prohibited; penalties.

1. Except as otherwise provided in this section, a person shall not use an electronic stun device on another person for any purpose other than self-defense.

2. Except as otherwise provided in this section, a person shall not have in his possession or under his custody or control any electronic stun device if he:

(a) Has been convicted of a felony in this State or any other state, or in any political subdivision thereof, or of a felony in violation of the laws of the United States of America, unless he has received a pardon and the pardon does not restrict his right to bear arms;

(b) Is a fugitive from justice;

(c) Has been adjudicated as mentally ill or has been committed to any mental health facility; or

(d) Is illegally or unlawfully in the United States.

3. A child under 18 years of age shall not have in his possession or under his custody or control any electronic stun device.

4. Except as otherwise provided in this section, a person within this State shall not sell, give or otherwise provide an electronic stun device to another person if he has actual knowledge that the other person:

(a) Is a child under 18 years of age;

(b) Has been convicted of a felony in this State or any other state, or in any political subdivision thereof, or of a felony in violation of the laws of the United States of America, unless he has received a pardon and the pardon does not restrict his right to bear arms;

(c) Is a fugitive from justice;

(d) Has been adjudicated as mentally ill or has been committed to any mental health facility; or

(e) Is illegally or unlawfully in the United States.

5. A person who violates the provisions of:

(a) Subsection 1 or paragraph (a) or (b) of subsection 2 is guilty of a category B felony and shall be punished by imprisonment in the state prison for a minimum term of not less than 1 year and a maximum term of not more than 6 years, and may be further punished by a fine of not more than $5,000.

(b) Paragraph (c) or (d) of subsection 2 is guilty of a category D felony and shall be punished as provided in NRS 193.130.

6. A child who violates subsection 3 commits a delinquent act and the court may order the detention of the child in the same manner as if the child had committed an act that would have been a felony if committed by an adult.

7. A person who violates the provisions of subsection 4 is guilty of a category D felony and shall be punished as provided in NRS 193.130.

8. The provisions of subsections 1, 2 and 4 do not apply to a peace officer who possesses or uses or sells, gives or otherwise provides to another person an electronic stun device within the scope of his duties.

9. As used in this section, "electronic stun device" means a device that:

(a) Emits an electrical charge or current that is transmitted by projectile, physical contact or other means; and

(b) Is designed to disable a person or animal temporarily or permanently.

202.360 Ownership or possession of firearm by certain persons prohibited; penalties.

1. A person shall not own or have in his possession or under his custody or control any firearm if he:

(a) Has been convicted of a felony in this or any other state, or in any political subdivision thereof, or of a felony in violation of the laws of the United States of America, unless he has received a pardon and the pardon does not restrict his right to bear arms;

(b) Is a fugitive from justice; or

(c) Is an unlawful user of, or addicted to, any controlled substance.

A person who violates the provisions of this subsection is guilty of a category B felony and shall be punished by imprisonment in the state prison for a minimum term of not less than 1 year and a maximum term of not more than 6 years, and may be further punished by a fine of not more than $5,000.

2. A person shall not own or have in his possession or under his custody or control any firearm if he:

(a) Has been adjudicated as mentally ill or has been committed to any mental health facility; or

(b) Is illegally or unlawfully in the United States.

A person who violates the provisions of this subsection is guilty of a category D felony and shall be punished as provided in NRS 193.130.

3. As used in this section:

(a) "Controlled substance" has the meaning ascribed to it in 21 U.S.C. § 802(6).

(b) "Firearm" includes any firearm that is loaded or unloaded and operable or inoperable.

202.362. Sale or disposal of firearm or ammunition to certain persons prohibited; penalty; exceptions.

1. Except as otherwise provided in subsection 3, a person within this state shall not sell or otherwise dispose of any firearm or ammunition to another person if he has actual knowledge that the other person:

(a) Is under indictment for, or has been convicted of, a felony in this or any other state, or in any political subdivision thereof, or of a felony in violation of the laws of the United States of America, unless he has received a pardon and the pardon does not restrict his right to bear arms;

(b) Is a fugitive from justice;

(c) Has been adjudicated as mentally ill or has been committed to any mental health facility; or

(d) Is illegally or unlawfully in the United States.

2. A person who violates the provisions of subsection 1 is guilty of a category B felony and shall be punished by imprisonment in the state prison for a minimum term of not less than 1 year and a maximum term of not more than 10 years, and may be further punished by a fine of not more than $10,000.

3. This section does not apply to a person who sells or disposes of any firearm or ammunition to:

(a) A licensed importer, licensed manufacturer, licensed dealer or licensed collector who, pursuant to 18 U.S.C. § 925(b), is not precluded from dealing in firearms or ammunition; or

(b) A person who has been granted relief from the disabilities imposed by federal laws pursuant to 18 U.S.C. § 925(c).

Concealed Firearms

202.3653. Definitions. As used in NRS 202.3653 to 202.369, inclusive, unless the context otherwise requires:

1. "Concealed firearm" means a loaded or unloaded pistol, revolver or other firearm which is carried upon a person in such a manner as not to be discernible by ordinary observation.

2. "Department" means the department of public safety.

3. "Permit" means a permit to carry a concealed firearm issued pursuant to the provisions of NRS 202.3653 to 202.369, inclusive, and Sections 2 and 3 of this act.

4. "Revolver" means a firearm that has a revolving cylinder with several chambers, which, by pulling the trigger or setting the hammer, are aligned with the barrel, placing the bullet in a position to be fired. The term includes, without limitation, a single or double derringer.

5. "Semiautomatic firearm" means a firearm which:

(a) Uses the energy of the explosive in a fixed cartridge to extract a fixed cartridge and chamber a fresh cartridge with each single pull of the trigger; and

(b) Requires the release of the trigger and another pull of the trigger for each successive shot

202.3657. Application for permit; eligibility; denial or revocation of permit.

1. Any person who is a resident of this state may apply to the sheriff of the county in which he resides for a permit on a form prescribed by regulation of the department. Any person who is not a resident of this state may apply to the sheriff of any county in this state for a permit on a form prescribed by regulation of the department. Application forms for permits must be furnished by the sheriff of each county upon request.

2. Except as otherwise provided in this section, the sheriff shall issue a permit for revolvers, one or more specific semiautomatic firearms , or for revolvers and one or more specific semiautomatic firearms, as applicable, to any person who is qualified to possess the firearm or firearms to which the application pertains] under state and federal law, who submits an application in accordance with the provisions of this section and who:

(a) Is 21 years of age or older;

(b) Is not prohibited from possessing a firearm pursuant to NRS 202.360; and

(c) Demonstrates competence with revolvers, each specific semiautomatic firearm to which the application pertains, or revolvers and each such semiautomatic firearm, as applicable, by presenting a certificate or other documentation to the sheriff which shows that he:

(1) Successfully completed a course in firearm safety approved by a sheriff in this State; or

(2) Successfully completed a course in firearm safety offered by a federal, state or local law enforcement agency, community college, university or national organization that certifies instructors in firearm safety. Such a course must include instruction in the use of revolvers, each semiautomatic firearm to which the application pertains, or revolvers and each such semiautomatic firearm and in the laws of this State relating to the use of a firearm. A sheriff may not approve a course in firearm safety pursuant to subparagraph (1) unless he determines that the course meets any standards that are established by the Nevada Sheriffs' and Chiefs' Association or, if the Nevada Sheriffs' and Chiefs' Association ceases to exist, its legal successor.

3. The sheriff shall deny an application or revoke a permit if he determines that the applicant or permittee:

(a) Has an outstanding warrant for his arrest.

(b) Has been judicially declared incompetent or insane.

(c) Has been voluntarily or involuntarily admitted to a mental health facility during the immediately preceding 5 years.

(d) Has habitually used intoxicating liquor or a controlled substance to the extent that his normal faculties are impaired. For the purposes of this paragraph, it is presumed that a person has so used intoxicating liquor or a controlled substance if, during the immediately preceding 5 years, he has been:

(1) Convicted of violating the provisions of NRS 484.379; or

(2) Committed for treatment pursuant to NRS 458.290 to 458.350, inclusive.

(e) Has been convicted of a crime involving the use or threatened use of force or violence punishable as a misdemeanor under the laws of this or any other state, or a territory or possession of the United States at any time during the immediately preceding 3 years.

(f) Has been convicted of a felony in this State or under the laws of any state, territory or possession of the United States.

(g) Has been convicted of a crime involving domestic violence or stalking, or is currently subject to a restraining order, injunction or other order for protection against domestic violence.

(h) Is currently on parole or probation from a conviction obtained in this State or in any other state or territory or possession of the United States.

(i) Has, within the immediately preceding 5 years, been subject to any requirements imposed by a court of this State or of any other state or territory or possession of the United States, as a condition to the court's:

(1) Withholding of the entry of judgment for his conviction of a felony; or

(2) Suspension of his sentence for the conviction of a felony.

(j) Has made a false statement on any application for a permit or for the renewal of a permit.

4. The sheriff may deny an application or revoke a permit if he receives a sworn affidavit stating articulable facts based upon personal knowledge from any natural person who is 18 years of age or older that the applicant or permittee has or may have committed an offense or engaged in any other activity specified in subsection 3 which would preclude the issuance of a permit to the applicant or require the revocation of a permit pursuant to this section.

5. If the sheriff receives notification submitted by a court or law enforcement agency of this or any other state, the United States or a territory or possession of the United States that a permittee or an applicant for a permit has been charged with a crime involving the use or threatened use of force or violence, the conviction for which would require the revocation of a permit or preclude the issuance of a permit to the applicant pursuant to this section, the sheriff shall suspend the person's permit or the processing of his application until the final disposition of the charges against him. If a permittee is acquitted of the charges against him, or if the charges are dropped, the sheriff shall restore his permit without imposing a fee.

6. An application submitted pursuant to this section must be completed and signed under oath by the applicant. The applicant's signature must be witnessed by an employee of the sheriff or notarized by a notary public. The application must include:

(a) The name, address, place and date of birth, social security number, occupation and employer of the applicant and any other names used by the applicant;

(b) A complete set of the applicant's fingerprints taken by the sheriff or his agent;

(c) A front-view colored photograph of the applicant taken by the sheriff or his agent;

(d) If the applicant is a resident of this State, the driver's license number or identification card number of the applicant issued by the Department of Motor Vehicles;

(e) If the applicant is not a resident of this State, the driver's license number or identification card number of the applicant issued by another state or jurisdiction;

(f) The make, model and caliber of each semiautomatic firearm to which the application pertains, if any;

(g) Whether the application pertains to revolvers;

(h) A nonrefundable fee in the amount necessary to obtain the report required pursuant to subsection 1 of NRS 202.366; and

(i) A nonrefundable fee set by the sheriff not to exceed $60.

202.366. Investigation of applicant for permit; issuance or denial of permit; expiration of permit.

1. Upon receipt by a sheriff of an application for a permit, the sheriff shall conduct an investigation of the applicant to determine if he is eligible for a permit. In conducting the investigation, the sheriff shall forward a complete set of the applicant's fingerprints to the Central Repository for Nevada Records of Criminal History for submission to the Federal Bureau of Investigation for its report concerning the criminal history of the applicant. The sheriff shall issue a permit to the applicant unless he is not qualified to possess a handgun pursuant to state or federal law or is not otherwise qualified to obtain a permit pursuant to NRS 202.3653 to 202.369, inclusive, and sections 2 and 3 of this act, or the regulations adopted pursuant thereto.

2. To assist the sheriff in conducting his investigation, any local law enforcement agency, including the sheriff of any county, may voluntarily submit to the sheriff a report or other information concerning the criminal history of an applicant.

3. Within 120 days after a complete application for a permit is submitted, the sheriff to whom the application is submitted shall grant or deny the application. If the application is denied, the sheriff shall send the applicant written notification setting forth the reasons for the denial. If the application is granted, the sheriff shall provide the applicant with a permit containing a colored photograph of the applicant and containing such other information as may be prescribed by the Department. The permit must be in substantially the following form:

NEVADA CONCEALED FIREARM PERMIT
County . Permit Number ..
Expires .... Date of Birth ....
Height .... Weight ....
Name .... Address ....
City .... Zip ....
Photograph
Signature
Issued by ....
Date of Issue ....
Make, model and caliber of each authorized semiautomatic firearm, If any ....
revolvers authorized ..yes .. no

4. Unless suspended or revoked by the sheriff who issued the permit, a permit expires 5 years after the date on which it is issued.

202.3667 Permittee to carry permit and proper identification when in possession of concealed firearm; penalty.

1. Each permittee shall carry the permit, or a duplicate issued pursuant to the provisions of NRS 202.367, together with proper identification whenever the permittee is in actual possession of a concealed firearm. Both the permit and proper identification must be presented if requested by a peace officer.

2. A permittee who violates the provisions of this section is subject to a civil penalty of $25 for each violation.

202.367 Duplicate permit; notification to sheriff of recovered permit; penalty.

1. A permittee shall notify the sheriff who issued his permit in writing within 30 days if:

(a) His permanent address changes; or

(b) His permit is lost, stolen or destroyed. ...

4. A permittee who fails to notify a sheriff pursuant to the provisions of this section is subject to a civil penalty of $25.

202.369. Regulations. The department may adopt such regulations as are necessary to carry out the provisions of NRS 202.3653 to 202.369, inclusive, and section 1 of this act.

Tear Gas Bombs and Weapons

202.370. Definitions. As used in NRS 202.370 to 202.440, inclusive:

1. "Shell," "cartridge" or "bomb" includes all shells, cartridges or bombs capable of being discharged or exploded, when such discharge or explosions will cause or permit the release or emission of tear gas.

2. "Tear gas" includes all liquid, gaseous or solid substances intended to produce temporary physical discomfort or permanent injury through being vaporized or otherwise dispersed in the air. The term does not include a liquid, gaseous or solid substance whose active ingredient is composed of natural substances or products derived from natural substances which cause no permanent injury through being vaporized or otherwise dispersed in the air.

3. "Weapon designed for the use of such shell, cartridge or bomb" includes all revolvers, pistols, fountain pen guns, billies, riot guns or other form of device, portable or fixed, intended for the projection or release of tear gas except those regularly manufactured and sold for use with firearm ammunition.

202.375. Applicability of NRS 202.370 to 202.440, inclusive, to small weapons containing "CS" tear gas and to certain law enforcement, correctional and military personnel.

1. The provisions of NRS 202.370 to 202.440, inclusive, do not apply to the sale or purchase by any adult, or the possession or use by any person, including a minor but not including a convicted person as defined in NRS 179C.010, of any form of:

(a) Cartridge which contains not more than 2 fluid ounces in volume of "CS" tear gas that may be propelled by air or another gas, but not an explosive, in the form of an aerosol spray; or

(b) Weapon designed for the use of such a cartridge which does not exceed that size, and which is designed and intended for use as an instrument of self-defense.

2. A seller, before delivering to a purchaser a cartridge or weapon which may be sold pursuant to subsection 1, must record and maintain for not less than 2 years the name and address of the purchaser and the brand name, model number or type, and serial number if there is one, of the weapon or cartridge, or both.

3. The provisions of NRS 202.370 to 202.440, inclusive, do not prohibit police departments or regular salaried peace officers thereof, sheriffs and their regular salaried deputies, the director, deputy director and superintendents of, and guards employed by, the department of corrections, personnel of the Nevada highway patrol or the military or naval forces of this state or of the United States from purchasing, possessing or transporting any shells, cartridges, bombs or weapons for official use in the discharge of their duties.

4. As used in this section, "CS" tear gas means a crystalline powder containing ortho-chlorobenzalmalononitrile.

202.380. Sale or possession of tear gas bombs or weapons which are not permitted under NRS 202.370 to 202.440, inclusive; penalties.

1. A person, other than a convicted person, who within this state knowingly sells or offers for sale, possesses or transports any form of shell, cartridge or bomb containing or capable of emitting tear gas, or any weapon designed for the use of such shell, cartridge or bomb, except as permitted under the provisions of NRS 202.370 to 202.440, inclusive, is guilty of a gross misdemeanor.

2. A convicted person who owns or has in his possession or under his custody or control any form of shell, cartridge or bomb containing or capable of emitting tear gas, or any weapon designed for the use of such a shell, cartridge or bomb, is guilty of a category B felony and shall be punished by imprisonment in the state prison for a minimum term of not less than 1 year and a maximum term of not more than 6 years, or by a fine of not more than $5,000, or by both fine and imprisonment.

3. As used in this section, the term "convicted person" has the meaning ascribed to it in NRS 179C.010.

202.390. Weapon to bear name of manufacturer and serial number; penalty for removal.

1. Each tear gas weapon sold, transported or possessed under the authority of NRS 202.370 to 202.440, inclusive, shall bear the name of the manufacturer and a serial number applied by him.

2. No person shall change, alter, remove or obliterate the name of the manufacturer, the serial number or any other mark of identification on any tear gas weapon. Possession of any such weapon upon which the same shall have been changed, altered, removed or obliterated, shall be presumptive evidence that such possessor has changed, altered, removed or obliterated the same.

3. Any person who violates any of the provisions of this section is guilty of a gross misdemeanor.

202.400. Permit for possession, transportation and use in protective system to be issued by sheriff.

1. The sheriff of any county may issue a permit for the possession and transportation of such shells, cartridges, bombs or weapons to any applicant who submits proof that good cause exists for issuance of the permit.

2. The permit may also allow the applicant to install, maintain and operate a protective system involving the use of such shells, cartridges, bombs or weapons in any place which is accurately and completely described in the application for the permit.

202.410. Applications for permits: Contents.

1. All applications for such permits shall:

(a) Be filed in writing;

(b) Be signed by the applicant if an individual, or by a member or officer qualified to sign if the applicant is a firm or corporation; and

(c) State the name, business in which engaged, business address, a full description of the place or vehicle in which such shells, cartridges, bombs or weapons are to be transported, kept, installed or maintained.

2. If such shells, cartridges, bombs or weapons are to be used in connection with or to constitute a protective system, the application shall also contain the name of the person who is to install such protective system.

202.420. Inspection of permits.

Every person, firm or corporation to whom a permit is issued shall either carry the same upon his person or keep the same in the place described in the permit. The permit shall be open to inspection by any peace officer or other person designated by the authority issuing the permit.

202.430. Revocation of permits.

Permits issued in accordance with NRS 202.370 to 202.440, inclusive, may be revoked by the issuing authority at any time when it shall appear that the need for the possession or transportation of such shells, cartridges, bombs, weapons, or protective system involving the use of the same, has ceased, or that the holder of the permit has engaged in an unlawful business or occupation or has wrongfully made use of such shells, cartridges, bombs or weapons or the permit issued.

202.440. License for retail sale of bombs or weapons; conditions. The sheriff of any county may also grant licenses in a form to be prescribed by him, effective for not more than 1 year from the date of issuance, to permit the sale at retail, at the place specified in the license, of such shells, cartridges, bombs or weapons, and to permit the installation and maintenance of protective systems involving the use of such shells, cartridges, bombs or weapons, subject to the following conditions, upon breach of any of which the license shall be subject to forfeiture:

1. Such business shall be carried on only in the building designated in the license.

2. Such license or certified copy thereof must be displayed on the premises in a place where it may easily be read.

3. No such shell, cartridge, bomb or weapon shall be delivered to any person not authorized to possess or transport the same under the provisions of NRS 202.370 to 202.440, inclusive. No protective system involving the use of such shells, cartridges, bombs or weapons shall be installed, nor shall supplies be sold for the maintenance of such system, unless the licensee has personal knowledge of the existence of a valid permit for the operation and maintenance of such system.

4. A complete record must be kept of sales made under the authority of the license, showing the name and address of the purchaser, the quantity and description of the articles purchased, together with the serial number, if any, the number and date of issue of the purchaser's permit, and the signature of the purchaser or purchasing agent. No sale shall be made unless the permit authorizing possession and transportation of shells, cartridges, bombs or weapons is displayed to the seller and the information herein required is copied therefrom. This record shall be open to the inspection of any peace officer or other person designated by the sheriff.

Title 42. Protection from Fire; Explosives

Chapter 476. Explosives and Inflammable Materials

476.010. Sales and records of dealers in explosives; penalty.

1. Except in the due course of trade or as otherwise provided in subsection 9, it is unlawful for any dealer in dynamite, nitroglycerine, gunpowder or any other explosive to distribute to any person, in any unusual manner, an excessive amount of such commodities

2. It is unlawful for any person to manufacture, import, purchase or distribute any explosive without creating and maintaining a written record that includes the information required pursuant to this section.

3. If a person involved in a transaction is not a business or governmental entity or an agent of a business or governmental entity, the written record required pursuant to subsection 2 must include all the following information:

(a) the name of the person.

(b) the signature of the person.

(c) the driver's license number of the person.

(d) the residential address of the person

(e) the date of the transaction.

(f) a statement of intended use.

4. If a person involved in a transaction is a business or governmental entity or an agent of a business or governmental entity, the written record required pursuant to subsection 2 must include all the following information with respect to the business or governmental entity and the agent of the business or governmental entity, if appropriate:

(a) the name of the business or governmental entity.

(b) the taxpayer identification number of the business or governmental entity.

(c) the principal and local addresses of the business or governmental entity.

(d) the name and any other appropriate personal identifying information that is sufficient to identify the agent authorized to act for the business or governmental entity.

(e) the date of the transaction.

(f) a statement of intended use.

5. It is unlawful for any person to knowingly and intentionally:

(a) make any false or misleading entry in a written record required pursuant to subsection 2; or

(b) fail to make an entry in a written record required pursuant to subsection 2.

6. Any person who keeps any explosive for any purpose shall do so in conformity with the regulations governing the storage of explosives promulgated by the attorney general of the United States pursuant to 18 U.S.C. Section 842 and set forth in 27 C.R.R. sections 555.201 et seq.

7. Any person who stores any explosive shall, within 24 hours after beginning to store the explosive, notify the local law enforcement agency and local fire department in whose jurisdiction the explosive is stored of:

(a) the type of explosive that is being stored; and

(b) the location of the site where the explosive is stored.

8. Any person who violates the provisions of this section is guilty of a gross misdemeanor.

9. The provisions of this section do not apply with respect to a person who is acting in his official capacity as an owner, officer or employee of a company, corporation or partnership engaged in the business of mining.

10. As used in this section:

(a) "distribute" means to sell, issue, give, transfer or otherwise dispose of an explosive.

(b) "person" means any of the following:

(1) a natural person.

(2) any form of business or social organization and any other nongovernmental legal entity, including, without limitation, a corporation, partnership, association, trust or unincorporated organization.

(3) a government, a political subdivision of a government or an agency or instrumentality of a government or a political subdivision of a government.

647.018. "Secondhand dealer" defined.

1. "Secondhand dealer" means any person engaged in whole or in part in the business of buying and selling metal junk, melted metals or secondhand personal property, other than used books, including, without limitation, antiques, coins and collectibles.

2. The term does not include a person who engages in the business of buying or selling secondhand firearms or any antique parts, accessories or other equipment relating to those firearms if:

(a) The person engages in that business at a show that:

(1) Is held at:

(I) A convention facility which is owned or operated by and located on the premises of a resort hotel; or

(II) A recreational facility which is owned or operated by a county fair and recreation board; and

(2) Is conducted for not more than 7 days during any 6-month period; and

(b) The person has been issued a license as a manufacturer, importer, dealer or collector pursuant to the provisions of 18 U.S.C. § 923.

[Current through the 74th (2007) Session and 24th Special (2008) Session]

Clark County Code

Title 6. Business Licenses

Chapter 6.12. Fees and Related Matters

6.12.557. Gunsmith shop. Any person who repairs firearms or who fits special barrels, stocks, or trigger mechanisms to firearms shall pay a semiannual license fee determined by the amount of semiannual gross revenue as contained in the schedule of Section 6.12.995 of this chapter. This includes the sale of parts and accessories necessary for conducting this business. This license does not permit the business to be conducted as a home occupation or from a warehouse location.

6.12.930. Sporting goods store. Any person who sells sports related wares and merchandise at retail as a regular course of business, maintaining an inventory commensurate with sales, established in a location with not less than one thousand square feet dedicated to the display of merchandise, excluding the warehouse and office areas, shall pay a semiannual license fee determined by the amount of semiannual gross revenue as contained in the schedule in Section 6.12.995 of this chapter. This definition does not include a person who makes occasional sales, exchanges, or purchases of sporting goods for the enhancement of a personal collection or for a hobby. Any business operating with a home occupation permit or as a mail order business is prohibited from taking orders for or selling any weapon as defined in Title 29 [sic] of this code.

Title 12. Public Peace, Safety and Morals

Chapter 12.04. Firearms and Air Guns

12.04.010. Definitions. In this chapter, unless the context clearly requires otherwise, the following definitions shall apply and be effective:

(1) Dealer means any person engaged in the business of buying or selling pistols at wholesale or retail, or of accepting pledges of pistols as security for loans.

(2) "Firearm capable of being concealed" means any gun, pistol, revolver or other firearm, the barrel of which does not exceed twelve inches in length, from which a dangerous projectile may be propelled by explosives, springs, gas or air as a means of propulsion.

(3) "Firearm" means any weapon with a caliber of .177 inches or greater, from which a projectile may be propelled by means of gunpowder.

(4) "Manager" means the natural person authorized by a dealer to approve and finalize all pistol sales, purchases or pledges and the person whose signature is required on all documents in connection therewith.

(5) "Pistol" means a firearm, the barrel of which does not exceed twelve inches, intended to be aimed and fired with one hand.

12.04.020. Dealing in pistols without permit unlawful.

(A) It is unlawful within the unincorporated area of Clark County, for any person, except a dealer having a pistol permit issued under the provisions of this chapter, to engage in the business of buying or selling pistols at retail or wholesale, or of accepting pledges of pistols as security for loans, and no such permit shall authorize the leasing of any pistol.

(B) Each natural person authorized or employed as a manager to buy, sell or accept pledges of pistols as security for loans must apply for and receive such pistol permit.

12.04.030. Business to which pistol permits may be issued. Applications for pistol permits to buy, sell or accept pistols as security for loans under this chapter, shall be accepted only from hardware stores, gunsmith shops, sporting goods stores, antique gun dealers and pawn shops, licensed by Clark County.

12.04.040. Application and investigation. Any person desiring a pistol permit to engage in the business of buying or selling pistols, or of accepting pledges of pistols as security for loans, in the unincorporated area of Clark County, or managing such business under the provisions of this chapter, shall first make written application to the board of county commissioners, on forms of application prescribed therefor by the office of the sheriff, setting out his name, address and other information required, and the business (other than dealing in pistols) intended to be carried on, and expressly requesting the issuance of a pistol permit to deal in pistols at such location. Such application will then be referred to the sheriff for investigation and recommendation, and then returned to the board of county commissioners, which may (after considering same in any manner that may appear to them proper), approve or disapprove such application. If disapproved, no pistol permit shall be issued. If approved, the office of the sheriff shall issue a pistol permit to such applicant to deal in pistols at the place named or described, and under the present business license of such applicant or applicant's employer.

12.04.050. Persons prohibited from purchasing or owning pistols. It is unlawful in the unincorporated area of Clark County, for any person to receive from another by loan, gift, purchase, or in any manner, or attempt to obtain in any manner or have in his possession or control, a pistol as defined by this chapter who:

(a) Has been convicted of a felony in the state of Nevada, or in any state or territory of the United States or any political subdivision thereof;

(b) Is a fugitive from justice;

(c) Is an unlawful user or seller of narcotics;

(d) Is a habitual drunkard;

(e) Is a member of an organization advocating criminal syndicalism as defined in NRS 203.160 through 203.190;

(f) Has been adjudged insane in the state or any other state and has not subsequently been adjudged sane;

(g) Is seventeen years of age or under.

12.04.060. Prohibited sales and transfer of a pistol. It shall be unlawful for any person or a dealer in pistols to sell, lease, lend, or otherwise transfer a pistol to any person whom he knows or has reasonable cause to believe:

(a) Is under the influence of intoxicating beverages;

(b) Is mentally ill or disturbed;

(c) Is seventeen years of age or under;

(d) Is a person prohibited from purchasing a pistol as provided in Section 12.04.050.

12.04.070. Certain persons prohibited from purchasing or owning pistols. It shall be unlawful for any person who is included in any one or more of the categories set forth in Section 12.04.050 of this chapter, to receive from another by loan, gift, purchase or in any other manner, or in any manner whatsoever, to attempt to obtain, or have in his or her possession or control, any pistol; provided, however, that a person under the age of eighteen years may have a pistol in his or her possession, or under his or her control, while accompanied by or under the immediate charge of his or her parent, guardian, or other responsible adult, and while engaged in hunting or target practice or other lawful purpose.

12.04.080. Time between sale and delivery of pistol. When any sale of a pistol is made by a dealer under this chapter, seventy-two hours must elapse between the time of sale and the time of delivery to the purchaser. When delivered, all pistols must be securely wrapped and be unloaded, and must be accompanied by a receipt signed by the dealer, setting forth the name, address, and description of the purchaser or transferee, a complete description of the pistol (including the manufacturer, model and manufacturer's serial number thereof), the date and time of sale, and the date and time of delivery, of such pistol, and advice to the purchaser or transferee if a resident of the county that the pistol must be registered with the sheriff within seventy-two hours.

12.04.090. Exceptions to Section 12.04.080. The aforesaid seventy-two hour waiting period shall not apply to the sale of a pistol to any person who, at the time of such sale, produces bona fide documentary evidence that he is a member of a federal law enforcement agency; or a peace officer of the state of Nevada or any political subdivision thereof, regularly employed for pay by the state or such subdivision; or that he currently owns a pistol which is duly registered in his name with any law enforcement agency in Clark County; or that the pistol is an unfinished kit which is designed to fire black powder of the cap and ball or flintlock variety. And said waiting period shall not apply to any person who requires the use of a pistol in connection with his employment, and who receives written permission from the sheriff with express waiver of the waiting period; however, all of the other provisions herein shall apply to any such sales.

12.04.100. Sheriff to be notified. A dealer making a sale of a pistol shall, within twenty-four hours thereof, furnish the sheriff with a duplicate copy of the receipt referred to in Section 12.04.080 of this chapter, signed by the dealer and by the purchaser or transferee of such pistol.

12.04.110. Registration of pistols within twenty-four hours. Any person receiving title to a pistol, whether by purchase, gift, or any other transfer, and whether from a dealer or from any other person, shall, within twenty-four hours of such receipt, personally appear at the county sheriff's office, together with the pistol, for the purpose of registering the same with the sheriff. It shall be the duty of the sheriff to register the pistol, and he may, and is hereby authorized to cooperate in any manner he sees fit with other law enforcement agencies, and with licensed dealers, relative to registration of pistols, so that efficient registration shall be secured at minimum cost and duplication.

12.04.120. Sale or possession of certain pistols prohibited. It shall be unlawful to purchase, sell, or in any manner to transfer, or have in possession or control, any pistol on which the name of the manufacturer, or the model, or the manufacturer's serial number has been removed, obliterated, concealed, or altered.

12.04.130. Window displays of pistols and ammunition prohibited. No pistols or ammunition shall be openly displayed in any store or business window, except only during such hours when said store is open for business and under the supervision of the licensee or adult authorized personnel.

12.04.140. Sale of foreign made pistols with ammunition restricted. No pistol of foreign make shall be sold with ammunition for the same unless the ammunition is expressly designed for use with such pistol.

12.04.150. Sale of homemade pistols prohibited. No homemade pistols shall be bought or sold except as provided by law under the federal firearms act.

12.04.160. Condition of pistols sold. All secondhand and used pistols, except antique pieces, sold or purchased, shall be in a safe and operable condition.

12.04.170. Unlawful to permit persons under eighteen years to have possession or control of pistol. It shall be unlawful within the unincorporated area of Clark County, for any person to aid, or knowingly permit, another under eighteen years of age to handle, or have in his or her possession, or under his or her control, any pistol, except while accompanied, by or under the immediate charge of, his or her parent, guardian, or other responsible adult and while engaged in hunting, target practice or other lawful purpose.

12.04.260. Unlawful use of electronic stun device. It is unlawful, within the unincorporated area of Clark County, for any person, other than a law enforcement officer, to use an electronic stun device (device which is designed to deliver an electric shock to the body or person of another) for any purpose other than self defense. It is further unlawful for any person who:

(a) Has been convicted of a felony in the United States or the state of Nevada, or in any state or territory of the United States or any public subdivision thereof; or

(b) Is a fugitive from justice; or

(c) Is an unlawful user or seller of narcotics; or

(d) Any minors;

to possess such electronic stun device.

12.04.270. Investigation and permit fee. All applicants for a pistol permit shall pay to the Las Vegas metropolitan police department a fee which reimburses the police department the actual cost of issuing the permit and investigation as set out in Section 6.04.015.

[Clark County Code current through Ord. 3657 passed June 3, 2008]

Las Vegas Municipal Code

Title 10. Public Peace, Morals and Welfare

Division IX. Weapons

Chapter 10.66. Pistols

10.66.010. Definitions.
As used in this Chapter:
"Dealer" means any person engaged in the business of buying or selling pistols at wholesale or retail, or of accepting pledges of pistols as security for loans.
"Firearm" means any device designed to be used as a weapon from which a projectile may be expelled through the barrel by the force of any explosion or other form of combustion.
"Firearm capable of being concealed" includes all firearms having a barrel less than twelve inches in length.
"Pistol" means a firearm capable of being concealed that is intended to be aimed and fired with one hand.

10.66.020. Dealer permit - Required - Leasing prohibited. It shall be unlawful for any person, except a dealer having a pistol permit issued under the provisions of this Chapter, to engage in the business of buying or selling pistols at retail or wholesale, or of accepting pledges of pistols as security for loans, and no such permit shall authorize the leasing of any pistol.

10.66.030. Dealer permit - Eligible firms. Applications for pistol permits shall be accepted only from duly licensed hardware stores, gunsmith shops, sporting goods stores, stores which outfit and equip police officers and security guards, Class IV secondhand dealers and pawnshops where pistols may be sold or purchased in compliance with this Chapter.

10.66.040. Dealer permit - Application - Investigation - Issuance.

(A) Any person desiring a pistol permit to engage in the business of buying or selling pistols, or of accepting pledges of pistols as security for loans, under the provisions of this Chapter, shall first apply to the City Council setting out his name, address, and other information as may be required and the business (other than dealing in pistols) intended to be carried on and stating a request for a pistol permit to deal in pistols at such place.

(B) The application will then be referred to the Sheriff of the Metropolitan Police Department, or his designee, for his investigation and his recommendation, then back to the Council which may, after considering same in any manner that may appear to it proper, approve or disapprove such application. If disapproved, then no pistol permit shall be issued. If approved, the Sheriff or his designee shall issue a pistol permit to such applicant to deal in pistols at the place named and under the present business license of such applicant.

10.66.050. Disqualifications for sale or transfer. It shall be unlawful for a dealer or any other person to sell, lease, rent or otherwise transfer a pistol to any person whom he knows or has reasonable cause to believe is included within any one or more of the following categories:

(A) Who is under the age of eighteen years;

(B) Who is under the influence of intoxicating beverages;

(C) Who is a fugitive from justice;

(D) Who has been convicted of a felony in the state or any state or territory of the United States or any political subdivision thereof;

(E) Who is an unlawful user or seller of narcotics;

(F) Who is a member of an association advocating criminal syndicalism as defined in NRS 203.160 through 203.190;

(G) Who has been adjudged insane in this State or elsewhere and has not subsequently been adjudged Sane by a court of competent jurisdiction.

10.66.060. Sale-delivery waiting period - Manner of delivery.

(A) When any sale of a pistol is made by a dealer under this Chapter, seventy-two hours must elapse between the time of sale and time of delivery to the purchaser, and when delivered, all pistols must be securely wrapped, must be unloaded and must be accompanied by a receipt, signed by dealer, setting forth:

(1) The name, address and description of the purchaser or transferee;

(2) A complete description of the pistol, including the manufacturer, model and manufacturer's serial number thereof;

(3) The date and time of sale and the date and time of delivery of such pistol; and

(4) A statement notifying the purchaser or transferee that the pistol must be registered with the Sheriff or his designee within seventy-two hours if the purchaser or transferee is a resident of the City.

(B) The seventy-two-hour waiting period before delivery shall not apply to the sale of a pistol to any person who, at the time of such sale, produces bona fide documentary evidence that he is a member of a Federal law enforcement agency, that he is a peace officer of the State or any political subdivision thereof who is regularly employed for pay by the State or such subdivision, or that he currently owns a pistol which is duly registered in his name with any law enforcement agency in the County, nor shall said waiting period apply to any person who requires the use of a pistol in his employment and receives written permission from the Sheriff or his designee to waive said waiting period; provided, however, that all of the other provisions shall apply to any of such sales.

10.66.070. Notice to Police of delivery A dealer making a sale of a pistol shall, within twenty-four hours after the delivery thereof, deliver to the Sheriff of the Metropolitan Police Department or his designee a duplicate copy of the receipt referred to in Section 10.66.060, signed by the dealer and by the purchaser or transferee of such pistol.

10.66.080. Altered or removed identification marks. It shall be unlawful to purchase, sell, or to transfer in any manner or to have in possession or control any pistol on which the name of the manufacturer, model or manufacturer's serial number has been removed or altered.

10.66.090. Window display. No pistols or ammunition shall be displayed in any store or business window.

10.66.100. Ammunition compatibility. No pistol of foreign make shall be sold with ammunition for the same unless said ammunition is expressly made for the pistol.

10.66.110. Home-made pistols. No home-made pistols shall be bought or sold except as provided by law under the Federal Firearms Act.

10.66.120. Safe and first-class condition. All secondhand and used pistols, except antique pieces, sold or purchased shall be in a safe and first-class condition.

10.66.130. Unlawful possession - Persons under age eighteen. It shall be unlawful for any person who is included within any one or more of the categories set forth in Section 10.66.050 to receive from another by loan, gift, purchase or in any manner, or to attempt to obtain in any manner, or have in his or her possession or control, any pistol; provided, however, that a person under the age of eighteen years may have a pistol in his or her possession or under his or her control while accompanied by or under the immediate charge of his or her parent, guardian, or responsible adult and while engaged in hunting or target practice or other lawful purpose.

10.66.140. Registration of ownership required.

(A) Any resident of the City receiving title to a pistol, whether by purchase, gift or other transfer, and whether from a dealer or any other person, shall, within seventy-two hours of such receipt, personally appear, together with such pistol, and register the same with the Sheriff of the Metropolitan Police Department or his designee. It shall be unlawful to possess a pistol which is not so registered.

(B) It shall be the duty of the Sheriff or his designee to register said pistol and he is authorized to cooperate with other law enforcement agencies and licensed dealers in effecting registration of pistols to the end that efficient registration will be secured at a minimum of cost and duplication.

10.66.150. Permitting possession by those under eighteen. It shall be unlawful for any person to aid or knowingly permit another under the age of eighteen years to handle or have in his or her possession or under his or her control any pistol except while accompanied by or under the immediate charge of his or her parent, guardian or responsible adult and while engaged in hunting, target practice or other lawful purpose.

[Las Vegas Municipal Code current through Ord. 5994 passed July 16, 2008]

Administrative Code, City of Reno

Title 5. Privileged Licenses, Permits and Franchises
Chapter 5.15. Street Vendors

5.15.100. Restrictions.

(a) Vendors shall not: ...

(7) Be allowed to sell alcoholic beverages, used goods, any controlled substance or paraphernalia, dangerous weapons, or pyrotechnics; ...

[Reno Municipal Code current through Ord. No. 6060, adopted Oct. 8, 2008]

NEW HAMPSHIRE
N.H. REV. STAT.

Title XII. Public Safety and Welfare

Chapter 159. Pistols and Revolvers

159:1. Definition. Pistol or revolver, as used herein, means any firearm with barrel less than 16 inches in length. It does not include antique pistols, gun canes, or revolvers. An antique pistol, gun cane, or revolver, for the purposes of this chapter means any pistol, gun cane, or revolver utilizing an early type of ignition, includeing, but not limited to, flintlocks, wheel locks, matchlocks, percussions and pin-fire, but no pistol, gun cane, or revolver which utilizes readily available center fire or rim-fire cartridges which are in common, current use shall be deemed to be an antique pistol, gun cane, or revolver. Nothing in this section shall prevent antique pistols, gun canes, or revolvers from being owned or transferred by museums, antique or arms collectors, or licensed gun dealers at auctions, gun shows, or private premises provided such ownership or transfer does not conflict with federal statutes.

159:3. Convicted Felons.

I. A person is guilty of a class B felony if he:

(a) Owns or has in his possession or under his control, a pistol, revolver, or other firearm, or slungshot, metallic knuckles, billies, stiletto, switchblade knife, sword cane, pistol cane, blackjack, dagger, dirk-knife, or other deadly weapon as defined in RSA 625:11, V; and

(b) Has been convicted in either a state or federal court in this or any other state, the District of Columbia, the Commonwealth of Puerto Rico, or any territory or possession of the United States of:

(1) A felony against the person or property of another; or

(2) A felony under RSA 318-B; or

(3) A felony violation of the laws of any other state, the District of Columbia, the United States, the Commonwealth of Puerto Rico or any territory or possession of the United States relating to controlled drugs as defined in RSA 318-B.

I-a. A person is guilty of a class B felony if such person completes and signs an application for purchase of a firearm and the person is a convicted felon under the provisions of paragraph I.

II. The state shall confiscate to the use of the state the weapon or weapons of persons convicted under this section.

III. It is an affirmative defense to a charge under this section that a felony of which a defendant has been convicted in another jurisdiction would not have constituted a felony in the state of New Hampshire at the time such felony was committed.

159:3-a. Armed Career Criminals.

I. No person who has been convicted of any combination of 3 or more felonies in this state or any other state under homicide, assault, sexual assault, arson, burglary, robbery, extortion, child pornography, or controlled drug laws, shall own or have in his possession or under his control, a pistol, revolver, rifle, shotgun, or any other firearm.

II. Any person who violates paragraph I shall be guilty of a felony and, notwithstanding RSA 651:2, II, shall be sentenced to a minimum mandatory term of 10 years imprisonment and a maximum term of imprisonment of not more than 40 years and shall be fined not more than $25,000.

III. Notwithstanding any other provision of law, neither the whole, nor any part of the minimum mandatory sentence provided under paragraph II shall be served concurrently with any other term, nor shall the whole or any part of such additional term of imprisonment be suspended or deferred. No action brought to enforce sentencing under this section shall be continued for sentencing, nor shall the provisions of RSA 651:20 relative to suspensions or RSA 651-A relative to parole apply to any sentence of imprisonment imposed.

159:4. Carrying Without License. No person shall carry a loaded pistol or revolver in any vehicle or concealed upon his person, except in his dwelling, house or place of business, without a valid license therefor as hereinafter provided. A loaded pistol or revolver shall include any pistol or revolver with a magazine, cylinder, chamber or clip in which there are loaded cartridges. Whoever violates the provisions of this section shall, for the first such offense, be guilty of a misdemeanor. For the second and for each subsequent violation of the provisions of this section, such person shall be guilty of a class B felony, provided such second or subsequent violation has occurred within 7 years of the previous conviction.

159:5. Exceptions. The provisions of RSA 159:3 and 4 shall not apply to marshals, sheriffs, policemen or other duly appointed peace and other law enforcement officers, or bailiffs and court officers responsible for court security; nor to the regular and ordinary transportation of pistols or revolvers as merchandise, nor to members of the armed services of the United States when on duty; nor to the national guard when on duty; nor to organizations by law authorized to purchase or receive such weapons; nor to duly authorized military or civil organizations when parading, or the members thereof when at, or going to or from, their customary places of assembly.

159:5-a Exceptions and Exemptions Not Required to be Negated.

In any complaint, information, or indictment, and in any action or proceeding brought for the enforcement of any provision of this chapter, it shall not be necessary to negate any exception, excuse, proviso, or exemption contained herein, and the burden of proof of any such exception, excuse, proviso or exemption shall be upon the defendant.

159:6. License to Carry.

I. The selectmen of a town or the mayor or chief of police of a city or some full-time police officer designated by them respectively, upon application of any resident of such town or city, or the director of state police, or some person designated by such director, upon application of a nonresident, shall issue a license to such applicant authorizing the applicant to carry a loaded pistol or revolver in this state for not less than 4 years from the date of issue, if it appears that the applicant has good reason to fear injury to the applicant's person or property or has any proper purpose, and that the applicant is a suitable person to be licensed. Hunting, target shooting, or self-defense shall be considered a proper purpose. The license shall be valid for all allowable purposes regardless of the purpose for which it was originally issued. The license shall be in duplicate and shall bear the name, address, description, and signature of the licensee. The original shall be delivered to the licensee and the duplicate shall be pre-served by the people issuing the same for 4 years. When required, license renewal shall take place within the month of the fourth anniversary of the license holder's date of birth following the date of issuance. The license shall be issued within 14 days after application, and, if such application is denied, the reason for such denial shall be stated in writing, the original of which such writing shall be delivered to the applicant, and a copy kept in the office of the person to whom the application was made. The fee for licenses issued to residents of the state shall be $10, which fee shall be for the use of the law enforcement department of the town granting said licenses; the fee for licenses granted to out-of-state residents shall be $20, which fee shall be for the use of the state. The director of state police is hereby authorized and directed to prepare forms for the licenses required under this chapter and forms for the application for such licenses and to supply the same to officials of the cities and towns authorized to issue the licenses. No other forms shall be used by officials of cities and towns. The cost of the forms shall be paid out of the fees received from nonresident licenses.

II. No photograph or fingerprint shall be required or used as a basis to grant, deny, or renew a license to carry for a resident or nonresident, unless requested by the applicant.

159:6-a Confidentiality of Licenses. Notwithstanding the provisions of RSA 91-A:4 or any other provision of law to the contrary, all papers and records, including applications, pertaining to the issuance of licenses pursuant to RSA 159:6 and all licenses issued pursuant to said section are subject to inspection only by law enforcement officials of the state or any political subdivision thereof or of the federal government while in the performance of official duties or upon written consent, for good cause shown, of the superior court in the county where said license was issued.

159:6-b Suspension or Revocation of License.

I. The issuing authority may order a license to carry a loaded pistol or revolver issued to any person pursuant to RSA 159:6 to be suspended or revoked for just cause, provided written notice of the suspension or revocation and the reason therefore is given to the licensee. A licensee whose license has been suspended or revoked shall be permitted a hearing on such suspension or revocation if a hearing is requested by the licensee to the issuing authority within 7 days of the suspension or revocation.

II. When the licensee hereunder ceases to be a resident of the community in which the license was issued he shall notify in writing the issuing authority at his new place of residence that he has a current license. Such license shall remain in effect until it expires pursuant to RSA 159:6.

159:6-c Appeal From Denial, Suspension, or Revocation.

Any person whose application for a license to carry a loaded pistol or revolver has been denied pursuant to RSA 159:6 or whose license to carry a loaded pistol or revolver has been suspended or revoked pursuant to RSA 159:6-b may within 30 days thereafter, petition the district or municipal court in the jurisdiction in which such person resides to determine whether the petitioner is entitled to a license. The court shall conduct a hearing within 14 days after receipt of the petition. During this hearing the burden shall be upon the issuing authority to demonstrate by clear and convincing proof why any denial, suspension, or revocation was justified, failing which the court shall enter an order directing the issuing authority to grant or reinstate the petitioner's license. The court shall issue its decision not later than 14 days after the hearing on whether the petitioner is entitled to a license.

159:6-d. Full Faith and Credit for Licenses From Other States; Reciprocity. Notwithstanding the provisions of RSA 159:6, no nonresident holding a current and valid license to carry a loaded pistol or revolver in the state in which he resides or who is a peace officer in the state in which he resides, shall be required to obtain a license to carry a loaded pistol or revolver within this state if:

I. Such nonresident carries upon his person the license held from the state in which he resides; and

II. The state in which such person is a resident provides a reciprocal privilege for residents of this state.

159:6-e Violation.

Any person aggrieved by a violation of the licensing sections of this chapter by a licensing entity may petition the superior court of the county in which the alleged violation occurred for injunctive relief. The court shall give proceedings under this chapter priority on the court calendar. Such a petitioner may appear with or without counsel. The petition shall be deemed sufficient if it states facts constituting a violation of the licensing sections of this chapter by the licensing entity, and may be filed by the petitioner or the petitioner's counsel with the clerk of court or the justice. The clerk of court or any justice shall order service by copy of the petition on the licensing entity or a person employed by the entity. If the justice finds that time is of the essence, the justice may order notice by any reasonable means, and shall have authority to issue an order ex parte when the justice reasonably deems such an order necessary to insure compliance with the provisions of this chapter.

159:6-f Remedies.

I. If any licensing entity or employee or member of the city council or board of selectmen, in violation of the provisions of this chapter, refuses to comply with this chapter, such entity or person shall be liable for reasonable attorney's fees and costs incurred in a lawsuit under this chapter to enforce the terms of this chapter, provided that the court finds that such lawsuit was necessary in order to obtain compliance with this chapter by the licensing authority. Fees shall not be awarded unless the court finds that the entity or person knew or should have known that the conduct engaged in was a violation of this chapter or when the parties, by agreement, provide that no such fees shall be paid. In any case in which fees are awarded under this chapter, upon a finding that an employee or other official of a licensing entity has acted in bad faith in refusing to comply with this chapter, the court may award such fees personally against such employee or other official.

II. The court may invalidate an action of a licensing entity taken in violation of the provisions of this chapter, if the circumstances justify such invalidation, and may require the licensing entity to issue a license or otherwise comply with the provisions of this chapter.

III. In addition to any other relief awarded pursuant to this chapter, the court may issue an order to enjoin future violations of this chapter.

159:7. Sales to Felons. No person shall sell, deliver, or otherwise transfer a pistol, revolver or any other firearm, to a person who has been convicted, in any jurisdiction, of a felony. Whoever violates the provisions of this section shall be guilty of a class B felony.

159:8. License to Sell. The selectmen of a town and the chief of police of a city may grant licenses, the form of which shall be prescribed by the director of the division of state police, effective for not more than 3 years from date of issue, permitting the licensee to sell at retail pistols and revolvers subject to the following conditions, for breach of any of which the licensee shall be subject to forfeiture:

I. The business shall be carried on only in the building designated in the license or at any organized sporting show or arms collectors' meeting sponsored by a chartered club or organization.

II. The license or a copy thereof, certified by the issuing authority, shall be displayed on the premises where it can easily be read.

III. No pistol, revolver, or other firearm shall be delivered to a purchaser not personally known to the seller or who does not present clear evidence of his identity; nor to a person who has been convicted of a felony.

159:8-a. Sales to Nonresidents; Attorney General. No person holding a license issued under the provisions of RSA 159:8 shall sell a pistol or revolver to a nonresident unless such nonresident has authority under the laws of the state of his residence, to purchase a pistol or revolver in the state of his residence, or unless the director of the division of state police, for good cause shown, has issued to such non-resident a permit for the purchase of a pistol or revolver. The attorney general shall, at least once annually, file with the secretary of state a summary of the laws of each state of the United States relative to the purchase of pistols and revolvers in such states; and a licensee may rely upon such summary in determining if a non-resident offering to purchase a pistol or revolver has authority to make such purchase under the laws of the state of his residence.

159:8-b. Penalties. If a licensee shall in any court be found guilty of a violation of any of the provisions of RSA 159:8-a, such court shall, for each such violation, order the suspension of his license for a period of 3 months, and may, in addition, impose a fine not in excess of $100.

159:10. Sale Without License. Any person who, without being licensed as herein provided, sells, advertises or exposes for sale, or has in his possession with intent to sell, pistols or revolvers shall be guilty of a class B felony if a natural person, or guilty of a felony if any other person.

159:11. False Information. Any person who, in purchasing or otherwise securing delivery of a pistol, revolver or other firearm, gives false information or offers false evidence of his identity, shall be guilty of a misdemeanor for the first offense, and be guilty of a class B felony for any subsequent offense.

159:12. Sale to Minors.

I. Any person who shall sell, barter, hire, lend or give to any minor any pistol or revolver shall be guilty of a misdemeanor.

II. This section shall not apply to:

(a) Fathers, mothers, grandparents, guardians, administrators or executors who give a revolver to their children or wards or to heirs to an estate.

(b) Individuals instructing minors in the safe use of firearms during a supervised firearms training program, provided the minor's parent or legal guardian has granted the minor permission to participate in such program.

(c) Licensed hunters accompanying a minor while lawfully taking wildlife.

(d) Individuals supervising minors using firearms during a lawful shooting event or activity.

159:13. Changing Marks. No person shall change, alter, remove or obliterate the name of the maker, model, manufacturer's number or other mark of identification on any pistol or revolver. Possession of any such firearms upon which the same shall have been changed, altered, removed or obliterated shall be presumptive evidence that such possessor has changed, altered, removed or obliterated the same. Any person who violates the provisions of this section shall be guilty of a misdemeanor.

159:14. Exemption. None of the provisions of this chapter shall prohibit an individual not licensed under the provisions thereof who is not engaged in the business of selling pistols or revolvers from selling a pistol or revolver to a person licensed under this chapter or to a person personally known to him.

State Jurisdiction

159:26. Firearms and Ammunition; Authority of the State.

I. To the extent consistent with federal law, the state of New Hampshire shall have authority and jurisdiction over the sale, purchase, ownership, use, possession, transportation, licensing, permitting, taxation, or other matter pertaining to firearms, firearms components, ammunition, or firearms supplies in the state. Except as otherwise specifically provided by statute, no ordinance or regulation of a political subdivision may regulate the sale, purchase, ownership, use, possession, transportation, licensing, permitting, taxation, or other matter pertaining to firearms, firearms components, ammunition, or firearms supplies in the state. Nothing in this section shall be construed as affecting a political subdivision's right to adopt zoning ordinances for the purpose of regulating firearms businesses in the same manner as other businesses or to take any action allowed under RSA 207:59.

II. Upon the effective date of this section, all municipal ordinances and regulations not authorized under paragraph I relative to the sale, purchase, ownership, use, possession, transportation, licensing, permitting, taxation, or other matter pertaining to firearms, firearm components, ammunition, or firearms supplies shall be null and void.

Chapter 159-D. Criminal Background Checks

159-D:1. Sale of Firearms; Criminal History Record and Protective Order Check. The department of safety may become the point of contact for the federal government for the purposes of the National Instant Criminal Background Check System (NICS).

159-D:2. Confidentiality.

I. If the department of safety conducts criminal background checks under RSA 159-D:1, any records containing information pertaining to a potential buyer or transferee who is not found to be prohibited from receipt or transfer of a firearm by reason of state or federal law, which are created by the department of safety to conduct the criminal background check, shall be confidential and may not be disclosed by the department or any officers or employees to any person or to another agency. The department shall destroy any such records after it communicates the corresponding approval number to the licensee and, in any event, such records shall be destroyed within one day after the day of the receipt of the licensee's request.

II. The department shall retain records containing any information pertaining to a potential buyer or transferee who is prohibited from receipt or transfer of a firearm for 3 years.

III. Notwithstanding the provisions of this section, the department may maintain only a log of dates of requests for criminal background checks and unique approval numbers corresponding to such dates for an indefinite period.

IV. Nothing in this section shall be construed to allow the department to maintain records containing the names of licensees who receive unique approval numbers or to maintain records of firearm transactions, including the names or other identification of licensees and potential buyers or transferees, including persons not otherwise prohibited by law from the receipt or possession of firearms.

159-D:3. Penalties for Attempts to Purchase Firearms Illegally. A person who completes and signs an application for purchase of a firearm and who knows that such purchase is illegal because he or she is subject to a protective order shall be guilty of a class A misdemeanor for a first offense and a class B felony for a second or subsequent offense.

Chapter 173-B. Protection of Persons From Domestic Violence

173-B:1. Definitions. In this chapter: ...

VII. "Deadly weapon" means "deadly weapon" as defined in RSA 625:11, V. ...

XI. "Firearm" means any weapon, including a starter gun, which will or is designed to or may readily be converted to expel a projectile by force of gunpowder. ...

173-B:4. Temporary Relief.

I. Upon a showing of an immediate and present danger of abuse, the court may enter temporary orders to protect the plaintiff with or without actual notice to defendant. The court may issue such temporary orders by telephone or facsimile. Such telephonically issued orders shall be made by a district or superior court judge to a law enforcement officer, shall be valid in any jurisdiction in the state, and shall be effective until the close of the next regular court business day. Such orders shall be returnable to the district court where the plaintiff resides or to which the plaintiff has fled, unless otherwise ordered by the issuing justice. If non-telephonic temporary orders are made ex parte, the party against whom such relief is issued may file a written request with the clerk of the court and request a hearing on such orders. Such hearing shall be held no less than 3 business days and no more than 5 business days after the request is received by the clerk. Such hearings may constitute the final hearing described in RSA 173-B:3, VII. Such temporary relief may direct the defendant to relinquish to a peace officer any and all firearms and ammunition in the control, ownership, or possession of the defendant, or any other person on behalf of the defendant for the duration of the protective order. Other temporary relief may include:

(a) Protective orders: ...

(9) Directing the defendant to relinquish to the peace officer, in addition to the relief specified in RSA 173-B:4, I, any and all deadly weapons specified in the protective order that are in the control, ownership, or possession of the defendant, or any other person on behalf of the defendant, for the duration of the protective order. ...

II. The defendant may be prohibited from purchasing, receiving, or possessing any deadly weapons and any and all firearms and ammunition for the duration of the order. The court may subsequently issue a search warrant authorizing the peace officer to seize any deadly weapons specified in the protective order and any and all firearms and ammunition, if there is probable cause to believe such firearms and ammunition and specified deadly weapons are kept on the premises or curtilage of the defendant and if the court has reason to believe that all such firearms and ammunition and specified deadly weapons have not been relinquished by the defendant.

173-B:5. Relief.

I. A finding of abuse shall mean the defendant represents a credible threat to the safety of the plaintiff. Upon a showing of abuse of the plaintiff by a preponderance of the evidence, the court shall grant such relief as is necessary to bring about a cessation of abuse. Such relief shall direct the defendant to relinquish to the peace officer any and all firearms and ammunition in the control, ownership, or possession of the defendant, or any other person on behalf of the defendant for the duration of the protective order. Other relief may include:

(a) Protective orders: ...

(6) Directing the defendant to relinquish to the peace officer, in addition to the relief specified in RSA 173-B:5, I, any and all deadly weapons specified in the protective order that are in the control, ownership, or possession of the defendant, or any other person on behalf of the defendant. ...

II. The defendant shall be prohibited from purchasing, receiving, or possessing any deadly weapons and any and all firearms and ammunition for the duration of the order. The court may subsequently issue a search warrant authorizing a peace officer to seize any deadly weapons specified in the protective order and any and all firearms and ammunition, if there is probable cause to believe such firearms and ammunition and specified deadly weapons are kept on the premises or curtilage of the defendant. ...

IX-a. If a criminal records check conducted by the department of safety indicates that a potential buyer or transferee is prohibited from receipt or possession of a firearm pursuant to a protective order issued under this chapter, the department of safety shall notify the administrative office of the courts of the denial. The administrative office of the courts shall immediately notify the plaintiff that the defendant has attempted to purchase or obtain a firearm in violation of the protective order.

X. (a) Within 15 days prior to the expiration of the protective orders, the defendant may request, by motion to the court, the return of any and all firearms and ammunition and specified deadly weapons held by the law enforcement agency while the protective order was in effect. Upon receipt of such a motion, the court shall schedule a hearing no later than 15 days after the expiration of the order. The court shall provide written notice to the plaintiff who shall have the right to appear and be heard, and to the law enforcement agency which has control of the firearms, ammunition, and specified deadly weapons. The scope of the hearing shall be limited to:

(1) Establishing whether the defendant is subject to any state or federal law or court order that precludes the defendant from owning or possessing a firearm; and

(2) Under circumstances where the plaintiff has requested an extension of the protective order, whether the plaintiff has established by a preponderance of the evidence that the defendant continues to represent a credible threat to the safety of the plaintiff.

(b) If the court finds that the defendant is not subject to any state or federal law or court order precluding the ownership or possession of firearms, or if the court denies the plaintiff's request to extend the protective order, the court shall issue a written order directing the law enforcement agency to return the requested firearms, ammunition, or deadly weapon to the defendant.

(c) Law enforcement agencies shall not release firearms and ammunition and specified deadly weapons without a court order granting such release. The law enforcement agency may charge the defendant a reasonable fee for the storage of any firearms and ammunition and specified deadly weapons taken pursuant to a protective order. The fee shall not exceed the actual cost incurred by the law enforcement agency for the storage of the firearms and ammunition and specified deadly weapons. The defendant may make alternative arrangements with a federally licensed firearms dealer for the storage of firearms, at the defendant's own expense, upon approval of the court. Such firearms shall be turned over to the appropriate law enforcement agency for transfer to the storage facility. Retrieval of such firearms shall be through the law enforcement agency responsible for their transfer to the storage facility pursuant to a court order as prescribed in this paragraph.

(d) No law enforcement agency shall be held liable for alleged damage or deterioration due to storage or transportation to any firearms and ammunition and specified deadly weapons held by a law enforcement agency, so long as due care is used.

Title LXII. Criminal Code

Chapter 625. Preliminary

625:11. General Definitions. The following definitions apply to this code. ...

V. "Deadly weapon" means any firearm, knife or other substance or thing which, in the manner it is used, intended to be used, or threatened to be used, is known to be capable of producing death or serious bodily injury. ...

Chapter 650-C. Negligent Storage of Firearms

650-C:1. Negligent Storage of Firearms.

I. Nothing in this section shall be construed to reduce or limit any existing right to purchase and own firearms or ammunition, or both, or to provide authority to any state or local agency to infringe upon the privacy of any family, home or business except by lawful warrant.

II. As used in this section, "child," "juvenile" or "youth" shall mean any person under 16 years of age.

III. Any person who stores or leaves on premises under that person's control a loaded firearm, and who knows or reasonably should know that a child is likely to gain access to the firearm without the permission of the child's parent or guardian, is guilty of a violation if a child gains access to a firearm and:

(a) The firearm is used in a reckless or threatening manner;

(b) The firearm is used during the commission of any misdemeanor or felony; or

(c) The firearm is negligently or recklessly discharged.

IV. Any person who violates paragraph III shall be fined not more than $1,000.

V. This section shall not apply whenever any of the following occurs:

(a) The child has completed firearm safety instructions by a certified firearms safety instructor or has successfully completed a certified hunter safety course.

(b) The firearm is kept secured in a locked box, gun safe, or other secure locked space, or in a location which a reasonable person would believe to be secure, or is secured with a trigger lock or similar device that prevents the firearm from discharging.

(c) The firearm is carried on the person or within such a close proximity thereto so that the individual can readily retrieve and use the firearm as if carried on the person.

(d) The child obtains or obtains and discharges the firearm in a lawful act of self-defense or defense of another person.

(e) The person who keeps a loaded firearm on any premises which are under such person's custody or control has no reasonable expectation, based on objective facts and circumstances, that a child is likely to be present on the premises.

(f) The child obtains the firearm as a result of an illegal entry of any premises by any person or an illegal taking of the firearm from the premises of the owner without permission of the owner.

VI. A parent or guardian of a child who is injured or who dies of an accidental shooting shall be prosecuted under this section only in those instances in which the parent or guardian behaved in a grossly negligent manner.

VII. Licensees shall conspicuously post at each purchase counter the following warning in bold type not less than one inch in height: "IT IS IMPORTANT THAT THE OWNER OF A FIREARM SEEK FIREARM SAFETY INSTRUCTIONS FROM A CERTIFIED FIREARMS INSTRUCTOR AND KEEP FIREARMS SECURED FROM UNAUTHORIZED USE." A licensee failing to display this warning to the purchaser of a firearm shall be guilty of a violation.

[Current through Chapter 1 of the 2008 Special Session]

NEW JERSEY

N.J. Rev. Stat.

Title 2C. The New Jersey Code of Criminal Justice

Chapter 1. Preliminary

2C:1-5. Abolition of common law crimes; all offenses defined by statute; application of general provisions of the code; limitation of local government laws

a. Common law crimes are abolished and no conduct constitutes an offense unless the offense is defined by this code or another statute of this State.

b. The provisions of subtitle 1 of the code are applicable to offenses defined by other statutes. The provisions of subtitle 3 are applicable to offenses defined by other statutes but the maximum penalties applicable to such offenses, if specifically provided in the statute defining such offenses, shall be as provided therein, rather than as provided in this code, except that if the non-code offense is a misdemeanor with a maximum penalty of more than 18 months imprisonment, the provisions of section 2C:43-1b shall apply.

c. This section does not affect the power to punish for contempt, either summarily or after indictment, or to employ any sanction authorized by law for the enforcement of an order or a civil judgment or decree.

d. Notwithstanding any other provision of law, the local governmental units of this State may neither enact nor enforce any ordinance or other local law or regulation conflicting with, or preempted by, any provision of this code or with any policy of this State expressed by this code, whether that policy be expressed by inclusion of a provision in the code or by exclusion of that subject from the code.

Chapter 39. Firearms, Other Dangerous Weapons, and Instruments of Crime

2C:39-1. Definitions The following definitions apply to this chapter and to chapter 58:

a. "Antique firearm" means any rifle or shotgun and "antique cannon" means a destructive device defined in paragraph (3) of subsection c. of this section, if the rifle, shotgun or destructive device, as the case may be, is incapable of being fired or discharged, or which does not fire fixed ammunition, regardless of date of manufacture, or was manufactured before 1898 for which cartridge ammunition is not commercially available, and is possessed as a curiosity or ornament or for its historical significance or value.

b. "Deface" means to remove, deface, cover, alter or destroy the name of the maker, model designation, manufacturer's serial number or any other distinguishing identification mark or number on any firearm.

c. "Destructive device" means any device, instrument or object designed to explode or produce uncontrolled combustion, including (1) any explosive or incendiary bomb, mine or grenade; (2) any rocket having a propellant charge of more than four ounces or any missile having an explosive or incendiary charge of more than one-quarter of an ounce; (3) any weapon capable of firing a projectile of a caliber greater than 60 caliber, except a shotgun or shotgun ammunition generally recognized as suitable for sporting purposes; (4) any Molotov cocktail or other device consisting of a breakable container containing flammable liquid and having a wick or similar device capable of being ignited. The term does not include any device manufactured for the purpose of illumination, distress signaling, line-throwing, safety or similar purposes. ...

e. "Explosive" means any chemical compound or mixture that is commonly used or is possessed for the purpose of producing an explosion and which contains any oxidizing and combustible materials or other ingredients in such proportions, quantities or packing that an ignition by fire, by friction, by concussion or by detonation of any part of the compound or mixture may cause such a sudden generation of highly heated gases that the resultant gaseous pressures are capable of producing destructive effects on contiguous objects. The term shall not include small arms ammunition, or explosives in the form prescribed by the official United States Pharmacopoeia.

f. "Firearm" means any handgun, rifle, shotgun, machine gun, automatic or semiautomatic rifle, or any gun, device or instrument in the nature of a weapon from which may be fired or ejected any solid projectable ball, slug, pellet, missile or bullet, or any gas, vapor or other noxious thing, by means of a cartridge or shell or by the action of an explosive or the igniting of flammable or explosive substances. It shall also include, without limitation, any firearm which is in the nature of an air gun, spring gun or pistol or other weapon of a similar nature in which the propelling force is a spring, elastic band, carbon dioxide, compressed or other gas or vapor, air or compressed air, or is ignited by compressed air, and ejecting a bullet or missile smaller than three-eighths of an inch in diameter, with sufficient force to injure a person.

g. "Firearm silencer" means any instrument, attachment, weapon or appliance for causing the firing of any gun, revolver, pistol or other firearm to be silent, or intended to lessen or muffle the noise of the firing of any gun, revolver, pistol or other firearm. ...

i. "Machine gun" means any firearm, mechanism or instrument not requiring that the trigger be pressed for each shot and having a reservoir, belt or other means of storing and carrying ammunition which can be loaded into the firearm, mechanism or instrument and fired therefrom.

j. "Manufacturer" means any person who receives or obtains raw materials or parts and processes them into firearms or finished parts of firearms, except a person who exclusively processes grips, stocks and other nonmetal parts of firearms. The term does not include a person who repairs existing firearms or receives new and used raw materials or parts solely for the repair of existing firearms.

k. "Handgun" means any pistol, revolver or other firearm originally designed or manufactured to be fired by the use of a single hand.

l. "Retail dealer" means any person including a gunsmith, except a manufacturer or a wholesale dealer, who sells, transfers or assigns for a fee or profit any firearm or parts of firearms or ammunition which he has purchased or obtained with the intention, or for the purpose, of reselling or reassigning to persons who are reasonably understood to be the ultimate consumers, and includes any person who is engaged in the business of repairing firearms or who sells any firearm to satisfy a debt secured by the pledge of a firearm.

m. "Rifle" means any firearm designed to be fired from the shoulder and using the energy of the explosive in a fixed metallic cartridge to fire a single projectile through a rifled bore for each single pull of the trigger.

n. "Shotgun" means any firearm designed to be fired from the shoulder and using the energy of the explosive in a fixed shotgun shell to fire through a smooth bore either a number of ball shots or a single projectile for each pull of the trigger, or any firearm designed to be fired from the shoulder which does not fire fixed ammunition.

o. "Sawed-off shotgun" means any shotgun having a barrel or barrels of less than 18 inches in length measured from the breech to the muzzle, or a rifle having a barrel or barrels of less than 16 inches in length measured from the breech to the muzzle, or any firearm made from a rifle or a shotgun, whether by alteration, or otherwise, if such firearm as modified has an overall length of less than 26 inches. ...

r. "Weapon" means anything readily capable of lethal use or of inflicting serious bodily injury. The term includes, but is not limited to, all (1) firearms, even though not loaded or lacking a clip or other component to render them immediately operable; (2) components which can be readily assembled into a weapon; (3) gravity knives, switchblade knives, daggers, dirks, stilettos, or other dangerous knives, billies, blackjacks, bludgeons, metal knuckles, sand-clubs, slingshots, cesti or similar leather bands studded with metal filings or razor blades imbedded in wood; and (4) stun guns; and any weapon or other device which projects, releases, or emits tear gas or any other substance intended to produce temporary physical discomfort or permanent injury through being vaporized or otherwise dispensed in the air.

s. "Wholesale dealer" means any person, except a manufacturer, who sells, transfers, or assigns firearms, or parts of firearms, to persons who are reasonably understood not to be the ultimate consumers, and includes persons who receive finished parts of firearms and assemble them into completed or partially completed firearms, in furtherance of such purpose, except that it shall not include those persons dealing exclusively in grips, stocks and other nonmetal parts of firearms.

t. "Stun gun" means any weapon or other device which emits an electrical charge or current intended to temporarily or permanently disable a person.

v. "Imitation firearm" means an object or device reasonably capable of being mistaken for a firearm.

w. "Assault firearm" means:

(1) The following firearms:

Algimec AGM1 type

Any shotgun with a revolving cylinder such as the "Street Sweeper" or "Striker 12"

Armalite AR-180 type

Australian Automatic Arms SAR

Avtomat Kalashnikov type semi-automatic firearms

Beretta AR-70 and BM59 semi-automatic firearms

Bushmaster Assault Rifle

Calico M-900 Assault carbine and M-900

CETME G3

Chartered Industries of Singapore SR-88 type

Colt AR-15 and CAR-15 series

Daewoo K-1, K-2, Max 1 and Max 2, AR 100 types

Demro TAC-1 carbine type

Encom MP-9 and MP-45 carbine types

FAMAS MAS223 types

FN-FAL, FN-LAR, or FN-FNC type semi-automatic firearms

Franchi SPAS 12 and LAW 12 shotguns

G3SA type

Galil type Heckler and Koch HK91, HK93, HK94, MP5, PSG-1

Intratec TEC 9 and 22 semi-automatic firearms

M1 carbine type

M14S type

MAC 10, MAC 11, MAC 11-9mm carbine type firearms

PJK M-68 carbine type

Plainfield Machine Company Carbine

Ruger K-Mini-14/5F and Mini-14/5RF

SIG AMT, SIG 550SP, SIG 551SP, SIG PE-57 types

SKS with detachable magazine type

Spectre Auto carbine type

Springfield Armory BM59 and SAR-48 type

Sterling MK-6, MK-7 and SAR types

Steyr A.U.G. semi-automatic firearms

USAS 12 semi-automatic type shotgun

Uzi type semi-automatic firearms

Valmet M62, M71S, M76, or M78 type semi-automatic firearms

Weaver Arm Nighthawk.

(2) Any firearm manufactured under any designation which is substantially identical to any of the firearms listed above.

(3) A semi-automatic shotgun with either a magazine capacity exceeding six rounds, a pistol grip, or a folding stock.

(4) A semi-automatic rifle with a fixed magazine capacity exceeding 15 rounds.

(5) A part or combination of parts designed or intended to convert a firearm into an assault firearm, or any combination of parts from which an assault firearm may be readily assembled if those parts are in the possession or under the control of the same person.

x. "Semi-automatic" means a firearm which fires a single projectile for each single pull of the trigger and is self-reloading or automatically chambers a round, cartridge, or bullet.

y. "Large capacity ammunition magazine" means a box, drum, tube or other container which is capable of holding more than 15 rounds of ammunition to be fed continuously and directly therefrom into a semi-automatic firearm.

z. "Pistol grip" means a well-defined handle, similar to that found on a handgun, that protrudes conspicuously beneath the action of the weapon, and which permits the shotgun to be held and fired with one hand.

aa. "Antique handgun" means a handgun manufactured before 1898, or a replica thereof, which is recognized as being historical in nature or of historical significance and either (1) utilizes a match, friction, flint, or percussion ignition, or which utilizes a pin-fire cartridge in which the pin is part of the cartridge or (2) does not fire fixed ammunition or for which cartridge ammunition is not commercially available.

bb. "Trigger lock" means a commercially available device approved by the Superintendent of State Police which is operated with a key or combination lock that prevents a firearm from being discharged while the device is attached to the firearm. It may include, but need not be limited to, devices that obstruct the barrel or cylinder of the firearm, as well as devices that immobilize the trigger.

cc. "Trigger locking device" means a device that, if installed on a firearm and secured by means of a key or mechanically, electronically or electromechanically operated combination lock, prevents the firearm from being discharged without first deactivating or removing the device by means of a key or mechanically, electronically or electromechanically operated combination lock.

dd. "Personalized handgun" means a handgun which incorporates within its design, and as part of its original manufacture, technology which automatically limits its operational use and which cannot be readily deactivated, so that it may only be fired by an authorized or recognized user. The technology limiting the handgun's operational use may include, but not be limited to: radio frequency tagging, touch memory, remote control, fingerprint, magnetic encoding and other automatic user identification systems utilizing biometric, mechanical or electronic systems. No make or model of a handgun shall be deemed to be a "personalized handgun" unless the Attorney General has determined, through testing or other reasonable means, that the handgun meets any reliability standards that the manufacturer may require for its commercially available handguns that are not personalized or, if the manufacturer has no such reliability standards, the handgun meets the reliability standards generally used in the industry for commercially available handguns.

2C:39-2. Presumptions

a. Possession of firearms, weapons, destructive devices, silencers, or explosives in a vehicle. When a firearm, weapon, destructive device, silencer, or explosive described in this chapter is found in a vehicle, it is presumed to be in the possession of the occupant if there is but one. If there is more than one occupant in the vehicle, it shall be presumed to be in the possession of all, except under the following circumstances:

(1) When it is found upon the person of one of the occupants, it shall be presumed to be in the possession of that occupant alone;

(2) When the vehicle is not a stolen one and the weapon or other instrument is found out of view in a glove compartment, trunk or other enclosed customary depository, it shall be presumed to be in the possession of the occupant or occupants who own or have authority to operate the vehicle; and

(3) When the vehicle is a taxicab and a weapon or other instrument is found in the passenger's portion of the vehicle, it shall be presumed to be in the possession of all the passengers, if there are any, and if not, in the possession of the driver.

b. Licenses and permits. When the legality of a person's conduct under this chapter depends on his possession of a license or permit or on his having registered with or given notice to a particular person or agency, it shall be presumed that he does not possess such a license or permit or has not registered or given the required notice, until he establishes the contrary.

2C:39-3. Prohibited weapons and devices

a. Destructive devices. Any person who knowingly has in his possession any destructive device is guilty of a crime of the third degree.

b. Sawed-off shotguns. Any person who knowingly has in his possession any sawed-off shotgun is guilty of a crime of the third degree.

c. Silencers. Any person who knowingly has in his possession any firearm silencer is guilty of a crime of the fourth degree.

d. Defaced firearms. Any person who knowingly has in his possession any firearm which has been defaced, except an antique firearm or an antique handgun, is guilty of a crime of the fourth degree.

e. Certain weapons. Any person who knowingly has in his possession any gravity knife, switchblade knife, dagger, dirk, stiletto, billy, blackjack, metal knuckle, sandclub, sling-shot, cestus or similar leather band studded with metal filings or razor blades imbedded in wood, ballistic knife, without any explainable lawful purpose, is guilty of a crime of the fourth degree.

f. Dum-dum or body armor penetrating bullets.

(1) Any person, other than a law enforcement officer or persons engaged in activities pursuant to subsection f. of N.J.S.2C:39-6, who knowingly has in his possession any hollow nose or dum-dum bullet, or

(2) any person, other than a collector of firearms or ammunition as curios or relics as defined in Title 18, United States Code, section 921 (a) (13) and has in his possession a valid Collector of Curios and Relics License issued by the Bureau of Alcohol, Tobacco and Firearms, who knowingly has in his possession any body armor breaching or penetrating ammunition, which means:

(a) ammunition primarily designed for use in a handgun, and

(b) which is comprised of a bullet whose core or jacket, if the jacket is thicker than.025 of an inch, is made of tungsten carbide, or hard bronze, or other material which is harder than a rating of 72 or greater on the Rockwell B. Hardness Scale, and

(c) is therefore capable of breaching or penetrating body armor, is guilty of a crime of the fourth degree. For purposes of this section, a collector may possess not more than three examples of each distinctive variation of the ammunition described above. A distinctive variation includes a different head stamp, composition, design, or color.

g. Exceptions.

(1) Nothing in subsection a., b., c., d., e., f., j. or k. of this section shall apply to any member of the Armed Forces of the United States or the National Guard, or except as otherwise provided, to any law enforcement officer while actually on duty or traveling to or from an authorized place of duty, provided that his possession of the prohibited weapon or device has been duly authorized under the applicable laws, regulations or military or law enforcement orders. Nothing in subsection h. of this section shall apply to any law enforcement officer who is exempted from the provisions of that subsection by the Attorney General. Nothing in this section shall apply to the possession of any weapon or device by a law enforcement officer who has confiscated, seized or otherwise taken possession of said weapon or device as evidence of the commission of a crime or because he believed it to be possessed illegally by the person from whom it was taken, provided that said law enforcement officer promptly notifies his superiors of his possession of such prohibited weapon or device.

(2)a. Nothing in subsection f.(1) shall be construed to prevent a person from keeping such ammunition at his dwelling, premises or other land owned or possessed by him, or from carrying such ammunition from the place of purchase to said dwelling or land, nor shall subsection f.(1) be construed to prevent any licensed retail or wholesale firearms dealer from possessing such ammunition at its licensed premises, provided that the seller of any such ammunition shall maintain a record of the name, age and place of residence of any purchaser who is not a li-censed dealer, together with the date of sale and quantity of ammunition sold.

b. Nothing in subsection f.(1) shall be construed to prevent a designated employee or designated licensed agent for a nuclear power plant under the license of the Nuclear Regulatory Commission from possessing hollow nose ammunition while in the actual performance of his official duties, if the federal licensee certifies that the designated employee or designated licensed agent is assigned to perform site protection, guard, armed response or armed escort duties and is appropriately trained and qualified, as prescribed by federal regulation, to perform those duties.

(3) Nothing in paragraph (2) of subsection f. or in subsection j. shall be construed to prevent any licensed retail or wholesale firearms dealer from possessing that ammunition or large capacity ammunition magazine at its licensed premises for sale or disposition to another licensed dealer, the Armed Forces of the United States or the National Guard, or to a law enforcement agency, provided that the seller maintains a record of any sale or disposition to a law enforcement agency. The record shall include the name of the purchasing agency, together with written authorization of the chief of police or highest ranking official of the agency, the name and rank of the purchasing law enforcement officer, if applicable, and the date, time and amount of ammunition sold or otherwise disposed A copy of this record shall be forwarded by the seller to the Superintendent of the Division of State Police within 48 hours of the sale or disposition.

(4) Nothing in subsection a. of this section shall be construed to apply to antique cannons as exempted in subsection d. of N.J.S.2C:39-6.

(5) Nothing in subsection c. of this section shall be construed to apply to any person who is specifically identified in a special deer management permit issued by the Division of Fish and Wildlife to utilize a firearm silencer as part of an alternative deer control method implemented in accordance with a special deer management permit issued pursuant to section 4 of P.L.2000, c.46 (C. 23:4-42.6), while the person is in the actual performance of the permitted alternative deer control method and while going to and from the place where the permitted alternative deer control method is being utilized. This exception shall not, however, otherwise apply to any person to authorize the purchase or possession of a firearm silencer.

h. Stun guns. Any person who knowingly has in his possession any stun gun is guilty of a crime of the fourth degree.

i. Nothing in subsection e. of this section shall be construed to prevent any guard in the employ of a private security company, who is licensed to carry a firearm, from the possession of a nightstick when in the actual performance of his official duties, provided that he has satisfactorily completed a training course approved by the Police Training Commission in the use of a nightstick.

j. Any person who knowingly has in his possession a large capacity ammunition magazine is guilty of a crime of the fourth degree unless the person has registered an assault firearm pursuant to section 11 of P.L. 1990, c. 32 (C. 2C:58-12) and the magazine is maintained and used in connection with participation in competitive shooting matches sanctioned by the Director of Civilian Marksmanship of the United States Department of the Army.

k. Handcuffs. Any person who knowingly has in his possession handcuffs as defined in P.L. 1991, c. 437 (C. 2C:39-9.2), under circumstances not manifestly appropriate for such lawful uses as handcuffs may have, is guilty of a disorderly persons offense. A law enforcement officer shall confiscate handcuffs possessed in violation of the law.

2C:39-4. Possession of weapons for unlawful purposes

a. Firearms. (1) Any person who has in his possession any firearm with a purpose to use it unlawfully against the person or property of another is guilty of a crime of the second degree.

(2) Any person who possesses, receives or transfers a community gun is guilty of a crime of the second degree and shall be sentenced to a term of imprisonment by the court. The term of imprisonment shall include the imposition of a minimum term. The minimum term shall be fixed at one-half of the sentence imposed by the court or three years, whichever is greater and during which the defendant shall be ineligible for parole. As used in this paragraph, "community gun" means a firearm that is transferred among, between or within any association of two or more persons who, while possessing that firearm, engage in criminal activity or use it unlawfully against the person or property of another.

b. Explosives. Any person who has in his possession or carries any explosive substance with a purpose to use it unlawfully against the person or property of another is guilty of a crime of the second degree.

c. Destructive devices. Any person who has in his possession any destructive device with a purpose to use it unlawfully against the person or property of another is guilty of a crime of the second degree.

d. Other weapons. Any person who has in his possession any weapon, except a firearm, with a purpose to use it unlawfully against the person or property of another is guilty of a crime of the third degree.

e. Imitation firearms. Any person who has in his possession an imitation firearm under circumstances that would lead an observer to reasonably believe that it is possessed for an unlawful purpose is guilty of a crime of the fourth degree.

2C:39-5. Unlawful Possession of Weapons

a. Machine guns. Any person who knowingly has in his possession a machine gun or any instrument or device adaptable for use as a machine gun, without being licensed to do so as provided in N.J.S.2C:58-5, is guilty of a crime of the third degree.

b. Handguns. Any person who knowingly has in his possession any handgun, including any antique handgun without first having obtained a permit to carry the same as provided in N.J.S.2C:58-4, is guilty of a crime of the third degree if the handgun is in the nature of an air gun, spring gun or pistol or other weapon of a similar nature in which the propelling force is a spring, elastic band, carbon dioxide, compressed or other gas or vapor, air or compressed air, or is ignited by compressed air, and ejecting a bullet or missile smaller than three-eighths of an inch in diameter, with sufficient force to injure a person. Otherwise it is a crime of the second degree.

c. Rifles and shotguns.

(1) Any person who knowingly has in his possession any rifle or shotgun without having first obtained a firearms purchaser identification card in accordance with the provisions of N.J.S.2C:58-3, is guilty of a crime of the third degree.

(2) Unless otherwise permitted by law, any person who knowingly has in his possession any loaded rifle or shotgun is guilty of a crime of the third degree.

d. Other weapons. Any person who knowingly has in his possession any other weapon under circumstances not manifestly appropriate for such lawful uses as it may have is guilty of a crime of the fourth degree.

e. Firearms or other weapons in educational institutions.

(1) Any person who knowingly has in his possession any firearm in or upon any part of the buildings or grounds of any school, college, university or other educational institution, without the written authorization of the governing officer of the institution, is guilty of a crime of the third degree, irrespective of whether he possesses a valid permit to carry the firearm or a valid firearms purchaser identification card.

(2) Any person who knowingly possesses any weapon enumerated in paragraphs (3) and (4) of subsection r. of N.J.S.2C:39-1 or any components which can readily be assembled into a firearm or other weapon enumerated in subsection r. of N.J.S.2C:39-1 or any other weapon under circumstances not manifestly appropriate for such lawful use as it may have, while in or upon any part of the buildings or grounds of any school, college, university or other educational institution without the written authorization of the governing officer of the institution is guilty of a crime of the fourth degree.

(3) Any person who knowingly has in his possession any imitation firearm in or upon any part of the buildings or grounds of any school, college, university or other educational institution, without the written authorization of the governing officer of the institution, or while on any school bus is a disorderly person, irrespective of whether he possesses a valid permit to carry a firearm or a valid firearms purchaser identification card.

f. Assault firearms. Any person who knowingly has in his possession an assault firearm is guilty of a crime of the third degree except if the assault firearm is licensed pursuant to N.J.S.2C:58-5; registered pursuant to section 11 of P.L.1990, c.32 (C.2C:58-12) or rendered inoperable pursuant to section 12 of P.L.1990, c.32 (C.2C:58-13).

g. (1) The temporary possession of a handgun, rifle or shotgun by a person receiving, possessing, carrying or using the handgun, rifle, or shotgun under the provisions of section 1 of P.L.1992, c.74 (C.2C:58-3.1) shall not be considered unlawful possession under the provisions of subsection b. or c. of this section.

(2) The temporary possession of a firearm by a person receiving, possessing, carrying or using the firearm under the provisions of section 1 of P.L.1997, c.375 (C.2C:58-3.2) shall not be considered unlawful possession under the provisions of this section.

h. A person who is convicted of a crime under subsection a., b. or f. of this section shall be ineligible for participation in any program of intensive supervision; provided, however, that this provision shall not apply to a crime under subsection b. involving only a handgun which is in the nature of an air gun, spring gun or pistol or other weapon of a similar nature in which the propelling force is a spring, elastic band, carbon dioxide, compressed or other gas or vapor, air or compressed air, or is ignited by compressed air, and ejecting a bullet or missile smaller than three-eighths of an inch in diameter, with sufficient force to injure a person.

i. A person convicted of violating subsection a., b. or f. of this section shall be sentenced by the court to a term of imprisonment, which shall include the imposition of a minimum term during which the defendant shall be ineligible for parole, if the court finds that the aggravating circumstance set forth in paragraph (5) of subsection a. of N.J.S.2C:44-1 applies. The minimum term of parole ineligibility shall be fixed at five years. The sentencing court shall make a finding on the record as to whether the aggravating circumstance set forth in paragraph (5) of subsection a. of N.J.S.2C:44-1 applies, and the court shall presume that there is a substantial likelihood that the defendant is involved in organized criminal activity if there is a substantial likelihood that the defendant is a member of an organization or group that engages in criminal activity. The prosecution at the sentencing hearing shall have the initial burden of producing evidence or information concerning the defendant's membership in such an organization or group.

2C:39-6. Exemptions

a. Provided a person complies with the requirements of subsection j. of this section, N.J.S.2C:39-5 does not apply to:

(1) Members of the Armed Forces of the United States or of the National Guard while actually on duty, or while traveling between places of duty and carrying authorized weapons in the manner prescribed by the appropriate military authorities;

(2) Federal law enforcement officers, and any other federal officers and employees required to carry firearms in the performance of their official duties;

(3) Members of the State Police and, under conditions prescribed by the superintendent, members of the Marine Law Enforcement Bureau of the Division of State Police;

(4) A sheriff, undersheriff, sheriff's officer, county prosecutor, assistant prosecutor, prosecutor's detective or investigator, deputy attorney general or State investigator employed by the Division of Criminal Justice of the Department of Law and Public Safety, investigator employed by the State Commission of Investigation, inspector of the Alcoholic Beverage Control Enforcement Bureau of the Division of State Police in the Department of Law and Public Safety authorized to carry such weapons by the Superintendent of State Police, State park police officer, or State conservation officer;

(5) Except as hereinafter provided, a prison or jail warden of any penal institution in this State or his deputies, or an employee of the Department of Corrections engaged in the interstate transportation of convicted offenders, while in the performance of his duties, and when required to possess the weapon by his superior officer, or a corrections officer or keeper of a penal institution in this State at all times while in the State of New Jersey, provided he annually passes an examination approved by the superintendent testing his proficiency in the handling of firearms;

(6) A civilian employee of the United States Government under the supervision of the commanding officer of any post, camp, station, base or other military or naval installation located in this State who is required, in the performance of his official duties, to carry firearms, and who is authorized to carry such firearms by said commanding officer, while in the actual performance of his official duties;

(7)(a) A regularly employed member, including a detective, of the police department of any county or municipality, or of any State, interstate, municipal or county park police force or boulevard police force, at all times while in the State of New Jersey;

(b) A special law enforcement officer authorized to carry a weapon as provided in subsection b. of section 7 of P.L.1985, c.439 (C.40A:14-146.14);

(c) An airport security officer or a special law enforcement officer appointed by the governing body of any county or municipality, except as provided in subsection (b) of this section, or by the commission, board or other body having control of a county park or airport or boulevard police force, while engaged in the actual performance of his official duties and when specifically authorized by the governing body to carry weapons;

(8) A full-time, paid member of a paid or part-paid fire department or force of any municipality who is assigned full-time or part-time to an arson investigation unit created pursuant to section 1 of P.L.1981, c.409 (C.40A:14-7.1) or to the county arson investigation unit in the county prosecutor's office, while either engaged in the actual performance of arson investigation duties or while actually on call to perform arson investigation duties and when specifically authorized by the governing body or the county prosecutor, as the case may be, to carry weapons. Prior to being permitted to carry a firearm, such a member shall take and successfully complete a firearms training course administered by the Police Training Commission pursuant to P.L.1961, c.56 (C.52:17B-66 et seq.), and shall annually qualify in the use of a revolver or similar weapon prior to being permitted to carry a firearm;

(9) A juvenile corrections officer in the employment of the Juvenile Justice Commission established pursuant to section 2 of P.L.1995, c.284 (C.52:17B-170) subject to the regulations promulgated by the commission;

(10) A designated employee or designated licensed agent for a nuclear power plant under license of the Nuclear Regulatory Commission, while in the actual performance of his official duties, if the federal licensee certifies that the designated employee or designated licensed agent is assigned to perform site protection, guard, armed response or armed escort duties and is appropriately trained and qualified, as prescribed by federal regulation, to perform those duties. Any firearm utilized by an employee or agent for a nuclear power plant pursuant to this paragraph shall be returned each day at the end of the employee's or agent's authorized official duties to the employee's or agent's supervisor. All firearms returned each day pursuant to this paragraph shall be stored in locked containers located in a secure area;

(11) A county corrections officer at all times while in the State of New Jersey, provided he annually passes an examination approved by the superintendent testing his proficiency in the handling of firearms.

b. Subsections a., b. and c. of N.J.S.2C:39-5 do not apply to:

(1) A law enforcement officer employed by a governmental agency outside of the State of New Jersey while actually engaged in his official duties, provided, however, that he has first notified the superintendent or the chief law enforcement officer of the municipality or the prosecutor of the county in which he is engaged; or

(2) A licensed dealer in firearms and his registered employees during the course of their normal business while traveling to and from their place of business and other places for the purpose of demonstration, exhibition or delivery in connection with a sale, provided, however, that the weapon is carried in the manner specified in subsection g. of this section.

c. Provided a person complies with the requirements of subsection j. of this section, subsections b. and c. of N.J.S.2C:39-5 do not apply to:

(1) A special agent of the Division of Taxation who has passed an examination in an approved police training program testing proficiency in the handling of any firearm which he may be required to carry, while in the actual performance of his official duties and while going to or from his place of duty, or any other police officer, while in the actual performance of his official duties;

(2) A State deputy conservation officer or a full-time employee of the Division of Parks and Forestry having the power of arrest and authorized to carry weapons, while in the actual performance of his official duties;

(3) (Deleted by amendment, P.L.1986, c.150.)

(4) A court attendant serving as such under appointment by the sheriff of the county or by the judge of any municipal court or other court of this State, while in the actual performance of his official duties;

(5) A guard in the employ of any railway express company, banking or building and loan or savings and loan institution of this State, while in the actual performance of his official duties;

(6) A member of a legally recognized military organization while actually under orders or while going to or from the prescribed place of meeting and carrying the weapons prescribed for drill, exercise or parade;

(7) A humane law enforcement officer of the New Jersey Society for the Prevention of Cruelty to Animals or of a county society for the prevention of cruelty to animals, while in the actual performance of his duties;

(8) An employee of a public utilities corporation actually engaged in the transportation of explosives;

(9) A railway policeman, except a transit police officer of the New Jersey Transit Police Department, at all times while in the State of New Jersey, provided that he has passed an approved police academy training program consisting of at least 280 hours. The training program shall include, but need not be limited to, the handling of firearms, community relations, and juvenile relations;

(10) A campus police officer appointed under P.L.1970, c.211 (C.18A:6-4.2 et seq.) at all times. Prior to being permitted to carry a firearm, a campus police officer shall take and successfully complete a firearms training course administered by the Police Training Commission, pursuant to P.L.1961, c.56 (C.52:17B-66 et seq.), and shall annually qualify in the use of a revolver or similar weapon prior to being permitted to carry a firearm;

(11) (Deleted by amendment, P.L.2003, c.168).

(12) A transit police officer of the New Jersey Transit Police Department, at all times while in the State of New Jersey, provided the officer has satisfied the training requirements of the Police Training Commission, pursuant to subsection c. of section 2 of P.L.1989, c.291 (C.27:25-15.1)

(13) A parole officer employed by the State Parole Board at all times. Prior to being permitted to carry a firearm, a parole officer shall take and successfully complete a basic course for regular police officer training administered by the Police Training Commission, pursuant to P.L.1961, c.56 (C.52:17B-66 et seq.), and shall annually qualify in the use of a revolver or similar weapon prior to being permitted to carry a firearm;

(14) A Human Services police officer at all times while in the State of New Jersey, as authorized by the Commissioner of Human Services;

(15) A person or employee of any person who, pursuant to and as required by a contract with a governmental entity, supervises or transports persons charged with or convicted of an offense;

(16) A housing authority police officer appointed under P.L.1997, c.210 (C.40A:14-146.19 et al.) at all times while in the State of New Jersey; or

(17) A probation officer assigned to the "Probation Officer Community Safety Unit" created by section 2 of P.L.2001, c.362 (C.2B:10A-2) while in the actual performance of the probation officer's official duties. Prior to being permitted to carry a firearm, a probation officer shall take and successfully complete a basic course for regular police officer training administered by the Police Training Commission, pursuant to P.L.1961, c.56 (C.52:17B-66 et seq.), and shall annually qualify in the use of a revolver or similar weapon prior to being permitted to carry a firearm.

d. (1) Subsections c. and d. of N.J.S.2C:39-5 do not apply to antique firearms, provided that such antique firearms are unloaded or are being fired for the purposes of exhibition or demonstration at an authorized target range or in such other manner as has been approved in writing by the chief law enforcement officer of the municipality in which the exhibition or demonstration is held, or if not held on property under the control of a particular municipality, the superintendent

(2) Subsection a. of N.J.S.2C:39-3 and subsection d. of N.J.S.2C:39-5 do not apply to an antique cannon that is capable of being fired but that is unloaded and immobile, provided that the antique cannon is possessed by (a) a scholastic institution, a museum, a municipality, a county or the State, or (b) a person who obtained a firearms purchaser identification card as specified in N.J.S.2C:58-3.

(3) Subsection a. of N.J.S.2C:39-3 and subsection d. of N.J.S.2C:39-5 do not apply to an unloaded antique cannon that is being transported by one eligible to possess it, in compliance with regulations the superintendent may promulgate, between its permanent location and place of purchase or repair.

(4) Subsection a. of N.J.S.2C:39-3 and subsection d. of N.J.S.2C:39-5 do not apply to antique cannons that are being loaded or fired by one eligible to possess an antique cannon, for purposes of exhibition or demonstration at an authorized target range or in the manner as has been approved in writing by the chief law enforcement officer of the municipality in which the exhibition or demonstration is held, or if not held on property under the control of a particular municipality, the superintendent, provided that performer has given at least 30 days' notice to the superintendent.

(5) Subsection a. of N.J.S.2C:39-3 and subsection d. of N.J.S.2C:39-5 do not apply to the transportation of unloaded antique cannons directly to or from exhibitions or demonstrations authorized under paragraph (4) of subsection d. of this section, provided that the transportation is in compliance with safety regulations the superintendent may promulgate. Nor do those subsections apply to transportation directly to or from exhibitions or demonstrations authorized under the law of another jurisdiction, provided that the superintendent has been given 30 days' notice and that the transportation is in compliance with safety regulations the superintendent may promulgate.

e. Nothing in subsections b., c. and d. of N.J.S.2C:39-5 shall be construed to prevent a person keeping or carrying about his place of business, residence, premises or other land owned or possessed by him, any firearm, or from carrying the same, in the manner specified in subsection g. of this section, from any place of purchase to his residence or place of business, between his dwelling and his place of business, between one place of business or residence and another when moving, or between his dwelling or place of business and place where such firearms are repaired, for the purpose of repair. For the purposes of this section, a place of business shall be deemed to be a fixed location.

f. Nothing in subsections b., c. and d. of N.J.S.2C:39-5 shall be construed to prevent:

(1) A member of any rifle or pistol club organized in accordance with the rules prescribed by the National Board for the Promotion of Rifle Practice, in going to or from a place of target practice, carrying such firearms as are necessary for said target practice, provided that the club has filed a copy of its charter with the superintendent and annually submits a list of its members to the superintendent and provided further that the firearms are carried in the manner specified in subsection g. of this section;

(2) A person carrying a firearm or knife in the woods or fields or upon the waters of this State for the purpose of hunting, target practice or fishing, provided that the firearm or knife is legal and appropriate for hunting or fishing purposes in this State and he has in his possession a valid hunting license, or, with respect to fresh water fishing, a valid fishing license;

(3) A person transporting any firearm or knife while traveling:

(a) Directly to or from any place for the purpose of hunting or fishing, provided the person has in his possession a valid hunting or fishing license; or

(b) Directly to or from any target range, or other authorized place for the purpose of practice, match, target, trap or skeet shooting exhibitions, provided in all cases that during the course of the travel all firearms are carried in the manner specified in subsection g. of this section and the person has complied with all the provisions and requirements of Title 23 of the Revised Statutes and any amendments thereto and all rules and regulations promulgated thereunder; or

(c) In the case of a firearm, directly to or from any exhibition or display of firearms which is sponsored by any law enforcement agency, any rifle or pistol club, or any firearms collectors club, for the purpose of displaying the firearms to the public or to the members of the organization or club, provided, however, that not less than 30 days prior to the exhibition or display, notice of the exhibition or display shall be given to the Superintendent of the State Police by the sponsoring organization or club, and the sponsor has complied with such reasonable safety regulations as the superintendent may promulgate. Any firearms transported pursuant to this section shall be transported in the manner specified in subsection g. of this section;

(4) A person from keeping or carrying about a private or commercial aircraft or any boat, or from transporting to or from such vessel for the purpose of installation or repair a visual distress signaling device approved by the United States Coast Guard.

g. All weapons being transported under paragraph (2) of subsection b., subsection e., or paragraph (1) or (3) of subsection f. of this section shall be carried unloaded and contained in a closed and fastened case, gunbox, securely tied package, or locked in the trunk of the automobile in which it is being transported, and in the course of travel shall include only such deviations as are reasonably necessary under the circumstances.

h. Nothing in subsection d. of N.J.S.2C:39-5 shall be construed to prevent any employee of a public utility, as defined in R.S.48:2-13, doing business in this State or any United States Postal Service employee, while in the actual performance of duties which specifically require regular and frequent visits to private premises, from possessing, carrying or using any device which projects, releases or emits any substance specified as being noninjurious to canines or other animals by the Commissioner of Health and Senior Services and which immobilizes only on a temporary basis and produces only temporary physical discomfort through being vaporized or otherwise dispensed in the air for the sole purpose of repelling canine or other animal attacks.

The device shall be used solely to repel only those canine or other animal attacks when the canines or other animals are not restrained in a fashion sufficient to allow the employee to properly perform his duties.

Any device used pursuant to this act shall be selected from a list of products, which consist of active and inert ingredients, permitted by the Commissioner of Health and Senior Services.

i. Nothing in N.J.S.2C:39-5 shall be construed to prevent any person who is 18 years of age or older and who has not been convicted of a felony, from possession for the purpose of personal self-defense of one pocket-sized device which contains and releases not more than three-quarters of an ounce of chemical substance not ordinarily capable of lethal use or of inflicting serious bodily injury, but rather, is intended to produce temporary physical discomfort or disability through being vaporized or otherwise dispensed in the air. Any person in possession of any device in violation of this subsection shall be deemed and adjudged to be a disorderly person, and upon conviction thereof, shall be punished by a fine of not less than $ 100.00.

j. A person shall qualify for an exemption from the provisions of N.J.S.2C:39-5, as specified under subsections a. and c. of this section, if the person has satisfactorily completed a firearms training course approved by the Police Training Commission.

Such exempt person shall not possess or carry a firearm until the person has satisfactorily completed a firearms training course and shall annually qualify in the use of a revolver or similar weapon. For purposes of this subsection, a "firearms training course" means a course of instruction in the safe use, maintenance and storage of firearms which is approved by the Police Training Commission. The commission shall approve a firearms training course if the requirements of the course are substantially equivalent to the requirements for firearms training provided by police training courses which are certified under section 6 of P.L.1961, c.56 (C.52:17B-71). A person who is specified in paragraph (1), (2), (3) or (6) of subsection a. of this section shall be exempt from the requirements of this subsection.

k. Nothing in subsection d. of N.J.S.2C:39-5 shall be construed to prevent any financial institution, or any duly authorized personnel of the institution, from possessing, carrying or using for the protection of money or property, any device which projects, releases or emits tear gas or other substances intended to produce temporary physical discomfort or temporary identification.

l. Nothing in subsection b. of N.J.S.2C:39-5 shall be construed to prevent a law enforcement officer who retired in good standing, including a retirement because of a disability pursuant to section 6 of P.L.1944, c.255 (C.43:16A-6), section 7 of P.L.1944, c.255 (C.43:16A-7), section 1 of P.L.1989, c.103 (C.43:16A-6.1) or any substantially similar statute governing the disability retirement of federal law enforcement officers, provided the officer was a regularly employed, full-time law enforcement officer for an aggregate of four or more years prior to his disability retirement and further provided that the disability which constituted the basis for the officer's retirement did not involve a certification that the officer was mentally incapacitated for the performance of his usual law enforcement duties and any other available duty in the department which his employer was willing to assign to him or does not subject that retired officer to any of the disabilities set forth in subsection c. of N.J.S.2C:58-3 which would disqualify the retired officer from possessing or carrying a firearm, who semi-annually qualifies in the use of the handgun he is permitted to carry in accordance with the requirements and procedures established by the Attorney General pursuant to subsection j. of this section and pays the actual costs associated with those semi-annual qualifications, who is 75 years of age or younger, and who was regularly employed as a full-time member of the State Police; a full-time member of an interstate police force; a full-time member of a county or municipal police department in this State; a full-time member of a State law enforcement agency; a full-time sheriff, undersheriff or sheriff's officer of a county of this State; a full-time State or county corrections officer; a full-time county park police officer; a full-time county prosecutor's detective or investigator; a full-time federal law enforcement officer; or is a qualified retired law enforcement officer as used in the federal "Law Enforcement Officers Safety Act of 2004," Pub. L. 108-277 [18 U.S.C.S. §§ 926B, 926C], domiciled in this State from carrying a handgun in the same manner as law enforcement officers exempted under paragraph (7) of subsection a. of this section under the conditions provided herein:

(1) The retired law enforcement officer shall make application in writing to the Superintendent of State Police for approval to carry a handgun for one year. An application for annual renewal shall be submitted in the same manner.

(2) Upon receipt of the written application of the retired law enforcement officer, the superintendent shall request a verification of service from the chief law enforcement officer of the organization in which the retired officer was last regularly employed as a full-time law enforcement officer prior to retiring. The verification of service shall include:

(a) The name and address of the retired officer;

(b) The date that the retired officer was hired and the date that the officer retired;

(c) A list of all handguns known to be registered to that officer;

(d) A statement that, to the reasonable knowledge of the chief law enforcement officer, the retired officer is not subject to any of the restrictions set forth in subsection c. of N.J.S.2C:58-3; and

(e) A statement that the officer retired in good standing.

(3) If the superintendent approves a retired officer's application or reapplication to carry a handgun pursuant to the provisions of this subsection, the superintendent shall notify in writing the chief law enforcement officer of the municipality wherein that retired officer resides. In the event the retired officer resides in a municipality which has no chief law enforcement officer or law enforcement agency, the superintendent shall maintain a record of the approval.

(4) The superintendent shall issue to an approved retired officer an identification card permitting the retired officer to carry a handgun pursuant to this subsection. This identification card shall be valid for one year from the date of issuance and shall be valid throughout the State. The identification card shall not be transferable to any other person. The identification card shall be carried at all times on the person of the retired officer while the retired officer is carrying a handgun. The retired officer shall produce the identification card for review on the demand of any law enforcement officer or authority.

(5) Any person aggrieved by the denial of the superintendent of approval for a permit to carry a handgun pursuant to this subsection may request a hearing in the Superior Court of New Jersey in the county in which he resides by filing a written request for such a hearing within 30 days of the denial. Copies of the request shall be served upon the superintendent and the county prosecutor. The hearing shall be held within 30 days of the filing of the request, and no formal pleading or filing fee shall be required. Appeals from the determination of such a hearing shall be in accordance with law and the rules governing the courts of this State.

(6) A judge of the Superior Court may revoke a retired officer's privilege to carry a handgun pursuant to this subsection for good cause shown on the application of any interested person. A person who becomes subject to any of the disabilities set forth in subsection c. of N.J.S.2C:58-3 shall surrender, as prescribed by the superintendent, his identification card issued under paragraph (4) of this subsection to the chief law enforcement officer of the municipality wherein he resides or the superintendent, and shall be permanently disqualified to carry a handgun under this subsection.

(7) The superintendent may charge a reasonable application fee to retired officers to offset any costs associated with administering the application process set forth in this subsection.

m. Nothing in subsection d. of N.J.S.2C:39-5 shall be construed to prevent duly authorized personnel of the New Jersey Division of Fish and Wildlife, while in the actual performance of duties, from possessing, transporting or using any device that projects, releases or emits any substance specified as being non-injurious to wildlife by the Director of the Division of Animal Health in the Department of Agriculture, and which may immobilize wildlife and produces only temporary physical discomfort through being vaporized or otherwise dispensed in the air for the purpose of repelling bear or other animal attacks or for the aversive conditioning of wildlife.

n. Nothing in subsection b., c., d. or e. of N.J.S.2C:39-5 shall be construed to prevent duly authorized personnel of the New Jersey Division of Fish and Wildlife, while in the actual performance of duties, from possessing, transporting or using hand held pistol-like devices, rifles or shotguns that launch pyrotechnic missiles for the sole purpose of frightening, hazing or aversive conditioning of nuisance or depredating wildlife; from possessing, transporting or using rifles, pistols or similar devices for the sole purpose of chemically immobilizing wild or non-domestic animals; or, provided the duly authorized person complies with the requirements of subsection j. of this section, from possessing, transporting or using rifles or shotguns, upon completion of a Police Training Commission approved training course, in order to dispatch injured or dangerous animals or for non-lethal use for the purpose of frightening, hazing or aversive conditioning of nuisance or depredating wildlife.

2C:39-7. Certain persons not to have weapons

a. Except as provided in subsection b. of this section, any person, having been convicted in this State or elsewhere of the crime of aggravated assault, arson, burglary, escape, extortion, homicide, kidnapping, robbery, aggravated sexual assault, sexual assault, bias intimidation in violation of N.J.S.2C:16-1 or endangering the welfare of a child pursuant to N.J.S.2C:24-4, whether or not armed with or having in his possession any weapon enumerated in subsection r. of N.J.S.2C:39-1, or any person convicted of a crime pursuant to the provisions of N.J.S.2C: 39-3, N.J.S.2C:39-4 or N.J.S.2C:39-9, or any person who has ever been committed for a mental disorder to any hospital, mental institution or sanitarium unless he possesses a certificate of a medical doctor or psychiatrist licensed to practice in New Jersey or other satisfactory proof that he is no longer suffering from a mental disorder which interferes with or handicaps him in the handling of a firearm, or any person who has been convicted of other than a disorderly persons or petty disorderly persons offense for the unlawful use, possession or sale of a controlled dangerous substance as defined in N.J.S. 2C: 35-2 who purchases, owns, possesses or controls any of the said weapons is guilty of a crime of the fourth degree.

b.(1) A person having been convicted in this State or elsewhere of the crime of aggravated assault, arson, burglary, escape, extortion, homicide, kidnapping, robbery, aggravated sexual assault, sexual assault, bias intimidation in violation of N.J.S.2C:16-1, endangering the welfare of a child pursuant to N.J.S.2C:24-4, stalking pursuant to P.L.1992, c.209 (C.2C:12-10) or a crime involving domestic violence as defined in section 3 of P.L.1991, c.261 (C.2C:25-19), whether or not armed with or having in his possession a weapon enumerated in subsection r. of N.J.S.2C:39-1, or a person having been convicted of a crime pursuant to the provisions of N.J.S.2C:35-3 through N.J.S.2C:35-6, inclusive; section 1 of P.L.1987, c.101 (C.2C:35-7); N.J.S. 2C:35-11; N.J.S.2C:39-3; N.J.S.2C:39-4; or N.J.S.2C:39-9 who purchases, owns, possesses or controls a firearm is guilty of a crime of the second degree and upon conviction thereof, the person shall be sentenced to a term of imprisonment by the court. The term of imprisonment shall include the imposition of a minimum term, which shall be fixed at five years, during which the defendant shall be ineligible for parole. If the defendant is sentenced to an extended term of imprisonment pursuant to N.J.S.2C:43-7, the extended term of imprisonment shall include the imposition of a minimum term, which shall be fixed at, or between, one-third and one-half of the sentence imposed by the court or five years, whichever is greater, during which the defendant shall be ineligible for parole.

(2) A person having been convicted in this State or elsewhere of a disorderly persons offense involving domestic violence, whether or not armed with or having in his possession a weapon enumerated in subsection r. of N.J.S. 2C:39-1, who purchases, owns, possesses or controls a firearm is guilty of a crime of the third degree.

(3) A person whose firearm is seized pursuant to the "Prevention of Domestic Violence Act of 1991," P.L.1991, c.261 (C.2C:25-17 et seq.) and whose firearm has not been returned, or who is subject to a court order prohibiting the possession of firearms issued pursuant to the "Prevention of Domestic Violence Act of 1991," P.L. 1991, c.261 (C.2C:25-17 et seq.) who purchases, owns, possesses or controls a firearm is guilty of a crime of the third degree, except that the provisions of this paragraph shall not apply to any law enforcement officer while actually on duty, or to any member of the Armed Forces of the United States or member of the National Guard while actually on duty or traveling to or from an authorized place of duty.

c. Whenever any person shall have been convicted in another state, territory, commonwealth or other jurisdiction of the United States, or any country in the world, in a court of competent jurisdiction, of a crime which in said other jurisdiction or country is comparable to one of the crimes enumerated in subsection a. or b. of this section, then that person shall be subject to the provisions of this section.

2C:39-9. Manufacture, transport, disposition and defacement of weapons and dangerous instruments and appliances

a. Machine guns. Any person who manufactures, causes to be manufactured, transports, ships, sells or disposes of any machine gun without being registered or licensed to do so as provided in chapter 58 is guilty of a crime of the third degree.

b. Sawed-off shotguns. Any person who manufactures, causes to be manufactured, transports, ships, sells or disposes of any sawed-off shotgun is guilty of a crime of the third degree.

c. Firearm silencers. Any person who manufactures, causes to be manufactured, transports, ships, sells or disposes of any firearm silencer is guilty of a crime of the fourth degree.

d. Weapons. Any person who manufactures, causes to be manufactured, transports, ships, sells or disposes of any weapon, including gravity knives, switchblade knives, ballistic knives, daggers, dirks, stilettos, billies, blackjacks, metal knuckles, sandclubs, slingshots, cestí or similar leather bands studded with metal filings, or, except as otherwise provided in subsection i. of this section, in the case of firearms if he is not licensed or registered to do so as provided in chapter 58, is guilty of a crime of the fourth degree. Any person who manufactures, causes to be manufactured, transports, ships, sells or disposes of any weapon or other device which projects, releases or emits tear gas or other substances intended to produce temporary physical discomfort or permanent injury through being vaporized or otherwise dispensed in the air, which is intended to be used for any purpose other than for authorized military or law enforcement purposes by duly authorized military or law enforcement personnel or the device is for the purpose of personal self-defense, is pocket-sized and contains not more than three-quarters of an ounce of chemical substance not ordinarily capable of lethal use or of inflicting serious bodily injury, or other than to be used by any person permitted to possess such weapon or device under the provisions of subsection d. of N.J.S.2C:39-5, which is intended for use by financial and other business institutions as part of an integrated security system, placed at fixed locations, for the protection of money and property, by the duly authorized personnel of those institutions, is guilty of a crime of the fourth degree.

e. Defaced firearms. Any person who defaces any firearm is guilty of a crime of the third degree. Any person who knowingly buys, receives, disposes of or conceals a defaced firearm, except an antique firearm or an antique handgun, is guilty of a crime of the fourth degree.

f.(1) Any person who manufactures, causes to be manufactured, transports, ships, sells, or disposes of any bullet, which is primarily designed for use in a handgun, and which is comprised of a bullet whose core or jacket, if the jacket is thicker than .025 of an inch, is made of tungsten carbide, or hard bronze, or other material which is harder than a rating of 72 or greater on the Rockwell B. Hardness Scale, and is therefore capable of breaching or penetrating body armor and which is intended to be used for any purpose other than for authorized military or law enforcement purposes by duly authorized military or law enforcement personnel, is guilty of a crime of the fourth degree.

(2) Nothing in this subsection shall be construed to prevent a licensed collector of ammunition as defined in paragraph (2) of subsection f. of N.J.S.2C:39-3 from transporting the bullets defined in paragraph (1) of this subsection from (a) any licensed retail or wholesale firearms dealer's place of business to the collector's dwelling, premises, or other land owned or possessed by him, or (b) to or from the collector's dwelling, premises or other land owned or possessed by him to any gun show for the purposes of display, sale, trade, or transfer between collectors, or (c) to or from the collector's dwelling, premises or other land owned or possessed by him to any rifle or pistol club organized in accordance with the rules prescribed by the National Board for the Promotion of Rifle Practice; provided that the club has filed a copy of its charter with the superintendent of the State Police and annually submits a list of its members to the superintendent, and provided further that the ammunition being transported shall be carried not loaded in any firearm and contained in a closed and fastened case, gun box, or locked in the trunk of the automobile in which it is being transported, and the course of travel shall include only such deviations as are reasonably necessary under the circumstances.

g. Assault firearms. Any person who manufactures, causes to be manufactured, transports, ships, sells or disposes of an assault firearm without being registered or licensed to do so pursuant to N.J.S.2C:58-1 et seq. is guilty of a crime of the third degree.

h. Large capacity ammunition magazines. Any person who manufactures, causes to be manufactured, transports, ships, sells or disposes of a large capacity ammunition magazine which is intended to be used for any purpose other than for authorized military or law enforcement purposes by duly authorized military or law enforcement personnel is guilty of a crime of the fourth degree.

i. Transporting firearms into this State for an unlawful sale or transfer. Any person who knowingly transports, ships or otherwise brings into this State any firearm for the purpose of unlawfully selling, transferring, giving, assigning or otherwise disposing of that firearm to another individual is guilty of a crime of the second degree. The temporary transfer of a firearm while hunting or target shooting, the transfer of any firearm that uses air or carbon dioxide to expel a projectile, or the transfer of an antique firearm shall not constitute a violation of this subsection.

2C:39-10. Violation of the regulatory provisions relating to firearms; false representation in applications

a.(1) Except as otherwise provided in paragraph (2) of this subsection, any person who knowingly violates the regulatory provisions relating to manufacturing or wholesaling of firearms (section 2C:58-1), retailing of firearms (section 2C:58-2), permits to purchase certain firearms (section 2C:58-3), permits to carry certain firearms (section 2C:58-4), licenses to procure machine guns or assault firearms (section 2C:58-5), or incendiary or tracer ammunition (section 2C:58-10), except acts which are punishable under section 2C:39-5 or section 2C:39-9, is guilty of a crime of the fourth degree.

(2) A licensed dealer who knowingly violates the provisions of subparagraph (d) of paragraph (5) of subsection a. of N.J.S.2C:58-2 is a disorderly person.

b. Any person who knowingly violates the regulatory provisions relating to notifying the authorities of possessing certain items of explosives (section 2C:58-7), or of certain wounds (section 2C:58-8) is a disorderly person.

c. Any person who gives or causes to be given any false information, or signs a fictitious name or address, in applying for a firearms purchaser identification card, a permit to purchase a handgun, a permit to carry a handgun, a permit to possess a machine gun, a permit to possess an assault firearm, or in completing the certificate or any other instrument required by law in purchasing or otherwise acquiring delivery of any rifle, shotgun, handgun, machine gun, or assault firearm or any other firearm, is guilty of a crime of the third degree.

d. Any person who gives or causes to be given any false information in registering an assault firearm pursuant to section 11 of P.L. 1990, c. 32 (C. 2C:58-12) or in certifying that an assault firearm was rendered inoperable pursuant to section 12 of P.L. 1990, c. 32 (C. 2C:58-13) commits a crime of the fourth degree.

e. Any person who knowingly sells, gives, transfers, assigns or otherwise disposes of a firearm to a person who is under the age of 18 years, except as permitted in section 14 of P.L. 1979, c. 179 (C. 2C:58-6.1), is guilty of a crime of the third degree. Notwithstanding any other provision of law to the contrary, the sentence imposed for a conviction under this subsection shall include a mandatory minimum three-year term of imprisonment, during which the defendant shall be ineligible for parole.

f. Unless the recipient is authorized to possess the handgun in connection with the performance of official duties under the provisions of N.J.S.2C:39-6, any person who knowingly sells, gives, transfers, assigns or otherwise disposes of a handgun to a person who is under the age of 21 years, except as permitted in section 14 of P.L. 1979, c. 179 (C. 2C:58-6.1), is guilty of a crime of the third degree.

2C:39-11. Pawnbrokers; loaning on firearms

a. Any pawnbroker who sells, offers to sell or to lend or to give away any weapon, destructive device or explosive is guilty of a crime of the third degree.

b. Any person who loans money, the security for which is any handgun, rifle or shotgun is guilty of a disorderly persons offense.

2C:39-12. Voluntary surrender No person shall be convicted of an offense under this chapter for possessing any firearms, weapons, destructive devices, silencers or explosives, if after giving written notice of his intention to do so, including the proposed date and time of surrender, he voluntarily surrendered the weapon, device, instrument or substance in question to the superintendent or to the chief of police in the municipality in which he resides, provided that the required notice is received by the superintendent or chief of police before any charges have been made or complaints filed against such person for the unlawful possession of the weapon, device, instrument or substance in question and before any investigation has been commenced by any law enforcement agency concerning the unlawful possession. Nothing in this section shall be construed as granting immunity from prosecution for any crime or offense except that of the unlawful possession of such weapons, devices, instruments or substances surrendered as herein provided.

2C:39-14. 2nd degree crimes

a. Any person who teaches or demonstrates to any other person the use, application, or making of any firearm, explosive or destructive device, or technique capable of causing injury or death to a person, knowing or having reason to know or intending that it will be employed for use in, or in furtherance of, an illegal activity is guilty of a crime of the second degree.

b. Any person who assembles with one or more persons for the purpose of training with, practicing with, or being instructed in the use of any firearm, explosive or destructive device, or technique capable of causing injury or death to a person, intending to unlawfully employ it for use in, or in furtherance of, an illegal activity is guilty of a crime of the second degree.

2C:39-15. Gun advertising requirement Any person who offers to sell a machine gun, semiautomatic rifle, or assault firearm by means of an advertisement published in a newspaper circulating within this State, which advertisement does not specify that the purchaser shall hold a valid license to purchase and possess a machine gun or assault firearm, or a valid firearms identification card to purchase and possess an automatic or semi-automatic rifle, is a disorderly person.

2C:39-16. Leader of a firearms trafficking network defined; first degree crime; fines; sentencing A person is a leader of a firearms trafficking network if he conspires with others as an organizer, supervisor, financier or manager, to engage for profit in a scheme or course of conduct to unlawfully manufacture, transport, ship, sell or dispose of any firearm. Leader of firearms trafficking network is a crime of the first degree.

As used in this section: "leader of a firearms trafficking network" means a person who occupies a position of authority or control over other persons in a scheme or organization of illegal firearms manufacturing, transporting, shipping or selling and who exercises that authority or control over others involved in the scheme or organization.

Notwithstanding the provisions of subsection a. of N.J.S.2C:43-3, the court may also impose a fine not to exceed $500,000.00 or five times the value of the firearms involved, whichever is greater.

Notwithstanding the provisions of N.J.S.2C:1-8, a conviction of leader of firearms trafficking network shall not merge with the conviction for any offense which is the object of the conspiracy. Nothing contained in this section shall prohibit the court from imposing an extended term pursuant to N.J.S.2C:43-7; nor shall this section be construed in any way to preclude or limit the prosecution or conviction of any person for conspiracy under N.J.S.2C:5-2, or any prosecution or conviction for weapons offenses under the provisions of chapter 39 of Title 2C of the New Jersey Statutes, N.J.S.2C:41-2 (racketeering activities) or subsection g. of N.J.S.2C:5-2 (leader of organized crime).

It shall not be necessary in any prosecution under this section for the State to prove that any intended profit was actually realized. The trier of fact may infer that a particular scheme or course of conduct was undertaken for profit from all of the attendant circumstances, including but not limited to the number of persons involved in the scheme or course of conduct, the actor's net worth and his expenditures in relation to his legitimate sources of income, the amount of firearms involved, or the amount of cash or currency involved.

It shall not be a defense to a prosecution under this section that the firearms were brought into or transported in this State solely for ultimate distribution or dispensing in another jurisdiction; nor shall it be a defense that any profit was intended to be made in another jurisdiction.

Chapter 58. Licensing and Other Provisions Relating to Firearms

2C:58-1. Registration of manufacturers and wholesale dealers of firearms

a. Registration. Every manufacturer and wholesale dealer of firearms shall register with the superintendent as provided in this section. No person shall engage in the business of, or act as a manufacturer or wholesale dealer of firearms, or manufacture or sell at wholesale any firearm, until he has so registered.

Applications for registration shall be made on such forms as shall be prescribed by the superintendent, and the applicant shall furnish such information and other particulars as may be prescribed by law or by any rules or regulations promulgated by the superintendent. Each application for registration or renewal shall be accompanied by a fee of $150.00.

The superintendent shall prescribe standards and qualifications for the registration of manufacturers and wholesalers of firearms, for the protection of the public safety, health and welfare. He shall refuse to register any applicant for registration unless he is satisfied that the applicant can be permitted to engage in business as a manufacturer or wholesale dealer of firearms without any danger to the public safety, health or welfare.

The superintendent shall issue a certificate of registration to every person registered under this section, and such certificate shall be valid for a period of 3 years from the date of issuance.

b. Wholesale dealer's agent. Every registered wholesale dealer of firearms shall cause each of his agents or employees actively engaged in the purchase or sale of firearms to be licensed with the superintendent as a wholesale dealer's agent. Applications for agents' licenses shall be submitted on such forms as shall be prescribed by the superintendent, and shall be signed by the registered wholesale dealer and by the agent. Each application shall be accompanied by a fee of $5.00, and each license shall be valid for so long as the agent or employee remains in the employ of the wholesale dealer and the wholesale dealer remains validly registered under this section. The superintendent shall prescribe standards and qualifications for licensed wholesale dealers' agents, for the protection of the public safety, health and welfare.

c. Revocation of certificate of registration or license. The superintendent may, after reasonable notice to all affected parties and a hearing if requested, revoke any certificate of registration or agent's license if he finds that the registered or licensed person is no longer engaged in the business of manufacturing or wholesaling firearms in this State or that he can no longer be permitted to carry on such business without endangering the public safety, health or welfare. A certificate or license may be cancelled at any time at the request of the registered or licensed person.

d. Appeals. Any person aggrieved by the refusal of the superintendent to register him as a manufacturer or wholesale dealer or a wholesale dealer's agent, or by revocation of his certificate or license, may appeal to the Appellate Division of the Superior Court.

e. Records of Sales. Every manufacturer and wholesale dealer shall keep a detailed record of each firearm sold by him. The record shall include the date of sale, the name and address of the purchaser, a description of each firearm and the serial number thereof. The records shall be available for inspection at all reasonable times by any law enforcement officer.

2C:58-2. Retailing of firearms; licensing of dealers and their employees

a. Licensing of retail dealers and their employees. No retail dealer of firearms nor any employee of a retail dealer shall sell or expose for sale, or possess with the intent of selling, any firearm unless licensed to do so as hereinafter provided. The superintendent shall prescribe standards and qualifications for retail dealers of firearms and their employees for the protection of the public safety, health and welfare.

Applications shall be made in the form prescribed by the superintendent, accompanied by a fee of $50.00 payable to the superintendent, and shall be made to a judge of the Superior Court in the county where the applicant maintains his place of business. The judge shall grant a license to an applicant if he finds that the applicant meets the standards and qualifications established by the superintendent and that the applicant can be permitted to engage in business as a retail dealer of firearms or employee thereof without any danger to the public safety, health and welfare. Each license shall be valid for a period of three years from the date of issuance, and shall authorize the holder to sell firearms at retail in a specified municipality.

In addition, every retail dealer shall pay a fee of $5.00 for each employee actively engaged in the sale or purchase of firearms. The superintendent shall issue a license for each employee for whom said fee has been paid, which license shall be valid for so long as the employee remains in the employ of said retail dealer.

No license shall be granted to any retail dealer under the age of 21 years or to any employee of a retail dealer under the age of 18 or to any person who could not qualify to obtain a permit to purchase a handgun or a firearms purchaser identification card, or to any corporation, partnership or other business organization in which the actual or equitable controlling interest is held or possessed by such an ineligible person.

All licenses shall be granted subject to the following conditions, for breach of any of which the license shall be subject to revocation on the application of any law enforcement officer and after notice and hearing by the issuing court:

(1) The business shall be carried on only in the building or buildings designated in the license, provided that repairs may be made by the dealer or his employees outside of such premises.

(2) The license or a copy certified by the issuing authority shall be displayed at all times in a conspicuous place on the business premises where it can be easily read.

(3) No firearm or imitation thereof shall be placed in any window or in any other part of the premises where it can be readily seen from the outside.

(4) No rifle or shotgun, except antique rifles or shotguns, shall be delivered to any person unless such person possesses and exhibits a valid firearms purchaser identification card and furnishes the seller, on the form prescribed by the superintendent, a certification signed by him setting forth his name, permanent address, firearms purchaser identification card number and such other information as the superintendent may by rule or regulation require. The certification shall be retained by the dealer and shall be made available for inspection by any law enforcement officer at any reasonable time.

(5) No handgun shall be delivered to any person unless:

(a) Such person possesses and exhibits a valid permit to purchase a firearm and at least seven days have elapsed since the date of application for the permit;

(b) The person is personally known to the seller or presents evidence of his identity;

(c) The handgun is unloaded and securely wrapped;

(d) Except as otherwise provided in subparagraph (e) of this paragraph, the handgun is accompanied by a trigger lock or a locked case, gun box, container or other secure facility; provided, however, this provision shall not apply to antique handguns. The exemption afforded under this subparagraph for antique handguns shall be narrowly construed, limited solely to the requirements set forth herein and shall not be deemed to afford or authorize any other exemption from the regulatory provisions governing firearms set forth in chapter 39 and chapter 58 of Title 2C of the New Jersey Statutes; and

(e) On and after the first day of the sixth month following the date on which the list of personalized handguns is prepared and delivered pursuant to section 3 of P.L.2002, c.130 (C.2C:58-2.4), the handgun is identified as a personalized handgun and included on that list or is an antique handgun. The provisions of subparagraph (d) of this section shall not apply to the delivery of a personalized handgun.

(6) The dealer shall keep a true record of every handgun sold, given or otherwise delivered or disposed of, in accordance with the provisions of subsections b. through e. of this section and the record shall note that a trigger lock, locked case, gun box, container or other secure facility was delivered along with the handgun.

b. Records. Every person engaged in the retail business of selling, leasing or otherwise transferring a handgun, as a retail dealer or otherwise, shall keep a register in which shall be entered the time of the sale, lease or other transfer, the date thereof, the name, age, date of birth, complexion, occupation, residence and a physical description including distinguishing physical characteristics, if any, of the purchaser, lessee or transferee, the name and permanent home address of the person making the sale, lease or transfer, the place of the transaction, and the make, model, manufacturer's number, caliber and other marks of identification on such handgun and such other information as the superintendent shall deem necessary for the proper enforcement of this chapter. The register shall be retained by the dealer and shall be made available at all reasonable hours for inspection by any law enforcement officer.

c. Forms of register. The superintendent shall prepare the form of the register as described in subsection b. of this section and furnish the same in triplicate to each person licensed to be engaged in the business of selling, leasing or otherwise transferring firearms.

d. Signatures on register. The purchaser, lessee or transferee of any handgun shall sign, and the dealer shall require him to sign his name to the register, in triplicate, and the person making the sale, lease or transfer shall affix his name, in triplicate, as a witness to the signature. The signatures shall constitute a representation of the accuracy of the information contained in the register.

e. Copies of register entries; delivery to chief of police or county clerk. Within 5 days of the date of the sale, assignment or transfer, the dealer shall deliver or mail by certified mail, return receipt requested, legible copies of the register forms to the office of the chief of police of the municipality in which the purchaser resides, or to the office of the captain of the precinct of the municipality in which the purchaser resides, and to the superintendent. If hand delivered a receipt shall be given to the dealer therefor.where a sale, assignment or transfer is made to a purchaser who resides in a municipality having no chief of police, the dealer shall, within 5 days of the transaction, mail a duplicate copy of the register sheet to the clerk of the county within which the purchaser resides.

2C:58-2.1. Guidelines for delivery of handguns

The Superintendent of State Police, in consultation with the Attorney General, shall promulgate guidelines to effectuate the purposes of P.L. 1999, c. 233.

2C:58-2.3. Reports as to availability of personalized handguns.

a. On the first day of the sixth month following the effective date of P.L.2002, c.130 (C.2C:58-2.2 et al.), the Attorney General shall report to the Governor and the Legislature as to the availability of personalized handguns for retail sales purposes. If the Attorney General determines that personalized handguns are not available for retail sales purposes, the Attorney General, every six months thereafter, shall report to the Governor and the Legislature as to the availability of personalized handguns for retail sales purposes until such time as the Attorney General shall deem that personalized handguns are available for retail sales purposes and so report to the Governor and the Legislature. In making this determination, the Attorney General may consult with any other neutral and detached public or private entity that may have useful information and expertise to assist in determining whether, through performance and other relevant indicators, a handgun meets the statutory definition of a personalized handgun set forth in N.J.S.2C:39-1.

b. For the purposes of this section, personalized handguns shall be deemed to be available for retail sales purposes if at least one manufacturer has delivered at least one production model of a personalized handgun to a registered or licensed wholesale or retail dealer in New Jersey or any other state. As used in this subsection, the term "production model" shall mean a handgun which is the product of a regular manufacturing process that produces multiple copies of the same handgun model, and shall not include a prototype or other unique specimen that is offered for sale.

2C:58-2.4. List of personalized handguns

a. On the first day of the 24th month following the date on which the Attorney General reports that personalized handguns are available for retail sales purposes pursuant to section 2 of P.L. 2002, c.130 (C.2C:58-2.3), the Attorney General shall direct the Superintendent of State Police to promulgate a list of personalized handguns that may be sold in the State. This list shall identify those handguns by manufacturer, model and caliber.

b. The list required under subsection a. of this section shall be prepared within six months of the Attorney General's directive to the superintendent and a copy thereof made available to registered and licensed firearms dealers in this State. Whenever a handgun is determined to meet the statutory definition of a personalized handgun as set forth in N.J.S.2C:39-1, the Attorney General shall report that determination in writing to the Governor and the Legislature within 60 days. The superintendent shall promptly amend and supplement the list to include handguns which meet the statutory definition of a personalized handgun as set forth in N.J.S.2C:39-1 or to remove previously listed handguns, if appropriate. Registered and licensed retail firearms dealers in this State shall be notified forthwith of any such changes in the list. The notice shall be given in a manner prescribed by rule and regulation. The Attorney General shall promulgate rules and regulations establishing a process for handgun manufacturers to demonstrate that their handguns meet the statutory definition of a personalized handgun set forth in N.J.S.2C:39-1 and request that their handgun be added to this list. These rules and regulations may require that the handgun manufacturer:

(1) deliver a handgun or handguns to the Attorney General or his designee for testing;

(2) pay a reasonable application fee; and

(3) pay any reasonable costs incurred in, or associated with, the testing and independent scientific analysis of the handgun, including any analysis of the technology the manufacturer has incorporated within the handgun's design to limit its operational use, that is conducted to determine whether the handgun meets the statutory definition of a personalized handgun set forth in N.J.S.2C:39-1.

2C:58-2.5. Sale of personalized handguns, inapplicability

a. On and after the first day of the sixth month following the preparation and delivery of the list of personalized handguns which may be sold in the State pursuant to section 3 of P.L.2002, c.130 (C.2C:58-2.4), no person registered or licensed by the superintendent as a manufacturer, wholesale dealer of firearms, retail dealer of firearms or agent or employee of a wholesale or retail dealer of firearms pursuant to the provisions of N.J.S.2C:58-1 or N.J.S.2C:58-2 shall transport into this State, sell, expose for sale, possess with the intent of selling, assign or otherwise transfer any handgun unless it is a personalized handgun or an antique handgun.

b. The provisions of this section shall not apply to handguns to be sold, transferred, assigned and delivered for official use to: (1) State and local law enforcement officers of this State; (2) federal law enforcement officers and any other federal officers and employees required to carry firearms in the performance of their official duties and (3) members of the Armed Forces of the United States or of the National Guard.

c. The provisions of this section also shall not apply to handguns to be sold, transferred, assigned and delivered solely for use in competitive shooting matches sanctioned by the Civilian Marksmanship Program, the International Olympic Committee or USA Shooting. The Attorney General may promulgate rules and regulations governing the scope and application of the exemption afforded under this section. The Attorney General, by rule and regulation, may require, at a minimum, that a person acquiring a handgun pursuant to this section submit valid proof of participation in these sanctioned shooting matches.

d. No later than 30 days after the preparation and delivery of the list of personalized handguns which may be sold in the State pursuant to section 3 of P.L.2002, c.130 (C.2C:58-2.4), there shall be established a seven-member commission in the Department of Law and Public Safety that shall meet at least once a year to determine whether personalized handguns qualify for use by State and local law enforcement officers. The Governor shall appoint the following six members of the commission: a county sheriff; a county law enforcement officer; a county prosecutor; one local law enforcement officer who shall be an active member of the New Jersey Fraternal Order of Police; one local law enforcement officer who shall be an active member of the New Jersey State Policemen's Benevolent Association; and an experienced firearms instructor qualified to teach a firearms training course approved by the Police Training Commission. The seventh member of the commission shall be the Superintendent of State Police.

The commission shall issue a report to the Attorney General upon its determination that personalized handguns qualify for use by State and local law enforcement officers. In making this determination, the commission shall consider any advantages and disadvantages to using these weapons in the performance of the official duties of law enforcement officers and shall give due regard to the safety of law enforcement officers and others. The commission shall expire thereafter. The Attorney General shall be authorized to promulgate rules and regulations that apply the provisions of this section to handguns to be sold, transferred, assigned and delivered for official use to State and local law enforcement officers upon a determination by the commission that personalized handguns qualify for use by State and local law enforcement officers.

e. A person who knowingly violates the provisions of this section is guilty of a crime of the fourth degree.

2C:58-2.6. Rules, regulations The Attorney General, in accordance with the provisions of the "Administrative Procedure Act," P.L.1968, c.410 (C.52:14B-1 et seq.), shall promulgate rules and regulations to effectuate the purposes of this act.

2C:58-3. Purchase of firearms

a. Permit to purchase a handgun. No person shall sell, give, transfer, assign or otherwise dispose of, nor receive, purchase, or otherwise acquire a handgun unless the purchaser, assignee, donee, receiver or holder is licensed as a dealer under this chapter or has first secured a permit to purchase a handgun as provided by this section.

b. Firearms purchaser identification card. No person shall sell, give, transfer, assign or otherwise dispose of nor receive, purchase or otherwise acquire an antique cannon or a rifle or shotgun, other than an antique rifle or shotgun, unless the purchaser, assignee, donee, receiver or holder is licensed as a dealer under this chapter or possesses a valid firearms purchaser identification card, and first exhibits said card to the seller, donor, transferor or assignor, and unless the purchaser, assignee, donee, receiver or holder signs a written certification, on a form prescribed by the superintendent, which shall indicate that he presently complies with the requirements of subsection c. of this section and shall contain his name, address and firearms purchaser identification card number or dealer's registration number. The said certification shall be retained by the seller, as provided in section 2C:58-2a., or, in the case of a person who is not a dealer, it may be filed with the chief of police of the municipality in which he resides or with the superintendent.

c. Who may obtain. No person of good character and good repute in the community in which he lives, and who is not subject to any of the disabilities set forth in this section or other sections of this chapter, shall be denied a permit to purchase a handgun or a firearms purchaser identification card, except as hereinafter set forth. No handgun purchase permit or firearms purchaser identification card shall be issued:

(1) To any person who has been convicted of any crime, or a disorderly persons offense involving an act of domestic violence as defined in section 3 of P.L.1991, c.261 (C.2C:25-19), whether or not armed with or possessing a weapon at the time of such offense;

(2) To any drug dependent person as defined in section 2 of P.L. 1970, c. 226 (C. 24:21-2), to any person who is confined for a mental disorder to a hospital, mental institution or sanitarium, or to any person who is presently an habitual drunkard;

(3) To any person who suffers from a physical defect or disease which would make it unsafe for him to handle firearms, to any person who has ever been confined for a mental disorder, or to any alcoholic unless any of the foregoing persons produces a certificate of a medical doctor or psychiatrist licensed in New Jersey, or other satisfactory proof, that he is no longer suffering from that particular disability in such a manner that would interfere with or handicap him in the handling of firearms; to any person who knowingly falsifies any information on the application form for a handgun purchase permit or firearms purchaser identification card;

(4) To any person under the age of 18 years for a firearms purchaser identification card and to any person under the age of 21 years for a permit to purchase a handgun;

(5) To any person where the issuance would not be in the interest of the public health, safety or welfare;

(6) To any person who is subject to a restraining order issued pursuant to the "Prevention of Domestic Violence Act of 1991," P.L.1991, c.261 (C.2C:25-17 et seq.) prohibiting the person from possessing any firearm;

(7) To any person who as a juvenile was adjudicated delinquent for an offense which, if committed by an adult, would constitute a crime and the offense involved the unlawful use or possession of a weapon, explosive or destructive device or is enumerated in subsection d. of section 2 of P.L.1997, c.117 (C:2C:43-7.2); or

(8) To any person whose firearm is seized pursuant to the "Prevention of Domestic Violence Act of 1991," P.L.1991, c.261 (C.2C:25-17 et seq.) and whose firearm has not been returned.

d. Issuance. The chief of police of an organized full-time police department of the municipality where the applicant resides or the superintendent, in all other cases, shall upon application, issue to any person qualified under the provisions of subsection c. of this section a permit to purchase a handgun or a firearms purchaser identification card.

Any person aggrieved by the denial of a permit or identification card may request a hearing in the Superior Court of the county in which he resides if he is a resident of New Jersey or in the Superior Court of the county in which his application was filed if he is a nonresident. The request for a hearing shall be made in writing within 30 days of the denial of the application for a permit or identification card. The applicant shall serve a copy of his request for a hearing upon the chief of police of the municipality in which he resides, if he is a resident of New Jersey, and upon the superintendent in all cases. The hearing shall be held and a record made thereof within 30 days of the receipt of the application for such hearing by the judge of the Superior Court. No formal pleading and no filing fee shall be required as a preliminary to such hearing. Appeals from the results of such hearing shall be in accordance with law.

e. Applications. Applications for permits to purchase a handgun and for firearms purchaser identification cards shall be in the form prescribed by the superintendent and shall set forth the name, residence, place of business, age, date of birth, occupation, sex and physical description, including distinguishing physical characteristics, if any, of the applicant, and shall state whether the applicant is a citizen, whether he is an alcoholic, habitual drunkard, drug dependent person as defined in section 2 of P.L. 1970, c. 226 (C.24:21-2), whether he has ever been confined or committed to a mental institution or hospital for treatment or observation of a mental or psychiatric condition on a temporary, interim or permanent basis, giving the name and location of the institution or hospital and the dates of such confinement or commitment, whether he has been attended, treated or observed by any doctor or psychiatrist or at any hospital or mental institution on an inpatient or outpatient basis for any mental or psychiatric condition, giving the name and location of the doctor, psychiatrist, hospital or institution and the dates of such occurrence, whether he presently or ever has been a member of any organization which advocates or approves the commission of acts of force and violence to overthrow the Government of the United States or of this State, or which seeks to deny others their rights under the Constitution of either the United States or the State of New Jersey, whether he has ever been convicted of a crime or disorderly persons offense, whether the person is subject to a restraining order issued pursuant to the "Prevention of Domestic Violence Act of 1991," P.L.1991, c.261 (C.2C:25-17 et. seq.) prohibiting the person from possessing any firearm, and such other information as the superintendent shall deem necessary for the proper enforcement of this chapter. For the purpose of complying with this subsection, the applicant shall waive any statutory or other right of confidentiality relating to institutional confinement. The application shall be signed by the applicant and shall contain as references the names and addresses of two reputable citizens personally acquainted with him.

Application blanks shall be obtainable from the superintendent, from any other officer authorized to grant such permit or identification card, and from licensed retail dealers.

The chief police officer or the superintendent shall obtain the fingerprints of the applicant and shall have them compared with any and all records of fingerprints in the municipality and county in which the applicant resides and also the records of the State Bureau of Identification and the Federal Bureau of Investigation, provided that an applicant for a handgun purchase permit who possesses a valid firearms purchaser identification card, or who has previously obtained a handgun purchase permit from the same licensing authority for which he was previously fingerprinted, and who provides other reasonably satisfactory proof of his identity, need not be fingerprinted again; however, the chief police officer or the superintendent shall proceed to investigate the application to determine whether or not the applicant has become subject to any of the disabilities set forth in this chapter.

f. Granting of permit or identification card; fee; term; renewal; revocation. The application for the permit to purchase a handgun together with a fee of $2.00, or the application for the firearms purchaser identification card together with a fee of $5.00, shall be delivered or forwarded to the licensing authority who shall investigate the same and, unless good cause for the denial thereof appears, shall grant the permit or the identification card, or both, if application has been made therefor, within 30 days from the date of receipt of the application for residents of this State and within 45 days for nonresident applicants. A permit to purchase a handgun shall be valid for a period of 90 days from the date of issuance and may be renewed by the issuing authority for good cause for an additional 90 days. A firearms purchaser identification card shall be valid until such time as the holder becomes subject to any of the disabilities set forth in subsection c. of this section, whereupon the card shall be void and shall be returned within five days by the holder to the superintendent, who shall then advise the licensing authority. Failure of the holder to return the firearms purchaser identification card to the superintendent within the said five days shall be an offense under section 2C:39-10a. Any firearms purchaser identification card may be revoked by the Superior Court of the county wherein the card was issued, after hearing upon notice, upon a finding that the holder thereof no longer qualifies for the issuance of such permit. The county prosecutor of any county, the chief police officer of any municipality or any citizen may apply to such court at any time for the revocation of such card.

There shall be no conditions or requirements added to the form or content of the application, or required by the licensing authority for the issuance of a permit or identification card, other than those that are specifically set forth in this chapter.

g. Disposition of fees. All fees for permits shall be paid to the State Treasury if the permit is issued by the superintendent, to the municipality if issued by the chief of police, and to the county treasurer if issued by the judge of the Superior Court.

h. Form of permit; quadruplicate; disposition of copies. The permit shall be in the form prescribed by the superintendent and shall be issued to the applicant in quadruplicate. Prior to the time he receives the handgun from the seller, the applicant shall deliver to the seller the permit in quadruplicate and the seller shall complete all of the information required on the form. Within five days of the date of the sale, the seller shall forward the original copy to the superintendent and the second copy to the chief of police of the municipality in which the purchaser resides, except that in a municipality having no chief of police, such copy shall be forwarded to the superintendent. The third copy shall then be returned to the purchaser with the pistol or revolver and the fourth copy shall be kept by the seller as a permanent record.

i. Restriction on number of firearms person may purchase. Only one handgun shall be purchased or delivered on each permit, but a person shall not be restricted as to the number of rifles or shotguns he may purchase, provided he possesses a valid firearms purchaser identification card and provided further that he signs the certification required in subsection b. of this section for each transaction.

j. Firearms passing to heirs or legatees. Notwithstanding any other provision of this section concerning the transfer, receipt or acquisition of a firearm, a permit to purchase or a firearms purchaser identification card shall not be required for the passing of a firearm upon the death of an owner thereof to his heir or legatee, whether the same be by testamentary bequest or by the laws of intestacy. The person who shall so receive, or acquire said firearm shall, however, be subject to all other provisions of this chapter. If the heir or legatee of such firearm does not qualify to possess or carry it, he may retain ownership of the firearm for the purpose of sale for a period not exceeding 180 days, or for such further limited period as may be approved by the chief law enforcement officer of the municipality in which the heir or legatee resides or the superintendent, provided that such firearm is in the custody of the chief law enforcement officer of the municipality or the superintendent during such period.

k. Sawed-off shotguns. Nothing in this section shall be construed to authorize the purchase or possession of any sawed-off shotgun.

l. Nothing in this section and in N.J.S. 2C:58-2 shall apply to the sale or purchase of a visual distress signaling device approved by the United States Coast Guard, solely for possession on a private or commercial aircraft or any boat; provided, however, that no person under the age of 18 years shall purchase nor shall any person sell to a person under the age of 18 years such a visual distress signaling device.

2C:58-3.1. Temporary transfer of firearms

a. Notwithstanding the provisions of N.J.S. 2C:39-9, N.J.S.2C:58-2, N.J.S.2C:58-3 or any other statute to the contrary concerning the transfer or disposition of firearms, the legal owner, or a dealer licensed under N.J.S.2C:58-2, may temporarily transfer a handgun, rifle or shotgun to another person who is 18 years of age or older, whether or not the person receiving the firearm holds a firearms purchaser identification card or a permit to carry a handgun. The person to whom a handgun, rifle or shotgun is temporarily transferred by the legal owner of the firearm or a licensed dealer may receive, possess, carry and use that handgun, rifle or shotgun, if the transfer is made upon a firing range operated by a licensed dealer, by a law enforcement agency, a legally recognized military organization or a rifle or pistol club which has filed a copy of its charter with the superintendent and annually submits to the superintendent a list of its members and if the firearm is received, possessed, carried and used for the sole purpose of target practice, trap or skeet shooting, or competition upon that firing range or instruction and training at any location.

A transfer under this subsection shall be for not more than eight consecutive hours in any 24-hour period and may be made for a set fee or an hourly charge.

The firearm shall be handled and used by the person to whom it is temporarily transferred only in the actual presence or under the direct supervision of the legal owner of the firearm, the dealer who transferred the firearm or any other person competent to supervise the handling and use of firearms and authorized to act for that purpose by the legal owner or licensed dealer. The legal owner of the firearm or the licensed dealer shall be on the premises or the property of the firing range during the entire time that the firearm is in the possession of the person to whom it is temporarily transferred.

The term "legal owner" as used in this subsection means a natural person and does not include an organization, commercial enterprise, or a licensed manufacturer, wholesaler or dealer of firearms.

b. Notwithstanding the provisions of N.J.S. 2C:39-9, N.J.S.2C:58-2, N.J.S.2C:58-3 or any other statute to the contrary concerning the transfer and disposition of firearms, a legal owner of a shotgun or a rifle may temporarily transfer that firearm to another person who is 18 years of age or older, whether or not the person receiving the firearm holds a firearms purchaser identification card. The person to whom a shotgun or rifle is temporarily transferred by the legal owner may receive, possess, carry and use that shotgun or rifle in the woods or fields or upon the waters of this State for the purposes of hunting if the transfer is made in the woods or fields or upon the waters of this State, the shot-gun or rifle is legal and appropriate for hunting and the person to whom the firearm is temporarily transferred possesses a valid license to hunt with a firearm, and a valid rifle permit if the firearm is a rifle, obtained in accordance with the provisions of chapter 3 of Title 23 of the Revised Statutes.

The transfer of a firearm under this subsection shall be for not more than eight consecutive hours in any 24-hour period and no fee shall be charged for the transfer.

The legal owner of the firearm which is temporarily transferred shall remain in the actual presence or in the vicinity of the person to whom it was transferred during the entire time that the firearm is in that person's possession.

The term "legal owner" as used in this subsection means a natural person and does not include an organization, commercial enterprise, or a licensed manufacturer, wholesaler or dealer of firearms.

c. No firearm shall be temporarily transferred or received under the provisions of subsections a. or b. of this section for the purposes described in section 1 of P.L. 1983, c. 229 (C. 2C:39-14).

d. An owner or dealer shall not transfer a firearm to any person pursuant to the provisions of this section if the owner or dealer knows the person does not meet the qualifications set forth in subsection c. of N.J.S.2C:58-3 for obtaining or holding a firearms purchaser identification card or a handgun purchase permit. A person shall not receive, possess, carry or use a firearm pursuant to the provisions of this section if the person knows he does not meet the qualifications set forth in subsection c. of N.J.S.2C:58-3 for obtaining or holding a firearms purchaser identification card or a handgun purchase permit.

2C:58-3.2. Temporary transfer of firearm for training purposes

a. Notwithstanding the provisions of N.J.S. 2C:39-9, N.J.S.2C:58-2, N.J.S.2C:58-3 or any other statute to the contrary, a person who is certified as an instructor in the use, handling and maintenance of firearms by the Police Training Commission, the Division of Fish, Game and Wildlife and the State Park Service in the Department of Environmental Protection [now, Department of Environmental Protection and Energy], the Director of Civilian Marksmanship of the United States Department of the Army or by a recognized rifle or pistol association that certifies instructors may transfer a firearm temporarily in accordance with the terms of this section to a person participating in a training course for the use, handling and maintenance of firearms by the Police Training Commission, the Division of Fish, Game and Wildlife, the Director of Civilian Marksmanship or by a recognized rifle or pistol association that certifies instructors. The person to whom a firearm is transferred by a certified instructor in accordance with the terms of this section may receive, possess, carry and use the firearm temporarily during the sessions of the course for the purpose of training and participating in the course.

b. A transfer of a firearm under this section may be made only if:

(1) the transfer is made upon a firearms range or, if the firearm is unloaded, in an area designated and appropriate for the training;

(2) the transfer is made during the sessions of the firearms course for the sole purpose of participating in the course;

(3) the transfer is made for not more than eight consecutive hours in any 24-hour period; and

(4) the transferred firearm is used and handled only in the actual presence and under the direct supervision of the instructor.

c. The transfer permitted by this section may be made whether or not the person participating in the course holds a firearms license, firearms purchaser identification card or a handgun purchase permit. However, an instructor shall not knowingly transfer a firearm under the terms of this section to a person who does not meet the qualifications set forth in subsection c. of N.J.S. 2C:58-3 for obtaining or holding a firearms purchaser identification card or a handgun purchase permit, and a person who knows that he does not meet such qualifications shall not receive the transferred firearm under the terms of this section.

d. No firearm shall be transferred or received under the provisions of this section for purposes described in section 1 of P.L. 1983, c. 229 (C. 2C:39-14).

2C:58-3.3. "Handgun ammunition" defined; sale, purchase, etc., regulated; violation, fourth degree crime

a. As used in this act, "handgun ammunition" means ammunition specifically designed to be used only in a handgun. "Handgun ammunition" shall not include blank ammunition, air gun pellets, flare gun ammunition, nail gun ammunition, paint ball ammunition, or any non-fixed ammunition.

b. No person shall sell, give, transfer, assign or otherwise dispose of, or receive, purchase, or otherwise acquire handgun ammunition unless the purchaser, assignee, donee, receiver or holder is licensed as a manufacturer, wholesaler, or dealer under this chapter or is the holder of and possesses a valid firearms purchaser identification card, a valid copy of a permit to purchase a handgun, or a valid permit to carry a handgun and first exhibits such card or permit to the seller, donor, transferor or assignor.

c. No person shall sell, give, transfer, assign or otherwise dispose of handgun ammunition to a person who is under 21 years of age.

d. The provisions of this section shall not apply to a collector of firearms or ammunition as curios or relics who purchases, receives, acquires, possesses, or transfers handgun ammunition which is recognized as being historical in nature or of historical significance.

e. A person who violates this section shall be guilty of a crime of the fourth degree, except that nothing contained herein shall be construed to prohibit the sale, transfer, assignment or disposition of handgun ammunition to or the purchase, receipt or acceptance of ammunition by a law enforcement agency or law enforcement official for law enforcement purposes.

f. Nothing in this section shall be construed to prohibit the transfer of ammunition for use in a lawfully transferred firearm in accordance with the provisions of section 1 of P.L.1992, c.74 (C.2C:58-3.1), section 1 of P.L.1997, c.375 (C.2C:58-3.2) or section 14 of P.L.1979, c.179 (C.2C:58-6.1).

g. Nothing in this section shall be construed to prohibit the sale of a de minimis amount of handgun ammunition at a firearms range operated by a licensed dealer; a law enforcement agency; a legally recognized military organization; or a rifle or pistol club which has filed a copy of its charter with the superintendent for immediate use at that range.

2C:58-4. Permits to carry handguns

a. Scope and duration of authority. Any person who holds a valid permit to carry a handgun issued pursuant to this section shall be authorized to carry a handgun in all parts of this State, except as prohibited by section 2C:39-5e. One permit shall be sufficient for all handguns owned by the holder thereof, but the permit shall apply only to a handgun carried by the actual and legal holder of the permit.

All permits to carry handguns shall expire 2 years from the date of issuance or, in the case of an employee of an armored car company, upon termination of his employment by the company occurring prior thereto whichever is earlier in time, and they may thereafter be renewed every 2 years in the same manner and subject to the same conditions as in the case of original applications.

b. Application forms. All applications for permits to carry handguns, and all applications for renewal of such permits, shall be made on the forms prescribed by the superintendent. Each application shall set forth the full name, date of birth, sex, residence, occupation, place of business or employment, and physical description of the applicant, and such other information as the superintendent may prescribe for the determination of the applicant's eligibility for a permit and for the proper enforcement of this chapter. The application shall be signed by the applicant under oath, and shall be indorsed by three reputable persons who have known the applicant for at least 3 years preceding the date of application, and who shall certify thereon that the applicant is a person of good moral character and behavior.

c. Investigation and approval. Each application shall in the first instance be submitted to the chief police officer of the municipality in which the applicant resides, or to the superintendent,

(1) if the applicant is an employee of an armored car company, or

(2) if there is no chief police officer in the municipality where the applicant resides, or

(3) if the applicant does not reside in this State. The chief police officer, or the superintendent, as the case may be, shall cause the fingerprints of the applicant to be taken and compared with any and all records maintained by the municipality, the county in which it is located, the State Bureau of Identification and the Federal Bureau of Identification. He shall also determine and record a complete description of each handgun the applicant intends to carry.

No application shall be approved by the chief police officer or the superintendent unless the applicant demonstrates that he is not subject to any of the disabilities set forth in 2C:58-3c., that he is thoroughly familiar with the safe handling and use of handguns, and that he has a justifiable need to carry a handgun. If the application is not approved by the chief police officer or the superintendent within 60 days of filing, it shall be deemed to have been approved, unless the applicant agrees to an extension of time in writing.

d. Issuance by Superior Court; fee. If the application has been approved by the chief police officer or the superintendent, as the case may be, the applicant shall forthwith present it to the Superior Court of the county in which the applicant resides, or to the Superior Court in any county where he intends to carry a handgun, in the case of a nonresident or employee of an armored car company. The court shall issue the permit to the applicant if, but only if, it is satisfied that the applicant is a person of good character who is not subject to any of the disabilities set forth in section 2C:58-3c., that he is thoroughly familiar with the safe handling and use of handguns, and that he has a justifiable need to carry a handgun. The court may at its discretion issue a limited-type permit which would restrict the applicant as to the types of handguns he may carry and where and for what purposes such handguns may be carried. At the time of issuance, the applicant shall pay to the county clerk of the county where the permit was issued a permit fee of $20.00.

e. Appeals from denial of applications. Any person aggrieved by the denial by the chief police officer or the superintendent of approval for a permit to carry a handgun may request a hearing in the Superior Court of the county in which he resides or in any county in which he intends to carry a handgun, in the case of a nonresident, by filing a written request for such a hearing within 30 days of the denial. Copies of the request shall be served upon the superintendent, the county prosecutor and the chief police officer of the municipality where the applicant resides, if he is a resident of this State. The hearing shall be held within 30 days of the filing of the request, and no formal pleading or filing fee shall be required. Appeals from the determination at such a hearing shall be in accordance with law and the rules governing the courts of this State.

If the superintendent or chief police officer approves an application and the Superior Court denies the application and refuses to issue a permit, the applicant may appeal such denial in accordance with law and the rules governing the courts of this State.

f. Revocation of permits. Any permit issued under this section shall be void at such time as the holder thereof becomes subject to any of the disabilities set forth in section 2C:58-3c, and the holder of such a void permit shall immediately surrender the permit to the superintendent who shall give notice to the licensing authority.

Any permit may be revoked by the Superior Court, after hearing upon notice to the holder, if the court finds that the holder is no longer qualified for the issuance of such a permit. The county prosecutor of any county, the chief police officer of any municipality, the superintendent or any citizen may apply to the court at any time for the revocation of any permit issued pursuant to this section.

2C:58-4.1. Employee of armored car company; application; letter from chief executive officer In addition to the requirements of N.J.S. 2C:58-4 any application to carry a handgun by an employee of an armored car company shall be accompanied by a letter from the chief executive officer of the armored car company verifying employment of the applicant; endorsing approval of the application; and agreeing to notify the superintendent forthwith upon the termination of the employee of any person to whom a permit is issued and to obtain from the employee the permit which shall thereupon be surrendered to the superintendent.

2C:58-5. Licenses to possess and carry machine guns and assault firearms

a. Any person who desires to purchase, possess and carry a machine gun or assault firearm in this State may apply for a license to do so by filing in the Superior Court in the county in which he resides, or conducts his business if a nonresident, a written application setting forth in detail his reasons for desiring such a license. The Superior Court shall refer the application to the county prosecutor for investigation and recommendation. A copy of the prosecutor's report, together with a copy of the notice of the hearing on the application, shall be served upon the superintendent and the chief police officer of every municipality in which the applicant intends to carry the machine gun or assault firearm, unless, for good cause shown, the court orders notice to be given wholly or in part by publication.

b. No license shall be issued to any person who would not qualify for a permit to carry a handgun under section 2C:58-4, and no license shall be issued unless the court finds that the public safety and welfare so require. Any person aggrieved by the decision of the court in granting or denying an application, including the applicant, the prosecutor, or any law enforcement officer entitled to notice under subsection a. who appeared in opposition to the application, may appeal said decision in accordance with law and the rules governing the courts of this State.

c. Upon the issuance of any license under this section, true copies of such license shall be filed with the superintendent and the chief police officer of the municipality where the licensee resides or has his place of business.

d. In issuing any license under this section, the court shall attach thereto such conditions and limitations as it deems to be in the public interest. Unless otherwise provided by court order at the time of issuance, each license shall expire 1 year from the date of issuance, and may be renewed in the same manner and under the same conditions as apply to original applications.

e. Any license may be revoked by the Superior Court, after a hearing upon notice to the holder thereof, if the court finds that the holder is no longer qualified for the issuance of such a license or that revocation is necessary for the public safety and welfare. Any citizen may apply to the court for revocation of a license issued under this section.

f. A filing fee of $75.00 shall be required for each application filed pursuant to the provisions of this section. Of this filing fee, $25.00 shall be forwarded to the State Treasury for deposit in the account used by the Violent Crimes Compensation Board in satisfying claims and for related administrative costs pursuant to the provisions of the "Criminal Injuries Compensation Act of 1971," P.L. 1971, c. 317 (C.52:4B-1 et seq.).

g. Any license granted pursuant to the provisions of this section shall expire two years from the date of issuance and may be renewed in the same manner and under the same conditions as apply to original applications. If the holder of a license dies, the holder's heirs or estate shall have 90 days to dispose of that firearm as provided in section 12 of P.L. 1990, c. 32 (C.2C:58-13).

h. If an assault firearm licensed pursuant to the provisions of this section is used in the commission of a crime, the holder of the license for that assault firearm shall be civilly liable for any damages resulting from that crime. The liability imposed by this subsection shall not apply if the assault firearm used in the commission of the crime was stolen and the license holder reported the theft of the firearm to law enforcement authorities within 24 hours of the license holder's knowledge of the theft.

i. Nothing in P.L. 1990, c. 32 (C.2C:58-12 et al.) shall be construed to abridge any exemptions provided under N.J.S.2C:39-6.

2C:58-6.1. Possession of firearms by minors; exceptions

a. No person under the age of 18 years shall purchase, barter or otherwise acquire a firearm and no person under the age of 21 years shall purchase, barter or otherwise acquire a handgun, unless the person is authorized to possess the handgun in connection with the performance of official duties under the provisions of N.J.S. 2C:39-6.

b. No person under the age of 18 years shall possess, carry, fire or use a firearm except as provided under paragraphs (1), (2), (3) and (4) of this subsection; and, unless authorized in connection with the performance of official duties under the provisions of N.J.S.2C:39-6, no person under the age of 21 years shall possess, carry, fire or use a handgun except under the following circumstances:

(1) In the actual presence or under the direct supervision of his father, mother or guardian, or some other person who holds a permit to carry a handgun or a firearms purchaser identification card, as the case may be; or

(2) For the purpose of military drill under the auspices of a legally recognized military organization and under competent supervision; or

(3) For the purpose of competition or target practice in and upon a firing range approved by the governing body of the municipality in which the range is located or the National Rifle Association and which is under competent supervision at the time of such supervision or target practice or instruction and training at any location; or

(4) For the purpose of hunting during the regularly designated hunting season, provided that he possesses a valid hunting license and has successfully completed a hunter's safety course taught by a qualified instructor or conservation officer and possesses a certificate indicating the successful completion of such a course.

c. A person who violates this section shall be guilty of a crime of the fourth degree. For purposes of this section the fact that the act would not constitute a crime if committed by an adult shall not be deemed to prohibit or require waiver of family court jurisdiction pursuant to N.J.S. 2C:4-11 or to preclude a finding of delinquency under the "New Jersey Code of Juvenile Justice," P.L. 1982, c. 77 (C.2A:4A-20 et seq.), P.L. 1982, c. 79 (C.2A:4A-60 et seq.), P.L. 1982, c. 80 (C.2A:4A-76 et seq.) and P.L. 1982, c. 81 (C. 2A:4A-70 et seq.).

2C:58-7. Persons possessing explosives or destructive devices to notify police

a. Any person who becomes the possessor of any explosive, destructive device, or ammunition therefor, which is or may be loaded or otherwise dangerous, except such as is possessed for any lawful commercial or other purpose in connection with which the use of explosives is authorized or as is authorized in subsection d. of N.J.S. 2C:39-6, shall within 15 days notify the police authorities of the municipality in which he resides or the State Police that the same is in his possession and shall present the same to them for inspection.

b. When any such ammunition, explosive or destructive device is presented for inspection it shall be inspected to ascertain whether or not it is loaded or of a dangerous character, and if it is found to be loaded or of dangerous character, it shall be destroyed or be unloaded or so processed as to remove its dangerous character before being returned to the possessor.

c. Any police officer having reasonable cause to believe that any person is possessed of any such ammunition, explosive, or destructive device shall investigate, under a proper search warrant when necessary, and shall seize the same for the purpose of inspection, unloading, processing or destruction, as provided in this section, and the same shall not be returned to the possessor thereof until it has been unloaded or so processed.

2C:58-12. Registration of assault firearms

a. Within 90 days of the effective date of P.L. 1990, c. 32 (C.2C:58-12 et al.), the Attorney General shall promulgate a list by trade name of any assault firearm which the Attorney General determines is an assault firearm which is used for legitimate target-shooting purposes. This list shall include, but need not be limited to, the Colt AR-15 and any other assault firearm used in competitive shooting matches sanctioned by the Director of Civilian Marksmanship of the United States Department of the Army.

b. The owner of an assault firearm purchased on or before May 1, 1990 which is on the list of assault firearms determined by the Attorney General to be legitimate for target-shooting purposes shall have one year from the effective date of P.L. 1990, c. 32 (C.2C:58-12 et al.) to register that firearm. In order to register an assault firearm, the owner shall:

(1) Complete an assault firearm registration statement, in the form to be prescribed by the Superintendent of the State Police;

(2) Pay a registration fee of $50.00 per each assault firearm;

(3) Produce for inspection a valid firearms purchaser identification card, a valid permit to carry handguns, or a copy of the permit to purchase a handgun which was used to purchase the assault firearm which is being registered; and

(4) Submit valid proof that the person is a member of a rifle or pistol club in existence prior to the effective date of P.L. 1990, c. 32 (C.2C: 58-12 et al.).

Membership in a rifle or pistol club shall not be considered valid unless the person joined the club no later than 210 days after the effective date of P.L. 1990, c. 32 (C.2C:58-12 et al.) and unless the rifle or pistol club files its charter with the Superintendent no later than 180 days following the effective date of P.L. 1990, c. 32 (C. 2C:58-12 et al). The rifle or pistol club charter shall contain the name and address of the club's headquarters and the name of the club's officers.

The information to be provided in the registration statement shall include, but shall not be limited to: the name and address of the registrant; the number or numbers on the registrant's firearms purchaser identification card, permit to carry handguns, or permit to purchase a handgun; the name, address, and telephone number of the rifle or pistol club in which the registrant is a member; and the make, model, and serial number of the assault firearm being registered. Each registration statement shall be signed by the registrant, and the signature shall constitute a representation of the accuracy of the information contained in the registration statement.

c. For an applicant who resides in a municipality with an organized full-time police department, the registration shall take place at the main office of the police department. For all other applicants, the registration shall take place at any State Police station.

d. Within 60 days of the effective date of P.L. 1990, c. 32 (C.2C-58-12 et al.), the Superintendent shall prepare the form of registration statement as described in subsection b. of this section and shall provide a suitable supply of statements to each organized full-time municipal police department and each State Police station.

e. One copy of the completed assault firearms registration statement shall be returned to the registrant, a second copy shall be sent to the Superintendent, and, if the registration takes place at a municipal police department, a third copy shall be retained by that municipal police department.

f. If the owner of an assault firearm which has been registered pursuant to this section dies, the owner's heirs or estate shall have 90 days to dispose of that firearm in accordance with section 12 of P.L. 1990, c. 32 (C.2C:58-13).

g. If an assault firearm registered pursuant to the provisions of this section is used in the commission of a crime, the registrant of that assault firearm shall be civilly liable for any damages resulting from that crime. The liability imposed by this subsection shall not apply if the assault firearm used in the commission of the crime was stolen and the registrant reported the theft of the firearm to law enforcement authorities within 24 hours of the registrant's knowledge of the theft.

h. Of the registration fee required pursuant to subsection b. of this section, $20.00 shall be forwarded to the State Treasury for deposit in the account used by the Violent Crimes Compensation Board in satisfying claims and for related administrative costs pursuant to the provisions of the "Criminal Injuries Compensation Act of 1971," P.L. 1971, c. 317 (C.52:4B-1 et seq.).

2C:58-13. Transfer of assault firearm to another; rendering inoperable; voluntarily surrendering

a. Any person who legally owns an assault firearm on the effective date of this act and who is unable to register or chooses not to register the firearm pursuant to section 11 of P.L. 1990, c. 32 (C.2C:58-12) may retain possession of that firearm for a period not to exceed one year from the effective date of this act. During this time period, the owner of the assault firearm shall either:

(1) Transfer the assault firearm to any person or firm lawfully entitled to own or possess such firearm;

(2) Render the assault firearm inoperable; or

(3) Voluntarily surrender the assault firearm pursuant to the provisions of N.J.S.2C:39-12.

b. If the owner of an assault firearm elects to render the firearm inoperable, the owner shall file a certification on a form prescribed by the Superintendent of the State Police indicating the date on which the firearm was rendered inoperable. This certification shall be filed with either the chief law enforcement officer of the municipality in which the owner resides or, in the case of an owner who resides outside this State but stores or possesses an assault firearm in this State, with the Superintendent of the State Police.

c. As used in this section, "inoperable" means that the firearm is altered in such a manner that it cannot be immediately fired and that the owner or possessor of the firearm does not possess or have control over the parts necessary to make the firearm operable.

2C:58-15. Minor's access to a loaded firearm; penalty, conditions

a. A person who knows or reasonably should know that a minor is likely to gain access to a loaded firearm at a premises under the person's control commits a disorderly persons offense if a minor gains access to the firearm, unless the person:

(1) Stores the firearm in a securely locked box or container;

(2) Stores the firearm in a location which a reasonable person would believe to be secure; or

(3) Secures the firearm with a trigger lock.

b. This section shall not apply:

(1) To activities authorized by section 14 of P.L. 1979, c. 179, (C.2C:58-6.1), concerning the lawful use of a firearm by a minor; or

(2) Under circumstances where a minor obtained a firearm as a result of an unlawful entry by any person.

c. As used in this act, "minor" means a person under the age of 16.

2C:58-16. Retailer's written warnings; wholesaler's warning; violation, penalty

a. Upon the retail sale or transfer of any firearm, the retail dealer or his employee shall deliver to the purchaser or transferee the following written warning, printed in block letters not less than one-fourth of an inch in height:

"IT IS A CRIMINAL OFFENSE, PUNISHABLE BY A FINE AND IMPRISONMENT, FOR AN ADULT TO LEAVE A LOADED FIREARM WITHIN EASY ACCESS OF A MINOR."

b. Every wholesale and retail dealer of firearms shall conspicuously post at each purchase counter the following warning, printed in block letters not less than one inch in height:

"IT IS A CRIMINAL OFFENSE TO LEAVE A LOADED FIREARM WITHIN EASY ACCESS OF A MINOR."

c. Violation of this section by any retail or wholesale dealer of firearms is a petty disorderly persons offense.

2C:58-17. KeepSafe program established

a. There is established a "KeepSafe" program to encourage and stimulate the safe storage of firearms in the State of New Jersey by providing instant rebates to firearms purchasers who purchase trigger locking devices.

Under the program, a person who purchases a firearm from a retail dealer licensed under the provisions of N.J.S.2C:58-2 shall be eligible for a $5 instant rebate when a compatible trigger locking device is purchased along with that firearm. The licensed retail dealer shall deduct the rebate from the price of the compatible locking device in order to reduce by $5 the cost of the device for the purchaser.

b. The Superintendent of State Police, in conjunction with the Attorney General, shall adopt guidelines in accordance with the Administrative Procedure Act, P.L. 1968, c. 410 (C.52:14B-1 et seq.), to effectuate the purposes of this act.

In addition, the superintendent shall prepare and deliver to each licensed retail firearms dealer in the State the forms necessary to record and report participation in the program. The forms, which shall set forth the name, address, telephone number, State tax number and State license number of the retail firearms dealer, the name of the firearms purchaser and his firearms purchaser identification card number or permit to purchase a handgun number, the make and model number of the compatible trigger locking device purchased and the date of the sale, shall be in duplicate. One copy shall be retained by the retail dealer for his records. The other shall be submitted to the Attorney General for reimbursement. The reimbursement copies shall be submitted monthly at a time prescribed by the superintendent. The submitting retail dealer shall be entitled to a reimbursement of $5 for each trigger locking device sold as part of the KeepSafe program. To help defray any administrative costs, each participating retail dealer shall receive, in addition to the reimbursement, $0.50 for each valid reimbursement copy submitted.

The superintendent also shall provide each licensed retail firearms dealer with a sign to be prominently displayed at a conspicuous place on the dealer's business premises where firearms are offered for sale. The sign shall state substantially the following:

"KEEP NEW JERSEY FIREARMS SAFE!!

TO ENCOURAGE NEW JERSEY GUN OWNERS TO STORE THEIR FIREARMS SAFELY, THE STATE IS OFFERING A $5 INSTANT REBATE WHEN YOU PURCHASE A COMPATIBLE TRIGGER LOCK ALONG WITH YOUR FIREARM. REMEMBER - THE USE OF A TRIGGER LOCK IS ONLY ONE ASPECT OF RESPONSIBLE FIREARM STORAGE. FIREARMS SHOULD BE STORED, UNLOADED AND LOCKED IN A LOCATION THAT IS BOTH SEPARATE FROM THEIR AMMUNITION AND INACCESSIBLE TO CHILDREN.

NEW JERSEY'S FAMILIES AND CHILDREN ARE PRECIOUS - KEEP THEM SAFE!"

2C:58-19. Report of loss, theft of firearm within 36 hours; violations, penalties

The legal owner of a firearm, upon discovering that the firearm is lost or stolen, shall report the loss or theft within 36 hours to the chief law enforcement officer of the municipality in which the loss or theft occurred or, if the municipality does not have a local police force, to the Superintendent of State Police.

A person who violates the provisions of this section shall be liable to a civil penalty of not less than $ 500 for a first offense, and not less than $ 1,000 for any second or subsequent offense. The civil penalty shall be collected pursuant to the "Penalty Enforcement Law of 1999," P.L.1999, c.274 (C.2A:58-10 et seq.).

[Current through New Jersey 213th Legislature, First Annual Session (P.L. 2008 ch. 97 & J.R. 13)]

NEW MEXICO
N.M. STAT.

Chapter 30. Criminal Offenses

Article 7. Weapons and Explosives

30-7-1. "Carrying a deadly weapon" "Carrying a deadly weapon" means being armed with a deadly weapon by having it on the person, or in close proximity thereto, so that the weapon is readily accessible for use.

30-7-2. Unlawful carrying of a deadly weapon

A. Unlawful carrying of a deadly weapon consists of carrying a concealed loaded firearm or any other type of deadly weapon anywhere, except in the following cases:

(1) in the person's residence or on real property belonging to him as owner, lessee, tenant or licensee;

(2) in a private automobile or other private means of conveyance, for lawful protection of the person's or another's person or property;

(3) by a peace officer in accordance with the policies of his law enforcement agency who is certified pursuant to the Law Enforcement Training Act [29-7-1 NMSA 1978];

(4) by a peace officer in accordance with the policies of his law enforcement agency who is employed on a temporary basis by that agency and who has successfully completed a course of firearms instruction prescribed by the New Mexico law enforcement academy or provided by a certified firearms instructor who is employed on a permanent basis by a law enforcement agency; or

(5) by a person in possession of a valid concealed handgun license issued to him by the department of public safety pursuant to the provisions of the Concealed Handgun Carry Act [29-19-1 NMSA 1978].

B. Nothing in this section shall be construed to prevent the carrying of any unloaded firearm.

C. Whoever commits unlawful carrying of a deadly weapon is guilty of a petty misdemeanor.

30-7-2.1. Unlawful carrying of a deadly weapon on school premises

A. Unlawful carrying of a deadly weapon on school premises consists of carrying a deadly weapon on school premises except by:

(1) a peace officer;

(2) school security personnel;

(3) a student, instructor or other school-authorized personnel engaged in army, navy, marine corps or air force reserve officer training corps programs or state-authorized hunter safety training instruction;

(4) a person conducting or participating in a school-approved program, class or other activity involving the carrying of a deadly weapon; or

(5) a person older than nineteen years of age on school premises in a private automobile or other private means of conveyance, for lawful protection of the person's or another's person or property.

B. As used in this section, "school premises" means:

(1) the buildings and grounds, including playgrounds, playing fields and parking areas and any school bus of any public elementary, secondary, junior high or high school in or on which school or school-related activities are being operated under the supervision of a local school board; or

(2) any other public buildings or grounds, including playing fields and parking areas that are not public school property, in or on which public school-related and sanctioned activities are being performed.

C. Whoever commits unlawful carrying of a deadly weapon on school premises is guilty of a fourth degree felony.

30-7-2.2. Unlawful possession of a handgun by a person; exceptions; penalty **A.** Unlawful possession of a handgun by a person consists of a person knowingly having a handgun in his possession or knowingly transporting a handgun, except when the person is:

(1) in attendance at a hunter's safety course or a handgun safety course;

(2) engaging in the use of a handgun for target shooting at an established range authorized by the governing body of the jurisdiction in which the range is located or in an area where the discharge of a handgun without legal justification is not prohibited by law;

(3) engaging in an organized competition involving the use of a handgun;

(4) participating in or practicing for a performance by an organization that has been granted exemption from federal income tax by the United States commissioner of internal revenue as an organization described in Section 501(c)(3) of the United States Internal Revenue Code of 1954, as amended or renumbered;

(5) legal hunting or trapping activities;

(6) traveling, with an unloaded handgun in his possession, to or from an activity described in Paragraph (1), (2), (3), (4) or (5) of this subsection; or

(7) on real property under the control of the person's parent, grandparent or legal guardian and the person is being supervised by his parent, grandparent or legal guardian.

B. A person who commits unlawful possession of a handgun by a person is guilty of a misdemeanor.

C. As used in this section:

(1) "person" means an individual who is less than nineteen years old; and

(2) "handgun" means a loaded or unloaded pistol, revolver or firearm which will or is designed to or may readily be converted to expel a projectile by the action of an explosion and the barrel length of which, not including a revolving, detachable or magazine breech, does not exceed twelve inches.

30-7-2.3. Seizure and forfeiture of a handgun possessed or transported by a person in violation of unlawful possession of a handgun by a person

A. A handgun is subject to seizure and forfeiture by a law enforcement agency when the handgun is possessed or transported by a person in violation of the offense of unlawful possession of a handgun by a person.

B. The provisions of the Forfeiture Act [31-27-1 NMSA 1978] apply to the seizure, forfeiture and disposal of a handgun subject to forfeiture pursuant to Subsection A of this section.

30-7-2.4 Unlawful carrying of a firearm on university premises; notice; penalty

A. Unlawful carrying of a firearm on university premises consists of carrying a firearm on university premises except by:

(1) a peace officer;

(2) university security personnel;

(3) a student, instructor or other university-authorized personnel who are engaged in army, navy, marine corps or air force reserve officer training corps programs or a state-authorized hunter safety training program;

(4) a person conducting or participating in a university-approved program, class or other activity involving the carrying of a firearm; or

(5) a person older than nineteen years of age on university premises in a private automobile or other private means of conveyance, for lawful protection of the person's or another's person or property.

B. A university shall conspicuously post notices on university premises that state that it is unlawful to carry a firearm on university premises.

C. As used in this section:

(1) "university" means a baccalaureate degree-granting post-secondary educational institution, a community college, a branch community college, a technical-vocational institute and an area vocational school; and

(2) "university premises" means:

(a) the buildings and grounds of a university, including playing fields and parking areas of a university, in or on which university or university-related activities are conducted; or

(b) any other public buildings or grounds, including playing fields and parking areas that are not university property, in or on which university-related and sanctioned activities are performed.

D. Whoever commits unlawful carrying of a firearm on university premises is guilty of a petty misdemeanor.

30-7-4. Negligent use of a deadly weapon

A. Negligent use of a deadly weapon consists of: **...**

(2) carrying a firearm while under the influence of an intoxicant or narcotic;

(3) endangering the safety of another by handling or using a firearm or other deadly weapon in a negligent manner; or **...**

B. The provisions of Paragraphs (1), (3) and (4) of Subsection A of this section shall not apply to a peace officer or other public employee who is required or authorized by law to carry or use a firearm in the course of his employment and who carries, handles, uses or discharges a firearm while lawfully engaged in carrying out the duties of his office or employment.

C. The exceptions from criminal liability provided for in Subsection B of this section shall not preclude or affect civil liability for the same conduct.

Whoever commits negligent use of a deadly weapon is guilty of a petty misdemeanor.

30-7-7. Unlawful sale, possession or transportation of explosives Unlawful sale, possession or transportation of explosives consists of:

A. knowingly selling or possessing any explosive or causing such explosive to be transported without having plainly marked in large letters in a conspicuous place on the box or package containing such explosive the name and explosive character thereof and the date of manufacture. For the purpose of this subsection, the term "explosive" is as defined in Section 2 [30-7-18 NMSA 1978] of the Explosives Act, but shall not include:

(1) explosive materials in medicine and medicinal agents in the forms prescribed by the official United States pharmacopoeia or the national formulary;

(2) small arms ammunition and components thereof;

(3) commercially manufactured black powder in quantities not to exceed fifty pounds, percussion caps, safety and pyrotechnic fuses, quills, quick and slow matches and friction primers intended to be used solely for sporting, recreational or cultural purposes as defined in Section 921(a)(16) [§ 921(a)(4)] of Title 18 of the United States Code, or in antique devices as exempted from the term "destructive device" in Section 921(a)(4) [§ 921(a)(16)] of Title 18 of the United States Code; or

(4) explosive materials transported in compliance with the regulations of the United States department of transportation and agencies thereof; or

B. knowingly transporting or taking any explosive upon or into any vehicle belonging to a common carrier transporting passengers. For the purpose of this subsection, the term "explosives" is as defined in Section 2 [30-7-18 NMSA 1978] of the Explosives Act, but shall not include:

(1) explosive materials in medicines and medicinal agents in the forms prescribed by the official United States pharmacopoeia or the national formulary;

(2) small arms ammunition or components thereof; or

(3) explosive materials transported in compliance with the regulations of the United States department of transportation and agencies thereof.

Whoever commits unlawful sale, possession or transportation of explosives as set forth in Subsection A of this section is guilty of a petty misdemeanor.

Whoever commits unlawful transportation of explosives as set forth in Subsection B of this section is guilty of a fourth degree felony.

30-7-9. Firearms; sale and purchase Residents of states contiguous to New Mexico may purchase firearms in New Mexico. Residents of New Mexico may purchase firearms in states contiguous to New Mexico.

30-7-16. Firearms or destructive devices; receipt, transportation or possession by a felon; penalty

A. It is unlawful for a felon to receive, transport or possess any firearm or destructive device in this state.

B. Any person violating the provisions of this section shall be guilty of a fourth degree felony and shall be sentenced in accordance with the provisions of the Criminal Sentencing Act [31-18-12 NMSA 1978].

C. As used in this section:

(1) "destructive device" means:

(a) any explosive, incendiary or poison gas: 1) bomb; 2) grenade; 3) rocket having a propellant charge of more than four ounces; 4) missile having an explosive or incendiary charge of more than one-fourth ounce; 5) mine; or 6) similar device;

(b) any type of weapon by whatever name known that will, or that may be readily converted to, expel a projectile by the action of an explosive or other propellant, the barrel or barrels of which have a bore of more than one-half inch in diameter, except a shotgun or shotgun shell that is generally recognized as particularly suitable for sporting purposes; and

(c) any combination of parts either designed or intended for use in converting any device into a destructive device as defined in this paragraph and from which a destructive device may be readily assembled.

The term "destructive device" does not include any device that is neither designed nor redesigned for use as a weapon or any device, although originally designed for use as a weapon, that is redesigned for use as a signaling, pyrotechnic, line throwing, safety or similar device;

(2) "felon" means a person convicted of a felony offense by a court of the United States or of any state or political subdivision thereof and:

(a) less than ten years have passed since the person completed serving his sentence or period of probation for the felony conviction, whichever is later;

(b) the person has not been pardoned for the felony conviction by the proper authority; and

(c) the person has not received a deferred sentence; and

(3) "firearm" means any weapon that will or is designed to or may readily be converted to expel a projectile by the action of an explosion; the frame or receiver of any such weapon; or any firearm muffler or firearm silencer. "Firearm" includes any handgun, rifle or shotgun.

30-7-18. Definitions

As used in the Explosives Act [30-7-17 NMSA 1978]:

A. "explosive" means any chemical compound or mixture or device, the primary or common purpose of which is to explode and includes but is not limited to dynamite and other high explosives, black powder, pellet powder, initiating explosives, detonators, safety fuses, squibs, detonating cord, igniter cord and igniters; and

B. "explosive device" or "incendiary device" means:

(1) any explosive bomb, grenade, missile or similar device;

(2) any device or mechanism used or created to start a fire or explosion with or without a timing mechanism except cigarette lighters and matches; or

(3) any incendiary bomb or grenade, fire bomb or similar device or any device which includes a flammable liquid or compound and a wick or igniting agent composed of any material which is capable of igniting the flammable liquid or compound.

30-7-19. Possession of explosives

A. Possession of explosives consists of knowingly possessing, manufacturing or transporting any explosive and either intending to use the explosive in the commission of any felony or knowing or reasonably believing that another intends to use the explosive to commit any felony.

B. Any person who commits possession of explosives is guilty of a fourth degree felony.

30-7-19.1. Possession of explosive device or incendiary device

A. Possession of an explosive device or incendiary device consists of knowingly possessing, manufacturing or transporting any explosive device or incendiary device or complete combination of parts thereof necessary to make an explosive device or incendiary device. This subsection shall not apply to any fireworks as defined in Section 60-2C-2 NMSA 1978 or any lawfully acquired household, commercial, industrial or sporting device or compound included in the definition of explosive device or incendiary device in Section 30-7-18 NMSA 1978 that has legitimate and lawful commercial, industrial or sporting purposes or that is lawfully possessed under Section 30-7-7 NMSA 1978.

B. Any person who commits possession of an explosive device or incendiary device is guilty of a fourth degree felony.

Chapter 32a. Children's Code

Article 2. Delinquency

32A-2-33. Child in possession of a firearm on school premises; detention; hearing

A. If a public school administrator or employee has reasonable cause to believe that a child is in possession of or has been in possession of a firearm on school premises in violation of Section 30-7-2.1 NMSA 1978, the administrator or employee shall immediately report the child's actions to a law enforcement agency and the children, youth and families department.

B. Upon receipt of a report pursuant to Subsection A of this section, the law enforcement agency may conduct an investigation to determine if there is probable cause to believe that the child possessed a firearm on school premises.

C. If the law enforcement agency determines there is probable cause to believe that the child possessed a firearm on school premises, the law enforcement agency may take the child into custody and deliver the child to a detention facility licensed by the department. After the child is delivered to a detention facility, the department shall comply with the notification provisions set forth in Subsection C of Section 32A-2-10 NMSA 1978. The child shall be detained in the detention facility, pending a detention hearing pursuant to the provisions of Section 32A-2-13 NMSA 1978.

D. As used in this section, "firearm" means any weapon that will or is designed to or may readily be converted to expel a projectile by the action of an explosion; the frame or receiver of any such weapon; or any firearm muffler or firearm silencer. "Firearm" includes any handgun, rifle or shotgun.

[Current through 2nd Special Session of the 48th Legislature (2008)]

Revised Ordinances of Albuquerque

Chapter 12. Criminal Code

Article 2. Offenses Relating to Public Order and Safety

12-2-9. Negligent Use of Weapons.

(A) Negligent use of a weapon consists of either: ...

(2) Carrying or having within one's reach or immediate grasp, a deadly weapon while under the influence of an intoxicating liquor or other drug; or

(3) Endangering the safety of another or his property by handling or using a firearm or other deadly weapon in a negligent manner; or

(4) Selling, loaning or furnishing any deadly weapon to a person with knowledge that the person is under the influence of alcohol or other drug or that the person is incompetent.

(B) Exception for peace officers. The provisions set forth above shall not be construed to forbid peace officers from carrying, wearing or discharging such weapons as shall be necessary in the proper discharge of their duties.

12-2-10. Unlawful Possession, Transfer or Sale of Weapons. ...

(B) Unlawful sale of a deadly weapon consists of giving, selling, trading, bartering, or exchanging for anything of value any deadly weapon or ammunition for any firearm to any person under the age of 18 years; provided, however, that nothing herein contained shall be construed to prevent any parent or legal guardian from purchasing such a weapon or ammunition for his child or ward.

[Revised Ordinances of Albuquerque current through Ord. 35-2008 passed Sept. 23, 2008]

Las Cruces Municipal Code

Chapter 16. Licenses, Taxation and Miscellaneous Business Regulations

Article V. Flea Markets

16-267. Responsibilities of vendors. No flea market vendor shall sell, exchange, or display for the purpose of sale or exchange any firearm, live ammunition, fresh meat, or any item of contraband as such is defined by federal or state law. ...

Chapter 19. Offenses and Miscellaneous Provisions

Article III. Offenses Against Public Peace, Order and Safety

Division 3. Weapons

19-164. Prohibited use of weapons and firearms.

(a) It shall be unlawful for any person to: ...

(4) Carry a firearm while under the influence of an intoxicant or narcotic.

(5) Endanger the safety or security of another person or his property by handling or using a firearm or other deadly weapon in a negligent manner.

[Las Cruces Municipal Code current through Ord. 2292, enacted Apr. 3, 2006]

NEW YORK

N.Y. Consolidated Laws

General Business Law

Article 26. Miscellaneous

396-ee. Sale of certain weapons; locking devices therefor

(1) No person, firm or corporation engaged in the retail business of selling rifles, shotguns or firearms, as such terms are defined in section 265.00 of the penal law, shall sell, deliver or transfer any such rifle, shotgun or firearm to another person unless the transferee is provided at the time of sale, delivery or transfer with a gun locking device and a label containing the quoted language specified in subdivision two of this section is either affixed to such rifle, shotgun or firearm or placed in the container in which such rifle, shotgun or firearm is sold, delivered or transferred. For the purposes of this section, the term "gun locking device" shall mean an integrated design feature or an attachable accessory that is resistant to tampering and is effective in preventing the discharge of such rifle, shotgun or firearm by a person who does not have access to the key, combination or other mechanism used to disengage the device. The division of state police shall develop and promulgate rules and regulations setting forth the specific devices or the minimum standards and criteria therefor which constitute an effective gun locking device.

(2) Every person, firm or corporation engaged in the retail business of selling rifles, shotguns or firearms, as such terms are defined in section 265.00 of the penal law, shall, in the place where such rifles, shotguns or firearms are displayed or transferred to the purchaser, post a notice conspicuously stating in bold print that: "The use of a locking device or safety lock is only one aspect of responsible firearm storage. For increased safety firearms should be stored unloaded and locked in a location that is both separate from their ammunition and inaccessible to children and any other unauthorized person."

(3) Any person, firm or corporation who fails to comply with the provisions of this section shall be guilty of a violation punishable as provided in the penal law. Any person, firm, or corporation who fails to comply with the provisions of this section after having been previously convicted of a violation of this section shall be guilty of a class A misdemeanor, punishable as provided in the penal law.

396-ff. Pistol and revolver ballistic identification databank

(1) For the purposes of this section, the following terms shall have the following meanings:

(a) "Manufacturer" means any person, firm or corporation possessing a valid federal license that permits such person, firm or corporation to engage in the business of manufacturing pistols or revolvers or ammunition therefor for the purpose of sale or distribution.

(b) "Shell casing" means that part of ammunition capable of being used in a pistol or revolver that contains the primer and propellant powder to discharge the bullet or projectile.

(2) On and after March first, two thousand one, any manufacturer that ships, transports or delivers a pistol or revolver to any person in this state shall, in accordance with rules and regulations promulgated by the division of state police, include in the container with such pistol or revolver a separate sealed container that encloses:

(a) a shell casing of a bullet or projectile discharged from such pistol or revolver; and

(b) any additional information that identifies such pistol or revolver and shell casing as required by such rules and regulations.

(3) A gunsmith or dealer in firearms licensed in this state shall, within ten days of the receipt of any pistol or revolver from a manufacturer that fails to comply with the provisions of this section, either (a) return such pistol or revolver to such manufacturer, or (b) notify the division of state police of such noncompliance and thereafter obtain a substitute sealed container through participation in a program operated by the state police as provided in subdivision four of this section.

(4) The division of state police shall no later than October first, two thousand, promulgate rules and regulations for the operation of a program which provides a gunsmith or a dealer in firearms licensed in this state with a sealed container enclosing the items specified in subdivision two of this section. The program shall at a minimum:

(a) be operational by January first, two thousand one;

(b) operate in at least five regional locations within the state; and

(c) specify procedures by which such gunsmith or dealer is to deliver a pistol or revolver to the regional program location closest to his or her place of business for testing and prompt return of such pistol or revolver.

(5) On and after March first, two thousand one, a gunsmith or dealer in firearms licensed in this state shall, within ten days of delivering to any person a pistol or revolver received by such gunsmith or dealer in firearms on or after such date, forward to the division of state police, along with the original transaction report required by subdivision twelve of section 400.00 of the penal law, the sealed container enclosing the shell casing from such pistol or revolver either (a) received from the manufacturer, or (b) obtained through participation in the program operated by the division of state police in accordance with subdivision four of this section.

(6) Upon receipt of the sealed container, the division of state police shall cause to be entered in an automated electronic databank pertinent data and other ballistic information relevant to identification of the shell casing and to the pistol or revolver from which it was discharged. The automated electronic databank will be operated and maintained by the division of state police, in accordance with its rules and regulations adopted after consultation with the Federal Bureau of Investigation and the United States Department of Treasury, Bureau of Alcohol, Tobacco and Firearms to ensure compatibility with national ballistic technology.

(7) Any person, firm or corporation who knowingly violates any of the provisions of this section shall be guilty of a violation, punishable as provided in the penal law. Any person, firm or corporation who knowingly violates any of the provisions of this section after having been previously convicted of a violation of this section shall be guilty of a class A misdemeanor, punishable as provided in the penal law.

399-r. Sale of paint pellet guns

(a) No person, firm, or corporation shall sell or offer to sell a paint pellet gun to any person under sixteen years of age.

(b) For purposes of this subdivision, the term "paint pellet gun" means a gun, air gun, pistol, rifle, or like device in appearance or function, capable of and designed for discharging and propelling through the air to a target a small quantity of paint enclosed within a pellet or pellet-like device or capsule or capsule-like device that breaks upon impact with the target, overspreading the target with paint.

(c) Any person, firm, or corporation who violates the provisions of this section shall be subject to a civil penalty of not more than five hundred dollars.

399-s. Posting of notice when air rifles or guns are sold

1. In each mercantile establishment in this state where air rifles or air guns are sold, a notice must be posted in a conspicuous place, and in close proximity to any air rifle or air gun displayed for sale, that in legible format states: "Except under supervision at a shooting range or for hunting pursuant to license, Section 265.05 of the New York State Penal Law makes it unlawful for any person under the age of sixteen years to possess an air rifle or air gun."

2. Any violation of this section shall be punishable by a civil penalty not to exceed one hundred dollars.

Article 39-B. Imitation Weapons

871. Definitions

1. "Firearm" as used in this section shall have the same meaning as that term is defined by subdivision three of section 265.00 of the penal law and shall include machine guns, rifles and shotguns as those terms are defined by subdivisions one, eleven and twelve, respectively, of section 265.00 of the penal law.

2. "Imitation weapon" means any device or object made of plastic, wood, metal or any other material which substantially duplicates or can reasonably be perceived to be an actual firearm, air rifle, pellet gun, or "B-B" gun; unless such imitation weapon (a) is colored other than black, blue, silver or aluminum, (b) is marked with a non-removable orange stripe which is at least one inch in width and runs the entire length of the barrel on each side and the front end of the barrel, and (c) has a barrel at least one inch in diameter that is closed for a distance of not less than one-half inch from the front-end of its barrel with the same material of which the imitation weapon is made. "Imitation weapon" does not include any nonfiring replica of an antique firearm, the original of which was designed, manufactured and produced prior to eighteen hundred ninety-eight.

872. Prohibitions No person, firm, corporation or agent or employee thereof shall import, manufacture, sell, hold for sale or distribute within the state any imitation weapon unless such device is imported, manufactured, sold, held for sale and distributed:

1. solely for subsequent transportation in interstate commerce; or

2. solely for lawful use in a theatrical production, including a motion picture, television or stage production.

873. Enforcement Whenever the attorney general shall believe from evidence satisfactory to him that any person, firm, corporation or association or agent or employee thereof has violated any provision of this article, he may bring an action or special proceeding in the supreme court for a judgment enjoining the continuance of such violation and for a civil penalty of not more than one thousand dollars for each violation. If it shall appear to the satisfaction of the court or justice that the defendant has violated any provision of this article, no proof shall be required that any person has been injured thereby nor that the defendant knowingly or intentionally violated such provision. In such action preliminary relief may be granted under article sixty-three of the civil practice law and rules. In connection with any such proposed application, the attorney general is authorized to take proof, issue subpoenas and administer oaths in the manner provided in the civil practice law and rules.

Article 39-DD. Sale of Firearms, Rifles or Shotguns at Gun Shows

895. Definitions For the purposes of this article:

1. "Gun show" means an event sponsored, whether for profit or not, by an individual, national, state or local organization, association or other entity devoted to the collection, competitive use, sporting use, or any other legal use of firearms, rifles or shotguns, or an event at which (a) twenty percent or more of the total number of exhibitors are firearm exhibitors or (b) ten or more firearm exhibitors are participating or (c) a total of twenty-five or more pistols or revolvers are offered for sale or transfer or (d) a total of fifty or more firearms, rifles or shotguns are offered for sale or transfer. The term gun show shall include any building, structure or facility where firearms, rifles or shotguns are offered for sale or transfer and any grounds used in connection with the event.

2. "Firearm exhibitor" means any person, firm, partnership, corporation or company that exhibits, sells, offers for sale, transfers, or exchanges firearms, rifles or shotguns at a gun show.

3. "Gun show operator" means any person, firm, partnership, corporation or company that organizes, produces, sponsors or operates a gun show.

4. "Firearm" has the same meaning as that term is defined in 18 U.S.C. 921(a)(3), but shall not include an "antique firearm" as that term is defined in 18 U.S.C. 921(a)(16).

5. "Rifle" has the same meaning as that term is defined in 18 U.S.C. 921(a)(7).

6. "Shotgun" has the same meaning as that term is defined in 18 U.S.C. 921(a)(5).

896. Operation of a gun show

1. A gun show operator shall:

(a) at all times during such show conspicuously post and maintain signs stating "A National Instant Criminal Background Check must be completed prior to all firearm sales or transfers, including sales or transfers of rifles or shotguns". Signs must be posted at all entrances to the gun show, at all places where admission tickets to the gun show are sold and not less than four additional locations within the grounds of the gun show;

(b) notify all firearm exhibitors in writing that a national instant criminal background check must be completed prior to all firearm sales or transfers, including sales or transfers of rifles or shotguns; and

(c) provide access at the gun show to a firearm dealer licensed under federal law who is authorized to perform a national instant criminal background check where the seller or transferor of a firearm, rifle or shotgun is not authorized to conduct such a check by (i) requiring firearm exhibitors who are firearm dealers licensed under federal law and who are authorized to conduct a national instant criminal background check to provide such a check at cost or (ii) designating a specific location at the gun show where a firearm dealer licensed under federal law who is authorized to conduct a national instant criminal background check will be present to perform such a check at cost. Any firearm dealer licensed under federal law who performs a national instant criminal background check pursuant to this paragraph shall provide the seller or transferor of the firearm, rifle or shotgun with a copy of the United States Department of Treasury, Bureau of Alcohol, Tobacco and Firearms Form ATF F 4473 and such dealer shall maintain such form and make such form available for inspection by law enforcement agencies for a period of ten years thereafter.

2. Whenever the attorney general shall believe from evidence satisfactory to him or her that a gun show operator has violated any of the provisions of this section, the attorney general may bring an action or special proceeding in the supreme court for a judgment enjoining the continuance of such violation and for a civil penalty in an amount not to exceed ten thousand dollars. If it shall appear to the satisfaction of the court or justice that the defendant has violated any provisions of this section, no proof shall be required that any person has been injured thereby nor that the defendant intentionally violated such provision. In such action preliminary relief may be granted under article sixty-three of the civil practice law and rules. In connection with any such proposed application, the attorney general is authorized to take proof, issue subpoenas and administer oaths in the manner provided in the civil practice law and rules.

897. Sale of a firearm, rifle or shotgun at a gun show

1. A national instant criminal background check shall be conducted and no person shall sell or transfer a firearm, rifle or shotgun at a gun show, except in accordance with the provisions of 18 U.S.C. 922(t).

2. No person shall offer or agree to sell or transfer a firearm, rifle or shotgun to another person at a gun show and transfer or deliver such firearm, rifle or shotgun to such person or person acting on his or her behalf thereafter at a location other than the gun show for the purpose of evading or avoiding compliance with 18 U.S.C. 922(t).

3. Any person who knowingly violates any of the provisions of this section shall be guilty of a class A misdemeanor punishable as provided for in the penal law.

Penal Law

Article 265. Firearms and Other Dangerous Weapons

265.00. Definitions As used in this article and in article four hundred, the following terms shall mean and include:

1. "Machine-gun" means a weapon of any description, irrespective of size, by whatever name known, loaded or unloaded, from which a number of shots or bullets may be rapidly or automatically discharged from a magazine with one continuous pull of the trigger and includes a submachine gun.

2. "Firearm silencer" means any instrument, attachment, weapon or appliance for causing the firing of any gun, revolver, pistol or other firearms to be silent, or intended to lessen or muffle the noise of the firing of any gun, revolver, pistol or other firearms.

3. "Firearm" means (a) any pistol or revolver; or (b) a shotgun having one or more barrels less than eighteen inches in length; or (c) a rifle having one or more barrels less than sixteen inches in length; or (d) any weapon made from a shotgun or rifle whether by alteration, modification, or otherwise if such weapon as altered, modified, or otherwise has an overall length of less than twenty-six inches; or (e) an assault weapon. For the purpose of this subdivision the length of the barrel on a shotgun or rifle shall be determined by measuring the distance between the muzzle and the face of the bolt, breech, or breechlock when closed and when the shotgun or rifle is cocked; the overall length of a weapon made from a shotgun or rifle is the distance between the extreme ends of the weapon measured along a line parallel to the center line of the bore. Firearm does not include an antique firearm. ...

6. "Dispose of" means to dispose of, give, give away, lease-loan, keep for sale, offer offer for sale, sell, transfer and otherwise dispose of.

7. "Deface" means to remove, deface, cover, alter or destroy the manufacturer's serial number or any other distinguishing number or identification mark.

8. "Gunsmith" means any person, firm, partnership, corporation or company who engages in the business of repairing, altering, assembling, manufacturing, cleaning, polishing, engraving or trueing, or who performs any mechanical operation on, any firearm, large capacity ammunition feeding device or machine-gun.

9. "Dealer in firearms" means any person, firm, partnership, corporation or company who engages in the business of purchasing, selling, keeping for sale, loaning, leasing, or in any manner disposing of, any assault weapon, large capacity ammunition feeding device, pistol or revolver.

10. "Licensing officer" means in the city of New York the police commissioner of that city; in the county of Nassau the commissioner of police of that county; in the county of Suffolk the sheriff of that county except in the towns of Babylon, Brookhaven, Huntington, Islip and Smithtown, the commissioner of police of that county; for the purposes of section 400.01 of this chapter the superintendent of state police; and elsewhere in the state a judge or justice of a court of record having his office in the county of issuance.

11. "Rifle" means a weapon designed or redesigned, made or remade, and intended to be fired from the shoulder and designed or redesigned and made or remade to use the energy of the explosive in a fixed metallic cartridge to fire only a single projectile through a rifled bore for each single pull of the trigger.

12. "Shotgun" means a weapon designed or redesigned, made or remade, and intended to be fired from the shoulder and designed or redesigned and made or remade to use the energy of the explosive in a fixed shotgun shell to fire through a smooth bore either a number of ball shot or a single projectile for each single pull of the trigger. ...

14. "Antique firearm" means: Any unloaded muzzle loading pistol or revolver with a matchlock, flintlock, percussion cap, or similar type of ignition system, or a pistol or revolver which uses fixed cartridges which are no longer available in the ordinary channels of commercial trade. ...

15. "Loaded firearm" means any firearm loaded with ammunition or any firearm which is possessed by one who, at the same time, possesses a quantity of ammunition which may be used to discharge such firearm.

15-a. "Electronic dart gun" means any device designed primarily as a weapon, the purpose of which is to momentarily stun, knock out or paralyze a person by passing an electrical shock to such person by means of a dart or projectile. ...

16. "Certified not suitable to possess a self-defense spray device, a rifle or shotgun" means that the director or physician in charge of any hospital or institution for mental illness, public or private, has certified to the superintendent of state police or to any organized police department of a county, city, town or village of this state, that a person who has been judicially adjudicated incompetent, or who has been confined to such institution for mental illness pursuant to judicial authority, is not suitable to possess a self-defense spray device, as defined in section 265.20 of this article, or a rifle or shotgun.

17. "Serious offense" means

(a) any of the following offenses defined in the former penal law as in force and effect immediately prior to September first, nineteen hundred sixty-seven: illegally using, carrying or possessing a pistol or other dangerous weapon; making or possessing burglar's instruments; buying or receiving stolen property; unlawful entry of a building; aiding escape from prison; that kind of disorderly conduct defined in subdivisions six and eight of section seven hundred twenty-two of such former penal law; violations of sections four hundred eighty-three, four hundred eighty-three-b, four hundred eighty-four-h and article one hundred six of such former penal law; that kind of criminal sexual act or rape which was designated as a misdemeanor; violation of section seventeen hundred forty-seven-d and seventeen hundred forty-seven-e of such former penal law; any violation of any provision of article thirty-three of the public health law relating to narcotic drugs which was defined as a misdemeanor by section seventeen hundred fifty-one-a of such former penal law, and any violation of any provision of article thirty-three-A of the public health law relating to depressant and stimulant drugs which was defined as a misdemeanor by section seventeen hundred forty-seven-b of such former penal law.

(b) **[**See also par. (b) below.**]** any of the following offenses defined in the penal law: illegally using, carrying or possessing a pistol or other dangerous weapon; possession of burglar's tools; criminal possession of stolen property in the third degree; escape in the third degree; jostling; fraudulent accosting; that kind of loitering defined in subdivision three of section 240.35; endangering the welfare of a child; the offenses defined in article two hundred thirty-five; issuing abortional articles; permitting prostitution; promoting prostitution in the third degree; stalking in the fourth degree; stalking in the third degree; the offenses defined in article one hundred thirty; the offenses defined in article two hundred twenty.

(b) **[**See also par. (b) above.**]** any of the following offenses defined in the penal law: illegally using, carrying or possessing a pistol or other dangerous weapon; possession of burglar's tools; criminal possession of stolen property in the third degree; escape in the third degree; jostling; fraudulent accosting; that kind of loitering defined in subdivision three of section 240.35; endangering the welfare of a child; the offenses defined in article two hundred thirty-five; issuing abortional articles; permitting prostitution; promoting prostitution in the third degree; stalking in the third degree; stalking in the fourth degree; the offenses defined in article one hundred thirty; the offenses defined in article two hundred twenty.

18. "Armor piercing ammunition" means any ammunition capable of being used in pistols or revolvers containing a projectile or projectile core, or a projectile or projectile core for use in such ammunition, that is constructed entirely (excluding the presence of traces of other substances) from one or a combination of any of the following: tungsten alloys, steel, iron, brass, bronze, beryllium copper, or uranium.

19. "Duly authorized instructor" means (a) a duly commissioned officer of the United States army, navy, marine corps or coast guard, or of the national guard of the state of New York; or (b) a duly qualified adult citizen of the United States who has been granted a certificate as an instructor in small arms practice issued by the United States army, navy or marine corps, or by the adjutant general of this state, or by the national rifle association of America, a not-for-profit corporation duly organized under the laws of this state; or (c) by a person duly qualified and designated by the department of environmental conservation under paragraph d of subdivision six of section 11-0713 of the environmental conservation law as its agent in the giving of instruction and the making of certifications of qualification in responsible hunting practices.

20. "Disguised gun" means any weapon or device capable of being concealed on the person from which a shot can be discharged through the energy of an explosive and is designed and intended to appear to be something other than a gun.

21. "Semiautomatic" means any repeating rifle, shotgun or pistol, regardless of barrel or overall length, which utilizes a portion of the energy of a firing cartridge or shell to extract the fired cartridge case or spent shell and chamber the next round, and which requires a separate pull of the trigger to fire each cartridge or shell.

22. "Assault weapon" means

(a) a semiautomatic rifle that has an ability to accept a detachable magazine and has at least two of the following characteristics:

(i) a folding or telescoping stock;

(ii) a pistol grip that protrudes conspicuously beneath the action of the weapon;

(iii) a bayonet mount;

(iv) a flash suppressor or threaded barrel designed to accommodate a flash suppressor;

(v) a grenade launcher; or

(b) a semiautomatic shotgun that has at least two of the following characteristics:

(i) a folding or telescoping stock;

(ii) a pistol grip that protrudes conspicuously beneath the action of the weapon;

(iii) a fixed magazine capacity in excess of five rounds;

(iv) an ability to accept a detachable magazine; or

(c) a semiautomatic pistol that has an ability to accept a detachable magazine and has at least two of the following characteristics:

(i) an ammunition magazine that attaches to the pistol outside of the pistol grip;

(ii) a threaded barrel capable of accepting a barrel extender, flash suppressor, forward handgrip, or silencer;

(iii) a shroud that is attached to, or partially or completely encircles, the barrel and that permits the shooter to hold the firearm with the nontrigger hand without being burned;

(iv) a manufactured weight of fifty ounces or more when the pistol is unloaded;

(v) a semiautomatic version of an automatic rifle, shotgun or firearm; or

(d) any of the weapons, or functioning frames or receivers of such weapons, or copies or duplicates of such weapons, in any caliber, known as:

(i) Norinco, Mitchell, and Poly Technologies Avtomat Kalashnikovs (all models);

(ii) Action Arms Israeli Military Industries UZI and Galil;

(iii) Beretta Ar70 (SC-70);

(iv) Colt AR-15;

(v) Fabrique National FN/FAL, FN/LAR, and FNC;

(vi) SWD M-10, M-11, M-11/9, and M-12;

(vii) Steyr AUG;

(viii) INTRATEC TEC-9, TEC-DC9 and TEC-22; and

(ix) revolving cylinder shotguns, such as (or similar to) the Street Sweeper and Striker 12;

(e) provided, however, that such term does not include:

(i) any rifle, shotgun or pistol that (A) is manually operated by bolt, pump, lever or slide action; (B) has been rendered permanently inoperable; or (C) is an antique firearm as defined in 18 U.S.C. 921(a)(16);

(ii) a semiautomatic rifle that cannot accept a detachable magazine that holds more than five rounds of ammunition;

(iii) a semiautomatic shotgun that cannot hold more than five rounds of ammunition in a fixed or detachable magazine;

(iv) a rifle, shotgun or pistol, or a replica or a duplicate thereof, specified in Appendix A to section 922 of 18 U.S.C. as such weapon was manufactured on October first, nineteen hundred ninety-three. The mere fact that a weapon is not listed in Appendix A shall not be construed to mean that such weapon is an assault weapon; or

(v) a semiautomatic rifle, a semiautomatic shotgun or a semiautomatic pistol or any of the weapons defined in paragraph (d) of this subdivision lawfully possessed prior to September fourteenth, nineteen hundred ninety-four.

23. "Large capacity ammunition feeding device" means a magazine, belt, drum, feed strip, or similar device, manufactured after September thirteenth, nineteen hundred ninety-four, that has a capacity of, or that can be readily restored or converted to accept, more than ten rounds of ammunition; provided, however, that such term does not include an attached tubular device designed to accept, and capable of operating only with, .22 caliber rimfire ammunition.

265.01. Criminal possession of a weapon in the fourth degree A person is guilty of criminal possession of a weapon in the fourth degree when:

(1) He or she possesses any firearm, electronic dart gun, electronic stun gun, gravity knife, switch-blade knife, pilum ballistic knife, metal knuckle knife, cane sword, billy, blackjack, bludgeon, plastic knuckles, metal knuckles, chuka stick, sand bag, sand-club, wrist-brace type slingshot or slungshot, shirken or "Kung Fu star"; or

(2) He possesses any dagger, dangerous knife, dirk, razor, stiletto, imitation pistol, or any other dangerous or deadly instrument or weapon with intent to use the same unlawfully against another; or

(3) He knowingly has in his possession a rifle, shotgun or firearm in or upon a building or grounds, used for educational purposes, of any school, college or university, except the forestry lands, wherever located, owned and maintained by the State University of New York college of environmental science and forestry, without the written authorization of such educational institution; or

(4) He possesses a rifle or shotgun and has been convicted of a felony or serious offense; or

(5) He possesses any dangerous or deadly weapon and is not a citizen of the United States; or

(6) He is a person who has been certified not suitable to possess a rifle or shotgun, as defined in subdivision sixteen of section 265.00, and refuses to yield possession of such rifle or shotgun upon the demand of a police officer. Whenever a person is certified not suitable to possess a rifle or shotgun, a member of the police department to which such certification is made, or of the state police, shall forthwith seize any rifle or shotgun possessed by such person. A rifle or shotgun seized as herein provided shall not be destroyed, but shall be delivered to the headquarters of such police department, or state police, and there retained until the aforesaid certificate has been rescinded by the director or physician in charge, or other disposition of such rifle or shotgun has been ordered or authorized by a court of competent jurisdiction.

(7) He knowingly possesses a bullet containing an explosive substance designed to detonate upon impact.

(8) He possesses any armor piercing ammunition with intent to use the same unlawfully against another.

Criminal possession of a weapon in the fourth degree is a class A misdemeanor.

265.02. Criminal possession of a weapon in the third degree A person is guilty of criminal possession of a weapon in the third degree when:

(1) Such person commits the crime of criminal possession of a weapon in the fourth degree as defined in subdivision one, two, three or five of section 265.01, and has been previously convicted of any crime; or

(2) Such person possesses any explosive or incendiary bomb, bombshell, firearm silencer, machine-gun or any other firearm or weapon simulating a machine-gun and which is adaptable for such use; or

(3) Such person knowingly possesses a machine-gun, firearm, rifle or shotgun which has been defaced for the purpose of concealment or prevention of the detection of a crime or misrepresenting the identity of such machine-gun, firearm, rifle or shotgun; or

(4) Such person possesses any loaded firearm. Such possession shall not, except as provided in subdivision one or seven, constitute a violation of this section if such possession takes place in such person's home or place of business; or

(5)(i) Such person possesses three or more firearms; or

(ii) such person possesses a firearm and has been previously convicted of a felony or a class A misdemeanor defined in this chapter within the five years immediately preceding the commission of the offense and such possession did not take place in the person's home or place of business; or

(6) Such person knowingly possesses any disguised gun; or

(7) Such person possesses an assault weapon; or

(8) Such person possesses a large capacity ammunition feeding device.

Criminal possession of a weapon in the third degree is a class D felony.

265.03. Criminal possession of a weapon in the second degree A person is guilty of criminal possession of a weapon in the second degree when:

(1) with intent to use the same unlawfully against another, such person:

(a) possesses a machine-gun; or

(b) possesses a loaded firearm; or

(c) possesses a disguised gun; or

(2) such person possesses five or more firearms.

Criminal possession of a weapon in the second degree is a class C felony.

265.04. Criminal possession of a weapon in the first degree A person is guilty of criminal possession of a weapon in the first degree when such person:

(1) possesses any explosive substance with intent to use the same unlawfully against the person or property of another; or

(2) possesses ten or more firearms.

Criminal possession of a weapon in the first degree is a class B felony.

265.05. Unlawful possession of weapons by persons under sixteen It shall be unlawful for any person under the age of sixteen to possess any air-gun, spring-gun or other instrument or weapon in which the propelling force is a spring or air, or any gun or any instrument or weapon in or upon which any loaded or blank cartridges may be used, or any loaded or blank cartridges or ammunition therefor, or any dangerous knife; provided that the possession of rifle or shotgun or ammunition therefor by the holder of a hunting license or permit issued pursuant to article eleven of the environmental conservation law and used in accordance with said law shall not be governed by this section.

A person who violates the provisions of this section shall be adjudged a juvenile delinquent.

265.06. Unlawful possession of a weapon upon school grounds It shall be unlawful for any person age sixteen or older to knowingly possess any air-gun, spring-gun or other instrument or weapon in which the propelling force is a spring, air, piston or CO2 cartridge in or upon a building or grounds, used for educational purposes, of any school, college or university, without the written authorization of such educational institution. Unlawful possession of a weapon upon school grounds is a violation.

265.08. Criminal use of a firearm in the second degree A person is guilty of criminal use of a firearm in the second degree when he commits any class C violent felony offense as defined in paragraph (b) of subdivision one of section 70.02 and he either:

(1) possesses a deadly weapon, if the weapon is a loaded weapon from which a shot, readily capable of producing death or other serious injury may be discharged; or

(2) displays what appears to be a pistol, revolver, rifle, shotgun, machine gun or other firearm.

Criminal use of a firearm in the second degree is a class C felony.

265.09. Criminal use of a firearm in the first degree

(1) A person is guilty of criminal use of a firearm in the first degree when he commits any class B violent felony offense as defined in paragraph (a) of subdivision one of section 70.02 and he either:

(a) possesses a deadly weapon, if the weapon is a loaded weapon from which a shot, readily capable of producing death or other serious injury may be discharged; or

(b) displays what appears to be a pistol, revolver, rifle, shotgun, machine gun or other firearm.

Criminal use of a firearm in the first degree is a class B felony.

(2) Sentencing. Notwithstanding any other provision of law to the contrary, when a person is convicted of criminal use of a firearm in the first degree as defined in subdivision one of this section, the court shall impose an additional consecutive sentence of five years to the minimum term of an indeterminate sentence imposed on the underlying class B violent felony offense where the person convicted of such crime displays a loaded weapon from which a shot, readily capable of producing death or other serious injury may be discharged, in furtherance of the commission of such crime, provided, however, that such additional sentence shall not be imposed if the court, having regard to the nature and circumstances of the crime and to the history and character of the defendant, finds on the record that such additional consecutive sentence would be unduly harsh and that not imposing such sentence would be consistent with the public safety and would not deprecate the seriousness of the crime. Notwithstanding any other provision of law to the contrary, the aggregate of the five year consecutive term imposed pursuant to this subdivision and the minimum term of the indeterminate sentence imposed on the underlying class B violent felony shall constitute the new aggregate minimum term of imprisonment, and a person subject to such term shall be required to serve the entire aggregate minimum term and shall not be eligible for release on parole or conditional release during such term. This subdivision shall not apply where the defendant's criminal liability for displaying a loaded weapon from which a shot, readily capable of producing death or other serious injury may be discharged, in furtherance of the commission of crime is based on the conduct of another pursuant to section 20.00 of the penal law.

265.10. Manufacture, transport, disposition and defacement of weapons and dangerous instruments and appliances

1. Any person who manufactures or causes to be manufactured any machine-gun, assault weapon, large capacity ammunition feeding device or disguised gun is guilty of a class D felony. Any person who manufactures or causes to be manufactured any switchblade knife, gravity knife, pilum ballistic knife, metal knuckle knife, billy, blackjack, bludgeon, plastic knuckles, metal knuckles, Kung Fu star, chuka stick, sandbag, sandclub or slungshot is guilty of a class A misdemeanor.

2. Any person who transports or ships any machine-gun, firearm silencer, assault weapon or large capacity ammunition feeding device or disguised gun, or who transports or ships as merchandise five or more firearms, is guilty of a class D felony. Any person who transports or ships as merchandise any firearm, other than an assault weapon, switchblade knife, gravity knife, pilum ballistic knife, billy, blackjack, bludgeon, plastic knuckles, metal knuckles, Kung Fu star, chuka stick, sand-bag or slungshot is guilty of a class A misde-meanor.

3. Any person who disposes of any machine-gun, assault weapon, large capacity ammunition feeding device or firearm silencer is guilty of a class D felony. Any person who knowingly buys, receives, disposes of, or conceals a machine-gun, firearm, large capacity ammunition feeding device, rifle or shotgun which has been defaced for the purpose of concealment or prevention of the detection of a crime or misrepresenting the identity of such machine-gun, firearm, large capacity ammunition feeding device, rifle or shotgun is guilty of a class D felony.

4. Any person who disposes of any of the weapons, instruments or appliances specified in subdivision one of section 265.01, except a firearm, is guilty of a class A misdemeanor, and he is guilty of a class D felony if he has previously been convicted of any crime.

5. Any person who disposes of any of the weapons, instruments, appliances or substances specified in section 265.05 to any other person under the age of sixteen years is guilty of a class A misdemeanor.

6. Any person who willfully defaces any machine-gun, large capacity ammunition feeding device or firearm is guilty of a class D felony.

7. Any person, other than a wholesale dealer, or gunsmith or dealer in firearms duly licensed pursuant to section 400.00, lawfully in possession of a firearm, who disposes of the same without first notifying in writing the licensing officer in the city of New York and counties of Nassau and Suffolk and elsewhere in the state the executive department, division of state police, Albany, is guilty of a class A misdemeanor.

265.11. Criminal sale of a firearm in the third degree A person is guilty of criminal sale of a firearm in the third degree when such person is not authorized pursuant to law to possess a firearm and such person unlawfully either:

(1) sells, exchanges, gives or disposes of a firearm or large capacity ammunition feeding device to another person; or

(2) possesses a firearm with the intent to sell it.

Criminal sale of a firearm in the third degree is a class D felony.

265.12. Criminal sale of a firearm in the second degree A person is guilty of criminal sale of a firearm in the second degree when such person:

(1) unlawfully sells, exchanges, gives or disposes of to another five or more firearms; or

(2) unlawfully sells, exchanges, gives or disposes of to another person or persons a total of five or more firearms in a period of not more than one year.

Criminal sale of a firearm in the second degree is a class C felony.

265.13. Criminal sale of a firearm in the first degree A person is guilty of a criminal sale of a firearm in the first degree when such person:

(1) unlawfully sells, exchanges, gives or disposes of to another ten or more firearms; or

(2) unlawfully sells, exchanges, gives or disposes of to another person or persons a total of ten or more firearms in a period of not more than one year.

Criminal sale of a firearm in the first degree is a class B felony.

265.14. Criminal sale of a firearm with the aid of a minor A person over the age of eighteen years of age is guilty of criminal sale of a weapon with the aid of a minor when a person under sixteen years of age knowingly and unlawfully sells, exchanges, gives or disposes of a firearm in violation of this article, and such person over the age of eighteen years of age, acting with the mental culpability required for the commission thereof, solicits, requests, commands, importunes or intentionally aids such person under sixteen years of age to engage in such conduct.

Criminal sale of a firearm with the aid of a minor is a class C felony.

265.15. Presumptions of possession, unlawful intent and defacement

1. The presence in any room, dwelling, structure or vehicle of any machine-gun is presumptive evidence of its unlawful possession by all persons occupying the place where such machine-gun is found.

2. The presence in any stolen vehicle of any weapon, instrument, appliance or substance specified in sections 265.01, 265.02, 265.03, 265.04 and 265.05 is presumptive evidence of its possession by all persons occupying such vehicle at the time such weapon, instrument, appliance or substance is found.

3. The presence in an automobile, other than a stolen one or a public omnibus, of any firearm, large capacity ammunition feeding device, defaced firearm, defaced rifle or shotgun, defaced large capacity ammunition feeding device, firearm silencer, explosive or incendiary bomb, bombshell, gravity knife, switchblade knife, pilum ballistic knife, metal knuckle knife, dagger, dirk, stiletto, billy, blackjack, plastic knuckles, metal knuckles, chuka stick, sandbag, sandclub or slungshot is presumptive evidence of its possession by all per-sons occupying such automobile at the time such weapon, instrument or appliance is found, except under the following circumstances: (a) if such weapon, instrument or appliance is found upon the person of one of the occupants therein; (b) if such weapon, instrument or appliance is found in an automobile which is being operated for hire by a duly licensed driver in the due, law-ful and proper pursuit of his or her trade, then such presumption shall not apply to the driver; or (c) if the weapon so found is a pistol or revolver and one of the occupants, not present under duress, has in his possession a valid license to have and carry concealed the same.

4. The possession by any person of the substance as specified in section 265.04 is presumptive evidence of possessing such substance with intent to use the same unlawfully against the person or property of another if such person is not licensed or otherwise authorized to possess such substance. The possession by any person of any dagger, dirk, stiletto, dangerous knife or any other weapon, instrument, appliance or substance designed, made or adapted for use primarily as a weapon, is presumptive evidence of intent to use the same unlawfully against another.

5. The possession by any person of a defaced machine-gun, firearm, rifle or shotgun is presumptive evidence that such person defaced the same.

6. The possession of five or more firearms by any person is presumptive evidence that such person possessed the firearms with the intent to sell same.

265.16. Criminal sale of a firearm to a minor A person is guilty of criminal sale of a firearm to a minor when he is not authorized pursuant to law to possess a firearm and he unlawfully sells, exchanges, gives or disposes of a firearm to another person who is or reasonably appears to be less than nineteen years of age who is not licensed pursuant to law to possess a firearm.

Criminal sale of a firearm to a minor is a class C felony.

265.17. Criminal purchase of a weapon A person is guilty of criminal purchase of a weapon when:

1. Knowing that he or she is prohibited by law from possessing a firearm, rifle or shotgun because of a prior conviction or because of some other disability which would render him or her ineligible to lawfully possess a firearm, rifle or shotgun in this state, such person attempts to purchase a firearm, rifle or shotgun from another person; or

2. Knowing that it would be unlawful for another person to possess a firearm, rifle or shotgun, he or she purchases a firearm, rifle or shotgun for, on behalf of, or for the use of such other person.

Criminal purchase of a weapon is a class A misdemeanor.

265.20. Exemptions

a. Sections 265.01, 265.02, 265.03, 265.04, 265.05, 265.10, 265.11, 265.12, 265.13, 265.15 and 270.05 shall not apply to:

1. Possession of any of the weapons, instruments, appliances or substances specified in sections 265.01, 265.02, 265.03, 265.04, 265.05 and 270.05 by the following:

(a) Persons in the military service of the state of New York when duly authorized by regulations issued by the adjutant general to possess the same.

(b) Police officers as defined in subdivision thirty-four of section 1.20 of the criminal procedure law.

(c) Peace officers as defined by section 2.10 of the criminal procedure law.

(d) Persons in the military or other service of the United States, in pursuit of official duty or when duly authorized by federal law, regulation or order to possess the same.

(e) Persons employed in fulfilling defense contracts with the government of the United States or agencies thereof when possession of the same is necessary for manufacture, transport, installation and testing under the requirements of such contract.

(f) A person voluntarily surrendering such weapon, instrument, appliance or substance, provided that such surrender shall be made to the superintendent of the division of state police or a member thereof designated by such superintendent, or to the sheriff of the county in which such person resides, or in the county of Nassau or in the towns of Babylon, Brookhaven, Huntington, Islip and Smithtown in the county of Suffolk to the commissioner of police or a member of the police department thereof designated by such commissioner, or if such person resides in a city, town other than one named in this subparagraph, or village to the police commissioner or head of the police force or department thereof or to a member of the force or department designated by such commissioner or head; and provided, further, that the same shall be surrendered by such person in accordance with such terms and conditions as may be established by such superintendent, sheriff, police force or department. Nothing in this paragraph shall be construed as granting immunity from prosecution for any crime or offense except that of unlawful possession of such weapons, instruments, appliances or substances surrendered as herein provided. A person who possesses any such weapon, instrument, appliance or substance as an executor or administrator or any other lawful possessor of such property of a decedent may continue to possess such property for a period not over fifteen days. If such property is not lawfully disposed of within such period the possessor shall deliver it to an appropriate official described in this paragraph or such property may be delivered to the superintendent of state police. Such officer shall hold it and shall thereafter deliver it on the written request of such executor, administrator or other lawful possessor of such property to a named person, provided such named person is licensed to or is otherwise lawfully permitted to possess the same. If no request to deliver the property is received by such official within two years of the delivery of such property, such official shall dispose of it in accordance with the provisions of section 400.05 of this chapter.

2. Possession of a machine-gun, firearm, large capacity ammunition feeding device, switchblade knife, gravity knife, pilum ballistic knife, billy or blackjack by a warden, superintendent, headkeeper or deputy of a state prison, penitentiary, workhouse, county jail or other institution for the detention of persons convicted or accused of crime or detained as witnesses in criminal cases, in pursuit of official duty or when duly authorized by regulation or order to possess the same.

3. Possession of a pistol or revolver by a person to whom a license therefor has been issued as provided under section 400.00 or 400.01 of this chapter; provided, that such a license shall not preclude a conviction for the offense defined in subdivision three of section 265.01 of this article.

4. Possession of a rifle, shotgun or longbow for use while hunting, trapping or fishing, by a person, not a citizen of the United States, carrying a valid license issued pursuant to section 11-0713 of the environmental conservation law.

5. Possession of a rifle or shotgun by a person other than a person who has been convicted of a class A-I felony or a violent felony offense, as defined in subdivision one of section 70.02 of this chapter, who has been convicted as specified in subdivision four of section 265.01 to whom a certificate of good conduct has been issued pursuant to section seven hundred three-b of the correction law.

6. Possession of a switchblade or gravity knife for use while hunting, trapping or fishing by a person carrying a valid license issued to him pursuant to section 11-0713 of the environmental conservation law.

7. Possession, at an indoor or outdoor shooting range for the purpose of loading and firing, of a rifle or shotgun, the propelling force of which is gunpowder by a person under sixteen years of age but not under twelve, under the immediate supervision, guidance and instruction of (a) a duly commissioned officer of the United States army, navy, air force, marine corps or coast guard, or of the national guard of the state of New York; or (b) a duly qualified adult citizen of the United States who has been granted a certificate as an instructor in small arms practice issued by the United States army, navy, air force or marine corps, or by the adjutant general of this state, or by the national rifle association of America, a not-for-profit corporation duly organized under the laws of this state; or (c) a parent, guardian, or a person over the age of eighteen designated in writing by such parent or guardian who shall have a certificate of qualification in responsible hunting, including safety, ethics, and landowner relations-hunter relations, issued or honored by the department of environmental conservation; or (d) an agent of the department of environmental conservation appointed to conduct courses in responsible hunting practices pursuant to article eleven of the environmental conservation law.

7-a. Possession and use, at an indoor or outdoor pistol range located in or on premises owned or occupied by a duly incorporated organization organized for conservation purposes or to foster proficiency in small arms or at a target pistol shooting competition under the auspices of or approved by the national rifle association for the purpose of loading and firing the same, by a person duly licensed to possess a pistol or revolver pursuant to section 400.00 or 400.01 of this chapter of a pistol or revolver duly so licensed to another person who is present at the time.

7-b. Possession and use, at an indoor or outdoor pistol range located in or on premises owned or occupied by a duly incorporated organization organized for conservation purposes or to foster proficiency in small arms or at a target pistol shooting competition under the auspices of or approved by the national rifle association for the purpose of loading and firing the same, by a person who has applied for a license to possess a pistol or revolver and pre-license possession of same pursuant to section 400.00 or 400.01 of this chapter, who has not been previously denied a license, been previously convicted of a felony or serious offense, and who does not appear to be, or pose a threat to be, a danger to himself or to others, and who has been approved for possession and use herein in accordance with section 400.00 or 400.01 of this chapter; provided however, that such possession shall be of a pistol or revolver duly licensed to and shall be used under the supervision, guidance and instruction of, a person specified in paragraph seven of this subdivision and provided further that such possession and use be within the jurisdiction of the licensing officer with whom the person has made application therefor or within the jurisdiction of the superintendent of state police in the case of a retired sworn member of the division of state police who has made an application pursuant to section 400.01 of this chapter.

7-c. Possession for the purpose of loading and firing, of a rifle, pistol or shotgun, the propelling force of which may be either air, compressed gas or springs, by a person under sixteen years of age but not under twelve, under the immediate supervision, guidance and instruction of (a) a duly commissioned officer of the United States army, navy, marine corps or coast guard, or of the national guard of the state of New York; or (b) a duly qualified adult citizen of the United States who has been granted a certificate as an instructor in small arms practice issued by the United States army, navy or marine corps, or by the adjutant general of this state, or by the national rifle association of America, a not-for-profit corporation duly organized under the laws of this state; or (c) a parent, guardian, or a person over the age of eighteen designated in writing by such parent or guardian who shall have a certificate of qualification in responsible hunting, including safety, ethics, and landowner relations-hunter relations, issued or honored by the department of environmental conservation.

7-d. Possession, at an indoor or outdoor shooting range for the purpose of loading and firing, of a rifle, pistol or shotgun, the propelling force of which may be either air, compressed gas or springs, by a person under twelve years of age, under the immediate supervision, guidance and instruction of (a) a duly commissioned officer of the United States army, navy, marine corps or coast guard, or of the national guard of the state of New York; or (b) a duly qualified adult citizen of the United States who has been granted a certificate as an instructor in small arms practice issued by the United States army, navy or marine corps, or by the adjutant general of this state, or by the national rifle association of America, a not-for-profit corporation duly organized under the laws of this state; or (c) a parent, guardian, or a person over the age of eighteen designated in writing by such parent or guardian who shall have a certificate of qualification in responsible hunting, including safety, ethics, and landowner relations-hunter relations, issued or honored by the department of environmental conservation.

7-e. Possession and use of a pistol or revolver, at an indoor or outdoor pistol range located in or on premises owned or occupied by a duly incorporated organization organized for conservation purposes or to foster proficiency in small arms or at a target pistol shooting competition under the auspices of or approved by an association or organization described in paragraph 7-a of this subdivision for the purpose of loading and firing the same by a person at least eighteen years of age but under the age of twenty-one who has not been previously convicted of a felony or serious offense, and who does not appear to be, or pose a threat to be, a danger to himself or to others; provided however, that such possession shall be of a pistol or revolver duly licensed to and shall be used under the immediate supervision, guidance and instruction of, a person specified in paragraph seven of this subdivision.

8. The manufacturer of machine-guns, assault weapons, large capacity ammunition feeding devices, disguised guns, pilum ballistic knives, switchblade or gravity knives, billies or blackjacks as merchandise and the disposal and shipment thereof direct to a regularly constituted or appointed state or municipal police department, sheriff, policeman or other peace officer, or to a state prison, penitentiary, workhouse, county jail or other institution for the detention of persons convicted or accused of crime or held as witnesses in criminal cases, or to the military service of this state or of the United States.

9. The regular and ordinary transport of firearms as merchandise, provided that the person transporting such firearms, where he knows or has reasonable means of ascertaining what he is transporting, notifies in writing the police commissioner, police chief or other law enforcement officer performing such functions at the place of delivery, of the name and address of the consignee and the place of delivery, and withholds delivery to the consignee for such reasonable period of time designated in writing by such police commissioner, police chief or other law enforcement officer as such official may deem necessary for investigation as to whether the consignee may lawfully receive and possess such firearms.

9-a.a. Except as provided in subdivision b hereof, the regular and ordinary transport of pistols or revolvers by a manufacturer of firearms to whom a license as a dealer in firearms has been issued pursuant to section 400.00 of this chapter, or by an agent or employee of such manufacturer of firearms who is otherwise duly licensed to carry a pistol or revolver and who is duly authorized in writing by such manufacturer of firearms to transport pistols or revolvers on the date or dates specified, directly between places where the manufacturer of firearms regularly conducts business provided such pistols or revolvers are transported unloaded, in a locked opaque container. For purposes of this subdivision, places where the manufacturer of firearms regularly conducts business includes, but is not limited to places where the manufacturer of firearms regularly or customarily conducts development or design of pistols or revolvers, or regularly or customarily conducts tests on pistols or revolvers, or regularly or customarily participates in the exposition of firearms to the public.

b. The transportation of such pistols or revolvers into, out of or within the city of New York may be done only with the consent of the police commissioner of the city of New York. To obtain such consent, the manufacturer must notify the police commissioner in writing of the name and address of the transporting manufacturer, or agent or employee of the manufacturer who is authorized in writing by such manufacturer to transport pistols or revolvers, the number, make and model number of the firearms to be transported and the place where the manufacturer regularly conducts business within the city of New York and such other information as the commissioner may deem necessary. The manufacturer must not transport such pistols and revolvers between the designated places of business for such reasonable period of time designated in writing by the police commissioner as such official may deem necessary for investigation and to give consent. The police commissioner may not unreasonably withhold his consent.

10. Engaging in the business of gunsmith or dealer in firearms by a person to whom a valid license therefor has been issued pursuant to section 400.00.

11. Possession of a firearm or large capacity ammunition feeding device by a police officer or swom peace officer of another state while conducting official business within the state of New York.

12. Possession of a pistol or revolver by a person who is a member or coach of an accredited college or university target pistol team while transporting the pistol or revolver into or through New York state to participate in a collegiate, olympic or target pistol shooting competition under the auspices of or approved by the national rifle association, provided such pistol or revolver is unloaded and carried in a locked carrying case and the ammunition therefor is carried in a separate locked container.

13. Possession of pistols and revolvers by a person who is a nonresident of this state while attending or traveling to or from, an organized competitive pistol match or league competition under auspices of, or approved by, the National Rifle Association and in which he is a competitor, within forty-eight hours of such event or by a person who is a non-resident of the state while attending or traveling to or from an organized match sanctioned by the International Handgun Metallic Silhouette Association and in which he is a competitor, within forty-eight hours of such event, provided that he has not been previously convicted of a felony or a crime which, if committed in New York, would constitute a felony, and further provided that the pistols or revolvers are transported unloaded in a locked opaque container together with a copy of the match program, match schedule or match registration card. Such documentation shall constitute prima facie evidence of exemption, providing that such person also has in his possession a pistol license or firearms registration card issued in accordance with the laws of his place of residence. For purposes of this subdivision, a person licensed in a jurisdiction which does not authorize such license by a person who has been previously convicted of a felony shall be presumed to have no prior conviction. The superintendent of state police shall annually review the laws of jurisdictions within the United States and Canada with respect to the applicable requirements for licensing or registration of firearms and shall publish a list of those jurisdictions which prohibit possession of a firearm by a person previously convicted of a felony or crimes which if committed in New York state would constitute a felony.

13-a. Except in cities not wholly contained within a single county of the state, possession of pistols and revolvers by a person who is a nonresident of this state while attending or traveling to or from, an organized convention or exhibition for the display of or education about firearms, which is conducted under auspices of, or approved by, the National Rifle Association and in which he is a registered participant, within forty-eight hours of such event, provided that he has not been previously convicted of a felony or a crime which, if committed in New York, would constitute a felony, and further provided that the pistols or revolvers are transported unloaded in a locked opaque container together with a copy of the convention or exhibition program, convention or exhibition schedule or convention or exhibition registration card. Such documentation shall constitute prima facie evidence of exemption, providing that such person also has in his possession a pistol license or firearms registration card issued in accordance with the laws of his place of residence. For purposes of this paragraph, a person licensed in a jurisdiction which does not authorize such license by a person who has been previously convicted of a felony shall be presumed to have no prior conviction. The superintendent of state police shall annually review the laws of jurisdictions within the United States and Canada with respect to the applicable requirements for licensing or registration of firearms and shall publish a list of those jurisdictions which prohibit possession of a firearm by a person previously convicted of a felony or crimes which if committed in New York state would constitute a felony.

14. Possession in accordance with the provisions of this paragraph of a self-defense spray device as defined herein for the protection of a person or property and use of such self-defense spray device under circumstances which would justify the use of physical force pursuant to article thirty-five of this chapter.

(a) As used in this section "self-defense spray device" shall mean a pocket sized spray device which contains and releases a chemical or organic substance which is intended to produce temporary physical discomfort or disability through being vaporized or otherwise dispensed in the air or any like device containing tear gas, pepper or similar disabling agent.

(b) The exemption under this paragraph shall not apply to a person who:

(i) is less than eighteen years of age; or

(ii) has been previously convicted in this state of a felony or any assault; or

(iii) has been convicted of a crime outside the state of New York which if committed in New York would constitute a felony or any assault crime.

(c) The department of health, with the cooperation of the division of criminal justice services and the superintendent of state police, shall develop standards and promulgate regulations regarding the type of self-defense spray device which may lawfully be purchased, possessed and used pursuant to this paragraph. The regulations shall include a requirement that every self-defense spray device which may be lawfully purchased, possessed or used pursuant to this paragraph have a label which states: "WARNING: The use of this substance or device for any purpose other than self-defense is a criminal offense under the law. The contents are dangerous - use with care. This device shall not be sold by anyone other than a licensed or authorized dealer. Possession of this device by any person under the age of eighteen or by anyone who has been convicted of a felony or assault is illegal. Violators may be prosecuted under the law."

15. Possession and sale of a self-defense spray device as defined in paragraph fourteen of this subdivision by a dealer in firearms licensed pursuant to section 400.00 of this chapter, a pharmacist licensed pursuant to article one hundred thirty-seven of the education law or by such other vendor as may be authorized and approved by the superintendent of state police.

(a) Every self-defense spray device shall be accompanied by an insert or inserts which include directions for use, first aid information, safety and storage information and which shall also contain a toll free telephone number for the purpose of allowing any purchaser to call and receive additional information regarding the availability of local courses in self-defense training and safety in the use of a self-defense spray device.

(b) Before delivering a self-defense spray device to any person, the licensed or authorized dealer shall require proof of age and a sworn statement on a form approved by the superintendent of state police that such person has not been convicted of a felony or any crime involving an assault. Such forms shall be forwarded to the division of state police at such intervals as directed by the superintendent of state police. Absent any such direction the forms shall be maintained on the premises of the vendor and shall be open at all reasonable hours for inspection by any peace officer or police officer, acting pursuant to his or her special duties. No more than two self-defense spray devices may be sold at any one time to a single purchaser.

16. The terms "rifle," "shotgun," "pistol," "revolver," and "firearm" as used in paragraphs three, four, five, seven, seven-a, seven-b, nine, nine-a, ten, twelve, thirteen and thirteen-a of this subdivision shall not include a disguised gun or an assault weapon.

b. Section 265.01 shall not apply to possession of that type of billy commonly known as a "police baton" which is twenty-four to twenty-six inches in length and no more than one and one-quarter inches in thickness by members of an auxiliary police force of a city with a population in excess of one million persons or the county of Suffolk when duly authorized by regulation or order issued by the police commissioner of such city or such county respectively. Such regulations shall require training in the use of the police baton including but not limited to the defensive use of the baton and instruction in the legal use of deadly physical force pursuant to article thirty-five of this chapter. Notwithstanding the provisions of this section or any other provision of law, possession of such baton shall not be authorized when used intentionally to strike another person except in those situations when the use of deadly physical force is authorized by such article thirty-five. ...

265.40 Purchase of rifles and/or shotguns in contiguous states

Definitions. As used in this act:

1. "Contiguous state" shall mean any state having any portion of its border in common with a portion of the border of the state of New York;

2. All other terms herein shall be given the meaning prescribed in Public Law 90-618 known as the "Gun Control Act of l968"(18 U.S.C. 921).

It shall be lawful for a person or persons residing in this state, to purchase or otherwise obtain a rifle and/or shotgun in a contiguous state, and to receive or transport such rifle and/ or shotgun into this state; provided, however, such person is otherwise eligible to possess a rifle and/or shotgun under the laws of this state.

Article 400. Licensing and Other Provisions Relating to Firearms

400.00. Licenses to carry, possess, repair and dispose of firearms

1. Eligibility. No license shall be issued or renewed pursuant to this section except by the licensing officer, and then only after investigation and finding that all statements in a proper application for a license are true. No license shall be issued or renewed except for an applicant (a) twenty-one years of age or older, provided, however, that where such applicant has been honorably discharged from the United States army, navy, marine corps, air force or coast guard, or the national guard of the state of New York, no such age restriction shall apply; (b) of good moral character; (c) who has not been convicted anywhere of a felony or a serious offense; (d) who has stated whether he has ever suffered any mental illness or been confined to any hospital or institution, public or private, for mental illness; (e) who has not had a license revoked or who is not under a suspension or ineligibility order issued pursuant to the provisions of section 530.14 of the criminal procedure law or section eight hundred forty-two-a of the family court act; (f) in the county of Westchester, who has successfully completed a firearms safety course and test as evidenced by a certificate of completion issued in his or her name and endorsed and affirmed under the penalties of perjury by a duly authorized instructor, except that: (i) persons who are honorably discharged from the United States army, navy, marine corps or coast guard, or of the national guard of the state of New York, and produce evidence of official qualification in firearms during the term of service are not required to have completed those hours of a firearms safety course pertaining to the safe use, carrying, possession, maintenance and storage of a firearm; and (ii) persons who were licensed to possess a pistol or revolver prior to the effective date of this paragraph are not required to have completed a firearms safety course and test; and (g) concerning whom no good cause exists for the denial of the license. No person shall engage in the business of gunsmith or dealer in firearms unless licensed pursuant to this section. An applicant to engage in such business shall also be a citizen of the United States, more than twenty-one years of age and maintain a place of business in the city or county where the license is issued. For such business, if the applicant is a firm or partnership, each member thereof shall comply with all of the requirements set forth in this subdivision and if the applicant is a corporation, each officer thereof shall so comply.

2. Types of licenses. A license for gunsmith or dealer in firearms shall be issued to engage in such business. A license for a pistol or revolver, other than an assault weapon or a disguised gun, shall be issued to (a) have and possess in his dwelling by a house-holder; (b) have and possess in his place of business by a merchant or storekeeper; (c) have and carry concealed while so employed by a messenger employed by a banking institution or express company; (d) have and carry concealed by a justice of the supreme court in the first or second judicial departments, or by a judge of the New York city civil court or the New York city criminal court; (e) have and carry concealed while so employed by a regular employee of an institution of the state, or of any county, city, town or village, under control of a commissioner of correction of the city or any warden, superintendent or head keeper of any state prison, penitentiary, workhouse, county jail or other institution for the detention of persons convicted or accused of crime or held as witnesses in criminal cases, provided that application is made therefor by such commissioner, warden, superintendent or head keeper; (f) have and carry concealed, without regard to employment or place of possession, by any person when proper cause exists for the issuance thereof; and (g) have, possess, collect and carry antique pistols which are defined as follows: (i) any single shot, muzzle loading pistol with a matchlock, flintlock, percussion cap, or similar type of ignition system manufactured in or before 1898, which is not designed for using rimfire or conventional centerfire fixed ammunition; and (ii) any replica of any pistol described in clause (i) hereof if such replica -

(1) is not designed or redesigned for using rimfire or conventional centerfire fixed ammunition, or

(2) uses rimfire or conventional centerfire fixed ammunition which is no longer manufactured in the United States and which is not readily available in the ordinary channels of commercial trade.

3. Applications.

(a) Applications shall be made and renewed, in the case of a license to carry or possess a pistol or revolver, to the licensing officer in the city or county, as the case may be, where the applicant resides, is principally employed or has his principal place of business as merchant or storekeeper; and, in the case of a license as gunsmith or dealer in firearms, to the licensing officer where such place of business is located. Blank applications shall, except in the city of New York, be approved as to form by the superintendent of state police. An application shall state the full name, date of birth, residence, present occupation of each person or individual signing the same, whether or not he is a citizen of the United States, whether or not he complies with each requirement for eligibility specified in subdivision one of this section and such other facts as may be required to show the good character, competency and integrity of each per-son or individual signing the application. An application shall be signed and verified by the applicant. Each individual signing an application shall submit one photograph of himself and a duplicate for each required copy of the application. Such photographs shall have been taken within thirty days prior to filing the application. In case of a license as gunsmith or dealer in firearms, the photographs submitted shall be two inches square, and the application shall also state the previous occupation of each individual signing the same and the location of the place of such business, or of the bureau, agency, subagency, office or branch office for which the license is sought, specifying the name of the city, town or village, indicating the street and number and otherwise giving such apt description as to point out reasonably the location thereof. In such case, if the applicant is a firm, partnership or corporation, its name, date and place of formation, and principal place of business shall be stated. For such firm or partnership, the application shall be signed and verified by each individual composing or intending to compose the same, and for such corporation, by each officer thereof.

(b) Application for an exemption under paragraph seven-b of subdivision a of section 265.20 of this chapter. Each applicant desiring to obtain the exemption set forth in paragraph seven-b of subdivision a of section 265.20 of this chapter shall make such request in writing of the licensing officer with whom his application for a license is filed, at the time of filing such application. Such request shall include a signed and verified statement by the person authorized to instruct and supervise the applicant, that has met with the applicant and that he has determined that, in his judgment, said applicant does not appear to be or poses a threat to be, a danger to himself or to others. He shall include a copy of his certificate as an instructor in small arms, if he is required to be certified, and state his address and telephone number. He shall specify the exact location by name, address and telephone number where such instruction will take place. Such licensing officer shall, no later than ten business days after such filing, request the duly constituted police authorities of the locality where such application is made to investigate and ascertain any previous criminal record of the applicant pursuant to subdivision four of this section. Upon completion of this investigation, the police authority shall report the results to the licensing officer without unnecessary delay. The licensing officer shall no later than ten business days after the receipt of such investigation, determine if the applicant has been previously denied a license, been convicted of a felony, or been convicted of a serious offense, and either approve or disapprove the applicant for exemption purposes based upon such determinations. If the applicant is approved for the exemption, the licensing officer shall notify the appropriate duly constituted police authorities and the applicant. Such exemption shall terminate if the application for the license is denied, or at any earlier time based upon any information obtained by the licensing officer or the appropriate police authorities which would cause the license to be denied. The applicant and appropriate police authorities shall be notified of any such terminations.

4. Investigation. Before a license is issued or renewed, there shall be an investigation of all statements required in the application by the duly constituted police authorities of the locality where such application is made. For that purpose, the records of the appropriate office of the department of mental hygiene concerning previous or present mental illness of the applicant shall be available for inspection by the investigating officer of the police authority. In order to ascertain any previous criminal record, the investigating officer shall take the fingerprints and physical descriptive data in quadruplicate of each individual by whom the application is signed and verified. Two copies of such fingerprints shall be taken on standard fingerprint cards eight inches square, and one copy may be taken on a card supplied for that purpose by the federal bureau of investigation; provided, however, that in the case of a corporate applicant that has already been issued a dealer in firearms license and seeks to operate a firearm dealership at a second or subsequent location, the original fingerprints on file may be used to ascertain any criminal record in the second or subsequent application unless any of the corporate officers have changed since the prior application, in which case the new corporate officer shall comply with procedures governing an initial application for such license. When completed, one standard card shall be forwarded to and retained by the division of criminal justice services in the executive department, at Albany. A search of the files of such division and written notification of the results of the search to the investigating officer shall be made without unnecessary delay. Thereafter, such division shall notify the licensing officer and the executive department, division of state police, Albany, of any criminal record of the applicant filed therein subsequent to the search of its files. A second standard card, or the one supplied by the federal bureau of investigation, as the case may be, shall be forwarded to that bureau at Washington with a request that the files of the bureau be searched and notification of the results of the search be made to the investigating police authority. The failure or refusal of the federal bureau of investigation to make the fingerprint check provided for in this section shall not constitute the sole basis for refusal to issue a permit pursuant to the provisions of this section. Of the remaining two fingerprint cards, one shall be filed with the executive department, division of state police, Albany, within ten days after issuance of the license, and the other remain on file with the investigating police authority. No such fingerprints may be inspected by any person other than a peace officer, who is acting pursuant to his special duties, or a police officer, except on order of a judge or justice of a court of record either upon notice to the licensee or without notice, as the judge or justice may deem appropriate. Upon completion of the investigation, the police authority shall report the results to the licensing officer without unnecessary delay.

4-a. Processing of license applications. Applications for licenses shall be accepted for processing by the licensing officer at the time of presentment. Except upon written notice to the applicant specifically stating the reasons for any delay, in each case the licensing officer shall act upon any application for a license pursuant to this section within six months of the date of presentment of such an application to the appropriate authority. Such delay may only be for good cause and with respect to the applicant. In acting upon an application, the licensing officer shall either deny the application for reasons specifically and concisely stated in writing or grant the application and issue the license applied for.

4-b. Westchester county firearms safety course certificate. In the county of Westchester, at the time of application, the licensing officer to which the license application is made shall provide a copy of the safety course booklet to each license applicant. Before such license is issued, such licensing officer shall require that the applicant submit a certificate of successful completion of a firearms safety course and test issued in his or her name and endorsed and affirmed under the penalties of perjury by a duly authorized instructor.

5. Filing of approved applications. The application for any license, if granted, shall be filed by the licensing officer with the clerk of the county of issuance, except that in the city of New York and, in the counties of Nassau and Suffolk, the licensing officer shall designate the place of filing in the appropriate division, bureau or unit of the police department thereof, and in the county of Suffolk the county clerk is hereby authorized to transfer all records or applications relating to firearms to the licensing authority of that county. The name and address of any person to whom an application for any license has been granted shall be a public record. Upon application by a licensee who has changed his place of residence such records or applications shall be transferred to the appropriate officer at the licensee's new place of residence. A duplicate copy of such application shall be filed by the licensing officer in the executive department, division of state police, Albany, within ten days after issuance of the license. Nothing in this subdivision shall be construed to change the expiration date or term of such licenses if otherwise provided for in law.

6. License: validity. Any license issued pursuant to this section shall be valid notwithstanding the provisions of any local law or ordinance. No license shall be transferable to any other person or premises. A license to carry or possess a pistol or revolver, not otherwise limited as to place or time of possession shall be effective throughout the state, except that the same shall not be valid within the city of New York unless a special permit granting validity is issued by the police commissioner of that city. Such license to carry or possess shall be valid within the city of New York in the absence of a permit issued by the police commissioner of that city, provided that (a) the firearms covered by such license have been purchased from a licensed dealer within the city of New York and are being transported out of said city forthwith and immediately from said dealer by the licensee in a locked container during a continuous and uninterrupted trip; or provided that (b) the firearms covered by such license are being transported by the licensee in a locked container and the trip through the city of New York is continuous and uninterrupted; or provided that (c) the firearms covered by such license are carried by armored car security guards transporting money or other valuables, in, to, or from motor vehicles commonly known as armored cars, during the course of their employment; or provided that (d) the licensee is a retired police officer as police officer is defined pursuant to subdivision thirty-four of section 1.20 of the criminal procedure law or a retired federal law enforcement officer, as defined in section 2.15 of the criminal procedure law, who has been issued a license by an authorized licensing officer as defined in subdivision ten of section 265.00 of this chapter; provided, further, however, that if such license was not issued in the city of New York it must be marked "Retired Police Officer" or "Retired Federal Law Enforcement Officer", as the case may be, and, in the case of a retired officer the license shall be deemed to permit only police or federal law enforcement regulations weapons; or provided that (e) the licensee is a peace officer described in subdivision four of section 2.10 of the criminal procedure law and the license, if issued by other than the city of New York, is marked "New York State Tax Department Peace Officer" and in such case the exemption shall apply only to the firearm issued to such licensee by the department of taxation and finance. A license as gunsmith or dealer in firearms shall not be valid outside the city or county, as the case may be, where issued.

7. License: form. Any license issued pursuant to this section shall, except in the city of New York, be approved as to form by the superintendent of state police. A license to carry or possess a pistol or revolver shall have attached the licensee's photograph, and a coupon which shall be removed and retained by any person disposing of a firearm to the licensee. Such license shall specify the weapon covered by caliber, make, model, manufacturer's name and serial number, or if none, by any other distinguishing number or identification mark, and shall indicate whether issued to carry on the person or possess on the premises, and if on the premises shall also specify the place where the licensee shall possess the same. If such license is issued to an alien, or to a person not a citizen of and usually a resident in the state, the licensing officer shall state in the license the particular reason for the issuance and the names of the persons certifying to the good character of the applicant. Any license as gunsmith or dealer in firearms shall mention and describe the premises for which it is issued and shall be valid only for such premises.

8. License: exhibition and display. Every licensee while carrying a pistol or revolver shall have on his or her person a license to carry the same. Every person licensed to possess a pistol or revolver on particular premises shall have the license for the same on such premises. Upon demand, the license shall be exhibited for inspection to any peace officer, who is acting pursuant to his or her special duties, or police officer. A license as gunsmith or dealer in firearms shall be prominently displayed on the licensed premises. A gunsmith or dealer of firearms may conduct business temporarily at a location other than the location specified on the license if such temporary location is the location for a gun show or event sponsored by any national, state, or local organization, or any affiliate of any such organization devoted to the collection, competitive use or other sporting use of firearms. Any sale or transfer at a gun show must also comply with the provisions of article thirty-nine-DD of the general business law. Records of receipt and disposition of firearms transactions conducted at such temporary location shall include the location of the sale or other disposition and shall be entered in the permanent records of the gunsmith or dealer of firearms and retained on the location specified on the license. Nothing in this section shall authorize any licensee to conduct business from any motorized or towed vehicle. A separate fee shall not be required of a licensee with respect to business conducted under this subdivision. Any inspection or examination of inventory or records under this section at such temporary location shall be limited to inventory consisting of, or records related to, firearms held or disposed at such temporary locations. Failure of any licensee to so exhibit or display his or her license, as the case may be, shall be presumptive evidence that he or she is not duly licensed.

9. License: amendment. Elsewhere than in the city of New York, a person licensed to carry or possess a pistol or revolver may apply at any time to his licensing officer for amendment of his license to include one or more such weapons or to cancel weapons held under license. If granted, a record of the amendment describing the weapons involved shall be filed by the licensing officer in the executive department, division of state police, Albany. Notification of any change of residence shall be made in writing by any licensee within ten days after such change occurs, and a record of such change shall be inscribed by such licensee on the reverse side of his license. Elsewhere than in the city of New York, and in the counties of Nassau and Suffolk, such notification shall be made to the executive department, division of state police, Albany, and in the city of New York to the police commissioner of that city, and in the county of Nassau to the police commissioner of that county, and in the county of Suffolk to the licensing officer of that county, who shall, within ten days after such notification shall be received by him, give notice in writing of such change to the executive department, division of state police, at Albany.

10. License: expiration, certification and renewal. Any license for gunsmith or dealer in firearms and, in the city of New York, any license to carry or possess a pistol or revolver, issued at any time pursuant to this section or prior to the first day of July, nineteen hundred sixty-three and not limited to expire on an earlier date fixed in the license, shall expire not more than three years after the date of issuance. In the counties of Nassau, Suffolk and Westchester, any license to carry or possess a pistol or revolver, issued at any time pursuant to this section or prior to the first day of July, nineteen hundred sixty-three and not limited to expire on an earlier date fixed in the license, shall expire not more than five years after the date of issuance; however, in the county of Westchester, any such license shall be certified prior to the first day of April, two thousand, in accordance with a schedule to be contained in regulations promulgated by the commissioner of the division of criminal justice services, and every such license shall be recertified every five years thereafter. For purposes of this section certification shall mean that the licensee shall provide to the licensing officer the following information only: current name, date of birth, current address, and the make, model, caliber and serial number of all firearms currently possessed. Such certification information shall be filed by the licensing officer in the same manner as an amendment. Elsewhere than in the city of New York and the counties of Nassau, Suffolk and Westchester, any license to carry or possess a pistol or revolver, issued at any time pursuant to this section or prior to the first day of July, nineteen hundred sixty-three and not previously revoked or cancelled, shall be in force and effect until revoked as herein provided. Any license not previously cancelled or revoked shall remain in full force and effect for thirty days beyond the stated expiration date on such license. Any application to renew a license that has not previously expired, been revoked or cancelled shall thereby extend the term of the license until disposition of the application by the licensing officer. In the case of a license for gunsmith or dealer in firearms, in counties having a population of less than two hundred thousand inhabitants, photographs and fingerprints shall be submitted on original applications and upon renewal thereafter only at six year intervals. Upon satisfactory proof that a currently valid original license has been despoiled, lost or otherwise removed from the possession of the licensee and upon application containing an additional photograph of the licensee, the licensing officer shall issue a duplicate license.

11. License: revocation and suspension. The conviction of a licensee anywhere of a felony or serious offense shall operate as a revocation of the license. A license may be revoked or suspended as provided for in section 530.14 of the criminal procedure law or section eight hundred forty-two-a of the family court act. A license may be revoked and cancelled at any time in the city of New York, and in the counties of Nassau and Suffolk, by the licensing officer, and elsewhere than in the city of New York by any judge or justice of a court of record. The official revoking a license shall give written notice thereof without unnecessary delay to the executive department, division of state police, Albany, and shall also notify immediately the duly constituted police authorities of the locality.

12. Records required of gunsmiths and dealers in firearms. Any person licensed as gunsmith or dealer in firearms shall keep a record book approved as to form, except in the city of New York, by the superintendent of state police. In the record book shall be entered at the time of every transaction involving a firearm the date, name, age, occupation and residence of any person from whom a firearm is received or to whom a firearm is delivered, and the caliber, make, model, manufacturer's name and serial number, or if none, any other distinguishing number or identification mark on such firearm. Before delivering a firearm to any person, the licensee shall require him to produce either a license valid under this section to carry or possess the same, or proof of lawful authority as an exempt person pursuant to section 265.20. In addition, before delivering a firearm to a peace officer, the licensee shall verify that person's status as a peace officer with the division of state police. After completing the foregoing, the licensee shall remove and retain the attached coupon and enter in the record book the date of such license, number, if any, and name of the licensing officer, in the case of the holder of a license to carry or possess, or the shield or other number, if any, assignment and department, unit or agency, in the case of an exempt person. The original transaction report shall be forwarded to the division of state police within ten days of delivering a firearm to any person, and a duplicate copy shall be kept by the licensee. The record book shall be maintained on the premises mentioned and described in the license and shall be open at all reasonable hours for inspection by any peace officer, acting pursuant to his special duties, or police officer. In the event of cancellation or revocation of the license for gunsmith or dealer in firearms, or discontinuance of business by a licensee, such record book shall be immediately surrendered to the licensing officer in the city of New York, and in the counties of Nassau and Suffolk, and elsewhere in the state to the executive department, division of state police

12-a. State police regulations applicable to licensed gunsmiths engaged in the business of assembling or manufacturing firearms. The superintendent of state police is hereby authorized to issue such rules and regulations as he deems reasonably necessary to prevent the manufacture and assembly of unsafe firearms in the state. Such rules and regulations shall establish safety standards in regard to the manufacture and assembly of firearms in the state including specifications as to materials and parts used, the proper storage and shipment of firearms, and minimum standards of quality control. Regulations issued by the state police pursuant to this subdivision shall apply to any person licensed as a gunsmith under this section engaged in the business of manufacturing or assembling firearms, and any violation thereof shall subject the licensee to revocation of license pursuant to subdivision eleven of this section. ...

12-c. Firearms records.

(a) Every employee of a state or local agency, unit of local government, state or local commission, or public or private organization who possesses a firearm or machine-gun under an exemption to the licensing requirements under this chapter, shall promptly report in writing to his employer the make, model, caliber and serial number of each such firearm or machine-gun. Thereafter, within ten days of the acquisition or disposition of any such weapon, he shall furnish such information to his employer, including the name and address of the person from whom the weapon was acquired or to whom it was disposed.

(b) Every head of a state or local agency, unit of local government, state or local commission, public authority or public or private organization to whom an employee has submitted a report pursuant to paragraph (a) of this subdivision shall promptly forward such report to the superintendent of state police.

(c) Every head of a state or local agency, unit of local government, state or local commission, public authority, or any other agency, firm or corporation that employs persons who may lawfully possess firearms or machine-guns without the requirement of a license therefor, or that employs persons licensed to possess fire-arms or machine-guns, shall promptly report to the superintendent of state police, in the manner prescribed by him, the make, model, caliber and serial number of every firearm or machine-gun possessed by it on the effective date of this act for the use of such employees or for any other use. Thereafter, within ten days of the acquisition or disposition of any such weapon, such head shall report such information to the superintendent of the state police, including the name and address of the person from whom the weapon was acquired or to whom it was disposed.

13. Expenses. The expense of providing a licensing officer with blank applications, licenses and record books for carrying out the provisions of this section shall be a charge against the county, and in the city of New York against the city.

14. Fees. In the city of New York and the county of Nassau, the annual license fee shall be twenty-five dollars for gunsmiths and fifty dollars for dealers in firearms. In such city, the city council and in the county of Nassau the Board of Supervisors shall fix the fee to be charged for a license to carry or possess a pistol or revolver and provide for the disposition of such fees. Elsewhere in the state, the licensing officer shall collect and pay into the county treasury the following fees: for each license to carry or possess a pistol or revolver, not less than three dollars nor more than ten dollars as may be determined by the legislative body of the county; for each amendment thereto, three dollars, and five dollars in the county of Suffolk; and for each license issued to a gunsmith or dealer in firearms, ten dollars. The fee for a duplicate license shall be five dollars. The fee for processing a license transfer between counties shall be five dollars. The fee for processing a license or renewal thereof for a qualified retired police officer as defined under subdivision thirty-four of section 1.20 of the criminal procedure law, or a qualified retired sheriff, undersheriff, or deputy sheriff of the city of New York as defined under subdivision two of section 2.10 of the criminal procedure law, or a qualified retired bridge and tunnel officer, sergeant or lieutenant of the triborough bridge and tunnel authority as defined under subdivision twenty of section 2.10 of the criminal procedure law, or a qualified retired uniformed court officer in the unified court system, or a qualified retired court clerk in the unified court system in the first and second judicial departments, as defined in paragraphs a and b of subdivision twenty-one of section 2.10 of the criminal procedure law or a retired correction officer as defined in subdivision twenty-five of section 2.10 of the criminal procedure law shall be waived in all counties throughout the state.

15. Any violation by any person of any provision of this section is a class A misdemeanor.

16. Unlawful disposal. No person shall except as otherwise authorized pursuant to law dispose of any firearm unless he is licensed as gunsmith or dealer in firearms.

17. Applicability of section. The provisions of article two hundred sixty-five relating to illegal possession of a firearm, shall not apply to an offense which also constitutes a violation of this section by a person holding an otherwise valid license under the provisions of this section and such offense shall only be punishable as a class A misdemeanor pursuant to this section. In addition, the provisions of such article two hundred sixty-five shall not apply to the possession of a firearm in a place not authorized by law, by a person who holds an otherwise valid license or possession of a firearm by a person within a one year period after the stated expiration date of an otherwise valid license which has not been previously cancelled or revoked shall only be punishable as a class A misdemeanor pursuant to this section.

400.05 Disposition of weapons and dangerous instruments, appliances and substances

1. Any weapon, instrument, appliance or substance specified in article two hundred sixty-five, when unlawfully possessed, manufactured, transported or disposed of, or when utilized in the commission of an offense, is hereby declared a nuisance. When the same shall come into the possession of any police officer or peace officer, it shall be surrendered immediately to the official mentioned in paragraph (f) of subdivision one of section 265.20, except that such weapon, instrument, appliance or substance coming into the possession of the state police shall be surrendered to the superintendent of state police. ...

400.10. Report of theft or loss of a firearm, rifle or shotgun

1.(a) Any owner or other person lawfully in possession of a firearm, rifle or shotgun who suffers the loss or theft of said weapon shall within twenty-four hours of the discovery of the loss or theft report the facts and circumstances of the loss or theft to a police department or sheriff's office.

(b) Whenever a person reports the theft or loss of a firearm, rifle or shotgun to any police department or sheriff's office, the officer or department receiving such report shall forward notice of such theft or loss to the division of state police via the New York Statewide Police Information Network. The notice shall contain information in compliance with the New York Statewide Police Information Network Operating Manual, including the caliber, make, model, manufacturer's name and serial number, if any, and any other distinguishing number or identification mark on the weapon.

2. The division of state police shall receive, collect and file the information referred to in subdivision one of this section. The division shall cooperate, and undertake to furnish or make available to law enforcement agencies this information, for the purpose of coordinating law enforcement efforts to locate such weapons.

3. Notwithstanding any other provision of law, a violation of paragraph (a) of subdivision one of this section shall be punishable only by a fine not to exceed one hundred dollars.

[Current through Ch. 564, September 4, 2008]

Code of the City of Albany

Chapter 193. Firearms and Ammunition

Article I. Carrying of Loaded Firearms

193-1. Unlawful to carry loaded firearms; exception. Any person, other than a peace officer, who shall in any public street, highway or place within the City of Albany have upon his person or carry any loaded pistol, revolver or other firearm without having been authorized as provided by law to carry the same shall be guilty of a misdemeanor, punishable by a fine not exceeding $1,000 or by imprisonment not exceeding one year, or by both such fine and imprisonment.

193-2. Commissioner of Police authorized to issue permit to carry loaded firearms. Any person, except as provided in this Article, who has occasion to carry a loaded revolver, pistol or firearm for his protection may apply to the Commissioner of Police, and such officer, if satisfied that the applicant is a proper and law-abiding person, shall give said person a permit allowing him to carry such loaded firearm for such period of time as he may deem proper. Any non-resident who does business in the City of Albany and has occasion to carry a loaded pistol, revolver or firearm while in said city must make application for permission to do so, to the Commissioner of Police, in the same manner as is required of residents of said city and shall be subject to the same conditions and restrictions.

Article II. Firearms and Ammunition

193-6. Purpose. The Common Council hereby finds that the storage, possession and display of firearms and the storage, possession and display and sale of ammunition in the City of Albany require the regulation thereof in the public interest, convenience and necessity.

193-7. Definitions. Unless otherwise expressly stated, whenever used in this Article, the following words shall have the meanings given to them by this section. When used herein:

Ammunition - Any projectile commonly known as a "bullet" or "shell."

Firearm - Any weapon from which a shot is discharged by force of an explosive, or a weapon which acts by force of gunpowder, and shall include any weapon capable of being loaded with powder, ball or ammunition, and shall include any weapon which is otherwise capable of being loaded with powder, ball or ammunition, whether completed, assembled or from which any part or piece has been removed therefrom.

Person - Any person, firm, partnership, association, corporation, company or organization of any kind.

193-8. Storage.

(A) Firearms and ammunition for sale shall be kept secured at all times when the place in which they are held for sale is not open for business. The manner in which the firearms and ammunition are to be secured must be approved by both the Chief of Police and the Chief of the Fire Department upon receipt of appropriate application for such approval. Such approval shall in no case be unreasonably withheld.

(B) The regulations provided for herein shall not apply to personal possession, use or ownership of firearms or ammunition therefor.

193-9. Sale of ammunition. Ammunition shall only be sold to persons possessing a valid firearms license, when a license is required for the type of firearm for which the ammunition is used. When purchasing any other type of ammunition, where a firearms license would not be required for the type of ammunition being purchased, a driver's license shall be required. A record shall be kept by the dealer of each sale of ammunition which shall show the type, caliber and quantity of ammunition sold, the name and address of the person receiving the same, the caliber, make, model, manufacturer's name and serial number of the firearm for which the purchaser is purchasing ammunition, the date and time of the transaction, and the identifying number of the firearms or driver's license. Such record shall be maintained for 10 years and shall be made available to state and local enforcement agencies upon request.

193-10. Report of theft or loss to Police Department. Any theft or loss of ammunition or firearms from a storage vault or otherwise shall be reported immediately to the Police Department of the City of Albany.

193-12. Penalties for offenses. Any person who shall violate or shall neglect or refuse to comply with this Article or any provisions thereof shall be guilty of a misdemeanor and, upon conviction thereof, shall be punishable by a fine not exceeding $1,000 or by imprisonment not exceeding one year, or by both such fine and imprisonment, and each day of violation shall constitute a separate offense and be subject to punishment as such.

Article III. Firearms

193-13. Findings and declaration. The Common Council of the City of Albany finds and declares that the proliferation and use of assault weapons pose and present a threat to the health, safety, welfare and security to the citizens of Albany.

193-14. Scope and intent. As used herein, "assault weapon" generally means a firearm of such a nature and with such a high rate of fire and/or capacity for firepower that its function as a legitimate sports or recreational firearm is substantially outweighed by the danger that it can be used to kill and injure human beings. While it is the intent, then, of this legislation to restrict the use and possession of assault weapons, it is not the intent of the Common Council to place restrictions on the use of those firearms which are primarily designed and intended for hunting, target practice or other legitimate sports or recreational activities.

193-15. Assault weapons.

A. As used herein, "assault weapon" includes any of the following or their copies:

(1) Pistols:

Bushmaster Firearms Company auto pistol
Calico Model 100-P auto pistol
Federal Engineering Corporation XP 450, XP 900
Holmes Firearms MP-83, MP-22 assault pistols
Intratec Tec-9 Auto, Tec-9M, Scorpion auto pistols
Israeli Military Industries UZI pistol
Iver Johnson Enforcer Model 300 auto
Ingram or S.W.D. MAC 10, MAC 11
Spectre double-action auto pistol
Ruger Mini 14

(2) Rifles and shotguns by manufacturers in the United States:

Auto-Ordinance Corporation Thompson M1, Mix, 27 A-1
Thompson A-1 Thompson Deluxe, 1927 A5
Colt AR-15A2-Delta H-Bar, AR-15A2 H-Bar
Federal Engineering Corporation XC-900, XC-450 auto carbines
Springfield Armory Inc. BM-59, SAR-48
S.W.D. Street Sweeper shotgun
Weaver Arms Corporation nighthawk

(3) Rifles and shotguns by manufacturers outside the United States:

Avtomat Kalashnikovs manufactured or imported by American Arms, Bushmaster Firearms, Daewoo, Mitchell, Norinco and Poly Technologies
Beretta AR-70 Sporter rifle
Fabrique National FN, FNC rifle
Franchi LAW 12 auto, SPAS 12 pump/auto shotguns
Heckler and Koch HK 91, 93, 94 rifles
Israeli Military Industries Galil Rifles, UZI carbines
Steyr Daimler-Pusch A.U.G. autoloading rifle
Striker 12 shotgun
Valmet M-76 standard rifle, M78 semi-auto

(4) Any other weapon to be subsequently designated by law.

B. A copy of any of the weapons listed in Subsection A(1), (2) and (3) of this section shall include any other model by the same manufacturer with the same action design having slight modifications or enhancements, provided that such weapon as modified or enhanced employs only ammunition of more than .22 caliber rimfire; any weapon with an action design identical or nearly identical to any of the listed weapons which has been designed from, renamed, renumbered or patterned after any of the listed weapons regardless of the manufacturer or country of origin, provided that such weapon employs only ammunition of more than .22 caliber rimfire; or any weapon which has been manufactured and sold by another company under a licensing agreement to manufacture or sell a weapon identical or nearly identical to any of the listed weapons regardless of the company or production or country of origin, provided that such weapon employs only ammunition of more than .22 caliber rimfire.

C. "Assault weapon" does not include weapons that do not use cartridges, manually operated bolt-action weapons, lever-action weapons, slide-action weapons, single-shot weapons, multiple-barrel weapons, revolving-cylinder weapons except shotguns, semiautomatic weapons that use exclusively a rotary Mannlicher-style magazine, any antique firearm as defined in Article 265 of the Penal Law or any assault weapon which has been modified either to render it permanently inoperable or permanently make it a device no longer defined as an assault weapon.

193-16. Prohibition; penalty; registration of existing weapons.

A. Any person, firm or corporation who or which shall, on or after the effective date of this Article, manufacture, import, keep, offer or expose for sale, give, lend or possess any assault weapon in the City of Albany, except as provided for herein, shall be guilty of a misdemeanor and, upon conviction, shall be subject to a fine not exceeding one thousand dollars ($1,000) and/or to imprisonment not exceeding one (1) year.

B. Any individual who lawfully acquired and possessed an assault weapon, as defined herein, prior to the effective date of this Article shall register the firearm with the Chief of the Albany Police Department pursuant to the provisions of § 193-17 herein. The registration shall be made within ninety (90) days of the effective date hereof and shall contain a description of the firearm that identifies it uniquely, including all identification marks and the full name, address, date of birth and fingerprints of the owner, together with such other information as the Chief of Police may deem appropriate. Any change of address of the owner must be registered with the Department within ninety (90) days of the change. The Department may charge a fee for registration not to exceed the actual processing costs of the Department. Such registration shall allow the possession of the firearm on the registrant's property or for the uses specified in § 193-17.

C. On and after the date fixed herein for the registration of assault weapons, no such weapon possessed pursuant hereto shall be sold or transferred to anyone within the City of Albany other than at a licensed gun dealer as defined in the Penal Law. Any individual who obtains title to an assault weapon registered under this section by bequest or intestate succession must, within ninety (90) days, either render the weapon inoperable in accordance with § 193-15, apply for a license pursuant to § 193-17 herein, sell the weapon to a licensed gun dealer or remove the weapon from the city.

193-17. Possession for competitive use; licensing and transportation.

A. A license for an assault weapon may be issued as provided in Subsection B to possess such weapon for the purpose of loading and firing the same while participating as a competitor in organized competitive matches or league competition under the auspices of or approval by a law enforcement agency or a nationally or state recognized organization that fosters proficiency in or promotes education about firearms or to possess such weapon for the purpose of loading and firing the same at a range located on premises owned and occupied by such agency or organization. Such license shall not permit use for any other purpose and shall not be transferable.

B. Application for such license as hereinbefore specified shall be made to the Chief of the Albany Police Department on forms prescribed by him for that purpose. Every person so licensed shall have such license on his or her person when engaged in such activity and while transporting such weapon to or from such competition or range.

C. All such weapons shall be transported in a case, together with a copy of the match program, match schedule or match registration. Where such person is transporting such weapon to or from a range, it shall be transported in a case. Magazines and ammunition for such weapon shall be carried and transported in a locked container separate from the weapon.

D. A licensed gun dealer, as defined in the Penal Law of the State of New York, who lawfully possessed an assault weapon prior to the effective date of this legislation or who gains possession thereof pursuant to this legislation may transport the weapon between dealers or out of the city, display it at any gun show licensed by a state or local governmental entity or sell it to a person licensed under this section. Any transporting allowed herein shall be accomplished as required in Subsection C.

193-18. Applicability to police and military forces. The provisions of this Article shall not apply to the sale to, purchase by or the use or possession of firearms by the Police Department, Sheriff's office, Marshal's office, corrections officers and the like or the military forces of this state while engaged in the discharge of their official duties or as otherwise permitted by regulation or law.

[Code of the City of Albany current as of Sept. 15, 2008]

Chapter 180. Firearms, Arrows and Other Weapons

Article I. Firearms and Arrows

180-1. Firearms, shotguns, rifles and other dangerous weapons.

A. Purpose and intent. The City of Buffalo finds that violent crime is a serious problem in the city and firearms and other dangerous weapons are frequently used in the commission of crimes, particularly homicides and assaults. The possession of such weapons also often leads to accidental deaths and injuries. The possession and use of assault weapons and ammunition-feeding devices for criminal purposes is increasing and poses a serious danger to public safety. The use of weapons by persons under the influence of drugs and/or alcohol can readily lead to serious injury or death. The possession of weapons in public facilities and places also poses a serious danger to public safety. The possession of toy or imitation weapons which substantially duplicate actual weapons poses a danger to the person possessing the weapon and to others. In order to promote and protect the health, safety and welfare of the public, the City of Buffalo finds it necessary to place restrictions upon the possession and use of such weapons. The restrictions imposed by this section are intended to be in addition to restrictions found in state law and are not intended to conflict with state law provisions.

B. Definitions. As used in this section, the following terms shall have the meanings indicated:

Ammunition - Explosives suitable to be fired from a firearm, machine gun, pistol, revolver, rifle, shotgun, assault weapon or other dangerous weapon.

Ammunition-Feeding Device - Magazines, belts, feedstrips, drums or clips capable of being attached to or utilized with any centerfire rifle, shotgun or pistol which employs the force of the expanding gases from a discharging cartridge to chamber a fresh round after each single pull of the trigger which, in the case of a rifle or shotgun, holds in excess of five cartridges, or in the case of a pistol holds in excess of 17 cartridges.

Assault Weapon:

(1) Any centerfire rifle or shotgun which employs the force of the expanding gases from a discharging cartridge to chamber a fresh round after each single pull of the trigger, and which is loaded or capable of being loaded with a combination of more than six cartridges in the ammunition-feeding device and chamber combined. For the purposes of this section, a weapon is capable of being loaded if it is possessed by one who, at the same time, possesses:

(a) In the case of a rifle, a fixed or detachable ammunition feeding device which is attached to or utilized with or capable of being attached to or utilized with such rifle and which has a capacity of more than five cartridges; or

(b) In the case of a shotgun, an ammunition-feeding device which is attached to or utilized with or capable of being attached to or utilized with such shotgun and which has a capacity of more than five cartridges.

(2) A center-fire rifle or shotgun which employs the force of expanding gases from a discharging cartridge to chamber a fresh round after each single pull of the trigger, and which has:

(a) A flash suppressor attached to the weapon reducing muzzle flash;

(b) A grenade launcher;

(c) A sighting device making a target visible at night;

(d) A barrel jacket surrounding all or a portion of the barrel, to dissipate heat therefrom; or

(e) A multi-burst trigger activator.

(3) Any stockless pistol grip shotgun.

(4) For purposes of this section, the term "assault weapon" shall not include any of the following:

(a) Any weapon which has been modified to render it permanently inoperable or permanently make it a device no longer defined as an "assault" weapon.

(b) Weapons that do not use cartridges or shells.

(c) Manually operated bolt-action weapons, lever-action weapons, slide-action weapons or single-shot weapons.

(d) Multiple-barrel weapons, revolving-cylinder weapons except shotguns, weapons that use exclusively a rotary Mannlicher-style magazine.

(e) Any antique firearm as defined in § 265.00 of the New York State Penal Law or any curio or relic as defined under United States law which is possessed by a licensed collector in accordance with United States law.

Dispose Of - To dispose of, give away, give, lease, loan, keep for sale, offer, offer for sale, sell, transfer or otherwise dispose of.

Drug - Any substance listed in § 3306 of the Public Health Law of the State of New York.

Dwelling - One or more rooms, in a building or portion thereof designed or used exclusively as the residence or sleeping place of one or more persons, designed for occupancy by one family for cooking, living and sleeping purposes, as defined in Chapter 511, Zoning, of this Code.

Firearm - Any pistol or revolver, or a shotgun having one or more barrels less than 18 inches in length or any weapon made from a shotgun (whether by alteration, modification or otherwise) if such weapon, as modified, has an overall length of less than 26 inches; or a rifle having one or more barrels less than 16 inches in length or any weapon made from a rifle (whether by alteration, modification or otherwise) if such weapon, as modified, has an overall length of less than 26 inches. For purposes of this definition, the length of a barrel on a shotgun or rifle shall be determined by measuring the distance between the muzzle and the face of the bolt, breech or breechlock when closed and when the shotgun or rifle is cocked; the overall length of a weapon made from a shotgun or rifle is the distance between the extreme ends of the weapon measured along a line parallel to the center line of the bore. Such definition, except as otherwise indicated, shall include both loaded and unloaded firearms, except that it shall not include any antique firearm as defined in federal or New York State law or any curio or relic as defined under United States law which is possessed by a licensed collector in accordance with United States law.

Locking Device - A design adaptation or attachable accessory that will prevent the use of the firearm, rifle or shotgun by an unauthorized user, and includes but is not limited to a trigger lock, which prevents the pulling of the trigger without the use of a key, or a combination handle, which prevents the use of the firearm, rifle or shotgun without the alignment of the combination tumblers.

Possess - Have physical possession or otherwise to exercise dominion or control over. The presence in an automobile of any firearm, rifle or shotgun which is openly visible is presumptive evidence of its possession by all persons occupying such automobile at the time such firearm, rifle or shotgun is found, except if such firearm, rifle or shotgun is found in a vehicle for hire.

Public Place - Any street, including the sidewalk portion thereof, park, playground, recreation area, cemetery or lot owned, leased, operated or controlled by or on behalf of any government, municipality or public authority or corporation within the boundaries of the city, which is generally accessible to the public, except grounds used for educational purposes.

Public Facility - Any building or facility owned, leased, operated or controlled by or on behalf of any government, municipality or public authority or corporation within the boundaries of the city, except buildings or facilities used for educational purposes.

Rifle - A weapon designed or redesigned, made or remade and intended to be fired from the shoulder and designed or redesigned and made or remade to use the energy of the explosive in a fixed metallic cartridge to fire only a single projectile through a rifled bore for each single pull of the trigger.

Shotgun - A weapon designed or redesigned, made or remade and intended to be fired from the shoulder and designed or redesigned and made or remade to use the energy of the explosive in a fixed shotgun shell to fire through a smooth bore either a number of ball shot or a single projectile for each single pull of the trigger.

C. No person shall possess a loaded or unloaded firearm, rifle, shotgun or air gun, or a dagger, dangerous knife, dirk, razor or stiletto, in a public place or public facility in the city. This prohibition shall not apply to:

(1) A police officer or peace officer authorized to possess the same.

(2) A government employee or licensed security guard authorized or required by employment or office to possess the same while acting within the scope of such employment.

(3) A person in the military service of the State of New York or the United States when duly authorized to possess the same.

(4) A person transporting a rifle or shotgun in a motor vehicle in the city in accordance with the provisions of § 11-0931, Subdivision 2, of the New York State Environmental Conservation Law, or otherwise transporting an unloaded rifle, shotgun or air gun in the city, provided that the same is completely enclosed or contained in a nontransparent carrying case and either:

(a) Said carrying case is locked; or

(b) A locking device is attached to the weapon and locked in a manner so as to prevent the weapon from being fired.

(5) An authorized person who, for the purpose of shooting practice, possesses a weapon at an established target range in a public place other than a park or public facility.

(6) A person voluntarily surrendering the same in accordance with the provisions of Section 265.20 of the Penal law.

(7) Possession of a firearm by a person licensed to carry a firearm pursuant to § 400.00 of the Penal Law or possession or transportation by a gunsmith or dealer in firearms in accordance with a license issued by the State of New York or the United States, except that this subsection shall not apply in a park or a public facility other than a parking garage.

D. No person shall store a firearm, rifle, shotgun or air gun in a dwelling in the city unless said firearm, rifle, shotgun or air gun is completely enclosed or contained in a nontransparent locked carrying case or in a locked gun rack, cabinet, closet or safe, or a locking device is attached to the weapon and locked in a manner so as to prevent the weapon from being fired. This requirement shall not apply to a rifle, shotgun or licensed firearm carried on the body of the owner or within such close proximity of the owner that the owner can retrieve it as quickly and easily as if it were carried on the owner's body.

E. No person shall dispose of any firearm, rifle, shotgun, air gun or ammunition in the city. This prohibition shall not apply to:

(1) A gunsmith or dealer in firearms duly licensed by the State of New York or the United States.

(2) A person disposing of the same to a gunsmith or dealer in firearms duly licensed by the State of New York or the United States.

(3) A person voluntarily surrendering the same in accordance with the provisions of § 265.20 of the Penal Law.

(4) A person disposing of a licensed firearm in accordance with law.

(5) Disposition by intestate or testamentary bequest.

(6) A person disposing of a rifle, shotgun, air gun or ammunition to a family member.

F. No person shall possess an assault weapon or an ammunition-feeding device in the city. This prohibition shall not apply to:

(1) A police officer or peace officer authorized to possess the same.

(2) A person in the military service of the State of New York or the United States when duly authorized to possess the same.

(3) A person voluntarily surrendering the same in accordance with the provisions of § 265.20 of the Penal law.

(4) A gunsmith or dealer in firearms duly licensed by the State of New York or the United States for weapons to be used by police officers or persons in the military service or for delivery outside of the city.

G. No person shall dispose of an assault weapon or ammunition-feeding device in the city. This prohibition shall not apply to:

(1) A person voluntarily surrendering the same in accordance with the provisions of § 265.20 of the Penal Law.

(2) A gunsmith or dealer in firearms duly licensed by the State of New York or the United States for weapons to be used by police offices or persons in the military service or for delivery outside of the city.

H. No person shall carry a firearm, shotgun, rifle or air gun in the city while such person has 1/10 of 1% or more by weight of alcohol in the person's blood as shown by chemical analysis of the person's blood, breath, urine or saliva.

I. No person shall carry a firearm, shotgun, rifle or air gun in the city while in an intoxicated condition.

J. No person shall carry a firearm, shotgun, rifle or air gun in the city while the person's ability to safely carry such weapon is impaired by the use of a drug.

K. Any person who carries a firearm, shotgun, rifle or air gun in the city shall be deemed to have given consent to a breath test and a chemical test of the person's breath, blood, urine or saliva for the purpose of determining the alcoholic or drug content of the person's blood, provided that any test is administered at the direction of a police officer having reasonable grounds therefor. ...

L. Discharge of weapons; permits. ...

M. Penalties for offenses. A violation of any provision of this section shall be punishable by a fine not to exceed $1,500 or by imprisonment not to exceed 15 days, or by both such fine and imprisonment. ...

180-2. Carrying loaded shotgun or rifle; exceptions. Nothing herein contained shall be construed to prohibit the possession, custody or use of said shotgun or rifle for the lawful purpose of self-defense or the defense of others.

[Code of the City of Buffalo current as of July 15, 2008]

Administrative Code of the City of New York

Title 10. Public Safety

Chapter 1. Public Safety

10-131 Firearms.

a. Pistols or revolvers, keeping or carrying.

1. The police commissioner shall grant and issue licenses hereunder pursuant to the provisions of article four hundred of the penal law. Unless they indicate otherwise, such licenses and permits shall expire on the first day of the second January after the date of issuance.

2. Every license to carry or possess a pistol or revolver in the city may be issued for a term of no less than one or more than three years. Every applicant for a license to carry or possess a pistol or revolver in the city shall pay therefor, a fee of three hundred forty dollars for each original or renewal application for a three year license period or part thereof, a fee often dollars for each replacement application of a lost license.

3. Every applicant to whom a license has been issued by any person other than the police commissioner, except as provided in paragraph five of this subdivision, for a special permit from the commissioner granting it validity within the city of New York, shall pay for such permit a fee of three hundred forty dollars, for each renewal a fee of three hundred forty dollars, for each replacement of a lost permit a fee of ten dollars.

4. Fees paid as provided herein shall not be refunded in the event that an original or renewal application, or a special validation permit application, is denied by the police commissioner.

5. A fee shall not be charged or collected for a license to have and carry concealed a pistol or revolver which shall be issued upon the application of the commissioner of correction or the warden or superintendent of any prison, penitentiary, workhouse or other institution for the detention of persons convicted or accused of crime or offense, or held as witnesses in criminal cases in the city.

6. The fees prescribed by this subdivision shall be collected by the police commissioner. ...

b. Air pistols and air rifles; selling or possessing.

1. It shall be unlawful for any person to sell, offer to sell or have in such person's possession any air pistol or air rifle or similar instrument in which the propelling force is a spring or air, except that the sale of such instruments if accompanied by delivery to a point without the city, and possession for such purpose, shall not be unlawful if such person shall have secured an annual license from the police commissioner of the city authorizing such sale and possession. The sale and delivery of such instruments within the city from one licensee to another licensee, and the use of such instruments in connection with an amusement licensed by the department of consumer affairs or at rifle or pistol ranges duly authorized by law shall not be considered a violation of this subdivision.

2. All persons dealing in such instruments referred to in this subdivision, shall keep a record showing the name and address of each person purchasing such instrument or instruments, together with place of delivery and said record shall be open to inspection during regular business hours by the officers of the police department of the city.

3. Every person to whom a license shall be granted to sell, possess and deliver the instruments described in this subdivision shall pay therefor an annual fee of ten dollars.

c. Discharge of small-arms. It shall be unlawful for any person to fire or discharge any gun, pistol, rifle, fowling-piece or other firearms in the city; provided that the provisions hereof shall not apply to premises designated by the police commissioner, a list of which shall be filed with the city clerk and published in the City Record.

d. Sale of toy-pistols. It shall be unlawful for any person to sell or dispose of to a minor any toy-pistol or pistol that can be loaded with powder and ball or blank cartridge to be exploded by means of metal caps; but nothing herein contained shall apply to the sale or disposal of what are known as firecracker pistols, torpedo pistols or such pistols as are used for the explosion of paper caps.

e. Tear gas; sale or possession of; fees for permits. ...

f. Violations. Except as is otherwise specifically provided in this section, any person who shall willfully violate any of the provisions of this section shall, upon conviction, be punishable by a fine of not more than fifty dollars, or by imprisonment not exceeding thirty days, or by both such fine and imprisonment.

g.1. It shall be unlawful for any person to sell or offer for sell, possess or use or attempt to use or give away, any toy or imitation firearm which substantially duplicates or can reasonably be perceived to be an actual firearm unless:

(a) the entire exterior surface of such toy or imitation firearm is colored white, bright red, bright orange, bright yellow, bright green, bright blue, bright pink or bright purple, either singly or as the predominant color in combination with other colors in any pattern; or

(b) such toy or imitation firearm is constructed entirely of transparent or translucent materials which permits unmistakable observation of the imitation or toy firearm's complete contents and

(c) the barrel of such toy or imitation firearm, other than the barrel of any such toy or imitation firearm that is a water gun, is closed with the same material of which the toy or imitation firearm is made for a distance of not less than one-half inch from the front end of said barrel, and;

(d) such toy or imitation firearm has legibly stamped thereon, the name of the manufacturer or some trade name, mark or brand by which the manufacturer can be readily identified; and

(e) such toy or imitation or firearm does not have attached thereto a laser pointer, as defined in paragraph one of subdivision a of section 10-134.2 of this code.

2. Paragraph one of this subdivision shall not apply to:

(a) the possession or display of toy or imitation firearms by a manufacturer or dealer solely for purposes of sales that are accompanied by delivery to a point without the city;

(b) any toy or imitation firearm that will be used only for or in the production of television programs or theatrical or motion picture presentations, provided, however, that such use of any toy or imitation firearm complies with all applicable laws, rules or regulations concerning request and receipt of waivers authorizing such use;

(c) non-firing collector replica antique firearms, which look authentic and may be a scale model but are not intended as toys modeled on real firearms designed, manufactured and produced prior to 1898;

(d) decorative, ornamental, and miniature objects having the appearance, shape and/or configuration of a firearm, including those intended to be displayed on a desk or worn on items such as bracelets, necklaces and key chains, provided that the objects measure no more than thirty-eight (38) millimeters in height by seventy (70) millimeters in length, the length measurement excluding any gun stock length measurement.

3. Authorized agents and employees of the department of consumer affairs, and of any other agency designated by the mayor, shall have the authority to enforce the provisions of this subdivision. A proceeding to recover any civil penalty pursuant to this subdivision shall be commenced by service of a notice of hearing that shall be returnable to the administrative tribunal of the department of consumer affairs. The administrative tribunal of such department shall have the power to impose civil penalties for a violation of this subdivision of not more than one thousand dollars ($1000).

4. Any person who shall violate this subdivision shall be guilty of a misdemeanor, punishable by a fine of not more than one thousand dollars ($1000) or imprisonment not exceeding one (1) year or both.

h. Rifles and shotguns; carrying or possessing.

1. It shall be unlawful for any person to carry or possess a loaded rifle or shotgun in public within the city limits. Any person who shall violate this paragraph shall be guilty of a misdemeanor punishable by a fine of not more than one thousand dollars, or imprisonment not exceeding one year, or by both such fine and imprisonment.

2. It shall be unlawful for any person to carry or possess an unloaded rifle or shotgun in public within the city limits unless such rifle or shotgun is completely enclosed, or contained, in a non-transparent carrying case. Any person who shall violate this paragraph shall be guilty of an offense punishable by a fine of not more than fifty dollars or by imprisonment not exceeding thirty days, or by both such fine and imprisonment.

3. The above provisions shall not apply to persons in the military service of the state of New York when duly authorized by regulations issued by the chief of staff to the governor to possess same, police officers and peace officers as defined in the criminal procedure law, or to participants in special events when authorized by the police commissioner.

i.1. It shall be unlawful for any person, except as otherwise authorized pursuant to law, to dispose of any ammunition or any ammunition feeding devise, as defined in section 10-301, designed for use in a firearm, rifle or shotgun, unless he or she is a dealer in firearms or a dealer in rifles and shotguns and such disposition is in accordance with law, provided that a person in lawful possession of such ammunition or ammunition feeding device may dispose of such ammunition or ammunition feeding device to a dealer in firearms who is authorized, or a dealer in rifles and shotguns who is authorized, to possess such ammunition or ammunition feeding device.

2. It shall be unlawful for any dealer in firearms or dealer in rifles and shotguns to dispose of any pistol or revolver ammunition of a particular caliber to any person not authorized to possess a pistol or revolver of such caliber within the city of New York.

3. It shall be unlawful for any person not authorized to possess a pistol or revolver within the city of New York to possess pistol or revolver ammunition, provided that a dealer in rifles and shotguns may possess such ammunition.

4. It shall be unlawful for any person authorized to possess a pistol or revolver of a particular caliber within the city of New York to possess pistol or revolver ammunition of a different caliber.

5. Notwithstanding the provisions of paragraphs two, three and four of this subdivision, any person authorized to possess a rifle within the city of New York may possess ammunition suitable for use in such rifle and a dealer in firearms or dealer in rifles and shotguns may dispose of such ammunition to such person pursuant to section 10-306.

6. It shall be unlawful for any person to possess any ammunition feeding device designed for use in a firearm except as provided in subparagraphs (a), (b), (c), (d) and (e) of this paragraph.

(a) Any pistol or revolver licensee or permittee may possess an ammunition feeding device designed for use in the pistol or revolver such licensee or permittee is authorized to possess, provided that such ammunition feeding device is not capable of holding more than seventeen rounds of ammunition and provided further that such ammunition feeding device does not extend below the grip of the pistol or revolver.

(b) Any person who is exempt pursuant to section 265.20 of the penal law from provisions of the penal law relating to possession of a firearm and who is authorized pursuant to any provision of law to possess a firearm without a license or permit therefor, may possess an ammunition feeding device suitable for use in such firearm, subject to the same conditions as apply with respect to such person's possession of such firearm.

(c) Any dealer in firearms may possess such ammunition feeding devices for the purpose of disposition authorized pursuant to paragraph seven of this subdivision.

(d) Any person who leases a firearm that has been certified by the commissioner as deactivated, from a dealer in firearms or a special theatrical dealer, for use during the course of any television, movie, stage or other similar theatrical production, or any professional photographer who leases a firearm that has been certified by the commissioner as deactivated, from a dealer in firearms or a special theatrical dealer, for use in the pursuance of his or her profession, may possess an ammunition feeding device suitable for use in such firearm, subject to the same conditions as apply with respect to such person's possession of such firearm.

(e) Any special theatrical dealer may possess such ammunition feeding devices exclusively for the purpose of leasing such ammunition feeding devices to such persons as are described in subparagraph (d) of this paragraph.

7. It shall be unlawful for any person to dispose of to another person any ammunition feeding device designed for use in a firearm, provided that a dealer in firearms may dispose of, to such persons as are described in subparagraphs (a) and (b) of paragraph six of this subdivision, such ammunition feeding devices as may be possessed by such persons and provided further that a person in lawful possession of such ammunition feeding devices may dispose of such ammunition feeding devices to a dealer in firearms. In addition, a dealer in firearms or a special theatrical dealer may lease, to such persons as are described in subparagraph (d) of paragraph six of this subdivision, such ammunition feeding devices as may be possessed by such persons.

8. Notwithstanding the provisions of paragraphs six and seven of this subdivision any person may, within ninety days of the effective date of this local law, dispose of an ammunition feeding device designed for use in a firearm to a dealer in firearms.

9. The regular and ordinary transport of ammunition or ammunition feeding devices as merchandise shall not be limited by this subdivision, provided that the person transporting such ammunition or ammunition feeding devices, where he or she knows or has reasonable means of ascertaining what he or she is transporting, notifies, in writing, the police commissioner of the name and address of the consignee and the place of delivery, and withholds delivery to the consignee for such reasonable period of time designated in writing by the police commissioner as the police commissioner may deem necessary for investigation as to whether the consignee may lawfully receive and possess such ammunition or ammunition feeding devices.

10. The regular and ordinary transport of ammunition or ammunition feeding devices by a manufacturer of ammunition or ammunition feeding devices, or by an agent or employee of such manufacturer who is duly authorized in writing by such manufacturer to transport ammunition or ammunition feeding devices on the date or dates specified, directly between places where the manufacturer regularly conducts business, provided such ammunition or ammunition feeding devices are transported in a locked opaque container, shall not be limited by this subdivision, provided that transportation of such ammunition or ammunition feeding devices into, out of or within the city of New York may be done only with the consent of the police commissioner of the city of New York. To obtain such consent, the manufacturer must notify the police commissioner in writing of the name and address of the transporting manufacturer, or agent or employee of the manufacturer who is authorized in writing by such manufacturer to transport ammunition or ammunition feeding devices, the quantity, caliber and type of ammunition or ammunition feeding devices to be transported and the place where the manufacturer regularly conducts business within the city of New York and such other information as the commissioner may deem necessary. The manufacturer shall not transport such ammunition or ammunition feeding devices between the designated places of business for such reasonable period of time designated in writing by the police commissioner as such official may deem necessary for investigation and to give consent. The police commissioner may not unreasonably withhold his or her consent. For the purposes of this paragraph, places where the manufacturer regularly conducts business include, but are not limited to, places where the manufacturer regularly or customarily conducts development or design of ammunition or ammunition feeding devices, or regularly or customarily conducts tests on ammunition or ammunition feeding devices.

11. A person shall be deemed authorized to possess a pistol or revolver within the city of New York if such person is authorized to possess a pistol or revolver within the city of New York pursuant to this section, section 10-302 or section 400.00 of the penal law, or is exempt pursuant to section 265.20 of the penal law from provisions of the penal law relating to possession of a firearm and is authorized pursuant to any provision of law to possess a pistol or revolver within the city of New York without a license or permit therefor. A person shall be deemed authorized to possess a rifle within the city of New York if such person is authorized to possess a rifle within the city of New York pursuant to section 10-303, or is a person permitted pursuant to section 10-305 to possess a rifle without a permit therefor.

12. No pistol or revolver ammunition or ammunition feeding device shall be disposed of to any person pursuant to this subdivision unless such person exhibits the license or permit authorizing him or her to possess a pistol or revolver within the city of New York or exhibits proof that he or she is exempt pursuant to section 265.20 of the penal law from provisions of the penal law relat-ing to possession of a firearm and proof that he or she is authorized pursuant to any provision of law to possess a pistol or revolver within the city of New York without a license or permit therefor.

13. A record shall be kept by the dealer of each disposition of ammunition or ammunition feeding devices under this subdivision which shall show the quantity, caliber and type of ammunition or ammunition feeding devices disposed of, the name and address of the person receiving same, the date and time of the transaction, and the number of the license or permit exhibited or description of the proof of status as a person not required to have a license or permit as required by paragraph twelve of this subdivision.

14. Any person who shall violate this subdivision shall be guilty of a misdemeanor punishable by a fine of not more than one thousand dollars or by imprisonment not exceeding one year, or by both such fine and imprisonment.

15. Any person who shall violate this subdivision shall be liable for a civil penalty of not more than one thousand dollars, to be recovered in a civil action brought by the corporation counsel in the name of the city in any court of competent jurisdiction.

16. The provisions of paragraph three, four and six of this subdivision shall not apply to a person voluntarily surrendering ammunition or ammunition feeding devices, provided that such surrender shall be made to the police commissioner or the commissioner's designee; and provided, further, that the same shall be surrendered by such person only after he or she gives notice in writing to the police commissioner or the commissioner's designee, stating his or her name, address, the nature of the property to be surrendered, and the approximate time of day and the place where such surrender shall take place. Such notice shall be acknowledged immediately upon receipt thereof by such authority. Nothing in this paragraph shall be construed as granting immunity from prosecution for any crime or offense except that of unlawful possession of such ammunition or ammunition feeding devices surrendered as herein provided. A person who possesses any such ammunition or ammunition feeding devices as an executor or administrator or any other lawful possessor of such property of a decedent may continue to possess such property for a period of not over fifteen days. If such property is not lawfully disposed of within such period, the possessor shall deliver it to the police commissioner or the commissioner's designee or such property may be delivered to the superintendent of state police. When such property is delivered to the police commissioner or the commissioner's designee, such officer shall hold it and shall thereafter deliver it on the written request of such executor, administrator or other lawful possessor of such property to a named person, provided such named person is licensed to or is otherwise lawfully permitted to possess the same. If no request to deliver the property is received within two years of the delivery of such property to such official, he or she shall dispose of it in accordance with the provisions of section 400.05 of the penal law.

10-135. Prohibition on sale and possession of electronic stun guns.

a. As used in this section, "electronic stun gun" shall mean any device designed primarily as a weapon, the purpose of which is to stun, render unconscious or paralyze a person by passing an electronic shock to such person, but shall not include an "electronic dart gun" as such term is defined in section 265.00 of the penal law.

b. It shall be unlawful for any person to sell or offer for sale or to have in his or her possession within the jurisdiction of the city any electronic stun gun.

c. Violation of this section shall be a class A misdemeanor.

d. The provisions of this section prohibiting the possession of electronic stun gun shall not apply to police officers as defined in the criminal procedure law, who are operating under regular department procedure or operation guidelines established by their department.

e. The provisions of this section shall not apply to manufacturers of electronic stun guns or importers and exporters or merchants of electronic stun guns, when such guns are scheduled to travel in the course of international, interstate, or intrastate commerce to a point outside the city. Such bulk shipments shall remain in their original shipping package, unopened, except for inspection and possible subdivision for further movement in interstate or intrastate commerce to a point outside the city.

Chapter 3. Firearms

10-301. Control and regulation of the disposition, purchase and possession of firearms, rifles, shotguns, and assault weapons.

Definitions. Whenever used in this chapter the following terms shall mean and include:

1. "Firearm."

(a) Any pistol or revolver;

(b) a shotgun having one or more barrels less than eighteen inches in length; or

(c) a rifle having one or more barrels less than sixteen inches in length;

(d) any weapon made from a shotgun or rifle whether by alteration, modification, or otherwise if such weapon as altered, modified, or otherwise has an overall length of less than twenty-six inches. For the purpose of this subdivision the length of the barrel on a shotgun or rifle shall be determined by measuring the distance between the muzzle and the face of the bolt, breech, or breechlock when closed and when the shotgun or rifle is cocked; the overall length of a weapon made from a shotgun or rifle is the distance between the extreme ends of the weapon measured along a line parallel to the center line of the bore. Firearm does not include an antique firearm. The provisions of this chapter relating to firearms shall not apply to assault weapons except as specifically provided.

2. "Rifle." A weapon designed or redesigned, made or remade, and intended to be fired from the shoulder, and, even if not designed or redesigned, made or remade, and intended to be fired from the shoulder, is not a firearm as defined in subdivision one of this section, and designed or redesigned and made or remade to use the energy of the explosive in a fixed metallic cartridge to fire only a single projectile through a rifled bore for each pull of the trigger. The provisions of this chapter relating to rifles shall not apply to assault weapons except as specifically provided.

3. "Shotgun." A weapon designed or redesigned, made or remade, and intended to be fired from the shoulder, and even if not designed or redesigned, made or remade, and intended to be fired from shoulder, is not a firearm as defined in subdivision one of this section, and designed or redesigned and made or remade to, use the energy of the explosive in a fixed shotgun shell, to fire through a smooth bore either a number of ball shot or a single projectile for each single pull of the trigger. The provisions of this chapter relating to shotguns shall not apply to assault weapons except as specifically provided.

4. "Gunsmith." Any person, firm, partnership, corporation, or company who engages in the business of repairing, altering, assembling, manufacturing, cleaning, polishing, engraving, or trueing, or who in the course of such business performs any mechanical operation on any rifle, shotgun, firearm, assault weapon or machinegun.

5. "Dealer in firearms." Any person, firm, partnership, corporation or company who engages in the business of purchasing, selling, keeping for sale, loaning, leasing, or in any manner disposing of any pistol or revolver or other firearms which may be concealed upon the person. Dealer in firearms shall not include a wholesale dealer.

6. "Dealer in rifles and shotguns." Any person, firm, partnership, corporation or company who engages in the business of purchasing, selling, keeping for sale, loaning, leasing, or in any manner disposing of any rifle, or shotgun. Dealer in rifles and shotguns shall not include a wholesale dealer.

7. "Ammunition." Explosives suitable to be fired from a firearm, machinegun, pistol, revolver, rifle, shotgun, assault weapon or other dangerous weapon.

8. "Dispose of." To dispose of, give away, give, lease, loan, keep for sale, offer, offer for sale, sell, transfer and otherwise dispose of.

9. "Deface." To remove, deface, cover, alter, or destroy the manufacturer's serial number or any other distinguishing number or identification mark.

10. "Commissioner." The police commissioner of the city of New York, or the commissioner's designee.

11. "Permit." The permit for purchase and possession of rifles and shotguns issued by the commissioner.

12. "Certificate." The certificate of registration for possession of rifles and shotguns.

13. "Serious offense." A serious offense as defined in subdivision seventeen of section 265.00 of the penal law.

14. "Business enterprise." Any proprietorship, company, partnership, corporation, association, cooperative, nonprofit organization or other entity engaged or seeking to engage in the activities regulated pursuant to section 10-302 of this chapter.

15. "Semiautomatic." Any firearm, rifle or shotgun that uses part of the energy of a fired cartridge to expel the case of the fired cartridge and load another cartridge into the firing chamber, and which requires a separate pull of the trigger to fire each cartridge.

16. "Assault weapon."

(a) Any semi-automatic centerfire or rimfire rifle or semi-automatic shotgun which has one or more of the following features:

1. folding or telescoping stock or no stock;

2. pistol grip that protrudes conspicuously beneath the action of the weapon;

3. bayonet mount;

4. flash suppressor or threaded barrel designed to accommodate a flash suppressor;

5. barrel shroud;

6. grenade launcher; or

7. modifications of such features, or other features, determined by rule of the commissioner to be particularly suitable for military and not sporting purposes. In addition, the commissioner shall, by rule, designate specific semi-automatic centerfire or rimfire rifles or semiautomatic shotguns, identified by make, model and/or manufacturer's name, as within the definition of assault weapon, if the commissioner determines that such weapons are particularly suitable for military and not sporting purposes. The commissioner shall inspect such specific designated semiautomatic centerfire or rimfire rifles or semi-automatic shotguns at least three times per year, and shall revise or update such designations as he or she deems appropriate.

(b) Any shotgun with a revolving-cylinder magazine.

(c) Any part, or combination of parts, designed or redesigned or intended to readily convert a rifle or shotgun into an assault weapon.

(d) "Assault weapon" shall not include any rifle or shotgun modified to render it permanently inoperative.

17. "Ammunition feeding device." Maga-zines, belts, feedstrips, drums or clips capable of being attached to or utilized with firearms, rifles, shotguns or assault weapons.

18. "Antique firearm." Any unloaded muzzle loading pistol or revolver with a matchlock, flintlock, percussion cap, or similar type of ignition system, or a pistol or revolver which uses fixed cartridges which are no longer available in the ordinary channels of commercial trade.

19. "Special theatrical dealer." Any person, firm, partnership, corporation or company who possesses assault weapons exclusively for the purpose of leasing such assault weapons to special theatrical permittees within the city and for theatrical purposes outside the city.

20. "Acquire." To gain possession of or title to a weapon through purchase, gift, lease, loan, or otherwise.

10-302. Licensing of gunsmiths, of whole-sale manufacturers of firearms, of assemblers of firearms, dealers in firearms, dealers in rifles and shotguns, and special theatrical dealers.

a. It shall be unlawful for any person or business enterprise to engage in the business of gunsmith, wholesale manufacturer of firearms, assembler of firearms, dealer in firearms, dealer in rifles and shotguns, or special theatrical dealer, unless such person, or business enterprise, has obtained a license to engage in such business in the manner prescribed in this section. No person shall engage in the business of dealer in rifles and shotguns or special theatrical dealer unless he or she has been issued a permit for the possession of shotguns and rifles pursuant to the provisions of this chapter. No person or business enterprise shall be eligible to apply for or to hold a license as a special theatrical dealer unless such person or business enterprise (1) possesses both a license as a dealer in firearms and a license as a dealer in rifles and shotguns and (2) has possessed such licenses and engaged in such business for at least one year.

b. No license shall be issued or renewed pursuant to this section except by the police commissioner, and then only after investigation and finding that all statements in a proper application for a license or renewal are true. No license shall be issued or renewed except for any applicant:

(1) of good moral character;

(2) who has not been convicted anywhere of a felony or of any serious offense;

(3) who has stated whether he or she has ever suffered any mental illness or been confined to any hospital or institution, public or private, for mental illness and who is free from any mental disorder, defects or diseases that would impair the ability safely to possess or use a firearm, rifle or shotgun;

(4) who has not been convicted of violating section 10-303.1 of this chapter; and

(5) concerning whom no good cause exists for the denial of a license.

c. (1) An applicant to engage in such business shall also be a citizen of the United States, more than twenty-one years of age and maintain a place of business in the city.

(2) Each applicant to engage in such business shall comply with all the requirements set forth in this section. If the applicant is a partnership, each partner thereof shall comply with all the requirements set forth in this section and if the applicant is a corporation, each officer thereof shall so comply.

(3) No applicant for a special theatrical dealer's license or renewal thereof shall be issued such license or renewal unless the applicant submits proof, in such form as the commissioner may require, that at least ten percent of the gross income earned by the applicant as a dealer in firearms, dealer in rifles and shotguns and, in the case of an application for renewal, special theatrical dealer, in the year preceding the application for such license or renewal, was earned from the lease for theatrical purposes of such ammunition feeding devices, firearms, rifles, shotguns or assault weapons as the applicant was authorized to lease. No applicant for a special theatrical dealer's license or renewal thereof shall be issued such license or renewal unless the gross income earned by the applicant as a dealer in firearms, dealer in rifles and shotguns and, in the case of an application for renewal, special theatrical dealer, in the year preceding the application for such license or renewal, exceeded fifty thousand dollars.

d. An application for a license or renewal thereof shall be made to the police commissioner. An application shall include the full name, date of birth, residence, present occupation of each person or individual signing the same, whether he or she is a citizen of the United States, whether he or she complies with each requirement for eligibility specified in subdivision b of this section and such other facts as may be required to show the good character, competency and integrity of each person or individual signing the application. An application shall be signed and affirmed by the applicant. Each individual signing an application shall submit one photograph of himself or herself and a duplicate for each required copy of the application. Such photographs shall have been taken within thirty days prior to filing the application. The photographs submitted shall be two inches square, and the application shall also state the previous occupation of each individual signing the same and the location of the place of such business, or of the bureau, agency, subagency, office or branch office for which the license is sought, indicating the street and number and otherwise giving such apt description as to point out reasonably the location thereof. In such case, if the applicant is a business enterprise, its name, date and place of formation, and principal place of business shall be stated. For a partnership, the application shall be signed and affirmed by each partner, and for a corporation, by each officer thereof. With respect to an application for a license as a dealer in rifles and shotguns or as a special theatrical dealer, a written statement shall be submitted by the individual applicant, or in the case of a business enterprise a responsible officer or agent thereof, stating (a) the identification number of the shotgun and rifle permit issued to the individual applicant or the responsible officer or agent in the case of a business enterprise, (b) the address of a regular place of business maintained by the applicant, (c) that since the issuance of the rifle and shotgun permit the individual applicant or responsible officer or agent has not become disqualified for issuance of such a permit, (d) that he or she undertakes to supervise the acts of his or her, or in the case of a business enterprise, its employees, (e) that the applicant has not previously been refused a license as a dealer in shotguns and rifles or as a special theatrical dealer and that no such license issued to the applicant has ever been revoked.

e. Before a license is issued or renewed, there shall be an investigation of all statements required in the application by the police department. For that purpose, the records of the department of mental hygiene concerning previous or present mental illness of the applicant shall be available for inspection by the investigating officer of the police department. In order to ascertain any previous criminal record, the investigating officer shall take the fingerprints and physical descriptive data in quadruplicate of each individual by whom the application is signed. Two copies of such fingerprints shall be taken on standard fingerprint cards eight inches square, and one copy may be taken on a card supplied for that purpose by the federal bureau of investigation. When completed, one standard card shall be promptly submitted to the division of criminal justice services where it shall be appropriately processed. A second standard card, or the one supplied by the federal bureau of investigation, as the case may be, shall be forwarded to that bureau at Washington with a request that the files of the bureau be searched and notification of the results of the search be made to the police department. The failure or refusal of the federal bureau of investigation to make the fingerprint check provided for in this section shall not constitute the sole basis for refusal to issue a license pursuant to the provisions of this section. Of the remaining two fingerprint cards, one shall be filed with the executive department, division of state police, Albany, within ten days after issuance of the license, and the other remain on file with the police department. No such fingerprints may be inspected by any person other than a peace officer, when acting pursuant to his or her special duties, or a police officer except on order of a judge or justice of a court of record either upon notice to the licensee or without notice, as the judge or justice may deem appropriate. Upon completion of the investigation, the police department shall report the results to the commissioner without unnecessary delay.

f. Applications shall indicate and licenses shall be issued for a gunsmith, a wholesale manufacturer of firearms, an assembler of firearms, a dealer in firearms, a dealer in rifles and shotguns or a special theatrical dealer.

g. The application for any license, if granted, shall be a public record. Such application shall be kept on file in the office of the police commissioner and, within ten days after the issuance of a license, a duplicate copy shall be filed in the executive department, division of state police, Albany.

h. No license shall be transferable to any other person or premises. The license shall mention and describe the premises for which it is issued and shall be valid only for such premises.

i. A license issued pursuant to this section shall be prominently displayed on the licensed premises. Failure of any person or business enterprise to so exhibit or display such license shall be presumptive evidence that the person or business enterprise is not duly licensed.

j. Any license or renewal issued pursuant to this section shall expire on the first day of the second January following the date of issuance. Any application to renew a license that has not previously expired, been revoked or cancelled shall thereby extend the term of the license until disposition is made of the application by the police commissioner.

k. The conviction of a licensee anywhere of a felony or a serious offense shall operate as a revocation of the license. Written notice of such revocation shall be forwarded to the executive division of state police in Albany.

l. The police commissioner, upon evidence of any disqualification set forth in subdivision b of this section, may at any time suspend or revoke and cancel a license issued pursuant to this section. Written notice of such suspension or revocation shall be forwarded to the executive department, division of state police, Albany.

m. No license shall be issued pursuant to this section unless the applicant therefor possesses and exhibits all licenses required under any federal or state law.

n. Records.

(1) Any person licensed under this section shall keep an accurate book record of every transaction involving a firearm, machinegun, rifle, shotgun or assault weapon. Such record shall be kept in the manner prescribed and contain the information required by the police commissioner.

(2) The records required by this section shall be subject to inspection at all times by members of the police department. Such records shall be maintained on the premises mentioned and described in the license, and preserved for record. In the event of suspension, cancellation or revocation of a license, or discontinuance of business by a licensee, such record shall be immediately surrendered to the police commissioner.

o. Rules and regulations.

(1) The police commissioner may make and promulgate such rules and regulations regarding the issuance and renewal of such licenses and prescribe such forms as are necessary to carry out the provisions of this section.

(2) Such rules and regulations shall prescribe reasonable standards and conditions under which firearms, component parts of firearms, rifles, shotguns, assault weapons and ammunition shall be kept at the store or premises of gunsmiths, including store and plant security, employment, record keeping and product quality control for the protection of the public safety, health and welfare. The foregoing enumeration shall not be construed as a limitation of the police commissioner's authority to promulgate rules and regulations hereunder.

(3) The violation of such rules and regulations shall be triable by a judge of the criminal court of the city of New York and punishable by not more than thirty days imprisonment or by a fine of not more than fifty dollars, or both.

p. The annual license fee for a license issued pursuant to this section shall be twenty-five dollars for gunsmiths, wholesale manufacturers and assemblers of firearms, fifty dollars for dealers in firearms, and one hundred fifty dollars for dealers in rifles and shotguns and special theatrical dealers.

q. Failure to obtain a license, by any person, firm, partnership, corporation or company, as required by the provisions of this section shall be punishable as a class A misdemeanor.

r. No dealer in rifles and shotguns may dispose of a rifle or shotgun to any person unless such person produces a valid rifle and shotgun permit, or proof of lawful authority as a police or peace officer, or is an exempt person as provided in this chapter.

s. Any suspension, denial or revocation of a license may be appealed by the applicant or licensee pursuant to procedures established by the police commissioner for administrative review.

10-302.1 [Effective July 17, 2005] Preventing the diversion of rifles and shotguns to criminals.

a. No dealer in rifles and shotguns shall: (i) sell more than one rifle or shotgun to any individual as part of the same sales transaction; or (ii) sell a rifle or shotgun to any individual if the dealer knows or should know that such individual has purchased a firearm, rifle or shotgun within the prior ninety days.

b. The provisions of this section shall not apply to the sale of firearms, rifles or shotguns to a police officer, as such term is defined in section 1.20 of the criminal procedure law, or to a federal law enforcement officer, as such term is defined in section 2.15 of the criminal procedure law, or to a public agency in furtherance of official business.

c. Before disposing of any firearm, rifle or shotgun to a person licensed by the commissioner to possess firearms, rifles or shotguns, any dealer in firearms, dealer in rifles and shotguns or other person shall contact the police department to ensure compliance with the requirements of this section.

d. Any dealer in firearms, dealer in rifles and shotguns or other person who disposes of any firearm, rifle or shotgun to a person licensed by an authority other than the commissioner to possess firearms, rifles or shotguns shall make reasonable efforts to contact such licensing authority and to ascertain the most recent date of acquisition by such licensee of a firearm, in the case of disposition of a firearm, or of a rifle or shotgun, in the case of disposition of a rifle or shotgun.

e. Any dealer in firearms or dealer in rifles or shotguns who disposes of any firearm, rifle or shotgun shall, before or at the time of disposing of such firearm, rifle or shotgun, record, in the record book required to be kept by subdivision n

f section ⇑10–302, the efforts made by such dealer to ensure compliance with the requirements of this section, any exception or exemption set forth in this section that such dealer reasonably believes would authorize the disposal of such firearm, rifle or shotgun, and the grounds for such dealer's belief that such exception or exemption applies.

f. Exceptions. The provisions of this section shall not apply to the sale of firearms, rifles or shotguns to (i) a police officer, as such term is defined in section 1.20 of the criminal procedure law, (ii) a federal law enforcement officer, as such term is defined in section 2.15 of the criminal procedure law, (iii) a public agency in furtherance of official business, (iv) persons in the military service of the state of New York, when duly authorized by regulations issued by the adjutant general to possess such weapons, (v) persons in the military or other service of the United States, in pursuit of official duty or when duly authorized by federal law, regulation or order to possess such weapons, (vi) persons employed in fulfilling defense contracts with the government of the United States or agencies thereof when possession of such weapons is necessary for manufacture, transport, installation and testing under the requirements of such contract, (vii) peace officers as defined in section 2.10 of the criminal procedure law, provided that such peace officers are authorized pursuant to law or regulation of the state or city of New York to possess a firearm, rifle or shotgun within the city of New York without a license or permit therefor, and are authorized by their employer to possess such firearm, rifle or shotgun, (viii) persons licensed as dealers, manufacturers or importers of firearms pursuant to chapter 44 of title 18 of the United States Code, (ix) any motion picture, television or video production company or entertainment or theatrical company whose production involves the use of firearms, rifles or shotguns, provided that such weapons shall be properly registered and a special theatrical permit shall have been issued for such weapons pursuant to rules established by the commissioner, (x) with respect to the sale of firearms only, persons licensed by the commissioner as gun custodians or special gun custodians, as those terms are used in title thirty-eight of the rules of the city of New York, and (xi) with respect to the sale of rifles and shotguns only, organizations possessing an organization registration certificate, as that term is used in title thirty-eight of the rules of the city of New York.

g. Exempt transactions. The requirements of this section shall not apply to: (i) any transaction in which a person acquires a firearm, rifle or shotgun by operation of law, or because of the death of another person for whom such person is an executor or administrator of an estate or a trustee of a trust created in a will, provided that within fifteen days such person surrenders such firearm, rifle or shotgun to the commissioner until it can be reacquired without violation of this section or other applicable law. If a firearm, rifle or shotgun is surrendered pursuant to this subdivision but no written request to reacquire it is received by the commissioner within two years of such surrender, the commissioner shall dispose of such firearm in accordance with the provisions of section 400.05 of the penal law;

(ii) the exchange of a firearm, rifle or shotgun by a dealer in firearms or a dealer in rifles and shotguns for another firearm, rifle or shotgun previously purchased from such dealer by the person requesting such exchange, provided that such exchange takes place within thirty days of such request;

(iii) the acquisition or disposal of an antique firearm, rifle or shotgun which is incapable of being fired or discharged or which does not fire fixed ammunition, or a firearm, rifle or shotgun manufactured prior to eighteen hundred ninety-four or whose design was patented and whose commercial manufacture commenced prior to eighteen hundred ninety-four and whose manufacture continued after such year without any substantial alteration in design or function, and for which cartridge ammunition is not commercially available and is possessed as a curiosity or ornament or for its historical significance and value;

(iv) the acquisition or disposal of a firearm at an indoor or outdoor pistol range when such acquisition or disposal begins a period of possession or use of the firearm that is authorized by paragraphs 7-a, 7-b, or 7-e of subdivision a of section 265.20 of the penal law;

(v) the sale of a firearm by a dealer in firearms to a person whose firearm is stolen or irretrievably lost, provided that: (1) such person has complied with any legal requirement to report the loss or theft, including but not limited to the applicable provisions of title thirty-eight of the rules of the city of New York and section 400.10 of the penal law;

(2) such person provides to such dealer a copy of a police report of the loss or theft or of any report made pursuant to the applicable provisions of title thirty-eight of the rules of the city of New York and section 400.10 of the penal law, which copy the dealer shall attach to the record book required to be kept by subdivision n of section 10–302;

(3) the copy provided pursuant to subparagraph two of this paragraph contains the name and address of the regulated firearm owner, a description of the regulated firearm, the location of the loss or theft, if known, the date of the loss or theft, if known, and the date when the loss or theft was reported to the law enforcement agency; and

(4) such person's attempt to replace the regulated firearm occurs within thirty days of the loss or theft of such firearm, if known, or, if such date is not known, within thirty days of the date when the loss or theft was reported to the law enforcement agency, as reflected by the information recorded on the police report; and

(vi) any other transaction authorized in advance in writing by the commissioner.

h. Penalties.

(i) In addition to the penalties specified in section ⇑10–310, any act or omission that constitutes or would constitute a violation of this section or of rules and regulations issued by the commissioner pursuant thereto shall be grounds for the revocation of a license to deal in firearms, deal in rifles and shotguns, possess firearms, or possess a rifle or shotgun.

(ii) Any firearm disposed of or acquired in violation of this section shall be a nuisance subject to surrender and forfeiture in accordance with the procedures specified in section 400.05 of the penal law.

i. The commissioner may make and promulgate such rules and regulations as are necessary to carry out the provisions of this section. Such rules and regulations may address, but need not be limited to:

(i) procedures for implementation of this section by the commissioner;

(ii) establishment of a database of firearm, rifle and shotgun purchases for the purpose of enforcing the requirements of this chapter; and

(iii) the specification of reasonable efforts required to comply with subdivision d of this section.

10-303. Permits for possession and purchase of rifles and shotguns. It shall be unlawful to dispose of any rifle or shotgun to any person unless said person is the holder of a permit for possession and purchase of rifles and shotguns; it shall be unlawful for any person to have in his or her possession any rifle or shotgun unless said person is the holder of a permit for the possession and purchase of rifles and shotguns.

The disposition of a rifle or shotgun, by any licensed dealer in rifles and shotguns, to any person presenting a valid rifle and shotgun permit issued to such person, shall be conclusive proof of the legality of such disposition by the dealer.

a. Requirements. No person shall be denied a permit to purchase and possess a rifle or shotgun unless the applicant:

(1) is under the age of twenty-one; or

(2) is not of good moral character; or

(3) has been convicted anywhere of a felony; of a serious offense as defined in § 265.00 (17) of the New York State Penal Law; of a misdemeanor crime of domestic violence as defined in § 921 (a) of title 18, United States Code; of a misdemeanor crime of assault as defined in the penal law where the applicant was convicted of such assault within the ten years preceding the submission of the application; or of any three misdemeanors as defined in local, state or federal law, however nothing in this paragraph shall preclude the denial of a permit to an applicant with fewer than three misdemeanor convictions; or

(4) has not stated whether he or she has ever suffered any mental illness or been confined to any hospital or institution, public or private, for mental illness; or

(5) is not now free from any mental disorders, defects or diseases that would impair the ability safely to possess or use a rifle or shotgun; or

(6) has been the subject of a suspension or ineligibility order issued pursuant to § 530.14 of the New York State Criminal Procedure Law or § 842-a of the New York State Family Court Act; or

(7) who is subject to a court order that

(a) was issued after a hearing of which such person received actual notice, and at which such person had the opportunity to participate;

(b) restrains such person from harassing, stalking, or threatening an intimate partner of such person or child of such intimate partner or person, or engaging in other conduct that would place an intimate partner in reasonable fear of bodily injury to the partner or child; and

(c)(i) includes a finding that such person represents a credible threat to the physical safety of such intimate partner or child; or

(ii) by its terms explicitly prohibits the use, attempted use, or threatened use of physical force against such intimate partner or child that would reasonably be expected to cause bodily injury;

(d) For purposes of this section only, "intimate partner" means, with respect to a person, the spouse of the person, a former spouse of the person, an individual who is a parent of a child of the person, and an individual who cohabitates or has cohabited with the person; or

(8) has been convicted of violating section 10-303.1 of this chapter; or

(9) unless good cause exists for the denial of the permit.

b. Application. Application for a rifle and shotgun permit shall be made to the police commissioner, shall be signed and affirmed by the applicant and shall state his or her full name, date of birth, residence, physical condition, occupation and whether he or she complies with each requirement specified in subdivision a of this section, and any other information required by the police commissioner to process the application. Each applicant shall submit with his or her application a photograph of himself or herself in duplicate, which shall have been taken within thirty days prior to the filing of the application. Any willful or material omission or false statement shall be a violation of this section and grounds for denial of the application.

c. Before a permit is issued or renewed, the police department shall investigate all statements required in the application. For that purpose, the records of the department of mental hygiene concerning previous or present mental illness of the applicant shall be available for inspection by the investigating officer of the police department. In order to ascertain any previous criminal record, the investigating officer shall take the fingerprints and physical descriptive data in quadruplicate of each individual by whom the application is signed. Two copies of such fingerprints shall be taken on standard fingerprint cards eight inches square, and one copy may be taken on a card supplied for that purpose by the federal bureau of investigation. When completed, one standard card shall be promptly submitted to the division of criminal justice services where it shall be appropriately processed. A second standard card, or the one supplied by the federal bureau of investigation, as the case may be, shall be forwarded to that bureau at Washington with a request that the files of the bureau be searched and notification of the results of the search be made to the police department. The failure or refusal of the federal bureau of investigation to make the fingerprint check provided for in this section shall not constitute the sole basis for refusal to issue a permit pursuant to the provisions of this section. Of the remaining two fingerprint cards, one shall be filed with the executive department, division of state police, Albany, within ten days after issuance of the permit, and the other remain on file with the police department. No such fingerprints may be inspected by any person other than a peace officer, when acting pursuant to his or her special duties, or a police officer, except on order of a justice of a court of record either upon notice to the permittee or without notice, as the judge or justice may deem appropriate. Upon completion of the investigation, the police department shall report the results to the police commissioner without unnecessary delay.

d. Fees. The fee for an application for a rifle and shotgun permit or renewal thereof shall be one hundred forty dollars.

e. Issuance.

(1) Upon completion of the investigation, and in no event later than thirty days from the submission of the application, unless the police commissioner determines more time is needed for an investigation and then it shall not exceed sixty days, the commissioner shall issue the permit or shall notify the applicant of the denial of the application and the reason or reasons therefor. The applicant shall have the right to appeal said denial pursuant to procedures established by the police commissioner for administrative review.

(2) Any person holding a valid license to carry a concealed weapon in accordance with the provisions of the penal law, shall be issued such permit upon filing an application and upon paying the established fee therefor, without the necessity of any further investigation, affidavits or fingerprinting, unless the police commissioner has reason to believe that the status of the applicant has changed since the issuance of the prior license.

f. Validity. Any person to whom a rifle and shotgun permit has been validly issued pursuant to this chapter may possess a rifle or shotgun. No permit shall be transferred to any other person. Every person carrying a rifle or shotgun shall have on his or her person a permit which shall be exhibited for inspection to any peace officer or police officer upon demand. Failure of any such person to so exhibit his or her permit shall be presumptive evidence that he or she is not duly authorized to possess a rifle or shotgun and the same may be considered by the police commissioner as cause for revocation or suspension of such permit. A permit shall be valid for three (3) years and shall be subject to automatic renewal, upon sworn application, and without investigation, unless the police commissioner has reason to believe that the status of the applicant has changed since the previous application.

g. Revocation or suspension. A permit shall be revoked upon the conviction in this state, or elsewhere, of a person holding a rifle or shotgun permit, of a felony or a serious offense. A permit may be revoked or suspended at any time upon evidence of any other disqualification set forth in subdivision a of this section. Upon revocation or suspension of a permit for any reason, the police commissioner shall immediately notify the New York state division of criminal justice services. The police commissioner shall from time to time send a notice and supplemental report hereof, containing the names, addresses and permit numbers of each person whose rifle and shotgun permit has been revoked or suspended to all licensed dealers in rifles and shotguns throughout the city for the purpose of notifying such dealers that no rifles or shotguns may be issued or sold or in any way disposed of to any such persons. The police commissioner or any police officer acting at the police commissioner's direction shall forthwith seize any rifle and shotgun permit which has been revoked or suspended hereunder and shall seize any rifle or shotgun possessed by such person, provided that the person whose rifle or shotgun permit has been revoked or suspended, or such person's appointee or legal representative, shall have the right at any time up to one year after such seizure to dispose of such rifle or shotgun to any licensed dealer or any other person legally permitted to purchase or take possession of such rifle or shotgun. The permittee shall have the right to appeal any suspension or revocation pursuant to procedures established by the commissioner for administrative review.

h. Non-residents. Non-residents of the city of New York may apply for a rifle or shotgun permit subject to the same conditions, regulations and requirements as residents of the city of New York.

10-303.1. Prohibition of the possession or disposition of assault weapons.

a. It shall be unlawful for any person to possess or dispose of any assault weapon within the city of New York, except as provided in subdivision d, e or f of this section or section 10-305. A person who peaceably surrenders an assault weapon to the commissioner pursuant to subdivision d, e or f of this section or subdivision f of section 10-305 shall not be subject to the criminal or civil penalties set forth in this section.

b. Criminal penalty. Any person who shall violate subdivision a of this section shall be guilty of an unclassified misdemeanor punishable by a fine of not more than five thousand dollars or by imprisonment of up to one year, or by both fine and imprisonment, for each assault weapon disposed of or possessed, provided that the first violation of subdivision a of this section involving possession of an assault weapon as defined in paragraph c of subdivision 16 of section 10-301 shall be an offense punishable by a fine of not more than three hundred dollars or imprisonment of not more than fifteen days, or both, on condition that (1) such first violation is not in conjunction with the commission of a crime and (2) the possessor has not been previously convicted of a felony or a serious offense.

c. Civil penalty. In addition to the penalties prescribed in subdivision b of this section, any person who shall violate subdivision a of this section shall be liable for a civil penalty of not more than twenty-five thousand dollars for each assault weapon disposed of or possessed, to be recovered in a civil action brought by the corporation counsel in the name of the city in any court of competent jurisdiction, provided that the first violation by any person of subdivision a of this section involving possession of an assault weapon as defined in paragraph c of subdivision sixteen of section 10-301 shall subject such person to a civil penalty of not more than five thousand dollars on condition that (1) such first violation is not in conjunction with the commission of a crime and (2) the possessor has not been previously convicted of a felony or a serious offense.

d. Disposition of assault weapons by permittees, licensees and previously exempt persons. Any person who, on or after the effective date of this local law, shall possess an assault weapon and a valid permit for possession and purchase of rifles and shotguns and a certificate of registration for such assault weapon, and any licensed dealer in firearms or licensed dealer in rifles and shotguns who is not licensed as a special theatrical dealer and who, on or after the effective date of this local law, shall possess an assault weapon, and any police officer or peace officer who, before the effective date of this law was exempt from the sections of the administrative code requiring rifle and shotgun permits and certificates, and who, upon the effective date of this local law, is not exempt from the sections of the administrative code prohibiting the possession or disposition of assault weapons, and who, on or after the effective date of this local law, shall possess an assault weapon, shall, within ninety days of the effective date of rules promulgated by the commissioner pursuant to subparagraph 7 of paragraph a of subdivision 16 of section 10-301, either:

(1) peaceably surrender his or her assault weapon pursuant to subdivision f of section 10-305 for the purpose of destruction of such weapon by the commissioner, provided that the commissioner may authorize the use of such weapon by the department; or

(2) lawfully remove such assault weapon from the city of New York. All assault weapons possessed by such permittees, licensees and previously exempt persons shall be subject to the provisions of this subdivision, whether defined as assault weapons in subdivision 16 of section 10-301 or in rules promulgated by the commissioner pursuant to subparagraph 7 of paragraph a of subdivision 16 of section 10-301.

e. Disposition of assault weapons by non-permittees. Any person who, on or after the effective date of this local law, shall possess an assault weapon and who is not the holder of a valid permit for possession and purchase of rifles and shotguns and a certificate of registration for such assault weapon, shall peaceably surrender his or her assault weapon pursuant to subdivision f of section 10-305 for the purpose of destruction of such weapon by the commissioner, provided that the commissioner may authorize the use of such weapon by the department, and provided further that heirs and legatees may dispose of assault weapons pursuant to subdivision f of this section.

f. Disposition of assault weapons by heirs and legatees. Any person who acquires an assault weapon on or after the effective date of this local law by the laws of intestacy or by testamentary bequest shall, within ninety days of such acquisition, either: (1) peaceably surrender such assault weapon pursuant to subdivision f of section 10-305 for the purpose of destruction of such weapon by the commissioner, provided that the commissioner may authorize the use of such weapon by the department; or (2) lawfully remove such assault weapon from the city of New York.

g. Within thirty days of the effective date of rules promulgated by the commissioner pursuant to subparagraph 7 of paragraph a of subdivision 16 of section 10-301, the commissioner shall send by regular mail to every person who has been issued a permit to possess a rifle or shotgun and whose rifle or shotgun the commissioner reasonably believes to be an assault weapon as defined in subdivision 16 of section 10-301 or as defined in such rules, a written notice setting forth the requirements and procedures relating to the disposition of such weapons, and the criminal and civil penalties that may be imposed upon the permittee for unlawful possession or disposition of such weapons. Failure by the commissioner to send, or the permittee to receive, such notice, shall not excuse such permittee for unlawful possession or disposition of such weapons.

h. Surrender of firearms. At the discretion of the police commissioner, any person convicted of violating this section may be subject to immediate surrender of all firearms in his or her possession.

10-303.2 Civil penalty; firearms dealers and manufacturers.

a. Definitions. For purposes of this section, the terms "firearm," "handgun," "dealer," "collector," and "manufacturer" shall have the meanings set forth in 18 U.S.C. § 921, as such section may be amended from time to time, or any successor provision thereto. The term "transfer" shall be deemed to include any sale, assignment, pledge, lease, loan, gift or other disposition. References to "statutes, laws or regulations" shall be deemed to include federal, state and local statutes, laws, local laws, ordinances, rules and regulations.

b. Manufacturer and Dealer Liability. A manufacturer or dealer shall be liable for any injury or death caused by a firearm that it has transferred, if (i) such injury or death results from the use of such firearm by an individual not authorized by law to possess such firearm in the city of New York, and (ii) such manufacturer or dealer, or any other individual or entity acting subsequent to such manufacturer or dealer, unlawfully transferred such firearm at any time prior to such injury or death. Such liability also includes the possible imposition of punitive damages. Liability under this section does not extend to any manufacturer or dealer that has complied with the following standards during a period of one year immediately preceding and including the transfer of such firearm:

(1) The manufacturer or dealer executes no transfers or agreements to transfer at gun shows except for gun shows that maintain a practice of performing instant criminal background checks consistent with 18 U.S.C. § 922(t), as such subsection may be amended from time to time and any successor provision thereto, on all transfers, whether by licensed or an unlicensed sellers.

(2) Any place of business operated by the manufacturer or dealer is located at a fixed address where:

(a) a record is maintained, as may be required by any statute, law or regulation, of the make, model, caliber or gauge, and serial number of all firearms held in inventory or offered for sale; and

(b) a record is maintained, as may be required by any statute, law or regulation, of the make, model, caliber or gauge, and serial number of all firearms sold, and of any identifying information required by any such statute, law or regulation to be obtained from purchasers;

(3) The manufacturer or dealer provides access to the aforementioned records to officers, employees and agents of public agencies conducting inspections, to the full extent required by applicable statutes, laws and regulations;

(4) The manufacturer or dealer limits transfers to any individual or entity to one handgun in any given thirty-day period, provided that this paragraph shall not apply to lawful transfers to (a) public agencies in furtherance of official business; (b) law enforcement officers employed by public agencies; (c) private security firms, holding any permits or licenses required by applicable statutes, laws and regulations, for the use of their agents and employees; (d) private operators of state and local correctional facilities, for the use of their agents and employees; or (e) licensed manufacturers, licensed dealers or licensed collectors, as those terms are defined by 18 U.S.C. § 921, as such section may be amended from time to time, or any successor provision thereto;

(5) The manufacturer or dealer has complied with all applicable statutes, laws and regulations governing the transfer of firearms; and

(6) The manufacturer or dealer has not transferred a firearm to any other manufacturer or dealer in circumstances in which the manufacturer or dealer transferring such firearm knew or should have known that such manufacturer or dealer had not complied with the standards set forth in this subdivision.

c. Exceptions.

(1) No action may be commenced pursuant to this section by any person injured or killed by the discharge of a firearm that is lawfully possessed by a law enforcement official employed by a public agency.

(2) This section shall not limit in scope any cause of action, other than that provided by this section, available to a person injured by or killed by a firearm.

(3) Nothing in this section shall prevent a manufacturer or dealer from seeking whole or partial indemnity or contribution for any liability incurred under this section from any third party wholly or partially responsible for the injury or death.

(4) Notwithstanding the provisions of subdivision b, there shall be no basis for liability under this section if the manufacturer or dealer proves by a preponderance of the evidence that: (i) the person injured or killed by the discharge of a firearm was committing or attempting to commit a crime (whether or not such crime is actually charged); (ii) the unlawful transfer or possession of the firearm is solely a result of the failure of the owner of the firearm to renew a license, permit or registration within six months of the date such renewal is required; or (iii) prior to the injury or death caused by the firearm, a lawful possessor of the firearm has reported its theft to a federal, state or local law enforcement agency, or reported its loss to an appropriate public agency.

(5) Notwithstanding the provisions of subdivision b, there shall be no basis for liability under this section if the manufacturer or dealer proves by a preponderance of the evidence that the manufacturer or dealer lawfully transferred the firearm to: (i) a public agency in furtherance of official business; (ii) a law enforcement officer employed by a public agency; (iii) a private security firm, holding any permits or licenses required by applicable statutes, laws and regulations, for the use of its agents and employees; or (iv) a private operator of a state or local correctional facility for the use of its agents and employees.

(6) Notwithstanding the provisions of subdivision b, there shall be no basis for liability under this section if the manufacturer or dealer proves by a preponderance of the evidence that the injury or death is not directly or indirectly related to any act or omission by such manufacturer or dealer, including but not limited to any failure by the manufacturer or dealer to comply with the standards set forth in subdivision b of this section.

10-304. Certificates of registration.

a. It shall be unlawful for any person to have in his or her possession any rifle or shotgun unless said person is the holder of a certificate of registration for such rifle or shotgun.

b. It shall be unlawful for any person who is not a licensed dealer to dispose of any rifle or shotgun for which such person does not have a certificate of registration unless such person files with the police commissioner a declaration in duplicate, signed and affirmed by the declarant which shall list by caliber, make, model, manufacturer's name and serial number, or if none, any other distinguishing number or identification mark, of each rifle and shotgun possessed by the declarant. Upon receipt of acknowledgement of said declaration by the police commissioner, the declarant may lawfully sell, transfer, or otherwise dispose of such rifles or shotguns to a licensed dealer or any other person legally permitted to purchase or take possession of such rifles or shotguns.

Any willful or material omission or false statement shall be a violation of this section.

c. Exhibition of certificate. Every person carrying a rifle or shotgun shall have on his or her person a certificate of registration valid for such weapon. Upon demand, the appropriate certificate shall be exhibited for inspection to any peace officer or police officer. Failure of any person to so exhibit his or her certificate shall be presumptive evidence that he or she is not duly authorized to possess such rifle or shotgun and may be considered by the police commissioner as cause for revocation or suspension of such person's permit.

d. Revocation. The revocation of a rifle or shotgun permit shall automatically be deemed to be a revocation of all certificates of registration for rifles and shotguns held by the person whose permit has been revoked.

e. Disposition of rifles and shotguns. No person lawfully in possession of a rifle or shotgun shall dispose of same except to a licensed dealer in firearms, licensed dealer in rifles and shotguns, the holder of a valid rifle and shotgun permit, an exempt person as enumerated in this chapter, or a non-resident of the city of New York not subject to the permit requirements of this chapter.

Any person so disposing of a rifle or shotgun shall report the disposition on forms provided by the commissioner setting forth the rifle and shotgun permit number of both seller and purchaser, the make, caliber, type, model and serial number, if any, and if the seller is a licensed dealer the certificate of registration number, of all such rifles and shotguns. Such form shall be signed by both seller and purchaser and the original shall be forwarded to the police commissioner within 72 hours of the disposition, one copy shall be retained by the seller, another by the purchaser.

1. If the seller is a licensed dealer, he or she shall at the time of the sale issue a certificate of registration to the purchaser provided to the dealer for that purpose by the police commissioner and shall forward to the police commissioner the duplicate thereof, together with the report of disposition.

2. If the seller is not a licensed dealer, the police commissioner shall, if the purchaser's rifle permit is valid, issue the certificate of registration within ten days of the receipt by the police commissioner of the report of disposition. Pending receipt of the certificate, but in no event for any longer than fourteen days from the date of purchase, the copy of the report of disposition shall serve in lieu of the purchaser's certificate of registration.

f. No fee shall be charged for a certificate of registration.

g. Notwithstanding any other provision of this section concerning the transfer, receipt, acquisition, or any other disposition of a rifle or shotgun, a rifle and shotgun permit shall not be required for the passing of a rifle or shotgun upon the death of an owner, to his or her heir or legatee, whether the same be by testamentary bequest or by the laws of intestacy, except that the person who shall so receive or acquire said rifle or shotgun shall be subject to all other provisions of this chapter, provided further that if the heir or legatee of the owner of such rifle or shotgun does not qualify to possess same under this chapter, the rifle or shotgun may be possessed by the heir or legatee for the purpose of sale as otherwise provided herein for a period not exceeding one hundred eighty days or for such further limited period beyond the one hundred eighty days as may be approved by the commissioner, said extensions in no event to exceed a total of ninety days.

10-305. Exemptions. The sections requiring rifle and shotgun permits and certificates and prohibiting the possession or disposition of assault weapons shall not apply as follows:

a. Minors. Any person under the age of twenty-one years may carry, fire, or use any rifle or shotgun in the actual presence or under the direct supervision of any person who is a holder of a rifle or shotgun permit, or for the purpose of military drill under the auspices of a legally recognized organization and under competent supervision or for the purpose of competition or target practice in and upon a firing range approved by the police commissioner or any other governmental agency authorized to provide such approval, or the national rifle association, which is under competent supervision at the time of such competition or target practice, provided that the rifle or shotgun is otherwise properly registered or exempt from registration by virtue of some other provision of this chapter. This exemption shall not apply to assault weapons.

b. Antiques and ornaments. The provisions of this chapter shall not apply to antique rifles and shotguns which are incapable of being fired or discharged or which do not fire fixed ammunition, or those weapons manufactured prior to eighteen hundred ninety-four and those weapons whose design was patented and whose commercial manufacture commenced prior to eighteen hundred ninety-four and whose manufacture continued after eighteen hundred ninety-four without any substantial alteration in design or function, and for which cartridge ammunition is not commercially available and are possessed as curiosities or ornaments or for their historical significance and value. This exemption shall not apply to assault weapons.

c. Persons in the military service in the state of New York, when duly authorized by regulations issued by the chief of staff to the governor to possess the same, and police officers, provided that such police officers shall not be exempt from the sections prohibiting the possession or disposition of assault weapons except during the performance of their duties as police officers, and other peace officers as defined in section 2.10 of the criminal procedure law, provided that such peace officers (1) are authorized pursuant to law or regulation of the state or city of New York to possess either (a) a firearm within the city of New York without a license or permit therefor, or (b) a rifle, shotgun or assault weapon within the city of New York without a permit therefor; and (2) are authorized by their employer to possess such rifle, shotgun or assault weapon; and (3) shall not possess such rifle, shotgun or assault weapon except during the performance of their duties as peace officers.

d. Persons in the military or other service of the United States, in pursuit of official duty or when duly authorized by federal law, regulation or order to possess the same.

e. Persons employed in fulfilling defense contracts with the government of the United States or agencies thereof when possession of the same is necessary for manufacture, transport, installation and testing under the requirements of such contract.

Any such person exempted by subdivisions c, d and e above, may purchase a rifle or shotgun only from a licensed dealer, and must submit to the dealer full and clear proof of identification, including shield number, serial number, military or governmental order or authorization, and military or other official identification. Any dealer who disposes of a rifle or shotgun to any exempt person without securing such identification shall be in violation of these sections.

f. A person may voluntarily surrender a rifle, shogun or assault weapon to the police commissioner, provided, that the same shall be surrendered by such person only after he or she gives notice in writing to the police commissioner, stating such person's name, address, the type of gun to be surrendered, and the approximate time of day and the place where such surrender shall take place and such time of day and place have been approved in writing by the police commissioner. Nothing in this subdivision shall be construed as granting immunity from prosecution for any crime or offense except that of unlawful possession of such rifle, shotgun or assault weapon.

g. The regular and ordinary transport of rifles, shotguns and assault weapons as merchandise provided that the person transporting such rifles, shotguns and assault weapons where he or she knows or has reasonable means of ascertaining what such person is transporting, notifies, in writing, the police commissioner of the name and address of the consignee and the place of delivery, and withholds delivery to the consignee for such reasonable period of time designated in writing by the police commissioner as the police commissioner may deem necessary for investigation as to whether the consignee may lawfully receive and possess such rifles, shotguns or assault weapons.

h. Possession by retail customers for the purpose of firing at duly licensed rifle target concessions at amusement parks, piers, and similar locations provided that the rifles to be so used be firmly chained or affixed to the counter and that the individual rifles are registered by the proprietor and that the proprietor is in possession of a rifle and shotgun permit. This exemption shall not apply to assault weapons.

i. (1) Non-residents in transit. Any other provision of this chapter to the contrary notwithstanding, a non-resident of the city of New York who, without a rifle and shotgun permit issued hereunder, enters the city of New York possessing a rifle or shotgun in the course of transit to a destination outside the city of New York, or a non-resident of the city of New York who enters the city of New York possessing an assault weapon in the course of transit to a destination outside the city of New York, shall have a period of twenty-four hours subsequent to such entering to be exempt from penalty under this chapter for the unlawful possession of a rifle, shotgun or assault weapon, provided that such rifle, shotgun or assault weapon shall at all times be unloaded and in a locked case, or locked automobile trunk, and that said non-resident is lawfully in possession of said rifle or shotgun according to the laws of his or her place of residence.

(2) Non-residents purchasing a rifle or shotgun from a licensed dealer. Any other provision of this chapter notwithstanding, a non-resident of the city of New York may purchase a rifle or shotgun from a licensed dealer provided that he or she presents the dealer with documentary evidence of his or her identity and place of residence, and the rifle or shotgun purchased is either personally delivered to the purchaser or transmitted by the dealer directly to the purchaser's residence. In the event the purchaser is traveling from the city by rail, ship or plane, the dealer is hereby authorized to deliver such rifle or shotgun at the appropriate terminal to a representative of the railroad, airline or shipping company, for placement aboard such train, plane or ship. If the rifle or shotgun is personally delivered to the non-resident purchaser within the city of New York, the purchaser shall have the rifle or shotgun removed from the city no later than twenty-four hours after the time of purchase. This exemption shall not apply to assault weapons.

j. Nothing herein contained shall be construed to be a prohibition of the conduct of business by manufacturers, wholesale dealers, interstate shippers, or any other individuals or firms properly licensed by the federal government, where such prohibition would be preempted by federal law.

k. Special theatrical permit. Nothing herein contained is intended to prevent the possession or utilization of any rifle, shotgun or assault weapon during the course of any television, movie, stage or other similar theatrical production, or by a professional photographer in the pursuance of his or her profession, provided however, that the rifle or shotgun so used shall be properly registered and a special theatrical permit shall have been issued for the rifle, shotgun or assault weapon pursuant to regulations established by the commissioner.

l. Persons in possession of, using or transporting rifles which have been issued by the director of civilian marksmanship of the department of the army, pursuant to the provisions of ten U.S.C., section 4307-4309, unto a civilian rifle club, or unto a rifle team representing an educational institution, provided that such persons are members in good standing of an accredited civilian rifle club, or are connected as students or coaches with such educational institution, shall not be required to obtain a certificate of registration for such rifle. This exemption shall not apply to assault weapons.

m. Any resident of the city of New York acquiring a rifle or shotgun outside the city of New York shall within seventy-two hours after bringing such weapon into the city make application for a rifle and shotgun permit, if such person does not already possess such permit, and for a certificate of registration.

Pending the issuance of such permit and/or certificate of registration such resident shall deposit such weapon with a designated officer, at the police precinct in which such person resides, who shall issue a receipt therefor and said weapon shall be retained at the precinct until the resident shall produce the proper permit and registration certificate. This exemption shall not apply to assault weapons.

n. The provisions of section 10-303 of this chapter shall not apply to persons who are members of units of war veterans organizations, which organizations are duly recognized by the veterans administration, pursuant to § 3402 of title 38 of the United States Code, and who are specifically designated to carry rifles or shotguns by the commanders of said units, while actually participating in going to or returning from, special events authorized by the commissioner. Said rifles or shotguns, to be carried, must be the property of the unit of the war veterans organization, must be registered with the police commissioner pursuant to section 10-304 of this chapter and must be kept at the unit's headquarters or some central place as registered.

o. Nothing herein, shall exempt a member of a unit of a war veterans organization from possessing a permit issued pursuant to section 10-303, to carry rifles or shotguns which are not the property of a war veterans organization; nor shall that member be exempt from registering such rifles or shotguns, pursuant to section 10-304, which said member may personally own, possess or purchase.

p. Any gunsmith licensed pursuant to section 10-302 may engage in the business of gunsmith as authorized by such license.

q. Notwithstanding the provisions of this chapter prohibiting the possession or disposition of assault weapons, a special theatrical dealer may possess such weapons exclusively for the purpose of leasing such weapons to special theatrical permittees within the city and for theatrical purposes outside the city and may, in addition, with the written approval of the commissioner, permanently remove one or more assault weapons from the city.

10-306. Disposition, purchase and possession of ammunition and ammunition feeding devices.

a. No person, except a dealer in rifles and shotguns, may dispose of to another person an ammunition feeding device which is designed for use in a rifle or shotgun and which is capable of holding more than five rounds or rifle or shotgun ammunition, except in the manner provided in this chapter for the disposition of assault weapons, provided that a person in lawful possession of such ammunition feeding devices may dispose of such ammunition feeding devices to a dealer in rifles and shotguns. No dealer in rifles and shotguns may dispose of such ammunition feeding devices except to a person who is exempt from subdivision a of section 10-303.1 pursuant to section 10- 305.

b. No person may possess an ammunition feeding device which is designed for use in a rifle or shotgun and which is capable of holding more than five rounds of rifle or shotgun ammunition, unless such person is exempt form subdivision a of section 10-303.1 pursuant to section 10-305, provided that a dealer in rifles and shotguns may possess such ammunition feeding devices for the purpose of disposition authorized pursuant to subdivision a of this section.

c. No ammunition suitable for use in a rifle of any caliber or for any shotgun or ammunition feeding device which is designed for use in a rifle or shotgun and which is capable of holding no more than five rounds of rifle or shotgun ammunition shall be disposed of to any person who has not been issued a rifle and shotgun permit and a certificate of registration and who does not exhibit same to the dealer at the time of the purchase. In no event shall rifle or shotgun ammunition be disposed of to or possessed by any such person except for a shotgun, or for the specific caliber of rifle, for which the certificate of registration has been issued. No ammunition feeding device which is designed for use in a rifle or shotgun and which is capable of holding more than five rounds of rifle or shotgun ammunition shall be disposed of by a dealer in rifles and shotguns to any person who does not exhibit proof that he or she is exempt from subdivision a of section 10-303.1 pursuant to section 10-305.

d. It shall þe unlawful for any person who is required to have a permit in order to possess a rifle or shotgun and who has not been issued such permit to possess rifle or shotgun ammunition or an ammunition feeding device which is designed for use in a rifle or shotgun.

e. A record shall be kept by the dealer of each disposition of ammunition or ammunition feeding devices under this section which shall show the type, caliber and quantity of ammunition or ammunition feeding devices disposed of, the name and address of the person receiving same, the caliber, make, model, manufacturer's name and serial number of the rifle or shotgun for which the purchaser is purchasing ammunition, the date and time of the transaction, and the number of the permit and certificate exhibited or description of the proof of exemption exhibited as required by this section. Such information shall be made available to all law enforcement agencies.

f. Notwithstanding any other provision of this section, ammunition and ammunition feeding devices which are designed for use in rifles or shotguns and which are capable of holding no more than five rounds of rifle or shotgun ammunition may be disposed of or possessed in the same manner and pursuant to the same requirements, rules and exemptions as apply to disposal or possession of rifles, shotguns or assault weapons under this chapter, provided that a special theatrical permittee may not possess live ammunition suitable for use in the rifle, shotgun or assault weapon such permittee is authorized to possess. Ammunition feeding devices which are designed for use in rifles or shotguns and which are capable of holding more than five rounds of rifle or shotgun ammunition may only be disposed of or possessed in the manner provided in this section.

g. Notwithstanding any other provision of this section, any person authorized to possess a pistol or revolver within the city of New York may possess ammunition suitable for use in such pistol or revolver and a dealer in firearms or dealer in rifles and shotguns may dispose of such ammunition to such person pursuant to subdivision i of section 10-131.

h. Dealers in rifles and shotguns and special theatrical dealers may lease ammunition feeding devices which are designed for use in rifles or shotguns to special theatrical permittees. Special theatrical permittees may possess such ammunition feeding devices subject to the same conditions as apply with respect to such permittee's possession of rifles, shotguns and assault weapons.

10-307. Supply of forms. The commissioner shall provide all dealers in rifles and shotguns with adequate supplies of all forms including applications for permits as required by this chapter, without charge.

10-308. Vehicles, rooms, dwellings or structures; possession therein. The presence of a rifle, or shotgun, or rifle or shotgun ammunition, in a vehicle, room, dwelling or structure, without a rifle and shotgun permit therefor and a certificate of registration therefor, or the presence of an assault weapon in a vehicle, room, dwelling or structure, shall be presumptive evidence of possession thereof by all persons occupying the vehicle, room, dwelling or structure at the time.

10-309. Identifying marks.

a. Defacing. Any person who alters, changes, removes, disfigures, obliterates or defaces the name of the maker, model, manufacturer's or serial number of a rifle, shotgun or assault weapon shall be in violation of this section.

b. Any rifle or shotgun sold or otherwise disposed of by a licensed dealer, which does not contain a manufacturer's or serial number, must have imbedded into the metal portion of such rifle or shotgun a dealer's number. Failure to so mark and identify any rifle or shotgun shall be a violation of this section.

10-310. Violation. Except as is otherwise provided in sections 10-302 and 10-303.1, violation of sections 10-301 through 10-309 shall be a misdemeanor punishable by a fine of not more than one thousand dollars or imprisonment of not more than one year or both, provided that the first violation of such sections involving possession of an unregistered rifle or shotgun or rifle or shotgun ammunition or an ammunition feeding device which is designed for use in a rifle or shotgun and which is capable of holding no more than five rounds of rifle or shotgun ammunition shall be an offense punishable by a fine of not more than three hundred dollars or imprisonment of not more than sixteen days, or both on condition that (a) the first violation of possession of an unregistered rifle or shotgun or rifle or shotgun ammunition or an ammunition feeding device which is designed for use in a rifle or shotgun and which is capable of holding no more than five rounds of rifle or shotgun ammunition is not in conjunction with the commission of a crime and (b) the possessor has not been previously convicted of a felony or a serious offense and (c) the possessor has not previously applied for and been denied a permit for such possession.

10-311. Sale of weapons without safety locking device prohibited.

a. It shall be unlawful for any person or business enterprise to dispose of any weapon which does not contain a safety locking device. For the purposes of this section and section 10-312: (1) weapon shall mean a firearm, rifle, shotgun, or assault weapon, as such terms are defined in section 10-301; or a machine gun, as defined in the penal law; and (2) a safety locking device shall mean a design adaptation or attachable accessory that will prevent the use of the weapon by an unauthorized user, and includes, but is not limited to, a trigger lock, which prevents the pulling of the trigger without the use of a key, or a combination handle, which prevents the use of the weapon without the alignment of the combination tumblers.

b. It shall be unlawful for any licensed manufacturer, licensed importer, or licensed dealer to dispose of any weapon in the city of New York unless it is accompanied by the following warning, which shall appear in conspicuous and legible type in capital letters, and which shall be printed on a label affixed to the weapon and on a separate sheet of paper included within the packaging enclosing the weapon:

"THE USE OF A LOCKING DEVICE OR SAFETY LOCK IS ONLY ONE ASPECT OF RESPONSIBLE FIREARM STORAGE. FIREARMS SHOULD BE STORED UNLOADED AND

LOCKED IN A LOCATION THAT IS BOTH SEPARATE FROM THEIR AMMUNITION AND INACCESSIBLE TO CHILDREN AND OTHER UNAUTHORIZED PERSONS."

c. Any person who applies for and obtains authorization to purchase a weapon or otherwise lawfully obtains a weapon pursuant to chapters one or three of title ten of this code shall be required to purchase or obtain a safety locking device at the time he or she purchases or obtains the weapon.

d. (1) The police commissioner is authorized to promulgate rules setting forth the types of safety locking devices which will comply with this section in accordance with subdivision a of this section. The city of New York and its agencies, officers or employees shall not be liable to any party by reason of any incident involving, or the use or misuse of, a safety locking device that may have been purchased in compliance with such rules promulgated by the commissioner.

(2) The police commissioner shall provide written notice of the requirements of this section and section 10-312 to all persons who receive an official authorization to purchase a weapon and all persons applying for renewal of a license or permit issued pursuant to chapters one or three of title ten, including any rules promulgated under this subdivision. All persons applying for a license or permit or applying for the renewal of a license or permit pursuant to chapters one or three of title ten of this code, shall receive from the commissioner information concerning the importance of using a safety locking device while a weapon is not in use, and a warning that weapons should be stored unloaded and locked in a location that is both separate from their ammunition and inaccessible to children and any other unauthorized persons.

e. Any violation of subdivisions a or b of this section or any rule promulgated thereunder shall be a misdemeanor and triable by a judge of the criminal court of the city of New York and punishable by imprisonment of not more than thirty days or by a fine of not more than five hundred dollars, or both.

10-312. Use of safety locking device required under certain circumstances.

a. It shall be unlawful for any person who is the lawful owner or lawful custodian of a weapon, as that term is defined in section 10-311, to store or otherwise place or leave such weapon in such a manner or under circumstances that it is out of his or her immediate possession or control, without having rendered such weapon inoperable by employing a safety locking device. Any person who violates this subdivision shall be guilty of a violation, punishable by imprisonment of not more than ten days or by a fine of not more than two hundred fifty dollars, or both.

b. Any person who violates subdivision a of this section having previously been found guilty of a violation of such subdivision, or under circumstances which create a substantial risk of physical injury to another person, shall be guilty of a misdemeanor punishable by imprisonment of not more than thirty days or by a fine of not more than one thousand dollars, or both.

c. The provisions of this section shall not apply to weapons owned or lawfully possessed by a police officer, as such term is defined in section 1.20 of the criminal procedure law, or a federal law enforcement officer, as such term is defined in section 2.15 of the criminal procedure law.

[NYC Administrative Code current through June 30, 2007]

Code of the City of Rochester

Chapter 47. Dangerous Articles

47-4. Storage and display of firearms, ammunition and explosives.

A. Purpose and intent. The Council finds that it is necessary to regulate the commercial storage, possession and display of firearms, ammunition or explosives pursuant to § 139-d of the General Municipal Law in order to provide for the public health, safety and welfare of all persons in the City of Rochester. The Council finds that the location of such activities close to residential uses is not compatible with residential uses and can pose a danger to residents through fire or explosion or as a result of burglaries at such locations. The Council therefore intends to regulate the location of such activities and to place additional regulations upon those activities in order to assure that such activities are conducted in a safe manner. The restrictions found herein shall be in addition to restrictions found Chapter 120 of the Municipal Code, Zoning Code, and whichever regulations are more restrictive shall be applicable to any potential location where such activities are to be conducted.

B. Location. The storage, possession or display of firearms, ammunition or explosives within a building occupied by a residential use, or within a building located within one hundred (100) feet of any residential use, which distance shall be measured from the closest point of the building, or portion thereof, used for the storage, possession or display of firearms, ammunition or explosives to the nearest point of the lot line of the property with a residential use, is hereby prohibited.

C. Standards of design, construction and maintenance of buildings and structures in which firearms, ammunition or explosives are stored. ...

D. Visibility of interior to be maintained at all times. The interior of any building or structure used for the storage, possession and display of firearms, ammunition or explosives shall be visible through any windows at all times when open for business, and no drapes or blinds should be used that would block the view of police or passersby who might observe unusual activity within the premises. The exterior of the premises shall be illuminated at night and during the hours when business is not conducted within.

E. Combustible materials. Combustible materials shall not be stored in any building or structure or that portion thereof used for the storage, possession and display of firearms, ammunition or explosives.

F. Fire-extinguishing equipment. Fully operable listed fire-extinguishing equipment shall be maintained in any building or structure used for the storage, possession and display of firearms, ammunition or explosives and made easily accessible.

G. Smoking and open flames prohibited. Smoking, matches, spark-producing devices and open flames shall be prohibited in any building or structure or that portion thereof used for the storage, possession and display of firearms, ammunition or explosives.

H. Standards of security for storage of firearms, ammunition or explosives.

(1) Storage of ammunition and explosives. All ammunition and explosives shall be stored in compliance with 9 NYCRR 1176 et seq. and 12 NYCRR 39 et seq. Further, all ammunition when being displayed shall be kept in locked cases or behind the counter in an area not accessible to the public.

(2) Storage of firearms when open for business.

(a) No firearms shall be stored, exhibited or displayed in windows of the premises.

(b) Firearms storage or inventory areas shall be physically separated from counter and display areas and access to these areas shall be carefully controlled.

(c) All firearm display cases shall be kept locked and secured at all times and not readily accessible to the public. All keys to such display cases shall not leave the control of authorized personnel.

(d) Trigger locks which disable firearms and prevent them from functioning must be locked to each firearm at all times, or the firearms must be secured in a locked case or be otherwise locked, or the firearms must be dispensed in an area behind the counter that is not accessible to the public. These requirements shall not apply to a firearm being shown to a customer, being repaired, or otherwise being worked on.

(3) Storage of firearms when not open for business. When not open for business, all firearms shall be stored in accordance with one of the following:

(a) All firearms shall be stored in a locked fireproof safe or vault located in the business premises;

(b) All firearms must be secured by a hardened steel rod or cable of at least one-eighth (1/8) inch in diameter through the trigger guard of the firearm. The steel cord or cable shall be secured with a hardened steel lock that has a shackle. The lock and shackle shall be protected or shielded from the use of a bolt cutter and the rod or cable shall be anchored in a manner that prevents the ready removal of the firearms from the premises; or

(c) All firearms shall be secured in a manner that prevents the ready removal of the firearms from the premises, as approved by the Chief of Police or the Chief's designee.

I. The regulations provided for herein shall not apply to the personal possession, use or ownership of firearms or ammunition therefor.

47-5. Firearms, shotguns, rifles and other dangerous weapons.

A. Purpose and intent. The Council finds that violent crime is a serious problem in the city and firearms and other dangerous weapons are frequently used in the commission of crimes, particularly homicides and assaults. The possession of such weapons also often leads to accidental deaths and injuries. The possession and use of assault weapons and ammunition feeding devices for criminal purposes is increasing and poses a serious danger to public safety. The use of weapons by persons under the influence of drugs and/or alcohol can readily lead to serious injury or death. The possession of weapons in public facilities and places also poses a serious danger to public safety. The possession of toy or imitation weapons which substantially duplicate actual weapons poses a danger to the person possessing the weapon and to others. In order to promote and protect the health, safety and welfare of the public, the Council finds it necessary to place restrictions upon the possession and use of such weapons. The restrictions imposed by this section are intended to be in addition to restrictions found in state law and are not intended to conflict with state law provisions.

B. As used in this section, the following terms shall have the meanings indicated:

Air Gun - [Note: This section was found unconstitutional.]

Ammunition - Explosives suitable to be fired from a firearm, machine gun, pistol, revolver, rifle, shotgun, assault weapon or other dangerous weapon.

Ammunition Feeding Device - Magazines, belts, feedstrips, drums or clips capable of being attached to or utilized with any center-fire rifle, shotgun or pistol which employs the force of the expanding gases from a discharging cart-ridge to chamber a fresh round after each single pull of the trigger which, in the case of a rifle or shotgun holds in excess of five (5) cartridges, or in the case of a pistol holds in excess of seventeen (17) cartridges.

Assault weapon:

(1) Any center-fire rifle or shotgun which employs the force of the expanding gases from a discharging cartridge to chamber a fresh round after each single pull of the trigger, and which is loaded or capable of being loaded with a combination of more than six (6) cartridges in the ammunition feeding device and chamber combined. For the purposes of this section, a weapon is capable of being loaded if it is possessed by one who, at the same time, possesses:

(a) In the case of a rifle, a fixed or detachable ammunition feeding device which is attached to or utilized with or capable of being attached to or utilized with such rifle and which has a capacity of more than five (5) cartridges; or

(b) In the case of a shotgun, an ammunition feeding device which is attached to or utilized with or capable of being attached to or utilized with such shotgun and which has a capacity of more than five (5) cartridges.

(2) A center-fire rifle or shotgun which employs the force of expanding gases from a discharging cartridge to chamber a fresh round after each single pull of the trigger, and which has:

(a) A flash suppressor attached to the weapon reducing muzzle flash;

(b) A grenade launcher;

(c) A sighting device making a target visible at night;

(d) A barrel jacket surrounding all or a portion of the barrel to dissipate heat therefrom; or

(e) A multi-burst trigger activator.

(3) Any stockless pistol grip shotgun.

(4) The following weapons manufactured prior to the effective date of this section. [Note: This section was found unconstitutional.]

(5) For purposes of this section, the term "assault weapon" shall not include any of the following:

(a) Any weapon which has been modified to render it permanently inoperable or permanently make it a device no longer defined as an "assault weapon";

(b) Weapons that do not use cartridges or shells;

(c) Manually operated bolt-action weapons, lever-action weapons, slide-action weapons or single-shot weapons;

(d) Multiple-barrel weapons, revolving-cylinder weapons except shotguns, weapons that use exclusively a rotary Mannlicher-style magazine; or

(e) Any antique firearm as defined in § 265.00 of the New York State Penal Law or any curio or relic as defined under United States law which is possessed by a licensed collector in accordance with United States Law.

Dispose Of - To dispose of, give away, give, lease, loan, keep for sale, offer, offer for sale, sell, transfer or otherwise dispose of.

Drug - Any substance listed in § 3306 of the Public Health Law of the State of New York.

Dwelling - As defined in Chapter 120 of the Municipal Code, Zoning Code.

Firearm - Any pistol or revolver; or a shotgun having one (1) or more barrels less than eighteen (18) inches in length or any weapon made from a shotgun (whether by alteration, modification or otherwise) if such weapon as modified has an overall length of less than twenty-six (26) inches; or a rifle having one (1) or more barrels less than sixteen (16) inches in length or any weapon made from a rifle (whether by alteration, modification or otherwise) if such weapon as modified has an overall length of less than twenty-six (26) inches. For purposes of this definition, the length of the barrel on a shotgun or rifle shall be determined by measuring the distance between the muzzle and the face of the bolt, breech or breechlock when closed and when the shotgun or rifle is cocked; the overall length of a weapon made from a shotgun or rifle is the distance between the extreme ends of the weapon measured along a line parallel to the center line of the bore. Such definition, except as otherwise indicated, shall include both loaded and unloaded firearms, except that it shall not include any antique firearm as defined in federal or New York State law or any curio or relic as defined under United States law which is possessed by a licensed collector in accordance with United States law.

Park - As defined in § 79-1 of the Municipal Code.

Possess - Have physical possession or otherwise to exercise dominion or control over. The presence in an automobile of any firearm, rifle or shotgun which is openly visible is presumptive evidence of its possession by all persons occupying such automobile at the time such firearm, rifle or shotgun is found, except if such firearm, rifle or shotgun is found in a vehicle for hire.

Public Place - Any street, including the sidewalk portion thereof, park, playground, recreation area, cemetery or lot owned, leased, operated or controlled by or on behalf of any government, municipality or public authority or corporation within the boundaries of the city, which is generally accessible to the public, except grounds used for educational purposes.

Public Facility - Any building or facility owned, leased, operated or controlled by or on behalf of any government, municipality or public authority or corporation within the boundaries of the city, except buildings or facilities used for educational purposes.

Rifle - A weapon designed or redesigned, made or remade and intended to be fired from the shoulder and designed or redesigned and made or remade to use the energy of the explosive in a fixed metallic cartridge to fire only a single projectile through a rifled bore for each single pull of the trigger.

Shotgun - A weapon designed or redesigned, made or remade and intended to be fired from the shoulder and designed or redesigned and made or remade to use the energy of the explosive in a fixed shotgun shell to fire through a smooth bore either a number of ball shot or a single projectile for each single pull of the trigger.

C. No person shall possess a loaded or unloaded firearm, rifle, shotgun or air gun, or a dagger, dangerous knife, dirk, razor or stiletto, in a public place or public facility in the city. This prohibition shall not apply to:

(1) A police officer or peace officer authorized to possess the same;

(2) A government employee or licensed security guard authorized or required by employment or office to possess the same while acting within the scope of such employment;

(3) A person in the military service of the State of New York or the United States when duly authorized to possess the same;

(4) A person transporting a rifle or shotgun in a motor vehicle in the city in accordance with the provisions of § 11-0931, Subdivision 2, of the New York State Environmental Conservation Law, or otherwise transporting an unloaded rifle, shotgun or air gun in the city, provided that the same is completely enclosed or contained in a nontransparent carrying case and either:

(a) Said carrying case is locked; or

(b) A locking device is attached to the weapon and locked in a manner so as to prevent the weapon from being fired;

(5) An authorized person who, for the purpose of shooting practice, possesses a weapon at an established target range in a public place other than a park or public facility;

(6) A person voluntarily surrendering the same in accordance with the provisions of § 265.20 of the Penal Law; or

(7) Possession of a firearm by a person licensed to carry a firearm pursuant to § 400.00 of the Penal Law or possession or transportation by a gunsmith or dealer in firearms in accordance with a license issued by the State of New York or the United States, except that this subsection shall not apply in a park or a public facility other than a parking garage.

D. No person shall store a firearm, rifle, shotgun or air gun in a dwelling in the city unless said firearm, rifle, shotgun or air gun is completely enclosed or contained in a nontransparent locked carrying case or in a locked gun rack, cabinet, closet or safe, or a locking device is attached to the weapon and locked in a manner so as to prevent the weapon from being fired. This requirement shall not apply to a rifle, shotgun or licensed firearm carried on the body of the owner or within such close proximity of the owner that the owner can retrieve it as quickly and easily as if it were carried on the owner's body.

E. No person shall dispose of any firearm, rifle, shotgun, air gun or ammunition in the city. This prohibition shall not apply to:

(1) A gunsmith or dealer in firearms duly licensed by the State of New York or the United States;

(2) A person disposing of the same to a gunsmith or dealer in firearms duly licensed by the State of New York or the United States;

(3) A person voluntarily surrendering the same in accordance with the provisions of § 265.20 of the Penal Law;

(4) A person disposing of a licensed firearm in accordance with law;

(5) Disposition by intestate or testamentary bequest; or

(6) A person disposing of a rifle, shotgun, air gun or ammunition to a family member.

F. No person shall possess an assault weapon or an ammunition feeding device in the city. This prohibition shall not apply to:

(1) A police officer or peace officer authorized to possess the same;

(2) A person in the military service of the State of New York or the United States when duly authorized to possess the same;

(3) A person voluntarily surrendering the same in accordance with the provisions of § 265.20 of the Penal Law; or

(4) A gunsmith or dealer in firearms duly licensed by the State of New York or the United States for weapons to be used by police officers or persons in the military service or for delivery outside of the city.

G. No person shall dispose of an assault weapon or ammunition feeding device in the city. This prohibition shall not apply to:

(1) A person voluntarily surrendering the same in accordance with the provisions of § 265.20 of the Penal Law; or

(2) A gunsmith or dealer in firearms duly licensed by the State of New York or the United States for weapons to be used by police officers or persons in the military service or for delivery outside of the city.

H. No person shall carry a firearm, shotgun, rifle or air gun in the city while such person has one-tenth of one per centum (1/10 of 1%) or more by weight of alcohol in the person's blood as shown by chemical analysis of the person's blood, breath, urine or saliva.

I. No person shall carry a firearm, shotgun, rifle or air gun in the city while in an intoxicated condition.

J. No person shall carry a firearm, shotgun, rifle or air gun in the city while the person's ability to safely carry such weapon is impaired by the use of a drug.

K. Any person who carries a firearm, shotgun, rifle or air gun in this city shall be deemed to have given consent to a breath test and a chemical test of the person's breath, blood, urine or saliva for the purpose of determining the alcoholic or drug content of the person's blood, provided that any test is administered at the direction of a police officer having reasonable grounds therefor. ...

L. [Note: This section was found unconstitutional.]

M. Discharge of weapons; permits. ...

N. The owner of a firearm, shotgun, rifle, assault weapon, machine gun or submachine gun, which becomes lost or stolen, shall report the loss or theft to the Rochester Police Department within 24 hours after the loss or theft is discovered or reasonably should be discovered. The owner of such a weapon shall store the weapon in a safe and secure manner as required in Subsection D of this section and shall check such weapon at least once each week, or immediately upon returning to the city if the owner is absent from the city for more than one week. Failure to perform such a check shall not be a defense to a prosecution for a violation of this subsection.

O. Notwithstanding the penalties contained in § 47-8, a violation of any provision of this section shall be punishable by a fine not to exceed one thousand dollars ($1,000.) or by imprisonment not to exceed one hundred eighty (180) days, or by both such fine and imprisonment. ...

47-8. Penalties. Any person or corporation violating any of the provisions of this chapter shall, upon conviction be punishable by a fine not exceeding $150, or by imprisonment not exceeding 15 days, or by both such fine and imprisonment, or by a penalty of not less than $5 nor more than $500 to be recovered by the City of Rochester in a civil action.

[Code of the City of Rochester current as of November 1, 2008]

Laws of Suffolk County

Chapter 233. Bullets, Armor-Piercing

233-1. Purpose. It is the intent of the Suffolk County Legislature to restrict the possession, disposition and use of certain handgun bullets that are designed primarily for the purpose of armor penetration so as to give law enforcement personnel a reasonable degree of protection from penetration of body armor. This chapter is not intended to restrict the availability of ammunition for personal defense, sporting or hunting purposes.

233-2. Definitions. As used in this chapter, the following terms shall have the meanings indicated:

Body Armor - Commercially available soft, bullet-resistant apparel with a penetration resistance equal to or greater than body armor classified as Threat Level Class II that complies with New York State Industrial Code Rule 55 or National Institute of Law Enforcement and Criminal Justice Standard 0101.01, dated May 1978.

Dealer - Any person engaged in the business of selling ammunition at wholesale or retail who is licensed under Chapter 44 of Title 18 of the United States Code.

Dispose Of - Dispose of, give, give away, loan, offer for sale, sell, transfer and otherwise dispose of, except to police or military units.

Handgun - A firearm originally designed to be fired by the use of a single hand.

Person - Natural person, firm, partnership, corporation or company.

Restricted Handgun Bullet - A handgun projectile that is capable of penetrating armor, including body armor as defined herein, and is comprised of a projectile whose composition includes any components having a hardness of 70 or greater on the Rockwell B hardness scale, specifically the Czechoslovakian manufactured 9-millimeter; and all KTW teflon-coated projectiles.

233-3. Prohibited acts.

A. Any person who, with intent to injure or kill, or whoever, during and in relation to the commission of a crime of violence for which he may be prosecuted in court, including a felony which provides for an enhanced punishment if committed by the use of a dangerous weapon or device, uses or carries any handgun loaded with armor-piercing ammunition, as defined herein, shall, in addition to the punishment provided for the commission of such felony, be sentenced to a term of imprisonment for not less than six months.

B. No dealer shall dispose of any restricted handgun bullet or bullets in Suffolk County, except to police or military units.

233-4. Penalties for offenses. Any person convicted of violating this chapter shall be deemed guilty of a misdemeanor punishable by a fine not exceeding $1,000 or by imprisonment for not more than six months, or by both such fine and imprisonment. Notwithstanding any other provision of law, the court shall not suspend the sentence of any person convicted of a violation of this chapter nor place him on probation, nor shall the term of imprisonment run concurrently with any other term of imprisonment, including that imposed for the felony in which the armor-piercing handgun ammunition was used or carried. No person sentenced under this section shall be eligible for parole.

Chapter 345. Licensed Occupations

Article V. Dealers in Secondhand Articles

345-47. Definitions. As used in this article, the following terms shall have the meanings indicated:

Antique Firearms - Any unloaded muzzle-loading pistol or revolver with a matchlock, flintlock, percussion cap or similar type of ignition system, or a pistol or revolver which uses fixed cartridges which are no longer available in the ordinary channels of commercial trade.

Business Day - Any calendar day except Sunday or any County holiday.

Dealer in Secondhand Articles:

A. Any person, corporation, partnership or other entity and its employees that, as a business, transacts more than five deals in the purchase or sale of the following articles within a twelve-month period:

(1) Antique firearms.

(2) Rifles.

(3) Shotguns. ...

B. Exempted operations:

(1) "Dealer in secondhand articles" shall not include any organizations formed for charitable purposes, which accept donations of secondhand articles and resell them to raise funds for purposes consistent with the formation of the organization, nor any persons, corporations, partnerships or other entities and their employees that, as a business, are principally engaged in the service and/or repair of electronic equipment or component parts thereof and who, from time to time, sell electronic equipment or component parts thereof, which have been left for repair and later abandoned.

(2) The sale of secondhand goods at events commonly known as "garage sales," "yard sales," or "estate sales" provided:

(a) The sale is held on noncommercial property; **(b)** None of the items offered for sale have been purchased for resale; **(c)** The owner of the property receives all proceeds; and **(d)** The sale period is no longer than 72 hours. At no time shall the property owner be permitted to conduct more than two events within a twelve-month period. **(3)** Secondhand books, comic books, magazines, post cards, and postage stamps.

Department - The Office of Consumer Affairs.

Proper Identification - Identification documents that contain the person's name and either a photograph or a physical description of said person. Social security cards, draft registration cards, voter registration cards and comparable documents shall not be considered sufficient identification for the purpose of this article.

Secondhand Article - An article or object which:

A. Has been previously sold at retail; or

B. Has been previously used or is not in a new condition.

345-48. License required; display

A. No person shall engage in any business as a dealer in secondhand articles without obtaining a license therefor from the office in accordance with and subject to the provisions of this article and Article I.

B. A licensee shall display the license obtained pursuant to this article in the establishment. If a licensee shall maintain more than one establishment within the County of Suffolk, he shall obtain duplicate licenses. The fee for duplicate licenses are set forth in § 345-49B.

C. No applicant for a license renewal shall have any outstanding judgment for child support against him or her, or be in arrears in child-support payments as determined by official court records or official government records, at the time an application is filed for such license renewal. ...

345-49. Fees.

A. An application fee of $25 shall accompany each application for a dealer in secondhand articles license.

B. The fee for a duplicate license for an additional dealer in a secondhand articles establishment shall be $50 per annum. It may be renewed biennially for a fee of $100.

C. The fee for a dealer in secondhand articles license or renewal thereof shall be $200 per annum.

345-50. Required records.

A. Each licensee shall keep records, legibly written in English, in a bound book. All entries shall be made in ink at the time of each transaction and shall include the computer transaction number. Each transaction shall also include:

(1) An accurate account and description of the article or thing bought, including but not limited to the make, model, color and serial number when present.

(2) The amount paid for the article.

(3) The date of transaction.

(4) The name, home residence, date of birth, sex and race of the person selling the items.

(5) The type of identification produced by the person selling the article.

(6) The signature of the person identified in the transaction.

B. No alterations or erasures are to be made to records. Erroneous entries are to have a simple line drawn through them and the corrected entry made on the next line.

C. The records shall be retained in the possession of the licensee for at least three years.

D. A written receipt shall be issued to the seller with the serial number of the transaction and the information required in Subsections A, B and C of this section.

345-51. Inspection of records and books. All records required to be kept pursuant to this article shall be open for inspection by the Office of Consumer Affairs, the Police Commissioner, a Chief of Police or any officer or employee duly authorized by them.

345-52. Prohibited acts.

A. No article purchased by a dealer in secondhand articles shall be sold or otherwise disposed of until the expiration of at least 21 business days from the date of purchase.

B. All articles subject to this holding period shall be available for inspection by the Director of the Office of Consumer Affairs, the Police Commissioner, the Chief of Police or any officer duly authorized by them.

C. Purchases or sales between licensed secondhand dealers shall be exempt from the provisions of this section only if evidence of full compliance with all provisions and conditions set forth in this article is obtained by the purchasing secondhand dealer from the selling secondhand dealer in the form of a receipt. This receipt shall be retained by the purchasing secondhand dealer for the period required by § 345-50C hereof.

345-52.1. Police order to hold property.

A. Investigative hold. Whenever a law enforcement or consumer affairs official notifies a licensee not to sell an item, the item shall not be sold or removed from the premises. The investigative hold shall be confirmed in writing by the originating agency within 72 hours and shall remain in effect for 15 days from the date of initial notification or until the investigative order is cancelled, or until an order to hold is issued, whichever comes first.

B. Order to hold. Whenever the Commissioner of Police, a Police Chief, or Director of Consumer Affairs notifies a licensee not to sell an item, the item shall not be sold or removed from the licensed premises until authorized to be released by the aforementioned individuals or their designees. The order to hold shall expire 90 days from the date it is placed. **C.** When an item is placed on hold, the person doing so shall provide identification and provide the licensee with the name and phone number of the holding agency and the case number related to the hold order. **D.** When an order to hold is no longer necessary, the licensee shall be notified in writing by the requesting agency. **345-52.2. Prohibited acts.** **A.** No article shall be purchased from a person who is unable to produce proper identification.

B. The acts enumerated in this section are not exclusive. Article I of this chapter also applies. **C.** No article shall be purchased from a person under the age of 18 without the written consent of his parent or guardian. **D.** No article shall be purchased that possesses an altered or obliterated serial number or any item that has had its serial number removed. **E.** No article shall be purchased from a person who appears to be intoxicated or of unsound mind.

345-53. Required reports.

A. Every person so licensed as a dealer in secondhand articles shall report to the Police Commissioner or Chief of Police, on or before the end of each business day in a form approved and supplied by the Commissioner of Police, the information described in § 345-50 of this article.

B. On or after March 1, 2004, any required report shall be filed electronically.

345-54. Temporary license pending issuance of permanent license.

A. The Office shall issue a temporary license to any applicant for a dealer in secondhand articles license if the Office has not, within 30 days after receipt of the application for such license, approved or disapproved such application.

B. The temporary license shall be valid for a period of 90 days. If within such ninety-day period the Office shall have failed to approve or disapprove the application, the Office shall then issue a regular license to the applicant. If the application is approved during the term of the temporary license, the Office shall then issue a regular license to the applicant.

C. The fee for a temporary license shall be $50. The fee for a regular license issued to replace a temporary license shall be $150.

D. A regular license issued pursuant to the provisions of Subsection B shall expire on the last day of the 24th month following the issuance of the temporary license.

345-54.1. Required bond. Notwithstanding the provisions of § 345-11A(4) herein, every applicant for a dealer in secondhand-articles license shall submit a five-thousand-dollar bond, or for renewal of a license, evidence of a bond issued in favor of the licensee. This bond shall be for the purpose of guaranteeing payments up to the face amount of the bond for bank drafts or other negotiable instruments issued by the licensee in exchange for the purchase of secondhand articles. All bonds shall be conditioned that the licensee will observe all laws in relation to precious metal dealers and will conduct business in conformity thereto. Such bond shall remain in full force during the entire period for which the license is valid.

[Laws of Suffolk County current as of Oct. 15, 2008]

Code of the City of Yonkers

Chapter 59. Fire Code

Article IV. Explosives

59-54. Definitions. Terms used in this article shall be defined as follows: ...

Explosive - Any chemical compound, mixture or device, the primary or common purpose of which is to function by explosion. The term "explosive" shall include all material which is classified as Class A, Class B or Class C explosives by the Interstate Commerce Commission and includes but is not limited to dynamite, black blasting powder, pellet powders, initiating explosives, blasting caps, electric blasting caps, safety fuse, fuse igniters, fuse lighters, squibs, cordeau detonant fuse, instantaneous fuse, igniter cord and igniters. ...

Magazine - Any building or structure approved for the storage of explosives. ...

Small Arms Ammunition - Includes any shotgun, rifle, pistol or revolver cartridges, percussion caps and primers. ...

59-55. General requirements.

A. The manufacture of explosives and small arms ammunition is prohibited in the City of Yonkers.

B. No person shall possess, keep, store, sell, offer for sale, give away, use, discharge, transport or dispose of in any manner any explosives within the City of Yonkers, except by the authority of a written license as provided in this article.

D. No person shall discharge any explosives, except small-arms ammunition and construction devices such as explosive rivets and explosive-driven pins or studs, for purposes other than blasting or demolition operations.

E. No person shall sell or give away any explosive, except small-arms ammunition and construction devices such as explosive rivets and explosive-driven pins or studs, to any person not in possession of a license to either possess, transport or use explosives as required by this article.

F. Encasement. Except while blasting, no person shall possess or store explosives, unless such explosives are completely enclosed or encased in tight metal, wooden or fiber containers. A person having possession or control of an explosive shall under no circumstances permit any grains or particles of it to remain on the outside of or about its container. Every container shall be plainly marked with the name of the explosive contained therein.

Article V. Ammunition

59-90. Definitions. As used in this article, terms shall be defined as follows:

Ammunition - A metal or other shell containing a fulminate or containing black or smokeless powder for the purpose of propelling projectiles or shot. The term shall also include black or smokeless powder packed for use as a propelling charge or for saluting purposes.

Small-Arms Ammunition - Any shotgun, rifle, pistol or revolver cartridges, and shall include percussion caps and primers.

59-91. Manufacture prohibited. It shall be unlawful to manufacture ammunition or small-arms ammunition in the City of Yonkers.

59-92. Power machinery prohibited. It shall be unlawful for any person to load ammunition or small-arms ammunition by power machinery in the city.

59-93. Permit for storage and sale.

A. It shall be unlawful for any person to store or sell or offer for sale any ammunition or small-arms ammunition within the city without a permit from the Fire Commissioner.

B. The permit shall state the location of the premises, the maximum amount of ammunition to be stored at any time and such other information as the Fire Commissioner may deem necessary.

59-94. Application for permit. Permits for the storage and sale of ammunition or small-arms ammunition may be issued by the Fire Commissioner upon written application giving in detail the following information:

A. The name and address of the applicant.

B. The location of the premises.

C. The nature of construction of the building.

D. Other purposes for which the building is used.

E. The nature of the business in which the applicant is engaged in such building.

F. The maximum quantity and description of the ammunition to be stored and sold.

59-95. Permit fees. For a permit allowing the storage and sale of ammunition or small-arms ammunition as provided in this article, the applicant shall pay an annual permit fee.

59-96. Storage and sale in designated locations prohibited. No permit for the storage and sale of ammunition or small-arms ammunition shall be issued for any of the following premises:

A. Premises which are occupied as a tenement house, school, theater or other place of public amusement or assembly.

B. Premises which are used as a drugstore, paint store, pawnshop or stationery store.

C. Premises which are artificially lighted by any means other than electricity.

D. Premises where cigars, cigarettes or tobacco are stored or kept for sale.

E. Premises where liquors are sold.

F. Premises where dry goods or other materials of a highly flammable nature are manufactured, stored or kept for sale.

59-97. Storage of ammunition.

A. The Fire Commissioner shall fix the maximum quantity of ammunition and small-arms ammunition to be stored in any premises for which a permit is issued, which quantity shall be stated in the permit. No permit shall be issued for the storage of ammunition or small-arms ammunition greater than the following:

(1) One hundred thousand (100,000) loaded shells containing shot for shotguns not exceeding No. 8 gauge.

(2) Two hundred thousand (200,000) cartridges for pistols.

(3) Two hundred thousand (200,000) cartridges for rifles, of a caliber not larger than fifty hundredths (.50) of an inch.

(4) One million (1,000,000) percussion caps or primers without anvils.

B. Ammunition and small-arms ammunition shall be kept out of the reach of the public. All ammunition and small-arms ammunition that is in excess of what is allowed to be kept on shelves or showcases shall be stored in an approved metal cabinet or vault, the location of which shall be approved by the Fire Commissioner.

59-98. Display and storage of ammunition restricted. It shall be unlawful for the holder of a permit for the storage and sale of ammunition or small-arms ammunition to store or exhibit in the windows or doors of the premises covered by the permit any cartridges or shells containing explosives.

59-99. Permit fee for use of blank cartridges. The Fire Commissioner may issue a permit allowing the use of blank cartridges in connection with performances in duly authorized theaters or places of amusement or for saluting purposes. The applicant shall pay an annual permit fee, except that no fee shall be charged for a permit for saluting purposes issued to the Grand Army or any other military or patriotic body.

Chapter 74. Firearms

74-1. License required to sell air guns. It shall be unlawful for any person to sell, offer to sell or have in his possession any air pistol or air rifle or similar instrument in which the propelling force is a spring or air, except that the sale of such instruments, if accompanied by delivery to a point without the city, and possession for such purpose shall not be unlawful if such person shall have secured an annual license from the Police Commissioner of the City of Yonkers authorizing such sale and possession.

74-2. Record of sales. All persons dealing in such instruments referred to in this chapter shall keep a record showing the name and address of each person purchasing such instrument or instruments, together with the place of delivery, and said record shall be open to inspection during regular business hours by the officers of the Police Department of the city.

74-3. Issuance of license. The Police Commissioner is hereby authorized to issue, in his discretion, upon payment of a license fee in the amount of ten dollars ($10.), an annual license authorizing the sale and possession of such instruments for delivery to a point without the city.

74-4. Pistol permit; registration; fee. The Police Commissioner may also issue a permit for a spring-, gas- or air-operated pistol to any person who holds a current New York State pistol permit. Said pellet gun shall be registered in the same manner as all other pistols, and the cost for said permit shall be one dollar ($1.).

74-6. Rifles and shotguns; penalties for offenses.

A. It shall be unlawful for any person to carry or possess a loaded rifle or shotgun, as those terms are defined in Penal Law § 265.00, in public within the city limits. Any violation of this subsection shall constitute a Class I offense.

B. It shall be unlawful for any person to carry or possess an unloaded rifle or shotgun, as those terms are defined in Penal Law § 265.00, in public within the city limits unless such rifle or shotgun is completely enclosed or contained in a nontransparent carrying case or a cover. Any violation of this subsection shall constitute a Class II offense.

C. The above provisions shall not apply to persons in the military service of the State of New York, when duly authorized by regulations issued by the Chief of Staff to the Governor to possess the same, or to peace officers, as defined in § 1.20, Subdivision 33, of the Criminal Procedure Law, or to participants in special events when authorized by the Police Commissioner.

D. The above provisions shall not apply to persons possessing or carrying a rifle or shotgun at a rifle range for which a license has been obtained from the Police Commissioner or to persons possessing or carrying a rifle or shotgun on their own premises.

74-7. Possession of imitation or inoperable guns, firearms and weapons prohibited.

A. The City Council finds and declares that there exists a danger to life, person and property in the City of Yonkers, especially on city-owned and -operated property and in the city public schools, with the possession and/or use of guns, firearms and weapons which are a replica of and cannot be easily distinguished from actual guns, firearms and weapons. The possession and, at times, use by simply displaying one of these imitation and/or inoperable guns, firearms and weapons has caused damage to life, limb and property in the City of Yonkers since the Penal Law of the State of New York does not extend its coverage to this area. Through the exercise of the police power granted to municipal governments and in order to afford the residents of the City of Yonkers greater safety and to protect their general welfare from individuals, both adults and juveniles are prohibited from using imitation, toy and inoperable pistols, revolvers, guns and other weapons of any type that may be readily mistaken for real guns, pistols, revolvers or weapons which are used or could be usable for nefarious purposes or as threats or potential threats to life, limb and property.

B. This section is not intended to forbid or restrict the sale, possession or use of true and actual toy pistols, guns, revolvers or other weapons, provided that the same are not substantial duplicates of actual pistols, guns, revolvers or weapons in appearance.

C. It shall be unlawful for any person to possess or use or attempt to use any imitation, toy or inoperable pistol, revolver, gun, firearm or any other weapon which substantially duplicates an actual pistol, revolver, gun, firearm or other weapon unless said imitation, toy or inoperable item or instrument shall be colored in colors other than blue, black, grays, silver or aluminum, and further provided, if resembling a gun or other firearm, that the barrel of said item shall be closed with the same material of which the item itself is made for a distance of not less than one-half (1/2) inch from the front end of the barrel of said item.

D. The provisions of Subsection C shall not apply to possession or display of such an instrument by a licensed manufacturer or dealer of the same, used solely in connection with his or her business. Further, any said instrumentalities used in theatrical productions licensed in advance by the City of Yonkers shall also be exempted when possessed, displayed and used in connection with said licensed theatrical productions.

E. Any violation of this section shall constitute a Class I offense.

74-8. Unlawful to possess bows in public; exceptions.

A. It shall be unlawful for any person to carry or possess a crossbow of any type, configuration or manufacture in public anywhere in the City of Yonkers. It shall be unlawful for any person to carry or possess any recurve bow or compound bow with a draw weight exceeding 35 pounds in a public place anywhere in the City of Yonkers.

B. The above provision shall not apply to persons carrying a longbow. The above provision shall not apply to persons carrying or possessing recurve bows or compound bows in a carrying case or carrying or possessing recurve bows or compound bows to, from, or at a firing range or competition where authorized or operated by the City of Yonkers or where such bow shooting is permitted. The above provision shall not apply to persons carrying or possessing a recurve or compound bow on their premises or private property.

74-9. Penalties for offenses.

Except as otherwise provided, any violation of this chapter shall constitute a Class II offense. Any violation of § 74-8 wherein it is alleged that the person or persons committing the offense caused injury to a person or an animal through the use of a crossbow, recurve bow or compound bow shall be a Class I offense.

[Code of the City of Yonkers current as of Aug. 15, 2008]

NORTH CAROLINA

N.C. GEN. STAT.

Chapter 14. Criminal Law

Article 23. Trespasses to Personal Property

14-160.1. Alteration, destruction or removal of permanent identification marks from personal property

(a) It shall be unlawful for any person to alter, deface, destroy or remove the permanent serial number, manufacturer's identification plate or other permanent, distinguishing number or identification mark from any item of personal property with the intent thereby to conceal or misrepresent the identity of said item.

(b) It shall be unlawful for any person knowingly to sell, buy or be in possession of any item of personal property, not his own, on which the permanent serial number, manufacturer's identification plate or other permanent, distinguishing number or identification mark has been altered, defaced, destroyed or removed for the purpose of concealing or misrepresenting the identity of said item.

(c) A violation of any of the provisions of this section shall be a Class 1 misdemeanor.

(d) This section shall not in any way affect the provisions of G.S. 20-108, 20-109(a) or 20-109(b).

Article 35. Offenses against the Public Peace

14-269.1. Confiscation and disposition of deadly weapons Upon conviction of any person for violation of G.S. 14-269, G.S. 14-269.7, or any other offense involving the use of a deadly weapon of a type referred to in G.S. 14-269, the deadly weapon with reference to which the defendant shall have been convicted shall be ordered confiscated and disposed of by the presiding judge at the trial in one of the following ways in the discretion of the presiding judge.

(1) By ordering the weapon returned to its rightful owner, but only when such owner is a person other than the defendant and has filed a petition for the recovery of such weapon with the presiding judge at the time of the defendant's conviction, and upon a finding by the presiding judge that petitioner is entitled to possession of same and that he was unlawfully deprived of the same without his consent.

(2) & (3) [Repealed]

(4) By ordering such weapon turned over to the sheriff of the county in which the trial is held or his duly authorized agent to be destroyed. The sheriff shall maintain a record of the destruction thereof.

(4a) [Repealed]

(4b) By ordering the weapon turned over to a law enforcement agency in the county of trial for (i) the official use of the agency or (ii) sale, trade, or exchange by the agency to a federally licensed firearm dealer in accordance with all applicable State and federal firearm laws. The court may order a disposition of the firearm pursuant to this subdivision only upon the written request of the head or chief of the law enforcement agency and only if the firearm has a legible, unique identification number. If the law enforcement agency sells the firearm, then the proceeds of the sale shall be remitted to the appropriate county finance officer as provided by G.S. 115C-452 to be used to maintain free public schools. The receiving law enforcement agency shall maintain a record and inventory of all firearms received pursuant to this subdivision.

(5) By ordering such weapon turned over to the North Carolina State Bureau of Investigation's Crime Laboratory Weapons Reference Library for official use by that agency. The State Bureau of Investigation shall maintain a record and inventory of all such weapons received.

(6) By ordering such weapons turned over to the North Carolina Justice Academy for official use by that agency. The North Carolina Justice Academy shall maintain a record and inventory of all such weapons received.

14-269.2. Weapons on campus or other educational property

(a) The following definitions apply to this section:

(1) Educational property. Any school building or bus, school campus, grounds, recreational area, athletic field, or other property owned, used, or operated by any board of education or school board of trustees, or directors for the administration of any school.

(1a) Employee. A person employed by a local board of education or school whether the person is an adult or a minor.

(1b) School. A public or private school, community college, college, or university.

(2) Student. A person enrolled in a school or a person who has been suspended or expelled within the last five years from a school, whether the person is an adult or a minor.

(3) Switchblade knife. A knife containing a blade that opens automatically by the release of a spring or a similar contrivance.

(4) Weapon. Any device enumerated in subsection (b), (b1), or (d) of this section.

(b) It shall be a Class I felony for any person to possess or carry, whether openly or concealed, any gun, rifle, pistol, or other firearm of any kind on educational property or to a curricular or extracurricular activity sponsored by a school. Unless the conduct is covered under some other provision of law providing greater punishment, any person who willfully discharges a firearm of any kind on educational property is guilty of a Class F felony. However, this subsection does not apply to a BB gun, stun gun, air rifle, or air pistol.

(b1) It shall be a Class G felony for any person to possess or carry, whether openly or concealed, any dynamite cartridge, bomb, grenade, mine, or powerful explosive as defined in G.S. 14-284.1, on educational property or to a curricular or extracurricular activity sponsored by a school. This subsection shall not apply to fireworks.

(c) It shall be a Class I felony for any person to cause, encourage, or aid a minor who is less than 18 years old to possess or carry, whether openly or concealed, any gun, rifle, pistol, or other firearm of any kind on educational property. However, this subsection does not apply to a BB gun, stun gun, air rifle, or air pistol.

(c1) It shall be a Class G felony for any person to cause, encourage, or aid a minor who is less than 18 years old to possess or carry, whether openly or concealed, any dynamite cartridge, bomb, grenade, mine, or powerful explosive as defined in G.S. 14-284.1 on educational property. This subsection shall not apply to fireworks.

(d) It shall be a Class 1 misdemeanor for any person to possess or carry, whether openly or concealed, any BB gun, stun gun, air rifle, air pistol, bowie knife, dirk, dagger, slungshot, leaded cane, switchblade knife, blackjack, metallic knuckles, razors and razor blades (except solely for personal shaving), firework, or any sharp-pointed or edged instrument except instructional supplies, unaltered nail files and clips and tools used solely for preparation of food, instruction, and maintenance, on educational property.

(e) It shall be a Class 1 misdemeanor for any person to cause, encourage, or aid a minor who is less than 18 years old to possess or carry, whether openly or concealed, any BB gun, stun gun, air rifle, air pistol, bowie knife, dirk, dagger, slungshot, leaded cane, switchblade knife, blackjack, metallic knuckles, razors and razor blades (except solely for personal shaving), firework, or any sharp-pointed or edged instrument except instructional supplies, unaltered nail files and clips and tools used solely for preparation of food, instruction, and maintenance, on educational property.

(f) Notwithstanding subsection (b) of this section it shall be a Class 1 misdemeanor rather than a Class I felony for any person to possess or carry, whether openly or concealed, any gun, rifle, pistol, or other firearm of any kind, on educational property or to a curricular or extracurricular activity sponsored by a school if:

(1) The person is not a student attending school on the educational property or an employee employed by the school working on the educational property; and

(1a) The person is not a student attending a curricular or extracurricular activity sponsored by the school at which the student is enrolled or an employee attending a curricular or extracurricular activity sponsored by the school at which the employee is employed; and

(2) [Repealed]

(3) The firearm is not loaded, is in a motor vehicle, and is in a locked container or a locked firearm rack.

(4) [Repealed]

(g) This section shall not apply to any of the following:

(1) A weapon used solely for educational or school-sanctioned ceremonial purposes, or used in a school-approved program conducted under the supervision of an adult whose supervision has been approved by the school authority.

(1a) A person exempted by the provisions of G.S. 14-269(b).

(2) Firefighters, emergency service personnel, and North Carolina Forest Service personnel, and any private police employed by a school, when acting in the discharge of their official duties.

(3) Home schools as defined in G.S. 115C-563(a).

(4) Weapons used for hunting purposes on the Howell Woods Nature Center property in Johnston County owned by Johnston Community College when used with the written permission of Johnston Community College or for hunting purposes on other educational property when used with the written permission of the governing body of the school that controls the educational property.

(5) A person registered under chapter 74c of the general statutes as an armed armored car service guard or an armed courier service guard when acting in the discharge of the guard's duties and with the permission of the college or university.

(6) A person registered under chapter 74c of the general statutes as an armed security guard while on the premises of a hospital or health care facility located on educational property when acting in the discharge of the guard's duties with the permission of the college or university.

(h) No person shall be guilty of a criminal violation of this section with regard to the possession or carrying of a weapon so long as both of the following apply:

(1) The person comes into possession of a weapon by taking or receiving the weapon from another person or by finding the weapon.

(2) The person delivers the weapon, directly or indirectly, as soon as practical to law enforcement authorities.

14-269.3. Carrying weapons into assemblies and establishments where alcoholic beverages are sold and consumed

(a) It shall be unlawful for any person to carry any gun, rifle, or pistol into any assembly where a fee has been charged for admission thereto, or into any establishment in which alcoholic beverages are sold and consumed. Any person violating the provisions of this section shall be guilty of a Class 1 misdemeanor.

(b) This section shall not apply to the following:

(1) A person exempted from the provisions of G.S. 14-269;

(2) The owner or lessee of the premises or business establishment;

(3) A person participating in the event, if he is carrying a gun, rifle, or pistol with the permission of the owner, lessee, or person or organization sponsoring the event; and

(4) A person registered or hired as a security guard by the owner, lessee, or person or organization sponsoring the event.

14-269.4. Weapons on State property and in courthouses It shall be unlawful for any person to possess, or carry, whether openly or concealed, any deadly weapon, not used solely for instructional or officially sanctioned ceremonial purposes in the State Capitol Building, the Executive Mansion, the Western Residence of the Governor, or on the grounds of any of these buildings, and in any building housing any court of the General Court of Justice. If a court is housed in a building containing nonpublic uses in addition to the court, then this prohibition shall apply only to that portion of the building used for court purposes while the building is being used for court purposes.

This section shall not apply to:

(1) Repealed by S.L. 1997-238, s. 3.

(1a) A person exempted by the provisions of G.S. 14-269(b),

(2) through (4) Repealed by S.L. 1997-238, s. 3.

(4a) Any person in a building housing a court of the General Court of Justice in possession of a weapon for evidentiary purposes, to deliver it to a law enforcement agency, or for purposes of registration,

(4b) Firearms in a courthouse, carried by detention officers employed by and authorized by the sheriff to carry firearms.

(5) State-owned rest areas, rest stops along the highways, and State-owned hunting and fishing reservations.

Any person violating the provisions of this section shall be guilty of a Class 1 misdemeanor.

14-269.7. Prohibitions on handguns for minors

(a) Any minor who possesses or carries a handgun is guilty of a Class 2 misdemeanor.

(b) This section does not apply:

(1) To officers and enlisted personnel of the armed forces of the United States when in discharge of their official duties or acting under orders requiring them to carry handguns.

(2) To a minor who possesses a handgun for educational or recreational purposes while the minor is supervised by an adult who is present.

(3) To an emancipated minor who possesses such handgun inside his or her residence.

(4) To a minor who possesses a handgun while hunting or trapping outside the limits of an incorporated municipality if he has on his person written permission from a parent, guardian, or other person standing in loco parentis.

(c) The following definitions apply in this section:

(1) Handgun. A firearm that has a short stock and is designed to be fired by the use of a single hand, or any combination of parts from which such a firearm can be assembled.

(2) Minor. Any person under 18 years of age.

14-269.8. Purchase or possession of firearms by person subject to domestic violence order prohibited

(a) In accordance with G.S. 50B-3.1, it is unlawful for any person to purchase or attempt to own, possess, purchase, or receive or attempt to own, possess, purchase, or receive a firearm, as defined in G.S. 14-409.39(2), machine gun, ammunition, or permits to purchase or carry concealed firearms if ordered by the court for so long as that protective order or any successive protective order entered against that person pursuant to Chapter 50B of the General Statutes is in effect.

(b) Any person violating the provisions of this section shall be guilty of a Class H felony.

Article 36A. Riots and Civil Disorders

14-288.1. Definitions Unless the context clearly requires otherwise, the definitions in this section apply throughout this Article: ...

(2) "Dangerous weapon or substance": Any deadly weapon, ammunition, explosive, incendiary device, radioactive material or device, as defined in G.S. 14-288.8(c)(5), or any instrument or substance designed for a use that carries a threat of serious bodily injury or destruction of property; or any instrument or substance that is capable of being used to inflict serious bodily injury, when the circumstances indicate a probability that such instrument or substance will be so used; or any part or ingredient in any instrument or substance included above, when the circumstances indicate a probability that such part or ingredient will be so used.

(3) "Declared state of emergency": A state of emergency found and proclaimed by the Governor under the authority of G.S. 14- 288.15, by any mayor or other municipal official or officials under the authority of G.S. 14-288.12, by any chairman of the board of commissioners of any county or other county official or officials under the authority of G.S. 14-288.13, by any chairman of the board of county commissioners acting under the authority of G.S. 14-288.14, by any chief executive official or acting chief executive official of any county or municipality acting under the authority of any other applicable statute or provision of the common law to preserve the public peace in a state of emergency, or by any executive official or military commanding officer of the United States or the State of North Carolina who becomes primarily responsible under applicable law for the preservation of the public peace within any part of North Carolina. ...

(9) "Riot": As defined in G.S. 14-288.2(a).

(10) "State of emergency": The condition that exists whenever, during times of public crisis, disaster, rioting, catastrophe, or similar public emergency, public safety authorities are unable to maintain public order or afford adequate protection for lives or property, or whenever the occurrence of any such condition is imminent.

14-288.7. Transporting dangerous weapon or substance during emergency; possessing off premises; exceptions

(a) Except as otherwise provided in this section, it is unlawful for any person to transport or possess off his own premises any dangerous weapon or substance in any area:

(1) In which a declared state of emergency exists; or

(2) Within the immediate vicinity of which a riot is occurring.

(b) This section does not apply to persons exempted from the provisions of G.S. 14-269 with respect to any activities lawfully engaged in while carrying out their duties.

(c) Any person who violates any provision of this section is guilty of a Class 1 misdemeanor.

14-288.8. Manufacture, assembly, possession, storage, transportation, sale, purchase, delivery, or acquisition of weapon of mass death and destruction; exceptions

(a) Except as otherwise provided in this section, it is unlawful for any person to manufacture, assemble, possess, store, transport, sell, offer to sell, purchase, offer to purchase, deliver or give to another, or acquire any weapon of mass death and destruction.

(b) This section does not apply to:

(1) Persons exempted from the provisions of G.S. 14-269 with respect to any activities lawfully engaged in while carrying out their duties.

(2) Importers, manufacturers, dealers, and collectors of firearms, ammunition, or destructive devices validly licensed under the laws of the United States or the State of North Carolina, while lawfully engaged in activities authorized under their licenses.

(3) Persons under contract with the United States, the State of North Carolina, or any agency of either government, with respect to any activities lawfully engaged in under their contracts.

(4) Inventors, designers, ordnance consultants and researchers, chemists, physicists, and other persons lawfully engaged in pursuits designed to enlarge knowledge or to facilitate the creation, development, or manufacture of weapons of mass death and destruction intended for use in a manner consistent with the laws of the United States and the State of North Carolina.

(c) The term "weapon of mass death and destruction" includes:

(1) Any explosive or incendiary:

a. Bomb; or

b. Grenade; or

c. Rocket having a propellant charge of more than four ounces; or

d. Missile having an explosive or incendiary charge of more than one-quarter ounce; or

e. Mine; or

f. Device similar to any of the devices described above; or

(2) Any type of weapon (other than a shotgun or a shotgun shell of a type particularly suitable for sporting purposes) which will, or which may be readily converted to, expel a projectile by the action of an explosive or other propellant, and which has any barrel with a bore of more than one-half inch in diameter; or

(3) Any firearm capable of fully automatic fire, any shotgun with a barrel or barrels of less than 18 inches in length or an overall length of less than 26 inches, any rifle with a barrel or barrels of less than 16 inches in length or an overall length of less than 26 inches, any muffler or silencer for any firearm, whether or not such firearm is included within this definition. For the purposes of this section, rifle is defined as a weapon designed or redesigned, made or remade, and intended to be fired from the shoulder; or

(4) Any combination of parts either designed or intended for use in converting any device into any weapon described above and from which a weapon of mass death and destruction may readily be assembled.

The term "weapon of mass death and destruction" does not include any device which is neither designed nor redesigned for use as a weapon; any device, although originally designed for use as a weapon, which is redesigned for use as a signaling, pyrotechnic, line-throwing, safety, or similar device; surplus ordnance sold, loaned, or given by the Secretary of the Army pursuant to the provisions of section 4684(2), 4685, or 4686 of Title 10 of the United States Code; or any other device which the Secretary of the Treasury finds is not likely to be used as a weapon, is an antique, or is a rifle which the owner intends to use solely for sporting purposes, in accordance with Chapter 44 of Title 18 of the United States Code.

(d) Any person who violates any provision of this section is guilty of a Class F felony.

14-288.20. Certain weapons at civil disorders

(a) The definitions in G.S. 14-288.1 do not apply to this section. As used in this section:

(1) The term "civil disorder" means any public disturbance involving acts or violence by assemblages of three or more persons, which causes an immediate danger of damage or injury to the property or person of any other individual or results in damage or injury to the property or person of any other individual.

(2) The term "firearm" means any weapon which is designed to or may readily be converted to expel any projectile by the action of an explosive; or the frame or receiver of such a weapon.

(3) The term "explosive or incendiary device" means (i) dynamite and all other forms of high explosives, (ii) any explosive bomb, grenade, missile, or similar device, and (iii) any incendiary bomb or grenade, fire bomb, or similar device, including any device which (i) consists of or includes a breakable container including a flammable liquid or compound, and a wick composed of any material which, when ignited, is capable of igniting that flammable liquid or compound, and (ii) can be carried or thrown by one individual acting alone.

(4) The term "law-enforcement officer" means any officer of the United States, any state, any political subdivision of a state, or the District of Columbia charged with the execution of the laws thereof; civil officers of the United States; officers and soldiers of the organized militia and state guard of any state or territory of the United States, the Commonwealth of Puerto Rico, or the District of Columbia; and members of the armed forces of the United States.

(b) A person is guilty of a Class H felony, if he:

(1) Teaches or demonstrates to any other person the use, application, or making of any firearm, explosive or incendiary device, or technique capable of causing injury or death to persons, knowing or having reason to know or intending that the same will be unlawfully employed for use in, or in furtherance of, a civil disorder; or

(2) Assembles with one or more persons for the purpose of training with, practicing with, or being instructed in the use of any firearm, explosive or incendiary device, or technique capable of causing injury or death to persons, intending to employ unlawfully the training, practicing, instruction, or technique for use in, or in furtherance of, a civil disorder.

(c) Nothing contained in this section shall make unlawful any act of any law-enforcement officer which is performed in the lawful performance of his official duties.

Article 39. Protection of Minors

14-315. Selling or giving weapons to minors

(a) Sale of Weapons Other Than Handguns. If a person sells, offers for sale, gives, or in any way transfers to a minor any pistol cartridge, brass knucks, bowie knife, dirk, shurikin, leaded cane, or slungshot, the person is guilty of a Class 1 misdemeanor and, in addition, shall forfeit the proceeds of any sale made in violation of this section.

(a1) Sale of Handguns. If a person sells, offers for sale, gives, or in any way transfers to a minor any handgun as defined in G.S. 14-269.7, the person is guilty of a Class H felony and, in addition, shall forfeit the proceeds of any sale made in violation of this section. This section does not apply in any of the following circumstances:

(1) The handgun is lent to a minor for temporary use if the minor's possession of the handgun is lawful under G.S. 14-269.7 and G.S. 14-316 and is not otherwise unlawful.

(2) The handgun is transferred to an adult custodian pursuant to Chapter 33A of the General Statutes, and the minor does not take possession of the handgun except that the adult custodian may allow the minor temporary possession of the handgun in circumstances in which the minor's possession of the handgun is lawful under G.S. 14-269.7 and G.S. 14-316 and is not otherwise unlawful.

(3) The handgun is a devise or legacy and is distributed to a parent or guardian under G.S. 28A-22-7, and the minor does not take possession of the handgun except that the parent or guardian may allow the minor temporary possession of the handgun in circumstances in which the minor's possession of the handgun is lawful under G.S. 14-269.7 and G.S. 14- 316 and is not otherwise unlawful.

(b) [Repealed]

(b1) Defense. It shall be a defense to a violation of this section if all of the following conditions are met:

(1) The person shows that the minor produced an apparently valid permit to receive the weapon, if such a permit would be required under G.S. 14- 402 or G.S. 14-409.1 for transfer of the weapon to an adult.

(2) The person reasonably believed that the minor was not a minor.

(3) The person either:

a. Shows that the minor produced a drivers license, a special identification card issued under G.S. 20-37.7, a military identification card, or a passport, showing the minor's age to be at least the required age for purchase and bearing a physical description of the person named on the card reasonably describing the minor; or

b. Produces evidence of other facts that reasonably indicated at the time of sale that the minor was at least the required age.

14-315.1. Storage of firearms to protect minors

(a) Any person who resides in the same premises as a minor, owns or possesses a firearm, and stores or leaves the firearm (i) in a condition that the firearm can be discharged and (ii) in a manner that the person knew or should have known that an unsupervised minor would be able to gain access to the firearm, is guilty of a Class 1 misdemeanor if a minor gains access to the firearm without the lawful permission of the minor's parents or a person having charge of the minor and the minor:

(1) Possesses it in violation of G.S. 14-269.2(b);

(2) Exhibits it in a public place in a careless, angry, or threatening manner;

(3) Causes personal injury or death with it not in self defense; or

(4) Uses it in the commission of a crime.

(b) Nothing in this section shall prohibit a person from carrying a firearm on his or her body, or placed in such close proximity that it can be used as easily and quickly as if carried on the body.

(c) This section shall not apply if the minor obtained the firearm as a result of an unlawful entry by any person.

(d) "Minor" as used in this section means a person under 18 years of age who is not emancipated.

14-315.2. Warning upon sale or transfer of firearm to protect minor

(a) Upon the retail commercial sale or transfer of any firearm, the seller or transferor shall deliver a written copy of G.S. 14-315.1 to the purchaser or transferee.

(b) Any retail or wholesale store, shop, or sales outlet that sells firearms shall conspicuously post at each purchase counter the following warning in block letters not less than one inch in height the phrase: "IT IS UNLAWFUL TO STORE OR LEAVE A FIREARM THAT CAN BE DISCHARGED IN A MANNER THAT A REASONABLE PERSON SHOULD KNOW IS ACCESSIBLE TO A MINOR."

(c) A violation of subsection (a) or (b) of this section is a Class 1 misdemeanor.

14-316. Permitting young children to use dangerous firearms

(a) It shall be unlawful for any parent, guardian, or person standing in loco parentis, to knowingly permit his child under the age of 12 years to have the possession, custody or use in any manner whatever, any gun, pistol or other dangerous firearm, whether such weapon be loaded or unloaded, except when such child is under the supervision of the parent, guardian or person standing in loco parentis. It shall be unlawful for any other person to knowingly furnish such child any weapon enumerated herein. Any person violating the provisions of this section shall be guilty of a Class 2 misdemeanor.

(b) Air rifles, air pistols, and BB guns shall not be deemed "dangerous firearms" within the meaning of subsection (a) of this section except in the following counties: Anson, Caldwell, Caswell, Chowan, Cleveland, Cumberland, Durham, Forsyth, Gaston, Harnett, Haywood, Mecklenburg, Stanly, Stokes, Surry, Union, Vance.

Article 52A. Sale of Weapons in Certain Counties

14-402. Sale of certain weapons without permit forbidden

(a) It is unlawful for any person, firm, or corporation in this State to sell, give away, or transfer, or to purchase or receive, at any place within this State from any other place within or without the State any pistol or crossbow unless: (i) a license or permit is first obtained under this Article by the purchaser or receiver from the sheriff of the county in which the purchaser or receiver resides; or (ii) a valid North Carolina concealed handgun permit is held under Article 54B of this Chapter by the purchaser or receiver who must be a resident of the State at the time of the purchase.

It is unlawful for any person or persons to receive from any postmaster, postal clerk, employee in the parcel post department, rural mail carrier, express agent or employee, railroad agent or employee within the State of North Carolina any pistol or crossbow without having in his or their possession and without exhibiting at the time of the delivery of the same and to the person delivering the same the permit from the sheriff as provided in G.S. 14-403. Any person violating the provisions of this section is guilty of a Class 2 misdemeanor.

(b) This section does not apply to an antique firearm or an historic edged weapon.

(c) The following definitions apply in this section:

(1) Antique firearm. Defined in G.S. 14-409.11.

(2) Bolt. A projectile made to be discharged from a crossbow. The bolt differs from an arrow in that the bolt is heavier and shorter than an arrow.

(3) Crossbow. A mechanical device consisting of, but not limited to, strings, cables, and prods transversely mounted on either a shoulder or hand-held stock. This device is mechanically held at full or partial draw and released by a trigger or similar mechanism that is incorporated into a stock or handle. When operated, the crossbow discharges a projectile known as a bolt.

(4) Historic edged weapon. Defined in G.S. 14-409.12.

14-403. Permit issued by sheriff; form of permit; expiration of permit The sheriffs of any and all counties of this State shall issue to any person, firm, or corporation in any county a license or permit to purchase or receive any weapon mentioned in this Article from any person, firm, or corporation offering to sell or dispose of the weapon. The license or permit shall expire five years from the date of issuance. The license or permit shall be in the following form:

North Carolina, _________________ County.

I, __________, Sheriff of said County, do hereby certify that I have conducted a criminal background check of the applicant, __________ whose place of residence is ____ in ____ (or) in _____Township, _____ County, North Carolina, and have received no information to indicate that it would be a violation of State or federal law for the applicant to purchase, transfer, receive, or possess a handgun. The applicant has further satisfied me as to his, her (or) their good moral character. Therefore, a license or permit is issued to ________ to purchase one pistol from any person, firm or corporation authorized to dispose of the same.

This license or permit expires five years from its date of issuance.

This _____ day of _________, _____

____

Sheriff.

14-404. Issuance or refusal of permit; appeal from refusal; grounds for refusal; sheriff's fee

(a) Upon application, the sheriff shall issue the license or permit to a resident of that county unless the purpose of the permit is for collecting, in which case a sheriff can issue a permit to a nonresident when the sheriff has done all of the following:

(1) Verified, before the issuance of a permit, by a criminal history background investigation that it is not a violation of State or federal law for the applicant to purchase, transfer, receive, or possess a handgun. The sheriff shall determine the criminal and background history of any applicant by accessing computerized criminal history records as maintained by the State Bureau of Investigation and the Federal Bureau of Investigation, by conducting a national criminal history records check, by conducting a check through the National Instant Criminal Background Check System (NICS), and by conducting a criminal history check through the Administrative Office of the Courts.

(2) Fully satisfied himself or herself by affidavits, oral evidence, or otherwise, as to the good moral character of the applicant.

(3) Fully satisfied himself or herself that the applicant desires the possession of the weapon mentioned for (i) the protection of the home, business, person, family or property, (ii) target shooting, (iii) collecting, or (iv) hunting.

(b) If the sheriff is not fully satisfied, the sheriff may, for good cause shown, decline to issue the license or permit and shall provide to the appli-cant within seven days of the refusal a written statement of the reason(s) for the refusal. An appeal from the refusal shall lie by way of petition to the chief judge of the district court for the dis-trict in which the application was filed. The deter-mination by the court, on appeal, shall be upon the facts, the law, and the reasonableness of the sheriff's refusal, and shall be final.

(c) A permit may not be issued to the following persons:

(1) One who is under an indictment or information for or has been convicted in any state, or in any court of the United States, of a felony (other than an offense pertaining to antitrust violations, unfair trade practices, or restraints of trade). However, a person who has been convicted of a felony in a court of any state or in a court of the United States and who is later pardoned may obtain a permit, if the purchase or receipt of a pistol or crossbow permitted in this Article does not violate a condition of the pardon.

(2) One who is a fugitive from justice.

(3) One who is an unlawful user of or addicted to marijuana or any depressant, stimulant, or narcotic drug (as defined in 21 U.S.C. section 802).

(4) One who has been adjudicated mentally incompetent or has been committed to any mental institution.

(5) One who is an alien illegally or unlawfully in the United States.

(6) One who has been discharged from the armed forces under dishonorable conditions.

(7) One who, having been a citizen of the United States, has renounced his or her citizenship.

(8) One who is subject to a court order that:

a. Was issued after a hearing of which the person received actual notice, and at which the person had an opportunity to participate;

b. Restrains the person from harassing, stalking, or threatening an intimate partner of the person or child of the intimate partner of the person, or engaging in other conduct that would place an intimate partner in reasonable fear of bodily injury to the partner or child; and

c. Includes a finding that the person represents a credible threat to the physical safety of the intimate partner or child; or by its terms explicitly prohibits the use, attempted use, or threatened use of physical force against the intimate partner or child that would reasonably be expected to cause bodily injury.

(d) Nothing in this Article shall apply to officers authorized by law to carry firearms if the officers identify themselves to the vendor or donor as being officers authorized by law to carry firearms and state that the purpose for the purchase of the firearms is directly related to the law officers' official duties.

(e) The sheriff shall charge for the sheriff's services upon issuing the license or permit a fee of five dollars ($5.00).

(f) Each applicant for a license or permit shall be informed by the sheriff within 30 days of the date of the application whether the license or permit will be granted or denied and, if granted, the license or permit shall be immediately issued to the applicant.

(g) An applicant shall not be ineligible to receive a permit under subsection (4) of subsection (c) of this section because of involuntary commitment to mental health services if the individual's rights have been restored under G.S. 122c-54.1.

14-405. Record of permits kept by sheriff The sheriff shall keep a book, to be provided by the board of commissioners of each county, in which he shall keep a record of all licenses or permits issued under this article, including the name, date, place of residence, age, former place of residence, etc., of each such person, firm, or corporation to whom or which a license or permit is issued.

14-406. Dealer to keep record of sales Every dealer in pistols, and other weapons mentioned in this Article shall keep an accurate record of all sales thereof, including the name, place of residence, date of sale, etc., of each person, firm, or corporation to whom or which such sales are made, which record shall be open to the inspection of any duly constituted State, county or police officer, within this State.

14-407.1. Sale of blank cartridge pistols The provisions of G.S. 14-402 and 14-405 to 14-406 shall apply to the sale of pistols suitable for firing blank cartridges. The sheriffs of the superior courts of all the counties of this State are authorized and may in their discretion issue to any person, firm or corporation, in any such county, a license or permit to purchase or receive any pistol suitable for firing blank cartridges from any person, firm or corporation offering to sell or dispose of the same, which said permit shall be in substantially the following form:

North Carolina, ________________ County.
I, __________, sheriff of said county, do hereby certify that __________, whose place of residence is __________ Street in __________ (or) in __________ Township in __________ County, North Carolina, having this day satisfied me that the possession of a pistol suitable for firing blank cartridges will be used only for lawful purposes, a permit is therefore given said __________ to purchase said pistol from any person, firm or corporation authorized to dispose of the same, this ________ day of __________, ________.

____
Sheriff

The sheriff shall charge for the sheriff's services, upon issuing such permit, a fee of fifty cents (50¢).

14-408. Violation of § 14-406 a misdemeanor Any person, firm, or corporation violating any of the provisions of G.S. 14-406 shall be guilty of a Class 2 misdemeanor.

14-409. Machine guns and other like weapons

(a) As used in this section, "machine gun" or "submachine gun" means any weapon which shoots, is designed to shoot, or can be readily restored to shoot, automatically more than one shot, without manual reloading, by a single function of the trigger. The term shall also include the frame or receiver of any such weapon, any combination of parts designed and intended for use in converting a weapon into a machine gun, and any combination of parts from which a machine gun can be assembled if such parts are in the possession or under the control of a person.

(b) It shall be unlawful for any person, firm or corporation to manufacture, sell, give away, dispose of, use or possess machine guns, submachine guns, or other like weapons as defined by subsection (a) of this section: Provided, however, that this subsection shall not apply to the following:

Banks, merchants, and recognized business establishments for use in their respective places of business, who shall first apply to and receive from the sheriff of the county in which said business is located, a permit to possess the said weapons for the purpose of defending the said business; officers and soldiers of the United States Army, when in discharge of their official duties, officers and soldiers of the militia when called into actual service, officers of the State, or of any county, city or town, charged with the execution of the laws of the State, when acting in the discharge of their official duties; the manufacture, use or possession of such weapons for scientific or experimental purposes when such manufacture, use or possession is lawful under federal laws and the weapon is registered with a federal agency, and when a permit to manufacture, use or possess the weapon is issued by the sheriff of the county in which the weapon is located. Provided, further, that any bona fide resident of this State who now owns a machine gun used in former wars, as a relic or souvenir, may retain and keep same as his or her property without violating the provisions of this section upon his reporting said ownership to the sheriff of the county in which said person lives.

(c) Any person violating any of the provisions of this section shall be guilty of a Class I felony.

Article 53A. Other Firearms

14-409.10. Purchase of rifles and shotguns out of State It shall be lawful for citizens of this State to purchase rifles and shotguns and ammunition therefor in states contiguous to this State.

14-409.11. "Antique firearm" defined

(a) The term "antique firearm" means any of the following:

(1) Any firearm (including any firearm with a matchlock, flintlock, percussion cap, or similar type of ignition system) manufactured on or before 1898.

(2) Any replica of any firearm described in subdivision (1) of this subsection if the replica is not designed ore redesigned for using rimfire or conventional centerfire fixed ammunition.

(3) Any muzzle loading rifle, muzzle loading shotgun, or muzzle loading pistol, which is designed to use black powder substitute, and which cannot use fixed ammunition.

(b) For purposes of this section, the term "antique firearm" shall not include any weapon which:

(1) Incorporates a firearm frame or receiver.

(2) Is converted into a muzzle loading weapon.

(3) Is a muzzle loading weapon that can be readily converted to fire fixed ammunition by replacing the barrel, bolt, breechblock, or any combination thereof.

14-409.12. "Historic edged weapons" defined The term "historic edged weapon" means any bayonet, trench knife, sword or dagger manufactured during or prior to World War II but in no event later than January 1, 1946.

Article 53B. Firearm Regulation

14-409.39. Definitions The following definitions apply in this Article:

(1) Dealer. Any person licensed as a dealer pursuant to 18 U.S.C. § 921, et seq., or G.S. 105-80.

(2) Firearm. A handgun, shotgun, or rifle which expels a projectile by action of an explosion.

(3) Handgun. A pistol, revolver, or other gun that has a short stock and is designed to be held and fired by the use of a single hand.

14-409.40. Statewide uniformity of local regulation

(a) It is declared by the General Assembly that the regulation of firearms is properly an issue of general, statewide concern, and that the entire field of regulation of firearms is preempted from regulation by local governments except as provided by this section. ...

(b) Unless otherwise permitted by statute, no county or municipality, by ordinance, resolution, or other enactment, shall regulate in any manner the possession, ownership, storage, transfer, sale, purchase, licensing, or registration of firearms, firearms ammunition, components of firearms, dealers in firearms, or dealers in handgun components or parts.

(c) Notwithstanding subsection (b) of this section, a county or municipality, by zoning or other ordinance, may regulate or prohibit the sale of firearms at a location only if there is a lawful, general, similar regulation or prohibition of commercial activities at that location. Nothing in this subsection shall restrict the right of a county or municipality to adopt a general zoning plan that prohibits any commercial activity within a fixed distance of a school or other educational institution except with a special use permit issued for a commercial activity found not to pose a danger to the health, safety, or general welfare of persons attending the school or educational institution within the fixed distance.

(d) No county or municipality, by zoning or other ordinance, shall regulate in any manner firearms shows with regulations more stringent than those applying to shows of other types of items.

(e) A county or municipality may regulate the transport, carrying, or possession of firearms by employees of the local unit of government in the course of their employment with that local unit of government.

(f) Nothing contained in this section prohibits municipalities or counties from application of their authority under G.S. 153A-129, 160A-189, 14-269, 14-269.2, 14-269.3, 14- 269.4, 14-277.2, 14-415.11, 14-415.23, including prohibiting the possession of firearms in public-owned buildings, on the grounds or parking areas of those buildings, or in public parks or recreation areas, except nothing in this subsection shall prohibit a person from storing a firearm within a motor vehicle while the vehicle is on these grounds or areas. Nothing contained in this section prohibits municipalities or counties from exercising powers provided by law in declared states of emergency under Article 36A of this Chapter. ...

Article 54A. The Felony Firearms Act

14-415.1. Possession of firearms, etc., by felon prohibited

(a) It shall be unlawful for any person who has been convicted of a felony to purchase, own, possess, or have in his custody, care, or control any firearm or any weapon of mass death and destruction as defined in G.S. 14-288.8(c). For the purposes of this section, a firearm is (i) any weapon, including a starter gun, which will or is designed to or may readily be converted to expel a projectile by the action of an explosive, or its frame or receiver, or (ii) any firearm muffler or firearm silencer. This section does not apply to an antique firearm, as defined in G.S. 14-409.11.

Every person violating the provisions of this section shall be punished as a Class G felon.

(b) Prior convictions which cause disentitlement under this section shall only include:

(1) Felony convictions in North Carolina that occur before, on, or after December 1, 1995; and

(2) [Repealed]

(3) Violations of criminal laws of other states or of the United States that occur before, on, or after December 1, 1995, and that are substantially similar to the crimes covered in subdivision (1) which are punishable where committed by imprisonment for a term exceeding one year.

When a person is charged under this section, records of prior convictions of any offense, whether in the courts of this State, or in the courts of any other state or of the United States, shall be admissible in evidence for the purpose of proving a violation of this section. The term "conviction" is defined as a final judgment in any case in which felony punishment, or imprisonment for a term exceeding one year, as the case may be, is permissible, without regard to the plea entered or to the sentence imposed. A judgment of a conviction of the defendant or a plea of guilty by the defendant to such an offense certified to a superior court of this State from the custodian of records of any state or federal court shall be prima facie evidence of the facts so certified.

(c) The indictment charging the defendant under the terms of this section shall be separate from any indictment charging him with other offenses related to or giving rise to a charge under this section. An indictment which charges the person with violation of this section must set forth the date that the prior offense was committed, the type of offense and the penalty therefor, and the date that the defendant was convicted or plead guilty to such offense, the identity of the court in which the conviction or plea of guilty took place and the verdict and judgment rendered therein.

14-415.3. Possession of a firearm or weapon of mass destruction by persons acquitted of certain crimes by reason of insanity or persons determined to be incapable to proceed prohibited

(a) It is unlawful for the following persons to purchase, own, possess, or have in the person's custody, care, or control, any firearm or any weapon of mass death and destruction as defined by G.S. 14-288.8(c):

(1) A person who has been acquitted by reason of insanity of any crime set out in G.S. 14-415.1(b) or any violation of G.S. 14-33(b)(1), 14-33(b)(8), or 14-34.

(2) A person who has been determined to lack capacity to proceed as provided in G.S. 15A-1002 for any crime set out in G.S. 14-415.1(b) or any violation of G.S. 14-33(b)(1), 14-33(b)(8), or 14-34.

(b) A violation of this section is a Class H felony. Any firearm or weapon of mass death and destruction lawfully seized for a violation of this section shall be forfeited to the State and disposed of as provided in G.S. 15-11.1.

Article 54B. Concealed Handgun Permit

14-415.10. Definitions The following definitions apply to this Article:

(1) Carry a concealed handgun. The term includes possession of a concealed handgun.

(1a) Deployed or deployment. Any military duty that removes a military permittee from the permittee's county of residence during which time the permittee's permit expires or will expire.

(2) Handgun. A firearm that has a short stock and is designed to be held and fired by the use of a single hand.

(2a) Military permittee. A person who holds a permit who is also a member of the armed forces of the United States, the armed forces reserves of the United States, the North Carolina Army National Guard, or the North Carolina Air National Guard.

(3) Permit. A concealed handgun permit issued in accordance with the provisions of this Article.

(3a) Proof of deployment. A copy of the military permittee's deployment orders or other written notification from the permittee's command indicating the start and end date of deployment and that orders the permittee to travel outside the permittee's county of residence.

(4) Qualified former sworn law enforcement officer. An individual who retired from service as a law enforcement officer with a local, State, or company police agency in North Carolina, other than for reasons of mental disability, who has been retired as a sworn law enforcement officer two years or less from the date of the permit application, and who satisfies all of the following:

a. Immediately before retirement, the individual was a qualified law enforcement officer with a local, State, campus police, or company police agency in North Carolina.

b. The individual has a nonforfeitable right to benefits under the retirement plan of the local, State, or company police agency as a law enforcement officer or has 20 or more aggregate years of law enforcement service and has retired from a company police agency that does not have a retirement plan.

c. The individual is not prohibited by State or federal law from receiving a firearm.

(5) Qualified sworn law enforcement officer. A law enforcement officer employed by a local, State, campus police, or company police agency in North Carolina who satisfies all of the following:

a. The individual is authorized by the agency to carry a handgun in the course of duty.

b. The individual is not the subject of a disciplinary action by the agency that prevents the carrying of a handgun.

c. The individual meets the requirements established by the agency regarding handguns.

(6) Qualified retired law enforcement officer. An individual who meets all of the following qualifications:

a. Retired in good standing from service with a public agency located in the United States as a law enforcement officer, other than for reasons of mental instability.

b. Prior to retirement, was authorized by law to engage in or supervise the prevention, detection, investigation, or prosecution of, or the incarceration of, any person for any violation of law, and had statutory powers of arrest.

c. Prior to retirement, was regularly employed as a law enforcement officer for a total of 15 years or more, or retired after completing probationary periods of service due to a service-connected disability, as determined by the agency.

d. Has a vested right to benefits under the retirement plan of the agency.

[Current through 2008 Regular Session]

NORTH DAKOTA

N.D. CENT. CODE

Title 12.1. Criminal Code

Chapter 12.1-25. Riot

12.1-25-02. Arming rioters.

1. A person is guilty of a class C felony if he:

a. Knowingly supplies a firearm, dangerous weapon, or destructive device for use in a riot;

b. Teaches another to prepare or use a firearm, dangerous weapon, or destructive device with intent that any such thing be used in a riot; or

c. While engaging in a riot, is knowingly armed with a firearm, dangerous weapon, or destructive device.

2. "Riot" has the meaning prescribed in section 12.1-25-01.

Chapter 12.1-32 - Penalties and Sentencing

12.1-32-07. Supervision of probationer - Conditions of probation - Revocation. ...

3. The court shall provide as an explicit condition of every probation that the defendant may not possess a firearm, destructive device, or other dangerous weapon while the defendant is on probation. Except when the offense is a misdemeanor offense under Section 12.1-17-01, 12.1-17-01.1, 12.1-17-05, or 12.1-17-07.1, or chapter 14-07.1, the court may waive this condition of probation if the defendant has pled guilty to, or has been found guilty of, a misdemeanor or infraction offense, the misdemeanor or infraction is the defendant's first offense, and the court has made a specific finding on the record before imposition of a sentence or a probation that there is good cause to waive the condition. The court may not waive this condition of probation if the court places the defendant under the supervision and management of the department of corrections and rehabilitation. The court shall provide as an explicit condition of probation that the defendant may not willfully defraud a urine test administered as a condition of probation. Unless waived on the record by the court, the court shall also provide as a condition of probation that the defendant undergo various agreed-to community constraints and conditions as intermediate measures of the department of corrections and rehabilitation to avoid revocation, which may include:

a. Community service;
b. Day reporting;
c. Curfew;
d. Home confinement;
e. House arrest;
f. Electronic monitoring;
g. Residential halfway house; or
h. Intensive supervision program.

Title 62.1. Weapons

Chapter 62.1-01. Definitions - General Provisions

62.1-01-01. General definitions. As used in this title, unless the context otherwise requires:

1. "Dangerous weapon" includes any switchblade or gravity knife, machete, scimitar, stiletto, sword, dagger, or knife with a blade of five inches or more; any throwing star, nunchaku, or other martial arts weapon; any billy, black-jack, sap, bludgeon, cudgel, metal knuckles, or sand club; any slungshot; any bow and arrow, crossbow, or spear; any stun gun; any weapon that will expel, or is readily capable of expelling, a projectile by the action of a spring, compressed air, or compressed gas including any such weapon, loaded or unloaded, commonly referred to as a BB gun, air rifle, or CO2 gun; and any projector of a bomb or any object containing or capable of producing and emitting any noxious liquid, gas, or substance.

2. "Direct supervision of an adult" means that an adult is present in such close proximity so as to be capable of observing and directing the actions of the individual supervised.

3. "Firearm" or "weapon" means any device which will expel, or is readily capable of expelling, a projectile by the action of an explosive and includes any such device, loaded or unloaded, commonly referred to as a pistol, revolver, rifle, gun, machine gun, shotgun, bazooka, or cannon.

4. "Gaming site" means any room or premises licensed by the attorney general or by a city or county governing body to conduct legal gaming operations.

5. "Government building" means a building which is owned, possessed, or used by or leased to the state of North Dakota, or any of its political subdivisions.

6. "Handgun" means any firearm that is not designed to be fired from the shoulder, which has a barrel less than sixteen inches [40.64 centimeters] long, and which is capable of firing, by the energy of an explosive in a fixed metallic cartridge, an exposed projectile through a rifled bore. The term includes the Thompson contender forty-five caliber single-shot center-fire with a pistol grip or similar firearm, if in compliance with the National Firearms Act [26 U.S.C. 5801-5872].

7. "Law enforcement officer" means a public servant authorized by law or by a government agency or branch to enforce the law and to conduct or engage in investigations or prosecutions for violations of law.

8. "Machine gun, submachine gun, or fully automatic rifle" means a firearm, mechanism, or instrument not requiring that the trigger be pressed for each shot, and having a reservoir, belt, or other means of storing and carrying ammunition which can be loaded into the firearm, mechanism, or instrument and fired therefrom at a rate of five or more shots to the second.

9. "Plain view" means the handgun is placed in such a location or carried in such a position as to be easily discernible by the ordinary observation of a passerby. In a motor vehicle, this includes being placed on the seat, dashboard, or in a gunrack as long as the handgun is not covered or is in any other way concealed from view.

10. "Rifle" means any firearm designed or redesigned, made or remade, and intended to be fired from the shoulder and using the energy of the explosive in a fixed metallic cartridge to fire only a single projectile through a rifled bore for each pull of the trigger.

11. "Secured" means the firearm is closed into the trunk or nonpassenger part of the vehicle; placed into a closed and secure carrying device; rendered inoperative by the use of a trigger, hammer, cylinder, slide, or barrel-locking device that renders the firearm incapable of firing until the device is unlocked and removed; or so disassembled or disabled as to be rendered incapable of firing.

12. "Short-barreled rifle" means a rifle having one or more barrels less than sixteen inches [40.64 centimeters] in length and any firearm made from a rifle, whether by alteration, modification, or otherwise, if the firearm, as modified, has an overall length of less than twenty-six inches.

13. "Short-barreled shotgun" means a shotgun having one or more barrels less than eighteen inches [45.72 centimeters] in length and any firearm made from a shotgun, whether by alteration, modification, or otherwise, if the firearm, as modified, has an overall length of less than twenty-six inches.

14. "Shotgun" means a firearm designed or redesigned, made or remade, and intended to be fired with one hand below or behind and one hand in front of the breach, which uses the energy of the explosive in a fixed shotgun shell to fire through a smooth or a rifled bore either a number of ball shot or a single projectile for each single pull of the trigger.

15. "Silencer" means any device for or attached to any firearm which will silence or deaden the sound or natural report of the firearm when it is discharged.

16. "Unloaded" means the chamber of the firearm does not contain a loaded shell. If the firearm is a revolver, then none of the chambers in the cylinder may contain a loaded shell. Handguns with a removable magazine or clip must have the magazine or clip removed from the firearm if the magazine or clip contains any loaded shells.

62.1-01-02. Forfeiture of dangerous weapon or firearm by person arrested and convicted of crime. Any firearm or dangerous weapon used or possessed while in the commission of a felony or a misdemeanor involving violence or intimidation must bed seized and, upon conviction and by motion, forfeited to the jurisdiction in which the arrest was made or the jurisdiction in which the charge arose. Except as provided in chapter 29-01 for stolen property, the forfeited firearm or dangerous weapon may be, pursuant to court order, sold at public auction, sold or traded to other law enforcement agencies or authorized firearm dealers, retained for use, or destroyed.

62.1-01-03. Limitation on authority of political subdivision regarding firearms. No political subdivision, including home rule cities or counties, may enact any ordinance relating to the purchase, sale, ownership, transfer of ownership, registration, or licensure of firearms and ammunition which is more restrictive than state law. All such existing ordinances are void.

Chapter 62.1-02. Possession of Weapons

62.1-02-01. Persons who are not to possess firearms - Penalty.

1. A person who has been convicted anywhere of a felony offense involving violence or intimidation in violation of chapters 12.1-16 through 12.1-25 or an equivalent felony offense of another state or the federal government is prohibited from owning a firearm or having one in possession or under control from the date of conviction and continuing for a period of ten years after the date of conviction or the date of release from incarceration, parole, or probation, whichever is latest.

2. A person who has been convicted anywhere of a felony offense of this or another state or the federal government not provided for in subsection 1 or who has been convicted of a class A misdemeanor offense involving violence or intimidation in violation of chapters 12.1-16 through 12.1-25 or an equivalent offense of another state or the federal government and the offense was committed while using or possessing a firearm, a dangerous weapon, or, as defined in subsections 7 and 8 of section 12.1-01-04, a destructive device or an explosive, is prohibited from owning a firearm or having one in possession or under control from the date of conviction and continuing for a period of five years after the date of conviction or the date of release from incarceration, parole, or probation, whichever is latest.

3. A person who is or has ever been diagnosed and confined or committed to a hospital or other institution in this state or elsewhere by a court of competent jurisdiction, other than a person who has had the petition that provided the basis for the diagnosis, confinement, or commitment dismissed under section 25-03.1-17, 25-03.1-18, or 25-03.1-19, or equivalent statutes of another jurisdiction, as a mentally ill person as defined in section 25-03.1-02, or as a mentally deficient person as defined in section 25-01-01, is prohibited from purchasing a firearm or having one in possession or under control. This limitation does not apply to a person who has not suffered from the disability for the previous three years.

4. A person under the age of eighteen years may not possess a handgun except that such a person, while under the direct supervision of an adult, may possess a handgun for the purposes of firearm safety training, target shooting, or hunting.

A person who violates subsection 1 or 2 is guilty of a class c felony, and a person who violates subsection 3 or 4 is guilty of a class a misdemeanor. For the purposes of this section, "conviction" means a determination that the person committed one of the above-mentioned crimes upon a verdict of guilt, a plea of guilty, or a plea of nolo contendere even though:

1. The court suspended execution of sentence in accordance with subsection 3 of section 12.1-32-02;

2. The court deferred imposition of sentence in accordance with subsection 4 of section 12.1-32-02;

3. The court placed the person on probation;

4. The person's conviction has been reduced in accordance with subsection 9 of section 12.1-32-02 or section 12.1-32-07.1;

5. Sentence dispositions, sentence reductions, or offense determinations equivalent to this section were imposed or granted by a court, board, agency, or law of another state or the federal government; or

6. The person committed an offense equivalent to an offense described in subsection 1 or 2 when that person was subject to juvenile adjudication or proceedings and a determination of a court under chapter 27-20 or of a court of another state or the federal government was made that the person committed the delinquent act or offense.

62.1-02-02. Sale of handgun regulated - Penalty. No person may transfer a handgun to any person who the transferor knows or has reasonable cause to believe is a person prohibited by section 62.1-02-01 from possessing a firearm. Any person who violates this section is guilty of a class A misdemeanor.

62.1-02-03. Possession or sale of short - barreled rifle or shotgun - Penalty - Application. A person who possesses, obtains, receives, sells, or uses a short-barreled rifle or a short-barreled shotgun is guilty of a class C felony. This section does not apply to a law enforcement officer who possesses, obtains, receives, sells, or uses a short-barreled rifle or a short-barreled shotgun in the course of or in connection with the officer's official duties, to a member of the armed forces of the United States or national guard, organized reserves, state defense forces, or state guard organizations who possesses or uses a short-barreled rifle or short-barreled shotgun issued to the member by that organization and while on official duty, or to any person who complies with the National Firearms Act [26 U.S.C. 5801-5872].

62.1-02-05. Possession of a firearm at a public gathering - Penalty - Application.

1. A person who possesses a firearm at a public gathering is guilty of a class B misdemeanor. For the purpose of this section, "public gathering" includes athletic or sporting events, schools or school functions, churches or church functions, political rallies or functions, musical concerts, and individuals in publicly owned parks where hunting is not allowed by proclamation and publicly owned or operated buildings.

2. This section does not apply to law enforcement officers; members of the armed forces of the United States or national guard, organized reserves, state defense forces, or state guard organizations, when on duty; competitors participating in organized sport shooting events; gun and antique shows; participants using blank cartridge firearms at sporting or theatrical events; any firearms carried in a temporary residence or motor vehicle; students and instructors at hunter safety classes; or private security personnel while on duty. In addition, a municipal court judge licensed to practice law in this state, a district court judge, and a retired North Dakota law enforcement officer are exempt from the prohibition and penalty in subsection 1 if the individual is otherwise licensed to carry a firearm under section 62.1-04-03 and maintains the same level of firearms proficiency as is required by the peace officers standards and training board for law enforcement officers. A local law enforcement agency shall issue a certificate of compliance under this section to an individual who is proficient.

3. This section does not prevent any political subdivision from enacting an ordinance which is less restrictive than this section relating to the possession of firearms at a public gathering. Such an ordinance supersedes this section within the jurisdiction of the political subdivision.

62.1-02-07. Use of firearm by certain minors prohibited - Penalty. Any parent, guardian, or other person having charge or custody of any minor under fifteen years of age who permits that minor to carry or use in public any firearm of any description loaded with powder and projectile, except when the minor is under the direct supervision of the parent, guardian, or other person authorized by the parent or guardian, is guilty of a class B misdemeanor.

62.1-02-08. Illegal firearms, ammunition, or explosive materials business.

1. A person is guilty of an offense if the person supplies a firearm, ammunition, or explosive material to, or procures or receives a firearm, ammunition, or explosive material for, a person prohibited by this title from receiving it if the transferor knows or has reasonable cause to believe that such person is prohibited by section 62.1-02-01 from receiving or possessing it.

2. The offense is a class C felony if the actor:

a. Was not licensed or otherwise authorized by law to handle, transfer, or engage in transactions with respect to the firearm, destructive device, or explosive material; or

b. Engaged in the forbidden transaction under circumstances manifesting the actor's readiness to supply or procure on other occasions in disregard of lawful restrictions.

Otherwise the offense is a class A misdemeanor.

Chapter 62.1-03. Handguns

62.1-03-02. Selling handgun to minors prohibited - Penalty. Any person who sells, barters, hires, lends, or gives any handgun to any minor is guilty of a class A misdemeanor. This section does not prohibit a person from lending or giving a handgun to a minor if the minor will be using the handgun under the direct supervision of an adult and for the purpose of firearm safety training, target shooting, or hunting.

62.1-03-03. Copy of federal license submitted to law enforcement officials. A retail dealer licensed to sell handguns by the federal government shall send a copy of the license, within seven days after receiving it, to the chief of police of the city and the sheriff of the county in which the dealer is licensed to sell handguns.

62.1-03-04. False information prohibited. No person, in purchasing or otherwise securing delivery of a handgun or in applying for a license to carry the handgun concealed, may give false information or offer false evidence of the person's identity.

62.1-03-05. Prohibited alterations in handgun. No person may change, alter, remove, or obliterate any mark of identification on a handgun, such as the name of the maker, model, or manufacturer's number or knowingly possess a handgun on which such alterations have been made. Possession of any handgun upon which any such identification mark has been changed, altered, removed, or obliterated creates a rebuttable presumption that the possessor made the alterations.

62.1-03-06. General penalty. Any person who violates any provision of this chapter, for which another penalty is not specifically provided, is guilty of a class A misdemeanor.

Chapter 62.1-05. Machineguns, Automatic Rifles, Silencers, Bombs

62.1-05-01. Possession and sale of machine guns, automatic rifles, silencers, and bombs - Penalty - Forfeiture. No person may purchase, sell, have, or possess a machine gun, fully automatic rifle, silencer, or bomb loaded with explosives or poisonous or dangerous gases, or any other federally licensed firearm or dangerous weapon unless that person has complied with the National Firearms Act [26 U.S.C. 5801-5872].

Any federal licensee who purchases, sells, has, or possesses those items for the licensee's protection or for sale must forward a copy of the licensee's federal license along with the required weapons transfer form to the licensee's local county sheriff and to the chief of the bureau of criminal investigation within five days of the receipt of those forms.

A person who violates this section is guilty of a class C felony. Upon arrest of that person, the firearm or dangerous weapon must be seized. Upon conviction of the person and motion to the court in which the conviction occurred, the firearm or dangerous weapon must be forfeited to the jurisdiction in which the arrest was made. The firearm or dangerous weapon may be sold at public auction, retained for use, or destroyed pursuant to the court's order. If a qualified local program as defined under section 12.1-32-02.2 has paid a reward for information that resulted in forfeiture of the item and the item has been sold, the jurisdiction shall, after payment of expenses for forfeiture and sale, repay the qualified local program for the reward that it has paid.

62.1-05-02. Persons exempt from chapter. This chapter does not apply to:

1. The authorized agent and a servant of a person who has a license to purchase, sell, have, or possess a machine gun, submachine gun, fully automatic rifle, silencer, or a bomb loaded with explosives or poisonous or dangerous gases.

2. Any officer or member of a duly authorized military organization while on official duty and using the firearm or dangerous weapon issued to the officer or member by that organization.

3. A North Dakota law enforcement officer.

4. Any federal officer authorized by the federal government to have or possess a machine gun, submachine gun, fully automatic rifle, silencer, or bomb loaded with explosives or poisonous or dangerous gases.

[Current through General Election of November 4, 2008]

NORTHERN MARIANA ISLANDS

N. Mar. I. Code

Title 6. Crimes and Criminal Procedure

General Provisions

Chapter 1. Preliminary Provisions

102. Definitions. The definitions in this and the following section apply throughout this title, unless otherwise specified or a different meaning is plainly required:

(a) "Automatic weapon" means a weapon of any description irrespective of size by whatever name designated or known, loaded or unloaded, from which may be repeatedly or automatically discharged a number of bullets contained in a magazine, ribbon or other receptacle, by one continued movement of the trigger or firing mechanism.

(b) "Carry" means having on one's person or in a motor vehicle or other conveyance. ...

(e) "Dangerous device" means any explosive, incendiary or poison gas bomb, grenade, mine or similar device, ... any instrument designed or redesigned for use as a weapon, or any other instrument which can be used for the purpose of inflicting bodily harm and which under the circumstances of its possession serves no lawful purpose.

(f) "Dangerous weapon" means any automatic weapon, dangerous device, firearm, gun, handgun, long gun, semiautomatic weapon, knife, machete, or other thing by which a fatal wound or injury may be inflicted. ...

(k) "Firearm" means any device, by whatever name known, which is designed or may be converted to expel or hurl a projectile or projectiles by the action of an explosion, a release, or an expansion of gas, including but not limited to guns and automatic and semiautomatic weapons, except a device designed or redesigned for use solely as a signaling, linethrowing, spearfishing, or industrial device, or a device which hurls a projectile by means of the release or expansion of carbon dioxide or air. Bang-sticks are firearms, provided that bang-sticks used for scientific purposes are not firearms if their temporary use is approved by the Governor.

(l) "Gun" means a handgun or long gun.

(m) "Handgun" means a pistol or revolver with an overall length less than 26 inches. ...

(q) "Long gun" means a rifle with one or more barrels of more than 18 inches in length. ...

103. Further Definitions. The definitions in this also apply throughout this title, unless otherwise specified or a different meaning is plainly required: ...

(n) "Semiautomatic weapon" means a weapon of any description irrespective of size, by whatever name designated or known, loaded or unloaded, from which may be repeatedly or automatically discharged a number of bullets contained in a magazine, ribbon or other receptacle by a like number of movements of the trigger or firing mechanism without recocking or resetting the trigger or firing mechanism. ...

Division 2. Contraband Offenses

Chapter 2. Weapons Control

Article 1. Weapons Control Act

2201. Short title. This article may be cited as the Commonwealth Weapons Control Act.

2202. Manufacture, Sale or Possession of Firearms and Dangerous Devices. No person may manufacture, purchase, sell, possess or carry any firearm, dangerous device or ammunition other than as provided in this article.

2203. Exemptions from Provisions of this Article. This article does not apply to:

(a) Law enforcement officers while engaged in official duty except to the extent that particular provisions are expressly made applicable to them.

(b) Firearms which are in unserviceable condition and which are incapable of being fired or discharged and which are kept as curios, ornaments or for their historical significance or value.

(c) Weapons or other dangerous devices which are not firearms and which are kept as ornaments, curios, or objects of historical or archeological interest; provided, that the article or articles are kept or displayed only in private homes, museums, or in connection with public exhibitions.

(d) Persons in the armed forces of the Untied States, whenever such persons are engaged in official duty except to the extent that particular provisions of this article are expressly made applicable to them.

(e) Patrons of shooting galleries, while on the premises of the shooting gallery.

(f) Parole officers while engaged in official duty except to the extent that particular provisions are expressly made applicable to them. Parole officers shall comply with all laws and regulations applicable to Department of Public Safety law enforcement officers relating to the training, carrying, and use of firearms.

2204. Identification Cards.

(a) No person may acquire or possess any firearm, dangerous device or ammunition unless the person holds an identification card issued pursuant to this article. The identification card is evidence of the holder's eligibility to possess and use or carry firearms, dangerous devices, or ammunition.

(b) Identification cards are issued only by the Department of Public Safety pursuant to regulations made by the Department of Public Safety in the manner which is or may be provided by law. The identification card shall have on its face all of the following:

(1) The name and address of the holder.

(2) The sex, height and weight of the holder.

(3) The birth date of the holder.

(4) The date of expiration for the card, which shall be two years from the date of issue.

(5) A photograph of the holder taken within 10 days prior to issuance.

(6) An endorsement setting forth the extent of the holder's eligibility to possess, use and carry firearms, dangerous devices, or ammunition.

(7) The number of the identification card.

(8) The manufacturer, model, type and serial number of the firearm.

(c) An applicant for an identification card shall make application for it on a form approved by the Department of Public Safety and supply any information necessary to afford the issuing agency reasonable opportunity to ascertain the facts required to appear on the face of the identification card, and to determine whether the applicant complies with all requirements of this article to possess, use, or carry firearms, dangerous devices or ammunition, as the case may be.

(d) No identification card may be issued until 15 days after application is made, and unless the issuing agency is satisfied that the applicant may lawfully possess and use or carry firearms, dangerous devices, or ammunition of the type or types enumerated on the identification card. Unless the application for use and possession is denied, the identification card shall issue within 60 days from the date of application.

(e) No person may be issued an identification card if the person has been:

(1) Acquitted of any criminal charge by reason of insanity.

(2) Adjudicated mentally incompetent.

(3) Treated in a hospital for mental illness, drug addiction or alcoholism.

(4) Convicted of a crime of which actual or attempted personal injury or death is an element.

(5) Convicted of a crime in connection with which firearms or dangerous devices were used or found in his or her possession.

(6) Convicted of a crime of which the use, possession or sale of narcotics or dangerous drugs is an element.

(f) No person may be issued an identification card if the person has a physical condition or impairment which makes the person unable to use a firearm or dangerous device with proper control.

(g) Any person suffering from a physical or mental defect, condition, illness or impairment which would make him ineligible for an identification card pursuant to this section may submit the certificate of a physician licensed to practice in the Commonwealth to the issuing agency or officer. If the certificate states that it is the subscribing physician's best opinion that the defect, condition, illness or impairment does not make the applicant incapable of possessing and using a firearm or dangerous device without danger to the public safety, the identification card may be issued. No such card may be valid for a period longer than six months.

(h) Any person who is ineligible for an identification card by reason of conviction of crime may be issued a card if:

(1) The most recent discharge from probation or parole or the termination of the most recent sentence, whichever is later, is more than 10 years prior to the time of application for the identification card; and

(2) The issuing agency finds that the person's record, taken as a whole, indicates that the possession, use, or carrying of a firearm or dangerous device, as the same may be, is not likely to constitute a special danger to the public safety.

(i) The holder of an identification card shall have it on or about his or her person at all times when carrying or using a firearm or dangerous device and shall display the card upon the request of any law enforcement official.

(j) A duplicate identification card may be issued to the holder of a lost, destroyed or defaced identification card upon proof of the loss, destruction or defacement, as the Department of Public Safety may require, upon payment of the fee required by 6 CMC § 2229 and upon surrender of any remaining portion of the original card. Notice shall be given the Department of Public Safety by the holder within 48 hours of the discovery of the loss, defacement or destruction. The holder shall notify the Department of Public Safety of any change of name or address from those appearing upon the identification card within 48 hours of such change.

(k) Only a person who is a permanent resident of the Commonwealth or a United States citizen or a United States national and a bona fide resident of the Commonwealth is eligible for an identification card or for renewal thereof. A bona fide resident of the Commonwealth means a person whose place of general abode is in the Commonwealth of the Northern Mariana Islands. The place of general abode of a person means his principal, actual dwelling place in fact, without regard to intent. This exemption does not apply to tourists or non-resident alien contract workers. Immediate relatives will be accorded the same status as their sponsor for the purposes of this section. In order to be issued a weapons identification card for use of .223 caliber rifles, special conditions, as promulgated by the Department of Public Safety, must be met.

(l)(1) Any person holding a firearms identification card pursuant to this section who desires renewal of the identification card shall submit an application for renewal at the Department of Public Safety 30 days prior to the expiration date of the person's current firearms identification card.

(2) The application forms shall contain a formal request for a renewal of two years and shall also contain a certification that the holder of the identification card has not been subjected to any of those conditions set forth in subsection (e) of this section.

(3) The identification card shall be automatically renewed upon its expiration date for a period of two years, provided that the requirements of subsections (l)(1) and (l)(2) of this section are complied with, and, provided, further, that the Department of Public Safety does not have any cause, pursuant to rules and regulations adopted under this article, to disapprove the renewal.

(4) If the time period for renewal stated in subsection (l)(1) of this section is not complied with, the identification card holder shall be required to follow the procedures for an original application.

2205. Identification Card Prerequisite to Purchase, Possession, and Use; Prima Facie Evidence of Possession.

(a) No person may purchase, possess or use a firearm, dangerous device, or ammunition unless he is the holder of an identification card issued pursuant to this article evidencing the eligibility of such person to purchase, possess and use a firearm, dangerous device, or ammunition. That person shall be at least 21 years of age. ...

2206. Carrying Firearms. No person may carry a firearm unless the person has in his or her immediate possession a valid identification card, and is carrying the firearm unloaded in a closed case or other securely wrapped or closed package or container, or locked in the trunk of a vehicle while en route to or from a target range, or area where the person hunts, or takes part in other sports involving firearms, or carries the firearm in plain sight on his or her person while actively engaged in hunting or sports involving the use of firearms.

2207. New Residents, Temporary Residents and Visitors to the Commonwealth. Visitors, new residents, and temporary residents in the Commonwealth may not import, transport, purchase, use or possess any firearm, dangerous device or ammunition in the Commonwealth without an identification card issued pursuant to this article. Any person who possesses any firearm, dangerous device, or ammunition shall, before or immediately upon his entrance into the Commonwealth, turn it in to the Department of Public Safety or the chief of police. The firearm, dangerous device, or ammunition shall be returned to the person upon his or her being issued an identification card pursuant to the provisions of this article or upon his departure from the Commonwealth.

2208. Law Enforcement Officers.

(a) Possession, use and carriage of firearms, ammunition and dangerous devices by law enforcement officers derives from the laws governing the powers, functions, and organization of the police and other organized forces of peace officers. Eligibility of law enforcement officers to possess, use and carry firearms, ammunition or dangerous devices while on duty is not subject to the holding of identification cards or any other qualifications prescribed in this article or in regulations pursuant thereto.

(b) Transfer of any firearm from or to a law enforcement officer or agency shall, except as provided in subsection (a) of this section, be subject to the provisions of this article and regulations made pursuant thereto.

(c) The head of a law enforcement agency of the Commonwealth shall furnish to the Department of Public Safety the names, ranks and badge numbers or similar identification of each person on his force who is authorized to possess, use and carry firearms in the course of official duty. Upon the occurrence of any changes in personnel to whom this subsection applies, the head of the law enforcement agency shall inform the Department of Public Safety promptly of the change.

(d) Whenever a law enforcement officer is not engaged in official duties, this article shall be applicable to him in the same manner and to the same extent as to any other person. ...

2209. License for Transfer.

(a) No dealer, manufacturer or wholesaler may transfer firearms, dangerous devices or ammunition except pursuant to a license as provided in this section.

(b) Any person, firm, corporation, or other entity proposing to engage in the business of selling firearms, ammunition, and dangerous devices at retail may apply for a dealer's license. The application shall be on a form approved by the Department of Public Safety and shall contain the following information:

(1) The name and address of the applicant, including the address of each separate location within the Commonwealth at which the applicant purposes to do business pursuant to the license;

(2) If the applicant is a partnership or association, the names and addresses of the partners or associates, or if the applicant is a corporation, the names and addresses of the officers and directors; and

(3) Such other information bearing on the applicant's ability to operate the business in a manner consonant with the public safety as the Department of Public Safety may require.

2210. Licenses for Transfer; Issuance and Renewal of Dealer's License.

(a) Upon receipt of a proper application and payment of the prescribed fee, the Department of Public Safety shall within 60 days issue a dealer's license to an applicant, if the applicant is found to be eligible for it pursuant to this article and any applicable regulations of the Department of Public Safety. The regulations shall place a reasonable limit on the number of dealers. The license shall list the types of firearms, ammunition, and dangerous devices which the dealer is authorized to offer for sale.

(b) A license issued pursuant to this section is valid for one year from the date of its issuance, unless sooner canceled, suspended or revoked. A license shall bear its expiration date on its face.

(c) A license issued pursuant to this section may be renewed annually upon application by the holder made on a form approved by the Department of Public Safety. Eligibility for renewal shall be the same terms and conditions as for an original license, except that renewal also may be denied on account of a violation of this article or regulations of the Department of Public Safety made pursuant thereto or for any conduct in the operation of the applicant's business which gives the Department of Public Safety grounds to believe that the applicant will no longer operate in a manner consonant with the public safety.

2211. License for Transfer; Display; Conduct of Dealer's Business. The holder of a dealer's license shall:

(a) Display his or her license in a conspicuous place at all times at the establishment described in the license. If a dealer has more than one place of business at which firearms, dangerous devices or ammunition are sold, the dealer shall display in the same manner a certified copy of the license at each additional place of business.

(b) Keep the records and file the reports required by this article and regulations made pursuant thereto.

(c) Display no firearms, dangerous devices, or ammunition in any place where they can be seen from outside the premises.

(d) Keep all firearms, dangerous devices and ammunition in a securely locked place at all times except when they are actually being shown to a customer or prospective customer or when actually being repaired or otherwise worked on.

(e) Permit only employees who are holder of identification cards making the eligible to purchase, possess and use firearms, dangerous devices or ammunition to have access to firearms, dangerous devices or ammunition.

2212. Records and Reports by Dealers.

(a) Every licensed dealer shall maintain records containing an inventory of firearms, dangerous devices, and ammunition or any of them received together with the name and address of the person from whom received, and the manufacturer, type and serial number of each firearm and dangerous device, the name and address of the person to whom transferred, the identification card number of that person, the manufacturer, type and serial number of the gun or dangerous device transferred and the date of transfer. These records shall be available for inspection at all reasonable times by the Department of Public Safety and the Department of Public Safety's duly designated representatives. These records shall be retained at least five years.

(b) Every dealer, at the time of any transfer of any firearm or dangerous device to any person other than a licensed dealer shall, within 24 hours of the transfer, supply the following information to the Department of Public Safety on a form approved by it:

(1) The name address and license number of the dealer.

(2) The manufacturer, type and serial number of the firearm or dangerous device transferred. No firearm may be transferred which does not have a serial number or from which the serial number has been removed, defaced, or altered.

(3) The name, address and identification number of the tranferee.

2213. Repair of Firearms.

(a) No person other than a dealer or manufacturer licensed pursuant to this article shall repair firearms or accept the same for repair.

(b) No person may accept any firearms for repair unless the person is shown an identification card evidencing eligibility of the holder to possess and use a firearm of the type offered for repair. Prior to returning any firearm, the manufacturer or dealer shall make and keep a record identical with that required for the purchase of a firearm pursuant to 6 CMC § 2212, and shall maintain the record for at least one year.

(c) Nothing in this section shall be construed to prohibit the repair or maintenance of a firearm by its owner.

2214. Transfer or Sale of Ammunition.

(a) No person may transfer ammunition, unless the person is a manufacturer, wholesaler or dealer licensed pursuant to this article. If the transfer is other than to another manufacturer, wholesaler or dealer, the transfer shall not be made until the transferor has ascertained that the transferee is the holder of an identification card evidencing eligibility to possess and use a firearm of the type for which the ammunition is suited. Upon transfer the transferor shall record the quantity, type and caliber or gauge transferred, the name and address of the transferee and the number of the identification card.

(b) No transferee of ammunition may transfer it to any person other than a dealer licensed pursuant to this article. Upon receipt of ammunition, the dealer shall make keep records with respect to the ammunition in the manner required by this section for ammunition sold by him.

2215. Private Sales or Transfers. No person other than a manufacturer, wholesaler or dealer licensed pursuant to this article may transfer a firearm or dangerous device to any person other than a manufacturer, wholesaler or dealer without first ascertaining that the transferee is a holder of an identification card issued pursuant to this article. Prior to any transfer, the transferor shall furnish the Department of Public Safety in person or by registered or certified mail, return receipt requested, a properly completed form approved by the Department of Public Safety providing information equivalent to that required to be furnished by a dealer upon the transfer by him of a firearm or dangerous device.

2216. Receipt or Use as Security.

(a) No person, other than a licensed dealer, may receive a firearm as a pledge or pawn, or in any other manner as security.

(b) A dealer receiving a firearm as a pledge, pawn or otherwise, as security, shall record promptly:

(1) The date of receipt;

(2) The full description of the item or items received including the manufacturer, type and serial number or numbers, if any;

(3) The name and address of the person making the pledge, pawn, or other deposit as security; and

(4) The number of the person's identification card.

No dealer may accept the pledge, pawn, or other deposit as security unless the person making the same exhibits an identification card evidencing an entitlement to possess and use a gun of the type involved.

(c) Upon the return or other disposition of the firearm in the dealer's possession pursuant to this section, the dealer shall make a record of the return or other disposition, including the date and the name and address of the person to whom the firearm was returned or disposed. No firearm may be returned or disposed of to any person who, at the time of the return or disposition, does not exhibit a valid identification card issued in the person's own name and entitling that person to possess and use the firearm involved.

2217. Manufacturers and Wholesalers.

(a) No person shall manufacture or deal in firearms, dangerous devices or ammunition at wholesale unless the person is the holder of:

(1) A dealer's license issued pursuant to 6 CMC § 2210; or

(2) A license issued pursuant to this section.

(b) Any person proposing to manufacture or deal at wholesale in firearms, dangerous devices or ammunition, who is not the holder of a dealer's license, may make application for a manufacturer's or wholesaler's license. The application shall contain the same information required for a dealer's license, and any addition information required by the Department of Public Safety as may be appropriate to administer this article. No manufacturer's license or wholesaler's license may authorize transfer or delivery within the Commonwealth except to a licensed dealer, manufacturer or wholesaler or to a political subdivision of the Commonwealth or, subject to applicable laws of the Commonwealth, for export.

(c) The Department of Public Safety shall issue, renew, cancel, deny, suspend or revoke manufacturer's and wholesaler's licenses on the same terms and subject to the same conditions as provided for dealer's licenses.

(d) Every manufacturer shall assign a unique serial number to each firearm manufactured and shall inscribe the number in or on the firearm in a manner that will resist removal, alteration, defacement or obliteration. The Department of Public Safety may make regulations for the style of the serial numbers and for the manner of the inscription.

2218. Registry of Firearms and Ammunition.

(a) The Department of Public Safety shall maintain a registry of firearms. The records in the registry shall be kept permanently unless there is a record of the destruction of the gun.

(b) Records kept in the registry shall include all records required to be filed with the Department of Public Safety pursuant to this article, copies of all records filed with an agency or officer of local government pursuant to this article, and any records deposited with the Department of Public Safety pursuant to subsection (c) of this section.

(c) Any dealer, manufacturer or wholesaler licensed pursuant to this article, upon discontinuance of the licensed business or activity, shall transmit all records kept pursuant to this article to the Department of Public Safety.

(d) Records relating to the repair of firearms shall be kept by the Department of Public Safety for a period of at least five years after transmittal.

(e) Records in the registry shall not be public records. They shall be made available only to law enforcement officers of the Commonwealth or its subdivisions, or at the discretion of the Department of Public Safety, to law enforcement officers and agencies of foreign governments.

2219. Cancellation, Denial, Suspension and Revocation of Licenses.

(a) Any license issued pursuant to this article shall be surrendered for cancellation immediately on the discontinuance or termination of business or upon the holder's discontinuing the manufacturing, selling, acquisition for sale or repair of firearms and the sale of ammunition.

(b) The issuing officer or agency may deny, suspend or revoke and identification card or a license issued pursuant to this chapter for failure of the applicant or holder to meet or continue to meet any of the requirements for eligibility therefore, or for any violation of this article or regulations in force pursuant to it. ...

2220. Shipment and Delivery of Firearms, Dangerous Devices, and Ammunition.

(a) No person may ship, transport or deliver any firearm, dangerous device or ammunition to anyone other than a licensed manufacturer, wholesaler, dealer, or person who possesses a valid identification card.

(b) Any person who ship, transports or delivers firearms or dangerous devices to a manufacturer, wholesaler, dealer or person possessing an identification card in the Commonwealth shall, before delivery, furnish to the Department of Public Safety an invoice listing the person's name and address, the name and address of the manufacturer, wholesaler, dealer or person possessing the identification card to whom the firearms or dangerous devices are to be delivered, the place of origin of the shipment, the number of firearms and dangerous devices of each type and the manufacturer and serial number of each firearm and dangerous device in the shipment.

(c) Any person who ships, transports or delivers ammunition to a manufacturer, wholesaler, dealer or person possessing an identification card in the Commonwealth shall, before delivery, furnish to the Department of Public Safety an invoice listing the person's name and address, the name and address of the manufacturer, wholesaler, dealer or person possessing the identification card to whom the ammunition is to be delivered, the place of origin of the shipment and the quantity if ammunition to each type in the shipment.

(d) If the shipment is by common carrier, a copy of the invoice required by subsections (b) and (c) of this section shall also be delivered to the common carrier. The common carrier shall deliver the invoice and any shipment to the chief of police who will verify the accuracy of the shipment and compliance with this article, before delivery to the manufacturer, wholesaler, dealer or person possessing an identification card. A copy of the invoice shall be left with the manufacturer, wholesaler, dealer or person possessing an identification card at the time delivery.

(e) If shipment is by other than common carrier, a copy of the invoice shall be furnished to the manufacturer, wholesaler, dealer or person possessing an identification card at the time delivery.

2221. Loss, Destruction or Theft of Firearms or Dangerous Devices.

Whoever owns or possesses a firearm or dangerous device shall, within 24 hours of discovery, notify the Department of Public Safety of the loss, theft or destruction of any firearm or dangerous device and, after giving notice, of recovery thereof.

2222. Prohibited Acts. No person shall:

(a) Knowingly remove, obliterate or alter the importer's or manufacturer's serial number of any firearm.

(b) Knowingly deface, alter or destroy an identification card.

(c) Acquire, possess or use any firearm silencer or muffler.

(d) Carry any gun or dangerous device while under the influence of alcohol or narcotic or other disabling drug.

(e) Import, sell, transfer, give away, purchase, possess or use any handgun, automatic weapon or ammunition other than:

(i) All .22 caliber rimfire cartridges and all regular .22 caliber rimfire cartridges.

(ii) All .22 caliber centerfire cartridges and .22 caliber rifles.

(iii) All .223 caliber centerfire cartridges and .223 caliber centerfire rifles. These require a special weapons identification card.

(iv) All .410 gauge shotgun shells and .410 gauge shotguns.

These firearms and ammunition are as defined by the Gun Data Book published by Harper and Row and the NRA Firearms Book published by the National Rifle Association of America. This section is subject to the law enforcement exemption provided at 6 CMC § 2203(a).

(f) Import, sell, transfer, give away, purchase, possess or use any explosives, ammunition or other projectiles other than those in subsection (e) of this section. This section is subject to the law enforcement exemption provided at 6 CMC § 2301(a)(4). ...

(i) In any manner or to any degree alter or modify any firearm from its original state of manufacture, including but not limited to alteration or modification of firing pins or chamber, or alteration or modification that creates automatic or semiautomatic weapons from single fire weapons, or to acquire, possess or use such altered or modified firearm.

2223. Forfeiture of Unlawful Item. All firearms, dangerous devices or ammunition unlawfully possessed, carried, used, shipped, transported or delivered into the Commonwealth are declared to be inimical to the public safety and are forfeited to the Commonwealth. When such forfeited articles are taken from any person, they shall be surrendered to the Department of Public Safety.

2224.Closing of Establishments During Emergencies.

In case of emergency concerning the public safety declared by the Governor, all establishments dealing in guns, dangerous devices or ammunition may be ordered closed by the official and be required to remain closed during the continuance of the emergency. During any such closure, any and all guns, dangerous devices and ammunition belonging to or in the keeping of a closed establishment may be impounded.

2225. Registration of Weapons Possessed on Effective Date of Article.

(a) Any person having in his or her possession a firearm or dangerous device on the effective date of this article shall, within 90 days of the effective date, furnish on a form approved by the Department of Public Safety to the agency or officer authorized to receive information concerning the transfer of firearms or dangerous devices pursuant to this article, equivalent information concerning any firearm or dangerous device in his or her possession.

(b) If, prior to the expiration of the 90 day period provided in subsection (a) of this section, the firearm is transferred, the transferor shall comply with the provisions of this article for furnishing of information on transfer and need not comply with subsection (a) of this section.

2226. Surrender of and Compensation for Weapons Held on Effective Date by Ineligible Persons.

Any person who possessed any firearm or dangerous device in the Commonwealth prior to the effective date of this article, and who is determined to be ineligible to possess or is prohibited from possessing the firearm or dangerous device under this article, shall tender such firearm or dangerous device to the Department of Public Safety or his delegate within 90 days of the effective date of this article and be reasonably compensated for it.

2227. Local laws. Nothing in this article shall be deemed to prevent any local government from further restricting, by local law or ordinance, the transfer, possession, use or carriage of firearms, ammunition or dangerous devices. This chapter shall supersede all district laws and municipal ordinances in conflict with this article.

2228. Authority of the Department of Public Safety to Promulgate Regulations. The Department of Public Safety may issue, amend and repeal regulations implementing this article in the manner which is or may be provided by law, as may be required by the public interest, safety and welfare.

2229. Fees for Licensing and Identification Cards. The fees for issuance and renewal of annual licenses and identification cards as required by this article shall be as follows:

(a) For the identification card, $25;

(b) For a dealer's license, $300;

(c) For a manufacturer's license, $1,000;

(d) For a wholesaler's license, $1,000;

(e) For replacement of lost, destroyed, or defaced identification card, $10. ...

2230. Penalties.

(a) Any person who, being a holder of a valid identification card fails to comply with 6 CMC § 2206 shall be guilty of a misdemeanor, and upon conviction thereof shall be fined not more than $100 or imprisoned not more than three months, or both.

(b) Any person who violates any other provision of this article or any regulation issued pursuant thereto shall be guilty of a felony, and upon conviction thereof, may be fined not more than $5,000 or imprisoned not more than five years, or both, and may be subject to confiscation of any firearm, dangerous device, or ammunition without compensation, involved in a violation of this article. The holder of any dealer's license or the manager or supervisor of employees of any establishment so licensed, or both, shall be liable for any violation of this article by his or her employee or agent committed in the course of the dealer's business, to the same extent as such employee or agent.

Chapter 3. Custom Violations

2301. Importation of Contraband.

(a) It shall be unlawful for any person to knowingly enter the Commonwealth with the intent to import, or attempt to cause another to bring or import into the Commonwealth, or conceal for the purpose of importation, any of the following items:

(1) Controlled substances as defined in this title.

(2) Currency, coin, travelers checks, money orders, and/or negotiable instruments of a total of more than $5,000 unless the same is reported to the Division of Customs in a signed customs declaration form prior to entry into the Commonwealth.

(3) Firearms and ammunition other than those rifles listed below or shotguns of .410 gauge and with normal factory-produced stocks and barrels unless the importer is authorized by the Director of the Department of Public Safety to import the same for law enforcement purposes. The following are not contraband:

(i) All .22 caliber rimfire cartridges and all regular .22 caliber rimfire cartridges.

(ii) All .22 center-fire cartridges and .22 caliber rifles.

(iii) All .223 caliber center-fire cartridges and .223 caliber center-fire rifles. These require a special weapons identification card.

(iv) All .410 gauge shotgun shells and .410 gauge shotguns.

(4) Ammunition other than regular long rifle .22 caliber rimfire cartridges or .410 gauge shotgun shells or dangerous devices as described in this title unless the importer has been authorized to import and possess the same by the Director of the Department of Public Safety or the Chief of the Customs Service for law enforcement purposes....

2302. Enforcement.

(a) The Customs Service, a division of the Department of Finance, shall have the primary responsibility and authority to enforce the provisions of this chapter. This authority shall be concurrent with the authority of any other law enforcement agency as provided by law.

(b) Any officer who is authorized by the Customs Service to enforce the provisions of this chapter may:

(1) Arrest any person, if there exists probable cause to believe that such person committed an act in violation of this chapter;

(2) Seize any evidence related to any violation of any provision of this chapter;

(3) Execute any warrant or other process issued by a court of competent jurisdiction.

2303. Forfeiture.

(a) All items of contraband as defined by 6 CMC § 2301 shall be subject to forfeiture pursuant to 6 CMC § 2150.

(b) Only firearms and ammunition of the .22, .22 magnum, and .223 caliber as referenced in 6 CMC § 2222(e) will be returned to their former lawful owners. No firearm shall be returned to the rightful owner unless that person has a firearm identification card and has reimbursed the Commonwealth government the amount compensated by the government for surrendering the firearm to the Department of Public Safety.

[Current through all Sessions to August 2007]

OHIO

Ohio Rev. Code

(A) The Ohio peace officer training commission shall prescribe, and shall make available to sheriffs, all of the following:

(1) An application form that is to be used under section 2923.125 of the Revised Code by a person who applies for a license to carry a concealed handgun or for the renewal of a license of that nature and that conforms substantially to the form prescribed in section 2923.1210 of the Revised Code; ...

(B)(1) The Ohio peace officer training commission, in consultation with the attorney general, shall prepare a pamphlet that does all of the following, in everyday language:

(a) Explains the firearms laws of this state;

(b) Instructs the reader in dispute resolution and explains the laws of this state related to that matter;

(c) Provides information to the reader regarding all aspects of the use of deadly force with a firearm, including, but not limited to, the steps that should be taken before contemplating the use of, or using, deadly force with a firearm, possible alternatives to using deadly force with a firearm, and the law governing the use of deadly force with a firearm.

(2) The attorney general shall consult with and assist the commission in the preparation of the pamphlet described in division (B)(1) of this section and, as necessary, shall recommend to the commission changes in the pamphlet to reflect changes in the law that are relevant to it. The commission shall make copies of the pamphlet available to any person, public entity, or private entity that operates or teaches a training course, class, or program described in division (B)(3)(a), (b), (c), and (e) of section 2923. 125 of the Revised Code and requests copies for distribution to persons who take the course, class, or program, and to sheriffs for distribution to applicants under section 2923.125 of the Revised Code for a license to carry a concealed handgun and applicants under that section for the renewal of a license to carry a concealed handgun. ...

(E) As used in this section, "handgun" has the same meaning as in section 2923.11 of the Revised Code.

Title LV. Roads--Highways--Bridges

Chapter 5502. Public Safety Department; Criminal Justice Services

5502.63. Poster and brochure describing safe firearms practices The division of criminal justice services in the department of public safety shall prepare a poster and a brochure that describe safe firearms practices. The poster and brochure shall contain typeface that is at least one-quarter inch tall. The division shall furnish copies of the poster and brochure free of charge to each federally licensed firearms dealer in this state.

As used in this section, "federally licensed firearms dealer" means an importer, manufacturer, or dealer having a license to deal in destructive devices or their ammunition, issued and in effect pursuant to the federal "Gun Control Act of 1968," 82 Stat. 1213, 18 U.S.C. 923 et seq., and any amendments or additions to that act or reen-actments of that act.

Title XXIX. Crimes - Procedure

Chapter 2921. Offenses Against Justice and Public Administration

Perjury

2921.13. Falsification; in theft offense; to purchase firearm

(A) No person shall knowingly make a false statement, or knowingly swear or affirm the truth of a false statement previously made, when any of the following applies: ...

(12) The statement is made in connection with the purchase of a firearm, as defined in section 2923.11 of the Revised Code, and in conjunction with the furnishing to the seller of the firearm of a fictitious or altered driver's or commercial driver's license or permit, a fictitious or altered identification card, or any other document that contains false information about the purchaser's identity.

(15) The statement is made in an application filed with a county sheriff pursuant to section 2923.125 of the Revised Code in order to obtain or renew a license to carry a concealed handgun or is made in an affidavit submitted to a county sheriff to obtain a temporary emergency license to carry a concealed handgun under section 2923.1213 of the Revised Code.

(B) No person, in connection with the purchase of a firearm, as defined in section 2923.11 of the Revised Code, shall knowingly furnish to the seller of the firearm a fictitious or altered driver's or commercial driver's license or permit, a fictitious or altered identification card, or any other document that contains false information about the purchaser's identity.

(C) No person, in an attempt to obtain a license to carry a concealed handgun under section 2923.125 of the Revised Code, shall knowingly present to a sheriff a fictitious or altered document that purports to be certification of the person's competence in handling a handgun as described in division (B)(3) of section 2923.125 of the Revised Code. ...

(F)(3) Whoever violates division (A)(12) or (B) of this section is guilty of falsification to purchase a firearm, a felony of the fifth degree.

(4) Whoever violates division (A)(15) or (C) of this section is guilty of falsification to obtain a concealed handgun license, a felony of the fourth degree.

(G) A person who violates this section is liable in a civil action to any person harmed by the violation for injury, death, or loss to person or property incurred as a result of the commission of the offense and for reasonable attorney's fees, court costs, and other expenses incurred as a result of prosecuting the civil action commenced under this division. A civil action under this division is not the exclusive remedy of a person who incurs injury, death, or loss to person or property as a result of a violation of this section.

Chapter 2923. Conspiracy, Attempt, and Complicity; Weapons Control; Corrupt Activity

Weapons Control

2923.11. Definitions As used in sections 2923.11 to 2923.24 of the Revised Code:

(A) "Deadly weapon" means any instrument, device, or thing capable of inflicting death, and designed or specially adapted for use as a weapon, or possessed, carried, or used as a weapon.

(B)(1) "Firearm" means any deadly weapon capable of expelling or propelling one or more projectiles by the action of an explosive or combustible propellant. "Firearm" includes an unloaded firearm, and any firearm that is inoperable but that can readily be rendered operable.

(2) When determining whether a firearm is capable of expelling or propelling one or more projectiles by the action of an explosive or combustible propellant, the trier of fact may rely upon circumstantial evidence, including, but not limited to, the representations and actions of the individual exercising control over the firearm.

(C) "Handgun" means any of the following:

(1) Any firearm that has a short stock and is designed to be held and fired by the use of a single hand;

(2) Any combination of parts from which a firearm of a type described in division (C)(1) of this section can be assembled.

(D) "Semi-automatic firearm" means any firearm designed or specially adapted to fire a single cartridge and automatically chamber a succeeding cartridge ready to fire, with a single function of the trigger.

(E) "Automatic firearm" means any firearm designed or specially adapted to fire a succession of cartridges with a single function of the trigger. "Automatic firearm" also means any semi-automatic firearm designed or specially adapted to fire more than thirty-one cartridges without reloading, other than a firearm chambering only .22 caliber short, long, or long-rifle cartridges.

(F) "Sawed-off firearm" means a shotgun with a barrel less than eighteen inches long, or a rifle with a barrel less than sixteen inches long, or a shotgun or rifle less than twenty-six inches long overall.

(G) "Zip-gun" means any of the following:

(1) Any firearm of crude and extemporized manufacture;

(2) Any device, including without limitation a starter's pistol, that is not designed as a firearm, but that is specially adapted for use as a firearm;

(3) Any industrial tool, signaling device, or safety device, that is not designed as a firearm, but that as designed is capable of use as such, when possessed, carried, or used as a firearm.

(H) "Explosive device" means any device designed or specially adapted to cause physical harm to persons or property by means of an explosion, and consisting of an explosive substance or agency and a means to detonate it. "Explosive device" includes without limitation any bomb, any explosive demolition device, any blasting cap or detonator containing an explosive charge, and any pressure vessel that has been knowingly tampered with or arranged so as to explode.

(I) "Incendiary device" means any firebomb, and any device designed or specially adapted to cause physical harm to persons or property by means of fire, and consisting of an incendiary substance or agency and a means to ignite it.

(J) "Ballistic knife" means a knife with a detachable blade that is propelled by a spring-operated mechanism.

(K) "Dangerous ordnance" means any of the following, except as provided in division (L) of this section:

(1) Any automatic or sawed-off firearm, zip-gun, or ballistic knife;

(2) Any explosive device or incendiary device;

(3) Nitroglycerin, nitrocellulose, nitrostarch, PETN, cyclonite, TNT, picric acid, and other high explosives; amatol, tritonal, tetrytol, pentolite, pecretol, cyclotol, and other high explosive compositions; plastic explosives; dynamite, blasting gelatin, gelatin dynamite, sensitized ammonium nitrate, liquid-oxygen blasting explosives, blasting powder, and other blasting agents; and any other explosive substance having sufficient brisance or power to be particularly suitable for use as a military explosive, or for use in mining, quarrying, excavating, or demolitions;

(4) Any firearm, rocket launcher, mortar, artillery piece, grenade, mine, bomb, torpedo, or similar weapon, designed and manufactured for military purposes, and the ammunition for that weapon;

(5) Any firearm muffler or silencer;

(6) Any combination of parts that is intended by the owner for use in converting any firearm or other device into a dangerous ordnance.

(L) "Dangerous ordnance" does not include any of the following:

(1) Any firearm, including a military weapon and the ammunition for that weapon, and regardless of its actual age, that employs a percussion cap or other obsolete ignition system, or that is designed and safe for use only with black powder;

(2) Any pistol, rifle, or shotgun, designed or suitable for sporting purposes, including a military weapon as issued or as modified, and the ammunition for that weapon, unless the firearm is an automatic or sawed-off firearm;

(3) Any cannon or other artillery piece that, regardless of its actual age, is of a type in accepted use prior to 1887, has no mechanical, hydraulic, pneumatic, or other system for absorbing recoil and returning the tube into battery without displacing the carriage, and is designed and safe for use only with black powder;

(4) Black powder, priming quills, and percussion caps possessed and lawfully used to fire a cannon of a type defined in division (L)(3) of this section during displays, celebrations, organized matches or shoots, and target practice, and smokeless and black powder, primers, and percussion caps possessed and lawfully used as a propellant or ignition device in small-arms or small-arms ammunition;

(5) Dangerous ordnance that is inoperable or inert and cannot readily be rendered operable or activated, and that is kept as a trophy, souvenir, curio, or museum piece.

(6) Any device that is expressly excepted from the definition of a destructive device pursuant to the "Gun Control Act of 1968," 82 Stat. 1213, 18 U.S.C. 921(a)(4), as amended, and regulations issued under that act.

(M) "Explosive" means any chemical compound, mixture, or device, the primary or common purpose of which is to function by explosion. "Explosive" includes all materials that have been classified as Division 1.1, Division 1.2, Division 1.3, or Division 1.4 explosives by the United States department of transportation in its regulations and includes, but is not limited to, dynamite, black powder, pellet powders, initiating explosives, blasting caps, electric blasting caps, safety fuses, fuse igniters, squibs, cordeau detonant fuses, instantaneous fuses, and igniter cords and igniters. "Explosive" does not include "fireworks," as defined in section 3743.01 of the Revised code, or any substance or material otherwise meeting the definition of explosive set forth in this section that is manufactured, sold, possessed, transported, stored, or used in any activity described in section 3743.80 of the Revised Code, provided the activity is conducted in accordance with all applicable laws, rules, and regulations, including, but not limited to, the provisions of section 3743.80 of the Revised Code and the rules of the fire marshal adopted pursuant to section 3737.82 of the Revised Code.

2923.122. Illegal conveyance or possession of deadly weapon or dangerous ordnance or illegal possession of object indistinguishable from firearm in school safety zone

(A) No person shall knowingly convey, or attempt to convey, a deadly weapon or dangerous ordnance into a school safety zone.

(B) No person shall knowingly possess a deadly weapon or dangerous ordnance in a school safety zone.

(C) No person shall knowingly possess an object in a school safety zone if both of the following apply:

(1) The object is indistinguishable from a firearm, whether or not the object is capable of being fired.

(2) The person indicates that the person possesses the object and that it is a firearm, or the person knowingly displays or brandishes the object and indicates that it is a firearm.

(D)(1) This section does not apply to any of the following:

(a) An officer, agents, or employee of this or any other state or the United States, or a law enforcement officer, authorized to carry deadly weapons or dangerous ordnance and acting within the scope of their duties, a security officer employed by a board of education or governing body of a school during the time that the security officer is on duty pursuant to that contract of employment, or any other person who has written authorization from the board of education or governing body of a school to convey deadly weapons or dangerous ordnance into a school safety zone or to possess a deadly weapon or dangerous ordnance in a school safety zone and who conveys or possesses the deadly weapon or dangerous ordnance in accordance with that authorization;

(b) Any person who is employed in this state, who is authorized to carry deadly weapons or dangerous ordnance, and who is subject to and in compliance with the requirements of section 109.801 of the revised code, unless the appointing authority of the person has expressly specified that the exemption provided in division (d)(1)(b) of this section does not apply to the person.

(2) Division (C) of this section does not apply to premises upon which home schooling is conducted. Division (C) of this section also does not apply to a school administrator, teacher, or employee who possesses an object that is indistinguishable from a firearm for legitimate school purposes during the course of employment, a student who uses an object that is indistinguishable from a firearm under the direction of a school administrator, teacher, or employee, or any other person who with the express prior approval of a school administrator possesses an object that is indistinguishable from a firearm for a legitimate purpose, including the use of the object in a ceremonial activity, a play, reenactment, or other dramatic presentation, or a ROTC activity or another similar use of the object.

(3) This section does not apply to a person who conveys or attempts to convey a handgun into, or possesses a handgun in, a school safety zone if, at the time of that conveyance, attempted conveyance, or possession of the handgun, all of the following apply:

(a) The person does not enter into a school building or onto school premises and is not at a school activity.

(b) The person is carrying a valid license or temporary emergency license to carry a concealed handgun issued to the person under section 2923. 125 or 2923.1213 of the Revised Code or a license to carry a concealed handgun that was issued by another state with which the attorney general has entered into a reciprocity agreement under section 109.69 of the Revised Code.

(c) The person is in the school safety zone in accordance with 18 U.S.C. 922(q)(2)(B).

(d) The person is not knowingly in a place described in division (B)(1) or (B)(3) to (10) of section 2923.126 of the Revised Code.

(4) This section does not apply to a person who conveys or attempts to convey a handgun into, or possesses a handgun in, a school safety zone if at the time of that conveyance, attempted conveyance, or possession of the handgun all of the following apply:

(a) The person is carrying a valid license or temporary emergency license to carry a concealed handgun issued to the person under section 2923.125 or 2923.1213 of the Revised code or a license to carry a concealed handgun that was issued by another state with which the attorney general has entered into a reciprocity agreement under section 109.69 of the Revised code.

(b) The person is the driver or passenger in a motor vehicle and is in the school safety zone while immediately in the process of picking up or dropping off a child.

(c) The person is not in violation of section 2923.16 of the Revised Code. (E)(1) Whoever violates division (A) or (B) of this section is guilty of illegal conveyance or possession of a deadly weapon or dangerous ordnance in a school safety zone. Except as other-wise provided in this division, illegal conveyance or possession of a deadly weapon or dangerous ordnance in a school safety zone is a felony of the fifth degree. If the offender previously has been convicted of a violation of this section, illegal conveyance or possession of a deadly weapon or dangerous ordnance in a school safety zone is a felony of the fourth degree.

(2) Whoever violates division (C) of this section is guilty of illegal possession of an object indistinguishable from a firearm in a school safety zone. Except as otherwise provided in this division, illegal possession of an object indistinguishable from a firearm in a school safety zone is a misdemeanor of the first degree. If the offender previously has been convicted of a violation of this section, illegal possession of an object indistinguishable from a firearm in a school safety zone is a felony of the fifth degree.

(F)(1) In addition to any other penalty imposed upon a person who is convicted of or pleads guilty to a violation of this section and subject to division (F)(2) of this section, if the offender has not attained nineteen years of age, regardless of whether the offender is attending or is enrolled in a school operated by a board of education or for which the state board of education prescribes minimum standards under section 3301.07 of the Revised Code, the court shall impose upon the offender a class four suspension of the offender's probationary driver's license, restricted license, driver's license, commercial driver's license, temporary instruction permit, or probationary commercial driver's license that then is in effect from the range specified in division (A)(4) of section 4510.02 of the Revised Code and shall deny the offender the issuance of any permit or license of that type during the period of the suspension. If the offender is not a resident of this state, the court shall impose a class four suspension of the nonresident operating privilege of the offender from the range specified in division (A)(4) of section 4510.02 of the Revised Code.

(2) If the offender shows good cause why the court should not suspend one of the types of licenses, permits, or privileges specified in division (F)(1) of this section or deny the issuance of one of the temporary instruction permits specified in that division, the court in its discretion may choose not to impose the suspension, revocation, or denial required in that division.

(G) As used in this section, "object that is indistinguishable from a firearm" means an object made, constructed, or altered so that, to a reasonable person without specialized training in firearms, the object appears to be a firearm.

2923.13. Having weapons while under disability

(A) Unless relieved from disability as provided in section 2923.14 of the Revised Code, no person shall knowingly acquire, have, carry, or use any firearm or dangerous ordnance, if any of the following apply:

(1) The person is a fugitive from justice.

(2) The person is under indictment for or has been convicted of any felony offense of violence, or has been adjudicated a delinquent child for the commission of an offense that, if committed by an adult, would have been a felony offense of violence.

(3) The person is under indictment for or has been convicted of any offense involving the illegal possession, use, sale, administration, distribution, or trafficking in any drug of abuse or has been adjudicated a delinquent child for the commission of an offense that, if committed by an adult, would have been an offense involving the illegal possession, use, sale, administration, distribution, or trafficking in any drug of abuse.

(4) The person is drug dependent, in danger of drug dependence, or a chronic alcoholic.

(5) The person is under adjudication of mental incompetence, has been adjudicated as a mental defective, has been committed to a mental institution, has been found by a court to be a mentally ill person subject to hospitalization by court order, or is an involuntary patient other than one who is a patient only for purposes of observation. As used in this division, "mentally ill person subject to hospitalization by court order" and "patient" have the same meanings as in section 5122.01 of the Revised Code.

(B) Whoever violates this section is guilty of having weapons while under disability, a felony of the third degree…

2923.14. Relief from disability

(A) Any person who, solely by reason of the person's disability under division (A)(2) or (3) of section 2923.13 of the Revised Code, is prohibited from acquiring, having, carrying, or using firearms, may apply to the court of common pleas in the county in which the person resides for relief from such prohibition.

(B) The application shall recite the following:

(1) All indictments, convictions, or adjudications upon which the applicant's disability is based, the sentence imposed and served, and any release granted under a community control sanction, post-release control sanction, or parole, any partial or conditional pardon granted, or other disposition of each case;

(2) Facts showing the applicant to be a fit subject for relief under this section.

(C) A copy of the application shall be served on the county prosecutor. The county prosecutor shall cause the matter to be investigated and shall raise before the court any objections to granting relief that the investigation reveals.

(D) Upon hearing, the court may grant the applicant relief pursuant to this section, if all of the following apply:

(1) The applicant has been fully discharged from imprisonment, community control, post-release control, and parole, or, if the applicant is under indictment, has been released on bail or recognizance.

(2) The applicant has led a law-abiding life since discharge or release, and appears likely to continue to do so.

(3) The applicant is not otherwise prohibited by law from acquiring, having, or using firearms.

(E) Costs of the proceeding shall be charged as in other civil cases, and taxed to the applicant.

(F) Relief from disability granted pursuant to this section:

(1) Applies only with respect to indictments, convictions, or adjudications recited in the application;

(2) Applies only with respect to firearms lawfully acquired, possessed, carried, or used by the applicant;

(3) Does not apply with respect to dangerous ordnance;

(4) May be revoked by the court at any time for good cause shown and upon notice to the applicant;

(5) Is automatically void upon commission by the applicant of any offense set forth in division (A)(2) or (3) of section 2923.13 of the Revised Code, or upon the applicant's becoming one of the class of persons named in division (A)(1), (4), or (5) of that section.

(G) As used in this section:

(1) "Community control sanction" has the same meaning as in section 2929.01 of the Revised Code.

(2) "Post-release control" and "post-release control sanction" have the same meanings as in section 2967.01 of the Revised Code.

2923.15. Using weapons while intoxicated

(A) No person, while under the influence of alcohol or any drug of abuse, shall carry or use any firearm or dangerous ordnance.

(B) Whoever violates this section is guilty of using weapons while intoxicated, a misdemeanor of the first degree.

2923.16 Improper handling firearms in a motor vehicle.

(A) No person shall knowingly discharge a firearm while in or on a motor vehicle.

(B) No person shall knowingly transport or have a loaded firearm in a motor vehicle in such a manner that the firearm is accessible to the operator or any passenger without leaving the vehicle.

(C) No person shall knowingly transport or have a firearm in a motor vehicle, unless it is unloaded and is carried in one of the following ways:

(1) In a closed package, box, or case;

(2) In a compartment that can be reached only by leaving the vehicle;

(3) In plain sight and secured in a rack or holder made for the purpose;

(4) In plain sight with the action open or the weapon stripped, or, if the firearm is of a type on which the action will not stay open or which cannot easily be stripped, in plain sight.

(D) No person shall knowingly transport or have a loaded handgun in a motor vehicle if, at the time of that transportation or possession, any of the following applies:

(1) The person is under the influence of alcohol, a drug of abuse, or a combination of them.

(2) The person's whole blood, blood serum or plasma, breath, or urine contains a concentration of alcohol prohibited for persons operating a vehicle, as specified in division (A) of section 4511.19 of the Revised Code, regardless of whether the person at the time of the transportation or possession as described in this division is the operator of or a passenger in the motor vehicle.

(E) No person who has been issued a license or temporary emergency license to carry a concealed handgun under section 2923.125 or 2923.1213 of the Revised Code shall do any of the following:

(1) Knowingly transport or have a loaded handgun in a motor vehicle unless one of the following applies:

(a) The loaded handgun is in a holster on the person's person.

(b) The loaded handgun is in a closed case, bag, box, or other container that is in plain sight and that has a lid, a cover, or a closing mechanism with a zipper, snap, or buckle, which lid, cover, or closing mechanism must be opened for a person to gain access to the handgun.

(c) The loaded handgun is securely encased by being stored in a closed, locked glove compartment or in a case that is locked.

(2) If the person is transporting or has a loaded handgun in a motor vehicle in a manner authorized under division (E)(1) of this section, knowingly remove or attempt to remove the loaded handgun from the holster, case, bag, box, container, or glove compartment, knowingly grasp or hold the loaded handgun, or knowingly have contact with the loaded handgun by touching it with the person's hands or fingers while the motor vehicle is being operated on a street, highway, or public property unless the person removes, attempts to remove, grasps, holds, or has the contact with the loaded handgun pursuant to and in accordance with directions given by a law enforcement officer;

(3) If the person is the driver or an occupant of a motor vehicle that is stopped as a result of a traffic stop or a stop for another law enforcement purpose or is the driver or an occupant of a commercial motor vehicle that is stopped by an employee of the motor carrier enforcement unit for the purposes defined in section 5503.34 of the revised code, and if the person is transporting or has a loaded handgun in the motor vehicle or commercial motor vehicle in any manner, fail to do any of the following that is applicable:

(A) If the person is the driver or an occupant of a motor vehicle stopped as a result of a traffic stop or a stop for another law enforcement purpose, promptly inform any law enforcement officer who approaches the vehicle while stopped that the person has been issued a license or temporary emergency license to carry a concealed handgun and that the person then possesses or has a loaded handgun in the motor vehicle;

(b) If the person is the driver or an occupant of a commercial motor vehicle stopped by an employee of the motor carrier enforcement unit for any of the defined purposes, fail to promptly inform the employee of the unit who approaches the vehicle while stopped that the person has been issued a license or temporary emergency license to carry a concealed handgun and that the person then possesses or has a loaded handgun in the commercial motor vehicle.

(4) If the person is the driver or an occupant of a motor vehicle that is stopped as a result of a traffic stop or a stop for another law enforcement purpose and if the person is transporting or has a loaded handgun in the motor vehicle in any manner, knowingly fail to remain in the motor vehicle while stopped or knowingly fail to keep the person's hands in plain sight at any time after any law enforcement officer begins approaching the person while stopped and before the law enforcement officer leaves, unless, regarding a failure to remain in the motor vehicle or to keep the person's hands in plain sight, the failure is pursuant to and in accordance with directions given by a law enforcement officer;

(5) If the person is the driver or an occupant of a motor vehicle that is stopped as a result of a traffic stop or a stop for another law enforcement purpose, if the person is transporting or has a loaded handgun in the motor vehicle in a manner authorized under division (E)(1) of this section, and if the person is approached by any law enforcement officer while stopped, knowingly remove or attempt to remove the loaded handgun from the holster, case, bag, box, container, or glove compartment, knowingly grasp or hold the loaded handgun, or knowingly have contact with the loaded handgun by touching it with the person's hands or fingers in the motor vehicle at any time after the law enforcement officer begins approaching and before the law enforcement officer leaves, unless the person removes, attempts to remove, grasps, holds, or has contact with the loaded handgun pursuant to and in accordance with directions given by the law enforcement officer;

(6) If the person is the driver or an occupant of a motor vehicle that is stopped as a result of a traffic stop or a stop for another law enforcement purpose and if the person is transporting or has a loaded handgun in the motor vehicle in any manner, knowingly disregards or fails to comply with any lawful order of any law enforcement officer given while the motor vehicle is stopped, including, but not limited to, a specific order to the person to keep the person's hands in plain sight.

(f)(1) Divisions (a), (b), (c), and (e) of this section do not apply to officers any of the following:

(a) An officer agent , or employee of this or any other state or the United States, a law enforcement officer, when authorized to carry or have loaded or accessible firearms in motor vehicles and acting within the scope of the officer's, agent's, or employee's duties;

(b) any person who is employed in this state, who is authorized to carry or have loaded or accessible firearms in motor vehicles, and who is subject to and in compliance with the requirements of section 109.801 of the revised code, unless the appointing authority of the person has expressly specified that the exemption provided in division (f)(1)(b) of this section does not apply to the person ...

(3) Divisions (B) and (C) of this section do not apply to a person if all of the following circumstances apply:

(a) At the time of the alleged violation of either of those divisions, the person is the operator of or a passenger in a motor vehicle.

(b) The motor vehicle is on real property that is located in an unincorporated area of a township and that either is zoned for agriculture or is used for agriculture.

(c) The person owns the real property described in division (D)(3)(b) of this section, is the spouse or a child of another person who owns that real property, is a tenant of another person who owns that real property, or is the spouse or a child of a tenant of another person who owns that real property.

(d) The person, prior to arriving at the real property described in division (D)(3)(b) of this section, did not transport or possess a firearm in the motor vehicle in a manner prohibited by division (B) or (C) of this section while the motor vehicle was being operated on a street, highway, or other public or private property used by the public for vehicular traffic or parking.

(4) Divisions (B) and (C) of this section do not apply to a person who transports or possesses a handgun in a motor vehicle if, at the time of that transportation or possession, all of the following apply:

(a) The person transporting or possessing the handgun is carrying a valid license or temporary emergency license to carry a concealed handgun issued to the person under section 2923.125 or 2923.1213 of the Revised Code or a license to carry a concealed handgun that was issued by another state with which the attorney general has entered into a reciprocity agreement under section 109.69 of the Revised Code.

(b) The person transporting or possessing the handgun is not knowingly in a place described in division (B) of section 2923.126 of the Revised Code.

(c) Either one of the following applies:

(I) The handgun is in a holster on the person's person.

(II) The handgun is in a closed case, bag, box, or other container that is in plain sight and that has a lid, a cover, or a closing mechanism with a zipper, snap, or buckle, which lid, cover, or closing mechanism must be opened for a person to gain access to the handgun.

(III) The handgun is securely encased by being stored in a closed, locked glove compartment or in a case that is locked.

(G)(1) The affirmative defenses authorized in divisions (D)(1) and (2) of section 2923.12 of the Revised Code are affirmative defenses to a charge under division (B) or (C) of this section that involves a firearm other than a handgun.

(2) It is an affirmative defense to a charge under division (B) or (C) of this section of improperly handling firearms in a motor vehicle that the actor transported or had the firearm in the motor vehicle for any lawful purpose and while the motor vehicle was on the actor's own property, provided that this affirmative defense is not available unless the person, prior to arriving at the actor's own property, did not transport or possess the firearm in a motor vehicle in a manner prohibited by division (B) or (C) of this section while the motor vehicle was being operated on a street, highway, or other public or private property used by the public for vehicular traffic.

(H) No person who is charged with a violation of division (B), (C), or (D) of this section shall be required to obtain a license or temporary emergency license to carry a concealed handgun under section 2923.125 or 2923.1213 of the Revised Code as a condition for the dismissal of the charge.

(I) Whoever violates this section is guilty of improperly handling firearms in a motor vehicle. Violation of division (a) of this section is a felony of the fourth degree. Violation of division (c) of this section is a misdemeanor of the fourth degree. A violation of division (d) of this section is a felony of the fifth degree or, if the loaded handgun is concealed on the person's person, a felony of the fourth degree. A violation of division (e)(3) of this section is a misdemeanor of the first degree, and, in addition to any other penalty or sanction imposed for the violation, the offender's license or temporary emergency license to carry a concealed handgun shall be suspended pursuant to division (a)(2) of section 2923.128 of the revised code. A violation of division (e)(1), (2), or (5) of this section is a felony of the fifth degree. A violation of division (e)(4) or (6) of this section is a misdemeanor of the first degree or, if the offender previously has been convicted of or pleaded guilty to a violation of division (e)(4) or (6) of this section, a felony of the fifth degree. In addition to any other penalty or sanction imposed for a misdemeanor violation of division (e)(4) or (6) of this section, the offender's license or temporary emergency license to carry a concealed handgun shall be suspended pursuant to division (a)(2) of section 2923.128 of the revised code. A violation of division (b) of this section is whichever of the following is applicable:

(1) If, at the time of the transportation or possession in violation of division (B) of this section, the offender was carrying a valid license or temporary emergency license to carry a concealed handgun issued to the offender under section 2923.125 or 2923.1213 of the Revised Code or a license to carry a concealed handgun that was issued by another state with which the attorney general has entered into a reciprocity agreement under section 109.69 of the Revised Code and the offender was not knowingly in a place described in division (B) of section 2923.126 of the Revised Code, the violation is a misdemeanor of the first degree or, if the offender previously has been convicted of or pleaded guilty to a violation of division (B) of this section, a felony of the fourth degree.

(2) If division (I)(1) of this section does not apply, a felony of the fourth degree.

(J) If a law enforcement officer stops a motor vehicle for a traffic stop or any other purpose, if any person in the motor vehicle surrenders a firearm to the officer, either voluntarily or pursuant to a request or demand of the officer, and if the officer does not charge the person with a violation of this section or arrest the person for any offense, the person is not otherwise prohibited by law from possessing the firearm, and the firearm is not contraband, the officer shall return the firearm to the person at the termination of the stop.

(K) As used in this section:

(1) "Motor vehicle," "street," and "highway" have the same meanings as in section 4511.01 of the Revised Code.

(2) "Occupied structure" has the same meaning as in section 2909.01 of the Revised Code.

(3) "Agriculture" has the same meaning as in section 519.01 of the Revised Code.

(4) "Tenant" has the same meaning as in section 1531.01 of the Revised Code.

(5) "Unloaded" means, with respect to a firearm employing a percussion cap, flintlock, or other obsolete ignition system, when the weapon is uncapped or when the priming charge is removed from the pan.

(6) "Commercial motor vehicle" has the same meaning as in division (a) of section 4506.25 of the Revised Code.

(7) "Motor carrier enforcement unit" means the motor carrier enforcement unit in the department of public safety, division of state highway patrol, that is created by section 5503.34 of the Revised Code.

2923.17. Unlawful possession of dangerous ordnance; illegally manufacturing or processing explosives

(A) No person shall knowingly acquire have, carry, or use any dangerous ordnance.

(B) No person shall manufacture or process an explosive at any location in this state unless the person first has been issued a license certificate of registration, or permit to do so from a fire official of a political subdivision of this state or from the office of the fire marshal.

(C) Division (A) of this section does not apply to:

(1) Officers, agents, or employees of this or any other state or the United States, members of the armed forces of the United States or the organized militia of this or any other state, and law enforcement officers, to the extent that any such person is authorized to acquire, have, carry, or use dangerous ordnance and is acting within the scope of the person's duties;

(2) Importers, manufacturers, dealers, and users of explosives, having a license or user permit issued and in effect pursuant to the "Organized Crime Control Act of 1970," 84 Stat. 952, 18 U.S.C. 843, and any amendments or additions thereto or reenactments thereof, with respect to explosives and explosive devices lawfully acquired, possessed, carried, or used under the laws of this state and applicable federal law;

(3) Importers, manufacturers, and dealers having a license to deal in destructive devices or their ammunition, issued and in effect pursuant to the "Gun Control Act of 1968," 82 Stat. 1213, 18 U.S.C. 923, and any amendments or additions thereto or reenactments thereof, with respect to dangerous ordnance lawfully acquired, possessed, carried, or used under the laws of this state and applicable federal law;

(4) Persons to whom surplus ordnance has been sold, loaned, or given by the secretary of the army pursuant to 70A Stat. 262 and 263, 10 U.S.C. 4684, 4685, and 4686, and any amendments or additions thereto or reenactments thereof, with respect to dangerous ordnance when lawfully possessed and used for the purposes specified in such section;

(5) Owners of dangerous ordnance registered in the national firearms registration and transfer record pursuant to the act of October 22, 1968, 82 Stat. 1229, 26 U.S.C. 5841, and any amendments or additions thereto or reenactments thereof, and regulations issued thereunder.

(6) Carriers, warehousemen, and others engaged in the business of transporting or storing goods for hire, with respect to dangerous ordnance lawfully transported or stored in the usual course of their business and in compliance with the laws of this state and applicable federal law;

(7) The holders of a license or temporary permit issued and in effect pursuant to section 2923.18 of the Revised Code, with respect to dangerous ordnance lawfully acquired, possessed, carried, or used for the purposes and in the manner specified in such license or permit.

(D) Whoever violates division (A) of this section is guilty of unlawful possession of dangerous ordnance, a felony of the fifth degree.

(E) Whoever violates division (B) of this section is guilty of illegally manufacturing or processing explosives, a felony of the second degree.

2923.18. License or permit to possess dangerous ordnance

(A) Upon application to the sheriff of the county or safety director or police chief of the municipality where the applicant resides or has his principal place of business, and upon payment of the fee specified in division (B) of this section, a license or temporary permit shall be issued to qualified applicants to acquire, possess, carry, or use dangerous ordnance, for the following purposes:

(1) Contractors, wreckers, quarrymen, mine operators, and other persons regularly employing explosives in the course of a legitimate business, with respect to explosives and explosive devices acquired, possessed, carried, or used in the course of such business;

(2) Farmers, with respect to explosives and explosive devices acquired, possessed, carried, or used for agricultural purposes on lands farmed by them;

(3) Scientists, engineers, and instructors, with respect to dangerous ordnance acquired, possessed, carried, or used in the course of bona fide research or instruction;

(4) Financial institution and armored car company guards, with respect to automatic firearms lawfully acquired, possessed, carried, or used by any such person while acting within the scope of his duties;

(5) In the discretion of the issuing authority, any responsible person, with respect to dangerous ordnance lawfully acquired, possessed, carried, or used for a legitimate research, scientific, educational, industrial, or other proper purpose.

(B) Application for a license or temporary permit under this section shall be in writing under oath to the sheriff of the county or safety director or police chief of the municipality where the applicant resides or has his principal place of business. The application shall be accompanied by an application fee of fifty dollars when the application is for a license, and an application fee of five dollars when the application is for a temporary permit. The fees shall be paid into the general revenue fund of the county or municipality. The application shall contain the following information:

(1) The name, age, address, occupation, and business address of the applicant, if he is a natural person, or the name, address, and principal place of business of the applicant, if the applicant is a corporation;

(2) A description of the dangerous ordnance for which a permit is requested;

(3) A description of the place or places where and the manner in which the dangerous ordnance is to be kept, carried, and used;

(4) A statement of the purposes for which the dangerous ordnance is to be acquired, possessed, carried, or used;

(5) Such other information as the issuing authority may require in giving effect to this section.

(C) Upon investigation, the issuing authority shall issue a license or temporary permit only if all of the following apply:

(1) The applicant is not otherwise prohibited by law from acquiring, having, carrying or using dangerous ordnance;

(2) The applicant is age twenty-one or over, if he is a natural person;

(3) It appears that the applicant has sufficient competence to safely acquire, possess, carry, or use the dangerous ordnance, and that proper precautions will be taken to protect the security of the dangerous ordnance and ensure the safety of persons and property;

(4) It appears that the dangerous ordnance will be lawfully acquired, possessed, carried, and used by the applicant for a legitimate purpose.

(D) The license or temporary permit shall identify the person to whom it is issued, identify the dangerous ordnance involved and state the purposes for which the license or temporary permit is issued, state the expiration date, if any, and list such restrictions on the acquisition, possession, carriage, or use of the dangerous ordnance as the issuing authority considers advisable to protect the security of the dangerous ordnance and ensure the safety of persons and property.

(E) A temporary permit shall be issued for the casual use of explosives and explosive devices, and other consumable dangerous ordnance, and shall expire within thirty days of its issuance. A license shall be issued for the regular use of consumable dangerous ordnance, or for any nonconsumable dangerous ordnance, which license need not specify an expiration date, but the issuing authority may specify such expiration date, not earlier than one year from the date of issuance, as it considers advisable in view of the nature of the dangerous ordnance and the purposes for which the license is issued.

(F) The dangerous ordnance specified in a license or temporary permit may be obtained by the holder anywhere in the state. The holder of a license may use such dangerous ordnance anywhere in the state. The holder of a temporary permit may use such dangerous ordnance only within the territorial jurisdiction of the issuing authority.

(G) The issuing authority shall forward to the state fire marshal a copy of each license or temporary permit issued pursuant to this section, and a copy of each record of a transaction in dangerous ordnance and of each report of lost or stolen dangerous ordnance, given to the local law enforcement authority as required by divisions (A)(4) and (5) of section 2923.20 of the Revised Code. The state fire marshal shall keep a permanent file of all licenses and temporary permits issued pursuant to this section, and of all records of transactions in, and losses or thefts of dangerous ordnance forwarded by local law enforcement authorities pursuant to this section.

2923.19. Failure to secure dangerous ordnance

(A) No person, in acquiring, possessing, carrying, or using any dangerous ordnance, shall negligently fail to take proper precautions:

(1) To secure the dangerous ordnance against theft, or against its acquisition or use by any unauthorized or incompetent person;

(2) To insure the safety of persons and property.

(B) Whoever violates this section is guilty of failure to secure dangerous ordnance, a misdemeanor of the second degree.

2923.20. Unlawful transactions in weapons.

(A) No person shall:

(1) Recklessly sell, lend, give, or furnish any firearm to any person prohibited by section 2923.13 or 2923.15 of the Revised Code from acquiring or using any firearm, or recklessly sell, lend, give, or furnish any dangerous ordnance to any person prohibited by section 2923.13, 2923.15, or 2923.17 of the Revised Code from acquiring or using any dangerous ordnance;

(2) Possess any firearm or dangerous ordinance with purpose to dispose of it in violation of division (A) of this section;

(3) Manufacture, possess for sale, sell, or furnish to any person other than a law enforcement agency for authorized use in police work, any brass knuckles, cestus, billy, blackjack, sandbag, switchblade knife, springblade knife, gravity knife, or similar weapon;

(4) When transferring any dangerous ordinance to another, negligently fail to require the transferee to exhibit such identification, license, or permit showing him to be authorized to acquire dangerous ordnance pursuant to section 2923.17 of the Revised Code, or negligently fail to take a complete record of the transaction and forthwith forward a copy of that record to the sheriff of the county or safety director or police chief of the municipality where the transaction takes place;

(5) Knowingly fail to report to law enforcement authorities forthwith the loss or theft of any firearm or dangerous ordnance in the person's possession or under the person's control.

(B) Whoever violates this section is guilty of unlawful transactions in weapons. A violation of division (A)(1) or (2) of this section is a felony of the fourth degree. A violation of division (A)(3) or (4) of this section is a misdemeanor of the second degree. A violation of division (A)(5) of this section is a misdemeanor of the fourth degree.

2923.201. Defacing identification marks of firearm

(A) No person shall do either of the following:

(1) Change, alter, remove, or obliterate the name of the manufacturer, model, manufacturer's serial number, or other mark of identification on a firearm.

(2) Possess a firearm knowing or having reasonable cause to believe that the name of the manufacturer, model, manufacturer's serial number, or other mark of identification on the firearm has been changed, altered, removed, or obliterated.

(B) (1) Whoever violates division (A)(1) of this section is guilty of defacing identification marks of a firearm. Except as otherwise provided in this division, defacing identification marks of a firearm is a misdemeanor of the first degree. If the offender previously has been convicted of or pleaded guilty to a violation of division (A)(1) of this section, defacing identification marks of a firearm is a felony of the fourth degree.

(2) Whoever violates division (A)(2) of this section is guilty of possessing a defaced firearm. Except as otherwise provided in this division, possessing a defaced firearm is a misdemeanor of the first degree. If the offender previously has been convicted of or pleaded guilty to a violation of division (A)(2) of this section, possessing a defaced firearm is a felony of the fourth degree.

(C) Division (A) of this section does not apply to any firearm on which no manufacturer's serial number was inscribed at the time of its manufacture.

2923.21. Improperly furnishing firearms to a minor

(A) No person shall do any of the following:

(1) Sell any firearm to a person who is under eighteen years of age;

(2) Subject to division (B) of this section, sell any handgun to a person who is under twenty-one years of age;

(3) Furnish any firearm to a person who is under eighteen years of age or, subject to division (B) of this section, furnish any handgun to a person who is under twenty-one years of age, except for lawful hunting, sporting, or educational purposes, including, but not limited to, instruction in firearms or handgun safety, care, handling, or marksmanship under the supervision or control of a responsible adult;

(4) Sell or furnish a firearm to a person who is eighteen years of age or older if the seller or furnisher knows, or has reason to know, that the person is purchasing or receiving the firearm for the purpose of selling the firearm in violation of division (A)(1) of this section to a person who is under eighteen years of age or for the purpose of furnishing the firearm in violation of division (A)(3) of this section to a person who is under eighteen years of age;

(5) Sell or furnish a handgun to a person who is twenty-one years of age or older if the seller or furnisher knows, or has reason to know, that the person is purchasing or receiving the handgun for the purpose of selling the handgun in violation of division (A)(2) of this section to a person who is under twenty-one years of age or for the purpose of furnishing the handgun in violation of division (A)(3) of this section to a person who is under twenty-one years of age;

(6) Purchase or attempt to purchase any firearm with the intent to sell the firearm in violation of division (A)(1) of this section to a person who is under eighteen years of age or with the intent to furnish the firearm in violation of division (A)(3) of this section to a person who is under eighteen years of age;

(7) Purchase or attempt to purchase any handgun with the intent to sell the handgun in violation of division (A)(2) of this section to a person who is under twenty-one years of age or with the intent to furnish the handgun in violation of division (A)(3) of this section to a person who is under twenty-one years of age.

(B) Divisions (A)(1) and (2) of this section do not apply to the sale or furnishing of a handgun to a person eighteen years of age or older and under twenty-one years of age if the person eighteen years of age or older and under twenty-one years of age is a law enforcement officer who is properly appointed or employed as a law enforcement officer and has received firearms training approved by the Ohio peace officer training council or equivalent firearms training.

(C) Whoever violates this section is guilty of improperly furnishing firearms to a minor, a felony of the fifth degree.

2923.211. Underage purchase of firearm or handgun

(A) No person under eighteen years of age shall purchase or attempt to purchase a firearm.

(B) No person under twenty-one years of age shall purchase or attempt to purchase a handgun, provided that this division does not apply to the purchase or attempted purchase of a handgun by a person eighteen years of age or older and under twenty-one years of age if the person eighteen years of age or older and under twenty-one years of age is a law enforcement officer who is properly appointed or employed as a law enforcement officer and has received firearms training approved by the Ohio peace officer training council or equivalent firearms training.

(C) Whoever violates division (A) of this section is guilty of underage purchase of a firearm, a delinquent act that would be a felony of the fourth degree if it could be committed by an adult. Whoever violates division (B) of this section is guilty of underage purchase of a handgun, a misdemeanor of the second degree.

2923.22. Permitted interstate transactions in firearms

(A) Any resident of Ohio age eighteen or over, and not prohibited by section 2923.13 or 2923.15 of the Revised Code or any applicable law of another state or the Revised Code from acquiring or using firearms, may purchase or obtain a rifle, shotgun, or ammunition therefor in Indiana, Kentucky, Michigan, Pennsylvania, or West Virginia.

(B) Any resident of Indiana, Kentucky, Michigan, Pennsylvania, or West Virginia, age eighteen or over, and not prohibited by section 2923.13 or 2923.15 of the Revised Code or the laws of his domicile or the United States from acquiring or using firearms, may purchase or obtain a rifle, shotgun, or ammunition therefor in Ohio.

(C) Any purchase and sale pursuant to this section shall be for such purposes and under such circumstances and upon such conditions as are prescribed by the "Gun Control Act of 1968," 82 Stat. 1213, 18 U.S.C. 922 (b)(3), and any amendments or additions thereto or reenactments thereof.

2923.23. Immunity from prosecution

(A) No person who acquires, possesses, or carries a firearm or dangerous ordnance in violation of section 2923.13 or 2923.17 of the Revised Code shall be prosecuted for such violation, if he reports his possession of firearms or dangerous ordnance to any law enforcement authority, describes the firearms of [or] dangerous ordnance in his possession and where they may be found, and voluntarily surrenders the firearms or dangerous ordnance to the law enforcement authority. A surrender is not voluntary if it occurs when the person is taken into custody or during a pursuit or attempt to take the person into custody under circumstances indicating that the surrender is made under threat of force.

(B) No person in violation of section 2923.13 of the Revised Code solely by reason of his being under indictment shall be prosecuted for such violation if, within ten days after service of the indictment he voluntarily surrenders the firearms and dangerous ordnance in his possession to any law enforcement authority pursuant to division (A) of this section, for safekeeping pending disposition of the indictment or of an application for relief under section 2923.14 of the Revised Code.

(C) Evidence obtained from or by reason of an application or proceeding under section 2923.14 of the Revised Code for relief from disability, shall not be used in a prosecution of the applicant for any violation of section 2923.13 of the Revised Code.

(D) Evidence obtained from or by reason of an application under section 2923.18 of the Revised Code for a permit to possess dangerous ordnance, shall not be used in a prosecution of the applicant for any violation of section 2923.13 or 2923.17 of the Revised Code.

2923.25 Trigger lock or gun locking device to be offered for sale Each federally licensed firearms dealer who sells any firearm, at the time of the sale of the firearm, shall offer for sale to the purchaser of the firearm a trigger lock, gun lock, or gun locking device that is appropriate for that firearm. Each federally licensed firearms dealer shall post in a conspicuous location in the dealer's place of business the poster furnished to the dealer pursuant to section 5502.63 of the Revised Code and shall make available to all purchasers of firearms from the dealer the brochure furnished to the dealer pursuant to that section.

As used in this section, "federally licensed firearms dealer" has the same meaning as in section 5502.63 of the Revised Code.

[Current through legislation passed by the 127th Ohio General Assembly and filed with the Secretary of State through November 3, 2008]

City of Akron Code

Title 13. General Offenses
Chapter 137. Weapons Control

Weapons Control Generally

137.01. Definitions. For the purpose of this chapter the following definitions shall apply unless the context clearly indicates or requires a different meaning.

"Airgun." Any air pistol, air rifle, BB gun, pump gun, pellet gun, CO-2 gun, or similar instrument or device capable of discharging ammunition by means of air pressure or spring action.

"Ammunition." Any leaden or metallic projectile, any pellet or any other substance capable of inflicting injuries to persons or property when used in an airgun.

"Automatic firearm." Any firearm designed or specially adapted to fire a succession of cartridges with a single function of the trigger. "Automatic Firearm" also means any semi-automatic firearm designed or specially adapted to fire more than thirty-one cartridges without reloading, other than a firearm chambering only .22 caliber short, long or long-rifle cartridges. ...

"Dangerous ordnance."

A. Any of the following, except as provided in subsection B of this section:

1. Any automatic or sawed-off firearm, or zip-gun;

2. Any explosive device or incendiary device;

3. Nitroglycerin, nitrocellulose, nitrostarch, PETN, cyclonite, TNT, picric acid and other high explosives; amatol, tritonal, tetrytol, pentolite, pecretol, cyclotol and other high explosive compositions; plastic explosives; dynamite, blasting gelatin, gelatin dynamite, sensitized ammonium nitrate, liquid-oxygen blasting explosives, blasting powder and other blasting agents; and any other explosive substance having sufficient brisance or power to be particularly suitable for use as a military explosive, or for use in mining, quarrying, excavating or demolitions;

4. Any firearm, rocket launcher, mortar, artillery piece, grenade, mine, bomb, torpedo or similar weapon, designed and manufactured for military purposes, and the ammunition for that weapon.

5. Any firearm muffler or silencer;

6. Any combination of parts that is intended by the owner for use in converting any firearm or other device into a dangerous ordnance;

7. A Taser or any device which shoots a dart-like object charged with volts of electricity.

B. "Dangerous Ordnance" does not include any of the following:

1. Any firearm, including a military weapon and the ammunition for that weapon, and regardless of its actual age, which employs a percussion cap or other obsolete ignition system, or which is designed and safe for use only with black powder;

2. Any pistol, rifle or shotgun, designed or suitable for sporting purposes, including a military weapon as issued or as modified, and the ammunition for that weapon, unless the firearm is an automatic or sawed-off firearm;

3. Any cannon or other artillery piece which, regardless of its actual age, is of a type in accepted use prior to 1887, has no mechanical, hydraulic, pneumatic or other system for absorbing recoil and returning the tube into battery without displacing the carriage and is designed and safe for use only with black powder;

4. Black powder, priming quills, and percussion caps possessed and lawfully used to fire a cannon of a type defined in subsection (B)(3) of this section during displays, celebrations, organized matches or shoots and target practice, and smokeless and black powder, primers and percussion caps possessed and lawfully used as a propellant or ignition device in small-arms or small-arms ammunition;

5. Dangerous ordnance which is inoperable or inert and cannot readily be rendered operable or activated, and which is kept as a trophy, souvenir, curio, or museum piece;

6. Any device which is expressly excepted from the definition of a destructive device pursuant to the "Gun Control Act of 1968," 82 Stat. 1213, 18 U.S.C. 921 (A) (4), and regulations issued under that act.

"Deadly weapon." Any instrument, device or thing capable of inflicting death and designed or specially adapted for use as a weapon, or possessed, carried or used as a weapon.

"Explosive device." Any device designed or specially adapted to cause physical harm to persons or property by means of an explosion, and consisting of an explosive substance or agency and a means to detonate it. "Explosive Device" includes without limitation any bomb, any explosive demolition device, any blasting cap or detonator containing an explosive charge and any pressure vessel which has been knowingly tampered with or arranged so as to explode.

"Firearm."

1. Any deadly weapon capable of expelling or propelling one or more projectiles by the action of an explosive or combustible propellant. "Firearm" includes an unloaded firearm, and any firearm which is inoperable but which can readily be rendered operable.

2. When determining whether a firearm is capable of expelling or propelling one or more projectiles by the action of an explosive or combustible propellant, the trier of fact may rely upon circumstantial evidence, including, but not limited to, the representations and actions of the individual exercising control over the firearm.

"Handgun." Any of the following: (1) any firearm that has a short stock and is designed to be held and fired by the use of a single hand; (2) any combination of parts from which a firearm described in the preceding definition can be assembled.

"Incendiary device." Any firebomb and any device designed or specially adapted to cause physical harm to persons or property by means of fire and consisting of an incendiary substance or agency and a means to ignite it.

"Pistol." Any firearm with a barrel less than twelve inches in length.

"Sawed-off firearm." A shotgun with a barrel less than eighteen inches long, or a rifle with a barrel less than sixteen inches long, or a shotgun or rifle less than twenty-six inches long overall.

"Semi-automatic firearm." Any firearm designed or specially adapted to fire a single cartridge and automatically chamber a succeeding cartridge ready to fire, with a single function of the trigger.

"Zip-gun." Any of the following:

(1) Any firearm of crude and extemporized manufacture;

(2) Any device, including without limitation a starter's pistol, not designed as a firearm, but which is specially adapted for use as a firearm;

(3) Any industrial tool, signaling device or safety device, not designed as a firearm, but which as designed is capable of use as such, when possessed, carried or used as a firearm.

137.02. Carrying weapons - Concealed weapons.

A. Weapons generally. Except as otherwise permitted by law, no person shall carry on or about his person a pistol, a knife having a blade two and one-half inches in length or longer, knuckles, a billy or other dangerous weapon. "Proper justification" includes, but is not limited to, the right of law enforcement officers and other persons specifically authorized by law to be armed within the scope of his or her duties. This section does not apply to a person who is engaged in a lawful business or pursuit justifying possession of such an item and the person did not use or intend to use the item as a weapon. It shall be an affirmative defense to a violation of this section that the defendant was at the time engaged in a lawful business, calling, employment or occupation and the circumstances in which he was placed justified a prudent man in possessing such a weapon for the defense of his person, property or family.

B.1. Except as otherwise permitted by law, no person shall knowingly carry or have, concealed on his person or concealed ready at hand, any deadly weapon or dangerous ordnance.

2. This section does not apply to officers, agents, or employees of this or any other state or of the United States, or to law enforcement officers, authorized to carry concealed weapons or dangerous ordnances, and acting within the scope of their duties.

3. It is an affirmative defense to a charge under this section of carrying or having control of a weapon other than dangerous ordnance, that the actor was not otherwise prohibited by law from having the weapon, and that any of the following apply:

a. The weapon was carried or kept ready at hand by the actor for defensive purposes, while he was engaged in or was going to or from his lawful business or occupation, which business or occupation was of such character or was necessarily carried on in such manner or at such a time or place as to render the actor particularly susceptible to criminal attack, that would justify a prudent man in going armed.

b. The weapon was carried or kept ready at hand by the actor for defensive purposes, while he was engaged in a lawful activity, and had reasonable cause to fear a criminal attack upon himself or a member of his family, or upon his home, such as would justify a prudent man in going armed.

c. The weapon was carried or kept ready at hand by the actor for any lawful purpose and while in his own home.

d. The weapon was being transported in a motor vehicle for any lawful purpose, and was not on the actor's person, and, if the weapon was a firearm, was carried in compliance with the applicable requirements of Ohio Revised Code §2923.16.

C. Penalty.

1. Whoever violates subsection A of this section is guilty of a misdemeanor of the third degree.

2. Whoever violates subsection B of this section is guilty of carrying concealed weapons, a misdemeanor of the first degree. If the offender previously has been convicted of a violation of this section or a like offense or of any offense of violence, if the weapon involved is a firearm and the violation of this section is committed at premises for which a D permit has been issued under R.C. Chapter 4303, if the weapon involved is a firearm which is either loaded or for which the offender has ammunition ready at hand, or if the weapon involved is a dangerous ordnance, then carrying concealed weapons is a felony of the third degree and shall be prosecuted under appropriate state law. If the offense is committed aboard an aircraft, or with purpose to carry a concealed weapon aboard an aircraft, regardless of the weapon involved, carrying concealed weapons is a felony and shall be prosecuted under appropriate state law.

137.03. Using weapons while intoxicated.

A. No person, while under the influence of alcohol or any drug of abuse, shall carry or use any firearm or dangerous ordnance.

B. Whoever violates this section is guilty of using weapons while intoxicated, a misdemeanor of the first degree.

137.04. Improperly handling firearms in a motor vehicle.

A. For the purpose of this section, "unloaded" means, with respect to a firearm employing a percussion cap, flintlock, or other obsolete ignition system, that the weapon is uncapped, or that the priming charge is removed from the pan.

B. Except as otherwise permitted by law, no person shall knowingly discharge a firearm while in or on a motor vehicle.

C. Except as otherwise permitted by law, no person shall knowingly transport or have a loaded firearm in a motor vehicle, in such manner that the firearm is accessible to the operator or any passenger without leaving the vehicle.

D. Except as otherwise permitted by law, no person shall knowingly transport or have a firearm in a motor vehicle, unless it is unloaded, and is carried in one of the following ways:

1. In a closed package, box or case;

2. In a compartment which can be reached only by leaving the vehicle;

3. In plain sight and secured in a rack or holder made for the purpose;

4. In plain sight with the action open or the weapon stripped, or, if the firearm is of a type on which the action will not stay open or which cannot easily be stripped, in plain sight.

E. This section does not apply to officers, agents, or employees of this or any other state or of the United States, or to law enforcement officers, authorized to carry or have loaded or accessible firearms in motor vehicles, and acting within the scope of their duties.

F. The affirmative defenses contained in §137.02(B)(3)(a) and (b) are affirmative defenses to a charge under subsection C or D of this section.

G. Whoever violates this section is guilty of improperly handling firearms in a motor vehicle. Violation of subsection B or C of this section is a misdemeanor of the first degree. Violation of subsection D of this section is a misdemeanor of the fourth degree. Penalty, see §130.99.

137.05. Failure to secure dangerous ordnance.

A. No person, in acquiring, possessing, carrying, or using any dangerous ordnance, shall negligently fail to take proper precautions:

1. To secure the dangerous ordnance against theft, or against its acquisition or use by any unauthorized or incompetent person;

2. To insure the safety of persons and property.

B. Whoever violates this section is guilty of failure to secure dangerous ordnance, a misdemeanor of the second degree.

137.06. License to possess dangerous ordnance.

A. Upon application to the Safety Director or Police Chief, and upon payment of the fee specified in subsection B of this section, a license or temporary permit shall be issued to qualified applicants to acquire, possess, carry, or use dangerous ordnance, for the following purposes:

1. Contractors, wreckers, quarrymen, mine operators, and other persons regularly employing explosives in the course of a legitimate business, with respect to explosives and explosive devices acquired, possessed, carried, or used in the course of such business;

2. Farmers, with respect to explosives and explosive devices acquired, possessed, carried, or used for agricultural purposes on lands farmed by them;

3. Scientists, engineers, and instructors, with respect to dangerous ordnance acquired, possessed, carried, or used in the course of bona fide research or instruction;

4. Financial institution and armored car company guards, with respect to automatic firearms lawfully acquired, possessed, carried, or used by any such person while acting within the scope of his duties;

5. In the discretion of the Safety Director or Police Chief, any responsible person, with respect to dangerous ordnance lawfully acquired, possessed, carried, or used for a legitimate research, scientific, educational, industrial, or other proper purpose.

B. Application for a license or temporary permit under this section shall be in writing under oath to the Safety Director or Police Chief. The application shall be accompanied by an application fee of fifty dollars when the application is for a license, and an application fee of five dollars when the application is for a temporary permit. The fees shall be paid into the General Fund of the municipality. The application shall contain the following information:

1. The name, age, address, occupation, and business address of the applicant, if he is a natural person, or the name, address and principal place of business of the applicant, if the applicant is a corporation;

2. A description of the dangerous ordnance for which a permit is requested;

3. A description of the place or places where and the manner in which the dangerous ordnance is to be kept, carried, and used;

4. A statement of the purposes for which the dangerous ordnance is to be acquired, possessed, carried, or used;

5. Such other information as the Safety Director or Police Chief may require in giving effect to this section.

C. Upon investigation, the Safety Director or Police Chief shall issue a license or temporary permit only if all of the following apply;

1. The applicant is not otherwise prohibited by law from acquiring, having, carrying, or using dangerous ordnance;

2. The applicant is age twenty-one or over, if he is a natural person;

3. It appears that the applicant has sufficient competence to safely acquire, possess, carry, or use the dangerous ordnance, and that proper precautions will be taken to protect the security of the dangerous ordnance and insure the safety of persons and property;

4. It appears that the dangerous ordnance will be lawfully acquired, possessed, carried, and used by the applicant for a legitimate purpose.

D. The license or temporary permit shall identify the person to whom it is issued, identify the dangerous ordnance involved, and state the purposes for which the license or temporary permit is issued, state the expiration date, if any, and list such restrictions on the acquisition, possession, carriage, or use of the dangerous ordnance as the Safety Director or Police Chief considers advisable to protect the security of the dangerous ordnance and insure the safety of persons and property.

E. A temporary permit shall be issued for the casual use of explosives and explosive devices and other consumable dangerous ordnance, and shall expire within thirty days of its issuance. A license shall he issued for the regular use of consumable dangerous ordnance or for any nonconsumable dangerous ordnance, which license need not specify an expiration date, but the Safety Director or Police Chief may specify such expiration date, not earlier than one year from the date of issuance, as he considers advisable in view of the nature of the dangerous ordnance and the purposes for which the license is issued.

F. The dangerous ordnance specified in a license or temporary permit may be obtained by the holder anywhere in the state. The holder of a license may use such dangerous ordnance anywhere in the state. The holder of a temporary permit may use such dangerous ordnance only within the territorial jurisdiction of the municipality.

G. The Safety Director or Police Chief shall forward to the State Fire Marshal a copy of each license or temporary permit issued pursuant to this section, and a copy of each record of a transaction in dangerous ordnance and of each report of lost or stolen dangerous ordnance, given to the Police Department as required by R.C. §§2923.20(A)(4) and (5).

137.07. Immunity from prosecution.

A. No person who acquires, possesses, or carries a firearm or dangerous ordnance in violation of R.C. §§2923.13 or 2923.17 shall be prosecuted for such violation if he reports his possession of firearms or dangerous ordnance to any law enforcement authority, describes the firearms or dangerous ordnance in his possession and where they may be found, and voluntarily surrenders the firearms or dangerous ordnance to the law enforcement authority. A surrender is not voluntary if it occurs when the person is taken into custody or during a pursuit or attempt to take the person into custody, under circumstances indicating that the surrender is made under threat of force.

B. Evidence obtained from or by reason of an application under §137.06 for a permit to possess dangerous ordnance, shall not be used in a prosecution of the applicant for any violation of R.C. §2923.17.

137.10. Certain handguns prohibited.

A For purposes of this section, "handgun" means a firearm having a barrel and firing mechanism designed to eject or propel a projectile by the action of an explosive or combustible propellant, and having a barrel not exceeding 12 inches in length, measured by the insertion therein of a rod with the receiver or slide closed, but does not include inoperable firearms which cannot be rendered operable, or handguns designed and safe only for use with black powder.

(B) No person shall possess, sell, transfer, give, deliver, or furnish a handgun having a retail value of $50 or less and having a barrel, slide, frame, or receiver which is a die casting of zinc alloy or any other nonhomogeneous metal which will melt or deform at a temperature of less than 800 degrees Fahrenheit, either assembled or disassembled.

(C) Such handguns are hereby declared to be illegal contraband, which may be confiscated and destroyed by the Police Division.

(D) Whoever violates this section is guilty of a misdemeanor of the first degree.

137.11. Use and possession of airguns and ammunition by minors.

A. For the purpose of this section the following definitions shall apply unless the context clearly indicates or requires a different meaning.

"Airgun." Any air pistol, air rifle, BB gun, pump gun, pellet gun, CO-2 gun, or similar instrument or device capable of discharging ammunition by means of air pressure or spring action.

"Ammunition." Any leaden or metallic projectile, any pellet, or any other substance capable of inflicting injuries to persons or property when used in an airgun.

B. It shall be unlawful for any person under the age of eighteen to use or have in his possession any airgun or ammunition.

C. The provisions of this section do not prohibit or render it unlawful to use or possess any airgun or ammunition for purposes of instruction in firearm safety, care, handling, or marksmanship under the supervision or control of a responsible adult.

D. It shall be unlawful for the parent or guardian of any such person under the age of eighteen in his charge or custody to knowingly permit any such person under the age of eighteen to use or have in his possession any airgun or ammunition unless such use or possession falls within the exception of subsection C of this section.

E. Whoever violates this section shall be guilty of a misdemeanor in the third degree.

137.13. Furnishing airguns or similar instruments to a minor.

A. No individual, partnership, association, corporation, or any other entity shall sell, offer for sale, give away, distribute, or furnish any airgun or ammunition to any person under the age of eighteen years.

B. The provisions of this section do not prohibit or render it unlawful for the parent or guardian of any such person under the age of eighteen in his charge or custody to furnish such person under the age of eighteen any airgun or ammunition for purposes of instruction in firearm safety, care, handling, or marksmanship under the supervision or control of a responsible adult.

C. Whoever violates this section is guilty of a misdemeanor in the third degree.

137.15. Facsimile firearms.

A. Definitions.

1. "Firearm" shall have the same meaning as used in §137.01 of this chapter.

2. "Replica or facsimile of a firearm" shall mean any device or object made of plastic, wood, metal or any other material which is a replica, facsimile or toy version of, or is otherwise recognizable as, a pistol, revolver, shotgun, sawed-off shotgun, rifle, machine, rocket launcher or any other firearm. As used in this section, "replica or facsimile of a firearm" shall include, but is not limited to, toy guns, movie props, hobby models (either in kit form or fully assembled), starter pistols, air guns, inoperative firearms or any other device which might reasonably be perceived to be a real firearm.

B. No person shall display, market for sale or sell any replica or facsimile of a firearm in the city. The provisions of this subsection shall not apply to any replica or facsimile firearm which, because of its distinct color, exaggerated size, or other design feature, cannot reasonably be perceived to be a real firearm.

C. Except in self-defense, no person shall draw, exhibit or brandish a replica or facsimile of a firearm or simulate a firearm in a rude, angry or threatening manner, with the intent to frighten, vex, harass or annoy or with the intent to commit an act which is a crime under the laws of the city, state or federal government against any other person.

D. No person shall draw, exhibit or brandish a replica or facsimile of a firearm or simulate a firearm in the presence of a law enforcement officer, fire fighter, emergency medical technician or paramedic engaged in the performance of his or her duties, when the person committing such brandishing knows or has reason to know that such law enforcement officer, fire fighter, emergency medical technician or paramedic is engaged in the performance of his or her duties.

E. Penalties.

1. Whoever violates §137.15(B) is guilty of unlawful sale of a replica firearm, a misdemeanor of the third degree.

2. Whoever violates §137.15(C) is guilty of brandishing a replica firearm, a misdemeanor of the first degree.

3. Whoever violates §137.15(D) is guilty of brandishing a replica firearm in the presence of a public safety officer, a misdemeanor of the first degree.

F. If any provision of this section is held to be invalid by a court of competent jurisdiction, such invalidity shall not affect the validity or enforceability of any other provision.

Sale of Weapons and Firearms

137.20. Definitions. For the purpose of this article the following definitions shall apply unless the context clearly indicates or requires a different meaning.

"Firearm." Any deadly weapon capable of expelling or propelling one or more projectiles by the action of an explosive or combustible propellant. "Firearm" includes an unloaded firearm, and any firearm which is inoperable but which can readily be rendered operable.

"Fugitive from justice." Any person who has fled or is fleeing from any law enforcement officer to avoid prosecution or incarceration for a felony or to avoid giving testimony in any criminal proceeding.

"Manufacturer" or "dealer." Any person engaged in the business of manufacturing, repairing, or selling firearms at wholesale or retail.

137.21. Possession restricted.

A. Except as permitted by law, no person who has been convicted of a felony in any court of the United States, the several states, territories, possessions, commonwealth countries or the District of Columbia, or who is a fugitive from justice or of unsound mind or a drug addict or an habitual drunkard, shall possess a firearm within the city.

B. No person under the age of eighteen years shall possess a pistol. The provisions of this division shall not apply to the issue of pistols to members of the armed forces of the United States, active or reserve, state militia or ROTC, when on duty or training, or to the temporary loan of pistols for instruction under the immediate supervision of a parent, guardian or adult instructor.

C. No person shall sell, lease, lend, or otherwise transfer a pistol to any person under eighteen years of age except as provided in subsection B.

D. Whoever violates this section is guilty of a misdemeanor of the first degree.

137.22. Removal of serial numbers.

A. No person shall receive, possess, sell, lease, or otherwise transfer any pistol from which the manufacturer's serial numbers or letters have been removed.

B. Whoever violates this section is guilty of a misdemeanor of the third degree.

137.23. Manufacturers' and dealers' licenses.

A. Application for a manufacturer's or dealer's license shall be made to the City License Clerk on forms prescribed and furnished by the city. The application shall indicate the business name of the applicant, the location where the business is to be conducted, the name and home address of the proprietor (proprietors, if a partnership) or the president and secretary, if a corporation, and such other pertinent data as may be required, and shall be accompanied by a license fee of $50 for each business location payable to the City License Clerk. This fee which is paid at the time of application is not refundable in any case. The license shall expire annually on October 31.

B. Licenses issued under this chapter shall not be transferable to any other person and the business may be conducted only at the location for which the license is issued.

137.24. Sales, transfers, and displays.

A. No manufacturer or dealer, except a manufacturer or dealer having a license issued, under the provisions of this chapter, shall sell any firearm at wholesale or retail.

B. No person shall sell, lease, lend, or otherwise transfer a firearm to any person who he knows or has reasonable cause to believe has been convicted of a felony, or is a fugitive from justice or of unsound mind or a drug addict or an habitual drunkard.

C. When delivered, all pistols must be securely wrapped and must be unloaded.

D. No dealer shall display any pistol in any part of the premises where it can readily be seen from the outside.

E. Whoever violates this section is guilty of a misdemeanor of the third degree.

137.241. Trigger lock requirement for handguns.

A. No manufacturer or dealer shall sell a handgun at retail unless at the time of the sale the dealer or manufacturer also sells to the purchaser of the handgun a trigger lock or other child-safety lock that is appropriate for the handgun or unless a trigger lock is an integral component of the handgun.

B. No person shall purchase a handgun at retail from a manufacturer or dealer unless at the time of the sale the person also purchases from the dealer a trigger lock or other child-safety lock that is appropriate for the handgun or unless a trigger lock is an integral component of the handgun.

C. It is an affirmative defense to a charge under this section that at the time of a sale of a handgun at retail the purchaser of the handgun demonstrated to the dealer who sold the handgun that the purchaser possessed a trigger lock or other child-safety lock that is appropriate for the handgun that was the subject of the sale.

D.1. Whoever violates Division (A) of this section is guilty of selling a handgun without a trigger lock or other child safety lock, a misdemeanor of the fourth degree.

2. Whoever violates subsection B of this section is guilty of purchasing a handgun without a trigger lock or child-safety lock, a misdemeanor of the fourth degree.

137.25. Manufacturers' and dealers' records.

A. Each licensed manufacturer or dealer shall maintain at his place of business, complete and adequate records of all firearm disposed of in the course of his business, whether manufactured by himself or acquired from other manufacturers or dealers. The records shall show and include:

1. The number of firearms of each type, together with a full and adequate description thereof, including the serial numbers if such weapons are numbered;

2. The name and address of each person from whom the firearms, if not the manufacturer's own product, were acquired and the date of acquisition; and

3. The disposition made of the firearms, including the name and principal address of each transferee, the address to which they are delivered and the date of disposition.

B. Each licensed dealer shall maintain similarly complete and adequate records at each store or place where firearms are sold, of all firearms acquired and disposed of in the course of his business at such store or place.

C. In addition to the records required by this section, every licensee under the provisions of this chapter shall at the time of every purchase, enter with typewritten or printed letters, in ink, on a blank form to be furnished by the Police Division, such information as may be called for by such blank form. The information shall be printed on the card by the licensee or his employee, except that the seller shall sign his name where required on the card. No entry on the card shall be erased, obliterated, altered, or defaced.

(1) Every licensee under the provisions of this chapter shall mail by noon of the following weekday to the Chief of Police or his authorized representative, to the address the Chief of Police or his authorized representative shall designate, the blanks furnished by the Police Division properly filled in and signed by the seller, in accordance with the provisions of this section, for all transactions of the preceding business day.

2. However, any licensee licensed under the provisions of this chapter for the first time must, for the first forty-five days he conducts business with the city, hand-deliver such blanks to the Chief of Police or his authorized representative, properly filled in and signed by the seller, on every weekday before the hours of 12:00 noon for all transactions of the preceding business day.

D. Whoever violates this section is guilty of a misdemeanor of the third degree.

137.26. False Information.

A. No person in applying for a license as a manufacturer or dealer shall give false information or offer false evidence of his identity.

B. No person shall give false information concerning the matters referred to in §137.21(A) or concerning his age or his name and address, or offer false evidence of his identity when purchasing a pistol.

C. Whoever violates this section is guilty of a misdemeanor of the third degree.

137.27. Obtaining weapons by theft or fraud.

A. No person shall procure or attempt to procure any firearm, regardless of dollar value, by theft, fraud, violence, or threat of violence.

B. Whoever violates this section is guilty of a misdemeanor of the third degree.

137.28. Unlawful transactions.

A. No person shall:

1. Manufacture, possess for sale, sell, or furnish to any person other than a law enforcement agency for authorized use in police work, any brass knuckles, cestus, billy, blackjack, sandbag, switchblade knife, springblade knife, gravity knife, or similar weapon;

2. When transferring any dangerous ordnance to another, negligently fail to require the transferee to exhibit such identification, license, or permit showing him to be authorized to acquire dangerous ordnance pursuant to R.C. §2923.17, or negligently fail to take a complete record of the transaction and forthwith forward a copy of the record to the sheriff of the county or safety director or police chief of the municipality where the transaction takes place;

3. Knowingly fail to report to law enforcement authorities forthwith the loss or theft of any firearm or dangerous ordnance in the person's possession and under his control.

B. Whoever violates this section is guilty of unlawful transactions in weapons. Violation of subsection (A)(1) or (2) of this section is a misdemeanor of the second degree. Violation of subsection (A)(3) of this section is a misdemeanor of the fourth degree.

137.291. Access to firearms by minors.

A. For the purposes of this section the term "minor" shall mean any person under the age of eighteen:

B. Except as provided in subsection C of this section, no person, including but not limited to a parent or legal guardian, shall store or leave a loaded or unloaded firearm in any place where the person knows, or reasonably should know based on the totality of the circumstances, that a minor is able to gain access to it.

C. Subsection B of this section shall not apply when:

1. A minor's access to a firearm is under the supervision or control of a responsible adult for purposes of lawful hunting or instruction in firearms safety, care, handling or marksmanship;

2. A minor has access to a firearm as a result of an unlawful entry into the place in which the firearm was found;

3. A minor obtains a firearm in a lawful act of self-defense or defense of another person or persons within a domicile;

4. A minor who, without permission of the lawful possessor of a firearm, obtains the firearm from the possessor's body.

5. A firearm is in a locked gun cabinet or similar locked location, or is secured with a trigger lock or other similar device which prevents the firearm from discharging ammunition.

D. Any person who violates subsection B of this section shall be guilty of a misdemeanor of the second degree.

137.30 Sale of explosives to minors.

A. No person shall sell, give away, or otherwise dispose of or deliver to any person under eighteen years of age any explosives, as defined in R.C. §3743.01(A), whether such person is acting for himself or for any other person.

B. Whoever violates this section is guilty of a misdemeanor of the third degree. Penalty, see §130.99.

[City of Akron Code current through Ord. 477-2008, passed October 27, 2008]

Cincinnati Municipal Code

Title VII. General Regulations

Chapter 708. Dangerous Weapons and Firearms

708-1. Definitions. For purposes of this chapter, the following words in Section 708-1-A - 708-1-E shall have the following meanings.

708-1-A. Deadly Weapon. "Deadly weapon" shall mean any instrument, device, or thing capable of inflicting death, and designed or specifically adapted for use as a weapon, or possessed, carried, or used as a weapon.

708-1-B. Firearm. "Firearm" shall mean any deadly weapon capable of expelling or propelling one or more projectiles by the action of an explosive or combustible propellant. "Firearm" includes an unloaded firearm, and any firearm which is inoperable but which can readily be rendered operable.

708-1-C. Handgun. "Handgun" shall mean any firearm designed to be fired while being held in one hand, or any combination of parts from which it is possible to assemble a handgun.

708-1-D. Dangerous Weapon. "Dangerous Weapon" shall mean any instrument, device or thing capable of inflicting death or serious physical injury and designed or specifically adapted for use as a weapon, including an air gun, BB gun, musket, rifle, shotgun, revolver, pistol or ammunition therefor, bowie knife, dirk or other device passed, carried or used as a weapon.

708-1-E. Person. "Person" shall mean any individual, organization, company or corporation.

708-1-F. Retail Dealer. "Retail dealer" means any person engaged in selling, exchanging, bartering or keeping a thing with the intention of selling, bartering, or exchanging directly it to or with a consumer.

708-3. Prevention of Distribution of Weapons and Firearms to Minors.

(A) No person shall knowingly and/or recklessly:

(1) Sell, barter, lend, provide or give to a minor under the age of 18 years a deadly weapon, dangerous weapon, firearm, or handgun.

(2) Sell, barter, lend, provide or give to a person under the age of 21 years a handgun.

(3) Furnish any firearm to a person under the age of 18 years, except for purposes of lawful hunting, or for purposes of instruction in firearms safety, care, handling, or marksmanship under the supervision or control of a parent or legal guardian.

(B) No parent or legal guardian of a minor:

(1) Who knows such minor possesses or has been provided a deadly weapon, dangerous weapon, firearm, or handgun shall fail to remove the deadly weapon, dangerous weapon, firearm, or handgun from the minor's possession or control; or if the deadly weapon, dangerous weapon, firearm, or handgun cannot be removed without jeopardizing the safety of the parent or legal guardian fail to notify the Cincinnati Police Division that the minor possesses the deadly weapon, dangerous weapon, firearm, or handgun.

(C) No person shall:

(1) Store, control or possess any firearm within or upon any premises of which that person has an ownership interest, control or custody of, in such a manner that the person knows or should know that a minor is likely to gain possession of the firearm, and in fact, the minor does obtain possession of the firearm.

(2) This provision shall not apply:

(a) where the minor obtains the firearm as a result of an illegal entry onto the premises; or

(b) where the firearm is kept in a locked container, access to which is not available to a minor; or

(c) the firearm is equipped with, and has in place, a locking device, the means of removal of which is not available to a minor; or

(d) the minor obtains the firearm in a lawful act of self-defense, or defense of a minor's home or property.

(D) It shall be unlawful for the owner to possess a firearm without obtaining liability insurance for the negligent use of the firearm by a minor or the negligent entrustment of the firearm to a minor and covering all potential victims of incidents committed by a minor in which the firearm is used. Proof of such liability insurance shall be required after an incident of negligent or accidental discharge in which the firearm was used by a minor has been committed.

(E) No minor under the age of 18 years shall:

(1) Buy, barter, hire, borrow, receive or possess a deadly weapon, dangerous weapon, firearm, or handgun except as otherwise provided at subsection (A)(3).

(F) For purposes of Section 708-3, the following words shall have the following meanings.

(1) "Deadly Weapon", "Dangerous Weapon", "Firearm", and "Handgun" shall have the same meaning as defined in Cincinnati Municipal Code §§709-1-A, 708-1-B, 708-1-C and 708-1-D.

(2) "Locked Container" shall mean a secure container which is enclosed on all sides and locked by a padlock, key lock, combination lock or similar device.

(3) "Locking Device" shall mean a trigger lock or similar device, which, when applied to a firearm, temporarily prevents the firearm from functioning.

(G) Whoever violates any provision of this chapter shall commit the offense of illegal distribution of firearms to a minor and shall be guilty of a misdemeanor of the first degree.

708-5. Display Advertising of Weapons Prohibited. No person shall exhibit for sale in show cases or show windows any dangerous weapon or firearm or any accessory therefor, or handgun ammunition, or display any sign, poster, carton or display card which suggests the sale of any dangerous weapon or firearm or accessory therefor, or handgun ammunition. Nothing herein shall apply to firearms having a barrel of 12 inches or more in length in establish-ments licensed to sell same.

708-7. Obliterating Identification Marks Prohibited. No person shall change, alter, remove or obliterate the name of the maker, model, manufacturer's number or other mark of identification on any dangerous weapon or firearm.

No person shall keep in such person's possession or under such person's control any dangerous weapon or firearm upon which the name of the maker, model, manufacturer's number or other mark of identification has been to such person's knowledge changed, altered, removed or obliterated.

708-9. Dealers in Weapons; License Fees. No person shall carry on the business of retail dealer in firearms, ammunition for firearms, or dangerous weapons without first obtaining a license therefor for a fee of $457 per year; and such license may be granted by the city manager upon written application of any such retail dealer. For a retail dealer only in shotguns and rifles of regulation size only, manufactured for hunting and sporting purposes only, and ammunition therefor, the annual license fee shall be the sum of $24. The application shall set forth the place of business of the applicant, the nature of the business, the names of all the partners, if a partnership, and the names of all officers, if a corporation. The city manager may authorize the city treasurer to issue a license to the applicant if the city manager is satisfied that the applicant is of good repute in the conduct of such business and that the conduct of such business will not be detrimental to the public safety and welfare.

The city manager may suspend or, after hearing, revoke a dealer's license whenever the city manager finds that the holder thereof has failed to comply with any of the provisions of this chapter. Upon suspending or revoking any such license, the city manager shall require the holder thereof to surrender immediately all licenses issued to such holder and the holder shall surrender promptly all such licenses required. Whenever the city manager suspends a dealer's license, the holder shall be notified immediately and afforded a hearing if no hearing has already been afforded and one is desired. After such hearing, the city manager shall either rescind the order of suspension or upon good cause appearing therefor shall continue the suspension or revoke the license.

708-11. Restriction on Pawnbrokers. A dealer licensed under the provisions of Section 708-9 who carries on the business of pawnbroker or secondhand dealer shall not sell any firearms or dangerous weapons except shotguns and rifles of regulation size manufactured for hunting and sporting purposes only.

708-13. Form of Application to Transfer Firearms. The city manager shall prescribe a form of application covering the purchase or transfer of firearms. Such form shall provide for the recording of the following facts:

(a) The name, address, and date of birth appearing on a state driver's license or state personal identification card of the transferee that is valid on its face and contains a photograph of the transferee, and a description of the identification card used;

(b) If the transferee is not personally known to the transferor, the name and address of the person by whom the transferee is identified;

(c) The purpose for which the transferee desires to use the firearm;

(d) An accurate description of the firearm, including the name of the manufacturer, model, caliber, and serial number;

(e) A description of the transferee, which shall include height, weight, age, color of hair, color of eyes, all visible marks and scars on face and hands, any deformity, nationality, and race.

There shall be printed on the face or the back of the application such extracts from the city ordinances and from the laws of the state of Ohio, including penalty provisions, as the city manager may deem desirable.

The city manager shall furnish sets of such application blanks through the chief of police.

708-15. Restriction on Transfer of Firearms. No person shall transfer ownership of a firearm to any transferee unless such person has first verified the identity of the transferee by personally reviewing the transferee's valid driver's license and has required the transferee to fill out in full and sign an application blank as provided for in Section 708-13. No person shall transfer such firearm to any transferee with whom the transferor is not personally acquainted unless such transferee is identified by a person who is personally known to the transferor.

The transferor shall mail the original of each application to the chief of police. One copy shall be delivered to the transferee and one preserved by the transferor. The transferor's copies of used application blanks, and all unused application blanks, shall be kept by the transferor so as to be open for inspection by city police officers at all times.

The provisions of this section and of Sections 708-13 and 708-33 shall not apply to sales at wholesale nor to sales of firearms which have a barrel over 12 inches in length.

708-17. Sale of Handgun Ammunition. No person shall sell handgun ammunition to any purchaser with whom the seller is not personally acquainted unless such purchaser is identified by a person with whom the seller is personally acquainted.

708-19. False Statements in Application. No person shall give any false information or make any false statements in any application or identification document or in any record or report provided for in Section 708-9 or 708-13.

708-21. Gunpowder Magazines and Vendor's License Fees. All vendors of gun-powder shall pay a license fee of $122 per year. All keepers or owners of gunpowder magazines shall pay a license fee of $696 per year.

708-29. Confiscation of Firearms. When any police officer shall discover any person in the act of violating any provision of Section 708-27 or 708-33, it shall be his duty to seize the firearms or cannon so unlawfully used, and to cause the arrest of such person. Upon conviction of the person so arrested, the chief of police shall or-der the firearms or cannon so seized destroyed.

708-31. Ammunition, Sale to Minor. No person shall sell to a minor any cap or other explosive device used in the discharge of toy pistols and torpedo canes, or any blank cartridge, whether for use in handguns, revolvers or cannons.

708-33. Waiting Period for Transfer of Firearms. Any person desiring to purchase, rent, lease or transfer ownership of a firearm from another person shall make application on forms as provided for in Section 708-13, to be distributed by the chief of police, which shall be signed by the applicant and a copy forwarded by certified mail to the chief of police by the transferor.

The chief of police shall cause a postal card to be mailed to the transferor on the same day the application is received in his office stating the time and date the application was received.

Unless the chief of police to whom the application is sent determines that a transaction would violate federal, state, or municipal code, the application shall be destroyed within 60 days after the date the application was received by the police division.

No person shall knowingly sell, deliver, rent, lease or transfer ownership and/or possession of a firearm to any person who is not otherwise prohibited by law from acquiring or possessing a firearm, until 15 consecutive days have elapsed from the time the application is received by the chief of police.

This section does not apply to the sale, delivery or transfer of a firearm, in accordance with applicable law or rule, to any law enforcement officer or to any officer, agent, or employee of this or any other state or the United States if the acquisition or possession of a firearm is within the scope of his official duties.

Whoever violates any part of this section is guilty of a misdemeanor of the first degree.

708-35. Retail Sales of Firearms, Location Restrictions.

(a) No person shall sell firearms, guns, ammunition, hunting knives or other dangerous weapons at any retail establishment within 1,000 feet of the boundary of a parcel of real estate upon which a school is located.

As used herein the word "school" means an accredited public or private educational institution for one or more of the grades kindergarten through the twelfth grade.

(b) The Safety Director may issue a permit for the sale of sporting firearms and ammunition therefor by an otherwise eligible permanent retail establishment contrary to (a), above. Permits will be good for a seventy-two (72) hour period and no more than one permit per month will be issued to any particular retail establishment. Such a permit may be issued only upon a determination by the Safety Director that all of the following requirements have been met:

(1) the sporting firearms displayed and offered for sale pursuant to this section shall be strictly limited to shotguns and rifles primarily suitable for hunting or target shooting purposes;

(2) sale of ammunition shall be strictly limited to that ammunition suitable only for those sporting firearms authorized for display or sale pursuant to (b)(1) above;

(3) no more than 20 sporting firearms may be kept, displayed or offered for sale on the premises at any one time nor shall any such firearms be actually sold at a price of less than two thousand dollars ($2,000.00) each;

(4) such sporting firearms shall not be exposed to public view from outside the permit premises and must be kept in a separately secured area on the premises continuously monitored by a security system approved by the Safety Director; and

(5) ammunition shall be secured on the premises separately from any firearms.

Compliance with the foregoing requirements shall be monitored by the Safety Director through periodic, on-site inspections. Said permit shall be revoked for noncompliance.

708-37. Possession or Sale of Semiautomatic Firearms Prohibited.

(a) No person shall sell, deliver, rent, lease, offer, or display for sale, or transfer ownership of, acquire or possess a semiautomatic firearm.

(b) No person shall sell, deliver, rent, lease, offer or display for sale, or transfer ownership of, acquire or possess any detachable magazine with the following capacities:

(1) More than ten rifle or carbine rounds;

(2) More than five shotgun rounds;

(3) More than 15 handgun rounds.

This section does not apply to any detachable magazine purchased or acquired for use with semiautomatic firearms registered pursuant to division (e)(3) of this section.

(c) No person shall sell, deliver, rent, lease, offer or display for sale, or transfer ownership of, acquire or possess any extension tube, drum, cylinder, or magazine capable of increasing the capacity of a fixed or detachable magazine.

(d) This section does not apply to law enforcement officers, agents, or employees of this or any other state or the United States, members of the Armed Forces of the United States or the organized militia of this or any other state, if the acquisition or possession of a semiautomatic firearm is authorized and within the scope of his official duties.

(e) Any person who is legally in possession of a semiautomatic firearm that was legally purchased or acquired before April 1, 1989, shall within 30 days from the effective date hereof:

(1) Remove said semiautomatic firearm from within the city limits of Cincinnati; or

(2) Modify said semiautomatic firearm to either render it permanently inoperable or to permanently make it a device no longer defined as a semiautomatic firearm; or

(3) Register the firearm with the chief of police pursuant to such procedures as the city manager may establish. The registration shall contain a description of the firearm including all identification marks, the full name, address, date of birth and such other information as the city manager may deem appropriate. The information will be provided on forms prescribed by the city manager. Any change of address of the owner must be registered with the chief of police within 14 days of the change.

The city manager may establish a fee for registration not to exceed the actual processing costs of the city.

(f) No semiautomatic firearm possessed pursuant to division (e)(3) above may be sold or transferred after the effective date of this section to anyone within the city of Cincinnati.

(g) Any person who obtains title to a semi-automatic firearm by bequest or intestate succession must within 90 days comply with the requirements of division (e) of this section.

(h) The term "semiautomatic" means any firearm designed or specially adapted to fire a single cartridge and automatically chamber a succeeding cartridge and ready to fire, with a single function of the trigger.

(i) For the purpose of this section, "semiautomatic firearm" shall have the following meanings:

(1) Any semiautomatic rifle or carbine that was originally designed with or has a fixed magazine or detachable magazine with a capacity of more than ten rounds;

(2) Any semiautomatic shotgun that was originally designed with or has a fixed magazine or detachable magazine with a capacity of more than five rounds;

(3) Any semiautomatic handgun that was originally designed with or has a fixed magazine or detachable magazine with a capacity of more than 15 rounds;

(4) Any semiautomatic handgun that is a modification of a rifle, carbine, or shotgun as described in (1) and (2) above;

(5) Any firearm which may be readily restored to an operable semiautomatic firearm, as defined in (1) through (4) above;

(6) Any part, or combination of parts, designed or intended to convert a weapon into a semiautomatic firearm, as defined in (1) through (4) above or any combination of parts from which a semiautomatic firearm may be readily assembled.

(j) Rifles or carbines which fire .22 caliber rimfire ammunition and have fixed, tubular magazines are exempt from this section.

Any person who violates any provision of this section guilty of a misdemeanor of the first degree.

708-39. Dangerous Weapons on School Property.

(a) No person shall knowingly possess, carry or use a deadly or dangerous weapon, dangerous ordnance or ammunition in a school building, on any school property or other facility used for school purposes or a school-sponsored activity.

(b) This section does not apply to officers, agents, or employees of this or any other state or the United States, or to law enforcement officers, authorized to carry firearms and acting within the scope of their duties.

(c) Whoever violates this section is guilty of having a weapon while on school property, a misdemeanor of the first degree.

(d) As used herein the word "school" shall mean an accredited public or private educational institution for one or more of the grades kindergarten through the twelfth grade.

708-99. Penalties. Violations of the provisions of this chapter shall be punishable as set forth in the sections hereunder and other applicable penalty sections of this code.

708-99-A. Violation of Section 708-3, 708-5, 708-7, 708-9, 708-11, 708-15, 708-17, 708-19, 708-27, 708-33, or 708-35. Whoever violates any provision of Section 708-3, 708-5, 708-7, 708-9, 708-11, 708-15, 708-17, 708-19, 708-27, 708-33, or 708-35 shall be fined not more than $1,000, or imprisoned not more than six months, or both.

708-99-B. Violation of Section 708-31. Whoever violates any provision of Section 708-31 shall be fined not more than $250, or imprisoned not more than 30 days, or both.

[Cincinnati Municipal Code current through Ord. 268-2008 effective Sept. 6, 2008]

Codified Ordinances of Cleveland

Part Six. Offenses and Business Activities Code

Title I. General Offenses

Chapter 627. Weapons and Explosives

627.01. Definitions. As used in this chapter:

(a) "Deadly weapon" means any instrument, device or thing capable of inflicting death, and designed or specially adapted for use as a weapon, or possessed, carried or used as a weapon.

(b)(1) "Firearm" means any deadly weapon capable of expelling or propelling one or more projectiles by the action of an explosive or combustible propellant or by pump action or by compressed gas. "Firearm" includes an unloaded firearm, and any firearm which is inoperable but which can readily be rendered operable.

(2) When determining whether a firearm is capable of expelling or propelling one or more projectiles by the action of an explosive or combustible propellant, the trier of fact may rely upon circumstantial evidence, including, but not limited to, the representations and actions of the individual exercising control over the firearm.

(c) "Handgun" means any firearm designed to be fired while being held in one hand.

(d) "Semi-automatic firearm" means any firearm designed or specially adapted to fire a single cartridge and automatically chamber a succeeding cartridge ready to fire, with a single function of the trigger.

(e) "Automatic firearm" means any firearm designed or specially adapted to fire a succession of cartridges with a single function of the trigger. "Automatic firearm" also means any semi-automatic firearm designed or specially adapted to fire more than thirty-one cartridges without reloading, other than a firearm chambering only .22 caliber short, long or long-rifle cartridges.

(f) "Sawed-off firearm" means a shotgun with a barrel less than eighteen inches long, or a rifle with a barrel less than sixteen inches long, or a shotgun or rifle less than twenty-six inches long overall.

(g) "Zip-gun" means any of the following:

(1) Any firearm of crude and extemporized manufacture;

(2) Any device, including without limitation a starter's pistol, not designed as a firearm, but which is specially adapted for use as a firearm;

(3) Any industrial tool, signaling device or safety device not designed as a firearm, but which as designed is capable of use as such, when possessed, carried or used as a firearm.

(h) "Explosive device" means any device designed or specially adapted to cause physical harm to persons or property by means of an explosion, and consisting of an explosive substance or agency and a means to detonate it.

"Explosive device" includes without limitation any bomb, any explosive demolition device, any blasting cap or detonator containing an explosive charge, and any pressure vessel which has been knowingly tampered with or arranged so as to explode.

(i) "Incendiary device" means any firebomb, and any device designed or specially adapted to cause physical harm to persons or property by means of fire, and consisting of an incendiary substance or agency and a means to ignite it.

(j) "Ballistic knife" means a knife with a detachable blade that is propelled by a spring-operated mechanism.

(k) "Dangerous ordnance" means any of the following, except as provided in division (l) of this section:

(1) Any automatic or sawed-off firearm, zip-gun or ballistic knife;

(2) Any explosive device or incendiary device;

(3) Nitroglycerin, nitrocellulose, nitrostarch, PETN, cyclonite, TNT, picric acid and other high explosives; amatol, tritonal, tetrytol, pentolite, pecretol, cyclotol, and other high explosive compositions; plastic explosives; dynamite, blasting gelatin, gelatin dynamite, sensitized ammonium nitrate, liquid-oxygen blasting explosives, blasting powder and other blasting agents; and any other explosive substance having sufficient brisance or power to be particularly suitable for use as a military explosive, or for use in mining, quarrying, excavating or demolitions;

(4) Any firearm, rocket launcher, mortar, artillery piece, grenade, mine, bomb, torpedo or similar weapon, designed and manufactured for military purposes, and the ammunition for that weapon;

(5) Any firearm muffler or silencer;

(6) Any combination of parts that is intended by the owner for use in converting any firearm or other device into a dangerous ordnance.

(l) "Dangerous ordnance" does not include any of the following:

(1) Any firearm, including a military weapon and the ammunition for that weapon, and regardless of its actual age, which employs a percussion cap or other obsolete ignition system, or which is designed and safe for use only with black powder;

(2) Any pistol, rifle or shotgun, designed or suitable for sporting purposes, including a military weapon as issued or as modified, and the ammunition for that weapon, unless such firearm is an automatic or sawed-off firearm;

(3) Any cannon or other artillery piece which, regardless of its actual age, is of a type in accepted use prior to 1887, has no mechanical, hydraulic, pneumatic or other system for absorbing recoil and returning the tube into battery without displacing the carriage, and is designed and safe for use only with black powder;

(4) Black powder, priming quills and percussion caps possessed and lawfully used to fire a cannon of a type defined in division (l)(3) of this section during displays, celebrations, organized matches or shoots, and target practice, and smokeless and black powder, primers and percussion caps possessed and lawfully used as a propellant or ignition device in small-arms or small-arms ammunition;

(5) Dangerous ordnance which is inoperable or inert and cannot readily be rendered operable or activated, and which is kept as a trophy, souvenir, curio or museum piece.

(6) Any device which is expressly excepted from the definition of a destructive device pursuant to the ``Gun Control Act of 1968," 82 Stat. 1213, 18 U.S.C. 921(a)(4), and any amendments or additions thereto or reenactments thereof, and regulations issued thereunder.

627.03. Using Weapons While Intoxicated.

(a) No person, while under the influence of alcohol or any drug of abuse, shall carry or use any firearm or dangerous ordnance.

(b) Notwithstanding the provisions of Sections 601.13 and 601.99(a), whoever violates this section is guilty of using weapons while intoxicated, and shall be fined not less than three hundred dollars ($300.00) nor more than one thousand dollars ($1,000), and imprisoned for not less than three days, nor more than six months. No part of this sentence shall, in any case whatsoever, be suspended or otherwise reduced.

627.04 Improperly Handling Firearms in a Motor Vehicle.

(a) No person shall knowingly discharge a firearm while in or on a motor vehicle.

(b) No person shall knowingly transport or have a loaded firearm in a motor vehicle, in such manner that the firearm is accessible to the operator or any passenger without leaving the vehicle.

(c) No person shall knowingly transport or have a firearm in a motor vehicle, unless it is unloaded, and is carried in one of the following ways:

(1) In a closed package, box or case;

(2) In a compartment which can be reached only by leaving the vehicle;

(3) In plain sight and secured in a rack or holder made for the purpose;

(4) In plain sight with the action open or the weapon stripped, or, if the firearm is of a type on which the action will not stay open or which cannot easily be stripped, in plain sight.

(d) This section does not apply to officers, agents or employees of this or any other state or the United States, or to law enforcement officers authorized to carry or have loaded or accessible firearms in motor vehicles, and acting within the scope of their duties.

(e) The affirmative defenses contained in Sec-tion 627.02(c)(1) and (2) are affirmative defenses to a charge under subsection (b) or (c) hereof.

(f) Notwithstanding the provisions of Sections 601.13 or 601.99(a), whoever violates this section is guilty of improperly handling firearms in a motor vehicle, and shall be fined not less than three hundred dollars ($300.00), nor more than one thousand dollars ($1,000), and imprisoned for not less than three days, nor more than six months. No part of this sentence shall, in any case whatsoever, be suspended or otherwise reduced.

(g) As used in this section, "unloaded" means, with respect to a firearm employing a percussion cap, flintlock or other obsolete ignition system, when the weapon is uncapped, or when the priming charge is removed from the pan.

627.05. Failure to Secure Dangerous Ordnance.

(a) No person, in acquiring, possessing, carrying or using any dangerous ordnance shall negligently fail to take proper precautions:

(1) To secure the dangerous ordnance against theft or against its acquisition or use by any unauthorized or incompetent person;

(2) To insure the safety of persons and property.

(b) Whoever violates this section is guilty of failure to secure dangerous ordnance, a misdemeanor of the second degree.

627.08. Possession of Firearms by Minors.

(a) No minor shall purchase, own, possess, receive, have on or about his person or use any firearm except pursuant to Section 627.07(a)(3).

(b) A juvenile who violates this section shall be adjudged an unruly child, with such disposition of the case as may be appropriate under RC Chapter 2151.

627.081 Prohibited Weapon Defined

For purposes of Section 627.082, a "prohibited weapon" is any weapon defined, described, or listed in any division of Section 627.01 other than division (a) or (1) of that section, division (a) or (b) of Section 627.06, division (a) of Section 627.14, division (a) or (b) of Section 627.15, division (a) of Section 627.17, or division (a) of Section 627.18.

(Ord. No. 1130-92. Passed 5-11-92, eff. 5-15-92)

627.082 Prohibited Weapons on School Property; Duty to Notify Police

(a) Subject to division (b) of this section, any school official or employee who discovers a prohibited weapon, as defined in Section 627.081, upon school grounds or in a school building shall immediately notify the Chief of Police or his designee and request the assistance of the Division of Police.

....**(b)** Division (a) of this section shall not apply:

(1) To any school official or employee who has personal knowledge that the notification required by division (a) has already been given with respect to a particular discovery of a particular prohibited weapon; or

(2) When the prohibited weapon is in the possession of an officer, agent or employee of this or any other state or the United States, members of the armed forces of the United States or the organized militia of this or any other state, or law enforcement officers, as defined in division (k) of Section 601.01, to the extent that any such person is authorized to possess a prohibited weapon and is acting within the scope of his or her duties.

(c) Any school official or employee who fails to comply with division (a) of this section is guilty of a misdemeanor of the first degree.

627.09. Possessing Deadly Weapons on Public Property.

(a) No person shall knowingly have in his possession or ready at hand any deadly weapon while on public property or in a public building.

(b) For the purpose of this section, public property and public buildings shall include, but not be limited to parks, playgrounds, beaches, marinas, courthouses, auditoriums, stadiums, office buildings, jails, storage areas and yards, greenhouses, plants and works and any other property, building or structure owned, leased or rented by a governmental unit, to schools, colleges, and other learning institutions, whether public, private or parochial, and to churches, synagogues and other places of worship.

(c) This section does not apply to officers, agents or employees of this or any other state or the United States, to law enforcement officers authorized to carry or possess deadly weapons or to persons with private or special police commissions, and acting within the scope of their duties, or if the deadly weapon was part of a public weapons display, show or exhibition or was in the possession of a person participating in an organized match, competition or practice session on public property, or in a public building.

(d) Notwithstanding the provisions of Sections 601.13 and 601.99(a), whoever violates this section is guilty of possessing deadly weapons on public property, and shall be fined not less than three hundred dollars ($300.00), nor more than one thousand dollars ($1,000), and imprisoned for not less than three days, nor more than six months. No part of this sentence shall, in any case whatsoever, be suspended or otherwise reduced.

627.11. Seizure and Confiscation of Deadly Weapons.

(a) In any situation where a deadly weapon is present and a person has been drinking or disturbing the peace, threatening bodily harm or causing or threatening a disturbance or violence, and there is reasonable cause for the investigating police officer to believe that such deadly weapon may be used to cause bodily harm, such deadly weapon may be seized by the police and kept in the custody of the Chief of Police until released by an order of a court of competent jurisdiction.

(b) Any deadly weapon seized by a police officer upon the arrest of any person, firm or corporation charged with a violation of any of the provisions of this chapter, or any felony or misdemeanor involving the use of a deadly weapon or the use of force or violence or the threat of the use of force or violence against the person of another, shall be confiscated by the Division of Police for disposal. However, any deadly weapon seized which has been reported stolen shall be returned to the owner thereof, unless possession by the owner would constitute a violation of any provision of these Codified Ordinances or of State or Federal law.

627.12 Voluntary Disposition of Handguns

(a) Any handgun, declared to be illegal under the provisions of Section 627.06(a)(4) may be disposed of by presenting the handgun by the person owning or possessing the gun, at any district police station in the City of Cleveland, at the Central Police Station or the Detective Bureau of the Division of Police.

(b) No person disposing of a handgun in the manner and at the places herein designated, shall be required to make any written or oral statement or report concerning the handgun or the circumstances surrounding its acquisition, possession or present or past ownership.
Receipt can be made available upon request.

627.16. Unlawful Display of Weapons.

(a) No person, firm or corporation shall exhibit for sale in showcases or show windows any revolvers, daggers, stilettos, brass or iron knuckles and billies, or display any signs, posters, cartoons or display cards, suggesting the sale of such weapons.

(b) Whoever violates this section is guilty of unlawful display of weapons, a misdemeanor of the first degree.

627.18 Tear Gas Guns

(a) No person not being a law enforcement officer acting in line of duty or a person engaged in repelling robbers, thieves, murderers or other law violators in the defense and protection of his home or place of business, shall aim and discharge at any person a weapon or device of any kind which impels by compressed air, spring release or other means a projectile containing any liquid or gas which is dangerous to the safety or health of such person, or which otherwise discharges any such liquid or gas upon the person of another.

627.23. Facsimile Firearms.

(a)(1) "Firearm" shall have the same meaning as used in Section 627.01(b) of this Chapter.

(2) "Replica or facsimile of a firearm" shall mean any device or object made of plastic, wood, metal or any other material which is a replica, facsimile or toy version of, or is otherwise recognizable as, a pistol, revolver, shotgun, sawed-off shotgun, rifle, machine gun, rocket launcher or any other firearm. As used in this section, "replica or facsimile of a firearm" shall include, but is not limited to, toy guns, movie props, hobby models (either in kit form or fully assembled), starter pistols, air guns, inoperative firearms or any other device which might reasonably be perceived to be a real firearm.

(b) No person shall display, market for sale or sell any replica or facsimile of a firearm in the City. The provisions of this subsection shall not apply to any replica or facsimile firearm which, because of its distinct color, exaggerated size, or other design feature, cannot reasonably be perceived to be a real firearm.

(c) Except in self-defense, no person shall draw, exhibit or brandish a replica or facsimile of a firearm or simulate a firearm in a rude, angry or threatening manner, with the intent to frighten, vex, harass or annoy or with the intent to commit an act which is a crime under the laws of the City, State or Federal government against any other person.

(d) No person shall draw, exhibit or brandish a replica or facsimile of a firearm or simulate a firearm in the presence of a law enforcement officer, fire fighter, emergency medical technician or paramedic engaged in the performance of his or her duties, when the person committing such brandishing knows or has reason to know that such law enforcement officer, fire fighter, emergency medical technician or paramedic is engaged in the performance of his or her duties.

(e)(1) Whoever violates Section 627.23(b) is guilty of unlawful sale of a replica firearm, a misdemeanor of the third degree.

(2) Whoever violates Section 627.23(c) is guilty of brandishing a replica firearm, a misdemeanor of the first degree.

(3) Whoever violates Section 627.23(d) is guilty of brandishing a replica firearm in the presence of a public safety officer, a misdemeanor of the first degree.

Chapter 627A. Possession of Firearms by Children

627A.01.Definitions. For purposes of this chapter:

(a) "Firearms" means any deadly weapon capable of expelling or propelling one or more projectiles by the action of an explosive or combustible propellant. "Firearm" includes an unloaded firearm, and any firearm that is inoperable but that can readily be rendered operable.

(b) "Ammunition" means any ammunition cartridge, shell or other device containing explosive or incendiary material and designed and intended for use in any firearm.

(c) "Child" means any person under the age of eighteen (18) years, and includes any person between the ages of eighteen (18) and twenty-one (21) years who is of sufficient mental incompetence as to have had a legal guardian appointed by the Probate Court.

627A.02. Access to Firearms.

(a) Except as provided in division (b) of this section, no person, including but not limited to a parent or legal guardian, shall store or leave a loaded or unloaded firearm in any place where the person knows, or reasonably should know based on the totality of the circumstances, that a child is able to gain access to it.

(b) Division (a) of this section shall not apply when:

(1) A child's access to a firearm is under the supervision or control of a responsible adult for purposes of lawful hunting or instruction in firearms safety, care, handling, or marksmanship.

(2) A child has access to a firearm as a result of an unlawful entry into the place in which the firearm was found;

(3) A child obtains a firearm in a lawful act of self-defense or defense of another person or persons within a domicile;

(4) A child who, without permission of the lawful possessor of a firearm, obtains the firearm from the possessor's body.

627A.03. Responsibility of Firearms Dealer.

(a) When selling any firearm, a licensed firearms dealer shall offer to sell or give the purchaser a trigger lock or similar device which prevents the firearm from discharging.

(b) At every purchase counter in every store, shop or sales outlet, licensed firearms dealers shall conspicuously post the following warning in block letters not less than one inch (1") in height: "IT IS UNLAWFUL TO STORE OR LEAVE A FIREARM WHERE CHILDREN CAN OBTAIN ACCESS." The same warning shall be distributed to each firearm purchaser at the time of the sale of a firearm.

627A.99. Penalties.

(a) Whoever violates the provisions of Section 627A.02 shall be guilty of a misdemeanor of the first degree.

(b) Whoever violates the provisions of Section 627A.03 shall be guilty of a misdemeanor of the third degree.

Chapter 628. Possession or Sale of Assault Weapons Prohibited

628.02. Definitions. For the purpose of this chapter:

(a) "Assault weapon" means:

(1) any semiautomatic action, center fire rifle or carbine that accepts a detachable magazine with a capacity of 20 rounds or more;

(2) any semiautomatic shotgun with a magazine capacity of more than six rounds;

(3) any semi-automatic handgun that is:

A. a modification of a rifle described in division (a) (1), or a modification of an automatic firearm; or

B. originally designed to accept a detachable magazine with a capacity of more than 20 rounds.

(4) any firearm which may be restored to an operable assault weapon as defined in divisions (a) (1), (a) (2) or (a) (3).

(5) any part, or combination of parts, designed or intended to convert a firearm into an assault weapon as defined in divisions (a) (1), (a) (2) or (a) (3), or any combination of parts from which an assault weapon as defined in divisions (a) (1), (a) (2) or (a) (3), may be readily assembled if those parts are in the possession or under the control of the same person.

(b) Assault weapon does not include any of the following:

(1) any firearm that uses .22 caliber rimfire ammunition with a detachable magazine with a capacity of 30 rounds or less.

(2) any assault weapon which has been modified to either render it permanently inoperable or to permanently make it a device no longer defined as an assault weapon.

(c) "Automatic firearm" means any firearm designed or specially adapted to fire a succession of cartridges with a single function of the trigger.

(d) "Deadly weapon" means any instrument, device or thing capable of inflicting death, and designed or specially adapted for use as a weapon, or possessed, carried or used as a weapon.

(e) "Firearm" means any deadly weapon capable of expelling or propelling one or more projectiles by the action of an explosive or combustible propellant. "Firearm" includes an unloaded firearm, and any firearm which is inoperable but which can readily be rendered operable.

(f) "Handgun" means any firearm designed to be fired while being held in one hand.

(g) "Person" means any individual, corporation, company, association, firm, partnership, club or society, including wholesale and retail gun dealers.

(h) "Rifle" means a weapon designed or redesigned, made or remade, and intended to be fired from the shoulder and designed or redesigned and made or remade to use the energy of the explosive in a fixed metallic cartridge to fire only a single projectile through a rifled bore for each single pull of the trigger.

(i) "Semiautomatic firearm" means any firearm designed or specially adapted to fire a single cartridge and automatically chamber a succeeding cartridge ready to fire, with a single function of the trigger.

(j) "Shotgun" means a firearm, whether or not it is intended to be fired from the shoulder, that is designed or redesigned, made or remade, to fire a fixed shotgun shell.

628.03. Unlawful Conduct.

(a) No person shall sell, offer or display for sale, give, lend or transfer ownership of, acquire or possess any assault weapon.

(b) This section shall not apply to any officer, agent, or employee of this or any other state or the United States, members of the armed forces of the United States or the organized militia of this or any other state, and law enforcement officers as defined in division (k) of Section 601.01, to the extent that any such person is authorized to acquire or possess an assault weapon and is acting within the scope of his duties. Further, this section shall not apply to the transportation of firearms through the City of Cleveland in accordance with federal law.

628.04. Seizure and Destruction of Assault Weapons. Any assault weapon is hereby declared to be contraband and shall be seized and disposed of in accordance with RC 2933.43.

628.99. Penalty. Whoever violates Section 628.03 is guilty of unlawful possession of an assault weapon, and shall be sentenced to six months imprisonment and fined one thousand dollars ($1,000.00) and no part of said sentence shall be reduced in any manner.

Title VII. Business Regulation

Chapter 674. Handgun Possession and Sale

674.01. Definitions. As used in this chapter:

(a) "Handgun" means any pistol, revolver or other firearm, having a barrel not exceeding twelve inches in length, measured by the insertion of a rod with the receiver or slide closed, or the barrel, receiver or any part of the firing mechanism of such weapon, which is designed to eject or propel a projectile by the action of an explosive or combustible propellant, but does not include inoperable handguns which cannot be rendered operable, curios, relics or antique handguns as presently or hereafter defined in Title 27, Part 178 of the Code of Federal Regulations, Commerce in Firearms.

(b) "Dealer" means any person, firm or corporation engaged in the business of selling or trading handguns at wholesale or retail within the limits of the City, whether as the principal business of such person, firm or corporation, or in addition thereto.

(c) "Resident" means any person who has a place of residence within the limits of the City of Cleveland.

(d) "Nonresident" means any person who does not have a place of residence within the limits of the City of Cleveland.

(e) "Identification card" means a handgun owner's identification card issued pursuant to Section 674.04.

(f) "Registration card" means a handgun registration card issued pursuant to Section 674.05.

674.02. General Prohibitions.

(a) No person shall sell, purchase, own, possess, transfer, give, deliver, receive or have on or about his person or use any handgun which does not contain an engraved serial number or other numerical identification. However, this prohibition shall not apply to any person who is in possession of such handgun on the effective date of this section, and who within a period of thirty days thereafter presents such handgun to the Division of Police, which shall inscribe thereon a serial number pursuant to law.

(b) No person shall receive or possess any handgun unless such person has an identification card issued to him and in effect, and a registration card for each such handgun, except to the extent that he is exempt pursuant to Section 674.06 from such requirement.

(c) No person shall have on or about his person or use any handgun unless such person has in his possession an identification card issued to him and in effect, and a registration card for each such handgun, except to the extent that he is exempt pursuant to Section 674.06 from such requirement.

(d) No person shall deliver, transfer or furnish any handgun to any person unless the transferee exhibits at the time of transfer an identification card valid on its face and issued to the transferee, and a registration card for each such handgun, except to the extent that he is exempt pursuant to Section 674.06 from such requirement.

(e) No person shall use or attempt to use his handgun owner's identification card to obtain any handgun for any person who does not have a handgun owner's identification card issued to him and in effect pursuant to this chapter. No person shall knowingly permit his identification card to be used by another to purchase, own, possess, receive, have on or about the person or use any handgun or handgun ammunition.

674.03. Handguns Confiscated. Any handgun which any person owns or possesses in violation of any of the provisions of this chapter shall, upon conviction of such person, be confiscated and destroyed by the Division of Police, except that any such handgun which has been reported stolen pursuant to Section 674.10 shall be returned to the owner thereof unless possession thereof by the owner would constitute a violation of any provision contained in this chapter.

674.04. Handgun Owner's Identification Card; Application; Prohibitions.

(a) Application for an identification card shall be made in writing, and shall be accompanied by an application fee of fifteen dollars ($15.00) which shall be paid into the treasury of the City, with a separate accounting made for it. If it does not appear upon investigation that the applicant is prohibited by this section from being issued an identification card, the Chief of Police shall issue an identification card to the applicant no sooner than seven (7) days and no more than sixty (60) days after the date of application.

For purposes of procuring an identification card, any and all forms and applications required shall be available, and may be filled out at the Central Police Station or any district police station.

(b) All handgun identification cards issued under this section shall be entitled "City of Cleveland, Ohio, Handgun Owner's Identification Card," be serially numbered according to a system devised by the Chief of Police; bear the date of issue, the date of expiration, the name of the Chief of Police, and the applicant's name, home address, birth date, physical description, and full face photograph; and shall be signed by the holder. A copy of each identification card shall be retained by the Chief, together with a copy of the application, which documents shall be maintained on permanent file by the Chief and shall be confidential and shall not be deemed a public record nor be disclosed to unauthorized persons.

(c) An identification card shall expire three (3) years after the birthday of the applicant that follows the date of issuance.

(d) An identification card shall not be issued to:

(1) A person now or subsequent prohibited by Section 2923.13 of the Revised Code from knowingly acquiring, having, carrying, or using any firearm or dangerous ordnance;

(2) A person under twenty-one (21) years of age;

(3) A person convicted of an illegal use or possession of narcotics;

(4) A person with more than one conviction of being drunk and disorderly or driving a motor vehicle while intoxicated, either of which occurs within one (1) year before the date of application;

(5) Any person with more than one conviction of a misdemeanor involving the use of force and violence, or the threat of the use of force and violence against the person of another within two (2) years before the application for the identification card.

(e) No person shall knowingly give any false information in making application for an identification card, and no person shall use or attempt to use an identification card to purchase, own, possess, receive, have on or about his person, or use any handgun, knowing the identification card belongs to another, or knowing it was obtained by means of false information, or when it is void by reason of the holder becoming a member of the class of persons prohibited by division (d) of this section from being issued an identification card.

(f) A possessor of an identification card shall become ineligible to possess such a card if he becomes one of that class of persons to whom an identification card cannot be issued under division (d) of this section, then that person shall immediately forfeit the card and return it to the Chief of Police. Any and all handguns owned by that person or in their possession shall be confiscated and disposed of by the Division of Police as provided in Section 674.03.

674.05. Registration of Handguns; Application; Fee.

(a) Application for a handgun registration card shall be made in writing by the person claiming to be the owner of the handgun to be registered at any office where identification cards may be issued under Section 674.04. The application shall be accompanied by an application fee of two dollars ($2.00) for each handgun to be registered, which shall be paid into the Treasury of the City, with separate accounting made for it When satisfied that the applicant holds a valid identification card and is not in that class of persons prohibited from holding the same, a registration card shall be issued to the applicant no sooner than three (3) days and no more than sixty (60) days after the date of application.

(b) All registration cards issued under this section shall be entitled "City of Cleveland, Ohio, Handgun Registration Card;" be serially numbered according to a system devised by the Chief of Police; bear date of issue, the name of the Chief of Police, the applicant's name, home address, identification card number, the signature of the applicant; and contain the name, type, caliber, and serial number of the handgun. A copy of each registration card shall be retained by the Chief, together with a copy of the application, which documents shall be maintained on permanent file by the Chief and shall not be deemed a public record nor be disclosed to unauthorized persons.

(c) Any person who sells or otherwise transfers possession of a registered handgun shall, within five days of the date of transfer of possession of the handgun, surrender the registration card for the handgun with the name, address, or social security number, and identification card number, if required by law, of the buyer endorsed on it, to any office where identification cards are issued, and obtain a receipt for it. The office receiving the same shall immediately cancel the registration card.

674.06. Exemptions.

(a) The following shall be exempt from the provisions of this chapter:

(1) Any state of the United States, the United States or any political subdivision, department or agency of either;

(2) An officer or agent of any state of the United States, or any agency of either, a member of the organized militia of any state or the armed forces of the United States, or a law enforcement officer of any political subdivision, but only to the extent that his official duties require him to purchase, own, possess, receive, carry or use handguns and not for the personal use of such individual.

(b) The requirements of an identification card and registration card contained in Sections 674.04 and 674.05 shall not apply to:

(1) A holder of a dealer's license issued pursuant to Section 674.07.

(2) A nonresident of the City who holds and exhibits upon request a valid permit, authorization or identification issued by the chief of police or the sheriff of his place of residence, to purchase, own, possess, receive, carry or use any handgun. It is provided further that handguns in the possession of or under the control of nonresidents shall at all times be unloaded and encased, except while on a suitable firing range, or while being used for lawful hunting, or while unloaded at a public firearms display, show or exhibition.

(3) A new resident of the City, during a period not exceeding thirty days after he becomes a resident. However, such person shall possess and exhibit upon request such valid permit, authorization or identification as may be required by the laws of his former domicile to own, possess, receive or have on or about the person any handgun, or proof of former residence in a jurisdiction which does not require the same.

(4) An executor, administrator, guardian, receiver, trustee in bankruptcy, or other fiduciary duly qualified and appointed by a court of competent jurisdiction, when acting in his fiduciary capacity, and an attorney for such a person, when such handgun constitutes property of the estate or trust, but not for the personal use by such individual.

(5) The next of kin or legatee of a decedent acquiring a handgun through distribution of an estate, during a period not exceeding thirty days from the date of his coming into possession thereof, but in any event not more than fifteen days after the appointment of an executor, administrator or other fiduciary.

(6) Carriers, warehousemen and others engaged in the business of transportation and/or storage and their employees, to the extent that the possession, receipt or having on or about the person of any handgun is in the ordinary course of business and in conformity with State or Federal laws, but not for the personal use of any such person.

(7) A person owning or possessing any handgun on February 9, 1976, who files an application for an identification card during a period not to exceed 100 days from such date, until such application is granted or denied. During any 100-day period, any person required to register a handgun shall cause the handgun to be brought, unloaded and securely wrapped, to the Central Police Station or to a district police station for verification of its type, caliber and serial number.

(8) A Federally licensed manufacturer of handguns.

674.07. Handgun Dealer's License; Fee.

(a) Every dealer in the City shall procure a license to engage in such business. Any dealer engaged in such business at more than one location in the City shall procure a separate license for each such location.

Application for a dealer's license shall be made in writing at the office of the Chief of Police and shall be accompanied by an application fee of one hundred dollars ($100.00), which shall be paid into the City treasury, with a separate accounting made therefor. If it appears upon investigation that the applicant holds a Federal dealer's license, the Chief of Police shall issue a dealer's license to the applicant within not less than seven days nor more than sixty days from the date of application. A copy of each dealer's license shall be retained by the Chief, together with a copy of the application, which documents shall be maintained on permanent file by the Chief.

(b) A dealer's license shall remain in effect for one year or until such time as it is revoked, surrendered or the dealer discontinues business as a dealer. Licenses issued pursuant to the provisions of this section are not transferable.

(c) Notwithstanding the provisions of division (a) hereof, no dealer's license shall be issued to any applicant whose dealer's license has been revoked, nor to any applicant who has been convicted of a violation of any provision of this chapter, until the expiration of five years from the effective date of such revocation or five years from the date of conviction. The license issued pursuant to this section shall be prominently displayed by every dealer at his place of business.

(d) The issuance of a dealer's license shall be restricted to locations within a general retail district or a semi-industrial district. Further, a dealer's license shall not be issued to an applicant whose location is within 1,000 feet of a school, church, day care center, liquor establishment or another handgun dealer.

(e) Ninety days after February 9, 1976, no person, firm or corporation shall engage in business as a dealer in the City without a dealer's license. Each day of continuing violation of this section shall be deemed a separate offense.

(f) No dealer, or any agent or employee of a dealer, shall fail to comply with this section or knowingly cause any false information to be entered on a record of purchase or sale of any handgun. No purchaser of any handgun shall give any false information for entry on such record of sale.

674.08. License Revocation.

(a) When any licensed dealer has been convicted of a violation of any of the provisions of this chapter, the Chief of Police shall revoke the license of such dealer. A certified copy of the order of revocation shall be sent forthwith to the dealer, who may appeal such order or revocation to the Director of Public Safety within ten days from the receipt of the order. The Director of Public Safety shall conduct a hearing upon such appeal within ten days from the receipt of a written notice of appeal from such order or revocation.

(b) No dealer shall employ, or enter into any partnership or corporation with any person, firm or corporation whose dealer's license has been revoked.

674.09. Handgun Dealer's Records.

(a) Every dealer shall maintain records of the purchase and sale of handguns. Such records shall be maintained on forms prescribed by the Chief of Police; contain the name of the dealer; and identify each handgun by manufacturer, model number or name, type, caliber and serial number. If such weapon is other than a standard model, or has been modified and improved, the record of sale shall give a brief description of the weapon including such information as may be necessary to identify it.

(b) The record of purchase shall also contain the name of the person from whom each handgun was purchased and the serial number of the seller's identification card, if any, and the serial number of the registration card of each handgun, if any.

(c) The record of sales shall also contain the place and date of the sale; the name, address, age and Social Security number of the purchaser; the serial number of the identification card of the purchaser, if any, or if the purchaser is a nonresident of Cleveland who is exempt from the requirement of an identification card pursuant to Section 674.06(b)(2), then such information as is contained in a valid and effective permit, authorization or identification issued at the purchaser's place of residence. The record of the sale shall also include the serial number of the registration card issued to the purchaser with respect to the handgun purchased. In the case of a sale from one dealer to another, the license number of the dealer shall be recorded instead of the serial numbers of the identification card and registration card. The record of any sale re-quired to be made under this section shall be turned over to the Chief within twenty-four hours of the sale.

(d) The Chief shall keep permanent records of all sales and transfers of handguns as required in this chapter, together with all copies of and applications for identification cards and registration cards. Such records shall be kept in such manner as to record all of the handguns in the ownership and/or possession of all persons who have been issued an identification card or registration card. Such records shall not be deemed public records and shall not be disclosed to unauthorized persons.

674.10. Stolen or Lost Handguns; Reports. Whenever any handgun is stolen or lost, the person losing possession thereof shall, immediately upon discovery of such theft or loss, make a report thereof to the Chief of Police showing the following:

(a) Name, address and Social Security number of the person owning or having possession of such handgun;

(b) Kind of handgun;

(c) Serial number of handgun;

(d) Model;

(e) Caliber;

(f) Manufacturer of handgun;

(g) Handgun registration card number, if any;

(h) Date and place of theft or loss;

(i) A complete statement of the facts and circumstances surrounding such theft or loss.

674.11. Construction and Interpretation. Wherever possible the provisions of this chapter shall not be deemed to be in conflict with the provisions of Chapter 627, but wherever possible shall be construed so as to give effect to both chapters. However, when a provision of this chapter conflicts or duplicates a provision of Chapter 627, the more restrictive provision and/or penalties shall control and govern.

674.99. Penalty. Whoever violates any of the provisions of this chapter shall be fined not less than one hundred dollars ($100.00) nor more than one thousand dollars ($1,000), or imprisoned not more than twelve months, or both. Each day such violation is continued or permitted to continue constitutes a separate offense.

[Codified Ordinances of Cleveland current through June. 30, 2008]

Columbus City Codes

Title 5. Business Regulation and Licensing Code

Article V. Businesses

Chapter 545. Weapons Sales

545.01. Definitions.

(a) "Weapon" means any instrument, device or thing, capable of inflicting death, and designed or specifically adapted for use as a weapon.

(b) "Firearm" means any deadly weapon capable of expelling or propelling one or more projectiles by the action of an explosive or combust-ible propellant. "Firearm" includes an unloaded firearm, and any firearm which is inoperable but which can readily be rendered operable.

(c) "Section" means License Section, Department of Public Safety, and is further defined in Sections 501.02 and 501.03, C.C.C.

(d) "Law enforcement officer" means any of the following:

(1) A sheriff, deputy sheriff, constable, marshal, deputy marshal, municipal police officer, or State highway patrolman;

(2) An officer, agent, or employee of the State or Federal government, or any of their agencies, instrumentalities or political subdivisions, upon whom, by statute, a duty to conserve the peace or to enforce all or certain laws is imposed and the authority to arrest violators is conferred.

(e) "Antique firearm" means any firearm manufactured prior to 1898.

(f) "Curio or collectible firearm" means any firearm manufactured fifty (50) or more years ago.

545.02. Weapon dealer license required.

(a) No person shall engage in the business of dealing weapons as defined in Section 545.01 (a), Columbus City Code (C.C.C.) without first procuring from the License section, hereafter known as the Section, a weapon dealer license.

(b) A weapon dealer that deals exclusively in transactions involving firearms that are antiques, curios or collectibles is exempt from the requirement to obtain a weapon dealer license.

545.03. Weapon dealer license application; license term; fee.

(a) Application for a weapon dealer license shall be made on forms provided by the Section, shall be made under oath and shall provide the following information and material:

(1) Full name, residence address, residence telephone number, date of birth, sex, race and social security number of applicant;

(2) Business name (D.B.A.), address, telephone number if different from above;

(3) If dealing in firearms, a copy of federal firearm license shall accompany such application.

(b) All weapon dealer licenses shall expire one year from date of issue.

(c) The annual license fee shall be one hundred dollars ($100.00) and shall be made payable at time of application. Any refunds shall be made in accordance with Section 501.09(A) and (S), C.C.C.

545.04. Dealer license prohibitions, conditions.

(a) No weapon dealer shall purchase, sell, barter, trade, give away or take possession of an "assault weapon" as defined in Section 2323.01(I), C.C.C. except as provided in Section 2323.05(C).

(b) No weapon dealer shall sell, barter, trade or give away any lawful weapon unless the individual so receiving such weapon is first in possession of a valid Weapon Transaction Permit.

(c) Every person who is licensed to deal in weapons described in C.C.C. 545.02 shall make out and deliver to the Division of Police, everyday before the hour of 12:00 noon, a legible and correct report of every sale, gift, or other transaction made under authority of such license, during the preceding twenty-four hours. The report shall contain the date of such transaction; the type of the weapon as described in C.C.C. 545.02; the full name of purchaser, or recipient with (his/her) address and age; dealer name and dealer license number; the serial number, kind, description and price of such weapon; the weapon transaction permit number; and the purpose given by such person for the acquisition of such weapon. The report shall be substantially in the following form:

Permit number, date of issue;

Full name, residence of permittee;

Date of birth, social security number of permittee;

Dealer name and dealer license number;

Weapon serial number, description;

For what purpose acquired.

(d) Display of license. Each weapon dealer license shall be prominently displayed at the place of business shown thereon, in full view of patrons.

545.05. Refusal of dealer license application.

(a) The Section may refuse to issue a dealer license if any of the following is found to be true:

(1) The applicant has made a false statement as to any matter in the application;

(2) The applicant has not complied with all applicable regulations of this chapter;

(3) Any other reason for which a license may be refused as provided in Chapter 501, C.C.C.

545.06. Weapon transaction permit required.

(a) No person shall purchase or take possession of any lawful weapon without first procuring from the License Section a valid weapon transaction permit.

(b) No weapon transaction permit shall be issued until at least seven (7) days have elapsed from the date of the permit application.

(c) This section does not apply to transactions involving firearms defined as antiques, curios or collectibles.

545.07. Weapon transaction permit application; permit term, fee.

(a) Application for a weapon transaction permit shall be made on forms provided by the section, shall be made under oath and shall provide the following information and material:

(1) Full name, residence address, residence telephone number, date of birth, sex, race, and social security number of applicant;

(2) A current criminal background review, performed by Columbus Police personnel;

(3) Fingerprints;

(4) Type of weapon to be transacted;

(5) Any other information or material required by the rules and regulations promulgated by the safety director under this section.

(b) All transaction permits expire thirty (30) days from date of issue.

(c) The transaction permit fee shall be ten dollars ($10.00) and shall be made payable at time of application. Any refunds shall be made in accordance with Section 501.09(A) and (B), C.C.C.

(d) Sections (a)(2) and (c) above shall not apply to law enforcement officers as defined in Section 545.01, C.C.C.

545.08. Refusal of weapon transaction permit.

(a) The section shall refuse a weapon transaction permit to the following individuals:

(1) Those having made any false statement as to any matter in their application;

(2) Those under the age of eighteen (18) years;

(3) Those under twenty-one (21) years of age, unless such individual is at least eighteen (18) years of age and wishes to purchase a shotgun or rifle only;

(4) Those convicted of any "offense of violence" as defined in Section 2901.01(I) of Ohio Revised Code, or any felony offense, within the last ten (10) years;

(5) Those under disability as defined in 2923.13 ORC, who have not been relieved from such disability as provided in ORC 2923.14;

(6) Any other reason for which a license may be refused as provided in Chapter 501, C.C.C.

545.09. License, permit revocation.

(a) Licenses and permits issued under this chapter may be revoked by the License Supervisor for any one or more of the following reasons:

(1) Fraud, misrepresentation or bribery in securing a license or during the course of business;

(2) Violation of any provision of this or any other applicable City, State or Federal statute, rule or regulation;

(3) And, any of the reasons which could have been grounds for refusing to issue the original license or permit as provided in this chapter or Chapter 501, C.C.C.

545.10. Application investigation. The Section shall examine all applications filed under this chapter and shall make, or cause to be made, any further investigation as is deemed necessary in order to perform duties prescribed by this chapter, as well as by Chapter 501, C.C.C.

545.11. License, permit appeals. Any individual or organization who has been refused a license or permit, or renewal of a license or permit under this chapter, or has had a license or permit issued under this chapter suspended or revoked, may appeal such decision as provided by Chapter 505, C.C.C.

545.12. Transfer of license, permit. No license or permit issued under this chapter shall be assigned or transferred by the named licensee or permittee to any other individual or organization.

545.13. Exemptions. This chapter shall not apply to:

(a) Any individual redeeming or removing from deposit or pledge, with proper proof of ownership, their own weapon.

545.99. Penalties.

(a) Whoever violates or fails to comply with Section 545.04(a), (b), (c); 545.06(a), (b); or 545.12 shall be deemed guilty of a misdemeanor of the first degree.

(b) Whoever violates or fails to comply with Section 545.04(d) shall be deemed guilty of a misdemeanor of the fourth degree.

Title 23. General Offenses Code

Chapter 2323. Attempt; Complicity; Weapons

2323.11 Definitions. As used in Chapter 2323 of the Columbus City Codes:

(A) "Deadly weapon" means any instrument, device, or thing capable of inflicting death, and designed or specially adapted for use as a weapon, or possessed, carried, or used as a weapon.

(B)(1) "Firearm" means any deadly weapon capable of expelling or propelling one or more projectiles by the action of an explosive or combustible propellant. "Firearm" includes an unloaded firearm, and any firearm which is inoperable but which can readily be rendered operable.

(2) When determining whether a firearm is capable of expelling or propelling one or more projectiles by the action of an explosive or combustible propellant, the trier of fact may rely upon circumstantial evidence, including, but not limited to, the representations and actions of the individual exercising control over the firearm.

(C) "Handgun" means any firearm designed to be fired while being held in one hand.

(D) "Semi-automatic firearm" means any firearm designed or specially adapted to fire a single cartridge and automatically chamber a succeeding cartridge ready to fire, with a single function of the trigger.

(E) "Automatic firearm" means any firearm designed or specially adapted to fire a succession of cartridges with a single function of the trigger. "Automatic firearm" also means any semiautomatic firearm designed or specially adapted to fire more than thirty-one cartridges without reloading, other than a firearm chambering only .22 caliber short, long, or long-rifle cartridges.

(F) "Large capacity magazine" means a box, drum, clip or other container which holds more than twenty rounds of ammunition to be fed continuously into a semiautomatic firearm, except a magazine designed to hold only .22 caliber rimfire cartridges.

(G) "Detachable magazine" means any ammunition feeding device, the function of which is to deliver one (1) or more ammunition cartridges into the firing chamber, and which can be removed from the firearm manually without the use of any tool or device.

(H) "Receiver" means that part of a firearm also referred to as the frame, which houses the firing mechanism and which is usually threaded at its forward portion to receive the barrel.

(I) "Thumbhole stock" means a stock with a hole through it to accommodate the thumb of the trigger hand.

(J) "Muzzle brake" means a device attached to the muzzle of a firearm that utilizes escaping gas to reduce recoil.

(K) "Muzzle compensator" means a device attached to the muzzle of a firearm that utilizes escaping gas to control muzzle movement.

(L) "Assault weapon" means any:

(1) Semi-automatic rifle that has the capacity to accept a detachable magazine and has one or more of the following:

(a) A pistol grip that protrudes conspicuously beneath the receiver of the weapon;

(b) Any feature capable of functioning as a protruding grip that can be held by the non-trigger hand;

(c) A folding, telescoping or thumbhole stock;

(d) A shroud attached to the barrel, or that partially or completely encircles the barrel, allowing the bearer to hold the firearm with the non-trigger hand without being burned, but excluding a slide that encloses the barrel; or

(e) A muzzle brake or muzzle compensator;

(2) Semi-automatic pistol with a fixed magazine, or any semi-automatic, centerfire with a fixed magazine, that has the capacity to accept more than ten rounds of ammunition;

(3) Semi-automatic pistol that has the capacity to accept a detachable magazine and has one or more of the following:

(a) Any feature capable of functioning as a protruding grip than can be held by the non-trigger hand;

(b) A folding, telescoping or thumbhold stock;

(c) A shroud attached to the barrel, or that partially or completely encircles the barrel, allowing the bearer to hold the firearm with the non-trigger hand without being burned, but excluding a slide that encloses the barrel;

(d) A muzzle brake or muzzle compensate-or; or

(e) The capacity to accept a detachable magazine at any location outside of the pistol grip;

(4) Semi-automatic shotgun that has two or more of the following:

(a) A pistol grip that protrudes conspicuously beneath the receiver of the weapon;

(b) A folding, telescoping or thumbhole stock;

(c) A fixed magazine capacity in excess of five standard two and three quarters inch, or longer, rounds; or

(d) An ability to accept a detachable magazine;

(5) Shotgun with a revolving cylinder;

(6) Conversion kit or combination of parts from which an assault weapon can be assembled if those parts are in the possession or under the control of the same person.

(M) Assault weapon does not include any antique firearm or any firearm that has been modified to either render it permanently inoperable or to permanently make it a device no longer defined as an assault weapon.

(N) "Antique firearm" means any firearm manufactured prior to 1898.

(O) "Curio or collectible firearm" means any firearm manufactured fifty (50) or more years ago.

2323.15. Using weapons while intoxicated.

(A) No person, while under the influence of alcohol or any drug of abuse, shall carry or use any firearm or dangerous ordnance as defined in Section 2923.11 of the Revised Code.

(B) Whoever violates this section is guilty of using weapons while intoxicated, a misdemeanor of the first degree.

2323.19 Failure to secure dangerous ordnance.

(A) No person, in acquiring, possessing, carrying, or using any dangerous ordnance, shall negligently fail to take proper precautions:

(1) To secure the dangerous ordnance against theft, or against its acquisition or use by an unauthorized or incompetent person;

(2) To insure the safety of persons and property.

(B) Whoever violates this section is guilty of failure to secure dangerous ordnance, a misdemeanor of the second degree.

(C) As used in this section, dangerous ordnance has the same meaning as used in Section 2923.11 of the Revised Code.

2323.20 Unlawful transactions in weapons.

(A) No person shall:

(1) Knowingly exhibit for sale in a show window any firearm.

(2) Sell a firearm to another person unless a permit has been issued pursuant to Chapter 545 of the Columbus City Codes.

(3) Manufacture, possess for sale, sell, or furnish to any person other than a law enforcement agency for authorized use in police work, any brass knuckles, cestus, billy, blackjack, sandbag, switchblade knife, springblade knife, gravity knife, or similar weapon;

(4) Knowingly fail to report to law enforcement authorities forthwith the loss or theft of any firearm in such person's possession or under his control.

(B) The Safety Director shall have the power to promulgate rules and regulations, including permit fees, regarding application for, and issuance of, firearm purchase permits.

(C) This section does not apply to transactions involving firearms defined as antiques, curios, or collectibles.

(D) Whoever violates any provision of this section is guilty of unlawful transactions in weapons. Violation of Division (A)(1) or (A)(4) of this section is a misdemeanor of the fourth degree. Violation of Division (A)(3) of this section is misdemeanor of the second degree. Violation of Division (A)(2) of this section is a misdemeanor of the first degree.

2323.21 Improperly furnishing firearms to a minor.

(A) No person shall:

(1) Sell any firearm to a person under age eighteen;

(2) Sell any handgun to a person under age twenty-one;

(3) Furnish any firearm to a person under age eighteen, except for purposes of lawful hunting, or for purposes of instruction in firearms safety, care, handling, or marksmanship under the supervision or control of a responsible adult.

(B) Whoever violates this section is guilty of improperly furnishing firearms to a minor, a misdemeanor of the second degree.

2323.31 Unlawful possession of assault weapons.

(A) No person shall sell, offer or display for sale, give, lend or transfer ownership of, acquire or possess any assault weapon.

(B) This section does not apply:

(1) To law enforcement officers of the United States, this State, this City, and members of the armed forces of the United States or this state if such person is authorized to acquire or possess an assault weapon and is acting within the scope of his duties;

(2) To the transportation of firearms through the City of Columbus in accordance with federal law;

(3) To the possession of an assault weapon by any person who lawfully owned and possessed that assault weapon prior to the effective date of the ordinance enacting this paragraph, provided that the person complies with all of the following:

(a) Within ninety (90) days of the effective date of the ordinance enacting this paragraph, the owner of the assault weapon must file a completed registration form with the License Section of the department of public safety, which form shall be prescribed by the director and which shall require the following information be provided:

(i) The owner's name and address,

(ii) The make, model, and serial number of the assault weapon(s) being registered, and

(iii) A statement by the owner, sworn to before a notary public, that the information provided on the form is complete and accurate and that the owner is in compliance with all applicable federal and state of Ohio statutes and regulations regarding the ownership and possession of the firearm(s) being registered;

(b) Except as otherwise provided in this section, the owner of an assault weapon registered pursuant to this paragraph shall not sell, give, lend or transfer ownership of a registered assault weapon;

(4) To the transfer to and possession of an assault weapon by a licensed gunsmith for purposes of service to or repair of the firearm, and the transfer of the assault weapon from the gunsmith to the assault weapon's lawful owner;

(5) To the sale or transfer of an assault weapon by a firearms dealer that is properly licensed under federal, state and local laws to any branch of the armed forces of the United States, or to a law enforcement agency for use by that agency or its employees for law enforcement purposes, and to the acquisition and possession of an assault weapon by the licensed firearms dealer for the purpose of such sale or transfer;

(6) To the possession of an assault weapon by any person while the person is either:

(a) Lawfully engaged in shooting while at a shooting range that is operating in accordance with all applicable laws and requirements and that has obtained all licenses, permits, and insurance coverage required for that operation; or

(b) Lawfully participating in a competitive sporting event that is officially sanctioned by USA Shooting, which is the national governing body for Olympic shooting competition in the United States, or by the Civilian Marksmanship Program, which is the program that was created by the United States Congress and that is supervised and controlled by the Corporation for the Promotion of Rifle Practice and Firearms Safety pursuant to 36 U.S.C. Sec. 40701, et seq.;

(7) To any person who receives an assault weapon as the result of the lawful distribution of a decedent's property by will or intestate succession, provided that, within ninety (90) days of acquiring possession of the weapon, that person shall either:

(a) Remove the assault weapon from the city of Columbus;

(b) Render the assault weapon permanently inoperable or to permanently make it a device no longer defined as an assault weapon; or

(c) Relinquish the assault weapon to the Columbus division of police for destruction.

(C) Whoever violates this section is guilty of unlawful possession of assault weapons, a misdemeanor of the first degree. If the offender previously has been convicted under this section, upon the second conviction the Court shall impose a sentence of imprisonment of at least thirty days, which sentence is mandatory and may not be suspended, modified, or subject to probation during that thirty day period. If the offender has twice previously been convicted under this section, upon the third or more conviction the Court shall impose a sentence of imprisonment of at least ninety days, which sentence is mandatory and may not be suspended, modified, or subject to probation during that ninety day period.

2323.32 Unlawful possession of a large capacity magazine.

(A) No person shall knowingly possess a large capacity magazine.

(B)(1) This section does not apply to law enforcement officers of the United States, this State, or City, and members of the armed forces of the United States or this state, if such person is authorized to possess such large capacity magazine and is acting within the scope of his duties.

(2) This section does not apply to a large capacity magazine which belongs to a firearm or which is possessed by the owner of a firearm which is registered with federal authorities under the National Firearms Act (26 U.S.C.A. Secs. 5801-5871), or if the large capacity magazine belongs to or is a part of an assault weapon that is lawfully possessed by the person under Section 2323.31(B) or has been rendered totally inoperable or inert and the firearm cannot be readily rendered operable or activated and which is kept as a trophy, souvenir, curio or museum piece.

(C) Whoever violates this section is guilty of unlawful possession of a large capacity magazine, a misdemeanor of the first degree.

[Columbus City Codes current through Ord. 1987-2008, passed December 15, 2008]

Code of Ordinances, City of Dayton

Title XIII. General Offenses

Chapter 138. Weapons and Explosives

Division 1. General Provisions

138.01. Definitions. For the purpose of §§ 138.01 to 138.09, the following words and phrases shall have the following meanings ascribed to them respectively:

Automatic firearm. Any firearm designed or specially adapted to fire a succession of cartridges with a single function of the trigger. Also any semiautomatic firearm designed or specially adapted to fire more than 18 cartridges without reloading, other than a firearm chambering only .22 caliber short, long, or long-rifle cartridges.

Dangerous ordnance. Any of the following:

(1) Any automatic or sawed-off firearm, or zip gun;

(2) Any explosive device or incendiary device;

(3) Nitroglycerin, nitrocellulose, nitrostarch, PETN, cyclonite, TNT, picric acid, and other high explosives; amatol, tritonal, tetrytol, pentolite, pecretol, cyclotol, and other high explosive compositions; plastic explosives; dynamite, blasting gelatin, gelatin dynamite, sensitized ammonium nitrate, liquid oxygen blasting explosives, blasting powder, and other blasting agents; and any other explosive substance having sufficient brisance or power to be particularly suitable for use as a military explosive, or for use in mining, quarrying, excavating, or demolitions;

(4) Any firearm, rocket launcher, mortar, artillery piece, grenade, mine, bomb, torpedo, or similar weapon, designed and manufactured for military purposes, and the ammunition therefor.

Dangerous ordnance. Does not include any of the following:

(1) Any firearm, including a military weapon and the ammunition therefor, and regardless of its actual age, which employs a percussion cap or other obsolete ignition system, or which is designed and safe for use only with black powder;

(2) Any pistol, rifle, or shotgun, designed or suitable for sporting purposes, including a military weapon as issued or as modified, and the ammunition therefor unless such firearm is an automatic or sawed-off firearm;

(3) Any cannon or other artillery piece which, regardless of its actual age, is of a type in accepted use prior to 1887, has no mechanical, hydraulic, pneumatic, or other system for absorbing recoil and returning the tube into battery without displacing the carriage, and is designed and safe for use only with black powder;

(4) Black powder, priming quills and percussion caps possessed and lawfully used to fire a cannon of a type defined in subsection (C)(3) of this section during displays, celebrations, organized matches or shoots, and target practice, and smokeless and black powder, primers, and percussion caps possessed and lawfully used as a propellant or ignition device in small arms or small arms ammunition;

(5) Dangerous ordnance which is inoperable or inert and cannot readily be rendered operable or activated, and which is kept as a trophy, souvenir, curio, or museum piece;

(6) Any device which is expressly excepted from the definition of a destructive device pursuant to the "Gun Control Act of 1968," 82 Stat. 1213, 18 U.S.C. 921 (A)(4), and any amendments or additions thereto or reenactments thereof, and regulations issued thereunder.

Deadly weapon. Any instrument, device, or thing capable of inflicting death, and designed or specially adapted for use as a weapon, or possessed, carried, or used as a weapon.

Explosive device. Any device designed or specially adapted to cause physical harm to persons or property by means of an explosion, and consisting of an explosive substance or agency and a means to detonate it. "Explosive device" includes without limitation any bomb, any explosive demolition device, any blasting cap or detonator containing an explosive charge, and any pressure vessel which has been knowingly tampered with or arranged so as to explode.

Firearm. Any deadly weapon capable of expelling or propelling one or more projectiles by the action of an explosive or combustible propellant. "Firearm" includes an unloaded firearm, and any firearm which is inoperable but which can readily be rendered operable.

Handgun. Any firearm designed to be fired while being held in one hand.

Incendiary device. Any firebomb, and any device designed or specially adapted to cause physical harm to persons or property by means of fire, and consisting of an incendiary substance or agency and a means to ignite it.

Sawed-off firearm. A shotgun with a barrel less than 18 inches long, or a rifle with a barrel less than 16 inches long, or a shotgun or rifle less than 26 inches long overall.

Semiautomatic firearm. Any firearm designed or specially adapted to fire a single cartridge and automatically chamber a succeeding cartridge ready to fire, with a single function of the trigger.

Zip gun. Any of the following:

(1) Any firearm of crude and extemporized manufacture;

(2) Any device, including, without limitation, a starter's pistol, not designed as a firearm, but which is specially adapted for use as such;

(3) Any industrial tool, signaling device, or safety device, not designed as a firearm, but which as designed is capable of use as such, when possessed, carried, or used as a firearm.

138.03. Using weapons while intoxicated.

(A) No person, while under the influence of alcohol or any drug of abuse, shall carry or use any firearm or dangerous ordnance.

(B) Whoever violates this section is guilty of using weapons while intoxicated, a misdemeanor of the first degree.

138.05. Failure to secure dangerous ordnance.

(A) No person, in acquiring, possessing, carrying, or using any dangerous ordnance shall negligently fail to take proper precautions:

(1) To secure the dangerous ordnance against theft, or against its acquisition or use by any unauthorized or incompetent person;

(2) To insure the safety of persons and property.

(B) Whoever violates this section is guilty of failure to secure dangerous ordnance, a misdemeanor of the second degree.

138.06. Unlawful transactions in weapons.

(A) No person shall:

(1) Manufacture, possess for sale, sell or furnish to any person other than a law enforcement agency for authorized use in police work, any brass knuckles, cestus, billy, blackjack, sandbag, switchblade knife, springblade knife, gravity knife, or similar weapon;

(2) When transferring any dangerous ordnance to another, negligently fail to require the transferee to exhibit such identification, license, or permit showing him to be authorized to acquire dangerous ordnance pursuant to § 138.05 of this Code or R.C. § 2923.17, or negligently fail to take a complete record of the transaction and forthwith forward a copy of such record to the sheriff of the county or safety director or police chief of the municipality where the transaction takes place;

(3) Knowingly fail to report to law enforcement authorities forthwith the loss or theft of any firearm or dangerous ordnance in such person's possession or under his control.

(B) Whoever violates this section is guilty of unlawful transactions in weapons. Violation of subsection (A)(1) or (2) of this section is a misdemeanor of the second degree. Violation of subsection (A)(3) of this section is a misdemeanor of the fourth degree.

138.07. Improperly furnishing firearms to a minor.

(A) No person shall:

(1) Sell any firearm to a person under age 18;

(2) Sell any handgun to a person under age 21;

(3) Furnish any firearm to a person under age 18, except for purposes of lawful hunting, or for purposes of instruction in firearms safety, care, handling, or marksmanship under the supervision or control of a responsible adult.

(B) Whoever violates this section is guilty of improperly furnishing firearms to a minor, a misdemeanor of the second degree.

Division 2. Firearm Owner's Identification Card

138.10. Definitions. For the purpose of §§ 138.10 to 138.23, the following words and phrases shall have the following meanings ascribed to them respectively:

Firearm. Any deadly weapon capable of expelling or propelling one or more projectiles by the action of an explosive or combustible propellant. "Firearm" includes an unloaded firearm, and any firearm which is inoperable but which can be easily restored to an operable condition.

Nonresident. Any person who does not have an actual place of residence in the city.

Offense of violence. Any violation as defined by R.C. § 2901.01(I)(1)-(4).

Possess. To knowingly carry or have a firearm on the person or ready at hand.

Resident. Any person who has an actual place of residence in the city.

138.11. Firearm owner's identification card required.

(A) No person shall possess any firearm, unless such person has a firearm owner's identification card issued to him/her and in effect under § 138.13, or is exempt from the requirement of an identification card under § 138.22, or is on a duly licensed and suitable firing range.

(B) No person shall sell, deliver, transfer, or furnish any firearm to any person, unless the transferee exhibits a firearm owner's identification card valid on its face and issued to the transferee under § 138.13, or unless the transferee exhibits evidence that he/she is exempt from the requirement under § 138.22.

(C) No person shall sell, deliver, transfer, or furnish any firearm to any person without first having notified the Director and Chief of Police of his/her intention to do so not less than five days prior thereto. Such notification shall be in writing, and shall contain the name, home address, social security number, and firearm owner's identification card number of the intended transferee. The Director and Chief of Police shall prescribe the form for such notification, and shall make them available upon request therefor.

(D) Every person regularly engaged in the business of selling or trading firearms at wholesale or retail whether as the principal business of such person or in addition thereto, having sold, delivered, or furnished a firearm to any person pursuant to the provisions of subsection (C) above, shall within five days thereafter, furnish to the Director and Chief of Police the details of such transaction, including the name and address of the transferee, his/her firearm owner's identification card number, and a description of the firearm obtained, including the manufacturer's brand name, caliber, and serial number.

(E) No person shall use or attempt to use a firearm owner's identification card to obtain any firearm, knowing such identification card to belong to another, or knowing it was obtained by means of false information, or when it is void by reason of the holder becoming a member of the class of persons prohibited by § 138.14 from being issued a firearm owner's identification card.

(F) No person shall use or attempt to use his/her firearm owner's identification card to obtain any firearm for any person who does not have a firearm owner's identification card issued to him/her and in effect under § 138.13 and is not exempt from the requirement of an identification card pursuant to § 138.22, and no person shall knowingly permit his/her firearm owner's identification card to be used by another to obtain any firearm.

(G) No person shall knowingly furnish false information when making application for a firearm owner's identification card.

138.12. Application for card; contents.

(A) Application for a firearm owner's identification card shall be made in writing to the Director and Chief of Police, and shall be accompanied by an application fee of $25.00, none of which shall be refunded if such application is denied.

(B) The application shall contain the name, home address, race, sex, ethnicity, birth date, height, weight, hair and eye color, social security number, full face photograph of the applicant, and shall be signed by the applicant.

(C) Applicants shall present corroborating evidence of their identity to the Director and Chief of Police or his/her agent with the initial application.

(D) Applications for a firearm owner's identification card shall be endorsed with the serial number and date of issuance of the card, or if such card is not issued, with the reasons therefor, and the date.

138.13. Issuance of card; forfeiture.

(A) If the applicant for a firearm owner's identification card is not a person prohibited by § 138.14 from being issued such card, the Director and Chief of Police shall notify the applicant of the approval of the issuance of such card not less than five nor more than 20 days from the date of the application by certified mail, return receipt requested, mailed to the applicant at the address stated in the application.

(B) The applicant must personally appear for the issuance of the firearm owner's identification card within 90 days after receipt of notice of approval at the location specified in such notice. Failure to appear within the 90-day period shall void the application, but shall not bar a subsequent application.

(C) If a person to whom a firearm owner's identification card has been issued becomes a prohibited person enumerated in § 138.14, the card is thereupon void and shall be forfeited to the Director and Chief of Police. The Director and Chief of Police shall cause a notice of forfeiture, requiring the return of the firearm owner's identification card within ten days of receipt of such notice, to be sent to the holder's recorded last known address by certified mail. The notice of forfeiture shall set forth the reasons therefor.

(D) No person shall fail to surrender in person to the Director and Chief of Police his/her firearm owner's identification card after receiving notice of forfeiture pursuant to subsection (C) of this section. If a person fails to surrender his/her firearm owner's identification card within 15 days after notice is sent, it may be seized by police officers and any firearm as defined by § 138.10(A) of this chapter possessed by the person at the time may be seized.

138.14. Persons not eligible for firearm owner's identification cards. A firearm owner's identification card shall not be issued to:

(A) Any person who is prohibited by 18 U.S.C. Section 922 from obtaining, possessing, carrying or using a firearm, and has not had such disability removed according to 18 U.S.C. Section 925;

(B) Any person who is prohibited by R.C. § 2923.13 from obtaining, possessing, carrying, or using a firearm, and has not had such disability removed according to R.C. § 2923.14;

(C) A person with more than one conviction of any offense involving drunkenness within one year prior to his/her application for firearm owner's identification card;

(D) A person with more than one conviction of disorderly conduct, or the state equivalent of such offense, within two years prior to his/her application for firearm owner's identification card;

(E) A person who has been charged with and/or indicted for or who has been convicted of any offense of violence, as defined in R.C. § 2901.01 or other offense involving the use of force or violence or the threat of the use of force or violence against the person of another, not constituting a felony, or has been adjudicated a juvenile delinquent for commission of any such offense;

(F) Any person who is under the age of 18 years;

(G) A person who has been charged with or convicted of any offense in violation of §§ 138.02, 138.03, 138.04, 138.05, 138.06, 138.07, 138.08, 138.09, or the state equivalent of such offenses.

(H) A nonresident of the City of Dayton, Ohio.

138.16. Contents of card. All cards issued under this section shall be entitled "City of Dayton Firearm Owner's Identification Card." They shall be serially numbered, shall bear the date of issue, the name of the issuing authority, and the name, home address, race, sex, ethnicity, birth date, height, weight, hair and eye color, social security number, a full face photograph of the holder, and signature of the holder. The firearm owner's identification card shall state that it is not a permit to carry a firearm. The card will also state that no person shall sell a handgun to the holder of such card if the holder is under the age of 21 years.

138.17. Expiration and renewal of card.

(A) A firearm owner's identification card expires three years after the holder's first birth date after issuance, and may be renewed for subsequent three-year periods by following the application procedure. The fee for a renewal application shall be $15.00, none of which shall be refunded if such application is denied.

(B) Any holder of a current, valid, "Handgun Owner's Identification Card" shall be granted all rights and privileges set forth in § 138.11, "Firearm Owner's Identification Card," and upon expiration of the handgun owner's identification card may renew the application, pay the above fee, and be issued a new firearm owner's identification card provided he/she is not ineligible as stated in § 138.14.

138.18. Form of application and card. The Director and Chief of Police shall design and prescribe the form for the application and for the firearm owner's identification card consistent with §§ 138.12 and 138.16, and shall make available a supply of such forms to each person requesting an application.

138.21. Director and Chief of Police authorized to promulgate procedure. The Director and Chief of Police, with the approval of the City Manager is authorized to promulgate such procedure as he/she deems appropriate to accomplish the orderly application for and issuance of firearm owner's identification cards, and for hearings conducted pursuant to § 138.15, and for the filing of applications by invalids and handicapped persons.

138.22. Persons exempt. The requirements of a firearm owner's identification card contained in §§ 138.10 to 138.23 do not apply to the following:

(A) Officers or agents of the United States or any state or any political subdivision, department, or agency, or members of the organized militia of any state, or the Armed Forces of the United States, or any law enforcement officer, to the extent that the official duties of any such person require him/her to obtain, possess, carry, or use firearms;

(B) Any private investigator or security guard that:

(1) Has successfully completed a basic firearm training program at a training school approved by the Ohio Peace Officer Training Council, and;

(2) Has requested, and received, a class A, B, or C license from the Ohio Department of Commerce, and;

(3)(a) Has a notation on his/her Private Investigator/Security Guard Provider Identification Card, issued by the Ohio Department of Commerce, Division of Licensing, that he/she is a firearm-bearer (FAB) specifically enumerating the type of firearm he/she is qualified to carry and carries the identification card whenever he/she carries a firearm in the course of engaging in the business of private investigation and/or security services, and;

(b) If the firearm-bearer (FAB) date has expired, carries his/her certificate of satisfactory requalification that is in effect while engaging in the business of private investigation and/or security services.

(4) No private investigator or security guard shall be considered a law enforcement officer for any purpose.

(5) Nothing in this section shall be construed as granting the right to carry a concealed weapon.

(C) Importers, manufacturers, and dealers licensed under federal law, to the extent that the acquisition, possession, carriage, or use of any firearm is in the ordinary course of business;

(D) Carriers, warehousemen, and others engaged in the business of transporting or storing firearms, and their agents or employees, to the extent that possessing any firearm by such persons is in the ordinary course of business and in conformity with the laws of the State of Ohio or the United States;

(E) Executors, administrators, guardians, receivers, trustees in bankruptcy, and other fiduciaries duly qualified and appointed by a court of competent jurisdiction, when acting in their official capacity, and their attorneys;

(F) Heirs and legatees coming into possession of any firearm through distribution of an estate and other persons other than those persons named in subsection (E) acquiring any firearm through operation of law, during a period not exceeding 30 days from the date of their coming into possession of such firearm;

(G) Nonresidents of the city moving to the city for the purpose of establishing residence therein, during a period not to exceed 30 days of their entry into the city for such purpose;

(H) Nonresidents of the city who enter the city for a period not to exceed 30 days with a firearm or firearms for the purpose of exhibiting or trading a firearm or firearms at a public firearm display, show, or exhibition. However, the firearm or firearms shall be unloaded at all times and must be kept encased except when at such display, show, or exhibition;

(I) Nonresidents of the city who enter the city for a period not to exceed 30 days for the purpose of participating in an organized and sanctioned firearm match; provided, that any firearm shall be unloaded and encased except while participating in such firearm match.

138.23. Nonapplicability of §§ 138.10 to 138.22. The provisions of §§ 138.10 to 138.22 shall not apply to the following:

(A) Any firearm which, regardless of its actual age, is designed and safe for use only with black powder, and uses a flintlock, percussion cap, or other obsolete ignition system, or with respect to any firearm which uses fixed ammunition of a type in use prior to 1899, which ammunition is no longer commercially manufactured in the United States and is not readily available in the ordinary channels of trade;

(B) Any device used exclusively for signaling or safety and recommended by the United States Coast Guard or the Interstate Commerce Commission;

(C) Any device used exclusively for firing of stud cartridges, rivets, or nails, and designed for industrial applications;

(D) Any device used exclusively as a starter gun for the purpose of firing blank cartridges as a means of controlling sporting events, or for use in theatrical productions; provided, that such gun is so constructed as not to be readily convertible to the firing or projectiles.

Division 3. Assault Weapons

138.24. Definitions. For the purposes of §§ 138.24 to 138.27 the following words and phrases shall have the following meanings ascribed to them respectively:

Assault weapon means:

(1) Any rifle or carbine:

(a) With a detachable magazine capable of containing more than ten rounds of ammunition, either inserted into the weapon or readily accessible, or;

(b) With a permanent magazine capable of containing more than ten rounds of ammunition, or;

(c) Which is a modification of a rifle or carbine as defined in (A)(1)(a) or (b), above, so as to be able to be fired while being held in one hand.

(2) Any shotgun:

(a) With a detachable magazine capable of containing more than six rounds of ammunition, either inserted into the weapon or readily accessible, or;

(b) With a permanent magazine capable of containing more than six rounds of ammunition.

(3) Any handgun:

(a) With a detachable magazine capable of containing more than 20 rounds of ammunition, either inserted into the weapon or readily accessible, or;

(b) With a permanent magazine capable of containing more than 20 rounds of ammunition.

(4) Any firearm which may be easily restored to an operable assault weapon as defined in subsection (A)(1), (A)(2), or (A)(3);

(5) A rifle/carbine magazine, capable of containing more than ten rounds of ammunition, or;

(a) A handgun magazine, capable of containing more than 20 rounds of ammunition, or;

(b) A shotgun magazine capable of containing more than six rounds of ammunition.

(6) Concerning sections (A)(1), (A)(2), (A)(3), (A)(4) and (A)(5); magazines need not be loaded to be in violation of this chapter.

(B) Assault weapon does not include any of the following:

(1) Any firearm that uses .22 caliber rimfire ammunition with a detachable or fixed magazine capable of containing 20 or fewer rounds of ammunition.

(2) Any assault weapon which has been modified to either render it permanently inoperable or to permanently make it a device no longer defined as an assault weapon.

Automatic firearm. Any firearm designed or specifically adapted to fire a succession of cartridges with a single function of the trigger.

Deadly weapon. Any instrument, device, or thing capable of inflicting death, and designed or specifically adapted for use as a weapon, or possessed, carried, or used as a weapon.

Easily restored. Any modification that can be done by hand or with the use of ordinary hand tools.

Firearm. Any deadly weapon capable of expelling or propelling one or more projectiles by the action of an explosive or combustible propellant. "Firearm" includes an unloaded firearm, and any firearm which is inoperable but which can be easily restored to an operable condition.

Handgun. Any firearm designed to be fired while being held in one hand.

Person. Any individual, corporation, company, association, firm, partnership, club, or society, including wholesale and retail gun dealers.

Rifle and/or carbine. Any firearm designed and intended to be fired from the shoulder and designed to use the energy of the explosive in a fixed cartridge to fire a single projectile through a rifle bore for each single pull of the trigger.

Semi-automatic firearm. Any firearm designed or specially adapted to fire a single cartridge and automatically chamber a succeeding cartridge ready to fire, with a single function of the trigger.

Shotgun. Any firearm, whether or not it is intended to be fired from the shoulder, originally designed to fire a fixed shotgun shell.

138.25. Prohibitions. No person shall sell, offer, or display for sale, give, lend or transfer ownership of, acquire or possess any assault weapon.

138.25.1. Persons exempt. Nothing in this section shall apply to any officer, agent, or employee of this or any other state or the United States, members of the armed forces of the United States or the organized militia of this or any other state, and law enforcement officers as defined in R.C. § 2901.01(K) to the extent that any such person is authorized to acquire or possess an assault weapon and is acting within the scope of his/her duties.

138.25.2. Safe passage. The prohibition against possessing an assault weapon(s) shall not apply when it is being transported through the City of Dayton, if such transportation is being made in accordance with 18 U.S.C. § 926A.

(A) Nothing in this section shall be construed as granting the right to carry a concealed weapon. See § 138.02.

(B) Nothing in this section shall be construed as granting the right to transport a loaded firearm in a motor vehicle, or in a manner that the firearm is accessible to the operator or any passenger without leaving the vehicle. See § 138.04.

138.26. Seizure and destruction of assault weapons. Any assault weapon is hereby declared to be contraband and shall be seized and disposed of as authorized by law.

138.27. Penalty. Any person who violates any provision of § 138.25 shall be guilty of a misdemeanor of the first degree. Each calendar day that any violation of § 138.25 is repeated or continued shall constitute a separate offense. If any person who violates any provision of § 136.25 was, at the time of the commission of such offense, also committing any drug offense, he or she shall be sentenced to serve the maximum term of imprisonment, to wit: 180 days. In addition, the offender shall be fined the maximum fine permitted, to wit: $1,000.00. These terms of imprisonment and fine are mandatory, shall not be suspended, and must be imposed by the court.

138.99. Penalty. Whoever violates §§ 138.11 or 138.13(D) shall be guilty of a misdemeanor of the first degree.

[City of Dayton Revised Code current through Ord. 30822-09 passed Jan. 7, 2009]

Toledo Municipal Code

Part 5. General Offenses Code

Chapter 549. Weapons and Explosives

549.01. Definitions. As used in this chapter:

(a) "Deadly weapon" means any instrument, device or thing capable of inflicting death, and designed or specially adapted for use as a weapon, or possessed, carried or used as a weapon.

(b) "Firearm" means any deadly weapon capable of expelling or propelling one or more projectiles by the action of an explosive or combustible propellant. "Firearm" includes an unloaded firearm, and any firearm which is inoperable but which can readily be rendered operable.

(c) "Handgun" means any firearm designed to be fired while being held in one hand.

(d) "Semi-automatic firearm" means any firearm designed or specially adapted to fire a single cartridge and automatically chamber a succeeding cartridge ready to fire, with a single function of the trigger.

(e) "Automatic firearm" means any firearm designed or specially adapted to fire a succession of cartridges with a single function of the trigger. "Automatic firearm" also means any semi-automatic firearm designed or specially adapted to fire more than twenty-one cartridges without reloading, other than a firearm chambering only .22 caliber short, long or long-rifle cartridges.

(f) "Sawed-off firearm" means a shotgun with a barrel less than eighteen inches long, or a rifle with a barrel less than sixteen inches long, or a shotgun or rifle less than twenty-six inches long overall.

(g) "Zip-gun" means any of the following:

(1) Any firearm of crude and extemporized manufacture,

(2) Any device, including without limitation a starter's pistol, not designed as a firearm, but which is specially adapted for use as such;

(3) Any industrial tool, signaling device or safety device, not designed as a firearm, but which as designed is, capable of use as such, when possessed, carried or used as a firearm.

(h) "Explosive device" means any device designed or specially adapted to cause physical harm to persons or property by means of an explosion, and consisting of an explosive substance or agency and a means to detonate it. "Explosive device" includes without limitation any bomb, any explosive demolition device, any blasting cap or detonator containing an explosive charge, and any pressure vessel which has been knowingly tampered with or arranged so as to explode.

(i) "Incendiary device" means any firebomb, and any device designed or specially adapted to cause physical harm to persons or property by means of fire, and consisting of an incendiary substance or agency and a means to ignite it.

(j) "Dangerous ordnance" means any of the following, except as provided in subsection (k) hereof:

(1) Any automatic or sawed-off firearm, or zip-gun;

(2) Any explosive device or incendiary device;

(3) Nitroglycerin, nitrocellulose, nitrostarch, PETN, cyclonite, TNT, picric acid and other high explosives; amatol, tritonal, tetrytol, pentolite, pecretol, cyclotol and other high explosive compositions; plastic explosives; dynamite, blasting gelatin, gelatin dynamite, sensitized ammonium nitrate, liquid-oxygen blasting explosives, blasting powder and other blasting agents; and any other explosive substance having sufficient brisance or power to be particularly suitable for use as a military explosive, or for use in mining, quarrying, excavating or demolitions;

(4) Any firearm, rocket launcher, mortar artillery piece, grenade, mine, bomb, torpedo or similar weapon, designed and manufactured for military purposes, and the ammunition therefor;

(5) Any firearm muffler or silencer;

(6) Any combination of parts that is intended by the owner for use in converting any firearm or other device into a dangerous ordnance.

(k) "Dangerous ordnance" does not include any of the following:

(1) Any firearm, including a military weapon and the ammunition therefor, and regardless of its actual age, which employs a percussion cap or other obsolete ignition system, or which is designed and safe for use only with black powder;

(2) Any pistol, rifle or shotgun, designed or suitable for sporting purposes, including a military weapon as issued or as modified, and the ammunition therefor unless such firearm is an automatic or sawed-off firearm;

(3) Any cannon or other artillery piece which, regardless of its actual age, is of a type in accepted use prior to 1887, has no mechanical, hydraulic, pneumatic or other system for absorbing recoil and returning the tube into battery without displacing the carriage, and is designed and safe for use only with black powder;

(4) Black powder, priming quills and percussion caps possessed and lawfully used to fire a cannon of a type defined in subsection (k)(3) hereof during displays, celebrations, organized matches or shoots, and target practice, and smokeless and black powder, primers and percussion caps possessed and lawfully used as a propellant or ignition device in small-arms or small arms ammunition;

(5) Dangerous ordnance which is inoperable or inert and cannot readily be rendered operable or activated, and which is kept as a trophy, souvenir, curio or museum piece.

(6) Any device which is expressly excepted from the definition of a destructive device pursuant to the "Gun Control Act of 1968," 82 Stat. 1213, 18 U.S.C. 921(a)(4), and any amendments or additions thereto or reenactments thereof, and regulations issued thereunder.

(l) "Shotgun" means any weapon designed or redesigned, made or remade, and intended to be fired from the shoulder and designed or redesigned, and made or remade to use the energy of the explosive in a fixed shotgun shell to fire through a smooth bore either a number of ball shot or a single projectile for each single pull of the trigger.

(m) "Rifle" means any weapon designed or redesigned, made or remade, and intended to be fired from the shoulder, and designed or redesigned, and made or remade to use the energy of the explosive in a fixed metallic cartridge to fire a single projectile through a rifled bore for each single pull of the trigger.

(n) "Fixed place of business" means an established location of a commercial or industrial enterprise owned, managed, or operated by a person, firm or corporation.

(o) "Unencased" means not enclosed in a case, container or receptacle designed and constructed specifically for the purpose of encasing a firearm, but shall not include a holster for a pistol or revolver.

(p) "Loaded" means any pistol, revolver, rifle, shotgun or any other weapon, by whatever name known, designed to expel a projectile or projectiles by the action of expanding gases, which contains a round or rounds of ammunition in the chamber, slide, receiver, barrel or cylinder or clip attached thereto.

(q) "Unloaded" means any pistol, revolver, rifle, shotgun or any other weapon, by whatever name known, designed to expel a projectile or projectiles by the action of expanding gases, which does not contain a round or rounds of ammunition in the chamber, receiver, slide, barrel, cylinder or clip attached thereto.

(r) "Firearms dealer" means any person, firm or corporation engaged in the business of selling or trading firearms or ammunition at wholesale or retail within the limits of the City, whether as the principal business of such person, firm or corporation, or in addition thereto.

(s) "Fugitive from justice" means a person who flees, escapes from custody, conceals or attempts to use any other unlawful means to avoid prosecution or punishment for a felony under the laws of this or any other state, the United States or any of its territories or possessions, the District of Columbia or any foreign country with which the United States has a treaty of extradition in effect.

(t) "Resident" means any person who has a factual place of residence within the limits of the City.

(u) "Nonresident" means any person who does not have a factual place of residence within the limits of the City.

(v) "Minor" means any person under the age of twenty-one years.

(w) "Director of Public Safety" means the Director of Public Safety or his or her designee.

(x) "Semi-automatic firearm" means any firearm designed or specially adapted to fire a single cartridge and automatically chamber a succeeding cartridge ready to fire, with a single function of the trigger.

"Certain semi-automatic firearms" means any of the following:

(A) A semi-automatic rifle less than forty-two inches in overall length that has an ability to accept a detachable magazine and has at least two of:

(i) a folding or telescoping stock;

(ii) a pistol grip that protrudes conspicuously beneath the action of the weapon.

(iii) a bayonet mount;

(iv) a flash suppressor or threaded barrel designed to accommodate a flash suppressor; and

(v) a grenade launcher;

(B) a semi-automatic pistol that has an ability to accept a detachable magazine and has at least two of:

(i) an ammunition magazine that attaches to the pistol outside of the pistol grip;

(ii) a threaded barrel capable of accepting a barrel extender, flash suppressor, forward handgrip, or silencer;

(iii) a shroud that is attached to, or partially or completely encircles, the barrel and that permits the shooter to hold the firearm with the non-trigger hand without being burned;

(iv) a manufactured weight of 50 ounces or more when the pistol is unloaded; and

(v) is known by the person charged at the time of the offense to be a semi-automatic version of an automatic firearm.

(C) A semi-automatic shotgun that has at least two of:

(i) a folding or telescoping stock;

(ii) a pistol grip that protrudes conspicuously beneath the action of the weapon;

(iii) a fixed magazine capacity in excess of 5 of the 2 3/4 inch standard rounds; and

(iv) an ability to accept detachable magazine.

549.03. Using weapons while intoxicated.

(a) No person, while under the influence of alcohol or any drug of abuse, shall carry or use any firearm or dangerous ordnance.

(b) Whoever violates this section is guilty of using weapons while intoxicated, a misdemeanor of the first degree.

549.05. Failure to secure dangerous ordnance.

(a) No person, in acquiring, possessing, carrying or using any dangerous ordnance or deadly weapon shall negligently fail to take proper precautions:

(1) To secure the dangerous ordnance or deadly weapon against theft, or against its acquisition or use by any unauthorized or incompetent person;

(2) To insure the safety of persons and property.

(b) Whoever violates this section is guilty of failure to secure dangerous ordnance, a misdemeanor of the second degree.

549.06. Unlawful transactions in weapons.

(a) No person shall:

(1) Manufacture, possess for sale, sell or furnish to any person other than a law enforcement agency for authorized use in police work, any brass knuckles, cestus, billy, blackjack, sandbag, switchblade knife, springblade knife, gravity knife or similar weapon;

(2) When transferring any dangerous ordnance to another, negligently fail to require the transferee to exhibit such identification, license or permit showing him to be authorized to acquire dangerous ordnance pursuant to Ohio R.C. 2923.17, or negligently fail to take a complete record of the transaction and forthwith forward a copy of such record to the sheriff of the county or safety director or police chief of the municipality where the transaction takes place;

(3) Knowingly fail to report to law enforcement authorities forthwith the loss or theft of any firearm or dangerous ordnance in such person's possession or under his control.

(b) Whoever violates this section is guilty of unlawful transactions in weapons. Violation of subsections (a)(1) or (2) hereof is a misdemeanor of the second degree. Violation of subsection (a)(3) hereof is a misdemeanor of the fourth degree.

549.07. Improperly furnishing firearms to a minor.

(a) No person shall:

(1) Sell any firearm to a person under age eighteen;

(2) Sell any handgun to a person under age twenty-one;

(3) Furnish any firearm to a person under age eighteen, except for purposes of lawful hunting, or for purposes of instruction in firearms safety, care, handling or marksmanship under the supervision or control of a responsible adult.

(b) Whoever violates this section is guilty of improperly furnishing firearms to a minor, a misdemeanor of the second degree.

549.08. Firearms confiscated. Any firearm seized by a police officer upon the arrest of any person, firm or corporation charged with a violation of any provision of this chapter shall, upon conviction of such person, firm or corporation, be ordered confiscated by the Municipal Court of the City, and such firearm shall be turned over to the Department of Police Operations for disposal, except that any firearm so seized which has been reported stolen shall be returned to the lawful owner thereof.

549.09. Handgun identification card; fee.

(a) No person shall purchase, own, possess, receive, have on or about his person, or use any handgun except upon a suitable firing range, unless such person has a handgun owner's identification card issued to him and in effect pursuant to this section, or unless such person is exempt from the requirement of an identification card pursuant to Section 549.10.

(b) Application for a handgun owner's identification card shall be made in writing under oath to the Chief of Police of the City, and shall be accompanied by an application fee of fifteen dollars every three years which shall be paid into the City Treasury. If it appears upon investigation that the applicant is a resident of the City and is not prohibited by this section from being issued an identification card, the Chief of Police shall issue such card to the applicant within not less than seven nor more than fourteen days from the date of application. A copy of such card shall be retained by the Chief of Police together with a copy of the application, which documents shall be maintained on permanent file by the Chief of Police. Handgun owner's identification cards issued pursuant to this section expire on the thirtieth day following the holder's third birthday after the date of issuance of such card, and are void at any time prior to such date upon the holder becoming a member of the class of persons prohibited by this section from being issued handgun owner's identification cards.

(c) All identification cards issued pursuant to this section shall be entitled "City of Toledo, Ohio, Handgun Owner's Identification Card", shall be serially numbered according to a system devised by the Chief of Police, shall bear the date of issue, the date of expiration, the name of the Chief of Police, and the applicant's name, home address, birth date, physical description, full face photograph and Social Security number and shall be signed by the holder. If, within ninety days after receipt of notice of approval of the application for a hand-gun owner's identification card, the applicant fails to appear for the processing and issuance of his identification card, such application shall be considered void but shall not bar the applicant from reapplying for such identification card.

(d) A handgun owner's identification card shall not be issued to any person who is:

(1) A person prohibited by Ohio R.C. 2923.13 from purchasing, owning, possessing, receiving, having on or about his person or using any firearm or ammunition;

(2) A minor;

(3) A person convicted of an illegal use or possession of narcotics, drugs of abuse or any drug dependent person;

(4) A person with more than one conviction or disorderly conduct involving intoxication within one year prior to his application for such identification card;

(5) Any person with more than one conviction or a misdemeanor involving the use of force and violence, or the threat of the use of force and violence against the person of another within one year prior to his application for such identification card;

(6) Any person convicted of a violation of any provision of this chapter, or Ohio R.C. 2923.11 to 2923.24 or any other comparable Municipal, State or Federal law;

(7) Any person who is otherwise prohibited by Federal law from owning, possessing, shipping, transporting, or receiving firearms or ammunition.

(e) No person shall knowingly give any false information in making application for a handgun owner's identification card, and no person shall use or attempt to use a handgun owner's identification card to purchase, own, possess, receive, have on or about his person, or use any handgun, knowing such identification card to belong to another, or knowing it was obtained by means of false information, or when it is void by reason of the class of persons prohibited by this section from being issued a handgun owner's identification card.

(f) No person shall use or attempt to use his handgun owner's identification card to obtain any handgun for any person who does not have a handgun owner's identification card issued to him and in effect pursuant to this section and is not exempt from the requirement of an identification card pursuant to Section 549. 10, and no person shall knowingly permit his identification card to be used by another to purchase, own, possess, receive, have on or about the person or use any handgun or handgun ammunition.

(g) If for reasons set forth in this chapter, an applicant is denied a handgun owner's identification card, such applicant may appeal such order or refusal to the Director of Public Safety within ten days from the receipt of a notice of such refusal. The Director of Public Safety shall conduct a hearing upon such appeal within thirty days from the receipt of a notice of appeal in writing from such order of refusal. If for reasons set forth in this chapter, an applicant is denied a handgun owner's identification card who, at the time of making the application is the owner or possessor of a handgun, the applicant shall, either within thirty days from the date of denial of such handgun owner's identification card, or if an appeal is filed by such applicant to the Director of Public Safety, within thirty days from the date of receipt of an affirmance of such order of refusal by the Director of Public Safety, dispose of such handgun pursuant to the provisions of Section 549.11.

(h) A possessor of a handgun owner's identification card shall become ineligible to possess such a card if he becomes one of that class of persons set forth in subsection (d) hereof, and he shall thereupon immediately forfeit such card and return the same to the Chief of Police, and he shall within thirty days from the date of forfeiture dispose of any handguns in his possession pursuant to the provisions of Section 549.11.

(i) A nonresident of the City who owns, manages or operates a business within the City, or whose terms of employment require him to own, possess or have on or about his person a handgun during the hours of his employment within the City, may make application for and obtain a handgun owner's identification card pursuant to the provisions of this section.

(j) Whoever violates this section is guilty of a misdemeanor of the first degree. Each day such violation is continued or permitted to continue shall constitute a separate offense and shall be punishable as such hereunder.

549.10. Handgun identification card exemptions.

(a) The requirements of a handgun owner's identification card contained in Section 549.09 do not apply to the purchase, ownership, possession, receipt, having on or about the person or use of any handgun or handgun ammunition by:

(1) Any state of the United States, or any political subdivision, department or agency of either;

(2) Officers or agents of any state or the United States or any political subdivision, department or agency of either, members of the organized militia of any state or the armed forces of the United States, or law enforcement officers of any political subdivision to the extent that the official duties of any such person require him to purchase, own, possess, receive, carry or use handguns;

(3) Licensed manufacturers and dealers, and their agents and servants, to the extent that the purchase, ownership, possession, receipt or use of any handgun is in the ordinary course of business;

(4) Nonresidents of the City who are residents of the State; provided that such persons are not prohibited by the provisions of Section 549.09 and Ohio R.C. 2923.13, and possess and exhibit upon request such valid permit, authorization or identification issued by the Chief of Police or the sheriff of their place of residence to purchase, own, possess, receive, carry or use any handgun and provided, further, that handguns in the possession of or under the control of nonresidents shall at all times be unloaded and encased, except while on a suitable firing range, or while being used for lawful hunting, or while unloaded at a public firearms display, show or exhibition. The requirements of a valid permit, authorization or identification as required by this subsection shall not apply to a nonresident who enters the City with firearms for the sole and only purpose of exhibiting or trading such fire-arms at a public firearms display, show or exhibition, or for the sole and only purpose of participating in an organized and sanctioned pistol match. The permit, authorization or identification provided for in this subsection shall be valid for thirty days only from the date thereof;

(5) Nonresidents of the City, moving to this City for the purpose of establishing residence herein, during a period not exceeding thirty days from the date of their entry into this City for such purpose; provided that such persons shall possess and exhibit upon request such valid permit, authorization or identification as may be required by the laws of their former domicile to own, possess, receive or have on or about the person any handgun;

(6) Executors, administrators, guardians, receivers, trustees in bankruptcy and other fiduciaries duly qualified and appointed by a court of competent jurisdiction, when acting in their fiduciary capacity, and their attorneys;

(7) Heirs and legatees acquiring any handgun or handgun ammunition through distribution of an estate, and other persons other than fiduciaries acquiring any handgun through operation of law, during a period not exceeding sixty days from the date of their coming into possession of any handgun;

(8) Carriers, warehousemen and others engaged in the business of transporting or storing firearms, handguns and their agents, servants, to the extent that the ownership, possession, receipt or having on or about the person of any handgun by such persons is in the ordinary course of business and in conformity with the laws of this State or the United States.

(b) No person shall purchase or otherwise obtain, or attempt to purchase or otherwise obtain, any handgun by claiming an exemption pursuant to this section from the requirement of a handgun owner's identification card contained in Section 549.09 knowing such claim of exemption to be false.

(c) Whoever violates this section is guilty of a misdemeanor of the first degree.

549.11. Handgun transfer restrictions.

(a) No person in this City shall purchase or otherwise obtain ownership or possession of, or as security for a loan, any handgun from any person unless the transferor exhibits a handgun owner's identification card valid on its face, issued to the transferor pursuant to Section 549. 09, or unless the transferor exhibits evidence that he is exempt from the requirement of an identification card pursuant to Section 549.10

(b) No person in this City shall sell, deliver, transfer or furnish any handgun to any person in this City, unless the transferee exhibits a handgun owner's identification card valid on its face and issued to the transferee pursuant to Section 549.09, or unless the transferee exhibits evidence that he is exempt from the requirement of an identification card pursuant to Section 549. 10.

(c) Any person in possession of a handgun who is denied a handgun owner's identification card, or as a possessor of a handgun owner's identification card becomes ineligible to possess such card, shall dispose of the handgun by selling or transferring it to a qualified transferee or by physically destroying the handgun.

(d) Whoever violates this section is guilty of a misdemeanor of the first degree. Each day such violation is continued or permitted to continue shall constitute a separate offense and shall be punishable as such hereunder.

549.12. Dealer's license.

(a) Every firearms dealer in this City shall procure a license to engage in such business. Any firearms dealer engaged in such business at more than one permanent location in this City shall procure a separate license for each such permanent location. Application for such license shall be made under oath to the Chief of Police.

(b) Upon application and payment of the fee prescribed in Section 549.13 the Chief of Police shall authorize the issuance of a license to engage in business as a firearms dealer. No license shall be issued to any applicant whose license has been revoked nor to any applicant who has been convicted of a violation of any section of this chapter, until the expiration of five years from the effective date of such revocation or five years from the date of conviction. No person, firm or corporation shall be issued a firearms dealer's license who does not meet the qualifications for a handgun owner's identification card. The license issued pursuant to this section shall be prominently displayed by every firearms dealer at his place of business. ...

(d) Whoever violates this section is guilty of a misdemeanor of the first degree. Each day such violation is continued or permitted to continue shall constitute a separate offense and shall be punishable as such hereunder.

549.13. Dealer's license fee. The fee for the license required by Section 549.12 shall be fifty dollars ($50.00) renewable annually. The license so issued shall be good and considered to be in full force and effect until such time as the license is revoked, surrendered or the licensee discontinues the business of being a gun dealer. Licenses issued pursuant to the provisions of this chapter are not transferable.

549.14. Records of firearms dealers.

(a) Every firearms dealer shall maintain a record of the sale of handguns. Such record of sale shall contain the name of the firearms dealer; the place and date of the sale; the name, address, age and Social Security number of the purchaser; and shall identify such handgun by manufacturer, model, number or name, type, caliber and serial number. If such weapon does not bear a serial number or is other than a standard model, or has been modified and improved, the record of sale shall give a brief description of the weapon including such information as may be necessary to identify it. The record of sale shall include the serial number of the identification card of the purchaser issued pursuant to Section 549.09, or if the purchaser is a nonresident of Toledo, then such information as is contained in a valid permit, authorization or identification as may be required by the laws of the residence of the purchaser, or valid permit, authorization or identification issued by the Chief of Police or the sheriff of the place of residence of such purchaser. Each such record of sale shall be signed by the purchaser and the salesman and no sale shall be made unless such identification card or permit is exhibited by the purchaser. In the instance of sales from a wholesale distributor to a dealer the license number of the dealer shall be recorded in lieu of the aforesaid identification card.

(b) Every firearms dealer shall, on or before the fifth day of each month, furnish to the Chief of Police a copy of the record of sales of handguns made during the preceding month, and shall make available at all times to the Chief of Police or his duly authorized agents the record of sales of all firearms required to be kept pursuant to the provisions of the Federal Firearms Act. In the case of a wholesale distributor, its records shall be made available at all times in lieu of the foregoing.

(c) No firearms dealer or any agent or employee of such dealer shall fail to comply with this section or knowingly cause any false information to be entered on a record of sale of any handgun, and no purchaser of any handgun shall give any false information for entry on such record of sale.

(d) As used in this section the term "wholesale distributor" shall be an individual, partnership or corporation, selling only to licensed dealers and not at retail or otherwise to individual consumers.

(e) Whoever violates this section is guilty of a misdemeanor of the first degree. Each day such violation is continued or permitted to continue shall constitute a separate offense and shall be punishable as such hereunder.

549.15. Revocation of license.

(a) When any licensed firearms dealer has been convicted of a violation of any provision of this chapter, the Chief of Police shall revoke the license of such firearms dealer. A certified copy of the order of revocation shall be sent forthwith to the firearms dealer, who may appeal such order of revocation to the Director of Public Safety within ten days from the receipt of the order. The Director of Public Safety shall conduct a hearing upon such appeal within thirty days from the receipt of a notice of appeal in writing from such order of revocation.

(b) No firearms dealer whose license has been revoked pursuant to this section shall engage in business as a firearms dealer in this City for a period of five years from the effective date of such revocation. Each day of continuing violation shall be deemed a separate offense.

(c) No firearms dealer shall employ, or enter into any partnership or corporation with any person, firm or corporation, whose firearms dealer's license has been revoked pursuant to the provisions of this section.

(d) Whoever violates this section is guilty of a misdemeanor of the first degree. Each day such violation is continued or permitted to continue shall constitute a separate offense and shall be punishable as such hereunder.

549.16. Carrying handguns in vehicles restricted.

(a) No person, unless exempted by any provision of this chapter, shall transport in any vehicle in or upon the streets or other public places of the City, any handgun without having in his possession an identification card required by the provisions of this chapter, or if a nonresident, an identification card, permit or authorization issued by the proper authorities of his residence, and provided further that the handgun shall at all times be unloaded and encased.

(b) Whoever violates this section is guilty of a misdemeanor of the first degree. Each day such violation is continued or permitted to continue shall constitute a separate offense and shall be punishable as such hereunder.

549.17. Stolen or lost firearms.

(a) Whenever any firearm is stolen or lost, the person losing possession thereof shall, immediately upon discovery of such theft or loss, make a report thereof to the Chief of Police showing the following:

(1) Name and address of the person owning or having possession of such firearm.

(2) Kind of firearm.

(3) Serial number.

(4) Model.

(5) Caliber.

(6) Manufacturer of firearm.

(7) Date and place of theft or loss.

(8) A complete statement of the facts and circumstances surrounding such theft or loss.

(b) Whoever violates this section is guilty of a misdemeanor of the first degree.

549.21. Exemptions.

(a) The provisions of this chapter shall not apply to officers or agents of any State or of the United States, or any political subdivision, division or agency of either, members of the organized militia of any state or the armed forces of the United States, or law enforcement officers of any political subdivision to the extent that the official duties of any such person require him to carry, be in possession of, or have on or about his person any such firearm or weapon, nor to a person whose terms of employment require him to carry, possess or have on or about his person during the hours of his employment such firearm or weapon; nor to the members of a color guard, military unit or marching unit, while such color guard, military unit or marching unit is participating in a duly authorized parade, athletic event, funeral or while so participating in such other civic events and functions in which the flag of the United States is displayed, provided, however, that any firearms borne by members of any such color guard, military unit or marching unit while so participating in such events shall at all times remain unloaded.

(b) Nothing in this chapter shall be construed to limit, restrict or modify the legitimate use of firearms as provided by the laws of the United States, or of the State of Ohio, nor shall it be construed to restrict, limit, prohibit or modify the conduct of legitimate business with manufacturers, wholesale dealers and retail dealers of firearms licensed by the Federal Government, or retail dealers of firearms licensed by the City, or with or by persons, firms or corporations engaged in the business of repairing, altering, assembling, cleaning, polishing, engraving or trueing or who perform any mechanical operation upon any firearm.

549.22. Failure to secure firearms.

(a) Except as provided in this section, no person shall knowingly store or leave a loaded firearm, or an unloaded firearm accompanied by ammunition, in any location where the person knows or reasonably should know that any person under eighteen years of age is likely to gain possession of such firearm.

(b) Section (a) shall not apply to:

(1) Any firearm which has been secured with a locking device which was designed to prevent the firearm from discharging;

(2) Any firearm which has been stored securely in a manner that could reasonably be expected to prevent persons under eighteen years of age from gaining possession of the firearm;

(3) Any firearm possessed by a person under eighteen years of age for target shooting, hunting or other lawful purpose if the person under eighteen years of age is, at all times that such person possesses the firearm, supervised by a person older than eighteen years of age;

(4) Any firearm controlled by a person between sixteen and eighteen years of age, by or with the consent of such person's parent or guardian, for the purpose of lawful hunting, or target shooting, if the person giving control of such firearm to the person between sixteen and eighteen years of age intends to use the firearm solely for the purpose of lawful hunting or target shooting;

(5) A firearm which a person under eighteen years of age possessed by entering property in violation of law and removing the firearm in violation of law;

(6) A firearm which a person under eighteen years of age possessed by removing it in violation of law from the body of a person who was carrying such firearm.

(7) A firearm which the owner of the firearm has ready at hand based on the owner's reasonable cause to believe a threat of criminal attack exists to the person, property or family of the firearm owner.

(8) Any firearm possessed by a person under age eighteen when possessed by such person in the lawful defense of his person, property or family.

(c) A dealer of firearms shall post in a conspicuous position on the premises where the dealer conducts business a sign that contains the following warning in block letters not less than one inch in height: "IT IS UNLAWFUL TO STORE OR LEAVE AN UNLOCKED FIREARM WHERE A PERSON UNDER 18 IS LIKELY TO OBTAIN POSSESSION OF IT."

(d)(1) Whoever violates Subsection (a) hereof is guilty of a misdemeanor of the first degree.

(2) Whoever violates subsection (c) hereof is guilty of a minor misdemeanor.

549.23. Possession of certain semi-automatic firearms prohibited, large capacity magazines prohibited, Acknowledgement Form.

(a) No person shall knowingly sell, deliver, rent, lease, display for sale, transfer ownership or possess certain semi-automatic firearms in the City of Toledo.

(b) No person shall knowingly possess a detachable ammunition magazine loaded with more than ten (10) rounds of ammunition.

(c) This section shall not apply to any assault weapons or any detachable ammunition magazines loaded with more than ten (10) rounds possessed or used by a law enforcement officer or law enforcement agency as provided for in Section 549.21, or any detachable ammunition magazine loaded with more than ten (10) rounds which was purchased by a Toledo Police Patrol Officer pursuant to Section 2129.90 or a Toledo Police Command Officer pursuant to Section 2109.88.

(d) It shall be an affirmative defense to prosecution under this section that either:

(1) At the time of violation the offender possessed a Toledo Certain Semi-Automatic Firearms Prohibited Acknowledgement Form describing by make, model and serial number the firearm(s) which would otherwise be unlawful to possess under this section and which form was signed by the person charged with the violation and acknowledgement before a Notary Public on or before the effective date of this section;

(2) That the offender provides a copy of the federal form 4473 showing proof of purchase date prior to the enactment of this ordinance and signed by the current possessor.

(3) That the proof of date of ownership or purchase provisions of (d)(1) or (d)(2) apply and the offender is an immediate family member of the person indicated as the owner of the firearm and is not otherwise prohibited from possessing firearms.

(e) Any person who acquires ownership of a prohibited semi-automatic firearm by inheritance or otherwise by operation of law shall not be in violation of this ordinance if the person takes possession of such prohibited semi-automatic firearm solely for the purpose of disposing of it, rendering it permanently inoperable, or permanently making it a devise no longer classified as a prohibited semi-automatic firearm provided that within 60 days of acquiring ownership the person in fact disposes of it, renders it permanently inoperable, or permanently makes it a device no longer classified as a prohibited semi-automatic firearm.

(f) Any person who otherwise is not prohibited from firearms ownership and who acquires ownership of a certain semi automatic firearm that is "grandfathered" by the provisions of section 549.23(d)(1) or (d)(2) by inheritance from an immediate family member (grandparent, parent, spouse, son, daughter, brother or sister) shall not be in violation of this ordinance so long as they complete a Toledo Certain Semi-Automatic Firearms Prohibited Acknowledgement Form within 60 days of acquiring ownership.

(g) Whoever violates this section shall be guilty of a misdemeanor of the first degree.

549.25. Certain handguns prohibited.

(A) The possession or sale of handguns with the following characteristics is prohibited:

(1) A revolver or single shot handguns which:

(a) Has an overall frame length with conventional grips, not measured diagonally of less than 4½"; or

(b) Has a barrel length of less than 3": or

(c) Fails to pass a safety test as follows: A Double Action Revolver must have a safety mechanism which automatically (or in a Single Action Revolver by manual operation) causes the hammer to retract to a point where the firing pin does not rest upon the primer of the cartridge or when not in lawful use an external locking or safety device which prevents the firearm from discharging; or

(d) Fails to achieve a score of forty-five (45) points as set forth in Appendix "A" of this chapter;

(2) A semiautomatic handgun which

(a) Does not have a positive manually operated safety device or a double action trigger pull; or when not in lawful use an external locking or safety device which prevents the firearm from discharging, or;

(b) Has a combined length and height of less than 10" with the height (right angle measurement to barrel without magazine or extension) being at least 4" and the length being at least 6"; or

(c) Fails to achieve a score of seventy-five (75) points as set forth in Appendix "A" of this chapter.

(B) No person shall sell, transfer, give, offer for sale or possess any handgun described in "A" above.

(1) Whoever violates this section is guilty of a misdemeanor of the first degree.

(C) Exceptions.

(1) Whether or not they are prohibited by "A" above, the following handguns are exempted from this section:

(a) Any handgun which is an antique or relic firearm, acquired primarily for display or investment, and not for the purpose of firing.

(b) Any handgun which has been modified to either render it permanently inoperable or to make it a device no longer prohibited by "A" above.

(c) Any handgun possessed or used by a law enforcement officer or law enforcement agency or a handgun purchased by a retired Toledo Police patrol officer pursuant to Section 2129.90 or Toledo Police Command Officer pursuant to Section 2109.88.

(d) Any person who acquires ownership of a handgun described in "A" above, by inheritance or otherwise by operation of law shall not be in violation of this section if the person takes possession of such handgun solely for the purpose of disposing of it, rendering it permanently inoperable, or permanently making it a device no longer described in this section, provided that within 30 days of acquiring ownership the person in fact disposes of it, renders it permanently inoperable, or permanently converts it into a device no longer prohibited by this section.

(e) No person shall be subject to prosecution under any section of Chapter 549 who voluntarily presents a handgun for inspection to the Toledo Police Department or who is transporting the handgun(s) for the purpose of having the handgun inspected or turning the handgun in. Any gun confiscated by the police pursuant to such an inspection will be held for at least 10 days for the owner to provide an address out-side of the city where such gun can be shipped.

(f) Any handgun manufactured 50 or more years ago, or any handgun determined by the U.S.B.A.T.F. to be a relic or curio, or replica of a relic or curio or any firearm that does not fire fixed ammunition. (e.g. Black Powder Guns)

(g) Any person who acquires ownership of a "grandfathered" prohibited handgun (handguns that had previously been in compliance by virtue of having been properly listed on a Handgun Acknowledgement form by an owner who had a valid Handgun Identification Card) by inheritance or otherwise by operation of law shall not be in violation of this section, so long as they obtain a Handgun Identification Card and complete a Toledo Handgun Acknowledgement Form within 30 days of acquiring ownership.

(D) It shall be an affirmative defense to prosecution under this section that at the time of the violation the offender possessed:

(1) a valid handgun owner's identification card; and

(2) a Toledo Handgun Acknowledgement form describing by make, model and serial number the handgun(s) which would otherwise be unlawful to possess under this section and which form was signed by the person charged with the violation and acknowledged before a Notary Public on or before the effective date of this section;

(E) The Toledo Handgun Acknowledgment form, Notice, and Appendix A shall be mailed to all persons who possess a Handgun Identification Permit. ...

[Toledo Municipal Code, current through January 1, 2009]

OKLAHOMA

OKLA. STAT.

Title 21. Crimes and Punishments

Chapter 53. Manufacturing, Selling and Wearing Weapons

1271.1. Detention or arrest of person under 18 years--Confiscation and forfeiture of prohibited weapons and firearms--Disposition of forfeited weapons and firearms

A. Whenever a person under eighteen (18) years of age is detained or arrested by a law enforcement officer and is carrying any weapon or firearm prohibited by Section 1272 of this title, each such prohibited weapon and firearm may be confiscated and forfeited to the State of Oklahoma by the law enforcement authority. Such confiscation and forfeiture shall not require that criminal charges be filed against the minor.

B. However, when a weapon or firearm confiscated pursuant to the provisions of this section has been taken by a minor without the permission of the owner, the weapon or firearm shall be returned to the owner pursuant to the procedures provided in Section 1321 of Title 22 of the Oklahoma Statutes, provided the possession of such weapon or firearm by the owner is not otherwise prohibited by law.

C. Any weapon or firearm confiscated and forfeited by any law enforcement authority may be sold at public auction, or when no longer needed as evidence in the criminal proceeding the confiscating authority may lease any firearm confiscated and forfeited by law pursuant to this section to any law enforcement agency for a period of one (1) year. Such lease may be renewed each year thereafter at the discretion of such authority to assist in the enforcement of the laws of this state or its political subdivisions. Any weapon or firearm deemed by the confiscating authority to be inappropriate for lease or sale shall be destroyed.

D. For purposes of this section, the term "confiscate" shall not be construed to prohibit any parent, guardian or other adult person from removing or otherwise seizing from any minor any weapon or firearm in the minor's possession. Provided however, no school authority shall return any weapon or firearm removed or otherwise seized from any minor to any person, and shall immediately deliver such weapon or firearm to a law enforcement authority for prosecution and forfeiture.

1272. Unlawful Carry

A. It shall be unlawful for any person to carry upon or about his or her person, or in a purse or other container belonging to the person, any pistol, revolver, shotgun or rifle whether loaded or unloaded ... or any other offensive weapon, whether such weapon be concealed or unconcealed, except this section shall not prohibit:

1. The proper use of guns and knives for hunting, fishing, educational or recreational purposes;

2. The carrying or use of weapons in a manner otherwise permitted by statute or authorized by the Oklahoma Self-Defense Act;

3. The carrying, possession and use of any weapon by a peace officer or other person authorized by law to carry a weapon in the performance of official duties and in compliance with the rules of the employing agency;

4. The carrying or use of weapons in a courthouse by a district judge, associate district judge or special district judge within this state, who is in possession of a valid concealed handgun license issued pursuant to the provisions of the Oklahoma Self-Defense Act and whose name appears on a list maintained by the Administrative Director of the Courts; or

5. The carrying and use of firearms and other weapons provided in this subsection when used for the purpose of living history reenactment. For purposes of this paragraph, "living history reenactment" means depiction of historical characters, scenes, historical life or events for entertainment, education, or historical documentation through the wearing or use of period, historical, antique or vintage clothing, accessories, firearms, weapons, and other implements of the historical period.

B. Any person convicted of violating the foregoing provision shall be guilty of a misdemeanor punishable as provided in Section 1276 of this title.

1273. Allowing minors to possess firearms

A. It shall be unlawful for any person within this state to sell or give to any child any of the arms or weapons designated in Section 1272 of this title; provided, the provisions of this section shall not prohibit a parent from giving his or her child a rifle or shotgun for participation in hunting animals or fowl, hunter safety classes, target shooting, skeet, trap or other recognized sporting events, except as provided in subsection B of this section.

B. It shall be unlawful for any parent or guardian to intentionally, knowingly, or recklessly permit his or her child to possess any of the arms or weapons designated in Section 1272 of this title, including any rifle or shotgun, if such parent is aware of a substantial risk that the child will use the weapon to commit a criminal offense or if the child has either been adjudicated a delinquent or has been convicted as an adult for any criminal offense.

C. It shall be unlawful for any child to possess any of the arms or weapons designated in Section 1272 of this title, except rifles or shotguns used for participation in hunting animals or fowl, hunter safety classes, target shooting, skeet, trap or other recognized sporting event. Provided, the possession of rifles or shotguns authorized by this section shall not authorize the possession of such weapons by any person who is subject to the provisions of Section 1283 of this title.

D. Any person violating the provisions of this section shall, upon conviction, be punished as provided in Section 1276 of this title, and, any child violating the provisions of this section shall be subject to adjudication as a delinquent. In addition, any person violating the provisions of this section shall be liable for civil damages for any injury or death to any person and for any damage to property resulting from any discharge of a firearm or use of any other weapon as provided in Section 10 of Title 23 of the Oklahoma Statutes. Any person convicted of violating the provisions of this section after having been issued a concealed handgun license pursuant to the provisions of the Oklahoma Self-Defense Act may be liable for an administrative violation as provided in Section 1276 of this title.

E. As used in this section, "child" means a person under eighteen (18) years of age.

1276. Penalty for 1272 and 1273

Any person violating the provisions of Section 1272 or 1273 shall, upon a first conviction, be adjudged guilty of a misdemeanor and the party offending shall be punished by a fine of not less than One Hundred Dollars ($100.00) nor more than Two Hundred Fifty Dollars ($250.00), or by imprisonment in the county jail for a period not to exceed thirty (30) days or both such fine and imprisonment. On the second and every subsequent violation, the party offending shall, upon conviction, be punished by a fine of not less than Two Hundred Fifty Dollars ($250.00) nor more than Five Hundred Dollars ($500.00), or by imprisonment in the county jail for a period not less than thirty (30) days nor more than three (3) months, or by both such fine and imprisonment.

Any person convicted of violating the provisions of Section 1272 or 1273 after having been issued a concealed handgun license pursuant to the provisions of the Oklahoma Self-Defense Act, Sections 1 through 25 of this act, shall have the license suspended for a period of six (6) months and shall be liable for an administrative fine of Fifty Dollars ($50.00) upon a hearing and determination by the Oklahoma State Bureau of Investigation that the person is in violation of the provisions of this section.

1277. Unlawful carry in certain places

A. It shall be unlawful for any person in possession of a valid concealed handgun license issued pursuant to the provisions of the Oklahoma Self-Defense Act to carry any concealed handgun into any of the following places:...

4. Any elementary or secondary school, or technology center school property;...

1279. Misdemeanor pointing a firearm
Except for an act of self-defense, it shall be unlawful for any person to point any pistol or any other deadly weapon whether loaded or not, at any other person or persons. Any person violating the provisions of this section shall, upon conviction, be guilty of a misdemeanor punishable as provided in Section 1280 of this title.

Any person convicted of violating the provisions of this section after having been issued a concealed handgun license pursuant to the provisions of the Oklahoma Self-Defense Act, Sections 1 through 25 of this act, may be subject to an administrative violation as provided in Section 1280 of this title.

1280. Penalty for 1279

Any person violating the provisions of Section 1279 of this title, upon conviction, shall be guilty of a misdemeanor. The person offending shall be punished by a fine of not less than One Hundred Dollars ($100.00) nor more than One Thousand Dollars ($1,000.00) and shall be imprisoned in the county jail for a period not less than three (3) nor more than twelve (12) months. Any person convicted of violating the provisions of Section 1279 of this title after having been issued a handgun license pursuant to the provisions of the Oklahoma Self-Defense Act, Sections I through 25 of this act, shall have the handgun license permanently revoked and shall be liable for an administrative fine of Fifty Dollars ($50.00) upon a hearing and determination by the Oklahoma State Bureau of Investigation that the person is in violation of the provisions of this section.

1280.1. Possession of firearm on school property

A. It shall be unlawful for any person to have in his or her possession on any public or private school property or while in any school bus or vehicle used by any school for transportation of students or teachers any firearm or weapon designated in Section 1272 of this title, except as provided in subsection C of this section or as otherwise authorized by law.

B. "School property" means any publicly or privately owned property held for purposes of elementary, secondary or vocational-technical education, and shall not include property owned by public school districts or private educational entities where such property is leased or rented to an individual or corporation and used for purposes other than educational.

C. Firearms and weapons are allowed on school property and deemed not in violation of subsection A of this section as follows:

1. A gun or knife designed for hunting or fishing purposes kept in a privately owned vehicle and properly displayed or stored as required by law, or a handgun carried in a vehicle pursuant to a valid handgun license authorized by the Oklahoma Self-Defense Act, provided such vehicle containing said gun or knife is driven onto school property only to transport a student to and from school and such vehicle does not remain unattended on school property;

2. A gun or knife used for the purposes of participating in the Oklahoma Department of Wildlife Conservation certified hunter training education course or any other hunting, fishing, safety or firearms training courses, or a recognized firearms sports event, team shooting program or competition, or living history reenactment, provided the course or event is approved by the principal or chief administrator of the school where the course or event is offered, and provided the weapon is properly displayed or stored as required by law pending participation in the course, event, program or competition; and

3. Weapons in the possession of any peace officer or other person authorized by law to possess a weapon in the performance of their duties and responsibilities.

D. Any person violating the provisions of this section shall, upon conviction, be guilty of a felony punishable by a fine not to exceed Five Thousand Dollars ($5,000.00), and imprisonment for not more than two (2) years. Any person convicted of violating the provisions of this section after having been issued a concealed handgun license pursuant to the provisions of the Oklahoma Self-Defense Act shall have the license permanently revoked and shall be liable for an administrative fine of One Hundred Dollars ($100.00) upon a hearing and determination by the Oklahoma State Bureau of Investigation that the person is in violation of the provisions of this section.

1281. Manufacturing slung shot

Any person who manufactures or causes to be manufactured, or sells or offers or keeps for sale or disposes of any instrument or weapon of the kind usually known as slung shot, or of any similar kind, is guilty of a misdemeanor.

1283. Convicted felons and delinquents

A. Except as provided in subsection B of this section, it shall be unlawful for any person convicted of any felony in any court of this state or of another state or of the United States to have in his or her possession or under his or her immediate control, or in any vehicle which the person is operating, or in which the person is riding as a passenger, or at the residence where the convicted person resides, any pistol, imitation or homemade pistol, altered air or toy pistol, machine gun, sawed-off shotgun or rifle, or any other dangerous or deadly firearm.

B. Any person who has previously been convicted of a nonviolent felony in any court of this state or of another state or of the United States, and who has received a full and complete pardon from the proper authority and has not been convicted of any other felony offense which has not been pardoned, shall have restored the right to possess any firearm or other weapon prohibited by subsection A of this section, the right to apply for and carry a concealed handgun pursuant to the Oklahoma Self-Defense Act, and the right to perform the duties of a peace officer, gunsmith, or for firearms repair.

C. It shall be unlawful for any person supervised by the Department of Corrections or any division thereof to have in his or her possession or under his or her immediate control, or at his or her residence, or in any passenger vehicle which the supervised person is operating or is riding as a passenger, any pistol, shotgun or rifle, including any imitation or homemade pistol, altered air or toy pistol, shotgun or rifle, while such person is subject to supervision, probation, parole or inmate status.

D. It shall be unlawful for any person previously adjudicated as a delinquent child or a youthful offender for the commission of an offense, which would have constituted a felony offense if committed by an adult, to have in said person's possession or under the person's immediate control, or have in any vehicle which he or she is driving or in which said person is riding as a passenger, or at the person's residence, any pistol, imitation or homemade pistol, altered air or toy pistol, machine gun, sawed-off shotgun or rifle, or any other dangerous or deadly firearm within ten (10) years after such adjudication.

E. Any person having been issued a concealed handgun license pursuant to the provisions of the Oklahoma Self-Defense Act and who thereafter knowingly or intentionally allows a convicted felon or adjudicated delinquent or a youthful offender as prohibited by the provisions of subsection A, C, or D of this section to possess or have control of any pistol authorized by the Oklahoma Self-Defense Act shall, upon conviction, be guilty of a felony punishable by a fine not to exceed Five Thousand Dollars ($5,000.00). In addition, the person shall have the handgun license revoked by the Oklahoma State Bureau of Investigation after a hearing and determination that the person has violated the provisions of this section.

F. Any convicted or adjudicated person violating the provisions of this section shall, upon conviction, be guilty of a felony punishable as provided in Section 1284 of this title.

G. For purposes of this section, "sawed-off shotgun or rifle" shall mean any shotgun or rifle which has been shortened to any length.

H. For purposes of this section, "altered toy pistol" shall mean any toy weapon which has been altered from its original manufactured state to resemble a real weapon.

I. For purposes of this section, "altered air pistol" shall mean any air pistol manufactured to propel projectiles by air pressure which has been altered from its original manufactured state.

1284. Penalty for 1283

Any previously convicted or adjudicated person who violates any provision of Section 1283 of this title shall be guilty of a felony and, upon conviction thereof, shall be punished by imprisonment in the State Penitentiary for a period not less than one (1) year nor more than ten (10) years.

1288. Purchases of firearms, ammunition and equipment in contiguous states by Oklahoma residents - Purchases in Oklahoma by residents of contiguous states

(a) Residents of the State of Oklahoma may purchase rifles, shotguns, ammunition, cartridge and shotgun shell handloading components and equipment in a state contiguous to the State of Oklahoma, provided that such residents conform to the applicable provisions of the Federal Gun Control Act of 1968, and regulations thereunder, as administered by the United States Secretary of the Treasury, and provided further that such residents conform to the provisions of law applicable to such purchase in the State of Oklahoma and in the contiguous state in which the purchase is made.

(b) Residents of a state contiguous to the State of Oklahoma may purchase rifles, shotguns, ammunition, cartridge and shotgun shell handloading components and equipment in the State of Oklahoma, provided that such residents conform to the applicable provisions of the Gun Control Act of 1968, and regulations thereunder, as administered by the United States Secretary of the Treasury and provided further that such residents conform to the provisions of law applicable to such purchase in the State of Oklahoma and in the state in which such persons reside.

Oklahoma Firearms Act of 1971

1289.1. Oklahoma Firearms Act of 1971 Sections 1289.1 through 1289.17 of this title may be known and cited as the "Oklahoma Firearms Act of 1971."

1289.3. Definitions for Firearms Act -- Pistols "Pistols" as used in the Oklahoma Firearms Act of 1971, Sections 1289.1 through 1289.17 of this title, shall mean any firearm capable of discharging a projectile composed of any material which may reasonably be expected to be able to cause lethal injury, with a barrel or barrels less than sixteen (16) inches in length, and using either gunpowder, gas or any means of rocket propulsion, but not to include flare guns, underwater fishing guns or blank pistols.

1289.4. Definitions for Firearms Act -- Rifles "Rifles" as used in the Oklahoma Firearms Act of 1971, Sections 1289.1 through 1289.17 of this title, shall mean any firearm capable of discharging a projectile composed of any material which may reasonably be expected to be able to cause lethal injury, with a barrel or barrels more than sixteen (16) inches in length, and using either gunpowder, gas or any means of rocket propulsion, but not to include archery equipment, flare guns or underwater fishing guns. In addition, any rifle capable of firing "shot" but primarily designed to fire single projectiles will be regarded as a "rifle".

1289.5. Definitions for Firearms Act -- Shotguns "Shotguns" as used in the Oklahoma Firearms Act of 1971, Sections 1289.1 through 1289.17 of this title, shall mean any firearm capable of discharging a series of projectiles of any material which may reasonably be expected to be able to cause lethal injury, with a barrel or barrels more than eighteen (18) inches in length, and using either gunpowder, gas or any means of rocket propulsion, but not to include any weapon so designed with a barrel less than eighteen (18) inches in length. In addition, any "shotgun" capable of firing single projectiles but primarily designed to fire multiple projectiles such as "shot" will be regarded as a "shotgun".

1289.7a. Transporting or storing firearms in locked motor vehicle on private premises—Prohibition proscribed--Liability--Enforcement

A. No person, property owner, tenant, employer, or business entity shall maintain, establish, or enforce any policy or rule that has the effect of prohibiting any person, except a convicted felon, from transporting and storing firearms in a locked motor vehicle, or from transporting and storing firearms locked in or locked to a motor vehicle on any property set aside for any motor vehicle.

B. No person, property owner, tenant, employer, or business entity shall be liable in any civil action for occurrences which result from the storing of firearms in a locked motor vehicle on any property set aside for any motor vehicle, unless the person, property owner, tenant, employer, or owner of the business entity commits a criminal act involving the use of the firearms. The provisions of this subsection shall not apply to claims pursuant to the Workers' Compensation Act.

C. An individual may bring a civil action to enforce this section. If a plaintiff prevails in a civil action related to the personnel manual against a person, property owner, tenant, employer or business for a violation of this section, the court shall award actual damages, enjoin further violations of this section, and award court costs and attorney fees to the prevailing plaintiff.

D. As used in this section, "motor vehicle" means any automobile, truck, minivan, sports utility vehicle, motorcycle, motor scooter, and any other vehicle required to be registered under the Oklahoma Vehicle License and Registration Act.

1289.10. Furnishing firearms to incompetent persons It shall be unlawful for any person to knowingly transmit, transfer, sell, lend or furnish any shotgun, rifle or pistol to any person who is under an adjudication of mental incompetency, or to any person who is mentally deficient or of unsound mind. Any person convicted of a violation of the provisions of this section shall be punished as provided in Section 1289.15 of this title.

Any person convicted of a violation of the provisions of this section after having been issued a concealed handgun license pursuant to the provisions of the Oklahoma Self-Defense Act, Section 1290.1 et seq. of this title, shall have the license suspended for a term of six (6) months and shall be subject to an administrative fine of Fifty Dollars ($50.00), upon a hearing and determination by the Oklahoma State Bureau of Investigation that the person is in violation of the provisions of this section.

1289.11. Reckless conduct It shall be unlawful for any person to engage in reckless conduct while having in his or her possession any shotgun, rifle or pistol, such actions consisting of creating a situation of unreasonable risk and probability of death or great bodily harm to another, and demonstrating a conscious disregard for the safety of another person. Any person convicted of violating the provisions of this section shall be punished as provided in Section 1289.15 of this title.

Any person convicted of a violation of the provisions of this section after having been issued a concealed handgun license pursuant to the Oklahoma Self-Defense Act, Sections 1 through 25 of this act, shall have the license revoked and shall be subject to an administrative fine of One Thousand Dollars ($1,000.00), upon a hearing and determination by the Oklahoma State Bureau of Investigation that the person is in violation of the provisions of this section.

1289.12. Giving firearms to convicted persons It shall be unlawful for any person within this state to knowingly sell, trade, give, transmit or otherwise cause the transfer of rifles, shotguns or pistols to any convicted felon or an adjudicated delinquent, and it shall be unlawful for any person within this state to knowingly sell, trade, give, transmit or otherwise cause the transfer of any shotgun, rifle or pistol to any individual who is under the influence of alcohol or drugs or is mentally or emotionally unbalanced or disturbed. All persons who engage in selling, trading or otherwise transferring firearms will display this section prominently in full view at or near the point of normal firearms sale, trade or transfer. Any person convicted of violating the provisions of this section shall be punished as provided in Section 1289.15 of this title.

Any person convicted of a violation of this section after having been issued a concealed handgun license pursuant to the Oklahoma Self-Defense Act, Sections 1 through 25 of this act, shall have the license suspended for six (6) months and shall be liable for an administrative fine of Fifty Dollars ($50.00), upon a hearing and determination by the Oklahoma State Bureau of Investigation that the person is in violation of the provisions of this section.

1289.13. Transporting a loaded firearm Except as otherwise provided by the provisions of the Oklahoma Self-Defense Act or another provision of law, it shall be unlawful to transport a loaded pistol, rifle or shotgun in a landborne motor vehicle over a public highway or roadway. However, a rifle or shotgun may be transported clip or magazine loaded and not chamber loaded when transported in an exterior locked compartment of the vehicle or trunk of the vehicle or in the interior compartment of the vehicle notwithstanding the provisions of Section 1289.7 of this title when the person is in possession of a valid handgun license pursuant to the Oklahoma Self-Defense Act.

Any person convicted of a violation of this section shall be punished as provided in Section 1289.15 of this title.

Any person who is the operator of a vehicle or is a passenger in any vehicle wherein another person who is licensed pursuant to the Oklahoma Self-Defense Act to carry a concealed handgun and is carrying a concealed handgun or has concealed a handgun or rifle or shotgun in such vehicle shall not be deemed in violation of the provisions of this section provided the licensee is in or near the vehicle.

1289.13A. Improper transportation of firearm--Fine and court costs--Confiscation of firearm

A. Notwithstanding the provisions of Section 1272 or 1289.13 of this title, any person stopped pursuant to a moving traffic violation who is transporting a loaded pistol in the motor vehicle without a valid concealed handgun permit authorized by the Oklahoma Self-Defense Act or valid license from another state, whether the loaded firearm is concealed or open in the vehicle, shall be issued a traffic citation in the amount of Seventy Dollars ($70.00), plus court costs for transporting a firearm improperly. In addition to the traffic citation provided in this section, the person may also be arrested for any other violation of law.

B. When the arresting officer determines that a valid handgun license exists, pursuant to the Oklahoma Self-Defense Act or any provision of law from another state, for any person in the stopped vehicle, any firearms permitted to be carried pursuant to that license shall not be confiscated, unless:

1. The person is arrested for violating another provision of law other than a violation of subsection A of this section; provided, however, if the person is never charged with an offense pursuant to this paragraph or if the charges are dismissed or the person is acquitted, the weapon shall be returned to the person; or

2. The officer has probable cause to believe the weapon is:

a. contraband, or

b. a firearm used in the commission of a crime other than a violation of subsection A of this section.

C. Nothing in this section shall be construed to require confiscation of any firearm.

1289.15. Penalties for Firearms Act of 1971 Any person adjudged guilty of violating any provision of Section 1289.9, 1289.10, 1289.11, 1289.12 or 1289.13 of this title shall, upon conviction, be punished by a fine of not less than Fifty Dollars ($50.00) nor more than Five Hundred Dollars ($500.00), or imprisonment in the county jail for not less than ten (10) days nor more than six (6) months, or by both such fine and imprisonment.

1289.16. Felony pointing firearms It shall be unlawful for any person to willfully or without lawful cause point a shotgun, rifle or pistol, or any deadly weapon, whether loaded or not, at any person or persons for the purpose of threatening or with the intention of discharging the firearm or with any malice or for any purpose of injuring, either through physical injury or mental or emotional intimidation or for purposes of whimsy, humor or prank, or in anger or otherwise, but not to include the pointing of shotguns, rifles or pistols by law enforcement authorities in the performance of their duties, members of the state military forces in the performance of their duties, members of the federal military reserve and active military components in the performance of their duties, or any federal government law enforcement officer in the performance of any duty, or in the performance of a play on stage, rodeo, television or on film, or in defense of any person, one's home or property. Any person convicted of a violation of the provisions of this section shall be punished as provided in Section 1289.17 of this title.

Any person convicted of a violation of the provisions of this section after having been issued a concealed handgun license pursuant to the Oklahoma Self-Defense Act, Sections 1 through 25 of this act, shall have the license revoked and shall be subject to an administrative fine of One Thousand Dollars ($1,000.00), upon a hearing and determination by the Oklahoma State Bureau of Investigation that the person is in violation of the provisions of this section.

1289.17. Penalties for 1289.16 Any violation of Section 1289.16 of this title shall constitute a felony, for which a person convicted thereof shall be sentenced to imprisonment in the State Penitentiary for not less than one (1) year nor more than ten (10) years.

1289.18. Definitions

A. "Sawed-off shotgun" shall mean any firearm capable of discharging a series of projectiles of any material which may reasonably be expected to be able to cause lethal injury, with a barrel or barrels less than eighteen (18) inches in length, and using either gunpowder, gas or any means of rocket propulsion.

B. "Sawed-off rifle" shall mean any rifle having a barrel or barrels of less than sixteen (16) inches in length or any weapon made from a rifle (whether by alteration, modification, or otherwise) if such a weapon as modified has an overall length of less than twenty-six (26) inches in length, including the stock portion.

C. Every person who has in his possession or under his immediate control a sawed-off shotgun or a sawed-off rifle, whether concealed or not, shall upon conviction be guilty of a felony for the possession of such device, and shall be punishable by a fine not to exceed One Thousand Dollars ($1,000.00), or imprisonment in the State Penitentiary for a period not to exceed two (2) years, or both such fine and imprisonment.

D. It is a defense to prosecution under this section, if the approved application form that authorized the making or transfer of the particular firearm to the defendant, which indicates the registration of the firearm to said defendant pursuant to the National Firearm's Act, is introduced.

1289.19. Restricted bullet and body armor defined As used in Sections 1289.20 through 1289.22 of this title and Section 2 of this act:

1. "Restricted bullet" means a round or elongated missile with a core of less than sixty percent (60%) lead and having a fluorocarbon coating, which is designed to travel at a high velocity and is capable of penetrating body armor; and

2. "Body armor" means a vest or shirt of ten (10) plies or more of bullet resistant material as defined by the Office of Development, Testing and Dissemination, a division of the United States Department of Justice.

1289.20. Manufacture of restricted bullets

A. Except for the purpose of public safety or national security, it shall be unlawful to manufacture, cause to be manufactured, import, advertise for sale or sell within this state any restricted bullet as defined in Section 1289.19 of this title.

B. Any person convicted of violating subsection A of this section shall be guilty of a felony and shall be punished by a fine of not less than Five Hundred Dollars ($500.00) nor more than Ten Thousand Dollars ($10,000.00), or by imprisonment in the State Penitentiary for not more than ten (10) years, or by both such fine and imprisonment.

1289.21. Possession or use of restricted bullets

A. It shall be unlawful for any person to possess, carry upon his person, use or attempt to use against another person any restricted bullet as defined in Section 1289.19 of this title.

B. Any person convicted of violating subsection A of this section shall be guilty of a felony and shall be punished by imprisonment in the State Penitentiary for not less than two (2) years nor more than ten (10) years. The sentence so imposed shall not be suspended.

1289.22. Exemptions The prohibition of possessing or using a restricted bullet shall not apply to law enforcement agencies when such bullet is used for testing, training or demonstration.

1289.24. Firearm regulation - State preemption

A.1. The State Legislature hereby occupies and preempts the entire field of legislation in this state touching in any way firearms, components, ammunition, and supplies to the complete exclusion of any order, ordinance, or regulation by any municipality or other political subdivision of this state. Any existing or future orders, ordinances, or regulations in this field, except as provided for in paragraph 2 of this subsection and subsection C of this section, are null and void.

2. A municipality may adopt any ordinance:

a. relating to the discharge of firearms within the jurisdiction of the municipality, and

b. allowing the municipality to issue a traffic citation for transporting a firearm improperly as provided for in Section 1289.13A of this title, provided however, that penalties contained for violation of any ordinance enacted pursuant to the provisions of this subparagraph shall not exceed the penalties established in the Oklahoma Self-Defense Act.

B. No municipality or other political subdivision of this state shall adopt any order, ordinance, or regulation concerning in any way the sale, purchase, purchase delay, transfer, ownership, use, keeping, possession, carrying, bearing, transportation, licensing, permit, registration, taxation other than sales and compensating use taxes, or other controls on firearms, components, ammunition, and supplies.

C. Except as hereinafter provided, this section shall not prohibit any order, ordinance, or regulation by any municipality concerning the confiscation of property used in violation of the ordinances of the municipality as provided for in Section 28-121 of Title 11 of the Oklahoma Statutes. Provided, however, no municipal ordinance relating to transporting a firearm improperly may include a provision for confiscation of property.

D. When a person's rights pursuant to the protection of the preemption provisions of this section have been violated, the person shall have the right to bring a civil action against the persons, municipality, and political subdivision jointly and severally for injunctive relief or monetary damages or both.

1289.24a. Lawsuits against gun manufacturers

1. The State Legislature declares that the lawful design, marketing, manufacturing, or sale of firearms or ammunition to the public is not unreasonably dangerous activity and does not constitute a nuisance.

2. The authority to bring suit and right to recover against any firearms or ammunition manufacturer, trade association, or dealer by or on behalf of any governmental unit created by or pursuant to an act of the Legislature or the Constitution, or any department, agency, or authority thereof, for damages, abatement, or injunctive relief resulting from or relating to the lawful design, manufacturing, marketing, or sale of firearms or ammunition to the public shall be reserved exclusively to the state. This paragraph shall not prohibit a political subdivision or local government authority from bringing an action against a firearms or ammunition manufacturer or dealer for breach of contract or warranty as to firearms or ammunition purchased by the political subdivision or local government authority. This bill shall not be construed to prohibit an individual from bringing a cause of action based upon an existing recognized theory of law.

1289.26. Use of body armor

Any person who commits or attempts to commit a felony while wearing body armor as defined in Section 1289.19 of this title, in addition to the penalty provided by statute for the felony committed or attempted, upon conviction shall be guilty of a felony for wearing such body armor, which shall be a separate offense from the felony committed or attempted, and shall be punishable by imprisonment in the State Penitentiary for a period of not more than ten (10) years for the first offense, and for a period of not more than twenty (20) years for any second or subsequent offense.

1368. Possession of explosives by convicted felons--Penalty

A. Any person who has been convicted of a felony under the laws of this or any other state or the laws of the United States who, with an unlawful intent, is in possession of any explosives, upon conviction, shall be guilty of a felony and shall be punished by a fine of not to exceed Five Thousand Dollars ($ 5,000.00), or by imprisonment in the State Penitentiary for a term not to exceed ten (10) years, or by both such fine and imprisonment.

B. For purposes of this section, the term "explosive" shall have the same definition as the term "explosive" as defined by Chapter 8 of Title 63 of the Oklahoma Statutes.

1767.5. Possession, manufacture, storage, or use of explosive without permit

A. Any person who shall possess, manufacture, store, or use any explosive, as defined in Section 121.1 of Title 63 of the Oklahoma Statutes, without having in the possession of the person a permit, or a copy thereof, issued pursuant to the Oklahoma Explosives and Blasting Regulation Act, shal be deemed guilty of a misdemeanor.

B. This section shall not be construed to:

1. Apply to any person or activity expressly exempted from the Oklahoma Explosives and Blasting Regulation Act;

2. Apply to, or repeal any laws pertaining to, the acts of mischief of juveniles involving noninjurious firecrackers or devices commonly called "stink bombs";

3. Apply to explosives while in transit in, into, or through this state, if the operator of the vehicle transporting the explosives carries in the vehicle the shipping papers required by 49 C.F.R., Section 172.200 et seq., and displays such papers to any law enforcement officer upon request;

4. Apply to any person who may possess, store or use gunpowder in a quantity reasonably calculated to be necessary for hunting or shooting purposes; or

5. Apply to any certified bomb technician employed by a federally accredited bomb squad of an agency of the federal government, this state, or any political subdivision of this state.

[Current through Emergency Legislation through Chapter 2 of the First Regular Session of the 52nd Legislature]

OREGON

OR. REV. STAT.

Title 16. Crimes and Punishments

Chapter 166. Offenses Against Public Order; Firearms and Other Weapons; Racketeering

Authority to Regulate Firearms

166.170 State preemption.

(1) Except as expressly authorized by state statute, the authority to regulate in any matter whatsoever the sale, acquisition, transfer, ownership, possession, storage, transportation or use of firearms or any element relating to firearms and components thereof, including ammunition, is vested solely in the Legislative Assembly.

(2) Except as expressly authorized by state statute, no county, city or other municipal corporation or district may enact civil or criminal ordinances, including but not limited to zoning ordinances, to regulate, restrict or prohibit the sale, acquisition, transfer, ownership, possession, storage, transportation or use of firearms or any element relating to firearms and components thereof, including ammunition. Ordinances that are contrary to this subsection are void.

166.173. Authority of city or county to regulate possession of loaded firearms in public places.

(1) A city or county may adopt ordinances to regulate, restrict or prohibit the possession of loaded firearms in public places as defined in ORS 161.015.

(2) Ordinances adopted under subsection (1) of this section do not apply to or affect:

(a) A law enforcement officer in the performance of official duty.

(b) A member of the military in the performance of official duty.

(c) A person licensed to carry a concealed handgun.

(d) A person authorized to possess a loaded firearm while in or on a public building or court facility under ORS 166.370.

166.174. City, county, municipal corporation or district regulation of possession or sale of firearms. Notwithstanding any other provision of law, a city, county or other municipal corporation or district may not adopt ordinances that regulate, restrict or prohibit the possession or sale of firearms in a public building that is rented or leased to a person during the term of the lease.

166.175. City regulation of purchase of used firearms.

(1) Notwithstanding any other provision of law, a city may continue to regulate the purchase of used firearms by pawnshops and secondhand stores.

(2) As used in this section, "secondhand store" means a store or business whose primary source of revenue is the sale of used merchandise.

166.176. Exception to preemption.

(1) Nothing in ORS 166.170 or 166.171 is intended to preempt, invalidate or in any way affect the operation of any provision of a county ordinance that was in effect on November 2, 1995, to the extent that the provision:

(a) Established a procedure for regulating, restricting or prohibiting the discharge of firearms; or

(b) Regulated, restricted or prohibited the discharge of firearms.

(2) Subsection (1) of this section does not apply to ordinances regulating, restricting or prohibiting the discharge of firearms on a shooting range or in a shooting gallery or other area designed and built for the purpose of target shooting.

Possession and Use of Weapons

166.190. Pointing firearm at another; courts having jurisdiction over offense. Any person over the age of 12 years who, with or without malice, purposely points or aims any loaded or empty pistol, gun, revolver or other firearm, at or toward any other person within range of the fire-arm, except in self-defense, shall be fined upon conviction in any sum not less than $10 nor more than $500, or be imprisoned in the county jail not less than 10 days nor more than six months, or both. Justice courts have jurisdiction concurrent with the circuit court of the trial of violations of this section. When any person is charged before a justice court with violation of this section, the court shall, upon motion of the district attorney, at any time before trial, act as a committing magistrate, and if probable cause be established, hold such person to the grand jury

166.210. Definitions. As used in ORS 166.250 to 166.270, 166.291 to 166.295 and 166.410 to 166.470:

(1) "Antique firearm" means:

(a) Any firearm, including any firearm with a matchlock, flintlock, percussion cap or similar type of ignition system, manufactured in or before 1898; and

(b) Any replica of any firearm described in paragraph (a) of this subsection if the replica:

(A) Is not designed or redesigned for using rimfire or conventional centerfire fixed ammunition; or

(B) Uses rimfire or conventional centerfire fixed ammunition that is no longer manufactured in the United States and that is not readily available in the ordinary channels of commercial trade.

(2) 'Corrections officer' has the meaning given that term in ORS 181.610.

(3) "Firearm" means a weapon, by whatever name known, which is designed to expel a projectile by the action of powder and which is readily capable of use as a weapon.

(4) "Firearms silencer" means any device for silencing, muffling or diminishing the report of a firearm.

(5) "Handgun" means any pistol or revolver using a fixed cartridge containing a propellant charge, primer and projectile, and designed to be aimed or fired otherwise than from the shoulder.

(6) "Machine gun" means a weapon of any description by whatever name known, loaded or unloaded, which is designed or modified to allow two or more shots to be fired by a single pressure on the trigger device.

(7) "Minor" means a person under 18 years of age.

(8) 'Offense' has the meaning given that term in ORS 161.505.

(9) "Parole and probation officer" has the meaning given that term in ORS 181.610.

(10) 'Peace officer' has the meaning given that term in ORS 133.005.

(11) "Short-barreled rifle" means a rifle having one or more barrels less than 16 inches in length and any weapon made from a rifle if the weapon has an overall length of less than 26 inches.

(12) "Short-barreled shotgun" means a shotgun having one or more barrels less than 18 inches in length and any weapon made from a shotgun if the weapon has an overall length of less than 26 inches.

166.220. Unlawful use of weapon.

(1) A person commits the crime of unlawful use of a weapon if the person:

(a) Attempts to use unlawfully against another, or carries or possesses with intent to use unlawfully against another, any dangerous or deadly weapon as defined in ORS 161.015; or

(b) Intentionally discharges a firearm, blowgun, bow and arrow, crossbow or explosive device within the city limits of any city or within residential areas within urban growth boundaries at or in the direction of any person, building, structure or vehicle within the range of the weapon without having legal authority for such discharge.

(2) This section does not apply to:

(a) Police officers or military personnel in the lawful performance of their official duties;

(b) Persons lawfully defending life or property as provided in ORS 161.219;

(c) Persons discharging firearms, blowguns, bows and arrows, crossbows or explosive devices upon public or private shooting ranges, shooting galleries or other areas designated and built for the purpose of target shooting; or

(d) Persons lawfully engaged in hunting in compliance with rules and regulations adopted by the State Department of Fish and Wildlife.

(3) Unlawful use of a weapon is a Class C felony.

166.250. Unlawful possession of firearms.

(1) Except as otherwise provided in this section or ORS 166.260, 166.270, 166.274, 166.291, 166.292 or 166.410 to 166.470, a person commits the crime of unlawful possession of a firearm if the person knowingly:

(a) Carries any firearm concealed upon the person;

(b) Possesses a handgun that is concealed and readily accessible to the person within any vehicle; or

(c) Possesses a firearm and:

(A) Is under 18 years of age;

(B)(i) While a minor, was found to be within the jurisdiction of the juvenile court for having committed an act which, if committed by an adult, would constitute a felony or a misdemeanor involving violence, as defined in ORS 166.470; and

(ii) Was discharged from the jurisdiction of the juvenile court within four years prior to being charged under this section;

(C) Has been convicted of a felony or found guilty, except for insanity under ORS 161.295, of a felony;

(D) Was committed to the Department of Human Services under ORS 426.130; or

(E) Was found to be mentally ill and subject to an order under ORS 426.130 that the person be prohibited from purchasing or possessing a firearm as a result of that mental illness.

(2) This section does not prohibit:

(a) A minor, who is not otherwise prohibited under subsection (1)(c) of this section, from possessing a firearm:

(A) Other than a handgun, if the firearm was transferred to the minor by the minor's parent or guardian or by another person with the consent of the minor's parent or guardian; or

(B) Temporarily for hunting, target practice or any other lawful purpose; or

(b) Any citizen of the United States over the age of 18 years who resides in or is temporarily sojourning within this state, and who is not within the excepted classes prescribed by ORS 166.270 and subsection (1) of this section, from owning, possessing or keeping within the person's place of residence or place of business any handgun, and no permit or license to purchase, own, possess or keep any such firearm at the person's place of residence or place of business is required of any such citizen. As used in this subsection, "residence" includes a recreational vessel or recreational vehicle while used, for whatever period of time, as residential quarters.

(3) Firearms carried openly in belt holsters are not concealed within the meaning of this section.

(4) Unlawful possession of a firearm is a Class A misdemeanor.

166.260. Persons not affected by ORS 166.250.

(1) ORS 166.250 does not apply to or affect:

(a) Sheriffs, constables, marshals, police officers, whether active or honorably retired, parole and probation officers or other duly appointed peace officers.

(b) Any person summoned by any such officer to assist in making arrests or preserving the peace, while said person so summoned is actually engaged in assisting the officer.

(c) The possession or transportation by any merchant of unloaded firearms as merchandise.

(d) Active or reserve members of the Army, Navy, Air Force, Coast Guard or Marine Corps of the United States, or of the National Guard, when on duty.

(e) Organizations which are by law authorized to purchase or receive weapons described in ORS 166.250 from the United States, or from this state.

(f) Duly authorized military or civil organizations while parading, or the members thereof when going to and from the places of meeting of their organization.

(g) A corrections officer while transporting or accompanying an individual convicted of or arrested for an offense and confined in a place of incarceration or detention while outside the confines of the place of incarceration or detention.

(h) A person who is licensed under ORS 166.291 and 166.292 to carry a concealed handgun.

(2) Except for persons who are otherwise prohibited from possessing a firearm under ORS 166.250 (1)(c) or 166.270, ORS 166.250 does not apply to or affect:

(a) Members of any club or organization, for the purpose of practicing shooting at targets upon the established target ranges, whether public or private, while such members are using any of the firearms referred to in ORS 166.250 upon such target ranges, or while going to and from such ranges.

(b) Licensed hunters or fishermen while engaged in hunting or fishing, or while going to or returning from a hunting or fishing expedition.

(3) The exceptions listed in subsection (1)(b) to (h) of this section constitute affirmative defenses to a charge of violating ORS 166.250.

166.262. Limitation on peace officer's authority to arrest for violating ORS 166.250 or 166.370. A peace officer may not arrest or charge a person for violating ORS 166.250(1)(a) or (b) or 166.370 (1) if the person has in the person's immediate possession a valid license to carry a firearm as provided in ORS 166.291 and 166.292.

166.270. Possession of weapons by certain felons.

(1) Any person who has been convicted of a felony under the law of this state or any other state, or who has been convicted of a felony under the laws of the Government of the United States, who owns or has in the person's possession or under the person's custody or control any firearm commits the crime of felon in possession of a firearm.

(2) Any person who has been convicted of a felony under the law of this state or any other state, or who has been convicted of a felony under the laws of the Government of the United States, who owns or has in the person's possession or under the person's custody or control any instrument or weapon having a blade that projects or swings into position by force of a spring or by centrifugal force or any blackjack, slungshot, sandclub, sandbag, sap glove or metal knuckles, or who carries a dirk, dagger or stiletto, commits the crime of felon in possession of a restricted weapon.

(3) For the purposes of this section, a person "has been convicted of a felony" if, at the time of conviction for an offense, that offense was a felony under the law of the jurisdiction in which it was committed. Such conviction shall not be deemed a conviction of a felony if:

(a) The court declared the conviction to be a misdemeanor at the time of judgment; or

(b) The offense was possession of marijuana and the conviction was prior to January 1, 1972.

(4) Subsection (1) of this section does not apply to any person who has been:

(a) Convicted of only one felony under the law of this state or any other state, or who has been convicted of only one felony under the laws of the United States, which felony did not involve criminal homicide, as defined in ORS 163.005, or the possession or use of a firearm or a weapon having a blade that projects or swings into position by force of a spring or by centrifugal force, and who has been discharged from imprisonment, parole or probation for said offense for a period of 15 years prior to the date of alleged violation of subsection (1) of this section; or

(b) Granted relief from the disability under 18 U.S.C. 925(c) or has had the person's record expunged under the laws of this state or equivalent laws of another jurisdiction.

(5) Felon in possession of a firearm is a Class C felony. Felon in possession of a restricted weapon is a Class A misdemeanor.

166.272. Possession of machine guns, short-barreled firearms and firearms silencers.

(1) A person commits the crime of unlawful possession of a machine gun, short-barreled rifle, short-barreled shotgun or firearms silencer if the person knowingly possesses any machine gun, short-barreled rifle, short-barreled shotgun or firearms silencer.

(2) Unlawful possession of a machine gun, short-barreled rifle, short-barreled shotgun or firearms silencer is a Class B felony.

(3) A peace officer may not arrest or charge a person for violating subsection (1) of this section if the person has in the person's immediate possession documentation showing that the machine gun, short-barreled rifle, short-barreled shotgun or firearms silencer is registered as required under federal law.

(4) It is an affirmative defense to a charge of violating subsection (1) of this section that the machine gun, short-barreled rifle, short-barreled shotgun or firearms silencer was registered as required under federal law.

166.274. Relief from prohibition against possessing or purchasing firearm.

(1) A person barred from possessing a firearm under ORS 166.250 (1)(c)(A), (B), (D) or (E) or barred from purchasing a firearm under ORS 166.470 may file a petition for relief from the bar in:

(a) A justice court in the petitioner's county of residence that is reasonably accessible to the petitioner; or

(b) If no justice court is reasonably accessible, the circuit court.

(2) A person may apply once per calendar year for relief under the provisions of this section.

(3)(a) A person petitioning for relief under this section shall serve a copy of the petition on:

(A) The city chief of police if the court in which the petition is filed is located in a city; or

(B) The sheriff of the county in which the court is located.

(b) The copy of the petition shall be served on the chief of police or sheriff at the same time the petition is filed at the court.

(4)(a) When a petition is denied, the judge shall cause that information to be entered into the Department of State Police computerized criminal history files.

(b) When a petition is granted, the judge shall cause that information and a fingerprint card of the petitioner to be entered into the Department of State Police computerized criminal history files. If, after a petition is granted, the petitioner is arrested and convicted of a crime that would disqualify the petitioner from purchasing or possessing a firearm, the Department of State Police shall notify the court that granted relief under this section. The court shall review the order granting relief and determine whether to rescind the order. The Department of State Police may charge a reasonable fee, under ORS 192.440, for the entry and maintenance of information under this section.

(5) Notwithstanding the provisions of ORS 9.320, a corporation, the state or any city, county, district or other political subdivision or public corporation in this state, without appearance by attorney, may appear as a party to an action under this section.

(6) If the petitioner seeks relief from the bar on possessing or purchasing a firearm, relief shall be granted when the petitioner demonstrates, by clear and convincing evidence, that the petitioner does not pose a threat to the safety of the public or the petitioner.

(7) A person barred from possessing or purchasing a firearm because the person, while a minor, was found to be within the jurisdiction of the juvenile court for committing an act which, if committed by an adult, would have constituted a felony or a misdemeanor involving violence, is not eligible to petition for relief under this section until more than four years have passed since the person was discharged from the jurisdiction of the juvenile court.

(8) Petitions filed under this section shall be heard and disposed of within 15 judicial days of filing or as soon as is practicable thereafter, but not more than 30 days thereafter. The judge shall then make findings and conclusions and issue a judgment based on the findings and conclusions in accordance with the requirements of law.

(9) Filing fees shall be as for any civil action filed in the court.

(10)(a) Initial appeals of petitions shall be heard de novo.

(b) Any party to a judgment under this subsection may appeal to the Court of Appeals in the same manner as for any other civil action.

(c) If the governmental entity files an appeal under this subsection and does not prevail, it shall be ordered to pay the attorney fees for the prevailing party.

166.275. Possession of weapons by inmates of institutions. Any person committed to any institution who, while under the jurisdiction of any institution or while being conveyed to or from any institution, possesses or carries upon the person, or has under the custody or control of the person any dangerous instrument, or any weapon including but not limited to any blackjack, slingshot, billy, sand club, metal knuckles, explosive substance, dirk, dagger, sharp instrument, pistol, revolver or other firearm without lawful authority, is guilty of a felony and upon conviction thereof shall be punished by imprisonment in the custody of the Department of Corrections for a term not more than 20 years.

166.300. Killing or injuring another.

(1) Any person who has committed, with firearms of any kind or description, murder in any degree, or manslaughter, either voluntary or involuntary, or who in a careless or reckless manner, kills or injures another with firearms, and who, at any time after committing murder or manslaughter or after said careless or reckless killing or injury of another, carries or bears firearms of any kind or description within this state, shall be punished upon conviction by a fine of not more than $500, or by imprisonment in the county jail not to exceed one year, or both.

(2) Subsection (1) of this section does not deprive the people of this state of the right to bear arms for the defense of themselves and the state, and does not apply to any peace officer in the discharge of official duties or to a member of any regularly constituted military organization while on duty with such military organization.

(3) Justice courts, county courts and all other courts having jurisdiction as justice courts, shall have concurrent jurisdiction with the circuit courts of all prosecutions under subsection (1) of this section.

166.350. Unlawful possession of armor piercing ammunition.

(1) A person commits the crime of unlawful possession of armor piercing ammunition if the person:

(a) Makes, sells, buys or possesses any handgun ammunition the bullet or projectile of which is coated with Teflon or any chemical compound with properties similar to Teflon and which is intended to penetrate soft body armor, such person having the intent that the ammunition be used in the commission of a felony; or

(b) Carries any ammunition described in paragraph (a) of this subsection while committing any felony during which the person or any accomplice of the person is armed with a firearm.

(2) As used in this section, "handgun ammunition" means ammunition principally for use in pistols or revolvers notwithstanding that the ammunition can be used in some rifles.

(3) Unlawful possession of armor piercing ammunition is a Class A misdemeanor.

Possession of Weapon or Destructive Device in Public Building or Court Facility

166.360. Definitions for ORS 166.360 to 166.380. As used in ORS 166.360 to 166.380, unless the context requires otherwise: ...

(3) "Loaded firearm" means:

(a) A breech-loading firearm in which there is an unexpended cartridge or shell in or attached to the firearm including but not limited to, in a chamber, magazine or clip which is attached to the firearm.

(b) A muzzle-loading firearm which is capped or primed and has a powder charge and ball, shot or projectile in the barrel or cylinder.

(4) "Public building" means a hospital, a capitol building, a public or private school, as defined in ORS 339.315, a college or university, a city hall or the residence of any state official elected by the state at large, and the grounds adjacent to each such building. The term also includes that portion of any other building occupied by an agency of the state or a municipal corporation, as defined in ORS 297.405, other than a court facility.

(5) "Weapon" means:

(a) A firearm;

(b) Any dirk, dagger, ice pick, slingshot, metal knuckles or any similar instrument or a knife other than an ordinary pocket knife, the use of which could inflict injury upon a person or property;

(c) Mace, tear gas, pepper mace or any similar deleterious agent as defined in ORS 163.211;

(d) An electrical stun gun or any similar instrument;

(e) A tear gas weapon as defined in ORS 163.211;

(f) A club, bat, baton, billy club, bludgeon, knobkerrie, nunchaku, nightstick, truncheon or any similar instrument, the use of which could inflict injury upon a person or property; or

(g) A dangerous or deadly weapon as those terms are defined in ORS 161.015.

166.382. Possession of destructive device prohibited; exceptions.

(1) A person commits the crime of unlawful possession of a destructive device if the person possesses:

(a) Any of the following devices with an explosive, incendiary or poison gas component:

(A) Bomb;

(B) Grenade;

(C) Rocket having a propellant charge of more than four ounces;

(D) Missile having an explosive or incendiary charge of more than one-quarter ounce; or

(E) Mine; or

(b) Any combination of parts either designed or intended for use in converting any device into any destructive device described in paragraph (a) of this subsection and from which a destructive device may be readily assembled.

(2) As used in this section:

(a) "Destructive device" does not include any device which is designed primarily or redesigned primarily for use as a signaling, pyrotechnic, line throwing, safety or similar device.

(b) "Possess" has the meaning given that term in ORS 161.015.

(3) This section does not apply to:

(a) Persons who possess explosives as provided in ORS 480.200 to 480.290.

(b) The possession of an explosive by a member of the Armed Forces of the United States while on active duty and engaged in the performance of official duties or by a member of a regularly organized fire or police department of a public agency while engaged in the performance of official duties.

(c) The possession of an explosive in the course of transportation by way of railroad, water, highway or air while under the jurisdiction of, or in conformity with, regulations adopted by the United States Department of Transportation.

(d) The possession, sale, transfer or manufacture of an explosive by a person acting in accordance with the provisions of any applicable federal law or regulation that provides substantially the same requirements as the comparable provisions of ORS 480.200 to 480.290.

(4) Possession of a destructive device is a Class C felony.

166.384. Unlawful manufacture of destructive device.

(1) A person commits the crime of unlawful manufacture of a destructive device if the person assembles, produces or otherwise manufactures:

(a) A destructive device, as defined in ORS 166.382; or

(b) A pyrotechnic device containing two or more grains of pyrotechnic charge in violation of chapter 10, Title 18 of the United States Code.

(2) Unlawful manufacture of a destructive device is a Class C felony.

166.385. Possession of hoax destructive device.

(1) A person commits the crime of possession of a hoax destructive device if the person knowingly places another person in fear of serious physical injury by:

(a) Possessing, manufacturing, selling, delivering, placing or causing to be placed a hoax destructive device; or

(b) Sending a hoax destructive device to another person.

(2) Possession of a hoax destructive device is a Class A misdemeanor.

(3) Notwithstanding subsection (2) of this section, possession of a hoax destructive device is a Class C felony if a person possesses, or threatens to use, a hoax destructive device while the person is committing or attempting to commit a felony.

(4) As used in this section, "hoax destructive device" means an object that reasonably appears, under the circumstances:

(a) To be a destructive device, as described in ORS 166.382 (1)(a), or an explosive, as defined in ORS 166.660, but is an inoperative imitation of a destructive device or explosive; or

(b) To contain a destructive device, as described in ORS 166.382 (1)(a), or an explosive, as defined in ORS 166.660.

Sale or Transfer of Firearms

166.410. Manufacture, importation or sale of firearms. Any person who manufactures or causes to be manufactured within this state, or who imports into this state, or offers, exposes for sale, or sells or transfers a handgun, short-barreled rifle, short-barreled shotgun, firearms silencer or machine gun, otherwise than in accordance with ORS 166.250, 166.260, 166.270, 166.291, 166.292 166.425, 166.450, 166.460 and 166.470, is guilty of a Class B felony.

166.412. Definitions; firearms transaction record; criminal record check.

(1) As used in this section:

(a) "Antique firearm" has the meaning given that term in 18 U.S.C. 921;

(b) "Department" means the Department of State Police;

(c) "Firearm" has the meaning given that term in ORS 166.210, except that it does not include an antique firearm;

(d) "Firearms transaction record" means the firearms transaction record required by 18 U.S.C. 921 to 929;

(e) "Firearms transaction thumbprint form" means a form provided by the department under subsection (12) of this section;

(f) "Gun dealer" means a person engaged in the business, as defined in 18 U.S.C. 921, of selling, leasing or otherwise transferring a firearm, whether the person is a retail dealer, pawnbroker or otherwise;

(g) "Handgun" has the meaning given that term in ORS 166.210; and

(h) "Purchaser" means a person who buys, leases or otherwise receives a firearm from a gun dealer.

(2) Except as provided in subsections (3)(c) and (13) of this section, a gun dealer shall comply with the following before a handgun is delivered to a purchaser:

(a) The purchaser shall present to the dealer current identification meeting the requirements of subsection (4) of this section.

(b) The gun dealer shall complete the firearms transaction record and obtain the signature of the purchaser on the record.

(c) The gun dealer shall obtain the thumbprints of the purchaser on the firearms transaction thumbprint form and attach the form to the gun dealer's copy of the firearms transaction record to be filed with that copy.

(d) The gun dealer shall request by telephone that the department conduct a criminal history record check on the purchaser and shall provide the following information to the department:

(A) The federal firearms license number of the gun dealer;

(B) The business name of the gun dealer;

(C) The place of transfer;

(D) The name of the person making the transfer;

(E) The make, model, caliber and manufacturer's number of the handgun being transferred;

(F) The name and date of birth of the purchaser;

(G) The social security number of the purchaser if the purchaser voluntarily provides this number to the gun dealer; and

(H) The type, issuer and identification number of the identification presented by the purchaser.

(e) The gun dealer shall receive a unique approval number for the transfer from the department and record the approval number on the firearms transaction record and on the firearms transaction thumbprint form.

(f) The gun dealer may destroy the firearms transaction thumbprint form five years after the completion of the firearms transaction thumbprint form.

(3)(a) Upon receipt of a request of the gun dealer for a criminal history record check, the department shall immediately, during the gun dealer's telephone call or by return call:

(A) Determine, from criminal records and other information available to it, whether the purchaser is disqualified under ORS 166.470 from completing the purchase; and

(B) Notify the dealer when a purchaser is disqualified from completing the transfer or provide the dealer with a unique approval number indicating that the purchaser is qualified to complete the transfer.

(b) If the department is unable to determine if the purchaser is qualified or disqualified from completing the transfer within 30 minutes, the department shall notify the dealer and provide the dealer with an estimate of the time when the department will provide the requested information.

(c) If the department fails to provide a unique approval number to a gun dealer or to notify the gun dealer that the purchaser is disqualified under paragraph (a) of this subsection before the close of the gun dealer's next business day following the request by the dealer for a criminal history record check, the dealer may deliver the handgun to the purchaser.

(4)(a) Identification required of the purchaser under subsection (2) of this section shall include one piece of current identification bearing a photograph and the date of birth of the purchaser that:

(A) Is issued under the authority of the United States government, a state, a political subdivision of a state, a foreign government, a political subdivision of a foreign government, an international governmental organization or an international quasi-governmental organization; and

(B) Is intended to be used for identification of an individual or is commonly accepted for the purpose of identification of an individual.

(b) If the identification presented by the purchaser under paragraph (a) of this subsection does not include the current address of the purchaser, the purchaser shall present a second piece of current identification that contains the current address of the purchaser. The Superintendent of State Police may specify by rule the type of identification that may be presented under this paragraph.

(c) The department may require that the dealer verify the identification of the purchaser if that identity is in question by sending the thumbprints of the purchaser to the department.

(5) The department shall establish a telephone number that shall be operational seven days a week between the hours of 8 a.m. and 10 p.m. for the purpose of responding to inquiries from dealers for a criminal history record check under this section.

(6) No public employee, official or agency shall be held criminally or civilly liable for performing the investigations required by this section provided the employee, official or agency acts in good faith and without malice.

(7)(a) The department may retain a record of the information obtained during a request for a criminal records check for no more than five years.

(b) The record of the information obtained during a request for a criminal records check by a gun dealer is exempt from disclosure under public records law.

(8) The Department of Human Services shall provide the Department of State Police with direct electronic access to information from the Department of Human Services' database of information identifying persons meeting the criteria in ORS 166.470 (1)(e) and (f) who were committed or subject to an order under ORS 426.130. The Department of State Police and the Department of Human Services shall enter into an agreement describing the access to information under this subsection.

(9) A law enforcement agency may inspect the records of a gun dealer relating to transfers of handguns with the consent of a gun dealer in the course of a reasonable inquiry during a criminal investigation or under the authority of a properly authorized subpoena or search warrant.

(10) When a handgun is delivered, it shall be unloaded.

(11) In accordance with applicable provisions of ORS 183.310 to 183.550, the Superintendent of State Police may adopt rules necessary for:

(a) The design of the firearms transaction thumbprint form;

(b) The maintenance of a procedure to correct errors in the criminal records of the department;

(c) The provision of a security system to identify dealers who request a criminal history record check under subsection (2) of this section; and

(d) The creation and maintenance of a database of the business hours of gun dealers.

(12) The department shall publish the firearms transaction thumbprint form and shall furnish the form to gun dealers on application at cost.

(13) This section does not apply to transactions between persons licensed as dealers under 18 U.S.C. 923.

166.414. Fees for conducting criminal history record checks.

(1) The Department of State Police may adopt a fee schedule for criminal history record checks required under ORS 166.412 and collect a fee for each criminal history record check requested. The fee schedule shall be calculated to recover the cost of performing criminal history record checks required under ORS 166.412, but may not exceed $10 per record check.

(2) Fees collected under this section shall be paid into the State Treasury and deposited in the General Fund to the credit of the State Police Account.

166.416. Providing false information in connection with a transfer of a firearm.

(1) A person commits the crime of providing false information in connection with a transfer of a firearm if the person knowingly provides a false name or false information or presents false identification in connection with a purchase or transfer of a firearm.

(2) Providing false information in connection with a transfer of a firearm is a Class A misdemeanor.

166.418. Improperly transferring a firearm.

(1) A person commits the crime of improperly transferring a firearm if the person is a gun dealer as defined in ORS 166.412 and sells, leases or otherwise transfers a firearm and intentionally violates ORS 166.412 or 166.434.

(2) Improperly transferring a firearm is a Class A misdemeanor.

166.421. Telephone request for firearms check. The Department of State Police may respond to a telephone request from any person requesting that the department determine if department records show that a fire-arm is stolen. No public employee, official or agency shall be held criminally or civilly liable for performing the investigation allowed by this section provided that the employee, official or agency acts in good faith and without malice.

166.422. Enforcement of ORS 166.412. Where appropriate, a person may enforce the legal duties imposed by ORS 166.412 (7), by the provisions of ORS 30.260 to 30.300 and ORS chapter 183.

166.425. Unlawful purchase of firearm.

(1) A person commits the crime of unlawfully purchasing a firearm if the person, knowing that the person is prohibited by state or federal law from owning or possessing the firearm or having the firearm under the person's custody or control, purchases or attempts to purchase the firearm.

(2) Unlawfully purchasing a firearm is a Class A misdemeanor.

166.427. Register of transfers of used firearms.

(1) Whenever a person engaged in the business, as defined in 18 U.S.C. 921, of selling, leasing or otherwise transferring a firearm, whether the person is a retail dealer, pawnbroker or otherwise, buys or accepts in trade, a used firearm, the person shall enter in a register the time, date and place of purchase or trade, the name of the person selling or trading the firearm, the number of the identification documentation presented by the person and the make, model and manufacturer's number of the firearm. The register shall be obtained from and furnished by the Department of State Police to the dealer on application at cost.

(2) The duplicate sheet of the register shall, on the day of purchase or trade, be hand delivered or mailed to the local law enforcement authority.

(3) Violation of this section by any person engaged in the business of selling, leasing or otherwise transferring a firearm is a Class C misdemeanor.

166.429. Firearms used in felony.

Any person who, with intent to commit a felony or who knows or reasonably should know that a felony will be committed with the firearm, ships, transports, receives, sells or otherwise furnishes any firearm in the furtherance of the felony is guilty of a Class B felony.

166.432. Definitions for ORS 166.412 and 166.433 to 166.441.

(1) As used in ORS 166.412, 166.433, 166.434, 166.436 and 166.438, "criminal background check" or "criminal history record check" means determining the eligibility of a person to purchase or possess a firearm by reviewing state and federal databases including, but not limited to, the:

(a) Oregon computerized criminal history system;

(b) Oregon mental health data system;

(c) Law Enforcement Data System;

(d) National Instant Criminal Background Check System; and

(e) Stolen guns system.

(2) As used in ORS 166.433, 166.434, 166.436, 166.438 and 166.441:

(a) "Gun dealer" has the meaning given that term in ORS 166.412.

(b) "Gun show" means an event at which more than 25 firearms are on site and available for transfer.

166.433. Findings regarding transfers of firearms.
The people of this state find that:

(1) The laws of Oregon regulating the sale of firearms contain a loophole that allows people other than gun dealers to sell firearms at gun shows without first conducting criminal background checks;

(2) It is necessary for the safety of the people of Oregon that any person who transfers a firearm at a gun show be required to request a criminal background check before completing the transfer of the firearm; and

(3) It is in the best interests of the people of Oregon that any person who transfers a firearm at any location other than a gun show be allowed to voluntarily request a criminal background check before completing the transfer of the firearm.

166.434. Firearms transfers by gun dealers; criminal background check.

(1) Notwithstanding the fact that ORS 166.412 requires a gun dealer to request a criminal history record check only when transferring a handgun, a gun dealer shall comply with the requirements of ORS 166.412 before transferring any firearm to a purchaser. The provisions of ORS 166.412 apply to the transfer of firearms other than handguns to the same extent that they apply to the transfer of handguns.

(2) In addition to the determination required by ORS 166.412 (3)(a)(A), in conducting a criminal background check or criminal history record check, the Department of State Police shall also determine whether the recipient is otherwise prohibited by state or federal law from possessing a firearm.

(3) Notwithstanding ORS 166.412 (5), the department is not required to operate the telephone number established under ORS 166.412 (5) on Thanksgiving Day or Christmas Day.

(4)(a) The department may charge a fee, not to exceed the amount authorized under ORS 166.414, for criminal background checks required under this section or ORS 166.436.

(b) The department shall establish a reduced fee for subsequent criminal background checks on the same recipient that are performed during the same day between the hours of 8 a.m. and 10 p.m.

166.436. Firearm transfers by persons other than gun dealers; criminal background checks; liability.

(1) The Department of State Police shall make the telephone number established under ORS 166.412 (5) available for requests from persons other than gun dealers for criminal background checks under this section.

(2) Prior to transferring a firearm, a transferor other than a gun dealer may request by telephone that the department conduct a criminal background check on the recipient and shall provide the following information to the department:

(a) The name, address and telephone number of the transferor;

(b) The make, model, caliber and manufacturer's number of the firearm being transferred;

(c) The name, date of birth, race, sex and address of the recipient;

(d) The social security number of the recipient if the recipient voluntarily provides that number;

(e) The address of the place where the transfer is occurring; and

(f) The type, issuer and identification number of a current piece of identification bearing a recent photograph of the recipient presented by the recipient. The identification presented by the recipient must meet the requirements of ORS 166.412 (4)(a).

(3)(a) Upon receipt of a request for a criminal background check under this section, the department shall immediately, during the telephone call or by return call:

(A) Determine from criminal records and other information available to it whether the recipient is disqualified under ORS 166.470 from completing the transfer or is otherwise prohibited by state or federal law from possessing a firearm; and

(B) Notify the transferor when a recipient is disqualified from completing the transfer or provide the transferor with a unique approval number indicating that the recipient is qualified to complete the transfer. The unique approval number is a permit valid for 24 hours for the requested transfer. If the firearm is not transferred from the transferor to the recipient within 24 hours after receipt of the unique approval number, a new request must be made by the transferor.

(b) If the department is unable to determine whether the recipient is qualified for or disqualified from completing the transfer within 30 minutes of receiving the request, the department shall notify the transferor and provide the transferor with an estimate of the time when the department will provide the requested information.

(4) A public employee or public agency incurs no criminal or civil liability for performing the criminal background checks required by this section, provided the employee or agency acts in good faith and without malice.

(5)(a) The department may retain a record of the information obtained during a request for a criminal background check under this section for the period of time provided in ORS 166.412 (7).

(b) The record of the information obtained during a request for a criminal background check under this section is exempt from disclosure under public records law.

(6) The recipient of the firearm must be present when the transferor requests a criminal background check under this section.

(7)(a) Except as otherwise provided in paragraphs (b) and (c) of this subsection, a transferor who receives notification under this section that the recipient is qualified to complete the transfer of a firearm is immune from civil liability for any use of the firearm from the time of the transfer unless the transferor knows, or reasonably should know, that the recipient is likely to commit an unlawful act involving the firearm.

(b) If the transferor is required to request a criminal background check under ORS 166.438, the immunity provided by paragraph (a) of this subsection applies only if, in addition to receiving the notification required by this section, the transferor has the recipient fill out the form required by ORS 166.438 (1)(a) and retains the form as required by ORS 166.438 (2).

(c) The immunity provided by paragraph (a) of this subsection does not apply:

(A) If the transferor knows, or reasonably should know, that the recipient of the firearm intends to deliver the firearm to a third person who the transferor knows, or reasonably should know, may not lawfully possess the firearm; or

(B) In any product liability civil action under ORS 30.900 to 30.920.

166.438. Transfer of firearms at gun shows; persons other than gun dealers.

(1) A transferor other than a gun dealer may not transfer a firearm at a gun show unless the transferor:

(a)(A) Requests a criminal background check under ORS 166.436 prior to completing the transfer;

(B) Receives notification that the recipient is qualified to complete the transfer; and

(C) Has the recipient complete the form described in ORS 166.441; or

(b) Completes the transfer through a gun dealer.

(2) The transferor shall retain the completed form referred to in subsection (1) of this section for at least five years and shall make the completed form available to law enforcement agencies for the purpose of criminal investigations.

(3) A person who organizes a gun show shall post in a prominent place at the gun show a notice explaining the requirements of subsections (1) and (2) of this section. The person shall provide the form required by subsection (1) of this section to any person transferring a firearm at the gun show.

(4) Subsection (1) of this section does not apply if the transferee is licensed as a dealer under 18 U.S.C. 923.

(5)(a) Failure to comply with the requirements of subsection (1), (2) or (3) of this section is a Class A misdemeanor.

(b) Notwithstanding paragraph (a) of this subsection, failure to comply with the requirements of subsection (1), (2) or (3) of this section is a Class C felony if the person has two or more previous convictions under this section.

(6) It is an affirmative defense to a charge of violating subsection (1) or (3) of this section that the person did not know, or reasonably could not know, that more than 25 firearms were at the site and available for transfer.

166.441. Form for transfer of firearm at gun show.

(1) The Department of State Police shall develop a form to be completed by a person seeking to obtain a firearm at a gun show from a transferor other than a gun dealer. The department shall consider including in the form all of the requirements for disclosure of information that are required by federal law for over-the-counter firearms transactions.

(2) The department shall make the form available to the public at no cost.

166.445. Short title. ORS 166.432 to 166.445 and the amendments to ORS 166.416, 166.418 and 166.460 by sections 9, 10 and 11, chapter 1, Oregon Laws 2001, shall be known as the Gun Violence Prevention Act.

166.450. Alteration, removal, or obliteration of firearm identification number.

Any person who intentionally alters, removes or obliterates the identification number of any firearm for an unlawful purpose, shall be punished upon conviction by imprisonment in the custody of the Department of Corrections for not more than five years. Possession of any such firearm is presumptive evidence that the possessor has altered, removed or obliterated the identification number.

166.460. Antique firearms excepted.

(1) ORS 166.250, 166.260, 166.291 to 166.295, 166.410, 166.412, 166.425, 166.434, 166.438 and 166.450 do not apply to antique firearms.

(2) Notwithstanding the provisions of subsection (1) of this section, possession of an antique firearm by a person described in ORS 166.250 (1)(c)(B), (C) or (D) constitutes a violation of ORS 166.250.

166.470. Limitations on sale of firearms.

(1) Unless relief has been granted under ORS 166.274, 18 U.S.C. 925(c) or the expunction laws of this state or an equivalent law of another jurisdiction, a person may not intentionally sell, deliver or otherwise transfer any firearm when the transferor knows or reasonably should know that the recipient:

(a) Is under 18 years of age;

(b) Has been convicted of a felony or found guilty, except for insanity under ORS 161.295, of a felony;

(c) Has any outstanding felony warrants for arrest;

(d) Is free on any form of pretrial release for a felony;

(e) Was committed to the Department of Human Services under ORS 426.130;

(f) After January 1, 1990, was found to be mentally ill and subject to an order under ORS 426.130 that the person be prohibited from purchasing or possessing a firearm as a result of that mental illness; or

(g) Has been convicted of a misdemeanor involving violence or found guilty, except for insanity under ORS 161.295, of a misdemeanor involving violence within the previous four years. As used in this paragraph, "misdemeanor involving violence" means a misdemeanor described in ORS 163.160, 163.187, 163.190, 163.195 or 166.155 (1)(b).

(2) A person may not sell, deliver or otherwise transfer any firearm that the person knows or reasonably should know is stolen.

(3) Subsection (1)(a) of this section does not prohibit:

(a) The parent or guardian, or another person with the consent of the parent or guardian, of a minor from transferring to the minor a firearm, other than a handgun; or

(b) The temporary transfer of any firearm to a minor for hunting, target practice or any other lawful purpose.

(4) Violation of this section is a Class A misdemeanor.

166.480. Sale or gift of explosives to children. Any person who sells, exchanges, barters or gives to any child, under the age of 14 years, any explosive article or substance, other than an ordinary firecracker containing not more than 10 grains of gunpowder or who sells, exchanges, barters or gives to any such child, any instrument or apparatus, the chief utility of which is the fact that it is used, or is ordinarily capable of being used, as an article or device to increase the force or intensity of any explosive, or to direct or control the discharge of any such explosive, is guilty of a misdemeanor.

166.490. Out-of-state purchase of firearms.

(1) As used in this section, unless the context requires otherwise:

(a) "Contiguous state" means California, Idaho, Nevada or Washington.

(b) "Resident" includes an individual or a corporation or other business entity that maintains a place of business in this state.

(2) A resident of this state may purchase or otherwise obtain a rifle or shotgun in a contiguous state and receive in this state or transport into this state such rifle or shotgun, unless the purchase or transfer violates the law of this state, the state in which the purchase or transfer is made or the United States.

(3) This section does not apply to the purchase, receipt or transportation of rifles and shotguns by federally licensed firearms manufacturers, importers, dealers or collectors.

(4) This section expires and stands repealed upon the date that section 922(b) (3) of the Gun Control Act of 1968 (18 U.S.C. 922(b) (3)) and regulations pursuant thereto are repealed or rescinded.

Title 38. Protection from Fire

Chapter 480. Explosives; Flammable Materials; Pressure Vessels

480.070. Possession of fire bombs prohibited.

(1) No person shall possess a fire bomb. For the purpose of this section a "fire bomb" is a breakable container containing a flammable liquid with a flash point of 150 degrees Fahrenheit or less, having an integral wick or similar device capable of being ignited, but no device commercially manufactured primarily for the purpose of illuminating shall be deemed to be a fire bomb for the purpose of this section.

(2) This section shall not prohibit the authorized use or possession of such fire bomb by a member of the Armed Forces of the United States or by any member of a regularly organized public fire or police department.

[Current through the 2007 Regular Session of the 74th Legislative Assembly]

PENNSYLVANIA
PA. CONS. STAT.

Title 18. Crimes and Offenses

Chapter 61. Firearms and Other Dangerous Articles

Subchapter A. Uniform Firearms Act

6101. Short title of subchapter This subchapter shall be known and may be cited as the Pennsylvania Uniform Firearms Act of 1995.

6102. Definitions Subject to additional definitions contained in subsequent provisions of this subchapter which are applicable to specific provisions of this subchapter, the following words and phrases, when used in this subchapter shall have, unless the context clearly indicates otherwise, the meanings given to them in this section:

"Commissioner." The Commissioner of the Pennsylvania State Police.

"Commonwealth Photo Imaging Network." The computer network administered by the Commonwealth and used to record and store digital photographs of an individual's face and any scars, marks, tattoos or other unique features of the individual.

"Conviction." A conviction, a finding of guilty or the entering of a plea of guilty or nolo contendere, whether or not judgment of sentence has been imposed, as determined by the law of the jurisdiction in which the prosecution was held. The term does not include a conviction which has been expunged or overturned or for which an individual has been pardoned unless the pardon expressly provides that the individual may not possess or transport firearms.

"County treasurer." The county treasurer or, in home rule or optional plan counties, the person whose duties encompass those of a county treasurer.

"Crime punishable by imprisonment exceeding one year." The term does not include any of the following:

(1) Federal or State offenses pertaining to antitrust, unfair trade practices, restraints on trade or regulation of business.

(2) State offenses classified as misdemeanors and punishable by a term of imprisonment not to exceed two years.

"Firearm." Any pistol or revolver with a barrel length less than 15 inches, any shotgun with a barrel length less than 18 inches or any rifle with a barrel length less than 16 inches, or any pistol, revolver, rifle or shotgun with an overall length of less than 26 inches. The barrel length of a firearm shall be determined by measuring from the muzzle of the barrel to the face of the closed action, bolt or cylinder, whichever is applicable.

"Fund." The Firearm Ownership Fund established in section 6111.3 (relating to Firearm Ownership Fund).

"Law enforcement officer." Any person employed by any police department or organization of the Commonwealth or political subdivision thereof who is empowered to effect an arrest with or without warrant and who is authorized to carry a firearm in the performance of that person's duties.

"Loaded." A firearm is loaded if the firing chamber, the nondetachable magazine or, in the case of a revolver, any of the chambers of the cylinder contain ammunition capable of being fired. In the case of a firearm which utilizes a detachable magazine, the term shall mean a magazine suitable for use in said firearm which magazine contains such ammunition and has been inserted in the firearm or is in the same container or, where the container has multiple compartments, the same compartment thereof as the firearm.

"Pennsylvania Sheriffs' Association." The State association of sheriffs authorized by the act of June 14, 1923 (P.L.774, No.305), entitled 'An act authorizing the sheriffs of the several counties of this Commonwealth to organize themselves into a State Association, for the purpose of holding annual meetings, to secure more uniformity and cooperation in the conduct of their offices, and providing for the payment of certain expenses in connection with such meetings by the various counties.'

"Safekeeping permit." As defined in 23 Pa.C.S. § 6102 (relating to definitions).

"Sheriff."

(1) Except as provided in paragraph (2), the sheriff of the county.

(2) In a city of the first class, the chief or head of the police department.

"State." When used in reference to different parts of the United States, includes the District of Columbia, the Commonwealth of Puerto Rico and territories and possessions of the United States.

6103. Crimes committed with firearms If any person commits or attempts to commit a crime enumerated in section 6105 (relating to persons not to possess, use, manufacture, control, sell or transfer firearms) when armed with a firearm contrary to the provisions of this subchapter, that person may, in addition to the punishment provided for the crime, also be punished as provided by this subchapter.

6104. Evidence of intent In the trial of a person for committing or attempting to commit a crime enumerated in section 6105 (relating to persons not to possess, use, manufacture, control, sell or transfer firearms), the fact that that person was armed with a firearm, used or attempted to be used, and had no license to carry the same, shall be evidence of that person's intention to commit the offense.

6105. Persons not to possess, use, manufacture, control, sell or transfer firearms.

(a) Offense defined

(1) A person who has been convicted of an offense enumerated in subsection (b), within or without this Commonwealth, regardless of the length of sentence or whose conduct meets the criteria in subsection (c) shall not possess, use, control, sell, transfer or manufacture or obtain a license to possess, use, control, sell, transfer or manufacture a firearm in this Commonwealth.

(2)(i) A person who is prohibited from possessing, using, controlling, selling, transferring or manufacturing a firearm under paragraph (1) or subsection (b) or (c) shall have a reasonable period of time, not to exceed 60 days from the date of the imposition of the disability under this subsection, in which to sell or transfer that person's firearms to another eligible person who is not a member of the prohibited person's household.

(ii) This paragraph shall not apply to any person whose disability is imposed pursuant to subsection (c)(6).

(a.1) Penalty.

(1) A person convicted of a felony enumerated under subsection (b) or a felony under the act of April 14, 1972 (P.L.233, No.64), known as The Controlled Substance, Drug, Device and Cosmetic Act, or any equivalent Federal statute or equivalent statute of any other state, who violates subsection (a) commits a felony of the second degree.

(2) A person who is the subject of an active protection from abuse order issued pursuant to 23 Pa.C.S. § 6108 (relating to relief), which order provided for the relinquishment of firearms, other weapons or ammunition during the period of time the order is in effect, commits a misdemeanor of the first degree if he intentionally or knowingly fails to relinquish a firearm, other weapon or ammunition to the sheriff as required by the order unless, in lieu of relinquishment, he provides an affidavit which lists the firearms, other weapons or ammunition to the sheriff in accordance with either 23 Pa.C.S. § 6108(a)(7)(i)(B), 6108.2 (relating to relinquishment for consignment sale, lawful transfer or safekeeping) or 6108.3 (relating to relinquishment to third party for safekeeping).

(3)(i) A person commits a misdemeanor of the third degree if he intentionally or knowingly accepts possession of a firearm, other weapon or ammunition from a person he knows is the subject of an active protection from abuse order issued pursuant to 23 Pa.C.S. § 6108, which order provided for the relinquishment of the firearm, other weapon or ammunition during the period of time the order is in effect.

(ii) This paragraph shall not apply to:

(A) a third party who accepts possession of a firearm, other weapon or ammunition relinquished pursuant to 23 Pa.C.S. § 6108.3; or

(B) a dealer licensed pursuant to section 6113 (relating to licensing of dealers) or subsequent purchaser from a dealer licensed pursuant to section 6113, who accepts possession of a firearm, other weapon or ammunition relinquished pursuant to 23 Pa.C.S. § 6108.2.

(4) It shall be an affirmative defense to any prosecution under paragraph (3) that the person accepting possession of a firearm, other weapon or ammunition in violation of paragraph (3):

(i) notified the sheriff as soon as practicable that he has taken possession; and

(ii) relinquished possession of any firearm, other weapon or ammunition possessed in violation of paragraph (3) as directed by the sheriff.

(5) A person who has accepted possession of a firearm, other weapon or ammunition pursuant to 23 Pa.C.S. § 6108.3 commits a misdemeanor of the first degree if he intentionally or knowingly returns a firearm, other weapon or ammunition to a defendant or intentionally or knowingly allows a defendant to have access to the firearm, other weapon or ammunition prior to either of the following:

(i) The sheriff accepts return of the safekeeping permit issued to the party pursuant to 23 Pa.C.S. § 6108.3(d)(1)(i).

(ii) The issuance of a court order pursuant to subsection (f)(2) or 23 Pa.C.S. § 6108.1(b) (relating to return of relinquished firearms, other weapons and ammunition and additional relief) which modifies a valid protection from abuse order issued pursuant to 23 Pa.C.S. § 6108, which order provided for the relinquishment of the firearm, other weapon or ammunition by allowing the defendant to take possession of the firearm, other weapon or ammunition that had previously been ordered relinquished.

(b) Enumerated offenses. The following offenses shall apply to subsection (a):

Section 908 (relating to prohibited offensive weapons).

Section 911 (relating to corrupt organizations).

Section 912 (relating to possession of weapon on school property).

Section 2502 (relating to murder).

Section 2503 (relating to voluntary manslaughter).

Section 2504 (relating to involuntary manslaughter) if the offense is based on the reckless use of a firearm.

Section 2702 (relating to aggravated assault).

Section 2703 (relating to assault by prisoner).

Section 2704 (relating to assault by life prisoner).

Section 2709.1 (relating to stalking).

Section 2716 (relating to weapons of mass destruction).

Section 2901 (relating to kidnapping).

Section 2902 (relating to unlawful restraint).

Section 2910 (relating to luring a child into a motor vehicle or structure).

Section 3121 (relating to rape).

Section 3123 (relating to involuntary deviate sexual intercourse).

Section 3125 (relating to aggravated indecent assault).

Section 3301 (relating to arson and related offenses).

Section 3302 (relating to causing or risking catastrophe).

Section 3502 (relating to burglary).

Section 3503 (relating to criminal trespass) if the offense is graded a felony of the second degree or higher.

Section 3701 (relating to robbery).

Section 3702 (relating to robbery of motor vehicle).

Section 3921 (relating to theft by unlawful taking or disposition) upon conviction of the second felony offense.

Section 3923 (relating to theft by extortion) when the offense is accompanied by threats of violence.

Section 3925 (relating to receiving stolen property) upon conviction of the second felony offense.

Section 4912 (relating to impersonating a public servant) if the person is impersonating a law enforcement officer.

Section 4952 (relating to intimidation of witnesses or victims).

Section 4953 (relating to retaliation against witness or victim).

Section 5121 (relating to escape).

Section 5122 (relating to weapons or implements for escape).

Section 5501(3) (relating to riot).

Section 5515 (relating to prohibiting of paramilitary training).

Section 5516 (relating to facsimile weapons of mass destruction).

Section 6110.1 (relating to possession of firearm by minor).

Section 6301 (relating to corruption of minors).

Section 6302 (relating to sale or lease of weapons and explosives).

Any offense equivalent to any of the above-enumerated offenses under the prior laws of this Commonwealth or any offense equivalent to any of the above-enumerated offenses under the statutes of any other state or of the United States.

(c) Other persons. In addition to any person who has been convicted of any offense listed under subsection (b), the following persons shall be subject to the prohibition of subsection (a):

(1) A person who is a fugitive from justice. This paragraph does not apply to an individual whose fugitive status is based upon a nonmoving or moving summary offense under Title 75 (relating to vehicles).

(2) A person who has been convicted of an offense under the act of April 14, 1972 (P.L.233, No.64), known as The Controlled Substance,

Drug, Device and Cosmetic Act, or any equivalent Federal statute or equivalent statute of any other state, that may be punishable by a term of imprisonment exceeding two years.

(3) A person who has been convicted of driving under the influence of alcohol or controlled substance as provided in 75 Pa.C.S. § 3802 (relating to driving under influence of alcohol or controlled substance) or the former 75 Pa.C.S. § 3731, on three or more separate occasions within a five-year period. For the purposes of this paragraph only, the prohibition of subsection (a) shall only apply to transfers or purchases of firearms after the third conviction.

(4) A person who has been adjudicated as an incompetent or who has been involuntarily committed to a mental institution for inpatient care and treatment under section 302, 303 or 304 of the provisions of the act of July 9, 1976 (P.L. 817, No.143), known as the Mental Health Procedures Act. This paragraph shall not apply to any proceeding under section 302 of the Mental Health Procedures Act unless the examining physician has issued a certification that inpatient care was necessary or that the person was committable.

(5) A person who, being an alien, is illegally or unlawfully in the United States.

(6) A person who is the subject of an active protection from abuse order issued pursuant to 23 Pa.C.S. § 6108, which order provided for the relinquishment of firearms during the period of time the order is in effect. This prohibition shall terminate upon the expiration or vacation of an active protection from abuse order or portion thereof relating to the confiscation of firearms.

(7) A person who was adjudicated delinquent by a court pursuant to 42 Pa.C.S. § 6341 (relating to adjudication) or under any equivalent Federal statute or statute of any other state as a result of conduct which if committed by an adult would constitute an offense under sections 2502, 2503, 2702, 2703 (relating to assault by prisoner), 2704, 2901, 3121, 3123, 3301, 3502, 3701 and 3923.

(8) A person who was adjudicated delinquent by a court pursuant to 42 Pa.C.S. § 6341 or under any equivalent Federal statute or statute of any other state as a result of conduct which if committed by an adult would constitute an offense enumerated in subsection (b) with the exception of those crimes set forth in paragraph (7). This prohibition shall terminate 15 years after the last applicable delinquent adjudication or upon the person reaching the age of 30, whichever is earlier.

(9) A person who is prohibited from possessing or acquiring a firearm under 18 U.S.C. § 922(g)(9) (relating to unlawful acts). If the offense which resulted in the prohibition under 18 U.S.C. § 922(g)(9) was committed, as provided in 18 U.S.C. § 921(a)(33)(A)(ii)(relating to definitions), by a person in any of the following relationships:

(i) The current or former spouse, parent or guardian of the victim;

(ii) A person with whom the victim shares a child in common;

(iii) A person who cohabits with or has cohabited with the victim as a spouse, parent or guardian; or

(iv) A person similarly situated to a spouse, parent or guardian of the victim; then the relationship need not be an element of the offense to meet the requirements of this paragraph.

(d) Exemption. A person who has been convicted of a crime specified in subsection (a) or (b) or a person whose conduct meets the criteria in subsection (c)(1), (2), (5), (7) or (9) may make application to the court of common pleas of the county where the principal residence of the applicant is situated for relief from the disability imposed by this section upon the possession, transfer or control of a firearm. The court shall grant such relief if it determines that any of the following apply:

(1) The conviction has been vacated under circumstances where all appeals have been exhausted or where the right to appeal has expired.

(2) The conviction has been the subject of a full pardon by the Governor.

(3) Each of the following conditions is met:

(i) The Secretary of the Treasury of the United States has relieved the applicant of an applicable disability imposed by Federal law upon the possession, ownership or control of a firearm as a result of the applicant's prior conviction, except that the court may waive this condition if the court determines that the Congress of the United States has not appropriated sufficient funds to enable the Secretary of the Treasury to grant relief to applicants eligible for the relief.

(ii) A period of ten years, not including any time spent in incarceration, has elapsed since the most recent conviction of the applicant of a crime enumerated in subsection (b) or a felony violation of The Controlled Substance, Drug, Device and Cosmetic Act of the offense which resulted in the prohibition under 18 U.S.C. § 922(g)(9).

(e) Proceedings.

(1) If a person convicted of an offense under subsection (a), (b) or (c)(1), (2), (5), (7) or (9) makes application to the court, a hearing shall be held in open court to determine whether the requirements of this section have been met. The com-missioner and the district attorney of the county where the application is filed and any victim or survivor of a victim of the offense upon which the disability is based may be parties to the proceeding.

(2) Upon application to the court of common pleas pursuant to paragraph (1) by an applicant who is subject to the prohibition under subsection (c)(3), the court shall grant such relief if a period of ten years, not including any time spent in incarceration, has passed since the applicant's most recent conviction under subsection (c)(3).

(f) Other exemptions and proceedings.

(1) Upon application to the court of common pleas under this subsection by an applicant subject to the prohibitions under subsection (c)(4), the court may grant such relief as it deems appropriate if the court determines that the applicant may possess a firearm without risk to the applicant or any other person.

(2) If application is made under this subsection for relief from the disability imposed under subsection (c)(6), notice of such application shall be given to the person who had petitioned for the protection from abuse order, and such person shall be a party to the proceedings. Notice of any court order or amendment to a court order restoring firearms possession or control shall be given to the person who had petitioned for the protection from abuse order, to the sheriff and to the Pennsylvania State Police. The application and any proceedings on the application shall comply with 23 Pa.C.S. Ch. 61 (relating to protection from abuse).

(3) All hearings conducted under this subsection shall be closed unless otherwise requested to be open by the applicant.

(4)(i) The owner of any seized or confiscated firearms or of any firearms ordered relinquished under 23 Pa.C.S. § 6108 shall be provided with a sign and dated written receipt by the appropriate law enforcement agency. This receipt shall include, but not limited to, a detailed identifying description indicating the serial number and condition of the firearm. In addition, the appropriate law enforcement agency shall be liable to the lawful owner of said confiscated, seized or relinquished firearm for any loss, damage or substantial decrease in value of said firearm that is a direct result of a lack of reasonable care by the appropriate law enforcement agency.

(ii) Firearms shall not be engraved or permanently marked in any manner, including, but not limited to, engraving of evidence or other identification numbers. Unless reasonable suspicion exists to believe that a particular firearm has been used in the commission of a crime, no firearm shall be test fired. Any reduction in the value of a firearm due to test firing, engraving or permanently marking in violation of this paragraph shall be considered damage, and the law enforcement agency shall be liable to the lawful owner of the firearm for the reduction in value caused by the test firing, engraving or permanently marking.

(iii) For purposes of this paragraph, the term "firearm" shall include any scope, sight, bipod, sling, light, magazine, clip, ammunition or other firearm accessory attached to or seized, confiscated or relinquished with a firearm.

(g) Other restrictions. Nothing in this section shall exempt a person from a disability in relation to the possession or control of a firearm which is imposed as a condition of probation or parole or which is imposed pursuant to the provision of any law other than this section.

(h) License prohibition. Any person who is prohibited from possessing, using, controlling, selling, purchasing, transferring or manufacturing any firearm under this section shall not be eligible for or permitted to obtain a license to carry a firearm under section 6109 (relating to licenses).

(i) Firearm. As used in this section only, the term "firearm" shall include any weapons which are designed to or may readily be converted to expel any projectile by the action of an explosive or the frame or receiver of any such weapon.

(j) Copy of order to State Police. If the court grants relief from the disabilities imposed under this section, a copy of the order shall be sent by the prothonotary within ten days of the entry of the order to the Pennsylvania State Police and shall include the name, date of birth and Social Security number of the individual.

[Publisher's Note: Persons, who have been granted an exemption under section 6105(d)(3), may still be prohibited from possessing firearms under federal law. This relates to the fact that a person convicted of "a crime punishable by more than one year" in Pennsylvania loses the right to serve on a jury. Only the Governor can restore this right. Therefore, unless such individuals receive a pardon (or federal relief from disabilities), they will remain federally prohibited even after the state court grants them an exemption under Pennsylvania's firearms laws.**]**

6105.1. Restoration of firearm rights for offenses under prior laws of this Commonwealth

(a) Restoration. A person convicted of a disabling offense may make application to the court of common pleas in the county where the principal residence of the applicant is situated for restoration of firearms rights. The court shall grant restoration of firearms rights after a hearing in open court to determine whether the requirements of this section have been met unless:

(1) the applicant has been convicted of any other offense specified in section 6105(a) or (b) (relating to persons not to possess, use, manufacture, control, sell or transfer firearms) or the applicant's conduct meets the criteria in section 6105(c)(1), (2), (3), (4), (5), (6) or (7);

(2) the applicant has been convicted of any other crime punishable by imprisonment exceeding one year as defined in section 6102 (relating to definitions); or

(3) the applicant's character and reputation is such that the applicant would be likely to act in a manner dangerous to public safety.

(b) Notice and standing.

(1) Notice of an application for restoration of firearms rights shall be provided to the Pennsylvania state police, the district attorney of the county where the disabling offense occurred and the district attorney of the county where the application is filed. The district attorney of the county where the application is filed, the district attorney of the county where the disabling offense occurred and the Pennsylvania State Police may, at their option, be parties to the proceeding.

(2) Notwithstanding paragraph (1), the standing of the Pennsylvania State Police as a party to a proceeding under this section shall be limited to determinations of whether the offense meets the definition of the phrase "disabling offense" or whether the provisions of subsection (a)(1) and (2) have been satisfied.

(c) Copy of order to Pennsylvania State Police. If the court grants restoration of firearms rights to an applicant, a copy of the order shall be sent by the Prothonotary within ten days of the entry of the order to the district attorneys and the Pennsylvania State Police, firearms division, and shall include the name, date of birth and social security number of the applicant.

(d) Expungement and pardon. A restoration of firearms rights under this section shall not result in the expungement of any criminal history record information nor will it constitute a gubernatorial pardon.

(e) Definitions. As used in this section, the following words and phrases shall have the meanings given to them in this subsection:

"Disabling offense." A conviction for any offense which:

(1) resulted in a federal firearms disability and is substantially similar to either an offense currently graded as a crime punishable by a term of imprisonment for not more than two years or conduct which no longer constitutes a violation of law; and

(2) was a violation of either of the following:

(i) the former act of May 1, 1929 (P.L.905, No.403), known as the Vehicle Code, or the former act of April 29, 1959 (P.L.58, No.32), known as the Vehicle Code; or

(ii) the former act of June 24, 1939 (P.L.872, No.375), known as the Penal Code.

The definition shall not include any offense which, if committed under contemporary standards, would constitute a misdemeanor of the second degree or greater under section 2701 (relating to simple assault) and was committed by a current or former spouse, parent or guardian of the victim, by a person with whom the victim shares a child in common, by a person who is cohabitating with or has cohabitated with the victim as a spouse, parent or guardian, or by a person similarly situated to a spouse, parent or guardian of the victim.

"Restoration of firearms rights." Relieving any and all disabilities with respect to a person's right to own, possess, use, control, sell, purchase, transfer, manufacture, receive, ship or transport firearms, including any disabilities imposed pursuant to this subchapter. The phrase shall also mean the restoration of the right to vote, to hold public office and to serve on a jury.

6110.1. Possession of firearm by minor

(a) Firearm. Except as provided in subsection (b), a person under 18 years of age shall not possess or transport a firearm anywhere in this Commonwealth.

(b) Exception. Subsection (a) shall not apply to a person under 18 years of age:

(1) who is under the supervision of a parent, grandparent, legal guardian or an adult acting with the expressed consent of the minor's custodial parent or legal guardian and the minor is engaged in lawful activity, including safety training, lawful target shooting, engaging in an organized competition involving the use of a firearm or the firearm is unloaded and the minor is transporting it for a lawful purpose; or

(2) who is lawfully hunting or trapping in accordance with 34 Pa.C.S. (relating to game).

(c) Responsibility of adult. Any person who knowingly and intentionally delivers or provides to the minor a firearm in violation of subsection (a) commits a felony of the third degree.

(d) Forfeiture. Any firearm in the possession of a person under 18 years of age in violation of this section shall be promptly seized by the arresting law enforcement officer and upon conviction or adjudication of delinquency shall be forfeited or, if stolen, returned to the lawful owner.

6110.2. Possession of firearm with altered manufacturer's number

(a) General rule. No person shall possess a firearm which has had the manufacturer's number integral to the frame or receiver altered, changed, removed or obliterated.

(b) Penalty. A person who violates this section commits a felony of the second degree.

(c) Definition. As used in this section, the term "firearm" shall have the same meaning as that term is defined in section 6105(i) (relating to persons not to possess, use, manufacture, control, sell or transfer firearms), except that the term shall not include antique firearms as defined in section 6118 (relating to antique firearms).

6111. Sale or transfer of firearms

(a) Time and manner of delivery.

(1) Except as provided in paragraph (2), no seller shall deliver a firearm to the purchaser or transferee thereof until 48 hours shall have elapsed from the time of the application for the purchase thereof, and, when delivered, the firearm shall be securely wrapped and shall be unloaded.

(2) Thirty days after publication in the Pennsylvania Bulletin that the Instantaneous Criminal History Records Check System has been established in accordance with the Brady Handgun Violence Prevention Act (Public Law 103-159, 18 U.S.C. 921 et seq.), no seller shall deliver a firearm to the purchaser thereof until the provisions of this section have been satisfied, and, when delivered, the firearm shall be securely wrapped and shall be unloaded.

(b) Duty of seller. No licensed importer, licensed manufacturer or licensed dealer shall sell or deliver any firearm to another person, other than a licensed importer, licensed manufacturer, licensed dealer or licensed collector, until the conditions of subsection (a) have been satisfied and until he has:

(1) For purposes of a firearm as defined in section 6102 (relating to definitions), obtained a completed application/record of sale from the potential buyer or transferee to be filled out in triplicate, the original copy to be sent to the Pennsylvania State Police, postmarked via first class mail, within 14 days of the sale, one copy to be retained by the licensed importer, licensed manufacturer or licensed dealer for a period of 20 years and one copy to be provided to the purchaser or transferee. The form of this application/record of sale shall be no more than one page in length and shall be promulgated by the Pennsylvania State Police and provided by the licensed importer, licensed manufacturer or licensed dealer. The application/record of sale shall include the name, address, birth date, gender, race, physical description and Social Security number of the purchaser or transferee, the date of the application and the caliber, length of barrel, make, model and manufacturer's number of the firearm to be purchased or transferred. the application/record of sale shall also contain the following question:

Are you the actual buyer of the firearm(s), as defined under 18 Pa.C.S. Section 6102, listed on this application/record of sale? Warning: you are not the actual buyer if you are acquiring the firearm(s) on behalf of another person, unless you are legitimately acquiring the firearm as a gift for any of the following individuals who are legally eligible to own a firearm:

(1) spouse;

(2) parent;

(3) child;

(4) grandparent; or

(5) grandchild.

(1.1) On the date of publication in the Pennsylvania Bulletin of a notice by the Pennsylvania State Police that the instantaneous records check has been implemented, all of the following shall apply:

(i) In the event of an electronic failure under section 6111.1(b)(2) (relating to Pennsylvania State Police) for purposes of a firearm which exceeds the barrel and related lengths set forth in section 6102, obtained a completed application/record of sale from the potential buyer or transferee to be filled out in triplicate, the original copy to be sent to the Pennsylvania State Police, postmarked via first class mail, within 14 days of sale, one copy to be retained by the licensed importer, licensed manufacturer or licensed dealer for a period of 20 years and one copy to be provided to the purchaser or transferee.

(ii) The form of the application/record of sale shall be no more than one page in length and shall be promulgated by the Pennsylvania State Police and provided by the licensed importer, licensed manufacturer or licensed dealer.

(iii) For purposes of conducting the criminal history, juvenile delinquency and mental health records background check which shall be completed within ten days of receipt of the information from the dealer, the application/ record of sale shall include the name, address, birthdate, gender, race, physical description and Social Security number of the purchaser or transferee and the date of application.

(iv) No information regarding the type of firearm need be included other than an indication that the firearm exceeds the barrel lengths set forth in section 6102.

(v) Unless it has been discovered pursuant to a criminal history, juvenile delinquency and mental health records background check that the potential purchaser or transferee is prohibited from possessing a firearm pursuant to section 6105 (relating to persons not to possess, use, manufacture, control, sell or transfer firearms), no information on the application/ record of sale provided pursuant to this subsection shall be retained as precluded by section 6111.4 (relating to registration of firearms) by the Pennsylvania State Police either through retention of the application/record of sale or by entering the information onto a computer, and, further, an application/record of sale received by the Pennsylvania State Police pursuant to this subsection shall be destroyed within 72 hours of the completion of the criminal history, juvenile delinquency and mental health records background check.

(1.2) Fees collected under paragraph (3) and section 6111.2 (relating to firearm sales surcharge) shall be transmitted to the Pennsylvania State Police within 14 days of collection.

(1.3) In addition to the criminal penalty under section 6119 (relating to violation penalty), any person who knowingly and intentionally maintains or fails to destroy any information submitted to the Pennsylvania State Police for purposes of a background check pursuant to paragraphs (1.1) and (1.4) or violates §6111.4 (relating to registration of firearms) shall be subject to a civil penalty of $250 per violation, entry or failure to destroy.

(1.4) Following implementation of the instantaneous records check by the Pennsylvania State Police on or before December 31, 1999, no application/record of sale shall be completed for the purchase or transfer of a firearm which exceeds the barrel lengths set forth in section 6102. A statement shall be submitted by the dealer to the Pennsylvania State Police, postmarked via first class mail, within 14 days of the sale, containing the number of firearms sold which exceed the barrel and related lengths set forth in section 6102, the amount of surcharge and other fees remitted and a list of the unique approval numbers given pursuant to paragraph (4), together with a statement that the background checks have been performed on the firearms contained in the statement. The form of the statement relating to performance of background checks shall be promulgated by the Pennsylvania State Police.

(2) Inspected photo identification of the potential purchaser or transferee, including, but not limited to, a driver's license, official Pennsylvania photo identification card or official government photo identification card. In the case of a potential buyer or transferee who is a member of a recognized religious sect or community whose tenets forbid or discourage the taking of photographs of members of that sect or community, a seller shall accept a valid-without-photo driver's license or a combination of documents, as prescribed by the Pennsylvania State Police, containing the applicant's name, address, date of birth and the signature of the applicant.

(3) Requested by means of a telephone call that the Pennsylvania State Police conduct a criminal history, juvenile delinquency history and a mental health record check. The purchaser and the licensed dealer shall provide such information as is necessary to accurately identify the purchaser. The requester shall be charged a fee equivalent to the cost of providing the service but not to exceed $2 per buyer or transferee.

(4) Received a unique approval number for that inquiry from the Pennsylvania State Police and recorded the date and the number on the application/record of sale form.

(5) Issued a receipt containing the information from paragraph (4), including the unique approval number of the purchaser. This receipt shall be prima facie evidence of the purchaser's or transferee's compliance with the provisions of this section.

(6) Unless it has been discovered pursuant to a criminal history, juvenile delinquency and mental health records background check that the potential purchaser or transferee is prohibited from possessing a firearm pursuant to section 6105, no information received via telephone following the implementation of the instantaneous background check system from a purchaser or transferee who has received a unique approval number shall be retained by the Pennsylvania State Police.

(7) For purposes of the enforcement of 18 United States Code § 922(d)(9), (g)(1) and (s)(1) (relating to unlawful acts), in the event the criminal history or juvenile delinquency background check indicates a conviction for a misdemeanor that the Pennsylvania State Police cannot determine is or is not related to an act of domestic violence, the Pennsylvania State Police shall issue a temporary delay of the approval of the purchase or transfer. During the temporary delay, the Pennsylvania State Police shall conduct a review or investigation of the conviction with courts, local police departments, district attorneys and other law enforcement or related institutions as necessary to determine whether or not the misdemeanor conviction involved an act of domestic violence. The Pennsylvania State Police shall conduct the review or investigation as expeditiously as possible. No firearm may be transferred by the dealer to the purchaser who is the subject of the investigation during the temporary delay. The Pennsylvania State Police shall notify the dealer of the termination of the temporary delay and either deny the sale or provide the unique approval number under paragraph (4).

(c) Duty of other persons. Any person who is not a licensed importer, manufacturer or dealer and who desires to sell or transfer a firearm to another unlicensed person shall do so only upon the place of business of a licensed importer, manufacturer, dealer or county sheriff's office, the latter of whom shall follow the procedure set forth in this section as if he were the seller of the firearm. The provisions of this section shall not apply to transfers between spouses or to transfers between a parent and child or to transfers between grandparent and grandchild.

(d) Defense. Compliance with the provisions of this section shall be a defense to any criminal complaint under the laws of this Commonwealth or other claim or cause of action arising under this chapter arising from the sale or transfer of any firearm.

(e) Nonapplicability of section. This section shall not apply to the following:

(1) Any firearm manufactured on or before 1898.

(2) Any firearm with a matchlock, flintlock or percussion cap type of ignition system.

(3) Any replica of any firearm described in paragraph (1) if the replica:

(i) is not designed or redesigned to use rimfire or conventional center fire fixed ammunition; or

(ii) uses rimfire or conventional center fire fixed ammunition which is no longer manufactured in the United States and which is not readily available in the ordinary channels of commercial trade.

(f) Application of section.

(1) For the purposes of this section only, except as provided by paragraph (2), "firearm" shall mean any weapon which is designed to or may readily be converted to expel any projectile by the action of an explosive or the frame or receiver of any such weapon.

(2) The provisions contained in subsections (a) and (c) shall only apply to pistols or revolvers with a barrel length of less than 15 inches, any shotgun with a barrel length of less than 18 inches, any rifle with a barrel length of less than 16 inches or any firearm with an overall length of less than 26 inches.

(3) The provisions contained in subsection (a) shall not apply to any law enforcement officer whose current identification as a law enforcement officer shall be construed as a valid license to carry a firearm or any person who possesses a valid license to carry a firearm under section 6109 (relating to licenses).

(4)(i) The provisions of subsection (a) shall not apply to any person who presents to the seller or transferor a written statement issued by the official described in subparagraph (iii) during the ten-day period ending on the date of the most recent proposal of such transfer or sale by the transferee or purchaser stating that the transferee or purchaser requires access to a firearm because of a threat to the life of the transferee or purchaser or any member of the household of that transferee or purchaser.

(ii) The issuing official shall notify the applicant's local police authority that such a statement has been issued. In counties of the first class the chief of police shall notify the police station or substation closest to the applicant's residence.

(iii) The statement issued under subparagraph (ii) shall be issued by the district attorney, or his designee, of the county of residence if the transferee or purchaser resides in a municipality where there is no chief of police. Otherwise, the statement shall be issued by the chief of police in the municipality in which the purchaser or transferee resides.

(g) Penalties.

(1) Any person, licensed dealer, licensed manufacturer or licensed importer who knowingly or intentionally sells, delivers or transfers a firearm in violation of this section commits a misdemeanor of the second degree.

(2) Any person, licensed dealer, licensed manufacturer or licensed importer who knowingly or intentionally sells, delivers or transfers a firearm under circumstances intended to provide a firearm to any person, purchaser or transferee who is unqualified or ineligible to control, possess or use a firearm under this chapter commits a felony of the third degree and shall in addition be subject to revocation of the license to sell firearms for a period of three years.

(3) Any person, licensed dealer, licensed manufacturer or licensed importer who knowingly and intentionally requests a criminal history, juvenile delinquency or mental health record check or other confidential information from the Pennsylvania State Police under this chapter for any purpose other than compliance with this chapter or knowingly and intentionally disseminates any criminal history, juvenile delinquency or mental health record or other confidential information to any person other than the subject of the information commits a felony of the third degree.

(3.1) Any person, licensed dealer, licensed manufacturer or licensed importer who knowingly and intentionally obtains or furnishes information collected or maintained pursuant to section 6109 for any purpose other than compliance with this chapter or who knowingly or intentionally disseminates, publishes or otherwise makes available such information to any person other than the subject of the information commits a felony of the third degree.

(4) Any person, purchaser or transferee commits a felony of the third degree if, in connection with the purchase, delivery or transfer of a firearm under this chapter, he knowingly and intentionally:

(I) Makes any materially false oral statement;

(II) Makes any materially false written statement, including a statement on any form promulgated by federal or state agencies; or

(III) Willfully furnishes or exhibits any false identification intended or likely to deceive the seller, licensed dealer or licensed manufacturer.

(5) Notwithstanding section 306 (relating to liability for conduct of another; complicity) or any other statute to the contrary, any person, licensed importer, licensed dealer or licensed manufacturer who knowingly and intentionally sells, delivers or transfers a firearm in violation of this chapter who has reason to believe that the firearm is intended to be used in the commission of a crime or attempt to commit a crime shall be criminally liable for such crime or attempted crime.

(6) Notwithstanding any act or statute to the contrary, any person, licensed importer, licensed manufacturer or licensed dealer who knowingly and intentionally sells or delivers a firearm in violation of this chapter who has reason to believe that the firearm is intended to be used in the commission of a crime or attempt to commit a crime shall be liable in the amount of the civil judgment for injuries suffered by any person so injured by such crime or attempted crime.

(h) Subsequent violation penalty.

(1) A second or subsequent violation of this section shall be a felony of the second degree and shall be punishable by a mandatory minimum sentence of imprisonment of five years. A second or subsequent offense shall also result in permanent revocation of any license to sell, import or manufacturer a firearm.

(2) Notice of the applicability of this subsection to the defendant and reasonable notice of the Commonwealth's intention to proceed under this section shall be provided prior to trial. The applicability of this section shall be determined at sentencing. The court shall consider evidence presented at trial, shall afford the Commonwealth and the defendant an opportunity to present necessary additional evidence and shall determine by a preponderance of the evidence if this section is applicable.

(3) There shall be no authority for a court to impose on a defendant to which this subsection is applicable a lesser sentence than provided for in paragraph (1), to place the defendant on probation or to suspend sentence. Nothing in this section shall prevent the sentencing court from imposing a sentence greater than that provided in this section. Sentencing guidelines promulgated by the Pennsylvania Commission on Sentencing shall not supersede the mandatory sentences provided in this section.

(4) If a sentencing court refuses to apply this subsection where applicable, the Commonwealth shall have the right to appellate review of the action of the sentencing court. The appellate court shall vacate the sentence and remand the case to the sentencing court for imposition of a sentence in accordance with this section if it finds that the sentence was imposed in violation of this subsection.

(i) Confidentiality. All information provided by the potential purchaser, transferee or applicant, including, but not limited to, the potential purchaser, transferee or applicant's name or identity, furnished by a potential purchaser or transferee under this section or any applicant for a license to carry a firearm as provided by section 6109 shall be confidential and not subject to public disclosure. In addition to any other sanction or penalty imposed by this chapter, any person, licensed dealer, State or local governmental agency or department that violates this subsection shall be liable in civil damages in the amount of $1,000 per occurrence or three times the actual damages incurred as a result of the violation, whichever is greater, as well as reasonable attorney fees.

(j) Exemption. The provisions of subsections (a) and (b) shall not apply to sales between Federal firearms licensees.

(1) The provisions of subsections (a) and (b) shall not apply to:

(I) Sales between Federal firearms licensees; or

(II) The purchase of firearms by a chief law enforcement officer, or his designee, for the official use of law enforcement officers.

(2) For the purposes of this subsection, the term "chief law enforcement officer" shall include the commissioner of the Pennsylvania state police, the chief or head of a police department, a county sheriff or any equivalent law enforcement official.

6111.1. Pennsylvania State Police

(a) Administration. The Pennsylvania State Police shall have the responsibility to administer the provisions of this chapter.

(b) Duty of Pennsylvania State Police.

(1) Upon receipt of a request for a criminal history, juvenile delinquency history and mental health record check of the potential purchaser or transferee, the Pennsylvania State Police shall immediately during the licensee's call or by return call forthwith:

(i) review the Pennsylvania State Police criminal history and fingerprint records to determine if the potential purchaser or transferee is prohibited from receipt or possession of a firearm under Federal or State law;

(ii) review the juvenile delinquency and mental health records of the Pennsylvania State Police to determine whether the potential purchaser or transferee is prohibited from receipt or possession of a firearm under Federal or State law; and

(iii) inform the licensee making the inquiry either:

(A) that the potential purchase or transfer is prohibited; or

(B) provide the licensee with a unique approval number.

(2) In the event of electronic failure, scheduled computer downtime or similar event beyond the control of the Pennsylvania State Police, the Pennsylvania State Police shall immediately notify the requesting licensee of the reason for and estimated length of the delay. If the failure or event lasts for a period exceeding 48 hours, the dealer shall not be subject to any penalty for completing a transaction absent the completion of an instantaneous records check for the remainder of the failure or similar event, but the dealer shall obtain a completed application/record of sale following the provisions of section 6111(b)(1) and (1.1) (relating to sale or transfer of firearms) as if an instantaneous records check has not been established for any sale or transfer of a firearm for the purpose of a subsequent background check.

(3) The Pennsylvania State Police shall fully comply, execute and enforce the directives of this section as follows:

(i) The instantaneous background check for firearms as defined in section 6102 (relating to definitions) shall begin in July 1, 1998.

(ii) The instantaneous background check for firearms that exceed the barrel lengths set forth in section 6102 shall begin on the later of:

(A) the date of publication of the notice under section 6111(a)(2); or

(B) December 31, 1998.

(4) The Pennsylvania State Police and any local law enforcement agency shall make all reasonable efforts to determine the lawful owner of any firearm confiscated by the Pennsylvania State Police or any local law enforcement agency and return said firearm to its lawful owner if the owner is not otherwise prohibited from possessing the firearm. When a court of law has determined that the Pennsylvania State Police or any local law enforcement agency have failed to exercise the duty under this subsection, reasonable attorney fees shall be awarded to any lawful owner of said firearm who has sought judicial enforcement of this subsection.

(c) Establish a telephone number. The Pennsylvania State Police shall establish a telephone number which shall be operational seven days a week between the hours of 8 a.m. and 10 p.m. local time for purposes of responding to inquiries as described in this section from licensed manufacturers, licensed importers and licensed dealers. The Pennsylvania State Police shall employ and train such personnel as are necessary to administer expeditiously the provisions of this section.

(d) Distribution. The Pennsylvania State Police shall provide, without charge, summaries of uniform firearm laws and firearm safety brochures pursuant to section 6125 (relating to distribution of uniform firearm laws and firearm safe-ty brochures).

(e) Challenge to records.

(1) Any person who is denied the right to receive, sell, transfer, possess, carry, manufacture or purchase a firearm as a result of the procedures established by this section may challenge the accuracy of that person's criminal history, juvenile delinquency history or mental health record pursuant to a denial by the instantaneous records check by submitting a challenge to the Pennsylvania state police within 30 days from the date of the denial.

(2) The Pennsylvania state police shall conduct a review of the accuracy of the information forming the basis for the denial, and shall have the burden of proving the accuracy of the record. Within 20 days after receiving a challenge, the Pennsylvania state police shall notify the challenger of the basis for the denial, including, but not limited to, the jurisdiction and docket number of any relevant court decision and provide the challenger an opportunity to provide additional information for the purposes of the review. The Pennsylvania state police shall communicate its final decision to the challenger within 60 days of the receipt of the challenge. The decision of the Pennsylvania state police shall include all information which formed a basis for the decision.

(3) If the challenge is ruled invalid, the person shall have the right to appeal the decision to the attorney general within 30 days of the decision. The attorney general shall conduct a hearing de novo in accordance with the administrative agency law. The burden of proof shall be upon the commonwealth.

(4) The decision of the attorney general may be appealed to the commonwealth court by an aggrieved party.

(f) Notification of mental health adjudication, treatment, commitment, drug use or addiction.

....**(1)** Notwithstanding any statute to the contrary, judges of the courts of common pleas shall notify the Pennsylvania state police, on a form developed by the Pennsylvania state police, of:

(i) The identity of any individual who has been adjudicated as an incompetent or as a mental defective or who has been involuntarily committed to a mental institution under the act of July 9, 1976 (p.l.817, no.143), known as the Mental Health Procedures Act, or who has been involuntarily treated as described in section 6105(c)(4) (relating to persons not to possess, use, manufacture, control, sell or transfer firearms) or as described in 18 U.S.C. section 922(g)(4) (relating to unlawful acts) and its implementing federal regulations; and

(ii) Any finding of fact or court order related to any person described in 18 U.S.C. section 922(g)(3).

(2) The notification shall be transmitted by the judge to the Pennsylvania state police within seven days of the adjudication, commitment or treatment.

(3) Notwithstanding any law to the contrary, the Pennsylvania state police may disclose, electronically or otherwise, to the united states attorney general or a designee, any record relevant to a determination of whether a person is disqualified from possessing or receiving a firearm under 18 U.S.C. Section 922 (g)(3) or (4) or an applicable state statute.

(g) Review by court.

(1) Upon receipt of a copy of the order of a court of competent jurisdiction which vacates a final order or an involuntary certification issued by a mental health review officer, the Pennsylvania State Police shall expunge all records of the involuntary treatment received under subsection (f).

(2) A person who is involuntarily committed pursuant to section 302 of the Mental Health Procedures Act may petition the court to review the sufficiency of the evidence upon which the commitment was based. If the court determines that the evidence upon which the involuntary commitment was based was insufficient, the court shall order that the record of the commitment submitted to the Pennsylvania State Police be expunged. A petition filed under this subsection shall toll the 60-day period set forth under section 6105(a)(2).

(3) The Pennsylvania State Police shall expunge all records of an involuntary commitment of an individual who is discharged from a mental health facility based upon the initial review by the physician occurring within two hours of arrival under section 302(b) of the Mental Health Procedures Act and the physician's determination that no severe mental disability existed pursuant to section 302(b) of the Mental Health Procedures Act. The physician shall provide signed confirmation of the determination of the lack of severe mental disability following the initial examination under section 302(b) of the Mental Health Procedures Act to the Pennsylvania State Police.

(h) Juvenile registry.

(1) The contents of law enforcement records and files compiled under 42 Pa.C.S. 6308 (relating to law enforcement records) concerning a child shall not be disclosed to the public except if the child is 14 years of age or older at the time of the alleged conduct and if any of the following apply:

(i) The child has been adjudicated delinquent by a court as a result of an act or acts which constitute any offense enumerated in section 6105.

(ii) A petition alleging delinquency has been filed by a law enforcement agency alleging that the child has committed an act or acts which constitute an offense enumerated in section 6105 and the child previously has been adjudicated delinquent by a court as a result of an act or acts which included the elements of one of such crimes.

(2) Notwithstanding any provision of this subsection, the contents of law enforcement records and files concerning any child adjudicated delinquent for the commission of any criminal activity described in paragraph (1) shall be recorded in the registry of the Pennsylvania State Police for the limited purposes of this chapter.

(i) Reports. The Pennsylvania State Police shall annually compile and report to the General Assembly, on or before December 31, the following information for the previous year:

(1) number of firearm sales, including the types of firearms;

(2) number of applications for sale of firearms denied, number of challenges of the denials and number of final reversals of initial denials;

(3) summary of the Pennsylvania State Police's activities, including the average time taken to complete a criminal history, juvenile delinquency history or mental health record check; and

(4) uniform crime reporting statistics compiled by the Pennsylvania State Police based on the National Incident-based Reporting System.

(j) Other criminal information. The Pennsylvania State Police shall be authorized to obtain any crime statistics necessary for the purposes of this chapter from any local law enforcement agency.

(j.1) Delinquency and mental health rec-ords. The provisions of this section which relate to juvenile delinquency and mental health rec-ords checks shall be applicable when the data has been made available to the Pennsylvania State Police but not later than October 11, 1999.

(j.2) Records check. The provisions of this section which relate to the instantaneous records check conducted by telephone shall be applicable 30 days following notice by the Pennsylvania State Police pursuant to section 6111(a)(2).

(j.3) Immunity. The Pennsylvania State Police and its employees shall be immune from actions for damages for the use of a firearm by a purchaser or for the unlawful transfer of a firearm by a dealer unless the act of the Pennsylvania State Police or its employees constitutes a crime, actual fraud, actual malice or willful misconduct.

(k) Definitions. As used in this section, the following words and phrases shall have the meanings given to them in this subsection:

"Firearm." The term shall have the same meaning as in section 6111.2 (relating to firearm sales surcharge).

"Physician." Any licensed psychiatrist or clin-ical psychologist as defined in the act of July 9, 1976 (P.L. 817, No. 143), known as the Mental Health Procedures Act.

6111.2. Firearm sales surcharge

(a) Surcharge imposed. There is hereby imposed on each sale of a firearm subject to tax under Article II of the act of March 4, 1971 (P.L. 6, No.2), known as the Tax Reform Code of 1971, an additional surcharge of $3. This shall be referred to as the Firearm Sale Surcharge. All moneys received from this surcharge shall be deposited in the Firearm Instant Records Check Fund.

(b) Increases or decreases. Five years from the effective date of this subsection, and every five years thereafter, the Pennsylvania State Police shall provide such information as necessary to the Legislative Budget and Finance Committee for the purpose of reviewing the need to increase or decrease the instant check fee. The committee shall issue a report of its findings and recommendations to the General Assembly for a statutory change in the fee.

(c) Revenue sources. Funds received under the provisions of this section and section 6111 (b)(3) (relating to sale or transfer of firearms) as estimated and certified by the Secretary of Revenue, shall be deposited within five days of the end of each quarter into the fund.

(d) Definition. As used in this section only, the term "firearm" shall mean any weapon which is designed to or may readily be converted to expel any projectile by the action of an explosion or the frame or receiver of any such weapon.

6111.3. Firearm Records Check Fund

(a) Establishment. The Firearm Records Check Fund is hereby established as a restricted account in the State Treasury, separate and apart from all other public money or funds of the Commonwealth, to be appropriated annually by the General Assembly, for use in carrying out the provisions of section 6111 (relating to firearm ownership). The moneys in the fund on June 1, 1998, are hereby appropriated to the Pennsylvania State Police.

(b) Source. The source of the fund shall be moneys collected and transferred under section 6111.2 (relating to firearm sales surcharge) and moneys collected and transferred under section 6111(b)(3).

6111.4. Registration of firearms Notwithstanding any section of this chapter to the contrary, nothing in this chapter shall be construed to allow any government or law enforcement agency or any agent thereof to create, maintain or operate any registry of firearm ownership within this Commonwealth. For the purposes of this section only, the term "firearm" shall include any weapon that is designed to or may readily be converted to expel any projectile by the action of an explosive or the frame or receiver of any such weapon.

6111.5. Rules and regulations The Pennsylvania State Police shall in the manner provided by law promulgate the rules and regulations necessary to carry out this chapter, including regulations to ensure the identity, confidentiality and security of all records and data provided pursuant hereto.

6112. Retail dealer required to be licensed No retail dealer shall sell, or otherwise transfer or expose for sale or transfer, or have in his possession with intent to sell or transfer, any firearm as defined in section 6113(d) (relating to licensing of dealers) without being licensed as provided in this chapter.

6113. Licensing of dealers

(a) General rule. The chief or head of any police force or police department of a city, and, elsewhere, the sheriff of the county, shall grant to reputable applicants licenses, in form prescribed by the Pennsylvania State Police, effective for three years from date of issue, permitting the licensee to sell firearms direct to the consumer, subject to the following conditions in addition to those specified in section 6111 (relating to sale or transfer of firearms), for breach of any of which the license shall be forfeited and the licensee subject to punishment as provided in this subchapter:

(1) The business shall be carried on only upon the premises designated in the license or at a lawful gun show or meet.

(2) The license, or a copy thereof, certified by the issuing authority, shall be displayed on the premises where it can easily be read.

(3) No firearm shall be sold in violation of any provision of this subchapter.

(4) No firearm shall be sold under any circumstances unless the purchaser is personally known to the seller or shall present clear evidence of the purchaser's identity.

(5) A true record in triplicate shall be made of every firearm sold, in a book kept for the purpose, the form of which may be prescribed by the Pennsylvania State Police, and shall be personally signed by the purchaser and by the person effecting the sale, each in the presence of the other, and shall contain the information required by section 6111. The record shall be maintained by the licensee for a period of 20 years.

(6) No firearm as defined in section 6102 (relating to definitions) shall be displayed in any part of any premises where it can readily be seen from the outside. In the event that the Commissioner of the Pennsylvania State Police shall find a clear and present danger to public safety within this Commonwealth or any area thereof, firearms shall be stored and safeguarded pursuant to regulations to be established by the Pennsylvania State Police by the licensee during the hours when the licensee is closed for business.

(7) The dealer shall possess all applicable current revenue licenses.

(b) Fee. The fee for issuing said license shall be $30, which fee shall be paid into the county treasury.

(c) Revocation. Any license granted under subsection (a) of this section may be revoked for cause by the person issuing the same, upon written notice to the holder thereof.

(d) Definitions. For the purposes of this section and section 6112 (relating to retail dealer required to be licensed) only unless otherwise specifically provided, the term "firearm" shall include any weapon that is designed to or may readily be converted to expel any projectile by the action of an explosive or the frame or receiver of any such weapon.

6114. Judicial review The action of the chief of police, sheriff, county treasurer or other officer under this subchapter shall be subject to judicial review in the manner and within the time provided by 2 Pa.C.S. Ch. 7 Subch. B (relating to judicial review of local agency action). A judgment sustaining a refusal to grant a license shall not bar, after one year, a new application; nor shall a judgment in favor of the petitioner prevent the defendant from thereafter revoking or refusing to renew such license for any proper cause which may thereafter occur. The court shall have full power to dispose of all costs.

6115. Loans on, or lending or giving firearms prohibited

(a) Offense defined. No person shall make any loan secured by mortgage, deposit or pledge of a firearm, nor, except as provided in subsection (b), shall any person lend or give a firearm to another or otherwise deliver a firearm contrary to the provisions of this subchapter.

(b) Exception.

(1) Subsection (a) shall not apply if any of the following apply:

(i) The person who receives the firearm is licensed to carry a firearm under section 6109 (relating to licenses).

(ii) The person who receives the firearm is exempt from licensing.

(iii) The person who receives the firearm is engaged in a hunter safety program certified by the Pennsylvania Game Commission or a firearm training program or competition sanctioned or approved by the National Rifle Association.

(iv) The person who receives the firearm meets all of the following:

(A) Is under 18 years of age.

(B) Pursuant to section 6110.1 (relating to possession of firearm by minor) is under the supervision, guidance and instruction of a responsible individual who:

(I) is 21 years of age or older; and

(II) is not prohibited from owning or possessing a firearm under section 6105 (relating to persons not to possess, use, manufacture, control, sell or transfer firearms).

(v) The person who receives the firearm is lawfully hunting or trapping and is in compliance with the provisions of Title 34 (relating to game).

(vi) A bank or other chartered lending institution is able to adequately secure firearms in its possession.

(2) Nothing in this section shall be construed to prohibit the transfer of a firearm under 20 Pa. C.S. Ch. 21 (relating to intestate succession) or by bequest if the individual receiving the firearm is not precluded from owning or possessing a firearm under section 6105.

(3) Nothing in this section shall be construed to prohibit the loaning or giving of a firearm to another in one's dwelling or place of business if the firearm is retained within the dwelling or place of business.

(4) Nothing in this section shall prohibit the relinquishment of firearms to a third party in accordance with 23 Pa.C.S. § 6108.3 (relating to relinquishment to third party for safekeeping).

6116. False evidence of identity In addition to any other penalty provided in this chapter, the furnishing of false information or offering false evidence of identity is a violation of section 4904 (relating to unsworn falsification to authorities).

6117. Altering or obliterating marks of identification

(a) Offense defined. No person shall change, alter, remove, or obliterate the manufacturer's number integral to the frame or receiver of any firearm which shall have the same meaning as provided in section 6105 (relating to persons not to possess, use, manufacture, control, sell or transfer firearms).

(b) Deleted by 2008, Oct. 17, P.L. 1628, No. 131, § 7, effective in 60 days [Dec. 16, 2008].

(c) Penalty. A violation of this section constitutes a felony of the second degree.

(d) Deleted by 2008, Oct. 17, P.L. 1628, No. 131, § 7, effective in 60 days [Dec. 16, 2008].

6118. Antique firearms

(a) General rule. This subchapter shall not apply to antique firearms.

(b) Exception. Subsection (a) shall not apply to the extent that such antique firearms, reproductions or replicas of firearms are concealed weapons as provided in section 6106 (relating to firearms not to be carried without a license), nor shall it apply to the provisions of section 6105 (relating to persons not to possess, use, manufacture, control, sell or transfer firearms) if such antique firearms, reproductions or replicas of firearms are suitable for use.

(c) Definition. As used in this section, the term "antique firearm" means:

(1) Any firearm with a matchlock, flintlock or percussion cap type of ignition system.

(2) Any firearm manufactured on or before 1898.

(3) Any replica of any firearm described in paragraph (2) if such replica:

(i) is not designed or redesigned for using rimfire or conventional center fire fixed ammunition; or

(ii) uses rimfire or conventional center fire fixed ammunition which is no longer manufactured in the United States and which is not readily available in the ordinary channels of commercial trade.

6119. Violation penalty Except as otherwise specifically provided, an offense under this subchapter constitutes a misdemeanor of the first degree.

6120. Limitation on the regulation of firearms and ammunition

(a) General rule. No county, municipality or township may in any manner regulate the lawful ownership, possession, transfer or transportation of firearms, ammunition or ammunition components when carried or transported for purposes not prohibited by the laws of this Commonwealth. ...

(b) Definitions. As used in this section, the following words and phrases shall have the meanings given to them in this subsection:

"Dealer." The term shall include any person engaged in the business of selling at wholesale or retail a firearm or ammunition.

"Firearms." This term shall have the meaning given to it in section 5515 (relating to prohibiting of paramilitary training) but shall not include air rifles as that term is defined in section 6304 (relating to sale and use of air rifles).

"Political subdivision." The term shall in-clude any home rule charter municipality, coun-ty, city, borough, incorporated town, town-ship or school district.

6121. Certain bullets prohibited

(a) Offense defined. It is unlawful for any person to possess, use or attempt to use a KTW teflon-coated bullet or other armor-piercing ammunition while committing or attempting to commit a crime of violence as defined in section 6102 (relating to definitions).

(b) Grading. An offense under this section constitutes a felony of the third degree.

(c) Sentencing. Any person who is convicted in any court of this Commonwealth of a crime of violence and who uses or carries, in the commission of that crime, a firearm loaded with KTW ammunition or any person who violates this section shall, in addition to the punishment provided for the commission of the crime, be sentenced to a term of imprisonment for not less than five years. Notwithstanding any other provision of law, the court shall not suspend the sentence of any person convicted of a crime subject to this subsection nor place him on probation nor shall the term of imprisonment run concurrently with any other term of imprisonment including that imposed for the crime in which the KTW ammunition was being used or carried. No person sentenced under this subsection shall be eligible for parole.

(d) Definition. As used in this section the term "armor-piercing ammunition" means ammunition which, when or if fired from any firearm as defined in section 6102 that is used or attempted to be used in violation of subsection (a) under the test procedure of the National Institute of Law Enforcement and Criminal Justice Standard for the Ballistics Resistance of Police Body Armor promulgated December 1978, is determined to be capable of penetrating bullet-resistant apparel or body armor meeting the requirements of Type IIA of Standard NILECJ-STD-0101.01 as formulated by the United States Department of Justice and published in December of 1978.

6122. Proof of license and exception

(a) General rule. When carrying a firearm concealed on or about one's person or in a vehicle, an individual licensed to carry a firearm shall, upon lawful demand of a law enforcement officer, produce the license for inspection. Failure to produce such license either at the time of arrest or at the preliminary hearing shall create a rebuttable presumption of nonlicensure.

(b) Exception. An individual carrying a firearm on or about his person or in a vehicle and claiming an exception under section 6106(b) (relating to firearms not to be carried without a license) shall, upon lawful demand of a law enforcement officer, produce satisfactory evidence of qualification for exception.

6123. Waiver of disability or pardons A waiver of disability from Federal authorities as provided for in 18 U.S.C. 925 (relating to exceptions; relief from disabilities), a full pardon from the Governor or an overturning of a conviction shall remove any corresponding disability under this subchapter except the disability under section 6105 (relating to persons not to possess, use, manufacture, control, sell or transfer firearms).

6124. Administrative regulations The commissioner may establish form specifications and regulations, consistent with section 6109(c) (relating to licenses), with respect to uniform forms control, including the following:

(1) License to carry firearms.

(2) Firearm registration.

(3) Dealer's license.

(4) Application for purchase of a firearm.

(5) Record of sale of firearms.

6125. Distribution of uniform firearm laws and firearm safety brochures It shall be the duty of the Pennsylvania State Police beginning January 1, 1996, to distribute to every licensed firearm dealer in this Commonwealth firearms safety brochures at no cost to the dealer. The brochures shall be written by the Pennsylvania State Police, with the cooperation of the Pennsylvania Game Commission, and shall include a summary of the major provisions of this subchapter, including, but not limited to, the duties of the sellers and purchasers and the transferees of firearms. The brochure or a copy thereof shall be provided without charge to each purchaser.

6127. Firearm tracing

(A) ILLEGAL POSSESSION. Upon confiscating or recovering a firearm from the possession of anyone who is not permitted by federal or state law to possess a firearm, a local law enforcement agency shall use the best available information, including a firearms trace where necessary, to determine how and from where the person gained possession of the firearm.

(b) TRACING.-- Local law enforcement shall use the National Tracing Center of the Federal Bureau of Alcohol, Tobacco, Firearms and Explosives in complying with subsection (a).

(c) NOTIFICATION.-- Local law enforcement agencies shall advise the Pennsylvania State Police of all firearms that are recovered in accordance with this section.

Subchapter B. Firearms Generally

6141.1. Purchase of rifles and shotguns outside this Commonwealth Nothing in this chapter shall be construed to prohibit a person in this Commonwealth who may lawfully purchase, use, control, sell, transfer or manufacture a firearm which exceeds the barrel and related lengths set forth in section 6102 (relating to definitions) from lawfully purchasing or otherwise obtaining such a firearm in a jurisdiction outside the Commonwealth.

6142. Locking device for firearms

(a) Offense defined. It shall be unlawful for any licensee to sell, deliver or transfer any firearm as defined in section 6102 (relating to definitions), other than an antique firearm as defined in section 6118 (relating to antique firearms), to any other person, other than another licensee, unless the transferee is provided with or purchases a locking device for that firearm or the design of the firearm incorporates a locking device.

(b) Exceptions. Firearms for transfer to or possession by any law enforcement officer employed by any Federal, State or local government entity or rail police employed and certified by a rail carrier as a police officer are not subject to the provisions of this section.

(c) Penalties. A violation of the provisions of this section shall be a summary offense.

(d) Good faith compliance. A licensee who in good faith complies with this section shall not be civilly liable as a result of such compliance with this section, except for any acts or omissions intentionally designed to harm or for grossly negligent acts or omissions which result in harm.

(e) Admissibility of evidence. A transferee's purchase or receipt of a locking device in conjunction with the purchase of a firearm pursuant to this section shall not be admissible as evidence in any civil action brought against the transferee.

(f) Definitions. As used in this section, the following words and phrases shall have the meanings given to them in this subsection:

"Licensee." Any licensed manufacturer, importer or dealer of firearms.

"Locking device." Either of the following:

(1) a device that, when installed on a firearm, is designed to prevent the firearm from being operated without first deactivating the device; or

(2) a device that is incorporated into the design of a firearm and that is designed to prevent the operation of the firearm by anyone not having access to the device.

Chapter 63. Minors

6302. Sale or lease of weapons or explosives

(a) Offense defined. A person is guilty of a misdemeanor of the first degree if he sells or causes to be sold or leases to any person under 18 years of age any deadly weapon, cartridge, gunpowder, or other similar dangerous explosive substance.

(b) Exception. The provisions of subsection (a) shall not prohibit hunting by minors under 18 years of age permitted under Title 34 (relating to game).

6303. Sale of starter pistols

(a) Offense defined. A person is guilty of a misdemeanor of the first degree if he sells, causes to be sold, gives or furnishes to any person under the age of 18 years, or if he, being under the age of 18 years, purchases, accepts, receives or possesses, any pistol commonly referred to as "starter pistol" specially designed to receive and discharge blank cartridges only or similar pistol.

(b) Exception. Nothing in this section shall prohibit the use of starter pistols for the purpose of starting or officiating at athletic events, use in dramatic productions, or other similar events.

6304. Sale and use of air rifles

(a) Sale or transfer of air rifles.

(1) It shall be unlawful for any dealer to sell, lend, rent, give, or otherwise transfer an air rifle to any person under the age of 18 years, where the dealer knows, or has reasonable cause to believe, the person to be under 18 years of age, or where such dealer has failed to make reasonable inquiry relative to the age of such person, and such person is under 18 years of age.

(2) It shall be unlawful for any person to sell, give, lend, or otherwise transfer any air rifle to any person under 18 years of age, except where the relationship of parent and child, guardian and ward or adult instructor and pupil exists between such person and the person under 18 years of age.

(b) Carrying or discharging air rifles.

(1) It shall be unlawful for any person under 18 years of age to carry any air rifle on the highways or public lands unless accompanied by an adult, except that a person under 18 years of age may carry such rifle unloaded in a suitable case or securely wrapped.

(2) It shall be unlawful for any person to discharge any air rifle from or across any highway or public land or any public place, except on a properly constructed target range.

(c) Exceptions.

(1) Nothing in this section shall make it unlawful for any person under 18 years of age to have in his possession any air rifle, if it is:

(i) kept within his domicile;

(ii) used by the person under 18 years of age and he is a duly enrolled member of any club, team or society organized for educational purposes and maintaining as part of its facilities or having written permission to use an indoor or outdoor rifle range under the supervision, guidance and instruction of a responsible adult, and then only, if said air rifle is actually being used in connection with the activities of said club, team or society under the supervision of a responsible adult; or

(iii) used in or on any private grounds or residence under circumstances when such air rifle can be fired, discharged or operated in such a manner as not to endanger persons or property, and then only, if it is used in such manner as to prevent the projectile from transversing any grounds or space outside the limits of such grounds or residence.

(2) Nothing in this section shall prohibit sales of air rifles:

(i) By wholesale dealers or jobbers.

(ii) To be shipped out of this Commonwealth.

(iii) To be used at a target range operated in accordance with paragraph (1) of this subsection or by members of the armed services of the United States or veterans' organizations.

(d) Seizure. Any law enforcement officer may seize, take, remove or cause to be removed, at the expense of the owner, all air rifles used or offered for sale in violation of this section.

(e) No preemption. The provisions of any ordinance enacted by any political subdivision which impose greater restrictions or limitations in respect to the sale and purchase, use or possession of air rifles, than is imposed by this section, shall not be invalidated or affected by this section.

(f) Grading. Any dealer violating the provisions of paragraph (a)(1) of this section shall be guilty of a misdemeanor of the third degree. Any person violating any other provision of this section shall be guilty of a summary offense.

(g) Definitions. As used in this section the following words and phrases shall have the meanings given to them in this subsection:

"Air rifle." Any air gun, air pistol, spring gun, spring pistol, B-B gun, or any implement that is not a firearm, which impels a pellet of any kind with a force that can reasonably be expected to cause bodily harm. The term does not include a paintball gun or paintball marker as defined in section 2707.2 (relating to paintball guns and paintball markers).

"Dealer." Any person engaged in the business of selling at retail or renting any air rifles.

Chapter 73. Trade and Commerce

7306. Incendiary devices

(a) Offense defined. A person is guilty of a misdemeanor of the first degree if he owns, manufactures, sells, transfers, uses or possesses any incendiary device or similar device or parts thereof, including but not limited to a "Molotov cocktail."

(b) Exception. The provisions of subsection (a) of this section shall not apply to authorized personnel of the United States, the Commonwealth or any political subdivision, who use incendiary devices as part of their duties.

(c) Definition. As used in this section the phrase "incendiary device" means any inflammable liquid enclosed in a readily breakable container that can be equipped with an igniter of any type.

[Current through the end of the 2008 Regular Session and the 2008 Special Sessions]

PUERTO RICO

P.R. Laws

Title 24. Health and Sanitation

Chapter 111. Controlled Substances Act of Puerto Rico

2516. Suspension or revocation of license to drive motor vehicles and license to carry firearms No person who has been convicted of a felony under this chapter or of any law of the United States or of any state, in connection with narcotic drugs, marihuana, depressant or stimulant substances, as well as of any foreign country and no person who has been declared addicted to narcotic drugs, may obtain a license from the corresponding authority to drive any kind of motor vehicle nor to hold, possess or carry firearms for a term of five (5) years from and after the serving of the sentence imposed by said conviction or after the date of the declaration. Officers or public employees in charge of the issuance of said licenses shall be prevented from extending them, whenever any of the circumstances expressed above may concur in the applicant [sic] for the license and any such licenses which have been issued prior to the conviction or declaration that the person is addicted to narcotic drugs, shall immediately be cancelled by the corresponding authority.

Title 25. Internal Security

Subtitle 1. Generally

Part V. Regulation of Firearms, Explosives, and Other Dangerous Devices

Chapter 51A. Puerto Rico Arms Act of 2000

Subchapter I. Preliminary Provisions

455. Definitions For the purposes of this chapter, the following terms shall have the meaning stated below:

(a) Law enforcement officer. Means any member or officer of the Government of Puerto Rico or of the United States of America, as well as any other political subdivision of Puerto Rico or the United States, among whose duties are to make arrests, including, but without being limited to the members of the Rangers Corps of the Department of Natural and Environmental Resources, the Puerto Rico Police, Deputy Police, Municipal Police, investigating agents of the Special Investigations Bureau of the Department of Justice, custodial officials of the Corrections Administration, custodial officials of the Pretrial Services Office, the National Guard while in official duty or practice, the custodial officials of the Juvenile Institutions Administration, the Internal Security Corps of the Ports Authority, the Director of the Drugs and Narcotics Control Division and the controlled substances inspectors of the Mental Health and Addiction Services Administration, the investigating agents of the Deputy Secretary Investigations Office of the Corrections System of the Department of Corrections and Rehabilitation, and the inspectors of the Public Service Commission, as well as the marshals of the General Court of Justice of Puerto Rico and of the federal court with jurisdiction throughout Puerto Rico, and the internal revenue inspectors of the Department of the Treasury.

(b) "Machine gun" or "automatic weapon". Means a weapon of any description, regardless of its size and of the name by which it is designated or known, either loaded or unloaded, that is capable of firing a rapid and repeated or automatic stream of bullets contained in a magazine, ammunition belt or other receptacle, by a single pull of the trigger. The term "machine gun" also includes a submachine gun, as well as any other firearm provided with a magazine to automatically fire all or part of the bullets or ammunition contained in the magazine, belt or any combination of the parts of a firearm, destined to, and with the intention of converting, modifying or altering said weapon to make it a machine gun.

(c) Weapon. Means any firearm, blade, or any other type of weapon regardless of its denomination.

(d) Sidearm. Means a stabbing, cutting or thrusting weapon (cold steel) that can be used as an instrument of aggression capable of inflicting grave bodily injury.

(e) Firearm. Means any weapon, regardless of the name by which it is known, designed to be or which may easily be converted to be, or which is capable of firing a round or rounds of ammunition by an explosive charge. This definition does not include those working tools such as, but not limited to, nail guns, when used for work, in the arts or a trade.

(f) Rifle. Means any shotgun, rifle or firearm designed to be fired from the shoulder.

(g) Pneumatic weapon. Means any weapon, regardless of the name by which it is known, that through the discharge of gas or a mixture of compressed gasses, is capable of firing one (1) or more projectiles.

(h) Antique firearm. Means:

(1) any firearm with a rifle matchlock, flintlock, or percussion cap mechanism made in or before 1898; or

(2) any replica of a firearm described in clause (1) above, if said replica:

(A) is not designed or redesigned to use rimfire or conventional centerfire ammunition, or

(B)uses rimfire or conventional centerfire ammunition no longer made in the United States and which cannot be found through normal and ordinary business channels; or

(C) any muzzle loading rifle, muzzle loading shotgun or muzzle loading pistol designed to be used with black gunpowder, or a substitute of black gunpowder, and which is unable to fire fixed ammunition. For the purposes of this clause, the term Antique Firearm shall not include any firearm that includes a frame or receiver, any firearm converted into a muzzle loading weapon, or any muzzle loading weapon that may be converted to be capable of firing fixed ammunition by means replacing the barrel, the bolt, the breech lock, or any combination thereof.

(D) The term fixed ammunition shall mean that which is completely assembled, meaning it is equipped with casing, gun powder, primer and slug.

(i) Gunsmith. Means any natural or juridical person who, on his/her own or through agents or employees, purchases, or introduces for sale, changes, exchanges, offers for sale or displays for sale, or has for sale in his/her business establishment, any firearm or ammunition, or who performs any mechanical or cosmetic work for a third party on any firearm or ammunitions.

(j) Armor piercing. Means a projectile that can be used in a handgun that is constructed entirely (excluding the presence or traces of other substances) or of a combination of an alloy of tungsten, steel, iron, tin, bronze, cupric beryllium or degraded uranium; or a fully-shielded, greater than twenty-two (22) caliber bullet, designed and intended to be used in a hand gun and whose shielding weighs more than twenty-five percent of its total weight. It excludes the shotgun ammunition required by federal or state environmental laws or hunting regulations for such purposes, a disintegrating bullet designed for target shooting, a projectile whose primary use determined by the Secretary of the Treasury of the United States for sports purposes, or any other projectile or nucleus of a projectile in which said Secretary finds that its main use is for industrial purposes, including a charge used in the digging of oil or gas wells.

(k) Home. Is the part of a building that is used or occupied by a single person or family.

(l) Committee. Means the Interagency Committee to Fight the Illegal Trafficking of Weapons, established in this chapter.

(m) Shotgun. Means a long-barreled firearm with one or more smooth bores, designed to be fired from the shoulder, which can fire cartridges of one (1) or more shots. It may be fed manually, or by a magazine or receptacle, and may be fired manually, automatically or semiautomatically. This definition includes sawed-off shotguns with barrels of less than 18 inches in length.

(n) Sports shooting federation. Means any federation attached to the Puerto Rico Olympic Committee that represents the Olympic target-shooting sport.

(o) Shooting license. Is the license issued by the Superintendent that authorizes the concessionaire to possess, carry and transport arms and ammunition, and, depending on their category, to carry firearms, target shooting or hunting.

(p) Ammunition. Means any bullet, cartridge, projectile, pellet, or any load that is placed or can be placed in a firearm to be fired.

(q) Pistol. Is any firearm that does not have a cylinder, which is fed manually or by a magazine, is not designed to be fired from the shoulder, and is capable of being fired semi-automatically or one shot at a time, depending on its class.

(r) Police. Means the Puerto Rico Police.

(s) Carry. Means the immediate possession or physical holding of a weapon, loaded or unloaded, on the person of the carrier, it being also understood when a weapon is not being transported pursuant to the provisions of this chapter.

(t) Revolver. Means any firearm that has a revolving cylinder with several chambers, which, by pulling the trigger or setting the hammer, are aligned with the barrel, placing the bullet in a position to be fired.

(u) Rifle. Means any firearm designed to be fired from the shoulder, which fires one or more projectiles. It may be fed manually or automatically by a magazine or removable receptacle and fired manually or semiautomatically. The word "rifle" also includes the word "carbine".

(v) Secretary. Means the Secretary of the Sports and Recreation Department.

(w) Superintendent. Means the Puerto Rico Police Superintendent.

(x) Transportation. Means the mediate or immediate possession of a weapon for the purpose of taking it from one place to another. Said transportation must be carried out by a person with a current weapons license and the weapon must be unloaded and transported inside a closed case whose contents are not visible and which may not be in plain sight.

(y) Vehicle. Is any means for the transport of persons or goods on land, sea or air.

(z) Category Change. Means to incorporate permits to a firearms license, regardless of its category, carrying, hunting or target shooting.

Subchapter II. Licenses and Regulations

456. Electronic register The Superintendent shall issue weapons and/or gunsmith licenses pursuant to the provisions of this chapter, which shall be designed to expedite the electronic registration of all transactions involving firearms and ammunition by the person who holds either of the above. It shall pertain to the Superintendent to provide, through regulations, the manner in which the Electronic Register system shall operate, and to ensure that the system that is designed directly informs the Police of every transaction made by a license holder. The Puerto Rico Police is granted a term of six (6) months from the date of effectiveness of this law to install this Register.

The weapons license shall be issued on an identification card, sufficiently small to be carried in a commonly used wallet, and shall contain, at least, a photograph of the petitioner, his/her complete name, date of birth, personal data and his/her weapons license number. It shall also contain the date of issue and expiration of the license, as provided below. It shall also contain the mechanisms to access the police electronic register system to ascertain its veracity and other relevant data, such as identification of its scope according to the categories of bearing, carrying the weapon, target shooting, hunting, or all categories. The license shall not contain the address of the petitioner, nor shall it mention the weapons or ammunition authorized for purchase, but the police electronic register shall contain and furnish said information to its users.

Provided, that until the police installs and makes the electronic register system available to gunsmiths, the Superintendent shall issue to every licensee a provisional identification card that contains, at least, a photograph of the licensee, his/her full name, date of birth, personal data, the license number and the caliber corresponding to the ammunition he/she is authorized to buy. It shall also contain the date of issue and expiration of the license, as provided below. The official identification card issued according to the provisions of this chapter shall be the only document providing the legal authority to conduct the authorized activities. Once the electronic register system is duly implemented, the Superintendent may only issue the electronic identification card. If the system is not available at the time a transaction takes place, said transaction shall be conducted according to the procedure the Superintendent provides through regulations.

456a. Weapons license

(a) The Superintendent shall issue a weapons license to any petitioner who meets the following requirements:

(1) Has reached the age of twenty-one (21) years.

(2) Has a negative criminal record certificate issued not later than thirty (30) days prior to the date of the application and has not been accused of, nor is pending or in the process of trial for any of the crimes listed in § 456j of this title or its equivalent, in Puerto Rico, the United States or abroad.

(3) Is not habitually inebriated or addicted to controlled substances.

(4) Has not been declared mentally incompetent by a court.

(5) Has not incurred or belonged to organizations involved in acts of violence, or directed to the overthrow of the constituted government.

(6) Has not been dishonorably discharged from the armed forces, or removed from any of the law enforcement agencies of the Government of Puerto Rico or its municipalities.

(7) Is not under a court order which prohibits harassing, stalking, threatening, or proximity to an intimate partner, the children of the latter or to any person, and who does not have a record of violence.

(8) Is a citizen of the United States, or a legal resident of Puerto Rico.

(9) Is not a person who, having been a citizen of the United States at a given time, has renounced said citizenship.

(10) Has submitted a sworn statement attesting to compliance with fiscal laws; it being established that it shall be grounds to deny the issue of the license applied for or to revoke the same if the petitioner has failed to comply with the fiscal laws of the Commonwealth of Puerto Rico.

(11) Has purchased a one hundred (100) dollar internal revenue stamp payable to the Puerto Rico Police; provided, that in those cases whereby the license is denied, the amount paid shall not be reimbursed.

(12) Has submitted, together with the application, one (1) sworn statement from three (3) persons that are not related by consanguinity or affinity to the petitioner and that under penalty of perjury, attest to the fact that the petitioner enjoys a good reputation in the community, that he/she does not have a tendency to commit acts of violence and that therefore they have no objection to the petitioner owning firearms. This statement shall be made in the form provided by the Superintendent along with the application for a weapons license.

(13) Has submitted the completed application, under oath, before a notary, accompanied by an impression of his/her fingerprints, taken by a technician of the Puerto Rico Police or a competent federal or state government agency, and that includes two (2) color photographs, two (2) inches by two (2) inches in size, sufficiently recent as to depict the petitioner in his/her true aspect at the time of the application.

(14) Has submitted a negative certification of debt from the Child Support Administration, issued not later than thirty (30) days prior to the date of the application.

(b) Every duly completed application, in duplicate, and, together with the documents and voucher indicated above, shall be filed with the General Police Headquarters or the law enforcement agency of the area in which the applicant resides. The applicant shall retain a sealed copy for his/her records. Within the term of five (5) working days, the Superintendent shall issue a certification stating that the application and all the required documents have been submitted, or shall require compliance with the requirements for the application in order to issue the certification. After said certificate is issued, the Superintendent shall determine and certify in writing, within a term that shall not exceed one hundred and twenty (120) calendar days, that the applicant meets the requirements established in this chapter for granting a weapons license. This may be achieved through an investigation in the files of any government agency in Puerto Rico, the United States, or abroad, to which he/she may have access (including the archives of the National Crime Information Center and the National Instant Criminal Background Check System, among others). Should the Superintendent's investigation result in a finding indicating that the person fails to meet all the requirements established in this chapter, the weapons license shall not be granted, but without it constituting an impairment, the applicant shall not be prevented from requesting said license again in the future. Should the Superintendent fail to issue a determination within the above mentioned term of one hundred and twenty (120) days, he/she shall be bound to issue a special provisional permit in favor of the applicant, within a term of ten (10) calendar days. Said special provisional permit shall grant all the rights, privileges and prerogatives of a regular license to carry weapons, for a term of sixty (60) calendar days, during which period the Superintendent must have reached a decision. If upon the conclusion of the effectiveness of said provisional permit, the Superintendent has not yet reached a decision as to the suitability of the applicant, said provisional permit shall automatically become a regular weapons license.

(c) The Superintendent, discretionally and unobtrusively, without disrupting the peace and tranquility of the person under investigation or breaching the privacy of his/her home, may conduct as many investigations as he/she deems pertinent after remitting the license to the applicant; provided, that regardless of whether the investigations are conducted or not, this shall not be an impediment for the license to be remitted under the terms indicated above. If after the pertinent investigation has been conducted by the Superintendent, it is found that the applicant has knowingly furnished false information in his/her application or does not meet the requirements established in this chapter, the license shall be immediately revoked and seized, as well as all the firearms and ammunition held by the applicant, who shall be subject to prosecution for the crime of perjury and for the corresponding violations of this chapter.

Any citizen to whom a license and/or permit is issued shall be responsible for the use of the licenses and the handling of the weapons, releasing the Commonwealth of Puerto Rico, its departments, agencies and municipalities from any liability for said individual use, except when these are vicariously liable for the acts of their employees or agents.

(d) The weapons license established in this Section empowers the licensee to be the owner of a maximum of two (2) firearms, except as provided below, regarding acquisitions through inheritance, or when the licensee holds a target shooting or hunting permit, in which case there shall be no established limit. Provided that any licensee who owns fifteen (15) weapons or more shall be bound to maintain eighty (80) percent of these locked in a safe place fixed to the real property in such a manner that the weapons cannot be easily removed. Any licensee under the obligation to fulfill the safety requirement shall submit to the Superintendent a sworn statement attesting to the fact that he/she is in compliance with the safety requirement. The Superintendent shall impose an administrative fine of one thousand ($1,000) dollars for each weapon property of the licensee that is removed due to noncompliance with the safety measures established herein. These safety requirements and the corresponding fine shall apply to any person who has more than fifteen (15) weapons in his/her possession: all licensees shall display in a conspicuous area visible to the clientele a clearly legible announcement notifying of this requirement. The license also empowers the licensee to acquire, buy, sell, donate, transfer, assign, hold, possess, have custody of and transport, bear and carry firearms, ammunition and any pertinent accessory anywhere under the jurisdiction of the Commonwealth of Puerto Rico; provided:

(1) That the firearms may be borne, carried, and transported in a hidden or unobtrusive manner, and the Superintendent shall provide by regulations, the procedure for any law enforcement officer as described herein, to issue a courtesy ticket which shall be prepared to such effect, which shall be remitted to the Puerto Rico Police Weapons License Office, to be filed with the record of the licensee. The Superintendent may impose a fine of up to two hundred ($200) dollars for recidivism in carrying or transporting arms in an ostentatious or conspicuous manner, as well as the procedure for the revision of the same, at the request of the interested party upon whom the fine is imposed. The Superintendent shall hold an administrative hearing within a term of not more than forty-five (45) days to uphold, review, modify or eliminate the imposed fine;

(2) Unless the licensee also holds a permit to carry weapons, the weapon may carried by the licensee on his/her person; and that to be able to transport said weapon without a permit to carry, the weapon must be unloaded and transported inside a closed case whose contents are not visible and which may not be in plain sight. Provided, that in the case of private security guards who hold a permit to carry, are uniformed and in the exercise of their duties, they may carry the weapon in plain sight.

(3) That the firearms may only be donated, sold, transferred, assigned, left in the custody, or otherwise conveyed by the transfer of the control or command of the same to persons who own a weapons or gunsmith license or to any of the persons mentioned in § 456c of this title.

(4) The licensee may only transport one (1) firearm at a time, except those licensees who also hold target shooting or hunting permits, who shall not be limited as to the number of firearms to be carried on their person while on the premises of an authorized gun club or in those places in which the hunting sport is practiced, pursuant to the applicable laws.

(5) The licensee shall only purchase ammunition of the caliber used by the weapons that he/she owns registered to his/her name.

(6) That the weapons license does not authorize the licensee to engage in the business of purchasing and selling firearms or ammunition, limiting the purchase and sale of the same to his/her personal firearms and ammunition; and

(7) The licensee may visit an authorized gun club once a year, and pay the required fees, to train in the use and handling of his/her weapons. Provided, that for this purpose, the Superintendent shall authorize the purchase of fifty (50) rounds of ammunition in addition to those allowed by this chapter, which must be used in their totality at the gun club which the licensee has attended.

(e) Within the term of forty-five (45) days from the date the Weapons License is received, which may be extended for thirty (30) additional days if requested within the original term, any licensee shall file a certification issued by an authorized official of a gun club authorized in Puerto Rico, if he/she has not filed it before with the Puerto Rico Police General Headquarters, personally or by certified mail with acknowledgment of receipt, to the effect that the petitioner has approved a course in the correct and safe use and handling of firearms. If the applicant fails to do so, he/she shall incur an administrative fault and pay a fine of one hundred ($100) dollars for each month in arrears, for up to a maximum of six (6) months, at the end of which the license shall be revoked and seized, as well as any weapon and ammunition the petitioner has acquired. For these purposes, the Superintendent shall authorize the purchase of up to a maximum of five hundred (500) rounds of ammunition in addition to those allowed by this chapter. Said ammunition shall have to be used completely by the licensee during the training for certification. The provisions of this paragraph shall not impair the provisions of clause (7) of the above paragraph.

The Superintendent shall handle any claim from the persons who for health reasons or special circumstances beyond their control are unable to comply with the certification duty imposed herein. In any case in which the Superintendent grants an extension, the new term for compliance with the requirements shall begin ten (10) days after the circumstances that led to the extension are no longer extant.

(f) The Puerto Rico Police shall issue the duplicates of weapons license cards that a licensee may want within the term of thirty (30) calendar days upon request after having paid fifty (50) dollars in internal revenue vouchers per duplicate. In case of a change in license category, the cost for the category change shall be twenty (20) dollars.

Every weapons license shall have the date on which it must be updated, which shall be five (5) years after its issue, and no person may make any transaction of a firearm or ammunition, nor shoot in a gun club, nor hunt, bear, carry, or transport weapons, if he/she has not applied for the updating thereof as indicated in this chapter, under penalty that the weapons license be revoked and an administrative fine of five hundred (500) dollars be imposed for shooting in a gun club or hunting, bearing, carrying or transporting weapons. After six (6) months have elapsed from the due date, he/she may only sell his/her firearms to a person with a dealer license.

Every five (5) years, on the fifth anniversary of the date of issue of the weapons license, the petitioner shall be bound to renew his/her license, by executing a sworn statement addressed to the Superintendent of Police, upon payment of the Internal Revenue vouchers in the denomination provided in [this section], stating that the circumstances that gave rise to the original granting are the same, or indicating the manner they have changed. This renewal may be made within six (6) months before, or thirty (30) days after the date the weapons license has expired. Failure to renew said license after the above mentioned thirty (30) days, shall entail an administrative fine of fifty ($50) dollars per month up to a maximum of six (6) months, amount which must be paid as a requirement for renewal. If the license is not renewed within six (6) months, the Superintendent shall revoke the same and seize the weapons and ammunition, but the licensee may renew and reinstate his/her license within up to six (6) additional months after the revocation or seizure, whichever is later, through the payment of twice the amount of the accrued fine. None of the above shall prevent a person whose license has been revoked through failure to act, from de novo applying for another license which shall be granted, provided he/she has paid any fines pending, in which case, he/she may recover the seized weapons if the Superintendent has not yet disposed of them.

It is hereby provided, that in the event the licensee is residing outside of Puerto Rico on the date the renewal of the license is due, or during the renewal period indicated above, the license shall not expire within the thirty (30) days of the licensee's return to Puerto Rico.

It is further provided that if the licensee happens to be a woman, and if she were pregnant on the date the renewal of the license is due or during the period of renewal mentioned above, said period shall not expire until ninety (90) days have elapsed after the birth. In any case in which a pregnant women is bound, under the provisions of this chapter, to submit a training certificate, the term to submit the same shall begin ninety (90) days after the birth.

The Superintendent shall notify every licensee by mail to his/her mailing address, the date on which his/her license must be updated, six (6) months prior to its expiration date. The Superintendent shall make available through the area police stations, the gunsmiths, and the Internet, all the forms needed to execute the renewal. Once the license has been renewed, the Superintendent shall issue the new license upon payment of the renewal fee, within the following thirty (30) calendar days, unless there is just cause to delay doing so.

Every licensee shall inform the Superintendent of his/her change of residential or mailing address within thirty (30) days of the change, under penalty of an administrative fine of two hundred (200) dollars, which shall be paid as a requirement for renewing the license.

(g) A person may at any time surrender his/her weapons license to the Police for its cancellation, and also surrender his/her weapons to the Police, or transfer the same to another person who owns a weapons or gunsmith license in force.

(h) It shall not be a requirement to own any firearms to obtain a weapons license and its categories.

456c. Procedure for issuing Weapons Licenses to certain government officials and former police officers The Governor, legislators, mayors, secretaries, directors and heads of agencies of the Government of Puerto Rico, Commonwealth and Federal judges, Commonwealth and Federal prosecutors, minors' advocates, the Superintendent, members of the police force, officials, agents and employees of the government of Puerto Rico who because of their office and the duties they perform are required to carry a weapon, and every law enforcement officer, may carry weapons. In addition, former governors, former legislators, former superintendents, former Commonwealth and federal judges, former Commonwealth and federal prosecutors, former minors' advocates, former mayors of Puerto Rico and former law enforcement officers, as long as they were honorably retired are not restricted by this chapter from owning firearms and in the case of former law enforcement officers, have served in such a capacity for not less than ten (10) years. The members of the United States Armed Forces and the Puerto Rico National Guard may also carry the weapons assigned to them by said organizations without a license while engaged in the official duties of their office. To such ends, the Superintendent shall establish an expedited procedure whereby the above mentioned officials, with the exception of the law enforcement officers and the Superintendent him/herself, shall be granted a weapons license with the corresponding permit to carry a weapon.

Those law enforcement officers, government officials and employees authorized to carry and train with firearms belonging to the Commonwealth of Puerto Rico or to the Federal Government, shall register the caliber of their official weapon in order to purchase and use ammunition on their weapons license with a permit to carry, with the prior authorization of the head or director of the agency and pursuant to the provisions of this chapter.

456d. Permits to carry weapons issued by the court

(a) The competent Part of the Court of First Instance shall grant authorization to the Superintendent to include in the petitioner's identification card a permit to bear, transport and carry, without identifying any particular weapon, unless there is just cause to deny said permit any legally owned pistol or revolver, unless there is just cause to deny said permit, with prior notification to the Department of Justice and upon a hearing before the same if so required by the latter, to any person who holds a weapons license and demonstrates that he/she fears for his/her safety. The petitioner shall include together with his/her application for a permit to carry firearms, a two hundred and fifty ($250) dollar internal revenue voucher drawn to the order of the Superintendent, which voucher shall have been previously submitted to the Superintendent, together with a certification issued by an authorized official of a gun club in Puerto Rico, stating that the petitioner has passed a course in the correct and safe use and handling of firearms.

The requirements for issuing a weapons license provided in § 456a of this title shall be considered by the court when evaluating the granting of the permit to carry a weapon.

The permit to carry a weapon issued by the Court may be renewed concurrently with the weapons license renewal by presenting before the Superintendent a one hundred ($100) dollar voucher drawn to the order of the Superintendent, and a sworn petition indicating that the circumstances that gave rise to the original granting of the license still prevail at the time the application is presented. In the event there is any change, said change shall be justified before the renewal is granted. The Superintendent shall notify the Court of the renewal of the permit to carry a weapon within thirty (30) days.

A sworn statement must be included to the effect that all the requirements established in § 456a of this title are met and that the entire contents of the application are true and correct.

(b) The permit to carry weapons granted herein shall have a term subject to the effectiveness of the weapons license and may be renewed for consecutive terms of five (5) years together with the weapons license. In the event the permit is denied, the amounts paid in vouchers shall not be reimbursable.

(c) As part of the application for the renewal of a weapons license and the permit to carry weapons, a person shall submit to the Superintendent a new certificate on the use, handling and safety measures of firearms, certified by a gun club. For such a purpose, the Superintendent shall authorize the purchase of a maximum of up to two hundred and fifty (250) rounds of ammunition in addition to those allowed by this chapter, which must be completely used up in the gun club that the licensee has attended during his/her training for certification.

Every law enforcement officer who because of his/her duties is assigned a weapon shall receive an annual training on the use and handling of said weapon offered by officials of or personnel contracted by the agencies that employ them and who are qualified to certify the use, handling and safety measures of a firearm. It shall be the duty of the agency that employs said official to submit a certification to the Superintendent indicating that the training herein established has taken place.

Those licensees who fail to comply with the requirement of the aforementioned certification may not carry a weapon until they are certified, under penalty of an administrative fine of five hundred (500) dollars; in the event of a second violation of the provisions of this paragraph, the Superintendent shall also revoke the permit to carry weapons without the need of a court authorization. In the case of law enforcement officers, after a grace period of 60 days has concluded, the agencies may not be able to use uncertified personnel pursuant to this section in functions that require using and/or carrying weapons.

(d) The permit to carry weapons shall be incorporated by the Superintendent to the weapons license of the licensee, stating the category to carry weapons as established in § 456a(f) of this title within ten (10) calendar days following the presentation of the authorization of the court by the licensee.

456f. Felony charges; seizure of weapons Upon a finding of probable cause for the arrest of any person to whom a license to carry weapons has been granted, for the commission of any of the crimes specified in § 456j of this title, or of violations of the provisions of this chapter, the court shall provisionally suspend the license until the final determination of the criminal procedure. Provided, further, that the court shall order the immediate seizure of all weapons and ammunition of the licensee for their custody in the police weapons and ammunition depot. Upon a final and binding determination that the accused is not guilty, the judge shall order the immediate return of the weapons license and all weapons and ammunition. All weapons and ammunition returned must be in the same condition they were when seized. The licensee shall be exempted from the payment of the storage fees. Should the judicial action result in a final and binding conviction, the Superintendent shall permanently revoke the license and shall seize all weapons and ammunition.

456g. Gunsmith license; report of transactions

(a) No person may be engaged in business as a gunsmith or trader of firearms and ammunition, unless he/she holds a gunsmith license issued by the Secretary of the Department of the Treasury. Said licenses shall expire one (1) year after their date of issue and shall again be subject to the formalities and application requirements of this chapter. Gunsmith licenses shall be subject to approval and certification by the Police, upon prior inspection of the safety measures required in the building where the establishment is located. The application for the renewal of a license shall be submitted thirty (30) days prior to its expiration date.

(b) Each transaction regarding the introduction of weapons into Puerto Rico by dealers, or to the sale of weapons and ammunition between dealers, must be informed through the electronic system established in this chapter. If said system is not available at the time of the transaction, the same shall be reported to the Superintendent on a form provided by the latter, which shall include the name, domicile, place of business and the specifics of the license of the vendor and the buyer, as well as the number, amount and description of the weapons or ammunition, including the serial numbers, subject to each transaction, as required by the Superintendent.

(c) A gunsmith who holds a license issued pursuant to this chapter, may acquire a weapon registered under the provisions of this chapter in the weapons register, by purchasing said weapon from the person who has registered the same in his/ her name, provided said person holds a weapons license issued pursuant to this chapter. When any sale of firearms or ammunition is made, said transaction must be reported through the electronic system established by this chapter. If said system is not available at the time of the transaction, the vendor and the buyer must notify this fact to the Superintendent in writing with return receipt requested, both on the same form provided by the latter for such purposes. In the case of a multiple weapons sale, more than one (1) at a time or more than one (1) weapon to the same person within a period of thirty (30) consecutive days, and if the electronic system is unavailable, the gunsmith, within twenty four (24) hours after delivering the weapons, shall notify the Superintendent by facsimile and by telephone, and shall register in his/her books the name and identification number of the person who received the information. Likewise, the same procedure shall be followed when carrying out any unitary sale of over six hundred (600) units of ammunition to any person with a weapons license for target shooting or hunting. If communication via facsimile and telephone is not achieved, the notification shall be made by certified mail with return receipt requested or in person.

(d) When any gunsmith, to his/her best judgment, detects abnormalities in the identification card of a licensee, or if the weapons delivery is denied or prohibited by federal provisions of law, said gunsmith shall immediately notify the Superintendent, or the person who is duly designated by the same, by facsimile and telephone and notify the gunsmiths. The Superintendent shall immediately proceed to investigate the licensee to determine whether the cancellation of the license and the filing of criminal charges are warranted.

Any infringement against the provisions set forth in paragraph (A) above shall constitute a felony punished with imprisonment for a fixed term of six (6) years. If there are aggravating circumstances, the established fixed penalty may be increased to up to a maximum of twelve (12) years; if there are extenuating circumstances, it may be reduced to a minimum of three (3) years. Provided that mechanical or cosmetic adjustment work between persons who are weapons license holders shall constitute no felony at all.

Any infringement against the provisions set forth in subsections (c) and (d) above shall be sanctioned with an administrative fine that shall not exceed five hundred ($500) dollars for failing to notify on the first infringement, and two thousand ($2,000) dollars for subsequent infringements, provided that should a third infringement arise, the Superintendent may decide to impose the fine or to initiate the procedure for the cancellation of the gunsmith license and, should another offense occur, the Superintendent shall then proceed to cancel the gunsmith license.

456h. Requirements

(a) Every person who wants to obtain or transfer the location of a weapons license shall file an application before the Secretary of the Department of the Treasury, sworn to before a notary, on the form, which shall be provided by the Secretary of the Treasury for such purpose, together with an internal revenue voucher of five hundred (500) dollars. After the application is approved by the Secretary of the Treasury, it shall be remitted to the Puerto Rico Police Superintendent. No license shall be issued pursuant to this section if it is shown that the petitioner, who must be a United States citizen, does not meet the requirements established in subsection (a) of § 456a of this title.

(b) No dealer license shall be issued under this section unless the police has investigated all the statements in the application beforehand, and unless the police archives and other accessible archives (including the archives of the National Crime Information Center and the National Instant Criminal Background Check System) have been examined in order to determine if the petitioner has any prior conviction. No license whatsoever shall be issued unless all the provisions of this section are complied with, or if the statements in the application are not true.

(c) If the petitioner is a corporation or a partnership, the application must be signed by the president, the secretary and the treasurer of the corporation; or by all the executive directors of the partnership; it shall indicate the name of the corporation or the partnership, the place and date of its incorporation or constitution, the site of its main office or domicile, the name of the city or town, street and number where the business, agency, subagency, office or branch for which the license is needed. A license issued under the provisions of this section shall be valid solely for the business mentioned and described in the license. Said license shall not be transferred to any other business or person, and shall be cancelled automatically when the corporation or partnership is dissolved, or any of the officials of the corporation who signed the application is replaced, or if a new director of the partnership is included, in the case of a partnership, although said license may be renewed as soon as the provisions of this section with regard to the new official or the new partner are complied with. In these cases, the Secretary of the Treasury shall issue a provisional license while the reorganization process is carried out.

(d) When the petitioner is a corporation or partnership, no license whatsoever shall be issued if any official of the corporation or a director of the partnership, who must be a United States citizen, does not meet the requirements established in subsection (a) of § 456a of this title.

(e) In every case, the dealer license must be issued within one hundred and twenty (120) days after the application has been filed in person or remitted by certified mail to the Secretary of the Department of the Treasury, without prejudice that the Superintendent may subsequently continue his/her investigation and revoke the license if there were legal grounds to do so.

456i.Conditions for operations of gunsmiths; evidence of transactions A person, partnership or corporation to which a gunsmith license has been issued may engage in the sale of arms and ammunition, or work as a gunsmith under the following conditions:

(a) The business may only be operated in the place designated in the license. Those dealers that have not been certified by the police to have met the safety measures pursuant to this chapter, shall not initiate operations until they have met them, nor can they keep weapons and ammunition in said place, other than those that the dealer is authorized to own and carry pursuant to the provisions of this chapter. Any infraction of this subsection by the dealer shall constitute a misdemeanor, which shall be punished by a fine of not less than five thousand ($5,000) dollars nor more than fifty thousand ($50,000) dollars, at the discretion of the Superintendent. It shall also bring about the revocation of the license by the Superintendent, who shall register any modification in the electronic register.

(b) No dealer shall receive any weapon to be repaired, modified, cleaned, engraved, polished or to have any other mechanical work performed, without first having been shown the weapons license, nor shall he/she accept any firearm whatsoever under any condition, that has a mutilated serial number. Any infraction of this subsection by the dealer shall constitute a misdemeanor and shall be sanctioned by a fine of ten thousand ($10,000) dollars. Failure to comply with this requirement shall bring about the revocation of the license by the Superintendent.

(c) The license of the dealer or a certified copy thereof shall be displayed in the establishment so that it can easily be read. Failure to comply with this requirement shall bring about the imposition of an administrative fine of five thousand ($5,000) dollars.

(d) Every dealer must place the following warning in a place that is visible to the buyer:

"The use of a locking device or safety lock on a firearm is advisable. Every loaded weapon as well as its ammunition must be kept out of the reach of minors or persons who are not authorized to use them. It is advisable to keep your weapons apart from the ammunition."

Failure to comply with this requirement shall bring about the imposition of an administrative fine of five thousand ($5,000) dollars.

(e) A record shall be kept of each weapon sold and each sale of ammunition in books provided for this purpose that shall be printed in the manner prescribed by the Superintendent, who shall furnish these books to the gunsmiths upon payment of the corresponding costs, as provided through regulations. The record of the sale shall be signed in person by the buyer and by the person who made the sale, in the presence of each other; and said record shall state the date, day and hour of the sale; caliber, make, model and manufacturer's number of the weapon, the caliber, make and amount of ammunition, and the name and number on the weapons license. The vendor shall record the description of the ammunition, the amount sold, and the date, day and hour of the sale, on the form provided by the Superintendent. Likewise, the electronic register shall contain documentary evidence of any weapon or munitions sold. The Superintendent must provide access to the electronic register to any person, partnership, society or corporation that has been issued a gunsmith license for the sole purpose of allowing the registration of the transactions to be conducted and ascertain that they are carried out according to the provisions of this chapter. The Superintendent shall have the obligation of keeping the register organized in a manner that will, at any time, expedite corroboration of the amount of ammunition acquired by each license holder and may not authorize the sale of calibers different from those registered in the name of the licensee.

(f) When the ammunition sold is as described in the second paragraph of § 459(a) of this title, the vendor shall keep a special register thereof in the books and forms intended for this purpose, that shall be printed in the manner prescribed by the Superintendent, which is also provided in the above subsection, showing the name of the buyer, the description of the ammunition and the date, day and hour of the sale. Said register shall also contain the following:

(1) A description of each weapon, including: (A) the manufacturer thereof; (B) the serial number that is engraved thereon; (C) the caliber of the weapon; and (D) the model and type of the weapon. In the case of the wholesale sale of weapons of the same caliber, model and type, the dealer may group said sales in his records, provided that they are made on the same date and to a single buyer.

(2) The name and address of each person from whom the weapon was received for sale in the armory, along with the date of acquisition.

(3) The name, license number and address of the natural or juridical person to whom it was sold, and the date of delivery.

The use of the electronic registry system shall not constitute an exemption from compliance with the provisions of this section.

(g) The documents and books shall be kept in the indicated establishment and described on the license, and shall be available during working hours for inspection by any public official or law enforcement agent. In those cases that the license is revoked as prescribed in this section, or the establishment goes out of business, said books and documents shall be immediately delivered to the Superintendent.

(h) No weapons, ammunition or imitations thereof shall be displayed in any part of a business establishment dedicated to the sale of weapons, where they can be seen from outside of the establishment. Failure to comply with this requirement may bring about the imposition of an administrative fine of five thousand ($5,000) dollars. Provided, that in cases of failure by the dealers to observe safety measures, or the measures provided in this section on two (2) or more occasions, the Superintendent, upon written notice, may revoke the license. If the person does not agree, he/she may file an action for review pursuant to §§ 2101 et seq. of Title 3.

(i) Any dealer who has been issued a license under the provisions of this section, who fails to keep the documents and books required herein, shall be guilty of a felony, and upon conviction, shall be punished with a penalty of imprisonment for a fixed term of twelve (12) years, and a penalty of a fine that shall not exceed one hundred thousand ($100,000) dollars. If there were aggravating circumstances, the fixed penalty established herein may be increased to a maximum of twenty-four (24) years; if there were mitigating circumstances, it can be reduced to a minimum of six (6) years. In addition, the Superintendent shall revoke any weapons or dealer license held by this person.

456j. Denial of license The Superintendent shall not issue a weapons license nor shall the Secretary of the Department of the Treasury issue a dealer license, or if they have been issued, they shall be revoked and the Superintendent shall attach the license and the arms and ammunitions of any person who has been convicted in or outside of Puerto Rico, for any felony or attempted felony, for conduct constituting domestic violence as typified in §§ 601 et seq. of Title 8; Provided, further, That no license shall be issued to any person who has a mental illness that disqualifies him/her to possess a weapon; a habitual drunkard, or one addicted to the use of narcotics or drugs; or any person who has renounced his/her American citizenship, or has been dishonorably discharged from the Armed Forces of the United States, or removed from any agency of law and order of the Government of Puerto Rico, or any person who has been convicted of any violation of the provisions of this chapter or former §§ 411-454 of this title.

456k. Registry of weapons; loss and surrender of firearms; death of the license holder

(a) The registry of weapons created in the general police headquarters, shall be adjusted in its organization and performance, to the provisions of this chapter, and shall be kept in a computerized, systematic and orderly fashion so that the search for information can be expedited. This registry must be duly safeguarded.

(b) Any legally owned firearm after this act becomes effective, shall be registered in the registry of weapons, in the event that it has not previously been registered. The Superintendent shall hand the affiant a certificate of said registration.

(c) Any person who owns or has a legally authorized weapon or ammunition under his/her control and loses the same, or the same disappears, or is stolen or is illegally appropriated, shall notify this fact by filing a complaint at the Police district or precinct in which he/she resides, or at the nearest Police station, immediately upon becoming aware of the loss, disappearance, theft or illegal appropriation. If he/she does not comply with this obligation, he/she shall be guilty of a misdemeanor, and, if convicted, shall be punished with a fine of up to a maximum of five thousand ($5,000) dollars per weapon or per every five hundred (500) rounds of ammunition, or fraction thereof, that he/she failed to report.

The Superintendent shall investigate every loss, disappearance, theft or illegal appropriation report, and shall keep a detailed register of the result of the same for the purpose of gathering statistics on the loss, disappearance, theft or illegal appropriation of weapons or ammunition reports. If the information submitted by the complainant is intentionally false, the Superintendent shall notify this fact to the Department of Justice for the filing of criminal charges.

(d) When a person who is duly authorized to possess weapons, dies, it shall be he duty of every administrator, executor, or trustee, or any of these who acts as such in Puerto Rico, and of any assistant administrator, agent or person who is legally authorized to administer the estate, to notify the Superintendent of the demise within thirty (30) days from the date of death. The notification shall state the name, residence and personal details of the deceased. Failure to notify this fact shall constitute a misdemeanor, and shall be punished with the payment of a fine that shall not exceed five hundred ($500) dollars. The Superintendent shall establish what is needed to receive, store, or provide custody of said weapons, which may be done by a person with a weapons license or a gun shop designated by the administrator, executor or trustee and/or disposal of said weapons, while the inheritance is distributed. If the weapons are adjudicated to an heir who is eligible to obtain a weapons license and said license is issued to him/her, said weapon or weapons shall be given to him/her; provided, that if said heir owns the maximum number of weapons permitted in this chapter, the Superintendent shall grant an authorization to possess the weapon acquired though inheritance, according to the form established by the Superintendent through regulations. If said license is denied, or the sale of said weapon at public auction is indicated, it may only be acquired by a person with a weapons license in force, by public auction, or by a gunsmith duly authorized by this chapter, and if it is not thus acquired, said weapon shall be delivered to the Superintendent to be confiscated as provided in this chapter. Also provided, that the Superintendent shall not deliver any weapon that, prior to the death of its owner, was not duly registered pursuant to subsection (b) of this section.

(e) Any acquisition, purchase, sale, donation, assignment or any other means of transferring the ownership of a firearm or ammunition, must be conducted before a person with a gunsmith license so as to be properly registered in the electronic register and in the books of the latter, pursuant to the provisions of § 456i of this title. The aforementioned means of ownership transfer may also be conducted between licensees through the weapons transfer forms provided by the Superintendent within the five (5) days following the issue thereof, so it may be duly noted and corrected in the electronic register.

456l. Grounds to empower law enforcement officers to seize weapons Any law enforcement officer shall seize the license, weapon and ammunition owned by a licensee when he/she has grounds to believe that the licensee has made, or shall make illegal use of said weapon and ammunition to harm other persons; has threatened to commit a crime; has stated the intent to commit suicide; has repeatedly demonstrated negligence or carelessness in handling the weapon; when it is deemed that the holder has a mental condition, is considered to be a habitual drunkard, or is addicted to controlled substances; or in any other situation of serious risk or danger that justifies this emergency measure. A law enforcement officer shall also seize the license, weapons and ammunition when the holder thereof is arrested for committing a felony or a misdemeanor that implies violence. Upon petition of the party from whom the weapon is seized, filed within the fifteen (15) work days after the weapon is seized, the Superintendent shall hold an administrative hearing within a term of not more than forty-five (45) days to sustain, revise or modify the seizing performed by the law enforcement. The Superintendent shall issue his decision within a term of not more than forty-five (45) days from the date of said formal administrative hearing, and if the Superintendent's decision is favorable to the injured party, he shall order the immediate return of the seized weapon or weapons.

456m. Semi-automatic assault weapons; manufacture, importing, distribution, possession and transfer

(a) No semiautomatic assault weapon shall be manufactured, or caused to be manufactured, offered, sold, rented, loaned, owned, used, transferred or mported. However, this prohibition shall not be applicable to:

(1) The possession, use, transfer in Puerto Rico or the importing from the United States territory by persons whose license contains the category of target shooting, hunting or who hold a dealer license, of those assault weapons that exist legally within the nation of the United States of America, on the date this act becomes effective.

(2) The manufacture, importing, sale or delivery by licensed dealers, to be used by the police officers of the government[s] of Puerto Rico or the United States, or for the use of the armed forces of the United Sates government or of Puerto Rico.

(b) The semiautomatic assault weapons referred to in this section are the following:

(1) Norinco, Mitchell, and Poly Technologies Avtomat Kalashnikovs (all AK models).

(2) Action Arms Israeli Military Industries UZI and Galil.

(3) Beretta Ar70 (SC-70).

(4) Colt AR-15.

(5) Fabrique National FN/FAL, FN/LAR, and FNC.

(6) SWD M-10, M-11, M-11/9, and M-12.

(7) Steyr AUG.

(8) INTRATEC tec-9, tec-dc9 AND tec-22.

(9) Revolving cylinder shotguns such as (or similar to) the Street Sweeper and Striker 12.

(c) Also considered as a semiautomatic assault weapon, is:

(1) A semiautomatic rifle that is retrofed by a magazine or removable receptacle which contains two (2) or more of the following characteristics:

(A) Folding or telescopic butt.

(B) Pistol grip that clearly overlaps the trigger action.

(C) Bayonet mount.

(D) Flash suppressor.

(E) Grenade launcher, excluding flare launchers.

(2) A semiautomatic pistol that can be backloaded by a magazine or removable receptacle which has more than two (2) of the following characteristics:

(A) A magazine or clip that is fixed on the pistol outside of the pistol grip.

(B) A barrel with spiral grooves on the front which is capable of accepting an extension to the barrel, a flash suppressor, a hand grip in front of the weapon or a silencer.

(C) A cover that can be fixed covering all or part of the barrel, allowing whoever fires the weapon to hold it with the hand that is not pulling the trigger, without getting burned.

(D) An unloaded manufacturing weight in excess of fifty (50) ounces.

(E) A semiautomatic version of an automatic weapon.

(3) A semiautomatic shotgun that has two (2) or more of the following characteristics:

(A) A collapsible or telescopic breech.

(B) A pistol grip that clearly overlaps the trigger action.

(C) A fixed ammunition magazine or clip that holds more than five (5) cartridges.

(D) Is capable of receiving a removable magazine or clip.

(d) Any person who violates the provisions of this Section shall be guilty of a felony, and upon conviction, shall be punished by imprisonment for a fixed term of twenty-four (24) years, without the right to suspended sentence, to parole, or to enjoy the benefit of a diversion program or to benefits or alternatives to imprisonment acknowledged in this jurisdiction, having to serve in calendar years the total amount of the penalty imposed. Should there be aggravating circumstances, the fixed penalty established may be increased up to a maximum of thirty-six (36) years; should there be extenuating circumstances, it may be reduced to a minimum of eighteen (18) years.

The possession or use of these weapons by members of the police and those other officers of the law that are duly authorized to bear arms in compliance of their duties, as established in this chapter, shall not constitute a crime.

The application of the provisions of this section shall be prospective as of the approval of this act.

Subchapter III. Target Shooting Permit

457c. Target-shooting permits

(a) Any person who holds a weapons license issued pursuant to this chapter, may request the Superintendent to issue a target-shooting license. He/she shall provide all the information required in the application forms prepared for these purposes by the Superintendent, under oath before a notary, which shall at least require an internal revenue voucher of twenty-five ($25) dollars, a two (2) inch by two (2) inch photograph, and a stamp from a sports shooting federation. The Superintendent shall issue the requested permit within the term of thirty (30) working days after receiving the application unless there is just cause for its denied.

(b) No target shooting permit shall be issued to any person whosoever who is not a member of a gun club or organization and a shooting federation duly recognized by the Secretary. ...

Subchapter IV. Security Agencies That Transport Valuables in Armored Vehicles

...457f. Special licenses The Superintendent may issue special licenses to those security agencies devoted to the transportation of valuables in armored vehicles that so request and that are duly authorized to operate as such, authorizing them to purchase, possess and dispose of weapons and maintain at their place of business a long barreled non-automatic weapons depot and an ammunition dump to be used solely and exclusively by the security agents employed by said agencies assigned to the transportation of valuables in armored vehicles while in the performance of their duties.

457j. Long barreled weapons depot and ammunition dump Any application for a special license to purchase, possess and dispose of weapons and maintain at the place of business a long barreled weapons depot and ammunition dump must be accompanied by attesting proof of the fact that the security agency employs five (5) persons or more for such a purpose.The applicant for a special license to purchase, possess and dispose of weapons and maintain at the place of business a long barreled weapons depot shall also comply with all provisions and security requirements for gunsmith licenses, as well as any other requirements provided by the Superintendent through regulations. Once the Superintendent has certified that the place of business of the applicant meets the obligatory safety requirements for a gunsmith license, the special license applied for shall be issued. The business of the applicant shall operate only in the designated premises and be subject to inspection by any police officer or agent of the Special Investigations Bureau of the Department of Justice and its license shall be displayed in a visible place in the applicant's office. No weapon whatsoever may be kept in said premises other than those whose possession has been authorized according to the provisions of this chapter.

457k. Limit to the number of weapons The special license to possess and maintain at the place of business a long barreled weapons depot shall allow the security agency to have under its control and care a specific number of long barreled weapons such as semiautomatic shotguns and rifles registered in its name in the registers of the Superintendent. The agency may only acquire two (2) long barreled weapons over the number of armored vehicles owned by the agency and devoted to the transportation of valuables as certified by the Public Service Commission.

457r. Ammunition Those security agencies that obtain the special license provided in this subchapter are hereby authorized to purchase a reasonable amount of ammunition for the weapons authorized for them by the Superintendent in said license. The security agency shall maintain a perpetual inventory of the authorized weapons and ammunition as well as a register of their daily movement. These registers shall be subject to inspection by the Puerto Rico Police. Also Provided, That the acquisition, purchase, sale, donation, assignment or any means of transferring the ownership of the firearms, munitions or accessories conducted by virtue of this special license must have been duly registered in the electronic register by the means provided in this chapter.

Subchapter V. Weapons

458. Manufacture, importation, sale and distribution of weapons It shall be necessary to hold a license issued pursuant to the requirements set forth in this chapter to manufacture, import, offer, sell or have available for sale, rent, or transfer any firearms or ammunition or that portion or part of a firearm on which the manufacturer of the same places the serial number of the firearm. Any infraction of this section shall constitute a felony and be punished with imprisonment for a fixed term of fifteen (15) years, without the right to suspended sentence, to parole, or to enjoy the benefits of a diversion program or to benefits or alternatives to imprisonment acknowledged in this jurisdiction, having to serve in calendar years the total amount of the penalty imposed. Should there be aggravating circumstances, the fixed penalty established may be increased up to a maximum of twenty-five (25) years; should there be extenuating circumstances, it may be reduced to a minimum of ten (10) years.

458a. Prohibition of the sale to unlicensed persons. No dealer shall deliver a weapon to a buyer unless he/she can show a weapons license in effect. When the buyer of the weapon is a hunter or shooter authorized to own firearms, the sale and delivery thereof shall be made in the same manner that is indicated in this chapter.

A dealer who knowingly sells firearms to a person without a license, shall be guilty of a felony, and upon conviction, shall be sanctioned with a penalty of imprisonment for a fixed term of fifteen (15) years. If there were aggravating circumstances, the penalty established may be increased to a maximum of twenty-five (25) years; if there were mitigating circumstances, it could be reduced to a minimum of ten (10) years.

A conviction under this section shall entail the automatic cancellation of the dealer license.

458b. Trade in automatic firearms Any person who sells or has for sale, or offers, delivers, rents, lends or otherwise disposes of any firearms that may be fired automatically, regardless of whether said weapon is classified as a machine gun or otherwise, shall be guilty of a felony and upon conviction shall be punished with a penalty of imprisonment for a fixed term of twenty-four (24) years, without the right to suspended sentence, to parole, or to enjoy the benefits of a diversion program or to benefits or alternatives to imprisonment acknowledged in this jurisdiction, having to serve in calendar years the total amount of the penalty imposed. Should there be aggravating circumstances, the fixed penalty established may be increased to a maximum of thirty-six (36) years; should there be extenuating circumstances, it may be reduced to a minimum of eighteen (18) years.

This penalty shall not apply to the sale or delivery of a machine gun or any other firearm that can be fired automatically, to be used by the Police and other law enforcement agents.

458e. Possession of weapons without license Any person who has or owns, but is not carrying a firearm without a license to do so, shall be guilty of a felony and upon conviction shall be punished with a penalty of imprisonment for a fixed term of five (5) years. Should there be aggravating circumstances, the penalty established may be increased to a maximum of ten (10) years; should there be mitigating circumstances, it may be reduced to a minimum of one (1) year.

Notwithstanding the above, when a person incurs conduct prohibited in this section without the intention of committing a crime with his/her own unlicensed firearm, and the person has never been convicted for a violation of this chapter, Act No. 348 of December 21, 1999, the provisions in Act No. 17 of March 11, 1915, or Act No. 27 of January 10, 2002, or any crimes provided under Section 456j of this title, and the weapon has not been reported as stolen or illegally appropriated, he/she shall incur a misdemeanor and upon conviction, shall be punished with a penalty of imprisonment for a term of not more than six (6) months, a fine that shall not exceed five thousand ($5,000) dollars or both penalties at the discretion of the court. The court, at its discretion, may impose the penalty of rendering community services in lieu of the established penalty of imprisonment.

In the event that the owner of the weapon shows satisfactory evidence that he/she owns a weapons license, although expired, and that he/she has requested its renewal within the term provided by this chapter, he/she shall not be guilty of any crime. If he/she has not requested its renewal within the established term, he/shall incur an administrative fault and shall have to pay three (3) times the accrued cost of the renewal fee.

458f. Illegal use or possession of automatic or semiautomatic long-barreled weapons or sawed-off shotguns Any person who , without the authorization of this chapter, carries, owns or uses a semiautomatic long-barreled weapon, a machine gun, carbine or rifle, or any modification thereof, or any other weapon that can be fired automatically, or a sawed-off shotgun of less of eighteen (18) inches, and which can cause grave bodily harm, shall incur a felony, and upon conviction shall be punished with imprisonment for a fixed term of twenty-four (24) years, without the right to a suspended sentence, to be released on parole, or enjoy the benefits of any diversion program or option to the term of imprisonment recognized in this jurisdiction, having to serve in calendar years the total amount of the penalty imposed.

Should there were aggravating circumstances, the fixed penalty established herein may be increased to a maximum of thirty-six (36) years; should there were mitigating circumstances, it may be reduced to a minimum of eighteen (18) years.

The possession or use of these weapons by the police and those other duly authorized law enforcement agents in the performance of their duty shall not constitute a crime.

458g. Possession or sale of silencing devices Any person who owns, sells, has for sale, lends, offers, delivers or provides any instrument, device, artifact or accessory that silences or reduces the noise of the shot fired from any firearm, shall be guilty of a felony, and upon conviction, shall be punished with imprisonment for a fixed term of twelve (12) years. If there were aggravating circumstances, the fixed penalty thus established may be increased to a maximum of twenty-four (24) years; if there were mitigating circumstances, it may be reduced to a minimum of six (6) years.

The provisions of this section shall not apply to duly authorized law enforcement officers in the performance of their duty.

458h. Furnishing of arms to third parties Any person who with criminal intent, furnishes or makes available to another person any firearm that has been under his/her custody or control, whether or not he/she is the proprietor thereof, shall be guilty of a felony, and upon conviction, shall be punished with imprisonment for a fixed term of twelve (12) years. If there were aggravating circumstances, the fixed penalty thus established may be increased to a maximum of twenty-four (24) years; if there were mitigating circumstances, it may be reduced to a minimum of six (6) years.

458i. Serial number or name of owner on firearm; removal or mutilation Every weapon shall have the name of the dealer or trademark or the name of the importer under which it shall be sold, adhered in such a way that it cannot be easily altered or erased, in addition to a serial number or the full name of its owner.

A person shall incur a felony and be punished with imprisonment for a fixed term of twelve (12) years, who:

(a) Willfully removes, mutilates, permanently covers, alters or erases the serial number or the name of the owner of any weapon.

(b) Knowingly buys, sells, receives, alienates, transfers, bears or holds any weapon whose serial number or the name of the owner has been removed, mutilated, permanently covered, altered or erased.

(c) Being a dealer or an agent or representative of said dealer, knowingly purchases, sells, receives, delivers, alienates, transfers, bears or has any weapon in his/her possession whose serial number or the name of its owner has been removed, mutilated, permanently covered, altered or erased.

If there were aggravating circumstances, the fixed penalty can be increased to a maximum of twenty-four (24) years; if there were mitigating circumstances, it may be reduced to a minimum of six (6) years.

458j. Presumptions The possession of a firearm by a person who does not have a weapons license shall be deemed as prima facie evidence of the fact that said person possesses the weapon with the intention of committing a crime.

The act of carrying a firearm by a person who does not have a weapons license with permit to carry, shall be deemed as prima facie evidence of the fact that said person carried the weapon with the intention of committing a crime.

The possession by any person of a weapon whose serial number or the name of the owner, has been removed, mutilated, permanently covered, altered or erased, shall be deemed prima facie evidence of the fact that said person removed, mutilated, covered, altered or erased said serial number or the name of the owner.

The possession by any person of a weapon whose serial number or the name of the owner, has been removed, mutilated, permanently covered, altered or erased, shall be deemed as prima facie evidence of the fact that said person possesses the weapon with the intention of committing a crime.

The possession of a weapon by any person at the time he/she commits, or attempts to commit a crime, shall be deemed prima facie evidence of the fact that said weapon was loaded at the time the crime was committed or intended to be committed.

The presence of three (3) or more firearms in a room, house, residence, establishment, office, structure or vehicle, shall constitute prima facie evidence of the fact that the owner or possessor of said room, house, residence, establishment, office, structure or vehicle, or those persons who occupy the room, house, residence, establishment, office or structure, traffic in or facilitate firearms illegally, provided said persons do not have a weapons license, a gunsmith license or a target shooting or hunting club license.

The presence of a machine gun or any other automatic weapon or any armor piercing ammunition in any room, house, residence, establishment, office, structure or vehicle, shall constitute prima facie evidence of its illegal possession by its owner or possessor of said building or vehicle, and by those persons who occupy the room, house, building, or structure in which said machine gun, automatic weapon or sawed-off shotgun is found, and who have the mediate or immediate possession thereof. This presumption shall not be applicable in those cases in which a public service vehicle is involved which at that time had been transporting paying passengers or when it is shown that incidental or emergency transportation is involved.

The presence of a machine gun or any other automatic weapon or any armor piercing ammunition in any room, house, residence, establishment, office, structure or vehicle, shall constitute prima facie evidence of the fact that the owner or possessor of said building or vehicle possesses the weapon or the ammunition with the intention of committing a crime.

The presence of a firearm or ammunition in any stolen vehicle shall constitute prima facie evidence of its illegal possession by all persons traveling in said vehicle at the time said weapon or ammunition is found.

The provisions of this section shall not apply to the law enforcement officers in the compliance of their official duties.

458k. Notice by carrier, warehouseman, or depository for the receipt of weapons; penalties Any sea, air or land carrier, and every warehouseman or depository who knowingly receives firearms, accessories or parts thereof, or ammunition, for delivery in Puerto Rico, shall not deliver said merchandise to the consignee until he shows his weapons or dealer license. Within five (5) working days of the delivery, the carrier, warehouseman, or depository shall notify the Superintendent of the name, address and license number of the consignee, and the number of firearms or ammunition including the caliber, that are delivered, personally or by certified mail with receipt requested, as well as any other information required by the Superintendent, by reg-ulations.

When the consignee does not have a weapons or dealer license, the carrier, warehouseman, or depository shall immediately notify the Superintendent of said fact, with the name and address of the consignee and the number of firearms or ammunition for delivery. Neither shall he deliver said merchandise to the consignee until he holds an authorization issued by the Superintendent to such effects.

The violation of any obligation established herein shall constitute a felony which shall be punished with imprisonment for a fixed term of twelve (12) years, and a fine of not less than two thousand ($2,000) dollars nor more than ten thousand ($10,000) dollars. If there were aggravating circumstances, the fixed penalty thus established can be raised to a maximum of twenty-four (24) years, if there were mitigating circumstances, it can be reduced to a minimum of six (6) years.

458l. Storage and custody of firearms in weapons and ammunitions depots Every dealer shall be bound to implement the safety measures required by the Police by regulations for the storage or custody of firearms and ammunition. The Police shall examine the dealers' places of business every three (3) months, which, if they do not meet the required safety measures, shall have thirty (30) days to comply with them, or if not, they shall deposit the firearms and ammunition they have for sale, for their custody and storage in the vault of another dealer or in the Weapons and Ammunition Depot of the Puerto Rico Police, within the term determined by the Superintendent, until the deficiency is corrected.

The dealers who in order to correct deficiencies, use the Weapons and Ammunition Depot, shall pay a monthly sum for the storage and custody of their arms and ammunitions which shall be determined by regulations. When establishing the cost of storage and custody, the operating costs of the Weapons and Ammunition Depot, and the handling of the receipt, classification, custody and delivery of the arms and ammunitions, shall be taken into consideration. The costs to be charged to the users of the Weapons and Ammunition Depot shall, under no circumstances, exceed the real and reasonable cost of the service rendered.

The Superintendent, or the person in charge of the Weapons and Ammunition Depot, shall remit an invoice to the dealers, as provided by regulations, in which the cost of storage and custody of their weapons shall be indicated, according to the dealer's use of the Weapons and Ammunition Depot during said month. Failure of the dealer to pay it shall be sufficient motive for the Superintendent, after a formal hearing, to revoke the license that he has issued to him/her.

The firearms of those citizens who have a weapons license, who want their arms to be kept temporarily in the Weapons and Ammunition Depot as a safety measure, may be stored by the payment of a fee, without impairing said citizen's option to keep their firearms in the private establishments of the dealers.

458n. Firing or pointing weapons

(a) Any person shall be found guilty of a felony, and punished with a penalty of imprisonment for a fixed term of five (5) years, who, except in cases of self defense or defense of third parties, or actions in the performance of official duties or legitimate sports activities:

(1) Willfully fires any weapon in a public place or any other place where there is any person who could be harmed, even though he/she causes no harm whatsoever to any person, or

(2) Intentionally, although without malice aforethought, points a weapon towards a person, even though he/she causes no harm whatsoever to any person.

If there were aggravating circumstances, the penalty thus established could be increased to a maximum of ten (10) years, if there were mitigating circumstances, it could be reduced to a minimum of one (1) year.

(b) Every person who, except in case of self-defense, or of third parties, or of acts in the performance of official functions or legitimate sports activities, incurs in any of the acts described above, using a pneumatic weapon, shall be guilty of a felony with a penalty of imprisonment for a fixed term of three (3) years. If there were aggravating circumstances, the penalty thus established may be increased to a maximum of six (6) years; of there were mitigating circumstances, it can be reduced to a minimum of six (6) months and one (1) day.

458o. Receipt, custody of weapons voluntarily deposited with Police; destruction The Superintendent shall establish, through regulations, that which regards the receipt, custody and disposal of those weapons that are voluntarily deposited by persons who have a license, or seized by the Police; or relinquished upon the death of a holder of a license; or upon the canceling of the grantee's license.

The Superintendent is authorized to sell, exchange, donate or cede the weapons to federal, state or municipal law enforcement agencies, or to other jurisdictions. He/she may also sell the weapons to dealers or a person with a weapons license issued under the provisions of this chapter, as provided by regulations.

The weapons or instruments seized pursuant to this section shall be stored by the Superintendent in the Weapons and Ammunitions Depot of the Police.

However, every machine gun, sawed-off shotgun, or any other weapon or instrument specified in § 458d of this title, that is carried, possessed or transported illegally, shall be deemed to be a public nuisance.

When any of these weapons or instruments is seized, it shall be delivered to the Superintendent who shall deal with its disposal and destruction, pursuant to the regulations promulgated to such effects.

458p. Weapons collections None of the provisions of this Act shall impede the preservation and keeping of collections of private weapons possessed by the owners thereof as a decoration or curios or the collections of weapons as antiques. In order to keep any type of weapon included in this Section, it shall first be necessary for the collector to obtain a target shooting or hunting permit, under the provisions of this chapter.

Antique firearms, as defined in this chapter, that lack a manufacturer's serial number shall be exempted from the requirement of registration, as defined in this chapter, but their existence shall be reported to the Weapons Register of the Puerto Rico Police along with three (3) different photographs detailing their particulars for the corresponding notation of their existence in the record of the license holder with a weapons license and a target shooting permit. Provided that, if the antique firearm is used in the commission of a crime it shall be considered as an unregistered firearm. It is also provided that under no circumstances shall it be required to mark, modify or alter the antique firearm in any way.

458q. Transporting of forbidden weapons; seizure The Secretary of Justice shall seize any property, as this term is defined in §§ 1723 et seq. of Title 34, known as the "Uniform Seizure Act of 1988", wherein any weapon is stored, loaded, unloaded, transported, carried or transferred, or in which it is found to be stored, loaded, unloaded, transported, taken or transferred in violation of this chapter.

The procedure established by §§ 1783 et seq. of Title 34 shall be followed for the seizure and disposal thereof.

458r. Weapons within reach of minors Any person who negligently leaves a firearms or automatic weapon within reach of a person who is under eighteen (18) years of age who does not hold a target shooting or hunting license, and who takes the weapon and harms another person or him/herself, shall be guilty of a felony and, upon conviction, shall be punished by imprisonment for a fixed term of two (2) years. If there were aggravating circumstances, the fixed penalty thus established, shall be increased to a maximum of five (5) years; if there were mitigating circumstances, it can be reduced to a minimum of six (6) months and one (1) day.

458s. Illegal appropriations of weapons or ammunition, theft Any person who intentionally, regardless of the means used for said purpose, illegally appropriates a firearm or ammunition, shall be guilty of a felony, and if convicted, shall be punished with imprisonment for a fixed term of ten (10) years, without the right to a suspended sentence, to be released on parole, or enjoy the benefits of any diversion program, benefits or option for the term of imprisonment acknowledged in this jurisdiction, having to serve in calendar years the total amount of the penalty imposed. Should there be aggravating circumstances, the fixed penalty established may be increased to a maximum of twenty (20) years; should there be extenuating circumstances, it may be reduced to a minimum of five (5) years.

If the person illegally appropriates, regardless of the means used for said purpose, more than one firearm, or if the person has a criminal record for having been convicted of a felony, the penalty shall be doubled.

Subchapter VI. Ammunition

459. Manufacture, distribution, possession and use It shall be necessary to hold a weapons, target shooting, hunting or gunsmith license, as the case may be, to manufacture, request the manufacture of, import, offer, purchase, sell or have for sale, keep, store, deliver, lend, transfer or otherwise dispose of or, own, use, carry or transport ammunition according to the requirements of this chapter. It shall likewise be necessary to have a permit issued by the police to purchase gunpowder. Any infraction of this section shall constitute a felony, and shall be sanctioned with a fixed term of imprisonment of six (6) years. Should there be aggravating circumstances, the fixed penalty established hereby shall be increased to a maximum of twelve (12) years; should there be mitigating circumstances, it may be reduced to a minimum of three (3) years.

It shall be deemed as an aggravating circumstance at the time of imposing the sentence to incur any of the acts described in this section without holding the corresponding license or permit to purchase gunpowder when the ammunition is of the kind commonly known as armor piercing. It shall not be deemed as a crime to manufacture, sell or deliver the ammunition described for the use of the Police and other law enforcement officers of the Government of Puerto Rico or the United States or for the use of the United States Armed Forces.

459a. Sale of ammunition to persons without a license; limit in the amount of ammunition A person with a weapons or gunsmith license may not sell ammunition to persons who do not show a weapons license or the permits set forth in this chapter. The sale of ammunition shall be limited exclusively to the type of ammunition used by the weapon or weapons that the purchaser has registered to his name.

Any violation of the provisions of the preceding paragraph shall constitute a felony and shall be punished by imprisonment with a fixed term of five (5) years. Should there be aggravating circumstances, the fixed penalty thus established may be increased to a maximum of eight (8) years; should there be extenuating circumstances, it may be reduced to a minimum of three (3) years and one (1) day.

A person who holds a weapons license, except in the target shooting or hunting categories, may only own a maximum of fifty (50) rounds per calendar year for each weapon he/she owns. If said person wishes to exchange the ammunition, whether by replacing the same or the acquiring new ammunition due to having used or lost some of it, he/she shall resort to the district or police precinct where he/she resides. The Police shall grant him/her an authorization to replace the ammunition in the amount established in this paragraph. In those cases in which the person wishes to acquire new ammunition because he/she has used or lost some of it, he/she shall report the circumstances under which he/she used or lost said ammunition. In order for the replacement of the ammunition to be granted, the circumstances under which it is to be used shall involve activities that are allowed and legitimate under our body of laws and the provisions of this chapter. The ammunition surrendered shall be impounded by the police.

Any infraction of the provisions of the preceding paragraph shall constitute a misdemeanor and be sanctioned with a term of imprisonment not to exceed six (6) months, a fine not to exceed five thousand ($5,000) dollars, or both penalties at the discretion of the court. The court, at its discretion, may impose the penalty of rendering community services in lieu of the established penalty of imprisonment.

459b. Purchase of a different caliber. Any person who having a valid weapons license, purchases ammunition of a caliber that is different to those that can be used in the firearms registered to his/her name, shall incur a felony, and upon conviction, shall be punished with a penalty of imprisonment for a fixed term of six (6) years. If there were aggravating circumstances, the fixed penalty thus established may be increased to a maximum of twelve (12) years; if there were mitigating circumstances, it may be reduced to a minimum of three (3) years.

Subchapter VII. Final Provisions

460. Hunting licensees All matters that refer to the licensing, regulating and control of the sport of hunting shall be governed by the provisions of §§ 107 et seq. of Title 12, known as the New Wildlife Act of Puerto Rico. However, the Superintendent shall see to the registration of all weapons and ammunition transactions of the holders of hunting licenses in the electronic register, pursuant to the provisions of this chapter.

460c. Conversion of licenses; terms; fees; rights; pending licenses

(a) Any license in force to have and possess a firearm, to practice target shooting, or to carry weapons, or any public official license shall be converted to a weapons license under its corresponding category, if any, pursuant to the provisions of this chapter, on or before the end of a term of six (6) months from the date this Act becomes effective. The conversion of the hunting license is voluntary on the part of its holder and may be done at any time according to the provisions of this chapter. Target shooting licenses held by minors under twenty-one (21) years of age shall be converted to target shooting licenses for minors and said conversion shall be free of cost. Provided, that until they are converted, they shall be governed by the provisions of the Acts under which they were issued.

It is hereby provided that once the conversion of a license to a weapons license is applied for, the original license shall not expire until the weapons license is granted, or is canceled because the grantee who requests the conversion does not qualify, pursuant to this chapter.

It is further provided that every weapon previously registered under another license, shall automatically be registered under the weapons license, when said license is converted to a weapons license.

(b) Every application for conversion of a license filed with the Superintendent, as established in this chapter, shall include an revenue voucher in the amount of fifty ($50) dollars, plus one ($1) dollar for each legally registered weapon held by the petitioner.

(c) In the case of licenses pending investigation that have been applied for by persons but not yet issued:

(1) License to have and possess. The Superintendent shall hand the documents of the application for the petitioner to submit his application pursuant to the provisions of this chapter. Provided, That the if the petitioner paid the corresponding fees according to Act No. 17 of January 19, 1951, as amended, he/she shall not have to pay the fees established in this chapter.

(2) Target-shooting license. The Superintendent shall hand the documents of the application for the petitioner to submit his/her application pursuant to the provisions of this chapter. Provided, that the petitioner shall submit an internal revenue voucher for the difference in the amount between the application fee provided in this chapter and those established in Act No. 75 of June 13, 1993, as amended.

(3) Hunting license. Will be governed by the provisions of §§ 107 et seq. of Title 12, known as the New Wildlife Act of Puerto Rico.

(d) Any special license issued to a security agency engaged in the transportation of valuables in armored vehicles or to its agents, issued pursuant to the provisions of Act No. 348 of December 21, 1999, shall retain its validity until its expiration, when it must be renewed according to the provisions of this chapter.

[Current through 2008 Sessions]

Leyes de Puerto Rico

Título 24. Salud Y Sanidad

Parte V. Sustancias Controladas

Capítulo 111. Ley de Sustancias Controladas de Puerto Rico

Subcapítulo 5. Disposiciones de Carácter Administrativo Y Otras

2516. Suspensión o revocación de licencia de conducir vehículos de motor y de licencia de portación de armas Ninguna persona que hubiere sido convicta por algún delito de este capítulo, o de cualquier ley de los Estados Unidos o de cualquier estado, relacionada con drogas narcóticas, marihuana, sustancias deprimentes o estimulantes, así como de cualquier país extranjero, y ninguna persona que haya sido declarada adicta a drogas narcóticas, podrá obtener licencia de la autoridad correspondiente para la conducción de ninguna clase de vehículo de motor ni de tenencia, posesión o portación de armas de fuego por un término de cinco (5) años a partir de la extinción de la sentencia por dicha convicción o a partir de la declaración. Los funcionarios o empleados públicos a cargo de expedir dichas licencias estarán impedidos de extenderlas cuando concurra alguna de las circunstancias y señaladas en el solicitante de la licencia y cualquiera de tales licencias que hubiere sido expedida con anterioridad a la convicción o declaración de que la persona es adicta a drogas narcóticas, será inmediatamente cancelada por la autoridad correspondiente. ...

Título 25. Seguridad Interna

Subtítulo 1. En General

Parte V. Reglamentación de Armas de Fuego, Explosivos Y Otros Artefactos Peligrosos

Capítulo 51A. Ley de Armas del 2000

Subcapítulo 1. Disposiciones Preliminares

455. Definiciones Para efectos de este capitulo, los siguientes términos tendrán el significado que a continuación se expresa:

(a) Agente del orden publico. Significa cualquier miembro u oficial del Gobierno de Puerto Rico o de los Estados Unidos de America, asi como cualquier subdivisión política de Puerto Rico o de Estados Unidos, entre cuyos deberes se encuentra el efectuar arrestos, incluyendo pero sin limitarse a los miembros del Cuerpo de Vigilantes del Departamento de Recursos Naturales y Ambientales, Policía de Puerto Rico, Policías Auxiliares, Policía Municipal, los agentes investigadores del Negociado de Investigaciones Especiales del Departamento de Justicia, los oficiales de custodia de la Administración de Corrección, los oficiales de custodia de la Oficina de Servicios con Antelación al Juicio, Guardia Nacional mientras se encuentren en funciones o ejercicios oficiales, los oficiales de custodia de la Administración de Instituciones Juveniles, el cuerpo de seguridad interna de la Autoridad de los Puertos, el Director de la División para el Control de Drogas y Narcóticos y los Inspectores de Sustancias Controladas de la Administración de Servicios de Salud Mental y contra la Adicción, los agentes investigadores de la Secretaria Auxiliar de Investigaciones del Sistema Correccional del Departamento de Corrección y Rehabilitación, y los Inspectores de la Comisión de Servicio Publico, así como los alguaciles del Tribunal General de Justicia de Puerto Rico y de los Tribunales Federales con jurisdicción en todo Puerto Rico, y los inspectores de rentas internas del Departamento de Hacienda.

(b) Ametralladora o arma automática. Significa un arma, de cualquier descripcion, independientemente de su tamaño y sin importar por que nombre se le designe o conozca, cargada o descargada, que pueda disparar repetida o automáticamente un numero de balas contenidas en un abastecedor, cinta u otro receptáculo, mediante una sola presión del gatillo. El termino "ametralladora" incluye también una subametralladora, así como cualquier otra arma de fuego provista de un dispositivo para disparar automáticamente la totalidad o parte de las balas o municiones contenidas en el abastecedor, o cualquier combinación de las partes de un arma de fuego destinada y con la intención de convertir, modificar o alterar dicha arma en una ametralladora.

(c) Arma. Se entenderá como toda arma de fuego, arma blanca o cualquier otro tipo de arma, independientemente de su denominación.

(d) Arma blanca. Significa un objeto punzante, cortante o contundente que pueda ser utilizado como un instrumento de agresión, capaz de infligir grave daño corporal.

(e) Arma de fuego. Significa cualquier arma, sin importar el nombre por el cual se conozca, diseñada o que pueda ser fácilmente convertida para ser o que sea capaz de lanzar una munición o municiones por la acción de una explosión. Esta definición no incluye aquellos artefactos de trabajo tales como, pero sin limitarse a, las pistolas de clavos utilizadas en la construcción, mientras se utilicen con fines de trabajo, arte u oficio.

(f) Arma larga. Significa cualquier escopeta, rifle o arma de fuego diseñada para ser disparada desde el hombro.

(g) Arma neumática. Significa cualquier arma, sin importar el nombre por el cual se conozca, que mediante la liberacion de gas o mezcla de gases comprimidos es capaz de impulsar uno o mas proyectiles.

(h) Arma de fuego antigua. Significa:

(1) cualquier arma de fuego con un mecanismo de tipo escopeta o fusil de mecha (matchlock), escopeta o fusil de chispa (flintlock), copo de percusión (percussion cap) manufacturado en o antes de 1898, o

(2) cualquier réplica de un arma de fuego descrita en la cláusula (1) de este inciso si dicha réplica:

(A) No está diseñada o rediseñada para utilizar munición de fuego anular (rim fire) o munición de tipo convencional de fuego central fijo (center fire).

(B) Utiliza munición de fuego anular (rim fire) o munición de tipo convencional de fuego central (center fire) que ya no es manufacturada en los Estados Unidos y que no se consigue por los canales normales y ordinarios de comercio.

(C) cualquier rifle de carga por el cañón (muzzle loading rifle), escopeta de carga por el cañón (muzzle loading shotgun) o pistola de carga por el cañón (muzzle loading pistol) que esté diseñada para utilizarse con pólvora negra, o un sustituto de pólvora negra, y que no pueda utilizar munición de tipo fijo. Para los propósitos de esta cláusula, el término "arma de fuego antigua" no incluirá cualquier arma que incorpore un armazón (frame) o recibidor (receiver), cualquier arma que sea convertida en un arma de carga por el cañón (muzzle loading weapon), o cualquier arma de carga por el cañón (muzzle loading weapon) que pueda ser convertida para ser capaz de disparar munición de tipo fijo mediante el reemplazo del cañón (barrel), cerrojo (bolt), anima (breech lock), o cualquier combinación de éstas.

(D) El término "munición de tipo fijo" significará aquella que está completamente ensamblada, entiéndase con casquillo, pólvora, fulminante y proyectil.

(i) Armero. Significa cualquier persona natural o jurídica que, por sí o por medio de sus agentes o empleados, compre o introduzca para la venta, cambie, permute, ofrezca en venta o exponga a la venta, o tenga a la venta en su establecimiento comercial cualquier arma de fuego o municiones, o que realice cualquier trabajo mecánico o cosmético para un tercero en cualquier arma de fuego o municiones.

(j) Armor piercing. Significa un proyectil que pueda ser usado en arma corta y que este construido enteramente (excluyendo la presencia o trazas de otras sustancias) o una combinación de aleación de tungsteno, acero, hierro, latón, bronce, berilio cuprico o uranio degradado; o un proyectil de cubierta completa mayor de calibre veintidós (22), diseñado e intencionado para usarse en arma corta y cuya cubierta tenga un peso de mas de veinticinco por ciento (25%) de su peso total. Excluye la munición de escopeta requerida por ley federal o estatal ambiental o reglamentación de caza para esos propósitos, un proyectil desintegrable diseñado para tiro al blanco, un proyectil en que se determine por el Secretario del Tesoro de los Estados Unidos que su uso primario es para propósito deportivo, o cualquier otro proyectil o núcleo del proyectil en cual dicho Secretario encuentre que su uso primordial es para fines industriales, incluyendo una carga usada en equipos de perforación de pozos de petróleo o de gas.

(k) Casa. Significa la parte de una edificación que es utilizada u ocupada por una sola persona o una sola familia.

(l) Comité. Significa el Comité Interagencial para Combatir el Trafico Ilegal de Armas, establecido en este capitulo.

(m) Escopeta. Significa un arma de fuego de canon largo con uno o mas cañones con interiores lisos, diseñada para ser disparada desde el hombro, la cual puede disparar cartuchos de uno o mas proyectiles. Puede ser alimentada manualmente o por abastecedor o receptáculo, y se puede disparar de manera manual, automática o semiautomática. Esta definición incluirá las escopetas con el canon cortado a menos de dieciocho (18) pulgadas.

(n) Federación de tiro. Significa cualquier federación adscrita al Comité Olímpico de Puerto Rico que represente el deporte olímpico de tiro al blanco.

(o) Licencia de armas. Significa aquella licencia concedida por el Superintendente que autorice al concesionario para tener, poseer y transportar armas, sus municiones, y dependiendo de su categoría, portar armas de fuego, tirar al blanco o cazar.

(p) Municiones. Significa cualquier bala, cartucho, proyectil, perdigón o cualquier carga que se ponga o pueda ponerse en un arma de fuego para ser disparada.

(q) Pistola. Significa cualquier arma de fuego que no tenga cilindro, la cual se carga manualmente o por un abastecedor, no diseñado para ser disparado del hombro, capaz de ser disparada en forma semiautomática o un disparo a la vez, dependiendo de su clase.

(r) Policía. Significa la Policía de Puerto Rico.

(s) Portacion. Significa la posesión inmediata o la tenencia física de un arma, cargada o descargada, sobre la persona del portador, entendiéndose también cuando no se este transportando un arma de conformidad a como se dispone en este capitulo.

(t) Revolver. Significa cualquier arma de fuego que contenga un cilindro giratorio con varias cámaras que, con la acción de apretar el gatillo o montar el martillo del arma, se alinea con el canon, poniendo la bala en posición de ser disparada.

(u) Rifle. Significa cualquier arma de fuego diseñada para ser disparada desde el hombro, que dispara uno o tres proyectiles. Puede ser alimentada manual o automáticamente por un abastecedor o receptáculo y se puede disparar de manera manual o semiautomática. El termino "rifle" incluye el termino "carabina."

(v) Secretario. Significa el Secretario del Departamento de Recreación y Deportes.

(w) Superintendente. Significa el Superintendente de la Policía de Puerto Rico.

(x) Transportación. Significa la posesión mediata o inmediata de un arma, con el fin de trasladarla de un lugar a otro. Dicha transportación deberá realizarse por una persona con licencia de armas vigente, y el arma deberá estar descargada y ser transportada dentro de un estuche cerrado que no refleje su contenido, y el cual a su vez no podrá estar a simple vista.

(y) Vehículo. Significa cualquier medio que sirva para transportar personas o cosas por tierra, mar o aire.

(z) Cambio de categoría. Significa incorporar permisos a una licencia de armas de fuego, independientemente de su categoría, portación, caza, o tiro al blanco.

Subcapítulo 2. Licencia Y Reglamentación

456. Registro electrónico El Superintendente expedirá licencias de armas y/o de armeros de conformidad con las disposiciones de este capítulo, las cuales facilitarán la inscripción electrónica de todas las transacciones de armas de fuego y municiones por parte de la persona tenedora de una de éstas. Corresponderá al Superintendente disponer mediante reglamentación la forma en que funcionará el sistema de registro electrónico, y éste se asegurará que el sistema diseñado haga llegar directamente a la Policía toda transacción que efectúe un tenedor de licencia. Se le concede a la Policía de Puerto Rico el término de seis (6) meses contados a partir de la vigencia de esta ley para instalar este registro.

La licencia de armas será representada por un carne lo suficientemente pequeño como para ser portado en billeteras de uso ordinario, y el cual contendrá al menos una fotografía del peticionario, el nombre completo de este, su fecha de nacimiento, sus senas personales y su numero de la licencia de armas. Contendrá, también, la fecha de expedición de la licencia y la fecha de su vencimiento, como mas adelante se dispone. Además, contendrá los mecanismos para lograr acceso al sistema de registro electrónico de la Policía para constatar su veracidad y otros datos pertinentes, tales como identificar el alcance del mismo mediante las categorías de portacion, tiro, caza o todas las categorías. El carne no contendrá la dirección del peticionario ni mención de sus armas o municiones autorizadas a comprar, pero el registro electrónico de la Policía contendrá y suministrara a sus usuarios tal información.

Disponiéndose, que mientras la Policía implementa y hace disponible a los armeros el sistema de registro electrónico, el Superintendente expedirá a cada concesionario de licencias de armas un carne provisional que al menos contenga una fotografía del concesionario y exprese su nombre completo, su fecha de nacimiento, sus senas personales, el numero de licencia, y los calibres correspondientes a las municiones que esta autorizado a comprar. Contendrá, también, la fecha de expedición de la licencia y la fecha de su vencimiento, como mas adelante se dispone. El carne oficial expedido de conformidad con las disposiciones de este titulo será el único documento acreditativo de autoridad legal para realizar las actividades autorizadas. Una vez este debidamente [implantado] el sistema de Registro Electrónico, el Superintendente únicamente podrá expedir el carne electrónico. De no estar disponible dicho sistema al momento de realizar alguna transacción, la misma se realizara según el procedimiento que el Superintendente disponga mediante reglamento.

456a. Licencia de armas

(a) El Superintendente expedirá una licencia de armas a cualquier peticionario que cumpla con los siguientes requisitos:

(1) Haber cumplido veintiún (21) anos de edad.

(2) Tener un certificado negativo de antecedentes penales expedido no mas de treinta (30) días previo a la fecha de la solicitud y no encontrarse acusado y pendiente o en proceso de juicio por algunos de los delitos enumerados en la sec. 456j de este titulo o sus equivalentes, tanto en Puerto Rico, los Estados Unidos o el extranjero.

(3) No ser ebrio habitual o adicto a sustancias controladas.

(4) No estar declarado incapaz mental por un tribunal.

(5) No incurrir ni pertenecer a organizaciones que incurran en actos de violencia o dirigidos al derrocamiento del gobierno construido.

(6) No haber sido separado de las Fuerzas Armadas bajo condiciones deshonrosas, o destituido de alguna de las agencias del orden publico del Gobierno de Puerto Rico o sus municipios.

(7) No estar bajo una orden del tribunal que le prohíba acosar, espiar, amenazar o acercarse a un compañero intimo, alguno de los niños de ese compañero o a persona alguna, y no tener un historial de violencia.

(8) Ser ciudadano de los Estados Unidos de America o residente legal de Puerto Rico.

(9) No ser persona que, habiendo sido ciudadano de los Estados Unidos alguna vez, renuncio a esa ciudadanía.

(10) Someter una declaración jurada atestiguando el cumplimiento con las leyes fiscales; estableciéndose que será razón para denegar la expedición de la licencia solicitada o para revocar esta el que el peticionario haya incumplido con las leyes fiscales del Estado Libre Asociado de Puerto Rico.

(11) Cancelar un comprobante de rentas internas de cien (100) dólares a favor de la Policía de Puerto Rico; Disponiéndose, que en los casos en que se deniegue la licencia, la cantidad pagada no será reembolsable.

(12) Someter en su solicitud una (1) declaración jurada de tres (3) personas que no tengan relación de consanguinidad o afinidad con el peticionario y que so pena de perjurio, atestigüen que el peticionario goza de buena reputación en su comunidad, y que no es propenso a cometer actos de violencia, por lo que no tienen objeción a que tenga armas de fuego. Esta declaración será en el formulario provisto por el Superintendente junto a la solicitud de licencia de armas.

(13) Someter su solicitud cumplimentada bajo juramento ante notario, acompañada de una muestra de sus huellas digitales, tomada por un técnico de la Policía de Puerto Rico o agencia gubernamental estatal o federal competente, y acompañada de dos (2) fotografías de dos (2) pulgadas por dos (2) pulgadas de tamaño, a colores, suficientemente reciente como para mostrar al peticionario en su apariencia real al momento de la solicitud.

(14) Someter una certificación negativa de deuda de la Administración para el Sustento de Menores, expedida no más de treinta (30) días previo a la fecha de la solicitud.

(b) Toda solicitud, en duplicado y debidamente cumplimentada, junto a los documentos y el comprobante arriba indicados, se radicaran en el cuartel general de la Policía o en las comandancias de[l] área donde resida el peticionario, reteniendo este la copia sellada para su constancia. Dentro de un termino de cinco (5) días laborables, el Superintendente expedirá una certificación de que la solicitud y todos los documentos requeridos han sido entregados, o requerirá la cumplimentacion de los requisitos de la solicitud para poder emitir la certificación. A partir de que se expida la mencionada certificación, el Superintendente, dentro de un término que no excederá de ciento veinte (120) días naturales, determinara y certificara por escrito si el peticionario cumple con los requisitos establecidos en este capitulo para la concesión de la licencia de armas. Esto podrá lograrse mediante una investigación en los archivos de cualquier agencia gubernamental de Puerto Rico, Estados Unidos o el exterior a la que pueda tener acceso (incluyendo los archivos del National Crime Information Center y del National Instant Criminal Background Check System, entre otros). De resultar la investigación del Superintendente en una determinación de que la persona no cumple con todos los requisitos establecidos en este capitulo, no le será concedida la licencia de armas, pero sin menoscabo a que el peticionario pueda solicitarla nuevamente en un futuro. Si el Superintendente no emite una determinación dentro del plazo antes mencionado de ciento veinte (120) días, este tendrá la obligación de expedir un permiso especial con carácter provisional a favor del peticionario, en un término [de] diez (10) días naturales. Dicho permiso especial con carácter provisional concederá todos los derechos, privilegios y prerrogativas de una licencia de armas ordinaria, durante una vigencia de sesenta (60) días naturales, periodo dentro del cual el Superintendente deberá alcanzar una determinación. Si al concluir la vigencia de dicho permiso con carácter provisional el Superintendente aun no hubiere alcanzado una determinación sobre la idoneidad del peticionario, dicho permiso con carácter provisional advendrá automáticamente a ser una licencia de armas ordinaria.

(c) El Superintendente podrá, discrecionalmente y de forma pasiva, sin perturbar la paz y tranquilidad del investigado o interrumpir la privacidad del hogar, realizar cuantas investigaciones estime pertinentes después de remitirse la licencia al peticionario; Disponiéndose, que el hecho de que se estén haciendo o no se hayan hecho las investigaciones no podrá ser impedimento para que se remita la licencia dentro de los términos antes indicados. Si después de realizada la investigación pertinente por el Superintendente resultare que el peticionario ha dado información falsa a sabiendas en su solicitud o no cumple con los requisitos establecidos en este capítulo, se procederá de inmediato a la revocación e incautación de la licencia ya la incautación de todas las armas de fuego y municiones que tuviera el peticionario, quedando éste sujeto a ser procesado por el delito de perjurio y por las correspondientes violaciones a este capítulo.

Todo ciudadano a quien se le otorgue una licencia y/o permiso, será responsable del uso de las licencias y del manejo de las armas, quedando libre de responsabilidad por dicho uso individual el Estado Libre Asociado de Puerto Rico, sus departamentos, agencias y municipios, excepto cuando éstos tengan responsabilidad vicaria por los actos de sus empleados o agentes.

(d) La licencia de armas que en esta sección se establece faculta al concesionario a ser propietario de un máximo de dos (2) armas de fuego, salvo lo dispuesto más adelante sobre adquisiciones por vía de herencia o que el concesionario posea un permiso de tiro al blanco o de caza, en cuyo caso no habrá límite establecido. Disponiéndose, que todo concesionario que posea quince (15) o más armas vendrá obligado a mantener el ochenta (80) por ciento de éstas en un lugar seguro, bajo llave y fijado al inmueble de forma que las armas no puedan ser sustraídas fácilmente. Todo concesionario obligado a cumplir con el requisito de seguridad, deberá someter al Superintendente una declaración jurada atestiguando que cumple con el requisito de seguridad. El Superintendente impondrá multa administrativa de mil (1,000) dólares por cada arma que le sea sustraída al concesionario de su propiedad que no cumpla con las medidas de seguridad aquí establecidas. Estos requisitos de seguridad y la multa correspondiente aplicarán a toda persona que tenga en su poder más de quince (15) armas, todo concesionario deberá exhibir en un área prominente y visible a la clientela un aviso claramente legible en que se informe de este requisito. La licencia, además, faculta al concesionario a adquirir, comprar, vender, donar, traspasar, ceder, tener, poseer, custodiar y transportar, conducir armas de fuego, municiones y cualquier accesorio pertinente, en todo lugar sujeto a la jurisdicción del Estado Libre Asociado de Puerto Rico; Disponiéndose, que:

(1) Las armas de fuego se podrán portar, conducir y transportar de forma oculta o no ostentosa, y el Superintendente dispondrá mediante reglamento, el procedimiento para que cualquier agente del orden público según descrito en este capítulo pueda expedir un boleto de cortesía que será preparado a esos efectos, el cual será remitido a la Oficina de Licencia de Armas de la Policía de Puerto Rico, donde se archivará en el expediente del tenedor de licencia. El Superintendente podrá imponer una multa de hasta doscientos (200) dólares por reincidir en portar, conducir o transportar armas de forma ostentosa o no oculta, así como el procedimiento de revisión de la misma, a solicitud de la parte interesada a quien se le impone la multa. El Superintendente celebrará una vista administrativa en un término no mayor de cuarenta y cinco (45) días para sostener, revisar, modificar o eliminar la multa impuesta.

(2) Salvo que se posea además un permiso de portación, el arma no se podrá portar en la persona del concesionario; y que para poder transportarla sin permiso de portación, se tendrá que hacer con el arma descargada, dentro de un estuche cerrado que no refleje su contenido, el cual no podrá estar a simple vista. Disponiéndose, que cuando se trate de un guardia de seguridad privado, con permiso de portar, uniformado y en el ejercicio de sus funciones, éstos podrán portar el arma a simple vista.

(3) Las armas de fuego o municiones sólo se podrán donar, vender, traspasar, ceder, dejar bajo la custodia o cualquier otra forma de traspaso de controlo de dominio, a personas que posean licencia de armas o de armero o una de las personas mencionadas en la sec. 456c de este título.

(4) El concesionario sólo podrá transportar un arma de fuego a la vez, salvo los concesionarios que posean a su vez permisos de tiro al blanco o de caza, quienes no tendrán limitación de cantidad para portar armas de fuego en su persona mientras se encuentren en los predios de un club de tiro autorizado o en aquellos lugares donde se practique el deporte de caza, en conformidad a las leyes aplicables.

(5) El concesionario sólo podrá comprar municiones de los calibres que puedan ser utilizados por las armas que posee inscritas a su nombre.

(6) Esta licencia de armas no autoriza al concesionario a dedicarse al negocio de compra y venta de armas de fuego o municiones, limitándose la compra y venta de éstas a sus armas y municiones personales.

(7) El concesionario podrá acudir a un club de tiro autorizado una vez al ano, pagando la cuota que allí se le requiera, para entrenarse en el uso y manejo de sus armas; Disponiéndose, que para este propósito el Superintendente autorizara la compra de cincuenta (50) municiones adicionales a las permitidas por este capitulo, las cuales tendrán que ser consumidas en su totalidad en el club de tiro al que el concesionario haya asistido.

(e) Dentro del término de cuarenta y cinco (45) días de recibir su Licencia de Armas, prorrogable por treinta (30) días más si así se solicita dentro del término original, todo concesionario deberá radicar, de no haberla radicado antes, en el Cuartel General de la Policía de Puerto Rico personalmente o por correo certificado con acuse de recibo, una certificación expedida por un oficial autorizado de un club de tiro autorizado en Puerto Rico al efecto de que el peticionario ha aprobado un curso en el uso y manejo correcto y seguro de armas de fuego conforme a esta Ley. De no hacerlo, incurrirá en una falta administrativa de cien (100) dólares por cada mes de atraso, hasta un máximo de seis (6) meses, al cabo de los cuales se revocará su licencia y se incautará la misma, así como toda arma y municiones que el peticionario haya adquirido. El Superintendente, para estos propósitos autorizará la compra de hasta un máximo de quinientas (500) municiones adicionales a las permitidas por esta Ley. Dichas municiones tendrán que ser consumidas en su totalidad por el concesionario durante el entrenamiento para la certificación. Lo dispuesto en este inciso no menoscabará lo dispuesto en el acápite (7) del inciso anterior.

El Superintendente habrá de atender cualquier reclamo de las personas que por razones de salud o circunstancia especiales, fuera de su control, no puedan cumplir con la obligación de certificación aquí impuesta. En todo caso en el que el Superintendente conceda una prórroga, el nuevo término para cumplir con los requisitos comenzará a partir de los diez (10) días después de resueltas las circunstancias que dieron lugar a la prórroga.

(f) La Policía de Puerto Rico expedirá los duplicados de carnés de licencia de armas que interese un peticionario dentro del término de treinta (30) días naturales de serle solicitado previo el pago de cincuenta (50) dólares en un comprobante de rentas internas por cada duplicado. En el caso de cambio de categoría de la licencia, el costo para el cambio de categoría será de veinte (20) dólares.

Todo carné de la licencia de armas tendrá la fecha en la cual deberá ser actualizado, que será cinco (5) años de expedido, y ninguna persona podrá hacer transacciones de armas de fuego o municiones, ni tirar en un club de tiro, ni cazar, ni portar, conducir o transportar armas, si no hubiese solicitado su actualización como se indica en este capítulo, so pena de que se revoque la licencia de armas y se imponga multa administrativa de quinientos (500) dólares por tirar en un club de tiro, cazar, portar, conducir o transportar armas. Transcurridos seis (6) meses de su fecha [de] vencimiento, sólo podrá vender sus armas de fuego a una persona con licencia de armero.

Cada cinco (5) años, en el quinto aniversario de la fecha de expedición de la licencia de armas, el peticionario vendrá obligado a renovar la misma cumplimentando una declaración jurada dirigida al Superintendente de la Policía, previo el pago del comprobante de rentas internas dispuesto en el Artículo 2.02 de esta Ley, haciendo constar que las circunstancias que dieron base al otorgamiento original se mantienen de igual forma o indicando de qué forma han cambiado. Dicha renovación se podrá realizar dentro de seis (6) meses antes o treinta (30) días después de la fecha de vencimiento de la licencia de armas. La no renovación de la licencia de armas transcurridos los treinta (30) días antes mencionados conllevará una multa administrativa de cincuenta (50) dólares por mes hasta un máximo de seis (6) meses, cantidad que deberá ser satisfecha como requisito a la renovación. Si pasados seis (6) meses no renueva la licencia de armas, el Superintendente revocará la misma e incautará las armas y municiones; Disponiéndose, que el concesionario podrá renovar y reinstalar su licencia hasta seis (6) meses más después de la revocación o la incautación, lo que fuese posterior, mediante el pago del doble de la multa acumulada. Nada de lo anterior impide que una persona a quien se le ha revocado su licencia de armas por su inacción solicite de novo otra licencia y se le conceda, siempre que hubiese pagado cualquier multa pendiente, en cuyo caso podrá recobrar las armas incautadas, si el Superintendente no hubiese dispuesto de ellas.

Se dispone que en el caso de que el concesionario estuviere residiendo fuera de Puerto Rico a la fecha aniversario de la renovación de la licencia o durante el período de renovación antes indicado, éste no vencerá hasta treinta (30) días de regresar el concesionario a Puerto Rico.

Se dispone, además, que en el caso de que el concesionario fuese una mujer y ésta estuviere en estado de gestación a la fecha aniversario de la renovación de la licencia o durante el período de renovación antes indicado, dicho período no vencerá hasta noventa (90) días después del alumbramiento. En todo caso que una mujer en estado de gestación venga obligada, bajo las disposiciones de este capítulo, a someter un certificado de entrenamiento, el término para radicarlo comenzará a partir de los noventa (90) días después del alumbramiento.

El Superintendente notificará a todo concesionario por correo dirigido a su dirección, seis (6) meses antes del vencimiento de la licencia de armas, la fecha en que ésta deberá ser renovada. El Superintendente pondrá a la disposición, a través de los cuarteles de área de la Policía, de los armeros y del Internet todos los formularios necesarios para llevar a cabo la renovación. Renovada la licencia, el Superintendente emitirá, previo satisfacción de derechos de renovación, el nuevo carné dentro de los próximos treinta (30) días naturales, a menos que tenga causa justificada para demorarlo.

Todo concesionario deberá informar al Superintendente su cambio de dirección residencial o postal dentro de treinta (30) días de realizarse el cambio, so pena de multa administrativa de doscientos (200) dólares, que deberá pagarse como requisito a la renovación de la licencia.

(g) En cualquier momento, una persona podrá entregar su licencia de armas a la Policía para su cancelación, y conjuntamente entregará sus armas a la Policía o las traspasará a otra persona con licencia de armas vigente o de armero.

(h) No será requisito poseer arma de fuego alguna para poder obtener licencia de armas y sus categorías.

456c. Funcionarios y ex policías, procedimiento de expedición de licencia

El Gobernador, los legisladores, los alcaldes, los secretarios, directores y jefes de agencias del Gobierno de Puerto Rico, los jueces estatales y federales, los fiscales estatales y federales y los procuradores de menores, el Superintendente, los miembros de la Policía, los funcionarios, agentes y empleados del Gobierno de Puerto Rico que por razón del cargo y las funciones que desempeñan vienen requeridos a portar armas, y todo agente del orden público, podrán portar armas de fuego. Podrán portar armas de fuego, además, los ex gobernadores, ex legisladores, ex superintendentes, ex jueces estatales y federales, ex fiscales estatales y federales, ex procuradores de menores, ex alcaldes de Puerto Rico y los ex agentes del orden público, siempre que su retiro haya sido honorable, que no estén impedidos por este capítulo de poseer armas de fuego y que, en el caso de ex agentes del orden público, hayan servido en dicha capacidad por no menos de diez (10) años. Además, los miembros de las Fuerzas Armadas de Estados Unidos y de la Guardia Nacional de Puerto Rico podrán portar sin licencia las armas que le asignen dichos cuerpos mientras se encuentren en funciones oficiales de su cargo. A esos fines, el Superintendente establecerá un procedimiento expedito mediante el cual otorgará a los funcionarios antes mencionados, salvo a los agentes del orden público y al propio Superintendente, una licencia de armas con el correspondiente permiso de portar.

Arguellos agentes del orden público, funcionarios y empleados gubernamentales autorizados a portar y entrenar con armas pertenecientes al Estado Libre Asociado de Puerto Rico o al gobierno federal, podrán inscribir el calibre de su arma oficial para poder comprar y utilizar municiones en su licencia de armas con permiso de portar, previa autorización del jefe o director de la agencia y en armonía con las disposiciones de este capítulo.

456d. Permisos de portación de armas expedidos por el tribunal

(a) La sala con competencia del Tribunal de Primera Instancia concederá, de no existir causa justificable para denegarlo, autorización al Superintendente para incluir en el carné del peticionario un permiso para portar, transportar y conducir, sin identificar arma en particular alguna, cualquier pistola o revólver legalmente poseído, previa notificación al Ministerio Público, y audiencia de éste así requerirlo, a toda persona poseedora de una licencia de armas que demostrare temer por su seguridad. El peticionario deberá radicar junto a su solicitud de portación, el recibo de un comprobante de rentas internas por la cantidad de doscientos cincuenta (250) dólares a favor del Superintendente, cuyo comprobante deberá haber sido presentado previamente al Superintendente, y una certificación expedida por un oficial autorizado de un club de tiro en Puerto Rico, al efecto de que el peticionario ha aprobado un curso en el uso y manejo correcto y seguro de armas de fuego.

Los requisitos exigidos para la expedición de una licencia de armas dispuestos en la sec. 456a de este título serán considerados por el tribunal al momento de evaluar la concesión del permiso de portación.

El permiso para portar armas expedido por el tribunal podrá renovarse concurrentemente con el procedimiento de renovación de la licencia de armas, mediante la presentación al Superintendente de un comprobante de cien (100) dólares a favor del Superintendente y una petición jurada en la que se haga constar que las circunstancias que dieron lugar a la concesión original de la licencia aún prevalecen al momento de presentarse la solicitud. En el caso de existir algún cambio, el mismo deberá ser justificado previo a la concesión de la renovación. El Superintendente notificará la renovación del permiso de portar armas al tribunal dentro de un término de treinta (30) días.

Deberá acompañarse una declaración jurada a los efectos que cumple con todos los requisitos establecidos en la sec. 456a de este título y que todo el contenido de la solicitud es correcto y cierto.

(b) El permiso de portacion aquí otorgado tendrá una duración sujeta a la vigencia de la licencia de armas, y podrá renovarse por términos consecutivos de cinco (5) anos, junto a la licencia de armas. En los casos en que se deniegue el permiso, las cantidades pagadas mediante comprobantes no serán reembolsables.

(c) Como parte de la solicitud de renovación de la licencia de armas y del permiso de portar, la persona deberá someter al Superintendente una nueva certificación en el uso, manejo y medidas de seguridad de armas de fuego, certificada por un club de tiro. El Superintendente, para estos propósitos, autorizará la compra de hasta un máximo de doscientas cincuenta (250) municiones adicionales a las permitidas en este capítulo, las cuales tendrán que ser consumidas en su totalidad en el club de tiro al que el concesionario haya asistido durante el entrenamiento de certificación.

Todo agente del orden público a quien por razón de sus funciones se le asigne un arma será adiestrado anualmente en el uso y manejo de dicha arma por funcionarios o contratistas de las agencias que los emplean que estén cualificados para certificar el uso, manejo y medidas de seguridad de armas de fuego. Será deber de la agencia que emplea a dicho agente someter una certificación al Superintendente de que el adiestramiento aquí establecido se ha llevado a cabo.

Los concesionarios que no cumplan con el requisito de certificación antes descrito no podrán portar un arma hasta tanto sean certificados, so pena de multa administrativa de quinientos (500) dólares; en caso de una segunda infracción a lo dispuesto en este inciso, el Superintendente, además, revocará el permiso de portación, sin mediar autorización del tribunal. En el caso de agentes del orden público, concluido un período de gracia de 60 días, las agencias no podrán utilizar personal no certificado de conformidad con esta sección en funciones que requieran el uso y/o portación de armas.

(d) El permiso de portacion será incorporado por el Superintendente a la licencia de armas del concesionario, haciendo constar la categoría de portar, según lo establecido en la sec. 456a(f) de este titulo, dentro de los diez (10) días naturales siguientes de haber entregado el concesionario la autorización del tribunal.

456f. Acusación por delito grave; ocupación de armas Luego de una determinación de causa probable para el arresto de cualquier persona a la cual se le haya otorgado una licencia de armas, por la comisión de cualquiera de los delitos especificados en la sec. 456j de este titulo o de violaciones a las disposiciones de este capitulo, el tribunal suspenderá provisionalmente la licencia hasta la determinación final del procedimiento criminal. Disponiéndose, además, que el tribunal ordenara la ocupación inmediata de la totalidad de las armas y municiones del concesionario para su custodia en el deposito de armas y municiones de la Policía. De resultar el acusado con una determinación de no culpabilidad, final y firme, el juez ordenara la inmediata devolución de su licencia de armas y de las armas y municiones. Toda arma y municiones así devueltas deberán entregarse en las mismas condiciones en que se ocuparon. El concesionario estará exento del pago por deposito. De resultar la acción judicial en una de culpabilidad, final y firme, el Superintendente revocara la licencia permanentemente y se incautara finalmente de todas sus armas y municiones.

456g. Licencia de armero; informe de transacciones

(a) Ninguna persona podrá dedicarse al negocio de armero o comerciante en armas de fuego y municiones, sin poseer una licencia de armero expedida por el Secretario del Departamento de Hacienda. Dichas licencias vencerán a partir de un (1) año desde la fecha de su expedición y estarán nuevamente sujetas a las formalidades y requisitos de solicitud de este capítulo. Las licencias de armeros estarán sujetas a la aprobación y certificación de la Policía, previa inspección, sobre las medidas de seguridad exigidas en la edificación donde esté ubicado el establecimiento. La solicitud para renovación de una licencia deberá radicarse con treinta (30) días de antelación a la fecha de su vencimiento.

(b) Cada transacción referente a la introducción de armas a Puerto Rico por armeros o a la venta de armas y municiones entre armeros deberá ser informada mediante el sistema electrónico establecido por este capitulo. De no estar disponible dicho sistema al momento de la transacción, será informada al Superintendente en formulario que proveerá este, debiendo expresar en el mismo el nombre, domicilio, sitio del negocio y particulares de la licencia, tanto del vendedor como del comprador, así como la cantidad y descripción, incluyendo el numero de serie de las armas o municiones objeto de cada transacción, según lo requiera el Superintendente.

(c) Un armero que posea una licencia expedida de acuerdo con este capítulo, podrá adquirir un arma que esté inscrita en el registro de armas bajo las disposiciones de este capítulo por compra de la persona que la haya inscrito a su nombre, siempre que tal persona tenga una licencia de armas expedida de acuerdo con este capítulo. Al efectuarse cualquier venta de armas de fuego o municiones, la transacción deberá ser informada mediante el sistema electrónico establecido por este capítulo. De no estar disponible dicho sistema al momento de la transacción el vendedor y comprador deberán notificarlo por escrito y con acuse de recibo al Superintendente, ambos mediante un mismo formulario que proveerá éste a esos fines. En el caso de venta múltiple de armas, más de una (1) a la vez o más de un (1) arma a la misma persona dentro de un período de treinta (30) días consecutivos, y de no estar disponible el sistema electrónico, el armero, dentro de veinticuatro (24) horas luego de entregar las armas, notificará mediante facsímil y por teléfono al Superintendente y anotará en sus libros el nombre y número de identificación de la persona que recibió la información. De igual forma se procederá cuando se lleve a cabo cualquier venta unitaria mayor a seiscientas (600) municiones a persona con licencia de armas con permiso de tiro al blanco o caza. De no lograr comunicación vía facsímil y telefónica, la notificación será mediante correo certificado con acuse de recibo, o personalmente.

(d) Cuando el armero, a su mejor juicio, detecte anormalidad en el carné de un concesionario o la entrega de armas sea negada o prohibida por disposición de ley federal, notificará de inmediato, vía facsímil y teléfono al Superintendente o a la persona que fehacientemente éste designe y notifique a los armeros. El Superintendente procederá de inmediato a investigar al concesionario para determinar si procede la cancelación de la licencia y la formulación de cargos criminales.

Toda infracción a lo dispuesto en el inciso (a) de esta sección constituirá delito grave, y será sancionada con pena de reclusión por un término fijo de seis (6) años. De mediar circunstancias agravantes, la pena fija establecida podrá ser aumentada hasta un máximo de doce (12) años; de mediar circunstancias atenuantes, podrá ser reducida hasta un mínimo de tres (3) años. Disponiéndose, que los trabajos de ajustes, mecánicos o cosméticos entre personas con licencia de armas no constituirán delito alguno.

Toda infracción a lo dispuesto en este inciso y el inciso (c) de esta sección será sancionada con pena de multa administrativa que no excederá de quinientos (500) dólares por no notificar en la primera infracción, y dos mil (2,000) dólares por infracciones subsiguientes; Disponiéndose, que de surgir una tercera falta, el Superintendente podrá optar por imponer la multa o iniciar el procedimiento para cancelar la licencia de armero y de surgir otra falta, el

Superintendente procederá a cancelar la licencia de armero.

456h. Requisitos de un peticionario para licencia de armero

(a) Toda persona que desee obtener o trasladar de local una licencia de armero radicara ante el Secretario del Departamento de Hacienda una solicitud jurada ante notario, acompañada de un comprobante de rentas internas de quinientos dólares ($500), en el formulario que proveerá el Secretario de Hacienda para estos propósitos. Luego de la aprobación de la solicitud por el Secretario de Hacienda, la misma será remitida al Superintendente de la Policía de Puerto Rico, y ninguna licencia será expedida de conformidad con esta sección si se demostrare que el peticionario no cumple con los requisitos establecidos en el inciso (a) de la sec. 456a de este titulo, excepto que tendrá que ser ciudadano de los Estados Unidos.

(b) Ninguna licencia de armero será expedida de conformidad con esta sección, sin que previamente se haya practicado por la Policía una investigación de todas las declaraciones contenidas en la solicitud y sin que los archivos de la Policía y los demás archivos accesibles (incluyendo los archivos del *National Crime Information Center* y del *National Instant Criminal Background Check System*) hayan sido examinados a los fines de determinar cualquier convicción anterior del peticionario. No se excederá licencia alguna si no se cumplieran con todas las disposiciones de esta sección, o si las declaraciones contenidas en la solicitud no resultaren ciertas.

(c) Si el peticionario es una corporación o una sociedad, la solicitud deberá estar firmada y jurada por el presidente, el secretario y el tesorero de la corporación, o por todos los socios directores de la sociedad; deberá indicar el nombre de la corporación o de la sociedad, sitio y fecha de su incorporación o constitución, sitio de su oficina principal o domicilio, nombre de la ciudad o pueblo, calle y numero donde será establecido el negocio, agencia, subagencia, oficina o sucursal para la cual se interese la licencia. Una licencia expedida bajo las disposiciones de esta sección será valida solamente para los negocios mencionados y descritos en la licencia. Dicha licencia no podrá traspasarse a ningún otro negocio ni a ninguna otra persona, y quedara automáticamente cancelada al disolverse la corporación o sociedad o sustituirse cualquiera de los oficiales de la corporación que suscribieron la solicitud, o al ingresar algún nuevo socio director en el caso de una sociedad, aunque dicha licencia podrá ser renovada tan pronto se cumpla con lo dispuesto en esta sección en relación con el nuevo oficial o el nuevo socio. En estos casos, el Secretario de Hacienda expedirá una licencia provisional mientras se efectúa el tramite de renovación.

(d) Cuando el peticionario es una corporación o sociedad, no se expedirá licencia alguna si cualquier oficial de la corporación no cumple con los requisitos establecidos en el inciso (a) de la sec. 456a de este titulo, excepto que tendrá que ser ciudadano de los Estados Unidos.

(e) En todo caso, la licencia de armero deberá ser expedida dentro de ciento veinte (120) días de radicada la solicitud personalmente o por correo certificado al Secretario del Departamento de Hacienda, sin perjuicio de que el Superintendente pueda continuar su investigación posteriormente y revocar la licencia si hubiera causa legal para ello.

456i. Condiciones para operaciones de armeros; constancias de transacciones Una persona, sociedad o corporación a la cual se le hubiera expedido una licencia de armero podrá dedicarse a la venta de armas y municiones, o al negocio de armero bajo las siguientes condiciones:

(a) El negocio se explotara solamente en el local designado en la licencia. Los armeros a quienes la Policía no les hubiese certificado que han cumplido con las medidas de seguridad de acuerdo con este capitulo, no podrán iniciar operaciones hasta cumplir con las mismas, ni podrán mantener en tal local armas o municiones que no sean aquellas que se este autorizado a poseer y portar por el armero de acuerdo con las disposiciones de este capitulo. La infracción [a] este inciso por el armero o aspirante a armero constituirá falta administrativa que será sancionada con multa no menor de cinco mil dólares ($5,000) ni mayor de cincuenta mil dólares ($50,000), a discreción del Superintendente. Además, conllevara que el Superintendente revoque la licencia. Este deberá inscribir cualquier modificación en el registro electrónico.

(b) Ningún armero recibirá arma alguna para su reparación, modificación, limpieza, grabación, pulimiento, o para efectuar cualquier otro trabajo mecánico, sin que se le muestre previamente la licencia de armas, ni aceptara un arma de fuego bajo condición alguna que tenga su numero de serie mutilado. La infracción de este inciso por parte del armero constituirá falta administrativa, y será sancionada con multa de diez mil dólares ($10,000). No cumplir con este requisito conllevara la revocación de la licencia por el Superintendente.

(c) La licencia de armero o copia certificada de la misma deberá colocarse en el establecimiento, de modo que pueda leerse con facilidad. No cumplir con este requisito podrá conllevar imposición de multa administrativa de cinco mil dólares ($5,000).

(d) Todo armero deberá tener en algún lugar visible al comprador o cliente la siguiente advertencia:

"El uso de un dispositivo de seguridad (locking device o safety lock) es recomendable para un arma de fuego. Toda arma cargada, así como sus municiones, deberán mantenerse fuera del alcance de menores o personas no autorizadas a utilizarlas. Es aconsejable guardar sus armas separadas de las municiones."

No cumplir con este requisito conllevara la imposición de una multa administrativa de cinco mil dólares ($5,000).

(e) Se llevara constancia de cada arma vendida y de cada venta de municiones, en libros destinados a este fin que serán impresos en la forma que prescriba el Superintendente, quien suministrara estos libros a los armeros, previo el pago por estos de los costos correspondientes, según se disponga mediante reglamento. La constancia de cada venta será firmada personalmente por el comprador y por la persona que efectúe la venta, haciéndolo cada uno en presencia del otro; y dicha constancia expresara la fecha, día y hora de la venta, calibre, fabricación, modelo y numero de fabrica del arma, calibre, marca y cantidad de municiones, y el nombre y numero de licencia de armas. El vendedor anotara la descripción de las municiones, la cantidad que vende y la fecha, día y hora de la venta en el formulario que le proveerá el Superintendente. De igual forma, en el registro electrónico se llevara constancia de cualquier arma o municiones vendidas. El Superintendente tendrá que dar acceso al registro electrónico a la persona, sociedad o corporación a la cual se le hubiera expedido una licencia de armero, a los únicos fines de poder registrar las transacciones a realizarse y que las mismas son conformes a las disposiciones de este capitulo. El Superintendente tendrá la obligación de mantener el registro organizado de forma que facilite comprobar en cualquier momento la cantidad de municiones que adquiere cada tenedor de licencia, y no autorizara la venta de calibres distintos a los que estén inscritos a favor del concesionario.

(f) Cuando las municiones vendidas sean de las descritas en el segundo párrafo de la sec. 459 de este título, el vendedor llevará un registro especial en libros y formularios destinados a este fin que serán impresos en la forma que prescriba el Superintendente, también suministrado de conformidad al inciso (e) de esta sección, en el que aparecerá el nombre del comprador, la descripción de las municiones, y la fecha, día y hora de la venta. Además, dicho registro contendrá lo siguiente:

(1) Una descripción completa de cada arma, incluyendo:

(A) El fabricante de la misma;

(B) el número de serie que aparezca grabado en la misma;

(C) el calibre del arma, y

(D) el modelo y tipo del arma.

En el caso de ventas al por mayor de armas del mismo calibre, modelo y tipo, el armero puede acumular dichas ventas en sus records, siempre y cuando las mismas sean hechas en una misma fecha y a un solo comprador.

(2) El nombre y dirección de cada persona de la cual se recibió el arma para la venta en la armería, conjuntamente con la fecha de adquisición.

(3) El nombre, numero de licencia y dirección de la persona natural o jurídica, a quien se vendió y la fecha de la entrega.

La utilización del sistema de registro electrónico no eximirá del cumplimiento de las disposiciones de esta sección.

(g) Los documentos y libros deberán mantenerse en el negocio indicado y descrito en la licencia, y deberán estar disponibles durante horas laborables para su inspección por cualquier funcionario del Ministerio Publico o agente del orden publico. En los casos de revocación de la licencia según se prescribe en esta sección o del cese de operaciones del negocio, dichos libros y constancias deberán ser entregados inmediatamente al Superintendente.

(h) No se exhibirán armas, municiones o imitaciones de los mismos en ningún lugar de un establecimiento comercial dedicado a la venta de armas, donde puedan ser vistas desde el exterior del negocio. No cumplir con este requisito podrá conllevar la imposición de una multa administrativa de cinco mil dólares ($5,000). Disponiéndose, que en casos de inobservancia de las medidas de seguridad o de las medidas dispuestas en esta sección en dos (2) o mas ocasiones, por parte de los armeros, el Superintendente, previa notificación escrita, podrá revocar la licencia. De la persona no estar de acuerdo, podrá llevar una acción de revisión, en conformidad a las secs. 2101 et seq. del Titulo 3.

(i) Todo armero a quien se le haya expedido una licencia bajo las disposiciones de esta sección, que deje de llevar las constancias y libros que aquí se exigen, incurrirá en delito grave y convicto que fuere, será sancionado con pena de reclusión por un termino fijo de doce (12) anos y pena de multa que no excederá de cien mil dólares ($100,000). De mediar circunstancias agravantes, la pena fija establecida podrá ser aumentada hasta un máximo de veinticuatro (24) anos; de mediar circunstancias atenuantes, podrá ser reducida hasta un mínimo de seis (6) anos. Además, el Superintendente revocara cualquier licencia de armas o de armero que esta persona posea.

456j. Fundamentos para rehusar expedir licencia El Superintendente no expedirá licencia de armas ni el Secretario del Departamento de Hacienda expedirá licencia de armero, o de haberse expedido se revocaran y el Superintendente se incautara de la licencia y de las armas y municiones de cualquier persona que haya sido convicta, en o fuera de Puerto Rico, de cualquier delito grave o su tentativa, por conducta constitutiva de violencia domestica según tipificada en las secs. 601 et seq. del Titulo 8, por conducta constitutiva de acecho según tipificada en las secs. 4013 a 4026 del Titulo 33, ni por conducta constitutiva de maltrato de menores según tipificada en las secs. 441 et seq. del Titulo 8. Disponiéndose, además, que tampoco se expedirá licencia alguna a una persona con un padecimiento mental que lo incapacite para poseer un arma, un ebrio habitual o adicto al uso de narcóticos o drogas, ni a persona alguna que haya renunciado a la ciudadanía americana o que haya sido separad[a] bajo condiciones deshonrosas de las Fuerzas Amadas de los Estados Unidos o destituido de alguna agencia del orden publico del Gobierno de Puerto Rico, ni a ninguna persona que haya sido convicta por alguna violación a las disposiciones de este capitulo o de las anteriores secs. 411 a 454 de este titulo.

456k. Registro de armas; pérdida y entrega de arma de fuego; muerte del poseedor de licencia

(a) El registro de armas creado en el cuartel general de la Policía se ajustara en su organización y funcionamiento a las disposiciones de este capitulo y será llevado en forma computadorizada, sistemática y ordenada de manera que se facilite la búsqueda de información. Este registro deberá estar debidamente custodiado.

(b) Toda arma de fuego legalmente poseída después de entrar en vigor esta ley deberá ser inscrita en el registro de armas, en caso de que no estuviere previamente inscrita. El Superintendente entregara al declarante una constancia de dicha inscripción.

(c) Toda persona que posea o tenga bajo su dominio un arma o municiones debidamente autorizadas por ley y las pierda, se le desaparezcan, se las roben o sean sustraídas mediante apropiación ilegal, deberá notificarlo mediante la presentación de una querella al distrito o precinto Policíaco donde éste resida o en el cuartel de la Policía más cercano, inmediatamente advenga en conocimiento de su pérdida, desaparición, robo o apropiación. Si no cumple con tal obligación, será culpable de delito menos grave y convicta que fuere, será sancionada con pena de multa hasta un máximo de cinco mil (5,000) dólares por cada arma o cada quinientas (500) o fracción de municiones dejadas de informar.

El Superintendente investigará todo informe de pérdida, desaparición, robo o apropiación ilegal, y llevará un registro detallado del resultado de éstos a los fines de levantar estadísticas sobre informes de pérdida, desaparición, robo o apropiación ilegal de armas o municiones. De ser intencionalmente falsa la información prestada por el querellante, el Superintendente notificará el hecho al Ministerio Público para que se presenten cargos criminales.

(d) Cuando falleciere una persona debidamente autorizada para la tenencia de armas, será deber de todo administrador, albacea o fideicomisario, o de cualquiera de ellos que actúe en Puerto Rico, y de cualquier subadministrador, agente o persona autorizada legalmente para administrar los bienes, notificar su fallecimiento al Superintendente dentro de los treinta (30) días siguientes a la fecha del fallecimiento. La notificación expresará el nombre, residencia y circunstancias personales del fallecido. No notificar este hecho constituirá delito menos grave que será sancionado con pena de multa no mayor de quinientos (500) dólares. El Superintendente dispondrá lo necesario para el recibo, almacenamiento, custodia, que podrá ser por una persona con licencia de armas o una armería designada por el administrador, albacea o fideicomisario y/o disposición de dichas armas, mientras se distribuye la herencia. Si las armas fueren adjudicadas a un heredero que sea elegible para obtener una licencia de armas, y se le expidiere tal licencia, dicha arma o armas le será entregada; Disponiéndose, que si dicho heredero ya fuera dueño del número máximo de armas permitido en este capítulo, el Superintendente concederá una autorización para la tenencia de las armas adquiridas por disposición de herencia, según el formulario que establezca éste mediante reglamento. De serle denegada tal licencia, o de disponerse la venta de dicha arma en pública subasta, la misma podrá ser adquirida únicamente por una persona con licencia de armas vigente, mediante subasta o por un armero debidamente autorizado por este capítulo y, de no ser así adquirida, dicha arma será entregada para su decomiso al Superintendente, tal como se dispone en este capítulo. Disponiéndose, además, que el Superintendente no entregará ninguna arma que, previo al fallecimiento de su dueño, no hubiese sido debidamente inscrita a tenor con el inciso (b) de esta sección.

(e) Cualquier adquisición, compra, venta, donación, cesión o cualquier forma de traspaso de titularidad de un arma de fuego y municiones deberá ser realizada ante una persona con licencia de armero para su correspondiente inscripción en el registro electrónico y en sus libros, en conformidad a lo dispuesto en la sec. 456i de este titulo. También podrán realizarse los mencionados medios de traspaso de titularidad entre concesionarios de licencia de armas mediante los formularios de traspaso de armas que provea el Superintendente, dentro de los cinco (5) días siguientes al otorgamiento, para la debida anotación y corrección en el registro electrónico.

456l. Motivos fundados para facultar a los agentes del orden publico a ocupar armas Cualquier agente del orden público ocupará la licencia, arma y municiones que posea un concesionario cuando tuviese motivos fundados para entender que el tenedor de la licencia hizo o hará uso ilegal de las armas y municiones, para causar daño a otras personas; por haber proferido amenazas de cometer un delito; por haber expresado su intención de suicidarse; cuando haya demostrado reiteradamente negligencia o descuido en el manejo del arma; cuando se estime que el tenedor padece de una condición mental, se le considere ebrio habitual o es adicto a sustancias controladas; o en cualquier otra situación de grave riesgo o peligro que justifique esta medida de emergencia. Un agente del orden público también ocupará la licencia, armas y municiones cuando se arreste al tenedor de la misma por la comisión de un delito grave o delito menos grave que implique violencia. A solicitud de la parte a quien se le ocupó el arma, hecha dentro de los quince (15) días laborables luego de la ocupación del arma, el Superintendente celebrará una vista administrativa en un término no mayor de cuarenta y cinco (45) días para sostener, revisar o modificar la ocupación del agente del orden público. El Superintendente deberá emitir su decisión en un plazo no mayor de cuarenta y cinco (45) días a partir de la celebración de dicha vista administrativa formal y de resultar favorable a la parte afectada la determinación de Superintendente, éste ordenará la devolución inmediata del arma o armas ocupadas.

456m. Armas de asalto semiautomáticas; fabricación, importación, distribución, posesión y transferencia

(a) No se podrá fabricar o hacer fabricar, ofrecer, vender, alquilar, prestar, poseer, usar, traspasar o importar un arma de asalto semiautomática. No obstante, esta prohibición no será de aplicación a:

(1) La posesión, uso, transferencia, en Puerto Rico o importación desde el territorio de los Estados Unidos, por personas cuya licencia contenga la categoría de tiro al blanco, de caza o posea licencia de armero, de aquellas armas de asalto legalmente existentes en la nación de los Estados Unidos de Norte América, a la fecha de entrar en vigor esta ley, o

(2) la fabricación, importación, venta o entrega, por personas con licencia de armero, para uso de estas armas en el cumplimiento del deber por los agentes del orden público, del Gobierno de Puerto Rico o de los Estados Unidos, o para el uso de las fuerzas armadas del Gobierno de los Estados Unidos o de Puerto Rico.

(b) Las armas de asalto semiautomáticas a que se refiere esta sección son las siguientes:

(1) *Norinco* , *Mitchell* , y *Poly Technologies Avtomat Kalashnikovs* (todos los modelos de AK;

(2) *Action Arms Israeli Military Industries UZI* y *Galil*;

(3) *Beretta* Ar70 (SC-70);

(4) *Colt* AR-15;

(5) *Fabrique National* FN/FAL, FN/LAR y FNC;

(6) *SWD M-10, M-11, M-11/9, y M-12*;

(7) *Steyr AUG*;

(8) *INTRATEC* TEC-9, TEC-DC9 y TEC-22, y

(9) escopetas *revolving cylinder*, tales como (o similares a) la *Street Sweeper* y el *Striker* 12.

(c) Además, será considerada como un arma de asalto semiautomática:

(1) Un rifle semiautomático que pueda ser alimentado mediante retroalimentación por un abastecedor o receptáculo removible y que contenga más de dos (2) de las siguientes características:

(A) Culata plegadiza o telescópica;

(B) empuñadura de pistola (*pistol grip*) que sobresale manifiestamente por debajo de la acción del arma;

(C) montura para bayoneta;

(D) supresor de fuego o rosca para acomodar un supresor de fuego (*flash suppressor*), o

(E) lanzador de granadas, excluyendo los lanzadores de bengalas.

(2) Una pistola semiautomática que pueda ser alimentada mediante retroalimentación por un abastecedor o receptáculo removible y que contenga más de dos (2) de las siguientes características:

(A) Un abastecedor o receptáculo de municiones que se fija a la pistola por fuera de la empuñadura de la pistola (*pistol grip*);

(B) un cañón con rosca en su punta delantera capaz de aceptar una extensión al cañón, supresor de fuego (*flash suppressor*), agarre para la mano al frente del arma o un silenciador;

(C) una cubierta que se puede fijar cubriendo parcial o total el cañón permitiendo a quien dispara el arma, sujetarla con la mano que no está oprimiendo el gatillo y no quemarse;

(D) un peso de manufactura en exceso a cincuenta (50) onzas descargada, o

(E) una versión semiautomática de un arma automática.

(3) Una escopeta semiautomática que contenga dos (2) o más de las siguientes características:

(A) Culata plegadiza o;

(B) empuñadura de pistola (*pistol grip*) que sobresale manifiestamente por debajo de la acción del arma;

(C) abastecedor o receptáculo de municiones fijo con capacidad para más de cinco (5) cartuchos, o

(D) capaz de recibir un abastecedor o receptáculo de municiones removible.

(d) Toda persona que viole las disposiciones de esta sección incurrirá en delito grave, y convicta que fuere será sancionada con pena de reclusión por un término fijo de veinticuatro (24) años, sin derecho a sentencia suspendida, a salir en libertad bajo palabra, o a disfrutar de los beneficios de algún programa de desvío, bonificaciones o alternativa a la reclusión reconocida en esta jurisdicción, debiendo cumplir en años naturales la totalidad de la pena impuesta. De mediar circunstancias agravantes, la pena fija establecida podrá ser aumentada hasta un máximo de treinta y seis (36) años; de mediar circunstancias atenuantes, podrá ser reducida hasta un mínimo de dieciocho (18) años.

No constituirá delito la posesión o uso de estas armas en el cumplimiento del deber por los miembros de la Policía, y aquellos otros agentes del orden público debidamente autorizados a portar armas de fuego conforme se establece en este capitulo.

La aplicación de las disposiciones de esta sección será prospectiva a partir de la aprobación de esta ley.

Subcapítulo 3. Permisos de Tiro Al Blanco

457c. Permisos de tiro al blanco

(a) Toda persona que tenga una licencia de armas expedida de conformidad con este capitulo podrá solicitar al Superintendente un permiso de tiro al blanco. Proveerá bajo juramento ante notario toda la información requerida en los formularios de solicitud preparados a esos efectos por el Superintendente, los cuales requerirán al menos un comprobante de rentas internas de veinticinco (25) dólares, un retrato [de] dos (2) pulgadas por dos (2) pulgadas, y un sello de la federación de tiro. El Superintendente, dentro del termino de treinta (30) días laborables de recibida la solicitud, expedirá el permiso solicitado, salvo que exista causa justificable para la denegación.

(b) No se expedirá permiso de tirador a persona alguna que no sea miembro de un club u organización de tiro al blanco y una federación de tiro, debidamente reconocidos por el Secretario. ...

Subcapítulo 4. Agencias de Seguridad Que Transporten Valores En Vehículos Blindados

457f. Licencias especiales El Superintendente podrá expedir una licencia especial a las agencias de seguridad que se dediquen al transporte de valores en vehículos blindados que así lo soliciten y que estén debidamente autorizadas a operar como tales; autorizándolas a comprar, poseer, disponer y mantener en su lugar de negocio un deposito para armas largas que no sean automáticas y municiones para ser utilizadas única y exclusivamente por los agentes de seguridad empleados por esta que estén asignados al transporte de valores en vehículos blindados y mientras estén en funciones de su empleo.

457j. Deposito de armas largas y municiones

Toda solicitud para obtener la licencia especial para comprar, poseer, disponer y mantener en el lugar de negocio un deposito para armas largas y municiones deberá acompañarse con prueba fehaciente de que la agencia de seguridad emplea cinco (5) personas o mas para tales fines.

El solicitante de una licencia especial de comprar, poseer, disponer y mantener en su lugar de negocio un deposito de armas largas cumplirá también con todas las disposiciones y requisitos de seguridad exigidos para las licencias de armeros, así como cualquier otro requisito que disponga el Superintendente mediante reglamento. Una vez el Superintendente certifique que el local del solicitante cumple con los requisitos de seguridad exigidos, se le expedirá la licencia especial solicitada. El negocio del solicitante operara únicamente en el local designado, estará sujeto a inspección por cualquier agente de la Policía o del Negociado de Investigaciones Especiales del Departamento de Justicia, y mantendrá la licencia en un sitio visible en su oficina. No podrá mantenerse en dicho local arma alguna que no sean aquellas que se este autorizando a poseer de acuerdo a las disposiciones de este capitulo.

457k. Limite de armas La licencia especial para poseer y mantener en su lugar de negocio un deposito de armas largas permitirá a la agencia de seguridad tener bajo su control y cuidado un numero especifico de armas largas tales como escopetas y rifles semiautomáticas, registradas a su nombre en los registros del Superintendente. La agencia solo podrá adquirir dos (2) armas largas en exceso del numero de vehículos blindados que posea la agencia, según certifique la Comisión de Servicio Publico, y que se dediquen al transporte de valores.

457r. Municiones Se autoriza a las agencias de seguridad que obtengan la licencia especial que dispone este capitulo a comprar una cantidad razonable de municiones para las armas que le han sido autorizadas por el Superintendente en la licencia. La agencia de seguridad mantendrá un inventario perpetuo de las armas y municiones autorizadas, así como un registro del movimiento diario de estas. Estos registros estarán sujetos a inspección por la Policía de Puerto Rico. Disponiéndose, además, que la adquisición, compra, venta, donación, cesión o cualquier forma de traspaso de titularidad de armas de fuego, municiones o accesorios, realizados en virtud de esta licencia especial, deben haber sido inscritos debidamente en el registro electrónico por los medios que dispone este capitulo.

Subcapítulo 5. Armas

458. Fabricación, importación, venta y distribución Se necesitará una licencia expedida conforme a los requisitos exigidos por este capítulo para fabricar, importar, ofrecer, vender o tener para la venta, alquilar o traspasar cualquier arma de fuego, municiones o aquella parte o pieza de un arma de fuego donde el fabricante de la misma coloca el número de serie del arma. Toda infracción a esta sección constituirá delito grave y será sancionada con pena de reclusión por un término fijo de quince (15) años, sin derecho a sentencia suspendida, a salir en libertad bajo palabra, o a disfrutar de los beneficios de algún programa de desvío, bonificaciones o alternativa a la reclusión reconocida en esta jurisdicción, debiendo cumplir en años naturales la totalidad de la pena impuesta. De mediar circunstancias agravantes, la pena fija establecida podrá ser aumentada hasta un máximo de veinticinco (25) años; de mediar circunstancias atenuantes, podrá ser reducida hasta un mínimo de diez (10) años.

458a. Prohibición vender armas a personas sin licencia Ningún armero entregara un arma de fuego a un comprador sin que este le muestre una licencia de arma vigente. Cuando el comprador del arma sea un cazador o tirador autorizado a poseer armas de fuego, la venta y entrega del arma se efectuara de la misma manera que se señala en este capitulo.

El armero que a sabiendas venda armas de fuego a una persona sin licencia, incurrirá en delito grave y convicto que fuere, será sancionado con pena de reclusión por un término fijo de quince (15) anos. De mediar circunstancias agravantes, la pena fija establecida podrá ser aumentada hasta un máximo de veinticinco (25) anos, de mediar circunstancias atenuantes, podrá ser reducida hasta un mínimo de diez (10) anos.

Una convicción bajo esta sección conllevara la cancelación automática de la licencia del armero.

458b. Comercio de armas de fuego automaticas Toda persona que venda o tenga para la venta, ofrezca, entregue, alquile, preste o en cualquier otra forma disponga de cualquier arma de fuego que pueda ser disparada automáticamente, independientemente de que dicha arma se denomine ametralladora o de otra manera, incurrirá en delito grave, y convicta que fuere será sancionada con pena de reclusión por un término fijo de veinticuatro (24) años, sin derecho a sentencia suspendida, a salir en libertad bajo palabra, o a disfrutar de los beneficios de algún programa de desvío, bonificaciones o alternativa a la reclusión reconocida en esta jurisdicción, debiendo cumplir en años naturales la totalidad de la pena impuesta. De mediar circunstancias agravantes, la pena fija establecida podrá ser aumentada hasta un máximo de treinta y seis (36) años; de mediar circunstancias atenuantes, podrá ser reducida hasta un mínimo de dieciocho (18) años.

Este delito no aplicará a la venta o entrega de una ametralladora o cualquier otra arma de fuego que pueda ser disparada automáticamente para uso de la Policía y otros agentes del orden público.

458e. Posesión sin licencia Toda persona que tenga o posea, pero que no esté portando, un arma de fuego sin tener licencia para ello, incurrirá en delito grave, y convicta que fuere será sancionada con pena de reclusión por un término fijo de cinco (5) años. De mediar circunstancias agravantes, la pena establecida podrá ser aumentada hasta un máximo de diez (10) años; de mediar circunstancias atenuantes, podrá ser reducida hasta un mínimo de un año.

No obstante todo lo anterior, cuando una persona incurra en las conductas prohibidas por esta sección sin la intención de cometer un delito con el arma de fuego poseída sin licencia, y se tratare de una persona que nunca haya sido convicta por violación a este capítulo, la Ley Núm. 348 de 21 de diciembre de 1999, las disposiciones de las secs. 1476 a 1482 del Título 32, o la Ley Núm. 27 de 10 de enero de 2002, o alguno de los delitos enumerados en la sec. 456j de este título, y el arma no sea una que ha sido reportada robada o apropiada ilegalmente, incurrirá en delito menos grave, y convicta que fuere será sancionada con pena de reclusión por un término no mayor de seis (6) meses, pena de multa que no excederá de cinco mil (5,000) dólares, o ambas penas a discreción del tribunal. El tribunal, a su discreción, podrá imponer la pena de prestación de servicios en la comunidad en lugar de la pena de reclusión establecida.

En caso de que el poseedor del arma demuestre con prueba fehaciente que posee una licencia de armas, aunque vencida, y que solicitó su renovación dentro del término provisto por este capítulo, no será culpable de delito alguno. Si no ha solicitado su renovación dentro del término provisto incurrirá en falta administrativa y tendrá que pagar el triple de los costos acumulados de los derechos de renovación.

458f. Posesión o uso ilegal de armas largas semiautomáticas, automáticas o escopeta de cañón cortado Toda persona que porte, posea o use sin autorización de este capítulo un arma larga semiautomática, una ametralladora, carabina, rifle, así como cualquier modificación de éstas o cualquiera otra arma que pueda ser disparada automáticamente o escopeta de cañón cortado a menos de dieciocho (18) pulgadas, y que pueda causar grave daño corporal, incurrirá en delito grave, y convicta que fuere será sancionada con pena de reclusión por un término fijo de veinticuatro (24) años, sin derecho a sentencia suspendida, a salir en libertad bajo palabra, o a disfrutar de los beneficios de algún programa de desvío, bonificaciones o alternativa a la reclusión reconocida en esta jurisdicción, debiendo cumplir en años naturales la totalidad de la pena impuesta.

De mediar circunstancias agravantes, la pena fija establecida podrá ser aumentada hasta un máximo de treinta y seis (36) años; de mediar circunstancias atenuantes, podrá ser reducida hasta un mínimo de dieciocho (18).

No constituirá delito la posesión o uso de estas armas en el cumplimiento del deber por los miembros de la Policía, y aquellos otros agentes del orden público debidamente autorizados.

458g. Posesión o venta de accesorios para silenciar. Toda persona que tenga en su posesión, venda, tenga para la venta, preste, ofrezca, entregue o disponga de cualquier instrumento, dispositivo, artefacto o accesorio, que silencie, reduzca el ruido del disparo de cualquier arma de fuego, incurrirá en delito grave y convicta que fuere, será sancionada con pena de reclusión por un termino fijo de doce (12) anos. De mediar circunstancias agravantes, la pena fija establecida podrá ser aumentada hasta un máximo de veinticuatro (24) anos; de mediar circunstancias atenuantes, podrá ser reducida hasta un mínimo de seis (6) anos.

Las disposiciones de esta sección no serán aplicables a los agentes del orden publico debidamente autorizados y en cumplimiento de sus funciones oficiales.

458h. Facilitación a terceros Toda persona que con intención criminal facilite o ponga a la disposición de otra persona cualquier arma de fuego que haya estado bajo su custodia o control, sea o no propietaria de la misma, incurrirá en delito grave y, convicta que fuere, será sancionada con pena de reclusión por un termino fijo de doce (12) anos. De mediar circunstancias agravantes, la pena fija establecida podrá ser aumentada hasta un máximo de veinticuatro (24) anos; de mediar circunstancias atenuantes, podrá ser reducida hasta un mínimo de seis (6) anos.

458i. Numero de serie o nombre de dueño en arma de fuego; remoción o mutilación Toda arma deberá llevar, en forma tal que no pueda ser fácilmente alterado o borrado, el nombre del armero o marca de fabrica bajo la cual se venderá el arma o el nombre del importador y, además, un numero de serie o el nombre completo de su poseedor grabado en la misma.

Incurrirá en delito grave y sancionada con pena de reclusión por un término fijo de doce (12) anos, toda persona que:

(a) Voluntariamente remueva, mutile, cubra permanentemente, altere o borre el número de serie o el nombre de su poseedor en cualquier arma;

(b) a sabiendas compre, venda, reciba, enajene, traspase, porte o tenga en su posesión, cualquier arma a la cual se le haya removido, mutilado, cubierto permanentemente, alterado o borrado el numero de serie o el nombre de su poseedor, o

(c) siendo un armero o un agente o representante de dicho armero, a sabiendas compre, venda, reciba, entregue, enajene, traspase, porte o tenga en su posesion, cualquier arma a la cual se le haya removido, mutilado, cubierto permanentemente, alterado o borrado su numero de serie o el nombre de su poseedor.

De mediar circunstancias agravantes, la pena fija establecida podra ser aumentada hasta un maximo de veinticuatro (24) anos; de mediar circunstancias atenuantes, podra ser reducida hasta un minimo de seis (6) anos.

458j. Presunciones La posesion de un arma de fuego por una persona que no posea una licencia de armas se considerara evidencia prima facie de que dicha persona posee el arma con la intencion de cometer delito.

La portacion de un arma de fuego por una persona que no posea una licencia de armas con permiso para portar, se considerara evidencia prima facie de que dicha persona portaba el arma con la intención de cometer delito.

La posesión por cualquier persona de un arma a la cual se le haya removido, mutilado, cubierto permanentemente, alterado o borrado su numero de serie o el nombre de su poseedor, se considerara evidencia prima facie de que dicha persona removió, mutilo, cubrió, altero o borro dicho numero de serie o el nombre de su poseedor.

La posesión por cualquier persona de un arma a la cual se le haya removido, mutilada, cubierto permanentemente, alterado o borrado su numero de serie o el nombre de su poseedor se considerara evidencia prima facie de que dicha persona posee el arma con la intención de cometer un delito.

La posesión por cualquier persona de un arma al momento de cometer o intentar cometer un delito, se considerara evidencia prima facie de que dicha arma estaba cargada al momento de cometer o intentar cometer el delito.

La presencia de tres (3) o mas armas de fuego en una habitación, casa, residencia, establecimiento, oficina, estructura o vehiculo, constituirá evidencia prima facie de que el dueño o poseedor de dicha habitación, casa, residencia, establecimiento, oficina, estructura o vehiculo, o aquellas personas que ocupen la habitación, casa, residencia, establecimiento, oficina o estructura, trafican y facilitan armas de fuego ilegalmente, siempre que estas personas no tengan una licencia de armas, de armero, de club de tiro o coto de caza.

La presencia de una ametralladora o cualquier otra arma de funcionamiento automático o de las municiones armor piercing en cualquier habitación, casa, residencia, establecimiento, oficina, estructura o vehiculo, constituirá evidencia prima facie de su posesión ilegal por el dueño o poseedor de dicha edificación o vehiculo, y por aquellas personas que ocupen la habitación, casa, edificio o estructura donde se encontrare tal ametralladora, arma de funcionamiento automático o escopeta de canon cortado, y que tengan la posesión mediata o inmediata de la misma. Esta presunción no será de aplicación en los casos que se trate de un vehiculo de servicio publico que en ese momento estuviere transportando pasajeros mediante paga, o que se demuestre que se trata de una transportación incidental o de emergencia.

La presencia de una ametralladora o cualquier otra arma de funcionamiento automático o de las municiones armour piercing en cualquier habitación, casa, residencia, establecimiento, oficina, estructura o vehiculo constituirá evidencia prima facie de que el dueño o poseedor de dicha edificación o vehiculo posee el arma o las municiones con la intención de cometer un delito.

La presencia de un arma de fuego o de municiones en cualquier vehiculo robado o hurtado constituirá evidencia prima facie de su posesión ilegal por todas las personas que viajaren en tal vehiculo al momento que dicha arma o municiones sean encontradas.

Las disposiciones de esta sección no aplicaran a los agentes del orden público en el cumplimiento de sus funciones oficiales.

458k. Notificación por porteador, almacenista o depositario de recibo de armas; penalidades Todo porteador marítimo, aéreo o terrestre, y todo almacenista o depositario que a sabiendas reciba armas de fuego, accesorios o partes de estas o municiones para entrega en Puerto Rico, no entregara dicha mercancía al consignatario hasta que este le muestre su licencia de armas o de armero. Después de cinco (5) días laborables de la entrega, el porteador, almacenista o depositario notificara al Superintendente, dirigiendo la notificación personalmente o por correo certificado con acuse de recibo, el nombre, dirección y numero de licencia del consignatario y el numero de armas de fuego o municiones, incluyendo el calibre, entregadas, así como cualquier otra información que requiera el Superintendente mediante reglamento.

Cuando el consignatario no tuviere licencia de armas o de armero, el porteador, almacenista o depositario notificara al Superintendente inmediatamente de tal hecho, el nombre y dirección del consignatario, y el número de armas de fuego o municiones para entrega. Además, no entregara dicha mercancía a tal consignatario hasta tener autorización al efecto, expedida por el Superintendente.

La violación de cualquier obligación aquí establecida constituirá un delito grave que será sancionada con pena de reclusión por un termino fijo de doce (12) anos y pena de multa no menor de dos mil dólares ($2,000) ni mayor de diez mil dólares ($10,000). De mediar circunstancias agravantes, la pena fija establecida podrá ser aumentada hasta un máximo de veinticuatro (24) anos; de mediar circunstancias atenuantes, podrá ser reducida hasta un mínimo de seis (6) anos.

458l. Almacenamiento y custodia de armas en depósitos de armas y municiones Todo armero vendrá obligado a implantar las medidas de seguridad exigidas por la Policía mediante reglamento para el almacenamiento o custodia de las armas y municiones. La Policía examinara cada tres (3) meses los locales de los armeros, los cuales de no cumplir con las medidas de seguridad exigidas, tendrán treinta (30) días para cumplir con las mismas o de lo contrario, deberán depositar las armas y municiones que posean para la venta, para su almacenamiento y custodia en la bóveda de otro armero o en el deposito de armas y municiones de la Policía de Puerto Rico, dentro del termino que determine el Superintendente, en lo que corrigen la deficiencia.

Los armeros que, para corregir deficiencias, utilicen el deposito de armas y municiones, pagaran por el almacenamiento y custodia de sus armas y municiones una mensualidad que se determinara mediante reglamento. Al establecer el costo de almacenamiento y custodia, se tomaran en consideración los costos de operación del depósito de armas y municiones y el manejo de las armas y municiones para efectos de recibo, clasificación, custodia y entrega de las mismas. Los costos a cargarse a los usuarios del depósito de armas y municiones bajo ningún concepto excederán los costos reales y razonables por concepto del servicio prestado.

El Superintendente o el encargado del deposito de armas y municiones enviara periódicamente a los armeros, según se disponga por reglamento, una factura en la que se indicara el costo del almacenamiento y custodia de sus armas, de acuerdo a la utilización del deposito de armas y municiones que durante dicho mes haya hecho el armero. La falta de pago por un armero será motivo suficiente para que el Superintendente, previa la celebración de una vista formal, pueda revocarle la licencia que hubiere expedido.

En el depósito de armas y municiones se almacenaran igualmente, mediante paga, las armas de aquellos ciudadanos con licencia de armas que interesen, como medida de seguridad, que sus armas sean guardadas temporeramente, sin menoscabo de que dichos ciudadanos puedan optar por guardar sus armas en negocios privados de armeros.

458n. Disparar o apuntar

(a) Incurrirá en delito grave con pena de reclusión por un término fijo de cinco (5) anos, toda persona que, salvo en casos de defensa propia o de terceros, o de actuaciones en el desempeño de funciones oficiales o actividades legítimas de deportes:

(1) Voluntariamente dispare cualquier arma en un sitio publico o en cualquier otro sitio donde haya alguna persona que pueda sufrir daño, aunque no le cause daño a persona alguna, o

(2) intencionalmente, aunque sin malicia, apunte hacia alguna persona con un arma, aunque no le cause daño a persona alguna.

De mediar circunstancias agravantes, la pena establecida podrá ser aumentada hasta un máximo de diez (10) anos; de mediar circunstancias atenuantes, podrá ser reducida hasta un mínimo de un ano.

(b) Sera culpable de delito grave con pena de reclusión por un termino fijo de tres (3) anos, toda persona que, salvo en casos de defensa propia o de terceros o de actuaciones en el desempeño de funciones oficiales o actividades legitimas de deportes, incurra en cualquiera de los actos descritos anteriormente utilizando un arma neumática. De mediar circunstancias agravantes, la pena establecida podrá ser aumentada hasta un máximo de seis (6) anos; de mediar circunstancias atenuantes, podrá ser reducida hasta un mínimo de seis (6) meses y un día.

458o. Recibo, custodia y disposición voluntaria por Policía; destrucción El Superintendente establecerá mediante reglamentación lo relacionado al recibo, custodia y disposición de aquellas armas que sean ocupadas o depositadas voluntariamente en la Policía por personas que tengan licencias; o fueren entregadas a la muerte del poseedor de una licencia; o por haberse cancelado la licencia al concesionario.

Se autoriza al Superintendente a vender, permutar, donar o ceder las armas a agencias del orden público federales, estatales o municipales u otras jurisdicciones. Además, podrá vender las armas a armeros o a una persona con licencia de armas expedida a tenor con lo dispuesto [por] este capitulo, según disponga mediante reglamento.

Las armas o instrumentos ocupados de acuerdo con esta sección serán almacenadas por el Superintendente en el deposito de armas y municiones de la Policía.

No obstante, toda ametralladora, escopeta de canon cortado o cualquier otra arma o instrumento especificado en la sec. 458d de este titulo, que se porte, posea o transporte ilegalmente, se considerara como un estorbo publico.

Cuando alguna de dichas armas o instrumentos sea ocupada la misma será entregada al Superintendente para que este se encargue de su disposición y destrucción, mediante la reglamentación promulgada al efecto.

458p. Colecciones. Nada de lo dispuesto en este capítulo impedirá que se conserven y mantengan colecciones privadas de armas y sus dueños las posean como adorno o materia de curiosidad, ni que se mantengan colecciones de armas como reliquias. Para la conservación de toda arma de las incluidas en esta sección, será necesario que el coleccionista primero obtenga un permiso de tiro al blanco o de caza, bajo las disposiciones de este capítulo.

Las armas de fuego antiguas, según se definen en este capítulo, que no estén provistas de número de serie por su manufacturero estarán exentas del requisito de registración, según definido en este capítulo, pero su existencia deberá ser informada al Registro de Armas de la Policía de Puerto Rico acompañada de tres (3) fotografías distintas que detallen sus particularidades para la correspondiente anotación de su existencia en el expediente del concesionario con licencia de armas y permiso de tiro. Disponiéndose, que de ser utilizada el arma de fuego antigua para cometer delito entonces se considerarán como arma de fuego no inscrita. Se dispone, además, que bajo ningún concepto se podrá requerir marcar, modificar o alterar de forma alguna el arma de fuego antigua.

458q. Transportación prohibida; confiscación El Secretario de Justicia confiscara cualquier propiedad, según este termino es definido en las secs. 1723 et seq. del Titulo 34, conocidas como "Ley Uniforme de Confiscaciones de 1988", en que se almacene, cargue, descargue, transporte, lleve o traslade o en el que se sorprenda almacenando, cargando, descargando, transportando, llevando o trasladando cualquier arma en violación [a] este capitulo.

Para la confiscación y disposición se seguirá el procedimiento establecido por las dichas secs. 1723 et seq. del Titulo 34.

458r. Menores Toda persona que negligentemente dejare un arma de fuego o arma neumática al alcance de una persona menor de dieciocho (18) anos que no tuviere un permiso para tiro al blanco o caza, y este se apodere del arma y causare daño a otra persona o a si mismo, cometerá delito grave y convicta que fuere, será sancionada con pena de reclusión por un termino fijo de dos (2) anos. De mediar circunstancias agravantes, la pena fija establecida podrá ser aumentada hasta un máximo de cinco (5) anos; de mediar [circunstancias] atenuantes, podrá ser reducida hasta un mínimo de seis (6) meses y un día.

458s. Apropiación ilegal de armas o municiones, robo Toda persona que intencionalmente, independientemente de los medios que utilice para ello, se apropie ilegalmente de un arma de fuego o municiones, incurrirá en delito grave y convicta que fuere, será sancionada con pena de reclusión por un término fijo de diez (10) años, sin derecho a sentencia suspendida, a salir en libertad bajo palabra, o a disfrutar de los beneficios de algún programa de desvío, bonificaciones o alternativa a la reclusión reconocida en esta jurisdicción, debiendo cumplir en años naturales la totalidad de la pena impuesta. De mediar circunstancias agravantes, la pena fija establecida podrá ser aumentada hasta un máximo de veinte (20) años; de mediar circunstancias atenuantes, podrá ser reducida hasta un mínimo de cinco (5) años.

En el caso que la persona se apropie ilegalmente, independientemente de los medios que utilice para ello, más de un arma de fuego o si la persona tuviere antecedentes penales por haber sido convicto por delito grave, la pena se duplicará.

Subcapítulo 6. Municiones

459. Fabricación, distribución, posesión y uso Se necesitará una licencia de armas, de tiro al blanco, de caza o de armero, según sea el caso, para fabricar, solicitar que se fabrique, importar, ofrecer, comprar, vender o tener para la venta, guardar, almacenar, entregar, prestar, traspasar, o en cualquier otra forma disponer de, poseer, usar, portar o transportar municiones, conforme a los requisitos exigidos por este capítulo. Asimismo, se necesitará un permiso expedido por la Policía para comprar pólvora. Toda infracción a este artículo constituirá delito grave, y será sancionada con pena de reclusión por un término fijo de seis (6) años. De mediar circunstancias agravantes, la pena fija establecida podrá ser aumentada hasta un máximo de doce (12) años; de mediar circunstancias atenuantes, podrá ser reducida hasta un mínimo de tres (3) años.

Será considerado como circunstancia agravante al momento de fijarse la sentencia, incurrir en cualquiera de las conductas descritas en este artículo sin la licencia o el permiso correspondiente para comprar pólvora, cuando las municiones sean de las comúnmente conocidas como armor piercing. No se constituirá delito la fabricación, venta o entrega de las municiones antes descritas para uso de la Policía y otros agentes del orden público del Gobierno de Puerto Rico o de los Estados Unidos o para el uso de las Fuerzas Armadas de los Estados.

459a. Venta de municiones a personas sin licencia; límite en el número de municiones Una persona con licencia de armas o de armero no podrá vender municiones a personas que no presenten una licencia de armas o los permisos contemplados en este capítulo. La venta de municiones se limitará exclusivamente al tipo de munición utilizada por el arma o las armas que el comprador tenga inscritas a su nombre.

Toda infracción a lo dispuesto en el primer párrafo de esta sección constituirá delito grave y será sancionada con pena de reclusión por un término fijo de cinco (5) años. De mediar circunstancias agravantes, la pena fija establecida podrá ser aumentada hasta un máximo de ocho (8) años; de mediar circunstancias atenuantes, podrá ser reducida hasta un mínimo de tres (3) años y un (1) día.

Una persona con licencia de armas, salvo las categorías de tiro al blanco o de caza, sólo podrá poseer como máximo cincuenta (50) balas por año natural por arma que posea. Si dicha persona deseare sustituir las municiones, ya sea mediante reemplazo o adquisición de nuevas municiones por haber utilizado o perdido alguna de las mismas, deberá acudir al distrito o precinto policíaco donde reside. La Policía le concederá una autorización para reemplazar las municiones manteniendo la cantidad establecida en este párrafo. En los casos donde la persona desee adquirir nuevas municiones por haber utilizado o perdido alguna de éstas, deberá informar las circunstancias en que utilizó o perdió las mismas. Para que se conceda el reemplazo de las municiones, las circunstancias en que se utilicen deberán ser actividades permitidas y legítimas al amparo de nuestro ordenamiento jurídico y lo dispuesto en este capítulo. Las municiones entregadas deberán ser decomisadas por la Policía.

Toda infracción a lo dispuesto en el párrafo anterior constituirá delito menos grave y será sancionad[a] con pena de reclusión que no excederá de seis (6) meses, pena de multa que no excederá cinco mil (5000) dólares, o ambas penas a discreción del tribunal. El tribunal, a su discreción, podrá imponer la pena de prestación de servicios en la comunidad en lugar de la pena de reclusión establecida.

459b. Compra de calibre distinto. Toda persona que, teniendo una licencia de armas valida, compre municiones de un calibre distinto a los que pueden ser utilizados en las armas de fuego inscritas a su nombre, incurrirá en delito grave y, convicta que fuere, será sancionada con pena de reclusión por un termino fijo de seis (6) anos. De mediar circunstancias agravantes, la pena fija establecida podrá ser aumentada hasta un máximo de doce (12) anos; de mediar circunstancias atenuantes, podrá ser reducida hasta un mínimo de tres (3) anos.

Subcapítulo 7. Disposiciones Finales

460. Licencias de caza Todo lo referente al licenciamiento, reglamentación y control del deporte de caza se regirá por lo dispuesto en las secs. 107 et seq. del Título 12, conocidas como la "Nueva Ley de Vida Silvestre de Puerto Rico". No obstante, el Superintendente facilitará la inscripción en el registro electrónico de las transacciones de armas y municiones de los tenedores de las licencias de caza, de conformidad a este capítulo.

460c Conversión de licencias; término; derechos; licencias en proceso de trámite

(a). Toda licencia vigente para tener y poseer un arma de fuego, de tiro al blanco, de portar o licencia de funcionario público, deberá ser convertida en una licencia de armas con su correspondiente categoría, si alguna, en conformidad a las disposiciones de este capítulo, en o antes de concluido un plazo de seis (6) meses a partir de la fecha en que comience a regir esta ley. La conversión de la licencia de caza es voluntaria por parte del tenedor de la misma, y podrá llevarse a cabo en cualquier momento de conformidad a las disposiciones de este capítulo. Las licencias de tiro al blanco poseídas por menores de veintiún (21) años serán convertidas a permisos de tiro para menores, y dicha conversión será libre de costo. Disponiéndose, que hasta tanto sean convertidas, éstas se regirán bajo las disposiciones de las leyes bajo las cuales fueron emitidas.

Se dispone que una vez que se solicite la conversión de otra licencia a licencia de armas, la licencia original no vencerá hasta que se conceda la licencia de armas o se cancele por no cualificar, en armonía con este capítulo, el concesionario que solicita la conversión.

Se dispone también que toda arma previamente inscrita bajo cualquier otra licencia, al convertirse la licencia previa en una licencia de armas, advendrán automáticamente a quedar inscritas bajo la licencia de armas.

(b) Toda solicitud de conversión de licencia ante el Superintendente, según establecido en este capitulo, deberá ser acompañada de un comprobante de rentas internas por la cantidad de cincuenta dólares ($50), mas un dólar ($1) por cada arma que el peticionario tenga inscrita legalmente.

(c) En el caso de las licencias pendientes de investigación, las cuales hayan sido solicitadas aunque aun no expedidas, por personas que habían solicitado:

(1) Licencia de tener y poseer. El Superintendente le entregara los documentos de la solicitud para que el peticionario someta su solicitud conforme a las disposiciones de este capitulo. Disponiéndose, que si el peticionario satisfizo el pago de los derechos correspondientes de conformidad con la Ley Num. 17 de 19 de enero de 1951, según enmendada, no tendrá que pagar los derechos establecidos en este capitulo.

(2) Licencia de tiro al blanco El Superintendente le entregara los documentos de la solicitud para que el peticionario someta su solicitud de licencia de armas conforme a las disposiciones de este capitulo. Disponiéndose, que el peticionario someterá un comprobante de rentas internas por la diferencia en la cantidad entre los derechos de solicitud dispuestos en este capitulo y los establecidos en la Ley Num. 75 de 13 de junio de 1953, según enmendada.

(3) Licencia de caza. Se regirá por las disposiciones de las secs. 107 et seq. del Titulo 12, conocidas como la "Nueva Ley de Vida Silvestre de Puerto Rico".

(d) Toda licencia especial expedida a una agencia de seguridad que se dedique al transporte de valores en vehículos blindados o a sus agentes, expedida al amparo de las secs. 1071 et seq. de este titulo, continuara con su validez hasta la conclusión de su vigencia, cuando deberá ser renovada de conformidad con las disposiciones de este capitulo.

[Contenido completo hasta la sesión del 2008]

RHODE ISLAND

R.I. GEN. LAWS

Title 11. Criminal Offense

Chapter 47. Weapons

11-47-1. Short title This chapter may be cited as the "Firearms Act".

11-47-2. Definitions When used in this chapter, the following words and phrases are construed as follows:

(1) "Antique firearm" is defined as that term is defined under the provisions of 18 U.S.C. § 921.

(2) "Crime of violence" means and includes any of the following crimes or an attempt to commit any of them; murder, manslaughter, rape, first or second degree sexual assault, first or second degree child molestation, kidnapping, first and second degree arson, mayhem, robbery, burglary, breaking and entering, any felony violation involving the illegal manufacture, sale, or delivery of a controlled substance, or possession with intent to manufacture, sell, or deliver a controlled substance classified in schedule I or schedule II of § 21-28-2.08, any violation of § 21-28-4.01.1 or 21-28-4.01.2 or conspiracy to commit any violation of these statutes, assault with a dangerous weapon, assault or battery involving grave bodily injury, and/or assault with intent to commit any offense punishable as a felony; upon any conviction of an offense punishable as a felony offense under § 12-29-5.

(3) "Firearm" includes any machine gun, pistol, rifle, air rifle, air pistol, "blank gun," "BB gun," or other instrument from which steel or metal projectiles are propelled, or which may readily be converted to expel a projectile, except recurve, compound, or longbows, and except instruments propelling projectiles which are designed or normally used for a primary purpose other than as a weapon. The frame or receiver of the weapon shall be construed as a firearm under the provisions of this section.

(4) "Fugitive from justice" means any person who has fled from any state, territory, the District of Columbia, or possession of the United States to avoid prosecution for a crime of violence or to avoid giving testimony in any criminal proceed-ing.

(5) "Licensing authorities" means the board of police commissioners of a city or town where the board has been instituted, the chief of police or superintendent of police of other cities and towns having a regular organized police force, and, in towns where there is no chief of police or superintendent of police, it means the town clerk who may issue licenses upon the recommendation of the town sergeant, and it also means any other person or body duly authorized by the city or town charter or by state law.

(6) "Machine gun" means any weapon which shoots, is designed to shoot, or can be readily restored to shoot, automatically more than one shot, without manual reloading, by a single function of the trigger. The term also includes the frame or receiver of any such weapon, any combination of parts designed and intended for use in converting a weapon into a machine gun, and any combination of parts from which a machine gun can be assembled if the parts are in the possession or under the control of a person.

(7) "Person" includes individual, partnership, firm, association, or corporation.

(8) "Pistol" includes any pistol or revolver, and any shotgun, rifle, or similar weapon with overall length less than twenty-six inches (26"), but does not include any pistol or revolver designed for the use of blank cartridges only.

(9) "Sawed-off rifle" means any rifle with overall length of less than twenty-six inches (26") and/or barrel length of less than sixteen inches (16").

(10) "Sawed-off shotgun" means any shotgun with overall length of less than twenty-six inches (26") and/or barrel length of less than eighteen inches (18"), or both.

(11) "Sell" includes let or hire, give, lend, and transfer, and the word "purchase" shall include hire, accept, and borrow, and the expression "purchasing" shall be construed accordingly.

11-47-5. Possession of arms by person convicted of crime of violence or who is a fugitive from justice

(a) No person who has been convicted in this state or elsewhere of a crime of violence or who is a fugitive from justice shall purchase, own, carry, transport, or have in his or her possession any firearm.

(b) Notwithstanding the provisions of subsection (a) of this section, no person convicted of an offense punishable as a felony offense under § 12-29-5 shall purchase, own, carry, transport, or have in his or her possession any firearm, for a period of two (2) years following the date of that conviction.

(c) No person who is in community confinement pursuant to the provisions of § 42-56-20.2 or who is otherwise subject to electronic surveillance or monitoring devices as a condition of parole shall purchase, carry, transport, or have in his or her possession any firearm. This subsection shall not apply to any person who has not been convicted of (or pleaded guilty or nolo contendere to) a crime of violence in a court of competent jurisdiction.

(d) Every person violating the provisions of this section shall, upon conviction, be punished by imprisonment for not less than two (2) nor more than ten (10) years; and for penalties provided in this section he or she shall not be afforded the benefit of suspension or deferment of sentence nor of probation.

11-47-5.1. Larceny of a firearm

(a) Every person who shall steal any firearm shall be deemed guilty of larceny. "Firearm", as utilized in this section only, shall not apply to an air rifle, air pistol, "blank gun," or "BB gun." Every person violating the provisions of this section shall be sentenced, upon conviction, to not less than one year nor more than ten (10) years.

(b) No person shall steal any firearm as defined in this section and then sell, lend, or transfer the firearm or firearms. Any person convicted of violating the provisions of this subsection shall be punished by imprisonment for not less than ten (10) years nor more than twenty (20) years, and the sentence shall be consecutive to any other sentence he or she may receive or is serving.

11-47-6. Mental incompetents, drug addicts, and drunkards prohibited from possession No person who is under guardianship or treatment or confinement by virtue of being a mental incompetent, or who has been adjudicated or is under treatment or confinement as a drug addict, or who has been adjudicated or is under treatment or confinement as an habitual drunkard, shall purchase, own, carry, transport, or have in his or her possession or under his or her control any firearm. Any person affected by the provisions of this section, other than a person who has been pronounced criminally insane by competent medical authority, after the lapse of a period of five (5) years from the date of being pronounced cured by competent medical authority, may, upon presentation of an affidavit issued by competent medical authority to the effect that he or she is a mentally stable person and a proper person to possess firearms, make application for the purchase of the firearm(s). Any person affected by the provisions of this section, in making application for the purchase of firearms and in executing the application, voluntarily waives his or her right to refuse or refrain from disclosing any confidential information, including, but not limited to, any information arising from the physician-patient relationship, pertinent to a determination by the proper authorities regarding the approval or disapproval of this application. Any person affected by the provisions of this section, in making application for the purchase of firearms and in executing the application, further agrees to allow the proper authorities to investigate any and all medical records of the applicant pertinent to a determination by the authorities regarding the approval or disapproval of this application. In the event that the application is approved, and if the person has no other disqualifying record, he or she will be allowed to purchase and possess firearms.

11-47-7. Possession of firearm by alien

(a) No unnaturalized foreign born person who entered the United States in violation of the laws of the United States or, having legally entered the United States in a lawful manner, but now remains in the United States in violation of the laws of the United States, shall purchase, own, carry, transport, or have in his or her possession or under his or her control any firearm.

(b) When any person is charged under this section, the law enforcement agency bringing the charge shall, prior to arraignment, notify the United States Office of Immigration and Naturalization of the charge and further notify the court, at arraignment, of the alleged status of the person so charged.

11-47-8. License or permit required for carrying pistol - Possession of machine gun

(a) No person shall, without a license or permit, issued as provided in §§ 11-47-11, 11-47-12 and 11-47-18, carry a pistol or revolver in any vehicle or conveyance or on or about his or her person whether visible or concealed, except in his or her dwelling house or place of business or on land possessed by him or her or as provided in §§ 11-47-9 and 11-47-10. The provisions of the above section shall not apply to any person who is the holder of a valid license or permit issued by the licensing authority of another state, or territory of the United States, or political subdivision thereof, allowing him or her to carry a pistol or revolver in any vehicle or conveyance or on or about his or her person whether visible or concealed, provided the person is merely transporting the firearm through the state in a vehicle or other conveyance without any intent on the part of said person to detain him or herself or remain within the state of Rhode Island. No person shall manufacture, sell, purchase, or possess a machine gun except as otherwise provided in this chapter. Every person violating the provision of this section shall, upon conviction, be punished by imprisonment for not less than one nor more than ten (10) years or by a fine up to ten thousand dollars ($10,000), or both, and except for a first conviction under this section shall not be afforded the provisions of suspension or deferment of sentence, nor a probation.

(b) No person shall have in his or her possession or under his or her control any sawed-off shotgun or sawed-off rifle as defined in § 11-47-2. Any person convicted of violating this subsection shall be punished by imprisonment for up to ten (10) years or by a fine of up to five thousand dollars ($5,000), or both.

(c) No person shall have in his or her possession or under his or her control any firearm while the person delivers, possesses with intent to deliver, or manufactures a controlled substance. Any person convicted of violating this subsection shall be punished by imprisonment for not less than two (2) years nor more than twenty (20) years; and further provided, that the sentence shall be consecutive to any sentence the person may receive for the delivery, possession with intent to deliver, or the manufacture of the controlled substance. It shall not be a defense to a violation of this subsection that a person has a license or permit to carry or possess a firearm.

11-47-9. Persons exempt from restrictions

The provisions of section 11-47-8 shall not apply to sheriffs, deputy sheriffs, the superintendent and members of the state police, members of the Rhode Island airport police department, members of the Rhode Island state marshals, Rhode Island state fire marshal, chief deputy state fire marshals, deputy state fire marshals assigned to the bomb squad, and those assigned to the investigation unit, correctional officers, all within the department of corrections, members of the city or town police force, capitol police investigators of the department of attorney general appointed pursuant to section 42-9-8.1, the witness protection coordinator for the witness protection review board as set forth in chapter 30 of title 12 and subject to the minimum qualifications of section 42-9-8.1, the director, assistant director, and other inspectors and agents at the Rhode Island state fugitive task force appointed pursuant to section 12-6-7.2, railroad police while traveling to and from official assignments or while on assignments, conservation officers, or other duly appointed law enforcement officers, nor to members of the Army, Navy, Air Force, and Marine Corps of the United States, the National Guard, or organized reserves, when on duty, nor to members of organizations by law authorized to purchase or receive firearms from the United States or this state, provided these members are at or going to or from their places of assembly or target practice, nor to officers or employees of the United States authorized by law to carry a concealed firearm, nor to any civilian guard or criminal investigator carrying sidearms or a concealed firearm in the performance of his or her official duties under the authority of the commanding officer of the military establishment in the state of Rhode Island where he or she is employed by the United States, nor to any civilian guard carrying sidearms or a concealed firearm in the performance of his or her official duties under the authority of the adjutant general where he or she is employed guarding a national guard facility, provided, that the commanding officer of the military establishment shall have on file with the attorney general of this state a list of the names and addresses of all civilian guards and criminal investigators so authorized, nor to duly authorized military organizations when on duty, nor to members when at or going to or from their customary places of assembly, nor to any individual employed in the capacity of warden, associate warden, major, captain, lieutenant, sergeant, correctional officer or investigator at any project owned or operated by a municipal detention facility corporation, including the Donald W. Wyatt Detention Facility, nor to the regular and/or ordinary transportation of pistols or revolvers as merchandise, nor to any person while transporting a pistol, or revolvers, unloaded from the place of purchase to their residence, or place of business, from their residence to their place of business or from their place of business to their residence, or to a federal firearms licensee for the purpose of sale, to or from a bona fide gunsmith, or firearms repair facility, to any police station or other location designated as a site of a bona fide "gun buy-back" program but only if said pistol or revolver is unloaded and any ammunition for said pistol or revolver is not readily or directly accessible from the passenger compartment of such vehicle while transporting same and further provided that in the case of a vehicle without a compartment separate from the passenger compartment the firearm or the ammunition shall be stored in a locked container. Persons exempted by the provisions of this section from the provisions of section 11-47-8 shall have the right to carry concealed firearms everywhere within this state; provided, that this shall not be construed as giving the right to carry concealed firearms to a person transporting firearms as merchandise or as household or business goods.

11-47-9.1. Additional exemptions

The provisions of §§ 11-47-8 and 11-47-11 shall not apply to members of the state police, members of city or town police forces, and members of the Rhode Island airport police department. Persons exempted by the provisions of this section from the provisions of § 11-47-8 shall have the right to carry concealed firearms everywhere within this state; provided, that this shall not be construed as giving the right to carry concealed firearms to a person transporting firearms as merchandise or as household or business goods.

11-47-19. Machine gun manufacturers' licenses or permits The attorney general may issue to any person, firm, or corporation, engaged in manufacturing in this state, a license or permit to manufacture and sell machine guns and any or all machine gun parts under any regulations that the attorney general may prescribe.

11-47-20. Sale or possession of silencers It shall be unlawful within this state to manufacture, sell, purchase, or possess any muffler, silencer, or device for deadening or muffling the sound of a firearm when discharged. Violations of this section shall be punished by imprisonment for not less than one year and one day.

11-47-20.1. Armor-piercing bullets It shall be unlawful within this state for any person to import, manufacture, sell, purchase, or otherwise transfer any bullets which have steel inner cores or cores of equivalent hardness and truncated cones and which are designed for use in pistols as armor-piercing or metal-piercing bullets. Any person who violates the provisions of this section shall be punished by imprisonment for not more than three (3) years, or a fine of not more than five thousand dollars ($5,000), or both. This section shall not apply to the purchase of those bullets by the Rhode Island state police, by any city or town police department of the state of Rhode Island, or by the department of environmental management for display as a part of a firearms training course under its auspices.

11-47-21. Restrictions on possession or carrying of explosives or noxious substances

Any person, except a member of the state police, the sheriff or the sheriff 's deputies, a member of the police force of any city or town, or a member of the Army, Navy, Air Force, or Marine Corps of the United States, or of the National Guard or organized reserves when on duty, who possesses, or carries on or about his or her person or in a vehicle, a bomb or bombshell, except for blasting or other commercial use, or who, with intent to use it unlawfully against the person or property of another, possesses or carries any explosive substance, or any noxious liquid, gas, or substance, shall be guilty of a violation of this chapter and punished as provided in § 11-47-26.

11-47-22. Forfeiture and destruction of unlawful firearms

(a) No property right shall exist in any firearm unlawfully possessed, carried, or used, and all unlawful firearms are hereby declared to be nuisances and forfeited to the state.

(b) When a firearm is lawfully seized, confiscated from or turned in by any person, it shall be placed in the custody of the superintendent of state police or the chief of police in the city or town in which it was seized, confiscated or turned in to. The officer who takes custody of the firearm shall promptly ascertain, using available record keeping systems, including, but not limited to, the National Crime Information Center, whether the firearm has been reported stolen and if stolen shall notify the reporting law enforcement agency of the recovery of said firearm. If the police department in the city or town in which the firearm was seized or confiscated has not been notified by a justice of the superior court or the attorney general that the firearm is necessary as evidence in a criminal or civil matter, it shall be returned to the lawful owner. However, any owner of a firearm who shall knowingly fail to report the loss or theft of the firearm to the proper law enforcement authorities shall not be entitled to its return.

(c) If a firearm is found not to be stolen and the owner cannot be readily ascertained within ninety (90) days subsequent to the seizure or confiscation of said firearm and the firearm is no longer necessary as evidence in a criminal or civil matter, the police department having custody of the firearm shall have the option of either

(1) destroying said firearm by rendering it permanently and irretrievably inoperable;

(2) transferring custody of said firearms to the state crime laboratory for the purpose of criminal investigation; or

(3) holding an auction of those firearms seized, confiscated or turned in to said police department; provided, however, any firearms to be transferred at auction shall be limited to antique firearms or curios or relics. For purposes of this section, an antique firearm shall be defined as any firearm (including any firearm with a matchlock, flintlock, percussion cap, or similar type of ignition system) manufactured on or before 1899, and any replica of such firearm if such replica is not designed or redesigned for using rim-fire or conventional center fire fixed ammunition, or uses rim-fire or conventional center fire fixed ammunition which is no longer manufactured in the United States and which is not readily available in the ordinary channels of commercial trade. Curios or relics shall be defined as firearms which are of special interest to collectors by reason of some quality other than is associated with firearms intended for sporting use or as offensive or defensive weapons. To be recognized as curios or relics, firearms must be manufactured at least fifty (50) years prior to the current date, but not including replicas thereof, and firearms which derive a substantial part of their monetary value from the fact that they are novel, rare, bizarre, or because of their association with some historical figure, period, or event.

(d) In the event that an auction is held, bidders shall be limited to bona fide holders of a valid Federal Firearms License for Retail Sale or a Federal Firearms Collector License. Any auction shall be advertised at least once a week for a period of three (3) weeks preceding the date of the auction in a newspaper of general circulation, said notice clearly stating the time, location and terms of said auction. All funds realized shall be used to purchase and provide necessary safety equipment, including, but not limited to, bulletproof vests, for the police department holding the auction and shall not revert to any general fund of the state, city or town, as the case may be.

(e) All firearms received by any police department in any manner shall be entered in the department's permanent records and listed by make, model, caliber and serial number and the manner in which said firearm was disposed of and, if by auction, the name and federal license number of the buyer. A copy of said record shall be forwarded to the office of the Rhode Island Attorney General and the Bureau of Alcohol, Tobacco and Firearms of the United States Treasury Department on at least an annual basis.

11-47-23. False information in securing firearm or license No person shall, in purchasing or otherwise securing delivery of a shotgun, rifle, pistol, or revolver, or in applying for a license to carry it, give false information or offer false evidence of his or her identity. Violation of the provisions of this section may be punished by imprisonment for not more than five (5) years.

11-47-24. Alteration of marks of identification on firearms No person shall change, alter, remove, or obliterate the name of the maker, model, manufacturer's number, or other mark of identification on any firearm. Possession of any firearm upon which any of these marks shall have been changed, altered, removed, or obliterated shall be prima facie evidence that the possessor has changed, altered, removed, or obliterated it. Violation of the provisions of this section may be punished by imprisonment for not more than five (5) years.

11-47-25. Antique firearms and collections This chapter shall not apply to antique firearms unsuitable for use, nor to collections of firearms utilized and maintained for educational, scientific, or any similar purpose without intent to use the firearms.

11-47-26. Penalties for violations than a machine gun, shotgun, rifle, or pistol, from which steel or other metal projectiles are propelled, shall for the first offense be punished by a fine of not more than Unless otherwise specified, any violation of any provi-sion of this chapter shall be punished by a fine of not more than one thousand dollars ($1,000), or imprisonment for not more than five (5) years, or both; provided, that a violation of any of the provisions of §§ 11-47-1 – 11-47-34 with relation to air rifle, air pistol, "blank gun," "BB gun," or other instrument other fifty dollars ($50.00) and for the second or any subsequent offense by a fine of not more than one hundred dollars ($100), or by imprisonment for not more than thirty (30) days, or both; and provided, further, that the provisions of chapter 1 of title 14 shall apply in the case of any person under the age of eighteen (18) years.

11-47-30. Sale, transfer or delivery of firearms to minors

(a) It shall be unlawful within this state for any person to sell, transfer, give, convey, or cause to be sold, transferred, given or conveyed any firearm to any person under eighteen (18) years of age, when the person knows or has reason to know that the recipient is under eighteen (18) years of age, except for the limited purposes set forth in §§ 11-47-33 and 11-47-34 and with the prior approval or consent of the parent or legal guardian of the minor.

(b) Every person violating this section shall be punished, upon conviction, by imprisonment for not less than ten (10) years and not more than twenty (20) years. The prohibitions of this section shall not apply to any federally and state licensed retail dealer who makes reasonable efforts to verify a purchaser's age and shall not apply to the sale of an air rifle, air pistol, "blank gun" or "BB gun."

11-47-31. Sale, transfer or delivery of ammunition to minors

(a) It shall be unlawful within this state for any person to sell, transfer, give, convey, or cause to be sold, transferred, given or conveyed any ammunition, including any priming charge of powder, propelling charge of powder, or any form of missile or projectile to be ejected from a firearm to any person under eighteen (18) years of age when the person knows or has reason to know that the recipient is under eighteen (18) years of age, except for the limited purposes set forth in §§ 11-47-33 and 11-47-34 and with the prior approval or consent of the parent or legal guardian of the minor.

(b) Every person violating this section shall be punished, upon conviction, by imprisonment for a term not to exceed ten (10) years. The prohibitions of this section shall not apply to any federally and state licensed retail dealer who makes reasonable efforts to verify a purchaser's age and shall not apply to the sale of ammunition for an air rifle, air pistol, "blank gun" or "BB gun."

11-47-32. Possession of ammunition by minor Except as provided in § 11-47-33, it shall be unlawful within this state for any person under eighteen (18) years of age to possess and use ammunition, including any priming charge of powder, propelling charge of powder, or any form of missile or projectile to be ejected from a firearm.

11-47-33. Possession of firearms by minors

(a) It shall be unlawful within this state for any person under eighteen (18) years of age to possess and use any firearm unless he or she shall hold a permit as provided in § 11-47-34, and unless the person is in the presence of a parent or guardian or supervising adult at any regular and recognized camp or rifle range approved by the Rhode Island state police or by the chief of police of the city or town in which the camp or rifle range is located; provided, that this provision shall not apply to minors engaged in lawful hunting activity under the supervision of a parent or guardian or qualified adult, minors participating in Reserve Officer Training Corps programs, ceremonial parade activities, competitive and target shooting, participants in state militia activities and minors participating in a basic firearms education program; provided, further, that a person under eighteen (18) years of age may carry a firearm, unloaded, in a suitable case to and from his or her home and the camp or range and from the camp or range to other camp or range when accompanied by a parent, guardian or supervising adult.

(b) For purposes of this section only, "qualified adult" means any person twenty-one (21) years of age or older and permitted by law to possess and use the firearm.

11-47-34. Firearms permits to minors The Rhode Island state police or the chief of police of the city or town in which the person resides shall issue permits to any person under eighteen (18) years of age only upon satisfactory proof of being engaged in a course of training in the use of firearms at a regular and recognized camp or rifle range, and provided that the person has the written consent of a parent or guardian. Valid membership cards of junior gun clubs or of junior divisions of senior gun clubs incorporated in the state of Rhode Island shall be prima facie evidence of the person under eighteen (18) years of age being engaged in a course of training in the use of firearms at a regular and recognized camp or rifle range.

11-47-35. Sale of concealable weapons – Safety courses and tests - Review board – Issuance of permits to certain government officers

(a)(1) No person shall deliver a pistol or revolver to a purchaser until seven (7) days shall have elapsed from twelve o'clock (12:00) noon of the day following the day of application for the purchase, and when delivered, the pistol or revolver shall be unloaded and securely wrapped, with the bill of sale to be enclosed within the wrapper with the pistol or revolver. Any citizen of the United States and/or lawful resident of this state who is twenty-one (21) years of age or older, and any nonresident member of the armed forces of the United States who is stationed in this state and who is twenty-one (21) years of age or older, may upon application purchase or acquire a pistol or revolver. At the time of applying for the purchase of a concealable firearm, the purchaser shall: (i) complete and sign in triplicate and deliver to the person selling the pistol or revolver the application form described in this section, and in no case shall it contain the serial number of the pistol or revolver; and (ii) shall present to the person selling the pistol or revolver a pistol/revolver safety certificate issued by the department of environmental management. The certificate shall be retained in the possession of the buyer. The pistol/revolver safety certificate shall certify that the purchaser has completed a basic pistol/revolver safety course as shall be administered by the department of environmental management.

(Face of application form)

Application to Purchase Pistol or Revolver

Date . . . . Hour A.M. P.M.
Name
(Street and number) (City or town) (State)
Date of Birth . . . . .Place of Birth
Height . . . . . Weight . . . . . Color hair
Color eyes
Scars
Tattoos
Other identifying marks
Are you a citizen of the United States
Are you a citizen of Rhode Island
How long
Where stationed
(Armed Forces only)
Have you ever been convicted of a crime of violence
(See § 11-47-2)
Have you ever been adjudicated or under confinement as addicted to a controlled substance
Have you ever been adjudicated or under confinement for alcoholism
Have you ever been confined or treated for mental illness
From whom is pistol or revolver being purchased
Seller's address
Seller's signature
Applicant's signature
(See § 11-47-23 for penalty for false information on this application)

(Reverse side of application form)

AFFIDAVIT: I certify that I have read and am familiar with the provisions of §§ 11-47-1 – 11-47-55, inclusive, of the general laws of the State of Rhode Island and Providence Plantations, and that I am aware of the penalties for violation of the provisions of the cited sections. I further certify that I have completed the required basic pistol/revolver safety course.
Signed
(over)
County of
State of Rhode Island
Subscribed and sworn before me this . . day of . . . . A.D. 20. .

Notary Public

(2) The person selling the pistol or revolver shall on the date of application sign and forward by registered mail or by delivery in person the original and duplicate copies of the application to the superintendent of the Rhode Island state police or the chief of police in the city or town in which the person has his or her residence or place of business. The superintendent of the Rhode Island state police or the chief of police in the city or town in which the person has his or her residence or place of business shall mark or stamp the original copy of the application form with the date and the time of receipt and return it by the most expeditious means to the person who is selling the pistol or revolver. The triplicate copy duly signed by the person who is selling the pistol or revolver shall within seven (7) days be sent by him or her by registered mail to the attorney general. The person who is selling the pistol or revolver shall retain the original copy duly receipted by the police authority to whom sent or delivered for a period of six (6) years with other records of the sale. It shall be the duty of the police authority to whom the duplicate copy of the application form is sent or delivered to make a background check of the applicant to ascertain whether he or she falls under the provisions of § 11-47-5, 11-47-6, 11-47-7, or 11-47-23. If, after the lapse of seven (7) days from twelve o'clock (12:00) noon of the day following application, no disqualifying information has been received from the investigating police authority by the person who is selling the pistol or revolver, he or she will deliver the firearm applied for to the applicant. Upon the finding of no disqualifying information under the provisions of the above cited sections of this chapter, and in no case later than thirty (30) days after the date of application, the duplicate and triplicate copies of the application will be destroyed. Retention of the duplicate and triplicate copies in violation of this section or any unauthorized use of the information contained in the copies by a person or agency shall be punishable by a fine of not more than one thousand dollars ($1,000). The provisions of this section shall not apply to bona fide sales at wholesale to duly licensed retail dealers, nor to purchases by retail dealers duly licensed under the provisions of § 11-47-39.

(b)(1) The department of environmental management shall establish the basic pistol/revolver safety course required by this section. The safety course shall consist of not less than two (2) hours of instruction in the safe use and handling of pistols and revolvers and the course shall be available to buyers continually throughout the year at convenient times and places but at least monthly at locations throughout the state, or more frequently as required. Proficiency in the use of pistols or revolvers shall not be prerequisite to the issuance of the safety certificate. No person shall be required to complete the course more than once; provided, that any person completing the course who is unable to produce the safety certificate issued by the department of environmental management shall be required to take the course again unless the person provides evidence to the department that he or she has successfully completed the course.

(2) The administration of the basic pistol/revolver safety course required by this section shall not exceed the cost of thirty-five thousand dollars ($35,000) in any fiscal year.

(c) Proof of passage of the department of environmental management's basic hunter safety course will be equivalent to the pistol/revolver safety certificate mandated by this section.

(d) Any person who has reason to believe that he or she does not need the required handgun safety course may apply by any written means to the department of environmental management to take an objective test on the subject of matter of the handgun safety course. The test shall be prepared, as well as an instruction manual upon which the test shall be based, by the department. The manual shall be made available by any means to the applicant who may, within the time limits for application, take the objective test at the department or at any location where the handgun safety course is being given. Any person receiving a passing grade on the test shall be issued a pistol/revolver safety certificate by the department. ...

(f) The following persons shall be issued basic pistol/revolver permits by the department of environmental management: sheriffs, deputy sheriffs, the superintendent and members of the state police, prison or jail wardens or their deputies, members of the city or town police force, members of the park police, conservation officers, members of the airport police, and officers of the United States government authorized by law to carry a concealed firearm and, at the discretion of the department of environmental management, any person who can satisfactorily establish that he or she formerly held one of these offices or were so authorized.

(g) Any person who is serving in the Army, Navy, Air Force, Marine Corps or Coast Guard on active duty shall not be required to obtain a basic pistol/revolver safety certificate or basic pistol/revolver permit under this section so long as he or she remains on active duty.

(h) Any person who is serving in the active reserve components of the Army, Navy, Air Force, Marine Corps or Coast Guard, or any person in an active duty paid status in the Rhode Island National Guard, shall not be required to obtain a basic pistol/revolver safety certificate under this section so long as he or she remains in active status.

11-47-35.1. Persons exempt from § 11-47-35 The provisions of § 11-47-35 shall not apply to full-time members of the state police, or full-time members of the state marshal's office, or full-time members of city or town police departments, or state marshals or correctional officers or persons licensed under § 11-47-11.

11-47-35.2. Sale of rifles/shotguns

(a) No person shall deliver a rifle or shotgun to a purchaser until seven (7) days shall have elapsed from twelve o'clock (12:00) noon of the day following the day of application for the purchase, and when delivered, the rifle or shotgun shall be unloaded and securely wrapped, with the bill of sale for it to be enclosed within the wrapper with the rifle or shotgun. Any citizen of the United States and/or lawful resident of this state who is eighteen (18) years of age or older, and any non-resident member of the armed forces of the United States who is stationed in this state and who is eighteen (18) years of age or older, may, upon application, purchase or acquire a rifle or shotgun. At the time of applying for the purchase of a shotgun or rifle the purchaser shall complete and sign in triplicate and deliver to the seller the application form described in this section, and in no case shall it contain the serial number of the rifle or shotgun.

(Face of application form)

Application to Purchase Shotgun or Rifle

Date . . . . Hour A.M. P.M.

Name

Address

(Street and number) (City or town) (State)

Date of Birth . . . . .Place of Birth

Height . . . . . Weight . . . . . Color hair

Color eyes

Scars

Tattoos

Other identifying marks

Are you a citizen of the United States

Are you a citizen of Rhode Island

How long

Where stationed

(Armed Forces only)

Have you ever been convicted of a crime of violence

(See § 11-47-2 General Laws of Rhode Island)

Have you ever been adjudicated or under confinement as addicted to a controlled substance

Have you ever been adjudicated or under confinement for alcoholism

Have you ever been confined or treated for mental illness

From whom is shotgun or rifle being purchased

Seller's address

Seller's signature

Applicant's signature

(See § 11-47-23 for penalty for false information on this application)

(Reverse side of application form)

AFFIDAVIT: I certify that I have read and am familiar with the provisions of §§ 11-47-1 – 11-47-59 of the general laws of the State of Rhode Island and Providence Plantations, and that I am aware of the penalties for violation of the provisions of the cited sections.

Signed

County of

State of Rhode Island

Subscribed and sworn before me this . . day of . . . . A.D. 20. . .

Notary Public

(b) The person who is selling the rifle or shotgun shall, on the date of application, sign and forward by registered mail or by delivery in person, the original and duplicate copies of the application to the superintendent of the Rhode Island state police or the chief of police in the city or town in which the seller has his or her residence or place of business. The superintendent of the Rhode Island state police or the chief of police in the city or town in which the person has his or her residence or place of business shall mark or stamp the original copy of the application form with the date and time of receipt and return it by the most expeditious means to the seller. The triplicate copy duly signed by the seller shall within seven (7) days be sent by him or her by registered mail to the attorney general. The person shall retain the original copy duly receipted by the police authority to whom sent or delivered for a period of six (6) years with other records of the sale. It shall be the duty of the police authority to whom the duplicate copy of the application form is sent or delivered to make a background check of the applicant to ascertain whether he or she falls under the provisions of § 11-47-5, 11-47-6, 11-47-7, or 11-47-23. If, after the lapse of seven (7) days from twelve o'clock (12:00) noon of the day following application, no disqualifying information has been received from the investigating police authority by the person who is selling the rifle or shotgun, he or she will deliver the firearm applied for to the applicant. Upon the finding of no disqualifying information under the provisions of the above cited sections of this chapter, and in no case later than thirty (30) days after the date of application, the duplicate and triplicate copies of the application will be destroyed. Retention of the duplicate and triplicate copies in violation of this chapter or any unauthorized use of the information contained in them by a person or agency shall be punishable by a fine of not more than one thousand dollars ($1,000). The provisions of this section shall not apply to bona fide sales at wholesale to duly licensed retail dealers, nor to purchases by retail dealers duly licensed under the provisions of § 11-47-39.

(c) The provisions of this section shall not apply to full-time members of the state police, full-time members of city or town police departments, persons licensed under §§ 11-47-9 and 11-47-11, or to sales of air rifles or " BB guns" or to sales of antique firearms as defined in § 11-47-2.

11-47-36. Purchase of concealable fire-arms from out of state dealers No citizen of this state shall purchase any concealable fire-arm outside of the state of Rhode Island unless he or she has duly executed the application form prescribed in § 11-47-35, the application form to be obtained by the purchaser from the city or town clerk of the city or town in which he or she resides or has his place of business. The original and duplicate copies of the application shall be delivered in person, duly executed, by the purchaser to the superintendent of the Rhode Island state police or to the chief of police of the city or town in which the purchaser resides. The purchaser shall send the triplicate copy of the application by registered mail to the attorney general within twenty-four (24) hours of the time of filing with the appropriate police authority. It shall be the duty of the police authority to whom the original and duplicate copies of the application are delivered to check the applicant's record to ascertain whether he or she falls under the provisions of § 11-47-5, 11-47-6, 11-47-7, or 11-47-23. If, after the lapse of seventy-two (72) hours from twelve o'clock (12:00) noon of the day following the date of application, no disqualifying record has been found by the investigating police authority, the original and duplicate copies of the application marked or stamped "approved" and signed by the investigating police authority will be returned to the applicant by the most expeditious means. The approved duplicate copy of the application shall be sent by the purchaser to the out-of-state dealer as proof of lawful purchase, and the original shall be retained by the purchaser along with the bill of sale for the firearm purchased for a period of six (6) years as proof of lawful purchase. The triplicate copy of the application shall be retained by the attorney general for a period which shall in no case exceed ninety (90) days, provided that no evidence of the nature as would disqualify the applicant has been found.

11-47-37. Sale to minors and others forbidden No person shall sell a pistol or revolver to any person under the age of twenty-one (21) or to one who he or she has reasonable cause to believe falls under the provisions of §§ 11-47-5, 11-47-6, 11-47-7, or 11-47-23.

11-47-38. Dealers to be licensed No retail dealer shall sell or otherwise transfer, or expose for sale or transfer, or have in his or her possession with intent to sell or otherwise transfer, any pistol, revolver, or other firearm without being licensed as provided in this chapter.

11-47-39. Issuance and conditions of dealer's license The duly constituted licensing authorities of any city, town, or political subdivision of this state may grant licenses in form prescribed by the attorney general effective for not more than one year from date of issue permitting the licensee to sell pistols and revolvers at retail within this state, subject to the following conditions in addition to those specified in §§ 11-47-35 and 11-47-36, for breach of any of which the license shall be forfeited and the licensee subject to punishment as provided in this chapter:

(1) The business shall be carried on only in the building designated in the license.

(2) The license or a copy of it, certified by the issuing authority, shall be displayed on the premises where it can easily be read.

(3) No pistol or revolver may be sold in violation of any provision of this chapter, nor shall a pistol or revolver be sold under any circumstances unless the purchaser is personally known to the seller or shall present clear evidence of his or her identity.

(4) The fee for issuing the license shall be five dollars ($5.00). The fee charged for the issuing of the license shall be applied for the use and benefit of the city or town.

11-47-40. Register of sales of firearms - Display of firearms

(a) Every person, firm, or corporation selling a pistol, revolver, or other firearm whether the seller is a retail dealer, pawnbroker, or otherwise shall keep a register in which shall be entered at the time of sale the date of sale, name, age, and residence of every purchaser of the a pistol, revolver, or other firearm, together with the caliber, make, model, manufacturer's number, or other mark of identification on the pistol, revolver, or other firearm. Every person, firm, or corporation who shall fail to keep a register and to enter the acts required by this section shall, upon conviction, be punished as provided in this chapter. The register shall be open at all reasonable hours for the mandatory monthly inspection of licensed firearm dealers to be conducted by state and/or local police officials.

(b) This section shall not apply to wholesale dealers' bona fide sales at wholesale to duly licensed retail dealers. It shall be unlawful for any person, firm, or corporation dealing in firearms to display any pistol, revolver, or imitation, or any firearm of a size which may be concealed upon the person, or placard advertising the sale of one, in any part of the premises of the person, firm, or corporation where it can be readily seen from the outside. "Firearm" as utilized in this section only does not apply to an air rifle, air pistol, "blank gun," or "BB gun."

11-47-41. Government firearm registration prohibited No government agency of this state or its political subdivisions shall keep or cause to be kept any list or register of privately owned firearms or any list or register of the owners of those firearms; provided, that the provisions of this section shall not apply to firearms which have been used in committing any crime of violence, nor to any person who has been convicted of a crime of violence.

11-47-42. Weapons other than firearms prohibited

(a)(1) No person shall carry or possess or attempt to use against another any instrument or weapon of the kind commonly known as a blackjack, slingshot, billy, sandclub, sandbag, metal knuckles, slap glove, bludgeon, stun-gun, or the so called "Kung-Fu" weapons, nor shall any person, with intent to use unlawfully against another, carry or possess a dagger, dirk, stiletto, sword-in-cane, bowie knife, or other similar weapon designed to cut and stab another, nor shall any person wear or carry concealed upon his person, any of the above-mentioned instruments or weapons, or any razor, or knife of any description having a blade of more than three (3) inches in length measuring from the end of the handle where the blade is attached to the end of the blade, or other weapon of like kind or description. Any person violating the provisions of this subsection shall be punished by a fine of not more than one thousand dollars ($1,000) or by imprisonment for not more than one year, or both, and the weapon so found shall be confiscated.

(2) Any person violating the provisions of this subsection while he or she is incarcerated within the confines of the adult correctional institutions shall be punished by a fine of not less than one thousand dollars ($1,000) nor more than three thousand dollars ($3,000), or by imprisonment for not less than one year nor more than five (5) years, or both, and the weapon so found shall be confiscated.

(b) No person shall sell to a person under eighteen (18) years of age, without the written authorization of the minor's parent or legal guardian, any stink bomb, blackjack, slingshot, bill, sandclub, sandbag, metal knuckles, slap glove, bludgeon, stungun, paint ball gun, so called "kung-fu" weapons, dagger, dirk, stiletto, sword-in-cane, bowie knife, razor, or knife of any description having a blade of more than three inches (3") in length as described in subsection (a) of this section, or any multi-pronged star with sharpened edges designed to be used as a weapon and commonly known as a Chinese throwing star, except that an individual who is actually engaged in the instruction of martial arts and licensed under § 5-43-1 may carry and possess any multi-pronged star with sharpened edges for the sole purpose of instructional use. Any person violating the provisions of this subsection shall be punished by a fine of not less than one thousand dollars ($1,000) nor more than three thousand dollars ($3,000), or by imprisonment for not less than one year nor more than five (5) years, or both, and the weapons so found shall be confiscated.

11-47-43. Collectors and police officers exempt from section 11-47-42 The provisions of § 11-47-42, so far as they forbid the possession of certain instruments or weapons, shall not apply to any person who possesses or is making a collection of the weapons as curios or for educational, professional, scientific, or any other lawful purpose, without intent to use the instrument or weapon unlawfully. Nor shall the provisions of § 11-47-42, so far as they relate to the possession or carrying of any billy, apply to sheriffs, constables, police, or other officers or guards whose duties require them to arrest or to keep and guard prisoners or property, nor to any person summoned by those officers to aid them in the discharge of their duties while actually engaged in their duties.

11-47-46. Sections 11-47-42 - 11-47-45 inapplicable to firearms The provisions of sections 11-47-42 - 11-47-45 shall not be construed to relate to the possession and carrying of air guns, pistols or firearms of any description.

11-47-47. Display of weapons No person, firm, or corporation shall display in a place of business by means of a window display any pistol, revolver, or other firearm, as defined in § 11-47-2, or any dagger, dirk, bowie knife, stiletto, metal knuckles, or blackjack. However, dealers in sporting goods may include in a window display pistols or revolvers upon a permit issued by the chief of police or town sergeant of any city or town. Any person, firm, or corporation violating the provisions of this section shall be punished by a fine not exceeding twenty-five dollars ($25.00) for the first offense and one hundred dollars ($100) for every subsequent offense.

11-47-48.1. Report of lost or stolen weapons Every person who owns a firearm shall report the loss or theft of their firearm to the local law enforcement agency within twenty-four (24) hours of the discovery of the loss or theft. Whoever knowingly violates this section shall be punished by a fine of not less than fifty dollars ($50.00) nor more than one hundred dollars ($100).

11-47-51. Loaded weapons in vehicles It is unlawful for any person to have in his or her possession a loaded rifle or loaded shotgun or a rifle or shotgun from the magazine of which all shells and cartridges have not been removed in or on any vehicle or conveyance or its attachments while upon or along any public highway, road, lane, or trail within this state; provided, that the provisions of this section shall not apply to sheriffs, deputy sheriffs, the superintendent and members of the state police, prison or jail wardens or their deputies, members of the city or town police force, investigators of the department of attorney general appointed pursuant to § 42-9-8.1, the director, assistant director and other inspectors and agents at the Rhode Island state fugitive task force appointed pursuant to § 12-6-7.2, nor to other duly appointed law enforcement officers, including conservation officers, nor to members of the Army, Navy, Air force, or Marine Corps of the United States, or the National Guard or organized reserves, when on duty, nor to officers or employees of the United States authorized by law to carry a concealed firearm, nor to any civilian guard or criminal investigator carrying sidearms or a concealed firearm in the performance of his or her official duties under the authority of the commanding officer of the military establishment in the state of Rhode Island where he or she is employed by the United States.

11-47-52. Carrying of weapon while under the influence of liquor or drugs It is unlawful to carry or transport any firearm in this state when intoxicated or under the influence of intoxicating liquor or narcotic drugs.

11-47-55. Enforcement of chapter Sheriffs, deputy sheriffs, the superintendent and members of the state police, members of the city or town police force, or other duly appointed law enforcement officers, including conservation officers, shall have the power to enforce the provisions of this chapter.

11-47-58. Firearms - State preemption The control of firearms, ammunition, or their component parts regarding their ownership, possession, transportation, carrying, transfer, sale, purchase, purchase delay, licensing, registration, and taxation shall rest solely with the state, except as otherwise provided in this chapter.

11-47-60. Possession of firearms on school grounds

(a)(1) No person shall have in his or her possession any firearm or other weapons on school grounds.

(2) For the purposes of this section, "school grounds" means the property of a public or private elementary or secondary school or in those portions of any building, stadium, or other structure on school grounds which were, at the time of the violation, being used for an activity sponsored by or through a school in this state or while riding school provided transportation.

(3) Every person violating the provisions of this section shall, upon conviction, be sentenced to imprisonment for not less than one year nor more than five (5) years, or shall be fined not less than five hundred dollars ($500) nor more than five thousand dollars ($5,000).

(4) Any juvenile adjudicated delinquent pursuant to this statute shall, in addition to whatever other penalties are imposed by the family court, lose his or her license to operate a motor vehicle for up to six (6) months. If the juvenile has not yet obtained the necessary age to obtain a license, the court may impose as part of its sentence a delay in his or her right to obtain the license when eligible to do so, for a period of up to six (6) months.

(b) The provisions of this section shall not apply to any person who shall be exempt pursuant to the provisions of §§ 11-47-9, 11-47-11, and 11-47-18 or to the following activities when the activities are officially recognized and sanctioned by the educational institution:

(1) Firearm instruction and/or safety courses;

(2) Government-sponsored military-related programs such as ROTC;

(3) Interscholastic shooting and/or marksmanship events;

(4) Military history and firearms collection courses and/or programs; and

(5) The use of blank guns in theatrical and/or athletic events.

(c) The provisions of this section shall not apply to colleges, universities, or junior colleges.

11-47-60.1. Safe storage

(a) Nothing in this section shall be construed to reduce or limit any existing right to purchase and own firearms and/or ammunition or to provide authority to any state or local agency to infringe upon the privacy of any family, home or business except by lawful warrant.

(b) A person who stores or leaves on premises under his or her control a loaded firearm and who knows or reasonably should know that a child is likely to gain access to the firearm without the permission of the child's parent or guardian, and the child obtains access to the firearm and causes injury to himself or herself or any other person with the firearm, is guilty of the crime of criminal storage of a firearm and, upon conviction, shall be fined not more than one thousand dollars ($1,000) or imprisoned for not more than one year, or both. For purposes of this section, a "child" is defined as any person who has not attained the age of sixteen (16) years.

(c) The provisions of subsection (b) of this section shall not apply whenever any of the following occurs:

(1) The child obtains the firearm as a result of an illegal entry of any premises by any person or an illegal taking of the firearm from the premises of the owner without permission of the owner;

(2) The firearm is kept in a locked container or in a location which a reasonable person would believe to be secured;

(3) The firearm is carried on the person or within such a close proximity so that the individual can readily retrieve and use the firearm as if carried on the person;

(4) The firearm is locked with a locking device;

(5) The child obtains or obtains and discharges the firearm in a lawful act of self-defense or defense of another person;

(6) The person who keeps a loaded firearm on any premises which is under his or her custody or control has no reasonable expectations, based on objective facts and circumstances, that a child is likely to be present on the premises.

(d) If the person who allegedly violated this section is the parent or guardian of a child who is injured or who dies as the result of an accidental shooting, the attorney general's department shall consider among other factors, the impact of the injury or death on the person who has allegedly violated this section when deciding whether to prosecute an alleged violation.

(2) It is the intent of the general assembly that a parent or guardian of a child who is injured or who dies of an accidental shooting shall be prosecuted only in those instances in which the parent or guardian behaved in a grossly negligent manner.

11-47-60.2. Possession of weapons on school grounds – Notification

(a) If a student is found to be carrying a weapon, as defined in § 11-47-42, a firearm or replica of a firearm, or commits an aggravated assault on school grounds as defined in § 11-47-60, the principal or designee shall immediately notify the student's parents and the local police and turn the weapon over, if any, to the local enforcement agency.

(b) Any person who has reasonable cause to know that any person is in violation of this statute shall notify the principal or designee. The principal or designee shall immediately notify the student's parents and the local police. Any person acting in good faith who makes a report under this section shall have immunity from any civil liability that might otherwise be incurred or imposed as a result of making the report.

(c) School superintendents shall receive notice from the clerk of the family court regarding the disposition of all cases involving juveniles from their school districts adjudged pursuant to this statute. This information shall remain confidential and be shared with school officials who deal directly with the student.

(d) The provisions of this section should not apply to the following activities when the activities are officially recognized and sanctioned by the educational institution:

(1) Firearm instructed and/or safety course;

(2) Government-sponsored military-related programs such as ROTC;

(3) Interscholastic shooting and/or marksmanship events;

(4) Military history and firearms collection courses and/or programs; and

(5) The use of blank guns in theatrical and/or athletic events.

(e) The provisions of this section shall not apply to colleges, universities or junior colleges.

11-47-60.3. Trigger lock required No licensed retail dealer shall deliver any pistol to any purchaser without providing a trigger lock or other safety device designed to prevent an unauthorized user from operating the pistol.

[Current through January 2008 Session]

Title 16. Crimes and Offenses

Chapter 23. Offenses Involving Weapons

Article 1. Pistols

16-23-10. Definitions. When used in this article:

(1) "Handgun" means any firearm designed to expel a projectile and designed to be fired from the hand, but shall not include any firearm gen-erally recognized or classified as an antique, curiosity, or collector's item, or any that does not fire fixed cartridges.

(2) "Dealer" means any person engaged in the business of selling firearms at retail or any person who is a pawnbroker.

(3) "Crime of violence" means murder, man-slaughter (except negligent manslaughter arising out of traffic accidents), rape, mayhem, kidnap-ping, burglary, robbery, housebreaking, assault with intent to kill, commit rape, or rob, assault with a dangerous weapon, or assault with intent to commit any offense punishable by imprison-ment for more than one year.

(4) "Fugitive from justice" means any person who has fled from or is fleeing from any law en-forcement officer to avoid prosecution or impris-onment for a crime of violence.

(5) "Subversive organization" means any group, committee, club, league, society, associ-ation or combination of individuals the purpose of which, or one of the purposes of which, is the establishment, control, conduct, seizure or over-throw of the government of the United States or any state or political subdivision thereof, by the use of force, violence, espionage, sabotage, or threats or attempts of any of the foregoing.

(6) "Conviction" as used herein shall include pleas of guilty, pleas of nolo contendere and for-feiture of bail.

(7) "Division" means the State Law Enforce-ment Division.

(8) The terms "purchase" or "sell" means to knowingly buy, offer to buy, receive, lease, rent, barter, exchange, pawn or accept in pawn.

(9) "Person" means any individual, corpora-tion, company, association, firm, partnership, so-ciety or joint stock company.

(10) "Luggage compartment" means the trunk of a motor vehicle which has a trunk; however, with respect to a motor vehicle which does not have a trunk, the term "luggage compartment" refers to the area of the motor vehicle in which the manufacturer designed that luggage be car-ried or to the area of the motor vehicle in which luggage is customarily carried. In a station wag-on, van, hatchback vehicle, or sport utility vehi-cle, the term "luggage compartment" refers to the area behind, but not under, the rearmost seat. In a truck, the term 'luggage compartment' refers to the area behind the rearmost seat, but not under the front seat.

16-23-30. Sale or delivery of handgun to and possession by certain persons unlawful; stolen handguns.

(A) It is unlawful for a person to knowingly sell, offer to sell, deliver, lease, rent, barter, exchange, or transport for sale into this State any handgun to:

(1) a person who has been convicted of a crime of violence in any court of the United States, the several states, commonwealths, territories, possessions, or the District of Columbia or who is a fugitive from justice or a habitual drunkard or a drug addict or who has been adjudicated mentally incompetent;

(2) a person who is a member of a subversive organization;

(3) a person under the age of eighteen, but this shall not apply to the issue of handguns to members of the Armed Forces of the United States, active or reserve, National Guard, State Militia, or R. O. T. C., when on duty or training or the temporary loan of handguns for instructions under the immediate supervision of a parent or adult instructor; or

(4) a person who by order of a circuit judge or county court judge of this State has been adjudged unfit to carry or possess a firearm, such adjudication to be made upon application by any police officer, or by any prosecuting officer of this State, or sua sponte, by the court, but a person who is the subject of such an application is entitled to reasonable notice and a proper hearing prior to any such adjudication."

(B) It is unlawful for a person enumerated in subsection (A) to possess or acquire handguns within this State.

(C) A person shall not knowingly buy, sell, transport, pawn, receive or possess any stolen handgun or one from which the original serial number has been removed or obliterated.

16-23-50. Penalties; disposition of fines; forfeiture and disposition of handguns.

(A)(1) A person, including a dealer, who vio-lates the provisions of this article, except Section 16-23-20, is guilty of a felony and, upon convic-tion, must be fined not more than two thousand dollars or imprisoned not more than five years, or both.

(2) A person violating the provisions of Sec-tion 16-23-20 is guilty of a misdemeanor and, upon conviction, must be fined not more than one thousand dollars or imprisoned not more than one year, or both.

(B) In addition to the penalty provided in this section, the handgun involved in the violation of this article must be confiscated. The handgun must be delivered to the chief of police of the municipality or to the sheriff of the county if the violation occurred outside the corporate limits of a municipality. The law enforcement agency that receives the confiscated handgun may use it within the agency, transfer it to another law en-forcement agency for the lawful use of that agency, trade it with a retail dealer licensed to sell handgun in this State for a handgun or any other equipment approved by the agency, or de-stroy it. A weapon must not be disposed of in any manner until the results of any legal pro-ceeding in which it may be involved are finally determined. If the State Law Enforcement Divi-sion seized the handgun, the division may keep the handgun for use by its forensic laboratory. Records must be kept of all confiscated hand-gun received by the law enforcement agencies under the provisions of this article.

16-23-55. Procedure for returning found handgun.

(A) A handgun that is found and turned over to a law enforcement agency must be held for a period of ninety days. During that period, the agency shall make a diligent effort to determine:

(1) if the handgun is stolen;

(2) if the handgun has been used in the commission of a crime; and (3) the true owner of the handgun.

(B) At least twice during the ninety-day holding period, the agency shall advertise the handgun with its full description in a newspaper having general circulation in the county where the handgun was found.

(C) After the ninety days have elapsed from publication of the first advertisement, and upon request of the individual who found and turned over the handgun, the agency shall return the handgun to this person if the individual fully completes the application process as described in Section 23-31-140 and in federal law, and pays all advertising and other costs incidental to returning the handgun. No handgun may be returned until the individual fully completes the application.

(D) Upon proper completion of the application, the law enforcement agency shall provide copies of the application in compliance with Section 23-31-140.

Article 3. Machineguns, Sawed-off Shotguns and Rifles

16-23-210. Definitions. When used in this article:

(a) "Machine gun" applies to and includes any weapon which shoots, is designed to shoot, or can be readily restored to shoot, automatically more than one shot, without manual reloading, by a single function of the trigger. The term shall also include the frame or receiver of any such weapon, any combination or parts designed and intended for use in converting a weapon into a machine gun, and any combination of parts from which a machine gun can be assembled if such parts are in the possession or under the control of a person.

(b) "Sawed-off shotgun" means a shotgun having a barrel or barrels of less than eighteen inches in length or a weapon made from a shot-gun which as modified has an overall length of less than twenty-six inches or a barrel or barrels of less than eighteen inches in length.

(c) "Shotgun" means a weapon designed or redesigned, made or remade, and intended to be fired from the shoulder and designed or rede-signed and made or remade to use the energy of the explosive in a fixed shotgun shell to fire through a smooth bore either a number of ball shot or a single projectile for each pull of the trig-ger. The term includes any such weapon which may be readily restored to fire a fixed shotgun shell but does not include an antique firearm as defined in this section.

(d) "Sawed-off rifle" means a rifle having a barrel or barrels of less than sixteen inches in length or a weapon made from a rifle which as modified has an overall length of less than twenty-six inches or a barrel or barrels of less than sixteen inches in length.

(e) "Rifle" means a weapon designed or re-designed, made or remade, and intended to be fired from the shoulder and designed or rede-signed and made or remade to use the energy of the explosive in a fixed cartridge to fire only a single projectile through a rifled bore for each single pull of the trigger. The term includes any such weapon which may be readily restored to fire a fixed cartridge but does not include an antique firearm as described in this section.

(f) "Antique firearm" means any firearm not designed or redesigned for using rim fire or conventional center fire ignition with fixed ammunition and manufactured in or before 1898 (including any matchlock, flintlock, percussion cap, or similar type of ignition system or replica thereof, whether actually manufactured before or after the year 1898) and also any firearm using fixed ammunition manufactured in or before 1898, for which ammunition is no longer manufactured in the United States and is not readily available in the ordinary channels of commercial trade.

(g) "Military firearm" means any military weap-on, firearm, or destructive device, other than a machine gun, that is manufactured for military use by a firm licensed by the federal government pursuant to a contract with the federal govern-ment and does not include a pistol, rifle, or shot-gun which fires only one shot for each pull of the trigger.

16-23-220. Unlawful transportation of machine gun, military firearm, or sawed-off shotgun or rifle within State. It is unlawful for a person to transport from one place to another in this State or for any railroad company, express company, or other common carrier or any officer, agent, or employee of any of them or other person acting in their behalf knowingly to ship or to transport from one place to another in this State a machine gun or firearm commonly known as a machine gun, military firearm, sawed-off shotgun, or sawed-off rifle, except as provided in Sections 16-23-250 and 23-31-330.

A person who violates the provisions of this section, upon conviction, must be punished pursuant to Section 16-23-260.

16-23-230. Unlawful storing, keeping, or possessing of machine gun, military firearm, or sawed-off shotgun or rifle. It is unlawful for a person to store, keep, possess, or have in possession or permit another to store, keep, possess, or have in possession a machine gun or firearm commonly known as a machine gun, military firearm, sawed-off shotgun, or sawed-off rifle, except as provided in Sections 16-23-250 and 23-31-330.

A person who violates the provisions of this section, upon conviction, must be punished pursuant to Section 16-23-260.

16-23-240. Unlawful sale, rental, or giving away of machine gun, military firearm, or sawed-off shotgun or rifle; exceptions. It is unlawful for a person to sell, rent, give away, or participate in any manner, directly or indirectly, in the sale, renting, giving away, or otherwise disposing of a machine gun, or firearm commonly known as a machine gun, military firearm, sawed-off shotgun, or sawed-off rifle, except as provided in Sections 16-23-250 and 23-31-330.

A person who violates the provisions of this section, upon conviction, must be punished pursuant to Section 16-23-260.

16-23-250. Exceptions to application of article. The provisions of this article do not apply to the Army, Navy, or Air Force of the United States, the National Guard, and organizations authorized by law to purchase or receive machine guns, military firearms, or sawed-off shotguns or sawed-off rifles, from the United States or from this State and the members of these organizations. Any peace officer of the State or of a county or other political subdivision, state constable, member of the highway patrol, railway policeman or warden, superintendent, head keeper or deputy of a state prison, correction facility, workhouse, county jail, city jail, or other institution for the detention of persons convicted or accused of crime or held as witnesses in criminal cases or persons on duty in the postal service of the United States or a common carrier while transporting direct to a police department, military, or naval organization or person authorized by law to possess or use a machine gun, or sawed-off shotgun, or sawed-off rifle, may possess machine guns, or sawed-off shotguns, or sawed-off rifles, when required in the performance of their duties. The provisions of this section must not be construed to apply to machine guns, or sawed-off shotguns, or sawed-off rifles kept for display as relics and which are rendered harmless and not usable.

The provisions of this article do not apply to a manufacturer of machine guns or military firearms licensed pursuant to the provisions of 18 U. S. C. Section 921 et seq., a person authorized to possess these weapons by the United States Department of the Treasury, the Bureau of Alcohol, Tobacco and Firearms, or any other federal agency empowered to grant this authorization, a common or contract carrier transporting or shipping any machine gun or military firearm to or from the manufacturer if the transportation or shipment is not prohibited by federal law, or persons licensed pursuant to Section 23-31-370.

16-23-260. Penalties. A person violating the provisions of this article is guilty of a felony and, upon conviction, must be fined not more than ten thousand dollars or imprisoned not more than ten years, or both.

16-23-270. Article not applicable to antique firearms. The provisions of this article shall not apply to antique firearms.

16-23-280. Manufacture and sale of machine guns by licensed manufacturer. Notwithstanding the provisions of this article, machine guns or military firearms manufactured by a firm licensed by the federal government and subject to the Federal Gun Control Act may be legally manufactured, transported, possessed, and sold within the State by the manufacturer thereof.

Article 5. Miscellaneous Offenses

16-23-405. Definition of "weapon"; confiscation and disposition of weapons used in commission or in furtherance of crime.

(A) Except for the provisions relating to rifles and shotguns in Section 16-23-460, as used in this chapter, 'weapon' means firearm (rifle, shotgun, pistol, or similar device that propels a projectile through the energy of an explosive), a blackjack, a metal pipe or pole, or any other type of device, or object which may be used to inflict bodily injury or death.

(B) A person convicted of a crime, in addition to a penalty, shall have a weapon used in the commission or in furtherance of the crime confiscated. Each weapon must be delivered to the chief of police of the municipality or to the sheriff of the county if the violation occurred outside the corporate limits of a municipality. The law enforcement agency that receives the confiscated weapon may use it within the agency, transfer it to another law enforcement agency for the lawful use of that agency, trade it with a retail dealer licensed to sell pistols in this State for a pistol or other equipment approved by the agency, or destroy it. A weapon may not be disposed of until the results of all legal proceedings in which it may be involved are finally determined. A firearm seized by the State Law Enforcement Division may be kept by the division for use by its forensic laboratory.

16-23-410. Pointing firearm at any person

It is unlawful for a person to present or point at another person a loaded or unloaded firearm.

A person who violates the provisions of this section is guilty of a felony and, upon conviction, must be fined in the discretion of the court or imprisoned not more than five years. This section must not be construed to abridge the right of self-defense or to apply to theatricals or like performances.

16-23-420. Carrying or displaying firearms in public buildings or areas adjacent thereto.

(A) It is unlawful for a person to possess a firearm of any kind any premises or property owned, operated, or controlled by a private or public school, college, university, technical college, other post-secondary institution, or in any publicly-owned building, without the express permission of the authorities in charge of the premises or property.

(B) It is unlawful for a person to enter the premises or property described in subsection (A) and to display, brandish, or threaten others with a firearm.

(C) A person who violates the provisions of this section is guilty of a felony and, upon conviction, must be fined not more than five thousand dollars or imprisoned not more than five years, or both.

(D) This section does not apply to a guard, law enforcement officer, or member of the armed forces, or student of military science. A married student residing in an apartment provided by the private or public school whose presence with a weapon in or around a particular building is authorized by persons legally responsible for the security of the buildings is also exempted from the provisions of this section.

(E) For purposes of this section, the terms 'premises' and 'property' do not include state or locally owned or maintained roads, streets or rights-of-way of them, running through or adjacent to premises or property owned, operated, or controlled by a private or public school, college, university, technical college, or other post-secondary institution, which are open full time to public vehicular traffic.

(F) This section does not apply to a person who is authorized to carry concealed weapons pursuant to Article 4, Chapter 31 of Title 23 when upon any premises, property, or building that is part of an interstate highway rest area facility.

16-23-430. Carrying weapons on school property.

(1) It shall be unlawful for any person, except State, county or municipal law-enforcement officers or personnel authorized by school officials, to carry on his person, while on any elementary or secondary school property, a knife, with a blade over two inches long, a blackjack, a metal pipe or pole, firearms or any other type of weapon, device or object which may be used to inflict bodily injury or death.

(2) A person who violates the provisions of this section is guilty of a felony and, upon conviction, must be fined not more than one thousand dollars or imprisoned not more than five years, or both. Any weapon or object used in violation of this section may be confiscated by the law enforcement division making the arrest.

16-23-450. Placing loaded trap gun, spring gun or like device. It shall be unlawful for any person to construct, set or place a loaded trap gun, spring gun or any like device in any manner in any building or in any place within this State, and any violation of the provisions of this section shall constitute a misdemeanor and be punished by a fine of not less than one hundred dollars nor more than five hundred dollars or by imprisonment of not less than thirty days nor more than one year or by both fine and imprisonment, in the discretion of the court.

16-23-470. Tear-gas gun or ammunition.

(A) It is unlawful for anyone except an authorized law enforcement officer to possess, use, transport, sell, or buy a tear-gas machine or gun, or its parts, or any ammunition, shells, or equipment that may be used in a tear-gas gun or machine. It is lawful for a person for self-defense purposes only to possess, use, transport, sell, or buy a tear-gas machine or gun, or its parts, or ammunition, shells, or equipment for a tear-gas machine or gun, but the capacity of a tear-gas cartridge, shell, or container shall not exceed fifty cubic centimeters nor shall a tear-gas machine or gun have the capability of shooting a cartridge, shell, or container of more than fifty cubic centimeters.

(B) A person who violates the provisions of this section is guilty of a misdemeanor and, upon conviction, must be imprisoned not more than three years or fined not more than five thousand dollars, or both.

(C) Except as permitted above, nothing in this section prohibits the purchase, sale, transportation, or use of tear gas for the destruction of insects or rodents if tear gas is not in containers or shells suitable for use in a tear-gas gun, equipment, or machine and if the purchaser has written authority for the purchase and use of tear gas from the county agent of the county in which he resides.

16-23-480. Manufacture or possession of article designed to cause damage by fire or other means. It is unlawful for a person to manufacture, cause to be manufactured, or possess any object or article which is designed to cause damage by fire or any other means to person or property either by ignition, detonation, or other means. It is unlawful for a person to possess any object or article solely for the purpose of causing damage by fire or other means to person or property either by ignition, detonation, or other means.

A person who violates the provisions of this section is guilty of a felony and, upon conviction, must be fined in the discretion of the court or imprisoned not more than five years, or both.

16-23-490. Additional punishment for possession of firearm or knife during commission of, or attempt to commit, violent crime.

(A) If a person is in possession of a firearm or visibly displays what appears to be a firearm or visibly displays a knife during the commission of a violent crime and is convicted of committing or attempting to commit a violent crime as defined in Section 16-1-60, he must be imprisoned five years, in addition to the punishment provided for the principal crime. This five-year sentence does not apply in cases where the death penalty or a life sentence without parole is imposed for the violent crime.

(B) Service of the five-year sentence is mandatory unless a longer mandatory minimum term of imprisonment is provided by law for the violent crime. The court may impose this mandatory five-year sentence to run consecutively or concurrently.

(C) The person sentenced under this section is not eligible during this five-year period for parole, work release, or extended work release. The five years may not be suspended and the person may not complete his term of imprisonment in less than five years pursuant to good-time credits or work credits, but may earn credits during this period.

(D) As used in this section, "firearm" means any machine gun, automatic rifle, revolver, pistol, or any weapon which will, or is designed to, or may readily be converted to expel a projectile; "knife" means an instrument or tool consisting of a sharp cutting blade whether or not fastened to a handle which is capable of being used to inflict a cut, slash, or wound.

(E) The additional punishment may not be imposed unless the indictment alleged as a separate count that the person was in possession of a firearm or visibly displayed what appeared to be a firearm or visibly displays a knife during the commission of the violent crime and conviction was had upon this count in the indictment. The penalties prescribed in this section may not be imposed unless the person convicted was at the same time indicted and convicted of a violent crime as defined in Section 16-1-60.

16-23-520. Use, transportation, manufacture, possession, purchase, or sale of teflon-coated ammunition. It is unlawful for a person to use, transport, manufacture, possess, distribute, sell, or buy any ammunition or shells that are coated with polytetrafluoroethylene (teflon).

A person who violates the provisions of this section is guilty of a felony and, upon conviction, must be imprisoned not more than five years or fined not more than five thousand dollars, or both.

Article 7. Bombs, Destructive Devices, and Weapons of Mass Destruction

16-23-710. Definitions. For purposes of this article:

(1) "Bacteriological weapon" and "biological weapon" mean devices which are designed in a manner as to permit the intentional release into the population or environment of microbiological or other biological materials, toxins, or agents, whatever their origin or method of production, in a manner not authorized by law, or any device, the development, production, or stockpiling of which is prohibited pursuant to the "Convention of the Prohibition of the Development, Production and Stockpiling of Bacteriological (Biological) and Toxin Weapons and their Destruction", 26 U.S.T. 583, TIAS 8063.

(2) "Bomb" includes a destructive device cap-able of being detonated, triggered, or set off to release any substance or material that is destructive, irritating, odoriferous, or otherwise harmful to one or more organisms including, but not limited to, human beings, livestock, animals, crops or vegetation, or to earth, air, water, or any other material or substance necessary or required to sustain human or any other individual form of life, or to real or personal property. ...

(4) "Building" means any structure, vehicle, watercraft, or aircraft:

(a) where any person lodges or lives; or

(b) where people assemble for purposes of business, government, education, religion, entertainment, public transportation, or public use or where goods are stored. Where a building consists of two or more units separately occupied or secured, each unit is considered both a separate building in itself and a part of the main building.

(5) "Device" means an object, contrivance, in-strument, technique, or any thing that is designed, manufactured, assembled, or capable of serving any purpose in a bomb, destructive device, explosive, incendiary, or weapon of mass destruction.

(6) "Detonate" means to explode or cause to explode.

(7) "Destructive device" means:

(a) a bomb, incendiary device, or any thing that can detonate, explode, or burn by mechanical, chemical, or nuclear means, or that contains an explosive, incendiary, poisonous gas, or toxic substance (chemical, biological, or nuclear materials) including, but not limited to, an incendiary or over-pressure device, or any other device capable of causing damage, injury, or death;

(b) a bacteriological weapon or biological weapon; or

(c) a combination of any parts, components, chemical compounds, or other substances, either designed or intended for use in converting any device into a destructive device which has been or can be assembled to cause damage, injury, or death.

(8) "Detonator" means a device containing a detonating charge used to initiate detonation in an explosive or any device capable of triggering or setting off an explosion or explosive charge including, but not limited to, impact or an impact device, a timing mechanism, electricity, a primer, primer or detonating cord, a detonating cap or device of any kind, detonating waves, electric blasting caps, blasting caps for use with safety fuses, shock tube initiator, and detonating cord delay connectors, or any other device capable of detonating or exploding a bomb, weapon of mass destruction, or destructive device.

(9) "Distribute" means the actual or construc-tive delivery or the attempted transfer from one person to another.

(10) "Explosive" means a chemical compound or other substance or a mechanical system intended for the purpose of producing an explosion capable of causing injury, death, or damage to property or an explosive containing oxidizing and combustible units or other ingredients in such proportions or quantities that ignition, fire, friction, concussion, percussion, or detonation may produce an explosion capable of causing injury, death, or damage to property. Explosives include, but are not limited to, the list of explosive materials published and periodically up-dated by the Bureau of Alcohol, Tobacco and Firearms.

(11) "Hoax device" or "replica" means a device or object which has the appearance of a destructive device.

(12) "Incendiary" means any material that:

(a) causes, or is capable of causing, fire when it is lit or ignited; and

(b) is used to ignite a flammable liquid or compound in an unlawful manner.

(13) "Incendiary device" means a destructive device, however possessed or delivered, and by whatever name called, containing or holding a flammable liquid or compound, which is capable of being ignited by any means possible. Incendiary device includes, but is not limited to, any form of explosive, explosive bomb, grenade, missile, or similar device, whether capable of be-ing carried or thrown by a person acting alone or with one or more persons, but does not include a device manufactured or produced for the pri-mary purpose of illumination or for marking de-tours, obstructions, defective paving, or other hazards on streets, roads, highways, or bridges, when used in a lawful manner.

(14) "Over-pressure device" means a contain-er filled with an explosive gas or expanding gas or liquid which is designed or constructed so as to cause the container to break, fracture, or rup-ture in a manner capable of causing death, injury, or property damage, and includes, but is not limited to, a chemical reaction bomb, an acid bomb, a caustic bomb, or a dry ice bomb.

(15) "Parts" means a combination of parts, components, chemical compounds, or other substances, designed or intended for use in converting any device into a destructive device.

(16) "Poisonous gases" mean a toxic chemical or its precursors that through its chemical action or properties on life processes, causes death or injury to human beings or other living organisms. However, the term does not include:

(a) riot control agents, smoke and obscuration materials, or medical products which are manufactured, possessed, transported, or used in accordance with the laws of this State or the United States;

(b) tear gas devices designed to be carried on or about the person which contain not more than fifty cubic centimeters of the chemical; or

(c) pesticides, as used in agriculture and household products.

(17) "Property" means real or personal property of any kind including money, choses in action, and other similar interest in property. ...

(19) "Weapon of mass destruction" means:

(a) any destructive device as defined in item (7);

(b) any weapon that is designed or intended to cause death or serious bodily injury through the release, dissemination, or impact of toxic or poisonous chemicals, or their precursors;

(c) any weapon involving a disease organism; or

(d) any weapon that is designed to release radiation or radioactivity at a level dangerous to human life.

16-23-715. Possession, threatened or attempted use of weapon of mass destruction for act of terrorism; penalty. A person who, without lawful authority, possesses, uses, threatens, or attempts or conspires to possess or use a weapon of mass destruction in furtherance of an act of terrorism is guilty of a felony and upon conviction:

(1) in cases resulting in the death of another person, must be punished by death or by imprisonment for life; or

(2) in cases which do not result in the death of another person, must be punished by imprisonment for not less than twenty-five years nor more than life.

16-23-720. Intentional detonation of or aiding, counseling or procuring detonation of destructive device; transporting of or possessing components of destructive device; penalties.

(C) A person who knowingly possesses, manufactures, transports, distributes, possesses with the intent to distribute a destructive device or any explosive, incendiary device, or over-pressure device or toxic substance or material which has been configured to cause damage, injury, or death, or a person who possesses parts, components, or materials which when assembled constitute a destructive device is guilty of a felony and, upon conviction, must be imprisoned for not less than two years nor more than fifteen years. ...

6-23-730. Hoax device or replica of destructive device or detonator; manufacture, possession or transport of; threat to use; penalties. A person who knowingly manufactures, possesses, transports, distributes, uses or aids, or counsels or conspires with another in the use of a hoax device or replica of a destructive device or detonator which causes any person reasonably to believe that the hoax device or replica is a destructive device or detonator is guilty of a misdemeanor and, upon conviction, must be imprisoned for not more than one year or fined not more than ten thousand dollars, or both.

16-23-770. Forfeiture of property used or intended for use in violation of article; storage and destruction; exceptions.

(A) All property used or intended for use in violation of this article and all proceeds derived from, realized from, or traced back to property used or intended for use in violation of this article is contraband and subject to forfeiture. Property subject to forfeiture must be seized by a law enforcement agency and forfeited to the State, a political subdivision of the State, or the seizing law enforcement agency. ...

(D) The provisions of this article do not apply to the lawful use of:

(1) fertilizers, propellant activated devices, or propellant activated industrial tools manufactured, imported, distributed, or used for their intended purposes;

(2) pesticides which are manufactured, stored, transported, distributed, possessed, or used in accordance with Chapter 7, Title 2, the federal Insecticide, Fungicide, and Rodenticide Act and the Environmental Pesticide Control Act of 1972;

(3) explosives, blasting agents, detonators, and other objects regulated and controlled by the South Carolina Explosives Control Act;

(4) ammunition for small arms and firearms;

(5) components of ammunition for small arms and firearms;

(6) ammunition reloading equipment;

(7) the use of small arms propellant when used in war reenactments;

(8) firearms, as defined in Section 16-8-10; or

(9) fireworks and explosives which are permitted to be sold, possessed, or used under Chapter 35 of Title 23.

(E) The provisions of this article do not apply to the military or naval forces of the United States, to the duly organized military force of a state or territory, or to police or fire departments in this State when they are acting within their official capacities and in performance of their duties.

Title 23. Law Enforcement and Public Safety

Chapter 31. Firearms

Article 1. Purchase of Rifles and Shotguns

23-31-10. Purchase of rifle or shotgun in contiguous state. Any resident of this State including a corporation or other business entity maintaining a place of business in this State, who may lawfully purchase and receive delivery of a rifle or shotgun in this State, may purchase a rifle or shotgun in a contiguous state and transport or receive the same in this State; provided, that the sale meets the lawful requirements of each such state, meets all lawful requirements of any Federal statute, and is made by a licensed importer, licensed manufacturer, licensed dealer, or licensed collector.

23-31-20. Purchase of rifle or shotgun in this State by resident of contiguous state. A resident of any state may purchase rifles and shotguns in this State if the resident conforms to applicable provisions of statutes and regulations of this State, the United States, and of the state in which the person resides.

Article 3. Regulation of Pistols

23-31-110. Definitions. When used in this article:

(a) "Pistol" means any firearm designed to expel a projectile and designed to be fired from the hand, but shall not include any firearm generally recognized or classified as an antique, curiosity, or collector's item, or any that does not fire fixed cartridges.

(b) The term "dealer" means any person engaged in the business of selling firearms at retail or any person who is a pawnbroker.

(c) The term "crime of violence" means murder, manslaughter (except negligent manslaughter arising out of traffic accidents), rape, mayhem, kidnapping, burglary, robbery, housebreaking assault with intent to kill, commit rape, or rob, assault with a dangerous weapon, or assault with intent to commit any offense punishable by imprisonment for more than one year.

(d) The term "fugitive from justice" means any person who has fled from or is fleeing from any law enforcement officer to avoid prosecution or imprisonment for a crime of violence.

(e) The term "subversive organization" means any group, committee, club, league, society, association or combination of individuals the purpose of which, or one of the purposes of which, is the establishment, control, conduct, seizure or overthrow of the government of the United States or any state or political subdivision there-of, by the use of force, violence, espionage, sabotage, or threats or attempts of any of the fore-going.

(f) The term "conviction" as used herein shall include pleas of guilty, pleas of nolo contendere and forfeiture of bail.

(g) The term "division" shall mean the State Law Enforcement Division.

(h) The terms "purchase" or "sell" mean to knowingly buy, offer to buy, receive, lease, rent, barter, exchange, pawn or accept in pawn.

(i) The term "person" shall mean any individual, corporation, company, association, firm, partnership, society or joint stock company.

23-31-130. Retail dealers shall be licensed. No retail dealer shall sell or otherwise transfer, or expose for sale or transfer, or have in his possession with intent to sell, or otherwise transfer, any pistol without being licensed as hereinafter provided.

23-31-140. Completion and contents of application prior to purchase of pistol; further restrictions on purchase; persons on active military duty.

(A) A person may not purchase a pistol from a dealer unless he is a resident of this State. For the purpose of this article, the possession of a valid South Carolina driver's license or Department of Motor Vehicles identification card constitutes proof of residency. However, residency is not required of a person who is on active duty in the United States military and who is in possession of a current United States military identification card.

(B) For purposes of this section, the purchase of a pistol does not include the redeeming of a pistol by its owner after it has been pledged to secure a loan.

(C) No person may purchase a pistol from a dealer unless he is a resident of this State. For the purpose of this article, the possession of a valid South Carolina driver's license or Department of Public Safety identification card constitutes proof of residency. However, residency is not required of a person who is on active duty in the United States military and who is in possession of a current United States military identification card.

(D) Upon proper completion of the application the dealer must submit the original application to the division, retain a copy for his records, and give a copy to the applicant upon his purchase of a pistol. The application to be submitted to the division must be accompanied by a firearm transaction record properly completed by the purchaser and the dealer.

For purposes of this section, the purchase of a pistol does not include the redeeming of a pistol by its owner after it has been pledged to secure a loan.

23-31-150. Issuance, duration, conditions and forfeiture of retail dealer's license. The Division shall grant a license to any person doing business in the State not ineligible to purchase, acquire or possess a pistol or be licensed as a dealer under the provisions of this article. Licenses shall be issued on a form furnished by the Division and be effective for two years from the date of issuance. Licensees shall be authorized to sell pistols at retail as dealers within this State subject to the following conditions, for breach of any of which the license shall be forfeited:

(a) The license or a copy thereof, authenticated by the issuing authority, shall at all times and places of sale be available for inspection or displayed where it can easily be read.

(b) No pistol shall knowingly be sold in violation of any provision of this article nor shall any pistol be sold without clear evidence as to the identity of the purchaser being furnished to the dealer.

(c) The fee for the issuance of such license shall be one hundred dollars and for renewal one hundred dollars every two years. The license fees shall be retained by the division for purposes of defraying the costs of administering the provisions of this article.

(d) Each applicant for a license shall post with the Division a bond in favor of the State with surety in the amount of ten thousand dollars. No bond shall be accepted for filing unless it is with a surety company authorized to do business in this State and conditioned that the principal named therein shall not do any act meriting suspension or revocation of his license under provisions of this article. In lieu of a bond, a cash deposit or a deposit of other securities acceptable to the Division of a value of ten thousand dollars shall be accepted. Any person aggrieved by any act of the principal named in such bond may in an action against the principal or surety therein, or both, recover damages. The aggregate liability of the surety for all breaches of the conditions of the bond shall, in no event, exceed the amount of such bond. The surety on the bond shall have the right to cancel such bond giving thirty days' notice to the Division and thereafter shall be relieved of liability for any breach of condition after the effective date of the cancellation.

(e) In order to insure compliance with the provisions of this article, dealers shall make available for inspection by the chief of the division or his agents, during normal business hours, all pistols in their possession.

(f) Each applicant for a license shall furnish to the Division a current federal firearms license and is required to maintain that federal firearms license in good standing as a condition of holding a retail dealer license issued under this section.

(g) A breach of any of the above conditions or violations of any provisions of this article by a dealer shall result in forfeiture of license, but the licensee is entitled to reasonable notice and proper hearing in the circuit court of the county in which he is licensed.

23-31-160. False information or evidence shall not be given. No person shall give false information or evidence regarding any information or evidence required herein.

23-31-170. Mortgage, deposit or pledge of pistol. Only a licensed dealer shall make a loan secured by a mortgage, deposit or pledge of a pistol and the dealer shall keep such records as are required herein. A licensed dealer may mortgage any pistol or stock of pistols but shall not deposit the same with any other person.

23-31-180. Certain pistols declared to be contraband; forfeiture, seizure, and destruction; disposal restrictions; use for display.

No licensed retail dealer may hold, store, handle, sell, offer for sale, or otherwise possess in his place of business a pistol or other handgun which has a die-cast, metal alloy frame or receiver which melts at a temperature of less than eight hundred degrees Fahrenheit.

A pistol or other handgun possessed or sold by a dealer in violation of this article is declared to be contraband and must be forfeited to or seized by the law enforcement agency in the municipality where forfeited or seized or to the law enforcement agency in the county where forfeited or seized if forfeited or seized outside a municipality. The weapon must be destroyed by the law enforcement agency which seized the weapon or the law enforcement agency to which the weapon is forfeited. A weapon must not be disposed of in any manner until the results of any legal proceeding in which it may be involved are finally determined.

However, a law enforcement agency may use the weapon for display purposes after the weapon has been rendered inoperable.

23-31-190. Penalties; disposition of fines; forfeiture and disposition of pistols. A person, including a dealer who violates the provisions of this article is guilty of a felony and, upon conviction, must be fined not more than two thousand dollars or imprisoned not more than five years, or both.

In addition to the penalty provided in this section the pistol involved in the violation must be confiscated. The pistol must be delivered to the chief of police of the municipality or to the sheriff of the county, if the violation occurred outside the corporate limits of a municipality. The law enforcement agencies that receive the confiscated pistols may use them within their department, transfer them to another law enforcement agency, or destroy them. Records must be kept of all confiscated pistols received by the law enforcement agencies under the provisions of this article.

Article 4. Concealed Weapons Permit

23-31-205. Name. This article may be cited as the "Law Abiding Citizens Self-Defense Act of 1996".

23-31-210. Definitions. As used in this article:

(1) "Resident" means an individual who is present in South Carolina with the intention of making a permanent home in South Carolina or military personnel on permanent change of station orders.

(2) "Picture identification" means:

(a) a valid South Carolina driver's license; or

(b) an official photographic identification card issued by the Department of Revenue, a federal or state law enforcement agency, an agency of the United States Department of Defense, or United States Department of State.

(3) "Proof of residence" means a person's cur-rent address on the original or certified copy of:

(a) a valid South Carolina driver's license;

(b) an official identification card issued by the Department of Revenue, a federal or state law enforcement agency, an agency of the United States Department of Defense, or United States Department of State;

(c) a voter registration card; or

(d) another document that SLED may determine that fulfills this requirement.

(4) "Proof of training" means an original document or certified copy of the document supplied by an applicant that certifies that he is either:

(a) a person who, within three years before filing an application, has successfully completed a basic or advanced handgun education course offered by a state, county, or municipal law enforcement agency or a nationally recognized organization that promotes gun safety. This education course must be a minimum of eight hours and must include, but is not limited to:

(i) information on the statutory and case law of this State relating to handguns and to the use of deadly force;

(ii) information on handgun use and safety;

(iii) information on the proper storage practice for handguns with an emphasis on storage practices that reduces the possibility of accidental injury to a child; and

(iv) the actual firing of the handgun in the presence of the instructor;

(b) an instructor certified by the National Rifle Association or another SLED-approved competent national organization that promotes the safe use of handguns;

(c) a person who can demonstrate to the Director of SLED or his designee that he has a proficiency in both the use of handguns and state laws pertaining to handguns;

(d) an active duty police handgun instructor;

(e) a person who has a SLED-certified or approved competitive handgun shooting classification; or

(f) a member of the active or reserve military, or a member of the National Guard who has had handgun training in the previous three years.

SLED shall promulgate regulations containing general guidelines for courses and qualifications for instructors which would satisfy the requirements of this item. For purposes of subitems (a) and (b), 'proof of training' is not satisfied unless the organization and its instructors meet or exceed the guidelines and qualifications contained in the regulations promulgated by SLED pursuant to this item.

(5) "Concealable weapon" means a firearm having a length of less than twelve inches measured along its greatest dimension that must be carried in a manner that is hidden from public view in normal wear of clothing except when needed for self-defense, defense of others, and the protection of real or personal property.

23-31-215. Issuance of permits.

(A) Notwithstanding any other provision of law, except subject to subsection (B) of this section, SLED must issue a permit, which is no larger than three and one-half inches by three inches in size, to carry a concealable weapon to a resident who is at least twenty-one years of age and who is not prohibited by state law from possessing the weapon upon submission of:

(1) a completed application signed by the person;

(2) one current full face color photograph of the person, not smaller than one inch by one inch nor larger than three inches by five inches;

....**(3)** proof of residence or if the person is a qualified nonresident, proof of ownership of real property in this State;

(4) proof of actual or corrected vision rated at 20/40 within six months of the date of application or, in the case of a person licensed to operate a motor vehicle in this State, presentation of a valid driver's license;

(5) proof of training;

(6) payment of a fifty-dollar application fee. This fee must be waived for disabled veterans and retired law enforcement officers; and

(7) a complete set of fingerprints unless, because of a medical condition verified in writing by a licensed medical doctor, a complete set of fingerprints is impossible to submit. In lieu of the submission of fingerprints, the applicant must submit the written statement from a licensed medical doctor specifying the reason or reasons why the applicant's fingerprints may not be taken. If all other qualifications are met, the Chief of SLED may waive the fingerprint requirements of this item. The statement of medical limitation must be attached to the copy of the application retained by SLED. A law enforcement agency may charge a fee not to exceed five dollars for fingerprinting an applicant.

(B) Upon submission of the items required by subsection (A) of this section, SLED must conduct or facilitate a local, state, and federal fingerprint review of the applicant. SLED must also conduct a background check of the applicant through notification to and input from the sheriff of the county where the applicant resides or if the applicant is a qualified nonresident, where the applicant owns real property in this State. The sheriff within ten working days after notification by SLED, must submit a recommendation on an application. Before making a determination whether or not to issue a permit under this article, SLED must consider the recommendation provided pursuant to this subsection. The failure of the sheriff to submit a recommendation within the ten-day period constitutes a favorable recommendation for the issuance of the permit to the applicant. If the fingerprint review and background check are favorable, SLED must issue the permit.

(C) SLED shall issue a written statement to an unqualified applicant specifying its reasons for denying the application within ninety days from the date the application was received; otherwise, SLED shall issue a concealable weapon permit. If an applicant is unable to comply with the provisions of Section 23-31-210(4), SLED shall offer the applicant a handgun training course that satisfies the requirements of Section 23-31-210(4)(a). The course shall cost fifty dollars. SLED shall use the proceeds to defray the training course's operating costs. If a permit is granted by operation of law because an applicant was not notified of a denial within the ninety-day notification period, the permit may be revoked upon written notification from SLED that sufficient grounds exist for revocation or initial denial.

(D) Denial of an application may be appealed. The appeal must be in writing and state the basis for the appeal. The appeal must be submitted to the Chief of SLED within thirty days from the date the denial notice is received. The chief shall issue a written decision within ten days from the date the appeal is received. An adverse decision shall specify the reasons for upholding the denial and may be reviewed by the Administrative Law Judge Division pursuant to Article 5, Chapter 23 of Title 1, upon a petition filed by an applicant within thirty days from the date of delivery of the division's decision.

(E) SLED must make permit application forms available to the public. A permit application form shall require an applicant to supply:

(1) name, including maiden name if applicable;

(2) date and place of birth;

(3) sex;

(4) race;

(5) height;

(6) weight;

(7) eye and hair color;

(8) current residence address, or if the applicant is a qualified nonresident, current residence address and where the applicant owns real property in this State; and

(9) all residence addresses for the three years preceding the application date.

(F) The permit application form shall require the applicant to certify that:

(1) he is not a person prohibited under state law from possessing a weapon;

(2) he understands the permit is revoked and must be surrendered immediately to SLED if the permit holder becomes a person prohibited under state law from possessing a weapon;

(3) he is a resident of this State, is military personnel on permanent change of station orders, or is a qualified nonresident; and

(4) all information contained in his application is true and correct to the best of his knowledge.

(G) Medical personnel, law enforcement agencies, organizations offering handgun education courses pursuant to Section 23-31-210(4)(a), and their personnel, who in good faith provide information regarding a person's application, must be exempt from liability that may arise from issuance of a permit; provided, however, a weapons instructor must meet the requirements established in Section 23-31-210(4)(b), (c), (d), (e), or (f) in order to be exempt from liability under this subsection.

(H) A permit application must be submitted in person or by mail to SLED headquarters which shall verify the legibility and accuracy of the required documents.

(I) SLED must maintain a list of all permit holders and the current status of each permit. SLED may release the list of permit holders or verify an individual's permit status only if the request is made by a law enforcement agency to aid in an official investigation, or if the list is required to be released pursuant to a subpoena or court order. SLED may charge a fee not to exceed its costs in releasing the information under this subsection. Except as otherwise provided in this subsection, a person in possession of a list of permit holders obtained from SLED must destroy the list.

(J) A permit is valid statewide unless revoked because the person has:

(1) become a person prohibited under state law from possessing a weapon;

(2) moved his permanent residence to another state;

(3) voluntarily surrendered the permit; or

(4) been charged with an offense that upon conviction, would prohibit the person from possessing a firearm. However, if the person subsequently is found not guilty of the offense, then his permit must be reinstated at no charge

Once a permit is revoked, it must be surrendered to a sheriff, police department, a SLED agent, or by certified mail to the Chief of SLED. A person who fails to surrender his permit in accordance with this subsection is guilty of a misdemeanor and, upon conviction, must be fined twenty-five dollars.

(K) A permit holder must have his permit identification card in his possession whenever he carries a concealable weapon. When carrying a concealable weapon pursuant to Article 4 of Chapter 31 of Title 23, a permit holder must inform a law enforcement officer of the fact that he is a permit holder and present the permit identification card when an officer (1) identifies himself as a law enforcement officer and (2) requests identification or a driver's license from a permit holder. A permit holder immediately must report the loss or theft of a permit identification card to SLED headquarters. A person who violates the provisions of this subsection is guilty of a misdemeanor and upon conviction, must be fined twenty-five dollars.

(L) SLED shall issue a replacement for lost, stolen, damaged, or destroyed permit identification cards after the permit holder has updated all information required in the original application and the payment of a five-dollar replacement fee. Any change of permanent address must be communicated in writing to SLED within ten days of the change accompanied by the payment of a fee of five dollars to defray the cost of issuance of a new permit. SLED shall then issue a new permit with the new address. A permit holder's failure to notify SLED in accordance with this subsection constitutes a misdemeanor punishable by a twenty-five dollar fine. The original permit shall remain in force until receipt of the corrected permit identification card by the permit holder, at which time the original permit must be returned to SLED.

(M) A permit issued pursuant to this section does not authorize a permit holder to carry a concealable weapon into a:

(1) police sheriff, or highway patrol station or any other law enforcement office or facility;

(2) detention facility, prison, or jail or any other correctional facility or office;

(3) courthouse or courtroom;

(4) polling place on election days;

(5) office of or the business meeting of the governing body of a county, public school district, municipality, or special purpose district;

(6) school or college athletic event not related to firearms;

(7) daycare facility or pre-school facility;

(8) place where the carrying of firearms is prohibited by federal law;

(9) church or other established religious sanctuary unless express permission is given by the appropriate church official or governing body; or

(10) hospital, medical clinic, doctor's office, or any other facility where medical services or procedures are performed unless expressly authorized by the employer.

A person who willfully violates a provision of this subsection is guilty of a misdemeanor and, upon conviction, must be fined not less than one thousand dollars or imprisoned not more than one year, or both, at the discretion of the court and have his permit revoked for five years.

Nothing contained herein may be construed to alter or affect the provisions of Sections 10-11-320, 16-23-420, 16-23-430, 16-23-465, 44-23-1080, 44-52-165, 50-9-830, and 51-3-145.

(N) Valid out-of-state permits to carry concealable weapons held by a resident of a reciprocal state must be honored by this State, provided, that the reciprocal state requires an applicant to successfully pass a criminal background check and a course in firearm training and safety. A resident of a reciprocal state carrying a concealable weapon in South Carolina is subject to and must abide by the laws of South Carolina regarding concealable weapons. SLED shall maintain and publish a list of those states as the states with which South Carolina has reciprocity.

(O) A permit issued pursuant to this article is not required for a person:

(1) specified in Section 16-23-20, items (1) through (5) and items (7) through (11);

(2) carrying a self-defense device generally considered to be nonlethal including the substance commonly referred to as 'pepper gas';

(3) carrying a concealable weapon in a manner not prohibited by law.

(P) A permit issued pursuant to this article is valid for four years. Subject to subsection (Q) of this section, SLED shall renew a permit upon:

(1) payment of a fifty-dollar renewal fee by the applicant. This fee must be waived for disabled veterans and retired law enforcement officers;

(2) completion of the renewal application; and

(3) submission of a photocopy of the applicant's valid South Carolina driver's license or South Carolina identification card.

(Q) Upon submission of the items required by subsection (P) of this section, SLED must conduct or facilitate a local, state, and federal fingerprint review of the applicant. If the background check is favorable, SLED must renew the permit.

(R) No provision contained within this article shall expand, diminish, or affect the duty of care owed by and liability accruing to, as may exist at law immediately before the effective date of this article, the owner of or individual in legal possession of real property for the injury or death of an invitee, licensee, or trespasser caused by the use or misuse by a third party of a concealable weapon. Absence of a sign prohibiting concealable weapons shall not constitute negligence or establish a lack of duty of care.

(S) Once a concealed weapon permit holder is no longer a resident of this State or is no longer a qualified nonresident, his concealed weapon permit is void, and immediately must be surrendered to SLED.

(T) During the first quarter of each calendar year, SLED must publish a report of the following information regarding the previous calendar year:

(1) the number of permits;

(2) the number of permits that were issued;

(3) the number of permit applications that were denied;

(4) the number of permits that were renewed;

(5) the number of permit renewals that were denied;

(6) the number of permits that were suspended or revoked; and

(7) the name, address, and county of a person whose permit was revoked, including the reason for the revocation under Section 23-31-215(J)(1).

The report must include a breakdown of such information by county.

23-31-217. Effect on Section 16-23-20. Nothing in this article shall affect the provisions of Section 16-23-20.

Article 5. Use and Possession of Machineguns, Sawed-off Shotguns and Rifles

23-31-310. Definitions. When used in this article:

(a) "Machine gun" applies to and includes any weapon which shoots, is designed to shoot, or can be readily restored to shoot, automatically more than one shot, without manual reloading, by a single function of the trigger. The term shall also include the frame or receiver of any such weapon, any combination or parts designed and intended for use in converting a weapon into a machine gun, and any combination of parts from which a machine gun can be assembled if such parts are in the possession or under the control of a person.

(b) "Sawed-off shotgun" means a shotgun having a barrel or barrels of less than eighteen inches in length or a weapon made from a shotgun which as modified has an overall length of less than twenty-six inches or a barrel or barrels of less than eighteen inches in length.

(c) "Shotgun" means a weapon designed or redesigned, made or remade, and intended to be fired from the shoulder and designed or redesigned and made or remade to use the energy of the explosive in a fixed shotgun shell to fire through a smooth bore either a number of ball shot or a single projectile for each pull of the trigger. The term includes any such weapon which may be readily restored to fire a fixed shotgun shell but does not include an antique firearm as defined in this section.

(d) "Sawed-off rifle" means a rifle having a barrel or barrels of less than sixteen inches in length or a weapon made from a rifle which as modified has an overall length of less than twenty-six inches or a barrel or barrels of less than sixteen inches in length.

(e) "Rifle" means a weapon designed or redesigned, made or remade, and intended to be fired from the shoulder and designed or redesigned and made or remade to use the energy of the explosive in a fixed cartridge to fire only a single projectile through a rifled bore for each single pull of the trigger. The term includes any such weapon which may be readily restored to fire a fixed cartridge but does not include an antique firearm as described in this section.

(f) "Antique firearm" means any firearm not designed or redesigned for using rim fire or conventional center fire ignition with fixed ammunition and manufactured in or before 1898 (including any matchlock, flintlock, percussion cap, or similar type of ignition system or replica thereof, whether actually manufactured before or after the year 1898) and also any firearm using fixed ammunition manufactured in or before 1898, for which ammunition is no longer manufactured in the United States and is not readily available in the ordinary channels of commercial trade.

(g) "Military firearm" means any military weap-on, firearm, or destructive device, other than a machine gun, that is manufactured for military use by a firm licensed by the federal government pursuant to a contract with the federal govern-ment and does not include a pistol, rifle, or shot-gun which fires only one shot for each pull of the trigger.

23-31-320. Exceptions to application of article. The provisions of this article shall not apply to the Army, Navy, or Air Force of the United States, the National Guard, and organizations authorized by law to purchase or receive machine guns, military firearms, or sawed-off shotguns or sawed-off rifles, from the United States or from this State and the members of such organizations. Any peace officer of the State or of any county or other political subdivision thereof, state constable, member of the highway patrol, railway policeman or warden, superintendent, head keeper or deputy of any state prison, penitentiary, work-house, county jail, city jail, or other institution for the detention of persons convicted or accused of crime or held as witnesses in criminal cases or person on duty in the postal service of the United States or any common carrier while transporting direct to any police department, military, or naval organization or person authorized by law to possess or use a machine gun, or sawed-off shotgun or sawed-off rifle, may possess machine guns, or sawed-off shotguns or sawed-off rifles, when required in the performance of their duties. Nor shall the provisions hereof be construed to apply to machine guns, or sawed-off shotguns or sawed-off rifles, kept for display as relics and which are rendered harmless and not usable.

The provisions of this article shall not apply to any manufacturer of machine guns or military firearms licensed pursuant to the provisions of 18 U. S. C. Section 921 et seq., nor to any common or contract carrier transporting or shipping any machine guns or military firearms to or from such manufacturer if the transportation or shipment is not prohibited by federal law. Any such manufacturer shall furnish to the South Carolina Law Enforcement Division the serial numbers of all machine guns or military firearms manufactured by it within thirty days of such manufacture and shall be subject to the penalties provided in Section 23-31-340 for noncompliance.

23-31-330. Application and registration of person allowed to possess machine gun or sawed-off shotgun or rifle.

(A) Every person permitted by Section 23-31-320 to possess a machine gun or sawed-off shotgun or sawed-off rifle, and any person elected or appointed to any office or position which entitles the person to possess a machine gun or sawed-off shotgun or sawed-off rifle, upon taking office, shall file with the State Law Enforcement Division on a blank to be supplied by the division on request an application which is properly sworn. The application must be approved by the sheriff of the county in which the applicant resides or has his principal place of business and include the applicant's name, residence and business address, physical description, whether or not ever charged or convicted of any crime, municipal, state, or otherwise, and where, if charged, and when it was disposed of. The applicant shall also give a description including the serial number and make of the machine gun or sawed-off shotgun or sawed-off rifle which he possesses or desires to possess. The State Law Enforcement Division shall file the application in its office. The division shall register the applicant together with the information required in the application in a book or index to be kept for that purpose, assign to him a number, and issue to him a card which shall bear the signature of the applicant and which he shall keep with him while he has the machine gun or sawed-off shotgun or sawed-off rifle in his possession. This registration must be made on the date application is received and filed with the division. The registration expires on December thirty-first of the year in which the license is issued.

(B) No permit or registration required by the provisions of this section is required where weapons are possessed by a governmental entity which has a significant public safety responsibility for the protection of life or property.

23-31-340. Penalties. A person who violates the provisions of this article is guilty of a felony and, upon conviction, must be fined not more than ten thousand dollars or imprisoned not more than ten years, or both.

23-31-350. Article not applicable to antique firearms. The provisions of this article shall not apply to antique firearms.

23-31-360. Unregistered possession of machine guns or military firearms by licensed manufacturer. Machine guns or military firearms manufactured by a firm licensed by the federal government and subject to the Federal Gun Control Act may be legally possessed by the manufacturer without being registered with the State Law Enforcement Division. The manufacturing firm shall furnish to SLED the serial numbers of all machine guns or military firearms manufactured by it within thirty days of their manufacture and it is subject to the penalties provided in Section 23-31-340 for noncompliance.

23-31-370. Special limited license for possession, transportation, and sale of machine guns; violations and penalties.

(a) The South Carolina Law Enforcement Division may issue a special limited license for the possession, transportation, and sale of machine guns in this State to persons: (1) who are authorized representatives of a machine gun manufacturer or dealer engaged in demonstrating and selling them to agencies authorized by law to possess them, or (2) who are engaged in professional movie-making or providing services to professional movie-makers who use machine guns as regulated by this article in the course of creating movie "special effects".

(b) Applications for the special license authorized by this section must be on a form prescribed by the division, duly sworn to, containing the applicant's name, business and residence address, a record of any criminal charges filed against the applicant in the United States for other than traffic law violations and the disposition of the charges, a description of the machine guns to be possessed, transported, or sold in this State, including their make and serial numbers, the sites within the State to which the machine guns will be transported, and such other information the division considers necessary to implement this section.

(c) The division may issue a special license pursuant to this section if it determines that the applicant has not been convicted of any offense other than traffic violations and the applicant clearly qualifies under item (1) or (2) of subsection (a). The special license is valid for a specified period not to exceed six months which must be stated on the license.

(d) Any person who knowingly and willfully makes any false statement for the purpose of obtaining the special license or who violates its terms, in addition to any other penalty provided by law, is guilty of a misdemeanor and, upon conviction, must be fined not more than five thousand dollars or imprisoned for not more than two years, or both.

Article 6. Using A Firearm while Under the Influence of Alcohol or A Controlled Substance

23-31-400. Definitions; unlawful use of firearm; violations.

(A) As used in this article:

(1) "Use a firearm" means to discharge a firearm.

(2) "Serious bodily injury" means a physical condition which creates a substantial risk of death, serious personal disfigurement, or protracted loss or impairment of the function of a bodily member or organ.

(B) It is unlawful for a person who is under the influence of alcohol or a controlled substance to use a firearm in this State.

(C) A person who violates the provisions of subsection (B) is guilty of a misdemeanor and, upon conviction, must be fined not less than two thousand dollars or imprisoned not more than two years.

(D) This article does not apply to persons lawfully defending themselves or their property.

Article 7. Local Regulations

23-31-510. Prohibition against regulation of certain matters. No governing body of any county, municipality, or other political subdivision in the State may enact or promulgate any regulation or ordinance that regulates or attempts to regulate:

(1) the transfer, ownership, possession, carrying, or transportation of firearms, ammunition, components of firearms, or any combination of these things; or

(2) a landowner discharging a firearm on the landowner's property to protect the landowner's family, employees, the general public, or the landowner's property from animals that the landowner reasonably believes pose a direct threat or danger to the landowner's property, people on the landowner's property, or the general public. For purposes of this item, the landowner's property must be a parcel of land comprised of at least twenty-five contiguous acres. Any ordinance regulating the discharge of firearms that does not specifically provide for an exclusion pursuant to this item is unenforceable as it pertains to an incident described in this item; otherwise, the ordinance is enforceable.

23-31-520. Matters subject to regulation. This article does not affect the authority of any county, municipality, or political subdivision to regulate the careless or negligent discharge or public brandishment of firearms, nor does it prevent the regulation of public brandishment of firearms during the times of or a demonstrated potential for insurrection, invasions, riots or natural disasters. This article denies any county, municipality, or political subdivision the power to confiscate a firearm or ammunition unless incident to an arrest.

23-31-600. Retired personnel; identification cards; qualification for carrying concealed weapon.

(A) For purposes of this section:

(1) "Identification card" is a photographic identification card complying with 18 U.S.C. Section 926C(d).

(2) "Qualified retired law enforcement officer" means any retired law enforcement officer as defined in 18 U.S.C. Section 926C(c) who at the time of his retirement was certified as a law enforcement officer in this State and who was trained and qualified to carry firearms in the performance of his duties.

(B) An agency or department within this State must comply with Section 3 of the Law Enforcement Officers Safety Act of 2004, 18 U.S.C. Section 926C, by issuing an identification card to any person who retired from that agency or department and who is a qualified retired law enforcement officer. If the agency or department currently issues credentials to active law enforcement officers, then the agency or department may comply with the requirements of this section by issuing the same credentials to retired law enforcement officers. If the same credentials are issued, then the agency or department must stamp the credentials with the word "RETIRED."

(C)(1) Subject to the limitations of subsection (E), a qualified retired law enforcement officer may carry a concealed weapon in this State if he possesses an identification card issued pursuant to subsection (C) along with a certification that he has, not less recently than one year before the date the individual is carrying the firearm, met the standards established by the agency for training and qualification for active law enforcement officers to carry a firearm of the same type as the concealed firearm.

(2) The firearms certification required by this subsection may be reflected on the identification card or may be in a separate document carried with the identification card.

(D) The restrictions contained in Sections 23-31-220 and 23-31-225 are applicable to a person carrying a concealed weapon pursuant to this section.

(E) The agency or department may charge the retired law enforcement officer a reasonable fee for issuing the identification card and must provide the retired officer with the opportunity to qualify to carry a firearm under the same standards for training and qualification for active law enforcement officers to carry firearms. However, the agency or department, as provided in 18 U. S. C. Section 926C(c)(5), may require the retired officer to pay the actual expenses of the training and qualification.

[Current through 2007 Regular Session]

SOUTH DAKOTA
S.D. CODIFIED LAWS

Title 7. Counties

7-18A-36. Ordinances restricting firearms, ammunition, or their component prohibited and void No county may pass any ordinance that restricts possession, transportation, sale, transfer, ownership, manufacture or repair of firearms or ammunition or their components. Any ordinances prohibited by this section are null and void.

Title 8. Townships

8-5-13 Ordinances regarding firearms and ammunition prohibited -- Ordinances null and void No township may pass any ordinance that restricts possession, transportation, sale, transfer, ownership, manufacture or repair of firearms or ammunition or their components. Any ordinances prohibited by this section are null and void.

Title 9. Municipal Government

9-19-20. Firearm restrictions prohibited No municipality may pass any ordinance that restricts possession, transportation, sale, transfer, ownership, manufacture or repair of firearms or ammunition or their components. Any ordinances prohibited by this section are null and void.

Title 22. Crimes

Chapter 1. Definitions and General Provisions

22-1-1. Construction The rule of the common law that penal statutes are to be strictly construed has no application to this title. All its criminal and penal provisions and all penal statutes shall be construed according to the fair import of their terms, with a view to effect their objects and promote justice.

22-1-2. Definitions Terms used in this title mean: ...

(2) "Actor," the person who takes the active part in a transaction; ...

(4) "Antique firearm," any firearm, including any firearm with a matchlock, flintlock, percussion cap or similar type of ignition system, manufactured before 1899, and any replica of any firearm described in this section if such replica is not designed or redesigned for using rimfire or conventional centerfire fixed ammunition or if it uses rimfire or conventional centerfire fixed ammunition which is no longer manufactured in the United States and which is not readily available in the ordinary channels of commercial trade; ...

(6) "Concealed," any firearm that is totally hidden from view. If any part of the firearm is capable of being seen, it is not concealed; ...

(8) "Controlled weapon" includes a firearm silencer, machine gun, or short shotgun, as those terms are defined in subdivisions (17), (23), and (46) of this section;

(9) "Crime of violence," any of the following crimes or an attempt to commit, or a conspiracy to commit, any of the following crimes: murder, manslaughter, rape, aggravated assault, riot, robbery, burglary in the first or second degree, arson, kidnapping, felony sexual contact as defined in §§ 22-22-7 and 22-22-19.1, felony child abuse as defined in § 26-10-1, or any other felony in the commission of which the perpetrator used force, or was armed with a dangerous weapon, or used any explosive or destructive device;

(10) "Dangerous weapon" or "deadly weapon," any firearm, knife, stun gun, knife, or device, instrument, material or substance, whether animate or inanimate, which is calculated or designed to inflict death or serious bodily harm, or by the manner in which it is used is likely to inflict death or serious bodily harm; ...

(13) "Destructive device,"

(a) Any bomb, grenade, explosive missile, or similar device or any launching device therefor; or

(b) Any breakable container which contains a flammable liquid with a flashpoint of one hundred and fifty degrees Fahrenheit or less and has a wick or similar device capable of being ignited;

(c) The term does not include "permissible fireworks," defined by § 34-37-5; any device which is neither designed nor redesigned for use as a weapon; any device, although originally designed for use as a weapon, which is redesigned for use as a signaling, pyrotechnic, line throwing, safety or similar device; surplus ordnance sold, loaned or given by the secretary of the army pursuant to the provisions of 10 U.S.C. §§ 4684(2), 4685, or 4686; or any other device which is an antique or is a rifle which the owner intends to use solely for sporting purposes;

(14) "Explosive," any substance, or combination of substances, that is used for the purpose of detonation and which, upon exposure to any external or internal force or condition, is capable of a relatively instantaneous release of gas and heat. The term does not include "permissible fireworks," as defined by § 34-37-5; ...

(16) "Firearm," any weapon from which a projectile or projectiles may be discharged by gunpowder. As used in this subdivision, the word "gunpowder" includes any propellant that upon oxidization emits heat and light and is commonly used in firearms cartridges;

(17) "Firearm silencer," any instrument, attachment, weapon or appliance for causing the firing of any gun, revolver, pistol, or other firearm to be silent, or intended to lessen or muffle the noise of the firing of any such weapon; ...

(23) "Machine gun," any firearm, whatever its size and usual designation, that automatically discharges two or more cartridges by a single function of the firing device; ...

(32) "Pistol," any firearm with a barrel less than sixteen inches in length, designed to expel a projectile or projectiles by the action of an explosive; ...

(44) "Seller," any person or employee engaged in the business of selling pistols at retail;

(45) "Short rifle," any rifle having a barrel less than sixteen inches long, or an overall length of less than twenty-six inches;

(46) "Short shotgun," a shotgun having a barrel less than eighteen inches long or an overall length of less than twenty-six inches; ...

Chapter 14. Unlawful Use of Weapons

22-14-5 Possession of firearm with altered serial number -- Penalty -- Exception Any person who possesses any firearm on which the manufacturer's serial number has been changed, altered, removed or obliterated is guilty of a Class 6 felony.

The provisions of this section do not apply to persons who have applied for a new serial number pursuant to § 23-7-43.

22-14-15. Possession of firearm by person convicted of violent crime or felony -- Penalty -- Time limit conviction as felony -- Fifteen-year period No person who has been convicted in this state or elsewhere of a crime of violence or a felony pursuant to § § 22-42-2, 22-42-3, 22-42-4, 22-42-7, 22-42-8, 22-42-9, 22-42-10 or 22-42-19, may possess or have control of a firearm. A violation of this section is a Class 6 felony. This section does not apply to any person who was last discharged from prison, jail, probation, or parole more than fifteen years prior to the commission of the principal offense.

22-14-15.1 Possession of firearm by person convicted of possession of controlled substance -- Penalty -- Time limit No person who has been convicted of a felony under chapter 22-42 or of a felony for a crime with the same elements in another state may possess or have control of a firearm. A violation of this section is a Class 6 felony. This section does not apply to any person who was last discharged from prison, jail, probation, or parole, for a felony under chapter 22-42 more than five years prior to the commission of the principal offense and is not subject to the restrictions in § 22-14-15.

22-14-15.2 Possession of firearm after conviction of crime involving domestic violence -- Misdemeanor -- Loss of right to possess firearm -- Restoration of rights No person who has been convicted of any misdemeanor crime involving an act of domestic violence may possess or have control of a firearm for a period of one year from the date of conviction. Any violation of this section is a Class 1 misdemeanor. At the end of the one year period, any civil rights lost as a result of this provision shall be restored. Any person who has lost their right to possess or have control of a firearm as a result of a misdemeanor conviction involving an act of domestic violence, prior to the date of the effectiveness of this Act, shall be restored to those civil rights one year after the effective date of this Act. This section shall be repealed on the date when any federal law restricting the right to possess firearms for misdemeanor domestic violence convictions is repealed.

Once eligible under the statute, a person convicted under this section may petition the convicting court for an order reflecting the restoration of any firearm rights lost, if the person has not been convicted within the prior year of a crime for which firearm rights have been lost. A petition filed under this section shall be verified by the petitioner and served upon the states attorney in the county where the conviction occurred. Thirty days after service upon the states attorney, the court shall enter the order, if the court finds that the petitioner is eligible for relief under this section.

EDITOR'S NOTE.

Pursuant to its own terms, this section is repealed on the date when any federal law restricting the right to possess firearms for misdemeanor domestic violence convictions is repealed.

22-14-16. Giving of firearm to person convicted of violent crime or felony – Penalty Any person who knows that another person is prohibited by § 22-14-15 or 22-14-15.1 from possessing a firearm, and who knowingly gives, loans, or sells a firearm to that person is guilty of a Class 6 felony.

22-14-17 Applicability of chapter to firearm incapable of discharge
The provisions of this chapter do not apply to any firearm which has been permanently altered so it is incapable of being discharged.

Chapter 14A. Explosives and Destructive Devices

22-14A-4 Sale or possession of destructive device – Penalty Any person who knowingly sells, offers for sale, transports or possesses any destructive device is guilty of a Class 4 felony. If such person has been previously convicted of a crime of violence in this state or elsewhere, the offense is a Class 3 felony.

22-14A-6. Possession of explosive or destructive device with intent to injure -- Penalty

Any person who has in his or her possession any explosive or destructive device under circumstances not described in § 22-14A-5, with intent to injure, intimidate, or terrify any person, or with intent to wrongfully injure or destroy any property, is guilty of a Class 3 felony.

22-14A-11. Explosion or ignition of destructive device -- Penalty

Any person who explodes or ignites any destructive device or explosive with intent to cause serious bodily injury and which results in serious bodily injury is guilty of a Class 2 felony.

22-14A-13 Possession of materials to make destructive device -- Penalty Any person who possesses any substance, material, or any combination of substances or materials, with the intent to make a destructive device without first obtaining a permit from the department of public safety to make such device, is guilty of a Class 5 felony.

22-14A-16. Chapter not applicable to military, law enforcement, and permit holders This chapter shall not apply to the armed forces of the United States, the National Guard, any law enforcement agency or any officer, agent, employee or member thereof acting in a lawful capacity and any person possessing a valid seller's permit or user's permit from the United States federal government for explosive and destructive devices.

22-14A-21 Possession of registered or licensed destructive device Any person may possess destructive devices that are registered with, or licensed by, the state or federal government pursuant to law.

Title 23. Law Enforcement

Chapter 7. Firearms Control

23-7-1. Definitions Terms used in this chapter, unless the context otherwise requires, mean:

(1) "Antique firearm," a firearm as defined in subdivision 22-1-2(4);

(2) "Concealed," a firearm as defined in subdivision 22-1-2(6);

(3) "Crime of violence," an action as defined in subdivision 22-1-2(9);

(4) "Pistol," a firearm as defined in subdivision 22-1-2(32);

(5) "Seller," a person as defined in subdivision 22-1-2(44).

23-7-1.1. Exception for antique or nondischargeable firearms This chapter shall not apply to antique firearms or to firearms which have been permanently altered so they are incapable of being discharged.

23-7-7 Permit to carry concealed pistol -- Valid statewide -- Criminal history check A permit to carry a concealed pistol shall be issued to any person by the sheriff of the county in which the applicant resides. The permit shall be valid throughout the state and shall be issued pursuant to § 23-7-7.1. Prior to issuing the permit, the sheriff shall execute a background investigation, including a criminal history check, of every applicant for the purposes of verifying the qualifications of the applicant pursuant to the requirements of § 23-7-7.1. For the purposes of this section, a background investigation is defined as a computer check of available online records.

23-7-7.1 Temporary permit to carry concealed pistol -- Requirements A temporary permit to carry a concealed pistol shall be issued within five days of application to a person if the applicant:

(1) Is eighteen years of age or older;

(2) Has never pled guilty to, nolo contendere to, or been convicted of a felony or a crime of violence;

(3) Is not habitually in an intoxicated or drugged condition;

(4) Has no history of violence;

(5) Has not been found in the previous ten years to be a "danger to others" or a "danger to self" as defined in § 27A-1-1 or is not currently adjudged mentally incompetent;

(6) Has been a resident of the county or municipality where the application is being made for at least thirty days;

(7) Has had no violations of chapter 23-7, 22-14, or 22-42 constituting a felony or misdemeanor in the two five years preceding the date of application or is not currently charged under indictment or information for such an offense; and

(8) Is a citizen of the United States or has been in the United States legally for at least two years; and

(9) Is not a fugitive from justice.

A person denied a permit may appeal to the circuit court pursuant to chapter 1-26.

23-7-7.3. Reciprocity -- Conditions The attorney general shall compare South Dakota permit issuance statutes with the permit issuance statutes in states with which reciprocity is sought or requested in order to determine whether the laws of the other state meet or exceed the requirements of this chapter for the issuance of a permit. The secretary of state may enter into reciprocity agreements with other states after the attorney general has notified the secretary of state that the other states' laws meet or exceed the provisions of this chapter.

23-7-9. Sale of pistol -- Waiting period -- Exemption -- Penalty No federal firearms licensee may deliver a pistol to a person who has purchased a pistol until forty-eight hours have elapsed from the time of the sale of the pistol. A person who has in his or her possession a valid permit to carry a concealed pistol is exempt from the forty-eight hour waiting period. When a pistol is delivered, the pistol shall be securely wrapped and shall be unloaded. A pistol that is securely wrapped and delivered to a purchaser pursuant to this section is not a concealed weapon under § 22-14-9. A violation of this section is a Class 1 misdemeanor.

23-7-10. Application to purchase pistol -- Form -- Requirements -- Retention At the time of purchase of a pistol, any person who does not have in his or her possession a valid permit to carry a concealed pistol shall complete an application to purchase a pistol. The form of the application to purchase a pistol shall be prescribed by the secretary of state. The application shall require the applicant's complete name, address, occupation, place and date of birth, physical description, the date and hour of application, a statement that the applicant has never pled guilty to, nolo contendere to, or been convicted of a crime of violence, a statement that the information on the application is true and correct, and his signature. The federal firearms licensee shall, within six hours, send by first class mail or hand deliver the completed application to the chief of police of the municipality or the sheriff of the county of which the purchaser is a resident. The duplicate shall be returned to the federal firearms licensee by the chief of police or sheriff and the original shall be retained by for a period of one year.

23-7-11. Wholesale sales -- Inapplicable Sections 23-7-7 to 23-7-12, inclusive, do not apply to sales at wholesale.

23-7-12. False information or identity -- Penalty. No person, in purchasing or otherwise securing delivery of a pistol or in applying for a permit to carry a concealed pistol, may give false information or offer false evidence of his identity. A violation of this section is a Class 6 felony.

23-7-18. Sale -- Purchaser identity determined -- Penalty No pistol shall be sold in violation of any provisions of this chapter, nor shall a pistol be sold under any circumstances unless the purchaser is personally known to the seller or shall present clear evidence of his identity. A violation of this section is a Class 1 misdemeanor.

23-7-40. Permissive firearms sales and delivery -- Contiguous states -- Federal law The state of South Dakota herewith permits residents of contiguous states and residents of this state, not otherwise precluded by any applicable laws, to purchase, sell, trade, convey, deliver, or transport rifles, shotguns, ammunition, reloading components or firearms accessories in South Dakota and in states contiguous to South Dakota. This authorization is enacted to implement for this state the permissive firearms sales and delivery provisions in section 922 (b), (3) (A) of Public Law 90-618 of the 90th United States Congress, second session.

23-7-41. Repeal of federal restrictions -- No effect upon permissive firearms sales and delivery -- Contiguous states In the event that presently enacted federal restrictions on the purchase, sale, trade, delivery, transportation or conveyance of firearms accessories are repealed by the United States Congress or set aside by courts of competent jurisdiction, § 23-7-40 shall in no way be interpreted to prohibit or restrict the purchase, sale, trade, delivery, transportation or conveyance of shotguns, rifles, ammunition, reloading components, or firearms accessories by residents of this state or of contiguous states, or of any other states who are otherwise competent to purchase, sell, trade, deliver, transport or convey rifles, shotguns, ammunition, reloading components or firearms accessories in this state, in contiguous states or other states.

23-7-43. New serial number required where original serial has been changed, altered, removed or obliterated Upon application by an owner of a firearm, the director of the division of criminal investigation shall engrave or stamp a new serial number on any firearm on which the manufacturer's serial number has been changed, altered, removed or obliterated.

23-7-44 Possession of pistol by minor prohibited -- Penalty No person under the age of eighteen years may knowingly possess a pistol. A violation of this section is a Class 1 misdemeanor.

23-7-45. Possession of pistol by minor prohibited -- Exceptions The provisions of § 23-7-44 or to a criminal prosecution brought after transfer pursuant to chapter 26-11, do not apply to any minor who has the consent of the minor's parent or guardian to possess such pistol, and:

(1) That the minor was in the presence of the minor's parent or guardian;

(2) That the minor was on premises owned or leased by the minor or the minor's parent, guardian, or immediate family member;

(3) That the minor was in the presence of a licensed or accredited gun safety instructor; or

(4) That the pistol was being used for farming, ranching, hunting, trapping, target shooting, or gun safety instruction.

23-7-46. Transfer of firearm or ammunition to minor -- Penalty No person may sell, transfer, give, loan, furnish, or deliver a firearm or firearm ammunition to any person under the age of eighteen years if such person knows or reasonably believes that the minor recipient of the transfer intended, at the time of transfer, to use the firearm or ammunition in the commission or attempted commission of a crime of violence as defined in subdivision 22-1-2 (9). The affirmative defenses contained in chapter 23-7 do not apply to a prosecution under this section. A violation of this section is a Class 5 felony.

Title 37. Trade Regulation

Chapter 17. Sale Of Serially Numbered Appliances And Equipment

37-17-1. Sale of goods with serial numbers -- Alteration of serial number -- Penalty Any person who knowingly sells or offers for sale any ... firearm ... or any mechanical or electrical device, appliance, contrivance, material, piece of apparatus, or equipment, which is identified by a serial number placed thereon by the manufacturer, the original serial number of which has been destroyed, removed, altered, covered, or defaced, is guilty of a Class 2 misdemeanor if the value of the property is four hundred dollars or less. If the value of the property is more than four hundred dollars, and less than one thousand dollars, such person is guilty of a Class 1 misdemeanor. If the value of the property is one thousand dollars or greater, such person is guilty of a Class 4 felony.

[Current through all 2008 Legislation passed at the 83rd Regular Sessions including Supreme Court Rule 08-09 and the November 2008 General Election]

TENNESSEE
TENN. CODE

Title 39. Criminal Offenses

Chapter 14 Offenses Against Property
Part 7 --Criminal Instruments

39-14-702. Possession of explosive components.

(a) A person commits an offense who unlawfully possesses any component part of an explosive including, but not limited to, a fuse cap, detonator or wiring, with the intent to produce or manufacture an explosive device.

(b) A violation of this section is a Class A misdemeanor.

Chapter 17. Offenses Against Public Health, Safety and Welfare

Part 13. Weapons

39-17-1301. Part Definitions. As used in this part, unless the context otherwise requires: ...

(2) "Crime of violence" includes any degree of murder, voluntary manslaughter, aggravated rape, rape, especially aggravated robbery, aggravated robbery, burglary, aggravated assault or aggravated kidnapping;

(3)(A) "Explosive weapon" means any explosive, incendiary or poisonous gas:

(i) Bomb;

(ii) Grenade;

(iii) Rocket;

(iv) Mine; or

(v) Shell, missile or projectile that is designed, made or adapted for the purpose of inflicting serious bodily injury, death or substantial property damage;

(B) "Explosive weapon" also means:

(i) Any breakable container which contains a flammable liquid with a flashpoint of one hundred fifty degrees (150°) Fahrenheit or less and has a wick or similar device capable of being ignited, other than a device which is commercially manufactured primarily for purposes of illumination; or

(ii) Any sealed device containing dry ice or other chemically reactive substances for the purposes of causing an explosion by a chemical reaction.

(4) "Firearm silencer" means any device designed, made or adapted to muffle the report of a firearm;

(5) "Hoax device" means any device that reasonably appears to be or is purported to be an explosive or incendiary device and is intended to cause alarm or reaction of any type by an official of a public safety agency or a volunteer agency organized to deal with emergencies.

(6) "Immediate vicinity" refers to the area within the person's immediate control within which the person has ready access to the ammunition;...

(9) "Machine gun" means any firearm that is capable of shooting more than two (2) shots automatically, without manual reloading, by a single function of the trigger;

(10) "Restricted firearm ammunition" means any cartridge containing a bullet coated with a plastic substance with other than a lead or lead alloy core or a jacketed bullet with other than a lead or lead alloy core or a cartridge of which the bullet itself is wholly composed of a metal or metal alloy other than lead. "Restricted firearm ammunition" does not include shotgun shells or solid plastic bullets;

(11) "Rifle" means any firearm designed, made or adapted to be fired from the shoulder and to use the energy of the explosive in a fixed metallic cartridge to fire a projectile through a rifled bore by a single function of the trigger;

(12) "Short barrel" means a barrel length of less than sixteen inches (16") for a rifle and eighteen inches (18") for a shotgun, or an overall firearm length of less than twenty-six inches (26");

(13) "Shotgun" means any firearm designed, made or adapted to be fired from the shoulder and to use the energy of the explosive in a fixed shotgun shell to fire through a smooth-bore barrel either a number of ball shot or a single projectile by a single function of the trigger; ...

(15) "Unloaded" means the rifle, shotgun or handgun does not have ammunition in the chamber, cylinder, clip or magazine, and no clip or magazine is in the immediate vicinity of the weapon.

39-17-1302. Prohibited weapons.

(a) A person commits an offense who intentionally or knowingly possesses, manufactures, transports, repairs or sells:

(1) An explosive or an explosive weapon;

(2) A device principally designed, made or adapted for delivering or shooting an explosive weapon;

(3) A machine gun;

(4) A short-barrel rifle or shotgun;

(5) A firearm silencer;

(6) Hoax device;

(7) A switchblade knife or knuckles; or

(8) Any other implement for infliction of serious bodily injury or death which has no common lawful purpose.

(b) It is a defense to prosecution under this section that the person's conduct:

(1) Was incident to the performance of official duty and pursuant to military regulations in the army, navy, air force, coast guard or marine service of the United States or the Tennessee national guard, or was incident to the performance of official duty in a governmental law enforcement agency or a penal institution;

(2) Was incident to engaging in a lawful commercial or business transaction with an organization identified in subdivision (b)(1);

(3) Was incident to using an explosive or an explosive weapon in a manner reasonably related to a lawful industrial or commercial enterprise;

(4) Was incident to using the weapon in a manner reasonably related to a lawful dramatic performance or scientific research;

(5) Was incident to displaying the weapon in a public museum or exhibition;

(6) Was licensed by the state of Tennessee as a manufacturer, importer or dealer in weapons; provided, that the manufacture, import, purchase, possession, sale or disposition of weapons is authorized and incident to carrying on the business for which licensed and is for scientific or research purposes or sale or disposition to the organization designated in subdivision (b)(1);

(7) Involved acquisition or possession of a sawed-off shotgun, sawed-off rifle, machine gun or firearm silencer which is validly registered to the person under federal law in the National Firearms Registration and Transfer Records. A person who acquires or possesses a firearm registered as required by this subdivision shall retain proof of registration; or

(8) Involved the manufacture and sale of an automatic knife; provided, that the sale of such knife was limited to:

(A) Retail establishments that represent in writing under oath before a notary public that they only sell the knives to law enforcement officers, military personnel and emergency medical technicians;

(B) Law enforcement officers;

(C) Military personnel; or

(D) Emergency medical technicians.

(c) It is an affirmative defense to prosecution under this section which the person must prove by a preponderance of the evidence that:

(1) The person's conduct was relative to dealing with the weapon solely as a curio, ornament or keepsake, and if the weapon is a type described in subdivisions (a)(1)-(5), that it was in a nonfunctioning condition and could not readily be made operable; or

(2) The possession was brief and occurred as a consequence of having found the weapon or taken it from an aggressor.

(d)(1) An offense under subdivision (a)(1) is a Class B felony.

(2) An offense under subdivisions (a)(2)-(5) is a class E felony.

(3) An offense under subdivision (a)(6) is a class C felony.

(4) An offense under subdivisions (a)(7)-(8) is a class A misdemeanor.

39-17-1303. Unlawful sale, loan or gift of firearm.

(a) A person commits an offense who:

(1) Intentionally, knowingly or recklessly sells, loans or makes a gift of a firearm or switchblade knife to a minor;

(2) Intentionally, knowingly or recklessly sells a firearm or ammunition for a firearm to a person who is intoxicated; or

(3) Intentionally, knowingly, recklessly or with criminal negligence violates the provisions of § 39-17-1316.

(b) It is a defense to prosecution under subdivision (a)(1) that:

(1) A firearm was loaned or given to a minor for the purposes of hunting, trapping, fishing, camping, sport shooting or any other lawful sporting activity; and

(2) The person is not required to obtain a license under § 39-17-1316.

(c) For purposes of this section, "intoxicated" means substantial impairment of mental or physical capacity resulting from introduction of any substance into the body.

(d) An offense under this section is a Class A misdemeanor.

39-17-1304. Restrictions on firearm ammunition.

(a) It is an offense for any person to possess, use or attempt to use restricted firearm ammunition while committing or attempting to commit a crime of violence. A violation of this section constitutes a separate and distinct felony.

(b) It is an offense for any person or corporation to manufacture, sell, offer for sale, display for sale or use in this state any ammunition cartridge, metallic or otherwise, containing a bullet with a hollow-nose cavity which is filled with an explosive material and designed to detonate upon impact; provided, that the provisions of this section shall not apply to any state or federal military unit or personnel for use in the performance of its duties.

(c)(1) A violation of subsection (a) by possession of restricted firearm ammunition is a Class E felony.

(2) A violation of subsection (a) by use or attempted use of restricted firearm ammunition is a Class D felony.

(3) A violation of subsection (b) is a Class E felony.

39-17-1307. Unlawful carrying or possession of a weapon.

(a)(1) A person commits an offense who carries with the intent to go armed a firearm, a knife with a blade length exceeding four inches (4"), or a club.

(2)(A) The first violation of subdivision (a)(1) is a Class C misdemeanor, and, in addition to possible imprisonment as provided by law, may be punished by a fine not to exceed five hundred dollars ($500).

(B) A second or subsequent violation of subdivision (a)(1) is a Class B misdemeanor.

(C) A violation of subdivision (a)(1) is a Class A misdemeanor if the person's carrying of a handgun occurred at a place open to the public where one (1) or more persons were present.

(b)(1) A person commits an offense who possesses a firearm, as defined in Section 39-11-106, and:

(A) Has been convicted of a felony involving the use or attempted use of force, violence or a deadly weapon; or

(B) Has been convicted of a felony drug offense.

(2) An offense under subdivision (b)(1) is a Class E felony.

(c)(1) A person commits an offense who possesses a handgun and has been convicted of a felony.

(2) An offense under subdivision (c)(1) is a Class E felony.

(d)(1) A person commits an offense who possesses a deadly weapon other than a firearm with the intent to employ it during the commission of, attempt to commit, or escape from a dangerous offense as defined in § 39-17-1324.

(2) A person commits an offense who possesses any deadly weapon with the intent to employ it during the commission of, attempt to commit, or escape from any offense not defined as a dangerous offense by § 39-17-1324.

(3) A violation of this subsection (d) is a Class E felony.

39-17-1308. Defenses to unlawful possession or carrying of a weapon.

(a) It is a defense to the application of § 39-17-1307 if the possession or carrying was:

(1) Of an unloaded rifle, shotgun or handgun not concealed on or about the person and the ammunition for the weapon was not in the immediate vicinity of the person or weapon;

(2) By a person authorized to possess or carry a firearm pursuant to § 39-17-1315 or § 39-17-1351;

(3) At the person's:

(A) Place of residence;

(B) Place of business; or

(C) Premises;

(4) Incident to lawful hunting, trapping, fishing, camping, sport shooting or other lawful activity;

(5) By a person possessing a rifle or shotgun while engaged in the lawful protection of livestock from predatory animals;

(6) By a Tennessee valley authority officer who holds a valid commission from the commissioner of safety pursuant to this part while such officer is in the performance of the officer's official duties;

(7) By a state, county or municipal judge or any federal judge or any federal or county magistrate; ...

(10) By any out-of-state, full-time, commissioned law enforcement officer who holds a valid commission card from the appropriate out-of-state law enforcement agency and a photo identification; provided, that if no such valid commission card and photo identification are retained, then it shall be unlawful for such officer to carry firearms in this state and the provisions of this section shall not apply. The defense provided by this subdivision (a)(10) shall only be applicable if the state where the out-of-state officer is employed has entered into a reciprocity agreement with this state that allows a full-time, commissioned law enforcement officer in Tennessee to lawfully car-ry or possess a weapon in such other state.

(b) The defenses described in this section are not available to persons described in § 39-17-1307(b)(1).

39-17-1309. Carrying weapons on school property.

(a) As used in this section, "weapon of like kind" includes razors and razor blades, except those used solely for personal shaving, and any sharp pointed or edged instrument, except unaltered nail files and clips and tools used solely for preparation of food, instruction and maintenance.

(b)(1) It is an offense for any person to possess or carry, whether openly or concealed, with the intent to go armed, any firearm, explosive, explosive weapon, bowie knife, hawk bill knife, ice pick, dagger, slingshot, leaded cane, switchblade knife, blackjack, knuckles or any other weapon of like kind, not used solely for instructional or school-sanctioned ceremonial purposes, in any public or private school building or bus, on any public or private school campus, grounds, recreation area, athletic field or any other property owned, used or operated by any board of education, school, college or university board of trustees, regents or directors for the administration of any public or private educational institution.

(2) A violation of this subsection (b) is a Class E felony.

(c)(1) It is an offense for any person to possess or carry, whether openly or concealed, any firearm, not used solely for instructional or school-sanctioned ceremonial purposes, in any public or private school building or bus, on any public or private school campus, grounds, recreation area, athletic field or any other property owned, used or operated by any board of education, school, college or university board of trustees, regents or directors for the administration of any public or private educational institution. It is not an offense under this subsection (c) for a nonstudent adult to possess a firearm, if the firearm is contained within a private vehicle operated by the adult and is not handled by the adult, or by any other person acting with the expressed or implied consent of such adult, while such vehicle is on school property.

(2) A violation of this subsection (c) is a Class B misdemeanor.

(d)(1) Each chief administrator of a public or private school shall display in prominent locations about the school a sign, at least six inches (6") high and fourteen inches (14") wide, stating: FELONY. STATE LAW PRESCRIBES A MAXIMUM PENALTY OF SIX (6) YEARS IMPRISONMENT AND A FINE NOT TO EXCEED THREE THOUSAND DOLLARS ($3,000) FOR CARRYING WEAPONS ON SCHOOL PROPERTY.

(2) As used in this subsection (d), "prominent locations about a school" includes, but is not limited to, sports arenas, gymnasiums, stadiums and cafeterias.

(e) The provisions of subsections (b) and (c) do not apply to the following persons:

(1) Persons employed in the army, air force, navy, coast guard or marine service of the United States or any member of the Tennessee national guard when in discharge of their official duties and acting under orders requiring them to carry arms or weapons;

(2) Civil officers of the United States in the discharge of their official duties;

(3) Officers and soldiers of the militia and the national guard when called into actual service;

(4) Officers of the state, or of any county, city or town, charged with the enforcement of the laws of the state, when in the discharge of their official duties;

(5) Any pupils who are members of the reserve officers training corps or pupils enrolled in a course of instruction or members of a club or team, and who are required to carry arms or weapons in the discharge of their official class or team duties;

(6) Any private police employed by the administration or board of trustees of any public or private institution of higher education in the discharge of their duties; and

(7) Any registered security officer/guard who meets the requirements of title 62, chapter 35, and who is discharging such officer's official duties.

39-17-1310. Affirmative defense to carrying weapons on school property. It is an affirmative defense to prosecution under § 39-17-1309 (a)-(d) that the person's behavior was in strict compliance with the requirements of one (1) of the following classifications:

(1) A person hunting during the lawful hunting season on lands owned by any public or private educational institution and designated as open to hunting by the administrator of the educational institution;

(2) A person possessing unloaded hunting weapons while transversing the grounds of any public or private educational institution for the purpose of gaining access to public or private lands open to hunting with the intent to hunt on the public or private lands unless the lands of the educational institution are posted prohibiting such entry;

(3) A person possessing guns or knives when conducting or attending "gun and knife shows" and the program has been approved by the administrator of the educational institution; or

(4) A person entering the property for the sole purpose of delivering or picking up passengers and who does not remove, utilize or allow to be removed or utilized any weapon from the vehicle.

39-17-1312. Inaction by persons eighteen (18) years of age or older, including parents or guardians, knowing a minor or student illegally possessing a firearm.

(a) It is an offense if a person eighteen (18) years of age or older, including a parent or other legal guardian, knows that a minor or student is in illegal possession of a firearm in or upon the premises of a public or private school, in or on such school's athletic stadium or other facility or building where school sponsored athletic events are conducted, or public park, playground or civic center, and the person, parent or guardian fails to prevent the possession or fails to report it to the appropriate school or law enforcement officials.

(b) A violation of this section is a Class A misdemeanor.

39-17-1314. Local regulation of firearms and ammunition preempted by state regulation - Actions against firearms or ammunition manufacturers, trade associations or dealers.

(a) No city, county, or metropolitan government shall occupy any part of the field of regulation of the transfer, ownership, possession or transportation of firearms, ammunition or components of firearms or combinations thereof; provided, that the provisions of this section shall be prospective only and shall not affect the validity of any ordinance or resolution lawfully enacted before April 8, 1986.

(b) The general assembly declares that the lawful design, marketing, manufacture and sale of firearms and ammunition to the public are not unreasonably dangerous activities and do not constitute a nuisance per se. ...

39-17-1315. Written directive and permit to carry handguns.

(a)(1) Any law enforcement officer, police officer, bonded and sworn deputy sheriff, director, commissioner, county magistrate or retired law enforcement officer who is bonded and who, at the time of receiving the written directive, has successfully completed and, except for a law enforcement officer who has retired in good standing as certified by the chief law enforcement officer of the organization from which the officer retired, continues to successfully complete on an annual basis a firearm training program of at least eight (8) hours duration, director or full-time employee of the Tennessee emergency management agency in the performance of the director's or employee's duty, any duly authorized representative or full-time employee of the board of probation and parole who has been specifically designated by the board to execute warrants issued pursuant to § 40-28-121 or § 40-35-311 or to perform such other duties as specifically designated by the board, or any other officer or person authorized to carry handguns by this, or any other law of this state, may carry handguns at all times pursuant to a written directive by the executive supervisor of the organization to which the person is or was attached or employed, regardless of the person's regular duty hours or assignments; however, a copy of the written directive shall be retained as a portion of the records of the particular law enforcement agency that shall issue the directive. Nothing in this subdivision (a)(1) shall prevent federal officers from carrying firearms as prescribed by federal law.

(2) Any duly elected and sworn constable in any county having a population of not less than eleven thousand one hundred (11,100) nor more than eleven thousand two hundred (11,200) according to the 1970 federal census or any subsequent federal census, and being a county in which such constables retain law enforcement powers and duties under the provisions of §§ 8-10-108, 40-6-210, 55-8-152, 57-5-202 and 57-9-101, are authorized to and may carry handguns at all times and may equip their vehicles with blue and red lights and sirens. The sheriff of such county shall issue a written directive or permit authorizing such constables to carry a handgun; provided, that each such constable has completed the same eight-hour annual firearm training program as is required by this subsection (a).

(3) The county commission may, by a two-thirds (2/3) vote, require any such constable to have in effect a liability policy or a corporate surety bond in an amount of not less than fifty thousand dollars ($50,000).

(b)(1) An individual, corporation or business entity is authorized to prohibit the possession of weapons by employees otherwise authorized by this subsection (b) on premises owned, operated or managed by such individual, corporation or busi-ness entity. Notice of such prohibition there-under shall be posted or otherwise noticed to all affected employees.

(2) An individual, corporation, business entity or governmental entity or agent thereof is authorized to prohibit possession of weapons by any person otherwise authorized by this subsection (b), at meetings conducted by, or on premises owned, operated, managed or under control of such individual, corporation, business entity or governmental entity. Notice of such prohibition shall be posted or announced.

39-17-1316. Sales of dangerous weapons – Certification of purchaser - Exceptions - Licensing of dealers - Definitions.

(a)(1) Any person appropriately licensed by the federal government may stock and sell firearms to persons desiring them; however, sales to persons who have been convicted of the offense of stalking as prohibited by § 39-17-315, who are addicted to alcohol, and who are ineligible to receive them under 18 U.S.C. § 922 are prohibited. For purposes of this subdivision (a)(1), the offense of violation of a protective order as prohibited by § 39-13-113 shall be considered a "misdemeanor crime of domestic violence" for purposes of 18 U.S.C. sec. 921.

(2) The provisions of this subsection (a) prohibiting the sale of a firearm to a person convicted of a felony shall not apply if:

(A) The person was pardoned for the offense;

(B) The conviction has been expunged or set aside; or

(C) The person's civil rights have been restored pursuant to title 40, chapter 29; and

(D) The person is not prohibited from possessing a firearm by the provisions of § 39-17-1307.

(b)(1) As used in this section, "firearm" has the meaning as defined in § 39-11-106, including handguns, long guns, and all other weapons which meet the definition except "antique firearms" as defined in 18 U.S.C. § 921.

(2) As used in this section, "gun dealer" means a person engaged in the business, as defined in 18 U.S.C. § 921, of selling, leasing, or otherwise transferring a firearm, whether the person is a retail dealer, pawnbroker, or otherwise.

(c) Except with respect to transactions between persons licensed as dealers under 18 U.S.C. § 923, a gun dealer shall comply with the following before a firearm is delivered to a purchaser:

(1) The purchaser shall present to the dealer current identification meeting the requirements of subsection (f);

(2) The gun dealer shall complete a firearms transaction record as required by 18 U.S.C. §§ 921 to 929, and obtain the signature of the purchaser on the record;

(3) The gun dealer shall obtain the thumbprints of the purchaser on the firearms transaction thumbprint form and attach the form to the gun dealer's copy of the firearms transaction record to be filed with that copy;

(4) The gun dealer shall request by telephone that the Tennessee bureau of investigation conduct a criminal history record check on the purchaser and shall provide the following information to the bureau:

(A) The federal firearms license number of the gun dealer;

(B) The business name of the gun dealer;

(C) The place of transfer;

(D) The name of the person making the transfer;

(E) The make, model, caliber and manufacturer's number of the firearm being transferred;

(F) The name, gender, race, and date of birth of the purchaser;

(G) The social security number of the purchaser, if one has been assigned; and

(H) The type, issuer and identification number of the identification presented by the purchaser.

(5) The gun dealer shall receive a unique approval number for the transfer from the bureau and record the approval number on the firearms transaction record and on the firearms transaction thumbprint form; and

(6) The gun dealer may destroy the firearms transaction thumbprint form one (1) year after the completion of the firearms transaction thumbprint form.

(d) Upon receipt of a request of the gun dealer for a criminal history record check, the Tennessee bureau of investigation shall immediately, during the gun dealer's telephone call or by return call:

(1) Determine, from criminal records and other information available to it, whether the purchaser is disqualified under the provisions of subdivision (a)(1) from completing the purchase; and

(2) Notify the dealer when a purchaser is disqualified from completing the transfer or provide the dealer with a unique approval number indicating that the purchaser is qualified to complete the transfer.

(e)(1) The Tennessee bureau of investigation may charge a reasonable fee, not to exceed ten dollars ($10.00), for conducting background checks and other costs incurred under the provisions of this section, and shall be empowered to bill gun dealers for checks run. ...

(f)(1) Identification required of the purchaser under subsection (c) shall include one (1) piece of current, valid identification bearing a photograph and the date of birth of the purchaser that:

(A) Is issued under the authority of the United States government, a state, a political subdivision of a state, a foreign government, a political subdivision of a foreign government, an international governmental organization or an international quasi-governmental organization; and

(B) Is intended to be used for identification of an individual or is commonly accepted for the purpose of identification of an individual.

(2) If the identification presented by the purchaser under subdivision (f)(1)(A) does not include the current address of the purchaser, the purchaser shall present a second piece of current identification that contains the current address of the purchaser.

(g) The Tennessee bureau of investigation may require that the dealer verify the identification of the purchaser if that identity is in question by sending the thumbprints of the purchaser to the bureau.

(h) The Tennessee bureau of investigation shall establish a telephone number that shall be operational seven (7) days a week between the hours of eight o'clock a.m. and ten o'clock p.m. Central Standard Time (8:00 a.m.-10:00 p.m. (CST)), except Christmas Day, Thanksgiving Day, and Independence Day, for the purpose of responding to inquiries from dealers for a criminal history record check under this section.

(i) No public employee, official or agency shall be held criminally or civilly liable for performing the investigations required by this section; provided the employee, official or agency acts in good faith and without malice.

(j) Upon the determination that receipt of a firearm by a particular individual would not violate this section, and after the issuance of a unique identifying number for the transaction, the Tennessee bureau of investigation shall destroy all records (except the unique identifying number and the date that it was assigned) associating a particular individual with a particular purchase of firearms.

(k) A law enforcement agency may inspect the records of a gun dealer relating to transfers of firearms in the course of a reasonable inquiry during a criminal investigation or under the authority of a properly authorized subpoena or search warrant.

(l) The Tennessee bureau of investigation shall publish the firearms transaction thumbprint form and shall furnish the form to gun dealers on application at cost.

(m)(1) The background check does not apply to transactions between licensed importers, licensed manufacturers, licensed dealers, or licensed collectors who meet the requirements of subsection (b) and certify prior to the transaction the legal and licensed status of both parties. The burden shall fall upon the transferor to determine the legality of the transaction in progress.

(2) The background check does not apply to transactions or transfers between a licensed importer, licensed manufacturer, or licensed dealer and a bona fide law enforcement agency or such agency's personnel. However, all other provisions and requirements of subsection (b) must be observed. The burden of proof of the legality of the transactions or transfers shall rest upon the transferor.

(3) The background check does not apply to any person eligible to purchase a firearm as set out in this section who wishes to make an occasional sale of a used or second-hand firearm legally purchased by the seller.

(n) The director of the Tennessee bureau of investigation is hereby authorized to make and issue all rules and regulations necessary to carry out the provisions of this section.

(o) In addition to the other grounds for denial, the bureau shall deny the transfer of a firearm if the background check reveals information indicating that the purchaser has been charged with a crime for which the purchaser, if convicted, would be prohibited under state or federal law from purchasing, receiving, or possessing a firearm; and, either there has been no final disposition of the case, or the final disposition is not noted.

(p) Upon receipt of the criminal history challenge form indicating a purchaser's request for review of the denial, the bureau shall proceed with efforts to obtain the final disposition information. The purchaser may attempt to assist the bureau in obtaining the final disposition information. If neither the purchaser nor the bureau is able to obtain the final disposition information within fifteen (15) calendar days of the bureau's receipt of the criminal history challenge form, the bureau shall immediately notify the federal firearms licensee that the transaction that was initially denied is now a "conditional proceed". A "conditional proceed" means that the federal firearms licensee may lawfully transfer the firearm to the purchaser.

(q) In any case in which the transfer has been denied pursuant to subsection (o), the inability of the bureau to obtain the final disposition of a case shall not constitute the basis for the continued denial of the transfer as long as the bureau receives written notice, signed and verified by the clerk of the court or the clerk's designee, that indicates that no final disposition information is available. Upon receipt of such letter by the bureau, the bureau shall immediately reverse the denial.

39-17-1319. Handgun possession prohibited - Exceptions.

(a) As used in this section and § 39-17-1320, unless the context otherwise requires:

(1) "Handgun" means a pistol, revolver or other firearm of any description, loaded or unloaded, from which any shot, bullet, or other missile can be discharged, the length of the barrel of which, not including any revolving, detachable, or magazine breech, does not exceed twelve inches (12"); and

(2) "Juvenile" means any person less than eighteen (18) years of age.

(b) Except as provided in this section, it is an offense for a juvenile to knowingly possess a handgun.

(c)(1) Illegal possession of a handgun by a juvenile is a delinquent act and, in addition to any other disposition authorized by law, such juvenile may be required to perform not more than one hundred (100) hours of community service work to be specified by the judge, and such juvenile's driving privileges shall be suspended for a period of one (1) year in accordance with the procedure set out in title 55, chapter 10, part 7.

(2) A second or subsequent violation of this section is a delinquent act and, in addition to any other disposition authorized by law, such juvenile may be required to perform not less than one hundred (100) nor more than two hundred (200) hours of community service work to be specified by the judge, and such juvenile's driving privileges shall be suspended for a period of two (2) years in accordance with the procedure set out in title 55, chapter 10, part 7.

(3) Any handgun illegally possessed in violation of this section shall be confiscated and disposed of in accordance with the provisions of § 39-17-1317.

(d)(1) It is a defense to prosecution under this section that the juvenile is:

(A) In attendance at a hunter's safety course or a firearms safety course;

(B) Engaging in practice in the use of a firearm or target shooting at an established range authorized by the governing body of the jurisdiction in which such range is located or any other area where the discharge of a firearm is not prohibited;

(C) Engaging in an organized competition involving the use of a firearm, or participating in or practicing for a performance by an organized group which is exempt from federal income taxation under § 501(c)(3) of the Internal Revenue Code of 1986 (26 U.S.C. § 501(c)(3)), as amended, and which uses firearms as part of such performance;

(D) Hunting or trapping pursuant to a valid license issued to such juvenile pursuant to title 70;

(E) Accompanied by such juvenile's parent or guardian and is being instructed by such adult or guardian in the use of the handgun possessed by the juvenile;

(F) On real property which is under the control of an adult and has the permission of that adult and the juvenile's parent or legal guardian to possess a handgun;

(G) Traveling to or from any activity described in subsection (d)(1) with an unloaded gun; or

(H) At the juvenile's residence and with the permission of the juvenile's parent or legal guardian, possesses a handgun and is justified in using physical force or deadly force.

(2) For purposes of subsection (d)(1)(G) a handgun is "unloaded" if:

(A) There is not a cartridge in the chamber of the handgun;

(B) There is not a cartridge in the cylinder of the handgun if the handgun is a revolver; or

(C) The handgun, and the ammunition for such handgun, are not carried on the person of a juvenile or are not in such close proximity to such juvenile that the juvenile could readily gain access to the handgun and the ammunition and load the handgun.

(e) Notwithstanding any other provision of this part to the contrary, the provisions of this section shall govern a juvenile who possesses a handgun.

39-17-1320. Providing handguns to juveniles - Penalties.

(a) It is an offense for a person intentionally, knowingly or recklessly to provide a handgun with or without remuneration to any person that the person providing the handgun knows or has reason to believe is a juvenile in violation of § 39-17-1319.

(b) It is an offense for a parent or guardian intentionally, knowingly or recklessly to provide a handgun to a juvenile or permit a juvenile to possess a handgun, if such parent or guardian knows of a substantial risk that such juvenile will use a handgun to commit a felony.

(c) Unlawfully providing or permitting a juvenile to possess a handgun in violation of subsection (a) is a Class A misdemeanor and in violation of subsection (b) is a Class D felony.

39-17-1321. Possession of handgun while under influence - Penalty.

(a) Notwithstanding whether a person has a permit issued pursuant to § 39-17-1315 or § 39-17-1351, it is an offense for a person to possess a handgun while under the influence of alcohol or any controlled substance.

(b) A violation of this section is a Class A misdemeanor.

39-17-1350. Law enforcement officers permitted to carry firearms at all times.

(a) Notwithstanding any provision of law to the contrary, any law enforcement officer may carry firearms at all times and in all places within Tennessee, on-duty or off-duty, regardless of the officer's regular duty hours or assignments, except as provided by subsection (c), federal law, lawful orders of court or the written directives of the executive supervisor of the employing agency.

(b) The authority conferred by this section is expressly intended to and shall supersede restrictions placed upon law enforcement officers' authority to carry firearms by other sections within this part.

(c) The authority conferred by this section shall not extend to a law enforcement officer:

(1) Who carries a firearm onto school grounds or inside a school building during regular school hours unless such officer immediately informs the principal that such officer will be present on school grounds or inside the school building and in possession of a firearm. If the principal is unavailable, the notice may be given to an appropriate administrative staff person in the principal's office;

(2) Who is consuming beer or an alcoholic beverage or who is under the influence of beer, an alcoholic beverage, or a controlled substance;

(3) Who is not engaged in the actual discharge of official duties as a law enforcement officer while within the confines of an establishment where beer or alcoholic beverages are sold for consumption on-the-premises; or

(4) Who is not engaged in the actual discharge of official duties as a law enforcement officer while attending a judicial proceeding.

(d) For purposes of this section, "law enforcement officer" means a person who is a full-time employee of the state in a position authorized by the laws of this state to carry a firearm and to make arrests for violations of some or all of the laws of this state, or a full-time police officer who has been certified by the peace officer standards and training commission, or a vested correctional officer employed by the department of correction.

(e) In counties having a population of not less than thirty thousand two hundred (30,200) nor more than thirty thousand four hundred seventy-five (30,475) or not less than one hundred eighteen thousand four hundred (118,400) nor more than one hundred eighteen thousand seven hundred (118,700) according to the 1990 federal census or any subsequent federal census, the authority conferred by this section shall only apply to law enforcement officers who are law enforcement officers for those counties or law enforcement officers for municipalities located therein.

(f)(1) The secretary of state shall, in consultation with the commissioner of the department of correction, design and issue to each requesting correction officer who is vested and employed by the department of correction, a state identification card certifying that the officer is authorized to carry a firearm pursuant to this section.

(2) Any correction officer desiring such an identification card shall notify the secretary of state and shall provide the officer's full name and residential address. Upon receipt of such a request, the secretary of state shall notify the commissioner of correction of the request. The commissioner of correction shall verify to the secretary of state whether the requesting officer is vested and employed by the department of correction and shall so certify in a letter to be maintained by the secretary.

(3) If the secretary of state receives certification that a requesting officer is vested and employed by the department, the secretary shall issue such officer an identification card so certifying. Such card shall be valid for as long as the officer remains vested and in the employment of the department of correction.

(4) A corrections officer issued a card pursuant to this subsection shall carry such card at all times the officer is carrying a firearm. Such card shall be sufficient proof that the corrections officer is authorized to carry a firearm pursuant to this section.

(5) If a vested corrections officer employed by the department resigns, is terminated, or is otherwise no longer employed by the department, the commissioner shall, within ten (10) days, so notify the secretary of state. Upon receiving such notice, the secretary of state shall revoke the identification card and send a letter of revocation to the corrections officer at such officer's last known address.

(6)(A) A person who is no longer a vested corrections officer employed by the department of correction but who still has an identification card issued by the secretary of state shall have ten (10) days from receipt of the letter of revocation from the secretary of state to return such card to the secretary.

(B) It is a Class C misdemeanor punishable by fine only of $50.00 for a person to knowingly fail to return an identification card as required by this subpart.

39-17-1351. Handgun carry permits.

(a) The citizens of this state have a right to keep and bear arms for their common defense; but the general assembly has the power, by law, to regulate the wearing of arms with a view to prevent crime.

(b) Except as provided in subsection (r), any resident of Tennessee who is a United States citizen or permanent lawful resident, as defined by § 55-50-102(30), who has reached twenty-one (21) years of age, may apply to the department of safety for a handgun carry permit. If the applicant is not prohibited from purchasing or possessing a firearm in this state pursuant to § 39-17-1316 or § 39-17-1307(b), 18 U.S.C. § 922(g), or any other state or federal law, and the applicant otherwise meets all of the requirements of this section, the department shall issue a permit to the applicant.

(c) The application for a permit shall be on a standard form developed by the department. The application shall clearly state in bold face type directly above the signature line that an applicant who, with intent to deceive, makes any false statement on such application commits the felony offense of perjury pursuant to § 39-16-702. The following are eligibility requirements for obtaining a handgun carry permit and the application shall require the applicant to disclose and confirm compliance with, under oath, the following information concerning the applicant and the eligibility requirements:

(1) Full legal name and any aliases;

(2) Addresses for the last five (5) years;

(3) Date of birth;

(4) Social security number;

(5) Physical description (height, weight, race, sex, hair color and eye color);

(6) That the applicant has not been convicted of a criminal offense that is designated as a felony, or that is one of the disqualifying misdemeanors set out in (c)(11), (c)(16), or (c)(18), with the exception of any federal or state offenses pertaining to antitrust violations, unfair trade practices, restraints of trade or other similar offenses relating to the regulations of business practices;

(7) That the applicant is not currently under indictment or information for any criminal offense that is designated as a felony, or that is one of the disqualifying misdemeanors set out in (c)(11), (c)(16), or (c)(18), with the exception of any federal or state offenses pertaining to antitrust violations, unfair trade practices, restraints of trade or other similar offenses relating to the regulations of business practices;

(8) That the applicant is not currently subject to any order of protection and, if so, the applicant shall provide a copy of the order;

(9) That the applicant is not a fugitive from justice;

(10) That the applicant is not an unlawful user of or addicted to alcohol or any controlled substance and the applicant has not been a patient in a rehabilitation program or hospitalized for alcohol or controlled substance abuse or addiction within ten (10) years from the date of application;

(11) That the applicant has not been convicted of the offense of driving under the influence of an intoxicant in this or any other state two (2) or more times within ten (10) years from the date of the application and that none of such convictions has occurred within five (5) years from the date of application or renewal;

(12) That the applicant has not been adjudicated as a mental defective; has not been committed to or hospitalized in a mental institution; has not had a court appoint a conservator for the applicant by reason of a mental defect; has not been judicially determined to be disabled by reason of mental illness, developmental disability or other mental incapacity; and has not, within seven (7) years from the date of application, been found by a court to pose an immediate substantial likelihood of serious harm, as defined in title 33, chapter 6, part 5, because of mental illness;

(13) That the applicant is not an alien and is not illegally or unlawfully in the United States;

(14) That the applicant has not been discharged from the Armed Forces under dishonorable conditions;

(15) That the applicant has not renounced the applicant's United States citizenship;

(16) That the applicant has not been convicted of a misdemeanor crime of domestic violence as defined in 18 U.S.C. § 921(33);

(17) That the applicant is not receiving social security disability benefits by reason of alcohol dependence, drug dependence or mental disability; and

(18) That the applicant has not been convicted of the offense of stalking...

(g)(1) Upon receipt of a permit application, the department shall:

(A) Forward two (2) full sets of fingerprints of the applicant to the Tennessee bureau of investigation; and

(B) Send a copy of the application to the sheriff of the county in which the applicant resides.

(2) Within thirty (30) days of receiving an application, the sheriff shall provide the department with any information concerning the truthfulness of the applicant's answers to the eligibility requirements of subsection (c) that is within the knowledge of the sheriff.

(h) Upon receipt of the fingerprints from the department, the Tennessee bureau of investigation shall:

(1) Within thirty (30) days from receipt of the fingerprints, conduct computer searches to determine the applicant's eligibility for a permit under subsection (c) as are available to the bureau based solely upon the applicant's name, date of birth and social security number and send the results of the searches to the department;

(2) Conduct a criminal history record check based upon one (1) set of the fingerprints received and send the results to the department; and

(3) Send one (1) set of the fingerprints received from the department to the federal bureau of investigation, request a federal criminal history record check based upon such fingerprints, as long as such service is available, and send the results of such check to the department.

(i) The department shall deny a permit application if it determines from information contained in the criminal history record checks conducted by the Tennessee and federal bureaus of investigation pursuant to subsection (h), or from other information that comes to the attention of the department, that the applicant does not meet the eligibility requirements of this section. The department shall not be required to confirm the applicant's eligibility for a permit beyond the information received from the Tennessee and federal bureaus of investigation and the sheriffs, if any...

(n)(1) Except as provided in (n)(2), a permit issued pursuant to this section shall be good for four (4) years and shall entitle the permit holder to carry any handgun or handguns that the permit holder legally owns or possesses. The permit holder shall have the permit in the holder's immediate possession at all times when carrying a handgun and shall display the permit on demand of a law enforcement officer.

(2) A Tennessee permit issued pursuant to this section to a person who is in or who enters into the United States armed forces shall continue in effect for so long as the person's service continues and such person is stationed outside this state, notwithstanding the fact that the person may be temporarily in this state on furlough, leave, or delay en route, and for a period not to exceed sixty (60) days following the date on which the person is honorably discharged or separated from service or returns to this state on reassignment to a duty station in this state, unless the permit is sooner suspended, cancelled or revoked for cause as provided by law. The permit is valid only when in the immediate possession of the permit holder and the permit holder has in the holder's immediate possession the holder's discharge or separation papers, if the permit holder has been discharged or separated from the service. ...

(q)(1) Prior to the expiration of a permit, a permit holder may apply to the department for the renewal of such permit by submitting, under oath, a renewal application with a renewal fee of fifty dollars ($50.00). The renewal application shall be on a standard form developed by the department of safety and shall require the applicant to disclose, under oath, the information concerning the applicant as set forth in subsection (c), and shall require the applicant to certify that the applicant still satisfies all the eligibility requirements of this section for the issuance of a permit. In the event the permit expires prior to the department's approval or issuance of notice of denial regarding the renewal application, the permit holder shall be entitled to continue to use the expired permit; provided, however, that the permit holder shall also be required to prove by displaying a receipt for the renewal application fee that the renewal application was delivered to the department prior to the expiration date of the permit.

(2) Any person whose handgun carry permit expires and who applies for a renewal of such handgun carry permit within six (6) months from the date of expiration shall only be required to comply with the renewal provisions of subdivision (q)(1). If the renewal application is filed six (6) months or more from the date of expiration, the person shall, for all purposes, be considered a new applicant.

(3) If a person whose handgun carry permit remained valid pursuant to subsection (n)(2) because such person was in the United States armed forces applies for a renewal of such permit within six (6) months of the expiration of the sixty (60) day period following discharge, separation, or return to this state on reassignment to a duty station in this state as provided in subdivision (n)(2), such person shall only be required to comply with the renewal provisions of subdivision (q)(1). If the renewal application is filed six (6) months or more from expiration of the sixty (60) day period following the date of honorable discharge, separation, or return to this state on reassignment to a duty station in this state, the person shall, for all purposes, be considered a new applicant.

(r)(1) A facially valid handgun permit, firearms permit, weapons permit or license issued by another state shall be valid in this state according to its terms and shall be treated as if it is a handgun permit issued by this state; provided, however, the provisions of this subsection (r) shall not be construed to authorize the holder of any out-of-state permit or license to carry, in this state, any firearm or weapon other than a handgun.

(2) For a person to lawfully carry a handgun in this state based upon a permit or license issued in another state, the person must be in possession of the permit or license at all times such person carries a handgun in this state.

(3)(A) The commissioner of safety shall enter into written reciprocity agreements with other states that require the execution of the agreements. The commissioner of safety shall prepare and publicly publish a current list of states honoring permits issued by the state of Tennessee and shall make the list available to anyone upon request. The commissioner of safety shall also prepare and publicly publish a current list of states who, after inquiry by the commissioner, refuse to enter into a reciprocity agreement with this state or honor handgun carry permits issued by this state. To the extent that any state may impose conditions in the reciprocity agreements, the commissioner of safety shall publish those conditions as part of the list. If another state imposes conditions on Tennessee permit holders in a reciprocity agreement, the conditions shall also become a part of the agreement and apply to the other state's permit holders when they carry a handgun in this state.

(B) If a person with a handgun permit from another state decides to become a resident of Tennessee, the person must obtain a Tennessee handgun permit within six (6) months of establishing residency in Tennessee. The permit may be issued based on the person having a permit from another state provided the other state has substantially similar permit eligibility requirements as this state. However, if during such six (6) month period the person applies for a handgun permit in this state and the application is denied, the person shall not be allowed to carry a handgun in this state based upon the other state's permit.

(C)(i) If a person who is a resident of and handgun permit holder in another state is employed in this state on a regular basis and desires to carry a handgun in this state, the person shall have six (6) months from the last day of the sixth month of regular employment in this state to obtain a Tennessee handgun carry permit. The permit may be issued based on the person having a permit from another state provided the other state has substantially similar permit eligibility requirements as this state. However, if during the six (6) month period the person applies for a handgun permit in this state and the application is denied, the person shall not be allowed to carry a handgun in this state based upon the other state's permit.

(ii) The provisions of this subdivision (r)(3)(C) shall not apply if the state of residence of the person employed in Tennessee has entered into a handgun permit reciprocity agreement with this state pursuant to this subsection (r).

(iii) As used in this subdivision (r)(3)(C), "employed in this state on a regular basis" means a person has been gainfully employed in this state for at least thirty (30) hours a week for six (6) consecutive months not counting any absence from employment caused by the employee's use of sick leave, annual leave, administrative leave or compensatory time. ...

(v) Any permit issued pursuant to this section shall be deemed a "license" within the meaning of title 36, chapter 5, part 7, dealing with the enforcement of child support obligations through license denial and revocation.

39-17-1360. Rules and regulations. The department of safety is authorized to promulgate rules and regulations pursuant to the Uniform Administrative Procedures Act, compiled in title

4, chapter 5, to implement the provisions of §§ 39-17-1351 - 39-17-1360.

[Current through 2008 Regular Session]

TEXAS
Tex. Code

Texas Family Code

Title 4. Protective Orders and Family Violence

Chapter 71. Definitions

71.003. Family "Family" includes individuals related by consanguinity or affinity, as determined under Sections 573.022 and 573.024, Government Code, individuals who are former spouses of each other, individuals who are the parents of the same child, without regard to marriage, and a foster child and foster parent, without regard to whether those individuals reside together.

71.005. Household "Household" means a unit composed of persons living together in the same dwelling, without regard to whether they are related to each other.

71.006. Member of a Household "Member of a household" includes a person who previously lived in a household.

Subtitle B. Protective Orders

Chapter 85. Issuance of Protective Order

Subchapter B. Contents of Protective Order

85.026. Warning on Protective Order

(a) Each protective order issued under this subtitle, including a temporary ex parte order, must contain the following prominently displayed statements in boldfaced type, capital letters, or underlined:

"A PERSON WHO VIOLATES THIS ORDER MAY BE PUNISHED FOR CONTEMPT OF COURT BY A FINE OF AS MUCH AS $ 500 OR BY CONFINEMENT IN JAIL FOR AS LONG AS SIX MONTHS, OR BOTH."

"NO PERSON, INCLUDING A PERSON WHO IS PROTECTED BY THIS ORDER, MAY GIVE PERMISSION TO ANYONE TO IGNORE OR VIOLATE ANY PROVISION OF THIS ORDER. DURING THE TIME IN WHICH THIS ORDER IS VALID, EVERY PROVISION OF THIS ORDER IS IN FULL FORCE AND EFFECT UNLESS A COURT CHANGES THE ORDER."

"IT IS UNLAWFUL FOR ANY PERSON, OTHER THAN A PEACE OFFICER, AS DEFINED BY SECTION 1.07, PENAL CODE, ACTIVELY ENGAGED IN EMPLOYMENT AS A SWORN, FULL-TIME PAID EMPLOYEE OF A STATE AGENCY OR POLITICAL SUBDIVISION, WHO IS SUBJECT TO A PROTECTIVE ORDER TO POSSESS A FIREARM OR AMMUNITION."

(b) Each protective order issued under this subtitle, except for a temporary ex parte order, must contain the following prominently displayed statement in boldfaced type, capital letters, or underlined:

"A VIOLATION OF THIS ORDER BY COMMISSION OF AN ACT PROHIBITED BY THE ORDER MAY BE PUNISHABLE BY A FINE OF AS MUCH AS $ 4,000 OR BY CONFINEMENT IN JAIL FOR AS LONG AS ONE YEAR, OR BOTH. AN ACT THAT RESULTS IN FAMILY VIOLENCE MAY BE PROSECUTED AS A SEPARATE MISDEMEANOR OR FELONY OFFENSE. IF THE ACT IS PROSECUTED AS A SEPARATE FELONY OFFENSE, IT IS PUNISHABLE BY CONFINEMENT IN PRISON FOR AT LEAST TWO YEARS."

(c) Each protective order issued under this subtitle, including a temporary ex parte order, must contain the following prominently displayed statement in boldfaced type, capital letters, or underlined:

"NO PERSON, INCLUDING A PERSON WHO IS PROTECTED BY THIS ORDER, MAY GIVE PERMISSION TO ANYONE TO IGNORE OR VIOLATE ANY PROVISION OF THIS ORDER. DURING THE TIME IN WHICH THIS ORDER IS VALID, EVERY PROVISION OF THIS ORDER IS IN FULL FORCE AND EFFECT UNLESS A COURT CHANGES THE ORDER."

Chapter 86. Law Enforcement Duties Relating to Protective Orders

86.002. Duty to Provide Information to Firearms Dealers

(a) On receipt of a request for a law enforcement information system record check of a prospective transferee by a licensed firearms dealer under the Brady Handgun Violence Prevention Act, 18 U.S.C. Section 922, the chief law enforcement officer shall determine whether the Department of Public Safety has in the department's law enforcement information system a record indicating the existence of an active protective order directed to the prospective transferee.

(b) If the department's law enforcement information system indicates the existence of an active protective order directed to the prospective transferee, the chief law enforcement officer shall immediately advise the dealer that the transfer is prohibited.

Texas Local Government Code

Title 7. Regulation of Land Use, Structures, Businesses, and Related Activities

Chapter 229. Miscellaneous Regulatory Authority of Municipalities

229.001. Firearms; Explosives

(a) A municipality may not adopt regulations relating to the transfer, private ownership, keeping, transportation, licensing, or registration of firearms, ammunition, or firearm supplies.

(b) Subsection (a) does not affect the authority a municipality has under another law to:

(1) require residents or public employees to be armed for personal or national defense, law enforcement, or another lawful purpose;

(2) regulate the discharge of firearms within the limits of the municipality;

(3) regulate the use of property, the location of a business, or uses at a business under the municipality's fire code, zoning ordinance, or land-use regulations as long as the code, ordinance, or regulations are not used to circumvent the intent of Subsection (a) or Subdivision (5) of this subsection;

(4) regulate the use of firearms in the case of an insurrection, riot, or natural disaster if the municipality finds the regulations necessary to protect public health and safety;

(5) regulate the storage or transportation of explosives to protect public health and safety, except that 25 pounds or less of black powder for each private residence and 50 pounds or less of black powder for each retail dealer are not subject to regulation; or

(6) regulate the carrying of a firearm by a person other than a person licensed to carry a concealed handgun under Subchapter H, Chapter 411, Government Code, at a:

(A) public park;

(B) public meeting of a municipality, county, or other governmental body;

(C) political rally, parade, or official political meeting; or

(D) nonfirearms-related school, college, or professional athletic event.

(c) The exception provided by Subsection (b)(6) does not apply if the firearm is in or is carried to or from an area designated for use in a lawful hunting, fishing, or other sporting event and the firearm is of the type commonly used in the activity.

(d) The exception provided by subsection (b)(4) does not authorize the seizure or confiscation of any firearm or ammunition from an individual who is lawfully carrying or possessing the firearm or ammunition.

Chapter 235. County Regulation of Matters Relating to Explosives and Weapons

Subchapter B. Firearms

235.021. Subdivisions Covered by Subchapter This subchapter applies only to a subdivision all or a part of which is located in the unincorporated area of a county and for which a plat is required to be prepared and filed under Chapter 232.

235.022. Authority to Regulate To promote the public safety, the commissioners court of a county by order may prohibit or otherwise regulate the discharge of firearms on lots that are 10 acres or smaller and are located in the unincorporated area of the county in a subdivision.

235.023. Prohibited Regulations This subchapter does not authorize the commissioners court to regulate the transfer, ownership, possession, or transportation of firearms and does not authorize the court to require the registration of firearms.

235.024. Injunction Any person is entitled to appropriate injunctive relief to prevent a violation or threatened violation of a prohibition or other regulation adopted under this subchapter from continuing or occurring.

Texas Penal Code

Title 10. Offenses Against Public Health, Safety, and Morals

Chapter 46. Weapons

46.01. Definitions In this chapter:...

(2) "Explosive weapon" means any explosive or incendiary bomb, grenade, rocket, or mine, that is designed, made, or adapted for the purpose of inflicting serious bodily injury, death, or substantial property damage, or for the principal purpose of causing such a loud report as to cause undue public alarm or terror, and includes a device designed, made, or adapted for delivery or shooting an explosive weapon.

(3) "Firearm" means any device designed, made, or adapted to expel a projectile through a barrel by using the energy generated by an explosion or burning substance or any device readily convertible to that use. Firearm does not include a firearm that may have, as an integral part, a folding knife blade or other characteristics of weapons made illegal by this chapter and that is:

(A) an antique or curio firearm manufactured before 1899; or

(B) a replica of an antique or curio firearm manufactured before 1899, but only if the replica does not use rim fire or center fire ammunition.

(4) "Firearm silencer" means any device designed, made, or adapted to muffle the report of a firearm.

(5) "Handgun" means any firearm that is designed, made, or adapted to be fired with one hand. ...

(9) "Machine gun" means any firearm that is capable of shooting more than two shots automatically, without manual reloading, by a single function of the trigger.

(10) "Short-barrel firearm" means a rifle with a barrel length of less than 16 inches or a shotgun with a barrel length of less than 18 inches, or any weapon made from a shotgun or rifle if, as altered, it has an overall length of less than 26 inches. ...

(12) "Armor-piercing ammunition" means handgun ammunition that is designed primarily for the purpose of penetrating metal or body armor and to be used principally in pistols and revolvers.

(13) "Hoax bomb" means a device that:

(A) reasonably appears to be an explosive or incendiary device; or

(B) by its design causes alarm or reaction of any type by an official of a public safety agency or a volunteer agency organized to deal with emergencies. ...

(16) "Zip gun" means a device or combination of devices that was not originally a firearm and is adapted to expel a projectile through a smooth-bore or rifled-bore barrel by using the energy generated by an explosion or burning substance.

46.03. Places Weapons Prohibited

(a) A person commits an offense if the person intentionally, knowingly, or recklessly possesses or goes with a firearm, illegal knife, club, or prohibited weapon listed in Section 46.05(a):

(1) on the physical premises of a school or educational institution, any grounds or building on which an activity sponsored by a school or educational institution is being conducted, or a passenger transportation vehicle of a school or educational institution, whether the school or educational institution is public or private, unless pursuant to written regulations or written authorization of the institution;...

46.04. Unlawful Possession of Firearm

(a) A person who has been convicted of a felony commits an offense if he possesses a firearm:

(1) after conviction and before the fifth anniversary of the person's release from confinement following conviction of the felony or the person's release from supervision under community supervision, parole, or mandatory supervision, whichever date is later; or

(2) after the period described by Subdivision (1), at any location other than the premises at which the person lives.

(b) A person who has been convicted of an offense under Section 22.01, punishable as a Class A misdemeanor and involving a member of the person's family or household, commits an offense if the person possesses a firearm before the fifth anniversary of the later of:

(1) the date of the person's release from confinement following conviction of the misdemeanor; or

(2) the date of the person's release from community supervision following conviction of the misdemeanor.

(c) A person, other than a peace officer, as defined by Section 1.07, actively engaged in employment as a sworn, full-time paid employee of a state agency or political subdivision, who is subject to an order issued under Section 6.504 or Chapter 85, Family Code, under Article 17.292 or Chapter 7A, Code of Criminal Procedure, or by another jurisdiction as provided by Chapter 88, Family Code, commits an offense if the person possesses a firearm after receiving notice of the order and before expiration of the order.

(d) In this section, "family," "household," and "member of a household" have the meanings assigned by Chapter 71, Family Code.

(e) An offense under Subsection (a) is a felony of the third degree. An offense under Subsection (b) or (c) is a Class A misdemeanor.

46.041. Unlawful Possession of Metal or Body Armor by Felon

(a) In this section, "metal or body armor" means any body covering manifestly designed, made, or adapted for the purpose of protecting a person against gunfire.

(b) A person who has been convicted of a felony commits an offense if after the conviction the person possesses metal or body armor.

(c) An offense under this section is a felony of the third degree.

46.05. Prohibited Weapons

(a) A person commits an offense if he intentionally or knowingly possesses, manufactures, transports, repairs, or sells:

(1) an explosive weapon;

(2) a machine gun;

(3) a short-barrel firearm;

(4) a firearm silencer;

(5) a switchblade knife;

(6) knuckles;

(7) armor-piercing ammunition;

(8) a chemical dispensing device; or

(9) a zip gun.

(b) It is a defense to prosecution under this section that the actor's conduct was incidental to the performance of official duty by the armed forces or national guard, a governmental law enforcement agency, or a correctional facility.

(c) It is a defense to prosecution under this section that the actor's possession was pursuant to registration pursuant to the National Firearms Act, as amended.

(d) It is an affirmative defense to prosecution under this section that the actor's conduct:

(1) was incidental to dealing with a switchblade knife, springblade knife, or short-barrel firearm solely as an antique or curio; or

(2) was incidental to dealing with armor-piercing ammunition solely for the purpose of making the ammunition available to an organization, agency, or institution listed in Subsection (b).

(e) An offense under this section is a felony of the third degree unless it is committed under Subsection (a)(5) or (a)(6), in which event, it is a Class A misdemeanor. ...

46.06. Unlawful Transfer of Certain Weapons

(a) A person commits an offense if the person:

(1) sells, rents, leases, loans, or gives a handgun to any person knowing that the person to whom the handgun is to be delivered intends to use it unlawfully or in the commission of an unlawful act;

(2) intentionally or knowingly sells, rents, leases, or gives or offers to sell, rent, lease, or give to any child younger than 18 years any firearm, club, or illegal knife;

(3) intentionally, knowingly, or recklessly sells a firearm or ammunition for a firearm to any person who is intoxicated;

(4) knowingly sells a firearm or ammunition for a firearm to any person who has been convicted of a felony before the fifth anniversary of the later of the following dates:

(A) the person's release from confinement following conviction of the felony; or

(B) the person's release from supervision under community supervision, parole, or mandatory supervision following conviction of the felony;

(5) sells, rents, leases, loans, or gives a handgun to any person knowing that an active protective order is directed to the person to whom the handgun is to be delivered; or

(6) knowingly purchases, rents, leases, or receives as a loan or gift from another a handgun while an active protective order is directed to the actor.

(b) In this section:

(1) "Intoxicated" means substantial impairment of mental or physical capacity resulting from introduction of any substance into the body.

(2) "Active protective order" means a protective order issued under Title 4, Family Code, that is in effect. The term does not include a temporary protective order issued before the court holds a hearing on the matter.

(c) It is an affirmative defense to prosecution under Subsection (a)(2) that the transfer was to a minor whose parent or the person having legal custody of the minor had given written permission for the sale or, if the transfer was other than a sale, the parent or person having legal custody had given effective consent.

(d) An offense under this section is a Class A misdemeanor, except that an offense under Subsection (a)(2) is a state jail felony if the weapon that is the subject of the offense is a handgun.

46.07. Interstate Purchase A resident of this state may, if not otherwise precluded by law, purchase firearms, ammunition, reloading components, or firearm accessories in contiguous states. This authorization is enacted in conformance with Section 922(b)(3)(A), Public Law 90-618, 90th Congress.

46.08. Hoax Bombs

(a) A person commits an offense if the person knowingly manufactures, sells, purchases, transports, or possesses a hoax bomb with intent to use the hoax bomb to:

(1) make another believe that the hoax bomb is an explosive or incendiary device; or

(2) cause alarm or reaction of any type by an official of a public safety agency or volunteer agency organized to deal with emergencies.

(b) An offense under this section is a Class A misdemeanor.

46.09. Components of Explosives

(a) A person commits an offense if the person knowingly possesses components of an explosive weapon with the intent to combine the components into an explosive weapon for use in a criminal endeavor.

(b) An offense under this section is a felony of the third degree.

46.11. Penalty If Offense Committed Within Weapon-Free School Zone

(a) Except as provided by Subsection (b), the punishment prescribed for an offense under this chapter is increased to the punishment prescribed for the next highest category of offense if it is shown beyond a reasonable doubt on the trial of the offense that the actor committed the offense in a place that the actor knew was:

(1) within 300 feet of the premises of a school; or

(2) on premises where:

(A) an official school function is taking place; or

(B) an event sponsored or sanctioned by the University Interscholastic League is taking place.

(b) This section does not apply to an offense under Section 46.03(a)(1).

(c) In this section:

(1) "Institution of higher education" and "premises" have the meanings assigned by Section 481.134, Health and Safety Code.

(2) "School" means a private or public elementary or secondary school.

46.13. Making a Firearm Accessible to a Child

(a) In this section:

(1) "Child" means a person younger than 17 years of age.

(2) "Readily dischargeable firearm" means a firearm that is loaded with ammunition, whether or not a round is in the chamber.

(3) "Secure" means to take steps that a reasonable person would take to prevent the access to a readily dischargeable firearm by a child, including but not limited to placing a firearm in a locked container or temporarily rendering the firearm inoperable by a trigger lock or other means.

(b) A person commits an offense if a child gains access to a readily dischargeable firearm and the person with criminal negligence:

(1) failed to secure the firearm; or

(2) left the firearm in a place to which the person knew or should have known the child would gain access.

(c) It is an affirmative defense to prosecution under this section that the child's access to the firearm:

(1) was supervised by a person older than 18 years of age and was for hunting, sporting, or other lawful purposes;

(2) consisted of lawful defense by the child of people or property;

(3) was gained by entering property in violation of this code; or

(4) occurred during a time when the actor was engaged in an agricultural enterprise.

(d) Except as provided by Subsection (e), an offense under this section is a Class C misdemeanor.

(e) An offense under this section is a Class A misdemeanor if the child discharges the firearm and causes death or serious bodily injury to himself or another person.

(f) A peace officer or other person may not arrest the actor before the seventh day after the date on which the offense is committed if:

(1) the actor is a member of the family, as defined by Section 71.003, Family Code, of the child who discharged the firearm; and

(2) the child in discharging the firearm caused the death of or serious injury to the child.

(g) A dealer of firearms shall post in a conspicuous position on the premises where the dealer conducts business a sign that contains the following warning in block letters not less than one inch in height:

"IT IS UNLAWFUL TO STORE, TRANSPORT, OR ABANDON AN UNSECURED FIREARM IN A PLACE WHERE CHILDREN ARE LIKELY TO BE AND CAN OBTAIN ACCESS TO THE FIREARM."

46.15. Nonapplicability

(a) Sections 46.02 and 46.03 do not apply to:

(1) peace officers or special investigators under Article 2.122, Code of Criminal Procedure, and neither section prohibits a peace officer or special investigator from carrying a weapon in this state, including in an establishment in this state serving the public, regardless of whether the peace officer or special investigator is engaged in the actual discharge of the officer's or investigator's duties while carrying the weapon;

(2) parole officers and neither section prohibits an officer from carrying a weapon in this state if the officer is:

(A) engaged in the actual discharge of the officer's duties while carrying the weapon; and

(B) in compliance with policies and procedures adopted by the Texas Department of Criminal Justice regarding the possession of a weapon by an officer while on duty;

(3) community supervision and corrections department officers appointed or employed under Section 76.004, Government Code, and neither section prohibits an officer from carrying a weapon in this state if the officer is:

(A) engaged in the actual discharge of the officer's duties while carrying the weapon; and

(B) authorized to carry a weapon under Section 76.0051, Government Code;

(4) a judge or justice of a federal court, the supreme court, the court of criminal appeals, a court of appeals, a district court, a criminal district court, a constitutional county court, a statutory county court, a justice court, or a municipal court who is licensed to carry a concealed handgun under Subchapter H, Chapter 411, Government Code;

(5) an honorably retired peace officer or federal criminal investigator who holds a certificate of proficiency issued under Section 1701.357, Occupations Code, and is carrying a photo identification that:

(A) verifies that the officer honorably retired after not less than 15 years of service as a commissioned officer; and

(B) is issued by a state or local law enforcement agency;

(6) a district attorney, criminal district attorney, municipal attorney, or county attorney who is licensed to carry a concealed handgun under Subchapter H, Chapter 411, Government Code; or

(7) [2 Versions: As amended by Acts 2007, 80th Leg., ch. 1214] a bailiff designated by an active judicial officer as defined by Section 411.201, Government Code, who is:

(A) licensed to carry a concealed handgun under Chapter 411, Government Code; and

(B) engaged in escorting the judicial officer.

(7) [2 Versions: As Amended by Acts 2007, 80th Leg., ch. 1222] an assistant district attorney, assistant criminal district attorney, or assistant county attorney who is licensed to carry a concealed handgun under Subchapter H, Chapter 411, Government Code.

(b) [3 Versions: As amended by Acts 2007, 80th Leg., ch. 647] Section 46.02 does not apply to a person who:

(1) is in the actual discharge of official duties as a member of the armed forces or state military forces as defined by Section 431.001, Government Code, or as a guard employed by a penal institution;

(2) is on the person's own premises or premises under the person's control unless the person is an employee or agent of the owner of the premises and the person's primary responsibility is to act in the capacity of a security guard to protect persons or property, in which event the person must comply with Subdivision (5);

(3) is traveling;

(4) is engaging in lawful hunting, fishing, or other sporting activity on the immediate premises where the activity is conducted, or is en route between the premises and the actor's residence, if the weapon is a type commonly used in the activity;

(5) holds a security officer commission issued by the Texas Private Security Board, if:

(A) the person is engaged in the performance of the person's duties as a security officer or traveling to and from the person's place of assignment;

(B) the person is wearing a distinctive uniform; and

(C) the weapon is in plain view;

(6) is carrying a concealed handgun and a valid license issued under Subchapter H, Chapter 411, Government Code, to carry a concealed handgun of the same category as the handgun the person is carrying;

(7) holds a security officer commission and a personal protection officer authorization issued by the Texas Private Security Board and is providing personal protection under Chapter 1702, Occupations Code;

(8) holds an alcoholic beverage permit or license or is an employee of a holder of an alcoholic beverage permit or license if the person is supervising the operation of the permitted or licensed premises; or

(9) is a student in a law enforcement class engaging in an activity required as part of the class, if the weapon is a type commonly used in the activity and the person is:

(A) on the immediate premises where the activity is conducted; or

(B) en route between those premises and the person's residence and is carrying the weapon unloaded.

(b) [3 Versions: As amended by Acts 2007, 80th Leg., ch. 693] Section 46.02 does not apply to a person who:

(1) is in the actual discharge of official duties as a member of the armed forces or state military forces as defined by Section 431.001, Government Code, or as a guard employed by a penal institution;

(2) is traveling;

(3) is engaging in lawful hunting, fishing, or other sporting activity on the immediate premises where the activity is conducted, or is en route between the premises and the actor's residence or motor vehicle, if the weapon is a type commonly used in the activity;

(4) holds a security officer commission issued by the Texas Private Security Board, if:

(A) the person is engaged in the performance of the person's duties as a security officer or traveling to and from the person's place of assignment;

(B) the person is wearing a distinctive uniform; and

(C) the weapon is in plain view;

(5) is carrying a concealed handgun and a valid license issued under Subchapter H, Chapter 411, Government Code, to carry a concealed handgun of the same category as the handgun the person is carrying;

(6) holds a security officer commission and a personal protection officer authorization issued by the Texas Private Security Board and is providing personal protection under Chapter 1702, Occupations Code; or

(7) holds an alcoholic beverage permit or license or is an employee of a holder of an alcoholic beverage permit or license if the person is supervising the operation of the permitted or licensed premises.

(b) [3 Versions: As amended by Acts 2007, 80th Leg., ch. 1048] Section 46.02 does not apply to a person who:

(1) is in the actual discharge of official duties as a member of the armed forces or state military forces as defined by Section 431.001, Government Code, or as a guard employed by a penal institution;

(2) is on the person's own premises or premises under the person's control unless the person is an employee or agent of the owner of the premises and the person's primary responsibility is to act in the capacity of a security guard to protect persons or property, in which event the person must comply with Subdivision (5);

(3) is traveling;

(4) is engaging in lawful hunting, fishing, or other sporting activity on the immediate premises where the activity is conducted, or is en route between the premises and the actor's residence, if the weapon is a type commonly used in the activity;

(5) holds a security officer commission issued by the Texas Private Security Board, if the person:

(A) is engaged in the performance of the person's duties as an officer commissioned under Chapter 1702, Occupations Code, or is traveling to or from the person's place of assignment; and

(B) is either:

(i) wearing the officer's uniform and carrying the officer's weapon in plain view; or

(ii) acting as a personal protection officer and carrying the person's security officer commission and personal protection officer authorization;

(6) is carrying a concealed handgun and a valid license issued under Subchapter H, Chapter 411, Government Code, to carry a concealed handgun of the same category as the handgun the person is carrying; or

(7) holds an alcoholic beverage permit or license or is an employee of a holder of an alcoholic beverage permit or license if the person is supervising the operation of the permitted or licensed premises.

(c) The provision of Section 46.02 prohibiting the carrying of a club does not apply to a noncommissioned security guard at an institution of higher education who carries a nightstick or similar club, and who has undergone 15 hours of training in the proper use of the club, including at least seven hours of training in the use of the club for nonviolent restraint. For the purposes of this subsection, "nonviolent restraint" means the use of reasonable force, not intended and not likely to inflict bodily injury.

(d) The provisions of Section 46.02 prohibiting the carrying of a firearm or carrying of a club do not apply to a public security officer employed by the adjutant general under Section 431.029, Government Code, in performance of official duties or while traveling to or from a place of duty.

(e) The provisions of Section 46.02 prohibiting the carrying of an illegal knife do not apply to an individual carrying a bowie knife or a sword used in a historical demonstration or in a ceremony in which the knife or sword is significant to the performance of the ceremony.

(f) Section 46.03(a)(6) does not apply to a person who possesses a firearm or club while in the actual discharge of official duties as:

(1) a member of the armed forces or state military forces, as defined by Section 431.001, Government Code; or

(2) an employee of a penal institution.

(g) [Repealed by Acts 2005, 79th Leg., chs. 1093 (H.B. 2110), § 4, effective Sept. 1, 2005, and Acts 2005, 79th Leg., ch. 1179, (S.B. 578), § 3, effective September 1, 2005.]

(h) [Repealed by Acts 2007, 80th Leg., ch. 693 (H.B. 1815), § 3(1), effective September 1, 2007.]

(i) [Repealed by Acts 2007, 80th Leg., ch. 693 (H.B. 1815), § 3(2), effective September 1, 2007.]

(j) The provisions of Section 46.02 prohibiting the carrying of a handgun do not apply to an individual who carries a handgun as a participant in a historical reenactment performed in accordance with the rules of the Texas Alcoholic Beverage Commission.

[Current through 2007 Regular Session]

UTAH

UTAH CODE

Title 76. Utah Criminal Code

Chapter 3. Punishments

76-3-203.2. Definitions - Use of dangerous weapon in offenses committed on or about school premises - Enhanced penalties

(1)(a) As used in this section and Section 76-10-505.5, "on or about school premises" means any of the following:

(i) in a public or private elementary, secondary, or on the grounds of any of those schools;

(ii) in a public or private vocational school or postsecondary institution or on the grounds of any of those schools or institutions;

(iii) in those portions of any building, park, stadium, or other structure or grounds which are, at the time of the act, being used for an activity sponsored by or through a school or institution under Subsections (1)(a)(i) and (ii);

(iv) in or on the grounds of a preschool or child-care facility; and

(v) within 1,000 feet of any structure, facility, or grounds included in Subsections (1)(a)(i), (ii), (iii), and (iv).

Chapter 10. Offenses Against Public Health, Safety, Welfare, and Morals

76-10-500. Uniform law

(1) The individual right to keep and bear arms being a constitutionally protected right, the Legislature finds the need to provide uniform laws throughout the state. Except as specifically provided by state law, a citizen of the United States or a lawfully admitted alien shall not be:

(a) prohibited from owning, possessing, purchasing, selling, transferring, transporting, or keeping any firearm at his place of residence, property, business, or in any vehicle lawfully in his possession or lawfully under his control; or

(b) required to have a permit or license to purchase, own, possess, transport, or keep a firearm.

(2) This part is uniformly applicable throughout this state and in all its political subdivisions and municipalities. All authority to regulate firearms shall be reserved to the state except where the Legislature specifically delegates responsibility to local authorities or state entities. Unless specifically authorized by the Legislature by statute, a local authority or state entity may not enact or enforce any ordinance, regulation, or rule pertaining to firearms.

76-10-501 Definitions As used in this part:

(1)(a) "Antique firearm" means any firearm:

(i)(A) with a matchlock, flintlock, percussion cap, or similar type of ignition system; and

(B) that was manufactured in or before 1898; or

(ii) that is a replica of any firearm described in this Subsection (1)(a), if the replica:

(A) is not designed or redesigned for using rimfire or conventional centerfire fixed ammunition; or

(B) uses rimfire or centerfire fixed ammunition which is:

(I) no longer manufactured in the United States; and

(II) is not readily available in ordinary channels of commercial trade; or

(iii) (A) that is a muzzle loading rifle, shotgun, or pistol; and

(B) is designed to use black powder, or a black powder substitute, and cannot use fixed ammunition.

(b) "Antique firearm" does not include:

(i) any weapon that incorporates a firearm frame or receiver;

(ii) any firearm that is converted into a muzzle loading weapon; or

(iii) any muzzle loading weapon that can be readily converted to fire fixed ammunition by replacing the:

(A) barrel;

(B) bolt;

(C) breechblock; or

(D) any combination of Subsection (1)(b)(iii) (A), (B), or (C).

(2)(a) "Concealed dangerous weapon" means a dangerous weapon that is covered, hidden, or secreted in a manner that the public would not be aware of its presence and is readily accessible for immediate use.

(b) A dangerous weapon shall not be considered a concealed dangerous weapon if it is a firearm which is unloaded and is securely encased.

(3) "Criminal history background check" means a criminal background check conducted by a licensed firearms dealer on every purchaser of a handgun through the division or the local law enforcement agency where the firearms dealer conducts business.

(4) "Curio or relic firearm" means any firearm that:

(a) is of special interest to a collector because of a quality that is not associated with firearms intended for:

(i) sporting use;

(ii) use as an offensive weapon; or

(iii) use as a defensive weapon;

(b)(i) was manufactured at least 50 years prior to the current date; and

(ii) is not a replica of a firearm described in Subsection (4)(b)(i);

(c) is certified by the curator of a municipal, state, or federal museum that exhibits firearms to be a curio or relic of museum interest;

(d) derives a substantial part of its monetary value:

(i) from the fact that the firearm is:

(A) novel;

(B) rare; or

(C) bizarre; or

(ii) because of the firearm's association with an historical:

(A) figure;

(B) period; or

(C) event; and

(e) has been designated as a curio or relic firearm by the director of the United States Treasury Department Bureau of Alcohol, Tobacco, and Firearms under 27 C.F.R. Sec. 178.11.

(5)(a) "Dangerous weapon" means any item that in the manner of its use or intended use is capable of causing death or serious bodily injury. The following factors shall be used in determining whether a knife, or any other item, object, or thing not commonly known as a dangerous weapon is a dangerous weapon:

(i) the character of the instrument, object, or thing;

(ii) the character of the wound produced, if any;

(iii) the manner in which the instrument, object, or thing was used; and

(iv) the other lawful purposes for which the instrument, object, or thing may be used.

(b) "Dangerous weapon" does not include any explosive, chemical, or incendiary device as defined by Section 76-10-306.

(6) "Dealer" means every person who is licensed under crimes and criminal procedure, 18 U.S.C. 923 and engaged in the business of selling, leasing, or otherwise transferring a handgun, whether the person is a retail or wholesale dealer, pawnbroker, or otherwise.

(7) "Division" means the Criminal Investigations and Technical Services Division of the Department of Public Safety, created in Section 53-10-103.

(8) "Enter" means intrusion of the entire body.

(9)(a) "Firearm" means a pistol, revolver, shotgun, sawed-off shotgun, rifle or sawed-off rifle, or any device that could be used as a dangerous weapon from which is expelled a projectile by action of an explosive.

(b) As used in Sections 76-10-526 and 76-10-527, "firearm" does not include an antique firearm.

(10) "Firearms transaction record form" means a form created by the division to be completed by a person purchasing, selling, or transferring a handgun from a dealer in the state.

(11) "Fully automatic weapon" means any firearm which fires, is designed to fire, or can be readily restored to fire, automatically more than one shot without manual reloading by a single function of the trigger.

(12)(a) "Handgun" means a pistol, revolver, or other firearm of any description, loaded or unloaded, from which any shot, bullet, or other missile can be discharged, the length of which, not including any revolving, detachable, or magazine breech, does not exceed 12 inches.

(b) As used in Sections 76-10-520, 76-10-521, and 76-10-522, "handgun" and "pistol or revolver" do not include an antique firearm.

(13) "House of worship" means a church, temple, synagogue, mosque, or other building set apart primarily for the purpose of worship in which religious services are held and the main body of which is kept for that use and not put to any other use inconsistent with its primary purpose.

(14) "Prohibited area" means any place where it is unlawful to discharge a firearm.

(15) "Readily accessible for immediate use" means that a firearm or other dangerous weapon is carried on the person or within such close proximity and in such a manner that it can be retrieved and used as readily as if carried on the person.

(16) "Residence" means an improvement to real property used or occupied as a primary or secondary residence.

(17) "Sawed-off shotgun" or "sawed-off rifle" means a shotgun having a barrel or barrels of fewer than 18 inches in length, or in the case of a rifle, having a barrel or barrels of fewer than 16 inches in length, or any dangerous weapon made from a rifle or shotgun by alteration, modification, or otherwise, if the weapon as modified has an overall length of fewer than 26 inches.

(18) "Securely encased" means not readily accessible for immediate use, such as held in a gun rack, or in a closed case or container, whether or not locked, or in a trunk or other storage area of a motor vehicle, not including a glove box or console box.

(19) "State entity" means each department, commission, board, council, agency, institution, officer, corporation, fund, division, office, committee, authority, laboratory, library, unit, bureau, panel, or other administrative unit of the state.

(20) "Violent felony" means the same as defined in Section 76-3-203.5.

76-10-502. When weapon deemed loaded

(1) For the purpose of this chapter, any pistol, revolver, shotgun, rifle, or other weapon described in this part shall be deemed to be loaded when there is an unexpended cartridge, shell, or projectile in the firing position.

(2) Pistols and revolvers shall also be deemed to be loaded when an unexpended cartridge, shell, or projectile is in a position whereby the manual operation of any mechanism once would cause the unexpended cartridge, shell, or projectile to be fired.

(3) A muzzle loading firearm shall be deemed to be loaded when it is capped or primed and has a powder charge and ball or shot in the barrel or cylinders.

76-10-503. Restrictions on possession, purchase, transfer, and ownership of dangerous weapons by certain persons

(1) For purposes of this section:

(a) A Category I restricted person is a person who:

(i) has been convicted of any violent felony as defined in Section 76-3-203.5;

(ii) is on probation or parole for any felony;

(iii) is on parole from a secure facility as defined in Section 62A-7-101; or

(iv) within the last ten years has been adjudicated delinquent for an offense which if committed by an adult would have been a violent felony as defined in Section 76-3-203.5.

(b) A Category II restricted person is a person who:

(i) has been convicted of or is under indictment for any felony;

(ii) within the last seven years has been adjudicated delinquent for an offense which if committed by an adult would have been a felony;

(iii) is an unlawful user of a controlled substance as defined in Section 58-37-2;

(iv) is in possession of a dangerous weapon and is knowingly and intentionally in unlawful possession of a Schedule I or II controlled substance as defined in Section 58-37-2;

(v) has been found not guilty by reason of insanity for a felony offense;

(vi) has been found mentally incompetent to stand trial for a felony offense;

(vii) has been adjudicated as mentally defective as provided in the Brady Handgun Violence Prevention Act, Pub. L. No. 103-159, 107 Stat. 1536 (1993), or has been committed to a mental institution;

(viii) is an alien who is illegally or unlawfully in the United States;

(ix) has been dishonorably discharged from the armed forces; or

(x) has renounced his citizenship after having been a citizen of the United States.

(2) A Category I restricted person who intentionally or knowingly agrees, consents, offers, or arranges to purchase, transfer, possess, use, or have under his custody or control, or who intentionally or knowingly purchases, transfers, possesses, uses, or has under his custody or control:

(a) any firearm is guilty of a second degree felony; or

(b) any dangerous weapon other than a firearm is guilty of a third degree felony.

(3) A Category II restricted person who purchases, transfers, possesses, uses, or has under his custody or control:

(a) any firearm is guilty of a third degree felony; or

(b) any dangerous weapon other than a firearm is guilty of a class A misdemeanor.

(4) A person may be subject to the restrictions of both categories at the same time.

(5) If a higher penalty than is prescribed in this section is provided in another section for one who purchases, transfers, possesses, uses, or has under this custody or control any dangerous weapon, the penalties of that section control.

(6) It is an affirmative defense to a charge based on the definition in Subsection (1)(b)(iv) that the person was:

(a) in possession of a controlled substance pursuant to a lawful order of a practitioner for use of a member of the person's household or for administration to an animal owned by the person or a member of the person's household; or

(b) otherwise authorized by law to possess the substance.

76-10-505.5. Possession of a dangerous weapon, firearm, or sawed-off shotgun on or about school premises - Penalties

(1) A person may not possess any dangerous weapon, firearm, or sawed-off shotgun, as those terms are defined in Section 76-10-501, at a place that the person knows, or has reasonable cause to believe, is on or about school premises as defined in Subsection 76-3-203.2 (1).

(2)(a) Possession of a dangerous weapon on or about school premises is a class B misdemeanor.

(b) Possession of a firearm or sawed-off shotgun on or about school premises is a class A misdemeanor.

(3) This section does not apply if:

(a) the person is authorized to possess a firearm as provided under Section 53-5-704, 53-5-705, 76-10-511, or 76-10-523 or as otherwise authorized by law;

(b) the possession is approved by the responsible school administrator;

(c) the item is present or to be used in connection with a lawful, approved activity and is in the possession or under the control of the person responsible for its possession or use; or

(d) the possession is:

(i) at the person's place of residence or on the person's property;

(ii) in any vehicle lawfully under the person's control, other than a vehicle owned by the school or used by the school to transport students; or

(iii) at the person's place of business which is not located in the areas described in Subsection 76-3-203.2 (1)(a)(i), (ii), or (iv).

(4) This section does not prohibit prosecution of a more serious weapons offense that may occur on or about school premises.

76-10-509. Possession of dangerous weapon by minor

(1) A minor under 18 years of age may not possess a dangerous weapon unless he:

(a) has the permission of his parent or guardian to have the weapon; or

(b) is accompanied by a parent or guardian while he has the weapon in his possession.

(2) Any minor under 14 years of age in possession of a dangerous weapon shall be accompanied by a responsible adult.

(3) Any person who violates this section is guilty of:

(a) a class B misdemeanor upon the first offense; and

(b) a class A misdemeanor for each subsequent offense.

76-10-509.4. Prohibition of possession of certain weapons by minors

(1) A minor under 18 years of age may not possess a handgun.

(2) Except as provided by federal law, a minor under 18 years of age may not possess the following:

(a) a sawed-off rifle or sawed-off shotgun; or

(b) a fully automatic weapon.

(3) Any person who violates Subsection (1) is guilty of:

(a) a class B misdemeanor upon the first offense; and

(b) a class A misdemeanor for each subsequent offense.

(4) Any person who violates Subsection (2) is guilty of a third degree felony.

76-10-509.5. Penalties for providing certain weapons to a minor

(1) Any person who provides a handgun to a minor when the possession of the handgun by the minor is a violation of Section 76-10-509.4 is guilty of:

(a) a class B misdemeanor upon the first offense; and

(b) a class A misdemeanor for each subsequent offense.

(2) Any person who transfers in violation of applicable state or federal law a sawed-off rifle, sawed-off shotgun, or fully automatic weapon to a minor is guilty of a third degree felony.

76-10-509.6. Parent or guardian providing firearm to violent minor

(1) A parent or guardian may not intentionally or knowingly provide a firearm to, or permit the possession of a firearm by, any minor who has been convicted of a violent felony as defined in Section 76-3-203.5 or any minor who has been adjudicated in juvenile court for an offense which would constitute a violent felony if the minor were an adult.

(2) Any person who violates this section is guilty of:

(a) a class A misdemeanor upon the first offense; and

(b) a third degree felony for each subsequent offense.

76-10-509.7. Parent or guardian knowing of minor's possession of dangerous weapon Any parent or guardian of a minor who knows that the minor is in possession of a dangerous weapon in violation of Section 76-10-509 or a firearm in violation of Section 76-10-509.4 and fails to make reasonable efforts to remove the firearm from the minor's possession is guilty of a class B misdemeanor.

76-10-509.9. Sales of firearms to juveniles

(1) A person may not sell any firearm to a minor under 18 years of age unless the minor is accompanied by a parent or guardian.

(2) Any person who violates this section is guilty of a third degree felony.

76-10-511. Possession of loaded weapon at residence authorized Except for persons described in Section 76-10-503, a person may have a loaded firearm at his place of residence, including any temporary residence or camp.

76-10-520. Number or mark assigned to pistol or revolver by Department of Public Safety The Department of Public Safety upon request may assign a distinguishing number or mark of identification to any pistol or revolver whenever it is without a manufacturer's number, or other mark of identification or whenever the manufacturer's number or other mark of identification or the distinguishing number or mark assigned by the Department of Public Safety has been destroyed or obliterated.

76-10-521. Unlawful marking of pistol or revolver

(1) Any person who places or stamps on any pistol or revolver any number except one assigned to it by the Department of Public Safety is guilty of a class A misdemeanor.

(2) This section does not prohibit restoration by the owner of the name of the maker, model, or of the original manufacturer's number or other mark of identification when the restoration is authorized by the Department of Public Safety, nor prevent any manufacturer from placing in the ordinary course of business the name of the make, model, manufacturer's number, or other mark of identification upon a new pistol or revolver.

76-10-522. Alteration of number or mark on pistol or revolver Any person who changes, alters, removes, or obliterates the name of the maker, the model, manufacturer's number, or other mark of identification, including any distinguishing number or mark assigned by the Department of Public Safety, on any pistol or revolver, without first having secured written permission from the Department of Public Safety to make the change, alteration, or removal, is guilty of a class A misdemeanor.

76-10-523. Persons exempt from weapons laws

(1) This part and Title 53, Chapter 5, Part 7, Concealed Weapon Act, do not apply to any of the following:

(a) a United States marshal;

(b) a federal official required to carry a firearm;

(c) a peace officer of this or any other jurisdiction;

(d) a law enforcement official as defined and qualified under Section 53-5-711;

(e) a judge as defined and qualified under Section 53-5-711;

(f) a common carrier while engaged in the regular and ordinary transport of firearms as merchandise; or

(g) a nonresident traveling in or through the state, provided that any firearm is:

(i) unloaded; and

(ii) securely encased as defined in Section 76-10-501.

(2) The provisions of Subsections 76-10-504(1)(a), (1)(b), and Section 76-10-505 do not apply to any person to whom a permit to carry a concealed firearm has been issued:

(a) pursuant to Section 53-5-704; or

(b) by another state or county.

76-10-524. Purchase of firearms in contiguous states pursuant to federal law This part will allow purchases of firearms and ammunition pursuant to U.S.C. Title 18 Chapter 44 Sec. 922b(3).

76-10-526. Criminal background check prior to purchase of a firearm - Fee - Exemption for concealed firearm permit holders

(1) For purposes of this section, "valid permit to carry a concealed firearm" does not include a temporary permit issued pursuant to Section 53-5-705.

(2)(a) To establish personal identification and residence in this state for purposes of this part, a dealer shall require an individual receiving a firearm to present one photo identification on a form issued by a governmental agency of the state.

(b) A dealer may not accept a driving privilege card issued in accordance with section 53-3-207 as proof of identification for the purpose of establishing personal identification and residence in this state as required under this subsection (2).

(3) A criminal history background check is required for the sale of a firearm by a licensed firearm dealer in the state.

(4)(a) An individual, except a dealer, purchasing a firearm from a dealer shall consent in writing to a criminal background check, on a form provided by the division.

(b) The form shall contain the following information:

(i) the dealer identification number;

(ii) the name and address of the individual receiving the firearm;

(iii) the date of birth, height, weight, eye color, and hair color of the individual receiving the firearm; and

(iv) the Social Security number or any other identification number of the individual receiving the firearm.

(5)(a) The dealer shall send the form required by Subsection (4) to the division immediately upon its completion.

(b) No dealer shall sell or transfer any firearm to an individual until the dealer has provided the division with the information in Subsection (4) and has received approval from the division under Subsection (7).

(6) The dealer shall make a request for criminal history background information by telephone or other electronic means to the division and shall receive approval or denial of the inquiry by telephone or other electronic means.

(7) When the dealer calls for or requests a criminal history background check, the division shall:

(a) review the criminal history files, including juvenile court records, to determine if the individual is prohibited from purchasing, possessing, or transferring a firearm by state or federal law;

(b) inform the dealer that:

(i) the records indicate the individual is so prohibited; or

(ii) the individual is approved for purchasing, possessing, or transferring a firearm;

(c) provide the dealer with a unique transaction number for that inquiry; and

(d) provide a response to the requesting dealer during the call for a criminal background, or by return call, or other electronic means, without delay, except in case of electronic failure or other circumstances beyond the control of the division, the division shall advise the dealer of the reason for the delay and give the dealer an estimate of the length of the delay.

(8)(a) The division shall not maintain any records of the criminal history background check longer than 20 days from the date of the dealer's request if the division determines that the individual receiving the gun is not prohibited from purchasing, possessing, or transferring the firearm under state or federal law.

(b) However, the division shall maintain a log of requests containing the dealer's federal firearms number, the transaction number, and the transaction date for a period of 12 months.

(9) If the criminal history background check discloses information indicating that the individual attempting to purchase the firearm is prohibited from purchasing, possessing, or transferring a firearm, the division shall inform the law enforcement agency in the jurisdiction where the person resides.

(10) If an individual is denied the right to purchase a firearm under this section, the individual may review the individual's criminal history information and may challenge or amend the information as provided in Section 53-10-108.

(11) The division shall make rules as provided in Title 63g, Chapter 3, Utah Administrative Rulemaking Act, to ensure the identity, confidentiality, and security of all records provided by the division pursuant to this part are in conformance with the requirements of the Brady Handgun Violence Prevention Act, Pub. L. No. 103-159, 107 Stat. 1536 (1993).

(12)(a)(i) All dealers shall collect a criminal history background check fee which is $7.50.

(ii) This fee remains in effect until changed by the division through the process under Section 63J-1- 303.

(b)(i) The dealer shall forward at one time all fees collected for criminal history background checks performed during the month to the division by the last day of the month following the sale of a firearm.

(ii) The division shall deposit the fees in the General Fund as dedicated credits to cover the cost of administering and conducting the criminal history background check program.

(13) An individual with a concealed firearm permit issued pursuant to Title 53, Chapter 5, Part 7, Concealed Weapon Act, shall be exempt from the background check and corresponding fee required in this section for the purchase of a firearm if:

(a) the individual presents the individual's concealed firearm permit to the dealer prior to purchase of the firearm; and

(b) the dealer verifies with the division that the individual's concealed firearm permit is valid.

76-10-527. Penalties

(1) This section shall apply only to a handgun until federal law requires the background check in Section 76-10-526 to extend to other firearms at which time this section shall also apply to those firearms.

(2) A dealer is guilty of a class A misdemeanor who willfully and intentionally:

(a) requests, obtains, or seeks to obtain criminal history background information under false pretenses; or

(b) disseminates criminal history background information.

(3) A person who purchases or transfers a firearm is guilty of a felony of the third degree who willfully and intentionally makes a false statement of the information required for a criminal background check in Section 76-10-526.

(4) A dealer is guilty of a felony of the third degree if the dealer willfully and intentionally sells or transfers a firearm in violation of this part.

(5) A person is guilty of a felony of the third degree who purchases a firearm with the intent to:

(a) resell or otherwise provide a firearm to any person who is ineligible to purchase or receive from a dealer a firearm; or

(b) transport a firearm out of this state to be resold to an ineligible person.

[Current through the 2008 Second Special Session]

VERMONT

VT. STAT.

Title 13. Crimes and Criminal Procedure

Chapter 37. Explosives

1603. Definitions For the purposes of this chapter:

(1) "Destructive device" means any:

(A) explosive, incendiary or poison gas bomb; or

(B) explosive, incendiary or poison gas grenade; or

(C) explosive, incendiary or poison gas rocket having a propellant charge of more than four ounces; or

(D) explosive, incendiary or poison gas missile having an explosive or incendiary charge of more than one-quarter ounce; or

(E) explosive, incendiary or poison gas mine; or

(F) device which consists of or includes a breakable container including a flammable liquid or compound, and a wick composed of any material which, when ignited, is capable of igniting the flammable liquid or compound, and can be carried or thrown by one individual acting alone; or

(G) device similar to those devices enumerated in paragraphs (1), (A)-(E) of this section.

A destructive device does not include a firearm or ammunition therefor.

(2) "Explosive" means dynamite, or any explosive compound of which nitroglycerin forms a part, or fulminate in bulk or dry condition, or blasting caps, or detonating fuses, or blasting powder or any other similar explosive. The term does not include a firearm or ammunition therefor or any components of ammunition for a firearm including primers, smokeless powder or black gunpowder.

(3) "Hoax device" means any device so designed, assembled, fabricated or manufactured as to convey the physical appearance of an explosive or incendiary bomb or the physical appearance of any of the devices enumerated in subdivisions (A)-(F) of division (1) of this section which is lacking an explosive or incendiary charge.

1604. Possession of destructive devices A person who manufactures, possesses, stores or transports a destructive device or a hoax device shall be imprisoned for not more than 10 years or fined not more than $5,000.00, or both.

1606. Possession and use of explosives A person who possesses, purchases, stores, uses or transports an explosive without a license as provided in chapter 177, subchapter 2, division 2 of Title 20 shall be imprisoned for not more than 5 years or fined not more than $1,000.00 or both.

1607. Sale of explosives A person who gives, transfers or sells an explosive to another who does not hold a license issued under chapter 177, subchapter 2, division 2 of Title 20 shall be imprisoned for not more than 5 years or fined not more than $1,000.00 or both.

1609. Record of sale

(a) A person may not give, transfer or sell an explosive to another unless the purchaser exhibits a valid license issued under chapter 177, subchapter 2, division 2 of Title 20.

(b) A person who gives, transfers or sells an explosive to another shall record the name and address of the purchaser, the license number of the purchaser, the date of sale, the type and quantity of explosives sold, the serial or lot number of the explosives, if any, and the purpose for which the explosive is to be used on forms provided by the commissioner of public safety. The purchaser holding a license shall keep a record of each purchase made and the disposition of the explosives, giving a full report without delay but in no event later than twenty-four hours after discovery of the loss or theft of any of such explosives to the commissioner of public safety. The records shall be kept by the seller and the purchaser for a period of two years and shall be open to inspection by any law enforcement officer.

(c) A person who violates a provision of this section shall be imprisoned for not more than five years or fined not more than $ 1,000.00 or both.

1610. Purchase in contiguous states

Any person holding a valid license under chapter 177, subchapter 2, division 2 of Title 20 may purchase explosives in any state contiguous to this state and transport them into this state, provided that he furnishes a record of each purchase to the commissioner of public safety within fifteen days of the transportation of the explosives into this state, and that he comply with both the laws applicable to the purchases in the contiguous state, and the pertinent statutes of the United States government.

1611. Exemptions

(a) Nothing contained in this chapter shall apply to the armed forces of the United States, the duly authorized militia of the state, the fire or police departments of this state, or to the state or any subdivision thereof.

(b) Nothing contained in this chapter shall apply to destructive devices or explosives while being transported upon vessels, motor vehicles or railroad cars in conformity with the regulations adopted by the interstate commerce commission.

(c) The provisions of section 1604 of this title do not apply to a person who holds a valid license issued under Title 18 of the United States Code, chapter 44, to manufacture, possess, use, store or transport a destructive device provided he is complying with the terms of the license.

1612. Placing a hoax device A person who willfully and maliciously puts, places or installs a hoax device in any building, house, facility of public transport, vehicular conveyance, ship, boat, aircraft, dam or reservoir for storing water shall be imprisoned for not more than 10 years or fined not more than $5,000.00, or both.

Chapter 76. Weapons of Mass Destruction

3501. Definitions

(a) As used in this chapter:

(1) "Chemical warfare agents" means:

(A) Any weaponized toxic or poisonous chemical, including the following agents or any analog of the following agents:

(i) Nerve agents, including Tabun (GA), Sarin (GB), Soman (GD), GF, and VX.

(ii) Choking agents, including Phosgene (CG) and Diphosgene (DP).

(iii) Blood agents, including HydrogenCyanide (AC), Cyanogen Chloride (CK), and Arsine (SA).

(iv) Blister agents, including mustards (H,HD (sulfur mustard), HN-1, HN-2, HN-3 (nitrogen mustard)), arsenicals, such as Lewisite (L), urticants, such as CX, and incapacitating agents, such as BZ.

(B) A dangerous chemical or hazardous material generally utilized in an industrial or commercial process when a person knowingly and intentionally utilizes the material with the intent to cause harm, and the use places persons at risk of serious bodily injury or death, or endangers the environment.

(2) "Health care provider" means a person, partnership, corporation, facility or institution, licensed, certified or authorized, by law, to provide professional health care service in this state to an individual during that individual's medical care, treatment or confinement.

(3) "Hoax weapon" means any substance, compound, or other item intended to convey the physical appearance or chemical properties of a weapon of mass destruction or asserted to contain a weapon of mass destruction, which is not a weapon of mass destruction or does not contain a weapon of mass destruction.

(4) "Law enforcement agency" means:

(A) A federal law enforcement agency, including the Bureau of Alcohol, Tobacco and Firearms, the Federal Bureau of Investigation, Military Police or Military Criminal Investigative Division, United States Marshals Service, Secret Service, Federal Emergency Management Agency, or the Department of Defense Threat Reduction Agency.

(B) One of the following Vermont law enforcement agencies:

(i) The department of public safety.

(ii) A municipal police department.

(iii) A sheriff's department.

(iv) The attorney general's office.

(v) A state's attorney's office.

(vi) The capitol police department.

(5) "Nuclear or radiological agents" means any improvised nuclear device (IND) which is any explosive device designed to cause a nuclear yield, any radiological dispersal device (RDD) which is any explosive device utilized to spread radioactive material, or a simple radiological dispersal device (SRDD) which is any container designed to release radiological material as a weapon without an explosion.

(6) "Vector" means a living organism or a molecule, including a recombinant molecule, or a biological product that may be engineered as a result of biotechnology, that is capable of carrying a biological agent or toxin to a host.

(7) "Weapon of mass destruction" means a chemical warfare agent, weaponized biological or biologic warfare agent, nuclear agent, or radiological agent.

(8) "Weaponization" means the deliberate processing, preparation, packaging, or synthesis of any substance or agent for use as a weapon or munition. "Weaponized agents" means those agents or substances that have been prepared for dissemination through any explosive, thermal, pneumatic, mechanical or other means.

(9) "Weaponized biological or biologic warfare agents" means:

(A) weaponized pathogens, including bacteria, viruses, rickettsia, yeasts, or fungi;

(B) genetically-engineered pathogens;

(C) weaponized toxins;

(D) weaponized vectors; and

(E) weaponized endogenous biological regulators (EBRs).

(b) The lawful use of chemicals for legitimate mineral extraction, industrial, agricultural, or commercial purposes is not proscribed by this chapter.

3502. Possession and use of weapons of mass destruction

(a) A person who knowingly and without lawful authority possesses, develops, manufactures, produces, transfers, acquires, or stockpiles any weapon of mass destruction shall be imprisoned not more than 20 years or fined not more than $ 100,000.00, or both.

(b) A person who uses or directly employs against other persons a weapon of mass destruction in a form that may cause disabling illness or injury in human beings shall be imprisoned not less than 20 years nor more than life and fined not more than $ 250,000.00.

(c) A person who uses a weapon of mass destruction in a form that may cause widespread damage to or disruption of water or food supplies shall be imprisoned not less than five years nor more than 30 years and fined not more than $ 250,000.00.

(d) A person who uses a weapon of mass destruction against livestock or crops with the intent to cause widespread and substantial damage to livestock or crops shall be imprisoned not more than 30 years and fined not more than $ 250,000.00.

(e) A person who uses a weapon of mass destruction in a form that may cause widespread and significant damage to public or private property shall be imprisoned not more than 30 years and fined not more than $ 250,000.00.

(f) A person who uses recombinant technology or any other biological advance to create new pathogens or more virulent forms of existing pathogens for the purpose of creating a weapon of mass destruction shall be imprisoned not more than 20 years or fined not more than $ 250,000.00, or both.

(g) A person who knowingly and intentionally places a hoax weapon in any public place, building, house, residence, facility of public transport, vehicular conveyance, train, ship, boat, aircraft, dam or reservoir for storing water, shall be imprisoned not more than five years or fined not more than $ 10,000.00, or both.

(h) No university, research institution, private company, individual, or hospital engaged in scientific or public health research and, as required, registered with the Centers for Disease Control and Prevention (CDC) pursuant to part 113 (commencing with Section 113.1) of subchapter E of chapter 1 of Title 9 or pursuant to Part 72 (commencing with Section 72.1) of Subchapter E of Chapter 1 of Title 42 of the Code of Federal Regulations, or any successor provisions, shall be subject to this section.

(i) Nothing in this section shall be construed to limit or restrict prosecution under any other applicable laws.

Chapter 85. Weapons

4004. Possession of dangerous or deadly weapon in a school bus or school building or on school property

(a) No person shall knowingly possess a firearm or a dangerous or deadly weapon while within a school building or on a school bus. A person who violates this section shall, for the first offense, be imprisoned not more than one year or fined not more than $1,000.00, or both, and for a second or subsequent offense shall be imprisoned not more than three years or fined not more than $5,000.00, or both.

(b) No person shall knowingly possess a firearm or a dangerous or deadly weapon on any school property with the intent to injure another person. A person who violates this section shall, for the first offense, be imprisoned not more than two years or fined not more than $1,000.00, or both, and for a second or subsequent offense shall be imprisoned not more than three years or fined not more than $5,000.00, or both.

(c) This section shall not apply to:

(1) A law enforcement officer while engaged in law enforcement duties.

(2) Possession and use of firearms or dangerous or deadly weapons if the board of school directors, or the superintendent or principal if delegated authority to do so by the board, authorizes possession or use for specific occasions or for instructional or other specific purposes.

(d) As used in this section:

(1) "School property" means any property owned by a school, including motor vehicles.

(2) "Owned by the school" means owned, leased, controlled or subcontracted by the school.

(3) "Dangerous or deadly weapon" has the meaning defined in section 4016 of this title.

(4) "Firearm" has the meaning defined in section 4016 of this title.

(5) "Law enforcement officer" has the meaning defined in section 4016 of this title.

(e) The provisions of this section shall not limit or restrict any prosecution for any other offense, including simple assault or aggravated assault.

4006. Record of firearm sales All pawnbrokers and retail merchants dealing in firearms shall keep a record book in which they shall record the sale by them of all revolvers and pistols, and the purchase by them of all second-hand revolvers and pistols. Such record shall include the date of the transaction, the marks of identification of the firearm, including the manufacturer's name, the caliber, model and manufacturer's number of the firearm, the name, address, birthplace, occupation, age, height, weight and color of eyes and hair of the purchaser or seller. Such purchaser or seller shall sign his name to the record and the pawnbroker or merchant shall preserve such record book for six years after the date of last entry and shall permit all enforcement officers to inspect the same at all reasonable times. A person, partnership or corporation who violates a provision of this section shall be fined not more than $100.00.

4007. Furnishing firearms to children A person, firm or corporation, other than a parent or guardian, who sells or furnishes to a minor under the age of sixteen years a firearm or other dangerous weapon or ammunition for firearms shall be fined not more than $50.00 nor less than $10.00. This section shall not apply to an instructor or teacher who furnishes firearms to pupils for instruction and drill.

4008. Possession of firearms by children

A child under the age of sixteen years shall not, without the consent of his parents or guardian, have in his possession or control a pistol or revolver constructed or designed for the use of gunpowder or other explosive substance with leaden ball or shot. A child who violates a provision of this section shall be deemed a delinquent child under the provisions of chapter 55 of Title 33.

4009. Negligent use of gun

A person who carelessly or negligently wounds another person by gunshot shall be imprisoned not more than five years or fined not more than $ 1,000.00, or both.

4010. Gun silencers A person who manufactures, sells or uses or possesses with intent to sell or use, an appliance known as or used for a gun silencer shall be fined $25.00 for each offense. The provisions of this section shall not prevent the use or possession of gun silencers for military purposes when so used or possessed under proper military authority and restriction.

4011. Aiming gun at another Any person who shall intentionally point or aim any gun, pistol or other firearm at or towards another, except in self-defense or in the lawful discharge of official duty, shall be punished by fine not exceeding $50.00. Any person who shall discharge any such firearm so intentionally aimed or pointed shall be punished by imprisonment for not more than one year or fined not more than $100.00, or both.

4013. Zip guns; switchblade knives A person who possesses, sells or offers for sale a weapon commonly known as a "zip" gun, or a weapon commonly known as a switchblade knife, the blade of which is three inches or more in length, shall be imprisoned not more than ninety days or fined not more than $100.00, or both.

4014. Purchase of firearms in contiguous states Residents of the state of Vermont may purchase rifles and shotguns in a state contiguous to the state of Vermont provided that such residents conform to the applicable provisions of the Gun Control Act of 1968, and regulations thereunder, as administered by the United States Secretary of the Treasury, and provided further that such residents conform to the provisions of law applicable to such purchase in the state of Vermont and in the contiguous state in which the purchase is made.

4015. Purchase of firearms by nonresidents Residents of a state contiguous to the state of Vermont may purchase rifles and shotguns in the state of Vermont, provided that such residents conform to the applicable provisions of the Gun Control Act of 1968, and regulations thereunder, as administered by the United States Secretary of the Treasury, and provided further that such residents conform to the provisions of law applicable to such purchase in the state of Vermont and in the state in which such persons reside.

Title 24. Municipal and County Governments

Part 2. Municipalities

Chapter 61. Regulatory Provisions; Police Power of Municipalities

Subchapter 11. Miscellaneous Regulatory Powers

2295. Authority of municipal and county governments to regulate firearms, ammunition, hunting, fishing and trapping Except as otherwise provided by law, no town, city or incorporated village, by ordinance, resolution or other enactment, shall directly regulate hunting, fishing and trapping or the possession, ownership, transportation, transfer, sale, purchase, carrying, licensing or registration of traps, firearms, ammunition or components of firearms or ammunition. This section shall not limit the powers conferred upon a town, city or incorporated village under section 2291(8) of this title. The provisions of this section shall supersede any inconsistent provisions of a municipal charter.

[Current through the 2007 Adjourned Session (2008)]

VIRGINIA

VA. CODE

Title 15.2. Counties, Cities and Towns

Chapter 9. General Powers of Local Governments

15.2-915. Control of firearms; applicability to authorities and local governmental agencies

A. No locality shall adopt or enforce any ordinance, resolution or motion, as permitted by § 15.2-1425, and no agent of such locality shall take any administrative action, governing the purchase, possession, transfer, ownership, carrying, storage or transporting of firearms, ammunition, or components or combination thereof other than those expressly authorized by statute. For purposes of this section, a statute that does not refer to firearms, ammunition, or components or combination thereof, shall not be construed to provide express authorization.

Nothing in this section shall prohibit a locality from adopting workplace rules relating to terms and conditions of employment of the workforce. Nothing in this section shall prohibit a law-enforcement officer, as defined in § 9.1-101 from acting within the scope of his duties.

The provisions of this section applicable to a locality shall also apply to any authority or to a local governmental entity, including a department or agency, but not including any local or regional jail or juvenile detention facility.

B. Any local ordinance, resolution or motion adopted prior to the effective date of this act governing the purchase, possession, transfer, ownership, carrying or transporting of firearms, ammunition, or components or combination thereof, other than those expressly authorized by statute, is invalid.

15.2-915.2. Regulation of transportation of a loaded rifle or shotgun The governing body of any county or city may by ordinance make it unlawful for any person to transport, possess or carry a loaded shotgun or loaded rifle in any vehicle on any public street, road, or highway within such locality. Any violation of such ordinance shall be punishable by a fine of not more than $100. Game wardens, sheriffs and all other law-enforcement officers shall enforce the provisions of this section. No ordinance adopted pursuant to this section shall be enforceable unless the governing body adopting such ordinance so notifies the Director of the Department of Game and Inland Fisheries by registered mail prior to May 1 of the year in which such ordinance is to take effect.

The provisions of this section shall not apply to duly authorized law-enforcement officers or military personnel in the performance of their lawful duties, nor to any person who reasonably believes that a loaded rifle or shotgun is necessary for his personal safety in the course of his employment or business.

15.2-915.3. Requiring fingerprinting for concealed handgun permit Notwithstanding § 15.2-915, a county or city shall by ordinance require any applicant for a concealed handgun permit to submit to fingerprinting for the purpose of obtaining the applicant's state or national criminal history record; however, such ordinance may not require fingerprinting for the renewal of an existing permit pursuant to Subsection I of § 18.2-308.

15.2-915.4. Counties, cities and towns authorized to regulate use of pneumatic guns

A. A locality may prohibit, by ordinance, the shooting of pneumatic guns in any areas of the locality that are in the opinion of the governing body so heavily populated as to make such conduct dangerous to the inhabitants thereof, and may require supervision by a parent, guardian, or other adult supervisor approved by a parent or guardian of any minor below the age of 16 in all uses of pneumatic guns on private or public property. The ordinance may specify that minors above the age of 16 may, with the written consent of a parent or guardian, use a pneumatic gun at any place designated for such use by the local governing body or on private property with the consent of the owner. The ordinance may specify that any minor, whether permitted by a parent or guardian to use a pneumatic gun or not, shall be responsible for obeying all laws, regulations and restrictions governing such use. Any penalty for a pneumatic gun offense set forth in such an ordinance shall not exceed a Class 3 misdemeanor.

B. No such ordinance authorized by subsection A shall prohibit the use of pneumatic guns at facilities approved for shooting ranges or on other property where firearms may be discharged.

C. Training of minors in the use of pneumatic guns shall be done only under direct supervision of a parent, guardian, Junior Reserve Officers Training Corps instructor, or a certified instructor. Training of minors above the age of 16 may also be done without direct supervision if approved by the minor's instructor, with the permission of and under the responsibility of a parent or guardian, and in compliance with all requirements of this section. Ranges and instructors may be certified by the National Rifle Association, a state or federal agency that has developed a certification program, any service of the Department of Defense, or any person authorized by these authorities to certify ranges and instructors.

D. Commercial or private areas designated for use of pneumatic paintball guns may be established and operated for recreational use. Equipment designed to protect the face and ears shall be provided to participants at such recreational areas, and signs must be posted to warn against entry into the paintball area by persons who are unprotected or unaware that paintball guns are in use.

E. As used in this section, "pneumatic gun" means any implement, designed as a gun, that will expel a BB or a pellet by action of pneumatic pressure. "Pneumatic gun" includes a paintball gun that expels by action of pneumatic pressure plastic balls filled with paint for the purpose of marking the point of impact.

Chapter 12. General Powers and Procedures of Counties

15.2-1206. Pistols and revolvers; license tax on dealers The governing body of any county may impose a license tax of not more than twenty-five dollars on persons engaged in the business of selling pistols and revolvers to the public.

15.2-1207. Same; reports of sales The governing body of any county may require sellers of pistols and revolvers to furnish the clerk of the circuit court of the county, within ten days after sale of any such weapon, with the name and address of the purchaser, the date of purchase, and the number, make and caliber of the weapon sold. The clerk shall keep a record of the reports.

15.2-1208. Same; in certain counties Chapter 297 of the Acts of 1944, approved March 29, 1944, requiring permits to sell or purchase pistols or revolvers in any county having a density of population of more than 1,000 a square mile, is repealed. Any records or copies thereof that were created pursuant to this section that are in the custody of any county shall be destroyed no later than July 31, 2004. Upon destroying the records, the county shall certify to the circuit court that such destruction has been completed.

15.2-1209.1. Counties may regulate carrying of loaded firearms on public highways The governing body of any county is hereby empowered to adopt ordinances making it unlawful for any person to carry or have in his possession, for the purpose of hunting, while on any part of a public highway within such county a loaded firearm when such person is not authorized to hunt on the private property on both sides of the highway along which he is standing or walking; and to provide a penalty for violation of such ordinance not to exceed a fine of $100. The provisions of this section shall not apply to persons carrying loaded firearms in moving vehicles or for purposes other than hunting, or to persons acting at the time in defense of persons or property.

Title 18.2. Crimes and Offenses Generally

Chapter 5. Crimes Against Property

Article 1. Arson and Related Crimes

18.2-85. Manufacture, possession, use, etc., of fire bombs or explosive materials or devices; penalties For the purpose of this section:

"Device" means any instrument, apparatus or contrivance, including its component parts, that is capable of producing or intended to produce an explosion but shall not include fireworks as defined in § 27-95.

"Explosive material" means any chemical compound, mechanical mixture or device that is commonly used or can be used for the purpose of producing an explosion and which contains any oxidizing and combustive agents or other ingredients in such proportions, quantities or packaging that an ignition by fire, friction, concussion, percussion, detonation or by any part of the compound or mixture may cause a sudden generation of highly heated gases. These materials include, but are not limited to, gunpowder, powders for blasting, high explosives, blasting materials, fuses (other than electric circuit breakers), detonators, and other detonating agents and smokeless powder.

"Fire bomb" means any container of a flammable material such as gasoline, kerosene, fuel oil, or other chemical compound, having a wick composed of any material or device or other substance which, if set or ignited, is capable of igniting such flammable material or chemical compound but does not include a similar device commercially manufactured and used solely for the purpose of illumination or cooking.

"Hoax explosive device" means any device which by its design, construction, content or characteristics appears to be or to contain a bomb or other destructive device or explosive but which is an imitation of any such device or explosive.

Any person who (i) possesses materials with which fire bombs or explosive materials or devices can be made with the intent to manufacture fire bombs or explosive materials or devices or, (ii) manufactures, transports, distributes, possesses or uses a fire bomb or explosive materials or devices shall be guilty of a Class 5 felony. Any person who constructs, uses, places, sends, or causes to be sent any hoax explosive device so as to intentionally cause another person to believe that such device is a bomb or explosive shall be guilty of a Class 6 felony.

Nothing in this section shall prohibit the authorized manufacture, transportation, distribution, use or possession of any material, substance, or device by a member of the armed forces of the United States, fire fighters or law-enforcement officers, nor shall it prohibit the manufacture, transportation, distribution, use or possession of any material, substance or device to be used solely for scientific research, educational purposes or for any lawful purpose, subject to the provisions of §§ 27-97 and 27-97.2.

Chapter 7. Crimes Involving Health And Safety

Article 4. Dangerous Use of Firearms or Other Weapons

18.2-284. Selling or giving toy firearms No person shall sell, barter, exchange, furnish or dispose of by purchase, gift or in any other manner any toy gun, pistol, rifle or other toy firearm, if the same shall, by action of an explosion of a combustible material, discharge blank or ball charges. Any person violating the provisions of this section shall be guilty of a Class 4 misdemeanor. Each sale of any of the articles hereinbefore specified to any person shall constitute a separate offense.

Nothing in this section shall be construed as preventing the sale of what are commonly known as cap pistols.

Article 5. Uniform Machinegun Act

18.2-288. Definitions When used in this article:

(1) "Machine gun" applies to any weapon which shoots or is designed to shoot automatically more than one shot, without manual reloading, by a single function of the trigger.

(2) "Crime of violence" applies to and includes any of the following crimes or an attempt to commit any of the same, namely, murder, manslaughter, kidnapping, rape, mayhem, assault with intent to maim, disable, disfigure or kill, robbery, burglary housebreaking, breaking and entering and larceny.

(3) "Person" applies to and includes firm, partnership, association or corporation.

18.2-289. Use of machine gun for crime of violence Possession or use of a machine gun in the perpetration or attempted perpetration of a crime of violence is hereby declared to be a Class 2 felony.

18.2-290. Use of machine gun for aggressive purpose Unlawful possession or use of a machine gun for an offensive or aggressive purpose is hereby declared to be a Class 4 felony.

18.2-291. What constitutes aggressive purpose Possession or use of a machine gun shall be presumed to be for an offensive or aggressive purpose:

(1) When the machine gun is on premises not owned or rented for bona fide permanent residence or business occupancy by the person in whose possession the machine gun may be found;

(2) When the machine gun is in the possession of, or used by, a person who has been convicted of a crime of violence in any court of record, state or federal, of the United States of America, its territories or insular possessions;

(3) When the machine gun has not been registered as required in § 18.2-295; or

(4) When empty or loaded shells which have been or are susceptible of use in the machine gun are found in the immediate vicinity thereof.

18.2-292. Presence prima facie evidence of use The presence of a machine gun in any room, boat or vehicle shall be prima facie evidence of the possession or use of the machine gun by each person occupying the room, boat, or vehicle where the weapon is found.

18.2-293. What article does not apply to The provisions of this article shall not be applicable to:

(1) The manufacture for, and sale of, machine guns to the armed forces or law-enforcement officers of the United States or of any state or of any political subdivision thereof, or the transportation required for that purpose; and

(2) Machine guns and automatic arms issued to the national guard of Virginia by the United States or such arms used by the United States army or navy or in the hands of troops of the national guards of other states or territories of the United States passing through Virginia, or such arms as may be provided for the officers of the State Police or officers of penal institutions.

18.2-293.1. What article does not prohibit Nothing contained in this article shall prohibit or interfere with:

(1) The possession of a machine gun for scientific purposes, or the possession of a machine gun not usable as a weapon and possessed as a curiosity, ornament, or keepsake; and

(2) The possession of a machine gun for a purpose manifestly not aggressive or offensive.

Provided, however, that possession of such machine guns shall be subject to the provisions of § 18.2-295.

18.2-294. Manufacturer's and dealer's register; inspection of stock Every manufacturer or dealer shall keep a register of all machine guns manufactured or handled by him. This register shall show the model and serial number, date of manufacture, sale, loan, gift, delivery or receipt of every machine gun, the name, address, and occupation of the person to whom the machine gun was sold, loaned, given or delivered, or from whom it was received. Upon demand every manufacturer or dealer shall permit any marshal, sheriff or police officer to inspect his entire stock of machine guns, parts, and supplies therefor, and shall produce the register, herein required, for inspection. A violation of any provisions of this section shall be punish-able as a Class 3 misdemeanor.

18.2-295. Registration of machine guns Every machine gun in this Commonwealth shall be registered with the Department of State Police within twenty-four hours after its acquisition or, in the case of semi-automatic weapons which are converted, modified or otherwise altered to become machine guns, within twenty-four hours of the conversion, modification or alteration. Blanks for registration shall be prepared by the Superintendent of State Police, and furnished upon application. To comply with this section the application as filed shall be notarized and shall show the model and serial number of the gun, the name, address and occupation of the person in possession, and from whom and the purpose for which, the gun was acquired or altered. The Superintendent of State Police shall upon registration required in this section forthwith furnish the registrant with a certificate of registration, which shall be valid as long as the registrant remains the same. Certificates of registration shall be retained by the registrant and produced by him upon demand by any peace officer. Failure to keep or produce such certificate for inspection shall be a Class 3 misdemeanor, and any peace officer, may without warrant, seize the machine gun and apply for its confiscation as provided in § 18.2-296. Upon transferring a registered machine gun, the transferor shall forthwith notify the Superintendent in writing, setting forth the date of transfer and name and address of the transferee. Failure to give the required notification shall constitute a Class 3 misdemeanor. Registration data shall not be subject to inspection by the public.

18.2-297. How article construed This article shall be so interpreted and construed as to effectuate its general purpose to make uniform the law of those states which enact it.

18.2-298. Short title of article This article may be cited as the "Uniform Machine Gun Act."

Article 6. "Sawed-off" Shotgun and "Sawed-off" Rifle Act"

18.2-299. Definitions When used in this article:

"Sawed-off shotgun" means any weapon, loaded or unloaded, originally designed as a shoulder weapon, utilizing a self-contained cartridge from which a number of ball shot pellets or projectiles may be fired simultaneously from a smooth or rifled bore by a single function of the firing device and which has a barrel length of less than 18 inches for smooth bore weapons and 16 inches for rifled weapons. Weapons of less than .225 caliber shall not be included.

"Sawed-off rifle" means a rifle of any caliber, loaded or unloaded, which expels a projectile by action of an explosion of a combustible material and is designed as a shoulder weapon with a barrel or barrels length of less than 16 inches or which has been modified to an overall length of less than 26 inches.

"Crime of violence" applies to and includes any of the following crimes or an attempt to commit any of the same, namely, murder, manslaughter, kidnapping, rape, mayhem, assault with intent to maim, disable, disfigure or kill, robbery, burglary, housebreaking, breaking and entering and larceny.

"Person" applies to and includes firm, partnership, association or corporation.

18.2-300. Possession or use of "sawed-off" shotgun or rifle

A. Possession or use of a "sawed-off" shotgun or "sawed-off" rifle in the perpetration or attempted perpetration of a crime of violence is a Class 2 felony.

B. Possession or use of a "sawed-off" shotgun or "sawed-off" rifle for any other purpose, except as permitted by this article and official use by those persons permitted possession by § 18.2-303, is a Class 4 felony.

18.2-303. What article does not apply to The provisions of this article shall not be applicable to:

(1) The manufacture for, and sale of, "sawed-off" shotguns or "sawed-off" rifles to the armed forces or law-enforcement officers of the United States or of any state or of any political subdivision thereof, or the transportation required for that purpose; and

(2) "Sawed-off" shotguns, "sawed-off" rifles and automatic arms issued to the National Guard of Virginia by the United States or such arms used by the United States Army or Navy or in the hands of troops of the national guards of other states or territories of the United States passing through Virginia, or such arms as may be provided for the officers of the State Police or officers of penal institutions.

18.2-303.1. What article does not prohibit Nothing contained in this article shall prohibit or interfere with the possession of a "sawed-off" shotgun or "sawed-off" rifle for scientific purposes, the possession of a "sawed-off" shotgun or "sawed-off" rifle possessed in compliance with federal law or the possession of a "sawed-off" shotgun or "sawed-off" rifle not usable as a firing weapon and possessed as a curiosity, ornament, or keepsake.

18.2-304. Manufacturer's and dealer's register; inspection of stock Every manufacturer or dealer shall keep a register of all "sawed-off" shotguns and "sawed-off" rifles manufactured or handled by him. This register shall show the model and serial number, date of manufacture, sale, loan, gift, delivery or receipt of every "sawed-off" shotgun and "sawed-off" rifle, the name, address, and occupation of the person to whom the "sawed-off" shotgun or "sawed-off" rifle was sold, loaned, given or delivered, or from whom it was received. Upon demand every manufacturer or dealer shall permit any marshal, sheriff or police officer to inspect his entire stock of "sawed-off" shotguns and "sawed-off" rifles, and "sawed-off" shotgun or "sawed-off" rifle barrels, and shall produce the register, herein required, for inspection. A violation of any provision of this section shall be punishable as a Class 3 misdemeanor.

18.2-307. Short title of article This article may be cited as the "Sawed-Off Shotgun and Sawed-Off Rifle Act."

Article 7. Other Illegal Weapons

18.2-308.1. Possession of firearm, stun weapon, or other weapon on school property prohibited

A. If any person possesses any (i) stun weapon as defined in this section; (ii) knife, except a pocket knife having a folding metal blade of less than three inches; or (iii) weapon, including a weapon of like kind, designated in subsection A of § 18.2-308, other than a firearm; upon (a) the property of any public, private or religious elementary, middle or high school, including buildings and grounds; (b) that portion of any property open to the public and then exclusively used for school-sponsored functions or extracurricular activities while such functions or activities are taking place; or (c) any school bus owned or operated by any such school, he shall be guilty of a Class 1 misdemeanor.

B. If any person possesses any firearm designed or intended to expel a projectile by action of an explosion of a combustible material while such person is upon (i) any public, private or religious elementary, middle or high school, including buildings and grounds; (ii) that portion of any property open to the public and then exclusively used for school-sponsored functions or extracurricular activities while such functions or activities are taking place; or (iii) any school bus owned or operated by any such school, he shall be guilty of a Class 6 felony; however, if the person possesses any firearm within a public, private or religious elementary, middle or high school building and intends to use, or attempts to use, such firearm, or displays such weapon in a threatening manner, such person shall be sentenced to a mandatory minimum term of imprisonment of five years to be served consecutively with any other sentence.

The exemptions set out in § 18.2-308 shall apply, mutatis mutandis, to the provisions of this section. The provisions of this section shall not apply to (i) persons who possess such weapon or weapons as a part of the school's curriculum or activities; (ii) a person possessing a knife customarily used for food preparation or service and using it for such purpose; (iii) persons who possess such weapon or weapons as a part of any program sponsored or facilitated by either the school or any organization authorized by the school to conduct its programs either on or off the school premises; (iv) any law-enforcement officer; (v) any person who possesses a knife or blade which he uses customarily in his trade; (vi) a person who possesses an unloaded firearm that is in a closed container, or a knife having a metal blade, in or upon a motor vehicle, or an unloaded shotgun or rifle in a firearms rack in or upon a motor vehicle; or (vii) a person who has a valid concealed handgun permit and possesses a concealed handgun while in a motor vehicle in a parking lot, traffic circle, or other means of vehicular ingress or egress to the school. For the purposes of this paragraph, "weapon" includes a knife having a metal blade of three inches or longer and "closed container" includes a locked vehicle trunk.

As used in this section:

"Stun weapon" means any device that emits a momentary or pulsed output, which is electrical, audible, optical or electromagnetic in nature and which is designed to temporarily incapacitate a person.

18.2-308.1:1. Possession or transportation of firearms by persons acquitted by reason of insanity; penalty; permit

A. It shall be unlawful for any person acquitted by reason of insanity and committed to the custody of the Commissioner of Mental Health, Mental Retardation and Substance Abuse Services, pursuant to § 19.2-181, on a charge of treason, any felony or any offense punishable as a misdemeanor under Title 54.1 or a Class 1 or Class 2 misdemeanor under this title, except those misdemeanor violations of (i) Article 2 (§ 18.2-266 et seq.) of Chapter 7 of this title, (ii) Article 2 (§ 18.2-415 et seq.) of Chapter 9 of this title, or (iii) § 18.2-119, or (iv) an ordinance of any county, city, or town similar to the offenses specified in (i), (ii), or (iii), to knowingly and intentionally possess or transport any firearm. A violation of this section shall be punishable as a Class 1 misdemeanor.

B. Any person so acquitted may, upon discharge from the custody of the Commissioner, petition the general district court in which he resides for a permit to possess or carry a firearm. If the court determines that the circumstances regarding the disability referred to in subsection A and the person's criminal history, treatment record, and reputation are such that the person will not be likely to act in a manner dangerous to public safety and that the granting of the relief would not be contrary to the public interest, the court shall grant the petition. Any person denied relief by the general district court may petition the circuit court for a de novo review of the denial. Upon a grant of relief in any court, the court shall enter a written order granting the petition and issue a permit, in which event the provisions of subsection A do not apply. The clerk of court shall certify and forward forthwith to the Central Criminal Records Exchange, on a form provided by the Exchange, a copy of any such order.

18.2-308.1:2. Purchase, possession or transportation of firearm by persons adjudicated legally incompetent or mentally incapacitated; penalty It shall be unlawful for any person who has been adjudicated (i) legally incompetent pursuant to former § 37.1-128.02 or former § 37.1-134, (ii) mentally incapacitated pursuant to former § 37.1-128.1 or former § 37.1-132 or (iii) incapacitated pursuant to Article 1.1 (§ 37.1-134.6 et seq.) of Chapter 4 of Title 37.1 and whose competency or capacity has not been restored pursuant to former § 37.1-134.1 or § 37.1-134. 16, to purchase, possess, or transport any firearm. A violation of this section shall be punishable as a Class 1 misdemeanor.

18.2-308.1:3. Purchase, possession or transportation of firearm by persons involuntarily committed; penalty

A. It shall be unlawful for any person involuntarily admitted to a facility or ordered to mandatory outpatient treatment pursuant to § 19.2-169.2, involuntarily admitted to a facility or ordered to mandatory outpatient treatment as the result of a commitment hearing pursuant to Article 5 (§ 37.2-814 et seq.) of Chapter 8 of Title 37.2, or who was the subject of a temporary detention order pursuant to § 37.2-809 and subsequently agreed to voluntary admission pursuant to § 37.2-805 to purchase, possess or transport a firearm. A violation of this subsection shall be punishable as a Class 1 misdemeanor.

B. Any person prohibited from purchasing, possessing or transporting firearms under this section may, at any time following his release from involuntary admission to a facility, his release from an order of mandatory outpatient treatment, or his release from voluntary admission pursuant to § 37.2-805 following the issuance of a temporary detention order, petition the general district court in the city or county in which he resides to restore his right to purchase, possess or transport a firearm. If the court determines that the circumstances regarding the disabilities referred to in subsection A and the person's criminal history, treatment record, and reputation are such that the person will not likely act in a manner dangerous to public safety and that granting the relief would not be contrary to the public interest, the court shall grant the petition. Any person denied relief by the general district court may petition the circuit court for a de novo review of the denial. Upon a grant of relief in any court, the court shall enter a written order granting the petition, in which event the provisions of subsection A shall no longer apply. The clerk of court shall certify and forward forthwith to the Central Criminal Records Exchange, on a form provided by the Exchange, a copy of any such order.

18.2-308.1:4. Purchase or transportation of firearm by persons subject to protective orders; penalty It shall be unlawful for any person who is subject to (i) a protective order entered pursuant to §§ 16.1-253, 16.1-253.1, 16.1-253.4, 16.1-279.1, 19.2-152.8, 19.2-152.9, or § 19.2-152.10; (ii) an order issued pursuant to subsection B of § 20-103; (iii) an order entered pursuant to subsection D of § 18.2-60.3; or (iv) an order issued by a tribunal of another state, the United States or any of its territories, possessions or commonwealths, or the District of Columbia pursuant to a statute that is substantially similar to those cited in clauses (i), (ii), or (iii) to purchase or transport any firearm while the order is in effect. Any person with a concealed handgun permit shall be prohibited from carrying any concealed firearm, and shall surrender his permit to the court entering the order, for the duration of any protective order referred to herein. A violation of this section is a Class 1 misdemeanor.

18.2-308.1:5. Purchase or transportation of firearm by persons convicted of certain drug offenses prohibited Any person who, within a thirty-six consecutive month period, has been convicted of two misdemeanor offenses under § 18.2-250 or § 18.2-250.1 shall be ineligible to purchase or transport a handgun. However, upon expiration of a period of five years from the date of the second conviction and provided the person has not been convicted of any such offense within that period, the ineligibility shall be removed.

18.2-308.2. Possession or transportation of firearms , stun weapons, tasers or concealed weapons by convicted felons; penalties; petition for permit; when issued

A. It shall be unlawful for (i) any person who has been convicted of a felony; (ii) any person adjudicated delinquent as a juvenile 14 years of age or older at the time of the offense of murder in violation of § 18.2-31 or 18.2-32, kidnapping in violation of § 18.2-47, robbery by the threat or presentation of firearms in violation of § 18.2-58, or rape in violation of § 18.2-61; or (iii) any person under the age of 29 who was adjudicated delinquent as a juvenile 14 years of age or older at the time of the offense of a delinquent act which would be a felony if committed by an adult, other than those felonies set forth in clause (ii), whether such conviction or adjudication occurred under the laws of the Commonwealth, or any other state, the District of Columbia, the United States or any territory thereof, to knowingly and intentionally possess or transport any firearm or ammunition for a firearm, any stun weapon as defined by § 18.2-308.1, or any explosive material, or to knowingly and intentionally carry about his person, hidden from common observation, any weapon described in subsection A of § 18.2-308. However, such person may possess in his residence or the curtilage thereof a stun weapon as defined by § 18.2-308.1. Any person who violates this section shall be guilty of a Class 6 felony. However, any person who violates this section by knowingly and intentionally possessing or transporting any firearm and who was previously convicted of a violent felony as defined in § 17.1-805 shall be sentenced to a mandatory minimum term of imprisonment of five years. Any person who violates this section by knowingly and intentionally possessing or transporting any firearm and who was previously convicted of any other felony within the prior 10 years shall be sentenced to a mandatory minimum term of imprisonment of two years. The mandatory minimum terms of imprisonment prescribed for violations of this section shall be served consecutively with any other sentence.

B. The prohibitions of subsection A shall not apply to (i) any person who possesses a firearm, explosive material or other weapon while carrying out his duties as a member of the

Armed Forces of the United States or of the National Guard of Virginia or of any other state, (ii) any law-enforcement officer in the performance of his duties, or (iii) any person who has been pardoned or whose political disabilities have been removed pursuant to Article V, Section 12 of the Constitution of Virginia provided the Governor, in the document granting the pardon or removing the person's political disabilities, may expressly place conditions upon the reinstatement of the person's right to ship, transport, possess or receive firearms.

C. Any person prohibited from possessing, transporting or carrying a firearm or stun weapon under subsection A, may petition the circuit court of the jurisdiction in which he resides for a permit to possess or carry a firearm or stun weapon; however, no person who has been convicted of a felony shall be qualified to petition for such a permit unless his civil rights have been restored by the Governor or other appropriate authority. The court may, in its discretion and for good cause shown, grant such petition and issue a permit. The provisions of this section relating to firearms and stun weapons shall not apply to any person who has been granted a permit pursuant to this subsection.

C1. Any person who was prohibited from possessing, transporting or carrying explosive material under subsection A may possess, transport or carry such explosive material if his right to possess, transport or carry explosive material has been restored pursuant to federal law.

D. For the purpose of this section, "explosive material" means any chemical compound mixture, or device, the primary or common purpose of which is to function by explosion; the term includes, but is not limited to, dynamite and other high explosives, black powder, pellet powder, smokeless gun powder, detonators, blasting caps and detonating cord but shall not include fireworks or permissible fireworks as defined in § 27-95.

18.2-308.2:01. Possession or transportation of certain firearms by aliens

A. It shall be unlawful for any person who is not a citizen of the United States or who is not a person lawfully admitted for permanent residence to knowingly and intentionally possess or transport any assault firearm or to knowingly and intentionally carry about his person, hidden from common observation, an assault firearm.

B. It shall be unlawful for any person who is not a citizen of the United States and who is not lawfully present in the United States to knowingly and intentionally possess or transport any firearm or to knowingly and intentionally carry about his person, hidden from common observation, any firearm. A violation of this section shall be punishable as a Class 6 felony.

C. For purposes of this section, "assault firearm" means any semi-automatic center-fire rifle or pistol that expels single or multiple projectiles by action of an explosion of a combustible material and is equipped at the time of the offense with a magazine which will hold more than 20 rounds of ammunition or designed by the manufacturer to accommodate a silencer or equipped with a folding stock.

18.2-308.2:1. Prohibiting the selling, etc., of firearms to certain persons Any person who sells, barters, gives or furnishes, or has in his possession or under his control with the intent of selling, bartering, giving or furnishing, any firearm to any person he knows is prohibited from possessing or transporting a firearm pursuant to § 18.2-308.1:1, 18.2-308.2, subsection B of § 18.2-308.2:01, or § 18.2-308.7 shall be guilty of a Class 6 felony. However, this prohibition shall not be applicable when the person convicted of the felony, adjudicated delinquent or acquitted by reason of insanity has (i) been issued a permit pursuant to subsection C of § 18.2-308.2 or subsection B of § 18.2-308.1:1, (ii) been pardoned or had his political disabilities removed in accordance with subsection B of § 18.2-308.2 or (iii) obtained a permit to ship, transport, possess or receive firearms pursuant to the laws of the United States.

18.2-308.2:2. Criminal history record information check required for the transfer of certain firearms

A. Any person purchasing from a dealer a firearm as herein defined shall consent in writing, on a form to be provided by the Department of State Police, to have the dealer obtain criminal history record information. Such form shall include only the written consent; the name, birth date, gender, race, citizenship, and social security number and/or any other identification number; the number of firearms by category intended to be sold, rented, traded, or transferred; and answers by the applicant to the following questions: (i) has the applicant been convicted of a felony offense or found guilty or adjudicated delinquent as a juvenile 14 years of age or older at the time of the offense of a delinquent act that would be a felony if committed by an adult; (ii) is the applicant subject to a court order restraining the applicant from harassing, stalking, or threatening the applicant's child or intimate partner, or a child of such partner, or is the applicant subject to a protective order; and (iii) has the applicant ever been acquitted by reason of insanity and prohibited from purchasing, possessing or transporting a firearm pursuant to § 18.2-308.1:1 or any substantially similar law of any other jurisdiction, been adjudicated legally incompetent, mentally incapacitated or adjudicated an incapacitated person and prohibited from purchasing a firearm pursuant to § 18.2-308.1:2 or any substantially similar law of any other jurisdiction, or been involuntarily admitted to an inpatient facility or involuntarily ordered to outpatient mental health treatment and prohibited from purchasing a firearm pursuant to § 18.2-308.1:3 or any substantially similar law of any other jurisdiction.

B. 1. No dealer shall sell, rent, trade or transfer from his inventory any such firearm to any other person who is a resident of Virginia until he has (i) obtained written consent and the other information on the consent form specified in subsection A, and provided the Department of State Police with the name, birth date, gender, race, citizenship, and social security and/or any other identification number and the number of firearms by category intended to be sold, rented, traded or transferred and (ii) requested criminal history record information by a telephone call to or other communication authorized by the State Police and is authorized by subdivision 2 of this subsection to complete the sale or other such transfer. To establish personal identification and residence in Virginia for purposes of this section, a dealer must require any prospective purchaser to present one photo-identification form issued by a governmental agency of the Commonwealth or by the United States Department of Defense, and other documentation of residence. Except where the photo-identification was issued by the United States Department of Defense, the other documentation of residence shall show an address identical to that shown on the photo-identification form, such as evidence of currently paid personal property tax or real estate tax, or a current (a) lease, (b) utility or telephone bill, (c) voter registration card, (d) bank check, (e) passport, (f) automobile registration, or (g) hunting or fishing license; other current identification allowed as evidence of residency by Part 178.124 of Title 27 of the Code of Federal Regulations and ATF Ruling 2001-5; or other documentation of residence determined to be acceptable by the Department of Criminal Justice Services, that corroborates that the prospective purchaser currently resides in Virginia. Where the photo-identification was issued by the Department of Defense, permanent orders assigning the purchaser to a duty post in Virginia shall be the only other required documentation of residence. For the purposes of this section and establishment of residency for firearm purchase, residency shall be deemed to be the permanent duty post of a member of the armed forces. When the photo-identification presented to a dealer by the prospective purchaser is a driver's license or other photo-identification issued by the Department of Motor Vehicles, and such identification form contains a date of issue, the dealer shall not, except for a renewed driver's license or other photo-identification issued by the Department of Motor Vehicles, sell or otherwise transfer a firearm to the prospective purchaser until 30 days after the date of issue of an original or duplicate driver's license unless the prospective purchaser also presents a copy of his Virginia Department of Motor Vehicles driver's record showing that the original date of issue of the driver's license was more than 30 days prior to the attempted purchase.

In addition, no dealer shall sell, rent, trade or transfer from his inventory any assault firearm to any person who is not a citizen of the United States or who is not a person lawfully admitted for permanent residence. To establish citizenship or lawful admission for a permanent residence for purposes of purchasing an assault firearm, a dealer shall require a prospective purchaser to present a certified birth certificate or a certificate of birth abroad issued by the United States State Department, a certificate of citizenship or a certificate of naturalization issued by the United States Citizenship and Immigration Services, an unexpired U.S. passport, a United States citizen identification card, a current voter registration card, a current selective service registration card, or an immigrant visa or other documentation of status as a person lawfully admitted for permanent residence issued by the United States Citizenship and Immigration Services.

Upon receipt of the request for a criminal history record information check, the State Police shall (1) review its criminal history record information to determine if the buyer or transferee is prohibited from possessing or transporting a firearm by state or federal law, (2) inform the dealer if its record indicates that the buyer or transferee is so prohibited, and (3) provide the dealer with a unique reference number for that inquiry.

2. The State Police shall provide its response to the requesting dealer during the dealer's request, or by return call without delay. If the criminal history record information check indicates the prospective purchaser or transferee has a disqualifying criminal record or has been acquitted by reason of insanity and committed to the custody of the Commissioner of Behavioral Health and Developmental Services, the State Police shall have until the end of the dealer's next business day to advise the dealer if its records indicate the buyer or transferee is prohibited from possessing or transporting a firearm by state or federal law. If not so advised by the end of the dealer's next business day, a dealer who has fulfilled the requirements of subdivision 1 of this subsection may immediately complete the sale or transfer and shall not be deemed in violation of this section with respect to such sale or transfer. In case of electronic failure or other circumstances beyond the control of the State Police, the dealer shall be advised immediately of the reason for such delay and be given an estimate of the length of such delay. After such notification, the State Police shall, as soon as possible but in no event later than the end of the dealer's next business day, inform the requesting dealer if its records indicate the buyer or transferee is prohibited from possessing or transporting a firearm by state or federal law. A dealer who fulfills the requirements of subdivision 1 of this subsection and is told by the State Police that a response will not be available by the end of the dealer's next business day may immediately complete the sale or transfer and shall not be deemed in violation of this section with respect to such sale or transfer.

3. Except as required by subsection D of § 9.1-132, the State Police shall not maintain records longer than 30 days, except for multiple handgun transactions for which records shall be maintained for 12 months, from any dealer's request for a criminal history record information check pertaining to a buyer or transferee who is not found to be prohibited from possessing and transporting a firearm under state or federal law. However, the log on requests made may be maintained for a period of 12 months, and such log shall consist of the name of the purchaser, the dealer identification number, the unique approval number and the transaction date.

4. On the last day of the week following the sale or transfer of any firearm, the dealer shall mail or deliver the written consent form required by subsection A to the Department of State Police. The State Police shall immediately initiate a search of all available criminal history record information to determine if the purchaser is prohibited from possessing or transporting a firearm under state or federal law. If the search discloses information indicating that the buyer or transferee is so prohibited from possessing or transporting a firearm, the State Police shall inform the chief law-enforcement officer in the jurisdiction where the sale or transfer occurred and the dealer without delay.

5. Notwithstanding any other provisions of this section, rifles and shotguns may be purchased by persons who are citizens of the United States or persons lawfully admitted for permanent residence but residents of other states under the terms of subsections A and B upon furnishing the dealer with proof of citizenship or status as a person lawfully admitted for permanent residence and one photo-identification form issued by a governmental agency of the person's state of residence and one other form of identification determined to be acceptable by the Department of Criminal Justice Services.

6. For the purposes of this subsection, the phrase "dealer's next business day" shall not include December 25.

C. No dealer shall sell, rent, trade or transfer from his inventory any firearm, except when the transaction involves a rifle or a shotgun and can be accomplished pursuant to the provisions of subdivision B 5 to any person who is not a resident of Virginia unless he has first obtained from the Department of State Police a report indicating that a search of all available criminal history record information has not disclosed that the person is prohibited from possessing or transporting a firearm under state or federal law. The dealer shall obtain the required report by mailing or delivering the written consent form required under subsection A to the State Police within 24 hours of its execution. If the dealer has complied with the provisions of this subsection and has not received the required report from the State Police within 10 days from the date the written consent form was mailed to the Department of State Police, he shall not be deemed in violation of this section for thereafter completing the sale or transfer.

D. Nothing herein shall prevent a resident of the Commonwealth, at his option, from buying, renting or receiving a firearm from a dealer in Virginia by obtaining a criminal history record information check through the dealer as provided in subsection C.

E. If any buyer or transferee is denied the right to purchase a firearm under this section, he may exercise his right of access to and review and correction of criminal history record information under § 9.1-132 or institute a civil action as provided in § 9.1-135, provided any such action is initiated within 30 days of such denial.

F. Any dealer who willfully and intentionally requests, obtains, or seeks to obtain criminal history record information under false pretenses, or who willfully and intentionally disseminates or seeks to disseminate criminal history record information except as authorized in this section shall be guilty of a Class 2 misdemeanor.

G. For purposes of this section:

"Actual buyer" means a person who executes the consent form required in subsection B or C, or other such firearm transaction records as may be required by federal law.

"Antique firearm":

1. Any firearm (including any firearm with a matchlock, flintlock, percussion cap, or similar type of ignition system) manufactured in or before 1898;

2. Any replica of any firearm described in subdivision 1 of this definition if such replica (i) is not designed or redesigned for using rimfire or conventional centerfire fixed ammunition or (ii) uses rimfire or conventional centerfire fixed ammunition that is no longer manufactured in the United States and that is not readily available in the ordinary channels of commercial trade;

3. Any muzzle-loading rifle, muzzle-loading shotgun, or muzzle-loading pistol that is designed to use black powder, or a black powder substitute, and that cannot use fixed ammunition. For purposes of this subdivision, the term "antique firearm" shall not include any weapon that incorporates a firearm frame or receiver, any firearm that is converted into a muzzle-loading weapon, or any muzzle-loading weapon that can be readily converted to fire fixed ammunition by replacing the barrel, bolt, breech-block, or any combination thereof; or

4. Any curio or relic as defined in this subsection.

"Assault firearm" means any semi-automatic center-fire rifle or pistol which expels single or multiple projectiles by action of an explosion of a combustible material and is equipped at the time of the offense with a magazine which will hold more than 20 rounds of ammunition or designed by the manufacturer to accommodate a silencer or equipped with a folding stock.

"Curios or relics" means firearms that are of special interest to collectors by reason of some quality other than is associated with firearms intended for sporting use or as offensive or defensive weapons. To be recognized as curios or relics, firearms must fall within one of the following categories:

1. Firearms that were manufactured at least 50 years prior to the current date, which use rimfire or conventional centerfire fixed ammunition that is no longer manufactured in the United States and that is not readily available in the ordinary channels of commercial trade, but not including replicas thereof;

2. Firearms that are certified by the curator of a municipal, state, or federal museum that exhibits firearms to be curios or relics of museum interest; and

3. Any other firearms that derive a substantial part of their monetary value from the fact that they are novel, rare, bizarre, or because of their association with some historical figure, period, or event. Proof of qualification of a particular firearm under this category may be established by evidence of present value and evidence that like firearms are not available except as collectors' items, or that the value of like firearms available in ordinary commercial channels is substantially less.

"Dealer" means any person licensed as a dealer pursuant to 18 U.S.C. § 921 et seq.

"Firearm" means any handgun, shotgun, or rifle that will or is designed to or may readily be converted to expel single or multiple projectiles by action of an explosion of a combustible material.

"Handgun" means any pistol or revolver or other firearm originally designed, made and intended to fire single or multiple projectiles by means of an explosion of a combustible material from one or more barrels when held in one hand.

"Lawfully admitted for permanent residence" means the status of having been lawfully accorded the privilege of residing permanently in the United States as an immigrant in accordance with the immigration laws, such status not having changed.

H. The Department of Criminal Justice Services shall promulgate regulations to ensure the identity, confidentiality and security of all records and data provided by the Department of State Police pursuant to this section.

I. The provisions of this section shall not apply to (i) transactions between persons who are licensed as firearms importers or collectors, manufacturers or dealers pursuant to 18 U.S.C. § 921 et seq.; (ii) purchases by or sales to any law-enforcement officer or agent of the United States, the Commonwealth or any local government; or (iii) antique firearms, curios or relics.

J. The provisions of this section shall not apply to restrict purchase, trade or transfer of firearms by a resident of Virginia when the resident of Virginia makes such purchase, trade or transfer in another state, in which case the laws and regulations of that state and the United States governing the purchase, trade or transfer of firearms shall apply. A National Instant Criminal Background Check System (NICS) check shall be performed prior to such purchase, trade or transfer of firearms.

J1. All licensed firearms dealers shall collect a fee of $2 for every transaction for which a criminal history record information check is required pursuant to this section, except that a fee of $5 shall be collected for every transaction involving an out-of-state resident. Such fee shall be transmitted to the Department of State Police by the last day of the month following the sale for deposit in a special fund for use by the State Police to offset the cost of conducting criminal history record information checks under the provisions of this section.

K. Any person willfully and intentionally making a materially false statement on the consent form required in subsection B or C or on such firearm transaction records as may be required by federal law, shall be guilty of a Class 5 felony.

L. Except as provided in § 18.2-308.2:1, any dealer who willfully and intentionally sells, rents, trades or transfers a firearm in violation of this section shall be guilty of a Class 6 felony.

L1. Any person who attempts to solicit, persuade, encourage, or entice any dealer to transfer or otherwise convey a firearm other than to the actual buyer, as well as any other person who willfully and intentionally aids or abets such person, shall be guilty of a Class 6 felony. This subsection shall not apply to a federal law-enforcement officer or a law-enforcement officer as defined in § 9.1-101, in the performance of his official duties, or other person under his direct supervision.

M. Any person who purchases a firearm with the intent to (i) resell or otherwise provide such firearm to any person who he knows or has reason to believe is ineligible to purchase or otherwise receive from a dealer a firearm for whatever reason or (ii) transport such firearm out of the Commonwealth to be resold or otherwise provided to another person who the transferor knows is ineligible to purchase or otherwise receive a firearm, shall be guilty of a Class 5 felony. However, if the violation of this subsection involves such a transfer of more than one firearm, the person shall be sentenced to a mandatory minimum term of imprisonment of five years.

N. Any person who is ineligible to purchase or otherwise receive or possess a firearm in the Commonwealth who solicits, employs or assists any person in violating subsection M shall be guilty of a Class 5 felony and shall be sentenced to a mandatory minimum term of imprisonment of five years.

O. All driver's licenses issued on or after July 1, 1994, shall carry a letter designation indicating whether the driver's license is an original, duplicate or renewed driver's license.

P. Except as provided in subdivisions 1, 2 and 3 of this subsection, it shall be unlawful for any person who is not a licensed firearms dealer to purchase more than one handgun within any 30-day period. A violation of this subsection shall be punishable as a Class 1 misdemeanor.

1. Purchases in excess of one handgun within a 30-day period may be made upon completion of an enhanced background check, as described herein, by special application to the Department of State Police listing the number and type of handguns to be purchased and transferred for lawful business or personal use, in a collector series, for collections, as a bulk purchase from estate sales and for similar purposes. Such applications shall be signed under oath by the applicant on forms provided by the Department of State Police, shall state the purpose for the purchase above the limit, and shall require satisfactory proof of residency and identity. Such application shall be in addition to the firearms sales report required by the Bureau of Alcohol, Tobacco and Firearms (ATF). The Superintendent of State Police shall promulgate regulations, pursuant to the Administrative Process (§ 2.2-4000 et seq.), for the implementation of an application process for purchases of handguns above the limit.

Upon being satisfied that these requirements have been met, the Department of State Police shall forthwith issue to the applicant a nontransferable certificate which shall be valid for seven days from the date of issue. The certificate shall be surrendered to the dealer by the prospective purchaser prior to the consummation of such sale and shall be kept on file at the dealer's place of business for inspection as provided in § 54.1-4201 for a period of not less than two years. Upon request of any local law-enforcement agency, and pursuant to its regulations, the Department of State Police may certify such local law-enforcement agency to serve as its agent to receive applications and, upon authorization by the Department of State Police, issue certificates forthwith pursuant to this subsection. Applications and certificates issued under this subsection shall be maintained as records as provided in subdivision B 3 . The Department of State Police shall make available to local law-enforcement agencies all records concerning certificates issued pursuant to this subsection and all records provided for in subdivision B 3.

2. The provisions of this subsection shall not apply to:

a. A law-enforcement agency;

b. An agency duly authorized to perform law-enforcement duties;

c. State and local correctional facilities;

d. A private security company licensed to do business within the Commonwealth;

e. The purchase of antique firearms as herein defined;

f. A person whose handgun is stolen or irretrievably lost who deems it essential that such handgun be replaced immediately. Such person may purchase another handgun, even if the person has previously purchased a handgun within a 30-day period, provided (i) the person provides the firearms dealer with a copy of the official police report or a summary thereof, on forms provided by the Department of State Police, from the law-enforcement agency that took the report of the lost or stolen handgun; (ii) the official police report or summary thereof contains the name and address of the handgun owner, the description of the handgun, the location of the loss or theft, the date of the loss or theft, and the date the loss or theft was reported to the law-enforcement agency; and (iii) the date of the loss or theft as reflected on the official police re-port or summary thereof occurred within 30 days of the person's attempt to replace the handgun. The firearms dealer shall attach a copy of the official police report or summary thereof to the original copy of the Virginia firearms transaction report completed for the transaction and retain it for the period prescribed by the Department of State Police;

g. A person who trades in a handgun at the same time he makes a handgun purchase and as a part of the same transaction, provided that no more than one transaction of this nature is completed per day;

h. A person who holds a valid Virginia permit to carry a concealed handgun;

i. A person who purchases a handgun in a private sale. For purposes of this subdivision, a private sale means purchase from a person who makes occasional sales, exchanges or purchases of firearms for the enhancement of a personal collection of curios or relics as herein defined, or who sells all or part of such collection of curios and relics; or

j. A law-enforcement officer. For purposes of this subdivision, a law-enforcement officer means any employee of a police department or sheriff's office that is part of or administered by the Commonwealth or any political subdivision thereof, and who is responsible for the prevention and detection of crime and the enforcement of the penal, traffic or highway laws of the Commonwealth.

3. For the purposes of this subsection, "purchase" shall not include the exchange or replacement of a handgun by a seller for a handgun purchased from such seller by the same person seeking the exchange or replacement within the 30-day period immediately preceding the date of exchange or replacement.

18.2-308.2:3. Criminal background check required for employees of a gun dealer to transfer firearms; exemptions; penalties

A. No person, corporation or proprietorship licensed as a firearms dealer pursuant to 18 U.S.C. § 921 et seq. shall employ any person to act as a seller, whether full-time or part-time, permanent, temporary, paid or unpaid, for the transfer of firearms under § 18.2-308.2:2, if such employee would be prohibited from possessing a firearm under §§ 18.2-308.1:1, 18.2-308.1:2, 18.2-308.1:3, 18.2-308.2, or 18.2-308.2:01 or is an illegal alien, or is prohibited from purchasing or transporting a firearm pursuant to § 18.2-308.1:4 or § 18.2-308.1:5.

B. Prior to permitting an applicant to begin employment, the dealer shall obtain a written statement or affirmation from the applicant that he is not disqualified from possessing a firearm and shall submit the applicant's fingerprints and personal descriptive information to the Central Criminal Records Exchange to be forwarded to the Federal Bureau of Investigation (FBI) for the purpose of obtaining national criminal history record information regarding the applicant.

C. Prior to August 1, 2000, the dealer shall obtain written statements or affirmations from persons employed before July 1, 2000, to act as a seller under § 18.2-308.2:2 that they are not disqualified from possessing a firearm. With-in five working days of the employee's next birthday, after August 1, 2000, the dealer shall submit the employee's fingerprints and personal descriptive information to the Central Criminal Records Exchange to be forwarded to the Federal Bureau of Investigation (FBI) for the purpose of obtaining national criminal history record information regarding the request.

C1. In lieu of submitting fingerprints pursuant to this section, any dealer holding a valid federal firearms license (FFL) issued by the Bureau of Alcohol, Tobacco and Firearms (ATF) may submit a sworn and notarized affidavit to the Department of State Police on a form provided by the Department, stating that the dealer has been subjected to a record check prior to the issuance and that the FFL was issued by the ATF. The affidavit may also contain the names of any employees that have been subjected to a record check and approved by the ATF. This exemption shall apply regardless of whether the FFL was issued in the name of the dealer or in the name of the business. The affidavit shall contain the valid FFL number, state the name of each person requesting the exemption, together with each person's identifying information, including their social security number and the following statement: "I hereby swear, under the penalty of perjury, that as a condition of obtaining a federal firearms license, each person requesting an exemption in this affidavit has been subjected to a fingerprint identification check by the Bureau of Alcohol, Tobacco and Firearms and the Bureau of Alcohol, Tobacco and Firearms subsequently determined that each person satisfied the requirements of 18 U.S.C. § 921 et seq. I understand that any person convicted of making a false statement in this affidavit is guilty of a Class 5 felony and that in addition to any other penalties imposed by law, a conviction under this section shall result in the forfeiture of my federal firearms license."

D. The Department of State Police, upon receipt of an individual's record or notification that no record exists, shall submit an eligibility report to the requesting dealer within 30 days of the applicant beginning his duties for new employees or within 30 days of the applicant's birthday for a person employed prior to July 1, 2000.

E. If any applicant is denied employment because of information appearing on the criminal history record and the applicant disputes the information upon which the denial was based, the Central Criminal Records Exchange shall, upon written request, furnish to the applicant the procedures for obtaining a copy of the criminal history record from the Federal Bureau of Investigation. The information provided to the dealer shall not be disseminated except as provided in this section.

F. The applicant shall bear the cost of obtaining the criminal history record unless the dealer, at his option, decides to pay such cost.

G. Upon receipt of the request for a criminal history record information check, the State Police shall establish a unique number for that firearm seller. Beginning September 1, 2001, the firearm seller's signature, firearm seller's number and the dealer's identification number shall be on all firearm transaction forms. The State Police shall void the firearm seller's number when a disqualifying record is discovered. The State Police may suspend a firearm seller's identification number upon the arrest of the firearm seller for a potentially disqualifying crime.

H. This section shall not restrict the transfer of a firearm at any place other than at a dealership or at any event required to be registered as a gun show.

I. Any person who willfully and intentionally requests, obtains, or seeks to obtain criminal history record information under false pretenses, or who willfully and intentionally disseminates or seeks to disseminate criminal history record information except as authorized by this section and § 18.2-308.2:2, shall be guilty of a Class 2 misdemeanor.

J. Any person willfully and intentionally making a materially false statement on the personal descriptive information required in this section shall be guilty of a Class 5 felony. Any person who offers for transfer any firearm in violation of this section shall be guilty of a Class 1 misdemeanor. Any dealer who willfully and knowingly employs or permits a person to act as a firearm seller in violation of this section shall be guilty of a Class 1 misdemeanor.

K. There is no civil liability for any seller for the actions of any purchaser or subsequent transferee of a firearm lawfully transferred pursuant to this section.

L. The provisions of this section requiring a seller's background check shall not apply to a licensed dealer.

M. Any person who willfully and intentionally makes a false statement in the affidavit as set out in subdivision C 1 shall be guilty of a Class 5 felony.

N. For purposes of this section:

"Dealer" means any person, corporation or proprietorship licensed as a dealer pursuant to 18 U.S.C. § 921 et seq.

"Firearm" means any handgun, shotgun, or rifle that will or is designed to or may readily be converted to expel single or multiple projectiles by action of an explosion of a combustible material.

"Place of business" means any place or premises where a dealer may lawfully transfer firearms.

"Seller" means for the purpose of any single sale of a firearm any person who is a dealer or an agent of a dealer, who may lawfully transfer firearms and who actually performs the criminal background check in accordance with the provisions of § 18.2-308.2:2.

"Transfer" means any act performed with intent to sell, rent, barter, trade or otherwise transfer ownership or permanent possession of a firearm at the place of business of a dealer.

18.2-308.3. Use or attempted use of restricted ammunition in commission or attempted commission of crimes prohibited; penalty

A. When used in this section:

"Restricted firearm ammunition" applies to bullets, projectiles or other types of ammunition that are: (i) coated with or contain, in whole or in part, polytetrafluorethylene or a similar product, (ii) commonly known as "KTW" bullets or "French Arcanes," or (iii) any cartridges containing bullets coated with a plastic substance with other than lead or lead alloy cores, jacketed bullets with other than lead or lead alloy cores, or cartridges of which the bullet itself is wholly comprised of a metal or metal alloy other than lead. This definition shall not be construed to include shotgun shells or solid plastic bullets.

B. It shall be unlawful for any person to knowingly use or attempt to use restricted firearm ammunition while committing or attempting to commit a crime. Violation of this section shall constitute a separate and distinct felony and any person found guilty thereof shall be guilty of a Class 5 felony.

18.2-308.4. Possession of firearms while in possession of certain controlled substances

A. It shall be unlawful for any person unlawfully in possession of a controlled substance classified in Schedule I or II of the Drug Control Act (§ 54.1-3400 et seq.) of Title 54.1 to simultaneously with knowledge and intent possess any firearm. A violation of this subsection is a Class 6 felony and constitutes a separate and distinct felony.

B. It shall be unlawful for any person unlawfully in possession of a controlled substance classified in Schedule I or II of the Drug Control Act (§ 54.1-3400 et seq.) to simultaneously with knowledge and intent possess any firearm on or about his person. A violation of this subsection is a Class 6 felony and constitutes a separate and distinct felony and any person convicted hereunder shall be sentenced to a mandatory minimum term of imprisonment of two years. Such punishment shall be separate and apart from, and shall be made to run consecutively with, any punishment received for the commission of the primary felony.

C. It shall be unlawful for any person to possess, use, or attempt to use any pistol, shotgun, rifle, or other firearm or display such weapon in a threatening manner while committing or attempting to commit the illegal manufacture, sale, distribution, or the possession with the intent to manufacture, sell, or distribute a controlled substance classified in Schedule I or Schedule II of the Drug Control Act (§ 54.1-3400 et seq.) of Title 54.1 or more than one pound of marijuana. Violation of this subsection is a Class 6 felony, and constitutes a separate and distinct felony and any person convicted hereunder shall be sentenced to a mandatory minimum term of imprisonment of five years. Such punishment shall be separate and apart from, and shall be made to run consecutively with, any punishment received for the commission of the primary felony.

18.2-308.5. Manufacture, import, sale, transfer or possession of plastic firearm prohibited It shall be unlawful for any person to manufacture, import, sell, transfer or possess any plastic firearm. As used in this section "plastic firearm" means any firearm, including machine guns and sawed-off shotguns as defined in this chapter, containing less than 3.7 ounces of electromagnetically detectable metal in the barrel, slide, cylinder, frame or receiver of which, when subjected to inspection by X-ray machines commonly used at airports, does not generate an image that accurately depicts its shape. A violation of this section shall be punishable as a Class 5 felony.

18.2-308.6. Possession of unregistered firearm mufflers or silencers prohibited; penalty It shall be unlawful for any person to possess any firearm muffler or firearm silencer which is not registered to him in the National Firearms Registration and Transfer Record. A violation of this section shall be punishable as a Class 6 felony.

18.2-308.7. Possession or transportation of certain firearms by persons under the age of eighteen; penalty It shall be unlawful for any person under 18 years of age to knowingly and intentionally possess or transport a handgun or assault firearm anywhere in the Commonwealth. For the purposes of this section, "handgun" means any pistol or revolver or other firearm originally designed, made and intended to fire single or multiple projectiles by means of an explosion of a combustible material from one or more barrels when held in one hand and "assault firearm" means any (i) semi-automatic centerfire rifle or pistol which expels single or multiple projectiles by action of an explosion of a combustible material and is equipped at the time of the offense with a magazine which will hold more than 20 rounds of ammunition or designed by the manufacturer to accommodate a silencer or equipped with a folding stock or (ii) shotgun with a magazine which will hold more than seven rounds of the longest ammunition for which it is chambered. A violation of this section shall be a Class 1 misdemeanor.

This section shall not apply to:

1. Any person (i) while in his home or on his property; (ii) while in the home or on the property of his parent, grandparent, or legal guardian, or (iii) while on the property of another who has provided prior permission, and with the prior permission of his parent or legal guardian if the person has the landowner's written permission on his person while on such property;

2. Any person who, while accompanied by an adult, is at, or going to and from, a lawful shooting range or firearms educational class, provided that the weapons are unloaded while being transported;

3. Any person actually engaged in lawful hunting or going to and from a hunting area or preserve, provided that the weapons are unloaded while being transported; and

4. Any person while carrying out his duties in the armed forces of the United States or the National Guard of this Commonwealth or any other state.

18.2-308.8. Importation, sale, possession or transfer of Striker 12's prohibited; penalty It shall be unlawful for any person to import, sell, possess or transfer the following firearms: the Striker 12, commonly called a "streetsweeper," or any semi-automatic folding stock shotgun of like kind with a spring tension drum magazine capable of holding twelve shotgun shells. A violation of this section shall be punishable as a Class 6 felony.

18.2-309. Furnishing certain weapons to minors; penalty

A. If any person sells, barters, gives or furnishes, or causes to be sold, bartered, given or furnished, to any minor a dirk, switchblade knife or bowie knife, having good cause to believe him to be a minor, such person shall be guilty of a Class 1 misdemeanor.

B. If any person sells, barters, gives or furnishes, or causes to be sold, bartered, given or furnished, to any minor a handgun, having good cause to believe him to be a minor, such person shall be guilty of a Class 6 felony. This subsection shall not apply to any transfer made between family members or for the purpose of engaging in a sporting event or activity.

18.2-311.1. Removing, altering, etc., serial number or other identification on firearm Any person, firm, association or corporation who or which intentionally removes, defaces, alters, changes, destroys or obliterates in any manner or way or who or which causes to be removed, defaced, altered, changed, destroyed or obliterated in any manner or way the name of the maker, model, manufacturer's or serial number, or any other mark or identification on any pistol, shotgun, rifle, machine gun or any other firearm shall be guilty of a Class 1 misdemeanor.

18.2-474.1. Delivery of drugs, firearms, explosives, etc., to prisoners

Notwithstanding the provisions of § 18.2-474, any person who shall willfully in any manner deliver, attempt to deliver, or conspire with another to deliver to any prisoner confined under authority of the Commonwealth of Virginia, or of any political subdivision thereof, any drug which is a controlled substance regulated by the Drug Control Act in Chapter 34 of Title 54.1 or marijuana, shall be guilty of a Class 5 felony. Any person who shall willfully in any manner so deliver or attempt to deliver or conspire to deliver to any such prisoner, firearms, ammunitions, or explosives of any nature shall be guilty of a Class 3 felony.

Nothing herein contained shall be construed to repeal or amend § 18.2-473.

Title 19.2. Criminal Procedure

Chapter 22.2. Miscellaneous Forfeiture Provisions

19.2-386.28. Forfeiture of weapons that are concealed, possessed, transported or carried in violation of law Any firearm, stun weapon or taser as defined by § 18.2-308.1, or any weapon concealed, possessed, transported or carried in violation of §§ 18.2-283.1, 18.2-287.4, 18.2-308.1:2, 18.2-308.1:3, 18.2-308.1:4, 18.2-308.2, 18.2-308.2:01, 18.2-308.2:1, 18.2-308.4, 18.2-308.5, 18.2-308.7, or § 18.2-308.8 shall be forfeited to the Commonwealth and disposed of as provided in § 19.2-386.29.

19.2-386.29. Forfeiture of certain weapons used in commission of criminal offense All pistols, shotguns, rifles, dirks, bowie knives, switchblade knives, ballistic knives, razors, slingshots, brass or metal knucks, blackjacks, stun weapons, and other weapons used by any person in the commission of a criminal offense, shall, upon conviction of such person, be forfeited to the Commonwealth by order of the court trying the case. The court shall dispose of such weapons as it deems proper by entry of an order of record. Such disposition may include the destruction of the weapons or, subject to any registration requirements of federal law, sale of the firearms to a licensed dealer in such firearms in accordance with the provisions of Chapter 22 (§ 19.2-369 et seq.) of this title regarding sale of property forfeited to the Commonwealth.

The proceeds of any sale of such weapon shall be paid in accordance with the provisions of Article VIII, Section 8 of the Constitution of Virginia. In addition, the court may authorize the seizing law-enforcement agency to use the weapon for a period of time as specified in the order. When the seizing agency ceases to so use the weapon, it shall be disposed of as otherwise provided in this section.

However, upon petition to the court and notice to the attorney for the Commonwealth, the court, upon good cause shown, shall return any such weapon to its lawful owner after conclusion of all relevant proceedings if such owner (i) did not know and had no reason to know of the conduct giving rise to the forfeiture and (ii) is not otherwise prohibited by law from possessing the weapon. The owner shall acknowledge in a sworn affidavit to be filed with the record in the case or cases that he has retaken possession of the weapon involved.

Title 52. Police (State)

Chapter 1. Department of State Police

52-4.4. Duties relating to criminal history record information checks required by licensed firearms dealers The Superintendent of the Department of State Police shall establish a toll-free telephone number which shall be operational seven days a week between the hours of 8:00 a.m. and 10:00 p.m., except December 25, for purposes of responding to inquiries from licensed firearms dealers, as such term is defined in 18 U.S.C. § 921 et seq., pursuant to the provisions of § 18.2-308.2:2. The Department shall hire and train such personnel as are necessary to administer the provisions of this section.

52-8.4:1. Regulations for firearms shows The Superintendent of State Police shall provide a form for use by promoters of firearms shows for the purpose of notifying the State Police and the chief of police, or the sheriff in localities without police departments, of their intent to conduct a firearms show pursuant to § 54.1-4201.1.

Title 54.1. Professions and Occupations

Chapter 42. Dealers in Firearms

54.1-4200. Definitions For the purpose of this chapter, unless the context requires a different meaning:

"Dealer in firearms" means (i) any person, firm, partnership, or corporation engaged in the business of selling, trading or transferring firearms at wholesale or retail; (ii) any person, firm, partnership, or corporation engaged in the business of making or fitting special barrels, stocks, or trigger mechanisms to firearms; or (iii) any person, firm, partnership, or corporation that is a pawnbroker.

"Engaged in business" means as applied to a dealer in firearms a person, firm, partnership, or corporation that devotes time, attention, and labor to dealing in firearms as a regular course of trade or business with the principal objective of livelihood and profit through repetitive purchase or resale of firearms, but such term shall not involve a person who makes occasional sales, exchanges, or purchases of firearms for the enhancement of a personal collection or for a hobby, or who sells all or part of his personal collection of firearms.

"Firearms show" means any gathering or exhibition, open to the public, not occurring on the permanent premises of a dealer in firearms, conducted principally for the purposes of exchanging, selling or trading firearms as defined in § 18.2-308.2:2.

54.1-4201. Inspection of records

A. Every dealer in firearms shall keep at his place of business, for not less than a period of two years, the original consent form required to be completed by § 18.2-308.2:2 for each firearm sale.

B. Every dealer in firearms shall admit to his place of business during regular business hours the chief law enforcement officer, or his designee, of the jurisdiction in which the dealer is located, or any law-enforcement official of the Commonwealth, and shall permit such law-enforcement officer, in the course of a bona fide criminal investigation, to examine and copy those federal and state records related to the acquisition or disposition of a particular firearm required by this section. This section shall not be construed to authorize the seizure of any records.

54.1-4201.1. Notification by sponsor of firearms show to State Police and local law-enforcement authorities required; records; penalty

A. No promoter of a firearms show shall hold such show without giving notice at least thirty days prior to the show to the State Police and the sheriff or chief of police of the locality in which the firearms show will be held. The notice shall be given on a form provided by the State Police. A separate notice shall be required for each firearms show.

"Promoter" means every person, firm, corporation, club, association, or organization holding a firearms show in the Commonwealth.

The promoter shall maintain for the duration of the show a list of all vendors or exhibitors in the show for immediate inspection by any law-enforcement authorities, and within five days after the conclusion of the show, by mail, by hand, by e-mail or by fax, transmit a copy of the complete vendor or exhibitor list to the law-enforcement authorities to which the 30-day prior notice was required. The vendor or exhibitor list shall contain the full name and residence address and the business name and address, if any, of the vendors or exhibitors.

B. A willful violation of this section shall be a Class 3 misdemeanor.

C. The provisions of this section shall not apply to firearms shows held in any town with a population of not less than 1,995 and not more than 2,010, according to the 1990 United States census.

54.1-4202. Penalties for violation of the provisions of this chapter Any person convicted of a first offense for willfully violating the provisions of this chapter shall be guilty of a Class 2 misdemeanor. Any person convicted of a second or subsequent offense under the provisions of this chapter shall be guilty of a Class 1 misdemeanor.

Title 59.1. Trade and Commerce

Chapter 10. Explosives

59.1-139. Persons possessing explosives to give notice of theft

Any person having in his possession any explosives covered by this chapter shall immediately notify the sheriff of the county or the police officials of the city in which any such explosives are being stored or used in the event that any such explosives are stolen.

Chapter 11.1. Firearms

59.1-148.4. Sale of firearms by law-enforcement agencies prohibited; exception A law-enforcement agency of this Commonwealth shall not sell or trade any firearm owned and used or otherwise lawfully in its possession except (i) to another law-enforcement agency of the Commonwealth, (ii) to a licensed firearms dealer, (iii) to the persons as provided in § 59.1-148.3 or (iv) as authorized by a court in accordance with § 19.2-386.29.

[Current through the 2008 Regular Session, Acts 2008, cc. 1 to 884, 2008 Special Session I, cc. 1 and 2, 2008 Special Session II, cc. 1 to 11]

VIRGIN ISLANDS

V.I. CODE

Title 14. Crimes

Chapter 113. Weapons

2253. Carrying of firearms; openly or concealed; evidence of intent to commit crime of violence; definitions

(a) Whoever, unless otherwise authorized by law, has, possesses, bears, transports or carries either, actually or constructively, openly or concealed any firearm, as defined in Title 23, section 451(d) of this code, loaded or unloaded, may be arrested without a warrant, and shall be sentenced to imprisonment of not less than one year nor more than five years and shall be fined not less than $5,000 nor more than $ 15,000 or both the fine and imprisonment, except that if such person shall have been convicted of a felony in any state, territory, or federal court of the United States, or if such firearm or an imitation thereof was had, possessed, borne, transported or carried by or under the proximate control of such person during the commission or attempted commission of a crime of violence, as defined in subsection (d) hereof, then such person shall be fined $25,000 and imprisoned not less than fifteen (15) years nor more than twenty (20) years. The foregoing applicable penalties provided for violation of this section shall be in addition to the penalty provided for the commission of, or attempt to commit, the felony or crime of violence.

(b) Whoever, unless otherwise authorized by law, has, possesses, bears, transports or carries either, actually or constructively, openly or concealed any machine gun or sawed-off shotgun, as defined in subsection (d)(2) and (3) of this section, loaded or unloaded, may be arrested without a warrant, and shall be sentenced to imprisonment of not less than two years nor more than seven years and shall be fined $25,000, except that if such person shall have been convicted of a felony in any state, territory or federal court of the United States, or if such machine gun or sawed-off shotgun or an imitation thereof was held, possessed, borne, transported by or under the proximate control of such person during the commission or attempted commission of a crime of violence, as herein defined, then such person shall be fined $ 50,000 and imprisoned not less than fifteen (15) years nor more than twenty (20) years. The foregoing applicable penalties provided for violation of this section shall be in addition to the penalty provided for the commission of, or attempt to commit, the crime of violence.

(c) In the trial of a person for committing or attempting to commit a crime of violence, as herein defined, the fact that he was armed with a firearm, used or attempted to be used, and had no license to carry the same, as required in Title 23, chapter 5 of the Code, shall be evidence of his intention to commit said crime of violence.

(d) As used in this chapter –

(1) "Crime of violence" shall have the same definition as that contained in Title 23, section 451(e) of this Code.

(2) "Machine gun" means any firearm, as defined in Title 23, section 451(d) of this Code, which shoots automatically more than 12 shots without reloading.

(3) "Sawed-off shotgun" means any firearm, as defined in Title 23, section 451(d) of this Code, designed to fire through a smooth bore either a number of ball shot or a single projectile, the barrel of which is less than 20 inches in length.

(4) The term "possession" as used in this section means both actual and constructive possession.

(5) "Constructive possession" means having the power and the intention at any given time to exercise dominion or actual control over the firearm either directly or through another person.

Title 23. Internal Security and Public Order

Chapter 5. Control of Firearms and Ammunition

451. Definitions As used in this chapter, unless the context clearly requires otherwise-

(a) "Ammunition" means any bullet, cartridge, projectile, buckshot, or any load placed or which may be placed in a firearm to be discharged.

(b) "Commissioner" means the Police Commissioner of the Virgin Islands.

(c) "Department" means the U.S. Virgin Islands Police Department (V.I.P.D.).

(d) "Firearm" means any device by whatever name known, capable of discharging ammunition by means of gas generated from an explosive composition, including any air, gas, or spring gun or any "B.B." pistols or "B.B." guns that have been adapted or modified to discharge projectiles as a firearm.

(e) "Crime of Violence" means the crime of, or the attempt to commit, murder in any degree, voluntary manslaughter, rape, arson, discharging or aiming firearms, mayhem, kidnapping, assault in the first degree, assault in the second degree, assault in the third degree, robbery, burglary, unlawful entry or larceny.

(f) "Dealer in firearms and/or ammunition" means any person engaged in the business of selling firearms and/or ammunition, for a profit or gain.

(g) "Gunsmith" means any person who engages in the business of repairing, altering, cleaning, polishing, engraving, bluing, or performing any mechanical operation on any firearm on an individual order basis.

452. Applicability of chapter No person shall have, possess, bear, transport or carry a firearm within the Virgin Islands, or engage in the business of dealer in firearms and/or ammunition or the business of gunsmith, except in compliance with the provisions of this chapter.

453. Persons who may lawfully carry firearms

(a) The following persons, in the discharge of their official duties, and in accordance with and subject to the conditions and restrictions imposed by the laws and regulations applicable to their conduct, may lawfully have, possess, bear, transport and carry firearms in the Virgin Islands:

(1) Members of the Armed Forces of the United States or of the organized reserves.

(2) Officers and employees of the United States duly authorized by Federal law to carry firearms.

(3) Persons employed in fulfilling defense contracts with the United States Government or agencies thereof where possession or use of firearms is necessary under the provisions of such contracts.

(4) Members of the police force of the Virgin Islands, marshals, or other duly authorized peace officers.

(5) Penitentiary and jail wardens and guards.

(b) The persons authorized by subsection (a) of this section lawfully to have, possess, bear, transport and carry firearms shall obtain such weapons and ammunition therefor only through the duly authorized officers or heads of their respective services or departments.

454. Persons who may be licensed to carry firearms. A firearm may be lawfully had, possessed, borne, transported or carried in the Virgin Islands by the following persons, provided a license for such purpose has been issued by the Commissioner in accordance with the provisions of this chapter:

(1) An officer or employee of the Government of the Virgin Islands in cases where such license, in the judgment of the Commissioner, should be issued to such officer or employee by reason of the duties of his position;

(2) An agent, messenger or other employee of a common carrier, bank or business firm, whose duties require him to protect money, valuables or other property in the discharge of his duties; And provided, That the employer of such person shall have justified to the satisfaction of the Commissioner the need for the issuance of the license;

(3) A person having a bona fide residence or place of business within the Virgin Islands, who established to the satisfaction of the Commissioner that he has good reason to fear death or great injury to his person or property, or who establishes any other proper reason for carrying a firearm, and the circumstances of the case, established by affidavit of the applicant and of at least two credible persons, demonstrate the need for such license;

(4) A person licensed to and actively engaged in the business of manufacturing, repairing or dealing in firearms in the Virgin Islands, or the agents or representatives of any such person, having necessity to handle or use firearms in the usual or ordinary course of business;

(5) With respect to a rifle or a shotgun a person possessing a valid and current Virgin Islands hunting license.

455. Application for license; form, oath; fees

(a) Every application for a license to have and possess a firearm shall be made under oath and on forms which the Commissioner shall prepare for such purpose. For the purposes of the enforcement of the provisions of this chapter, the applicant shall furnish all information as may be required of him by the Commissioner.

(b) The initial fee for a license under section 454 of this chapter shall be $75.00. The license may be renewed every three years for a fee of $150.00.

(c) Upon renewal of a license to have and possess a firearm, the receipt from the U.S. Virgin Islands Police Department (V.I.P.D.) for the renewal fee will serve as a temporary license until the official license can be provided to the licensee by the Commissioner.

(d) The Commissioner shall ensure that the renewal license is presented to the licensee within forty-five (45) days of receipt of payment for the renewal fee.

(e) Notwithstanding the provisions of this section, no person shall be charged with possession of an unlicensed firearm if the subject weapon had been previously licensed and said license has expired not more than ninety (90) days prior to arrest; Provided, however, that this subsection shall not apply to persons who possess, bear, transport, carry or have under their control in any vehicle, any firearm during the commission or attempted commission of a crime of violence, as defined in subsection (d) of section 2253, Title 14, Virgin Islands Code.

456. Qualifications of applicant

(a) The Commissioner shall not issue a license for firearms under section 454 of this chapter until all the circumstances and facts set forth in the application have been investigated, and the records of the Department and other available records have been examined, and unless such investigation establishes to the satisfaction of the Commission:

(1) the truth of such circumstances and facts;

(2) that the applicant is a resident of the Virgin Islands, including with respect to shotguns or rifles a minor not under 16 years of age, or a nonresident who holds a current and valid license to hunt in the Virgin Islands, or an alien bonded under applicable Federal and Virgin Islands statutes for employment with a person, firm, corporation or other business entity duly licensed in the Virgin Islands to carry on the business of providing security, guard, patrol and private detective services; Provided, however, That in the case of any such bonded alien the license shall be issued to the business entity by which he is employed;

(3) that the applicant is a person of good moral character;

(4) that the applicant's fingerprints have been duly taken and/or checked with the records of the Department or other appropriate sources; and

(5) that no proper reason exists to deny such application.

456a. Persons ineligible to possess or carry firearms or ammunition

(a) The following persons are ineligible for a license to possess or carry a firearm or ammunition as provided in this chapter:

(1) a person who has been convicted in any court for a crime punishable by imprisonment for a term exceeding one year;

(2) a person who is a fugitive from justice;

(3) a person who is an unlawful user of or addicted to any controlled substance as defined in title 19, section 593(6) of the Virgin Islands Code;

(4) a person who has been adjudicated as a mental defective or who has been committed to a mental institution:

(A) For the purpose of this section the phrase "committed to a mental institution" includes commitment to a mental institution involuntarily, but does not include a person held in a mental institution for observation.

(B) For the purpose of this section, the phrase "mental institution" includes mental health facilities, mental hospitals, sanitariums, psychiatric facilities and other facilities that provide diagnoses by licensed professionals for mental retardation or mental illness, including a psychiatric ward in a public or private hospital.

(5) a person who, being an alien, is illegally or unlawfully in the United States;

(6) a person who has been discharged from the United States Armed Forces under dishonorable conditions;

(7) a person who, having been a citizen of the United States, has renounced his citizenship;

(8) a person who is subject to a court order that-

(A) was issued after a hearing of which the person received actual notice, and at which the person had an opportunity to participate;

(B) restrains the person from harassing, stalking, or threatening an intimate partner of such person or child of such intimate partner or person, or engaging in other conduct that would place an intimate partner in reasonable fear of bodily injury to the partner or child; and

(C)(i) includes a finding that the person represents a credible threat to the physical safety of such intimate partner or child; or

(ii) by its terms explicitly prohibits the use, attempted use, or threatened use of physical force against such intimate partner or child that would reasonably be expected to cause bodily injury; or

(9) a person who has been convicted in any court of a misdemeanor crime of domestic violence.

458. Grounds for refusing to issue license

(a) The Commissioner shall not issue a license to have and possess a firearm to any person convicted in or outside the Virgin Islands of any crime of violence; or of any violation of a narcotic or "harmful drug" law; nor to any person who is mentally incompetent, or a habitual drunkard or narcotic or drug addict; nor to any person convicted for the violation of the provisions of this chapter; nor to any person who for justifiable reasons is deemed to be an improper person by the Commissioner.

460. Reciprocal recognition of out-of-state licenses Unless otherwise prohibited by any state or federal law, a license to possess or to carry firearms, issued by any competent authority of any state or territory of the United States and in accordance with the same or similar requirements as set forth in the preceding sections pertaining to the applicant's eligibility, and the establishment of his reputation through fingerprints, shall be recognized as valid within the Virgin Islands and shall allow the holder thereof to exercise all of the privileges in connection therewith, while said licensee is a visitor or transient resident herein.

Any marshal, sheriff, constable, police or other peace officer, of any state or territory of the United States, whose duty it is to serve process and make arrests, may, while travelling through or in the Virgin Islands on official business, carry such weapons or equipment as has been authorized by his appointing authority.

461. License to sell firearms and/or ammunition; gunsmiths; report of transactions; private transfer sales to minors or aliens

(a) No person may engage in the business of dealer in firearms and/or ammunition or as a gunsmith without holding a license therefor issued by the Commissioner of Licensing and Consumer Affairs upon favorable report of the Police Commissioner.

(b) Each transaction referring to the importation, or to the sale of firearms and ammunition between dealers, shall be reported to the Commissioner on forms which he shall provide, and the name, domicile, place of business, and the number of the license of the vendor and vendee, as well as the quantities and descriptions of the firearms or ammunition which are the subject of each transaction, shall be set forth therein, as required by the Commissioner.

(c) Any person, not otherwise engaged in the business of a dealer in firearms and/or ammunition, may transfer a firearm to another by sale, gift, exchange or otherwise only upon prior reporting to the Commissioner of the details of the proposed transaction. The Commissioner shall thereupon determine the eligibility of the proposed transferee for a license to possess firearms, and shall make such further determination as may be necessary in the circumstances.

(d) No person licensed or otherwise, may sell or furnish firearms or ammunition to a minor, except that a shotgun or rifle of such type or caliber as the Commissioner may prescribe or ammunition therefor, may be sold or furnished by a licensed dealer to a minor who displays a hunting or sporting license issued him in accordance with the laws of the Virgin Islands, and who further displays the written consent of his parent, guardian, or other responsible person acting in their absence and interest, in which such sale or delivery has been authorized.

462. Application for dealer's or gunsmith's license; form and content; term; fee; renewals

(a) Any person wishing to obtain a license to engage in the business of dealer in firearms and/or ammunition or the business of gunsmith, or both, shall file with the Commissioner of Licensing and Consumer Affairs a sworn application in the manner provided by the said Commissioner of Licensing and Consumer Affairs in consultation with the Police Commissioner which application shall contain all the information necessary so that the license may be issued in accordance with the provisions of this chapter. Such application shall be transmitted to the Police Commissioner and no license shall be issued under this section without the latter Commissioner first making an investigation of all the statements set forth in the application and without the records of the Department first being examined for the purpose of determining any prior convictions of the applicant, and only when such an investigation establishes that the statements of the applicant are true, and that the applicant has not been convicted of any crime of violence and the Commissioner so certifies, recommending that the license be issued. Where the applicant is a corporation or partnership, no license shall be issued if any officer of the corporation, or partner of the partnership, has been convicted of any crime of violence.

(b) Licenses hereunder shall be issued for a term of one year, expiring in any event on the 15th day of January succeeding the date of issuance.

(c) The annual fee for a license as a dealer of firearms and/or ammunition shall be $100, for a license as gunsmith, $100, and for a license as both, $150.

463. Qualifications of dealer or gunsmith No license to engage in the business of gunsmith or of dealer in firearms and/or ammunition shall be issued to a person not over 21 years of age, and not a resident of the Virgin Islands and a citizen of the United States.

When the applicant is a corporation, it shall be organized under the laws of the Virgin Islands, and if a partnership, all partners shall be residents of the Virgin Islands and citizens of the United States.

464. Corporation or partnership application for license Where the applicant is a corporation or a partnership, the application shall be signed and sworn to by the president, the secretary and the treasurer of the corporation or by all the managing partners of the partnership; and it shall set forth the name of the corporation or of the partnership, place and date of the incorporation or organization thereof, and the principal place of business. The requirements of subparagraphs (3) and (4) of section 456 of this chapter shall be applied to the president, the secretary, and the treasurer of the corporation and, in the proper case, to all partners of the partnership. A license issued under the provisions of this section shall be valid only for the business establishments named and described in the license. Said license may not be transferred to any other business establishment or to any other person and shall be automatically cancelled upon the dissolution of the corporation or partnership, or the replacement of any of the officers of the corporation signing the application, or upon the admission of any new partner in the case of a partnership, even though such license may be renewed as soon as the provisions of subparagraphs (3) and (4) of section 456 of this chapter in connection with the new officer or the new partner are complied with. In these cases the Commissioner may issue a provisional issue for a term of not more than thirty days while the renewal is being processed.

465. Conditions for dealers' operations; Records of transactions Any person, to whom a license has been issued under section 461 of this chapter may engage in the business of gunsmith or of dealer in firearms and/or ammunition under the following conditions:

(1) The business shall be operated only on the premises specified in the license.

(2) The license, or a copy thereof, certified by the authority issuing the same, shall be posted in the establishment so that it may be easily read.

(3) Under no circumstances shall a weapon or shall ammunition be sold unless the vendor is personally acquainted with the buyer or the latter clearly establishes his identity.

(4) A record in triplicate shall be kept of each firearm sold and of each sale of ammunition, on books devoted to this purpose which shall be printed in the manner prescribed by the Commissioner and the record of each sale shall be personally signed by the buyer and by the person making the sale, each in the presence of the other; and such record shall set forth the day and hour of the sale, caliber, manufacture, model and factory number of the weapon, caliber mark, and quantity of ammunition, the name, birthplace, address and occupation of the buyer. Said record shall also contain a thumbprint of the buyer and shall also set forth whether the buyer is personally known to the vendor, and in case he should not be the manner in which the buyer established his identity. The vendor shall transmit a copy of such record by registered mail, within 24 hours following the sale, to the Commissioner; he shall send the duplicate within 48 hours following the sale to the Attorney General and shall keep the triplicate for six years.

(5) A licensee, if limited to the business of a gunsmith, shall not be required to forward reports of the work performed under said license, but shall maintain at his place of business an accurate and legible accounting of the nature and type of the jobs or work performed, together with the name and address of the customers, a description of the firearm, including the serial number which accounting shall be available as required by the Commissioner.

466. Sales of weapons and ammunition without licenses prohibited; Sales slips No dealer in firearms or ammunition shall deliver a firearm to a purchaser without the latter's handing over to him a license to have and possess a firearm, duly issued in accordance with the provisions of this chapter, and unless said license contains an authorization for the purchase of such firearm, and said dealer shall not sell to such purchaser any other weapon than the one described on the said license. The dealer in firearms and ammunition shall separate from such license and keeps the dealer's coupon, and shall return the license to the purchaser. No weapon shall be sold to the holder of a license from which the coupon has been removed, and in no case shall a pistol, revolver, or other firearm be delivered to the purchaser thereof until after forty-eight (48) hours have elapsed from the time he applies for his purchase and the same shall be delivered to him unloaded and securely wrapped.

No dealer in firearms or ammunition shall sell any quantity of ammunition to any person failing to present a firearm license.

467. Selling firearms and ammunition without a license Whoever, without being licensed in accordance with the provisions of this chapter, sells, has in his possession with the intent to sell, exposes for sale, or advertises for sale, any firearm or ammunition, or without being so licensed engages in the business of a gunsmith, shall be fined a minimum of $5,000 or five (5) years imprisonment or both; and a maximum of not more than $10,000 or ten (10) years imprisonment or both.

467a. Importation of firearms without a license Except as provided in section 476 of this chapter, whoever, without being licensed in accordance with this chapter, imports, carries, or brings any firearm or ammunition into the Virgin Islands and fails to report to the Commissioner as required by section 470 of this chapter shall be punished by a fine of not less than $ 50,000 or imprisonment for a minimum of 25 years or both fine.

470. Report of firearms purchased outside or brought into the Virgin Islands; fees; penalty

(a) Any person other than a licensed dealer, who purchases or otherwise obtains any firearm or ammunition from any source within or outside the Virgin Islands shall report such fact in writing or in person to the Commissioner immediately after receipt of the firearm or ammunition, furnishing a complete description of the firearm or ammunition purchased or otherwise obtained. He shall also furnish his own name, address, date of birth and occupation.

(b) Any person upon entering the Virgin Islands bringing with him any firearm or ammunition shall report in writing or in person to the Commissioner immediately of his arrival, furnishing a complete description of the firearm or ammunition brought into the Virgin Islands. He shall also furnish his own name, address, date of birth and occupation.

(c) In the event the person reporting under subsections (a) or (b), above, is qualified for a license to carry firearms in the Virgin Islands, the Commissioner shall issue the same, upon payment of the proper fee, and the firearm shall be registered in the Weapons Register provided for in section 469 of this chapter. If the person is not qualified for a license then the Commissioner shall retain the firearms or ammunition for disposition in accordance with the provisions of section 475 of this chapter, but no prosecution shall lie against the person for unlawful possession of the firearm or ammunition.

(d) Any person who fails to comply with this section shall be punished as provided in section 484 of this chapter.

471. Report of loss of firearm Every per-son possessing or having under his control a firearm, and who loses the same or finds that it has disappeared, shall report the same to the Commissioner within 10 days, following discovery of such loss or disappearance, and in the event of his failing to do [so] such person shall be fined no more than $100.

473. Firearms Register

(a) The Commissioner shall establish a Firearms Register within the Department and maintain the same in a systematic and orderly manner, so that the names of the persons licensed as dealers in firearms, gunsmiths or to carry firearms in the Virgin Islands as well as the essential details concerning the firearms registered, may be easily found.

(b) Every firearm authorized to be licensed under section 454 which is duly licensed after this chapter takes effect, shall be registered in the Firearms Register provided for in the preceding subsection. If such firearm does not bear a serial number, or if the same is illegible, a nongovernmental licensee shall have his full name engraved on the butt or the stock of the weapon, and shall so set forth in his declaration. The Commissioner shall deliver to the declarant a record of such registration.

(c) The firearms authorized to be had, possessed, or carried under paragraphs (4) and (5) of subsection (a) of section 453 and under subparagraph (1) of section 454 of this chapter, shall also be registered in the Firearms Register above provided for.

474. Death of licensee Upon the death of a person leaving in his estate a firearm it shall be the duty of his administrator, executor or any other person legally authorized to represent the estate to report to the Commissioner the brand, caliber and serial number if any, of the firearm in question. Upon such report, or upon information otherwise verified, the Commissioner shall make the necessary provisions for the custody of said firearm by an interested party or by its deposit with the Department, pending the settlement of the estate. If the firearm is assigned to an heir or distributee who is eligible to obtain a license for a firearm, and such license is issued, said firearm shall be delivered to such heir or distributee. If the heir or distributee is not eligible for a firearms license, then the firearm shall be disposed of as part of the estate through public or private sale as the District Court of the Virgin Islands may direct; provided that the firearm may be acquired only by a duly licensed dealer, a person eligible for a license to possess or carry firearms in the Virgin Islands, or a nonresident of the Virgin Islands in compliance with the laws of the buyer's state.

475. Deposit of firearms in Department of Public Safety, disposition

(a) All firearms surrendered, seized or confiscated under the provisions of this chapter, shall be disposed of as provided in this section.

(b) Any firearm unlawfully borne, possessed, transported or carried, is hereby declared a public nuisance and when any such weapon or instrument is seized by the police the same shall be delivered to the Commissioner. In the cases where criminal actions are filed, the judgment convicting the defendant shall, in addition to the penalties herein prescribed, carry with it the forfeiture in favor of the Government of the Virgin Islands of the firearm seized, and it shall be the duty of the court or judge to deliver the firearm so forfeited to the Commissioner who shall direct the disposition thereof.

(c) Those firearms which are in good condition may be preserved by the Commissioner for governmental uses.

(d) The Commissioner shall, whenever he may deem it advisable, and at least once a year, destroy and render completely unserviceable the firearms delivered to him under the provisions of this section, except when a certification is presented to him from a court, or the United States Attorney or the Attorney General to the effect that the preservation thereof, or of any of them, is necessary or indispensable for the ends of justice, and except when the Commissioner himself directs that they, or any of them, be retained in any police or detective laboratory to the end that they may be examined, compared, identified, or be the object of any other experiment tending to discover or prevent crime.

(e) Before disposing of a firearm, as provided in this section, the Commissioner shall direct that the class, make, number and caliber of such firearms be carefully recorded and filed.

476. Collections of antique firearms; Certificates of uselessness No provision hereof shall prevent that private collections of antique firearms, which may not be used as weapons, be preserved and maintained and that their owners possess them as ornaments or as matters of curiosity, nor the collections of firearms kept as relics, but for the preservation of any weapon of those included in this section the prior inspection thereof and approval therefor by the Commissioner shall be necessary and he shall render such firearms useless, so that the same may not be used as such. The Commissioner shall issue a certificate of uselessness of all the weapons possessed under the provisions of this section, and the possession of any firearm not included in said certificate shall be subject to all the provisions hereof.

479. Discharging or aiming firearms Any person who, otherwise than in self-defense or in the discharge of official duty:

(b) intentionally although without malice aims a revolver, pistol or other firearm toward any person; or

(d) causes physical injury to any person by the discharge of a firearm while aiming intentionally, although without malice, at any person, shall be guilty of a felony.

481. Alteration of identifying marks of weapons prohibited

(a) No person shall within the Virgin Islands change, alter, remove or obliterate the name of the maker, model, manufacturer's number, or other mark or identification on any pistol, machinegun, or sawed-off shotgun. Possession of any pistol, machinegun, or sawed-off shotgun upon which any such mark shall have been changed, altered, removed, or obliterated shall be prima facie evidence that the possessor has changed, altered, removed or obliterated the same within the Virgin Islands: Provided, however, that nothing contained in this section shall apply to any officer or agent of the United States or the Government of the Virgin Islands engaged in experimental work.

(b) Whoever, unless otherwise authorized by law, has, possesses, bears, transports or carries either, actually or constructively, openly or concealed, any firearm, as defined in section 451(d) of this title, loaded or unloaded, with altered or obliterated identification marks, in a public place, a residential area, a vehicle or any place where persons are gathered shall be imprisoned for not less than fifteen (15) years without parole.

482. Illegal use of license; penalty Whoever knowingly allows the use of his license, issued under this chapter or any prior law of the Virgin Islands, by another person for the purpose of obtaining or transporting firearms shall be punished as provided in section 484 of this chapter.

484. General penalty section Any person who violates the provisions of this chapter shall, except when otherwise specifically provided herein, be fined not more than $5,000 or imprisoned not more than three years or both; provided that if the violation occurs after such person has been convicted in the Virgin Islands of a violation of this chapter, or of a crime of violence, either in the Virgin Islands or in another jurisdiction, such person shall be fined not more than $10,000 or imprisoned not more than ten years or both.

487. Seizure and forfeiture

(a) The following items of property shall be subject to seizure and forfeiture, and, upon forfeiture, no property right shall exist in them:

1. Any firearm being worn, borne, or transported by any person not authorized pursuant to section 454 of this title, and, therefore, in violation of section 484 of this title.

2. All ammunition or other parts of or appurtenances to any such firearm worn, carried, or transported by such person or found in the immediate vicinity of such firearm.

(b) Any property subject to seizure under subsection (a) of this section may be seized by any duly authorized officer, as an incident to an arrest or search and seizure. Any such officer seizing such property under this section shall either place the property under seal or remove the same to a location designated by the Police Commissioner. Such officer shall, at the time of seizure or as soon as possible thereafter, execute and deliver to the possessor a signed and dated receipt for the item seized.

(c) Upon the seizure of property pursuant to this section, the Attorney General shall be notified thereof by the U.S. Virgin Islands Police Department (V.I.P.D.), which Department shall also furnish the name and address of the owner thereof, if known. The Attorney General shall notify the owner by certified mail of the seizure, if the registered owner is not the person from whom the item was seized, and the Attorney General's determination of whether the owner knew or should have known that the property was worn, borne, transported or used without lawful authority under section 454 of this title.

1. If the Attorney General determines that the owner neither knew nor should have known of the unauthorized use or intended unauthorized use of the property, he shall surrender the property upon request to the owner unless he determines that the property is needed as evidence in a pending criminal case, in which event he shall return the property upon the final conclusion of the case or cases in which the property is needed as evidence.

2. If the Attorney General determines that the property should be forfeited to the Government, he shall petition the Territorial Court in the name of the Government of the Virgin Islands against the property as designated by make, model, year and serial number or other identifying characteristic. The petition shall allege the seizure and set forth in general terms the causes or grounds of forfeiture. It shall also pray that the property be condemned as forfeited to the Government and disposed of according to law.

3. If the owner or owners of the property are unknown or cannot be found, notice of the seizure and intended forfeiture proceedings shall be made by publication in one or more newspapers published in the District in which the action is brought. The notice shall state the substance and object of the original petition and give notice of the intended forfeiture proceedings.

4. Within 30 days after service of the notice of seizure and intended forfeiture proceedings or within 30 days after the date of publication, the owner of the property seized may file an answer under oath to the petition.

5. The court shall retain custody of the seized property pending prosecution of the person accused of violating section 454 of this title and in case such person be found guilty, the property shall remain in the custody of the court until the hearing on the forfeiture is held. The hearing shall be scheduled no more than 30 days after conviction of the defendant, and reasonable notice shall be given to those parties filing an answer to the petition.

6. If no timely answer is filed, the Court shall hear evidence upon the unauthorized use of the property and shall upon satisfactory proof thereof, order the property forfeited to the Government of the Virgin Islands.

7. At the scheduled hearing, any owner who filed a timely answer may show by competent evidence that the property was not in fact used in violation of section 454 of this title or that he neither knew nor should have known that the property was being, or was to be so used. Upon the determination that the property was not so used, the Court shall order that the property be released to the owner.

8. If after a full hearing the Court decides that the property was used in violation of section 454 of this title or that the owner knew or should have known that the property was being, or was to be so used, the Court shall order that the property be forfeited to the Government of the Virgin Islands.

(d) Whenever property is forfeited under this section, it shall be turned over to the Police Commissioner for immediate destruction in the manner he deems appropriate.

(e) Before disposing of a firearm, as provided in this chapter, the Commissioner shall direct that the class, make, number and caliber of such firearm, as well as the time of receipt of such firearm by the Commissioner and the date of destruction, if any, of same, shall be carefully recorded and filed.

489. Registration of firearms upon purchase from dealer; registration of firearms transferred from non-dealer

(a) At the time that any firearm is purchased from a licensed firearms dealer, the dealer and the purchaser shall complete such registration documents as required by the Commissioner of Police before the firearm may be removed from the premises. The firearms dealer shall forward said documents to the Firearms Division of the Police Department by the end of the next business day.

(b) Any person other than a firearms dealer wishing to transfer a firearm to another person shall effect the transfer at the Firearms Division of the Virgin Islands Police Department or at the place of business of a licensed firearms dealer within the Virgin Islands. Both individuals involved in the transfer must present the appropriate firearm licenses to the Firearms Division or the firearms dealer before such transfer may be effected. A record of transfer, as provided by the Virgin Islands Police Department, shall be completed at the time of transfer, and no transfer of a firearm shall be valid absent the completion of registration and transfer documents as required by this section. The firearms dealer shall forward said documents to the Firearms Division of the Police Department by the end of the next business day.

(c) The completion of registration and transfer documents as required by this section shall constitute compliance with the requirements of Section 470, subsection (a) of this chapter

489a. Safe storage of firearms; penalties

(a) Every person who owns, possesses, purchases, or acquires a firearm, as that word is defined at Section 451(d) of this chapter, shall be responsible for the safe storage of the firearm.

(b) Any person who owns, possesses, purchases, or acquires a firearm, as that word is defined in section 451(d) of this chapter, whose failure to lock or safely store the firearm directly results in a person not licensed to own or possess the firearm to gain access thereto, and said unlicensed person injures or kills himself or another person with said firearm, may be, upon conviction, punished by a fine not to exceed $2,500, imprisonment not to exceed two years, or both, and be subject to forfeiture of his license and firearm. This section shall not be construed as precluding the charge or conviction of any other appropriate violation of law.

(c) For purposes of this section, "safe storage" means the storage of a firearm in a locked manner so as to prevent discharge or storage of a firearm in a safe location that is inaccessible to all except the licensed owner of the firearm. For purposes of this section a firearm is locked when the device installed on it or incorporated into its design is activated or set to prevent the firearm from being discharged.

[Current through Act 7023 of the 2008 Regular Session]

WASHINGTON

WASH. REV. CODE

Title 9. Crimes and Punishments

Chapter 9.41. Firearms and Dangerous Weapons

9.41.010. Terms defined Unless the context clearly requires otherwise, the definitions in this section apply throughout this chapter.

(1) "Firearm" means a weapon or device from which a projectile or projectiles may be fired by an explosive such as gunpowder.

(2) "Pistol" means any firearm with a barrel less than sixteen inches in length, or is designed to be held and fired by the use of a single hand.

(3) "Rifle" means a weapon designed or redesigned, made or remade, and intended to be fired from the shoulder and designed or redesigned, made or remade, and intended to use the energy of the explosive in a fixed metallic cartridge to fire only a single projectile through a rifled bore for each single pull of the trigger.

(4) "Short-barreled rifle" means a rifle having one or more barrels less than sixteen inches in length and any weapon made from a rifle by any means of modification if such modified weapon has an overall length of less than twenty-six inches.

(5) "Shotgun" means a weapon with one or more barrels, designed or redesigned, made or remade, and intended to be fired from the shoulder and designed or redesigned, made or remade, and intended to use the energy of the explosive in a fixed shotgun shell to fire through a smooth bore either a number of ball shot or a single projectile for each single pull of the trigger.

(6) "Short-barreled shotgun" means a shotgun having one or more barrels less than eighteen inches in length and any weapon made from a shotgun by any means of modification if such modified weapon has an overall length of less than twenty-six inches.

(7) "Machine gun" means any firearm known as a machine gun, mechanical rifle, submachine gun, or any other mechanism or instrument not requiring that the trigger be pressed for each shot and having a reservoir clip, disc, drum, belt, or other separable mechanical device for storing, carrying, or supplying ammunition which can be loaded into the firearm, mechanism, or instrument, and fired therefrom at the rate of five or more shots per second.

(8) "Antique firearm" means a firearm or replica of a firearm not designed or redesigned for using rim fire or conventional center fire ignition with fixed ammunition and manufactured in or before 1898, including any matchlock, flintlock, percussion cap, or similar type of ignition system and also any firearm using fixed ammunition manufactured in or before 1898, for which ammunition is no longer manufactured in the United States and is not readily available in the ordinary channels of commercial trade.

(9) "Loaded" means:

(a) There is a cartridge in the chamber of the firearm;

(b) Cartridges are in a clip that is locked in place in the firearm;

(c) There is a cartridge in the cylinder of the firearm, if the firearm is a revolver;

(d) There is a cartridge in the tube or magazine that is inserted in the action; or

(e) There is a ball in the barrel and the firearm is capped or primed if the firearm is a muzzle loader.

(10) "Dealer" means a person engaged in the business of selling firearms at wholesale or retail who has, or is required to have, a federal firearms license under 18 U.S.C. Sec. 923(a). A person who does not have, and is not required to have, a federal firearms license under 18 U.S.C. Sec. 923(a), is not a dealer if that person makes only occasional sales, exchanges, or purchases of firearms for the enhancement of a personal collection or for a hobby, or sells all or part of his or her personal collection of firearms.

(11) "Crime of violence" means:

(a) Any of the following felonies, as now existing or hereafter amended: Any felony defined under any law as a class A felony or an attempt to commit a class A felony, criminal solicitation of or criminal conspiracy to commit a class A felony, manslaughter in the first degree, manslaughter in the second degree, indecent liberties if committed by forcible compulsion, kidnapping in the second degree, arson in the second degree, assault in the second degree, assault of a child in the second degree, extortion in the first degree, burglary in the second degree, residential burglary, and robbery in the second degree;

(b) Any conviction for a felony offense in effect at any time prior to June 6, 1996, which is comparable to a felony classified as a crime of violence in (a) of this subsection; and

(c) Any federal or out-of-state conviction for an offense comparable to a felony classified as a crime of violence under (a) or (b) of this subsection.

(12) "Serious offense" means any of the following felonies or a felony attempt to commit any of the following felonies, as now existing or hereafter amended:

(a) Any crime of violence;

(b) Any felony violation of the uniform controlled substances act, chapter 69.50 RCW, that is classified as a class B felony or that has a maximum term of imprisonment of at least ten years;

(c) Child molestation in the second degree;

(d) Incest when committed against a child under age fourteen;

(e) Indecent liberties;

(f) Leading organized crime;

(g) Promoting prostitution in the first degree;

(h) Rape in the third degree;

(i) Drive-by shooting;

(j) Sexual exploitation;

(k) Vehicular assault, when caused by the operation or driving of a vehicle by a person while under the influence of intoxicating liquor or any drug or by the operation or driving of a vehicle in a reckless manner;

(l) Vehicular homicide, when proximately caused by the driving of any vehicle by any person while under the influence of intoxicating liquor or any drug as defined by RCW 46.61.502, or by the operation of any vehicle in a reckless manner;

(m) Any other class B felony offense with a finding of sexual motivation, as "sexual motivation" is defined under RCW 9.94A.030;

(n) Any other felony with a deadly weapon verdict under RCW 9.94A.602; or

(o) Any felony offense in effect at any time prior to June 6, 1996, that is comparable to a serious offense, or any federal or out-of-state conviction for an offense that under the laws of this state would be a felony classified as a serious offense.

(13) "Law enforcement officer" includes a general authority Washington peace officer as defined in RCW 10.93.020, or a specially commissioned Washington peace officer as defined in RCW 10.93.020. "Law enforcement officer" also includes a limited authority Washington peace officer as defined in RCW 10.93.020 if such officer is duly authorized by his or her employer to carry a concealed pistol.

(14) "Felony" means any felony offense under the laws of this state or any federal or out-of-state offense comparable to a felony offense under the laws of this state.

(15) "Sell" refers to the actual approval of the delivery of a firearm in consideration of payment or promise of payment of a certain price in money.

(16) "Barrel length" means the distance from the bolt face of a closed action down the length of the axis of the bore to the crown of the muzzle, or in the case of a barrel with attachments to the end of any legal device permanently attached to the end of the muzzle.

(17) "Family or household member" means "family" or "household member" as used in RCW 10.99.020.

9.41.040. Unlawful possession of firearms - Ownership, possession by certain persons -- Penalties

(1)(a) A person, whether an adult or juvenile, is guilty of the crime of unlawful possession of a firearm in the first degree, if the person owns, has in his or her possession, or has in his or her control any firearm after having previously been convicted or found not guilty by reason of insanity in this state or elsewhere of any serious offense as defined in this chapter.

(b) Unlawful possession of a firearm in the first degree is a class B felony punishable according to chapter 9A.20 RCW.

(2)(a) A person, whether an adult or juvenile, is guilty of the crime of unlawful possession of a firearm in the second degree, if the person does not qualify under subsection (1) of this section for the crime of unlawful possession of a firearm in the first degree and the person owns, has in his or her possession, or has in his or her control any firearm:

(i) After having previously been convicted or found not guilty by reason of insanity in this state or elsewhere of any felony not specifically listed as prohibiting firearm possession under subsection (1) of this section, or any of the following crimes when committed by one family or household member against another, committed on or after July 1, 1993: Assault in the fourth degree, coercion, stalking, reckless endangerment, criminal trespass in the first degree, or violation of the provisions of a protection order or no-contact order restraining the person or excluding the person from a residence (RCW 26.50.060, 26.50.070, 26.50.130, or 10.99.040);

(ii) After having previously been involuntarily committed for mental health treatment under RCW 71.05.320, 71.34.090, chapter 10.77 RCW, or equivalent statutes of another jurisdiction, unless his or her right to possess a firearm has been restored as provided in RCW 9.41.047;

(iii) If the person is under eighteen years of age, except as provided in RCW 9.41.042; and/or

(iv) If the person is free on bond or personal recognizance pending trial, appeal, or sentencing for a serious offense as defined in RCW 9.41.010.

(b) Unlawful possession of a firearm in the second degree is a 8 class C felony punishable according to chapter 9A.20 RCW.

(3) Notwithstanding RCW 9.41.047 or any other provisions of law, as used in this chapter, a person has been "convicted", whether in an adult court or adjudicated in a juvenile court, at such time as a plea of guilty has been accepted, or a verdict of guilty has been filed, notwithstanding the pendency of any future proceedings including but not limited to sentencing or disposition, post-trial or post-factfinding motions, and appeals. Conviction includes a dismissal entered after a period of probation, suspension or deferral of sentence, and also includes equivalent dispositions by courts in jurisdictions other than Washington state. A person shall not be precluded from possession of a firearm if the conviction has been the subject of a pardon, annulment, certificate of rehabilitation, or other equivalent procedure based on a finding of the rehabilitation of the person convicted or the conviction or disposition has been the subject of a pardon, annulment, or other equivalent procedure based on a finding of innocence. Where no record of the court's disposition of the charges can be found, there shall be a rebuttable presumption that the person was not convicted of the charge.

(4) Notwithstanding subsection (1) or (2) of this section, a person convicted or found not guilty by reason of insanity of an offense prohibiting the possession of a firearm under this section other than murder, manslaughter, robbery, rape, indecent liberties, arson, assault, kidnapping, extortion, burglary, or violations with respect to controlled substances under RCW 69.50.401(a) and 69.50.410, who received a probationary sentence under RCW 9.95.200, and who received a dismissal of the charge under RCW 9.95.240, shall not be precluded from possession of a firearm as a result of the conviction or finding of not guilty by reason of insanity. Notwithstanding any other provisions of this section, if a person is prohibited from possession of a firearm under subsection (1) or (2) of this section and has not previously been convicted or found not guilty by reason of insanity of a sex offense prohibiting firearm ownership under subsection (1) or (2) of this section and/or any felony defined under any law as a class A felony or with a maximum sentence of at least twenty years, or both, the individual may petition a court of record to have his or her right to possess a firearm restored:

(a) Under RCW 9.41.047; and/or

(b)(i) If the conviction or finding of not guilty by reason of insanity was for a felony offense, after five or more consecutive years in the community without being convicted or found not guilty by reason of insanity or currently charged with any felony, gross misdemeanor, or misdemeanor crimes, if the individual has no prior felony convictions that prohibit the possession of a firearm counted as part of the offender score under RCW 9.94A.525; or

(ii) If the conviction or finding of not guilty by reason of insanity was for a nonfelony offense, after three or more consecutive years in the community without being convicted or found not guilty by reason of insanity or currently charged with any felony, gross misdemeanor, or misdemeanor crimes, if the individual has no prior felony convictions that prohibit the possession of a firearm counted as part of the offender score under RCW 9.94A.525 and the individual has completed all conditions of the sentence.

(5) In addition to any other penalty provided for by law, if a person under the age of eighteen years is found by a court to have possessed a firearm in a vehicle in violation of subsection (1) of this section or to have committed an offense while armed with a firearm during which offense a motor vehicle served an integral function, the court shall notify the department of licensing within twenty-four hours and the person's privilege to drive shall be revoked under RCW 46.20.265.

(6) Nothing in chapter 129, Laws of 1995 shall ever be construed or interpreted as preventing an offender from being charged and subsequently convicted for the separate felony crimes of theft of a firearm or possession of a stolen firearm, or both, in addition to being charged and subsequently convicted under this section for unlawful possession of a firearm in the first or second degree. Notwithstanding any other law, if the offender is convicted under this section for unlawful possession of a firearm in the first or second degree and for the felony crimes of theft of a firearm or possession of a stolen firearm, or both, then the offender shall serve consecutive sentences for each of the felony crimes of conviction listed in this subsection.

(7) Each firearm unlawfully possessed under this section shall be a separate offense.

9.41.042. Children - Permissible firearm possession RCW 9.41.040(2)(a)(iii) shall not apply to any person under the age of eighteen years who is:

(1) In attendance at a hunter's safety course or a firearms safety course;

(2) Engaging in practice in the use of a firearm or target shooting at an established range authorized by the governing body of the jurisdiction in which such range is located or any other area where the discharge of a firearm is not prohibited;

(3) Engaging in an organized competition involving the use of a firearm, or participating in or practicing for a performance by an organized group that uses firearms as a part of the performance;

(4) Hunting or trapping under a valid license issued to the person under Title 77 RCW;

(5) In an area where the discharge of a firearm is permitted, is not trespassing, and the person either: (a) Is at least fourteen years of age, has been issued a hunter safety certificate, and is using a lawful firearm other than a pistol; or (b) is under the supervision of a parent, guardian, or other adult approved for the purpose by the parent or guardian;

(6) Traveling with any unloaded firearm in the person's possession to or from any activity described in subsection (1), (2), (3), (4), or (5) of this section;

(7) On real property under the control of his or her parent, other relative, or legal guardian and who has the permission of the parent or legal guardian to possess a firearm;

(8) At his or her residence and who, with the permission of his or her parent or legal guardian, possesses a firearm for the purpose of exercising the rights specified in RCW 9A.16.020(3); or

(9) Is a member of the armed forces of the United States, national guard, or organized reserves, when on duty.

9.41.045. Possession by offenders As a sentence condition and requirement, offenders under the supervision of the department of corrections pursuant to chapter 9.94A RCW shall not own, use, or possess firearms or ammunition. In addition to any penalty imposed pursuant to RCW 9.41.040 when applicable, offenders found to be in actual or constructive possession of firearms or ammunition shall be subject to the appropriate violation process and sanctions as provided for in RCW 9.94A.634. Firearms or ammunition owned, used, or possessed by offenders may be confiscated by community corrections officers and turned over to the Washington state patrol for disposal as provided in RCW 9.41.098.

9.41.047. Restoration of possession rights

(1) At the time a person is convicted or found guilty by reason of insanity of an offense making the person ineligible to possess a firearm, or at the time a person is committed by court order under RCW 71.05.320, 71.34.090, or chapter 10.77 RCW for mental health treatment, the convicting or committing court shall notify the person, orally and in writing, that the person must immediately surrender any concealed pistol license and that the person may not possess a firearm unless his or her right to do so is restored by a court of record. For purposes of this section a convicting court includes a court in which a person has been found not guilty by reason of insanity.

The convicting or committing court also shall forward a copy of the person's driver's license or identicard, or comparable information, to the department of licensing, along with the date of conviction or commitment.

(2) Upon receipt of the information provided for by subsection (1) of this section, the department of licensing shall determine if the convicted or committed person has a concealed pistol license. If the person does have a concealed pistol license, the department of licensing shall immediately notify the license-issuing authority which, upon receipt of such notification, shall immediately revoke the license.

(3)(a) A person who is prohibited from possessing a firearm, by reason of having been involuntarily committed for mental health treatment under RCW 71.05.320, 71.34.090, chapter 10.77 RCW, or equivalent statutes of another jurisdiction may, upon discharge, petition a court of record to have his or her right to possess a firearm restored. At the time of commitment, the court shall specifically state to the person that he or she is barred from possession of firearms.

(b) The secretary of social and health services shall develop appropriate rules to create an approval process under this subsection. The rules must provide for the restoration of the right to possess a firearm upon a showing in a court of competent jurisdiction that the person is no longer required to participate in an inpatient or outpatient treatment program, is no longer required to take medication to treat any condition related to the commitment, and does not present a substantial danger to himself or herself, others, or the public. Unlawful possession of a firearm under this subsection shall be punished as a class C felony under chapter 9A.20 RCW.

(c) A person petitioning the court under this subsection (3) shall bear the burden of proving by a preponderance of the evidence that the circumstances resulting in the commitment no longer exist and are not reasonably likely to recur. If a preponderance of the evidence in the record supports a finding that the person petitioning the court has engaged in violence and that it is more likely than not that the person will engage in violence after his or her right to possess a firearm is restored, the person shall bear the burden of proving by clear, cogent, and convincing evidence that he or she does not present a substantial danger to the safety of others.

(4) No person who has been found not guilty by reason of insanity may petition a court for restoration of the right to possess a firearm unless the person meets the requirements for the restoration of the right to possess a firearm under RCW 9.41.040(4).

9.41.080. Delivery to ineligible persons No person may deliver a firearm to any person whom he or she has reasonable cause to believe is ineligible under RCW 9.41.040 to possess a firearm. Any person violating this section is guilty of a class C felony, punishable under chapter 9A.20 RCW.

9.41.090. Dealer deliveries regulated - Hold on delivery

(1) In addition to the other requirements of this chapter, no dealer may deliver a pistol to the purchaser thereof until:

(a) The purchaser produces a valid concealed pistol license and the dealer has recorded the purchaser's name, license number, and issuing agency, such record to be made in triplicate and processed as provided in subsection (5) of this section. For purposes of this subsection (1)(a), a "valid concealed pistol license" does not include a temporary emergency license, and does not include any license issued before July 1, 1996, unless the issuing agency conducted a records search for disqualifying crimes under RCW 9.41.070 at the time of issuance;

(b) The dealer is notified in writing by the chief of police or the sheriff of the jurisdiction in which the purchaser resides that the purchaser is eligible to possess a pistol under RCW 9.41.040 and that the application to purchase is approved by the chief of police or sheriff; or

(c) Five business days, meaning days on which state offices are open, have elapsed from the time of receipt of the application for the purchase thereof as provided herein by the chief of police or sheriff designated in subsection (5) of this section, and, when delivered, the pistol shall be securely wrapped and shall be unloaded. However, if the purchaser does not have a valid permanent Washington driver's license or state identification card or has not been a resident of the state for the previous consecutive ninety days, the waiting period under this subsection (1)(c) shall be up to sixty days.

(2)(a) Except as provided in (b) of this subsection, in determining whether the purchaser meets the requirements of RCW 9.41.040, the chief of police or sheriff, or the designee of either, shall check with the national crime information center, the Washington state patrol electronic data base, the department of social and health services electronic data base, and with other agencies or resources as appropriate, to determine whether the applicant is ineligible under RCW 9.41.040 to possess a firearm.

(b) Once the system is established, a dealer shall use the state system and national instant criminal background check system, provided for by the Brady Handgun Violence Prevention Act (18 U.S.C. Sec. 921 et seq.), to make criminal background checks of applicants to purchase firearms. However, a chief of police or sheriff, or a designee of either, shall continue to check the department of social and health services' electronic data base and with other agencies or resources as appropriate, to determine whether applicants are ineligible under RCW 9.41.040 to possess a firearm.

(3) In any case under subsection (1)(c) of this section where the applicant has an outstanding warrant for his or her arrest from any court of competent jurisdiction for a felony or misdemeanor, the dealer shall hold the delivery of the pistol until the warrant for arrest is served and satisfied by appropriate court appearance. The local jurisdiction for purposes of the sale shall confirm the existence of outstanding warrants within seventy-two hours after notification of the application to purchase a pistol is received. The local jurisdiction shall also immediately confirm the satisfaction of the warrant on request of the dealer so that the hold may be released if the warrant was for an offense other than an offense making a person ineligible under RCW 9.41.040 to possess a pistol.

(4) In any case where the chief or sheriff of the local jurisdiction has reasonable grounds based on the following circumstances: (a) Open criminal charges, (b) pending criminal proceedings, (c) pending commitment proceedings, (d) an outstanding warrant for an offense making a person ineligible under RCW 9.41.040 to possess a pistol, or (e) an arrest for an offense making a person ineligible under RCW 9.41.040 to possess a pistol, if the records of disposition have not yet been reported or entered sufficiently to determine eligibility to purchase a pistol, the local jurisdiction may hold the sale and delivery of the pistol beyond five days up to thirty days in order to confirm existing records in this state or elsewhere. After thirty days, the hold will be lifted unless an extension of the thirty days is approved by a local district court or municipal court for good cause shown. A dealer shall be notified of each hold placed on the sale by local law enforcement and of any application to the court for additional hold period to confirm records or confirm the identity of the applicant.

(5) At the time of applying for the purchase of a pistol, the purchaser shall sign in triplicate and deliver to the dealer an application containing his or her full name, residential address, date and place of birth, race, and gender; the date and hour of the application; the applicant's driver's license number or state identification card number; a description of the pistol including the make, model, caliber and manufacturer's number if available at the time of applying for the purchase of a pistol. If the manufacturer's number is not available, the application may be processed, but delivery of the pistol to the purchaser may not occur unless the manufacturer's number is recorded on the application by the dealer and transmitted to the chief of police of the municipality or the sheriff of the county in which the purchaser resides; and a statement that the purchaser is eligible to possess a pistol under RCW 9.41.040.

The application shall contain a warning substantially as follows:

CAUTION: Although state and local laws do not differ, federal law and state law on the possession of firearms differ. If you are prohibited by federal law from possessing a firearm, you may be prosecuted in federal court. State permission to purchase a firearm is not a defense to a federal prosecution.

The purchaser shall be given a copy of the department of fish and wildlife pamphlet on the legal limits of the use of firearms, firearms safety, and the fact that local laws and ordinances on firearms are preempted by state law and must be consistent with state law.

The dealer shall, by the end of the business day, sign and attach his or her address and deliver a copy of the application and such other documentation as required under subsection (1) of this section to the chief of police of the municipality or the sheriff of the county of which the purchaser is a resident. The triplicate shall be retained by the dealer for six years. The dealer shall deliver the pistol to the purchaser following the period of time specified in this section unless the dealer is notified of an investigative hold under subsection (4) of this section in writing by the chief of police of the municipality or the sheriff of the county, whichever is applicable, denying the purchaser's application to purchase and the grounds thereof. The application shall not be denied unless the purchaser is not eligible to possess a pistol under RCW 9.41.040 or 9.41.045, or federal law.

The chief of police of the municipality or the sheriff of the county shall retain or destroy applications to purchase a pistol in accordance with the requirements of 18 U.S.C. Sec. 922.

(6) A person who knowingly makes a false statement regarding identity or eligibility requirements on the application to purchase a pistol is guilty of false swearing under RCW 9A.72.040.

(7) This section does not apply to sales to licensed dealers for resale or to the sale of antique firearms.

9.41.094. Waiver of confidentiality A signed application to purchase a pistol shall constitute a waiver of confidentiality and written request that the department of social and health services, mental health institutions, and other health care facilities release, to an inquiring court or law enforcement agency, information relevant to the applicant's eligibility to purchase a pistol to an inquiring court or law enforcement agency.

9.41.097. Supplying information on persons purchasing pistols or applying for concealed pistol licenses

(1) The department of social and health services, mental health institutions, and other health care facilities shall, upon request of a court or law enforcement agency, supply such relevant information as is necessary to determine the eligibility of a person to possess a pistol or to be issued a concealed pistol license under RCW 9.41.070 or to purchase a pistol under RCW 9.41.090.

(2) Mental health information received by: (a) The department of licensing pursuant to RCW 9.41.047 or 9.41.170; (b) an issuing authority pursuant to RCW 9.41.047 or 9.41.070; (c) a chief of police or sheriff pursuant to RCW 9.41.090 or 9.41.170; (d) a court or law enforcement agency pursuant to subsection (1) of this section, shall not be disclosed except as provided in RCW 42.17.318.

9.41.0975. Officials and agencies - Immunity, writ of mandamus

(1) The state, local governmental entities, any public or private agency, and the employees of any state or local governmental entity or public or private agency, acting in good faith, are immune from liability:

(a) For failure to prevent the sale or transfer of a firearm to a person whose receipt or possession of the firearm is unlawful;

(b) For preventing the sale or transfer of a firearm to a person who may lawfully receive or possess a firearm;

(c) For issuing a concealed pistol license to a person ineligible for such a license;

(d) For failing to issue a concealed pistol license to a person eligible for such a license;

(e) For revoking or failing to revoke an issued concealed pistol license;

(f) For errors in preparing or transmitting information as part of determining a person's eligibility to receive or possess a firearm, or eligibility for a concealed pistol license;

(g) For issuing a dealer's license to a person ineligible for such a license; or

(h) For failing to issue a dealer's license to a person eligible for such a license.

(2) An application may be made to a court of competent jurisdiction for a writ of mandamus:

(a) Directing an issuing agency to issue a concealed pistol license wrongfully refused;

(b) Directing a law enforcement agency to approve an application to purchase wrongfully denied;

(c) Directing that erroneous information resulting either in the wrongful refusal to issue a concealed pistol license or in the wrongful denial of a purchase application be corrected; or

(d) Directing a law enforcement agency to approve a dealer's license wrongfully denied.

The application for the writ may be made in the county in which the application for a concealed pistol license or to purchase a pistol was made, or in Thurston County, at the discretion of the petitioner. A court shall provide an expedited hearing for an application brought under this subsection (2) for a writ of mandamus. A person granted a writ of mandamus under this subsection (2) shall be awarded reasonable attorneys' fees and costs.

9.41.098. Forfeiture of firearms - Disposition - Confiscation

(1) The superior courts and the courts of limited jurisdiction of the state may order forfeiture of a firearm which is proven to be:

(a) Found concealed on a person not authorized by RCW 9.41.060 or 9.41.070 to carry a concealed pistol: PROVIDED, That it is an absolute defense to forfeiture if the person possessed a valid Washington concealed pistol license within the preceding two years and has not become ineligible for a concealed pistol license in the interim. Before the firearm may be returned, the person must pay the past due renewal fee and the current renewal fee;

(b) Commercially sold to any person without an application as required by RCW 9.41.090;

(c) In the possession of a person prohibited from possessing the firearm under RCW 9.41.040 or 9.41.045;

(d) In the possession or under the control of a person at the time the person committed or was arrested for committing a felony or committing a nonfelony crime in which a firearm was used or displayed;

(e) In the possession of a person who is in any place in which a concealed pistol license is required, and who is under the influence of any drug or under the influence of intoxicating liquor, as defined in chapter 46.61 RCW;

(f) In the possession of a person free on bail or personal recognizance pending trial, appeal, or sentencing for a felony or for a nonfelony crime in which a firearm was used or displayed, except that violations of Title 77 RCW shall not result in forfeiture under this section;

(g) In the possession of a person found to have been mentally incompetent while in possession of a firearm when apprehended or who is thereafter committed pursuant to chapter 10.77 or 71.05 RCW;

(h) Used or displayed by a person in the violation of a proper written order of a court of general jurisdiction; or

(i) Used in the commission of a felony or of a nonfelony crime in which a firearm was used or displayed.

(2) Upon order of forfeiture, the court in its discretion may order destruction of any forfeited firearm. A court may temporarily retain forfeited firearms needed for evidence.

(a) Except as provided in (b), (c), and (d) of this subsection, firearms that are: (i) Judicially forfeited and no longer needed for evidence; or (ii) forfeited due to a failure to make a claim under RCW 63.32.010 or 63.40.010; may be disposed of in any manner determined by the local legislative authority. Any proceeds of an auction or trade may be retained by the legislative authority. This subsection (2)(a) applies only to firearms that come into the possession of the law enforcement agency after June 30, 1993. ...

(3) The court shall order the firearm returned to the owner upon a showing that there is no probable cause to believe a violation of subsection (1) of this section existed or the firearm was stolen from the owner or the owner neither had knowledge of nor consented to the act or omission involving the firearm which resulted in its forfeiture.

(4) A law enforcement officer of the state or of any county or municipality may confiscate a firearm found to be in the possession of a person under circumstances specified in subsection (1) of this section. After confiscation, the firearm shall not be surrendered except: (a) To the prosecuting attorney for use in subsequent legal proceedings; (b) for disposition according to an order of a court having jurisdiction as provided in subsection (1) of this section; or (c) to the owner if the proceedings are dismissed or as directed in subsection (3) of this section.

9.41.100. Dealer licensing and registration required Every dealer shall be licensed as provided in RCW 9.41.110 and shall register with the department of revenue as provided in chapters 82.04 and 82.32 RCW.

9.41.110. Dealer's licenses, by whom granted, conditions, fees - Employees, fingerprinting and background checks - Wholesale sales excepted - Permits prohibited

(1) No dealer may sell or otherwise transfer, or expose for sale or transfer, or have in his or her possession with intent to sell, or otherwise transfer, any pistol without being licensed as provided in this section.

(2) No dealer may sell or otherwise transfer, or expose for sale or transfer, or have in his or her possession with intent to sell, or otherwise transfer, any firearm other than a pistol without being licensed as provided in this section.

(3) No dealer may sell or otherwise transfer, or expose for sale or transfer, or have in his or her possession with intent to sell, or otherwise transfer, any ammunition without being licensed as provided in this section.

(4) The duly constituted licensing authorities of any city, town, or political subdivision of this state shall grant licenses in forms prescribed by the director of licensing effective for not more than one year from the date of issue permitting the licensee to sell firearms within this state subject to the following conditions, for breach of any of which the license shall be forfeited and the licensee subject to punishment as provided in RCW 9.41.010 through 9.41.810. A licensing authority shall forward a copy of each license granted to the department of licensing. The department of licensing shall notify the department of revenue of the name and address of each dealer licensed under this section.

(5)(a) A licensing authority shall, within thirty days after the filing of an application of any person for a dealer's license, determine whether to grant the license. However, if the applicant does not have a valid permanent Washington driver's license or Washington state identification card, or has not been a resident of the state for the previous consecutive ninety days, the licensing authority shall have up to sixty days to determine whether to issue a license. No person shall qualify for a license under this section without first receiving a federal firearms license and undergoing fingerprinting and a background check. In addition, no person ineligible to possess a firearm under RCW 9.41.040 or ineligible for a concealed pistol license under RCW 9.41.070 shall qualify for a dealer's license.

(b) A dealer shall require every employee who may sell a firearm in the course of his or her employment to undergo fingerprinting and a background check. An employee must be eligible to possess a firearm, and must not have been convicted of a crime that would make the person ineligible for a concealed pistol license, before being permitted to sell a firearm. Every employee shall comply with requirements concerning purchase applications and restrictions on delivery of pistols that are applicable to dealers.

(6)(a) Except as otherwise provided in (b) of this subsection, the business shall be carried on only in the building designated in the license. For the purpose of this section, advertising firearms for sale shall not be considered the carrying on of business.

(b) A dealer may conduct business temporarily at a location other than the building designated in the license, if the temporary location is within Washington state and is the location of a gun show sponsored by a national, state, or local organization, or an affiliate of any such organization, devoted to the collection, competitive use, or other sporting use of firearms in the community. Nothing in this subsection (6)(b) authorizes a dealer to conduct business in or from a motorized or towed vehicle.

In conducting business temporarily at a location other than the building designated in the license, the dealer shall comply with all other requirements imposed on dealers by RCW 9.41.090, 9.41.100, and 9.41.110. The license of a dealer who fails to comply with the requirements of RCW 9.41.080 and 9.41.090 and subsection (8) of this section while conducting business at a temporary location shall be revoked, and the dealer shall be permanently ineligible for a dealer's license.

(7) The license or a copy thereof, certified by the issuing authority, shall be displayed on the premises in the area where firearms are sold, or at the temporary location, where it can easily be read.

(8)(a) No pistol may be sold: (i) In violation of any provisions of RCW 9.41.010 through 9.41.810; nor (ii) may a pistol be sold under any circumstances unless the purchaser is personally known to the dealer or shall present clear evidence of his or her identity.

(b) A dealer who sells or delivers any firearm in violation of RCW 9.41.080 is guilty of a class C felony. In addition to any other penalty provided for by law, the dealer is subject to mandatory permanent revocation of his or her dealer's license and permanent ineligibility for a dealer's license.

(c) The license fee for pistols shall be one hundred twenty-five dollars. The license fee for firearms other than pistols shall be one hundred twenty-five dollars. The license fee for ammunition shall be one hundred twenty-five dollars. Any dealer who obtains any license under subsection (1), (2), or (3) of this section may also obtain the remaining licenses without payment of any fee. The fees received under this section shall be deposited in the account under RCW 69.50.520.

(9)(a) A true record in triplicate shall be made of every pistol sold, in a book kept for the purpose, the form of which may be prescribed by the director of licensing and shall be personally signed by the purchaser and by the person effecting the sale, each in the presence of the other, and shall contain the date of sale, the caliber, make, model and manufacturer's number of the weapon, the name, address, occupation, and place of birth of the purchaser and a statement signed by the purchaser that he or she is not ineligible under RCW 9.41.040 to possess a firearm.

(b) One copy shall within six hours be sent by certified mail to the chief of police of the municipality or the sheriff of the county of which the purchaser is a resident; the duplicate the dealer shall within seven days send to the director of licensing; the triplicate the dealer shall retain for six years.

(10) Subsections (2) through (9) of this section shall not apply to sales at wholesale.

(11) The dealer's licenses authorized to be issued by this section are general licenses covering all sales by the licensee within the effective period of the licenses. The department shall provide a single application form for dealer's licenses and a single license form which shall indicate the type or types of licenses granted.

(12) Except as provided in RCW 9.41.090, every city, town, and political subdivision of this state is prohibited from requiring the purchaser to secure a permit to purchase or from requiring the dealer to secure an individual permit for each sale.

9.41.120. Firearms as loan security No person other than a duly licensed dealer shall make any loan secured by a mortgage, deposit or pledge of a pistol. Any licensed dealer receiving a pistol as a deposit or pledge for a loan shall keep such records and make such reports as are provided by law for pawnbrokers and secondhand dealers in cities of the first class. A duly licensed dealer may mortgage any pistol or stock of pistols but shall not deposit or pledge the same with any other person.

9.41.122. Out-of-state purchasing Residents of Washington may purchase rifles and shotguns in a state other than Washington: PROVIDED, That such residents conform to the applicable provisions of the federal Gun Control Act of 1968, Title IV, Pub. L. 90-351 as administered by the United States secretary of the treasury: AND PROVIDED FURTHER, That such residents are eligible to purchase or possess such weapons in Washington and in the state in which such purchase is made.

9.41.124. Purchasing by nonresidents Residents of a state other than Washington may purchase rifles and shotguns in Washington: PROVIDED, That such residents conform to the applicable provisions of the federal Gun Control Act of 1968, Title IV, Pub. L. 90-351 as administered by the United States secretary of the treasury: AND PROVIDED FURTHER, That such residents are eligible to purchase or possess such weapons in Washington and in the state in which such persons reside.

9.41.129. Recordkeeping requirements The department of licensing may keep copies or records of applications for concealed pistol licenses provided for in RCW 9.41.070, copies or records of applications for alien firearm licenses, copies or records of applications to purchase pistols provided for in RCW 9.41.090 and copies or records of pistol transfers provided for in RCW 9.41.110. The copies and records shall not be disclosed except as provided in RCW 42.56.240(4).

9.41.135. Verification of licenses and registration - Notice to federal government

(1) At least once every twelve months, the department of licensing shall obtain a list of dealers licensed under 18 U.S.C. Sec. 923(a) with business premises in the state of Washington from the United States bureau of alcohol, tobacco, and firearms. The department of licensing shall verify that all dealers on the list provided by the bureau of alcohol, tobacco, and firearms are licensed and registered as required by RCW 9.41.100.

(2) At least once every twelve months, the department of licensing shall obtain from the department of revenue and the department of revenue shall transmit to the department of licensing a list of dealers registered with the department of revenue, and a list of dealers whose names and addresses were forwarded to the department of revenue by the department of licensing under RCW 9.41.110, who failed to register with the department of revenue as required by RCW 9.41.100.

(3) At least once every twelve months, the department of licensing shall notify the bureau of alcohol, tobacco, and firearms of all dealers licensed under 18 U.S.C. Sec. 923(a) with business premises in the state of Washington who have not complied with the licensing or registration requirements of RCW 9.41.100. In notifying the bureau of alcohol, tobacco, and firearms, the department of licensing shall not specify whether a particular dealer has failed to comply with licensing requirements or has failed to comply with registration requirements.

9.41.140. Alteration of identifying marks - Exceptions No person may change, alter, remove, or obliterate the name of the maker, model, manufacturer's number, or other mark of identification on any firearm. Possession of any firearm upon which any such mark shall have been changed, altered, removed, or obliterated, shall be prima facie evidence that the possessor has changed, altered, removed, or obliterated the same. This section shall not apply to replacement barrels in old firearms, which barrels are produced by current manufacturers and therefor do not have the markings on the barrels of the original manufacturers who are no longer in business. This section also shall not apply if the changes do not make the firearm illegal for the person to possess under state or federal law.

9.41.170. Alien's license to carry firearms - Exception

(1) It is a class C felony for any person who is not a citizen of the United States to carry or possess any firearm, without first having obtained an alien firearm license from the director of licensing. In order to be eligible for a license, an alien must provide proof that he or she is lawfully present in the United States, which the director of licensing shall verify through the appropriate authorities. Except as provided in subsection (2)(a) of this section, and subject to the additional requirements of subsection (2)(b) of this section, the director of licensing may issue an alien firearm license only upon receiving from the consul domiciled in this state representing the country of the alien, a certified copy of the alien's criminal history in the alien's country indicating the alien is not ineligible under RCW 9.41.040 to own, possess, or control a firearm, and the consul's attestation that the alien is a responsible person.

(2)(a) Subject to the additional requirements of (b) of this subsection, the director of licensing may issue an alien firearm license without a certified copy of the alien's criminal history or the consul's attestation required by subsection (1) of this section, if the alien has been a resident of this state for at least two years and: (i) The alien is from a country without a consul domiciled within this state, or (ii) the consul has failed to provide, within ninety days after a request by the alien, the criminal history or attestation required by subsection (1) of this section.

(b) Before issuing an alien firearm license under subsection (1) of this section or this subsection (2), the director of licensing shall ask the local law enforcement agency of the jurisdiction in which the alien resides to complete a background and fingerprint check to determine the alien's eligibility under RCW 9.41.040 to own, possess, or control a firearm. The law enforcement agency shall complete a background check within thirty days after the request, unless the alien does not have a valid Washington driver's license or Washington state identification card. In the latter case, the law enforcement agency shall complete the background check within sixty days after the request.

A signed application for an alien firearm license shall constitute a waiver of confidentiality and written request that the department of social and health services, mental health institutions, and other health care facilities release information relevant to the applicant's eligibility for an alien firearm license to an inquiring law enforcement agency.

(3) The alien firearm license shall be valid for five years from the date of issue so long as the alien is lawfully present in the United States. The nonrefundable fee, paid upon application, for the five-year license shall be fifty-five dollars plus additional charges imposed by the Federal Bureau of Investigation that are passed on to the applicant. The fee shall be distributed as follows:

(a) Fifteen dollars shall be paid to the department of licensing;

(b) Twenty-five dollars shall be paid to the Washington state patrol; and

(c) Fifteen dollars shall be paid to the local law enforcement agency conducting the background check.

(4) This section shall not apply to Canadian citizens resident in a province which has an enactment or public policy providing substantially similar privilege to residents of the state of Washington and who are carrying or possessing weapons for the purpose of using them in the hunting of game while such persons are in the act of hunting, or while on a hunting trip, or while such persons are competing in a bona fide trap or skeet shoot or any other organized contest where rifles, pistols, or shotguns are used. Nothing in this section shall be construed to allow aliens to hunt or fish in this state without first having obtained a regular hunting or fishing license.

9.41.185. Coyote getters The use of "coyote getters" or similar spring-triggered shell devices shall not constitute a violation of any of the laws of the state of Washington when the use of such "coyote getters" is authorized by the state department of agriculture and/or the state department of fish and wildlife in cooperative programs with the United States Fish and Wildlife Service, for the purpose of controlling or eliminating coyotes harmful to livestock and game animals on range land or forest areas.

9.41.190. Unlawful firearms - Exceptions

(1) It is unlawful for any person to manufacture, own, buy, sell, loan, furnish, transport, or have in possession or under control, any machine gun, short-barreled shotgun, or short-barreled rifle; or any part designed and intended solely and exclusively for use in a machine gun, short-barreled shotgun, or short-barreled rifle, or in converting a weapon into a machine gun, short-barreled shotgun, or short-barreled rifle; or to assemble or repair any machine gun, short-barreled shotgun, or short-barreled rifle.

(2) This section shall not apply to:

(a) Any peace officer in the discharge of official duty or traveling to or from official duty, or to any officer or member of the armed forces of the United States or the state of Washington in the discharge of official duty or traveling to or from official duty; or

(b) A person, including an employee of such person if the employee has undergone fingerprinting and a background check, who or which is exempt from or licensed under federal law, and engaged in the production, manufacture, repair, or testing of machine guns, short-barreled shotguns, or short-barreled rifles:

(i) To be used or purchased by the armed forces of the United States;

(ii) To be used or purchased by federal, state, county, or municipal law enforcement agencies; or

(iii) For exportation in compliance with all applicable federal laws and regulations.

(3) It shall be an affirmative defense to a prosecution brought under this section that the machine gun, short-barreled shotgun, or short-barreled rifle was acquired prior to July 1, 1994, and is possessed in compliance with federal law.

(4) Any person violating this section is guilty of a class C felony.

9.41.220. Unlawful firearms and parts contraband All machine guns, short-barreled shotguns, or short-barreled rifles, or any part designed and intended solely and exclusively for use in a machine gun, short-barreled shotgun, or short-barreled rifle, or in converting a weapon into a machine gun, short-barreled shotgun, or short-barreled rifle, illegally held or illegally possessed are hereby declared to be contraband, and it shall be the duty of all peace officers, and/or any officer or member of the armed forces of the United States or the state of Washington, to seize said machine gun, short-barreled shotgun, or short-barreled rifle, or parts thereof, wherever and whenever found.

9.41.240. Possession of pistol by person from eighteen to twenty-one Unless an exception under RCW 9.41.042, 9.41.050, or 9.41.060 applies, a person at least eighteen years of age, but less than twenty-one years of age, may possess a pistol only:

(1) In the person's place of abode;

(2) At the person's fixed place of business; or

(3) On real property under his or her control.

9.41.280. Possessing dangerous weapons on school facilities - Penalty - Exceptions

(1) It is unlawful for a person to carry onto, or to possess on, public or private elementary or secondary school premises, school-provided transportation, or areas of facilities while being used exclusively by public or private schools:

(a) Any firearm;

(b) Any other dangerous weapon as defined in RCW 9.41.250;

(c) Any device commonly known as "nun-chu-ka sticks", consisting of two or more lengths of wood, metal, plastic, or similar substance connected with wire, rope, or other means;

(d) Any device, commonly known as "throwing stars", which are multi-pointed, metal objects designed to embed upon impact from any aspect; or

(e) Any air gun, including any air pistol or air rifle, designed to propel a BB, pellet, or other projectile by the discharge of compressed air, carbon dioxide, or other gas.

(2) Any such person violating subsection (1) of this section is guilty of a gross misdemeanor. If any person is convicted of a violation of subsection (1)(a) of this section, the person shall have his or her concealed pistol license, if any revoked for a period of three years. Anyone convicted under this subsection is prohibited from applying for a concealed pistol license for a period of three years. The court shall send notice of the revocation to the department of licensing, and the city, town, or county which issued the license.

Any violation of subsection (1) of this section by elementary or secondary school students constitutes grounds for expulsion from the state's public schools in accordance with RCW 28A.600.010. An appropriate school authority shall promptly notify law enforcement and the student's parent or guardian regarding any allegation or indication of such violation.

Upon the arrest of a person at least twelve years of age and not more than twenty-one years of age for violating subsection (1)(a) of this section, the person shall be detained or confined in a juvenile or adult facility for up to seventy-two hours. The person shall not be released within the seventy-two hours until after the person has been examined and evaluated by the county-designated mental health professional unless the court in its discretion releases the person sooner after a determination regarding probable cause or on probation bond or bail.

Within twenty-four hours of the arrest, the arresting law enforcement agency shall refer the person to the county-designated mental health professional for examination and evaluation under chapter 71.05 or 71.34 RCW and inform a parent or guardian of the person of the arrest, detention, and examination. The county-designated mental health professional shall examine and evaluate the person subject to the provisions of chapter 71.05 or 71.34 RCW. The examination shall occur at the facility in which the person is detained or confined. If the person has been released on probation, bond, or bail, the examination shall occur wherever is appropriate.

The county-designated mental health professional may determine whether to refer the person to the county-designated chemical dependency specialist for examination and evaluation in accordance with chapter 70.96A RCW. The county-designated chemical dependency specialist shall examine the person subject to the provisions of chapter 70.96A RCW. The examination shall occur at the facility in which the person is detained or confined. If the person has been released on probation, bond, or bail, the examination shall occur wherever is appropriate.

Upon completion of any examination by the county-designated mental health professional or the county-designated chemical dependency specialist, the results of the examination shall be sent to the court, and the court shall consider those results in making any determination about the person.

The county-designated mental health professional and county-designated chemical dependency specialist shall, to the extent permitted by law, notify a parent or guardian of the person that an examination and evaluation has taken place and the results of the examination. Nothing in this subsection prohibits the delivery of additional, appropriate mental health examinations to the person while the person is detained or confined.

If the county-designated mental health professional determines it is appropriate, the county-designated mental health professional may refer the person to the local regional support network for follow-up services or the department of social and health services or other community providers for other services to the family and individual.

(3) Subsection (1) of this section does not apply to:

(a) Any student or employee of a private military academy when on the property of the academy;

(b) Any person engaged in military, law enforcement, or school district security activities;

(c) Any person who is involved in a convention, showing, demonstration, lecture, or firearms safety course authorized by school authorities in which the firearms of collectors or instructors are handled or displayed;

(d) Any person while the person is participating in a firearms or air gun competition approved by the school or school district;

(e) Any person in possession of a pistol who has been issued a license under RCW 9.41.070, or is exempt from the licensing requirement by RCW 9.41.060, while picking up or dropping off a student;

(f) Any nonstudent at least eighteen years of age legally in possession of a firearm or dangerous weapon that is secured within an attended vehicle or concealed from view within a locked unattended vehicle while conducting legitimate business at the school;

(g) Any nonstudent at least eighteen years of age who is in lawful possession of an unloaded firearm, secured in a vehicle while conducting legitimate business at the school; or

(h) Any law enforcement officer of the federal, state, or local government agency.

(4) Subsections (1)(c) and (d) of this section do not apply to any person who possesses nunchu-ka sticks, throwing stars, or other dangerous weapons to be used in martial arts classes authorized to be conducted on the school premises.

(5) Except as provided in subsection (3)(b), (c), (f), and (h) of this section, firearms are not permitted in a public or private school building.

(6) "GUN-FREE ZONE" signs shall be posted around school facilities giving warning of the prohibition of the possession of firearms on school grounds.

9.41.290. State preemption The state of Washington hereby fully occupies and preempts the entire field of firearms regulation within the boundaries of the state, including the registration, licensing, possession, purchase, sale, acquisition, transfer, discharge, and transportation of firearms, or any other element relating to firearms or parts thereof, including ammunition and reloader components. Cities, towns, and counties or other municipalities may enact only those laws and ordinances relating to firearms that are specifically authorized by state law, as in RCW 9.41.300, and are consistent with this chapter. Such local ordinances shall have the same penalty as provided for by state law. Local laws and ordinances that are inconsistent with, more restrictive than, or exceed the requirements of state law shall not be enacted and are preempted and repealed, regardless of the nature of the code, charter, or home rule status of such city, town, county, or municipality.

9.41.300. Weapons prohibited in certain places - Local laws and ordinances – Exceptions - Penalty

(1) It is unlawful for any person to enter the following places when he or she knowingly possesses or knowingly has under his or her control a weapon:

(a) The restricted access areas of a jail, or of a law enforcement facility, or any place used for the confinement of a person (i) arrested for, charged with, or convicted of an offense, (ii) held for extradition or as a material witness, or (iii) otherwise confined pursuant to an order of a court, except an order under chapter 13.32A or 13.34 RCW. Restricted access areas do not include common areas of egress or ingress open to the general public;

(b) Those areas in any building which are used in connection with court proceedings, including courtrooms, jury rooms, judge's chambers, offices and areas used to conduct court business, waiting areas, and corridors adjacent to areas used in connection with court proceedings. The restricted areas do not include common areas of ingress and egress to the building that is used in connection with court proceedings, when it is possible to protect court areas without restricting ingress and egress to the building. The restricted areas shall be the minimum necessary to fulfill the objective of this subsection (1)(b).

For purposes of this subsection (1)(b), "weapon" means any firearm, explosive as defined in RCW 70.74.010, or any weapon of the kind usually known as slung shot, sand club, or metal knuckles, or any knife, dagger, dirk, or other similar weapon that is capable of causing death or bodily injury and is commonly used with the intent to cause death or bodily injury.

In addition, the local legislative authority shall provide either a stationary locked box sufficient in size for pistols and key to a weapon owner for weapon storage, or shall designate an official to receive weapons for safekeeping, during the owner's visit to restricted areas of the building. The locked box or designated official shall be located within the same building used in connection with court proceedings. The local legislative authority shall be liable for any negligence causing damage to or loss of a weapon either placed in a locked box or left with an official during the owner's visit to restricted areas of the building.

The local judicial authority shall designate and clearly mark those areas where weapons are prohibited, and shall post notices at each entrance to the building of the prohibition against weapons in the restricted areas;

(c) The restricted access areas of a public mental health facility certified by the department of social and health services for inpatient hospital care and state institutions for the care of the mentally ill, excluding those facilities solely for evaluation and treatment. Restricted access areas do not include common areas of egress and ingress open to the general public;

(d) That portion of an establishment classified by the state liquor control board as off-limits to persons under twenty-one years of age; or

(e) The restricted access areas of a commercial service airport designated in the airport security plan approved by the federal transportation security administration, including passenger screening checkpoints at or beyond the point at which a passenger initiates the screening process. These areas do not include airport drives, general parking areas and walkways, and shops and areas of the terminal that are outside the screening checkpoints and that are normally open to unscreened passengers or visitors to the airport. Any restricted access area shall be clearly indicated by prominent signs indicating that firearms and other weapons are prohibited in the area.

(2) Cities, towns, counties, and other municipalities may enact laws and ordinances:

(a) Restricting the discharge of firearms in any portion of their respective jurisdictions where there is a reasonable likelihood that humans, domestic animals, or property will be jeopardized. Such laws and ordinances shall not abridge the right of the individual guaranteed by Article I, section 24 of the state Constitution to bear arms in defense of self or others; and

(b) Restricting the possession of firearms in any stadium or convention center, operated by a city, town, county, or other municipality, except that such restrictions shall not apply to:

(i) Any pistol in the possession of a person licensed under RCW 9.41.070 or exempt from the licensing requirement by RCW 9.41.060; or

(ii) Any showing, demonstration, or lecture involving the exhibition of firearms.

(3)(a) Cities, towns, and counties may enact ordinances restricting the areas in their respective jurisdictions in which firearms may be sold, but, except as provided in (b) of this subsection, a business selling firearms may not be treated more restrictively than other businesses located within the same zone. An ordinance requiring the cessation of business within a zone shall not have a shorter grandfather period for businesses selling firearms than for any other businesses within the zone.

(b) Cities, towns, and counties may restrict the location of a business selling firearms to not less than five hundred feet from primary or secondary school grounds, if the business has a storefront, has hours during which it is open for business, and posts advertisements or signs observable to passersby that firearms are available for sale. A business selling firearms that exists as of the date a restriction is enacted under this subsection (3)(b) shall be grandfathered according to existing law.

(4) Violations of local ordinances adopted under subsection (2) of this section must have the same penalty as provided for by state law.

(5) The perimeter of the premises of any specific location covered by subsection (1) of this section shall be posted at reasonable intervals to alert the public as to the existence of any law restricting the possession of firearms on the premises.

(6) Subsection (1) of this section does not apply to:

(a) A person engaged in military activities sponsored by the federal or state governments, while engaged in official duties;

(b) Law enforcement personnel, except that subsection (1)(b) of this section does apply to a law enforcement officer who is present at a courthouse building as a party to an action under chapter 10.14, 10.99, or 26.50 RCW, or an action under Title 26 RCW where any party has alleged the existence of domestic violence as defined in RCW 26.50.010; or

(c) Security personnel while engaged in official duties.

(7) Subsection (1)(a) of this section does not apply to a person licensed pursuant to RCW 9.41.070 who, upon entering the place or facility, directly and promptly proceeds to the administrator of the facility or the administrator's designee and obtains written permission to possess the firearm while on the premises or checks his or her firearm. The person may reclaim the firearms upon leaving but must immediately and directly depart from the place or facility.

(8) Subsection (1)(c) of this section does not apply to any administrator or employee of the facility or to any person who, upon entering the place or facility, directly and promptly proceeds to the administrator of the facility or the administrator's designee and obtains written permission to possess the firearm while on the premises.

(9) Subsection (1)(d) of this section does not apply to the proprietor of the premises or his or her employees while engaged in their employment.

(10) Any person violating subsection (1) of this section is guilty of a gross misdemeanor.

(11) "Weapon" as used in this section means any firearm, explosive as defined in RCW 70.74.010, or instrument or weapon listed in RCW 9.41.250.

9.41.310. Information pamphlet After a public hearing, the department of fish and wildlife shall publish a pamphlet on firearms safety and the legal limits of the use of firearms. The pamphlet shall include current information on firearms laws and regulations and state preemption of local firearms laws. This pamphlet may be used in the department's hunter safety education program and shall be provided to the department of licensing for distribution to firearms dealers and persons authorized to issue concealed pistol licenses. The department of fish and wildlife shall reimburse the department of licensing for costs associated with distribution of the pamphlet.

9.41.800. Surrender of weapons or licenses - Prohibition on future possession or licensing

(1) Any court when entering an order authorized under RCW 9A.46.080, 10.14.080, 10.99.040, 10.99.045, 26.09.050, 26.09.060, 26.10.040, 26.10.115, 26.26.130, 26.50.060, 26.50.070, or 26.26.590 shall, upon a showing by clear and convincing evidence, that a party has: Used, displayed, or threatened to use a firearm or other dangerous weapon in a felony, or previously committed any offense that makes him or her ineligible to possess a firearm under the provisions of RCW 9.41.040:

(a) Require the party to surrender any firearm or other dangerous weapon;

(b) Require the party to surrender any concealed pistol license issued under RCW 9.41.070;

(c) Prohibit the party from obtaining or possessing a firearm or other dangerous weapon;

(d) Prohibit the party from obtaining or possessing a concealed pistol license.

(2) Any court when entering an order authorized under RCW 9A.46.080, 10.14.080, 10.99.040, 10.99.045, 26.09.050, 26.09.060, 26.10.040, 26.10.115, 26.26.130, 26.50.060, 26.50.070, or 26.26.590 may, upon a showing by a preponderance of the evidence but not by clear and convincing evidence, that a party has: Used, displayed, or threatened to use a firearm or other dangerous weapon in a felony, or previously committed any offense that makes him or her ineligible to possess a pistol under the provisions of RCW 9.41.040:

(a) Require the party to surrender any firearm or other dangerous weapon;

(b) Require the party to surrender a concealed pistol license issued under RCW 9.41.070;

(c) Prohibit the party from obtaining or possessing a firearm or other dangerous weapon;

(d) Prohibit the party from obtaining or possessing a concealed pistol license.

(3) The court may order temporary surrender of a firearm or other dangerous weapon without notice to the other party if it finds, on the basis of the moving affidavit or other evidence, that irreparable injury could result if an order is not issued until the time for response has elapsed.

(4) In addition to the provisions of subsections (1), (2), and (3) of this section, the court may enter an order requiring a party to comply with the provisions in subsection (1) of this section if it finds that the possession of a firearm or other dangerous weapon by any party presents a serious and imminent threat to public health or safety, or to the health or safety of any individual.

(5) The requirements of subsections (1), (2), and (4) of this section may be for a period of time less than the duration of the order.

(6) The court may require the party to surrender any firearm or other dangerous weapon in his or her immediate possession or control or subject to his or her immediate possession or control to the sheriff of the county having jurisdiction of the proceeding, the chief of police of the municipality having jurisdiction, or to the restrained or enjoined party's counsel or to any person designated by the court.

9.41.810. Penalty Any violation of any provision of this chapter, except as otherwise provided, shall be a misdemeanor and punishable accordingly.

Title 9A. Washington Criminal Code

Chapter 9A.56. Theft and Robbery

9A.56.300. Theft of a firearm

(1) A person is guilty of theft of a firearm if he or she commits a theft of any firearm.

(2) This section applies regardless of the value of the firearm taken in the theft.

(3) Each firearm taken in the theft under this section is a separate offense.

(4) The definition of "theft" and the defense allowed against the prosecution for theft under RCW 9A.56.020 shall apply to the crime of theft of a firearm.

(5) As used in this section, "firearm" means any firearm as defined in RCW 9.41.010.

(6) Theft of a firearm is a class B felony.

9A.56.310. Possessing a stolen firearm

(1) A person is guilty of possessing a stolen firearm if he or she possesses, carries, delivers, sells, or is in control of a stolen firearm.

(2) This section applies regardless of the stolen firearm's value.

(3) Each stolen firearm possessed under this section is a separate offense.

(4) The definition of "possessing stolen property" and the defense allowed against the prosecution for possessing stolen property under RCW 9A.56.140 shall apply to the crime of possessing a stolen firearm.

(5) As used in this section, "firearm" means any firearm as defined in RCW 9.41.010.

(6) Possessing a stolen firearm is a class B felony.

[Current for all Legislation through 2008]

WEST VIRGINIA
W. VA. CODE

Chapter 8. Municipal Corporations

8-12-5. General powers of every municipality and the governing body thereof. In addition to the powers and authority granted by: (i) The constitution of this state; (ii) other provisions of this chapter; (iii) other general law; and (iv) any charter, and to the extent not inconsistent or in conflict with any of the foregoing except special legislative charters, every municipality and the governing body thereof shall have plenary power and authority therein by ordinance or resolution, as the case may require, and by appropriate action based thereon: **...**

(16) To arrest, convict and punish any individual for carrying about his person any revolver or other pistol, dirk, bowie knife, razor, slingshot, billy, metallic or other false knuckles, or any other dangerous or other deadly weapon of like kind or character;

8-12-5a. Limitations upon municipalities' power to restrict the purchase, possession, transfer, ownership, carrying, transport, sale and storage of certain weapons and ammunition. The provisions of section five [§ 8-12-5] of this article notwithstanding, neither a municipality nor the governing body of any municipality may limit the right of any person to purchase, possess, transfer, own, carry, transport, sell or store any revolver, pistol, rifle or shotgun or any ammunition or ammunition components to be used therewith nor to so regulate the keeping of gunpowder so as to directly or indirectly prohibit the ownership of the ammunition. Nothing herein shall in any way impair the authority of any municipality, or the governing body thereof, to enact any ordinance or resolution respecting the power to arrest, convict and punish any individual under the provisions of subdivision (16), section five [§ 8-12-5(16)] of this article or from enforcing any such ordinance or resolution: Provided, that any municipal ordinance in place as of the effective date of this section shall be excepted from the provisions of this section: Provided, however, that no provision in this section may be construed to limit the authority of a municipality to restrict the commercial use of real estate in designated areas through planning or zoning ordinances.

Chapter 27. Mentally Ill Persons

27-12-3. Miscellaneous offenses. If any person shall entice any patient from any state hospital who has been legally committed thereto, or attempt to do so; or shall counsel, cause, influence or assist, or attempt to do so, any such patient to escape or attempt to escape therefrom, or harbor or conceal any such patient who has escaped therefrom; or shall, without the permission of the superintendent of any such hospital, give or sell to any such patient, whether on the premises thereof or elsewhere, any money, firearms, drugs, cigarettes, tobacco, or any other article whatsoever; or shall receive from the hands of any such patient anything of value, whether belonging to the state or not; or shall cause or influence, or attempt to cause or influence, any such patient to violate any rule or to rebel against the government or discipline of such hospital; or shall tease, pester, annoy, or molest any such patient, he shall be guilty of a misdemeanor, and, on conviction thereof, shall be fined not less than ten nor more than one hundred dollars, or imprisoned not exceeding six months, or, in the discretion of the court, both fined and imprisoned. If any person shall aid or abet the commission of any of the foregoing offenses or aid or abet an attempt to commit the same, he shall be guilty of the same as if he were the principal, and be punished as above provided. In the trial of an indictment for committing any of the above-named offenses, the accused may be found guilty of an attempt to commit the same, or of aiding or abetting another in committing or in an attempt to commit the same. If any person, not her husband, shall have sexual intercourse with any female patient who is a patient of any of said state hospitals, he shall be guilty of a felony, and, on conviction thereof, shall be confined in the penitentiary not less than ten nor more than fifteen years; and if such female patient be under sixteen years of age, he shall be imprisoned not less than ten nor more than twenty years.

Chapter 28. State Correctional and Penal Institutions

28-1-8. Offenses relating to youth facilities; penalties; escape; arrest and return. If any person shall entice or attempt to entice away from any youth facility any youth legally committed to the same, or shall aid or abet any youth to escape therefrom, or shall harbor, conceal or aid or abet in harboring or concealing, any youth who shall have escaped therefrom, or shall, without the permission of the superintendent, give or sell, or aid or abet any other person to give or sell, to any youth in the youth facility, whether on the premises of such institution or otherwise, any ... firearms, ... or other articles whatsoever, or shall in any way cause or influence, or attempt to cause or influence or aid or abet therein, any youth in the youth facility to violate any rule of the institution or to rebel against the government of said facility in any particular, or shall receive by the hands of any such youth anything of value, whether belonging to the state or otherwise, such person shall be guilty of a misdemeanor, and, upon conviction thereof, shall be fined not less than ten, nor more than one hundred dollars, or be confined not more than twelve months in the county jail, or, in the discretion of the court, both fined and imprisoned. The superintendent, or any of his assistants, or anyone authorized in writing by him, or any sheriff, constable, policeman or other peace officer, shall have power, and it is hereby made his duty, to arrest any youth, when in his power to do so, who shall have escaped from said facility, and return him thereto.

Chapter 61. Crimes and Their Punishment

Article 3E. Offenses Involving Explosives.

61-3E-1. Definitions. As used in this article, unless the context otherwise requires:

(a) "Destructive device" means any bomb, grenade, mine, rocket, missile, pipe bomb or similar device containing an explosive, incendiary, explosive gas or expanding gas which is designed or so constructed as to explode by such filler and is capable of causing bodily harm or property damage; any combination of parts, either designed or intended for use in converting any device into a destructive device and from which a destructive device may be readily assembled.

"Destructive device" does not include a firearm as such is defined in section two, article seven of this chapter or model rockets and their components as defined in section twenty-three, article three, chapter twenty-nine of this code.

(b) "Explosive material" means any chemical compound, mechanical mixture or device that is commonly used or can be used for the purpose of producing an explosion and which contains any oxidizing and combustive units or other ingredients in such proportions, quantities or packaging that an ignition by fire, by friction, by concussion, by percussion, by detonator or by any part of the compound or mixture may cause a sudden generation of highly heated gases. These materials include, but are not limited to, powders for blasting, high or low explosives, blasting materials, blasting agents, blasting emulsions, blasting fuses other than electric circuit breakers, detonators, blasting caps and other detonating agents and black or smokeless powders not manufactured or used for lawful sporting purposes or fireworks defined in section twenty-three, article three, chapter twenty-nine of this code which are not used in violation of this article. Also included are all explosive materials listed annually by the office of the state fire marshal and published in the state register, said publication being hereby mandated.

(c) "Hoax bomb" means any device or object that by its design, construction, content or characteristics appears to be, or is represented to be or to contain a destructive device, explosive material or incendiary device as defined in this section, but is, in fact, an inoperative facsimile or imitation of such a destructive device, explosive material or incendiary device.

(d) "Incendiary device" means a container containing gasoline, kerosene, fuel oil, or derivative thereof, or other flammable or combustible material, having a wick or other substance or device which, if set or ignited, is capable of igniting such gasoline, kerosene, fuel oil, or derivative thereof, or other flammable or combustible material: Provided, That no similar device commercially manufactured and used solely for the purpose of illumination shall be deemed to be an incendiary device.

(e) "Legal authority" means that right as expressly stated by statute or law.

(f) "Person" shall mean an individual, corporation, company, association, firm, partnership, society or joint stock company.

(g) "Storage magazine" is defined to mean any building or structure, other than an explosives manufacturing building, approved by the legal authority for the storage of explosive materials.

61-3E-3. Illegal possession of destructive devices, explosive materials or incendiary devices; penalty. Any person who possesses or manufactures any explosive material without first obtaining a permit to use explosives from the office of the state fire marshal or who possesses or manufacturers any destructive device or incendiary device shall be guilty of a felony and, upon conviction thereof, shall be committed to the custody of the Division of Corrections for not less than one nor more than ten years or fined not more than five thousand dollars, or both.

61-3E-7. Manufacture, purchase, sale, advertising for sale, transporting or possess on or use of a hoax bomb; possession or use in commission of a felony; penalty.

(a) Any person who knowingly manufactures, purchases, sells, advertises for sale, transports or possesses a hoax bomb with intent to violate any provision of this code shall be guilty of a misdemeanor. Any person convicted of a violation of this section shall be incarcerated in a county or regional jail for not less than six months nor more than one year, or fined five thousand dollars, or both.

(b) Notwithstanding the provisions of subsection (a) of this section, any person who possesses or uses a hoax bomb to commit or attempt to commit any felony shall be guilty of a felony and, upon conviction thereof, shall be committed to the custody of the Division of Corrections for not less than one nor more than ten years or fined not more than ten thousand dollars, or both.

61-3E-8. Theft of explosive material from storage magazines or buildings; penalty.

Any person who breaks and enters or shall enter without breaking any storage magazine, shop, office, storehouse, warehouse or any other building or out-house adjoining thereto, any railcar, boat, vessel or motor vehicle within the jurisdiction of any county within this state where explosive material is stored, with the intent to commit larceny shall be guilty of a felony and, upon conviction thereof, shall be committed to the custody of the Division of Corrections for not less than one nor more than ten years or fined not more than ten thousand dollars, or both.

61-3E-9. Receipt, possession, storage, sale or transportation of stolen explosive material; penalty.

Any person who receives, conceals, transports, ships, stores, barters, sells or disposes of any explosive material knowing or have reason to know that such materials is stolen is guilty of a felony and, upon conviction thereof, shall be committed to the custody of the Division of Corrections for not less than one nor more than ten years or fined not more than ten thousand dollars, or both.

61-3E-11. Exemptions.

(a) Unless specifically prohibited by any provision of this code or the laws of the United States, nothing in this article shall prohibit the authorized manufacture, sale, transportation, distribution, use or possession of any explosive material by any person holding a permit for such issued by the office of the state fire marshal. Any person performing a lawful activity pursuant to or regulated by the terms of a permit issued by the division of environmental protection, or any office thereof, shall be exempt from the provisions of this article.

(b) Unless specifically prohibited by any other provision of this code or the laws of the United States, nothing in this section shall prohibit the authorized manufacture, transportation, distribution, use or possession of any explosive, destructive device or incendiary device by a member of the armed forces or law enforcement officers whenever such persons are acting lawfully and in the line of duty; nor shall it prohibit the manufacture, transportation, distribution, use or possession of any explosive material, destructive device or incendiary device to be used solely for lawful scientific research or lawful educational purposes. Any person engaged in otherwise lawful blasting activities failing to obtain a permit or in possession of an expired permit issued by the office of the state fire marshal shall not be construed to be in violation of the article.

61-3E-12. Contraband, seizure, forfeiture. Any destructive device, explosive material, incendiary device or hoax bomb possessed, involved in, used or intended to be used in a violation of this article or any violation of any criminal law or regulation of this state are hereby declared to be contraband and any property interest therein shall be vested in the state of West Virginia. Said contraband may be seized by the office of the state fire marshal or other law enforcement agency conducting said investigation and upon application to the circuit court of the county in which said contraband is seized be forfeited to the state of West Virginia for destruction or for training purposes by the office of the state fire marshal or other law enforcement agency.

Article 7. Dangerous Weapons

61-7-1. Legislative findings. The Legislature finds that the overwhelming support of the citizens of West Virginia for article three, section twenty-two of the constitution of this state, commonly known as the "Right to Keep and Bear Arms Amendment", combined with the obligation of the state to reasonably regulate the right of persons to keep and bear arms for self-defense requires the reenactment of this article.

61-7-2. Definitions. As used in this article, unless the context otherwise requires: ...

(7) "Pistol" means a short firearm having a chamber which is integral with the barrel, designed to be aimed and fired by the use of a single hand.

(8) "Revolver" means a short firearm having a cylinder of several chambers that are brought successively into line with the barrel to be discharged, designed to be aimed and fired by the use of a single hand.

(9) "Deadly weapon" means an instrument which is designed to be used to produce serious bodily injury or death, or is readily adaptable to such use. The term "deadly weapon" shall include, but not be limited to, the instruments defined in subdivisions (1) through (8), inclusive, of this section, or other deadly weapons of like kind or character which may be easily concealed on or about the person. For the purposes of section one-a [§ 18A-5-1a], article five, chapter eighteen-a of this code and section eleven-a [§ 61-7-11a], article seven of this chapter, in addition to the definition of "knife" set forth in subdivision (3) of this section, the term "deadly weapon" also includes any instrument included within the definition of "knife" with a blade of three and one-half inches or less in length. Additionally, for the purposes of section one-a, article five, chapter eighteen-a of this code and section eleven-a, article seven of this chapter, the term "deadly weapon" includes explosive, chemical, biological and radiological materials. Notwithstanding any other provision of this section, the term "deadly weapon" does not include any item or material owned by the school or county board, intended for curricular use, and used by the student at the time of the alleged offense solely for curricular purposes.

(10) "Concealed" means hidden from ordinary observation so as to prevent disclosure or recognition. A deadly weapon is concealed when it is carried on or about the person in such a manner that another person in the ordinary course of events would not be placed on notice that the deadly weapon was being carried.

(11) "Firearm" means any weapon which will expel a projectile by action of an explosion. ...

61-7-7. Persons prohibited from possessing firearms; classifications; reinstatement of rights to possess; offenses; penalties.

(a) Except as provided in this section, no person shall possess a firearm, as such is defined in section two of this article, who:

(1) Has been convicted in any court of a crime punishable by imprisonment for a term exceeding one year;

(2) Is habitually addicted to alcohol;

(3) Is an unlawful user of or habitually addicted to any controlled substance;

(4) Has been adjudicated as a mental defective or who has been involuntarily committed to a mental institution pursuant to the provisions of chapter twenty-seven of this code: Provided, That once an individual has been adjudicated as a mental defective or involuntarily committed to a mental institution, he or she shall be duly notified that they are to immediately surrender any firearms in their ownership or possession: Provided, however, That the mental hygiene commissioner or circuit judge shall first make a determination of the appropriate public or private individual or entity to act as conservator for the surrendered property;

(5) Is an alien illegally or unlawfully in the United States;

(6) Has been discharged from the armed forces under dishonorable conditions;

(7) Is subject to a domestic violence protective order that:

(A) Was issued after a hearing of which such person received actual notice and at which such person had an opportunity to participate;

(B) Restrains such person from harassing, stalking or threatening an intimate partner of such person or child of such intimate partner or person, or engaging in other conduct that would place an intimate partner in reasonable fear of bodily injury to the partner or child; and

(C)(i) Includes a finding that such person represents a credible threat to the physical safety of such intimate partner or child; or

(ii) By its terms explicitly prohibits the use, attempted use or threatened use of physical force against such intimate partner or child that would reasonably be expected to cause bodily injury; or

(8) Has been convicted of a misdemeanor offense of assault or battery either under the provisions of section twenty-eight [§ 61-2-28], article two of this chapter or the provisions of subsection (b) or (c), section nine [§ 61-2-9] of said article in which the victim was a current or former spouse, current or former sexual or intimate partner, person with whom the defendant has a child in common, person with whom the defendant cohabits or has cohabited, a parent or guardian, the defendant's child or ward or a member of the defendant's household at the time of the offense or has been convicted in any court of any jurisdiction of a comparable misdemeanor crime of domestic violence.

Any person who violates the provisions of this subsection shall be guilty of a misdemeanor and, upon conviction thereof, shall be fined not less than one hundred dollars nor more than one thousand dollars or confined in the county jail for not less than ninety days nor more than one year, or both.

(b) Notwithstanding the provisions of subsection (a) of this section, any person:

(1) Who has been convicted in this state or any other jurisdiction of a felony crime of violence against the person of another or of a felony sexual offense; or

(2) Who has been convicted in this state or any other jurisdiction of a felony controlled substance offense involving a Schedule I controlled substance other than marijuana, a Schedule II or a Schedule III controlled substance as such are defined in sections two hundred four [§ 60A-2-204], two hundred five [§ 60A-2-205] and two hundred six [§ 60A-2-206], article two, chapter sixty-a of this code and who possesses a firearm as such is defined in section two of this article shall be guilty of a felony and, upon conviction thereof, shall be confined in a state correctional facility for not more than five years or fined not more than five thousand dollars, or both. The provisions of subsection (c) of this section shall not apply to persons convicted of offenses referred to in this subsection or to persons convicted of a violation of this subsection.

(c) Any person prohibited from possessing a firearm by the provisions of subsection (a) of this section may petition the circuit court of the county in which he or she resides to regain the ability to possess a firearm and if the court finds by clear and convincing evidence that the person is competent and capable of exercising the responsibility concomitant with the possession of a firearm, the court may enter an order allowing the person to possess a firearm if such possession would not violate any federal law: Provided, That a person prohibited from possessing a firearm by the provisions of subdivision (4), subsection (a) of this section may petition to regain the ability to possess a firearm in accordance with the provisions of section five [§ 61-7A-5], article seven-a of this chapter.

61-7-8. Possession of deadly weapons by minors; prohibitions. Notwithstanding any other provision of this article to the contrary, a person under the age of eighteen years who is not married or otherwise emancipated shall not possess or carry concealed or openly any deadly weapon: Provided, that a minor may possess a firearm upon premises owned by said minor or his family or on the premises of another with the permission of his or her parent or guardian and in the case of property other than his or her own or that of his family, with the permission of the owner or lessee of such property: Provided, however, that nothing in this section shall prohibit a minor from possessing a firearm while hunting in a lawful manner or while traveling from a place where he or she may lawfully possess a deadly weapon, to a hunting site, and returning to a place where he or she may lawfully possess such weapon.

A violation of this section by a person under the age of eighteen years shall subject the child to the jurisdiction of the circuit court under the provisions of article five [§§ 49-5-1 et seq.], chapter forty-nine of this code, and such minor may be proceeded against in the same manner as if he or she had committed an act which if committed by an adult would be a crime, and may be adjudicated delinquent.

61-7-9. Possession of machine guns, penalties. It shall be unlawful for any person to carry, transport, or have in his possession, any machine gun, submachine gun, or any other fully automatic weapon unless he or she has fully complied with applicable federal statutes and all applicable rules and regulations of the Secretary of the Treasury of the United States relating to such firearms.

Any person who violates the provision of this section shall be guilty of a misdemeanor, and, upon conviction thereof, shall be fined not less than one thousand dollars nor more than five thousand dollars, or shall be confined in the county jail for not less than ninety days, or more than one year, or both.

61-7-10. Display of deadly weapons for sale or hire; sale to prohibited persons; penalties.

(a)(1) It shall be unlawful for any person to publicly display and offer for rent or sale, or, where the person is other than a natural person, to knowingly permit an employee thereof to publicly display and offer for rent or sale, to any passersby on any street, road or alley, any deadly weapon, machine gun, submachine gun or other fully automatic weapon, any rifle, shotgun or ammunition for same.

(2) Any person violating the provisions of this subsection shall be guilty of a misdemeanor, and, upon conviction thereof, shall be fined not more than five thousand dollars or shall be confined in the county jail for not more than one year, or both fined and confined, except that where the person violating the provisions of this subsection is other than a natural person, such person shall be fined not more than ten thousand dollars.

(b)(1) It shall be unlawful for any person to knowingly sell, rent, give or lend, or, where the person is other than a natural person, to knowingly permit an employee thereof to knowingly sell, rent, give or lend, any deadly weapon to a person prohibited from possessing same by any provision of this article.

(2) Any person violating the provisions of this subsection shall be guilty of a felony, and, upon conviction thereof, shall be fined not more than twenty-five thousand dollars or shall be imprisoned in the penitentiary of this state for a definite term of years of not less than three years nor more than ten years, or both fined and imprisoned, except that where the person violating the provisions of this subsection is other than a natural person, such person shall be fined not more than fifty thousand dollars.

61-7-11a. Possessing deadly weapons on premises of educational facilities; reports by school principals; suspension of driver license; possessing deadly weapons on premises housing courts of law and in offices of family law master.

(a) The Legislature hereby finds that the safety and welfare of the citizens of this state are inextricably dependent upon assurances of safety for children attending, and the persons employed by, schools in this state and for those persons employed with the judicial department of this state. It is for the purpose of providing such assurances of safety, therefore, that subsections (b), (g) and (h) of this section are enacted as a reasonable regulation of the manner in which citizens may exercise those rights accorded to them pursuant to section twenty-two, article three of the Constitution of the state of West Virginia.

(b)(1) It shall be unlawful for any person to possess any firearm or any other deadly weapon on any school bus as defined in section one [§ 17A-1-1], article one, chapter seventeen-a of this code, or in or on any public or private primary or secondary education building, structure, facility or grounds thereof, including any vocational education building, structure, facility or grounds thereof where secondary vocational education programs are conducted or at any school-sponsored function.

(2) This subsection shall not apply to:

(A) A law enforcement officer acting in his or her official capacity;

(B) A person specifically authorized by the board of education of the county or principal of the school where the property is located to conduct programs with valid educational purposes;

(C) A person who, as otherwise permitted by the provisions of this article, possesses an unloaded firearm or deadly weapon in a motor vehicle, or leaves an unloaded firearm or deadly weapon in a locked motor vehicle;

(D) Programs or raffles conducted with the approval of the county board of education or school which include the display of unloaded firearms; or

(E) The official mascot of West Virginia University, commonly known as "The Mountaineer", acting in his or her official capacity.

(3) Any person violating this subsection shall be guilty of a felony, and, upon conviction thereof, shall be imprisoned in the penitentiary of this state for a definite term of years of not less than two years nor more than ten years, or fined not more than five thousand dollars, or both.

(c) It shall be the duty of the principal of each school subject to the authority of the state board of education to report any violation of subsection (b) of this section discovered by such principal to the state superintendent of schools within seventy-two hours after such violation occurs. The state board of education shall keep and maintain such reports and may prescribe rules establishing policy and procedures for the making and delivery of the same as required by this subsection. In addition, it shall be the duty of the principal of each school subject to the authority of the state board of education to report any violation of subsection (b) of this section discovered by such principal to the appropriate local office of the division of public safety within seventy-two hours after such violation occurs.

(d) In addition to the methods of disposition provided by article five [§§ 49-5-1 et seq.], chapter forty-nine of this code, any court which adjudicates a person who is fourteen years of age or older as delinquent for a violation of subsection (b) of this section may, in its discretion, order the Division of Motor Vehicles to suspend any driver's license or instruction permit issued to such person for such period of time as the court may deem appropriate, such suspension, however, not to extend beyond such person's nineteenth birthday; or, where such person has not been issued a driver's license or instruction permit by this state, order the division of motor vehicles to deny such person's application for the same for such period of time as the court may deem appropriate, such denial, however, not to extend beyond such person's nineteenth birthday. Any suspension ordered by the court pursuant to this subsection shall be effective upon the date of entry of such order. Where the court orders the suspension of a driver's license or instruction permit pursuant to this subsection, the court shall confiscate any driver's license or instruction permit in the adjudicated person's possession and forward the same to the division of motor vehicles.

(e)(1) If a person eighteen years of age or older is convicted of violating subsection (b) of this section, and if such person does not act to appeal such conviction within the time periods described in subdivision (2) of this subsection, such person's license or privilege to operate a motor vehicle in this state shall be revoked in accordance with the provisions of this section.

(2) The clerk of the court in which the person is convicted as described in subdivision (1) of this subsection shall forward to the commissioner a transcript of the judgment of conviction. If the conviction is the judgment of a magistrate court, the magistrate court clerk shall forward such transcript when the person convicted has not requested an appeal within twenty days of the sentencing for such conviction. If the conviction is the judgment of a circuit court, the circuit clerk shall forward such transcript when the person convicted has not filed a notice of intent to file a petition for appeal or writ of error within thirty days after the judgment was entered.

(3) If, upon examination of the transcript of the judgment of conviction, the commissioner shall determine that the person was convicted as described in subdivision (1) of this subsection, the commissioner shall make and enter an order revoking such person's license or privilege to operate a motor vehicle in this state for a period of one year, or, in the event the person is a student enrolled in a secondary school, for a period of one year or until the person's twentieth birthday, whichever is the greater period. The order shall contain the reasons for the revocation and the revocation period. The order of suspension shall advise the person that because of the receipt of the court's transcript, a presumption exists that the person named in the order of suspension is the same person named in the transcript. The commissioner may grant an administrative hearing which substantially complies with the requirements of the provisions of section two [§ 17C-5A-2], article five-a, chapter seventeen-c of this code upon a preliminary showing that a possibility exists that the person named in the notice of conviction is not the same person whose license is being suspended. Such request for hearing shall be made within ten days after receipt of a copy of the order of suspension. The sole purpose of this hearing shall be for the person requesting the hearing to present evidence that he or she is not the person named in the notice. In the event the commissioner grants an administrative hearing, the commissioner shall stay the license suspension pending the commissioner's order resulting from the hearing.

(4) For the purposes of this subsection, a person is convicted when such person enters a plea of guilty or is found guilty by a court or jury.

(f)(1) It shall be unlawful for any parent(s), guardian(s) or custodian(s) of a person less than eighteen years of age who knows that said person is in violation of subsection (b) of this section, or who has reasonable cause to believe that said person's violation of said subsection is imminent, to fail to immediately report such knowledge or belief to the appropriate school or law-enforcement officials.

(2) Any person violating this subsection shall be guilty of a misdemeanor, and, upon conviction thereof, shall be fined not more than one thousand dollars, or shall be confined in jail not more than one year, or both.

(g)(1) It shall be unlawful for any person to possess any firearm or any other deadly weapon on any premises which houses a court of law or in the offices of a family law master.

(2) This subsection shall not apply to:

(A) A law enforcement officer acting in his or her official capacity; and

(B) A person exempted from the provisions of this subsection by order of record entered by a court with jurisdiction over such premises or offices.

(3) Any person violating this subsection shall be guilty of a misdemeanor, and, upon conviction thereof, shall be fined not more than one thousand dollars, or shall be confined in jail not more than one year, or both.

(h)(1) It shall be unlawful for any person to possess any firearm or any other deadly weapon on any premises which houses a court of law or in the offices of a family law master with the intent to commit a crime.

(2) Any person violating this subsection shall be guilty of a felony, and, upon conviction thereof, shall be imprisoned in the penitentiary of this state for a definite term of years of not less than two years nor more than ten years, or fined not more than five thousand dollars, or both.

(i) Nothing in this section may be construed to be in conflict with the provisions of federal law.

61-7-14. Right of certain persons to limit possession of firearms on premises.

Notwithstanding the provisions of this article, any owner, lessee or other person charged with the care, custody and control of real property may prohibit the carrying openly or concealing of any firearm or deadly weapon on property under his or her domain: Provided, That for purposes of this section "person" means an individual or any entity which may acquire title to real property.

Any person carrying or possessing a firearm or other deadly weapon on the property of another who refuses to temporarily relinquish possession of such firearm or other deadly weapon, upon being requested to do so, or to leave such premises, while in possession of such firearm or other deadly weapon, shall be guilty of a misdemeanor, and, upon conviction thereof, shall be fined not more than one thousand dollars or confined in the county jail not more than six months, or both: Provided, That the provisions of this section shall not apply to those persons set forth in subsections (3) through (6), section six [§ 61-7-6] of this code while such persons are acting in an official capacity: Provided, however, That under no circumstances may any person possess or carry or cause the possession or carrying of any firearm or other deadly weapon on the premises of any primary or secondary educational facility in this state unless such person is a law-enforcement officer or he or she has the express written permission of the county school superintendent.

Article 7A. State Mental Health Registry; Reporting of Persons Proscribed from Firearm Possession Due to Mental Condition to the National Instant Criminal Background Check System; Legislative Findings; Definitions; Reporting Requirements; Reinstatement of Rights Procedures.

61-7A-1. Legislative intent.

It is the intention of the Legislature in the enactment of this article to clarify the persons whom it intends to proscribe from the possession of firearms due to substance abuse or mental illness; establish a process in reporting the names of persons proscribed from possession of firearms due to mental illness to the Central State Mental Health Registry; authorize reporting by registry to the National Instant Criminal Background Check System; and to prescribe a means for reinstating one's ability to lawfully possess a firearm.

61-7A-2. Definitions.

As used in this article and as the terms are deemed to mean in 18 U. S. C. § 922(g) and section seven [§ 61-7-7], article seven of this chapter as each exists as of the thirty-first day of January, two thousand eight:

(1) "A person adjudicated as a mental defective" means a person who has been determined by a duly authorized court, tribunal, board or other entity to be mentally ill to the point where he or she has been found to be incompetent to stand trial due to mental illness or insanity, has been found not guilty in a criminal proceeding by reason of mental illness or insanity or has been determined to be unable to handle his or her own affairs due to mental illness or insanity.

(2) "Committed to a mental institution" means to have been involuntarily committed for treatment pursuant to the provisions of chapter twenty-seven [§§ 27-1-1 et seq.] of this code.

(3) "Mental institution" means any facility or part of a facility used for the treatment of persons committed for treatment of mental illness or addiction.

61-7A-3. Persons whose names are to be supplied to the central state mental health registry.

(a) The Superintendent of the West Virginia State Police and the Secretary of the Department of Health and Human Resources, or their designees, shall cooperate with the circuit clerk of each county and Administrator of the West Virginia Supreme Court of Appeals in compiling and maintaining a database containing the names and identifying information of persons who have been adjudicated to be mentally defective or who have been committed for treatment of a mental illness pursuant to the provisions of chapter twenty-seven [§§ 27-1-1 et seq.] of this code. The registry shall be maintained by the Administrator of the Supreme Court of Appeals or the Superintendent of the West Virginia State Police.

(b) The name of any person who has been adjudicated to be mentally defective or who has been committed for treatment of a mental illness pursuant to the provisions of chapter twenty-seven [§§ 27-1-1 et seq.] of this code which shall be provided to the Superintendent of the West Virginia State Police for inclusion in the Central State Mental Health Registry. Upon receipt of the information being received by the Central State Mental Health Registry it may be transmitted to the National Instant Criminal Background Check System and to county sheriffs;

(c) The Secretary of Department of Human Resources and the circuit clerk of each county shall, as soon as practicable after the effective date of this article, supply to the Superintendent of the West Virginia State Police for inclusion in the Central State Mental Health Registry the name and identifying information required by the provisions of subsection (d) of this section of all persons covered by the provisions of this article and shall on an ongoing basis continue to provide such information as it is developed;

(d) The central state mental health registry shall contain the name, address at the time of commitment or adjudication, date of birth, date of commitment or adjudication and of all persons who have been adjudicated to be mentally defective or who have been committed for treatment of a mental illness pursuant to the provisions of chapter twenty-seven [§§ 27-1-1 et seq.] of this code.

(e) The Central State Mental Health Registry shall provide only such information about a person on the registry to county sheriffs and the National Instant Criminal Background Check System as is necessary to identify registrants; and

(f) On or before the first day of January, two thousand ten, the Central State Mental Health Registry shall contain the name, address at the time of commitment or adjudication, date of birth, date of commitment or adjudication and any other identifying characteristics of all persons who have been adjudicated to be mentally defective or who have been committed for treatment of a mental illness pursuant to the provisions of chapter twenty-seven [§§ 27-1-1 et seq.] of this code. Under no circumstances shall the registry contain information relating to any diagnosis or treatment provided.

61-7A-4. Confidentiality; limits on use of registry information.

(a) Notwithstanding any provision of this code to the contrary, the Superintendent of the State Police, the Secretary of the Department of Health and Human Resources and the circuit clerks and the Administrator of the Supreme Court of Appeals may provide notice to the Central State Mental Health Registry and the National Instant Criminal Background Check System established pursuant to Section 103(d) of the Brady Handgun Violence Protection Act, 18 U. S. C. § 922, that a person: (i) Has been involuntarily committed as provided in chapter twenty-seven [§§ 27-1-1 et seq.] of this code; (ii) has been adjudicated mentally incompetent in a proceeding under article six-a [§§ 61-6A-1 et seq.] of this chapter; or (iii) has regained the ability to possess a firearm by order of a circuit court in a proceeding under section five [§ 61-7A-5] of this article.

(b) The information contained in the Central State Mental Health Registry is to be used solely for the purpose of records checks related to firearms purchases and for eligibility for a state license or permit to possess or carry a concealed firearm.

(c) Whenever a person's name and other identifying information has been added to the Central State Mental Health Registry, a review of the state concealed handgun registry shall be undertaken and if such review reveals that the person possesses a current concealed handgun license, the sheriff of the county issuing the concealed handgun license shall be informed of the person's change in status.

61-7A-5. Petition to regain right to possess firearms.

(a) Any person who is prohibited from possessing a firearm pursuant to the provisions of section seven [§ 61-7-7], article seven of this chapter or by provisions of federal law by virtue solely of having previously been adjudicated to be mentally defective or to having a prior involuntary commitment to a mental institution pursuant to chapter twenty-seven [§§ 27-1-1 et seq.] of this code may petition the circuit court of the county of his or her residence to regain the ability to lawfully possess a firearm. If the court finds by clear and convincing evidence that the person is competent and capable of exercising the responsibilities concomitant with the possession of a firearm, the court may enter an order allowing the petitioner to possess a firearm.

(b) The circuit clerk of each county shall provide the Superintendent of the West Virginia State Police or his or her designee with a certified copy of any order entered pursuant to the provisions of this section. If the order restores the petitioner's ability to possess a firearm, petitioner's name shall be promptly removed from the Central State Mental Health Registry and the superintendent shall forthwith inform the Federal Bureau of Investigation or other federal entity operating the National Instant Criminal Background Check System of the court action.

[Current through 2008 Regular and 1st and 2nd Extraordinary Sessions]

WISCONSIN
WIS. STAT.

Chapter 66. General Municipality Law

66.0409. Local regulation of firearms.

(1) In this section:

(a) "Firearm" has the meaning given in s. 167.31(1)(c).

(b) "Political subdivision" means a city, village, town or county.

(c) "Sport shooting range" means an area designed and operated for the practice of weapons used in hunting, skeet shooting and similar sport shooting.

(2) Except as provided in subs. (3) and (4), no political subdivision may enact an ordinance or adopt a resolution that regulates the sale, purchase, purchase delay, transfer, ownership, use, keeping, possession, bearing, transportation, licensing, permitting, registration or taxation of any firearm or part of a firearm, including ammunition and reloader components, unless the ordinance or resolution is the same as or similar to, and no more stringent than, a state statute.

(3)(a) Nothing in this section prohibits a county from imposing a sales tax or use tax under subchapter V of chapter 77 on any firearm or part of a firearm, including ammunition and reloader components, sold in the county.

(b) Nothing in this section prohibits a city, village or town that is authorized to exercise village powers under s. 60.22(3) from enacting an ordinance or adopting a resolution that restricts the discharge of a firearm.

(4)(a) Nothing in this section prohibits a political subdivision from continuing to enforce an ordinance or resolution that is in effect on November 18, 1995, and that regulates the sale, purchase, transfer, ownership, use, keeping, possession, bearing, transportation, licensing, permitting, registration or taxation of any firearm or part of a firearm, including ammunition and reloader components, if the ordinance or resolution is the same as or similar to, and no more stringent than, a state statute.

(am) Nothing in this section prohibits a political subdivision from continuing to enforce until November 30, 1998, an ordinance or resolution that is in effect on November 18, 1995, and that requires a waiting period of not more than 7 days for the purchase of a handgun.

(b) If a political subdivision has in effect on November 17, 1995, an ordinance or resolution that regulates the sale, purchase, transfer, ownership, use, keeping, possession, bearing, transportation, licensing, permitting, registration or taxation of any firearm or part of a firearm, including ammunition and reloader components, and the ordinance or resolution is not the same as or similar to a state statute, the ordinance or resolution shall have no legal effect and the political subdivision may not enforce the ordinance or resolution on or after November 18, 1995.

(c) Nothing in this section prohibits a political subdivision from enacting and enforcing a zoning ordinance that regulates the new construction of a sport shooting range or when the expansion of an existing sport shooting range would impact public health and safety.

(5) A county ordinance that is enacted or a county resolution that is adopted by a county under sub. (2) or a county ordinance or resolution that remains in effect under sub. (4)(a) or (am) applies only in those towns in the county that have not enacted an ordinance or adopted a resolution under sub. (2) or that continue to enforce an ordinance or resolution under sub. (4)(a) or (am), except that this subsection does not apply to a sales or use tax that is imposed under subchapter V of chapter 77.

Chapter 167. Safeguards of Persons and Property

167.31. Safe use and transportation of firearms and bows.

(1) Definitions. In this section:

(a) "Aircraft" has the meaning given under s. 114.002(3).

(b) "Encased" means enclosed in a case that is expressly made for the purpose of containing a firearm and that is completely zipped, snapped, buckled, tied or otherwise fastened with no part of the firearm exposed.

(bg) "Family member of the landowner" means a person who is related to the landowner as a parent, child, spouse, or sibling.

(bn) "Farm tractor" has the meaning given in s. 340.01(16).

(c) "Firearm" means a weapon that acts by force of gunpowder.

(d) "Highway" has the meaning given under s. 340.01(22).

(dm) "Implement of husbandry" has the meaning given in s. 340.01(24).

(e) "Motorboat" has the meaning given under s. 30.50 (6).

(em) "Peace officer" has the meaning given in s. 939.22(22).

(f) "Roadway" has the meaning given under s. 340.01(54).

(fm) "Street" means a highway that is within the corporate limits of a city or village.

(g) "Unloaded" means any of the following:

1. Having no shell or cartridge in the chamber of a firearm or in the magazine attached to a firearm.

2. In the case of a cap lock muzzle-loading firearm, having the cap removed.

3. In the case of a flint lock muzzle-loading firearm, having the flashpan cleaned of powder.

(h) "Vehicle" has the meaning given in s. 340.01(74), and includes a snowmobile, as defined in s. 340.01(58a), and an electric personal assistive mobility device, as defined in s. 340.01 (15pm), except that for purposes of subs. (4)(c) and (cg) and (4m) "vehicle" has the meaning given for "motor vehicle" in s. 29.001(57).

(2) Prohibitions; motorboats and vehicles; highways and roadways.

(a) Except as provided in sub. (4), no person may place, possess or transport a firearm, bow or crossbow in or on a motorboat with the motor running, unless the firearm is unloaded or unless the bow or crossbow is unstrung or is enclosed in a carrying case.

(b) Except as provided in sub. (4), no person may place, possess or transport a firearm, bow or crossbow in or on a vehicle, unless the firearm is unloaded and encased or unless the bow or crossbow is unstrung or is enclosed in a carrying case.

(c) Except as provided in sub. (4), no person may load or discharge a firearm or shoot a bolt or an arrow from a bow or crossbow in or from a vehicle.

(d) Except as provided in sub. (4) (a), (bg), (cg), (e), and (g), no person may discharge a firearm or shoot a bolt or an arrow from a bow or crossbow from or across a highway or within 50 feet of the center of a roadway.

(e) A person who violates pars. (a) to (d) is subject to a forfeiture of not more than $100.

(3) Prohibitions; aircraft.

(a) Except as provided in sub. (4), no person may place, possess or transport a firearm, bow or crossbow in or on an aircraft, unless the firearm is unloaded and encased or unless the bow or crossbow is unstrung or is enclosed in a carrying case.

(b) Except as provided in sub. (4), no person may load or discharge a firearm or shoot a bolt or an arrow from a bow or crossbow in or from an aircraft.

(c) A person who violates par. (a) or (b) shall be fined not more than $1,000 or imprisoned not more than 90 days or both.

(4) Exceptions.

(a) Subsections (2) and (3) do not apply to any of the following who, in the line of duty, place, possess, transport, load or discharge a firearm in, on or from a vehicle, motorboat or aircraft or discharge a firearm from or across a highway or within 50 feet of the center of a roadway:

2. A member of the U.S. armed forces.

3. A member of the national guard.

4. A private security person who meets all of the following requirements:

a. He or she holds either a private detective license issued under s. 440.26(2)(a)2. or a private security permit issued under s. 440.26(5).

b. He or she holds a certificate of proficiency to carry a firearm issued by the department of regulation and licensing.

c. He or she is performing his or her assigned duties or responsibilities.

d. He or she is wearing a uniform that clearly identifies him or her as a private security person.

e. His or her firearm is in plain view, as defined by rule by the department of regulation and licensing.

(am)1. Subsections (2)(a), (c) and (d) and (3)(a) and (b) do not apply to a peace officer who, in the line of duty, loads or discharges a firearm in, on or from a vehicle, motorboat or aircraft or discharges a firearm from or across a highway or within 50 feet of the center of a roadway.

2. Subsection (2)(b) does not apply to a peace officer who places, possesses or transports a firearm in or on a vehicle, motorboat or aircraft while in the line of duty.

3. Subsection (2)(b) does not apply to a person employed as a peace officer who places, possesses or transports a firearm in or on a vehicle while traveling in the vehicle from his or her residence to his or her place of employment as a peace officer.

(b) Subsections (2)(a), (b) and (c) and (3)(a) and (b), and (3m) do not apply to the holder of a scientific research license under s. 169.25 or a scientific collector permit under s. 29.614 who is using a net gun or tranquilizer gun in an activity related to the purpose for which the license or permit was issued. ...

(5) Weapons surcharge.

(a) If a court imposes a fine or forfeiture for a violation of this section, the court shall also impose a weapons surcharge under ch. 814 equal to 75% of the amount of the fine or forfeiture.

(b) If a fine or forfeiture is suspended in whole or in part, the weapons surcharge shall be reduced in proportion to the suspension.

(c) If any deposit is made for an offense to which this subsection applies, the person making the deposit shall also deposit a sufficient amount to include the weapons surcharge under this subsection. If the deposit is forfeited, the amount of the weapons surcharge shall be transmitted to the secretary of administration under par. (d). If the deposit is returned, the amount of the weapons surcharge shall also be returned. ...

Chapter 175. Miscellaneous Police Provisions

175.30. Purchase of firearms in contiguous states permitted. It is lawful for a resident of this state or a corporation or other business entity maintaining a place of business in this state to purchase or otherwise obtain a rifle or shotgun in a state contiguous to this state if the transfer complies with federal law and the laws of both states.

175.35. Waiting period for purchase of handguns.

(1) In this section:

(ag) "Criminal history record" includes information reported to the department under s. 938.396(2(g)(n)) that indicates a person was adjudicated delinquent for an act that if committed by an adult in this state would be a felony.

(ar) "Firearms dealer" means any person engaged in the business of importing, manufacturing or dealing in firearms and having a license as an importer, manufacturer or dealer issued by the U.S. department of the treasury.

(at) "Firearms restrictions record search" means a search of department of justice records to determine whether a person seeking to purchase a handgun is prohibited from possessing a firearm under s. 941.29. "Firearms restriction record search" includes a criminal history record search, a search to determine whether a person is prohibited from possessing a firearm under s. 51.20(13)(cv), a search to determine whether the person is subject to an injunction under s. 813.12 or 813.122, or a tribal injunction, as defined in s. 813.12(1)(e), issued by a court established by any federally recognized Wisconsin Indian tribe or band, except the Menominee Indian tribe of Wisconsin, that includes notice to the respondent that he or she is subject to the requirements and penalties under s. 941.29 and that has been filed with the circuit court under s. 806.247(3), and a search to determine whether the person is prohibited from possessing a firearm under s. 813.125(4m).

(b) "Handgun" means any weapon designed or redesigned, or made or remade, and intended to be fired while held in one hand and to use the energy of an explosive to expel a projectile through a smooth or rifled bore.

(c) "Working day" means each day except Saturday, Sunday or a legal holiday under s. 995.20.

(2) When a firearms dealer sells a handgun, he or she may not transfer possession of that handgun to any other person until all of the following have occurred:

(a) The transferee has provided identification as required by rule under sub. (2g)(a).

(b) The transferee has completed the notification form described in sub. (2g)(b).

(c) The firearms dealer has conveyed the information from the completed notification form to the department of justice as required by rule under sub. (2g)(b) and requested a firearms restrictions record search.

(d) Forty-eight hours, subject to extension under sub. (2g)(c)4.c., have elapsed from the time that the firearms dealer has received a confirmation number regarding the firearms restrictions record search under sub. (2g)(c) from the department of justice and the firearms dealer has not been notified that the transfer would be in violation of s. 941.29.

(2e) When a transferee completes the notification form described in sub. (2g)(b), the transferee shall provide truthful information.

(2f) When a firearms dealer requests that the department of justice provide a firearms restrictions record search under sub. (2g), he or she shall provide truthful information about his or her status as a firearms dealer and shall provide an accurate firearms dealer identification number obtained under sub. (2h). A person may request that the department provide a firearms restrictions record search under sub. (2g) only if he or she is a firearms dealer.

(2g)(a) The department of justice shall promulgate rules prescribing procedures for a transferee to provide and a firearms dealer to inspect identification containing a photograph of the transferee.

(b) The department of justice shall promulgate rules prescribing a notification form for use under sub. (2) requiring the transferee to provide his or her name, date of birth, gender, race and social security number and other identification necessary to permit an accurate firearms restrictions record search under par. (c)3. and the required notification under par. (c)4. The department of justice shall make the forms available at locations throughout the state.

(c) The department of justice shall promulgate rules for firearms restrictions record searches regarding transferees under sub. (2), including procedures for all of the following:

1. A firearms dealer to convey the information from a completed notification form to the department using a toll-free telephone number provided by the department.

2. The department to provide the firearms dealer with a confirmation number confirming the receipt of the information under subdivision 1.

3. The department to conduct the firearms restrictions record search regarding the transferee. The rules shall include, but not be limited to, a requirement that the department use the transaction information for management of enforcement system and the national crime information center system.

4. The department to notify the dealer, either during the initial telephone call or as soon thereafter as practicable, of the results of the firearms restrictions record search as follows:

a. If the search indicates that the transferee is prohibited from possessing a firearm under s. 941.29, the department shall provide the firearms dealer with a unique nonapproval number. The department may not disclose to the firearms dealer the reason the transferee is prohibited from possessing a firearm under s. 941.29.

b. If the search indicates that the transferee is not prohibited from possessing a firearm under s. 941.29, the department shall provide the firearms dealer with a unique approval number.

c. If the search indicates a felony charge without a recorded disposition, the deadline under sub. (2)(d) is extended to the end of the 3rd complete working day commencing after the day on which the finding is made. The department shall notify the firearms dealer of the extension as soon as practicable. During the extended period, the department shall make every reasonable effort to determine the disposition of the charge and notify the firearms dealer of the results as soon as practicable.

(2h) Upon the request of any firearms dealer, the department of justice shall provide that firearms dealer with a unique firearms dealer identification number for use under this section.

(2i) The department shall charge a firearms dealer an $8 fee for each firearms restrictions record search that the firearms dealer requests under sub. (2)(c). The firearms dealer may collect the fee from the transferee. The department may refuse to conduct firearms restrictions record searches for any firearms dealer who fails to pay any fee under this subsection within 30 days after billing by the department.

(2j) A firearms dealer shall maintain the original record of all completed notification forms and a record of all confirmation numbers and corresponding approval or nonapproval numbers that he or she receives regarding firearms restrictions record searches under sub. (2g). The firearms dealer shall mail the duplicate copy of each completed notification form to the department of justice.

(2k)(ag) In this subsection:

1. "Law enforcement agency of a physically adjacent state" has the meaning given in s. 175.46(1)(b).

2. "Wisconsin law enforcement agency" means a governmental unit of one or more persons employed by this state or a political subdivision of this state for the purpose of preventing and detecting crime and enforcing state laws or local ordinances, employees of which unit are authorized to make arrests for crimes while acting within the scope of their authority.

(ar) Except as provided in pars. (b) to (j) and as necessary to administer this section, the department of justice shall do all of the following:

1. Deny access to any record kept under this section.

2. Check each duplicate notification form received under sub. (2j) against the information recorded by the department regarding the corresponding request for a firearms restrictions record search under sub. (2g). If the department previously provided a unique approval number regarding the request and nothing in the duplicate completed notification form indicates that the transferee is prohibited from possessing a firearm under s. 941.29, the department shall destroy all records regarding that firearms restrictions record search within 30 days after receiving the duplicate form.

(b) Notwithstanding par. (ar), the department of justice may maintain all of the following:

1. Records necessary to comply with federal law.

2.a. Except as provided in subd. 2.b., a log of dates of requests for firearms restrictions record searches under sub. (2g) together with confirmation numbers, unique approval and nonapproval numbers and firearms dealer identification numbers corresponding to those dates.

b. Within 3 years after the department issues a unique approval number, the department shall destroy all corresponding information contained in the log under subd. 2.a.

3. Records necessary to administer this section.

(c) Notwithstanding par. (ar), the department of justice shall provide access to any record under this section under all of the following circumstances:

1. The department of justice receives a record request that is submitted in writing by a Wisconsin law enforcement agency.

2. The request submitted under subd. 1. appears on the Wisconsin law enforcement agency's letterhead and contains all of the following:

a. A statement that the Wisconsin law enforcement agency is conducting an investigation of a crime in which a handgun was used or was attempted to be used or was unlawfully possessed.

b. A statement by a division commander or higher authority within the Wisconsin law enforcement agency that he or she has a reasonable suspicion that the person who is the subject of the information request has obtained or is attempting to obtain a handgun.

c. The signature of a division commander or higher authority within the Wisconsin law enforcement agency.

(d) Whenever a Wisconsin law enforcement agency makes a request for information under par. (c), the agency shall report to the subject of the information request the fact that a request has been made and the name of the Wisconsin law enforcement agency that made the request. The agency shall make the report whenever the earliest of the following occurs:

1. The person who is the subject of the information request under par. (c)2.b. is no longer material to the investigation conducted under par. (c)2.a.

2. The Wisconsin law enforcement agency has completed its investigation under par. (c)2.a.

3. One year after the date that the Wisconsin law enforcement agency made the request under par. (c).

(e) A Wisconsin law enforcement agency may disclose information that is provided by the department of justice under par. (c) to another law enforcement agency. If there is a request for information from a requester other than a law enforcement agency, the Wisconsin law enforcement agency shall not disclose information to the requester that is provided by the department of justice under par. (c). If there is a request by a requester other than a law enforcement agency to copy or inspect any record of the Wisconsin law enforcement agency that contains that information, the agency, acting under s. 19.36(6), shall delete any portion of the record that relates to that information before release.

(f) A Wisconsin law enforcement agency that is provided access to a record under par. (c) shall destroy all corresponding information contained in the record when the earliest of the following occurs:

1. The person who is the subject of the information request under par. (c)2.b. is no longer material to the investigation conducted under par. (c)2.a.

2. The Wisconsin law enforcement agency has completed its investigation under par. (c)2.a.

3. One year after the date the Wisconsin law enforcement agency made the request under par. (c).

(g) If a search conducted under sub. (2g) indicates that the transferee is prohibited from possessing a firearm under s. 941.29, the attorney general or his or her designee may disclose to a law enforcement agency that the transferee has attempted to obtain a handgun.

(h) If a search conducted under sub. (2g) indicates a felony charge without a recorded disposition and the attorney general or his or her designee has reasonable grounds to believe the transferee may pose a danger to himself, herself or another, the attorney general or his or her designee may disclose to a law enforcement agency that the transferee has obtained or has attempted to obtain a handgun.

(i) The department of justice may not charge a fee for any services that the department provides under pars. (c) to (j).

(j) If a law enforcement agency of a physically adjacent state makes a request under par. (c), the department shall comply with the request under all of the following circumstances:

1. The law enforcement agency of the physically adjacent state agrees to comply with all the requirements under this subsection.

2. The physically adjacent state allows Wisconsin law enforcement agencies similar or greater access to similar information from that physically adjacent state.

(2L) The department of justice shall promulgate rules providing for the review of nonapprovals under sub. (2g)(c)4.a. Any person who is denied the right to purchase a handgun because the firearms dealer received a nonapproval number under sub. (2g)(c)4.a. may request a firearms restrictions record search review under those rules. If the person disagrees with the results of that review, the person may file an appeal under rules promulgated by the department.

(2t) This section does not apply to any of the following:

(a) Transfers of any handgun classified as an antique by regulations of the U.S. department of the treasury.

(b) Transfers of any handgun between firearms dealers or between wholesalers and dealers.

(c) Transfers of any handgun to law enforcement or armed services agencies.

(3) Any person who intentionally violates sub. (2), (2e), (2f) or (2j) shall be fined not less than $500 nor more than $10,000 and may be imprisoned for not more than 9 months.

175.37. Warning whenever transferring a firearm.

(1) Upon the retail commercial sale or retail commercial transfer of any firearm, the seller or transferor shall provide to the buyer or transferee the following written warning in block letters not less than one-fourth inch in height: "IF YOU LEAVE A LOADED FIREARM WITHIN THE REACH OR EASY ACCESS OF A CHILD YOU MAY BE FINED OR IMPRISONED OR BOTH IF THE CHILD IMPROPERLY DISCHARGES, POSSESSES OR EXHIBITS THE FIREARM."

(2) Any person who violates sub. (1) may be fined not more than $500 or imprisoned for not more than 30 days or both.

Chapter 813. Injunctions, Ne Exeat and Receivers

813.12. Domestic abuse restraining orders and injunctions. ...

(4m) Notice of restriction on firearm possession; surrender of firearms.

(a) An injunction issued under sub. (4) shall do all of the following:

1. Inform the respondent named in the petition of the requirements and penalties under s. 941.29.

2. Except as provided in par. (ag), require the respondent to surrender any firearms that he or she owns or has in his or her possession to the sheriff of the county in which the action under this section was commenced, to the sheriff of the county in which the respondent resides or to another person designated by the respondent and approved by the judge or circuit court commissioner. The judge or circuit court commissioner shall approve the person designated by the respondent unless the judge or circuit court commissioner finds that the person is inappropriate and places the reasons for the finding on the record. If a firearm is surrendered to a person designated by the respondent and approved by the judge or circuit court commissioner, the judge or circuit court commissioner shall inform the person to whom the firearm is surrendered of the requirements and penalties under s. 941.29(4).

(ag) If the respondent is a peace officer, an injunction issued under sub. (4) may not require the respondent to surrender a firearm that he or she is required, as a condition of employment, to possess whether or not he or she is on duty.

(am)1. When a respondent surrenders a firearm under par. (a)2. to a sheriff, the sheriff who is receiving the firearm shall prepare a receipt for each firearm surrendered to him or her. The receipt shall include the manufacturer, model and serial number of the firearm surrendered to the sheriff and shall be signed by the respondent and by the sheriff to whom the firearm is surrendered.

2. The sheriff shall keep the original of a receipt prepared under subd. 1. and shall provide an exact copy of the receipt to the respondent. When the firearm covered by the receipt is returned to the respondent under par. (b), the sheriff shall surrender to the respondent the original receipt and all of his or her copies of the receipt.

3. A receipt prepared under subd. 1. is conclusive proof that the respondent owns the firearm for purposes of returning the firearm covered by the receipt to the respondent under par. (b).

4. The sheriff may not enter any information contained on a receipt prepared under subd. 1. into any computerized or direct electronic data transfer system in order to store the information or disseminate or provide access to the information.

(aw) A sheriff may store a firearm surrendered to him or her under par. (a)2. in a warehouse that is operated by a public warehouse keeper licensed under ch. 99. If a sheriff stores a firearm at a warehouse under this paragraph, the respondent shall pay the costs charged by the warehouse for storing that firearm.

(b) A firearm surrendered under par. (a)2. may not be returned to the respondent until a judge or circuit court commissioner determines all of the following:

1. That the injunction issued under sub. (4) has been vacated or has expired and not been extended.

2. That the person is not prohibited from possessing a firearm under any state or federal law or by the order of any federal court or state court, other than an order from which the judge or circuit court commissioner is competent to grant relief.

(c) If a respondent surrenders a firearm under par. (a)2. that is owned by a person other than the respondent, the person who owns the firearm may apply for its return to the circuit court for the county in which the person to whom the firearm was surrendered is located. The court shall order such notice as it considers adequate to be given to all persons who have or may have an interest in the firearm and shall hold a hearing to hear all claims to its true ownership. If the right to possession is proved to the court's satisfaction, it shall order the firearm returned. If the court returns a firearm under this paragraph, the court shall inform the person to whom the firearm is returned of the requirements and penalties under s. 941.29 (4). ...

813.122. Child abuse restraining orders and injunctions....

(5m) Notice of restriction on firearm possession; surrender of firearms.

(a) An injunction issued under sub. (5) shall do all of the following:

1. Inform the respondent named in the petition of the requirements and penalties under s. 941.29

2. Except as provided in par. (ag), require the respondent to surrender any firearms that he or she owns or has in his or her possession to the sheriff of the county in which the action under this section was commenced, to the sheriff of the county in which the respondent resides or to another person designated by the respondent and approved by the judge or circuit court commissioner. The judge or circuit court commissioner shall approve the person designated by the respondent unless the judge or circuit court commissioner finds that the person is inappropriate and places the reasons for the finding on the record. If a firearm is surrendered to a person designated by the respondent and approved by the judge or circuit court commissioner, the judge or circuit court commissioner shall inform the person to whom the firearm is surrendered of the requirements and penalties under s. 941.29 (4)

(ag) If the respondent is a peace officer, an injunction issued under sub. (5) may not require the respondent to surrender a firearm that he or she is required, as a condition of employment, to possess whether or not he or she is on duty.

813.125. Harassment restraining orders and injunctions. ...

(4m) Restriction on firearm possession; surrender of firearms.

(a) If a judge or circuit court commissioner issues an injunction under sub. (4) and the judge or circuit court commissioner determines, based on clear and convincing evidence presented at the hearing on the issuance of the injunction, that the respondent may use a firearm to cause physical harm to another or to endanger public safety, the judge or circuit court commissioner may prohibit the respondent from possessing a firearm.

(b) An order prohibiting a respondent from possessing a firearm issued under par. (a) remains in effect until the expiration of the injunction issued under sub. (4).

(c) An order issued under par. (a) that prohibits a respondent from possessing a firearm shall do all of the following:

1. Inform the respondent named in the petition of the requirements and penalties under s. 941.29.

2. Except as provided in par. (cg), require the respondent to surrender any firearms that he or she owns or has in his or her possession to the sheriff of the county in which the action under this section was commenced, to the sheriff of the county in which the respondent resides or to another person designated by the respondent and approved by the judge or circuit court commissioner. The judge or circuit court commissioner shall approve the person designated by the respondent unless the judge or circuit court commissioner finds that the person is inappropriate and places the reasons for the finding on the record. If a firearm is surrendered to a person designated by the respondent and approved by the judge or circuit court commissioner, the judge or circuit court commissioner shall inform the person to whom the firearm is surrendered of the requirements and penalties under s. 941.29(4).

(cg) If the respondent is a peace officer, an order issued under par. (a) may not require the respondent to surrender a firearm that he or she is required, as a condition of employment, to possess whether or not he or she is on duty. ...

(d) A firearm surrendered under par. (c)2. may not be returned to the respondent until a judge or circuit court commissioner determines all of the following:

1. That the injunction issued under sub. (4) has been vacated or has expired.

2. That the person is not prohibited from possessing a firearm under any state or federal law or by the order of any federal court or state court, other than an order from which the judge or circuit court commissioner is competent to grant relief.

(e) If a respondent surrenders a firearm under par. (c)2. that is owned by a person other than the respondent, the person who owns the firearm may apply for its return to the circuit court for the county in which the person to whom the firearm was surrendered is located. The court shall order such notice as it considers adequate to be given to all persons who have or may have an interest in the firearm and shall hold a hearing to hear all claims to its true ownership. If the right to possession is proved to the court's satisfaction, it shall order the firearm returned. If the court returns a firearm under this paragraph, the court shall inform the person to whom the firearm is returned of the requirements and penalties under s. 941.29(4).

Chapter 939. Crimes - General Provisions

Subchapter I. Preliminary Provisions

939.22. Words and phrases defined. In chs. 939 to 948 and 951, the following words and phrases have the designated meanings unless the context of a specific section manifestly requires a different construction or the word or phrase is defined in s. 948.01 for purposes of ch. 948:

(2) "Airgun" means a weapon which expels a missile by the expansion of compressed air or other gas. ...

(10) "Dangerous weapon" means any firearm, whether loaded or unloaded; any device designed as a weapon and capable of producing death or great bodily harm; any ligature or other instrumentality used on the throat, neck, nose, or mouth of another person to impede, partially or completely, breathing or circulation of blood; any electric weapon, as defined in s. 941.295(4); or any other device or instrumentality which, in the manner it is used or intended to be used, is calculated or likely to produce death or great bodily harm. ...

Chapter 940. Crimes Against Life and Bodily Security

Subchapter II. Bodily Security

940.24. Injury by negligent handling of dangerous weapon, explosives or fire.

(1) Whoever causes bodily harm to another by the negligent operation or handling of a dangerous weapon, explosives or fire is guilty of a Class I felony.

(2) Whoever causes bodily harm to an unborn child by the negligent operation or handling of a dangerous weapon, explosives or fire is guilty of a Class I felony.

Chapter 941. Crimes against Public Health and Safety

Subchapter III. Weapons

941.20. Endangering safety by use of dangerous weapon.

(1) Whoever does any of the following is guilty of a Class A misdemeanor:

(a) Endangers another's safety by the negligent operation or handling of a dangerous weapon; or

(b) Operates or goes armed with a firearm while he or she is under the influence of an intoxicant; or

(bm) Operates or goes armed with a firearm while he or she has a detectable amount of a restricted controlled substance in his or her blood. A defendant has a defense to any action under this paragraph that is based on the defendant allegedly having a detectable amount of methamphetamine, gamma-hydroxybutyric acid, or delta-9-tetrahydrocannabinol in his or her blood, if he or she proves by a preponderance of the evidence that at the time of the incident or occurrence he or she had a valid prescription for methamphetamine or one of its metabolic precursors, gamma-hydroxybutyric acid, or delta-9-tetrahydrocannabinol.

(c) Except as provided in sub. (1m), intentionally points a firearm at or toward another. ...

(b) Whoever intentionally points a firearm at or towards a law enforcement officer, a fire fighter, an emergency medical technician, a first responder, or an ambulance driver who is acting in an official capacity and who the person knows or has reason to know is a law enforcement officer, a fire fighter, an emergency medical technician, a first responder, or an ambulance driver, is guilty of a Class H felony.

941.25. Manufacturer to register machine guns. Every manufacturer shall keep a register of all machine guns manufactured or handled by him or her. This register shall show the model and serial number, date of manufacture, sale, loan, gift, delivery or receipt, of every machine gun, the name, address, and occupation of the person to whom the machine gun was sold, loaned, given or delivered, or from whom it was received; and the purpose for which it was acquired by the person to whom the machine gun was sold, loaned, given or delivered, or from whom received. Upon demand every manufacturer shall permit any marshal, sheriff or police officer to inspect his or her entire stock of machine guns, parts, and supplies therefor, and shall produce the register required under this section for inspection. Whoever violates any provision of this section is subject to a Class B forfeiture.

941.26. Machine guns and other weapons; use in certain cases; penalty.

(1)(a) No person may sell, possess, use or transport any machine gun or other full automatic firearm.

(b) Except as provided in sub. (4), no person may sell, possess, use or transport any tear gas bomb, hand grenade, projectile or shell or any other container of any kind or character into which tear gas or any similar substance is used or placed for use to cause bodily discomfort, panic, or damage to property.

(1m) No person may take a firearm that is not designed to shoot more than one shot, without manual reloading, by a single function of the trigger and modify the firearm so that it does shoot more than one shot, without manual reloading, by a single function of the trigger.

(2)(a) Any person violating sub. (1)(a) is guilty of a Class H felony.

(b) Any person violating sub. (1m) is guilty of a Class F felony.

(c) Except as provided in par. (d), any person who violates sub. (1)(b) regarding the possession, noncommercial transportation or use of the bomb, grenade, projectile, shell or container under sub. (1)(b) is guilty of a Class A misdemeanor.

(d) Any person who violates sub. (1)(b) regarding the possession, noncommercial transportation or use of the bomb, grenade, projectile, shell or container under sub. (1)(b) in self-defense or defense of another, as allowed under s. 939.48, is subject to a Class D forfeiture.

(e) Any person who violates sub. (1)(b) regarding the sale or commercial transportation of the bomb, grenade, projectile, shell or container under sub. (1)(b) is guilty of a Class H felony.

(f) Any person who violates sub. (1)(b) regarding the use of the bomb, grenade, projectile, shell or container under sub. (1)(b) to cause bodily harm or bodily discomfort to a person who the actor knows, or has reason to know, is a peace officer who is acting in an official capacity is guilty of a Class H felony.

(g) Any person who violates sub. (1)(b) regarding the use of the bomb, grenade, projectile, shell or container under sub. (1)(b) during his or her commission of another crime to cause bodily harm or bodily discomfort to another or who threatens to use the bomb, grenade, projectile, shell or container during his or her commission of another crime to incapacitate another person is guilty of a Class H felony.

(3) This section does not apply to the sale, possession, modification, use or transportation of any weapons or containers under sub. (1) or (1m) to or by any armed forces or national guard personnel in the line of duty, any civil enforcement officer of the state or of any city or county. This section does not apply to the sale, possession, modification, use or transportation of weapons under sub. (1)(a) or (1m) to or by any person duly authorized by the chief of police of any city or the sheriff of any county. This section does not apply to the restoration of any weapon under sub. (1)(a) or (1m) by a person having a license to collect firearms as curios or relics issued by the U.S. department of the treasury. The restriction on transportation contained in this section does not apply to common carriers.

(4)(a) Subsections (1) to (3) do not apply to any device or container that contains a combination of oleoresin of capsicum and inert ingredients but does not contain any other gas or substance that will cause bodily discomfort.

(b) Whoever intentionally uses a device or container described under par. (a) to cause bodily harm or bodily discomfort to another is guilty of a Class A misdemeanor.

(c) Paragraph (b) does not apply to any of the following:

1. Any person acting in self-defense or defense of another, as allowed under s. 939.48.

2. Any peace officer acting in his or her official capacity. Notwithstanding section 939.22 (22), for purposes of this subdivision, peace officer does not include a commission warden who is not a state-certified commission warden.

3. Any armed forces or national guard personnel acting in the line of duty.

(d) Whoever intentionally uses a device or container described under par. (a) to cause bodily harm or bodily discomfort to a person who the actor knows, or has reason to know, is a peace officer who is acting in an official capacity is guilty of a Class H felony.

(e) Whoever uses a device or container described under par. (a) during his or her commission of another crime to cause bodily harm or bodily discomfort to another or who threatens to use the device or container during his or her commission of another crime to incapacitate another person is guilty of a Class H felony.

(f) Any person who offers for sale a device or container described under par. (a) and who leaves in his or her place of business an unsold device or container in a place where customers have ready access to the device or container is subject to a Class C forfeiture.

(g) 1. Any person who sells or distributes a device or container described under par. (a) to a person who has not attained 18 years of age is subject to a Class C forfeiture.

2. A person who proves all of the following by a preponderance of the evidence has a defense to prosecution under subd. 1.

a. That the purchaser or distributee falsely represented that he or she had attained the age of 18 and presented an identification card.

b. That the appearance of the purchaser or distributee was such that an ordinary and prudent person would believe that the purchaser or distributee had attained the age of 18.

c. That the sale was made in good faith, in reasonable reliance on the identification card and appearance of the purchaser or distributee and in the belief that the purchaser or distributee had attained the age of 18.

(h) Any person who intentionally offers for sale a device or container in a place where customers have direct access to the device or container is guilty of a Class A misdemeanor.

(i)1. Whoever intentionally sells a device or container described under par. (a) that does not meet the safety criteria provided in rules promulgated under subd. 2. is guilty of a Class A misdemeanor.

2. The department of justice shall promulgate rules providing safety criteria for devices or containers described under par. (a). In promulgating the rules, the department shall do all of the following:

a. Consider recommendations of law enforcement agencies, as defined in s. 165.83(1)(b), and manufacturers of devices or containers described under par. (a).

b. Provide allowable amounts of oleoresin of capsicum, inert ingredients and total ingredients for a device or container described under par. (a).

c. Provide a maximum effective range for a device or container described under par. (a).

d. Provide other requirements to ensure that a device or container described under par. (a) is effective and appropriate for self-defense purposes.

3. Subdivisions 1. and 2. do not apply to sales of devices or containers described under par. (a) for use by peace officers or armed forces or national guard personnel.

(j)1. Whoever intentionally sells a device or container described under par. (a) without providing the purchaser with all of the following is guilty of a Class A misdemeanor:

a. A proper label on the device or container.

b. Written safety instructions for using the device or container.

c. A package that contains a clear, highlighted message to the purchaser cautioning him or her to read and follow the safety instructions.

2. The department of justice shall promulgate rules providing the requirements for labeling, packaging and written safety instructions under subd. 1.

(k) Any person who has not attained the age of 18 years and who possesses a device or container described under par. (a) is subject to a Class E forfeiture.

(l) Any person who has been convicted of a felony in this state or has been convicted of a crime elsewhere that would be a felony if committed in this state who possesses a device or container described under par. (a) is subject to a Class A misdemeanor. This paragraph does not apply if the person has received a pardon for the felony or crime.

941.27. Machine guns.

(1) Definition. In ss. 941.25 and 941.26, "machine gun" means any of the following:

(a) Any weapon that shoots, is designed to shoot or can be readily restored to shoot, automatically more than one shot, without manual reloading, by a single function of the trigger.

(b) The frame or receiver of any weapon described under par. (a) or any part designed and intended solely and exclusively, or combination of parts designed and intended, for use in converting a weapon into a weapon described under par. (a).

(c) Any combination of parts from which a weapon described under par. (a) can be assembled if those parts are in the possession or under the control of a person.

(2) Exceptions. Sections 941.25 and 941.26 shall not prohibit or interfere with the manufacture for, and sale of, machine guns to the military forces or the peace officers of the United States or of any political subdivision thereof, or the transportation required for that purpose; the possession of a machine gun for scientific purpose, or the possession of a machine gun not usable as a weapon and possessed as a curiosity, ornament or keepsake; or the possession of a machine gun other than one adapted to use pistol cartridges for a purpose manifestly not aggressive or offensive.

941.28. Possession of short-barreled shotgun or short-barreled rifle.

(1) In this section:

(a) "Rifle" means a firearm designed or redesigned, made or remade, and intended to be fired from the shoulder or hip and designed or redesigned and made or remade to use the energy of a propellant in a metallic cartridge to fire through a rifled barrel a single projectile for each pull of the trigger.

(b) "Short-barreled rifle" means a rifle having one or more barrels having a length of less than 16 inches measured from closed breech or bolt face to muzzle or a rifle having an overall length of less than 26 inches.

(c) "Short-barreled shotgun" means a shotgun having one or more barrels having a length of less than 18 inches measured from closed breech or bolt face to muzzle or a shotgun having an overall length of less than 26 inches.

(d) "Shotgun" means a weapon designed or redesigned, made or remade, and intended to be fired from the shoulder or hip and designed or redesigned and made or remade to use the energy of a propellant in a fixed shotgun shell to fire through a smooth bore either a number of ball shot or a single projectile for each single pull of the trigger.

(2) No person may sell or offer to sell, transport, purchase, possess or go armed with a short-barreled shotgun or short-barreled rifle.

(3) Any person violating this section is guilty of a Class H felony.

(4) This section does not apply to the sale, purchase, possession, use or transportation of a short-barreled shotgun or short-barreled rifle to or by any armed forces or national guard personnel in line of duty, any peace officer of the United States or of any political subdivision of the United States or any person who has complied with the licensing and registration requirements under 26 USC 5801 to 5872. This section does not apply to the manufacture of short-barreled shotguns or short-barreled rifles for any person or group authorized to possess these weapons. The restriction on transportation contained in this section does not apply to common carriers. This section shall not apply to any firearm that may be lawfully possessed under federal law, or any firearm that could have been lawfully registered at the time of the enactment of the national firearms act of 1968.

(5) Any firearm seized under this section is subject to s. 968.20(3) and is presumed to be contraband.

941.29. Possession of a firearm.

(1) A person is subject to the requirements and penalties of this section if he or she has been:

(a) Convicted of a felony in this state.

(b) Convicted of a crime elsewhere that would be a felony if committed in this state.

(bm) Adjudicated delinquent for an act committed on or after April 21, 1994, that if committed by an adult in this state would be a felony.

(c) Found not guilty of a felony in this state by reason of mental disease or defect.

(d) Found not guilty of or not responsible for a crime elsewhere that would be a felony in this state by reason of insanity or mental disease, defect or illness.

(e) Committed for treatment under s. 51.20 (13)(a) and ordered not to possess a firearm under s. 51.20(13)(cv).

(f) Enjoined under an injunction issued under s. 813.12 or 813.122 or under a tribal injunction, as defined in s. 813.12(1)(e), issued by a court established by any federally recognized Wisconsin Indian tribe or band, except the Menominee Indian tribe of Wisconsin, that includes notice to the respondent that he or she is subject to the requirements and penalties under s. 941.29 and that has been filed under s. 806.247(3).

(g) Ordered not to possess a firearm under s. 813.125(4m).

(2) A person specified in sub. (1) is guilty of a Class G felony if he or she possesses a firearm under any of the following circumstances:

(a) The person possesses a firearm subsequent to the conviction for the felony or other crime, as specified in sub. (1)(a) or (b).

(b) The person possesses a firearm subsequent to the adjudication, as specified in sub. (1)(bm).

(c) The person possesses a firearm subsequent to the finding of not guilty or not responsible by reason of insanity or mental disease, defect or illness as specified in sub. (1)(c) or (d).

(d) The person possesses a firearm while subject to the court order, as specified in sub. (1)(e) or (g).

(e) The person possesses a firearm while the injunction, as specified in sub. (1)(f), is in effect.

(3) Any firearm involved in an offense under sub. (2) is subject to s. 968.20(3).

(4) A person is concerned with the commission of a crime, as specified in s. 939.05(2)(b), in violation of this section if he or she knowingly furnishes a person with a firearm in violation of sub. (2).

(5) This section does not apply to any person specified in sub. (1) who:

(a) Has received a pardon with respect to the crime or felony specified in sub. (1) and has been expressly authorized to possess a firearm under 18 USC app. 1203; or

(b) Has obtained relief from disabilities under 18 USC 925(c).

(6) The prohibition against firearm possession under this section does not apply to any correctional officer employed before May 1, 1982, who is required to possess a firearm as a condition of employment. This exemption applies if the officer is eligible to possess a firearm under any federal law and applies while the officer is acting in an official capacity.

(7) This section does not apply to any person who has been found not guilty or not responsible by reason of insanity or mental disease, defect or illness if a court subsequently determines both of the following:

(a) The person is no longer insane or no longer has a mental disease, defect or illness.

(b) The person is not likely to act in a manner dangerous to public safety.

(8) This section does not apply to any person specified in sub. (1)(bm) if a court subsequently determines that the person is not likely to act in a manner dangerous to public safety. In any action or proceeding regarding this determination, the person has the burden of proving by a preponderance of the evidence that he or she is not likely to act in a manner dangerous to public safety.

(9) This section does not apply to a person specified in sub. (1)(e) if the prohibition under s. 51.20(13)(cv)1. has been canceled under s. 51.20(13)(cv)2. or (16)(gm).

(10) The prohibition against firearm possession under this section does not apply to a person specified in sub. (1)(f) if the person satisfies any of the following:

(a) The person is a peace officer and the person possesses a firearm while in the line of duty or, if required to do so as a condition of employment, while off duty. Notwithstanding section 939.22 (22), for purposes of this paragraph, peace officer does not include a commission warden who is not a state-certified commission warden.

(b) The person is a member of the U.S. armed forces or national guard and the person possesses a firearm while in the line of duty.

941.2965 Restrictions on use of facsimile firearms.

(1) In this section, "facsimile firearm" means any replica, toy, starter pistol or other object that bears a reasonable resemblance to or that reasonably can be perceived to be an actual firearm. "Facsimile firearm" does not include any actual firearm.

(2) No person may carry or display a facsimile firearm in a manner that could reasonably be expected to alarm, intimidate, threaten or terrify another person. Whoever violates this section is subject to a Class C forfeiture.

(3) Subsection (2) does not apply to any of the following:

(a) Any peace officer acting in the discharge of his or her official duties. Notwithstanding section 939.22 (22), this paragraph does not apply to a commission warden.

(b) Any person engaged in military activities, sponsored by the state or federal government, acting in the discharge of his or her official duties.

(c) Any person who is on his or her own real property, in his or her own home or at his or her own fixed place of business.

(d) Any person who is on real property and acting with the consent of the owner of that property.

941.297 Sale or distribution of imitation firearms.

(1) In this section, "look-alike firearm" means any imitation of any original firearm that was manufactured, designed and produced after December 31, 1897, including and limited to toy guns, water guns, replica nonguns, and air-soft guns firing nonmetallic projectiles. "Look-alike firearm" does not include any imitation, nonfiring, collector replica of an antique firearm developed prior to 1898, or any traditional beebee, paintball or pellet-firing air gun that expels a projectile through the force of air pressure.

(2) Beginning November 1, 1992, no person may sell or distribute any look-alike firearm. Whoever violates this subsection is subject to a Class A forfeiture.

(3) This section does not apply to the sale or distribution of a look-alike firearm that complies with the marking or waiver requirements under 15 USC 5001 (b).

941.298 Firearm silencers.

(1) In this section, "firearm silencer" means any device for silencing, muffling or diminishing the report of a portable firearm, including any combination of parts, designed or redesigned, and intended for use in assembling or fabricating such a device, and any part intended only for use in that assembly or fabrication.

(2) Whoever sells, delivers or possesses a firearm silencer is guilty of a Class H felony.

(3) Subsection (2) does not apply to sales or deliveries of firearm silencers to or possession of firearm silencers by any of the following:

(a) Any peace officer who is acting in compliance with the written policies of the officer's department or agency. This paragraph does not apply to any officer whose department or agency does not have such a policy. Notwithstanding section 939.22 (22), this paragraph does not apply to a commission warden.

(b) Any armed forces or national guard personnel, while in the line of duty.

(c) Any person who has complied with the licensing and registration requirements under 26 USC 5801 to 5872.

Subchapter IV. Other Dangerous Instrumentalities and Practices

941.31 Possession of explosives.

(1) Whoever makes, buys, transports, possesses, or transfers any explosive compound or offers to do the same, either with intent to use such explosive to commit a crime or knowing that another intends to use it to commit a crime, is guilty of a Class F felony.

(2)(a) In this subsection, "improvised explosive device" means a destructive explosive device capable of causing bodily harm, great bodily harm, death or property damage; with some type of explosive material and a means of detonating the explosive material, directly, remotely, or with a timer either present or readily capable of being inserted or attached; which may include a pipe or similar casing, with the ends of the pipe or casing capped, plugged or crimped, and a fuse or similar object sticking out of the pipe or casing; and made by a person not engaged in the legitimate manufacture or legitimate use of explosives, or otherwise authorized by law to do so. "Improvised explosive device" does not include ammunition for any rifle, pistol or shotgun.

(b) Whoever makes, buys, sells, transports, possesses, uses or transfers any improvised explosive device, or possesses materials or components with intent to assemble any improvised explosive device, is guilty of a Class H felony.

(c) This subsection does not apply to the transportation, possession, use, or transfer of any improvised explosive device by any armed forces or national guard personnel or to any peace officer in the line of duty or as part of a duty-related function or exercise. The restriction on transportation in this subsection does not apply to common carriers. Notwithstanding section 939.22 (22), this paragraph does not apply to a commission warden.

Chapter 943. Crimes against Property

Subchapter I. Damage

943.06. Molotov cocktails.

(1) As used in this section, "fire bomb" means a breakable container containing a flammable liquid with a flash point of 150 degrees Fahrenheit or less, having a wick or similar device capable of being ignited, but does not mean a device commercially manufactured primarily for the purpose of illumination.

(2) Whoever possesses, manufactures, sells, offers for sale, gives or transfers a fire bomb is guilty of a Class H felony.

(3) This section shall not prohibit the authorized use or possession of any such device by a member of the armed forces or by fire fighters or law enforcement officers.

Chapter 948. Crimes against Children

948.55 Leaving or storing a loaded firearm within the reach or easy access of a child.

(1) In this section, "child" means a person who has not attained the age of 14 years.

(2) Whoever recklessly stores or leaves a loaded firearm within the reach or easy access of a child is guilty of a Class A misdemeanor if all of the following occur:

(a) A child obtains the firearm without the lawful permission of his or her parent or guardian or the person having charge of the child.

(b) The child under par. (a) discharges the firearm and the discharge causes bodily harm or death to himself, herself or another.

(3) Whoever recklessly stores or leaves a loaded firearm within the reach or easy access of a child is guilty of a Class C misdemeanor if all of the following occur:

(a) A child obtains the firearm without the lawful permission of his or her parent or guardian or the person having charge of the child.

(b) The child under par. (a) possesses or exhibits the firearm in a public place or in violation of s. 941.20.

(4) Subsections (2) and (3) do not apply under any of the following circumstances:

(a) The firearm is stored or left in a securely locked box or container or in a location that a reasonable person would believe to be secure.

(b) The firearm is securely locked with a trigger lock.

(c) The firearm is left on the person's body or in such proximity to the person's body that he or she could retrieve it as easily and quickly as if carried on his or her body.

(d) The person is a peace officer or a member of the armed forces or national guard and the child obtains the firearm during or incidental to the performance of the person's duties. Notwithstanding section 939.22 (22), for purposes of this paragraph, peace officer does not include a commission warden who is not a state-certified commission warden.

(e) The child obtains the firearm as a result of an illegal entry by any person.

(f) The child gains access to a loaded firearm and uses it in the lawful exercise of a privilege under s. 939.48.

(g) The person who stores or leaves a loaded firearm reasonably believes that a child is not likely to be present where the firearm is stored or left.

(h) The firearm is rendered inoperable by the removal of an essential component of the firing mechanism such as the bolt in a breech-loading firearm.

(5) Subsection (2) does not apply if the bodily harm or death resulted from an accident that occurs while the child is using the firearm in accordance with s. 29.304 or 948.60(3).

948.60. Possession of a dangerous weapon by a person under 18.

(1) In this section, "dangerous weapon" means any firearm, loaded or unloaded; any electric weapon, as defined in s. 941.295(4); metallic knuckles or knuckles of any substance which could be put to the same use with the same or similar effect as metallic knuckles; a nunchaku or any similar weapon consisting of 2 sticks of wood, plastic or metal connected at one end by a length of rope, chain, wire or leather; a cestus or similar material weighted with metal or other substance and worn on the hand; a shuriken or any similar pointed star-like object intended to injure a person when thrown; or a manrikigusari or similar length of chain having weighted ends.

(2)(a) Any person under 18 years of age who possesses or goes armed with a dangerous weapon is guilty of a Class A misdemeanor.

(b) Except as provided in par. (c), any person who intentionally sells, loans or gives a dangerous weapon to a person under 18 years of age is guilty of a Class I felony.

(c) Whoever violates par. (b) is guilty of a Class H felony if the person under 18 years of age under par. (b) discharges the firearm and the discharge causes death to himself, herself or another.

(d) A person under 17 years of age who has violated this subsection is subject to the provisions of ch. 938 unless jurisdiction is waived under s. 938.18 or the person is subject to the jurisdiction of a court of criminal jurisdiction under s. 938.183.

(3)(a) This section does not apply to a person under 18 years of age who possesses or is armed with a dangerous weapon when the dangerous weapon is being used in target practice under the supervision of an adult or in a course of instruction in the traditional and proper use of the dangerous weapon under the supervision of an adult. This section does not apply to an adult who transfers a dangerous weapon to a person under 18 years of age for use only in target practice under the adult's supervision or in a course of instruction in the traditional and proper use of the dangerous weapon under the adult's supervision.

(b) This section does not apply to a person under 18 years of age who is a member of the armed forces or national guard and who possesses or is armed with a dangerous weapon in the line of duty. This section does not apply to an adult who is a member of the armed forces or national guard and who transfers a dangerous weapon to a person under 18 years of age in the line of duty.

(c) This section applies only to a person under 18 years of age who possesses or is armed with a rifle or a shotgun if the person is in violation of s. 941.28 or is not in compliance with ss. 29.304 and 29.593. This section applies only to an adult who transfers a firearm to a person under 18 years of age if the person under 18 years of age is not in compliance with ss. 29.304 and 29.593 or to an adult who is in violation of s. 941.28.

948.605. Gun-free school zones.

(1) Definitions. In this section:

(a) "Encased" has the meaning given in s. 167.31(1)(b).

(ac) "Firearm" does not include any beebee or pellet-firing gun that expels a projectile through the force of air pressure or any starter pistol.

(am) "Motor vehicle" has the meaning given in s. 340.01(35).

(b) "School" has the meaning given in s. 948.61(1)(b).

(c) "School zone" means any of the following:

1. In or on the grounds of a school.

2. Within 1,000 feet from the grounds of a school.

(2) Possession of firearm in school zone.

(a) Any individual who knowingly possesses a firearm at a place that the individual knows, or has reasonable cause to believe, is a school zone is guilty of a Class I felony.

(b) Paragraph (a) does not apply to the possession of a firearm:

1. On private property not part of school grounds;

2. If the individual possessing the firearm is licensed to do so by a political subdivision of the state or bureau of alcohol, tobacco and firearms in which political subdivision the school zone is located, and the law of the political subdivision requires that, before an individual may obtain such a license, the law enforcement authorities of the political subdivision must verify that the individual is qualified under law to receive the license;

3. That is not loaded and is:

a. Encased; or

b. In a locked firearms rack that is on a motor vehicle;

4. By an individual for use in a program approved by a school in the school zone;

5. By an individual in accordance with a contract entered into between a school in the school zone and the individual or an employer of the individual;

6. By a law enforcement officer or state-certified commission warden acting in his or her official capacity; or

7. That is unloaded and is possessed by an individual while traversing school grounds for the purpose of gaining access to public or private lands open to hunting, if the entry on school grounds is authorized by school authorities.

8. By a person who is legally hunting in a school forest if the school board has decided that hunting may be allowed in the school forest under s. 120.13(38).

(3) Discharge of firearm in a school zone.

(a) Any individual who knowingly, or with reckless disregard for the safety of another, discharges or attempts to discharge a firearm at a place the individual knows is a school zone is guilty of a Class G felony.

(b) Paragraph (a) does not apply to the discharge of, or the attempt to discharge, a firearm:

1. On private property not part of school grounds;

2. As part of a program approved by a school in the school zone, by an individual who is participating in the program;

3. By an individual in accordance with a contract entered into between a school in a school zone and the individual or an employer of the individual; or

4. By a law enforcement officer acting in his or her official capacity.

948.61. Dangerous weapons other than firearms on school premises.

(1) In this section:

(a) "Dangerous weapon" has the meaning specified in s. 939.22(10), except "dangerous weapon" does not include any firearm and does include any beebee or pellet-firing gun that expels a projectile through the force of air pressure or any starter pistol.

(b) "School" means a public, parochial or private school which provides an educational program for one or more grades between grades 1 and 12 and which is commonly known as an elementary school, middle school, junior high school, senior high school or high school.

(c) "School premises" means any school building, grounds, recreation area or athletic field or any other property owned, used or operated for school administration.

(2) Any person who knowingly possesses or goes armed with a dangerous weapon on school premises is guilty of:

(a) A Class A misdemeanor.

(b) A Class I felony, if the violation is the person's 2nd or subsequent violation of this section within a 5-year period, as measured from the dates the violations occurred.

(3) This section does not apply to any person who:

(a) Uses a weapon solely for school-sanctioned purposes.

(b) Engages in military activities, sponsored by the federal or state government, when acting in the discharge of his or her official duties.

(c) Is a law enforcement officer or state-certified commission warden acting in the discharge of his or her official duties.

(d) Participates in a convocation authorized by school authorities in which weapons of collectors or instructors are handled or displayed.

(e) Drives a motor vehicle in which a dangerous weapon is located onto school premises for school-sanctioned purposes or for the purpose of delivering or picking up passengers or property. The weapon may not be removed from the vehicle or be used in any manner.

(f) Possess or uses a bow and arrow or knife while legally hunting in a school forest if the school board has decided that hunting may be allowed in the school forest under s. 120.13(38).

(4) A person under 17 years of age who has violated this section is subject to the provisions of ch. 938, unless jurisdiction is waived under s. 938.18 or the person is subject to the jurisdiction of a court of criminal jurisdiction under s. 938.183.

[Current through 2007 Wis. Act 41; Dec. 18, 2007]

WYOMING

WYO. STAT.

Title 6 Crimes and Offenses

Chapter 3 Offenses Against Property

Article 1. Arson and Related Offenses

6-3-111. Possession, manufacture, transportation and sale of explosives, improvised explosive device, or incendiary apparatus with unlawful intent prohibited; penalties; definition; exception.

(a) As used in this section:

(i) "Explosive" means any chemical or mechanical compound, substance or mixture that is commonly used or intended to cause an explosion and which contains any oxidizing and combustive units or other ingredients in such proportions, quantities or packing that an ignition by fire, friction, concussion, percussion or by detonation of any part of the compound or mixture is likely to cause such a sudden generation of heated gases that the resultant gaseous pressures are capable of producing destructive effects on nearby objects, or of destroying life or limb;

(ii) "Improvised explosive device" means any device, not commercially manufactured in the ordinary course of interstate commerce, which contains explosives as defined by paragraph (i) of this subsection;

(iii) "Incendiary apparatus" means any fuse, accelerant, time delay ignition apparatus, mechanism, device or material or combination of materials designed, devised or reasonably calculated to cause, spread or accelerate the rate of burning of a fire, or to cause additional damage at or by a fire, or to cause an explosion in connection with a fire;

(iv) The terms "explosive," "improvised explosive device" and "incendiary apparatus" shall be construed to include and refer to any explosive, incendiary, bomb, grenade, rocket having a propellant charge of more than four (4) ounces, missile having an explosive or incendiary charge of more than one-quarter (1/4) ounce, mine or similar device containing or represented to contain any poison gas, nerve gas, biological agent or other chemical or substance capable of causing death or serious physical injury.

(b) Any person who possesses, manufactures, transports, sells or delivers to another person any explosive, improvised explosive device, or incendiary apparatus, with the intent unlawfully to endanger the life or physical well being of another, to commit assault or battery or to inflict bodily harm or injury upon the person of another, or with the intent to assist another person to do the same, is guilty of a felony. Upon conviction, he shall be punished by imprisonment for not more than ten (10) years, a fine of not more than ten thousand dollars ($10,000.00), or both.

(c) Any person who possesses, manufactures, transports, sells or delivers to another person any explosive, improvised explosive device, or incendiary apparatus, with the intent to cause injury or damage to the property of another as defined in W.S. 6-3-103(c), or with the intent to assist another person to do the same, is guilty of a felony. Upon conviction, he shall be punished by imprisonment for not more than five (5) years, a fine of not more than five thousand dollars ($5,000.00), or both.

(d) A person is guilty of intimidation by explosive device if he knowingly, and with the intent to threaten, intimidate or terrorize another person, uses any object or material and represents it to be an explosive, improvised explosive device or incendiary apparatus, and thereby places another person in reasonable fear of imminent physical harm. Upon conviction he shall be punished by imprisonment for not more than five (5) years, a fine of not more than ten thousand dollars ($10,000.00), or both.

(e) Nothing contained in this section shall be construed to apply to any law enforcement officer if performed in the lawful performance of his official duties, nor to any person customarily engaged in the lawful business of manufacture, transportation, sale or use of such materials and devices, if performed in the ordinary course of business and without the criminal intent described in this section, nor to any person actually and lawfully engaged in demolition activity on a ranch, farm or construction site with the authority of the owner thereof, and acting without the criminal intent described in this section.

Title 6. Crimes and Offenses

Chapter 8. Weapons

Article 1. Weapons Offenses

6-8-102. Use or possession of firearm by person convicted of certain felony offenses; penalties. Any person who has previously pleaded guilty to or been convicted of committing or attempting to commit a violent felony or a felony under W.S. 6-5-204(b), and has not been pardoned and who uses or knowingly possesses any firearm is guilty of a felony punishable by imprisonment for not more than three (3) years, a fine of not more than five thousand dollars ($5,000.00), or both.

6-8-103. Possession, manufacture or disposition of deadly weapon with unlawful intent; penalties.

A person who knowingly possesses, manufactures, transports, repairs or sells a deadly weapon with intent to unlawfully threaten the life or physical well-being of another or to commit assault or inflict bodily injury on another is guilty of a felony punishable by imprisonment for not more than five (5) years, a fine of not more than one thousand dollars ($1,000.00), or both.

6-8-104. Wearing or carrying concealed weapons; penalties; exceptions; permits.

(a) A person who wears or carries a concealed deadly weapon is guilty of a misdemeanor punishable by a fine of not more than seven hundred fifty dollars ($750.00), imprisonment in the county jail for not more than six (6) months, or both, unless:

(i) The person is a peace officer;

(ii) The person possesses a permit under this section; or

(iii) The person holds a valid permit authorizing him to carry a concealed firearm authorized and issued by a governmental agency or entity in another state that recognizes Wyoming permits, is a valid statewide permit, and the state has laws similar to the provisions of this section, as determined by the attorney general, including a proper background check of the permit holder.

(b) The attorney general is authorized to issue permits to carry a concealed firearm to persons qualified as provided by this subsection. The attorney general shall promulgate rules necessary to carry out this section no later than October 1, 1994. Applications for a permit to carry a concealed firearm shall be made available and distributed by the division of criminal investigation and local law enforcement agencies. The permit shall be valid throughout the state for a period of five (5) years from the date of issuance. The permittee shall carry the permit, together with valid identification at all times when the permittee is carrying a concealed firearm and shall display both the permit and proper identification upon request of any peace officer. ...

(o) Within thirty (30) days after the changing of a permanent address, or within thirty (30) days after the loss or destruction of a permit, the permittee, including any permittee under paragraph (a)(iii) of this section, shall so notify the division. Violation of this subsection may result in cancellation or revocation of the permit.

(p) In the event that a permit is lost or destroyed, the permit shall be automatically invalid, and the person to whom the same was issued may, upon payment of a five dollar ($5.00) fee to the division, obtain a duplicate, upon furnishing a notarized statement to the division that the permit has been lost or destroyed. ...

(t) No permit issued pursuant to this section or any permit issued from any other state shall authorize any person to carry a concealed firearm into: ...

(vi) Any school, college or professional athletic event not related to firearms; ...

(ix) Any elementary or secondary school facility;

(x) Any college or university facility without the written consent of the security service of the college or university; or

(xi) Any place where the carrying of firearms is prohibited by federal law or regulation or state law. ...

(y) As used in this section:

(i) "Division" means the division of criminal investigation within the office of the attorney general;

(ii) "Firearm" means any pistol, revolver or derringer, designed to be fired by the use of a single hand. ...

Article 2. Firearms Regulation

6-8-203. Firearms information to be kept in place of business; inspection by peace officer. The information required by federal law to be maintained on firearms shall be kept by every wholesaler, retailer, pawnbroker and dealer in firearms in the place of business of the wholesaler, retailer, pawnbroker or dealer, and shall be subject to inspection by any peace officer at all reasonable times.

Article 3. Rifles and Shotguns [Repealed]

Article 4. Regulation by State

6-8-401. Firearm, weapon and ammunition regulation and prohibition by state.

(a) The sale, transfer, purchase, delivery, taxation, manufacture, ownership, transportation, storage, use and possession of firearms, weapons and ammunition shall be authorized, regulated and prohibited by the state, and regulation thereof is preempted by the state. Except as authorized by W.S. 15-1-103(a)(xviii), no city, town or county shall authorize, regulate or prohibit the sale, transfer, purchase, delivery, taxation, manufacture, ownership, transportation, storage, use or possession of firearms, weapons and ammunition except as specifically provided by this chapter. This section shall not affect zoning or other ordinances which encompass firearms businesses along with other businesses. Zoning and other ordinances which are designed for the purpose of restricting or prohibiting the sale, purchase, transfer or manufacture of firearms or ammunition a method of regulating firearms or ammunition as a method of regulating firearms or ammunition are in conflict with this section and are prohibited.

(b) As used in this article, "firearm" means any weapon which will or is designed to expel any projectile by the action of an explosive.

Title 23 Game and Fish

Chapter 3 General Regulatory Provisions

Article 1. Bird and Animal Provisions

23-3-112. Firearms; automatic weapon or silencer prohibited.

(a) No person shall take into or possess in the game fields or forests of Wyoming any fully automatic weapon or device designed to silence or muffle the report of any firearm.

(b) Violation of this section constitutes a high misdemeanor punishable as provided in W.S. 23-6-202(a)(ii).

Title 36 State Lands

Chapter 8 State Parks and Reserves

Article 1. In General

36-8-104. Use of firearms, fireworks, explosives and weapons prohibited; exceptions.

(a) The use of firearms, fireworks, explosives and weapons of all kinds is prohibited in all state parks, campgrounds, recreational grounds, historic landmarks or historic sites providing, however, that:

(i) Firearms and weapons that otherwise comply with state law may be used at such times and in such places as the Wyoming game and fish commission may designate; and

(ii) A supervised public fireworks display may be held in a state park that is not subject to an open fire ban if the fireworks operator has obtained both a special use permit from the superintendent of the park, pursuant to department rules and regulations, and a permit for the display pursuant to W.S. 35-10-203.

36-8-105. Penalty for violating W.S. 36-8-103 through 36-8-105.

Any person violating the provisions of this law shall be guilty of a misdemeanor and upon the conviction thereof shall be fined not more than one hundred dollars ($100.00). Each and every county and state law enforcement officer, game wardens, deputy game wardens of the Wyoming game and fish commission and the director of the department of state parks and cultural resources shall enforce the provisions of this act.

[Current through all 2008 Legislation]

Made in the USA
Monee, IL
05 November 2026

67811393R00331